Meteorites, Comets and Planets

Citations
Please use the following example for citations:
Truran Jr J.W. and Heger A. (2003) Origin of the elements, pp. 1–15. In *Meteorites, Comets, and Planets* (ed. A.M. Davis) Vol. 1 *Treatise on Geochemistry* (eds. H.D. Holland and K.K. Turekian), Elsevier–Pergamon, Oxford.

Cover photo: Solar system. Artwork of the solar system, showing the paths (blue lines) of the nine planets as they orbit around the Sun. The four inner planets are (from inner to outer): Mercury, Venus, Earth and Mars. Around the inner solar system is a broad belt containing thousands of minor planets (asteroids). The five outer planets are (inner to outer): Jupiter (lower right), Saturn (lower left), Uranus (upper left), Neptune (upper right) and Pluto (far right). The star clouds of our galaxy, the Milky Way, appear as a broad luminous band at upper right. (DETLEV VAN RAVENSWAAY / SCIENCE PHOTO LIBRARY)

Meteorites, Comets and Planets

Edited by

A. M. Davis
University of Chicago, IL, USA

TREATISE ON GEOCHEMISTRY
Volume 1

Executive Editors

H. D. Holland
Harvard University, Cambridge, MA, USA

and

K. K. Turekian
Yale University, New Haven, CT, USA

ELSEVIER

2005

AMSTERDAM – BOSTON – HEIDELBERG – LONDON – NEW YORK – OXFORD
PARIS – SAN DIEGO – SAN FRANCISCO – SINGAPORE – SYDNEY – TOKYO

Library
Quest University Canada
3200 University Boulevard
Squamish, BC V8B 0N8

ELSEVIER B.V.	ELSEVIER Inc.	**ELSEVIER Ltd**	ELSEVIER Ltd
Radarweg 29	525 B Street, Suite 1900	**The Boulevard, Langford Lane**	84 Theobalds Road
P.O. Box 211, 1000 AE Amsterdam	San Diego, CA 92101-4495	**Kidlington, Oxford OX5 1GB**	London WC1X 8RR
The Netherlands	USA	**UK**	UK

© 2005 Elsevier Ltd. All rights reserved.

This work is protected under copyright by Elsevier Ltd., and the following terms and conditions apply to its use:

Photocopying
Single photocopies of single chapters may be made for personal use as allowed by national copyright laws. Permission of the Publisher and payment of a fee is required for all other photocopying, including multiple or systematic copying, copying for advertising or promotional purposes, resale, and all forms of document delivery. Special rates are available for educational institutions that wish to make photocopies for non-profit educational classroom use.

Permissions may be sought directly from Elsevier's Rights Department in Oxford, UK: phone (+44) 1865 843830, fax (+44) 1865 853333, e-mail: permissions@elsevier.com. Requests may also be completed on-line via the Elsevier homepage (http://www.elsevier.com/locate/permissions).

In the USA, users may clear permissions and make payments through the Copyright Clearance Center, Inc., 222 Rosewood Drive, Danvers, MA 01923, USA; phone: (+1) (978) 7508400, fax: (+1) (978) 7504744, and in the UK through the Copyright Licensing Agency Rapid Clearance Service (CLARCS), 90 Tottenham Court Road, London W1P 0LP, UK; phone: (+44) 20 7631 5555; fax: (+44) 20 7631 5500. Other countries may have a local reprographic rights agency for payments.

Derivative Works
Tables of contents may be reproduced for internal circulation, but permission of the Publisher is required for external resale or distribution of such material. Permission of the Publisher is required for all other derivative works, including compilations and translations.

Electronic Storage or Usage
Permission of the Publisher is required to store or use electronically any material contained in this work, including any chapter or part of a chapter.

Except as outlined above, no part of this work may be reproduced, stored in a retrieval system or transmitted in any form or by any means, electronic, mechanical, photocopying, recording or otherwise, without prior written permission of the Publisher.
Address permissions requests to: Elsevier's Rights Department, at the fax and e-mail addresses noted above.

Notice
No responsibility is assumed by the Publisher for any injury and/or damage to persons or property as a matter of products liability, negligence or otherwise, or from any use or operation of any methods, products, instructions or ideas contained in the material herein. Because of rapid advances in the medical sciences, in particular, independent verification of diagnoses and drug dosages should be made.

First edition 2005

Library of Congress Cataloging in Publication Data
A catalog record is available from the Library of Congress.

British Library Cataloguing in Publication Data
A catalogue record is available from the British Library.

ISBN: 0-08-044720-1 (Paperback)

The following chapters are US Government works in the public domain and not subject to copyright:
    Calcium–Aluminum-rich Inclusions in Chondritic Meteorites
    Achondrites

∞ The paper used in this publication meets the requirements of ANSI/NISO Z39.48-1992 (Permanence of Paper).
Printed in Italy

DEDICATED TO

HAROLD C. UREY
(1893–1981)

Photograph provided by University of Chicago

Contents

Executive Editors' Foreword	ix
Contributors to Volume 1	xiii
Volume Editor's Introduction	xv
1.01 Origin of the Elements J. W. TRURAN, Jr. and A. HEGER	1
1.02 Presolar Grains E. K. ZINNER	17
1.03 Solar System Abundances of the Elements H. PALME and A. JONES	41
1.04 The Solar Nebula A. P. BOSS	63
1.05 Classification of Meteorites A. N. KROT, K. KEIL, C. A. GOODRICH, E. R. D. SCOTT and M. K. WEISBERG	83
1.06 Oxygen Isotopes in Meteorites R. N. CLAYTON	129
1.07 Chondrites and their Components E. R. D. SCOTT and A. N. KROT	143
1.08 Calcium–Aluminum-rich Inclusions in Chondritic Meteorites G. J. MacPHERSON	201
1.09 Nebular versus Parent-body Processing A. J. BREARLEY	247
1.10 Structural and Isotopic Analysis of Organic Matter in Carbonaceous Chondrites I. GILMOUR	269
1.11 Achondrites D. W. MITTLEFEHLDT	291
1.12 Iron and Stony-iron Meteorites H. HAACK and T. J. McCOY	325
1.13 Cosmic-ray Exposure Ages of Meteorites G. F. HERZOG	347
1.14 Noble Gases F. A. PODOSEK	381
1.15 Condensation and Evaporation of Solar System Materials A. M. DAVIS and F. M. RICHTER	407
1.16 Early Solar System Chronology K. D. McKEEGAN and A. M. DAVIS	431
1.17 Planet Formation J. E. CHAMBERS	461
1.18 Mercury G. J. TAYLOR and E. R. D. SCOTT	477

1.19	Venus B. FEGLEY, Jr.	487
1.20	The Origin and Earliest History of the Earth A. N. HALLIDAY	509
1.21	The Moon P. H. WARREN	559
1.22	Mars H. Y. McSWEEN, Jr.	601
1.23	Giant Planets J. I. LUNINE	623
1.24	Major Satellites of the Giant Planets T. V. JOHNSON	637
1.25	Comets D. E. BROWNLEE	663
1.26	Interplanetary Dust Particles J. P. BRADLEY	689

Subject Index   713

Executive Editors' Foreword

H. D. Holland and
Harvard University, Cambridge, MA, USA
and
K. K. Turekian
Yale University, New Haven, CT, USA

Geochemistry has deep roots. Its beginnings can be traced back to antiquity, but many of the discoveries that are basic to the science were made between 1800 and 1910. The periodic table of elements was assembled, radioactivity was discovered, and the thermodynamics of heterogeneous systems was developed. The solar spectrum was used to determine the composition of the Sun. This information, together with chemical analyses of meteorites, provided an entry to a larger view of the universe.

During the first half of the twentieth century, a large number of scientists used a variety of methods to determine the major-element composition of the Earth's crust, and the geochemistries of many of the minor elements were defined by V. M. Goldschmidt and his associates using the then new technique of emission spectrography. V. I. Vernadsky founded biogeochemistry. The crystal structures of most minerals were determined by X-ray diffraction techniques. Isotope geochemistry was born, and age determinations based on radiometric techniques began to define the absolute geologic timescale. The intense scientific efforts during World War II yielded new analytical tools and a group of people who trained a new generation of geochemists at a number of universities. But the field grew slowly. In the 1950s, a few journals were able to report all of the important developments in trace-element geochemistry, isotopic geochronometry, the exploration of paleoclimatology and biogeochemistry with light stable isotopes, and studies of phase equilibria. At the meetings of the American Geophysical Union, geochemical sessions were few, none were concurrent, and they all ranged across the entire field.

Since then the developments in instrumentation and the increases in computing power have been spectacular. The education of geochemists has been broadened beyond the old, rather narrowly defined areas. Atmospheric and marine geochemistry have become integrated into solid Earth geochemistry; cosmochemistry and biogeochemistry have contributed greatly to our understanding of the history of our planet. The study of Earth has evolved into "Earth System Science," whose progress since the 1940s has been truly dramatic.

Major ocean expeditions have shown how and how fast the oceans mix; they have demonstrated the connections between the biologic pump, marine biology, physical oceanography, and marine sedimentation. The discovery of hydrothermal vents has shown how oceanography is related to economic geology. It has revealed formerly unknown oceanic biotas, and has clarified the factors that today control, and in the past have controlled the composition of seawater.

Seafloor spreading, continental drift and plate tectonics have permeated geochemistry. We finally understand the fate of sediments and oceanic crust in subduction zones, their burial and their exhumation. New experimental techniques at temperatures and pressures of the deep Earth interior have clarified the three-dimensional structure of the mantle and the generation of magmas.

Moon rocks, the treasure trove of photographs of the planets and their moons, and the successful search for planets in other solar systems have all revolutionized our understanding of Earth and the universe in which we are embedded.

Geochemistry has also been propelled into the arena of local, regional, and global anthropogenic problems. The discovery of the ozone hole came as a great, unpleasant surprise, an object lesson for optimists and a source of major new insights into the photochemistry and dynamics of the atmosphere. The rise of the CO_2 content of the atmosphere due to the burning of fossil fuels and deforestation has been and will continue to be at the center of the global change controversy, and will yield new insights into the coupling of atmospheric chemistry to the biosphere, the crust, and the oceans.

The rush of scientific progress in geochemistry since World War II has been matched by organizational innovations. The first issue of *Geochimica et Cosmochimica Acta* appeared in June 1950. The Geochemical Society was founded in 1955 and adopted *Geochimica et Cosmochimica Acta* as its official publication in 1957. The International Association of Geochemistry and Cosmochemistry was founded in 1966, and its journal, *Applied Geochemistry*, began publication in 1986. *Chemical Geology* became the journal of the European Association for Geochemistry.

The Goldschmidt Conferences were inaugurated in 1991 and have become large international meetings. Geochemistry has become a major force in the Geological Society of America and in the American Geophysical Union. Needless to say, medals and other awards now recognize outstanding achievements in geochemistry in a number of scientific societies.

During the phenomenal growth of the science since the end of World War II an admirable number of books on various aspects of geochemistry were published. Of these only three attempted to cover the whole field. The excellent *Geochemistry* by K. Rankama and Th.G. Sahama was published in 1950. V. M. Goldschmidt's book with the same title was started by the author in the 1940s. Sadly, his health suffered during the German occupation of his native Norway, and he died in England before the book was completed. Alex Muir and several of Goldschmidt's friends wrote the missing chapters of this classic volume, which was finally published in 1954.

Between 1969 and 1978 K. H. Wedepohl together with a board of editors (C. W. Correns, D. M. Shaw, K. K. Turekian and J. Zeman) and a large number of individual authors assembled the *Handbook of Geochemistry*. This and the other two major works on geochemistry begin with integrating chapters followed by chapters devoted to the geochemistry of one or a small group of elements. All three are now out of date, because major innovations in instrumentation and the expansion of the number of practitioners in the field have produced valuable sets of high-quality data, which have led to many new insights into fundamental geochemical problems.

At the Goldschmidt Conference at Harvard in 1999, Elsevier proposed to the Executive Editors that it was time to prepare a new, reasonably comprehensive, integrated summary of geochemistry. We decided to approach our task somewhat differently from our predecessors. We divided geochemistry into nine parts. As shown below, each part was assigned a volume, and a distinguished editor was chosen for each volume. A tenth volume was reserved for a comprehensive index:

(i) *Meteorites, Comets, and Planets*: Andrew M. Davis

(ii) *Geochemistry of the Mantle and Core*: Richard Carlson

(iii) *The Earth's Crust*: Roberta L. Rudnick

(iv) *Atmospheric Geochemistry*: Ralph F. Keeling

(v) *Freshwater Geochemistry, Weathering, and Soils*: James I. Drever

(vi) *The Oceans and Marine Geochemistry*: Harry Elderfield

(vii) *Sediments, Diagenesis, and Sedimentary Rocks*: Fred T. Mackenzie

(viii) *Biogeochemistry*: William H. Schlesinger

(ix) *Environmental Geochemistry*: Barbara Sherwood Lollar

(x) *Indexes*

The editor of each volume was asked to assemble a group of authors to write a series of chapters that together summarize the part of the field covered by the volume. The volume editors and chapter authors joined the team enthusiastically. Altogether there are 155 chapters and 9 introductory essays in the Treatise. Naming the work proved to be somewhat problematic. It is clearly not meant to be an encyclopedia. The titles *Comprehensive Geochemistry* and *Handbook of Geochemistry* were finally abandoned in favor of *Treatise on Geochemistry*.

The major features of the Treatise were shaped at a meeting in Edinburgh during a conference on Earth System Processes sponsored by the Geological Society of America and the Geological Society of London in June 2001. The fact that the Treatise is being published in 2003 is due to a great deal of hard work on the part of the editors, the authors, Mabel Peterson (the Managing Editor), Angela Greenwell (the former Head of Major Reference Works), Diana Calvert (Developmental Editor, Major Reference Works),

Bob Donaldson (Developmental Manager), Jerome Michalczyk and Rob Webb (Production Editors), and Friso Veenstra (Senior Publishing Editor). We extend our warm thanks to all of them. May their efforts be rewarded by a distinguished journey for the Treatise.

Finally, we would like to express our thanks to J. Laurence Kulp, our advisor as graduate students at Columbia University. He introduced us to the excitement of doing science and convinced us that all of the sciences are really subdivisions of geochemistry.

Contributors to Volume 1

A. P. Boss
Carnegie Institution of Washington, DC, USA

J. P. Bradley
Institute of Geophysics and Planetary Physics, Lawrence Livermore, National Laboratory, CA, USA

A. J. Brearley
University of New Mexico, Albuquerque, NM, USA

D. E. Brownlee
University of Washington, WA, USA

J. E. Chambers
The SETI Institute, Mountain View, CA, USA

R. N. Clayton
University of Chicago, IL, USA

A. M. Davis
The University of Chicago, IL, USA

B. Fegley Jr.
Washington University, St. Louis, MO, USA

I. Gilmour
The Open University, Milton Keynes, UK

C. A. Goodrich
University of Hawaii at Manoa, Honolulu, HI, USA

H. Haack
University of Copenhagen, Denmark

A. N. Halliday
Eidgenössische Technische Hochschule, Zürich, Switzerland

A. Heger
University of Chicago, IL, USA

G. F. Herzog
Rutgers University, Piscataway, NJ, USA

T. V. Johnson
California Institute of Technology, Pasadena, CA, USA

A. Jones
Université Paris Sud, France

K. Keil
University of Hawaii at Manoa, Honolulu, HI, USA

A. N. Krot
University of Hawaii at Manoa, Honolulu, HI, USA

J. I. Lunine
The University of Arizona, Tucson, AZ, USA

G. J. MacPherson
Smithsonian Institution, Washington, DC, USA

T. J. McCoy
Smithsonian Institution, Washington, DC, USA

K. D. McKeegan
University of California, Los Angeles, CA, USA

H. Y. McSween Jr.
University of Tennessee, Knoxville, TN, USA

D. W. Mittlefehldt
NASA/Johnson Space Center, Houston, TX, USA

H. Palme
Universität zu Köln, Germany

F. A. Podosek
Washington University, St. Louis, MO, USA

F. M. Richter
University of Chicago, IL, USA

E. R. D. Scott
University of Hawaii at Manoa, Honolulu, HI, USA

G. J. Taylor
University of Hawaii, Honolulu, HI, USA

J. W. Truran Jr.
University of Chicago, IL, USA

P. H. Warren
University of California, Los Angeles, CA, USA

M. K. Weisberg
Kingsborough College of the City University of New York, Brooklyn, NY, USA

E. K. Zinner
Washington University, St. Louis, MO, USA

Volume Editor's Introduction

A. M. Davis

University of Chicago, Chicago, IL, USA

The first volume of the *Treatise on Geochemistry* covers the chemistry of everything except the Earth as well as the early history of the Earth. We have learned a great deal about the present-day solar system from remote sensing observations of planets, asteroids, and comets with telescopes and spacecraft. Through astronomical observations of young stars, we have learned about what the early stages of the solar system might have been like, and from observations of extra-solar planets, whether our solar system is typical or not. The discoveries of giant planets in close orbits around stars suggest that our solar system may be unusual, but Earth-sized planets remain difficult to detect around other stars. Meteorites and their constituents provide unique witnesses to planet formation, the formational stages of the solar nebula and even stars that lived their entire lives before the solar system was born.

Cosmochemistry has made tremendous progress since the 1960s, largely because of improvements in analytical technology. The Apollo program to bring back samples of the Moon was initially driven by competition with the Soviet Union, but the large investment in laboratories to analyze the returned sample provided a critical boost to cosmochemistry in the late 1960s. The development of analytical instrumentation capable of extreme isotopic precision (thermal ionization and gas source mass spectrometry), precise microbeam chemical analysis (electron probe microanalysis), and sensitive trace-element analysis (neutron activation analysis) at this time coupled with the serendipitous falls of the primitive carbonaceous chondrites Allende and Murchison in 1969 led to considerable growth in our knowledge of the early solar system.

The discovery of calcium-aluminum-rich refractory inclusions (CAIs) in Allende led to the paradigm of the high-temperature condensation sequence and realization that meteorites contained objects that were the oldest in the solar system. Thirty years later, we know that the CAIs are the oldest objects in the solar system, but they have experienced complex thermal and chemical histories both in the solar nebula and in meteorite parent bodies.

It was in CAIs that the great discovery of non-mass-dependent oxygen isotopic fractionation (expressed as ^{16}O excesses in CAIs) in the solar system was made in 1973 by Clayton and co-workers. Oxygen isotopes have proved to be extraordinarily valuable for studying a wide variety of nebular and parent-body processes and for classification of solar-system objects. The oxygen-isotopic anomalies in CAIs seemed clearly to be of nucleosynthetic origin when they were discovered and this discovery provoked searches for other isotopic anomalies, including evidence of the former existence of short-lived radionuclides such as ^{26}Al. It is ironic that it now seems more likely that the non-mass-dependent oxygen isotopic anomalies are of chemical origin.

In the late 1980s, a truly astounding discovery was made: that primitive meteorites like Murchison contain grains of genuine stardust. These grains are only a few nanometers to a few micrometers across. They formed in outflows from red giant stars and supernovae, and survived potentially destructive processes in the interstellar medium, in the solar nebula, in meteorite parent bodies, and during atmospheric entry.

Each stardust grain is a sample of a single star. The ion microprobe, which was developed in the 1970s, made isotopic analyses of single stardust grains possible and showed that presolar silicon carbide and graphite grains have $^{12}C/^{13}C$ and $^{14}N/^{15}N$ ratios that span factors of 10^4! The small size and high yield of interesting information from stardust has strongly pushed technology and led to incredibly capable new instruments such as the NanoSIMS ion microprobe, which can routinely analyze submicron grains, and microbeam laser resonance mass spectrometry, which can actually count 30% of the atoms in a grain and provide isotopic analyses of ppm-level trace elements in single presolar grains.

Another important discovery was that it appears that some meteorites, the so-called SNC meteorites, come from Mars. This idea arose after the Viking landers showed that Mars has isotopically heavy nitrogen in its atmosphere. A similar isotopic pattern was seen in some unusual achondrites that had young radiometric ages of only 200 Ma (most meteorites have ages about the same as that of the solar system, ~4.5 Ga). The idea that there are martian meteorites was also given a boost from the discovery of lunar meteorites among meteorites returned from Antarctic ice fields. Although there are only ~20 martian meteorites known, they have taught us much of what we know about Mars. However, the many spacecraft that have observed Mars, both from orbit and from landers, have shown how limited a sample of Mars is provided by martian meteorites.

Antarctica has proved to be a wonderful place to collect meteorites since the realization in the early 1970s that there are natural meteorite concentration mechanisms at work there. There are now annual collection expeditions to many ice fields in Antarctica that have nearly doubled the number of known individual meteorites in the world. Meteorites can also be concentrated in deserts and many meteorites have been recovered from North Africa and Australia. A number of new kinds of meteorites have been discovered in Antarctica and desert locations and there has been a lot of activity in meteorite classification.

The start of the twenty-first century is an exciting time in cosmochemistry. New instruments, such as the ICP mass spectrometer, are allowing new isotopic systems such as ^{182}Hf-^{182}W to be explored. For the first time since Apollo, more than a generation ago, there is the prospect of samples being returned to Earth from spacecraft: Genesis will bring back samples of the Sun next year; Stardust will return cometary and genuine interstellar dust in 2006; MUSES-C will return material from an asteroid in 2008; and sometime in the next decade, samples will be brought back from Mars. Further sample return missions are in the planning stage. All these missions are driving instrumentation to allow us to learn more and more from smaller and smaller samples. This is the dawn of a new era of exploration of the solar system.

The aim of this volume is to bring to the reader some of the excitement of this field while trying to cover the chemistry of the entire solar system in a few hundred pages. This volume begins with chapters on the origin of the elements (see Chapter 1.01) and the use of presolar stardust grains (see Chapter 1.02) to learn about processes in other stars. Chapter 1.03 discusses elemental abundances in the solar system. Bulk solar system elemental composition provides a standard of reference for the many geochemical processes that fractionate the elements from one another. Chapter 1.04 deals with the properties of the solar nebula.

Meteorites provide our primary probes of the solar system and their classification is detailed and abundantly illustrated in Chapter 1.05 Oxygen isotopes are a key to understanding many cosmochemical processes and are essential for meteorite classification (Chapter 1.06). The most abundant kind of meteorites, the chondrites, as well as the components within them, are dealt with in Chapter 1.07. A particularly important component, calcium-aluminum-rich refractory inclusions, often called CAIs, were witnesses to the earliest high-temperature chemical processing in the solar system and merit their own chapter, Chapter 1.08. The question of whether some chemical processing occurs in the solar nebula or in meteorite parent bodies is often hotly debated and is discussed in Chapter 1.09. Some primitive chondritic meteorites are quite rich in organic compounds and these many have implications for the origin of life on Earth. Organics in meteorites are detailed in Chapter 1.10. The stony achondrite meteorites are explored in Chapter 1.11 and the iron and stony-iron meteorites, some of which represent cores and core-mantle boundaries of asteroids, are discussed in Chapter 1.12. The histories of meteorites and the lunar surface through exposure to cosmic rays are the subject of Chapter 1.13. Noble gases in meteorites provide a unique window on solar system processes (Chapter 1.14). High-temperature gas-solid fractionation effects due to condensation and evaporation are dealt with from both an experimental and theoretical point of view in Chapter 1.15. The final chapter on meteorites, Chapter 1.16, discusses early solar system chronology through the use of short-lived and long-lived chronometers.

The next section of this volume begins with a general discussion of planet formation processes

(Chapter 1.17) and moves on to discuss Mercury (Chapter 1.18), Venus (Chapter 1.19), the early Earth (Chapter 1.20), the Moon (Chapter 1.21), Mars (Chapter 1.22), the giant planets (Chapter 1.23) and the major satellites of the giant planets (Chapter 1.24). The volume finishes with comets (Chapter 1.25) and interplanetary dust particles (Chapter 1.26).

This volume is dedicated to Harold C. Urey, who is widely regarded as the father of cosmochemistry. Urey won the Nobel Prize in Chemistry in 1934 for the discovery of deuterium. Urey's fascination with the solar system came after his move to the University of Chicago in 1945. Particularly influential were his book *The Planets* (Urey, 1952) and the following papers: Craig and Urey (1953), in which the chemical and redox relationships between different kinds of chondrites were first recognized; Suess and Urey (1957), an improved table of solar system abundances of the elements, which had a profound influence on the seminal work on the origin of the elements by Burbidge *et al.* (1957) and Cameron (1957); and Miller and Urey (1959), which was a major milestone in studies of the origin of life.

REFERENCES

Burbidge E. M., Burbidge G. R., Fowler W. A., and Hoyle F. (1957) Synthesis of the elements in stars. *Rev. Mod. Phys.* **29**, 547–650.

Cameron A. G. W. (1957) *Stellar Evolution, Nuclear Astrophysics and Nucleogenesis*. Chalk River Report CRL-41, Atomic Energy Canada, Ltd.; *Pubs. Astron. Soc. Pacific* **69**, 201–222.

Miller S. L. and Urey H. C. (1959) Organic compound synthesis on the primitive Earth. *Science* **130**, 245–251.

Suess H. E. and Urey H. C. (1957) Abundances of the elements. *Rev. Mod. Phys.* **28**, 53–74.

Urey H. C. (1952) *The Planets*. Yale University Press, New Haven.

Urey H. C. and Craig H. (1953) The composition of stony meteorites and the origin of the meteorites. *Geochim. Cosmochim. Acta* **4**, 36–82.

1.01
Origin of the Elements

J. W. Truran, Jr. and A. Heger
University of Chicago, IL, USA

1.01.1 INTRODUCTION	1
1.01.2 ABUNDANCES AND NUCLEOSYNTHESIS	2
1.01.3 INTERMEDIATE MASS STARS: EVOLUTION AND NUCLEOSYNTHESIS	3
1.01.3.1 Shell Helium Burning and ^{12}C Production	4
1.01.3.2 s-Process Synthesis in Red Giants	4
1.01.4 MASSIVE-STAR EVOLUTION AND NUCLEOSYNTHESIS	5
1.01.4.1 Nucleosynthesis in Massive Stars	7
1.01.4.1.1 Hydrogen burning	7
1.01.4.1.2 Helium burning and the s-process	7
1.01.4.1.3 Hydrogen and helium shell burning	7
1.01.4.1.4 Carbon burning	8
1.01.4.1.5 Neon and oxygen burning	8
1.01.4.1.6 Silicon burning	8
1.01.4.1.7 Explosive nucleosynthesis	8
1.01.4.1.8 The p-process	9
1.01.4.1.9 The r-process	9
1.01.5 TYPE Ia SUPERNOVAE: PROGENITORS AND NUCLEOSYNTHESIS	9
1.01.6 NUCLEOSYNTHESIS AND GALACTIC CHEMICAL EVOLUTION	12
REFERENCES	14

1.01.1 INTRODUCTION

Nucleosynthesis is the study of the nuclear processes responsible for the formation of the elements which constitute the baryonic matter of the Universe. The elements of which the Universe is composed indeed have a quite complicated nucleosynthesis history, which extends from the first three minutes of the Big Bang through to the present. Contemporary nucleosynthesis theory associates the production of certain elements/isotopes or groups of elements with a number of specific astrophysical settings, the most significant of which are: (i) the cosmological Big Bang, (ii) stars, and (iii) supernovae.

Cosmological nucleosynthesis studies predict that the conditions characterizing the Big Bang are consistent with the synthesis only of the lightest elements: ^{1}H, ^{2}H, ^{3}He, ^{4}He, and ^{7}Li (Burles *et al.*, 2001; Cyburt *et al.*, 2002). These contributions define the primordial compositions both of galaxies and of the first stars formed therein. Within galaxies, stars and supernovae play the dominant role both in synthesizing the elements from carbon to uranium and in returning heavy-element-enriched matter to the interstellar gas from which new stars are formed. The mass fraction of our solar system (formed ~4.6 Gyr ago) in the form of heavy elements is ~1.8%, and stars formed today in our galaxy can be a factor 2 or 3 more enriched (Edvardsson *et al.*, 1993). It is the processes of nucleosynthesis operating in stars and supernovae that we will review in this chapter. We will confine our attention to three broad categories of stellar and supernova site with which specific nucleosynthesis products are understood to be identified: (i) intermediate mass stars, (ii) massive stars and associated type II supernovae, and (iii) type Ia supernovae. The first two of these sites are the straightforward consequence of the evolution of single stars, while type Ia supernovae are understood to result from binary stellar evolution.

Stellar nucleosynthesis resulting from the evolution of single stars is a strong function of stellar mass (Woosley *et al.*, 2002). Following phases of hydrogen and helium burning, all stars consist of a carbon–oxygen core. In the mass range of the so-called "intermediate mass" stars ($1 \lesssim M/M_\odot \lesssim 10$), the temperatures realized in their degenerate cores never reach levels at which carbon ignition can occur. Substantial element production occurs in such stars during the asymptotic giant branch (AGB) phase of evolution, accompanied by significant mass loss, and they evolve to white dwarfs of carbon–oxygen (or, less commonly, oxygen–neon) composition. In contrast, the increased pressures that are experienced in the cores of stars of masses $M \gtrsim 10 M_\odot$ yield higher core temperatures that enable subsequent phases of carbon, neon, oxygen, and silicon burning to proceed. Collapse of an iron core devoid of further nuclear energy then gives rise to a type II supernova and the formation of a neutron star or black hole remnant (Heger *et al.*, 2003). The ejecta of type IIs contain the ashes of nuclear burning of the entire life of the star, but are also modified by the explosion itself. They are the source of most material (by mass) heavier than helium.

Observations reveal that binary stellar systems comprise roughly half of all stars in our galaxy. Single star evolution, as noted above, can leave in its wake compact stellar remnants: white dwarfs, neutron stars, and black holes. Indeed, we have evidence for the occurrence of all three types of condensed remnant in binaries. In close binary systems, mass transfer can take place from an evolving companion onto a compact object. This naturally gives rise to a variety of interesting phenomena: classical novae (involving hydrogen thermonuclear runaways in accreted shells on white dwarfs (Gehrz *et al.*, 1998)), X-ray bursts (hydrogen/helium thermonuclear runaways on neutron stars (Strohmayer and Bildsten, 2003)), and X-ray binaries (accretion onto black holes). For some range of conditions, accretion onto carbon–oxygen white dwarfs will permit growth of the CO core to the Chandrasekhar limit $M_{Ch} = 1.4 M_\odot$, and a thermonuclear runaway in to core leads to a type Ia supernova.

In this chapter, we will review the characteristics of thermonuclear processing in the three environments we have identified: (i) intermediate-mass stars; (ii) massive stars and type II supernovae; and (iii) type Ia supernovae. This will be followed by a brief discussion of galactic chemical evolution, which illustrates how the contributions from each of these environments are first introduced into the interstellar media of galaxies. Reviews of nucleosynthesis processes include those by Arnett (1995), Trimble (1975), Truran (1984), Wallerstein *et al.* (1997), and Woosley *et al.* (2002). An overview of galactic chemical evolution is presented by Tinsley (1980).

1.01.2 ABUNDANCES AND NUCLEOSYNTHESIS

The ultimate goal of nucleosynthesis theory is, of course, to explain the composition of the Universe, as reflected, for example, in the stellar and gas components of galaxies. Significant progress has been achieved in this regard as a consequence of a wealth of new information of cosmic abundances—spectroscopic properties of stars in our galaxy and of gas clouds and galaxies at high redshifts—pouring in from new ground- and space-based observatories. Given that, it remains true that the most significant clues to nucleosynthesis are those provided by our detailed knowledge of the elemental and isotopic composition of solar system matter. The mass fractions of the stable isotopes in the solar are displayed in Figure 1. Key features that reflect the nature of the nuclear processes by which the heavy elements are formed include: (i) the large abundances of ^{12}C and ^{16}O, the main products of stellar helium burning; (ii) the dominance of the α-particle nuclei through calcium (^{20}Ne, ^{24}Mg, ^{28}Si, ^{32}S, ^{36}Ar, and ^{40}Ca); (iii) the "nuclear statistical equilibrium" peak at the position of ^{56}Fe; and (iv) the abundance peaks in the region past iron at the neutron closed shell positions (zirconium, barium, and lead), confirming the occurrence of processes of neutron-capture synthesis. The solar system abundance patterns associated specifically with the slow (s-process) and fast (r-process) processes of neutron capture synthesis are shown in Figure 2.

It is important here to call attention to the revised determinations of the oxygen and carbon abundances in the Sun. Allende Prieto *et al.* (2001) derived an accurate oxygen abundance for the Sun of $\log \varepsilon(O) = 8.69 \pm 0.05$ dex, a value approximately a factor of 2 below that quoted by Anders and Grevesse (1989). Subsequently, Allende Prieto *et al.* (2002) determined the solar carbon abundance to be $\log \varepsilon(C) = 8.39 \pm 0.04$ dex, and the ratio C/O = 0.5 ± 0.07. The bottom line here is a reduction in the abundances of the two most abundant heavy elements in the Sun, relative to hydrogen and helium, by a factor ~ 2. The implications of these results for stellar evolution, nucleosynthesis, the formation of carbon stars, and galactic chemical evolution remain to be explored.

Guided by early compilations of the "cosmic abundances" as reflected in solar system material (e.g., Suess and Urey, 1956), Burbidge *et al.* (1957) and Cameron (1957) identified the nuclear processes by which element formation occurs in stellar and supernova environments: (i) *hydrogen burning,* which powers stars for $\sim 90\%$ of their lifetimes; (ii) *helium burning,* which is responsible for the production of ^{12}C and ^{16}O, the two most abundant elements heavier than helium; (iii) the *α-process,* which we now understand as a combination of

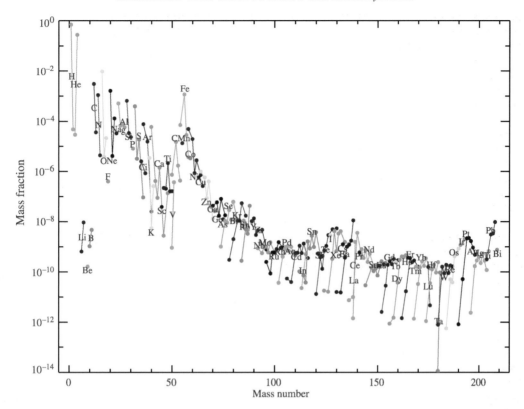

Figure 1 The abundances of the isotopes present in solar system matter are plotted as a function of mass number A (the solar system abundances for the heavy elements are those compiled by Palme and Jones (see Chapter 1.03).

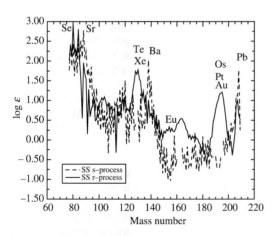

Figure 2 The s-process and r-process abundances in solar system matter (based upon the work by Käppeler et al., 1989). Note the distinctive s-process signature at masses $A \sim 88$, 138, and 208 and the corresponding r-process signatures at $A \sim 130$ and 195, all attributable to closed-shell effects on neutron capture cross-sections. It is the r-process pattern thus extracted from solar system abundances that can be compared with the observed heavy element patterns in extremely metal-deficient stars (the total solar system abundances for the heavy elements are those compiled by Anders and Grevesse, 1989), which are very similar to those from the compilation of Palme and Jones (see Chapter 1.03).

carbon, neon, and oxygen burning; (iv) the *equilibrium process*, by which silicon burning proceeds to the formation of a nuclear statistical equilibrium abundance peak centered on mass $A = 56$; (v) the slow (*s-process*) and rapid (*r-process*) mechanisms of neutron capture synthesis of the heaviest elements ($A \gtrsim 60-70$); and (vi) the *p-process*, a combination of the γ-*process* and the ν-*process*, which we understand to be responsible for the synthesis of a number of stable isotopes of nuclei on the proton-rich side of the valley of beta stability. Our subsequent discussions will identify the astrophysical environments in which these diverse processes are now understood to occur.

1.01.3 INTERMEDIATE MASS STARS: EVOLUTION AND NUCLEOSYNTHESIS

Intermediate-mass red giant stars are understood to be the primary source both of ^{12}C and of the heavy s-process (slow neutron capture) elements, as well as a significant source of ^{14}N and other less abundant CNO isotopes. Their contributions to galactic nucleosynthesis are a consequence of the occurrence of nuclear reactions in helium shell thermal pulses on the AGB, the subsequent dredge-up of matter into the hydrogen-rich envelope by convection, and mass loss. This is a very complicated evolution. Current stellar models, reviewed by Busso et al. (1999), allow the formation of low mass ($\sim 1.5 M_\odot$) carbon stars, which represent the main source of s-process nuclei. A detailed review of the nucleosynthesis products (chemical yields) for low- and intermediate-mass stars is provided by Marigo (2001).

Red giant stars have played a significant role in the historical development of nucleosynthesis theory. While the pivotal role played by nuclear reactions in stars in providing an energy source sufficient to power stars like the Sun over billions of years was established in the late 1930s, it remained to be demonstrated that nuclear processes in stellar interiors might play a role in the synthesis of heavy nuclei. The recognition that heavy-element synthesis is an ongoing process in stellar interiors followed the discovery by Merrill (1952) of the presence of the element technetium in red giant stars. Since technetium has no stable isotopes, and the longest-lived isotope has a half-life $\tau_{1/2} \sim 4.6$ Myr, its presence in red-giant atmospheres indicates its formation in these stars. This confirmed that the products of nuclear reactions operating at high temperatures and densities in the deep interior can be transported by convection to the outermost regions of the stellar envelope.

The role of such convective "dredge-up" of matter in the red-giant phase of evolution of $1-10 M_\odot$ stars is now understood to be an extremely complex process (Busso et al., 1999). On the first ascent of the giant branch (prior to helium ignition), convection can bring the products of CNO cycle burning (e.g., ^{13}C, ^{17}O, and ^{14}N) to the surface. A second dredge-up phase occurs following the termination of core helium burning. The critical third dredge up, occurring in the aftermath of thermal pulses in the helium shells of these AGB stars, is responsible for the transport of both ^{12}C and s-process nuclei (e.g., technetium) to the surface. The subsequent loss of this enriched envelope matter by winds and planetary nebula formation serves to enrich the interstellar media of galaxies, from which new stars are born. A brief review of the mechanisms of production of ^{12}C and the "main" component of the s-process of neutron capture nucleosynthesis is presented in the following sections.

1.01.3.1 Shell Helium Burning and ^{12}C Production

Stellar helium burning proceeds by means of the "triple-alpha" reaction in which three ^{4}He nuclei are converted into ^{12}C, followed by the ^{12}C$(\alpha, \gamma)^{16}$O reaction, which forms ^{16}O at the expense of ^{12}C. Core helium burning in massive stars ($M \gtrsim 10 M_\odot$) occurs at high temperatures, which increases the rate of the ^{12}C$(\alpha, \gamma)^{16}$O reaction and favors the production of oxygen. Typically, the ^{16}O/^{12}C ratio in the oxygen-rich mantles of massive stars prior to collapse is a factor $\sim 2-3$ higher than the solar values. Massive stars are thus the major source of oxygen, while low- and intermediate-mass stars dominate the production of carbon.

The advantage of the helium shells of low- and intermediate-mass stars for ^{12}C production arises from the fact that, for conditions of incomplete helium burning, the ^{12}C/^{16}O ratio is high. Following a thermal pulse in the helium shell, convective dredge-up of matter from the helium shell brings helium, s-process elements, and a significant mass of ^{12}C to the surface. It is the surface enrichment associated with this source of ^{12}C that leads to the condition that the envelope ^{12}C/^{16}O ratio exceeds 1, such that "carbon star" is born. Calculations of galactic chemical evolution indicate that this source of carbon is sufficient to account for the level of ^{12}C in galactic matter. The levels of production of ^{14}N, ^{13}C, and other CNO isotopes in this environment are significantly more difficult to estimate, and thus the corresponding contributions of AGB stars to the galactic abundances of these isotopes remain uncertain.

1.01.3.2 s-Process Synthesis in Red Giants

The formation of most of the heavy elements occurs in one of two processes of neutron capture: the s-process or the r-process. These two broad divisions are distinguished on the basis of the relative lifetimes for neutron captures (τ_n) and electron decays (τ_β). The condition that $\tau_n > \tau_\beta$, where τ_β is a characteristic lifetime for β-unstable nuclei near the valley of β-stability, ensures that as captures proceed the neutron-capture path will itself remain close to the valley of β-stability. This defines the s-process. In contrast, when $\tau_n < \tau_\beta$, it follows that successive neutron captures will proceed into the neutron-rich regions off the β-stable valley. Following the exhaustion of the neutron flux, the capture products approach the position of the valley of β-stability by β-decay, forming the r-process nuclei. The s-process and r-process patterns in solar system matter are those shown in Figure 2.

The environment provided by thermal pulses in the helium shells of intermediate-mass stars on the AGB provides conditions consistent with the synthesis of the bulk of the heavy s-process isotopes through bismuth. Neutron captures in AGB stars are driven by a combination of neutron sources: the ^{13}C$(\alpha, n)^{16}$O reaction provides the bulk of the neutron budget at low-neutron densities, while the ^{22}Ne(α,n)^{25}Mg operating at high temperatures helps to set the timescale for critical reaction branches. This s-process site (the main s-process component) is understood to operate in low-mass AGB stars ($M \sim 1\text{--}3M_\odot$) and to be responsible for the synthesis of the s-process nuclei in the mass range $A \gtrsim 90$. Calculations reviewed by Busso *et al.* (1999) indicate a great sensitivity both to the characteristics of the ^{13}C "pocket" in which neutron production occurs and to the initial metallicity of the star. In their view, this implies that the solar system abundances are not the result of a unique s-process but rather the consequence of a complicated galactic chemical evolutionary history which witnessed mixing of the products of s-processing in stars of different metallicity and a range of ^{13}C pockets. We can hope that observations of the s-process abundance patterns in stars as a function of metallicity will ultimately be better able to guide and to constrain such theoretical models.

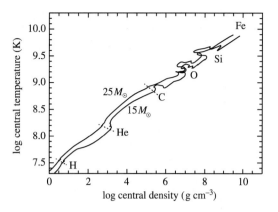

Figure 3 Evolution of the central temperature and density in stars of $15M_\odot$ and $25M_\odot$ from birth as hydrogen burning stars until iron core collapse (Table 1). In general, the trajectories follow a line of $\rho \propto T^3$, but with some deviation downward (towards higher ρ at a given T) due to the decreasing entropy of the core. Nonmonotonic behavior is observed when nuclear fuels are ignited and this is exacerbated in the $15M_\odot$ model by partial degeneracy of the gas (source Woosley *et al.*, 2002).

1.01.4 MASSIVE-STAR EVOLUTION AND NUCLEOSYNTHESIS

Generally speaking, the evolution of a massive star follows a well-understood path of contraction to increasing central density and temperature. The contraction is caused by the energy loss of the star, due to light radiated from the surface and neutrino losses (see below). The released potential energy is in part converted into internal energy of the gas (Virial theorem). This path of contraction is interrupted by nuclear fusion—first hydrogen is burned to helium, then helium to carbon and oxygen. This is followed by stages of carbon, neon, oxygen, and silicon burning, until finally a core of iron is produced, from which no more energy can be extracted by nuclear burning. The onsets of these burning phases as the star evolves through the temperature–density plane are shown in Figure 3, for stars of masses $15M_\odot$ and $25M_\odot$. Each fuel burns first in the center of the star, then in one or more shells (Figure 4). Most burning stages proceed convectively: i.e., the energy production rate by the burning is so large and centrally concentrated that the energy cannot be transported by radiation (heat diffusion) alone, and convective motions dominate the heat transport. The reason for this is the high-temperature sensitivity of nuclear reaction rates: for hydrogen burning in massive stars, nuclear energy generation has a $\propto T^{18}$ dependence, and the dependence is even stronger for later burning stages. The important consequence is that, due to the efficient mixing caused by the convection, the entire unstable region evolves essentially chemically homogeneously—replenishing the fuel at the bottom of burning region (central or shell burning) and depleting the fuel elsewhere in that region at the same time. As a result, shell burning of this fuel then commences outside that region.

Table 1 summarizes the burning stages and their durations for a $20M_\odot$ star. The timescale for helium burning is ~ 10 times shorter than that of hydrogen burning, mostly because of the lower energy release per unit mass. The timescale of the burning stages and contraction beyond central helium burning is greatly reduced by thermal neutrino losses that carry away energy *in situ*, instead of requiring that it be transported to the stellar surface by diffusion or convection. These losses increase with temperature (as $\propto T^9$). When the star has built up a large-enough iron core, exceeding its effective Chandrasekhar mass (the maximum mass for which such a core can be stable), the core collapses to form a neutron star or a black hole (see Woosley *et al.* (2002) for a more extended review). A supernova explosion may result (e.g., Colgate and White, 1966) that ejects most of the layers outside the iron core, including many of the ashes from the preceding burning phase. However, when the supernova shock front travels outward, for a brief time peak temperatures are reached, that exceed the maximum temperatures that have been reached in each region in the preceding hydrostatic burning stages (Table 2). This defines the transient stage of "explosive nucleosynthesis" that is critical to the formation of an equilibrium peak dominated by ^{56}Ni (Truran *et al.*, 1967).

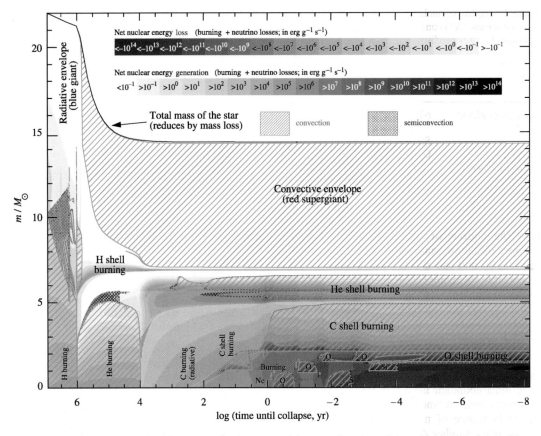

Figure 4 Interior structure of a $22M_\odot$ star of solar composition as a function of time (logarithm of time till core collapse) and enclosed mass. *Green hatching* and *red cross hatching* indicate convective and semiconvective regions. Convective regions are typically well mixed and evolve chemically homogeneously. *Blue shading* indicates energy generation and *pink shading* energy loss. Both take into account the sum of nuclear and neutrino loss contributions. The *thick black* line at the top indicates the total mass of the star, being reduced by mass loss due to stellar winds. Note that the mass loss rate actually increases at late times of the stellar evolution. The decreasing slope of the total mass of the star in the figure is due to the logarithmic scale chosen for the time axis.

Table 1 Hydrostatic nuclear burning stages in massive stars. The table gives burning stages, main and secondary products (ashes), typical temperatures and burning timescales for a $20M_\odot$ star, and the main nuclear reactions. An ellipsis ($\cdots$) indicates more than one product of the double carbon and double oxygen reactions, and a chain of reactions leading to the buildup of iron group elements for silicon burning.

Fuel	Main products	Secondary products	T (10^9 K)	Duration (yr)	Main reaction
H	He	^{14}N	0.037	8.1×10^6	$4\text{H} \rightarrow {}^4\text{He}$ (CNO cycle)
He	O, C	^{18}O, ^{22}Ne s-Process	0.19	1.2×10^6	$3{}^4\text{He} \rightarrow {}^{12}\text{C}$; $^{12}\text{C} + {}^4\text{He} \rightarrow {}^{16}\text{O}$
C	Ne, Mg	Na	0.87	9.8×10^2	$^{12}\text{C} + {}^{12}\text{C} \rightarrow \cdots$
Ne	O, Mg	Al, P	1.6	0.60	$^{20}\text{Ne} \rightarrow {}^{16}\text{O} + {}^4\text{He}$; $^{20}\text{Ne} + {}^4\text{He} \rightarrow {}^{24}\text{Mg}$
O	Si, S	Cl, Ar, K, Ca	2.0	1.3	$^{16}\text{O} + {}^{16}\text{O} \rightarrow \cdots$
Si	Fe	Ti, V, Cr, Mn, Co, Ni	3.3	0.031	$^{28}\text{Si} \rightarrow {}^{24}\text{Mg} + {}^4\text{He} \cdots$; $^{28}\text{Si} + {}^4\text{He} \rightarrow {}^{24}\text{Mg} \cdots$

Massive stars build up most of the heavy elements from oxygen through the iron group from the initial hydrogen and helium of which they are formed. They also make most of the s-process heavy elements up to atomic mass numbers 80–90 from initial iron, converting initial carbon, oxygen, and nitrogen into ^{22}Ne, thus providing a neutron source for the s-process. Massive stars are probably

Table 2 Explosive nucleosynthesis in supernovae. Similar to Table 1, but for the explosion of the star. The "(A, B)" notation means "A" is on the *ingoing* channel and "B" is on the *outgoing* channel. An "α" is same as ^{4}He, "γ" denotes a photon, i.e., a photodisintegration reaction when on the ingoing channel, "n" is a neutron, β^- shows β-decay, and ν indicates neutrino-induced reactions.

Fuel	Main products	Secondary products	T (10^9 K)	Duration (s)	Main reaction
"Innermost ejecta"	r-Process (low Y_e)		>10	1	(n, γ), β^-
Si, O	^{56}Ni	Iron group	>4	0.1	(α, γ)
O	Si, S	Cl, Ar, K, Ca	3–4	1	^{16}O + ^{16}O
Ne, O	O, Mg, Ne	Na, Al, P	2–3	5	(γ, α)
"Heavy elements"		p-Process, ν-process:	2–3	5	(γ, n)
		^{11}B, ^{19}F, ^{138}La, ^{180}Ta		5	(ν, ν'), (ν_e, e$^-$)

also the source of most of the proton-rich isotopes of atomic mass number greater than 100 (p-process) and the site of the r-process that is responsible for many of the neutron-rich isotopes from barium through uranium and thorium.

1.01.4.1 Nucleosynthesis in Massive Stars

The most abundant product of the evolution of massive stars is oxygen, ^{16}O in particular—the third most abundant isotope in the Universe and the most abundant "metal." Massive stars are also the main source of most heavy elements up to atomic mass number $A \sim 80$, of some of the rare proton-rich nuclei, and of the r-process nuclei from barium to uranium. In the following, we will briefly review the burning stages and nuclear processes that characterize the evolution of massive stars and the resulting core collapse supernovae.

1.01.4.1.1 Hydrogen burning

The first stage of stellar burning and energy generation is hydrogen burning, during which hydrogen is transformed to helium by the so-called CNO cycle. An initial abundance of carbon, nitrogen, and oxygen (of which ^{16}O is the most abundant) is collected into ^{14}N—the proton capture on this isotope is the slowest reaction in this cycle. The total number of CNO isotopes remains unchanged—they operate only as nuclear catalysts in the conversion of hydrogen into helium.

1.01.4.1.2 Helium burning and the s-process

After hydrogen is depleted, the star contracts towards helium burning. Before helium burning is ignited, the ^{14}N is converted into ^{18}O by ^{14}N(α, γ)^{18}F(β^+)^{18}O and then burned to ^{22}Ne by ^{18}O(α, γ)^{22}Ne, where α stands for a ^{4}He nucleus and the bracket notation means that the first particle is on the ingoing channel and the second on the outgoing channel. The first stage of helium burning involves the interaction of three α-particles to form carbon (^{4}He(α, γ)^{8}Be*(α, γ)^{12}C; "3α reaction"). As helium becomes depleted, the α-capture on ^{12}C takes over and produces ^{16}O. Indeed, as helium is entirely depleted in the central regions of the star, most of the ashes are ^{16}O, while ^{12}C only comprises 10–20%, decreasing with increasing stellar mass.

Towards the end of the central helium burning phase, the temperature becomes high enough for the ^{22}Ne(α, n)^{25}Mg reaction to proceed, which provides a neutron source for the so-called "weak" component of the s-process. In massive stars, this process builds up elements of atomic mass number of up to 80–90, starting with neutron capture on original ^{56}Fe present in the interstellar medium from which the star was born. In this environment, the timescale for neutron capture is slow compared to the weak (β^-) decay of radioactive nuclei produced by the capture. The s-process path thus proceeds along the neutron-rich side of the valley of stable nuclei. The main competitor reaction for the neutron source is the destruction channel for ^{22}Ne that does not produce neutrons, ^{22}Ne(α, γ)^{26}Mg. The main neutron sink or "poison" for the s-process in this environment is neutron captures on the progeny of the ^{22}Ne(α, n) reaction, among which ^{25}Mg is itself a major neutron poison.

1.01.4.1.3 Hydrogen and helium shell burning

During central helium burning, hydrogen continues to burn in a shell at the outer edge of the helium core, leaving the ashes (helium) at the base of the shell and increasing the size of the helium core. This growth of the helium core by the continued addition of fresh helium fuel contributes to the destruction of carbon at the end of helium burning.

At the completion of core helium burning, helium continues to burn in a shell overlying the helium-free core. It first burns radiatively, but then becomes convective, especially in more massive stars. Since helium is not completely consumed until the death of the star, some s-processing may start here as well, as a consequence of the higher temperatures characterizing the shell burning phase. By the time the stable burning phases of evolution are completed and the iron core is on the verge of collapse, this shell consists of a mixture of helium, carbon, and oxygen, each of which comprises several tens of percent of the matter.

1.01.4.1.4 Carbon burning

After a contraction phase of several 10^4 yr carbon burning starts in the center. This produces mostly ^{20}Ne (via ^{12}C(^{12}C, α)^{20}Ne), but also some ^{24}Mg (via ^{12}C(^{12}C, γ)^{24}Mg) and ^{23}Na (via ^{12}C(^{12}C, p)^{23}Na). There is also some continuation of the s-process from "unburned" ^{22}Ne after the end of central helium burning, operating with neutrons released during carbon burning via the reaction ^{12}C(^{12}C, n)^{23}Mg.

1.01.4.1.5 Neon and oxygen burning

Neon burning is induced by photodisintegration of ^{20}Ne into an α-particle and ^{16}O. The α-particle is then captured by another ^{20}Ne nucleus to make ^{24}Mg or by ^{24}Mg to make ^{28}Si. Secondary products include ^{27}Al, ^{31}P, and ^{32}S.

This is closely followed by oxygen burning, for which the dominant reaction is ^{16}O + ^{16}O. The significant products are the α-particle nuclei ^{28}Si, ^{32}S, ^{36}Ar, and ^{40}Ca, together with such isotopes of odd-Z nuclei such as ^{35}Cl, ^{37}Cl, and ^{39}Cl.

1.01.4.1.6 Silicon burning

Finally, ^{28}Si (and ^{32}S) is burned in a similar way as ^{20}Ne: photodisintegration of ^{28}Si and resulting lighter isotopes accompanies the gradual buildup of iron peak nuclei with higher binding energies per nucleon. At the same time weak processes—positron decaying and electron capture—also become important, producing an neutron excess and allowing the formation of nuclei along the valley of stability. The ultimate result is the formation of an iron "equilibrium" peak of nuclei of maximal nuclear binding energies, from which no further energy can be gained by means of nuclear fusion. The extent in mass of convective central silicon burning is typically $\sim 1.05 M_\odot$. Several brief stages of shell silicon burning follow until the star has built a core of iron group elements ("iron core") large enough to collapse, typically $1.3 M_\odot$ to $\gtrsim 2 M_\odot$. The core had been held up, in part, by electron degeneracy pressure, but when their Fermi energy becomes sufficiently large they can be captured on protons. This serves both to neutronize the core and to reduce pressure support, ultimately leading to the collapse of the core to a neutron star or black hole. It follows that virtually all of the iron-peak products of this phase of quasi-hydrostatic silicon burning do not escape from the star and thus do not contribute to galactic nucleosynthesis. It is the overlying regions of the core that will generally be ejected.

1.01.4.1.7 Explosive nucleosynthesis

Following the collapse of the core a supernova shock front, driven by energy deposition by neutrinos from the hot protoneutron star, travels outwards through the star, setting the stage for a brief phase of "explosive nucleosynthesis." In the inner regions, in silicon- and oxygen-rich layers, the photodisintegration of silicon into α-particles and free nucleons is accompanied by the capture of these particles onto silicon and heavier nuclei, leading to the formation of a nuclear statistical equilibrium peak centered on mass $A = 56$. Since the material here is characterized by a small excess of neutrons over protons, it follows that the final products of these explosive burning episodes must lie along or very near to the $Z = N$ line. The dominant species in situ therefore include the nuclei ^{44}Ti, ^{48}Cr, ^{52}Fe, ^{56}Ni, and ^{60}Zn. Following decay, these contribute to the production of the most proton-rich stable isotopes at these mass numbers: ^{44}Ca, ^{48}Ti, ^{52}Cr, ^{56}Fe, and ^{60}Ni. The neutron enrichments characteristic of matter of solar composition also allow the production of the odd-Z species ^{51}V (formed as ^{51}Mn), ^{55}Mn (formed as ^{55}Co), and ^{59}Co (formed as ^{59}Cu).

Further out, in the oxygen, neon, and carbon shells, explosive burning can act to modify somewhat the abundance patterns resulting from earlier hydrostatic burning phases. The final nucleosynthesis products of the evolution of massive stars and associated type II core collapse supernovae have been discussed by a number of authors (Woosley and Weaver, 1995; Thielemann et al., 1996; Nomoto et al., 1997; Limongi et al., 2000; Woosley et al., 2002). Most of the elements and isotopes in the region from ^{16}O to ^{40}Ca are formed in relative proportions consistent with solar system matter. The same may be said of the nuclei in the iron-peak region from approximately ^{48}Ti to ^{64}Zn. A unique signature of such massive star/type II supernova nucleosynthesis, however, is the fact that nuclei in the ^{16}O to ^{40}Ca region are overproduced by a factor $\sim 2-3$ with respect to nuclei in the ^{48}Ti to ^{64}Zn region—relative to solar system abundances. We will see how this signature, reflected in the compositions of the oldest stars in our galaxy, provides important constraints on models of galactic chemical evolution.

1.01.4.1.8 The p-process

The "p-process" generally serves to include all possible mechanisms that can contribute to the formation of proton-rich isotopes of heavy nuclei. It was previously believed that these nuclei might be formed by proton captures. We now understand that, for the conditions that obtain in massive stars and accompanying supernovae, proton captures are not sufficient. The two dominant processes making proton-rich nuclei in massive stars are the γ-process and the ν-process.

The γ-process involves the photodisintegration of heavy elements. Obviously, this process is not very efficient, as the effective "seed" nuclei are the primordial heavy-element constituents of the star. Most of the heavy p-elements with atomic mass number greater than 100 that are produced in massive stars (see Figure 5) are formed by this mechanism. It is primarily for this reason that these proton-rich heavy isotopes are rare in nature.

The rarest of these proton-rich isotopes have a different and more exotic sources: the immense neutrino flux from the forming hot neutron star can either convert a neutron into a proton (making, e.g., ^{138}La, ^{180}Ta) or knock out a nucleon (e.g., ^{11}B, ^{19}F). Since the neutrino cross-sections are quite small, it is clear that the "parent" nucleus must have an abundance that is at least several thousand times greater than that of the neutrino-produced "daughter" if this process is to contribute significantly to nucleosynthesis. This is indeed true for the cases of ^{11}B (parent ^{12}C) and ^{19}F (parent ^{20}Ne). Production of ^{138}La and ^{180}Ta, the two rarest stable isotopes in the Universe, is provided by neutrino interactions with ^{138}Ba and ^{180}Hf, respectively.

1.01.4.1.9 The r-process

It is generally accepted that the r-process synthesis of the heavy neutron capture elements in the mass regime $A \gtrsim 130-140$ occurs in an environment associated with massive stars. This results from two factors: (i) the two most promising mechanisms for r-process synthesis—a neutrino heated "hot bubble" and neutron star mergers—are both tied to environments associated with core collapse supernovae; and (ii) observations of old stars (discussed in Section 1.01.6) confirm the early entry of r-process isotopes into galactic matter.

- The r-process model that has received the greatest study in recent years involves a high-entropy (neutrino-driven) wind from a core collapse supernova (Woosley et al., 1994; Takahashi et al., 1994). The attractive features of this model include the facts that it may be a natural consequence of the neutrino emission that must accompany core collapse in collapse events, that it would appear to be quite robust, and that it is indeed associated with massive stars of short lifetime. Recent calculations have, however, called attention to a significant problem associated with this mechanism: the entropy values predicted by current type II supernova models are too low to yield the correct levels of production of both the lighter and heavier r-process nuclei.

- The conditions estimated to characterize the decompressed ejecta from neutron star mergers (Lattimer et al., 1977; Rosswog et al., 1999) may also be compatible with the production of an r-process abundance pattern generally consistent with solar system matter. The most recent numerical study of r-process nucleosynthesis in matter ejected in such mergers (Freiburghaus et al., 1999) show specifically that the r-process heavy nuclei in the mass range $A \gtrsim 130-140$ are produced in solar proportions. Here again, the association with a massive star/core collapse super-nova environment is consistent with the early appearance of r-process nuclei in the galaxy and the mechanism seems quite robust.

Massive stars may also contribute to the abundances of the lighter r-process isotopes ($A \lesssim 130-140$). The helium and carbon shells of massive stars undergoing supernova explosions can give rise to neutron production via such reactions as ^{13}C$(\alpha, n)^{16}$O, ^{18}O$(\alpha, n)^{21}$Ne, and ^{22}Ne$(\alpha, n)^{25}$Mg, involving residues of hydrostatic burning phases. Early studies of this problem (Hillebrandt et al., 1981; Truran et al., 1978) indicated that these conditions might allow the production of at least the light ($A \lesssim 130-140$) r-process isotopes (see also Truran and Cowan (2000)). A recent analysis of nucleosynthesis in massive stars by Rauscher et al. (2002) has, however, found that this yields only a slight redistribution of heavy-mass nuclei at the base of the helium shell and no significant production of r-process nuclei above mass $A = 100$. This environment may, nevertheless, provide a source of r-process-like anomalies in grains.

1.01.5 TYPE Ia SUPERNOVAE: PROGENITORS AND NUCLEOSYNTHESIS

Two broad classes of supernovae are observed to occur in the Universe: type I and type II. We have learned from our discussion in the previous section that type II (core collapse) supernovae are products of the evolution of massive

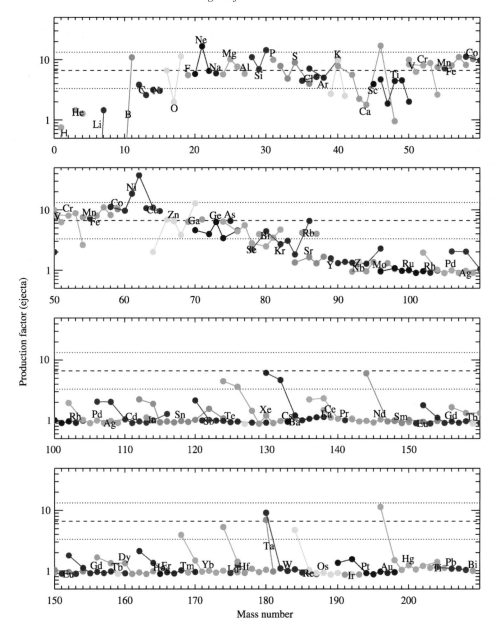

Figure 5 Production factor in the ejecta of a $15M_\odot$ star from Rauscher *et al.* (2002). The production factor is defined as the average abundance of the isotope in the stellar ejecta divided by its solar abundance. To guide the eye, we mark the production factor of ^{16}O by a *dashed* line and a range of "acceptable co-production," i.e., within a factor 2 of ^{16}O, by *dotted* lines. The absolute value of the production factors does not matter so much, as the ejecta will be diluted with the interstellar material after the supernova explosion. However, to reproduce a solar *abundance pattern* for certain isotopes in massive stars, they should be closely co-produced with the most abundant product of massive stars, ^{16}O.

stars ($M \gtrsim 10M_\odot$). Observationally, the critical distinguishing feature of type I supernovae is the absence of hydrogen features in their spectra at maximum light. Theoretical studies focus attention on models for type I events involving either exploding white dwarfs (type Ia) or the explosions of massive stars (similar to those discussed previously) which have, via wind-driven mass loss or binary effects, shed virtually their entire hydrogen envelopes prior to (iron) core collapse (e.g., types Ib and Ic). We are concerned here with supernovae of type Ia, a subclass of the type Is, which are understood to be associated with the explosion of a white dwarf in a close binary system. The standard model for SNe Ia involves specifically the growth of a carbon–oxygen white dwarf to the Chandrasekhar-limit mass in a close binary system and its subsequent incineration (see, e.g., the review of type Ia progenitor models by Livio, 2000).

The iron-peak nuclei observed in nature had their origin in supernova explosions. Type II and type Ia supernovae provide the dominant sites in which this "explosive nucleosynthesis" mechanism is known to operate. These two supernova sites operate on distinctively different timescales and eject different amounts of iron. As we shall see, an understanding of the detailed nucleosynthesis patterns emerging from these two classes of events provides important insights into the star formation histories of galaxies. Type II (core collapse) supernovae produce both the intermediate-mass nuclei from oxygen to calcium- and iron-peak nuclei. Calculations of charged-particle nucleosynthesis in massive stars and type II supernovae (Woosley and Weaver, 1995; Thielemann et al., 1996; Nomoto et al., 1997; Limongi et al., 2000) reveal one particularly significant distinguishing feature of the emerging abundance patterns: the elements from oxygen through calcium (and titanium) are overproduced relative to iron (peak nuclei) by a factor $\sim$2–3. This means that SNe II produce only $\sim$1/3–1/2 of the iron in galactic matter. (We note that, while all such models necessarily make use of an artificially induced shock wave via thermal energy deposition (Thielemann et al., 1996) or a piston (Woosley and Weaver, 1995), the general trends obtained from such nucleosynthesis studies are expected to be valid.) As we shall see in our discussion of chemical evolution, these trends are reflected in the abundance patterns of metal deficient stars in the halo of our galaxy (Wheeler et al., 1989; McWilliam, 1997). This leaves to SNe Ia the need to produce the $\sim$1/3–1/2 of the iron-peak nuclei from titanium to zinc (Ti–V–Cr–Mn–Fe–Co–Ni–Cu–Zn).

Calculations of explosive nucleosynthesis associated with carbon deflagration models for type Ia events (Thielemann et al., 1986; Iwamoto et al., 1999; Dominguez et al., 2001) predict that sufficient iron-peak nuclei $\sim$(0.6–0.8$M_\odot$ of ^{56}Fe in the form of ^{56}Ni) are synthesized to explain both the powering of the light curves due to the decays of ^{56}Ni and ^{56}Co and the observed mass fraction of ^{56}Fe in galactic matter. Estimates of the timescale for first entry of the ejecta of SNe Ia into the interstellar medium of our galaxy yield $\sim$2 × 10^9 yr. at a metallicity [Fe/H] $\sim$ −1 (Kobayashi et al., 2000; Goswami and Prantzos, 2000).

A representative nucleosynthesis calculation for such a type Ia supernova event is shown in Figure 6. Note particularly the region of mass fraction between $\sim$0.2$M_\odot$ and 0.8$M_\odot$, which is dominated by the presence of ^{56}Ni, is in nuclear statistical equilibrium. It is this nickel mass that is responsible—as a consequence of the decay of ^{56}Ni through ^{56}Co to ^{56}Fe—for the bulk of the luminosity of type Ia supernovae at maximum

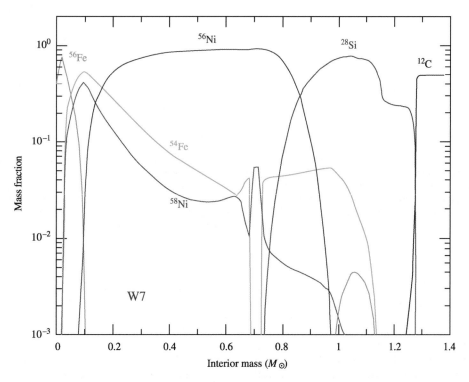

Figure 6 The composition of the core of a type Ia supernova as a function of interior mass. Note the region of $\sim$0.6$M_\odot$ within which the dominant product is ^{56}Ni (source Timmes et al., 2003).

light. This is a critical factor in making SNe Ia the tool of choice for the determination of the cosmological distance scale—as they are the brightest stellar objects known. From the point of view of nucleosynthesis, we then understand that it is the $\sim 0.6 M_\odot$ of iron-peak nuclei ejected per event by type Ia supernovae that represents the bulk of ^{56}Fe in galactic matter. The remaining mass, which is converted into nuclei from ^{16}O to ^{40}Ca, does not make a significant contribution to the synthesis of these elements.

1.01.6 NUCLEOSYNTHESIS AND GALACTIC CHEMICAL EVOLUTION

We have concentrated in this review on three broad categories of stellar and supernova nucleosynthesis sites: (i) the mass range $1 \lesssim M/M_\odot \lesssim 10$ of "intermediate"-mass stars, for which substantial element production occurs during the AGB phase of their evolution; (ii) the mass range $M \gtrsim 10 M_\odot$, corresponding to the massive star progenitors of type II ("core collapse") supernovae; and (iii) type Ia supernovae, which are understood to arise as a consequence of the evolution of intermediate mass stars in close binary systems.

In the context of models of galactic chemical evolution, it is extremely important to know as well the production timescales for each of these sites—i.e., the effective timescales for the return of a star's nucleosynthesis yields to the interstellar gas. The lifetime of a $10 M_\odot$ star is $\sim 5 \times 10^7$ yr. We can thus expect all massive stars $M \gtrsim 10 M_\odot$ to evolve on timescales $\tau_{SNII} < 10^8$ yr, and to represent the first sources of heavy-element enrichment of stellar populations. In contrast, intermediate-mass stars evolve on timescales $\tau_{IMS} \gtrsim 10^8 - 10^9$ yr. Finally, the timescale for SNe Ia product enrichment is a complicated function of the binary history of type Ia progenitors (see, e.g., Livio, 2000). Observations and theory suggest a timescale $\tau_{SNIa} \gtrsim (1.5-2) \times 10^9$ yr.

Many significant features of the evolution of our galaxy follow from a knowledge of the nuclear ashes and evolutionary timescales for the three sites we have surveyed. The primordial composition of the galaxy was that which it inherited from cosmological nucleosynthesis—primarily hydrogen and helium. The characteristics of the first stellar contributions to nucleosynthesis, whether associated with population II or population III, reflect (as might be expected) the nucleosynthesis products of the evolution of massive stars ($M \gtrsim 10 M_\odot$) of short lifetimes ($\tau \lesssim 10^8$ yr). The significant trends in galactic chemical evolution of concern here are those involving the timing of first entry of the products of the other two broad classifications of nucleosynthesis contributors that we have identified: low-mass stars (s-process) and type Ia supernovae (iron-peak nuclei).

Spectroscopic abundance studies of the oldest stars in our galaxy, down to metallicities

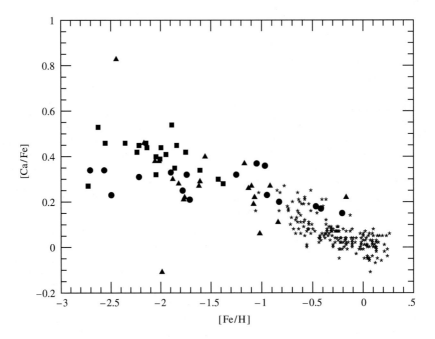

Figure 7 Observed evolution of the calcium to iron abundance ratio with metallicity (▲: Hartmann and Gehren (1998); ■: Zhao and Magain (1990); ●: Gratton and Sneden (1991); ★: Edvardsson *et al.*, (1993)).

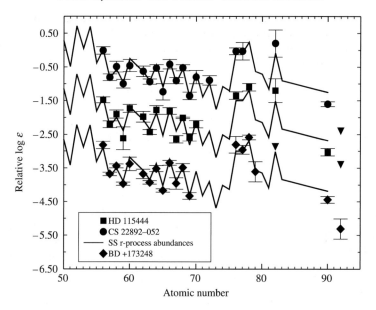

Figure 8 The heavy-element abundance patterns for the three stars CS 22892-052, HD 155444, and BD+17°3248 are compared with the scaled solar system r-process abundance distribution (solid line) (Sneden *et al.*, 2003; Westin *et al.*, 2000; Cowan *et al.*, 2002). Upper limits are indicated by inverted triangles (source Truran *et al.*, 2003).

[Fe/H] ∼−4 to −3, have been reviewed by Wheeler *et al.* (1989). Such studies typically reveal abundance trends which can best be understood as reflecting the nucleosynthesis products of the massive stars and associated type II supernovae that can be expected to evolve and to enrich the interstellar media of galaxies on rapid timescales ($\lesssim 10^8$ yr). Metal deficient stars ($-1.5 \lesssim$ [Fe/H] $\lesssim -3$) in our own galaxy's halo show two significant variations with respect to solar abundances: the elements in the mass range from oxygen to calcium are overabundant—relative to iron-peak nuclei—by a factor ∼2–3, and the abundance pattern in the heavy element region $A \gtrsim 60$–70 closely reflects the r-process abundance distribution that is characteristic of solar system matter, with no evidence for an s-process nucleosynthesis contribution. For purposes of illustration, the trends in [Ca/Fe] as a function of metallicity [Fe/H] are shown in Figure 7. Note the factor 2–3 overabundance of calcium with respect to iron at metallicities below [Fe/H] ∼−1. We also display in Figure 8 the detailed agreement of metal-poor star heavy-element abundance pattern with that of solar system r-process abundances, for three metal-poor but r-process-rich stars: CS 22892-052 (Sneden *et al.*, 2003), HD 115444 (Westin *et al.*, 2000), and BD+17°3248 (Cowan *et al.*, 2002). Both of these features are entirely consistent with nucleosynthesis expectations for massive stars and associated type II supernovae.

The signatures of an increasing s-process contamination first appears at an [Fe/H] ∼−2.5 to −2.0.

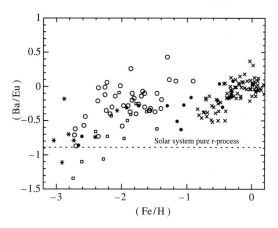

Figure 9 The history of the [Ba/Eu] ratio is shown as a function of metallicity [Fe/H]. This ratio reflects to a good approximation the ratio of s-process to r-process elemental abundances, and thus measures the histories of the contributions from these two nucleosynthesis processes to galactic matter (○: Burris *et al.* (2000); ×: Woolf *et al.* (1995) and Edvardsson *et al.* (1993); ●: Gratton and Sneden (1994); ∗: McWilliam (1997); □: Zhao and Magain (1990)).

The evidence for this is provided by an increase in the ratio of barium (the abundance of which in galactic matter is dominated by s-process contributions) to europium (almost exclusively an r-process product). The ratio [Ba/Eu] is shown in Figure 9 as a function of [Fe/H] for a large sample of halo and disk stars. Note that at the lowest metallicities the [Ba/Eu] ratio clusters around the "pure" r-process value ([Ba/Eu]$_{\text{r-process}}$ ∼−0.9); at a metallicity [Fe/H]∼−2.5, the Ba/Eu ratio shows a gradual increase. In the context of our earlier review of nucleosynthesis sites, this provides observational evidence for the first input from AGB stars, on timescales perhaps approaching $\tau_{IMS} \gtrsim 10^9$ yr.

In contrast, evidence for entry of the iron-rich ejecta of SNe Ia is seen first to appear at a metallicity [Fe/H] ~ -1.5 to -1.0. This may be seen reflected in the abundance histories of [Ca/Fe] with [Fe/H], in Figure 7. This implies input from supernovae Ia on timescales $\tau_{SNIa} \gtrsim (1-2) \times 10^9$ yr. It may be of interest that this seems to appear at approximately the transition from halo to (thick) disk stars.

These observed abundance histories confirm theoretical expectations for the nucleosynthesis contributions from different stellar and supernova sites and the timescales on which enrichment occurs. They provide extremely important tests of numerical simulations and illustrate how the interplay of theory and observation can provide important constraints on the star formation and nucleosynthesis history of our galaxy.

REFERENCES

Allende Prieto C., Lambert D. L., and Asplund M. (2001) The forbidden abundance of oxygen in the Sun. *Astrophys. J.* **556**, L63–L66.

Allende Prieto C., Lambert D. L., and Asplund M. (2002) A reappraisal of the solar photospheric C/O ratio. *Astrophys. J.* **573**, L137–L140.

Anders E. and Grevesse N. (1989) Abundances of the elements: meteoritic and solar. *Geochim. Cosmochim. Acta* **53**, 197–214.

Arnett W. D. (1995) Explosive nucleosynthesis revisited: yields. *Ann. Rev. Astron. Astrophys.* **33**, 115–132.

Burbidge E. M., Burbidge G. R., Fowler W. A., and Hoyle F. (1957) Synthesis of the elements in stars. *Rev. Mod. Phys.* **29**, 547–650.

Burles S., Nollett K. M., and Turner M. S. (2001) Big Bang nucleosynthesis predictions for precision cosmology. *Astrophys. J.* **552**, L1–L6.

Burris D. L., Pilachowski C. A., Armandroff T. A., Sneden C., Cowan J. J., and Roe H. (2000) Neutron-capture elements in the early galaxy: insights from a large sample of metal-poor giants. *Astrophys. J.* **544**, 302–319.

Busso M., Gallino R., and Wasserburg G. J. (1999) Nucleosynthesis in asymptotic giant branch stars: relevance for galactic enrichment and solar system formation. *Ann. Rev. Astron. Astrophys.* **37**, 239–309.

Cameron A. G. W. (1957) *Stellar Evolution Nuclear Astrophysics and Nucleogenesis*. Chalk River Report, AELC (Atomic Energy Canada), CRL-41.

Colgate S. A. and White R. H. (1966) The hydrodynamic behavior of supernovae explosions. *Astrophys. J.* **143**, 626–681.

Cowan J. J., Sneden C., Burles S., Ivans I. I., Beers T. C., Truran J. W., Lawler J. E., Primas F., Fuller G. M., Pfeiffer B., and Kratz K.-L. (2002) The chemical composition and age of the metal-poor halo star BD + 17°3248. *Astrophys. J.* **572**, 861–879.

Cyburt R. H., Fields B. D., and Olive K. A. (2002) Primordial nucleosynthesis with CMB inputs: probing the early universe and light element astrophysics. *Astropart. Phys.* **17**, 87–100.

Domínguez I., Höflich P., and Straniero O. (2001) Constraints on the progenitors of Type Ia supernovae and implications for the cosmological equation of state. *Astrophys. J.* **557**, 279–291.

Edvardsson B., Andersen J., Gustafsson B., Lambert D. L., Nissen P. E., and Tomkin J. (1993) The chemical evolution of the galactic disk I. Analysis and results. *Astron. Astrophys.* **275**, 101–152.

Freiburghaus C., Rosswog S., and Thielemann F.-K. (1999) r-Process in neutron star mergers. *Astrophys. J.* **525**, L121–L124.

Gehrz R. D., Truran J. W., Williams R. E., and Starrfield S. (1998) Nucleosynthesis in classical novae and its contribution to the interstellar medium. *Proc. Astron. Soc. Pacific* **110**, 3–26.

Goswami A. and Prantzos N. (2000) Abundance evolution of intermediate mass elements (C to Zn) in the Milky Way halo and disk. *Astron. Astrophys.* **359**, 191–212.

Gratton R. G. and Sneden C. (1991) Abundances of elements of the Fe-group in metal-poor stars. *Astron. Astrophys.* **241**, 501–525.

Gratton R. G. and Sneden C. (1994) Abundances of neutron-capture elements in metal-poor stars. *Astron. Astrophys.* **287**, 927–946.

Hartmann K. and Gehren T. (1989) Metal-poor subdwarfs and early galactic nucleosynthesis. *Astron. Astrophys.* **199**, 269–270.

Heger A., Fryer C. L., Woosley S. E., Langer N., and Hartmann D. H. (2003) How massive single stars end their life. *Astrophys. J.* **591**, 288–300.

Hillebrandt W., Klapdor H. V., Oda T., and Thielemann F.-K. (1981) The r-process during explosive helium burning in supernovae. *Astron. Astrophys.* **99**, 195–198.

Iwamoto K., Brachwitz F., Nomoto K., Kishimoto N., Hix R., and Thielemann F.-K. (1999) Nucleosynthesis in Chandrasekhar mass models for Type Ia supernovae and constraints on progenitor systems and burning-front propagation. *Astrophy. J. Suppl.* **125**, 439–462.

Käppeler F., Beer H., and Wisshak K. (1989) s-Process nucleosynthesis–nuclear physics and the classical model. *Rev. Prog. Phys.* **52**, 945–1013.

Kobayashi C., Tsujimoto T., and Nomoto K. (2000) The history of the cosmic supernova rate derived from the evolution of the host galaxies. *Astrophys. J.* **539**, 26–38.

Lattimer J. M., Mackie F., Ravenhall D. G., and Schramm D. N. (1977) The decompression of cold neutron star matter. *Astrophys. J.* **213**, 225–233.

Limongi M., Straniero O., and Chieffi A. (2000) Massive stars in the range 13–25$M_\odot$: evolution and nucleosynthesis: II. The solar metallicity models. *Astrophys. J. Suppl.* **125**, 625–644.

Livio M. (2000) The progenitors of Type Ia supernovae. In *Type Ia Supernovae: Theory and Cosmology* (eds. J. C. Niemeyer and J. W. Truran). Cambridge University Press, Cambridge, pp. 33.

Marigo P. (2001) Chemical yields from low- and intermediate-mass stars: model predictions and basic observational constraints. *Astron. Astrophys.* **370**, 194–217.

McWilliam A. (1997) Abundance ratios and galactic chemical evolution. *Ann. Rev. Astron. Astrophys.* **35**, 503–556.

Merrill P. (1952) Technetium in Red Giant Stars. *Science* **115**, 484–485.

Nomoto K., Hashimoto M., Tsujimoto T., Thielemann F.-K., Kishimoto N., Kubo Y., and Nakasato N. (1997) Nucleosynthesis in type II supernovae. *Nucl. Phys.* **616A**, 79c–91c.

Rauscher T., Heger A., Hoffman R. D., and Woosley S. E. (2002) Nucleosynthesis in massive stars with improved nuclear and stellar physics. *Astrophys. J.* **576**, 323–348.

Rosswog S., Liebendörfer M., Thielemann F.-K., Davies M. B., Benz W., and Piran T. (1999) Mass ejection in neutron star mergers. *Astron. Astrophys.* **341**, 499–526.

Sneden C., Cowan J. J., Lawler J. E., Ivans I. I., Burles S., Beers T. C., Primas F., Hill V., Truran J. W., Fuller G. M., Pfeiffer B., and Kratz K.-L. (2003) The extremely metal-poor, neutron-capture-rich star CS 22892-052:

a comprehensive abundance analysis. *Astrophys. J.* (in press).

Suess H. E. and Urey H. C. (1956) Abundances of the elements. *Rev. Mod. Phys.* **28**, 53–74.

Takahashi K., Witti J., and Janka H.-T. (1994) Nucleosynthesis in neutrino-driven winds from protoneutron stars: II. The r-process. *Astron. Astrophys.* **286**, 857–869.

Thielemann F.-K., Nomoto K., and Yokoi K. (1986) Explosive nucleosynthesis in carbon deflagration models of Type I supernovae. *Astron. Astrophys.* **158**, 17–33.

Thielemann F.-K., Nomoto K., and Hashimoto M. (1996) Core-collapse supernovae and their ejecta. *Astrophys. J.* **460**, 408–436.

Timmes F. X., Brown E. F., and Truran J. W. (2003) On variations in the peak luminosity of Type Ia supernovae. *Astrophys. J.* (in press).

Tinsley B. (1980) Evolution of the stars and gas in galaxies. *Fundament. Cosmic Phys.* **5**, 287–388.

Trimble V. (1975) The origin and abundances of the chemical elements. *Rev. Mod. Phys.* **47**, 877–976.

Truran J. W. (1984) Nucleosynthesis. *Ann. Rev. Nucl. Part. Sci.* **34**, 53–97.

Truran J. W. and Cowan J. J. (2000) On the site of the weak r-process component. In *Proceedings Ringberg Workshop on Nuclear Astrophysics*. Max Planck Publication, MPA-P12, 64p.

Truran J. W., Arnett W. D., and Cameron A. G. W. (1967) Nucleosynthesis in supernova shock waves. *Can. J. Phys.* **45**, 2315–2332.

Truran J. W., Cowan J. J., and Cameron A. G. W. (1978) The helium-driven r-process in supernovae. *Astrophys. J.* **222**, L63–L67.

Truran J. W., Cowan J. J., Pilachowski C. A., and Sneden C. (2003) Probing the neutron-capture nucleosynthesis history of galactic matter. *Proc. Astron. Soc. Pacific* **114**, 1293–1308.

Wallerstein G., Iben I., Jr., Parker P., Boesgaard A. M., Hale G. M., Champagne A. E., Barnes C. A., Käppeler F., Smith V. V., Hoffman R. D., Timmes F. X., Sneden C., Boyd R. N., Meyer B. S., and Lambert D. L. (1997) Synthesis of the elements in stars: forty years of progress. *Rev. Mod. Phys.* **69**, 995–1084.

Westin J., Sneden C., Gustafsson B., and Cowan J. J. (2000) The r-process-enriched low-metallicity giant HD 115444. *Astrophys. J.* **530**, 783–799.

Wheeler J. C., Sneden C., and Truran J. W. (1989) Abundance ratios as a function of metallicity. *Ann. Rev. Astron. Astrophys.* **27**, 279–349.

Woolf V. M., Tomkin J., and Lambert D. L. (1995) The r-process element europium in galactic disk F and G dwarf stars. *Astrophys. J.* **453**, 660–672.

Woosley S. A. and Weaver T. A. (1995) The evolution and explosion of massive stars: II. Explosive hydrodynamics and nucleosynthesis. *Astrophys. J. Suppl.* **101**, 181–235.

Woosley S. E., Wilson J. R., Mathews G. J., Hoffman R. D., and Meyer B. S. (1994) The r-process and neutrino-heated supernova ejecta. *Astrophys. J.* **433**, 229–246.

Woosley S. E., Heger A., and Weaver T. A. (2002) The evolution and explosion of massive stars. *Rev. Mod. Phys.* **74**, 1015–1071.

Zhao G. and Magain P. (1990) The chemical composition of the extreme halo stars: II. Green spectra of 20 dwarfs. *Astron. Astrophys.* **238**, 242–248.

1.02
Presolar Grains

E. K. Zinner

Washington University, St. Louis, MO, USA

1.02.1	INTRODUCTION	17
1.02.2	HISTORICAL BACKGROUND	18
1.02.3	TYPES OF PRESOLAR GRAINS	19
1.02.4	ANALYSIS TECHNIQUES	19
1.02.5	ASTROPHYSICAL IMPLICATIONS OF THE STUDY OF PRESOLAR GRAINS	20
1.02.6	SILICON CARBIDE	20
	1.02.6.1 Mainstream Grains	22
	1.02.6.2 Type Y and Z Grains	25
	1.02.6.3 Type A + B Grains	26
	1.02.6.4 Type X Grains	26
	1.02.6.5 Nova Grains	28
	1.02.6.6 Grain Size Effect	28
1.02.7	SILICON NITRIDE	29
1.02.8	GRAPHITE	29
	1.02.8.1 Physical Properties	29
	1.02.8.2 Isotopic Compositions	30
1.02.9	OXIDE GRAINS	32
1.02.10	DIAMOND	33
1.02.11	CONCLUSION AND FUTURE PROSPECTS	34
ACKNOWLEDGMENTS		34
REFERENCES		34

1.02.1 INTRODUCTION

Traditionally, astronomers have studied the stars by using, with rare exception, electromagnetic radiation received by telescopes on and above the Earth. Since the mid-1980s, an additional observational window has been opened in the form of microscopic presolar grains found in primitive meteorites. These grains had apparently formed in stellar outflows of late-type stars and in the ejecta of stellar explosions and had survived the formation of the solar system. They can be located in and extracted from their parent meteorites and studied in detail in the laboratory. Their stellar origin is recognized by their isotopic compositions, which are completely different from those of the solar system and, for some elements, cover extremely wide ranges, leaving little doubt that the grains are ancient stardust.

By the 1950s it had been conclusively established that the elements from carbon on up are produced by nuclear reactions in stars and the classic papers by Burbidge *et al.* (1957) and Cameron (1957) provided a theoretical framework for stellar nucleosynthesis. According to these authors, nuclear processes produce elements with very different isotopic compositions, depending on the specific stellar source. The newly produced elements are injected into the interstellar medium (ISM) by stellar winds or as supernova (SN) ejecta, enriching the galaxy in "metals" (all elements heavier than helium) and after a long galactic history the solar system is believed to have formed from a mix of this material. In fact, the original work by Burbidge *et al.* and Cameron was stimulated by the observation of regularities in the abundance of the nuclides in the solar system as obtained by the study of meteorites (Suess and Urey, 1956). Although providing only a grand average of many stellar sources, the solar system abundances of the elements and isotopes (Anders and Grevesse, 1989; Grevesse et al., 1996; see Chapter 1.03; Lodders, 2003) remained an important test for nucleosynthesis theory (e.g., Timmes et al., 1995).

In contrast, the study of stellar grains permits information to be obtained about individual stars, complementing astronomical observations of elemental and isotopic abundances in stars (e.g., Lambert, 1991), by extending measurements to elements that cannot be measured astronomically. In addition to nucleosynthesis and stellar evolution, presolar grains provide information about galactic chemical evolution, physical properties in stellar atmospheres, mixing of SN ejecta and conditions in the parent bodies of the meteorites in which the grains are found.

This new field of astronomy has grown to an extent that not all aspects of presolar grains can be treated in detail in this chapter. The interested reader is therefore referred to some recent reviews (Anders and Zinner, 1993; Ott, 1993; Zinner, 1998a,b; Hoppe and Zinner, 2000; Nittler, 2003) and to the compilation of papers found in Bernatowicz and Zinner (1997). The latter not only contains several detailed review papers on presolar dust grains but also a series of chapters on stellar nucleosynthesis. Further information on nucleosynthesis can be obtained from the textbooks by Clayton (1983b) and Arnett (1996), and from reviews by Käppeler et al. (1989), Meyer (1994), and Wallerstein et al. (1997).

1.02.2 HISTORICAL BACKGROUND

Although the work by Burbidge et al. (1957) and Cameron (1957), and subsequent work by nuclear astrophysicists made it clear that many different stellar sources must have contributed to the material that formed the solar system and although astronomical observations indicate that some of this material was in the form of interstellar (IS) grains (e. g., Mathis, 1990), it was generally believed that it had been thoroughly homogenized in a hot solar nebula (Cameron, 1962). The uniform isotopic composition of all available solar system material seemed to confirm this opinion.

The first evidence for isotopic heterogeneity of the solar nebula and a hint of the survival of presolar grains came from hydrogen (Boato, 1954) and the noble gases xenon (Reynolds and Turner, 1964) and neon (Black and Pepin, 1969; Black, 1972) but it was only after the discovery of anomalies in oxygen, a rock-forming element (Clayton et al., 1973), that the concept of survival of presolar material in primitive meteorites was widely accepted. The finding of ^{16}O excesses was followed by the detection of isotopic anomalies in other elements such as magnesium, calcium, titanium, chromium, and barium in refractory inclusions (CAIs for calcium- and aluminum-rich inclusions) (Wasserburg, 1987; Clayton et al., 1988; Lee, 1988). Also large anomalies in carbon (Halbout et al., 1986) and nitrogen (Lewis et al., 1983) indicated the presence of presolar grains. However, it was the pursuit of the carriers of the "exotic" (i.e., isotopically anomalous) noble gas components of neon and xenon (Figure 1) by Ed Anders and his colleagues at the University of Chicago that led to their ultimate isolation (see Anders and Zinner, 1993). The approach taken by these scientists, "burning down the haystack to find the needle," consisted of tracking the noble gas carriers through a series of increasingly harsher chemical dissolution and physical separation steps (Tang and Anders, 1988b; Amari et al., 1994). Their effort culminated in the isolation and identification of diamond, the carrier of Xe-HL (Lewis et al., 1987), silicon carbide, the carrier of Ne-E(H) and Xe-S

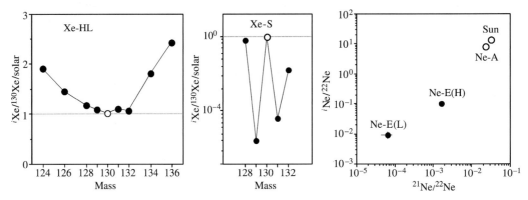

Figure 1 Exotic noble gas components present in presolar carbonaceous grains. Diamond is the carrier of Xe-HL, SiC the carrier of Xe-S and Ne-E(H), and graphite the carrier of Ne-E(L) (source Anders and Zinner, 1993).

(Bernatowicz et al., 1987; Tang and Anders, 1988b), and graphite, the carrier of Ne-E(L) (Amari et al., 1990).

Once isolated, SiC and graphite (for diamond, see below) were found to be anomalous in *all* their isotopic ratios, and it is this feature that identifies them as presolar grains. This distinguishes them from other materials in meteorites such as CAIs that also carry isotopic anomalies in some elements but, in contrast to bona fide stardust, formed in the solar system. They apparently inherited their anomalies from incompletely homogenized presolar material. Another distinguishing feature is that anomalies in presolar grains are up to several orders of magnitude larger than those in CAIs and match those expected for stellar atmospheres (Zinner, 1997).

1.02.3 TYPES OF PRESOLAR GRAINS

Table 1 shows the types of presolar grains identified as of early 2000s. It also lists the sizes, approximate abundances, and stellar sources. In addition to the three carbonaceous phases that were discovered because they carry exotic noble gas components (Figure 1) and which can be isolated from meteorites in almost pure form by chemical and physical processing, presolar oxide, silicon nitride (Si_3N_4), and silicates were identified by isotopic measurements in the ion microprobe and the number of such grains available for study is much smaller than for the carbonaceous phases. Most oxide grains are spinel ($MgAl_2O_4$) and corundum (Al_2O_3), but hibonite ($CaAl_{12}O_{19}$) and possibly titanium oxide have also been found (Hutcheon et al., 1994; Nittler et al., 1994; Nittler and Alexander, 1999; Choi et al., 1998; Zinner et al., 2003). While all these grains, as well as presolar Si_3N_4 (Nittler et al., 1995), were located by single grain analysis of acid residues, presolar silicates were discovered by isotopic imaging of chemically untreated interplanetary dust particles (IDPs) (Messenger et al., 2003).

Finally, titanium-, zirconium-, and molybdenum-rich carbides, cohenite (($Fe,Ni)_3C$), kamacite (Fe–Ni), and elemental iron were found as tiny subgrains inside of graphite spheres (Bernatowicz et al., 1991, 1996, 1999; Croat et al., 2003). While TiC inside of an SiC grain (Bernatowicz et al., 1992) could have formed by exsolution, there can be little doubt that interior grains in graphite must have formed prior to the condensation of the spherules.

1.02.4 ANALYSIS TECHNIQUES

Although the abundance of carbonaceous presolar grains in meteorites is low, once they are identified, almost pure samples can be prepared and studied in detail. Enough material of these phases can be obtained for "bulk" analysis, i.e., analysis of collections of large numbers of grains either by gas mass spectrometry (GMS) of carbon, nitrogen, and the noble gases (Lewis et al., 1994; Russell et al., 1996, 1997), by thermal ionization mass spectrometry (TIMS) of strontium, barium, neodymium, samarium, dysprosium (Ott and Begemann, 1990; Prombo et al., 1993; Richter et al., 1993, 1994; Podosek et al., 2003) or secondary ion mass spectrometry (SIMS) (Zinner et al., 1991; Amari et al., 2000). While only averages over many grains are obtained in this way, it allows the measurement of trace elements such as the noble gases and heavy elements that cannot be analyzed otherwise.

However, because presolar grains come from different stellar sources, information on individual stars is obtained by the study of single grains. This challenge has been successfully taken up by the application of a series of microanalytical techniques. For isotopic analysis, the ion microprobe has become the instrument of choice. While most SIMS measurements have been on grains 1 μm in size or larger, a new type of ion probe, the NanoSIMS, allows measurements of grains an order of magnitude smaller (e.g., Zinner et al., 2003). Ion probe analysis has led to the discovery of new types of presolar grains such as corundum (Hutcheon et al., 1994; Nittler et al., 1994) and silicon nitride (Nittler et al., 1995). It also has led to the identification of rare subpopulations of presolar dust such as SiC grains

Table 1 Types of presolar grains in primitive meteorites and IDPs.

Grain type	Noble-gas components	Size	Abundance	Stellar sources
Diamond	Xe-HL	2 nm	1,000 ppm	Supernovae?
Silicon carbide	Ne-E(H), Xe-S	0.1–20 μm	10 ppm	AGB, SNe, J-stars, novae
Graphite	Ne-E(L)	1–20 μm	1–2 ppm	SNe, AGB
Oxides		0.15–3 μm	1 ppm	RG, AGB, SNe
Silicon nitride		0.3–1 μm	~3 ppb	SNe, AGB
Ti-, Fe-, Zr-, Mo-carbides		10–200 nm		SNe
Kamacite, iron		~10–20 nm		SNe
Olivine		0.1–0.3 μm		

of type X (Amari *et al.*, 1992) and type Y (Hoppe *et al.*, 1994). Searches for presolar oxide grains and rare subpopulations of SiC profited from the application of isotopic imaging in the ion probe, which allows the rapid analysis of a large number of grains (Nittler *et al.*, 1997). Laser ablation and resonant ionization mass spectrometry (RIMS) (Savina *et al.*, 2003b) has been successfully applied to isotopic analysis of the heavy elements strontium, zirconium, molybdenum, and barium in individual SiC and graphite grains (Nicolussi *et al.*, 1997, 1998a,b,c; Savina *et al.*, 2003a). Single grain measurements of helium and neon have been made by laser heating and gas mass spectrometry (Nichols *et al.*, 2003).

The surface morphology of grains has been studied by secondary electron microscopy (SEM) (Hoppe *et al.*, 1995). Such studies have been especially useful for pristine SiC grains that have not been subjected to any chemical treatment (Bernatowicz *et al.*, 2003). Finally, the transmission electron microscope (TEM) played an important role in the discovery of presolar SiC (Bernatowicz *et al.*, 1987) and internal TiC and other subgrains in graphite (Bernatowicz *et al.*, 1991). It has also been successfully applied to the study of diamonds (Daulton *et al.*, 1996) and of polytypes of SiC (Daulton *et al.*, 2002, 2003).

1.02.5 ASTROPHYSICAL IMPLICATIONS OF THE STUDY OF PRESOLAR GRAINS

There are many stages in the long history of presolar grains from their stellar birth to their incorporation into primitive meteorites and, in principle, the study of the grains can provide information on all of them.

The isotopic composition of a given circumstellar grain reflects that of the stellar atmosphere from which the grain condensed. The atmosphere's composition in turn is determined by several factors: (i) by the galactic history of the material from which the star itself formed, (ii) by nucleosynthetic processes in the star's interior, and (iii) by mixing episodes in which newly synthesized material is dredged from the interior into the star's envelope. In supernovae, mixing of different layers with different nucleosynthetic history accompanies the explosion and the ejection of material. The isotopic compositions of grains provide information on these processes.

Grain formation occurs when temperatures in the expanding envelope of red giants (RGs) or in SN ejecta are low enough for the condensation of minerals. Many late-type stars are observed to be surrounded by dust shells of grains whose mineral compositions reflect the major chemistry of the gas (e.g., Little-Marenin, 1986). The study of morphological features of pristine grains, of internal grains, and of trace-element abundances can give information on the physical and chemical properties of stellar atmospheres (Bernatowicz *et al.*, 1996; Amari *et al.*, 1995a; Lodders and Fegley, 1998; Kashiv *et al.*, 2001, 2002; Croat *et al.*, 2003).

After their formation as circumstellar grains or as SN condensates, grains enter a long journey through the ISM. They should be distinguished from true IS grains that form in the ISM, e.g., in dense molecular clouds. Grains of stellar origin are most likely to be covered by mantles of IS cloud material. During their IS history, grains are subjected to a variety of destructive processes, such as evaporation in SN shocks and sputtering by shocks and stellar winds. They are also exposed to galactic cosmic rays that leave a record in the form of cosmogenic nuclides (Tang and Anders, 1988a; Ott and Begemann, 2000).

Grains might go in and out of IS clouds before some were finally incorporated into the dense molecular cloud from which our solar system formed. The final step in the complex history of stellar grains is the formation of planetesimals and of the parent bodies of the meteorites in which we find these presolar fossils. By far the largest fraction of the solids, even in primitive meteorites, formed in the solar system and the fraction of surviving presolar grains is small (see Table 1). Primitive meteorites experienced varying degrees of metamorphism on their parent bodies and these metamorphic processes affected different types of presolar grains in different ways. The abundance of different grain types can thus give information about conditions in the solar nebula and about parent-body processes (Huss and Lewis, 1995; Mendybaev *et al.*, 2002).

1.02.6 SILICON CARBIDE

Silicon carbide is the best-studied presolar grain type. It has been found in carbonaceous, unequilibrated ordinary, and enstatite chondrites with concentrations ranging up to ~10 ppm (Huss and Lewis, 1995). Most SiC grains are less than 0.5 μm in diameter. Murchison is an exception in that grain sizes are, on average, much larger than those in other meteorites (Amari *et al.*, 1994; Huss *et al.*, 1997; Russell *et al.*, 1997). This difference is still not understood but it, and the fact that plenty of Murchison is available, is the reason that by far most measurements have been made on Murchison SiC. Many SiC grains show euhedral crystal features (Figure 2) but there are large variations. Morphological studies by high-resolution SEM (Bernatowicz *et al.*, 2003) reveal detailed crystallographic features that give information about growth conditions. Such information is also obtained from TEM studies that show that only the cubic (3C) (~80%) and hexagonal (2H) polytypes are present, indicating low pressures and condensation temperatures in stellar outflows (Bernatowicz et al., 1987; Daulton et al., 2002, 2003). A preponderance of cubic SiC has been observed astronomically in carbon stars (Speck et al., 1999).

The availability of >1 μm SiC grains and relatively high concentrations of trace elements (Amari et al., 1995a) allow the isotopic analysis of the major and of many trace elements in individual grains. In addition to the major elements, carbon and silicon, isotopic data are available for the diagnostic (in terms of nucleosynthesis and stellar origin) elements nitrogen, magnesium, calcium, titanium, the noble gases, and the heavy refractory elements strontium, zirconium, molybdenum, barium, neodymium, samarium, and dysprosium. Carbon, nitrogen, and silicon isotopic as well as inferred $^{26}Al/^{27}Al$ ratios in a large number of individual grains (Figures 3–5) have led to the classification into different populations (Hoppe and Ott, 1997): mainstream grains (~93% of the total), and the minor subtypes A, B, X, Y, Z, and nova grains.

Most of presolar SiC is believed to have originated from carbon stars, late-type stars of low mass ($1-3M_\odot$) in the thermally pulsing (TP) asymptotic giant branch (AGB) phase of evolution (Iben and Renzini, 1983). Dust from such stars has been proposed already one decade prior to identification of SiC to be a minor constituent of primitive meteorites (Clayton and Ward, 1978; Srinivasan and Anders, 1978; Clayton, 1983a). Several pieces of evidence point to such an origin. Mainstream grains have $^{12}C/^{13}C$ ratios similar to those found in carbon stars (Figure 6), which are considered to be the most prolific injectors of carbonaceous dust grains into the ISM (Tielens, 1990). Many carbon stars show the 11.3 μm

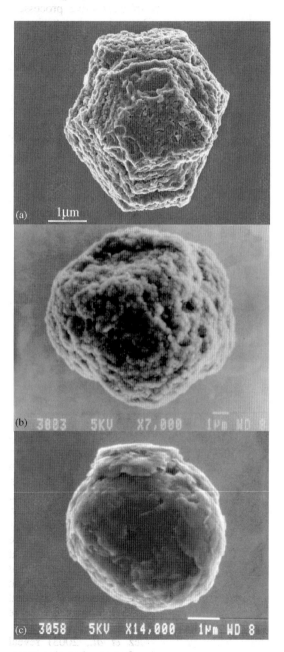

Figure 2 Secondary electron micrographs of (a) presolar SiC, (b) presolar graphite (cauliflower type), and (c) presolar graphite (onion type). Photographs courtesy of Sachiko Amari and Scott Messenger.

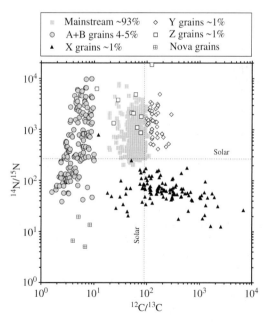

Figure 3 Nitrogen and carbon isotopic ratios of individual presolar SiC grains. Because rare grain types were located by automatic ion imaging, the number of grains of different types do not correspond to their abundances in the meteorites; these abundances are given in the legend (sources Alexander, 1993; Hoppe et al., 1994, 1996a; Nittler et al., 1995; Huss et al., 1997; Amari et al., 2001a,b,c).

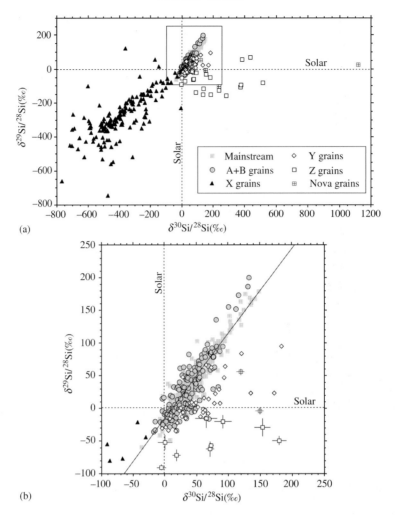

Figure 4 Si isotopic ratios of different types of presolar SiC grains plotted as δ-values, deviations in per mil (‰) from the solar ratios: $\delta^i\text{Si}/^{28}\text{Si} = [(^i\text{Si}/^{28}\text{Si})_{\text{meas}}/(^i\text{Si}/^{28}\text{Si})_{\odot-1}] \times 1{,}000$. Mainstream grains plot along a line of slope 1.4 (solid line). Symbols are the same as those in Figure 3. Sources same as in Figure 3.

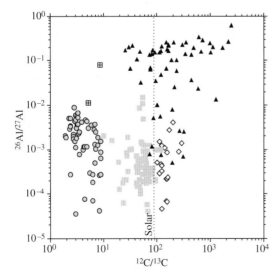

Figure 5 Aluminum and carbon isotopic ratios of individual presolar SiC grains. Symbols and sources for the data are the same as those for Figure 3.

emission feature typical of SiC (Treffers and Cohen, 1974; Speck et al., 1997). Finally, AGB stars are believed to be the main source of the s-process (slow neutron capture nucleosynthesis) elements (e.g., Busso et al., 2001), and the s-process isotopic patterns of the heavy elements exhibited by mainstream SiC provide the most convincing argument for their origin in carbon stars (see below).

1.02.6.1 Mainstream Grains

Mainstream grains have $^{12}\text{C}/^{13}\text{C}$ ratios between 10 and 100 (Figure 2). They have carbon and nitrogen isotopic compositions (Zinner et al., 1989; Stone et al., 1991; Virag et al., 1992; Alexander, 1993; Hoppe et al., 1994, 1996a; Nittler et al., 1995; Huss et al., 1997; Amari et al., 2002; Nittler and Alexander, 2003) that are roughly in agreement with an AGB origin.

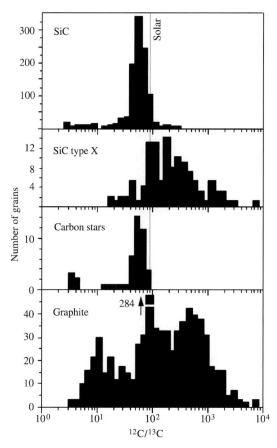

Figure 6 The distributions of carbon isotopic ratios measured in presolar SiC (Hoppe et al., 1994; Nittler et al., 1995) and graphite grains (Hoppe et al., 1995) from the Murchison meteorite are compared to astronomical measurements of the atmospheres of carbon stars (Lambert et al., 1986).

Carbon-13 and ^{15}N excesses relative to solar are the signature of hydrogen burning via the CNO cycle that occurred during the main sequence phase of the stars. This material is brought to the star's surface by the first (and second) dredge-up. The carbon isotopic ratios are also affected by shell helium burning and the third dredge-up during the TP-AGB phase (Busso et al., 1999). This process adds ^{12}C to the envelope, increases the ^{12}C/^{13}C ratio from the low values resulting from the first dredge-up, and, by making C > O, causes the star to become a carbon star.

Envelope ^{12}C/^{13}C ratios predicted by canonical stellar evolution models range from ~20 after first dredge-up in the RG phase to ~300 in the late TP-AGB phases (El Eid, 1994; Gallino et al., 1994; Amari et al., 2001b). Predicted ^{14}N/^{15}N ratios are 600–1,600 (Becker and Iben, 1979; El Eid, 1994), falling short of the range observed in the grains. However, the assumption of deep mixing ("cool bottom processing") of envelope material to deep hot regions in $M < 2.5 M_\odot$ stars during their RG and AGB phases (Charbonnel, 1995; Wasserburg et al., 1995; Langer et al., 1999; Nollett et al., 2003) results in partial hydrogen burning, with higher ^{14}N/^{15}N and lower ^{12}C/^{13}C ratios in the envelope than in canonical models (see also Huss et al., 1997).

Two other isotopes that are a signature for AGB stars are ^{26}Al and ^{22}Ne. Figure 5 shows inferred ^{26}Al/^{27}Al ratios in different types of SiC grains. The existence of the short-lived radioisotope ^{26}Al ($T_{1/2} = 7.3 \times 10^5$ yr) is inferred from large ^{26}Mg excesses. Aluminiun-26 is produced in the hydrogen shell by proton capture on ^{25}Mg and mixed to the surface by the third dredge-up (Forestini et al., 1991). It can also be produced during "hot bottom burning" (Lattanzio et al., 1997), but this process is believed to prevent carbon-star formation (Frost and Lattanzio, 1996). Neon-22, the main component in Ne-E, is produced in the helium shell by ^{14}N + 2α. The neon isotopic ratios measured in SiC bulk samples (Lewis et al., 1990, 1994) are very close to those expected for helium shell material (Gallino et al., 1990). In contrast to krypton and xenon and heavy refractory elements, neon as well as helium and argon show very little dilution of helium shell material with envelope material, indicating a special implantation mechanism by an ionized wind. Another piece of evidence that the Ne-E(H) component originated from the helium shell of AGB stars and not from the decay of ^{22}Na (Clayton, 1975) is the fact that in individual grains, of which only ~5% carry ^{22}Ne, it is always accompanied by ^{4}He (Nichols et al., 2003). Excesses in ^{21}Ne in SiC relative to the predicted helium-shell composition have been interpreted as being due to spallation by galactic cosmic rays (Tang and Anders, 1988a; Lewis et al., 1990, 1994), which allows the determination of grain lifetimes in the IS medium. Inferred exposure ages depend on grain size and range from 10 Myr to 130 Myr (Lewis et al., 1994). However, this interpretation has recently been challenged (Ott and Begemann, 2000) and the question of IS ages of SiC is not settled.

The silicon isotopic compositions of most mainstream grains are characterized by enrichments in the heavy silicon isotopes of up to 200‰ relative to their solar abundances (Figure 4). In a silicon three-isotope plot the data fall along a line with slope 1.4, which is shifted slightly to the right of the solar system composition. In contrast to the light elements carbon, nitrogen, neon, and aluminum and the heavy elements (see below), the silicon isotopic ratios of mainstream grains cannot be explained by nuclear processes taking place within their parent stars. In AGB stars the silicon isotopes are affected by neutron capture in the helium shell leading to excesses in ^{29}Si and ^{30}Si along a slope 0.2–0.5

line in a δ-value silicon three-isotope plot (Gallino et al., 1990, 1994; Brown and Clayton, 1992; Lugaro et al., 1999; Amari et al., 2001b). Predicted excesses are only on the order of 20‰ in low-mass AGB stars of close-to-solar metallicity (metallicity is the abundance of all elements heavier than helium). This led to the proposal that many stars with varying initial silicon isotopic compositions contributed SiC grains to the solar system (Clayton et al., 1991; Alexander, 1993) and that neutron-capture nucleosynthesis in these stars only plays a secondary role in modifying these compositions. Several explanations have been given for the initial silicon ratios in the parent stars, which in their late stages of evolution became the carbon stars that produced the SiC. One is the evolution of the silicon isotopic ratios through galactic history as different generations of supernovae produced silicon with increasing ratios of the secondary isotopes ^{29}Si and ^{30}Si to the primary ^{28}Si (Gallino et al., 1994; Timmes and Clayton, 1996; Clayton and Timmes, 1997a,b). Clayton (1997) addressed the problem that most SiC grains have higher than solar ^{29}Si/^{28}Si and ^{30}Si/^{28}Si ratios by considering the possibility that the mainstream grains originated from stars that were born in central, more metal-rich regions of the galaxy and moved to the molecular cloud from which our sun formed. Alexander and Nittler (1999), alternatively, suggested that the Sun has an atypical silicon isotopic composition. Lugaro et al. (1999) explained the spread in the isotopic compositions of the parent stars by local heterogeneities in the galaxy caused by the stochastic nature of the admixture of the ejecta from supernovae of varying type and mass.

Titanium isotopic ratios in single grains (Ireland et al., 1991; Hoppe et al., 1994; Alexander and Nittler, 1999) and in bulk samples (Amari et al., 2000) show excesses in all isotopes relative to ^{48}Ti, a result expected of neutron capture in AGB stars. However, as for silicon, theoretical models (Lugaro et al., 1999) cannot explain the range of ratios observed in single grains. Furthermore, titanium ratios are correlated with those of silicon, also indicating that the titanium isotopic compositions are dominated by galactic evolution effects (Alexander and Nittler, 1999). Excesses of ^{42}Ca and ^{43}Ca relative to ^{40}Ca observed in bulk samples (Amari et al., 2000) agree with predictions for neutron capture. Large ^{44}Ca excesses are apparently due to the presence of type X grains (see below). Iron isotopic ratios have been measured by RIMS in single grains (Davis et al., 2002; Tripa et al., 2002). Depletions in ^{54}Fe are much larger than predicted by neutron capture in AGB stars and ^{57}Fe does not show the predicted excesses. While it is quite possible that, as for silicon and titanium, galactic evolution effects dominate and while admixture of SN ejecta results in the observed ^{54}Fe depletions, the SN mixing model also predicts substantial ^{57}Fe excesses, which are not observed (see also Clayton et al., 2002).

All heavy elements measured so far show the signature of the s-process (Figure 7, see also Figure 10). They include the noble gases krypton and xenon (Lewis et al., 1990, 1994) but also the heavy elements strontium (Podosek et al., 2003), barium (Ott and Begemann, 1990; Zinner et al., 1991; Prombo et al., 1993), neodymium and samarium (Zinner et al., 1991; Richter et al., 1993), and dysprosium (Richter et al., 1994). Although most measurements were made on bulk samples, it is clear that mainstream grains dominate. Single-grain measurements of strontium (Nicolussi et al., 1998b), zirconium (Nicolussi et al., 1997), molybdenum (Nicolussi et al., 1998a), and barium (Savina et al., 2003a) have been made with RIMS. Large enrichments of certain heavy elements such as yttrium, zirconium, barium, and cerium in single mainstream grains also indicate large overabundances of s-process elements in the parent stars (Amari et al., 1995a). For all the isotopic compositions of the elements listed above except for dysprosium there is good agreement with theoretical models of the s-process in low-mass AGB stars (Gallino et al., 1993, 1997; Lugaro et al., 2003). Discrepancies with earlier model calculations were caused by incorrect nuclear cross-sections and could be resolved by improved experimental determinations (e.g., Guber et al., 1997; Wisshak et al., 1997; Koehler et al., 1998).

The s-process isotopic patterns observed in grains allow the determination of different parameters affecting the s-process such as neutron exposure, temperature, and neutron density (Hoppe and Ott, 1997). Since these parameters depend in turn on stellar mass and metallicity as well as on the neutron source operating in AGB stars, they allow information to be obtained about the parent stars of the grains. For example, the barium isotopic ratios indicate a neutron exposure that is only half of that inferred for the solar system (Ott and Begemann, 1990; Gallino et al., 1993). Another example is provided by the abundance of ^{96}Zr in single grains, which is sensitive to neutron density because of the relatively short half-life of ^{95}Zr (~64 d). While the ^{13}C(α,n) source with its low neutron density destroys ^{96}Zr, activation of the ^{22}Ne(α,n) source during later thermal pulses in AGB stars restores some of this isotope, whose abundance thus varies with pulse number. Some grains have essentially no ^{96}Zr, indicating that the ^{22}Ne(α,n) source was weak in their parent stars, pointing to low-mass AGB stars as the source of mainstream grains (Lugaro et al., 2003).

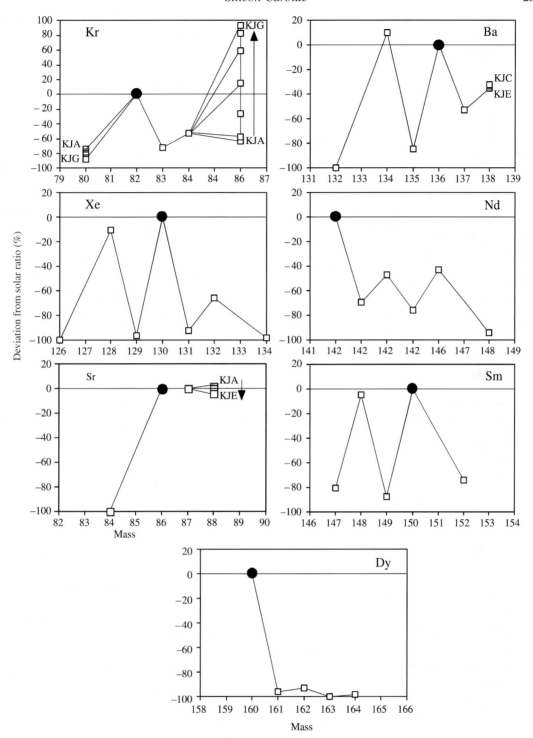

Figure 7 Isotopic patterns measured in bulk samples of SiC extracted from the Murchison meteorite. Isotopic ratios are relative to the reference isotope plotted as a solid circle and are normalized to the solar isotopic ratios. Data are from Lewis *et al.* (1994) (Kr and Xe), Podosek *et al.* (2003) (Sr), Prombo *et al.* (1993) (Ba), Richter *et al.* (1993) (Nd and Sm), and Richter *et al.* (1994) (Dy).

1.02.6.2 Type Y and Z Grains

Type Y grains have $^{12}C/^{13}C > 100$ and silicon isotopic compositions that lie to the right of the mainstream correlation line (Figures 3 and 4(b)) (Hoppe *et al.*, 1994; Amari *et al.*, 2001b). Type Z grains have even larger ^{30}Si excesses relative to ^{29}Si and, on average, lower $\delta^{29}Si$ values than Y grains. However, they are distinguished from Y grains by having $^{12}C/^{13}C < 100$

(Alexander, 1993; Hoppe et al., 1997). Comparison of the carbon, silicon, and titanium isotopic ratios of Y grains with models of nucleosynthesis indicates an origin in low-to-intermediate-mass AGB stars with approximately half the solar metallicity (Amari et al., 2001b). Such stars dredge up more ^{12}C, and silicon and titanium that experienced neutron capture, from the helium shell (see also Lugaro et al., 1999). According to their silicon isotopic ratios, Z grains came from low-mass stars of even lower (~one-third solar) metallicity (Hoppe et al., 1997). In order to achieve the relatively low ^{12}C/^{13}C ratios of these grains, the parent stars must have experienced cool bottom processing (Wasserburg et al., 1995; Nollett et al., 2003) during their RG and AGB phase. From the theoretically inferred metallicities and average silicon isotopic ratios of mainstream Y and Z grains, Zinner et al. (2001) derived the galactic evolution of the silicon isotopic ratios as a function of metallicity. This evolution differs from the results of galactic evolution models based on the yields of supernovae (Timmes and Clayton, 1996) and has important implications concerning the relative contributions from type II and type Ia supernovae during the history of our galaxy.

1.02.6.3 Type A + B Grains

Grains of type A + B have ^{12}C/^{13}C < 10, but their silicon isotopic ratios plot along the mainstream line (Figures 3 and 4). In contrast to mainstream grains, many A + B grains have lower than solar ^{14}N/^{15}N ratios (Hoppe et al., 1995, 1996a; Huss et al., 1997; Amari et al., 2001c). While the isotopic ratios of mainstream, Y and Z grains find an explanation in nucleosynthetic models of AGB stars, a satisfactory explanation of the data in terms of stellar nucleosynthesis is more elusive for the A + B grains. The low ^{12}C/^{13}C ratios of these grains combined with the requirement for a carbon-rich environment during their formation indicate helium burning followed by limited hydrogen burning in their stellar sources. However, the astrophysical sites for this process are not well known. There might be two different kinds of A + B grains with corresponding different stellar sources (Amari et al., 2001c). Grains with no s-process enhancements (Amari et al., 1995a; Pellin et al., 2000b; Savina et al., 2003c) probably come from J-type carbon stars that also have low ^{12}C/^{13}C ratios (Lambert et al., 1986). Unfortunately, J stars are not well understood and there are no astronomical observations of nitrogen isotopic ratios in such stars. Furthermore, the low ^{14}N/^{15}N ratios observed in some of the grains as well as the carbon-rich nature of their parent stars appear to be incompatible with the consequences of hydrogen burning in the CNO cycle, which seems to be responsible for the low ^{12}C/^{13}C ratios of J stars and the grains. A + B grains with s-process enhancements might come from post-AGB stars that undergo a very late thermal pulse. An example of such a star is Sakurai's object (e.g., Asplund et al., 1999; Herwig, 2001). However, grains with low ^{14}N/^{15}N ratios pose a problem. Huss et al. (1997) proposed that the currently used ^{18}O(p,α)^{15}N reaction rate is too low by a factor of 1,000. This would result in low ^{12}C/^{13}C and ^{14}N/^{15}N ratios if an appropriate level of cool bottom processing is considered.

1.02.6.4 Type X Grains

Although SiC grains of type X account for only 1% of presolar SiC, a fairly large number can be located by ion imaging (Nittler et al., 1997; Hoppe et al., 1996b, 2000; Lin et al., 2002; Besmehn and Hoppe, 2003). X grains are characterized by mostly ^{12}C and ^{15}N excesses relative to solar (Figures 3 and 6), excesses in ^{28}Si (Figure 4) and very large ^{26}Al/^{27}Al ratios, ranging up to 0.6 (Figure 5). About 10–20% of the grains show large ^{44}Ca excesses, which must come from the decay of short-lived ^{44}Ti ($T_{1/2}$ = 60 yr) (Amari et al., 1992; Hoppe et al., 1996b, 2000; Nittler et al., 1996; Besmehn and Hoppe, 2003). Inferred ^{44}Ti/^{48}Ti ratios range up to 0.6 (Figure 8). In contrast to presolar graphite, which contains subgrains of TiC, titanium in SiC seems to occur in solid solution and radiogenic ^{44}Ca is uniformly distributed in most of the grains. Only in one X grains a pronounced heterogeneity points to a titanium-rich subgrain (Besmehn and Hoppe, 2003). Because ^{44}Ti can only be produced in SN explosions (Timmes et al., 1996), grains with evidence for ^{44}Ti, and by implications all X grains, must have an SN origin. In type II supernovae ^{44}Ti is produced in the nickel- and silicon-rich inner zones (see Figure 9) (Woosley and Weaver, 1995; Timmes et al., 1996). Silicon in the Si/S zone consists of almost pure ^{28}Si. Also the other isotopic signatures of X grains are compatible with an origin in type II supernovae: high ^{12}C/^{13}C and low ^{14}N/^{15}N ratios are the signature of helium burning (Figure 9) and high ^{26}Al/^{27}Al ratios can be reached in the He/N zone by hydrogen burning.

However, these isotopic signatures occur in massive stars in very different stellar zones, which experienced different stages of nuclear burning before the SN explosion (Figure 9) (e.g., Woosley and Weaver, 1995; Rauscher et al., 2002). The isotopic signatures of the X grains suggest deep and inhomogeneous mixing of matter from these different zones in the SN ejecta. While the titanium and silicon isotopic signature of the X

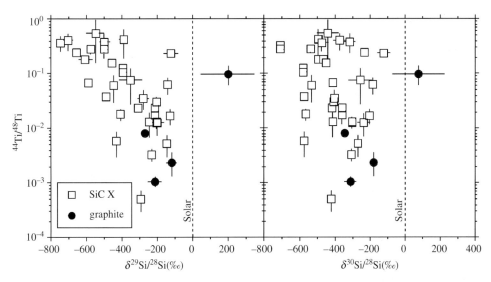

Figure 8 $^{44}Ti/^{48}Ti$ ratios inferred from ^{44}Ca excesses in SiC grains of type X and graphite grains are plotted against Si isotopic ratios. Except for one graphite, all grains with evidence for ^{44}Ti have ^{28}Si excesses (sources Amari et al., 1992, unpublished; Hoppe et al., 1994, 1996b, 2000; Nittler et al., 1996; Besmehn and Hoppe, 2003).

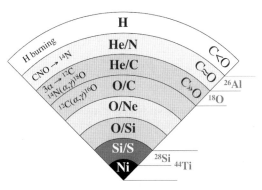

Figure 9 Schematic structure of a massive star before its explosion as a type II supernova (source Woosley and Weaver, 1995). Such a star consists of different layers, which are labeled according to their most abundant elements that experienced different stages of nucleosynthesis. Indicated are dominant nuclear reactions in some layers and the layers in which isotopes abundant in grains of an inferred SN origin are produced.

grains requires contributions from the Ni, O/Si and Si/S zones, which experienced silicon-, neon-, and oxygen-burning, significant contributions must also come from the He/N and He/C zones that experienced hydrogen and incomplete helium burning in order to achieve C > O, the condition for SiC condensation (Larimer and Bartholomay, 1979; Lodders and Fegley, 1997). Furthermore, addition of material from the intermediate oxygen-rich layers must be severely limited. Astronomical observations indicate extensive mixing of SN ejecta (e.g., Ebisuzaki and Shibazaki, 1988; Hughes et al., 2000) and hydrodynamic models of SN explosions predict mixing in the ejecta initiated by the formation of Rayleigh–Taylor instabilities (e.g., Herant et al., 1994). However, it still has to be seen whether mixing can occur on a microscopic scale and whether these instabilities allow mixing of matter from nonneighboring zones while excluding large contributions from the intermediate oxygen-rich zones. Clayton et al. (1999) and Deneault et al. (2003) suggested condensation of carbonaceous phases in type II SN ejecta even while C < O because of the destruction of CO in the high-radiation environment of the ejecta. While this might work for graphite, it is doubtful whether SiC can condense from a gas with C < O (Ebel and Grossman, 2001). Even for graphite, the presence of subgrains of elemental iron inside of graphite grains whose isotopic signatures indicate an SN origin argues against formation in an oxygen-rich environment (Croat et al., 2003).

Although multizone mixing models can qualitatively reproduce the isotopic signatures of X grains, several ratios, in particular the large ^{15}N excesses and excesses of ^{29}Si over ^{30}Si found in most grains, cannot be explained quantitatively and indicate deficiencies in the existing models. The latter is a long-standing problem: SN models cannot account for the solar $^{29}Si/^{30}Si$ ratio (Timmes and Clayton, 1996). Studies of SiC X grains isolated from the Qingzhen enstatite chondrite (Lin et al., 2002) suggest that there are two population of X grains with different trends in the silicon isotopic ratios, the minor population having lower-than-solar $^{29}Si/^{30}Si$ ratios. Recently, Clayton et al. (2002) and Deneault et al. (2003) have tried to account for isotopic signatures from different SN zones by considering implantation into newly condensed grains as they pass through different regions of the ejecta, specifically through zones with reverse shocks.

Some SiC X grains also show large excesses in ^{49}Ti (Amari et al., 1992; Nittler et al., 1996; Hoppe and Besmehn, 2002). The correlation of these excesses with the V/Ti ratio (Hoppe and Besmehn, 2002) indicates that they come from the decay of short-lived ^{49}V ($T_{1/2} = 330$ d) and that the grains must have formed within a few months of the explosion. ^{49}V is produced in the Si/S zone, which contains almost pure ^{28}Si. RIMS isotopic measurements of iron, strontium, zirconium, molybdenum, and barium have been made on X grains (Pellin et al., 1999, 2000a; Davis et al., 2002). The most complete and interesting are the molybdenum measurements, which reveal large excesses in ^{95}Mo and ^{97}Mo. Figure 10 shows the molybdenum isotopic patterns of a mainstream and an X grain. The mainstream grain has a typical s-process pattern, in agreement with bulk measurements of other heavy elements such as xenon, barium, and neodymium (Figure 7). The molybdenum pattern of the X grain is completely different and indicates neutron capture at much higher neutron densities. While it does not agree with the pattern expected for the r-process, it is successfully explained by a neutron-burst model (Meyer et al., 2000). In the type II SN models by Rauscher et al. (2002) an intense neutron burst is predicted to occur in the oxygen layer just below the He/C zone, accounting for the molybdenum isotopic patterns observed in X grains.

Type Ia supernovae offer an alternative explanation for the isotopic signature of X grains. In the model by Clayton et al. (1997) nucleosynthesis takes place by explosive helium burning of a helium cap on top of a white dwarf. This process produces most of the isotopic signatures of the SN grains. The isotopes ^{12}C, ^{15}N, ^{26}Al, ^{28}Si, and ^{44}Ti are all made by helium burning during the explosion, which makes the transport of ^{28}Si and ^{44}Ti through the massive oxygen-rich zone into the overlying carbon-rich zones of a type II SN unnecessary. Mixing is limited to material from helium burning and to matter that experienced CNO processing. The best match with the X grain data, however, is achieved for mixing scenarios that yield O > C (Amari et al., 1998). Other problems include the questions whether high enough gas densities can be achieved in the ejecta for the condensation of micrometer-sized grains and whether type Ia supernovae can generate a neutron burst necessary for the molybdenum isotopic pattern. More work is needed to decide whether a type Ia SN origin for X grains is a realistic alternative.

1.02.6.5 Nova Grains

A few grains have isotopic ratios that are best explained by a nova origin (Amari et al., 2001a). These grains have low ^{12}C/^{13}C and ^{14}N/^{15}N ratios (Figure 3), large ^{30}Si excesses (Figure 4), and high ^{26}Al/^{27}Al ratios (Figure 5). All these features are predicted to be produced by explosive hydrogen burning taking place in classical novae (e.g., Kovetz and Prialnik, 1997; Starrfield et al., 1998; José et al., 1999) but the predicted anomalies are much larger than those found in the grains, and the nova ejecta have to be mixed with material of close-to-solar isotopic compositions. A comparison of the data with the models implicates ONe novae with a white dwarf mass of at least 1.25 $M_\odot$ as the most likely sources (Amari et al., 2001a).

1.02.6.6 Grain Size Effect

Grain size distributions of SiC have been determined for several meteorites and while grain sizes vary from 0.1 μm to 20 μm, the distributions are different for different meteorites. Murchison appears to have, on average, the largest grains (Amari et al., 1994), while SiC from Indarch (Russell et al., 1997) and Orgueil (Huss et al., 1997) is much finer grained. Various isotopic and other properties vary with grain size. Both the ^{22}Ne-E(H)/^{130}Xe-S and the ^{86}Kr/^{82}Kr ratios increase with grain size (Lewis et al., 1994)

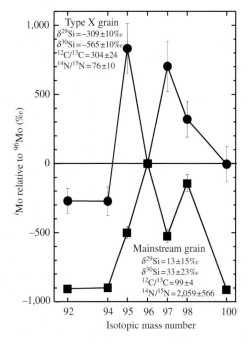

Figure 10 Molybdenum isotopic patterns measured by RIMS in a type X and a mainstream SiC grain (source Pellin et al., 1999).

and the first ratio has been used as a measure for the average grain size in meteorites for which no detailed size distributions have been determined (Russell et al., 1997). The $^{86}Kr/^{82}Kr$ ratio is a function of neutron exposure and the data indicate that exposure decreases with increasing grains size. The $^{88}Sr/^{86}Sr$ and $^{138}Ba/^{136}Ba$ ratios also depend on grain size but the dependence of neutron exposure on grain size inferred from these isotopic ratios is just the opposite of that inferred from the $^{86}Kr/^{82}Kr$ ratio. This puzzle has not been resolved. A possible explanation is a different trapping mechanism for noble gases and refractory elements, respectively (Zinner et al., 1991), or different populations of carrier grains if, as for neon (Nichols et al., 2003), only a small fraction of the grains carry krypton. Excesses in ^{21}Ne relative to the predicted helium-shell composition, interpreted as being due to spallation by galactic cosmic rays, increase with grain size (Tang and Anders, 1988a; Lewis et al., 1990, 1994). However, the correlation of the $^{21}Ne/^{22}Ne$ ratio with the s-process $^{86}Kr/^{82}Kr$ ratio (Hoppe and Ott, 1997) and the recent determination of spallation recoil ranges (Ott and Begemann, 2000) cast doubt on a chronological interpretation. Other grain size effects are, on average, larger $^{14}N/^{15}N$ ratios for smaller grains (Hoppe et al., 1996a) and an increasing abundance of Z grains among smaller SiC grains (Hoppe et al., 1996a, 1997). There are also differences in the distribution of different grain types in SiC from different meteorites: whereas the abundance of X grains in SiC from Murchison and other carbonaceous chondrites is ~1%, it is only ~0.1% in SiC from the enstatite chondrites Indarch and Qingzhen (Besmehn and Hoppe, 2001; Lin et al., 2002).

1.02.7 SILICON NITRIDE

Presolar silicon nitride (Si_3N_4) grains are extremely rare (in Murchison SiC-rich separates ~5% of SiC of type X), but automatic ion imaging has been successfully used to detect those with large ^{28}Si excesses (Nittler et al., 1995; Besmehn and Hoppe, 2001; Lin et al., 2002; Nittler and Alexander, 2003). The carbon, nitrogen, aluminum, and silicon isotopic signatures of these grains are the same as those of SiC grains of type X, i.e., large ^{15}N and ^{28}Si excesses and high $^{26}Al/^{27}Al$ ratios (Figure 12). Although so far no resolvable ^{44}Ca excesses have been detected (Besmehn and Hoppe, 2001), the similarity with X grains implies an SN origin for these grains. While Si_3N_4 grains in SiC-rich residues from Murchison are extremely rare and, if present, are of type X, enstatite chondrites contain much higher abundances of Si_3N_4 (Alexander et al., 1994; Besmehn and Hoppe, 2001; Amari et al., 2002).

Most of them have normal isotopic compositions. Recent measurements of small (0.25–0.65 μm) grains from Indarch revealed several Si_3N_4 grains with carbon and nitrogen isotopic ratios similar to those of mainstream SiC grains, but contamination from attached SiC grains cannot be excluded (Amari et al., 2002).

1.02.8 GRAPHITE

Graphite, the third type of carbonaceous presolar grains, was isolated because it is the carrier of Ne-E(L) (Amari et al., 1990; Amari et al., 1995b). Subsequent isotopic measurements of individual grains revealed anomalies in many different elements.

1.02.8.1 Physical Properties

Only grains ≥1 μm in diameter carry Ne-E(L) and only round grains, which range up to 20 μm in size, appear to be of presolar origin (Amari et al., 1990; Zinner et al., 1995). Presolar graphite has a range in density (1.6–2.2 g cm^{-3}) and four different density fractions have been isolated (Amari et al., 1994). Average grain sizes decrease with increasing density, and density fractions differ in the distribution of their carbon and noble gas isotopic compositions (Amari et al., 1995b; Hoppe et al., 1995). SEM studies revealed two basic morphologies (Hoppe et al., 1995): dense aggregates of small scales ("cauliflowers," Figure 2(b)) and grains with smooth or shell-like platy surfaces ("onions," Figure 2(c)). TEM analysis of microtomed sections of graphite spherules (Bernatowicz et al., 1991, 1996) found the surface morphology reflected in the internal structure of the grains. Cauliflowers consist of concentrically packed scales of poorly crystallized carbon, whereas onions either consist of well-crystallized graphite throughout or of a core of tightly packed graphene sheets of only several atomic layers surrounded by a mantle of well-crystallized graphite. Most graphite spherules contain small (20–500 nm) internal grains of mostly titanium carbide (TiC) (Bernatowicz et al., 1991); however, also zirconium- and molybdenum-rich carbides have been found (Bernatowicz et al., 1996). Recent studies of graphite spherules whose oxygen and silicon isotopic compositions indicated an SN origin (see below) did not detect Zr–Mo-rich carbides but revealed internal kamacite, cohenite, and iron grains in addition to TiC (Bernatowicz et al., 1999; Croat et al., 2003). Both cauliflowers and onions contain internal grains, which must have condensed before the graphite, and were apparently captured and included by the growing

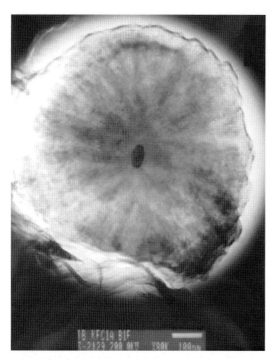

Figure 11 Transmission electron micrograph of a slice through a presolar graphite grain (onion). The grain in the center is TiC and apparently acted as condensation nucleus for the growth of the graphite spherule. Photo courtesy of Thomas Bernatowicz.

spherules. Some onions show TiC grains at their center that apparently acted as condensation nuclei for the graphite (Figure 11). Sizes of internal grains and graphite spherules and their relationship and chemical compositions provide information about physical properties such as pressure, temperature, and C/O ratio in the gas from which the grains condensed (Bernatowicz et al., 1996; Croat et al., 2003).

1.02.8.2 Isotopic Compositions

Noble gas measurements were made on bulk samples of four density fractions (Amari et al., 1995b). In contrast to SiC, a substantial fraction of Ne-E in graphite seems to come from the decay of short-lived ($T_{1/2} = 2.6$ yr) ^{22}Na (Clayton, 1975). This is supported by the low ^{4}He/^{22}Ne ratios measured in individual grains (Nichols et al., 2003). Krypton in graphite has two s-process components with apparent different neutron exposures, residing in different density fractions (Amari et al., 1995b).

Ion microprobe analyses of single grains revealed the same range of ^{12}C/^{13}C ratios as in SiC grains but the distribution is quite different (Figure 6). Most anomalous grains have ^{12}C excesses, similar to SiC X grains. A substantial fraction has low ^{12}C/^{13}C ratios like SiC A + B grains. Most graphite grains have close-to-solar nitrogen isotopic ratios (Hoppe et al., 1995; Zinner et al., 1995). In view of the enormous range in carbon isotopic ratios these normal nitrogen ratios cannot be intrinsic and most likely are the result of isotopic equilibration, either on the meteorite parent body or in the laboratory. Apparently, elements such as nitrogen are much more mobile in graphite than in SiC. An exception are graphite grains of low density (≤ 2.05 g cm^{-3}), which have anomalous nitrogen (Figure 12). Low-density (LD) graphite grains have in general higher trace-element concentrations than those with higher densities and for this reason have been studied for their isotopic compositions in detail (Travaglio et al., 1999). Those with nitrogen anomalies have ^{15}N excesses (Figure 12). Many LD grains have large ^{18}O excesses (Amari et al., 1995c) and high ^{26}Al/^{27}Al ratios that almost reach those of SiC X grains (Figure 12) and are much higher than those of mainstream SiC grains (Figure 5). ^{18}O excesses are correlated with ^{12}C/^{13}C ratios. Many grains for which silicon isotopic ratios could be determined with sufficient precision show ^{28}Si excesses, although large ^{29}Si and ^{30}Si excesses are also seen. The similarities of the isotopic signatures with those of SiC X point to an SN origin of LD graphite grains. The ^{18}O excesses are compatible with such an origin. Helium burning produces ^{18}O from ^{14}N, which dominates the CNO isotopes in material that had undergone hydrogen burning via the CNO cycle. As a consequence, the H/C zone in pre-SNII massive stars (see Figure 9), which experienced partial helium burning, has a high ^{18}O abundance (Woosley and Weaver, 1995). Wolf-Rayet stars during the WN–WC transitions are predicted to also show ^{12}C, ^{15}N, and ^{18}O excesses and high ^{26}Al/^{27}Al ratios (Arnould et al., 1997) but also large excesses in ^{29}Si and ^{30}Si and are therefore excluded for LD graphite grains with ^{28}Si excesses.

There are additional features that indicate an SN origin of LD graphite grains. A few grains show evidence for ^{44}Ti (Nittler et al., 1996), others have large excesses of ^{41}K, which must be due to the decay of the radioisotope ^{41}Ca ($T_{1/2} = 1.05 \times 10^{5}$ yr) (Amari et al., 1996). Inferred ^{41}Ca/^{40}Ca ratios are much higher (0.001–0.01) than those predicted for the envelopes of AGB stars (Wasserburg et al., 1994) but are in the range expected for the carbon- and oxygen-rich zones of type II supernovae, where neutron capture leads to the production of ^{41}Ca (Woosley and Weaver, 1995). Measurements of calcium isotopic ratios in grains without evidence for ^{44}Ti show excesses in ^{42}Ca, ^{43}Ca, and ^{44}Ca, with ^{43}Ca having the largest excess (Amari et al., 1996; Travaglio et al., 1999). This pattern is best

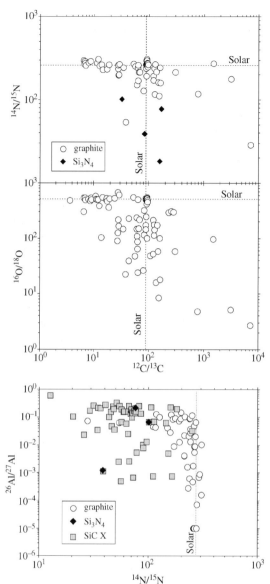

Figure 12 Nitrogen, oxygen, carbon, and aluminum isotopic ratios measured in individual low-density graphite grains. Also shown are data for presolar Si$_3$N$_4$ and SiC grains of type X (source Zinner, 1998a).

explained by neutron capture in the He/C and O/C zones (Figure 9) of type II supernovae. In cases where titanium isotopic ratios have been measured (Amari et al., 1996; Nittler et al., 1996; Travaglio et al., 1999), they show large excesses in ^{49}Ti and smaller ones in ^{50}Ti. This pattern also indicates neutron capture and is well matched by predictions for the He/C zone (Amari et al., 1996). However, large ^{49}Ti excesses in grains with relatively low (10–100) ^{12}C/^{13}C ratios can only be explained if contributions from the decay of ^{49}V are considered (Travaglio et al., 1999). Nicolussi et al. (1998c) have reported RIMS measurements of zirconium and molybdenum isotopic ratios in graphite grains from the highest density fraction (2.15–2.20 g cm^{-3}), but no other isotopic ratios had been measured. Several grains show s-process patterns for zirconium and molybdenum, similar to those exhibited by mainstream SiC grains, although two grains with a distinct s-process pattern for zirconium have normal molybdenum. Two grains have extreme ^{96}Zr excesses, indicating an SN origin, but the molybdenum isotopes in one are almost normal. Molybdenum, like nitrogen, might have suffered isotopic equilibration in graphite. High-density graphite grains apparently come from AGB stars as previously indicated by the krypton data (Amari et al., 1995b) and from supernovae. It remains to be seen whether also LD grains have multiple stellar sources.

In order to obtain better constraints on theoretical models of SN nucleosynthesis, Travaglio et al. (1999) tried to match the isotopic compositions of LD graphite grains by performing mixing calculations of different type II SN layers (Woosley and Weaver, 1995). While the results reproduce the principal isotopic signatures of the grains, there remain several problems. The models do not produce enough ^{15}N and yield too low ^{29}Si/^{30}Si ratios. The models also cannot explain the magnitude of ^{26}Al/^{27}Al, especially if SiC X grains are also considered, and give the wrong sign in the correlation of this ratio with the ^{14}N/^{15}N ratio. Furthermore, large neutron-capture effects observed in calcium and titanium can be only achieved in a mix with O > C. Clayton et al. (1999) proposed a kinetic condensation model that allows formation of graphite in the high-radiation environment of SN ejecta even when O > C, which relaxes the chemical constraint on mixing. However, it remains to be seen whether SiC and Si$_3$N$_4$ can also form under oxidizing conditions. Additional information about the formation environment of presolar graphite is, in principle, provided by the presence of indigenous polycyclic aromatic hydrocarbons (PAHs) (Messenger et al., 1998). PAHs with anomalous carbon ratios show different mass envelopes, which indicate different formation conditions.

A few graphite grains appear to come from novae. Laser extraction GMS of single grains show that, like SiC grains, only a small fraction contains evidence for Ne-E. Two of these grains have ^{20}Ne/^{22}Ne ratios that are lower than ratios predicted to result from helium burning in any known stellar sources, implying decay of ^{22}Na (Nichols et al., 2003). Furthermore, their ^{22}Ne is not accompanied by ^{4}He, expected if neon was implanted. The ^{12}C/^{13}C of these two grains are 4 and 10, in the range of SiC grains with a putative nova origin. Another graphite grain with ^{12}C/^{13}C = 8.5 has a large ^{30}Si excess of 760‰ (Amari et al., 2001a).

In summary, although a few graphite grains have the isotopic signature expected for condensates from AGB stars (isotopically heavy carbon and light nitrogen, enhanced abundances of s-process elements) and fewer still those of nova grains, the majority seems to have an SN origin (see also Figure 6). This remains an unsolved puzzle because stars that produce SiC are also expected to form graphite and the apparent underabundance of graphite from AGB stars points to deficiencies in the current understanding of the condensation of carbonaceous phases in carbon-rich stellar atmospheres and the survival of different grain types in the ISM.

1.02.9 OXIDE GRAINS

In contrast to the carbonaceous presolar phases, presolar oxide grains apparently do not carry any "exotic" noble gas component. They have been identified by ion microprobe oxygen isotopic measurements of single grains from acid residues free of silicates. In contrast to SiC, essentially all of which is of presolar origin, most oxide grains found in meteorites formed in the solar system and only a small fraction is presolar. The oxygen isotopic compositions of the presolar oxide minerals identified so far are plotted in Figure 13. They include 198 corundum or likely corundum grains (Huss et al., 1994; Hutcheon et al., 1994; Nittler et al., 1994, 1997, 1998; Nittler and Alexander, 1999; Strebel et al., 1996; Choi et al., 1998, 1999; Krestina et al., 2002), 41 spinel grains (Nittler et al., 1997; Choi et al., 1998; Zinner et al., 2003), and 5 hibonite grains (Choi et al., 1999; Krestina et al., 2002). These numbers, however, cannot be used to infer relative abundances of these mineral phases. Most presolar corundum and hibonite grains are ≥ 1 μm. In contrast, most presolar spinels were found by NanoSIMS analysis of grains down to 0.15 μm in diameter. At this size, the abundance of presolar spinel among all spinel grains is ~3%, whereas it is $\leq 0.5\%$ among >1 μm spinel grains. Furthermore, searches for presolar oxide grains have been made in different types of residues, some containing spinel, others not. Another complication is that more than half of all presolar corundum grains have been found by automatic direct $^{18}O/^{16}O$ imaging searches in the ion microprobe (Nittler et al., 1997), a method that does not detect grains with anomalies in the $^{17}O/^{16}O$ ratio but with close-to-normal $^{18}O/^{16}O$. The oxygen isotopic distribution of corundum in Figure 13, therefore, does not reflect the true distribution. NanoSIMS oxygen isotopic raster imaging of tightly packed submicron grains from the Murray CM2 chondrite led to the identification of an additional 252 presolar spinel and 32 presolar corundum grains (Nguyen et al., 2003).

Nittler et al. (1997) have classified presolar oxide grains into four different groups according to their oxygen isotopic ratios. Grains with $^{17}O/^{16}O >$ solar (3.82×10^{-4}) and $0.001 < ^{18}O/^{16}O <$ solar (2.01×10^{-3}), comprising group 1, have oxygen isotopic ratios similar to those observed in RG and AGB stars (Harris and Lambert, 1984; Harris et al., 1987; Smith and Lambert, 1990), indicating such an origin also for the grains. These compositions can be explained by hydrogen burning in the core of low-to-intermediate-mass stars followed by mixing of core material into the envelope during the first dredge-up (also second dredge-up in low-metallicity stars with $M > 3M_\odot$) (Boothroyd et al., 1994; Boothroyd and Sackmann, 1999). Variations in $^{17}O/^{16}O$ ratios mainly correspond to differences in stellar mass, while those in $^{18}O/^{16}O$ can be explained by assuming that stars with different metallicities contributed oxide grains to the solar system. According to galactic chemical evolution models, $^{17}O/^{16}O$ and $^{18}O/^{16}O$ ratios are expected to increase as a function of stellar metallicity (Timmes et al., 1995). Grains with depletions in both ^{17}O and ^{18}O (group 3) could thus come from low-mass stars (producing only small ^{17}O enrichments) with lower than solar metallicity (originally having lower than solar $^{17}O/^{16}O$ and $^{18}O/^{16}O$ ratios). The oxygen isotopic ratios of group 3 grains have been used to obtain an estimate of the age of the galaxy (Nittler and Cowsik, 1997). Group 2 grains have ^{17}O excesses and large ^{18}O depletions ($^{18}O/^{16}O < 0.001$). Such depletions cannot be produced by the first and second dredge-up but have been successfully explained by extra mixing (cool bottom processing) of low-mass ($M < 1.65M_\odot$) stars during the AGB phase

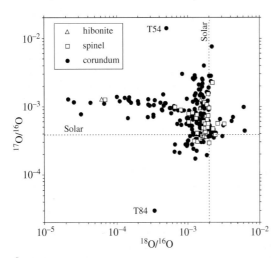

Figure 13 Oxygen isotopic ratios in individual oxide grains (sources Nittler et al., 1997, 1998; Chai et al., 1998, 1999; Krestina et al., 2002, unpublished data; Zinner et al., 2003).

that circulates material from the envelope through regions close to the hydrogen-burning shell (Wasserburg et al., 1995; Denissenkov and Weiss, 1996; Nollett et al., 2003). Group 4 grains have both ^{17}O and ^{18}O excesses. If they originated from AGB stars they could either come from low-mass stars, in which ^{18}O produced by helium burning of ^{14}N during early pulses was mixed into the envelope by third dredge-up (Boothroyd and Sackmann, 1988) or from stars with high metallicity. More likely for the grains with the largest ^{18}O excesses is an SN origin as suggested by Choi et al. (1998) if ^{18}O-rich material from the He/C zone can be admixed to material from oxygen-rich zones.

The only grain that has the typical isotopic signature expected for SN condensates, namely, a large ^{16}O excess, is grain T84 (Figure 13) (Nittler et al., 1998). All oxygen-rich zones (O/C, O/Ne, O/Si—see Figure 9) are dominated by ^{16}O (Woosley and Weaver, 1995; Thielemann et al., 1996; Rauscher et al., 2002). The paucity of such grains, whose abundance is expected to dominate that of carbonaceous phases with an SN origin, remains an unsolved mystery. It has been suggested that oxide grains from supernovae are smaller than those from red giant stars, but recent measurements of submicron grains have not uncovered any additional oxides with large ^{16}O excesses (Zinner et al., 2003; Nguyen et al., 2003). Another grain that does not fit into the four groups is T54 (Nittler et al., 1997). This grain could come from a star with >$5M_\odot$ that experienced hot bottom burning, a condition during which the convective envelope extends into the hydrogen-burning shell (Boothroyd et al., 1995; Lattanzio et al., 1997).

Some but not all grains in the four groups show evidence for initial ^{26}Al (Nittler et al., 1997; Choi et al., 1998, 1999; Krestina et al., 2002). Because ^{26}Al is produced in the hydrogen-burning shell (Forestini et al., 1991), dredge-up of material during the TP AGB phase is required, and grains without ^{26}Al must have formed before their parent stars reached this evolutionary stage. Initial $^{26}Al/^{27}Al$ ratios are highest in group 2 grains (Nittler et al., 1997; Choi et al., 1998), which is explained if these grains formed in the later stages of the AGB phase when more ^{18}O had been destroyed by cool bottom processing and more ^{26}Al dredged up (Choi et al., 1998). Titanium isotopic ratios have been determined in a few presolar corundum grains (Choi et al., 1998). The observed ^{50}Ti excesses agree with those predicted to result from neutron capture in AGB stars. The depletions in all titanium isotopes relative to ^{48}Ti found in one grain indicate that, just as for SiC grains, galactic evolution affects the isotopic compositions of the parent stars of oxide grains. One hibonite grain was found to have a large ^{41}K excess, corresponding to a $^{41}Ca/^{40}Ca$ ratio of 1.5×10^{-4}, within the range of values predicted for the envelope of AGB stars (Wasserburg et al., 1994).

To date, all attempts to identify presolar silicates in primitive meteorites have been unsuccessful (Nittler et al., 1997; Messenger and Bernatowicz, 2000). However, Messenger et al. (2003) have discovered presolar grains, including silicates, in IDPs. These grains are 0.2–1 μm in size and the discovery was made possible by the high spatial resolution of the NanoSIMS. It remains to be seen whether meteorites also contain presolar silicates in this size range (previous searches have been made mostly on ≥1 μm grains) or whether such grains were preserved only in IDPs.

1.02.10 DIAMOND

Although diamond is the most abundant presolar grain species (~500 ppm) and was the first to be isolated (Lewis et al., 1987), it remains the least understood. The only presolar isotopic signatures (indicating an SN origin) are those of Xe-HL and tellurium (Richter et al., 1998), to a marginal extent also those of strontium and barium (Lewis et al., 1991). However, the carbon isotopic composition of bulk diamonds is essentially the same as that of the solar system (Russell et al., 1991, 1996) and diamonds are too small (the average size is ~2.6 nm—hence nanodiamonds) to be analyzed as single grains. At present, it is not known whether or not this normal carbon isotopic composition is the result of averaging over grains that have large carbon isotopic anomalies, and whether all nanodiamonds are of presolar origin. Nitrogen shows a ^{15}N depletion of 343‰, but isotopically light nitrogen is produced by the CN cycle in all stars and is therefore not very diagnostic. More recent measurements have shown that the nitrogen isotopic ratio of Jupiter (Owen et al., 2001) is very similar to that of the nanodiamonds, which therefore is not necessarily a presolar signature. Furthermore, the concentration of Xe-HL is such that only one diamond grain in a million contains a xenon atom. As of early 2000s, all attempts to separate different, isotopically distinct, components among nanodiamonds have met with only limited success. Stepped pyrolysis indicates that nitrogen and the noble gas components Xe-HL and Ar-HL are decoupled, with nitrogen being released at lower temperature (Verchovsky et al., 1993a,b), and it is likely that nitrogen and the exotic gases are located in different carriers. A solar origin of a large fraction of the nanodiamonds remains a distinct possibility (Dai et al., 2002).

The light and heavy isotope enrichment in Xe-HL has been interpreted as being due to the p- and r-process, and thus requires an SN origin (Heymann and Dziczkaniec, 1979, 1980; Clayton, 1989). In one model, Xe-H is made by a short neutron burst, with neutron densities intermediate between those characteristic for the r- and s-processes (Clayton, 1989; Howard et al., 1992). Ott (1996) kept the standard r-process but proposed that xenon is separated from iodine and tellurium precursors on a timescale of a few hours after their production. Measurements of tellurium isotopes in nanodiamonds show almost complete absence of the isotopes ^{120}Te, $^{122-126}$Te, and a slight excess of ^{128}Te relative to ^{130}Te (Richter et al., 1998). This pattern agrees much better with a standard r-process and early element separation than with the neutron burst model. Clayton and co-workers (Clayton, 1989; Clayton et al., 1995) have also tried to attribute the diamonds and their carbon and nitrogen isotopic compositions to a type II supernova. This requires mixing of contributions from different SN zones. In contrast, Jørgensen (1988) proposed that diamond and Xe-HL were produced by different members of a binary system of low-mass $(1-2M_\odot)$ stars, diamond in the winds of one member, a carbon star, while Xe-HL by the other, which exploded as a type Ia supernova. However, at present we do not have an unambiguous identification of the origin of the Xe-HL and tellurium, and of the diamonds (in case they have a different origin).

1.02.11 CONCLUSION AND FUTURE PROSPECTS

The study of presolar grains has provided a wealth of information on galactic evolution, stellar nucleosynthesis, physical properties of stellar atmospheres, and conditions in the solar nebula and on meteoritic parent bodies. However, there are still many features that are not well understood with existing models of nucleosynthesis and stellar evolution and stellar structure. Examples are the carbon and nitrogen isotopic compositions of SiC A + B grains, ^{15}N and ^{29}Si excesses in SN grains, and the paucity of oxide grains from supernovae. The grain data, especially correlated isotopic ratios of many elements, thus provide a challenge to nuclear astrophysicist in tightening constraints on theoretical models.

Continuing instrumental developments allow us to make new and more measurements on the grains and likely lead to new discoveries. For example, the NanoSIMS features high spatial resolution and sensitivity, making isotopic analysis of small grains possible, and this capability has already resulted in the discovery of presolar silicate grains in IDPs (Messenger et al., 2003) and the identification of a large number of presolar spinel grains (Zinner et al., 2003; Nguyen et al., 2003). The NanoSIMS will also make it possible to analyze internal grains that have been studied in detail in the TEM (Stadermann et al., 2003). Another example is the application of RIMS to grain studies. As the number of elements that can be analyzed is being expanded (e.g., to the rare-earth elements), unexpected discoveries such as the molybdenum isotopic patterns in SiC X grains (Pellin et al., 1999) will probably result. RIMS measurements can also be made on grains, such as graphites, for which the isotopic ratios of many elements are measured with the ion microprobe.

It is clear that the discovery of presolar grains and their detailed study have opened a new and fruitful field of astrophysical research.

ACKNOWLEDGMENTS

I thank Sachiko Amari, Peter Hoppe, Natasha Krestina, Larry Nittler, and Roger Strebel for providing unpublished data, and Sachiko Amari, Tom Bernatowicz, and Scott Messenger for providing micrographs. I am grateful for help from and discussions with Sachiko Amari, Peter Hoppe, Gary Huss, and Larry Nittler. Andy Davis provided useful comments on the manuscript.

REFERENCES

Alexander C. M. O'D. (1993) Presolar SiC in chondrites: how variable and how many sources? *Geochim. Cosmochim. Acta* **57**, 2869–2888.

Alexander C. M. O'D. and Nittler L. R. (1999) The galactic evolution of Si, Ti, and O isotopic ratios. *Astrophys. J.* **519**, 222–235.

Alexander C. M. O'D., Swan P., and Prombo C. A. (1994) Occurrence and implications of silicon nitride in enstatite chondrites. *Meteoritics* **29**, 79–85.

Amari S., Anders E., Virag A., and Zinner E. (1990) Interstellar graphite in meteorites. *Nature* **345**, 238–240.

Amari S., Hoppe P., Zinner E., and Lewis R. S. (1992) Interstellar SiC with unusual isotopic compositions: grains from a supernova? *Astrophys. J.* **394**, L43–L46.

Amari S., Lewis R. S., and Anders E. (1994) Interstellar grains in meteorites: I. Isolation of SiC, graphite, and diamond: size distributions of SiC and graphite. *Geochim. Cosmochim. Acta* **58**, 459–470.

Amari S., Hoppe P., Zinner E., and Lewis R. S. (1995a) Trace-element concentrations in single circumstellar silicon carbide grains from the Murchison meteorite. *Meteoritics* **30**, 679–693.

Amari S., Lewis R. S., and Anders E. (1995b) Interstellar grains in meteorites: III. Graphite and its noble gases. *Geochim. Cosmochim. Acta* **59**, 1411–1426.

Amari S., Zinner E., and Lewis R. S. (1995c) Large ^{18}O excesses in interstellar graphite grains from the Murchison meteorite indicate a massive star origin. *Astrophys. J.* **447**, L147–L150.

Amari S., Zinner E., and Lewis R. S. (1996) ^{41}Ca in presolar graphite of supernova origin. *Astrophys. J.* **470**, L101–L104.

Amari S., Zinner E., Clayton D. D., and Meyer B. S. (1998) Presolar grains from supernovae: the case for a type Ia SN source. *Meteorit. Planet. Sci.* **33**, A10.

Amari S., Zinner E., and Lewis R. S. (2000) Isotopic compositions of different presolar silicon carbide size fractions from the Murchison meteorite. *Meteorit. Planet. Sci.* **35**, 997–1014.

Amari S., Gao X., Nittler L., Zinner E., José J., Hernanz M., and Lewis R. S. (2001a) Presolar grains from novae. *Astrophys. J.* **551**, 1065–1072.

Amari S., Nittler L. R., Zinner E., Gallino R., Lugaro M., and Lewis R. S. (2001b) Presolar SiC grains of type Y: origin from low-metallicity AGB stars. *Astrophys. J.* **546**, 248–266.

Amari S., Nittler L. R., Zinner E., Lodders K., and Lewis R. S. (2001c) Presolar SiC grains of type A and B: their isotopic compositions and stellar origins. *Astrophys. J.* **559**, 463–483.

Amari S., Jennings C., Nguyen A., Stadermann F. J., Zinner E., and Lewis R. S. (2002) NanoSIMS isotopic analysis of small presolar SiC grains from the Murchison and Indarch meteorites. In *Lunar Planet. Sci.* **XXXIII**, #1205. The Lunar and Planetary Institute, Houston (CD-ROM).

Anders E. and Grevesse N. (1989) Abundances of the elements: meteoritic and solar. *Geochim. Cosmochim. Acta* **53**, 197–214.

Anders E. and Zinner E. (1993) Interstellar grains in primitive meteorites: diamond, silicon carbide, and graphite. *Meteoritics* **28**, 490–514.

Arnett D. (1996) *Supernovae and Nucleosynthesis*. Princeton University Press, Princeton, 598p.

Arnould M., Meynet G., and Paulus G. (1997) Wolf-Rayet stars and meteoritic anomalies. In *Astrophysical Implications of the Laboratory Study of Presolar Materials* (eds. T. J. Bernatowicz and E. Zinner). AIP, New York, pp. 179–202.

Asplund M., Lambert D. L., Kipper T., Pollacco D., and Shetrone M. D. (1999) The rapid evolution of the born-again giant Sakurai's object. *Astron. Astrophys.* **343**, 507–518.

Becker S. A. and Iben I., Jr. (1979) The asymptotic giant branch evolution of intermediate-mass stars as a function of mass and composition: I. Through the second dredge-up phase. *Astrophys. J.* **232**, 831–853.

Bernatowicz T. J. and Zinner E. (eds.) (1997) *Astrophysical Implications of the Laboratory Study of Presolar Materials*. AIP, New York, 750p.

Bernatowicz T., Fraundorf G., Tang M., Anders E., Wopenka B., Zinner E., and Fraundorf P. (1987) Evidence for interstellar SiC in the Murray carbonaceous meteorite. *Nature* **330**, 728–730.

Bernatowicz T. J., Amari S., Zinner E., and Lewis R. S. (1991) Interstellar grains within interstellar grains. *Astrophys. J.* **373**, L73–L76.

Bernatowicz T. J., Amari S., and Lewis R. S. (1992) TEM studies of a circumstellar rock. In *Lunar Planet. Sci.* **XXIII**. The Lunar and Planetary Institute, Houston (CD-ROM), pp. 91–91.

Bernatowicz T. J., Cowsik R., Gibbons P. C., Lodders K., Fegley B., Jr., Amari S., and Lewis R. S. (1996) Constraints on stellar grain formation from presolar graphite in the Murchison meteorite. *Astrophys. J.* **472**, 760–782.

Bernatowicz T., Bradley J., Amari S., Messenger S., and Lewis R. (1999) New kinds of massive star condensates in a presolar graphite from Murchison. In *Lunar Planet. Sci.* **XXX**, #1392. The Lunar and Planetary Institute, Houston (CD-ROM).

Bernatowicz T. J., Messenger S., Pravdivtseva O., Swan P., and Walker R. M. (2003) Pristine presolar silicon carbide. *Geochim. Cosmochim. Acta* (in press).

Besmehn A. and Hoppe P. (2001) Silicon- and calcium-isotopic compositions of presolar silicon nitride grains from the Indarch enstatite chondrite. In *Lunar Planet. Sci.* **XXXII**, #1188. The Lunar and Planetary Institute, Houston (CD-ROM).

Besmehn A. and Hoppe P. (2003) A NanoSIMS study of Si- and Ca–Ti-isotopic compositions of presolar silicon carbide grains from supernovae. *Geochim. Cosmochim. Acta* (in press).

Black D. C. (1972) On the origins of trapped helium, neon, and argon isotopic variations in meteorites: II. Carbonaceous meteorites. *Geochim. Cosmochim. Acta* **36**, 377–394.

Black D. C. and Pepin R. O. (1969) Trapped neon in meteorites: II. *Earth Planet. Sci. Lett.* **6**, 395–405.

Boato G. (1954) The isotopic composition of hydrogen and carbon in the carbonaceous chondrites. *Geochim. Cosmochim. Acta* **6**, 209–220.

Boothroyd A. I. and Sackmann I.-J. (1988) Low-mass stars: III. Low-mass stars with steady mass loss: up to the asymptotic giant branch and through the final thermal pulses. *Astrophys. J.* **328**, 653–670.

Boothroyd A. I. and Sackmann I.-J. (1999) The CNO isotopes: deep circulation in red giants and first and second dredge-up. *Astrophys. J.* **510**, 232–250.

Boothroyd A. I., Sackmann I.-J., and Wasserburg G. J. (1994) Predictions of oxygen isotope ratios in stars and of oxygen-rich interstellar grains in meteorites. *Astrophys. J.* **430**, L77–L80.

Boothroyd A. I., Sackmann I.-J., and Wasserburg G. J. (1995) Hot bottom burning in asymptotic giant branch stars and its effect on oxygen isotopic abundances. *Astrophys. J.* **442**, L21–L24.

Brown L. E. and Clayton D. D. (1992) SiC particles from asymptotic giant branch stars: Mg burning and the s-process. *Astrophys. J.* **392**, L79–L82.

Burbidge E. M., Burbidge G. R., Fowler W. A., and Hoyle F. (1957) Synthesis of the elements in stars. *Rev. Mod. Phys.* **29**, 547–650.

Busso M., Gallino R., and Wasserburg G. J. (1999) Nucleosynthesis in AGB stars: relevance for Galactic enrichment and solar system formation. *Ann. Rev. Astron. Astrophys.* **37**, 239–309.

Busso M., Gallino R., Lambert D. L., Travaglio C., and Smith V. V. (2001) Nucleosynthesis and mixing on the asymptotic giant branch: III. Predicted and observed s-process abundances. *Astrophys. J.* **557**, 802–821.

Cameron A. G. W. (1957) Chalk River Report CRL-41, Atomic Energy Canada. *Pubs. Astron. Soc. Pacific* **69**, 201–222.

Cameron A. G. W. (1962) The formation of the sun and planets. *Icarus* **1**, 13–69.

Charbonnel C. (1995) A consistent explanation for $^{12}C/^{13}C$, ^{7}Li, and ^{3}He anomalies in red giant stars. *Astrophys. J.* **453**, L41–L44.

Choi B.-G., Huss G. R., Wasserburg G. J., and Gallino R. (1998) Presolar corundum and spinel in ordinary chondrites: origins from AGB stars and a supernova. *Science* **282**, 1284–1289.

Choi B.-G., Wasserburg G. J., and Huss G. R. (1999) Circumstellar hibonite and corundum and nucleosynthesis in asymptotic giant branch stars. *Astrophys. J.* **522**, L133–L136.

Clayton D. D. (1975) Na-22, Ne-E, extinct radioactive anomalies and unsupported Ar-40. *Nature* **257**, 36–37.

Clayton D. D. (1983a) Discovery of s-process Nd in Allende residue. *Astrophys. J.* **271**, L107–L109.

Clayton D. D. (1983b) *Principles of Stellar Evolution and Nucleosynthesis*. University of Chicago Press, Chicago and London, 612p.

Clayton D. D. (1989) Origin of heavy xenon in meteoritic diamonds. *Astrophys. J.* **340**, 613–619.

Clayton D. D. (1997) Placing the sun and mainstream SiC particles in galactic chemodynamic evolution. *Astrophys. J.* **484**, L67–L70.

Clayton D. D. and Timmes F. X. (1997a) Implications of presolar grains for Galactic chemical evolution. In *Astrophysical Implications of the Laboratory Study of Presolar Materials* (eds. T. J. Bernatowicz and E. Zinner). AIP, New York, pp. 237–264.

Clayton D. D. and Timmes F. X. (1997b) Placing the Sun in galactic chemical evolution: mainstream SiC particles. *Astrophys. J.* **483**, 220–227.

Clayton D. D. and Ward R. A. (1978) s-Process studies: xenon and krypton isotopic abundances. *Astrophys. J.* **224**, 1000–1006.

Clayton D. D., Obradovic M., Guha S., and Brown L. E. (1991) Silicon and titanium isotopes in SiC from AGB stars. In *Lunar Planet. Sci.* **XXII**. The Lunar and Planetary Institute, Houston (CD-ROM), pp. 221–222.

Clayton D. D., Meyer B. S., Sanderson C. I., Russell S. S., and Pillinger C. T. (1995) Carbon and nitrogen isotopes in type II supernova diamonds. *Astrophys. J.* **447**, 894–905.

Clayton D. D., Arnett W. D., Kane J., and Meyer B. S. (1997) Type X silicon carbide presolar grains: type Ia supernova condensates? *Astrophys. J.* **486**, 824–834.

Clayton D. D., Liu W., and Dalgarno A. (1999) Condensation of carbon in radioactive supernova gas. *Science* **283**, 1290–1292.

Clayton D. D., Meyer B. S., The L.-S., and El Eid M. F. (2002) Iron implantation in presolar supernova grains. *Astrophys. J.* **578**, L83–L86.

Clayton R. N., Grossman L., and Mayeda T. K. (1973) A component of primitive nuclear composition in carbonaceous meteorites. *Science* **182**, 485–488.

Clayton R. N., Hinton R. W., and Davis A. M. (1988) Isotopic variations in the rock-forming elements in meteorites. *Phil. Trans. Roy. Soc. London A* **325**, 483–501.

Croat T. K., Bernatowicz T., Amari S., Messenger S., and Stadermann F. J. (2003) Structural, chemical, and isotopic microanalytical investigations of graphite from supernovae. *Geochim. Cosmochim. Acta* (in press).

Dai Z. R., Bradley J. P., Joswiak D. J., Brownlee D. E., Hill H. G. M., and Genge M. J. (2002) Possible *in situ* formation of meteoritic nanodiamonds in the early solar system. *Nature* **418**, 157–159.

Daulton T. L., Eisenhour D. D., Bernatowicz T. J., Lewis R. S., and Buseck P. (1996) Genesis of presolar diamonds: comparative high-resolution transmission electron microscopy study of meteoritic and terrestrial nano-diamonds. *Geochim. Cosmochim. Acta* **60**, 4853–4872.

Daulton T. L., Bernatowicz T. J., Lewis R. S., Messenger S., Stadermann F. J., and Amari S. (2002) Polytype distribution in circumstellar silicon carbide. *Science* **296**, 1852–1855.

Daulton T. L., Bernatowicz T. J., Lewis R. S., Messenger S., Stadermann F. J., and Amari S. (2003) Polytype distribution of circumstellar silicon carbide: microstructural characterization by transmission electron microscopy. *Geochim. Cosmochim. Acta* (in press).

Davis A. M., Gallino R., Lugaro M., Tripa C. E., Savina M. R., Pellin M. J., and Lewis R. S. (2002) Presolar grains and the nucleosynthesis of iron isotopes. In *Lunar Planet. Sci.* **XXXIII**, #2018. The Lunar and Planetary Institute, Houston (CD-ROM).

Deneault E. A.-N., Clayton D. D., and Heger A. (2003) Supernova reverse shocks and SiC growth. *Astrophys. J.* **594**, 312–325.

Denissenkov P. A. and Weiss A. (1996) Deep diffusive mixing in globular-cluster red giants. *Astron. Astrophys.* **308**, 773–784.

Ebel D. S. and Grossman L. (2001) Condensation from supernova gas made of free atoms. *Geochim. Cosmochim. Acta* **65**, 469–477.

Ebisuzaki T. and Shibazaki N. (1988) The effects of mixing of the ejecta on the hard x-ray emissions from SN 1987A. *Astrophys. J.* **327**, L5–L8.

El Eid M. (1994) CNO isotopes in red giants: theory versus observations. *Astron. Astrophys.* **285**, 915–928.

Forestini M., Paulus G., and Arnould M. (1991) On the production of ^{26}Al in AGB stars. *Astron. Astrophys.* **252**, 597–604.

Frost C. A. and Lattanzio J. C. (1996) AGB stars: what should be done? In *Stellar Evolution: What should be done; 32nd Liège Int. Astrophys. Coll.* (eds. A. Noel, D. Fraipont-Caro, M. Gabriel, N. Grevesse, and P. Demarque). Université de Liège, Liège, Belgium, pp. 307–325.

Gallino R., Busso M., and Lugaro M. (1997) Neutron capture nucleosynthesis in AGB stars. In *Astrophysical Implications of the Laboratory Study of Presolar Materials* (eds. T. J. Bernatowicz and E. Zinner). AIP, New York, pp. 115–153.

Gallino R., Busso M., Picchio G., and Raiteri C. M. (1990) On the astrophysical interpretation of isotope anomalies in meteoritic SiC grains. *Nature* **348**, 298–302.

Gallino R., Raiteri C. M., and Busso M. (1993) Carbon stars and isotopic Ba anomalies in meteoritic SiC grains. *Astrophys. J.* **410**, 400–411.

Gallino R., Raiteri C. M., Busso M., and Matteucci F. (1994) The puzzle of silicon, titanium, and magnesium anomalies in meteoritic silicon carbide grains. *Astrophys. J.* **430**, 858–869.

Grevesse N., Noels A., and Sauval A. J. (1996) Standard abundances. In *Cosmic Abundances* (eds. S. S. Holt and G. Sonneborn). BookCrafters, San Francisco, pp. 117–126.

Guber K. H., Spencer R. R., Koehler P. E., and Winters R. R. (1997) New ^{142,144}Nd (n, γ) cross-sections and the s-process origin of the Nd anomalies in presolar meteoritic silicon carbide grains. *Phys. Rev. Lett.* **78**, 2704–2707.

Halbout J., Mayeda T. K., and Clayton R. N. (1986) Carbon isotopes and light element abundances in carbonaceous chondrites. *Earth Planet. Sci. Lett.* **80**, 1–18.

Harris M. J. and Lambert D. L. (1984) Oxygen isotopic abundances in the atmospheres of seven red giant stars. *Astrophys. J.* **285**, 674–682.

Harris M. J., Lambert D. L., Hinkle K. H., Gustafsson B., and Eriksson K. (1987) Oxygen isotopic abundances in evolved stars: III. 26 carbon stars. *Astrophys. J.* **316**, 294–304.

Herant M., Benz W., Hix W. R., Fryer C. L., and Colgate S. A. (1994) Inside the supernova: a powerful convective engine. *Astrophys. J.* **435**, 339–361.

Herwig F. (2001) The evolutionary timescale of Sakurai's object: a test of convection theory? *Astrophys. J.* **554**, L71–L74.

Heymann D. and Dziczkaniec M. (1979) Xenon from intermediate zones of supernovae. In *Lunar Planet. Sci.* **X**. The Lunar and Planetary Institute, Houston (CD-ROM), pp. 1943–1959.

Heymann D. and Dziczkaniec M. (1980) A first roadmap for kryptology. In *Lunar Planet. Sci.* **XI**. The Lunar and Planetary Institute, Houston (CD-ROM), pp. 1179–1212.

Hoppe P. and Besmehn A. (2002) Evidence for extinct vanadium-49 in presolar silicon carbide grains from supernovae. *Astrophys. J.* **576**, L69–L72.

Hoppe P. and Ott U. (1997) Mainstream silicon carbide grains from meteorites. In *Astrophysical Implications of the Laboratory Study of Presolar Materials* (eds. T. J. Bernatowicz and E. Zinner). AIP, New York, pp. 27–58.

Hoppe P. and Zinner E. (2000) Presolar dust grains from meteorites and their stellar sources. *J. Geophys. Res.* **105**, 10371–10385.

Hoppe P., Amari S., Zinner E., Ireland T., and Lewis R. S. (1994) Carbon, nitrogen, magnesium, silicon, and titanium isotopic compositions of single interstellar silicon carbide grains from the Murchison carbonaceous chondrite. *Astrophys. J.* **430**, 870–890.

Hoppe P., Amari S., Zinner E., and Lewis R. S. (1995) Isotopic compositions of C, N, O, Mg and Si, trace element abundances, and morphologies of single circumstellar graphite grains in four density fractions from the Murchison meteorite. *Geochim. Cosmochim. Acta* **59**, 4029–4056.

Hoppe P., Strebel R., Eberhardt P., Amari S., and Lewis R. S. (1996a) Small SiC grains and a nitride grain of circumstellar origin from the Murchison meteorite: implications for stellar evolution and nucleosynthesis. *Geochim. Cosmochim. Acta* **60**, 883–907.

Hoppe P., Strebel R., Eberhardt P., Amari S., and Lewis R. S. (1996b) Type II supernova matter in a silicon carbide grain from the Murchison meteorite. *Science* **272**, 1314–1316.

Hoppe P., Annen P., Strebel R., Eberhardt P., Gallino R., Lugaro M., Amari S., and Lewis R. S. (1997) Meteoritic silicon carbide grains with unusual Si-isotopic compositions: evidence for an origin in low-mass metallicity asymptotic giant branch stars. *Astrophys. J.* **487**, L101–L104.

Hoppe P., Strebel R., Eberhardt P., Amari S., and Lewis R. S. (2000) Isotopic properties of silicon carbide X grains from the Murchison meteorite in the size range 0.5–1.5 μm. *Meteorit. Planet. Sci.* **35**, 1157–1176.

Howard W. M., Meyer B. S., and Clayton D. D. (1992) Heavy-element abundances from a neutron burst that produces Xe–H. *Meteoritics* **27**, 404–412.

Hughes J. P., Rakowski C. E., Burrows D. N., and Slane P. O. (2000) Nucleosynthesis and mixing in Cassiopeia A. *Astrophys. J.* **528**, L109–L113.

Huss G. R. and Lewis R. S. (1995) Presolar diamond, SiC, and graphite in primitive chondrites: abundances as a function of meteorite class and petrologic type. *Geochim. Cosmochim. Acta* **59**, 115–160.

Huss G. R., Fahey A. J., Gallino R., and Wasserburg G. J. (1994) Oxygen isotopes in circumstellar Al_2O_3 grains from meteorites and stellar nucleosynthesis. *Astrophys. J.* **430**, L81–L84.

Huss G. R., Hutcheon I. D., and Wasserburg G. J. (1997) Isotopic systematics of presolar silicon carbide from the Orgueil (CI) carbonaceous chondrite: implications for solar system formation and stellar nucleosynthesis. *Geochim. Cosmochim. Acta* **61**, 5117–5148.

Hutcheon I. D., Huss G. R., Fahey A. J., and Wasserburg G. J. (1994) Extreme ^{26}Mg and ^{17}O enrichments in an Orgueil corundum: identification of a presolar oxide grain. *Astrophys. J.* **425**, L97–L100.

Iben I., Jr. and Renzini A. (1983) Asymptotic giant branch evolution and beyond. *Ann. Rev. Astron. Astrophys.* **21**, 271–342.

Ireland T. R., Zinner E. K., and Amari S. (1991) Isotopically anomalous Ti in presolar SiC from the Murchison meteorite. *Astrophys. J.* **376**, L53–L56.

Jørgensen U. G. (1988) Formation of Xe–Hl-enriched diamond grains in stellar environments. *Nature* **332**, 702–705.

José J., Coc A., and Hernanz M. (1999) Nuclear uncertainties in the NeNa–MgAl cycles and production of ^{22}Na and ^{26}Al during nova outbursts. *Astrophys. J.* **520**, 347–360.

Käppeler F., Beer H., and Wisshak K. (1989) s-Process nucleosynthesis–nuclear physics and the classic model. *Rep. Prog. Phys.* **52**, 945–1013.

Kashiv Y., Cai Z., Lai B., Sutton S. R., Lewis R. S., Davis A. M., and Clayton R. N. (2001) Synchroton x-ray fluorescence: a new approach for determining trace element concentrations in individual presolar grains. In *Lunar Planet. Sci.* **XXXII**, #2192. The Lunar and Planetary Institute, Houston (CD-ROM).

Kashiv Y., Cai Z., Lai B., Sutton S. R., Lewis R. S., Davis A. M., Clayton R. N., and Pellin M. J. (2002) Condensation of trace elements into presolar SiC stardust grains. In *Lunar Planet. Sci.* **XXXIII**, #2056. The Lunar and Planetary Institute, Houston (CD-ROM).

Koehler P. E., Spencer R. R., Guber K. H., Winters R. R., Raman S., Harvey J. A., Hill N. W., Blackmon J. C., Bardayan D. W., Larson D. C., Lewis T. A., Pierce D. E., and Smith M. S. (1998) High resolution neutron capture and transmission measurement on ^{137}Ba and their impact on the interpretation of meteoritic barium anomalies. *Phys. Rev. C* **57**, R1558–R1561.

Kovetz A. and Prialnik D. (1997) The composition of nova ejecta from multicycle evolution models. *Astrophys. J.* **477**, 356–367.

Krestina N., Hsu W., and Wasserburg G. J. (2002) Circumstellar oxide grains in ordinary chondrites and their origin. In *Lunar Planet. Sci.* **XXXIII**, #1425. The Lunar and Planetary Institute, Houston (CD-ROM).

Lambert D. L. (1991) The abundance connection—the view from the trenches. In *Evolution of Stars: The Photospheric Abundance Connection* (eds. G. Michaud and A. Tutukov). Kluwer, Dordrecht, pp. 451–460.

Lambert D. L., Gustafsson B., Eriksson K., and Hinkle K. H. (1986) The chemical composition of carbon stars: I. Carbon, nitrogen, and oxygen in 30 cool carbon stars in the galactic disk. *Astrophys. J. Suppl.* **62**, 373–425.

Langer N., Heger A., Wellstein S., and Herwig F. (1999) Mixing and nucleosynthesis in rotating TP-AGB stars. *Astron. Astrophys.* **346**, L37–L40.

Larimer J. W. and Bartholomay M. (1979) The role of carbon and oxygen in cosmic gases: some applications to the chemistry and mineralogy of enstatite chondrites. *Geochim. Cosmochim. Acta* **43**, 1455–1466.

Lattanzio J. C., Frost C. A., Cannon R. C., and Wood P. R. (1997) Hot bottom burning nucleosynthesis in $6M_\odot$ stellar models. *Nuclear Phys.* **A621**, 435c–438c.

Lee T. (1988) Implications of isotopic anomalies for nucleosynthesis. In *Meteorites and the Early Solar System* (eds. J. F. Kerridge and M. S. Matthews). University of Arizona Press, Tucson, pp. 1063–1089.

Lewis R. S., Anders E., Wright I. P., Norris S. J., and Pillinger C. T. (1983) Isotopically anomalous nitrogen in primitive meteorites. *Nature* **305**, 767–771.

Lewis R. S., Tang M., Wacker J. F., Anders E., and Steel E. (1987) Interstellar diamonds in meteorites. *Nature* **326**, 160–162.

Lewis R. S., Amari S., and Anders E. (1990) Meteoritic silicon carbide: pristine material from carbon stars. *Nature* **348**, 293–298.

Lewis R. S., Huss G. R., and Lugmair G. (1991) Finally, Ba and Sr accompanying Xe-HL in diamonds form Allende. In *Lunar Planet. Sci.* **XXII**. The Lunar and Planetary Institute, Houston (CD-ROM), pp. 887–888.

Lewis R. S., Amari S., and Anders E. (1994) Interstellar grains in meteorites: II. SiC and its noble gases. *Geochim. Cosmochim. Acta* **58**, 471–494.

Lin Y., Amari S., and Pravdivtseva O. (2002) Presolar grains from the Qingzhen (EH3) meteorite. *Astrophys. J.* **575**, 257–263.

Little-Marenin I. R. (1986) Carbon stars with silicate dust in their circumstellar shells. *Astrophys. J.* **307**, L15–L19.

Lodders K. (2003) Solar system abundances and condensation temperatures of the elements. *Astrophys. J.* **591**, 1220–1247.

Lodders K. and Fegley B., Jr. (1997) Condensation chemistry of carbon stars. In *Astrophysical Implications of the Laboratory Study of Presolar Materials* (eds. T. J. Bernatowicz and E. Zinner). AIP, New York, pp. 391–423.

Lodders K. and Fegley B., Jr. (1998) Presolar silicon carbide grains and their parent stars. *Meteorit. Planet. Sci.* **33**, 871–880.

Lugaro M., Zinner E., Gallino R., and Amari S. (1999) Si isotopic ratios in mainstream presolar SiC grains revisited. *Astrophys. J.* **527**, 369–394.

Lugaro M., Davis A. M., Gallino R., Pellin M. J., Straniero O., and Käppeler F. (2003) Isotopic compositions of strontium, zirconium, molybdenum, and barium in single presolar SiC grains and asymptotic giant branch stars. *Astrophys. J.* **593**, 486–508.

Mathis J. S. (1990) Interstellar dust and extinction. *Ann. Rev. Astron. Astrophys.* **28**, 37–70.

Mendybaev R. A., Beckett J. R., Grossman L., Stolper E., Cooper R. F., and Bradley J. P. (2002) Volatilization kinetics of silicon carbide in reducing gases: an experimental study with applications to the survival of presolar grains in the solar nebula. *Geochim. Cosmochim. Acta* **66**, 661–682.

Messenger S. and Bernatowicz T. J. (2000) Search for presolar silicates in Acfer 094. *Meteorit. Planet. Sci.* **35**, A109.

Messenger S., Amari S., Gao X., Walker R. M., Clemett S., Chillier X. D. F., Zare R. N., and Lewis R. S. (1998) Indigenous polycyclic aromatic hydrocarbons in circumstellar graphite grains from primitive meteorites. *Astrophys. J.* **502**, 284–295.

Messenger S., Keller L. P., Stadermann F. J., Walker R. M., and Zinner E. (2003) Samples of stars beyond the solar system: silicate grains in interplanetary dust. *Science* **300**, 105–108.

Meyer B. S. (1994) The r-, s-, and p-processes in nucleosynthesis. *Ann. Rev. Astron. Astrophys.* **32**, 153–190.

Meyer B. S., Clayton D. D., and The L.-S. (2000) Molybdenum and zirconium isotopes from a supernova neutron burst. *Astrophys. J.* **540**, L49–L52.

Nguyen A., Zinner E., and Lewis R. S. (2003) Identification of small presolar spinel and corundum grains by isotopic raster imaging. *Proc. 6th Torino Workshop, Publ. Astron. Soc. Australia.* (in press).

Nichols R. H., Jr., Kehm K., Hohenberg C. M., Amari S., and Lewis R. S. (2003) Neon and helium in single interstellar SiC and graphite grains: asymptotic giant branch, Wolf-Rayet, supernova and nova sources. *Geochim. Cosmochim. Acta* (submitted).

Nicolussi G. K., Davis A. M., Pellin M. J., Lewis R. S., Clayton R. N., and Amari S. (1997) s-Process zirconium in presolar silicon carbide grains. *Science* **277**, 1281–1283.

Nicolussi G. K., Pellin M. J., Lewis R. S., Davis A. M., Amari S., and Clayton R. N. (1998a) Molybdenum isotopic composition of individual presolar silicon carbide grains from the Murchison meteorite. *Geochim. Cosmochim. Acta* **62**, 1093–1104.

Nicolussi G. K., Pellin M. J., Lewis R. S., Davis A. M., Clayton R. N., and Amari S. (1998b) Strontium isotopic composition in individual circumstellar silicon carbide grains: a record of s-process nucleosynthesis. *Phys. Rev. Lett.* **81**, 3583–3586.

Nicolussi G. K., Pellin M. J., Lewis R. S., Davis A. M., Clayton R. N., and Amari S. (1998c) Zirconium and molybdenum in individual circumstellar graphite grains: new isotopic data on the nucleosynthesis of heavy elements. *Astrophys. J.* **504**, 492–499.

Nittler L. R. (2003) Presolar stardust in meteorites: recent advances and scientific frontiers. *Earth Planet. Sci. Lett.* **209**, 259–273.

Nittler L. R. and Alexander C. M. O'D. (1999) Automatic identification of presolar Al- and Ti-rich oxide grains from ordinary chondrites. In *Lunar Planet. Sci.* **XXX**, #2041. The Lunar and Planetary Institute, Houston (CD-ROM).

Nittler L. R. and Alexander C. M. O'D. (2003) Automated isotopic measurements of micron-sized dust: application to meteoritic presolar silicon carbide. *Geochim. Cosmochim. Acta* (in press).

Nittler L. R. and Cowsik R. (1997) Galactic age estimates from O-rich stardust in meteorites. *Phys. Rev. Lett.* **78**, 175–178.

Nittler L. R., Alexander C. M. O'D., Gao X., Walker R. M., and Zinner E. K. (1994) Interstellar oxide grains from the Tieschitz ordinary chondrite. *Nature* **370**, 443–446.

Nittler L. R., Hoppe P., Alexander C. M. O'D., Amari S., Eberhardt P., Gao X., Lewis R. S., Strebel R., Walker R. M., and Zinner E. (1995) Silicon nitride from supernovae. *Astrophys. J.* **453**, L25–L28.

Nittler L. R., Amari S., Zinner E., Woosley S. E., and Lewis R. S. (1996) Extinct ^{44}Ti in presolar graphite and SiC: proof of a supernova origin. *Astrophys. J.* **462**, L31–L34.

Nittler L. R., Alexander C. M. O'D., Gao X., Walker R. M., and Zinner E. (1997) Stellar sapphires: the properties and origins of presolar Al_2O_3 in meteorites. *Astrophys. J.* **483**, 475–495.

Nittler L. R., Alexander C. M. O'D., Wang J., and Gao X. (1998) Meteoritic oxide grain from supernova found. *Nature* **393**, 222.

Nollett K. M., Busso M., and Wasserburg G. J. (2003) Cool bottom processes on the thermally pulsing asymptotic giant branch and the isotopic composition of circumstellar dust grains. *Astrophys. J.* **582**, 1036–1058.

Ott U. (1993) Interstellar grains in meteorites. *Nature* **364**, 25–33.

Ott U. (1996) Interstellar diamond xenon and timescales of supernova ejecta. *Astrophys. J.* **463**, 344–348.

Ott U. and Begemann F. (1990) Discovery of s-process barium in the Murchison meteorite. *Astrophys. J.* **353**, L57–L60.

Ott U. and Begemann F. (2000) Spallation recoil and age of presolar grains in meteorites. *Meteorit. Planet. Sci.* **35**, 53–63.

Owen T., Mahaffy P. R., Niemann H. B., Atreya S., and Wong M. (2001) Protosolar nitrogen. *Astrophys. J.* **553**, L77–L79.

Pellin M. J., Davis A. M., Lewis R. S., Amari S., and Clayton R. N. (1999) Molybdenum isotopic composition of single silicon carbide grains from supernovae. In *Lunar Planet. Sci.* **XXX**, #1969. The Lunar and Planetary Institute, Houston (CD-ROM).

Pellin M. J., Calaway W. F., Davis A. M., Lewis R. S., Amari S., and Clayton R. N. (2000a) Toward complete isotopic analysis of individual presolar silicon carbide grains: C, N, Si, Sr, Zr, Mo, and Ba in single grains of type X. In *Lunar Planet. Sci.* **XXXI**, #1917. The Lunar and Planetary Institute, Houston (CD-ROM).

Pellin M. J., Davis A. M., Calaway W. F., Lewis R. S., Clayton R. N., and Amari S. (2000b) Zr and Mo isotopic constraints on the origin of unusual types of presolar SiC grains. In *Lunar Planet. Sci.* **XXXI**, #1934. The Lunar and Planetary Institute, Houston (CD-ROM).

Podosek F. A., Prombo C. A., Amari S., and Lewis R. S. (2003) s-Process Sr isotopic compositions in presolar SiC from the Murchison meteorite. *Astrophys. J.* (in press).

Prombo C. A., Podosek F. A., Amari S., and Lewis R. S. (1993) s-Process Ba isotopic compositions in presolar SiC from the Murchison meteorite. *Astrophys. J.* **410**, 393–399.

Rauscher T., Heger A., Hoffman R. D., and Woosley S. E. (2002) Nucleosynthesis in massive stars with improved nuclear and stellar physics. *Astrophys. J.* **576**, 323–348.

Reynolds J. H. and Turner G. (1964) Rare gases in the chondrite Renazzo. *J. Geophys. Res.* **69**, 3263–3281.

Richter S., Ott U., and Begemann F. (1993) s-Process isotope abundance anomalies in meteoritic silicon carbide: new data. In *Nuclei in the Cosmos* (eds. F. Käppeler and K. Wisshak). Institute of Physics Publishing, Bristol and Philadelphia, pp. 127–132.

Richter S., Ott U., and Begemann F. (1994) s-Process isotope abundance anomalies in meteoritic silicon carbide: data for Dy. In *Proc. European Workshop on Heavy Element Nucleosynthesis* (eds. E. Somorjai and Z. Fülöp). Inst. Nucl. Res. Hungarian Acad. Sci., Debrecen, pp. 44–46.

Richter S., Ott U., and Begemann F. (1998) Tellurium in presolar diamonds as an indicator for rapid separation of supernova ejecta. *Nature* **391**, 261–263.

Russell S. S., Arden J. W., and Pillinger C. T. (1991) Evidence for multiple sources of diamond from primitive chondrites. *Science* **254**, 1188–1191.

Russell S. S., Arden J. W., and Pillinger C. T. (1996) A carbon and nitrogen isotope study of diamond from primitive chondrites. *Meteorit. Planet. Sci.* **31**, 343–355.

Russell S. S., Ott U., Alexander C. M. O'D., Zinner E. K., Arden J. W., and Pillinger C. T. (1997) Presolar silicon carbide from the Indarch (EH4) meteorite: comparison with silicon carbide populations from other meteorite classes. *Meteorit. Planet. Sci.* **32**, 719–732.

Savina M. R., Davis A. M., Tripa C. E., Pellin M. J., Clayton R. N., Lewis R. S., Amari S., Gallino R., and Lugaro M. (2003a) Barium isotopes in individual presolar silicon carbide grains from the Murchison meteorite. *Geochim. Cosmochim. Acta* **67**, 3201–3214.

Savina M. R., Pellin M. J., Tripa C. E., Veryovkin I. V., Calaway W. F., and Davis A. M. (2003b) Analyzing individual presolar grains with CHARISMA. *Geochim. Cosmochim. Acta* **67**, 3215–3225.

Savina M. R., Tripa C. E., Pellin M. J., Davis A. M., Clayton R. N., Lewis R. S., and Amari S. (2003c) Isotopic composition of molybdenum and barium in single presolar silicon carbide grains of type A + B. In *Lunar Planet. Sci.* **XXXIV**, #2079. The Lunar and Planetary Institute, Houston (CD-ROM).

Smith V. V. and Lambert D. L. (1990) The chemical composition of red giants: III. Further CNO isotopic and s-process abundances in thermally pulsing asymptotic giant branch stars. *Astrophys. J. Suppl.* **72**, 387–416.

Speck A. K., Barlow M. J., and Skinner C. J. (1997) The nature of silicon carbide in carbon star outflows. *Mon. Not. Roy. Astron. Soc.* **234**, 79–84.

Speck A. K., Hofmeister A. M., and Barlow M. J. (1999) The SiC problem: astronomical and meteoritic evidence. *Astrophys. J.* **513**, L87–L90.

Srinivasan B. and Anders E. (1978) Noble gases in the Murchison meteorite: possible relics of s-process nucleosynthesis. *Science* **201**, 51–56.

Stadermann F. J., Bernatowicz T., Croat T. K., Zinner E., Messenger S., and Amari S. (2003) Titanium and oxygen isotopic compositions of sub-micrometer TiC crystals within presolar graphite. In *Lunar Planet. Sci.* **XXXIV**, #1627. The Lunar and Planetary Institute, Houston (CD-ROM).

Starrfield S., Truran J. W., Wiescher M. C., and Sparks W. M. (1998) Evolutionary sequences for Nova V1974 Cygni using new nuclear reaction rates and opacities. *Mon. Not. Roy. Astron. Soc.* **296**, 502–522.

Stone J., Hutcheon I. D., Epstein S., and Wasserburg G. J. (1991) Correlated Si isotope anomalies and large ^{13}C enrichments in a family of exotic SiC grains. *Earth Planet. Sci. Lett.* **107**, 570–581.

Strebel R., Hoppe P., and Eberhardt P. (1996) A circumstellar Al- and Mg-rich oxide grain from the Orgueil meteorite. *Meteoritics* **31**, A136.

Suess H. E. and Urey H. C. (1956) Abundances of the elements. *Rev. Mod. Phys.* **28**, 53–74.

Tang M. and Anders E. (1988a) Interstellar silicon carbide: how much older than the solar system? *Astrophys. J.* **335**, L31–L34.

Tang M. and Anders E. (1988b) Isotopic anomalies of Ne, Xe, and C in meteorites: II. Interstellar diamond and SiC: carriers of exotic noble gases. *Geochim. Cosmochim. Acta* **52**, 1235–1244.

Thielemann F.-K., Nomoto K., and Hashimoto M.-A. (1996) Core-collapse supernovae and their ejecta. *Astrophys. J.* **460**, 408–436.

Tielens A. G. G. M. (1990) Carbon stardust: from soot to diamonds. In *Carbon in the Galaxy: Studies from Earth and Space*, Conf. Publ. 3061 (eds. J. C. Tarter, S. Chang, and D. J. deFrees). NASA, pp. 59–111.

Timmes F. X. and Clayton D. D. (1996) Galactic evolution of silicon isotopes: application to presolar SiC grains from meteorites. *Astrophys. J.* **472**, 723–741.

Timmes F. X., Woosley S. E., and Weaver T. A. (1995) Galactic chemical evolution: hydrogen through zinc. *Astrophys. J. Suppl.* **98**, 617–658.

Timmes F. X., Woosley S. E., Hartmann D. H., and Hoffman R. D. (1996) The production of ^{44}Ti and ^{60}Co in supernovae. *Astrophys. J.* **464**, 332–341.

Travaglio C., Gallino R., Amari S., Zinner E., Woosley S., and Lewis R. S. (1999) Low-density graphite grains and mixing in type II supernovae. *Astrophys. J.* **510**, 325–354.

Treffers R. and Cohen M. (1974) High-resolution spectra of cold stars in the 10- and 20-micron regions. *Astrophys. J.* **188**, 545–552.

Tripa C. E., Pellin M. J., Savina M. R., Davis A. M., Lewis R. S., and Clayton R. N. (2002) Fe isotopic composition of presolar SiC mainstream grains. In *Lunar Planet. Sci.* **XXXIII**, #1975. The Lunar and Planetary Institute, Houston (CD-ROM).

Verchovsky A. B., Franchi I. A., Arden J. W., Fisenko A. V., Semionova L. F., and Pillinger C. T. (1993a) Conjoint release of N, C, He, and Ar from Cδ by stepped pyrolysis: implications for the identification of their carriers. *Meteoritics* **28**, 452–453.

Verchovsky A. B., Russell S. S., Pillinger C. T., Fisenko A. V., and Shukolyukov Y. A. (1993b) Are the Cδ light nitrogen and noble gases are located in the same carrier? In *Lunar Planet. Sci.* **XXIV**. The Lunar and Planetary Institute, Houston (CD-ROM), pp. 1461–1462.

Virag A., Wopenka B., Amari S., Zinner E., Anders E., and Lewis R. S. (1992) Isotopic, optical, and trace element properties of large single SiC grains from the Murchison meteorite. *Geochim. Cosmochim. Acta* **56**, 1715–1733.

Wallerstein G., Iben I., Jr., Parker P., Boesgaard A. M., Hale G. M., Champagne A. E., Barnes C. A., Käppeler F., Smith V. V., Hoffman R. D., Timmes F. X., Sneden C., Boyd R. N., Meyer B. S., and Lambert D. L. (1997) Synthesis of the elements in stars: forty years of progress. *Rev. Mod. Phys.* **69**, 995–1084.

Wasserburg G. J. (1987) Isotopic abundances: inferences on solar system and planetary evolution. *Earth Planet. Sci. Lett.* **86**, 129–173.

Wasserburg G. J., Busso M., Gallino R., and Raiteri C. M. (1994) Asymptotic giant branch stars as a source of short-lived radioactive nuclei in the solar nebula. *Astrophys. J.* **424**, 412–428.

Wasserburg G. J., Boothroyd A. I., and Sackmann I.-J. (1995) Deep circulation in red giant stars: a solution to the carbon and oxygen isotope puzzles? *Astrophys. J.* **447**, L37–L40.

Wisshak K., Voss F., Käppeler F., and Kazakov L. (1997) Neutron capture in neodymium isotopes: implications for the s-process. *Nuclear Phys.* **A621**, 270c–273c.

Woosley S. E. and Weaver T. A. (1995) The evolution and explosion of massive stars: II. Explosive hydrodynamics and nucleosynthesis. *Astrophys. J. Suppl.* **101**, 181–235.

Zinner E. (1997) Presolar material in meteorites: an overview. In *Astrophysical Implications of the Laboratory Study of Presolar Materials* (eds. T. J. Bernatowicz and E. Zinner). AIP, New York, pp. 3–26.

Zinner E. (1998a) Stellar nucleosynthesis and the isotopic composition of presolar grains from primitive meteorites. *Ann. Rev. Earth Planet. Sci.* **26**, 147–188.

Zinner E. (1998b) Trends in the study of presolar dust grains from primitive meteorites. *Meteorit. Planet. Sci.* **33**, 549–564.

Zinner E., Tang M., and Anders E. (1989) Interstellar SiC in the Murchison and Murray meteorites: isotopic composition of Ne, Xe, Si, C, and N. *Geochim. Cosmochim. Acta* **53**, 3273–3290.

Zinner E., Amari S., and Lewis R. S. (1991) s-Process Ba, Nd, and Sm in presolar SiC from the Murchison meteorite. *Astrophys. J.* **382**, L47–L50.

Zinner E., Amari S., Wopenka B., and Lewis R. S. (1995) Interstellar graphite in meteorites: isotopic compositions and structural properties of single graphite grains from Murchison. *Meteoritics* **30**, 209–226.

Zinner E., Amari S., Gallino R., and Lugaro M. (2001) Evidence for a range of metallicities in the parent stars of presolar SiC grains. *Nuclear Phys. A* **A688**, 102c–105c.

Zinner E., Amari S., Guinness R., Nguyen A., Stadermann F., Walker R. M., and Lewis R. S. (2003) Presolar spinel grains from the Murray and Murchison carbonaceous chondrites. *Geochim. Cosmochim. Acta* (in press).

1.03
Solar System Abundances of the Elements

H. Palme
Universität zu Köln, Germany

and

A. Jones
Université Paris Sud, France

1.03.1 ABUNDANCES OF THE ELEMENTS IN THE SOLAR NEBULA	41
1.03.1.1 Historical Remarks	41
1.03.1.2 Solar System Abundances of the Elements	42
1.03.1.2.1 Is the elemental and isotopic composition of the solar nebula uniform?	42
1.03.1.2.2 The composition of the solar photosphere	43
1.03.1.3 Abundances of Elements in Meteorites	45
1.03.1.3.1 Undifferentiated and differentiated meteorites	45
1.03.1.3.2 Cosmochemical classification of elements	45
1.03.1.4 CI Chondrites as Standard for Solar Abundances	47
1.03.1.4.1 Chemical variations among chondritic meteorites	47
1.03.1.4.2 CI chondrites	51
1.03.1.4.3 The CI abundance table	51
1.03.1.4.4 Comparison with Anders and Grevesse abundance table	53
1.03.1.5 Solar System Abundances of the Elements	53
1.03.1.5.1 Comparison of meteorite and solar abundances	53
1.03.1.5.2 Solar system abundances versus mass number	54
1.03.1.5.3 Other sources for solar system abundances	55
1.03.2 THE ABUNDANCES OF THE ELEMENTS IN THE ISM	55
1.03.2.1 Introduction	55
1.03.2.2 The Nature of the ISM	56
1.03.2.3 The Chemical Composition of the ISM	57
1.03.2.3.1 The composition of the interstellar gas and elemental depletions	57
1.03.2.3.2 The composition of interstellar dust	57
1.03.2.3.3 Did the solar system inherit the depletion of volatile elements from the ISM?	59
1.03.2.3.4 The ISM oxygen problem	59
1.03.3 SUMMARY	60
REFERENCES	60

1.03.1 ABUNDANCES OF THE ELEMENTS IN THE SOLAR NEBULA

1.03.1.1 Historical Remarks

At the beginning of the twentieth century attempts were made to define the average composition of cosmic matter, utilizing compositional data on the Earth's crust and meteorites. This led to the discovery by Harkins (1917) that elements with even atomic numbers are more abundant than those with odd atomic numbers, the so-called Oddo–Harkins rule, best exemplified for the rare earth elements (REEs). During the

1920s and 1930s Victor Moritz Goldschmidt and his colleagues in Göttingen, and later in Oslo, measured and compiled a wealth of chemical data on terrestrial rocks, meteorites, and individual phases of meteorites. On the basis of these data Goldschmidt (1938) set up a cosmic abundance table which he published in 1938 in the ninth volume of his *Geochemische Verteilungsgesetze der Elemente* (*The Geochemical Laws of the Distribution of the Elements*) entitled *Die Mengenverhältnisse der Elemente und der Atom-Arten* (*The Proportions of the Elements and the Various Kinds of Atoms*). Goldschmidt believed that meteorites would provide the average composition of cosmic matter. He used the word "cosmic" because, in citing contemporaneous astronomers, he thought that meteorites represent interstellar material from outside the solar system. In his book he mentioned a second reason for using meteorite data. Most meteorites will be representative of average cosmic matter, because they have not been affected by physicochemical processes (e.g., melting and crystallization), although chondrules within them have experienced melting and crystallization, the meteorites as a whole have not. In contrast, the crust of the Earth, which formed by the melting of the mantle, provides only a very biased sampling of elemental abundances in the bulk Earth. Goldschmidt calculated the average concentrations of elements in cosmic matter by using a weighted mean of element abundances in meteorite phases: metal (two parts), sulfide (one part), and silicates (10 parts). In this way he obtained the cosmic abundances of 66 elements.

It was during the same time that astronomers began to extract quantitative information about elemental abundances in the Sun by solar absorption spectroscopy and it was soon realized that the compositions of the Sun and the whole Earth are similar, except for hydrogen and other extremely volatile elements (see Russell, 1941).

Almost 20 years after Goldschmidt, Suess and Urey (1956) published a new abundance table, which in part relied on solar abundances. In addition, Suess and Urey (1956) introduced arguments based on nucleosynthesis. Their so-called semiempirical abundance rules, primarily the smooth abundance variation of odd-mass nuclei with increasing mass number, were applied to estimate abundances for elements for which analytical data from meteorites were not available or had large errors. The Suess and Urey compilation was very influential for theories of nucleosynthesis and for the development of nuclear astrophysics in general. Later compilations by Cameron (1973), Anders and Grevesse (1989), Palme and Beer (1993), and others took into account improved analytical data on meteorites and the more accurate determination of elemental abundances in the solar photosphere. Over the years there has been a continuous convergence of abundances derived from meteorites and those obtained from solar absorption line spectroscopy. The agreement is now better than ±10% for most elements, as described below.

1.03.1.2 Solar System Abundances of the Elements

1.03.1.2.1 Is the elemental and isotopic composition of the solar nebula uniform?

In the past it was assumed that the Sun, the planets and all other objects of the solar system formed from a gaseous nebula with well-defined chemical and isotopic composition. The discovery of comparatively large and widespread variations in oxygen isotopic compositions has cast doubt upon this assumption (see Begemann, 1980 and references therein). In addition, evidence for incomplete mixing in the primordial solar nebula is provided by isotopic anomalies for a variety of elements in the refractory inclusions of carbonaceous chondrites and by detection of the huge isotope anomalies of carbon, nitrogen, silicon, and some heavy elements in tiny grains of meteorites, such as silicon carbide, nanodiamond, and graphite grains (e.g., Anders and Zinner, 1993; Chapter 1.02). A good example for such isotope anomalies is given in Figure 1, where the unusual isotopic composition of neodymium in an aggregate of SiC grains from the Murchison carbonaceous chondrite is shown (Richter *et al.*, 1992). The neodymium isotopic compositions of all other

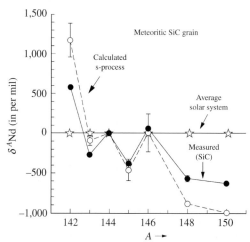

Figure 1 Nd isotopes in an aggregate of meteoritic SiC from Murchison. The deviation of the Nd isotopic composition from normal is given in per mil (δ). All ratios are normalized to ^{144}Nd. Full symbols are measured ratios. Error bars are in most cases smaller than symbol sizes. Calculated s-process productions are indicated. All previous analyses of Nd isotopes in terrestrial, lunar, or meteorite samples fall along the line marked "average solar system," which is used for normalization (source Richter *et al.*, 1992).

solar system materials analyzed (i.e., terrestrial, lunar, and meteoritic samples) are indistinguishable within the scale of Figure 1 and these other materials would fall on the line designated "average solar system." Such s-process components have also been found for other elements: For example, Nicolussi et al. (1998) identified nearly pure s-process molybdenum in some SiC grains (see Chapter 1.02, figure 10). These findings confirm the presence of material of distinct nucleosynthetic origins at the time of accretion of meteorite parent bodies. However, such isotope anomalies are confined to a very small fraction (a few ppm) of the bulk of a meteorite, i.e., this material is truly exotic. Moreover, it is likely that the more widespread oxygen isotope anomalies are not of nucleosynthetic origin but were produced by fractionation processes within the solar nebula or a precursor molecular cloud (Chapter 1.06), it is still a reasonable working hypothesis that the bulk of the matter of the solar system formed from a chemically and isotopically uniform reservoir, the primordial solar nebula. The composition of this nebula, the average solar system composition, is well known and carries the signatures of a variety of nucleosynthetic processes in stellar environments.

Although the elemental composition of the solar system is roughly similar to that of many other stars, in particular with respect to the relative abundances of the nongaseous elements, there are, in detail, compositional differences among stars and there are, in addition, truly exotic stars that make the term "cosmic abundances of elements" questionable. We will therefore use the term "solar system abundances of the elements" in this chapter.

1.03.1.2.2 The composition of the solar photosphere

The quantitative determination of elemental abundances in the Sun involves three steps: (i) construction of a numerical model atmosphere; (ii) calculation of an emitted spectrum based on the model atmosphere; and (iii) comparison of this spectrum with the observed spectrum (Cowley, 1995). Another assumption usually made in calculating solar abundances is that of local thermodynamic equilibrium (LTE), i.e., "the quantum-mechanical states of atoms, ions, and molecules are populated according to the relations of Boltzmann and Saha, valid strictly in thermodynamic equilibrium" (Holweger, 2001). It is not clear if, in the highly inhomogeneous and dynamic plasma permeated by an intense, anisotropic radiation, the LTE assumption is justified. In recent calculations effects of NLTE (nonthermal local equilibrium thermodynamics) and of photospheric granulation are taken into account (Holweger, 2001). Another important factor in the accuracy of solar abundance determinations are transition probabilities determined in laboratory experiments. The main need for improving solar abundance data is more accurate transition probabilities (Grevesse and Sauval, 1998).

In Table 1, the composition of the solar photosphere as obtained by absorption spectroscopy is given. Abundances are normalized to 10^{12} H atoms, the usual practice in astronomy. Most of the data are from Grevesse and Sauval (1998), which is an update of the photospheric abundance table by Anders and Grevesse (1989). For nitrogen, magnesium, silicon, and iron new photospheric abundances from Holweger (2001) were used. These data are marked H in Table 1. Their uncertainties range from about 30% for nitrogen to 12% for silicon. The accuracy of iron is given as 20%. The standard deviations listed by Holweger (2001) are, in all cases except silicon, larger than those given by Grevesse and Sauval (1998). Holweger ascribes this to his more conservative procedure for calculating errors. A new determination of the solar lead is included (Biemont et al., 2000) and marked B in Table 1. The oxygen abundance was taken from a paper by Allende Prieto et al. (2001), marked A1 in Table 1. The solar oxygen abundance has gone down considerably, from 8.93 ± 0.35 (Anders and Grevesse, 1989) and 8.83 ± 0.06 (Grevesse and Sauval, 1998) to 8.736 ± 0.078 (Holweger, 2001) and 8.69 ± 0.05 (Allende Prieto et al., 2001). This 50% decrease is important because, based on the old value, it was thought that the interstellar medium (ISM) must have had a different H/O ratio than the Sun (see below). The carbon abundance has also been revised downward, as indicated in Table 1 (Allende Prieto et al., 2002). The new carbon and oxygen lead to a higher C/O ratio of 0.50 ± 0.07 compared to the earlier ratio of 0.43 ± 0.06 (Anders and Grevesse, 1989), although both values overlap within error bars. This implies a somewhat more reducing nebular gas, as all C is present as CO, and the higher C/O ratio thus reduces the number of O_2 molecules.

Rare gases have no appropriate lines in the solar spectrum. The helium abundance of Grevesse and Sauval (1998) is derived from standard solar models. The helium abundance in the outer layers of the Sun seems to have decreased over the lifetime of the Sun, from $N_{He}/N_H = 0.098$ at the beginning of the solar system to the present value of $N_{He}/N_H = 0.085$ corresponding to an abundance of 10.93 ± 0.004 (Grevesse and Sauval, 1998), which is given in Table 1. A detailed discussion of the solar helium abundance can be found in Lodders (2003).

The abundance of neon was calculated from an Ne/Mg abundance ratio of 3.16 ± 0.07 derived from emerging magnetic flux regions observed in Skylab spectroheliograms. This value is thought to be representative of the Ne/Mg abundance

Table 1 Solar photospheric abundances and meteorite derived solar system abundances (log abundance $a(H) = 12$).

	Element	Solar photosphere	SD		Meteorite (CI)	SD	Sun/meteorite
1	H	12.00					
2	He	10.99[a]	0.02	G[b]			
3	Li	1.10	0.10	G	3.30	0.04	0.006
4	Be	1.40	0.09	G	1.41	0.04	0.98
5	B	2.70	−0.12, +0.21	C	2.77	0.04	0.74
6	C	8.39	0.04	A2	7.39	0.04	9.90
7	N	7.93	0.11	H	6.32	0.04	40.6
8	O	8.69	0.05	A1	8.43	0.04	1.82
9	F	4.56[a]	0.3	G	4.45	0.06	1.29
10	Ne	8.00[a]	0.07	H			
11	Na	6.33	0.03	G	6.30	0.02	1.07
12	Mg	7.54	0.06	H	7.56	0.01	0.94
13	Al	6.47	0.07	H	6.46	0.01	1.02
14	Si	7.54	0.05	H	7.55	0.01	0.99
15	P	5.45	(0.04)	G	5.44	0.04	1.02
16	S	7.33	0.11	G	7.19	0.04	1.37
17	Cl	5.5[a]	0.3	G	5.26	0.06	1.74
18	Ar	6.40[a]	0.06	G			
19	K	5.12	0.13	G	5.11	0.02	1.03
20	Ca	6.36	0.02	G	6.33	0.01	1.07
21	Sc	3.17	0.10	G	3.08	0.01	1.22
22	Ti	5.02	0.06	G	4.95	0.04	1.18
23	V	4.00	0.02	G	3.99	0.02	1.02
24	Cr	5.67	0.03	G	5.67	0.01	0.99
25	Mn	5.39	0.03	G	5.51	0.01	0.75
26	Fe	7.45	0.08	H	7.49	0.01	0.92
27	Co	4.92	0.04	G	4.90	0.01	1.05
28	Ni	6.25	0.04	G	6.23	0.02	1.05
29	Cu	4.21	0.04	G	4.28	0.04	0.85
30	Zn	4.60	0.08	G	4.66	0.04	0.87
31	Ga	2.88	(0.10)	G	3.11	0.02	0.59
32	Ge	3.41	0.14	G	3.62	0.04	0.62
33	As			G	2.35	0.02	
34	Se			G	3.40	0.04	
35	Br			G	2.61	0.04	
36	Kr	3.30[a]	0.06	P			
37	Rb	2.60	(0.15)	G	2.40	0.04	1.59
38	Sr	2.97	0.07	G	2.88	0.04	1.22
39	Y	2.24	0.03	G	2.21	0.04	1.06
40	Zr	2.60	0.02	G	2.59	0.04	1.02
41	Nb	1.42	0.06	G	1.39	0.04	1.07
42	Mo	1.92	0.05	G	1.95	0.04	0.93
44	Ru	1.84	0.07	G	1.80	0.04	1.11
45	Rh	1.12	0.12	G	1.10	0.08	1.05
46	Pd	1.69	0.04	G	1.68	0.04	1.01
47	Ag	(0.94	0.25)	G	1.23	0.04	
48	Cd	1.77	0.11	G	1.75	0.04	1.05
49	In	(1.66	0.15)	G	0.80	0.04	
50	Sn	2.00	(0.3)	G	2.12	0.04	0.76
51	Sb	1.00	(0.3)	G	1.00	0.04	0.99
52	Te			G	2.22	0.04	
53	I			G	1.50	0.08	
54	Xe	2.16[a]	0.09	P			
55	Cs			G	1.12	0.04	
56	Ba	2.13	0.05	G	2.21	0.04	0.83
57	La	1.17	0.07	G	1.21	0.02	0.91
58	Ce	1.58	0.09	G	1.62	0.02	0.90
59	Pr	0.71	0.08	G	0.80	0.04	0.81
60	Nd	1.50	0.06	G	1.48	0.02	1.04
62	Sm	1.01	0.06	G	0.98	0.02	1.08
63	Eu	0.51	0.08	G	0.55	0.02	0.92

(continued)

Table 1 (continued).

	Element	Solar photosphere	SD		Meteorite (CI)	SD	Sun/meteorite
64	Gd	1.12	0.04	G	1.08	0.02	1.10
65	Tb	(−0.1	0.3)	G	0.34	0.04	
66	Dy	1.14	0.08	G	1.16	0.02	0.96
67	Ho	(0.26	0.16)	G	0.50	0.04	
68	Er	0.93	0.06	G	0.96	0.02	0.93
69	Tm	(0.00	0.15)	G	0.15	0.04	
70	Yb	1.08	(0.15)	G	0.95	0.02	
71	Lu	0.06	0.10	G	0.13	0.04	0.85
72	Hf	0.88	(0.08)	G	0.74	0.04	1.37
73	Ta			G	−0.14	0.06	
74	W	(1.11	0.15)	G	0.66	0.04	
75	Re			G	0.29	0.04	
76	Os	1.45	0.10	G	1.38	0.04	1.18
77	Ir	1.35	(0.10)	G	1.36	0.02	0.97
78	Pt	1.80	0.30	G	1.67	0.06	1.36
79	Au	(1.01	0.15)	G	0.84	0.02	
80	Hg			G	1.15	0.08	
81	Tl	(0.9	0.2)	G	0.81	0.04	
82	Pb	2.00	0.06	B	2.05	0.04	0.89
83	Bi			G	0.69	0.06	
90	Th				0.07	0.04	
92	U	(<−0.47)		G	−0.52	0.04	

SD—standard deviation in dex: 0.1—12%, 0.2—60%, 0.3—100%. Meteorite data: log a_E + 1.546; a_E abundances relative to 10^6 Si atoms (see Table 3). Solar data: A1—Allende Prieto et al. (2001); A2—Allende Prieto et al. (2002); B—Biemont et al. (2000); C—Cunha and Smith (1999); G—Grevesse and Sauval (1998); H—Holweger (2001); P—Palme and Beer (1993).
[a] Abundances are not derived from the photosphere. [b] Average Sun; outer layers of Sun are 10% lower in He; values in parenthesis are defined as uncertain by Grevesse and Sauval (1998).

ratio in the solar photosphere (Reames, 1998). Following Holweger (2001) we adopt a value of log A_{Ne} = 8.001 ± 0.069 calculated from the Ne/Mg and Ne/O ratios of Reames (1998). The argon value is based on data from coronal spectra and a more precise SEP (solar energetic particles) value by Reames (1998), as discussed in Grevesse and Sauval (1998).

For krypton and xenon abundances were derived from computer fits of σN (neutron capture cross-section times abundance) versus mass number. Nuclei that are shielded from the r-process, so-called s-only nuclei, were used for the fit and the abundances of ^{82}Kr and ^{128}Xe were calculated. From these data, and the isotopic composition of the solar wind, the krypton and xenon elemental abundances were calculated (Palme and Beer, 1993) and are listed in Table 1. The meteorite data given in Table 1 will be discussed in a later section.

1.03.1.3 Abundances of Elements in Meteorites

1.03.1.3.1 Undifferentiated and differentiated meteorites

There are two different groups of meteorites—undifferentiated and differentiated. The undifferentiated meteorites are pieces of planetesimals that have never been heated to melting temperatures. Their chemical and isotopic composition should be representative of the bulk parent planetesimal from which the meteorites were derived. Differentiated meteorites are pieces of planetesimals that were molten and differentiated into core, mantle, and crust. A sample from such a body will not be representative of the bulk planet and it is not a trivial task to derive, from the samples available, the bulk composition of the parent planet. Undifferentiated meteorites reflect, at least to some degree, the composition of the solar nebula from which they formed. There are, however, variabilities in the composition of undifferentiated meteorites, which must reflect inhomogeneities in the solar nebula or disequilibrium during formation of solids from gas, or both. A more comprehensive discussion of meteorite classification is given by Krot et al. (Chapter 1.05).

1.03.1.3.2 Cosmochemical classification of elements

Many of the processes that are responsible for the variable chemical composition of primitive meteorites are related to the temperature of formation of meteoritic components. Although it is difficult to unambiguously ascertain the condensation origin of any single meteoritic component, it is clear from the elemental patterns that condensation processes must have occurred in the early solar system and that the volatilities of the elements in the solar nebula environment were important in establishing the various meteorite compositions. Condensation temperatures provide a convenient measure of volatility. These temperatures are calculated by assuming thermodynamic equilibrium between solids and a cooling gas of solar composition. Major elements condense as minerals while minor and trace elements condense in solid solution with the major phases. The temperature where 50% of an element is in the solid phase is called the 50% condensation temperature (Wasson, 1985; Lodders, 2003). Within this framework five more or less well-defined components that account for the variations in the elemental abundances in primitive meteorites may be defined (Table 2). In addition, the state of oxidation of a meteorite is an important parameter that adds to the complex textural variability among chondritic meteorites. According to the condensation temperatures the following components are distinguished:

(i) *Refractory component*. The first phases to condense from a cooling gas of solar composition are calcium, aluminum oxides, and silicates associated with a comparatively large number of trace elements, such as REEs, zirconium, hafnium, and scandium. These elements are often named refractory lithophile elements (RLEs), in contrast to the refractory siderophile elements (RSEs) comprising metals with low vapor pressures, e.g., tungsten, osmium, and iridium condensing at similarly high temperatures as multicomponent metal alloys. Both RLEs and RSEs are enriched in Ca–Al-rich inclusions (Chapter 1.08) by a factor of 20 on average, reflecting the fact that the refractory component makes up ~5% of the total condensible matter (Grossman and Larimer, 1974). Variations in Al/Si, Ca/Si, etc., ratios of bulk chondritic meteorites may be ascribed to the incorporation of variable amounts of an early condensed refractory phase.

(ii) *Magnesium silicates*. The major fraction of condensible matter is associated with the three most abundant elements heavier than oxygen—silicon, magnesium, and iron. In the reducing environment of the solar nebula iron condenses almost entirely as metal, while magnesium and silicon form forsterite (Mg_2SiO_4), which is, to a large extent, converted to enstatite ($MgSiO_3$) at lower temperatures by reaction with gaseous SiO. As forsterite has an atomic Mg/Si ratio twice the solar system ratio, loss or gain of a forsterite component is the most simple way for producing variations in Mg/Si ratios. Thus, variations in Mg/Si ratios of bulk meteorites are produced by the incorporation of various amounts of early-formed forsterite.

(iii) *Metallic iron*. Metal condenses as an Fe–Ni alloy at about the same temperature as forsterite, the sequence depending on pressure. At pressures above 10^{-4} bar iron metal condenses before forsterite and at lower pressures forsterite condenses ahead of metal (Grossman and Larimer, 1974). Variations in the concentrations of iron and other siderophile elements in meteorites are produced by the incorporation of variable fractions of metal.

(iv) *Moderately volatile elements*. These have condensation temperatures between those of magnesium silicates and FeS (troilite). The most abundant of the moderately volatile elements is sulfur, which condenses by reaction of gaseous sulfur with solid iron at 710 K, independent of pressure. Other moderately volatile elements condense in solid solution with major phases. Moderately volatile elements are distributed among sulfides, silicates, and metals.

Table 2 Cosmochemical classification of the elements.

	Elements	
	Lithophile (silicate)	Siderophile + chalcophile (sulfide + metal)
Refractory	$T_c = 1,850–1,400$ K Al, Ca, Ti, Be, Ba, Sc, V, Sr, Y, Zr, Nb, Ba, REE, Hf, Ta, Th, U, Pu	Re, Os, W, Mo, Ru, Rh, Ir, Pt, Rh
Main component	$T_c = 1,350–1,250$ K Mg, Si, Cr, Li	Fe, Ni, Co, Pd
Moderately volatile	$T_c = 1,230–640$ K Mn, P, Na, B, Rb, K, F, Zn	Au, Cu, Ag, Ga, Sb, Ge, Sn, Se, Te, S
Highly volatile	$T_c < 640$ K Cl, Br, I, Cs, Tl, H, C, N, O, He, Ne, Ar, Kr, Xe	In, Bi, Pb, Hg

T_c—Condensation temperatures at a pressure of 10^{-4} bar (Wasson, 1985; for B, Lauretta and Lodders, 1997).

Their abundances are in most cases below solar, i.e., they have lower element/silicon ratios than the Sun or CI chondrites, they are depleted (see below). In Figure 2, abundances of moderately volatile elements in CV3 meteorites relative to those in CI meteorites are plotted. Increasing depletions correlate with decreasing condensation temperatures but are independent of the geochemical properties of the elements. Depletions of moderately volatile elements in meteorites are produced by incomplete condensation. The amount and the relative abundances of these elements in meteorites are probably the result of removal of volatiles during condensation (Palme et al., 1988).

(v) *Highly volatile elements.* These have condensation temperatures below that of FeS (Table 1). The group of highly volatile elements comprises elements with very different geochemical affinity, such as the chalcophile lead and the atmophile elements nitrogen and rare gases. Similar processes as those invoked for the depletion of moderately volatile elements are responsible for variations in these elements. In addition, heating on small parent bodies may lead to loss of highly volatile elements.

(vi) *Oxygen fugacity and oxygen isotopic composition.* The oxygen fugacities recorded in meteoritic minerals are extremely variable, from the high oxygen fugacity recorded in the magnetite of carbonaceous chondrites to the extremely reducing conditions in enstatite chondrites, reflected in the presence of substantial amounts of metallic silicon dissolved in FeNi. The oxygen fugacity of a meteorite is, however, not well defined. Large variations in oxygen fugacity are often recorded in individual components of a single meteorite. The various components of primitive meteorites apparently represent extreme disequilibrium. Oxygen isotopes give a similar picture (Clayton, 1993). It has been suggested that variations in $\Delta^{17}O$ were produced by reaction of ^{16}O-rich material with a gas rich in ^{17}O and ^{18}O (Clayton, 1993). The gas phase may be considered an additional independent component of meteorites. Thus, the extent of the gas–solid reaction at various temperatures, and possibly also fluid–solid reactions, on a parent body determine the degree of oxidation and the oxygen fugacity of meteoritic components and bulk meteorites (Chapter 1.06).

1.03.1.4 CI Chondrites as Standard for Solar Abundances

1.03.1.4.1 Chemical variations among chondritic meteorites

In Figure 3 the variations of selected element ratios in the different groups of chondritic meteorites are shown. All ratios are normalized to the best estimate of the average solar system ratios, the CI ratios (taken from Table 3, see below). Meteorite groups are arranged in the sequence of decreasing bulk oxygen contents, i.e., decreasing average oxygen fugacity. Element ratios in the solar photosphere determined by absorption line spectroscopy are shown for comparison (Table 1).

In Figure 3, aluminum is representative of refractory elements in general and the Al/Si ratios indicate the size of the refractory component relative to the major fraction of the meteorite. It is clear from this figure that the Al/Si ratio of CI meteorites agrees best with the solar ratio, although the ratios in CM (Type 2 carbonaceous chondrites) and even OC (ordinary chondrites) are almost within the error bar of the solar ratio. The errors of the meteorite ratios are below 10%, in many cases below 5%. A very similar pattern as for aluminum would be obtained for other refractory elements (calcium, titanium, scandium, REEs, etc.), as ratios among refractory elements in meteorites are constant in all classes of chondritic meteorites, at least within ~5–10%. The average Sun/CI meteorite ratio of 19 refractory lithophile elements (Al, Ca, Ti, V, Sr, Y, Zr, Nb, Ba, La, Ce, Pr, Nd, Sm, Eu, Gd, Dy, Er, Lu, see Table 2) is 1.004 with a standard deviation of 0.12. The accuracy of the solar abundance determination of each of these elements is better than 25% (see Table 1). Elements with larger errors were not considered. Calculating the standard error of the mean for the average refractory element/silicon ratio leads to an uncertainty of about ±0.03 in the

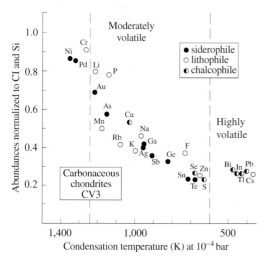

Figure 2 Abundances of volatile elements in CV3 chondrites (e.g., Allende) normalized to CI chondrites and Si. There is a continuous decrease of abundances with increasing volatility as measured by the condensation temperature. The sequence contains elements of very different geochemical character, indicating that volatility is the only relevant parameter in establishing this pattern (source Palme, 2000).

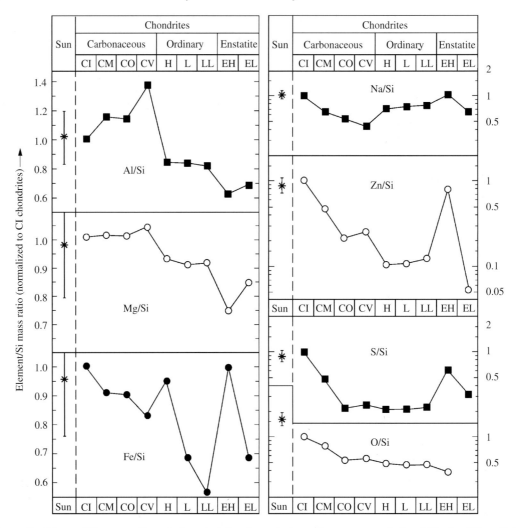

Figure 3 Element/Si mass ratios of characteristic elements in various groups of chondritic (undifferentiated) meteorites. Meteorite groups are arranged according to decreasing oxygen content. The best match between solar abundances and meteoritic abundances is with CI meteorites (see text for details).

ratio of refractory elements between CI meteorites and the Sun. Thus the absolute level of refractory elements (measured by refractory element/silicon ratios) and ratios among refractory elements are the same in CI meteorites and the Sun. The level of refractory elements in other chondritic meteorites is higher in CM (by 13%) and in CV (by 25%) and lower in H chondrites (by 10%) and enstatite chondrites (by 20%). Thus, the agreement between refractory elements in the Sun and in CI meteorites is statistically significant and all other groups of meteorites will not match solar refractory element abundances. The CR meteorites (Renazzo type meteorites) have CI ratios of Al/Si (Bischoff et al., 1993) (not shown in Figure 3). However, these meteorites are depleted in volatile elements relative to CI meteorites, disqualifying them as a solar system standard.

The Mg/Si ratios of CI chondrites also match with the solar abundance ratio (Figure 3). This is, however, less diagnostic, as all groups of carbonaceous chondrites have the same Mg/Si ratio (Wolf and Palme, 2001). The OC and the EC have significantly less magnesium. The error of the solar ratio is ~19% (combining the errors of silicon and magnesium, Table 1) and thus covers the range of all classes of chondrites except EH chondrites.

Until recently, the range of the bulk iron content in chondritic meteorites varied by about a factor of 2, with most meteorite groups being depleted in iron (Figure 3). There are new, recently discovered subgroups of carbonaceous chondrites, with large excesses of iron (see Chapter 1.05). These CH and CB chondrites indicate that metal behaves as an independent component, some groups of chondritic meteorites are enriched in iron and other siderophile elements, others are depleted. The excellent agreement of CI meteorite abundance ratios with solar abundance ratios is

Table 3 Solar system abundances based on CI meteorites.

		Palme and Beer (1993, updated)				Anders and Grevesse (1989)	
	Element	Mean CI abundance (by weight)	Lit.	Estimated accuracy (%)	Atoms per 10^6 atoms of Si	Mean CI abundance (by weight)	Atoms per 10^6 atoms of Si
1	H	2.02 (%)	*	10	5.27×10^6	2.02 (%)	
2	He	56 (nL g^{-1})	*			56 (nL g^{-1})	
3	Li	1.49 (ppm)		10	56.5	1.50 (ppm)	57.1
4	Be	0.0249 (ppm)		10	0.727	0.0249 (ppm)	0.73
5	B	0.69 (ppm)	Z	13	16.8	0.870 (ppm)	21.2
6	C	3.22 (%)	*	10	7.05×10^5	3.45 (%)	1.01×10^7
7	N	3,180 (ppm)	*	10	5.97×10^4	0.318 (%)	3.13×10^6
8	O	46.5 (%)	*	10	7.64×10^6	46.4 (%)	2.38×10^7
9	F	58.2 (ppm)		15	806	60.7 (ppm)	843
10	Ne	203 (pL g^{-1})	*			203 (pL g^{-1})	
11	Na	4,982 (ppm)		5	5.70×10^4	5,000 (ppm)	5.74×10^4
12	Mg	9.61 (%)		3	1.04×10^6	9.89 (%)	1.074×10^6
13	Al	8,490 (ppm)	W	3	8.27×10^4	8,680 (ppm)	8.49×10^4
14	Si	10.68 (%)		3	$\equiv 10^6$	10.64 (%)	$\equiv 10^6$
15	P	926 (ppm)	W	7	7.86×10^3	1,220 (ppm)	1.04×10^4
16	S	5.41 (%)	D	5	4.44×10^5	6.25 (%)	5.15×10^5
17	Cl	698 (%)		15	5.18×10^3	704 (ppm)	5.24×10^3
18	Ar	751 (pL g^{-1})	*			751 (pL g^{-1})	
19	K	544 (ppm)		5	3.66×10^3	558 (ppm)	3.77×10^3
20	Ca	9,320 (ppm)	W	3	6.12×10^4	9,280 (ppm)	6.11×10^4
21	Sc	5.90 (ppm)		3	34.5	5.82 (ppm)	34.2
22	Ti	458 (ppm)	W	4	2.52×10^3	436 (ppm)	2.40×10^3
23	V	54.3 (ppm)		5	280	56.5 (ppm)	293
24	Cr	2,646 (ppm)		3	1.34×10^4	2,660 (ppm)	1.35×10^4
25	Mn	1,933 (ppm)		3	9.25×10^3	1,990 (ppm)	9.55×10^3
26	Fe	18.43 (%)		3	8.68×10^5	19.04 (%)	9.00×10^5
27	Co	506 (ppm)		3	2.26×10^3	502 (ppm)	2.25×10^3
28	Ni	1.077 (%)		3	4.82×10^4	1.10 (%)	4.93×10^4
29	Cu	131 (ppm)		10	542	126 (ppm)	522
30	Zn	323 (ppm)		10	1.30×10^3	312 (ppm)	1.26×10^3
31	Ga	9.71 (ppm)		5	36.6	10.0 (ppm)	37.8
32	Ge	32.6 (ppm)		10	118	32.7 (ppm)	119
33	As	1.81 (ppm)		5	6.35	1.86 (ppm)	6.56
34	Se	21.4 (ppm)	D	5	71.3	18.6 (ppm)	62.1
35	Br	3.50 (ppm)		10	11.5	3.57 (ppm)	11.8
36	Kr	8.7 (pL g^{-1})				8.7 (pL g^{-1})	
37	Rb	2.32 (ppm)		5	7.14	2.30 (ppm)	7.09
38	Sr	7.26 (ppm)		5	21.8	7.80 (ppm)	23.5
39	Y	1.56 (ppm)	J2	3	4.61	1.56 (ppm)	4.64
40	Zr	3.86 (ppm)	J2	2	11.1	3.94 (ppm)	11.4
41	Nb	0.247 (ppm)	J2	3	0.699	0.246 (ppm)	0.698
42	Mo	0.928 (ppm)		5	2.54	0.928 (ppm)	2.55
44	Ru	0.683 (ppm)	J1	3	1.78	0.712 (ppm)	1.86
45	Rh	0.140 (ppm)	J1	3	0.358	0.134 (ppm)	0.344
46	Pd	0.556 (ppm)		10	1.37	0.560 (ppm)	1.39
47	Ag	0.197 (ppm)		10	0.480	0.199 (ppm)	0.486
48	Cd	0.680 (ppm)		10	1.59	0.686 (ppm)	1.61
49	In	0.0780 (ppm)		10	0.178	0.080 (ppm)	0.184
50	Sn	1.68 (ppm)		10	3.72	1.720 (ppm)	3.82
51	Sb	0.133 (ppm)		10	0.287	0.142 (ppm)	0.309
52	Te	2.27 (ppm)		10	4.68	2.320 (ppm)	4.81
53	I	0.433 (ppm)		20	0.897	0.433 (ppm)	0.90
54	Xe	8.6 (pL g^{-1})				8.6 (pL g^{-1})	
55	Cs	0.188 (ppm)		5	0.372	0.187 (ppm)	0.372
56	Ba	2.41 (ppm)		10	4.61	2.340 (ppm)	4.49
57	La	0.245 (ppm)		5	0.464	0.2347 (ppm)	0.4460
58	Ce	0.638 (ppm)		5	1.20	0.6032 (ppm)	1.136

(continued)

Table 3 (continued).

	Element	Palme and Beer (1993, updated)				Anders and Grevesse (1989)	
		Mean CI abundance (by weight)	Lit.	Estimated accuracy (%)	Atoms per 10^6 atoms of Si	Mean CI abundance (by weight)	Atoms per 10^6 atoms of Si
59	Pr	0.0964 (ppm)		10	0.180	0.0891 (ppm)	0.1669
60	Nd	0.474 (ppm)		5	0.864	0.4524 (ppm)	0.8279
62	Sm	0.154 (ppm)		5	0.269	0.1471 (ppm)	0.2582
63	Eu	0.0580 (ppm)		5	0.100	0.0560 (ppm)	0.0973
64	Gd	0.204 (ppm)		5	0.341	0.1966 (ppm)	0.3300
65	Tb	0.0375 (ppm)		10	0.0621	0.0363 (ppm)	0.0603
66	Dy	0.254 (ppm)		5	0.411	0.2427 (ppm)	0.3942
67	Ho	0.0567 (ppm)		10	0.0904	0.0556 (ppm)	0.0889
68	Er	0.166 (ppm)		5	0.261	0.1589 (ppm)	0.2508
69	Tm	0.0256 (ppm)		10	0.0399	0.0242 (ppm)	0.0378
70	Yb	0.165 (ppm)		5	0.251	0.1625 (ppm)	0.2479
71	Lu	0.0254 (ppm)		10	0.0382	0.0243 (ppm)	0.0367
72	Hf	0.107 (ppm)		5	0.158	0.104 (ppm)	0.154
73	Ta	0.0142 (ppm)	J2	6	0.0206	0.0142 (ppm)	0.0207
74	W	0.0903 (ppm)	J1	4	0.129	0.0926 (ppm)	0.133
75	Re	0.0395 (ppm)	J1	4	0.0558	0.0365 (ppm)	0.0517
76	Os	0.506 (ppm)	J1*	2	0.699	0.486 (ppm)	0.675
77	Ir	0.480 (ppm)	J1	4	0.657	0.481 (ppm)	0.661
78	Pt	0.982 (ppm)	J1	4	1.32	0.990 (ppm)	1.34
79	Au	0.148 (ppm)	J1	4	0.198	0.140 (ppm)	0.187
80	Hg	0.310 (ppm)		20	0.406	0.258 (ppm)	0.34
82	Tl	0.143 (ppm)		10	0.184	0.142 (ppm)	0.184
82	Pb	2.53 (ppm)		10	3.21	2.470 (ppm)	3.15
83	Bi	0.111 (ppm)		15	0.140	0.114 (ppm)	0.144
90	Th	0.0298 (ppm)		10	0.0338	0.0294 (ppm)	0.0335
92	U	0.00780 (ppm)		10	0.0086	0.0081 (ppm)	0.0090

Data from Palme and Beer (1993), except: rare gases in nL g^{-1} or pL g^{-1} at (STP) from Anders and Grevesse (1989), elements marked D (Dreibus et al., 1995), J1 (Jochum, 1996); J1* Os is calculated from the average carbonaceous chondrite ratio Ir/Os of 0.0949 from Jochum (1996); J2 (Jochum et al., 2000); W (Wolf and Palme, 2001); Z (Zhai and Shaw, 1994); * elements incompletely condensed in CI meteorites. Average CI abundances from Anders and Grevesse (1989), Table 1, columns 6 and 2, except C, N, O and rare gases He, Ne, Ar, Kr, Xe which are only from Orgueil.

obvious, although the formal error of the solar iron abundance is quite large (Table 1). However, the same argument that has been applied to refractory elements can be used here. The patterns of nickel and cobalt are identical to that of iron and the errors in the solar abundances of nickel and cobalt are only half of the error in the iron abundance determination (Table 1). The combined error associated with the solar/CI meteorite ratio for these three metals is then below 10%, assuming constant ratios between the three metals. Thus only CI, H, and EH chondrites have solar Fe/Si ratios. As H and EH chondrites have very different refractory element contents, they are not suitable meteorites for representing solar abundances.

In Figure 3, sodium, zinc, and sulfur are representative of the abundances of moderately volatile elements (Figure 2 and Table 2). Abundance variations reach a factor of 5 for sulfur and 10 for zinc. All three elements show excellent agreement of solar with CI abundances, in contrast to other groups of chondritic meteorites, except for the enstatite chondrites, which reach the level of CI abundances. However, enstatite chondrites cannot be representative of solar abundances because of their low refractory element contents and their fractionated Mg/Si ratios (Figure 3).

The new photospheric lead determinations (Table 1) show that there is now agreement between the CI lead and the solar lead abundance. Thus the excellent match of CI chondrites with the solar photosphere can be extended to some of the highly volatile elements.

The CI chondrites not only have the highest contents of volatile elements but also the highest content of oxygen (a large fraction in the form of water) of all chondritic meteorites. However, in contrast to other highly volatile elements, such as lead, the amount of oxygen contained in CI meteorites is still a factor of 2 below that of the solar photosphere (Figure 3), implying that water is not fully condensed. The concept of incomplete condensation of volatile elements in most meteorite groups is supported by the observation that there is no group of meteorites that is enriched in these elements relative to the average solar system abundances. There was no late redistribution of volatiles, for example, by reheating. In CI meteorites, the moderately volatile elements and some of the highly volatile elements are fully condensed; other groups of chondrites acquired lower fractions of volatiles because the solar gas dissipated during condensation (Palme et al., 1988).

In summary, there is only one group of meteorites, the CI chondrites, that closely match solar abundances for elements representing the various cosmochemical groups and excluding the extremely volatile elements such as the rare gases, hydrogen, carbon, oxygen, and nitrogen and also the element lithium, which will be discussed in Section 1.03.1.5.3. All other chondrite groups deviate from solar abundances and the deviations can be understood, at least in principle, by gas–solid fractionation processes in the early solar system.

1.03.1.4.2 CI chondrites

Among the more than twenty thousand recovered meteorites there are only five CI meteorites: Orgueil, Ivuna, Alais, Tonk, and Revelstoke. These meteorites are very fragile and are easily fragmented on atmospheric entry. In addition, their survival time on Earth is short. All five CI meteorites are observed falls. Most analyses have been performed on the Orgueil meteorite, simply because Orgueil is the largest CI meteorite and material is easily available for analysis. However, problems with sample size, sample preparation, and the mobility of some elements are reflected in the chemical inhomogeneities within the Orgueil meteorite, and often make a comparison of data obtained by different authors difficult. This contributes significantly to the uncertainties in CI abundances.

The chemical composition of CI chondrites as shown in Figure 3 is the basis for designating CI meteorites as primitive or unfractionated. Texturally and mineralogically they are far from being primitive. CI meteorites are microbreccias with millimeter to submillimeter clasts with variable composition. Late stage fractures filled with carbonates, hydrous calcium, and magnesium sulfate demonstrate that low temperature processes have affected the meteorite. The CI meteorites have no chondrules and thus consist almost entirely of extremely fine-grained hydrous silicates with some magnetite. High temperature phases such as olivine and pyroxene are frequently found (Dodd, 1981). Although the CI meteorites undoubtedly match solar abundances very closely, processes that occurred late on the Orgueil parent body have largely established their present texture and mineralogy. On a centimeter scale these processes must have been essentially isochemical, otherwise the composition of Orgueil would not be solar for so many elements.

1.03.1.4.3 The CI abundance table

The CI abundance table given here (Table 3) is largely based on the compilation by Palme and Beer (1993). Relevant data published after 1993 are incorporated as described below. In addition, the widely used CI data of Anders and Grevesse (1989) are listed for comparison.

In the 1993 compilation rhodium was the only element that had not been determined in the Orgueil meteorite. Jochum (1996) reported a value of 134 ppb for his Orgueil analysis, in good agreement with the value listed in the 1993 compilation, estimated from rhodium in other meteorites. The values for ruthenium, tungsten, rhenium, osmium, iridium, platinum, and gold were all taken from the paper by Jochum (1996) and are marked J1 in Table 3. The difference between the new data from Jochum (1996) and the old data listed in the Palme and Beer (1993) compilation is in all cases below 5%.

New sulfur and selenium data for CI chondrites were obtained by Dreibus et al. (1995) in a study of sulfur and selenium in chondritic meteorites. With the measured CI contents of 5.41% for sulfur and 21.4 ppm for selenium in Orgueil a CI S/Se ratio of 2,540 is obtained. A nearly identical S/Se ratio of 2,560 ± 150 was calculated as average of carbonaceous chondrites, considering only observed meteorite falls. The average ratio of all meteorite falls analyzed by Dreibus et al. (1995) was 2,500 ± 270. These data suggest that the new sulfur and selenium content of Orgueil provides a reliable estimate for the solar system average. The new CI sulfur content is slightly higher than the sulfur content given in the Palme and Beer (1993) compilation and significantly lower than the sulfur content of 6.25% in the compilation of Anders and Grevesse (1989). The corresponding change in selenium is to 21.4 from 21.3 ppm in Palme and Beer (1993) and from 18.6 ppm in Anders and Grevesse (1989).

New phosphorus and titanium XRF data for CI chondrites were reported by Wolf and Palme (2001). The change in phosphorus is significant. The new CI abundance is 926 ppm, which is much lower than the values in the older compilations, 1,105 ppm in Palme and Beer (1993) and 1,200 ppm in Anders and Grevesse (1989), respectively. The new phosphorus contents are considered to be more reliable. The changes in titanium are small. Wolf and Palme (2001) also reported major element concentrations of CI meteorites and other carbonaceous chondrites. Their magnesium and silicon contents were almost identical to those given by Palme and Beer (1993): 10.69% versus 10.68% (for silicon) in Palme and Beer (1993) and 9.60% versus 9.61% (for magnesium) in Palme and Beer (1993). The aluminum, calcium, and iron concentrations of Palme and Beer (1993) were, however, slightly different. The average between the new data and those listed earlier in Palme and Beer (1993) was used in the present compilation.

Figure 4 is a plot of some refractory and moderately volatile elements in the various types of carbonaceous chondrites. Calculated condensation temperatures at a pressure of 10^{-4} bar (Wasson, 1985) are indicated to the right. There is a trend for increasing depletions (decreasing abundances) with decreasing condensation temperatures (see also Figure 2). However, the sequence of decreasing condensation temperatures for silicon, chromium, iron, and phosphorus does not match with the depletions observed in carbonaceous chondrites. The differences are small; however, the condensation temperatures of phosphorus and chromium are not well known and the condensation temperature of iron (condensing as metal) relative to silicon and other lithophile elements is dependent on nebular pressure (Grossman and Larimer, 1974). These minor discrepancies do not change the basic conclusion that these depletions are related to condensation temperatures. A number of conclusions with regard to the CI abundances can be drawn from Figure 4.

(i) The Mg/Si weight ratio, 0.90, is constant within one percent in bulk carbonaceous chondrites, except for a three percent depletion of magnesium in CV chondrites. There is no apparent trend that would suggest volatility related depletion of silicon, comparable to the depletions observed for chromium, manganese, zinc, etc.

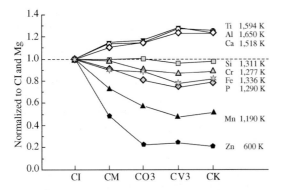

Figure 4 Abundances of refractory and moderately volatile elements in various groups of carbonaceous chondrites, normalized to CI and Mg. Refractory elements increase from CI to CV3 chondrites while Mg/Si ratios are constant in all groups of carbonaceous chondrites. Although the elements Cr, Fe, and P are significantly less depleted than Mn and Zn, they show a similar behavior, suggesting volatility related depletions of Cr, Fe, and P in carbonaceous chondrites of higher metamorphic grades (source Wolf and Palme, 2001).

(ii) There is a clear trend of decreasing Fe/Mg ratios with the increasing depletion of moderately volatile elements, suggesting that the variations in iron contents are probably related to the volatility of iron. Thus in defining the CI ratio for iron one cannot rely on other carbonaceous chondrites. There is an uncertainty of ~5% in the CI Fe/Mg ratio of 1.92 (by weight) given in Table 1. This uncertainty reflects variable iron contents in CI samples and it will be difficult to obtain more accurate CI iron contents.

(iii) A similar problem is encountered with refractory elements. Wolf and Palme (2001) noted a 20% variation in the calcium content of various CI meteorite samples. The corresponding aluminum contents are fortunately much more constant and the Al/Mg ratio of 0.0865 is estimated to be accurate to within 2%. As a result there is some uncertainty in the solar system Ca/Al ratio, from ~1.07 to 1.10. Members of the reduced subgroup in CV chondrites have Ca/Al ratios at least 10% below those of other carbonaceous chondrites (Wolf and Palme, 2001). In general, however, ratios among refractory elements in other chondritic meteorites may be used to improve the accuracy of chondritic refractory element ratios, which are variable in CI meteorites due to inhomogeneous distribution. An extreme example for CI variability is U. Rocholl and Jochum (1993) found Th/U ratios in CI chondrites varying from 1.06 to 3.79. The variability is primarily a result of low temperature mobilization of uranium under aqueous conditions. As other chondritic meteorites are also variable, although to a lesser extent, the CI uranium content is not too well determined. Rocholl and Jochum (1993) suggest a value of 3.9 ± 0.2. Other elements, such as barium, caesium, and antimony, may also be affected by alteration processes on parent bodies, although to a lesser extent (Rocholl and Jochum, 1993).

(iv) The geochemically similar, but cosmochemically dissimilar elements magnesium and chromium are fractionated in carbonaceous chondrites: chromium behaves as a slightly more volatile element than magnesium (Figure 4). The Fe/Cr ratio is apparently less variable than the Mg/Cr ratio reflecting the volatility related behavior of iron in carbonaceous chondrites (Figure 4). Wolf and Palme (2001) found an average Mg/Cr ratio in CI of 36.52 compared to 40.75 in CO chondrites, but Fe/Cr ratios of 70.89 for CI and 71.05 for CO. Much stronger variations are found for other moderately volatile elements, such as manganese, sodium, potassium, sulfur, zinc, selenium, etc. (see Figures 2–4). The concentrations of these elements in other carbonaceous chondrites cannot be used to infer or improve their CI abundances.

1.03.1.4.4 Comparison with Anders and Grevesse abundance table

A comparison with the Anders and Grevesse (1989) compilation (Table 3) shows that there are few elements for which the difference between this compilation and that of Anders and Grevesse (1989) exceeds five percent. This is not surprising as both compilations rely, at least in part, on the same sources of data. Differences above 10% are found for phosphorus, sulfur, and selenium. The new values listed, as discussed above, are more reliable. Differences between 5% and 10% are found for carbon, strontium, antimony, cerium, praseodymium, rhenium, and mercury. The rare earth abundaces in CI meteorites reported here are the same as those in Palme and Beer (1993), which are based on data by Evenson et al. (1978). There is an excellent agreement of the REE data of Evenson et al. (1978) with those in two Orgueil samples determined by Beer et al. (1984), except for a small difference of a few percent in the light REE. Because of this general agreement the hafnium, barium, strontium, rubidium, and caesium concentrations of Beer et al. (1984) were used in the present compilation. The Anders and Grevesse (1989) REE and strontium data were obtained by averaging data sets of various authors. The absolute concentrations of the REEs are slightly lower, exceeding 5% for strontium, cerium, and praseodymium. The differences in carbon and antimony concentrations between Palme and Beer (1993) and Anders and Grevesse (1989) are related to the selection of data literature. Mercury abundances in CI chondrites are extremely variable. The high contents of mercury in Orgueil may reflect contamination (Palme and Beer, 1993). A single analysis of Ivuna (0.31 ppm) was therefore used in the present compilation. The value of 0.31 ppm is in agreement with a mercury content inferred from nuclear abundance systematics (Palme and Beer, 1993).

1.03.1.5 Solar System Abundances of the Elements

1.03.1.5.1 Comparison of meteorite and solar abundances

In Table 1 the Si-normalized meteorite abundances of Table 3 ($\log A_{Si} = 6$) are converted to the H-normalized abundances ($\log A_H = 12$). The conversion factor between the two scales was calculated by dividing the H-normalized solar abundances by the Si-normalized meteorite abundances. The comparison was made for all elements with an error of the corresponding photospheric abundance of less than 0.1 dex, i.e., less than ~25%. Thirty-four elements qualified for this procedure, and the log of the average ratio of solar abundance per 10^{12} H atoms/meteorite abundance per 10^6 silicon atoms is 1.546 ± 0.045. As each of the solar and meteorite abundance measurements are independent the error of the mean may be used which gives 1.546 ± 0.008, corresponding to a ratio of 35.16 ± 0.65. Thus,

$$\log A_{ast} = \log A_{met} + 1.546$$

This yields a silicon abundance on the astronomical scale of $\log A_{ast}(Si) = 7.546$ and a hydrogen abundance on the meteoritic scale of $\log A_{met}(H) = 10.45$ or 2.84×10^{10} which is given in Table 3. Anders and Grevesse (1989) calculated a value of 1.554 for the ratio of solar to meteoritic abundances, which leads to a hydrogen abundance of 2.97×10^{10} on the meteoritic scale. Lodders (2003) used a conversion factor of 1.540 based on the ratio of photospheric and meteoritic silicon.

In column 6 of Table 1, the meteorite data are given on the astronomical scale and in column 8 the ratios of the photospheric abundances to the meteoritic abundances are listed. These ratios are displayed in Figure 5. The solar abundances of carbon, nitrogen, and oxygen are higher because these elements are incompletely condensed in CI meteorites. Although the new solar oxygen abundance is 50% lower than the old value, the solar oxygen abundance is still a factor of 2 above the meteorite value, which is, however, significantly less than the depletion of nitrogen (factor 41) and of carbon (factor 9.9). A major revision in photospheric abundances was found for beryllium which is now in good agreement with meteoritic abundances (Grevesse and Sauval, 1998). As the photospheric abundance of boron appears to fit with meteorite data, at least within a factor of 2, which is also the error assigned to the photospheric boron determination by Grevesse and Sauval (1998), lithium is the only light element that is strongly depleted in the Sun. "The Li–Be–B problem is now reduced to explaining how the Sun can deplete Li by a factor of 160 whereas Be and B are not destroyed" (Grevesse and Sauval, 1998).

The recent revision of the photospheric lead abundance brought this element into agreement with meteoritic lead abundances to within 10% (Biemont et al., 2000).

In comparing meteorite with solar abundances in Figure 5 only ratios are plotted where the error of the photospheric abundances is below 0.1 dex, i.e., below 25%. In the Figure 5 ±10% variation is indicated. The largest discrepancy between solar and meteoritic abundances is now found for sulfur, scandium, strontium, and manganese, the only elements where the difference between solar and meteoritic abundances exceeds 20%. The solar sulfur and scandium abundances had been revised by Grevesse and Sauval (1998). Their new values are higher than those of

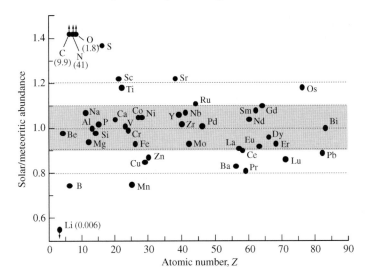

Figure 5 Comparison of solar and meteoritic abundances (see Table 1). The elements C, N, and O are incompletely condensed in meteorites. Li is consumed by fusion processes in the interior of the Sun, but not Be and B. Solar and meteoritic abundances agree in most cases within 10%. Only the four elements S, Mn, Sc, and Sr differ by more than 20% from CI abundances. The difference is below 10% for 27 elements. Only elements with uncertainties of less than 25% in the photosphere are plotted.

Anders and Grevesse (1989), which had agreed to within 5% with CI abundances. The error in the photospheric abundance of strontium is 17% according to Grevesse and Sauval (1998) and the difference is thus not significant. The only element that has consistently shown a major deviation within the stated error is manganese (Figure 5). In the compilation of Grevesse and Sauval (1998) an uncertainty of 7% is assigned to the manganese abundance. Recently Prochaska and McWilliam (2000) pointed out that there are so far unrecognized problems in the photospheric determination of manganese by incorrect treatment of hyperfine splitting. These authors also mention that similar problems may be involved with the photospheric abundance determination of scandium. The errors associated with the photospheric abundances of the other elements are all in excess of 10%, some are considerably higher, e.g., praseodymium with 20%.

The agreement between meteoritic and solar abundances must be considered excellent and there is not much room left for further improvements. Obvious candidates for redetermination of the solar abundances are manganese and sulfur.

1.03.1.5.2 Solar system abundances versus mass number

The isotopic compositions of the elements are not discussed in this chapter. The compilation by Palme and Beer (1993) contains a list of isotopes and their relative abundances. Lodders (2003) has prepared a new compilation. From these data the abundances for individual mass numbers can be calculated using the elemental abundances as given in Table 1. Figure 6 is a plot of abundances versus mass number. The generally higher abundance of even masses is apparent. Plots for even and odd mass numbers are more or less smooth, the latter forming a considerably smoother curve than the former. Historically, the so-called abundance rules, established by Suess (1947), postulating a smooth dependence of isotopic abundances on mass number A, especially of odd-A nuclei, played an important role in estimating unknown or badly determined abundances. Later this rule was modified and supplemented by two additional rules (Suess and Zeh, 1973) in order to make the concept applicable to the now more accurate abundance data. However, the smoothness of odd-A nuclei abundances itself has been questioned (Anders and Grevesse, 1989; Burnett and Woolum, 1990). Figure 7(a) is an enlargement of a part of Figure 6. The high abundance of ^{89}Y (Figure 7(a)), an apparent discontinuity (Burnett and Woolum, 1990), reflects the low neutron capture cross-section of a dominantly s-process nucleus with a magic neutron number (50). Also, there are major breaks in the abundance curve of odd-A nuclei in the region of molybdenum, rhodium, indium, tin, and antimony (Figure 7(a)). In detail there is therefore no smoothness of odd-A nuclei with mass number. Similar arguments apply to the abundances of odd-A isotopes of neodymium, samarium, and europium (Figure 7(b)). They do not follow a smooth trend with the lower abundance of samarium.

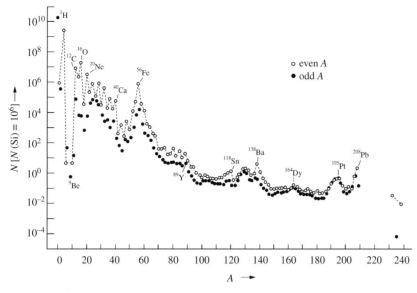

Figure 6 Solar system abundances by mass number. Atoms with even masses are more abundant than those with odd masses (Oddo–Harkins rule) (source Palme and Beer, 1993).

1.03.1.5.3 Other sources for solar system abundances

Emission spectroscopy of the solar corona, solar energetic particles (SEP) and the composition of the solar wind yield information on the composition of the Sun. Solar wind data were used for isotopic decomposition of rare gases. Coronal abundances are fractionated relative to photospheric abundances. Elements with high first ionization potential are depleted relative to the rest (see Anders and Grevesse, 1989 for details).

The composition of dust grains of comet Halley has been determined with impact ionization time-of-flight mass spectrometers on board the Vega-I, Vega-II, and Giotto spacecrafts. The abundances of 16 elements and magnesium, which is used for normalization, are on average CI chondritic to within a factor of 2–3, except for hydrogen, carbon, and nitrogen which are significantly higher in Halley dust, presumably due to the presence of organic compounds (Jessberger et al., 1988). There is no evidence for a clear enhancement of volatile elements relative to CI.

Many of the micron-sized interplanetary dust particles (IDPs) have approximately chondritic bulk composition (see Chapter 1.26 for details). Porous IDPs match the CI composition better than nonporous (smooth) IDPs. On an average, IDPs show some enhancement of moderately volatile and volatile elements (see Palme, 2000). Arndt et al. (1996) found similar enrichments in their suite of 44 chondritic particles (average size 17.2 ± 1.2 μm). The elements chlorine, copper, zinc, gallium, selenium, and rubidium were enriched by factors of 2.2–2.7. In addition, these authors also reported very high enrichments of bromine (29 × CI) and arsenic (7.4 × CI), perhaps acquired in the Earth's atmosphere.

These particles probably come from the asteroid belt (Flynn, 1994). They are brought to Earth by the action of the Poynting–Robertson effect. Perhaps they are derived from sources that contain uncondensed volatiles from the inner part of the nebula. This would be the only example of a clear enhancement of moderately volatile elements in solar system material.

1.03.2 THE ABUNDANCES OF THE ELEMENTS IN THE ISM

1.03.2.1 Introduction

The solar system was formed as the result of the collapse of a cloud of pre-existing interstellar gas and dust. We should therefore expect a close compositional relationship between the solar system and the interstellar material from which it formed. If we make the assumption that the composition of the ISM has remained unchanged since the formation of the solar system, we can use the local ISM as a measure of the original presolar composition. Differences between the solar system and current local ISM would imply that fractionation occurred during the formation of the solar system, that the local ISM composition changed after solar system formation or that the solar system formed in a different part of the galaxy and then migrated to its present location. Studies of solar system and local ISM composition are therefore fundamental to the formation of the

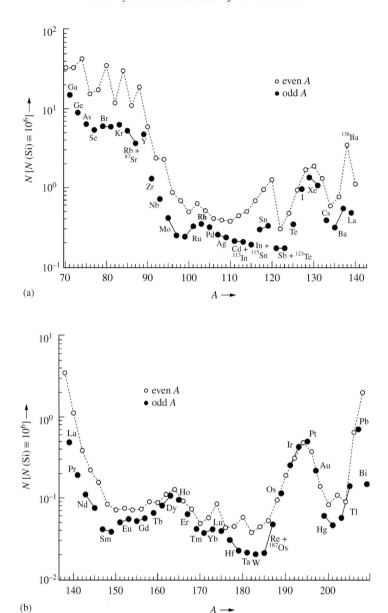

Figure 7 Enlarged parts of Figure 6: (a) mass range 70–140 and (b) mass range 138–209. The abundances of odd mass nuclei are not a smooth function of mass number, e.g., Y and Sn.

solar system, the nature of the local ISM and the general processes leading to low-mass star formation.

The discovery of presolar grains in meteorites has, for the first time, enabled the precise chemical and isotopic analysis of interstellar material (e.g., Anders and Zinner, 1993; Chapter 1.02). The huge variations in the isotopic compositions of all the elements analyzed in presolar grains is in stark contrast to the basically uniform isotopic composition of solar system materials (see Figure 1). This uniformity would have required an effective isotopic homogenization of all the material in the solar nebula, i.e., gas and dust, during the early stages of the formation of the solar system.

1.03.2.2 The Nature of the ISM

The ISM is the medium between the stars. For present purposes, we will also consider the media immediately surrounding stars (generally considered as circumstellar media) as part of the ISM. Most of the matter in the ISM is in the form of a very tenuous gas with densities of less than one hydrogen atom per cm^3 to perhaps a million hydrogen atoms per cm^3. For comparison the terrestrial atmosphere contains about 10^{19} hydrogen atoms per cm^3.

Interstellar matter is comprised of both gas and dust. The gas consists of atomic and polyatomic ions and radicals, and also of molecules. It is the form of hydrogen in the ISM that is used to describe its nature, i.e., ionized (H^+), atomic (H), or molecular (H_2), with increasing density from the ionized to the molecular gas. The dust is primarily composed of amorphous carbons and silicates and makes up ~1% of the mass of the ISM. The dust particles have sizes in the nanometer to micrometer range (e.g., Mathis, 1990). The total amount of absorption and scattering along any given line of sight is called the interstellar extinction. The degree of extinction depends upon the wavelength, the size of the dust grains and their composition.

1.03.2.3 The Chemical Composition of the ISM

1.03.2.3.1 The composition of the interstellar gas and elemental depletions

The elements incorporated into grains in the ISM cannot be observed directly. They are underrepresented in or depleted from the gas phase. Only the fraction of an element that remains in the gas phase can be detected in the ISM, provided that the element has accessible and observable transitions. Initial measurements of the abundances of elements in the gas phase date back to the mid-1960s, when with the advent of space missions, it became possible to eliminate the absorbing effects of the Earth's atmosphere (e.g., Morton and Spitzer, 1966). With the sensitive spectrographs onboard the International Ultraviolet Explorer (IUE) and the Hubble Space Telescope (HST, e.g., the Goddard High Resolution Spectrometer, GHRS, and more recently the Space Telescope Imaging Spectrograph, STIS) many lines of sight have now been studied for a large number of elements (e.g., Savage and Sembach, 1996; Howk et al., 1999; Cartledge et al., 2001).

Hydrogen and helium are the most abundant elements in the ISM gas phase. To date some 30–40 elements heavier than helium have been observed and their gas phase abundances determined. Based on the existing data (e.g., Savage and Sembach, 1996; Howk et al., 1999; Sofia and Meyer, 2001) the local ISM sampled out to a few kiloparsecs from the Sun appears to be rather uniform in chemical composition.

As an example we show the results of abundance determinations along the line of sight towards ζ Oph (ζ Ophiuchus), a moderately reddened star that is frequently used as standard for depletion studies. Molecules are observed along this line of sight and the material is a blend of cool diffuse clouds and a large cold cloud. The atomic hydrogen column density is log N(H) = 21.12 ± 0.10. In Table 4 all data are normalized to 10^{12} atoms of hydrogen and in the last column the ratios of the ζ Oph abundances to the solar abundances are given. In Figure 8 these ratios are plotted against condensation temperatures (see Savage and Sembach, 1996 for details). The abundances of many of the highly volatile and moderately volatile elements up to condensation temperatures of around 900 K (at 10^{-4} bar) are, within a factor of 2, the same in the ISM and in the Sun, independent of the condensation temperatures. This suggests that these elements predominantly reside in the gas phase in the ISM and that their elemental abundances in the ISM are similar to those of the solar system. At higher condensation temperatures a clear trend of increasing depletions with increasing condensation temperatures is seen. The more refractory elements are condensed into grains in the outflows of evolved stars or perhaps in the ejected remnants associated with supernovae explosions.

There are several elements whose abundances deviate from the general trend (e.g., phosphorus and arsenic in Figure 8). It will be important to find out during the course of future work whether these deviations reflect problems with the extremely complex analyses or whether they are true variations that indicate particular chemical processes in the ISM or are characteristic of condensation in stellar outflows.

It should be emphasized that the depletion pattern in the ISM does not reflect thermodynamic equilibrium between dust and gas. The temperature in the ISM is so low that virtually all elements, including the rare gases, should be condensed in grains if thermodynamic equilibrium between dust and gas is assumed. The depletion pattern rather reflects conditions at higher temperatures established during the condensation of minerals in the outflows of dying stars or supernovae explosions. This pattern is then frozen in the cold ISM.

1.03.2.3.2 The composition of interstellar dust

As concluded in the previous section, the dust in the ISM is primarily composed of the elements carbon, oxygen, magnesium, silicon, and iron. This argument is based on the elemental make-up of the solid phase in the ISM determined from the depletions. However, the exact chemical and mineralogical composition of the dust in the ISM can be determined through infrared observations of the absorption of starlight by cold dust ($T \approx 20$ K) along lines of sight toward distant stars, and also by the emission features from hot dust ($T \approx$ a few hundred kelvin) in the regions close to stars. Such observations reveal the spectral signatures of amorphous aliphatic and

Table 4 Abundances of elements in the gas phase of the ISM in the direction of ζ Ophiucus.

Element	T_c (K)	Solar system		ζ Ophiucus cool			ζ Ophiucus cool/solar
		log X	SD (%)	log X	SD (%)	Lit.	
Highly volatile elements							
Ar	25	6.40	5	6.08	45	(1)	0.48
Kr	25	3.30	15	2.97	15	(1)	0.47
C	75	8.39	30	8.14	35	(1)	0.56
N	120	7.93	30	7.90	15	(1)	0.93
O	180	8.69	20	8.48	15	(1)	0.62
Pb	427	2.05	15	1.34	40	(1)	0.19
Cd	429	1.77	30	1.67	10	(2)	0.79
Tl	448	0.81	60	1.27	30	(1)	2.9
Moderately volatile elements							
S	648	7.19	30	7.45	90	(1)	1.8
Zn	660	4.66	20	3.98	30	(1)	0.21
Te	680	2.22		<3.01		(1)	<6.2
Se	684	3.40		3.45	70	(1)	1.1
Sn	720	2.12	90	2.16	25	(1)	1.1
F	736	4.45	90	4.26	60	(3)	0.65
Ge	825	3.62	40	3.01	10	(1)	0.25
Cl	863	5.26	90	5.27	60	(1)	1.0
B	908	2.87	90	1.95	25	(1)	0.12
Ga	918	3.11	25	1.99	15	(1)	0.076
Na	970	6.30	5	5.36	25	(1)	0.11
K	1,000	5.11	35	4.04	80	(1)	0.085
Cu	1,037	4.28	10	2.92	5	(1)	0.044
As	1,135	2.35	??	2.16	25	(1)	0.65
P	1,151	5.44	10	5.07	75	(1)	0.43
Mn	1,190	5.51	5	4.08	5	(1)	0.037
Li	1,225	3.30	10	1.73	15	(1)	0.027
Mg-silicates and metallic FeNi							
Cr	1,301	5.67	5	3.4	5	(1)	5.3×10^{-3}
Si	1,311	7.55	15	6.24	5	(1)	4.9×10^{-2}
Fe	1,337	7.49	20	5.24	5	(1)	5.7×10^{-3}
Mg	1,340	7.56	15	6.33	5	(1)	5.9×10^{-2}
Ni	1,354	6.23	10	3.51	5	(1)	1.9×10^{-3}
Co	1,356	4.90	10	2.15	30	(1)	1.8×10^{-3}
Refractory elements							
V	1,455	3.99	5	<2.06		(1)	$<1.2 \times 10^{-2}$
Ca	1,518	6.33	5	2.61	15	(1)	1.9×10^{-4}
Ti	1,598	4.95	14	1.91	10	(1)	9.2×10^{-4}

T_c—condensation temperatures at 10^{-4} bar (Wasson, 1985), except B (Lauretta and Lodders, 1997); log X—log of abundances relative to 10^{12} atoms of H; SD's are given to the nearest 5%. (1) Savage and Sembach (1996); (2) Sofia *et al*. (1999); (3) Snow and York (1981).

aromatic hydrocarbons and amorphous silicates. In dense clouds (densities of the order of 10^3–10^5 hydrogen atoms per cm³), where the matter is well-shielded from the destructive effects of stellar UV-light, we observe molecular species such as H_2, H_2O, CO, CO_2, and CH_3OH.

The interstellar carbon grains contain both aliphatic and aromatic C—H and C—C bonds; both are observed in absorption and emission in the ISM, but beyond this their exact composition is not known. Given the seemingly uniform elemental abundances in our local ISM we might expect that the dust composition would also be chemically uniform. However, the inferred elemental composition of silicates in the ISM indicates that they have an olivine-type stoichiometry where the depletions are largest, and a mixed oxide/silicate stoichiometry where lower depletions indicate that some dust erosion has occurred (e.g., Savage and Sembach, 1996; Jones, 2000 and references therein). In the lower density regions above the galactic plane we clearly see a different dust stoichiometry (e.g., Savage and Sembach, 1996). This change in composition is presumably a reflection of the effects of shock waves that have lifted interstellar clouds high above the plane and that have, at the same time, eroded and destroyed some fraction of the dust incorporated into these clouds.

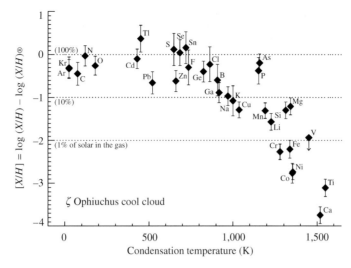

Figure 8 Abundances of elements along the line of sight towards ζ Oph (ζ Ophiuchus), a moderately reddened star that is frequently used as standard for depletion studies. The ratios of ζ Oph abundances to the solar abundances are plotted against condensation temperatures. The abundances of many of the highly volatile and moderately volatile elements up to condensation temperatures of around 900 K are, within a factor of 2, the same in the ISM and in the Sun. At higher condensation temperatures a clear trend of increasing depletions with increasing condensation temperatures is seen. It is usually assumed that the missing refractory elements are in grains (source Savage and Sembach, 1996).

1.03.2.3.3 Did the solar system inherit the depletion of volatile elements from the ISM?

Most meteorites are depleted in moderately volatile and highly volatile elements (see Figures 2–4). The terrestrial planets Earth, Moon, Mars, and the asteroid Vesta show similar or even stronger depletions (e.g., Palme et al., 1988; Palme, 2001). The depletion patterns in meteorites and in the inner planets are qualitatively similar to those in the ISM. It is thus possible that the material in the inner solar system inherited the depletions from the ISM by the preferential accretion of dust grains and the loss of gas during the collapse of the molecular cloud that led to the formation of the solar system. There is, however, little support for this hypothesis:

(i) The general uniformity of the isotopic compositions of solar system materials and the extreme variations in presolar grains suggest isotopic and elemental homogenization at the beginning of the solar system.

(ii) This applies likewise to all systems involving radioactive nuclei that are used for dating. For example, the extremely uniform strontium isotopic composition in all solar system materials at the beginning of the solar system, 4.566 billion years ago, indicates that the strontium isotopes were homogenized at that time, including ^{87}Sr, the decay product of ^{87}Rb ($T_{1/2} = 5 \times 10^{10}$ yr). In the ISM, the refractory strontium is in grains and the volatile rubidium in the gas phase. The separation of these two elements over hundreds of millions of years without thorough remixing at the beginning of the solar system, would produce a large range in initial ^{87}Sr/^{86}Sr ratios at the beginning of the solar system and should have been observed in solar system materials (see Palme, 2001).

1.03.2.3.4 The ISM oxygen problem

Snow and Witt (1996) and others argued that the composition of the ISM is different from the composition of the Sun. Based on stellar compositional data, these authors concluded that heavy elements in the Sun are only two-thirds of the solar composition, with the implication that the heavy elements either fractionated from hydrogen during formation of the Sun (Snow, 2000) or that the Sun was formed in a different place in the Milky Way where the heavy element abundances were higher. Another possibility was that there were additional "hidden" reservoirs. In particular, there was a problem of too much interstellar oxygen when using the solar oxygen abundance as standard. Where could that excess oxygen be stored? Some 20% of the solar oxygen abundance must have been combined with magnesium, silicon, and iron, etc. in the form of the amorphous silicates observed in the ISM. Another 40% was directly observed in the gas as atomic oxygen (Meyer et al., 1998). The "missing" 40% of oxygen remained elusive; it could not be in the form of molecules, e.g., H_2O and CO, because they were not observed in sufficient abundance. The recent re-evaluation of the solar oxygen abundance (Holweger, 2001; Allende Prieto et al., 2001), and of interstellar oxygen (Sofia and Meyer, 2001), has resolved this problem. The new solar oxygen is now only ~60% of its previous value and so there is, and indeed never was, a problem. Recent interstellar abundance determinations using observations of solar-type F and G stars in the galactic neighborhood (Sofia and Meyer, 2001) suggest that many elemental abundances in the ISM are close to solar abundances.

1.03.3 SUMMARY

Updated solar photospheric abundances are compared with meteoritic abundances. It is shown that only one group of chondritic meteorites, the CI chondrites, matches solar abundances in refractory lithophile, siderophile, and volatile elements. All other chondritic meteorites differ from CI chondrites. The agreement between solar and CI abundances for all elements heavier than oxygen and excluding rare gases has constantly improved since Goldschmidt (1938) published his first comprehensive table of cosmic abundances.

The abundances of 39 nongaseous elements in the Sun have assigned errors below 30%. Only the four elements sulfur, manganese, scandium, and strontium differ by more than 20% from CI abundances. The difference is below 10% for 27 of these elements. The agreement between meteoritic and solar abundances must therefore be considered excellent and there is not much room left for further improvements. Obvious candidates for redetermination of solar abundances are manganese and sulfur. The limiting factor in the accuracy of meteorite abundances is the inherent variability of CI chondrites, primarily the Orgueil meteorite.

The ISM from which the solar system formed has volatile and moderately volatile elements, within a factor of 2, the same composition as the Sun. The more refractory elements of the ISM are depleted from the gas and are concentrated in grains.

REFERENCES

Allende Prieto C., Lambert D. L., and Asplund M. (2001) The forbidden abundance of oxygen in the sun. *Astrophys. J.* **556**, L63–L66.

Allende Prieto C., Lambert D. L., and Asplund M. (2002) A reappraisal of the solar photospheric C/O ratio. *Astrophys. J.* **573**, L137–L140.

Anders E. and Grevesse N. (1989) Abundances of the elements: meteoritic and solar. *Geochim. Cosmochim. Acta* **53**, 197–214.

Anders E. and Zinner E. (1993) Interstellar grains in primitive meteorites: diamond, silicon carbide, and graphite. *Meteoritics* **28**, 490–514.

Arndt P., Bohsung J., Maetz M., and Jessberger E. (1996) The elemental abundances in interplanetary dust particles. *Meteoritics Planet. Sci.* **31**, 817–833.

Beer H., Walter G., Macklin R. L., and Patchett P. J. (1984) Neutron capture cross sections and solar abundances of ^{160,161}Dy, ^{170,171}Yb, ^{175,176}Lu, and ^{176,177}Hf for the s-process analysis of the radionuclide ^{176}Lu. *Phys. Rev.* **C30**, 464–478.

Begemann F. (1980) Isotope anomalies in meteorites. *Rep. Prog. Phys.* **43**, 1309–1356.

Biemont E., Garnir H. P., Palmeri P., Li Z. S., and Svanberg S. (2000) New f-values in neutral lead obtained by time-resolved laser spectroscopy, and astrophysical applications. *Mon. Not. Roy. Astronom. Soc.* **312**, 116–122.

Bischoff A., Palme H., Ash R. D., Clayton R. N., Schultz L., Herpers U., Stöffler D., Grady M. M., Pillinger C. T., Spettel B., Weber H., Grund T., Endreß M., and Weber D. (1993) Paired Renazzo-type (CR) carbonaceous chondrites from the Sahara. *Geochim. Cosmochim. Acta* **57**, 1587–1603.

Burnett D. S. and Woolum D. S. (1990) The interpretation of solar abundances at the $N = 50$ neutron shell. *Astron. Astrophys.* **228**, 253–259.

Cameron A. G. W. (1973) Abundances of the elements in the solar system. *Space Sci. Rev.* **15**, 121–146.

Cartledge S. I. B., Meyer D. M., and Lauroesch J. T. (2001) Space telescope imaging spectrograph observations on interstellar oxygen and krypton in translucent clouds. *Astrophys. J.* **562**, 394–399.

Clayton R. N. (1993) Oxygen isotopes in meteorites. *Ann. Rev. Earth Planet. Sci.* **21**, 115–149.

Cowley C. R. (1995) *An introduction to cosmochemistry*. Cambridge University Press, Cambridge, 480pp.

Cunha K. and Smith V. V. (1999) A determination of the solar photospheric boron abundance. *Astrophys. J.* **512**, 1006–1013.

Dodd R. T. (1981) *Meteorites, a petrologic-chemical synthesis*. Cambridge University Press, Cambridge, 368pp.

Dreibus G., Palme H., Spettel B., Zipfel J., and Wänke H. (1995) Sulfur and selenium in chondritic meteorites. *Meteoritics* **30**, 439–445.

Evenson N. M., Hamilton P. J., and O'Nions R. K. (1978) Rare-earth abundances in chondritic meteorites. *Geochim. Cosmochim. Acta* **42**, 1199–1212.

Flynn G. J. (1994) Interplanetary dust particles collected from the stratosphere: physical, chemical, and mineralogical properties and implications for their sources. *Planet. Space. Sci.* **42**, 1151–1161.

Goldschmidt V. M. (1938) Geochemische Verteilungsgestze der Elemente IX. Die Mengenverhältnisse der Elemente und der Atom-Arten. Skrifter Utgitt av Det Norske Vidensk. Akad. Skrifter I. Mat. Naturv. Kl. No. 4, Oslo, 1937, 148pp.

Grevesse N. and Sauval A. J. (1998) Standard solar composition. *Space Sci. Rev.* **85**, 161–174.

Grossman L. and Larimer J. W. (1974) Early chemical history of the solar system. *Rev. Geophys. Space Phys.* **12**, 71–101.

Harkins W. D. (1917) The evolution of the elements and the stability of complex atoms. *J. Am. Chem. Soc.* **39**, 856–879.

Holweger H. (2001) Photospheric abundances: problems, updates, implications. In *Solar and Galactic Composition* (ed. R. F. Wimmer-Schweinsgruber). American Institute of Physics, pp. 23–30.

Howk C., Savage B. D., and Fabian D. (1999) Abundances and physical conditions in the warm neutral medium toward μ Columbae. *Astrophys. J.* **525**, 253–293.

Jessberger E. K., Christoforidis A., and Kissel J. (1988) Aspect of the major element composition of Halley's dust. *Nature* **332**, 691–695.

Jochum K. P. (1996) Rhodium and other platinum-group elements in carbonaceous chondrites. *Geochim. Cosmochim. Acta* **60**, 3353–3357.

Jochum K. P., Stolz A. J., and McOrist G. (2000) Niobium and tantalum in carbonaceous chondrites: constraints on the solar system and primitive mantle niobium/tantalum, zirconium/ niobium, and niobium/uranium ratios. *Meteoritics Planet. Sci.* **35**, 229–235.

Jones A. P. (2000) Depletion patterns and dust evolution in the interstellar medium. *J. Geophys. Res.* **105**, 10257–10268.

Lauretta D. S. and Lodders K. (1997) The cosmochemical behavior of beryllium and boron. *Earth. Planet. Sci. Lett.* **146**, 315–327.

Lodders K. (2003) Solar system abundances and condensation temperatures of the elements. *Astrophys. J.* **591**, 1220–1247.

Mathis J. S. (1990) Interstellar dust and extinction. *Ann. Rev. Astron. Astrophys.* **28**, 37–70.

Meyer D. M., Jura M., and Cardelli J. A. (1998) The definitive abundance of interstellar oxygen. *Astrophys. J.* **493**, 222–229.

Morton D. C. and Spitzer L. (1966) Line spectra of delta and pi Scorpii in the far-ultraviolet. *Astrophys. J.* **144**, 1–12.

Nicolussi G. K., Pellin M. J., Lewis R. S., Davis A. M., Amari S., and Clayton R. N. (1998) Molybdenum isotopic composition of individual presolar silicon carbide grains from the Murchison meteorite. *Geochim. Cosmochim. Acta* **62**, 1093–1104.

Palme H. (2000) Are there chemical gradients in the inner solar system? *Space Sci. Rev.* **92**, 237–262.

Palme H. (2001) Chemical and isotopic heterogeneity in protosolar matter. *Phil. Trans. Roy. Soc. London* **A359**, 2061–2075.

Palme H. and Beer H. (1993) Abundances of the elements in the solar system. In Landolt-Börnstein, Group VI: *Astronomy and Astrophysics: Instruments; Methods; Solar System* (ed. H. H. Voigt). Springer, Berlin, vol. 3(a), pp. 196–221.

Palme H., Larimer J. W., and Lipschutz M. E. (1988) Moderately volatile elements. In *Meteorites and the Early Solar System* (eds. J. F. Kerridge and M. S. Matthews). University of Arizona Press, Tucson, pp. 436–461.

Prochaska J. X. and McWilliam A. (2000) On the perils of hyperfine splitting: a reanalysis of Mn and Sc abundance trends. *Astrophys. J.* **537**, L57–L60.

Reames D. V. (1998) Solar energetic particles: sampling coronal abundances. *Space Sci. Rev.* **85**, 327–340.

Richter S., Ott U., and Begemann F. (1992) S-process isotope anomalies: neodymium, samarium and a bit more strontium. *Lunar Planet. Sci.* **XXIII**, 1147–1148.

Rocholl A. and Jochum K. P. (1993) Th, U and other trace elements in carbonaceous chondrites: implications for the terrestrial and solar system Th/U ratios. *Earth. Planet. Sci. Lett.* **117**, 265–278.

Russell H. N. (1941) The cosmical abundance of the elements. *Science* **94**, 375–381.

Savage B. D. and Sembach K. R. (1996) Interstellar abundances from absorption-line observations with the Hubble Space Telescope. *Ann. Rev. Astron. Astrophys.* **34**, 279–329.

Snow T. P. (2000) Composition of interstellar gas and dust. *J. Geophys. Res.* **105**, 10239–10248.

Snow T. P. and Witt A. N. (1996) Interstellar depletions updated: where all the atoms went. *Astrophys. J.* **468**, L65–L68.

Snow T. P. and York D. G. (1981) The detection of interstellar fluorine in the line of sight toward Delta Scorpii. *Astrophys. J.* **247**, L39–L41.

Sofia U. J. and Meyer D. M. (2001) Interstellar abundance standards revisited. *Astrophys. J.* **554**, L221–L224.

Sofia U. J., Meyer D. M., and Cardelli J. A. (1999) The abundance of interstellar tin and cadmium. *Astrophys. J.* **522**, L137–L140.

Suess H. (1947) Über kosmische Kernhäufigkeiten I. Mitteilung: Einige Häufigkeitsregeln und ihre Anwendung bei der Abschätzung der Häufigkeitwerte für die mittelsschweren und schweren Elemente: II. Mitteilung: Einzelheiten in der Häufigkeitverteilung der mittelschweren und schweren Kerne. *Z. Naturforsch.* **2a**, 311–321, 604–608.

Suess H. E. and Urey H. C. (1956) Abundances of the elements. *Rev. Mod. Phys.* **28**, 53–74.

Suess H. E. and Zeh H. D. (1973) The abundances of the heavy elements. *Astrophys. Space Sci.* **23**, 173–187.

Wasson J. T. (1985) *Meteorites: Their Record of Early Solar-system History*. W. H. Freeman, New York, 267pp.

Wolf D. and Palme H. (2001) The solar system abundances of P and Ti and the nebular volatility of P. *Meteoritics Planet. Sci.* **36**, 559–571.

Zhai M. and Shaw D. M. (1994) Boron cosmochemistry: Part 1. Boron in meteorites. *Meteoritics* **29**, 607–615.

1.04
The Solar Nebula

A. P. Boss

Carnegie Institution of Washington, DC, USA

1.04.1	INTRODUCTION	63
1.04.2	FORMATION OF THE SOLAR NEBULA	64
	1.04.2.1 Observations of Precollapse Clouds	64
	1.04.2.1.1 Cloud properties	64
	1.04.2.1.2 Onset of collapse phase	65
	1.04.2.1.3 Outcome of collapse phase	66
	1.04.2.2 Observations of Star-forming Regions	67
	1.04.2.2.1 Protostellar phases: class −I, 0, I, II, III objects	67
	1.04.2.2.2 Ubiquity of bipolar outflows from earliest phases	67
1.04.3	ASTROPHYSICAL ANALOGUES FOR THE SOLAR NEBULA	68
	1.04.3.1 Optical and IR-wave Observations	69
	1.04.3.1.1 Mass accretion rates—episodicity and outbursts	69
	1.04.3.1.2 Temperature profiles from spectral energy distributions	69
	1.04.3.2 Millimeter-wave Observations	70
	1.04.3.2.1 Disk masses	70
	1.04.3.2.2 Keplerian rotation	70
1.04.4	NEBULA TRANSPORT AND EVOLUTION	70
	1.04.4.1 Angular Momentum Transport Mechanisms	70
	1.04.4.1.1 Hydrodynamic turbulence	71
	1.04.4.1.2 Magnetorotational-driven turbulence	72
	1.04.4.1.3 Gravitational torques in a marginally unstable disk	72
	1.04.4.2 Evolution of the Solar Nebula	73
	1.04.4.2.1 Viscous accretion disk models	74
	1.04.4.2.2 Volatility patterns and transport in inner solar system	74
	1.04.4.2.3 Turbulent and gas drag transport	75
	1.04.4.3 Clump Formation in a Marginally Gravitationally Unstable Disk	75
	1.04.4.3.1 Chondrule formation in nebular shock fronts	76
	1.04.4.3.2 Mixing processes in marginally gravitationally unstable disks	76
	1.04.4.4 X-wind Model for Processing Solids	77
	1.04.4.4.1 Local irradiation (^{10}Be)	77
	1.04.4.4.2 Thermal processing of solids and ^{26}Al formation	78
1.04.5	SOLAR NEBULA REMOVAL	78
	1.04.5.1 Observational Constraints on Disk Life Times	78
	1.04.5.2 Removal Mechanisms	79
	1.04.5.2.1 Inward flow onto protosun	79
	1.04.5.2.2 UV photoevaporation by protosun	79
	1.04.5.2.3 UV photoevaporation in Orion-like regions	79
	1.04.5.2.4 Final scouring by widened stellar outflow	79
1.04.6	SUMMARY	79
REFERENCES		80

1.04.1 INTRODUCTION

The solar nebula was the rotating, flattened disk of gas and dust from which the solar system originated ~4.6 Gyr ago. Much of the motivation for cosmochemical studies of meteorites, comets, and other primitive bodies stems from the desire to use the results to constrain or otherwise illuminate the physical and chemical conditions in the solar nebula, in the hope of learning more about the processes that led to the formation of the planets. In addition to cosmochemical studies, there are important lessons to be learned about the planet formation process from astrophysical observations of young stellar objects and their accompanying protoplanetary disks, from the discovery of other planetary systems, and from theoretical models. A key resource for learning more about these subjects is the compendium volume (Mannings *et al.*, 2000).

1.04.2 FORMATION OF THE SOLAR NEBULA

The protosun and solar nebula were formed by the self-gravitational collapse of a dense molecular cloud core, much as we see new stars being formed today in regions of active star formation. The formation of the solar nebula was largely an initial value problem, i.e., given detailed knowledge of the particular dense molecular cloud core which was the presolar cloud, one could in principle calculate the flow of gas and dust subject to the known laws of physics and predict the basic outcome. Specific details of the outcome cannot be predicted, however, as there appears to be an inevitable amount of stochastic, chaotic evolution involved, e.g., in the orbital motions of any ensemble of gravitationally interacting particles. Nevertheless, we expect that at least the gross features of the solar nebula should be predictable from theoretical models of cloud collapse, constrained by astronomical observations of young stellar objects. For this reason, the physical structure of likely precollapse clouds is of interest with regard to inferring the formation mechanism of the protosun and the structure of the accompanying solar nebula.

1.04.2.1 Observations of Precollapse Clouds

Astronomical observations at long wavelengths (e.g., millimeter) are able to probe deep within interstellar clouds of gas and dust, which are opaque at short wavelengths (e.g., visible wavelengths). These clouds are composed primarily of molecular hydrogen gas, helium, and molecules such as carbon monoxide, hence the term molecular clouds. About 1% by mass of these clouds is in the form of submicron-sized dust grains, with about another 1% composed of gaseous molecules and atoms of elements heavier than helium. Regions of active star formation are located within molecular clouds and complexes ranging in mass from a few solar masses to over a million. This association of young stars with molecular clouds is the most obvious manifestation of the fact that stars form from these clouds. Many of the densest regions of these clouds were found to contain embedded IR objects, i.e., newly formed stars whose light is scattered, absorbed, and re-emitted at IR wavelengths in the process of exiting the placental cloud core. Such cores have already succeeded in forming stars. Initial conditions for the collapse of the presolar cloud can be more profitably ascertained from observations of dense cloud cores which do not appear to contain embedded IR objects, i.e., precollapse cloud cores.

1.04.2.1.1 Cloud properties

Precollapse cloud cores are composed of cold molecular gas with temperatures in the range $\sim 7-15$ K, and with gas densities $\sim 10^3 - 10^5$ mol cm^{-3} (Figure 1). Some clouds may be denser yet, but this is hard to determine because of the limited density ranges for which suitable molecular tracers are abundant (typically isotopes of carbon monoxide and ammonia). Masses of these clouds range from roughly a solar mass to thousands of solar masses, with the distribution of clump masses fitting a power-law such that most of the clumps are of low mass, as is also true of stars in general. The cloud properties described below are used to constrain the initial conditions for hydrodynamic models of the collapse of cloud cores.

Density profiles. Large radio telescopes have enabled the high spatial resolution mapping of precollapse clouds and the determination of their interior density structure. While such clouds undoubtedly vary in all three spatial dimensions, typically the observations are averaged in angle to yield an equivalent, spherically symmetric density profile. These radial-density profiles have shown

Figure 1 The Barnard 86 dark cloud, seen in silhouette against a dense field of stars, is typical of the dense clouds of molecular gas and dust grains in which low-mass stars form in relative isolation from massive stars. (reproduced by permission of David Malin Images and the Anglo-Australian Observatory).

that precollapse clouds typically have flat density profiles near their centers, as is to be expected for a cloud which has not yet collapsed to form a star, surrounded by an envelope with a steeply declining profile that could be fit with a power law. The density profile thus resembles that of a Gaussian distribution, or more precisely, the profile of the Bonnor–Ebert sphere, which is the equilibrium configuration for an isothermal gas cloud (Alves et al., 2001).

Cloud shapes. While precollapse clouds often have a complicated appearance, attempts have been made to approximate their shapes with simple geometries. Triaxial spheroids seem to be required in general, though most lower-mass clouds appear to be more nearly oblate than prolate (Jones et al., 2001). On the larger scale, prolate shapes seem to give a better fit than oblate spheroids.

Angular momentum distributions. Precollapse clouds have significant interior velocity fields, which appear to be a mixture of turbulence derived from fast stellar winds and outflows, and magnetohydrodynamic (MHD) waves associated with the ambient magnetic field. In addition, there may be evidence for a systematic shift in velocities across one axis of the cloud, which can be interpreted as solid-body rotation. When estimated in this manner, typical rotation rates are found to be below the level needed for cloud support by centrifugal force, yet large enough to result in considerable rotational flattening once cloud collapse begins. Ratios of rotational to gravitational energy in dense cloud cores range from 0.002 to 1, with a typical value being 0.02 (Goodman et al., 1993).

1.04.2.1.2 Onset of collapse phase

Dense cloud cores are supported against their own self-gravity by a combination of turbulent motions, magnetic fields, thermal (gas) pressure, and centrifugal force, in decreasing order of importance. Turbulent motions inevitably die out over timescales which are comparable to a cloud's free-fall time (the time over which an idealized, pressureless sphere of gas initially of uniform density would collapse to form a star) once the source of turbulence is removed. For a dense cloud core, free-fall times are ~ 0.1 Myr. However, dense clouds do not collapse on this timescale, because once turbulence decays, magnetic fields provide support against self-gravity.

Ambipolar diffusion and loss of magnetic support. Magnetic field strengths in dense clouds are measured by Zeeman splitting of molecular lines, and found to be large enough ($\sim 10-1,000$ μG) to be capable of supporting clouds, when both static magnetic fields and MHD waves are present (Crutcher, 1999). Field strengths are found to depend on the density to roughly the 1/2 power, as is predicted to be the case if ambipolar diffusion controls the cloud's dynamics (Mouschovias, 1991). Ambipolar diffusion is the process of slippage of the primarily neutral gas molecules past the ions, to which the magnetic field lines are effectively attached. This process occurs over timescales of a few million years or more for dense cloud cores, and inevitably leads to the loss of sufficient magnetic field support such that the slow inward contraction of the cloud turns into a rapid, dynamic collapse phase when the magnetic field is no longer in control. This is generally believed to be the process through which stars in regions of low-mass star formation begin their life, the "standard model" of star formation (Shu et al., 1987). (Low-mass stars have masses no more than a few times that of the Sun; high-mass stars have masses more than 10 times the Sun's mass.)

Shock-triggered collapse. In regions of high-mass star formation (Figure 2), where the great majority of stars are believed to form, quiescent star formation of the type envisioned in the standard model occurs only until the phase when

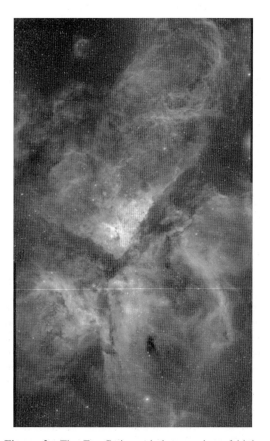

Figure 2 The Eta Carina nebula, a region of high-mass star formation. Most low-mass stars form in such regions, prior to the formation of massive OB stars, which disrupt subsequent star formation through their huge luminosities.

high-mass stars begin to form and evolve. The process of high-mass star formation is less well understood than that of low-mass stars, but observations make it clear that events such as the supernova explosions that terminate the life of massive stars can result in the triggering of star formation in neighboring molecular clouds that are swept up and compressed by the expanding supernova shock front (Preibisch and Zinnecker, 1999). Even strong protostellar outflows are capable of triggering the collapse of dense cloud cores (Foster and Boss, 1996). A supernova shock-triggered origin for the presolar cloud has been advanced as a likely source of the short-lived radioisotopes (e.g., ^{26}Al) that existed in the early solar nebula (Cameron and Truran, 1977). Detailed models of shock-triggered collapse have shown that injection of shock-front material into the collapsing protostellar cloud can occur, provided that the shock speed is ~ 25 km s^{-1} (Boss, 1995), as is appropriate for a moderately distant supernova, or for the wind from an evolved red giant star, which might also account for some short-lived isotopes (Goswami and Vanhala, 2000).

1.04.2.1.3 Outcome of collapse phase

Once a cloud begins to collapse as a result of ambipolar diffusion or triggering by a shock wave, supersonic inward motions develop and soon result in the formation of an optically thick first core, with a size of order 10 AU (1 AU = Earth–Sun distance). This central core is primarily supported by the thermal pressure of the molecular hydrogen gas, while the remainder of the cloud continues to fall onto the core. For a solar-mass cloud, this core has a mass of ~ 0.01 solar masses. Once the central temperature reaches $\sim 2,000$ K, thermal energy goes into dissociating the hydrogen molecules, lowering the thermal pressure, and leading to a second collapse phase, during which the first core disappears and a second, final core is formed at the center, with a radius a few times that of the Sun. This core then accretes mass from the infalling cloud over a timescale of ~ 1 Myr (Larson, 1969). In the presence of rotation or magnetic fields, however, the collapsing cloud becomes flattened into a pancake, and may then fragment into two or more protostars. At this point, we cannot reliably predict what sort of dense cloud core will form what sort of star or stellar system, much less what sorts of planetary systems will accompany them, but certain general trends are evident.

Fragmentation leading to ejection of single stars. The standard model pertains only to formation of single stars, whereas most stars are known to be members of binary or multiple star systems. There is growing observational evidence that multiple star formation may be the rule, rather than the exception (Reipurth, 2000). If so, then it may be that single stars, such as the Sun, are formed in multiple protostar systems, only to be ejected soon thereafter as a result of the decay of the multiple system into an orbitally stable configuration. In that case, the solar nebula would have been subject to strong tidal forces during the close encounters with other protostars prior to its ejection from the multiple system. This hypothesis has not been investigated in detail (but see Kobrick and Kaula, 1979). Detailed models of the collapse of magnetic cloud cores, starting from initial conditions defined by observations of molecular clouds, show that while initially prolate cores tend to fragment into binary protostars, initially oblate clouds form multiple protostar systems that are highly unstable and likely to eject single protostars and their disks (Boss, 2002a). Surprisingly, magnetic fields were found to enhance the tendency for a collapsing cloud to fragment by helping to prevent the formation of a single central mass concentration of the type assumed to form in the standard model of star formation.

Collapse leading to single star formation. In the case of nonmagnetic clouds, where thermal pressure and rotation dominate, single protostars can result from the collapse of dense cloud cores that are rotating slowly enough to avoid the formation of a large-scale protostellar disk that could then fragment into a binary system (Boss and Myhill, 1995). Alternatively, the collapse of an initially strongly centrally condensed (power-law), nonmagnetic cloud leads to the formation of a single central body (Yorke and Bodenheimer, 1999). However, considering that most cloud cores are believed to be supported to a significant extent by magnetic fields, the applicability of these results is uncertain. In the case of shock-triggered collapse, calculations have shown that weakly magnetic clouds seem to form single protostars when triggering occurs after the core has already contracted toward high central densities (Vanhala and Cameron, 1998). Binary systems can result when triggering occurs prior to this phase.

In the case of nonmagnetic collapse of a spherical cloud (Yorke and Bodenheimer, 1999), the protostar that forms is orbited by a protostellar disk with a similar mass. When angular momentum is transported outward by assumed gravitational torques, and therefore mass is transported inward onto the protostar, the amount of mass remaining in the disk is still so large that most of this matter must eventually be accreted by the protostar through other processes. Hence, the disk at this phase must still be considered a protostellar disk, not a relatively late phase, protoplanetary disk where any objects which form have some hope of survival in orbit. Thus, even in the relatively simple case of nonmagnetic clouds, it is not yet possible to compute the expected detailed structure of a protoplanetary disk, starting from the initial conditions of a dense cloud core. Calculations starting from less idealized initial conditions, such as a segment of an infinite sheet, suffer from the same limitations (Boss and Hartmann, 2001).

Because of the complications of multiple protostar formation, magnetic field support, possible shock-wave triggering, and angular momentum transport, among others, a definitive theoretical model for the collapse of the presolar cloud has not yet emerged.

1.04.2.2 Observations of Star-forming Regions

Observations of star-forming regions have advanced our understanding of the star-formation process considerably in the last few decades. We now can study examples of nearly all phases of the evolution of a dense molecular cloud core into a nearly fully formed star (i.e., the roughly solar-mass T Tauri stars). As a result, the theory of star formation is relatively mature, with future progress expected to center on defining the role played by binary and multiple stars and on refining observations of known phases of evolution.

1.04.2.2.1 Protostellar phases: class −I, 0, I, II, III objects

Protostellar evolution can be conveniently subdivided into six phases, which form a sequence in time. The usual starting point is the precollapse cloud, which collapses to form the first protostellar core, which is then defined to be a class −I object. The first core collapses to form the final, second core, or class 0 object, which has a core mass less than that of the infalling envelope. Class I, II, and III objects (Lada and Shu, 1990) are defined in terms of their spectral energy distributions at mid-IR wavelengths, where the emission is diagnostic of the amount of cold, circumstellar dust. Class I objects have mid-IR fluxes that increase with increasing wavelength, and are optically invisible, IR protostars with so much dust emission that the circumstellar gas mass is ~0.1 of a solar mass or more. Class II objects have excess mid-IR fluxes that decrease with increasing wavelength, implying less dust emission compared to class I objects, and a gas mass of ~0.01 solar masses. Class II objects are usually optically visible, T Tauri stars, where most of the circumstellar gas resides in a disk rather than in the surrounding envelope. Class III objects show no signs of mid-IR flux in excess of that of the blackbody emission from the star itself, and so are "naked" T Tauri stars, with only trace amounts of circumstellar gas and dust. While these classes imply a progression in time from objects with more to less gas emission, the time for this to occur for any given object is highly variable: some class III objects appear to be only 0.1 Myr old, while some class II objects have ages of several million years, based on theoretical models of the evolution of stellar luminosities and surface temperatures. Evidence for dust debris disks has been found around even older stars, such as Beta Pictoris, with an age of ~10 Myr, though its disk mass is much smaller than that of even class III objects. Debris disks appear to owe their existence to erosional collisions between the members of a swarm of unseen smaller bodies.

Multiple examples of all of these phases of protostellar evolution have been found, with the exception of the short-lived class −I objects. It is noteworthy that observations of protostars and young stars find a higher frequency of binary and multiple systems than is the case for mature stars, implying the orbital decay of many of these young systems (Reipurth, 2000; Smith et al., 2000).

1.04.2.2.2 Ubiquity of bipolar outflows from earliest phases

A remarkable aspect of young stellar objects is the presence of strong molecular outflows for essentially all young stellar objects, even the class 0 objects. This means that at the same time that matter is still accreting onto the protostar, it is also losing mass through a vigorous wind directed in a bipolar manner in both directions along the presumed rotation axis of the protostar/disk system. In fact, the energy needed to drive this wind appears to be derived from mass accretion by the protostar, as observed wind momenta are correlated with protostellar luminosities and with the amount of mass in the infalling envelope (Andre, 1997).

There are two competing mechanisms for driving bipolar outflows, both of which depend on magnetic fields to sling ionized gas outward and to remove angular momentum from the star/disk system (Shu et al., 2000; Konigl and Pudritz, 2000). One mechanism is the X-wind model, where coronal winds from the central star and from the inner edge of the accretion disk join together to form the magnetized X-wind, launched from an orbital radius of a few stellar radii. The other mechanism is a disk wind, launched from the surface of the disk over a much larger range of distances, from less than 1 AU to as far away as 100 AU or so. In both mechanisms, centrifugal support of the disk gas makes it easier to launch this material outward, and bipolar flows develop in the directions perpendicular to the disk, because the disk forces the outflow into these preferred directions. Because it derives from radii deeper within the star's gravitational potential well, an X-wind is energetically favored over a disk wind. However, observations show that during FU Orionis-type outbursts, mass is added onto the central star so rapidly that the X-wind region is probably crushed out of existence, implying that the strong outflows that still occur during these outbursts must be caused by an extended disk wind (Hartmann and Kenyon, 1996). All T Tauri stars are believed to experience FU Orionis outbursts, so disk winds may be the primary driver of bipolar outflows.

1.04.3 ASTROPHYSICAL ANALOGUES FOR THE SOLAR NEBULA

Considerable insight into the conditions likely to have accompanied the formation of our solar system can be gained by detailed observations of the circumstellar environments of young stars with masses similar to that of the Sun (Figure 3). The theory of stellar evolution is largely a two-parameter theory, where stellar properties depend primarily on the star's mass and its metallicity (composition). Because we know the Sun's mass and can estimate its composition fairly well through solar spectroscopy and analysis of CI chondrites (see Chapter 1.03; Anders and Grevesse, 1989), one might expect that we could find an exact analogue for the solar nebula. However, while gross stellar properties depend, to a minor extent, on internal rotation and magnetic fields, we can expect that the properties of protoplanetary disks will depend much more strongly on these factors, controlling as they do the approach to and evolution of a largely centrifugally supported disk. Hence, an exact analogue for the solar nebula may be difficult to locate, and we must consider the full range of protoplanetary disks around solar-mass young stars as likely suspects. Note that because there

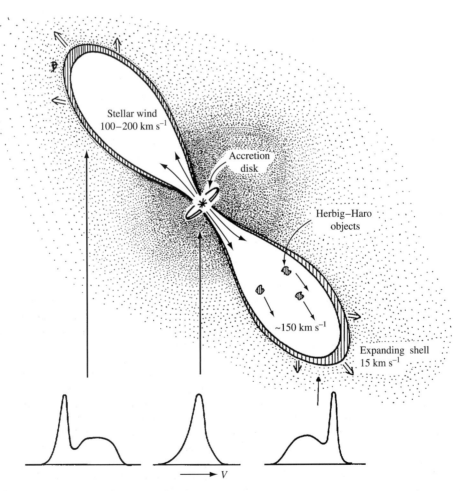

Figure 3 The prototypical young star with a bipolar outflow, L1551 IRS5 (Snell *et al.*, 1980), now known to harbor a binary protostar, with a separation of 45 AU, and where each protostar is orbited by a disk with a radius of ~10 AU and a mass of ~0.05 solar masses (Snell *et al.*, 1980) (reproduced by permission of University of Chicago Press and American Astronomical Society from *Astrophys. J.*, **1980**, *239*, L17–L22).

is as yet no conclusive proof for the existence of any protoplanet orbiting a young star, the very use of the term "protoplanetary" disk to describe the matter in orbit around these stars is a bit of an optimistic assumption.

1.04.3.1 Optical and IR-wave Observations

The prototypical young star of solar mass is T Tauri, an optically visible star located in the Taurus star-forming region. T Tauri stars are highly variable, with rapid, irregular changes in their luminosity, presumably caused by the irregular accretion of matter from their disks. T Tauri has been discovered to have a close binary star as a companion, and possibly other stellar companions as well (Koresko, 2000), so it may not be a good analogue for the early Sun. However, T Tauri's status as a multiple system is very much in step with the observational trend for single young stars to be rare (Reipurth, 2000).

Because of the limited spatial resolution of even the largest telescopes observing the closest young stars, much of what we know about these stars and their disks is inferred from indirect evidence, such as the distribution of radiation with wavelength emitted by these objects. Compared to the spectrum of single temperature black body, as expected for an isolated star, T Tauri shows evidence for both UV and IR excess emission (Kenyon and Hartmann, 1987). The excess UV emission is believed to arise from the boundary layer between the star's surface and the accretion disk, where the gravitational energy released by infall onto the star's surface heats the infalling gas to higher temperatures than the stellar surface. The excess IR emission is believed to be derived from warm dust grains orbiting close to the star, either in a centrifugally supported disk, or still infalling onto the disk in a thick envelope surrounding the disk. This emission is generally optically thick, meaning that these observations can only probe the surface of the emitting region and not deeper layers in the disk.

1.04.3.1.1 Mass accretion rates—episodicity and outbursts

Observations of UV excesses and emission lines in young stars are used to estimate the rate at which matter from an unseen disk is accreting onto the central protostar (Calvet et al., 2000). To fuel this accretion, matter from the initial cloud core infalls onto the protostar's disk. In general, these two mass accretion rates are not the same. During the earliest phases of evolution, the disk accretes mass from its infalling envelope at a rate of about a solar mass or less per 10^5 yr, with this disk mass accretion rate falling off steeply after a few hundred thousand years, as the placental cloud is drained. The stellar mass accretion rate after this initial period varies between a low rate of about a solar mass per 100 Myr to episodic peaks a factor of 1,000 or more times higher during FU Orionis outbursts. This behavior suggests while the disk may gain mass fairly steadily, the mass does not accrete steadily onto the protostar, but rather is stored temporarily in the disk and then is accreted rapidly by the star in episodic bursts. FU Orionis outbursts result in vigorous heating of the innermost regions of the disk (well inside an astronomical unit), and may occur up to ages of a million years or so. The stellar mass accretion rate declines and significant accretion ceases altogether within 10 Myr at the very latest (class III objects).

1.04.3.1.2 Temperature profiles from spectral energy distributions

The amount of energy being emitted at IR wavelengths can be used to estimate the surface temperatures of disks, with a typical surface temperature being ~200 K at a distance of 1 AU from the protostar (Boss, 1998; Woolum and Cassen, 1999). However, surface temperatures are of less interest for cosmochemists than midplane temperatures. Detailed models of the disk interior and thermodynamics, including detailed dust grain opacities, can be coupled with the observed spectral energy distributions to estimate midplane temperatures (D'Alessio et al., 2001) for disks being heated internally by viscous dissipation and externally by radiation from the central star. For typical T Tauri disks, the estimated midplane temperatures range from ~1,500 K inside ~0.2 AU, to ~300 K at 1 AU, to ~30 K at 10 AU (Figure 4). These values correspond to quiescent disks, when the mass accretion rate is quite low, and the assumed disk masses range from 0.02 to 0.1 of a solar mass. During periods of significantly higher mass accretion, midplane temperatures are expected to be correspondingly higher, e.g., between 1,000 K and 1,500 K at 1 AU, and greater than 500 K at 2.5 AU (Boss, 1998). During FU Orionis outbursts, midplane temperatures exceed 10^3 K, turning the disk gas into an ionized plasma, but these extreme temperatures are confined to the region within 0.1 AU of the star (Bell and Lin, 1994). The evolution of midplane temperatures with time for protoplanetary disks is only poorly understood, and probably consists of a gradually decreasing trend due to the moderation of infalling gas onto the disk and hence the lessening of internal heating, as well as reduced surface irradiation by the protostar as it approaches the main sequence,

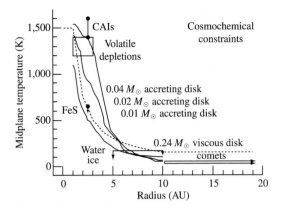

Figure 4 Midplane temperature as a function of heliocentric radius for a solar nebula with varying mass (inside 10 AU) undergoing mass accretion at a rate of a solar mass in ~0.1–1 Myr, compared to various cosmochemical constraints, and the results of a viscous accretion disk model (dashed line) with a mass of 0.24 solar masses (source Boss, 1998).

punctuated by short-lived bursts of heating in the innermost disk associated with FU Orionis outbursts (Calvet *et al.*, 2000).

1.04.3.2 Millimeter-wave Observations

Millimeter-wave emission often consists of contributions from both the infalling envelope and from any disk that has formed, making the separation of their respective contributions difficult. Millimeter-wave observations are also limited in spatial resolution (as of early 2000s) to the outermost regions of disks, at ~100 AU or so, and extrapolating these measurements inward to the planetary regions at 1–10 AU may not be fully warranted. Nevertheless, these measurements provide the best constraints available on the masses and physical conditions within protoplanetary disks.

1.04.3.2.1 Disk masses

Thermal emission from small dust particles in protoplanetary disks is optically thin at large distances from the star at millimeter wavelengths, allowing an estimate of the entire amount of dust present to be made, assuming a power-law distribution of the dust. The corresponding total amount of gas is typically assumed to be 100 times the dust mass. For T Tauri stars (class II objects), the inferred total gas and dust mass is ~0.01–0.1 solar mass, based on the dust emission coming from distances of 100 AU or more (Beckwith *et al.*, 1990). There is some evidence for higher disk masses in younger stars: a disk around a class I object with a mass of ~0.3 solar masses has a mass of at least 0.1 solar masses; inside 120 AU, the disk is optically thick at millimeter wavelengths and the amount of mass residing there can only be inferred (Launhardt and Sargent, 2001). Evidently, some disks have masses that are a significant fraction of their protostar's mass, at least at early phases.

1.04.3.2.2 Keplerian rotation

While dust continuum emission can be used to estimate disk masses, line emission from optically thin molecules (e.g., ^{13}CO) can be used to map the line-of-sight velocities in disks, using the Doppler shift of the moving gas. Evidence for Keplerian velocity profiles is typically found (e.g., Launhardt and Sargent, 2001), as is to be expected for gas in a stable orbit around a central protostar. These measurements only apply to the gas at considerable distances from the young star, however, typically at several hundred astronomical units or more. The situation inside these disks at planetary distances is not constrained by these observations, and even the outer disk measurements are subject to possible confusion with infalling gas in the envelopes and outflowing gas from stellar winds.

1.04.4 NEBULA TRANSPORT AND EVOLUTION

On theoretical grounds, even an initially highly centrally condensed (i.e., power-law density profile) cloud core is likely to collapse to form a protostar surrounded by a fairly massive protostellar disk. Observations of disks around young stars imply that at early ages, disk masses may be a significant fraction of the protostar's mass. As a result, the expectation is that protostellar disks must somehow transport most of their mass inward to be accreted by the protostar, eventually evolving into protoplanetary disks, where planetary bodies should be able to form and survive their subsequent interactions with the disk. The transition point from a protostellar disk to a protoplanetary disk is not clear, and the physical mechanisms responsible for disk evolution in either of these two phases remain uncertain, though progress seems to have been made in ruling out several proposed mechanisms.

1.04.4.1 Angular Momentum Transport Mechanisms

The basic theory of the evolution of an accretion disk can be derived by assuming that there is some physical mechanism operating that results in an effective viscosity of the gas. Because the intrinsic molecular viscosity of hydrogen gas is far too small to have an appreciable effect on disk evolution in a reasonable amount of time, theorists have sought other sources for an effective viscosity, such as turbulence. In a fully turbulent flow, the effective viscosity can be equal to the molecular viscosity multiplied by a large factor: the ratio of the Reynolds number of the disk (~10 billion) to the critical Reynolds number for the onset of turbulence (~1,000), or a factor of ~10 million. Under very general conditions, it can be shown (Lynden-Bell and Pringle, 1974) that a viscous disk will evolve in such a manner as to transport most of its mass inward, thereby becoming more tightly gravitationally bound, and minimizing the total energy of the system. In order to conserve angular momentum, this means that angular momentum must be transported outward along with a small fraction of the mass, so that the accretion disk expands outside some radius. The loss of significant angular momentum by centrifugally launched winds somewhat relieves this need for the accretion disk to expand; this additional angular momentum sink was not recognized when the theory was first developed. While the basic physics of a viscous accretion disk is fairly well developed, the physical mechanism(s) responsible for disk evolution remain contentious.

1.04.4.1.1 Hydrodynamic turbulence

Given the high Reynolds number ($Re = LV/v$, where $L =$ characteristic length scale, $V =$ characteristic velocity, and $v =$ kinematic viscosity) of a protoplanetary disk, one might expect that a turbulent cascade of energy would occur and result in an effective turbulent viscosity, which might be sufficient to drive disk evolution. However, because of the strong differential rotation in a Keplerian disk, a high Reynolds number is not a sufficient condition for fully developed turbulence. Instead, the Rayleigh criterion (which states that the radial gradient of the square of the specific angular momentum must be positive for stable flows), which applies to rotating fluids but is not strictly applicable to the solar nebula, suggests that Keplerian disks are stable with respect to turbulence.

Vertical convectively driven turbulence. While differential rotation may inhibit convective motions in the radial direction in a disk, motions parallel to the rotation axis are relatively unaffected by rotation. In a disk where heat is being generated near the midplane, and where dust grains are the dominant source of opacity, the disk is likely to be unstable to convective motions in the vertical direction, which carry the heat away from the disk's midplane and deposit it close to the disk's surface, where it can be radiated away. Convective instability was conjectured to lead to sufficiently robust turbulence for the resulting turbulent viscosity to be large enough to drive disk evolution (Lin and Papaloizou, 1980), a seemingly attractive, self-consistent scenario which has motivated much of the work on viscous evolution of the solar nebula. However, three-dimensional hydrodynamic models of vertically convectively unstable disks have shown that the convective cells that result are sheared by differential rotation to such an extent that the net transport of angular momentum is very small, and may even be in the wrong direction (Stone *et al.*, 2000). As a result, convectively driven disk evolution does not seem to be a major driver.

Rotational shear-induced turbulence. It has also been suggested that finite-amplitude (nonlinear) disturbances to Keplerian flow could result in a self-sustaining shear instability that would produce significant turbulence (Dubrulle, 1993). However, when three-dimensional hydrodynamic models were again used to investigate this possibility, it was found that the initially assumed turbulent motions decayed rather than grew (Stone *et al.*, 2000). Evidently purely hydrodynamic turbulence can neither grow spontaneously nor be self-sustained upon being excited by an external perturbation.

Global baroclinic instability-driven turbulence. In spite of these discouraging results for hydrodynamic turbulence, another possibility remains and is under investigation (Klahr and Bodenheimer, 2003), that of a global baroclinic instability. In this mechanism, turbulence results in essence from steep temperature gradients in the radial direction, which then battle centrifugal effects head-on. Three-dimensional hydrodynamic models imply that this mechanism can drive inward mass transport and outward angular momentum transport, as desired. However, a detailed stability analysis remains to be performed, and the global baroclinic instability has not yet been confirmed by other workers.

Rossby waves. These waves occur in planetary atmospheres as a result of shearing motions and can produce large-scale vortices such as the Great Red Spot on Jupiter. Rossby waves have been proposed to occur in the solar nebula as a result of Keplerian rotation coupled with a source of vortices. While prograde rotation (cyclonic) vortices are quickly dissipated by the background Keplerian flow, retrograde (anticyclonic) vortices are able to survive for longer periods of time (Godon and Livio, 1999). Rossby waves have been advanced as a significant source of angular momentum transport in the disk (Li *et al.*, 2001). Rossby vortices could serve as sites for concentrating dust particles, but the difficulty in forming the vortices in the first place, coupled with their eventual decay, makes this otherwise attractive idea somewhat dubious (Godon and Livio, 2000). In addition, the restriction of these numerical studies to thin, two-dimensional disk models, where refraction of the waves away from the midplane is not possible, suggests that in a fully three-dimensional calculation, Rossby waves may be less vigorous than in the thin disk calculations (Stone et al., 2000).

1.04.4.1.2 Magnetorotational-driven turbulence

While a purely hydrodynamic source for turbulence has not yet been demonstrated, the situation is much different when MHD effects are considered in a shearing, Keplerian disk. In this case, the Rayleigh criterion for stability can be shown to be irrelevant: provided only that the angular velocity of the disk decreases with radius, even an infinitesimal magnetic field will grow at the expense of the shear motions.

Balbus–Hawley instability. Balbus and Hawley (1991) pointed out that in the presence of rotational shear, even a small magnetic field will grow on a very short timescale. The basic reason is that magnetic field lines can act like rubber bands, linking two parcels of ionized gas. The parcel which is closer to the protosun will orbit faster than the other, increasing its distance from the other parcel. This leads to stretching of the magnetic field lines linking the parcel, and so to a retarding force on the forward motion of the inner parcel. This force transfers angular momentum from the inner parcel to the outer parcel, which means that the inner parcel must fall farther inward toward the protosun, increasing its angular velocity, and therefore leading to even more stretching of the field lines and increased magnetic forces. Because of this positive feedback, extremely rapid growth of an infinitesimal seed field occurs. Consequently, the magnetic field soon grows so large that its subsequent evolution must be computed with an MHD code.

Three-dimensional MHD models of a small region in the solar nebula (Hawley et al., 1995) have shown that, as expected, a tiny seed magnetic field soon grows and results in a turbulent disk where the turbulence is maintained by the magnetic instability. In addition, the magnetic turbulence results in a net outward flow of angular momentum, as desired. The magnetic field grows to a certain value and then oscillates about that mean value, depending on the assumed initial field geometry, which is large enough to result in relatively vigorous angular momentum transport. While promising, these studies of the magnetorotational instability (MRI) are presently restricted to small regions of the nebula, and the global response of the disk to this instability remains to be determined.

Ionization structure and layered accretion. MRI is a powerful phenomenon, but is limited to affecting nebula regions where there is sufficient ionization for the magnetic field, which is coupled only to the ions, to have an effect on the neutral atoms and molecules. The MRI studies described above all assume ideal MHD, i.e., a fully ionized plasma, where the magnetic field is frozen into the fluid. At the midplane of the solar nebula, however, the fractional ionization is expected to be quite low in the planetary region. Both ambipolar diffusion and resistivity (ohmic dissipation) are effective at limiting magnetic field strengths and suppressing MRI-driven turbulence, but a fractional ionization of only ~ 1 ion per 1,000 billion atoms is sufficient for MRI to proceed in spite of ambipolar diffusion and ohmic dissipation. Close to the protosun, disk temperatures are certainly high enough for thermal ionization to create an ionization fraction greater than this, and thus to maintain full-blown MRI turbulence. Given that a temperature of at least 1,400 K is necessary, MRI instability may be limited to the innermost 0.2 AU or so in quiescent phases, or as far out as ~ 1 AU during rapid mass accretion phases (Boss, 1998; Stone et al., 2000).

At greater distances, disk temperatures are too low for thermal ionization to be effective. Cosmic rays were thought to be able to ionize the outer regions of the nebula, but the fact that bipolar outflows are likely to be magnetically driven means that cosmic rays may have a difficult time reaching the disk midplane (Dolginov and Stepinski, 1994). However, the coronae of young stars are known to be prolific emitters of hard X-rays, which can penetrate the bipolar outflow and reach the disk surface at distances of ~ 1 AU or so, where they are attenuated (Glassgold et al., 1997). As a result, the solar nebula is likely to be a layered accretion disk (Gammie, 1996), where MRI turbulence results in inward mass transport within thin, lightly ionized surface layers, while the layers below the surface do not participate in MRI-driven transport. Thus, the bulk of the disk, from just below the surface to the midplane, is expected to be a magnetically dead zone. Layered accretion is thought to be capable of driving mass inflow at a rate of about a solar mass in 100 Myr, sufficient to account for observed mass accretion rates in quiescent T Tauri stars.

1.04.4.1.3 Gravitational torques in a marginally unstable disk

The remaining possibility for large-scale mass transport in the solar nebula is gravitational torques. The likelihood that much of the solar nebula was a magnetically dead zone where MRI transport was ineffective leads to the suggestion that there might be regions where inward MRI mass transport would cease, leading to a local pileup of mass, which might then cause at least a local gravitational instability of the disk (Gammie, 1996). In addition, there is observational and theoretical evidence that protostellar disks tend to start their lives with sufficient mass to be gravitationally unstable in their cooler regions, leading to the formation of nonaxisymmetric structure and hence the action of gravitational torques, and that these torques may be the dominant transport mechanism in early phases of evolution (Yorke and Bodenheimer, 1999).

In order for gravitational torques to be effective, a protostellar disk or the solar nebula must be significantly nonaxisymmetric, e.g., threaded by clumps of gas, or by spiral arms, much like a spiral galaxy. In that case, trailing spiral structures, which form inevitably as a result of Keplerian shear, will result in the desired outward transport of angular momentum. This is because in a Keplerian disk, an initial bar-shaped density perturbation will be sheared into a trailing spiral arm configuration. The inner end of the bar rotates faster than the outer end and therefore moves ahead of the outer end. Because of the gravitational attraction between the inner end and the outer end, the inner end will have a component of this gravitational force in the backward direction, while the outer end will feel an equal and opposite force in the forward direction. The inner end will thus lose orbital angular momentum, while the outer end gains this angular momentum. As a result, the inner end falls closer to the protosun, while the outer end moves farther away, with a net outward transport of angular momentum.

Rapid mass and angular momentum transport. Models of the growth of nonaxisymmetry during the collapse and formation of protostellar disks show that large-scale bars and spirals can form with the potential to transfer most of the disk angular momentum outward on timescales as short as 1,000 yr to 0.1 Myr (Boss, 1989), sufficiently fast to allow protostellar disks to transport the most of their mass inward onto the protostar and thereby evolve into protoplanetary disks.

Early numerical models of the evolution of a gravitationally unstable disk (e.g., Cassen *et al.*, 1981) suggested that a disk would have to be comparable in mass to the central protostar in order to be unstable, i.e., gravitational instability could occur in protostellar, but not in protoplanetary disks. Analytical work on the growth of spiral density waves implied that for a solar-mass star, gravitational instability could occur in a disk with a mass as low as 0.19 solar masses (Shu *et al.*, 1990). Newer three-dimensional hydrodynamic models have shown that vigorous gravitational instability can occur in a disk with a mass of 0.1 solar mass or even less, in orbit around a solar-mass star (Boss, 2000), because of the expected low midplane temperatures (~ 30 K) in the outer disk implied by cometary compositions (Kawakita *et al.*, 2001) and by observations of disks (D'Alessio *et al.*, 2001). Similar models with a complete thermodynamic treatment (Boss, 2002b), including convective transport and radiative transfer, show that a marginally gravitationally unstable solar nebula develops a robust pattern of clumps and spiral arms, persisting for many disk rotation periods, and resulting in episodic mass accretion rates onto the central protosun that vary between accreting a solar mass in 10 Myr to as short as 1,000 yr. The latter rates appear to be high enough to account for FU Orionis outbursts.

Global process versus local viscosity. Angular momentum transport by a strongly gravitationally unstable disk is rapid, so it is unlikely that protostellar or protoplanetary disks are ever strongly gravitationally unstable, because they can probably evolve away from such a strongly unstable state faster than they can be driven into it by, e.g., accretion of more mass from an infalling envelope or radiative cooling. As a result, it is much more likely that a disk will approach gravitational instability from a marginally unstable state (Cassen *et al.*, 1981). Accordingly, models of gravitationally unstable disk have focused almost exclusively on marginally gravitationally unstable disks (e.g., Boss, 2000), where the outer disk, beyond ~ 5 AU, primarily participates in the instability. Inside ~ 5 AU, disk temperatures appear to be too high for an instability to grow there, though these inner regions may still be subject to shock fronts driven by clumps and spiral arms in the gravitationally unstable, outer region. One-armed spiral density waves can propagate right down to the stellar surface.

Gravitational forces are intrinsically global in nature, and their effect on different regions of the nebula can be expected to be highly variable in both space and time. Alternatively, turbulent viscosity is a local process that is usually assumed to operate more or less equally efficiently throughout a disk. As a result, it is unclear if gravitational effects can faithfully be modeled as a single, effective viscosity capable of driving disk evolution in the manner envisioned by Lynden-Bell and Pringle (1974). Nevertheless, efforts have been made to try to quantify the expected strength of gravitational torques in this manner (Lin and Pringle, 1987). Three-dimensional models of marginally gravitationally unstable disks imply that such an effective viscosity is indeed large and comparable to that in MRI models (Laughlin and Bodenheimer, 1994).

1.04.4.2 Evolution of the Solar Nebula

Given an effective source of viscosity, in principle, the time evolution of the solar nebula can be calculated in great detail, at least in the context of the viscous accretion disk model. The strength of an effective viscosity is usually quantified by the alpha parameter. Alpha is often defined in various ways, but typically alpha is defined to be the constant that when multiplied by the sound speed and the vertical scale height of the disk (two convenient measures of a typical velocity and length scale) yields the effective viscosity of the disk (Lynden-Bell and Pringle, 1974). Three-dimensional MHD models of the MRI imply a typical MRI alpha of ~0.005–0.5 (Stone et al., 2000). Similarly, three-dimensional models of marginally gravitationally unstable disks imply an alpha of ~0.03 (Laughlin and Bodenheimer, 1994). Steady mass accretion at the low rates found in quiescent T Tauri stars requires an alpha of ~0.01 (Calvet et al., 2000), in rough agreement with these estimates.

Once planets have formed and become massive enough to open gaps in their surrounding disks, their orbital evolution becomes tied to that of the gas. As the gaseous disk is transported inward by viscous accretion, these planets must also migrate inward. The perils of orbital migration for planetary formation and evolution are covered in the Chapter 1.17. Here we limit ourselves to considering the evolution of dust and gas prior to the formation of planetary-size bodies.

1.04.4.2.1 Viscous accretion disk models

The generation of viscous accretion disk models was an active area of research during the period when convective instability was believed to be an effective source of viscosity. Ruden and Pollack (1991) constructed models where convective instabilty was assumed to control the evolution, so that in regions where the disk became optically thin and thus convectively stable, the effective viscosity vanished. Starting with an alpha of ~0.01, they found that disks evolved for about a million years before becoming optically thin, often leaving behind a disk with a mass of ~0.1 solar mass. Midplane temperatures at 1 AU dropped precipitously from ~1,500 K initially to ~20 K when convection ceased and the disk was optically thin at that radius. Similarly, dramatic temperature drops occur throughout the disk in these models, and the outer regions of the models eventually became gravitationally unstable as a result.

Given that convective instability is no longer considered to be a possible driver of disk evolution, the Ruden and Pollack (1991) models are interesting, but not likely to be applicable to the solar nebula. Unfortunately, little effort has gone into generating detailed viscous accretion models in the interim: the theoretical focus seems to have been more on the question of determining which mechanisms are contenders for disk evolution than on the question of the resulting disk evolution. In particular, the realization that the MRI mechanism is likely to have operated only in the magnetically active surface layers of the disk, and not in the magnetically dead bulk of the disk, presents a formidable technical challenge for viscous accretion disk models, which have usually been based on the assumption that the nebula can be represented by a thin, axisymmetric disk (e.g., Ruden and Pollack, 1991), greatly simplifying the numerical solution. The need for consideration of the vertical as well as the radial structure of the disk, and possibly the azimuthal (nonaxisymmetric) structure as well, points toward the requirement of a three-dimensional MHD calculation of the entire disk. As of the early 2000s, such a calculation has not been performed, and even the existing MRI calculations on small regions of a disk can only be carried forward in time for a small fraction of the expected lifetime of the disk.

Some progress has been made in two-dimensional hydrodynamic models of a thick disk evolving under the action of a globally defined alpha viscosity, representing the effects of torques in a marginally gravitationally unstable disk (Yorke and Bodenheimer, 1999), but in these models the evolution eventually slows down and leaves behind a fairly massive protostellar disk after 10 Myr.

1.04.4.2.2 Volatility patterns and transport in inner solar system

One aspect of particular interest about viscous accretion disk models is the evolution of solid particles, both in terms of their thermal processing and their transport in the nebula. Interstellar dust grains are small enough (submicron-sized) to remain well coupled to the gas, so they will move along with the gas. During this phase, the gas and dust may undergo trajectories that are outward at first, as the disk accretes matter from the infalling envelope and expands by outward angular momentum transport, followed by inward motion once accretion stops and the disk continues to accrete onto the protostar (Cassen, 1996). Once collisional coagulation gets underway and grain growth begins, solid particles begin to move with respect to the gas, suffering gas drag and additional radial migration as a result (Weidenschilling, 1988).

The bulk compositions of the bodies in the inner solar system show a marked depletion of volatile elements compared to the solar composition. Cassen (2001) has shown that the depletions of the moderately volatile elements (those with condensation temperatures above ~650 K) can be explained as a result of the condensation of hot gases and coagulation of the resulting refractory dust grains into solids that are decoupled from the gas through the rapid growth of kilometer-sized planetesimals. The moderately volatile elements remain in gaseous form at these temperatures, and so avoid being incorporated into the planetesimals that will eventually form the terrestrial planets and asteroids. In order for this process to work, significant regions of the nebula must have been hot enough at the midplane to keep moderately volatile elements in the gaseous form, a situation that would characterize the nebula when mass accretion rates were on the order of a solar mass in less than 10 Myr. The moderately volatile gases would then be removed from the terrestrial planet region along with the more volatile elements and molecules (e.g., hydrogen, helium) by viscous accretion onto the protosun. The postulated rapid growth from dust grains to kilometer-sized bodies required by this scenario appears to be possible (Woolum and Cassen, 1999).

1.04.4.2.3 Turbulent and gas drag transport

Cyr et al. (1998) studied the redistribution of water in the inner solar system by inward radial migration of icy solids caused by gas drag and outward transport of water vapor by gaseous diffusion to ~5 AU, where it is assumed to condense back into icy particles. Their models showed that the abundance of water vapor might be enhanced at distances of several astronomical units, and suggest that cycling of water vapor might be related to the spectral evidence for variations in the water of hydration in asteroids. Gail (2001) studied the radial mixing of gases and small particles by diffusion in a turbulent disk, and found that materials that were thermally processed inside 1 AU could be transported outward to 10 AU or beyond, possibly explaining observations of crystallized silicate dust grains in comets. Turbulent diffusion of water has also been invoked to explain the D/H ratio of LL3 meteorites and the ice giant planets, compared to that of the Sun (Hersant et al., 2001).

One means of speeding the growth of dust grains into kilometer-sized planetesimals envisions the vertical sedimentation of dust grains into a thin sub-disk of dust particles, sufficiently dense as to undergo a local gravitational instability that could rapidly form comet-sized bodies. However, in a turbulent nebula, the ability of dust grains to sediment to the midplane is countered by the tendency of the turbulent motions to loft the dust grains back upwards. Detailed models of the interactions of dust grains with turbulent flows suggest that a dust sub-disk gravitational instability is likely to be circumvented by these interactions (e.g., Dubrulle et al., 1995; Dobrovolskis et al., 1999).

A perennial problem in meteoritics is understanding what effect gas drag has on the orbital migration of solids. In a laminar disk where the gas pressure decreases monotonically with distance from the star, solids will always face a headwind from the gas (which orbits slower than the Keplerian rate, because the outward pressure force weakens the effective gravity of the protosun as felt by the gas). As a result, solids will spiral inward in such a disk (Weidenschilling, 1988), with centimeter-sized solids migrating ~20 AU within 0.1 Myr, and millimeter-sized solids traversing ~0.6 AU in the same time interval. Given that chondritic meteorites contain centimeter-sized and millimeter-sized solids (i.e., refractory inclusions and chondrules, respectively), as well as much smaller grains of matrix material, it has always been hard to understand how these components could have been processed locally and aggregated into a single rock, unless the aggregation took place on a timescale much less than 0.1 Myr. This seems unlikely, considering that the evidence for short-lived isotopes in these solids implies that most refractory inclusions formed several million years before most chondrules. One possible solution is that in a turbulent disk, the weak but finite coupling of centimeter-sized solids to the turbulent gas will still permit some centimeter-sized solids to resist inward migration caused by gas drag and to be transported outward to distances of several astronomical units or so (Cuzzi et al., 2003). Another possibility is that in a marginally gravitationally unstable disk (see next section), the local pressure maxima associated with spiral-arm-like structures will concentrate solids in the centers of the arms, because solids will face both headwinds and tailwinds in this situation, with their orbits converging at the location of the gas pressure maxima (Haghighipour and Boss, 2003). This process would prevent or at least slow down loss of the solids to the protosun.

1.04.4.3 Clump Formation in a Marginally Gravitationally Unstable Disk

Given the apparent limitation of MRI-driven accretion to the surfaces of protoplanetary disks, it would appear that gravitational torques might have to be responsible for the evolution in the bulk of the disk. In addition, there are strong theoretical reasons why gravitational torques may be effective, including the difficulty in forming the gas giant planets by the conventional means of core accretion. The standard model for Jupiter formation by core accretion envisions a nebula that has a surface density high enough for a solid core to form within

~8 Myr through runaway accretion (Pollack et al., 1996). However, such a nebula is likely to be marginally gravitationally unstable, a situation that could result in the rapid formation of gas giant planets in a few thousand years by the formation of self-gravitating clumps of gas and dust (Boss, 1997, 2000, 2002b).

In their pioneering study of a marginally unstable disk, Laughlin and Bodenheimer (1994) found strong spiral arm formation, but no clumps, presumably as a result of the limited spatial resolution that was computationally possible at the time (up to 2.5×10^4 particles). Boss (2000) has shown that when a million or more grid points are included, three-dimensional hydrodynamic models of marginally gravitationally unstable disks demonstrate the persistent formation of self-gravitating clumps, though even these models do not appear to have sufficient spatial resolution to follow the high-density clumps indefinitely in time. Regardless of whether or not such disk instability models can lead to gas giant planet formation, the likelihood that the solar nebula was at least episodically marginally gravitationally unstable has important implications for cosmochemistry.

1.04.4.3.1 Chondrule formation in nebular shock fronts

Perhaps the most well-known, unsolved problem in cosmochemistry is the question of the mechanism whereby dust grain aggregates were thermally processed to form chondrules and some rounded refractory inclusions. Chondrule compositions and textures require rapid heating and somewhat slower cooling for their explanation; a globally hot nebula is inconsistent with these requirements (Cassen, 2001). A wide variety of mechanisms has been proposed and generally discarded (see Desch and Cuzzi (2000) for an update, including the status of lightning as a flash heating mechanism likely to have occurred in the solar nebula), but the work of Desch and Connolly (2002) seems to have largely solved the problem. In a marginally gravitationally unstable nebula, clumps and spiral arms at ~8 AU will drive one-armed spiral arms into the inner nebula that at times result in shock fronts that are orientated roughly perpendicular to the orbits of bodies in the asteroidal region. Because of the tendency toward corotation in self-gravitating structures, this leads to solids encountering a shock front at speeds ~10 km s^{-1}, sufficiently high to result in postshock temperatures of ~3,000 K (Figure 5). Detailed one-dimensional models of heating and cooling processes in such a shock front have shown that shock speeds ~7 km s^{-1} are optimal for matching chondrule cooling rates and therefore textures (Desch and Connolly, 2002).

1.04.4.3.2 Mixing processes in marginally gravitationally unstable disks

If disk evolution near the midplane is largely controlled by gravitational torques rather than by a turbulent process such as MRI or convection, then mixing processes might be profoundly different as a result. Gravitational torques could potentially result in matter flowing through the disk without being rapidly homogenized through mixing by turbulence. As a result, spatially heterogeneous regions of the disk might persist for some amount of time, if they were formed in the first place by processes such as the triggered injection of shock-wave material

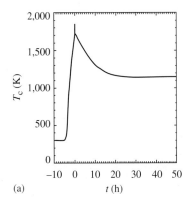
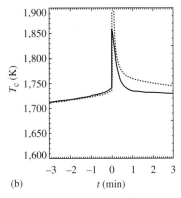

Figure 5 Thermal history of a chondrule (solid lines) passing through a 7.5 km s^{-1} shock wave, over timescales of: (a) hours and (b) minutes. Chondrules are heated for hours before the shock by radiation. At the shock front, chondrules must slow to the gas velocity; in 1 min it takes to do so, gas drag heating causes a spike in temperature. Afterwards, chondrules are heated for hours by radiation and thermal exchange with the hot gas (dashed line) (Desch and Connolly, 2002) (reproduced by permission of Sheridan Press and Meteoritical Society from *Meteorit. Planet. Sci.*, **2002**, *37*, 183–207).

(Vanhala and Boss, 2000) or the spraying and size sorting of solids processed by an X-wind onto the surface of the nebula (Shu et al., 1996). However, because convective motions appear to play an important part in cooling the disk midplane in recent models of disk instability (Boss, 2002b), it is unclear if gravitational torques could act in isolation without interference from convective motions or other sources of turbulence. At any rate, spatially heterogeneous regions might only last for short fraction of the nebular lifetime, requiring rapid coagulation and growth of kilometer-sized bodies if evidence of this phase is to be preserved, as seems to be the case for the ^{53}Mn/^{55}Mn ratio in the inner solar system (Lugmair and Shukolyukov, 1998), and possibly also for oxygen isotopes.

1.04.4.4 X-wind Model for Processing Solids

If an X-wind is responsible for driving bipolar outflows, then there are possibly important implications for the thermal processing of solids and the production of short-lived radioisotopes (Shu et al., 1996, 2001). The basic idea is that some of the solids that spiral inward and approach the boundary layer between the solar nebula and the protosun will be lifted upward by the same magnetically driven wind that powers bipolar outflows. While close to the protosun, these solids will be subject to heating by the solar radiation field and to spallation by particles from solar flares. Following this processing, the solids will be lofted onto size-sorted trajectories that return them to the surface of the solar nebula at several astronomical units or beyond (Figure 6). Note, however, that if a disk wind operates along with or instead of an X-wind, then any solids lofted from the inner region may be unable to return directly to the asteroid region ~2.5 AU, as the disk wind being launched at those same distances may prevent their infall onto the disk.

1.04.4.4.1 Local irradiation (^{10}Be)

While most of the short-lived isotopes found in primitive meteorites could be products of stellar nucleosynthesis (Goswami and Vanhala, 2000), evidence indicates that ^{10}Be may have also been alive in the early solar nebula (McKeegan et al., 2000). Beryllium-10 cannot be synthesized in stellar interiors, but can be formed by spallation reactions with energetic particles such as those that accompany stellar flares. The evidence for ^{10}Be thus suggests that some short-lived isotopes were formed by spallation reactions near the surface of the nebula or in the X-wind region (Gounelle et al., 2001). However, the fact that ^{10}Be does not correlate with the presence of evidence for ^{26}Al or ^{41}Ca (Marhas et al., 2002) implies that the mechanism that produced ^{10}Be did

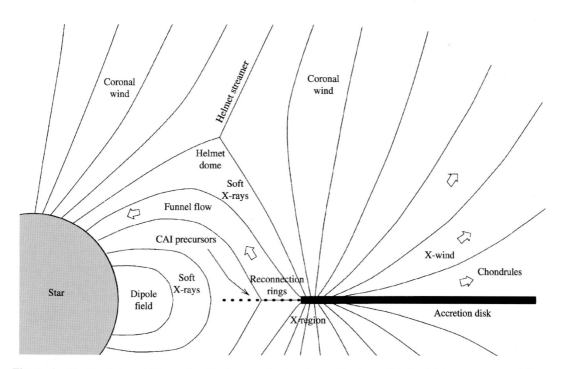

Figure 6 The X-wind model for meteoritical processing involves subjecting solids to high temperatures and fluxes of energetic particles from the early Sun during and after being lifted from the midplane of the solar nebula in a bipolar wind, and afterward falling back onto the nebula at greater distances (Shu et al., 2001) (reproduced by permission of University of Chicago Press and American Astronomical Society from Astrophys. J., **2001**, 548, 1029–1050).

not produce the ^{26}Al and ^{41}Ca. It has also been proposed that ^{10}Be could have been formed by spallation in the debris associated with a core-collapse supernova (Cameron, 2001). Galactic cosmic rays might also have produced the ^{10}Be in the presolar cloud (Desch et al., 2003). In addition, the short-lived radioisotope ^{60}Fe cannot be produced by spallation reactions, but is produced by stellar nucleosynthesis, so the isotopic evidence still seems to require some synthesis in a stellar source, ejection from the star, transport across the interstellar medium, injection into the presolar cloud, collapse of the presolar cloud, and formation of centimeter-sized particles, all within ~0.1 Myr. Such a scenario would seem to argue strongly in favor of a supernova trigger for the formation of the solar nebula (Cameron and Truran, 1977), perhaps in addition to whatever production occurred locally as a result of spallation reactions.

1.04.4.4.2 Thermal processing of solids and ^{26}Al formation

The X-wind model provides a ready explanation for the formation of meteoritic components such as refractory inclusions, which require access to relatively high temperatures, because the X-wind mechanism hypothesizes the cycling of solids through the hot boundary layer located between the inner edge of the nebula and the protosun and back outward to the asteroid region. Refractory inclusions might then also acquire ^{26}Al by spallation reactions close to the protosun, thereby explaining the high frequency of evidence for live ^{26}Al at the canonical ratio of ^{26}Al/^{27}Al of $\sim 5 \times 10^{-5}$ in refractory inclusions (Gounelle et al., 2001). However, given that stellar flares are known to occur with a wide range of intensities, it is unclear whether the X-wind scenario can explain the remarkably uniformity of the canonical ratio of ^{26}Al/^{27}Al in refractory inclusions. Unless these refractory inclusions are shielded by a mantle of less refractory material during particle irradiation (Gounelle et al., 2001), spallation reactions will produce an abundance of the short-lived isotope ^{41}Ca that will be much greater than is observed, if the same particle flux is used to explain the canonical ^{26}Al/^{27}Al ratio (Goswami et al., 2001), which seems to be required given the correlation between the presence of ^{41}Ca and ^{26}Al. It has been argued that hypothesizing such a less refractory mantle on refractory cores is inconsistent with petrologic constraints (Simon et al., 2002).

The X-wind model has also been advanced as a means for thermal processing of chondrules (Shu et al., 1996). While it has not yet been possible to calculate the detailed thermal history of chondrule precursors to see if the required impulsive heating and slower cooling rates can be matched, the fact that this mechanism implies size sorting of the particles as they are lofted upward and return to the nebula on ballistic trajectories, means that small dust particles and chondrule-sized particles that were thermally processed together in the X-wind region will eventually be separated upon their return to the nebula (Desch and Connolly, 2002). As a result, it is difficult to explain the fact that chondrules and the fine-grained matrix in which they reside are chemically complementary: their combined, bulk composition is roughly solar, though individually they are not. This seems to indicate thermal processing in a more closed system, such as the one that occurs with shock wave processing within the nebula, may be a better means to explain chondrule thermal processing.

1.04.5 SOLAR NEBULA REMOVAL

Once the planet formation process is finished, or at least once bodies large enough to form the terrestrial planets in the absence of any gas have formed, there remains the problem of how to remove any residual nebular gas and dust that might remain. A long-lived disk could represent a danger to planets, as inward migration with respect to the disk by Earth-mass planets or migration along with the disk for Jupiter-mass planets represent significant threats (see Chapter 1.17). The existence of short-period Jupiter-like planets in orbit around solar-type stars (Mayor and Queloz, 1995; Marcy and Butler, 1996) is best explained by inward orbital migration, so this danger must be taken seriously.

1.04.5.1 Observational Constraints on Disk Life Times

The most robust constraints on the timescale for disk removal come from astronomical observations of young stars. Because of the limited spatial resolution of interferometric arrays, molecular hydrogen gas or tracer species such as carbon monoxide cannot be mapped at scales of less than ~10 AU in most disks. Instead, the presence of the gaseous portion of a disk is inferred indirectly by the presence of ongoing mass accretion from the disk onto the star. This accretion leads to enhanced emission in the star's hydrogen alpha line, which is then a diagnostic of the presence of disk gas. Observations of hydrogen alpha emission in the Orion OB1 association have shown that hydrogen alpha emission drops toward zero once the stars reach an age of a few million years (Briceno et al., 2001). The presence of the dusty portion of disks is signaled by excess IR emission, derived from dust grains in the innermost region of the disk. Again, observations imply that the dust disk largely disappears by ages of a few million years (Briceno et al., 2001). In the Orion nebula cluster, ages are thought to be even shorter as a result of photoevaporation caused by UV irradiation by newly formed massive stars. However, at least portions of other disks last for as long as 10 Myr; Beta Pictoris is ~10 Myr old and has a remnant dust disk which seems to be replenished by collisions between the members of an unseen population of orbiting bodies.

1.04.5.2 Removal Mechanisms

Leftover gas and dust that is not accreted by the planetary system can be removed in either of only two ways: either it accretes onto the central protostar, or it is transported out of the system altogether.

1.04.5.2.1 Inward flow onto protosun

Continued accretion onto the protosun is the simplest, most efficient means for removing excess material, as this process liberates energy rather than requiring energy, and the protosun has an endless appetite. The volatile gases that were not accreted by the terrestrial planets (Cassen, 2001), e.g., were presumably largely accreted by the protosun. The principal obstacles to this removal mechanism may be the need to provide an ongoing source of viscosity which could drive the transport, and barriers (disk gaps) erected by massive planets, particularly in the giant planet region.

1.04.5.2.2 UV photoevaporation by protosun

While formation as a single star in an isolated, dense cloud core is usually imagined for the presolar cloud, in reality there are very few examples of isolated star formation. Most stars form in regions of high-mass star formation, similar to Orion, with a smaller fraction forming in smaller clusters of low-mass stars, like Taurus. The radiation environment differs considerably between these two extremes, with Taurus being relatively benign, and with Orion being flooded with UV radiation once massive stars begin to form. Even in Taurus, though, individual young stars emit UV and X-ray radiation at levels considerably greater than mature stars (Feigelson and Montmerle, 1999). UV radiation from the protosun has been suggested as a means of removing the residual gas from the outermost solar nebula through photoevaporation of hydrogen atoms (Shu et al., 1993), a process estimated to require ~10 Myr.

1.04.5.2.3 UV photoevaporation in Orion-like regions

UV radiation from a nearby massive star has been invoked as a means to photoevaporate more rapidly the gas in the outer solar nebula and then to form the ice giant planets by photoevaporating the gaseous envelopes of the outermost gas giant protoplanets (Boss et al., 2002), as may be happening in the protoplanetary disks seen in Hubble images of the Orion nebula cluster. This environment would lead to much shorter disk lifetimes than in regions like Taurus, yet may still be able to yield planetary systems similar to our own. In addition, this new scenario shortens the outer disk lifetime, leading to more rapid depletion of the inner disk, and so helps to prevent subsequent loss of planets by inward orbital migration driven by disk interactions.

1.04.5.2.4 Final scouring by widened stellar outflow

The simple picture of the solar nebula being removed by a spherically symmetric T Tauri wind has long since been supplanted by the realization that young stars have directed, bipolar outflows that do not sweep over most of the disk. However, mature stars such as the Sun do have approximately isotropic winds, so there must be some transition phase where the bipolar star/disk wind evolves into a more spherically symmetric stellar wind. Presumably this enhanced stellar wind would eventually scour any remaining gas and dust from the system. In addition, Poynting–Robertson drag and radiation pressure are able to remove the smaller dust grains around older stars, the former by orbital decay inward onto the star, and the latter by being driven outward.

1.04.6 SUMMARY

In conclusion, we first highlight a few issues of special interest to meteoriticists.

(i) The solar nebula gas was probably cooler than 650 K at 2.5 AU at the time of chondrule formation (e.g., Figure 4), but was shock-heated to considerably higher temperatures, over 2,000 K (Figure 5), as a result of shock fronts driven by massive clumps in a marginally gravitationally unstable nebula. Such an unstable nebula seems to be required in any model of gas giant planet formation (Pollack et al., 1996; Boss, 1997).

(ii) Calcium–aluminum-rich inclusions (CAIs) have refractory compositions that seem to require condensation from very high temperatures, 1,700 K or more. The high midplane temperatures close to the protosun provide a possible means for explaining the condensation of CAIs in the context of the X-wind model (Shu et al., 2001). However, an attractive alternative has emerged, where CAI and chondrule compositions are explained as a result of partial evaporation of CI composition dust (with gas-melt exchange) during flash heating events at 2–3 AU in the nebula (Alexander, 2003). In this model, CAI-like and chondrule-like elemental and isotopic compositions result from a few hours of heating at 1,800 K or so, consistent with the heating experienced by solids in nebular shock fronts (see Figure 5; Desch and Connolly, 2002).

(iii) Expected lifetimes for chondrules and CAIs (i.e., millimeter- to centimeter-sized solids) are expected to be shorter than the nebula lifetime, ~1 Myr or less, due to orbital decay into the protosun caused by gas drag (Weidenschilling, 1988). However, it has been shown that in the context of a marginally gravitationally unstable nebula, where the gas forms spiral arms, solids will not migrate monotonically toward the protosun, but rather toward the centers of the spiral arms, on timescales of 0.01–0.1 Myr (Haghighipour and Boss, 2003), where they are likely to coalesce.

While we have learned much in the last few decades, it is clear that we still have much to learn about protoplanetary disks and the solar nebula. Future progress will require a sustained effort in three areas: observation, theory, and laboratory work. For example, the Atacama Large Millimeter Array (ALMA) is anticipated to be able to provide millimeter-wave maps of protoplanetary disks at much higher spatial resolution than possible with available millimeter arrays, allowing structures on the 1 AU scale to be probed in the nearest star-forming regions, revealing for the first time conditions in the inner planetary region. Laboratory analyses of samples of comets, asteroids, solar wind, and interplanetary dust particles will continue to provide the detailed clues needed to decipher the origin of our solar system. Theorists must weave together all of the constraints produced by observations, laboratory work, and their own increasingly detailed models of nebular processes, in order to recreate the grand tapestry of planetary system formation.

REFERENCES

Alexander C. M. O.'D. (2003) Making CAIs and chondrules from CI dust in a canonical nebula. *Lunar Planet. Sci.* **XXXIV**, #1391. The Lunar and Planetary Institute (CD-ROM).

Alves J. F., Lada C. J., and Lada E. A. (2001) Internal structure of a cold dark molecular cloud inferred from the extinction of background starlight. *Nature* **409**, 159–161.

Anders E. and Grevesse N. (1989) Abundances of the elements: meteoritic and solar. *Geochim. Cosmochim. Acta* **53**, 197–214.

Andre P. (1997) The evolution of flows and protostars. In *Herbig–Haro Flows and the Birth of Low Mass Stars* (eds. B. Reipurth and C. Bertout). Kluwer, Dordrecht, pp. 483–494.

Balbus S. A. and Hawley J. F. (1991) A powerful local shear instability in weakly magnetized disks: I. Linear analysis. *Astrophys. J.* **376**, 214–222.

Beckwith S. V. W., Sargent A. I., Chini R. S., and Gusten R. (1990) A survey for circumstellar disks around young stellar objects. *Astron. J.* **99**, 924–945.

Bell K. R. and Lin D. N. C. (1994) Using FU Orionis outbursts to constrain self-regulated protostellar disks. *Astrophys. J.* **427**, 987–1004.

Boss A. P. (1989) Evolution of the solar nebula: I. Nonaxisymmetric structure during nebula formation. *Astrophys. J.* **345**, 554–571.

Boss A. P. (1995) Collapse and fragmentation of molecular cloud cores: II. Collapse induced by stellar shock waves. *Astrophys. J.* **439**, 224–236.

Boss A. P. (1997) Giant planet formation by gravitational instability. *Science* **276**, 1836–1839.

Boss A. P. (1998) Temperatures in protoplanetary disks. *Ann. Rev. Earth Planet. Sci.* **26**, 53–80.

Boss A. P. (2000) Possible rapid gas giant planet formation in the solar nebula and other protoplanetary disks. *Astrophys. J.* **536**, L101–L104.

Boss A. P. (2002a) Collapse and fragmentation of molecular cloud cores: VII. Magnetic fields and multiple protostar formation. *Astrophys. J.* **568**, 743–753.

Boss A. P. (2002b) Evolution of the solar nebula: V. Disk instabilities with varied thermodynamics. *Astrophys. J.* **576**, 462–472.

Boss A. P. and Hartmann L. W. (2001) Protostellar collapse in a rotating, self-gravitating sheet. *Astrophys. J.* **562**, 842–851.

Boss A. P. and Myhill E. A. (1995) Collapse and fragmentation of molecular cloud cores: III. Initial differential rotation. *Astrophys. J.* **451**, 218–224.

Boss A. P., Wetherill G. W., and Haghighipour N. (2002) Rapid formation of ice giant planets. *Icarus* **156**, 291–295.

Briceno C., Vivas A. K., Calvet N., Hartmann L., Pacheco R., Herrera D., Romero L., Berlind P., and Sanchez G. (2001) The CIDA-QUEST large-scale survey of Orion OB1: evidence for rapid disk dissipation in a dispersed stellar population. *Science* **291**, 93–96.

Calvet N., Hartmann L., and Strom S. E. (2000) Evolution of disk accretion. In *Protostars and Planets IV* (eds. V. Mannings, A. P. Boss, and S. S. Russell). University of Arizona Press, Tucson, pp. 377–400.

Cameron A. G. W. (2001) Some properties of r-process accretion disks and jets. *Astrophys. J.* **562**, 456–469.

Cameron A. G. W. and Truran J. W. (1977) The supernova trigger for formation of the solar system. *Icarus* **30**, 447–461.

Cassen P. (1996) Models for the fractionation of moderately volatile elements in the solar nebula. *Meteorit. Planet. Sci.* **31**, 793–806.

Cassen P. (2001) Nebula thermal evolution and the properties of primitive planetary materials. *Meteorit. Planet. Sci.* **36**, 671–700.

Cassen P. M., Smith B. F., Miller R., and Reynolds R. T. (1981) Numerical experiments on the stability of preplanetary disks. *Icarus* **48**, 377–392.

Crutcher R. M. (1999) Magnetic fields in molecular clouds: observations confront theory. *Astrophys. J.* **520**, 706–713.

Cuzzi J. N., Davis S. S., and Dobrovolskis A. R. (2003) Creation and distribution of CAIs in the protoplanetary nebula. *Lunar Planet. Sci.* **XXXIV**, #1749, The Lunar and Planetary Institute (CD-ROM).

Cyr K. E., Sears W. D., and Lunine J. I. (1998) Distribution and evolution of water ice in the solar nebula: implications for solar system body formation. *Icarus* **135**, 537–548.

D'Alessio P., Calvet N., and Hartmann L. (2001) Accretion disks around young objects: III. Grain growth. *Astrophys. J.* **553**, 321–334.

Desch S. J. and Connolly H. C. (2002) A model of the thermal processing of particles in solar nebula shocks: application to the cooling rates of chondrules. *Meteorit. Planet. Sci.* **37**, 183–207.

Desch S. J. and Cuzzi J. N. (2000) The generation of lightning in the solar nebula. *Icarus* **143**, 87–105.

Desch S. J., Srinivasan G., and Connolly H. C. (2003) An interstellar origin for the beryllium 10 in CAIs. *Lunar Planet. Sci.* **XXXIV**, #1394. The Lunar and Planetary Institute (CD-ROM).

Dobrovolskis A. R., Dacles-Mariani J. S., and Cuzzi J. N. (1999) Production and damping of turbulence by particles in the solar nebula. *J. Geophys. Res.* **104**, 30805–30815.

Dolginov A. Z. and Stepinski T. F. (1994) Are cosmic rays effective for ionization of protoplanetary disks? *Astrophys. J.* **427**, 377–383.

Dubrulle B. (1993) Differential rotation as a source of angular momentum transfer in the solar nebula. *Icarus* **106**, 59–76.

Dubrulle B., Morfill G., and Sterzik M. (1995) The dust subdisk in the protoplanetary nebula. *Icarus* **114**, 237–246.

Feigelson E. D. and Montmerle T. (1999) High energy processes in young stellar objects. *Ann. Rev. Astron. Astrophys.* **37**, 363–408.

Foster P. N. and Boss A. P. (1996) Triggering star formation with stellar ejecta. *Astrophys. J.* **468**, 784–796.

Gail H.-P. (2001) Radial mixing in protoplanetary accretion disks: I. Stationary disc models with annealing and carbon combustion. *Astron. Astrophys.* **378**, 192–213.

Gammie C. F. (1996) Layered accretion in T Tauri disks. *Astrophys. J.* **457**, 355–362.

Glassgold A. E., Najita J., and Igea J. (1997) X-ray ionization of protoplanetary disks. *Astrophys. J.* **480**, 344–350.

Godon P. and Livio M. (1999) On the nonlinear hydrodynamic stability of thin Keplerian disks. *Astrophys. J.* **521**, 319–327.

Godon P. and Livio M. (2000) The formation and role of vortices in protoplanetary disks. *Astrophys. J.* **537**, 396–404.

Goodman A. A., Benson P. J., Fuller G. A., and Myers P. C. (1993) Dense cores in dark clouds: VIII. Velocity gradients. *Astrophys. J.* **406**, 528–547.

Goswami J. N. and Vanhala H. A. T. (2000) Extinct radionuclides and the origin of the solar system. In *Protostars and Planets IV* (eds. V. Mannings, A. P. Boss, and S. S. Russell). University of Arizona Press, Tucson, pp. 963–994.

Goswami J. N., Marhas K. K., and Sahijpal S. (2001) Did solar energetic particles produce the short-lived nuclides present in the early solar system? *Astrophys. J.* **549**, 1151–1159.

Gounelle M., Shu F. H., Shang H., Glassgold A. E., Rehm K. E., and Lee T. (2001) Extinct radioactivities and protosolar cosmic rays: self-shielding and light elements. *Astrophys. J.* **548**, 1051–1070.

Haghighipour N. and Boss A. P. (2003) On pressure gradients and rapid migration of solids in a nonuniform solar nebula. *Astrophys. J.* **583**, 996–1003.

Hartmann L. and Kenyon S. J. (1996) The FU Orionis phenomenon. *Ann. Rev. Astron. Astrophys.* **34**, 207–240.

Hawley J. F., Gammie C. F., and Balbus S. A. (1995) Local three-dimensional magnetohydrodynamic simulations of accretion disks. *Astrophys. J.* **440**, 742–763.

Hersant F., Gautier D., and Hure J.-M. (2001) A two-dimensional model for the primordial nebula constrained by D/H measurements in the solar system: implications for the formation of giant planets. *Astrophys. J.* **554**, 391–407.

Jones C. E., Basu S., and Dubinski J. (2001) Intrinsic shapes of molecular cloud cores. *Astrophys. J.* **551**, 387–393.

Kawakita H., Watanabe J., Ando H., Aoki W., Fuse T., Honda S., Izumiura H., Kajino T., Kambe E., Kawanomoto S., Sato B., Takada-Hidai M., and Takeda Y. (2001) The spin temperature of NH_3 in comet C/1999S4 (LINEAR). *Science* **294**, 1089–1091.

Kenyon S. J. and Hartmann L. (1987) Spectral energy distributions of T Tauri stars: disk flaring and limits on accretion. *Astrophys. J.* **323**, 714–733.

Klahr H. H. and Bodenheimer P. (2003) Turbulence in accretion disks: vorticity generation and angular momentum transport via the global baroclinic instability. *Astrophys. J.* **582**, 869–892.

Kobrick M. and Kaula W. M. (1979) A tidal theory for the origin of the solar nebula. *Moon Planet.* **20**, 61–101.

Konigl A. and Pudritz R. E. (2000) Disk winds and the accretion-outflow connection. In *Protostars and Planets IV* (eds. V. Mannings, A. P. Boss, and S. S. Russell). University of Arizona Press, Tucson, pp. 759–788.

Koresko C. D. (2000) A third star in the T Tauri system. *Astrophys. J.* **531**, L147–L149.

Lada C. J. and Shu F. H. (1990) The formation of sunlike stars. *Science* **248**, 564–572.

Larson R. B. (1969) Numerical calculations of the dynamics of a collapsing proto-star. *Mon. Not. Roy. Astron. Soc.* **145**, 271–295.

Laughlin G. and Bodenheimer P. (1994) Nonaxisymmetric evolution in protostellar disks. *Astrophys. J.* **436**, 335–354.

Launhardt R. and Sargent A. I. (2001) A young protoplanetary disk in the Bok globule CB 26? *Astrophys. J.* **562**, L173–L175.

Li H., Colgate S. A., Wendroff B., and Liska R. (2001) Rossby wave instability of thin accretion disks: III. Nonlinear simulations. *Astrophys J.* **551**, 874–896.

Lin D. N. C. and Papaloizou J. (1980) On the structure and evolution of the primordial solar nebula. *Mon. Not. Roy. Astron. Soc.* **191**, 37–48.

Lin D. N. C. and Pringle J. E. (1987) A viscosity prescription for a self-gravitating accretion disk. *Mon. Not. Roy. Astron. Soc.* **225**, 607–613.

Lugmair G. W. and Shukolyukov A. (1998) Early solar system timescales according to $^{53}Mn-^{53}Cr$ systematics. *Geochim. Cosmochim. Acta* **62**, 2863–2886.

Lynden-Bell D. and Pringle J. E. (1974) The evolution of viscous disks and the origin of the nebular variables. *Mon. Not. Roy. Astron. Soc.* **168**, 603–637.

Mannings V., Boss A. P., and Russell S. S. (2000) *Protostars and Planets IV*. University of Arizona Press, Tucson.

Marcy G. W. and Butler R. P. (1996) A planetary companion to 70 Virginis. *Astrophys J. Lett.* **464**, L147–L151.

Marhas K. K., Goswami J. N., and Davis A. M. (2002) Short-lived nuclides in hibonite grains from Murchison: evidence for solar system evolution. *Science* **298**, 2182–2185.

Mayor M. and Queloz D. (1995) A Jupiter-mass companion to a solar-type star. *Nature* **378**, 355–359.

McKeegan K. D., Chaussidon M., and Robert F. (2000) Incorporation of short-lived ^{10}Be in a calcium–aluminum-rich inclusion from the Allende meteorite. *Science* **289**, 1334–1337.

Mouschovias T. Ch. (1991) Magnetic braking, ambipolar diffusion, cloud cores, and star formation: natural length scales and protostellar masses. *Astrophys. J.* **373**, 169–186.

Pollack J. B., Hubickyj O., Bodenheimer P., Lissauer J. J., Podolak M., and Greenzweig Y. (1996) Formation of the giant planets by concurrent accretion of solids and gas. *Icarus* **124**, 62–85.

Preibisch T. and Zinnecker H. (1999) The history of low-mass star formation in the upper Scorpius OB association. *Astron. J.* **117**, 2381–2397.

Reipurth B. (2000) Disintegrating multiple systems in early stellar evolution. *Astron. J.* **120**, 3177–3191.

Rodriguez L. F., D'Alessio P., Wilner D. J., Ho P. T. P., Torrelles J. M., Curiel S., Gomez Y., Lizano S., Pedlar A., Canto J., and Raga A. C. (1998) Compact protoplanetary disks around the stars of a young binary system. *Nature* **395**, 355–357.

Ruden S. P. and Pollack J. B. (1991) The dynamical evolution of the solar nebula. *Astrophys. J.* **375**, 740–760.

Shu F. H., Adams F. C., and Lizano S. (1987) Star formation in molecular clouds: observation and theory. *Ann. Rev. Astron. Astrophys.* **25**, 23–72.

Shu F. H., Tremaine S., Adams F. C., and Ruden S. P. (1990) Sling amplification and eccentric gravitational instabilities in gaseous disks. *Astrophys. J.* **358**, 495–514.

Shu F. H., Johnstone D., and Hollenbach D. (1993) Photoevaporation of the solar nebula and the formation of the giant planets. *Icarus* **106**, 92–101.

Shu F. H., Shang H., and Lee T. (1996) Toward an astrophysical theory of chondrites. *Science* **271**, 1545–1552.

Shu F. H., Najita J. R., Shang H., and Li Z.-Y. (2000) X-winds: theory and observations. In *Protostars and Planets IV* (eds. V. Mannings, A. P. Boss, and S. S. Russell). University of Arizona Press, Tucson, pp. 789–814.

Shu F. H., Shang H., Gounelle M., Glassgold A. E., and Lee T. (2001) The origin of chondrules and refractory inclusions in chondritic meteorites. *Astrophys. J.* **548**, 1029–1050.

Simon S. B., Davis A. M., Grossman L., and McKeegan K. D. (2002) A hibonite-corundum inclusion from Murchison: a first-generation condensate from the solar nebula. *Meteorit. Planet. Sci.* **37**, 533–548.

Smith K. W., Bonnell I. A., Emerson J. P., and Jenness T. (2000) NGC 1333/IRAS 4: a multiple star formation laboratory. *Mon. Not. Roy. Astron. Soc.* **319**, 991–1000.

Snell R. L., Loren R. B., and Plambeck R. L. (1980) Observations of CO in L1551—evidence for stellar wind driven shocks. *Astrophys. J.* **239**, L17–L22.

Stone J. M., Gammie C. F., Balbus S. A., and Hawley J. F. (2000) Transport processes in protostellar disks. In *Protostars and Planets IV* (eds. V. Mannings, A. P. Boss, and S. S. Russell). University of Arizona Press, Tucson, pp. 589–612.

Vanhala H. A. T. and Boss A. P. (2000) Injection of radioactivities into the presolar cloud: convergence testing. *Astrophys. J.* **538**, 911–921.

Vanhala H. A. T. and Cameron A. G. W. (1998) Numerical simulations of triggered star formation: I. Collapse of dense molecular cloud cores. *Astrophys. J.* **508**, 291–307.

Weidenschilling S. J. (1988) Formation processes and time scales for meteorite parent bodies. In *Meteorites and the Early Solar System* (eds. J. F. Kerridge and M. S. Matthews). University of Arizona Press, Tucson, pp. 348–371.

Woolum D. S. and Cassen P. (1999) Astronomical constraints on nebula temperatures: implications for planetesimal formation. *Meteorit. Planet. Sci.* **34**, 897–907.

Yorke H. W. and Bodenheimer P. (1999) The formation of protostellar disks: III. The influence of gravitationally induced angular momentum transport on disk structure and appearance. *Astrophys. J.* **525**, 330–342.

1.05
Classification of Meteorites

A. N. Krot, K. Keil, C. A. Goodrich, and E. R. D. Scott

University of Hawaii at Manoa, Honolulu, HI, USA

and

M. K. Weisberg

Kingsborough College of the City University of New York, Brooklyn, NY, USA

1.05.1 INTRODUCTION	84
1.05.2 CLASSIFICATION OF CHONDRITIC METEORITES	86
1.05.2.1 Taxonomy	86
1.05.2.2 Primary Classification Parameters	87
1.05.2.2.1 Bulk chemical compositions	87
1.05.2.2.2 Bulk oxygen isotopic compositions	88
1.05.2.2.3 Bulk nitrogen and carbon abundances and isotopic compositions	88
1.05.2.2.4 Oxidation state	88
1.05.2.3 Secondary Classification Parameters	89
1.05.2.3.1 Petrologic type	89
1.05.2.3.2 Shock metamorphism stages	90
1.05.2.3.3 Classification of breccias	90
1.05.2.3.4 Degree of terrestrial weathering	90
1.05.2.4 Mineralogical and Geochemical Characteristics of Chondrite Groups	91
1.05.2.4.1 Carbonaceous chondrites	91
1.05.2.4.2 Ordinary chondrites	100
1.05.2.4.3 Enstatite (EH, EL, and Lewis Cliff 87223) chondrites	101
1.05.2.4.4 K (Kakangari-like) chondrites	101
1.05.2.4.5 R (Rumuruti-like) chondrites	103
1.05.3 CLASSIFICATION OF IDPS	103
1.05.4 CLASSIFICATION OF NONCHONDRITIC METEORITES	103
1.05.4.1 Primitive Achondrites	104
1.05.4.1.1 Acapulcoites and lodranites	104
1.05.4.1.2 Winonaites and the silicate-bearing IAB and IIICD irons	106
1.05.4.1.3 Silicate-bearing IIE irons	107
1.05.4.2 Differentiated Achondrites	107
1.05.4.2.1 Angrites	107
1.05.4.2.2 Aubrites	108
1.05.4.2.3 Brachinites	108
1.05.4.2.4 Ureilites	108
1.05.4.2.5 HED meteorites: howardites, eucrites, diogenites	109
1.05.4.3 Mesosiderites	112
1.05.4.4 Pallasites	113
1.05.4.4.1 Main-group pallasites	113
1.05.4.4.2 Eagle Station pallasite grouplet	113
1.05.4.4.3 Pyroxene–pallasite grouplet	114
1.05.4.5 Irons	114
1.05.4.5.1 Chemical groups	114
1.05.4.5.2 Structural groups	115
1.05.4.6 Silicate-bearing IVA Irons	116

1.05.5 MARTIAN (SNC) METEORITES 116
  1.05.5.1 Shergottites 116
  1.05.5.2 Nakhlites (Clinopyroxenites/Wehrlites) 118
  1.05.5.3 Chassignite (Dunite) 118
  1.05.5.4 Allan Hills 84001 (Orthopyroxenite) 120
1.05.6 LUNAR METEORITES 120
ACKNOWLEDGMENTS 120
REFERENCES 120

1.05.1 INTRODUCTION

Solid bodies of extraterrestrial material that penetrate the atmosphere and reach the Earth's surface are called *meteorites*. Other extraterrestrial materials include micrometer-sized *interplanetary dust particles* (IDPs) collected in the lower stratosphere and polar ices. Most meteorites and IDPs are fragments of asteroids, but some IDPs may represent cometary material and some meteorites are fragments of the planets Mars and Earth's moon. Meteorites recovered following observed falls are called *falls*; those which cannot definitely be associated with observed falls are called *finds*. Meteorites are given names based on the location where they were recovered (e.g., the Allende meteorite fell in Allende, Mexico). Meteorites recovered in Antarctica and the deserts of Australia and northern Africa are given names and numbers, because numerous samples are found in the same locations. Fragments thought to be of the same meteorite fall, which, in Antarctica or hot deserts, may have different numbers or even names because they were found in different locations, are called *paired*. More than 1.7×10^4 meteorites are currently known; this number is rapidly growing through the discovery of large concentrations of meteorites in the cold and hot deserts.

Based on their bulk compositions and textures, meteorites can be divided into two major categories, *chondrites* and *nonchondritic meteorites*; the latter include the *primitive achondrites* and *igneously differentiated meteorites* (Figure 1).

They are further classified into *groups* using a classification scheme based on their oxygen isotopes, chemistry, mineralogy, and petrography (Table 1 and Figure 1). The goals of this classification scheme are to provide descriptive labels for classes of meteorites with similar origins or formation histories that could be derived from the same asteroidal or planetary body, and to reveal possible genetic links between various classes. In this chapter, we describe the existing meteorite classification, classification parameters, and general mineralogical and geochemical characteristics of the meteorite groups. Classification of IDPs and meteorite–asteroid connections are only briefly mentioned; interested readers can find additional

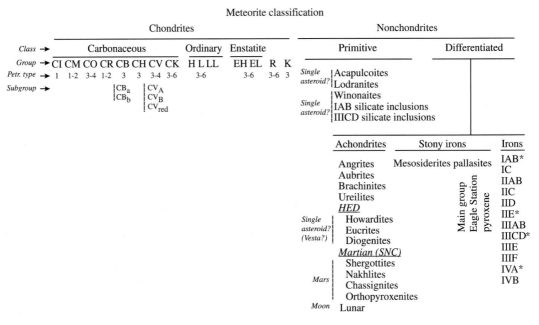

Figure 1 Classification of meteorites.

Table 1 Meteorite groups and numbers of their members.

	Falls	Total
Chondrites		
Carbonaceous		
CI (Ivuna-like)	5	5
CM (Mighei-like)	15	171
CR (Renazzo-like)	3	78
CO (Ornans-like)	5	85
CV (Vigarano-like)	6	49
CV-oxidized Allende-like		
CV-oxidized Bali-like		
CV-reduced		
CK (Karoonda-like)	2	73
CH (ALHA85005-like)	0	11
CB (Bencubbin-like)		
CB_a: Bencubbin, Weatherford, Gujba	0	3
CB_b: QUE94411, Hammadah al Hamra 237	0	2
Ordinary		
H	316	6962
L	350	6213
LL	72	1048
Enstatite		
EH	8	125
EL	7	38
R (Rumuruti-like)	1	19
K (Kakangari-like)	1	3
Ungrouped:		
Acfer 094		
Adelaide		
Belgica-7904, Yamato-86720, Yamato-82162, Dhofar-225		
Coolidge		
Loongana 001		
LEW85332		
MAC87300		
MAC88107		
Ningqiang		
Tagish Lake		
Nonchondrites		
Primitive		
Acapulcoites	1	12
Lodranites	1	14
Winonaites	1	11
Differentiated (planetary)		
Achondrites		
Angrites	1	4
Aubrites	9	46
Brachinites	0	7
HED meteorites		
Eucrites	25	200
Howardites	20	93
Diogenites	10	94
Ureilites	5	110
Stony-irons		
Pallasites	5	50
main group pallasites		45
Eagle Station pallasites		3
pyroxene-pallasites	0	2
Mesosiderites	7	66
Irons		
IAB (nonmagmatic, related to IIICD and winonaites)	5	131
IC	0	11
IIAB	6	103
IIC	0	8
IID	3	16
IIE (related to H chondrites)	1	18
IIF	1	5
IIIAB	11	230
IIICD (nonmagmatic, related to IAB and winonaites)	3	41
IIIE	0	13
IIIF	0	6
IVA	4	64
IVB	0	13
Ungrouped	8	111
Ungrouped nonchondrites		
ALHA77255		
Bocaiuva		
Deep Springs		
Divnoe		
Enon		
Guin		
LEW86211		
LEW86220		
LEW88763		
Mbosi		
Northwest Africa 011		
Northwest Africa 176		
Puente del Zacate		
QUE93148		
Sahara 00182		
Sombrerete		
Tucson		
Differentiated (planetary)		
Martian (SNC):	4	26
Shergottites	2	18
Nakhlites (clinopyroxenites/wehrlites)	1	6
Chassigny (dunite)	1	1
Orthopyroxenite (ALH84001)	0	1
Lunar	0	18

Number of meteorites are from Grady (2000).

information in Rietmeijer (1998) and Burbine *et al.* (2002).

Chondrites consist of four major components: chondrules, FeNi-metal, refractory inclusions (Ca–Al-rich inclusions (CAIs) and amoeboid olivine aggregates (AOAs)), and fine-grained matrix material. It is generally accepted that the refractory inclusions, chondrules, and FeNi-metal are formed in the solar nebula by high-temperature processes that included condensation and evaporation. Many CAIs and most chondrules and FeNi-metal were subsequently melted during multiple brief heating episodes. Matrix, some CAIs, and metal in some chondrites (e.g., CH and CB) appear to have escaped these high-temperature nebular events (see Chapter 1.07–1.09). Although most chondrites experienced thermal processing on their parent asteroids, such as aqueous alteration, thermal, and shock metamorphism, they did not experience melting and igneous differentiation, and thus largely preserve records of physical and chemical processes in the solar nebula. Deciphering these records is a primary goal of chondrite studies.

Nonchondritic meteorites lack chondritic textures and are formed by partial or complete melting and planetary differentiation of chondritic precursor asteroids or larger planetary bodies (Mars, Moon), and hence provide unique opportunities to study these processes on extraterrestrial bodies (see Chapter 1.11, 1.12, 1.21, and 1.22). Several groups of nonchondritic meteorites experienced only low degrees of melting, and have largely retained their chondritic bulk compositions. To emphasize the relatively unprocessed nature of these nonchondrites and their intermediate status relative to chondrites and highly differentiated meteorites, they are referred to as primitive achondrites and discussed as a separate category of nonchondritic meteorites.

1.05.2 CLASSIFICATION OF CHONDRITIC METEORITES

1.05.2.1 Taxonomy

Based on bulk chemistry (Figures 2 and 3), bulk oxygen isotopic compositions (Figure 4), mineralogy, petrology (Figures 8–13 and 15), and proportions of various chondritic components (Table 2), 14 chondrite groups have been recognized. Several other chondrites are mineralogically and/or chemically unique and defy classification into the existing chondrite groups; these are commonly called *ungrouped* chondrites (Table 1 and Figure 1).

Conventionally, a chondrite "group" is defined as having a minimum of five unpaired chondrites of similar mineralogy, petrography, bulk isotopic properties, and bulk chemical compositions in major, nonvolatile elements. According to this definition, K chondrites could be considered as a *grouplet*: there are only two known K chondrites (see K (Kakangari-like) chondrites).

Thirteen out of the 14 chondrite groups comprise three major *classes*: *carbonaceous* (C), *ordinary* (O), and *enstatite* (E), each of which contains distinct groups.

The term carbonaceous is somewhat of a misnomer, because only the CI, CM, and CR chondrites are significantly enriched in carbon relative to noncarbonaceous chondrites (Figure 5). The CI chondrites lack chondrules. Carbonaceous and noncarbonaceous chondrites can be resolved (with some exceptions) by several characteristics: (i) mean refractory lithophile/Si abundance ratios relative to CI chondrites (≥ 1.00 in carbonaceous chondrites, ≤ 0.95 in noncarbonaceous chondrites); (ii) oxygen isotopic composition ($\Delta^{17}O \leq -2‰$ for carbonaceous chondrites, except CI; $\Delta^{17}O \geq -1‰$ for noncarbonaceous chondrites); (iii) refractory inclusion abundances (≥ 0.1 vol.% in carbonaceous chondrites, except CI; ≤ 0.1 vol.% in noncarbonaceous chondrites); and (iv) matrix/chondrule modal abundance ratio (≥ 0.9 in carbonaceous chondrites, except CH and CB; ≤ 0.9 in noncarbonaceous chondrites, except K chondrites) (Kallemeyn et al., 1996).

There are eight well-resolved groups of carbonaceous chondrites: CI, CM, CR, CH, CB, CV, CK, and CO. The letters designating the groups refer to a typical chondrite (commonly fall) in the group: CI (Ivuna-like), CM (Mighei-like), CO (Ornans-like), CR (Renazzo-like), CH (ALH85085-like), CB (Bencubbin-like), CV (Vigarano-like), and CK (Karoonda-like). The exception to this rule is the CH group, in which the "H" refers to high metal abundance and high iron concentration (Bischoff et al., 1993b).

Ordinary chondrites are divided into three groups: H, L, and LL. The letters designating the groups refer to the bulk iron contents: H chondrites have high total iron contents, L chondrites have low total iron contents, and LL chondrites have low metallic iron relative to total iron, as well as low total iron contents.

Enstatite chondrites comprise two groups with different contents of metallic iron: EH and EL. Additionally, there is an ungrouped E chondrite, LEW87223 (Grossman et al., 1993).

R (Rumuruti-like) and K (Kakangari-like) chondrite groups are different from other chondrites and have been suggested to represent additional chondrite classes.

The term *clan* is used as a higher order of classification than group. It was originally defined to encompass chondrites that have chemical, mineralogical, and isotopic similarities and were believed to have formed at about the same time in the same small region of the solar nebula, in a narrow range of heliocentric distances (Kallemeyn and Wasson, 1981; Kallemeyn et al., 1996). Weisberg et al. (1995a) used the term for chondrites that have chemical, mineralogical, and isotopic similarities suggesting a petrogenetic kinship, but have petrologic and/or bulk chemical characteristics that challenge a group relationship. Four carbonaceous chondrite clans have been recognized: (i) the CR clan, which includes the CR, CH, CB, and a unique chondrite LEW85332; (ii) the CM–CO clan; (iii) the CV–CK clan; and (iv) the CI clan (Kallemeyn et al., 1996). The noncarbonaceous chondrites constitute three clans: O+R, E, and K.

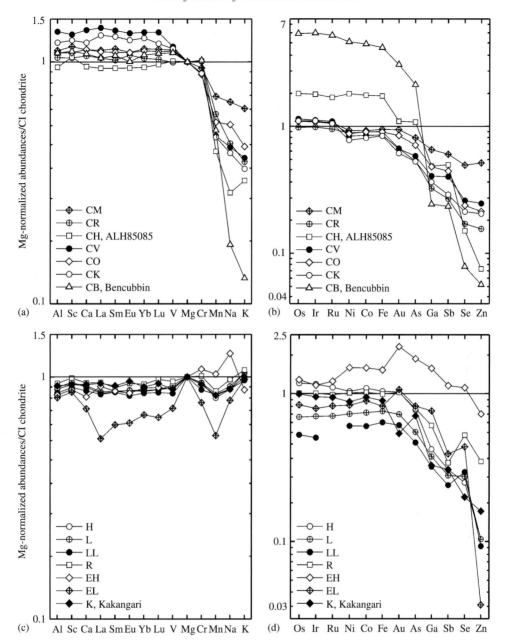

Figure 2 Magnesium- and CI-normalized bulk lithophile (a, c) and siderophile (b, d) element abundances of the carbonaceous (a, b) and noncarbonaceous (c, d) chondrite groups (sources Kallemeyn and Wasson, 1981, 1982, 1985; Kallemeyn et al., 1978, 1989, 1991, 1994, 1996; Kallemeyn, unpublished).

1.05.2.2 Primary Classification Parameters

1.05.2.2.1 Bulk chemical compositions

The bulk compositions of chondrites closely match the compositions of the solar photosphere, with the exception of a few highly volatile elements (hydrogen, carbon, nitrogen, helium) and lithium (see Chapter 1.03). The chondrite groups also show different levels of depletions in moderately volatile elements (e.g., manganese, sodium, potassium, gallium, antimony, selenium, and zinc). The elemental abundances of bulk Ivuna-like carbonaceous chondrites (CI) are viewed as a measure of average solar-system abundances and are used as a reference composition for many solar-system materials. Bulk compositions of other chondrite groups are similar to, but distinct from, CI chondrites, most non-volatile elements varying within a factor of 3 relative to CI abundances (Figure 2). Each chondrite group has a unique and well-defined elemental abundance pattern (the Bencubbin-like

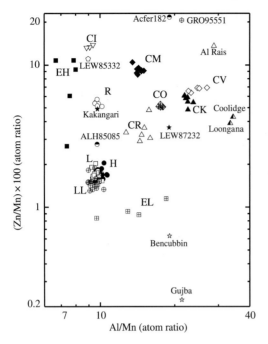

Figure 3 Al/Mn versus (Zn/Mn) × 100 atom ratios in chondritic meteorites (sources Kallemeyn and Wasson, 1981, 1982, 1985; Kallemeyn et al., 1978, 1989, 1991, 1994, 1996; Wasson and Kallemeyn, 1988; Kallemeyn, unpublished).

carbonaceous chondrite group (CB) is the only exception). Carbonaceous chondrites are enriched in refractory lithophile elements relative to CI, and depleted in volatile elements. Siderophile and chalcophile elements show volatility-controlled abundance patterns.

Elemental abundance patterns for ordinary, Rumuruti-like (R), and Kakangari-like (K) chondrites are fairly flat and enriched relative to CI for lithophile and refractory lithophile elements. Enstatite chondrites have the lowest abundance of refractory lithophile elements.

In addition to abundance patterns, other bulk compositional criteria for resolving the chondrite groups involve compositional diagrams of elements of different volatility, e.g., Sb/Ni versus Ir/Ni, Sc/Mg versus Ir/Ni, and Al/Mn versus Zn/Mn (Figure 3).

1.05.2.2.2 Bulk oxygen isotopic compositions

Most chondrite classes (ordinary, carbonaceous, and enstatite chondrites) plot in unique positions on a three-isotope oxygen plot (Figure 4). Carbonaceous and K chondrites fall below the terrestrial fractionation line (CI chondrites are the only exception), whereas ordinary, enstatite, and R chondrites each form well-defined clusters on or above the terrestrial fractionation line. However, there does not appear to be a resolvable difference between many groups within the chondrite classes. For example, bulk oxygen isotopic compositions of EH and EL enstatite chondrites completely overlap, as do those of the type-3 H, L, and LL ordinary chondrites. The CO, CV, and CK carbonaceous chondrites also have similar bulk oxygen isotopic compositions.

1.05.2.2.3 Bulk nitrogen and carbon abundances and isotopic compositions

Nitrogen abundances and isotopic compositions (Figure 6) can differentiate most of the existing chondrite groups (Kung and Clayton, 1978; Kerridge, 1985). Carbon abundance and isotopic compositions (Figure 5) can differentiate only CI and Mighei-like (CM) carbonaceous chondrite groups.

1.05.2.2.4 Oxidation state

The chondrite groups record a wide range of oxidation states that were probably established through a combination of nebular and asteroidal processes (Rubin et al., 1988a; Krot et al., 2000a). The oxidation state is reflected by the distribution of iron among its three common oxidation states: 0 (FeNi-metal and Fe-sulfides), +2 (silicates), and +3 (oxides). The oxidation states of chondrite groups can be summarized in a Urey–Craig diagram, where iron present in metal and sulfide phases is plotted versus iron present in silicate and oxide phases (Figure 7). Oxidation state generally increases in the order enstatite–ordinary–carbonaceous+R chondrite classes, with K chondrites being intermediate between ordinary and enstatite chondrites. Within the ordinary chondrites, the oxidation state increases from H to L to LL, as is reflected in an increase in the Fe/(Fe + Mg) ratios of their olivine and pyroxene. For carbonaceous chondrites, oxidation state increases in the order CB–CH–CR–CO–CV–CK–CM–CI.

We note, however, that this classification parameter does not generally reflect the oxidation state of individual chondritic components, e.g., CAIs in all chondrite groups formed under highly reducing conditions (see Chapter 1.08), and magnesian (type-I) and ferrous (type-II) chondrules (both of which occur in the same meteorites) require formation under different redox conditions (see Chapter 1.07). In addition, the role of nebular and asteroidal processes in establishing of the oxidation states of chondrites remains controversial (e.g., Krot et al., 2000a).

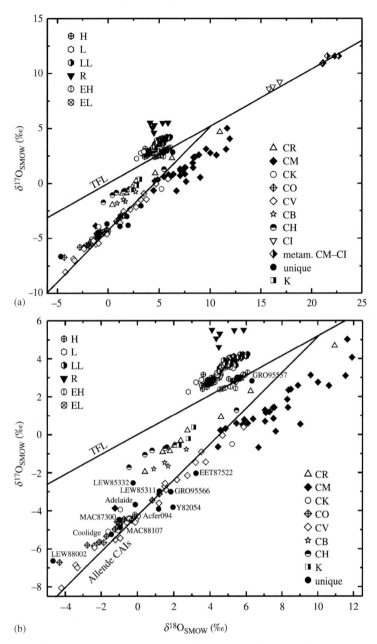

Figure 4 Bulk oxygen isotope compositions of chondrite groups and ungrouped chondrites (source Clayton and Mayeda, 1999).

1.05.2.3 Secondary Classification Parameters

1.05.2.3.1 Petrologic type

Van Schmus and Wood (1967) introduced a classification scheme that provides a guide to the degree of thermal and aqueous alteration experienced by a chondrite. According to this widely used scheme, chondrites are divided into petrologic (or petrographic) types 1–6. The sequence type 3 (commonly called *unequilibrated*) to type 6 (commonly called *equilibrated*) represents an increasing degree of chemical equilibrium and textural recrystallization, presumably due to thermal metamorphism. Type 1 represents higher degree of aqueous alteration compared to type 2, based largely on the abundance of hydrous silicates. Type 3 chondrites are widely considered the least modified by secondary processes (see also Chapter 1.07). We note, however, that this scheme cannot be applied to several CI and CM chondrites (e.g., Belgica-7904) that appear to have experienced aqueous alteration followed by thermal metamorphism and dehydration. Several ordinary and enstatite chondrites that experienced extensive recrystallization and possibly melting have been designated

Table 2 Summary of the average petrographic properties of the chondritic meteorites.

	Carbonaceous								Ordinary			Enstatite		Additional	
	CI	CM	CO	CR	CH	CB	CV	CK	H	L	LL	EH	EL	K	R
CAI + AOA (vol.%)	≪1	≪1	13	0.5	0.1	<0.1	10	4	<0.1	<0.1	<0.1	<0.1	<0.1	<0.1	<0.1
chd (vol.%)	≪1	20	48	50–60	~70	30–40	45	15	60–80	60–80	60–80	60–80	60–80	27	>40
matrix (vol.%)	>99	70	34	30–50	5	<5	40	75	10–15	10–15	10–15	<0.1	<0.1	60	30
metal (vol. %)	0	0.1	1–5	5–8	20	60–70	0–5	<0.01	8.4	4.1	2.0	10.1	10.2	7.4	<0.1
chd, mean diam. mm		0.3	0.15	0.7	0.02	0.1–20	1	0.7	0.3	0.7	0.9	0.2	0.6	0.6	0.4

as type 7, although some may be impact and not internally derived melts.

The type-3 ordinary, CO, and CV chondrites are commonly subdivided into 10 subtypes (3.0–3.9), of which 3.0 is the least metamorphosed (e.g., Sears et al., 1991). Table 3 provides a summary of the criteria used to define petrologic types.

1.05.2.3.2 Shock metamorphism stages

The degree of shock metamorphism (caused by impacts) recorded in a chondrite is determined from a variety of mineralogical and textural parameters (e.g., Stöffler et al., 1991; Scott et al., 1992). The classification scheme by Stöffler et al. (1991) is based on shock effects observed in olivine and plagioclase (Table 4). Since olivine is rare in enstatite chondrites, Rubin et al. (1997) extended this shock classification scheme to orthopyroxene (Table 4).

1.05.2.3.3 Classification of breccias

In addition to shock metamorphism, impacts between solar-system objects (asteroids and comets) result in formation of breccias—rocks composed of fragments derived from previous generations of rocks, cemented together to form a new lithology. Many chondrites are breccias. A breccia in which the clasts belong to the same group of meteorites is called *genomict*. A breccia in which the clasts and/or matrix belong to different meteorite groups is called *polymict*. The classification of chondritic breccias is summarized in Table 5.

1.05.2.3.4 Degree of terrestrial weathering

Degree of terrestrial weathering is an additional classification parameter commonly applied to meteorite finds. Two classification schemes are used: one for hand specimens of Antarctic meteorites (commonly used), and one for meteorites as they appear in polished sections (rarely used). Weathering categories for hand specimens are: A—minor rustiness; B—moderate rustiness; C—severe rustiness; and e—evaporite minerals visible to the naked eye (e.g., Grossman, 1994).

Wlotzka (1993) suggested the following progressive alteration stages for meteorites as they appear in polished sections: W0—no visible oxidation of metal or sulfides; W1—minor oxide veins and rims around metal and troilite; W2—moderate oxidation of ~20–60% of metal; W3—heavy oxidation of metal and troilite, 60–95% being replaced; W4—complete oxidation of metal and troilite, but no oxidation of silicates; W5—beginning alteration of mafic silicates,

Classification of Chondritic Meteorites

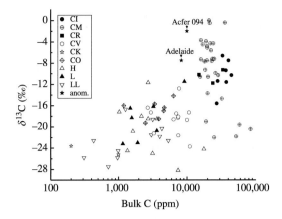

Figure 5 Bulk carbon contents (ppm) versus $\delta^{13}C$ (‰) in chondrites (sources Kerridge, 1985; Grady and Pillinger, 1986).

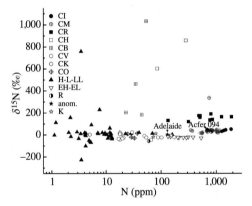

Figure 6 Bulk nitrogen contents (ppm) versus $\delta^{15}N$ (‰) in chondrites (sources Kerridge, 1985; Grady and Pillinger, 1990; Hashizume and Sugiura, 1995; Sugiura, 1995).

mainly along cracks; and W6—massive replacement of silicates by clay minerals and oxides.

1.05.2.4 Mineralogical and Geochemical Characteristics of Chondrite Groups

1.05.2.4.1 Carbonaceous chondrites

CI (Ivuna-like) chondrites. Although CI chondrites are compositionally the most primitive meteorites in that they provide the best match to the solar photosphere (see Chapter 1.03), their primary mineralogy and petrography were erased by extensive, multistage aqueous alteration on their parent asteroid(s) (McSween, 1979; Endress and Bischoff, 1993, 1996; Endress et al., 1996). All five known CI chondrites are regolith breccias consisting of various types of heavily hydrated lithic fragments (lithologies). The fragments are composed of a fine-grained phyllosilicate-rich matrix containing magnetite, sulfides, sulfates, and carbonates; they are devoid of chondrules and

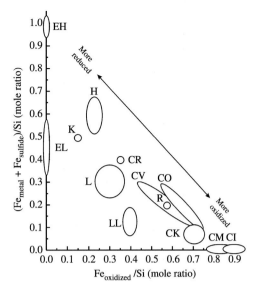

Figure 7 Urey–Craig diagram showing relative iron contents and oxidation states of the chondrite groups. Iron present in metal and sulfide phases is plotted versus iron present in silicate and oxide phases, for bulk chondrite compositions (after Brearley and Jones, 1998) (reproduced by permission of the Mineralogical Society of America from *Reviews in Mineralogy* **1998**, *36*, 1–398).

CAIs, but contain a few isolated olivine and pyroxene grains (e.g., McSween and Richardson, 1977; Endress and Bischoff, 1993, 1996; Leshin et al., 1997). The fragments are cemented by networks of secondary carbonate and calcium- and magnesium-sulfate veins; some or all of the sulfate veins could be of terrestrial origin (Gounelle and Zolensky, 2001).

CM (Mighei-like) and ungrouped CM/CI-like chondrites. The CM chondrites (Figure 8(a)) can be distinguished from other carbonaceous chondrite groups based on bulk chemical (Figures 2 and 3) and oxygen isotopic compositions (Figure 4). The major mineralogical and petrographic characteristics include: (i) abundant (~30 vol.%) relatively small (~300 μm) chondrules of various textural types (largely porphyritic) and chemical compositions (largely FeO-poor—type I) which are partially replaced by phyllosilicates; (ii) presence of fine-grained accretionary rims around many chondrules; (iii) nearly complete absence of FeNi-metal due to extensive aqueous alteration; (iv) common presence of CAIs and AOAs; and (v) high abundance (~70 vol.%) of matrix composed of phyllosilicates, tochilinite, carbonates, sulfides, and magnetite. The degree of aqueous alteration experienced by CM chondrites varies widely (e.g., McSween, 1979; Browning et al., 1996). Yamato-82042 and Elephant Moraine (EET)-83334 contain virtually no anhydrous silicates and can be classified as CM1 chondrites (Zolensky et al., 1997).

Table 3 Summary of criteria for petrologic types.

Criterion	Petrologic type					
	1	2	3	4	5	6
Homogeneity of olivine and low-Ca pyroxene compositions	—————— >5% mean deviations ——————			<5%	———— Homogeneous ————	
Structural state of low-Ca pyroxene	———————— Predominantly monoclinic ————————			>20% monoclinic	<20% monoclinic	Orthorhombic
Feldspar	———————— Minor primary grains only ————————			Secondary, <2 μm grains	Secondary, 2–50 μm grains	Secondary, >50 μm grains
Chondrule glass	Altered, mostly absent[a]		Clear, isotropic, variable abundance	———— Devitrified, absent ————		
Maximum Ni in metal	<20 wt.%; taenite minor or absent		———————— >20 wt.% Kamacite and taenite in exsolution relationship ————————			
Mean Ni in sulfides	>0.5 wt.%		———————————— <0.5 wt.% ————————————			
Matrix	All fine-grained, opaque	Mostly fine, opaque	Clastic, minor opaque	———— Transparent, recrystallized coarsening from 4 to 6 ————		
Chondrule–matrix integration	No chondrules	———— Chondrules very sharply defined ————		Chondrules well defined	Chondrules readily delineated	Chondrules poorly defined
Carbon (wt.%)	3–5	0.8–2.6	———————————— <1.5 ————————————			
Water (wt.%)	18–22	2–16	0.3–3	———————— <1.5 ————————		

After Van Schmus and Wood (1967), with modifications from Sears and Dodd (1988) and Brearley and Jones (1998).
[a] Chondrule glass is rare in CM2 chondrites, but is preserved in many CR2 chondrites.

Table 4 Classification scheme for shock metamorphism in chondrites.

Shock stage	Description	Effect resulting from equilibration peak shock pressure			Shock pressure (GPa)[a]
		Olivine	Plagioclase	Orthopyroxene	
S1	Unshocked		Sharp optical extinction, irregular fractures		<4–5
S2	Very weakly shocked	Undulatory extinction, irregular fractures	Undulatory extinction, irregular fractures	Undulatory extinction, irregular and some planar fractures	5–10
S3	Weakly shocked	Planar fractures, undulatory extinction, irregular fractures	Undulatory extinction	Clinoenstatite lamellae on (100), undulatory extinction, planar and irregular fractures	15–20
S4	Moderately shocked	Weak mosaicism, planar fractures	Undulatory extinction, partially isotropic, planar deformation features		30–35
S5	Strongly shocked	Strong mosaicism, planar fractures, planar deformation fractures	Maskelynite		45–55
S6	Very strongly shocked	Solid-state recrystallization and staining, ringwoodite, melting	Shock melted (normal glass)	Majorite, melting	75–90
	Shock melted	Whole-rock melting (impact melt rocks and melt breccias)			

Sources: Stöffler et al. (1991) and Rubin et al. (1997).
Shock effects in ordinary chondrites are characterized by effects in olivine and plagioclase; shock levels in carbonaceous chondrites are characterized by effects mostly in olivine (Scott et al., 1992); shock levels in enstatite chondrites are characterized by effects mostly in orthopyroxene. The prime shock criteria for each shock stage are in italics.
[a] Shock pressures for ordinary chondrites only.

Table 5 Classification of chondritic breccias.

Breccia type	Description
Lithic fragments	Xenolithic—clasts of different chemical class to host
	Cognate—impact melt fragments
Regolith breccias	Fragmental debris on surface of body (contain solar-wind gases, solar flare tracks, agglutinates, etc.), usually have light–dark structure
Fragmental breccias	Fragmental debris with no regolith properties
Impact melt breccias	Unmelted debris in igneous matrices
Granulitic breccias	Metamorphosed fragmental breccias
Primitive breccias	Type-3 ordinary chondrite breccias

After Keil (1982).

Tagish Lake (Figure 8(b)) is an ungrouped carbonaceous chondrite with CI and CM affinities that has higher abundances of carbon (5.8 wt.%) and presolar nanodiamonds than any other chondrite (Brown et al., 2000; Clayton and Mayeda, 2001; Grady et al., 2002; Mittlefehldt, 2002; Zolensky et al., 2002).

The ungrouped meteorites Belgica-7904, Yamato-86720, Yamato-82162, and Dhofar-225 have petrographic, mineralogical, and chemical affinities to the CM and CI carbonaceous chondrites (e.g., Tomeoka et al., 1989; Tomeoka, 1990; Paul and Lipschutz, 1990; Bischoff and Metzler, 1991; Ikeda, 1992; Hiroi et al., 1993a, 1996; Clayton and Mayeda, 1999; Ivanova et al., 2002), but have distinct oxygen isotopic compositions (Figure 4). In contrast to CI and CM chondrites, these meteorites experienced aqueous alteration followed by reheating and dehydration (e.g., Akai, 1988; Yanai et al., 1995). The reheating resulted in partial loss of carbon, nitrogen, and water, structural transformation of serpentine to olivine, and precipitation of abundant fine-grained troilite (possibly by decomposition of tochilinite).

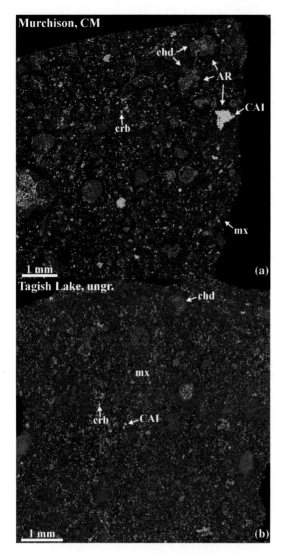
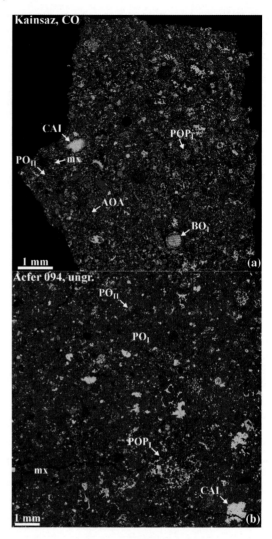

Figure 8 Combined elemental maps (Mg—red, Ca—green, Al—blue) of: (a) the CM carbonaceous chondrite Murchison and (b) ungrouped CM-like carbonaceous chondrite Tagish Lake. Murchison contains abundant magnesian chondrules (red) with porphyritic textures, CAIs (bluish), and heavily hydrated matrix (purple) containing relatively rare grains of carbonate (green). Most chondrules and CAIs are surrounded by fine-grained accretionary rims. Tagish Lake contains higher proportion of heavily hydrated matrix and carbonates, and fewer chondrules and CAIs. Many chondrules are surrounded by fine-grained accretionary rims. AR = accretionary rim; CAI = Ca–Al-rich inclusion; chd = chondrule; crb = carbonates; mx = matrix.

Figure 9 Combined elemental maps of: (a) the CO3.1 carbonaceous chondrite Kainsaz and (b) ungrouped CO/CM-like carbonaceous chondrite Acfer 094. Kainsaz contains abundant small chondrules, CAIs, and AOAs. Acfer 094 is texturally and mineralogically similar to CO chondrites, but contains higher abundance of matrix. AOA = amoeboid olivine aggregate; BO = barred olivine chondrule; PO(P)$_{I, II}$ = type I (II) porphyritic olivine (pyroxene) chondrule.

The nature of this reheating (its peak temperature and duration) remains poorly understood; two mechanisms discussed in the literature include impact shock heating and thermal metamorphism (e.g., Ikeda and Prinz, 1993).

CO (Ornans-like) and ungrouped CO/CM-like chondrites. Characteristics of the CO group (Figure 9(a)) include: (i) abundant (35–45 vol.%) relatively small (~150 μm), rounded chondrules (Rubin, 1989); (ii) a high proportion of metal-rich magnesian chondrules; ferroan (type II) chondrules are also common; (iii) abundant CAIs and AOAs (~10 vol.%); (iv) a relatively low abundance of matrix (30–45 vol.%); (v) a metamorphic sequence (3.0–3.7), in many ways analogous to that of type-3 ordinary chondrites (McSween, 1977a; Scott and Jones, 1990); and (vi) secondary alteration minerals in chondrules and CAIs that are similar to those in CV chondrites (e.g., nepheline, sodalite, ferrous olivine, hedenbergite, andradite, and ilmenite, with phyllosilicates being virtually absent; e.g., Keller and Buseck (1990a), Tomeoka *et al.*

(1992), Rubin et al. (1985), Rubin (1998), and Russell et al. (1998a)).

Acfer 094 (Figure 9(b)) is an ungrouped, type-3 carbonaceous chondrite breccia with mineralogical, petrological, nitrogen isotopic, and oxygen isotopic affinities to the CM and CO groups (Newton et al., 1995; Greshake, 1997). Its bulk chemical composition is similar to that of CM chondrites, and oxygen isotopic composition (Figure 4) and matrix modal abundance are similar to those of CO chondrites. In contrast to CM chondrites, Acfer 094 shows no evidence for aqueous alteration and thermal metamorphism and has different carbon isotope compositions. It has higher abundances of presolar SiC and diamonds than any other chondrite (Tagish Lake is the only exception).

Adelaide is an ungrouped, type-3 carbonaceous chondrite with affinities to the CM–CO clan, but appears to have escaped the thermal metamorphism and alteration commonly observed in the CM and CO carbonaceous chondrites (Fitzgerald and Jones, 1977; Davy et al., 1978; Kallemeyn and Wasson, 1982; Hutcheon and Steele, 1982; Kerridge, 1985; Brearley, 1991; Huss and Hutcheon, 1992; Krot et al., 2001b,c).

MAC88107 and MAC87300 comprise a grouplet of types 2 and 3 carbonaceous chondrites with bulk chemical compositions intermediate between CO and CM chondrites, and oxygen isotopic compositions similar to the CK chondrites (Clayton and Mayeda, 1999). In contrast to the CK chondrites, they show no evidence for thermal metamorphism (Krot et al., 2000d). Contrary to the CM chondrites, which have no detectable natural or induced thermoluminescence (TL), MAC87300 and MAC88107 have both (Sears et al., 1991). They experienced very minor aqueous alteration that appears to have occurred at higher temperatures ($\sim$120–220 °C) than that inferred for CM chondrites ($\sim$25 °C; Clayton and Mayeda, 1999), and resulted in formation of saponite, serpentine, magnetite, nickel-bearing sulfides, fayalite, and hedenbergite (Krot et al., 2000d).

CR (Renazzo-like) and ungrouped CR-like chondrites. Characteristics of the CR group (Figures 10(a) and 11(a), and Table 2) include: (i) abundant large FeNi-metal-rich, porphyritic, magnesian (type-I) chondrules, typically surrounded by multilayered coarse-grained, igneous rims that in many cases contain a silica phase; (ii) high ($\sim$0.5 wt.%) Cr_2O_3 content in chondrule olivines; (iii) abundant heavily hydrated matrix and matrix-like lithic clasts (commonly referred to in the literature as dark inclusions); (iv) abundant FeNi-metal with a positive nickel versus cobalt trend and a solar Ni/Co ratio; (v) low abundance of CAIs and AOAs; (vi) uniform ^{16}O-enrichment of AOAs and most CAIs; (vii) whole-rock refractory lithophile elemental abundances close to solar (Figure 2); (viii) whole-rock oxygen isotopic compositions that form a unique CR-mixing line (Figure 4); and (ix) bulk nitrogen isotopic compositions with large positive anomalies in $\delta^{15}N$ (Figure 6) (McSween, 1977b; Bischoff et al., 1993a; Weisberg et al., 1993; Kallemeyn et al., 1994; Krot et al., 2000b, 2002; Aléon et al., 2002).

Most of the CR chondrites are of petrologic type 2 and generally contain abundant phyllosilicates, carbonates, platelet and framboidal magnetite, and sulfides (Weisberg et al., 1993). GRO95577 contains no anhydrous silicates and is classified as CR1 (Weisberg and Prinz, 2000). The Al Rais meteorite is an anomalous CR chondrite due to a high abundance of matrix component (>70 vol.%). The Kaidun meteorite is a complex chondritic breccia that appears to consist mainly of CR material, but also contains clasts of enstatite, ordinary, and several carbonaceous chondrite groups (Ivanov, 1989; Clayton and Mayeda, 1999).

Sahara 00182 (SAH00182) is an ungrouped metal-rich C3 chondrite (Grossman and Zipfel, 2001; Weisberg, 2001). It contains large (up to 2 mm in diameter), metal-rich, multilayered chondrules that are texturally similar to those in the CR chondrites. However, SAH00182 differs from the CR chondrites in that its chondrules contain troilite (FeS), and the chondrules and matrix lack the hydrous phyllosilicates that are characteristic of many CR chondrites. Its whole-rock oxygen isotopic composition plots along the CCAM line (Grossman and Zipfel, 2001).

The recently described metal-rich Tafassasset meteorite has an oxygen isotopic composition similar to CR chondrites and an equigranular texture (Bourot-Denise et al., 2002). It is, however, characterized by uniformly ferrous olivine (Fa_{30}) and orthopyroxene (Fs_{25}), and the presence of albitic plagioclase (An_{24-45}), chromite, and merrilite; it may be the first metamorphosed CR chondrite.

Lewis Cliff 85332 (LEW85332) is a unique, metal-rich type-3 carbonaceous chondrite breccia with chemical, oxygen isotopic, and petrographic characteristics similar to those of the CR chondrites (Rubin and Kallemeyn, 1990; Prinz et al., 1992; Weisberg et al., 1995a; Brearley, 1997). It has a high abundance of heavily hydrated matrix ($\sim$30 vol.%), anhydrous chondrules and CAIs, and CR-like refractory lithophile element abundance ratios. Relative to the CR chondrites, however, it has smaller chondrules and higher abundances of manganese and most volatile siderophile elements.

CH (ALH85085-like) chondrites. The CH chondrites (Figures 10(b) and 11(b)) are characterized by: (i) abundant FeNi-metal ($\sim$20 vol.%)

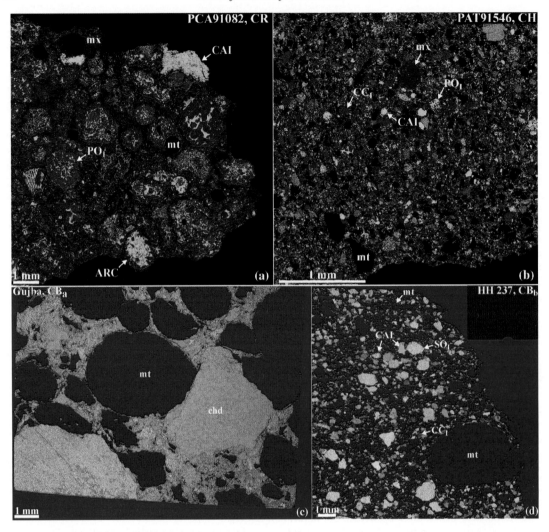

Figure 10 Combined elemental maps of the: (a) CR carbonaceous chondrite PCA91082, (b) CH carbonaceous chondrite PAT91546, (c) CB_a carbonaceous chondrite Gujba, and (d) CB_b carbonaceous chondrite Hammadah al Hamra 237. (a) The CR chondrite PCA91082 contains large porphyritic, metal-rich (mt), magnesian (type I) chondrules, heavily hydrated fine-grained matrix, and rare anorthite-rich chondrules (ARC) and CAIs. Many chondrules are surrounded by coarse-grained, igneous rims. (b) The CH chondrite PAT91546 contains abundant small chondrules and chondrule fragments, FeNi-metal grains, and CAIs. Interchondrule matrix material is virtually absent; heavily hydrated matrix lumps (mx) are present instead. Most chondrules are FeNi-metal-free and have cryptocrystalline (CC) and barred olivine textures; chondrules of porphyritic textures (PO) are rare. (c) The CB_a chondrite Gujba consists of large chondrule fragments with cryptocrystalline and very fine grained textures, and FeNi-metal grains. (d) The CB_b chondrite Hammadah al Hamra 237 contains abundant FeNi-metal, chondrules, and rare CAIs. Chondrules have either cryptocrystalline (reddish colors) or skeletal olivine (bluish colors) textures. Matrix material is absent.

with a positive nickel versus cobalt trend and a solar Ni/Co ratio; (ii) small chondrule and CAI sizes (average diameter ~20 μm; ~90 μm in the CH chondrite Acfer 182); (iii) dominance of cryptocrystalline (CC) chondrules; (iv) high (~0.5 wt.%) Cr_2O_3 contents of chondrule olivines; (v) lack of matrix and presence of heavily hydrated lithic clasts mineralogically similar to those in CB chondrites instead; (vi) nearly solar bulk refractory lithophile element abundances (Figure 2); (vii) large depletions in bulk volatile and moderately volatile element abundances (Figure 2); (viii) whole-rock oxygen isotopic compositions plotting along the CR-mixing line (Figure 4); (ix) large positive anomalies in $\delta^{15}N$ (Figure 6); and (x) presence of a unique component of subsolar rare gases (Grossman et al., 1988; Scott, 1988; Weisberg et al., 1988; Wasson and Kallemeyn, 1990; Bischoff et al., 1993b).

When the ALH85085 meteorite was first described, it sparked much controversy because

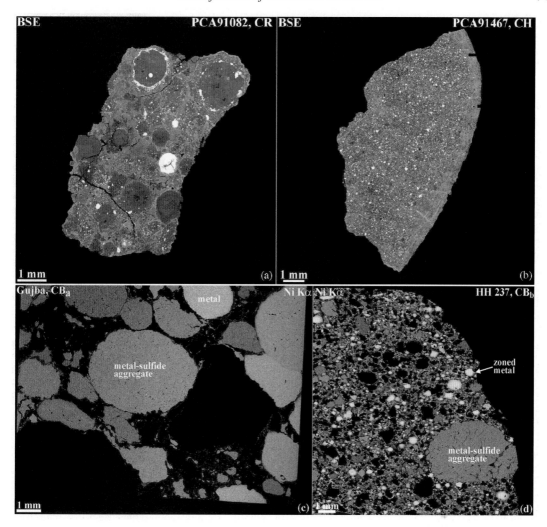

Figure 11 Backscattered electron images (a, b) and elemental maps in Ni Kα X-rays (c, d) of the: (a) CR chondrite PCA91082, (b) CH chondrite PCA91467, (c) CB$_a$ chondrite Gujba, and (d) CB$_b$ chondrite Hammadah al Hamra 237. (a) In the CR chondrite PCA91467, FeNi-metal occurs as nodules in chondrules, igneous rims around chondrules and in the matrix. (b) In the CH chondrite PAT91546, FeNi-metal occurs largely as irregularly shaped grains outside chondrules. (c) In the CB$_a$ chondrite Gujba, most FeNi-metal grains are texturally similar to metal-sulfide aggregates in the CB$_b$ chondrite Hammadah al Hamra 237. Although there are compositional variations between metal grains, each grain is compositionally uniform. (d) In Hammadah al Hamra 237, FeNi-metal occurs exclusively outside chondrules; ~20% of all metal grains are compositionally zoned, with cores enriched in Ni relative to edges. Some of the zoned metal grains contain silicate inclusions. Metal-sulfide aggregates are compositionally uniform.

of these unusual characteritics. Wasson and Kallemeyn (1990) suggested that it is a modified "subchondritic" meteorite formed as a result of a highly energetic asteroidal collision. More recently, it has been argued that the refractory inclusions, chondrules, and zoned metal grains in these meteorites are in fact pristine nebular products (Meibom et al., 1999; Krot et al., 2000c; McKeegan et al., 2003).

CB (Bencubbin-like) chondrites. The CB chondrites (Figures 10(c), (d), and 11(c), (d)) are a group of very metal rich (60–70 vol.%) meteorites including Bencubbin, Gujba, Weatherford, Hammadah al Hamra (HH) 237, and Queen Alexandra Range (QUE) 94411 which is paired with QUE94627 (Weisberg et al., 2001). Characteristics of the CB chondrites include: (i) very high metal/silicate ratios (Table 2); (ii) extreme depletion in moderately volatile lithophile elements (Figure 2); (iii) extreme enrichment in δ^{15}N (Figure 6); (iv) bulk oxygen isotopic compositions plotting near CR chondrites (Figure 4); (v) bulk refractory lithophile element abundances of $(1.1-1.5) \times$ CI (Figure 2(a)); (vi) FeNi-metal with a positive nickel versus cobalt trend and a solar Ni/Co ratio; (vii) magnesium-rich (Fa$_{\leq 4}$, Fs$_{\leq 4}$), FeNi-metal-free chondrules with only nonporphyritic (cryptocrystalline and skeletal-olivine) textures;

(viii) high (~0.5 wt.%) Cr_2O_3 contents in chondrule olivines; (ix) rarity and uniform ^{16}O-depletion of CAIs; and (x) virtual absence of fine-grained matrix material and presence of rare heavily hydrated lithic clasts instead (Lovering, 1962; Mason and Nelen, 1968; Kallemeyn et al., 1978; Newsom and Drake, 1979; Weisberg et al., 1990, 2001; Rubin et al., 2001; Krot et al., 2001a, 2002; Greshake et al., 2002).

Based on petrologic and chemical differences among the CB chondrites, they have been divided into two subgroups: CB_a, which includes Bencubbin, Weatherford, and Gujba; and CB_b, which includes HH 237 and QUE94411 (Weisberg et al., 2001). Differences between the subgroups include FeNi-metal abundances, CAI abundances, chondrule sizes, compositional range of FeNi-metal, and nitrogen isotopic compositions (Figures 6, 10(c), (d), and 11(c), (d)). The CB_a contains ~60 vol.% metal, centimeter-sized chondrules, and exceptionally rare CAIs (the only CAI was found in Gujba; Weisberg et al., 2002); FeNi-metal ranges from 5 wt.% Ni to 8 wt.% Ni, and $\delta^{15}N$ is up to ~1,000‰. The CB_b contains >70 vol.% metal, millimeter-sized chondrules; FeNi-metal ranges from 4 wt.% Ni to 15 wt.% Ni; refractory inclusions are common; and $\delta^{15}N$ is up to ~200‰.

CV (Vigarano-like) and ungrouped CV-like chondrites. The CV chondrites (Figure 12) are characterized by: (i) millimeter-sized chondrules with mostly porhyritic textures, most of which are magnesium rich and ~50% of which are surrounded by coarse-grained igneous rims; (ii) high matrix/chondrule ratios (0.5–1.2); (iii) unique presence of abundant salite-hedenbergite ± andradite nodules in the matrix (Krot et al., 1998); (iv) high abundance of millimeter-to-centimeter-sized CAIs and AOAs; and (v) common occurrence of igneous melilite-spinel-pyroxene ± anorthite (type-B) CAIs (MacPherson et al., 1988).

The CV chondrites are a diverse group of meteorites that was originally divided into oxidized (CV_{Ox}) and reduced (CV_R) subgroups, based principally on modal metal/magnetite ratios and nickel content of metal and sulfides (McSween, 1977b). Weisberg et al. (1997a) subdivided the oxidized CV chondrites into the Allende-like (CV_{OxA}) and Bali-like (CV_{OxB}) subgroups (Figure 13). Matrix/chondrule ratios increase in the order CV_R(0.5–0.6), CV_{OxA}(0.6–0.7), and $CV3_{OxB}$(0.7–1.2); metal/magnetite ratios decrease in the same order (McSween, 1977b; Simon et al., 1995). Oxygen isotopic compositions of the CV chondrites show a wide range along a slope ~1 line on the three isotope diagram, with the CV_{OxB} chondrites being slightly depleted in ^{16}O relative to the CV_R and CV_{OxA} chondrites (Weisberg et al., 1997a; Figure 1).

There are significant mineralogical differences between the CV subgroups. These differences are largely secondary and resulted from late-stage alteration that affected, to various degrees, most of the CV chondrites (Krot et al., 1995, 1998). For example, primary minerals in chondrules in CV_{OxB} chondrites are replaced by phyllosilicates, magnetite, nickel-rich sulfides, fayalite (Fa_{95-100}), and hedenbergite (Hua and Buseck, 1995; Krot et al., 1998). Chondrules in CV_{OxA} chondrites are replaced by magnetite, nickel-rich sulfides, fayalitic olivine (Fa_{30-60}), nepheline, and sodalite. Chondrules in CV_R chondrites are virtually unaltered (Kimura and Ikeda, 1998; Krot et al., 1995, 1998).

Although all of the CV chondrites have been classified as type 3, matrices of CV_{OxB} chondrites contain abundant phyllosilicates; phyllosilicates are rare in both the CV_R and CV_{OxA} chondrites (Tomeoka and Buseck, 1982, 1990; Keller and Buseck, 1990b; Keller and McKay, 1993; Keller et al., 1994; Zolensky et al., 1993).

Coolidge and Loongana 001 comprise a grouplet of carbonaceous chondrites with chemical, oxygen isotopic and petrographic characteristics similar to those of the CV chondrites: high matrix modal abundance (~20–30 vol.%), 1–2 vol.% refractory inclusions, and high refractory lithophile abundance ratios (~1.35 × CI). However, both meteorites are of higher petrologic type, 3.8–4, and have smaller chondrules and lower volatile element abundances than CV chondrites (Kallemeyn and Wasson, 1982; Scott and Taylor, 1985; Noguchi, 1994; Kallemeyn and Rubin, 1995).

CK (Karoonda-like) and CK/CV-like chondrites. Most of the CK chondrites (Figure 13(a)) are of high petrologic type (4–6) and are characterized by: (i) high abundance of matrix; (ii) large porphyritic chondrules (700–1,000 μm), with nonporphyritic chondrules and igneous rims around chondrules being virtually absent; (iii) high degree of oxidation indicated by high Fa contents of olivine (Fa_{29-33}), nearly complete absence of FeNi-metal (Figure 7), high nickel content in sulfides, and abundant magnetite which commonly contains ilmenite and spinel exsolution lamellae; (iv) rarity of CAIs and AOAs; (v) large compositional variations in plagioclase (An_{45-78}); (vi) bulk oxygen isotopic compositions plotting within the CO and CV fields (Figure 4); (vii) bulk refractory lithophile element abundances (~1.2 × CI) between those of CV and CO chondrites (Figure 2); and (viii) depletion in moderately volatile elements relative to CV and CO chondrites (Figure 2) (Kallemeyn et al., 1991).

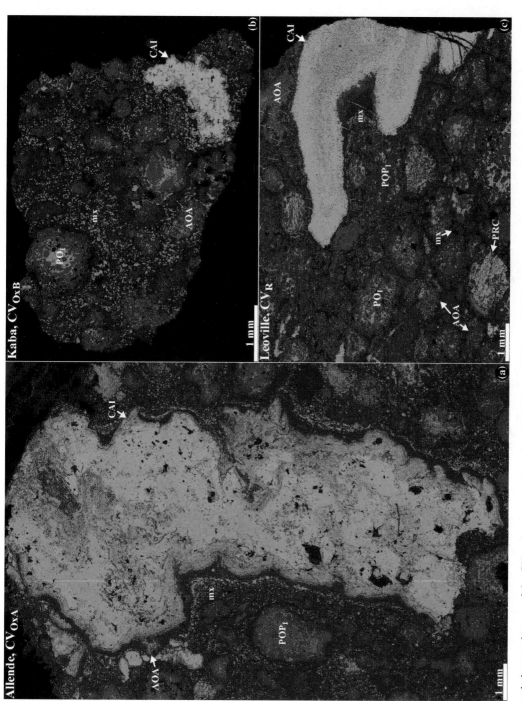

Figure 12 Combined elemental maps of the CV carbonaceous chondrites: (a) Allende (CV_{OxA}), (b) Kaba (CV_{OxB}), and (c) Leoville (CV_R). The CV chondrites contain large CAIs, AOAs, and chondrules, and fine-grained matrix. Most chondrules have porphyritic textures and magnesium-rich compositions; plagioclase-rich chondrules (PRCs) are relatively common. The CV matrices contain abundant secondary Ca-, Fe-rich pyroxenes (green spots). The Kaba matrix is hydrated; matrices in Leoville and Allende are anhydrous. Image of the Allende meteorite is not representative; large CAIs are relatively rare.

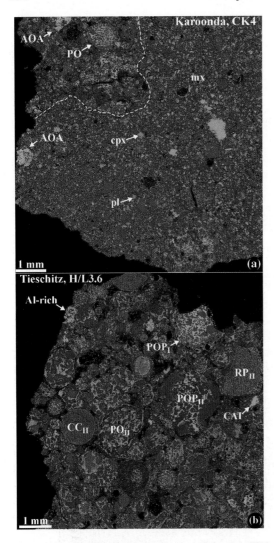

Figure 13 Combined elemental maps of the: (a) CK4 carbonaceous chondrite breccia Karoonda and (b) the H/L3.6 ordinary chondrite Tieschitz. Karoonda consists of coarse-grained matrix containing abundant plagioclase (pl) and high-Ca pyroxene (cpx) grains, and rare poorly recognized chondrules and AOAs. Outlined region contains rare plagioclase grains and well-recognized chondrules and may represent CK material of lower petrologic type. Tieschitz contains abundant type I and type II chondrules of porhyritic (PO, POP, PP), radial pyroxene (RP), and cryptocrystalline (CC) textures; Al-rich chondrules and CAIs are rare. Most chondrules are surrounded by Al-rich (bluish) fine-grained material composed of a nepheline-like phase resulted from *in situ* aqeuous alteration (source Hutchison *et al.*, 1998).

Ningqiang is an ungrouped carbonaceous chondrite that shares many petrologic, as well as oxygen isotopic and TL characteristics with CV and CK chondrites (Rubin *et al.*, 1988b; Kallemeyn *et al.*, 1991; Koeberl *et al.*, 1987; Mayeda *et al.*, 1988; Guimon *et al.*, 1995; Kallemeyn, 1996). As a result, it has been classified as a CV anomalous (Koeberl *et al.*, 1987; Rubin, 1988b; Weisberg *et al.*, 1996b), and CK anomalous (Kallemeyn, 1996).

1.05.2.4.2 Ordinary chondrites

H, L, and LL chondrites. Ordinary chondrites (Figure 13(b)) are characterized by: (i) magnesium-normalized refractory lithophile abundances $\sim 0.85 \times CI$ chondrites (Figure 2); (ii) oxygen isotopic compositions plotting above the terrestrial fractionation line (Figure 4); (iii) high abundance of chondrules with nonporphyritic and FeO-rich (type-II) chondrules being common and aluminum-rich chondrules being rare; (iv) rarity of CAIs and AOAs; and (v) a large range in degree of metamorphism (petrologic types 3–6), with several type 3.0–3.1 ordinary chondrites (e.g., Semarkona and Bishunpur) showing evidence for aqueous alteration (e.g., Hutchison *et al.*, 1987, 1998; Alexander *et al.*, 1989; Sears and Weeks, 1991; Sears *et al.*, 1980, 1991).

Abundance ratios (CI- and Mg-normalized) of the refractory lithophile elements and of selenium and zinc are very similar for the three ordinary chondrite groups; siderophile element abundances decrease and oxidation state increases through the sequence H–L–LL (Figures 2 and 7). Based on evidence for differences in impact history (there is an H-chondrite cluster of cosmic-ray exposure ages at ~ 7 Ma and an L-chondrite cluster of outgassing ages < 900 Ma; Keil, 1964; Heymann, 1967; Crabb and Schultz, 1981) and the scarcity of evidence for intergroup mixing in ordinary chondrite breccias, it is generally accepted that the H, L, and LL chondrites are formed in at least three separate parent asteroids (Kallemeyn *et al.*, 1989).

A number of taxonomic parameters are used to resolve the individual ordinary chondrite groups. Small, systematic textural differences among the ordinary chondrite groups are observed for metal abundances and chondrule size (Table 2). On a plot of cobalt concentration in kamacite versus Fa content in olivine, there is a hiatus between H and L, but no hiatus between L and LL chondrites (Figure 14). Olivine composition is not a reliable indicator of group for unequilibrated ordinary chondrites of low petrologic type (< 3.5) and siderophile element abundances are used instead (Kallemeyn *et al.*, 1989; Sears *et al.*, 1991). Several chondrites (Bjurböle, Cynthiana, Knyahinya, Qidong, Xi Ujimgin) fall between the main L and LL clusters (Figure 14), possibly indicating formation on a separate parent asteroid (Kallemeyn *et al.*, 1989). Based on abundance of siderophile elements and compositions of olivine and kamacite (Figure 14) that are intermediate between H and L chondrites, Tieschitz (Figure 13(b)) and

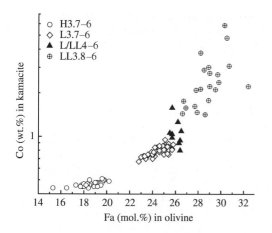

Figure 14 Fayalite (Fa) content (mol.%) in olivine versus Co content (wt.%) in FeNi-metal from ordinary chondrites (source Rubin, 1990).

Bremervörde are classified as H/L chondrites (Kallemeyn et al., 1989).

Low-FeO ordinary chondrites. The chemical properties of the Burnwell, Willaroy, Suwahib (Buwah), Moorabie, Cerro los Calvos, and "Wray (a)" chondrites plot on extensions of the H–L–LL trends of ordinary chondrites towards more reducing compositions. These meteorites have generally lower fayalite in olivine (Fa_{13-15}), ferrosilite in orthopyroxene, cobalt in kamacite (0.30–0.45 wt.%), FeO in the bulk chemical compositions, and $\Delta^{17}O$, compared to that observed in equilibrated ordinary chondrites. They have, however, ordinary chondrite abundances of refractory lithophile elements, zinc and other taxonomic elements, and matrix/chondrule abundance ratios typical of ordinary chondrites (Wasson et al., 1993; McCoy et al., 1994; Russell et al., 1998b). All of them are of petrologic type between 3.8 and 4. Wasson et al. (1993) argue that the low-FeO chondrites were originally normal H or L chondrites that have been altered by metamorphism in a highly reducing regolith. McCoy et al. (1994) and Russell et al. (1998b) conclude that reduction did not play a role in establishing their mineral compositions and suggest that these meteorites sampled other than H, L, or LL parent asteroid(s).

1.05.2.4.3 Enstatite (EH, EL, and Lewis Cliff 87223) chondrites

Enstatite chondrites (Figures 15(a) and (b)) are characterized by: (i) unique mineralogy (e.g., oldhamite (CaS), niningerite ((Mg,Fe,Mn)S), alabandite ((Mn,Fe)S), osbornite (TiN), sinoite (Si_2N_2O), silicon-rich kamacite, daubreelite ($FeCr_2S_4$), caswellsilverite ($NaCrS_2$), and perryite ($(Ni,Fe)_x(Si,P)$)) indicating formation under extremely reducing conditions (e.g., Keil, 1968; Brearley and Jones, 1998); (ii) bulk oxygen isotopic compositions that plot along the terrestrial fractionation line (Figure 4), close to the oxygen composition of the Earth and Moon (Clayton et al., 1984); (iii) abundant enstatite-rich chondrules having cryptocrystalline and porphyritic textures with olivine-bearing chondrules being rare (absent in types 4–6 enstatite chondites); (iv) rarity of CAIs (Guan et al., 2000; Fagan et al., 2001); and (v) very low abundance of fine-grained matrix and fine-grained chondrule rims.

Based on mineralogy and bulk chemistry, enstatite chondrites are divided into EH and EL groups (e.g., Sears et al., 1982). The higher abundance of opaque minerals and the occurrence of niningerite and various alkali sulfides (e.g., caswellsilverite, djerfisherite) are diagnostic criteria for EH chondrites, while alabandite is characteristic for EL chondrites (Lin and El Goresy, 2002). In addition, EH chondrites are characterized by enrichments of silicon in kamacite and the occurrence of perryite. The EH and EL chondrites range in petrologic type from EH3 to EH5 and from EL3 to EL6. EH3 chondrites differ from EL3 in having higher modal abundances of sulfides (7–16 vol.% versus 7–10 vol.%), lower abundances of enstatite (56–63 vol.% versus 64–66 vol.%), and more silicon-rich (1.6–4.9 wt.% versus 0.2–1.2 wt.%) and nickel-poor (2–4.5 wt.% versus 3.6–8.7 wt.%) metal compositions. LEW87223 differs from both EH3 and EL3 chondrites in having the highest abundance of metal (~17 vol.%) and FeO-rich enstatite, and the lowest abundance (~3 vol.%) of sulfides (Weisberg et al., 1995b).

Some E chondrites are classified as melt rocks (McCoy et al., 1995; Weisberg et al., 1997b; Rubin and Scott, 1997). They have homogeneous, in some cases coarse-grained (millimeter-sized enstatite), achondritic textures and are chemically and oxygen-isotopically similar to E chondrites; some are metal rich (up to 25 vol.%). These E-chondrite melt rocks have been interpreted to be a result of impact melting on the E-chondrite-like parent bodies (McCoy et al., 1995; Rubin and Scott, 1997) or internally derived melts (Olsen et al., 1977; Patzer et al., 2001; Weisberg et al., 1997b). Itqiy is an E-chondrite melt rock that is depleted in plagioclase and has a fractionated rare earth element (REE) pattern suggesting that it may be the product of partial melting (Patzer et al., 2001).

1.05.2.4.4 K (Kakangari-like) chondrites

A grouplet of two chondrites, Kakangari and Lewis Cliff 87232 (Figure 15(c)), is chemically and isotopically distinct from the ordinary, enstatite,

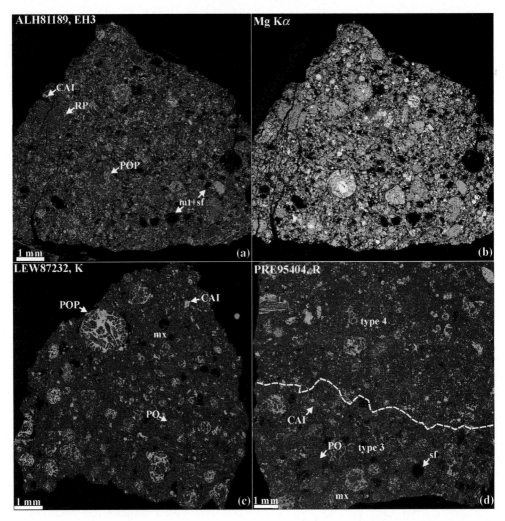

Figure 15 Combined elemental maps (a, c, d) and Mg Kα X-ray map (b) of the (a, b) EH3 enstatite chondrite ALH81189, (c) the Kakangari-like chondrite LEW87232, and (d) the R3-5 chondrite breccia PRE95404. The EH3 chondrite ALH81189 (a, b) is composed of enstatite-rich chondrules and metal-sulfide nodules; forsterite-bearing (bright grains in (b)) chondrules are rare; CAIs are very rare; matrix material is very minor (difficult to recognize). (c) The Kakangari-like meteorite LEW87232 consists of crystalline matrix largely composed of magnesian pyroxenes and olivine, magnesian pyroxene-rich chondrules, and rare CAIs. (d) The R3-5 chondrite breccia PRE95404 contains two lithologically different protions. The lower part of the section consists of type-3 material; it contains abundant chondrules, large sulfides nodules (sf), and very rare CAIs surrounded by a fine-grained matrix material. The rest of the section represents type-4 material; it contains poorly defined chondrules and relatively coarse-grained matrix. Ferromagnesian silicates in chondrules and matrix have similar ferroan compositions; magnesium-rich chondrules and chondrule fragments are rare.

and carbonaceous classes, although these meteorites have properties related to each (Srinivasan and Anders, 1977; Davis et al., 1977; Brearley, 1989; Weisberg et al., 1996a). The K chondrites have: (i) high matrix abundance (70–77 vol.%) like carbonaceous chondrites (Table 2); (ii) metal abundances (6–9 vol.%) similar to H chondrites; (iii) average olivine (Fa$_2$) and enstatite (Fs$_4$) compositions indicating an oxidation state intermediate between H and enstatite chondrites; (iv) mineralogically unique matrix composed largely of enstatite (Fs$_3$); (v) refractory lithophile element abundances ($<1 \times$ CI) similar to those of the ordinary chondrites (Figure 2); (vi) volatile lithophile element abundances similar to ordinary chondrites; (vii) chalcophile element abundances between those of H and enstatite chondrites (Figure 2); (viii) whole-rock oxygen isotopic compositions that are below the terrestrial fractionation line (Figure 4), near the CR, CB, and CH chondrites (CR-mixing line); and (ix) abundance and isotopic compositions of nitrogen and carbon within the range of CV chondrites (Figures 5 and 6).

Lea County 002 was originally classified as a Kakangari-like chondrite, although Weisberg et al. (1996a) reported differences in matrix abundances and chondrule sizes by comparison to Kakangari/Lewis Cliff 87232 (33 vol.% versus 70–77 vol.% and 1.0–1.25 mm versus 0.25–0.5 mm, respectively). CAIs in Kakangari and Lea County 002 appear to be different as well (Prinz et al., 1991a). Mineralogical observations (Krot, unpublished) and oxygen isotopic compositions (Figure 4) suggest instead that Lea County 002 is a CR chondrite. For more on IDPs, see Chapter 1.26.

1.05.2.4.5 R (Rumuruti-like) chondrites

The R chondrites (Figures 16(d)) are characterized by: (i) a high abundance of matrix (~50 vol.%); (ii) high oxidation states, reflected in the occurrence of abundant NiO-bearing olivine with Fa_{37-40} (in equilibrated R chondrites) and nearly complete absence of FeNi-metal; (iii) refractory lithophile and moderately volatile element abundances similar to those in ordinary chondrites (~0.95 × CI); (iv) absence of significant depletions in manganese and sodium and enrichment in volatile elements such as gallium, selenium, sulfur, and zinc, which distinguish R chondrites from ordinary chondrites (Figure 2); (v) whole-rock $\Delta^{17}O$ values higher than for any other chondrite group (Figure 4); and (vi) extreme rarity of CAIs (Weisberg et al., 1991; Bischoff et al., 1994; Rubin and Kallemeyn, 1994; Schulze et al., 1994; Kallemeyn et al., 1996; Russell, 1998). Most R chondrites are metamorphosed (petrologic type >3.6) and brecciated (Figure 15(d)); they have typical light/dark structure and solar-wind implanted rare gases and are best described as R3–6 regolith breccias (Bischoff, 2000).

1.05.3 CLASSIFICATION OF IDPS

Based on the bulk chemistry, IDPs are divided into two groups: (i) micrometer-sized *chondritic particles* and (ii) micrometer-sized *nonchondritic particles*. A particle is defined as chondritic when magnesium, aluminum, silicon, sulfur, calcium, titanium, chromium, manganese, iron, and nickel occur in relative proportions similar (within a factor of 2) to their solar element abundances, as represented by the CI carbonaceous chondrite composition (Brownlee et al., 1976). Chondritic IDPs differ significantly in form and texture from the components of known carbonaceous chondrite groups and are highly enriched in carbon relative to the most carbon-rich CI carbonaceous chondrites (Rietmeijer, 1992; Thomas et al., 1996; Rietmeijer, 1998, 2002).

The chondritic IDPs occur as aggregate particles (make up ~90% of chondritic IDPs) and nonaggregate particles. A subdivision of chondritic IDPs includes: (i) particles that form highly porous (*chondritic porous (CP) IDPs*) and somewhat porous aggregates (*chondritic filled (CF) IDPs*); (ii) smooth particles (*chondritic smooth (CS) IDPs*); and (iii) compact particles with a rough surface (*chondritic rough (CR) IDPs*) (MacKinnon and Rietmeijer, 1987; Rietmeijer and Warren, 1994; Rietmeijer, 1998, 2002).

Three groups of chondritic IDPs have been distinguished using infrared spectroscopy: (i) olivine-rich particles + carbon phases; (ii) pyroxene-rich particles + carbon phases; and (iii) layered lattice silicate-rich particles + carbon phases (Sandford and Walker, 1985; MacKinnon and Rietmeijer, 1987). The layered lattice silicate-rich particles are further divided into serpentine-rich and smectite-rich particles (Rietmeijer, 1998).

In an attempt to reconcile various chondritic IDP classifications, Rietmeijer (1994) proposed to classify IDPs on the basis of properties of identifiable textural entities in these particles and their porosity as *aggregate IDPs* and *collapsed aggregate IDPs*. In highly porous aggregates the constituents are loosely bound (CP IDPs). Most aggregates are more compact with little pore space (CF IDPs). The collapsed aggregate particles typically have a smooth surface (CS IDPs).

The nonchondritic IDPs consist mostly of magnesium-rich olivines and pyroxenes, and FeNi-sulfides (Rietmeijer, 1998). Chondritic materials are often attached to the surface of nonchondritic IDPs, which suggests that the groups have a common origin. For more on IDPs, see Chapter 1.26.

1.05.4 CLASSIFICATION OF NONCHONDRITIC METEORITES

The nonchondritic meteorites contain virtually none of the components found in chondrites and were derived from chondritic materials by planetary melting, and fractionation caused their bulk compositions to deviate to various degrees from chondritic materials. The degrees of melting that these rocks experienced are highly variable and, thus, these meteorites can be divided into two major categories: *primitive* and *differentiated* (Figure 1). There is no clear-cut boundary between these categories, however.

The differentiated meteorites were derived from parent bodies that experienced large-scale partial melting, isotopic homogenization (ureilites are the only exception), and subsequent differentiation. Based on abundance of FeNi-metal, these meteorites are commonly divided into three types: *achondrites* (metal-poor), *stony irons*, and *irons*; each of the types contains several meteorite groups and ungrouped members.

The primitive nonchondritic meteorites have approximately chondritic bulk compositions, but their textures are igneous or metamorphic. They are generally thought to be ultrametamorphosed chondrites or residues of very low degrees of partial melting.

Several groups of achondrites and iron meteorites are genetically related and were possibly derived from single asteroids or planetary bodies (acapulcoites and lodranites; winonaites, silicate inclusions in IAB and IIICD irons; howardites, eucrites and diogenites; shergottites, nakhlites, Chassigny and ALH84001).

1.05.4.1 Primitive Achondrites

Several groups of stony (silicate-rich) meteorites that have essentially bulk chondritic compositions but achondritic textures formed by low degrees of melting and have collectively been referred to as "primitive achondrites" (Prinz et al., 1983b). These include the acapulcoites, lodranites, and winonaites. Two groups of silicate-bearing iron meteorites, IAB and IIICD, may have originated from the same parent body as the winonaites. Some of the silicate inclusions in IIE irons also resemble primitive achondrites.

1.05.4.1.1 Acapulcoites and lodranites

The acapulcoites and lodranites are fine- to medium-grained equigranular rocks composed of orthopyroxene, olivine, Cr-diopside, Na-plagioclase, FeNi-metal, schreibersite, troilite, whitlockite, Cl-apatite, chromite, and graphite (Figures 16(a) and (b)). Although this mineral assemblage is rather similar to that of ordinary chondrites, and rare relict chondrules have been reported in several acapulcoites (e.g., Schultz et al., 1982; McCoy et al., 1996), their mineral compositions, mineral abundances, and grain size differ from ordinary chondrites (Table 6).

There is no clear-cut distinction between acapulcoites and lodranites. Typically, acapulcoites (Figure 16(a)) are fine grained (150–230 μm), with approximately chondritic abundances of olivine, pyroxene, plagioclase, metal, and troilite, whereas lodranites (Figure 16(b)) are coarse grained (540–700 μm) and depleted in troilite and plagioclase (Nagahara, 1992; McCoy et al., 1996; Mittlefehldt et al., 1998). The bulk oxygen isotopic compositions of acapulcoites and lodranites are similar and form a cluster with significant dispersion in $\Delta^{17}O$ (Figure 17).

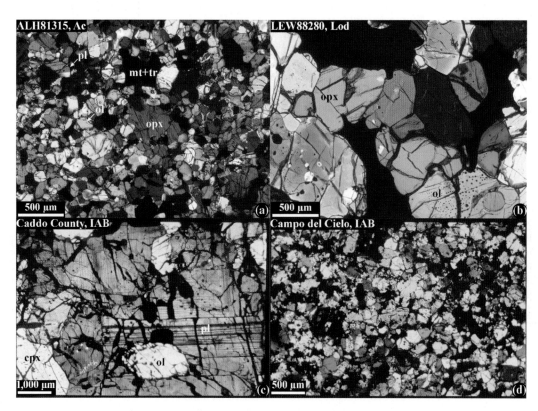

Figure 16 Photomicrographs of primitive achondrites in transmitted light with partially crossed nicols. (a) Acapulcoite ALH81315 and (b) lodranite LEW88280 have similar mineralogy (mostly orthopyroxene, olivine, Cr-diopside, Na-plagioclase, FeNi-metal, and troilite), but different grain size. (c) Silicate inclusion in IAB iron Caddo County is a basaltic clast composed of plagioclase (pl, gray), clinopyroxene (cpx, whitish) and olivine (ol, pink). (d) Silicate inclusion in IAB iron Campo del Cielo has chondritic mineralogy ((a, b)—photograph courtesy of D. Mittlefehldt; (c, d)—photograph courtesy of G. Benedix).

Table 6 Synopsis of petrologic characteristics of the asteroidal, nonchondrite groups, and models of their origin.

Group	Mineralogy	Texture	Origin
Acapulcoites–lodranites	Olivine (Fo_{87-97}), orthopyroxene (En_{86-92}), clinopyroxene ($En_{50-54}Wo_{43-46}$), metal, ±troilite, ±Na-plagioclase (An_{12-31})	Fine- to coarse-grained equigranular	Ultrametamorphism, low-degree partial melting ± melt migration
IAB–IIICD silicate inclusions, winonaites	Olivine (Fa_{1-8}), orthopyroxene (Fs_{1-9}), clinopyroxene ($Fs_{2-4}Wo_{44-45}$), Na-plagioclase ($An_{11-22}Ab_{76-85}$), troilite, metal, daubreilite, schreibersite, graphite	Fine-grained equigranular	Ultrametamorphism, low-degree partial melting, impact mixing
IIE silicate inclusions	Olivine (Fo_{79-86}), orthopyroxene (Fs_{14-17}), clinopyroxene ($Fs_{6-18}Wo_{36-45}$), Na-plagioclase ($An_{0-14}Ab_{9-93}$), troilite, metal	Chondritic to gabbroic	Metamorphism, low-degree partial melting, melt migration, shock
Nonmagmatic irons IAB, IIE, IIICD metal phase	Metal, troilite, schreibersite, ±graphite	Coarse-grained, generally exsolved	Impact melting (?)
Angrites	Al–Ti-diopside (TiO_2, 1–5%, Al_2O_3, 4–12%; Fs_{12-50}), anorthite ($An_{86-100}Ab_{0-13}$), Ca-olivine (Fo_{32-89}), kirschsteinite	Fine- to coarse-grained igneous	Crystallized melts, cumulates
Aubrites	Enstatite, pigeonite, diopside, forsterite, Si-bearing kamacite, Na-plagioclase ($An_{2-24}Ab_{75-94}$), Fe, Mn, Mg, Cr, Ti, Na, K-sulfides	Breccias of igneous lithologies	Melted and crystallized residues from partial melting
Brachinites	Olivine (Fo_{65-70}), augite ($En_{44-48}Wo_{39-47}$), ±orthopyroxene (En_{69-72}), Fe-sulfide, ±Na-plagioclase ($An_{22-33}Ab_{68-76}$)	Medium- to coarse-grained equigranular	Metamorphosed chondrite, partial melting residue
Ureilites	Olivine (Fo_{74-95}), orthopyroxene (En_{75-87}), pigeonite ($En_{68-87}Wo_{2-16}$), augite ($En_{52-60}Wo_{32-39}$)	Coarse-grained granular	Partial melting residue
HED (howardites, eucrites, diogenites)	Orthopyroxene (En_{14-77}), olivine (Fo_{56-73}), chromite, pigeonite, Ca-plagioclase ($An_{73-96}Ab_{5-23}$), silica	Breccias (howardites), fine- to coarse-grained basalts (eucrites), coarse-grained orthopyroxenites (diogenites)	Cumulates, crystallized melts, impact mixing
Mesosiderites	Orthopyroxene (En_{41-77}), olivine (Fo_{63-92}), chromite, pigeonite, Ca-plagioclase ($An_{91-93}Ab_{7-9}$), silica, phosphates	Breccias of igneous lithologies, impact melts	Impact melted crustal igneous lithologies
Silicates and oxides in pallasites	Olivine (Fo_{80-90}), chromite, ±low-Ca pyroxene (En_{83-91})	Coarse-grained, rounded to angular	Cumulates from core–mantle boundary
Pallasite and mesosiderite metal phase	Metal, troilite, schreibersite, ±graphite	Coarse-grained, generally exsolved	Fractional crystallization of core
Magmatic irons	Metal, troilite, schreibersite, ±graphite	Coarse-grained, generally exsolved	Fractional crystallization of core

After Mittlefehldt et al. (1998).

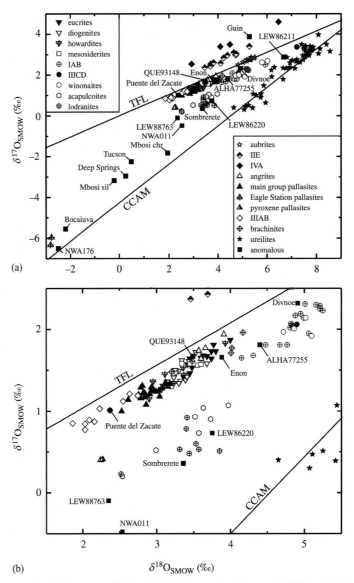

Figure 17 Bulk oxygen isotopic compositions of primitive achondrites and differentiated meteorites (sources Clayton and Mayeda, 1996; Yamaguchi et al., 2002).

EET84302 is texturally and mineralogically intermediate between acapulcoites and lodranites (Takeda et al., 1994; Mittlefehldt et al., 1996; McCoy et al., 1997a,b).

LEW86220 is related to acapulcoites and lodranites through both its oxygen isotopic (Figure 17) and mineral compositions (Clayton and Mayeda, 1996; McCoy et al., 1997b). This meteorite contains two lithologies: one is coarse grained and has mineralogy similar to acapulcoites, and the other is very coarse grained (up to 9 mm) and consists of FeNi-metal, troilite, plagioclase, and augite (Mittlefehldt et al.,1998).

Cosmic-ray exposure ages are between ~5.5 Ma and 7 Ma for all acapulcoites and most of the lodranites, possibly indicating sampling from a single impact event on a common parent body (McCoy et al., 1996, 1997a,b).

1.05.4.1.2 Winonaites and the silicate-bearing IAB and IIICD irons

Winonaites have roughly chondritic mineralogy and chemical composition (Table 6), but achondritic, recrystallized textures (Figures 16(c) and (d)). They are fine- to medium grained, mostly equigranular rocks but some (Pontlyfni and Mount Morris (Wisconsin)) contain what appear to be relic chondrules. Their mineral compositions are intermediate between those of enstatite and H chondrites, and FeNi–FeS veins that constitute the first partial melts of a chondritic precursor material are common (Benedix et al., 1998).

Silicate inclusions in IAB irons are linked through oxygen isotopic (Figure 17) and mineral compositions to the winonaites and may be related to IIICD irons (Kracher, 1982; Kallemeyn and Wasson, 1985; Palme *et al.*, 1991; Kimura *et al.*, 1992; McCoy *et al.*, 1993; Choi *et al.*, 1995; Yugami *et al.*, 1997; Takeda *et al.*, 1997; Clayton and Mayeda, 1996; Benedix *et al.*, 1998, 2000). These inclusions contain variable amounts of low-calcium pyroxene, high-calcium pyroxene, olivine, plagioclase, troilite, graphite, phosphates and FeNi-metal, and minor amounts of daubreelite and chromite (e.g., Mittlefehldt *et al.*, 1998). The abundance of silicates varies significantly between different IAB and IIICD irons.

Bunch *et al.* (1970) proposed a dual classification scheme (*Odessa type* and *Copiapo type*) of the IAB irons based on modes, mineralogy, and petrography of their silicate inclusions. Benedix *et al.* (2000) showed that, on the basis of the composition of their inclusions, IAB irons fall broadly into five types: (i) sulfide-rich inclusions with abundant silicates (e.g., Mundrabilla); (ii) nonchondritic, silicate-rich inclusions (e.g., Caddo County and Ocotillo) and in some cases (e.g., Winona) coarse-grained, olivine-rich inclusions; (iii) angular, chondritic silicate inclusions (e.g., Campo del Cielo; Copiapo type of Bunch *et al.*, 1970); (iv) rounded, often graphite-rich inclusions (e.g., Toluca; Odessa type of Bunch *et al.*, 1970); and (v) phosphate-rich inclusions (e.g., San Cristobal, Carlton).

The observed mineralogical and oxygen isotopic similarities (Figure 17) between the silicate-bearing inclusions and winonaites suggest that they formed on a common parent asteroid. It is less clear whether IAB and IIICD irons sample a common parent body (McCoy *et al.*, 1993; Choi *et al.*, 1995; Benedix *et al.*, 1998, 2000).

1.05.4.1.3 Silicate-bearing IIE irons

Silicate-bearing IIE iron meteorites have a large diversity of inclusion types, which range from clasts with recognizable chondrules (primitive), to elongate blebs of quenched basaltic melts (differentiated). Their mineralogy (Table 6) is principally olivine, orthopyroxene, clinopyroxene, plagioclase-tridymite glass, and phosphate (Bunch and Olsen, 1968; Bence and Burnett, 1969; Bunch *et al.*, 1970; Olsen and Jarosewich, 1971; Bild and Wasson, 1977; Osadchii *et al.*, 1981; Prinz *et al.*, 1983a; Olsen *et al.*, 1994; Casanova *et al.*, 1995; McCoy, 1995; Ikeda and Prinz, 1996; Ikeda *et al.*, 1997).

Based on the mineralogy and petrography of their silicate inclusions, IIE irons are divided into five groups, corresponding to increasing degrees of silicate melting: (i) Netschaëvo (chondritic clasts); (ii) Techado (partially melted, but undifferentiated

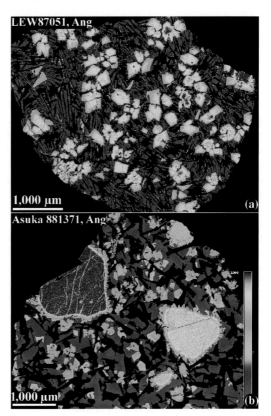

Figure 18 Elemental maps in Mg Kα X-rays of angrites: (a) LEW87051 and (b) Asuka 881371. These ophitic-textured rocks contain large, subhedral to euhedral grains of magnesian olivine (yellow and red) set in an ophitic textured groundmass of euhedral laths of anorthite (black) intergrown with euhedral to sunhedral, highly zoned Al–Ti-diopside (purple) and ferroan olivine (blue and green) (photograph courtesy of G. McKay).

clasts); (iii) Watson (totally melted inclusions that have lost metal and troilite); (iv) Weekeroo Station and Miles (plagioclase–orthopyroxene–clinopyroxene "basaltic" partial melts); and (v) Colomera, Kodaikanal, and Elga (plagioclase–clinopyroxene partial melts) (Mittlefehldt *et al.*, 1998). The bulk oxygen isotopic compositions of silicates in IIE irons are similar to those of H chondrites (Figures 4 and 18). A genetic link between IIE silicate-bearing irons and H chondrites has been inferred by Olsen *et al.* (1994), Casanova *et al.* (1995), and Clayton and Mayeda (1996), but questioned by Bogard *et al.* (2000).

1.05.4.2 Differentiated Achondrites

1.05.4.2.1 Angrites

The angrites are medium- to coarse-grained (up to 2–3 mm) igneous rocks of generally basaltic composition and consist mainly of Ca–Al–Ti-rich pyroxene, calcium-rich olivine, and anorthitic plagioclase (Table 6 and Figure 18); spinel, troilite, kirschsteinite, whitlockite, titanomagnetite, and FeNi-metal are accessory (e.g., Prinz et al., 1977; McKay et al., 1988, 1990; Yanai, 1994; Mittlefehldt et al., 1998). Although oxygen isotopic compositions of angrites are indistinguishable from those of the HED meteorites, brachinites, and mesosiderites (Figure 17), the unusual mineralogies and compositions of the angrites suggest that they are not related to any of these other meteorite groups. Angrites are the most highly alkali-depleted basalts in the solar system and have low abundances of the moderately volatile element gallium. However, they are not notably depleted in the highly volatile elements—bromium, selenium, zinc, indium, and cadmium—compared to lunar basalts and basaltic eucrites. All angrites have superchondritic Ca/Al ratios, which contrasts with the chondritic Ca/Al ratios of eucritic basalts (Mittlefehldt et al., 1998).

1.05.4.2.2 Aubrites

The aubrites (or enstatite achondrites) are highly reduced enstatite pyroxenites. Except Shallowater, which has an igneous texture (Keil et al., 1989), all are fragmental (Figure 19) or, less commonly, regolith breccias, as indicated by their high contents of solar-wind-implanted noble gases (e.g., Keil, 1989). They consist mostly (~75–95 vol.%) of nearly FeO-free enstatite, but also contain lesser amounts of albitic plagioclase and nearly FeO-free diopside and forsterite (e.g., Watters and Prinz, 1979). Clasts of both igneous and impact-melt origin are common, and the precursors to the breccias were mostly coarse-grained (probably plutonic) orthopyroxenites (Okada et al., 1988). Other phases include accessory silicon-bearing FeNi-metal

Figure 19 Hand-specimen photograph of the Bustee aubrite showing the brecciated texture of the rock (9 cm in horizontal dimension) (photograph courtesy of the Smithsonian Institution).

(e.g., Casanova et al., 1993) and a host of unusual sulfides in which elements that are normally lithophile are chalcophile, such as titanium in heideite ($FeTi_2S_4$) (Keil and Brett, 1974) and calcium in oldhamite (CaS) (Wheelock et al., 1994).

Similarities in mineralogy and oxygen isotopic compositions (Figures 4 and 17) suggest that aubrites are related to enstatite chondrites, although they are not thought to have formed on the same parent bodies. Arguments have been presented that the brecciated aubrites, the igneous Shallowater aubrite, and the H- and L-group enstatite chondrites represent samples from four different asteroidal parent bodies (Keil, 1989).

1.05.4.2.3 Brachinites

The brachinites are medium- to coarse-grained (0.1–2.7 mm) equigranular dunitic wehrlites, consisting dominantly of olivine (74–98%); they also contain augite (4–15%), plagioclase (0–10%), orthopyroxene (traces), chromite (0.5–2%), Fe-sulfides (3–7%), minor phosphates, and FeNi-metal (Table 6 and Figure 20). Lithophile element abundances in Brachina are close to chondritic and are unfractionated, whereas other brachinites show depletions in aluminum, calcium, rubidium, potassium, and sodium (Mittlefehldt et al., 1998). Siderophile element abundances in brachinites are near-chondritic (~(0.1–1) × CI). Trapped noble gases show planetary-type patterns. Oxygen isotopic compositions are indistinguishable from those of HED meteorites (Figure 17), but brachinites are clearly not from the HED parent body.

The question of whether brachinites are primitive, partial melt residues, or igneous cumulates (Warren and Kallemeyn, 1989; Nehru et al., 1996) remains open. Petrologic and geochemical differences between the different brachinites suggest that they may not have all formed by the same process (Mittlefehldt et al., 1998; Mittlefehldt and Berkley, 2002).

1.05.4.2.4 Ureilites

The ureilites (Figure 21) are carbon-bearing ultramafic rocks composed of olivine, pyroxene, and <10% dark interstitial material consisting of varying amounts of carbon, metal, sulfides, and minor fine-grained silicates (Mittlefehldt et al., 1998). The characteristic features of ureilites that distinguish them from other achondrite groups include: (i) high CaO contents in olivine and pigeonite; (ii) high Cr_2O_3 contents in olivine; (iii) high (up to ~5 wt.%) carbon contents; (iv) presence of reduced rims on olivine (and sometimes pyroxene) where in contact with graphite; and (v) large range in oxygen isotopic compositions (Figure 17) which plot along

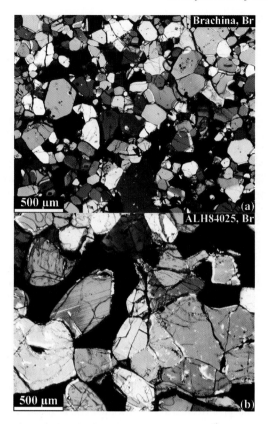

Figure 20 Photomicrographs of brachinites—(a) Brachina and (b) ALH84025—in transmitted light with partially crossed nicols. Brachina has an equigranular texture of olivine, augite, and plagioclase, with minor chromite, FeNi-metal, and sulfides. ALH84025 has an equigranular texture of olivine and augite with fine twin lamellae and interstitial troilite (black). Note coarse grain size (compared to Brachina) and absence of plagioclase (photograph courtesy of M. Wadhwa).

the line with slope ~1 nearly identical to the carbonaceous chondrite anhydrous mineral line (Goodrich, 1992; Clayton and Mayeda, 1996). Based on the mineralogy and petrology, ureilites are divided into three groups (lithological types): (i) *olivine–pigeonite ureilites*, (ii) *olivine–orthopyroxene–(augite) ureilites*, and (iii) *polymict ureilites*.

The olivine–pigeonite ureilites (Figure 21(a)), which constitute >90% of all ureilites, are coarse grained (average grain size ~1 mm), highly equilibrated assemblages of ~80 vol.% of olivine ranging in composition from Fo_{76} to Fo_{95} (olivine compositions within each ureilite are homogeneous), and uninverted pigeonite. Many have pronounced fabric, defined by both a lineation and a foliation (Figure 21(a)). Graphite occurs mainly along grains boundaries (Figure 21(c)).

Most of the olivine–orthopyroxene–(augite) ureilites have poikilitic textures, consisting of large (up to ~5 mm) oikocrysts of low-calcium pyroxene (either orthopyroxene or inverted pigeonite) enclosing rounded chadocrysts of olivine and augite (Figure 21(b)); all these minerals may contain magmatic inclusions (Figure 21(d)), which are absent in olivine–pigeonite ureilites (Goodrich *et al.*, 2001). META78008, one of the olivine–orthopyroxene–(augite) ureilites, contains some pigeonite (Berkley and Goodrich, 2001).

The polymict ureilites are fragmental breccias and regolith breccias containing solar-wind-implanted gases. They consist almost entirely of mineral and lithic clasts of olivine–pigeonite and olivine–orthopyroxene–(augite) ureilite materials. In addition, there are plagioclase clasts, feldspathic melts clasts, Ca–Al–Ti-rich clasts that resemble angrites, heavily hydrated clasts of CI-like materials, and chondritic fragments (Prinz *et al.*, 1986, 1987, 1988; Ikeda and Prinz, 2000; Ikeda *et al.*, 2000; Kita *et al.*, 2000).

The olivine–pigeonite ureilites are now accepted to be residues from at least 15% melting, the minimum required to eliminate plagioclase from a generally chondritic parent body (Warren and Kallemeyn, 1992; Scott *et al.*, 1993; Goodrich *et al.*, 2002), whereas the olivine–augite–orthopyroxene ureilites appear to be cumulates (Goodrich *et al.*, 2001). The small percentage of feldspathic material in polymict ureilites may sample the missing basaltic crust of the ureilite parent body.

1.05.4.2.5 HED meteorites: howardites, eucrites, diogenites

The HED meteorites have traditionally been classified into one group, because there is strong evidence that they originated on the same asteroidal parent body (Drake, 2001). This evidence includes their identical oxygen isotopic compositions (Figure 17), similarities in Fe/Mn ratios of pyroxenes (e.g., Papike, 1998), the occurrence of polymict breccias consisting of materials of eucritic and diogenitic parentage (e.g., the howardites; Fredriksson and Keil (1963)), and the existence of rocks intermediate between diogenites and cumulate eucrites (Takeda and Mori, 1985). Similarities in the mineralogical compositions of the HEDs and the surface mineralogy of asteroid 4 Vesta, as determined by ground-based visible and near-infrared spectroscopy, suggest that the HEDs are impact ejecta off 4 Vesta (McCord *et al.*, 1970). This proposal has been strengthened by the discovery of the so-called "Vestoids" of the Vesta family, which are small (<10 km diameter) asteroids that are dynamically linked to Vesta (e.g., Binzel and Xu, 1993).

The HEDs are discussed in order of increasing depth in their parent body, as suggested by Takeda

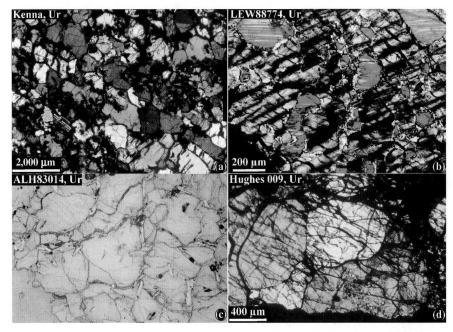

Figure 21 Photomicrographs in transmitted light with crossed polars (a, b, d) and reflected light (c) of ureilites Kenna (a), LEW88774 (b), ALH83014 (c), and Hughes 009 (d). (a) Kenna contains coarse-grained olivine and pigeonite in equilibrated (note abundant triple junctions), foliated texture. Abundant black material along olivine grain boundaries and in crosscutting veins consists of fine-grained graphite and reduction rims/veins characteristic of ureilites. (b) LEW88774 has large oikocryst of low-Ca pyroxene (now orthopyroxene with augite exsolution lamellae, oriented diagonally upper right to lower left) enclosing irregularly shaped, embayed and corroded grains of augite (blue) which are in optical continuity with one another (as can be seen in the similar orientation of shock-produced twins) and appear to be remnants of once single crystals of primary augite which reacted with the liquid from which the oikocryst crystallized. (c) In ALH83014, ureilite with very low shock level, graphite grains occur along grain boundaries and within olivine grains as large, euhedral crystals rather than the fine-grained masses seen in most ureilites. (d) The olivine–augite–orthopyroxene ureilite Hughes 009 shows highly equilibrated texture typical of olivine–pigeonite ureilites. Olivine contains magmatic inclusions (center of crystal, bottom).

(1979, 1997) in his layered crust model of the HED parent body.

Howardites. These are lithified regolith from the surface of the HED parent body, as indicated by high contents of solar-wind-implanted noble gases in their fine-grained, clastic, impact-produced matrices (e.g., Suess *et al.*, 1964) (Figure 22(a)). They are polymict breccias and consist mostly of eucritic and diogenitic material with >10 vol.% of orthopyroxene as well as some olivine. In addition, they contain abundant impact-melt clasts and breccia fragments, and occasionally carbonaceous chondrite xenoliths (e.g., Metzler *et al.*, 1995; Pun *et al.*, 1998).

Eucrites. These are pyroxene–plagioclase basalts and are subdivided into three major subclasses: the *noncumulate* eucrites (*basaltic* eucrites), the *cumulate* eucrites, and the *polymict* eucrites.

Polymict eucrites (Figure 22(b)) are polymict breccias consisting mostly of eucritic material, but they also contain <10 vol.% of a diogenitic component in the form of orthopyroxene.

Noncumulate eucrites originally formed as quickly cooled surface lava flows (*unequilibrated* noncumulate eucrites), but most were subsequently metamorphosed (*metamorphosed* noncumulate eucrites). Unequilibrated noncumulate eucrites, also referred to as the *unmetamorphosed* or *least-metamorphosed* noncumulate eucrites or *Pasamonte-type* eucrites, are surface lava flows (Figure 22(c)) that cooled quickly (Walker *et al.*, 1978). As a result of their fast cooling, their pyroxenes (pigeonite of Mg# ~70–20) are zoned, and exsolution lamellae are only visible by TEM. These rocks have experienced only minor metamorphism (e.g., Takeda and Graham, 1991).

Metamorphosed (equilibrated) noncumulate eucrites are also collectively referred to as the *ordinary* eucrites. They include the *Juvinas type* (main group) and the *Stannern* and *Nuevo Laredo types*. They are unbrecciated or monomict-brecciated, metamorphosed basalts (Figure 22(d)) and contain homogeneous low-calcium pigeonite (Mg# ~42–30) with fine exsolution lamellae of high-calcium pyroxene. The high abundance

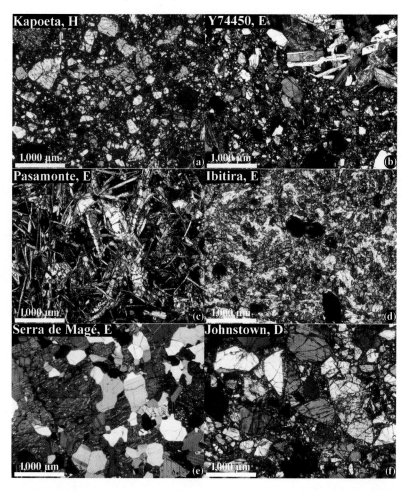

Figure 22 Thin section photomicrographs in transmitted light with crossed polars of typical HED meteorite types. (a) Howardite Kapoeta consisting of mineral fragments of highly variable grain sizes. (b) Polymict eucrite Y74450 consisting of basaltic rock (upper right) and mineral fragments. (c) Unequilibrated noncumulate eucrite Pasamonte showing the typical basaltic texture of plagioclase (light) and pyroxene (colored). (d) Metamorphosed (equilibrated) noncumulate eucrite Ibitira showing a recrystallized texture, with plagioclase (white) and pyroxene (colored). The round, dark areas in the center, bottom, and the left of image are vesicles. (e) Cumulate eucrite Serra de Magé. The rock consists of large crystals of plagioclase (lighter material, with straight twin lamellae) and mostly orthopyroxene with complex augite exsolution lamellae (darker material, with irregular, sometimes worm-like exsolution lamellae). (f) Diogenite Johnstown illustrating the highly brecciated nature of the rock that consists essentially entirely of orthopyroxene (after Keil, 2002).

of metamorphosed noncumulate eucrites suggests that the HED parent body experienced global metamorphism and that most of the original unequilibrated noncumulate eucrites have been extensively metamorphosed and transformed into the metamorphosed noncumulate eucrites (e.g., Yamaguchi et al., 1996, 1997). That many noncumulate eucrites had been extensively thermally metamorphosed was recognized early, and prompted Takeda and Graham (1991) to establish a classification of increasing metamorphism from type 1 to type 6. It should be noted that this scale is mainly based on the compositions and textures of the pyroxenes of these rocks and, thus, is different from the well-known metamorphic (petrologic type) scale of ordinary chondrites established by Van Schmus and Wood (1967).

Cumulate eucrites include the *Binda type* and *Moore County type*. They are coarse-grained gabbros, and many are unbrecciated (Figure 22(e)). They contain orthopyroxene inverted from low-calcium clinopyroxene (Mg# ~67–58) and orthopyroxene inverted from pigeonite (Mg# ~57–45).

Northwest Africa 011 (NWA 011) is a basaltic eucrite (Yamaguchi et al., 2002) composed of coarse-grained (~1 mm in size) anhedral clinopyroxene (pigeonite and relict augite) and a fine-grained to stubby (50–100 μm in size) plagioclase ($Or_{0.3-0.5}$ $Ab_{11.0-22.1}$). The oxygen isotopic composition of NWA 001 is similar to CR chondrites (Yamaguchi et al. (2000); Figure 17).

Diogenites. These are coarse-grained, cumulate orthopyroxenites that are usually highly brecciated and comminuted (Figure 22(f)). On average, they consist mostly of orthopyroxene, but also contain minor olivine, clinopyroxene, chromite, plagioclase, FeNi-metal, troilite, and a silica phase (Bowman et al., 1997).

1.05.4.3 Mesosiderites

Mesosiderites are breccias composed of roughly equal proportions of silicates and FeNi-metal plus troilite (Figure 23). Their silicate fraction consists of mineral and lithic clasts in a fine-grained fragmental or igneous matrix (e.g., Floran, 1978; Hewins, 1983; Mittlefehldt et al., 1998). There is essentially a continuum in particle sizes between lithic and mineral clasts and fine-grained matrix grains, making the distinction between matrix and clast arbitrary. The lithic clasts are largely basalts, gabbros, and pyroxenites with minor amounts of dunite and rarer anorthosite (Scott et al., 2001). Mineral clasts consist of coarse-grained orthopyroxene, olivine, and plagioclase. FeNi-metal in mesosiderites is mostly in the form of millimeter or submillimeter grains that are intimately mixed with similarly sized silicate grains (Figure 23).

The mesosiderites are divided into three petrologic classes (Hewins, 1984, 1988) based on orthopyroxene content (Figure 24): the abundance of orthopyroxene increases from A (basaltic in composition) to B (more ultramafic) to C (orthopyroxenite).

Mesosiderites have been further subdivided into four subtypes based on silicate textures, which was originally thought to reflect increasing metamorphic equilibration of the silicates (Powell, 1971). The lowest metamorphic grade 1 is characterized by a fine-grained fragmental matrix; grades 2 and 3 are characterized by recrystallized matrix, while the highest grade 4 mesosiderites are melt–matrix breccias (Powell, 1971; Floran et al., 1978). The criteria for the mesosiderite subtypes are summarized in Table 7. There are several problems with this textural classification scheme: (i) it is often difficult to distinguish between true matrix material and fine-grained breccia or impact melt clasts; (ii) textures are variable in some mesosiderites (Hewins, 1984); (iii) several grade 1 mesosiderites contain some igneous textured matrix material (Floran, 1978); and (iv) Hewins (1984) reinterpreted the type-3B mesosiderite plagioclase poikilitic matrix to have an igneous texture.

Chaunskij is an unusual, highly metamorphosed, heavily shocked, metal-rich, cordierite-bearing mesosiderite (Petaev et al., 2000).

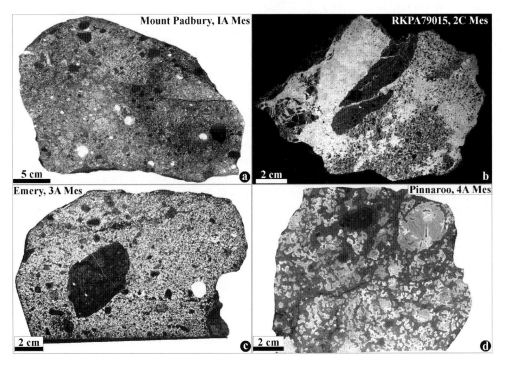

Figure 23 (a) The type-1A mesosiderite Mount Padbury having numerous centimeter-sized silicate and metal clasts dispersed in a finely divided metal–silicate matrix. (b) The type-2C mesosiderite RKPA79015 having large silicate-free metal regions. (c) The type-3A mesosiderite Emery showing a pyroxene poikiloblastic texture. (d) Polished and etched slab of Pinnaroo, a type-4A mesosiderite, showing coarse segregation of metal and silicate and large silicate clast (upper center) (photograph courtesy of the Smithsonian Institution).

1.05.4.4 Pallasites

Pallasites are stony irons composed of roughly equal amounts of silicate (dominantly olivine), metal, and troilite (Figure 25). There are three separate pallasite types that are distinguished by differences in silicate mineralogy and composition (Table 6), metal, and oxygen isotopic composition (Figure 17): (i) the main group; (ii) the Eagle Station grouplet; and (iii) the pyroxene–pallasite grouplet (e.g., Mittlefehldt et al., 1998). These differences suggest that pallasites originated on at least three separate asteroidal bodies.

1.05.4.4.1 Main-group pallasites

The main-group pallasites are composed dominantly of olivine, with minor amounts of low-calcium pyroxene, chromite, phosphates (farringtonite $(Mg,Fe)_3(PO_4)_2$, stanfieldite, whitlockite), troilite, and schreibersite (Buseck, 1977). Most contain olivine of $\sim Fo_{88\pm1}$, but a few have ferroan olivines (down to Fo_{82}). Springwater is an anomalous main-group pallasite based on its olivine composition and greater abundance of phosphates (Buseck, 1977). Metal is close in composition to that of nickel-rich IIIAB irons, possibly indicating formation on the same parent body (Scott, 1977). Although oxygen isotopic compositions of main-group pallasites are similar to those of the HEDs (Figure 17), it is unlikely that they formed on the same parent asteroid, because pallasites are thought to represent the core–mantle boundary(ies) of their parent asteroid(s), and the HED parent body (Vesta?) is probably still intact with only its crust sampled (Drake, 2001).

1.05.4.4.2 Eagle Station pallasite grouplet

The Eagle Station grouplet is mineralogically similar to the main-group pallasites, but has more ferroan and calcium-rich olivine (Buseck, 1977; Davis and Olsen, 1991) (Table 6). Their metal is

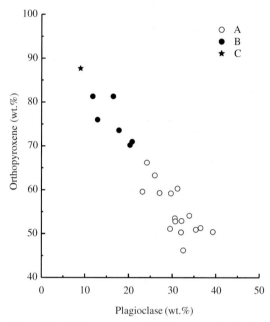

Figure 24 Orthopyroxene (wt.%) versus plagioclase (wt.%) for mesosiderites showing the generally greater basaltic component, indicated by plagioclase, in the compositional class-A mesosiderites compared to the more ultramafic-rich class-B mesosiderites (source Mittlefehldt et al., 1998) (reproduced by permission of the Mineralogical Society of America from *Reviews in Mineralogy* **1998**, *36*, 1–495).

Table 7 Summary of criteria for textural types of mesosiderites.

Grade 1	Grade 2	Grade 3
Pronounced cataclastic texture; marked angularity in small and large silicate fragments	Recognizable cataclastic texture in thin section; angularity limited to larger silicate fragments	Brecciation recognizable only macroscopically; no cataclastic angularity in thin section
Abundant extremely fine-grained comminuted silicates; many grains <10 μm	Fine-grained material coarser than in grade 1; most grains >10 μm	No extremely fine-grained silicates; most grains >100 μm
Grain boundaries between silicate fragments not intergrown; saturated contacts occur only within lithic fragments	Saturated contacts may occur between smaller silicate grains; some intergrowth visible between small matrix grains and large fragments	Silicate grains of all sizes show interlocking boundaries; rims of large pyroxene grains have ophitic textures indistinguishable from the matrix
Abundant pigeonite showing little or no inversion to orthopyroxene	Minor pigeonite showing partial inversion to orthopyroxene	Accessory pigeonite showing ample evidence of inversion to orthopyroxene; augite exsolution in orthopyroxene often shows two orientations
Partial replacement of olivine by pyroxene		Extensive/complete replacement of olivine by pyroxene

After Powell (1971).

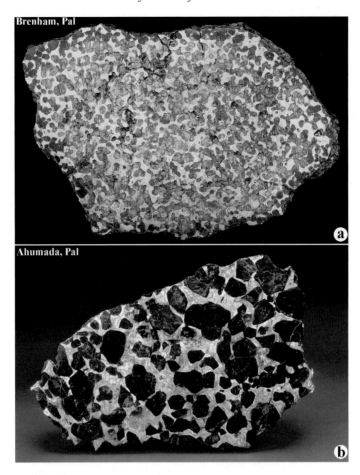

Figure 25 Polished slabs of the main-group pallasites: (a) Brenham and (b) Ahumada. Brenham (30 cm in horizontal dimension) contains well-rounded olivines dispersed in metal. Ahumada (15 cm in horizontal dimension) contains angular and subangular olivine grains in metal. Photograph courtesy of the Smithsonian Institution.

close in composition to IIF irons and has higher nickel and iridium contents than those of the main group pallasites (Scott, 1977). Kracher et al. (1980), therefore, suggested that Eagle Station grouplet pallasites and IIF irons formed in a close proximity in the solar nebula region but not on the same parent asteroid. Eagle Station pallasites have oxygen isotopic compositions quite similar to those of CV chondrites (Figures 4 and 17), suggesting another linkage (Clayton and Mayeda, 1996).

1.05.4.4.3 Pyroxene–pallasite grouplet

The pyroxene–pallasite grouplet consists of Vermillion and Yamato-8451 (Boesenberg et al., 2000). These meteorites contain ~14–63 vol.% olivine, 30–43 vol.% metal, 0.7–3 vol.% pyroxene, 0–1 vol.% troilite, and minor whitlockite. The occurrence of millimeter-sized pyroxenes distinguishes these pallasites from main-group and Eage Station grouplet pallasites (Hiroi et al., 1993b; Boesenberg et al., 2000; Yanai and Kojima, 1995).

The pyroxene pallasites have metal of different composition from other pallasites, possibly indicating formation on a different parent asteroid (e.g., Mittlefehldt et al., 1998). Metal of Y-8451 has same nickel and iridium content as the Eagle Station grouplet, but the gold content is like that of the main group pallasites (Wasson et al., 1998). Metal in Vermillion has nickel, iridium, and gold contents similar to main group pallasites, but has different contents of moderately volatile siderophile elements (e.g., gallium and germanium) than any other group of pallasites (e.g., Mittlefehldt et al., 1998). Oxygen isotopes of pyroxene pallasites are also distinct from those of main-group pallasites (Figure 17).

1.05.4.5 Irons

Iron meteorites are commonly classified on the basis of chemical and structural properties.

1.05.4.5.1 Chemical groups

On plots of log (E) versus log (Ni), where E denotes some well-determined trace element,

~85% of iron meteorites fall into one of 13 chemical groups: *IAB, IC, IIAB, IIC, IID, IIE, IIF, IIIAB, IIICD, IIIE, IIIF, IVA,* and *IVB*. With two exceptions, these groups fall within restricted areas on a log (Ge) versus log (Ni) plot (Figure 26(a)), whereas on log (Ir) versus log (Ni) plot the same samples form narrow, elongated fields spanning a considerable range in iridium contents (Figure 26(b)). It is customary to refer to these clusters as groups as if they contain at least five members. Individual samples that fall outside defined chemical groups are called ungrouped or anomalous iron meteorites, and some of these form grouplets of two to four individuals.

The taxonomic significance of gallium and germanium derives from the fact that they are the most volatile siderophile elements (Wasson and Wai, 1976; Wai and Wasson, 1979) and tend to be strongly fractionated between different iron groups. Early determinations of these elements (Lovering *et al.*, 1957) resolved only four groups, designated by Roman numerals I–IV in order of decreasing content of gallium and germanium. Letters were added later to distinguish additional groups. Not all of these groups proved to be independent, leading to the current (confusing) nomenclature, in which some groups have a combination of two letters after their Roman numeral.

Chemical evidence indicates that magmatic iron meteorite groups formed by fractional crystallization, most likely in the cores of differentiated asteroids.

Of the 13 major iron meteorite groups, only the IAB and IIICD groups have broad ranges in nickel and some trace elements. These ranges are difficult to explain by simple fractional crystallization or a metallic core. In addition, many members of these groups contain silicate inclusions, which are roughly chondritic in mineralogy and chemical compositions, but the textures are achondritic, exhibiting recrystallized textures. Silicate inclusions in IAB irons are linked through oxygen isotopic compositions and mineral compositions to winonaites and may be related to IIICD irons (Figure 17).

About 15% of all known iron meteorites do not belong to one of the above defined 13 chemical groups (Wasson and Kallemeyn, 1990). If each ungrouped individual or grouplet represents a separate parent body, then iron meteorites sample between 30 and 50 asteroids (Scott, 1979; Wasson, 1990). Mineralogy of several ungrouped iron meteorites with silicate inclusions and known oxygen isotopic compositions are listed in Table 8, which also contains information about several ungrouped nonchondritic meteorites.

1.05.4.5.2 Structural groups

The metallic phase of many iron meteorites shows a texture called the Widmanstätten pattern, an oriented intergrowth of body-centered cubic α-FeNi (kamacite), and high-nickel regions composed of several phases (a tetragonal FeNi phase (tetrataenite), a body-centered cubic FeNi phase ($α_2$-FeNi), and awaruite ($FeNi_3$)). Kamacite forms lamellae with a nickel content of <6%. The width of the kamacite lamellae is a function of the cooling rate and, hence, is approximately constant in a given iron meteorite, but varies widely between different meteorites. Because kamacite lamellae are oriented along octahedralplanes, irons which show this structure are called *octahedrites* (Figures 27(a)–(e)). According to the width of the kamacite lamellae, octahedrites are further subdivided into coarsest (*Ogg*, bandwidths >3.3 mm), coarse (*Og*, 1.3–3.3 mm), medium (*Om*, 0.5–1.3 mm), fine (*Of*, 0.2–0.5 mm), and finest (*Off*, <0.2 mm). Plessitic octahedrites (*Opl*) are transitional to ataxites (*D*).

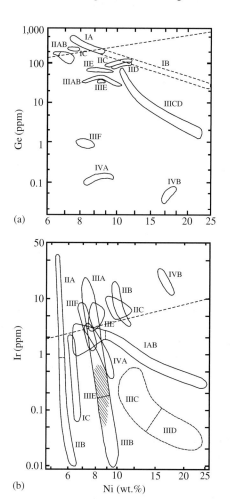

Figure 26 Plots of: (a) Ni versus Ge and (b) Ni versus Ir showing the fields for the 13 iron meteorite groups (source Scott and Wasson, 1975).

Table 8 Major characteristics of some ungrouped nonchondritic meteorites.

Meteorite	Silicate mineralogy	Classification	Refs.
ALHA77255	Silica	Anomalous iron	[1]
Bocaiuva	Ol (Fa$_8$), Opx (Fs$_8$), Pl (An$_{49}$), Cpx (Fs$_5$Wo$_{42}$)	Silicate-bearing iron, *Off*	[2]
Deep Springs		Silicate-bearing iron	[3]
Divnoe	Ol (Fa$_{20-28}$), Px (Fs$_{20-28}$Wo$_{0.5-2.5}$), Pl (An$_{45-32}$)	Ol-rich primitive achondrite	[4–6]
Enon	Ol (Fo$_{91}$), Opx (Fs$_{11}$Wo$_2$), Cpx (Fs$_4$Wo$_{44}$), Pl (An$_{15}$Ab$_{79}$), Chr, Wt, Tr, Schr	Stony iron	[7,8]
Guin		Iron with affinities to IIE, IAB, and IIICD	[9]
LEW86211	Troilite–metal intergrowth with olivine–pyroxene inclusions	Anomalous	[1,10]
LEW88763	Ol (71 vol.%; Fo$_{63-64}$), Px (7%; Fs$_{29}$Wo$_4$ + Aug (Fs$_{16}$Wo$_{38}$), Pl (10%; Ab$_{55-74}$An$_{19-44}$), Met + Tr + Chr + Ilm (6%)	Primitive achondrite	[11]
Mbosi	Pl, Px, Chr, Tr, Schr, Sil	Silicate-bearing iron	[12]
NWA 176	Ol (20.9 vol.%, Fa$_{11}$), Opx (17.5%, Fs$_{11}$Wo$_3$), Cpx (1.3%, Fs$_6$Wo$_{42}$), Pl (3.9%, An$_{50}$Ab$_{47}$), Mer (0.2%), Chr (0.1%)	Silicate-bearing iron	[13]
Puente del Zacate	Ol (23 wt.%, Fa$_4$), Px (14%, Fs$_6$Wo$_1$), Cr-Di (15%, Fs$_3$Wo$_{47}$), Pl (15%, An$_{14}$Ab$_{82}$), Grph (27%), Tr, Chr, Daub, Met	Silicate-bearing iron	[14]
QUE93148	Ol (65 vol.%), Opx (13%), Met (22%), Aug (<0.1%)	Olivine-rich achondrite	[15]
Sombrerete	Opx (14.7%, En$_{68}$), albitic glass (66.7%), Pl (9%), Cl-Apt (8%), Ks, Trid, Chr, Ilm, Rt	Silicate-bearing iron	[16,17]
Tucson	Ol (66 vol.%, Fo$_{99-100}$), En (30%, 0.5–21 wt.% Al$_2$O$_3$), Di (3%, 5–18 wt.% Al$_2$O$_3$), Pl, Sp, Brz	Silicate-bearing iron	[18]

Brz = brezinaite; Chr = chromite; Cl-Apt = Cl-apatite; Cpx = clinopyroxene; Daub = daubreelite; Di = diopside; Grph = graphite; Ilm = ilmenite; Ks = kaersutite; Mer = merrillite; Met = FeNi-metal; Ol = olivine; Opx = orthopyroxene; Pl = plagioclase; Px = pyroxene; Rt = rutile; Schr = schreibersite; Tr = troilite; Trid = tridymite; Wt = whitlockite.
[1]—Clayton and Mayedsa (1996); [2]—Malvin *et al*. (1985); [3]—Schaudy *et al*. (1972); [4]—Petaev *et al*. (1994); [5]—McCoy *et al*. (1992a); [6]—Weigel *et al*. (1996); [7]—Bunch *et al*. (1970); [8]—Kallemeyn and Wasson (1985); [9]—Rubin *et al*. (1986); [10]—Prinz *et al*. (1991b); [11]—Swindle *et al*. (1998); [12]—Olsen *et al*. (1996a); [13]—Liu *et al*. (2001); [14]—Olsen *et al*. (1996b); [15]—Goodrich and Righter (2000); [16]—Prinz *et al*. (1982); [17]—Malvin *et al*. (1984); [18]—Nehru *et al*. (1982).

Some irons with Ni contents <6% consist almost entirely of kamacite and show no Widmanstätten pattern; they are called *hexahedrites* (H) (Figure 27(f)).

1.05.4.6 Silicate-bearing IVA Irons

The IVA iron meteorite group contains four silicate-bearing members: Bishop Canyon, Gibeon, Steinbach, and São João Nepomuceno (Mittlefehldt *et al*., 1998). Silicates in IVA irons are nonchondritic in their mineralogy (Table 6) and consist of orthobronzite–clinobronzite–tridymite-rich inclusions in Steinbach and São João Nepomuceno and of individual SiO$_2$ grains in Bishop Canyon and Gibeon (e.g., Ulff-Möller *et al*., 1995; Scott *et al*., 1996; Haack *et al*., 1996; Marvin *et al*., 1997).

1.05.5 MARTIAN (SNC) METEORITES

Twenty-six meteorites are believed to be martian rocks. This group was formerly referred to by the acronym SNC (for *Shergottites, Nakhlites,* and *Chassigny*). However, this designation is no longer comprehensive, and the simple term "martian meteorites" is recommended instead (Treiman *et al*., 2000). The martian meteorites are volcanic or subvolcanic and plutonic rocks. Their diverse, highly fractionated compositions, and young crystallization ages (1.3 Ga and possibly ~180 Ma), suggest that they are derived from a large, planet-sized body (e.g. Wood and Ashwal, 1981; Ashwal *et al*., 1982; Jones, 1986). Their unique oxygen isotopic compositions (Figure 28) and FeO/MnO ratios (Figure 29) indicate that this body is not Earth or Moon. Similarities between the isotopic compositions of nitrogen and noble gases of the martian atmosphere (as determined by the Viking Landers) and those trapped in impact-produced glasses in some shergottites suggest that it is Mars (e.g., McSween and Treiman, 1998; Bogard *et al*., 2000; Nyquist *et al*., 2001; McSween, 2002 and references therein; see Chapter 1.22).

1.05.5.1 Shergottites

The shergottites are the most abundant (18 out of 26) and the most diverse of the martian meteorite subgroups. They are commonly divided into two types: basaltic and lherzolitic.

Figure 27 Polished and etched slabs of iron meteorites. (a) The fine octahedrite IIICD iron Carlton. (b) The medium octahedrite IIIAB iron Casas Grandes. (c) The coarse octahedrite IIE silicate-bearing iron Weekeroo Station. Silicate inclusions in (c) consist of plagioclase, orthopyroxene, and clinopyroxene. (d) The coarse octahedrite IAB silicate-bearing iron Campo del Cielo. (e) The coarsest octahedrite IIAB iron Santa Luzia. (f) The hexahedrite IIAB iron Bennett County (photograph courtesy of the Smithsonian Institution).

The *basaltic shergottites* consist predominantly of clinopyroxene (pigeonite and augite) and plagioclase (now shock-produced glass or maskelynite), and have basaltic or diabasic textures. The absence of olivine in these rocks and their low Mg#s (23–52) indicate that they crystallized from fractionated magmas (Stolper and McSween, 1979). Many of them (including

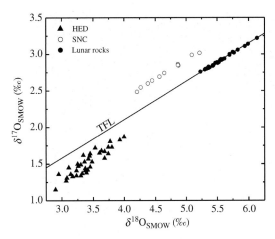

Figure 28 Bulk oxygen isotopic compositions of SNC meteorites, lunar meteorites, and HED meteorites (source Clayton and Mayeda, 1996).

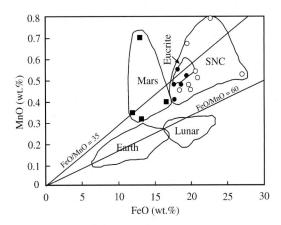

Figure 29 Weight percentages of FeO and MnO in solar-system basalts and martian meteorites. Filled circles are SNC basalts, open circles are other martian meteorites, and filled squares are Mars' surface materials. Martian meteorites have essentially the same FeO/MnO ratio as do Mars surface materials and eucrite basalts. Only Chassigny has a comparable FeO/MnO ratio to terrestrial basalts (source Treiman et al., 2000).

the type shergottite, Shergotty) contain cumulus pyroxene, and have strongly foliated textures suggesting crystal accumulation in near-surface dikes or lava flows (Stolper and McSween, 1979; McCoy et al., 1992b). However, some (e.g., QUE94201 and Los Angeles) have higher plagioclase contents and may represent magma compositions (McSween et al., 1996; McKay et al., 2002; Rubin et al., 2000).

The *lherzolitic shergottites* are magnesian (Mg# ~ 70) olivine–clinopyroxene–chromite cumulates, dominated by coarse-grained poikilitic pigeonite enclosing rounded olivine (and chromite) crystals, with interstitial areas of finer-grained FeO-rich olivine, pigeonite, augite, maskelynite, and other late phases (McSween et al., 1979a,b; Treiman et al., 1994; Ikeda, 1997). McSween et al. (1979a,b) noted that the dominant mineralogy of these rocks is consistent with early crystallization from magmas having the crystallization sequence inferred for basaltic shergottites such as Shergotty. This led to their classification as shergottites. Some authors, however, feel that they should be referred to simply as martian lherzolites (e.g., Treiman et al., 1994).

EET79001 is a unique shergottite consisting of two lithologies (designated A and B) separated by an obvious contact (Figure 30(a)). Lithology B is a clinopyroxene–plagioclase rock resembling the basaltic shergottites. Lithology A, however, is distinct from either the basaltic or lherzolitic shergottites. It has a porphyritic texture consisting of megacrysts of olivine, orthopyroxene, and chromite, in a finer-grained pigeonite–plagioclase groundmass. Four new shergottites (e.g., Dar al Gani 476 and Sayh al Uhaymir 005 (Figure 30(b))) consist entirely of olivine–porphyritic lithologies resembling EET79001 lithology A. These meteorites have been referred to by various terms, including "basaltic shergottite" (on the basis of plagioclase content), "*transitional shergottite*," or "*mixed shergottite*." We recommend the term "olivine-phyric shergottite," which emphasizes their differences from basaltic shergottites such as Shergotty or QUE94201, and does not have genetic implications.

1.05.5.2 Nakhlites (Clinopyroxenites/Wehrlites)

Nakhlites are clinopyroxenites or wehrlites, dominated by augite with lesser olivine. Their coarse-grained textures (Figure 30(c)) and the presence of exsolution lamellae in augite (which require slow cooling) suggest that they are cumulates. They also contain interstitial material consisting of a microcrystalline groundmass of radiating crystalline plagioclase, pigeonite, ferroaugite, titanomagnetite, pyrite, troilite, chlorapatite, and rare silica-rich glass (e.g., Berkley et al., 1980; Treiman, 1986, 1990, 1993).

1.05.5.3 Chassignite (Dunite)

Chassigny is the only known martian dunite. It is an olivine–chromite cumulate consisting of 90% olivine (Fa$_{\sim 32}$), 5% pyroxene, 2% feldspar (An$_{\sim 20}$) that has been transformed into maskelynite, and 3% accessory minerals including chromite (Figure 30(d)). Melt inclusions in olivine contain hydrous amphibole,

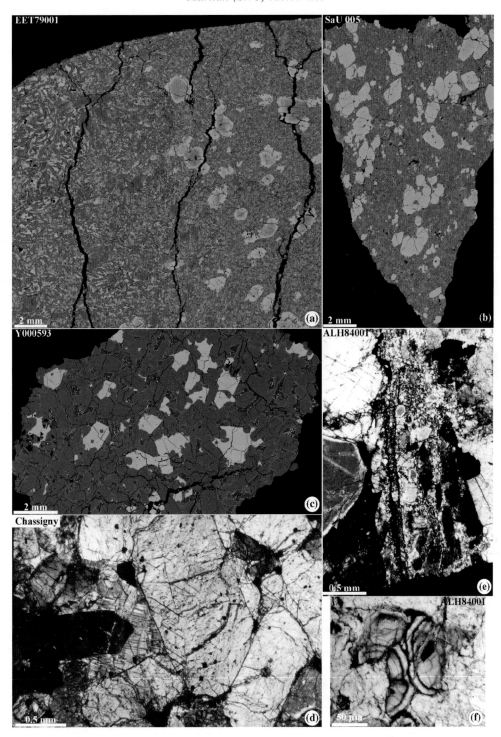

Figure 30 Combined elemental maps (a–c) in Al (red), Fe (green) and Ca Kα (blue) X-rays and thin section photomicrographs in transmitted light with crossed polars (d–f) of martian meteorites: (a) Elephant Moraine (EET) 79001, (b) Sayh al Uhaymir (SaU) 005, (c) Yamato (Y) 000593, (d) Chassigny, and (e, f) ALH84001. Image of EET79001 shergottite shows a contact between lithology A (right) and lithology B (left). Lithology A is an olivine-phyric basalt containing megacrysts of olivine in a finer grained groundmass consisting principally of pigeonite (green-blues) and maskelynite (magenta). Lithology B is a pyroxene–plagioclase (maskelynite) basalt. It lacks olivine, is coarser-grained than the groundmass of lithology A, and has a higher augite content (some of the blues). SaU 005 is an olivine-phyric basalt mineralogically similar to lithology A of EET79001. Olivine megacrysts is a finer-grained groundmass consisting principally of pigeonite (green-blues) and maskelynite (magenta). Y000593 is a clinopyroxenite containing augite (blue), olivine (green), and interstitial areas (red) consisting principally of radiating sprays of plagioclase and alkali feldspars. Chassigny is an olivine-chromite cumulate. Olivine grains contain hydrous amphibole-bearing melt inclusions. ALH84001 is a coarse-grained orthopyroxenite containing abundant carbonates.

Figure 31 Lunar meteorite North West Africa (NWA) 773 consists of two distinct lithologies: cumulate olivine norite and regolith breccia. The cumulate portion is composed of olivine, pigeonite, augite, feldspar, and opaques (troilite, chromite, Fe-metal). The breccia portion contains fragments of cumulate portion as well as silica glass, hedenbergitic pyroxene, volcanic rocks, and unusual lithic clasts with fayalite + Ba-rich K-feldspar + silica + plagioclase (photograph courtesy of M. Killgore).

suggesting that the rock formed under relatively highly oxidizing conditions (Floran et al., 1978).

1.05.5.4 Allan Hills 84001 (Orthopyroxenite)

The notorious ALH84001 meteorite that has been postulated to contain nanofossils of martian origin (McKay et al., 1996) is the only known martian orthopyroxenite. Originally classified as a diogenite (i.e., a pyroxenite of the HED clan), it was shown by Mittlefehldt (1994) to be a unique martian cumulate orthopyroxenite. While all other SNCs have crystallization ages of a few hundred Ma, the age of ALH84001 is ~4.5 Ga (e.g., Nyquist et al., 2001) and thus is a fragment of the ancient martian crust. This coarse-grained rock (Figures 30(e) and (f)) consists mainly of large (up to 6 mm long) orthopyroxene crystals (96%), with 2% chromite, 1% plagioclase (maskelynite), 0.15% phosphate, and accessory phases such as olivine, augite, pyrite, and carbonates (Mittlefehldt, 1994).

1.05.6 LUNAR METEORITES

ALHA81005 was the first meteorite to be recognized, by comparison with Apollo samples, as a lunar rock (e.g., Warren et al., 1983). Another lunar meteorite, Yamato-791197, was actually found in Antarctica two years prior to ALHA81005, but was not identified as lunar until after the revelation of ALHA81005's identity. As of this writing, ~26 lunar meteorites have been recovered from Antarctica and hot deserts (Korotev, 2002), representing a wide range of different lunar rock types (for a description of lunar rock types, see Papike et al., 1998). Specifically, three of the lunar meteorites are *unbrecciated, crystalline mare basalts*, three are *mare basalt breccias* (Figure 31), four are *mixed mare/highlands breccias*, 13 or 14 are *highlands regolith breccias*, and two or three are *highlands impact-melt breccias*. Note that the mare basalts and breccias consist mostly of pyroxene, plagioclase, and olivine, with minor or accessory amounts of ilmenite, chromite, troilite, and traces of metallic iron, whereas the highlands rocks are pronounced anorthositic and are dominated by plagioclase of anorthite composition.

ACKNOWLEDGMENTS

The authors thank G. Benedix, B. Cowen, M. Killgore, R. Korotev, G. McKay, T. McCoy, D. Mittlefehldt, M. Petaev, A. Treiman, M. Wadhwa, and M. Zolensky for providing images and thin sections. This work was supported in part by NASA Grants NAG5-10610 (A. Krot, P.I.), NAG 5-11591 (K. Keil, P.I.), and NAG5-11546 (M. Weisberg, P.I.). This is Hawai'i Institute of Geophysics and Planetology Publication No. HIGP 1299 and School of Ocean and Earth Science and Technology Publication No. SOEST 6216.

REFERENCES

Akai J. (1988) Incompletely transformed serpentine-type phyllosilicates in the matrix of Antarctic CM chondrites. *Geochim. Cosmochim. Acta* **52**, 1593–1599.

Aléon J., Krot A. N., and McKeegan K. D. (2002) Ca–Al-rich inclusions and amoeboid olivine aggregates from the CR carbonaceous chondrites. *Meteorit. Planet. Sci.* **37**, 1729–1755.

Alexander C. M. O., Hutchison R., and Barber D. J. (1989) Origin of chondrule rims and interchondrule matrices in unequilibrated ordinary chondrites. *Earth Planet. Sci.* **95**, 187–207.

Ashwal L. D., Warner J. L., and Wood C. A. (1982) SNC meteorites: evidence against an asteroidal origin. *Proc. Lunar Planet. Sci. Conf.* **13**(suppl.) *J. Geophys. Rev.* **87**, A393–A400.

Bence A. E. and Burnett D. S. (1969) Chemistry and mineralogy of the silicates and metal of the Kodaikanal meteorite. *Geochim. Cosmochim. Acta* **33**, 387–407.

Benedix G. K., McCoy T. J., Keil K., Bogard D. D., and Garrison D. H. (1998) A petrologic and isotopic study of winonaites: evidence for early partial melting, brecciation, and metamorphism. *Geochim. Cosmochim. Acta* **62**, 2535–2553.

Benedix G. K., McCoy T. J., Keil K., and Love S. G. (2000) A petrologic study of the IAB iron meteorites: constraints on the formation of the IAB-winonaite parent body. *Meteorit. Planet. Sci.* **35**, 1127–1141.

Berkley J. L. and Goodrich C. A. (2001) Evidence for multi-episodic igneous events in ureilite MET 78008. *Meteorit. Planet. Sci.* **36**, A18–A19.

Berkley J. L., Keil K., and Prinz M. (1980) Comparative petrology and origin of Governador Valadares and other nakhlites. *Proc. Lunar. Planet. Sci. Conf.* **11**, 1089–1102.

Bild R. W. and Wasson J. T. (1977) Netchaëvo: a new class of chondritic meteorite. *Science* **197**, 58–62.

Binzel R. P. and Xu S. (1993) Chips off of asteroid 4 Vesta: evidence for the parent body of basaltic achondrite meteorites. *Science* **260**, 186–191.

Bischoff A. (2000) Mineralogical characterization of primitive, type-3 lithologies in Rumuruti chondrites. *Meteorit. Planet. Sci.* **35**, 699–706.

Bischoff A. and Metzler K. (1991) Mineralogy and petrography of the anomalous carbonaceous chondrites Yamato-86720, Yamato-82162, and Belgica-7904. *Proc. NIPR Symp. Antarct. Meteorit.* **4**, 226–246.

Bischoff A., Palme H., Ash R. D., Clayton R. N., Schultz L., Herpers U., Stöffler D., Grady M. M., Pillinger C. T., Spettel B., Weber H., Grund T., Endreβ M., and Weber D. (1993a) Paired Renazzo-type (CR) carbonaceous chondrites from the Sahara. *Geochim. Cosmochim. Acta* **57**, 1587–1603.

Bischoff A., Palme H., Schultz L., Weber D., Weber H., and Spettel B. (1993b) Acfer 182 and paired samples, an iron-rich carbonaceous chondrite: similarities with ALH85085 and relationship to CR chondrites. *Geochim. Cosmochim. Acta* **57**, 2631–2648.

Bischoff A., Geiger T., Palme H., Spettel B., Schultz L., Scherer P., Loeken T., Bland P., Clayton R. N., Mayeda T. K., Herpers U., Meltzow B., Michel R., and Dittrich-Hannen B. (1994) Acfer 217: a new member of the Rumuruti chondrite group (R). *Meteoritics* **29**, 264–274.

Boesenberg J. S., Davis A. M., Prinz M., Weisberg M. K., Clayton R. N., and Mayeda T. K. (2000) The pyroxene pallasites Vermillion and Yamato 8451: not quite a couple. *Meteoritics Planet. Sci.* **35**, 757–769.

Bogard D. D., Garrison D. H., and McCoy T. J. (2000) Chronology and petrology of silicates from IIE iron meteorites: evidence of a complex parent body evolution. *Geochim. Cosmochim. Acta* **64**, 2133–2154.

Bourot-Denise M., Zanda B., and Javoy M. (2002) Tafassesset: an equilibrated CR chondrite. In *Lunar Planet. Sci.* **XXXIII**, #1611. The Lunar and Planetary Institute, Houston (CD-ROM).

Bowman L. E., Spilde M. N., and Papike J. J. (1997) Automated energy dispersive spectrometer modal analysis applied to diogenites. *Meteorit. Planet. Sci.* **32**, 869–875.

Brearley A. J. (1989) Nature and origin of matrix in the unique type 3 chondrite, Kakangari. *Geochim. Cosmochim. Acta* **53**, 197–214.

Brearley A. J. (1991) Mineralogical and chemical studies of matrix in the Adelaide meteorite, a unique carbonaceous chondrite with affinities to ALH A77307 (CO3). In *Lunar Planet. Sci.* **XXII**. The Lunar and Planetary Institute, Houston, pp. 133–134.

Brearley A. J. (1997) Phyllosilicates in the matrix of the unique carbonaceous chondrite, LEW 85332 and possible implications for the aqueous alteration of CI chondrites. *Meteorit. Planet. Sci.* **32**, 377–388.

Brearley A. J. and Jones R. H. (1998) Chondritic meteorites. In *Planetary Materials*, Reviews in Mineralogy (ed. J. J. Papike). Mineralogical Society of America, Washington, DC, vol. 36, chap. 3, pp. 3-1–3-398.

Brown P. G., Hildebrand A. R., Zolensky M. E., Grady M., Clayton R. N., Mayeda T. K., Tagliaferri E., Spalding R., MacRae N. D., Hoffman E. L., Mittlefehldt D. W., Wacker J. F., Bird J. A., Campbell M. D., Carpenter R., Gingerich H., Glatiotis M., Greiner E., Mazur M. J., McCausland P. J., Plotkin H., and Mazur T. R. (2000) The fall, recovery, orbit, and composition of the Tagish Lake meteorite: a new type of carbonaceous chondrite. *Science* **290**, 320–325.

Browning L. B., McSween H. Y. Jr., and Zolensky M. E. (1996) Correlated alteration effects in CM carbonaceous chondrites. *Geochim. Cosmochim. Acta* **60**, 2621–2633.

Brownlee D. E., Tomandl D., and Hodge P. W. (1976) Extraterrestrial particles in the stratosphere. In *Interplanetary Dust and the Zodiacal Light* (eds. H. Elsasser and H. Fechtig). Springer, New York, pp. 279–284.

Bunch T. E. and Olsen E. (1968) Potassium feldspar in Weekeroo Station, Kodaikanal, and Colomera iron meteorites. *Science* **160**, 1223–1225.

Bunch T. E., Keil K., and Olsen E. (1970) Mineralogy and petrology of silicate inclusions in iron meteorites. *Contrib. Mineral. Petrol.* **25**, 297–340.

Burbine T. H., McCoy T. J., Meibom A., Gladman B., and Keil K. (2002) Meteoritic parent bodies: their number and identification. In *Asteroids III* (eds. W. Bottke, A. Cellino, P. Paolicchi, and R. P. Binzel). University of Arizona Press, Tucson, pp. 653–669.

Buseck P. R. (1977) Pallasite meteorites mineralogy, petrology and geochemistry. *Geochim. Cosmochim. Acta* **41**, 711–740.

Casanova I., Keil K., and Newsom H. E. (1993) Composition of metal in aubrites: constraints on core formation. *Geochim. Cosmochim. Acta* **57**, 675–682.

Casanova I., Graf T., and Marti K. (1995) Discovery of an unmelted H-chondrite inclusion in iron meteorite. *Science* **268**, 540–542.

Choi B.-G., Quyang X., and Wasson J. T. (1995) Classification and origin of IAB and IIICD iron meteorites. *Geochim. Cosmochim. Acta* **59**, 593–612.

Clayton R. N. and Mayeda T. K. (1996) Oxygen-isotope studies of achondrites. *Geochim. Cosmochim. Acta* **60**, 1999–2018.

Clayton R. N. and Mayeda T. K. (1999) Oxygen isotope studies of carbonaceous chondrites. *Geochim. Cosmochim. Acta* **63**, 2089–2104.

Clayton R. N. and Mayeda T. K. (2001) Oxygen isotope composition of the Tagish Lake carbonaceous chondrite. In *Lunar Planet. Sci.* **XXXII**, #1885. The Lunar and Planetary Institute, Houston, (CD-ROM).

Clayton R. N., Mayeda T. K., and Rubin A. E. (1984) Oxygen isotopic compositions of enstatite chondrites and aubrites. *Proc. 15th Lunar Planet. Sci. Part 1: J. Geophys. Res.* **B89**, C245–C249.

Crabb J. and Schultz L. (1981) Cosmic-ray exposure ages of the ordinary chondrites and their significance for parent body stratigraphy. *Geochim. Cosmochim. Acta* **45**, 2151–2160.

Davis A. M. and Olsen E. J. (1991) Phosphates in pallasite meteorites as probes of mantle processes in small planetary bodies. *Nature* **353**, 637–640.

Davis A. M., Grossman L., and Ganapathy R. (1977) Yes, Kakangari is a unique chondrite. *Nature* **265**, 230–232.

Davy R., Whitehead S. G., and Pitt G. (1978) The Adelaide meteorite. *Meteoritics* **13**, 121–139.

Drake M. J. (2001) The eucrite/Vesta story. *Meteoritics Planet. Sci.* **36**, 501–513.

Endress M. and Bischoff A. (1993) Mineralogy, degree of brecciation, and aqueous alteration of CI chondrites Orgueil, Ivuna, and Alais. *Meteoritics* **28**, 345–346.

Endress M. and Bischoff A. (1996) Carbonates in CI chondrites: clues to parent body evolution. *Geochim. Cosmochim. Acta* **60**, 489–507.

Endress M., Zinner E. K., and Bischoff A. (1996) Early aqueous activity on primitive meteorite parent bodies. *Nature* **379**, 701–703.

Fagan T. J., McKeegan K. D., Krot A. N., and Keil K. (2001) Calcium, aluminum-rich inclusions in enstatite chondrites (2): oxygen isotopes. *Meteorit. Planet. Sci.* **36**, 223–230.

Fitzgerald M. J. and Jones J. B. (1977) Adelaide and Bench Crater—members of a new subgroup of the carbonaceous chondrites. *Meteoritics* **12**, 443–458.

Floran R. J. (1978) Silicate petrography, classification, and origin of the mesosiderites: review and new observations. *Proc. Lunar Planet. Sci. Conf.* **9**, 1053–1081.

Floran R. J., Caulfield J. B. D., Harlow G. E., and Prinz M. (1978) Impact origin for the Simondium, Pinnaroo, and Hainholz mesosiderites: implications for impact processes beyond the Earth–Moon system. *Proc. Lunar Planet. Sci. Conf.* **9**, 1083–1114.

Fredriksson K. and Keil K. (1963) The light-dark structure in the Pantar and Kapoeta stone meteorites. *Geochim. Cosmochim. Acta* **27**, 717–739.

Goodrich C. A. (1992) Ureilites: a critical review. *Meteoritics* **27**, 327–352.

Goodrich C. A. and Righter K. (2000) Petrology of unique achondrite Queen Alexandra Range 93148: a piece of the pallasite (howardite-eucrite-diogenite?) parent body? *Meteorit. Planet. Sci.* **35**, 521–535.

Goodrich C. A., Fioretti A. M., Tribaudino M., and Molin G. (2001) Primary trapped melt inclusions in olivine in the olivine-augite-orthopyroxene ureilite Hughes 009. *Geochim. Cosmochim. Acta* **65**, 621–652.

Goodrich C. A., Krot A. N., Scott E. R. D., Taylor G. J., Fioretti A. M., and Keil K. (2002) Origin and evolution of the ureilite parent body and its offspring. In *Lunar Planet Sci.* **XXXIII**, #1379. The Lunar and Planetary Institute, Houston (CD-ROM).

Gounelle M. and Zolensky M. E. (2001) A terrestrial origin for sulfate veins in CI1 chondrites. *Meteorit. Planet. Sci.* **36**, 1321–1329.

Grady M. M. (2000) *Catalogue of Meteorites*. Cambridge University Press, London, 689 p.

Grady M. M. and Pillinger C. T. (1986) Carbon isotope relationships in winonaites and forsterite chondrites. *Geochim. Cosmochim. Acta* **50**, 255–263.

Grady M. M. and Pillinger C. T. (1990) ALH85085: nitrogen isotope analysis of a highly unusual primitive chondrite. *Earth Planet. Sci. Lett.* **97**, 29–40.

Grady M. M., Verchovsky A. B., Franchi I. A., Wright I. P., and Pillinger C. T. (2002) Light element geochemistry of the Tagish Lake CI2 chondrite: comparison with CI1 and CM2 meteorites. *Meteorit. Planet. Sci.* **37**, 713–735.

Greshake A. (1997) The primitive matrix components of the unique carbonaceous chondrite Acfer 094: a TEM study. *Geochim. Cosmochim. Acta* **61**, 437–452.

Greshake A., Krot A. N., Meibom A., Weisberg M. K., and Keil K. (2002) Heavily-hydrated matrix lumps in the CH and metal-rich chondrites QUE 94411 and Hammadah al Hamra 237. *Meteorit. Planet. Sci.* **37**, 281–294.

Grossman J. N. (1994) The Meteoritical Bulletin, No. 76: the US Antarctic Meteorite Collection. *Meteoritics* **29**, 100–143.

Grossman J. N. and Zipfel J. (2001) The Meteoritical Bulletin, No. 85. *Meteorit. Planet. Sci.* **36**, A293–A322.

Grossman J. N., Rubin A. E., and MacPherson G. J. (1988) ALH85085: a unique volatile-poor carbonaceous chondrite with possible implications for nebular fractionation processes. *Earth Planet. Sci. Lett.* **91**, 33–54.

Grossman J. N., MacPherson G. J., and Crozaz G. (1993) LEW 87223: a unique E chondrite with possible links to H chondrites. *Meteoritics* **28**, 358.

Guan Y., Huss G. R., MacPherson G. J., and Wasserburg G. J. (2000) Calcium–aluminum-rich inclusions from enstatite chondrites: indigenous or foreign? *Science* **289**, 1330–1333.

Guimon R. K., Symes S. J. K., Sears D. W. G., and Benoit P. H. (1995) Chemical and physical studies of type 3 chondrites XII: the metamorphic history of CV chondrites and their components. *Meteorit. Planet. Sci.* **30**, 704–714.

Hashizume K. and Sugiura N. (1995) Nitrogen isotopes in bulk ordinary chondrites. *Geochim. Cosmochim. Acta* **59**, 4057–4070.

Haack H., Scott E. R. D., Love S. G., Brearley A. J., and McCoy T. J. (1996) Thermal histories of IVA stony-iron and iron meteorites: evidence for asteroid fragmentation and reacrretion. *Geochim. Cosmochim. Acta* **60**, 3103–3113.

Hewins R. H. (1983) Impact versus internal origins for mesosiderites. *Proc. 14th Lunar Planet. Sci. Conf.: Part I. J. Geophys. Res.* **88**, B257–B266.

Hewins R. H. (1984) The case for a melt matrix in plagioclase-POIK mesosiderites. *Proc. 15th Lunar Planet. Sci. Conf.: Part I. J. Geophys. Res.* **89**, C289–C297.

Hewins R. H. (1988) Petrology and pairing of mesosiderites from Victoria Land, Antarctica. *Meteoritics* **23**, 123–129.

Heymann D. (1967) On the origin of hypersthene chondrites: ages and shock effects of black meteorites. *Icarus* **6**, 189–221.

Hiroi T., Pieters C. M., Zolensky M. E., and Lipschutz M. E. (1993a) Evidence of thermal metamorphism on the C, G, B, and F asteroids. *Science* **261**, 1016–1018.

Hiroi T., Bell J. F., Takeda H., and Pieters C. M. (1993b) Spectral comparison between olivine-rich asteroids and pallasites. *Proc. NIPR Symp. Antarct. Meteorit.* **6**, 234–245.

Hiroi T., Pieters C. M., Zolensky M. E., and Prinz M. (1996) Reflectance spectra (UV-3 μm) of heated Ivuna (CI) meteorite and newly identified thermally metamorphosed CM chondrites. In *Lunar Planet. Sci.* **XXVII**. The Lunar and Planetary Institute, Houston, pp. 551–552.

Hua X. and Buseck P. R. (1995) Fayalite in the Kaba and Mokoia carbonaceous chondrites. *Geochim. Cosmochim. Acta* **59**, 563–587.

Huss G. R. and Hutcheon I. D. (1992) Abundant ^{26}Mg* in Adelaide refractory inclusions. *Meteoritics* **27**, 236.

Hutcheon I. D. and Steele I. M. (1982) Refractory inclusions in the Adelaide carbonaceous chondrite. In *Lunar Planet. Sci.* **XXIII**. The Lunar and Planetary Institute, Houston, pp. 352–353.

Hutchison R., Alexander C. M. O., and Barber D. J. (1987) The Semarkona meteorite: first recorded occurrence of smectite in an ordinary chondrite, and its implications. *Geochim. Cosmochim. Acta* **31**, 1103–1106.

Hutchison R., Alexander C. M. O'D., and Bridges J. C. (1998) Elemental redistribution in Tieschitz and the origin of white matrix. *Meteorit. Planet. Sci.* **33**, 1169–1179.

Ikeda Y. (1992) An overview of the research consortium, "Antarctic carbonaceous chondrites with CI affinities, Yamato-86720, Yamato-82162, and Belgica-7904". *Proc. NIPR Symp. Antarct. Meteorit.* **5**, 49–73.

Ikeda Y. (1997) Petrology of the Yamato 793695 lherzolitic shergottite. *Proc. NIPR Symp. Antarct. Meteorit.* **7**, 9–29.

Ikeda Y. and Prinz M. (1993) Petrologic study of the Belgica 7904 carbonaceous chondrite: hydrous alteration, thermal metamorphism, and relationship to CM and CI chondrites. *Geochim. Cosmochim. Acta* **57**, 439–452.

Ikeda Y. and Prinz M. (1996) Petrology of silcate inclusions in the Miles IIE iron. *Proc. NIPR Symp. Antarct. Meteorit.* **9**, 143–173.

Ikeda Y. and Prinz M. (2000) Magmatic inclusions and felsic clasts in the Dar al Gani 319 polymict ureilite. *Meteorit. Planet. Sci.* **36**, 481–500.

Ikeda Y., Ebihara M., and Prinz M. (1997) Petrology and chemistry of the Miles IIE iron: I. Description and petrology of twenty new silicate inclusions. *Antarct. Meteorit. Res.* **10**, 355–372.

Ikeda Y., Prinz M., and Nehru C. E. (2000) Lithic and mineral clasts in the Dar al Gani (DAG) 319 polymict ureilite. *Antarct. Meteorit. Res.* **13**, 177–221.

Ivanov A. V. (1989) The meteorite Kaidun: composition and history of formation. *Geokhimiya* **2**, 259–266.

Ivanova M. A., Taylor L. A., Clayton R. N., Mayeda T. K., Nazarov M. A., Brandstätter F., and Kurat G. (2002) Dhofar 225 vs. the CM clan: metamorphosed or new type of carbonaceous chondrite? In *Lunar Planet. Sci.* **XXXIII**, #1437. The Lunar and Planetary Institute, Houston (CD-ROM).

Jones J. H. (1986) A discussion of isotopic systematics and mineral zoning in the shergottites: evidence for a 180 m. y. igneous crystallization age. *Geochim. Cosmochim. Acta* **50**, 969–977.

Kallemeyn G. W. (1996) The classificational wanderings of the Ningqiang chondrite. In *Lunar Planet. Sci.* **XXVII**. The Lunar and Planetary Institute, Houston, pp. 635–636.

Kallemeyn G. W. and Rubin A. E. (1995) Coolidge and Loongana 001: a new carbonaceous chondrite grouplet. *Meteoritics* **30**, 20–27.

Kallemeyn G. W. and Wasson J. T. (1981) The compositional classification of chondrites: I. The carbonaceous chondrite groups. *Geochim. Cosmochim. Acta* **45**, 1217–1230.

Kallemeyn G. W. and Wasson J. T. (1982) The compositional classification of chondrites: III. Ungrouped carbonaceous chondrites. *Geochim. Cosmochim. Acta* **46**, 2217–2228.

Kallemeyn G. W. and Wasson J. T. (1985) The compositional classification of chondrites: IV. Ungrouped chondritic meteorites and clasts. *Geochim. Cosmochim. Acta* **49**, 261–270.

Kallemeyn G. W., Boynton W. V., Willis J., and Wasson J. T. (1978) Formation of the Bencubbin polymict meteoritic breccia. *Geochim. Cosmochim. Acta* **42**, 507–515.

Kallemeyn G. W., Rubin A. E., Wang D., and Wasson J. T. (1989) Ordinary chondrites: bulk compositions, classification, lithophile-element fractionations, and composition-petrographic type relationships. *Geochim. Cosmochim. Acta* **53**, 2747–2767.

Kallemeyn G. W., Rubin A. E., and Wasson J. T. (1991) The compositional classification of chondrites: V. The Karoonda (CK) group of carbonaceous chondrites. *Geochim. Cosmochim. Acta* **55**, 881–892.

Kallemeyn G. W., Rubin A. E., and Wasson J. T. (1994) The compositional classification of chondrites: VI. The CR carbonaceous chondrite group. *Geochim. Cosmochim. Acta* **58**, 2873–2888.

Kallemeyn G. W., Rubin A. E., and Wasson J. T. (1996) The compositional classification of chondrites: VII. The R chondrite group. *Geochim. Cosmochim. Acta* **60**, 2243–2256.

Keil K. (1964) Possible correlation between classifications and potassium–argon ages of chondrites. *Nature* **203**, 511.

Keil K. (1968) Mineralogical and chemical relationships among enstatite chondrites. *J. Geophys. Res.* **73**, 6945–6976.

Keil K. (1982) Composition and origin of chondritic breccias. In *Workshop on Lunar Breccias and Soils and their Meteoritic Analogs*, LPI Technical Report 82-02 (eds. G. J. Taylor and L. L. Wilkening). The Lunar and Planetary Institute, Houston, pp. 65–83.

Keil K. (1989) Enstatite meteorites and their parent bodies. *Meteoritics* **24**, 195–208.

Keil K. (2002) Geological history of asteroid 4 Vesta: The smallest terrestrial planet. In *Asteroids III* (eds. W. F. Bottke Jr., A. Cellino, and P. Paolicchi, R. P. Binzel). University of Arizona Press, Tucson, pp. 573–589.

Keil K. and Brett R. (1974) Heideite (Fe, Cr)$_{1+x}$(Ti, Fe)$_2$S$_4$, a new mineral in the Bustee enstatite achondrite. *Am. Mineral.* **59**, 465–470.

Keil K., Ntaflos Th., Taylor G. J., Brearley A. J., Newsom H. E., and Romig A. D. Jr. (1989) The Shallowater aubrite: evidence for origin by planetesimal impacts. *Geochim. Cosmochim. Acta* **53**, 3291–3307.

Keller L. P. and Buseck P. R. (1990a) Matrix mineralogy of the Lance CO3 carbonaceous chondrite. *Geochim. Cosmochim. Acta* **54**, 1155–1163.

Keller L. P. and Buseck P. R. (1990b) Aqueous alteration in the Kaba CV3 carbonaceous chondrite. *Geochim. Cosmochim. Acta* **54**, 2113–2120.

Keller L. P. and McKay D. S. (1993) Aqueous alteration of the Grosnaja CV3 carbonaceous chondrite. *Meteoritics* **28**, 378.

Keller L. P., Thomas K. L., Clayton R. N., Mayeda T. K., DeHart J. M., and McKay D. S. (1994) Aqueous alteration of the Bali CV3 chondrite: evidence from mineralogy, mineral chemistry, and oxygen isotopic compositions. *Geochim. Cosmochim. Acta* **58**, 5589–5598.

Kerridge J. F. (1985) Carbon, hydrogen, and nitrogen in carbonaceous chondrites: abundances and isotopic compositions in bulk samples. *Geochim. Cosmochim. Acta* **49**, 1707–1714.

Kimura M. and Ikeda Y. (1998) Hydrous and anhydrous alterations of chondrules in Kaba and Mokoia CV chondrites. *Meteorit. Planet. Sci.* **33**, 1139–1146.

Kimura M., Tsuchiayama A., Fukuoka T., and Iimura Y. (1992) Antarctic primitive achondrites, Yamato-74025, -75300, and -75305: their mineralogy, thermal history, and the relevance to winonaites. *Proc. NIPR Symp. Antarct. Meteorit.* **5**, 165–190.

Kita N. T., Liu Y. Z., Ikeda Y., Prinz M., and Morishita Y. (2000) Identification of a variety of clasts in the Dar al Gani 319 polymict ureilite using secondary ion mass spectrometer oxygen-isotopic analyses. *Meteorit. Planet. Sci.* **35**, A88–A89.

Koeberl C., Ntaflos T., Kurat G., and Chai C. F. (1987) Petrology and Geochemistry of the Ningqiang (CV3) chondrite. In *Lunar Planet. Sci.* **XVIII**. The Lunar and Planetary Institute, Houston, pp. 499–500.

Korotev R. L. (2002) Lunar meteorites. Website, http://epsc.wustl.edu/admin/resources/moon_meteorites.html.

Kracher A. (1982) Crystallization of a S-saturated Fe, Ni-melt, and the origin of the iron meteorite groups IAB and IIICD. *Geophys. Res. Lett.* **9**, 412–415.

Kracher A., Willis J., and Wasson J. T. (1980) Chemical classification of iron meteorites: IX. A new group (IIF), revision of IAB and IIICD, and data on 57 additional irons. *Geochim. Cosmochim. Acta* **44**, 773–787.

Krot A. N., Scott E. R. D., and Zolensky M. E. (1995) Mineralogic and chemical variations among CV3 chondrites and their components: nebular and asteroidal processing. *Meteoiritics* **30**, 748–775.

Krot A. N., Petaev M. I., Scott E. R. D., Choi B.-G., Zolensky M. E., and Keil K. (1998) Progressive alteration in CV3 chondrites: more evidence for asteroidal alteration. *Meteor. Planet. Sci.* **33**, 1065–1085.

Krot A. N., Fegley B., Palme H., and Lodders K. (2000a) Meteoritical and astrophysical constraints on the oxidation state of the solar nebula. In *Protostars and Planets IV* (eds. A. P. Boss, V. Manning, and S. S. Russell). University of Arizona Press, Tucson, pp. 1019–1055.

Krot A. N., Weisberg M. K., Petaev M. I., Keil K., and Scott E. R. D. (2000b) High-temperature condensation signatures in Type I chondrules from CR carbonaceous chondrites. In *Lunar Planet. Sci.* **XXXI**, #1470. The Lunar and Planetary Institute, Houston (CD-ROM).

Krot A. N., Meibom A., and Keil K. (2000c) Volatile-poor chondrules in CH carbonaceous chondrites: formation at high ambient nebular temperature. In *Lunar Planet. Sci.* **XXXI**, #1481. The Lunar and Planetary Institute, Houston (CD-ROM).

Krot A. N., Brearley A. J., Petaev M. I., Kallemeyn G. W., Sears D. W. G., Benoit P. H., Hutcheon I. D., Zolensky M. E., and Keil K. (2000d) Evidence for *in situ* growth of fayalite and hedenbergite in MacAlpine Hills 88107, ungrouped carbonaceous chondrite related to CM–CO clan. *Meteorit. Planet. Sci.* **35**, 1365–1387.

Krot A. N., McKeegan K. D., Russell S. S., Meibom A., Weisberg M. K., Zipfel J., Krot T. V., Fagan T. J., and Keil K. (2001a) Refractory Ca, Al-rich inclusions and

Al-diopside-rich chondrules in the metal-rich chondrites Hammadah al Hamra 237 and QUE 94411. *Meteorit. Planet. Sci.* **36**, 1189–1217.

Krot A. N., Huss G. R., and Hutcheon I. D. (2001b) Corundum-hibonite refractory inclusions from Adelaide: condensation or crystallization from melt? *Meteorit. Planet. Sci.* **36**, A105.

Krot A. N., Hutcheon I. D., and Huss G. R. (2001c) Aluminum-rich chondrules and associated refractory inclusions in the unique carbonaceous chondrite Adelaide. *Meteorit. Planet. Sci.* **36**, A105–A106.

Krot A. N., Aléon J., and McKeegan K. D. (2002) Mineralogy, petrography and oxygen-isotopic compositions of Ca, Al-rich inclusions and amoeboid olivine aggregates in the CR carbonaceous chondrites. In *Lunar Planet. Sci.* **XXXIII**, #1412. The Lunar and Planetary Institute, Houston (CD-ROM).

Kung C. C. and Clayton R. N. (1978) Nitrogen abundances and isotopic compositions in stony meteorites. *Earth Planet. Sci. Lett.* **38**, 421–435.

Leshin L. A., Rubin A. E., and McKeegan K. D. (1997) The oxygen isotopic composition of olivine and pyroxene from CI chondrites. *Geochim. Cosmochim. Acta* **61**, 835–845.

Lin Y. and El Goresy A. (2002) A compartive study of opaque phases in Qingzhen (EH3) and MacAlpine Hills 88136 (EL3): representatives of EH and EL parent bodies. *Meteorit. Planet. Sci.* **37**, 577–599.

Liu M., Scott E. R. D., Keil K., Wasson J. T., Clayton R. N., Mayeda T. K., Eugster O., Crozaz G., and Floss C. (2001) Northwest Africa 176, a unique iron meteorite with silicate inclusions related to Bocaiuva. In *Lunar Planet. Sci.* **XXXII**, #2152. The Lunar and Planetary Institute, Houston (CD-ROM).

Lovering J. F. (1962) The evolution of the meteorites—evidence for the co-existence of chondritic, achondritic, and iron meteorites in a typical parent meteorite body. In *Researches on Meteorites* (ed. C. B. Moore). Wiley, New York, pp. 179–197.

Lovering J. F., Nichiporuk W., Chodos A., and Brown H. (1957) The distribution of gallium, germanium, cobalt, chromium, and copper in iron and stony-iron meteorites in relation to nickel content and structure. *Geochim. Cosmochim. Acta* **11**, 263–278.

MacKinnon I. D. R. and Rietmeijer F. J. M. (1987) Mineralogy of chondritic interplanetary dust particles. *Rev. Geophys.* **25**, 1527–1553.

MacPherson G. J., Wark D. A., and Armstrong J. T. (1988) Primitive material surviving in chondrites: refractory inclusions. In *Meteorites and the Early Solar System* (eds. J. F. Kerridge and M. S. Matthews). University of Arizona Press, Tucson, pp. 746–807.

Malvin D. J., Wang D., and Wasson J. T. (1984) Chemical classification of iron meteorites: X. Multielement studies of 43 irons, resolution of group IIIE from IIIAB, and evaluation of Cu as a taxonomic parameter. *Geochim. Cosmcohim. Acta* **48**, 785–804.

Malvin D. J., Wasson J. T., Clayton R. N., Mayeda T. K., and Curvello W. S. (1985) Bocaiuva—a silicate-inclusion bearing iron meteorite related to the Eagle-Station pallasites. *Meteortics* **20**, 259–273.

Marvin U. B., Petaev M. I., Croft W. J., and Kilgore M. (1997) Silica minerals in the Gibeon IVA iron meteorite. In *Lunar Planet. Sci.* **XXVIII**. The Lunar and Planetary Institute, Houston, pp. 879–880.

Mason B. and Nelen J. (1968) The Weatherford meteorite. *Geochim. Cosmochim. Acta* **32**, 661–664.

Mayeda K., Clayton R. N., Krung D. A., and Davis M. (1988) Oxygen, silicon and magnesium isotopes in Ningqiang chondrules. *Meteoritics* **23**, 288.

McCoy T. J. (1995) Silicate-bearing IIE irons: early mixing and differentiation in a core-mantle environment and shock resetting of ages. *Meteoritics* **30**, 542–543.

McCoy T. J., Keil K., and Clayton R. N. (1992a) Petrogenesis of the lodranite–acapulcoite parent body. *Meteoritics* **27**, 258–259.

McCoy T. J., Taylor G. J., and Keil K. (1992b) Zagami: product of a two-stage magmatic history. *Geochim. Cosmochim. Acta* **56**, 3571–3582.

McCoy T. J., Scott E. R. D., and Haack H. (1993) Genesis of the IIICD iron meteorites: evidence from silicate-bearing inclusions. *Meteoritics* **28**, 552–560.

McCoy T. J., Keil K., Scott E. R. D., Benedix G. K., Ehlmann A. J., Mayeda T. K., and Clayton R. N. (1994) Low FeO ordinary chondrites: a nebular origin and a new chondrite parent body. In *Lunar Planet. Sci.* **XXV**. The Lunar and Planetary Institute, Houston, pp. 865–866.

McCoy T. J., Keil K., Bogard D. D., Garrison D. H., Casanova I., Lindstrom M. M., Brearley A. J., Kehm K., Nichols R. H. Jr., and Hohenberg C. M. (1995) Origin and history of impact-melt rocks of enstatite chondrite parentage. *Geochim. Cosmochim. Acta* **59**, 161–175.

McCoy T. J., Keil K., Clayton R. N., Mayeda T. K., Bogard D. D., Garrison D. H., Huss G. R., Hutcheon I. D., and Wieler R. (1996) A petrologic, chemical and isotopic study of Monument Draw and comparison with other acapulcoites: evidence for formation by incipient partial melting. *Geochim. Cosmochim. Acta* **60**, 2681–2708.

McCoy T. J., Keil K., Clayton R. N., Mayeda T. K., Bogard D. D., Garrison D. H., and Wieler R. (1997a) A petrologic and isotopic study of lodranites: evidence for early formation as partial melt residues from heterogeneous precursors. *Geochim. Cosmochim. Acta* **61**, 623–637.

McCoy T. J., Keil K., Muenow D. W., and Wilson L. (1997b) Partial melting and melt migration in the acapulcoite–lodranite parent body. *Geochim. Cosmochim. Acta* **61**, 639–650.

McCord T. B., Adams J. B., and Johnson T. V. (1970) Asteroid Vesta: spectral reflectivity and compositional implications. *Science* **168**, 1445–1447.

McKay D. S., Gibson E. K., Thomas-Keprta K. L., Vali H., Romanek C. S., Clement S. J., Chillier X. D. F., Maechling C. R., and Zare R. N. (1996) Search for past life on Mars: possible relic biogenic activity in martian meteorite ALH84001. *Science* **273**, 924–930.

McKay G., Lindstrom D., Yang S.-R., and Wagstaff J. (1988) Petrology of a unique achondrite Lewis Cliff 87051. In *Lunar Planet. Sci.* **XIX**. The Lunar and Planetary Institute, Houston, pp. 762–763.

McKay G., Crozaz G., Wagstaff J., Yang S.-R., and Lundberg L. (1990) A petrographic, electron microprobe, and ion microprobe study of mini-angrite Lewis Cliff 86010. In *Lunar Planet. Sci.* **XXI**. The Lunar and Planetary Institute, Houston, pp. 771–772.

McKay G., Koizumi E., Mikouchi T., Le L., and Schwandt C. (2002) Crystallization of shergottite QUE 94201: an experimental study. In *Lunar Planet. Sci.* **XXXIII**, #2051. The Lunar and Planetary Institute, Houston (CD-ROM).

McKeegan K. D., Sahijpal S., Krot A. N., Weber D., and Ulyanov A. A. (2003) Preservation of primary oxygen isotopic compositions in Ca, Al-rich inclusions from CH chondrites. *Earth Planet. Sci. Lett.* (submitted).

McSween H. Y. Jr. (1977a) Carbonaceous chondrites of the Ornans type: a metamorphic sequence. *Geochim. Cosmochim. Acta* **41**, 479–491.

McSween H. Y. Jr. (1977b) Petrographic variations among carbonaceous chondrites of the Vigarano type. *Geochim. Cosmochim. Acta* **41**, 1777–1790.

McSween H. Y. Jr. (1979) Are carbonaceous chondrites primitive or processed? A review. *Rev. Geophys. Space Phys.* **17**, 1059–1078.

McSween H. Y. Jr. (2002) The rocks of Mars, from far and near. *Meteorit. Planet. Sci.* **37**, 7–25.

McSween H. Y. Jr. and Richardson S. M. (1977) The composition of carbonaceous chondrite matrix. *Geochim. Cosmochim. Acta* **41**, 1145–1161.

McSween H. Y. Jr. and Treiman A. H. (1998) Martian meteorites. In *Planetary Materials,* Reviews in Mineralogy (ed. J. J. Papike). Mineralogical Society of America, Washington, DC, vol. 36, chap. 6, pp. 6-1–6-53.

McSween H. Y. Jr., Taylor L. A., and Stolper E. M. (1979a) Allan Hills 77005: a new meteorite type found in Antarctica. *Science* **204**, 1201–1203.

McSween H. Y. Jr., Stolper E. M., Taylor L. A., Muntean R. A., O'Kelly G. D., Eldridge J. S., Biswas S., Ngo H. T., and Lipschutz M. E. (1979b) Petrogenetic relationship between Allan Hills 77005 and other achondrites. *Earth Planet. Sci. Lett.* **45**, 275–284.

McSween H. Y. Jr., Eisenhour D. D., Taylor L. A., Wadhwa M., and Crozaz G. (1996) QUE94201 shergotite: crystallization of a martian basaltic magma. *Geochim. Cosmochim. Acta* **60**, 4563–4569.

Meibom A., Petaev M. I., Krot A. N., Wood J. A., and Keil K. (1999) Primitive FeNi metal grains in CH carbonaceous chondrites formed by condensation from a gas of solar composition. *J. Geophys. Res.* **104**, 22053–22059.

Metzler K., Bobe K.-D., Palme H., Spettel B., and Stöffler D. (1995) Thermal and impact metamorphism of the HED parent body. *Planet. Space Sci.* **43**, 499–525.

Mittlefehldt D. W. (1994) ALH84001, a cumulate orthopyroxenite member of the martian meteorite clan. *Meteoritics* **29**, 214–221.

Mittlefehldt D. W. (2002) Geochemistry of the ungrouped carbonaceous chondrite Tagish Lake, the anomalous CM chondrite Bells, and comparison with CI and CM chondrites. *Meteorit. Planet. Sci.* **37**, 703–712.

Mittlefehldt D. W. and Berkley J. L. (2002) Petrology and geochemistry of paired brachinites EET 99402 and EET 99407. In *Lunar Planet. Sci.* **XXXIII**, #1008. The Lunar and Planetary Institute, Houston (CD-ROM).

Mittlefehldt D. W., Lindstrom M. M., Bogard D. D., Garrison D. H., and Field S. W. (1996) Acapulco- and Lodran-like achondrites: petrology, geochemistry, chronology and origin. *Geochim. Cosmochim. Acta* **60**, 867–882.

Mittlefehldt D. W., McCoy T. J., Goodrich C. A., and Kracher A. (1998) Non-chondritic meteorites from asteroidal bodies. In *Planetary Materials,* Reviews in Mineralogy (ed. J. J. Papike). Mineralogical Society of America, Washington, DC, vol. 36, chap. 4, pp. 4-1–4-495.

Nagahara H. (1992) Yamato-8002: partial melting residue on the "unique" chondrite parent body. *Proc. NIPR Symp. Antarct. Meteorit.* **5**, 191–223.

Nehru C. E., Prinz M., and Delaney J. S. (1982) The Tucson iron and its relationship to enstatite meteorites. *Proc. 13th Lunar Planet. Sci. Conf.: J. Geophys. Res.* **B57**(suppl. 1), A3675–AA373.

Nehru C. E., Prinz M., Weisberg M. K., Ebihara M. E., Clayton R. E., and Mayeda T. K. (1996) Brachinites: a new primitive achondrite group. *Meteoritics* **27**, 267.

Newsom H. E. and Drake M. J. (1979) The origin of metal clasts in the Bencubbin meteoritic breccia. *Geochim. Cosmochim. Acta* **43**, 689–707.

Newton J., Bischoff A., Arden J. W., Franchi I. A., Geiger T., Greshake A., and Pillinger C. T. (1995) Acfer 094, a uniquely primitive carbonaceous chondrite from the Sagara. *Meteoritics* **30**, 47–56.

Noguchi T. (1994) Petrology and mineralogy of the Coolidge meteorite (CV4). *Proc. 7th NIPR Symp. Antarct. Meteorit.* 42–72.

Nyquist L. E., Bogard D. D., Shih C.-Y., Greshake A., Stöffler D., and Eugster O. (2001) Ages and geologic histories of martian meteorites. *Space Sci. Rev.* **96**, 105–164.

Okada A., Keil K., Taylor G. J., and Newsom H. (1988) Igneous history of the aubrite parent asteroid: evidence from the Norton County enstatite achondrite. *Meteoritics* **23**, 59–74.

Olsen E. and Jarosewich E. (1971) Chondrules: first occurrence in an iron meteorite. *Science* **174**, 583–585.

Olsen E., Bunch T. E., Jarosewich E., Noonan A. F., and Huss G. I. (1977) Happy Canyon: a new type of enstatite chondrite. *Meteoritics* **12**, 109–123.

Olsen E., Davis A., Clarke R. S. Jr., Schultz L., Weber H. W., Clayton R. N., Mayeda T. K., Jarosewich E., Sylvester P., Grossman L., Wang M.-S., Lipschutz M. E., Steele I. M., and Schwade J. (1994) Watson: a new link in the IIE iron chain. *Meteoritics* **29**, 200–213.

Olsen E. J., Clayton R. N., Mayeda T. K., Davis A. M., Clarke R. S. Jr., and Wasson J. T. (1996a) Mbosi: an anomalous iron with unique silicate inclusions. *Meteoritics Planet. Sci.* **31**, 633–639.

Olsen E. J., Davis A. M., Clayton R. N., Mayeda T. K., Moore C. B., and Steele I. M. (1996b) A silicate inclusion in Puente Del Zacate: a IIIA iron meteorite. *Science* **273**, 1365–1367.

Osadchii E. G., Baryshnikova G. V., and Novikov G. V. (1981) The Elga meteorite: silicate inclusions and shock metamorphism. *Proc. Lunar Planet. Sci. Conf.* **12B**, 1049–1068.

Palme H., Hutcheon I. D., Kennedy A. K., Sheng Y. J., and Spettel B. (1991) Trace element distributions in minerals from a silicate inclusion in the Caddo IAB-iron meteorite. In *Lunar Planet. Sci.* **XXII**. The Lunar and Planetary Institute, Houston, pp. 1015–1016.

Papike J. J. (1998) Comparative planetary mineralogy: chemistry of melt-derived pyroxene, feldspar, and olivine. In *Planetary Materials,* Reviews in Mineralogy (ed. J. J. Papike). Mineralogical Society of America, Washington, DC, vol. 36, chap. 7, pp. 7-10–7-11.

Papike J. J., Ryder G., and Shearer C. K. (1998) Lunar samples. In *Planetary Materials,* Reviews in Mineralogy (ed. J. J. Papike). Mineralogical Society of America, Washington, DC, vol. 36, chap. 5, pp. 5-1–5-234.

Patzer A., Hill D. H., and Boynton W. V. (2001) Itqiy: a metal-rich enstatite meteorite with achondritic texture. *Meteorit. Planet. Sci.* **36**, 1495–1505.

Paul R. L. and Lipschutz M. E. (1990) Consortium study of labile trace elements in some Antarctic carbonaceous chondrites: Antarctic and non-Antarctic meteorite comparisons. *Proc. NIPR Symp. Antarct. Meteorit.* **3**, 80–95.

Petaev M. I., Barsukova L. D., Lipschutz M. E., Ariskin A. A., Clayton R. N., and Mayeda T. K. (1994) The Divnoe meteorite: petrology, chemistry, oxygen isotopes and origin. *Meteorit. Planet. Sci.* **29**, 183–199.

Petaev M. I., Clarke R. S. Jr., Jarosewich E., Zaslavskaya N. I., Kononkova N. N., Wang M.-S., Lipschutz M. I., Olsen E. J., Davis A. M., Steele I. M., Clayton R. N., Mayeda T. K., and Kallemeyn G. W. (2000) The Chaunskij anomalous mesosiderite: petrology, chemistry, oxygen isotopes, classification and origin. *Geochem. Int.* **38**, S322–S350.

Powell B. N. (1971) Petrology and chemistry of mesosiderites: II. Silicate textures and compositions and metal-silciate relationships. *Geochim. Cosmochim. Acta* **35**, 5–34.

Prinz M., Keil K., Hlava P. F., Berkley J. L., Gomes C. B., and Curvello W. S. (1977) Studies of Brazilian meteorites: III. Origin and history of the Angra dos Reis achondrite. *Earth Planet. Sci. Lett.* **35**, 317–330.

Prinz M., Nehru C. E., and Delaney J. S. (1982) Sombrerete: an iron with highly fractionated amphibole-bearing Na–P-rich silicate inclusion. In *Lunar Planet. Sci.* **XIII**. The Lunar and Planetary Institute, Houston, pp. 634–635.

Prinz M., Nehru C. E., Delaney J. S., Weisberg M., and Olsen E. (1983a) Globular silicate inclusions in IIE irons and Sombrerete: highly fractionated minimum melts. In *Lunar Planet Sci.* **XIV**, The Lunar and Planetary Institute, Houston, pp. 618–619.

Prinz M., Nehru C. E., Delaney J. S., and Weisberg M. (1983b) Silicates in IAB and IIICD irons, winonaites, lodranites, and Brachina: a primitive and modified primitive group. In *Lunar Planet. Sci.* **XIV**. The Lunar and Planetary Institute, Houston, pp. 616–617.

Prinz M., Weisberg M. K., and Nehru C. E. (1986) North Haig and Nilpena: paired polymict ureilites with Angra dos

Reis-related and other clasts. In *Lunar Planet. Sci.* **XVII**. The Lunar and Planetary Institute, Houston, pp. 681–682.

Prinz M., Weisberg M. K., Nehru C. E., and Delaney J. S. (1987) EET83309, a polymict ureilite: recognition of a new group. In *Lunar Planet. Sci.* **XVIII**. The Lunar and Planetary Institute, Houston, pp. 802–803.

Prinz M., Weisberg M. K., and Nehru C. E. (1988) Feldspathic components in polymict ureilites. In *Lunar Planet. Sci.* **XIX**. The Lunar and Planetary Institute, Houston, pp. 947–948.

Prinz M., Chatterjee N., Weisberg M. K., Clayton R. N., and Mayeda T. K. (1991a) Lea County 002: a second Kakangari-type chondrite. In *Lunar Planet. Sci.* **XXII**. The Lunar and Planetary Institute, Houston, pp. 1097–1098.

Prinz M., Chatterjee N., Weisberg M. K., Clayton R. N., and Mayeda T. K. (1991b) Silicate inclusions in Antarctic irons. In *Lunar Planet. Sci.* **XXII**. The Lunar and Planetary Institute, Houston, pp. 1101–1102.

Prinz M., Weisberg M. K., Brearley A., Grady M. M., Pillinger C. T., Clayton R. N., and Mayeda T. K. (1992) LEW85332: a C2 chondrite in the CR clan. *Meteoritics* **27**, 278–279.

Pun A., Keil K., Taylor G. J., and Wieler R. (1998) The Kapoeta howardite: implications for the regolith evolution of the howardite–eucrite–diogenite parent body. *Meteorit. Planet. Sci.* **33**, 835–851.

Rietmeijer F. J. M. (1992) Pregraphitic and poorly graphitised carbons in porous chondritic micrometeorites. *Geochim. Cosmochim. Acta* **56**, 1665–1671.

Rietmeijer F. J. M. (1994) A proposal for a petrological classification scheme of carbonaceous chondritic micrometeorites. In *Analysis of Interplanetary Dust*, AIP Conf. Proc. (eds. M. E. Zolensky, T. L. Wilson, F. J. M. Rietmeijer, and G. J. Flynn). American Institute of Physics, NY, vol. 310, pp. 231–240.

Rietmeijer F. J. M. (1998) Interplanetary dust particles. In *Planetary Materials*, Reviews in Mineralogy (ed. J. J. Papike). Mineralogical Society of America, Washington, DC, vol. 36, chap. 2, pp. 2-1–2-95.

Rietmeijer F. J. M. (2002) The earliest chemical dust evolution in the solar nebula. *Chemie der Erde* **62**, 1–45.

Rietmeijer F. J. M. and Warren J. L. (1994) Windows of opportunity in the NASA Johnson Space Center Cosmic Dust Collection. In *Analysis of Interplanetary Dust*, AIP Conf. Proc. (eds. M. E. Zolensky, T. L. Wilson, F. J. M. Rietmeijer, and G. J. Flynn). American Institute of Physics, NY, vol. 310, pp. 255–275.

Rubin A. E. (1989) Size-frequency distributions of chondrules in CO_3 chondrites. *Meteoritics* **24**, 179–189.

Rubin A. E. (1990) Kamacite and olivine in ordinary chondrites: intergroup and intragroup relationships. *Geochim. Cosmochim. Acta* **54**, 1217–1232.

Rubin A. E. (1998) Correlated petrologic and geochemical characteristics of CO3 chondrites. *Meteorit. Planet. Sci.* **33**, 385–391.

Rubin A. E. and Kallemeyn G. W. (1990) Lewis Cliff 85332: a unique carbonaceous chondrite. *Meteoritics* **25**, 215–225.

Rubin A. E. and Kallemeyn G. W. (1994) Pecora Escarpment 91002: a member of the Rumuruti (R) chondrite group. *Meteoritics* **29**, 255–264.

Rubin A. E. and Scott E. R. D. (1997) Abee and related EH chondrite impact-melt breccias. *Geochim. Cosmcohim. Acta* **61**, 425–435.

Rubin A. E., James J. A., Keck B. D., Weeks K. S., Sears D. W. G., and Jarosewich E. (1985) The Colony meteorite and variations in CO3 chondrite properties. *Meteoritics* **20**, 175–197.

Rubin A. E., Jerde E. A., Zong P., Wasson J. T., Westcott J. W., Mayeda T. K., and Clayton R. N. (1986) Properties of the Guin ungrouped iron meteorite: the origin of Guin and of group-IIE irons. *Earth Planet. Sci. Lett.* **76**, 209–226.

Rubin A. E., Fegley B., and Brett R. (1988a) Oxidation state in chondrites. In *Meteorites and the Early Solar System* (eds. J. F. Kerridge and M. S. Matthews). University of Arizona Press, Tucson, pp. 488–511.

Rubin A. E., Wang D., Kallemeyn G. W., and Wasson J. T. (1988b) The Ningqiang meteorite; classification and petrology of an anomalous CV chondrite. *Meteoritics* **23**, 13–23.

Rubin A. E., Keil K., and Scott E. R. D. (1997) Shock metamorphism of enstatite chondrites. *Geochim. Cosmochim. Acta* **61**, 847–858.

Rubin A. E., Warren P. H., Greenwood J. P., Verish R. S., Leshin L. A., Hervig R. L., Clayton R. N., and Mayeda T. K. (2000) Los Angeles: the most differentiated basaltic martian meteorite. *Geology* **28**, 1011–1014.

Rubin A. E., Kallemeyn G. W., Wasson J. T., Clayton R. N., Mayeda T. K., Grady M. M., and Verchovsky A. B. (2001) Gujba: a new Bencubbin-like meteorite fall from Nigeria. In *Lunar Planet. Sci.* **XXXII**, #1779. The Lunar and Planetary Institute, Houston (CD-ROM).

Russell S. S. (1998) A survey of calcium–aluminum-rich inclusions from Rumuritiite chondrites: implications for relationships between meteorite groups. *Meteorit. Planet. Sci.* **33**, A131–A132.

Russell S. S., Huss G. R., Fahey A. J., Greenwood R. C., Hutchison R., and Wasserburg G. J. (1998a) An isotopic and petrologic study of calcium–aluminum-rich inclusions from CO3 chondrites. *Geochim. Cosmochim. Acta* **62**, 689–714.

Russell S. S., McCoy T. J., Jarosewich E., and Ash R. D. (1998b) The Burnwell, Kentucky, low iron oxide chondrite fall: description, classification and origin. *Meteorit. Planet. Sci.* **33**, 853–856.

Sandford S. A. and Walker R. M. (1985) Laboratory infrared transmission spectra of individual interplanetary dust particles from 2.5 to 25 microns. *Astrophys. J.* **291**, 838–851.

Schaudy R., Wasson J. T., and Buchwald V. F. (1972) The chemical classification of iron meteorites: VI. A reinvestigation of irons with Ge concentrations lower than 1 ppm. *Icarus* **17**, 174–192.

Schultz L., Palme H., Spettel B., Weber H. W., Wanke H., Christophe Michel-Levy M., and Lorin J. C. (1982) Allan Hills A77081—an unusual stony meteorite. *Earth Planet. Sci. Lett.* **61**, 23–31.

Schulze H., Bischoff A., Palme H., Spettel B., Dreibus G., and Otto J. (1994) Mineralogy an chemistry of Rumuruti: the first meteorite fall of the new R chondrite group. *Meteorit. Planet. Sci.* **29**, 275–286.

Scott E. R. D. (1977) Geochemical relationship between some pallasites and iron meteorites. *Mineral. Mag.* **41**, 262–275.

Scott E. R. D. (1979) Origin of anomalous iron meteorites. *Mineral. Mag.* **43**, 415–421.

Scott E. R. D. (1988) A new kind of primitive chondrite, Allan Hills 85085. *Earth Planet. Sci. Lett.* **91**, 1–18.

Scott E. R. D. and Jones R. H. (1990) Disentangling nebular and asteroidal features of CO3 carbonaceous chondrites. *Geochim. Cosmochim. Acta* **54**, 2485–2502.

Scott E. R. D. and Taylor G. J. (1985) Petrology of types 4–6 carbonaceous chondrites. *Proc. 15th Lunar Planet. Sci. Conf. Part 2: J. Geophys. Res.* **B90**, C699–C709.

Scott E. R. D. and Wasson J. T. (1975) Classification and properties of iron meteorites. *Rev. Geophys. Space Phys.* **13**, 527–546.

Scott E. R. D., Keil K., and Stöffler D. (1992) Shock metamorphism of carbonaceous chondrites. *Geochim. Cosmochim. Acta* **56**, 4281–4293.

Scott E. R. D., Taylor G. J., and Keil K. (1993) Origin of ureilite meteorites and implications for planetary accretion. *Geophys. Res. Lett.* **20**, 415–418.

Scott E. R. D., Haack H., and McCoy T. J. (1996) Core crystallization and silicate-metal mixing in the parent body of the IVA iron and stony-iron meteorites. *Geochim. Cosmochim. Acta* **60**, 1615–1631.

Scott E. R. D., Haack H., and Love S. G. (2001) Formation of mesosiderites by fragmentation and reaccretion of a large differentiated asteroid. *Meteorit. Planet. Sci.* **36**, 869–881.

Sears D. W. G. and Dodd R. T. (1988) Overview and classification of meteorites. In *Meteorites and Early Solar System* (eds. J. F. Kerridge and M. S. Matthews). University of Arizona Press, Tucson, pp. 3–31.

Sears D. W. G. and Weeks K. S. (1991) Chemical and physical studies of type 3 chondrites: 2. Thermoluminescence of sixteen type 3 ordinary chondrites and relationships with oxygen isotopes. *Proc. 14th Lunar Planet. Sci. Conf. Part 1: J. Geophys. Res.* **88**, B301–B311.

Sears D. W. G., Grossman J. N., Melcher C. L., Ross L. M., and Mills A. A. (1980) Measuring the metamorphic history of unequilibrated ordinary chondrites. *Nature* **287**, 791–795.

Sears D. W. G., Kallemeyn G. W., and Wasson J. T. (1982) The compositional classification of chondrites: II. The enstatite chondrite groups. *Geochim. Cosmochim. Acta* **46**, 597–608.

Sears D. W. G., Hasan F. A., Batchelor J. D., and Lu J. (1991) Chemical and physical studies of type 3 chondrites: XI. Metamorphism, pairing, and brecciation of ordinary chondrites. *Proc. Lunar Planet. Sci. Conf.* **21**, 493–512.

Simon S. B., Grossman L., Casanova I., Symes S., Benoit P., Sears D. W. G., and Wacker J. F. (1995) Axtell, a new CV3 chondrite find from Texas. *Meteoritics* **30**, 42–46.

Srinivasan B. and Anders E. (1977) Noble gases in the unique chondrite, Kakangari. *Meteoritics* **12**, 417–424.

Stöffler D., Keil K., and Scott E. R. D. (1991) Shock metamorphism in ordinary chondrites. *Geochim. Cosmochim. Acta* **55**, 3845–3867.

Stolper E. and McSween H. Y. Jr. (1979) Petrology and origin of the shergottite meteorites. *Geochim. Cosmochim. Acta* **43**, 1475–1498.

Suess H. E., Wänke H., and Wlotzka F. (1964) On the origin of gas-rich meteorites. *Geochim. Cosmochim. Acta* **28**, 595–607.

Sugiura N. (1995) Isotopic composition of nitrogen in the PCA 91002 R chondrite group. *Proc. NIPR Symp. Antarct. Meteorit.* **8**, 273–285.

Swindle T. D., Kring D. A., Burkland M. K., Hill D. H., and Boynton W. V. (1998) Noble gases, bulk chemistry, and petrography of olivine-rich achondrites Eagles Nest and Lewis Cliff 88763: comparison to brachinites. *Meteorit. Planet Sci.* **33**, 31–48.

Takeda H. (1979) A layered-crust model of a howardite parent body. *Icarus* **40**, 455–470.

Takeda H. (1997) Mineralogical records of early planetary processes on the howardite, eucrite, diogenite parent body with reference to Vesta. *Meteorit. Planet. Sci.* **32**, 841–853.

Takeda H. and Graham A. L. (1991) Degree of equilibration of eucritic pyroxenes and thermal metamorphism of the earliest planetary crust. *Meteoritics* **26**, 129–134.

Takeda H. and Mori H. (1985) The diogenite–eucrite links and the crystallization history of a crust of their parent body. *Proc. Lunar Planet. Sci. Conf. Part 2: J. Geophys. Res.* **90**(suppl.), C626–C648.

Takeda H., Mori H., Hiroi T., and Saito J. (1994) Mineralogy of new Antarctic achondrites with affinity to Lodran and a model of their evolution in an asteroid. *Meteoritics* **29**, 830–842.

Takeda H., Yugami K., Bogard D. D., and Miyamoto M. (1997) Plagioclase-augite-rich gabbro in the Caddo County IAB iron and the missing basalts associated with iron meteorites. In *Lunar Planet. Sci.* **XXVIII**. The Lunar and Planetary Institute, Houston, pp. 1409–1410.

Thomas K. L., Keller L. P., and McKay D. S. (1996) A comprehensive study of major, minor, and light element abundances in over 100 interplanetary dust particles. In *Physics, Chemistry, and Dynamics of Interplanetary Dust*, Astron. Soc. Pacific Conf. Ser. (eds. B. S. Gustafson and M. S. Hanner). vol. 104, pp. 283–286.

Tomeoka K. (1990) Mineralogy and petrology of Belgica-7904: a new kind of carbonaceous chondrite from Antarctica. *Proc. NIPR Symp. Antarct. Meteorit.* **3**, 40–54.

Tomeoka K. and Buseck P. R. (1982) Intergrown mica and montmorillonite in the Allende carbonaceous chondrite. *Nature* **299**, 326–327.

Tomeoka K. and Buseck P. R. (1990) Phyllosilicates in the Mokoia CV carbonaceous chondrite: evidence for aqueous alteration in an oxidizing condition. *Geochim. Cosmochim. Acta* **54**, 1745–1754.

Tomeoka K., Kojima H., and Yanai K. (1989) Yamato-86720: a CM carbonaceous chondrite having experienced extensive aqueous alteration and thermal metamorphism. *Proc. NIPR Symp. Antarct. Meteorit.* **2**, 55–74.

Tomeoka K., Nomura K., and Takeda H. (1992) Na-bearing Ca–Al-rich inclusions in the Yamato-791717 CO carbonaceous chondrite. *Meteoritics* **27**, 136–143.

Treiman A. H. (1986) The parental magma of the Nakhla achondirte: ultrabasic volcanism on the shergottite parent body. *Geochim. Cosmochim. Acta* **50**, 1061–1070.

Treiman A. H. (1990) Complex petrogenesis of the Nakhla (SNC) meteorite: evidence from petrography and mineral chemistry. *Proc. Lunar Planet. Sci. Conf.* **20**, 273–280.

Treiman A. H. (1993) The parent magma of the Nakhla (SNC) meteorite, inferred from magmatic inclusions. *Geochim. Cosmochim. Acta* **57**, 4753–4767.

Treiman A. H., McKay G. A., Bogard D. D., Mittlefehldt D. W., Wang M.-S., Keller L., Lipschutz M. E., Lindstrom M. M., and Garrison D. (1994) Comparison of the LEW88516 and ALH77005 meteorites: similar but distinct. *Meteoritics* **29**, 581–592.

Treiman A. H., Gleason J. D., and Bogard D. D. (2000) The SNC meteorites are from Mars. *Planet. Space Sci.* **48**, 1213–1230.

Ulff-Møller F., Rasnussen K. L., Prinz M., Palme H., Spettel B., and Kallemeyn G. W. (1995) Magmatic activity on the IVA parent body: evidence from silicate-bearing iron meteorites. *Geochim. Cosmochim. Acta* **59**, 4713–4728.

Van Schmus W. R. and Wood J. A. (1967) A chemical-petrologic classification for the chondritic meteorites. *Geochim. Cosmochim. Acta* **31**, 747–765.

Walker D., Powell M. A., Lofgren G. E., and Hays J. F. (1978) Dynamic crystallization of a eucrite basalt. *Proc. 9th Lunar Planet. Sci. Conf.* 1369–1391.

Wai C. M. and Wasson J. T. (1979) Nebular condensation of Ga, Ge and Sb and chemical classification of iron meteorites. *Nature*, **282**, 790–793.

Warren P. H. and Kallemeyn G. W. (1989) Allan Hills 84025: the second Brachinite, far more differentiated than Brachina and an ultramafic achondritic clasts from L chondrite Yamato 95097. *Proc. Lunar Planet Sci. Conf.* **19**, 475–486.

Warren P. H. and Kallemeyn G. W. (1992) Explosive volcanism and the graphite-oxygen fucacity buffer on the parent asteroid(s) of the ureilite meteorites. *Icarus* **100**, 110–126.

Warren P. H., Taylor J. G., and Keil K. (1983) Regolith breccia Allan Hills A81005: evidence of lunar origin, and petrography of pristine and nonpristine clasts. *Geophys. Res. Lett.* **10**, 779–782.

Wasson J. T. (1990) Ungrouped iron meteorites in Antarctica: Origin of anomalously high abundance. *Science* **249**, 900–902.

Wasson J. T. and Kallemeyn G. W. (1988) Composition of chondrites. *Phil. Trans. Roy. Soc. London* **A325**, 535–544.

Wasson J. T. and Kallemeyn G. W. (1990) Allan Hills 85085: a subchondritic meteorite of mixed nebular and regolithic heritage. *Earth Planet. Sci. Lett.* **101**, 148–161.

Wasson J. T., Rubin A. E., and Kallemeyn G. W. (1993) Reduction during metamorphism of four ordinary chondrites. *Geochim. Cosmochim. Acta* **57**, 1867–1869.

Wasson J. T., Choi B.-G., Jerde E. A., and Ulf-Møller F. (1998) Chemical classification of iron meteorites: XII. New members of the magmatic groups. *Geochim. Cosmochim. Acta* **62**, 715–724.

Wasson J. T. and Wai C. M. (1976) Explanation for a very low Ga and Ge concentrations in some iron meteorite groups. *Nature* **261**, 114–116.

Watters T. R. and Prinz M. (1979) Aubrites: their origin and relationship to enstatite chondrites. *Proc. 10th Lunar Planet. Sci. Conf.* **10**, 1073–1093.

Weigel A., Eugster O., Koeberl C., and Krähenbühl U. (1996) Primitive differentiated meteorite Divnoe and its relationship to brachinites. In *Lunar Planet. Sci.* **XXVII**. The Lunar and Planetary Institute, Houston, pp. 1403–1404.

Weisberg M. K. (2001) Sahara 00182, the first CR3 chondrite and formation of multi-layered chondrules. *Meteorit. Planet. Sci.* **36**, A222–A223.

Weisberg M. K. and Prinz M. (2000) The Grosvenor Mountains 95577 CR1 Chondrite and hydration of the CR chondrites. *Meteorit. Planet. Sci.* **35**, A168.

Weisberg M. K., Prinz M., and Nehru C. E. (1988) Petrology of ALH85085: a chondrite with unique characteristics. *Earth Planet. Sci. Lett.* **91**, 19–32.

Weisberg M. K., Prinz M., and Nehru C. E. (1990) The Bencubbin chondrite breccia and its relationship to CR chondrites and the ALH85085 chondrite. *Meteoritics* **25**, 269–279.

Weisberg M. K., Prinz M., Kojima H., Yanai K., Clayton R. N., and Mayeda T. K. (1991) The Carlisle Lakes-type chondrites: a new grouplet with high $\Delta^{17}O$ and evidence for nebular oxidation. *Geochim. Cosmochim. Acta* **55**, 2657–2669.

Weisberg M. K., Prinz M., Clayton R. N., and Mayeda T. K. (1993) The CR (Renazzo-type) carbonaceous chondrite group and its implications. *Geochim. Cosmochim. Acta* **57**, 1567–1586.

Weisberg M. K., Prinz M., Clayton R. N., Mayeda T. K., Grady M. M., and Pillinger C. T. (1995a) The CR chondrite clan. *Proc. NIPR Symp. Antarct. Meteorit.* **8**, 11–32.

Weisberg M. K., Boesenberg J. S., Kozhushko G., Prinz M., Clayton R. N., and Mayeda T. K. (1995b) EH3 and EL3 chondrites: a petrologic-oxygen isotopic study. In *Lunar Planet. Sci.* **XXVI**. The Lunar and Planetary Institute, Houston, pp. 1481–1482.

Weisberg M. K., Prinz M., Clayton R. N., Mayeda T. K., Grady M. M., Franchi I., Pillinger C. T., and Kallemeyn G. W. (1996a) The K (Kakangari) chondrite grouplet. *Geochim. Cosmochim. Acta* **60**, 4253–4263.

Weisberg M. K., Prinz M., Zolensky M. E., Clayton R. N., Mayeda T. K., and Ebihara M. (1996b) Ningqiang and its relationship to oxidized CV3 chondrites. *Meteorit. Planet. Sci.* **31**, 150–151.

Weisberg M. K., Prinz M., Clayton R. N., and Mayeda T. K. (1997a) CV3 chondrites: three subgroups, not two. *Meteorit. Planet. Sci.* **32**, 138–139.

Weisberg M. K., Prinz M., and Nehru C. E. (1997b) QUE94204: an EH-chondritic melt rock. In *Lunar Planet. Sci.* **XXVIII**. The Lunar and Planetary Institute, Houston, pp. 1525–1526.

Weisberg M. K., Prinz M., Clayton R. N., Mayeda T. K., Sugiura N., Zashu S., and Ebihara M. (2001) A new metal-rich chondrite grouplet. *Meteorit. Planet. Sci.* **36**, 401–418.

Weisberg M. K., Boesenberg J. S., and Ebel D. S. (2002) Gujba and the origin of the CB chondrites. In *Lunar Planet. Sci.* **XXXIII**, #1551. The Lunar and Planetary Institute, Houston (CD-ROM).

Wheelock M. M., Keil K., Floss C., Taylor G. J., and Crozaz G. (1994) REE geochemistry of oldhamite-dominated clasts from the Norton County aubrite: igneous origin of oldhamite. *Geochim. Cosmochim. Acta* **58**, 449–458.

Wlotzka F. (1993) A weathering scale for the ordinary chondrites. *Meteoritics* **28**, 460.

Wood C. A. and Ashwal L. D. (1981) SNC meteorites: igneous rocks from Mars? *Proc. Lunar Planet. Sci. Conf.* **12B**, 1359–1376.

Yamaguchi A., Taylor G. J., and Keil K. (1996) Global crustal metamorphism of the eucrite parent body. *Icarus* **124**, 97–112.

Yamaguchi A., Taylor G. J., and Keil K. (1997) Metamorphic history of the eucritic crust of 4 Vesta. *J. Geophys. Res.* **102**, 13381–13386.

Yamaguchi A., Clayton R. N., Mayeda T. K., Ebihara M., Oura Y., Miura Y. N., Haramura H., Misawa K., Kojima H., and Nagao K. (2002) A new source of basaltic meteorites inferred from Northwest Africa 011. *Science* **296**, 334–336.

Yanai K. (1994) Angrite Asuka-881371: preliminary examination of a unique meteorite in the Japanese collection of antarctic meteorites. *Proc. NIPR Symp. Antarct. Meteorit.* **7**, 30–41.

Yanai K. and Kojima H. (1995) Yamato-8451: a newly identified pyroxene-bearing pallasite. *Proc. NIPR Symp. Antarct. Meteorit.* **8**, 1–10.

Yanai K., Kojima H., and Haramura H. (1995) *Catalog of the Antarctic Meteorites*. NIPR, Tokyo.

Yugami K., Takeda H., Kojima H., and Miyamoto M. (1997) Modal abundances of primitive achondrites and the end-member mineral assemblage of the differentiated trend. *Symp. Antarct. Meteorit.* **22**, 220–222.

Zolensky M. E., Barrett R. A., and Browning L. B. (1993) Mineralogy and composition of matrix and chondrule rims in carbonaceous chondrites. *Geochim. Cosmochim. Acta* **57**, 3123–3148.

Zolensky M. E., Mittlefehldt D. W., Lipschutz M. E., Wang M.-S., Clayton R. N., Mayeda T. K., Grady M. M., Pillinger C. T., and Barber D. (1997) CM chondrites exhibit the complete petrologic range from type 2 to 1. *Geochim. Cosmochim. Acta* **61**, 5099–5115.

Zolensky M. E., Nakamura K., Gounelle M., Mikouchi T., Kasama T., Tachikawa O., and Tonui E. (2002) Mineralogy of Tagish Lake: an ungrouped type 2 carbonaceous chondrite. *Meteorit. Planet. Sci.* **37**, 737–761.

1.06
Oxygen Isotopes in Meteorites
R. N. Clayton
University of Chicago, IL, USA

1.06.1 INTRODUCTION	129
1.06.2 ISOTOPIC ABUNDANCES IN THE SUN AND SOLAR NEBULA	131
1.06.2.1 Isotopes of Light Elements in the Sun	131
1.06.2.2 Oxygen Isotopic Composition of the Solar Nebula	131
1.06.2.2.1 Primitive nebular materials	132
1.06.2.2.2 Sources of oxygen isotopic heterogeneity in the solar nebula	134
1.06.3 OXYGEN ISOTOPES IN CHONDRITES	135
1.06.3.1 Thermal Metamorphism of Chondrites	137
1.06.3.1.1 Ordinary chondrites	137
1.06.3.1.2 CO chondrites	138
1.06.3.1.3 CM and CI chondrites	138
1.06.4 OXYGEN ISOTOPES IN ACHONDRITES	138
1.06.4.1 Differentiated (Evolved) Achondrites	138
1.06.4.2 Undifferentiated (Primitive) Achondrites	140
1.06.5 SUMMARY AND CONCLUSIONS	140
REFERENCES	141

1.06.1 INTRODUCTION

Oxygen isotope abundance variations in meteorites are very useful in elucidating chemical and physical processes that occurred during the formation of the solar system (Clayton, 1993). On Earth, the mean abundances of the three stable isotopes are ^{16}O: 99.76%, ^{17}O: 0.039%, and ^{18}O: 0.202%. It is conventional to express variations in abundances of the isotopes in terms of isotopic ratios, relative to an arbitrary standard, called SMOW (for standard mean ocean water), as follows:

$$\delta^{18}O = \left[\frac{(^{18}O/^{16}O)_{sample}}{(^{18}O/^{16}O)_{SMOW}} - 1\right] \times 1,000$$

$$\delta^{17}O = \left[\frac{(^{17}O/^{16}O)_{sample}}{(^{17}O/^{16}O)_{SMOW}} - 1\right] \times 1,000$$

The isotopic composition of any sample can then be represented by one point on a "three-isotope plot," a graph of $\delta^{17}O$ versus $\delta^{18}O$. It will be seen that such plots are invaluable in interpreting meteoritic data. Figure 1 shows schematically the effect of various processes on an initial composition at the center of the diagram. Almost all terrestrial materials lie along a "fractionation" trend; most meteoritic materials lie near a line of "^{16}O addition" (or subtraction).

The three isotopes of oxygen are produced by nucleosynthesis in stars, but by different nuclear processes in different stellar environments. The principal isotope, ^{16}O, is a primary isotope (capable of being produced from hydrogen and helium alone), formed in massive stars (>10 solar masses), and ejected by supernova explosions. The two rare isotopes are secondary nuclei (produced in stars from nuclei formed in an earlier generation of stars), with ^{17}O coming primarily from low- and intermediate-mass stars (<8 solar masses), and ^{18}O coming primarily from high-mass stars (Prantzos et al., 1996). These differences in type of stellar source result in large observable variations in stellar isotopic abundances as functions of age, size, metallicity, and galactic location (Prantzos et al., 1996). In their paper reporting the discovery of ^{18}O in the Earth's atmosphere, Giauque and

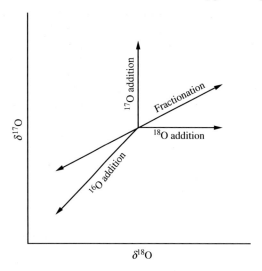

Figure 1 Schematic representation of various isotopic processes shown on an oxygen three-isotope plot. Almost all terrestrial materials plot along a line of "fractionation"; most primitive meteoritic materials plot near a line of "^{16}O addition."

Johnston (1929) refer to nonuniform distribution of oxygen isotopes as a "remote possibility," whereas Manian et al. (1934) sought to find variations in oxygen isotope abundances in meteorites as evidence for an origin outside the solar system.

In addition to the abundance variations due to nuclear processes, there are important isotopic variations produced within molecular clouds, the precursors to later star-formation. The most important process is isotopic self-shielding in the UV photodissociation of CO (van Dishoeck and Black, 1988). This process results from the large differences in abundance between $C^{16}O$, on the one hand, and $C^{17}O$ and $C^{18}O$ on the other. Photolysis of CO occurs by absorption of stellar UV radiation in the wavelength range 90–100 nm. The reaction proceeds by a predissociation mechanism, in which the excited electronic state lives long enough to have well-defined vibrational and rotational energy levels. As a consequence, the three isotopic species—$C^{16}O$, $C^{17}O$, and $C^{18}O$—absorb at different wavelengths, corresponding to the isotope shift in vibrational frequencies. Because of their different number densities, the abundant $C^{16}O$ becomes optically thick in the outermost part of the cloud (nearest to the external source of UV radiation), while the rare $C^{17}O$ and $C^{18}O$ remain optically thin, and hence dissociate at a greater rate in the cloud interior. The differences in chemical reactivity between $C^{16}O$ *molecules* and ^{17}O and ^{18}O *atoms* may lead to isotopically selective reaction products. This scenario has been suggested to explain meteoritic isotope patterns, as discussed below (Yurimoto and Kuramoto, 2002).

Stable isotope abundances in meteoritic material provide an opportunity to evaluate the thoroughness of mixing of isotopes of diverse stellar sources. Molybdenum presents a good test case: it has seven stable isotopes, derived from at least three types of stellar sources, corresponding to the r-process, s-process, and p-process. Presolar silicon carbide grains, extracted from primitive meteorites, contain molybdenum that has been subject to s-process neutron capture in red-giant stars, resulting in large enrichments of isotopes at masses 95, 96, 97, 98, and severe depletions (up to 100%) of isotopes at masses 92 and 94 (p-process) and 100 (r-process) (Nicolussi et al., 1998). Complementary patterns have been found in whole-rock samples of several meteorites, with >1,000-fold smaller amplitude, suggesting the preservation of a small fraction of the initial isotopic heterogeneity (Yin et al., 2002; Dauphas et al., 2002). Oxygen is another element for which primordial isotopic heterogeneity might be preserved. This is discussed further below.

It would be highly desirable to have samples of oxygen-rich mineral grains that have formed in stellar atmospheres and have recorded the nucleosynthetic processes in individual stars. Similar samples are already available for carbon-rich grains, in the form of SiC and graphite, primarily from asymptotic giant branch (AGB) stars and supernovae (Anders and Zinner, 1993). These presolar grains have provided a wealth of detailed information concerning nucleosynthesis of carbon, nitrogen, silicon, calcium, titanium, and heavier elements (see Chapter 1.02). It is thought that such carbon-rich minerals should form only in environments with C/O > 1, as in the late stages of AGB evolution, or in carbon-rich layers of supernovae. By analogy, one would expect to form oxide and silicate minerals in environments with C/O < 1, as is common for most stars. Indeed there is evidence in infrared spectra for the formation of Al_2O_3 (corundum) and silicates, such as olivine (Speck et al., 2000) around evolved oxygen-rich stars. However, searches for such grains in meteorites have yielded only a very small population of corundum grains, a few grains of spinel and hibonite, and no silicates (Nittler et al., 1997). The observed oxygen isotopic compositions of presolar corundum grains show clear evidence of nuclear processes in red-giant stars, and have had significant impact on the theory of these stars (Boothroyd and Sackmann, 1999).

There are several possible reasons for the failure to recognize and analyze large populations of oxygen-rich presolar grains:

(i) they may not exist: oxygen ejected in supernova explosions may not condense into mineral grains on the short timescale available;

(ii) they may be smaller in size than can be detected by applicable techniques (~0.1 μm); and

(iii) they may be destroyed in the laboratory procedures used to isolate other types of presolar grains.

1.06.2 ISOTOPIC ABUNDANCES IN THE SUN AND SOLAR NEBULA

1.06.2.1 Isotopes of Light Elements in the Sun

For most of the chemical elements, the relative abundances of their stable isotopes in the Sun and solar nebula are well known, so that any departures from those values that may be found in meteorites and planetary materials can then be interpreted in terms of planet-forming processes. This is best illustrated for the noble gases: neon, argon, krypton, and xenon. The solar isotopic abundances are known through laboratory mass-spectrometric analysis of solar wind extracted from lunar soils (Eberhardt et al., 1970) and gas-rich meteorites. Noble gases in other meteorites and in the atmospheres of Earth and Mars show many substantial differences from the solar composition, due to a variety of nonsolar processes, e.g., excesses of ^{40}Ar and ^{129}Xe due to radioactive decay of ^{40}K and ^{129}I, excesses of ^{134,136}Xe due to fission of ^{244}Pu, excesses of ^{21,22}Ne and ^{38}Ar due to cosmic-ray spallation reactions, and mass-dependent isotopic fractionation due to gas loss in a gravitational field (Pepin, 1991). For many of the nonvolatile elements, such as magnesium and iron, it is thought that terrestrial and solar isotopic abundances are almost identical, so that small differences in meteoritic materials can serve as tracers of local solar nebular processes, such as evaporation of molten silicate droplets in space (Davis et al., 1990).

For a few very important elements, notably hydrogen, helium, carbon, nitrogen, and oxygen, neither of these procedures can be used to infer accurate values for the isotopic abundances in the Sun and solar nebula. Hydrogen is a special case, in that the Sun's initial complement of deuterium (^{2}H) has been destroyed early in solar history by nuclear reaction, so that today's solar-wind hydrogen is deuterium free (Epstein and Taylor, 1970). Attempts to infer the D/H ratio in the solar nebula, based on measurements in comets, meteorites, and planets, have left an uncertainty of a factor of 2 in the primordial ratio. Helium isotope ratios (^{4}He/^{3}He) are variable in the solar wind (Gloeckler and Geiss, 1998), and are strongly modified in solid solar system bodies by radioactive production of ^{4}He from uranium and thorium, and by cosmic-ray production of both isotopes.

Carbon isotope abundances (^{12}C/^{13}C ratios) are measurable in solar energetic particles (by spacecraft) (Leske et al., 2001), and in molecules in the Sun and comets (Wyckoff et al., 2000). All values are consistent with a mean terrestrial ratio of 89, but with analytical uncertainties of ~10%. The ^{12}C/^{13}C ratio in the present near-solar interstellar medium has been found to be 70 ± 10, interpreted as reflecting "galactic chemical evolution" over the age of the solar system (Prantzos et al., 1996). Isotopic measurement of solar-wind carbon in lunar soils has not produced definitive results due to lunar, meteoritic, and terrestrial background contamination.

Nitrogen isotope abundances (^{14}N/^{15}N ratios) in the Sun and solar nebula have presented problems of measurement and interpretation for decades (Kerridge, 1993). The terrestrial atmospheric ratio is 272, but meteoritic and lunar ratios vary by about a factor of 3 (Owen et al., 2001). Estimates of the nitrogen isotope ratio of the Sun and solar system range from ~30% lower than terrestrial, measured in the solar wind (Kallenbach et al., 1998) to ~30% higher than terrestrial, measured in ammonia in Jupiter (Owen et al., 2001). The lunar regolith should be an excellent collector of solar wind, since the indigenous nitrogen content of lunar rocks is very low. A range in ^{14}N/^{15}N of over 35% has been found for the implanted nitrogen, which correlates in concentration with solar-wind-derived noble gases. However, there is reason to believe that ~80–90% of this nitrogen has come from unidentified nonsolar sources (Hashizume et al., 2000). The net result is that the isotopic composition of nitrogen in the Sun and in the solar nebula is very poorly known.

The observational constraints on the solar isotopic abundances of oxygen are also poor. The only solar-wind measurement, by the Advanced Composition Explorer, yielded a ratio of ^{16}O/^{18}O consistent with the terrestrial value, with 20% uncertainty (Wimmer-Schweingruber et al., 2001). An earlier spectroscopic measurement of the solar photosphere gave a similar result (Harris et al., 1987). No information is available on the solar ^{17}O abundance. The very limited state of knowledge of the solar isotope abundances of carbon, nitrogen, and oxygen illustrates the importance of the NASA Genesis mission to collect a pure solar-wind sample and return it to Earth for laboratory measurement.

1.06.2.2 Oxygen Isotopic Composition of the Solar Nebula

In the absence of an accurate independent measurement of oxygen isotope abundances in the Sun, the original nebular composition must be inferred from measurements on meteorites and planets. Spectroscopic measurements of planetary atmospheres have revealed no departures from terrestrial values, within rather large analytical uncertainties. Meteorites have significant advantages over other solar system bodies for tracing early nebular processes: (i) laboratory isotopic analyses are ~1,000 times more precise than remote sensing methods, and (ii) meteorites contain components that have been unchanged since their formation in the nebula 4.5 billion years ago.

For isotopic studies, oxygen has two major advantages over other light elements, such as hydrogen, carbon, and nitrogen: (i) it has three stable isotopes, rather than two, and (ii) it is a major constituent of both the solid and gaseous components of the solar nebula. The latter point is illustrated by Table 1, showing the solar system abundances of the major rock-forming elements. Over a broad range of temperatures, below the temperature of condensation of silicates, and above the temperature of condensation of ices, ~25% of the oxygen in the nebula is in the form of solid anhydrous minerals, and 75% is in gaseous molecules, primarily CO and H_2O. A major question is whether or not the solid and gaseous reservoirs had the same oxygen isotopic composition. Evidence was presented earlier for thorough isotopic homogenization of heavy elements such as molybdenum, but it is not clear that any initial isotopic differences between gas and solid would also be erased. Thus, it is not yet established whether the oxygen isotopic heterogeneity discussed below was inherited by the nebula or was generated *within* the nebula by local processes.

1.06.2.2.1 Primitive nebular materials

Chondritic meteorites are sedimentary rocks composed primarily of chondrules, typically sub-millimeter-sized spherules believed to have been molten droplets in the solar nebula, formed by melting of dust in a brief, local heating event. During the high-temperature stage, with a duration of some hours, the droplets could interact chemically and undergo isotopic exchange with the surrounding gas. Thus, isotopic analyses of individual chondrules can provide information about both the dust and gas components of the nebula. Figure 2 shows oxygen isotopic compositions of chondrules from the three major groups—ordinary (O), carbonaceous (C), and enstatite (E) chondrites—and one minor group—Rumuruti-type (R)-chondrites. The three-isotope graph has two useful properties: (i) samples related to one another by ordinary mass-dependent isotopic fractionation lie on a line of slope = 0.52, like the line labeled terrestrial fractionation (TF), and (ii) samples that are two-component mixtures lie on a straight line connecting the compositions of the end-members.

Figure 2 shows that chondrule compositions do not lie on a mass-dependent fractionation line, thus indicating isotopic heterogeneity in the nebula. Chondrules from different chondrite classes occupy different regions of the diagram, and for each class, they form near-linear arrays that are considerably steeper than a mass-dependent fractionation line. For comparison, it can be noted that analogous three-isotope graphs for iron in various meteorite types are strictly mass dependent, and show no evidence for nebular heterogeneity (Zhu et al., 2001).

Another group of primitive objects with a direct link to the solar nebula are the calcium–aluminum-rich inclusions (CAIs), that range in size from a few μm to >1 cm (see Chapter 1.08). They are found in all types of primitive

Table 1 Atomic abundances of major elements (relative to silicon).[a]

H	28,840
He	2,455
C	7.08
N	2.45
O	14.13
Mg	1.02
Si	1.00
Fe	0.85

[a] Source: Palme and Jones (Treatise on Geochemistry, Chapter 1.03).

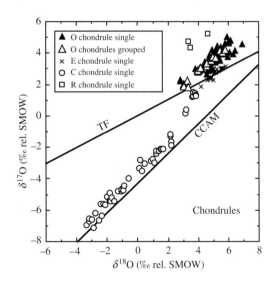

Figure 2 Oxygen isotopic compositions of chondrules from all classes of chondritic meteorites: ordinary (O), enstatite (E), carbonaceous (C), and Rumuruti-type (R). The TF line and carbonaceous chondrite anhydrous mineral (CCAM) line are shown for reference in this and many subsequent figures. Equations for these lines are: TF—$\delta^{17} = 0.52\delta^{18}$ and CCAM—$\delta^{17} = 0.941\delta^{18} - 4.00$ (sources Clayton et al., 1983, 1984, 1991; Weisberg et al., 1991).

chondrites but are rare in all but the CV carbonaceous chondrites. Their bulk chemical compositions correspond to the most refractory 5% of condensable solar matter (Grossman, 1973). They may represent direct condensates from the nebular gas, followed, in many cases, by further chemical and isotopic interaction with the gas. Their radiometric ages have been measured with high precision (Allègre et al., 1995), and indicate solidification earlier than any other solar system rocks (excluding the presolar dust grains). Thus, the oxygen isotope abundances in CAIs may provide the best guide to the composition of the nebular gas.

CAIs exhibit a specific, characteristic pattern of oxygen isotope abundances. *Within* an individual CAI, different minerals have different isotopic compositions, with all data points falling on a straight line in the three-isotope graph. This behavior is illustrated in Figure 3, based on analyses of physically separated minerals from the Allende (CV3) carbonaceous chondrite (Clayton et al., 1977). Each analysis represents a large number of grains. Figure 4 shows data obtained by ion microprobe analysis, where each point represents only one grain (Aléon et al., 2002; Fagan et al., 2002; Itoh et al., 2002; Jones et al., 2002; Krot et al., 2002). The line labeled "CCAM" is the same in Figures 3 and 4. Although the ion microprobe data have larger analytical uncertainties, leading to greater scatter in the data, it is clear that the same pattern exists at both the microscopic and macroscopic level.

Within individual CAIs, the sequence of isotopic composition, in terms of ^{16}O-enrichment, is spinel ≧ pyroxene > olivine > melilite = anorthite. A straight line on the three-isotope graph is indicative of some sort of two-component mixture. The fact that the range of variation in the individual-grain studies is almost the same as the range in the bulk-sample studies shows that the end-members do not lie much beyond the observed range of variation. All studies, as of early 2003, reveal an ^{16}O-rich end-member near $-45‰$ for both $\delta^{17}O$ and $\delta^{18}O$, frequently represented by spinel, the most refractory of the CAI phases. The most obvious interpretation is that this end-member represents the composition of the primary nebular gas, from which the CAIs originally condensed. Subsequent reaction and isotopic exchange with an isotopically modified gas could then yield the observed heterogeneities on a micrometer to millimeter scale (Clayton et al., 1977; Clayton, 2002).

Another argument that the ^{16}O-rich end-member was a ubiquitous component of primitive solids is that it is found in many different chemical forms (different minerals) in many classes of meteorites: CAIs and amoeboid olivine aggregates (AOAs) from Efremovka (CV3) (Aléon et al., 2002; Fagan et al., 2002), AOA from a CO chondrite, Y 81020 (Itoh et al., 2002), and CAI and AOA from CM and CR chondrites (Krot et al., 2002). This isotopic composition can also serve as an end-member for the chondrule mixing line in Figure 2.

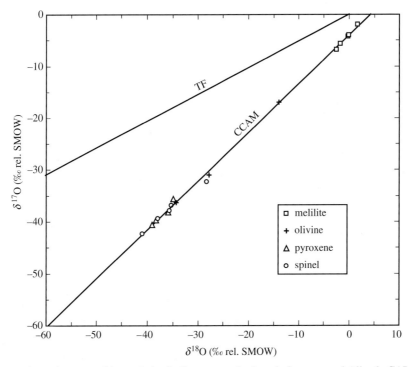

Figure 3 Oxygen isotopic compositions of physically separated minerals from several Allende CAIs. These points were used to define the CCAM line (source Clayton et al., 1977).

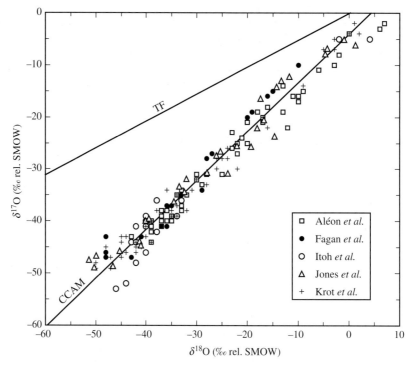

Figure 4 Ion microprobe oxygen isotope analyses of single grains in several carbonaceous chondrites. Analytical uncertainties are typically about ±2‰. The CCAM line is shown for reference sources: are noted in the figure.

If the ^{16}O-rich composition is indeed the isotopic composition of the primordial solar nebula, the consequences for solar system formation are profound. As noted above, materials with the ^{16}O-rich composition are ubiquitous, but they are also rather rare, never amounting to more than a few percent of the host meteorite. The implication is that all the other material in the inner solar system has undergone some process that changed its ^{16}O-abundance by 4–5%. This must have been a major chemical or physical process that must leave evidence in forms other than the isotopic composition of oxygen.

1.06.2.2.2 Sources of oxygen isotopic heterogeneity in the solar nebula

The oxygen isotopic patterns in primitive materials, shown in Figures 2–4, imply some sort of two-component mixing involving one component enriched in ^{16}O relative to the other. Several models have been proposed for the origin of these components. They can be subdivided into two categories: (i) reservoirs that were inherited from the molecular cloud from which the nebula was presumed to have come, and (ii) reservoirs that were generated within a nebula that was initially isotopically homogeneous. Two versions of the inheritance model have been proposed: (i) the dust and gas components had different proportions of supernova-produced ^{16}O, with a higher ^{16}O-abundance in the dust (Clayton et al., 1977), and (ii) the gas component was depleted in ^{16}O by photochemical processes in the molecular cloud, whereas the dust was not (Yurimoto and Kuramoto, 2002). Two versions of the local generation model have been proposed: (i) a gas-phase mass-independent fractionation reaction, as has been observed in the laboratory for synthesis of O_3 from O_2 (Heidenreich and Thiemens, 1983), and (ii) isotopic self-shielding in the photolysis of CO during the accretion of the Sun (Clayton, 2002).

On the basis of analyses of bulk CAIs or separated minerals, the inheritance of ^{16}O-rich condensates from supernovae appears plausible. However, the magnitude of the isotopic difference between the putative dust and gas reservoirs is a free parameter, so that the model has no predictive power. Furthermore, the observed oxygen isotope anomalies in presolar oxide grains are best understood as resulting from processing in red giant stars, and do not show ^{16}O-excesses.

Inheritance of ^{16}O-poor gas from the presolar molecular cloud is based on the well-known phenomenon of photochemical self-shielding, in which photolysis of CO in the cloud interior affects preferentially the less-abundant isotopic species: $C^{17}O$ and $C^{18}O$. Yurimoto and Kuramoto (2002) proposed that the ^{17}O and ^{18}O atoms thus formed can react with hydrogen to form water ice. When a portion of the cloud collapses to form the solar nebula, the ice evaporates to form a

^{16}O-depleted gas reservoir. This is a plausible scenario if the dust and ice components can maintain their isotopic distinction during nebular collapse and heating.

The mass-independent isotope fractionation in the gas-phase synthesis of O_3 from O_2 produces a slope-1 line on a three-isotope plot, with ozone being depleted in ^{16}O, and residual oxygen being enriched (Heidenreich and Thiemens, 1983). This process occurs naturally in the Earth's stratosphere, and the heavy-isotope excess can be chemically transferred to other atmospheric molecules (Thiemens, 1999). It is not clear how the isotopic anomalies could have been transmitted to meteoritic materials.

The process of photochemical self-shielding, which selects isotopes on the basis of their abundances, rather than their mass, might also have occurred within an initially homogeneous solar nebula, with the nascent Sun as the UV light source (Clayton, 2002). Because the density of the solar nebula was many orders of magnitude greater than the density of its precursor molecular cloud, the depth of penetration of solar UV was only a small fraction of 1 AU. The X-wind model of growth of the Sun from an accretion disk (Shu et al., 1996) provides a plausible setting for the self-shielding, in which accretionary matter is irradiated by the hot, early Sun at a distance of <0.1 AU, and a sizable fraction of the irradiated, chemically processed material is carried outward by the magnetically driven X-wind, to form the rocky parts of the inner solar system. The self-shielding mechanism can only act to increase the ^{17}O and ^{18}O abundances in the processed material, so that the primary solar system isotopic composition must have been at the ^{16}O-rich end of the mixing lines of Figures 3 and 4. In this picture, the ^{16}O-rich compositions at the lower end of the mixing line in Figure 4 represent condensates from the unaltered nebular gas, while the less-^{16}O-rich compositions are produced by chemical interaction with atomic oxygen, enriched in ^{17}O and ^{18}O by the photolysis of CO. The direct measurement of the oxygen isotope abundances in the solar wind, as sampled by the Genesis spacecraft, should allow a definitive choice from the proposed models.

1.06.3 OXYGEN ISOTOPES IN CHONDRITES

Chondritic meteorites, characterized by their relatively unfractionated chemical compositions, and usually consisting of chondrules and some matrix, can be subdivided into classes, as follows: carbonaceous chondrites—Vigarano-type (CV), Ornans-type (CO), Mighei-type (CM), Renazzo-type (CR), Karoonda-type (CK), Bencubbin-type (CB), and ALH 85085-type (CH) (see Chapter 1.05 for details); ordinary chondrites—high-iron (H), low-iron (L), and low-iron, low-metal (LL); and enstatite chondrites—high-iron (EH), low-iron (EL); R-chondrites, characterized by olivine with very high ferrous iron content.

Figure 5 shows schematically the locations and ranges of whole-rock oxygen isotopic compositions of the various chondrite classes. The distribution is intrinsically two dimensional,

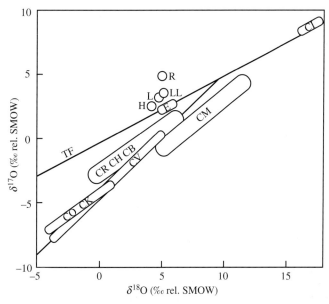

Figure 5 Schematic representation of the locations and ranges of whole-rock isotopic compositions of major chondrite classes. The TF and CCAM lines are shown for reference (CCAM mostly hidden by the CV balloon).

so that there is no single parameter, such as distance from the Sun, by which to order the groups. It is noteworthy that the ranges of compositions within each carbonaceous chondrite group are much larger than the ranges for the ordinary (O) and enstatite (E) groups. The large ranges for CM and CR–CH–CB groups are the result of varying degrees of low-temperature aqueous alteration, which produces phyllosilicates enriched in the heavy isotopes. It will be shown below that the CO and CM groups form a continuous trend on the three-isotope graph, and that the variations within the CO group may also be due to small degrees of aqueous alteration. The range of ~10‰ in the CV chondrites is not associated with phyllosilicate formation, but may represent internal isotopic heterogeneity due to the presence of ^{16}O-rich refractory phases. The whole-rock isotopic compositions of CV meteorites fall along the same CCAM line shown in Figure 3 for mineral analyses of refractory inclusions.

The isotopic composition of enstatite chondrites is indistinguishable from that of the Earth, which has led to suggestions that such material was a major "building block" for our planet (Javoy et al., 1986). It should be noted that the Earth does not occupy a special place on the diagram: it cannot be considered as "normal" or as an end-member of a mixing line.

The most abundant stony meteorites are the ordinary chondrites (H, L, and LL). Whole-rock oxygen isotopic compositions are similar, but resolvable, for the three iron groups (Clayton et al., 1991). Remarkably, analyses of individual chondrules from these meteorites show no grouping corresponding to the classification of the host meteorites (Figure 6). All ordinary chondrite chondrules appear to be derived from a single population. It is possible that the differences in whole-rock composition result from a size sorting of chondrules, as it is known that mean chondrule sizes increase in the order: H < L < LL (Rubin, 1989).

The isotopic compositions of the various carbonaceous chondrite groups imply some genetic relations among them. The CK data fall within the CO range, suggesting that CK are metamorphosed from CO-like precursors. The CR, CH, and CB meteorites have many properties in common, although the three groups differ in texture and metal content (Weisberg et al., 2001).

The characteristic feature of CM chondrules is the coexistence of roughly equal amounts of high-temperature anhydrous silicates (olivine and pyroxene) and low-temperature hydrous clay minerals. It is generally believed that the clay minerals were formed by aqueous alteration of the high-temperature phases, either in space or in the parent body. Figure 7 shows that the phyllosilicate matrix is systematically enriched in the heavy isotopes of oxygen, relative to the whole rock. The tie-lines between whole-rock compositions and matrix compositions have slopes of ~0.7, implying that the water reservoir had a composition with $\Delta^{17}O$ more positive than the silicate reservoir. Clayton and Mayeda (1999) showed that the observed patterns can be accounted for with a simple closed-system hydration reaction at temperatures near 0 °C,

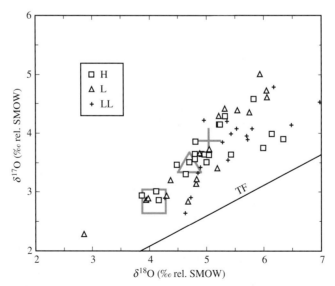

Figure 6 Oxygen isotopic compositions of individual chondrules from ordinary chondrites, with symbols showing the H, L, or LL group of the parent meteorite. The large gray symbols show the mean compositions of H, L, and LL whole rocks. There is no correlation between chondrule composition and parent composition, indicating that all ordinary chondrite chondrules are drawn from the same population. The TF line is shown for reference. The least-squares line fit to the chondrule data has a slope of 0.69 (source Clayton et al., 1991).

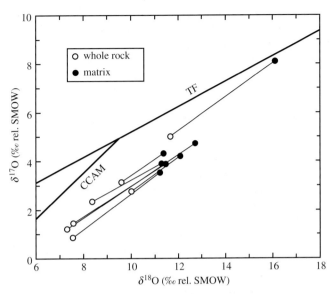

Figure 7 Whole-rock and separated matrix isotopic compositions for several CM chondrites. The matrix is always enriched in the heavier isotopes as a consequence of low-temperature aqueous alteration (source Clayton and Mayeda, 1999).

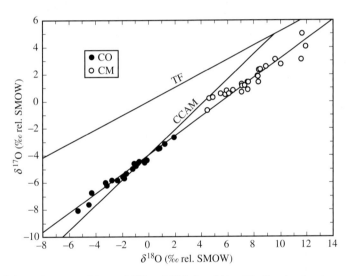

Figure 8 Whole-rock isotopic compositions of CO and CM chondrites. The line is a least-squares fit to all the data. The large range within each group reflects variations in water/rock ratio, which increases from lower left to upper right. The anhydrous precursors to both CO and CM groups should lie at the lower end of the CO group (source Clayton and Mayeda, 1999).

with water/rock ratios in the range 0.4–0.6 (in terms of oxygen atoms).

Since the phyllosilicates in CM chondrites are a few per mil enriched in heavy isotopes relative to the whole rock, material balance requires that the unanalyzed residual anhydrous silicates must be depleted in heavy isotopes by a comparable amount. This puts their composition into the range of CO chondrites. Figure 8 shows the relationship between CO and CM chondrites, which apparently represents different water/rock ratios in the aqueous alteration of a common CO-like precursor. The genetic association of CO and CM chondrites was discussed from a chemical and textural viewpoint by Rubin and Wasson (1986).

1.06.3.1 Thermal Metamorphism of Chondrites

1.06.3.1.1 Ordinary chondrites

Most ordinary chondrites have been thermally metamorphosed within their parent bodies, as evidenced by mineral recrystallization and chemical homogenization, from which a petrographic scale, from 3 (least metamorphosed) to 6 (most metamorphosed), has been constructed

(Van Schmus and Wood, 1967). This metamorphism occurred in an almost closed system with respect to oxygen, resulting in less than 0.5‰ variation in $\delta^{18}O$ for different metamorphic grades within each iron group (H, L, and LL) (Clayton et al., 1991). This observation is in accord with the inferred anhydrous state of the O-chondrite parent bodies.

During metamorphism and recrystallization, oxygen isotopes are redistributed among mineral phases, according to the mass-dependent equilibrium fractionations corresponding to the peak metamorphic temperature. The measured mineral-pair fractionations (usually for major minerals: olivine, pyroxene, and feldspar) can then be used for metamorphic thermometry, yielding temperatures of 600 °C for an L4 chondrite, and 850 ± 50 °C for several type-5 and type-6 chondrites (Clayton et al., 1991). Isotopic equilibration, even in type-6 chondrites, involves oxygen atom transport only over distances of a few millimeters (Olsen et al., 1981).

1.06.3.1.2 CO chondrites

The CO3 meteorites have been subdivided into metamorphic grades from 3.0 to 3.7, with peak temperatures in the range 450–600 °C (Rubin, 1998). In contrast to the metamorphism of the ordinary chondrites, the CO metamorphism probably occurred in the presence of water, and the system was not closed with respect to oxygen. The least metamorphosed CO meteorites, ALH 77307 and Y 81020, are the most ^{16}O-rich; Loongana 001 and HH 073, classified as 3.8 or 4, are the most ^{16}O-poor. There is not, however, a simple one-to-one correspondence between metamorphic grade and isotopic composition. Since most of the CO chondrites are finds, some may have been altered by terrestrial weathering.

1.06.3.1.3 CM and CI chondrites

Some CM and CI chondrites have been heated to temperatures sufficient to cause dehydration of phyllosilicates (Akai, 1990). Reflectance spectra of many asteroids suggest that metamorphosed carbonaceous chondrite material is common (Hiroi et al., 1993). Oxygen isotopic compositions of dehydrated CM and CI chondrites occupy two separate regions in the three-isotope graph (Figure 9), one falling near the terrestrial line with $\delta^{18}O$ and $\delta^{17}O$ values more positive than the unmetamorphosed CI chondrites, and the other falling near the CM chondrite region, but lower in $\delta^{17}O$ by ~1‰. Thus, neither group has isotopic compositions corresponding to those of any unmetamorphosed meteorites. The process of thermal dehydration itself probably introduces

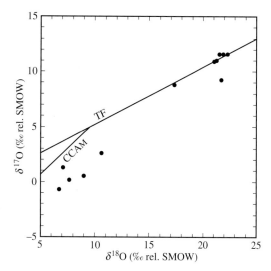

Figure 9 Oxygen isotopic compositions of metamorphosed CM- and CI-like materials, identified by their reflectance spectra, which show dehydration of phyllosilicates (Akai, 1990). The upper group may have had CI-like precursors, and the lower group may have had CM-like precursors, but the isotopic effects of metamorphism are uncertain (source Clayton and Mayeda, 1999).

some isotopic fractionation. However, in rocks as chemically complex as CM and CI chondrites, it is not possible to predict either the magnitude or direction of such fractionation effects (Clayton et al., 1997).

Another class of objects that have apparently gone through a cycle of hydration and dehydration are the dark inclusions commonly found in CV chondrites (Clayton and Mayeda, 1999). These have a wide range of isotopic compositions, enriched in the heavy oxygen isotopes, reflecting primarily the hydration step.

1.06.4 OXYGEN ISOTOPES IN ACHONDRITES

The achondritic meteorites can be subdivided into the differentiated achondrites: igneous rocks from parent bodies that were extensively melted, and the undifferentiated, or primitive, achondrites from parent bodies that underwent little melting.

1.06.4.1 Differentiated (Evolved) Achondrites

Oxygen isotope variations in differentiated bodies—such as Earth, Moon, Mars—and the parent body of the HED group (for howardites, eucrites, and diogenites) show characteristic mass-dependent fractionation lines, with a slope of 0.52 on the three-isotope plot

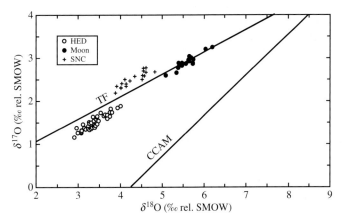

Figure 10 Oxygen isotopic compositions of whole-rock meteorites from differentiated bodies: HED, possibly asteroid Vesta; SNC, possibly Mars; and Moon. Each body produces a slope-1/2 mass-dependent fractionation line, with values of $\Delta^{17}O$ characteristic of the whole source planet. The isotopic compositions of lunar meteorites are indistinguishable from those of terrestrial mantle rocks. Note that HED and SNC lie on opposite sides of TF. This and Figure 11 are drawn to the same scale for comparison (source Clayton and Mayeda, 1996).

(Figure 10). In general, different bodies define separate, parallel trends with different $\Delta^{17}O$ ($\Delta^{17}O = \delta^{17}O - 0.52\ \delta^{18}O$), reflecting the different bulk isotopic composition of the parent body. These separate fractionation lines are useful in demonstrating that diverse lithologies are derived from a common source reservoir. For example, the SNC meteorites (for Shergotty, Nakhla, and Chassigny) range from basalts to peridotite, but all have the same value of $\Delta^{17}O = 0.30\textperthousand$, supporting other evidence for a common parent body. This same property identified ALH 84001 as a member of this group, even though its petrography and age are unique (Clayton and Mayeda, 1996). Similarly, the HED meteorites have a uniform value of $\Delta^{17}O = -0.25\textperthousand$, again implying a common parent body, perhaps the asteroid Vesta. Caution must be applied, however, in that accidental coincidences may occur. For example, Earth, Moon, and the aubrite (enstatite achondrite) parent body have indistinguishable values of $\Delta^{17}O = 0\textperthousand$ (Wiechert et al., 2001), but are obviously separate bodies.

Besides the HED group, other achondrites that also have $\Delta^{17}O$ close to $-0.25\textperthousand$ are the mesosiderites, main-group pallasites, and IIIAB iron meteorites. It remains a subject of controversy whether this similarity is an accident, whether it implies a single parent body, or whether it implies several related parent bodies derived from some common reservoir.

Differentiated, or evolved, achondrite types are rare. Other than the HED group, there are only two other known asteroidal sources, one for the angrites (six known meteorites) (Mittlefehldt et al., 2002) and one for the unique basalt NWA 011 (Yamaguchi et al., 2002). Oxygen isotopic compositions of angrites are barely resolvable from those of the HED group (Clayton and Mayeda, 1996), whereas that of NWA 011 is distinctly different, and falls near the CR chondrite group.

The pallasites, coarse-grained stony-iron meteorites, also form three oxygen isotope groups: the main group, the Eagle Station pallasites (three members), and the pyroxene pallasites (two members) (Clayton and Mayeda, 1996).

In contrast to the small number of differentiated parent bodies represented by evolved achondritic meteorites, the number of parent bodies inferred from the chemical compositions of iron meteorites may be as large as 50 (Wasson, 1990). Of the 13 major iron meteorite groups, 10 appear to be from cores of differentiated meteorites. Many additional cores are inferred from the "ungrouped" irons, which make up ~15% of iron meteorites. It is a puzzle why we appear to sample many more cores than mantles of these asteroids (see Chapter 1.12 for further discussion).

Some iron meteorites contain oxygen-bearing phases: chromite, phosphates, and silicates, which can be isotopically analyzed to search for genetic associations with stony meteorite groups. There is a clear isotopic and chemical connection between the group IAB irons and winonaites (one class of primitive achondrite, discussed below). An equally strong association is found for IIIAB irons and main-group pallasites (Clayton and Mayeda, 1996). Oxygen-isotope links are also found between IIE irons and H-group ordinary chondrites, and between group IVA irons and L- or LL-group ordinary chondrites. Several ungrouped irons have oxygen isotopes in the range of CV carbonaceous chondrites (Clayton and Mayeda, 1996).

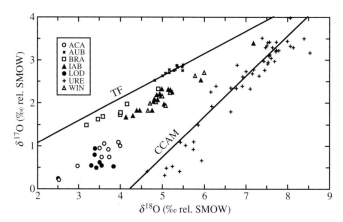

Figure 11 Oxygen isotopic compositions of whole-rock "primitive" achondrites. Data for a given class are much more scattered than data for differentiated achondrites shown in Figure 10, as a result of incomplete melting and homogenization. Several genetic associations are implied by the data: (i) aubrites and enstatite chondrites; (ii) acapulcoites and lodranites; (iii) IAB irons and winonaites; and (iv) ureilites and dark inclusions in carbonaceous chondrites. This and Figure 10 are drawn to the same scale for comparison (source Clayton and Mayeda, 1996).

The close association of some iron meteorites with chondrites suggests an origin as impact melts, rather than asteroidal cores.

1.06.4.2 Undifferentiated (Primitive) Achondrites

A combination of chemical, textural, and oxygen isotopic information permits recognition of several groups of achondrites that may have undergone a small degree of partial melting, but were never part of a differentiation process on an asteroidal scale (Goodrich and Delaney, 2000). They argue that the distinction between "evolved" achondrites and "primitive" achondrites is that the former are melts, while the latter are residues of partial melting. These residues may retain chemical and isotopic heterogeneities inherited from their unmelted precursors. By this criterion, the primitive achondrites include the aubrites, lodranites/acapulcoites, brachinites, winonaites, and ureilites. Their oxygen isotopic compositions are shown in Figure 11.

The aubrites (enstatite achondrites) are clearly closely related to the enstatite chondrites, discussed earlier. They also share the property of being highly reduced, and have identical oxygen isotopic compositions (Clayton et al., 1984). Experimental studies by McCoy et al. (1999) show that partial melting of an E-chondrite can yield aubritic material by removal of a basaltic melt and a metal-sulfide melt. This is the best known instance of a genetic connection between chondrites and achondrites.

The brachinites are a small group of olivine-rich residues, with oxygen isotopic compositions overlapping those of the HED achondrites, although no genetic association has been proposed (Clayton and Mayeda, 1996).

The lodranites and acapulcoites are almost certainly from the same parent body (McCoy et al., 1997). Both groups are residues of partial melting, with the lodranites having undergone more extensive melting and recrystallization. Their oxygen isotopic compositions overlap and form a blob on the three-isotope graph, rather than a fractionation line that would indicate complete melting. The mean isotopic composition of this group falls within the range of CR and CH chondrites, but no genetic association has been proposed.

The winonaites and silicates in IAB irons share the same range of oxygen isotopic composition, and probably have a common parent body. A model for generation of winonaites and IAB–IIICD irons in the same body (completely independent of oxygen isotope evidence) was presented by Kracher (1985).

The ureilites are the most remarkable of the primitive achondrites. They are coarse-grained, carbon-bearing ultramafic rocks, and are residues of extensive partial melting that has removed a basaltic component. Their oxygen isotopic compositions are highly variable, and extend along an extrapolation of the CCAM line. The isotopic variability is probably inherited from heterogeneous carbonaceous chondrite precursors (Clayton and Mayeda, 1988).

1.06.5 SUMMARY AND CONCLUSIONS

Oxygen isotope abundances display a remarkable variability on all scales studied, from micrometers to planetary dimensions. Two types of processes have been identified and associated with some of this variability: (i) a nebular interaction between condensates and the ambient gas, probably at a temperature sufficiently high that mass-dependent fractionation effects were relatively small, and (ii) low-temperature aqueous alteration, probably occurring within the parent bodies. The first process begins with ^{16}O-rich condensates, as are seen in CAIs, and enriches them in ^{17}O and ^{18}O along the CCAM line. The second process modifies these materials, especially in the carbonaceous chondrites, with further enrichment of the two heavier isotopes, but on lines of shallower slope as a consequence of large low-temperature fractionation effects. Although these two processes may be adequate to account for the observed variability in the chondritic meteorites, they do not seem to provide an explanation for the range of compositions of the achondrites. Some *ad hoc* explanations have been proposed, such as the existence of additional oxygen reservoirs or time-variable oxygen sources (Wasson, 2000), but these are basically untestable.

Many questions remain unanswered. What was the anhydrous precursor for the CR–CH–CB group? What were the precursors of the metamorphosed CM and CI meteorites? What was the relationship between ureilites and carbonaceous chondrites? Why are so few differentiated parent bodies represented by achondrites? Why is the isotopic composition of the Earth identical to that of the Moon, but different from that of Mars? What is the relationship (if any) between the Earth and the enstatite meteorites? Some of these questions may be successfully addressed once we have accurate fine-scaled chronology so as to put solar nebular events into the correct time sequence.

REFERENCES

Akai J. (1990) Mineralogical evidence of heating events in Antarctic carbonaceous chondrites. *Proc. NIPR Symp. Antarct. Meteorit.* **3**, 55–68.

Aléon J., Krot A. N., McKeegan K. D., MacPherson G. J., and Ulyanov A. A. (2002) Oxygen isotopic composition of fine-grained Ca–Al-rich inclusions in the reduced CV3 chondrite Efremovka. *Lunar Planet. Sci.* **XXXIII**, #1426. The Lunar and Planetary Institute, Houston (CD-ROM).

Allègre C. J., Manhès G., and Göpel C. (1995) The age of the Earth. *Geochim. Cosmochim. Acta* **59**, 1445–1446.

Anders E. and Zinner E. (1993) Interstellar grains in primitive meteorites: diamond, silicon carbide and graphite. *Meteoritics* **28**, 490–514.

Boothroyd A. I. and Sackmann I. J. (1999) The CNO isotopes: deep circulation in red giants and first and second dredge-up. *Astrophys. J.* **510**, 232–250.

Clayton R. N. (1993) Oxygen isotopes in meteorites. *Ann. Rev. Earth Planet. Sci.* **21**, 115–149.

Clayton R. N. (2002) Self-shielding in the solar nebula. *Nature* **415**, 860–861.

Clayton R. N. and Mayeda T. K. (1988) Formation of ureilites by nebular processes. *Geochim. Cosmochim. Acta* **52**, 1313–1318.

Clayton R. N. and Mayeda T. K. (1996) Oxygen isotope studies of achondrites. *Geochim. Cosmochim. Acta* **60**, 1999–2017.

Clayton R. N. and Mayeda T. K. (1999) Oxygen isotope studies of carbonaceous chondrites. *Geochim. Cosmochim. Acta* **63**, 2089–2104.

Clayton R. N., Onuma N., Grossman L., and Mayeda T. K. (1977) Distribution of the presolar component in Allende and other carbonaceous chondrites. *Earth Planet. Sci. Lett.* **34**, 209–224.

Clayton R. N., Onuma N., Ikeda Y., Mayeda T. K., Hutcheon I. D., Olsen E. J., and Molini-Velsko C. (1983) Oxygen isotopic compositions of chondrules in Allende and ordinary chondrites. In *Chondrules and their Origins* (ed. E. A. King). Lunar and Planetary Institute, Houston, pp. 37–43.

Clayton R. N., Mayeda T. K., and Rubin A. E. (1984) Oxygen isotopic compositions of enstatite chondrites and aubrites. *J. Geophys. Res.* **89**, C245–C249.

Clayton R. N., Mayeda T. K., Goswami J. N., and Olsen E. J. (1991) Oxygen isotope studies of ordinary chondrites. *Geochim. Cosmochim. Acta* **35**, 2317–2338.

Clayton R. N., Mayeda T. K., Hiroi T., Zolensky M. E., and Lipschutz M. E. (1997) Oxygen isotopes in laboratory-heated CI and CM chondrites. *Meteorit. Planet. Sci.* **32**, A30.

Dauphas N., Marty B., and Reisberg L. (2002) Molybdenum nucleosynthetic dichotomy revealed in primitive meteorites. *Astrophys. J.* **569**, L139–L142.

Davis A. M., Hashimoto A., Clayton R. N., and Mayeda T. K. (1990) Isotope mass fractionation during evaporation of Mg$_2$SiO$_4$. *Nature* **347**, 655–658.

Eberhardt P., Geiss J., Graf H., Grögler N., Krähenbühl U., Schwaller H., Schwarzmüller J., and Stettler A. (1970) Trapped solar wind noble gases, exposure age, and K/Ar-age in Apollo 11 lunar fine material. *Proc. Apollo 11 Lunar Sci. Conf.*, Pergamon, New York, pp. 1037–1070.

Epstein S. and Taylor H. P. (1970) The concentration and isotopic composition of hydrogen, carbon, and silicon in Apollo 11 lunar rocks and minerals. *Proc. Apollo 11 Lunar Sci. Conf.*, Pergamon, New York, pp. 1085–1096.

Fagan T. J., Yurimoto H., Krot A. N., and Keil K. (2002) Constraints on oxygen isotopic evolution from an amoeboid olivine aggregate and Ca, Al-rich inclusions from the CV3 Efremovka. *Lunar Planet. Sci.* **XXXIII**, #1507. The Lunar and Planetary Institute, Houston (CD-ROM).

Giauque W. F. and Johnston H. L. (1929) An isotope of oxygen, mass 18. Interpretation of the atmospheric absorption bands. *J. Am. Chem. Soc.* **51**, 1436–1441.

Gloeckler G. and Geiss J. (1998) Measurement of the abundance of helium-3 in the Sun and in the local interstellar cloud with SWICS on Ulysses. *Space Sci. Rev.* **84**, 275–284.

Goodrich C. A. and Delaney J. S. (2000) Fe/Mg–Fe/Mn relations of meteorites and primary heterogeneity of primitive achondrite parent bodies. *Geochim. Cosmochim. Acta* **64**, 149–160.

Grossman L. (1973) Refractory trace elements in Ca–Al-rich inclusions in the Allende meteorite. *Geochim. Cosmochim. Acta* **37**, 1119–1140.

Harris M. J., Lambert D. L., and Goldman A. (1987) The ^{12}C/^{13}C and ^{16}O/^{18}O ratios in the solar photosphere. *Mon. Not. Roy. Astron. Soc.* **224**, 237–255.

Hashizume K., Chaussidon M., Marty B., and Robert F. (2000) Solar wind record on the Moon: deciphering presolar from planetary nitrogen. *Science* **290**, 1142–1145.

Heidenreich J. E. and Thiemens M. H. (1983) A non-mass-dependent isotope effect in the production of ozone from molecular oxygen. *J. Chem. Phys.* **78**, 892–895.

Hiroi T., Pieters C. M., Zolensky M. E., and Lipschutz M. E. (1993) Evidence of thermal metamorphism in the C, G, B, and F asteroids. *Science* **261**, 1016–1018.

Itoh S., Rubin A. E., Kojima H., Wasson J. T., and Yurimoto H. (2002) Amoeboid olivine aggregates and AOA-bearing chondrule from Y 81020 CO 3.0 chondrite: distribution of oxygen and magnesium isotopes. *Lunar Planet. Sci.* **XXXIII**, #1490. The Lunar and Planetary Institute, Houston (CD-ROM).

Javoy M., Pineau F., and Delorme H. (1986) Carbon and nitrogen isotopes in the mantle. *Chem. Geol.* **57**, 41–62.

Jones R. H., Leshin L. A., and Guan Y. (2002) Heterogeneity and ^{16}O-enrichments in oxygen isotope ratios of olivine from chondrules in the Mokoia CV3 chondrite. *Lunar Planet. Sci.* **XXXIII**, #1571. The Lunar and Planetary Institute, Houston (CD-ROM).

Kallenbach R., Geiss J., Ipavich F. M., Gloeckler G., Bochsler P., Gliem F., Hefti S., Hilchenbach M., and Hovestadt D. (1998) Isotopic composition of solar-wind nitrogen: first *in situ* determination with the CELIAS/MTOF spectrometer on board SOHO. *Astrophys. J.* **507**, L185–L188.

Kerridge J. F. (1993) Long-term compositional variation in solar corpuscular radiation: evidence from nitrogen isotopes in the lunar regolith. *Rev. Geophys.* **31**, 423–437.

Kracher A. (1985) The evolution of partially differentiated planetesimals: evidence from iron meteorite groups IAB and IIICD. *J. Geophys. Res.* **90**, C689–C698.

Krot A. N., Aléon J., and McKeegan K. D. (2002) Mineralogy, petrography and oxygen-isotopic compositions of Ca–Al-rich inclusions and amoeboid olivine aggregates in the CR carbonaceous chondrites. *Lunar Planet. Sci.* **XXXIII**, #1412. The Lunar and Planetary Institute, Houston (CD-ROM).

Leske R. A., Mewaldt R. A., Cohen C. M. S., Christian E. R., Cummings A. C., Slocum P. L., Stone E. C., von Rosenvinge T. T., and Wiedenbeck M. E. (2001) Isotopic abundances in the solar corona as inferred from ACE measurements of solar energetic particles. In *Solar and Galactic Composition* (ed. R. F. Wimmer-Schweingruber). AIP Conf. Proc. 598, Melville, NY, pp. 127–132.

Manian S. H., Urey H. C., and Bleakney W. (1934) An investigation of the relative abundance of the oxygen isotopes $O^{16}:O^{18}$ in stone meteorites. *J. Am. Chem. Soc.* **56**, 2601–2609.

McCoy T. J., Keil K., Clayton R. N., Mayeda T. K., Bogard D. D., Garrison D. H., and Wieler R. (1997) A petrologic and isotopic study of lodranites: evidence for early formation as partial melt residues from heterogeneous precursors. *Geochim. Cosmochim. Acta* **61**, 623–637.

McCoy T. J., Dickinson T. L., and Lofgren G. E. (1999) Partial melting of the Indarch (EH4) meteorite: a textural, chemical, and phase relations view of melting and melt migration. *Meteorit. Planet. Sci.* **34**, 735–746.

Mittlefehldt D. W., Kilgore M., and Lee M. T. (2002) Petrology and geochemistry of d'Orbigny, geochemistry of Sahara 99555, and the origin of angrites. *Meteorit. Planet. Sci.* **37**, 345–369.

Nicolussi G. K., Pellin M. J., Lewis R. S., Davis A. M., Amari S., and Clayton R. N. (1998) Molybdenum isotopic composition of individual presolar silicon carbide grains from the Murchison meteorite. *Geochim. Cosmochim. Acta* **62**, 1093–1104.

Nittler L. R., Alexander C. M. O'D., Gao X., Walker R. M., and Zinner E. (1997) Stellar sapphires: the properties and origins of presolar Al_2O_3 in meteorites. *Astrophys. J.* **483**, 475–495.

Olsen E. J., Mayeda T. K., and Clayton R. N. (1981) Cristobalite-pyroxene in an L6 chondrite: implications for metamorphism. *Earth Planet. Sci. Lett.* **56**, 82–88.

Owen T., Mahaffy P. R., Niemann H. B., Atreya S., and Wong M. (2001) Protosolar nitrogen. *Astrophys. J.* **553**, L77–L79.

Pepin R. O. (1991) On the origin and early evolution of terrestrial planet atmospheres and meteoritic volatiles. *Icarus* **92**, 2–79.

Prantzos N., Aubert O., and Audouze J. (1996) Evolution of carbon and oxygen isotopes in the Galaxy. *Astron. Astrophys.* **309**, 760–774.

Rubin A. E. (1989) Size-frequency distributions of chondrules in CO3 chondrites. *Meteoritics* **24**, 179–189.

Rubin A. E. (1998) Correlated petrologic and geochemical characteristics of CO3 chondrites. *Meteorit. Planet. Sci.* **33**, 385–391.

Rubin A. E. and Wasson J. T. (1986) Chondrules in the Murray CM2 meteorite and compositional differences between CM–CO and ordinary chondrite chondrules. *Geochim. Cosmochim. Acta* **50**, 307–315.

Shu F. H., Shang H., and Lee T. (1996) Toward an astrophysical theory of chondrites. *Science* **271**, 1545–1552.

Speck A. K., Barlow M. J., Sylvester R. J., and Hofmeister A. M. (2000) Dust features in the 10 μm infrared spectra of oxygen-rich evolved stars. *Astron. Astrophys.* **146**, 437–464.

Thiemens M. H. (1999) Atmospheric science—mass-independent isotope effects in planetary atmospheres and the early solar system. *Science* **283**, 341–345.

van Dishoeck E. F. and Black J. H. (1988) The photodissociation and chemistry of interstellar CO. *Astrophys. J.* **334**, 771–802.

Van Schmus W. R. and Wood J. A. (1967) A chemical-petrologic classification for the chondritic meteorites. *Geochim. Cosmochim. Acta* **31**, 747–765.

Wasson J. T. (1990) Ungrouped iron meteorites in Antarctica: origin of anomalously high abundance. *Science* **249**, 900–902.

Wasson J. T. (2000) Oxygen isotopic evolution of the solar nebula. *Rev. Geophys.* **38**, 491–512.

Weisberg M. K., Prinz M., Kojima H., Yanai K., Clayton R. N., and Mayeda T. K. (1991) The Carlisle Lakes-type chondrites: a new grouplet with high $\Delta^{17}O$ and evidence for nebular oxidation. *Geochim. Cosmochim. Acta* **55**, 2657–2669.

Weisberg M. K., Prinz M., Clayton R. N., Mayeda T. K., Sugiura N., Zashu S., and Ebihara M. (2001) A new metal-rich chondrite grouplet. *Meteorit. Planet. Sci.* **36**, 401–418.

Wiechert U., Halliday A. N., Lee D.-C., Snyder G. A., Taylor L. A., and Rumble D. (2001) Oxygen isotopes and the Moon-forming giant impact. *Science* **294**, 345–348.

Wimmer-Schweingruber R. F., Bochsler P., and Gloeckler G. (2001) The isotopic composition of oxygen in the fast solar wind. *Geophys. Res. Lett.* **28**, 2763–2766.

Wyckoff S., Kleine M., Peterson B. A., Wehinger P. A., and Ziurys L. M. (2000) Carbon isotope abundances in comets. *Astrophys. J.* **535**, 991–999.

Yamaguchi A., Clayton R. N., Mayeda T. K., Ebihara M., Oura Y., Miura Y. N., Haramura H., Misawa K., Kojima H., and Nagao K. (2002) A new source of basaltic meteorites inferred from Northwest Africa 011. *Science* **196**, 334–336.

Yin Q., Jacobsen S. B., and Yamashita K. (2002) Diverse supernova sources of pre-solar material inferred from molybdenum isotopes in meteorites. *Nature* **415**, 881–885.

Yurimoto H. and Kuramoto K. (2002) A possible scenario introducing heterogeneous oxygen isotopic distribution in protoplanetary disks. *Meteorit. Planet. Sci. Suppl.* **37**, A153.

Zhu X. K., Guo Y., O'Nions R. K., Young E. D., and Ash R. D. (2001) Isotopic homogeneity of iron in the early solar nebula. *Nature* **412**, 311–313.

1.07
Chondrites and their Components

E. R. D. Scott and A. N. Krot

University of Hawai'i at Manoa, Honolulu, HI, USA

1.07.1	INTRODUCTION	144
	1.07.1.1 What are Chondrites?	144
	1.07.1.2 Why Study Chondrites?	144
	1.07.1.3 Historical Views on Chondrite Origins	145
	1.07.1.4 Chondrites and the Solar Nebula	146
1.07.2	CLASSIFICATION AND PARENT BODIES OF CHONDRITES	146
	1.07.2.1 Chondrite Groups, Clans, and Parent Bodies	146
	1.07.2.2 Ordinary Chondrites	148
	1.07.2.3 Carbonaceous Chondrites	148
	1.07.2.4 Enstatite Chondrites	149
1.07.3	BULK COMPOSITION OF CHONDRITES	149
	1.07.3.1 Cosmochemical Classification of Elements	149
	1.07.3.2 Chemical Compositions of Chondrites	150
	1.07.3.3 Isotopic Compositions of Chondrites	151
	1.07.3.4 Oxygen Isotopic Compositions	151
1.07.4	METAMORPHISM, ALTERATION, AND IMPACT PROCESSING	152
	1.07.4.1 Petrologic Types	152
	1.07.4.1.1 Type 1–2 chondrites	152
	1.07.4.1.2 Type 3 chondrites	153
	1.07.4.1.3 Type 4–6 chondrites	154
	1.07.4.2 Thermal History and Modeling	154
	1.07.4.3 Impact Processing of Chondritic Asteroids	155
1.07.5	CHONDRITIC COMPONENTS	155
	1.07.5.1 Calcium- and Aluminum-rich Inclusions	156
	1.07.5.1.1 Comparison of CAIs from different chondrite groups	157
	1.07.5.1.2 Oxygen Isotopic compositions of CAIs	158
	1.07.5.2 Forsterite-rich Accretionary Rims Around CAIs	160
	1.07.5.3 Amoeboid Olivine Aggregates (AOAs)	160
	1.07.5.3.1 Mineralogy and petrology of AOAs	161
	1.07.5.3.2 Trace elements and isotopic composition of AOAs	162
	1.07.5.3.3 Origin of amoeboid olivine aggregates	162
	1.07.5.4 Aluminum-rich Chondrules	163
	1.07.5.4.1 Isotopic and trace element studies of aluminum-rich chondrules	165
	1.07.5.4.2 Origin of aluminum-rich chondrules	166
	1.07.5.4.3 Relict CAIs in ferromagnesian chondrules	168
	1.07.5.5 Chondrules	168
	1.07.5.5.1 Chondrule textures, types, and thermal histories	169
	1.07.5.5.2 Compositions of chondrules	171
	1.07.5.5.3 Compound chondrules, relict grains, and chondrule rims	172
	1.07.5.5.4 Closed-system crystallization	172
	1.07.5.5.5 Open-system crystallization	173
	1.07.5.5.6 Chondrules that formed on asteroids	175
	1.07.5.6 Metal and Troilite	175
	1.07.5.6.1 CR chondrite metallic Fe,Ni	176
	1.07.5.6.2 CH and CB chondrite metallic Fe,Ni	176
	1.07.5.6.3 Troilite	177
	1.07.5.7 Matrix Material	177
	1.07.5.7.1 CI1 chondrites	179
	1.07.5.7.2 CM2 chondrite matrices	180

1.07.5.7.3 CR2 chondrite matrices	180
1.07.5.7.4 CO3 chondrite matrices	180
1.07.5.7.5 CK3 chondrite matrices	182
1.07.5.7.6 CV3 chondrite matrices	182
1.07.5.7.7 Matrix of ungrouped C chondrites, Acfer 094, and Adelaide	184
1.07.5.7.8 H–L–LL3 chondrite matrices	184
1.07.5.7.9 K3 chondrite matrix (Kakangari)	184
1.07.5.7.10 Heavily altered, matrix-rich lithic clasts	185
1.07.5.7.11 Origins of matrix phases	186
1.07.6 FORMATION AND ACCRETION OF CHONDRITIC COMPONENTS	187
1.07.7 HEATING MECHANISMS IN THE EARLY SOLAR SYSTEM	188
1.07.7.1 Nebular Shocks	189
1.07.7.2 Jets and Outflows	189
1.07.7.3 Impacts on Planetesimals	189
ACKNOWLEDGMENTS	190
REFERENCES	190

1.07.1 INTRODUCTION

1.07.1.1 What are Chondrites?

Chondrites are meteorites that provide the best% clues to the origin of the solar system. They are the oldest known rocks—their components formed during the birth of the solar system ca. 4,567 Ma—and their abundances of nonvolatile elements are close to those in the solar photosphere. Chondrites are broadly ultramafic in composition, consisting largely of iron, magnesium, silicon, and oxygen. The most abundant constituents of chondrites are chondrules, which are igneous particles that crystallized rapidly in minutes to hours. They are composed largely of olivine and pyroxene, commonly contain metallic Fe,Ni and are 0.01–10 mm in size. Some chondrules are rounded as they were once entirely molten but many are irregular in shape because they were only partly melted or because they accreted other particles as they solidified. Chondrites themselves were never molten. The definition of a chondrite has expanded recently with the discovery in Antarctica and the Sahara Desert of extraordinary meteorites with chondrules 10–100 μm in size, and chondrites so rich in metallic Fe,Ni that they were initially classified as iron meteorites with silicate inclusions. Thus, in meteoritics, as in other fields of planetary science, new discoveries sometimes require definitions to be modified.

Chondrites are so diverse in their mineralogical and textural characteristics that it is not possible to describe a typical chondrite. We show one with diversely textured chondrules including prominent, aesthetically pleasing, rounded chondrules (Figure 1(a)), and another with more uniformly textured chondrules (Figure 1(b)). Owing to the high abundance of rounded or droplet chondrules in the abundant, so-called "ordinary" chondrites (Figure 1(a)), studies of the origin of chondrules have commonly been based on these chondrites.

Chondrites contain diverse proportions of three other components: refractory inclusions (0.01–10 vol.%), metallic Fe,Ni (<0.1–70%), and matrix material (1–80%). Refractory inclusions are tens of micrometers to centimeters in dimensions, lack volatile elements, and are the products of high-temperature processes including condensation, evaporation, and melting. Two types are recognized: calcium- and aluminum-rich inclusions or CAIs, and amoeboid olivine aggregates. CAIs are composed of minerals such as spinel, melilite, hibonite, perovskite and Al–Ti-diopside, which are absent in other chondritic components (see Chapter 1.08). Amoeboid olivine aggregates consist of fine-grained olivine, Fe,Ni metal, and a refractory component largely composed of aluminum-diopside, anorthite, spinel and rare melilite. Grains of metallic Fe,Ni occur inside and outside the chondrules as grains up to a millimeter in size and, like the chondrules and refractory inclusions, formed at high temperatures. Matrix material is volatile-rich, and fine-grained (5–10 μm) and forms rims on other components and fills the interstices between them. Chondrite matrices have diverse mineralogies: most are disequilibrium mixtures of hydrated and anhydrous silicates, oxides, metallic Fe,Ni, sulfides, and organic material and contain rare presolar grains.

1.07.1.2 Why Study Chondrites?

Goldschmidt, Suess, and Urey showed that chondrites provide the best estimates for the mean abundances of condensable elements in the solar system. These estimates were essential for developing theories for the formation of elements in evolved stars (see Chapters 1.01 and 1.03). Presolar grains (see Chapter 1.02) provide additional clues to nucleosynthesis and the subsequent growth of circumstellar grains. Chondrules, metal grains, refractory inclusions, and matrix materials formed under very diverse conditions in the solar system and appear to offer

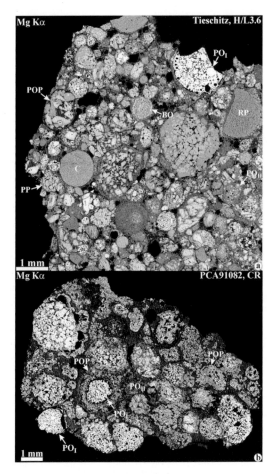

Figure 1 Maps showing magnesium concentrations in two chondrites: (a) PCA91082, a CR2 carbonaceous chondrite, and (b) Tieschitz, an H/L3.6 ordinary chondrite. In CR chondrites, as in most carbonaceous chondrites, nearly all chondrules have porphyritic textures and are composed largely of forsterite (white grains), enstatite (gray), and metallic Fe,Ni (black). The subscripts show type I chondrules, which are common, and type II, which are FeO-rich and rare in this chondrite. Tieschitz, like other ordinary chondrites, is composed of all kinds of chondrules with diverse FeO concentrations. Key to chondrule types: BO, barred olivine; C, cryptocrystalline, PO, porphyritic olivine; POP, porphyritic olivine-pyroxene; PP, porphyritic pyroxene; RP, radial pyroxene. These maps were made with an electron microprobe from Mg Kα X-rays.

insights into processes that occurred during the formation of the Sun and planets from a collapsing cloud of interstellar dust and gas. The rocks themselves provide clues to the geological processes including impact processes that affected asteroids over 4.5 Ga. Studies of chondrites help us to match chondrite groups with asteroid classes, to understand the origin and evolution of the asteroid belt, the nature of planetesimals that accreted into terrestrial planets, and the reason for the dearth of planetary material between Mars and Jupiter. Finally, chondrite studies help us to understand the physical and mineralogical structure of unmelted asteroids and to assess what should be done about rogue near-Earth objects that threaten Earth.

In this chapter, we focus on the insights that chondrites offer into the earliest stages in the formation of asteroids and planetesimals in the solar nebula—the protoplanetary disk of dust and gas that evolved into the planetary system. Other chapters review many other aspects of chondrite studies.

1.07.1.3 Historical Views on Chondrite Origins

At the beginning of the space age in the 1960s, there were two views about the origin of chondrules and the variations in the concentrations of the most volatile elements in chondrites. Ringwood, Urey, and others invoked volcanism, impacts and other planetary processes, whereas Wood, Larimer, and Anders invoked processes in the solar nebula (Wood, 1963; Larimer, 1967; Larimer and Anders, 1967). In the early 1970s, studies of CAIs by Grossman and others appeared to strongly favor an origin for chondritic components as condensates from the solar nebula (Grossman and Larimer, 1974). However, this simple picture was disrupted by discoveries of ubiquitous isotopic anomalies in CAIs and the complexity of the mineralogical and isotopic record in chondritic components, which seemed inconsistent with the standard model of a hot, monotonically cooling solar nebula. In addition, astronomical evidence and theoretical studies suggested that the solar nebula was not hot enough to vaporize silicates where chondrites formed. The solar nebula was envisaged as a place where there were dynamic, energetic, dust-rich zones with constantly changing dust/gas ratios and fluctuating high temperatures. "Mineral material caught in this maelstrom went through multiple cycles of melting, evaporation, recondensation, crystallization and aggregation" (Wood, 1988).

Since the mid-1980s, much progress has been made in understanding chondrites, though the origin of their components has not been resolved. Discoveries of new kinds of carbonaceous chondrites have shown that the Allende meteorite, which is the most-studied chondrite, is not as pristine as once believed. Its components were modified in asteroids (e.g., Krot et al., 1995, 1998a) or the solar nebula (e.g., Palme and Fegley, 1990; Weisberg and Prinz, 1998). Detailed isotopic and chemical data for CAIs and chondrules from ion microprobe studies have been invaluable in understanding how and where chondritic components were altered, but have not yet resolved their origin. Most authors now accept that chondrules and CAIs formed in some kind of solar nebula, and several promising models have been developed, but because of inconsistencies with chondrite data, some workers favor asteroidal models (e.g., Sanders, 1996; Lugmair and Shukolyukov, 2001; Hsu et al., 2000). To meteoriticists like Wood, who argued in 1963 that chondrites were aggregates of solar nebular condensates, it has seemed that chondrite research has stagnated and that little progress has been made in developing plausible theories for the origins of chondrules, CAIs and chondrites (Kerr, 2001). However, most find the lure of studying presolar grains, solar nebula particles, and annual batches of new types of meteorites from Antarctica to be irresistible (Bischoff, 2001).

1.07.1.4 Chondrites and the Solar Nebula

Chondritic components have generally been thought of as products of the "minimum mass" solar nebula, which contains just enough mass to make the planets. However, Wood (2000b) argues that planetary material began to form and accrete at an earlier stage during the most active and violent stages of protosun formation. In this view, localized heating events may have been more important for understanding chondrites and their components than nebula-wide thermal variations (Cassen, 2001a). Chondrules and CAIs may have formed at the inner edge of the disk prior to injection into the asteroid belt by bipolar outflows (Liffman and Brown, 1996; Shu et al., 1996, 2001). Alternatively, Alexander et al. (2001) argue that CAIs and chondrules formed in localized heating by shocks in the asteroid belt. The 7° difference between the planetary and solar angular momentum vectors (Hutchison et al., 2001), the multiplicity of protostellar systems, and the discovery of numerous "hot Jupiters" around other stars provide additional reminders of the complexity of disk evolution.

1.07.2 CLASSIFICATION AND PARENT BODIES OF CHONDRITES

1.07.2.1 Chondrite Groups, Clans, and Parent Bodies

Many thousands of chondrites have been classified into 15 groups: about 15 other chondrites do not fit comfortably into these groups and are called ungrouped (Table 1). Thirteen of these groups comprise three classes: carbonaceous (CI, CM, CO, CV, CR, CH, CB, and CK), ordinary (H, L, and LL), and enstatite (EH and EL). The K and R chondrites do not belong to the three classes (see Chapter 1.05). Chondrites are also classified into petrologic types 1–6, which indicate the extent of asteroidal processing (see table 2 of Chapter 1.05). Type 3 chondrites are the least metamorphosed and least altered; type 6 are the most metamorphosed. Type 1 chondrites, which lack chondrules, are composed almost entirely of minerals that formed during aqueous alteration. Type 2 chondrites are partly altered.

Elemental ratios are commonly used for comparing chondrite compositions instead of concentrations, as some chondrites have excessive amounts of one component like water or metallic Fe,Ni, for example, that dilute elements in other components. The elements silicon and magnesium are commonly used for normalizing lithophiles; nickel is used for comparing siderophile elements. CI chondrites, which have the highest concentrations of volatile elements, are used as the reference standard because they provide the best match for the composition of the solar photosphere, neglecting the incompletely condensed elements hydrogen, helium, carbon, nitrogen, oxygen, and the noble gases (see Chapter 1.03).

Ratios of refractory and moderately volatile lithophile (figure 3 in Chapter 1.05) or siderophile elements (Figure 2) provide the best criteria for classifying chondrites. On these plots, the range within each group is a small fraction of the total range shown by all chondrites. Each group has a unique chemical composition implying that its members formed in a separate location. The CB group, which is split into two subgroups with rather different properties, is an exception. Unfortunately, many newly discovered, unusual chondrites have not been chemically analyzed, so their classification and the compositional limits of newly discovered groups are not well defined. Two other properties are useful for classifying chondrites and understanding their compositional diversity and genetic relationships: bulk oxygen isotopic compositions (see Chapter 1.06) and the distribution of iron among silicate, metal, and sulfide phases (figure 7 of Chapter 1.05). The latter requires classical methods of rock analysis (Jarosewich, 1990), which are no longer applied to chondrite. Many chondrites groups can also be distinguished by other parameters such as the sizes and proportions of their components and the minerals in their chondrules, CAIs or other components. See Brearley and Jones (1998), a detailed account of chondrite mineralogy (hereafter abbreviated, B & J (1998).

The chemical, mineralogical, and isotopic properties of chondrites distinguish closely related groups, which constitute "clans." For example, chondrules in the H, L, and LL groups are chemically and isotopically similar, though their average properties are different, and these three groups form a clan. Four other clans are recognized: the EH–EL clan, the CM–CO clan, the CV–CK clan, and the CR–CH–CB clan.

Cosmic-ray exposure ages suggest that chondrites in many groups were exposed to space as

Table 1 Abundances of refractory inclusions, chondrules, metallic Fe,Ni, and matrix and other key properties of the chondrite groups.

Group	Refract. lith./Mg rel. CI[a]	CAI and AOA (vol.%)	Chondrule avg. dia. (mm)	Chondrules (vol.%)[b]	Metal (vol.%)	Matrix (vol.%)[c]	Fall frequency[d] (%)	Examples
Carbonaceous								
CI	1.00	<0.01		<5	<0.01	95	0.5	Ivuna, Orgueil
CM	1.15	5	0.3	20	0.1	70	1.6	Murchison
CO	1.13	13	0.15	40	1–5	30	0.5	Ornans
CV	1.35	10	1.0	45	0–5	40	0.6	Vigarano, Allende
CR	1.03	0.5	0.7	50–60	5–8	30–50	0.3	Renazzo
CH	1.00	0.1	0.02–0.09	~70	20	5	0	ALH 85085
CB$_a$	1.0	<0.1	~5	40	60	<5	0	Bencubbin
CB$_b$	1.4	<0.1	~0.5	30	70	<5	0	QUE 94411
CK	1.21	4	0.8	15	<0.01	75	0.2	Karoonda
Ordinary								
H	0.93	0.01–0.2	0.3	60–80	8	10–15	34.4	Dhajala
L	0.94	<0.1	0.5	60–80	3	10–15	38.1	Khohar
LL	0.90	<0.1	0.6	60–80	1.5	10–15	7.8	Semarkona
Enstatite								
EH	0.87	<0.1	0.2	60–80	8	<0.1–10	0.9	Qingzhen, Abee
EL	0.83	<0.1	0.6	60–80	15	<0.1–10	0.8	Hvittis
Other								
K	0.9	<0.1	0.6	20–30	6–9	70	0.1	Kakangari
R	0.95	<0.1	0.4	>40	<0.1	35	0.1	Rumuruti

Source: Scott *et al.* (1996); other data from Weisberg *et al.* (1996, 2001), Rubin (2000), Krot *et al.* (2002a), Kimura *et al.* (2002), Bischoff *et al.* (1993).
[a] Mean ratio of refractory lithophiles relative to magnesium, normalized to CI chondrites. [b] Includes chondrule fragments and silicates inferred to be fragments of chondrites. [c] Includes matrix-rich clasts, which account for all matrix in CH and CB$_b$ chondrites (Greshake *et al.*, 2002). [d] Fall frequencies based on 918 falls of differentiated meteorites and classified chondrites (Grady, 2000).

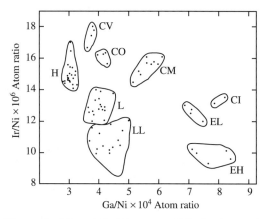

Figure 2 Plot of Ga/Ni versus Ir/Ni showing bulk compositions of chondrites in nine groups (Scott and Newsom, 1989). Each group is well resolved. The proportions of these siderophiles are not correlated with other chemical properties showing that the chemical variations in chondrites are complex (reproduced by permission of Verlag der Zeitschrift für Naturforschuang from *Z. Naturforsch.*, **1989**, *44a*, 927).

meter-sized objects in a limited number of impacts (Chapter 1.13). For example, about a third of the LL chondrites were exposed to cosmic rays 15 Ma and half the H chondrites have exposure ages of 7 Ma (Graf and Marti, 1994, 1995). Three or four large collisions appear to have generated two-thirds of all H chondrites (Wieler, 2002). Since all petrologic types are found in the 7 Ma and 33 Ma peaks, it is possible that all H chondrites come from a single body. Some other groups show distinctive impact ages or shock properties, suggesting that they were involved in a single impact. For example, many L chondrites, which are strongly shocked, have radiometric ages of ~0.5 Ga (Bogard, 1995). These features suggest that each group comes from one, or possibly a few bodies.

We do not know for certain that two chondrites groups did not originate from a single body but it is rather unlikely (except possibly for the EH and EL chondrites). Evidence from meteorites, impact modeling of asteroid collisions, and asteroid density determinations suggests that all asteroids, except the few largest, were well fragmented and mixed by collisions before meteoroids were removed. Impacts also weld rock fragments from different locations into coherent rocks that reach earth as meteorites. Thus, if CV and CO chondrites, for example, had formed in a single body, we might expect to find meteorites containing both types of material. Their absence makes it unlikely that one body ever contained both groups. Nevertheless, regolith breccias contain foreign clasts showing that material from different sources was mixed together. In addition, one EL6 chondrite had an orbit almost identical to that of an H5 chondrite (Spurny *et al.*, 2003).

Given the extraordinary diversity of chemical and mineralogical properties of the 15 chondrite groups and the rapid rate at which new groups are being identified (only nine were listed by Wasson, 1985), we should expect many more unusual types to be discovered. Our sampling of chondrites is also biased towards tough rocks that can survive the journey to Earth, and we cannot expect these to be representative samples. Thus, it is important to study components in all kinds of chondrites if we want to understand how they formed, not just those in the most common chondrite groups or those that are present in the largest falls, like Allende.

1.07.2.2 Ordinary Chondrites

Since ordinary chondrites account for ~80% of all meteorite falls (Table 1), it was once believed that their parent bodies were common in the main asteroid belt. However, spectral studies show that most asteroids are dark and featureless (C and related types), and most of the brighter S type differ from ordinary chondrites. H group chondrites probably come from one or more S-type asteroids, possibly 6 Hebe (Burbine *et al.*, 2002). Ordinary chondrites are probably rare in the main part of the asteroid belt (Meibom and Clark, 1999), though they may account for ~20% of the near-Earth objects (Binzel *et al.*, 2002). A few ordinary chondrites including Tieschitz (Figure 1(a)) do not fit comfortably into the H, L, and LL groups and may be derived from separate bodies.

1.07.2.3 Carbonaceous Chondrites

Some carbonaceous chondrites are rich in carbon (CI and CM chondrites have 1.5–6% carbon), but others are not. Carbonaceous chondrites are now defined on the basis of their refractory elemental abundances, which equal or exceed those in CI chondrites. Carbonaceous chondrites are derived from very diverse asteroids, which probably formed in very different locations. The parent bodies of CI and CM chondrites are highly altered, yet the parent bodies of CH and CB chondrites are less altered than all other chondrite bodies. Young *et al.* (1999) infer from oxygen isotopic compositional data that CI, CM, and CV chondrites could have been derived from different zones in a single, aqueously altered body. However, bulk chemical differences between these groups indicate fractionation during nebular processes, not aqueous alteration (see below), and the components in CM and CV chondrites are quite different.

Most carbonaceous chondrites are thought to come from the low-albedo, C-type asteroids, which are the most abundant type between 2.7 AU and 3.4 AU (Bell *et al.*, 1989), CM chondrites may be derived from an altered C-like asteroid called G-type (Burbine *et al.*, 2002). The ungrouped C chondrite, Tagish Lake, has been linked with D asteroids, which appear to dominate the asteroid population beyond 4 AU (Jones *et al.*, 1990; Hiroi *et al.*, 2001). CB chondrites have only very minor amounts of phyllosilicates and may come from W-type asteroids ("wet-M" asteroids). Lodders and Osborne (1999) discuss the possibility that CM and CI chondrites could be derived from a small fraction of comets that evolve into near-Earth objects after losing volatiles. However, chondritic aggregate interplanetary dust particles (IDPs), which probably do come from comets, have few phyllosilicates (Rietmeijer, 1998, 2002), unlike CI chondrites. Campins and Swindle (1998) infer that cometary meteorites, if they exist, should resemble C-rich, unaltered chondrites without chondrules.

1.07.2.4 Enstatite Chondrites

The genetic relationships among E chondrites are poorly understood, in part because of poor sampling but also because the cause of their chemical and mineralogical diversity is obscure. EH and EL chondrites may come from one (Kong *et al.*, 1997) or two bodies (Keil, 1989). Although there are compositional discontinuities between the EH and EL groups (e.g., in gold and aluminum), Kong *et al.* concluded that elemental concentrations appear to vary continuously through the sequence EH4,5, EH3, EL3, and EL6. Oxygen-isotopic compositions of EH and EL chondrites show a large overlap, but do not resolve the issue (Newton *et al.*, 2000). On the basis of their noble gas analyses, Patzer and Schultz (2002b) advanced an entirely different proposal that "EH3 and EL3 chondrites are derived from one body and EH4–6 and EL4–6 from a second." They found that E3 and E4–6 chondrites types have different trapped noble gas components (called "Q-gases"), and inferred that this difference could not be attributed to asteroidal processing. In addition, about a third of the E3 chondrites contain solar noble gases, which are absent among E4–6 chondrites. Unfortunately, the cosmic-ray exposure age data do not yet provide definitive constraints. The 60-odd E chondrites have a wide range of exposure ages, 0.07–70 Ma with broad peaks as their precision is ~20% (cf. 10% for ordinary chondrites) (Patzer and Schultz, 2002b; Okazaki *et al.*, 2000). However, the data are not inconsistent with a single E chondrite body, as Kong *et al.* (1997) proposed. E chondrites may be derived from one or more M-type asteroids (Burbine *et al.*, 2002), which are concentrated in the inner asteroid belt at 2–3 AU (Bell *et al.*, 1989).

1.07.3 BULK COMPOSITION OF CHONDRITES

1.07.3.1 Cosmochemical Classification of Elements

The chemical fractionations observed among chondrites and the compositions of many chondritic components are best understood in terms of quenched equilibrium between phases in a nebula of solar composition (Palme, 2001; Chapters 1.03 and 1.15). The equilibrium model assumes that minerals condensed from, or equilibrated with, a homogeneous solar nebula at diverse temperatures. Isotopic variations among chondrites and their components show that this assumption is not correct and detailed petrologic studies have identified relatively few chondritic components that resemble equilibrium nebular products. Nevertheless, the equilibrium model is invaluable for understanding the chemical composition of chondrites and their components as the solar nebular signature is etched deeply into the chemistry and mineralogy.

In a cooling nebula at a pressure of 10^{-3} atm, calcium, aluminum, and titanium condense as oxides in the temperature range 1,800–1,450 K along with refractory trace elements like the rare earth elements (REEs) and platinum group metals (figure 2 of Chapter 1.15). Between 1,450 K and 1,350 K, magnesium and silicon condense as forsterite and enstatite and iron condenses as metallic Fe,Ni along with cobalt. Between 1,250 K and 650 K, the alkalis and moderately volatile siderophiles condense and at 650 K, iron reacts to form FeS. Below 650 K, the highly volatile elements like the halogens and inert gases condense. The division of the elements into refractories (1,800–1,450 K), major elements (1,350–1,250 K), moderately volatile elements (1,250–650 K) and highly volatile elements (<650 K) is the key to understanding the chemical variations among chondrites and their components. Note that only three phases condense entirely from the gas phase—Al_2O_3, Mg_2SiO_4, and Fe,Ni—all other minerals form by reaction between solids and gas.

At 10^{-3} atm, FeO concentrations in silicates are minimal above 650 K and all condensed phases are solid. If the total pressure increases, condensation temperatures rise, but the sequence in which minerals condense does not change significantly except that silicate melts become stable at pressures above ~10^{-2} atm. If condensation occurs in systems that have been enriched in dust relative to gas, FeO concentrations in silicates become appreciable at high temperatures (up to ~1,100 K for 10^3-fold dust enrichments), and liquids are also stabilized (Wood and Hashimoto, 1993; Ebel and Grossman, 2000). Pressures and temperatures at the midplane of the nebula vary considerably as the solar nebula evolves, decreasing as the accretion rate falls (Wood, 2000a; Woolum and Cassen, 1999). For a detailed understanding of chondritic features, fractional condensation models are required (Petaev and Wood, 2000).

1.07.3.2 Chemical Compositions of Chondrites

A large number chondrites have been analyzed with instrumental and radiochemical neutron activation analysis using ~250–350 mg samples of chondrites (Kallemeyn and Wasson, 1981; Wasson and Kallemeyn, 1988; Kallemeyn et al., 1991, 1994, 1996). Samples of this size can also be analyzed for many elements with comparable accuracy using X-ray fluorescence analysis (Wolf and Palme, 2001). Mean compositions of chondrite groups are given by Lodders and Fegley (1998).

Chemical variations among chondrites are commonly attributed to the accretion of different proportions of five components: the refractory component, magnesium silicates, metallic Fe,Ni, moderately volatile elements, and highly volatile elements (see chapters in Kerridge and Matthews, 1988; Palme, 2000). In addition, variations in oxygen fugacity are invoked to explain the diverse distribution of iron among metallic and silicate phases (figure 7 of Chapter 1.05). In general, chemical fractionation patterns are simple: refractory elements, for example, are uniformly enriched relative to CI chondrites in carbonaceous chondrites by factors of 1.0–1.4, whereas they are depleted in ordinary and enstatite chondrites by factors of 0.8–0.95 (Figure 3). Moderately volatile elements decrease in abundance with decreasing condensation temperature. (Note that enstatite chondrites do not follow these rules as closely as the other groups.) Although the chemical fractionation patterns for most groups of chondrites are known (Wasson and Kallemeyn, 1988), their origins have proved to be surprisingly elusive. In particular, the relative importance of localized processes that made chondrules and CAIs and nebula-wide thermal gradients is controversial (see Cassen, 2001a).

(i) *Refractory elements*. All the elements that condense above 1,450 K are precisely those found in CAIs, and there is some correlation between the bulk concentrations of refractory elements and refractory inclusions (Table 1). This suggests that the fractionation of refractory elements among chondrites is simply caused by addition or loss of CAIs. However, this is incorrect as the proportion of the refractory elements that is present in CAIs is generally small, except for CVs, where it approaches ~50%. Chondrules, which are much more abundant and have approximately CI-like levels of refractories relative to silicon, account for most of the refractory elements in chondrites. Note that the depletions of refractory elements in E and O chondrites require that their chondrules formed from a condensate that was depleted in refractories.

(ii) *Magnesium silicates*. Forsterite and enstatite, which are the major minerals in chondrules, form in the temperature range 1,450–1,350 K, the latter by reaction between forsterite and

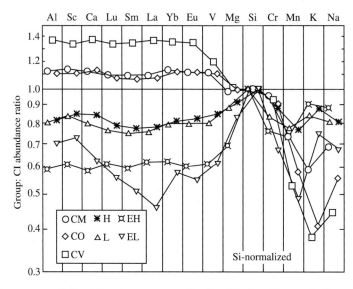

Figure 3 Mean abundances of lithophile elements normalized to CI chondrites and silicon arranged in order of increasing volatility in seven chondrite groups (Wasson and Kallemeyn, 1988). Refractories (elements condensing above V) are uniformly enriched in CO, CM, and CV chondrites and depleted in H, L, and EH chondrites. Moderately volatile elements, which condense below magnesium and silicon, are all depleted relative to CI chondrites. These fractionations are related in poorly understood ways to the formation of CAIs and chondrules (reproduced by permission of The Royal Society from *Phil. Trans. Roy. Soc. London*, **1988**, *A325*, p. 539).

gaseous SiO. Thus, equilibrium condensation can produce a range of molar enstatite/forsterite ratios of 0–1.2 (figure 2 of Chapter 1.15), provided that the condensates are sequestered from the gas over a range of temperatures. To reach the composition of E chondrites, which are almost pure enstatite, would require removal of forsterite condensate. A correlation between the Mg/Si and refractory/Si ratios of the chondrite groups suggests that refractories were partly associated with forsterite (Larimer and Wasson, 1988).

(iii) *Metallic Fe,Ni*, which condenses over the same temperature range as the magnesium silicates, is closely associated with them in chondrules. At 10^{-5} atm, Fe,Ni condenses at slightly lower temperatures than forsterite, but above 10^{-4} atm there is a reversal. Some groups like the CV3 chondrites did not suffer metal/silicate fractionation (figure 2 of Chapter 1.15), but the EH–EL, H–L–LL, and CR–CH–CB clans clearly did. Fe/Si ratios are $\sim(2-5) \times$ CI levels in CH and CB chondrites because metal was enriched relative to silicate; EL, L, and LL chondrites were depleted in metallic Fe,Ni.

(iv) *Moderately volatile elements*. The gradual depletion of moderately volatiles with decreasing condensation temperature can be attributed to loss of fine volatile-rich dust (Wasson and Kallemeyn, 1988), incomplete condensation due to isolation of condensed phases (Palme *et al.*, 1988; Cassen, 2001b), or to evaporation prior to or during chondrule formation (see, e.g., Young, 2000; Lugmair and Shukolyukov, 2001; see also Chapter 1.15).

(v) *Highly volatile elements* may have failed to accrete from the nebula because of high ambient temperatures during accretion or inefficient accretion of volatile-rich dust, or they may simply have been lost during asteroidal metamorphism.

1.07.3.3 Isotopic Compositions of Chondrites

Major and minor planetary bodies are essentially uniform isotopically (e.g., Palme, 2001). The great majority of elements are isotopically uniform to within $\sim 0.01\%$. The exceptions are found for oxygen (discussed below), effects due to short-lived isotopes (Chapter 1.16), slightly larger effects (per mil or less) due to anomalies in ^{54}Cr and ^{50}Ti in chondrites and their components (e.g., Podosek *et al.*, 1997; Niemeyer, 1988), and isotopically anomalous components in chondrites: interstellar grains, certain rare CAIs called FUN inclusions and some hibonite inclusions (see Chapter 1.08). The isotopic anomalies in interstellar grains are orders of magnitude larger than those in refractory grains in chondrites and the latter probably reflect incomplete homogenization of the former (e.g., Ott, 1993, 2001). Depending on one's viewpoint, the isotopic anomalies provide evidence that vaporization of presolar materials was limited (Taylor, 2001) or that isotopic homogenization was remarkably complete (Palme, 2001).

1.07.3.4 Oxygen Isotopic Compositions

The discovery of large and ubiquitous oxygen isotopic anomalies in refractory inclusions in carbonaceous chondrites and smaller anomalies in chondrules transformed the study of chondrites (see Clayton, 1993; Chapter 1.06). It galvanized the search for presolar grains, provided another way to classify meteorites, and offered numerous constraints on the origin of chondrites and their constituents and subsequent asteroidal processing. In particular, the oxygen isotopic anomalies appeared to eliminate the conviction that the meteorites and inner planets were derived from a hot, gaseous and homogeneous nebula (e.g., Begemann, 1980). However, the origin of the oxygen anomalies remains a mystery, presolar grains were not discovered by searching for the source of the anomalous oxygen (Ott, 1993), and the oxygen anomalies are now known to be associated with nebular condensates as well as evaporative residues (see Scott and Krot, 2001; Krot *et al.*, 2002b). The recent introduction of ion and laser microprobe techniques has produced an explosion of new oxygen isotopic analyses of minerals in CAIs and chondrules which have been especially valuable in distinguishing asteroidal and nebula processes (e.g., Choi *et al.*, 1998; Yurimoto *et al.*, 1998; Yurimoto and Wasson, 2002).

Differences between the bulk compositions of chondrites, planets and asteroids can be attributed to accretion from different batches of CAIs, chondrules and other components, which are spread along ^{16}O variation lines on the standard three-isotope plot. The preservation of oxygen isotopic anomalies shows that there were numerous oxygen reservoirs for the manufacture of chondrules and CAIs that were quite separate.

The source of the anomalies has been attributed to nucleosynthetic differences between batches of presolar materials (Wasson, 2000), mass-independent partitioning among oxygen-bearing molecules in the nebular gas (Thiemens, 1999), photochemical self-shielding of CO during accretion of the solar nebula (Clayton, 2002), and photochemical processes in the presolar molecular cloud (Yurimoto and Kuramoto, 2002). Whatever the source, the ^{16}O variations in CAIs and chondrules imply that the anomalies can be understood in terms of mixing of ^{16}O-rich and ^{16}O-poor components in the solar nebula. Since oxygen is present in the gas and the solid phases over a wide range of temperatures, the anomalies could have been preserved by enriching dust relative to gas prior to vaporization (Cassen, 2001b; Scott and Krot, 2001). Luck et al. (2003) inferred that ^{16}O excesses were correlated with ^{63}Cu-isotopic anomalies, but the reason for this is not clear. It is hoped that the NASA Genesis mission will determine the oxygen isotopic composition of the solar wind and establish the origin of the oxygen anomalies (Wiens et al., 1999; Clayton, 2002).

1.07.4 METAMORPHISM, ALTERATION, AND IMPACT PROCESSING

All chondrites were modified in some way by geological processes in asteroids operating over 4.5 Gyr. If we want to understand how the components in chondrites were formed, we must understand how chondritic materials were modified in asteroids. Three processes affected chondrites: aqueous and hydrothermal alteration, thermal metamorphism, and impacts.

The most important heat source for the alteration and metamorphism of chondrites was heat from the decay of ^{26}Al, and to a much lesser extent, ^{60}Fe (Chapter 1.16). ^{26}Al appears to have been widespread: excess ^{26}Mg from the decay of ^{26}Al has been measured in plagioclase separates from several rapidly cooled H4 chondrites (Zinner and Göpel, 2002) as well as in achondrites. Impacts also provide kinetic energy but for asteroids smaller than a few hundred kilometers in size, they are not an effective heat source for global metamorphism (Keil et al., 1997). For alternative views, see Rubin (1995), who favors heating by asteroidal impacts, and Kurat (1988), who argues for nebular heating processes. Electrical induction heating of asteroids by plasma outflow from the protosun is not currently favored (see Scott et al., 1989). Arguments against the concept of hot accretion without radiometric heating (Hutchison et al., 1979) were given by Haack et al. (1992); Rubin and Brearley (1996).

Alteration and metamorphism used to be considered as separate processes: the former affecting carbonaceous bodies and the latter operating on enstatite and ordinary chondrites. However, it is now clear that fluids of some kind were present during thermal processing of virtually all chondrite classes.

1.07.4.1 Petrologic Types

Chondrites are classified into petrologic (or petrographic) types 1–6 to provide a guide to the extent and nature of asteroidal processing. This classification scheme, which has been invaluable (Anders and Kerridge, 1988), nevertheless provides only a very rough guide as our ideas about asteroidal processes have changed significantly since the scheme was first established, and because one number cannot summarize all the effects of complex asteroidal processing. As a result, the classification is not rigorous and may be ambiguous or even misleading.

The original division of chondrites by Van Schmus and Wood (1967) into six petrologic types was based on two underlying assumptions that are now known to be incorrect: 1) that type 1 chondrites are the least altered chondrites that best represent the mineralogy of the original materials that accreted into asteroids, and 2) that types 2–6 represent increasing degrees of thermal metamorphism of material that mineralogically resembled type 1 chondrites. Although CI1 chondrites are *chemically* most pristine as they are compositionally similar to the solar composition, almost all of their *minerals* formed during aqueous alteration. Type 3 chondrites are closest mineralogically to the original materials that accreted into asteroids (McSween, 1979).

An additional problem with the division into six petrologic types is that it ignores the role played by impacts in modifying and mixing chondritic material. Many chondrites are breccias of materials with diverse alteration or metamorphic histories. Thus the petrologic type may provide no information about the metamorphic history of the whole chondrite, only an approximate guide to the history of most constituents.

The criteria used for assigning petrologic types were first given by Van Schmus and Wood (1967) and have not changed significantly since (table 3 of Chapter 1.05). Table 2 in this chapter shows the numbers of classified chondrites in petrologic types 1–6 within each group. The unrepresented cells in Table 2 may reflect poor sampling, e.g., in the poorly represented K group. However in many cases, it is likely that the missing types were volumetrically insignificant. It is very unlikely, for example, that the ordinary chondrite parent bodies ever contained significant amounts of type 1–2 material.

1.07.4.1.1 Type 1–2 chondrites

There is much ambiguity about the classification of type 1 and 2 chondrites that are not clearly associated with the major groups, CI and CM. Different workers favor different criteria, so that type 1, for example, may mean a chondrite with major minerals similar to those in CI1, or carbon-rich like CI1, or simply that all the chondrules have been almost completely altered. Since the original CI and CM chondrites were studied (see, Zolensky and McSween, 1988), CM1 chondrites have been described (Zolensky et al., 1997), as well as several CI or CM chondrites that

Table 2 Numbers of classified chondrites in petrologic types 1–6 by group.

	1	2	3	4	5	6	Total no.
Carbonaceous							
CI	5						5
CM	4	44					48
CO			31				31
CV		1	35				36
CR	1	13	1				15
CH			7				7
CB$_a$			3				3
CB$_b$			2				2
CK			2	13	6	1	23
Ordinary[a]							
H			187	1,371	3,319	1,784	6,661
L			316	415	1,220	4,053	6,004
LL			71	64	419	406	960
Enstatite							
EH			18	9	6	2	35
EL			8	0	2	19	29
Other							
K			2				2
R			3	2			14
			9 R3–6				

Source: Based largely on lists in Grady (2000) with additional data from Krot et al. (2002a) for CR, CH, and CB and Kallemeyn et al. (1996) and Bischoff (2000) for R chondrites. A small number of chondrites that are classified as intermediate types (e.g., 3–4, 4–5, etc.) and mixed breccias (e.g., type 3–5) were omitted except for the R chondrites, which are mostly breccias of type 3–5 or 3–6 material. [a] Numbers of H, L, and LL chondrites have not been corrected for probable pairings: the number of individual meteorites is probably ~2–4× smaller than the number of specimens.

were metamorphosed after alteration but have not been assigned a petrologic type (B & J, 1998, pp. 70–71, 240–244).

CR2 chondrites, which contain abundant Fe,Ni, are the least altered type 1–2 chondrites and offer exceptional insights into the nature of nebular components. Renazzo was the first to be recognized, but other CR2s including PCA 91082 (Figure 1(a)) appear to be the least altered (Wood, 1967; Krot et al., 2002a). GRO 95577, a type 1 chondrite, is the most altered CR (Weisberg and Prinz, 2000). Since there are only two known CR1 and CR3 chondrites, the term "CR chondrite" is commonly used as a synonym for CR2. Similarly, CM is used for CM2.

1.07.4.1.2 Type 3 chondrites

Type 3 chondrites show very diverse mineralogies and in several groups they have been subdivided. CO3, H3, L3, and LL3 chondrites, which are divided into subtypes 3.0–3.9, show correlated variations in mineralogy and concentrations of carbon and highly volatile elements that can largely be attributed to metamorphism. Subtypes were first defined for ordinary chondrites based largely on their thermoluminescence sensitivity, which is a measure of plagioclase growth in chondrule mesostases (Sears and Hasan, 1987; Sears et al., 1991, 1995a). Many other properties vary systematically with progressive

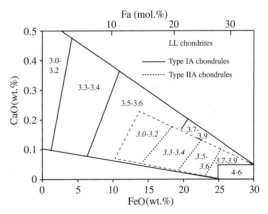

Figure 4 Plot showing effects of metamorphism on the mean CaO and FeO concentrations in olivine in two types of chondrules in LL3 chondrites (Scott et al., 1994). Type 3 chondrites can be readily subdivided into subtypes using these parameters or many others (e.g., Sears and Hasan, 1987) (reproduced by permission of Elsevier from *Geochim. Cosmochim. Acta*, **1994**, *58*, 1208).

metamorphism and have been used to assign subtypes, e.g., chemical changes in chondrule olivines (Figure 4), the chemical homogenization of olivines, FeO/(FeO + MgO) ratios in matrices (Huss et al., 1981), loss of trapped noble gases and loss of carbon (Sears and Hasan, 1987), chromium levels in metallic Fe,Ni grains (Scott and Jones, 1990), and growth of ferroan olivine in amoeboid–olivine aggregates (Chizmadia et al., 2002). Especially in low-subtype chondrites, there is evidence that hydrous fluids promoted chemical equilibration and alteration in chondrule mesostases (B & J, 1998), CAIs (Russell et al., 1998), and matrices (Section 1.07.5.7). With increasing metamorphism, fluids were lost from the parent bodies. For CV3 chondrites, the effects of alteration are more complex: subtypes based on thermoluminescence sensitivities have been assigned but are not in common use (Guimon et al., 1995; Krot et al., 1995).

Although type 3 chondrites are the least altered and metamorphosed chondrites, only a few of them are now thought to be relatively pristine samples, and all of these have components that were tainted by their asteroidal environment. Among the most pristine are the CB_b3 and CH3 chondrites, Vigarano and Leoville (CV3), ALHA77307 (CO3.0), Semarkona (LL3.0), Kakangari (K3), and two ungrouped carbonaceous chondrites, Acfer 094 and Adelaide. Chondrites in type 3 chondrites in the ordinary, K, CO, and CV groups were probably heated to around 400–600 °C for $>10^6$ yr, as they contain kamacite and taenite that equilibrated at these temperatures (Keil, 2000). However, the CH3 and CB3 chondrites contain martensite that has not exsolved to kamacite and taenite and has retained chemical zoning patterns that were established in the solar nebula (Meibom et al., 1999, 2000; Krot et al., 2002a). Electron microscopic studies of the metal show that CH3 chondrites were not heated above 300 °C for more than a year and probably experienced much lower peak temperatures (Reisener et al., 2000).

1.07.4.1.3 Type 4–6 chondrites

Type 4–6 ordinary chondrites, which are also called equilibrated chondrites, come closest to representing an isochemical, metamorphic sequence. Types 4–5 were heated to 500–800 °C and type 6 to 800–1000 °C (see Keil, 2000). The most noticeable metamorphic effects are due to grain coarsening and growth of feldspar, chromite and phosphate grains. Some type 4–6 chondrites are not simple metamorphic rocks: they are fragmental or regolith breccias of rock fragments that were metamorphosed prior to assembly (Rubin, 1990). In these cases, the petrologic type may only be a measure of the metamorphic history of most material in the rock. In addition, some seemingly unshocked type 6 chondrites were actually shocked and then metamorphosed so that they had a more complex recrystallization history (Rubin, 2002). Within each group of ordinary chondrites, there are small but systematic increases in the FeO concentrations of olivines and pyroxenes with increasing petrologic. These are probably due to oxidation of metal by traces of water vapor during metamorphism (McSween and Labotka, 1993). Fluid inclusions are present in halite in two H chondrite regolith breccias and may have been derived from asteroidal fluids or icy projectiles (see Rubin et al., 2002).

The criteria for classifying types 4–6 were based solely on the properties of H, L, and LL types 3–6 chondrites and most cannot be used for asteroidally processed carbonaceous and enstatite chondrites. For these chondrites, the type is based almost entirely on the degree to which the chondrules are delineated. If the chondrules have been obscured by recrystallization during prolonged heating, the petrologic type is a good measure of the metamorphic history. However, for the EH4–6 and EL4–6 chondrites, evidence for simple thermal metamorphism is lacking. Rapid cooling rates recorded by sulfides and numerous examples of impact heating suggest that impact processing may have been more important for enstatite chondrites than heating by decay of ^{26}Al (Zhang et al., 1996; Rubin et al., 1997; Lin and Kimura, 1998). R chondrites experienced ubiquitous brecciation and although some have been classified as R4 (or R3), all are probably breccias of type 3–5 or 3–6 material (Bischoff, 2000).

1.07.4.2 Thermal History and Modeling

To understand geological processes on chondritic bodies and to estimate their sizes we need to investigate the thermal history of unshocked or only lightly shocked samples. The simplest metamorphosed chondrites are the ordinary chondrites, for which there are three kinds of constraints: (i) mineral equilibration temperatures (e.g., McSween and Patchen, 1989); (ii) radiometric ages that date cooling below isotopic closure in minerals using isotope systems such as ^{207}Pb–^{206}Pb, ^{40}K–^{39}Ar, and ^{87}Rb–^{87}Sr (e.g., Trieloff et al., 2003); (iii) cooling-rate determinations based on annealing of Pu fission tracks (Lipschutz et al., 1989) and metallographic cooling rates (Taylor et al., 1987). The cooling rates determined by the metallographic technique, which are mostly $1-10^3$ °C Ma^{-1} in the temperature range of 400–500 °C, are compatible with cooling rates in the range 100–500 °C determined by the fission track technique. A third technique based on Fe–Mg ordering in orthopyroxene gives cooling rates in the range 340–480 °C that are systematically higher by several orders of magnitude (Folco et al., 1997), probably because metamorphism was insufficient to reequilibrate the ordering imposed during chondrule formation (Artioli and Davoli, 1994). Cooling rates can also be estimated from radiometric age data for mineral grains that close isotopically at different temperatures (Bogard, 1995; Ganguly and Tirone, 2001).

Metamorphic ages of ordinary chondrites from Pb–Pb ages of phosphates in seven H4–6 chondrites range from 4.50 Ga to 4.56 Ga (Göpel et al., 1994), while their Ar–Ar ages range from 4.45 Ga to 4.53 Ga (Trieloff et al., 2003). The H4 chondrites, which have metallographic cooling rates of 10^3 °C Myr^{-1}, have the oldest ages while the H6 chondrites with cooling rates of ~10 °C Myr^{-1} have younger ages. The negative correlation between age and metamorphic type for these seven chondrites is compatible with an onion-shell model in which burial depth correlates with petrologic type. However, Pb–Pb ages for two ordinary type 3 chondrites are significantly younger than the phosphate ages for the H6 chondrites and are not readily reconciled with this model (Göpel et al., 1994). Metallographic cooling rates for larger numbers of unshocked, ordinary chondrites show no correlation between cooling rate and petrologic type (Taylor et al., 1987). If the petrologic types were once arranged sequentially in layers, the bodies must have been catastrophically fragmented and reassembled by impacts during cooling from peak metamorphic temperatures (Grimm, 1985). Regolith breccias show extreme variations in the cooling rates of their metal grains from $1-10^3$ °C Myr^{-1} implying that the ordinary chondrite bodies were scrambled after metamorphism so that material from all depths could be combined together (Williams et al., 1999).

Estimates of the sizes of the ordinary chondrite parent bodies can be derived from thermal models as the timescale for cooling is controlled by conduction for dry asteroids >10–20 km in size (McSween et al., 2002). However, the thermal conductivity is very sensitive to the porosity, which can be reduced by sintering and increased by impacts. Layers of regolith <1 km thick can drastically reduce the cooling rate of asteroids and increase near-surface thermal gradients (Haack et al., 1990). Akridge et al. (1998) argue that H chondrites are derived from depths of <10 km from a 100 km radius asteroid and that material at greater depths has not been sampled. However, Bennett and McSween (1996) infer complete sampling of bodies with radii of ~80–95 km for the H and L chondrite parent bodies.

For volatile-rich carbonaceous chondrites like CI and CM chondrites, constraints on thermal histories are derived from carbonate ages and oxygen isotopic data. The former indicate that alteration began soon after CAI formation and lasted ~20 Myr (Endress et al., 1996; Brearley et al., 2001). Oxygen-isotopic compositions of carbonates provide model-dependent temperatures under 50 °C (see McSween et al., 2002). Modeling metamorphism of wet asteroids requires more complex models to track fluid flow and chemical and mineralogical changes during metamorphism (e.g., Wilson et al., 1999; Cohen and Coker, 2000; Young, 2001). McSween et al. (2002) suggest that thermal buffering by ice and water prevented high-temperature metamorphism in carbonaceous chondrite bodies. Buffering from ices probably saved comets from major volatile loss, but some CV and CK chondrites were clearly heated above 500 °C.

1.07.4.3 Impact Processing of Chondritic Asteroids

Impacts fragmented, mixed, modified, melted, devolatilized, and lithified chondritic material. Asteroids were impacted throughout their history: when they accreted, during alteration and metamorphism, and during the 4.5 Ga after their parent bodies were heated. For reviews of impact processing of ordinary chondrites, see Stöffler et al. (1991); carbonaceous chondrites, Scott et al. (1992); and enstatite chondrites, Rubin et al. (1997). Experimental studies of impact metamorphism of ordinary chondrites were made by Schmitt (2000); Langenhorst et al. (2002), CV chondrites by Nakamura et al. (2000), and CM chondrites by Tomeoka et al. (1999). Meteorite evidence for impact processing of chondritic and other asteroids was reviewed by Keil et al. (1994) and Scott (2002). High-pressure minerals in shock veins in chondrites also provide clues to mineral stability at depth in the Earth (Stöffler, 1997).

Advocates for impact heating of chondrites point to Portales Valley, a breccia of H6 clasts embedded in a matrix of once-molten metallic Fe,Ni, which cooled slowly to form a Widmanstätten pattern (Rubin et al., 2001). Gaffey and Gilbert (1998) suggest that molten metal on the H parent body was derived from impact-formed melt sheets or residues of metal-rich projectiles and that the IIE irons were also derived from this melt. However, the origin of Portales Valley and the link with IIE irons remain obscure as asteroids are thought to be too small for impacts to generate melt sheets (Keil et al., 1997).

1.07.5 CHONDRITIC COMPONENTS

Below we review the geochemical properties and possible origins of chondritic components focusing on what appear to be primary features in the least altered type 2–3 chondrites. The rich diversity of chondritic components even in unaltered chondrites and the vast differences between chondrite groups are key features of chondrites.

Few comprehensive reviews of the origin and properties of chondritic components have appeared since Kerridge and Matthews (1988). Exceptions include a comprehensive review of chondrite mineralogy (B & J, 1998), an edited book on chondrules and their origins (Hewins et al., 1996), and several reviews on chondrules and CAIs (Hewins, 1997; Rubin, 2000; Jones et al., 2000a; Ireland and Fegley, 2000).

First we review CAIs and their forsterite-rich accretionary rims, then amoeboid olivine inclusions and aluminum-rich chondrules, which are intermediate in composition between CAIs and chondrules, and finally matrix, which is a complex mixture of many ingredients. Although chondrules are volumetrically more important, we start with CAIs as these formed first and their origins are better constrained by chemical and isotopic data. Chemical and isotopic variations in chondrules are much more modest than those in CAIs but there are enough related features and components with intermediate compositions to suggest that chondrules cannot be understood in isolation from CAIs and other components.

1.07.5.1 Calcium- and Aluminum-rich Inclusions

There are two types of refractory inclusions: calcium- and aluminum-rich inclusions (this section) and amoeboid olivine aggregates (Section 1.07.5.3). Since the mineralogy, chemistry and isotope chemistry of refractory inclusions were reviewed by MacPherson et al. (1988), many new analyses have been made of CAIs in CV, CM, CO, CR, CH, CB, ordinary and enstatite chondrites that provide important constraints on physicochemical conditions, time, and place of CAI formation. CAIs are addressed in detail in Chapter 1.08, the role of condensation and evaporation in their formation in Chapter 1.15, and their clues to early solar system chronology in Chapter 1.16.

CAIs are the oldest objects in chondrites excluding presolar grains and probably formed during the most energetic phase of protosolar disk evolution (e.g., Ireland and Fegley, 2000; Wood, 2000b). They are composed of the minerals corundum, hibonite, grossite, perovskite, melilite, spinel, Al–Ti-diopside, anorthite, and forsterite, which are predicted to condense from a cooling gas of solar composition at temperatures >1,200–1,300 K and total pressure of 10^{-5} bar (e.g., Ireland and Fegley, 2000; Ebel and Grossman, 2000). Some CAIs are thought to be condensates because they have irregular shapes, fluffy textures, and minerals that match the sequence of mineral formation (Figure 5),

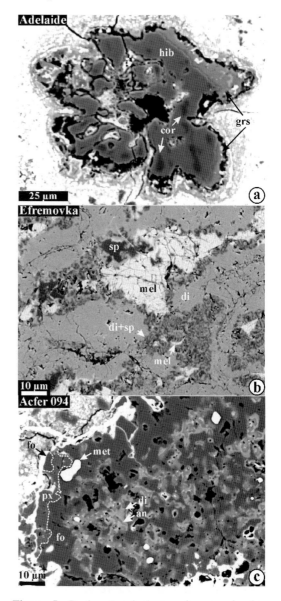

Figure 5 Backscattered electron images of refractory inclusions with clear condensation signatures: (a) CAI in Adelaide (unique carbonaceous chondrite), (b) CAI in Efremovka (CV3), and (c) AOA in Acfer 094 (unique carbonaceous chondrite). (a) Corundum (cor) is replaced by hibonite (hib), which is corroded by grossite (grs) (reproduced by permission of University of Arizona from Meteorit. Planet. Sci., 2001, 36, A105; © The Meteoritical Society). The inferred sequence of mineral crystallization is consistent with the predicted equilibrium condensation sequence at total pressure of $\leq 10^{-5}$ bar and CI chondritic dust/gas enrichment relative to solar composition of ~1,000 (Simon et al., 2002). (b) Replacement of melilite by a fine-grained mixture of spinel and diopside, suggesting the following gas–solid reaction: $Ca_2Al_2SiO_7 + 3SiO(g) + 3Mg(g) + 6H_2O(g) = 2CaMgSi_2O_6 + MgAl_2O_4 + 6H_2(g)$ (after Krot et al., 2003f). (c) Replacement of forsterite by low-calcium pyroxene (px), suggesting the following reaction $Mg_2SiO_4 + SiO(g) + H_2O(g) = Mg_2Si_2O_6 + H_2(g)$ (after Krot et al., 2003b).

and trace element chemistry (e.g., Simon et al., 2002). The strongest argument in favor of condensation origin of some CAIs comes from their group II REE pattern, which is a highly fractionated pattern with much lower abundances of the heavy REEs Gd–Er and Lu, exhibited by ~30% of all CAIs (Palme and Boynton, 1993; Ireland and Fegley, 2000). This volatility-controlled pattern requires the prior removal of the ultrarefractory elements. However, the preservation of large isotopic anomalies (e.g., ^{48}Ca, ^{50}Ti) in platy hibonites (PLACs) in CM chondrites and their lack of ^{26}Al, plus mass fractionation effects and smaller nuclear anomalies in FUN inclusions (see Chapter 1.08), favor evaporation as the predominant process for other CAIs. CAIs may represent mixtures of solar nebula condensates and refractory residue materials formed by evaporating material of presolar origin (Ireland and Fegley, 2000).

Because of the diversity of CAIs, there is no universal classification scheme for all groups. The best studied CAIs from CV chondrites are divided into type A (melilite–spinel-rich), type B (composed of pyroxene, melilite, spinel, ±plagioclase), forsterite-bearing type B, type C (melilite-poor, pyroxene–plagioclase-rich), and fine-grained spinel-rich (e.g., MacPherson et al., 1988). Type A CAIs are subdivided into fluffy and compact subtypes. In other chondrite groups, CAIs are typically classified according to their modal mineralogy (e.g., corundum-rich, grossite-rich, hibonite-rich, etc.). In CV chondrites, most coarse-grained CAIs were probably once molten. However, in other chondrite groups, CAIs were probably not molten, though definitive compositional data for comparison with melt–solid distribution coefficients are typically not available.

1.07.5.1.1 Comparison of CAIs from different chondrite groups

A comparison of CAIs from carbonaceous, ordinary and enstatite groups (Table 3) shows that most inclusion types can be found in many chondrite groups. However, the relative proportions of CAI types and their mean sizes, which are correlated with chondrule size (May et al., 1999), differ among the chondrite groups. For example, type B CAIs are almost exclusively found in CV chondrites; PLACs occur mostly in CM chondrites; grossite-rich, hibonite-rich and aluminum-diopside spherules are common only in CH chondrites; CAIs in enstatite chondrites are rich in spinel and, in many cases, hibonite. There are also mineralogical and isotopic differences between CAIs of the same mineralogical type: for example, (i) melilite in compact type A CAIs from CV chondrites is more åkermanitic than that in compact melilite-rich CAIs from CO and CR chondrites (Weber and Bischoff, 1997; Russell et al., 1998; Aléon et al., 2002a); (ii) the grossite-rich and hibonite-rich CAIs in CR chondrites are ^{16}O-rich and have a canonical ^{26}Al/^{27}Al ratio (Marhas et al., 2002; Aléon et al., 2002a), whereas those in CH chondrites are ^{16}O-rich, but generally lack radiogenic ^{26}Mg (^{26}Mg*) (MacPherson et al., 1989; Kimura et al., 1993; Weber et al., 1995; McKeegan et al., 2000a). The grossite-rich and hibonite-rich CAIs in CB chondrites are ^{16}O-poor (Krot et al., 2001e); no Al–Mg-isotopic measurements have been published yet.

Table 3 Distribution of types of CAIs, AOAs, accretionary rims, and aluminum-rich chondrules in type 2 and 3 chondrites.

Object	CM	CO	CV	CR	CH	CB	CK	Acfer 094	Adelaide	H–L–LL	EH–EL
Melilite-rich (Type A)	r	c	c	c	c	c	—	c	c	r	—
Al,Ti-px-rich (Type B)	—	—	c	r	—	r	r	—	—	—	—
Fo-Type B CAIs	—	—	r	—	—	r	—	—	—	—	—
Plagioclase-pyroxene-rich (Type C)	—	r	c	r	—	—	—	r	—	—	—
Hibonite ± spinel-rich	c	c	c	r	c	c	—	r	c	r	r
Grossite-rich	—	r	—	r	c	r	—	c	—	—	—
Spinel ± pyroxene-rich	c	c	c	c	c	c	r	c	c	r	r
Pyx-hibonite spherules	r	r	—	—	c	c	—	r	—	—	r
Fo-rich accretionary rims	—	—	c	—	—	—	—	—	—	—	—
Amoeboid olivine aggregates (AOAs)	c	c	c	c	c	r	c	c	c	—	—
Aluminum-rich chondrules	c	c	c	c	c	c	c	c	c	c	c

Key: r, rare; c, common; —, absent. Source: MacPherson et al. (1988) with additional data: CHs, Kimura et al. (1993); COs, Russell et al. (1998); CMs, MacPherson et al. (1984), Simon et al. (1994); CRs, Aléon et al., (2002a), Weisberg et al. (1993), Kallemeyn et al. (1994); CKs, Weber and Bischoff (1997), McSween (1977), MacPherson and Delaney (1985), Kallemeyn et al. (1991), Keller et al. (1992), Noguchi, (1993); ECs, Bischoff et al. (1985), Guan et al. (2000), Fagan et al. (2000); CBs, Krot et al. (2001a); Acfer 094, Adelaide, Krot (unpubl.).

1.07.5.1.2 Oxygen Isotopic compositions of CAIs

In situ ion microprobe analyses of oxygen isotopic compositions of individual minerals in CAIs in the chondrite groups provide very important constraints on the origin of CAIs. CAIs from CV chondrites typically exhibit oxygen isotope heterogeneity within an individual inclusion, with spinel, hibonite, and pyroxene enriched and anorthite and melilite depleted in ^{16}O. At the same time, several coarse-grained igneous CAIs containing ^{16}O-rich and ^{16}O-poor melilite, anorthite and Al–Ti-diopside, and Al–Ti-diopside with oxygen isotope zoning have been described in the Allende and Efremovka meteorites (e.g., Kim et al., 1998; Yurimoto et al., 1998; Ito et al., 1999; Imai and Yurimoto, 2000). Based on these observations and experimental data on oxygen-isotope self-diffusion in CAI minerals (Ryerson and McKeegan, 1994), Yurimoto et al. (1998) concluded that coarse-grained CAIs in CV chondrites were initially ^{16}O-rich and subsequently experienced oxygen isotope exchange with ^{16}O-poor nebular gas while partly molten during reheating. ^{16}O-poor compositions of secondary minerals (nepheline, sodalite, andradite, and hedenbergite) in CV CAIs (Cosarinsky et al., 2003) and oxygen isotope heterogeneity in fine-grained spinel-rich CAIs that were probably not melted suggest that at least some of the oxygen isotopic exchange resulted from secondary alteration (Aléon et al., 2002b; Fagan et al., 2002a).

CAIs from CO chondrites (Figure 6) show a correlation between oxygen isotopic composition and degree of thermal metamorphism: spinel, hibonite, melilite, diopside, anorthite, and olivine in most CAIs from CO3.0 chondrites Colony and Y-81020 are uniformly ^{16}O-enriched, whereas melilite and secondary nepheline in CAIs from metamorphosed CO chondrites Kainsaz (3.2) and Ornans (3.2) are ^{16}O-depleted relative to spinel and diopside. These observations favor oxygen isotope exchange during fluid–rock interaction in an asteroidal setting (Itoh et al., 2000; Wasson et al., 2001). However, the ^{16}O-poor compositions of some CAIs in CO3.0 chondrites (e.g., Yurimoto et al., 2001) suggests that some isotopic exchange also occurred in the solar nebula in the presence of ^{16}O-poor gas.

CAIs in type 2–3 CR, CH, CB, CO, CM, ordinary and enstatite chondrites are typically isotopically uniform (within 3–4‰): both ^{16}O-rich and ^{16}O-poor CAIs have been found. Most CAIs in CR and CH chondrites are ^{16}O-rich (Figure 7); the ^{16}O-poor CAIs in CR, CH, and CB chondrites show petrographic evidence for extensive melting (Aléon et al., 2002a; Krot et al., 2001e; McKeegan et al., 2003). CAIs in CM, ordinary and enstatite chondrites are ^{16}O-rich

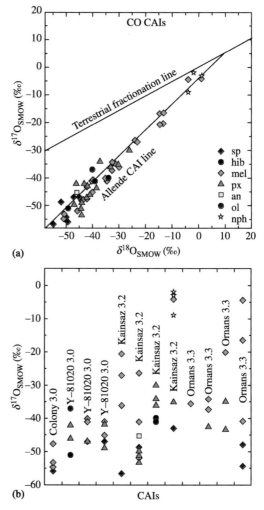

Figure 6 Oxygen-isotopic compositions of individual minerals in CAIs from the CO chondrites Y-81020, Colony, Kainsaz and Ornans. Primary minerals in CAIs from the least metamoprhosed CO chondrites Y-81020 (type 3.0) and Colony (3.0) are uniformly ^{16}O-enriched, whereas CAIs from Kainsaz (3.2) and Ornans (3.3) tend to show oxygen isotopic heterogeneity with spinel and high-calcium pyroxene enriched in ^{16}O and melilite and secondary nepheline depleted in ^{16}O. Based on these observations, Wasson et al. (2001) inferred that oxygen isotope exchange took place during thermal metamorphism and alteration in an asteroid (data from Itoh et al., 2000; Wasson et al., 2001).

(Δ^{17}O <-20‰). In CM chondrites, hibonites in spinel–hibonite spherules (SHIBs), which generally lack stable-isotopic anomalies in ^{48}Ca and ^{49,50}Ti, have canonical (^{26}Al/^{27}Al)$_0$ and (^{41}Ca/^{40}Ca)$_0$ ratios and group II REE patterns, whereas platy hibonites (PLACs), which generally have group III REE patterns, large (>10‰) stable-isotopic anomalies in ^{48}Ca and ^{50}Ti, and lack ^{26}Al and ^{41}Ca, are similarly ^{16}O-enriched. No obvious correlation between hibonite morphology and radiogenic ^{26}Mg, Ca (δ^{48}Ca) or Ti

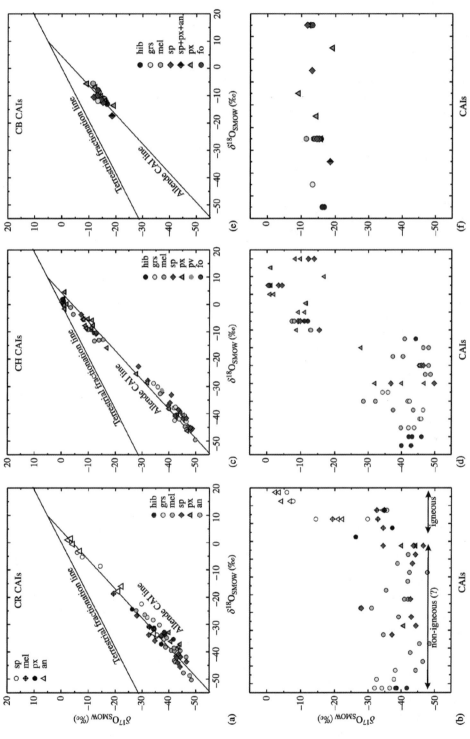

Figure 7 (a,b) Oxygen-isotopic compositions of individual minerals in CAIs from CR chondrites (data from Aléon et al., 2002a). Most CAIs are ^{16}O-rich and isotopically homogeneous; CAIs, which show petrographic evidence for extensive melting are less ^{16}O-rich. Secondary phyllosilicates replacing ^{16}O-rich melilite plot along terrestrial fractionation line. (c,d) Oxygen-isotopic compositions of individual minerals in CAIs from CH chondrites show a bi-modal distribution (data from McKeegan et al., 2003). The most refractory CAIs, which are composed of grossite, hibonite, spinel, perovskite, and melilite, are enriched in ^{16}O; aluminum-diopside rims around these CAIs, pyroxene-spinel CAIs and two melilite-rich CAIs surrounded by a forsterite layer are ^{16}O-poor. (e,f) Oxygen-isotopic compositions of individual minerals in CAIs from CB chondrites Hammadah al Hamra 237 and QUE 94411. The CAIs are rather homogeneous and ^{16}O-poor (data from Krot et al., 2001a).

(δ^{50}Ti) isotopic anomalies has been found (Sahijpal et al., 2000; Goswami et al., 2001).

To summarize, variations in oxygen isotopic compositions within a single CAI in CV and CO chondrites require isotopic exchange, however, the exact mechanism of exchange remains poorly understood. Proposed mechanisms include gas–solid and/or gas–melt exchange in the solar nebula (e.g., Clayton, 1993; Yurimoto et al., 1998, 2000, 2001), and solid–fluid exchange in an asteroidal setting (e.g., Wasson et al., 2001; Fagan et al., 2002a,b). The presence of ^{16}O-poor CAIs with igneous textures in primitive CR, CH, and CB chondrites suggests that in these cases oxygen isotopic exchange occurred during melting. It remains unclear whether this exchange occurred in the CAI- or chondrule-forming regions and whether this exchange was late or early.

The oxygen isotopic compositions of chondritic components are generally assumed to reflect isotopic variations among the different locations within the meteorite-forming regions of the solar nebula. However, since CAIs in most chondrite groups have similar ^{16}O-rich isotopic compositions, it was inferred that CAIs formed in a single, restricted nebular locale and were then unevenly distributed to the regions where chondrites accreted after sorting by size and type (McKeegan et al., 1998; Guan et al., 2000). Although formation in a single restricted region appears to be generally accepted, sorting by size and type cannot explain the observed variations in mineralogy and isotope chemistry of CAIs in different chondrite groups. We suggest instead that CAI formation occurred multiple times in the CAI-forming region under diverse physicochemical conditions (e.g., dust/gas ratio, peak heating temperature, cooling rate, number of recyclings) and that CAIs were subsequently isolated from the hot nebular gas at various ambient nebular temperatures. The isolation could be due to removal by stellar wind (e.g., Shu et al., 1996, 2001). The ^{16}O-rich and ^{16}O-poor CAIs reflect either temporal variations in oxygen isotopic compositions of the nebular gas in the CAI-forming region(s) (e.g., Yurimoto et al., 2001) or radial transport of some CAI precursors into an ^{16}O-poor gas prior to their melting (see also Scott and Krot, 2001; Aléon et al., 2002a; McKeegan et al., 2003).

1.07.5.2 Forsterite-rich Accretionary Rims Around CAIs

Olivine-rich, accretionary rims around CAIs, which have grain sizes >5–20 μm, were first described in the altered CV chondrite Allende by MacPherson et al. (1985). In this meteorite, the accretionary rims have four layers that differ in texture and mineralogy. The innermost layer consists of either pyroxene needles, olivine, clumps of hedenbergite and andradite or olivine "doughnuts" (i.e., crystals with central cavities). The next two layers outward both contain olivine plates and laths. The final layer separating accretionary rims from the Allende matrix contains clumps of andradite and hedenbergite surrounded by salitic pyroxene needles. Nepheline and Fe,Ni-sulfides are common constituents in all layers. Based on the observed disequilibrium mineral assemblages, abundant euhedral crystals with pore space between them, and rim thicknesses controlled by the underlying topography of their host inclusions, MacPherson et al. (1985) concluded that the layered accretionary rims in Allende are aggregates of gas–solid condensates which reflect significant fluctuations in physicochemical conditions in the solar nebula and grain/gas separation processes. Hiyagon (1998) analyzed the oxygen isotope compositions of minerals in accretionary rims around two type A CAIs from Allende. Olivines in both accretionary rims are characterized by ^{16}O-rich-isotopic compositions (δ^{18}O = −28‰, δ^{17}O = −30‰), whereas wollastonite, andradite and hedenbergite show much heavier oxygen isotopic compositions (δ^{18}O from +2‰ to +10‰, δ^{17}O from −3‰ to +4‰).

In the reduced CV chondrites, which are less altered than Allende, accretionary rims are not layered and consist of coarse-grained (20–40 μm), anhedral forsterite (Fa$_{1-8}$ versus Fa$_{5-50}$ in Allende), Fe,Ni-metal nodules, and fine-grained refractory components composed of aluminum-diopside, anorthite, spinel, and, in some cases, forsterite (Krot et al., 2001b; 2002b). These rims surround different types of CAIs: types A, B, and fine-grained spinel-rich (Figure 8). Forsterite grains in the accretionary rims are ^{16}O-enriched similar to spinel, aluminum-diopside, and hibonite of the host CAIs (Figure 9).

Krot et al. (2001b, 2002b) concluded that forsterite-rich rims formed by accretion of high-temperature condensates in the CAI-forming regions. The rims were emplaced after high-temperature events experienced by the CAIs, such as melting and formation of Wark–Lovering rims (see Wark and Boynton, 2001), which must therefore have occurred in the CAI-forming region as well. Andradite, wollastonite, hedenbergite, and fayalitic olivine in the in the Allende rims formed by alteration under oxidizing conditions in the presence of ^{16}O-poor fluid (Krot et al., 1995; 1998a,b,c,d).

1.07.5.3 Amoeboid Olivine Aggregates (AOAs)

Amoeboid olivine aggregates (AOAs) are irregularly shaped objects with fine grain sizes

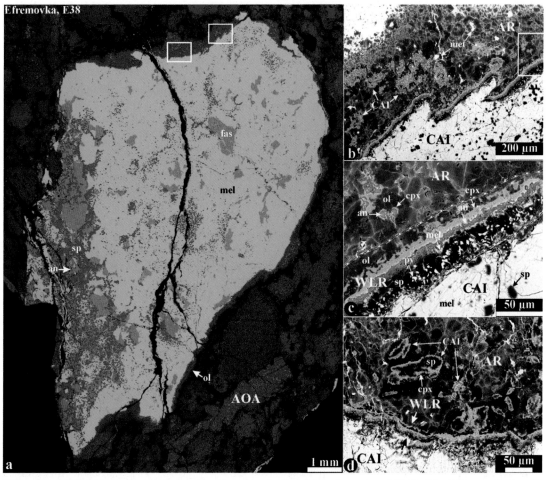

Figure 8 (a) Combined elemental map in magnesium (red), calcium (green) and Al K$_\alpha$ (blue) X-rays of a type B CAI E38 from Efremovka with an accretionary rim; (b)–(d) backscattered electron images from regions marked in elemental map. This CAI consists of melilite (mel), aluminum, titanium-diopside (fas), spinel (sp), and anorthite (an) and is surrounded by a multilayered Wark–Lovering rim composed of spinel, perovskite (pv), anorthite, aluminum-diopside, and forsterite, and a thick outer forsterite-rich accretionary rim (red in (a)). The accretionary rim consists of forsterite (ol) and refractory components composed of spinel, aluminum-diopside, and anorthite.

(5–20 μm) that occupy up to a few percent of type 2–3 carbonaceous chondrites (Table 3). AOAs are similar in size to co-accreted chondrules and CAIs being mostly 100–500 μm in size in CO (Chizmadia et al., 2002), up to 5 mm in size in CV (Komatsu et al., 2001), and typically <500 μm in CR chondrites (Aléon et al., 2002a). In the least altered chondrites, they are porous aggregates and their mineralogy matches that expected for high-temperature nebular condensates. Unlike CAIs and chondrules, AOAs do not appear to show mineralogical and isotopic differences between groups.

1.07.5.3.1 Mineralogy and petrology of AOAs

AOAs, which were first described in the heavily altered Allende CV3 chondrite (Grossman and Steele, 1976), are best observed in unaltered chondrites, CO3.0, CR2, CH, CB chondrites, Adelaide, and Acfer 094. Here they consist of anhedral, fine-grained (1–20 μm) forsterite (Fa$_{<1-3}$), Fe,Ni-metal, and a refractory component composed of aluminum-diopside, spinel, anorthite, and rare melilite, or CAIs of similar mineralogy. The refractory objects are characterized by significant variations in size (1–250 μm), shape (irregular, rounded), distribution (uniform, heterogeneous), modal mineralogy (spinel-rich, spinel-poor), and abundance (rare, abundant) (Figure 10). Forsterite grains typically contain numerous pores and tiny inclusions of aluminum-diopside; some AOAs show triple junctions between forsterite grains and coarse-grained shells surrounding finer-grained cores, indicating high-temperature annealing (Komatsu et al., 2001). About 10% of AOAs contain low-calcium pyroxene replacing forsterite (Figure 5(c)).

1.07.5.3.2 Trace elements and isotopic composition of AOAs

Trace elements concentrations have been measured in 10 AOAs from Allende (Grossman et al., 1979). Most AOAs have unfractionated abundances of refractory lithophile and siderophile elements (2–20 × CI); two AOAs analyzed have group II REE patterns.

Oxygen-isotopic analysis of primary minerals (forsterite, aluminum-diopside, spinel, anorthite) in AOAs from CM, CR, CB, and unmetamorphosed CO chondrites are uniformly enriched in ^{16}O, whereas anorthite, melilite and secondary minerals (nepheline and sodalite) in AOAs from CVs and metamorphosed COs are ^{16}O-depleted to various degrees (Hiyagon and Hashimoto, 1999a; Imai and Yurimoto, 2000, 2001; Itoh et al., 2002; Krot et al., 2001b, 2002b) (Figure 12). Melilite grains in AOAs from the reduced CV3 chondrite Efremovka are ^{16}O-enriched relative to those from in Allende, suggesting exchange during alteration (Fagan et al., 2002a). Imai and Yurimoto (2000, 2001) reported two generations of olivines in Allende AOAs: primary forsteritic olivines, which are ^{16}O-rich ($\Delta^{17}O \sim -20‰$), and secondary fayalitic olivines, which are ^{16}O-poor ($\Delta^{17}O \sim -5‰$) and probably grew on the parent asteroid in the presence of ^{16}O-poor fluid.

Three AOAs in the CO3.0 chondrite Y-81020 that were studied for Al–Mg systematics showed ^{26}Mg excesses corresponding to initial $^{26}Al/^{27}Al$ ratios of $(2.7 \pm 0.7) \times 10^{-5}$, $(2.9 \pm 1.5) \times 10^{-5}$ and $(3.2 \pm 1.5) \times 10^{-5}$. Comparable excesses of ^{26}Mg were not detected in the AOA-bearing chondrule (Itoh et al., 2002). The lower $^{26}Al/^{27}Al$ ratios in AOAs compared to a canonical value of $\sim 5 \times 10^{-5}$ found in most CAIs (e.g., MacPherson et al., 1995) suggest that either AOAs formed 0.1–0.5 Ma after CAIs or that ^{26}Al was heterogeneously distributed in the CAI-forming region (Itoh et al., 2002).

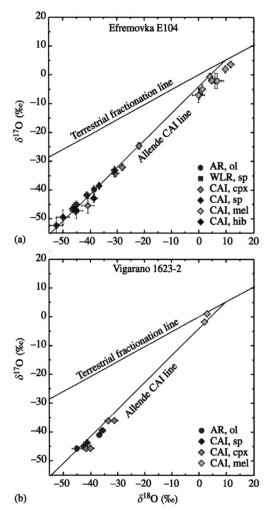

Figure 9 Oxygen-isotopic compositions of minerals in the Efremovka CAI E104 (a) and Vigarano CAI 1623-2 (b) and their forsterite-rich accretionary rims (AR) and Wark–Lovering rims (WLR) showing that both formed in an ^{16}O-rich reservoir (after Krot et al., 2002b).

Although AOAs show no clear evidence of being melted, some appear to contain CAIs that have been melted (Figure 10). AOAs in CO3.0 chondrites contain ~8 vol.% metallic Fe,Ni: troilite is rare or absent (Chizmadia et al., 2002).

Secondary nepheline, sodalite and ferrous olivine are common in AOAs in the altered and metamorphosed CV and CO chondrites, but these minerals are absent in AOAs from the unaltered chondrites. Olivines in AOAs from CV chondrites are generally more fayalitic than in the unaltered chondrites, with fayalite concentrations increasing from Leoville and Vigarano to Efremovka to Allende (Figure 11). This sequence correlates with the degree of secondary alteration and thermal metamorphism experienced by CV chondrites (e.g., Krot et al., 1995; 1998a,b,c,d; Komatsu et al., 2001). In CO3 chondrites, fayalite contents are well correlated with petrogic subtype (Chizmadia et al., 2002).

1.07.5.3.3 Origin of amoeboid Olivine aggregates

The mineralogy, petrology, ^{16}O-rich compositions, depletion in moderately volatile elements, such as manganese, Cr, and Na, and the general absence of low-calcium pyroxenes associated with AOAs strongly suggest that AOAs are aggregates of grains of forsterite, aluminum-diopside, spinel, anorthite and Fe,Ni-metal that condensed in the nebula from an ^{16}O-rich gaseous reservoir and aggregated with refractory objects. Mineralogical and chemical similarities between AOAs and forsterite-rich accretionary rims suggest that AOAs formed contemporaneously with these

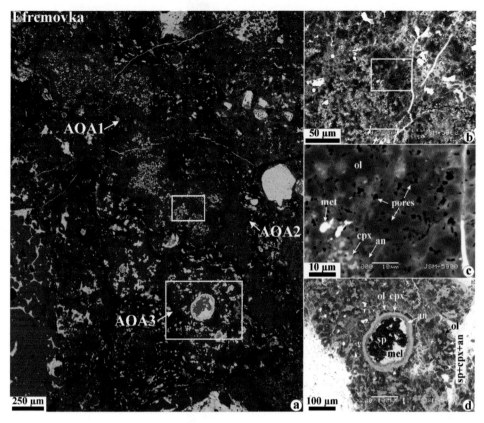

Figure 10 Combined elemental map in magnesium (red), calcium (green), and Al K_α (blue) X-rays (a) and BSE images (b)–(d) of amoeboid–olivine aggregates (AOAs) in the reduced CV chondrite Efremovka. They consist of fine-grained forsteritic olivines (ol), Fe,Ni-metal (met), and CAIs composed of aluminum-diopside (cpx), anorthite (an), spinel (sp), and rare melilite (mel); melilite is replaced by anorthite. Although AOAs show no clear evidence of being melted, some CAIs inside them appear to have experienced melting. Regions outlined in (a) and (b) are shown in detail in (b)–(d).

rims. Some refractory objects inside AOAs were melted prior to aggregation. AOAs subsequently experienced high-temperature annealing and solid-state recrystallization without substantial melting and nebular alteration resulting in formation of anorthite and, in some cases, low-calcium pyroxene. The AOAs formed in the CAI-forming region(s) and were largely absent from chondrule-forming region(s) when chondrules formed. They were either removed from the CAI-forming region prior to condensation of low-calcium pyroxene or else gas–solid condensation of low-calcium-pyroxene was kinetically inhibited.

1.07.5.4 Aluminum-rich Chondrules

Aluminum-rich chondrules are a diverse group of objects with igneous textures and >10 wt.% Al_2O_3 (Bischoff and Keil, 1984; Huss et al., 2001), which appear to have distinct origins from ferromagnesian chondrules (Section 1.07.5.5). Their mineralogy, bulk chemical compositions, oxygen and Al–Mg isotope systematics suggest a close link between CAIs and aluminum-rich chondrules. Using the $CaO-MgO-Al_2O_3-SiO_2$ phase diagram (Figure 13), Huss et al. (2001) subdivided aluminum-rich chondrules in ordinary chondrites according to their bulk composition and mineralogy into three groups: *plagioclase-rich*, which plot in the anorthite (+spinel) field on the liquidus and contain abundant euhedral calcic plagioclase crystals, *olivine-rich*, which plot in the forsterite (+spinel) field and commonly contain large euhedral olivine phenocrysts, and *glassy*, which plot near the corundum–forsterite side of the ternary and are characterized by abundant transparent Na–Al-rich and calcium-poor glass that encloses a phenocryst assemblage of olivine, pyroxene, and spinel. In carbonaceous chondrites, however, Krot and Keil (2002) and Krot et al. (2002c) found a continuum between the plagioclase-rich and olivine-rich aluminum-rich chondrules: glassy aluminum-rich chondrules are absent (Figure 14). Because aluminum-rich chondrules in carbonaceous chondrites all have similar mineralogy (magnesian low-calcium

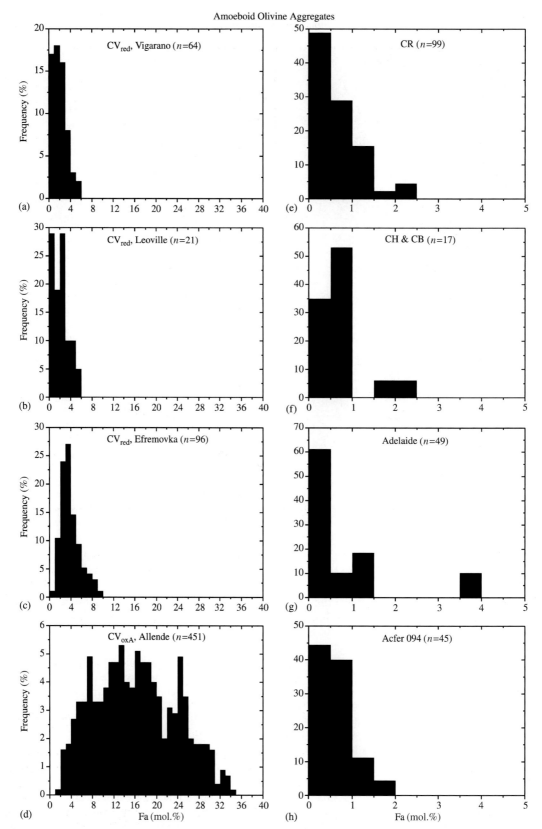

Figure 11 Compositions of olivine in AOAs from the reduced CV chondrites Vigarano (a), Leoville (b), Efremovka (c), oxidized CV chondrite Allende (d), CR chondrites (e), CH and CB chondrites (f), unique carbonaceous chondrites Adelaide (g), and Acfer 094 (h). Olivines in CRs, Adelaide, and Acfer 094 are magnesium-rich compared to olivines in CV AOAs. Fayalite contents in olivines from CV AOAs increase in the order Leoville and Vigarano, Efremovka, Allende; this is correlated with the degree of secondary alteration and thermal metamorphism experienced by CV chondrites. Data for Allende AOAs are from Hashimoto and Grossman (1987); and for Efremovka, Leoville, and Vigarano AOAs from Komatsu *et al.* (2001).

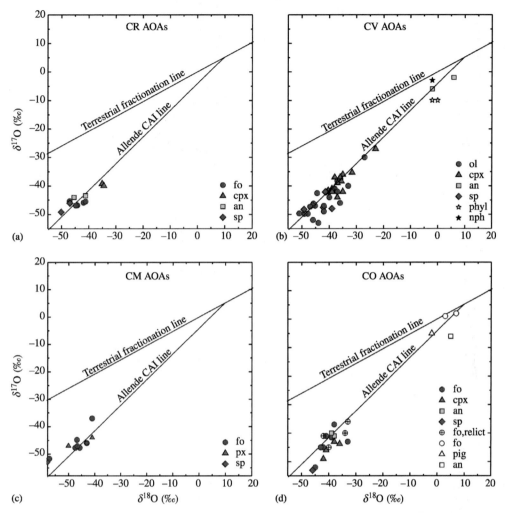

Figure 12 Ion microprobe analyses of oxygen-isotopic compositions of individual minerals in AOAs from CR2 (a), CV3 (b), CM2 (c), and CO3 (d) chondrites. Key: fo = forsterite, cpx = Al-diopside, an = anorthite, sp = spinel; fas = Al–Ti-diopside; phyl = phyllosilicates; nph = nepheline; fo relict = relict AOA forsterite in chondrules; pig = chondrule pigeonite (data for CR AOAs from Aléon et al. (2002a); CV AOAs are from Imai and Yurimoto (2000, 2001); Fagan et al. (2002b); Hiyagon and Hashimoto (1999a); Krot et al., 2002b); CM AOAs (Hiyagon and Hashimoto, 1999b); CO AOAs (Itoh et al., 2002); CB AOA (Krot et al., 2001a).

pyroxene and forsterite phenocrysts, Fe,Ni-metal nodules, interstitial anorthitic plagioclase, Al–Ti–Cr-rich low-calcium and high-calcium pyroxenes, and crystalline mesostasis composed of silica, anorthite and high-calcium pyroxene), Krot and Keil (2002) and Krot et al. (2002c) called them all plagioclase-rich chondrules, irrespective of modal mineralogy.

Because the aluminum-rich chondrules are compositionally most similar to CAIs and often contain aluminum-rich phases with high Al/Mg ratios (plagioclase and, occasionally, glass) which make them suitable for Al–Mg isotope studies, these chondrules received special attention (e.g., Kring and Boynton, 1990; Sheng et al., 1991; Hutcheon and Jones, 1995; Russell et al., 1996, 2000; Hutcheon et al., 1994, 2000; Marhas et al., 2002; Srinivasan et al., 2000a,b;

Huss et al., 2001; Krot and Keil, 2002; Krot et al., 2001c; 2002a,c).

1.07.5.4.1 Isotopic and trace element studies of aluminum-rich chondrules

Oxygen-isotopic data for seven aluminum-rich chondrules from ordinary chondrites (Russell et al., 2000) are continuous with the ordinary chondrite ferromagnesian chondrule field above the terrestrial line, but extend it to more ^{16}O-enriched values ($\delta^{18}O = -15.7 \pm 1.8‰$, $\delta^{17}O = -13.5 \pm 2.6‰$) along a mixing line of slope = 0.83 ± 0.09 (Figure 15(a)). Two chondrules exhibit significant internal-isotopic heterogeneity indicative of partial exchange with a gaseous reservoir (Russell et al., 2000). Porphyritic aluminum-rich chondrules are consistently

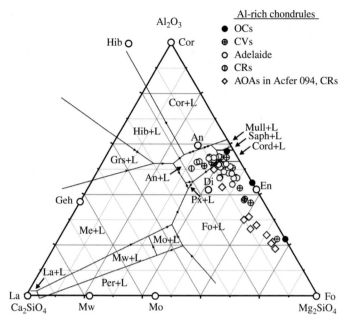

Figure 13 Bulk compositions of aluminum-rich chondrules plotted in the system $Al_2O_3-Mg_2SiO_4-Ca_2SiO_4$ and projected from spinel ($MgAl_2O_4$). Mineral abbreviations: An = anorthite, Grs = grossite, Cor = corundum, Cord = cordierite, Di = diopside, Fo = forsterite, Geh = gehlenite, Hib = hibonite, La = larnite, L = liquid, Mel = melilite solid solution, Mull = mullite, Mo = monticellite, Mw = merwinite, Per = periclase, Pyx = pyroxene solid solution, Saph = sapphirine (data from Bischoff and Keil (1984), Sheng et al. (1991), Huss et al. (2001), Krot and Keil (2002), Krot et al. (2002a, 2003g).

^{16}O-rich relative to nonporphyritic ones, suggesting that degree of melting was a key factor during nebular exchange (Russell et al., 2000).

Oxygen-isotopic compositions of aluminum-rich chondrules from CR and CH chondrites (Krot et al., 2003c) are plotted in Figure 15(b). Three out of six chondrules analyzed exhibit large internal-isotopic heterogeneity, whereas others are isotopically uniform; all chondrules have porphyritic textures. In contrast to aluminum-rich chondrules in ordinary chondrites, oxygen isotopic heterogeneity is due to the presence of relict CAIs (Krot and Keil, 2002).

Aluminum–magnesium isotope measurements of aluminum-rich chondrules in ordinary and carbonaceous chondrites were summarized by Huss et al. (2001). About 10–15% of all aluminum-rich chondrules studied show $^{26}Mg^*$. The inferred $(^{26}Al/^{27}Al)_0$ ratio of these chondrules is $\sim(0.5-2) \times 10^{-5}$, which is consistent with the values found in ferromagnesian chondrules from type 3.0–3.1 unequilibrated ordinary chondrites (Kita et al., 2000; McKeegan et al., 2000b; Mostefaoui et al., 2000; 2002).

The REEs in most aluminum-rich chondrules from carbonaceous chondrites studied are unfractionated (Kring and Boynton, 1990; Weisberg et al., 1991). However, some have highly fractionated REE patterns, similar to volatility-controlled group II and ultrarefractory patterns seen in refractory inclusions (Kring and Boynton, 1990; Misawa and Nakamura, 1988, 1996).

1.07.5.4.2 Origin of aluminum-rich chondrules

MacPherson and Russell (1997) discussed several possible models for the origin of aluminum-rich chondrules including impact melting of aluminous (evolved) parent body surface material, mixing of aluminous components with ferromagnesian chondrule precursors prior to chondrule formation, and volatilization. They concluded that (i) no single model appears capable of explaining the entire spectrum of aluminum-rich chondrule bulk compositions, (ii) aluminum-rich chondrules are not intermediate in an evolutionary sense between known CAIs and chondrules, and (iii) aluminum-rich chondrules are more closely related to normal chondrules, either by addition of an anorthite-like component to chondrule precursors or possibly, in some cases, by volatilization of chondrule or chondritic material of unusual oxygen isotopic composition. Russell et al. (2000) concluded that "if aluminum-rich chondrules were mixtures of ferromagnesian chondrules and CAI material, their bulk chemical compositions would require them to exhibit larger ^{16}O enrichments than we observe. Therefore, aluminum-rich chondrules are not simple mixtures of these two components."

However, Krot and Keil (2002) and Krot et al. (2001c, 2002c) showed that ~15% of aluminum-rich chondrules from different carbonaceous chondrites are associated with relict CAIs

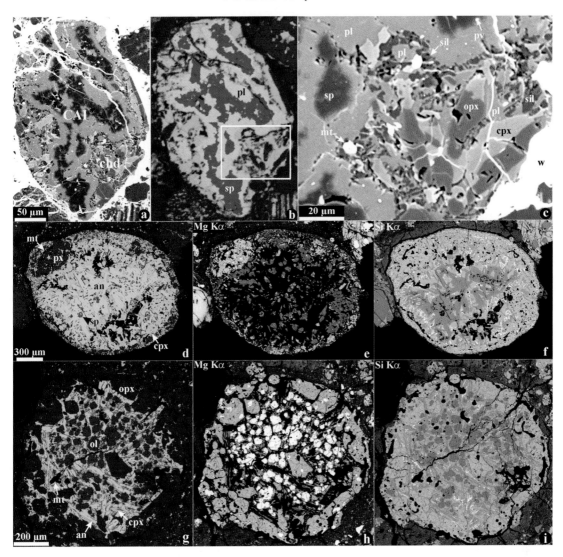

Figure 14 Backscattered electron images ((a), (c)), combined elemental map in magnesium (red), calcium (green) and Al K_α (blue) X-rays ((b), (d), (g)), and elemental maps in magnesium ((e), (h)) and Si K_α X-rays ((f), (i)) of the CAI-bearing aluminum-rich chondrule #2 in the CH chondrite Acfer 182 ((a)–(c)) and plagioclase-bearing chondrules #1 ((d)–(f)) and #17 ((g)–(i)) in the CR chondrites EET92042 and PCA91082, respectively. (a)–(c) Relict CAI consists of irregularly-shaped spinel (sp) core surrounded by anorthite (pl); tiny perovskite (pv) inclusions occur in spinel. The chondrule portion consists of low-calcium pyroxene (opx), high-calcium pyroxene (cpx), Fe,Ni-metal nodules (met) and silica-rich crystalline mesostasis composed of silica (sil), anorthite and high-calcium pyroxene. Region outlined in (b) is shown in detail in (c). White veins in (a) are veins of terrestrial weathering products. (d)–(f) Chondrule #1 consists of anorthite poikilitically enclosing spinel grains, which are relict, high-calcium pyroxene, low-calcium pyroxene, Fe,Ni-metal nodules, minor forsterite and silica-rich crystalline mesostasis composed of silica, anorthite and high-calcium pyroxene. (g)–(i) Chondrule #17 consists of forsteritic olivine, low-calcium pyroxene, anorthitic plagioclase, high-calcium pyroxene, Fe,Ni-metal nodules (black), and silica-rich crystalline mesostasis composed of anorthite, silica, and high-calcium pyroxene. Low-calcium pyroxene occurs as elongated grains overgrown by high-calcium pyroxene in the chondrule cores and as phenocrysts poikilitically enclosing forsterite grains in chondrule peripheries.

(Figure 14). These relict CAIs are mineralogically similar to CAIs in AOAs and consist of spinel, anorthite, ± aluminum-diopside, and ± forsterite; most of them were extensively melted during chondrule formation. Krot and colleagues inferred that aluminum-rich chondrules formed by melting of the ferromagnesian chondrule precursors (magnesian olivine and pyroxenes, Fe,Ni-metal) mixed with the refractory materials, including relict CAIs, composed of anorthite, spinel, high-calcium pyroxene and forsterite. This mixing explains the relative enrichment of anorthite-rich chondrules in ^{16}O compared to typical ferromagnesian chondrules (Figure 12), and the group II REE patterns of some of the aluminum-rich chondrules (Misawa and Nakamura, 1988, 1996; Kring and Boynton, 1990).

1.07.5.4.3 Relict CAIs in ferromagnesian chondrules

Mineralogical observations indicate that CAIs outside chondrules show no clear evidence for being melted during a chondrule-forming event (Krot et al., 2002a,b,c, 2003a). Relict CAIs inside ferromagnesian chondrules are exceptionally rare. Only six relict CAIs have been described in ferromagnesian chondrules: in the unique carbonaceous chondrites Acfer 094 and Adelaide, the CH chondrites Acfer 182 and Patuxent Range 91546, the CV chondrite Allende, and the H3.4 ordinary chondrite Sharps (Bischoff and Keil, 1984; Misawa and Fujita, 1994; Krot et al., 1999, 2003a). Relict CAIs are also present in AOAs (Itoh et al., 2002; Hiyagon, 2000). These relict CAIs are very refractory, unlike those in aluminum-rich chondrules, being composed of hibonite, grossite, spinel, melilite, perovskite, and platinum group metals (Figure 16). The relict CAIs in chondrules are corroded by host chondrule melts, but have largely preserved the ^{16}O-rich isotopic signatures of typical CAIs (Figure 15(c)). These observations indicate that some CAIs were present in the chondrule-forming regions and were melted together with chondrule precursors, suggesting that at least some CAIs formed before chondrules.

Published magnesium isotope measurements of the relict CAIs and CAI- and AOA-bearing chondrules show no resolvable ^{26}Mg* (Krot et al., 1999, 2002c; Itoh et al., 2002), implying upper limits for $(^{26}\text{Al}/^{27}\text{Al})_I$ of $<1 \times 10^{-5}$ and $<5 \times 10^{-6}$. These data suggest either these CAIs formed without ^{26}Al or their Al–Mg systems were reset during chondrule formation. If these CAIs formed initially with a canonical $(^{26}\text{Al}/^{26}\text{Mg})_I$ that was reset during chondrule melting, the host chondrules must have formed at least ~2 Myr after the CAIs.

1.07.5.5 Chondrules

Chondrules are igneous-textured particles composed mainly of olivine and low-calcium pyroxene crystals set in a feldspathic glass or microcrystalline matrix. (Chondrules with >10 wt.% Al_2O_3, which have closer links with CAIs are reviewed in Section 1.07.5.4.) Many chondrules with FeO-poor silicates also contain metallic Fe,Ni grains, often clustered near the periphery. Metal grains outside chondrules that are of comparable sizes to those inside have geochemical properties suggesting they were once ejected from or broken out of chondrules, or in the case of the CB group (where reduced chondrules lack metal), that metal and chondrules

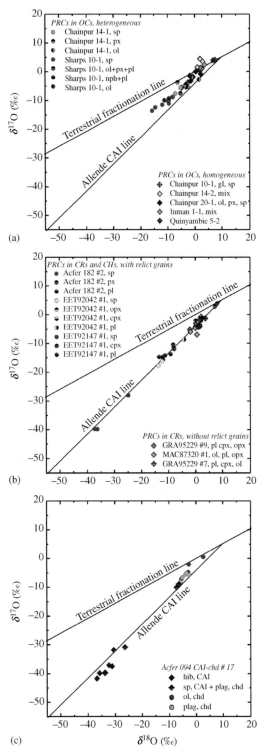

Figure 15 Oxygen-isotopic compositions of individual minerals in (a) aluminum-rich chondrules from ordinary chondrites (Russell et al., 2000) and (b) CR carbonaceous chondrites (Krot et al., 2002b). (c) Oxygen-isotopic compositions of individual minerals in the CAI-bearing chondrule #17 from Acfer 094 (data from Krot et al., 2003a) (Abbreviations: chd, chondrule; cpx, clinopyroxene; gl, glass; hib, hibonite; nph, nepheline; ol, olivine; opx, orthopyroxene; pl, plagioclase; px, pyroxene; sp, spinel).

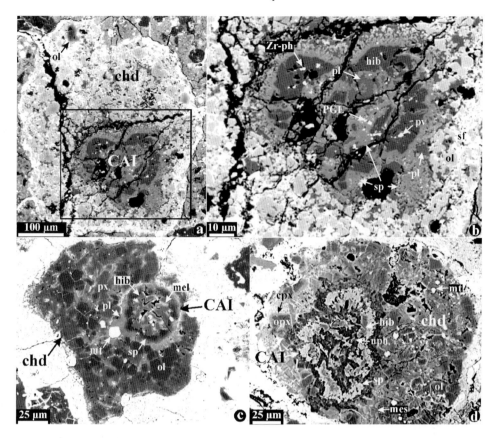

Figure 16 Backscattered electron images of the CAI-bearing chondrules #17 in Acfer 094 ((a), (b)), #9d in Adelaide (c), and #1 in Sharps (d). Relict CAI Acfer 094 ((a), (b)) consists of hibonite (hib) with inclusions of perovskite (pv), a Zr-bearing phase (Zr-ph), and Re-, Ir-, Os-bearing (PGE) nuggets. It is surrounded by a shell of fine-grained, iron-spinel (sp). The host chondrule consists of ferrous olivine (ol), plagioclase mesostasis (pl), Fe,Ni-sulfides (sf), Cr-spinel, and relict grains of forsteritic olivine (fo). The relict CAI in Adelaide (c) consists of hibonite, perovskite and melilite (mel); it is surrounded by a magnesium-spinel rim. Its host chondrule consists of forsteritic olivine, low-calcium pyroxene, glassy mesostasis, and kamacitic metal. The relict CAI in Sharps (d) consists of hibonite surrounded by iron-spinel and a nepheline-like phase (nph). Its host chondrule consists of forsteritic olivine, low-calcium pyroxene, high-calcium pyroxene, anorthitic mesostasis, and kamacitic metal (after Krot *et al.* (2003a)).

formed in the same environment. A close link between metal and reduced silicates is consistent with mineral–gas equilibration in the solar nebula as forsterite, enstatite, and metallic Fe,Ni are stable over similar temperature ranges (figure 2 of Chapter 1.15).

Links between asteroids and chondrites suggest that chondrules are abundant in the inner part of the asteroid belt and that the terrestrial planets probably formed from chondritic material. Thus, much of the mass that accreted in the inner solar system would have been derived from chondrules. In the absence of sticky organics, dry grains of silicate do not readily adhere so that chondrule formation may have been the process that triggered accretion in the inner solar system (e.g., Wood, 1996a; Cuzzi *et al.*, 2001; Scott, 2002). Understanding the origin of chondrules has been a major goal of meteorite studies since chondrites were first described over 200 years ago.

1.07.5.5.1 Chondrule textures, types, and thermal histories

Chondrule textures are conveniently divided into two types; porphyritic, which have large crystals of olivine and/or pyroxene in a fine-grained or glassy matrix, and nonporphyritic, which include those with cryptocrystalline, radial-pyroxene, and barred-olivine textures (Figures 1 and 17). Table 4 shows how the proportions of different types vary considerably among the groups. Carbonaceous chondrites tend to have the highest fraction of porphyritic chondrules, but CH and CB groups are exceptions. Mean sizes and abundances of chondrules in each group are given in Table 1.

Laboratory experiments constrain the initial conditions and thermal history required for each textural type, but there are too many variables (e.g., initial state of chondrule, peak temperature, time at peak temperature, and cooling rate) to

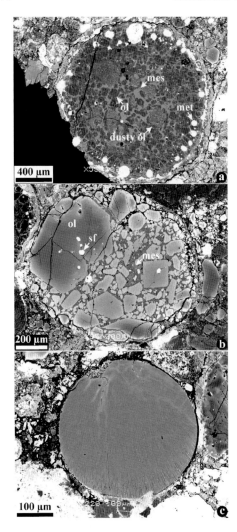

Figure 17 Backscattered electron images of three chondrules in Tieschitz (H/L3.6) chondrite. (a) Type IA porphyritic chondrule composed largely of forsteritic olivine (ol), mesostasis (mes), metallic Fe,Ni droplets (white). Dusty olivines are relict, FeO-rich olivines that crystallized elsewhere and formed tiny metallic Fe,Ni particles when heated in the chondrule melt. (b) Type IIA porphyritic chondrule containing large euhedral, FeO-bearing, olivine phenocrysts, dark mesostasis (mes), and white sulfide droplets (sf). (c) Nonporphyritic chondrule containing fine pyroxene crystals that appear to radiate from the upper edge of the chondrule.

make very quantitative statements (Lofgren, 1996; Hewins, 1997). Porphyritic chondrules crystallized from melts with many nuclei after incomplete melting of fine-grained, precursor materials (Lofgren, 1996; Connolly et al., 1998). Peak temperature and duration of heating are linked as porphyritic textures can be reproduced by heating fine-grained material to $\sim 100\,°C$ above the liquidus temperature (generally $1{,}350-1{,}800\,°C$) for 5 min, or just below the liquidus temperature for hours (Hewins and Connolly, 1996; Lofgren, 1996). Nonporphyritic chondrules crystallized from melts that were superheated above their liquidus for long enough to destroy all solids that could act as nuclei. For barred olivine and radial pyroxene chondrules, nucleation was probably induced by collision with solid grains (Connolly and Hewins, 1995). Radial textured chondrules crystallized while cooling at $5-3{,}000\,°C\,h^{-1}$, barred olivine chondrules at $500-3{,}000\,°C\,h^{-1}$, and porphyritic chondrules at $5-100\,°C\,h^{-1}$ (Desch and Connolly, 2002). Cooling rates above the liquidus may have been much higher; below the solidus, chondrules cooled more slowly at rates of $0.1-50\,°C\,h^{-1}$ (Weinbruch et al., 2001).

Porphyritic chondrules are classified into type I, which are FeO-poor in unmetamorphosed chondrites, and type II, which are FeO-rich. Type I chondrules account for >95% of chondrules in CO, CV, CR (Figure 1(a)), and CM chondrites and olivine-rich varieties were probably abundant in the material that accreted to form the Earth (Hewins and Herzberg, 1996). Types I and II are both subdivided into A, which are olivine-rich (>90 modal percent olivine), and B, which are pyroxene-rich (Jones, 1990; 1994, 1996a,b). Olivines in type IA chondrules in unmetamorphosed chondrites are mostly $Fa_{0.3-5}$ and Fa_{2-10} in IB; type II chondrule olivines are Fa_{10-20} in ordinary chondrites and up to Fa_{50} in carbonaceous chondrites. Rounded type I chondrules commonly have low-calcium pyroxene rims, glassy mesostases, and grains of metallic Fe,Ni, whereas type IIs lack these features, contain small chromites and commonly have larger olivines (Figure 17; Scott and Taylor, 1983). Type I and II chondrules can be identified from these features in many type 2–6 chondrites so that metamorphic and alteration effects can be tracked (McCoy et al., 1991; Hanowski and Brearley, 2001). Type I chondrules also occur as finer grained, irregularly shaped aggregates, which represent partly melted dust aggregates or aggregates of partly melted particles. In ordinary chondrites, which have very diverse textured chondrules, all chondrules having olivines with Fa <10 may be referred to as type I by some authors (Hewins, 1997).

Chondrules have also been classified using cathodoluminescence properties of the glass and olivine (Sears et al., 1995a). For unmetamorphosed chondrites, groups A and B in this scheme correspond to types I and II. During metamorphism, the cathodoluminescence properties and chondrule group change. The merits of the two schemes were reviewed by Scott et al. (1994) and Sears et al. (1995b).

Table 4 Proportions of chondrule types in chondrite groups.

Type/group	CM	CO	CV	CR	CH	CB$_b$	CK	H-L-LL	EH	EL	R
Porphyritic											
Olivine		8						23	0.1		
Oliv-pyx		69						48	4		
Pyroxene		18						10	77		
Granular oliv-pyx		<0.1	<0.1					3	1		
Total porphyritic	~95	95	94	96–98	20	~1	>99	84	82	~87	~92
Barred olivine		2	6		1			4	0.1		
Radial pyx.		2	0.2					7	13		
Cryptocrystalline		1	0.1		79[a]			5	5		
Total nonporph	~5	5	6.3	2–4	80	~99	<1	16	18	~13	~8

Source: Rubin (2000), Kallemeyn et al. (1994), Weisberg et al. (2001), Scott (1988).
[a] Includes radial pyroxene and glass chondrules.

1.07.5.5.2 Compositions of chondrules

Chemical compositions of chondrules have been determined from extracted samples using neutron activation analysis and by in situ analysis in polished sections using electron microprobe and ion probe analysis (see e.g., Gooding et al., 1980; Grossman et al., 1988; Alexander, 1995). Chondrules typically show flat refractory abundances that are relatively close to the mean chondrite value and abundances of moderately volatile elements that scatter more widely about the mean. Type II chondrules are relatively unfractionated with near-CI levels of refractories and moderately volatile elements. However, type I chondrules show systematic depletions of moderately volatile elements and a broader spread of refractory abundances with the silicon-rich type IB chondrules being poorer in refractories than the silicon-poor type IA chondrules (Figure 18). The source of the fractionations in type I chondrules is discussed below.

Oxygen-isotopic compositions of large individual chondrules suggest that chondrules formed from several different reservoirs that were poorer in ^{16}O than the CAI source (Figure 19). Chondrules in ordinary chondrites plot above the terrestrial fractionation line, enstatite chondrite chondrules plot on or near the line, and carbonaceous chondrite chondrules lie below. Ion probe analyses for oxygen isotopes in olivine show that most chondrules have homogeneous compositions and extend the bulk compositional fields previously defined (see references in Scott and Krot, 2001). Chondrules are isotopically homogeneous in comparison to CAIs: ^{50}Ti anomalies (see Niemeyer, 1988) and mass fractionation effects (e.g., Clayton et al., 1985; Misawa and Fujita, 2000; Alexander and Wang, 2001; Alexander et al., 2000) are much smaller than those in many CAIs. Inferred initial concentrations of ^{26}Al in chondrules suggest that chondrules formed ~2 Ma after CAIs- (Kita et al., 2000; Mostefaoui et al.,

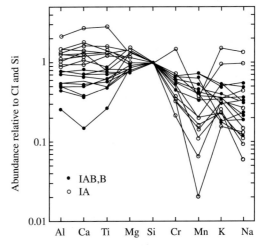

Figure 18 Elemental abundances in type IA chondrules, which are olivine-rich and type IB and IAB chondrules, which are pyroxene rich, from the Semarkona (LL3.0) chondrite normalized to CI chondrites and silicon. Type IA chondrules are enriched in refractories relative to IAB and IB. For moderately volatile elements, all type I chondrules are depleted, with most IA showing larger depletions than types IB and IAB. (Jones (1994); reproduced by permission of Elsevier from *Geochim. Cosmochim. Acta*, **1994**, 58, 5538).

2002), consistent with limited lead-isotopic ages (Amelin et al., 2002). See Gilmour (2002) and Chapter 1.16.

Effects of secondary processes in modifying volatile elements (Grossman et al., 2000, 2002) and oxygen isotopes in chondrules, especially in ordinary chondrites (Bridges et al., 1998; Franchi et al., 2001) require further study (see Chapter 1.09).

Trapped and cosmogenic noble gases in chondrules offer clues to their irradiation history and formation but not definitive constraints (see Chapter 1.14). Trapped noble gases are generally absent or rare in chondrules, but significant amounts of trapped ^{36}Ar, ^{84}Kr, and ^{132}Xe from a

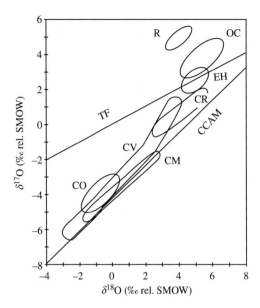

Figure 19 Oxygen-isotopic plot showing fields of analyses of large chondrules in chondrite groups. The data show that chondrules are derived from many isotopically distinct reservoirs (Rubin (2000); reproduced by permission of Elsevier from *Earth Sci. Rev.*, **2000**, *50*, p. 5).

so-called "subsolar" component were found in chondrules in an EH3/4 chondrite (Okazaki et al., 2001). This component is intermediate in composition between that of the Sun and the major trapped component in type 1–3 chondrites ("Q-type" gases) and occurs in E4–6 chondrites (Patzer and Schultz, 2002b), CR2, CH3, CB_a3 and K3 chondrites (see Bischoff et al., 1993). Okazaki et al. (2001) inferred that solar wind had been implanted into chondrule precursor material near the protosun and then fractionated during melting to leave a subsolar pattern. However, subsolar noble gases are present in several other sites suggesting that the subsolar gases were redistributed (Busemann et al., 2002). Further laser microprobe analyses of chondrules and other components in the most pristine chondrites are clearly needed to establish which chondrules, if any, acquired solar gases from the nebula or were irradiated by high-energy particles near the inner edge of the disk.

1.07.5.5.3 Compound chondrules, relict grains, and chondrule rims

Chondrules in all groups except CB and CH show evidence for collisions: with each other while partly molten and with solid grains while molten and solid. Compound chondrules form by collisions of partly molten chondrules and their abundance gives some constraint on the number density of chondrules in the region where they formed. Unfortunately, there are large uncertainties in chondrule speed and the provenance of the joined chondrules, which make estimates of the chondrule density very uncertain. Wasson et al. (1995) estimate that in ordinary chondrites, ~2.5% of all chondrules are compound. In over half of these, the chondrules that collided have such similar compositions that they could have been derived from a single large droplet. Many other compound chondrules may have formed by part melting of a porous aggregate containing an embedded chondrule: relatively few compound chondrules probably formed by random collisions.

Foreign grains in chondrules appear to have been derived from other chondrules (Jones, 1996b). Type I olivines can be found in type II chondrules and vice versa, indicating that chondrule formation was a repeated process and that material was recycled through the chondrule furnace.

Chondrules may have two types of rims: fine-grained matrix rims and coarse-grained igneous rims: both are missing in CH and CB chondrites. Matrix rims vary in composition but the variations are uncorrelated with chondrule composition (Section 1.07.5.7). However, the compositions of igneous rims are related to those of the host chondrule: type I chondrules have igneous rims that are FeO poor, while type II chondrules have FeO-rich rims (Krot and Wasson, 1995). The igneous rims formed by collisions between chondrules and smaller partly molten particles, or solid particles that were accreted and later melted.

From the abundances of igneous rims, compound chondrules, and relict grains and the dearth of unmelted, chondrule-sized dustballs, we infer that collisions between partly melted objects and remelting of aggregates from such collisions were important processes that enabled chondrules to form from fine particles.

1.07.5.5.4 Closed-system crystallization

Understanding whether chondrules acted as open or closed systems during chondrule formation is crucial for elucidating their origin (Sears et al., 1996). Type II chondrules appear to be good candidates for closed system crystallization, as chemical zoning profiles across their phenocrysts are consistent with fractional crystallization (Jones, 1990; Jones and Lofgren, 1993). Type II chondrules contain significant concentrations of sodium and other volatile elements that would have been readily lost if the chondrules had been molten for more than a few minutes (Yu and Hewins, 1998). Such chondrules are commonly considered as archetypal chondrules formed by closed-system melting. Although most workers infer that type II chondrules resulted from single heating events, Wasson and Rubin (2003) argue for multiple events with low degrees of melting, which would help to explain how volatiles were retained.

Many workers once argued that all chondrules formed by isochemical melting of precursor material and that the chemical fractionations among chondrules, except for metal-loss, result from processes that predate chondrule formation (e.g., Gooding et al., 1980). Few workers hold this view now, nevertheless it is commonly held that chemical fractionations prior to chondrule formation were much more important than any fractionation during chondrule formation, besides loss of metal (e.g., Grossman, 1996). Thus, the standard model for chondrule formation envisages that volatile loss from chondrules was minimized by flash melting of dust aggregates during melting events that lasted seconds to minutes, rather than the minutes to hours indicated by experiments (Connolly et al., 1998).

The major difficulty with closed-system models is that they do not explain why some chondrules are olivine-rich and others are pyroxene-rich. It is very difficult to see how fine-grained nebula dust could have been sorted into high Mg/Si dustballs for olivine chondrules and low Mg/Si dustballs for pyroxene-rich chondrules (e.g., Wood, 1996a). Although matrix contains both phases, they are not segregated into chondrule-sized volumes. The only plausible way to produce both olivine-rich and pyroxene-rich chondrules is by high-temperature reactions between condensed material and nebular gas. New evidence, especially from CR–CH–CB clan chondrites, supports this thesis.

1.07.5.5.5 Open-system crystallization

Unlike the archetypal chondrules, type I chondrules show a variety of evidence for volatility-controlled chemical fractionations that may have occurred during chondrule formation. Hewins et al. (1997) studied microporphyritic and cryptoporphyritic type I chondrules in Semarkona (LL3.0), which have irregularly shaped interstitial metal grains rather than the rounded droplets in mesostasis common in coarser type I chondrules. They found an inverse correlation between grain size and abundance for the moderately volatile elements, sodium, potassium, and sulfur, as well as iron and nickel (Figure 20). Hewins et al. inferred that volatilization increased with the extent of melting of a fine-grained precursor having a chondritic composition. Alternatively, these very fine-grained chondrules may have formed by aggregation of partly melted particles that scavenged volatiles with an efficiency that depended on their grain size.

Mesostases in some type IA chondrules in Semarkona show volatility-related chemical zoning (Matsunami et al., 1993; Nagahara et al., 1999). Mesostasis near the rims is higher in volatiles and lower in refractories than mesostasis near the core. Matsunami et al. invoked recondensation and reduction processes in the solar nebula. The chemical variations across type I chondrules and those in microporphyritic and cryptoporphyritic type I chondrules (Figure 20) mimic the fractionation patterns shown by the bulk chemical data for chondrules in the IA–IB sequence (Figure 18) suggesting that a single-process-controlled chemical variations within and among type I chondrules.

The concentration of pyroxene phenocrysts around the periphery of many type I chondrules was reproduced experimentally by Tissandier et al. (2002), who allowed melt to react with

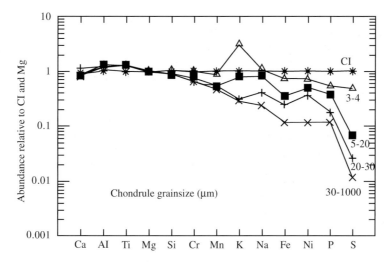

Figure 20 Elemental abundances in microporphyritic and cryptoporphyritic type I chondrules in Semarkona showing an inverse correlation between grain size (in μm) and abundance for the moderately volatile elements (chromium to sulfur). Hewins et al. (1997) attribute this to volatilization during melting (reproduced by permission of National Institute of Polar Research from Antarct. Meteorit. Res., 1997, 10, 281).

gaseous SiO during crystallization. They inferred that such gas–melt interactions were prevalent during chondrule formation, and invoked high dust–gas ratios to stabilize chondrule melts in the nebula (Ebel and Grossman, 2000). Further evidence for gas–melt interactions is provided by silicon-rich, igneous rims on many type I chondrules in CR2 chondrites (Krot et al., 2003d). These rims contain high-calcium and low-calcium pyroxene, glassy mesostasis, Fe,Ni, and silica, with high concentrations of chromium and manganese and low concentrations of aluminum and calcium in high-calcium pyroxene (Figure 21). Mesostasis zoning follows that in the ordinary chondrites with silicon, sodium, potassium, and manganese increasing and calcium, magnesium, aluminum, and chromium decreasing from the olivine-rich core, through the pyroxene-rich rim to the silicon-rich rim. FeO remains nearly constant. Krot et al. (2003d) infer that gas–melt fractional condensation or gas–solid condensation followed by remelting produced these chemical fractionations. Isolation of early-formed forsterite condensates may have resulted naturally from crystallization, as forsterite would have equilibrated with the nebula more slowly than the melt.

Strong evidence for chondrule formation by condensation of melts is provided by the chondrules in CB_b chondrites, which have unusual textures and compositions (Krot et al., 2001d, 2003e). The chondrules are cryptocrystalline or skeletal olivine, lack igneous or fine-grained rims and relict grains, and do not form compound objects (Figure 22). They clearly formed from total melts in a dust-free environment. Some cryptocrystalline chondrules are embedded in zoned metal grains, which have compositional profiles consistent with nebular condensation (Krot et al., 2001d, 2003e). Chondrules have uniform flat refractory element abundances but the variation $((0.02-3) \times CI)$ is far wider than in other chondrites (Figure 23). The depletions of refractory elements relative to silicon in the cryptocrystalline chondrules indicate that silicon or the refractory elements, or both, condensed into chondrules or their precursors. The near-CI bulk abundances of refractory elements in the chondrites favor closed-system fractional condensation. Moderately volatile elements in chondrules show large depletions (Figure 23) that are inversely correlated with condensation temperature, as for bulk chondrites (Figure 3) and type I chondrules

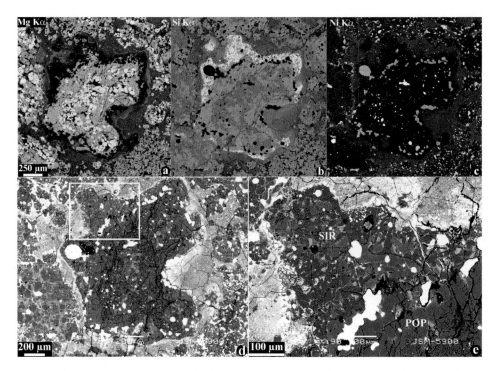

Figure 21 X-ray elemental scanning maps using Mg K_α (a), Si K_α (b) and Ni K_α (c), plus backscattered electron images ((d) and (e)) of a typical type I porphyritic chondrule in the QUE 99177 CR2 chondrite, which has a forsterite-rich core, a porphyritic olivine–pyroxene rim (POP) and an outermost silica-rich rim (SIR). The Ni K_α map (c) shows that metallic Fe,Ni grains in the chondrule core are richer in nickel than those in the outer portions. The zoned structure of the chondrule can be explained by gas–melt fractional condensation or gas–solid condensation followed by remelting (after Krot et al., 2003d).

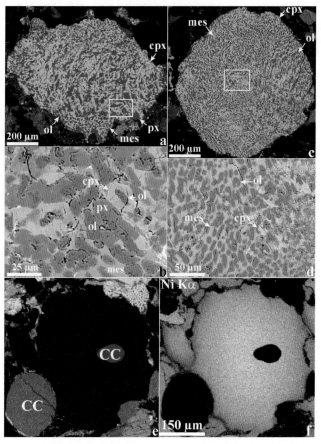

Figure 22 Combined elemental map in magnesium (red), calcium (green), and Al K_α (blue) X-rays ((a), (b)) and backscattered electron images ((b), (d)) of two barred or skeletal olivine chondrules, and a combined X-ray map (e) and NiKα scanning map (f) of a zoned Fe,Ni metal grain with an enclosed cryptocrystalline chondrule (CC) in the CB$_b$ chondrite HaH 237 (Krot et al., 2002a). The chondrules, which contain forsteritic olivine (ol), low-calcium pyroxene (px), high-calcium pyroxene (cpx) and mesostasis (mes), lack rims and relict grains and clearly formed in a dust-free environment from total melts (reproduced by permission of University of Arizona on behalf of The Meteoritical Society from *Meteorit. Planet. Sci.*, **2002**, *37*, p. 1451).

(Figure 18). Chondrule textures are consistent with condensation of melts.

1.07.5.5.6 Chondrules that formed on asteroids

Chondrites contain a small fraction of chondrules that may have formed on asteroids. One rare type of chondrule ($\sim$0.1%) in type 4–6 ordinary chondrite breccias is composed largely of plagioclase (or mesostasis of plagioclase composition) and chromite (Krot et al., 1993). Krot and Rubin (1993) suggest that these chondrules may have formed by impact melting as they find impact melts with similar compositions inside shocked ordinary chondrites and such chondrules are absent in type 3 chondrites. In addition, some chromite-rich chondrules contain chromite-rich aggregates, which appear to be fragments of equilibrated chondrites. Other possible impact products were described in LL chondrites, which are mostly breccias, by Wlotzka et al. (1983). They found microporphyritic potassium-rich clasts with highly variable K/Na ratios that are totally unlike chondrules in type 3 chondrites. Large chondrules and clasts in Julesberg (L3) may also be impact melt products according to Ruzicka et al. (1998).

1.07.5.6 Metal and Troilite

Two kinds of metal are found in chondrites: grains composed of refractory elements (iridium, osmium, ruthenium, molybdenum, tungsten, and rhenium), which condense along with the refractory oxides above $\sim$1,600 K at 10^{-4} atm, and grains composed predominantly of iron, cobalt, and nickel, which condense with forsterite and enstatite at $\sim$1,350–1,450 K. The former are associated with CAIs (Palme and Wlotzka 1976) and the latter with chondrules, typically type I or FeO-poor chondrules (B & J 1998, pp. 244–278). Unfortunately, few chondrites preserve a good record of the formation history

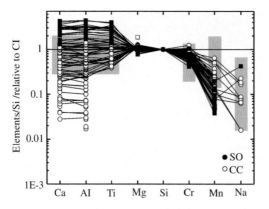

Figure 23 Bulk concentrations of lithophile elements normalized to CI chondrites and silicon in skeletal-olivine and cryptocrystalline chondrules including cryptocrystalline inclusions in metallic Fe,Ni in two CB_b chondrites (HH 237 and QUE 94411). Shaded regions show compositional range of chondrules in other carbonaceous chondrites, excluding the CH group chondrules that are closely related to CB chondrules. The wide range of refractory abundances in CB_b chondrules ((0.02–3) × CI levels) appears to reflect fractional condensation with skeletal-olivine chondrules condensing at higher temperatures than cryptocrystalline chondrules. Both types are more depleted in moderately volatile elements than other carbonaceous chondrites (after Krot et al., 2002a).

of their metal grains because subsequent low-temperature reactions commonly formed oxides and sulfides and thermal metamorphism allowed kamacite to exsolve from taenite. Most refractory nuggets have been studied in Allende CAI, where rampant oxidation and sulfurization and redistribution of volatile tungsten and molybdenum compounds have severely compromised the high-temperature nebular record (Palme et al., 1994).

Grains of metallic Fe,Ni in most unshocked type 3–6 chondrites provide a record of slow cooling at ~1–1,000 K Myr^{-1} through the temperature range ~550–350 °C, when kamacite and taenite ceased to equilibrate (Wood, 1967). In most type 2 and 3.0–3.3 chondrites, metallic Fe,Ni grains typically contain concentrations of 0.1–1% chromium, silicon, and phosphorus, which are not found in type 4–6 chondrites, and reflect high-temperature processing prior to accretion.

The most pristine metallic Fe,Ni grains are found in the CR–CH–CB clan (Krot et al., 2002a). Most metal grains with >8% Ni lack coexisting low-nickel kamacite and high-nickel taenite showing that these chondrites were not heated above ~300 °C for more than a year (Reisener et al., 2000). Metal grains are not present in most chondrules in CH chondrites, which are cryptocrystalline, and are entirely absent from chondrules in CB chondrites.

1.07.5.6.1 CR chondrite metallic Fe,Ni

CR2 chondrites contain 5–8 vol.% metallic Fe,Ni grains in type I chondrules and between them, which are not associated with troilite. Grains outside chondrules were probably once associated with chondrules. Metal grains have bulk concentrations of 4–15 wt.% Ni, solar Co/Ni ratios and up to ~1% chromium and phosphorus (Weisberg et al., 1993). Grains at chondrule cores have higher nickel concentrations than those at the rims (Figure 21(c)) and all grains are depleted in moderately volatile elements (Connolly et al., 2001). Kong and Palme (1999) noted that Renazzo metal grains are covered with layers of carbon, which appear to have exsolved at low temperatures. Two kinds of processes have been invoked to account for the origin and composition of the metal grains: condensation as nebular solids or recondensation during chondrule formation (Grossman and Olsen, 1974; Weisberg et al., 1993; Connolly et al., 2001) and silicate–metal equilibration and reduction during chondrule formation (Scott and Taylor, 1983; Zanda et al., 1994). Given that the silicate portions of type I chondrules in CR2 chondrites have retained a record of condensation processes (Section 1.07.5.5.5), it seems likely that the metallic portions have also, and that both kinds of processes were involved in making CR metal.

1.07.5.6.2 CH and CB chondrite metallic Fe,Ni

Between 0.01% and 1% of the metal grains in CH chondrites (Meibom et al., 1999, 2000) and all the zoned metal grains in CB_b chondrites (Meibom et al., 2001; Campbell et al., 2001) have radial chemical zoning patterns that are consistent with formation as solid nebular condensates above 1,200 K (Figure 24). Fractional condensation models suggest that grains in QUE 94411 formed under a variety of oxidation conditions with enhanced dust–gas ratios of 10–40 × solar prior to vaporization (Petaev et al., 2001). Kinetic growth models suggest growth timescales of 1–4 d at cooling rates of a few degrees per hour and removal from hot gas before temperatures fell below ~1,200 K (Meibom et al., 2001). The tiny fraction of appropriately zoned grains in CH chondrites might have condensed in convective updrafts in the nebula and been transported outwards to cooler regions by convection (Meibom et al., 2000). However, the large fraction of zoned grains in CB_b chondrites requires a more efficient transport mechanism such as an X-wind (Meibom et al., 2001; Shu et al., 2001).

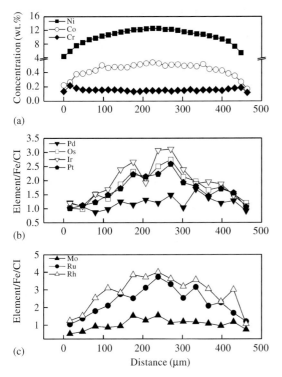

Figure 24 Concentration profiles of siderophile elements in a radially zoned Fe,Ni grain in the CB_b chondrite, QUE 94411: (a) electron microprobe data; (b) and (c) trace element data from laser ablation ICPMS (Campbell et al., 2001). The nickel, cobalt, and chromium profiles can be matched by nonequilibrium nebular condensation assuming an enhanced dust-gas ratio of $\sim 36 \times$ solar, partial condensation of chromium into silicates, and isolation of 4% of condensates per degree of cooling (Petaev et al., 2001). Concentrations of the refractory siderophile elements, osmium, iridium, platinum, ruthenium, and rhodium, are enriched at the center of the grain by factors of 2.5–3 relative to edge concentrations, which are near CI levels after normalization to iron (reproduced by permission of University of Arizona on behalf of The Meteoritical Society from *Meteorit. Planet. Sci.*, **2002**, *37*, pp. 1451–1490).

CB chondrites also contain relatively homogeneous metal grains with micrometer-sized troilite droplets. Large grains in CB_a chondrites have near-solar Co/Ni ratios and 0.1–1 wt.% Cr and P and were among the first metallic grains to be tagged as nebular condensates (Newsom and Drake, 1979; Weisberg et al., 1990). Moderately volatile siderophile elements gallium, germanium, arsenic, and antimony are depleted by factors of up to 10^3 from CI-levels (Campbell et al., 2002). Condensation models suggest that the precursor gas was enriched in metals by a factor of 10^7 above solar levels prior to condensation. Hypervelocity impacts between asteroids or disruption of molten planetesimals might provide an appropriate environment for the condensation of CB_a metal as molten sulfur-bearing droplets (Campbell et al., 2002). However, the presence of CAIs in CB_a chondrites and the intimate mixture of zoned and sulfur-bearing metallic Fe,Ni grains in CB_b chondrites argues against a unique origin for the sulfur-bearing grains.

1.07.5.6.3 Troilite

Troilite in chondrites may have condensed below ~ 650 K by reaction of nebular gas with metallic Fe,Ni, or crystallized from Fe–Ni–S melts in chondrules or during secondary processing in asteroids. Whether any metal grains in chondrites have preserved nebular signatures is questionable as sulfur is readily mobilized during metamorphism and shock (Lauretta et al., 1997). Criteria for identifying nebular sulfide condensates (Lauretta et al., 1998) and igneous sulfides (Rubin et al., 1999; Kong et al., 2000) have been proposed. Rubin et al. (1999) found that 13% of the chondrules in Semarkona (LL3.0) contain metal-troilite nodules with igneous textures inside phenocrysts. They inferred that these chondrules were heated only briefly in the nebula, otherwise sulfur would have been lost. Lauretta et al. (2001) infer that troilite and fayalite rims on metal grains in Bishunpur (LL3.1) formed in a totally different way at $\sim 1,200$ K in a dust-enriched nebula.

1.07.5.7 Matrix Material

Matrix material is best defined as the optically opaque mixture of mineral grains 10 nm to 5 μm in size that rims chondrules, CAIs, and other components and fills in the interstices between them (Scott et al., 1988). Matrix grains are generally distinguished from fragments of chondrules, CAIs and other components by their distinctive sizes, shapes, and textures. Minerals found in matrices include silicates, oxides, sulfides, metallic Fe,Ni, and especially in type 2 chondrites, phyllosilicates and carbonates (Table 5). Matrices are broadly chondritic in composition though richer in FeO than chondrules and have refractory abundances that deviate more from bulk chondrite values (McSween and Richardson, 1977; Brearley, 1996). Matrix typically accounts for 5–50 vol.% of the chondrite (Table 1).

Matrix rims, which are also called fine-grained rims, are similar in chemical composition and mineralogy to the interstitial matrix material, though matrix rims may be finer grained. Chemically, matrix rims and rimmed objects appear to be unrelated: rims on chondrules do not differ significantly from rims on CAIs if alteration effects are ignored. Rims are not uniform in width (Figure 25), but rim thickness is correlated with chondrule size within a chondrite group (Metzler et al., 1992; Paque and Cuzzi, 1997).

Table 5 Mineralogy of chondrite matrices.

Group	Phases	Ref.
Carbonaceous		
CI	Serpentine, saponite, ferrihydrite, magnetite, Ca–Mg carbonate, pyrrhotite	1
CM	Serpentine, tochilinite, pyrrhotite, amorphous phase, calcite	1
CR2	Olivine, serpentine, saponite, minor magnetite, FeS, pentlandite, pyrrhotite, calcite	2
CO3.0: ALHA77307	Amorphous silicate, Fo_{30-98}, low-calcium pyroxene, Fe,Ni metal, magnetite, sulfides	3
CO3.1–3.6	Fayalitic olivine, minor phyllosilicates and ferric oxide	4
CV3 reduced	Fayalitic olivine, low-calcium pyroxene, low-nickel metal, FeS	5
CV3 oxidized:		
Bali-like	Fayalitic olivine, phyllosilicate, fayalite, Ca–Fe pyroxene, pentlandite, magnetite	6
Allende-like	Fayalitic olivine, Ca–Fe pyroxene, nepheline, pentlandite	7
Ungrouped C:		
Acfer 094	Amorphous silicate, forsterite, enstatite, pyrrhotite, ferrihydrite, traces of phyllosilicate	8
Adelaide	Fayalitic olivine, amorphous silicate, enstatite, pentlandite, and magnetite	9
Ordinary		
LL3.0: Semarkona	Smectite, fayalitic olivine, forsterite, enstatite, calcite, magnetite	10
H, L, LL 3.1–3.6	Fayalitic olivine, amorphous material, pyroxene, albite, Fe,Ni metal	11
Other		
K3: Kakangari	Enstatite, forsterite, albite, ferrihydrite, troilite, Fe,Ni metal	12

References: 1, Zolensky *et al.* (1993); 2, Endress *et al.* (1994); 3. Brearley (1993); 4, Brearley and Jones (1998); 5. Lee *et al.*, (1996); 6. Keller *et al.*, (1994); 7. Scott *et al.* (1988); 8, Greshake (1997); 9, Brearley (1991); 10, Alexander *et al.* (1989a); 11, Alexander *et al.* (1989b); 12, Brearley (1989).

However, matrix rims are absent in the CH, CB, and K chondrite groups. Clearly, matrix rims around chondritic components were acquired after the components formed and before final lithification of these components into a rock. (Note that some rims around altered mineral grains may have formed in situ by cementation of interstitial material during aqueous alteration.) Most authors envisage that the chondritic components acquired their rims by accreting dust in the nebula prior to accretion into planetesimals (e.g., Morfill *et al.*, 1998; Liffman and Toscano, 2000; Cuzzi *et al.*, 2001), though some argue that rims form in asteroidal regoliths (e.g., Symes *et al.*, 1998; Tomeoka and Tanimura, 2000).

Millimeter-to-submillimeter lumps of matrix-rich material are present in many chondrites. Lumps that are mineralogically similar to nearby matrix occurrences are probably pieces of rims or aggregates of interstitial material. However, heavily altered matrix-rich lumps called dark clasts with distinctive mineralogies probably have a different origin (Section 5.7.10). In CH and CB chondrites, all the matrix material is present as dark clasts: there are no matrix rims or interstitial material.

According to Anders and Kerridge (1988), "controversy permeates every aspect of the study of matrix material, from definition of matrix to theories of its origin." Although controversy remains, major progress has been made in identifying matrix components and understanding their origin. Since matrix is commonly richer in volatiles than other components, it is commonly referred to as the "low-temperature component" of chondrites. But this is a misnomer as it is composed of diverse materials that formed under different conditions. Prior to detailed TEM studies and the acquisition of mineralogical and isotopic evidence for aqueous alteration on asteroids, all matrix minerals in carbonaceous chondrites including magnetite, carbonates and phyllosilicates were thought to have condensed in the solar nebula (see McSween, 1979; see Chapter 1.09). However, matrix minerals are now thought to be a complex mixture of presolar materials and nebular condensates that were mixed with fine chondrule fragments and experienced aqueous alteration and metamorphism (e.g., Scott *et al.*, 1988). However, nebular mechanisms for forming magnetite and phyllosilicates are still being studied (Hong and Fegley, 1998; Ciesla *et al.*, 2003). Matrix investigations are handicapped by the extremely fine-grained nature of the constituents and the difficulty of distinguishing primary and secondary mineralogical features (Brearley, 1996). The fine grain size (commonly as small as 50–100 nm), high porosity and permeability of matrix ensure that it is more susceptible to alteration by aqueous fluids and metamorphism in asteroids than other chondritic components.

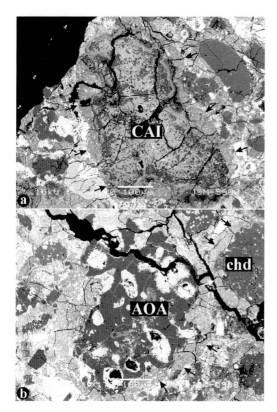

Figure 25 Backscattered electron images showing fine-grained matrix rims on (a) Ca–Al-rich inclusion (CAI), and (b) an amoeboid–olivine aggregate (AOA) in ALHA77307. The light-gray rims contain FeO-rich silicate material and are crossed by dark cracks; arrows mark the outer edge of the rims.

Most of the matrix data that we review are chemical and mineralogical as there is little isotopic information about individual matrix grains. Bulk oxygen isotopic compositions for matrix samples commonly differ from those of associated chondrules (e.g., Scott et al., 1988), but lack of oxygen-isotope data for samples of key chondrites and specific matrix components severely limits inferences about matrix origins.

Previous reviews of chondrite matrices have addressed aqueous alteration in the matrices of carbonaceous chondrites (Buseck and Hua, 1993), matrix mineralogies and origins (Scott et al., 1988), possible relations between matrix and chondrules (Brearley, 1996), and the diverse mineralogy of matrices in different chondrite groups (B & J 1998, pp. 191–244). Here we focus on geochemical and mineralogical features in type 1–3 chondrites that offer the best insights into the origin of the components of matrix and chondrites and the relationship between matrix and chondrules. Unfortunately, well characterized matrix material from E3 and R3 chondrites is lacking. Some important constituents of matrix material are discussed elsewhere: presolar grains (see Chapter 1.02), carbon and organic phases (see Chapter 1.10) and noble gases (see Chapter 1.14).

1.07.5.7.1 CI1 Chondrites

CI1 chondrites contain neither chondrules nor refractory inclusions and have generally been considered as virtually pure matrix material. They are composed almost entirely of phyllosilicates, oxides, sulfides, carbonates and other minerals (Table 5) that were once widely thought to be nebular condensates but are now known to be products of aqueous alteration (B & J 1998, p. 192). CI1 chondrites contain 65 wt.% of serpentine–saponite intergrowths, which are too fine-scale for accurate analysis, 10% magnetite, 7% pyrrhotite, 5% poorly crystalline ferrihydrite ($5Fe_2O_3 \cdot 9H_2O$) and smaller amounts of pentlandite and other phases (Bland et al., 2002; B & J 1998, p. 192). Carbonates constitute ~5% of the meteorites: the largest have been dated by $^{53}Mn-^{53}Cr$ isotope systematics at ~6–20 Ma after CAI formation (Endress et al., 1996).

CI1 chondrites also contain a small fraction of isolated olivine and pyroxene grains up to 400 μm in size with chemical, oxygen isotopic compositions, and rounded inclusions of metallic Fe,Ni and trapped melt indicating that they were derived from chondrules (Leshin et al., 1997). Brearley and Jones (1998) estimate the abundance of olivine and pyroxene at <1 vol.%, but X-ray diffraction studies by Bland et al. (2002) indicate 7 wt.% olivine. The difference probably reflects a high abundance of olivine crystallites embedded in phyllosilicates (P. A. Bland, private communication). CI1 chondrites also contain rare refractory grains with oxygen isotopic compositions comparable to those in CAIs (see Scott and Krot, 2001).

Why do the most highly altered chondrites provide the best match with the composition of the solar photosphere? We have yet to answer this question satisfactorily, but many factors are probably responsible. First, chondrules, which are deficient in volatiles relative to matrix material, were scarce in the CI precursor material. Second, the rocks appear to have been affected by closed-system alteration. The prominent sulfate veins in CI1 chondrites appear to have formed in fractures on Earth (Gounelle and Zolensky, 2001) and not in the parent body, so it is likely that aqueous solutions did not percolate through the rock. Third, CI1 chondrites are unique fine-grained breccias as they are entirely composed of diverse, altered lithic fragments less than a few hundred micrometers in size (Morlok et al., 2002). They appear to very representative samples of one or more asteroids that were thoroughly scrambled. Fourth, they probably formed further from the Sun than other chondrite groups and therefore acquired higher concentrations of carbonaceous, volatile-rich material.

1.07.5.7.2 CM2 chondrite matrices

In CM2 chondrites, the interstitial matrix material and the matrix rims around chondrules are very similar mineralogically (Zolensky et al., 1993). Matrices are dominated by phyllosilicates, which are also present in chondrules and CAIs, and formed as a result of aqueous alteration, but the location is controversial (see Chapter 1.09). Phyllosilicates are mostly 10–100 nm in size and are largely serpentines with diverse morphologies, degrees of crystallinity and compositions (Barber, 1981; Zolensky et al., 1993). Crystals of the Fe^{3+}-rich serpentine, cronstedite, reach up to 1×10 μm in size and may be intergrown with tochilinite, $(Fe,Mg)(OH)_2 \cdot (Fe,Ni)S$ (see B & J 1998, p. 202). Metzler et al. (1992), Bischoff (1998), and Lauretta et al. (2000) argue that alteration occurred mainly in small precursor planetesimals prior to accretion of the CM body. Hanowski and Brearley (2000, 2001) disagree pointing to the occurrence of cross-cutting veins of alteration minerals, loss of calcium from chondrule mesostases and deposition as carbonate in the surrounding matrix, and other features favoring *in situ* alteration on asteroids.

CM2 matrices also contain small olivine grains down to 0.1 μm in size, which are close to Fo_{100} in composition (Barber, 1981). These forsterites lack defects, are sometimes euhedral and appear to have formed before the phyllosilicates (Barber, 1981; B & J 1998, pp. 216). Euhedral Fa_{40-50} olivines are also embedded in phyllosilicate (Lauretta et al., 2000). Pyroxenes are rarer and are dominantly magnesium-rich, $Fs_{<2}$ (Brearley, 1995). Some olivines and pyroxenes are especially enriched in manganese (1–2 wt.% MnO) and are comparable in composition to the low iron, high manganese olivines and pyroxenes present in IDPs (Klöck et al., 1989). These were called LIME silicates (Low-Iron, Manganese Enriched) by Klöck et al., who argued that they are nebular condensates.

As much as 15% of CM matrices may be amorphous silicate, which is associated with comparable volumes of very poorly crystalline phyllosilicates (Barber, 1981). In ALH81002, 5 vol.% of the matrix rims are composed of amorphous regions a few micrometers across that are rich in silicon, aluminum, magnesium, iron, and presumably oxygen, and contain 5–40 nm grains of iron-rich olivines and metallic Fe,Ni (Lauretta et al., 2000). Chizmadia and Brearley (2003) found more abundant amorphous material in the CM2 chondrite, Y 791198. (Origins of matrix components are reviewed in Section 5.7.11.)

Rim thickness appears to be correlated with object size (Metzler et al., 1992). However, rims around grains of calcite, tochilinite and other alteration products may be artifacts of alteration rather than preaccretionary features. Hua et al. (2002) compared the concentrations of trace and minor elements in matrix rims around different types of chondrules and other objects in two CM chondrites and found no relationship between bulk compositions of rims and enclosed objects.

1.07.5.7.3 CR2 chondrite matrices

CR2 chondrite matrices are predominantly composed of olivine with grain sizes of 0.2–0.3 μm accompanied by intergrowths of serpentine and saponite with dimensions of 20–300 nm and minor sulfide and magnetite grains 0.1–25 μm in size (Zolensky et al., 1993; Endress et al., 1994). CR2 chondrites also contain ~3 vol.% of more altered matrix lumps mostly 0.1–1 mm in size, which are similar in bulk composition to the other matrix material (Endress et al., 1994). However, phyllosilicates, magnetite and chondrule fragments are more abundant in the matrix lumps than in interstitial material, and sulfides are richer in nickel suggesting that the lumps are chondritic clasts that formed in a totally different environment.

1.07.5.7.4 CO3 chondrite matrices

The matrix of the type 3.0 CO chondrite, ALHA77307, is very different mineralogically from matrices in CO3.1–3.6 chondrites (Table 5). It is largely composed of amorphous silicate material (Figure 26), which forms discrete regions 1–5 μm in size, and 30–40 vol.% of heterogeneously distributed tiny olivine crystals and crystal aggregates (Brearley, 1993). Matrix rims around chondrules and other components are fairly homogeneous (Figure 27) and mineralogically indistinguishable and nearly isochemical with interstitial material (Brearley et al., 1995). Rims are slightly lower in manganese and calcium (Table 6). As in other chondrites, matrix material is broadly chondritic in composition but aluminum, nickel, and potassium are enriched relative to silicon and CI chondrites, and calcium and titanium are depleted. Several moderately volatile elements are enriched relative to bulk CO chondrite by factors of 2–3 (Figure 27).

The amorphous phase in the ALHA77307 matrix is heterogeneous on the submicrometer scale but homogeneous on a 10 μm scale (Table 6). It is mostly composed of silicon, iron, magnesium, and oxygen with relatively high concentrations of aluminum, nickel, sulfur, and phosphorus. Its Ca/Al ratio is much lower than those in chondrule mesostases, which are closer to chondritic values, precluding an origin from chondrule glass (Brearley, 1993). Brearley's chemical and textural studies also suggest that the amorphous component was not formed by shock melting of matrix material, as Bunch et al. (1991) proposed for the matrix rims. In some

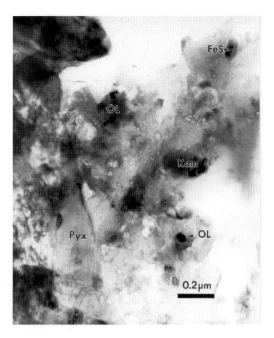

Figure 26 Transmission electron micrograph of the matrix in ALHA77307 (CO3.0) showing that it is largely composed of amorphous material with embedded crystallites of kamacite (Kam), pyrrhotite (FeS), olivine (OL) and low-calcium pyroxene (Pyx) (Brearley, 1993). Matrices of all carbonaceous chondrites probably resembled this matrix material prior to asteroidal alteration and metamorphism (reproduced by permission of Elsevier from *Geochim. Cosmochim. Acta*, **1993**, *57*, 1534).

areas, the amorphous phase appears to show incipient formation of phyllosilicates.

Three types of olivine occur in the amorphous phase of ALHA77307 (Brearley, 1993): (i) abundant irregularly shaped and poorly crystalline olivines <300 nm in size with Fa_{10-70}; (ii) well crystallized, subhedral, larger forsterite crystals 200 nm to >4 μm in size (Fa_{95-98}); and (iii) micrometer-sized, isolated crystals and similar-sized aggregates of 100 nm grains of well-crystallized manganese-rich forsterites, with up to 2 wt.% MnO. Pyroxene, which is not as abundant as olivine, is mostly poor in calcium and iron and also occurs in several varieties. Orthopyroxene forms angular crystals 1–1.5 μm in size and is intergrown with clinopyroxene in aggregates of 0.2–0.3 μm crystals. High-resolution TEM images of an intergrowth indicate cooling from the protopyroxene stability field above 1,000 °C at ~1,000–5,000 °C h^{-1} (Jones and Brearley, 1988). Manganese-rich enstatite crystals are ~0.1 μm in size, largely clinoenstatite, indicative of rapid cooling from above 1,000 °C at >1,000 °C h^{-1}, and are associated with manganese-rich olivines in micrometer sized aggregates.

Some amorphous regions in ALHA77307 contain rounded grains of kamacite with 4–5 %Ni and nickel-bearing sulfides, which are both <200 nm in size. Elsewhere, magnetite occurs as crystals as small as 50 nm and aggregates <10 μm

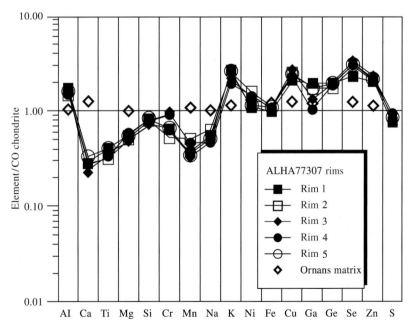

Figure 27 Elemental abundances in matrix material in CO3 chondrites normalized to bulk chondrite: data for rims on chondrules in ALHA77307 (type 3.0) obtained by synchrotron X-ray fluorescence microprobe analysis (Brearley *et al.*, 1995) and for bulk matrix in Ornans (type 3.3) analyzed by instrumental neutron activation (data from Rubin and Wasson, 1988). Chondrule rims in ALHA77307 are uniform in composition but deviate significantly from the bulk chondrite composition. Matrix in Ornans is relatively unfractionated (reproduced by permission of Elsevier from *Geochim. Cosmochim. Acta*, **1995**, *59*, 4310).

Table 6 Composition of interstitial matrix and matrix rims on chondrules in the CO3.0 chondrite, ALHA77307, and amorphous matrix phases in ALHA77307 and Acfer 094.

wt.%	Interstitial matrix 3 areas ALHA77307	Matrix rims 8 chondrules ALHA77307	Amorphous phase ALHA77307 matrix (CO3.0)	Amorphous phase Acfer 094 matrix (ungrouped)
No. anal.	36	106	8	6
SiO_2	33.6 (0.9)	33.3 (0.7)	53 (9)	47 (4)
TiO_2	0.07 (0.02)	0.06 (0.01)	<0.04	NA
Al_2O_3	4.4 (0.3)	4.7 (0.3)	5.2 (1.6)	3.8 (0.9)
Cr_2O_3	0.43 (0.04)	0.36 (0.02)	0.44 (0.18)	0.7 (0.2)
FeO	36.2 (2.5)	37.3 (2.1)	20 (6)	26 (3)
MnO	0.27 (0.04)	0.16 (0.05)	0.24 (0.20)	0.6 (~0.2)
MgO	16.7 (1.5)	15.8 (0.8)	8.3 (1.7)	10.1 (3)
CaO	1.1 (0.2)	0.76 (0.12)	0.41 (0.38)	1.4 (0.4)
Na_2O	0.32 (0.06)	0.25 (0.06)	NA	~0.7
K_2O	0.11 (0.02)	0.14 (0.05)	NA	NA
NiO	3.4 (0.4)	3.9 (0.7)	4.7 (2.6)	5.2 (2.2)
P_2O_5	0.32 (0.09)	0.25 (0.09)	3.6 (2.1)	NA
SO_3	3.4 (0.8)	3.0 (0.6)	4.5 (2.8)	4.0 (2.0)
Total	100	100	100	100

Source: Brearley (1993) and Greshake (1997). Matrix analyzed by electron microprobe: amorphous phase by analytical transmission electron microscopy. Totals are normalized to 100%. Figures in parentheses are 1σ of means of analyzed regions for matrix and 1σ of analyses for amorphous phase. NA, not analyzed.

in size. The close proximity of magnetite-rich and metal-sulfide-rich regions suggests that they formed separately under very different conditions in the nebula (Brearley, 1993). From a weak correlation between the Mg/(Mg + Fe) ratios of the olivines and the adjacent amorphous material, Brearley (1993) proposed that the olivines had crystallized from the amorphous phase during low-temperature annealing. However, the very different properties of the magnesian and ferroan olivines suggest they formed under different conditions. Presumably, the larger, highly crystalline forsterites crystallized at higher temperatures. Brearley (1993) argued against a condensation origin for the iron-poor, manganese-rich silicates as proposed by Klöck et al. (1989), and suggested that they formed by annealing of amorphous material.

Matrices in the more metamorphosed CO3.1–3.6 chondrites are unlike the ALHA77307 matrix as they are dominated by submicrometer grains of FeO-rich olivine Fo_{40-60} with minor pyroxene, spinels and metallic Fe,Ni (see B & J 1988, pp. 217–220). In Lancé and Ornans, the matrix, like the mesostasis in chondrules, also contains minor phyllosilicates and other alteration products. Lancé matrix contains magnesium-rich serpentine and poorly crystalline material between FeO-rich olivines (Keller and Buseck, 1990). Hydrous ferric oxides are also present on grain boundaries in Kainsaz and Warrenton indicating incipient alteration. Matrix olivines in type ≥ 3.4 chondrites have more equilibrated compositions than those in types 3.1–3.3 (B & J 1998, p. 220). Ornans matrix is close to bulk CO composition (Rubin and Wasson, 1988; Figure 27).

The evidence that CO3 chondrites form a metamorphic sequence (Scott and Jones, 1990; B & J 1998, p. 38) suggests that they all originally contained matrix material like that in ALHA77307. The amorphous material in ALHA77307, which shows only incipient formation of fayalitic olivine and phyllosilicates, could have been readily replaced by fayalitic olivine by further fluid-assisted metamorphism to produce the type 3.1–3.6 matrices. Additional studies are needed to see if the minor feldspathic component between FeO-rich olivines in types 3.1–3.6 (B & J 1998, p. 217) is related to the amorphous component in ALHA77307.

1.07.5.7.5 CK3 chondrite matrices

Matrices of CK4–6 chondrites are composed largely of olivine, low-calcium pyroxene and plagioclase grains that are mostly 20–100 μm in size with minor magnetite and sulfide (B & J 1998, p. 239). CK3 chondrite matrices appear to resemble those in CV3 chondrites, which are discussed below. Matrix textures in the Kobe CK4 chondrite were interpreted by Tomeoka et al. (2001) to result from partial melting during shock metamorphism, although Kobe itself is only classified as a shock stage S2.

1.07.5.7.6 CV3 chondrite matrices

Matrix material in CV3 chondrites occurs as rims on chondrules and other components as well as interstitial material, which is coarser grained. The volume of a matrix rim is approximately equal to the volume of the enclosed chondrule (Paque and Cuzzi, 1997). Matrices are mineralogically diverse but are dominated by iron-rich olivine, Fa_{30-60}, which is coarser than in other chondrite groups. It forms elongate grains up to 20 μm in length and subhedral finer grains <100 nm to 3 μm in the interstices (B & J 1998, pp. 220; Scott et al., 1988). Olivines contain planar defects and small inclusions of sulfide, metal or magnetite, and poorly graphitized carbon (Tomeoka and Buseck, 1990; Keller et al., 1994; Brearley, 1999). Reduced CV3 chondrites appear to be the least altered as they have abundant metallic Fe,Ni rather than magnetite: minor matrix minerals include low-calcium pyroxene, metallic, Fe,Ni, and FeS. In the Bali-like oxidized CV3 chondrites, matrices lack metal and contain abundant phyllosilicates, fayalite (Fa >95), calcium-rich pyroxene, pentlandite and magnetite in addition to iron-rich olivine (Figure 28(a)). In the Allende-like oxidized CV3s, hydrated silicates and fayalite are rare or absent, but Ca–Fe-rich pyroxene, nepheline, and pentlandite are abundant, and magnetite is present (Krot et al., 1995, 1998a,b,c,d; Brearley, 1999; B & J 1998, p.220–5) (Figure 28(b)).

Proponents of a nebular origin for the secondary minerals in oxidized CV3 chondrites argue that fayalitic olivine condensed in the solar nebula at high f_{O_2} following evaporation of chondritic dust, and coated preexisting chondrules (Nagahara et al., 1988; Palme and Fegley, 1990; Weinbruch et al., 1990). Kimura and Ikeda (1998) infer that anhydrous alteration minerals like nepheline in Allende formed at ~600–800 °C whereas estimates from zoning in olivine point to maximum temperatures of ~400–550 °C. They, therefore, conclude that the anhydrous alteration occurred prior to accretion of the chondritic components. Brenker et al. (2000) who studied aggregates of andradite and hedenbergite 20–50 μm in size in the Allende matrix, infer that the Ca–Fe-rich pyroxene cooled rapidly from >1,050 °C at >10 °C h^{-1}. They argue that a localized nebular environment is more compatible with this thermal history than an asteroidal environment.

Evidence favoring progressive alteration in asteroids includes fayalite veins that cross-cut chondrules and matrices, replacement of chondrules by fayalite-rich matrix material in dark inclusions, and comparable alteration minerals in chondrules and matrices (Krot et al., 1995, 1998a,b,c,d; Kimura and Ikeda, 1998). In Bali-like CV3s, saponite replaces calcium-rich minerals in chondrules and CAIs; magnetite is replaced by fayalite and calcium-rich pyroxene. In Allende-like CV3s, fayalitic olivine replaces low-calcium pyroxene; magnetite-sulfide grains in matrices and chondrules are replaced by fayalitic olivine and calcium-rich pyroxenes; nepheline and calcium-rich pyroxene form veins in and around CAIs and dark inclusions. Isotopic data also favor asteroidal sites for alteration. Coexisting magnetite and fayalite in Mokoia show large mass fractionation in oxygen isotopes, consistent with aqueous alteration (Choi et al., 2000). Nepheline lacks evidence for ^{26}Mg excesses indicating formation >3–5 Myr after CAI formation (MacPherson et al., 1995). $^{53}Mn-^{53}Cr$ isotope systematics suggest that fayalite formed ~7–16 Myr after CAIs (Hutcheon et al., 1998).

The mineralogy of CV3 chondrite matrices prior to alteration is not clear. The two reduced CV3 chondrites that have been studied by TEM are not ideal samples. In Leoville, which is moderately shocked, Nakamura et al. (1992) found aggregates of rounded, 10–100 nm olivine grains with interstitial amorphous material, but attributed this to shock. In Vigarano, which is a breccia of altered and unaltered material

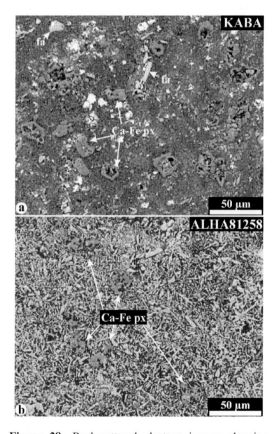

Figure 28 Backscattered electron images showing matrix material in the Bali-like oxidized CV3 chondrite, Kaba (a) and the Allende-like, oxidized CV3 chondrite, ALHA81258 (b). Kaba matrix contains fayalite (fa), Ca–Fe pyroxene, fine-grained, iron-rich olivine, pentlandite, magnetite and phyllosilicates. The ALHA81258 matrix lacks phyllosilicates and contains coarse-grained laths of fayalitic olivine (light-gray) and patches of Ca–Fe pyroxene plus nepheline and magnetite. Both matrices lack the amorphous silicate present in ALHA77307 and appear to be highly altered.

(Lee et al., 1996), oxygen isotopic mapping of the matrix revealed ^{16}O-rich forsterite grains 2 μm in size that may be fragments of AOA or related objects, as well as similar-sized ^{16}O-poor fayalite and enstatite grains (Kunihiro et al., 2002). Clearly this technique is essential for unraveling the origins of matrix components.

1.07.5.7.7 Matrix of ungrouped C chondrites, Acfer 094, and Adelaide

Acfer 094, which is closely linked to the CO and CM groups, has abundant matrix material and higher concentrations of presolar SiC than any other chondrite (Newton et al., 1995). Its matrix is composed of 30 vol.% forsterites 10–300 nm in size and Fa_{0-2} in composition, 40% amorphous material, and 20% enstatite with Fs_{0-3} and grain sizes of 200–300 nm (Greshake, 1997). The enstatites show intergrowths of ortho- and clino enstatite indicating rapid cooling at ~1,000 °C h^{-1}. Minor components included 5% sulfides, <1% phyllosilicates and <2% ferrihydrite.

The highly unusual combination of amorphous material and forsterite in Acfer 094 matches that in the CO3.0 chondrite, ALH77307. The amorphous phases in the two chondrites are compositionally similar (Table 6) and both contain submicrometer enstatites and forsterites with high concentrations of MnO (0.6–2 wt.%; Greshake, 1997). These concentrations are higher than in chondrules and comparable to those in high-manganese, low-iron silicates reported in IDPs by Klöck et al. (1989). Acfer 094, like ALH77307, contains <1% phyllosilicate in the amorphous material and 100–300 nm sulfide grains (pyrrhotite and minor pentlandite) as in ALHA77307, but it lacks matrix metal grains. Since ferrihydrite is present in the amorphous material in Acfer 094, metal may once have been present. Greshake (1997) proposed that the amorphous material was a presolar or solar condensate and that the FeO-poor silicates formed in the solar nebula by condensation or annealing of amorphous material.

Adelaide, which also has links to CO and CM groups, has a matrix that lacks phyllosilicates and is dominated by fayalitic olivine Fa_{31-84} occurring as aggregates of 100–200 nm crystals and platy and corroded single crystals (Brearley, 1991). Amorphous material, which is rich in iron and sulfur, is present but is less abundant than in ALHA77307. High-manganese enstatite forms 1–1.5 μm clusters of 100–500 nm crystals but forsterites are absent. Pentlandite forms aggregates of 10–30 nm crystals and magnetite occurs as 400–6,500 nm clusters of 100 nm grains.

1.07.5.7.8 H–L–LL3 chondrite matrices

Matrix material in type 3 ordinary chondrites forms rims on chondrules, metal grains and other components and occurs as lumps (B & J 1998, pp. 231–7). In the LL3.0 chondrite, Semarkona, matrix consists largely of phyllosilicates formed by aqueous alteration with fine-grained olivine Fa_{5-38} and low-calcium pyroxene, calcite, and magnetite (Alexander et al., 1989a). Magnetite grains show large mass fractionation of oxygen isotopes suggesting that a limited supply of aqueous fluid was largely consumed during asteroidal alteration (Choi et al., 1998). Klöck et al. (1989) identified an additional population of forsterites $Fa_{<1}$ and enstatites $Fs_{<2}$ having MnO concentrations of 0–1.4 wt.% that are correlated with FeO. The high-manganese silicates are comparable in composition to the iron-poor, manganese-rich silicates identified in IDPs (Klöck et al., 1989).

Matrices in type 3.1–3.6 chondrites are largely composed of FeO-rich olivines, which may have skeletal or elongate morphologies, and an amorphous feldspathic phase (Nagahara, 1984; B & J 1998, pp. 231–7). The origins of these phases are uncertain. Alexander et al. (1989b) argued that interchondrule matrix and matrix rims in ordinary chondrites were derived from broken chondrules, contrary to Brearley et al. (1989), who suggested an origin by annealing of amorphous material. Ikeda et al. (1981) identified two types of matrix in ordinary chondrites with abundant amorphous material, one contained Fe,Ni droplets 10–100 nm in size. They speculated that the amorphous material formed by shock. Fayalitic olivine (Fa_{54-94}) occurs in the matrices of Krymka (LL3.1) and Bishunpur where it rims metal grains (Weisberg et al., 1997; Lauretta et al., 2001). These authors argued that it formed in the solar nebula, not during aqueous or hydrothermal alteration. Strong evidence for asteroidal alteration is present in Tieschitz (H3.6), which contains nepheline and albite in veins between chondrules and in chondrule mesostases (Hutchison et al., 1998). This so-called "white matrix" formed when aqueous fluids attacked chondrule mesostases forming voids.

1.07.5.7.9 K3 chondrite matrix (Kakangari)

Kakangari has many unusual properties as chondrules and matrix both contain enstatite and forsterite, and matrix silicates are more magnesian than those in chondrules (Weisberg et al., 1996; B & J 1998, p. 82). Matrix and chondrules have similar bulk chemical compositions though matrix is richer in ^{16}O (Weisberg et al., 1996).

Kakangari matrix contains mostly euhedral to subhedral crystals with grain sizes of 200 nm to

1.5 μm, which occur as individual crystals and polycrystalline units or clusters 2–8 μm in size, some of which have well annealed textures. It is composed of 50 vol.% enstatite, Fs_{2-5}, ~20 vol.% forsterite, Fa_{0-3}, 10 vol.% albite crystals 250–1,000 nm in size, $Or_{<3}$, and <1 vol.% anorthite up to 500 nm in size. Minor phases include chromium-spinel, troilite grains 1–5 μm in size and ~10 vol.% ferrihydrite, which forms acicular aggregates up to 2 μm in length in pores. The lower FeO concentrations in olivine relative to coexisting pyroxene in matrix and chondrules as in EH3 chondrites, suggest that reduction occurred during cooling as olivine diffusion rates are faster than those in pyroxene (Scott and Taylor, 1983; B & J, 1998, pp. 72–74).

Olivine occurs mainly as inclusions in enstatite and is also found in albite, spinel, and troilite (Brearley, 1989). At least three kinds of polycrystalline clusters appear to have formed separately: (i) enstatite-rich units with forsterite inclusions, (ii) coarser versions with albite and olivine inclusions, and (iii) forsterite–anorthite units with no enstatite. Enstatite crystals are intergrowths of ortho- and clinopyroxene and their microstructures indicate cooling from >1,000 °C at ~1,000 °C h^{-1}. Brearley (1989) suggests that the Kakangari matrix formed from amorphous or partly crystalline particles <10 μm in size that were annealed at 1,100–1,200 °C or possibly higher, and then rapidly cooled in an hour. The chondrules in Kakangari could have formed from similar material that was heated to higher temperatures, partly melted, and quenched at comparable rates, provided that the chondrules acquired lower ^{16}O concentrations when molten.

The bulk composition of Kakangari suggests that the hypothesized precursor material for the chondrules and matrix had near-CI levels of moderately volatile elements that were not lost during high-temperature processing. Reduction probably occurred during this processing.

1.07.5.7.10 Heavily altered, matrix-rich lithic clasts

Heavily altered, matrix-rich chondritic clasts are very different from the most common lithic clasts in chondrites, which are fragments of chondrules and CAIs or lithic fragments resembling the host rock. They are also unlike rare foreign chondritic clasts that are shocked and appear to have impacted the parent body at high speed after accretion, e.g., an LL chondrite clast in an H chondrite (Lipschutz et al., 1989).

Altered, matrix-rich chondritic clasts are similar in size to chondrules and CAIs in the host chondrite, unshocked, and appear to have been gently incorporated into the host material at an early stage. Prior to incorporation, they were aqueously altered in a different location. They occupy a few vol.% of CR2 (see Section 1.07.5.7.3), CV3, CO3, CH, and CB chondrites. In CV3s they tend to have fewer and smaller chondrules than the host chondrites and some are rimmed by matrix material (Kracher et al., 1985; Johnson et al., 1990). Although Palme et al. (1989) and Weisberg and Prinz (1998) argued that some are unaltered aggregates of nebular condensates, there is now abundant evidence favoring formation by asteroidal alteration (Kojima et al., 1993; Kojima and Tomeoka, 1996; Krot et al., 1997, 1998a,b,c,d). I–Xe ages of some CV3 clasts confirm that they were altered before their hosts (Hohenberg et al., 2001). Dark clasts in CO3 chondrites are smaller than in CV3 chondrites (50–200 μm; cf. millimeter to centimeter in CV3s) (Itoh and Tomeoka, 2003). In CH3 chondrites and CB3 chondrites, matrix-rich chondritic clasts are the only matrix material and they are heavily altered to type 1 or 2 levels, unlike the chondrules and metal grains, which are pristine (Greshake et al., 2002). In CH chondrites these clasts are mostly 5–200 μm in size and have a size distribution comparable to that of the chondrules (Scott, 1988), suggesting that sizes of dark clasts are correlated with chondrule size.

Some authors infer that matrix-rich clasts are merely pieces of the body that supplied the host chondrite (e.g., Itoh and Tomeoka, 2003). But their unusual mineralogy, limited size ranges, which are correlated with chondrule size, and matrix rims on some clasts favor a separate origin. The rims suggest they were lofted after alteration into the dusty environment that formed rims on chondrules and other components. Most matrix-rich clasts may be fragments of bodies that formed before the parent body of the host chondrite, and were then accreted with another batch of chondrules and CAIs to form a later generation of chondritic asteroids. Such planetesimals have been involved by Metzler et al. (1992) and Bischoff (1998). Early formed, volatile-rich planetesimals may have been disrupted by explosive volatile release (Wilson et al., 1999) or impact fragmentation and dispersal.

The unique chondrite, Kaidun, appears to be a breccia composed almost entirely of millimeter and submillimeter sized fragments of diverse kinds of chondrites (Zolensky and Ivanov, 2001). Kaidun may be a clast-rich, chondrule-poor chondrite that formed in a nebula region where chondrules were much rarer than clasts (Scott, 2002). CI chondrites and the Tagish Lake chondrite also appear to have formed from sub-millimeter particles of altered chondritic material (Nakamura et al., 2003). Thus disaggregation of early-formed planetesimals may have been widespread.

1.07.5.7.11 Origins of matrix phases

Our survey suggests that unaltered carbonaceous chondrite matrices are composed largely of amorphous silicate and sub-micrometer forsterites and enstatites and micrometer-sized aggregates of such crystals. In addition, primitive matrices also contain up to a few volume percent of carbonaceous material, ppm levels of presolar grains (see Chapter 1.02), and refractory grains. Micrometer-sized corundum, spinel and hibonite grains of solar-system origin are much more common than presolar grains (e. g., Strebel et al., 2000; Nittler et al., 1994) and probably formed as isolated grains rather than fragments of CAIs. Not enough is known about unaltered matrices in ordinary and enstatite chondrites to infer their detailed mineralogy.

Amorphous ferromagnesian silicate is best preserved in the matrices of ALHA77307, Acfer 094, and Adelaide, though it is absent in Kakangari. It is more fractionated than bulk matrix with higher Al/Ca, Ni/Si, and P/Si than CI chondrites. Metallic Fe,Ni is abundant only in the ALHA77307 matrix but it may have been altered in the other chondrites. Related amorphous material may be present in CM, CO, and ordinary chondrites. The aluminum-rich nature of amorphous material may account for the high Al/Ca ratio of some matrix materials (Figure 27), the depletion of refractory elements with respect to aluminum in matrix materials in ordinary chondrites (Alexander, 1995), and the low Ca/Al ratio of certain chondrites like Adelaide and Kakangari (Brearley, 1991). The cause of the enrichment of aluminum, nickel and phosphorus in amorphous matrix is uncertain as these elements have different geochemical properties. This amorphous material did not form from quenched melt from chondrule mesostases, by impact in situ or elsewhere, or by dehydration of phyllosilicates. It appears to have accreted into chondrule rims as small particles (1–5 μm in size in ALHA77307).

Forsterites $Fa_{<5}$ and enstatites $Fa_{<5}$, which are 10–300 nm in size, are found in the matrices of CM chondrites, ALHA77307, Acfer 094, and probably also in Semarkona, for which grain sizes were not reported. Enstatites and forsterite grains with higher FeO concentrations tend to be MnO-rich (0.6–2 wt.%). Intergrowths of ortho- and clinoenstatite and forsterite–enstatite associations indicate that all magnesian silicates cooled from >1,300 K at $\sim 10^3$ to 10^4 °C h^{-1} as micrometer- and submicrometer-sized grains prior to accretion. Oxygen-isotopic mapping of grains 0.03–2 μm in size shows that Acfer 094 contains <20 ppm of presolar silicates (Messenger and Bernatowicz, 2000). Therefore, the abundant iron-poor matrix silicates must have formed in the solar nebula.

Primitive chondrite matrices are surprisingly similar to porous chondritic aggregate IDPs in that both contain amorphous material, FeO-poor silicates some of which are MnO-rich (Klöck et al., 1989), refractory grains, and presolar grains. Comets also contain amorphous and FeO-poor silicates (Hanner, 1999). ^{16}O-rich refractory micrograins are present in micrometeorites and chondritic aggregate IDPs suggesting additional links between chondrite matrices and particles from the outer solar system (Greshake et al., 1996; Engrand and Maurette, 1998; Engrand et al., 1999). The amorphous aggregates in ALHA77307 that are rich in metal and sulfide may be related to the glass with embedded metal and sulfide (GEMS) grains that are a major constituent of porous chondritic IDPs and resemble amorphous interstellar grains (Bradley et al., 1999; Rietmeijer, 1998, 2002). However GEMS are richer in SiO_2 and much smaller: 100–500 nm in size with kamacite and sulfide grains that are <5 nm in size, cf. <200 nm in ALH77307. GEMS may have formed by amorphization of interstellar grains by cosmic-ray irradiation (like amorphous rims on some lunar grains) or by shocks (Bradley, 1994; Bradley et al., 1999). However, nearly all GEMS have solar-like oxygen isotopic compositions and are more likely to be solar system products (Messenger et al., 2003). Amorphous material is known to condense around evolved stars (Hill et al., 2001), and the amorphous matrix material may be a solar system analogue.

Two high-temperature processes have been proposed for magnesian silicates in matrices: disequilibrium condensation (Klöck et al., 1989; Nagahara et al., 1988; Greshake, 1997) and annealing of amorphous condensate particles (Brearley, 1993; Greshake, 1997). Annealing experiments suggest that magnesium-rich silicates can crystallize from amorphous condensates in a few hours at 1,050 K, or a year or more below 1,000 K (Hallenbeck et al., 2000; Rietmeijer et al., 2002). If the magnesian silicates formed close to the protosun by annealing or condensation, they could have been lofted across the disk by outflows powered by reconnecting magnetic field lines, turbulence, or radiation pressure (Shu et al., 1996; 2001; Nuth et al., 2002). Alternatively, Harker and Desch (2002) suggest that magnesian silicates in comets formed at 5–10 AU by annealing caused by nebular shocks. We infer that magnesian silicates were ubiquitous in the solar nebula at 2–10 AU as a result of high-temperature processing and quenching in hours or days. The similarity of the cooling rates inferred for the magnesian silicates in the matrix and for chondrules suggests that matrix silicates and chondrules may have been formed in related events.

Submicrometer fayalitic olivines are heterogeneously distributed in the matrices of primitive chondrites. In Acfer 094 they are absent but the ALHA77307 matrix contains 30–40 vol.%, which probably formed in situ by low-temperature annealing of amorphous material (Brearley, 1993). Sub-micrometer sized fayalitic olivines are also present in matrices of CO3.1–3.6, type 3 ordinary chondrites and the interstices of larger olivines in CV3 chondrites. An origin for these olivines by annealing of amorphous material also appears plausible but the evidence is weak. In some instances (e.g., in ALH77307) annealing and crystallization of olivines may have started prior to accretion, as Kakangari matrix grains were extensively annealed before chondrules and matrix accreted. However, for the coarse, abundant fayalitic grains and pure fayalite in CV3 chondrites, an asteroidal origin is indicated as their formation required an aqueous fluid. Although fayalitic olivine is generally inferred to have been a major constituent of the solar nebula (e.g., Nagahara et al., 1994; Ozawa and Nagahara, 2001), we find no firm evidence from chondrite matrices that any *crystalline*, FeO-rich silicate formed in the solar nebula.

Chemical differences between matrix and chondrules suggest that all chondrules could not have formed from matrix material. For example, matrix material is typically richer in FeO than chondrules, and mean refractory elemental abundances deviate further from CI levels than for chondrules. Some authors argue from chemical differences that chondrules and matrices are chemically complementary and therefore formed in the same location. For example, Fe/Si ratios in chondrules and matrix are lower and higher respectively than CI ratios whereas the bulk compositions of CM, CO, CV, and other chondrites are close to solar implying that chondrules and matrix were chemically fractionated from one another (Wood, 1996a). The case for Fe/Si is not very convincing, however, as there is a wide range of CI-normalized Fe/Si ratios from 0.4 in LL chondrites to ≥ 2 in CH and CB chondrites. However, for refractory elements, a stronger case can be made. CI-normalized, mean Ti/Al ratios for matrix and chondrules in the Renazzo CR2 chondrite are 0.5 and 1.5, respectively, whereas bulk Ti/Al ratios for CR and other chondrites are within 10% of CI levels (Klerner and Palme, 2000). Palme and Klerner (2000) infer that chondrules and matrix formed from a single reservoir that behaved as a closed system during condensation and processing of refractory elements.

1.07.6 FORMATION AND ACCRETION OF CHONDRITIC COMPONENTS

Several common themes have emerged from this review of chondritic components. Type 3 chondrites, for example, are more affected by alteration than generally realized. Only a handful of chondrites meet the minimal requirements for pristine chondrites, and none of these escaped the imprint of asteroidal modification. Whether this alteration occurred in parent bodies, as we infer, or in the nebula, the alteration environment was quite different from that at the location where each component formed.

Mean sizes of chondrules, CAIs, amoeboid olivine inclusions, dark clasts, and possibly metallic Fe,Ni grains tend to be correlated among the chondrite groups. Thus, in groups with relatively small chondrules, all components tend to be relatively small. This sorting process occurred after all components formed, possibly during turbulent accretion triggered by chondrule formation (Cuzzi et al., 2001). Aerodynamic sorting should cause metal particles to accrete with larger chondrules, as particles are sorted according to the product of their radius and density (Kuebler et al., 1999). However, Akridge and Sears (1999) have also argued for aerodynamic sorting but they proposed that components were sorted in an asteroidal regolith during degassing.

Many different processes were involved in making each chondritic component. Unaltered chondrite matrices may contain at least six different types of micrometer-to-nanometer-sized components, which formed in diverse environments: amorphous FeO-rich silicate, forsterite and enstatite grains, refractory grains, presolar grains, carbonaceous material, and iron-rich olivine. Chondrules formed by several nebular processes (closed-system melting, condensation, and possibly evaporation) and at least one asteroidal process (impact melting in regoliths). CAIs may be condensates, residues or processed versions of both. An exception to this preference for complexity is provided by the amoeboid olivine inclusions: all AOAs could have formed by the same basic process: nebular condensation. Aluminum-rich chondrules may provide a second exception, at least within carbonaceous chondrites.

Each chondrite component appears to have been manufactured predominantly by brief high-temperature processes. Chondrules and igneous CAIs have textures indicating cooling in minutes or hours. Many enstatite grains in matrices contain ortho–clino intergrowths like those in chondrules, and cooled in less than a few hours. Metal grains in CH chondrites cooled from high temperatures in days, at most. However, the generation of group II REE pattern characteristic of many fine-grained CAIs is one process that required longer timescales of many years (Palme and Boynton, 1993).

The high-temperature processes that formed chondritic components typically involved gas–solid or gas–liquid exchange over part of the temperature range 1,200–2,000 K. Rapid cooling in the formation environment or removal to cooler localities ensured an extraordinary diversity of mineralogical and chemical compositions among chondritic components. Chemical fractionations during brief high-temperature processing were so ubiquitous and extensive that they may be largely responsible for at least four of the major chemical variations shown by chondrites: variations in refractory and moderately volatile elements, metal–silicate ratios, and forsterite–enstatite variations.

The search for nebular condensates among chondrites has generated several interesting but seemingly false leads including wollastonite needles (e.g., Kerridge, 1993), blue-luminescing enstatite rims (Weisberg et al., 1994; Hsu and Crozaz, 1998), and refractory forsterite grains (Weinbruch et al., 2000; Jones et al., 2000b). FeO-rich matrix silicates are also unlikely to be nebular condensates. They appear to have formed predominantly by asteroidal processes, though some may have formed by nebular annealing of FeO-rich amorphous material.

In the final section we review the heat sources that may have operated in the early solar system to produce such diverse chondritic components.

1.07.7 HEATING MECHANISMS IN THE EARLY SOLAR SYSTEM

Possible heating mechanisms for making chondrules have been reviewed by Cassen (1996), Boss (1996), and Rubin (2000), and for CAIs and chondrules by Jones et al. (2000a). Nearly all heating models have focused on chondrule origins, but Shu et al. (1996, 2001) address both CAI and chondrule formation close to the protosun; Hood (1998) and Desch and Connolly (2002) consider mechanisms for separately melting ferromagnesian and refractory aggregates to make chondrules and igneous CAIs.

Three kinds of processes have been invoked for forming chondrules: melting of dust aggregates (generally considered to be the standard model), condensation of melts, or melts and crystals (Ebel and Grossman, 2000; Krot et al., 2001d), and impacts into solid or partly molten asteroids (e.g., Symes et al., 1998; Sanders, 1996). As discussed above, we believe that all three processes formed chondrules, and that the standard model cannot account for olivine-rich and pyroxene-rich chondrules.

Rubin (2000) has reviewed models for forming chondrules and their consistency with various petrologic constraints (Table 7): condensation, exothermic chemical reactions, jetting during nebular collisions between particles, ablation in protoplanetary atmospheres, nebular lightning

Table 7 Review by Rubin (2000) of chondrule-formation models and their consistency with petrologic and geochemical constraints.

	Oxygen isotopes	Incomplete melting	Evaporation during long-term heating	Relict grains and moderate volatiles	Rapid cooling	Chondrule abundance	Chondrule size[a]	Multiple heating	Microchondrules in rims	Primary troilite
Condensation	−	−	/	−	−	+	−	−	−	+
Exothermic chemical reactions	?	?	+	?	−	−	−	−	−	−
Jetting	?	+	+	+	+	−	−	−	−	+
Meteor ablation	?	+	+	+	+	−	+	−	−	+
Nebular lightning	+	+	+	+	+	+	+	+	+	+
Supernova shock waves	+	+	+	+	+	−	−	−	−	+
Aerodynamic drag heating	+	+	+	+	+	?	−	+	−	+
Magnetic reconnection flares	?	+	+	+	+	+	+	+	+	+
Gas dynamic shock waves	?	+	+	+	+	+	+	+	?	+
Planetesimal bow shocks	−	−	−	−	+	+	?	−	−	−
FU Orionis outbursts	−	+	+	−	+	+	+	+	+	+
Bipolar outflows	+	+	+	+	+	+	?	+	−	+
Gamma-ray bursts	+	+	+	+	+	+	?	+	+	+
Radiative heating	−	−	−	−	−	+	+	−	+	−
Planetesimal collision	−	+	+	−	−	−	−	−	−	+
Impact-melting in parent-body regolith	−	+	+	+	+	−	−	?	−	+

+ = consistent; − = inconsistent; ? = possibly consistent after ad hoc modifications; / = not applicable.
[a] Chondrule size distribution within a group is much narrower than the total variation shown by all chondrites.

(Desch and Cuzzi, 2000), supernova shock waves, aerodynamic drag heating at the accretion edge of the disk, magnetic reconnection flares, gas dynamic shock waves (Section 1.07.7.1), bow shocks from early formed planetesimals (Hood, 1998; Weidenschilling et al., 1998), FU Orionis outbursts, bipolar outflows (Section 1.07.7.2), gamma-ray bursts, radiative heating (Eisenhour and Buseck, 1995), and impacts into asteroidal regoliths and molten planetesimals (Section 1.07.7.3). Note that the constraints listed in Table 7 apply to what we have called, "archetypal chondrules," which are most closely represented by type II chondrules. Chronological and magnetic constraints are not included in Table 7.

Below we review three models that are currently popular: nebular shocks, jets and outflows near the protosun, and impacts on planetesimals.

1.07.7.1 Nebular Shocks

Nebular shock waves are discontinuities between hot compressed gas moving faster than the local sound speed [$= 1.2 \times 10^5 \times (T/300)^{1/2}$ cm s^{-1}] and cooler, less dense, slowly moving gas (Cassen, 1996). Particles overtaken by shocks are suddenly enveloped in a blast of wind moving at several kilometers per second, which subjects the particles to frictional heating. Particles are also heated by radiation from the shock front before they are overtaken and by conduction from hot gas. Desch and Connolly (2002) find that, contrary to earlier studies, cooling rates in the range 10–1,000 °C h^{-1} can be produced without invoking shocks of unreasonable speeds. They studied a range of shock speeds (5–10 km s^{-1}) and gas densities and found conditions under which chondrules would partially melt, conditions that would cause dust to evaporate, and at the higher shock speeds, conditions that would evaporate chondrules. The shock model appears to be capable of forming droplets of the correct size (Connolly and Love, 1998; Susa and Nakamoto, 2002), it seems consistent with many petrologic constraints from archetypal chondrules (Table 7), and predicts a correlation between cooling rate and the concentration of chondrules, which appears consistent with compound chondrule statistics (Desch and Connolly, 2002).

A major problem for the shock model has been to find a source of powerful, pervasive, and repeatable shocks. Several have been proposed: clumpy material falling into the nebula (Tanaka et al., 1998), bow shocks from planetesimals scattered by Jupiter (Hood, 1998; Weidenschilling et al., 1998), and spiral-arm instabilities in the solar nebula (Wood, 1996b). Boss infers that clumps and spiral arms could generate ~10 km s^{-1} shocks in the asteroid belt (Chapter 1.04). Iida et al. (2001) and Ciesla and Hood (2002) also favor shock heating models. Genge (2000) proposed that shocks formed chondrules by surface melting and ablation of 1–500 m sized planetesimals.

1.07.7.2 Jets and Outflows

Two types of jet or outflow models have been proposed. Liffman and Brown (1996) suggested that chondrules formed in a bipolar outflow by ablation of planetesimals and were then injected into the asteroid belt. The observed correlation between rim thickness and chondrule size may have arisen when chondrules reentered the dusty nebula at hypersonic speeds (Liffman and Toscano, 2000).

Shu et al. (1996, 2001) have proposed that chondrules and CAIs formed at the inner edge of the protostellar disk. About 0.05 AU from the protosun, the rotation rate of the disk matches that of the protostar and the disk is terminated. Inflowing gas is ionized and either whipped into the protostar by the star's magnetic field, or ejected in the bipolar outflows, while rocks spiral inwards until they evaporate or are launched by the winds. CAIs are thought to form during quiescent conditions whereas chondrules form when the position of the inner edge of the disk fluctuates due to intense flares from the protosun. Shu and colleagues suggest that CAIs and chondrules are hurled outwards by the so-called X-wind, which they infer powers the bipolar outflows. CAIs and chondrules fall back onto the disk at several AU or beyond and are available for accretion into asteroids and planets or recycling back to the inner edge of the disk. Smaller particles fall back at greater distances from the protosun. Detailed thermal histories for chondrules and CAIs have not been calculated for this model.

1.07.7.3 Impacts on Planetesimals

Two kinds of impacts have been proposed for chondrule formation: impacts at ~5 km s^{-1} between solid bodies that melt regolith material (e.g., Symes et al., 1998; Sears, 1998; Bridges et al., 1998), and impacts onto already molten planetesimals at lower speeds that puncture the planet spewing forth molten drops of silicate and metal (Sanders, 1996). Many factors suggest that the vast majority of chondrules could not form by impact in regoliths (Taylor et al., 1983). Molten droplets are rare on the regolith of Vesta, the presumed parent of the eucrites, and meteorite evidence suggests that chondrule-free asteroids did not form in the inner part of the asteroid belt (Scott, 2002). (See Sears (1998) for a contrary view.) However, the chromite-rich chondrules present in brecciated type 4–6 ordinary chondrites (Krot and Rubin, 1993) and perhaps a few percent of other chondrules in H, L, and LL chondrites probably did form this way after the chondritic bodies had accreted. Since impact melt will be ejected at speeds of several kilometers per second, much faster than the escape velocity of the asteroid (less than a few hundred meters per second), most impact melt will not re-accrete to the target asteroid. Impact melt should be rarer on the surface of asteroids than on the Moon (Taylor et al., 1983).

Any asteroid more than ~10 km in radius that accreted <1.5 Ma after most CAIs formed would have melted <2.5 Ma later, assuming that ^{26}Al was relatively homogeneously distributed with ^{26}Al/^{27}Al ratio of ~5 × 10^{-5} when CAIs formed (e.g., Sanders and Hevey, 2002; Yoshino et al., 2003). The parent asteroids of differentiated asteroids appear to have formed in this interval and several suffered catastrophic impacts during or soon after igneous differentiation (Keil et al., 1994). Sanders (1996) argues that chondrules formed in impacts on molten planetesimals with thin crusts, before they crystallized. Sanders and Hevey (2002) suggest that convection was vigorous enough to prevent gravitational segregation of melted Fe,Ni droplets to the core (see also Yoshino et al., 2003). This model envisages that molten metal–silicate droplets were dispersed into the solar nebula and that they accreted long after they had cooled and mixed with impact droplets from other asteroidal impacts, nebula dust, CAIs, and material from the cool unmelted crust of the target asteroids. The advantages of this model are that abundant chondrules are generated, the model readily accommodates the several Ma interval between CAI and chondrule formation, and, provided gas loss occurred, it could explain why many chondrules lost volatiles (Lugmair and Shukolyukov, 2001). Disadvantages are summarized in Table 7.

ACKNOWLEDGMENTS

This research has made use of NASA's Astrophysics Data System and was supported in part by NASA grants NAG5-11591 (K. Keil, P. I.) and NAG5-10610 (A. Krot, P. I.). We thank A. Davis and an anonymous reviewer for many helpful comments. This is SOEST publication #6179 and HIGH publication #1292.

REFERENCES

Akridge D. G. and Sears D. W. G. (1999) The gravitational and aerodynamic sorting of meteoritic chondrules and metal: experimental results with implications for chondritic meteorites. *J. Geophys. Res.* **104**, 11853–11864.

Akridge G., Benoit P. H., and Sears D. W. G. (1998) Regolith and megaregolith formation of H-chondrites: thermal constraints on the parent body. *Icarus* **132**, 185–195.

Aléon J., Krot A. N., and McKeegan K. D. (2002a) Calcium–aluminum-rich inclusions and amoeboid olivine aggregates from the CR carbonaceous chondrites. *Meteorit. Planet. Sci.* **37**, 1729–1755.

Aléon J., Krot A. N., McKeegan K. D., MacPherson G. J., and Ulyanov A. A. (2002b) Oxygen isotopic composition of fine-grained Ca–Al-rich inclusions in the reduced CV3 chondrite Efremovka. In *Lunar Planet. Sci.* **XXXIII**, #1426. The Lunar and Planetary Institute, Houston (CD-ROM).

Alexander C. M. O'D. (1995) Trace element contents of chondrule rims and interchondrule matrix in ordinary chondrites. *Geochim. Cosmochim. Acta* **59**, 3247–3266.

Alexander C. M. O'D. and Wang J. (2001) Iron isotopes in chondrules: implications for the role of evaporation during chondrule formation. *Meteorit. Planet. Sci.* **36**, 419–428.

Alexander C. M. O'D., Barber D. J., and Hutchison R. (1989a) The microstructure of Semarkona and Bishunpur. *Geochim. Cosmochim. Acta* **53**, 3045–3057.

Alexander C. M. O'D., Hutchison R., and Barber D. J. (1989b) Origin of chondrule rims and interchondrule matrices in unequilibrated ordinary chondrites. *Earth Planet. Sci. Lett.* **95**, 187–207.

Alexander C. M. O'D., Grossman J. N., Wang J., Zanda B., Bourot-Denise M., and Hewins R. H. (2000) The lack of potassium-isotopic fractionation in Bishunpur chondrules. *Meteorit. Planet. Sci.* **35**, 859–868.

Alexander C. M. O'D., Boss A. P., and Carlson R. W. (2001) The early evolution of the inner solar system: a meteoritic perspective. *Science* **293**, 64–68.

Amelin Y., Krot A. N., Hutcheon I. D., and Ulyanov A. A. (2002) Lead isotopic ages of chondrules and calcium–aluminum-rich inclusions. *Science* **297**, 1678–1683.

Anders E. and Kerridge J. F. (1988) Future directions in meteorite research. In *Meteorites and the Early Solar System* (eds. J. F. Kerridge and M. S. Matthews). University of Arizona Press, Tucson, pp. 1155–1186.

Artioli G. and Davoli G. (1994) Low-Ca pyroxenes from LL group chondritic meteorites: crystal structural studies and implications for their thermal histories. *Earth Planet. Sci. Lett.* **128**, 469–478.

Barber D. J. (1981) Matrix phyllosilicates and associated minerals in C2M carbonaceous chondrites. *Geochim. Cosmochim. Acta* **45**, 945–970.

Begemann F. (1980) Isotopic anomalies in meteorites. *Rep. Prog. Phys.* **43**, 1309–1356.

Bell J. F., Davis D. R., Hartmann W. K., and Gaffey M. J. (1989) Asteroids: the big picture. In *Asteroids II* (eds. R. P. Binzel, T. Gehrels, and M. S. Matthews). University of Arizona Press, Tucson, pp. 921–945.

Bennett M. E. and McSween H. Y., Jr. (1996) Revised model calculations for the thermal histories of ordinary chondrite parent bodies. *Meteorit. Planet. Sci.* **31**, 783–792.

Binzel R. P., Lupishko D. F., Martino M. D., Whiteley R. J., and Hahn G. J. (2002) Physical properties of near-Earth objects. In *Asteroids III* (eds. W. F. Bottke, Jr., A. Cellino, P. Paolicchi, and R. P. Binzel). University of Arizona Press, Tucson, pp. 255–271.

Bischoff A. (1998) Aqueous alteration of carbonaceous chondrites—a review. *Meteorit. Planet. Sci.* **33**, 1113–1122.

Bischoff A. (2000) Mineralogical characterization of primitive, type 3 lithologies in Rumuruti chondrites. *Meteorit. Planet. Sci.* **35**, 699–706.

Bischoff A. (2001) Fantastic new chondrites, achondrites, and lunar meteorites as the result of recent meteorite search expeditions in hot and cold deserts. *Earth Moon Planet.* **85**, 87–97.

Bischoff A. and Keil K. (1984) Al-rich objects in ordinary chondrites: related origin of carbonaceous and ordinary chondrites and their constituents. *Geochim. Cosmochim. Acta* **48**, 693–709.

Bischoff A., Keil K., and Stöffler D. (1985) Perovskite–hibonite–spinel-bearing inclusions and Al-rich chondrules

and fragments in enstatite chondrites. *Chem. Erde* **44**, 97–106.
Bischoff A., Palme H., Schultz L., Weber D., Weber H. W., and Spettel B. (1993) Acfer 182 and paired samples, an iron-rich carbonaceous chondrite: similarities with ALH85085 and relationship to CR chondrites. *Geochim. Cosmochim. Acta* **57**, 2631–2648.
Bland P. A., Cressey G., Alard O., Rogers N. W., Forder S. D., and Gounelle M. (2002) Modal mineralogy of carbonaceous chondrites and chemical variation in chondrite matrix. In *Lunar Planet. Sci.* **XXXIII**, #1754. The Lunar and Planetary Institute, Houston (CD-ROM).
Bogard D. D. (1995) Impact ages of meteorites: a synthesis. *Meteoritics* **30**, 244–268.
Boss A. P. (1996) A concise guide to chondrule formation models. In *Chondrules and the Protoplanetary Disk* (eds. R. H. Hewins, R. H. Jones, and E. R. D. Scott). Cambridge University Press, Cambridge, pp. 257–263.
Bradley J. P. (1994) Chemically anomalous, preaccretionary irradiated grains in interplanetary dust particles from comets. *Science* **265**, 925–929.
Bradley J. P., Keller L. P., Snow T. P., Hanner M. S., Flynn G. J., Gezo J. C., Clemette S. J., Brownlee D. E., and Bowey J. E. (1999) An infrared spectral match between GEMS and interstellar grains. *Science* **285**, 1716–1718.
Brearley A. J. (1989) Nature and origin of matrix in the unique type 3 chondrite, Kakangari. *Geochim. Cosmochim. Acta* **53**, 2395–2411.
Brearley A. J. (1991) Mineralogical and chemical studies of matrix in the Adelaide meteorite: a unique carbonaceous chondrite with affinities to ALHA77307 (CO3). *Lunar Planet. Sci.* **22**, 133–134.
Brearley A. J. (1993) Matrix and fine-grained rims in the unequilibrated CO3 chondrite, ALHA77307: origins and evidence for diverse, primitive nebular dust components. *Geochim. Cosmochim. Acta* **57**, 1521–1550.
Brearley A. J. (1995) Aqueous alteration and brecciation in Bells, an unusual, saponite-bearing, CM chondrite. *Geochim. Cosmochim. Acta* **59**, 2291–2317.
Brearley A. J. (1996) Nature of matrix in unequilibrated chondrites and its possible relationship to chondrules. In *Chondrules and the Protoplanetary Disk* (eds. R. H. Hewins, R. H. Jones, and E. R. D. Scott). Cambridge University Press, Cambridge, pp. 137–151.
Brearley A. J. (1999) Origin of graphitic carbon and pentlandite inclusions in matrix olivines in the Allende meteorite. *Science* **285**, 1380–1382.
Brearley A. J. and Jones R. H. (1998) Chondritic meteorites. In *Planetary Materials*, Reviews in Mineralogy (ed. J. J. Papike). Mineralogical Society of America, Washington, DC, Volume 36, pp. 3-1–3-398.
Brearley A. J., Scott E. R. D., Keil K., Clayton R. N., Mayeda T. K., Boynton W. V., and Hill D. H. (1989) Chemical, isotopic and mineralogical evidence for the origin of matrix in ordinary chondrites. *Geochim. Cosmochim. Acta* **53**, 2081–2093.
Brearley A. J., Bajt S., and Sutton S. (1995) Distribution of moderately volatile trace elements in fine-grained chondrule rims in the unequilibrated CO3 chondrite, ALH A77307. *Geochim. Cosmochim. Acta* **59**, 4307–4316.
Brearley A. J., Hutcheon I. D., and Browning L. (2001) Compositional zoning and Mn–Cr systematics in carbonates from the Y791198 CM2 carbonaceous chondrite. In *Lunar Planet. Sci.* **XXXII**, #1458. The Lunar and Planetary Institute, Houston (CD-ROM).
Brenker F. E., Palme H., and Klerner S. (2000) Evidence for solar nebula signatures in the matrix of the Allende meteorite. *Earth Planet. Sci. Lett.* **178**, 185–194.
Bridges J. C., Franchi I. A., Hutchison R., Sexton A. S., and Pillinger C. T. (1998) Correlated mineralogy, chemical compositions, oxygen isotopic compositions and size of chondrules. *Earth Planet. Sci. Lett.* **155**, 183–196.

Bunch T. E., Schultz P., Cassen P., Brownlee D., Podolak M., Lissauer J., Reynolds R., and Chang S. (1991) Are some chondrule rims formed by impact processes? Observations and experiments. *Icarus* **91**, 76–92.
Burbine T. H., McCoy T. J., Meibom A., Gladman B., and Keil K. (2002) Meteoritic parent bodies: their number and identification. In *Asteroids III* (eds. W. F. Bottke, Jr., A. Cellino, P. Paolicchi, and R. P. Binzel). University of Arizona Press, Tucson, pp. 653–666.
Buseck P. R. and Hua X. (1993) Matrices of carbonaceous chondrite meteorites. *Ann. Rev. Earth Planet. Sci.* **21**, 255–305.
Busemann H., Bauer H., and Wieler R. (2002) Phase Q—a carrier for subsolar gases. In *Lunar Planet. Sci.* **XXXIII**, #1462. The Lunar and Planetary Institute, Houston (CD-ROM).
Campbell A. J., Humayun M., Meibom A., Krot A. N., and Keil K. (2001) Origin of zoned metal grains in the QUE94411 chondrite. *Geochim. Cosmochim. Acta* **65**, 163–180.
Campbell A. J., Humayun M., and Weisberg M. K. (2002) Siderophile element constraints on the formation of metal in the metal-rich chondrites Bencubbin, Weatherford, and Gujba. *Geochim. Cosmochim. Acta* **66**, 647–660.
Campins H. and Swindle T. D. (1998) Expected characteristics of cometary meteorites. *Meteorit. Planet. Sci.* **33**, 1201–1211.
Cassen P. (1996) Overview of models of the solar nebula: potential chondrule-forming events. In *Chondrules and the Protoplanetary Disk* (eds. R. H. Hewins, R. H. Jones, and E. R. D. Scott). Cambridge University Press, Cambridge, pp. 21–28.
Cassen P. (2001a) Unresolved questions regarding the origins of Solar System solids. *Phil. Trans. Roy. Soc. London A* **359**, 1935–1946.
Cassen P. (2001b) Nebular thermal evolution and the properties of primitive planetary materials. *Meteorit. Planet. Sci.* **36**, 671–700.
Chizmadia L. J. and Brearley A. J. (2003) Mineralogy and textural characteristics of fine-grained rims in Yamato 791198 CM2 carbonaceous chondrite: constraints on the location of aqueous alteration. In *Lunar Planet. Sci.* **XXXIV**, #1419. The Lunar and Planetary Institute, Houston (CD-ROM).
Chizmadia L. J., Rubin A. E., and Wasson J. T. (2002) Mineralogy and petrology of amoeboid olivine inclusions in CO3 chondrites: relationship to parent-body aqueous alteration. *Meteorit. Planet. Sci.* **37**, 1781–1796.
Choi B. G., McKeegan K. D., Krot A. N., and Wasson J. T. (1998) Extreme oxygen-isotope compositions in magnetite from unequilibrated ordinary chondrites. *Nature* **392**, 577–579.
Choi B. G., Krot A. N., and Wasson J. T. (2000) Oxygen isotopes in magnetite and fayalite in CV chondrites Kaba and Mokoia. *Meteorit. Planet. Sci.* **35**, 1239–1248.
Ciesla F. J. and Hood L. L. (2002) The nebular shock wave model for chondrule formation: shock processing in a particle-gas suspension. *Icarus* **158**, 281–293.
Ciesla F. J., Lauretta D. S., Cohen B. A., and Hood L. L. (2003) A nebular origin for chondritic fine-grained phyllosilicates. *Science* **299**, 549–552.
Clayton R. N. (1993) Oxygen isotopes in meteorites. *Ann. Rev. Earth Planet. Sci.* **21**, 115–149.
Clayton R. N. (2002) Self-shielding in the solar nebula. *Nature* **415**, 860–861.
Clayton R. N., Mayeda T. K., and Molini-Velsko C. A. (1985) Isotopic variations in solar system material: evaporation and condensation of silicates. In *Protostars and Planets II* (eds. D. C. Black and M. S. Matthews). Univeristy of Arizona Press, Tucson, pp. 755–771.
Cohen B. A. and Coker R. F. (2000) Modeling of liquid water on CM meteorite parent bodies and implications for amino acid racemization. *Icarus* **145**, 369–381.
Connolly H. C., Jr. and Hewins R. H. (1995) Chondrules as products of dust collisions with totally molten droplets

within a dust-rich nebular environment: an experimental investigation. *Geochim. Cosmochim. Acta* **59**, 3131–3246.

Connolly H. C., Jr. and Love S. G. (1998) The formation of chondrules: petrologic tests of the shock wave model. *Science* **280**, 62–67.

Connolly H. C., Jr., Jones B. D., and Hewins R. H. (1998) The flash melting of chondrules: an experimental investigation into the melting history and physical nature of chondrule precursors. *Geochim. Cosmochim. Acta* **62**, 2725–2735.

Connolly H. C., Jr., Huss G. R., and Wasserburg G. J. (2001) On the formation of Fe–Ni metal in Renazzo-like carbonaceous chondrites. *Geochim. Cosmochim. Acta* **65**, 4567–4588.

Cosarinsky M., Leshin L. A., MacPherson G. J., Krot A. N., and Guan Y. (2003) Oxygen isotopic composition of Ca–Fe-rich silicates in and around an Allende Ca–Al-rich inclusion. In *Lunar Planet. Sci.* **XXXIV**, #1043. The Lunar and Planetary Institute, Houston (CD-ROM).

Cuzzi J. N., Hogan R. C., Paque J. M., and Dobrovolskis A. R. (2001) Size-selective concentration of chondrules and other small particles in protoplanetary nebula turbulence. *Astrophys. J.* **546**, 496–508.

Desch S. J. and Connolly H. C., Jr. (2002) A model of the thermal processing of particles in solar nebula shocks: application to the cooling rates of chondrules. *Meteorit. Planet. Sci.* **37**, 183–207.

Desch S. J. and Cuzzi J. N. (2000) The generation of lightning in the solar nebula. *Icarus* **143**, 87–105.

Ebel D. S. and Grossman L. (2000) Condensation in dust enriched systems. *Geochim. Cosmochim. Acta* **64**, 339–366.

Eisenhour D. D. and Buseck P. R. (1995) Chondrule formation by radiative heating: a numerical model. *Icarus* **117**, 197–211.

Endress M., Keil K., Bischoff A., Spettel B., Clayton R. N., and Mayeda T. K. (1994) Origin of the dark clasts in the Acfer 059/El Djouf 001 CR2 chondrite. *Meteoritics* **29**, 26–40.

Endress M., Zinner E., and Bischoff A. (1996) Early aqueous activity on primitive meteorite bodies. *Nature* **379**, 701–703.

Engrand C. and Maurette M. (1998) Carbonaceous micrometeorites from Antarctica. *Meteorit. Planet. Sci.* **33**, 565–580.

Engrand C., McKeegan K. D., and Leshin L. A. (1999) Oxygen isotopic compositions of individual minerals in Antarctic micrometeorites: further links to carbonaceous chondrites. *Geochim. Cosmochim. Acta* **63**, 2623–2636.

Fagan T. J., Krot A. N., and Keil K. (2000) Calcium, aluminum-rich inclusions in enstatite chondrites (I): Mineralogy and textures. *Meteorit. Planet. Sci.* **35**, 771–783.

Fagan T. J., Yurimoto H., Krot A. N., and Keil K. (2002a) Contrasting oxygen isotopic evolution of fine and coarse refractory inclusions from the CV3 Efremovka. *NIPR Antarct. Symp. Meteorit.* **27**, 15–17.

Fagan T. J., Yurimoto H., Krot A. N., and Keil K. (2002b) Constraints on oxygen isotopic evolution from an amoeboid olivine aggregate and Ca, Al-rich inclusion from the CV3 Efremovka. In *Lunar Planet. Sci.* **XXXIII**, #1507. The Lunar and Planetary Institute, Houston (CD-ROM).

Folco L., Mellini M., and Pillinger C. T. (1997) Equilibrated ordinary chondrites: constraints for thermal history from iron-magnesium ordering in orthopyroxene. *Meteorit. Planet. Sci.* **32**, 567–575.

Franchi I. A., Baker l., Bridges J. C., Wright I. P., and Pillinger C. T. (2001) Oxygen isotopes and the early solar system. *Phil. Trans. Roy. Soc. London A* **359**, 2019–2035.

Gaffey M. J. and Gilbert S. L. (1998) Asteroid 6 Hebe: the probable parent body of the H-type ordinary chondrites and the IIE irons. *Meteorit. Planet. Sci.* **33**, 1281–1295.

Ganguly J. and Tirone M. (2001) Relationship between cooling rate and cooling age of a mineral: theory and applications to meteorites. *Meteorit. Planet. Sci.* **36**, 167–176.

Genge M. J. (2000) Chondrule formation by the ablation of small planetesimals. *Meteorit. Planet. Sci.* **35**, 1143–1150.

Gilmour J. D. (2002) The solar system's first clocks. *Science* **297**, 1658–1659.

Gooding J. L., Keil K., Fukuoka T., and Schmitt R. A. (1980) Elemental abundances in chondrules from unequilibrated chondrites: evidence for chondrule origin by melting of preexisting materials. *Earth Planet Sci. Lett.* **50**, 171–180.

Göpel C., Manhès G., and Allègre C. J. (1994) U–Pb systematics of phosphates from equilibrated ordinary chondrites. *Earth Planet. Sci. Lett.* **121**, 153–171.

Goswami J. N., Marhas K. K., and Sahijpal S. (2001) Did solar energetic particles produce the short-lived nuclides present in the early solar system? *Astrophys. J.* **549**, 1151–1159.

Gounelle M. and Zolensky M. E. (2001) A terrestrial origin for the sulfate veins in CI1 chondrites. *Meteorit. Planet. Sci.* **36**, 1321–1329.

Grady M. M. (2000) *Catalogue of Meteorites*. Cambridge University Press, Cambridge, England, 689pp.

Graf T. and Marti K. (1994) Collisional records in LL-chondrites. *Meteoritics* **29**, 643–648.

Graf T. and Marti K. (1995) Collisional history of H chondrites. *J. Geophys. Res.* **100**, 21247–21264.

Greshake A. (1997) The primitive matrix components of the unique carbonaceous chondrite Acfer 094: a TEM study. *Geochim. Cosmochim. Acta* **61**, 437–452.

Greshake A., Hoppe P., and Bischoff A. (1996) Mineralogy, chemistry, and oxygen isotopes of refractory inclusions from stratospheric interplanetary dust particles and micrometeorites. *Meteorit. Planet. Sci.* **31**, 739–748.

Greshake A., Krot A. N., Meibom A., Weisberg M. K., Zolensky M. E., and Keil K. (2002) Heavily-hydrated lithic clasts in CH chondrites and the related, metal-rich chondrites Queen Alexandra Range 94411 and Hammadah al Hamra 237. *Meteorit. Planet. Sci.* **37**, 281–293.

Grimm R. E. (1985) Penecontemporaneous metamorphism, fragmentation, and reassembly of ordinary chondrite asteroids. *J. Geophys. Res.* **90**, 2022–2028.

Grossman J. N. (1996) Chemical fractionations of chondrites. In *Chondrules and the Protoplanetary Disk* (eds. R. H. Hewins, R. H. Jones, and E. R. D. Scott). Cambridge University Press, Cambridge, pp. 243–253.

Grossman J. N., Rubin A. E., Nagahara H., and King E. A. (1988) Properties of chondrules. In *Meteorites and the Early Solar System* (eds. J. F. Kerridge and M. S. Matthews). University of Arizona Press, Tucson, pp. 619–659.

Grossman J. N., Alexander C. M. O'D., Wang J. H., and Brearley A. J. (2000) Bleached chondrules: evidence for widespread aqueous processes on the parent asteroids of ordinary chondrites. *Meteorit. Planet. Sci.* **35**, 467–486.

Grossman J. N., Alexander C. M. O'D., Wang J. H., and Brearley A. J. (2002) Zoned chondrules in Semarkona: evidence for high- and low-temperature processing. *Meteorit. Planet. Sci.* **37**, 49–73.

Grossman L. and Larimer J. W. (1974) Early chemical history of the solar system. *Rev. Geophys. Space Phys.* **12**, 71–101.

Grossman L. and Olsen E. (1974) Origin of high-temperature fraction of C2 chondrites. *Geochim. Cosmochim. Acta* **38**, 178–187.

Grossman L. and Steele I. M. (1976) Amoeboid olivine aggregates in the Allende meteorite. *Geochim. Cosmochim. Acta* **40**, 149–155.

Grossman L., Ganapathy R., Methot R. L., and Davis A. M. (1979) Trace elements in the Allende meteorite amoeboid olivine aggregates. *Geochim. Cosmochim. Acta* **43**, 817–829.

Guan Y., McKeegan K. D., and MacPherson G. J. (2000) Oxygen isotopes in calcium–aluminum-rich inclusions from enstatite chondrites: new evidence for a single CAI source in the solar nebula. *Earth Planet. Sci. Lett.* **181**, 271–277.

Guimon R. K., Symes S. J. K., Sears D. W. G., and Benoit P. H. (1995) Chemical and physical studies of type 3 chondrites XII: the metamorphic history of CV chondrites and their components. *Meteoritics* **30**, 704–714.

Haack H., Warren P. H., and Rasmussen K. L. (1990) Effects of regolith/megaregolith insulation on the cooling histories of differentiated asteroids. *J. Geophys. Res.* **95**, 5111–5124.

Haack H., Taylor G. J., Scott E. R. D., and Keil K. (1992) Thermal history of chondrites: hot accretion vs. metamorphic reheating. *Geophys. Res. Lett.* **19**, 2235–2238.

Hallenbeck S. L., Nuth J. A., and Nelson R. N. (2000) Evolving optical properties of annealing silicate grains: from amorphous condensate to crystalline material. *Astrophys. J.* **535**, 247–255.

Hanner M. S. (1999) The silicate material in comets. *Space Sci. Rev.* **90**, 99–108.

Hanowski N. P. and Brearley A. J. (2000) Iron-rich aureoles in the CM carbonaceous chondrites, Murray, Murchison, and Allan Hills 81002: evidence for *in situ* aqueous alteration. *Meteorit. Planet. Sci.* **35**, 1291–1308.

Hanowski N. P. and Brearley A. J. (2001) Aqueous alteration of chondrules in the CM carbonaceous chondrite, Allan Hills 81002: implications for parent body alteration. *Geochim. Cosmochim. Acta* **65**, 495–518.

Harker D. E. and Desch S. L. (2002) Annealing of silicate dust by nebular shocks at 10 AU. *Astrophys. J.* **565**, L109–L112.

Hashimoto A. and Grossman L. (1987) Alteration of Al-rich inclusions inside amoeboid olivine aggregates in the Allende meteorite. *Geochim. Cosmochim. Acta* **51**, 1685–1704.

Hewins R. H. (1997) Chondrules. *Ann. Rev. Earth Planet. Sci.* **25**, 61–83.

Hewins R. H. and Connolly H. C., Jr. (1996) Peak temperatures of flash-melted chondrules. In *Chondrules and the Protoplanetary Disk* (eds. R. H. Hewins, R. H. Jones, and E. R. D. Scott). Cambridge University Press, Cambridge, pp. 197–204.

Hewins R. H. and Herzberg C. T. (1996) Nebular turbulence, chondrule formation, and the composition of the Earth. *Earth Planet. Sci. Lett.* **144**, 1–7.

Hewins R. H., Jones R. H., and Scott E. R. D. (1996) *Chondrules and the Protoplanetary Disk*. Cambridge University Press, Cambridge, 346p.

Hewins R. H., Yu Y., Zanda B., and Bourot-Denise M. (1997) Do nebular fractionations, evaporative losses, or both, influence chondrule compositions? *Antarct. Meteorit. Res.* **10**, 275–298.

Hill H. G. M., Grady C. A., Nuth J. A., Hallenbeck S. L., and Sitko M. L. (2001) Constraints on nebular dynamics and chemistry based on observations of annealed magnesium silicate grains in comets and disks surrounding Herbig Ae/Be stars. *Proc. Natl. Acad. Sci.* **98**, 2182–2187.

Hiroi T., Zolensky M. E., and Pieters C. M. (2001) The Tagish Lake meteorite: a possible sample from a D-type asteroid. *Science* **293**, 2234–2236.

Hiyagon H. (1998) Distribution of oxygen isotopes in and around some refractory inclusions from Allende. In *Lunar Planet. Sci.* XXIX, #1582. The Lunar and Planetary Institute, Houston (CD-ROM).

Hiyagon H. (2000) An ion microprobe study of oxygen isotopes in some inclusions in Kainsaz and Y-81020 CO3 chondrites. *Symp. Antarct. Meteorit.* **25**, 19–21.

Hiyagon H. and Hashimoto A. (1999a) ^{16}O excesses in olivine inclusions in Yamato-86009 and Murchison chondrites and their relation to CAIs. *Science* **283**, 828–831.

Hiyagon H. and Hashimoto A. (1999b) An ion microprobe study of oxygen isotopes in various types of inclusions in Yamato-791717 (CO3) chondrite: oxygen isotopes vs. fayalite content in olivine. *Symp. Antarct. Meteorit.* **24**, 37–39.

Hohenberg C. M., Meshik A. P., Pravdivtseva O. V., and Krot A. N. (2001) I–Xe dating: dark inclusions from Allende CV3. *Meteorit. Planet. Sci.* **36**, A83.

Hong Y. and Fegley B., Jr. (1998) Experimental studies of magnetite formation in the solar nebula. *Meteorit. Planet. Sci.* **33**, 1101–1112.

Hood L. (1998) Thermal processing of chondrule precursors in planetesimal bow shocks. *Meteorit. Planet. Sci.* **33**, 97–107.

Hsu W. and Crozaz G. (1998) Mineral chemistry and the origin of enstatite in unequilibrated enstatite chondrites. *Geochim. Cosmochim. Acta* **62**, 1993–2004.

Hsu W., Wasserburg G. J., and Huss G. R. (2000) High time resolution by use of the ^{26}Al chronometer in the multi-stage formation of a CAI. *Earth Planet. Sci. Lett.* **182**, 15–29.

Hua X., Wang J., and Buseck P. R. (2002) Fine-grained rims in the Allan Hills 81002 and Lewis Cliff 90500 CM2 meteorites: their origin and modification. *Meteorit. Planet. Sci.* **37**, 229–244.

Huss G. R., Keil K., and Taylor G. J. (1981) The matrices of unequilibrated ordinary chondrites: implications for the origin and history of chondrites. *Geochim. Cosmochim. Acta* **45**, 33–51.

Huss G. R., MacPherson G. J., Wasserburg G. J., Russell S. S., and Srinivasan G. (2001) Aluminum-26 in calcium–aluminum-rich inclusions and chondrules from unequilibrated ordinary chondrites. *Meteorit. Planet. Sci.* **36**, 975–997.

Hutcheon I. D. and Jones R. H. (1995) The ^{26}Al–^{26}Mg record of chondrules: clues to nebular chronology. In *Lunar Planet. Sci.* XXVI. The Lunar and Planetary Institute, Houston, pp. 647–648.

Hutcheon I. D., Huss G. R., and Wasserburg G. J. (1994) A search for ^{26}Al in chondrites: chondrule formation time scales. In *Lunar Planet. Sci.* XXV. The Lunar and Planetary Institute, Houston, pp. 587–588.

Hutcheon I. D., Krot A. N., Keil K., Phinney D. L., and Scott E. R. D. (1998) ^{53}Mn–^{53}Cr dating of fayalite formation in the CV3 chondrite Mokoia: evidence for asteroidal alteration. *Science* **282**, 1865–1867.

Hutcheon I. D., Krot A. N., and Ulyanov A. A. (2000) ^{26}Al in anorthite-rich chondrules in primitive carbonaceous chondrites: evidence chondrules post-date CAI. In *Lunar Planet. Sci.* XXXI, #1869. The Lunar and Planetary Institute, Houston (CD-ROM).

Hutchison R., Bevan A. W. R., Agrell S. O., and Ashworth J. R. (1979) Accretion temperature of the Tieschitz, H3, chondritic material. *Nature* **280**, 116–119.

Hutchison R., Alexander C. M. O'D., and Bridges J. C. (1998) Elemental redistribution in Tieschitz and the origin of white matrix. *Meteorit. Planet. Sci.* **33**, 1169–1179.

Hutchison R., Williams I. P., and Russell S. S. (2001) Theories of planetary formation: constraints from the study of meteorites. *Phil. Trans. Roy. Soc. London A* **359**, 2077–2091.

Iida A., Nakamoto T., Susa H., and Nakagawa Y. (2001) A shock heating model for chondrule formation in a protoplanetary disk. *Icarus* **153**, 430–450.

Ikeda Y., Kimura M., Mori H., and Takeda H. (1981) Chemical compositions of matrices of unequilibrated ordinary chondrites. *Mem. Nat. Inst. Polar Res. Spec. Issue* **20**, 136–154.

Imai H. and Yurimoto H. (2000) Oxygen and magnesium isotopic distributions in a Type C CAI from the Allende meteorite. In *Lunar Planet. Sci.* XXXI, #1510. The Lunar and Planetary Institute, Houston (CD-ROM).

Imai H. and Yurimoto H. (2001) Two generations of olivine-growth in an amoeboid olivine aggregate from the Allende meteorite. In *Lunar Planet. Sci.* XXXII, #1580. The Lunar and Planetary Institute, Houston (CD-ROM).

Ireland T. R. and Fegley B., Jr. (2000) The solar system's earliest chemistry: systematics of refractory inclusions. *Int. Geol. Rev.* **42**, 865–894.

Ito M., Yurimoto H., and Nagasawa H. (1999) Oxygen isotope microdistribution *vs.* composition in melilite/fassaite in the Allende CAI 7R-19-1: a new evidence for multiple heating. In *Lunar Planet. Sci.* XXX, #1538. The Lunar and Planetary Institute, Houston (CD-ROM).

Itoh D. and Tomeoka K. (2003) Dark inclusions in CO3 chondrites: new indicators of parent-body processes. *Geochim. Cosmochim. Acta* **67**, 153–169.

Itoh S., Kojima H., and Yurimoto H. (2000) Petrography and oxygen isotope chemistry of calcium–aluminum-rich inclusions in CO chondrites. In *Lunar Planet. Sci.* **XXXI**, #1323. The Lunar and Planetary Institute, Houston (CD-ROM).

Itoh S., Rubin A. E., Kojima H., Wasson J. T., and Yurimoto H. (2002) Amoeboid olivine aggregates and AOA-bearing chondrule from Y-81020 CO 3.0 chondrite: distribution of oxygen and magnesium isotopes. In *Lunar Planet. Sci.* **XXXIII**, #1490. The Lunar and Planetary Institute, Houston (CD-ROM).

Jarosewich E. (1990) Chemical analyses of meteorites: a compilation of stony and iron meteorite analyses. *Meteoritics* **25**, 323–337.

Johnson C. A., Prinz M., Weisberg M. K., Clayton R. N., and Mayeda T. K. (1990) Dark inclusions in Allende, Leoville, and Vigarano: evidence for nebular oxidation of CV3 constituents. *Geochim. Cosmochim. Acta* **54**, 819–830.

Jones R. H. (1990) Petrology and mineralogy of type II, FeO-rich chondrules in Semarkona (LL3.0): origin by closed system fractional crystallization, with evidence for supercooling. *Geochim. Cosmochim. Acta* **54**, 1785–1802.

Jones R. H. (1994) Petrology of FeO-poor, porphyritic pyroxene chondrules in the Semarkona chondrite. *Geochim. Cosmochim. Acta* **58**, 5325–5340.

Jones R. H. (1996a) FeO-rich, porphyritic pyroxene chondrules in unequilibrated ordinary chondrites. *Geochim. Cosmochim. Acta* **60**, 3115–3138.

Jones R. H. (1996b) Relict grains in chondrules: evidence for chondrules recycling. In *Chondrules and the Protoplanetary Disk* (eds. R. H. Hewins, R. H. Jones, and E. R. D. Scott). Cambridge University Press, Cambridge, pp. 163–172.

Jones R. H. and Brearley A. J. (1988) Kinetics of the clinopyroxene–orthopyroxene transition: constraints on the thermal histories of chondrules and type 3–6 chondrites. *Meteoritics* **23**, 277–278.

Jones R. H. and Lofgren G. E. (1993) A comparison of FeO-rich, porphyritic olivine chondrules in unequilibrated chondrites and experimental analogues. *Meteoritics* **28**, 213–221.

Jones R. H., Lee T., Connolly H. C., Jr., Love S. G., and Shang H. (2000a) Formation of chondrules and CAIs: theory vs. observation. In *Protostars and Planets IV* (eds. V. Mannings, A. P. Boss, and S. S. Russell). University of Arizona Press, Tucson, pp. 927–962.

Jones R. H., Saxton J. M., Lyon I. C., and Turner G. (2000b) Oxygen isotopes in chondrule olivine and isolated olivine grains. *Meteorit. Planet. Sci.* **35**, 849–857.

Jones T. D., Lebovsky L. A., Lewis J. S., and Marley M. S. (1990) The composition and origin of the C, P, and D asteroids: water as a tracer of thermal evolution in the outer belt. *Icarus* **88**, 172–192.

Kallemeyn G. W. and Wasson J. T. (1981) The compositional classification of chondrites: I. The carbonaceous chondrite groups. *Geochim. Cosmochim. Acta* **45**, 1217–1230.

Kallemeyn G. W., Rubin A. E., and Wasson J. T. (1991) The compositional classification of chondrites: V. The Karoonda (CK) group of carbonaceous chondrites. *Geochim. Cosmochim. Acta* **55**, 881–892.

Kallemeyn G. W., Rubin A. E., and Wasson J. T. (1994) The compositional classification of chondrites: VI. The CR carbonaceous chondrite group. *Geochim. Cosmochim. Acta* **58**, 2873–2888.

Kallemeyn G. W., Rubin A. E., and Wasson J. T. (1996) The compositional classification of chondrites: VII. The R chondrite group. *Geochim. Cosmochim. Acta* **60**, 2243–2256.

Keil K. (1989) Enstatite meteorites and their parent bodies. *Meteoritics* **24**, 195–208.

Keil K. (2000) Thermal alteration of asteroids: evidence from meteorites. *Planet. Space Sci.* **48**, 887–903.

Keil K., Haack H., and Scott E. R. D. (1994) Catastrophic fragmentation of asteroids: evidence from meteorites. *Planet. Space Sci.* **42**, 1109–1122.

Keil K., Stöffler D., Love S. G., and Scott E. R. D. (1997) Constraints on the role of impact heating and melting in asteroids. *Meteorit. Planet. Sci.* **32**, 349–363.

Keller L. P. and Buseck P. R. (1990) Matrix mineralogy of the Lancé CO3 carbonaceous chondrite. *Geochim. Cosmochim. Acta* **54**, 1155–1163.

Keller L. P., Clark J. C., Lewis C. F., and Moore C. B. (1992) Maralinga, a metamorphosed carbonaceous chondrite found in Australia. *Meteoritics* **27**, 87–91.

Keller L. P., Thomas K. L., Clayton R. N., Mayeda T. K., DeHart J. M., and McKay D. S. (1994) Aqueous alteration of the Bali CV3 chondrite: evidence from mineralogy, mineral chemistry, and oxygen isotopic compositions. *Geochim. Cosmochim. Acta* **58**, 5589–5598.

Kerr R. A. (2001) A meteoriticist speaks out, his rocks remain mute. *Science* **293**, 1581–1584.

Kerridge J. F. (1993) What can meteorites tell us about nebular conditions and processes during planetesimal accretion? *Icarus* **106**, 135–150.

Kerridge J. F. and Matthews M. S. (1988) *Meteorites and the Early Solar System*. Arizona State University, Tucson, pp. 1269.

Kim G. L., Yurimoto H., and Sueno S. (1998) ^{16}O-enriched melilite and anorthite coexisting with ^{16}O-depleted melilite in a CAI. In *Lunar Planet. Sci.* **XXIX**, #1344. The Lunar and Planetary Institute, Houston (CD-ROM).

Kimura M. and Ikeda Y. (1998) Hydrous and anhydrous alteration in Kaba and Mokoia CV chondrites. *Meteorit. Planet. Sci.* **33**, 1139–1146.

Kimura M., El Goresy A., Palme H., and Zinner E. (1993) Ca-, Al-rich inclusions in the unique chondrite ALH85085: petrology, chemistry, and isotopic compositions. *Geochim. Cosmochim. Acta* **57**, 2329–2359.

Kimura M., Hiyagon H., Palme H., Spettel B., Wolf D., Clayton R. N., Mayeda T. K., Sato T., Suzuki A., and Kojima H. (2002) Yamato 792947, 793408 and 82038: the most primitive H chondrites, with abundant refractory inclusions. *Meteorit. Planet. Sci.* **37**, 1417–1434.

Kita N. T., Nagahara H., Togashi S., and Morishita Y. (2000) A short formation period of chondrules in the solar nebula. *Geochim. Cosmochim. Acta* **48**, 693–709.

Klerner S. and Palme H. (2000) Large titanium/aluminum fractionation between chondrules and matrix in Renazzo and other carbonaceous chondrites. *Meteorit. Planet. Sci.* **35**, A89–A89.

Klöck W., Thomas K. L., McKay D. S., and Palme H. (1989) Unusual olivine and pyroxene composition in interplanetary dust and unequilibrated ordinary chondrites. *Nature* **339**, 126–128.

Kojima T. and Tomeoka K. (1996) Indicators of aqueous alteration and thermal metamorphism on the CV parent body: microtextures of a dark inclusion. *Geochim. Cosmochim. Acta* **60**, 2651–2666.

Kojima T., Tomeoka K., and Takeda H. (1993) Unusual dark clasts in the Vigarano CV3 carbonaceous chondrite: record of the parent body process. *Meteoritics* **28**, 649–658.

Komatsu M., Krot A. N., Petaev M. I., Ulyanov A. A., Keil K., and Miyamoto M. (2001) Mineralogy and petrography of amoeboid olivine aggregates from the reduced CV3 chondrites Efremovka, Leoville, and Vigarano: products of nebular condensation, accretion, and annealing. *Meteorit. Planet. Sci.* **36**, 629–641.

Kong P. and Palme H. (1999) Compositional and genetic relationship between chondrules, chondrule rims, metal, and matrix in the Renazzo chondrite. *Geochim. Cosmochim. Acta* **63**, 3673–3682.

Kong P., Mori T., and Ebihara M. (1997) Compositional continuity of enstatite chondrites and implications for heterogeneous accretion of the enstatite chondrite parent body. *Geochim. Cosmochim. Acta* **61**, 4895–4914.

Kong P., Deloule E., and Palme H. (2000) REE-bearing sulfide in Bishunpur (LL3.1)—a highly unequilibrated ordinary chondrite. *Earth Planet. Sci. Lett.* **177**, 1–7.

Kracher A., Keil K., Kallemeyn G. W., Wasson J. T., Clayton R. N., and Huss G. I. (1985) The Leoville acccretionary breccia. *Proc. 16th Lunar Planet. Sci. Conf.: J. Geophys. Res.* **90**, D123–D135.

Kring D. A. and Boynton W. V. (1990) Trace-element compositions of Ca-rich chondrules from Allende: relationships between refractory inclusions and ferromagnesian chondrules. *Meteoritics* **25**, 377.

Krot A. N. and Keil K. (2002) Anorthite-rich chondrules in CR and CH carbonaceous chondrites: genetic link between Ca, Al-rich inclusions and ferromagnesian chondrules. *Meteorit. Planet. Sci.* **37**, 91–111.

Krot A. N. and Rubin A. E. (1993) Chromite-rich mafic silicate chondrules in ordinary chondrites: formation by impact melting. In *Lunar Planet. Sci.* **XXXIII**. The Lunar and Planetary Institute, Houston, pp. 827–828.

Krot A. N. and Wasson J. T. (1995) Igneous rims on low-FeO and high-FeO chondrules in ordinary chondrites. *Geochim. Cosmochim. Acta* **59**, 4951–4966.

Krot A., Ivanova M. A., and Wasson J. T. (1993) The origin of chromitic chondrules and the volatility of Cr under a range of nebular conditions. *Earth Planet. Sci. Lett.* **119**, 569–584.

Krot A. N., Scott E. R. D., and Zolensky M. E. (1995) Mineralogical and chemical modification of components in CV3 chondrites: nebular or asteroidal processing? *Meteoritics* **30**, 748–775.

Krot A. N., Scott E. R. D., and Zolensky M. E. (1997) Origin of fayalitic olivine rims and lath-shaped matrix olivine in the CV3 chondrite Allende and its dark inclusions. *Meteorit. Planet. Sci.* **32**, 31–49.

Krot A. N., Petaev M. I., Scott E. R. D., Choi B.-G., Zolensky M. E., and Keil K. (1998a) Progressive alteration in CV3 chondrites: more evidence for asteroidal alteration. *Meteorit. Planet. Sci.* **33**, 1065–1085.

Krot A. N., Petaev M. I., and Meibom A. (1998b) Mineralogy and petrography of Ca–Fe-rich silicate rims around heavily altered Allende dark inclusions. In *Lunar Planet. Sci.* **XXIX**, #1555. The Lunar and Planetary Institute, Houston (CD-ROM).

Krot A. N., Petaev M. I., Zolensky M. E., Keil K., Scott E. R. D., and Nakamura K. (1998c) Secondary calcium–iron-rich minerals in the Bali-like and Allende-like oxidized CV3 chondrites and Allende dark inclusions. *Meteorit. Planet. Sci.* **33**, 623–645.

Krot A. N., Brearley A. J., Ulyanov A. J., Biryukov V. V., Swindle T. D., Keil K., Mittlefehldt D. W., Scott E. R. D., and Nakamura K. (1998d) Mineralogy, petrography, bulk chemical, iodine–xenon, and oxygen isotopic compositions of dark inclusions in the reduced CV3 chondrite Efremovka. *Meteorit. Planet. Sci.* **34**, 67–89.

Krot A. N., Sahijpal S., McKeegan K. D., Weber D., Greshake A., Ulyanov A. A., Hutcheon I. D., and Keil K. (1999) Mineralogy, aluminum–magnesium and oxygen isotope studies of the relict calcium–aluminum-rich inclusions in chondrules. *Meteorit. Planet. Sci.* **34**, A68–A69.

Krot A. N., McKeegan K. D., Russell S. S., Meibom A., Weisberg M. K., Zipfel J., Krot T. V., Fagan T. J., and Keil K. (2001a) Refractory Ca, Al-rich inclusions and Al-diopside-rich chondrules in the metal-rich chondrites Hammadah al Hamra 237 and QUE 94411. *Meteorit. Planet. Sci.* **36**, 1189–1217.

Krot A. N., Ulyanov A. A., Meibom A., and Keil K. (2001b) Forsterite-rich accretionary rims around Ca, Al-rich inclusions from the reduced CV3 chondrite Efremovka. *Meteorit. Planet. Sci.* **36**, 611–628.

Krot A. N., Hutcheon I. D., and Huss G. R. (2001c) Aluminum-rich chondrules and associated refractory inclusions in the unique carbonaceous chondrite Adelaide. *Meteorit. Planet. Sci.* **36**, A105–A106.

Krot A. N., Meibom A., Russell S. S., Alexander C. M. O., Jeffries T. E., and Keil K. (2001d) A new astrophysical setting for chondrule formation. *Science* **291**, 1776–1779.

Krot A. N., Huss G. R., and Hutcheon I. D. (2001e) Corundum–hibonite refractory inclusions from Adelaide: condensation or crystallization from melt? *Meteorit. Planet. Sci.* **36**, A105.

Krot A. N., Meibom A., Weisberg M. K., and Keil K. (2002a) The CR chondrite clan: implications for early solar system processes. *Meteorit. Planet. Sci.* **37**, 1451–1490.

Krot A. N., McKeegan K. D., Leshin L. A., MacPherson G. J., and Scott E. R. D. (2002b) Existence of an ^{16}O-rich gaseous reservoir in the solar nebula. *Science* **295**, 1051–1054.

Krot A. N., Hutcheon I. D., and Keil K. (2002c) Anorthite-rich chondrules in the reduced CV chondrites: evidence for complex formation history and genetic links between CAIs and ferromagnesian chondrules. *Meteorit. Planet. Sci.* **37**, 155–182.

Krot A. N., McKeegan K. D., Huss G. R., Liffman K., Sahijpal S., Hutcheon I. D., Srinivasan G., Bischoff A., and Keil K. (2003a) Aluminum–magnesium and oxygen isotope study of relict Ca–Al-rich inclusions in chondrules. *Earth Planet. Sci. Lett.* (in preparation).

Krot A. N., Petaev M. I., and Yurimoto H. (2003b) Low-Ca pyroxene in amoeboid olivine aggregates in primitive carbonaceous chondrites. In *Lunar Planet. Sci.* **XXXIV**, #1441. The Lunar and Planetary Institute, Houston (CD-ROM).

Krot A. N., Libourel G., and Chaussidon M. (2003c) Oxygen isotopic compositions of chondrules in CR chondrites. *Meteorit. Planet. Sci.* (in preparation).

Krot A. N., Libourel G., Goodrich C. A., Petaev M. I., and Killgore M. (2003d) Silica-rich igneous rims around magnesian chondrules in CR carbonaceous chondrites: evidence for fractional condensation during chondrule formation. In *Lunar Planet. Sci.* **XXXIV**, #1451. The Lunar and Planetary Institute, Houston (CD-ROM).

Krot A. N., Meibom A., Aléon J., McKeegan K. D., Russell S. S., Hezel D. C., Jeffries E., and Keil K. (2003e) Chondrules in the CB_b metal-rich chondrites, Hammadah al Hamra 237 and QUE 94411. *Meteorit. Planet. Sci.* (submitted).

Krot A. N., MacPherson G. J., Ulyanov A. A., and Petaev M. I. (2003f) Fine-grained, spinel-rich inclusions from the reduced CV chondrites Efremovka and Leoville: I. Mineralogy, Petrology and bulk chemistry. *Meteorit. Planet. Sci.* (submitted).

Krot A. N., Fagan T. J., Keil K., McKeegan K. D., Sahijpal S., Hutcheon I. D., Petaev M. I., and Yurimoto H. (2003g) Ca, Al-rich inclusions, amoeboid olivine aggregates, and Al-rich chondrules from the unique carbonaceous chondrite Acfer 094: I. Mineralogy and petrology. *Geochim. Cosmochim Acta* (submitted).

Kuebler K. E., McSween Jr. H. Y., Carlson W. D., and Hirsch D. (1999) Sizes and masses of chondrules and metal-troilite grains in ordinary chondrites: possible implications for nebular sorting. *Icarus* **141**, 96–106.

Kunihiro T., Nagashima K., and Yurimoto H. (2002) Distribution of oxygen isotopes in matrix from the Vigarano CV3 meteorite. In *Lunar Planet. Sci.* **XXXIII**, #1549. The Lunar and Planetary Institute, Houston (CD-ROM).

Kurat G. (1988) Primitive meteorites: an attempt towards unification. *Proc Trans. Roy. Soc. London A* **325**, 459–482.

Langenhorst F., Poirier J.-P., Deutsch A., and Hornemann U. (2002) Experimental approach to generate shock veins in single-crystal olivine by shear melting. *Meteorit. Planet. Sci.* **37**, 1541–1553.

Larimer J. W. (1967) Chemical fractionations in meteorites: I. Condensation of the elements. *Geochim. Cosmochim. Acta* **31**, 1215–1238.

Larimer J. W. and Anders E. (1967) Chemical fractionations in meteorites: II. Abundance patterns and their interpretation. *Geochim. Cosmochim. Acta* **31**, 1239–1270.

Larimer J. W. and Wasson J. T. (1988) Refractory lithophile elements. In *Meteorites and the Early Solar System* (eds. J. F. Kerridge and M. S. Matthews). University of Arizona Press, Tucson, pp. 394–415.

Lauretta D. S., Lodders K., Fegley B., and Kremser D. T. (1997) The origin of sulfide-rimmed metal grains in ordinary chondrites. *Earth Planet. Sci. Lett.* **151**, 289–301.

Lauretta D. S., Lodders K., and Fegley B. (1998) Kamacite sulfurization in the solar nebula. *Meteorit. Planet. Sci.* **33**, 821–833.

Lauretta D. S., Hua X., and Buseck P. R. (2000) Mineralogy of fine-grained rims in the ALH 81002 CM chondrite. *Geochim. Cosmochim. Acta* **64**, 3263–3273.

Lauretta D. S., Buseck P. R., and Zega T. J. (2001) Opaque minerals in the matrix of the Bishunpur (LL3.1) chondrite: constraints on the chondrule formation environment. *Geochim. Cosmochim. Acta* **65**, 1337–1353.

Lee M. R., Hutchison R., and Graham A. L. (1996) Aqueous alteration in the matrix of the Vigarano (CV3) carbonaceous chondrite. *Meteorit. Planet. Sci.* **31**, 477–483.

Leshin L. A., Rubin A. E., and McKeegan K. D. (1997) The oxygen isotopic composition of olivine and pyroxene from CI chondrites. *Geochim. Cosmochim. Acta* **61**, 835–845.

Liffman K. and Brown M. (1996) The motion and size sorting of particles ejected from a protostellar accretion disk. *Icarus* **116**, 275–290.

Liffman K. and Toscano M. (2000) Chondrule fine-grained mantle formation by hypervelocity impact of chondrules with a dusty gas. *Icarus* **143**, 106–125.

Lin Y. and Kimura M. (1998) Petrographic and mineralogical study of new EH melt rocks and a new enstatite chondrite grouplet. *Meteorit. Planet. Sci.* **33**, 501–511.

Lipschutz M. E., Gaffey M. J., and Pellas P. (1989) Meteoritic parent bodies: nature, number, size and relationship to present-day asteroids. In *Asteroids II* (eds. R. P. Binzel, T. Gehrels, and M. S. Matthews). University of Arizona Press, Tucson, pp. 740–777.

Lodders K. and Fegley B., Jr. (1998) *The Planetary Scientist's Companion*. Oxford University Press, New York, 371p.

Lodders K. and Osborne R. (1999) Perspectives on the comet-asteroid-meteorite link. *Space Sci. Rev.* **90**, 289–297.

Lofgren G. (1996) A dynamic crystallization model for chondrules melts. In *Chondrules and the Protoplanetary Disk* (eds. R. H. Hewins, R. H. Jones, and E. R. D. Scott). Cambridge University Press, Cambridge, pp. 187–196.

Luck J. M., Ben-Othman D., Barrat J. A., and Albarede F. (2003) Coupled ^{63}Cu and ^{16}O excesses in chondrites. *Geochim. Cosmochim. Acta* **67**, 143–151.

Lugmair G. W. and Shukolyukov A. (2001) Early solar system events and timescales. *Meteorit. Planet. Sci.* **36**, 1017–1026.

MacPherson G. J. and Delaney J. S. (1985) A fassaite-two olivine–pleonaste-bearing refractory inclusion from Karoonda. In *Lunar Planet. Sci.* **XVI**. The Lunar and Planetary Institute, Houston, pp. 515–516.

MacPherson G. J. and Russell S. S. (1997) Origin of aluminum-rich chondrules: constraints from major-element chemistry. *Meteorit. Planet. Sci.* **32**, A83.

MacPherson G. J., Grossman L., Hashimoto A., Bar-Matthews M., and Tanaka T. (1984) Petrographic studies of refractory inclusions from the Murchison meteorite. *Proc. 15th Lunar Planet. Sci. Conf.: J. Geophys. Res. Suppl.* **89**, C299–C312.

MacPherson G. J., Hashimoto A., and Grossman L. (1985) Accretionary rims on inclusions in the Allende meteorite. *Geochim. Cosmochim. Acta* **49**, 2267–2279.

MacPherson G. J., Wark D. A., and Armstrong J. T. (1988) Primitive material surviving in chondrites: refractory inclusions. In *Meteorites and the Early Solar System* (eds. J. F. Kerridge and M. S. Matthews). University of Arizona Press, Tucson, pp. 746–807.

MacPherson G. J., Davis A. M., and Grossman J. N. (1989) Refractory inclusions in the unique chondrite ALH85085. *Meteoritics* **24**, 297.

MacPherson G. J., Davis A. M., and Zinner E. K. (1995) The distribution of aluminum-26 in the early solar system. *Meteoritics* **30**, 365–386.

Marhas K. K., Goswami J. N., and Davis A. M. (2002) Short-lived nuclides in hibonite: evidence for solar-system evolution. *Science* **298**, 2182–2185.

Matsunami S., Ninagawa K., Nishimura S., Kubono N., Yamamoto I., Kohata M., Wada T., Yamashita Y., Lu J., Sears D. W. G., and Nishimura H. (1993) Thermoluminescence and compositional zoning in the mesostasis of a Semarkona group A1 chondrule and new insights into the chondrule-forming process. *Meteoritics* **57**, 2101–2110.

May C., Russell S. S., and Grady M. M. (1999) Analysis of chondrule and CAI size and abundance in CO3 and CV3 chondrites: a preliminary study. In *Lunar Planet. Sci.* **XXX**, #1688. The Lunar and Planetary Institute, Houston (CD-ROM).

McCoy T. J., Scott E. R. D., Jones R. H., Keil K., and Taylor G. J. (1991) Composition of chondrule silicates in LL3-5 chondrites and implications for their nebular history and parent body metamorphism. *Geochim. Cosmoschim. Acta* **55**, 601–619.

McKeegan K. D., Leshin L. A., Russell S. S., and MacPherson G. J. (1998) Oxygen isotopic abundances in calcium–aluminum-rich inclusions from ordinary chondrites: implications for nebular heterogeneity. *Science* **280**, 414–418.

McKeegan K. D., Chaussidon M., and Robert F. (2000a) Incorporation of short-lived ^{10}Be in a calcium–aluminum-rich inclusion from the Allende meteorite. *Science* **289**, 1334–1337.

McKeegan K. D., Greenwood J. P., Leshin L., and Cosarinsky M. (2000b) Abundance of ^{26}Al in ferromagnesian chondrules of unequilibrated ordinary chondrites. In *Lunar Planet. Sci.* **XXX**, #2009. The Lunar and Planetary Institute, Houston (CD-ROM).

McKeegan K. D., Sahijpal S., Krot A. N., Weber D., and Ulyanov A. A. (2003) Preservation of primary oxygen isotopic compositions in Ca, Al-rich inclusions from CH chondrites. *Earth Planet. Sci. Lett.* (submitted).

McSween H. Y., Jr. (1977) Petrographic variations among carbonaceous chondrites of the Vigarano type. *Geochim. Cosmochim. Acta* **41**, 1777–1790.

McSween H. Y., Jr. (1979) Are carbonaceous chondrites primitive or processed—a review. *Rev. Geophys. Space Phys.* **17**, 1059–1078.

McSween H. Y. and Labotka T. C. (1993) Oxidation during metamorphism of the ordinary chondrites. *Geochim. Cosmochim. Acta* **57**, 1105–1114.

McSween H. Y., Jr. and Patchen A. D. (1989) Pyroxene thermobarometry in LL chondrites and implications for parent body metamorphism. *Meteoritics* **24**, 219–226.

McSween H. Y., Jr. and Richardson S. M. (1977) The composition of carbonaceous chondrite matrix. *Geochim. Cosmochim. Acta* **41**, 1145–1161.

McSween H. Y., Jr., Ghosh A., Grimm R. E., Wilson L., and Young E. D. (2002) Thermal evolution models of asteroids. In *Asteroids III* (eds. W. F. Bottke, Jr., A. Cellino, P. Paolicchi, and R. P. Binzel) pp. 559–571.

Meibom A. and Clark B. E. (1999) Evidence for the insignificance of ordinary chondrite material in the asteroid belt. *Meteorit. Planet. Sci.* **34**, 7–24.

Meibom A., Petaev M. I., Krot A. N., Wood J. A., and Keil K. (1999) Primitive FeNi metal grains in CH carbonaceous chondrites formed by condensation from a gas of solar composition. *J. Geophys. Res.* **104**, 22053–22059.

Meibom A., Desch S. J., Krot A. N., Cuzzi J. N., Petaev M. I., Wilson L., and Keil K. (2000) Large-scale thermal events in the solar nebula: evidence from Fe,Ni metal grains in primitive meteorites. *Science* **288**, 839–841.

Meibom A., Petaev M. I., Krot A. N., Keil K., and Wood J. A. (2001) Growth mechanism and additional constraints on FeNi condensation in the solar nebula. *J. Geophys. Res.* **106**, 32797–32801.

Messenger S. and Bernatowicz T. J. (2000) Search for presolar silicates in Acfer 094. *Meteorit. Planet. Sci.* **35**, A109–A110.

Messenger S., Keller L. P., Stadermann F. J., Walker R. M., and Zinner E. (2003) Samples of stars beyond the solar system: silicate grains in interplanetary dust. *Science* **300**, 105–108.

Metzler K., Bischoff A., and Stöffler D. (1992) Accretionary dust mantles in CM chondrites: evidence for solar nebula processes. *Geochim. Cosmochim. Acta* **56**, 2873–2897.

Misawa K. and Fujita T. (1994) A relict refractory inclusion in a ferromagnesian chondrule from the Allende meteorite. *Nature* **368**, 723–726.

Misawa K. and Fujita T. (2000) Magnesium isotopic fractionations in barred olivine chondrules from the Allende meteorite. *Meteorit. Planet. Sci.* **35**, 85–94.

Misawa K. and Nakamura N. (1988) Highly fractionated rare-earth elements in ferromagnesian chondrules from the Felix (CO3) meteorite. *Nature* **334**, 47–49.

Misawa K. and Nakamura N. (1996) Origin of refractory precursor components of chondrules from carbonaceous chondrites. In *Chondrules and the Protoplanetary Disk* (eds. R. H. Hewins, R. H. Jones, and E. R. D. Scott). Cambridge University Press, Cambridge, pp. 99–105.

Morfill G. E., Durisen R. H., and Turner G. W. (1998) An accretion rim constraint on chondrule formation theories. *Icarus* **134**, 180–184.

Morlok A., Floss C., Zinner E., Bischoff A., Henkel T., Rost D., Stephan T., and Jessberger E. K. (2002) Trace elements in CI chondrites: a heterogeneous distribution. In *Lunar Planet. Sci.* **XXXIII**, #1269. The Lunar and Planetary Institute, Houston (CD-ROM).

Mostefaoui S., Kita N. T., Tachibana S., Nagahara H., Togashi S., and Morishita Y. (2000) Correlated initial aluminum-26/aluminum-27 with olivine and pyroxene proportions in chondrules from highly unequilibrated ordinary chondrites. *Meteorit. Planet. Sci.* **35**, A114.

Mostefaoui S., Kita N. T., Togashi S., Tachibana S., Nagahara H., and Morishita Y. (2002) The relative formation ages of ferromagnesian chondrules inferred from their initial aluminum-26/aluminum-27 ratios. *Meteorit. Planet. Sci.* **37**, 421–438.

Nagahara H. (1984) Matrices of type 3 ordinary chondrites—primitive nebular records. *Geochim. Cosmochim. Acta* **48**, 2581–2595.

Nagahara H., Kushiro I., Mysen B., and Mori H. (1988) Experimental vaporization and condensation of olivine solid solution. *Nature* **331**, 516–518.

Nagahara H., Kushiro I., and Mysen B. (1994) Evaporation of olivine: low pressure phase relations of the olivine system and its implication for the origin of chondritic components in the solar nebula. *Geochim. Cosmochim. Acta* **58**, 1951–1963.

Nagahara H., Kita N. T., Ozawa K., and Morishita Y. (1999) Condensation during chondrule formation: elemental and Mg isotopic evidence. In *Lunar Planet. Sci.* **XXX**, #1342. The Lunar and Planetary Institute, Houston (CD-ROM).

Nakamura T., Tomeoka K., and Takeda H. (1992) Shock effects of the Leoville CV carbonaceous chondrite: a transmission electron microscope study. *Earth Planet. Sci. Lett.* **114**, 159–170.

Nakamura T., Tomeoka K., Takaoka N., Sekine T., and Takeda H. (2000) Impact-induced textural changes of CV carbonaceous chondrites: experimental reproduction. *Icarus* **146**, 289–300.

Nakamura T., Noguchi T., Zolensky M. E., and Tanaka M. (2003) Mineralogy and noble-gas signatures of the carbonate-rich lithology of the Tagish Lake carbonaceous chondrite: evidence for an accretionary breccia. *Earth Planet. Sci. Lett.* **207**, 83–101.

Newsom H. E. and Drake M. J. (1979) The origin of metal clasts in the Bencubbin breccia. *Geochim. Cosmochim. Acta* **43**, 689–707.

Newton J., Bischoff A., Arden J. W., Franchi I. A., Geiger T., Greshake A., and Pillinger C. T. (1995) Acfer 094, a uniquely primitive carbonaceous chondrite from the Sahara. *Meteorit. Planet. Sci.* **30**, 47–56.

Newton J., Franchi I. A., and Pillinger C. T. (2000) The oxygen isotopic record in enstatite meteorites. *Meteorit. Planet. Sci.* **35**, 689–698.

Niemeyer S. (1988) Isotopic diversity in nebular dust: the distribution of Ti isotopic anomalies. *Geochim. Cosmochim. Acta* **52**, 2941–2954.

Nittler L. R., Alexander C. M. O'D., Gao X., Walker R. M., and Zinner E. K. (1994) Interstellar grains from the Tieschitz ordinary chondrite. *Nature* **370**, 443–446.

Noguchi T. (1993) Petrology and mineralogy of CK chondrites: implications for the metamorphism of the CK chondrite parent body. *Proc. NIPR Symp. Antarct. Meteorit.* **6**, 204–233.

Nuth J. A., Reitmeijer F. J. M., and Hill H. G. M. (2002) Condensation processes in astrophysical environments: the composition and structure of cometary grains. *Meteorit. Planet. Sci.* **37**, 1579–1590.

Okazaki R., Takaoka N., Nakamura T., and Nagao K. (2000) Cosmic-ray exposure ages of enstatite chondrites. *Antarct. Meteorit. Res.* **13**, 153–169.

Okazaki R., Takaoka N., Nagao K., Sekiya M., and Nakamura T. (2001) Noble-gas-rich chondrules in an enstatite meteorite. *Nature* **412**, 795–798.

Ott U. (1993) Interstellar grains in meteorites. *Nature* **364**, 25–33.

Ott U. (2001) Presolar grains in meteorites: an overview and some implications. *Planet. Space Sci.* **49**, 763–767.

Ozawa K. and Nagahara H. (2001) Chemical and isotopic fractionations by evaporation and their cosmochemical implications. *Geochim. Cosmochim. Acta* **65**, 2171–2199.

Palme H. (2000) Are there chemical gradients in the inner solar system? *Space Sci. Rev.* **92**, 237–262.

Palme H. (2001) Chemical and isotopic heterogeneity in protosolar matter. *Phil. Trans. Roy. Soc. London A* **359**, 2061–2075.

Palme H. and Boynton W. V. (1993) Meteoritic constraints on conditions in the solar nebula. In *Protostars and Planets III* (eds. E. H. Levy and J. I. Lunine). University of Arizona Press, Tucson, AZ, pp. 979–1004.

Palme H. and Fegley B. (1990) High temperature condensation of iron-rich olivine in the solar nebula. *Earth Planet. Sci. Lett.* **101**, 180–195.

Palme H. and Klerner S. (2000) Formation of chondrules and matrix in carbonaceous chondrites. *Meteorit. Planet. Sci.* **35**, A124–A124.

Palme H. and Wlotzka F. (1976) A metal particle from a Ca, Al-rich refractory inclusion from the meteorite, Allende, and the condensation of refractory siderophile elements. *Earth Planet. Sci. Lett.* **33**, 45–60.

Palme H., Larimer J. W., and Lipschutz M. E. (1988) Moderately volatile elements. In *Meteorites and the Early Solar System* (eds. J. F. Kerridge and M. S. Matthews). University of Arizona Press, Tucson, pp. 436–461.

Palme H., Kurat G., Spettel B., and Burghele A. (1989) Chemical composition of an unusual xenolith of the Allende meteorite. *Z. Naturforsch.* **44a**, 1005–1014.

Palme H., Hutcheon I. D., and Spettel B. (1994) Composition and origin of refractory-metal-rich assemblages in a Ca,Al-rich Allende inclusion. *Geochim. Cosmochim. Acta* **58**, 495–513.

Paque J. M. and Cuzzi J. N. (1997) Physical characteristics of chondrules and rims, and aerodynamic sorting in the solar nebula. *Lunar Planet. Sci.* **28**, 1071–1072.

Patzer A. and Schultz L. (2002a) Noble gases in enstatite chondrites: I. exposure ages, pairing, and weathering effects. *Meteorit. Planet. Sci.* **36**, 947–961.

Patzer A. and Schultz L. (2002b) Noble gases in enstatite chondrites: II. *Meteorit. Planet. Sci.* **37**, 601–612.

Petaev M. I. and Wood J. A. (2000) The condensation with partial isolation (CWPI) model of condensation in the solar nebula. *Meteorit. Planet. Sci.* **33**, 1123–1137.

Petaev M. I., Meibom A., Krot A. N., Wood J. A., and Keil K. (2001) The condensation origin of zoned metal grains in

Queen Alexandra Range 94411: implications for the formation of the Bencubbin-like chondrites. *Meteorit. Planet. Sci.* **36**, 96–106.

Podosek F. A., Ott U., Brannon J. C., Neal C. R., Bernatowicz T. J., Swan P., and Mahan S. E. (1997) Thoroughly anomalous chromium in Orgueil. *Meteorit. Planet. Sci.* **32**, 617–627.

Reisener R., Meibom A., Krot A. N., Goldstein, J. I., and Keil K. (2000) Microstructure of condensate Fe–Ni metal particles in the CH chondrite PAT91546. In *Lunar Planet. Sci.* **XXXI**, #1445. The Lunar and Planetary Institute, Houston (CD-ROM).

Rietmeijer F. J. M. (1998) Interplanetary dust particles. In *Planetary Materials*, Reviews in Mineralogy (ed. J. J. Papike), Mineralogical Society of America, Washington, DC, vol. 36, pp. 2-1–2-95.

Rietmeijer F. J. M. (2002) The earliest chemical dust evolution in the solar nebula. *Chem. Erde* **62**, 1–45.

Rietmeijer F. J. M., Hallenbeck S. L., Nuth J. A., and Karner J. M. (2002) Amorphous magnesiosilicate smokes annealed in vacuum: the evolution of magnesium silicates in circumstellar and cometary dust. *Icarus* **156**, 269–286.

Rubin A. E. (1990) Kamacite and olivine in ordinary chondrites: intergroup and intragroup relationships. *Geochim. Cosmochim. Acta* **54**, 1217–1232.

Rubin A. E. (1995) Petrologic evidence for collisional heating of chondritic asteroids. *Icarus* **113**, 156–167.

Rubin A. E. (2000) Petrologic, geochemical and experimental constraints on models of chondrule formation. *Earth Sci. Rev.* **50**, 3–27.

Rubin A. E. (2002) Post-shock annealing of Miller Range 99301 (LL6): implications for impact heating of ordinary chondrites. *Geochim. Cosmochim. Acta* **66**, 3327–3337.

Rubin A. E. and Brearley A. J. (1996) A critical evaluation of the evidence for hot accretion. *Icarus* **124**, 86–96.

Rubin A. E. and Wasson J. T. (1988) Chondrules in the Ornans CO3 meteorite and the timing of chondrule formation relative to nebular fractionation events. *Geochim. Cosmochim. Acta* **52**, 425–432.

Rubin A. E., Scott E. R. D., and Keil K. (1997) Shock metamorphism of enstatite chondrites. *Geochim. Cosmochim. Acta* **61**, 847–858.

Rubin A. E., Sailer A. L., and Wasson J. T. (1999) Troilite in the chondrules of type 3 ordinary chondrites: implications for chondrule formation. *Geochim. Cosmochim. Acta* **63**, 2281–2298.

Rubin A. E., Ulff-Møller F., Wasson J. T., and Carlson W. D. (2001) The Portales Valley meteorite breccia: evidence for impact-induced melting and metamorphism of an ordinary chondrite. *Geochim. Cosmochim. Acta* **65**, 323–342.

Rubin A. E., Zolensky M. E., and Bodnar R. J. (2002) The halite-bearing Zag and Monahans (1998) meteorite breccias: shock metamorphism, thermal metamorphism and aqueous alteration on the H-chondrite parent body. *Meteorit. Planet. Sci.* **37**, 124–141.

Russell S. S., Srinivasan G., Huss G. R., Wasserburg G. J., and MacPherson G. J. (1996) Evidence for widespread ^{26}Al in the solar nebula and new constraints for nebula timescales. *Science* **273**, 757–762.

Russell S. S., Huss G. R., Fahey A. J., Greenwood R. C., Hutchison R., and Wasserburg G. J. (1998) An isotopic and petrologic study of calcium–aluminum-rich inclusions from CO3 meteorites. *Geochim. Cosmochim. Acta* **62**, 689–714.

Russell S. S., MacPherson G. J., Leshin L. A., and McKeegan K. D. (2000) ^{16}O enrichments in aluminum-rich chondrules from ordinary chondrites. *Earth Planet. Sci. Lett.* **184**, 57–74.

Ruzicka A., Snyder G. A., and Taylor L. A. (1998) Megachondrules and large, igneous-textured clasts in Julesberg (L3) and other ordinary chondrites: vapor-fractionation, shock-melting, and chondrule formation. *Geochim. Cosmochim. Acta* **62**, 1419–1442.

Ryerson F. J. and McKeegan K. D. (1994) Determination of oxygen self-diffusion in åkermanite, anorthite, diopside, and spinel: implications for oxygen isotopic anomalies and the thermal histories of Ca–Al-rich inclusions. *Geochim. Cosmochim. Acta* **58**, 3713–3734.

Sahijpal S., McKeegan K. D., Goswami J. N., and Davis A. M. (2000) Oxygen isotopic compositions of Murchison hibonites with wide-ranging radiogenic and neutron-rich stable isotope anomalies. In *Lunar Planet. Sci.* **XXXI**, #1502. The Lunar and Planetary Institute, Houston (CD-ROM).

Sanders I. S. (1996) A chondrule-forming scenario involving molten planetesimals. In *Chondrules and the Protoplanetary Disk* (eds. R. H. Hewins, R. H. Jones, and E. R. D. Scott). Cambridge University Press, Cambridge, pp. 327–334.

Sanders I. S. and Hevey P. J. (2002) ^{26}Al meltdown of planetesimals, and its implications for the age and chemistry of chondrules. *Meteorit. Planet. Sci.* **37**, A125–A125.

Schmitt R. T. (2000) Shock experiments with the H6 chondrite Kernouvé: pressure calibration of microscopic shock effects. *Meteorit. Planet. Sci.* **35**, 545–560.

Scott E. R. D. (1988) A new kind of primitive chondrite, Allan Hills 85085. *Earth Planet. Sci. Lett.* **91**, 1–18.

Scott E. R. D. (2002) Meteoritic evidence for the accretion and collisional evolution of asteroids. In *Asteroids III* (eds. W. F. Bottke Jr., A. Cellino, P. Paolicchi, R. P. Binzel). University of Arizona Press, Tucson, pp. 697–709.

Scott E. R. D. and Jones R. H. (1990) Disentangling nebular and asteroidal features of CO3 carbonaceous chondrites. *Geochim. Cosmochim. Acta* **54**, 2485–2502.

Scott E. R. D. and Krot A. N. (2001) Oxygen isotopic compositions and origins of calcium–aluminum-rich inclusions and chondrules. *Meteorit. Planet. Sci.* **36**, 1307–1319.

Scott E. R. D. and Newsom H. (1989) Planetary compositions—clues from meteorites and asteroids. *Z. Naturforsch.* **44a**, 924–934.

Scott E. R. D. and Taylor G. J. (1983) Chondrules and other components in C, O, and E chondrites: similarities in their properties and origins. *J. Geophys. Res.* **88**(suppl.), B275–B286.

Scott E. R. D., Barber D. J., Alexander C. M., Hutchison R., and Peck J. A. (1988) Primitive material surviving in chondrites: matrix. In *Meteorites and the Early Solar System* (eds. J. F. Kerridge and M. S. Matthews). University of Arizona Press, Tucson, pp. 718–745.

Scott E. R. D., Taylor G. J., Newsom H., Herbert F., Zolensky M., and Kerridge J. F. (1989) Chemical, thermal and impact processing of asteroids. In *Asteroids II* (eds. R. P. Binzel, T. Gehrels, and M. S. Matthews). University of Arizona Press, Tucson, pp. 701–739.

Scott E. R. D., Keil K., and Stöffler D. (1992) Shock metamorphism of carbonaceous chondrites. *Geochim. Cosmochim. Acta* **56**, 4281–4293.

Scott E. R. D., Jones R. H., and Rubin A. E. (1994) Classification, metamorphic history, and pre-metamorphic composition of chondrules. *Geochim. Cosmochim. Acta* **58**, 1203–1209.

Scott E. R. D., Love S. G., and Krot A. N. (1996) Formation of chondrules and chondrites in the protoplanetary nebula. In *Chondrules and the Protoplanetary Disk* (eds. R H. Hewins, R H. Jones, and E. R. D. Scott). Cambridge University Press, pp. 87–96.

Sears D. W. G. (1998) The case for rarity of chondrules and calcium–aluminum-rich inclusions in the early solar system and some implications for astrophysical models. *Astrophys. J.* **498**, 773–778.

Sears D. W. G. and Hasan F. A. (1987) The type three ordinary chondrites: a review. *Surv. Geophys.* **9**, 43–97.

Sears D. W. G., Batchelor D. J., Lu J., and Keck B. D. (1991) Metamorphism of CO and CO-like chondrites and comparisons with type 3 ordinary chondrites. *Proc. NIPR Symp. Antarct. Meteorit.* **4**, 319–343.

Sears D. W. G., Morse A. D., Hutchison R., Guimon R. K., Jie L., Alexander C. M. O'D., Benoit P. H., Wright I., Pillinger C., Xie T., and Lipschutz M. E. (1995a) Metamorphism and aqueous alteration in low petrographic type ordinary chondrites. *Meteoritics* **30**, 169–181.

Sears D. W. G., Huang S., and Benoit P. H. (1995b) Chondrule formation, metamorphism, brecciation, an important new primary chondrule group, and the classification of chondrules. *Earth Planet. Sci. Lett.* **131**, 27–39.

Sears D. W. G., Huang S., and Benoit P. H. (1996) Open-system behavior during chondrule formation. In *Chondrules and the Protoplanetary Disk* (eds. R. H. Hewins, R. H. Jones, and E. R. D. Scott). Cambridge University Press, Cambridge, pp. 221–231.

Sheng Y. J., Hutcheon I. D., and Wasserburg G. J. (1991) Origin of plagioclase-olivine inclusions in carbonaceous chondrites. *Geochim. Cosmochim. Acta* **55**, 581–599.

Shu F. H., Shang H., and Lee T. (1996) Toward an astrophysical theory of chondrites. *Science* **271**, 1545–1552.

Shu F. H., Shang H., Gounelle M., and Glassgold A. E. (2001) The origin of chondrules and refractory inclusions in chondritic meteorites. *Astrophys. J.* **548**, 1029–1050.

Simon S. B., Yoneda S., Grossman L., and Davis A. M. (1994) A $CaAl_4O_7$-bearing refractory spherule from Murchison: evidence for very high-temperature melting in the solar nebula. *Geochim. Cosmochim. Acta* **58**, 1937–1949.

Simon S. B., Davis A. M., Grossman L., and McKeegan K. D. (2002) A hibonite-corundum inclusion from Murchison: a first generation condensate from the solar nebula. *Meteorit. Planet. Sci.* **37**, 533–548.

Spurny P., Oberst J., and Heinlein D. (2003) Photographic observations of Neuschwanstein, a second meteorite from the orbit of the Pribram chondrite. *Nature* **423**, 151–153.

Srinivasan G., Huss G. R., and Wasserburg G. J. (2000a) A petrographic, chemical and isotopic study of calcium-aluminum-rich inclusions and aluminum-rich chondrules from the Axtell (CV3) chondrite. *Meteorit. Planet. Sci.* **35**, 1333–1354.

Srinivasan G., Krot A. N., and Ulyanov A. A. (2000b) Aluminum–magnesium systematics in anorthite-rich chondrules and calcium–aluminum-rich inclusions from the reduced CV chondrite Efremovka. *Meteorit. Planet. Sci.* **35**, A151–A152.

Stöffler D. (1997) Minerals in the deep Earth: a message from the asteroid belt. *Science* **278**, 1576–1577.

Stöffler D., Keil K., and Scott E. R. D. (1991) Shock metamorphism of ordinary chondrites. *Geochim. Cosmochim. Acta* **55**, 3845–3867.

Strebel R., Huth J., and Hoppe P. (2000) *In situ* location by cathodoluminescence and SIMS isotopic analyses of small corundum grains in the Krymka meteorite. In *Lunar Planet. Sci.* **XXXI**, #1585. The Lunar and Planetary Institute, Houston (CD-ROM).

Susa H. and Nakamoto T. (2002) On the maximal size of chondrules in shock wave heating model. *Astrophys. J.* **564**, L57–L60.

Symes S. J. K., Sears D. W. G., Huang S. X., and Benoit P. H. (1998) The crystalline lunar spherules: their formation and implications for the origin of meteoritic chondrules. *Meteorit. Planet. Sci.* **33**, 13–29.

Tanaka K. K., Tanaka H., Nakazawa K., and Nakagawa Y. (1998) Shock heating due to accretion of a clumpy cloud onto a protoplanetary disk. *Icarus* **134**, 137–154.

Taylor G. J., Scott E. R. D., and Keil K. (1983) Cosmic setting for chondrule formation. In *Chondrules and their Origin* (ed. E. A. King). The Lunar and Planetary Institute, Houston, pp. 262–278.

Taylor G. J., Maggiore P., Scott E. R. D., Rubin A. E., and Keil K. (1987) Original structures and fragmentation histories of asteroids: evidence from meteorites. *Icarus* **69**, 1–13.

Taylor S. R. (2001) *Solar System Evolution: A New Perspective*. Cambridge University Press, Cambridge, 460p.

Thiemens M. H. (1999) Mass-independent isotopic effects in planetary atmospheres and the early solar system. *Science* **283**, 341–345.

Tissandier L., Libourel G., and Robert F. (2002) Gas–melt interactions and their bearing on chondrule formation. *Meteorit. Planet. Sci.* **37**, 1377–1389.

Tomeoka K. and Buseck P. R. (1990) Phyllosilicates in the Mokoia CV carbonaceous chondrite: evidence for aqueous alteration in an oxidizing environment. *Geochim. Cosmochim. Acta* **54**, 1745–1754.

Tomeoka K. and Tanimura I. (2000) Phyllosilicate-rich chondrule rims in the Vigarano CV3 chondrite: evidence for parent-body processes. *Geochim. Cosmochim. Acta* **64**, 1971–1988.

Tomeoka K., Yamahana Y., and Sekine T. (1999) Experimental shock metamorphism of the Murchison CM carbonaceous chondrite. *Geochim. Cosmochim. Acta* **63**, 3683–3703.

Tomeoka K., Ohnishi I., and Nakamura N. (2001) Silicate darkening in the Kobe CK chondrite: evidence for shock metamorphism at high temperatures. *Meteorit. Planet. Sci.* **36**, 1535–1545.

Trieloff M., Jessberger E. K., Herrwerth I., Hopp J., Fiéni C., Ghélis, Bourot-Denise M., and Pellas P. (2003) Structure and thermal history of the H-chondrite asteroid revealed by thermochronometry. *Nature* **422**, 502–506.

Van Schmus W. R. and Wood J. A. (1967) A chemical-petrologic classification for the chondritic meteorites. *Geochim. Cosmochim. Acta* **31**, 747–765.

Wark D. and Boynton W. V. (2001) The formation of rims on calcium–aluminum-rich inclusions: Step I. Flash heating. *Meteorit. Planet. Sci.* **36**, 1135–1166.

Wasson J. T. (1985) *Meteorites: Their Record of Early Solar System History*. W. H. Freeman, New York.

Wasson J. T. (2000) Oxygen-isotopic evolution of the solar nebula. *Rev. Geophys. Space Phys.* **38**, 491–512.

Wasson J. T. and Kallemeyn G. W. (1988) Compositions of chondrites. *Phil. Trans. Roy. Soc. London A* **325**, 535–544.

Wasson J. T. and Rubin A. E. (2003) Ubiquitous low-FeO relict grains in type II chondrules and limited overgrowths on phenocrysts following the final melting events. *Geochim. Cosmochim. Acta* **67**, 2239–2250.

Wasson J. T., Krot A. N., Lee M. S., and Rubin A. E. (1995) Compound chondrules. *Geochim. Cosmochim. Acta* **59**, 1847–1869.

Wasson J. T., Yurimoto H., and Russell S. S. (2001) ^{16}O-rich melilite in CO3.0 chondrites: possible formation of common, ^{16}O-poor melilite by aqueous alteration. *Geochim. Cosmochim. Acta* **65**, 4539–4549.

Weber D. and Bischoff A. (1997) Refractory inclusions in the CR chondrite Acfer 059-El Djouf 001: petrology, chemical composition, and relationship to inclusion populations in other types of carbonaceous chondrites. *Chem. Erde* **57**, 1–24.

Weber D., Zinner E., and Bischoff A. (1995) Trace element abundances and magnesium, calcium, and titanium isotopic compositions of grossite-containing inclusions from the carbonaceous chondrite Acfer 182. *Geochim. Cosmochim. Acta* **59**, 803–823.

Weidenschilling S. J., Marzari F., and Hood L. L. (1998) The origin of chondrules at Jovian resonances. *Science* **279**, 681–684.

Weinbruch S., Palme H., Müller W. F., and El Goresy A. (1990) FeO-rich rims and veins in Allende forsterite: evidence for high-temperature condensation at oxidizing conditions. *Meteoritics* **25**, 115–125.

Weinbruch S., Palme H., and Spettel B. (2000) Refractory forsterite in primitive meteorites: condensates from the solar nebula?. *Meteorit Planet. Sci.* **35**, 161–171.

Weinbruch S., Müller W. F., and Hewins R. H. (2001) A transmission electron microscope study of exsolution and coarsening in iron-bearing clinopyroxene from synthetic analogues of chondrules. *Meteorit. Planet. Sci.* **36**, 1237–1248.

Weisberg M. K. and Prinz M. (1998) Fayalitic olivine in CV3 chondrite matrix and dark inclusions: a nebular origin. *Meteorit. Planet. Sci.* **33**, 1087–1099.

Weisberg M. K. and Prinz M. (2000) The Grosvenor Mountains 95577 CR1 chondrite and hydration of the CR chondrites. *Meteorit. Planet. Sci.* **35**, A168.

Weisberg M. K., Prinz M., and Nehru C. E. (1990) The Bencubbin chondrite breccia and its relationship to CR chondrites and the ALH85085 chondrite. *Meteoritics* **25**, 269–279.

Weisberg M. K., Prinz M., Kennedy A., and Hutcheon I. D. (1991) Trace elements in refractory-rich inclusions in CR2 chondrites. *Meteoritics* **26**, 407–408.

Weisberg M. K., Prinz M., Clayton R. N., and Mayeda T. K. (1993) The CR (Renazzo-type) carbonaceous chondrite group and its implications. *Geochim. Cosmochim. Acta* **57**, 1567–1586.

Weisberg M. K., Prinz M., and Fogel R. A. (1994) The evolution of enstatite and chondrules in unequilibrated enstatite chondrites. *Meteoritics* **29**, 362–373.

Weisberg M. K., Prinz M., Clayton R. N., Mayeda T. K., Grady M. M., Franchi I., Pillinger C. T., and Kallemeyn G. W. (1996) The K chondrite grouplet. *Geochim. Cosmochim. Acta* **60**, 4253–4263.

Weisberg M. K., Zolensky M. E., and Prinz M. (1997) Fayalitic olivine in matrix of the Krymka LL3.1 chondrite: vapor–solid growth in the solar nebula. *Meteorit. Planet. Sci.* **32**, 791–801.

Weisberg M. K., Prinz M., Clayton R. N., Mayeda T. K., Sugiura N., Zashu S., and Ebihara M. (2001) A new metal-rich chondrite grouplet. *Meteorit. Planet. Sci.* **36**, 401–418.

Wiens R. C., Huss G. R., and Burnett D. S. (1999) The solar oxygen isotopic composition: predictions and implications for solar nebula processes. *Meteorit. Planet. Sci.* **34**, 99–107.

Wieler R. (2002) Cosmic-ray-produced noble gases in meteorites. In *Noble Gases in Geochemistry and Cosmochemistry*, Rev. Mineral. Geochem. (eds. D. Porcelli, C. J. Ballentine, and R. Wieler) Mineralogical Society of America, Washington, DC, pp. 125–170.

Williams C. V., Keil K., Taylor G. J., and Scott E. R. D. (1999) Cooling rates of equilibrated clasts in ordinary chondrite regolith breccias: implications for parent body histories. *Chem. Erde* **59**, 287–305.

Wilson L., Keil K., Browning L. B., Krot A. N., and Bourcier W. (1999) Early aqueous alteration, explosive disruption, and reprocessing of asteroids. *Meteorit. Planet. Sci.* **34**, 541–557.

Wlotzka F., Palme H., Spettel B., Wänke H., Fredriksson K., and Noonan A. F. (1983) Alkali differentiation in LL chondrites. *Geochim. Cosmochim. Acta* **47**, 743–757.

Wolf D. and Palme H. (2001) The solar system abundance of phosphorus and titanium and the nebular volatility of phosphorus. *Meteorit. Planet. Sci.* **36**, 559–571.

Wood J. A. (1963) On the origin of chondrules and chondrites. *Icarus* **2**, 152–180.

Wood J. A. (1967) Chondrites: their metallic minerals, thermal histories, and parent planets. *Icarus* **6**, 1–49.

Wood J. A. (1988) Chondritic meteorites and the solar nebula. *Ann. Rev. Earth Planet. Sci.* **16**, 53–72.

Wood J. A. (1996a) Unresolved issues in the formation of chondrules and chondrites. In *Chondrules and the Protoplanetary Disk* (eds. R. H. Hewins, R. H. Jones, and E. R. D. Scott). Cambridge University Press, Cambridge, pp. 55–69.

Wood J. A. (1996b) Processing of chondritic and planetary material in spiral density waves in the nebula. *Meteorit. Planet. Sci.* **31**, 641–645.

Wood J. A. (2000a) Pressure and temperature profiles in the solar nebula. *Space Sci. Rev.* **92**, 87–93.

Wood J. A. (2000b) The beginning: swift and violent. *Space Sci. Rev.* **92**, 97–112.

Wood J. A. and Hashimoto A. (1993) Mineral equilibrium in fractionated nebula systems. *Geochim. Cosmochim. Acta* **57**, 2377–2388.

Woolum D. S. and Cassen P. (1999) Astronomical constraints on nebular temperatures: implications for planetesimal formation. *Meteorit. Planet. Sci.* **34**, 897–907.

Yoshino T., Walter M. J., and Katsura T. (2003) Core formation triggered by permeable flow. *Nature* **422**, 154–157.

Young E. D. (2000) Assessing the implications of K isotope cosmochemistry for evaporation in the preplanetary solar nebula. *Earth Planet. Sci. Lett.* **183**, 321–333.

Young E. D. (2001) The hydrology of carbonaceous chondrite parent bodies and the evolution of planet progenitors. *Phil. Trans. Roy. Soc. London A* **359**, 2095–2110.

Young E. D., Ash R. D., England P., and Rumble D., III (1999) Fluid flow in chondritic parent bodies: deciphering the compositions of planetesimals. *Science* **286**, 1331–1335.

Yu Y. and Hewins R. H. (1998) Transient heating and chondrule formation: evidence from sodium loss in flash heating simulation experiments. *Geochim. Cosmochim. Acta* **62**, 159–172.

Yurimoto H. and Kuramoto K. (2002) A possible scenario introducing heterogeneous oxygen isotopic distribution into protoplanetary disks. *Meteorit. Planet. Sci.* **37**, A153.

Yurimoto H. and Wasson J. T. (2002) Extremely rapid cooling of a carbonaceous-chondrite chondrule containing very ^{16}O-rich olivine and a ^{26}Mg excess. *Geochim. Cosmochim. Acta* **66**, 4355–4363.

Yurimoto H., Ito M., and Nagasawa H. (1998) Oxygen isotope exchange between refractory inclusion in Allende and solar nebula gas. *Science* **282**, 1874–1877.

Yurimoto H., Koike O., Nagahara H., Morioka M., and Nagasawa H. (2000) Heterogeneous distribution of Mg isotopes in anorthite single crystal from Type B CAIs in Allende meteorite. In *Lunar Planet. Sci. XXXI*, #1593. The Lunar and Planetary Institute, Houston (CD-ROM).

Yurimoto H., Rubin A. E., Itoh S., and Wasson J. T. (2001) Non-stoichiometric Al-rich spinel in an ultrarefractory inclusion from CO chondrite. In *Lunar Planet. Sci. XXXIII*, #1557. The Lunar and Planetary Institute, Houston (CD-ROM).

Zanda B., Bourot-Denise M., Perron C., and Hewins R. H. (1994) Origin and metamorphic redistribution of silicon, chromium, and phosphorus in the metal of chondrites. *Science* **265**, 1846–1849.

Zhang Y., Huang S., Schneider D., Benoit P. H., DeHart J. M., Lofgren G. E., and Sears D. W. G. (1996) Pyroxene structures, cathodoluminescence and the thermal history of the enstatite chondrites. *Meteorit. Planet. Sci.* **31**, 87–96.

Zinner E. and Göpel C. (2002) Aluminum-26 in H4 chondrites: implications for its production and its usefulness as a fine-scale chronometer for early solar system events. *Meteorit. Planet. Sci.* **37**, 1001–1013.

Zolensky M. E. and Ivanov A. V. (2001) Kaidun: a smorgasbord of new asteroid samples. *Meteorit. Planet. Sci.* **36**, A233.

Zolensky M. E. and McSween H. Y., Jr. (1988) Aqueous alteration. In *Meteorites and the Early Solar System* (eds. J. F. Kerridge and M. S. Matthews). University of Arizona Press, Tucson, pp. 114–143.

Zolensky M., Barrett R., and Browning L. (1993) Mineralogy and composition of matrix and chondrule rims in carbonaceous chondrites. *Geochim. Cosmochim. Acta* **57**, 3123–3148.

Zolensky M. E., Mittlefehldt D. W., Lipschutz M. E., Wang M.-S., Clayton R. N., Mayeda T. K., Grady M. M., Pillinger C., and Barber D. (1997) CM chondrites exhibit the complete petrologic range from type 2 to 1. *Geochim. Cosmochim. Acta* **61**, 5099–5115.

1.08
Calcium–Aluminum-rich Inclusions in Chondritic Meteorites

G. J. MacPherson

Smithsonian Institution, Washington, DC, USA

1.08.1	INTRODUCTION	201
1.08.2	SOME ESSENTIAL TERMINOLOGY: STRUCTURAL ELEMENTS OF A CAI	202
	1.08.2.1 The Generic CAI	202
	1.08.2.2 Comments on Primary and Secondary Mineralogy	203
	1.08.2.3 Rim Sequences	204
	1.08.2.4 Metal Grains and Fremdlinge	205
1.08.3	MINERALOGY AND MINERAL CHEMISTRY	206
	1.08.3.1 Melilite	206
	1.08.3.2 Pyroxene	209
	1.08.3.3 Hibonite	210
	1.08.3.4 Perovskite	210
	1.08.3.5 Grossite and $CaAl_2O_4$	210
	1.08.3.6 Anorthite	211
1.08.4	DIVERSITY AND MAJOR-ELEMENT BULK CHEMISTRY	211
	1.08.4.1 FUN CAIs	219
	1.08.4.2 Unresolved Problems	221
1.08.5	DISTRIBUTION AMONG CHONDRITE TYPES	222
1.08.6	AGES	229
1.08.7	TRACE ELEMENTS	230
1.08.8	OXYGEN ISOTOPES	232
	1.08.8.1 General	232
	1.08.8.2 The Data	232
	1.08.8.3 Synthesis of Existing Data, and Unresolved Issues	235
1.08.9	SHORT-LIVED RADIONUCLIDES	237
1.08.10	CAIs, CHONDRULES, CONDENSATION, AND MELT DISTILLATION	240
1.08.11	CONCLUSIONS AND REFLECTIONS: WHERE AND HOW DID CAIs FORM?	241
ACKNOWLEDGMENTS		241
REFERENCES		241

1.08.1 INTRODUCTION

Calcium–aluminum-rich inclusions (CAIs) are submillimeter- to centimeter-sized clasts in chondritic meteorites, whose ceramic-like chemistry and mineralogy set them apart from other chondrite components. Since their first descriptions more than 30 years ago (e.g., Christophe Michel-Lévy, 1968), they have been the objects of a vast amount of study. At first, interest centered on the close similarity of their mineralogy to the first phases predicted by thermodynamic calculations to condense out of a gas of solar composition during cooling from very high temperatures (e.g., Lord, 1965; Grossman, 1972). Immediately thereafter, CAIs were found to be extremely old (4.56 Ga) and to possess unusual isotopic compositions (in particular, in magnesium and oxygen) suggestive of a presolar dust component. In short, they appear to be the oldest and most primitive objects formed in the infant solar system.

In the late 1980s (e.g., MacPherson et al., 1988), the attention of most workers in the field was focused on understanding the petrogenesis and isotopic compositions of CAIs within a relatively restricted number of chondrite varieties. Much has changed since then. We now have extended our data sets beyond CV and CM chondrites to CAIs from ordinary, enstatite, and a wider range of carbonaceous chondrites. Out of this has emerged an ironic fact: the large centimeter-sized CAI "marbles" (the so-called type Bs; see below) that are so prominent in CV chondrites, and upon which so many of the original concepts were based owing to the abundance and availability of material from the Allende meteorite, turn out to be the exceptions rather than the norm. Indeed, we now know that the Allende parent body itself experienced so much postaccretion reprocessing that its CAIs reveal only a murky picture of the early solar nebula. Another profound change since 1988 has been the development of ion microprobe technology permitting microanalysis of oxygen isotopes within standard petrographic thin sections. One of the debates raging in 1988 centered on the difficulty of making CAIs at the distance of the asteroid belt where the host chondrites presumably accreted. Since then, a combination of isotopic (especially beryllium-boron and oxygen) evidence and theoretical modeling have suggested the interesting possibility that CAIs all formed very close to the infant sun and were later dispersed out to the respective chondrite accretion regions. Yet another difference from 1988 is that the science of meteoritics has matured past the point where workers study CAIs without consideration of other chondritic components, most notably chondrules. Bulk composition differences aside, CAIs and chondrules are both products of very high temperature events in the earliest solar system. Like chondrules, many CAIs apparently solidified from partially to completely molten droplets. Moreover, chondrules and CAIs both have far more complex histories than was generally recognized in the late 1980s: both may have experienced multiple melting episodes, followed by prolonged nebular and/or asteroidal modification due to gaseous or liquid interactions and reheating as a result of shock processes. In short, although CAIs are still recognized as being the oldest objects formed in our solar system and possessing isotopic traces of the presolar dust from which they ultimately formed, they are a long way from being the primitive condensates they were once thought to be. Rather than stressing the unusual isotopic characteristics of CAIs and the presolar implications, this chapter will instead treat CAIs as probes of the earliest solar system that reveal the nature of the high-temperature events, their chronology, and the likely locales where such events occurred.

This chapter is not, and does not pretend to be, an exhaustive review of the sum of knowledge about CAIs. Indeed, so much new data have been generated since 1988, especially isotopic data from an ever-growing number of sophisticated ion microprobe labs, that it would far beyond the space limitations of this chapter to attempt such a review. It should be regarded rather as a primer, or broad introduction. Rather than compiling a comprehensive database to illustrate the current knowledge of CAI properties, I have instead used representative data sets taken primarily from published papers rather than from abstracts.

1.08.2 SOME ESSENTIAL TERMINOLOGY: STRUCTURAL ELEMENTS OF A CAI

CAIs are complex objects. Once thought to be simple and pristine aggregates of high-temperature nebular condensate grains, they are now universally recognized to have experienced prolonged histories of incomplete back-reactions with a nebular gas, reheating that commonly led to one or multiple episodes of melting, impact-induced shock, and secondary mineralization in the nebula or on an asteroidal parent body or both. Some of this complexity is manifest in the basic structure of CAIs. In the specialized literature dealing with them, unwary readers can find themselves hopelessly confused by references to, e.g., multiple kinds of rims, multiple generations of individual phases, and primary versus secondary minerals, just to name a few.

These topics cannot be avoided in this review. Therefore, before delving into the details of CAI diversity, mineralogy, and chemistry (not to mention petrogenesis), some orientation is in order.

1.08.2.1 The Generic CAI

Figure 1 is a schematic drawing of a generic CAI, in this case a melilite-rich one such as might be found in the Allende CV3 meteorite. The interior contains the essential primary phases, consisting of large melilite crystals that enclose much smaller crystals of spinel and hibonite. The original grain boundaries between the melilite crystals are largely erased by pervasive fine-grained secondary minerals that replace the melilite along grain boundaries and cleavage planes. Closer examination of real CAIs shows that this fine-grained material actually corrodes the melilite, establishing, beyond doubt, a replacement relationship. Surrounding the entire CAI is a thin (typically ten to a few tens of micrometers thick) sequence of nearly monomineralic layers,

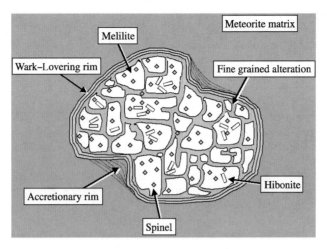

Figure 1 Cartoon illustrating the major structural components of a hypothetical CAI. This one is intentionally patterned after CAIs in the Allende CV3 chondrite in order to illustrate extensive secondary alteration. Here, as in most real CAIs, the secondary minerals preferentially replace primary melilite (as indicated by the irregular crystal shapes and embayed margins).

referred to collectively as a Wark–Lovering rim (after the investigators who first described such rims). The minerals in such rims commonly duplicate those minerals found in the CAI interior, such as melilite and spinel and pyroxene; yet, the rim minerals clearly represent a different generation formed by a separate event. A characteristic of the Wark–Lovering rim is that the rim and its individual layers retain a remarkably consistent thickness around the periphery of even the most irregularly shaped CAIs. Overlying the Wark–Lovering rim is an accretionary rim, which is completely different in mineralogy, texture, and origin from the former. Accretionary rims are layered sequences of very fine-grained olivine and pyroxene, very much like the enclosing meteorite matrices in composition and texture. Accretionary rims are highly variable in thickness, in a very predictable way: they preferentially fill in topographic pockets on the surfaces of the CAIs, like levee deposits, and are thin or absent around topographic protrusions. External to the accretionary rim is the meteorite matrix.

1.08.2.2 Comments on Primary and Secondary Mineralogy

The mineralogy of CAIs is commonly described in terms of what might be called "primary" and "secondary" minerals. Although useful, these terms are relative and potentially misleading. "Primary" refers to a phase that apparently formed when the inclusion itself first formed, for example, by direct condensation, melt solidification, or solid-state recrystallization. "Secondary" is a petrographic term for any phase that texturally appears to be replacing another phase; e.g., one that not only mantles but even corrodes that earlier phase. But two examples should serve to alert the reader to the potential pitfalls when reading or using these terms in the context of CAIs. First, the equilibrium condensation sequence of phases during cooling of a hot solar gas is itself a discontinuous reaction series: in all such thermodynamic calculations, most phases appear during the reaction and disappearance of an earlier, higher-temperature phase as a result of falling temperature and changing gas composition (e.g., Yoneda and Grossman, 1995). Even hibonite ($CaAl_{12}O_{19}$), one of the very highest temperature "primary" phases, condenses as a result of a reaction between corundum and the gas. Of course, no one ever refers to hibonite as a secondary or alteration phase. Consider as a second example the minerals anorthite and calcic pyroxene. Many unmelted melilite-rich CAIs commonly show a clear replacement of melilite by fine-grained anorthite and calcic pyroxene, and these occurrences are commonly referred to as secondary. Yet, there is an entire class of CAIs (the type Bs) that clearly crystallized from molten droplets in which two of the "primary" crystallizing phases were anorthite and calcic pyroxene. To add to the confusion, the primary melilite in these type B CAIs is itself commonly replaced by fine-grained anorthite and pyroxene, meaning that a single CAI can contain two separate generations of pyroxene and feldspar that both formed through high-temperature processes! Less confusion exists for some phases such as hydrous minerals and iron-bearing silicates that arguably formed through parent body (asteroidal) as opposed to nebular processes (although even that interpretation is debated). The point here is not to dissuade the reader from using the terms "primary" and "secondary." To the contrary, the terms have value when properly used. For example, the fine-grained secondary anorthite, nepheline, sodalite, and grossular garnet that replace melilite and spinel in some CAIs commonly have a fundamentally different magnesium isotopic composition than the primary phases they replace: the primary phases contain excess ^{26}Mg that can be attributed to decay of the short-lived radionuclide ^{26}Al, whereas the secondary phases lack such excesses. This isotopic difference suggests a significant age difference. The point of this discussion, therefore, is to create an awareness of the complexities—and, hence, potential confusion—arising from usage of the terms "primary" and "secondary."

1.08.2.3 Rim Sequences

Wark and Lovering (1977) first described in detail the very thin multilayered rims that surround CAIs in the Allende chondrite, and which have now been found to be nearly universal around CAIs in other chondrite types. Many workers consequently refer to all such rims as Wark–Lovering rims, wherever they occur, although Wark himself (see Wark and Boynton, 2001) advocates using the term only for those containing spinel + melilite + pyroxene (i.e., specifically excluding ones that consist mainly of pyroxene). They vary considerably in their complexity and mineralogy, but the essential structure is consistent (see Figure 2): (i) they consist of one or more thin layers, each of which is monomineralic or bimineralic; (ii) the layers maintain a remarkably consistent thickness even where the rims wrap around the convoluted exteriors of highly irregularly-shaped CAIs; and (iii) where the CAI is broken or fractured, the rims are as well, indicating that the rims did not form *in situ* within the meteorite matrices where they now reside.

Early detailed studies of rim sequences focused primarily on examples from Allende, which to some extent was misleading and unfortunate. Rims around Allende CAIs contain abundant nepheline, iron-rich spinel, hedenbergite and andradite; melilite is rare. Only later did studies of CAI rims in reduced CV chondrites reveal that some of the above properties are intrinsic to Allende (and its oxidized CV3 relatives) rather than being intrinsic to rims: the nepheline systematically replaces melilite and anorthite, spinel in the reduced CV rims contains far less oxidized iron, and andradite and hedenbergite are virtually restricted to the oxidized subgroup. The alkali and iron signatures of Allende and the oxidized subgroup CV3s have been interpreted to be the result of parent-body metasomatism (e.g., Krot *et al.*, 1995); in any case they are not intrinsic to the rim sequences. The original rim sequences on CV3 CAIs, as revealed in reduced CV3 chondrites, consist of: an innermost layer consisting of iron-poor spinel ± perovskite ± hibonite; a layer of aluminum-rich melilite, a layer of anorthite that partially replaces the melilite, and a layer of pyroxene that varies systematically in composition from highly titanium-, aluminum-rich fassaite outwards to aluminum-rich diopside. In some cases a layer of forsterite overlies the pyroxene (e.g., Figure 2). Rims on CAIs in other chondrite types can deviate significantly from this pattern, such as the absence

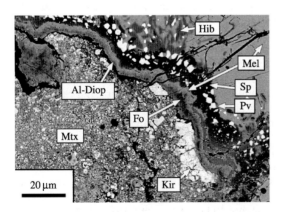

Figure 2 Backscattered electron photograph of a Wark–Lovering rim sequence surrounding a type A CAI in the Vigarano (CV3) chondrite. Some of the minerals in the rim duplicate minerals occurring in the interior of the CAI, notably melilite, spinel, perovskite, and hibonite. Abbreviations: Al-Diop—aluminum diopside; Fo—forsterite; Hib—hibonite; Kir—kirschsteinite; Mel—melilite; Mtx—meteorite matrix; Pv—perovskite; Sp—spinel.

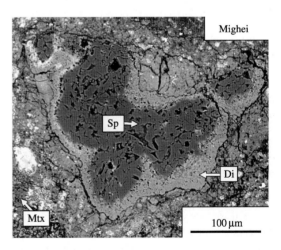

Figure 3 Backscattered electron photograph of a "nodular" spinel–pyroxene inclusion from the Mighei CM chondrite. Small bright patches locally separating the spinel interior from the diopside rim are iron-rich phyllosilicate (after MacPherson and Davis, 1994). (reproduced by permission of Elsevier from *Geochim. Cosmochim. Acta* **1994**, *58*, 5599–5625). Abbreviations as in Figure 1, except: Diop—diopside.

of melilite from rims on most CAIs in CM chondrites (Figure 3).

Advanced analytical techniques have permitted the trace element and isotopic compositions of individual layers within rim sequences to be determined, and these have been highly informative. The reader is referred to Wark and Boynton (2001) for a thorough summary of virtually all of the known measurements of rim sequences.

Wark and Boynton (2001; henceforth, WB) argue that rims formed in multiple stages: the earliest event involved flash heating and melt distillation to produce a calcium–aluminum-rich melt that solidified to aluminates (mainly hibonite)+perovskite+possibly glass. A second stage involved the reintroduction of magnesium and silicon from the gas onto the rim surface, setting up a diffusion-controlled process that resulted in formation of the thin monomineralic rim layers that are now observed. Introduction of alkalis and oxidized iron came last where it occurred at all. Their model is based primarily on their own and others' work showing that the refractory trace elements not only are enriched in rims relative to the CAI interiors but also tend to reflect the individual peculiar trace element fractionations that each CAI possesses. CAIs with distinctive highly fractionated rare-earth element (REE) signatures (known as group II; e.g., see Section 1.08.7 for an explanation of group II) tend to have rims with similar REE patterns. Their model has the advantage of explaining both the trace element enrichments of rims together with the multilayered aspect that cannot be the result of melt solidification (e.g., MacPherson *et al.*, 1981). Also, by reintroducing magnesium in order to form the spinel layer, it avoids the problem that intense distillation would severely deplete the residue in magnesium and hence spinel would not form. Nevertheless, significant difficulties remain with the WB model. First is the issue that melt distillation will tend to produce mass-dependent isotopic fractionation that in many cases is simply not observed in rims. WB cite numerous cases from the literature where distillation has been invoked to explain the magnesium isotopic properties of various CAIs, but none of those cases involve rim sequences. In fact, using the data compiled by WB, in only two out of 18 cases were actual rim phases enriched in heavy isotopes relative to any phases in the CAI interiors. If all of the magnesium now present in rims was reintroduced subsequent to rim formation, then a key test of the WB model will be to look for small degrees of isotopic fractionation in calcium (which was not all evaporated). Oxygen isotopes are an equally problematic issue for their model, as WB admit. Rim spinel and pyroxene have been analyzed by ion microprobe (see references cited in WB) and shown to have similar levels of ^{16}O-enrichment to the CAI interiors, with no evidence for mass-dependent fractionation except in the isotopically unusual objects known as FUN inclusions (see Section 1.08.4.1) where such an isotopic signature is intrinsic anyway. Finally, the rim sequences that WB exclude as not being "true" Wark–Lovering rims cannot be so easily dismissed. Pyroxene rims on spinel are common on CAIs in most chondrite varieties, including the CVs where WB focus most of their attention. How a pyroxene rim on a spinel grain would form by the WB model is (to this writer at least) not clear.

Lest the above discussion of the WB model for rims sound too negative, there are many strengths of their model. Rims must have formed in multiple stages, and the general trace element enrichment in rims over the host CAIs is an important constraint that must be addressed. It is also likely that the multilayered aspect must result from solid-state processes involving diffusion (eg., see MacPherson *et al.*, 1981).

1.08.2.4 Metal Grains and Fremdlinge

Many CAIs in different chondrites contain metal grains, ranging from iron–nickel alloys to alloys of nearly pure platinum-group elements (see references listed in MacPherson *et al.*, 1988; for examples of metal in CAIs from diverse chondrite varieties, see also Weber and Bischoff (1997) and Guan *et al.*, (2000a)). Fremdlinge, however, are a phenomenon restricted to CAIs in CV3 oxidized subgroup meteorites. They are complex assemblages including metal, sulfides, oxides, and phosphates, which were described in detail and illustrated by El Goresy *et al.* (1978; to which the reader is referred; see also Armstrong *et al.*, 1985). Once thought to possibly be presolar grains, Fremdlinge are now recognized to be products of secondary alteration and sulfidization of original simple metal grains that can be observed in their pristine form in the reduced-subgroup CV3s such as Vigarano and Leoville. Blum *et al.* (1988) reported experiments in which they were able to reproduce important features of the metal and oxide speciation of Fremdlinge simply by low-temperature oxidation of homogeneous metal. The tiny noble metal nuggets (noted above) that occur as dispersed grains within many kinds of CAI are not the result of such secondary processes, and appear to record high-temperature condensation of refractory siderophile elements into metallic alloys. Within an individual CAI, the compositions of such grains can vary widely (e.g. osmium-rich and platinum-rich separate grains). The problem represented by this observation of diverse metal grains, and also diverse Fremdlinge, is how assemblages of such highly fractionated individual phases can add up to bulk CAIs having essentially unfractionated trace element abundances (see discussion of this topic in Section 1.08.7; see also Sylvester et al., 1990).

1.08.3 MINERALOGY AND MINERAL CHEMISTRY

Brearley and Jones (1998) have given an exhaustive review of the long list of minerals that have been found in CAIs, along with all of the chemical variations found in those minerals, and there is no need to repeat their treatment here. However, a large number of the known CAI minerals are rare in occurrence or minor in abundance. In fact, the mineralogy of most CAIs in most chondrite types can be defined mainly in terms of relatively few essential minerals: spinel, melilite, perovskite, hibonite, calcic pyroxene, metal, anorthite, and olivine. Somewhat rarer but important phases are grossite and corundum. Table 1 lists the principal chemical variations of abundant and rarer phases along with selected comments on their occurrences. Table 2 lists the most common "secondary" minerals that are found in CAIs from a variety of chondrite types.

Other than metal, all of the primary phases in CAIs are oxides and silicates of calcium, aluminum, magnesium, and titanium. Although the relative abundance of different CAI types varies from chondrite type to chondrite type, the occurrence of the various primary phases is virtually universal: e.g., spinel-pyroxene-rich and melilite-rich CAIs are found in all but CI chondrites (even in enstatite chondrites, the significance of which is discussed below). Analyses by electron microprobe and ion microprobe universally demonstrate that these primary phases, in their pristine (unaltered) state, are highly depleted in volatile trace components such as alkalies or iron. The most abundant minor elements are vanadium and chromium. The single critical fact that first drew attention to CAIs in mid-1960s and subsequently is the remarkable similarity of the observed primary phases to those predicted by thermodynamic calculations (e.g., Lord, 1965; Grossman, 1972) to condense out of a hot gas of solar composition. This is what led to the initial speculation, noted above, that CAIs might actually be aggregates of solar nebula condensate grains. Although it is now recognized that very few CAIs are preserved condensate aggregates, there remains a high likelihood that they have been reprocessed from such aggregates. Without doubt, all of the primary CAI phases share in common a very high temperature origin.

The following sections highlight the most genetically significant aspects of the major phases only, in terms of both chemistry and modes of occurrence; it emphatically is not a review of all of the mineral chemical variations, and for that the reader is referred to Brearley and Jones (1998).

1.08.3.1 Melilite

Melilite in CAIs consists almost exclusively of the binary solid solution gehlenite [$Ca_2Al_2SiO_7$]–åkermanite [$Ca_2MgSi_2O_7$], and thus differs from terrestrial igneous melilite that generally contains significant amounts of iron and sodium in solid solution (but see below). Melilite in this simple system is a very diagnostic and informative mineral. The liquidus phase equilibria for the gehlenite–åkermanite system is well determined (e.g., Osborn and Schairer, 1941). There is continuous solid solution between the two end members, with a thermal minimum at ~72% (weight) åkermanite. On both sides of the minimum the composition of melilite crystallizing from a melt is a strong function of temperature, and on the aluminum-rich side the first-formed (highest temperature) melilite is relatively aluminum-rich and becomes progressively magnesium-rich with falling temperature. This is the compositional region of interest for most CAIs (except forsterite-bearing type B CAIs, or FoBs), and the expected result for an igneous melilite in this range (such as those in type Bs) is a compositionally zoned crystal having an aluminum-rich core and magnesium-rich rim. Fortuitously, the birefringence of melilite is such a strong function of composition that, in a standard thin section (30 μm thickness), it is trivially easy to recognize even complex zoning profiles within individual melilite crystals. With practice, one can visually estimate the compositions of individual portions of a melilite crystal within ~5 mol.% (expressed herein in terms of fraction of the åkermanite component, e.g., $Åk_{10}$). Figure 4 is a photomicrograph of part of a type B CAI taken in cross-polarized transmitted light, showing a tabular melilite crystal in contact with a pyroxene crystal. The aluminum-rich core of the melilite crystal has bright blue-white birefringence, which changes outward through the crystal to a thin very dark anomalous blue-purple zone (at ~$Åk_{50}$) and, finally, to the very magnesium-rich margin of the crystal (abutting the pyroxene) that has an olive color. The important point here is that this visually-apparent compositional zoning profile is precisely that expected (from phase equilibria) for melilite crystallizing from an aluminum-rich melt

Table 1 Important primary phases in CAIs.

Phase	Principal component(s)	Comments	Distribution
Most common phases			
Spinel	$MgAl_2O_4$ $FeAl_2O_4$	Mg-rich variety is a "primary" phase in many CAI varieties; Fe-rich ("secondary") in altered zones and in Wark–Lovering rims.	Common in CAIs in all unequilibrated chondrite types (except CI)
Melilite	$Ca_2Al_2SiO_7$ [gehlenite]– $Ca_2MgSi_2O_7$ [åkermanite] solid solution	A "primary" phase in many CAI varieties; common also in Wark–Lovering rims	Rare in CM chondrites
Perovskite	$CaTiO_3$	A "primary" phase in some CAI varieties; common also in Wark–Lovering rims	Common in CAIs in all unequilibrated chondrite types (except CI)
Hibonite	$CaAl_{12}O_{19}$ A significant substitution is: $Mg + Ti^{4+} \leftrightarrow 2Al$	A "primary" phase in some CAI varieties; common also in Wark–Lovering rims	Common in CAIs in all unequilibrated chondrite types (except CI)
Calcic pyroxene (commonly referred to by the useful, but officially discredited, name "fassaite")	$CaMgSi_2O_6$–$CaAl_2SiO_6$– $CaTi^{4+}Al_2O_6$– $CaTi^{3+}AlSiO_6$ solid solution	A "primary" phase only in type B CAIs; a "secondary" or rim phase in most CAI varieties	Common in CAIs in all unequilibrated chondrite types (except CI)
Anorthite	$CaAl_2Si_2O_8$	A "primary" phase only in type B CAIs; a "secondary" phase in many melilite-bearing CAI varieties	Common in CAIs in most unequilibrated chondrite types (except CI), as a rim phase; type B CAIs are restricted to CV3 chondrites
Forsterite	Mg_2SiO_4 Olivine in Forsterite-bearing type B (FoB) CAIs commonly contains a significant component of $CaMgSiO_4$	A "primary" phase only in type FoB CAIs; elsewhere, restricted to Wark–Lovering rims	Found in CAIs in most unequilibrated chondrite types (except CI), as a rim phase. Type FoB CAIs are restricted to CV3 chondrites
Less Common Phases			
Grossite	$CaAl_4O_7$ No significant chemical variation	Apparently restricted to CAIs with low Mg and Si	Common in CH chondrites; rare in all others
Corundum	Al_2O_3	Primary phase	Rare
NiFe metal	Taenite; kamacite; awaruite.	Primary phase	Common phase in CAIs from many chondrite types.
Rhönite	$Ca_4(Mg,Al,Ti)_{12}(Si,Al)_{12}O_{40}$	Primary phase	Rare
Noble (platinum-group) metal alloys		Primary phase	Common trace phase in CAIs from many chondrite types.
$CaAl_2O_4$	$CaAl_2O_4$ No significant chemical variation	Apparently restricted to CAIs with low Mg and Si	Only one known occurrence, in the CH chondrite NWA470; Ivanova et al. (2002)

Stolper (1982). Hence, it is called "normal" zoning. Complexities (oscillations) can occur in the last stages of crystallization once plagioclase and pyroxene finally appear in the crystallizing sequence (e.g., see MacPherson et al., 1984b), but the overall trend is from aluminum-rich to magnesium-rich melilite with falling temperature. This is an igneous melilite.

Table 2 Important secondary phases in CAIs.

Phase	Principal component	Occurrence and other comments
Nepheline	$NaAlSiO_4$	Most common in CO3 and oxidized subgroup CV3 chondrites
Sodalite	$NaAlSiO_4 \cdot NaCl$	Most common in oxidized subgroup CV3 chondrites
Wollastonite	$CaSiO_3$	Most common in CV3 chondrites
Hedenbergite	$CaFeSi_2O_6$	Most common in oxidized subgroup CV3 chondrites
Grossular	$Ca_3Al_2Si_3O_{10}$	Commonly replace primary melilite
Monticellite	$CaMgSiO_4$	
Andradite	$Ca_3Fe_2Si_3O_{12}$	Most common in oxidized subgroup CV3 chondrites
Calcite	$CaCO_3$	
Phyllosilicate		Most common in CM, CR chondrites
Serpentine	$(Mg,Fe)_6Si_4O_{10}(OH)_8$	
Cronstedite	$(Mg,Fe)_2Al_3Si_5AlO_{18}$	
Tochilinite	$6Fe_{1-x}S \cdot 5(Mg,Fe)(OH)_2$	

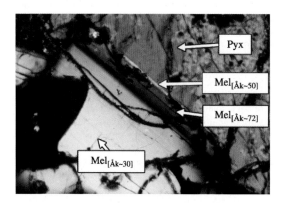

Figure 4 Transmitted light photomicrograph of a large-zoned melilite crystal in an Allende type B CAI, taken in cross-polarized light. The colors produced by the polarized light show clearly the chemical variations in the crystal, from its bright white aluminum-rich core (composition indicated as mole % of the magnesium end-member, åkermanite) out to the olive-colored magnesium-rich mantle. At the immediate contact with the large pyroxene crystal at upper right, the melilite has a very thin and bright white aluminum-rich rim that formed when the pyroxene began crystallizing. The zoning pattern exhibited by this crystal is exactly that expected during crystallization from an aluminum-rich melt. The melilite crystal is ~0.5 mm in length. Abbreviations as used previously except: Åk—åkermanite; Pyx—pyroxene.

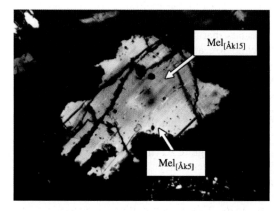

Figure 5 Transmitted light photomicrograph of a reversely zoned melilite crystal in an Allende type A CAI, taken in cross-polarized light. Unlike the crystal in Figure 4, this one has a magnesium-rich core surrounded by a bright aluminum-rich core mantle. This zoning pattern cannot be explained by melt crystallization. See Allen et al. (1978) for a detailed description of this CAI. The crystal is ~200 μm in length. The dark surrounding material (and in cross-cutting veins) is fine-grained secondary anorthite and nepheline that replaces the melilite. Abbreviations as used previously.

In contrast, the melilite crystal in Figure 5 is from a type A CAI. The composition of this crystal is more aluminum-rich than even the core of the crystal shown in the previous figure. The brightly birefringent rim of this crystal is more aluminum-rich than the darker core, exactly the opposite from that in the crystal in Figure 4 and thus counter to expectations for a crystal formed out of a silicate melt. This "reversely zoned" crystal is typical of melilite in most type A CAIs and represents a strong argument against such CAIs ever having been molten.

The interpretation that most type As were never melted and, by implication, represent preserved nebular condensates, was long ago recognized to be in conflict with the nearly universal property of type A CAIs that spinel is texturally enclosed within melilite and therefore appears to have formed earlier. Equilibrium condensation calculations are in universal agreement (e.g., Grossman, 1972; Kornacki and Fegley, 1984; Yoneda and Grossman, 1995; Petaev and Wood, 1998b; Ebel and Grossman, 2000) that melilite should begin to condense prior to spinel. This problem remains unresolved to this day, although many workers have used qualitative arguments to conclude that many type As really are condensates and the aberrant crystallization sequence results from kinetic factors (e.g., see Beckett and Stolper, 1994).

One other aspect of melilite chemistry that merits special comment is a small but persistent sodium content of melilite in type B CAIs. Consistent levels of several tenths of a weight percent (as Na_2O) have been measured by numerous investigators (see review by Brearley and Jones, 1998), and the sodium content correlates positively with the åkermanite content (Mg/Al ratio) of the melilite. Type B CAI melts formed at such high temperatures and low pressures that sodium is unexpected at any level. MacPherson and Davis (1993) showed that the sodium is crystal-chemically substituted into the melilite, incorporated when the melilite crystallized from its melt. Those authors argued, and Beckett and Stolper (2000), and Beckett et al. (2000) concurred that the most likely explanation for the elevated sodium contents in these melilites is a multistage CAI history in which melting followed secondary mineralization of the precursor material by sodium-bearing fluids or gas. The sodium was thus incorporated into the later melt. This is compelling evidence that the seemingly "simple" type B CAIs experienced multiple stages of melting and probably secondary alteration as well (since in many cases the sodium-bearing melilite has been re-altered).

1.08.3.2 Pyroxene

High-calcium, iron-free pyroxene with enrichment in aluminum and commonly titanium occurs in two fundamentally distinct settings in CAIs: it is a primary igneous phase in type B CAIs, and it occurs in the Wark–Lovering rim sequences of most kinds of CAIs in virtually every kind of CAI-bearing chondrite. Together with spinel, it is the most ubiquitous CAI phase. The extensive chemical variations of this pyroxene are reviewed in detail by Brearley and Jones (1998) and will not be repeated here. The petrogenetically essential property that requires comment here is the presence of both trivalent and quadrivalent titanium (Dowty and Clarke, 1973; Burns and Huggins, 1973). The pyroxene in type B CAIs has green to blue-green pleochroism in thin section owing to the titanium, and was termed fassaite by Clarke et al. (1970), Marvin et al. (1970), and Dowty and Clarke (1973) in analogy to a terrestrial calcium-saturated (one calcium per six oxygens) pyroxene having substantial aluminum and ferric iron. The CAI pyroxenes differ in having trivalent titanium instead of ferric iron. Beckett (1986) showed experimentally that the calculated Ti^{3+}/Ti^{4+} ratios of the natural type B1 pyroxenes (and their enclosing CAIs) require formation in an extremely reducing environment, roughly consistent with a hot hydrogen gas. A major control over oxygen fugacity in the solar nebula is generally believed to be the C/O ratio, and the type B1 CAI pyroxenes imply formation in a gas with C/O ~0.86 (Beckett, 1986).

The name fassaite remains embedded in the literature even though the International Mineralogical Association (Morimoto et al., 1988) recommended—wrongly in this writer's opinion—dropping that name for this unique and important pyroxene variety in favor of "subsilicic titanoan aluminian pyroxene."

Brearley and Jones (1998) reviewed how the composition of CAI pyroxene is highly variable, within individual inclusions (concentric and sector zoning), among differing CAI varieties and among differing petrographic settings. Among silicate-rich CAIs, the type A CAIs have only minor amounts of pyroxene but it is extremely titanium-rich (up to ~17 wt.% Ti calculated as TiO_2). Types B, FoB, and C CAIs have progressively less titanium-rich pyroxene. The pyroxene in Wark–Lovering rim sequences and spinel-pyroxene CAIs (including the fine-grained spinel-rich CAIs in CV3 chondrites) is predominantly an aluminum-rich diopside. In many rims, the pyroxene layer shows a progressive variation in composition across the thickness of the layer from diopside to fassaite (Wark and Lovering, 1977).

An interesting feature of many CAIs in oxidized subgroup CV3 meteorites is the presence of hedenbergite (along with andradite) along the outside of the Wark–Lovering rims sequences. The presence of pyroxene that contains oxidized iron, within a few tens of micrometers of a pyroxene layer in the rims that contains trivalent titanium, is testimony to the extreme disequilibrium nature of chondrite assemblages and to the very different conditions that led to formation of the two pyroxenes. Whereas the CAI fassaites formed in a hot and highly reducing solar nebular gas, the hedenbergites and associated minerals may well have formed by fluid deposition within an asteroidal parent body (e.g., Krot et al., 1995).

1.08.3.3 Hibonite

Although corundum is predicted to be the first major element condensate from a hot solar gas, that phase is quite rare in CAIs and instead hibonite is the most commonly encountered aluminate. From an isotopic point of view especially, hibonite is one of the most important of all CAI phases (e.g., Ireland, 1988; Ireland *et al.*, 1988; Hinton *et al.*, 1988).

Hibonite alone of the calcium aluminates found in CAIs shows substantial chemical variations, which are dominated by the coupled substitution

$$Ti^{4+} + Mg^{2+} \Leftrightarrow Al^{3+}$$

Meteoritic hibonite is commonly intensely colored, most often having a sapphire-blue pleochroism (Figure 6) that, in intensity, is proportional to the concentration of titanium. Like fassaite, some of the titanium in blue hibonite is trivalent. Ihinger and Stolper (1986) showed that such hibonite must have formed under highly reducing conditions that was, nevertheless, more oxidizing than those inferred from fassaite. Although the hibonite in reduced CV3 chondrites (Vigarano, Efremovka, Leoville) is blue, that in oxidized subgroup CV3 meteorites is generally orange. Ihinger and Stolper (1986) showed that such hibonite derives its color from the presence of vanadium and formation under conditions much more oxidizing than the blue hibonite. It is likely that originally blue hibonite was altered by the same late stage highly oxidizing processes that formed such phases as hedenbergite and andradite in Allende and its kindred CV3s.

1.08.3.4 Perovskite

Perovskite is generally stoichiometric $CaTiO_3$, and few studies report detailed analyses of this phase (in part also because it tends to be very fine grained). It is most notable for its capacity to accommodate significant amounts of large ion lithophile trace elements such as the rare earths and yttrium and high field strength elements such as zirconium; the latter two elements have been reported in concentrations of several tenths of a weight percent (as oxides) in electron microprobe analyses. Brearley and Jones (1998) give a thorough report of the occurrences and chemistry of this phase.

1.08.3.5 Grossite and $CaAl_2O_4$

Both of these phases exhibit virtually no chemical variations whatsoever, and are always close to theoretical stoichiometry (Weber and Bischoff, 1994; Aléon *et al.*, 2002a; Ivanova *et al.*, 2002). This is true even when coexisting hibonite (in the case of grossite) contains substantial titanium and magnesium. Ivanova *et al.* (2002) argued that this chemical purity is due to the fact that aluminum atoms in both grossite and $CaAl_2O_4$ are in tetrahedral rather than octahedral

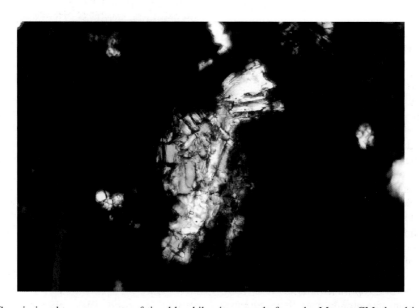

Figure 6 Cosmic jewels: an aggregate of tiny blue hibonite crystals from the Murray CM chondrite. This entire object is no more than ~100 μm in maximum size. The color of these sapphire-like gems is real and is due to the presence of substantial titanium in the crystal structure. The color variation is partly due to the crystals being compositionally zoned (their margins are more titanium-rich), and also to their being pleochroic under plane polarized light: the color is visible only when the crystals are in specific orientations relative to the polarizers on the microscope through which this photo was taken.

coordination, unlike hibonite, thus providing no suitable site for titanium and negating possible coupled substitution with magnesium.

The single most remarkable feature of grossite and $CaAl_2O_4$ is their extreme rarity except in CH chondrites. Weber and Bischoff (1994) originally argued that the critical factor in stabilizing grossite is a bulk composition with high Ca/Al. However, Beckett and Stolper (1994) and Simon et al. (1994) suggested that elevated TiO_2 contents tend to stabilize hibonite at the expense of grossite. More recently, Ivanova et al. (2002) suggested that depletion in bulk MgO and SiO_2 also might be important for stabilizing both grossite and $CaAl_2O_4$. In any case, grossite- and $CaAl_2O_4$-bearing CAIs are probably products of nebular condensation of their precursors: trace element fractionations suggest vapor–solid volatility control, and the absence of mass-dependent isotopic fractionation points to condensation rather than volatilization as the cause (see Weber et al., 1995). The problem is that equilibrium nebular condensation calculations by numerous workers have arrived at differing conclusions regarding the stability of these two aluminates. Primarily, the disparities arise from differing sources for thermodynamic input data (free energy of formation) used for the calculations. One common theme to more recent calculations is that enhanced nebular dust/gas ratios tend to stabilize both phases, suggesting that grossite- and $CaAl_2O_4$-bearing CAIs may be indicative of unusual nebular conditions in the regions where they formed (Ivanova et al., 2002).

1.08.3.6 Anorthite

Like calcic pyroxene, sodium-free plagioclase occurs in two fundamentally distinct settings in CAIs: it is a primary igneous phase in type B CAIs, and it occurs as a secondary phase that replaces melilite in Wark–Lovering rim sequences and in CAI interiors. The phase shows very little chemical variation (see Brearley and Jones, 1998).

Anorthite has an importance far out of proportion to its simple chemistry. First, because of its very low MgO content (hence, high Al/Mg ratio) and the fact that it occurs not just in many CAIs but also in aluminum-rich chondrules, it is a prime analytical target in the search for evidence of extinct ^{26}Al. Second, like calcic pyroxene, anorthite is one of the principal phases that CAIs share in common with aluminum-rich chondrules. In fact, MacPherson and Huss (2000) showed that the trends defined within CMAS ($CaO-MgO-Al_2O_3-SiO_2$) space by silicate-bearing CAI and aluminum-rich chondrule bulk compositions converge on a region intermediate between the compositions of anorthite and calcic pyroxene. In essence, anorthite and calcic pyroxene are the links between CAIs and Al-rich chondrules. Finally, the similar dual natures of anorthite and calcic pyroxene, occurring both as secondary phases in melilite-bearing CAIs (especially in type As) and as primary igneous phases in type B CAIs, suggests the interesting possibility that type B CAIs form by the melting of altered melilite-rich type As (MacPherson and Huss, 2000; see also Section 1.08.10).

1.08.4 DIVERSITY AND MAJOR-ELEMENT BULK CHEMISTRY

Our knowledge of the diversity of CAI varieties, and distribution of those varieties among the many kinds of chondrites, has increased significantly since the mid-1980s, thanks to an increasing number of dedicated investigators doing systematic studies with sophisticated new analytical instrumentation. Little was known in 1988 about the tiny and rare CAIs that occur in ordinary and enstatite chondrites, but detailed petrologic and isotopic data now exist for significant numbers of such objects. Newly identified classes of chondrites, such as CB and CH, have their own population of CAIs and these also have been studied in detail. With CAIs, as with chondrules, the comparison of populations and properties across meteorite classes is potentially very informative with regard to nebular heterogeneity/homogeneity.

The 1988 review (MacPherson et al., 1988) discussed at length the numerous, sometimes conflicting and confusing nomenclature systems for CAIs: some based on petrography, some based on trace element (especially rare earth) fractionation patterns, and some based on major element chemistry. The underlying cause of the confusion remains the same: a mountain of accumulated evidence demonstrates unequivocally that trace element fractionation patterns are largely decoupled from major element chemistry, and isotopic signatures are largely decoupled from both. A classification based on one set of properties does not make sense with respect to the other sets of properties.

This author's opinion on the matter is not only unchanged from 1988 but, if anything, even stronger. CAI nomenclature should be based primarily on a combination of mineralogy and bulk major-element chemistry, which are connected, plus some consideration of petrographic features. Although trace element fractionation patterns (and isotopic signatures) are profoundly important in understanding CAI petrogenesis, they should not be used to name or classify the CAIs themselves. Where two groups of CAIs overlap in bulk chemistry but differ texturally, one should consider the possibility that the two groups are petrogenetically somehow related. For example, the so-called fine-grained spinel-rich inclusions from CV3 chondrites look nothing like the igneous-textured type C CAIs, yet the two groups largely overlap in bulk composition; the former may well be the unmelted precursors of the latter.

Comparison of the observed mineralogy and textures of a particular object with those predicted by phase equilibria for its specific bulk composition can be a powerful way of determining whether melting occurred or not. In this review, therefore, the descriptions of CAI diversity and nomenclature will be based on the graphical portrayal of CAI bulk compositions using relevant phase diagrams. The choice of which particular *name* to ascribe to a specific kind of CAI is relatively unimportant except as a convenient shorthand means of conveying a sense of the mineralogy and chemistry of that particular class of objects. The critical point is to consistently associate *nomenclature* with *chemistry*. Except for some of the more newly recognized varieties of CAIs, the nomenclature used herein is based mostly on that developed at the University of Chicago in the early 1970s on the basis of textural and, especially, mineralogical differences. Those names have achieved common usage and are recognizable to most workers in the field. Most importantly, however, the various "Chicago" CAI types were defined primarily in terms of mineralogy, which reflects their differences in major element bulk chemistry. The reader can refer to types A and B CAIs by any preferred alternative names, but the two will always show systematic differences in chemistry and mineralogy.

The phase diagram of choice in the 1988 review was that developed by Stolper (1982) to portray the compositional and phase relationships of silicate-rich CAIs found in the CV3 chondrites. Most silicate-rich CAIs contain abundant spinel, and most melted ones arguably have spinel as the liquidus phase. Stolper advantageously projected from that common saturated phase onto an appropriate plane (forsterite–anorthite–gehlenite) in order to achieve accurate representation of the phase equilibria within a two-dimensional diagram. Stolper's diagram is ideally suited for its intended purpose, namely looking at the phase relationships of type B CAIs. However, it has the disadvantage of limited compositional scope: it cannot represent either silicate-poor CAIs or chondrules, thus making it difficult to look at the compositional relationships of a much wider range of objects than just silicate-rich CAIs.

Huss *et al.* (2001) introduced a more general diagram, based in principle on Stolper's diagram but with a greatly extended compositional scope. Shown in Figure 7, it is a projection from spinel onto the plane Al_2O_3–Mg_2SiO_4–Ca_2SiO_4. Note that Stolper's diagram is contained (in somewhat distorted form owing to difference in projection angle) within this one, and can be visualized by connecting the points (minerals) labeled gehlenite (Geh), forsterite (Fo), and anorthite (An). Figure 7(a) shows the compositions of types A, B and FoB CAIs, spinel–pyroxene spherules, hibonite–pyroxene spherules, and grossite-rich CAIs. Figure 7(b) shows the compositions of type C CAIs and fine-grained spinel-rich CAIs such as those found in the reduced CV3 chondrites Efremovka and Leoville. For no reasons other than historical precedent and personal choice, the CAI varieties will be described starting with the silicate-rich type A and B varieties first.

Grossman (1975) first used the term type A to refer to CAIs that consist mainly of aluminum-rich melilite. Both on Figure 7(a) and on Stolper's diagram, type A CAIs basically plot along a line connecting the melilite end-members gehlenite and åkermanite. Relative to their flashier igneous cousins the type Bs, type As have received less attention. Yet, unlike the type Bs, numerous studies have revealed that type As occur in nearly all chondrite varieties except CIs (of course) and enstatite chondrites. In most chondrites they are less than 1 mm in maximum size, but in the CV3 chondrites they can achieve sizes of 2 cm or more. Their essential structure consists of densely intergrown melilite that encloses spinel, hibonite, perovskite, and noble metal nuggets as the chief primary accessory phases; generally, there also is a small amount of very titanium–aluminum-rich pyroxene in the interiors as small grains, and the rare phase rhönite is known primarily from type As. Early attempts by MacPherson and Grossman (1979) to distinguish two variants of type A, eventually known as "fluffy" and "compact" and distinguished on the basis of differences in shape and internal textures, have evolved into the general recognition that although a relatively few type As experienced at least partial melting (the "compact" variety; see Simon *et al.* (1999) for a thorough study of compact type As) most type As were never melted (see MacPherson and Grossman, 1984; MacPherson *et al.*, 1988; Beckett and Stolper, 1994). Fluffy type As are distinguished by highly irregular shapes and often aggregate structures, highly aluminum-rich melilite ($Åk_{<25}$) that commonly is reversely zoned in a monotonic manner inconsistent with igneous crystallization, and hibonite (if present) enclosed within the CAI interior (MacPherson and Grossman, 1984). Compact type As tend to be more spheroidal in shape, have more magnesium-rich melilite that shows

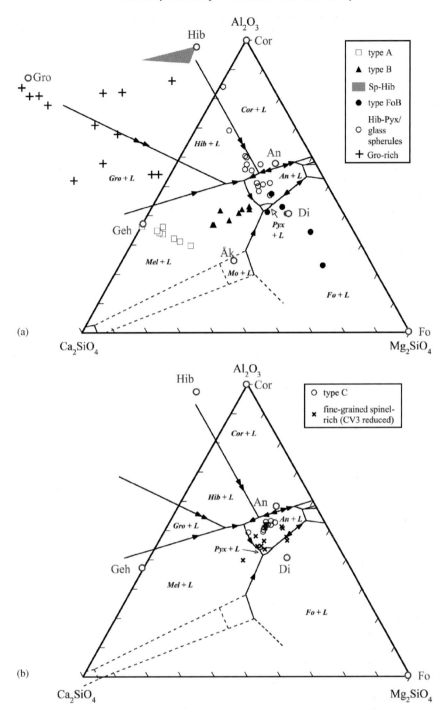

Figure 7 Bulk compositions of major CAI varieties, projected from spinel (MgAl$_2$O$_4$) onto the plane Al$_2$O$_3$–Mg$_2$SiO$_4$–Ca$_2$SiO$_4$. (a) Types A, B, FoB, as well as spinel–hibonite, hibonite–pyroxene spherules and grossite-rich. (b) Type C CAIs and the fine-grained spinel-rich inclusions common in CV3 (reduced subgroup) chondrites. The fine-grained CAIs have approximately the required compositions and mineralogy to be the unmelted progenitors of the type Cs. Details for calculating the plotting coordinates for this diagram are given in Huss et al. (2000). Mineral compositions are in red. Abbreviations: An—anorthite; Cor—corundum; Di—diopside; Fo—forsterite; Geh—gehlenite; Gro—grossite; Hib—hibonite; L—liquid; Mo—monticellite; Pyx—pyroxene.

complex zoning, and any hibonite tends to be concentrated around the CAI periphery just below the Wark–Lovering rim (Simon et al., 1999). Mineral chemical, trace element, and experimental petrologic evidence suggest that the relatively rare compact type As experienced partial melting (Beckett and Stolper, 1994; Simon et al., 1999) but the more common fluffy type As probably were never melted at all (MacPherson and Grossman, 1984; Beckett and Stolper, 1994).

Type As in general are more primitive than type Bs by virtue of their more refractory bulk chemistry (lower MgO, SiO_2), and fluffy type As are more primitive than even compact type As in having never been molten. fluffy type As are primitive isotopically as well: both their aluminum–magnesium and beryllium–boron isotope systems show less evidence for later disturbance than do those of type B CAIs (MacPherson et al., 1995, 2003). Nevertheless, fluffy type As are not the aggregates of individual nebula condensate melilite crystals they were once thought to be based on studies of Allende examples (MacPherson and Grossman, 1984). Subsequent studies of unaltered examples from Vigarano and other relatively pristine chondrites showed these objects to be aggregates of one or more irregular nodules, each consisting of densely intergrown melilite + spinel + hibonite + perovskite and definitely not having the expected properties of loose clusters of vapor-grown individual crystals. Rather, the internal texture within each nodule is consistent with solid-state recrystallization. These essential features can be seen in Figure 8, which shows part of a large fluffy type A CAI from Vigarano. Note the aggregate structure of the whole and dense textures of each separate nodule, and the fact that each nodule is individually surrounded by a Wark–Lovering rim sequence. Secondary mineralization of the kind found in Allende is largely missing from this Vigarano example. Although this CAI is a relatively large example (~1 cm in maximum size), the individual nodules resemble the smaller whole CAIs found in chondrites other than CV3s. (see Section 1.08.5).

True type B CAIs (Figure 9) are large (commonly centimeter-sized or even larger) and occur only in CV3 chondrites. They consist of coarse-grained melilite that ranges in composition from $\sim Åk_{10}$ to $Åk_{72}$, aluminum–titanium-rich calcic pyroxene, anorthite, and spinel. Wark and Lovering (1982) subdivided type Bs into two variants: type B1 CAIs have a nearly monomineralic mantle of melilite surrounding a core enriched in pyroxene, anorthite, spinel, nickel-iron metal in reduced CV3s (or Fremdlinge in oxidized CV3s), and noble metal nuggets. Type B2 CAIs lack a melilite-rich mantle and overall resemble the cores of the type B1s. Mineral chemical and experimental petrologic studies have shown that the basic features of type Bs are consistent with solidification from partially molten droplets (Clarke et al., 1970; Kurat et al., 1975; MacPherson and Grossman, 1981; Wark and Lovering, 1982; Stolper, 1982; Stolper and Paque, 1986; Beckett, 1986). In Figure 7(a), the type B CAIs plot within the melilite + spinel primary phase volume, and define a trend extending away from and at a high angle to that defined by the type As, toward a position intermediate between the minerals diopside and anorthite. The predicted crystallization sequence for type B melts is spinel followed by melilite, then anorthite or pyroxene as the third. This is broadly consistent with observed textural relationships, although

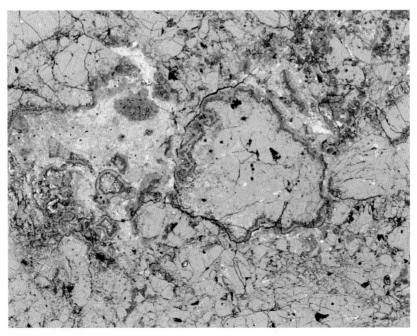

Figure 8 A backscattered electron image of Vigarano "fluffy" type A CAI 477-5. Melilite-rich nodules are each surrounded by Wark–Lovering rims. The field of view is 2 mm.

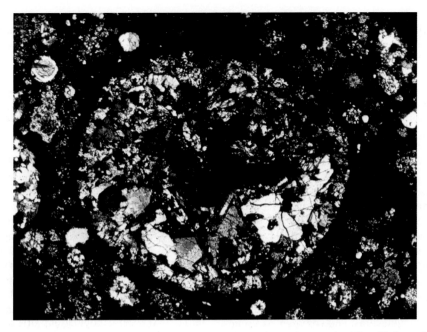

Figure 9 A type B1 inclusion from the Allende CV3 chondrite. This centimeter-sized "marble" consists mainly of melilite (bluish-white), titanium–aluminum-rich calcic pyroxene (bright colors), and anorthite and spinel (not readily visible in photo). Type B1 inclusions figure prominently in the early petrologic, trace element, and isotopic studies of CAIs, in part because of the richness of information about physicochemical histories available from petrologic, chemical and isotopic properties. Ironically, because type B inclusions occur only in CV3 chondrites, they are nonrepresentative of CAIs in general. Photograph taken in cross-polarized transmitted light. The colors are not the true colors of the crystals; they are artifacts of the polarized light.

kinetics can play a role in the relative order of anorthite and pyroxene (MacPherson *et al.*, 1984b). As discussed below, however, these are not the products of simple one-stage heating and cooling events that they were once imagined to be.

Weber and Bischoff (1997) described a fragment of a type-B-like CAI from the CR chondrite Acfer 059/El Djouf 001. Their brief description suggests that it has significant differences from the type Bs described above. In particular, the melilite is highly aluminous within a very narrow range, $Åk_{15}–Åk_{25}$, which is very different from type B melilite that extends all the way to the melilite binary minimum composition of $Åk_{72}$. Assuming that the narrow range reported by Weber and Bischoff (1997) is not an artifact of sampling by the small fragment they studied, this observation suggests a melt composition (if it was melted) far more aluminum-rich than type B melts.

The overwhelming importance of type B CAIs during the first 20 years of CAI literature derives from several factors. First, and has been noted many times by many authors, these objects are large, abundant, and eye-catching within a very large meteorite (Allende) that fell in 1969 and was immediately distributed to scientists around the world. Second, their size and large crystals (commonly 0.5–1 mm for phases other than spinel) made them easy to study, especially for investigations that required mineral separates (trace element partitioning, radiometric ages, internal oxygen isotope distributions, and the search for excess ^{26}Mg). Third, the recognition that these objects solidified from melts means that their phases arguably were in original isotopic equilibrium. To understand the importance of this, consider that such equilibrium was (and is) a requirement for the interpretation of ^{26}Mg excesses in different phases of type B CAIs as due to the *in situ* decay product of ^{26}Al (see Lee *et al.*, 1977). Finally, because they are igneous, they are amenable to experimental petrologic studies that lead to a relatively clear understanding of how they formed. In short, the answers to this paragraph's opening sentence can be summarized as: they were obvious, they were available, they were easy, and they were very interesting—and they still are.

The reason why type B CAIs seem to be confined largely (or exclusively, if the CR object noted above is not really a true type B) to CV3 chondrites remains an important unsolved mystery. The irony that type Bs may be the least representative CAIs could turn out to be doubly ironic if solving the mystery of their limited distribution reveals a fundamentally important truth about the early solar nebula.

Forsterite-bearing Type B (FoB) CAIs, also been named "Type B3" in the review by Wark *et al.* (1987), are an uncommon variant of type Bs that, as their name suggests, contain abundant large forsterite crystals as an essential phase in addition to pyroxene, spinel, and melilite. Anorthite is minor or absent altogether. Like normal type Bs, these are large: typically centimeter-sized. The first detailed description was published by Dominik et al. (1978), although Clayton et al. (1977) previously had not only analyzed but showed a photograph of one (see their figure 5). Others have been described by Clayton et al. (1984), Wark et al. (1987), and Davis et al. (1991). The bulk compositions of FoBs plot within the forsterite+spinel primary phase volume on Figure 7(a), meaning that spinel or forsterite will be the first crystallizing phases from melts of such compositions, followed by pyroxene or melilite. Virtually all workers agree that these objects did solidify from partial melts (see Wark et al., 1987, for a summary). FoBs are important for two reasons that are probably related. First, an unusual fraction of the known FoBs have magnesium and oxygen isotopic signatures that show evidence of mass-dependent fractionation such as might be expected from melt evaporation. Second, many FoBs show extreme mineral chemical disequilibrium between their interior and outer regions. Melilite in the interiors is commonly so magnesium-rich, upwards of Åk$_{75}$, that it plots on the magnesium-rich side of the melilite binary; melilite in the mantles of the CAIs is far more aluminum-rich, in some cases even Åk$_{12}$ (Wark et al., 1987). These melilite compositions cannot be reconciled with simple fractional crystallization, nor can the presence of hibonite in the outermost portions of one of these objects (Davis et al., 1991). Such observations, coupled with large degrees of isotopic mass fractionation, led Davis et al. (1991) to conclude that the example they studied must have experienced extreme melt evaporation from its outer surface prior to solidification. This is likely the case for some other FoBs as well. The combined petrologic and isotopic evidence from some of these objects makes them perfect test cases for understanding melt evaporation, and consequent isotopic fractionation, under actual solar nebular conditions. This has not been explored in sufficient detail (although see Davis et al., 1991); given the spate of recent experimental and theoretical studies on the role of volatilization in the formation and evolution of CAIs in general (e.g., Floss et al., 1996, 1998; Grossman et al., 2000, 2002; Richter et al., 2002), this oversight is surprising.

Hibonite-rich CAIs are best known from (but not unique to) the CM chondrites, where they are relatively abundant and have been extensively studied by (among others) Macdougall (1979, 1981); MacPherson et al. (1983, 1984a), Hinton et al. (1988), Fahey et al. (1987a), Ireland (1988, 1990), and Ireland et al. (1988). These mostly consist of hibonite alone, or hibonite with spinel ± perovskite. They range in form from single isolated crystals sitting in the meteorite matrices (the platy hibonite crystals (PLACs) of Ireland, 1988), to loose aggregates of crystals (Figure 6), to dense spherules that probably crystallized from melt droplets (see example in Figure 10). Rare examples contain corundum in addition to hibonite (Bar-Matthews et al., 1982; MacPherson et al., 1984a; Simon et al., 2002). The correlated isotopic–petrologic studies of CM hibonite-rich inclusions by Fahey et al. (1987a), Hinton et al. (1988), and Ireland and coworkers (Ireland, 1988, 1990; Ireland et al., 1988) are among the most detailed studies of large numbers of CAIs ever conducted. They demonstrated a huge diversity of trace element and isotopic patterns among seemingly similar objects, and showed that a significant fraction of CM hibonite CAIs contain large nuclear (nonradiogenic) isotope anomalies coupled with lack of evidence for initial ^{26}Al. These populations of grains, mostly formed directly as nebular condensates or derived from same, are powerful testimony to the isotopic heterogeneity of the solar nebula on a grain-to-grain scale.

Hibonite–silicate spherules are a relatively new addition to the CAI lexicon. The first reported description was given by Kurat (1975), but Ireland et al. (1991) and Simon et al. (1998) are more comprehensive studies of this class of objects (see also Beckett and Stolper, 1994; Grossman et al., 1988; Kimura et al., 1993; Russell et al., 1998). Hibonite–silicate spherules consist variously of hibonite laths within highly aluminous pyroxene, hibonite (±perovskite) embedded in an aluminous pyroxene-like glass (e.g., Figure 11), or more complex assemblages that can include melilite and grossite. These spherules are small, the largest being ~170 μm in diameter and others

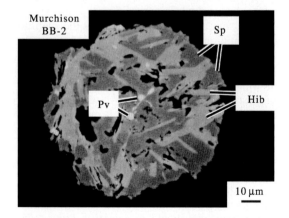

Figure 10 Backscattered electron photograph of a spinel–hibonite spherule from the Murchison CM chondrite (after MacPherson et al., 1983) (reproduced by permission of Elsevier from *Geochim. Cosmochim. Acta* **1983**, *47*, 823–839). Abbreviations as used previously.

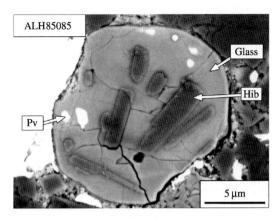

Figure 11 Backscattered electron photograph of a tiny hibonite-glass spherule from the CH chondrite ALH85085. Abbreviations as used previously.

being much smaller (such as the example in Figure 11).

Detailed isotopic studies by Ireland et al. (1991) on three hibonite–glass spherules showed a rare correlation between a CAI type and important isotopic properties. Specifically, they contain large nuclear isotope anomalies in calcium and titanium together with little evidence for initial live ^{26}Al. Russell et al. (1998) and Simon et al. (1998) analyzed hibonite–pyroxene spherules and also found similar magnesium isotopic signatures and, with one exception analyzed by Russell et al., large nuclear isotope anomalies. None of the spherules are FUN inclusions, however, because they lack large degrees of mass-dependent isotopic fractionation in magnesium (Ireland et al., 1991; Simon et al., 1998).

There is universal agreement that these spherules solidified from melt droplets. However, Beckett and Stolper (1994) showed that hibonite in some spherules is not even the liquidus phase (e.g., in Figure 7, not all plot within the hibonite + spinel field, and even some of those that do have spinel rather than hibonite as the liquidus phase). In such cases, the hibonite crystals may be unmelted relicts from the precursor assemblage. A relict origin is also indicated in one of the spherules studied by Simon et al. (1998), in which the hibonite and its enclosing pyroxene are not in isotopic equilibrium. Even for those compositions where hibonite is the expected liquidus phase (plotting in the hibonite phase volume on Figure 7), hibonite + pyroxene in the absence of anorthite and spinel is not expected for these bulk compositions. For these, Beckett and Stolper argued that rapid cooling could account for bypassing the nucleation of anorthite and spinel.

The hibonite–pyroxene and hibonite–glass spherules are important because they represent a rare example of isotopic properties correlating with CAI mineralogy and bulk composition. Many FUN inclusions, e.g., are indistinguishable from non-FUN CAIs in terms of mineralogy and bulk chemistry.

Type C inclusions consist mostly of spinel + calcic pyroxene + anorthite, with little or no melilite, and have distinctive ophitic or poikilitic textures that were immediately recognizable as igneous in origin. They were originally named "type I" by Grossman (1975) on the basis that their pyroxene had a composition apparently intermediate in composition between that of types A and B. Wark 1987, reviewed them and, noting that the bulk compositions of these CAIs are not in any way intermediate between those of types A and B, renamed them type C. Type C CAI bulk compositions plot exclusively within the anorthite (+spinel) primary phase volume on Figure 7(b), which is consistent with the textures indicating anorthite + spinel to be the first crystallizing phases. Type C CAIs are relatively rare, and their chief importance lies in their bulk compositions. Of all the silicate-rich CAIs, the type Cs have compositions that are most at variance with the predictions of equilibrium condensation calculations. Beckett and Grossman (1988) explored the possible thermodynamic reasons for this, and argued that reaction of an equilibrium condensate assemblage of melilite + spinel (i.e., a type A CAI) reacted with a silica-rich gas to produce anorthite + pyroxene. MacPherson and Huss (2000) argued that the key to understanding type Cs may be simpler: their compositions would not be at variance with the predictions of equilibrium condensation calculations if the relative condensation sequence of anorthite and forsterite were reversed. Petaev and Wood (1998a) showed that the order of appearance of these two phases is pressure dependent, with lower nebular pressures favoring anorthite before (at temperatures higher than) forsterite and higher pressures favoring the reverse.

In terms of sheer numbers, probably the most abundant variety of CAI in most chondrites consists largely of spinel and calcic pyroxene. Some of these contain lesser amounts of anorthite, melilite, hibonite, and perovskite, but pyroxene–spinel-rich assemblages are almost universal. These objects range in size from a few tens of micrometers up to 1–2 cm (in CV3 chondrites). Large or small, from whatever kind of chondrite, the essential structure consists of small spinel grains or dense nodules of spinel grains or even chains of spinels, enveloped in a continuous thin rim of aluminum diopside that binds the entire structure together.

The best-known spinel–pyroxene inclusions are the large (in size) very fine-grained objects. They were originally recognized in Allende because of their very distinctive pink color, arising from the presence of iron-rich spinel. They became known as fine-grained inclusions because of their grain size relative to the types A and B CAIs, but this really is no longer a useful term given that nearly all other CAIs in all other chondrite groups have comparable or smaller grain size. The examples in Allende are permeated with nepheline, sodalite, hedenbergite, and andradite, and show a pronounced concentric zoned structure defined by differing abundances of minerals (e.g., McGuire and Hashimoto, 1989). Melilite is virtually absent. However, work on examples from reduced-subgroup CV3 chondrites (e.g., Ulyanov, 1984; Boynton et al., 1986; Wark et al., 1986; MacPherson et al., 2002) shows unambiguously that the alkali- and iron-rich assemblage in Allende CAIs represents a late-stage secondary mineralization that largely obscures the original structure of the inclusions. The "progenitor" inclusions observed in Leoville and Efremovka lack the alkali- and iron-rich phases, containing instead anorthite and melilite. Zoned examples (Figures 12(a) and (b)) have melilite-free cores consisting mostly of spinel grains (±perovskite and hibonite) embedded in sequential rims of abundant anorthite and less pyroxene; the mantles contain abundant melilite surrounding the spinel, with minor anorthite replacing the melilite and pyroxene surrounding both. Unzoned examples tend to be smaller and contain little or no anorthite. The bulk compositions of these CV3 inclusions are important for two reasons.

First, it was recognized very early (Tanaka and Masuda, 1973; Grossman and Ganapathy, 1976b) that many of them have highly unusual REE fractionation patterns that are depleted both in the most refractory elements (the heavy rare earths) and the most volatile elements (europium and ytterbium). These patterns have only been successfully modeled in terms of fractional condensation (Boynton, 1975; Davis and Grossman, 1979), and CAIs containing such patterns are universally regarded as having originated as condensates. Specifically, they cannot be volatilization residues. Second, the bulk compositions of most of the Efremovka and Leoville spinel–pyroxene-rich CAIs plot within the anorthite (+spinel) primary phase volume on Figure 7(b) and overlap those of the type C CAIs. Mineralogically and chemically, the Leoville and Efremovka inclusions are suitable candidates for the precursor material that was melted to form the igneous type Cs.

Figures 3 and 13 show two spinel-pyroxene-rich CAIs from Mighei (CM) that are typical of the kinds of seen in many chondrite types, including CO, CR, ordinary, and enstatite chondrites. Besides being much smaller than their CV3 kindred, these CM examples typically are deficient in melilite and anorthite. Figure 3 shows what MacPherson et al. (1983) called a "nodular" spinel-pyroxene inclusion, in which the central spinel region consists of numerous spinel crystals in a dense overgrowth. In contrast, the CAI in Figure 13 is a "chain-like" inclusion (MacPherson and Davis, 1994) that consists of numerous small linear chains of spinel and even some individual spinel crystals. Each has its own individual pyroxene rim. Although clearly the ensemble is part of a single object, its individual components are separated from one another by up to 50 μm of intervening meteorite matrix.

Inclusions that contain grossite as a major phase were not known until the late 1980s–early 1990s,

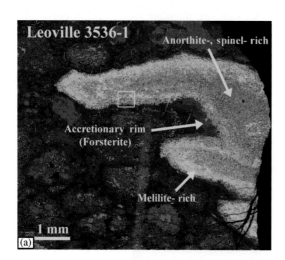
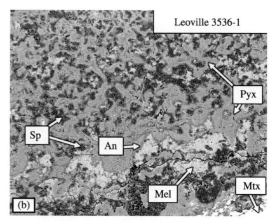

Figure 12 (a) False-color X-ray area map of a fine-grained spinel-rich CAI from the Leoville CV3 chondrite and (b) backscattered electron photograph showing details of the outer margin of the same CAI (area outlined by small rectangle in (a)). In (a), the bright aqua blue is melilite, yellow-green is diopside, dark blue and pale blue are spinel and anorthite. Bright red outside of the CAI is forsterite in chondrules, plus a thin accretionary rim immediately mantling much of the inclusion. Abbreviations as used previously.

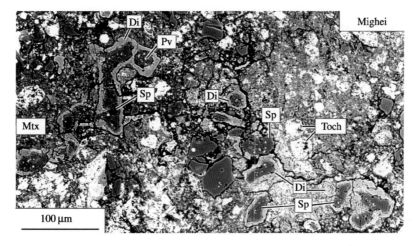

Figure 13 Backscattered electron photograph of a chain-like spinel–pyroxene inclusion from the Mighei CM chondrite. Bright patches in the surrounding matrix are tochilinite (Toch); other abbreviations as used previously (after MacPherson and Davis, 1994) (reproduced by permission of Elsevier from *Geochin. Cosmochin. Acta* **1994**, *58*, 5599–5625).

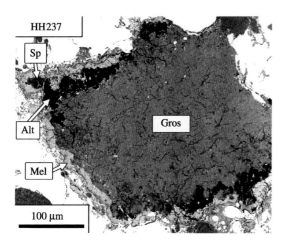

Figure 14 Backscattered electron photograph of a grossite-rich CAI from the CH chondrite HH 237 (image provided by Dr. A. N. Krot, University of Hawaii; after Krot *et al.*, 2002a) (reproduced by permission of the Meteoritical Society from *Meteorit. Planet. Sci.* **2002**, *37*, 1451–1490). Abbreviations as used previously, except: Alt—alteration; Gros—grossite.

when grossite-rich CAIs were found to be abundant in the chondrites ALH85085, Acfer 182, and Acfer 059-El Djouf 001 (Grossman *et al.*, 1988; Weber and Bischoff, 1994, 1997; and references therein). Grossite-rich CAIs are now recognized to be a distinctive property of the CR clan of meteorites, including CR, CH, and CB (Bencubbin-like; see Krot *et al.* (2002a), for a review of CR clan meteorites and their CAIs). An example of a CAI that consists mostly of grossite is shown in Figure 14. A grossite-rich CAI from another CR chondrite, NWA 470, was recently reported by Ivanova *et al.* (2002) to contain the first known natural occurrence of calcium monoaluminate, $CaAl_2O_4$.

Amoeboid olivine aggregates (hereafter, AOAs) are strictly speaking not themselves CAIs. As their name implies they consist mostly of (generally forsteritic) olivine, and in general have highly irregular shapes that suggest they are unmelted aggregates of grains (Grossman and Steele, 1976). AOAs occur in a wide range of chondrite types. Many AOAs do contain small nodules of refractory minerals including spinel, calcic pyroxene, anorthite, perovskite, and more rarely melilite (e.g., Grossman and Steele, 1976; Hashimoto and Grossman, 1987; Komatsu *et al.*, 2001). AOAs are interesting for two reasons. First, their bulk compositions straddle the gap between those of CAIs and chondrules, and indeed are close in composition to the predicted precursors for aluminum-rich chondrules although AOAs tend to be rather too olivine-rich (Komatsu *et al.*, 2001). Second, forsterite in AOAs are highly ^{16}O-rich, suggesting that they formed in the same or identical oxygen isotopic reservoir as did the primary minerals in most CAIs (e.g., Hiyagon and Hashimoto, 1999; Aléon *et al.*, 2002a; Krot *et al.*, 2002b). This in turn supports the idea that in at least some places it was the nebular gas, rather than presolar grains, that was ^{16}O-rich (see Section 1.08.8)

1.08.4.1 FUN CAIs

Wasserburg *et al.* (1977) defined a class of CAIs that have unusual isotopic properties even relative to other CAIs; these were named FUN CAIs in reference to their Fractionation and Unidentified Nuclear effects. The characteristic signatures included little or no excess of ^{26}Mg from the decay of ^{26}Al, large mass-dependent isotopic fractionation effects in both magnesium and oxygen, and relatively large nonradiogenic nuclear anomalies in a variety of elements including barium, titanium, calcium, and others. The definition of FUN became blurred with time, when other inclusions were recognized that have large nuclear anomalies ("UN" CAIs) in the absence of significant mass-dependent isotopic fractionation effects (e.g., Ireland, 1988). Still other CAIs have been found to have large mass-dependent fractionation effects but only small nuclear anomalies ("F" CAIs; e.g., inclusion "TE" reported by Clayton et al. (1984) and Niederer et al. (1985)). One feature unites FUN, UN, and F CAIs: they contained little or no live ^{26}Al at the time of their formation.

A hallmark of the two prototypical FUN inclusions (whose individual names are C1 and EK-1-4-1) is that their only distinguishing characteristics are their isotopic compositions. In all other properties they are identical to normal type B CAIs. To this day it remains true that there is no certain way to recognize a FUN CAI except by isotopic analysis. For example, the FUN inclusion AXCAI-2271 from Axtell (Srinivasan et al., 2000), shown in Figure 15, is a compact type A (melilite-rich) CAI whose ordinary appearance gives no hint of its peculiar isotopic properties. Nevertheless, certain kinds of CAIs do have a statistical propensity toward unusual isotopic properties. FoBs are relatively rare, yet four of them possess large degrees of mass-dependent isotopic fractionation (Clayton et al., 1984; Davis et al., 1991; MacPherson and Davis, 1992) in magnesium and (where analyzed) silicon and oxygen. However, only one of these, Vigarano 1623-5, has accompanying large nuclear anomalies (Loss et al., 1994). Papanastassiou and Brigham (1989) identified a type of spinel-rich inclusion (purple spinel inclusions (PSIs)) in Allende for which approximately one in five possesses nuclear isotope anomalies. Ireland (1988, 1990) showed that many hibonite-bearing CAIs in the CM chondrite Murchison have larger nuclear isotope anomalies even than those observed in the CV3 FUN inclusions. Like the FUN inclusions, those Murchison CAIs having large nuclear anomalies in titanium and calcium lack evidence for extinct ^{26}Al. Most of them differ from true FUN inclusions, however, in lacking significant mass-dependent isotopic fractionation.

The origin of mass-dependent isotopic fractionation in FUN CAIs is commonly (and somewhat casually) assumed to be the result of Rayleigh-type distillation, while the inclusions were molten. It is true that a strong case for distillation has been made in the case of the so-called HAL-type hibonites (see Section 1.08.7), based on trace element and isotopic properties (Lee et al., 1979, 1980; Davis et al., 1982; Ireland et al., 1992; Floss et al., 1996). Such an origin is problematic for other FUN CAIs, however, especially those that are otherwise identical in bulk composition to non-FUN CAIs. Most notably this is true of the FoBs that also happen to have F or FUN properties (Clayton et al., 1984; Davis et al., 1991). These objects are magnesium-rich relative to other CAIs, yet distillation experiments conducted on chondritic starting materials consistently show that

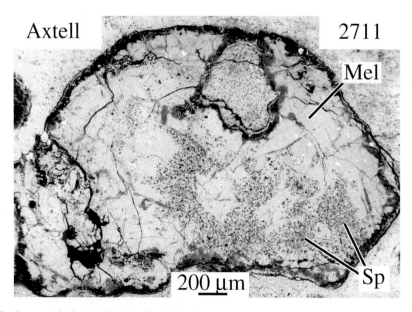

Figure 15 Backscattered electron image of a FUN CAI from the Axtell CV3 chondrite. Except for its isotopic properties, this object is an ordinary type A CAI (photo courtesy of Dr. Gary Huss, Arizona State University; after Srinivasan et al., 2000). Abbreviations as used previously.

magnesium evaporates rapidly once more volatile species such as alkalies and iron have evaporated (Notsu et al., 1978; Hashimoto, 1983; Floss et al., 1996). To produce the observed large mass-dependent fractionation in magnesium through evaporation would require the loss of 70% of the original magnesium in some cases (Davis et al., 1990, 1991), so the starting material for these still-magnesium-rich objects would have to be considerably enriched in magnesium relative to solar (CI chondritic). Moreover, most CAIs contain less magnesium than these and yet show far smaller degrees of isotopic mass fractionation. Also puzzling is the observation by Loss et al. (1994) that two FUN CAIs—one a forsterite-bearing inclusion from Vigarano (1623-5) and the other (C1) a type B from Allende—have nearly identical isotopic properties yet very different bulk major-element compositions. This coincidence requires either remarkable convergent evolution of the two CAIs or else their isotopic compositions reflect a nebular reservoir rather than distillation (Loss et al., 1994). In the case of the Vigarano FUN inclusion, Davis et al. (1991) showed that the magnesium-poor outer mantle almost certainly did experience melt evaporation during a rapid heating event, but the magnesium-rich CAI interior was untouched by this event and yet still shows significant mass-dependent isotopic fractionation. This isotopic signature thus predates at least the most recent melting/evaporation event, and Davis and coworkers discussed in detail the difficulties with producing the core isotopic composition by melt distillation (although they could not completely rule it out).

The isotopic properties of FUN inclusions relative to "normal" CAIs continue to pose major problems for virtually all global models for CAI formation and evolution. Models that can explain "normal" CAIs generally fail to explain FUN CAIs without extraordinary *ad hoc* contortions, and vice versa. Mainly, the difficulties center around ^{26}Al. As Clayton et al. (1988) pointed out, there is a remarkable mutual exclusivity between ^{26}Al and large nuclear isotope anomalies in CAIs. Figure 16, modified and updated from figure 7 of Clayton et al., shows that CAIs with normal inferred initial abundances of ^{26}Al (^{26}Al/^{27}Al $\sim 5 \times 10^{-5}$) have anomalies of ^{50}Ti (for example) of at most ~10 parts per thousand (per mil, ‰), whereas CAIs with larger titanium anomalies formed with little or no live ^{26}Al. The ^{26}Al and ^{50}Ti likely originated in different astrophysical sources and probably entered the solar system in different presolar grains. Yet, during the high-temperature processing in the solar nebula that formed the CAIs, the two isotopic reservoirs did not mix. Given that live ^{26}Al apparently was quite widespread in the early solar nebula at an initial abundance ratio (^{26}Al/^{27}Al)$_0 \sim 5 \times 10^{-5}$ (review

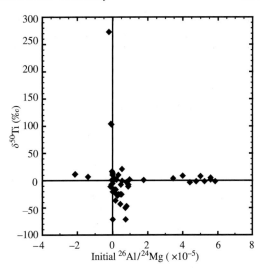

Figure 16 ^{50}Ti isotopic composition versus calculated initial ^{26}Al/^{27}Al for hibonite-bearing CAIs from CM chondrites. Titanium isotope anomalies larger than ±10‰ do not occur in CAIs with initial ^{26}Al/^{27}Al $\sim 5 \times 10^{-5}$, and vice versa. Diagram is updated from Clayton et al. (1988, figure 7). Only analyses having Al/Mg > 20 are shown (sources Fahey et al., 1987a; Hinton et al., 1984; Hutcheon et al., 1980; Ireland, 1988, 1990; Ireland and Compston, 1987; Ireland et al., 1991; Macdougall and Phinney, 1979).

by MacPherson et al., 1995; see also Russell et al., 1996; Guan et al., 2000a), how then can its absence from FUN CAIs be explained? If initial ^{26}Al/^{27}Al ratio was perfectly uniform throughout the nebula, then the magnesium isotopic differences between FUN and normal CAIs would suggest that the FUN CAIs formed or were reprocessed several million years later than the normal CAIs. This is difficult to reconcile with the observation that the FUN CAIs preserve such large presolar isotopic anomalies despite their long residence time in a violent and locally very hot solar nebula. The alternative that ^{26}Al was heterogeneous in the solar nebula requires near-complete physical separation of two contemporaneous and isotopically distinct reservoirs. A very different alternative (e.g., Wood, 1998) postulates that ^{26}Al was injected into the solar system by a very time-restricted event, and that FUN and non-FUN CAIs were formed before and after this event, respectively. In other words, exactly counter to the expectations based on initial ^{26}Al/^{27}Al differences, the FUN inclusions would actually be older than non-FUN CAIs.

1.08.4.2 Unresolved Problems

For forsterite-bearing FUN CAIs and other FUN CAIs that are not HAL-type hibonites, we still do not understand whether their isotopic fractionation is the result of melt evaporation or some other process and, if the former, what the precursor materials might have been. It is not even clear if "normal" and FUN CAIs formed by the same or completely separate mechanisms. If they did form the same way, there is a problem explaining why FUN CAIs show such large degrees of isotopic mass-dependent fractionation when normal CAIs do not even though the latter have equally or more refractory bulk compositions than FUN CAIs. If different formation mechanisms are involved, then remarkable convergent evolution of CAI populations is required. Much more work is needed in these areas.

A critically needed experiment is to measure with very high precision the absolute radiometric age of one or more FUN CAIs and thus determine whether FUN CAIs are older, younger, or contemporaneous with normal CAIs. Beyond just helping to better understand when and where FUN CAIs formed relative to "normal" CAIs, this experiment is critical for better understanding observed differences in initial $^{26}Al/^{27}Al$ among early solar system objects.

1.08.5 DISTRIBUTION AMONG CHONDRITE TYPES

The many and varied chondrite groups differ in bulk chemistry, grain size, mineral chemistry, and the relative abundance of components such as chondrule and CAI varieties. Yet all chondrites are believed to come from the asteroid belt, and the different chondrite "flavors" reflect in some fashion the heterogeneity of the asteroid belt. The asteroid belt of course is only a tiny fraction of the solar system, but to the extent that it is at all representative, heterogeneity in the asteroid belt implies heterogeneity in the rest of the solar system. From the perspective of CAIs, this heterogeneity is potentially very important. For example, CAIs are the principal carriers of evidence for some extinct radionuclides such as ^{26}Al and ^{41}Ca, but as of the early 1990s most of the work on isotopes in CAIs had been done on a limited variety of CAIs from the CV3 and CM2 carbonaceous chondrites. The situation for oxygen isotopes was similar. In order to utilize extinct short-lived radionuclide data as high-resolution chronometers, e.g., it is necessary to establish that isotopic differences between any two or more objects are due to differences in time of formation rather than nebular heterogeneity in the distribution of isotopes. For this reason, we have witnessed, since the early 1990s, an explosion of studies that compare the petrologic and isotopic properties of CAIs (especially) across the entire spectrum of chondrite groups: how do the CAI populations vary, are there systematic isotopic differences between CAI types, and are there systematic isotopic differences between CAIs from different chondrite groups? This section addresses the first question, of systematic differences in the abundance of CAI varieties across chondrite groups.

At the time of the 1988 review, little was known about CAIs in meteorites other than CV3 and CM chondrites. Since then, there have been extensive petrologic and isotopic studies of CAIs in ordinary and enstatite chondrites, CO3 carbonaceous chondrites, and the variants of the CR chondrite clan (CR, CB, CH). All of these chondrite groups contain small (<1 mm) melilite-rich, hibonite-rich, and spinel-pyroxene-rich CAIs as the most common varieties. However, they differ in containing different proportions of some CAI varieties. To give two examples, CH chondrites contain an exceptional abundance of grossite-rich CAIs that are only rarely found in other chondrites, and CM chondrites are notable for the rarity of melilite-bearing CAIs. And, again, the CV3 chondrites are now recognized as being highly atypical in terms of their CAI population (see below). Both the differences and similarities among the chondrite group CAI populations are informative. Consider the enstatite chondrites, which contain CAIs that are mineralogically similar to ones found in other chondrite groups. Equilibrium condensation calculations indicate that, if the characteristic enstatite chondrite mineralogy (e.g., oldhamite) formed as a result of nebular condensation, it did so under conditions so reducing that the most refractory elements would have condensed as nitrides and sulfides rather than silicates and oxides (e.g., Larimer and Bartholomay, 1979). To the extent that CAIs or their precursors may represent condensates, the observation that enstatite chondrites in fact contain "normal" silicate- and oxide-bearing CAIs suggests at least the possibility that all CAIs in all chondrite types share some common origin unrelated to the chondrites in which they reside. As will be discussed later (see Section 1.08.9), isotopic evidence (especially beryllium–boron and oxygen) suggests this same possibility. Following are general observations about CAI populations in each of the major chondrite types.

CV3 chondrites. Allende, Vigarano, and their kindred contain the greatest dynamic range of CAIs, in terms of both variety and size. The centimeter-sized CAIs in CV3 chondrites are an order of magnitude larger than even the largest (~1 mm) CAIs found in any other chondrite types. And, although the CV3s also contain abundant numbers of the same small CAI varieties that are found in other the chondrite types, the stereotypical type B CAIs and their rarer forsterite-bearing relatives (the FoBs) are virtually restricted to CV3s. The point cannot be stressed too highly: CV3 chondrites and their huge type B CAIs are the oddballs of the chondrite world, during the 1970s, they virtually defined how we understood the solar nebula.

The historic irony of the CV3s does not end there. The most-studied CV3 meteorite is Allende, and yet it is a very misleading representative of that chondrite group. Allende and its components are pervaded by alkali- and iron-rich secondary minerals that replace primary phases, and the CAIs are strongly affected. All of the oxidized subgroup CV3s show this alteration to some degree, which has greatly hindered deciphering the primary petrologic and isotopic properties of the CAIs. In contrast, the reduced subgroup CV3 meteorites—Vigarano, Leoville, Efremovka, and Arch—largely lack this secondary mineralization. Most recent CAI studies have focused on these meteorites. The fall of so much Allende material at such a fortuitous time (see introductory statement by Grossman, 1980) unquestionably had a profound positive influence on our understanding of the early solar system, but the path to progress would have been straighter and quicker if several tons of Vigarano or Efremovka had fallen instead!

CO3 chondrites. Russell *et al.* (1998, 2000) studied the petrologic and isotopic characteristics of more than 450 CAIs from 12 CO3 meteorites covering a range of petrologic subtypes, and including the CO3-like meteorites MAC87300 and MAC88107. All of the CAIs are less than ~1 mm in size, and most are less than ~0.5 mm. Russell *et al.* (1998) showed that a progression in petrologic subtype from 3.0 up through 3.7 correlates with a change in the CAI population, primarily as a result of secondary mineralization that changed not only the mineralogy but also mineral chemistry of the CAIs. The best estimate of the original CAI population comes from the least-metamorphosed (3.0) CO3 meteorites Colony and ALHA77307, which contain abundant melilite-rich (type A; one is shown in Figure 17), spinel–pyroxene-rich, and hibonite-rich CAIs that are little affected by secondary mineralization. Rarer types include hibonite–pyroxene spherules. The type A CAIs are basically smaller versions of their counterparts in CV3 chondrites, and the spinel–pyroxene inclusions are indistinguishable from the most abundant CAIs that occur in CM meteorites. With progressive increase of petrologic subtype comes systematic replacement of melilite by secondary minerals and transformation of iron-free spinel into iron-rich spinel. For example, Isna (3.7) contains no melilite-rich CAIs and spinel typically contains 50% or more of the hercynite ($FeAl_2O_4$) component.

CM chondrites. CM meteorites have perhaps the most restricted population of CAIs because, other than phases occurring in Wark–Lovering rims, "primary" silicates such as melilite are very rare. All CAIs in the CMs are small, only rarely achieving sizes even as large as 1 mm (the largest known to this author is 1.7 mm; MacPherson and Davis, 1994). Most are less than 0.5 mm.

The dominant CAIs are spinel–pyroxene inclusions, with centers that are rich in spinel, with or without accessory hibonite and perovskite, and Wark–Lovering rims consisting mainly of aluminous diopside and iron-rich phyllosilicate. These CAIs can take the form of compact objects (e.g., see figure 4 of MacPherson and Davis, 1994), porous aggregates, or even distended chains of spinel nodules (e.g., Figure 13). In some cases, the entire CAI and its rim are encased in an outer shell of forsterite or enstatite.

Hibonite-rich CAIs in CM chondrites are among the most-studied of all CAIs. They are easily recognized in thin section and in hand sample because of the sapphire-blue color of the hibonite. Because hibonite is expected to be one of the earliest (highest temperature) condensates formed in the solar nebula, numerous ion microprobe studies were conducted on hibonite-rich CAIs in the late 1980s and early 1990s by groups from the University of Chicago and Washington University. These studies revealed an unexpected diversity of trace element and isotopic signatures in this group of mineralogically similar objects (Ireland, 1988, 1990; Ireland *et al.*, 1988; Hinton *et al.*, 1988). Some are microspherules (e.g., Figure 10) that almost certainly solidified from melt droplets. Others are loose aggregates of crystals and, in some cases, even solitary hibonite crystals. Ireland (1988) classified hibonite-rich CM inclusions into (i) spinel–hibonite inclusions, SHIBs, that include the microspherules such as that in Figure 10; (ii) blue aggregates, BAGs, that are nearly monomineralic hibonite clusters; and (iii) platy crystals, PLACs, that are single crystals of hibonite. Corundum is more common as an accessory in CM hibonite-rich CAIs than it is in CAIs from any other chondrite types, although even here corundum is scarce (see Bar-Matthews *et al.*, 1982; MacPherson *et al.*, 1984a; Fahey, 1988; Simon *et al.*, 2002). It generally occurs together with hibonite and perovskite (eg., Figure 18).

CM CAIs are notable for the paucity of melilite. Although melilite has been reported as a minor accessory in several CM inclusions, only one melilite-rich object has ever been described (see figure 3 of MacPherson *et al.*, 1983).

Hibonite–pyroxene spherules and grossite-bearing CAIs have been reported from CM chondrites (Ireland *et al.*, 1991, and Simon *et al.*, 1994, respectively); however, grossite is rare in CM chondrites.

Secondary mineralization in CM CAIs is abundant, and is dominated by hydrous phases.

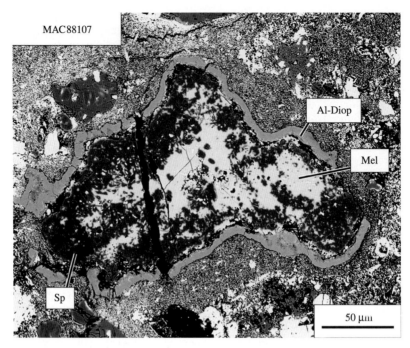

Figure 17 Backscattered electron image of a type A CAI in the CO3-like meteorite MAC88107. Abbreviations as used previously.

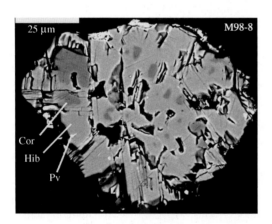

Figure 18 Backscattered electron image of a hibonite-rich, corundum-bearing CAI in the CM chondrite Murchison (photo courtesy of Simon et al., 2002) (reproduced by permission of the Meteoritical Society from Meteorit. Planet. Sci. **2002**, 37, 533–548). Abbreviations as used previously.

The principal secondary minerals in CM CAIs are diverse phyllosilicates (cronstedite, serpentine, and the Fe–Ni–S–O phase tochilinite; see review by Tomeoka et al., 1989) and calcite. Feldspathoids have only rarely been reported (MacPherson et al., 1983; El Goresy et al., 1984).

CR clan chondrites. The CR "clan" includes a wide diversity of chondritic meteorites that share sufficient mineralogic and isotopic properties in common to be considered genetically related (Weisberg et al., 1995); it includes the CR and CH groups, the unique meteorite LEW 85332, and the Bencubbin-like meteorites. Krot et al. (2002a) and Aléon et al. (2002a) have extensively reviewed the CR chondrites and their CAIs, respectively. The significant petrographic and mineralogic diversity that exists even within the individual groups extends to the CAIs.

CR-clan meteorites have become increasingly important among carbonaceous chondrites, because most of the meteorites (Al Rais and Renazzo are exceptions) show very little evidence for pervasive metamorphism or metasomatism that characterize most other carbonaceous chondrites. CR CAIs show only minor secondary mineralization at most, and the oxygen isotope signatures generally do not follow the pattern seen in CAIs from CV3 and CO3 chondrites of internal disequilibrium among the phases of each individual CAI. In short, pristine petrologic and isotopic characteristics are better preserved in CR CAIs than in those from other carbonaceous chondrites. CAIs from the CH group have also attracted particular attention because of the notable abundance of grossite-bearing and grossite-rich inclusions, a phase which is rare in all other chondrite groups.

CAIs from the CR group proper are generally less than 1 mm (most are less than 500 μm). The most common varieties are melilite-rich type As (e.g., Figures 19(a) and (b) and spinel-rich aggregates that range from objects indistinguishable from the fine-grained spinel+pyroxene+melilite inclusions (Figure 19(c)) like those in

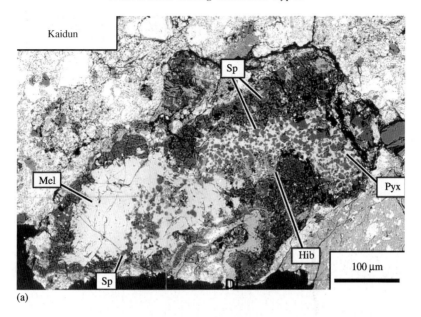

Figure 19 Backscattered electron images of: (a) a type A CAI in the CR lithology meteorite breccia Kaidun; (b) a type A CAI in the CR meteorite GRA95229; (c) a spinel-rich aggregate in PCA91082; and (d) a spinel nodule in the CR lithology of Kaidun. Abbreviations: Phyll—phyllosilicate; W-L Rim—Wark–Lovering rim sequence; others as used previously (photos 19(b) and 19(c) courtesy of Dr. A. N. Krot, University of Hawaii; after Aléon et al., 2002) (reproduced by permission of the Meteoritical Society from *Meteorit. Planet. Sci.* **2002**, *37*, 1729–1755).

reduced CV3 chondrites to small and simple spinel nodules (Figure 19(d)). Hibonite-rich and grossite-rich inclusions are relatively rare. The CR chondrite Acfer 059/El Djouf 001 is one of the only non-CV3 meteorites from which a CAI like type B has been described (Weber and Bischoff, 1997), although the small grain size (≤100 μm) and highly aluminous melilite composition of the Acfer object set it apart from the "true" type Bs (found in CV3s) that have millimeter-sized crystals and melilite containing up to ~72% of the magnesium-rich (åkermanite) component. Amoeboid olivine aggregates that enclose refractory nodules also occur. Secondary alteration of CAIs is minor to nonexistent in many CRs. Alkali-bearing phases apparently are very rare (Aléon et al., 2002a); calcite and phyllosilicates are more common.

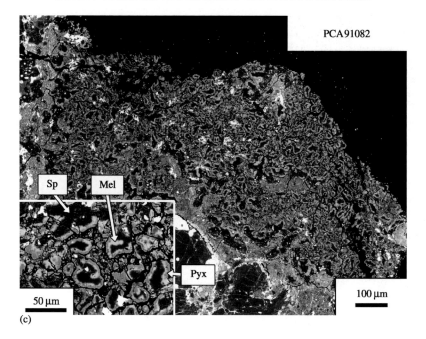

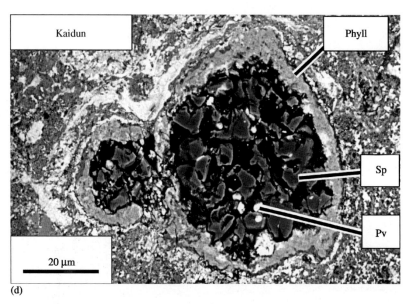

Figure 19 (continued).

Bischoff et al. (1993) first proposed the name CH for a small group of very metal-rich chondrites that otherwise are chemically and mineralogically similar to the CR chondrites. The best-studied CH meteorites are ALH85085 (and several similar Antarctic meteorites) and Acfer 182. The two are chemically similar but the components in Acfer 182 are, on average, larger than those in ALH85085.

The CAI population in CH meteorites is distinctive in several respects relative to other chondrites. First, the CAIs are very small. The largest CAI reported in Acfer 182 is 450 μm in size and most are less than 300 μm (Weber et al., 1995); those in ALH85085 are even smaller, with the largest reported CAI in this meteorite being 110 μm (Grossman et al., 1988). A second unusual feature is an exceptional abundance of not only grossite-bearing but grossite-rich CAIs. More than half of the CAIs studied by Weber and Bischoff (1994) in Acfer 182 contain grossite, and in more than one third it is a major phase. Grossite is common in the CAIs in ALH85085 as well (Grossman et al., 1988; Kimura et al., 1993), Figure 20(a). The only known reported natural occurrence of calcium monoaluminate, $CaAl_2O_4$, is in a grossite-rich CAI in the CH chondrite NWA 470 (Ivanova et al., 2002). The CAI population in

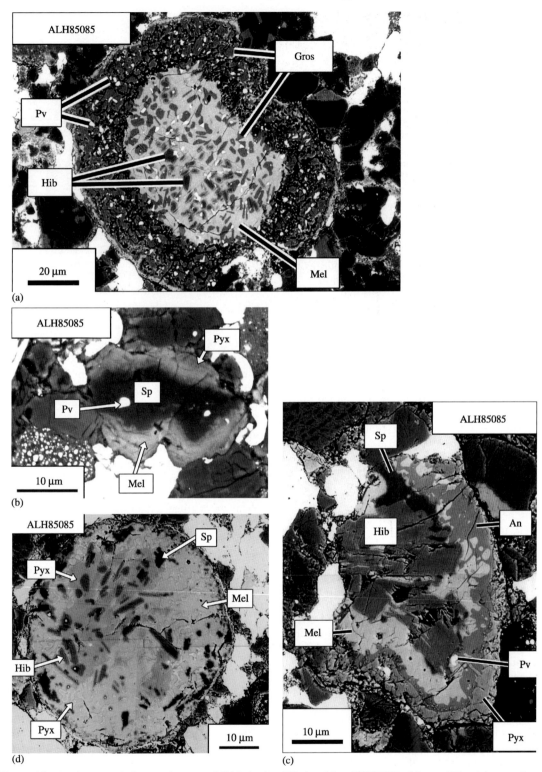

Figure 20 Backscattered electron images of CAIs in the CH chondrite ALH85085: (a) a melilite-rich, grossite-bearing CAI; (b) a spinel-rich CAI; (c) a hibonite-rich CAI, with anorthite replacing melilite; and (d) a hibonite-silicate spherule. Abbreviations as used previously (photos 20(a) and 20(b) modified from Grossman et al. (1988) (reproduced by permission of Elsevier from *Earth Planet. Sci. Lett.* **1988**, *91*, 33–54).

ALH85085 contains an unusual abundance of hibonite-rich microspherules (e.g., Figure 11) in which hibonite and grossite commonly occur as microphenocrysts; such objects apparently are less common in Acfer 182. Other CAI varieties in CH chondrites are similar to ones found in other chondrite types, and include spinel- and melilite-rich examples (e.g., Figures 20(b) and (d)).

The CAIs in CH chondrites show little secondary mineralization; rare anorthite replaces melilite in some CAIs (e.g., Figure 20(c)), but alkali-rich phases, grossular, and phyllosilicates have not been found. Wark–Lovering rims occur on some but not all CAIs; those on the microspherules in ALH85085 tend to consist mainly of a wispy layer of calcic pyroxene.

One CAI was reported (Weisberg et al., 2002) in the Bencubbin-like meteorite Gujba, which contains melilite + pyroxene + spinel. No photographs or detailed descriptions have been published; CAIs have not (as of this writing) been reported in Bencubbin or Weatherford.

Ordinary chondrites. CAIs in ordinary chondrites are small and rare. Until Bischoff and Keil (1984) conducted the first systematic search, only a smattering of CAIs in ordinary chondrites had ever been reported (see brief summary in MacPherson et al., 1988). Bischoff and Keil (1984) used optical microscopy to search 41 thin sections of various ordinary chondrites, and reported a variety of very small, mainly spinel-bearing objects that contain pyroxene as a common accessory and in many cases have secondary nepheline. They also described a hibonite–spinel CAI fragment that turned out to have unusual isotopic properties (Hinton and Bischoff, 1984) and one exceptional hibonite–melilite–spinel–perovskite CAI from Semarkona. Russell et al. (1996) and Huss et al. 2001a later used systematic X-ray area mapping of thin sections of unequilibrated ordinary chondrites (UOC) in the hopes of locating significant numbers of CAIs having sufficient size (>10 μm) and grain size to be suitable for their isotopic studies. The results were mixed: although numerous small and mostly spinel-bearing fragments were found, only three new larger objects were found suitable for analysis (in addition to the hibonite-rich Semarkona CAI noted above that had been found by Bischoff and Keil, 1984). More recently, Kimura et al. (2002) also used X-ray area mapping to locate 43 CAIs larger than 17 μm in size within sections of three unequilibrated H chondrites from the Japanese Antarctic meteorite collection. Of these, four were suitable for oxygen isotopic analysis. Taking the recent work together with that of Bischoff and Keil (1984) and others, the maximum dimension of ordinary chondrite CAIs is 480 μm and most are significantly smaller. Although spinel-rich ones are the most commonly encountered varieties, three others are now known to contain melilite (e.g., Figure 21) and three more contain hibonite. Nepheline, sodalite, and phyllosilicate are common secondary minerals, and their presence does not correlate with metamorphic grade of the host; e.g., the type A CAI from Quinyambie (3.4) shown in Figure 21 is completely unaltered,

Figure 21 Backscattered electron image of an unaltered type A inclusion in the unequilibrated ordinary chondrite Quinyambie. Abbreviations as used previously (photo modified from Russell et al., 1996) (reproduced by permission of the American Association for the Advancement of Science from *Science* **1996**, 273, 757–762).

whereas a Semarkona CAI (3.0) studied by Huss et al. (2001) is partially replaced by sodalite.

Enstatite chondrites. Surveys of CAIs in enstatite chondrites have been reported by Bischoff et al. (1985), Guan et al. (2000a,b), and Fagan et al. (2001). As was the case for the ordinary chondrite CAIs, the more recent studies have relied on X-ray area mapping of whole thin sections rather than optical searches in order to locate these rare and very small objects. None of the ~90 whole CAIs (not including mineral fragments) reported in the above studies are larger than ~140 μm, making them on average among the smallest of all CAI populations. These CAIs are very unevenly distributed among the different enstatite chondrites. Fifty of the 80 (>1/2) CAIs found by Guan et al. (2002) came from three thin sections of one Antarctic meteorite (EET87746) and two thin sections of one African meteorite (Sahara 97072), whereas two thin sections each of the meteorites MAC88136 and PCA91020 yielded a total of only six CAIs. Single thin sections of each of eight other enstatite chondrites yielded no CAIs whatever.

The mineralogy of enstatite chondrite CAIs presents instructive and very important contrasts with the mineralogy not only of CAIs from other chondrite types but also with their own host enstatite chondrites. For example, the essential primary mineralogy and structures of enstatite chondrite CAIs are similar to many CAIs seen in other chondrite types; however, there is a weak late-stage overprint of distinctive mineralogy within the enstatite chondrite CAIs unlike that found in other CAIs. Conversely, the characteristic enstatite chondrite phases oldhamite and osbornite (TiN) do not occur inside the CAIs even though the host chondrites contain these phases. Further, Guan et al. (2000a) also showed that the metal inside one CAI contains very low (0.2 wt.%) metallic silicon whereas the metal in the host chondrite contains ~2.5%. Therefore, the enstatite chondrite CAIs clearly formed under more oxidizing conditions than did their host chondrites, and then experienced a very late stage event that was moderately reducing and unlike late stage modification experienced by CAIs from other chondrite groups.

Spinel–hibonite (Figure 22(a)) and spinel–pyroxene inclusions are the most common varieties; melilite apparently is uncommon and has only been reported in two CAIs (e.g., Figure 22(b); Guan et al. 2000a; Fagan et al., 2000). Metal, perovskite, and plagioclase are all known accessories. Some rim pyroxene and even spinel inside the CAIs contain significant oxidized iron.

Many CAIs show a distinctive late-stage mineralogy that consists of titanium-bearing troilite and geikelite, $MgTiO_3$; the troilite fills in pore spaces and cracks, whereas the geikelite rims perovskite (Fagan et al., 2000; Guan et al. 2000a). Textural evidence suggests that the troilite and geikelite even postdate the Wark–Lovering rims. Secondary alteration of the primary phases is common.

The most remarkable facts about the enstatite chondrite CAIs are that (i) they exist at all, and (ii) they are not particularly remarkable as CAIs go. If enstatite chondrite CAIs originated by condensation under the highly reducing conditions considered necessary for forming the characteristic enstatite chondrite mineralogy, refractory lithophile elements like calcium and aluminum are expected to be largely locked up in sulfides, carbides, and nitrides (e.g., Lodders and Fegley, 1995). Phases such as melilite, hibonite, and spinel should not be present. The fact that enstatite chondrite CAIs do contain these phases has been used to argue (e.g., Fagan et al., 2000; Guan et al., 2000a) that those CAIs did not originate in the putative enstatite chondrite formation zone, and instead share a common origin with all other CAIs in a single restricted solar nebular region—possibly near the infant sun.

1.08.6 AGES

A variety of isotopic evidence demonstrates CAIs to be the oldest objects formed in the solar system.

The most precise absolute radiometric ages are based on lead isotopes, and recent work by Amelin et al. (2002) established the ages of two inclusions from Efremovka as being 4.567 Ga with a precision of better than 1 Myr. This result confirms, with substantially better precision, previous results by Allégre et al. (1994) and other earlier data (see review of pre-1979 lead and other isotopic age data in Grossman, 1980). Significantly, the improved precision of the techniques used by Amelin et al. allowed those authors to obtain highly precise whole-rock isochron ages of ferromagnesian chondrules from the CR chondrule Acfer 059. Their data demonstrated clearly, for the first time, that CAIs have absolute ages ~2 Myr older than at least some ferromagnesian chondrules, thus confirming results based on differences in initial abundances of ^{26}Al in the two kinds of objects. This important result goes a long way toward refuting the argument (e.g., Wood, 1996) that CAIs and chondrules necessarily formed at about the same time because a 1–2 Myr age difference is too long for CAIs to have remained in solar orbit without

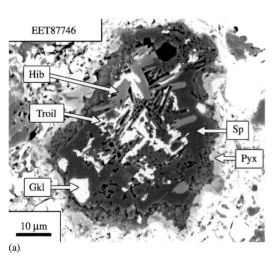
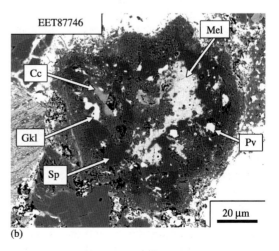

Figure 22 Backscattered electron images of CAI in the enstatite chondrite EET87746: (a) 4640-1, a spinel-hibonite inclusion and (b) 4631-1, one of the rare melilite-bearing inclusions. Abbreviations: Cc—calcite; Gkl—geikelite; Troil—troilite; others as used previously (photos modified from Guan et al., 2000a) (reproduced by permission of the American Association for the Advancement of Science from Science **2000**, 289, 1330–1333).

falling into the Sun owing to gas drag. Clearly, the physics of the early solar nebular disk is not well understood.

Another precise radiochronometric tool is initial $^{87}Sr/^{86}Sr$, especially in objects with low bulk rubidium. Gray et al. (1973) found one Allende CAI to have the lowest initial ratio yet seen in any solar system material: $(^{87}Sr/^{86}Sr)_0 = 0.69877$. This value has endured as a solar system benchmark for strontium isotopes, and was given the name ALL. Later work by Podosek et al. (1991) on a different set of Allende CAIs also found very low initial ratios: 0.698793–0.698865, which are, however, not quite as low as ALL. An interesting result of the study of Podosek et al. (1991) is that they reanalyzed the same CAI (D7; USNM 3898 of Podosek et al.) on which ALL is based but obtained a significantly higher initial ratio than did the Caltech group. This discrepancy remains unresolved. Nevertheless, the low initial strontium ratios of CAIs are consistent with their being the oldest objects formed in the solar system.

With the exception of some early potassium–argon ages of nearly 5 Ga, and which are believed to be analytical artifacts, all high quality radiometric ages of CAIs are no older than ~4.56–4.57 Ga. This observation suggests that whole CAIs are not themselves presolar objects, although it does not rule out the possibility that presolar CAI precursors were reprocessed (melted or otherwise completely recrystallized) in the earliest solar system. For more on CAI chronology, see Chapter 1.16.

1.08.7 TRACE ELEMENTS

The bulk trace element abundance patterns in CAIs are generally agreed to reflect element volatility, with the most refractory elements enriched relative to solar (CI chondrite) abundances, and volatile elements depleted.

The main fractionation patterns were recognized very early as a result of studies of Allende CAIs (e.g., Grossman, 1973; Grossman and Ganapathy, 1976a,b; Martin and Mason, 1974; Mason and Martin, 1977; Tanaka and Masuda, 1973) and later found to occur in CAIs from other chondrites and chondrite types as well (e.g., Fahey et al., 1987a; Hinton et al., 1988; Ireland, 1990; Ireland et al., 1988; Russell et al., 1998; Guan et al., 2002). Consider first the REEs. Mason and Martin (1977) classified five distinct REE patterns (a sixth applies only to ferromagnesian objects and is not considered here); examples of their groups I, III, V, and VI patterns are shown in Figure 23(a); these are essentially flat (unfractionated) patterns that are distinguished on the basis of enrichments or depletions

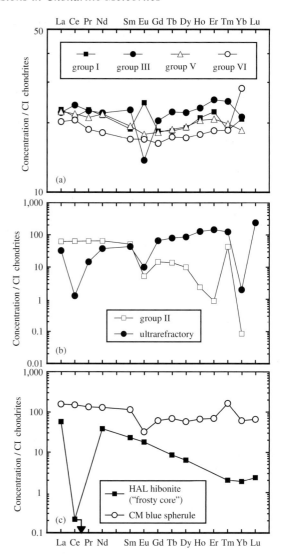

Figure 23 Examples of REE patterns observed in CAIs. CI abundances from Anders and Grevesse (1989) (sources Mason and Taylor, 1982; Ireland et al., 1988; Davis et al., 1982).

in the relatively volatile elements europium and ytterbium. The essential point is that, although enriched, most of the REEs are present in their cosmic proportions relative to one another. This effectively rules out any kind of planetary process (e.g., igneous differentiation), and thus most workers interpret these patterns in terms of complete condensation from a solar gas. Tanaka and Masuda (1973) first identified a very different kind of REE pattern, shown in Figure 23(b), which Mason and Martin labeled group II and in which the REEs are highly fractionated from one another. Such a pattern is depleted in both the most refractory (the heavy REEs, except for ytterbium) and most volatile REEs (europium and ytterbium). Boynton (1975) and later Davis and Grossman (1979) showed unequivocally that such a pattern can only arise through fractional condensation, in which a component containing the most refractory REEs was removed by condensation and then the remaining gas condensed all the remaining REEs except the most volatile elements (europium and ytterbium). The high-temperature fractionating phase that removed the most refractory REE was originally thought to be perovskite (e.g., Davis and Grossman, 1979) but later calculations showed in fact that hibonite must be the phase responsible (MacPherson and Davis, 1994). Group II patterns are dominant (although not universal) among the fine-grained spinel-rich CAIs in the CV3 chondrites, and are common among hibonite grains and hibonite inclusions in CM chondrites (e.g., Fahey et al., 1987a; Ireland, 1990). An example of a group-II-like REE pattern from a CM inclusion, a spinel–hibonite spherule, is shown in Figure 23(c). The complementary REE pattern to group II, the so-called ultrarefractory, is rarely observed (see review by Simon et al., 1996). This material represents the highest-temperature fraction removed prior to group II condensation. An example of an ultrarefractory REE pattern, from a platy single hibonite crystal (PLAC) from Murchison, is shown in Figure 23(b). Only recently have ultrarefractory REE patterns been identified in CAIs from a CV chondrite (El Goresy et al., 2002; Hiyagon et al., 2003), which is ironic considering that CVs are where group II REE patterns were first discovered. Finally, a small number of hibonite grains have been discovered to have REE fractionation patterns completely unlike those discussed above. An example is shown in Figure 23(c), from the FUN inclusion HAL that has become the prototype for REE patterns of this variety. It is characterized by a huge negative cerium anomaly superimposed on a pattern in which all the other REEs fall off in abundance smoothly with increasing atomic number. HAL and its related hibonites are also characterized by large degrees of mass-dependent isotopic fractionation in oxygen, magnesium, and calcium, together with low bulk concentrations of magnesium and titanium. Collectively, these properties are interpreted by most workers to mean that HAL-type hibonites formed by melt distillation (see Lee et al., 1979, 1980; Davis et al., 1982; Ireland et al., 1992; Floss et al., 1996; Wang et al., 2001). Because cerium is more volatile under oxidizing conditions whereas europium is more refractory, the negative cerium anomalies coupled with small or no europium anomalies in HAL-type hibonites were believed indicative of distillation under oxidizing conditions relative to those under which most CAIs originally formed. However, Floss et al. (1996) and Wang et al. (2001) experimentally showed that cerium anomalies could result from very localized oxidizing conditions at the surface of the distilling CAIs, not necessarily from nebular oxidation conditions.

Refractory siderophile elements are enriched but unfractionated (i.e., cosmic relative proportions) in most CAIs, similar to the case for refractory lithophile elements. The fractionated group II CAIs are depleted in refractory siderophiles. These observations suggest that the refractory siderophile abundances, like the lithophiles, are the result of volatility controlled processes. Unlike the refractory lithophiles, however, the refractory siderophiles are predicted by equilibrium condensation theory to condense as relatively simple metallic alloys rather than as trace components in oxides and silicates. The very wide range of alloy compositions observed in the tiny metal nuggets within CAIs led Sylvester et al., (1990) to propose that condensation of the various metals was controlled partly by the crystallographic structures of their high-temperature solid phases. Each metal alloyed mainly with other metals having similar structures, of three basic kinds: face-centered cubic (iridium, platinum, rhodium, gold), body-centered cubic (tungsten, molybdenum), and hexagonal close-packed (rhenium, osmium, ruthenium). It is interesting that, despite the highly variable compositions of individual nuggets within individual CAIs, the CAI bulk compositions nevertheless have unfractionated (except in group II) abundances of the different siderophile elements. Each CAI must have sampled a statistically large number of tiny grains in order to achieve the overall cosmic abundance patterns.

Some CAIs are slightly depleted in molybdenum and tungsten relative to other refractory elements. Because those two elements are more volatile during condensation under oxidizing conditions, Fegley and Palme (1985) concluded that CAIs showing such depletions must have formed under relatively oxidizing conditions. However, Sylvester et al. (1990) suggested that their three metal-phase condensation model (above) is better able to explain details of molybdenum and tungsten abundance patterns without resorting to unusual oxidation conditions.

Allende CAIs commonly contain alkali- and iron-rich phases, as a result of which their bulk compositions are enriched in those and other volatile elements. Early workers recognized the apparent incompatibility of enrichment in both highly refractory and highly volatile elements. In general, the volatiles were interpreted to be the result of secondary mineralization, possibly on the parent body, and unrelated to the high-temperature nebular conditions under which the CAIs first formed (e.g., Grossman and Ganapathy, 1975). However, observations of condensate-like nepheline needles in some Allende fine-grained spinel-rich CAIs led Grossman et al. (1975) to conclude that at least some of the nepheline in these particular inclusions did form in the nebula. Accordingly, Grossman and Ganapathy (1975) concluded that some of the volatile element concentrations in the fine-grained inclusions are indigenous. Much later work (Mao et al., 1990) on the bulk composition of a fine-grained spinel-rich CAI from the reduced CV3 Leoville revealed it to have extremely low concentrations of volatile elements such as gold and sodium. Conversely, a fine-grained CAI from another reduced subgroup CV chondrite, Vigarano, has elevated contents of these elements comparable to those of its Allende counterparts. These data are consistent with a parent-body metasomatic origin for the most volatile elements in these inclusions, in which the degree of metasomatism was controlled by the porosity of the host rocks (MacPherson and Krot, 2002. Leoville and its near-twin Efremovka are highly flattened and dense objects that probably experienced intense shock deformation (Cain et al., 1986), and in this respect they differ not only from Allende and other oxidized subgroup CV3s but also Vigarano.

1.08.8 OXYGEN ISOTOPES

1.08.8.1 General

The discovery by Clayton et al. (1973) that Allende CAIs contain non-mass-dependent enrichments in ^{16}O (relative to ^{17}O, ^{18}O) was in part responsible for setting both the excitement and tone of CAI research for the ensuing 30 years. This isotopic signature, together with evidence for extinct ^{26}Al and nonradiogenic isotope anomalies in elements such as titanium and calcium, was attributed to the isotopic compositions of the presolar precursor dust from which the CAIs formed. Hence, CAIs have astrophysical significance.

Clayton et al. (1977) showed in more detail that the Allende CAIs individually have isotopic disequilibrium, and the constituent minerals disperse along the same mixing line defined by analyses of multiple whole CAIs. This was later shown to be true for CAIs from Leoville and Vigarano CAIs as well (Clayton et al., 1986, 1987). In general, spinel was found to be the most ^{16}O-rich phase, with pyroxene, anorthite, and melilite having progressively lower enrichments in that order. A significant observation was that spinel almost always showed a maximum of ~4% enrichment in ^{16}O relative to terrestrial standards; in other words, there seemed to be a very consistent upper limit to ^{16}O-enrichment regardless of CAI. All of this early work was done by reacting bromine pentafluoride with either whole CAIs or mineral separates, followed by conventional gas mass spectrometric analysis. Even for whole CAIs, work was generally restricted to the large CAIs from the CV3 chondrites. This CV3 bias was, in retrospect, misleading as will be shown below.

Several significant events have taken place since this author last reviewed CAIs (MacPherson et al., 1988) that dramatically changed the perspective on oxygen isotopes in the early solar system, both in general and in CAIs in particular. The first of these events was the development of techniques allowing the microanalysis of oxygen isotopes *in situ* in insulating minerals within a standard petrographic thin section. This was originally attained with the (then) new Cameca ims-1270 large radius machine (eg., Leshin et al., 1996), and the techniques were then extended to other modern machines as well. The second (and without question the most profound) event since the 1988 review was the positive identification and detailed study of preserved presolar grains within meteorites. Most of the work on presolar grains has been on relatively exotic materials such as silicon carbide, microdiamonds, and graphite (e.g., review by Zinner, 1998; Chapter 1.02); Messenger et al. (2003) recently identified and analyzed presolar silicates. For both the carbon-rich phases and the silicates, the critical observation in the context of this review is that all but a rare few grains are enriched in ^{17}O or ^{18}O but *not* ^{16}O. There is precious little evidence for any ^{16}O-rich presolar solids in the earliest solar system, (Nittler et al., 1997) and no ^{16}O-rich "precursor grains" have ever been found preserved within a CAI. The presolar grain research in particular has caused the original discoverer of ^{16}O-rich anomalies in CAIs recently (Clayton, 2002) to abandon his long-held belief that the anomalies were due to presolar grains and instead attribute them to mass-independent fractionation processes within the solar nebula. Such a mechanism, albeit different in detail, was first proposed by Thiemens and Heidenreich (1983). The third important development since 1988, enabled by the new *in situ* capabilities of the ion microprobe, has been the analysis of oxygen isotopes in many different kinds of CAIs (and chondrules) too small to analyze by the more conventional method, and in many different kinds of chondrites other than CV3s. One result has been to reinforce the growing recognition that CAIs in CV3 chondrites in general, and Allende in particular, are non-representative of the whole population of CAIs.

1.08.8.2 The Data

Whole-CAI and mineral separates data for CAIs from the CV3 chondrites Allende, Leoville, Vigarano, and Efremovka are shown in simplified form in Figure 24(a). The data cluster tightly

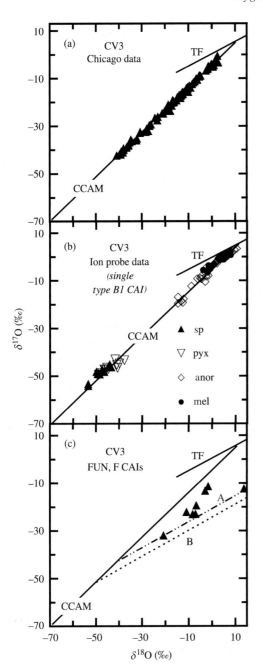

Figure 24 Oxygen isotopes in CV3 chondrite CAIs. Data in (a, c) taken from the summary table of Clayton et al. (1988) and references therein; ion probe data (b) from McKeegan et al. (1996).

along a line (slope ~0.95) known as the carbonaceous chondrite anhydrous minerals (CCAM) line, which in turn is very close to the line defined by Allende CAIs alone (Clayton et al., 1977). An example ion probe data set for a single type B1 CAI (McKeegan et al., 1996) is shown in Figure 24(b); except for greater scatter (relative to Figure 23(a)) owing to the lower precision of the ion probe data, the minerals cluster close to the same CCAM line. The ion probe data extend the dynamic range of CV3 CAI data. The early work by Clayton et al. (1977) used crushing and heavy liquids to separate individual minerals by density differences. It is very difficult to achieve perfect separation by this method, so the resulting isotopic analyses were to some degree cross-contaminated. No such difficulty exists with the ion probe, and CV3 CAI spinels analyzed by ion probe consistently show greater ^{16}O enrichments (δ^{18}O: −50‰ to −55‰) than revealed by the conventional techniques (δ^{18}O: −40‰ to −42‰). These ion probe data are consistent with the earlier data in showing that spinel and pyroxene are the most ^{16}O-rich phases, and anorthite and melilite are much less enriched. However, rare ^{16}O-rich melilite has been found in some CV3 CAIs. Yurimoto et al. (1994) reported ion probe data for an Allende type A CAI in which some of the melilite crystals have the same level of ^{16}O-enrichment as many CV3 spinels, with δ^{18}O nearly −40‰. Clayton et al. (1977) showed that secondary phases such as nepheline, sodalite, and iron-rich silicates (andradite, hedenbergite) show very little ^{16}O-enrichment. Recent ion microprobe work by Cosarinsky et al., (2003) has shown that the iron-rich silicates in particular, together with wollastonite, define a slope-1/2 line near the upper end of the CCAM line and just below the terrestrial fractionation line, which is suggestive of low temperature parent body processes.

Whole CAI data for CV3 FUN CAIs are shown in Figure 24(c), along with data for two CAIs that show strong mass-dependent oxygen isotopic fractionation without accompanying nuclear anomalies (so-called "F" CAIs). The data are consistent with an interpretation (e.g., Clayton et al., 1984) that FUN and F CAIs originally had a ^{16}O-rich composition similar to that defined by normal CAI spinels, which then evolved by melt distillation and Rayleigh-type isotope fractionation to produce a slope-1/2 array approximated by the hypothetical line "A" on the figure. Later exchange with a ^{16}O-poor gas caused the phases to evolve above the line "A." Ion microprobe analyses (Davis et al., 2000) of individual phases in a Vigarano forsterite-bearing FUN CAI showed that the olivine, pyroxene, and spinel do in fact define a line very similar to but somewhat below line "A"; this line, "B," intersects the CCAM line at approximately the composition of CAI spinels and pyroxenes as analyzed by ion probe (see Figure 24(b)). Thus, the ion probe work supports and refines the earlier hypothesis of Clayton et al. (1984).

Ion probe isotope data for CAIs from CM and CO3 meteorites are shown in Figure 25. The CM data (Figure 25(a); data from Fahey et al., 1987b), mainly for spinel and hibonite, are ^{16}O-rich and

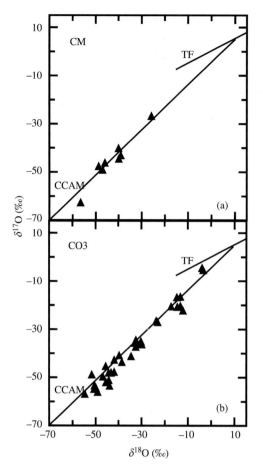

Figure 25 Oxygen isotopes in CM and CO chondrite CAIs. (sources Fahey *et al.*, 1987b; Wasson *et al.*, 2001).

cluster close to the CCAM line. With the exception of one point, the data do not extend below about $\delta^{18}O \sim -50‰$, consistent with CV3 spinels and pyroxenes. The exception is a hibonite in a corundum-bearing inclusion, and which has $\delta^{18}O \sim -57‰$. The CO3 data as a whole (Figure 25(b); data from Wasson *et al.*, 2001) define a line very similar to that defined by CV3 CAIs. However, what distinguishes the CO3 data is that within individual CAIs from the least metamorphosed meteorites, such as Kainsaz and Colony, the minerals exhibit a smaller range of composition than displayed by the data set as a whole. Although not explicitly set apart by different symbols on Figure 25, *even melilite and feldspar* are ^{16}O-rich relative to those phases in CV3 CAIs. CAIs from the more-metamorphosed CO3s, such as Ornans, show a wider range internally and commonly (not always) have ^{16}O-depleted melilite (Wasson *et al.*, 2001).

Data for CAIs from unequilibrated ordinary and enstatite chondrites, and from CB and CH carbonaceous chondrites, are shown in Figures 26(a)–(d). The data for UOC CAIs (Figure 26(a)) fall into two clusters, and each cluster consists of data from a single CAI (data from McKeegan *et al.*, 1998). These CAIs are both melilite-rich type As but contain very little secondary mineralization, which distinguishes them from CV3 and some CO3 CAIs. The lower (^{16}O-rich) cluster is from a melilite-rich CAI in Semarkona (LL3.0); none of the analyses represents a pure single-phase in this small and fine-grained object, but two of the points sampled ~50% melilite in addition to spinel and hibonite and perovskite. All of the data nevertheless plot at $\delta^{18}O \sim -40‰$, and there is no evidence of internal disequilibrium. The second CAI is from Quinyambie; all of the analyses sampled 70–80 vol.% melilite, and although the data are not as ^{16}O-rich as the Semarkona CAI they nonetheless are enriched relative to most CV3 melilite. The two critical facts that derive from this combined UOC data set are: (i) CAIs from UOCs sampled the same isotopic reservoirs as did CV3 CAIs; and (ii) UOC CAIs have more uniform internal isotopic compositions than do CV3 CAIs, and on average they are more ^{16}O-rich. The data for enstatite chondrite CAIs (Figure 26(c)) also fall into two clusters, representing in this case 14 CAIs (data from Guan *et al.*, 2000b; Fagan *et al.*, 2001). Data for 13 of those CAIs fall exclusively in the lower ^{16}O-rich cluster, with $\delta^{18}O \sim -30‰$ to $-50‰$. Several of the most ^{16}O-rich analyses sampled 70–80 vol.% melilite, and most of the rest are dominated by spinel and hibonite. Only one of the CAIs has a bimodal internal distribution of oxygen isotope compositions, and it is an unusual spinel-rich spherule that contains abundant sodalite (Fagan *et al.*, 2001). The spinel is ^{16}O-rich, and the sodalite is ^{16}O-depleted. The remaining CAI is uniformly ^{16}O-depleted (Fagan *et al.*, 2001), and consists largely of spinel, plagioclase, and troilite. The feldspar and spinel are both ^{16}O-depleted, the former being more so. Unlike most other CAIs from most other chondrites, there is no clear evidence that this object was ever ^{16}O-rich. With this one exception, however, the enstatite chondrite CAIs present a similar picture to the UOC CAIs in terms of oxygen isotopes: They sampled a ^{16}O-rich reservoir indistinguishable from that sampled by CV3 CAIs, but differ from the latter in generally preserving this original isotopic signature in an undisturbed form. Most show little or no evidence for isotopic disequilibrium. The existence, in an enstatite chondrite, of a CAI with an (apparent) primary isotopic signature that is ^{16}O-*poor* is significant (Fagan *et al.*, 2001), and as shown below it is not unique.

Data for CAIs from CH chondrites are shown in Figure 26(b) (unpublished data courtesy of K. D. McKeegan and A. N. Krot). The bimodal data reflect two apparent populations of CAIs and not variations within individual CAIs.

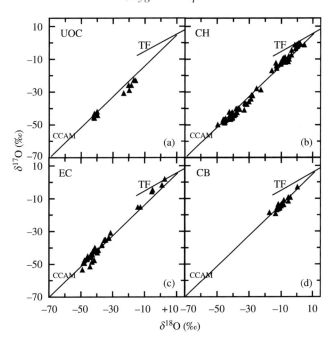

Figure 26 Oxygen isotopes in CB, CH, EC, and UOC. (sources Krot *et al.*, 2001; K. D. McKeegan unpublished data (for CH); McKeegan *et al.*, 1998; Guan *et al.*, 2000b; Fagan *et al.*, 2001).

Hibonite-, spinel-, and grossite-rich CAIs are ^{16}O-rich, and pyroxene-, plagioclase-, and melilite-rich CAIs are ^{16}O-poor. Unlike data for CAIs from the other chondrite types, the total data set for CH CAIs appear to define a slightly different line from CCAM. Weighted regression of the data gives a slope of ~1.02 (McKeegan and Krot, personal communication). Similar to the single enstatite chondrite CAI discussed above, there exists a population of CAIs in CH chondrites that are ^{16}O-poor and yet do not show significant internal isotopic heterogeneity within individual CAIs. The data for CB CAIs are even more extreme in this regard (Figure 26(d)): all of the data for all CB CAIs are relatively depleted in ^{16}O, even analyses dominated by hibonite, spinel, grossite, and pyroxene as well as those dominated by melilite (Krot *et al.*, 2001).

Individual mineral isotopic data for three CAIs from CR chondrites are shown in Figure 27, taken from the extensive ion probe data set of Aléon *et al.* (2002a). The reduced data set is intended to clearly demonstrate a general property of oxygen isotopes in the CR clan CAIs, namely that each individual CAI shows very little internal isotopic heterogeneity among its constituent minerals. For the three CAIs shown, spinel and melilite in each inclusion have very similar isotopic compositions; this is in marked contrast with the pattern generally seen in CV3s (e.g. Figure 24(b)). Such data suggest (Aléon *et al.*, 2002a) that the CR clan CAIs have experienced less secondary reprocessing than their CV kindred, and support the hypothesis that most CAIs originally had homogeneous internal isotopic compositions that in some cases (e.g., CVs) were greatly affected by secondary processes.

1.08.8.3 Synthesis of Existing Data, and Unresolved Issues

There is now general agreement that, despite isotopic differences between the various chondrite classes and even their chondrules, the majority of CAIs from most chondrite types sampled the same single ^{16}O-rich reservoir having a composition $\delta^{18}O \sim \delta^{17}O \sim -50$ to $-60‰$. Even those CAIs that are significantly depleted in ^{16}O, such as those from CB chondrites and some from CH and enstatite chondrites, plot along mixing lines that extrapolate through the same ^{16}O-rich composition. The many isotopic analyses now available, especially ion probe data obtained at very fine analytical scales, show no evidence of ^{16}O hot spots that might be preserved presolar grains. To the contrary, most data show a surprisingly consistent upper limit of ^{16}O-enrichment at the value noted above. Although it is possible that a mixture of nearly pure ^{16}O-bearing presolar grains ($\delta^{18}O \sim \delta^{17}O \sim -1,000‰$) together with ^{16}O-depleted material could have produced a large population of CAIs with such a restricted ^{16}O-rich composition at $\delta^{18}O \sim \delta^{17}O \sim -50$ to $-60‰$, it seems more likely that the result would be a large array of compositions in which some are far more ^{16}O-rich than any observed in the CAIs. This conclusion, based on the observations, is one

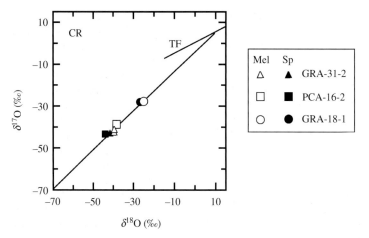

Figure 27 Oxygen isotopes in minerals within individual CR chondrite CAIs. Data taken from the much larger data set of Aléon et al. (2002) (reproduced by permission of the Meteoritical Society from *Meteorit. Planet. Sci.* **2002**, *37*, 1729–1755).

factor leading to the alternative idea that the carrier of the ^{16}O-rich isotopic signature was not presolar grains but rather the gas in which CAIs formed and equilibrated. Supporting evidence comes from analyses of chondritic materials that are interpreted to be primary nebular condensates, such as magnesian olivines in accretionary rims around CAIs (Krot et al., 2002b) and individual mineral phases in CAIs whose volatility-fractionated trace element patterns indicate them to be condensates (Aléon et al., 2002b). If the petrogenetic interpretation of such material is correct, they must have formed out of a ^{16}O-rich gas. Finally there is the evidence of known presolar grains that are (with very rare exceptions) enriched in either ^{18}O or ^{17}O but not ^{16}O.

The origin and location of this ^{16}O-rich gaseous reservoir is the subject of much current debate (see Chapter 1.06). Thiemens and Heidenreich (1983) originally showed that gaseous reactions involving asymmetric molecules can produce non-mass-dependent isotopic fractionation, with ^{16}O-rich and ^{16}O-rich components whose spread in compositions approximate the observed range along the CCAM line. This process has been demonstrated to occur naturally in Earth's upper atmosphere. More recent alternative models for the solar nebula involve self-shielding (Clayton, 2002; Young and Lyons, 2003). What has not been satisfactorily demonstrated by any of these models is how molecular fractionations can be isolated and trapped in *grains* having large ^{16}O-enrichments. An alternative model is that ^{16}O-rich solids did exist but were completely evaporated within limited nebular regions, and high temperature recondensation of this gas led to CAIs or their precursors. The latter model runs up against the problem of the paucity of ^{16}O-rich presolar grains in the population of such grains that have been analyzed.

Best fit lines through various oxygen isotopic data sets of whole-CAI or individual mineral sets generally give a slope of somewhat less than unity (typically ~0.95). However, Young and Russell (1998) used a UV-laser fluorination microprobe analyses to measure with high precision the individual mineral compositions within an individual Allende CAI and found that the primary minerals define a slope of precisely 1.00. They argued that slopes of less than unity are in part a result of the secondary alteration and reprocessing that many CAIs have experienced. This result is potentially very important in discriminating models for the origin of the ^{16}O-rich component in primitive solar system materials. If the component originated in ^{16}O-rich presolar grains, there is no particular expectation for any specific slope much less one of exactly 1.00. However, non-mass-dependent fractionation through photo-dissociation of gaseous asymmetric molecules will, in pure form, tend towards making products that disperse along a line of slope 1.00 (M. Thiemens, personal communication, 2003).

The origin of the CAI mixing line(s) within individual CAIs also remains unresolved. The observation cited above that many "pristine" CAIs (e.g., in CR clan chondrites) have relatively homogeneous and ^{16}O-rich internal isotopic compositions suggests that the extreme heterogeneity observed in CAIs from the CV chondrites is due to secondary processing such as aqueous alteration or alkali metasomatism. However, secondary processing alone does not explain the fact that some CAIs have homogeneous internal isotopic compositions that are not very ^{16}O-rich (Krot et al., 2001), unless they were homogenized (melted?) subsequent to the secondary reprocessing. Moreover, the details of this processing, and where it took place, are not clear. For example, the generally accepted model for internal isotopic heterogeneity in the CV CAIs is that originally ^{16}O-rich minerals exchanged to differing degrees with the altering medium (gas or liquid) owing to their different oxygen diffusion coefficients. Yet, laboratory measurements of diffusion coefficients predict relative exchange rates that are somewhat inconsistent with the compositions of natural phases (see Ryerson and McKeegan, 1994) without resorting to complex CAI petrogenesis (which may, however, be justified; *ibid.*).

1.08.9 SHORT-LIVED RADIONUCLIDES

A general review of short-lived nuclides in the early solar system is given in Chapter 1.16, so only brief comments specifically relevant to CAI chronology and possible constraints on where CAIs formed are given here.

Lee *et al.* (1976) demonstrated unambiguously that an Allende type B CAI contains excesses of ^{26}Mg relative to normal (terrestrial) values, and that the magnitude of the excesses in each mineral correlate positively with the bulk Al/Mg of that phase. In other words, the ^{26}Mg isotopic signature is largest in aluminum-rich phases. This was the first compelling evidence for the *in situ* decay of the short-lived radionuclide ^{26}Al ($t_{1/2} \sim 0.7$ Myr) at the time of CAI formation. The ^{26}Mg excesses observed in that CAI correspond to an initial abundance of ^{26}Al, relative to the stable isotope ^{27}Al, of ~ 50 ppm. Birck and Allègre (1985) found manganese-correlated excesses of ^{53}Cr in mineral separates from an Allende CAI, demonstrating the original presence of ^{53}Mn ($t_{1/2} \sim 3.7$ Myr). Since then, analogous data for the calcium–potassium and beryllium–boron isotopic systems have demonstrated the original presence in CAIs of two other short-lived radionuclides, ^{41}Ca ($t_{1/2} \sim 0.1$ Myr; Srinivasan *et al.*, 1994, 1996) and ^{10}Be ($t_{1/2} \sim 1.5$ Myr; McKeegan *et al.*, 2000). The importance of such data is threefold: (i) the CAIs (hence, the solar system) must have formed within a very short time of the production of the radionuclides themselves; (ii) ^{26}Al could have been potent heat source contributing to heating and melting of early-formed planetesimals within the solar system, if it was present at anything like the abundance inferred by Lee *et al.* (1976) and others; and (iii) short-lived nuclides in general can serve as high-resolution chronometers of early solar system events provided that their abundance ratios were uniform throughout the solar system.

In the same paper where they demonstrated the *in situ* decay of ^{26}Al in an Allende CAI, Lee *et al.* (1976) also showed that another Allende inclusion contains no resolvable excesses of ^{26}Mg and, hence, had little or no ^{26}Al at the time of its formation. This object is a FUN inclusion, and (as noted earlier) it is now known that FUN CAIs as a group are characterized by having had little or no ^{26}Al at the time of their formation. This result immediately suggested that at some scale the solar system was not isotopically homogeneous with respect to ^{26}Al and cast doubt on the use of ^{26}Al variations as a chronometer. Nevertheless, numerous studies over the ensuing 25+ years have shown chronologic information can be extracted from ^{26}Al variations among solar system materials. A critical enabling step in this direction was the development of reliable methods for the analysis of aluminum–magnesium isotopic systematics using an ion microprobe (e.g., Hutcheon, 1982) through which large numbers of CAIs from a wide range of chondrite types have been analyzed at the microscale in combination with careful petrographic and petrologic characterization. Although many CAIs were found to contain evidence for extinct ^{26}Al (e.g., see review by MacPherson *et al.*, 1995), the ion probe data rarely define perfect correlations between ^{26}Mg and ^{27}Al/^{24}Mg. For example, data for one type B inclusion studied by Podosek *et al.* (1991) are shown in Figure 28. On such diagrams for extinct radionuclides, the slope of the correlation line (fossil isochron) gives the initial ratio of radiogenic to nonradiogenic isotopes (e.g., ^{26}Al/^{27}Al) provided that the system formed at equilibrium. Although most of the data in Figure 28 plot within error of a best-fit line corresponding to a slope (initial ^{26}Al/^{27}Al) of 4×10^{-5}, some high Al/Mg points plot below the best-fit line and some low Al/Mg points plot above it. In this and the other CAIs they studied, Podosek and colleagues showed that such deviations from isochronism are correlated with petrographic evidence for secondary events such as

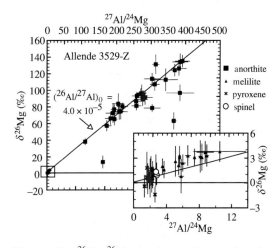

Figure 28 ^{26}Al–^{26}Mg isochron diagram for the Allende type B CAI 3529-Z. (source Podosek *et al.*, 1996).

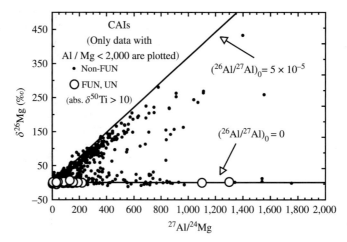

Figure 29 ^{26}Al–^{26}Mg isochron diagram for a large number of CAIs. Data taken from numerous sources listed in MacPherson et al. (1995)

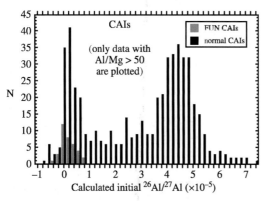

Figure 30 Histogram of calculated initial ^{26}Al/^{27}Al values, comparing normal and FUN CAIs. FUN CAIs rarely show evidence of live ^{26}Al, and when evidence is available (in the case of the hibonite-rich Allende CAI HAL), the initial value is very low. Data sources as in Figure 29.

reheating that could have partially reset the isotopic system in the observed systematic way (see Wasserburg and Papanastassiou, 1982; Podosek et al., 1991). A synthesis by MacPherson et al. (1995) of data available through 1996 showed that the initial isotopic signature $(^{26}Al/^{27}Al)_0 \sim 5 \times 10^{-5}$ was widespread (but not universal) throughout CAIs from a variety of chondrite types and, by extension, possibly widespread throughout the early solar system as well.

Representative data are plotted in Figure 29 on a magnesium isotope correlation diagram and in Figure 30 as a histogram of calculated initial ^{26}Al/^{27}Al. A majority of the points in Figure 29 lie close to or slightly below a line of slope 5×10^{-5}. In contrast to the situation that existed in 1995, we now know that CAIs from ordinary and enstatite chondrites also share this isotopic signature (Russell et al., 1996; Guan et al., 2000a). Points that plot above the line are confined to those having low Al/Mg ratios and these, together with points having high Al/Mg that plot between the line of slope 5×10^{-5} and the reference ("normal") line of zero slope, result mainly from later isotopic disturbance as noted above. Data for all FUN/UN CAIs plot essentially on the reference line of zero slope; so too do analyses of many secondary phases in the CAIs, such as nepheline and grossular. Taken at face value, the data would suggest that the FUN/UN CAIs and alteration phases formed significantly later than the original formation of the primary CAIs. Although reasonable with respect to the secondary alteration, as noted earlier this interpretation is difficult to reconcile with the other isotopic properties of the FUN/UN CAIs. The histogram in Figure 30 is constructed by calculating (see MacPherson et al., 1995) an initial ^{26}Al/^{27}Al value for every point in Figure 29. The dominant peak is actually centered on an initial ^{26}Al/^{27}Al of $\sim 4.5 \times 10^{-5}$, somewhat lower than the commonly quoted canonical value of 5×10^{-5}. The peak centered on ^{26}Al/^{26}Al ~ 0 should not really be considered a peak at all, as this isotopic system cannot distinguish between an object formed 4 Myr after time zero (^{26}Al/^{27}Al $\sim 4.5 \times 10^{-5}$) and something that formed 4 Gyr later. Nevertheless, at least some of the material that defines this "peak" (setting aside the FUN/UN CAIs) probably formed at least 1–2 Myr later than the normal CAIs. One of the most interesting developments in the past ten years has been the accumulation of a substantial body of Al–Mg isotopic data for chondrules. A histogram of initial ^{26}Al/^{27}Al values for chondrules from a variety of chondrite types (Figure 31) shows a peak at just under 1×10^{-5}. The data are taken from both aluminum-rich and normal ferromagnesian chondrules; there is no systematic isotopic difference

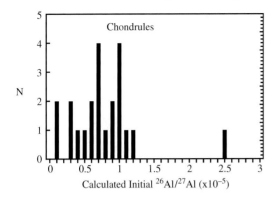

Figure 31 Histogram of calculated initial $^{26}Al/^{27}Al$ values in chondrules. Data from numerous sources as summarized in Table 2 of Huss *et al.* (2001)

accretion regions (e.g., McKeegan *et al.*, 1998; Guan *et al.*, 2000b; Fagan *et al.*, 2001). Shu *et al.* (1996) had independently proposed a mechanism by which CAIs formed very near the protosun and were ejected outward into the solar nebula (the "X-wind" model). The evidence from ^{10}Be adds increased support to such a model, which also has the further advantage of addressing the problem of the energy source that fed high temperature events such as refractory evaporation and condensation and CAI melting (e.g., Wood, 1996). This conclusion is elaborated further below, in Section 1.08.11.

1.08.10 CAIs, CHONDRULES, CONDENSATION, AND MELT DISTILLATION

It is now universally recognized that few CAIs are the simple aggregates of pristine nebula condensates that they were once thought to be. Most have had long and complex histories, including multiple episodes of partial melting interspersed in some cases with secondary alteration (e.g., MacPherson and Davis, 1993; Beckett *et al.*, 2000). However, there is still lingering debate over the relative roles of nebular condensation versus melt distillation in controlling the bulk compositions of many CAIs. Equally, the relationship between CAIs and chondrules is only beginning to be explored.

Although trace element patterns and the lack of *large* degrees of isotopic mass-dependent fractionation (except in HAL-type FUN CAIs, as noted earlier) tend to support the idea that CAI bulk chemistry was established primarily by condensation, it is well known (e.g., see discussions by Stolper (1982) and Beckett and Grossman (1988)) that the bulk compositions of types B and C CAIs differ slightly but systematically from the predictions of equilibrium thermodynamic calculations for condensation from a hot solar gas. At least one possibility for this discrepancy is distillation when the condensate precursors were melted, which then modified the bulk compositions. Niederer and Papanastassiou (1984) demonstrated that there are *small* degrees of mass-dependent isotopic fractionation in many CAIs, but not always in favor of the heavy isotopes. They concluded that both condensation and melt distillation played some role. This debate has recently been taken up anew in a series of empirical and theoretical studies at the University of Chicago and elsewhere. Grossman *et al.* (2000, 2002) and Richter *et al.* (2002) argued that the bulk chemical and isotopic properties of type B CAIs are consistent with varying degrees of distillation of melt droplets whose original compositions were

between the two. The critical points to note are: (1) the histogram peak clearly is not at zero; and (2) the ~5× difference in initial $^{26}Al/^{27}Al$ between chondrules and many CAIs implies an age difference of 1–2 Myr. This chondrule–CAI difference was hinted at by the very sparse data available at the time of the 1995 review paper by MacPherson *et al.*, but Wood (1996) and others regarded the difference as being nonrepresentative (most of the chondrule data were from aluminum-rich chondrules, whose relation, if any, to ferromagnesian chondrules was unknown) and probably nonchronologic in origin anyway. This must be rethought in light of the new data noted above and, especially, with the result of Amelin *et al.* (2002) establishing absolute age differences of 1–2 Myr between at least some chondrules and some CAIs.

The demonstration by McKeegan *et al.* (2000) that another short-lived nuclide, ^{10}Be, also existed in CAIs at the time of their formation, has very different significance from ^{26}Al. McKeegan *et al.* (2000) and others (e.g., MacPherson *et al.*, 2003) have argued convincingly that the ^{10}Be, unlike ^{26}Al, was largely produced by spallation reactions within the solar nebula because it is not made by stellar nucleosynthesis. The most likely location where this occurred was very near the protosun. In this context the overwhelming importance of the ^{10}Be is not for chronology but rather for *location*. So far only CAIs in CV3 chondrites have been demonstrated to contain excess ^{10}B from the decay of extinct ^{10}Be, and this result needs to be tested in CAIs from other chondrite types. However, there is a high degree of similarity in both oxygen and magnesium isotopes among CAIs from all chondrite types, which suggests that all CAIs from all chondrite types may have formed in a restricted nebular location and then were widely dispersed into the various chondrite derived from condensation. In all of these studies the need for distillation is driven not just by the small amounts of mass-dependent isotopic fractionation but also by the observation that type B CAIs are systematically magnesium-depleted relative to the predictions of equilibrium condensation. Collectively, the evidence is compelling that some melt distillation occurred; however, this may not be the whole story for why CAI bulk compositions deviate from the predictions of condensation. Although these studies recognize that silicon is also volatile during distillation, the evidence for silicon depletion in type B bulk compositions is weak at best (Richter et al., 2002). In fact, when type C CAIs are considered as part of the problem (Beckett and Grossman, 1988), the combined data suggest if anything that silica is enriched in type C and many type B CAIs relative to the predictions of equilibrium condensation. Figure 32 shows the bulk compositions of type A, B, and C CAIs relative to the predictions of the equilibrium condensation calculations of Yoneda and Grossman (1995), plotted on the same compositional diagram from Figure 7. It is clear that the type C CAIs plot on an extension of the same trend defined by the type Bs, and simple deficiency in magnesium cannot explain the difference between the observed and calculated trends. Moreover, the combined CAI composition trend converges with the trend defined by aluminum-rich chondrules, at a point intermediate between the compositions of the minerals anorthite and diopside. MacPherson and Huss (2000) used this evidence to argue that a simpler explanation for the trends lies within the condensation calculations themselves: if anorthite were to condense prior to forsterite in contrast to the result usually obtained from such calculations, the observed and calculated bulk composition trends come into much closer agreement. Because anorthite is one of the most abundant secondary phases in many CAIs (in rims, and replacing melilite), this is tantamount to proposing that alteration (followed by melting) is what defines the bulk composition trend that leads from type A CAIs (the primary precursors) in the direction of pyroxene and anorthite, directly through the observed fields for type B and C CAIs (Figure 32). This idea has theoretical support from Petaev and Wood (1998a), who showed that the calculated order of appearance of forsterite and anorthite is pressure dependent: lower nebular pressures favor anorthite condensing before (at higher temperatures than) forsterite, whereas higher pressures favor the reverse. The bonus from

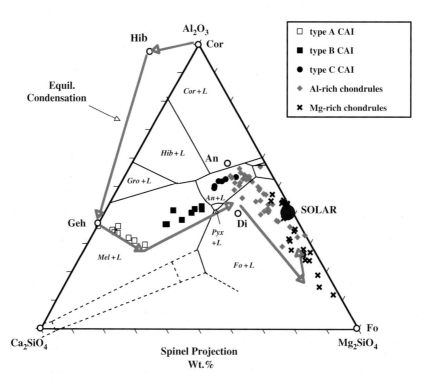

Figure 32 Bulk compositions of major CAI varieties, aluminum-rich chondrules, ferromagnesian chondrules, and the projected trend of total condensed solids as calculated for equilibrium condensation (Yoneda and Grossman, 1995), all projected (as in Figure 7) from spinel ($MgAl_2O_4$) onto the plane Al_2O_3–Mg_2SiO_4–Ca_2SiO_4. Data taken from numerous sources as listed in MacPherson and Huss (2000). Abbreviations: An—anorthite; Cor—corundum; Di—diopside; Fo—forsterite; Geh—gehlenite; Gro—grossite; L—liquid; Pyx—pyroxene.

this model is that it potentially provides a link between chondrules and CAIs, since this reversal of forsterite and anorthite condensation also brings the compositions of aluminum-rich chondrules into alignment with theoretical predictions, which was in fact the original point that MacPherson and Huss (2000) made.

It is very likely that postalteration melting as well as melt distillation played roles in governing the bulk compositions of many CAIs. Detailed additional studies, especially high-precision isotopic fractionation measurements, will be required to sort out the relative contributions.

1.08.11 CONCLUSIONS AND REFLECTIONS: WHERE AND HOW DID CAIs FORM?

One of the great unknowns about CAIs is in essence the same riddle that exists for chondrules and can in fact be called the "chondrule problem": what was the heating mechanism? Both kinds of objects formed at very high temperatures and, regardless of the ambient temperature of the regions where they formed, all true chondrules and most CAIs experienced transient higher temperatures that caused brief melting followed by constrained (i.e., not simply radiative) cooling. Some workers have suggested that the only possible solution will come from theory rather than observations. However, it is the opinion of this author that for CAIs at least, we may be converging on the answer because of the new isotopic evidence. The combined oxygen, aluminum–magnesium, and beryllium–boron isotopic evidence summarized at the end of Section 1.08.9 suggests that by being able to say *where* CAIs formed, near the protosun, we may be closing in on knowing *how*. This is an area that requires much additional study, both observational and theoretical.

ACKNOWLEDGMENTS

This chapter has benefited greatly from rewarding collaborations and/or discussions over many years with numerous valued colleagues, including most importantly John Beckett, Catherine Caillet, Bob Clayton, Ghislaine Crozaz, Andy Davis, Larry Grossman, Yunbin Guan, Gen Hashimoto, Gary Huss, Sasha Krot, Laurie Leshin, Kevin McKeegan, Frank Podosek, Sara Russell, Gerry Wasserburg, John Wood, and Ernst Zinner. The debates have been lively and enjoyable, even (sometimes especially) when we disagreed. None should be held accountable for any errors of thought or word in this review. The editorial patience and help of Andy Davis is appreciated more than I can express.

REFERENCES

Aléon J., Krot A. N., and McKeegan K. D. (2002a) Calcium–aluminum-rich inclusions and amoeboid olivine aggregates from the CR carbonaceous chondrites. *Meteorit. Planet. Sci.* **37**, 1729–1755.

Aléon J., Krot A.N., McKeegan K. D., MacPherson G. J., and Ulyanov A. A. (2002b) Oxygen isotopic composition of fine-grained Ca-Al-rich inclusions in the reduced CV3 chondrite Efremovka. In *Lunar Planet. Sci.* **XXXIII**, #1426. The Lunar and Planetary Institute, Houston (CD-ROM).

Allégre C. J., Manhés G., and Göpel C. (1994). The age of the Earth. *Geochim. Cosmochim. Acta* **59**, 1445–1456.

Allen J. M., Grossman L., Davis A. M., and Hutcheon I. D. (1978) Mineralogy, textures, and mode of formation of a hibonite-bearing Allende inclusion. *Proc. 9th Lunar Planet. Sci. Conf.*, 1209–1233.

Amelin Y., Krot A. N., Hutcheon I. D., and Ulyanov A. A. (2002) Lead isotopic ages of chondrules and calcium–aluminum-rich inclusions. *Science* **297**, 1678–1683.

Anders E. and Grevesse N. (1989) Abundances of the elements: meteoritic and solar. *Geochim. Cosmochim. Acta* **53**, 197–214.

Armstrong J. T., El Goresy A., and Wasserburg G. J. (1985) Willy: a prize noble Ur-Fremdling—its history and implications for the formation of Fremdlinge and CAI. *Geochim. Cosmochim. Acta* **49**, 1001–1022.

Bar-Matthews M., Hutcheon I. D., MacPherson G. J., and Grossman L. (1982) A corundum-rich inclusion in the Murchison carbonaceous chondrite. *Geochim. Cosmochim. Acta* **46**, 31–41.

Beckett J. R. (1986) The origin of calcium-, aluminum-rich inclusions from carbonaceous chondrites: an experimental study. PhD Dissertation, University of Chicago, 373 p.

Beckett J. R. and Grossman L. (1988) The origin of type C inclusions from carbonaceous chondrites. *Earth. Planet. Sci. Lett.* **89**, 1–14.

Beckett J. R. and Stolper E. (1994) The stability of hibonite, melilite and other aluminous phases in silicate melts: implications for the origin of hibonite-bearing inclusions from carbonaceous chondrites. *Meteoritics* **29**, 41–65.

Beckett J. R. and Stolper E. (2000) The partitioning of Na between melilite and liquid: Part I. The role of crystal chemistry and liquid composition. *Geochim. Cosmochim. Acta* **64**, 2509–2517.

Beckett J. R. Simon S. B., and Stolper E. (2000) The partitioning of Na between melilite and liquid: Part II. Applications to Type B inclusions from carbonaceous chondrites. *Geochim. Cosmochim. Acta* **64**, 2519–2534.

Birck J. L. and Allègre C. J. (1985) Evidence for the presence of ^{53}Mn in the early solar system. *Geophys. Res. Lett.* **11**, 745–748.

Bischoff A., and Keil K. (1984) Al-rich objects in ordinary chondrites: related origin of carbonaceous and ordinary chondrites and their constituents. *Geochim. Cosmochim. Acta* **48**, 693–709.

Bischoff A., Keil K., and Stöffler D. (1985) Perovskite–hibonite–spinel-bearing inclusions and Al-rich chondrules and fragments in enstatite chondrites. *Chem. Erde* **44**, 97–106.

Bischoff A., Palme H., Schultz L., Weber D., Weber H. W., and Spettel B. (1993) Acfer 182 and paired samples, an iron-rich carbonaceous chondrite: similarities with ALH 85085 and relationship to CR chondrites. *Geochim. Cosmochim. Acta* **57**, 2631–2648.

Blum J. D., Wasserburg G. J., Hutcheon I. D., Beckett J. R., and Stolper E. M. (1988) "Domestic" origin of opaque assemblages in refractory inclusions in meteorites. *Nature* **331**, 405–409.

Boynton W. V. (1975) Fractionation in the solar nebula: condensation of yttrium and the rare earth elements. *Geochim. Cosmochim. Acta* **39**, 569–584.

Boynton W. V., Wark D. A., and Ulyanov A. A. (1986) Trace elements in Efremovka fine-grained inclusion E14: evidence for high temperature, oxidizing fractionations in the solar nebula. In *Lunar Planet. Sci.* **XVII**. The Lunar and Planetary Institute, Houston, pp. 78–79.

Brearley A. J. and Jones R. H. (1998) Chondritic meteorites. In *Planetary Materials* (ed. J. J. Papike). *Rev. Mineral.* **36**, 3-01–3-398.

Burns R. G. and Huggins F. E. (1973) Visible-region absorption spectra of a Ti^{3+} fassaite from the Allende meteorite. *Am. Mineral.* **58**. 955–961.

Cain P. M., McSween H. Y. Jr., and Woodward N. B. (1986) Structural deformation of the Leoville meteorite. *Earth Planet. Sci. Lett.* **77**, 165–175.

Christophe Michel-Lévy M. (1968) Un chondre exceptionnel dans la météorite de Vigarano. *Bull. Soc. Fr. Minéral. Cristallogr.* **91**, 212–214.

Clarke R. S., Jr., Jarosewich E., Mason B., Nelen J., Gomez M., and Hyde J. R. (1970) The Allende, Mexico, meteorite shower. *Smithson. Contrib. Earth Sci.* **5**, 1–30.

Clayton R. N. (2002) Self-shielding in the solar nebula. *Nature* **415**, 860–861.

Clayton R. N., Grossman L., and Mayeda T. K. (1973) A component of primitive nuclear composition in carbonaceous chondrites. *Science* **182**, 485–488.

Clayton R. N., Onuma N., Grossman L., and Mayeda T. K. (1977) Distribution of the pre-solar component in Allende and other carbonaceous chondrites. *Earth Planet. Sci. Lett.* **34**, 209–224.

Clayton R. N., MacPherson G. J., Hutcheon I. D., Davis A. M., Grossman L., Mayeda T. K., Molini-Velsko C., and Allen J. M. (1984) Two forsterite-bearing FUN inclusions in the Allende meteorite. *Geochim. Cosmochim. Acta* **48**, 535–548.

Clayton R. N., Mayeda T. K., Palme H., and Laughlin J. (1986) Oxygen, silicon, and magnesium isotopes in Leoville refractory inclusions. In *Lunar Planet. Sci.* **XVII**. The Lunar and Planetary Institute, Houston, pp. 139–140.

Clayton R. N., Mayeda T. K., MacPherson G. J., and Grossman L. (1987) Oxygen and silicon isotopes in inclusions and chondrules from Vigarano. In *Lunar Planet. Sci.* **XVIII**. The Lunar and Planetary Institute, Houston, pp. 185–186.

Clayton R. N., Hinton R. W., and Davis A. M. (1988) Isotopic variations in the rock-forming elements in meteorites. *Phil. Trans. Roy. Soc. London* **325**, 483–501.

Cosarinsky M., Leshin L. A., MacPherson G. J., Krot A. N., and Guan Y. (2003) Oxygen isotopic composition of Ca–Fe-rich silicates in and around an Allende Ca–Al-rich inclusion. In *Lunar Planet. Sci.* **XXXIV**, #1043. The Lunar and Planetary Institute, Houston (CD-ROM).

Davis A. M. and Grossman L. (1979) Condensation and fractionation of rare earths in the solar nebula. *Geochim. Cosmochim. Acta* **43**, 1611–1632.

Davis A. M., Tanaka T., Grossman L., Lee T., and Wasserburg G. J. (1982) Chemical composition of HAL, an isotopically-unusual Allende inclusion. *Geochim. Cosmochim. Acta* **46**, 1627–1651.

Davis A. M., Hashimoto A., Clayton R. N., and Mayeda T. K. (1990) Isotope mass fractionation during evaporation of Mg_2SiO_4. *Nature* **347**, 655–658.

Davis A. M., MacPherson G. J., Clayton R. N., Mayeda T. K., Sylvester P., Grossman L., Hinton R. W., and Laughlin J. R. (1991) Melt solidification and late-stage evaporation in the evolution of a FUN inclusion from the Vigarano C3V chondrite. *Geochim. Cosmochim. Acta* **55**, 621–638.

Davis A. M., McKeegan K. D., and MacPherson G. J. (2000) Oxygen isotopic compositions of individual minerals from the FUN inclusion Vigarano 1623-5. *Meteorit. Planet. Sci.* **35**, A47.

Dominik B., Jessberger E. K., Staudacher Th., Nagel K., and El Goresy A. (1978) A new type of white inclusion in Allende: petrography, mineral chemistry, $^{40}Ar/^{39}Ar$ ages, and genetic implications. *Proc. 9th Lunar Planet. Sci. Conf.*, 1249–1266, Pergamon.

Dowty E. and Clarke R. S., Jr. (1973) Crystal structure refinement and optical properties of a Ti^{3+} fassaite from the Allende meteorite. *Am. Mineral.* **58**, 230–242.

Ebel D. S. and Grossman L. (2000) Condensation in dust-enriched systems. *Geochim. Cosmochim. Acta* **64**, 339–366.

El Goresy A., Nagel K., and Ramdohr P. (1978) Fremdlinge and their noble relatives. *Proc. 9th Lunar Planet. Sci. Conf.*, pp. 1279–1303.

El Goresy A., Palme H., Yabuki H., Nagel K., Herrwerth I., and Ramdohr P. (1984) A calcium–aluminum-rich inclusion from the Essebi (CM2) chondrite: evidence for captured spinel–hibonite spherules and for an ultra-refractory rimming sequence. *Geochim. Cosmochim. Acta* **48**, 2283–2298.

El Goresy A., Zinner E., Matsunami S., Palme H., Spettel B., Lin Y., and Nazarov M. (2002) Efremovka E 1.1.1: a CAI with ultrarefractory REE patterns and enormous enrichments of Sc, Zr, and Y in fassaite and perovskite. *Geochim. Cosmochim. Acta* **66**, 1459–1491.

Fagan T. J., Krot A. N., and Keil K. (2000) Calcium–aluminum-rich inclusions in enstatite chondrites: I. Mineralogy and textures. *Meteorit. Planet. Sci.* **35**, 771–781.

Fagan T. J., McKeegan K. D., Krot A. N., and Keil K. (2001) Calcium–aluminum-rich inclusions in enstatite chondrites: II. Oxygen isotopes. *Meteorit. Planet. Sci.* **36**, 223–230.

Fahey A. J. (1988) Ion microprobe measurements of Mg, Ca, Ti and Fe isotopic ratios and trace element abundances in hibonite-bearing inclusions from primitive meteorites. Ph.D. thesis, Washington University, 206 p.

Fahey A., Goswami J. N., McKeegan K. D., and Zinner E. (1987a) ^{26}Al, ^{244}Pu, ^{50}Ti, REE, and trace element abundances in hibonite grains from CM and CV meteorites. *Geochim. Cosmochim. Acta* **51**, 329–350.

Fahey A. J., Goswami J. N., McKeegan K. D., and Zinner E. K. (1987b) ^{16}O excesses in Murchison and Murray hibonite: a case against a late supernova injection origin of isotopic anomalies in O, Mg, Ca, and Ti. *Astrophys. J.* **323**, L91–L95.

Fegley B., Jr. and Palme H. (1985) Evidence for oxidizing conditions in the solar nebula from Mo and W depletions in refractory inclusions in carbonaceous chondrites. *Earth. Planet. Sci. Lett.* **72**, 311–326.

Floss C., El Goresy A., Zinner E., Kransel G., Rammensee W., and Palme H. (1996) Elemental and isotopic fractionations produced through evaporation of the Allende CV chondrite: implications for the origin of HAL-type hibonite inclusions. *Geochim. Cosmochim. Acta* **60**, 1975–1997.

Floss C., El Goresy A., Zinner E., Palme H., Weckwerth G., and Rammensee W. (1998) Corundum-bearing residues produced through evaporation of natural and synthetic hibonite. *Meteorit. Planet. Sci.* **33**, 191–206.

Gray C. M., Papanastassiou D. A., and Wasserburg G. J. (1973) The identification of early condensates from the solar nebula. *Icarus* **20**, 213–239.

Grossman J. N., Rubin A. E., and MacPherson G. J. (1988) Allan Hills 85085: a unique volatile-poor carbonaceous chondrite with implications for nebular agglomeration and fractionation processes. *Earth Planet. Sci. Lett.* **91**, 33–54.

Grossman L. (1972) Condensation in the primitive solar nebula. *Geochim. Cosmochim. Acta* **36**, 597–619.

Grossman L. (1973) Refractory trace elements in Ca–Al-rich inclusions in the Allende meteorite. *Geochim. Cosmochim. Acta* **37**, 1119–1140.

Grossman L. (1975) Petrography and mineral chemistry of Ca-rich inclusions in the Allende meteorite. *Geochim. Cosmochim. Acta* **39**, 433–454.

Grossman L. (1980) Refractory inclusions in the Allende meteorite. *Ann. Rev. Earth Planet. Sci.* **8**, 559–608.

Grossman L. and Ganapathy R. (1975) Volatile elements in Allende inclusions. *Proc. 6th Lunar Planet. Sci. Conf.*, 1729–1736.

Grossman L. and Ganapathy R. (1976a) Trace elements in the Allende meteorite: I. Coarse-grained, Ca-rich inclusions. *Geochim. Cosmochim. Acta* **40**, 331–344.

Grossman L. and Ganapathy R. (1976b) Trace elements in the Allende meteorite: II. Fine-grained, Ca-rich inclusions. *Geochim. Cosmochim. Acta* **40**, 967–977.

Grossman L. and Steele I. M. (1976) Amoeboid olivine aggregates in the Allende meteorite. *Geochim. Cosmochim. Acta* **40**, 149–155.

Grossman L., Ebel D. S., Simon S. B., Davis A. M., Richter F. M., and Parsad P. M. (2000) Major element chemical and isotopic compositions of refractory inclusions in C3 chondrites: the separate roles of condensation and evaporation. *Geochim. Cosmochim. Acta* **64**, 2879–2894.

Grossman L., Ebel D. S., and Simon S. B. (2002) Formation of refractory inclusions by evaporation of condensate precursors. *Geochim. Cosmochim. Acta* **66**, 145–161.

Grossman L., Fruland R. M., and McKay D. S. (1975) Scanning electron microscopy of a pink inclusion from the Allende meteorite. *Geophys. Res. Lett* **2**, 37–40.

Guan Y., Huss G. R., MacPherson G. J., and Wasserburg G. J. (2000a) Calcium–aluminum-rich inclusions from enstatite chondrites: indigenous or foreign? *Science* **289**, 1330–1333.

Guan Y., McKeegan K. D., and MacPherson G. J. (2000b) Oxygen isotopes in calcium–aluminum-rich inclusions from enstatite chondrites: new evidence for a single CAI source in the solar nebula. *Earth. Planet. Sci. Lett.* **181**, 271–277.

Guan Y., Huss G. R., MacPherson G. J., and Leshin L. A. (2002) Rare earth elements of calcium–aluminum-rich inclusions in unequilibrated enstatite chondrites: characteristics and implications. *Meteorit. Planet. Sci.* **37**, A59.

Hashimoto A. (1983) Evaporation metamorphism in the early solar nebula-evaporation experiments on the melt FeO–MgO–SiO$_2$–CaO–Al$_2$O$_3$ and chemical fractionations of primitive materials. *Geochem. J.* **17**, 111–145.

Hashimoto A. and Grossman L. (1987) Alteration of Al-rich inclusions inside amoeboid olivine aggregates in the Allende meteorite. *Geochim. Cosmochim. Acta* **51**, 1685–1704.

Hinton R. W. and Bischoff A. (1984) Ion microprobe magnesium isotope analysis of plagioclase and hibonite from ordinary chondrites. *Nature* **308**, 169–172.

Hinton R. W., Grossman L, and MacPherson G. J. (1984) Magnesium and calcium isotopes in hibonite-bearing CAIs. *Meteoritics* **19**, 240–241.

Hinton R. W., Davis A. M., Scatena-Wachel D. E., Grossman L., and Draus R. J. (1988) A chemical and isotopic study of hibonite-rich refractory inclusions in primitive meteorites. *Geochim. Cosmochim. Acta* **52**, 2573–2598.

Hiyagon H. and Hashimoto A. (1999) ^{16}O excesses in olivine inclusions in Yamato-86009 and Murchison chondrites and their relation to CAIs. *Science* **283**, 828–831.

Hiyagon H., Hashimoto A., Kimura M., and Ushikubo T. (2003) First discovery of an ultra-refractory nodule in an Allende fine-grained inclusion. In *Lunar Planet. Sci.* **XXXIV**, #1552. The Lunar and Planetary Institute, Houston (CD-ROM).

Huss G. R., MacPherson G. J., Wasserburg G. J., Russell S. S., and Srinivasan G. (2001) ^{26}Al in CAIs and Al-chondrules from unequilibrated ordinary chondrites. *Meteorit. Planet. Sci.* **36**, 975–997.

Hutcheon I. D. (1982) Ion probe magnesium isotopic measurements of Allende inclusions. *Am. Chem. Soc. Symp. Ser.* **176**, 95–128.

Hutcheon I. D., Bar-Matthews M., Tanaka T., MacPherson G. J., Grossman L., Kawabe I., and Olsen E. (1980) A Mg isotope study of hibonite-bearing Murchison inclusions. *Meteoritics* **15**, 306–307.

Ihinger P. D. and Stolper E. (1986) The color of meteoritic hibonite: an indicator of oxygen fugacity. *Earth Planet. Sci. Lett.* **78**, 67–79.

Ireland T. R. (1988) Correlated morphological, chemical, and isotopic characteristics of hibonites from the Murchison carbonaceous chondrite. *Geochim. Cosmochim. Acta* **52**, 2827–2839.

Ireland T. R. (1990) Presolar isotopic and chemical signatures in hibonite-bearing refractory inclusions from the Murchison carbonaceous chondrite. *Geochim. Cosmochim. Acta* **54**, 3219–3237.

Ireland T. R. and Compston W. (1987) Large heterogeneous ^{26}Mg excesses in a hibonite from the Murchison meteorite. *Nature* **327**, 689–692.

Ireland T. R., Fahey A. J., and Zinner E. K. (1988) Trace element abundances in hibonites from the Murchison carbonaceous chondrite: constraints on high-temperature processes in the solar nebula. *Geochim. Cosmochim. Acta* **52**, 2841–2854.

Ireland T. R., Fahey A. J., and Zinner E. K. (1991) Hibonite-bearing microspherules: a new type of refractory inclusion with large isotopic anomalies. *Geochim. Cosmochim. Acta* **55**, 367–379.

Ireland T. R., Zinner E. K., Fahey A. J., and Esat T. M. (1992) Evidence for distillation in the formation of HAL and related hibonite inclusions. *Geochim. Cosmochim. Acta* **56**, 2503–2520.

Ivanova M. A., Petaev M. I., MacPherson G. J., Nazarov M. A., Taylor L. A., and Wood J. A. (2002) The first known natural occurrence of CaAl$_2$O$_4$, in a Ca–Al-rich inclusion from the CH chondrite NWA470. *Meteorit. Planet. Sci.* **37**, 1337–1344.

Kimura M., El Goresy A., Palme H., and Zinner E. (1993) Ca-, Al-rich inclusions in the unique chondrite ALH85085: petrology, chemistry and isotopic compositions. *Geochim. Cosmochim. Acta* **57**, 2329–2359.

Kimura M., Hiyagon H., Palme H., Spettel B., Wolf D., Clayton R. N., Mayeda T. K., Sato T., Suzuki A., and Kojima H. (2002) Yamato 79247, 79348 and 82038: the most primitive H chondrites, with abundant refractory inclusions. *Meteorit. Planet. Sci.* **37**, 1417–1434.

Komatsu M., Krot A. N., Petaev M. I., Ulyanov A. A., Keil K., and A. Miyamoto M. (2001) Mineralogy and petrography of amoeboid olivine aggregates from the reduced CV3 chondrites Efremovka, Leoville, and Vigarano: Products of nebular condensation, accretion, and annealing. *Meteorit. Planet. Sci.* **36**, 629–641.

Kornacki A. S. and Fegley B., Jr. (1984) Origin of spinel-rich chondrules and inclusions in carbonaceous and ordinary chondrites. *Proc. 14th Lunar Planet. Sci. Conf.* B588–B596.

Krot A. N., Scott E. R. D., and Zolensky M. E. (1995) Mineralogical and chemical modification of components in CV3 chondrites: nebular or asteroidal processing? *Meteoritics* **30**, 748–775.

Krot A. N., McKeegan K. D., Russell S. S., Meibom A., Weisberg M. K., Zipfel J., Krot T. V., Fagan T. J., and Keil K. (2001) Refractory calcium–aluminum-rich inclusions and aluminum-diopside-rich chondrules in the metal-rich chondrites Hammadah al Hamra 237 and Queen Alexandra range 94411. *Meteorit. Planet. Sci.* **36**, 1189–1216.

Krot A. N., Meibom A., Weisberg M. K., and Keil K. (2002a) The CR chondrite clan: implications for early solar system processes. *Meteorit. Planet. Sci.* **37**, 1451–1490.

Krot A. N., McKeegan K. D., Leshin L. A., MacPherson G. J., and Scott E. R. D. (2002b) Existence of an ^{16}O-rich gaseous reservoir in the solar nebula. *Science* **295**, 1051–1054.

Kurat G. (1975) Der kohlige Chondrit Lancé: Eine petrologische Analyse der komplexen Genese eines Chondriten. *Tschermaks Min. Petr. Mitt.* **22**, 38–78.

Kurat G., Hoinkes G., and Fredriksson K (1975) Zoned Ca–Al-rich chondrule in Bali: new evidence against the primordial condensation model. *Earth Planet. Sci. Lett.* **26**, 140–144.

Larimer J. W. and Bartholomay M. (1979) The role of carbon and oxygen in cosmic gases: some applications to the chemistry and mineralogy of enstatite chondrites. *Geochim. Cosmochim. Acta* **43**, 1455–1466.

Lee T., Papanastassiou D. A., and Wasserburg G. J. (1976) Demonstration of ^{26}Mg excess in Allende and evidence for ^{26}Al. *Geophys. Res. Lett.* **3**, 109–112.

Lee T., Papanastassiou D. A., and Wasserburg G. J. (1977) Aluminum-26 in the early solar system: fossil or fuel? *Ap. J. Lett.* **211**, L107–L110.

Lee T., Russell W. A., and Wasserburg G. J. (1979) Ca isotopic anomalies and the lack of aluminum-26 in an unusual Allende inclusion. *Astrophys. J.* **228**, L93–L98.

Lee T., Mayeda T. K., and Clayton R. N. (1980) Oxygen isotopic anomalies in the Allende inclusion HAL. *Geophys. Res. Lett.* **7**, 493–496.

Leshin L. A., Rubin A. E., and McKeegan K. D. (1996) Oxygen isotopic compositions of olivine and pyroxene from CI chondrites. In *Lunar Planet. Sci.* **XXVII**. The Lunar and Planetary Institute, Houston, pp. 745–746.

Lodders K. and Fegley B., Jr. (1995) The origin of circumstellar silicon carbide grains found in meteorites. *Meteoritics* **30**, 661–678.

Lord H. C., III (1965) Molecular equilibria and condensation in a solar nebula and cool stellar atmospheres. *Icarus* **4**, 279–288.

Loss R. D., Lugmair G. W., Davis A. M., and MacPherson G. J. (1994) Isotopically distinct reservoirs in the solar nebula: isotope anomalies in Vigarano meteorite inclusions. *Astrophys. J.* **436**, L193–L196.

Macdougall J. D. (1979) Refractory-element-rich inclusions in CM meteorites. *Earth Planet. Sci. Lett.* **42**, 1–6.

Macdougall J. D. (1981) Refractory spherules in the Murchison meteorite: are they chondrules? *Geophys. Res. Lett.* **8**, 966–969.

Macdougall J. D. and Phinney D. (1979) Magnesium isotopic variations in hibonite from the Murchison meteorite: an ion microprobe study. *Geophys. Res. Lett.* **6**, 215–218.

MacPherson G. J. and Davis A. M. (1992) Evolution of a Vigarano forsterite-bearing CAI. *Meteoritics* **27**, 253.

MacPherson G. J. and Davis A. M. (1993) A petrologic and ion microprobe study of a Vigarano Type B2 refractory inclusion: evolution by multiple stages of melting and alteration. *Geochim. Cosmochim. Acta* **57**, 231–243.

MacPherson G. J. and Davis A. M. (1994) Refractory inclusions in the prototypical CM chondrite, Mighei. *Geochim. Cosmochim. Acta* **58**, 5599–5625.

MacPherson G. J. and Grossman L. (1979) Melted and non-melted coarse-grained Ca-, Al-rich inclusions in Allende. *Meteoritics* **14**, 479–480.

MacPherson G. J. and Grossman L. (1981) A once-molten, coarse-grained, Ca-rich inclusion in Allende. *Earth Planet. Sci. Lett.* **52**, 16–24.

MacPherson G. J. and Grossman L. (1984) Fluffy Type-A inclusions in the Allende meteorite. *Geochim. Cosmochim. Acta* **48**, 29–46.

MacPherson G. J. and Krot A. N. (2002) Distribution of Ca-Fe-silicates in CV3 chondrites: possible controls by parent-body compaction. *Meteorit. Planet. Sci.* **37**, A91.

MacPherson G. J. and Huss G. R. (2000) Convergent evolution of CAIs and chondrules: evidence from bulk compositions and a cosmochemical phase diagram. In *Lunar Planet. Sci.* **XXXI**, #1796, The Lunar and Planetary Institute, Houston (CD-ROM).

MacPherson G. J., Grossman L., Allen J. M., and Beckett J. R. (1981) Origin of rims on coarse-grained inclusions in the Allende meteorite. *Proc. 12th Lunar Planet. Sci. Conf.*, 1079–1091.

MacPherson G. J., Bar-Matthews M., Tanaka T., Olsen E., and Grossman L. (1983) Refractory inclusions in the Murchison meteorite. *Geochim. Cosmochim. Acta* **47**, 823–839.

MacPherson G. J., Grossman L., Hashimoto A., Bar-Matthews M., and Tanaka T. (1984a) Petrographic studies of refractory inclusions from the Murchison meteorite. *Proc. 15th Lunar Planet. Sci. Conf.*, C299–C312.

MacPherson G. J., Paque J. M., Stolper E., and Grossman L. (1984b) The origin and significance of reverse zoning in melilite from Allende Type B inclusions. *J. Geology* **92**, 289–305.

MacPherson G. J., Wark D. A., and Armstrong J. T. (1988) Primitive materials surviving in chondrites: refractory inclusions. In *Meteorites and the Early Solar System* (eds. J. F. Kerridge and M. S. Matthews). University of Arizona Press, Tucson, pp. 746–807.

MacPherson G. J., Davis A. M., and Zinner E. K. (1995) ^{26}Al in the early solar system: a reappraisal. *Meteoritics* **30**, 365–386.

MacPherson G. J., Krot A. N., Ulyanov A. A., and Hicks T. (2002) A comprehensive study of pristine, fine-grained, spinel-rich inclusions from the Leoville and Efremovka CV3 chondrites: I. Petrology. In *Lunar Planet. Sci.* **XXXIII**, #1526 The Lunar and Planetary Institute, Houston (CD-ROM).

MacPherson G. J., Huss G. R., and Davis A. M. (2003) Extinct ^{10}Be in Type A CAIs from CV chondrites. *Geochim. Cosmochim. Acta* **67**, 3165–3179.

Mao X.-Y., Ward B. J., Grossman L., and MacPherson G. J. (1990) Chemical compositions of refractory inclusions from the Vigarano and Leoville carbonaceous chondrites. *Geochim. Cosmochim. Acta* **54**, 2121–2132.

Martin P. M. and Mason B. (1974) Major and trace elements in the Allende meteorite. *Nature* **249**, 333–334.

Marvin U. B., Wood J. A., and Dickey J. S., Jr. (1970) Ca–Al rich phases in the Allende meteorite. *Earth Planet. Sci. Lett.* **7**, 346–350.

Mason B. and Martin P. M. (1977) Geochemical differences among components of the Allende meteorite. *Smithson. Contrib. Earth Sci.* Smithsonian Institution Press, Washington, DC, vol. 19, 84–95.

Mason B. and Taylor S. R. (1982) Inclusions in the Allende meteorite. *Smithson. Contrib. Earth Sci.* **25**, 1–30.

McGuire A. V. and Hashimoto A. (1989) Origin of zoned fine-grained inclusions in the Allende meteorite. *Geochim. Cosmochim. Acta* **53**, 1123–1133.

McKeegan K. D., Leshin L. A., Russell S. S., and MacPherson G. J. (1996) *In situ* measurement of O isotopic anomalies in a Type B Allende CAI. *Meteoritics* **31**, pp. A86–A87.

McKeegan K. D., Leshin L. A., Russell S. S., and MacPherson G. J. (1998) Oxygen isotopic abundances in calcium–aluminum-rich inclusions from ordinary chondrites: implications for nebular heterogeneity. *Science* **280**, 414–418.

McKeegan K. D., Chaussidon M., and Robert F. (2000) Incorporation of short-lived ^{10}Be in a calcium–aluminum-rich inclusion from the Allende meteorite. *Science* **289**, 1334–1337.

Messenger S., Keller L. P., Stadermann F. J., Walker R. M., and Zinner E. (2003) Samples of stars beyond the solar system: silicate grains in interplanetary dust. *Science* **300**, 105–108.

Morimoto N., Fabries J., Ferguson A. K., Ginzburg I. V., Ross M., Seifert F. A., Zussman J., Aoki K., and Gottardi G. (1988) Nomenclature of pyroxenes. *Am. Mineral.* **73**, 1123–1133.

Niederer F. R. and Papanastassiou D. A. (1984) Ca isotopes in refractory inclusions. *Geochim. Cosmochim. Acta* **48**, 1279–1294.

Niederer F. R., Papanastassiou D. A., and Wasserburg G. J. (1985) Absolute isotopic abundances of Ti in meteorites. *Geochim. Cosmochim. Acta* **49**, 835–851.

Nittler L. R., Alexander C. M. O.'D., Gao X., Walker R. M. and Zinner E. (1997) Stellar sapphires: the properties and origins of presolar Al_2O_3 in meteorites. *Astrophys. J.* **483**, 475–495.

Notsu K., Onuma N., Nishida N., and Nagasawa H. (1978) High temperature heating of the Allende meteorite. *Geochim. Cosmochim. Acta* **42**, 903–907.

Osborn E. F. and Schairer J. F. (1941) The ternary system pseudowollastonite–åkermanite–gehlenite. *Am. J. Sci.* **239**, 715–763.

Papanastassiou D. A. and Brigham C. A. (1989) The identification of meteorite inclusions with isotope anomalies. *Astrophys. J.* **338**, L37–L40.

Petaev M. I. and Wood J. A. (1998a) The CWPI model of nebular condensation: effects of pressure on the condensation sequence. *Meteorit. Planet. Sci.* **33**, A122.

Petaev M. I. and Wood J. A. (1998b) The condensation with partial isolation (CWPI) model of condensation in the solar nebula. *Meteorit. Planet. Sci.* **33**, 1123–1137.

Podosek F. A., Zinner E. K., MacPherson G. J., Lundberg L. L., Brannon J. C., and Fahey A. J. (1991) Correlated study of initial $^{87}Sr/^{86}Sr$ and Al/Mg isotopic systematics and petrologic properties in a suite of refractory inclusions from the Allende meteorite. *Geochim. Cosmochim. Acta* **55**, 1083–1110.

Richter F. M., Davis A. M., Ebel D. S., and Hashimoto A. (2002) Elemental and isotopic fractionation of Type B calcium-, aluminum-rich inclusions: experiments, theoretical considerations, and constraints on their thermal evolution. *Geochim. Cosmochim. Acta* **66**, 521–540.

Russell S. S., Srinivasan G., Huss G. R., Wasserburg G. J., and MacPherson G. J. (1996) Evidence for widespread ^{26}Al in the solar nebula and constraints for nebula timescales. *Science* **273**, 757–762.

Russell S. S., Huss G. R., Fahey A. J., Greenwood R. C., Hutchison R., and Wasserburg G. J. (1998) An isotopic and petrologic study of calcium–aluminum-rich inclusions from CO3 meteorites. *Geochim. Cosmochim. Acta* **62**, 689–714.

Russell S. S., Davis A. M., MacPherson G. J., Guan Y., and Huss G. R. (2000) Refractory inclusions from the ungrouped carbonaceous chondrites MAC87300 and MAC88107. *Meteorit. Planet. Sci.* **35**, 1051–1066.

Ryerson F. J. and McKeegan K. D. (1994) Determination of oxygen self-diffusion in åkermanite, anorthite, diopside, and spinel: implications for oxygen isotopic anomalies and the thermal histories of Ca–Al-rich inclusions. *Geochim. Cosmochim. Acta* **58**, 3713–3734.

Shu F. H., Shang H. and Lee T. (1996) Toward an astrophysical theory of chondrites. *Science* **271**, 1545–1552.

Simon S. B., Yoneda S., Grossman L., and Davis A. M. (1994) A $CaAl_4O_7$-bearing refractory spherule from Murchison: evidence for very high-temperature melting in the solar nebula. *Geochim. Cosmochim. Acta* **58**, 1937–1949.

Simon S. B., Davis A. M., and Grossman L. (1996) A unique ultrarefractory inclusion from the Murchison meteorite. *Meteorit. Planet. Sci.* **31**, 106–115.

Simon S. B., Davis A. M., Grossman L., and Zinner E. K. (1998) Origin of hibonite–pyroxene spherules found in carbonaceous chondrites. *Meteorit. Planet. Sci.* **33**, 411–424.

Simon S. B., Davis A. M., Grossman L., and McKeegan K. D. (1999) Origin of compact type A refractory inclusions from CV3 carbonaceous chondrites. *Geochim. Cosmochim. Acta* **63**, 1233–1248.

Simon S. B., Davis A. M., Grossman L., and McKeegan K. D. (2002) A hibonite–corundum inclusion from Murchison: a first-generation condensate from the solar nebula. *Meteorit. Planet. Sci.* **37**, 533–548.

Srinivasan G., Ulyanov A. A., and Goswami J. N. (1994) ^{41}Ca in the early solar system. *Astrophys. J.* **431**, L67–L70.

Srinivasan G., Sahijpal S., Ulyanov A. A., and Goswami J. N. (1996) Ion microprobe studies of Efremovka CAIs: II. Potassium isotope composition and ^{41}Ca in the early solar system. *Geochim. Cosmochim. Acta* **60**, 1823–1835.

Srinivasan G., Huss G. R., and Wasserburg G. J. (2000) A petrographic, chemical, and isotopic study of calcium–aluminum-rich inclusions and aluminum-rich chondrules from the Axtell (CV3) chondrite. *Meteorit. Planet. Sci.* **35**, 1333–1354.

Stolper E. (1982) Crystallization sequences of Ca–Al-rich inclusions from Allende: an experimental study. *Geochim. Cosmochim. Acta* **46**, 2159–2180.

Stolper E. and Paque J. (1986) Crystallization sequences of Ca–Al-rich inclusions from Allende: the effects of cooling rate and maximum temperature. *Geochim. Cosmochim. Acta* **50**, 1785–1806.

Sylvester P. J., Ward B. J., Grossman L. and Hutcheon I. D. (1990) Chemical compositions siderophile element-rich opaque assemblages in an Allende inclusion. *Geochim. Cosmochim. Acta* **54**, 3491–3508.

Tanaka T. and Masuda A. (1973) Rare-earth elements in matrix, inclusions, and chondrules of the Allende meteorite. *Icarus* **19**, 523–530.

Thiemens M. H. and Heidenreich J. E. (1983) The mass independent fractionation of oxygen: a novel isotope effect and its possible cosmochemical implications. *Science* **219**, 1073–1075.

Tomeoka K., McSween H. Y., Jr., and Buseck P. R. (1989) Mineralogical alteration of CM carbonaceous chondrites: a review. *Proc. NIPR Symp. Antarct. Meteorit.* **2**, 221–234.

Ulyanov A. A. (1984) On the origin of fine-grained Ca, Al-rich inclusions in the Efremovka carbonaceous chondrite. In *Lunar Planet. Sci.* **XV**. The Lunar and Planetary Institute, Houston, pp. 872–873.

Wang J., Davis A. M., Clayton R. N., Mayeda T. K., and Hashimoto A. (2001) Chemical and isotopic fractionation during the evaporation of the $FeO–MgO–SiO_2–CaO–Al_2O_3–TiO_2$–REE melt system. *Geochim. Cosmochim. Acta* **65**, 479–494.

Wark D. A. and Boynton W. V. (2001) The formation of rims on calcium–aluminum-rich inclusions: Step I. Flash heating. *Meteorit. Planet. Sci.* **36**, 1135–1166.

Wark D. A. (1987) Plagioclase-rich inclusions in carbonaceous chondrite meteorites: liquid condensates? *Geochim. Cosmochim. Acta* **51**, 221–242.

Wark D. A. and Lovering J. F. (1977) Marker events in the early solar system: evidence from rims on Ca–Al-rich inclusions in carbonaceous chondrites. *Proc. 8th Lunar Sci. Conf.*, 95–112.

Wark D. A., and Lovering J. F. (1982) The nature and origin of type B1 and B2 Ca–Al-rich inclusions in the Allende meteorite. *Geochim. Cosmochim. Acta* **46**, 2581–2594.

Wark D. A., Kornacki A. S., Boynton W. V., and Ulyanov A. A. (1986) Efremovka fine-grained inclusion E14: comparisons with Allende. In *Lunar Planet. Sci.* **XVII**. The Lunar and Planetary Institute, Houston, pp. 921–922.

Wark D. A., Boynton W. V., Keays R. R., and Palme H. (1987) Trace element and petrologic clues to the formation of forsterite-bearing Ca–Al-rich inclusions in the Allende meteorite. *Geochim. Cosmochim. Acta* **51**, 607–622.

Wasserburg G. J. and Papanastassiou D. A. (1982) Some short-lived nuclei in the early solar system—a connection with the placental ISM. In *Essays in Nuclear Astrophysics* (eds. C. A. Barnes, D. D. Clayton and D. N. Schramm). Cambridge University Press, Cambridge, pp. 77–140.

Wasserburg G. J., Lee T., and Papanastassiou D. A. (1977) Correlated O and Mg isotopic anomalies in Allende inclusions: II. Magnesium. *Geophys. Res. Lett.* **4**, 299–302.

Wasson J. T., Yurimoto H., and Russell S. S. (2001) ^{16}O-rich melilite in CO3.0 chondrites: possible formation of common, ^{16}O-poor melilite by aqueous alteration. *Geochim. Cosmochim. Acta* **65**, 4539–4549.

Weber D. and Bischoff A. (1994) The occurrence of grossite ($CaAl_4O_7$) in chondrites. *Geochim. Cosmochim. Acta* **58**, 3855–3877.

Weber D. and Bischoff A. (1997) Refractory inclusions in the CR chondrite Acfer 059-El Djouf 001: petrology, chemical composition, and relationship to inclusion populations in other types of carbonaceous chondrites. *Chem. Erde* **57**, 1–24.

Weber D., Zinner E., and Bischoff A. (1995) Trace element abundances and magnesium, calcium, and titanium isotopic compositions of grossite-containing inclusions from the carbonaceous chondrite Acfer 182. *Geochim. Cosmochim. Acta* **59**, 803–823.

Weisberg M. K., Prinz M., Clayton R. N., Mayeda T. K., Grady M. M., and Pillinger C. T. (1995) The CR chondrite clan. *Proc. NIPR Symp. Antarct. Meteorit.* **8**, 11–32.

Weisberg M. K., Boesenberg J. S., and Ebel D. S. (2002) Gujba and the evolution of the Bencubbin-like (CB) chondrites. In *Lunar Planet. Sci.* **XXXIII**, #1551. The Lunar and Planetary Institute, Houston (CD-ROM).

Wood J. A. (1996) Unresolved issues in the formation of chondrules and chondrites. In *Chondrules and the Protoplanetary Disk* (eds. R. H. Hewins, R. H. Jones, and E. R. D. Scott). Cambridge University Press, Cambridge, pp. 55–69.

Wood J. A. (1998) Meteoritic evidence for the infall of large interstellar dust aggregates during formation of the solar system. *Astrophys. J.* **503**, L101–L104.

Young E. D. and Lyons J. R. (2003) CO self-shielding in the outer solar nebula: an astrochemical explanation for the oxygen isotope slope-1 line. In *Lunar Planet. Sci.* **XXXIV**, #1923. The Lunar and Planetary Institute, Houston (CD-ROM).

Young E. D. and Russell S. S. (1998) Oxygen reservoirs in the early solar nebula inferred from an Allende CAI. *Science* **282**, 452–455.

Yoneda S. and Grossman L. (1995) Condensation of CaO–MgO–Al_2O_3–SiO_2 liquids from cosmic gases. *Geochim. Cosmochim. Acta* **59**, 3413–3444.

Yurimoto H., Ito M., and Nagasawa H. (1994) Oxygen isotope exchange between refractory inclusion in Allend and solar nebula gas. *Science* **282**, 1874–1877.

Zinner E. (1998) Stellar nucleosynthesis and the isotopic composition of presolar grains from primitive meteorites. *Ann. Rev. Earth Planet. Sci.* **26**, 147–188.

1.09
Nebular versus Parent-body Processing

A. J. Brearley

University of New Mexico, Albuquerque, NM, USA

1.09.1 INTRODUCTION	247
1.09.2 NEBULAR OR ASTEROIDAL PROCESSING: SOME CRITERIA	248
1.09.3 AQUEOUS ALTERATION	248
1.09.3.1 CI Carbonaceous Chondrites	248
1.09.3.1.1 Evidence for asteroidal alteration	249
1.09.3.1.2 Timing of alteration	249
1.09.3.1.3 Summary	249
1.09.3.2 CM Carbonaceous Chondrites	250
1.09.3.2.1 Timing of alteration	251
1.09.3.2.2 Appraisal of the evidence for pre-accretionary alteration	251
1.09.3.2.3 Evidence for asteroidal alteration	252
1.09.3.2.4 Summary	254
1.09.3.3 CR Carbonaceous Chondrites	254
1.09.3.4 CO Carbonaceous Chondrites	254
1.09.3.5 CV Carbonaceous Chondrites	255
1.09.3.5.1 Evidence for pre-accretionary alteration	256
1.09.3.5.2 Evidence for parent-body alteration	256
1.09.3.6 Unequilibrated Ordinary Chondrites	256
1.09.4 OXIDATION AND METASOMATISM	257
1.09.4.1 CV Carbonaceous Chondrites	257
1.09.4.2 Dark inclusions in CV Chondrites	258
1.09.4.3 Nebular and Parent-body Alteration Models for CV Chondrites	259
1.09.4.3.1 Nebular alteration	259
1.09.4.3.2 Parent-body alteration	259
1.09.4.3.3 Problems with the nebular-alteration model	259
1.09.4.3.4 Problems with the asteroidal alteration model	261
1.09.4.4 CO Carbonaceous Chondrites	261
1.09.4.4.1 Evidence for parent-body alteration of CO Chondrites	262
1.09.4.4.2 Evidence for nebular alteration of CO Chondrites	262
1.09.4.5 Unequilibrated Ordinary Chondrites	263
1.09.5 FUTURE WORK	263
ACKNOWLEDGMENTS	264
REFERENCES	264

1.09.1 INTRODUCTION

Chondritic meteorites contain a complex record of processes that occurred during the earliest stages of solar system evolution, from the formation of the earliest solids by condensation in the solar nebula to the accretion of asteroidal parent bodies and subsequent processing within these small planetesimals. Of key interest is the very earliest history of nebular solids, which provide important constraints on nebular conditions and processes that are the ground truth for astrophysical models of nebular evolution (e.g., Boss, 1996; Shu *et al.*, 1996; Jones *et al.*, 2000).

Although chondritic meteorites certainly contain a record of nebular processes, this very early record has been modified or obscured to different degrees by later processing. A major area of endeavor for researchers studying chondritic meteorites is to identify the nature of this secondary processing and establish where this processing occurred. This is a complex and challenging task, because secondary processing can occur in different environments and/or times and manifests itself in primitive chondrites in a number of different ways. These include, but are not limited to, aqueous alteration, oxidation, and metasomatism/metamorphism. This chapter reviews the cosmochemical evidence for secondary alteration within a framework of mineralogical observations and evaluates the evidence for nebular versus asteroidal alteration of chondritic materials.

1.09.2 NEBULAR OR ASTEROIDAL PROCESSING: SOME CRITERIA

To provide a framework for the discussion of alteration effects in chondritic meteorites, it is useful to outline criteria that may be used to distinguish between solar-nebular processes versus processes that could plausibly have occurred within asteroidal parent bodies. Several different criteria have been used in a rather *ad hoc* fashion in the literature, leading to confusion. In most cases, none of the criteria listed below provide exclusive proof of alteration in a particular environment, but generally support alteration in one environment or the other. Criteria that have been used are: (i) heterogeneity versus homogeneity of the mineralogical effects of alteration within different primitive chondritic components (e.g., CAIs, chondrules, matrix, etc.), i.e., heterogeneous alteration of components may be most consistent with pre-accretionary alteration; (ii) stable isotopic heterogeneity versus homogeneity of alteration products in altered chondrites as an indicator of alteration in different or the same isotopic reservoir; (iii) correlation of mineralogical alteration effects with stable-isotopic data; (iv) evidence of mass-independent fractionation (nebular) versus mass-dependent fractionation (asteroidal) in the oxygen-isotopic composition of bulk chondrites and individual components; (v) evidence (or lack of) bulk-compositional variations that might constrain the location of alteration, i.e., isochemical alteration is most consistent with alteration after accretion; (vi) constraints provided by the thermodynamic stabilities of mineral phases under solar nebular and asteroidal conditions; (vii) experimental and theoretical constraints on the kinetics of alteration reactions (i.e., gas–solid, liquid–solid, etc.); and (viii) timing of alteration, i.e., evidence of

the formation of alteration phases more than several million years after CAI formation is most consistent with asteroidal alteration, assuming nebular lifetimes of 5–10 Myr (Podosek and Cassen, 1994).

1.09.3 AQUEOUS ALTERATION

Of the secondary processes that have affected chondritic meteorites, aqueous alteration is among the most widespread. Evidence of varying degrees of aqueous alteration is present in all the major chondrite groups, with the exception of the enstatite chondrites. This alteration is typically indicated by the presence of hydrous phyllosilicates (principally serpentines and smectite clays), often associated with carbonates, sulfates, oxides (magnetite), and secondary sulfides. The variable alteration assemblages present in different chondrite groups are principally the result of alteration under different conditions (P, T, f_{O_2}, water/rock ratio) (e.g., Zolensky *et al.*, 1993).

Four principal models have been proposed to explain the presence of alteration phases in chondritic meteorites. These models can be summarized as follows: (i) reaction of anhydrous, high-temperature condensate phases with water vapor as the solar nebula cooled to temperatures below ~375 K (e.g., Grossman and Larimer, 1974]); (ii) hydration of anhydrous dust in icy regions of the nebula during the passage of shock waves (Ciesla *et al.*, 2003); (iii) alteration within small (tens of meters), ephemeral parent bodies that were subsequently disrupted and their altered components accreted with unaltered materials into the final asteroidal parent bodies (pre-accretionary alteration) (Metzler *et al.*, 1992; Bischoff, 1998); and (iv) alteration within asteroidal parent bodies (Kerridge and Bunch, 1979; Zolensky and McSween, 1988).

1.09.3.1 CI Carbonaceous Chondrites

The CI chondrites represent one of the most curious paradoxes of cosmochemistry. Despite their unfractionated compositions, the CI chondrites are the most altered of all chondrites, with water contents of ~19.5 wt.% (Nagy *et al.*, 1963). Anhydrous phases (olivines and pyroxenes) represent less than 1 vol.% of these meteorites (Leshin *et al.*, 1997). CI chondrites are complex meteorites that consist of a dark, fine-grained matrix comprised of phyllosilicates with magnetite, sulfides, carbonates, and sulfates embedded within it (e.g., DuFresne and Anders, 1962; Nagy, 1966). They have experienced extensive brecciation on their asteroidal parent bodies that caused mixing of compositionally distinct lithic clasts on the millimeter to submillimeter scale (Endress and Bischoff, 1996).

1.09.3.1.1 Evidence for asteroidal alteration

The CI chondrites have long been cited as the classic example of asteroidal aqueous alteration, because of the presence of ubiquitous sulfate veins (DuFresne and Anders, 1962; Richardson, 1978; Fredriksson and Kerridge, 1988). These veins crosscut the dark, fine-grained matrix and can extend across the entire meteorite sample or stone. These veins have commonly been attributed to the widespread movement of water within the CI parent body. However, Gounelle and Zolensky (2001) have reappraised the origin of these veins and concluded that they are terrestrial, not asteroidal, in origin. Their preferred interpretation is that the veins formed as a result of the dissolution, local transport, and precipitation of extraterrestrial sulfates by absorbed terrestrial water. Thus, one of the widely accepted lines of evidence to support parent-body alteration should now be treated with caution.

Other lines of textural evidence, however, still provide support for asteroidal alteration of the CI chondrites. These include: (i) crosscutting phyllosilicate-rich veins (Tomeoka, 1990) and (ii) the presence of carbonates that appear to represent fragments of an earlier generation of carbonate veins or resemble vein fillings (Richardson, 1978; Endress and Bischoff, 1996).

In addition, chemical and isotopic data provide support for asteroidal aqueous alteration. The highly unfractionated bulk composition of the CI chondrites indicates that alteration must have been essentially isochemical, with the exception of the most volatile elements such as hydrogen, carbon, oxygen, and nitrogen. These elements were probably lost as gaseous species during alteration (Wilson et al., 1999). Given the variable solubilities of refractory and moderately volatile elements such as calcium, rubidium, strontium, etc., it seems highly improbable that the unfractionated abundances of these elements could have been retained during open-system alteration in either the solar nebula or an asteroidal environment. This evidence implies that fluid flow was essentially zero and that alteration occurred under isochemical conditions, a scenario which is most compatible with asteroidal alteration.

Oxygen-isotopic data for CI chondrites provide some constraints on the location of alteration. The bulk compositions of CI chondrites lie close to the terrestrial fractionation line on a three-isotope plot, with a $\Delta^{17}O$ of $+0.38 \pm 0.09\permil$ ($2\sigma_m$) (Rowe et al., 1994; see Chapter 1.06 for a definition of $\Delta^{17}O$). Carbonates with $\Delta^{17}O = \sim 0.37\permil$ appear to be in isotopic equilibrium with the host rock and define a mass-dependent fractionation line with phyllosilicates that make up the bulk of these chondrites (Leshin et al., 2001). Phyllosilicate and carbonate compositions indicate that they equilibrated with the same fluid and that alteration took place within an asteroidal parent body (Clayton and Mayeda, 1999; Benedix et al., 2000). Magnetites from Alais, Ivuna, and Orgueil have $\Delta^{17}O = +1.70 \pm 0.05\permil$ and are clearly not at isotopic equilibrium with the host rock.

1.09.3.1.2 Timing of alteration

Constraints on the timing of aqueous alteration of CI chondrites come from both long- and short-lived isotope geochronology. Rubidium–strontium dating of Orgueil carbonates shows that they formed within 50 Myr of formation of the Orgueil parent body (Macdougall et al., 1984). This early formation age is supported by manganese-chromium and iodine-xenon dating of CI chondrites. Manganese-chromium isotopic data (Figure 1) from carbonates in Ivuna and Orgueil (Endress et al., 1996; Hutcheon and Phinney, 1996) yield formation ages of 16.5–18.3 Myr after CAI formation, based on inferred initial $^{53}Mn/^{55}Mn$ ratios between 1.42×10^{-6} to 1.99×10^{-6} and a solar system initial $^{53}Mn/^{55}Mn$ ratio of $(4.4 \pm 1.0) \times 10^{-5}$ (Lugmair and Shukolyukov, 2001).

Iodine-xenon dating of magnetite separates from Orgueil (Herzog et al., 1973; Lewis and Anders, 1975) yield closure ages of $\sim 2.0 \pm 1$ Myr after Vigarano, indicating that magnetite formation occurred very early and probably over a very restricted time period (Swindle, 1998). Such data are suggestive of a nebular origin for the magnetite. However, more recent data (Hohenberg et al., 2000; Pravdivtseva et al., 2003) indicate that Orgueil magnetite is actually 5 Myr younger than reported previously by Lewis and Anders (1975), more consistent with asteroidal formation.

1.09.3.1.3 Summary

These data suggest that alteration of the CI chondrites took place over an extended timescale, indicating an asteroidal environment for alteration (Hutcheon, 1997). Given that early-formed magnetite is almost certainly the product of aqueous activity (e.g., Kerridge and Bunch, 1979), alteration probably occurred, almost entirely, within an asteroidal environment. It is possible that precursor components of the CI chondrites experienced pre-accretionary alteration (Bischoff, 1998). However, it is clear that

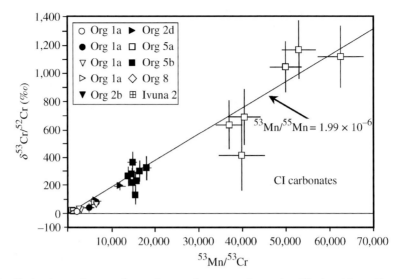

Figure 1 Mn–Cr isochron diagram for carbonate fragments from the CI chondrites, Orgueil, and Ivuna. ^{53}Cr excesses are found in all the fragments analyzed by ion microprobe, indicating the presence of live ^{53}Mn at the time of carbonate crystallization (after Endress *et al.*, 1996).

advanced parent-body alteration has completely obliterated any evidence of such processes.

1.09.3.2 CM Carbonaceous Chondrites

CM chondrites are a diverse, complex group of meteorites that have experienced aqueous alteration and brecciation to different degrees. The least-brecciated CM chondrites have a primary accretionary texture (Metzler *et al.*, 1992) that is characterized by the presence of well-developed fine-grained rims (Figure 2) that mantle all the macroscopic components of the chondrite (i.e., chondrules, etc.). Brecciation on the CM chondrite parent body caused progressive fragmentation of this primary texture and produced a clastic texture that is characteristic of many CM chondrites. Most CM chondrites are petrologic type 2, with fully hydrated matrices and variably altered chondrules, CAIs, etc. The alteration phases are dominated by phyllosilicates, principally serpentine, with associated carbonates, oxides, sulfides, and sulfates (see Brearley and Jones, 1998). A few CM chondrites exist which show essentially 100% hydration, and are termed CM1 chondrites (Zolensky *et al.*, 1997).

Within the CM2 chondrites, a progression in the degree of alteration has been recognized (McSween, 1979, 1987). A sequence of alteration has been established (Tomeoka and Buseck, 1985; Zolensky *et al.*, 1993; Browning *et al.*, 1996), based on the relative alteration susceptibilities of the primary phases (olivine, pyroxene, chondrule mesostasis, troilite metal) in aqueous fluids. Iron, nickel metal, troilite and chondrule glass are most susceptible to alteration, whereas olivine and low-calcium pyroxene are

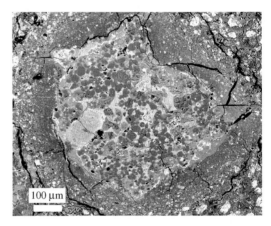

Figure 2 Backscattered electron image of fine-grained rim surrounding a highly altered chondrule in ALH81002, a moderately altered CM2 carbonaceous chondrite with a primary accretionary texture.

much more resistant. In the least-altered CM2s, matrix, chondrule metal, sulfide, and mesostasis glass have been hydrated, whereas olivine and pyroxene are completely unaltered. As alteration proceeds, replacement of chondrule olivines and pyroxenes become progressively more extensive (Hanowski and Brearley, 2001).

Oxygen-isotopic data for bulk CM chondrites and mineral separates (Clayton and Mayeda, 1999) lie along a mixing line with slope 0.7, rather than the slope 0.5 carbonaceous chondrite anhydrous mineral (CCAM) (see line Chapter 1.06). This line is attributable to the mixing of anhydrous ^{16}O-rich silicate material in chondrules and CAIs, etc. and hydrated ^{16}O-poor material, largely phyllosilicates that occurs in the matrix. The bulk δ^{18}O of CM chondrites is weakly correlated with the degree of aqueous alteration based on mineralogical criteria (Browning et al., 1996), moving towards more ^{16}O-poor compositions as alteration increases (Figure 3).

The compositions of individual components of CM chondrites suggest a complex alteration history. Calcite and phyllosilicates in Murchison fall on a mass-dependent fractionation line with $\Delta^{17}O = -1.4‰$ (Clayton and Mayeda, 1984), indicating that they formed at equilibrium from the same aqueous reservoir. Clayton and Mayeda (1984, 1999) have argued that this was most likely to have occurred within an asteroidal parent body consistent with the very high $\delta^{13}C$ of carbonates in CM chondrites (+40‰ to +70‰) (Clayton, 1963; Halbout et al., 1986). Analyses of carbonate from four splits from Murchison (Benedix et al., 2003) show a range of $\delta^{18}O$ values (+26.6‰ to +35.5‰) (Grady et al., 1988) indicative of carbonate formation at constant temperature from an evolving reservoir or the temperature varied during carbonate growth.

Carbonates and sulfates from other CM chondrites (Benedix et al., 2003; Airieau et al., 2001) have $\Delta^{17}O$ values that become progressively lower as a function of the degree of alteration of the host chondrite. Assuming that the altering fluid had a higher $\Delta^{17}O$ than the anhydrous-precursor phases (Clayton and Mayeda, 1999), these values indicate that the change in $\Delta^{17}O$ was due to precipitation of carbonates and sulfates from a fluid that became progressively more equilibrated with the host chondrite as alteration proceeded.

1.09.3.2.1 Timing of alteration

The chronology of alteration in CM chondrites is imperfectly understood. Iodine-xenon dating of mineral separates from Murchison, Murray, Mighei, and Cold Bokkeveld (Lewis and Anders, 1975; Niemeyer and Zaikowski, 1980) indicate an extended alteration history for the CMs. Relative to the formation of Murchison magnetite, Murray, and Cold Bokkeveld samples yield closure ages of 10.1 ± 3.2 Myr and at least 11 Myr, respectively, after CAI formation.

Manganese–chromium dating of carbonates in CM chondrites shows a similar spread in the duration of alteration to that obtained by iodine-xenon dating. Carbonates in Yamato 791198 (CM2) yield an initial $^{53}Mn/^{55}Mn$ ratio of $(1.3 \pm 0.6) \times 10^{-5}$ (Brearley et al., 2001), very similar to the ratio determined in carbonates in the unique chondrite Kaidun of $(9.4 \pm 1.6) \times 10^{-6}$ (Hutcheon et al., 1999). These data constrain carbonate crystallization to no more than 10 Myr, after CAI formation. Younger formation ages are indicated from dolomites in ALH84034 (CM1), with a calculated initial $^{53}Mn/^{55}Mn$ ratio of $(5.0 \pm 1.5) \times 10^{-6}$ (Brearley and Hutcheon, 2000), corresponding to a formation age ~14 Myr after CAI formation. These data show that alteration of CM chondrites extended for a significant period after CAI formation, and for periods longer than estimated for the lifespan of the solar nebula. These extended periods of alteration appear to be most consistent with alteration within an asteroidal parent body, especially for carbonates in the most highly altered CM chondrites.

1.09.3.2.2 Appraisal of the evidence for pre-accretionary alteration

The major controversy surrounding the CM chondrites is where aqueous alteration occurred. Most researchers have favored a post-accretion asteroidal scenario (e.g., Kerridge and Bunch, 1979; Bunch and Chang, 1980, Browning et al., 1996; Hanowski and Brearley, 2001). However, a significant body of evidence suggests that alteration of some components of CM chondrites occurred prior to accretion (Metzler et al., 1992; Bischoff, 1998). In this scenario, alteration occurred in small, ephemeral protoplanetary bodies that were disrupted by impacts and their altered fragments were later re-accreted into the final CM2 asteroid. Alternatively, Ciesla et al. (2003) have suggested that hydration of dust could have occurred during the passage of shock waves through ice-rich

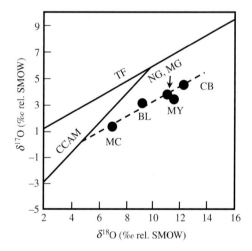

Figure 3 Oxygen-isotopic compositions of CM2 carbonaceous chondrites plotted in a three-isotope diagram. The CM2s form a linear array with a slope of 0.7, with enrichments in the heavy isotopes that show a general increase as a function of increasing alteration. Nogoya (NG) and Cold Bokkeveld (CB) are the most heavily altered CMs and Murchison (MC) is one of the least altered. Chondrites with intermediate degrees of alteration are not well resolved in this diagram (after Browning et al., 1996; source Clayton, 1993).

regions of the solar nebula. The pre-accretionary and asteroidal models are not necessarily mutually exclusive: pre-accretionary alteration may have occurred, but most CM chondrites were also altered in the final parent body (Metzler et al., 1992).

Bischoff (1998) reviewed the evidence for pre-accretionary alteration of CM chondrites. Most evidence centers on the relationship between fine-grained rims (Figure 2) and the objects (i.e., chondrules) they surround, as well as the mineral assemblages present in the rims themselves. A key assumption in the pre-accretionary model is that the rims formed by accretion of fine-grained dust onto chondrules within the solar nebula. Although this model has received considerable support, asteroidal formation mechanisms for fine-grained rims have also been proposed (e.g., Richardson, 1981; Sears et al., 1991, 1993). However, enrichments of primordial noble gases in fine-grained rims appear to be most consistent with a nebular model (Nakamura et al., 1999). The key lines of evidence for pre-accretionary alteration are as follows.

(i) *Unaltered chondrule glass juxtaposed against hydrated fine-grained rim material.* Metzler et al. (1992) have argued that alteration of fine-grained rims could not have occurred without affecting chondrule glass, which is highly susceptible to aqueous alteration. Fine-grained rim material must therefore have been altered prior to accretion onto the chondrule in the solar nebula; no further aqueous alteration could have occurred within the parent body.

(ii) *Fresh, unaltered fracture surfaces of chondrule olivines juxtaposed against hydrated fine-grained rim material.* These relationships have been interpreted as representing the following alteration sequence: (a) alteration of the chondrule prior to fracturing; (b) accretion of a fine-grained rim consisting of a mixture of anhydrous and hydrated dust in the solar nebula; and (c) accretion into the CM parent body where no subsequent aqueous alteration occurred. Asteroidal fluids would have altered the exposed surface of the fractured olivine grain in contact with fine-grained rim material. The absence of such alteration indicates that no asteroidal alteration occurred.

(iii) *Disequilibrium mineral assemblages within fine-grained rims.* Fine-grained rims in the weakly altered CM2, Yamato 791198 contain an intimate mixture of primary (Fe,Ni metal, troilite, olivine) and alteration phases (phyllosilicates, tochilinite, sulfides) (Figure 2) (Metzler et al., 1992). This complex assemblage has been interpreted as the result of mixing of unaltered and altered nebular dust prior to asteroidal accretion (Metzler et al., 1992; Bischoff, 1998). Asteroidal alteration should have caused oxidation of Fe,Ni metal and troilite to secondary alteration products.

(iv) *Alteration of CAIs.* Based on textural and compositional criteria, Armstrong et al. (1982) and MacPherson and Davis (1994) have argued that alteration of CAIs in Murchison and Mighei did not take place on the final meteorite parent body. Based on the stability of calcite in a hibonite-bearing inclusion, Armstrong et al. (1982) concluded that a period of alteration and metamorphism was required before incorporation into the final CM parent body. However, MacPherson and Davis (1994) have argued that alteration of CAIs in Mighei occurred in the solar nebula. Contrary to these conclusions, Greenwood et al. (1994) have argued that alteration of Cold Bokkeveld CAIs occurred within an asteroidal parent body.

1.09.3.2.3 Evidence for asteroidal alteration

Veining in CM chondrites. Rare examples of veins in CM chondrites have been documented in the literature. Lee (1993) described calcium sulfate veins crosscutting the matrix of Cold Bokkeveld, and Hanowski and Brearley (2000) found thin veins of iron oxyhydroxide associated with altered metal grains in chondrules in Murchison. These veins cut across fine-grained rims and extend into the matrix of the chondrite (Figure 4), demonstrating that metal-alteration occurred after accretion of the chondrule and its fine-grained rim into the parent body.

Iron-rich aureoles. In some CM chondrites, aureoles of iron enrichment are present surrounding

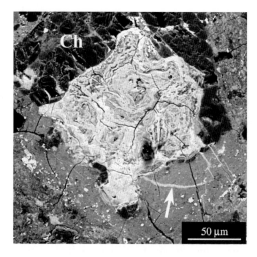

Figure 4 Backscattered electron image of a highly altered Fe,Ni metal nodule in a chondrule in the Murchison CM2 carbonaceous chondrite. Thin veins of iron oxyhydroxides are present extending from the metal grain through the fine-grained rim, into the clastic matrix of the chondrite (reproduced by permission of University of Arizona and The Meteoritical Society from *Meteorit. Planet. Sci.*, **2000**, *35*, 1300).

large, altered metal grains (Hanowski and Brearley, 2000). These aureoles incorporate matrix, chondrules, and mineral fragments, and have outer edges that sometimes crosscut fine-grained rims and interfinger between chondrules and mineral fragments. These observations demonstrate that alteration of the metal grains occurred after asteroidal accretion and that transport and precipitation of oxidized iron took place within partially consolidated chondritic materials.

Bulk compositional homogeneity of CM chondrites. CM chondrites exhibit a wide degree of aqueous alteration, but have quite homogeneous bulk compositions (Figure 5). This evidence indicates that, irrespective of the extent of alteration, elemental mass transfer occurred on a localized scale and alteration was isochemical for most elements. This is illustrated by the behavior of the highly soluble element calcium (Figure 5). In bulk CMs, calcium is unfractionated relative to other refractory elements, but during aqueous alteration calcium was extremely mobile, on a local scale (millimeter to centimeter). Calcium has been completely leached from chondrule mesostases, which have been replaced by iron-rich serpentine (Richardson, 1981; Hanowski and Brearley, 2001). In the parent-body alteration model, calcium precipitates as carbonates in the matrix. Maintaining a constant solar Ca/Al ratio in chondrites whose components have been altered to different degrees in different protoplanetary bodies, prior to final accretion, is not readily explained by the pre-accretionary model. This would require mixing of altered chondrules with depleted calcium contents with an exact proportion of fine-grained dust with enriched calcium contents, in order to retain the solar Ca/Al ratio of the final chondrite.

Elemental exchange between chondrules and matrix during progressive alteration. In addition to calcium, magnesium, and iron show evidence of elemental exchange, notably between chondrules and matrix, during alteration. McSween (1979, 1987) proposed a model for CM chondrite alteration involving the early alteration of iron-rich matrix materials followed by later, alteration of magnesium-rich olivine and pyroxene in chondrules. This model predicts that with progressive alteration, the matrix should become more magnesium-rich and chondrules magnesium-depleted. Although brecciation effects complicate this trend, the magnesium content of matrix does increase with progressive alteration (McSween, 1987). In addition, the iron-rich serpentine alteration products of chondrule silicates become more magnesium-rich as progressive alteration of chondrules occurs (Hanowski and Brearley, 1997).

Homogeneity of chondrule alteration. A potential criterion for differentiating between asteroidal and nebular alteration is heterogeneity in the degree of alteration of nebular materials such as chondrules and CAIs within the same chondrite. Provided the effects of primary bulk compositions, mineralogy, grain sizes, and textures of different objects (e.g., chondrules) can be constrained, similar degrees of alteration are most consistent with alteration of all components *in situ*, under the same conditions of temperature and pressure, etc. In ALH81002, a moderately altered, unbrecciated CM2 chondrite, Hanowski and Brearley (2001) found that individual chondrules show widely differing degrees of alteration. However, when chondrules of the same type are compared, the style and extent of alteration is consistent, as are the compositions of the serpentine alteration products. Collectively, these observations favor alteration of chondrules within the same environment where the alteration products in all chondrules equilibrated with the same fluid. This does not necessarily preclude pre-accretionary alteration. However, the probability of transporting groups of chondrules with the same degrees of alteration through a turbulent nebula as proposed in the pre-accretionary model, without mixing them with chondrules with different degrees of alteration, is small.

Oxygen isotopic compositions. The oxygen-isotopic data indicates that there are systematic relationships between the bulk isotopic composition of CM chondrites and the degree of aqueous alteration (Browning *et al.*, 1996). In addition, the compositions of individual phases (carbonates and sulfates) also evolve with

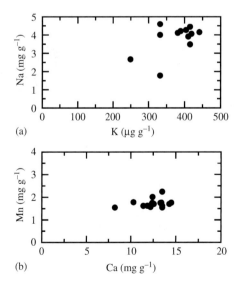

Figure 5 Concentrations of soluble elements in CM chondrites, illustrating the restricted compositional range of these elements in chondrites with variable degrees of alteration. (a) Sodium versus potassium and (b) manganese versus calcium (sources Kallemeyn and Wasson, 1981; Jarosewich, 1990).

increasing degree of alteration in a manner that is most consistent with formation from a fluid that is progressively equilibrating with the host rock (e.g., Benedix et al., 2003). These systematic relationships would be extremely difficult to maintain in a pre-accretionary alteration environment where disruption of planetesimals occurs prior to accretion of the final parent body.

Two parent body models have been proposed to explain the oxygen isotopic composition of carbonates in CM chondrites: (i) a closed system, two reservoir model (Clayton and Mayeda, 1984, 1999); and (ii) a fluid-flow model (Young et al., 1999; Young, 2001; Cohen and Coker, 2000). Current oxygen-isotopic data are generally most consistent with the closed-system model, but can also be reconciled with the fluid-flow model if the CM chondrites sample a restricted region of the CM asteroid (Benedix et al., 2003), just downstream of the model alteration front proposed by Young (2001).

1.09.3.2.4 Summary

In the least altered CM chondrites, there is textural evidence of alteration in a pre-accretionary environment. However, an asteroidal location appears to have been the prevalent alteration environment for most CM chondrites, which show a number of textural, chemical, and isotopic characteristics that are most consistent with alteration after accretion. Pre-accretionary alteration may have been important for all CM chondrites, but its effects have been overprinted by later parent-body alteration. For the CM chondrites, distinguishing these pre-accretionary effects from asteroidal alteration remains a significant challenge.

1.09.3.3 CR Carbonaceous Chondrites

Aqueous alteration in the CR chondrites has not been investigated in detail, although the key characteristics have been described (Weisberg et al., 1993; Zolensky et al., 1993). CR chondrites show variable degrees of aqueous alteration, indicated by a variation in the extent of alteration of chondrule mesostasis. In Renazzo and Al Rais, mesostasis is typically replaced by phyllosilicates, but other CRs contain unaltered glassy mesostasis. Matrices of CR chondrites contain phyllosilicates (serpentine and saponite), calcium carbonates, sulfides, and magnetite (Weisberg et al., 1993; Zolensky et al., 1993), with varying proportions of olivine. Weisberg et al. (1993) concluded that alteration of the components in CR chondrites occurred prior to the final lithification of the CR parent body, because chondrules show variable degrees of alteration, within the same chondrite. They argued that alteration occurred within the CR parent body and the components were mixed together by later brecciation. However, Weisberg et al. (1993) did not rule out alteration in the nebula. Ichikawa and Ikeda (1995) reported unaltered chondrule glass in contact with hydrous matrix in Yamato 8449 and also noted marked compositional differences between phyllosilicates in chondrules and those in the adjacent matrix. Based on these lines of evidence they argued that CR chondrites must have experienced aqueous alteration prior to accretion.

Bulk CR chondrites and separated components (chondrules, matrix, and mineral fragments) have oxygen-isotope compositions that lie along a mixing line of slope 0.59, distinct from similar mixing lines for the CI and CM chondrites (Weisberg et al., 1993; Clayton and Mayeda, 1999). Anhydrous silicates have the lightest isotopic compositions, whereas more hydrated components have much heavier oxygen-isotopic compositions. This mixing line is interpreted as being the result of mixing between at least two oxygen-isotopic reservoirs: a ^{16}O-rich anhydrous component and a ^{16}O-poor hydrous component. The range of δ^{17}O values of the CR chondrites appears to be related to the degree of hydration with the highest δ^{17}O values correlating with the highest degree of alteration, based on abundance of phyllosilicates and bulk water content. Magnetite and phyllosilicate-rich matrix separates from Renazzo have oxygen-isotopic compositions that lie on the terrestrial fractionation line, indicating that both materials formed from the same fluid (Clayton and Mayeda, 1977). This observation and the well-defined mixing line for the CR chondrites suggest they must have interacted with the same fluid, a scenario which is most compatible with alteration within an asteroidal environment (Clayton and Mayeda, 1999).

1.09.3.4 CO Carbonaceous Chondrites

CO chondrites have clearly experienced limited aqueous alteration during a relatively low-temperature event. However, the CO chondrites may have also experienced aqueous alteration contemporaneously with metamorphism (Rubin, 1998). This latter type of alteration is best described as fluid-assisted metamorphism, rather than aqueous alteration and shares some similarities with alteration in the oxidized CV chondrites. Hence, CO chondrites may have experienced two periods of aqueous alteration.

CO3 chondrites have been affected minimally by low-temperature aqueous alteration and evidence of incipient alteration is restricted to a few members of this group. The effects of alteration appear to be restricted largely to the matrix. Minor amounts of hydrous phyllosilicates (serpentine, saponite) have been described in ALH77307, Lancé and Ornans (Brearley, 1993; Kerridge, 1964; Keller and Buseck, 1990a). In Lancé and Ornans, Keller and Buseck (1990a) found that phyllosilicates occur interstitially to matrix olivines and sometimes within veins in the larger matrix olivines. Evidence of phyllosilicates replacing chondrules glass has also been reported in ALH77307 (Ikeda, 1983).

Evidence to constrain the location of alteration for the CO chondrites is limited. However, most data suggest a late-stage event that postdated metamorphism, implying that it was probably a parent-body process. Development of alteration phases in CO matrices has occurred interstitially to matrix phases and there is no evidence of distinct hydrous phases intermixed with unaltered phases that would support pre-accretionary alteration. However, formation of phyllosilicates in chondrules in ALH77307 may have occurred prior to accretion, based on the fact that the ALH77307 matrix shows essentially no evidence of aqueous alteration (Brearley, 1993).

1.09.3.5 CV Carbonaceous Chondrites

Aqueous alteration in the CV groups is variable in extent and complex. In the reduced CV chondrites, evidence for aqueous alteration appears to be minimal. Lee et al. (1996) found incipient aqueous alteration in the matrix of Vigarano in the form of ferrihydrite and rare, very fine-grained saponite. The ubiquitous presence of these alteration products interstitial to matrix olivines led Lee et al. (1996) to conclude that alteration occurred in an asteroidal environment.

In the oxidized CV chondrites, phyllosilicates have been widely reported in the Bali subgroup, and the range of alteration in this group varies from essential zero to regions of some chondrites where alteration is almost 50%. In Bali, Kaba and Mokoia, matrix, chondrules, and CAIs have all been extensively replaced by phyllosilicates (Cohen et al., 1983; Keller and Buseck, 1990b; Tomeoka and Buseck, 1990, Keller et al., 1994; Kimura and Ikeda, 1998). Matrix olivines have commonly been replaced by saponite in all these meteorites, and a variety of phyllosilicates, including iron-bearing saponite, sodium phlogopite, aluminum-rich serpentine, and sodium–potassium mica, are found in altered chondrules and CAIs. Elemental redistribution is extensive in the most altered regions of Bali, resulting in the formation of secondary phosphates, magnetite, and carbonates. Alteration veins are also present that appear to postdate formation of a prominent shock-produced foliation.

Members of the Bali subgroup have oxygen isotopic compositions that lie on the CCAM line (Figure 6), but are typically lower in ^{16}O than either the Allende subgroup or the reduced CV chondrites, an effect that may be due to aqueous alteration (Krot et al., 1995; Clayton and Mayeda, 1996). The isotopic effects of aqueous alteration are especially well developed in heavily altered regions of Bali; such regions show significant heavy isotope depletions consistent with low-temperature aqueous alteration.

Evidence for aqueous alteration in the Allende subgroup is very limited, in part because Allende itself is the only chondrite of this subgroup that has been studied in detail. Allende has an exceptionally low water content (<0.2 wt.%; Jarosewich et al., 1987), which reflects the fact that it shows no obvious signs of aqueous alteration. Allende apparently contains no hydrous phases within its fine-grained matrix (Brearley, 1999). Nevertheless, rare phyllosilicates such as phlogopite, montmorillonite, clintonite, margarite, saponite, and chlorite (Wark and Lovering, 1977; Hashimoto and Grossman, 1987; Keller and Buseck, 1991; Tomeoka and Buseck, 1982) occur in CAIs. In chondrules, Brearley (1997a) reported the rare, but ubiquitous presence of talc, amphibole, and disordered biopyriboles. Compared with the

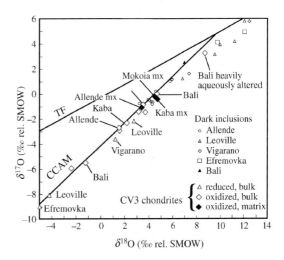

Figure 6 Oxygen-isotopic compositions of CV chondrites and their dark inclusions plotted on a three-isotope diagram. Bulk CV3 chondrites plot close to the CCAM line defined by the isotopic composition of anhydrous minerals from CAIs in Allende. The reduced CV chondrites are isotopically lighter than the oxidized CV chondrites such as Allende. Dark inclusions from Allende also plot on the CCAM line, whereas inclusions from the reduced CV chondrites, Vigarano, Leoville, and Efremovka have oxygen-isotopic compositions that show significant heavy isotope enrichments, indicative of aqueous alteration (source Krot et al., 1999).

hydrous phases observed in the Bali-subgroup, it is notable that with the exception of montmorillonite and saponite, the phases present in Allende all have higher thermal stabilities. The style of aqueous alteration in Allende is therefore quite different from the Bali-like oxidized CV chondrites.

The bulk compositions of CV chondrites exhibit a significant degree of compositional variability that is not apparent in other chondrite groups. This is particularly notable in the behavior of volatile elements, which are distinctly different between the oxidized and reduced subgroups. For example, sodium, manganese, and iron are all enriched in the oxidized group (Figure 7). The origin of these compositional variations appears to be strongly correlated with the fact that the oxidized subgroup has undergone extensive aqueous alteration and metasomatism, showing that alteration did not occur under closed-system conditions.

1.09.3.5.1 Evidence for pre-accretionary alteration

Most workers have argued that alteration of the Bali group CV chondrites occurred in an asteroidal environment, although Kimura and Ikeda (1998) did not rule out the possibility of pre-accretionary alteration. In the case of Allende, the very heterogeneous character of the phases present in CAIs, in addition to the fact that they are extremely rare and are heterogeneously distributed, has led most workers to favor formation of these hydrous phases in the solar nebula prior to accretion.

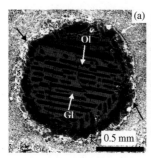
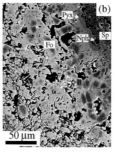

Figure 7 (a) Backscattered electron image of a barred olivine chondrule from Allende, showing the presence of a narrow fayalitic olivine rim that is a characteristic alteration effect in the oxidized CV chondrites. (b) Backscattered electron image of the outer zone of type A CAI from Allende that is surrounded by a coarse-grained, forsterite-rich accretionary rim. The outer zone of the CAI has experienced metasomatism and melilite been replaced by secondary phases including nepheline. Forsterite in the accretionary rim is replaced by fayalitic olivine ((a) reproduced by permission of American Association for the Advancement of Science from *Science*, **1997**, *278*, 76 and (b) courtesy of A. N. Krot).

Hashimoto and Grossman (1987) and Keller and Buseck (1991), for example, have argued that these phases formed by reaction of individual CAIs with a nebular gas, that also caused iron–alkali–halogen metasomatism (see below). The heterogeneous distribution of alteration phases can then be attributed to the fact that some CAIs interacted with the gas, but others did not. The interrelationship between aqueous alteration of the CV chondrites and metasomatism/oxidation will be discussed in detail below. The variable bulk compositions of the oxidized CV chondrites may be the result of nebular aqueous alteration, but could also be the result of open-system parent-body alteration as well.

1.09.3.5.2 Evidence for parent-body alteration

Parent-body alteration appears to be strongly indicated for the Bali group CV chondrites, based on the presence of crosscutting veins and the development of alteration phases interstitially to matrix phases. This is further indicated by the fact that all components in the most-altered chondrites have been affected, although in some chondrites (Bali in particular), alteration is heterogeneous and highly altered regions are mixed with less-altered regions. For the Allende group of CV chondrites, Brearley (1999) observed that the same rare, hydrous phases occur in several low-calcium pyroxene-bearing chondrules within a single thin section and the style and abundance of the phases was consistent from one chondrule to another. This observation suggests that all the chondrules were altered under similar conditions, for similar lengths of time. Whilst this does not rule out alteration in the solar nebula, it appears to be more compatible with alteration within a parent-body environment, in which all the chondrules interacted with the same aqueous fluid.

1.09.3.6 Unequilibrated Ordinary Chondrites

The effects of aqueous alteration are apparent in the least equilibrated ordinary chondrites such as Semarkona, Bishunpur, and Chainpur. The effects of alteration are most widely developed in the matrices of these chondrites (Hutchison et al., 1987; Alexander et al., 1989a,b), but alteration of chondrules has recently been recognized (Hutchison et al., 1998; Grossman et al., 2000, 2002). In these chondrites, alteration of the matrices is not complete and precursor phases are still present. The dominant alteration phase in the matrix is an amorphous phase and an iron-rich smectite phase replacing olivine, pyroxene, etc. Other minor alteration phases include maghemite, nickel-rich pyrrhotite, pentlandite, magnetite, and carbides (see below). Calcite also occurs sometimes in rims around chondrules.

Evidence for aqueous alteration of chondrules is indicated by the presence of so-called bleached chondrules that occur in several unequilibrated chondrites including Semarkona (LL3.0) Krymka (LL3.0), Bishunpur (LL3.1), and Chainpur (LL3.4) (Kurat, 1969; Christophe Michel-Lévy, 1976; Grossman et al., 2000). The outer portions of these radial pyroxene and cryptocrystalline chondrules have porous outer zones where mesostasis has been removed, probably by dissolution and is correlated with significant depletions in alkalis and aluminum. Deloule and Robert (1995) measured very elevated D/H ratios in the matrix of Semarkona which they attributed to the presence of interstellar water. Grossman et al. (2000) found D/H ratios in the bleached chondrules that lie in the range of values measured by Deloule and Robert (1995). However, these measurements are compromised to some extent by the fact that D/H ratios measured by ion probe are variable as a function of depth into the sample. This variability has been attributed to the exchange of terrestrial water with interlayer water in smectite. Despite some ambiguity, the elevated D/H ratios in both matrix and altered chondrules indicate that they were both exchanged with an isotopically similar reservoir, implying that alteration occurred within an asteroidal environment. This is also supported by the similar compositions of matrix and chondrule smectite replacement products. Textural evidence also supports the idea of asteroidal alteration. Grossman et al. (2000) described a chondrule that appears to have undergone one stage of alteration that produced a bleached zone around its exterior. This chondrule was later fragmented, probably by regolith processes and a second bleached zone developed along the surfaces exposed by the fracture.

1.09.4 OXIDATION AND METASOMATISM

Evidence of oxidation that may be related to metasomatic processes appears to be essentially restricted to the oxidized CV chondrites, some of the CO chondrites, and a few unequilibrated ordinary chondrites. This section will examine the evidence for these processes and will include a brief mineralogical overview of the evidence followed by a discussion of the available chemical and isotopic evidence for oxidation and metasomatism.

1.09.4.1 CV Carbonaceous Chondrites

The CV chondrites exhibit the most complex evidence of secondary oxidation and metasomatism of any of the chondrite groups; however, these effects are largely restricted to the oxidized subgroup. Additional insights into these complex processes have come from studies of dark inclusions, the angular lithic clasts that occur in many CV chondrites. These objects appear to represent highly processed lithologies derived from other locations within the CV parent asteroid. A brief summary of alteration effects in CV chondrites is presented here; Krot et al. (1995) reviewed this subject in detail.

The principal types of alteration effects in CV chondrites are: (i) iron–alkali–halogen metasomatism of CAIs, chondrules, and matrix; (ii) formation of ferrous olivine rims on chondrules, isolated olivine grains, etc.; (iii) oxidation of sulfidization of opaque assemblages in CAIs, chondrules, and matrix; and (iv) formation of phyllosilicates (discussed above). Additional possible effects of alteration noted by Krot et al. (1995) also include the formation of platy olivines, calcium–iron pyroxenes, and andradite in the matrix, and the formation of pure fayalite.

Iron–alkali–halogen metasomatism: Many components in the oxidized CV chondrites have been affected by a metasomatic event that has caused enrichments in chlorine, sodium, and potassium, or iron. Primary phases in CAIs, such as melilite, hibonite, spinel, aluminum-titanium-diopside, and perovskite, show evidence of replacement or modification that occurred at relatively low temperatures, probably <1,000 K (MacPherson et al., 1988; Brearley and Jones, 1998). The alteration is well developed in CAIs in the oxidized CVs (Grossman, 1980), but CAIs in the reduced subgroup are minimally affected (e.g., MacPherson et al., 1988). Secondary alteration products include nepheline, sodalite, grossular, wollastonite, hedenbergite, and ferroan spinel, which clearly corrode minerals or vein and fill cavities within CAIs (Figure 7(b)). This alteration affected the bulk compositions of CAIs resulting in enrichments in alkalis (sodium, potassium, rubidium), halogens (chlorine, iodine, bromine), and moderately volatile elements (manganese, iron, copper, zinc, chromium, gold, gallium, silicon).

Similar effects are also observed in chondrules (e.g., Ikeda and Kimura, 1995, Kimura and Ikeda, 1998) and matrix (Krot et al., 1998a). Ikeda and Kimura (1996) have emphasized the mineralogical similarities between alteration in chondrules and CAIs. In the case of chondrules, anorthite-normative mesostasis and plagioclase has been replaced by nepheline and sodalite with minor amounts of andradite, kirchsteinite, wollastonite, and hedenbergite, etc.

Fayalitic rims on chondrules, CAIs, etc. The presence of fayalitic rims on forsteritic chondrule phenocrysts and the partial-to-complete replacement of low-calcium pyroxene by fayalitic olivine are widespread alteration phenomena in the oxidized CV chondrites (Figure 7(a)). Veins of fayalitic olivine also occur along fractures within olivine grains. Fayalitic olivine rims (<15 μm in width) are ubiquitous around forsteritic fragments in the matrix and around CAIs.

Oxidation and sulfidization of opaque assemblages. Opaque assemblages (Fremdlinge) occur in CAIs, in chondrules, and in the matrix in CV chondrites. In the reduced subgroup, the primary opaque assemblage consists of kamacite, taenite, and troilite (McSween, 1977a). In the oxidized subgroup, magnetite is the dominant opaque phase and awaruite ($FeNi_3$) is the metallic phase. Pyrrhotite and pentlandite are the main sulfides present. Opaque assemblages in Fremdlinge are unusual occurrences that consist of platinum-group element-rich alloys, nickel–iron alloys, sulfides, and oxides that have complex histories (e.g., El Goresy et al., 1978; Blum et al., 1989).

Oxygen-isotopic compositions of bulk CV chondrites all lie along the CCAM line (Figure 6), with slope 1, defined by the isotopic compositions of separated minerals in Allende CAIs, indicating that both arrays are probably produced by the same process (Clayton and Mayeda, 1999). The Allende subgroup has compositions which overlap the reduced CV chondrites, suggesting that the bulk isotopic composition of these meteorites has not been affected by an alteration process, such as aqueous alteration which would have caused heavy isotope enrichments. However, ion microprobe studies of individual components in Allende indicate a complex alteration history that is not evident from the bulk isotopic data. For example, the oxygen-isotopic compositions of coexisting magnetite and olivine in Allende chondrules are not in isotopic equilibrium (Choi et al., 1998). Chondrule olivines fall on or are slightly above the CCAM line, whereas most magnetites form a cluster with similar $\delta^{18}O$, but smaller $\Delta^{17}O$ than the olivines. These data show that the magnetites were formed from a distinct isotopic reservoir.

1.09.4.2 Dark Inclusions in CV Chondrites

CV chondrites commonly contain dark, lithic clasts, termed dark inclusions, which are considered to have a close genetic relationship to the host CV chondrites (see Krot et al., 1995). Dark inclusions contain the same alteration features that occur in the oxidized CV chondrites, but the range of alteration in dark inclusions is much wider. Consequently, dark inclusions provide additional insights into the alteration processes that affected chondrites. Dark inclusions have been classified into two types: type A and type B (e.g., Johnson et al., 1990; Krot et al., 1995). Type A inclusions resemble the host CV chondrites, but have smaller chondrules and CAIs. Type B inclusions contain no chondrules and CAIs, but consist largely of aggregates with chondrule-like shapes consisting fine-grained FeO-rich olivine (Kojima and Tomeoka, 1996), set in a matrix of finer-grained olivine. In dark inclusions in the oxidized CV chondrites, the olivine aggregates contain platy olivine whereas in the reduced CVs, chondrules have been perfectly pseudomorphed by dense aggregates of fine-grained FeO-rich olivine (Kracher et al., 1985; Johnson et al., 1990; Krot et al., 1999). Most dark inclusions in Allende are intermediate in character between these two types and are referred to as type A/B. The bulk compositions of dark inclusions are typically significantly depleted in moderately volatile elements compared with their host chondrites (Figure 8).

Oxygen isotopic data for dark inclusions (Clayton and Mayeda, 1999) provide some important insights into their possible origins. Allende dark inclusions have compositions that are only very slightly displaced from the CCAM line to ^{16}O-rich composition (Figure 6). However, inclusions from Efremovka, Leoville, and Vigarano are displaced to the right of the CCAM line forming an array that extends towards the terrestrial fractionation line and shares a number of similarities with the arrays found in the CR and CM chondrites. Clayton and Mayeda (1999) have argued that large heavy isotope enrichments in these dark inclusions are likely to be the result of low-temperature aqueous alteration processes, although mineralogically there is little evidence of hydrous phases in these inclusions. For some inclusions (e.g., Efremovka,

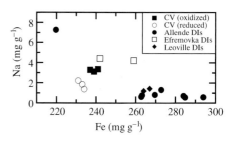

Figure 8 Bulk sodium versus iron contents of oxidized and reduced CV chondrites and dark inclusions. The enrichment in sodium and iron of the oxidized group appears to be the result of metasomatic processes (sources Kallemeyn and Wasson, 1981; Kracher et al., 1985; Bischoff et al., 1988; Palme et al., 1989; Jarosewich, 1990; Krot et al., 1999).

Krot et al., 1999), there is a clear correlation between degree of replacement of primary phases with heavy isotope enrichment, providing further support for such a model.

1.09.4.3 Nebular and Parent-body Alteration Models for CV chondrites

1.09.4.3.1 Nebular alteration

Most studies of oxidized CV chondrites have focused on alteration effects in one particular component, for example, CAIs or chondrules, usually treating them as unrelated phenomena. As a consequence, a rather incoherent picture emerged of alteration in which each type of component experienced alteration under different conditions. Palme and Wark (1988) and Palme et al. (1991) recognized that multiple alteration features in oxidized CV chondrites may be related by a common nebular process (Hua et al., 1988; Palme and Fegley, 1990; Weinbruch et al., 1990). In this model, alteration of all the CV3 components occurred at high temperature (~1,500 K) in the nebula in the presence of an oxidizing gas ($H_2O/H_2 \sim 1$) that was produced by the evaporation and recondensation of pre-existing material. During recondensation, rims of fayalitic olivine were formed on olivine fragments and CAI, and oxidation and sulfidization of opaque assemblages occurred. Platy matrix olivines are considered to be the products of nebular condensation (e.g., Weisberg and Prinz, 1998), as are the pure fayalitic olivines in some oxidized CV chondrites (Hua and Buseck, 1995). During this event, volatile elements condensed and produced the observed metasomatic enrichments in these elements. Components in the reduced CV chondrite group escaped this intense processing. The consistency of alteration effects in Allende requires that parent-body accretion took place very soon after the high-temperature event, in order to prevent mixing of altered and unaltered materials. Based on iodine-xenon data (Swindle, 1998) accretion may have occurred ~3.7 Myr after CAI formation (Palme et al., 1991).

In this model, type A and A/B dark inclusions are considered to be result of the same kind of processes that affected the components of the oxidized CV chondrites. For type B dark inclusions, Kurat et al. (1989) and Palme et al. (1989) have proposed an origin for the porous chondrule-shaped aggregates of fayalitic olivine that involves condensation and aggregation of primary magnesian olivine. Following condensation, these aggregates underwent metasomatic alteration reactions with an oxidizing nebula, to produce the iron-rich olivine compositions, sulfidization of metal grains, etc.

1.09.4.3.2 Parent-body alteration

Parent-body alteration models for the oxidized CV chondrites essentially treat all the alteration effects as being related, taking place contemporaneously under the same conditions. Housley and Cirlin (1983) first proposed such a model for Allende, in which alteration took place in the absence of water and involved vapor transport and catalysis by CO, HCl, H_2S, and SO_2 gases. A more complex alternative model has been proposed by Kojima and Tomeoka (1993, 1996) based on studies of dark inclusions. They proposed that dark inclusions underwent a period of hydration in an asteroidal environment followed by a period of thermal metamorphism, analogous to the process that has affected unusual carbonaceous chondrites such as Yamato 793321 (Tomeoka et al., 1989; Akai, 1990). During metamorphic heating, phyllosilicates were dehydrated and transformed into platy olivines that constitute the matrix of Allende and dark inclusions in Vigarano and Allende. In this model, the chondrule-shaped aggregates of platy olivine grains in type B dark inclusions result from complete replacement of chondrules.

Krot et al. (1995, 1997a, 1998a, 2000a) have extended this model to alteration of the host CV chondrites. Based on petrographic observations and thermodynamic modeling, they concluded that calcium-iron-rich pyroxene and fayalite–magnetite assemblages could be produced by metasomatic fluids in an asteroidal environment, at temperatures of 400–500 °C.

1.09.4.3.3 Problems with the nebular-alteration model

Both nebular and asteroidal models have strengths and weaknesses, but at present neither satisfy all the available data. Some of the weaknesses of both models are discussed below. Resolution of these issues may ultimately provide a more coherent model for alteration of this group of meteorites.

Origin of platy matrix olivine grains. The platy olivines in dark inclusions and Allende matrix are generally considered to be nebular condensates (e.g., Wiesberg and Prinz, 1998). However, these olivines contain multiple inclusions of poorly graphitized carbon, amorphous carbon, pentlandite, and chromite, which are problematic for the nebular condensation model (Brearley, 1999). Pyrrhotite rather than pentlandite can condense under oxidizing conditions in the nebular, but at temperatures below the condensation temperature of FeO-rich olivine even under extreme dust-to-gas enrichments (Ebel and Grossman, 2000). Pentlandite would therefore not have been a stable phase during the condensation of FeO-bearing olivine. Similarly, the organic precursor to the poorly graphitized carbon should not have been present during condensation, unless it represents a component of presolar dust that survived evaporation and recondensation. This is a possibility, but the selective inclusion of phases that are relatively rare in the solar nebula over more common refractory phases remains a problem for the nebular evaporation–condensation model.

Presence of veins in dark inclusions. Veins of fayalitic olivine, nepheline, calcium-iron pyroxene, andradite, and sulfides several to 100 μm s in thickness and several millimeters in length occur in Vigarano and Allende dark inclusions (Kojima and Tomeoka, 1993, 1996; Krot et al., 2000a). The veins crosscut chondrules, CAIs, and matrix and sometimes branch into smaller veins. These veins are interpreted as pathways along which fluids moved during aqueous alteration and vein-filling minerals precipitated from solution. The crosscutting relationships of the veins show that alteration must have occurred after accretion of the components, an association that is consistent with asteroidal alteration and difficult to explain by nebular processes (Kojima and Tomeoka, 1996).

Formation of calcium-iron pyroxenes. Aggregates of calcium-iron pyroxenes sometimes associated with andradite and wollastonite occur commonly in the matrices of oxidized CV chondrites (Krot et al., 1998a,b). Although these phases have been interpreted as having a high-temperature origin, possibly as condensates (e.g., Brenker et al., 2000), thermodynamic calculations show that it is not possible to condense these phases in the nebula under any oxidizing conditions (Ebel and Grossman, 2000). These calculations show that, even under extreme dust-to-gas enrichments (1,000 × CI, $P^{tot} = 10^{-3}$ bar), the maximum amount of FeO that can enter clinopyroxene is ~5.5 wt.%, significantly lower than that observed in clinopyroxene in CV matrices (the latter can extend to almost pure hedenbergite in composition).

Rims of calcium-iron-rich pyroxene around dark inclusions. Rims of calcium-iron pyroxenes are often present around dark inclusions where they come into contact with Allende host (Kurat et al., 1989; Johnson et al., 1990; Krot et al., 2000a). Interpretations of these rims have varied, and include nebular as well as asteroidal origins. However, a number of aspects of the rims suggest that they are likely to have formed within an asteroidal parent body (Krot et al., 2000a). These include the observation that the calcium-iron-rich phases in the rims are closely intergrown with Allende matrix olivines, indicating that growth of the rim occurred after emplacement into the Allende parent body.

Timing of alteration. The nebular model proposed by Palme et al. (1991) requires that accretion occurred very rapidly after the high-temperature oxidizing event in order to prevent mixing of altered and unaltered components in the nebula. However, iodine-xenon and manganese–chromium dating of alteration phases in CV chondrites and dark inclusions indicate that the formation of these phases spanned several million years. Iodine-xenon data from a variety of Allende samples including whole rock, chondrules, and assorted inclusions show that well-defined iodine-xenon isochrons are extremely rare in any of these objects (Swindle, 1998). However, in many cases, the ratio of iodine to trapped xenon is very high, enabling a precise closure time for the objects to be defined. The closure ages determined from different samples are rather variable, extending over several million years. Iodine is considered to be located in the secondary phase, sodalite, indicating that alteration occurred over an extended time interval or that the sodalite has undergone later disturbance that also took place over several million years. Data for two dark inclusions from Efremovka (Krot et al., 1999) yield well-defined isochrons with formation ages 3.2 ± 1.7 Myr and 8.9 ± 2.6 Myr, respectively, after Vigarano. Both inclusions show significant heavy isotope enrichments in their oxygen-isotopic compositions indicating that they have been altered by aqueous fluids.

Nearly pure fayalite with 1 wt.% MnO occurs associated with magnetite in the Kaba and Mokoia CV chondrites (Hua and Buseck, 1995). The fayalite has been interpreted as the product of replacement of magnetite by an oxidizing nebular gas with $H_2O/H_2 \sim 2 \times 10^3 - 10^4$ at 800–1,200 K. However, Krot et al. (1998) have argued for a low-temperature (<300 °C) asteroidal origin for the fayalite based on textural, isotopic, and thermodynamic grounds. This interpretation appears to be supported by manganese–chromium dating of fayalite grains from Mokoia (Hutcheon et al., 1998) which yield an initial $^{53}Mn/^{55}Mn$ ratio of $(2.32 \pm 0.18) \times 10^{-6}$. Depending on the choice of initial solar system ratio of $^{53}Mn/^{55}Mn$, this ratio indicates that formation of the fayalite occurred 7–16 Myr after CAI formation, strongly implying asteroidal formation.

While these data do not rule out alteration in the solar nebula, they indicate that the alteration history of CV chondrites and their dark inclusions was a process that extended over several million years. Maintaining separation of altered and unaltered objects within the nebula for this period of time is problematic. In models for asteroidal alteration, spatial separation of unaltered and altered regions appears to be much less problematical and also provides an environment where alteration can occur for extended periods of time.

1.09.4.3.4 Problems with the asteroidal alteration model

Oxygen isotopes. The oxygen-isotopic compositions of CV chondrites and dark inclusions provide important insights into the alteration processes that affected the CV chondrites. Dark inclusions in reduced CV chondrites (e.g., Efremovka, Leoville, and Vigarano) have heavy-isotope enrichments (Clayton and Mayeda, 1999) that are indicative of aqueous alteration, consistent with the model proposed by Kojima and Tomeoka (1996). However, Allende dark inclusions, even those which show extensive alteration, have a much more restricted range of oxygen-isotopic compositions (3‰), with minimal displacements from the CCAM line (Figure 6). This indicates that hydration/dehydration (Kojima et al., 1995) did not play a significant role in their formation. This is also true of Allende matrix, which is mineralogically and texturally very similar to Allende and Vigarano inclusions and also has an oxygen-isotopic composition that lies on the CCAM line (Clayton and Mayeda, 1999). However, laser ablation ICPMS data for chondrules and CAIs from Allende (Young and Russell, 1998) show some analyses with oxygen-isotopic compositions that lie off the CCAM line and show heavy isotope enrichments. These data provide some indication that aqueous alteration played at least some role in the alteration of Allende. The isotopic disequilibrium between magnetite and olivine in Allende (Choi et al., 1998) also suggests that Allende interacted with an isotopically distinct reservoir that may have been an aqueous fluid.

Formation of platy olivines. The matrices of the reduced and oxidized subgroups are texturally distinct. Matrix olivines in the reduced subgroup are fine-grained and usually anhedral in shape, whereas in the oxidized subgroup, matrix olivines have platy morphologies. This platy morphology is usually interpreted as the product of vapor–solid condensation (e.g., Weisberg and Prinz, 1998). In the parent-body model, the platy olivines are the product of dehydration of phyllosilicate phases as suggested by Kojima and Tomeoka (1996). However, Allende and dark inclusions from Allende and Vigarano inclusions, all of which contain platy olivines, have oxygen-isotopic compositions that are either on or are very close to the CCAM mixing line. These data are incompatible with an origin by dehydration because of the lack of any significant heavy isotope enrichment that would be expected if the precursors had been formed by a process that involved interaction with an aqueous fluid (Clayton and Mayeda, 1999).

Timing of metasomatism and aqueous alteration. In the asteroidal model, proposed by Kojima and Tomeoka (1996) and Krot et al. (1995), aqueous alteration should precede metamorphism. However, Kimura and Ikeda (1998) have reported textural evidence from the Bali subgroup of the CV chondrites that indicates that hydrous alteration occurred after higher temperature anhydrous alteration (metasomatism) rather than preceding it.

1.09.4.4 CO Carbonaceous Chondrites

Although not as extensively developed, there is evidence that the components of some CO chondrites were affected by iron–alkali metasomatism and oxidation processes similar to those experienced by CV chondrites.

Metasomatic and oxidation effects in chondrules in CO chondrites are limited in extent compared to CV chondrites. Metasomatic effects are best developed in rare plagioclase-rich chondrules and in primary anorthite in type I chondrules (Kurat and Kracher, 1980; Jones, 1997). In plagioclase-rich chondrules (Figure 9), which consist of abundant primary anorthite, orthopyroxene, and augite, plagioclase has been altered to nepheline. Ferrosalite pyroxene is present in the mesostasis, which seems to have formed as result of alteration of groundmass diopside, plagioclase, and silica. Replacement of

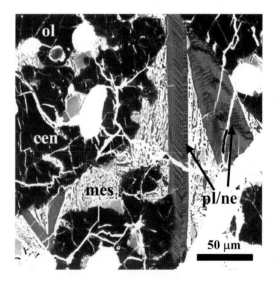

Figure 9 Backscattered electron image of a region of a metasomatized type I chondrule from the CO chondrite Kainsaz (3.1). The chondrule contains clinoenstatite (cen), olivine (ol) metal, and sulfide grains (bright) and fine-grained mesostasis (mes). Primary anorthite has been extensively replaced by nepheline, which occurs as dark lamellae within the plagioclase. In addition the mesostasis has been partially replaced by salitic pyroxene which is also considered to be a metasomatic alteration product (reproduced by permission of Mineralogical Society of America from *Planetary Materials*, **1998**, *36*, chap. 3, 3-1–3-398).

primary anorthite by nepheline in type I chondrules is also a common phenomenon in many CO3 chondrites (Ikeda, 1982).

Oxidation effects in CO chondrites are variably developed. Many CO chondrites contain unaltered kamacite, whereas in ALH77307, Ornans, and several Antarctic CO chondrites, magnetite is common, replacing metal (McSween, 1977b; Scott and Jones, 1990; Shibata, 1996). Pentlandite appears to have replaced troilite in ALH77307, whereas troilite still coexists with magnetite in Ornans.

Secondary alteration of CAIs in CO chondrites has been noted by a number of workers (Ikeda, 1982; Holmberg and Hashimoto, 1992; Tomeoka et al., 1992; Kojima et al., 1995; Russell et al., 1998). Nepheline is the most common replacement phase, with minor amounts of sodalite, ilmenite, and monticellite (Brearley and Jones, 1998). Russell et al. (1998) studied CAIs in a suite of CO3 chondrites ranging from petrologic type 3.0–3.7. They observed that CAIs showed significant effects of alteration, such that, in some cases, all the primary components of the CAIs have either been replaced or had their compositions modified. The key evidence of alteration includes increases in the FeO content of spinels in spinel–pyroxene inclusions, increased replacement of melilite by nepheline, sodalite, and diopside, and disturbed aluminum-magnesium systematics in some inclusions.

1.09.4.4.1 Evidence for parent-body alteration of CO chondrites

The CO chondrites show a range of degrees of thermal metamorphism that most workers agree occurred in an asteroidal environment (Scott and Jones, 1990; Sears et al., 1991). This conclusion is based on a number of correlated textural and mineralogical changes that occur in chondrules, matrix, and metal as a function of increasing petrologic type. Russell et al. (1998) found that many alteration effects in CAIs are also correlated with increasing petrologic type. These include a progressive increase in the FeO content of spinel, indicating that diffusive elemental exchange has occurred between matrix and CAIs, similar to the effects that are observed in olivine and pyroxene in chondrules (e.g., Scott and Jones, 1990; Jones, 1993). In addition, perovskite decreases and ilmenite increases in abundance in CAIs and melilite-bearing inclusions decrease significantly through the petrologic sequence. This trend is consistent with observations of Allende CAIs that demonstrate that melilite is more susceptible to alteration than other refractory phases (e.g., McGuire and Hashimoto, 1989).

Russell et al. (1998) also found CAIs with both disturbed and undisturbed aluminum-magnesium systematics. Above petrologic type 3.4 most CAIs exhibit disturbed isochrons, whereas below type 3.4, no clear correlation exists between the degree of disturbance and the petrologic type of the chondrite. This combination of mineralogical and isotopic effects, correlated with petrologic type, led Russell et al. (1998) to conclude that most of the metasomatic alteration of CAIs occurred within a parent-body environment. However, some CAIs may have experienced alteration in a pre-accretionary environment (see below).

Rubin (1998) has attempted to reconcile a number of diverse chemical and mineralogical characteristics of the CO chondrites in terms of a model that involves alteration in an asteroidal environment. This model has some elements which are comparable with that proposed for the parent-body alteration of the CV chondrites. Alteration by aqueous fluids is indicated by oxygen-isotopic data from melilite in CO chondrites (Wasson et al., 2001), which show progressive heavy isotope enrichments that are generally correlated with petrologic type. Wasson et al. argued that this is the result of oxygen exchange between melilite and H_2O during asteroidal aqueous alteration/metamorphism.

The ungrouped CM/CO chondrite MAC88107 also shows clear evidence of asteroidal alteration that is similar to that observed in the oxidized CV chondrites. For example, this meteorite shows evidence for in situ formation of hedenbergite and fayalite (Krot et al., 2000a), indicated by the occurrence of these phases in veins that originate at opaque nodules, crosscut fine-grained rims around chondrules and terminate at the boundaries with neighboring fine-grained rims. Manganese-chromium dating of the fayalite in MAC88107 (Krot et al., 2000a) yields a $^{53}Mn/^{55}Mn$ initial ratio of $(2.32 \pm 0.18) \times 10^{-6}$ indicating a formation age of ~ 18 Myr after CAI formation (using the solar system initial ratio of Lugmair and Shukolyukov (2001). This value is entirely consistent with formation of the fayalite in an asteroidal environment.

1.09.4.4.2 Evidence for nebular alteration of CO chondrites

The main evidence for nebular metasomatism comes from a few CAIs in low petrologic type CO chondrites. Although Russell et al. (1998) found that alteration in most CAIs correlated with petrologic type, they observed rare inclusions in low petrologic type CO chondrites (e.g., Colony, 3.0) that show evidence of alteration and enrichment in alkalis, that was inconsistent with the unaltered nature of other similar inclusions in the same chondrite. The advanced degree of alteration of these inclusions does not correlate with the low degree of alteration experienced by the host chondrite and therefore cannot be related to metamorphism. In these particular cases, it appears that the CAIs were altered prior to accretion and not within the parent body. Additional support for this conclusion also comes from the aluminum–magnesium systematics of inclusions from very low petrologic grade CO chondrites. In these chondrites, inclusions with both disturbed and undisturbed isochrons occur, showing that some inclusions were disturbed prior to accretion and not within the parent body. The nature of the process that disturbed these inclusions is unclear, but it may have been episodes of heating that involved evaporation or condensation in the solar nebula rather than nebular metasomatism.

1.09.4.5 Unequilibrated Ordinary Chondrites

The secondary effects of metasomatism and oxidation have been reported in a number of unequilibrated ordinary chondrites (UOCs), although are only mildly developed. The mineralogical effects of metasomatism in UOCs are manifested by the formation of feldspathoids and, to a lesser extent, scapolite, associated with enrichments in sodium and chlorine. The effects of metasomatism are best described in the Tieschitz (H3.6) chondrite and have focused on the presence of an unusual "White Matrix" (Christophe Michel-Lévy, 1976). This material is a matrix component that occurs interstitially to chondrules and consists of nepheline, albite, and an unidentified sodium-rich aluminosilicate (Ashworth, 1981; Alexander, 1989b). Some chondrules in Tieschitz are also enriched in aluminum, alkalis, barium, fluorine, and chlorine in regions of mesostasis that have clearly undergone replacement. Unaltered glassy mesostasis in the same chondrules is not enriched in these elements (Hutchison et al., 1998). In addition, Bridges et al. (1997) described a number of chondrules separated from Chainpur (LL3.4) and Parnallee (LL3.6) that contain mesostasis that is enriched in sodium and chlorine and contain microcrystalline sodalite, nepheline, and scapolite.

Several different mechanisms have been proposed to explain the origin of the alkali and chlorine enrichments in unequilibrated ordinary chondrites. In the case of Tieschitz, in which the effects of this process are manifested in both matrix and chondrules, most workers have invoked alteration by fluids within an asteroidal environment. However, Bridges et al. (1997) favored a pre-accretionary environment for the metasomatism of Chainpur and Parnallee chondrules. In contrast, Grossman et al. (2002) have argued that the metasomatic and aqueous alteration effects in the UOCs were all related effects that occurred within an asteroidal environment.

Carbide-magnetite assemblages. In some unequilibrated ordinary chondrites, such as Semarkona (LL3.0), assemblages consisting of magnetite, carbides (cohenite, haxonite), kamacite, troilite with minor taenite and pentlandite have been described (e.g., Taylor et al., 1981; Hutchison et al., 1987; Krot et al., 1997b; Keller, 1998). Carbide–magnetite assemblages occur in matrix and chondrules and their formation has been attributed to oxidation and carbidization of metallic Fe,Ni. Taylor et al. (1981) proposed a complex nebular model for the formation of these assemblages involving sequential steps of sulfidization, carbidization, and finally oxidation with progressively decreasing temperature. Krot et al. (1997b) and Keller (1998) have argued that an asteroidal setting for alteration is more likely and resulted from hydrothermal alteration of Fe,Ni metal by carbon-oxygen-hydrogen-bearing fluid. In the case of Semarkona, this model is supported by the presence of hydrated phases in the matrix (e.g., Hutchison et al., 1987) and by the oxygen-isotopic compositions of magnetite grains (Choi et al., 1998). The magnetite shows a range of $\delta^{18}O$ of 13‰ and has the highest $\Delta^{17}O$ of any measured solar system material. Formation of magnetite in a semi-infinite reservoir such as the solar nebula should have produced a limited range in the $\delta^{18}O$ values, because magnetite-water fractionation factor is relatively temperature insensitive at low temperatures. The large spread in $\delta^{18}O$ appears to be most indicative of Raleigh fractionation as a result of growth from a limited water reservoir, a scenario that is most consistent with an asteroidal setting.

1.09.5 FUTURE WORK

Differentiating between nebular and asteroidal processes is a complex, challenging, and controversial task. Since the 1980s, significant advances have been made in characterizing the types of secondary processes that have affected chondritic meteorites. These data have stimulated important debate and resulted in a rigorous appraisal of old models. Nevertheless, our understanding of the conditions of alteration and where they occurred remains imperfect. However, major advances are likely to come in the future with the application of modern microbeam techniques to understand the microstructural, trace-element and isotopic characteristics of individual phases. In particular, recent studies of oxygen isotopes at the grain scale have already contributed a significant body of data that allows differentiation between possible nebular and asteroidal alteration scenarios. Additional studies in the following areas are likely to improve our understanding of these complex issues and are likely to be exciting and important areas of research for some time to come.

(i) Studies of the aqueous alteration behavior of complex, heterogeneous assemblages of starting materials in the presence of carbon-oxygen-hydrogen fluids are essential to appraise the evidence for pre-accretionary environments and examine the possible role of microchemical environments in influencing alteration reactions.

(ii) Isotopic analyses of individual alteration components *in situ* to provide a complete alteration history of CI and CM chondrites and provide important constraints on the fluid composition and fluid evolution during alteration.

(iii) Further experimental studies to constrain the effects of aqueous alteration and dehydration on the oxygen-isotopic compositions of chondritic materials.

(iv) Experimental studies to constrain the origins of the platy morphology of matrix olivines in oxidized CV chondrites (dehydration versus condensation).

(v) Detailed petrologic and isotopic studies to improve our understanding of the relationship between aqueous alteration and metasomatism, relevant to the genesis of the CV chondrites. The interrelationship between these two processes is currently not well understood and is at the heart of the nebular- versus asteroidal-alteration debate.

(vi) Further studies on the chronology of alteration of chondritic meteorites are essential to clarify the timing and duration of alteration. In particular, an exact knowledge of the phases that are being dated by iodine-xenon techniques would be very beneficial, but is a major challenge because of the current lack of knowledge of the mineralogical siting of iodine. Careful SEM and TEM studies of mineral separates used for iodine-xenon dating would dramatically improve this situation.

ACKNOWLEDGMENTS

A. J. Brearley (PI) was supported by NASA grants NAG5-9798 and NAG5-11862.

REFERENCES

Airieau S. A., Farquhar J., Jackson T. L., Leshin L. A., Thiemens M. H., and Bao H. (2001) Oxygen isotope systematics of CI and CM chondrite sulfate: implications for evolution and mobility of water in planetesimals. *Lunar and Planetary Science* **XXXII**, #1744. Lunar and Planetary Institute, Houston (CD-ROM).

Akai J. (1990) Mineralogical evidence of heating events in Antarctic carbonaceous chondrites, Y-86720 and Y-82162. *Proc. NIPR Symp. Antarct. Meteorit.* **3**, 55–68.

Alexander C. M. O'D., Barber D. J., and Hutchison R. (1989a) The microstructure of Semarkona and Bishunpur. *Geochim. Cosmochim. Acta* **53**, 3045–3057.

Alexander C. M. O'D., Hutchison R., and Barber D. J. (1989b) Origin of chondrule rims and interchondrule matrices in unequilibrated ordinary chondrites. *Earth Planet. Sci. Lett.* **95**, 187–207.

Armstrong J. T., Meeker G. P., Huneke J. C., and Wasserburg G. J. (1982) The Blue Angel: I. The mineralogy and petrogenesis of a hibonite inclusion from the Murchison meteorite. *Geochim. Cosmochim. Acta* **46**, 575–595.

Ashworth J. R. (1981) Fine structure in H-group chondrites. *Proc. Roy. Soc. London* **A374**, 179–194.

Benedix S. A., Leshin L. A., Farquhar J., Jackson T. L., and Thiemens M. H. (2000) Carbonates in CM chondrites: oxygen isotope geochemistry and implications for alteration of the CM parent body. *Lunar Planet. Sci.* **XXXI**, #1840. Lunar and Planetary Institute, Houston (CD-ROM).

Benedix S. A., Leshin L. A., Farquhar J., Jackson T. L., and Thiemens M. H. (2003) Carbonates in CM chondrites: constraints on alteration conditions from oxygen isotopic compositions and petrographic observations. *Geochim. Cosmochim. Acta* **67**, 1577–1588.

Bischoff A. (1998) Aqueous alteration of carbonaceous chondrites: evidence for preaccretionary alteration—a review. *Meteorit. Planet. Sci.* **33**, 1113–1122.

Bischoff A., Palme H., Spettel B., Clayton R. N., and Mayeda T. K. (1988) The chemical composition of dark inclusions from the Allende meteorite. *Lunar Planet. Sci.* **XIX**, 88–89.

Blum J. D., Wasserburg G. J., Hutcheon I. D., Beckett J. R., and Stolper E. M. (1989) Origin of opaque assemblages in CV meteorites: implications for nebular and planetary processes. *Geochim. Cosmochim. Acta* **53**, 543–556.

Boss A. P. (1996) Large scale processes in the solar nebula. In *Chondrules and the Protoplanetary Disk* (eds. R. H. Hewins, R. H. Jones, and E. R. D. Scott). Cambridge University Press, Cambridge, UK, pp. 29–34.

Brearley A. J. (1993) Matrix and fine-grained rims in the unequilibrated CO3 chondrite, ALH A77307: origins and evidence for diverse, primitive nebular dust components. *Geochim. Cosmochim. Acta* **57**, 1521–1550.

Brearley A. J. (1997a) Disordered biopyriboles, amphibole, and talc in the Allende meteorite: products of nebular or parent body aqueous alteration? *Science* **276**, 1103–1105.

Brearley A. J. (1997b) Chondrites and the solar nebula. *Science* **278**, 76.

Brearley A. J. (1999) Origin of graphitic carbon and pentlandite inclusions in matrix olivines in the Allende meteorite. *Science* **285**, 1380–1382.

Brearley A. J. and Hutcheon I. D. (2000) Carbonates in the CM1 chondrite ALH84034: mineral chemistry, zoning and Mn–Cr systematics. *Lunar and Planetary Science* **XXXI**, #1407. Lunar and Planetary Institute, Houston (CD-ROM).

Brearley A. J. and Jones R. H. (1998) Chondritic meteorites. In *Planetary Materials*, Rev. Mineral. (ed. J. J. Papike). Mineralogical Society of America, vol. 36, pp. 3-1–3-398.

Brearley A. J., Hutcheon I. D., and Browning L. (2001) Compositional zoning and Mn–Cr systematics in carbonates from the Y791198 CM2 carbonaceous chondrite. *Lunar and Planetary Science* **XXXII**, #1458. Lunar and Planetary Institute, Houston (CD-ROM).

Brenker F. E., Palme H., and Klerner S. (2000) Evidence for solar nebula signatures in the matrix of the Allende meteorite. *Earth Planet. Sci. Lett.* **178**, 185–194.

Bridges J. C., Alexander C. M. O'D., Hutchison R., Franchi I. A., and Pillinger C. T. (1997) Sodium-, chlorine-rich mesostases in Chainpur (LL3) and Parnallee (LL3) chondrules. *Meteorit. Planet. Sci.* **32**, 555–565.

Browning L. B., McSween H. Y. J., and Zolensky M. E. (1996) Correlated alteration effects in CM carbonaceous chondrites. *Geochim. Cosmochim. Acta* **60**, 2621–2633.

Bunch T. E. and Chang S. (1980) Carbonaceous chondrites: II. Carbonaceous chondrite phyllosilicates and light element geochemistry as indicators of parent body processes and surface conditions. *Geochim. Cosmochim. Acta* **44**, 1543–1577.

Choi B.-G., McKeegan K. D., Leshin L. A., and Wasson J. T. (1997) Origin of magnetite in oxidized CV chondrites: *in situ* measurement of oxygen isotope compositions of Allende magnetite and olivine. *Earth Planet. Sci. Lett.* **146**, 337–349.

Choi B. G., McKeegan K. D., Krot A. N., and Wasson J. T. (1998) Extreme oxygen-isotope compositions in magnetite from unequilibrated ordinary chondrites. *Nature* **392**, 577–579.

Christophe Michel-Lévy M. (1976) La matrice noire et blanche de la chondrite de Tieschitz. *Earth Planet. Sci. Lett.* **30**, 143–150.

Ciesla F. J., Lauretta D. S., Cohen B. A., and Hood L. L. (2003) A nebular origin for chondritic fine-grained phyllosilicates. *Science* **299**, 549–552.

Clayton R. N. (1963) Carbon isotopes in meteoritic carbonates. *Science* **140**, 192–193.

Clayton R. N. (1993) Oxygen isotopes in meteorites. *Ann. Rev. Earth Planet. Sci.* **21**, 115–119.

Clayton R. N. and Mayeda T. K. (1977) Oxygen isotopic compositions of seperated fractions of the Leoville and Renazzo carbonaceous chondrites. *Meteoritics* **12**, A199.

Clayton R. N. and Mayeda T. K. (1984) The oxygen isotope record in Murchison and other carbonaceous chondrites. *Earth Planet. Sci. Lett.* **67**, 151–161.

Clayton R. N. and Mayeda T. K. (1996) Oxygen isotope relations among CO, CK, and CM chondrites and carbonaceous chondrite dark inclusions. *Meteorit. Planet. Sci.* **31**, A30.

Clayton R. N. and Mayeda T. K. (1999) Oxygen isotope studies of carbonaceous chondrites. *Geochim. Cosmochim. Acta* **63**, 2089–2104.

Cohen B. A. and Coker R. A. (2000) Modeling of liquid water on CM meteorite parent bodies and implications for amino acid racemizations. *Icarus* **145**, 369–381.

Cohen R. E., Kornacki A. S., and Wood J. A. (1983) Mineralogy and petrology of chondrules and inclusions in the Mokoia CV3 chondrite. *Geochim. Cosmochim. Acta* **47**, 1739–1757.

Deloule E. and Robert F. (1995) Interstellar water in meteorites? *Geochim. Cosmochim. Acta* **59**, 4695–4706.

DuFresne E. R. and Anders E. (1962) On the chemical evolution of the carbonaceous chondrites. *Geochim. Cosmochim. Acta* **26**, 1085–1114.

Ebel D. S. and Grossman L. (2000) Condensation in dust-enriched systems. *Geochim. Cosmochim. Acta* **64**, 339–366.

El Goresy A., Nagel K., and Ramdohr P. (1978) Fremdlinge and their noble relatives. *Proc. 9th Lunar Planet. Sci. Conf.* 1279–1303.

Endress M. and Bischoff A. (1996) Carbonates in CI chondrites: clues to parent body evolution. *Geochim. Cosmochim. Acta* **60**, 489–507.

Endress M., Zinner E., and Bischoff A. (1996) Early aqueous activity on primitive meteorite parent bodies. *Nature* **379**, 701–703.

Fredriksson K. and Kerridge J. F. (1988) Carbonates and sulfates in CI chondrites: formation by aqueous activity on the parent body. *Meteoritics* **23**, 35–44.

Gounelle M. and Zolensky M. E. (2001) A terrestrial origin for sulfate veins in CI1 chondrites. *Meteorit. Planet. Sci.* **35**, 1321–1329.

Greenwood R. C., Lee M. R., Hutchison R., and Barber D. J. (1994) Formation and alteration of CAIs in Cold Bokkeveld (CM2). *Geochim. Cosmochim. Acta* **58**, 1035–1913.

Grossman J. N., Alexander C. M. O'D., Wang J. H., and Brearley A. J. (2000) Bleached chondrules: evidence for widespread aqueous processes on the parent asteroids of ordinary chondrites. *Meteorit. Planet. Sci.* **35**, 467–486.

Grossman J. N., Alexander C. M. O'D., Wang J. H., and Brearley A. J. (2002) Zoned chondrules in Semarkona: evidence for high- and low-temperature processing. *Meteorit. Planet. Sci.* **37**, 49–73.

Grossman L. (1980) Refractory inclusions in the Allende meteorite. *Ann. Rev. Earth Planet. Sci.* **8**, 559–608.

Grossman L. and Larimer J. W. (1974) Early chemical history of the solar system. *Rev. Geophys. Space Phys.* **12**, 71–101.

Grady M. M., Wright I. P., Swart P. K., and Pillinger C. T. (1988) The carbon and oxygen isotopic composition of meteoritic carbonates. *Geochim. Cosmochim. Acta* **52**, 2855–2866.

Halbout J., Mayeda T. K., and Clayton R. N. (1986) Carbon isotopes and light element abundances in carbonaceous chondrites. *Earth Planet. Sci. Lett.* **80**, 1–8.

Hanowski N. P. and Brearley A. J. (2000) Iron-rich aureoles in the CM carbonaceous chondrites, Murray, Murchison and ALH81002: evidence for *in situ* alteration. *Meteorit. Planet. Sci.* **35**, 1291–1308.

Hanowski N. P. and Brearley A. J. (2001) Aqueous alteration of chondrules in the CM carbonaceous chondrites, Allan Hills 81002. *Geochim. Cosmochim. Acta* **65**, 495–518.

Hashimoto A. and Grossman L. (1987) Alteration of Al-rich inclusions inside amoeboid olivine aggregates in the Allende meteorite. *Geochim. Cosmochim. Acta* **51**, 1685–1704.

Herzog G. F., Anders E., Alexander E. C. J., Davis P. K., and Lewis R. S. (1973) Iodine-129/Xenon-129 age of magnetite from the Orgueil meteorite. *Science* **180**, 489–491.

Hohenberg C. M., Pravdivtseva O., and Meshik A. (2000) Reexamination of anomalous I–Xe ages: Orgueil and Murchison magnetites and Allegan feldspar. *Geochim. Cosmochim. Acta* **64**, 4257–4263.

Holmberg A. A. and Hashimoto A. (1992) A unique (almost) unaltered spinel-rich fine-grained inclusion in Kainsaz. *Meteoritics* **27**, 149–153.

Housley R. M. and Cirlin E. H. (1983) On the alteration of Allende chondrules and the formation of matrix. In *Chondrules and Their Origins* (ed. E. A. King). Lunar and Planetary Institute, Houston, pp. 145–161.

Hua X., Adam J., Palme H., and El Goresy A. (1988) Fayalite-rich rims, veins and haloes around and in forsteritic olivines in CAIs and chondrules in carbonaceous chondrites: types, compositional profiles and constraints on their formation. *Geochim. Cosmochim. Acta* **52**, 1389–1408.

Hua X. and Buseck P. R. (1995) Fayalite in Kaba and Mokoia carbonaceous chondrites. *Geochim. Cosmochim. Acta* **59**, 563–578.

Hutcheon I. D. (1997) Chronologic constraints on secondary alteration processes. In *Workshop on Parent-body and Nebular Modification of Chondritic Materials* (eds. M. E. Zolensky, A. N. Krot, and E. R. D. Scott). Lunar and Planetary Institute, Houston, p. 27.

Hutcheon I. D. and Phinney D. L. (1996) Radiogenic $^{53}Cr^*$ in Orgueil carbonates: chronology of aqueous activity on the CI parent body. *Lunar and Planetary Science* **XXVII**. Lunar and Planetary Institute, Houston, pp. 577–578.

Hutcheon I. D., Krot A. N., Keil K., Phinney D. L., and Scott E. R. D. (1998) $^{53}Mn-^{53}Cr$ dating of fayalite formation in the CV3 chondrite Mokoia: evidence for asteroidal alteration. *Science* **282**, 1865–1867.

Hutcheon I. D., Weisberg M. K., Phinney D. L., Zolensky M. E., Prinz M., and Ivanov A. V. (1999) Radiogenic ^{53}Cr in Kaidun carbonates: evidence for very early aqueous activity. *Lunar and Planetary Science* **XXX**, #1722. Lunar and Planetary Institute, Houston (CD-ROM).

Hutchison R., Alexander C. M. O., and Barber D. J. (1987) The Semarkona meteorite: first recorded occurrence of smectite in an ordinary chondrite, and its implications. *Geochim. Cosmochim. Acta* **51**, 1875–1882.

Hutchison R., Alexander C. M. O. D., and Bridges J. C. (1998) Elemental redistribution in Tieschitz and the origin of white matrix. *Meteorit. Planet. Sci.* **33**, 1169–1180.

Ichikawa O. and Ikeda Y. (1995) Petrology of the Yamato-8449 CR chondrite. *Proc. NIPR Symp. Antarct. Meteorit.* **8**, 63–78.

Ikeda Y. (1982) Petrology of the ALH-77003 chondrite (C3). *Proc. 7th Symp. Antarct. Meteorit.* 34–65.

Ikeda Y. (1983) Alteration of chondrules and matrices in the four Antarctic carbonaceous chondrites ALH 77307 (C3),

Y-790123 (C2), Y-75293(C2), and Y-74662(C2). *Proc. NIPR Symp. Antarct. Meteorit.* **8**, 93–108.

Ikeda Y. and Kimura M. (1995) Anhydrous alteration of Allende chondrules in the solar nebula: I. Description and alteration of chondrules with known oxygen-isotopic compositions. *Proc. NIPR Symp. Antarct. Meteorit.* **8**, 97–122.

Ikeda Y. and Kimura M. (1996) Anhydrous alteration of Allende chondrules in the solar nebula: III. Alkali-zoned chondrules and heating experiments for anhydrous alteration. *Proc. NIPR Symp. Antarct. Meteorit.* **9**, 51–68.

Jarosewich E. (1990) Chemical analyses of meteorites: a compilation of stony and iron meteorite analyses. *Meteoritics* **25**, 323–337.

Jarosewich E., Clarke R. S. J., and Barrows J. N. (1987) The Allende meteorite reference sample. *Smithson. Contrib. Earth Sci.* **27**.

Johnson C. A., Prinz M., Weisberg M. K., Clayton R. N., and Mayeda T. K. (1990) Dark inclusions in Allende, Leoville, and Vigarano: evidence for nebular oxidation of CV3 constituents. *Geochim. Cosmochim. Acta* **54**, 819–830.

Jones R. H. (1993) Effect of metamorphism on isolated olivine grains in CO3 chondrites. *Geochim. Cosmochim. Acta.* **57**, 2853–2867.

Jones R. H. (1997) Alteration of plagioclase-rich chondrules in CO3 chondrites: evidence for late-stage sodium and iron metasomatism in a nebular environment. In *Workshop on Parent-body and Nebular Modification of Chondritic Materials* (eds. M. E. Zolensky, A. N. Krot, and E. R. D. Scott). Lunar and Planetary Institute, Houston, pp. 30–31.

Jones R. H., Lee T., Connolly H. C., Jr., Love S. G., and Shang H. (2000) Formation of chondrules and CAIs: theory vs. observation. In *Protostars and Planets IV* (eds. V. Mannings, A. P. Boss, and S. S. Russell). University of Arizona Press, Tucson, pp. 927–962.

Kallemeyn G. W. and Wasson J. T. (1981) The compositional classification of chondrites: I. The carbonaceous chondrite groups. *Geochim. Cosmochim. Acta* **45**, 1217–1230.

Keller L. P. (1998) A transmission electron microscope study of iron-nickel carbides in the matrix of the Semakona unequilibrated ordinary chondrite. *Meteorit. Planet. Sci.* **33**, 913.

Keller L. P. and Buseck P. R. (1990a) Matrix mineralogy of the Lancé CO3 carbonaceous chondrite: a transmission electron microscope study. *Geochim. Cosmochim. Acta.* **54**, 1155–1163.

Keller L. P. and Buseck P. R. (1990b) Aqueous alteration in the Kaba CV3 carbonaceous chondrite. *Geochim. Cosmochim. Acta* **54**, 2113–2120.

Keller L. P. and Buseck P. R. (1991) Calcic micas in the Allende meteorite: evidence for hydration reactions in the early solar nebula. *Science* **252**, 946–949.

Keller L. P., Thomas K. L., Clayton R. N., Mayeda T. K., DeHart J. M., and McKay D. S. (1994) Aqueous alteration of the Bali CV3 chondrite: evidence from mineralogy, mineral chemistry, and oxygen isotopic compositions. *Geochim. Cosmochim. Acta* **58**, 5589–5598.

Kerridge J. F. (1964) Low-temperature minerals from the fine-grained matrix of some carbonaceous meteorites. *Ann. NY Acad. Sci.* **119**, 41–53.

Kerridge J. F. and Bunch T. E. (1979) Aqueous activity on asteroids: evidence from carbonaceous chondrites. In *Asteroids* (ed. T. Gehrels). University of Arizona Press, Tucson, pp. 745–764.

Kimura M. and Ikeda Y. (1998) Hydrous and anhydrous alterations of chondrules in Kaba and Mokoia CV chondrites. *Meteorit. Planet. Sci.* **33**, 1139–1146.

Kojima T. and Tomeoka K. (1993) Unusual dark clasts in the Vigarano CV3 carbonaceous chondrite: record of parent body processes. *Meteoritics* **28**, 649–658.

Kojima T. and Tomeoka K. (1996) Indicators of aqueous alteration and thermal metamorphism on the CV parent body: microtextures of a dark inclusion from Allende. *Geochim. Cosmochim. Acta* **60**, 2651–2666.

Kojima T., Yada S., and Tomeoka K. (1995) Ca–Al-rich inclusion in three Antarctic CO3 chondrites, Yamato 81020, Yamato-82050, and Yamato-790992: record of low temperature alteration processes. *Proc. NIPR Symp. Antarct. Meteorit.* **8**, 79–86.

Kracher A., Keil K., Kallemeyn G. W., Wasson J. T., Clayton R. N., and Huss G. I. (1985) The Leoville (CV3) accretionary breccia. *Proc. 16th Lunar Planet Sci. Conf.* 123–135.

Krot A. N., Scott E. R. D., and Zolensky M. E. (1995) Mineralogical and chemical modification of components in CV3 chondrites: nebular or asteroidal processing? *Meteoritics* **30**, 748–775.

Krot A. N., Scott E. R. D., and Zolensky M. E. (1997a) Origin of fayalitic olivine rims and lath-shaped matrix olivine in the CV3 chondrite Allende and its dark inclusions. *Meteoritics* **32**, 31–49.

Krot A. N., Zolensky M. E., Wasson J. T., Scott E. R. D., Keil K., and Ohsumi K. (1997b) Carbide-magnetite assemblages in type-3 ordinary chondrites. *Geochim. Cosmochim. Acta* **61**, 219–237.

Krot A. N., Petaev M. I., Scott E. R. D., Choi B.-G., Zolensky M. E., and Keil K. (1998a) Progressive alteration in CV3 chondrites: more evidence for asteroidal alteration. *Meteorit. Planet. Sci.* **33**, 1065–1085.

Krot A. N., Petaev M. I., Zolensky M. E., Keil K., Scott E. R. D., and Nakamura K. (1998b) Secondary calcium-iron minerals in the Bali-like and and Allende-like oxidized CV3 chondrites and Allende inclusions. *Meteorit. Planet. Sci.* **33**, 623–645.

Krot A. N., Brearley A. J., Ulyanov A. A., Biryukov V. V., Swindle T. D., Keil K., Mittlefehldt D. W., Scott E. R. D., Clayton R. N., and Mayeda T. K. (1999) Mineralogy, petrography, and bulk chemical, I-Xe, and oxygen isotopic compositions of dark inclusions in the reduced CV3 chondrite Efremovka. *Meteorit. Planet. Sci.* **34**, 67–90.

Krot A. N., Petaev M. I., Meibom A., and Keil K. (2000a) In situ growth of Ca-rich rims around Allende dark inclusions. *Geochem. Int.* **38**, S351–S368.

Krot A. N., Brearley A. J., Petaev M. I., Kallemeyn G. W., Sears D. W. G., Benoit P. H., Hutcheon I. D., Zolensky M. E., and Keil K. (2000b) Evidence for in situ growth of fayalite and hedenbergite in MacAlpine Hills 88107, ungrouped carbonaceous chondrite related to the CM–CO clan. *Meteorit. Planet. Sci.* **35**, 1365–1387.

Kurat G. (1969) The formation of chondrules and chondrites and some observations on chondrules from the Tieschitz meteorite. In *Meteorite Research* (ed. P. M. Millman). Reidel, Dordrecht, pp. 185–190.

Kurat G. and Kracher A. (1980) Basalts in the Lancé carbonaceous chondrite. *Z. Naturforsch* **35a**, 180–190.

Kurat G., Palme H., Brandstatter F., and Huth J. (1989) Allende Xenolith AF: undisturbed record of condensation and aggregation of matter in the Solar Nebula. *Z. Naturforsch* **44a**, 988–1004.

Lee M. (1993) The petrography, mineralogy, and origins of calcium sulphate within the cold Bokkeveld CM carbonaceous chondrite. *Meteoritics* **28**, 53–62.

Lee M. R., Hutchison R., and Graham A. L. (1996) Aqueous alteration in the matrix of the Vigarano (CV3) carbonaceous chondrite. *Meteorit. Planet. Sci.* **31**, 477–483.

Leshin L. A., Rubin A. E., and McKeegan K. D. (1997) The oxygen isotopic composition of olivines and pyroxenes from CI chondrites. *Geochim. Cosmochim. Acta* **61**, 835–845.

Leshin L. A., Farquhar J., Guan Y., Pizzarello S., Jackson T. L., and Thiemens M. H. (2001) Oxygen isotopic anatomy of Tagish Lake: relationship to primary and secondary minerals in CI and CM chondrites. *Lunar and Planetary Science*, **XXXII** #1843. Lunar and Planetary Institute, Houston (CD-ROM).

Lewis R. S. and Anders E. (1975) Condensation time of the solar nebula from extinct ^{129}I in primitive meteorites. *Proc. Natl. Acad. Sci.* **72**, 268–273.

Lugmair G. W. and Shukolyukov A. (2001) Early solar system events and timescales. *Meteorit. Planet. Sci.* **36**, 1017–1026.

Macdougall J. D., Lugmair G. W., and Kerridge J. F. (1984) Early solar system aqueous activity: Sr isotope evidence from the Orgueil CI meteorite. *Nature* **307**, 249–251.

MacPherson G. J. and Davis A. M. (1994) Refractory inclusions in the prototypical CM chondrite, Mighei. *Geochim. Cosmochim. Acta* **58**, 5599–5625.

MacPherson G. J., Wark D. A., and Armstrong J. T. (1988) Primitive material surviving in chondrites: refractory inclusions. In *Meteorites and the Early Solar System* (eds. J. F. Kerridge and M. S. Matthews). University of Arizona Press, Tucson, pp. 746–807.

McGuire A. V. and Hashimoto A. (1989) Origin of zoned fine-grained inclusions in the Allende meteorite. *Geochim. Cosmochim. Acta* **53**, 1123–1133.

McSween H. Y., Jr. (1977a) Petrographic variations among carbonaceous chondrites of the Vigarano type. *Geochim. Cosmochim. Acta* **41**, 1777–1790.

McSween H. Y., Jr. (1977b) Carbonaceous chondrites of the Ornans type: a metamorphic sequence. *Geochim. Cosmochim. Acta* **44**, 477–491.

McSween H. Y., Jr. (1979) Alteration in CM carbonaceous chondrites inferred from modal and chemical variations in matrix. *Geochim. Cosmochim. Acta* **43**, 1761–1770.

McSween H. Y., Jr. (1987) Aqueous alteration in carbonaceous chondrites: mass balance constraints on matrix mineralogy. *Geochim. Cosmochim. Acta* **51**, 2469–2477.

Metzler K., Bischoff A., and Stöffler D. (1992) Accretionary dust mantles in CM chondrites: evidence for solar nebula processes. *Geochim. Cosmochim. Acta* **56**, 2873–2897.

Nagy B. (1966) Investigation of the Orgueil carbonaceous meteorite. *Geol. Foren. Stockholm Forh.* **88**, 235–272.

Nagy B., Meinschein W. G., and Hennessy D. J. (1963) Aqueous low temperature environment of the Orgueil meteorite: parent body. *Ann. NY Acad. Sci.* **108**, 534–552.

Nakamura T., Nagao K., and Takaoka N. (1999) Microdistribution of primordial noble gases in CM chondrites determined by *in situ* laser microprobe analysis: decipherment of nebular processes. *Geochim. Cosmochim. Acta* **63**, 241–255.

Niemeyer S. and Zaikowski A. (1980) I-Xe age and trapped Xe components in the Murray (C-2) chondrite. *Earth Planet. Sci. Lett.* **48**, 335–347.

Palme H. and Fegley B., Jr. (1990) High-temperature condensation of iron-rich olivine in the solar nebula. *Earth Planet. Sci. Lett.* **101**, 180–195.

Palme H. and Wark D. (1988) CV-chondrites: high temperature gas-solid equilibrium vs. parent body metamorphism. *Lunar and Planetary Science* **XIX**. Lunar and Planetary Institute, Houston, pp. 897–898.

Palme H., Kurat G., Spettel B., and Burghele A. (1989) Chemical composition of an unusual xenolith of the Allende meteorite. *Z. Naturforsch.* **44a**, 1005–1014.

Palme H., Weinbruch S., and El Goresy A. (1991) Reheating of Allende components before accretion. *Meteoritics* **26**, 383.

Podosek F. A. and Cassen P. (1994) Theoretical, observational, and isotopic estimates of the lifetime of the solar nebula. *Meteoritics* **29**, 6–25.

Pravdivtseva O. V., Hohenberg C. M., and Meshik A. P. (2003) The I-Xe age of Orgueil magnetite: new results. *Lunar and Planetary Science* **XXXIV**, #1863. Lunar and Planetary Institute, Houston (CD-ROM).

Richardson S. M. (1978) Vein formation in the C1 carbonaceous chondrites. *Meteoritics* **13**, 141–159.

Richardson S. M. (1981) Alteration of mesostasis in chondrules and aggregates from three C2 carbonaceous chondrites. *Earth Planet. Sci. Lett.* **52**, 67–75.

Rowe M. W., Clayton R. N., and Mayeda T. K. (1994) Oxygen isotopes in separated components of CI and CM meteorites. *Geochim. Cosmochim. Acta* **58**, 5341–5348.

Rubin A. E. (1998) Correlated petrologic and geochemical characteristics of CO3 chondrites. *Meteorit. Planet. Sci.* **33**, 385–391.

Russell S. S., Huss G. R., Fahey A. J., Greenwood R. C., Hutchison R., and Wasserburg G. J. (1998) An isotopic and petrologic study of calcium–aluminum-rich inclusions from CO3 chondrites. *Geochim. Cosmochim. Acta* **62**, 689–714.

Scott E. R. D. and Jones R. H. (1990) Disentangling nebular and asteroidal features of CO3 carbonaceous chondrites. *Geochim. Cosmochim. Acta* **54**, 2485–2502.

Sears D. W. G., Batchelor D. J., Lu J., and Keck B. D. (1991) Metamorphism of CO and CO-like chondrites and comparisons with type 3 ordinary chondrites. *Proc. NIPR Symp. Antarct. Meteorit.* **4**, 319–343.

Sears D. W. G., Benoit P. H., and Lu J. (1993) Two chondrule groups each with distinctive rims in Murchison recognized by cathodoluminescence. *Meteoritics* **28**, 669–675.

Shibata Y. (1996) Opaque minerals in Antarctic CO3 carbonaceous chondrites, Yamato-74135, -790992, -791717, -81020, -81025, -82050 and Allan Hills-77307. *Proc. NIPR Symp. Antarct. Meteorit.* **9**, 79–96.

Shu F. H., Shang H., and Lee T. (1996) Toward an astrophysical theory of chondrites. *Science* **271**, 1545–1552.

Swindle T. (1998) Implications of iodine-xenon studies for the timing and location of secondary alteration. *Meteorit. Planet. Sci.* **33**, 1147–1156.

Taylor G. J., Okada A., Scott E. R. D., Rubin A. E., Huss G. R., and Keil K. (1981) The occurrence and implications of carbide-magnetite assemblages in unequilibrated ordinary chondrites. *Lunar and Planetary Science* **XII**. Lunar and Planetary Institute, Houston (CD-ROM), pp. 1076–1078.

Tomeoka K. (1990) Phyllosilicate veins in a CI meteorite: evidence for aqueous alteration on the parent body. *Nature* **345**, 138–140.

Tomeoka K. and Buseck P. R. (1982) Intergrown mica and montmorillonite in the Allende carbonaceous chondrite. *Nature* **299**, 326–327.

Tomeoka K. and Buseck P. R. (1985) Indicators of aqueous alteration in CM carbonaceous chondrites: microtextures of a layered mineral containing Fe, S, O, and Ni. *Geochim. Cosmochim. Acta* **49**, 2149–2163.

Tomeoka K. and Buseck P. R. (1990) Phyllosilicates in the Mokoia CV carbonaceous chondrite: evidence for aqueous alteration in an oxidizing condition. *Geochim. Cosmochim. Acta* **54**, 1787–1796.

Tomeoka K., Kojima H., and Yanai K. (1989) Yamato-86720: a CM carbonaceous chondrite having experienced extensive aqueous alteration and thermal metamorphism. *Proc. NIPR Symp. Antarct. Meteorit.* **2**, 55–74.

Tomeoka K., Nomura K., and Takeda H. (1992) Na-bearing Ca–Al-rich inclusions in the Yamato-791717 CO carbonaceous chondrite. *Meteoritics* **27**, 136–143.

Wark D. A. and Lovering J. F. (1977) Marker events in the early evolution of the solar system: evidence from rims on Ca–Al-rich inclusions from carbonaceous chondrites. *Proc. 8th Lunar Planet. Sci. Conf.* 95–112.

Wasson J. T., Yurimoto H., and Russell S. S. (2001) ^{16}O-rich melilite in CO3.0 chondrites: possible formation of common, ^{16}O-poor melilite by aqueous alteration. *Geochim. Cosmochim. Acta* **65**, 4539–4549.

Weinbruch S., Palme H., Muller W. F., and El Goresy A. (1990) FeO-rich rims and veins in Allende forsterite: evidence for high temperature condensation at oxidizing conditions. *Meteoritics* **25**, 115–125.

Weisberg M. K. and Prinz M. (1998) Fayalitic olivine in CV3 chondrite matrix and dark inclusions: a nebular origin. *Meteorit. Planet. Sci.* **33**, 1087–1099.

Weisberg M. K., Prinz M., Clayton R. N., and Mayeda T. K. (1993) The CR (Renazzo-type) carbonaceous chondrite group and its implications. *Geochim. Cosmochim. Acta* **57**, 1567–1586.

Wilson L., Keil K., Browning L. B., Krot A. N., and Bourcier W. (1999) Early aqueous alteration, explosive disruption, and reprocessing of asteroids. *Meteorit. Planet. Sci.* **34**, 541–557.

Young E. D. (2001) The hydrology of carbonaceous chondrite parent bodies and the evolution of planet progenitors. *Phil. Trans. Roy. Soc.* **359**, 2095–2110.

Young E. and Russell S. S. (1998) Oxygen reservoirs in the early solar nebula inferred from an Allende CAI. *Science* **282**, 452–455.

Young E. D., Ash R. D., England P., and Rumble D., III (1999) Fluid flow in chondritic parent bodies: deciphering the compositions of planetesimals. *Science* **286**, 1331–1335.

Zolensky M. and McSween H. Y., Jr. (1988) Aqueous alteration. In *Meteorites and the Early Solar System* (eds. J. F. Kerridge and M. S. Matthews). University of Arizona Press, Tucson, pp. 114–143.

Zolensky M. E., Barrett T., and Browning L. (1993) Mineralogy and composition of matrix and chondrule rims in carbonaceous chondrites. *Geochim. Cosmochim. Acta* **57**, 3123–3148.

Zolensky M. E., Mittlefehldt D. W., Lipschutz M. E., Wang M.-S., Clayton R. N., Mayeda T. K., Grady M. M., Pillinger C., and Barber D. (1997) CM chondrites exhibit the complete petrologic range from type 2 to 1. *Geochim. Cosmochim. Acta* **61**, 5099–5115.

1.10
Structural and Isotopic Analysis of Organic Matter in Carbonaceous Chondrites

I. Gilmour

The Open University, Milton Keynes, UK

1.10.1	INTRODUCTION	269
1.10.2	ORGANIC MATERIAL IN CARBONACEOUS CHONDRITES	270
1.10.3	EXTRACTABLE ORGANIC MATTER	270
	1.10.3.1 Abundance and Distribution of Extractable Compounds	270
	1.10.3.1.1 Carboxylic acids	270
	1.10.3.1.2 Sulfonic and phosphonic acids	272
	1.10.3.1.3 Amino acids	272
	1.10.3.1.4 Aromatic hydrocarbons	274
	1.10.3.1.5 Heterocyclic compounds	274
	1.10.3.1.6 Aliphatic hydrocarbons	275
	1.10.3.1.7 Amines and amides	275
	1.10.3.1.8 Alcohols, aldehydes, ketones, and sugar-related compounds	276
	1.10.3.2 Stable-isotopic Investigations of Classes of Organic Compounds	276
	1.10.3.3 Compound-specific Isotopic Studies	277
	1.10.3.3.1 Carbon	277
	1.10.3.3.2 Nitrogen	280
	1.10.3.3.3 Hydrogen	280
	1.10.3.3.4 Sulfur	280
1.10.4	MACROMOLECULAR MATERIAL	280
	1.10.4.1 Structural Studies Using Pyrolysis and Chemical Degradation	280
	1.10.4.2 NMR and Electron Spin Resonance Studies	281
	1.10.4.3 Stable-isotopic Studies	282
1.10.5	*IN SITU* EXAMINATION OF METEORITIC ORGANIC MATTER	284
1.10.6	ENVIRONMENTS OF FORMATION	285
REFERENCES		287

1.10.1 INTRODUCTION

The most ancient organic molecules available for study in the laboratory are those carried to Earth by infalling carbonaceous chondrite meteorites. All the classes of compounds normally considered to be of biological origin are represented in carbonaceous meteorites and, aside from some terrestrial contamination; it is safe to assume that these organic species were produced by nonbiological methods of synthesis. In effect, carbonaceous chondrites are a natural laboratory containing organic molecules that are the product of ancient chemical evolution. Understanding the sources of organic molecules in meteorites and the chemical processes that led to their formation has been the primary research goal. Circumstellar space, the solar nebulae, and asteroidal meteorite parent bodies have all been suggested as environments where organic matter may have been formed. Determination of the provenance of meteoritic organic matter requires detailed structural and isotopic information, and the fall of the Murchison CM2 chondrite in 1969 enabled the first systematic organic analyses to be performed on comparatively pristine samples of extraterrestrial organic material. Prior to that, extensive work had been undertaken on the organic matter in a range of meteorite samples galvanized, in part, by the controversial debate in the early 1960s on possible evidence for former life in the Orgueil carbonaceous chondrite (Fitch et al., 1962; Meinschein et al., 1963). It was eventually demonstrated that the suggested biogenic material was terrestrial contamination (Fitch and Anders, 1963; Anders et al., 1964); however, the difficulties created by contamination have posed a continuing problem in the analysis and interpretation of organic material in meteorites (e.g., Watson et al., 2003); this has significant implications for the return of extraterrestrial samples by space missions. Hayes (1967) extensively reviewed data acquired prior to the availability of Murchison samples.

Developments in the analysis of meteoritic organic matter have largely been driven by progress in analytical capabilities. The limited availability of samples, often restricted to a few grams at most, has presented a series of analytical challenges and significant advances were made in the late 1960s and early 1970s when the coupling of gas chromatography with electron impact mass spectrometry (GCMS) enabled detailed structural information to be obtained on individual compounds (e.g., Hayes and Biemann, 1968). Light-element stable-isotope measurements of meteoritic organic matter can provide important information on its origins and the potential of such measurements has long been recognized (Boato, 1954). Indeed, meteoritic research has led to significant improvements in stable-isotope-ratio mass spectrometry (see Pillinger (1984), for a review). The 1970s saw the start of extensive quantitative analysis of solvent extractable compounds from Murchison together with the first attempts to resolve isotopic heterogeneities in the stable isotopes of carbon, hydrogen and nitrogen and this work has been the subject of regular detailed reviews (Anders et al., 1973; Hayatsu and Anders, 1981; Mullie and Reisse, 1987; Cronin and Chang, 1993; Sephton and Gilmour, 2001b).

The principal focus of this review is on the analysis of the organic matter in the Murchison CM2 chondrite, together with data from other meteorites, where it can be shown that they have not been compromised by terrestrial contamination. It primarily covers work undertaken since 1980, a period that has seen the increasing use of stable-isotopic techniques to elucidate the sources of meteoritic organic matter, improved methods to study the structure of organic matter such as NMR, and the first in situ examinations of organic matter in meteorites; these approaches have provided significant advances in our understanding of the processes involved in the synthesis of extraterrestrial organic matter.

1.10.2 ORGANIC MATERIAL IN CARBONACEOUS CHONDRITES

The carbonaceous chondrites contain up to 5% carbon in a variety of forms, including organic matter, carbonates, and minor amounts of "exotic" presolar grain material such as diamond, graphite, and silicon carbide. For a review of the classification of presolar grains and carbonaceous meteorites, see Chapters 1.02 and 1.05. Less than 25% of the organic matter in carbonaceous chondrites is present as relatively low-molecular-weight "free" compounds which can be extracted with common organic solvents. The remaining 75% or so is present as high-molecular-weight macromolecular materials that persist after prolonged treatment of crushed samples of meteorites with organic solvents and acids such as HF–HCl that remove, or partially remove, silicates and other minerals (Hayatsu and Anders, 1981). These two components require different analytical strategies and are reviewed separately.

1.10.3 EXTRACTABLE ORGANIC MATTER

1.10.3.1 Abundance and Distribution of Extractable Compounds

1.10.3.1.1 *Carboxylic acids*

Early identifications of long-chain monocarboxylic acids (fatty acids) are generally believed to have been terrestrial contaminants (e.g., Smith and Kaplan, 1970). However, short-chain monocarboxylic acids ($<C_{10}$) constitute the most abundant class of extractable compounds in Murchison (Figure 1 and Table 1; Yuen and Kvenvolden, 1973; Lawless and Yuen, 1979; Yuen et al., 1984). They show complete structural diversity with all C_2–C_5 isomers present and with equal concentrations of both branched- and straight-chain isomers; there is a general decrease in the abundance of short-chain monocarboxylic acids with increasing carbon number.

The distribution of monocarboxylic acids has been investigated in other CM2s, most notably Antarctic samples (Shimoyama et al., 1986, 1989; Naraoka et al., 1987; Naraoka et al., 1999). Where present, the abundances of monocarboxylic acids ranged from 9.5 ppm (Yamato-791198) to 191 ppm (Asuka-881458). As with Murchison structural diversity is apparent, however, Yamato-74662 and Asuka-881458 showed an increasing predominance of straight chain over branched isomers (Shimoyama et al., 1989; Naraoka et al., 1999). Some Antarctic samples were devoid of

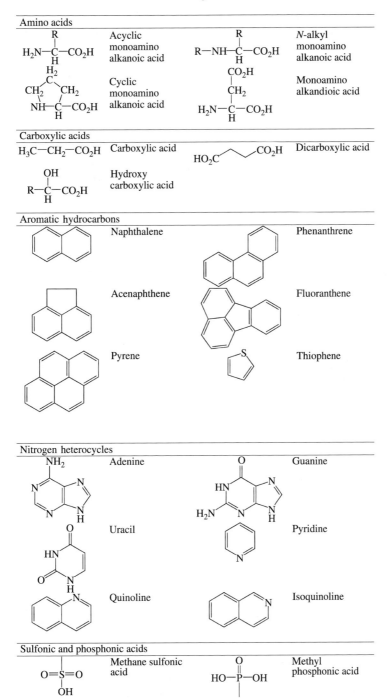

Figure 1 Representative structures of classes of organic compounds identified in carbonaceous chondrites.

monocarboxylic acids, which may indicate that they have lost carboxylic acids due to leaching by Antarctic ice.

A survey undertaken by Lawless et al. (1974) identified 17 dicarboxylic acids (Figure 1) in Murchison including 15 saturated and two unsaturated aliphatic compounds. As with monocarboxylic acids, complete structural diversity was observed. Cronin et al. (1993) examined the distribution of hydroxymonocarboxylic acids, dicarboxylic acids, and hydroxydicarboxylic acids in detail and identified at least 40 dicarboxylic acids with chain lengths up to C_9 and with most chain and substitution position isomers represented at each carbon number. Cronin and Chang (1993) suggested that the dicarboxylic acids are present in Murchison as carboxylate dianions, an interpretation supported by the presence of the calcium salt of oxalic acid (Lawless et al., 1974). Analyses of dicarboxylic

Table 1 Abundances of major classes of organic compounds in the Murchison (CM2) carbonaceous chondrite.

Compounds	Abundance (ppm)	References
Carbon dioxide	106	1
Carbon monoxide	0.06	1
Methane	0.14	1
Hydrocarbons		
Aliphatic	12–35	2
Aromatic	15–29	3
Carboxylic acids		
Monocarboxylic	332	4,1
Dicarboxylic	25.7	5
α-Hydroxycarboxylic	14.6	6
Amino acids	60	7
Alcohols	11	8
Aldehydes	11	8
Ketones	16	8
Sugars and related compounds	~60	9
Ammonia	19	10
Amines	8	11
Urea	25	12
Basic *N*-heterocycles	0.05–0.5	13
Pyridinecarboxylic acids	>7	14
Dicarboximides	>50	14
Pyrimidines	0.06	15
Purines	1.2	16
Benzothiophenes	0.3	17
Sulfonic acids	67	18
Phosphonic acids	1.5	19

References: 1. Yuen *et al.* (1984), 2. Kvenvolden *et al.* (1970), 3. Pering and Ponnamperuma (1971), 4. Lawless and Yuen (1979), 5. Lawless *et al.* (1974), 6. Peltzer *et al.* (1984), 7. Cronin *et al.* (1988), 8. Jungclaus *et al.* (1976b), 9. Cooper *et al.* (2001), 10. Pizzarello *et al.* (1994), 11. Jungclaus *et al.* (1976a), 12. Hayatsu *et al.* (1975), 13. Stoks and Schwartz (1982), 14. Pizzarello *et al.* (2001), 15. Stoks and Schwartz (1979), 16. Stoks and Schwartz (1981), 17. Shimoyama and Katsumata (2001), 18. Cooper *et al.* (1997), and 19. Cooper *et al.* (1992).

acids in the Tagish Lake CI chondrite revealed a homologous series of saturated and unsaturated C_3–C_{10} acids (Pizzarello and Huang, 2002). Linear saturated acids were predominant and showed decreasing abundance with increasing chain length. In all, Pizzarello and Huang (2002) found 44 dicarboxylic acids, with succinic acid the most abundant. The distribution of Tagish Lake dicarboxylic acids showed good compound-to-compound correspondence with those observed in Murchison.

Hydroxy acids (α-hydroxy-carboxylic acids, Figure 1) were first reported in Murchison in the late 1970s (Peltzer and Bada, 1978). These compounds are racemic and correspond structurally to the more abundant amino acids. The presence of hydroxy acids is significant in two respects: (i) They are useful thermometers and indicate that Murchison has not experienced temperatures that would cause the pyrolytic breakdown of hydroxyacids, many of which will decompose at temperatures of less than 120 °C. (ii) The presence of α-hydroxy acids along with the structurally similar α-amino acids suggests the formation of both classes of compounds by a Strecker synthesis, the aqueous-phase component of amino acid synthesis by a Miller–Urey process (Peltzer and Bada, 1978). The α-hydroxydicarboxylic acids reported by Cronin *et al.* (1993) had not been found in previous studies of hydroxy acids in meteorites (Peltzer and Bada, 1978). Each class of hydroxy acids is numerous with carbon chains up to C_8 or C_9 and many, if not all, chain and substitution position isomers represented at each carbon number. The α-hydroxycarboxylic acids and α-hydroxydicarboxylic acids correspond structurally to many of the known meteoritic α-aminocarboxylic acids and α-aminodicarboxylic acids, respectively, a fact that supports the proposal that a Strecker synthesis was involved in the formation of both classes of compounds.

1.10.3.1.2 Sulfonic and phosphonic acids

Cooper *et al.* (1992, 1997) identified a homologous series of C_1–C_4 alkyl phosphonic acids and alkanesulfonic acids in water extracts from Murchison. Isomeric diversity was apparent with five of the eight possible C_1–C_4 alkyl phosphonic and seven of the eight C_1–C_4 alkanesulfonic acids identified. The relative abundances of both classes of compound decreased exponentially with increasing carbon number.

1.10.3.1.3 Amino acids

In comparison to living organisms and terrestrial sedimentary organic matter, the distribution of amino acids (Figure 1) reported in meteorites is unusual. Their abundance and distribution has been extensively investigated in the Murchison CM2 meteorite and to date more than 70 amino acids have been identified in hot-water extracts and many others have been partially characterized (e.g., Kvenvolden *et al.*, 1970; Nagy, 1975; Cronin *et al.*, 1981; Engel and Nagy, 1982; Cronin *et al.*, 1985, 1988; Cronin and Pizzarello, 1986). Murchison amino acids include eight of the protein amino acids and 11 others that are biologically common. The total amino acid concentration reported for Murchison is around 60 ppm (Table 1, Cronin *et al.*, 1988). However, differences in the concentrations of amino acids (Table 2) have been observed for separate stones of the Murchison CM2 meteorite that have been attributed to varying levels of terrestrial contamination and to different extraction procedures. Shock and Schulte (1990) have suggested that most of the differences in amino acid distributions

Table 2 Carbon, nitrogen, and hydrogen stable isotope compositions and abundances of individual amino acids in the Murchison meteorite.

Compound(s)	Formula	Abundance (nmol g^{-1})	$\delta^{13}C$ (‰)	Refs.	$\delta^{15}N$ (‰)	Refs.	δD (‰)	Refs.
Glycine	$C_2H_5NO_2$	28.1–31.0	+22	1	+37	2		
D-alanine	$C_3H_7NO_2$	12.9–17.1	+30	1	+60	2		
L-alanine	$C_3H_7NO_2$		+27	1	+57	2		
β-alanine	$C_3H_7NO_2$	5.7–8.1			+61	2		
Glycine + alanine			+41	3			+1,072	3
Sarcosine	$C_3H_7NO_2$				+129	2		
α-aminoisobutyric acid	$C_4H_9NO_2$	15.0–19.0	+5	1	+184	2	+67	3
D,L-aspartic acid	$C_4H_7NO_4$	1.0–3.9	+4	3	+61	2	+214	3
L-glutamic acid	$C_5H_9NO_4$		+6	1	+58	2	+523	3
D-glutamic	$C_5H_9NO_4$	1.9–4.6			+60	2		
D,L-proline	$C_5H_9NO_2$				+50	2		
Isovaline	$C_5H_{11}NO_2$		+17	1				
Isovaline + valine		4.6–7.5	+30	3			+713	
L-leucine	$C_6H_{13}NO_2$	0.8–1.6			+60	2		

References: 1. Engel et al. (1990) and Silfer (1991), 2. Engel and Macko (1997), and 3. Pizzarello et al. (1991).

and abundances reported for Murchison can, based on solubility data, be explained by differences in extractions procedures (Bada et al., 1983; Cronin et al., 1988). This conclusion is supported by the observation that the use of acid hydrolysis and repeated extractions of a sample of Murchison enabled the recovery of additional amino acids (Cronin, 1976a,b). It is also not possible to exclude minor heterogeneities in amino acid distribution between stones or the possible decomposition of amino acids with prolonged residence time on Earth (Cronin et al., 1988). Nonetheless, there is reasonable agreement between studies of different stones and Table 2 reports the ranges in abundance for amino acids.

Amino acid abundances in several other CM chondrites have also been analyzed, though in less detail compared to Murchison (e.g., Lawless et al., 1972, 1973; Cronin and Moore, 1976; Cronin et al., 1981, 1995). Amino acid distribution in other CM chondrites is apparently similar to that of Murchison, although analysis of several samples is compromised by terrestrial contamination. Antarctic CM2s revealed an amino acid distribution and abundance similar to Murchison and are generally believed to have suffered less terrestrial biologically derived addition to the amino acid abundances, although low abundances in some samples may indicate that they have lost amino acids due to water leaching in Antarctic ice (Shimoyama et al., 1979, 1985; Shimoyama and Harada, 1984). There have been limited analyses of the amino acids distribution of CI chondrites. Orgueil (CI1) apparently contains 11 amino acids, six of which were nonprotein (Lawless et al., 1972). Analyses of both Orgueil and Ivuna (CI1) using high-performance liquid chromatography suggest that the amino acid distribution in the CIs is distinct from CM2s, notably β-alanine is relatively more abundant in Orgueil and Ivuna compared with Murchison while α-aminoisobutyric acid is only present in trace amounts (Ehrenfreund et al., 2001). Analysis of water-soluble extracts of Tagish Lake showed that amino acids were less abundant in this possible CI2 chondrite by almost three orders of magnitude (<0.1 ppm) compared with Murchison (Kminek et al., 2002; Pizzarello et al., 2001). Kminek et al. (2002) concluded that their extremely low abundance and similarities between amino acids in Tagish Lake and in its ice-melt water indicated that the amino acids detected in the lake were terrestrial contaminants.

Cronin et al. (1995) summarized the characteristics of the abundance, distribution, and structure of Murchison amino acids as follows:

(i) Two simple structural types of amino acids have been identified: monoamino monocarboxylic acids and monoamino dicarboxylic acids. Two variations on these structural types occur: n-alkyl secondary amino acids (e.g., sarcosine) and cyclic secondary amino acids.

(ii) There is complete structural diversity. All of the possible α-amino monocarboxylic acids through C_7 have been identified, as have all of the chain and amino position isomers through C_5.

(iii) Homologous series of amino acids show exponential declines in concentration with increasing carbon number, each addition of a carbon atom to the series corresponding to a decrease in concentration of around 70%.

(iv) Branched-chain isomers predominate. At each carbon number, branched isomers are more abundant than straight-chain isomers.

(v) Enantiomers occur in approximately equal amounts. The majority of protein and nonprotein amino acids occur as racemic mixtures. However, there is evidence that there may be slight L-excesses in some protein and nonprotein amino acids, although these results remain controversial.

(vi) The amino acids coexist with a closely matched set of hydroxy acids.

(vii) The amino acids have unusually high $\delta^{13}C$, $\delta^{15}N$, and δD values (see below).

(viii) Acid hydrolysis of aqueous extracts of Murchison indicates that a substantial fraction of the amino acids exist as acid-labile precursors.

Engel and Nagy (1982) detected L-excesses in five protein amino acids from Murchison (alanine, glutamic acid, proline, aspartic acid, and leucine) that they argued were indigenous to the meteorite. This conclusion was criticized, largely on the grounds that the sampling procedure had not completely excluded terrestrial contaminants (Bada et al., 1983). Earlier controversy over the possibility of enantiomeric excesses in meteorites had largely centered on the possible presence of enantiomeric excesses in Orgueil (Nagy et al., 1964), although these results had proved to be difficult to confirm (Hayatsu, 1965) and no consensus had been reached (Hayes, 1967). Subsequent stable-isotopic evidence (see below) strengthened the argument that the L-excesses in protein amino acids in Murchison were indigenous. However, the possibility of terrestrial contamination, which would preferentially lead to an apparent L-excess in protein amino acids, led to an examination of the enantiomeric distribution of nonprotein amino acids (Cronin and Pizzarello, 1997; Pizzarello and Cronin, 1998, 2000). Small excesses (2–9%) in the α-methyl-α-amino acids in Murchison were observed, together with smaller excesses (1–6%) in the same compounds in Murray (Cronin and Pizzarello, 1997; Pizzarello and Cronin, 2000). It was argued that these L-excesses were indigenous to Murchison on the grounds that the excesses were observed in four amino acids not known in nature, there was no apparent correlation between L-excess and potential terrestrial contaminants, and that the analytical procedures adopted reduced the risk of co-elution of other amino acids during the chromatographic analysis.

1.10.3.1.4 Aromatic hydrocarbons

Early investigations of the distribution of aromatic hydrocarbons in meteorites employed several analytical approaches including gas chromatography, spectroscopic techniques (IR, UV, and fluorescence) and mass spectrometry, and reported a wide range of aromatic and polycyclic aromatic molecules with higher molecular weights predominating (Hayes, 1967). However, the development of GCMS in the early 1970s led to a more systematic series of investigations employing both solvent and thermal extraction, particularly of Murchison (Oró et al., 1971; Pering and Ponnamperuma, 1971; Studier et al., 1972; Levy et al., 1973) although the relative abundances obtained for different aromatic compounds varied considerably. This almost certainly reflects differences in the analytical approach, in particular problems associated with the loss of more volatile aromatic species when solvent extracts are dried prior to analysis. More recent investigations of aromatic hydrocarbons identified the three-ring polycyclic aromatic hydrocarbons (PAHs) fluoranthene and pyrene as the most abundant compounds in solvent extracts of Murchison and a number of Antarctic CM2s (Basile et al., 1984; Naraoka et al., 1988; Shimoyama et al., 1989; Krishnamurthy et al., 1992; Gilmour and Pillinger, 1994). However, when Sephton et al. (1998) used supercritical CO_2 as an extraction solvent to minimize the loss of volatile material, they obtained a positive correlation between the abundance of aromatic compounds and their volatility. Higher-molecular-weight PAHs are apparently relatively minor components concentrated in previous studies as a result of the loss of more volatile species.

1.10.3.1.5 Heterocyclic compounds

A number of early studies reported the presence of urine and pyrimidine bases in water extracts of carbonaceous chondrites, although contamination was often suspected (Hayes, 1967). Investigations of nitrogen heterocycles in the late 1970s and early 1980s (Stoks and Schwartz, 1979, 1981, 1982) confirmed the presence of several nitrogen bases and extended earlier studies, at the same time identifying several potential terrestrial contaminants (Hayatsu, 1964; Hayatsu et al., 1968, 1975; van der Velden and Schwartz, 1977). Several classes of basic and neutral nitrogen heterocycles have since been identified in Murchison including:

(i) Purines: xanthine, hypoxanthine, guanine, and adenine (Stoks and Schwartz, 1981, 1982).

(ii) Pyrimidines: uracil (Stoks and Schwartz, 1979).

(iii) Quinolines and isoquinolines with 0–4 carbon atoms as methyl or ethyl side chains (Stoks and Schwartz, 1982).

(iv) Pyridines: C_3-alkyl pyridines and C_5-alkyl pyridines, carboxylated pyridines (Hayatsu et al., 1975; Stoks and Schwartz, 1982; Pizzarello et al., 2001).

The quinolines, isoquinolines, and pyridines are structurally diverse with a large number of alkyl-substituted isomers. Direct isotopic measurements have not been made on nitrogen heterocycles.

Modern GCMS analysis and sample preparation techniques have overcome many the analytical problems associated with the investigation of organic sulfur compounds associated with potential reactions between elemental sulfur and organic matter during analysis (Hayes, 1967). Shimoyama and Katsumata (2001) examined the distribution of aromatic thiophenes in Murchison detecting benzothiophene, dibenzothiophene, alky-substituted dibenzothiophenes, and benzonaphthothiophenes at concentrations levels of 0.3 ppm (Table 1) confirming earlier studies (Basile et al., 1984).

1.10.3.1.6 Aliphatic hydrocarbons

Considerable importance was attached to the presence of long-chain n-alkanes in carbonaceous chondrites in studies during the 1960s and 1970s. Analyses of the Orgueil and other meteorites in the early 1960s had identified the presence of saturated hydrocarbons (e.g., Nagy et al., 1961; Meinschein et al., 1963) and it was suggested, somewhat controversially, that these compounds were evidence for biogenic activity on the meteorites' parent bodies. However, when Oró et al. (1966) analyzed both interior and exterior portions of Orgueil, they observed a decrease in the abundance of alkanes away from the surface of the meteorite together with the presence of the biologically derived isoprenoid hydrocarbons pristane and phytane in a total of 19 meteorites. Such results were strongly suggestive that alkanes were the result of terrestrial contamination. The problem of contamination was demonstrated in 1969 when the Allende (CV3) chondrite was subjected to organic analysis within seven days of its fall. Both n-alkanes and isoprenoids were found concentrated at its surface (Han et al., 1969).

Although there was considerable evidence of a terrestrial source for n-alkanes, analysis of the Murchison meteorite revived arguments in favor of an extraterrestrial origin for these compounds. Analysis of Murchison revealed a similar distribution of hydrocarbons to those found in Orgueil and Murray (Studier et al., 1972). Studier et al. (1968) had identified n-alkanes as the principal hydrocarbon species above C_{10} in Orgueil and Murray, while below C_{10} aliphatic hydrocarbons were markedly deficient and benzene and alkylbenzenes were predominant. They obtained a similar hydrocarbon distribution using a Fischer–Tropsch type (FTT) synthesis from CO and H_2 in the presence of iron meteorite powder and suggested that catalytic reactions of this type may have occurred on a large scale in the solar nebula, converting CO to less volatile carbon compounds.

However, a key feature of catalyzed reactions such as Fischer–Tropsch is their structural selectivity. More recent analyses of hydrocarbons of interior and exterior portions of the Murchison, Murray, and Allende meteorites together with the highly metamorphosed New Concord chondrite have again indicated that n-alkanes are preferentially concentrated toward the surface of these meteorites strongly suggesting that they are terrestrial contaminants (Cronin and Pizzarello, 1990). However, indigenous branched and cyclic aliphatic hydrocarbons in Murchison appear to show complete structural diversity within homologous series arguing strongly against a Fischer–Tropsch origin (Cronin and Pizzarello, 1990). Furthermore, isotopic fractionations accompanying Fischer–Tropsch reactions in the laboratory (Yuen et al., 1990) do not explain the isotopic distributions observed in n-alkanes in meteoritic material (Sephton et al., 2001).

1.10.3.1.7 Amines and amides

Pizzarello et al. (1994) undertook a detailed investigation of nitrogenous compounds in Murchison isolating volatile bases and identifying a series of aliphatic amines, confirming and extending earlier work (Hayatsu et al., 1975; Jungclaus et al., 1976a). Significant ^{13}C, ^{15}N, and D-enrichments were observed in the isolated volatile bases, confirming their extraterrestrial origin and suggestive of an interstellar origin (Table 3). As with other groups of compounds in Murchison, the aliphatic amines showed decreasing abundance with increasing carbon number and almost complete structural diversity through C_5. Branched-chain isomers were more abundant than straight-chain isomers. Pizzarello et al. (1994) suggested that aliphatic amines could have originated by two possible routes: direct incorporation from molecular clouds, supported by the detection of methylamine in interstellar environments, or via the decarboxylation of α-amino acids, supported by the observation that most of the aliphatic amines could be produced by the decarboxylation of known α-amino acids in Murchison.

Cooper and Cronin (1995) detected a wide range of linear and cyclic amides in water extracts of Murchison, extending the previous positive detection of guanylurea (Hayatsu et al., 1968) These included many mono- and dicarboxylic acid amides, hydroxyacid amides, and other amides with no known terrestrial source. These compounds were characterized by a structural diversity of isomers up to C_8 and a decline in abundance with increasing carbon number.

Table 3 Carbon, nitrogen, and hydrogen stable isotope compositions of organic fractions from the Murchison meteorite.

Fraction	$\delta^{13}C$ (‰)	$\delta^{15}N$ (‰)	δD (‰)	References
Benzene-methanol extract	+5			1
Methanol extract	+7	+88	+406	2
	+4		+957	3
Volatile hydrocarbons				
Freeze thaw disaggregation	0		−92	3
Hot water extract	+6		+217	3
H_2SO_4 treatment	+17		+410	3
Isolated fractions				
Aliphatic fraction[a]	−11.5/−5		+264/+103	3
Aromatic fraction[a]	−5.5/−5		+407/+244	3
Polar fraction[a]	+6/+5		+946/+751	3
Amino acids		+102		4
	+23	+90	+1,370	5
	+26		+1,137	6
Monocarboxylic acids	+7	−1	+377	5
	−1		+652	3
Hydroxy acids	+4		+573	7
Dicarboxylic acids	+6		+357	7
Volatile bases (ammonia, amines)	+22	+93	+1,221	4
Neutral polyhydroxylated compounds	−6		+119	8

References: 1. Chang *et al.* (1978), 2. Becker and Epstein (1982), 3. Krishnamurthy *et al.* (1992), 4. Pizzarello *et al.* (1994), 5. Epstein *et al.* (1987), 6. Pizzarello *et al.* (1991), 7. Cronin *et al.* (1993), and 8. Cooper *et al.* (2001).
[a] Analyses performed on two samples held, respectively, at Chicago and Arizona State University.

1.10.3.1.8 Alcohols, aldehydes, ketones, and sugar-related compounds

Alcohols, aldehydes, and ketones have not been the subject of extensive investigation for 20–25 years since initial studies of their distribution and abundance in Murchison, Murray, and Orgueil (Meinschein *et al.*, 1963; Studier *et al.*, 1965; Hayes and Biemann, 1968; Jungclaus *et al.*, 1976b; Basile *et al.*, 1984). The abundances of these compounds in Murchison are reported in Table 1. Junglaus *et al.* (1976b) positively identified C_1–C_4 alcohols, and C_2, C_4, and C_5 carbonyl compounds. The aromatic ketones fluoren-9-one, anthracene-dione, phenanthrenedione, benzanthracen-7-one, and anthracen-9(10H)-one were identified in Murchison as part of an investigation of aromatic hydrocarbons (Basile *et al.*, 1984).

Sugar-related compounds were first reported in ethanol extracts of Murray and Orgueil (Degens and Bajor, 1962; Kaplan *et al.*, 1963); however, it has been only since early 2000s that these compounds have received further investigation. Cooper *et al.* (2001) identified polyhydroxylated (polyols) sugars, sugar alcohols, and sugar acids in Murchison at abundances similar to those for amino acids (Table 1). Their indigeneity was confirmed through stable-isotope measurements of an isolated polyol fraction, which gave a $\delta^{13}C$ value of −6‰ and a δD value of +119‰ (Table 3). Cooper *et al.* (2001) found marked decreases in the abundances of higher-molecular-weight polyols and almost complete isomeric diversity.

1.10.3.2 Stable-isotopic Investigations of Classes of Organic Compounds

Stable-isotope studies from the 1960s through the early 1980s focused on making isotopic measurements on broad groupings of meteoritic organic compounds, e.g., HF/HCl acid residues, which concentrated the macromolecular component of the organic matter, or on solvent extracts (Krouse and Modzeleski, 1970; Kvenvolden *et al.*, 1970; Smith and Kaplan, 1970; Becker and Epstein, 1982; Robert and Epstein, 1982; Yang and Epstein, 1983). Such studies were successful in establishing that meteoritic organic matter contained material of probable interstellar origin (Yang and Epstein, 1985). Subsequent work began the process of attempting to isolate chemically identifiable components; using chromatographic techniques it was possible to obtain fractions with broadly similar chemical structures. This led to isotopic measurements being obtained on assemblages of compounds, predominantly from the Murchison meteorite, including aliphatic and aromatic hydrocarbons (Krishnamurthy *et al.*, 1992), carboxylic acids (Epstein *et al.*, 1987; Pizzarello *et al.*, 1991; Krishnamurthy *et al.*, 1992; Cronin *et al.*, 1993), amino acids (Epstein *et al.*, 1987; Pizzarello *et al.*, 1991), and polar hydrocarbons (Pizzarello et al., 1994); the data for Murchison are summarized in Table 3. These investigations confirmed that a significant proportion of the solvent extractable organic matter was indigenous to meteorites since significant enrichments in ^{13}C, ^{15}N, and D were detected. The deuterium enrichments also provided strong evidence for the role of interstellar processes in the origin of meteoritic organic matter.

1.10.3.3 Compound-specific Isotopic Studies

The ability to measure the isotopic compositions of individual compounds in a complex mixture, termed compound-specific isotope analysis (CSIA), has been a long-held goal of stable-isotope mass spectrometry, and its potential to identify indigenous and contaminant material in meteorites was recognized well before analytical developments permitted such measurements on a routine basis (Pillinger, 1982). The determination of the stable-isotope compositions of individual compounds provides a powerful means of elucidating the reaction mechanisms and possible source environments from which the organic constituents have formed. Using preparative chromatography to isolate fractions containing simple mixtures or individual compounds, which are subsequently converted to a form suitable for isotope-ratio mass spectrometry, it has been possible to obtain δ^{13}C and δD measurements on C_2–C_5 amino acids (Table 2; Pizzarello et al., 1991) and δ^{13}C, δD, δ^{34}S, and δ^{33}S of C_1–C_3 sulfonic acids (Table 4; Cooper et al., 1997). However, it is the advent of combined gas chromatography isotope-ratio mass spectrometry (GC-IRMS) (Hayes et al., 1990) that has enabled the comparatively rapid determination of the isotopic compositions of individual compounds to high levels of precision. Since the approach relies on the ability of analytes to be separated chromatographically, exploiting gas–stationary phase partitioning, volatile meteoritic compounds with hydrocarbon skeletons and relatively few functional groups have been the main focus of study. As of early 2000s, work on Murchison and a limited number of other meteorites has provided carbon stable-isotopic compositions for CO, CO_2, C_2–C_5 aliphatic hydrocarbons (Table 5; Yuen et al., 1984), C_2–C_5 carboxylic acids (Table 5; Yuen et al., 1984), dicarboxylic acids (Table 5, Pizzarello et al., 2001; Pizzarello and Huang, 2002), C_6–C_{20} aromatic compounds (Table 6; Yuen et al., 1984; Naraoka et al., 2000; Gilmour and Pillinger, 1994; Sephton et al., 1998), and C_{12}–C_{18} n-alkanes (Sephton et al., 2001). GC-IRMS analyses of the less volatile polar compounds present analytical difficulties since derivatives of these compounds have to be chemically modified to increase their volatility and make them amenable to gas chromatographic analysis. It is then necessary to correct the δ^{13}C value obtained for the derivatized compound in order to determine the true δ^{13}C value of the molecule being studied. Such procedures have enabled carbon and nitrogen isotopic compositions to be determined on C_2–C_6 amino acids from the Murchison meteorite as trifluoroacetic acid-isopropyl ester derivatives (Table 2; Engel and Macko, 1997; Engel et al., 1990).

1.10.3.3.1 Carbon

Figure 2 shows that the C_1–C_5 aliphatic hydrocarbons, amino acids, carboxylic acids, and sulfonic acids from the Murchison meteorite appear to follow a common trend when their δ^{13}C values are plotted against carbon number. δ^{13}C values generally decrease as the amount of carbon in the molecules increases. This trend has been interpreted as the result of a kinetic isotope effect during the sequential formation of higher-molecular-weight compounds from simpler precursors (Yuen et al., 1984). The more reactive ^{12}C is preferentially added during the synthesis of the carbon skeleton of these compounds.

Extractable C_6–C_{22} aromatic compounds from the Murchison (CM2) and Asuka-881458 (CM2) meteorites also appear to follow a systematic trend when their δ^{13}C values are plotted against carbon number (Figure 3), although benzene and toluene from Murchison and naphthalene and biphenyl from Asuka-881458 are clear outliers. The δ^{13}C values of many of the aromatic compounds extracted (e.g., Asuka-881458 fluoranthene: $-8.3‰$) are markedly

Table 4 Carbon, hydrogen, and sulfur isotopic compositions of individual sulfonic acids from the Murchison (CM2) meteorite.

Compound	$\delta^{13}C$ (‰)	δD (‰)	$\delta^{33}S$ (‰)	$\delta^{34}S$ (‰)	$\delta^{36}S$ (‰)	$\Delta^{33}S$
Methanesulfonic acid	+29.8	+483	+7.63	+11.27	+22.5	+2
Ethanesulfonic acid	+9.1	+787	+0.33	+1.13	+0.8	−0.24
Propanesulfonic acid	−0.4	+536	+0.20	+1.20	+2.1	−0.40
1-Methylethanesulfonic acid	−0.9	+852	+0.32	+0.68	+2.9	−0.02

Source: Cooper et al. (1997).

Table 5 Carbon isotope compositions of CO, CO$_2$, carboxylic acids, dicarboxylic acids, and volatile hydrocarbons in the Murchison (CM2), Tagish Lake (CI), and Orgueil (CI1) meteorites together with hydrogen isotope compositions for dicarboxylic acids.

Compound	Murchison $\delta^{13}C$ (‰)	Tagish Lake $\delta^{13}C$ (‰)	Orgueil $\delta^{13}C$ (‰)	Murchison δD (‰)	Tagish Lake δD (‰)	Orgueil δD (‰)
CO[a]	−32.0 ± 2.0					
CO$_2$[a]	+29.1 ± 0.2					
Monocarboxylic acids[a]						
Acetic acid	2					
Propionic acid	+17.4 ± 0.2					
Isobutyric acid	+16.9 ± 0.2					
Butyric acid	+11.0 ± 0.3					
Isovaleric acid	+8.0 ± 4.5					
Valeric acid	+4.5 ± 0.2					
Dicarboxylic acids[b]						
Succinic acid	+22.5 ± 0.6	+22.5 ± 0.6	−23.8	+1,124	+1,116	+389
Methylsuccinic acid	+15.4 ± 3.0	+15.4 ± 3.0	−19.5	+1,106	+1,112	+1,225
Glutaric acid	+22.9 ± 1.5	+22.9 ± 1.5	−20.4	+1,387	+1,322	+795
2-Methyl glutaric	+27.9 ± 1.0	+18.6 ± 0.7	−10.3	+1,463	+1,263	+1,551
3-Methyl glutaric	+19.1	+12.6				
Adipic	+21.4	+5.5 ± 0.9				
Volatile hydrocarbons[a]						
Methane	+9.2 ± 1.0					
Ethane	+3.7 ± 0.1					
Ethene	+0.1 ± 0.4					
Propane	+1.2 ± 0.1					
Isobutane	+4.4 ± 0.1					
Butane	+2.4 ± 0.1					

[a] Yuen *et al.* (1984). [b] Pizzarello and Huang (2002).

Table 6 Carbon stable isotope compositions of solvent extractable aromatic compounds in carbonaceous chondrites.

Compound	Orgueil[a] $\delta^{13}C$ (‰)	Tagish Lake[b] $\delta^{13}C$ (‰)	Cold Bokkeveld[a] $\delta^{13}C$ (‰)	Murchison $\delta^{13}C$ (‰)	Asuka-991458[f] $\delta^{13}C$ (‰)
Benzene				−28.7 ± 0.2[c]	
Toluene			−24	−28.8 ± 1.1[d]	
Naphthalene				−12.6 ± 2.3[d]	−26.2 ± 1.2
1-Methylnaphthalene				−11.1[d]	
2-Methylnaphthalene				−5.8[d]	
Biphenyl					−25.9 ± 0.9
Phenanthrene	−23	−25	−27.0 ± 1.2	−7.5 ± 1.2[e]	−12.9 ± 0.9
Methyl-phenanthenes					−13.4 ± 0.8
Fluoranthene	−19			−5.9 ± 1.2[e]	−8.3 ± 0.5
Pyrene	−17			−13.1 ± 1.3[e]	−15.8 ± 0.8
Chrysene	−23			−14.5 ± 2.2[e]	
Benzo[ghi]fluoranthene	−15			−14.2 ± 2.2[e]	−15.8 ± 0.8
Benzanthracene	−15				
Benz[a]anthracene, Chrysene and triphenylene					−11.7 ± 2.4
Benzo[e]pyrene				−22.3 ± 4.1[e]	
Benzopyrenes and perylene					−19.1 ± 1.5
Benzo[f]fluoranthene				−15.4 ± 3.3[e]	
Benzofluoranthenes					−14.0 ± 0.5
Dibenzanthracenes	−17				
Benzo[ghi]perylene					−25.2 ± 0.5

[a] Sephton and Gilmour (2000). [b] Pizzarello *et al.* (2001). [c] Yuen *et al.* (1984). [d] Sephton *et al.* (1998).
[e] Gilmour and Pillinger (1994). [f] Naraoka *et al.* (2000).

more ^{13}C-enriched than typical terrestrial PAHs confirming the indigenous nature of these compounds. The predominant aromatic trend is consistent that observed for C_1–C_5 compounds from Murchison and indicates an origin by a synthetic process progressively adding ^{12}C to the carbon skeleton with kinetic isotopic fractionation determining the distribution of carbon isotopes between compounds rather than thermodynamic equilibrium (Gilmour and Pillinger, 1994; Naraoka et al., 2000). The δ^{13}C values obtained for extractable aromatic hydrocarbons in CM2 meteorites display a significant amount of isotopic heterogeneity with a range in δ^{13}C values of over 20‰. Compounds, with relatively high molecular weights, but which differ by only one or two carbon atoms also display significant differences in their δ^{13}C values. This has led to the suggestion that during the synthetic processes that led to bond formation, isotopic fractionation was at its most extreme, implying that synthesis took place in a low-temperature environment such as interstellar space (Sephton and Gilmour, 2000). The isotopic heterogeneity displayed by aromatic compounds in Murchison and Asuka-881458 may also contain evidence for different synthetic pathways. There is a 7.5‰ difference in δ^{13}C values between PAHs isomers containing a five-carbon ring (e.g., fluoranthene) and those without (e.g., pyrene), which has been interpreted as evidence of two possible pathways for the formation of PAHs (Gilmour and Pillinger, 1994; Naraoka et al., 2000).

The δ^{13}C values for the C_{12}–C_{26} n-alkanes from six chondrites are shown in Figure 4 (Sephton et al., 2001). None of the n-alkanes exhibit either the ^{13}C-enrichments or systematic isotopic trends that apparently characterize indigenous organic matter in meteorites. Most of the δ^{13}C values are similar both in value and in the trends shown within homologous series to δ^{13}C variations observed for terrestrial petroleum products or other terrestrial fossil hydrocarbons. These features confirm the long-held suspicion that these molecules are contaminants from the

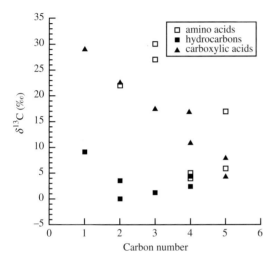

Figure 2 Carbon stable-isotope compositions of low-molecular-weight hydrocarbons, amino acids, and monocarboxylic acids from the Murchison meteorite plotted against carbon number. Carbon number 1 denotes methane and CO_2, 2 denotes ethane, ethanoic acid, glycine, etc. (source Yuen et al., 1984).

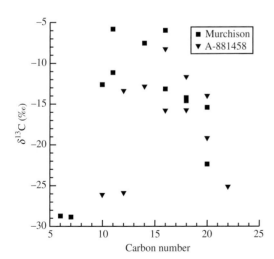

Figure 3 Carbon stable-isotope compositions of solvent extractable aromatic and PAHs plotted against carbon number from the Murchison and Asuka-881458 CM2 carbonaceous chondrites (sources Yuen et al., 1984; Gilmour and Pillinger, 1994; Sephton et al., 1998; Naraoka et al., 2000).

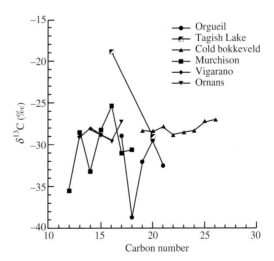

Figure 4 Carbon stable-isotope compositions of solvent extractable n-alkanes from the Orgueil (CI), Cold Bokkeveld (CM2), Murchison (CM2), Vigarano (CV3), Ornans (CO), and Tagish Lake carbonaceous chondrites plotted against carbon number (sources Sephton et al., 2001; Pizzarello et al., 2001).

terrestrial environment added to the meteorite following its fall to Earth (Cronin and Pizzarello, 1990; Sephton et al., 2001). It is interesting to note, however, that n-alkanes with negative $\delta^{13}C$ values were recorded in the Tagish Lake meteorite, despite samples being held at temperatures below freezing and, once collected, in clean conditions (Pizzarello et al., 2001).

Engel et al. (1990) determined the $\delta^{13}C$ values for individual amino acid enantiomers to attempt to ascertain whether the reported L-excess in Murchison amino acids (Engel and Nagy, 1982) was a consequence of terrestrial contamination, one of the original applications envisaged for CSIA of meteoritic organics (Pillinger, 1982). Similar, nonterrestrial $\delta^{13}C$ values were obtained for L- and D-alanine of +27‰ and +30‰, respectively (Table 2), apparently confirming the indigenous nature of these acids. They argued that the more negative $\delta^{13}C$ value for the L-enantiomer could not be explained by terrestrial contamination and that this excess was indigenous to the meteorite.

1.10.3.3.2 Nitrogen

The $\delta^{15}N$ values of individual amino acid enantiomers in Murchison have also been used to try and ascertain whether the reported L-excess is indigenous (Engel and Macko, 1997). Engel and Macko (1997) found that the L- and D-enantiomers of alanine and glutamic acid have significantly higher $\delta^{15}N$ values (ca. +60‰, Table 2) than their terrestrial counterparts (ca. −10‰ to +20‰) and argued that the excess of L-enantiomers over D-enantiomers is extraterrestrial in origin and not the result of terrestrial contamination. The observed ^{15}N-enrichments in the C_2-C_6 amino acids in Murchison (e.g., $\delta^{15}N = +184$‰ for α-aminoisobutyric acid, Table 2) suggest an interstellar source for these compounds or their precursors (Engel and Macko, 1997). However, Pizzarello and Cronin (1998) considered that other meteoritic amino acids may be co-eluting with L-alanine, thereby contributing to both the L-excess and $\delta^{15}N$ determinations. In parallel work, these authors examined the enantiomers of nonprotein amino acids and observed an L-excesses in the α-methyl-α-amino acids in Murchison and Murray reported above (Cronin and Pizzarello, 1997, 1999; Pizzarello and Cronin, 1998, 2000).

1.10.3.3.3 Hydrogen

Compound-specific hydrogen isotopic measurements of amino and sulfonic acids display significant deuterium enrichments suggesting an interstellar origin for these compounds or their precursors (Pizzarello et al., 1991; Cooper et al., 1997). Cooper et al. (1997) proposed that the relatively constant δD values for different sulfonic acids (Table 4) implies that the hydrogenation of their unsaturated precursors occurred within a pool of nearly uniform deuterium enrichment.

1.10.3.3.4 Sulfur

Cooper et al. (1997) determined the sulfur isotope compositions for a homologous series of sulfonic acids from Murchison (Table 4). They observed a nonmass-dependant enrichment in ^{33}S ($\delta^{33}S = +2$) in methanesulfonic acid that was attributed to the gas-phase UV-irradiation of CS_2 in an interstellar environment, prior to the production of the methanesulfonic acid precursor.

1.10.4 MACROMOLECULAR MATERIAL

Anders and Kerridge (1988) considered establishing the origins of the macromolecular organic matter in meteorites a daunting task, and studies of the macromolecular organic matter in carbonaceous chondrites have presented a series of analytical challenges. The elemental composition of the Murchison macromolecule has been determined as $C_{100}H_{71}N_3O_{12}S_2$ based on elemental analysis (Hayatsu et al., 1977) and revised to $C_{100}H_{48}N_{1.8}O_{12}S_2$ based on pyrolytic release studies (Zinner, 1988). Such large macromolecules amenable to direct study using NMR, while other analytical approaches attempt to break the structure down, using techniques such as pyrolysis or chemical degradation, into fragments that are easier to study.

1.10.4.1 Structural Studies Using Pyrolysis and Chemical Degradation

The majority of the carbon in the Murchison macromolecular material is present within aromatic ring systems. This aromatic nature has been revealed by a series of pyrolysis studies of meteorites such as Orgueil (CI1), Murchison (CM2), Murray (CM2), and Allende (CV3) in which the macromolecular material was thermally fragmented to produce benzene, toluene, alkylbenzenes, naphthalene, alkylnaphthalenes, and PAHs with molecular weights up to around 200–300 amu (Simmonds et al., 1969; Studier et al., 1972; Levy et al., 1973; Bandurski and Nagy, 1976; Holtzer and Oró, 1977; Murae, 1995; Kitajima et al., 2002). Further identification of the aromatic units in the Murchison macromolecular material was achieved by Hayatsu et al. (1977), who used sodium dichromate oxidation to selectively remove aliphatic side chains in the macromolecular material and thereby isolate and release the aromatic cores present. These studies have revealed a dominance of single-ring aromatic entities but also a significant amount of two- to four-ring aromatic cores bound to the macromolecular material by a number of aliphatic linkages.

Sephton et al. (1998, 1999, 2000) used hydrous pyrolysis followed by supercritical extraction to examine insoluble organic matter in Orgueil (CI1), Murchison (CM2), and Cold Bokkeveld (CM2). The hydrous pyrolysates obtained for the three meteorites displayed a remarkable degree of qualitative similarity suggesting that the macromolecular materials in different carbonaceous chondrites are apparently composed of essentially the same aromatic structural units, predominantly one to three ring alkyl-substituted aromatic structures. Significant quantitative differences were observed, however, and these were interpreted as indications of the different parent body histories of the three meteorites (Sephton et al., 2000).

Aliphatic hydrocarbon moieties are present in significant amounts within the Murchison macromolecular material and several pyrolysis studies have indicated that these entities exist within or around the aromatic network as hydroaromatic rings and short alkyl substituents or bridging groups (e.g., Hayatsu et al., 1977; Holtzer and Oró, 1977; Levy et al., 1973).

Several pyrolysis experiments have released oxygen-containing moieties such as phenols (Studier et al., 1972; Hayatsu et al., 1977; Sephton et al., 1998), benzene carboxylic acids (Studier et al., 1972), propanone (Levy et al., 1973; Biemann, 1974), and methylfuran (Holtzer and Oró, 1977) from the Murchison macromolecular material. Sodium dichromate oxidation of the macromolecular material liberated the cyclic diaryl ether dibenzofuran and aromatic ketones such as fluorenone, benzophenone, and anthraquinone. Each of these organic units appears to be bound into the macromolecular network by two to four aliphatic linkages (Hayatsu et al., 1977). Hayatsu et al. (1980) used alkaline cupric oxide to selectively cleave organic moieties incorporated into the Murchison macromolecular material by ether groups. In this way, these authors established the presence of significant amounts of phenolic species bound to the macromolecular material by ether linkages.

Thiophenes are common pyrolysis products of the Murchison macromolecular material and thiophene, methylthiophene, dimethylthiophene, and benzothiophene have been detected in this way (Biemann, 1974; Holtzer and Oró, 1977; Sephton et al., 1998). Sodium dichromate oxidation has revealed the presence of substituted benzothiophene and dibenzothiophene moieties within the macromolecular material (Hayatsu et al., 1980).

Pyrolysis has also led to the tentative identifications of the nitrogen heterocyclics, cyanuric acid (Studier et al., 1972), and alkylpyridines (Hayatsu et al., 1977) in the Murchison macromolecular material. Similar types of analyses have revealed acetonitrile and benzonitrile (Levy et al., 1973; Holtzer and Oró, 1977). Substituted pyridine, quinoline, and carbazole were observed in sodium dichromate oxidation of the macromolecular material (Hayatsu et al., 1977).

1.10.4.2 NMR and Electron Spin Resonance Studies

Cronin et al. (1987) first employed ^{13}C NMR spectroscopy to detect aromatic carbon within the Orgueil, Murchison, and Allende macromolecular materials. These studies indicated degrees of aromaticity of 47% and 40% for macromolecular material in Orgueil and Murchison, respectively; however, Cronin et al. concluded that they were underestimating the contribution of nonprotonated aromatic carbon and invoked the presence of extensive polycyclic aromatic sheets significantly larger than the one to four ring aromatic entities isolated by Hayatsu et al. (1977). Gardinier et al. (2000) revised the levels of aromatic carbon in Orgueil to between 69% and 78% and in Murchison to between 61% and 68% and estimated the function group abundance for the Murchison macromolecular material.

Cody et al. (2002) undertook an extensive NMR investigation using double- and single-resonance solid-state (^{1}H and ^{13}C) NMR to study Murchison macromolecular material. This indicated that it was a complex organic solid composed of a wide range of organic (aromatic and aliphatic) functional groups, including numerous oxygen-containing functional groups (Figure 5 and Table 7). They refined the estimates of aromatic carbon within the Murchison organic residue to 61–66%. Aside from the presence of interstellar diamond, they found no evidence for significant amounts of large polycyclic aromatic sheets, concluding that such structures comprised a maximum of 10%, and probably much less, of the macromolecular material. Using both single-pulse and cross-polarized spectra, Cody et al. (2002) concluded that the fraction of aromatic carbon directly bonded to hydrogen is low (~30%), indicating that the aromatic molecules in the Murchison organic residue are highly substituted and estimated the H/C ratio at 0.53–0.63, compared with and H/C ratio of 0.53 determined using elemental analysis (Zinner, 1988). NMR data suggest that the range of oxygen-containing organic functionality in Murchison is substantial with a wide range in O/C ratio possible, depending on whether various oxygen-containing organic functional groups exist as free acids and hydroxyls or are linked as esters (Gardinier et al., 2000; Cody et al., 2002). The lower O/C values are consistent with elemental analyses, requiring that oxygen-containing functional groups in the Murchison macromolecule are highly linked. NMR data also indicated a significant proportion of tertiary (methyne, CH) carbon, suggesting that aliphatic carbon chains within the Murchison organic macromolecule are highly branched (Cody et al., 2002). In contrast to Murchison and Orgueil macromolecular materials, NMR of insoluble organic matter in Tagish Lake (a possible CI2) is extremely aromatic (Pizzarello et al., 2001).

Binet et al. (2002) have undertaken an initial electron paramagnetic resonance study to examine the distribution of free radicals in Murchison and Orgueil macromolecular material. They suggest that there are radical-rich regions, which could represent regions of pristine interstellar organic matter preserved within the macromolecular material.

1.10.4.3 Stable-isotopic Studies

Measurements of bulk carbon, nitrogen, and hydrogen isotopic composition indicate that chondritic macromolecular material contains significant enrichments in some of the heavier isotopes. Kerridge (1985) reviewed the contents of carbon, hydrogen, and nitrogen and $\delta^{13}C$, $\delta^{15}N$, and δD for 25 whole-rock samples of carbonaceous chondrites, updating and reviewing many earlier studies (e.g., Boato, 1954; Smith and Kaplan, 1970; Kung and Clayton, 1978; Kolodny et al., 1980; Robert and Epstein, 1982; Yang and Epstein, 1983).

Two analytical approaches have been adopted in attempts to obtain stable-isotopic information on insoluble organic matter in carbonaceous chondrites: stepped-combustion analysis (e.g., Kerridge, 1983; Swart et al., 1983) and CSIA of pyrolysis products (e.g., Sephton et al., 1998). Stepped-combustion analysis has proved to be more successful in providing information on the major-elemental constituents of chondritic organic matter, i.e., carbon, hydrogen, nitrogen, and oxygen, whereas CSIA has started to yield detailed carbon isotopic and structural information.

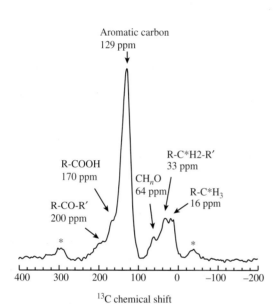

Figure 5 Cross polarization NMR spectrum of organic macromolecular material from the Murchison (CM2) chondrite. Prominent peaks and shoulders are assigned to probable functional groups identified by their respective chemical shifts in ppm (source Cody et al., 2002).

Table 7 Estimates of functional group abundance by cross-polarization NMR.

Carbon type	% of total	Hydrogen content[a] ($C_{100}H_n$)	Oxygen content[b] ($C_{100}O_m$)
CH_3	8.0	24	0
CH, CH_2	8.7	8.7–17.4	0
$CH_{n=1-2}$, $OH_{i=0.1}$	5.5	0–16.5	5.5 (max)
Aromatic (C-H, C-R)	53.7	0–53.7	0
Aromatic (C-O)	7.6	0–7.6	7.6 (max)
R-COOH,R′	7.4	0–7.4	14.8 (max)
R-CO-R′	9.1	0	9.1
	Aromatic fraction (F_a) = 0.61	$n = 32.7–126.6$	$m = 26.8–37.0$

Source: Cody et al. (2002).
[a] Uncertainties in identifying specific functional groups results in a range of n. The lower limit considers all aromatic carbon to be nonprotonated, the oxygen-substituted carbon to be a tertiary ether, the carboxylate to be an ester, and the carbonyl to be a ketone. The upper limit is derived by assuming that all the aromatic carbon is protonated, the oxygen-substituted carbon is a primary alcohol, and the carboxylate is an acid.
[b] The maximum value of m is derived by assuming that all oxygen-substituted alkyl carbon is in the form of alcohol, all the oxygen-substituted aromatic carbon is hydroxyl, and the carboxylate is in the form of free acid. The minimum value of m is derived by assuming that all carboxylate is linked to aromatic carbon via aromatic esters, the remainder of aromatic oxygen is linked to aliphatic carbon via alkyl aryl ethers, and that the remainder of the aliphatic oxygen linked as aliphatic ethers.

Kerridge et al. (1987) undertook a high-resolution stepped-combustion experiment of Murchison insoluble organic matter that indicated significant variations in $\delta^{13}C$, $\delta^{15}N$, and δD and attempted to relate stable-isotopic composition to structural moieties within the insoluble organic matter (summarized in Figure 6), based on the observation that aliphatic-rich organic matter combusts at lower temperatures than aromatic-rich. High-sensitivity stable-isotope measurements combined with stepped combustion (Pillinger, 1984, 1987) have been extensively applied to the study of presolar grains in meteorites, and Alexander et al. (1998) applied this technique to the study of nitrogen and carbon abundances and isotopic compositions of acid-insoluble carbonaceous material in 13 chondritic meteorites. They found a range in $\delta^{15}N$ values for organic material from $-40‰$ to $+260‰$, with the most anomalous nitrogen being associated with the petrologically most primitive meteorites. This suggested that two basic nitrogen-containing components were present within the organic material: one with $\delta^{15}N$ values of between ~0‰ and $-40‰$ and a second ^{15}N-enriched component with $\delta^{15}N$ values $>+260‰$. The ^{15}N-enriched component was interpreted as being presolar and comprising 40–70% of the total nitrogen released in the experiment.

The combined use of pyrolysis techniques followed by CSIA measurements (both offline hydrous pyrolysis and online pyrolysis-GC-IRMS) enables the isotopic compositions of fragments of the macromolecular material to be determined. This approach has been applied to the study of the stable carbon isotope distribution of insoluble organic matter in three carbonaceous chondrites: Orgueil (CI1), Murchison (CM2), and Cold Bokkeveld (CM2) (Sephton, 1998, 2000; Sephton and Gilmour, 2001a), and both isotopic and structural information has been obtained for the macromolecular material in these meteorites (Table 8 and Figure 7). Relatively large and systematic differences in $\delta^{13}C$ values are observed for molecules that differ by only one or two carbon atoms, with compounds from Cold Bokkeveld displaying the widest range in $\delta^{13}C$ values (~27‰). Systematic differences in $\delta^{13}C$ values are apparent with increasing carbon number: $\delta^{13}C$ values in Murchison become more positive with increasing carbon number, while $\delta^{13}C$ values in Cold Bokkeveld become more negative with increasing carbon number above C_8, a similar trend is observed in Orgueil. There are strong similarities in $\delta^{13}C$ values between free and macromolecular aromatic moieties in Murchison, and, to a lesser extent, Cold Bokkeveld, suggesting a genetic relationship between the two. Based on these data, Sephton et al. (2000) advocate that the free aromatic compounds in these meteorites were derived from macromolecular material via parent body processes. The systematic shift in $\delta^{13}C$ values with carbon number have been interpreted as resulting from kinetic isotope effects during bond formation and destruction in the aromatic carbon skeletons of these compounds with the relatively large differences in $\delta^{13}C$ values, indicating that significant isotopic fractionations have occurred such as might be expected in low-temperature interstellar environments (Sephton and Gilmour, 2000). Comparing data for Murchison in Figures 3 and 7, it is apparent that two trends are evident: $\delta^{13}C$ values for low-molecular-weight compounds

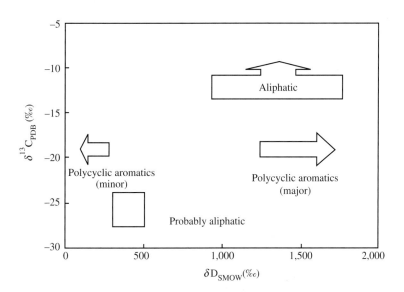

Figure 6 Carbon and hydrogen stable-isotope compositions of discrete moieties identified using stepped combustion of macromolecular material from the Murchison (CM2) chondrite (source Kerridge et al., 1987).

Table 8 Carbon stable isotope compositions of individual aromatic molecules released by the pyrolysis of macromolecular materials in carbonaceous chondrites.

Compound	Orgueil (CI1)[a] $\delta^{13}C_{PDB}$ (‰)	Murchison (CM2) $\delta^{13}C_{PDB}$ (‰)	Cold Bokkeveld (CM2)[b] $\delta^{13}C_{PDB}$ (‰)
Toluene	-25.0 ± 1.1	-24.6 ± 0.2[b]	-22.9 ± 2.3
		-1.3 ± 2.0[c]	
		-5.4[c]	
Ethylbenzene	-10.0 ± 0.6	-21.9 ± 1.4[b]	$+2.5 \pm 2.3$
m-Xylene	-17.0 ± 0.5	-19.6[b]	$+4.0 \pm 3.8$
		-21.7 ± 0.4[c]	
		-20.3[c]	
p-Xylene	-14.0 ± 1.5	-17.8[b]	$+2.4 \pm 0.3$
C_3-alkylbenzene	-22.5 ± 0.7	-18.1[b]	-1.0 ± 2.0
Benzaldehyde		-23.7 ± 0.6[c]	
		-24.0 ± 0.3[c]	
Phenol		-24.1[b]	-14.0 ± 1.5
C_3-alkylbenzene	-14.5 ± 1.8		-8.0 ± 0.4
C_4-alkylbenzene	-23.6 ± 1.1		-12.2 ± 1.1
2-Methylphenol	-12.8 ± 0.9	-10.3[b]	-3.9 ± 2.2
3-Methylphenol		-10.4[b]	-3.4
		-13.9 ± 0.9[c]	
		-18.5 ± 1.9[c]	
Naphthalene	-14.3 ± 1.4	-6.5 ± 2.5[b]	-6.1 ± 1.3
		-5.5 ± 0.5[c]	
		-6.4 ± 0.8[c]	
Benzothiophene		-15.8[b]	
2-Methylnaphthalene	-17.5 ± 1.5	-5.6 ± 2.1[b]	-12.1
		-6.4 ± 0.8[c]	
1-Methylnaphthalene	-18.7 ± 0.4	-7.2 ± 2.0[b]	-15.1
		-7.1[c]	
Acenaphthene		-5.9 ± 1.7[b]	

[a] Sephton et al. (2000). [b] Sephton et al. (1998). [c] Sephton and Gilmour (2001b).

become more positive with increasing carbon number, while $\delta^{13}C$ values for high-molecular-weight compounds become more negative, i.e., signatures of both bond formation (synthesis of higher homologs from lower ones) and bond destruction (cracking of higher homologs to form lower ones). Sephton and Gilmour (2000) suggest that these trends imply that the carbon skeletons were produced in the restrictive environment of the icy organic-rich mantles of interstellar grains where radiation-induced reactions simultaneously create and destroy organic matter to produce material with an intermediate level of complexity. Once formed, these complex organic residues would be available to participate in the formation of the carbonaceous chondrites following the collapse of the interstellar cloud.

1.10.5 *IN SITU* EXAMINATION OF METEORITIC ORGANIC MATTER

Early attempts at the examination of organic matter *in situ* in meteorites employed fluorescence and suggested that organic matter was coated on the surfaces of mineral grains (Alpern and Benkeiri, 1973). More recently, Pearson et al. (2002) used an organic labeling technique to map the distribution of organic matter in the Murchison (CM2), Ivuna (CI1), Orgueil (CI1), and Tagish Lake chondrites. A strong association was observed between the distribution of organic matter and hydrous clay minerals, suggesting that the production of clays by aqueous processes influenced the distribution of organic matter in meteorites.

The development of two-step laser mass spectrometry in which organic molecules are first desorbed and then ionized using a laser before detection in a mass spectrometer led to the first *in situ* identifications of PAHs in chondrites (Hahn et al., 1988; Kovalenko et al., 1991, 1992). The high spatial resolution of this approach has enabled its application to the study of organic matter in interplanetary dust particles collected by aircraft and in micrometeoritic material recovered from Antarctica. Clemett et al. (1993) examined eight stratospherically collected interplanetary dust particles and identified PAHs and their alkylated derivatives including high-mass PAHs not observed in similar studies of meteorites. The same approach has been used to recognize apparently indigenous PAHs in Antarctic micrometeorites (Clemett et al., 1998). The apparent lack of a spatial relationship between organic compounds and carbonate minerals in the martian meteorite ALH 84001 has been used to infer that PAHs present in that meteorite are terrestrial in origin and not a remnant of martian biogenic activity (Stephan et al., 2003).

Messenger et al. (1998) used two-step laser mass spectrometry to identify PAHs in individual circumstellar graphite grains extracted from the Murchison (CM2) and Acfer 094 meteorites. Some 70% of the grains studied had appreciable concentrations of PAHs (500–5,000 ppm), and in several cases correlated isotopic anomalies were observed between PAHs (phenanthrene, C_1 and C_2 alkyl phenanthrene, chrysene, and C_3 alkyl chrysene) and their parent grains. These correlations were most evident for ^{13}C-depleted grains. This isotopic linkage between specific molecules and the circumstellar grains they are associated with, suggests a genetic relationship between the two, and the authors suggested that PAH-like material was produced in the gas phase prior to the formation of graphite in a circumstellar environment.

1.10.6 ENVIRONMENTS OF FORMATION

Hypotheses for the origin of meteoritic organic matter must account for its molecular and isotopic composition and be consistent with models of meteorite petrogenesis; consequently, a number of potential environments have been considered (Table 9). Until the early 1990s, the favored hypothesis involved the catalytic hydrogenation of CO in the solar nebula. However, a characteristic of such catalytic reactions is their structural selectivity. FTT synthesis, in particular, produces a structurally selective suite of hydrocarbons and other compounds that, initially, were believed to

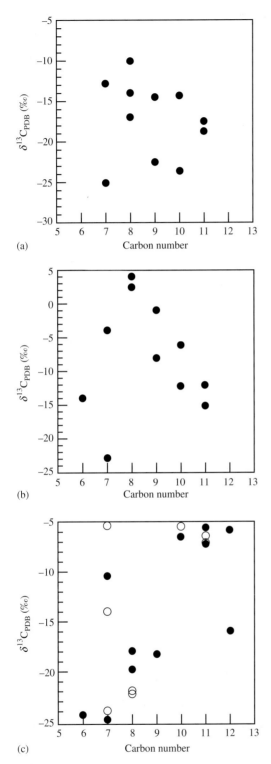

Figure 7 Carbon stable-isotope compositions of individual aromatic and polycyclic aromatic compounds, plotted against carbon number. Compounds were obtained by hydrous pyrolysis of macromolecular material from: (a) Orgueil (CI), (b) Cold Bokkeveld (CM2), and (c) by hydrous pyrolysis (closed symbols) and online pyrolysis GC-IRMS (open symbols) of Murchison (CM2) (sources (a) Sephton et al., 2000; (b) Sephton et al., 1998; (c) Sephton and Gilmour, 2001a).

Table 9 Sources and processes potentially involved in the production of meteoritic organic matter.

Ion–molecule reactions in interstellar clouds
Radiation chemistry in interstellar grain mantles
Condensation in stellar outflows
Equilibrium reactions in the solar nebula
Surface catalysis (Fischer–Tropsch) in the solar nebula
Kinetically controlled reactions in the solar nebula
Radiation chemistry (Miller–Urey) in the nebula
Photochemistry in nebular surface regions
Liquid-phase reactions on parent asteroid
Surface catalysis (Fischer–Tropsch) on asteroid
Radiation chemistry (Miller–Urey) in asteroid atmosphere
Thermal processing during asteroid metamorphism

Source: Kerridge (1999).

closely resemble those observed in meteorites (Studier et al., 1968; Hayatsu et al., 1971, 1972; Yoshino et al., 1971). However, the recognition of structural diversity in indigenous hydrocarbons in Murchison (Cronin and Pizzarello, 1990) together with inability of the isotopic fractionations associated with FTT reactions (Yuen et al., 1990) to explain the isotopic distributions observed in meteorites indicates that catalytic synthesis is not the primary process involved in the synthesis of meteoritic organics. Catalytic reactions such as FTT also have problems in accounting for the preservation of deuterium-rich interstellar material in meteorites, and petrographic evidence indicates that the catalysts necessary to trigger the reaction in the primitive solar nebula were formed much later on the meteorite parent body (Kerridge et al., 1979; Bunch and Chang, 1980). This all suggests that events in the solar nebula and on the meteorite parent body are more likely to amount to the secondary processing of pre-existing interstellar organic matter than primary synthesis. It is worth noting, however, that FTT reactions primarily fail to account for the properties of solvent-extractable organic matter in the Murchison meteorite, which does not exclude this process having operated elsewhere in the early solar system. Fischer–Tropsch is the only feasible thermally driven pathway available in the solar nebula to convert CO into other forms of carbon, and models suggest that the process may have been most efficient in the vicinity of the nebula equivalent to the present position of the asteroid belt (Kress and Tielens, 2001). Further studies of catalytic synthesis reactions using catalysts appropriate to specific nebular and asteroidal environments may indicate that such processes remain a significant mechanism for the production of extraterrestrial organic matter (Llorca and Casanova, 2000).

It has become increasingly apparent that the production of meteoritic organic matter must have involved a combination of different processes taking place in a wide range of environments. The detailed structural and isotopic analysis of meteoritic organic matter reviewed in this chapter has led to the development of new models for the origin of organic compounds in meteorites that envisage a distinctly different set of processes from the synthesis models proposed by earlier workers (e.g., Yoshino et al., 1971; Miller et al., 1976). Foremost has been the development of the so-called interstellar parent-body hypothesis. Petrographic evidence indicates that carbonaceous chondrites experienced a period of aqueous activity which, when combined with evidence for interstellar organics, led to the development of a model in which meteoritic organic matter is envisaged as being the product of parent body aqueous and thermal processing of reactive, volatile precursors such as water, HCN, NH_3, and ^{13}C-, ^{15}N-, and D-rich interstellar organics (Cronin and Chang, 1993). The model has superceded earlier ones that envisaged a primarily nebular origin for meteoritic organic matter on the strength of its ability to account for a number of the features common to meteoritic organics that have been reviewed in this chapter. These include:

(i) The proportion of amino to hydroxy acids observed in carbonaceous chondrites broadly matches that predicted by the Strecker cyanohydrin synthesis, a set of aqueous reactions in which cyanide, ammonia, and aldehydes and/or ketones are converted to amino acids and hydroxy acids (Peltzer et al., 1984). There is also a rough correlation between abundance and the extent of aqueous alteration (Cronin, 1989), although the abundances of amino acids in meteorites are much lower than those predicted by a Strecker model (Cronin and Chang, 1993).

(ii) The structural diversity observed in virtually all classes of organic compounds, a predominance of branched-chain isomers and an exponential decline in abundance with increasing molecular weight. These observations are not consistent with structurally selective syntheses, but rather suggest the random production of precursor compounds.

(iii) Systematic decreases in $\delta^{13}C$ values with increasing molecular weight indicative of the synthesis of higher-molecular-weight homologs from lower ones (i.e., carbon chains are constructed by the addition of C atoms). Large differences in $\delta^{13}C$ values between deuterium-enriched substantial molecules such as PAHs indicate that such species were synthesized under conditions, such as the low temperatures of dense interstellar cloud environments, where isotopic fractionations were maximized. (Sephton and Gilmour, 2000). Some PAHs also appear to have been derived from circumstellar environments (Messenger et al., 1998).

(iv) Deuterium enrichments in a number of the organic constituents, e.g., amino acids, volatile bases, and components of the macromolecular material. This deuterium enrichment is believed to a signature of ion–molecule reactions in the interstellar medium (Kolodny et al., 1980; Robert and Epstein, 1982; Kerridge, 1983). Ionization of simple gaseous compounds in interstellar clouds (e.g., CH_4, CH_2O, H_2O, N_2, and NH_3) by cosmic radiation leads to fragmentation and the production of ions that react with neutral molecules to produce deuterium-enriched products. Continued ion–molecule interactions are thought to produce increasingly deuterium-enriched and complex organic matter that condenses on to dust grains (Robert and Epstein, 1982). Within interstellar clouds simple gas-phase molecules condense as icy mantles around silicate dust grains (Greenburg, 1984). When these mantles are subjected to increased temperatures or periods of UV radiation, thermally or photolytically driven polymerization reactions occur, which results in complex organic products that may also be deuterium-enriched (Sandford, 1996). Once formed, these deuterium-enriched complex organic residues would be available to participate in the formation of the carbonaceous chondrites following the collapse of the interstellar cloud.

(v) The widespread distribution, in both our own and other galaxies, of a 3.4 μm absorption feature attributed to saturated aliphatic hydrocarbons, suggesting that the organic component of interstellar dust is available for incorporation in newly forming planetary systems (e.g., Pendleton et al., 1994; Pendleton, 1995).

(vi) A physical, not biological, enantiomeric excess. The cause of the chiral excess is not known, although the selective destruction of one enantiomer by UV circularly polarized light (UVCPL) from neutron stars in a presolar cloud has been invoked (Bonner and Rubenstein, 1987). The infall of icy/organic interstellar grains into the protosolar nebula would have resulted in their eventual incorporation into progressively larger planetesimals, culminating in the formation of the asteroidal parent bodies of the carbonaceous chondrites. The internal heating of the Murchison parent body by the decay of short-lived nuclides is thought to have initiated periods of aqueous alteration as liquid water was present at temperatures of less than 20 °C (Clayton and Mayeda, 1984). This would have resulted in the reaction of interstellar molecules with water and with each other to form more complex organics. Further organic synthesis could have occurred as hot fluids reacted on the surfaces of mineral catalysts during transport from the interior of the parent body. Continued heating of simpler organic molecules may then have resulted in the production of the macromolecular organic matter. Alternatively, it has also been suggested that the macromolecular material could have been produced in the solar nebula via the gas-phase pyrolysis of simple interstellar cloud-derived hydrocarbons (e.g., C_2H_2, CH_4) at temperatures of 900–1,100 K (Morgan et al., 1991) followed by polymerization to form of high-molecular-weight aromatic structures.

REFERENCES

Alexander C. M. O'D., Russell S. S., Arden J. W., Ash R. D., Grady M. M., and Pillinger C. T. (1998) The origin of chondritic macromolecular organic matter: a carbon and nitrogen isotope study. *Meteorit. Planet. Sci.* **33**, 603–622.

Alpern B. and Benkeiri Y. (1973) Distribution de la matière organique de la météorite d'Orgueil par microscopie en fluorescence. *Earth. Planet. Sci. Lett.* **19**, 422–428.

Anders E. and Kerridge J. F. (1988) Future directions in meteorite research. In *Meteorites and the Early Solar System* (eds. J. F. Kerridge and M. S. Matthews). University of Arizona Press, Tucson, pp. 1149–1154.

Anders E., DuFresne E. R., Hayatsu R., Cavaille A., DuFresne A., and Fitch F. W. (1964) Contaminated meteorite. *Science* **146**, 1157–1161.

Anders E., Hayatsu R., and Studier M. H. (1973) Organic compounds in meteorites. *Science* **182**, 781–789.

Bada J. L., Cronin J. R., Ho M. S., Kvenvolden K. A., Lawless J. G., Miller S. L., Oro J., and Steinberg S. (1983) On the reported optical activity of amino acids in the Murchison meteorite. *Nature* **301**, 494–496.

Bandurski E. L. and Nagy B. (1976) The polymer-like material in the Orgueil meteorite. *Geochim. Cosmochim. Acta* **40**, 1397–1406.

Basile B. P., Middleditch B. S., and Oró J. (1984) Polycyclic aromatic hydrocarbons in the Murchison meteorite. *Org. Geochem.* **5**, 211–216.

Becker R. H. and Epstein S. (1982) Carbon, hydrogen, and nitrogen isotopes in solvent-extractable organic matter from carbonaceous chondrites. *Geochim. Cosmochim. Acta* **46**, 97–103.

Biemann K. (1974) Test result on the Viking gas chromatograph-mass spectrometer experiment. *Origins Life* **5**, 417–430.

Binet L., Gourier D., Derenne S., and Robert F. (2002) Heterogeneous distribution of paramagnetic radicals in insoluble organic matter from the Orgueil and Murchison meteorites. *Geochim. Cosmochim. Acta* **66**, 4177–4186.

Boato G. (1954) The isotopic composition of hydrogen and carbon in the carbonaceous chondrites. *Geochim. Cosmochim. Acta* **6**, 209–220.

Bonner W. A. and Rubenstein E. (1987) Supernovae, neutron stars and biomolecular chirality. *Biosystems* **20**, 99–111.

Bunch T. E. and Chang S. (1980) Carbonaceous chondrites: II. Carbonaceous chondrite phyllosilicates and light element geochemistry as indicators of parent body processes and surface conditions. *Geochim. Cosmochim. Acta* **44**, 1543–1577.

Chang S., Mack R., and Lennon K. (1978) Carbon chemistry of separated phases of Murchison and Allende. In *Lunar Planet. Sci.* **IX**. The Lunar and Planetary Institute, Houston, pp. 157–159.

Clayton R. N. and Mayeda T. K. (1984) The oxygen isotope record in Murchison and other carbonaceous chondrites. *Earth Planet. Sci. Lett.* **67**, 151–161.

Clemett S. J., Maechling C. R., Zare R. N., Swan P. D., and Walker R. M. (1993) Identification of complex aromatic molecules in individual interplanetary dust particles. *Science* **262**, 721–725.

Clemett S. J., Chillier X. D. F., Gillette S., Zare R. N., Maurette M., Engrand C., and Kurat G. (1998) Observation of indigenous polycyclic aromatic hydrocarbons in 'giant' carbonaceous Antarctic micrometeorites. *Origins Life Evol. Biosphere* **28**, 425–448.

Cody G. D., Alexander C. M. O'D., and Tera F. (2002) Solid-state (^1H and ^{13}C) nuclear magnetic resonance spectroscopy of insoluble organic residue in the Murchison meteorite: a self-consistent quantitative analysis. *Geochim. Cosmochim. Acta* **66**, 1851–1865.

Cooper G. and Cronin J. R. (1995) Linear and cyclic aliphatic carboxamides of the Murchison meteorite; hydrolyzable derivatives of amino acids and other carboxylic acids. *Geochim. Cosmochim. Acta* **59**, 1003–1015.

Cooper G., Kimmich N., Belisle W., Sarinana J., Brabham K., and Garrel L. (2001) Carbonaceous meteorites as a source of sugar-related organic compounds for the early Earth. *Nature* **414**, 879–882.

Cooper G. W., Onwo W. M., and Cronin J. R. (1992) Alkyl phosphonic acids and sulfonic acids in the Murchison meteorite. *Geochim. Cosmochim. Acta* **56**, 4109–4115.

Cooper G. W., Thiemens M. H., Jackson T. L., and Chang S. (1997) Sulfur and hydrogen isotope anomalies in meteorite sulfonic acids. *Science* **277**, 1072–1074.

Cronin J. R. (1976a) Acid-labile amino acid precursors in the Murchison meteorite: I. Chromatographic fractionation. *Origins Life* **7**, 337–342.

Cronin J. R. (1976b) Acid-labile amino acid precursors in the Murchison meteorite: II. A search for peptides and amino acyl amindes. *Origins Life* **7**, 343–348.

Cronin J. R. (1989) Origin of organic compounds in carbonaceous chondrites. *Adv. Space. Res.* **9**, 59–64.

Cronin J. R. and Chang S. (1993) Organic matter in meteorites: molecular and isotopic analysis of the Murchison meteorite. In *The Chemistry of Life's Origins* (ed. J. M. Greenberg). Kluwer, Boston, pp. 209–258.

Cronin J. R. and Moore C. B. (1976) Amino acids of the Nogoya and Mokoia carbonaceous chondrites. *Geochim. Cosmochim. Acta* **40**, 853–857.

Cronin J. R. and Pizzarello S. (1986) Amino acids of the Murchison meteorite: 3. 7 carbon acyclic primary alpha-amino alkanoic acids. *Geochim. Cosmochim. Acta* **50**, 2419–2427.

Cronin J. R. and Pizzarello S. (1990) Aliphatic hydrocarbons of the Murchison meteorite. *Geochim. Cosmochim. Acta* **54**, 2859–2868.

Cronin J. R. and Pizzarello S. (1997) Enantiomeric excesses in meteoritic amino acids. *Science* **275**, 951–955.

Cronin J. R. and Pizzarello S. (1999) Amino acid enantiomer excesses in meteorites: origin and significance. *Adv. Space Res.* **23**, 293–299.

Cronin J. R., Gandy W. E., and Pizzarello S. (1981) Amino acids of the Murchison meteorite: 1. 6 carbon acyclic primary alpha-amino alkanoic acids. *J. Molec. Evol.* **17**, 265–272.

Cronin J. R., Pizzarello S., and Yuen G. U. (1985) Amino acids of the Murchison meteorite: 2. 5 carbon acyclic primary beta-amino, gamma-amino and delta-amino alkanoic acids. *Geochim. Cosmochim. Acta* **49**, 2259–2265.

Cronin J. R., Pizzarello S., and Fyre J. S. (1987) ^{13}C NMR spectroscopy of the insoluble carbon of carbonaceous chondrites. *Geochim. Cosmochim. Acta* **51**, 229–303.

Cronin J. R., Pizzarello S., and Cruikshank D. P. (1988) Organic matter in carbonaceous chondrites, planetary satellites, asteroids and comets. In *Meteorites and the Early Solar System* (eds. J. F. Kerridge and M. S. Matthews). University of Arizona Press, Tucson, pp. 819–857.

Cronin J. R., Pizzarello S., Epstein S., and Krishnamurthy R. V. (1993) Molecular and isotopic analyses of the hydroxy-acids, dicarboxylic-acids, and hydroxydicarboxylic acids of the Murchison meteorite. *Geochim. Cosmochim. Acta* **57**, 4745–4752.

Cronin J. R., Cooper G., and Pizzarello S. (1995) Characteristics and formation of amino acids and hydroxy acids of the Murchison meteorite. *Adv. Space Res.* **3**, 91–97.

Degens E. T. and Bajor M. (1962) Amino acids and sugars in the Brudeheim and Murray meteorites. *Naturwiss.* **49**, 605–606.

Ehrenfreund P., Glavin D. P., Botta O., Cooper G., and Bada J. L. (2001) Extraterrestrial amino acids in Orgueil and Ivuna: tracing the parent body of CI type carbonaceous chondrites. *Proc. Natl. Acad. Sci.* **98**, 2138–2141.

Engel M. H. and Macko S. A. (1997) Isotopic evidence for extraterrestrial non-racemic amino acids in the Murchison meteorite. *Nature* **389**, 265–267.

Engel M. H. and Nagy B. (1982) Distribution and enantiomeric composition of amino acids in the Murchison meteorite. *Nature* **296**, 837–840.

Engel M. H., Macko S. A., and Silfer J. A. (1990) Carbon isotope composition of individual amino acids in the Murchison meteorite. *Nature* **348**, 47–49.

Epstein S., Krishnamurthy R. V., Cronin J. R., Pizzarello S., and Yuen G. U. (1987) Unusual stable isotope ratios in amino acid and carboxylic acid extracts from the Murchison meteorite. *Nature* **326**, 477–479.

Fitch F. W. and Anders E. (1963) Observations on the nature of the organized elements in carbonaceous chondrites. *Ann. NY Acad. Sci.* **108**, 495–513.

Fitch F. W., Schwarcz H. P., and Anders E. (1962) 'Organized elements' in carbonaceous chondrites. *Nature* **193**, 1123–1125.

Gardinier A., Derenne S., Robert F., Behar F., Largeau C., and Maquet J. (2000) Solid state CP/MAS ^{13}C NMR of the insoluble organic matter of the Orgueil and Murchison meteorites: quantitative study. *Earth. Planet. Sci. Lett.* **184**, 9–21.

Gilmour I. and Pillinger C. T. (1994) Isotopic compositions of individual polycyclic aromatic hydrocarbons from the Murchison meteorite. *Mon. Not. Roy. Astron. Soc.* **269**, 235–240.

Greenburg J. M. (1984) Chemical evolution in space. *Origins Life* **14**, 25–36.

Hahn J. H., Zenobi R., Bada J. L., and Zare R. N. (1988) Application of two-step laser mass spectrometry to cosmogeochemistry: direct analysis of meteorites. *Science* **239**, 1523–1525.

Han J., Simoneit B. R., Burlingame A. L., and Calvin M. (1969) Organic analysis on the Pueblito de Allende meteorite. *Nature* **222**, 364–365.

Hayatsu R. (1964) Orgeuil meteorite: organic nitrogen contents. *Science* **146**, 1291–1293.

Hayatsu R. (1965) Optical activity in the Orgueil meteorite. *Science* **149**, 443–447.

Hayatsu R. and Anders E. (1981) Organic compounds in meteorites and their origins. *Top. Curr. Chem.* **99**, 1–37.

Hayatsu R., Studier M. H., Oda A., Fuse K., and Anders E. (1968) Origin of organic matter in early solar system: II. Nitrogen compounds. *Geochim. Cosmochim. Acta* **32**, 175–190.

Hayatsu R., Studier M. H., and Anders E. (1971) Origin of organic matter in early solar system: IV. Amino acids: confirmation of catalytic synthesis by mass spectrometry. *Geochim. Cosmochim. Acta* **35**, 939–951.

Hayatsu R., Studier M. H., Matsuoka S., and Anders E. (1972) Origin of organic matter in early solar system: VI. Catalytic synthesis of nitriles, nitrogen bases and porphyrin-like pigments. *Geochim. Cosmochim. Acta* **36**, 555–571.

Hayatsu R., Studier M. H., Moore L. P., and Anders E. (1975) Purines and triazines in the Murchison meteorite. *Geochim. Cosmochim. Acta* **39**, 471–488.

Hayatsu R., Matsuoka S., Scott R. G., Studier M. H., and Anders E. (1977) Origin of organic matter in the early solar system: VII. The organic polymer in carbonaceous chondrites. *Geochim. Cosmochim. Acta* **41**, 1325–1339.

Hayatsu R., Scott R. G., Studier M. H., Lewis R. S., and Anders E. (1980) Carbynes in meteorites: detection, low temperature origin and implications for interstellar molecules. *Science* **209**, 1515–1518.

Hayes J. M. (1967) Organic constituents of meteorites—a review. *Geochim. Cosmochim. Acta* **31**, 1395–1440.

Hayes J. M. and Biemann K. (1968) High resolution mass spectrometric investigations of the organic constituents of the Murray and Holbrook chondrites. *Geochim. Cosmochim. Acta* **32**, 239–269.

Hayes J. M., Freeman K. H., Popp B. N., and Hoham C. H. (1990) Compound-specific isotopic analyses—a novel tool for reconstruction of ancient biogeochemical processes. *Org. Geochem.* **16**, 1115–1128.

Holtzer G. and Oró J. (1977) Pyrolysis of organic compounds in the presence of ammonia: the Viking Mars Lander site alteration experiment. *Org. Geochem.* **1**, 37–52.

Jungclaus G., Cronin J. R., Moore C. B., and Yuen G. U. (1976a) Aliphatic amines in the Murchison meteorite. *Nature* **261**, 126–128.

Jungclaus G. A., Yuen G. U., Moore C. B., and Lawless J. G. (1976b) Evidence for the presence of low molecular weight alcohols and carbonyl compounds in the Murchison meteorite. *Meteoritics* **11**, 231–237.

Kaplan I. R., Degens E. T., and Reuter J. H. (1963) Organic compounds in stony meteorites. *Geochim. Cosmochim. Acta* **27**, 805–834.

Kerridge J. F. (1983) Isotopic composition of carbonaceous-chondrite kerogen: evidence for an interstellar origin of organic matter in meteorites. *Earth. Planet. Sci. Lett.* **64**, 186–200.

Kerridge J. F. (1985) Carbon, hydrogen, and nitrogen in carbonaceous chondrites—abundances and isotopic compositions in bulk samples. *Geochim. Cosmochim. Acta* **49**, 1707–1714.

Kerridge J. F. (1999) Formation and processing of organics in the early solar system. *Space Sci. Rev.* **90**, 275–288.

Kerridge J. F., Mackay A. L., and Boynton W. V. (1979) Magnetite in CI carbonaceous chondrites: origin by aqueous activity on a planetesimal surface. *Science* **205**, 395–397.

Kerridge J. F., Chang S., and Shipp R. (1987) Isotopic characterization of kerogen-like material in the Murchison carbonaceous chondrite. *Geochim. Cosmochim. Acta* **51**, 2527–2540.

Kitajima F., Nakamura T., Takaoka N., and Murae T. (2002) Evaluating the thermal metamorphism of CM chondrites by using the pyrolytic behavior of carbonaceous macromolecular matter. *Geochim. Cosmochim. Acta* **66**, 163–172.

Kminek G., Botta O., Glavin D. P., and Bada J. L. (2002) Amino acids in the Tagish Lake meteorite. *Meteorit. Planet. Sci.* **37**, 697–701.

Kolodny Y., Kerridge J. F., and Kaplan I. R. (1980) Deuterium in carbonaceous chondrites. *Earth Planet. Sci. Lett.* **46**, 149–158.

Kovalenko L. J., Philippoz J. M., Bucenell J. R., Zenobi R., and Zare R. N. (1991) Organic chemical analysis on a microscopic scale using two-step laser desorption/laser ionization mass spectrometry. *Space Sci. Rev.* **56**, 191–195.

Kovalenko L. J., Maechling C. R., Clemett S. J., Philippoz J. M., Zare R. N., and Alexaner C. M. O. D. (1992) Microscopic organic analysis using two step laser mass spectrometry: application to meteoritic acid residues. *Anal. Chem.* **64**, 682–690.

Kress M. E. and Tielens A. G. G. M. (2001) The role of Fischer-Tropsch catalysis in solar nebular chemistry. *Meteorit. Planet. Sci.* **36**, 75–91.

Krishnamurthy R. V., Epstein S., Cronin J. R., Pizzarello S., and Yuen G. U. (1992) Isotopic and molecular analyses of hydrocarbons and monocarboxylic acids of the Murchison meteorite. *Geochim. Cosmochim. Acta* **56**, 4045–4058.

Krouse H. R. and Modzeleski V. E. (1970) $^{13}C/^{12}C$ abundances in components of carbonaceous chondrites and terrestrial samples. *Geochim. Cosmochim. Acta* **34**, 459–474.

Kung C. C. and Clayton R. N. (1978) Nitrogen abundances and isotopic composition in stony meteorites. *Earth Planet. Sci. Lett.* **38**, 421–435.

Kvenvolden K., Lawless J., Peterson E., Flors J., Ponnamperuma C., Kaplan I. R., and Moore C. (1970) Evidence for extraterrestrial amino acids and hydrocarbons in the Murchison meteorite. *Nature* **228**, 923–936.

Lawless J. G. and Yuen G. U. (1979) Quantification of monocarboxylic acids in the Murchison carbonaceous meteorite. *Nature* **282**, 396–398.

Lawless J. G., Kvenvolden K. A., Peterson E., Ponnamperuma C., and Jarosewich E. (1972) Evidence for amino-acids of extraterrestrial origin in the Orgueil meteorite. *Nature* **236**, 66–67.

Lawless J. G., Peterson E., and Kvenvolden K. A. (1973) Amino acids in meteorites. *Astrophys. Space Sci. Lib.* **40**, 167–168.

Lawless J. G., Zeitman B., Pereira W. E., Summons R. E., and Duffield A. M. (1974) Dicarboxylic acids in the Murchison meteorite. *Nature* **251**, 40–42.

Levy R. L., Grayson M. A., and Wolf C. J. (1973) The organic analysis of the Murchison meteorite. *Geochim. Cosmochim. Acta* **37**, 467–483.

Llorca J. and Casanova I. (2000) Reaction between H_2, CO, and H_2S over Fe, Ni metal in the solar nebula: experimental evidence for the formation of sulfur bearing organic molecules and sulfides. *Meteorit. Planet. Sci.* **35**, 841–848.

Meinschein W. G., Nagy B., and Hennessy D. J. (1963) Evidence in meteorites of former life. *Ann. NY Acad. Sci.* **108**, 553–579.

Messenger S., Amari S., Gao X., Walker R. M., Clemett S. J., Chillier X. D. F., Zare R. N., and Lewis R. S. (1998) Indigenous polycyclic aromatic hydrocarbons in circumstellar graphite grains from primitive meteorites. *Astrophys. J.* **501**, 284–295.

Miller S. L., Urey H. C., and Oró J. (1976) Origin of organic compounds on the primitive Earth and in meteorites. *J. Molec. Evol.* **9**, 59–72.

Morgan W. A., Feigelson E. D., Wang H., and Frenklach M. (1991) A new mechanism for the formation of meteoritic kerogen-like material. *Science* **252**, 109–112.

Mullie F. and Reisse J. (1987) Organic matter in carbonaceous chondrites. *Top. Curr. Chem.* **137**, 83–117.

Murae T. (1995) Characterization of extraterrestrial high-molecular-weight organic-matter by pyrolysis-gas chromatography mass-spectrometry. *J. Anal. Appl. Pyrol.* **32**, 65–73.

Nagy B., Meinschein W. G., and Hennessy D. J. (1961) Mass spectrometric analysis of the Orgueil meteorite: evidence for biogenic hydrocarbons. *Ann. NY Acad. Sci.* **93**, 534–552.

Nagy B., Claus G., Colombo U., Gazzarrini F., Modzeleski V. E., Murphy M. T. J., and Rouser G. (1964) Optical activity in saponified organic matter isolated from the interior of the Orgueil meteorite. *Nature* **202**, 228–233.

Nagy B. J. (1975) *Carbonaceous Meteorites*. Elsevier, Amsterdam.

Naraoka H., Shimoyama A., Komiya M., Yamamoto H., and Harada K. (1987) Carboxylic acids and hydrocarbons in Antarctic carbonaceous chondrites. In *Twelfth Symposium on Antarctic Meteorites*. National Institute of Polar Research, Tokyo, pp. 9–11.

Naraoka H., Shimoyama A., Komiya M., Yamamoto H., and Harada K. (1988) Hydrocarbons in the Yamato-791198 carbonaceous chondrite from Antarctica. *Chem. Lett.* **1988**, 831–934.

Naraoka H., Shimoyama A., and Harada K. (1999) Molecular distribution of monocarboxylic acids in Asuka chondrites from Antarctica. *Origins Life Evol. Biosphere* **29**, 187–201.

Naraoka H., Shimoyama A., and Harada K. (2000) Isotopic evidence from an Antarctic carbonaceous chondrite for two reaction pathways of extraterrestrial PAH formation. *Earth Planet. Sci. Lett.* **184**, 1–7.

Oró J., Nooner D. W., Zlatkis A., and Wisktrom S. A. (1966) Paraffinic hydrocarbons in Orgueil, Murray, Mokoia, and other meteorites. *Life Sci. Space Res.* **4**, 63–100.

Oró J., Gibert J., Lichtenstein H., Wikstrom S., and Flory D. A. (1971) Amino-acids, aliphatic and aromatic hydrocarbons in the Murchison meteorite. *Nature* **230**, 105–106.

Pearson V. K., Sephton M. A., Kearsley A. T., Bland P. A., Franchi I. A., and Gilmour I. (2002) Clay mineral-organic matter relationships in the early solar system. *Meteorit. Planet. Sci.* **37**, 1829–1833.

Peltzer E. T. and Bada J. L. (1978) α-Hydroxycarboxylic acids in the Murchison meteorite. *Nature* **272**, 443–444.

Peltzer E. T., Bada J. L., Schlesinger G., and Miller S. L. (1984) The chemical conditions on the parent body of the Murchison meteorite: some conclusions based on amino, hydroxy and dicarboxylic acids. *Adv. Space Res.* **4**, 69–74.

Pendleton Y. J. (1995) Laboratory comparisons of organic materials to interstellar dust and the Murchison meteorite. *Planet. Space Sci.* **43**, 1359–1364.

Pendleton Y. J., Sandford S., Allamandola L., Tielens A. G. G. M., and Sellgren K. (1994) Near infrared absorption spectroscopy of interstellar hydrocarbon grains. *Astrophys. J.* **437**, 683–696.

Pering K. L. and Ponnamperuma C. (1971) Aromatic hydrocarbons in the Murchison meteorite. *Science* **173**, 237–239.

Pillinger C. (1982) Not quite full circle? Non-racemic amino acids in the Murchison meteorite. *Nature* **296**, 802.

Pillinger C. T. (1984) Light element stable isotopes in meteorites; from grams to picograms. *Geochim. Cosmochim. Acta* **48**, 2739–2766.

Pillinger C. T. (1987) Stable isotope measurements of meteorites and cosmic dust particles. *Phil. Trans. Roy. Soc. London A* **323**, 313–322.

Pizzarello S. and Cronin J. R. (1998) Alanine enantiomers in the Murchison Meteorite. *Nature* **394**, 236.

Pizzarello S. and Cronin J. R. (2000) Non-racemic amino acids in the Murray and Murchison meteorites. *Geochim. Cosmochim. Acta* **64**, 329–338.

Pizzarello S. and Huang Y. (2002) Molecular and isotopic analyses of Tagish Lake alkyl dicarboxylic acids. *Meteorit. Planet. Sci.* **37**, 687–696.

Pizzarello S., Krishnamurthy R. V., Epstein S., and Cronin J. R. (1991) Isotopic analyses of amino acids from the Murchison meteorite. *Geochim. Cosmochim. Acta* **55**, 905–910.

Pizzarello S., Feng X., Epstein S., and Cronin J. R. (1994) Isotopic analyses of nitrogenous compounds from the Murchison meteorite; ammonia, amines, amino acids, and polar hydrocarbons. *Geochim. Cosmochim. Acta* **58**, 5579–5587.

Pizzarello S., Huang Y. S., Becker L., Poreda R. J., Nieman R. A., Cooper G., and Williams M. (2001) The organic content of the Tagish Lake meteorite. *Science* **293**, 2236–2239.

Robert F. and Epstein S. (1982) The concentration and isotopic composition of hydrogen, carbon, and nitrogen in carbonaceous meteorites. *Geochim. Cosmochim. Acta* **46**, 81–95.

Sandford S. A. (1996) The inventory of interstellar materials available for the formation of the solar-system. *Meteorit. Planet. Sci.* **31**, 449–476.

Sephton M. A. and Gilmour I. (2000) Aromatic moieties in meteorites: relics of interstellar grain processes? *Astrophys. J.* **540**, 588–591.

Sephton M. and Gilmour I. (2001a) Pyrolysis-gas chromatography-isotope ratio mass spectrometry of macromolecular material in meteorites. *Planet. Space Sci.* **49**, 465–471.

Sephton M. A. and Gilmour I. (2001b) Compound specific isotope analysis of the organic constituents in carbonaceous chondrites. *Mass Spec. Rev.* **20**, 111–120.

Sephton M. A., Pillinger C. T., and Gilmour I. (1998) $\delta^{13}C$ of free and macromolecular polyaromatic structures in the Murchison meteorite. *Geochim. Cosmochim. Acta* **62**, 1821–1828.

Sephton M. A., Pillinger C. T., and Gilmour I. (1999) Small-scale hydrous pyrolysis of macromolecular material in meteorites. *Planet. Space Sci* **47**, 181–187.

Sephton M. A., Pillinger C. T., and Gilmour I. (2000) Aromatic moieties in meteoritic macromolecular materials: analysis by hydrous pyrolysis and $\delta^{13}C$ of individual compounds. *Geochim. Cosmochim. Acta* **64**, 321–328.

Sephton M. A., Pillinger C. T., and Gilmour I. (2001) Normal alkanes in meteorites: molecular $\delta^{13}C$ values indicate an origin by terrestrial contamination. *Precamb. Res.* **106**, 45.

Shimoyama A. and Harada K. (1984) Amino acid depleted carbonaceous chondrites (C2) from Antarctica. *Geochem. J.* **18**, 281–286.

Shimoyama A. and Katsumata H. (2001) Polynuclear aromatic thiophenes in the murchison carbonaceous chondrite. *Chem. Lett.* **2001**, 202–203.

Shimoyama A., Ponnamperuma C., and Yanai K. (1979) Amino acids in the Yamato carbonaceous chondrite from Antarctica. *Nature* **282**, 394–396.

Shimoyama A., Harada K., and Yanai K. (1985) Amino acids from the Yamato-791198 carbonaceous chondrite from Antarctica. *Chem. Lett.* **1985**, 1183–1186.

Shimoyama A., Naraoka H., Yamamoto H., and Harada K. (1986) Carboxylic acids in the Yamato-791198 carbonaceous chondrite from Antarctica. *Chem. Lett.* **1986**, 1561–1564.

Shimoyama A., Naraoka H., Komiya M., and Harada K. (1989) Analyses of carboxylic acids and hydrocarbons in Antarctic carbonaceous chondrites, Yamato-74662 and Yamato-793321. *Geochem. J.* **23**, 181–193.

Shock E. L. and Schulte M. D. (1990) Amino acid synthesis in carbonaceous meteorites by aqueous alteration of polycyclic aromatic hydrocarbons. *Nature* **343**, 728–731.

Silfer J. A. (1991) Stable carbon and nitrogen isotope signatures of amino acids as molecular probes in geologic systems. PhD Thesis, University of Oklahoma.

Simmonds P. G., Shulman G. P., and Stembridge C. H. (1969) Organic analysis by pyrolysis gas chromatography-mass spectrometry. A candidate experiment for the biological exploration of Mars. *J. Chromatog. Sci.* **7**, 36–41.

Smith J. W. and Kaplan I. R. (1970) Endogenous carbon in carbonaceous meteorites. *Science* **167**, 1367–1370.

Stephan T., Jessberger E. K., Heiss C. H., and Rost R. (2003) TOF-SIMS analysis of polycyclic aromatic hydrocarbons in Alan Hills 84001. *Meteorit. Planet. Sci.* **38**, 109–116.

Stoks P. G. and Schwartz A. W. (1979) Uracil in carbonaceous meteorites. *Nature* **282**, 709–710.

Stoks P. G. and Schwartz A. W. (1981) Nitrogen-heterocyclic compounds in meteorites: significance and mechanisms of formation. *Geochim. Cosmochim. Acta* **45**, 563–569.

Stoks P. G. and Schwartz A. W. (1982) Basic nitrogen-heterocyclic compounds in the Murchison Meteorite. *Geochim. Cosmochim. Acta* **46**, 309–315.

Studier M. H., Hayatsu R., and Anders E. (1965) Organic compounds in carbonaceous chondrites. *Science* **149**, 1455–1459.

Studier M. H., Hayatsu R., and Anders E. (1968) Origin of organic matter in early solar system—I. Hydrocarbons. *Geochim. Cosmochim. Acta* **32**, 151–173.

Studier M. H., Hayatsu R., and Anders E. (1972) Origin of organic matter in Early Solar System—V: Further studies of meteoritic hydrocarbons and discussion of their origin. *Geochim. Cosmochim. Acta* **36**, 189–215.

Swart P. K., Grady M. M., and Pillinger C. T. (1983) A method for the identification and elimination of contamination during carbon isotopic analyses of extraterrestrial samples. *Meteoritics* **18**, 137–154.

van der Velden W. and Schwartz A. W. (1977) Search for purines and pyrimidines in the Murchison meteorite. *Geochim. Cosmochim. Acta* **41**, 961–968.

Watson J. S., Pearson V. K., Gilmour I., and Sephton M. (2003) Contamination by sesquiterpenoid derivatives in the Orgueil carbonaceous chondrite. *Org. Geochem.* **34**, 37–47.

Yang J. and Epstein S. (1983) Interstellar organic matter in meteorites. *Geochim. Cosmochim. Acta* **47**, 2199–2216.

Yang J. and Epstein S. (1985) A search for presolar organic matter in meteorite. *Geophys. Res. Lett.* **12**, 73–76.

Yoshino D., Hayatsu R., and Anders E. (1971) Origin of organic matter in early solar system: III. Amino acids: catalytic synthesis. *Geochim. Cosmochim. Acta* **35**, 927–938.

Yuen G. U. and Kvenvolden K. A. (1973) Monocarboxylic acids in Murray and Murchison carbonaceous meteorites. *Nature* **246**, 301–303.

Yuen G., Blair N., Des Marais D. J., and Chang S. (1984) Carbon isotope composition of low molecular weight hydrocarbons and mono carboxylic acids from Murchison meteorite. *Nature* **307**, 252–254.

Yuen G. U., Pecore J. A., Kerridge J. F., Pinnavaia T. J., Rightor E. G., Flores J., K. M. W., Mariner R., DesMarais D. J., and Chang S. (1990) Carbon isotopic fractionation in Fischer-Tropsch type reactions. *Lunar. Planet. Sci.* **21**, 1367–1368.

Zinner E. (1988) Interstellar cloud material in meteorites. In *Meteorites and the Early Solar System* (eds. J. F. Kerridge and M. S. Mathews). University of Arizona Press, Tucson, pp. 956–983.

1.11
Achondrites

D. W. Mittlefehldt

NASA/Johnson Space Center, Houston, TX, USA

1.11.1 INTRODUCTION	291
1.11.2 PRIMITIVE ACHONDRITES	292
1.11.2.1 Acapulcoite–Lodranite Clan	294
1.11.2.2 Winonaite–IAB-iron Silicate Inclusion Clan	301
1.11.2.3 Zag (b)	302
1.11.3 DIFFERENTIATED ACHONDRITES	303
1.11.3.1 Angrites	303
1.11.3.2 Aubrites	305
1.11.3.3 Brachinites	307
1.11.3.4 Howardite–Eucrite–Diogenite Clan	309
1.11.3.5 Mesosiderite Silicates	311
1.11.3.6 Ureilites	312
1.11.3.7 Itqiy	315
1.11.3.8 Northwest Africa 011	315
1.11.4 UNCATEGORIZED ACHONDRITES	315
1.11.4.1 IIE Iron Silicates	315
1.11.5 SUMMARY	317
ACKNOWLEDGMENT	317
REFERENCES	317

1.11.1 INTRODUCTION

This chapter covers the major and minor achondrite groups, and three newly described unique achondrites. The discussion of other unique achondrites by Mittlefehldt *et al.* (1998) is still current. The silicates of the stony-iron mesosiderites show many similarities with the howardites, and the silicate inclusions in IAB irons are closely related to the stony winonaites. Therefore, these will be included here. Finally, some IIE irons contain nonchondritic silicate inclusions, and they will also be considered.

The meteorites discussed are all samples of asteroids, although the exact sources are generally not known more precisely than that. Current practice is to divide the achondrites into two broad categories—differentiated achondrites and primitive achondrites. The former generally have igneous textures and compositions far removed from those of nebular materials, while the latter have metamorphic textures and compositions less distinct from those of nebular materials. I will define these categories more precisely as:

Differentiated achondrites. They are achondrites that exhibit igneous textures or igneous textures modified by impact and/or thermal metamorphism, and that have compositions of lithophile, siderophile, chalcophile, and atmophile elements that are highly fractionated from the ranges of chondritic materials.

Primitive achondrites. They are achondrites that exhibit equilibrated, metamorphic textures, possibly modified by impact. Rare members have relict chondritic textures, but examples with classic, unequilibrated chondritic (i.e., type 3) textures are absent. They have compositions of lithophile, siderophile, chalcophile, and atmophile elements that are at most only moderately fractionated from the range of nebular materials. Some members may be quite fractionated, but the group as a whole is dominated by primitive materials.

The majority of silicate inclusions of IIE irons do not easily fit into either of these categories. These I will call uncategorized as:
Uncategorized achondrites. They are achondrites with textures and compositions that do not clearly denote mode of origin (igneous versus metamorphic), or that have not been sufficiently well characterized to permit unambiguous categorization.

Table 1 lists the meteorite groups discussed in this chapter. Typical textures of representative members of each group are shown in Chapter 1.05. Some of the meteorite groups have been gathered into clans when the weight of evidence suggests that they originate on a single parent body. This reduces the 13 groups to representing possibly as few as nine parent bodies. Table 2 contains compositional data for representative achondrites. Figure 1 is a plot of Na/Al versus FeO/MnO for most of the achondrite groups discussed here. Sodium is a moderately volatile lithophile element, aluminum is a refractory lithophile element, and both are concentrated in plagioclase. To first order, the Na/Al ratio is a marker of moderately volatile element depletions of parent bodies. Ferrous iron and MnO do not greatly fractionate during igneous processes and, to first order, the FeO/MnO ratio is a marker for the oxidation state of parent bodies.

The petrology and composition of achondrites is a broad topic to cover, and only a fraction of the relevant literature can be cited. Mittlefehldt *et al.* (1998) presented a more complete treatment of the topic and referencing. Sources of petrologic and chemical *data* will generally be cited at the start of each achondrite section, while sources of *ideas* or very specific information will be cited where they are discussed. Here mg# is molar $100 \times \text{MgO}/(\text{MgO} + \text{FeO})$ and cr# is molar $100 \times \text{Cr}_2\text{O}_3/(\text{Cr}_2\text{O}_3 + \text{Al}_2\text{O}_3)$.

1.11.2 PRIMITIVE ACHONDRITES

The terms primitive and modified primitive achondrites were coined by Prinz *et al.* (1983) in order to distinguish those meteorite groups that lacked members with classic chondritic textures, but yet were not highly differentiated. Three clans were included in this achondrite supergroup—the acapulcoite–lodranite clan, the winonaite–IAB-iron silicate inclusion clan, and the brachinites. Subsequent work on new brachinites shows that most, maybe all, of them are igneous rocks. For this reason, they are considered differentiated achondrites here. Figure 2 shows Mg- and CI-normalized abundance plots for representative primitive achondrites and other select meteorites.

Table 1 Synopsis of petrologic characteristics of meteorite groups and modes of origin.

Category/group	Rock type	Texture type	$\Delta^{17}O$	Origin
Primitive achondrites				
Acapulcoite–lodranite clan	Chondritic ultramafic	Equigranular, metamorphic	-0.85 to -1.49	Metamorphism $\pm$ partial melting
Winonaite–IAB-iron silicate clan	Chondritic ultramafic	Equigranular, metamorphic	-0.48 ± 0.10	Metamorphism $\pm$ partial melting
Zag (b)	Ultramafic	?	-0.46	Partial melt residue?
Differentiated achondrites				
Angrites	Mafic	Igneous	-0.15 ± 0.06	Melt crystallization
Aubrites	Nonchondritic ultramafic	Brecciated igneous	0.02 ± 0.04	Melt crystallization, melt residue (?)
Brachinites	Chondritic and nonchondritic ultramafic	Equigranular	-0.26 ± 0.09	Melt crystallization, metamorphosed chondrite (?)
Howardite–eucrite–diogenite clan	Nonchondritic ultramafic to mafic	Brecciated igneous	-0.25 ± 0.08	Melt crystallization, impact modified
Mesosiderite silicates	Nonchondritic ultramafic to mafic	Brecciated igneous	-0.24 ± 0.09	Melt crystallization, impact modified
Ureilites	Nonchondritic ultramafic	Granular	-0.23 to ± 2.45	Melt crystallization, melting residue
Itqiy	Nonchondritic ultramafic, metal-rich	Igneous		Melt crystallization, impact modified (?)
Northwest Africa 011	Mafic	Igneous	-1.80	Melt crystallization, impact modified (?)
Uncategorized achondrites				
IIE-iron silicates	Chondritic ultramafic, mafic and silicic	Metamorphic, igneous	$+0.59 \pm 0.07$	Metamorphism, partial melting (?), impact melting (?)

Oxygen isotope data from Clayton and Mayeda (1996), Delaney *et al.* (2000), and Yamaguchi *et al.* (2002).

Table 2 Compositional data on representative achondrites.

		Acapulcoite–lodranite						IAB–Winonaite				
		aca Acapulco	aca ALHA77081	trans EET 84302	lod Gibson	lod MAC 88177	lod Y-791491	IAB Udei Station	IAB Campo del Cielo	IAB Landes	win Pontlyfni	win Tierra Blanca
Reference(s)		1, 2									7, 8	5, 9
Na	mg g^{-1}	6.42	7.52	10.5	2.8	0.29	0.71	5.25	8.8	9.57	6.422	3.69
Mg	mg g^{-1}	153.8	157	210	188	221	212	197	159	153	119.2	147
Al	mg g^{-1}	11.2	12	1.1	3.4	2.3	3.1	10	14.3	14	13.1	7.1
Si	mg g^{-1}	176.5	191	180	221	222	179				144.8	177
P	mg g^{-1}	1.6									0.48	1.2
S	mg g^{-1}	21.1									70.4	
K	μg g^{-1}	530	690	980	190	17	150	614			700	359
Ca	mg g^{-1}	9.8	5.9	5.1	9.15	7	9.51	8.3	5.5	11.1	11.8	8.1
Sc	μg g^{-1}	8.5	10.3	7.35	10.5	8.97	6.07	12.6	5.9	9.3	5.79	8.03
Ti	mg g^{-1}	0.7		0.6	0.72	0.48	0.36			0.9	0.6	0.66
V	μg g^{-1}	92	88.7					68	29	48		56
Cr	mg g^{-1}	6.995	7.19	1.5	2.8	3.85	2.35	3.23	1.5	1.73	2.087	1.91
Mn	mg g^{-1}	3.26	3.03	4.1	4.2	4.1	2.2	2.37	2.3	2.19	2.02	2.04
Fe	mg g^{-1}	204.5	248	268	159	102	245	95	126	220	296	197
Co	μg g^{-1}	640	795	1,620	720	100	1,250	110	402	730	883	762
Ni	mg g^{-1}	13.3	15.6	20.5	8.1	2.16	12.9	2.75	4.48	12	23.4	12.2
Zn	μg g^{-1}	250	306	62	62	79	105	237	182	185	130	134
Ga	μg g^{-1}	7.78	10.4	3	8.1	0.6	4.4	5.3			16.6	10.6
As	μg g^{-1}	1.57	2.14	5.31	2.12	0.17	4.28	0.33			3.2	2.55
Se	μg g^{-1}	10.8	10.3	0.68	0.51	0.88	0.53	10.4			16	11.6
Br	ng g^{-1}		350	2,900	3,200	3,300	4,400	450				280
Rb	ng g^{-1}			2,800	2,600	1,100	1,400					
Sr	μg g^{-1}											
Zr	ng g^{-1}											
Sb	ng g^{-1}	61	65	77	150	25	130	36			390	114
Cs	ng g^{-1}			10	58	45	90					
Ba	μg g^{-1}			9.1		5						
La	ng g^{-1}	742	290	53	170	11	14	158	50	150	260	190
Ce	ng g^{-1}			400	700	100	100				700	
Pr	ng g^{-1}											
Nd	ng g^{-1}			500	550							
Sm	ng g^{-1}	250	200	220	210	21	30	92	16	180	103	92
Eu	ng g^{-1}	100	94	89	20	17	13	60	90	110	100	49
Gd	ng g^{-1}			310	260	130						
Tb	ng g^{-1}	62	80	65	50	35	13					
Dy	ng g^{-1}											

(continued)

Table 2 (continued).

		Acapulcoite–lodranite					IAB–Winonaite				
		aca Acapulco	aca ALHA77081	trans EET 84302	lod Gibson	lod MAC 88177	lod Y-791491	IAB Udei Station	IAB Campo del Cielo	IAB Landes	win Pontlyfni	win Tierra Blanca
Ho	ng g⁻¹											
Er	ng g⁻¹											
Yb	ng g⁻¹	300	300	250	270	160	120	147	110	190	90	159
Lu	ng g⁻¹	40	57	49	50	24	18	26	20	32	13	24
Hf	ng g⁻¹	120	210		50	70					110	
Ta	ng g⁻¹			10	14	14	11					
W	ng g⁻¹		50	510	190	140	540					
Re	ng g⁻¹							3				
Os	ng g⁻¹	60	60			31		50				105
Ir	ng g⁻¹	670	950	2,125	1,220	8.4	2,020	37	740	700	780	1,240
Au	ng g⁻¹	536	840	204	90	8	291	38	169	270	586	1,140
Th	ng g⁻¹	149	200	70	34		41				250	
U	ng g⁻¹	70		70	50	10						261

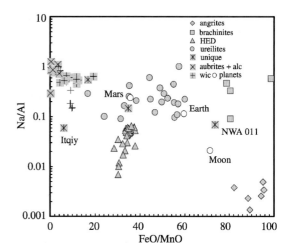

Figure 1 Na/Al versus FeO/MnO for achondrites. The FeO/MnO for acapulcoite–lodranite clan (alc), winonaite–IAB-iron silicate inclusion clan (wic) and aubrites are for orthopyroxene. All other data are bulk rock values. Meteorite data are from sources listed in the text, while planet data are from McDonough and Sun (1995), Taylor (1982), and Wänke and Dreibus (1994).

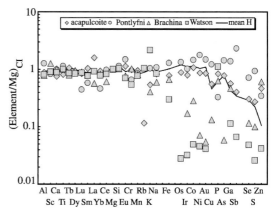

Figure 2 Nebular chemistry diagram for select achondrites compared to mean H-chondrite (Wasson and Kallemeyn, 1988). Data are arranged in order of increasing volatility within cosmochemical groups—lithophile (Al–Na), siderophile (Os–Sb), and chalcophile (Se–Zn). Data are from: mean acapulcoites—Lodders and Fegley (1998); winonaite Pontlyfni—Davis *et al.* (1977), Graham *et al.* (1977); Brachina—Johnson *et al.* (1977), Nehru *et al.* (1983); and IIE iron Watson—Olsen *et al.* (1994).

1.11.2.1 Acapulcoite–Lodranite Clan

The acapulcoite–lodranite clan is the prototypical primitive achondrite group. The acapulcoites have modal mineral abundances very much like those of ordinary chondrites, relict chondrules are reported from a few, and their bulk compositions are broadly chondritic (McCoy *et al.*, 1996; Mittlefehldt *et al.*, 1996, 1998). The lodranites are distinctly nonchondritic. They are generally depleted in

Table 2 (continued).

		Angrite			Aubrite		Brachinite		HED			
		Angra dos Reis	LEW 87051	D'Orbigny	Aubres	Peña Blanca Spring	Shallowater	Brachina	ALH 84025	euc Sioux County	euc Nuevo Laredo	euc Stannern	euc Ibitira
Reference		10	10	11	12, 13	13, 14	12, 15	16	17	18, 19	20	18, 19	9, 18
Na	mg g^{-1}	0.223	0.174	0.109	0.706	3.16	3.236	4.94	0.51	3.3	3.76	4.6	1.56
Mg	mg g^{-1}	65.1	117	39.6	232.7	238.4	201.5		184	42.9	34	42	43.9
Al	mg g^{-1}	49.5	48.7	65.6	2.4	2.9	5.1		0.9	67.97	63	65.28	68.7
Si	mg g^{-1}	204.3	188.9	179	273.7	267.5	232.4		170	229.2	231	232.3	228
P	mg g^{-1}			0.7	2		22	1.31		0.39		0.445	0.44
S	mg g^{-1}				2.83	2.6	3.2	5.17					
K	μg g^{-1}	163.7	77.2	69	517	240	380	370	18	295	400	550	200
Ca	mg g^{-1}	49.6	36	108	6.5	8.6	0.739	12	13.6	73.98	74	76.27	78.5
Sc	μg g^{-1}	12.3	4.4	40.3	0.185	9	6.8			34.5	33.3	29.4	29.1
Ti	mg g^{-1}			5.3		0.4	0.128		108	4.08	4.9	5.88	4.79
V	μg g^{-1}	1.5	1.09	0.262	0.241	6.4	1.1	88	3.94	1.98	64	2.26	2.12
Cr	mg g^{-1}	0.77	1.9	2.2	1.486	0.62	0.68	4.44	2.53	4.34	1.93	4.07	3.72
Mn	mg g^{-1}	73.1	148	193	4.1	1.94	150	2.65	250	144.7	4.5	138.2	141
Fe	mg g^{-1}	21.3	27.4	33	6	9.6	490	207.9	365	6.8	153	7.5	11.7
Co	μg g^{-1}	0.097	0.044	0.091	0.161	9.8	9.5	235	5.1	0.0017	2.17	0.003	0.01
Ni	mg g^{-1}				3.61	0.146	43	1.3	164	5.3	0.003	7.8	5.6
Zn	μg g^{-1}						41	313	2.1	1.27	1	1.73	1.19
Ga	μg g^{-1}			0.53			2	7.6	0.55				
As	μg g^{-1}			0.74	2.46	2.37	4.2	0.18	12.4				
Se	μg g^{-1}							3.5	330				
Br	ng g^{-1}				605			490					
Rb	ng g^{-1}	130	67	142		320				170	80	690	170
Sr	μg g^{-1}	160		66						76	71	89	81
Zr	μg g^{-1}				2.2	4.9			56	40.88		91.23	49.3
Sb	ng g^{-1}				122	220							
Cs	ng g^{-1}	36		52						9		16	8
Ba	μg g^{-1}	6,140	2,320	3,760	347	97.3	21	220	65	16.9	44	59.1	33.7
La	ng g^{-1}	19,200	6,200	9,180				600		1,820	3,790	5,210	2,150
Ce	ng g^{-1}									5,190	10,000	13,760	5,680
Pr	ng g^{-1}					56.8				838		2,070	868
Nd	ng g^{-1}	16,300	1,480	2,230	241		8.1	190	34	4,320	6,500	10,070	4,300
Sm	ng g^{-1}	5,760	536	876	16.7	2.24	27	70	33	1,540	2,330	3,160	1,450
Eu	ng g^{-1}	1,780								415	710	779	653
Gd	ng g^{-1}									2,070		4,010	1,970
Tb	ng g^{-1}	1,390	360	511	25.2	6.65		47		398	540	724	380
Dy	ng g^{-1}							380		2,850	3,300	4,950	2,630
Ho	ng g^{-1}							70		629		1,060	592

(continued)

Table 2 (continued).

		Angrite			Aubrite			Brachinite		HED			
		Angra dos Reis	LEW 87051	D'Orbigny	Aubres	Peña Blanca Spring	Shallowater	Brachina	ALH 84025	euc Sioux County	euc Nuevo Laredo	euc Stannern	euc Ibitira
Er	ng g^{-1}	4,820	1,520	2,130	101	41		220	98	1,830	2,410	3,000	1,730
Yb	ng g^{-1}	686	239	308				36	16	1,940		2,810	1,710
Lu	ng g^{-1}	2,790	1,170	1,480						287	350	421	261
Hf	ng g^{-1}	350	110	200	14.8	7.2				1,160	1,600	2,350	1,210
Ta	ng g^{-1}									147	178	368	153
W	ng g^{-1}						77	91.5		43		220	80
Re	ng g^{-1}				0.129	0.093					6.8		
Os	ng g^{-1}				0.843	0.45			250		80		
Ir	ng g^{-1}				0.822	0.31	380	135	123		0.083		
Au	ng g^{-1}				4.27	3.4		12.6	61		1.5		
Th	ng g^{-1}	640	220	434						287	430	680	248
U	ng g			116	3.17					74	140	172	73

a basaltic component, expressed in modal mineralogy by deficits in plagioclase and high-calcium pyroxene, and in bulk composition by depletions in highly incompatible elements (McCoy et al., 1997a; Mittlefehldt et al., 1996, 1998). Some members have transitional (e.g., EET 84302, GRA 95209) or unusual (LEW 86220) characteristics that make assignment to one type or the other controversial. All will be referred to as acapulcoite–lodranite clan achondrites.

Petrologic and chemical information on acapulcoite–lodranite clan meteorites was summarized from: Bild and Wasson (1976); Kallemeyn and Wasson (1985); McCoy et al. (1996, 1997a,b); Mittlefehldt et al. (1996); Nagahara (1992); Nagahara and Ozawa (1986); Palme et al. (1981); Schultz et al. (1982); Takeda et al. (1994a); Torigoye et al. (1993); Weigel et al. (1999); and Zipfel et al. (1995).

The mineralogy of the acapulcoite–lodranite clan is similar to that of ordinary chondrites, though mineral compositions are quite distinct. Unlike equilibrated ordinary chondrites, mineral compositions vary considerably within the clan. The acapulcoite–lodranite clan meteorites are composed of olivine, orthopyroxene, diopside, plagioclase, metal, troilite, chromite, apatite, whitlockite, schreibersite, and graphite. Not all members contain all of these phases, and abundances vary. Olivine compositions range from mg# 86 to mg# 97, and orthopyroxene compositions range from mg# 86 to mg# 96. Diopside is chromium rich, with Cr_2O_3 contents mostly in the range 0.9–1.9 wt.%. Plagioclase is sodic, ranging from An_{10} to An_{31}. However, some reported analyses are nonstoichiometric, suggesting that the data should be treated cautiously. The range for demonstrably stoichiometric analyses is An_{10} to An_{24}.

Acapulcoite–lodranite clan meteorites vary from chondritic compositions (acapulcoites) to compositions depleted in minimum melts in the silicate and metal–sulfide systems (lodranites). Lodders and Fegley (1998) have estimated the average bulk composition of the chondritic members of the clan. Figure 2 shows Mg- and CI-normalized abundances for this average, compared to an H-chondrite average (Wasson and Kallemeyn, 1988). The mean acapulcoite abundance pattern is very similar to that of ordinary chondrites—average refractory lithophile/Mg ratios are 0.95 versus 0.98, and moderately volatile and volatile elements show similar depletion patterns in the two groups. The lower average refractory siderophile/Mg ratio in the acapulcoite–lodranite clan (0.83) compared to H-chondrites (1.16) reflects in part metal/silicate fractionation—Fe/Mg ratios are 0.78 versus 1.02. However, part of the difference may be a sampling artifact. Acapulcoite–lodranite clan meteorites can be heterogeneous (e.g., Takeda et al., 1994a)

Table 2 (continued).

						HED						
		cum euc Binda	cum euc Serra de Magé	cum euc Moore County	dio Shalka	dio Johnstown	poly euc Macibini	poly euc Petersburg	poly euc Y-74450	how Malvern	how Frankfort	how Kapoeta
Reference		18, 19	18, 19	18, 19	21, 22	23	19, 24	25, 26			21, 27	
Na	mg g^{-1}	1.93	1.85	3.34	0.024	0.15	3.2	2.87	3.8	2.73	1.01	2.05
Mg	mg g^{-1}	107	68.44	56.7	155.5	153.6	50.5	67.78	45.7	72.6	126.3	95
Al	mg g^{-1}	36.8	67.18	66.97	3.2	7.9	64.16	57.28	60.5	52.4	22.6	44
Si	mg g^{-1}	235.7	226.4	225.9	241.1	245.4	230.6	229.9	225.7	230	237.9	235
P	mg g^{-1}	0.17	0.25	0.16	0.01	0.18	0.48			0.305	0.15	
S	mg g^{-1}					2.2		2.7	2.36	1.53		
K	µg g^{-1}	85	60	180	13	9.2	290	270	450	410	81	180
Ca	mg g^{-1}	41.6	64.9	67	5.2	13.1	71.3	60.7	71.1	58	26	37.2
Sc	µg g^{-1}	14.5	23.5	19.8	9.9	15.8	29.2	26.15	30	26.5	20.5	20.7
Ti	mg g^{-1}	1.02	1.02	2.58	0.37	0.63	4.3	3.4	5.4	2.7	1.36	1.8
V	µg g^{-1}					115			67.5	91.4		
Cr	mg g^{-1}	5.61	3.08	2.81	16.5	5.61	2.5	3.16	2.71	3.855	7.94	4.75
Mn	mg g^{-1}	3.72	4.88	3.49	4.28	3.75	4.17	4	4.11	4.06	4.34	3.83
Fe	mg g^{-1}	130.8	125.8	134	126.5	123.7	142.3	136.8	144.3	141.5	139.1	136
Co	µg g^{-1}	11.8	14.4	11.3	18	38.1	8.52	69	12	15.5	23.6	28
Ni	mg g^{-1}	0.0057	0.0118	0.0042	0.005	0.15	0.043	1.247		0.19	0.125	0.41
Zn	µg g^{-1}	6.1	1.6	3.4		0.65	2.4		1.44		2.17	4.2
Ga	µg g^{-1}	0.9	0.68	1.6		0.18			1.46		0.51	1.04
As	µg g^{-1}					0.032	0.6		0.0056		0.058	0.092
Se	ng g^{-1}				0.3		0.62		0.37		0.05	
Br	ng g^{-1}				120		250		47			
Rb	ng g^{-1}	130	30	60		51	2,000		450	170	100	
Sr	µg g^{-1}	33	23	70			77		85	59	230	
Zr	µg g^{-1}	6.36	9.91	15.38			54	47	68	38	24	
Sb	ng g^{-1}						18					
Cs	ng g^{-1}	6	1	1								
Ba	µg g^{-1}	6	1.5	20.6			32	27	19		9.5	
La	ng g^{-1}	382	146	1,280	9	63	3,110	2,480	45.4	2,270	7.9	1,390
Ce	ng g^{-1}	1,010	494	3,380			8,210	6,300	4,790		980	
Pr	ng g^{-1}	154	92	521		21			12,000		2,700	
Nd	ng g^{-1}	773	548	2,700			5,580	5,200	1,900	1,280	370	830
Sm	ng g^{-1}	261	256	906	4	59	1,860	1,599	8,800		1,700	540
Eu	ng g^{-1}	277	133	527	2	11	660	551	2,780	450	560	320
Gd	ng g^{-1}	355	407	1,260		240	2,300		700		180	1,250
Tb	ng g^{-1}	70.5	83.9	239			414	392	3,400	350	860	290
Dy	ng g^{-1}	507	631	1,690	210				660	2,200	150	1,260
Ho	ng g^{-1}	118	151	376	59				4,200		1,020	230
									970		200	

(continued)

Table 2 (continued).

		cum euc Binda	cum euc Serra de Magé	cum euc Moore County	dio Shalka	dio Johnstown	HED poly euc Macibini	poly euc Petersburg	poly euc Y-74450	how Malvern	how Frankfort	how Kapoeta
Er	ng g^{-1}	356	461	1,100		140	1,840	1,599	2,750	1,350	640	790
Yb	ng g^{-1}	387	510	1,040		170	290	232	2,610	200	620	890
Lu	ng g^{-1}	62.4	81.5	160		27			360		93	140
Hf	ng g^{-1}	180	190	530	30		1,440	1,170	2,080	900	490	600
Ta	ng g^{-1}	21	31	30	6		180	144	300	110	40	100
W	ng g^{-1}	20	14	18		3.5			85		126	36
Re	ng g^{-1}								700			
Os	ng g^{-1}											
Ir	ng g^{-1}					6.4	0.7			9	4.1	20
Au	ng g^{-1}	46	24	125		1.7	0.7		0.27		3.4	6.8
Th	ng g^{-1}	27	19	64		2.8	320	263	163		38	51
U	ng g^{-1}						70					

making representative sampling difficult. In spite of the similarities, the acapulcoite–lodranite clan is clearly distinct from ordinary chondrites. The acapulcoite–lodranite clan does not fit into the pattern of correlated siderophile element abundances with oxidation state shown by the ordinary chondrites (e.g., Wasson, 1972) and the latter show a distinctly different oxygen-isotopic mixing trend (see Chapter 1.06).

Some members of the acapulcoite–lodranite clan are depleted in minimum melts. The first minimum melt occurs in the metal–sulfide system. The resulting geochemical signature is increasing refractory siderophile element/Ni ratios with decreasing chalcophile element/siderophile element ratios as melts are lost. Metal–sulfide veins are present in many acapulcoite–lodranite clan meteorites, and these have been interpreted as evidence for melting (McCoy *et al.*, 1997b). Figure 3 shows Se/Co versus Ir/Ni for acapulcoite–lodranite clan meteorites—many have high Ir/Ni and low Se/Co, consistent with melt depletion in the metal–sulfide system (see Mittlefehldt *et al.*, 1996). A few plot in the low Ir/Ni, high Se/Co quadrant, indicating enrichment in a minimum-melt fraction of the metal–sulfide system. This suggests that melt from elsewhere in the parent body migrated into these stones.

The lithophile elements also show evidence for melting in the silicate–oxide system on the acapulcoite–lodranite clan parent body. Figure 4 shows normalized Na/Sc versus Sm/Sc for these meteorites. Loss of a partial melt in the silicate–oxide system will result in a decrease in both of these ratios (see Mittlefehldt *et al.*, 1996). Several members of the acapulcoite–lodranite clan have

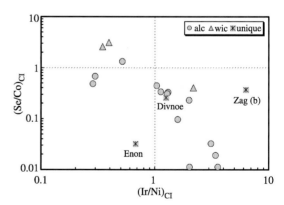

Figure 3 CI-normalized siderophile–chalcophile element ratios in acapulcoite–lodranite clan (alc), winonaite–IAB-iron silicate inclusion clan (wic) and some unique meteorites showing the effects of melting in the metal–sulfide system. Residual solid metal will have high Ir/Ni and low Se/Co, while metallic melt will have the opposite characteristics. Data from sources listed in the text, except for Divnoe (Petaev *et al.*, 1994; Weigel *et al.*, 1997) and Enon (Kallemeyn and Wasson, 1985).

Table 2 (continued).

		Mesosiderite					Ureilite					IIE	
		Barea	Estherville	Mincy	Veramin	Chaunskij	ALHA77257	ALHA81101	Haverö	META 78008	EET 83309	Watson	Miles	Miles
Reference		29, 30	30, 31	30, 31	30, 32	30, 33	9, 34, 35	34, 35						
Na	mg g^{-1}	1.54	1.83	0.79	0.292	2	0.183	0.174	0.242	0.58	1.45	38	39	39
Mg	mg g^{-1}	91.7	102.4	124	136.3	48.4	239.1	215	224	210	210	8.1	39	32.3
Al	mg g^{-1}	48.7	51.2	39.7	30.9	61.36	1.3		1.1			189.4	42	65.1
Si	mg g^{-1}	233	240.8	249	251.5	191.4	192.2	0.87	188	2.3	3.7	13.4	55.5	43.7
P	mg g^{-1}	5.28	2.4	6.07	3.2	8.16	0.3					212.7		
S	mg g^{-1}	13.6				39.4						2.6		
K	µg g^{-1}	184		100		110	100	8	35	13	238	3.1		
Ca	mg g^{-1}	41.8	52	42.4	26.9	52.3	7.3	6.9		9.3	8.1	2,300	44.7	5,600
Sc	µg g^{-1}	18.19	19.88	19.8	11.15	20.5	7.7	7.2	4.9	8.4	7.8	16.4	24.1	30.5
Ti	mg g^{-1}	1.52	1.9	1.8	1.3	2.1	0.2		0.36			9.972		25
V	µg g^{-1}						99	99		71	94	0.84	108	2.73
Cr	mg g^{-1}	5.42	6.08	5.03	6.68	3.83	4.8	4.9	4.09	3.23	4.85	2.173	4.07	113
Mn	mg g^{-1}	4.36		4.6	3.7	2.31	3.1	2.89	2.66	2.94	2.88	2.9	1.31	5.55
Fe	mg g^{-1}	128.9	93.11	67.6	88.21	181.51	105.5	149	150	155	134	110.2	74	2.07
Co	µg g^{-1}	50.5	39.5	138	108.5	311	89	71	101	123	135	41.87	302	90
Ni	mg g^{-1}	1.224	1.042	3.28	2.8	17.7	0.89	0.82	0.92	1.17	1.67	0.629	6.4	263
Zn	µg g^{-1}						243	159	235	280	284	16.4	14	9
Ga	µg g^{-1}					6.1	1.84	1.44	1.13	3.3	3	2.19		139
As	µg g^{-1}	3.3	2.5	0.27	0.26	2.52	0.19	0.112		0.24	0.39			
Se	µg g^{-1}			2.9	2.15	10.6		1.26		0.3	2.1	2.02		
Br	ng g^{-1}			150		5,740					220			
Rb	ng g^{-1}													
Sr	µg g^{-1}		96						0.7			4.54		
Zr	µg g^{-1}													
Sb	ng g^{-1}				30	31	20				19	24.5		
Cs	ng g^{-1}													
Ba	µg g^{-1}		19	19								1,110		
La	ng g^{-1}	5,760	1,780	424	164	1,155	13.8	9.9	70		185	414	590	
Ce	ng g^{-1}	12,100	4,500			2,180	39.2	23.4				953	2,400	
Pr	ng g^{-1}								19					
Nd	ng g^{-1}	4,000					16.2	10.6						
Sm	ng g^{-1}	369	703	295	75	684	6.4	3	14	14.6	62	216.6	730	200
Eu	ng g^{-1}	233	318	150	44	284	1.24	0.96	4.1	8	30	74.9	370	100
Gd	ng g^{-1}						12.4	10.4	25					
Tb	ng g^{-1}	101	146	58	31	206	4.1	2.1				59.8		
Dy	ng g^{-1}						40.6	22.5	22			391		
Ho	ng g^{-1}						10.6	5.5	5.4					
Er	ng g^{-1}								18					
Yb	ng g^{-1}	473	730	403	112	805	68.8	47.5	25	84	76	242	730	290
Lu	ng g^{-1}	68	105	63	27	137	12.7	9.4	8.5	30	14	40.6	138	64

(continued)

essentially chondritic ratios, and thus have not lost or gained a basaltic partial melt. Others have low Na/Sc and Sm/Sc ratios, indicating depletion in a basaltic melt fraction. Trace lithophile element data on pyroxenes, plagioclase, and calcium-phosphates in acapulcoite–lodranite clan meteorites also present evidence for loss of basaltic melts by some members of the clan (Floss, 2000). Acapulcoite–lodranite clan meteorites generally display no evidence for inclusion of basaltic melts, but coarse-grained diopside and plagioclase regions in LEW 86220 are interpreted as crystallized mafic melt (McCoy et al., 1997b).

An unusual characteristic of acapulcoite–lodranite clan meteorites is that some have relatively high trapped noble gas contents, considering their apparent high-temperature history (Bild and Wasson, 1976; Palme et al., 1981; Schultz et al., 1982). The chondritic members of the clan have trapped ^{36}Ar contents of $\sim (3-6) \times 10^{-8}$ cm^3 STP g^{-1} (Weigel et al., 1999), $\sim 3-10$ times the contents in many type-6 ordinary chondrites (e.g., Eugster et al., 1993, 1998). Among ordinary chondrites, volatile element contents are inversely correlated with metamorphic grade (e.g., Heymann and Mazor, 1968; Wasson, 1972). Thus, the relatively high trapped noble gas contents in some acapulcoite–lodranite clan meteorites seem anomalous. Figure 5 shows, however, that for most of them, the content of trapped ^{36}Ar is positively correlated with bulk rock Na/Sc—samples that have lost a basaltic melt have trapped ^{36}Ar contents depleted by up to a factor of ~ 100. Several members have

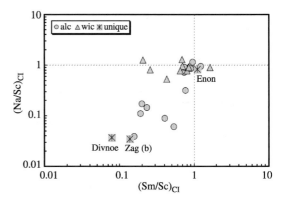

Figure 4 CI-normalized lithophile element ratios in acapulcoite–lodranite clan (alc), winonaite–IAB-iron silicate inclusion clan (wic) and some unique meteorites showing the effects of melting in the silicate–oxide system. Residual silicate source regions will have low Na/Sc and Sm/Sc, while mafic melts will have the opposite characteristics. Many acapulcoite–lodranite clan and winonaite–IAB-iron silicate inclusion clan meteorites have essentially chondritic lithophile element ratios. Data from sources listed in the text, except for Divnoe (Petaev et al., 1994; Weigel et al., 1997), and Enon (Kallemeyn and Wasson, 1985).

Table 2 (continued).

		Mesosiderite				Ureilite			IIE		
		Barea	Estherville	Mincy	Veramin	Chaunskij	ALHA77257	ALHA81101	Haverö	META 78008	EET 83309	Watson	Miles	Miles
Hf	ng g^{-1}	190	420	226	107	540								660
Ta	ng g^{-1}	54	69			52			180			8.9		
W	ng g^{-1}								30		16			
Re	ng g^{-1}													
Os	ng g^{-1}						112			460	270	25.6		
Ir	ng g^{-1}	21.4	10.4			18.8	184	37	240	340	249	29.2	36	30
Au	ng g^{-1}	13.4				92	20	11.6	24	21.4	32	10.9	25	22
Th	ng g^{-1}		67											
U	ng g^{-1}								6			13.5		

References: 1. Yanai and Kojima (1991); 2. Zipfel et al. (1995); 3. Schultz et al. (1982); 4. Weigel et al. (1999); 5. Kallemeyn and Wasson (1985); 6. Bild (1977); 7. Graham et al. (1977); 8. Davis et al. (1977); 9. Jarosewich (1990); 10. Mittlefehldt and Lindstrom (1990); 11. Mittlefehldt et al. (2002); 12. Easton (1985); 13. Wolf et al. (1983); 14. Lodders et al. (1993); 15. Keil et al. (1989); 16. Nehru et al. (1983); 17. Warren and Kallemeyn (1989a); 18. Barrat et al. (2000); 19. McCarthy et al. (1973); 20. Warren and Jerde (1987); 21. McCarthy (1972); 22. Mittlefehldt (1994); 23. Wänke et al. (1977); 24. Buchanan et al. (2000b); 25. Mason et al. (1979); 26. Buchanan and Reid (1996); 27. Palme et al. (1978); 28. Wänke et al. (1972); 29. Mason and Jarosewich (1973); 30. Mittlefehldt, unpublished; 31. Simpson and Ahrens (1977); 32. Powell (1971); 33. Petaev et al. (2000); 34. Warren and Kallemeyn (1992); 35. Spitz and Boynton (1991); 36. Wänke et al. (1972); 37. Warren and Kallemeyn (1989b); 38. Olsen et al. (1994); 39. Ebihara et al. (1997).

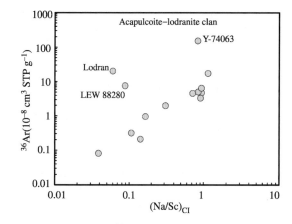

Figure 5 Trapped ^{36}Ar versus Na/Sc for acapulcoite–lodranite clan meteorites. For most acaplucoite–lodranite clan meteorites, there is a good correlation, indicating progressive gas loss with melting. Sodium and scandium data from sources listed in the text; trapped ^{36}Ar was taken from Weigel et al. (1999, table 8).

anomalously high trapped ^{36}Ar contents, indicating that simple thermal degassing of a uniform protolith cannot explain all the noble gas data.

A formation age has been determined only for Acapulco; its Sm–Nd and Pb–Pb ages are 4.60 ± 0.03 Ga and 4.557 ± 0.002 Ga (Prinzhofer et al., 1992; Göpel et al., 1992). Ages determined by ^{39}Ar–^{40}Ar for several acapulcoite–lodranite clan meteorites show that radiogenic argon retention began roughly at 4.51 Ga, about 40 Ma after formation (McCoy et al., 1996, 1997a; Mittlefehldt et al., 1996; Pellas et al., 1997).

The acapulcoite–lodranite clan is recognized as having formed by extensive high-temperature metamorphism and anatexis of primitive, reduced chondritic material (see McCoy et al., 1997b, 1996). A range in final temperatures was reached, resulting in rocks retaining their chondritic composition, rocks depleted in metal–sulfide system minimum melts, and rocks depleted in mafic partial melts.

1.11.2.2 Winonaite–IAB-iron Silicate Inclusion Clan

The winonaites and silicate inclusions in some IAB iron meteorites compose the other major primitive achondrite clan. Choi et al. (1995) compared metal compositions of IAB and IIICD irons and concluded that they represent a single meteorite group. Three IIICD meteorites contain silicate inclusions. McCoy et al. (1993) discussed some petrologic distinctions between them and inclusions in IAB irons, but Choi et al. (1995) did not find these to be compelling arguments against grouping. There are only very limited data available on inclusions in IIICD irons, and they will not be discussed. Y-75261 is unique—it is an impact melt breccia (Benedix et al., 1998), the only such winonaite, has more magnesian mafic silicates and more calcic plagioclase (Yanai and Kojima, 1995), but has an oxygen-isotopic composition identical to winonaites (Clayton and Mayeda, 1996). Y-75261 may not be a winonaite, and it will not be discussed further.

Petrologic and chemical information on winonaite–IAB-iron silicate inclusion clan meteorites was summarized from: Benedix et al. (1998, 2000); Bild (1977); Bunch et al. (1970); Davis et al. (1977); Fukuoka and Schmitt (1978); Kallemeyn and Wasson (1985); Kimura et al. (1992); Prinz et al. (1980); and Takeda et al. (2000).

The winonaite–IAB-iron silicate inclusion clan contains members with primitive chondritic compositions, basalt-depleted lithologies and lithologies containing mafic segregations. Modal plagioclase contents are like those of ordinary chondrites, and some winonaites contain relict chondrules. Texturally, winonaites show internal heterogeneity and substantial differences between members. The textures are typically metamorphic. Winonaites contain olivine, orthopyroxene, clinopyroxene, plagioclase, troilite, metal, chromite, daubreelite, schreibersite, graphite, alabandite, potassium feldspar, and apatite. The silicate inclusions in IAB iron meteorites are not as well described owing to the difficulty of sampling these heterogeneous inclusions. Benedix et al. (2000) divide all inclusions into five types: (i) angular, chondritic silicate; (ii) nonchondritic silicate-rich; (iii) sulfide-rich; (iv) rounded, often graphite-rich; and (v) phosphate-bearing. This nomenclature describes the total range of inclusions, not just the silicate inclusions. The graphite-rich inclusions only rarely contain silicates. Some sulfide-rich inclusions contain silicates that are apparently angular chondritic silicates. The mineralogy of silicate inclusions in IAB is more diverse, likely due to reaction with the host metal. They contain olivine, orthopyroxene, clinopyroxene, plagioclase, troilite, metal, chromite, daubreelite, graphite, and the phosphates apatite, whitlockite and rarely, brianite.

The silicates in the clan are magnesian. Olivine compositions range from mg# 92 to mg# 99—most winonaite olivines have mg# ≥ 98, while most IAB-iron silicate inclusions have olivines with mg# ≤ 97. Orthopyroxenes range from mg# 91 to mg# 99, and the winonaites again tend to be more magnesian. Diopside grains in IAB silicates mostly have Cr_2O_3 contents >0.6 wt.%, with Pine River being exceptional at 0.38 wt.% (Bunch et al., 1970; Benedix et al., 2000). Contrast this with winonaite diopside grains, many of which have lower Cr_2O_3 contents of <0.4 wt.%. However, winonaite QUE 94535 contains diopside grains with differing compositions, with Cr_2O_3 ranging from <0.01 wt.% to 1.02 wt.% (my analyses). Plagioclase is sodic; compositions range from An_9 to An_{26}.

There are few comprehensive bulk analyses of winonaite–IAB-iron silicate inclusion clan meteorites. Takeda et al. (2000) discussed average bulk compositions for winonaites and IAB-iron silicate inclusions, which within uncertainty show unfractionated lithophile and siderophile element patterns for refractory through moderately volatile elements (Figure 2). Poor sampling may explain the slight fractionations that are observed. Trace lithophile element data on winonaite-IAB-iron silicate inclusion clan meteorites show a much wider range in patterns than observed for the acapulcoite–lodranite clan meteorites. The rare earth elements (REEs) show three distinct patterns: (i) approximately chondritic (e.g., Y-74025, Kimura et al., 1992); (ii) enriched patterns with Eu depletions (e.g., Linwood IAB—Fukuoka and Schmitt, 1978); and (iii) depleted patterns with Eu enrichments (e.g., Campo del Cielo IAB—Bild, 1977). Patterns (ii) and (iii) have been determined on different splits of Winona (Prinz et al., 1980) testifying to compositional heterogeneity within individual meteorites. REE patterns similar to those observed for basalt-depleted acapulcoite–lodranite clan meteorites have not been found for winonaite–IAB-iron silicate inclusion clan meteorites.

A plagioclase–diopside-rich mafic segregation (gabbro) from a Caddo County IAB iron inclusion has an REE pattern like the enriched patterns determined on some winonaite–IAB-iron silicate inclusion clan meteorites (Takeda et al., 2000). Similar mafic segregations are present in other winonaite–IAB-iron silicate inclusion clan meteorites (Benedix et al., 1998, 2000). Kallemeyn and Wasson (1985) interpreted the depleted patterns in two winonaites as having resulted from loss of a phosphate phase, while Bild (1977) ascribed similar REE patterns in IAB silicates to variations in abundances of diopside and plagioclase. Winonaite–IAB silicates tend to have roughly chondritic Na/Sc but variable Sm/Sc (Figure 4), which favors the Kallemeyn and Wasson (1985) interpretation. Centimeter-scale modal heterogeneity observed in many winonaite–IAB-iron silicate inclusion clan meteorites (Benedix et al., 1998, 2000; Bunch et al., 1970; Takeda et al., 2000; Wlotzka and Jarosewich, 1977) suggests that the compositional variations plausibly reflect localized heterogeneity rather than parent body processes.

Few modern radiometric ages have been determined for winonaite–IAB-iron silicate inclusion clan meteorites. The IAB-iron silicate inclusions were formed very early in solar system history as I–Xe ages are within ±3 Ma of that of the Bjurböle chondrite (Niemeyer, 1979a), which has been calibrated to an absolute age of 4.566 ± 0.002 Ga (Brazzle et al., 1999). The Ar–Ar ages for several winonaites and IAB silicates range from 4.53 Ga to ≥4.40 Ga (Benedix et al., 1998; Niemeyer, 1979b; Takeda et al., 2000), indicating that cooling to argon closure took 40–170 Ma.

The formation of the winonaite–IAB-iron silicate inclusion clan meteorites is an unsettled question. There is as yet no consensus on how the metal phase was formed—impact melting (Choi et al., 1995), fractional melting with melt segregation (Kelly and Larimer, 1977), batch melting with melt separation (Kracher, 1982, 1985), and fractional crystallization (McCoy et al., 1993) are suggested mechanisms. Fractional crystallization of a core has serious difficulty explaining the entrapment of unmelted chondritic silicates, as does the batch melting-melt segregation model. Impact melting is believed by many to produce small quantities of melt that quench too quickly (Keil et al., 1997) to explain the IAB-iron suite. The fractional melting-melt segregation model requires solid metal/liquid metal partition coefficients that are not supported by data on the magmatic iron meteorite groups, or experiments. Detailed discussions of the models can be found in Benedix et al. (2000), Choi et al. (1995), Mittlefehldt et al. (1998), and Wasson et al. (1980). Benedix et al. (2000) have presented the most comprehensive model attempting to explain the origin of the winonaite–IAB-iron silicate inclusion clan through breakup and reassembly of a hot, partially differentiated parent body. The breakup and reassembly event effectively freezes the evolution of the parent body, and mixes different materials together. This model may explain the clan, but it will be difficult to quantitatively test it.

1.11.2.3 Zag (b)

Zag (b) is classified as a primitive achondrite. Petrologic and compositional information is from Delaney et al. (2000), and my unpublished analysis. Zag (b) is an unbrecciated, ultramafic rock containing 68% olivine, 8% orthopyroxene, 10% high-calcium clinopyroxene, 14% sulfide, metal and weathering products, and minor chromite and phosphate. However, Zag (b) is modally heterogeneous, and only a single thin section was studied. Olivine is homogeneous with mg# 80, but contains "channels" of orthopyroxene associated with sulfide, metal and/or weathering products. This orthopyroxene likely was produced by reduction of the olivine. It is more magnesian- and calcium-poor than coarse orthopyroxene, whose composition is $Wo_{2.3}En_{71.9}Fs_{25.8}$.

Olivine has a lower FeO/MnO than the coarse orthopyroxene, indicating that olivine has been reduced and is not in equilibrium with the orthopyroxene. The composition of clinopyroxene is $Wo_{43.7}En_{45.4}Fs_{10.9}$. Zag (b) is close to the unique achondrite Divnoe in oxygen-isotopic composition, but olivine is more magnesian and orthopyroxene is more ferroan in Zag (b) (Petaev et al., 1994). Zag (b) is depleted in highly incompatible elements—the La content is $0.25 \times CI$, La/Yb is $0.5 \times CI$, and Eu/Sm is $0.7 \times CI$—and is similar to basalt-depleted members of the acapulcoite–lodranite clan (Figure 4). The siderophile–chalcophile element contents of Zag (b) are relatively high—cobalt, nickel, gold, and selenium are at $0.2-0.6 \times CI$, while iron is $1.6 \times CI$, giving Zag (b) a very high Ir/Ni ratio (Figure 3). Because of heterogeneity, these characteristics may not be representative.

1.11.3 DIFFERENTIATED ACHONDRITES

Differentiated achondrites represent the products of classical igneous processes acting on the silicate–oxide system of asteroidal bodies—partial to complete melting, and magmatic crystallization. Iron meteorites represent the complementary metal–sulfide system products of this process.

1.11.3.1 Angrites

The angrite group consists of seven meteorites linked by identical oxygen-isotopic compositions (Clayton and Mayeda, 1996), similar unusual mineralogy (e.g., McKay et al., 1988, 1990; Mittlefehldt et al., 2002; Prinz et al., 1977; Yanai, 1994), and several distinctive geochemical characteristics (Mittlefehldt and Lindstrom, 1990; Mittlefehldt et al., 2002; Warren et al., 1995). Although some are petrologically anomalous, the preponderance of distinctive characteristics indicates that they plausibly originated on a common parent body. The angrites are mafic igneous rocks from the crust of a differentiated asteroid.

Petrologic and chemical information on angrites was taken from: Crozaz and McKay (1990); Delaney and Sutton (1988); Goodrich (1988); McKay et al. (1988); Mikouchi and McKay (2001); Mikouchi et al. (1995, 1996, 2000a,b); Mittlefehldt and Lindstrom (1990); Mittlefehldt et al. (2002); Prinz et al. (1977, 1988); Tera et al. (1970); Warren and Davis (1995); Warren et al. (1995); Yanai (1994); plus personal communication from G. A. McKay.

Angrites contain distinctive aluminian–titanian–diopside (formerly called fassaite), calcium-rich olivine, and kirschsteinite, and all but Angra dos Reis contain plagioclase. Aluminous spinel, troilite, whitlockite, ulvöspinel–magnetite solid solution, and nickel-rich metal are common minor or trace phases. Rare phases include celsian, baddeleyite, and a silico-phosphate phase. Angra dos Reis has an equilibrated texture, with groundmass olivine and pyroxene occurring as small xenomorphic, equidimensional grains joining at triple junctures, and larger, poikilitic pyroxene grains enclosing smaller grains. LEW 86010 has a hypidiomorphic–granular texture. Pyroxene grains are mostly anhedral, olivine grains are anhedral to subhedral, and plagioclase grains are subhedral to euhedral. Asuka 881371, D'Orbigny, LEW 87051 and Sahara 99555 have quenched, igneous textures—ophitic textures, common graphic intergrowths of olivine and plagioclase, and some are porphyritic with coarse subhedral to euhedral olivine grains.

Olivine and kirschsteinite compositions are uniform in Angra dos Reis and LEW 86010. Their olivines are more ferroan than the olivine cores of the quench-textured angrites, while the kirschsteinite grains are more magnesian. Olivines in Angra dos Reis and LEW 86010 have mg#s of 53 and 33, respectively, compared to most magnesian phenocrysts in LEW 87051, A-881371, D'Orbigny and Sahara 99555 with mg#s of 80, 67, 63, and 63. More magnesian olivine grains, interpreted to be xenocrystic, occur in LEW 87051, A-881371, and D'Orbigny, and can have mg#s of 85–91. Olivine in the quench-textured angrites are zoned to mg# ~ 0, and the CaO contents increase with decreasing mg#; reaching a maximum of ~ 20 mol.% in D'Orbigny. Kirschsteinite grains in Angra dos Reis and LEW 86010 have ~ 47 mol.% calcium, and mg#s of 38 and 23, respectively. Kirschsteinite in the quench-textured angrites is subcalcic, with calcium contents of $\sim 31-37$ mol.%, and mg#s of ~ 10 to 0.

The most magnesian pyroxenes are found in Angra dos Reis and LEW 86010, with compositions of $Wo_{55}En_{33}Fs_{12}$ and $Wo_{51}En_{36}Fs_{13}$, respectively. Pyroxenes in Angra dos Reis are homogeneous, while those in LEW 86010 exhibit moderate zoning from mg# of ~ 75 to ~ 50. Pyroxene cores have lower TiO_2 and Al_2O_3, and higher Cr_2O_3 contents than rims. Pyroxenes in the quench-textured angrites are highly zoned. The most magnesian pyroxene cores have mg#s in the range of 63–58, except for Sahara 99555 with mg# of ~ 50, and are zoned to mg# of ~ 0. The Al_2O_3 and TiO_2 contents are high and variable. The TiO_2 contents increase from 1–2 wt.% in the cores to 3–5 wt.% in the rims, while Al_2O_3 shows complex variation with mg#. Plagioclase grains are virtually end-member anorthite—the Na_2O contents are $\sim 0.02-0.04$ wt.%.

Two types of spinel are present in angrites: hercynitic spinel that crystallized relatively early, and ulvöspinel–magnetite solid solution that occurs in the mesostasis. The Cr_2O_3 contents are low in hercynitic spinels from the metamorphosed angrites—1.2–3.3 wt.%, and between 8.2 wt.% and 11.4 wt.% in the cores of grains in D'Orbigny. An unusual grain from A-881371 has a high Cr_2O_3 content of 20.9 wt.% (Mikouchi et al., 1996). A large spinel grain in D'Orbigny is interpreted to be xenocrystic based on its unusual size and composition, and a possible xenocrystic grain occurs in A-881371 (Mittlefehldt et al., 2002). The TiO_2 contents of the hercynitic spinels are generally <1 wt.%. The ulvöspinel–magnetite solid solution grains show little variation—their TiO_2 contents are 25–27 wt.%, and Al_2O_3 contents are 2.2–4.9 wt.%. These mesostasis spinels typically are 71–77 mol.% ulvöspinel, with 17–21 mol.% magnetite, calculated assuming stoichiometry. The high magnetite component indicates formation under more oxidizing conditions than experienced by most achondrites. This is supported by intrinsic oxygen fugacity measurements (Brett et al., 1977), experimental studies of europium partitioning (McKay et al., 1994), and the high-bulk rock FeO/MnO (Figure 1).

The angrites are basaltic in composition, and critically silica undersaturated—all have normative olivine, Ca_2SiO_4, and nepheline and lack orthopyroxene (Mittlefehldt et al., 2002). Angrites are also characterized by extreme depletions in moderately volatile elements (Figure 1). They have very low abundances of all the alkali elements, and have Ga/Al ratios lower than for any other achondrite, lunar sample, or martian meteorite (e.g., Warren et al., 1995). This extreme depletion in moderately volatile elements does not extend to the highly volatile/mobile elements—angrite data for these overlap the ranges for eucrites and lunar basalts (Mittlefehldt et al., 1998; Warren et al., 1995). An exception to this is lead—the $^{204}Pb/^{238}U$ ratios inferred for source regions based on chronological studies (Lugmair and Galer, 1992; Tera et al., 1997) are substantially lower for angrites than eucrites (Mittlefehldt et al., 1998). Thus, the angrite parent body was depleted in volatile lead relative to refractory lithophile elements as compared to the eucrite parent body (Lugmair and Galer, 1992).

Angrites are enriched in incompatible lithophile elements as expected for basalts (Figure 6), and all but Angra dos Reis have similar fractionated refractory lithophile element patterns (Mittlefehldt et al., 2002). All are depleted in aluminum relative to other incompatible refractory lithophile elements. They have Ca/Al ratios of ~ 1.47–$1.59 \times$ CI, with Angra dos Reis showing an extreme ratio of $\sim 3.06 \times$ CI—they are distinct from basaltic eucrites in this (Mittlefehldt et al., 2002). All but Angra dos Reis have higher Sc/Sm ratios for a given Yb/La ratio than do basaltic eucrites (Figure 7), indicating distinct fractionation paths (Mittlefehldt et al., 1998). The Sc/Sm ratio of Angra dos Reis is lower than those of all main-group and Nuevo Laredo-trend basaltic eucrites, but identical to that of the Stannern-trend basaltic eucrite bouvante (Mittlefehldt et al., 1998).

Chronological studies show that the angrites are essentially as old as the solar system. Isotopic anomalies in chromium due to the decay of

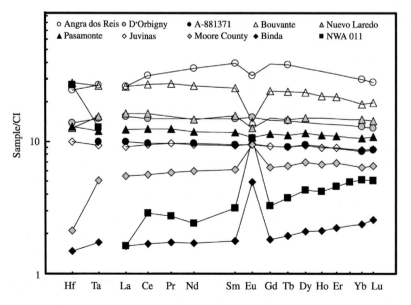

Figure 6 REE and hafnium, tantalum diagram for representative mafic differentiated achondrites. Data are taken from sources listed in the text.

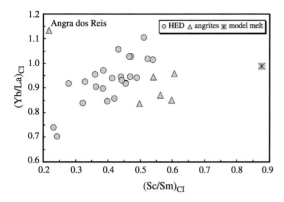

Figure 7 CI-normalized Yb/La versus Sc/Sm for mafic differentiated achondrites. Model melt is calculated to be in equilibrium with only olivine. The generally lower Sc/Sm ratios of the achondrites suggest that equilibration with, or fractionation of pyroxene also occurred. Data from sources listed in the text.

short-lived ^{53}Mn ($t_{1/2}$ 3.7 Ma) have been measured in mineral separates from LEW 86010 (Lugmair et al., 1992; Nyquist et al., 1994). These data show that LEW 86010 was formed within 15–20 Ma of the oldest known solar system materials (Nyquist et al., 1994). Angra dos Reis, A-881371, D'Orbigny, and LEW 86010 are enriched in xenon derived from fission of refractory lithophile ^{244}Pu ($t_{1/2}$ 81.8 Ma), but not in ^{129}Xe excess from the decay of volatile ^{129}I ($t_{1/2}$ 16 Ma) (Eugster et al., 1991, 2002; Hohenberg, 1970; Hohenberg et al., 1991; Lugmair and Marti, 1977; Wasserburg et al., 1977; Weigel et al., 1997). Finally, Angra dos Reis and LEW 86010 have anomalies in ^{142}Nd/^{144}Nd ratios consistent with decay of short-lived ^{146}Sm ($t_{1/2}$ 103 Ma) (Jacobsen and Wasserburg, 1984; Lugmair and Marti, 1977; Lugmair and Galer, 1992; Nyquist et al., 1994).

The Sm–Nd ages for Angra dos Reis and LEW 86010 are 4.53–4.56 Ga (Jacobsen and Wasserburg, 1984; Lugmair and Galer, 1992; Lugmair and Marti, 1977; Nyquist et al., 1994). The very low ^{204}Pb/^{238}U ratio of their parent body allows for very precise Pb–Pb ages of angrite formation. Lugmair and Galer (1992) determined Pb–Pb model ages for pyroxene separates from Angra dos Reis and LEW 86010 of 4.5578 Ga. Preliminary Pb–Pb ages for D'Orbigny and A-881371 are 4.559 Ga and 4.567 Ga, respectively (Jagoutz et al., 2002; Premo and Tatsumoto, 1995).

Angrites are believed to have formed as partial melts of primitive source materials under relatively oxidizing conditions (Longhi, 1999; Mittlefehldt and Lindstrom, 1990; Mittlefehldt et al., 2002). Angra dos Reis is distinct from the others—its major-element composition indicates a different fractionation path for its parent melt (Longhi, 1999). Key incompatible element ratios and contents (Ca/Al, Sc/Sm, Yb/La, REE; Figures 6 and 7) in Angra dos Reis are quite distinct from all other angrites, and support a different petrogenetic history for it (Mittlefehldt et al., 1998, 2002). All other angrites could have followed a similar partial melting/crystallization path (Longhi, 1999). In detail, the angrite suite does not form a single fractionation sequence; several parent melts are required (see Mittlefehldt et al., 2002).

1.11.3.2 Aubrites

Aubrites are the most reduced achondrites in our collections. Their silicates are essentially FeO free (Figure 1); they contain sulfides of calcium, chromium, manganese, titanium, and sodium—all normally lithophile elements—and silicon-bearing FeNi metal. They share a similar highly reduced nature, unusual mineralogy, and oxygen-isotopic composition with enstatite chondrites (Clayton et al., 1984; Keil, 1968; Watters and Prinz, 1979), and are generally considered to be related to enstatite chondrite-like parent bodies. Shallowater and Mt. Egerton may be related to the aubrites. They will be discussed here even though they show substantial differences from typical aubrites.

Petrologic and chemical information on aubrites was taken from: Biswas et al. (1980); Easton (1985); Floss and Crozaz (1993); Floss et al. (1990); Keil et al. (1989); Lodders et al. (1993); McCoy (1998); Okada and Keil (1982); Okada et al. (1988); Reid and Cohen (1967); Strait (1983); Watters and Prinz (1979); Wheelock et al. (1994); and Wolf et al. (1983).

Aubrites are brecciated orthopyroxenites consisting of 75–98 vol.% FeO-free enstatite (mg# >99), with variable contents of plagioclase, high-calcium pyroxene and forsterite, minor amounts of kamacite, troilite, oldhamite, daubreelite, and alabandite, and trace amounts of caswellsilverite, djerfisherite, heideite, niningerite, and schreibersite. The original enstatite grains were very coarse and exhibit remnant original igneous textures (Okada et al., 1988). High-calcium pyroxene is FeO free with Wo$_{40-46}$. Plagioclase modal abundances vary from 0.3 vol.% to 7.3 vol.%, with Bishopville having an anomalously high abundance (16.2 vol.%). They are sodic, typically Ab$_{88}$ to Ab$_{95}$, although more calcic plagioclase grains are also present. Forsterite is FeO free.

Aubrites contain an unusual suite of sulfides, many containing normally lithophile elements as their major cations. Oldhamite, a minor phase in aubrites, typically occurs as grains in the matrix, but also in sulfide-rich clasts. Oldhamite is the major REE carrier in aubrites, with abundances typically 100–1,000 × CI chondrites. REE patterns in oldhamite grains are highly variable.

There are fewer analyses of REE in alabandite, and these show less variability. Alabandite shows LREE-depleted patterns, with lanthanum abundances of ~0.1 × CI and lutetium abundances of ~10 × CI (Wheelock et al., 1994).

Silicon-bearing kamacite is a minor component of most aubrites, and contains 3.7–6.8 wt.% Ni and 0.1–2.4 wt.% Si. Metal nodules from several aubrites contain approximately chondritic abundances of refractory to moderately volatile siderophile elements (Casanova et al., 1993).

Shallowater and Mt. Egerton are unique meteorites that may be closely related to the aubrites. Shallowater is unbrecciated, contains higher abundances of both metal and troilite than other aubrites (Keil et al., 1989; Watters and Prinz, 1979), and the pyroxene is ordered orthopyroxene, rather than disordered pyroxene common to the group (Reid and Cohen, 1967). Shallowater also contains xenoliths composed of twinned clinoenstatite, forsterite, plagioclase, FeNi metal, and troilite that comprise ~20 vol.% of the rock (Keil et al., 1989). Mt. Egerton is also unbrecciated and composed of centimeter-sized enstatite crystals with ~21 wt.% interstitial FeNi metal.

Basaltic vitrophyre clasts have been found in two aubrites, Khor Temiki and LEW 87007 (Fogel, 1994, 1997). These clasts contain 60–90 vol.% enstatite, 10–40 vol.% of Na–Al–K–Ca-enriched glass, and minor diopside or olivine, troilite, metal.

Bulk analyses for major or trace elements exist for some aubrites, but the very coarse grain sizes of the rocks make truly representative sampling impossible. Aubrites are depleted in a basaltic component. They have CI-normalized Al/Si ratios of 0.08–0.54 and Ca/Si ratios of 0.04–0.37, lower than in any chondrite group (see Lodders and Fegley, 1998). Incompatible trace lithophile elements are usually $<1 \times$ CI (Figure 8). However, multiple analyses of individual aubrites can show differences in REE by a factor of 10, including samples with REE $>1 \times$ CI. Most aubrites exhibit depletions in europium. Exceptions are Bishopville, which has a much higher modal plagioclase content, Shallowater, and some samples of Peña Blanca Spring. Wolf et al. (1983) showed that a dark fraction of Khor Temiki was enriched in trace lithophile, siderophile, and chalcophile elements compared to light samples. Newsom et al. (1996) did a petrographic and chemical study of a number of dark clasts from Khor Temiki, and found a range of REE contents of ~0.4–20 × CI REE (Figure 8), and variable chalcophile and siderophile element contents. They concluded that the dark clasts are simply shock-darkened material containing differing amounts of sulfides.

Aubrites are depleted in siderophile elements relative to chondritic abundances. Excluding Shallowater, bulk rocks have CI-normalized Fe/Si ratios of 0.006–0.055. Shallowater, with a higher modal metal content (3.7% versus ≤0.7%; Watters and Prinz, 1979), has a CI-normalized Fe/Si of 0.33 (Easton, 1985). Trace siderophile elements are also depleted; for example, excluding dark samples, iridium varies from ~$10^{-5} \times$ CI to $10^{-2} \times$ CI. Dark clasts from Khor

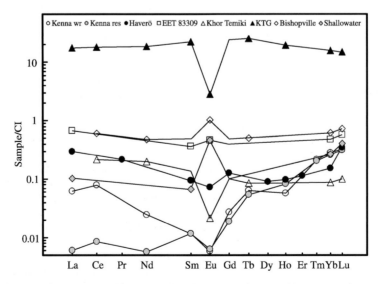

Figure 8 REE diagram for representative ultramafic differentiated achondrites. Kenna wr and Kenna res are whole-rock and acid-leached residues, illustrating the easily removed LREE-enriched component found in many ureilites. KTG is a sulfide-rich dark clast from Khor Temiki, illustrating that REEs are concentrated in sulfides in aubrites. Kenna, Haverö, and EET 83309 are ureilites, the others are aubrites. Data are taken from sources listed in the text.

Temiki have iridium abundances varying from $1.8 \times 10^{-3} \times$ CI to $1.7 \times 10^{-2} \times$ CI (Newsom et al., 1996), while Khor Temiki dark has an iridium abundance of $5.4 \times$ CI (Wolf et al., 1983). Shallowater samples have iridium abundances of $0.58-1.48 \times$ CI (Keil et al., 1989).

The chronology of aubrites is poorly constrained. Bogard et al. (1967) presented Rb–Sr and K–Ar ages for Norton County that are compatible with the age of the solar system, while Compston et al. (1965) determined an age of 3.7 Ga for Bishopville. Hohenberg (1967) determined that Shallowater began retaining ^{129}Xe from ^{129}I decay at roughly the same time as chondritic meteorites.

Some authors have argued that aubrites are nebular materials (Sears, 1980; Wasson and Wai, 1970), but the consensus, as of early 2000s, is that they are igneous rocks. Okada et al. (1988) argued that the textures, mineral compositions, and trace-element compositions of Norton County favored an origin involving extensive fractional crystallization of an ultramafic magma. By inference, the same holds for other aubrites. In contrast, Lodders et al. (1993) concluded that while the aubrite parent body was melted sufficiently to segregate metal and sulfide into a core, large-scale silicate fractionation did not occur. Keil et al. (1989) suggested that the anomalous aubrite Shallowater is an igneous rock from a distinct parent body formed when a solid enstatite chondrite-like asteroid crashed into a molten aubrite-like asteroid, and rapidly cooled and crystallized.

1.11.3.3 Brachinites

The brachinite group is small and of somewhat diverse petrology. Nehru et al. (1992) classified them as primitive achondrites, but recent studies have favored cumulate origin for several members (Mittlefehldt et al., 2003; Swindle et al., 1998; Warren and Kallemeyn, 1989a).

Petrologic and compositional details were taken from: Floran et al. (1978a); Johnson et al. (1977); Mittlefehldt et al. (2003); Nehru et al. (1983, 1996); Smith et al. (1983); Swindle et al. (1998); and Warren and Kallemeyn (1989a). For many, the only available information is in Grady (2000) or the Meteoritical Bulletin.

Brachinites consist dominantly of olivine, 79–93%, and all contain high-calcium pyroxene, 3–15%. Only some contain plagioclase—ALH 84025, Eagles Nest, and NWA 595 are plagioclase-free, while the remainder contain from a trace up to 9.9% for Brachina. Orthopyroxene is absent or at trace levels in most, but Hughes 026 contains 1.6%, while NWA 595 reportedly contains 10–15% orthopyroxene. Brachina contains orthopyroxene only in melt inclusions in olivine (Nehru et al., 1983). Chromite and iron-sulfide are minor components reported in most brachinites, and metal and phosphates are trace components in several. Brachinites typically have equigranular or xenomorphic granular textures, but with varying grain sizes. Some contain prismatic olivine grains, some showing preferred orientation (Mittlefehldt et al., 2003). High-calcium pyroxene, chromite, and plagioclase appear to be interstitial to olivine, while the textural setting of orthopyroxene is not well described.

Olivine grains are homogeneous but their compositions differ among brachinites from Fo_{64} to Fo_{71}. Their FeO/MnO ratios from 52 to 77, and they contain 0.09–0.27 wt.% CaO. High-calcium pyroxene is usually diopside, but is augite (Wo < 45) in ALH 84025 and Brachina. High-calcium pyroxene is more magnesian than olivine, with mg# of 79–82. Minor element contents of high-calcium pyroxenes exhibit the following ranges: Na_2O—0.36–0.61%; Al_2O_3—0.55–1.03%; TiO_2—0.1–0.4%; Cr_2O_3—0.68–1.04%. Orthopyroxene is $Wo_{2-4} En_{69-73} Fs_{25-28}$. Chromite has cr# of 73–84, and mg# of 18–30. The classification description of Reid 027 suggests that it contains very unusual chromites, with cr# of 93 and mg# of 9 (Grady, 2000). The TiO_2 contents of chromite are typically in the range 1.0–1.4 wt.%, but those in Brachina have ~ 2.5 wt.%. Sulfide is mainly troilite, with variable nickel contents from 0.04 wt.% to 2.0 wt.% in ALH 84025 and from 0.6 wt.% to 3.6 wt.% in Brachina.

Compositional data are available for ALH 84025, Brachina, Eagles Nest, and EET 99402 and EET 99407, which are paired. The brachinites are quite variable in composition. Brachina is approximately chondritic in lithophile element abundances—CI-normalized element/Mg ratios are between 0.76 and 1.04 for most refractory and moderately volatile lithophile elements (Figure 2). The two most volatile of the moderately volatile elements, sodium and zinc, have ratios of 0.60, while refractory elements thorium and lanthanum have ratios of 2.7 and 1.3, respectively. Thorium and lanthanum have the highest continental crust/CI ratios of the elements measured, so contamination of this find is a possible explanation. Eagles Nest shows enrichments in several elements concentrated in the continental crust indicating that contamination is pervasive. Nevertheless, CI-normalized Na/Mg is 0.043, consistent with the absence of plagioclase. ALH 84025 also has low CI-normalized Al/Mg and Na/Mg ratios of 0.055, consistent with the absence of plagioclase. ALH 84025 is strongly depleted in highly incompatible refractory lithophile elements with La/Mg and Sm/Mg ratios of 0.15 and 0.12. Curiously, Eu/Sm is $2.55 \times$ CI (Figure 9), suggesting that some plagioclase is present although it has not been observed. An average of the EET 99402–EET 99407 pair has

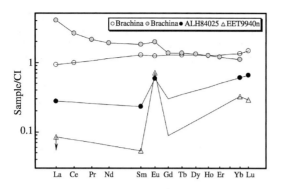

Figure 9 REE diagram for brachinites illustrating the wide range in REE contents of these achondrites. Two samples of Brachina show differing LREE contents, suggesting that terrestrial contamination may have occurred. Data are taken from sources listed in the text.

the most fractionated lithophile element pattern. Ratios of Al/Mg and Na/Mg are $0.6 \times$ CI and $0.3 \times$ CI, consistent with the presence of plagioclase, and Eu/Sm is $13.2 \times$ CI (Figure 9). The highly incompatible refractory lithophile elements are severely depleted, with La/Mg and Sm/Mg of ≤ 0.05 and $0.03 \times$ CI.

Siderophile element abundance patterns are variable, but brachinites are depleted in them—Ir/Mg ratios are $0.12-0.15 \times$ CI for ALH 84025, Brachina, and Eagles Nest, and $0.038 \times$ CI for EET 99402–EET 99407. All are enriched in cobalt relative to other siderophile elements, with Co/Ni ratios of $1.6-2.1 \times$ CI for ALH 84025, Brachina and Eagles Nest, but $6.2 \times$ CI for EET 99402–EET 99407. The high Co/Ni and Co/Mg ratios are plausibly due to substantial lithophile character for cobalt compared to other siderophile elements in these relatively oxidized meteorites (Mittlefehldt et al., 2003; Figure 1).

There are few published data for noble gases in brachinites, many only in abstracts. The trapped noble gas contents of ALH 84025, Brachina, and EET 99402 are relatively high, with $^{36}Ar_{trapped}$ of $(0.82-2.6) \times 10^{-8}$ cm^3 STP g^{-1}, and have isotopic ratios within the range of ureilites (Mittlefehldt et al., 2003; Ott et al., 1985, 1987).

Limited chronological information is available for brachinites. The short-lived nuclide systems indicate that the brachinites were formed early in solar system history. Excess ^{129}Xe from in situ decay of ^{129}I ($t_{1/2}$ 16 Ma) has been measured in Brachina, ALH 84025, and Eagles Nest. The Brachina data display a correlation of ^{129}Xe/^{132}Xe with ^{128}Xe/^{132}Xe, indicating that ^{129}I was present at the time of formation (Mittlefehldt et al., 2003), while the Eagles Nest data indicate that it began retaining xenon within ~ 50 Ma of primitive chondrites (Swindle et al., 1998).

Excess ^{53}Cr, correlated with the Mn/Cr ratio, indicates that ^{53}Mn ($t_{1/2}$ 3.7 Ma) was present in Brachina, and a formation age of 4.5637 ± 0.0009 Ga was calculated by comparison with the ^{53}Mn/^{55}Mn ratio and Pb–Pb age of angrite LEW 86010 (Wadhwa et al., 1998).

A ^{39}Ar–^{40}Ar spectrum of EET 99402 indicates that a degassing event occurred 4.13 Ga ago, assuming the event totally degassed one of the carrier phases for potassium (Mittlefehldt et al., 2003). Brachina shows a more complex age spectrum, but is compatible with a degassing event at 4.13 Ga (Mittlefehldt et al., 2003).

Interpretation of the origin of brachinites is in flux. Originally, Brachina was thought to be an igneous rock that crystallized from a melt of its own composition (Johnson et al., 1977; Floran et al., 1978a), but this cannot explain the fractionated brachinites discovered since. Currently, the major hypotheses are: (i) metamorphism and oxidation of chondritic material, possibly including anatexis (Nehru et al., 1983, 1992, 1996); or (ii) accumulation from a magma (Mittlefehldt et al., 2003; Swindle et al., 1998; Warren and Kallemeyn, 1989a).

Brachina has near-chondritic abundances of lithophile elements (Figure 2), and relatively high abundances of siderophile elements and trapped noble gas contents (Ott et al., 1985, 1987). These suggest that igneous processes have not significantly modified it, a defining characteristic of primitive achondrites. However, melt inclusions are present in its olivine (Floran et al., 1978a; Nehru et al., 1992), but have not been described from acapulcoites and winonaites (e.g., Benedix et al., 1998; McCoy et al., 1996), and its siderophile element contents are lower than those of acapulcoites and winonaites (Figure 2). These favor an igneous origin for Brachina. ALH 84025, Eagles Nest, and EET 99402–EET 99407 are distinctly fractionated from primitive compositions, with aluminum and sodium depletions, and fractionated, igneous REE patterns (Figure 9).

Nehru et al. (1992, 1996) posited that brachinites evolved from CI-like material oxidized during metamorphism. In some locations, intense heating causes partial melting and removal of basalts, while other areas only reached metamorphic temperatures. Warren and Kallemeyn (1989a) suggested that ALH 84025 is an olivine heteradcumulate, and that Brachina is an olivine orthocumulate. They found that the textures and compositions of these rocks are more like those of cumulates than metamorphic rocks. Mittlefehldt et al. (2003) and Swindle et al. (1998) argued for a cumulate origin for EET 99402–EET 99407 and Eagles Nest based on mineralogy, texture, and composition.

Significant petrologic and geochemical differences exist among brachinites and the same process may not have formed all of them. About half of the brachinites have not been subjected to detailed study. Thus, interpretation of their petrogenesis will likely undergo significant revision.

1.11.3.4 Howardite–Eucrite–Diogenite Clan

The Howardite–Eucrite–Diogenite (HED) clan is the most extensive suite of differentiated crustal rocks from an asteroid. They strongly influence ideas about asteroidal differentiation processes and the nature of the energy source responsible for heating asteroid-size bodies early in solar system history. The HED clan is also the only asteroidal meteorite group for which we have a strong candidate parent body. Circumstantial evidence makes a strong case that 4 Vesta may be the parent body of the HED suite. This asteroid is covered with rock having reflectance spectra like those of eucrites and howardites (McCord et al., 1970). Several kilometer-size asteroids with identical reflectance spectra and orbits similar to that of 4 Vesta have been identified (Binzel and Xu, 1993). A large crater is present on 4 Vesta that could be the ejection source for these small "Vestoids" (Thomas et al., 1997). Thus, the HED suite allows us, with some confidence, to associate specific types of igneous processes with a body of known size.

There is a very large body of literature on the petrology and composition of HED meteorites. Representative studies include: Barrat et al. (2000); Bowman et al. (1997); Buchanan and Reid (1996); Buchanan et al. (2000a,b); Delaney et al. (1984); Floran et al. (1981); Fowler et al. (1994, 1995); Fukuoka et al. (1977); Lovering (1975); Metzler et al. (1995); Mittlefehldt (1979, 1994, 2000); Mittlefehldt and Lindstrom (2003); Palme et al. (1978, 1988); Sack et al. (1991); Takeda and Graham (1991); Takeda et al. (1994b,c); Wänke et al. (1972, 1973, 1977); Warren and Jerde (1987); and Yamaguchi et al. (2001).

The HED suite is composed of mafic and ultramafic igneous rocks, most of which are breccias. The parent lithologies were mostly metamorphosed, which has obscured original igneous zoning in most cases. The suite contains three main igneous lithologies—basalt, cumulate gabbro, and orthopyroxenite.

Diogenites are orthopyroxenites. Most are monomict, fragmental breccias of original coarse-grained rock composed of >90% orthopyroxene, with minor chromite. Olivine is a minor phase, but reaches high modal abundances in some. Troilite, metal, and silica are accessory phases in many diogenites. Diopside occurs as exsolution products. Plagioclase occurs in several members, but often in the breccia matrix and may be foreign contaminants in these cases. Phosphates are rare accessory phases. Orthopyroxene is magnesian—most have orthopyroxene with mg#s of 72–77, and Wo_{1-2}. The most magnesian diogenite has an mg# of 84, while the most ferroan have mg#s of 66. The major element contents of orthopyroxene grains are very uniform, but the minor and trace incompatible elements show considerable variation in the suite and within some individuals. Diogenite olivines generally have mg#s from 73 to 70, with EETA79002 having more magnesian (mg# 76) and Peckelsheim having more ferroan (mg# 65) olivines. Chromite in diogenites is variable in mg# (13–29) and Al_2O_3 (6.1–22.1 wt.%), and low in TiO_2 (0.11–1.47 wt.%).

Most basaltic eucrites are fragmental breccias of fine to medium grained, subophitic to ophitic basalts composed of pigeonite and plagioclase, with minor silica, ilmenite, and chromite, and accessory phosphates, troilite, metal, fayalitic olivine, zircon, and baddeleyite. The few unbrecciated basaltic eucrites have all been highly metamorphosed. The pyroxenes were originally ferroan pigeonite, $\sim Wo_{6-15}En_{29-43}Fs_{48-58}$, which exsolved augite during metamorphism. In most eucrites, pyroxene Fe/Mg is uniform as a result of metamorphism; original igneous zoning is preserved in very few. Plagioclase is calcic, with most in the range An_{75-93}, and igneous zoning is commonly preserved. Chromite grains are ferroan (mg# 1.4–7.8) and enriched in TiO_2 (1.4–24 wt.%) compared to diogenite chromites, but have Al_2O_3 contents overlapping the low end of the diogenite range (3–9 wt.%).

Cumulate eucrites are intermediate in mg# between diogenites and basaltic eucrites. They are coarse-grained gabbros, many unbrecciated, composed of pigeonite, plagioclase, and minor chromite with silica, ilmenite, metal, troilite, and phosphate as trace accessory phases. The original igneous pyroxene was pigeonite, $\sim Wo_{7-16}En_{38-61}Fs_{32-46}$, which exsolved augite and, in some, inverted to orthopyroxene. Plagioclase is generally more calcic than typical for basaltic eucrites, with most in the range An_{91-95}. Chromite grains have mg#s of 4.0–11.5, TiO_2 contents of 3.2–10.1 wt.%, and Al_2O_3 between 5.7 wt.% and 8.3 wt.%.

The HED suite includes a wide variety of polymict rocks consisting of fragmental and melt–matrix breccias dominantly of material from the three igneous lithologies. Most of these polymict breccias—polymict diogenites, howardite, and polymict eucrites—are mixtures of diogenites and basaltic eucrites. However, some polymict eucrites and howardites contain substantial cumulate eucrite components, and polymict cumulate eucrites are also represented.

Diogenites are very uniform in bulk major and minor element composition, with the exception of

Cr$_2$O$_3$, which reflects variation in chromite content. The basaltic eucrites are also very uniform in composition, and have very limited ranges in mg#, CaO, Al$_2$O$_3$, TiO$_2$, and Cr$_2$O$_3$. The cumulate eucrites show a wider range in bulk major and minor element composition, but this is partially due to sampling problems for these coarse-grained, heterogeneous rocks. The mg# of the cumulate eucrites ranges from 65 to 44, while the Cr$_2$O$_3$ contents vary from 0.33 wt.% to 1.05 wt.%. Both the diogenites and cumulate eucrites show considerable variation in trace incompatible lithophile element contents from three causes: (i) variation in the trace-element content of the cumulus minerals; (ii) variation in the amount of a trapped melt component; and (iii) admixture of small amount of basaltic eucrite contaminants even in supposed monomict breccias. (Monomict breccias are defined as containing no foreign components (see Delaney et al., 1983), but in practice, this is impossible to establish. Hence, many of the so-called monomict breccias may well contain some small amount of foreign material.) As a result, the contents of highly incompatible elements, such as lanthanum, vary by a factor of 700 within the diogenite suite, and by a factor of ~20 among cumulate eucrites. In contrast, the lithophile element variations in the basaltic eucrite suite are small—about a factor of 2 for lanthanum.

Substantial variations in minor and trace incompatible lithophile element contents of diogenite cumulus orthopyroxenes are established by microbeam measurements. Strong correlations among incompatible lithophile elements, such as aluminum, titanium, and samarium, are consistent with igneous fraction, but they are not anticorrelated with mg# as would be expected. Major elements are thus decoupled from minor- and trace-element contents among diogenites. The trace-element contents of cumulate eucrite pyroxenes were redistributed during exsolution of augite, and igneous trends are difficult to establish. Hsu and Crozaz (1997) suggested that cumulate eucrites were formed from magmas with trace-element contents unlike those of basaltic eucrites, but this has been disputed (Treiman, 1997). Bulk rock trace element contents of basaltic eucrites show two distinct trends. One shows slightly increasing mg#, with decreasing incompatible lithophile element content (Stannern-trend), while the other shows a substantial decrease in mg#, with increasing incompatible lithophile element contents (Nuevo Laredo-trend). These are considered to be distinct magmatic trends, but their origin is controversial.

The HED suite is poor in moderately volatile and volatile elements, and very depleted in highly siderophile elements. The moderately volatile and volatile lithophile element content of the HED parent body is estimated to be similar to that of the Moon (see Figure 1). Chondrite-normalized ratios of sodium, potassium, rubidium, and caesium to incompatible refractory lithophile elements are ~0.04, ~0.04, ~0.01, and ~0.004 (Mittlefehldt, 1987). The innate contents of highly siderophile elements in HED igneous lithologies are difficult to establish because of extensive impact brecciation and contamination by chondritic debris (see Chou et al., 1976). Nevertheless, the iridium contents of a suite of Antarctic monomict eucrites are at $10^{-5}-10^{-6} \times$ CI abundances (Warren et al., 1996), indicating efficient separation of metal from silicate on the parent body. Diogenites and cumulate eucrites have iridium contents of $10^{-4}-10^{-6} \times$ CI. The contents of rhenium and osmium are similarly in the range of $10^{-4}-10^{-6} \times$ CI in igneous lithologies (Warren et al., 1996). Because it is more easily oxidizable, cobalt contents are much higher than for other siderophile elements; typical cobalt contents are 6–62 µg g^{-1} for diogenite orthopyroxene clasts, 6–10 µg g^{-1} for cumulate eucrites, and typically 3–8 µg g^{-1} for basaltic eucrites, $\sim 10^{-2} \times$ CI.

The bulk major-, minor-, and trace-element contents of the polymict breccias are intermediate in composition between those of the igneous lithologies that compose them. However, the siderophile element contents of polymict breccias can be higher than in the igneous lithologies because of admixed chondritic debris. The iridium content of howardites is typically on the order of $10^{-2} \times$ CI—two to four orders of magnitude higher than in the parent igneous lithologies.

Magmatism on the HED parent body was nearly isochronous with formation of the rocky planetesimals. A best estimate for the formation age of eucrites from Rb–Sr data is 4.548 ± 0.058 Ga (Smoliar, 1993), while combining Mn–Cr systematics on diogenites and basaltic eucrites with Mn–Cr and Pb–Pb data on the angrite LEW 86010 results in a time of differentiation of the HED parent body of 4.568 ± 0.0009 Ga (Lugmair and Shukolyukov, 1997). Unbrecciated cumulate eucrites are younger; they yield Pb–Pb ages in the range 4.484–4.399 Ga (Tera et al., 1997), but there is no consensus whether this reflects late formation or metamorphism. The Ar–Ar ages on brecciated HED meteorites were reset in the time period 4.1–3.4 Ga ago, which is suggested to be a period of impact modification of the HED parent body surface (Bogard, 1995). The unbrecciated, metamorphosed basaltic eucrites have Ar–Ar ages of ~4.5 Ga, between the ages of HED differentiation and impact modification (Bogard and Garrison, 1995, 2001, 2002; Yamaguchi et al., 2001).

There is no consensus on the mechanism by which the HED parent body differentiated.

Detailed models for HED petrogenesis include those of Barrat et al. (2000), Righter and Drake (1997), Ruzicka et al. (1997), Shearer et al. (1997), Stolper (1977), and Warren (1997). The contending models are: (i) partial melting of a primitive body; (ii) total melting followed by fractional crystallization of a magma ocean; (iii) total melting followed by equilibrium crystallization of a magma ocean; and (iv) *in situ* crystallization of a magma ocean. Each of these models has what appears to be a fatal flaw, suggesting that either not all HED meteorite types are from the same parent body, or that assumptions that go into the models are invalid (Mittlefehldt and Lindstrom, 2003). Partial melting models appear incapable of explaining the putative correlation between the moderately siderophile element tungsten and incompatible lithophile elements such as lanthanum, and the very low siderophile element contents. The equilibrium crystallization model seems incapable of explaining the extreme incompatible element fractionations observed in the diogenite suite orthopyroxenes, while the fractional crystallization and *in situ* crystallization models cannot explain both the Stannern- and Nuevo-Laredo-trends as arising from a common process.

1.11.3.5 Mesosiderite Silicates

Mesosiderites are stony irons in which the rocky material is a polymict breccia of crustal rock from a differentiated body—basalt, gabbro, pyroxenite, dunite, and anorthosite. The silicates are very similar to the HED suite meteorites. Thus, mesosiderite silicates are discussed here, but the metallic phase is not.

Petrologic and compositional information on mesosiderites was taken from: Delaney et al. (1980, 1981); Floran (1978); Floran et al. (1978b); Hewins (1984); Ikeda et al. (1990); Kimura et al. (1991); Mittlefehldt (1979, 1980, 1990); Mittlefehldt et al. (1979); Powell (1971); Rubin and Jerde (1987, 1988); Rubin and Mittlefehldt (1992); and Simpson and Ahrens (1977).

The silicates consist of mineral and lithic clasts set in a fine-grained fragmental to impact–melt matrix. The most common lithic clasts are basalts, gabbros, and orthopyroxenites, while dunites are minor and anorthosites are rare. The most common mineral clasts are centimeter-sized orthopyroxene and olivine fragments, while millimeter-sized plagioclase fragments are less common.

Olivine clasts are typically single-crystal fragments, varying in composition from Fo_{92} to Fo_{58}. Mesosiderites also contain fine-grained olivine. Most literature analyses are of unspecified grains, and metamorphic equilibration may have altered the compositions of smaller grains. Coarse-grained olivines have FeO/MnO of 40–45, within the range of main-group pallasite olivines (Mittlefehldt et al., 1998). Low-calcium pyroxene clasts include centimeter-sized orthopyroxene and millimeter-sized pigeonite. The orthopyroxene clasts are compositionally similar to diogenites and orthopyroxenite clasts in howardites. Compositional ranges are from about Fs_{20} to Fs_{40} for low-calcium pyroxenes, and orthopyroxenes more ferroan than about Fs_{30} were inverted from pigeonite (Powell, 1971). Plagioclase grains are calcic, like those of HED meteorites.

Basaltic and gabbroic clasts are generally similar to eucrites and mafic clasts in howardites. They are composed of ferroan pigeonite and calcic plagioclase, with minor to accessory silica, whitlockite, augite, chromite, and ilmenite. Metal and troilite are common, but these may have been added during brecciation. Many of the mafic clasts are distinguishable from similar HED meteorite materials in detail—they generally have higher modal proportions of tridymite and whitlockite (Nehru et al., 1980; Rubin and Mittlefehldt, 1992), and have chromite > ilmenite, opposite to what is found for basaltic eucrites (Delaney et al., 1984; Nehru et al., 1980). Pyroxene compositions in mafic clasts generally have lower FeO contents and lower FeO/MnO ratios than those of HED meteorites (Mittlefehldt, 1990; Mittlefehldt et al., 1998).

The mesosiderite matrix varies from cataclastic texture, with highly angular mineral fragments (texture grade 1) to igneous-textured matrix (texture grade 4) (Floran, 1978). Matrix pyroxene grains were affected by FeO-reduction and metamorphic equilibration. They typically have low FeO/MnO ratios and are relatively magnesian, irrespective of metamorphic grade—no matrix pyroxene grains have compositions as ferroan as pyroxenes in the basaltic clasts. This is unlike the case for howardites, where ferroan pigeonite is a common constituent of the matrix (e.g., Reid et al., 1990).

There are very few bulk analyses for coarse-grained olivine clasts from mesosiderites, and observed distinctions between them and pallasite olivines are mostly due to chromite and metal inclusions in the former (Mittlefehldt, 1980). Orthopyroxene clasts are generally similar to diogenites in major-, minor-, and trace-element composition (see Mittlefehldt et al., 1998).

The mesosiderite basalt and the gabbro clasts display wider ranges in trace-element compositions than do similar HED lithologies. All mesosiderite clasts have high nickel and cobalt contents and some have high selenium contents, undoubtedly due to impact mixing with matrix metal and troilite. Some mesosiderite basalts are identical in composition to basaltic eucrites.

A clast from Mount Padbury has a mg# of ~36, a molar FeO/MnO of ~36, and a flat REE pattern at 9–10 × CI chondrite abundances (Mittlefehldt, 1979)—all within the ranges for basaltic eucrites. However, many basaltic clasts are distinct in major element composition, with higher mg#s and lower molar FeO/MnO ratios than those of basaltic eucrites, and have LREE-depleted patterns and $(Eu/Sm)_{CI} > 1$—patterns unknown among unaltered basaltic eucrites. Some gabbro clasts are similar to cumulate eucrites in major- and trace-element contents, but many are distinct in having extreme depletions in the most incompatible elements (Mittlefehldt, 1979; Rubin and Mittlefehldt, 1992). In extreme cases, samarium abundances are only 0.02–0.03 × CI chondrites (Rubin and Jerde, 1987; Rubin and Mittlefehldt, 1992), much less than the 1–2 × CI typical of cumulate eucrites. These clasts have $(Eu/Sm)_{CI}$ of 220–260, the most extreme ratios known among solar system igneous rocks (Mittlefehldt et al., 1992).

Mesosiderite silicates are broadly similar to howardites in bulk composition, although there are distinctions. Major differences between mesosiderite silicates and howardites are that the former have generally lower FeO, systematically lower FeO/MnO and are enriched in P_2O_5 compared to howardites (Simpson and Ahrens, 1977). Mesosiderite silicates typically have lower incompatible trace-element contents, such as the LREE, than do howardites.

Age determinations have been done for many mesosiderites, but some of these are for bulk samples of polymict breccias and thus are difficult to interpret. Brouxel and Tatsumoto (1991) and Prinzhofer et al. (1992) showed that Estherville and Morristown were formed early in the history of the solar system and later metamorphosed. The U–Pb ages on zircons in a basaltic clast from Vaca Muerta yielded an age of 4.563 ± 0.015 Ga (Ireland and Wlotzka, 1992). The petrology of this clast was not given, so its origin is not known. Regardless, this demonstrates that magmatic processes occurred very early on the mesosiderite parent body. Stewart et al. (1994) studied three clasts from Vaca Muerta and one from Mount Padbury. The Sm–Nd isochron ages vary from 4.52 GPa to 4.48 Ga for three igneous clasts, and 4.42 Ga for an impact–melt clast.

Numerous $^{39}Ar-^{40}Ar$ ages have been done for a wide range of whole-rock and igneous clast samples (Bogard and Garrison, 1998; Bogard et al., 1990). Most have stepwise argon release profiles exhibiting modest increases in calculated age with extraction temperature. The $^{39}Ar-^{40}Ar$ ages for these samples are ~3.9 Ga, indicating either slow cooling or thermal resetting (Bogard and Garrison, 1998; Bogard et al., 1990).

The silicate fraction of mesosiderites is broadly similar to the HED meteorites, and the original formation of the silicates is generally believed to be the same as that of the HED suite. However, differences in detail between most mesosiderite lithic clasts, and eucrites and diogenites suggest that the later history of mesosiderites was different. Mittlefehldt (1979, 1990), Mittlefehldt et al. (1992), and Rubin and Mittlefehldt (1992) argued that many of the mesosiderite mafic clasts were remelted after metal–silicate mixing. This process formed magnesian basalts and gabbros with low FeO/MnO, high modal abundances of tridymite and/or whitlockite, and LREE-depleted patterns with high Eu/Sm. A large fraction of the mafic clasts are from remelted parent rocks, suggesting that a large fraction of the parent body crust was remelted after metal–silicate mixing.

The iron metal and troilite of mesosiderites are presumed to represent core materials of an asteroid. Mixing of this with crustal silicates requires an unusual formation process. Some have suggested that a naked molten core (a core with the silicate crust and mantle largely stripped off) impacted a differentiated asteroid at low velocity (Wasson and Rubin, 1985). Others have suggested that an impact disrupted the differentiated, mesosiderite parent body, which reaccreted. This process mixed materials from different portions of the parent body, with mesosiderites representing a location where the core and crust were mixed together (Haack et al., 1996; Scott et al., 2001).

Mesosiderites are polymict breccias, as are many HED meteorites, thus presenting evidence for impact mixing on the surfaces of their parent bodies. Impact mixing on asteroidal bodies is expected to be parent-body-wide (e.g., Housen et al., 1979), leading some researchers to argue that mesosiderites and HEDs were formed on different parent bodies (see Mittlefehldt et al., 1998; Rubin and Mittlefehldt, 1993).

1.11.3.6 Ureilites

The ureilite group is the second largest achondrite group, and exhibits some characteristics of both primitive and differentiated achondrites. However, the dominant silicates of ureilites seem to require a high-temperature, igneous origin, and no true primitive materials related to the ureilites are known (e.g., Goodrich, 1992).

Petrologic and compositional information on ureilites is from: Berkley and Jones (1982); Berkley et al. (1976, 1978, 1980); Binz et al. (1975); Boynton et al. (1976); Goodrich and Berkley (1986); Higuchi et al. (1976); Jaques and Fitzgerald (1982); Janssens et al. (1987); Sinha et al. (1997); Spitz and Boynton (1991);

Takeda (1987); Wang and Lipschutz (1995); Warren and Kallemeyn (1989b); and Wasson *et al*. (1976).

Ureilites are ultramafic rocks composed dominantly of olivine and pyroxene, with <10% interstitial material rich in elemental carbon. Modal abundances of the major silicates are quite variable, with pyroxene varying from 0% to 90%. Roughly one-half of ureilites contain pigeonite as the sole pyroxene, but in the others, pyroxene assemblages include various combinations of pigeonite, orthopyroxene, and/or augite.

Textures of ureilites are variable. Most contain millimeter-sized anhedral olivine and pyroxene grains with curved borders joining in triple junctions. Some have a poikilitic texture in which olivine grains contain rounded inclusions of pyroxene, and low-calcium pyroxene may contain inclusions of olivine and/or augite. Some contain elongated olivine and pyroxene grains showing foliation and lineation. Several have a mosaic texture of small, recrystallized grains. A few ureilites have a bimodal texture in which large, centimeter-sized crystals of low-calcium pyroxene poikilitically enclose regions with typical ureilite texture. Olivine grains typically have reduced rims where in contact with carbon-rich material. These rims consist of nearly end-member forsterite and/or enstatite containing small inclusions of low-nickel metal.

Olivine cores are very homogeneous in mg# within each ureilite, and range from ~74 to 95 with a peak at ~79 for the group, but Mn/Mg and Cr/Mg ratios are roughly constant (Mittlefehldt *et al*., 1998). They have high CaO (0.30–0.45 wt.%) and Cr_2O_3 (0.56–0.85%) contents, much higher than found for primitive achondrite olivines. Pyroxene core compositions show a similar range in mg#, and are also rich in Cr_2O_3, up to about 1.3 wt.%.

The carbon-rich material occurs mostly along silicate grain boundaries, but also intrudes silicates along fractures. Graphite is the most common polymorph, and is usually very fine-grained. In a few ureilites, graphite is present as large, millimeter-sized, euhedral crystals intergrown with metal and/or sulfide, and sometimes penetrates silicate grains. Diamond and lonsdaleite are also present in most, if not all, ureilites, typically as micrometer-sized grains within graphite.

Ureilites contain a variety of trace accessory phases. Metallic spherules composed of cohenite, metal, and sulfide are included in olivine and pigeonite of a few ureilites, and metal, phosphide, and sulfide are present as interstitial phases in most. The interstitial regions also commonly contain fine-grained silicates, including low-calcium pyroxene, augite and Si–Al–alkali glass.

Most ureilites are unbrecciated or are monomict, but a few are polymict. These latter fragmental breccias contain lithic clasts of normal monomict ureilite material, plus other materials, including clasts similar to enstatite chondrites or aubrites, chondritic clasts, feldspathic melt rocks, and clasts of a Ca–Al–Ti-rich assemblage similar to angrites. Most mineral fragments in polymict ureilites are identical to phases found in monomict ureilites. However, some compositionally distinct olivine grains are found, and plagioclase, which is absent in monomict ureilites, is common in polymict ureilites where it spans the entire range of An_{0-100}.

Ureilites show evidence for widely differing levels of shock. Some have virtually unmodified primary mineral grains indicative of very low shock levels. Shock damage in the silicates progressively varies from minor fracturing, undulatory extinction, and kink band development in olivine through complete shattering or mosaicism. Pyroxene shows the progressive development of melt–glass inclusions. Diamond and/or lonsdaleite are generally believed to have been produced from graphite by shock loading. The range of estimated shock pressures is from <20 GPa to at least 100 GPa.

The lithophile element contents of ureilites demonstrate that they are highly fractionated from chondritic abundances. They are enriched in scandium, vanadium, magnesium, chromium, and manganese, and depleted in aluminum and sodium as would be expected for plagioclase-free, olivine–pyroxene rocks. REE abundances are generally low compared to chondrites and most other types of achondrites. Many ureilites show a V-shaped REE pattern with extreme middle REE depletion (Figure 8); europium can be as low as $0.005 \times CI$. Many other ureilites have LREE-depleted patterns, with lanthanum varying in the range $0.005-0.1 \times CI$, and low Eu/Sm ratios. Both patterns have been found in individual ureilites. The V-shaped REE patterns result from addition of a LREE-enriched component to typical LREE-depleted ureilitic material (e.g., Boynton *et al*., 1976), but the nature of this component is obscure.

The refractory siderophile elements—rhenium, osmium, tungsten, and iridium—are relatively unfractionated within each ureilite, with abundance in the range $0.1-2 \times CI$ among ureilites. The moderately volatile siderophile elements have generally lower abundances, with Ni at $0.006-0.2 \times CI$, and they tend to decrease with nebular volatility, although gold and cobalt are reversed. Carbon-rich separates from Kenna and Haverö have siderophile element patterns similar to bulk ureilites, but at higher abundances. Moderately volatile siderophile elements show linear correlations with the refractory siderophile element iridium, suggesting that the siderophile elements are contained in two components, one refractory-rich, and the other refractory-poor. The refractory-rich component appears to be associated with the carbon-rich material (see Boynton et al., 1976). However, the siderophile element abundances are correlated with solid metal/liquid metal partition coefficients. Goodrich et al. (1987) suggested that they might be controlled by magmatic fractionation of metal rather than component mixing.

Ureilites have trapped noble gas contents at roughly chondritic abundances, with elemental and isotopic ratios distinctive of a planetary-type pattern as in CM chondrites. Gas contents are quite variable, with xenon contents varying by a factor of ~ 100. The noble gases are largely contained in the carbon polymorphs (Göbel et al., 1978; Weber et al., 1971, 1976). Diamond is the principal carrier and graphite is virtually free of trapped gases in diamond-bearing ureilites (Göbel et al., 1978; Weber et al., 1976). However, ALHA78019 is diamond-free yet has trapped noble gas abundances comparable to those of diamond-bearing ureilites (Wacker, 1986). The gases in ALHA78019 are contained in fine-grained, unidentified polymorph(s) of carbon; the coarse-grained graphite is gas free.

Ureilites exhibit a wide range of oxygen-isotopic compositions distinct from simple mass fractionation trends (Clayton and Mayeda, 1988). They plot roughly along the carbonaceous chondrite anhydrous mineral (CCAM) mixing line defined by materials from CV, CO, and CM chondrites, which is thought to be a nebular mixing line (Clayton et al., 1973; Clayton, 1993). This distribution is unique among achondrites, although acapulcoites and lodranites show somewhat similar isotopic heterogeneity but to a much lesser degree. The range in $\Delta^{17}O$ for ureilites is -0.23 to -2.45, compared to -0.85 to -1.49 for acapulcoites–lodranites (Clayton and Mayeda, 1996). Ureilite $\Delta^{17}O$ is correlated with olivine mg# (Mittlefehldt et al., 1998). Ureilites also show some dispersion across the CCAM line, consistent with mass-dependent fractionation.

Interpretation of chronological data on ureilites is contentious. There appear to be three groups in the Sm–Nd system. Several fall along a line with a slope consistent with an age of 3.79 Ga. A second set defines a line with a slope corresponding to an age of 4.23 Ga. A third set plots along a 4.55 Ga chondritic evolution line. The three lines intersect at $^{147}Sm/^{144}Nd \sim 0.51$. Torigoye-Kita et al. (1995a,b) argued that the 3.79 Ga line is a mixing line between 4.55 Ga ureilitic material and terrestrial contaminants. They noted that the initial neodymium-isotopic composition of the 3.79 Ga group is similar to that of average continental crust. (All the ureilites studied are finds, albeit from widely differing locations.) Goodrich and Lugmair (1995) and Goodrich et al. (1995) showed that acid residues and leached mineral separates fall along the 3.79 Ga or the 4.23 Ga line, rather than the chondritic evolution line. They believe this favors a chronological interpretation, and thus 3.79 Ga and 4.23 Ga date events on the ureilite parent body. Regardless, at least some ureilites were formed early in solar system history. There is significant terrestrial lead contamination in ureilites, but leaching residues containing the least amount of contamination yield a Pb–Pb isochron age of 4.563 ± 0.021 Ga (Torigoye-Kita et al., 1995a,c). Bogard and Garrison (1994) have done $^{39}Ar-^{40}Ar$ measurements on Kenna, Novo Urei, PCA 82506, and Haverö. They infer argon degassing events of ~ 3.3 Ga for Haverö, $\sim 3.3-3.7$ Ga for Novo Urei, ~ 4.1 Ga for Kenna, and 4.5–4.6 Ga for PCA 82506.

The ureilites exhibit a mix of primitive and differentiated features. The extreme oxygen-isotopic heterogeneity, the abundances of noble gases, and the relatively high siderophile element contents suggest that they are primitive rocks, but the refractory mineralogy and depletion in incompatible elements suggests a high-temperature origin. These disparate characteristics hamper development of petrogenetic models. There are two main models for ureilite genesis: (i) ultramafic igneous cumulates, or (ii) partial melt residues (see Goodrich, 1992). It is usually assumed that all ureilites originated on a single parent body, but the data do not require this.

The absence of plagioclase, extreme depletion in incompatible lithophile elements, and the textural and compositional equilibration of olivine and pyroxene are consistent with a partial-melt-residue model. These characteristics are also found in lodranites, which are believed to be partial-melt residues. Modeling suggests that at least 15% melting is needed to eliminate plagioclase, and as much as 20–30% melting is needed to explain the more extreme REE depletions and negative europium anomalies (Goodrich, 1997; Spitz and Goodrich, 1987; Warren and Kallemeyn, 1992).

The foliation and lineation of silicate minerals in ureilites argues for a cumulate model (Berkley and Jones, 1982; Berkley et al., 1976, 1980), and penetration of euhedral graphite grains into silicates is difficult to explain by a residue model (Treiman and Berkley, 1994). Goodrich et al. (1987) argued that the superchondritic Ca/Al ratios of olivine requires that the ureilite parent magmas were derived from previously plagioclase-depleted sources, implying a fractional fusion mechanism for the genesis of the parent magmas. Presently, few workers favor a cumulate model for ureilites. Warren and Kallemeyn (1989b) proposed a hybrid, paracumulate model, in which ureilites form as partial-melt residues in a zone rich in magma. The process of magma expulsion results in the textural characteristics that mimic those of a cumulate.

Regardless of mode of origin, the ureilite suite lacks correlations between mg# and bulk composition, modal mineralogy, mineral minor element composition, pyroxene type, or incompatible element depletion that would be expected if all ureilites were related by either partial melting of single source, or fractional crystallization of a single magma. This is consistent with the heterogeneous oxygen-isotopic composition of the ureilite suite (Clayton and Mayeda, 1988).

The high carbon contents of ureilites and the distribution of their oxygen-isotopic compositions along the CCAM line suggest that their parent body was originally similar to carbonaceous chondrites. However, no data associate ureilites with a specific carbonaceous chondrite group. The carbon-isotopic compositions of ureilites are distinct from those of any carbonaceous chondrite (Grady et al., 1985).

1.11.3.7 Itqiy

Itqiy is an unusual enstatite-rich meteorite described by Patzer et al. (2001). It is composed of ~78% silicate and 22% metal plus "rust" and a suite of sulfides of unspecified abundance. The silicates and metal are relatively coarse-grained with an equigranular, recrystallized texture. The silicate is nearly pure, very homogeneous enstatite, $Wo_{3.0}En_{96.8}Fs_{0.2}$—no other silicates are mentioned. The metal is kamacite with an average of 5.8 wt.% Ni and 3.1 wt.% Si. Sulfides include oldhamite, an Mg–Mn–Fe sulfide with variable composition, and an Fe–Cr sulfide intermediate between troilite and daubreelite. The sulfides occur in intergranular assemblages with Mg–Mn–Fe sulfide hosts containing oldhamite, Fe–Cr sulfide, and kamacite globules. The compositions of the Mg–Mn–Fe sulfides do not match those in EH or EL chondrites.

Itqiy is distinct from chondritic meteorites in bulk composition. Aluminum, LREE, europium, sodium, potassium, vanadium, chromium, and manganese are all depleted. Itqiy has La/Yb of $0.10 \times CI$, and Eu/Sm of $0.16 \times CI$. Refractory siderophile elements are enriched $\sim 2-3 \times CI$, while moderately volatile siderophile elements are at roughly CI abundances. The bulk rock Mg/Si and Fe/Si ratios are greater than those of EH or EL chondrites.

Patzer et al. (2001) concluded that Itqiy is generally related to the EH and EL chondrites, but from a distinct parent body. They infer that it was heated sufficiently to generate melts in the sulfide–metal and silicate systems. The basaltic and metal–sulfide partial melts were expelled from the rock, leaving Itqiy as a residue depleted in incompatible lithophile, siderophile, and chalcophile elements.

1.11.3.8 Northwest Africa 011

Northwest Africa 011 (NWA 011) has briefly been described by Afanasiev et al. (2000) and more extensively by Yamaguchi et al. (2002), the source of most information presented here. NWA 011 is composed of relatively coarse, anhedral pigeonite and augite, and fine-grained, mostly interstitial plagioclase with a recrystallized texture, interpreted to indicate that it is a recrystallized breccia. Minor phases are silica, chromite, ilmenite, calcium phosphate, ferroan olivine, troilite, and baddeleyite. Pyroxenes make up ~53% of the rock (Afanasiev et al., 2000). They are equilibrated in Fe/Mg, with approximate bulk compositions of $Wo_{16}En_{30}Fs_{54}$ for pigeonite and $Wo_{29}En_{27}Fs_{44}$ for augite. Pyroxene FeO/MnO is ~65. Relict plagioclase laths show igneous zoning from $Ab_{6.1}Or_{0.2}$ to $Ab_{35.6}Or_{1.5}$, completely overlapping the range for fine-grained, recrystallized plagioclase. Silica and calcium phosphate are commonly associated and heterogeneously distributed, while ilmenite and chromite ± ferroan olivine are associated.

NWA 011 is basaltic in composition, with enrichments in aluminum, calcium, and titanium relative to chondrites, and is ferroan, with an mg# of 36. NWA 011 has a cumulate–eucrite-like REE pattern (Figure 6). However, Afanasiev et al. (2000) report that NWA 011 has a negative europium anomaly and samarium content of $13.3 \times CI$, more like those expected for a mafic melt composition. Yamaguchi et al. (2002) suggest that the heterogeneous distribution of calcium phosphate is the likely explanation for this discrepancy. This is supported by their hafnium and tantalum analyses, which are at $12.8 \times CI$ and $26.8 \times CI$, abundances expected of a melt but not a pyroxene–plagioclase cumulate (Figure 6). The contents of cobalt, nickel, and iridium are 10–100 times higher than observed for basaltic eucrites, which Yamaguchi et al. (2002) suggest might indicate chondritic contamination. The oxygen-isotopic composition of NWA 011 is similar to that of some CR chondrites, and unlike those of any other achondrite.

1.11.4 UNCATEGORIZED ACHONDRITES

1.11.4.1 IIE Iron Silicates

The IIE iron group contains 18 members, of which 11 are reported to contain silicate inclusions (Grady, 2000). These inclusions, described for only eight IIE irons, are of widely differing types, from chondritic silicates to quenched mafic melts to centimeter-sized single crystals of alkali feldspar. Because of the wide range in inclusion types, including primitive and highly evolved compositions, they are considered as uncategorized achondrites here. The silicate inclusions are divided into three groups for discussion—chondritic, mafic, and silicic.

Petrologic and compositional information on IIE iron silicates was taken from: Bence and Burnett (1969); Bild and Wasson (1977); Bogard et al. (2000); Bunch and Olsen (1968); Bunch et al. (1970); Casanova et al. (1995); Ebihara et al. (1997); Ikeda and Prinz (1996); Ikeda et al. (1997); Olsen and Jarosewich (1970, 1971); Olsen et al. (1994); Osadchii et al. (1981); and Prinz et al. (1983b).

Chondritic inclusions are found in Netschaëvo and Techado. They contain olivine, orthopyroxene, diopside, sodic plagioclase, phosphates, FeNi metal, and troilite, and relict chondrules are present in Netschaëvo. Mineral compositions differ between Netschaëvo and Techado; olivines have mg#s of 85.9 and 83.6, orthopyroxenes are $Wo_{1.4}Fs_{13.6}$ and $Wo_{1.6}Fs_{15.3}$, and plagioclases are $Ab_{81.8}Or_{4.3}$ and $Ab_{78.9}Or_{6.0}$. The single silicate inclusion identified in Watson is also roughly chondritic in bulk composition. It contains olivine (mg# 79.4), orthopyroxene ($Wo_{3.8}Fs_{17.6}$), calcic pyroxene ($Wo_{41.1}Fs_{9.0}$), antiperthitic alkali feldspar with potassium feldspar lamellae ($Ab_{57.2}Or_{41.4}$) in an albite host ($Ab_{92.6}Or_{5.2}$), chromite, troilite, and metal. The texture of this inclusion is igneous—orthopyroxene crystals up to 1 mm in size poikilitically enclose olivine crystals, with olivine, calcic pyroxene, plagioclase, and troilite occurring interstitial to orthopyroxene grains. This texture is somewhat similar to that of PAT 91501, an impact–melt of an L-chondrite (Mittlefehldt and Lindstrom, 2001).

Mafic inclusions dominate the inclusion types in Weekeroo Station and Miles. They are composed of ~25% orthopyroxene, 25% clinopyroxene, and 50% feldspar, but modes are variable. Most of the inclusions in Miles are coarse-grained gabbros, but some fine-grained, cryptocrystalline inclusions are also present. Weekeroo Station contains coarse-grained pyroxene–plagioclase inclusions and inclusions with coarse pyroxene contained in fine-grained radiating plagioclase–tridymite groundmass. Orthopyroxene in Weekeroo Station is Fs_{22}, while that in Miles is $Fs_{19.9-23.2}$. These are more FeO rich than those in the chondritic inclusions. Both plagioclase and alkali feldspar are present in Weekeroo Station and Miles.

Silicic inclusions, common in Colomera, Kodaikanal, and Elga, are dominated by glass or cryptocrystalline material of plagioclase–tridymite composition and clinopyroxene in a ratio of ~2 : 1. The silicic inclusions have textures varying from radiating, fine-grained intergrowths of plagioclase, and tridymite to glassy inclusions. Mineral compositions within this group are somewhat more diverse. Olivine has a mg# of 78 in Elga and 79 in Kodaikanal, orthopyroxene is Wo_2Fs_{22} in Colomera and Wo_3Fs_{16} in Elga and Kodaikanal, while clinopyroxene is more variable, with $Wo_{40.5-46.4}Fs_{8.6-14.0}$ in Colomera, $Wo_{40.7-44.4}Fs_{8.6-11.6}$ in Elga, and $Wo_{37.1-42.5}Fs_{7.8-10.3}$ in Kodaikanal. Elga also contains very low calcium orthopyroxene, $Wo_{0.4}Fs_{14.8}$ (Osadchii et al., 1981). Colomera and Kodaikanal contain both plagioclase and potassium feldspar, while Elga contains alkali feldspar exhibiting a range of compositions.

Few comprehensive bulk compositional analyses are available for silicate inclusions from IIE irons, and many of them are of small samples. Netschaëvo silicates have magnesium-normalized abundances of refractory, moderately volatile, and volatile lithophile elements within the ranges of ordinary chondrites. The nickel-normalized abundances of refractory and moderately volatile siderophile elements are also similar to those of ordinary chondrites. The silicates have siderophile/Mg ratios of $(1.9-2.2) \times CI$ chondrites, however. The silicate inclusion in Watson has CI-normalized element/Mg ratios of ~0.86 for most refractory and moderately volatile lithophile elements (Figure 2). Siderophile elements are depleted, and show increasing abundance with increasing volatility (Olsen et al., 1994): $Os/Mg = 0.028 \times CI$ and $Sb/Mg = 0.066 \times CI$.

A composite of 12 inclusions from Weekeroo Station is broadly mafic in composition—depleted in MgO and enriched in SiO_2, Al_2O_3, and CaO compared to the chondritic inclusions in Netschaëvo and Watson. It is quartz normative and tridymite is observed in the inclusions (Olsen and Jarosewich, 1970). Six gabbroic and three cryptocrystalline inclusions from Miles show considerable compositional overlap in magnesium, aluminum, and calcium between the two types, but the analyzed masses were small, from 5.6 mg to 60.4 mg. The compatible and incompatible lithophile trace elements are also quite variable, but they generally show fractionated lithophile element abundances. The plagiophile elements (sodium, aluminum, and potassium) and the incompatible elements (titanium and hafnium) are enriched, while magnesium is depleted relative to CI chondrites. The REE abundances are generally elevated over CI values, although some of the cryptocrystalline clasts have LREE depletions. The cryptocrystalline clasts are more extremely fractionated (Ebihara et al., 1997). While clearly not chondritic in composition, these clasts are not obviously partial melts of a chondritic source either

(Ebihara et al., 1997). Because of sample heterogeneity, the analyzed clasts may not be representative of the bulk of the silicate material.

The IIE-iron silicate inclusions are unusual among meteorites in that they appear to have a range of formation ages. Bogard et al. (2000) summarized the existing radiometric age data on IIE-iron silicate inclusions. Colomera, Miles, Techado, and Weekeroo Station have Ar–Ar and/or Rb–Sr ages >4.3 Ga, while Kodaikanal, Netschaëvo, and Watson have Ar–Ar, Rb–Sr, and/or Pb–Pb ages of ~3.7 Ga. Weekeroo Station has an I–Xe formation age of 4.555 Ga (Niemeyer, 1980, calibrated to the Bjurböle absolute age determined by Brazzle et al. (1999), which is older than Rb–Sr isochron ages (Burnett and Wasserburg, 1967a; Evensen et al., 1979). The tungsten-isotopic composition of metal and silicate in Watson are different, indicating that these phases did not equilibrate (Snyder et al., 1998). Burnett and Wasserburg (1967b) argued that the relatively low initial $^{87}Sr/^{86}Sr$ ratio of Kodaikanal for its very high Rb/Sr is not compatible with simple metamorphic reequilibration, and requires Rb/Sr fractionation at ~3.7 Ga.

Bogard et al. (2000) reviewed models for the formation of IIE irons. The two main models are: (i) IIE irons were formed by endogenous igneous processes (e.g., Casanova et al., 1995), or (ii) exogenous, impact-driven processes (e.g., Wasson and Wang, 1986). The fine-grained textures of some inclusions argue for shock remelting (Bogard et al., 2000; Osadchii et al., 1981), but it is not clear that these inclusions were formed by this process, as opposed to being simply remelted. Young ages for some IIE irons also argue in favor of an impact process as internal heating of asteroids was long since dead by 3.8 Ga. Burnett and Wasserburg (1967b) showed that Rb–Sr fractionation occurred 3.8 Ga ago in the Kodaikanal inclusions, and this age does not correspond to a simple metamorphic re-equilibration event. However, Kodaikanal could be a special case in which the impact resulted in thorough remelting and chemical fractionation, while other IIE irons were simply shock heated to a lesser degree. In this case, the original formation of the silicates could have been by endogenous processes ~4.56 Ga ago. The textural heterogeneity, small size of many inclusions, and the paucity of detailed studies of many IIE irons hamper clear understanding of the formation of the silicates.

1.11.5 SUMMARY

Achondrites as a whole show one curious property when compared to the total range of meteorite types—chondrites, achondrites, stony irons, and irons. Many of the achondrite types occupy a very limited range of oxygen-isotopic compositions (Clayton and Mayeda, 1996), yet exhibit a wide range in other nebula-imposed properties, such as f_{O_2} and moderately volatile element content. For example, the mean oxygen-isotopic composition of angrites is $\delta^{18}O = 3.68$, $\Delta^{17}O = -0.15$, and of brachinites is $\delta^{18}O = 3.78$, $\Delta^{17}O = -0.26$ (Clayton and Mayeda, 1996), but the bulk rock Na/Al are different by a factor of ~120 (Figure 1). Compare this with the variation in Na/Al ratios among all chondrite groups of only a factor of <5 (Lodders and Fegley, 1998), but much wider range in oxygen-isotopic composition (see Chapter 1.06). Similarly, the winonaite–IAB-iron silicate inclusion clan has a mean oxygen-isotopic composition ($\delta^{18}O = 5.05$, $\Delta^{17}O = -0.48$; Clayton and Mayeda, 1996) close to that of the brachinites, but are much more reduced judging by FeO/MnO (Figure 1) or olivine mg# (92–99 versus 64–71). It is generally assumed that meteorites with similar oxygen-isotopic compositions were formed in a relatively limited region of the solar nebula. For the achondrites just mentioned, this would imply that moderately volatile element fractionations and oxidation state varied widely within a limited region, which is counterintuitive. Greater understanding of the nature of achondrite parent bodies and their thermal processing could thus contribute to answering fundamental questions on the chemical evolution of the solar nebula.

ACKNOWLEDGMENT

The NASA Cosmochemistry Program supported this work.

REFERENCES

Afanasiev S. V., Ivanova M. A., Korochantsev A. V., Kononkova N. N., and Nazarov M. A. (2000) Dhofar 007 and Northwest Africa 011: two new eucrites of different types. *Meteorit. Planet. Sci.* **35**, A19.

Barrat J. A., Blichert-Toft J., Gillet Ph., and Keller F. (2000) The differentiation of eucrites: the role of in situ crystallization. *Meteorit. Planet. Sci.* **35**, 1087–1100.

Bence A. E. and Burnett D. S. (1969) Chemistry and mineralogy of the silicates and metal of the Kodaikanal meteorite. *Geochim. Cosmochim. Acta* **33**, 387–407.

Benedix G. K., McCoy T. J., Keil K., Bogard D. D., and Garrison D. H. (1998) A petrologic and isotopic study of winonaites: evidence for early partial melting, brecciation, and metamorphism. *Geochim. Cosmochim. Acta* **62**, 2535–2553.

Benedix G. K., McCoy T. J., Keil K., and Love S. G. (2000) A petrologic study of the IAB iron meteorites: constraints on the formation of the IAB-winonaite parent body. *Meteorit. Planet. Sci.* **35**, 1127–1141.

Berkley J. L. and Jones J. H. (1982) Primary igneous carbon in ureilites: petrological implications. *Proc. 13th Lunar Planet. Sci. Conf.: J. Geophys. Res.* **87**, A353–A364.

Berkley J. L., Brown H. G., Keil K., Carter N. L., Mercier J.-C. C., and Huss G. (1976) The Kenna ureilite: an ultramafic rock with evidence for igneous, metamorphic and shock origin. *Geochim. Cosmochim. Acta* **40**, 1429–1437.

Berkley J. L., Taylor G. J., and Keil K. (1978) Fluorescent accessory phases in the carbonaceous matrix of ureilites. *Geophys. Res. Lett.* **5**, 1075–1078.

Berkley J. L., Taylor G. J., Keil K., Harlow G. E., and Prinz M. (1980) The nature and origin of ureilites. *Geochim. Cosmochim. Acta* **44**, 1579–1597.

Bild R. W. (1977) Silicate inclusions in group IAB irons and a relation to the anomalous stones Winona and Mt. Morris (Wis.). *Geochim. Cosmochim. Acta* **41**, 1439–1456.

Bild R. W. and Wasson J. T. (1976) The Lodran meteorite and its relationship to the ureilites. *Min. Mag.* **40**, 721–735.

Bild R. W. and Wasson J. T. (1977) Netschaëvo: a new class of chondritic meteorite. *Science* **197**, 58–62.

Binz C. M., Ikramuddin M., and Lipschutz M. E. (1975) Contents of eleven trace elements in ureilite achondrites. *Geochim. Cosmochim. Acta* **39**, 1576–1579.

Binzel R. P. and Xu S. (1993) Chips off of asteroid 4 Vesta: evidence for the parent body of basaltic achondrite meteorites. *Science* **260**, 186–191.

Biswas S., Walsh T., Bart G., and Lipschutz M. E. (1980) Thermal metamorphism of primitive meteorites: XI. The enstatite meteorites: origin and evolution of a parent body. *Geochim. Cosmochim. Acta* **44**, 2097–2109.

Bogard D. D. (1995) Impact ages of meteorites: a synthesis. *Meteorit. Planet. Sci.* **30**, 244–268.

Bogard D. D. and Garrison D. H. (1994) ^{39}Ar–^{40}Ar ages of four ureilites. In *Lunar Planet. Sci.* **XXV**. The Lunar and Planetary Institute, Houston, pp. 137–138.

Bogard D. D. and Garrison D. H. (1995) ^{39}Ar–^{40}Ar age of the Ibitira eucrite and constraints on the time of pyroxene equilibration. *Geochim. Cosmochim. Acta* **59**, 4317–4322.

Bogard D. D. and Garrison D. H. (1998) ^{39}Ar–^{40}Ar ages and thermal history of mesosiderites. *Geochim. Cosmochim. Acta* **62**, 1459–1468.

Bogard D. D. and Garrison D. H. (2001) Early thermal history of eucrites by ^{39}Ar–^{40}Ar. In *Lunar Planet. Sci.* **XXXII**, #1138. The Lunar and Planetary Institute, Houston (CD-ROM).

Bogard D. D. and Garrison D. H. (2002) Argon-39–Argon-40 ages of two meteorites. In *Lunar Planet. Sci.* **XXXIII**, #1212. The Lunar and Planetary Institute, Houston (CD-ROM).

Bogard D. D., Burnett D. S., Eberhardt P., and Wasserburg G. J. (1967) ^{87}Rb–^{87}Sr and ^{40}K–^{40}Ar ages of the Norton County achondrite. *Earth Planet. Sci. Lett.* **3**, 179–189.

Bogard D. D., Garrison D. H., Jordan J. L., and Mittlefehldt D. (1990) ^{39}Ar–^{40}Ar dating of mesosiderites: evidence for major parent body disruption <4 Ga. *Geochim. Cosmochim. Acta* **54**, 2549–2564.

Bogard D. D., Garrison D. H., and McCoy T. J. (2000) Chronology and petrology of silicates from IIE iron meteorites: evidence of a complex parent body evolution. *Geochim. Cosmochim. Acta* **64**, 2133–2154.

Bowman L. E., Spilde M. N., and Papike J. J. (1997) Automated energy dispersive spectrometer modal analysis applied to the diogenites. *Meteorit. Planet. Sci.* **32**, 869–875.

Boynton W. V., Starzyk P. M., and Schmitt R. A. (1976) Chemical evidence for the genesis of the ureilites, the achondrite Chassigny, and the nakhlites. *Geochim. Cosmochim. Acta* **40**, 1439–1447.

Brazzle R. H., Pravdivtseva O. V., Meshik A. P., and Hohenberg C. M. (1999) Verification and interpretation of the I-Xe chronometer. *Geochim. Cosmochim. Acta* **63**, 739–760.

Brett R., Huebner J. S., and Sato M. (1977) Measured oxygen fugacities of the Angra dos Reis achondrite as a function of temperature. *Earth Planet. Sci. Lett.* **35**, 363–368.

Brouxel M. and Tatsumoto M. (1991) The Estherville mesosiderite: U–Pb, Rb–Sr, and Sm–Nd isotopic study of a polymict breccia. *Geochim. Cosmochim. Acta* **55**, 1121–1133.

Buchanan P. C. and Reid A. M. (1996) Petrology of the polymict eucrite Petersburg. *Geochim. Cosmochim. Acta* **60**, 135–146.

Buchanan P. C., Mittlefehldt D. W., Hutchison R., Koeberl C., Lindstrom D. J., and Pandit M. K. (2000a) Petrology of the Indian eucrite Piplia Kalan. *Meteorit. Planet. Sci.* **35**, 609–615.

Buchanan P. C., Lindstrom D. J., Mittlefehldt D. W., Koeberl C., and Reimold W. U. (2000b) The South African polymict eucrite Macibini. *Meteorit. Planet. Sci.* **35**, 1321–1331.

Bunch T. E. and Olsen E. (1968) Potassium feldspar in Weekeroo Station, Kodaikanal, and Colomera iron meteorites. *Science* **160**, 1223–1225.

Bunch T. E., Keil K., and Olsen E. (1970) Mineralogy and petrology of silicate inclusions in iron meteorites. *Contrib. Mineral. Petrol.* **25**, 297–240.

Burnett D. S. and Wasserburg G. J. (1967a) ^{87}Rb–^{87}Sr ages of silicate inclusions in iron meteorites. *Earth Planet. Sci. Lett.* **2**, 397–408.

Burnett D. S. and Wasserburg G. J. (1967b) Evidence for the formation of an iron meteorite at 3.8×10^9 years. *Earth Planet. Sci. Lett.* **2**, 137–147.

Casanova I., Keil K., and Newsom H. E. (1993) Composition of metal in aubrites: constraints on core formation. *Geochim. Cosmochim. Acta* **57**, 675–682.

Casanova I., Graf T., and Marti K. (1995) Discovery of an unmelted H-chondrite inclusion in an iron meteorite. *Science* **268**, 540–542.

Choi B.-G., Ouyang X., and Wasson J. T. (1995) Classification and origin of IAB and IIICD iron meteorites. *Geochim. Cosmochim. Acta* **59**, 593–612.

Chou C.-L., Boynton W. V., Bild R. W., Kimberlin J., and Wasson J. T. (1976) Trace element evidence regarding a chondritic component in howardite meteorites. *Proc. 7th Lunar Sci. Conf.* 3501–3518.

Clayton R. N. (1993) Oxygen-isotopes in meteorites. *Ann. Rev. Earth Planet. Sci.* **21**, 115–149.

Clayton R. N. and Mayeda T. K. (1988) Formation of ureilites by nebular processes. *Geochim. Cosmochim. Acta* **52**, 1313–1318.

Clayton R. N. and Mayeda T. K. (1996) Oxygen-isotope studies of achondrites. *Geochim. Cosmochim. Acta* **60**, 1999–2018.

Clayton R. N., Grossman L., and Mayeda T. K. (1973) A component of primitive nuclear composition in carbonaceous chondrites. *Science* **181**, 485–487.

Clayton R. N., Mayeda T. K., and Rubin A. E. (1984) Oxygen-isotopic compositions of enstatite chondrites and aubrites. *Proc. 15th Lunar Planet. Sci. Conf.: J. Geophys. Res.* **89**(suppl.), C245–C249.

Compston W., Lovering J. F., and Vernon M. J. (1965) Rubidium–strontium age of the Bishopville aubrite and its component enstatite and feldspar. *Geochim. Cosmochim. Acta* **29**, 1085–1099.

Crozaz G. and McKay G. (1990) Rare earth elements in Angra dos Reis and Lewis Cliff 86010, two meteorites with similar but distinct magma evolutions. *Earth Planet. Sci. Lett.* **97**, 369–381.

Davis A. M., Ganapathy R., and Grossman L. (1977) Pontlyfni: a differentiated meteorite related to the group IAB irons. *Earth Planet. Sci. Lett.* **35**, 19–24.

Delaney J. S. and Sutton S. R. (1988) Lewis Cliff 86010, an ADORable Antarctican. In *Lunar Planet. Sci.* **XIX**. The Lunar and Planetary Institute, Houston, 265–266.

Delaney J. S., Nehru C. E., and Prinz M. (1980) Olivine clasts from mesosiderites and howardites: clues to the nature of achondritic parent bodies. *Proc. 11th Lunar Planet. Sci. Conf.* 1073–1087.

Delaney J. S., Nehru C. E., Prinz M., and Harlow G. E. (1981) Metamorphism in mesosiderites. *Proc. 12th Lunar Planet. Sci. Conf.* 1315–1342.

Delaney J. S., Takeda H., Prinz M., Nehru C. E., and Harlow G. E. (1983) The nomenclature of polymict basaltic achondrites. *Meteoritics* **18**, 103–111.

Delaney J. S., Prinz M., and Takeda H. (1984) The polymict eucrites. *Proc 15th Lunar Planet. Sci. Conf.: J. Geophys. Res.* **89** (suppl.), C251–C288.

Delaney J. S., Zanda B., Clayton R. N., and Mayeda T. (2000) Zag (b): a ferroan achondrite intermediate between brachinites and lodranites. In *Lunar Planet. Sci.* **XXXI**, #1745. The Lunar and Planetary Institute, Houston (CD-ROM).

Easton A. J. (1985) Seven new bulk chemical analyses of aubrites. *Meteoritics* **20**, 571–573.

Ebihara M., Ikeda Y., and Prinz M. (1997) Petrology and chemistry of the Miles IIE iron: II. Chemical characteristics of the Miles silicate inclusions. *Antarct. Meteorit. Res.* **10**, 373–388.

Eugster O., Michel Th., and Niedermann S. (1991) ^{244}Pu–Xe formation and gas retention age, exposure history, and terrestrial age of angrites LEW86010 and LEW87051: comparison with Angra dos Reis. *Geochim. Cosmochim. Acta* **55**, 2957–2964.

Eugster O., Michel Th., Niedermann S., Wang D., and Yi W. (1993) The record of cosmogenic, radiogenic, fissiogenic, and trapped noble gases in recently recovered Chinese and other chondrites. *Geochim. Cosmochim. Acta* **57**, 1115–1142.

Eugster O., Polnau E., and Terribilini D. (1998) Cosmic ray- and gas retention ages of newly recovered and unusual chondrites. *Earth Planet. Sci. Lett.* **164**, 511–519.

Eugster O., Busemann H., Kurat G., Lorenzetti S., and Varela M. E. (2002) Characterization of the noble gases and CRE age of the D'Orbigny angrite. *Meteorit. Planet. Sci.* **37**, A44.

Evensen N. M., Hamilton P. J., Harlow G. E., Klimentidis R., O'Nions R. K., and Prinz M. (1979) Silicate inclusions in Weekeroo Station: planetary differentiates in an iron meteorite. In *Lunar Planet. Sci.* **X**. The Lunar and Planetary Institute, Houston, pp. 376-378.

Floran R. J. (1978) Silicate petrography, classification, and origin of the mesosiderites: review and new observations. *Proc. 9th Lunar Planet. Sci. Conf.* 1053–1081.

Floran R. J., Prinz M., Hlava P. F., Keil K., Nehru C. E., and Hinthorne J. R. (1978a) The Chassigny meteorite: a cumulate dunite with hydrous amphibole-bearing melt inclusions. *Geochim. Cosmochim. Acta* **42**, 1213–1229.

Floran R. J., Caulfield J. B. D., Harlow G. E., and Prinz M. (1978b) Impact origin for the Simondium, Pinnaroo, and Hainholz mesosiderites: implications for impact processes beyond the Earth–Moon system. *Proc. 9th Lunar Planet. Sci. Conf.* 1083–1114.

Floran R. J., Prinz M., Hlava P. F., Keil K., Spettel B., and Wänke H. (1981) Mineralogy, petrology, and trace element geochemistry of the Johnstown meteorite: a brecciated orthopyroxenite with siderophile and REE-rich components. *Geochim. Cosmochim. Acta* **45**, 2385–2391.

Floss C. (2000) Complexities on the acapulcoite–lodranite parent body: evidence from trace element distributions in silicate minerals. *Meteorit. Planet. Sci.* **35**, 1073–1085.

Floss C. and Crozaz G. (1993) Heterogeneous REE patterns in oldhamite from the aubrites: their nature and origin. *Geochim. Cosmochim. Acta* **57**, 4039–4057.

Floss C., Strait M. M., and Crozaz G. (1990) Rare earth elements and the petrogenesis of aubrites. *Geochim. Cosmochim. Acta* **57**, 4039–4057.

Fogel R. A. (1994) Aubrite basalt vitrophyres: high sulfur silicate melts and a snapshot of aubrite formation. *Meteoritics* **29**, 466–467.

Fogel R. A. (1997) A new aubrite basalt vitrophyre from the LEW 87007 aubrite. In *Lunar Planet. Sci.* **XXVIII**. The Lunar and Planetary Institute, Houston, pp. 369–370.

Fowler G. W., Papike J. J., Spilde M. N., and Shearer C. K. (1994) Diogenites as asteroidal cumulates: insights from orthopyroxene major and minor element chemistry. *Geochim. Cosmochim. Acta* **58**, 3921–3929.

Fowler G. W., Shearer C. K., Papike J. J., and Layne G. D. (1995) Diogenites as asteroidal cumulates: insights from orthopyroxene trace element chemistry. *Geochim. Cosmochim. Acta* **59**, 3071–3084.

Fukuoka T. and Schmitt R. A. (1978) Chemical compositions of silicate inclusions in IAB iron meteorites. In *Lunar Planet. Sci.* **IX**. The Lunar and Planetary Institute, Houston, pp. 359–361.

Fukuoka T., Boynton W. V., Ma M.-S., and Schmitt R. A. (1977) Genesis of howardites, diogenites, and eucrites. *Proc. 8th Lunar. Sci. Conf.* 187–210.

Göbel R., Ott U., and Begemann F. (1978) On trapped noble gases in ureilites. *J. Geophys. Res.* **83**, 855–867.

Goodrich C. A. (1988) Petrology of the unique achondrite LEW86010. In *Lunar Planet. Sci.* **XIX**. The Lunar and Planetary Institute, Houston, pp. 399–400.

Goodrich C. A. (1992) Ureilites: a critical review. *Meteoritics* **27**, 327–252.

Goodrich C. A. (1997) Preservation of a nebular mg-Δ^{17}O correlation during partial melting of ureilites. In *Lunar Planet. Sci.* **XXVIII**. The Lunar and Planetary Institute, Houston, pp. 435–436.

Goodrich C. A. and Berkley J. L. (1986) Primary magmatic carbon in ureilites: evidence from cohenite-bearing spherules. *Geochim. Cosmochim. Acta* **50**, 681–691.

Goodrich C. A. and Lugmair G. W. (1995) Stalking the LREE-enriched component in ureilites. *Geochim. Cosmochim. Acta* **59**, 2609–2620.

Goodrich C. A., Keil K., Berkley J. L., Laul J. C., Smith M. R., Wacker J. F., Clayton R. N., and Mayeda T. K. (1987) Roosevelt County 027: a low-shock ureilite with interstitial silicates and high noble-gas concentration. *Meteoritics* **22**, 191–218.

Goodrich C. A., Lugmair G. W., Drake M. J., and Patchett P. J. (1995) Comment on Torigoye-Kita *et al.* (1995a). *Geochim. Cosmochim. Acta* **59**, 4083–4085.

Göpel Ch., Manhès G., and Allègre C. J. (1992) U–Pb study of the Acapulco meteorite. *Meteoritics* **27**, 226.

Grady M. M. (2000) *Catalogue of Meteorites*, 5th edn. Cambridge University Press, Cambridge, UK.

Grady M. M., Wright I. P., Swart P. K., and Pillinger C. T. (1985) The carbon and nitrogen isotopic composition of ureilites: implications for their genesis. *Geochim. Cosmochim. Acta* **49**, 903–915.

Graham A. L., Easton A. J., and Hutchison R. (1977) Forsterite chondrites: the meteorites Kakangari, Mount Morris (Wisconsin), Pontlyfni, and Winona. *Min. Mag.* **41**, 201–210.

Haack H., Scott E. R. D., and Rasmussen K. L. (1996) Thermal and shock history of mesosiderites and their large parent asteroid. *Geochim. Cosmochim. Acta* **60**, 2609–2619.

Hewins R. H. (1984) The case for a melt matrix in plagioclase-POIK mesosiderites. *Proc 15th Lunar Planet. Sci. Conf.: J. Geophys. Res.* **89** (suppl.), C289–C297.

Heymann D. and Mazor E. (1968) Noble gases in unequilibrated ordinary chondrites. *Geochim. Cosmochim. Acta* **32**, 1–19.

Higuchi H., Morgan J. W., Ganapathy R., and Anders E. (1976) Chemical variation in meteorites-X Ureilites. *Geochim. Cosmochim. Acta* **40**, 1563–1571.

Hohenberg C. M. (1967) I-Xe dating of the Shallowater achondrite. *Earth Planet. Sci. Lett.* **3**, 357–362.

Hohenberg C. M. (1970) Xenon from the Angra dos Reis meteorite. *Geochim. Cosmochim. Acta* **34**, 185–191.

Hohenberg C. M., Bernatowicz T. J., and Podosek F. A. (1991) Comparative xenology of two angrites. *Earth Planet. Sci. Lett.* **102**, 167–177.

Housen K. R., Wilkening L. L., Chapman C. R., and Greenberg R. J. (1979) Asteroidal regoliths. *Icarus* **39**, 317–352.

Hsu W. and Crozaz G. (1997) Mineral chemistry and the petrogenesis of eucrites: II. Cumulate eucrites. *Geochim. Cosmochim. Acta* **61**, 1293–1302.

Ikeda Y. and Prinz M. (1996) Petrology of silicate inclusions in the Miles IIE iron. *Proc. NIPR Symp. Antarct. Meteorit.* **9**, 143–173.

Ikeda Y., Ebihara M., and Prinz M. (1990) Enclaves in the Mt. Padbury and Vaca Muerta mesosiderites: magmatic and residue (or cumulate) rock types. *Proc. NIPR Symp. Antarct. Meteorit.* **3**, 99–131.

Ikeda Y., Ebihara M., and Prinz M. (1997) Petrology of and chemistry of the Miles IIE iron: I. Description and petrology of twenty new silicate inclusions. *Antarct. Meteorit. Res.* **10**, 355–372.

Ireland T. R. and Wlotzka F. (1992) The oldest zircons in the solar system. *Earth Planet. Sci. Lett.* **109**, 1–10.

Jacobsen S. B. and Wasserburg G. J. (1984) Sm–Nd isotopic evolution of chondrites and achondrites: II. *Earth Planet. Sci. Lett.* **67**, 137–150.

Jagoutz E., Jotter R., Varela M. E., Zartman R., Kurat G., and Lugmair G. W. (2002) Pb–U–Th isotopic evolution of the D'Orbigny angrite. In *Lunar Planet. Sci.* **XXXIII**, #1043. The Lunar and Planetary Institute, Houston (CD-ROM).

Janssens M.-J., Hertogen J., Wolf R., Ebihara M., and Anders E. (1987) Ureilites: trace element clues to their origin. *Geochim. Cosmochim. Acta* **51**, 2275–2283.

Jaques A. L. and Fitzgerald M. J. (1982) The Nilpena ureilite, an unusual polymict breccia: implications for origin. *Geochim. Cosmochim. Acta* **46**, 893–900.

Jarosewich E. (1990) Chemical analyses of meteorites: a compilation of stony and iron meteorite analyses. *Meteoritics* **25**, 323–337.

Johnson J. E., Scrymgour J. M., Jarosewich E., and Mason B. (1977) Brachina meteorite—a chassignite from South Australia. *Rec. S. Austral. Mus.* **17**, 309–319.

Kallemeyn G. W. and Wasson J. T. (1985) The compositional classification of chondrites: IV. Ungrouped chondritic meteorites and clasts. *Geochim. Cosmochim. Acta* **49**, 261–270.

Keil K. (1968) Mineralogical and chemical relationships among enstatite chondrites. *J. Geophys. Res.* **73**, 6945–6976.

Keil K., Ntaflos Th., Taylor G. J., Brearley A. J., Newsom H. E., and Romig A. D., Jr. (1989) The Shallowater aubrite: evidence for origin by planetesimal impact. *Geochim. Cosmochim. Acta* **53**, 3291–3307.

Keil K., Stöffler D., Love S. G., and Scott E. R. D. (1997) Constraints on the role of impact heating and melting in asteroids. *Meteorit. Planet. Sci.* **32**, 349–363.

Kelly W. R. and Larimer J. W. (1977) Chemical fractionation in meteorites: VII. Iron meteorites and the cosmochemical history of the metal phase. *Geochim. Cosmochim. Acta* **41**, 93–111.

Kimura M., Ikeda Y., Ebihara M., and Prinz M. (1991) New enclaves in the Vaca Muerta mesosiderite: petrogenesis and comparison with HED meteorites. *Proc. NIPR Symp. Antarct. Meteorit.* **4**, 263–306.

Kimura M., Tsuchiyama A., Fukuoka T., and Iimura Y. (1992) Antarctic primitive achondrites, Yamato-74025,-75300, and-75305: their mineralogy, thermal history, and the relevance to winonaite. *Proc. NIPR Symp. Antarct. Meteorit.* **5**, 165–190.

Kracher A. (1982) Crystallization of a S-saturated Fe, Ni-melt, and the origin of the iron meteorite groups IAB and IIICD. *Geophys. Res. Lett.* **9**, 412–415.

Kracher A. (1985) The evolution of partially differentiated planetesimal: evidence from iron meteorite groups IAB and IIICD. *Proc. 15th Lunar Planet. Sci. Conf. J. Geophys. Res.* **90**(suppl.), C689–C698.

Lodders K. and Fegley B., Jr. (1998) *The Planetary Scientist's Companion.* Oxford University Press, Oxford, England, 371pp.

Lodders K., Palme H., and Wlotzka F. (1993) Trace elements in mineral separates of the Peña Blanca Spring aubrite: implications for the evolution of the aubrite parent body. *Meteoritics* **28**, 538–551.

Longhi J. (1999) Phase equilibrium constraints on angrite petrogenesis. *Geochim. Cosmochim. Acta* **63**, 573–585.

Lovering J. F. (1975) The Moama eucrites—a pyroxene-plagioclase adcumulate. *Meteoritics* **10**, 101–114.

Lugmair G. W. and Galer S. J. G. (1992) Age and isotopic relationships among the angrites Lewis Cliff 86010 and Angra dos Reis. *Geochim. Cosmochim. Acta* **56**, 1673–1694.

Lugmair G. W. and Marti K. (1977) Sm–Nd–Pu timepieces in the Angra dos Reis meteorite. *Earth Planet. Sci. Lett.* **35**, 273–284.

Lugmair G. W. and Shukolyukov A. (1997) ^{53}Mn–^{53}Cr isotope systematics of the HED parent body. In *Lunar Planet. Sci.* **XXVIII**. The Lunar and Planetary Institute, Houston, pp. 851–852.

Lugmair G. W., MacIsaac C., and Shukolyukov A. (1992) The ^{53}Mn–^{53}Cr isotopic system and early planetary evolution. In *Lunar Planet. Sci.* **XXIII**. The Lunar and Planetary Institute, Houston, pp. 823–824.

Mason B. and Jarosewich E. (1973) The Barea, Dyarrl Island, and Emery meteorites, and a review of the mesosiderites. *Min. Mag.* **39**, 204–215.

Mason B., Jarosewich E., and Nelen J. A. (1979) The pyroxene-plagioclase achondrites. *Smithson. Contrib. Earth Sci.* **22**, 27–45.

McCarthy T. S., Ahrens L. H., and Erlank A. J. (1972) Further evidence in support of the mixing model for howardite origin. *Earth Planet. Sci. Lett.* **15**, 86–93.

McCarthy T. S., Erlank A. J., and Willis J. P. (1973) On the origin of eucrites and diogenites. *Earth Planet. Sci. Lett.* **18**, 433–442.

McCord T. B., Adams J. B., and Johnson T. V. (1970) Asteroid Vesta: spectral reflectivity and compositional implications. *Science* **168**, 1445–1447.

McCoy T. J. (1998) A pyroxene-oldhamite clast in Bustee: igneous aubritic oldhamite and a mechanism for the Ti enrichment in aubritic troilite. *Antarct. Meteorit. Res.* **11**, 32–48.

McCoy T. J., Keil K., Scott E. R. D., and Haack H. (1993) Genesis of the IIICD iron meteorites: evidence from silicate-bearing inclusions. *Meteoritics* **28**, 552–560.

McCoy T. J., Keil K., Clayton R. N., Mayeda T. K., Bogard D. D., Garrison D. H., Huss G. R., Hutcheon I. D., and Wieler R. (1996) A petrologic, chemical, and isotopic study of monument draw and comparison with other acapulcoites: evidence for formation by incipient partial melting. *Geochim. Cosmochim. Acta* **60**, 2681–2708.

McCoy T. J., Keil K., Clayton R. N., Mayeda T. K., Bogard D. D., Garrison D. H., and Wieler R. (1997a) A petrologic and isotopic study of lodranites: evidence for early formation as partial melt residues from heterogeneous precursors. *Geochim. Cosmochim. Acta* **61**, 623–637.

McCoy T. J., Keil K., Muenow D. W., and Wilson L. (1997b) Partial melting and melt migration in the acapulcoite–lodranite parent body. *Geochim. Cosmochim. Acta* **61**, 639–650.

McDonough W. F. and Sun S.-S. (1995) The composition of the Earth. *Chem. Geol.* **120**, 223–253.

McKay G., Lindstrom D., Yang S.-R., and Wagstaff J. (1988) Petrology of a unique achondrite Lewis Cliff 86010. In *Lunar Planet. Sci.* **XIX**. The Lunar and Planetary Institute, Houston, pp. 762–763.

McKay G., Crozaz G., Wagstaff J., Yang S.-R., and Lundberg L. (1990) A petrographic, electron microprobe, and ion microprobe study of mini-angrite Lewis Cliff 87051. In *Lunar Planet. Sci.* **XXI**. The Lunar and Planetary Institute, Houston, pp. 771–772.

McKay G., Le L., Wagstaff J., and Crozaz G. (1994) Experimental partitioning of rare earth elements and strontium: constraints on petrogenesis and redox conditions during crystallization of Antarctic angrite Lewis Cliff 86010. *Geochim. Cosmochim. Acta* **58**, 2911–2919.

Metzler K., Bobe K. D., Palme H., Spettel B., and Stöffler D. (1995) Thermal and impact metamorphism on the HED parent asteroid. *Planet. Space Sci.* **43**, 499–525.

Mikouchi T. and McKay G. (2001) Mineralogical investigation of D'Orbigny: a new angrite showing close affinities to Asuka 881371, Sahara 99555 and Lewis Cliff 87051. In *Lunar Planet. Sci.* **XXXII**, #1876. The Lunar and Planetary Institute, Houston (CD-ROM).

Mikouchi T., Takeda H., Miyamoto M., Ohsumi K., and McKay G. A. (1995) Exsolution lamellae of kirschsteinite in magnesium-iron olivine from an angrite meteorite. *Am. Mineral.* **80**, 585–592.

Mikouchi T., Miyamoto M., and McKay G. A. (1996) Mineralogical study of angrite Asuka-881371: its possible relation to angrite LEW87051. *Proc. NIPR Symp. Antarct. Meteorit.* **9**, 174–188.

Mikouchi T., McKay G., Le L., and Mittlefehldt D. W. (2000a) Preliminary examination of Sahara 99555: mineralogy and experimental studies of a new angrite. In *Lunar Planet. Sci.* **XXXI**, #1970. The Lunar and Planetary Institute, Houston (CD-ROM).

Mikouchi T., McKay G., and Le L. (2000b) A new angrite Sahara 99555: mineralogical comparison with Angra dos Reis, Lewis Cliff 86010, Lewis Cliff 87051, ad Asuka 881371 angrites. *Antarct. Meteorit.* **XXV**, 74–76.

Mittlefehldt D. W. (1979) Petrographic and chemical characterization of igneous lithic clasts from mesosiderites and howardites and comparison with eucrites and diogenites. *Geochim. Cosmochim. Acta* **43**, 1917–1935.

Mittlefehldt D. W. (1980) The composition of mesosiderite olivine clasts and implications for the origin of pallasites. *Earth Planet. Sci. Lett.* **51**, 29–40.

Mittlefehldt D. W. (1987) Volatile degassing of basaltic achondrite parent bodies: evidence from alkali elements and phosphorus. *Geochim. Cosmochim. Acta* **51**, 267–278.

Mittlefehldt D. W. (1990) Petrogenesis of mesosiderites: I. Origin of mafic lithologies and comparison with basaltic achondrites. *Geochim. Cosmochim. Acta* **54**, 1165–1173.

Mittlefehldt D. W. (1994) The genesis of diogenites and HED parent body petrogenesis. *Geochim. Cosmochim. Acta* **58**, 1537–1552.

Mittlefehldt D. W. (2000) Petrology and geochemistry of the Elephant Moraine A79002 diogenite: a genomict breccia containing a magnesian harzburgite component. *Meteorit. Planet. Sci.* **35**, 901–912.

Mittlefehldt D. W. and Lindstrom M. M. (1990) Geochemistry and genesis of the angrites. *Geochim. Cosmochim. Acta* **54**, 3209–3218.

Mittlefehldt D. W. and Lindstrom M. M. (2001) Petrology and geochemistry of Patuxent Range 91501, an impact melt from the L chondrite parent body, and Lewis Cliff 88663, an L7 chondrite. *Meteorit. Planet. Sci.* **36**, 439–457.

Mittlefehldt D. W. and Lindstrom M. M. (2003) Geochemistry of eucrites: genesis of basaltic eucrites, and Hf and Ta as petrogenetic indicators for altered antarctic eucrites. *Geochim. Cosmochim. Acta* **67**, 1911–1935.

Mittlefehldt D. W., Chou C.-L., and Wasson J. T. (1979) Mesosiderites and howardites: igneous formation and possible genetic relationships. *Geochim. Cosmochim. Acta* **43**, 673–688.

Mittlefehldt D. W., Rubin A. E., and Davis A. M. (1992) Mesosiderite clasts with the most extreme positive europium anomalies among solar system rocks. *Science* **257**, 1096–1099.

Mittlefehldt D. W., Lindstrom M. M., Bogard D. D., Garrison D. H., and Field S. W. (1996) Acapulco- and Lodran-like achondrites: petrology, geochemistry, chronology and origin. *Geochim. Cosmochim. Acta* **60**, 867–882.

Mittlefehldt D. W., McCoy T. J., Goodrich C. A., and Kracher A. (1998) Non-chondritic meteorites from asteroidal bodies. In *Planetary Materials*, Rev. Mineral., 36 (ed. J. J. Papike). Mineralogical Society of America, Washington, DC, pp. 4-1–4-195.

Mittlefehldt D. W., Killgore M., and Lee M. T. (2002) Petrology and geochemistry of D'Orbigny, geochemistry of Sahara 99555, and the genesis of angrites. *Meteorit. Planet. Sci.* **37**, 345–369.

Mittlefehldt D. W., Bogard D. D., Berkley J. L., and Garrison D. H. (2003) Brachinites—igneous rocks from a differentiated asteroid. *Meteorit. Planet. Sci.* (submitted).

Nagahara H. (1992) Yamato-8002: partial melting residue on the "unique" chondrite parent body. *Proc. NIPR Symp.: Antarct. Meteorit.* **5**, 191–223.

Nagahara H. and Ozawa K. (1986) Petrology of Yamato-791493, "lodranite": melting, crystallization, cooling history, and relationship to other meteorites. *Mem. NIPR Spec. Issue* **41**, 181–205.

Nehru C. E., Delaney J. S., Harlow G. E., and Prinz M. (1980) Mesosiderite basalts and the eucrites. *Meteoritics* **15**, 337–338.

Nehru C. E., Prinz M., Delaney J. S., Dreibus G., Palme H., Spettel B., and Wänke H. (1983) Brachina: a new type of meteorite, not a Chassignite. *Proc. 14th Lunar Planet. Sci. Conf.: J. Geophys. Res.* **88** (suppl.), B237–B244.

Nehru C. E., Prinz M., Weisberg M. K., Ebihara M. E., Clayton R. N., and Mayeda T. K. (1992) Brachinites: a new primitive achondrite group. *Meteoritics* **27**, 267.

Nehru C. E., Prinz M., Weisberg M. K., Ebihara M. E., Clayton R. N., and Mayeda T. K. (1996) A new brachinite and petrogenesis of the group. In *Lunar Planet. Sci.* **XXVII**. The Lunar and Planetary Institute, Houston, pp. 943–944.

Newsom H. E., Ntaflos Th., and Keil K. (1996) Dark clasts in the Khor Temiki aubrite: not basalts. *Meteorit. Planet. Sci.* **31**, 146–151.

Niemeyer S. (1979a) I-Xe dating of silicate and troilite from IAB iron meteorites. *Geochim. Cosmochim. Acta* **43**, 843–860.

Niemeyer S. (1979b) ^{40}Ar–^{39}Ar dating of inclusions from IAB iron meteorites. *Geochim. Cosmochim. Acta* **43**, 1829–1840.

Niemeyer S. (1980) I-Xe and ^{40}Ar–^{39}Ar dating of silicate from Weekeroo Station and Netschaëvo iron meteorites. *Geochim. Cosmochim. Acta* **44**, 33–44.

Nyquist L., Bansal B., Wiesmann H., and Shih C.-Y. (1994) Neodymium, strontium, and chromium isotopic studies of the LEW86010 and Angra dos Reis meteorites and the chronology of the angrite parent body. *Meteoritics* **29**, 872–885.

Okada A. and Keil K. (1982) Caswellsilverite, $NaCrS_2$: a new mineral in the Norton County enstatite achondrite. *Am. Mineral.* **67**, 132–136.

Okada A., Keil K., Taylor G. J., and Newsom H. (1988) Igneous history of the aubrite parent asteroid: evidence from the Norton County enstatite achondrite. *Meteoritics* **23**, 59–74.

Olsen E. and Jarosewich E. (1970) The chemical composition of the silicate inclusions in the Weekeroo Station iron meteorite. *Earth Planet. Sci. Lett.* **8**, 261–266.

Olsen E. and Jarosewich E. (1971) Chondrules: first occurrence in an iron meteorite. *Science* **174**, 583–585.

Olsen E., Davis A., Clarke R. S., Jr., Schultz L., Weber H. W., Clayton R., Mayeda T., Jarosewich E., Sylvester P., Grossman L., Wang M.-S., Lipschutz M. E., Steele I. M., and Schwade J. (1994) Watson: a new link in the IIE iron chain. *Meteoritics* **29**, 200–213.

Osadchii Eu. G., Baryshnikova G. V., and Novikov G. V. (1981) The Elga meteorite: silicate inclusions and shock metamorphism. *Proc. 12th Lunar Planet. Sci. Conf.* 1049–1068.

Ott U., Löhr H. P., and Begemann F. (1985) Noble gases and the classification of Brachina. *Meteoritics* **20**, 69–78.

Ott U., Begemann F., and Löhr H. P. (1987) Noble gases in ALH 84025: like Brachina, unlike Chassigny. *Meteoritics* **22**, 476–477.

Palme H., Baddenhausen H., Blum K., Cendales M., Dreibus G., Hofmeister H., Kruse H., Palme C., Spettel B., Vilcsek E., and Wänke H. (1978) New data on lunar samples and achondrites and a comparison of the least fractionated samples from the Earth, the Moon and the eucrite parent body. *Proc. 9th Lunar. Sci. Conf.* 25–57.

Palme H., Schultz L., Spettel B., Weber H. W., Wänke H., Christophe Michel-Levy M., and Lorin J. C. (1981) The Acapulco meteorite: chemistry, mineralogy and irradiation effects. *Geochim. Cosmochim. Acta* **45**, 727–752.

Palme H., Wlotzka F., Spettel B., Dreibus G., and Weber H. (1988) Camel Donga: a eucrite with high metal content. *Meteoritics* **23**, 49–57.

Patzer A., Hill D. H., and Boynton W. V. (2001) Itqiy: a metal-rich enstatite meteorite with achondritic texture. *Meteorit. Planet. Sci.* **36**, 1495–1505.

Pellas P., Fiéni C., Trieloff M., and Jessberger E. K. (1997) The cooling history of the Acapulco meteorite as recorded by the ^{244}Pu and ^{40}Ar–^{39}Ar chronometers. *Geochim. Cosmochim. Acta* **61**, 3477–3501.

Petaev M. I., Barsukova L. D., Lipschutz M. E., Wang M.-S., Ariskin A. A., Clayton R. N., and Mayeda T. K. (1994) The Divnoe meteorite: petrology, chemistry, oxygen isotopes and origin. *Meteoritics* **29**, 182–199.

Petaev M. I., Clarke R. S., Jr., Jarosewich E., Zaslavskaya N. I., Knonnkova N. N., Wang M.-S., Lipschutz M. E., Olsen E. J., Davis A. M., Steele I. M., Clayton R. N., Mayeda T. K., and Kallemeyn G. W. (2000) The Chaunskij anomalous mesosiderite: petrology, chemistry, oxygen isotopes, classification, and origin. *Geochem. Int.* **38**(3), S322–S350.

Powell B. N. (1971) Petrology and chemistry of mesosiderites: II. Silicate textures and compositions and metal-silicate relationships. *Geochim. Cosmochim. Acta* **35**, 5–34.

Premo W. R. and Tatsumoto M. (1995) Pb isotopic systematics of angrite Asuka-881371. *Antarct. Meteorit.* **XX**, 204–206.

Prinz M., Keil K., Hlava P. F., Berkley J. L., Gomes C. B., and Curvello W. S. (1977) Studies of Brazilian meteorites: III. Origin and history of the Angra dos Reis achondrite. *Earth Planet. Sci. Lett.* **35**, 317–330.

Prinz M., Waggoner D. G., and Hamilton P. J. (1980) Winonaites: a primitive achondritic group related to silicate inclusions in IAB irons. In *Lunar Planet. Sci.* **XI**. The Lunar and Planetary Institute, Houston, pp. 902–904.

Prinz M., Nehru C. E., Delaney J. S., and Weisberg M. (1983) Silicates in IAB and IIICD irons, winonaites, lodranites and Brachina: a primitive and modified-primitive group. In *Lunar Planet. Sci.* **XIV**. The Lunar and Planetary Institute, Houston, pp. 616–617.

Prinz M., Weisberg M. K., and Nehru C. E. (1988) LEW86010, a second angrite: relationship to CAI's and opaque matrix. In *Lunar Planet. Sci.* **XIX**. The Lunar and Planetary Institute, Houston, pp. 949–950.

Prinzhofer A., Papanastassiou D. A., and Wasserburg G. J. (1992) Samarium–neodymium evolution of meteorites. *Geochim. Cosmochim. Acta* **56**, 797–815.

Reid A. M. and Cohen A. J. (1967) Some characteristics of enstatite from enstatite achondrites. *Geochim. Cosmochim. Acta* **31**, 661–672.

Reid A. M., Buchanan P., Zolensky M. E., and Barrett R. A. (1990) The Bholghati howardite: petrography and mineral chemistry. *Geochim. Cosmochim. Acta* **54**, 2161–2166.

Righter K. and Drake M. J. (1997) A magma ocean on Vesta: core formation and petrogenesis of eucrites and diogenites. *Meteorit. Planet. Sci.* **32**, 929–944.

Rubin A. E. and Jerde E. A. (1987) Diverse eucritic pebbles in the Vaca Muerta mesosiderite. *Earth Planet. Sci. Lett.* **84**, 1–14.

Rubin A. E. and Jerde E. A. (1988) Compositional differences between basaltic and gabbroic clasts in mesosiderites. *Earth Planet. Sci. Lett.* **87**, 485–490.

Rubin A. E. and Mittlefehldt D. W. (1992) Classification of mafic clasts from mesosiderites: implications for endogenous igneous processes. *Geochim. Cosmochim. Acta* **56**, 827–840.

Rubin A. E. and Mittlefehldt D. W. (1993) Evolutionary history of the mesosiderite asteroid: a chronologic and petrologic synthesis. *Icarus* **101**, 201–212.

Ruzicka A., Snyder G. A., and Taylor L. A. (1997) Vesta as the howardite, eucrite and diogenite parent body: implications for the size of a core and for large-scale differentiation. *Meteorit. Planet. Sci.* **32**, 825–840.

Sack R. O., Azeredo W. J., and Lipschutz M. E. (1991) Olivine diogenites: the mantle of the eucrite parent body. *Geochim. Cosmochim. Acta* **55**, 1111–1120.

Schultz L., Palme H., Spettel B., Weber H. W., Wänke H., Christophe Michel-Levy M., and Lorin J. C. (1982) Allan Hills A77081—an unusual stony meteorite. *Earth Planet. Sci. Lett.* **61**, 23–31.

Scott E. R. D., Haack H., and Love S. G. (2001) Formation of mesosiderites by fragmentation and reaccretion of a large differentiated asteroid. *Meteorit. Planet. Sci.* **36**, 869–881.

Sears D. W. (1980) Formation of E-chondrites and aubrites—a thermodynamic model. *Icarus* **43**, 184–202.

Shearer C. K., Fowler G. W., and Papike J. J. (1997) Petrogenetic models for magmatism on the eucrite parent body: evidence from orthopyroxene in diogenites. *Meteorit. Planet. Sci.* **32**, 877–889.

Simpson A. B. and Ahrens L. H. (1977) The chemical relationship between howardites and the silicate fraction of mesosiderites. In *Comets, Asteroids, Meteorites—Interpretations, Evolution, and Origins* (ed. A. H. Delsemme). U. Toledo Press, Toledo, OH, pp. 445–450.

Sinha S. K., Sack R. O., and Lipschutz M. E. (1997) Ureilite meteorites: equilibration temperatures and smelting reactions. *Geochim. Cosmochim. Acta* **61**, 4235–4242.

Smith J. V., Steele I. M., and Leitch C. A. (1983) Mineral chemistry of the shergottites, nakhlites, Chassigny, Brachina, pallasites, and ureilites. *Proc. 14th Lunar Planet. Sci. Conf.: J. Geophys. Res.* **88** (suppl.), B229–B236.

Smoliar M. I. (1993) A survey of Rb–Sr systematics of eucrites. *Meteoritics* **28**, 105–113.

Snyder G. A., Lee D.-C., Ruzicka A. M., Taylor L. A., Halliday A. N., and Prinz M. (1998) Evidence of late impact fractionation and mixing of silicates on iron meteorite parent bodies: Hf–W, Sm–Nd, and Rb–Sr isotopic studies of silicate inclusions in IIE irons. In *Lunar Planet. Sci.* **XXIX**, #1142. The Lunar and Planetary Institute, Houston (CD-ROM).

Spitz A. H. and Boynton W. V. (1991) Trace element analysis of ureilites: new constraints on their petrogenesis. *Geochim. Cosmochim. Acta* **55**, 3417–3430.

Spitz A. H. and Goodrich C. A. (1987) Rare earth element tests of ureilite petrogenesis models. *Meteoritics* **21**, 515–516.

Stewart B. W., Papanastassiou D. A., and Wasserburg G. J. (1994) Sm–Nd chronology and petrogenesis of mesosiderites. *Geochim. Cosmochim. Acta* **58**, 3487–3509.

Stolper E. (1977) Experimental petrology of eucrite meteorites. *Geochim. Cosmochim. Acta* **41**, 587–611.

Strait M. M. (1983) Chemical variations in enstatite achondrites. PhD Thesis, Arizona State University.

Swindle T. D., Kring D. A., Burkland M. K., Hill D. H., and Boynton W. V. (1998) Noble gases, bulk chemistry, and petrography of olivine-rich achondrites Eagles Nest and Lewis Cliff 88763: comparison to brachinites. *Meteorit. Planet. Sci.* **33**, 31–48.

Takeda H. (1987) Mineralogy of Antarctic ureilites and a working hypothesis for their origin and evolution. *Earth Planet. Sci. Lett.* **81**, 358–370.

Takeda H. and Graham A. L. (1991) Degree of equilibration of eucritic pyroxenes and thermal metamorphism of the earliest planetary crust. *Meteoritics* **26**, 129–134.

Takeda H., Mori H., Hiroi T., and Saito J. (1994a) Mineralogy of new Antarctic achondrites with affinity to Lodran and a model of their evolution in an asteroid. *Meteoritics* **29**, 830–842.

Takeda H., Yamaguchi A., Nyquist L. E., and Bogard D. D. (1994b) A mineralogical study of the proposed paired eucrites Y-792769 and Y-793164 with reference to cratering events on their parent body. *Proc. NIPR Symp.: Antarct. Meteorit.* **7**, 73–93.

Takeda H., Mori H., and Bogard D. D. (1994c) Mineralogy and ^{39}Ar–^{40}Ar age of an old pristine basalt: thermal history of the HED parent body. *Earth Planet. Sci. Lett.* **122**, 183–194.

Takeda H., Bogard D. D., Mittlefehldt D. W., and Garrison D. H. (2000) Mineralogy, petrology, chemistry, and ^{39}Ar–^{40}Ar and exposure ages of the Caddo County IAB iron: evidence for early partial melt segregation of a gabbro area rich in plagioclase-diopside. *Geochim. Cosmochim. Acta* **64**, 1311–1327.

Taylor S. R. (1982) *Planetary Science: A Lunar Perspective.* Lunar Planetary Institute, Houston, TX.

Tera F., Eugster O., Burnett D. S., and Wasserburg G. J. (1970) Comparative study of Li, Na, K, Rb, Cs, Ca, Sr, and Ba abundances in achondrites and in Apollo 11 lunar samples. *Proc. Apollo. 11 Lunar Sci. Conf.* **11**, 1637–1657.

Tera F., Carlson R. W., and Boctor N. Z. (1997) Radiometric ages of basaltic achondrites and their relation to the early history of the solar system. *Geochim. Cosmochim. Acta* **61**, 1713–1731.

Thomas P. C., Binzel R. P., Gaffey M. J., Storrs A. D., Wells E. N., and Zellner B. H. (1997) Impact excavation on asteroid 4 Vesta: Hubble Space Telescope results. *Science* **277**, 1492–1495.

Torigoye N., Yamamoto K., Misawa K., and Nakamura N. (1993) Compositions of REE, K, Rb, Sr, Ba, Mg, Ca, Fe, and Sr isotopes in Antarctic "unique" meteorites. *Proc. NIPR Symp.: Antarct. Meteorit.* **6**, 100–119.

Torigoye-Kita N., Misawa K., and Tatsumoto M. (1995a) U–Th–Pb and Sm–Nd isotopic systematics of the Goalpara ureilite: resolution of terrestrial contamination. *Geochim. Cosmochim. Acta* **59**, 381–390.

Torigoye-Kita N., Misawa K., and Tatsumoto M., (1995b) Reply to the comment by Goodrich et al. (1995). *Geochim. Cosmochim. Acta* **59**, 4087–4091.

Torigoye-Kita N., Tatsumoto M., Meeker G. P., and Yanai K. (1995c) The 4.56 Ga age of the MET 78008 ureilite. *Geochim. Cosmochim. Acta* **59**, 2319–2329.

Treiman A. H. (1997) The parent magmas of the cumulate eucrites: a mass balance approach. *Meteorit. Planet. Sci.* **32**, 217–230.

Treiman A. H. and Berkley J. L. (1994) Igneous petrology of the new ureilites Nova 001 and Nullarbor 010. *Meteoritics* **29**, 843–848.

Wacker J. (1986) Noble gases in the diamond-free ureilite, ALHA78019: the roles of shock and nebular processes. *Geochim. Cosmochim. Acta* **50**, 633–642.

Wadhwa M., Shukolyukov A., and Lugmair G. W. (1998) ^{53}Mn–^{53}Cr systematics in Brachina: a record of one of the earliest phases of igneous activity on an asteroid. In *Lunar Planet. Sci.* **XXIX**, #1480. The Lunar and Planetary Institute, Houston (CD-ROM).

Wang M.-S. and Lipschutz M. E. (1995) Volatile trace elements in Antarctic ureilites. *Meteoritics* **30**, 319–324.

Wänke H. and Dreibus G. (1994) Chemistry and accretion history of Mars. *Phil. Trans. Roy. Soc. London* **A349**, 285–293.

Wänke H., Baddenhausen H., Balacescu A., Teschke F., Spettel B., Dreibus G., Palme H., Quijano-Rico M., Kruse H., Wlotzka F., and Begemann F. (1972) Multielement analyses of lunar samples and some implications of the results. *Proc. 3rd Lunar Sci. Conf.* 1251–1268.

Wänke H., Baddenhausen H., Dreibus G., Jagoutz E., Kruse H., Palme H., Spettel B., and Teschke F. (1973) Multielement analyses of Apollo 15, 16, and 17 samples and the bulk composition of the moon. *Proc. 4th Lunar Sci. Conf.* 1461–1481.

Wänke H., Baddenhausen H., Blum K., Cendales M., Dreibus G., Hofmeister H., Kruse H., Jagoutz E., Palme C., Spettel B., Thacker R., and Vilcsek E. (1977) On the chemistry of lunar samples and achondrites. Primary matter in the lunar highlands: a re-evaluation. *Proc. 8th Lunar Sci. Conf.* 2191–2213.

Warren P. H. (1997) Magnesium oxide-iron oxide mass balance constraints and a more detailed model for the relationship between eucrites and diogenites. *Meteorit. Planet. Sci.* **32**, 945–963.

Warren P. H. and Davis A. M. (1995) Consortium investigation of the Asuka-881371 angrite: petrographic, electron microprobe, and ion microprobe observations. *Antarct. Meteorit.* **XX**, 257–260.

Warren P. H. and Jerde E. A. (1987) Composition and origin of Nuevo Laredo trend eucrites. *Geochim. Cosmochim. Acta* **51**, 713–725.

Warren P. H. and Kallemeyn G. W. (1989a) Allan Hills 84025: the second Brachinite, far more differentiated than Brachina, and an ultramafic achondritic clast from L chondrite Yamato 75097. *Proc. 19th Lunar Planet. Sci. Conf.* 475–486.

Warren P. H. and Kallemeyn G. W. (1989b) Geochemistry of polymict ureilite EET83309: and a partially-disruptive impact model for ureilite origin. *Meteoritics* **24**, 233–246.

Warren P. H. and Kallemeyn G. W. (1992) Explosive volcanism and the graphite-oxygen fugacity buffer on the parent asteroid(s) of the ureilite meteorites. *Icarus* **100**, 110–126.

Warren P. H., Kallemeyn G. W., and Mayeda T. (1995) Consortium investigation of the Asuka-881371 angrite: bulk-rock geochemistry and oxygen-isotopes. *Antarct. Meteorit.* **XX**, 261–264.

Warren P. H., Kallemeyn G. W., Arai T., and Kaneda K. (1996) Compositional-petrologic investigations of eucrites and the QUE94201 shergottite. *Antarct. Meteorit.* **XXI**, 195–197.

Wasserburg G. J., Tera F., Papanastassiou D. A., and Huneke J. C. (1977) Isotopic and chemical investigations on Angra dos Reis. *Earth Planet. Sci. Lett.* **35**, 294–316.

Wasson J. T. (1972) Formation of ordinary chondrites. *Rev. Geophys. Space Phys.* **10**, 711–759.

Wasson J. T. and Kallemeyn G. W. (1988) Compositions of chondrites. *Phil. Trans. Roy. Soc. London* **A325**, 535–544.

Wasson J. T. and Rubin A. E. (1985) Formation of mesosiderites by low-velocity impacts as a natural consequence of planet formation. *Nature* **318**, 168–170.

Wasson J. T. and Wai C. M. (1970) Composition of the metal, schreibersite and perryite of enstatite achondrites and the origin of enstatite chondrites and achondrites. *Geochim. Cosmochim. Acta* **34**, 169–184.

Wasson J. T. and Wang J. (1986) A nonmagmatic origin of group-IIE iron meteorites. *Geochim. Cosmochim. Acta* **50**, 725–732.

Wasson J. T., Chou C.-L., Bild R. W., and Baedecker P. A. (1976) Classification of and elemental fractionation among ureilites. *Geochim. Cosmochim. Acta* **40**, 1449–1458.

Wasson J. T., Willis J., Wai C. M., and Kracher A. (1980) Origin of iron meteorite groups IAB and IIICD. *Z. Naturforsch.* **35a**, 781–795.

Watters T. R. and Prinz M. (1979) Aubrites: their origin and relationship to enstatite chondrites. *Proc. 10th lunar Planet. Sci. Conf.* 1073–1093.

Weber H. W., Hintenberger H., and Begemann F. (1971) Noble gases in the Haverö ureilite. *Earth Planet. Sci. Lett.* **13**, 205–209.

Weber H. W., Begemann F., and Hintenberger H. (1976) Primordial gases in graphite–diamond-kamacite inclusions from the Haverö ureilite. *Earth Planet. Sci. Lett.* **29**, 81–90.

Weigel A., Eugster O., Koeberl C., and Krähenbühl U. (1997) Differentiated achondrites Asuka 881371, an angrite, and Divnoe: noble gases, ages, chemical composition, and relation to other meteorites. *Geochim. Cosmochim. Acta* **61**, 239–248.

Weigel A., Eugster O., Koeberl C., Michel R., Krähenbühl U., and Neumann S. (1999) Relationships among lodranites and acapulcoites: noble gas isotopic abundances, chemical composition, cosmic-ray exposure ages, and solar cosmic ray effects. *Geochim. Cosmochim. Acta* **61**, 175–192.

Wheelock M. M., Keil K., Floss C., Taylor G. J., and Crozaz G. (1994) REE geochemistry of oldhamite-dominated clasts from the Norton County aubrite: igneous origin of oldhamite. *Geochim. Cosmochim. Acta* **58**, 449–458.

Wolf R., Ebihara M., Richter G. R., and Anders E. (1983) Aubrites and diogenites: trace element clues to their origin. *Geochim. Cosmochim. Acta* **47**, 2257–2270.

Wlotzka F. and Jarosewich E. (1977) Mineralogical and chemical compositions of silicate inclusions in the El Taco, Camp del Cielo, iron meteorite. *Smithson. Contrib. Earth Sci.* **19**, 104–125.

Yamaguchi A., Taylor G. J., Keil K., Floss C., Crozaz G., Nyquist L. E., Bogard D. D., Garrison D., Reese Y., Wiesman H., and Shih C.-Y. (2001) Post-crystallization reheating and partial melting of eucrite EET 90020 by impact into the hot crust of asteroid 4 Vesta ~4.5 Ga ago. *Geochim. Cosmochim. Acta* **65**, 3577–3599.

Yamaguchi A., Clayton R. N., Mayeda T. K., Ebihara M., Oura Y., Miura Y. N., Haramura H., Misawa K., Kojima H., and Nagao K. (2002) A new source of basaltic meteorites inferred from Northwest Africa 011. *Science* **296**, 334–336.

Yanai K. (1994) Angrite Asuka-881371: preliminary examination of a unique meteorite in the Japanese collection of Antarctic meteorites. *Proc. NIPR Symp.: Antarct. Meteorit.* **7**, 30–41.

Yanai K. and Kojima H. (1991) Yamato-74063: chondritic meteorite classified between E and H chondrite groups. *Proc. NIPR Symp.: Antarct. Meteorit.* **4**, 118–130.

Yanai K. and Kojima H. (1995) *Catalog of Antarctic Meteorites*. National Institute of Polar Research Tokyo, Japan.

Zipfel J., Palme H., Kennedy A. K., and Hutcheon I. D. (1995) Chemical composition and origin of the Acapulco meteorite. *Geochim. Cosmochim. Acta* **59**, 3607–3627.

1.12
Iron and Stony-iron Meteorites

H. Haack

University of Copenhagen, Denmark

and

T. J. McCoy

Smithsonian Institution, Washington, DC, USA

1.12.1	INTRODUCTION	325
1.12.2	CLASSIFICATION AND CHEMICAL COMPOSITION OF IRON METEORITES	327
	1.12.2.1 Group IIAB Iron Meteorites	327
	1.12.2.2 Group IIIAB Iron Meteorites	327
	1.12.2.3 Group IVA Iron Meteorites	328
	1.12.2.4 Group IVB Iron Meteorites	328
	1.12.2.5 Silicate-bearing Iron Meteorites	329
	1.12.2.6 Mesosiderites	329
	1.12.2.7 Ungrouped Iron Meteorites	329
1.12.3	ACCRETION AND DIFFERENCES IN BULK CHEMISTRY BETWEEN GROUPS OF IRON METEORITES	330
1.12.4	HEATING AND DIFFERENTIATION	332
1.12.5	FRACTIONAL CRYSTALLIZATION OF METAL CORES	333
	1.12.5.1 Imperfect Mixing during Crystallization	335
	1.12.5.2 Late-stage Crystallization and Immiscible Liquid	335
	1.12.5.3 The Missing Sulfur-rich Meteorites	336
1.12.6	COOLING RATES AND SIZES OF PARENT BODIES	336
1.12.7	PALLASITES	338
1.12.8	PARENT BODIES OF IRON AND STONY-IRON METEORITES	339
1.12.9	FUTURE RESEARCH DIRECTIONS	340
REFERENCES		341

1.12.1 INTRODUCTION

Without iron and stony-iron meteorites, our chances of ever sampling the deep interior of a differentiated planetary object would be next to nil. Although we live on a planet with a very substantial core, we will never be able to sample it. Fortunately, asteroid collisions provide us with a rich sampling of the deep interiors of differentiated asteroids.

Iron and stony-iron meteorites are fragments of a large number of asteroids that underwent significant geological processing in the early solar system. Parent bodies of iron and some stony-iron meteorites completed a geological evolution similar to that continuing on Earth—although on much smaller length- and timescales—with melting of the metal and silicates, differentiation into core, mantle, and crust, and probably extensive volcanism. Iron and stony-iron meteorites are our only available analogues to materials found in the deep interiors of Earth and other terrestrial planets. This fact has been recognized since the work of Chladni (1794), who argued that stony-iron meteorites must have originated in outer space and fallen during fireballs and that they provide our closest analogue to the material that comprises our own planet's core. This chapter deals with our current knowledge of these meteorites. How did they form? What can they tell us about the early evolution of the solar system and its solid bodies? How closely do they resemble the materials from planetary interiors? What do we know and don't we know?

Iron and stony-iron meteorites constitute ~6% of meteorite falls (Grady, 2000). Despite their scarcity among falls, iron meteorites are our only samples of ~75 of the ~135 asteroids from which meteorites originate (Keil et al., 1994; Scott, 1979; Meibom and Clark, 1999; see also Chapter 1.05), suggesting that both differentiated asteroids and the geologic processes that produced them were common.

Despite the highly evolved nature of iron and stony-iron meteorites, their chemistry provides important constraints on the processes operating in the solar nebula. Although most of them probably formed through similar mechanisms, their characteristics are diverse in terms of chemistry, mineralogy, and structure. Significant differences in bulk chemistry between iron meteorites from different cores as well as variations in chemistry between meteorites from the same core provide evidence of the complex chemical evolution of these evolved meteorites. Intergroup variations for volatile siderophile elements (e.g., gallium and germanium) extend more than three orders of magnitude, hinting that iron meteorite parent bodies formed under diverse conditions. These differences reflect both the nebular source material and geological processing in the parent bodies.

Can we be sure that the iron meteorites are indeed fragments of cores? Since no differentiated asteroid has yet been visited by a spacecraft, we rely on circumstantial evidence. Some M-type asteroids have spectral characteristics expected from exposed metallic cores (Tholen, 1989), while others exhibit basaltic surfaces, a hallmark of global differentiation. Although olivine-rich mantles should dominate the volume of differentiated asteroids, there is an enigmatic lack of olivine-rich asteroids (and meteorites) that could represent mantle material (Burbine et al., 1996). Until we visit an asteroid with parts of a core–mantle boundary exposed, our best evidence supporting a core origin is detailed studies of iron meteorites.

Iron–nickel alloys are expected in the cores of differentiated asteroids, but what other evidence supports the notion that iron meteorites sample the metallic cores of differentiated asteroids? What suggests that these asteroids were sufficiently heated to trigger core formation, and that iron meteorites sample cores rather than isolated pods of once molten metal? First and foremost, trace-element compositional trends in most groups of iron meteorites are consistent with fractional crystallization of a metallic melt (Scott, 1972), thus constraining peak temperatures. The temperatures required to form a metallic melt are sufficiently high to cause substantial melting of the associated silicates and trigger core formation (Taylor et al., 1993). In addition, meter-sized taenite crystals and slow metallographic cooling rates in some iron meteorites suggest crystallization and cooling at considerable depth. Due to the high thermal conductivity of solid metal compared with mantle and crust materials, metal cores are believed to be virtually isothermal during cooling (Wood, 1964). Iron meteorites from the same core should therefore have identical cooling rates. Metallographic cooling rates are different from group to group but are, with the exception of group IVA, relatively uniform within each group (Mittlefehldt et al., 1998). This is consistent with each group cooling in a separate core surrounded by a thermally insulating mantle.

If the iron and stony-iron meteorites came from fully differentiated asteroids, how did these asteroids heat to the point of partial melting and how did the metal segregate from the silicates? Unlike large planets, where potential energy release triggers core formation, small asteroids require an additional heat source. The heat source(s) for asteroidal melting produced a wide range of products, from unmetamorphosed chondrites to fully molten asteroids, as well as partially melted asteroids. Samples from these latter asteroids provide us with a rare opportunity to observe core formation—frozen in place.

Most iron and stony-iron meteorites came from asteroids that were sufficiently heated to completely differentiate. The better-sampled iron meteorite groups provide us with an unparalleled view of a crystallizing metallic magma. Except for an enigmatic paucity of samples representing the late stage Fe–S eutectic melt, the trends recorded by the larger groups of iron meteorites cover the entire crystallization sequence of their parent-body cores (Scott and Wasson, 1975). Within individual groups, the abundance of meteorites in each compositional range is also in good agreement with the expectations based on numerical models (Scott and Wasson, 1975).

For some asteroids, near-catastrophic impacts played an important role in their early geological evolution. Impact events ranging from local melting of the target area to complete disruption of the parent body are recorded in most groups of meteorites (Keil et al., 1994). Major impacts in the early solar system caused remixing of metal and silicate and large-scale redistribution of cold and hot material in the interior of some of the parent bodies. These processes had profound consequences for the geochemical and thermal evolution of the surviving material. Later impacts on the meteorite parent bodies dispersed fragments in the solar system, some of which have fallen on the Earth in the form of meteorites. These fragments provide us with an opportunity to study the geological evolution of a complete suite of rocks from differentiated asteroids in the laboratory.

1.12.2 CLASSIFICATION AND CHEMICAL COMPOSITION OF IRON METEORITES

To use meteorites as a guide to parent-body evolution, we need groupings that represent individual parent bodies. In this section, we briefly discuss the characteristics of iron and stony-iron meteorites used to discriminate between the different parent bodies. Historically, iron meteorites were classified on the basis of macroscopic structure, being divided into hexahedrites, octahedrites (of varying types based on kamacite bandwidth), and ataxites (see Buchwald, 1975 and references therein). Beginning in the 1950s, it was recognized that the chemical composition of iron meteorites varied, often in concert with the structure. Today, the chemical classification—based on bulk chemical analysis of the metal—is the standard for iron meteorite groupings (see Chapter 1.05), with structural and other data (e.g., cosmic-ray exposure ages) providing supporting information about iron meteorite history.

In a series of 12 papers beginning in 1967, John Wasson and co-workers analyzed the chemistry of the vast majority iron meteorites using INAA (Wasson, 1967, 1969, 1970, 1971; Wasson and Kimberlin, 1967; Wasson et al., 1989, 1998; Schaudy et al., 1972; Scott et al., 1973; Scott and Wasson, 1976; Kracher et al., 1980; Malvin et al., 1984), producing the definitive database on the chemistry and classification of iron meteorites. In the early papers, the meteorites were analyzed for nickel, iridium, gallium, and germanium, whereas the later papers include analyses for chromium, cobalt, nickel, copper, gallium, germanium, arsenic, antimony, tungsten, rhenium, iridium, platinum, and gold. Other papers by Wasson and co-workers that include analytical data have focused on individual groups or improved techniques (Scott, 1977a, 1978; Pernicka and Wasson, 1987; Rasmussen et al., 1984; Wasson and Wang, 1986; Choi et al., 1995; Wasson, 1999; Wasson and Richardson, 2001). Additional resources on the chemical composition of iron meteorites include Buchwald (1975), Moore et al. (1969) (nickel, cobalt, phosphorus, carbon, sulfur, and copper), and Hoashi et al. (1990, 1992, 1993a,b) (platinum group elements).

Although the original classification was entirely based on chemistry, additional data on structure, cooling rates, and mineralogy suggest that it is a genetic classification where each group represents one parent core (Scott and Wasson, 1975). The chemical compositions within each group form coherent trends generally consistent with fractional crystallization and the members have similar or uniformly varying structure and mineralogy. Despite significant overlap in the compositional clusters, multiple elements, along with structure and mineralogy, distinguish the groups. Most groups have uniform cooling rates, supporting the idea that each group represents an individual asteroid core. Cosmic-ray exposure ages further support a genetic classification in groups IIIAB and IVA, where most members have cosmic-ray exposure ages of 650 Ma and 400 Ma, respectively.

A detailed description of the individual iron meteorite groups is found in Scott and Wasson (1975). We include a brief synopsis of some of the important characteristics of the largest groups, particularly those with implications for the evolution of the parent bodies. Differences between these groups serve to illustrate some of the diverse characteristic of iron meteorites. The groups range from the volatile and silicate-rich IAB–IIICDs to the highly volatile depleted groups IVA and, in particular, IVB. Differences in the oxygen fugacities of the parent cores are reflected in the difference in mineralogy between the highly reduced IIABs and the more oxidized group IIIAB iron meteorites.

1.12.2.1 Group IIAB Iron Meteorites

Group IIAB iron meteorites (103 members) have the lowest nickel concentrations of any group of iron meteorites (5.3–6.6 wt.% Ni) (see figure 26, Chapter 1.05). The IIAB irons exhibit a distinct structural appearance, owing to their low bulk nickel concentration and, thus, predominance of kamacite. Structurally, they range from hexahedrites (former group IIA) to coarsest octahedrites (former group IIB). In hexahedrites, kamacite crystals are generally larger than the specimen, whereas coarsest octahedrites have kamacite lamellae wider than 3.3 mm (Buchwald, 1975). The low nickel concentration, absence of phosphates (Buchwald, 1975, 1984), and occurrence of graphite in IIAB iron meteorites suggest formation from highly reduced material.

1.12.2.2 Group IIIAB Iron Meteorites

With 230 classified meteorites, IIIAB is by far the largest group of iron meteorites. IIIAB irons are medium octahedrites, often heavily shocked

(Buchwald, 1975; Stöffler et al., 1988), and have uniform exposure ages and cooling rates. They have the by far highest abundance of phosphates, suggesting formation from relatively oxidized material (Kracher et al., 1977; Olsen and Fredriksson, 1966; Olsen et al., 1999). Silicate inclusions are documented in only two members (Haack and Scott, 1993; Kracher et al., 1977; Olsen et al., 1996). Observed compositional trends can be broadly matched by numerical models for fractional crystallization of a common core (Haack and Scott, 1993; Ulff-Møller, 1998; Chabot and Drake, 1999; Wasson, 1999; Chabot and Drake, 2000). Metallographic cooling rates of IIIAB iron meteorites are fairly uniform with an average of 50 K Myr^{-1} (Rasmussen, 1989), corresponding to a parent-body diameter of ~50 km (Haack et al., 1990). An interesting feature of IIIAB's is that 19 of 20 measured cosmic-ray exposure ages point to a breakup of the parent core 650 ± 100 Myr ago (Voshage, 1967; Voshage and Feldmann, 1979; see Chapter 1.13), suggesting catastrophic dispersal then (Keil et al., 1994). The large number of IIIAB iron meteorites and the almost complete absence of unusual or poorly understood features make group IIIAB iron meteorites the best available samples of a crystallized metal core from a differentiated body.

1.12.2.3 Group IVA Iron Meteorites

IVA irons are fine octahedrites (Buchwald, 1975). With 48 members, IVA is the third-largest group. They have very low volatile concentrations and an unusually low Ir/Au ratio. The chemical trends defined by group IVA are broadly consistent with numerical models of fractional crystallization (Scott et al., 1996; Wasson and Richardson, 2001). Several features of group IVA suggest a complex parent body evolution after core crystallization. Two IVA irons contain rare tridymite crystals and two others contain ~50 vol.% silicates mixed with metal on a centimeter scale (Haack et al., 1996a; Reid et al., 1974; Scott et al., 1996; Ulff-Møller et al., 1995). Although the metal in the latter two crystallized within the solid silicate matrix, it has normal structure, chemistry, and cooling rate for group IVA (Schaudy et al., 1972; Haack et al., 1996a; Scott et al., 1996). Metallographic cooling rates of group IVA irons are the subject of dispute between those who believe that an apparent cooling rate variation correlated with nickel concentration is an artifact (Willis and Wasson, 1978a,b; Wasson and Richardson, 2001) and those that favor a true variation (Moren and Goldstein, 1978, 1979; Rasmussen, 1982; Rasmussen et al., 1995; Haack et al., 1996a). Haack et al. (1996a) argued that the diverse, but slow, metallographic cooling rates and evidence of rapid cooling from 1,200 °C of the tridymite–pyroxene assemblage in one of the stony-irons, Steinbach, implied that the IVA parent body went through a breakup and reassembly event shortly after core crystallization. The model was disputed by Wasson and Richardson (2001), who argued that the apparent correlation between metallographic cooling rates and chemistry for IVA iron meteorites implies that the cooling rates are in error. Like IIIAB, they are often heavily shocked and have similar cosmic-ray exposure ages of ~400 ± 100 Ma (Voshage, 1967; Voshage and Feldmann, 1979; see Chapter 1.13) suggesting parent-body disruption at that time (Keil et al., 1994). A recent compilation of the chemical compositions of IVA iron meteorites may be found in Wasson and Richardson (2001).

1.12.2.4 Group IVB Iron Meteorites

Despite containing only 13 members, group IVB exhibits unique properties and deserves special mention. Group IVB irons have the lowest volatile concentrations of any group (Table 1; see figure 26, Chapter 1.05). IVB is also enriched in refractory elements, with iridium, rhenium, and osmium concentrations an order of magnitude higher than in any other group of irons (Scott, 1972; see also Table 1). No chondrite groups have siderophile compositions similar to group IVB iron meteorites (Rasmussen et al., 1984). Re–Os

Table 1 Average compositions of meteorites from the major iron and stony-iron meteorite groups.

Group	Re (ng g^{-1})	Ir (µg g^{-1})	Ni (wt.%)	Co (mg g^{-1})	Cu (µg g^{-1})	Au (µg g^{-1})	Ga (µg g^{-1})	Ge (µg g^{-1})	S (wt.%)
IAB	260	2.0	9.50	4.9	234	1.75	63.6	247	
IIAB	1,780 (250)	12.5 (1.3)	5.65	4.6	133	0.71 (1.0)	58.63	174	(17)
IIIAB	439 (200)	3.2 (5.0)	8.33	5.1	156	1.12 (0.7)	19.79	39.1	(12)
IVA	230 (150)	1.8 (1.8)	8.51	4.0	137	1.55 (1.6)	2.14	0.12	(3)
IVB	2,150 (3,500)	18.0 (22)	17.13	7.6	<9	0.14 (0.15)	0.23	0.055	(0)

Data are from the references listed in Section 1.12.2. Numbers in parentheses are the calculated initial liquid compositions of the core from Chabot and Drake (2000) (Re, S) and Jones and Drake (1983) (Ir, Au). Note the significant differences between the estimates of the initial liquid and the average meteorite compositions. For elements such as Re and Ir with distribution coefficients between solid and liquid metal that are far from unity, the average meteorite composition may be different from the initial composition of the liquid core (see also table 3, Mittlefehldt et al., 1998).

systematics (Smoliar et al., 1996) suggest that IVB irons, unlike other groups, require a nonchondritic osmium reservoir. High metallographic cooling rates of IVB irons (Rasmussen et al., 1984) suggest a small parent body.

1.12.2.5 Silicate-bearing Iron Meteorites

Group IAB, IIICD, and IIE irons differ in a number of properties from other iron groups. The compositional trends in these groups are unlike those in other irons (e.g., IIIAB) for which an origin by fractional crystallization of a common metallic core is inferred. Although sometimes termed "nonmagmatic," IAB, IIICD, and IIE irons clearly experienced melting during their history, although perhaps of a different type than, for example, group IIIAB. Unlike other groups, metals in IAB and IIICD irons show considerable ranges in nickel, gallium, and germanium concentrations but very restricted ranges for all other elements (see figure 26, Chapter 1.05). Iridium in these groups varies within a factor of 10, compared to more than three orders of magnitude in IIIAB. These trends cannot be explained by simple fractional crystallization (Scott, 1972; Scott and Wasson, 1975). IAB irons also have metal compositions close to cosmic abundances (Scott, 1972), suggesting limited parent-body processing.

Silicates provide further evidence for the unusual origin of IAB-IIICD. While differentiated silicates might be expected in association with iron meteorites, silicates in IAB-IIICD irons are broadly chondritic (Mittlefehldt et al., 1998; Benedix et al., 2000; see Chapter 1.11). Models for the origins of IAB-IIICD iron meteorites include crystallization of a sulfur- and carbon-rich core in a partially differentiated object (Kracher, 1985; McCoy et al., 1993), breakup and reassembly of a partially differentiated object at its peak temperature (Benedix et al., 2000), or crystal segregation in isolated impact melt pools on the surface of a porous chondritic body (Wasson and Kallemeyn, 2002). A recent compilation of the chemical compositions of IAB and IIICD iron meteorites may be found in Wasson and Kallemeyn (2002).

Group IIE is a much smaller group with very diverse characteristics in terms of metal textures and silicate mineralogy. The chemical composition of the metal is very restricted and inconsistent with fractional crystallization (Scott and Wasson, 1975; Wasson and Wang, 1986). The silicate inclusions range from metamorphosed chondrites (e.g., Netschaevo) to highly differentiated silicates (e.g., Kodaikanal, Weekeroo Station), with intermediate members present (e.g., Watson, Techado) (Bogard et al., 2000). The most primitive members of the group resemble H chondrites in both mineral chemistry and oxygen isotopic composition (Clayton and Mayeda, 1996; Ruzicka et al., 1999), although the match is not perfect (Bogard et al., 2000). The most perplexing feature of this group is the presence of silicate inclusions which give chronometric ages of ~3.8 Ga, suggesting formation ~0.8 Ga after the formation of the solar system. A range of models have been proposed in the past decade, including both impact-induced (Olsen et al., 1994; Ruzicka et al., 1999) and incipient parent-body melting (Bogard et al., 2000). Young chronometric ages would virtually require that impact played a role in the formation of some members of this group.

1.12.2.6 Mesosiderites

Mesosiderites are arguably the most enigmatic group of differentiated meteorites. Mesosiderites are breccias composed of roughly equal proportions of Fe-Ni metal and silicates. Unlike pallasites, where the silicates are consistent with a deep mantle origin, the silicates in the mesosiderites are basaltic, gabbroic, and pyroxenitic (Powell, 1971; see Chapter 1.11). Metal compositions of mesosiderites are almost uniform, suggesting that the metal was molten when mixed with the silicates (Hassanzadeh et al., 1989). Many innovative models have attempted to explain the enigmatic mixture of crustal materials with core metal (see review in Hewins, 1983). Some relatively new models include impacts of molten planetesimals onto the surface of a large differentiated asteroid (Rubin and Mittlefehldt, 1992; Wasson and Rubin, 1985) and breakup and reassembly of a large differentiated asteroid with a still molten core (Scott et al., 2001).

Another unusual characteristic of mesosiderites is their very slow metallographic cooling rates of less than 1 K Myr^{-1} (Powell, 1969; Haack et al., 1996b; Hopfe and Goldstein, 2001). These cooling rates are the slowest for any natural geological material, suggesting that the mesosiderite parent body must have been large. Mesosiderites have young Ar-Ar ages of ~3.95 Ga (Bogard and Garrison, 1998). The young Ar-Ar ages have been attributed to extended cooling within a large asteroid (Haack et al., 1996b; Bogard and Garrison, 1998) or impact resetting (Bogard et al., 1990; Rubin and Mittlefehldt, 1992).

1.12.2.7 Ungrouped Iron Meteorites

Iron meteorites require a large number of parent bodies to account for their diverse properties. Ungrouped irons alone require 40-50 different parent bodies (Scott, 1979; Wasson, 1990), a number unmatched by the types of meteorites that represent mantle and crust materials (Burbine et al., 1996). In addition, gallium and germanium concentrations of ungrouped irons are not randomly distributed as expected if they represent a large and continuous population of poorly sampled parent bodies (Scott, 1979). Ungrouped irons tend to have gallium and germanium concentrations in the same ranges as those defined by the original Ga–Ge groups I–IV. The origin of ungrouped irons is poorly understood. While many of them sample poorly known asteroidal cores, they include a variety of anomalous types, including highly reduced silicon-bearing irons (e.g., Horse Creek) and at least one that essentially quenched from a molten state (e.g., Nedagolla).

1.12.3 ACCRETION AND DIFFERENCES IN BULK CHEMISTRY BETWEEN GROUPS OF IRON METEORITES

The most striking differences in chemical composition between groups of iron meteorites are the differences in concentration of volatile elements (several orders of magnitude) and the smaller but important threefold variation in nickel concentration (Table 1). These bulk chemical differences cannot be attributed to fractional crystallization of asteroidal cores (Scott, 1972), but suggest that processes during asteroid accretion and core-formation produced metallic melt bodies with a range of compositions. These processes occurred exclusively in the nebula (condensation–evaporation), in both nebular and parent-body settings (oxidation–reduction) and exclusively on the parent body (metal–silicate segregation, degassing of volatiles and impacts). In some cases, it is possible to relate specific chemical characteristics of an iron meteorite group to a specific process, but in most cases some ambiguity remains. In this section, we discuss the diverse chemical compositions of iron meteorites and its possible origins.

Chondrites are generally considered representative of the material from which asteroids, including iron meteorite parent bodies, and planets formed. The heterogeneity of primitive chondrites shows that solids in the early solar system were, to some extent, chemically and isotopically diverse. Metal abundances range from zero in some carbonaceous chondrites (e.g., CI and CM) to more than 60 wt.% in some CH/CB chondrites (Campbell et al., 2001; see Chapter 1.05). The chemical composition of chondritic metal is also diverse, primarily reflecting oxidation–reduction processes, although in rare cases the metal formed as high temperature condensates (Kong and Ebihara, 1997; Kong et al., 1997; Campbell et al., 2001). Although ordinary chondrites are depleted in gallium and germanium by up to an order of magnitude relative to CI chondrites (Wasson and Wai, 1976), the variation in volatile concentrations observed among iron meteorite groups has no counterpart in chondrites.

The main processes that control the composition of nebular metal—condensation and fractionation from the gas at high temperatures and oxidation–reduction processes (Kelly and Larimer, 1977)—are illustrated in Figure 1. Oxidation of iron from the metal will enrich the remaining metal in elements less easily oxidized than iron. Differences in mineralogy and chemistry between iron meteorite groups show that oxidation–reduction processes were important. The oxygen fugacity of the molten core is reflected in the abundance of the oxygen-bearing phosphates found in some groups of iron meteorites (Olsen and Fredriksson, 1966; Olsen et al., 1999). Phosphate minerals are typical in group IIIAB but have never been observed in IIAB iron meteorites (Buchwald, 1975, 1984; Scott and Wasson, 1975). The higher nickel concentration in group IIIAB compared to IIAB (Table 1) is consistent with the latter forming from more reduced material. Whether this difference is a result of a nebula process or a parent-body process remains an open question.

The other important process in the nebula is condensation and fractionation from the gas. Condensation at high-temperatures results in refractory element enrichment and volatile depletion. High-temperature condensates may be preserved if isolated from the gas before more volatile elements condense (Meibom et al., 2000; Petaev et al., 2001). Processes operating within the parent bodies may mask this process. Volatile-element depletions could result from degassing of the parent-body during heating, partial melting, and impact (Rasmussen et al., 1984; Keil and Wilson, 1993). Parent-body degassing will not, however, result in significant refractory element enrichment. Thus, it may be possible to distinguish volatile depletions caused by high-temperature condensation and parent-body degassing.

The best evidence for the importance of condensation and fractionation is the composition of group IVB irons and ungrouped irons of similar composition (Figure 2). The bulk compositions of groups IVA and IVB irons are consistent with high-temperature condensation of the source material (Kelly and Larimer, 1977) (Figure 1). These groups are depleted in the volatile to moderately volatile elements sulfur, phosphorus, gallium, germanium, phosphorus, antimony, copper, and gold, with the nonmetals sulfur and phosphorus of particular importance, since they have significant effects on core crystallization and Widmanstätten pattern formation. While the depletion of volatile elements in group IVA can result from parent-body degassing (Keil and Wilson, 1993), enrichments in the refractory

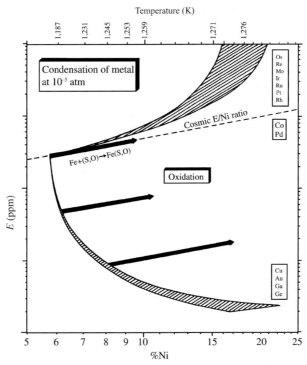

Figure 1 Compositional evolution of metal condensing at a pressure of 10^{-5} atm. The first metal to condense contains 19 ± 3 wt.% Ni and decreases to the cosmic value of 5.7 wt.% Ni as cooling commences. Elements more refractory than Ni are enriched in the early condensates, whereas elements more volatile than Ni are depleted. Oxidation of Fe shifts the composition of the metal in the direction of the heavy arrows (Kelly and Larimer, 1977) (reproduced by permission of Elsevier from *Geochim. Cosmochim. Acta*, **1977**, *41*, 93–111).

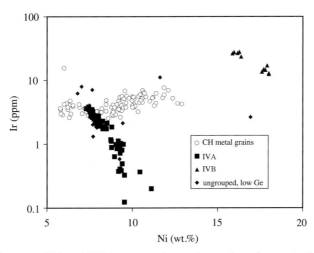

Figure 2 Ni versus Ir for group IVA and IVB iron meteorites and a number of ungrouped iron meteorites with low Ge concentrations and compositions intermediate between IVA and IVB. Also shown are compositions of zoned metal grains from CH-like chondrites (sources Campbell et al., 2001; Wasson and Richardson, 2001; Rasmussen et al., 1984).

elements rhenium, osmium, and iridium in group IVB cannot (Kelly and Larimer, 1977; Scott, 1972; Wasson and Wai, 1976). High-temperature condensation for group IVA might also explain the unusually high nickel concentration, although oxidation cannot be excluded (Figure 1).

The source material for IVB irons is not preserved among known chondrites (Rasmussen et al., 1984), although a few CB chondrites contain metal grains thought to sample high-temperature condensates (Meibom et al., 2000; Campbell et al., 2001; see Chapter 1.05). These metal grains are compositionally intermediate between IVA and IVB irons (Figure 2) and may be the first direct evidence that such materials were produced in the nebula. A close relationship between type IVA and IVB irons and CB chondrites is also suggested by the elevated $\delta^{15}N$ values in all three groups (Kerridge, 1985; Prombo and Clayton, 1985, 1993). Metal from IVB irons would have been isolated from a nebular gas above 1,500 K (Petaev et al., 2001), even higher than metal grains in CB chondrites. The postulated chondritic metal corresponding to IVB irons was either entirely incorporated into bodies that differentiated, or these chondrites remained unsampled.

1.12.4 HEATING AND DIFFERENTIATION

How did asteroidal cores come to exist in the first place? What were the physical processes? What were the heat sources? In this section, we address these fundamental issues by considering evidence from iron meteorites, primitive chondrites, achondrites, experiments, and numerical calculations. The simplest case for core formation is metal sinking through a silicate matrix that has experienced a high degree of partial melting. Numerous experimental studies (e.g., Takahashi, 1983; Walker and Agee, 1988; McCoy et al., 1998) demonstrate that metal and sulfide tend to form rounded globules, rather than an interconnected network, at moderate degrees of silicate partial melt. These globules then sink through silicate mush. Taylor (1992) calculated that at silicate fractions of 0.5, metal particles ~10–1,000 cm sink readily through a crystal mush, although it remains unclear how millimeter-sized metal particles in primitive chondritic meteorites attain these sizes. Settling would be rapid, with core formation requiring tens to thousands of years, depending on parent-body size and degree of silicate melting. Many large iron meteorites (e.g., Canyon Diablo, Hoba) and shower-producing irons (e.g., Gibeon) exceeded 1–10 m, supporting the idea that irons originated in central cores, rather than dispersed metallic masses.

A more interesting case is porous flow through interconnected networks in a largely solid silicate matrix. This scenario is of interest both because any fully melted body would have experienced an earlier stage of partial melting and because early partial melts are sulfur enriched. The Fe, Ni–FeS eutectic occurs at ~950 °C and contains ~85 wt.% FeS (Kullerud, 1963). Using Darcy's law, Taylor (1992) calculated flow velocities of 270 m yr^{-1} at 10% melting, suggesting core formation ~10^2–10^4 yr. Taylor (1992) did note considerable uncertainty about interfacial angles, which control melt migration, in Fe,Ni–FeS melts and that considerable experimental evidence argued against metal-sulfide melt migration at low degrees of partial melting under static conditions. In contrast, Urakawa et al. (1987) suggested that oxygen-rich melts at higher pressure might readily segregate. Rushmer et al. (2000) reviewed recent experimental evidence for metal-sulfide melt migration at pressures from 1 atm to 25 GPa (Herpfer and Larimer, 1993; Ballhaus and Ellis, 1996; Minarik et al., 1996; Shannon and Agee, 1996, 1998; Gaetani and Grove, 1999). For our purposes, many of these experiments are limited in their utility, since pressures in even the largest asteroids were only fractions of a GPa (Rushmer et al., 2000). These experiments demonstrate that only anion-dominated metallic melts exhibit dihedral angles less than 60° and form interconnected networks (Figure 3). This is consistent with the work of Taylor (1992) and suggests that even Fe–S eutectic melts at low pressure (anions/cations ~0.8) would not migrate to form a core under static conditions.

Perhaps dynamic processes played a role in the formation of asteroidal cores. Rushmer et al. (2000) found considerable evidence for iron sulfide melt mobility in the absence of silicate partial melting in experiments on an H6 ordinary chondrite. These authors suggest that metal-sulfide segregation may be possible, although the pressure–strain regime used may not be applicable to asteroidal-sized bodies. Keil and Wilson (1993) suggested that overpressure during Fe,Ni–FeS eutectic melting might create veins and volatiles might cause these veins to rise in small bodies.

These results suggest two very different types of cores result from partial melting. At low degrees of partial melting, cores may be sulfur-rich or even

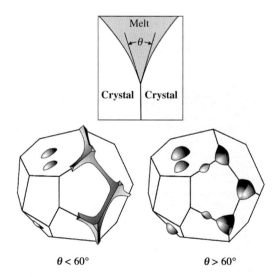

Figure 3 Illustration of the dihedral angle θ and a depiction of two end-member microstructures for static systems. If $\theta < 60°$, an interconnected network will form and melt migration can occur. If $\theta > 60°$, melt forms isolated pockets. In experimental systems, interconnectedness only occurs in anion-rich static systems.

sulfur-dominated. The small sulfur rich cores would coexist with a mantle that contains essentially chondritic silicate mineralogy and metal abundance. As near-total melting is achieved, metal drains efficiently to the center of the body, forming metal-dominated cores, which may be depleted in sulfur relative to the chondritic precursor as a result of explosive removal of the Fe,Ni–FeS eutectic melt (Keil and Wilson, 1993).

While most meteorites sample either primitive or fully differentiated asteroids, a few offer direct evidence for processes occurring during core formation. Acapulcoites and lodranites (McCoy et al., 1996, 1997a, b; Mittlefehldt et al., 1996) offer tantalizing clues to the nature of asteroidal core formation. In the acapulcoites, millimeter- to centimeter-long veins of Fe,Ni–FeS eutectic melts formed in the absence of silicate partial melting (Figure 4). The bulk composition of these samples is unchanged from chondritic, suggesting that only localized melt migration occurs. In contrast, lodranites are depleted in both plagioclase–pyroxene (basaltic) melts and the Fe,Ni–FeS eutectic melts, suggesting that silicate partial melting opened conduits for metal-sulfide melt migration. While sulfur-rich cores may have formed at low degrees of partial melting (e.g., <20%), they required silicate partial melting. Impact during core formation is a complicating factor. Mesosiderites and silicate-bearing IAB and IIICD irons may have formed by disruption of a partially to fully differentiated parent body prior to core crystallization and solidification (Benedix et al., 2000; Scott et al., 2001). The consequences of an impact during core formation/crystallization have not been fully explored.

Iron meteorite parent bodies experienced heating and melting at temperatures in excess of 1,500 °C (Taylor, 1992). What was the heat source? Melting occurred very early in solar system history. ^{182}Hf–^{182}W and ^{187}Re–^{187}Os systematics in iron meteorites (e.g., Horan et al., 1998) suggest core formation within 5 Myr of each other and formation of the first solids in the solar system. Early differentiation is also supported by the presence of excess ^{26}Mg from the decay of extinct ^{26}Al (half-life of 0.73 Ma) in the eucrites Piplia Kalan (Srinivasan et al., 1999) and Asuka-881394 (Nyquist et al., 2001).

Several heat sources for core formation can be ruled out. Early melting was not caused by decay of the long-lived radionuclides (e.g., uranium, potassium, and thorium) that contribute to the Earth's heat budget. Similarly, accretional heating—which could have caused planet-wide melting in the terrestrial planets—would have produced heating of no more than a few tens of degrees in asteroids. Keil et al. (1997) argued that the maximum energy released during accretion is equal to the gravitational binding energy of the asteroid after accretion. For a 100 km body, this equates to a temperature increase of only 6 °C. Finally, impacts sufficiently energetic to melt asteroids will also catastrophically disrupt the asteroid.

The most likely heat sources for melting of asteroidal parent bodies are electrical conduction heating by the T-Tauri solar wind from the pre-main-sequence Sun (e.g., Sonett et al., 1970) or short-lived radioactive isotopes (Keil, 2000). Melting of asteroids by ^{26}Al has gained broader acceptance since the discovery of excess ^{26}Mg in Piplia Kalan (Srinivasan et al., 1999). Taylor (1992) points out an interesting conundrum. In chondrites, ^{26}Al is concentrated in plagioclase—an early melting silicate. Loss of basaltic melts leaves plagioclase- and troilite-depleted residues akin to the lodranites (e.g., McCoy et al., 1997a), which are sufficiently depleted in ^{26}Al that no further heating occurs. Several authors (e.g., Shukolyukov and Lugmair, 1992) suggest ^{60}Fe as a heat source. Recent evidence for very high concentrations of ^{60}Fe in chondritic troilite (Mostefaoui et al., 2003; Tachibana and Huss, 2003) raises the possibility that ^{60}Fe could cause melting even in the absence of other heat sources. Occurring in both oxidized and reduced forms, ^{60}Fe would be retained throughout all layers of a body during differentiation and could provide the heat necessary for global melting and, ultimately, core formation in asteroids.

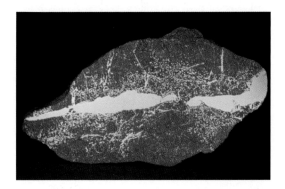

Figure 4 The Monument Draw acapulcoite contains centimeter scale veins of Fe,Ni metal and troilite formed during the first melting of asteroids. In the absence of silicate melting, these veins were unable to migrate substantial distances and, thus, would not have contributed to core formation. Length of specimen is ~7.5 cm (Smithsonian specimen USNM 7050).

1.12.5 FRACTIONAL CRYSTALLIZATION OF METAL CORES

Although iron meteorites provide a guide to understanding the ongoing crystallization of the Earth's core, important differences in the physical settings exist. First and foremost, the central pressures within the iron meteorite parent bodies did not exceed 0.1 GPa, compared to central pressures in excess of 350 GPa on Earth. The steep pressure gradient in the Earth's core, which causes the core to crystallize from the inside out, was absent in the asteroidal cores. Since the core is cooled through the mantle, Haack and Scott (1992) argued that the onset of core crystallization in asteroids was probably from the base of the mantle. Crystallization of the asteroidal cores was likely in the form of kilometer-sized dendrites as light buoyant liquid, enriched in the incompatible element sulfur inhibited crystallization of the outer core.

The fractional crystallization trends preserved in iron meteorites are unparalleled in terrestrial magmatic systems. Trace-element variations span more than three orders of magnitude within group IIIAB irons (Figure 5). The traditional way to display the bulk compositional data of iron meteorites is in a log nickel versus log element diagram. More recent work has used gold as the reference element (Wasson, 1999; Wasson and Richardson, 2001). Gold is a better choice because the distribution coefficient is further from unity giving a natural variation that far exceeds the analytical uncertainty (Haack and Scott, 1993). In a log E versus log Ni diagram, ideal fractional crystallization from a perfectly mixed liquid with constant distribution coefficients will result in straight chemical trends where the slope is given by $(D_E - 1)/(D_{Ni} - 1)$, where D_E and D_{Ni} are the liquid metal/solid metal distribution coefficients for the element and nickel, respectively. Although most trends define almost straight lines, differences in slope from group to group and curved trends for some elements such as gallium and germanium show that the distribution coefficients cannot be constant. The slopes of the Ni–Ir trends correlate with the volatile concentrations (including sulfur) for the different groups. Experimental work has shown that the distribution coefficients are functions of the phosphorus, sulfur, and carbon concentration in the liquid (Goldstein and Friel, 1978; Narayan and Goldstein, 1981, 1982; Willis and Goldstein, 1982; Jones and Drake, 1983; Malvin et al., 1986; Jones and Malvin, 1990; Chabot and Drake, 1997, 1999, 2000; Liu and Fleet, 2001; Chabot and Jones, 2003). Using the experimentally determined distribution coefficients it is possible to calculate fractional crystallization trends (Figure 5). Early numerical models assumed that the liquid remained perfectly mixed throughout crystallization (Willis and Goldstein, 1982; Jones and Drake, 1983), whereas later models have included the effects

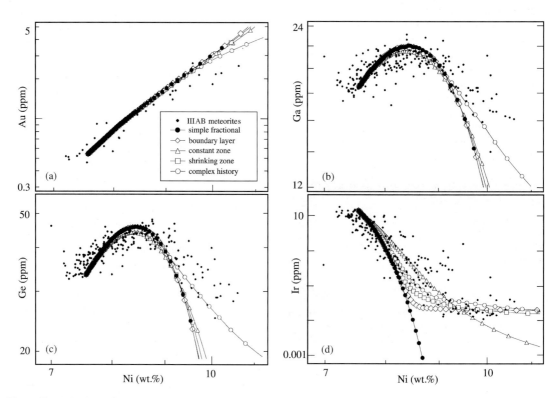

Figure 5 Calculated fractional crystallization trends for group IIIAB iron meteorites using several different types of models. The simple fractional crystallization models assume a perfectly mixed liquid whereas the other four models assume different types of imperfect mixing (reproduced by permission of the Meteoritical Society from *Meteorit. Planet. Sci.* **1999**, *34*, 235–246).

of assimilation (Malvin, 1988), imperfect mixing of the liquid (Chabot and Drake, 1999), liquid immiscibility (Ulff-Møller, 1998; Chabot and Drake, 2000), and trapping of sulfur-rich liquid (Haack and Scott, 1993; Scott et al., 1996; Wasson, 1999; Wasson and Richardson, 2001).

Numerical models of the fractional crystallization process are broadly consistent with the observed trends. There are, however, several features that remain poorly understood.

1.12.5.1 Imperfect Mixing during Crystallization

The compositional data scatter from the average trend for each group are significantly greater than the analytical uncertainty, in particular for compatible elements (Pernicka and Wasson, 1987; Haack and Scott, 1992, 1993; Scott et al., 1996; Wasson, 1999; Wasson and Richardson, 2001). The scatter shows that the assumption of metal crystallizing from a perfectly mixed liquid breaks down during the course of crystallization. An interesting example of this scatter is observed among and within the multiton fragments of the IIIAB iron meteorite Cape York (Esbensen and Buchwald, 1982; Esbensen et al., 1982). The compositional trend defined by the different Cape York fragments diverges from the general trend defined by IIIAB iron meteorites (Figure 6). The observation that the compositional variation within a single meteorite shower covers most of the scatter within the entire group suggests that the compositional scatter of group IIIAB is due to processes operating on a meter scale. Wasson (1999) noted that both the Cape York trends and the scatter in the IIIAB compositions tend to fall between the compositions of solid metal and its inferred coexisting liquid. Using modified distribution coefficients, he was able to model the Cape York trend as a mixing line between liquid and solid (Figure 6). He suggested that the observed scatter and the Cape York trend are caused by diffusional homogenization of trapped melt pools and solid metal.

1.12.5.2 Late-stage Crystallization and Immiscible Liquid

For most groups the modeled compositional trends tend to deviate from the observations toward the late stages of the crystallization (Figure 5). There are several possible reasons for these discrepancies. The distribution coefficients are functions of the concentration of the incompatible elements sulfur and phosphorus in the liquid. The phosphorus concentration of the melt is well constrained but the behavior of sulfur during crystallization remains poorly understood. Sulfur is almost insoluble in solid metal, and it is therefore not possible to estimate the liquid

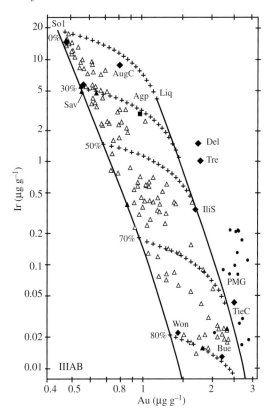

Figure 6 Solid (left) and liquid (right) tracks shown on a plot of Ir versus Au. Crosses at 0%, 30%, 50%, 70%, and 80% crystallization show solid–liquid mixing curves. The compositions of the Cape York irons Savik and Agpalilik fall close to the mixing line calculated for 30% crystallization. Also shown are the unusual IIIAB meteorites Augusta County, Buenaventura, Delegate, Ilinskaya Stanitza, Tieraco Creek, Treysa, and Wonyulgunna (source Wasson, 1999).

concentration by analyzing iron meteorites that represent the crystallized metal. Sulfur may only be determined indirectly by treating it as a free parameter in the fractional crystallization models and choose the sulfur composition that provides the best fit to the data (Jones and Drake, 1983; Haack and Scott, 1993; Chabot and Drake, 2000). The concentration of sulfur in the liquid as crystallization proceeds would lead to changes in the slope of the elemental trends that are not matched by observations. Apparently, the distribution coefficients are either incorrect (Wasson, 1999) or some mechanism prevented the concentration of sulfur in the late-stage liquid (Haack and Scott, 1993; Chabot and Drake, 1999).

As sulfur and phosphorus concentrations increase, liquid immiscibility may also play an increasingly important role. Experiments show that at high sulfur and phosphorus concentrations (and high oxygen fugacity), the liquid will enter a two-phase field (Jones and Drake, 1983; Chabot and Drake, 2000) (Figure 7). Ulff-Møller (1998)

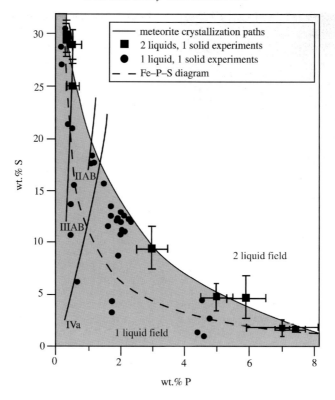

Figure 7 Two-liquid field in the Fe–S–P system as determined by Raghavan (1988) (hatched line) and Chabot and Drake (2000) (white area). Chabot and Drake performed their experiment at more oxidizing condition—assumed relevant for group IIIAB iron meteorites. Also shown are the trends followed by the major iron meteorite groups. Note that all groups eventually enter the two-liquid field (reproduced by permission of the Meteoritical Society from *Meteorit. Planet. Sci.*, **2000**, *35*, 807–816).

modeled the crystallization of the IIIAB core taking liquid immiscibility into account. He found that the effect of liquid immiscibility changes the compositional trends resulting from late-stage crystallization significantly.

1.12.5.3 The Missing Sulfur-rich Meteorites

Models predict that a considerable volume of Fe,Ni–FeS eutectic melt should be produced at the end of core crystallization. However, the number of sulfur-rich meteorites falls far short of the predicted abundances. For the sulfur- and phosphorus-rich group IIAB, the models predict that the unsampled volume of eutectic melt is ~70% (Chabot and Drake, 2000). Although arguments may be made that the sulfur-rich material is weaker and, therefore, more easily broken down in space and ablates more rapidly during atmospheric entry (Kracher and Wasson, 1982), it remains a mystery why the number of sulfur-rich meteorites is so low. It is, however, worth noting that although the mantle and crust probably comprised more than 85 vol.% of differentiated parent bodies, achondrites that could represent mantle and crust materials of the iron meteorite parent bodies are astonishingly rare. Clearly, processes that selectively remove more fragile materials en route to the Earth are of significant importance.

1.12.6 COOLING RATES AND SIZES OF PARENT BODIES

After the cores of the differentiated asteroids had crystallized, a slow cooling period commenced. During this period the most prominent feature of iron meteorites evolved—the Widmanstätten pattern (Figure 8). Several characteristics of the Widmanstätten pattern may be used to constrain the thermal evolution and the sizes of the iron meteorite parent bodies.

In a typical parent-body core with nickel concentrations in the range 7–15 wt.% Ni, the low-nickel phase kamacite will grow at the expense of the high-nickel phase taenite between 800 °C and 500 °C. Several other elements are partitioned between the two phases during growth of the kamacite phase (Rasmussen *et al.*, 1988). The kamacite grows as platelets in four possible orientations relative to the taenite host. Since most

Figure 8 Widmanstätten pattern in a polished and etched section of the IID iron meteorite, Carbo. 10 cm field of view. The brownish tear shaped troilite nodule to the left has acted as precipitation site for kamacite (Geological Museum, University of Copenhagen, specimen 1990.143).

Table 2 Metallographic cooling rates and corresponding parent-body radii of the main iron and stony-iron meteorite groups.

Group	Cooling rate (K Myr^{-1})	Parent-body radius (km)	Refs.
IAB	25	>33	a
IIAB	6–12	45–65	b
IIIAB	15–85	20–40	c
	7.5–15	42–58	a
IVA	19–3,400	>40	d
IVB	170–230	12–14	e
Pallasites	2.5–4	80–100	a
Mesosiderites	0.5	200	a

References: (a) Yang et al. (1997), (b) Saikumar and Goldstein (1988), (c) Rasmussen (1989), (d) Rasmussen et al. (1995), and (e) Rasmussen et al. (1984). A radius of the parent body can only be calculated for those groups where the meteorites are believed to have cooled in the core. For other groups a minimum parent-body size may be calculated.

parent taenite crystals had dimensions larger than typical meteorites, a continuous pattern of four different sets of kamacite plates may be observed on etched surfaces of typical iron meteorites. The limiting factor for the growth of kamacite is the slow diffusion of nickel through taenite. Nickel profiles across taenite lamellae may, therefore, be used to determine the so-called metallographic cooling rate of the parent-body core at ~500 °C (Wood, 1964; Rasmussen, 1981; Herpfer et al., 1994; Yang et al., 1997; Hopfe and Goldstein, 2001; Rasmussen et al., 2001). A number of revisions of the metallographic cooling rate method have been implemented since its original formulation (Wood, 1964; Goldstein and Ogilvie, 1965). Revisions have resulted from improved ternary (Fe–Ni–P) phase diagrams (Doan and Goldstein, 1970; Romig and Goldstein, 1980; Yang et al., 1996) and, in particular, the discovery that small amounts of phosphorus may increase the diffusion rate of nickel through taenite by an order of magnitude (Narayan and Goldstein, 1985; Dean and Goldstein, 1986).

Several other characteristics of the Widmanstätten pattern have provided additional information on the thermal history of the metal. Diffusion controlled growth of schreibersite has been used to determine cooling rates of hexahedrites (Randich and Goldstein, 1978). Although nickel diffusion through kamacite is much faster than through taenite, the decreasing temperature will eventually result in zoned kamacite as well. Zoned kamacite has been used to infer cooling rates at temperatures below 400 °C (Haack et al., 1996b; Rasmussen et al., 2001). Yang et al. (1997) showed that the size of the so-called island phase is inversely correlated with cooling rate and may be used to infer the cooling rate at temperatures ~320 °C. The island phase is high-nickel tetrataenite that forms irregular globules with dimensions up to 470 nm in the nickel-rich rims of taenite.

With the exception of group IVA iron meteorites, the metallographic cooling rates tend to be similar within each group but different from group to group (Table 2). This is consistent with the idea that each iron meteorite group cooled in its own separate metallic core surrounded by an insulating mantle. The cooling rates for the different groups of iron meteorites are generally in the range 10–100 °C Myr^{-1}.

The metallographic cooling rates combined with numerical models of the thermal evolution of the parent bodies may be used to constrain the sizes of the parent bodies (Wood, 1964; Goldstein and Short, 1967; Haack et al., 1990). The main uncertainties in the numerical models are the states of the mantle, surface, and crust during cooling. Observations of present-day asteroids show that they tend to be covered by a thick insulting regolith and that they may be heavily brecciated. A brecciated mantle and/or a highly porous regolith cover on the surface of an asteroid could potentially slow the cooling rate by a factor of 5–10 (Haack et al., 1990). An approximate relationship between metallographic cooling rates and parent-body radius for regolith covered asteroids was given by Haack et al. (1990): $R = 149 \times CR^{-0.465}$, where R is the radius in km and CR the cooling rate in K Myr^{-1}. Table 2 gives a compilation of cooling rates and corresponding parent-body sizes.

1.12.7 PALLASITES

Pallasites are the most abundant group of stony-irons. Their features are easily reconciled with simple models of asteroid differentiation, yet challenge some of our basic assumptions about core formation and crystallization (Mittlefehldt et al., 1998). The 50 known pallasites are divided into main group pallasites, the Eagle Station grouplet (three members) and the pyroxene pallasite grouplet (two members). These are distinguished based on oxygen isotopic, mineral, and metal compositions. Main group pallasites are comprised of subequal mixtures of forsteritic olivine and iron, nickel metal, often heterogeneously distributed. Clusters of olivine grains reach several centimeters (Ulff-Møller et al., 1998), while Brenham has metal cross-cutting zones with more typical pallasitic texture (Figure 9). Olivine morphology also differs significantly, from angular–subangular in Salta to rounded in Thiel mountains (Buseck, 1977; Scott, 1977b).

Olivines in the main group pallasites cluster around $Fo_{88\pm1}$, although some reach Fo_{82}, and exhibit core-to-rim zoning of aluminum, chromium, calcium, and manganese (Zhou and Steele, 1993; Hsu et al., 1997). Chromite, low-calcium pyroxene and a variety of phosphates comprise <1 vol.% each (Buseck, 1977; Ulff-Møller et al., 1998). Phosphates show REE patterns, inherited from the olivines, largely consistent with the olivines forming as cumulates at the base of the mantle (Davis and Olsen, 1991; Davis and Olsen, 1996). The metal composition is similar to high-nickel IIIAB irons, but with a greater scatter in nickel ($\sim$7–13 wt.%) and some other siderophile elements (Davis, 1977; Scott, 1977b). The pallasite-IIIAB iron link is supported by the similarity in oxygen isotopic compositions of their silicates (Clayton and Mayeda, 1996).

The Eagle Station trio (Eagle Station, Cold Bay, Itzawisis) is comprised dominantly of iron and nickel metal and olivine with lesser amounts of clinopyroxene, orthopyroxene, chromite, and phosphates. Olivines in the Eagle Station (Fo_{80-81}) are more ferroan than in the main group and differ substantially in their Fe/Mn ratio (Mittlefehldt et al., 1998). Metal in the Eagle Station pallasites has higher iridium and nickel compared to main group pallasites and is closer to metal in IIF irons (Kracher et al., 1980). They also differ dramatically in oxygen isotopic composition from main group pallasites (Clayton and Mayeda, 1996).

Pyroxene pallasites are represented by only two members: Vermillion and Yamato 8451. The two meteorites share the common feature of containing pyroxene, but differ substantially from one another. Yamato 8451 consists of $\sim$60% olivine, 35% metal, 2% pyroxene, and 1% troilite (Hiroi et al., 1993; Yanai and Kojima, 1995). In contrast, Vermillion contains around 14% olivine, and less than 1% each of orthopyroxene, chromite, and phosphates (Boesenberg et al., 2000). The iron concentration in olivine (Fo_{88-90}) is similar to main group pallasites, although pyroxene pallasites have lower Fe/Mn ratios (Mittlefehldt et al., 1998). The two pyroxene pallasites do not share a common metal composition with each other or, in detail, with other pallasites (Wasson et al., 1998).

Pallasites are both intriguing and perplexing. The two lithologies, Fe,Ni metal and olivine, are reasonable assemblages expected at the core–mantle boundary of a differentiated asteroid. The lower mantle should be primarily olivine, either a residue from high degrees of partial melting or an early cumulate phase from a global magma ocean. However, the marked density contrast between metal and silicates should lead to rapid separation. Further, some features of pallasitic olivine, such as the angular shapes and marked minor element zoning, seem inconsistent with formation at a relatively quiescent, deep-seated core–mantle boundary. As Mittlefehldt et al. (1998) note, the core–mantle boundary origin remains the most plausible, despite these difficulties.

If we assume a core–mantle boundary origin (see Mittlefehldt et al., 1998, for references suggesting alternative models), how can these

Figure 9 The Brenham pallasite contains areas of both olivine-free regions and areas more typical of pallasites. In this specimen, pallasitic material is cross-cut by a metallic region, suggesting that silicate–metal mixing at the core–mantle boundary was a dynamic process. Length of specimen is $\sim$13 cm (Smithsonian specimen USNM 266).

disparate features be reconciled and what are the implications for the nature of asteroidal cores during cooling and solidification. Scott (1977c) suggested that the main group pallasite metal is a reasonable crystallization product after ~80% crystallization of IIIAB metallic melt. However, asteroidal cores probably crystallized from the core–mantle boundary inwards and we should expect pallasites to have the signature of early-crystallizing, low-nickel melts, not late-crystallizing, high-nickel melts. Haack and Scott (1993) postulated that residual metallic melt migrated to the core–mantle boundary between dendrites and Ulff-Møller et al. (1997) proposed mixing of late-stage metallic melt with earlier solidified metal to produce the range of pallasite compositions. Late-stage intrusions might explain the presence of angular olivines, which were fractured during the intrusion (Scott, 1977b), and the heterogeneous textures seen in pallasites like Brenham. Ulff-Møller et al. (1997) suggested that high-pressure injection of metal, perhaps caused by impact, might provide the most plausible mechanism for producing pallasites. Finally, pallasites have experienced a long history of subsolidus cooling and annealing. Both minor-element zoning in olivine and rounded olivines might be best explained by subsolidus annealing and diffusion. The small Eagle Station and pyroxene-bearing pallasite grouplets remain enigmatic, but suggest that similar processes operated on other parent bodies and may have been common among highly differentiated asteroids.

1.12.8 PARENT BODIES OF IRON AND STONY-IRON METEORITES

While iron and stony-iron meteorites provide important snapshots of the origin and evolution of highly differentiated asteroids, they lack geologic context. From which type of asteroid do iron and stony-iron meteorites originate? Are there enough metallic asteroids to account for the enormous diversity among iron meteorites, particularly the ungrouped irons? Have we sampled corresponding abundances of crustal and mantle material in both the meteorite and asteroid populations? These are the questions that we explore in this section.

The major tool that allows us to relate classes of meteorites to classes of asteroids is spectral reflectance (Burbine et al., 2002). Unfortunately, iron meteorites, with their paucity of silicate phases, have relatively featureless spectra with red spectral slopes and moderate albedos (e.g., Cloutis et al., 1990) (Figure 10). There appears to be no simple correlation between nickel abundance and spectra redness and, thus, distinguishing different chemical groups (e.g., low-nickel IIAB from

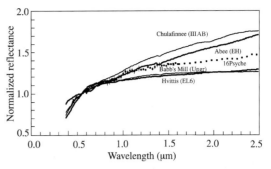

Figure 10 Normalized reflectance versus wavelength (μm) for M-type 16 Psyche versus iron and enstatite chondrite meteorites (lines) (Gaffey, 1976; Bell et al., 1988; Bus, 1999). All spectra are normalized to unity at 0.55 μm. Spectra for all three are relatively featureless with red spectral slopes and moderate albedos. While some M-class asteroids may be metallic core material, existing spectral and density data are inconsistent with this explanation for all M asteroids.

high-nickel IVB) is probably not possible. Historically, M-class asteroids, of which ~40 are known (Tholen, 1989), have been linked to iron meteorites. These asteroids exhibit moderate visual albedos, relatively featureless spectra and red spectral slopes, similar to that of iron meteorites. Radar albedos—an indirect measure of near-surface bulk density—have been used as supporting evidence for the link between iron meteorites and M-class asteroids, as M asteroids tend to have higher radar albedos than C or S asteroids (Magri et al., 1999) and the highest asteroid radar albedos are known from the M asteroids 6178 (1986 DA) and 216 Kleopatra (Ostro et al., 1991, 2000). There is reason to believe that not all (perhaps not many) M-class asteroids are either core fragments or largely intact stripped cores of differentiated asteroids. The enstatite chondrites also exhibit nearly featureless spectra (owing to the nearly Fe^{2+}-free composition of the enstatite; Keil, 1968) with red spectral slopes and moderate albedos (Gaffey, 1976) and have been suggested as possible meteoritic analogues to M asteroids (Gaffey and McCord, 1978) (Figure 10). Rivkin et al. (2000) found that more than one-third of observed M-class asteroids have 3 μm absorption features suggestive of hydrated silicates. Although this conclusion remains the subject of significant debate (Gaffey et al., 2002a; Rivkin et al., 2002), it would be inconsistent with core material. Finally, recent ground-based measurements of bulk densities for M asteroids 16 Psyche (Viateau, 2000; Britt et al., 2002) and 22 Kalliope (Margot and Brown, 2001) are ~2 g cm^{-3}, far below that expected for even highly fragmented core material.

Equally puzzling is the lack of olivine-rich asteroids corresponding to the large numbers of asteroidal cores sampled by iron meteorites. If we have sampled ~65 asteroidal cores, we might expect corresponding abundances of olivine-rich meteorites and A-type (olivine-rich) asteroids (Chapman, 1986; Bell *et al.*, 1989; Burbine *et al.*, 1996; Gaffey *et al.*, 2002b). The near absence of dunite meteorites and paucity of A-type asteroids has come to be known as "The Great Dunite Shortage." The most likely explanation centers on the age of disruption of differentiated parent bodies and the durability of metallic meteorites and asteroids relative to their stony counterparts. If disruption of differentiated bodies occurred relatively early in the history of the solar system, the dunitic mantle may have been fragmented to below the current observational limit of ~5 km. Dunitic meteorites may well have fallen to Earth as a consequence of this fragmentation of the mantles, but terrestrial ages for meteorites stretch back only ~2 Myr (Welten *et al.*, 1997). In contrast, core fragments may prove much more versatile and may still be sampled by impacts today. Indeed, iron meteorites are much more durable, with cosmic-ray exposure ages to more than 2 Gyr (e.g., Deep Springs, Voshage and Feldmann, 1979). Much slower Yarkovsky-driven transfer times from the asteroid belt for metallic objects also suggest that these sample impacts further back in time (Farinella *et al.*, 1998).

1.12.9 FUTURE RESEARCH DIRECTIONS

In this chapter, we have touched upon our current state of knowledge about iron and stony-iron meteorites, the processes that formed them and the places from which they originate. Our knowledge of all of these is far from complete. In this section, we briefly address the future of these fields. What questions remain unanswered? Which are most pressing to understand the origin of iron meteorites? What tools do we need to move forward?

As students of meteorites, our first inclination is to look to the meteorite record for missing types and new meteorite recoveries to fill those gaps. The substantial number of ungrouped iron meteorites suggests that the cores of differentiated bodies are poorly sampled. Additional recoveries may serve to establish new groups that include existing ungrouped irons. However, the number of ungrouped irons continues to grow much more rapidly than the number of new groups. A more fundamental recovery that would increase our understanding of the relative roles of late-stage melt trapping versus concentration would be the discovery of a large (several kilograms), sulfide meteorite. As we noted earlier, few meteorites exist that can reasonably be attributed to a eutectic melt in both phase proportions and compositions.

A second field ripe for future work is experimental studies designed to increase our understanding of the processes of core crystallization. Although we now have more than a rudimentary understanding of core crystallization, uncertainties in both partition coefficients (particularly as a function of sulfur, phosphorus, and carbon concentration) and liquid immiscibility in sulfur-, phosphorus-, and carbon-rich systems limit our understanding of the nature of late-stage core crystallization. Recent experimental studies have made significant headway in these areas, and parametrization of partitioning data allows firm predictions about partitioning over a much broader range of conditions, but experimental verification and determination are essential to further strengthen our understanding of these processes.

At the dawn of the twenty-first century, spacecraft missions hold the promise of finally placing meteorites in the context of the worlds from which they came. Although, to our knowledge, none are currently proposed, there is considerable reason to contemplate a mission to an iron meteorite parent body. A mission to a metal-dominated asteroid, particularly one that combines multispectral imaging with remote chemical analyses, may reveal the overall structure of the original body and provide answers to some of our long-standing questions. What is the nature of the core–mantle boundary? Are the pallasites typical examples of this boundary? Where is the olivine hiding? What is the structure of exposed metallic cores and what is the distribution of sulfur-rich material?

We already stand at the threshold of a better understanding of asteroids in general and differentiated asteroids in particular. The Galileo and NEAR missions gave us our first closeup looks at the S-class asteroids Gaspra, Ida, and Eros and the C-type asteroid Mathilde. While all are thought to be primitive chondritic bodies, these encounters revealed new insights into the geologic processes that have modified them during 4.5 Ga of impact bombardment. Our knowledge of differentiated asteroids is considerably less clear. The classification system described in Chapter 1.05 divides the meteorites into groups that came from the same parent body or at least from very similar parent bodies. In some cases we may infer the depth relationship between meteorites but we have no information on the horizontal distribution of the samples we have obtained. We cannot tell if they all came from the same fragment of the original asteroid or if they were more uniformly distributed across the parent body. At least some of these questions may be answered by NASA's DAWN mission, which will visit 4 Vesta by the end of the decade. With its complement of multi-spectral imagers and gamma-ray spectrometer, it should answer many of our questions about the relationship between surficial basaltic materials, deeper-seated (?) pyroxenitic material, and olivine-rich (?) mantle, including their lateral and horizontal relationships.

Asteroid spectroscopy has provided possible links between iron and stony-iron meteorites and their possible parent as discussed in the previous section. Iron meteorites provide evidence that numerous metal objects exist but the lack of spectral features and the surprising low density determined for some M-type asteroids makes it impossible to establish a firm link between iron meteorites and M-type asteroids. The high cosmic-ray exposure ages of iron meteorites show, not surprisingly, that iron–nickel alloys are much more resistant to impacts and space erosion than stony targets. We should, therefore, expect that the surfaces of stripped metal cores or fragments of metal cores display a dramatic cratered terrain unlike anything that we have ever seen before. Although the subtle compositional differences that occur within and between iron meteorite groups could likely not be discerned from orbit, the difference between metal-dominated lithologies, troilite-rich lithologies, and dunitic/pallasitic lithologies could be easily distinguished on the basis of both spectral absorption features (e.g., Britt et al., 1992) and, more readily, orbital X-ray or gamma-ray compositional mapping finally resolving questions about the existence and/or spatial distribution of these important lithologies. The distribution of sulfur-rich pockets, for example, can provide evidence on the core crystallization process (Haack and Scott, 1992) and the fate of the sulfur-rich material missing from our meteorite collections.

Such a mission might also resolve the role and extent of impact in the formation of asteroidal cores. Impact clearly played a role in the formation of both mesosiderites and silicate-bearing IAB–IIICD irons, but whether those impacts were global or local remains an open question. The slow cooling rates of mesosiderites suggest that the parent body was very large, probably several hundred kilometers in radius (Haack et al., 1996b). Since large asteroids are more difficult to destroy, this opens the possibility that the parent body may still exist. Davis et al. (1999) suggested that the largest M-type asteroid, 16 Psyche, could be the shattered remains of the mesosiderite parent body. A mission to Phyche may not only reveal the nature of M-type asteroids, it could also possibly provide evidence on one of the biggest impact events documented in our meteorite collections.

Finally, much has been written in recent years about the hazard presented to mankind by asteroid impact (Gehrels, 1994). While considerable effort has been put into understanding possible mitigation schemes, it is important to remember that energy scales linearly with mass. Thus, a metallic asteroid of comparable diameter would release ~2.5 times the energy of a stony asteroid. Further, its physical properties would certainly be quite different from that of a stony asteroid. The aforementioned mission might not only provide substantial insights into the geologic history of asteroids, but also provide essential data for understanding the physical properties (e.g., material distribution, global cracks) that would allow us to ensure the continuation of human history.

REFERENCES

Ballhaus C. and Ellis D. J. (1996) Mobility of core melts during Earth's accretion. *Earth Planet. Sci. Lett.* **143**, 137–145.

Bell J. F., Davis D. R., Hartmann W. K., and Gaffey M. J. (1989) Asteroids: the big picture. In *Asteroids II* (eds. R. P. Binzel, T. Gehrels, and M. S. Matthews). University Arizona Press, Tucson, pp. 921–945.

Bell J. F., Owensby P. D., Hawke B. R., and Gaffey M. J. (1988) The 52-color asteroid survey: final results and interpretation (abstract). *Lunar Planet. Sci.* **XIX**. The Lunar and Planetary Institute, Houston, pp. 57–58.

Benedix G. K., McCoy T. J., Keil K., and Love S. G. (2000) A petrologic study of the IAB iron meteorites: constraints on the formation of the IAB-winonaite parent body. *Meteorit. Planet. Sci.* **35**, 1127–1141.

Boesenberg J. S., Davis A. M., Prinz M., Weisberg M. K., Clayton R. N., and Mayeda T. K. (2000) The pyroxene pallasites, Vermillion and Yamato 8451: not quite a couple. *Meteorit. Planet. Sci.* **35**, 757–769.

Bogard D. D. and Garrison D. H. (1998) ^{39}Ar–^{40}Ar ages and thermal history of mesosiderites. *Geochim. Cosmochim. Acta* **62**, 1459–1468.

Bogard D. D., Garrison D. H., Jordan J. L., and Mittlefehldt D. (1990) ^{39}Ar–^{40}Ar dating of mesosiderites—evidence for major parent body disruption less than 4 Ga ago. *Geochim. Cosmochim. Acta* **54**, 2549–2564.

Bogard D. D., Garrison D. H., and McCoy T. J. (2000) Chronology and Petrology of silicates from IIE iron meteorites: evidence of a complex parent body evolution. *Geochim. Cosmochim. Acta* **64**, 2133–2154.

Britt D. T., Bell J. F., Haack H., and Scott E. R. D. (1992) The reflectance spectrum of troilite and the T-type asteroids. *Meteoritics* **27**, 207.

Britt D. T., Yeomans D., Housen K., and Consolmagno G. (2002) Asteroid density, porosity, and structure. In *Asteroids III* (eds. W. Bottke, A. Cellino, P. Paolicchi, and R. P. Binzel). University of Arizona Press, Tucson, pp. 485–500.

Buchwald V. F. (1975) *Handbook of Iron Meteorites*. University of California Press, Berkeley, 1426pp.

Buchwald V. F. (1984) Phosphate minerals in meteorites and lunar rocks. In *Phosphate Minerals* (eds. J. O. Nriagu and P. B. Moore). Springer, New York, pp. 199–214.

Burbine T. H., Meibom A., and Binzel R. P. (1996) Mantle material in the main belt: battered to bits? *Meteorit. Planet. Sci.* **31**, 607–620.

Burbine T. H., McCoy T. J., Meibom A., Gladman B., and Keil K. (2002) Meteoritic parent bodies: their number and identification. In *Asteroids III* (eds. W. Bottke, A. Cellino, P. Paolicchi, and R. P. Binzel). University of Arizona Press, Tucson, pp. 653–667.

Bus S. J. (1999) Compositional structure in the asteroid belt: results of a spectroscopic survey. PhD Thesis, Massachusetts Institute of Technology, 367p.

Buseck P. R. (1977) Pallasite meteorites—mineralogy, petrology, and geochemistry. *Geochim. Cosmochim. Acta* **41**, 711–740.

Campbell A. J., Humayun M., Meibom A., Krot A. N., and Keil K. (2001) Origin of zoned metal grains in the QUE94411 chondrite. *Geochim. Cosmochim. Acta* **65**, 163–180.

Chabot N. L. and Drake M. J. (1997) An experimental study of silver and palladium partitioning between solid and liquid metal, with applications to iron meteorites. *Meteorit. Planet. Sci.* **32**, 637–645.

Chabot N. L. and Drake M. J. (1999) Crystallization of magmatic iron meteorites: the role of mixing in the molten core. *Meteorit. Planet. Sci.* **34**, 235–246.

Chabot N. L. and Drake M. J. (2000) Crystallization of magmatic iron meteorites: the effects of phosphorus and liquid immiscibility. *Meteorit. Planet. Sci.* **35**, 807–816.

Chabot N. L. and Jones J. H. (2003) The parameterization of solid metal-liquid metal partitioning of siderophile elements. *Lunar Planet. Sci.* **XXXIV**, #1004. The Lunar and Planetary Institute, Houston (CD-ROM).

Chapman C. R. (1986) Implications of the inferred compositions of the asteroids for their collisional evolution. *Mem. Soc. Astron. Italiana* **57**, 103–114.

Chladni E. F. F. (1794) *Über den Ursprung der von Pallas Gefundenen und anderer ihr ähnlicher Eisenmassen, und Über Einige Damit in Verbindung stehende Naturerscheinungen*. Johan Friedrich Hartknoch, Riga, Latvia, 63p.

Choi B. G., Ouyang X. W., and Wasson J. T. (1995) Classification and origin of IAB and IIICD iron meteorites. *Geochim. Cosmochim. Acta* **59**, 593–612.

Clayton R. N. and Mayeda T. K. (1996) Oxygen isotope studies of achondrites. *Geochim. Cosmochim. Acta* **60**, 1999–2017.

Cloutis E. A., Gaffey M. J., Smith D. G. W., and Lambert R. S. J. (1990) Reflectance spectra of "featureless" materials and the surface mineralogies of M- and E-class asteroid. *J. Geophys. Res.* **95**, 281–293.

Davis A. M. (1977) The cosmochemical history of the pallasites. PhD Dissertation, Yale University, 285p.

Davis A. M. and Olsen E. J. (1991) Phosphates in pallasite meteorites as probes of mantle processes in small planetary bodies. *Nature* **353**, 637–640.

Davis A. M. and Olsen E. J. (1996) REE patterns in pallasite phosphates—a window on mantle differentiation in parent bodies. *Meteorit. Planet. Sci.* **31**, A34.

Davis D. R., Farinella P., and Marzari F. (1999) The missing Psyche family: collisionally eroded or never formed? *Icarus* **137**, 140–151.

Dean D. C. and Goldstein J. I. (1986) Determination of the interdiffusion coefficients in the Fe–Ni and Fe–Ni–P systems below 900 °C. *Metall. Trans.* **17A**, 1131–1138.

Doan A. S. and Goldstein J. I. (1970) The ternary phase diagram Fe–Ni–P. *Metall. Trans.* **1**, 1759–1767.

Esbensen K. H. and Buchwald V. F. (1982) Planet(oid) core crystallization and fractionation: evidence from the Cape York iron meteorite shower. *Phys. Earth Planet. Int.* **29**, 218–232.

Esbensen K. H., Buchwald V. F., Malvin D. J., and Wasson J. T. (1982) Systematic compositional variations in the Cape York iron meteorites. *Geochim. Cosmochim. Acta* **46**, 1913–1920.

Farinella P., Vokrouhlicky D., and Hartmann W. K. (1998) Meteorite delivery via Yarkovsky orbital drift. *Icarus* **132**, 378–387.

Gaetani G. A. and Grove T. L. (1999) Wetting of mantle olivine by sulfide melt: implications for Re/Os ratios in mantle peridotite and late-stage core formation. *Earth Planet. Sci. Lett.* **169**, 147–163.

Gaffey M. J. (1976) Spectral reflectance characteristics of the meteorite classes. *J. Geophys. Res.* **81**, 905–920.

Gaffey M. J. and McCord T. B. (1978) Asteroid surface materials: mineralogical characterizations from reflectance spectra. *Space Sci. Rev.* **21**, 555–628.

Gaffey M. J., Cloutis E. A., Kelley M. S., and Reed K. L. (2002a) Mineralogy of asteroids. In *Asteroids III* (eds. W. Bottke, A. Cellino, P. Paolicchi, and R. P. Binzel). University of Arizona Press, Tucson, pp. 183–204.

Gaffey M. J., Kelley M. S., and Hardersen P. S. (2002b) Meteoritic and asteroidal constraints on the identification and collisional evolution of asteroid families. *Lunar Planet. Sci.* **XXXIII**, #1506. The Lunar and Planetary Institute, Houston (CD-ROM).

Gehrels T. (1994) *Hazards due to Comets and Asteroids*. Tucson, University of Arizona Press, Tucson.

Goldstein J. I. and Friel J. J. (1978) Fractional crystallization of iron meteorites, an experimental study. *Proc. 9th Lunar Planet. Sci. Conf.* 1423–1435.

Goldstein J. I. and Ogilvie R. E. (1965) The growth of the Widmanstätten pattern in metallic meteorites. *Geochim. Cosmochim. Acta* **29**, 893–920.

Goldstein J. I. and Short J. M. (1967) The iron meteorites, their thermal history and parent bodies. *Geochim. Cosmochim. Acta* **31**, 1733–1770.

Grady M. (2000) *Catalogue of Meteorites*. Cambridge University Press, Cambridge.

Haack H. and Scott E. R. D. (1992) Asteroid core crystallization by inward dendritic growth. *J. Geophys. Res.* **97**, 14727–14734.

Haack H. and Scott E. R. D. (1993) Chemical fractionations in group IIIAB iron meteorites—origin by dendritic crystallization of an asteroidal core. *Geochim. Cosmochim. Acta* **57**, 3457–3472.

Haack H., Rasmussen K. L., and Warren P. H. (1990) Effects of regolith megaregolith insulation on the cooling histories of differentiated asteroids. *J. Geophys. Res.* **95**, 5111–5124.

Haack H., Scott E. R. D., Love S. G., Brearley A. J., and McCoy T. J. (1996a) Thermal histories of IVA stony-iron and iron meteorites: evidence for asteroid fragmentation and reaccretion. *Geochim. Cosmochim. Acta* **60**, 3103–3113.

Haack H., Scott E. R. D., and Rasmussen K. L. (1996b) Thermal and shock history of mesosiderites and their large parent asteroid. *Geochim. Cosmochim. Acta* **60**, 2609–2619.

Hassanzadeh J., Rubin A. E., and Wasson J. T. (1989) Large metal nodules in mesosiderites. *Meteoritics* **24** 276–276.

Herpfer M. A. and Larimer J. W. (1993) Core formation: an experimental study of metallic melt-silicate segregation. *Meteoritics* **28**, 362.

Herpfer M. A., Larimer J. W., and Goldstein J. I. (1994) A comparison of metallographic cooling rate methods used in meteorites. *Geochim. Cosmochim. Acta* **58**, 1353–1365.

Hewins R. H. (1983) Impact versus internal origins for mesosiderites. *J. Geophys. Res.* **88**, B257–B266.

Hiroi T., Bell J. F., Takeda H., and Pieters C. M. (1993) Spectral comparison between olivine-rich asteroids and pallasites. *Proc. NIPR Symp. Antarct. Meteorit.* **6**, 234–245.

Hoashi M., Brooks R. R., and Reeves R. D. (1990) The ruthenium content of iron meteorites. *Meteoritics* **25**, 371–372.

Hoashi M., Varelaalvarez H., Brooks R. R., Reeves R. D., Ryan D. E., and Holzbecher J. (1992) Revised classification of some iron meteorites by use of statistical procedures. *Chem. Geol.* **98**, 1–10.

Hoashi M., Brooks R. R., and Reeves R. D. (1993a) Palladium, platinum and ruthenium in iron meteorites and their taxonomic significance. *Chem. Geol.* **106**, 207–218.

Hoashi M., Brooks R. R., Ryan D. E., Holzbecher J., and Reeves R. D. (1993b) Chemical evidence for the pairing of some iron meteorites. *Geochem. J.* **27**, 163–169.

Hopfe W. D. and Goldstein J. I. (2001) The metallographic cooling rate method revised: application to iron meteorites and mesosiderites. *Meteorit. Planet. Sci.* **36**, 135–154.

Horan M. F., Smoliar M. I., and Walker R. J. (1998) ^{182}W and ^{187}Re–^{187}Os systematics of iron meteorites: chronology for melting, differentiation, and crystallization in asteroids. *Geochim. Cosmochim. Acta* **62**, 545–554 (Erratum, 1653).

Hsu W., Huss G. R., and Wasserburg G. J. (1997) Mn–Cr systematics of differentiated meteorites. *Lunar Planet. Sci.* **XXVIII**. The Lunar and Planetary Institute, Houston, pp. 609–610.

Jones J. H. and Drake M. J. (1983) Experimental investigations of trace-element fractionation in iron meteorites: 2, The influence of sulfur. *Geochim. Cosmochim. Acta* **47**, 1199–1209.

Jones J. H. and Malvin D. J. (1990) A nonmetal interaction-model for the segregation of trace-metals during solidification of Fe–Ni–S, Fe–Ni–P, and Fe–Ni–S–P alloys. *Metall. Trans.* **21B**, 697–706.

Keil K. (1968) Mineralogical and chemical relationships among enstatite chondrites. *J. Geophys. Res.* **73**, 6945–6976.

Keil K. (2000) Thermal alteration of asteroids: evidence from meteorites. *Planet. Space Sci.* **48**, 887–903.

Keil K. and Wilson L. (1993) Explosive volcanism and the compositions of cores of differentiated asteroids. *Earth Planet. Sci. Lett.* **117**, 111–124.

Keil K., Haack H., and Scott E. R. D. (1994) Catastrophic fragmentation of asteroids—evidence from meteorites. *Planet. Space Sci.* **42**, 1109–1122.

Keil K., Stöffler D., Love S. G., and Scott E. R. D. (1997) Constraints on the role of impact heating and melting in asteroids. *Meteorit. Planet. Sci.* **32**, 349–363.

Kelly W. R. and Larimer J. W. (1977) Chemical fractionations in meteorites: VIII. Iron meteorites and the cosmochemical history of the metal phase. *Geochim. Cosmochim. Acta* **41**, 93–111.

Kerridge J. F. (1985) Carbon, hydrogen and nitrogen in carbonaceous chondrites abundances and isotopic compositions in bulk samples. *Geochim. Cosmochim. Acta* **49**, 1707–1714.

Kong P. and Ebihara M. (1997) The origin and nebular history of the metal phase of ordinary chondrites. *Geochim. Cosmochim. Acta* **61**, 2317–2329.

Kong P., Mori T., and Ebihara M. (1997) Compositional continuity of enstatite chondrites and implications for heterogeneous accretion of the enstatite chondrite parent body. *Geochim. Cosmochim. Acta* **61**, 4895–4914.

Kracher A. (1985) The evolution of partially differentiated planetesimals: evidence from iron meteorite groups IAB and IIICD. *J. Geophys. Res.* **90**, C689–C698.

Kracher A. and Wasson J. T. (1982) The role of S in the evolution of the parental cores of the iron meteorites. *Geochim. Cosmochim. Acta* **46**, 2419–2426.

Kracher A., Kurat G., and Buchwald V. F. (1977) Cape York: the extraordinary mineralogy of an ordinary iron meteorite and its implication for the genesis of IIIAB irons. *Geochem. J.* **11**, 207–217.

Kracher A., Willis J., and Wasson J. T. (1980) Chemical classification of iron meteorites: IX. A new group (IIF), revision of IAB and IIICD, and data on 57 additional irons. *Geochim. Cosmochim. Acta* **44**, 773–787.

Kullerud G. (1963) The Fe–Ni–S system. *Ann. Rep. Geophys. Lab.* **67**, 4055–4061.

Liu M. H. and Fleet M. E. (2001) Partitioning of siderophile elements (W, Mo, As, Ag, Ge, Ga, and Sn) and Si in the Fe–S system and their fractionation in iron meteorites. *Geochim. Cosmochim. Acta* **65**, 671–682.

Magri C., Ostro S. J., Rosema K. D., Thomas M. L., Mitchell D. L., Campbell D. B., Chandler J. F., Shapiro I. I., Giorgini J. D., and Yeomans D. K. (1999) Mainbelt asteroids: results of Arecibo and Goldstone radar observations of 37 objects during 1980–1985. *Icarus* **140**, 379–407.

Malvin D. J. (1988) Assimilation-fractional crystallization of magmatic iron meteorites. *Lunar Planet. Sci.* **XIX**. Lunar and Planetary Institute, Houston, pp. 720–721.

Malvin D. J., Wang D., and Wasson J. T. (1984) Chemical classification of iron meteorites: X. Multielement studies of 43 irons, resolution of group-IIIE from group-IIIAB and evaluation of Cu as a taxonomic parameter. *Geochim. Cosmochim. Acta* **48**, 785–804.

Malvin D. J., Jones J. H., and Drake M. J. (1986) Experimental investigations of trace-element fractionation in iron meteorites: 3. Elemental partitioning in the system Fe–Ni–S–P. *Geochim. Cosmochim. Acta* **50**, 1221–1231.

Margot J. L. and Brown M. E. (2001) Discovery and characterization of binary asteroids 22 Kalliope and 87 Sylvia. *Bull. Am. Astron. Soc.* **33**, 1133.

McCoy T. J., Keil K., Scott E. R. D., and Haack H. (1993) Genesis of the IIICD iron meteorites: evidence from silicate-bearing inclusions. *Meteoritics* **28**, 552–560.

McCoy T. J., Keil K., Clayton R. N., Mayeda T. K., Bogard D. D., Garrison D. H., Huss G. R., Hutcheon I. D., and Wieler R. (1996) A petrologic, chemical, and isotopic study of monument draw and comparison with other acapulcoites: evidence for formation by incipient partial melting. *Geochim. Cosmochim. Acta* **60**, 2681–2708.

McCoy T. J., Keil K., Clayton R. N., Mayeda T. K., Bogard D. D., Garrison D. H., and Wieler R. (1997a) A petrologic and isotopic study of lodranites: evidence for early formation as partial melt residues from heterogeneous precursors. *Geochim. Cosmochim. Acta* **61**, 623–637.

McCoy T. J., Keil K., Muenow D. W., and Wilson L. (1997b) Partial melting and melt migration in the acapulcoite-lodranite parent body. *Geochim. Cosmochim. Acta* **61**, 639–650.

McCoy T. J., Dickinson T. L., and Lofgren G. E. (1998) Partial melting of the Indarch (EH4) meteorite: textural view of melting and melt migration. *Meteorit. Planet. Sci.* **33**, A100–A101.

Meibom A. and Clark B. E. (1999) Evidence for the insignificance of ordinary chondritic material in the asteroid belt. *Meteorit. Planet. Sci.* **34**, 7–24.

Meibom A., Desch S. J., Krot A. N., Cuzzi J. N., Petaev M. I., Wilson L., and Keil K. (2000) Large-scale thermal events in the solar nebula: evidence from Fe, Ni metal grains in primitive meteorites. *Science* **288**, 839–841.

Minarik W. G., Ryerson F. J., and Watson E. B. (1996) Textural entrapment of core-forming melts. *Science* **272**, 530–533.

Mittlefehldt D. W., Lindstrom M. M., Bogard D. D., Garrison D. H., and Field S. W. (1996) Acapulco- and Lodran-like achondrites: petrology, geochemistry, chronology, and origin. *Geochim. Cosmochim. Acta* **60**, 867–882.

Mittlefehldt D. W., McCoy T. J., Goodrich C. A., and Kracher A. (1998) Non-chondritic meteorites from asteroidal bodies. *Rev. Min.* **36**, D1–D195.

Moore C. B., Lewis C. F., and Nava D. (1969) Superior analysis of iron meteorites. In *Meteorite Research* (ed. P. M. Millman). Reidel, Dordrecht, pp. 738–748.

Moren A. E. and Goldstein J. I. (1978) Cooling rate variations of group-IVA iron meteorites. *Earth Planet. Sci. Lett.* **40**, 151–161.

Moren A. E. and Goldstein J. I. (1979) Cooling rates of group IVA iron meteorites determined from a ternary Fe–Ni–P model. *Earth Planet. Sci. Lett.* **43**, 182–196.

Mostefaoui S., Lugmair G. W., Hoppe P., and El Goresy A. (2003) Evidence for live iron-60 in Semarkona and Chervony Kut: a NanoSIMS atudy. *Lunar Planet. Sci. Conf.* **XXXIV**, #1585. The Lunar and Planetary Institute, Houston (CD-ROM).

Narayan C. and Goldstein J. I. (1981) Experimental determination of ternary partition-coefficients in Fe–Ni–X alloys. *Metall. Trans.* **12A**, 1883–1890.

Narayan C. and Goldstein J. I. (1982) A dendritic solidification model to explain Ge–Ni variations in iron meteorite chemical groups. *Geochim. Cosmochim. Acta* **46**, 259–268.

Narayan C. and Goldstein J. I. (1985) A major revision of iron meteorite cooling rates—an experimental study of the growth of the Widmanstätten pattern. *Geochim. Cosmochim. Acta* **49**, 397–410.

Nyquist L. E., Reese Y., Wiesmann H., Shih C. Y., and Takeda H. (2001) Dating eucrite formation and metamorphism. In *Antarctic Meteorites XXVI*. National Institute for Polar Research, Tokyo, pp. 113–115.

Olsen E. and Fredriksson K. (1966) Phosphates in iron and pallasite meteorites. *Geochim. Cosmochim. Acta* **30**, 459–470.

Olsen E., Davis A., Clarke R. S., Jr., Schultz L., Weber H. W., Clayton R., Mayeda T., Jarosewich E., Sylvester P., Grossman L., Wang M.-S., Lipschutz M. E., Steele I. M., and Schwade J. (1994) Watson: a new link in the IIE iron chain. *Meteoritics* **29**, 200–213.

Olsen E. J., Davis A. M., Clayton R. N., Mayeda T. K., Moore C. B., and Steele I. M. (1996) A silicate inclusion in Puente del Zacate, a IIIA iron meteorite. *Science* **273**, 1365–1367.

Olsen E. J., Kracher A., Davis A. M., Steele I. M., Hutcheon I. D., and Bunch T. E. (1999) The phosphates of IIIAB iron meteorites. *Meteorit. Planet. Sci.* **34**, 285–300.

Ostro S. J., Campbell D. B., Chandler J. F., Hine A. A., Hudson R. S., Rosema K. D., and Shapiro I. I. (1991) Asteroid 1986 DA: radar evidence for a metallic composition. *Science* **252**, 1399–1404.

Ostro S. J., Hudson R. S., Nolan M. C., Margot J. L., Scheeres D. J., Campbell D. B., Magri C., Giorgini J. D., and Yeomans D. K. (2000) Radar observations of asteroid 216 Kleopatra. *Science* **288**, 836–839.

Pernicka E. and Wasson J. T. (1987) Ru, Re, Os, Pt and Au in iron meteorites. *Geochim. Cosmochim. Acta* **51**, 1717–1726.

Petaev M. I., Meibom A., Krot A. N., Wood J. A., and Keil K. (2001) The condensation origin of zoned metal grains in Queen Alexandra range 94411. implications for the formation of the Bencubbin-like chondrites. *Meteorit. Planet. Sci.* **36**, 93–106.

Powell B. N. (1969) Petrology and chemistry of mesosiderites: I. Textures and composition of nickel–iron. *Geochim. Cosmochim. Acta* **33**, 789–810.

Powell B. N. (1971) Petrology and chemistry of mesosiderites: 2. Silicate textures and compositions and metal-silicate relationships. *Geochim. Cosmochim. Acta* **35**, 5–34.

Prombo C. A. and Clayton R. N. (1985) A striking nitrogen isotope anomaly in the Bencubbin and Weatherford meteorites. *Science* **230**, 935–937.

Prombo C. A. and Clayton R. N. (1993) Nitrogen isotopic compositions of iron meteorites. *Geochim. Cosmochim. Acta* **57**, 3749–3761.

Raghavan V. (1988) The Fe–P–S system. In *Phase Diagrams of Ternary Iron Alloys*. Indian National Scientific Documentation Centre, Vol. 2, pp. 209–217.

Randich E. and Goldstein J. I. (1978) Cooling rates of seven hexahedrites. *Geochim. Cosmochim. Acta* **42**, 221–234.

Rasmussen K. L. (1981) The cooling rates of iron meteorites—a new approach. *Icarus* **45**, 564–576.

Rasmussen K. L. (1982) Determination of the cooling rates and nucleation histories of 8 group IVA iron meteorites using local bulk Ni and P variation. *Icarus* **52**, 444–453.

Rasmussen K. L. (1989) Cooling rates of IIIAB iron meteorites. *Icarus* **80**, 315–325.

Rasmussen K. L., Malvin D. J., Buchwald V. F., and Wasson J. T. (1984) Compositional trends and cooling rates of group IVB iron meteorites. *Geochim. Cosmochim. Acta* **48**, 805–813.

Rasmussen K. L., Malvin D. J., and Wasson J. T. (1988) Trace-element partitioning between taenite and kamacite—relationship to the cooling rates of iron meteorites. *Meteoritics* **23**, 107–112.

Rasmussen K. L., Ulff-Møller F., and Haack H. (1995) The thermal evolution of IVA iron meteorites—evidence from metallographic cooling rates. *Geochim. Cosmochim. Acta* **59**, 3049–3059.

Rasmussen K. L., Haack H., and Ulff-Møller F. (2001) Metallographic cooling rates of group IIF iron meteorites. *Meteorit. Planet. Sci.* **36**, 883–896.

Reid A. M., Williams R. J., and Takeda H. (1974) Coexisting bronzite and clinobronzite and thermal evolution of the Steinbach meteorite. *Earth Planet. Sci. Lett.* **22**, 67–74.

Rivkin A. S., Howell E. S., Lebofsky L. A., Clark B. E., and Britt D. T. (2000) The nature of M-class asteroids from 3 μm observations. *Icarus* **145**, 351–368.

Rivkin A. S., Howell E. S., Vilas F., and Lebofsky L. A. (2002) Hydrated minerals on asteroids: the astronomical record. In *Asteroids III* (eds. W. Bottke, A. Cellino, P. Paolicchi, and R. P. Binzel). University of Arizona Press, Tucson, pp. 235–253.

Romig A. D. and Goldstein J. I. (1980) Determination of the Fe–Ni and Fe–Ni–P phase diagrams at low temperatures (700 to 300°C). *Metall. Trans.* **11A**, 1151–1159.

Rubin A. E. and Mittlefehldt D. W. (1992) Mesosiderites—a chronological and petrologic synthesis. *Meteoritics* **27**, 282–282.

Rushmer T., Minarik W. G., and Taylor G. J. (2000) Physical processes of core formation. In *Origin of the Earth and Moon* (eds. R. M. Canup and K. Righter). University of Arizona Press, Tucson, pp. 227–243.

Ruzicka A., Fowler G. W., Snyder G. A., Prinz M., Papike J. J., and Taylor L. A. (1999) Petrogenesis of silicate inclusions in the weekeroo station IIE iron meteorite: differentiation, remelting, and dynamic mixing. *Geochim. Cosmochim. Acta* **63**, 2123–2143.

Saikumar V. and Goldstein J. I. (1988) An evaluation of the methods to determine the cooling rates of iron meteorites. *Geochim. Cosmochim. Acta* **52**, 715–726.

Schaudy R., Wasson J. T., and Buchwald V. F. (1972) Chemical classification of iron meteorites: VI. Reinvestigation of irons with Ge concentrations lower than 1 ppm. *Icarus* **17**, 174–192.

Scott E. R. D. (1972) Chemical fractionation in iron meteorites and its interpretation. *Geochim. Cosmochim. Acta* **36**, 1205–1236.

Scott E. R. D. (1977a) Composition, mineralogy and origin of group IC iron meteorites. *Earth Planet. Sci. Lett.* **37**, 273–284.

Scott E. R. D. (1977b) Formation of olivine-metal textures in pallasite meteorites. *Geochim. Cosmochim. Acta* **41**, 693–710.

Scott E. R. D. (1977c) Geochemical relationships between some pallasites and iron meteorites. *Min. Mag.* **41**, 265–272.

Scott E. R. D. (1978) Tungsten in iron meteorites. *Earth Planet. Sci. Lett.* **39**, 363–370.

Scott E. R. D. (1979) Origin of anomalous iron meteorites. *Mineral. Mag.* **43**, 415–421.

Scott E. R. D. and Wasson J. T. (1975) Classification and properties of iron meteorites. *Rev. Geophys. Space Phys.* **13**, 527–546.

Scott E. R. D. and Wasson J. T. (1976) Chemical classification of iron meteorites: VIII. Groups IC, IIE, IIIF and 97 other irons. *Geochim. Cosmochim. Acta* **40**, 103–115.

Scott E. R. D., Wasson J. T., and Buchwald V. F. (1973) Chemical classification of iron meteorites: VII. Reinvestigation of irons with Ge concentrations between 25 and 80 ppm. *Geochim. Cosmochim. Acta* **37**, 1957–1983.

Scott E. R. D., Haack H., and McCoy T. J. (1996) Core crystallization and silicate-metal mixing in the parent body of the IVA iron and stony-iron meteorites. *Geochim. Cosmochim. Acta* **60**, 1615–1631.

Scott E. R. D., Haack H., and Love S. G. (2001) Formation of mesosiderites by fragmentation and reaccretion of a large differentiated asteroid. *Meteorit. Planet. Sci.* **36**, 869–881.

Shannon M. C. and Agee C. B. (1996) High pressure constraints on percolative core formation. *Geophys. Res. Lett.* **23**, 2717–2720.

Shannon M. C. and Agee C. B. (1998) Percolation of core melts at lower mantle conditions. *Science* **280**, 1059–1061.

Shukolyukov A. and Lugmair G. (1992) ^{60}Fe-Light my fire. *Meteoritics* **27**, 289.

Smoliar M. I., Walker R. J., and Morgan J. W. (1996) Re–Os ages of group IIA, IIIA, IVA, and IVB iron meteorites. *Science* **271**, 1099–1102.

Sonett C. P., Colburn D. S., Schwartz K., and Keil K. (1970) The melting of asteroidal-sized parent bodies by unipolar dynamo induction from a primitive T Tauri sun. *Astrophys. Space Sci.* **7**, 446–488.

Srinivasan G., Goswami J. N., and Bhandari N. (1999) ^{26}Al in eucrite Piplia Kalan: plausible heat source and formation chronology. *Science* **284**, 1348–1350.

Stöffler D., Bischoff A., Buchwald V. F., and Rubin A. E. (1988) Shock effects in meteorites. In *Meteorites and the Early Solar System* (eds. J. F. Kerridge and M. S. Matthews). University of Arizona Press, Tucson, pp. 165–204.

Tachibana S. and Huss G. I. (2003) The initial abundance of ^{60}Fe in the solar system. *Astrophys. J.* **588**, L41–L44.

Takahashi E. (1983) Melting of a Yamato L3 chondrite (Y-74191) up to 30 kbar. *Proc. 8th Symp. Antarct. Meteorit.*, 168–180.

Taylor G. J. (1992) Core formation in asteroids. *J. Geophys. Res.* **97**, 14717–14726.

Taylor G. J., Keil K., McCoy T., Haack H., and Scott E. R. D. (1993) Asteroid differentiation—pyroclastic volcanism to Magma Oceans. *Meteoritics* **28**, 34–52.

Tholen D. J. (1989) Asteroid taxonomic classification. In *Asteroids II* (eds. R. P. Binzel, T. Gehrels, and M. S. Matthews). University of Arizona Press, Tucson, pp. 1139–1150.

Ulff-Møller F. (1998) Effects of liquid immiscibility on trace element fractionation in magmatic iron meteorites: a case study of group IIIAB. *Meteorit. Planet. Sci.* **33**, 207–220.

Ulff-Møller F., Rasmussen K. L., Prinz M., Palme H., Spettel B., and Kallemeyn G. W. (1995) Magmatic activity on the IVA parent body—evidence from silicate-bearing iron meteorites. *Geochim. Cosmochim. Acta* **59**, 4713–4728.

Ulff-Møller F., Tran J., Choi B.-G., Haag R., Rubin A. E., and Wasson J. T. (1997) Esquel: implications for pallasite formation processes based on the petrography of a large slab. *Lunar Planet. Sci.* **XXVIII**. The Lunar and Planetary Institute, Houston, pp. 1465–1466.

Ulff-Møller F., Choi B. G., Rubin A. E., Tran J., and Wasson J. T. (1998) Paucity of sulfide in a large slab of Esquel: new perspectives on pallasite formation. *Meteorit. Planet. Sci.* **33**, 221–227.

Urakawa S., Kato M., and Kumazawa M. (1987) Experimental study on the phase relations in the system Fe–Ni–O–S up to 15 GPa. In *High Pressure Research in Mineral Physics.* TERRAPUB/AGU, pp. 95–111.

Viateau B. (2000) Mass and density of asteroids (16) psyche and (121) Hermione. *Astron. Astrophys.* **354**, 725–731.

Voshage H. (1967) Bestrahlungsalter und Herkunft der Eisenmeteorite. *Z. Naturforschg.* **22a**, 477–506.

Voshage H. and Feldmann H. (1979) Investigations on cosmic-ray-produced nuclides in iron meteorites: 3. Exposure ages, meteoroid sizes and sample depths determined by mass-spectrometric analyses of Potassium and rare gases. *Earth Planet. Sci. Lett.* **45**, 293–308.

Walker D. and Agee C. B. (1988) Ureilite compaction. *Meteoritics* **23**, 81–91.

Wasson J. T. (1967) Chemical classification of iron meteorites: I. A study of iron meteorites with low concentrations of gallium and germanium. *Geochim. Cosmochim. Acta* **31**, 161–180.

Wasson J. T. (1969) Chemical classification of iron meteorites: III. Hexahedrites and other irons with germanium concentrations between 80 ppm and 200 ppm. *Geochim. Cosmochim. Acta* **33**, 859–876.

Wasson J. T. (1970) Chemical classification of iron meteorites: IV. Irons with Ge concentrations greater than 190 ppm and other meteorites associated with group I. *Icarus* **12**, 407–423.

Wasson J. T. (1971) Chemical classification of iron meteorites: V. Groups IIIC and IIID and other irons with germanium concentrations between 1 and 25 ppm. *Icarus* **14**, 59–70.

Wasson J. T. (1990) Ungrouped iron meteorites in Antarctica—origin of anomalously high abundance. *Science* **249**, 900–902.

Wasson J. T. (1999) Trapped melt in IIIAB irons: solid/liquid elemental partitioning during the fractionation of the IIIAB magma. *Geochim. Cosmochim. Acta* **63**, 2875–2889.

Wasson J. T. and Kallemeyn G. W. (2002) The IAB iron-meteorite complex: a group, five subgroups, numerous grouplets, closely related, mainly formed by crystal segregation in rapidly cooling melts. *Geochim. Cosmochim. Acta* **66**, 2445–2473.

Wasson J. T. and Kimberlin J. (1967) Chemical classification of iron meteorites: II. Irons and pallasites with germanium concentrations between 8 and 100 ppm. *Geochim. Cosmochim. Acta* **31**, 2065–2093.

Wasson J. T. and Richardson J. W. (2001) Fractionation trends among IVA iron meteorites: contrasts with IIIAB trends. *Geochim. Cosmochim. Acta* **65**, 951–970.

Wasson J. T. and Rubin A. E. (1985) Formation of mesosiderites by low-velocity impacts as a natural consequence of planet formation. *Nature* **318**, 168–170.

Wasson J. T. and Wai C. M. (1976) Explanation for the very low Ga and Ge concentrations in some iron meteorite groups. *Nature* **261**, 114–116.

Wasson J. T. and Wang J. M. (1986) A nonmagmatic origin of group IIE iron meteorites. *Geochim. Cosmochim. Acta* **50**, 725–732.

Wasson J. T., Ouyang X. W., Wang J. M., and Jerde E. (1989) Chemical classification of iron meteorites: XI. Multi-element studies of 38 new irons and the high abundance of ungrouped irons from Antarctica. *Geochim. Cosmochim. Acta* **53**, 735–744.

Wasson J. T., Choi B. G., Jerde E. A., and Ulff-Møller F. (1998) Chemical classification of iron meteorites: XII. New members of the magmatic groups. *Geochim. Cosmochim. Acta* **62**, 715–724.

Welten K. C., Alderliesten C., Van der Borg K., Lindner L., Loeken T., and Schultz L. (1997) Lewis Cliff 86360: an Antarctic L-chondrite with a terrestrial age of 2.35 million years. *Meteorit. Planet. Sci.* **32**, 775–780.

Willis J. and Goldstein J. I. (1982) The effects of C, P, and S on trace-element partitioning during solidification in Fe–Ni alloys. *J. Geophys. Res.* **87**, A435–A445.

Willis J. and Wasson J. T. (1978a) Cooling rates of group-IVA iron meteorites. *Earth Planet. Sci. Lett.* **40**, 141–150.

Willis J. and Wasson J. T. (1978b) Core origin for group-IVA iron meteorites—reply. *Earth Planet. Sci. Lett.* **40**, 162–167.

Wood J. A. (1964) Cooling rates and parent planets of several iron meteorites. *Icarus* **3**, 429–459.

Yanai K. and Kojima H. (1995) Yamato-8451: a newly identified pyroxene-bearing pallasite. *Proc. NIPR Symp. Antarct. Meteorit.* **8**, 1–10.

Yang C. W., Williams D. B., and Goldstein J. I. (1996) A revision of the Fe–Ni phase diagram at low temperatures (<400 degrees C). *J. Phase Equil.* **17**, 522–531.

Yang C. W., Williams D. B., and Goldstein J. I. (1997) A new empirical cooling rate indicator for meteorites based on the size of the cloudy zone of the metallic phases. *Meteorit. Planet. Sci.* **32**, 423–429.

Zhou H. and Steele I. M. (1993) Chemical zoning and diffusion of Ca, Al, Mn, and Cr in olivine of springwater pallasite. *Lunar Planet. Sci.* **XXIV**. The Lunar and Planetary Institute, Houston, pp. 1573–1574.

1.13
Cosmic-ray Exposure Ages of Meteorites

G. F. Herzog
Rutgers University, Piscataway, NJ, USA

1.13.1 INTRODUCTION	348
1.13.2 CALCULATION OF EXPOSURE AGES	349
1.13.2.1 Basic Equations	349
1.13.2.2 Factors Influencing Production Rates	349
1.13.2.3 Measurement Units and Quantities	350
1.13.2.4 Calibration of Production Rates	351
1.13.2.4.1 ^{26}Al versus ^{21}Ne calibration	351
1.13.2.4.2 $^{83}Kr/^{81}Kr$ calibration	351
1.13.2.5 Equations for Calculating One-stage CRE Ages	352
1.13.2.5.1 $^{21}Ne-^{22}Ne/^{21}Ne$ ages	352
1.13.2.5.2 $^{38}Ar-^{22}Ne/^{21}Ne$ ages	353
1.13.2.5.3 ^{3}He ages	353
1.13.2.5.4 $^{36}Cl/^{36}Ar$ ages	354
1.13.2.5.5 $^{81}Kr/Kr$ ages	354
1.13.2.5.6 $^{40}K/K$ ages	354
1.13.2.6 The Importance of Half-lives	356
1.13.3 CARBONACEOUS CHONDRITES	356
1.13.3.1 CI, CM, CO, CV, and CK Chondrites	356
1.13.3.2 The CR Clan	357
1.13.4 H-CHONDRITES	357
1.13.5 L-CHONDRITES	358
1.13.6 LL CHONDRITES	359
1.13.7 E-CHONDRITES	359
1.13.8 R-CHONDRITES	360
1.13.9 LODRANITES AND ACAPULCOITES	360
1.13.10 LUNAR METEORITES	361
1.13.10.1 Overview	361
1.13.10.2 Construction of CRE Histories	361
1.13.10.3 Production Rate of Lunar Meteorites	364
1.13.11 HOWARDITE–EUCRITE–DIOGENITE (HED) METEORITES	364
1.13.11.1 Eucrites	364
1.13.11.2 Diogenites	365
1.13.11.3 Howardites	365
1.13.11.4 Kapoeta	366
1.13.12 ANGRITES	367
1.13.13 UREILITES	367
1.13.14 AUBRITES (ENSTATITE ACHONDRITES)	367
1.13.15 BRACHINITES	368
1.13.16 MARTIAN METEORITES	368
1.13.17 MESOSIDERITES	370
1.13.18 PALLASITES	371

1.13.19 IRONS 371
1.13.20 THE SMALLEST PARTICLES: MICROMETEORITES, INTERPLANETARY DUST PARTICLES,
AND INTERSTELLAR GRAINS 372
   1.13.20.1 Background 372
   1.13.20.2 Micrometeorites and IDPs 372
   1.13.20.3 Interstellar Grains 374
1.13.21 CONCLUSIONS 374
ACKNOWLEDGMENTS 375
REFERENCES 375

1.13.1 INTRODUCTION

The classic idea of a cosmic-ray exposure (CRE) age for a meteorite is based on a simple but useful picture of meteorite evolution, the one-stage irradiation model. The precursor rock starts out on a parent body, buried under a mantle of material many meters thick that screens out cosmic rays. At a time t_i, a collision excavates a precursor rock—a "meteoroid." The newly liberated meteoroid, now fully exposed to cosmic rays, orbits the Sun until a time t_f, when it strikes the Earth, where the overlying blanket of air (and possibly of water or ice) again shuts out almost all cosmic rays (cf. Masarik and Reedy, 1995). The quantity $t_f - t_i$ is called the CRE age, t. To obtain the CRE age of a meteorite, we measure the concentrations in it of one or more cosmogenic nuclides (Table 1), which are nuclides that cosmic rays produce by inducing nuclear reactions. Many shorter-lived radionuclides excluded from Table 1 such as ^{22}Na ($t_{1/2} = 2.6$ yr) and ^{60}Co ($t_{1/2} = 5.27$ yr) can also furnish valuable information, but can be measured only in meteorites that fell within the last few half-lives of those nuclides (see, e.g., Leya et al. (2001) and references therein).

Table 1 Cosmogenic nuclides used for calculating exposure ages.

Nuclide	Half-life[a] (Myr)
Radionuclides	
^{14}C	0.005730
^{59}Ni	0.076
^{41}Ca	0.1034
^{81}Kr	0.229
^{36}Cl	0.301
^{26}Al	0.717
^{10}Be	1.51
^{53}Mn	3.74
^{129}I	15.7
Stable nuclides	
^{3}He	
^{21}Ne	
^{38}Ar	
^{83}Kr	
^{126}Xe	

[a] http://www2.bnl.gov/ton.

CRE ages have implications for several interrelated questions. From how many different parent bodies do meteorites come? How well do meteorites represent the population of the asteroid belt? How many distinct collisions on each parent body have created the known meteorites of each type? How often do asteroids collide? How big and how energetic were the collisions that produced meteoroids? What factors control the CRE age of a meteorite and how do meteoroid orbits evolve through time? We will touch on these questions below as we examine the data.

By 1975, the CRE ages of hundreds of meteorites had been estimated from noble gas measurements. Histograms of the CRE age distributions pointed to several important observations.

(i) The CRE ages of meteorites increase in the order stones <stony irons <irons.

(ii) The CRE ages of stones rarely exceed 100 Myr; the average ages of stony irons are typically between 50 Myr and 200 Myr; the CRE ages of irons vary with group but more often than not exceed 200 Myr.

(iii) The CRE ages of stones and of irons are neither uniformly distributed nor tightly clustered.

These early conclusions imply first that meteoroid production does not take place uniformly through time, for if it did, then we ought to see a distribution of CRE ages without peaks. Second, they imply that mechanical toughness contributes to the survival ability of meteoroids, a hypothesis that helps explain the greater fraction of irons with high CRE ages and the much shorter CRE ages of, e.g., the relatively fragile carbonaceous chondrites. Third, comparisons of the CRE age distributions of different types of stones point to the importance of orbits. Although aubrites and CI carbonaceous chondrites, e.g., are both fairly fragile, aubrites have much larger CRE ages. This difference (along with dynamical calculations) suggested early on that the original orbit of the parent body affects CRE ages.

Since the early 1970s, several developments have brought the landscape of CRE ages into sharper focus. The number of meteorites available for analyses has increased greatly, by a factor of ~ 10, thanks to abundant finds in the Antarctic, northern Africa/Arabia, and Australia. With increased sampling, the statistical properties of

CRE age distributions have become more convincing. Further, the world's collection of meteorites collection has become more diverse. In this respect, the lunar and the martian meteorites take pride of place but leave ample room for R, CH, and CB chondrites, new angrites, and other unusual specimens. At the same time, better experimental methods have lowered detection limits for cosmogenic nuclides and the modeling calculations needed to interpret the measurements have improved.

With greater analytical power has come the ability to recognize and, increasingly, to characterize more complex irradiation histories. As it turns out, many meteorites retain the effects not only of recent irradiation but also of irradiations that took place at earlier times, in different settings.

(i) Collisions in space reduced the sizes and changed the shapes of some meteoroids. The cosmogenic nuclide inventories in such meteorites may record the two distinct periods of exposure.

(ii) Certain components of polymict meteoritic breccias (rocks that consist of unlike grains cemented together) spent time at the surfaces of their parent bodies before they were buried in parent bodies or perhaps in meteoroids. While at the parent-body surface these components must have been exposed directly not only to galactic cosmic rays (the high-energy particles from outside the solar system that are responsible for most of the production of cosmogenic nuclides), but also to lower-energy cosmic rays from the Sun.

(iii) Selected petrologic phases—the chondrules and the calcium- and aluminum-rich inclusions found in some meteorites—may have been irradiated just after forming in the very early solar system. One proposed mechanism is irradiation by the so-called X-wind, an intense outflow of nuclear-active particles hypothesized for the primitive Sun (Shu et al., 1996).

(iv) Interstellar grains isolated from certain meteorites retain cosmogenic nuclides made by irradiation in interstellar space or, perhaps, close to other stars, at a time predating the formation of the solar system. We will interweave a few examples of multistage exposures into the discussion, but our main emphasis will be on the most recent one.

Honda and Arnold (1967), Wasson (1974), Reedy et al. (1983), Caffee et al. (1988), Vogt et al. (1990), Tuniz et al. (1998), Wieler and Graf (2001), and Eugster (2003) have published general reviews of CRE ages.

1.13.2 CALCULATION OF EXPOSURE AGES

1.13.2.1 Basic Equations

During a one-stage exposure, cosmogenic nuclides accumulate in meteoroids much as rain collects in a bucket. We can determine the length of time the bucket is outside (its exposure age) by measuring the height of the water that collects. The calculation requires that the rain fell at a known rate, preferably constant. Further, the bucket must be empty initially; be of known size and shape; and lose water only in known and reproducible ways. In our imperfect analogy, the bucket plays the role of the meteoroid while the rain doubles as both cosmic rays and their products, the cosmogenic nuclides. Translated into the language of CRE ages, the standard calculations for a one-stage irradiation require: a constant flux of cosmic rays; and either (i) known production rates, P (and known loss rates if radioactive decay or diffusion matter) for the cosmogenic nuclides of interest; or (ii) known production rate ratios. When losses are negligible, we have for the concentration of a stable cosmogenic nuclide, s,

$$s = P_s t \qquad (1)$$

and for the concentration of a radioactive cosmogenic nuclide r, with decay constant λ,

$$r = \frac{P_r(1 - e^{-\lambda t})}{\lambda} \qquad (2)$$

On combining Equations (1) and (2), we have

$$\frac{s}{r} = \frac{P_s}{P_r} \frac{\lambda t}{(1 - e^{-\lambda t})} \qquad (3)$$

Given either s and P_s, or r and P_r, or the ratios s/r and P_s/P_r, one may calculate a CRE age. The problem is how to get the production rates or the ratio P_s/P_r.

1.13.2.2 Factors Influencing Production Rates

The production rate of each nuclide, i, depends on numerous factors unique to each meteorite. At any time t', the full expression for the production rate at a location with coordinates x, y, z in the meteoroid is given by

$$P_i(x, y, z, t', \text{composition})$$
$$= \sum_k \sum_j N_j \int_E \{\phi_k(E, x, y, z, t')\} \sigma_{j,k}(E)\, dE \qquad (4a)$$

where ϕ_k is the flux of nuclear-active particles of type k (mainly protons and neutrons, but also some alpha particles and pions) and energy E; $\sigma_{j,k}$ is the nuclear cross-section for the production of i from chemical element j by particle k; and N_j is the concentration of element j in the meteoroid (assumed constant). In the absence of information about $x, y,$ and z, one customarily assumes a spherical meteoroid with radius R. Under spherical symmetry and with uniform composition, the depth of the sample below the meteoroid surface, d, then sets the production rate:

$$P_i(d, R, t', \text{composition})$$
$$= \sum_k \sum_j N_j \int_E \{\phi_k(E, d, R, t')\}\sigma_{j,k}(E)dE. \quad (4b)$$

For a constant flux of galactic cosmic rays (for further discussion of this point, see Lavielle et al. (1999)), the time dependence vanishes. The parameters, size and depth, remain along with the elemental abundances and nuclear cross-sections; in principle, the elemental abundances and nuclear cross-sections can be measured directly.

During the 1990s teams led by Janet Sisterson at Harvard and by Rolf Michel at Hannover measured many nuclear cross-sections relevant to the calculation of production rates. Cross-sections for nuclear reactions induced by energetic neutrons, although not measured directly in most cases, have been inferred from thick target irradiations. Modern modeling calculations incorporate all these results. Given the pre-atmospheric size and shape of a meteoroid and the location of a sample within it, these calculations can now reproduce the production rates of many cosmogenic nuclides to within 15% with the introduction of just one universal parameter, a measure of the intensity of galactic cosmic rays (Leya et al., 2000; Masarik et al., 2001). Even these best estimates, however, ultimately derive from absolute CRE age calibrations that are based on radioactive cosmogenic nuclides.

Cosmogenic nuclides can be divided into three broad and imperfect categories according to the energies of the particles that produce them (e.g., Tuniz et al., 1998). At the highest energies, nuclear spallation reactions dominate. The concentrations of nuclides produced in such reactions—^{21}Ne in iron meteorites for example—tend to decrease with increasing depth. At intermediate energies, simpler two-body nuclear reactions and compound nucleus reactions are common. The nuclear particles with the energies in this range are for the most part *secondary* particles, themselves generated by reactions of higher-energy cosmic rays. The concentrations of the cosmogenic nuclides produced in reactions induced by these particles—^{21}Ne in stony meteoroids for example—typically increase and pass through a maximum as the depth in the meteoroid increases. At the lowest energies, we have thermal neutrons. Eventually, through inelastic collisions, secondary neutrons slow and may reach thermal velocities, $\sqrt{2kT/m}$, where k is Boltzmann's constant, T the temperature, and m the neutron mass. Special importance attaches to the fluxes of thermal neutrons, because certain isotopes have extraordinarily large cross-sections for their capture, and hence for making detectable concentrations of cosmogenic nuclides. Examples include ^{59}Co(n,γ)^{60}Co, ^{35}Cl(n,γ)^{36}Cl, ^{40}Ca(n,γ)^{41}Ca, ^{58}Ni(n,γ)^{59}Ni, ^{79}Br(n,γ)^{80}Br → ^{80}Kr, ^{149}Sm(n,γ)^{150}Sm, and ^{157}Gd(n,γ)^{158}Gd. Nuclear modeling calculations and direct measurements of depth profiles show that the production rates of these nuclides peak at greater depths in larger bodies relative to cosmogenic nuclides produced at higher energies (Eberhardt et al., 1963; Spergel et al., 1986).

1.13.2.3 Measurement Units and Quantities

The concentrations of (stable) ^{3}He, ^{21}Ne, and ^{38}Ar constitute by far the largest and best bank of data for the calculation of CRE ages (Schultz and Franke, 2002). Noble gas concentrations are often reported in 10^{-8} cm^3 STP (g sample)$^{-1}$, where STP refers to a temperature of 273.15 K and a pressure of 1 atm. To convert the concentration of an isotope from 10^{-8} cm^3 STP (g sample)$^{-1}$ to atoms per gram of sample, one multiplies by N_A/RT, where N_A is Avogadro's number and R is the gas constant, i.e., by 2.687×10^{11}:

$$s[\text{atom}(\text{g sample})^{-1}]$$
$$= 2.687 \times 10^{11} \, s[10^{-8} \, \text{cm}^3 \, \text{STP})$$
$$(\text{g sample})^{-1}] \quad (5)$$

To a reasonable first approximation, the production rate of ^{21}Ne in a very common type of meteorite, an H-chondrite, is 8×10^{10} atom g^{-1} Myr^{-1}. Thus, an irradiation lasting 7.5 Myr produces $\sim 6 \times 10^{11}$ atom per gram of ^{21}Ne, or $\sim 2 \times 10^{-8}$ cm^3 STP per gram of ^{21}Ne.

Radionuclide concentrations are normally reported as *activities*, A, in units of disintegrations per minute per kilogram (dpm kg^{-1}). Nishiizumi (1987) compiled a new and comprehensive source of data for ^{36}Cl, ^{26}Al, ^{10}Be, and ^{53}Mn in meteorites. Table 2 shows a few activities that are typical for stony meteorites of small to moderate size (pre-atmospheric radius less than 40 cm). To convert from dpm kg^{-1} to atom g^{-1}, one uses the radioactive decay law,

Table 2 Typical activities (dpm kg^{-1}) of selected cosmogenic radionuclides in stony meteorites.

Nuclide	Activity
^{14}C	50
^{41}Ca	24[a]
^{81}Kr	0.003
^{36}Cl	22[a]
^{26}Al	60
^{10}Be	20
^{53}Mn	400[a]
^{129}I	0.0003

[a] Metal phase.

$\lambda r = -dr/dt = A$. From the expression for the half-life, $t_{1/2} = \ln(2)/\lambda$, we have after conversion of units (1 yr = 365.25 d)

$$r(\text{atom g}^{-1}) = 7.588 \times 10^8 \times A \text{ (dpm kg}^{-1}) \times t_{1/2}(\text{My}) \quad (6)$$

As shown in the next section, we often want the ratios of noble gas *concentrations* (atom g^{-1}) to radionuclide *activities* (atom Myr^{-1} g^{-1}); after appropriate conversion of units, such ratios have dimensions of time and are closely related to CRE ages. In converting the units, the half-life does not enter and we have, for all noble gas–radionuclide pairs,

$$\frac{s(\text{atom g}^{-1})}{r(\text{atom My}^{-1} \text{ g}^{-1})} = 510.9 \times \frac{s(10^{-8} \text{ cm}^3 \text{ STP g}^{-1})}{r(\text{dpm kg}^{-1})} \quad (7)$$

1.13.2.4 Calibration of Production Rates

To infer CRE ages from cosmogenic noble gas contents, we must know their absolute production rates. We now discuss two examples that illustrate how to obtain this information.

1.13.2.4.1 ^{26}Al versus ^{21}Ne calibration

Inspection of Equation (2) shows that in the limit of CRE ages long compared to $1/\lambda$, the concentration of a cosmogenic radionuclide approaches the constant value P/λ. Hence, in this case, the activity at the time of a meteorite fall (or the collection of a lunar sample) is equal to the production rate. Once the production rate is known for one meteorite, one may analyze for the same radionuclide in a second meteorite with, perhaps, a shorter CRE age. If P was the same in the second meteorite (an important assumption), then we can solve Equation (2) for the CRE age. In a kind of bootstrapping process, measurements of other cosmogenic nuclides in the second meteorite then can serve as the basis for calculating *their* production rates, and so on.

Herzog and Anders (1971) tried to calibrate the ^{21}Ne production rate through measurements of the activity of ^{26}Al, $A(^{26}\text{Al})$, and the concentration of ^{21}Ne, [^{21}Ne], for a suite of meteorites. From Equation (1), we have $T = {}^{21}\text{Ne}/P_{21}$. We substitute this result in Equation (2), written for ^{26}Al, and rearrange

$$A(^{26}\text{Al}) = P_{26}\left[1 - \exp\left(-\lambda_{26} \frac{[^{21}\text{Ne}]}{P_{21}}\right)\right] \quad (8)$$

A nonlinear fitting routine applied to Equation (8) gives the parameters P_{26} and P_{21}. In practice, the important information about P_{21} comes from meteorites with low ^{21}Ne contents. Ironically, this method originally gave too large a value for P_{21}, because several of the meteorites with low ^{21}Ne contents had complex rather than one-stage exposure histories. In other words, the meteorites chosen for study contained ^{21}Ne but not much ^{26}Al left over from an earlier irradiation period, because the ^{26}Al created in the earlier period had largely decayed away.

The principle of the method is sound, however, and with careful sample selection and the addition of shielding adjustments the procedure works well for several pairs of cosmogenic nuclides. Once P_{21} and P_{26} are known, we can calculate the ratio P_{21}/P_{26}. On expanding the exponential in Equation (8), we then have, for any meteorite with $t \gg 1/\lambda_{26}$ the relation,

$$\frac{[^{21}\text{Ne}]}{A(^{26}\text{Al})} \approx \frac{P_{21}t}{P_{26}} \quad \left(t \gg \frac{1}{\lambda_{26}}\right) \quad (9)$$

where units are atom per unit mass or atom per unit mass per unit time. For L-chondrites, the value of the production rate ratio P_{21}/P_{26} is ~2.5 (atom ^{21}Ne/atom ^{26}Al), where the units of time cancel out, or 0.005 (10^{-8} cm^3 STP g^{-1} Myr^{-1})/(dpm kg^{-1}) and varies slightly with shielding.

1.13.2.4.2 ^{83}Kr/^{81}Kr calibration

Marti (1967) introduced the ^{81}Kr/Kr method for determining CRE ages. The ^{26}Al/^{21}Ne method sketched above rests only on analyses of meteorites; in contrast, the ^{81}Kr/Kr method also demands knowledge of the relative cross-sections for certain nuclear reactions. To see how the method works, we write Equation (1) for ^{83}Kr, a stable isotope produced mainly by spallation, and Equation (2) for ^{81}Kr, which has a half-life of 0.229 Myr, and then divide one by the other, obtaining

$$\frac{^{83}\text{Kr}}{^{81}\text{Kr}} = \frac{\lambda_{81} P_{83} t}{P_{81}(1 - e^{-\lambda_{81}t})} \quad (10)$$

Note that the appearance of λ_{81} here reflects the fact that it is the concentration ratio of the two nuclides, rather than a concentration/activity ratio that is measured. As CRE ages of less than a few half-lives of ^{81}Kr are rare, for the great majority of meteorites Equation (10) simplifies to

$$\frac{^{83}\text{Kr}}{^{81}\text{Kr}} \approx \frac{\lambda_{81} P_{83} t}{P_{81}} \quad (t \gg \lambda_{81}) \quad (11)$$

Provided that the ratio P_{83}/P_{81} is known, the calculation of a CRE age reduces to the measurement of the ^{83}Kr/^{81}Kr (atom/atom) ratio. In the earliest calculations, P_{81}/P_{83} was set by the relation

$$\frac{P_{81}}{P_{83}} = 0.95 \left(\frac{^{80}\text{Kr} + ^{82}\text{Kr}}{2 \times ^{83}\text{Kr}} \right)_{\text{spallogenic}} \quad (12)$$

The factor of 0.95 was chosen to take account of the isobaric yield, which, in turn, was based on cross-section measurements made for the element silver. Since then laboratory simulations of meteorite irradiation have furnished more relevant cross-section data (Gilabert et al., 2002). Although these authors recommend no changes to the calculation of ^{81}Kr/Kr ages, the new data may make it necessary to revise Equation (12).

The use of ^{80}Kr and ^{82}Kr to arrive at P_{81}/P_{83} (Equation (12)) is unsatisfactory for samples that contain appreciable amounts of Br, from which neutron capture produces ^{80}Kr and ^{82}Kr, but not ^{81}Kr. To avoid this problem, P_{81}/P_{83} has often been calculated from an empirical relation obtained by selecting samples with negligible neutron capture effects and then regressing calculated values of P_{81}/P_{83} (from Equation (12)) on directly measured values of ^{78}Kr/^{83}Kr (Marti and Lugmair, 1971). The regression gave

$$\frac{P_{81}}{P_{83}} = 1.262 \left(\frac{^{78}\text{Kr}}{^{83}\text{Kr}} \right)_{\text{spallogenic}} + 0.381 \quad (13)$$

Underlying this equation is the additional assumption that ratios of production rate ratios (e.g., $P_{81}/P_{83} \div P_{78}/P_{83}$) are insensitive to changes in composition and in shielding.

Eugster (1988) and Eugster and Michel (1995) have used ^{81}Kr/Kr CRE ages to derive equations for ^{3}He, ^{21}Ne, and ^{22}Ne production rates in chondrites. These empirical correlations are embodied in production rate equations for ^{3}He, ^{21}Ne, and ^{38}Ar that are now in wide use.

1.13.2.5 Equations for Calculating One-stage CRE Ages

Here we describe and comment critically on a few of the many methods used to calculate CRE ages.

1.13.2.5.1 ^{21}Ne–^{22}Ne/^{21}Ne ages

The age equation is a specialized form of Equation (1):

$$t = \frac{^{21}\text{Ne}}{P_{21}} \quad (14)$$

Eugster (1988) and Eugster and Michel (1995) give formulas for the numerical calculation of P_{21} in ordinary chondrites and certain achondrites, respectively. The absolute timescale is set by reference to the ^{81}Kr/Kr ages of selected meteorites. The formula for the production rate of ^{21}Ne has the general form

$$P = F \times S \times A \quad (15)$$

Here F (dimensionless) corrects for composition over a limited range; S (dimensionless) corrects for shielding through the cosmogenic ^{22}Ne/^{21}Ne ratio, also over a limited range; A is a normalizing constant with units of 10^{-8} cm^3 STP g^{-1} Myr^{-1}, or more specifically, the production rate for average L-chondrite composition (Avg L) and shielding. By definition, "average" shielding means the shielding associated with a cosmogenic ^{22}Ne/^{21}Ne ratio of 1.11. For the case of ordinary chondrites, we have

$$F = \frac{(1.63[\text{Mg}] + 0.6[\text{Al}] + 0.32[\text{Si}])_{\text{sample}}}{(1.63[\text{Mg}] + 0.6[\text{Al}] + 0.32[\text{Si}])_{\text{Avg L}}} \quad (16)$$

and

$$S = \left(\frac{1}{4.494 \frac{^{22}\text{Ne}}{^{21}\text{Ne}} - 3.988} \right) \quad (17)$$

where we normally express elemental concentrations as mass fractions, although mass percentages will do as well. Equation (17) may be used over a range of ^{22}Ne/^{21}Ne ratios extending from ~ 1.08 ($S = 1.16$) to 1.21 ($S = 0.69$), and perhaps for ^{22}Ne/^{21}Ne ratios outside this range, but with lower reliability (Masarik et al., 2001). The normalization constant (average production rate) is given by

$$A = 0.332 \times 10^{-8} \text{ cm}^3 \text{ STP g}^{-1} \text{ Myr}^{-1} \quad (18)$$

Measurements needed. Cosmogenic ^{21}Ne content and ^{22}Ne/^{21}Ne ratio of the sample. Although seldom made, direct measurements of the elemental composition of a sample aliquot are highly desirable.

Range of applicability. Stones and the stony portions of stony-iron meteorites: best for ^{22}Ne/^{21}Ne ratios between 1.09 and 1.18.

Limitations. (i) The presence of large amounts of trapped neon may prevent an accurate determination of the cosmogenic ^{22}Ne/^{21}Ne ratio, and, less frequently, of the ^{21}Ne content itself; (ii) for ratios less than ~ 1.09 (i.e., in the interiors of large meteorites), the ^{22}Ne/^{21}Ne ratio does not determine the ^{21}Ne production rate uniquely (cf. Masarik et al., 2001); (iii) the behavior of the ^{22}Ne/^{21}Ne ratio for samples containing unusually large concentrations of sodium is not known; (iv) reliability is not well established in samples with high ^{22}Ne/^{21}Ne ratios (>1.22); (v) while ^{21}Ne does not usually leak out of meteorites, it may do so in cases of exceptionally intense heating; and (vi) the ^{22}Ne/^{21}Ne ratio depends on the Mg/Si ratio. Corrections for variations in the Mg/Si ratio are possible, but in practice it is usually necessary to recalibrate the production rates (through ^{81}Kr or some other method).

1.13.2.5.2 ^{38}Ar–^{22}Ne/^{21}Ne ages

The age equation is again a specialized form of Equation (1):

$$t_{38} = \frac{^{38}\text{Ar}}{P_{38}} \quad (19)$$

and has been calibrated in the same way as Equation (14) for ^{21}Ne–^{22}Ne/^{21}Ne ages (Eugster, 1988; Eugster and Michel, 1995). The cosmogenic ^{38}Ar production rate has the same form as Equation (15) with

$$F = \frac{(1.58[\text{Ca}]+0.086[\text{Fe}+\text{Ni}]+0.33[\text{Ti}+\text{Cr}+\text{Mn}]+11[\text{K}])_{\text{sample}}}{(1.58[\text{Ca}]+0.086[\text{Fe}+\text{Ni}]+0.33[\text{Ti}+\text{Cr}+\text{Mn}]+11[\text{K}])^{\text{Avg L}}} \quad (20)$$

and

$$S = 2.706 - 1.537 \times \left(\frac{^{22}\text{Ne}}{^{21}\text{Ne}}\right)_{\text{cos mogenic}} \quad (21)$$

Note that for ^{38}Ar the (empirical) shielding correction varies linearly with the cosmogenic ^{22}Ne/^{21}Ne ratio, rather than inversely as for ^{21}Ne. The normalization constant is given by

$$A = 0.0462 \times 10^{-8} \text{ cm}^3 \text{ STP g}^{-1} \text{ Myr}^{-1} \quad (22)$$

Measurements needed. Cosmogenic ^{21}Ne, ^{22}Ne, and ^{38}Ar. Direct measurements of elemental composition are also desirable.

Range of applicability. Best for stony meteorites and the stony phases of stony-iron meteorites.

Limitations. (i) The remarks made above concerning the calculation of cosmogenic ^{22}Ne/^{21}Ne ratios for ^{21}Ne CRE ages apply. While with ^{21}Ne the corrections for trapped gases are usually small, with ^{38}Ar the corrections are often 20% or more. (ii) In the normal deconvolution of spallogenic ^{38}Ar, one assumes that the trapped argon component has a ^{36}Ar/^{38}Ar ratio between 5.32 and 5.36 and that the (high-energy) spallogenic component has a ratio of 0.63–0.65. In some meteorites, however, the argon may comprise a third component, namely, ^{36}Ar produced by (low-energy) neutron capture on ^{35}Cl (Göbel et al., 1982). This effect is restricted mainly to interior portions of large meteorites. Failure to account for any neutron-produced ^{36}Cl leads to an underestimate of the cosmogenic ^{38}Ar content. (iii) Compositional variations are a more serious problem. In stony meteorites, the ^{38}Ar production rate depends sensitively on the concentrations of potassium and calcium, which can vary and are rarely well known.

1.13.2.5.3 ^{3}He ages

Again, the age equation is a specialized form of Equation (1):

$$t_3 = \frac{^{3}\text{He}}{P_3} \quad (23)$$

Eugster (1988) and Eugster and Michel (1995) give formulas for the numerical calculation of P_3 in different kinds of stony meteorites. The approach is similar to the one taken for ^{21}Ne. For ^{3}He, however, we have

$$F = \frac{(2.66 - 0.0096[\text{Ti}+\text{Cr}+\text{Mn}+\text{Fe}+\text{Ni}])_{\text{sample}}}{(2.66 - 0.0096[\text{Ti}+\text{Cr}+\text{Mn}+\text{Fe}+\text{Ni}])^{\text{Avg L}}} \quad (24)$$

and

$$S = 1.00 \quad (25)$$

Here the ^{3}He production rate is assumed to be independent of shielding, a good assumption in some, but perhaps not all cases (Wright et al., 1973). The normalization constant is given by

$$A = 1.61 \times 10^{-8} \text{ cm}^3 \text{ STP g}^{-1} \text{ Myr}^{-1} \quad (26)$$

Range of applicability. All stones and the silicate phases of stony irons.

Measurements needed. When trapped corrections are small as is usually the case, we need only the total measured ^{3}He content to calculate t_3. Within fairly broad limits, the value of P_3 is not very sensitive to elemental composition.

Limitations. (i) Meteorites are susceptible to ^{3}He (and ^{3}H, a precursor of ^{3}He) losses on heating, which probably explains why so many ^{3}He CRE ages are smaller than the corresponding ^{21}Ne ages. Losses of ^{3}He can often, but not always, be identified by plotting ^{3}He/^{21}Ne ratios against ^{22}Ne/^{21}Ne concentrations and seeing whether the meteorite of interest conforms to a well-established general trend (Eberhardt et al., 1966) or falls well below it. The method does not infallibly identify diffusion losses because both complex exposure histories and very heavy shielding can also cause a data point to lie below the trend line (Masarik et al., 2001). It is fair to say, however, that if a sample appears to have a low ^{3}He/^{21}Ne ratio for any reason, the corresponding ^{3}He age should be regarded with caution. (ii) Contrary to the implication of Equation (25), with increasing meteoroid size and sample depth, there must come a point where shielding depresses P_3. We know from direct measurement that ^{3}He concentrations decrease with depth in several large iron meteorites. Alternatively, they decrease little or not at all in the L5 chondrite, Knyahinya (Graf et al., 1990a), which had a pre-atmospheric radius of 40–50 cm. We conclude that shielding corrections for ^{3}He are probably small in most stones but may be appreciable in stony irons and irons.

1.13.2.5.4 $^{36}Cl/^{36}Ar$ ages

A ^{36}Cl/^{36}Ar age may be calculated from Equation (3) with due allowance for the decay of ^{36}Cl into ^{36}Ar (cf. Begemann et al., 1976; Albrecht et al., 2000):

$$\frac{t}{1-e^{-\lambda t}} = (430 \pm 16) \times \frac{^{36}\text{Ar}[10^{-8}\,\text{cm}^3\,\text{STP}\,\text{g}^{-1}]}{^{36}\text{Cl}[\text{dpm}/(\text{kg metal})]} \quad (27)$$

Other authors give values of the constant between 425 and 433 (Lavielle et al., 1999; Terribilini et al., 2000a; Leya et al., 2000). For the few meteorites with very short exposure ages, this equation must be solved iteratively. Most meteoroids, however, orbit in space for times long compared to the half-life of ^{36}Cl, ~300 kyr. Thus, $e^{-\lambda t}$ is negligible for ages >1.5 Myr and we have

$$t(\text{Myr}) = 430\frac{^{36}\text{Ar}}{^{36}\text{Cl}} \quad (28)$$

where ^{36}Ar is in 10^{-8} cm^3 STP (g metal)$^{-1}$, ^{36}Cl in dpm (kg metal)$^{-1}$ and the constant 430 has the units necessary to convert the ratio of measured quantities to Myr.

Measurements needed. ^{36}Ar, ^{38}Ar, and ^{36}Cl in clean metal. A measurement of the ^{36}Ar/^{38}Ar ratio makes it possible to correct for trapped ^{36}Ar.

Range of applicability. Metal or magnetite under all shielding conditions; all falls and those finds for which the terrestrial age is known or known to be short compared to the half-life of ^{36}Cl.

Limitations. (i) As with cosmogenic ^{38}Ar, the deconvolution of the cosmogenic component of ^{36}Ar requires corrections for the presence of trapped ^{36}Ar. (ii) Long terrestrial ages will lower appreciably the measured ^{36}Cl contents. An independent measure of terrestrial age may be necessary to correct for this effect. (iii) The measurements are sensitive to the presence of calcium- and of rare potassium-bearing impurities in the metal phase.

1.13.2.5.5 $^{81}Kr/Kr$ ages

We presented in Section 1.13.2.4.2 the equations needed for calculating ^{81}Kr/Kr CRE ages.

Measurements needed. ^{81}Kr/^{83}Kr and either ^{78}Kr/^{83}Kr or ^{80}Kr/^{83}Kr and ^{82}Kr/^{83}Kr when neutron capture on bromine is negligible.

Range of applicability. Falls and finds for which terrestrial age is known. For exposures long compared to the half-life of ^{81}Kr, ~229 kyr, the term $1 - e^{-\lambda_{81} t}$ reduces to 1 and the age equation becomes easy to solve. For shorter ages the equation must be solved iteratively. Production from strontium dominates and explains why the early measurements of cosmic-ray krypton focused on eucrites and lunar samples, both of which are rich in strontium. In cases where the terrestrial age is not known and the exposure age is long, one may use the measured ^{81}Kr content to obtain a terrestrial age.

Limitations. (i) As with other radionuclide-based ages, the terrestrial age of the sample must be known. (ii) Concentrations of ^{81}Kr are quite low in most meteorites, typically just 5×10^5 atom g^{-1} in chondrites. For this reason, ^{81}Kr measurements are still scarce and their uncertainties can be relatively large, often ~20%. (iii) Production rates for krypton isotopes may vary with the abundances of rubidium, yttrium, and zirconium relative to strontium. It should be understood that the original basis for the calculation of P_{81}/P_{83} was a set of relative cross-section measurements for the production of krypton from silver (Marti, 1967).

1.13.2.5.6 $^{40}K/K$ ages

Cosmic rays produce stable ^{39}K and ^{41}K along with radioactive ^{40}K ($t_{1/2}$ = 1.27 Gyr) in meteoroids. At first glance, it would seem straightforward to write an equation similar to Equation (3) with ^{40}K playing the role of r, and, say ^{41}K playing the role of s. An immediate wrinkle is that virtually all meteoritic materials contain some nonspallogenic (native or primordial) potassium, and perhaps some potassium introduced by terrestrial contamination. Native potassium in the stony portions of meteorites overwhelms any spallogenic component. Even irons and the metal phases of stony irons contain enough native (or contaminant) potassium to make the necessary corrections significant, and especially so for relatively short CREs (i.e., <~100 Myr).

By itself, the need to correct for native potassium would not seem terribly problematic. After all, such corrections are routine in calculating CRE ages with the noble gases, e.g., ^{21}Ne and ^{38}Ar. In the case of potassium, however, an experimental constraint introduces a second wrinkle. Specifically, the mass spectrometry done to date for potassium does not yield absolute concentrations of the isotopes as it does for noble gases, but rather a pair of isotope *ratios*, conventionally ^{39}K/^{40}K and ^{41}K/^{40}K. Without the absolute concentrations of ^{x}K, the separate calculation of the ratio s/r becomes impossible and we must give up on any strict analogue to Equation (3).

Fortunately, as Voshage and co-workers were able to show, the available data do allow the construction algebraically of a mathematical expression that removes the native component and provides a measure of spallogenic potassium production. We begin with the relations below:

$$^{39}K = {}^{39}K_{sp} + {}^{39}K_p \quad (29a)$$

$$^{40}K = {}^{40}K_{sp} \quad (29b)$$

$$^{41}K = {}^{41}K_{sp} + {}^{41}K_p \quad (29c)$$

$$a = {}^{41}K_p/{}^{39}K_p \sim 0.07 \quad (30)$$

The subscripts sp and p denote spallogenic and primordial, respectively. The (primordial) ratio $^{41}K_p/{}^{39}K_p$ is assumed to have the terrestrial value. Note that Equation (29b) builds in two additional assumptions: that irradiation took place in one stage and that $^{40}K_p = 0$. The four equations can be combined to eliminate $^{39}K_p$ and $^{41}K_p$, yielding the result

$$\frac{{}^{41}K_{sp}}{{}^{40}K_{sp}} - a\frac{{}^{39}K_{sp}}{{}^{40}K_{sp}} = \frac{{}^{41}K}{{}^{40}K} - a\frac{{}^{39}K}{{}^{40}K} \quad (31)$$

Voshage refers to the *measurable* quantity on the right-hand side as M. To first order, we have $^{41}K_{sp}/{}^{40}K_{sp} \sim {}^{39}K_{sp}/{}^{40}K_{sp} \sim 1$. Because a (~0.07) is small compared to $^{x}K_{sp}/{}^{40}K_{sp}$ (~1), when the primordial correction is also small, M approaches $^{41}K_{sp}/{}^{40}K_{sp}$, i.e., the ratio s/r in the terminology of Equation (3). By general analogy with Equation (3), it remains to take account of production rates. To do so we write

$$^{39}K_{sp} = P_{39}t \quad (32a)$$

$$^{40}K_{sp} = P_{40}/\lambda_{40} \times (1 - e^{\lambda t}) \quad (32b)$$

$$^{41}K_{sp} = P_{41}t \quad (32c)$$

From these equations we can construct an expression equal to the left-hand side of Equation (31). Skipping the algebraic details, we find

$$\frac{{}^{41}K_{sp}}{{}^{40}K_{sp}} - a\frac{{}^{39}K_{sp}}{{}^{40}K_{sp}}$$
$$= \left(\frac{P_{41}}{P_{40}} - a\frac{P_{39}}{P_{40}}\right)\frac{\lambda t}{1 - e^{-\lambda t}} \quad (33)$$

where λ is the decay constant of ^{40}K, taken as 0.546 Gyr^{-1} by Voshage and co-workers. Voshage refers to the terms in parentheses on the right-hand side of Equation (33) as N. Finally, combining Equations (31) and (33), we obtain the equation for $^{40}K/K$ ages

$$\frac{{}^{41}K}{{}^{40}K} - a\frac{{}^{39}K}{{}^{40}K} = \left(\frac{P_{41}}{P_{40}} - a\frac{P_{39}}{P_{40}}\right)\frac{\lambda t}{1 - e^{-\lambda t}} \quad (34)$$

When the primordial correction is small, N approaches the production rate ratio P_{41}/P_{40} to within 10% or so.

The evaluation of N presents a last major hurdle. With the half-life of ^{40}K so long, with exposure ages of meteorites too short for ^{40}K to saturate (see below), and with the absolute concentrations of ^{40}K unknown, we cannot get directly at ^{40}K production rates from the available data as we *can* do for shorter lived nuclides such as ^{26}Al and ^{81}Kr. At the same time, the lack of measured cross-sections for the relevant reactions (e.g., Fe(p, X)^{40}K)—a lack that persists to this day—poses a problem for obtaining the production rates from modeling. To make matters still more complicated, N varies with irradiation hardness or shielding, so that it is not enough to evaluate N once and apply that result to numerous meteorites, as is done for $^{36}Cl/^{36}Ar$ dating.

Space does not permit a detailed discussion of the two basic methods used to address these difficulties. In brief, both of them start with measurements of $^{4}He/^{21}Ne$ ratios, which provide a measure of the irradiation hardness in metal (but not in stony phases). One method then relies on modeling calculations. Voshage and Hintenberger (1963) used those of Arnold et al. (1961), first to find the conditions needed to reproduce the observed spallogenic $^{4}He/^{21}Ne$ ratios, and then to obtain production rates for the potassium isotopes from *estimated* cross-sections. As it turned out, the calculated values of N correlate linearly with the $^{4}He/^{21}Ne$ ratios, thereby providing a basis for calculating N for any other iron with a known $^{4}He/^{21}Ne$ ratio. The values of N found in this way range from ~1.45 to ~1.57.

In discussing the second method, we pass over the work of Voshage and Hintenberger (1963) in favor of a more recent treatment by Lavielle et al. (1999). These authors made three assumptions: first, $^{40}K/K$ CRE ages should closely approximate $^{36}Cl/^{36}Ar$ CRE ages, at least for the last 500 Myr; second, $^{36}Cl/^{36}Ar$ CRE ages of irons can be written in the form $t_{36} = k\,{}^{36}Ar/{}^{36}Cl$, i.e., as in Equation (28), but with the constant k allowed temporarily to float; and third, N can be written in the form

$$N = a_1\left[1 + a_2\left(\exp\left\{a_3 \times \frac{{}^{4}He}{{}^{21}Ne}\right\}\right)\right] \quad (35)$$

They then assembled measured potassium isotope ratios, $^{36}Cl/^{36}Ar$ ratios, and $^{4}He/^{21}Ne$ ratios for selected iron meteorites and calculated the values of the parameters a_n and k that made the $^{40}K/K$

and ^{36}Cl/^{36}Ar CRE ages agree best. This method ultimately recovers to within a few percent the values of N estimated by Voshage and Hintenberger (1963). It also gives the result $k = 590$ (measurement units) that differs substantially from 430, the value shown in Equation (28). We will return to this point later.

Measurements needed. ^{39}K/^{40}K, ^{41}K/^{40}K, and ^{4}He/^{21}Ne ratios.

Range of applicability. Irons and the metal phases of other meteorites with CRE ages greater than 100 Myr and preferably greater than 200 Myr.

Limitations. The ^{40}K/K method of calculating CRE ages works poorly for ages less than 100 Myr because of the need for large corrections for native potassium. The absolute calibration of the method is indirect and rests on educated guesses and assumptions. Nickel content may influence the results but is not normally considered explicitly.

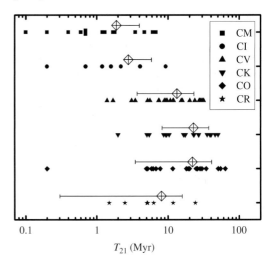

Figure 1 ^{21}Ne exposure ages (Myr) of carbonaceous chondrites. Large symbols show group averages.

1.13.2.6 The Importance of Half-lives

The half-lives (decay constants) that appear in many of the equations used to calculate CRE ages undergo revision from time to time. For example, at this writing (April, 2003), the half-lives of ^{10}Be and of ^{53}Mn are under active scrutiny. In all cases, the absolute production rates depend inversely on the half-life of the relevant radionuclide of interest. Thus, an increase of 10% in the half-life of ^{81}Kr would decrease by 10% any value of P_s based on ^{81}Kr measurements. It follows that all CRE ages would then shift upward by 10%. While such changes may be important for understanding the dynamics of meteoroid delivery to Earth, they do not affect the relative values of the CRE ages, and hence the characteristic shapes of CRE age distributions.

1.13.3 CARBONACEOUS CHONDRITES

1.13.3.1 CI, CM, CO, CV, and CK Chondrites

Mazor *et al.* (1970) carried out the first extensive set of CRE ages for carbonaceous chondrites, based mainly on ^{21}Ne. Figure 1 presents ^{21}Ne exposure ages for carbonaceous chondrites calculated from the compilation of (Schultz and Franke, 2002), assuming that the neon was in each case a mixture of solar and cosmogenic gas. Some cautions are in order for CI and CM chondrites in particular.

- CI and CM chondrites often have large concentrations of interfering noncosmogenic components. The process of stripping away these contributions to obtain cosmogenic ^{21}Ne adds to the uncertainty of the result.

- The presence of noncosmogenic components also undermines the one-stage irradiation scenario.
- The measurement uncertainties of the cosmogenic ^{22}Ne/^{21}Ne ratio, which is a durable indicator of shielding in ordinary chondrites, are increased in making corrections for noncosmogenic components.
- The CI and CM chondrites retain some cosmogenic noble gases from early, regolith irradiations.

For these reasons, we expect CRE ages of CI and CM meteorites to have a precision no better than ~ 30–40%. Carbonaceous chondrites belonging to the other classes generally have lower concentrations of noncosmogenic gas and may give better precision, perhaps 10–15%. Inspection of Figure 1 reveals several interesting features.

The CI and CM chondrites have unusually short CRE ages, many of them less than 1 Myr. The group averages are 1.8 ± 2.1 Myr and 2.8 ± 3.1 Myr, respectively. Except for lunar meteorites, no other group of meteorites has such a high proportion (1/4–1/2) of short-lived objects. Scherer and Schultz (2000) list three possible reasons why meteorites may have short exposure ages: the parent body orbited close to a resonance; the parent body was in an Earth crossing orbit when a collision released the meteoroid; and the meteoroids are so fragile that collisions destroy them if they fail to reach Earth quickly.

For CI and CM chondrites, cosmogenic radionuclides provide an excellent alternative way to calculate CRE ages. Based on ^{26}Al and ^{10}Be measurements, Caffee and Nishiizumi (1997) present a preliminary age distribution for selected C2 carbonaceous chondrites, which are, for the most part, CM chondrites. Remarkably, they find a strong peak at 0.2 Myr, a peak absent from the distribution of ^{21}Ne ages. Perhaps this difference merely reflects the large uncertainties of the noble gas CRE ages. It seems equally likely, however, that many of the CI and CM chondrites retain significant fractions of ^{21}Ne from earlier irradiations. Within the CM2 chondrites (e.g., Murchison, Murray, and Cold Bokkeveld), the concentrations of ^{21}Ne in individual "irradiated" grains (grains with track densities that show previous exposure in a regolith) vary by factors of 10–30 (Hohenberg et al., 1990). Even within the "unirradiated" grains (grains with low track densities, which presumably had minimal previous exposure in a regolith), ^{21}Ne ages vary by a factor of 2, from 1.4 Myr to 2.8 Myr.

The average ^{21}Ne ages for CV, CK, and CO chondrites, not corrected for pairing, are 13 ± 10 Myr, 23 ± 14 Myr, and 22 ± 18 Myr, respectively. These averages exceed the CRE ages of the CI and CM chondrites by a factor of ~10. The data hint but do not clearly establish that the CV chondrites have shorter CRE ages than the CK and CO chondrites. Goswami et al. (2001) have shown that at least one of the CK meteorites, Kobe, had a complex exposure history, for which the most recent stage lasted only ~1 Myr.

1.13.3.2 The CR Clan

The CR clan comprises ~20 meteorites subgrouped as CR, CH, or CB meteorites. Noble gas analyses have been reported for five of them: Acfer 094 (paired with El Djouf 001), Al Rais, GRA 95229, Loongana, and the type specimen, Renazzo (Scherer and Schultz, 2000; Mazor et al., 1970). Because of the presence of trapped gases, for only one of these meteorites, Loongana, can the noncosmogenic ^{22}Ne/^{21}Ne ratio be determined with sufficient precision to make a reliable shielding correction. Accordingly we have calculated the ^{21}Ne ages shown in Figure 1 by assuming average shielding, average CR composition (Lodders and Fegley, 1998), and solar neon isotope ratios. The results have large uncertainties—30% or so—in part because the isotopic composition of the trapped component is not entirely certain. Nishiizumi et al. (1996b) present ^{10}Be, ^{26}Al, and ^{36}Cl activities for the CR chondrites (EET 87770, MAC 87320, and PCA 91082). PCA 91082 has an unusually low ^{10}Be content, only 10.9 dpm kg^{-1}, which is consistent with an age of 1.5 Myr. The activities for the other meteorites appear to be saturated and give only lower bounds on the exposure ages. At this early stage, the age distribution of CR chondrites seems to be most like that of the CV chondrites both in terms of average age, ~8 Myr, and range, from 1 Myr to 25 Myr.

Table 3 CRE ages (Myr) of CH chondrites.

Meteorite	T_{Gas}	T_{Radio}[a]
Acfer 182 (+207 + 214)	12[b]	
ALH 85085	1.7 ± 0.8[c]	0.9
PAT 91546	>8[d]	>3.5
PCA 91328 (+452 + 467)	4.3[d]	1.1
RKP 92435	1.5[d]	0.6

[a] Nishiizumi et al. (1996a). [b] Bischoff et al. (1993).
[c] Eugster and Niedermann (1990). [d] Weber et al. (2001).

Noble gas analyses and CRE ages have been reported for all but one (NWA 470) of the six known CH chondrites: the CRE ages are shown in Table 3. Weber et al. (2001) note apparent systematic differences between CRE ages based on noble gases, T_{Gas}, or on cosmogenic radionuclides, T_{Radio}, and suggest that multiple stages of exposure may explain the differences. At first glance, it appears that the CRE ages of CH chondrites seem most similar to those of the CR and CV chondrites. This group needs more work.

Five bencubbinites (CB chondrites) are known—Bencubbin, Gujba, Weatherford, Hammada al Hamra 237, and QUE 94411 (paired with QUE 94627)—three of them found since 1994. Begemann et al. (1976) obtained a ^{36}Cl/^{36}Ar age of 36 ± 5 Myr for Bencubbin after making a substantial correction for trapped argon. From the noble gas contents of Gujba silicates, Rubin et al. (2003) calculate a CRE age of 26 ± 7 Myr with the unusually large uncertainty reflecting a lack of a compositional analysis and an unusually low ^{22}Ne/^{21}Ne ratio of ~1.05. An earlier noble gas analysis of Weatherford by Stauffer (1962) is difficult to interpret because of the lack of compositional information. In sum, the state of knowledge about CRE ages for this interesting group of meteorites is inadequate.

1.13.4 H-CHONDRITES

Among ordinary chondrites, ~50% (well over seven *thousand* stones) are classified as H-chondrites and 3/4 of those belong to just two petrologic groups, H5 or H6. Graf and Marti (1995) have reviewed noble gas data for more than 400 H-chondrites, extracted CRE ages from those data, and examined their distributions. The CRE ages rest ultimately on the production rates of Eugster (1988), which is to say on an absolute calibration based on the ^{81}Kr/Kr ages of ~20 ordinary chondrites. Graf et al. (2001) subsequently presented ^{36}Cl/^{36}Ar ages for 16 H-chondrites. Below we summarize some conclusions from these articles.

The CRE age distribution for all H-chondrites spans an apparent range from less than 1 Myr to

~80 Myr (Figure 2). For stones at the low end of the distribution, the possibility of complex irradiation looms large, and indeed, that possibility has been confirmed in several cases (Herzog et al., 1997). Graf and Marti (1995) conclude that "it appears there are very few H-chondrites with short exposure ages." As noted by Wieler (2002), this observation has two implications—first that collisions producing H-chondrites are rare events, occurring with a frequency of no more than a few per million years; and second, that the collisions take place so far away that the fragments take several million years to get to Earth.

An enormous peak between about 6 Myr and 10 Myr in the CRE age distribution encompasses nearly half the H-chondrites. The H5 and H4-chondrites populate the peak more heavily, both in an absolute and a relative sense, than do the H3 or H6 chondrites. Interestingly, the position of the maxima may differ slightly among the groups. The effect is not statistically significant at present, but, if verified, would signal two events close in time, one of them leading to the production of H5 meteorites and a second one to other petrologic types of H-chondrites. Graf and Marti (1995) also note other distinguishing features of the H5 chondrites. One subset has lower ^{3}He/^{38}Ar ratios, which may reflect greater heating in orbit, and a different distribution of fall times (see also Wieler and Graf (2001)).

A second peak in the CRE age distribution for all H-chondrites crops up at ~33 Myr and seems fairly robust, but the uncertainties are appreciable. The distribution for H6 chondrites has a unique peak at ~24 Myr.

What do these ages tell us about the history of the H-chondrites? First, as has been known for some time, the age distribution is not consistent with continuous delivery, but rather is dominated by a very small number of events, consistent with a correspondingly small number of parent bodies. Second, the existence of a peak at 33 Myr common to H-chondrites of all petrologic types suggests that most types of H-chondrites are present in at least some asteroids. As Graf and Marti (1995) note, material that was at the *surface* of the parent body for some period of time (as indicated by the presence of trapped noble gases) occurs about as often in one petrologic type as another. Thus, at the time of the major collision at 6–10 Myr, the source of H-chondrites was *not* an ancient layered object with, say, matter of higher petrographic type sheltered in the interior. In this context, it would be interesting to re-examine the meteorites with an age of 33 Myr to see whether trapped gas contents correlate with petrographic type.

1.13.5 L-CHONDRITES

Numbering well over 6,000, the L-chondrites are dominated by two petrographic types, the L5 (20%) and L6 (65%). Marti and Graf (1992) have reviewed L-chondrite exposure ages calculated from the light noble gases. The collective CRE age distribution looks at first to have an exponential envelope that rises from zero to a peak at or near 40 Myr with a fairly sharp decrease beyond (Figure 3). On closer examination, the CRE age distributions of the L5 and L6 chondrites appear to be neither completely monotonic nor the same for all petrologic types. The strongest peak occurs at ~40 Myr. With a little imagination one can make out a peak at 5 Myr and a broad hump running from

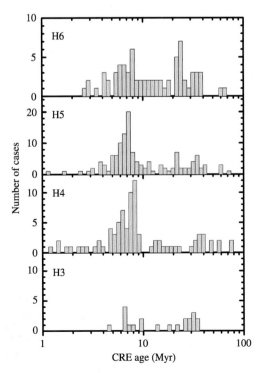

Figure 2 Exposure ages (Myr) of H chondrites (source Graf and Marti, 1995).

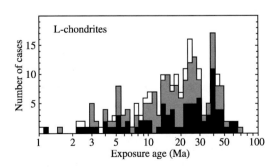

Figure 3 Exposure ages of L-chondrites. Different fills indicate varying degrees of precision (source Marti and Graf, 1992).

20 Myr to 30 Myr with a maximum somewhere in the vicinity of 28 Myr. At present we do not seem to have any L3 or L4 chondrites with exposure ages of 50 Myr or more.

A handful of L-chondrites have the low ^{21}Ne contents, $< \sim 0.4 \times 10^{-8}$ cm^3 STP g^{-1} associated with an exposure age of 1 Myr or less, namely, Farmington (L5), Pampa (L4), Shaw (L6-7), and Ladder Creek. Herzog et al. (1997) concluded that Ladder Creek had a complex history, but found no evidence for one in Shaw. Marti and Matthew (2002) conclude that Farmington, too, had a simple exposure history in the sense that the material was deeply shielded until ~25 kyr before the present.

The authors of many early studies of L-chondrites searched for trends relating CRE ages to losses of the radiogenic gases ^{40}Ar and ^{4}He. Interestingly, among L-chondrites with significant losses of ^{40}Ar, the peaks at 5 Myr and 28 Myr noted above emerge much more clearly (Marti and Graf, 1992) than they do in Figure 3. That we have samples from (at least) two events in heated portions of a parent body seems clear; it is not obvious whether the events took place on the same or different parent bodies.

1.13.6 LL CHONDRITES

Graf and Marti (1994) compiled light-noble-gas exposure ages for ~60 LL chondrites. We have recalculated ^{21}Ne ages for 77 LL chondrites by using the formula of Eugster (1988) without culling the data and simply averaging where analyses for more than one sample were available (Figure 4). The figure includes one LL chondrite with a short exposure age, Hunter, which was not in the data set of Graf and Marti (1994). In general, Figure 4 is similar to figure 1 of Graf and Marti (1994). The distribution runs from 0.5 Myr for Hunter to just under 80 Myr for Soko Banja. The strongest peak in the distribution, with 20% of the L-chondrites, occurs at ~15 Myr. Whether different petrologic types of LL chondrite populate the peak to different degrees seems to us debatable given the small numbers of meteorites in the LL3 category. In our larger data set, we have two meteorites, Richfield (LL 3.7) and Acfer 160 (LL3.8-6), with ^{21}Ne ages close to 15 Myr. A broad peak spreads from 27 Myr to 33 Myr, overlapping the 28 Myr peak in the L-chondrite CRE age distribution.

1.13.7 E-CHONDRITES

Patzer and Schultz (2001) calculated ^{21}Ne exposure ages of ~60 E-chondrites with shielding corrections based on ^{22}Ne/^{21}Ne ratios (Eugster, 1988) and adjustments for composition based on average compositions of one of the following groups: EH3; EH4,5; EL3; and EL5,6 (Kong et al., 1997). In shape, the distribution of ^{21}Ne ages (Figure 5) resembles most closely that of the L-chondrites, although it has relatively more members with short CRE ages and fewer with ages greater than 40 Myr. Similarities to the age distribution of the enstatite achondrites or aubrites (Figure 6) are limited.

Patzer and Schultz (2001) identified possible clusters of ages at about 3.5 Myr, 8 Myr, and 25 Myr with a well-advised caution that confirmation is needed. While the experimental data are not in doubt, the bellwether ^{21}Ne CRE ages do not agree particularly well with either ^{3}He or ^{38}Ar CRE ages, the typical spread being ~25%, with many larger excursions. Reasons for the disagreements may

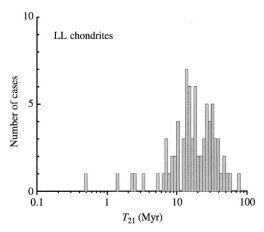

Figure 4 ^{21}Ne exposure ages of LL chondrites. ^{21}Ne ages recalculated with the formulas of Eugster (1988). Multiple analyses for meteorites were averaged. Meteorite finds from Antarctica and northern Africa with similar exposure ages and collected at the same site were treated as paired (source L. Schultz).

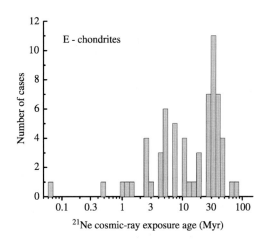

Figure 5 Cosmic-ray exposure ages of E-chondrites (sources Patzer and Schultz, 2001; Patzer et al., 2001; Okazaki et al., 2000).

include weathering, particularly of calcium-bearing sulfides, target element variations in the case of ^{38}Ar, and diffusion losses for ^{3}He.

1.13.8 R-CHONDRITES

Figure 7 shows CRE ages for a relatively new group of ordinary chondrites, the R-chondrites, named for the type specimen, Rumuruti. In tabulating ages, we have either quoted directly from the literature or, where possible, recalculated ^{21}Ne ages using the formalism of Eugster (1988) and the average compositional data of Lodders and Fegley (1998). The presence of large concentrations of trapped noble gases limits the accuracy of the shielding corrections in several cases. The CRE ages range from 0.2 Myr for Northwest Africa 053 to ~50 Myr for Hughes 030. Five of the 21 R-chondrites in Figure 7 have CRE ages close to 7.5 Myr: Carlisle Lakes, 6.9 ± 1.0; Northwest Africa 755, 8.0; Ouzina, 8.0; Y79357, 8.3 and 8.6 (shown separately in Figure 7); and Daral Gani 013, 9.5 ± 1.4. Although it would be premature to identify these five meteorites with a cluster, we cannot resist pointing out the coincidence with the major peaks in the respective CRE age distributions of the H-chondrites and of the acapulcoite/lodranite group (see below).

1.13.9 LODRANITES AND ACAPULCOITES

Several lines of evidence point to a common parent body for lodranites and acapulcoites: overlapping oxygen isotope composition, identical mineral constituents, complex thermal histories, and, as we will see, a peaked distribution of CRE ages (Clayton and Mayeda, 1996; McCoy et al., 1997). Terribilini et al. (2000a), Weigel et al. (1999), and McCoy et al. (1996) report analyses of noble gases in lodranites and acapulcoites and give references to earlier work. Xue et al. (1994) and Terribilini et al. (2000a) report analyses of cosmogenic radionuclides. The CRE ages of lodranites and acapulcoites fall in a fairly narrow range running from ~4 Myr to ~10 Myr (Figure 8). Terribilini et al. (2000a) note that H-chondrites also have a peak in this region. They do not suggest a common parent body for the two groups of meteorites but raise the possibility of unusually high collisional activity in the asteroid belt, due perhaps to the passage through the asteroid belt of a third party of impactors.

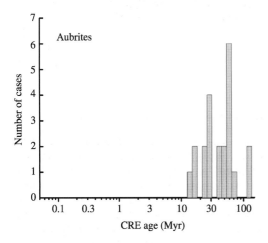

Figure 6 CRE ages of aubrites (for references, see Section 1.13.14).

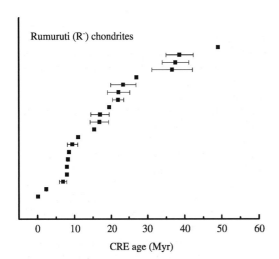

Figure 7 Cosmic-ray exposure ages of R-chondrites, based on noble gas contents (source Weber and Schultz (2001) and references therein).

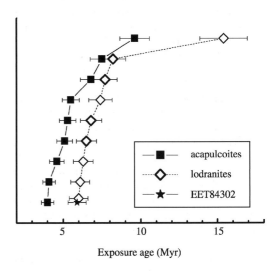

Figure 8 Top panel: CRE ages of lodranites Gibson, Y791491, LEW 88280, FRO 90011, GRA 95209, Y74357, Lodran, MAC 88177, and QUE93148. Bottom panel: acapulcoites ALH 77081, ALH 81261, FRO 95029, GRA 98028, ALH 81187, ALH 84190, Monument Draw, Acapulco, and Y74063 and of EET 84302.

The lodranites and acapulcoites present two kinds of problems for the calculation of CRE ages. First, the lodranites are coarse grained and many of them contain a large fraction of metal. Special attention to elemental composition is, therefore, desirable in interpreting the data. Second, as it turns out, many lodranites and acapulcoites have exceptionally high cosmogenic ^{22}Ne/^{21}Ne ratios, which normally indicate near-surface locations, probably in small meteoroids, where solar cosmic rays may have left their mark on the samples. The quantitation of effects due to solar cosmic rays introduces another level of difficulty in the calculation of CRE ages. The ages shown in Figure 8 were calculated in various ways: mostly from ^{21}Ne and ^{22}Ne/^{21}Ne ratios, but with a few ^{36}Cl/^{36}Ar ages, one ^{81}Kr/^{83}Kr age, and model-dependent estimates of cosmogenic nuclide contributions due to solar cosmic rays.

1.13.10 LUNAR METEORITES

1.13.10.1 Overview

We summarize results pertaining to the CRE histories of lunar meteorites in Table 4. The meanings of the column headers are discussed in the table notes. Eugster (1989) and Warren (1994, 2001) have written specialized reviews dealing with questions related to the exposure histories of lunar meteorites. Some general observations and conclusions from this work follow.

- Most lunar meteorites travel to Earth in less than 1 Myr and some in less than 0.1 Myr; other kinds of meteorites generally take longer. As the Moon is close by, the short transit times are intuitively appealing. They also fall naturally out of theoretical calculations, as does the expectation that a small fraction of the lunar meteorites will have longer exposure ages (cf. Gladman et al., 1995).
- Most lunar meteorites contain detectable concentrations of cosmogenic nuclides attributable to irradiation within a few meters of the lunar surface. This observation sets lunar meteorites apart from the general run of meteorites in which the signs of parent-body irradiation are rarer and generally harder to detect. The near-surface irradiations prior to launch also hint at the importance of "smaller" impact events in launching lunar meteorites. We will return to the meaning of "smaller" below.
- Many lunar meteorites are regolith breccias, meaning that over time at least some of their constituent grains resided not only *below* the surface waiting for launch as just described, but also *at* the very surface, where the grains could collect solar wind ions, and presumably at other depths as well. Such grains may have extremely complex exposure histories. For this reason, if considered alone, the concentrations of stable cosmogenic nuclides in lunar meteorites can seldom be interpreted uniquely. For lunar meteorites especially, the combination of radioactive and stable cosmogenic nuclides is important. Even with both kinds of measurements, however, the reconstruction of a grain-by-grain irradiation history is probably beyond reach. In any case, if detailed exposure histories on the Moon are of interest, it is probably simpler to study nonregolithic rocks that astronauts have brought back from the Moon.
- Most lunar meteorites are breccias with low porosity. Warren (2001) notes that low-porosity breccias constitute a larger fraction of lunar meteorites than they do of the lunar surface as a whole. The low porosity of lunar meteorites seems to correlate with high mechanical strength and by implication, with resistance to collisional destruction. By extension, this observation points strongly to the importance of the mechanical properties of the target rock in determining the kinds of material that we sample from asteroids and Mars.

1.13.10.2 Construction of CRE Histories

Sorting out the complex irradiation histories of lunar meteorite requires some art. As a first step, we usually assume that irradiation on the Moon has left no measurable traces and try to explain the observations with a simple, one-stage or 4π irradiation history in space. Here the modifier 4π refers to the assumption, almost always true, that the meteoroid was small enough in space for cosmic rays to reach interior material from all solid angles. In the idealized world of spherical meteoroids, a one-stage history has four important parameters: the size of the meteoroid, the depth of the sample within it, the duration of the irradiation in space, and the terrestrial age. In principle, during transit to Earth, *solar* cosmic rays could also affect matter within a few millimeters of the surface. Ablation losses usually remove such material, but in rare instances the surface material survives. When the one-stage irradiation model fails to explain the results for a lunar meteorite, we precede it with an earlier stage of irradiation on the Moon. In the approximation that the local topography of the lunar surface did not much affect the galactic cosmic ray (GCR) flux and that erosion of the lunar surface was negligible, the earlier stage introduces two more parameters—the depth of the sample while on the Moon and the duration of the irradiation there. Neither of these approximations probably holds in reality. Thus, the inferred conditions of early exposure will generally have large uncertainties.

Table 4 Exposure histories of lunar meteorites.

Meteorite	Pair	Mass (g)	$D_{2\pi}$ (g cm^{-2})	$T_{2\pi}$ (Myr)	$R_{4\pi}$ (g cm^{-2})	$T_{4\pi}$ (Myr)	T_{Terr} (kyr)	Notes	Refs.
Cumulate olivine norite with regolith breccia									
NWA 773		633		160					a
Feldspathic fragmental breccia									
Dhofar 081		174		100					a
Dhofar 081		174	190–210	>4		Short	200	(i)	b
Dhofar 280	081	251							
Dhofar 489		34.4							
Feldspathic fragmental/regolith breccia									
Yamato 82193		712	>1,000			9	~90		c
82192	86032								
Feldspathic impact melt breccia									
NWA 482		1,015			15–19	0.9 ± 0.2	60–120	(ii)	d
Dhofar 026		148	1,100–1,300	10		<~0.003			d
Dhofar 026		148				~0			d
Dhofar 026		148				<0.01			e
Dhofar 301		9							
Dhofar 302	025?	3.83							
Dhofar 303		4.15							
Feldspathic regolith breccia									
Dar al Gani 400		1,425		<3		<1			f
"1153" alleged									z
ALH 81005		31.4	164			0.0025	9		c
Dar al Gani 262		513	50	Long	>3	0.5	300		g
Dar al Gani 262		513	75–85	Long		Short	50–60		g
Dar al Gani 262		513	55–85		=10	<0.15			h
Dar al Gani 262		513	50–80	500–1000					i
Dhofar 025	025	751				4–20	500–600		j
MAC 88104	88105	724	370			~0.04	210–250		c
MAC 88104	88105	724	85	630 ± 200		<0.24	100–600	(iii)	k
MAC 88104	88105	724	360–400	>5		0.04–0.05	210–250		l

Meteorite							Ref
MAC 88104	88105	724	390–500	>5	0.04–0.11	100–190	m
QUE 93069	94269	24.5	~90	1,000 ± 400	0.15 ± 0.02	<15	n
QUE 93069	94269	24.5	65–90	>500	~0.02–0.05	5–10	o
Y 791197		52.4	5			30–90	c
Y 791197		52.4			<0.019		p
Y 983885		289		450			
Feldspathic/mare regolith breccia							
Calcalong Creek		19	40–50	>300			q
QUE 94281		23	270–320		<0.2	<70	r
QUE 94281		23		400 ± 60		150–200	s
Y 79374/981031		194.7	165				c
Y 79374/981031		194.7	40	510 ± 140	<0.02	<20	t
Y 79374/981031		194.7	150–190				u
Y 79374/981031		194.7	35 ± 15	700 ± 200	<0.12		v
Mare basalt							
NWA 032/479		~456	>1100		0.042	<80	b
Y 793169		6.1	Deep	<12	1.1 ± 0.2	<50	w
Y 793169		6.1	500				n
Dhofar 287		154		50 ± 10			
Asuka 881757		442	>3 m		0.9	<50	c
Mare polymict breccia							
EET 87521	96008	84	565	26	<0.1	15–50	x
EET 87521		84	540–600		<0.01	80 ± 30	y
EET 96008	87521	84	200–600	26			i
				(iv)			

Notes: The petrographic discriptions are from http://epsc.wustl.edu/admin/resources/meteorites/moon_meteorites_list.html
Pair refers paired meteorites from the same locality. Mass is the recovered mass. $D_{2\pi}$ is the depth at which irradiation on the Moon took place. $T_{2\pi}$ is the duration of the lunar irradiation. $R_{4\pi}$ is the radius of the meteoroid while in transit to Earth. $T_{4\pi}$ is the duration of transit to Earth. T_{Terr} is the terrestrial age. (i) Greshake et al. (2001) note similarities to MAC 88104/5. (ii) Assume density 2.7 g cm^3. (iii) $T_{2\pi}$ before compaction. (iv) Full model has three stages on Moon. References: (a) Eugster and Lorenzetti (2001). (b) Nishiizumi and Caffee (2001a). (c) Warren (1994). (d) Nishiizumi and Caffee (2001b). (e) Shukolyukov et al. (2001). (f) Scherer et al. (1998). (g) Nishiizumi et al. (1998). (h) Bischoff et al. (1998). (i) Eugster et al. (2000). (j) Nishiizumi and Caffee (2001b). (k) Eugster et al. (1991b). (l) Nishiizumi et al. (1991a). (m) Vogt et al. (1991). (n) Thalmann et al. (1996). (o) Nishiizumi et al. (1996b). (p) Ostertag et al. (1986). (q) Swindle et al. (1995). (r) Nishiizumi and Caffee (1996b). (s) Polnau and Eugster (1998). (t) Takaoka and Yoshida (1992). (u) Nishiizumi et al. (1991b). (v) Eugster et al. (1992). (w) Nishiizumi et al. (1992a). (x) Vogt et al. (1993). (y) Nishiizumi et al. (1999). (z) Yanai (2000).

As all known lunar meteorites are finds (and therefore have nonzero terrestrial ages), we need at least four measured quantities to determine the four parameters of a simple one-stage history. Similarly, for a simple two-stage history, we need at least six measured quantities. Typically the data set available comprises ^{3}He, ^{21}Ne, ^{22}Ne, ^{38}Ar, ^{36}Cl, ^{26}Al, and ^{10}Be. Occasionally we may have other information—the concentrations of spallogenic krypton isotopes, spallogenic xenon isotopes, ^{14}C, ^{41}Ca, and ^{53}Mn, the densities of nuclear tracks (tracks/unit area), and the concentrations of certain isotopes produced by thermal neutrons, e.g., ^{36}Ar (from ^{36}Cl) and ^{158}Gd.

In practice, the important information about terrestrial age derives mainly from ^{14}C, ^{36}Cl, or ^{81}Kr. Chlorine-36, although suboptimal for terrestrial ages less than 100 kyr, is the most widely used of the three and the potential of ^{41}Ca has yet to be realized. In the determination of the other parameters of the irradiation histories, the possibility of diffusion loss compromises ^{3}He with the consequence that it is respected when it confirms other results but rejected when it does not. As noted above (Section 1.13.2.5.2), the calculation of ^{38}Ar CRE ages leaves room for improvement and for this reason results based on ^{38}Ar may generate skepticism. Commonly, then, we end up with five pieces of information on which to build the exposure histories of lunar meteorites: ^{21}Ne, ^{22}Ne, ^{36}Cl, ^{26}Al, and ^{10}Be, and the hope, sometimes realized, that the elemental composition is known for the particular sample analyzed for cosmogenic nuclides. The importance of spallogenic ^{22}Ne arises mainly through the spallogenic ^{22}Ne/^{21}Ne ratio, which as noted above provides some measure of shielding conditions. In lunar samples, unfortunately, the presence of large concentrations of trapped (solar) gases increases the uncertainties of the calculated spallogenic ^{22}Ne contents. Problems aside, armed with this information, we can proceed by an iterative process to find the set of parameters that best matches the data, provided that we also have adequate knowledge of production rate systematics for the cosmogenic nuclides.

Space does not allow us to discuss production rate systematics in detail. In brief, several groups have studied this issue theoretically, with modeling calculations, two of them in ways that have proved accessible and useful for meteoriticists (cf. Leya et al., 2000; Masarik and Reedy, 1994). The calculated production rates of spallogenic nuclides are probably good to within ~15%. Both sets of calculations incorporate a parameter that, in effect, specifies the intensity of the GCR flux. This flux is smaller by perhaps 3–5% at 1 AU than at 2–5 AU (Reedy et al., 1993; Reedy and Masarik, 1994; Michel et al., 1996). Additional important information used to calibrate the production rates comes from another source: the empirical studies of depth profiles in meteorites and lunar samples. In daily practice, researchers are likely to rely on a combination of results from modeling calculations and from measured depth profiles.

1.13.10.3 Production Rate of Lunar Meteorites

Vogt et al. (1991) quote J. Melosh to the effect that an impactor striking the Moon must have a radius greater than 10 m in order to accelerate rocks to escape velocity. The cumulative rate of influx onto the Moon's surface for impactors of $R \geq 10$ m is ~ 30 Ma^{-1} (Melosh, 1989, p. 189). As Table 4 shows, of 18 ± 6 meteorites thought to have left the Moon independently and for which we have estimates of the transit time to Earth, 16 have transit times $(T_{4\pi}) < \sim 1$ Myr. If a similar proportion holds for the lunar meteorites not yet analyzed, then the arrival rate of lunar meteorites will have bumped up against the estimated production rate of 30 Ma^{-1} even though we have sampled effectively only a tiny fraction of the Earth's surface. Vogt et al. (1991) inferred as much on the basis of a much smaller sample set. They, therefore, suggested that the impact events on the Moon must typically launch not one or even a few meteoroids, but large numbers of them. The alternative explanation is that smaller impactors, for which fluxes are higher, also produce lunar meteorites. The common occurrence of lunar meteorites with signs of *lunar* irradiation would seem to point in this direction in the sense that smaller impacts probably excavate material from shallower depths.

1.13.11 HOWARDITE–EUCRITE–DIOGENITE (HED) METEORITES

Just two major impact events, one at ~20 Myr and the other at ~40 Myr, on one body could account for the great majority of the eucrites, howardites, and diogenites reaching Earth today. One or more events at earlier times also seem likely.

1.13.11.1 Eucrites

Among the various types of exposure ages, ^{81}Kr/Kr CRE ages appear to be the most reliable for eucrites (plagioclase–pigeonite achondrites). Eucrites contain strontium (~80 ppm), yttrium (~15 ppm), and zirconium (20–90 ppm) at levels considerably higher than those that occur in ordinary chondrites. The high concentrations of these three elements enhance the production of spallogenic krypton and so facilitate the determination of ^{81}Kr/Kr exposure ages. Argon-38 CRE ages may also be trustworthy when adjusted for shielding and sample composition (Aylmer et al., 1988). Most eucrites have ~7 wt.% of calcium, a prime target for production of ^{38}Ar. Unfortunately, disappointingly few reliable CRE ages can be calculated from well over 200(!) analyses of cosmogenic ^{3}He and ^{21}Ne in eucrites. Difficulties arise partly because of diffusion losses from feldspar (Megrue, 1966; Heymann et al., 1968). Feldspar typically constitutes 30–40% of eucrites (Kitts and Lodders, 1998), and any small, randomly chosen sample may contain more or less of that mineral. Further, low magnesium contents make the ^{22}Ne/^{21}Ne shielding monitor nearly useless.

Figure 9 summarizes the ^{81}Kr/Kr ages of 19 of the ~200 known eucrites (The 1977 *Appendix to the Catalog of Meteorites* lists only 31 eucrites!). All 19 of them have values between 4 Myr and 60 Myr. Eugster and Michel (1995) assign these ^{81}Kr/Kr CRE ages to five clusters at ages (Myr) of 6 ± 1, 12 ± 2, 21 ± 4, 38 ± 8, and 73 ± 3. We note also that a suite of Yamato meteorites that may be paired with each other may populate a separate cluster at 73 Myr (Miura et al., 1993). These ^{81}Kr/Kr CRE ages are uncertain because the terrestrial ages of the meteorites are not known. In independent work, Shukolyukov and Begemann (1996) reassigned essentially the same eucrites to five clusters at 7 ± 1 Myr, 10 ± 1 Myr, 14 ± 1 Myr, 22 ± 2 Myr, and 37 ± 1 Myr. Shukolyukov and Begemann have provided a muscular criterion for deciding when meteorites belong in a cluster. In our view, the clusters of five eucrites near 20 Myr (30% of the cases) and of four meteorites at 37 Myr (20% of the cases) are plausible; the identification of groupings below 20 Myr seems doubtful at this time. Recent estimates of the ^{38}Ar ages of the eucrites (Piplia Kalan and Vissannapeta) appear to place both meteorites in the cluster between 20 Myr and 25 Myr (Bhandari et al., 1998; Mahajan et al., 2000).

1.13.11.2 Diogenites

With strontium, yttrium, and zirconium concentrations much lower than in eucrites, the diogenites present a less attractive target for ^{81}Kr/Kr measurements and we are aware of only three such ages (Eugster and Michel, 1995). Fortunately, the higher magnesium contents in diogenites reinstate the value of the ^{22}Ne/^{21}Ne ratio as a shielding monitor. Welten et al. (1997, 2001a) present a careful assessment of the exposure ages of ~20 diogenites, based on results for light noble gases with shielding corrections from ^{22}Ne/^{21}Ne and compositional corrections based on same-sample elemental analyses.

The diogenites, as the eucrites, define a range of CRE ages that lie mostly between about 5 Myr and 60 Myr. Figure 10 lends unambiguous support to a cluster of ages at ~22 Myr (1/3–1/2 of the cases, depending on pairing), as suggested by Eugster and Michel (1995). Perhaps one-sixth of the diogenites have CRE ages near 37 Myr. With only five diogenites plotted for CRE ages between 4 Myr and 20 Myr and only one possible coincidence among those five (Aioun el Atrouss and TIL 82410 at 12 Myr), we regard the identification of clusters below 20 Myr as premature for diogenites.

1.13.11.3 Howardites

Relatively precise CRE ages are available for only a few of the more than 100 howardites

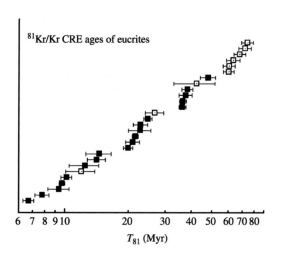

Figure 9 Eucrite exposure ages. Data shown as closed squares from Marti (1967); Hudson (1981); Freundel et al. (1986); Shukolyukov and Begemann (1996), and Miura et al. (1998). Data shown as open squares from Miura et al. (1993). Additional results may be found in Paetsch et al. (2001).

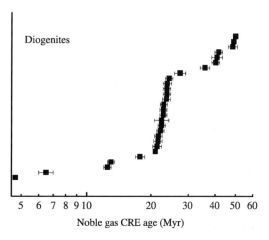

Figure 10 Diogenite exposure ages from Welten et al. (1997, 2001a). Other results from Miura and Nagao (2000) and from Paetsch et al. (2001) agree.

known. Welten et al. (1997) tabulate CRE ages from ^{21}Ne and ^{38}Ar for 19 howardites. In 10 cases the CRE ages agree within the uncertainties; in nine cases T_{21} and T_{38} disagree by more than 50%. With feldspar a major mineral and same-sample elemental analyses uncommon, these discrepancies are not surprising. Eugster and Michel (1995) recommend CRE ages for 17 howardites among which they include ^{81}Kr/Kr ages for two stones, namely, Bholghati (17 Myr) and Petersburg (21.5 Myr). Published ^{26}Al activities (Nishiizumi, 1987) suggest generally normal levels of shielding. Measurements of other cosmogenic radionuclides are scarce.

Figure 11 shows the CRE ages mentioned above along with recent data for Lohawat. All but two or three of the CRE ages may well belong to just two clusters at about 20 Myr and 40 Myr, consistent with the best-defined peaks in the CRE age distributions of the eucrites and of the diogenites. At present, it is hard to tell whether Luotolax (70 Myr) or Lohawat (110 Myr?) has the largest CRE age for a howardite. In either case, the value is larger than the maxima seen among diogenites, although some Yamato eucrites may have CRE ages as high as 70 Myr (see above). While it is easy to identify the howardite with the smallest CRE age (Chaves), it is hard to tell what, exactly, that CRE age may be: the error bars allow a range from 6 Myr to 19 Myr! In contrast, both the diogenites and the eucrites clearly include younger members: ~15% of the diogenites and up to 50% of the eucrites have CRE ages less than 19 Myr.

1.13.11.4 Kapoeta

An interesting controversy has arisen over the CRE history of Kapoeta. Inclusions in Kapoeta contain variable concentrations of noble gases produced by cosmic rays. Neither (i) shielding effects nor (ii) compositional variations can explain the variations in the cosmogenic gases inasmuch as (i) only a few centimeters separate the inclusions and (ii) the effects of elemental composition are explicitly taken into account. By elimination, the variations must record an earlier period of irradiation. To characterize this earlier irradiation we first remove the effects of the most recent stage of irradiation. From measurements of cosmogenic radionuclides, Caffee and Nishiizumi (2001) persuasively conclude that Kapoeta was last irradiated as a body with a diameter 20 cm for ~3 Myr prior to striking Earth in 1942. With this information and the application of modeling calculations, one can estimate the concentrations of noble gases attributable to the last 3 Myr of cosmic-ray irradiation and subtract them from the observed totals. The balance represents the contributions from earlier irradiation. It is the characterization of the early irradiation that has occasioned disagreement. Wieler et al. (2000) account for the early neon with a GCR-only irradiation lasting, perhaps, tens to hundreds of million years at a depth of several hundred g cm^{-2} in a large body (2π irradiation). Rao et al. (1997) do not agree, suggesting instead that *solar cosmic rays* (SCRs) produced as much as 80% of the early neon, with GCR producing the rest.

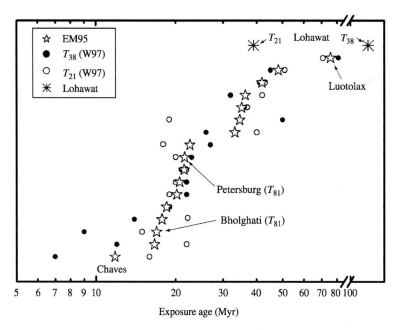

Figure 11 CRE ages of howardites. EM = best estimates from Eugster and Michel (1995). $T_{38} = {}^{38}$Ar CRE age from Welten et al. (1997). $T_{21} = {}^{21}$Ne CRE age from Welten et al. (1997). Data for Lohawat from Sisodia et al. (2001). Analytical uncertainties of the CRE ages are ~10%, although as shown the results of different methods may disagree by larger percentages.

In more modern (lunar) samples, Rao et al. argue that SCRs produce 40% or less of the cosmogenic neon. Among several possible explanations for the higher apparent proportion of SCR-produced neon in Kapoeta, Rao et al. favor an SCR flux enhanced by a factor of 10 or more, probably generated by an early active Sun (see also Hohenberg et al. (1990)). Wieler et al. (2000) observe that the divergence of views hinges on a technical point, namely, the deconvolution of SCR and GCR neon. The results reflect the (model-dependent) neon isotopic ratios adopted for SCR and GCR neon in the phases analyzed (feldspars and pyroxenes). We regard this argument as unsettled.

1.13.12 ANGRITES

Oxygen isotope systematics tie the angrites to the HED complex (Clayton and Mayeda, 1996), but CRE ages do not. CRE ages are known for five, and a very uncertain lower limit for one (LEW 87051) of the angrites (Figure 12). The better-defined ages range from 5.4 Myr for Asuka 881371 to 56 Myr for Angra dos Reis. As far as we know, the angrites show no signs of earlier irradiation as a small body or in a regolith.

1.13.13 UREILITES

Goodrich et al. (2002) suggest that the more than 110 ureilites known today were excavated by a single collision and then reassembled in a single parent body. Figure 13 shows the distribution of approximate ^{21}Ne ages of 26 ureilites calculated from the data compiled by Schultz and Franke, 2002 and a ^{21}Ne production rate of 0.412×10^{-8} cm^3 STP g^{-1} Myr^{-1}. Four of the samples have unusually short ages—less than 2 Myr. ALH 78019 is remarkable in having a CRE age of less than 100 kyr; such low values are common only

among lunar meteorites and, to a lesser degree, CM chondrites. Kenna has the maximum CRE age for a ureilite, ~35 Myr. The age distribution may have a cluster in the vicinity of 10 Myr. The peak sharpens for CRE ages based on ^{21}Ne/^{10}Be ratios (Graf et al., 1990b) and data from Aylmer et al. (1990). If real, a cluster would lend credibility to the idea that many ureilites come from one parent body (Goodrich et al., 2002).

1.13.14 AUBRITES (ENSTATITE ACHONDRITES)

Early on, Eberhardt et al. (1965) presented the irradiation histories of aubrites and identified the essential features of their CRE age distribution. More recently, Lorenzetti et al. (2003) compiled available data, reported new measurements (for Mt. Egerton, Mayo Belwa, and what are probably five distinct Antarctic meteorites), and updated the calculations of CRE ages. We show the results in Figure 14 (and in Figure 6).

Whether considered individually or on average, the aubrites have the largest CRE ages of all stony meteorites. For example, the Norton County fall has a CRE age of over 100 Myr. It is an often-remarked curiosity that this atypically large CRE age was also the first one ever reported for a stone (Begemann et al., 1957). Even after 40 years of analyses, Norton County, along with one other aubrite, Mayo Belwa, ranks among the stones with the longest CRE ages. At the other extreme, the smallest CRE age for aubrites, ~12 Myr, belongs to the type specimen for the group, the Aubres fall.

Eberhardt et al. (1965) noted an apparent cluster between 40 Myr and 50 Myr in the CRE age distribution of the aubrites. The original cluster comprised six meteorites—Bishopville,

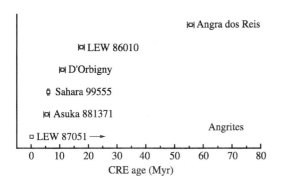

Figure 12 CRE ages of angrites. Sahara 99555, Bischoff et al. (2000); D'Orbigny, Kurat et al. (2001); Asuka 881371, Weigel et al. (1997); Angra dos Reis, Lugmair and Marti (1977); LEW 86010 and LEW 87051, Eugster et al. (1991a).

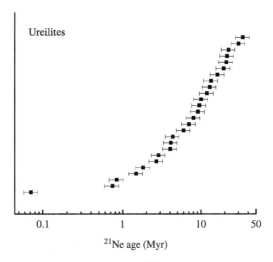

Figure 13 Approximate ^{21}Ne CRE ages of 26 ureilites.

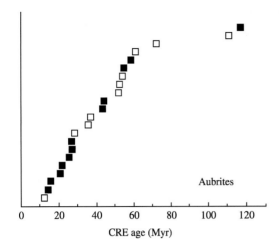

Figure 14 CRE ages of aubrites from Eberhardt et al. (1965) (open squares) and Lorenzetti et al. (2003) (closed squares). Analytical uncertainties of the ages are ~10%; the true uncertainties are larger because of the possibility that some aubrites had multistage exposures.

Bustee, Cumberland Falls, Khor Temiki, Peña Blanca Spring, and Pesyanoe. Lorenzetti et al. (2003) re-evaluated these CRE ages and in so doing increased them by ~20%, introduced more spread among them, and added a backdrop of five Antarctic aubrites with lower CRE ages. Thus, taken as a group, the aubrite ages no longer show any strong tendency to cluster, although we suspect that one collision produced both Norton County and Mayo Belwa because of their unusually large CRE ages. The large CRE ages of the aubrites are often attributed to storage in orbits free from objects likely to cause destructive collisions. Gaffey et al. (1992) have associated the aubrites with E-type Apollo asteroid 3103 based on their spectral properties.

Eberhardt et al. (1965) noted two features important in interpreting the irradiation histories of aubrites, namely the presence of solar noble gases and of isotopic anomalies associated with irradiation by thermal neutrons. Six of the aubrites contain solar noble gases (Lorenzetti et al., 2003), a fact that has significance for their CRE ages (Caffee et al., 1988). First, the presence of solar noble gases in interior samples of meteorites implies a two-stage (at least) irradiation by galactic cosmic rays. Second, the presence of solar wind noble gases implies irradiation *at a surface* by solar energetic particles or solar cosmic rays as well as by galactic cosmic rays. These observations raise interesting and largely unanswered questions about how to specify the duration and intensity of the surface irradiations.

The long CRE ages of aubrites make them good samples to search for the production of *stable* isotopes by thermal neutrons. At the same time, their compositions (low iron contents) allow the development of higher fluxes of thermal neutrons, which also support higher production rates of *radioactive* nuclides (Spergel et al., 1986). Hidaka et al. (1999) have shown that the neutron fluences (10^{16} n cm^{-2}) in aubrites vary from less than 0.01 in Happy Canyon to 1.19 in Norton County to 3.99 in Cumberland Falls. Working from the other direction, i.e., from published ^{41}Ca activities (Fink et al., 1992) and a CRE age of 100 Myr, Fink et al., 2002 have calculated similar neutron fluences in various Norton County samples.

1.13.15 BRACHINITES

At this time, seven brachinites and one close relation are known. CRE ages based on noble gases are available for three of these meteorites, but may be problematic for one reason or another: Brachina, 3 Myr (Ott et al., 1985); ALH 84025, 10 Myr (Ott et al., 1987); and Eagles Nest, 57 Myr (Swindle et al., 1998). Mittlefehldt and Berkley (2002) have noted that all the brachinites come from the southern hemisphere, either Australia or the Antarctic. Notwithstanding the geographical closeness of the find locations, if the CRE ages hold up, they suggest at least three independent falls.

1.13.16 MARTIAN METEORITES

Table 5 shows CRE ages of 24 meteorites identified as martian as of March 2002. The results are based on a critical review of the literature referenced in the notes to Table 5. The distribution of CRE ages for martian meteorites appears in Figure 15. The ages range from a low of ~0.6 Myr for EET 79001 to a high no greater than 20 Myr. Thus, the CRE ages of martian meteorites tend to be short relative to those of asteroidal meteorites. Most workers agree that one impact ~12 Myr ago produced five nakhlites and that the CRE age difference between the nakhlites and the orthopyroxenite ALH 84001 is big enough to infer an event. Opinion is divided as to whether Chassigny, with its distinct mineralogy, came out of the same martian crater as the nakhlites (Terribilini et al., 1998; Nyquist et al., 1998).

Much ink has flowed over the number of impacts required to produce the shergottites. The controversy arises partly out of technical issues concerning production rates and partly because of the apparent scarcity on Mars of target sites with rocks that appear likely to have the crystallization ages measured for the martian meteorites, which tend to be young, ~200 Myr (Nyquist et al., 1998, 2001). Clearly, if martian terrain with the right crystallization age is rare, then each new discovery of

Table 5 CRE ages of martian meteorites.

	Pairs	Total mass (kg)	Number	CRE age (Myr)
Shergottites/basalts			10	
Dar al Gani 476	489, 670, 735, 876	6.37		1.20 ± 0.15
Dhofar 019		1.06		17 ± 6
EETA 79001		7.90		0.6 ± 0.1
Los Angeles 001	002	0.70		3.0 ± 0.3
Northwest Africa 480		0.028		2.4 ± 0.2
Northwest Africa 856		0.320		
QUE 94201		0.01		2.5 ± 0.6
Sayh Al Uhaymir 005	008, 051, 194	10.51		1.3 ± 0.2
Shergotty		5		2.5 ± 0.6
Zagami		18.1		2.8 ± 0.2
Shergottites/unspecified			3	
Dhofar 378		0.015		
GRV 9927		0.010		
Northwest Africa 1068	1,110	0.772		
Shergottites/lherzolites			4	
ALHA 77005		0.48		3.3 ± 0.6
LEW 88516		0.013		4 ± 1
Y793605		0.016		4.5 ± 0.5
YA 1075		0.055		
Nakhlites/N-clinopyroxenites			5	
Governador Valadares		0.16		10.1 ± 22
Lafayette		0.8		11.4 ± 2.1
Nakhla		10		10.8 ± 0.4
Northwest Africa 817		0.104		11 ± 1
Y000593	Y000749	13.7		12 ± 1
Chassignite/Dunite			1	
Chassigny		4		12 ± 1
Orthopyroxenite			1	
ALH 84001		1.9		14 ± 1

References. **Dar al Gani 476**: Zipfel *et al.* (2000); Scherer and Schultz (1999); Park *et al.* (2001); Folco *et al.* (2000). **Dhofar 019**: Shukolyukov *et al.* (2000); Park *et al.* (2001); Nishiizumi *et al.* (2002). **EET 79001**: Eugster *et al.* (1997); Schnabel *et al.* (2001); Nishiizumi *et al.* (1986); Jull and Donahue (1988). **Los Angeles**: Garrison and Bogard (2000); Nishiizumi *et al.* (2000a). **Northwest Africa 480**: Marty *et al.* (2001). **QUE 94201**: Eugster *et al.* (1997); Garrison and Bogard (1998); Nishiizumi and Caffee (1996a); Schnabel *et al.* (2001). **Sayh al Uhaymir 005**: Paetsch *et al.* (2000); Park *et al.* (2001); Nishiizumi *et al.* (2001). **Shergotty**: Eugster *et al.* (1997); Garrison and Bogard (1998); Terribiliini *et al.* (2000b); Nishiizumi (1987); Nishiizumi and Caffee (1996a); **Zagami**: Eugster *et al.* (1997); Terribiliini *et al.* (2000b); Schnabel *et al.* (2001). **Allan Hills 77005**: Eugster *et al.* (1997); Garrison *et al.* (1995); Nyquist *et al.* (1998). Nishiizumi (1987); Nishiizumi *et al.* (1986, 1994); Schnabel *et al.* (2001); Schultz and Freundel (1984). **Lewis Cliffs 88516**: Eugster *et al.* (1997); Jull *et al.* (1994); Nishiizumi *et al.* (1992b); Schnabel *et al.* (2001). **Y793605**: Terribilini *et al.* (1998); Nagao *et al.* (1997, 1998); Nishiizumi and Caffee (1997). **Governador Valadares, Lafayette, and Nakhla**: Eugster *et al.* (1997); Bogard and Husain (1977); Terribilini *et al.* (2000b); Jull *et al.* (1999). **Northwest Africa 817**: Marty *et al.* (2001). **Yamato 000593**: Imae (2002). **Chassigny**: Eugster *et al.* (1997); Terribilini *et al.* (1998, 2000b); Matthew and Marti (2001). **Allan Hills 84001**: Eugster *et al.* (1997); Matthew and Marti (2001); Garrison and Bogard (1998); Jull *et al.* (1994); Nishiizumi *et al.* (1994).

a shergottite with that crystallization age makes it harder to imagine how separate impactors could have found their way to that terrain. The hypothesis of one impact dispels this problem. A sufficiently large impact might well sample several different lithologies, e.g., nakhlites and Chassigny. EET 79001 provides direct evidence that different lithologies may coexist at close quarters.

In our view the one-crater hypothesis for shergottites has two weaknesses. First, measured crystallization ages must be equated to the absolute ages of martian provinces, which are inferred from crater counts. Nyquist *et al.* (2001) discuss several difficulties associated with this exercise. Second, it requires a complex reinterpretation of the cosmogenic nuclide data, i.e., either the construction of multistage exposure histories where one-stage histories would seem to suffice or an *ad hoc* treatment of production rates. Based on inspection of Figure 15, we favor five distinct events and times for launching the shergottites: (i) one <1 Myr; (ii) two between 1 Myr and 2 Myr; (iii) five between 2 Myr and 3 Myr; (iv) three between 3 Myr and 4 Myr; and (v) one (Dhofar 019) >10 Myr. Even the observation that Dhofar 019 has a CRE age more than twice as large as does any other shergottite, however, will not end this debate until we can be certain of minimal pre-irradiation on Mars and have better calculations of production rates. Measurements of

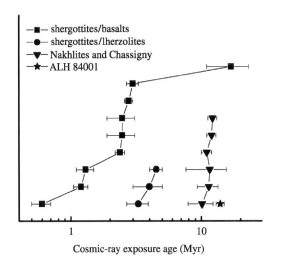

Figure 15 CRE ages of martian meteorites.

gadolinium and neodymium isotope systematics such as those of Hidaka *et al.* (2001) should help with respect to the former. We would note in this context that the number of shergottites with known CRE ages, 12, is still fairly small; no doubt, surprises are waiting.

Numerous authors (e.g., Warren, 1994; Wieler, 2002; Nyquist *et al.*, 2001) have contrasted the exposure histories and other properties of lunar and martian meteorites. On average, we would expect key systematic differences to relate to their respective distances from the Earth (or more precisely how easily their ejecta could attain Earth-crossing orbits), the respective depths of their gravitational wells, the mechanical properties of their regoliths, and the relative fluxes of impacting bodies.

- Martian meteorites have larger average recovered masses than lunar meteorites. According to Mileikowsky *et al.* (2000), ejecta size increases with the size of the impact event. With more energy required (and hence available) for launch from Mars than from the Moon, it seems reasonable to expect larger martian fragments. Once ejected, larger fragments would be more likely to survive collisional destruction en route to Earth.
- More martian than lunar meteorites appear to have come to Earth per crater. According to Mileikowsky *et al.* (2000), the total mass of the ejecta increases with the size of the impact event. With more energy required for launch from Mars, and with launches rare, a greater likelihood of the pairing of source craters for martian meteorites seems reasonable.
- The numbers of lunar and of martian meteorites are nearly equal. The near equality at first seems odd in light of the closeness and smaller gravitational field of the Moon. The absolute number of martian meteorites may be misleading, however. As suggested by the grouping of CRE ages above, that number is almost certainly larger than the number of events on Mars that produced the meteorites. Further, asteroids may strike Mars more often than they do than the Moon, producing more ejecta (Wieler, 2002).
- Many lunar meteorites retain a record of cosmic-ray irradiation on the Moon while martian meteorites show few obvious signs of pre-irradiation near the surface of Mars. The former result points to a launch depth within a few meters of the lunar surface. The latter observation suggests that martian meteorites came from depths of a few meters or more below the surface although further investigation may alter this view. Warren (1994) argues that the mechanical properties of the lunar regolith favor near-surface lunar objects while weathering on Mars may have destroyed smaller, suitable candidates near the martian surface. In considering closely related issues, Nyquist *et al.* (2001) conclude, "The launch mechanism for the Martian meteorites is sufficiently uncertain that a number of possible mechanisms should continue to be evaluated."

1.13.17 MESOSIDERITES

Begemann *et al.* (1976) presented the first comprehensive survey of CRE ages for mesosiderites. They calculated $^{36}Cl/^{36}Ar$ ages for metal phases and ^3He, ^{21}Ne, and ^{38}Ar ages for the silicates. Nagai *et al.* (1993) investigated production rate systematics of cosmogenic nuclides in mesosiderites. Since then, Terribilini *et al.* (2000c); Albrecht *et al.* (2000); Nishiizumi *et al.* (2000b), and Welten *et al.* (2001b) have reported new cosmogenic nuclide data for various members of the group.

The $^{36}Cl/^{36}Ar$ CRE ages for the metal phases of mesosiderites are probably the most reliable. Figure 16, therefore, shows the distribution of CRE ages for mesosiderites based mainly on metal-phase $^{36}Cl/^{36}Ar$, and bulk krypton isotopic measurements. The average, ~90 Myr, is intermediate between CRE ages typical for the generally younger stones and the generally older irons. Five of the mesosiderite ages surpass the CRE ages of the oldest stones. The oldest CRE age, 340 Myr for EET 87500, may be compared with CRE age of the oldest iron, ~2,500 Myr for Deep Springs. The age distribution does not have any strong clusters, although Welten *et al.* (2001b) note that three of 19 mesosiderites (Chinguetti, Crab Orchard, and Estherville) have CRE ages close to 70 Myr. Terribilini *et al.* (2000c) call attention to the dissimilarity of the CRE age distributions for mesosiderites and HED meteorites.

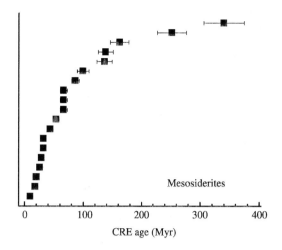

Figure 16 CRE ages of mesosiderites from Terribilini et al. (2000c); Albrecht et al. (2000), Welten et al. (2001b); Lavielle et al. (1998), and Nishiizumi et al. (2001).

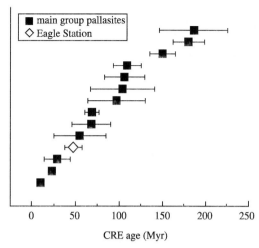

Figure 17 CRE ages of pallasites (sources Begemann et al., 01976; Honda et al., 2002; Megrue, 1968; Miura, 1995; Nagao et al., 1983; Schultz and Hintenberger, 1967; Shukolyukov and Petaev, 1992).

Despite their brecciated nature, only two mesosiderites—Veramin (Begemann et al., 1976) and Eltanin (Nishiizumi et al., 2000b)—seem to contain appreciable concentrations of solar wind or SEP gases. The meteorite Eltanin deserves special consideration. Fragments of this object were recovered from deep-ocean sediments beneath stormy southern seas. The meteoroid was probably enormous (Gersonde et al., 1997). Silicates dominate and metal is rare in the material recovered. The silicates seem to be intermediate in character between those of eucrites and of mesosiderites (Kyte et al., 2000). Eltanin's exposure age of 20 Myr lies close to the CRE ages of several eucrites.

1.13.18 PALLASITES

Most CRE ages and histories of pallasites rest shakily on noble gas analyses alone. We summarize the results in Figure 17. Pallasite exposure ages are longer than those of most stony meteorites and comparable to those of the other major stony iron group, the mesosiderites.

1.13.19 IRONS

Figure 18 shows: (i) the ^{40}K/K CRE ages of ~80 iron meteorites and (ii) the ^{38}Ar ages of ~160 iron meteorites based on shielding-corrected production rates calculated according to a method not discussed in detail here, that of Lavielle et al. (1985). These authors set the parameters of the ^{38}Ar production rate so that the ^{40}K/K and the ^{38}Ar ages would agree for as many irons as possible. Most of the ^{40}K/K CRE ages lie between 200 Myr and 1,000 Myr. The ^{38}Ar fill in nicely the low-age gaps in the ^{40}K/K CRE age distributions, gaps that partly reflect the limitations of the ^{40}K/K dating method. A few irons have CRE ages larger than 1,000 Myr with the championship held by Deep Springs at over 2,000 Myr. For group I irons, the ^{40}K/K ages define a cluster at ~900 Myr. For group III irons, both the ^{40}K/K and the ^{38}Ar-based CRE ages establish a cluster at ~650 Myr; in addition, the ^{38}Ar-based CRE ages suggest a second peak in the vicinity of 450 Myr (Lavielle et al., 1985). In considering these plots, it is worth noting that the chemical groupings/classifications of iron meteorites may change. For example, Wasson and Kallemeyn (2002) recommend reclassifying the IIIC and IIID subgroups and making them part of the IAB complex.

A comparison of the ^{40}K/K and ^{36}Cl/^{36}Ar ages of 17 irons (Lavielle et al., 1999) raises two unsettled and related issues. First, as succinctly summarized by Eugster (2003), "there is still some unresolved bias in the age scales for the different methods." In particular, CRE ages based on shorter-lived radionuclides including ^{36}Cl are ~30% lower than ^{40}K/K CRE ages. The standard explanation is that the intensity of the cosmic-ray flux increased at some time within the last 100 Myr or so. The increase raised the activities and hence the production rates for the short-lived radionuclides, but had a smaller effect on the longer-lived ^{40}K. Compared to the shorter-lived radionuclides, ^{40}K should give a better measure of the average cosmic-ray flux because its half-life is closer to the typical CRE ages of irons. If the cosmic-ray flux indeed changed, then the production rates (but not necessarily the total concentrations of cosmogenic nuclides!) should have increased by the same fractions and at the same time for all meteoroids.

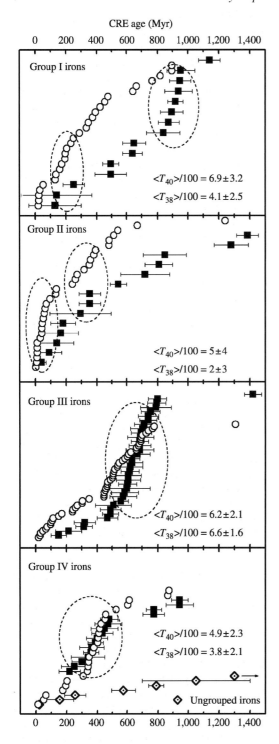

Figure 18 ^{40}K/K CRE ages of iron meteorites as closed squares (Voshage, 1967; Voshage and Feldmann, 1979; Voshage et al., 983), ^{38}Ar–^{4}He/^{21}Ne ages of iron meteorites as open circles after Lavielle et al. (1985) with noble gas data taken from the compilation of Schultz and Franke (2002). Possible clusters of ages are marked with dashed circles (see also Eugster, 2003). The angular brackets denote group averages of the two types of ages for groups I–IV. For the ungrouped irons $<T_{40}> = 732 \pm 292$ Myr and $T_{38} = 469 \pm 567$ Myr.

The related unresolved issue concerns the prevalence of multiple exposures. For example, Lavielle et al. (1999) identify Bendego as an iron that may have undergone multiple periods of exposure, presumably initiated by collisions, and there are others such as Canyon Diablo (Michlovich et al., 1994). In general, we would expect collisions to reduce the sizes of meteoroids and thereby increase production rates, i.e., to have the same qualitative effect as would an increase in the cosmic-ray flux. Such changes, however, should change production rates at random times and by random increments. The relative importance of multiple exposures in irons is difficult to assess from the available data.

1.13.20 THE SMALLEST PARTICLES: MICROMETEORITES, INTERPLANETARY DUST PARTICLES, AND INTERSTELLAR GRAINS

1.13.20.1 Background

We will use the term micrometeorite in the most general sense to include all particles with a maximum dimension less than 0.5 mm, i.e., those that melted (the cosmic spherules), those that did not (the "unmelted" micrometeorites), and the interplanetary dust particles (IDPs). According to theory, several kinds of physical forces act on small particles and should limit their expected transit times to Earth (Burns et al., 1979; Gustafson, 1994) and hence their exposure to cosmic rays. In the inner solar system, Poynting–Robinson drag slows the smaller (<100 μm) particles down to the point that they fall into the Sun. The timescale, <10^5 yr, for this process is much shorter than the one defined by CRE ages of asteroidal meteoroids. For larger particles, those in the 100–500 μm range, collisional destruction may limit the transit times to 50–60 kyr (cf. Kortenkamp and Dermott, 1998).

1.13.20.2 Micrometeorites and IDPs

At present, quantitative assessments of the CRE ages of small particles are few. The measurements, especially for the tiny IDPs, are difficult, and even heroic. *Solar* noble gases (solar wind and solar energetic particles) occur widely in micrometeorites in concentrations that are consistent with relatively short periods of exposure in space as small bodies (Nier, 1995). In contrast, what we know from the cosmogenic nuclides is either ambiguous or seems to require much longer periods of irradiation. Olinger et al. (1990) analyzed the neon in individual, unmelted micrometeorites in the size range between

0.1 mm and 0.5 mm. Then, with the aid of various assumptions, they estimated CRE ages, obtaining values from less than 0.5 Myr to 20 Myr. The *range* of CRE ages was well over a factor of 10 regardless of the method of calculation. Among the IDPs, which are smaller, several have $^3\mathrm{He}/^4\mathrm{He}$ ratios that indicate appreciable concentrations of cosmogenic $^3\mathrm{He}$ (Nier, 1995). Pepin *et al.* (2001) measured $^3\mathrm{He}$ concentrations in samples from a single IDP that correspond to a CRE age $\sim 10^9$ yr! To explain the long period of exposure, these authors consider the possibility that the samples came not from the inner solar system, but from much farther away, in the Edgeworth–Kuiper Belt (30–120 AU), where collision rates were lower and particle survival times longer. Transport of the IDPs as part of larger objects might preserve them from collisional destruction along the way. Pepin *et al.* also discuss several other imaginative explanations for the large concentrations of $^3\mathrm{He}$, including implantation by solar energetic particles, interstellar pickup, and an unusual primordial component rich in $^3\mathrm{He}$. The view that a parent-body irradiation produced $^3\mathrm{He}$ and $^{21}\mathrm{Ne}$ seems to us the most appealing.

Turning now to larger particles, Figure 19 shows the $^{10}\mathrm{Be}$ and $^{26}\mathrm{Al}$ activities of micrometeorites harvested from the deep sea and from the Antarctic (Raisbeck *et al.*, 1985a; Nishiizumi *et al.*, 1991c, 1995). The gray region of Figure 19 shows the "allowed" range of $^{26}\mathrm{Al}$ and $^{10}\mathrm{Be}$ activities expected for small particles that have undergone a one-stage irradiation by galactic cosmic rays and varying degrees of irradiation by solar cosmic rays. The lower curve close to the *x*-axis traces the expected growth of the two nuclides in small *metallic* particles based on production rates for iron meteorites (Albrecht *et al.*, 2000). The lower curve terminates at the point where the activities reach the production rates, $P_{10} \sim 6$ and $P_{26} \sim 4.2$. Contrary to expectations, several I-type spherules have $^{10}\mathrm{Be}$ activities larger than the maximum production rate of ~ 6 dpm kg^{-1}. These experimental results would have plotted at even higher values had we corrected for the terrestrial addition of oxygen to the samples. Solar cosmic rays offer no way to bring the straying data points into the "allowed" region for SCR do not produce appreciable quantities of $^{26}\mathrm{Al}$ or $^{10}\mathrm{Be}$ from iron and nickel (Reedy, 1987b). In sum, the high measured $^{10}\mathrm{Be}$ activities are not consistent with the presumed pre-atmospheric composition of the spherules—metallic—or with what we think we know about irradiation conditions in the solar system. Other than experimental error, possible explanations include an origin in a location where production rates were higher (Raisbeck *et al.*, 1985b; Pepin *et al.*, 2001); the presence of siliceous material in the pre-atmospheric object; or terrestrial addition of $^{10}\mathrm{Be}$ in an aqueous environment.

The upper curve in Figure 19 traces the expected growth of $^{10}\mathrm{Be}$ and $^{26}\mathrm{Al}$ in stony material exposed to both solar and galactic cosmic rays (Reedy, 1987a, 1990). Numerous data points for S-type spherules lie to the right of the allowed field. Here an explanation does not require much of a stretch, although the construction of unambiguous exposure histories is not yet possible. If the particles had short lifetimes in space, less than say 1 Myr, then they should preserve cosmogenic radionuclides produced if they lay within 50 cm or so of the surface of their precursor bodies. There, $^{10}\mathrm{Be}$ chondritic production rates of 20 dpm kg^{-1} or more could have prevailed. Unclear is what fraction of the observed $^{26}\mathrm{Al}$ derives from the earlier irradiation. The extraordinarily high $^{26}\mathrm{Al}$ activities (relative to chondritic values of ~ 50 dpm kg^{-1}) and $^{26}\mathrm{Al}/^{10}\mathrm{Be}$ ratios observed in virtually all the S-type particles require significant SCR exposure. In view of the prevalence of $^{26}\mathrm{Al}$ activities well above chondritic values, it seems likely that the S-type spherules in the "forbidden" field acquired most of their SCR-produced $^{26}\mathrm{Al}$ recently, as small bodies. We cannot yet rule out a contribution from earlier irradiation at the surface of a larger body, however. In this connection, it would be interesting to know the depth range in the S-type parent bodies from which collisions excavate the S-type particles. Although a few S-type spherules appear to the left of the "allowed" region, their error bars reach into it. We will assume experimental uncertainties account for these observations.

We turn finally to the allowed region itself. The simplest interpretation is that the particles follow intermediate curves of growth (not shown) where $^{26}\mathrm{Al}$ production from SCRs varied because of variations in particle size and geometry. In view of the fairly uniform distribution of data points, however, we think it more likely that most of these

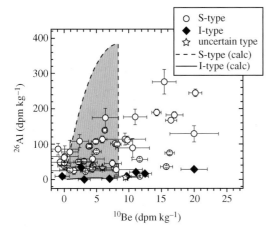

Figure 19 $^{10}\mathrm{Be}$ and $^{26}\mathrm{Al}$ activities of micrometeorites.

particles, like their siblings to the right of the allowed region, retain some record of earlier irradiation. Thus, the sorting out of their exposure histories will require deconvolution of effects due to SCR and GCR in at least two stages of irradiation. It will be an interesting task made worthwhile by the clues of high-flux irradiations at large perihelia.

1.13.20.3 Interstellar Grains

Several types of grains separated from carbonaceous chondrites have unusual isotopic compositions thought to reflect an origin outside the solar system. As these interstellar grains traveled through interstellar space, it seems certain that cosmic rays would have produced cosmogenic nuclides in them. Vogt et al. (1990) reviewed early treatments of this question. They concluded that the uncertainties of ^{21}Ne CRE ages, ~40 Myr, found for SiC grains were too large to permit meaningful comparisons with (much longer) theoretical estimates for the lifetimes of interstellar grains. In particular, the CRE age calculation (Equation (1)) requires: (i) a production rate, which depends on the unknown intensity cosmic-ray flux more than 4.5 Gyr ago in far away places; and (ii) an estimate of the cosmogenic ^{21}Ne content. To arrive at the cosmogenic ^{21}Ne content, one must separate it from trapped components (e.g., solar wind) and from other, local cosmogenic components (e.g., ^{21}Ne produced in an asteroidal regolith or later, during exposure as a meteoroid). Further, because of the small grain sizes—only a few micrometers—of the interstellar SiC grains, one must compensate explicitly for recoil losses.

Viewed in the light of new developments, the prospect of having reliable ^{21}Ne CRE ages for interstellar grains has dimmed further. Ott and Begemann (2000) measured recoil losses of ^{21}Ne from SiC in the laboratory. They found that the recoil losses were much larger (over 90%) for small (<1 μm) grains than originally estimated. Thus, analyses of ^{21}Ne in small grains provide little reliable information about total exposure. As if this result were not discouraging enough, Ott and Begemann reopened the question of how to distinguish the cosmogenic ^{21}Ne content of interstellar origin. They concluded that the ^{21}Ne previously classified as cosmogenic might, in fact, be nucleosynthetic, i.e., a special type of trapped component. On the bright side, Ott and Begemann suggest that cosmogenic ^{126}Xe could offer a more solid base for calculating exposure ages of interstellar grains. With various assumptions, they estimate presolar xenon ages for finer grains to be 100–300 Myr and infer that the larger grains are younger than the smaller ones.

1.13.21 CONCLUSIONS

By and large, the generalizations about CRE ages made in the early 1970s have stood up well against a tide of new data. The field has also progressed considerably since then. With a few notable exceptions, the exposure age distributions of the well-established meteorite groups are now known and it is clear that exposure ages differ systematically among different groups of meteorites. As Wieler and Graf (2001) conclude, the clumpiness of the CRE age distributions strongly suggests that a small number of collisions produced a very large fraction of the known meteorites. It follows, these authors continue, that we may have sampled only a tiny portion of the potential parent bodies in the asteroid belt. We have identified a few of those parent bodies and have hints about the identities of others. Figure 20 plots the average ages of lunar meteorites (0.2 ± 0.3 Myr excluding only Dhofar 025), of Martian meteorites (7 ± 5 Myr), and of HED meteorites (25 ± 8 Myr), which are tied to the asteroid Vesta and its kin, against aphelion of the respective parent bodies. Also shown are five ordinary chondrite falls with known orbital parameters. The scatter of the data for the H5 chondrites alerts us to the dangers of oversimplifying (see below). Nonetheless, it is hard to resist the conclusion that on average, the further away the source of a meteorite, the longer that meteorite takes to get to Earth.

Recently, theorists seem to have come to grips with all the factors needed to account for the principal features of CRE age distributions. A decade ago, in the reigning view, the Earth competed for meteoroids against collisional destruction and ejection from the solar system or incineration by the Sun (Greenberg and Nolan, 1989). Earth capture depended on injection of the meteoroid into an orbit close to or in a "chaotic resonance in the inner main belt"

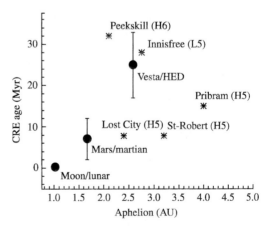

Figure 20 CRE ages of various meteorites versus aphelion of parent body or meteoroid.

(Bottke et al., 2000). This picture could accommodate qualitatively the relatively long exposure ages of stony irons and irons (resistance to collisional destruction) and depressed ^{3}He ages of many meteorites (^{3}He losses at small perihelion) and those elements of the model remain. The model failed, however, on several counts having to do with the observed properties of asteroids (Bottke et al., 2002). Further and of more direct relevance here, the calculations of Gladman et al. (1997) showed that meteoroid transport through the resonances took much less time than actually clocked by the CRE ages. Consideration of the Yarkovsky effect now seems to have resolved the order-of-magnitude discrepancies between theory and measured CRE ages (Farinella et al., 1998; Vokrouhlický et al., 2000; Spitale and Greenberg, 2002). The Yarkovsky effect refers to the asymmetric heating of meteoroids by solar radiation, which leads to small, instantaneous forces of acceleration that change the principal elements of the meteoroid orbit. The changes occur slowly enough that a meteoroid may spend considerable time—tens of millions of years or more—in the main belt before it reaches a resonance that brings it to Earth, always provided that collisions do not destroy the object first. Interestingly, the physical properties of irons give rise to a smaller Yarkovsky effect for irons allowing a longer random-walk passage through the main belt. While much work in modeling the Yarkovsky effect remains to be done (Spitale and Greenberg, 2002), the essential physical elements that control CRE age distributions now seem to be understood. We should not forget, however, that the creation of a meteoroid begins with a collision on a parent body. Crater counting and the general clumpiness of the CRE age distributions suggest that these meteoroid-producing events are rare and that chance is a crucial factor in determining the CRE ages.

Many interesting questions remain for future research. Cosmic rays have irradiated, albeit weakly at times, the stuff of meteorites ever since parent bodies formed 4.5 Gyr ago. In our view, the ultimate goal of CRE studies is to document the exposure histories of meteorites and their components over that entire period. We have tools for examining the earliest irradiations and the most recent. The period in between presents a major challenge. We conclude with a few questions addressable in the nearer term.

(i) What are the CRE age distributions of the CHs, the CBs, the pallasites, and the micrometeorites?

(ii) How tight are the clusters of CRE ages? Can we resolve CRE ages that differ by <5%?

(iii) Can we determine separately the CRE histories of chondrules and other small components of meteorites (e.g., Polnau et al., 2001)?

(iv) What is the overall likelihood that meteoroids experienced multiple, recent periods of irradiation?

(v) How has the flux of galactic cosmic rays varied in time?

(vi) Can we construct CRE histories from analyses of cosmogenic nuclides made aboard spacecraft?

(vii) Can we reduce the mass needed for analyzing cosmogenic nuclides so that it will be possible to construct CRE histories for materials returned from the surfaces of asteroids and comets?

ACKNOWLEDGMENTS

I thank Ludolf Schultz for making available his most recent compilation of noble gas data. We are grateful to Robert Reedy, Rainer Wieler, Donald Bogard, and Andrew Davis for taking the time to make many valuable comments.

REFERENCES

Albrecht A., Schnabel C., Vogt S., Xue S., Herzog G. F., Begemann F., Weber H. W., Middleton R., Fink D., and Klein J. (2000) Light noble gases and cosmogenic radionuclides in Estherville, Budulan and other mesosiderites: implications for exposure histories and production rates. Meteorit. Planet. Sci. 35, 975–986.

Arnold J. R., Honda M., and Lal D. (1961) Record of cosmic-ray intensity in meteorites. J. Geophys. Res. 66, 3519–3531.

Aylmer D., Herzog G. F., Klein J., and Middleton R. (1988) ^{10}Be and ^{26}Al contents of eucrites: implications for production rates and exposure ages. Geochim. Cosmochim. Acta 52, 1691–1698.

Aylmer D., Vogt S., Herzog G. F., Klein J., Fink D., and Middleton R. (1990) Low ^{10}Be and ^{26}Al contents of ureilites: production at meteoroid surfaces. Geochim. Cosmochim. Acta 54, 1775–1784.

Begemann F., Geiss J., and Hess D. C. (1957) Radiation age of a meteorite from cosmic-ray produced He^3 and H^3. Phys. Rev. 107, 540–542.

Begemann F., Weber H. W., Vilcsek E., and Hintenberger H. (1976) Rare gases and ^{36}Cl in stony-iron meteorites: cosmogenic elemental production rates, exposure ages, diffusion losses and thermal histories. Geochim. Cosmochim. Acta 40, 353–368.

Bhandari N., Murty S. V. S., Suthar K. M., Shukla A. D., Ballabh G. M., Sisodia M. S., and Vaya V. K. (1998) The orbit and exposure history of the Piplia Kalan eucrite. Meteorit. Planet. Sci. 33, 455–461.

Bischoff A., Palme H., Schultz L., Weber D., Weber H. W., and Spettel B. (1993) Acfer 182 and paired samples, an iron-rich carbonaceous chondrite: similarities with ALH 85085 and relationship to CR chondrites. Geochim. Cosmochim. Acta 57, 2631–2648.

Bischoff A., Weber D., Clayton R. N., Faestermann T., Franchi I. A., Herpers U., Knie K., Korschinek G., Kubik P. W., Mayeda T. K., Merchel S., Michel R., Neumann S., Palme H., Pillinger C. T., Schultz L., Sexton A. S., Spettel B., Verchovsky A. B., Weber H. W., Weckwerth G., and Wolf D. (1998) Petrology, chemistry and isotopic compositions of the lunar highland regolith breccia Dar al Gani 262. Meteorit. Planet. Sci. 33, 1243–1257.

Bischoff A., Clayton R. N., Markl G., Mayeda T. K., Palme H., Schultz L., Srinivasan G., Weber H. W., Weckwerth G., and

Wolf D. (2000) Mineralogy, chemistry, noble gases and oxygen- and magnesium-isotopic compositions of the angrite Sahara 99555. *Meteorit. Planet. Sci.* **35**, A27.

Bogard D. D. and Husain L. (1977) A new 1.3-aeon-young achondrite. *Geophys. Res. Lett.* **4**, 69–71.

Bottke W. F., Vokrouhlický D., Rubincam D., and Brož M. (2002) The effect of Yarkovsky thermal forces on the dynamical evolution of asteroids and meteoroids. In *Asteroids III* (eds. W. F. Bottke, A. Cellino, P. Paolicchi, and R. Binzel). University of Arizona Press, Tucson, pp. 395–408.

Bottke W. F., Rubincam D. P., and Burns J. A. (2000) Dynamical evolution of main belt meteoroids: numerical simulations incorporating planetary perturbations and Yarkovsky thermal forces. *Icarus* **145**, 301–331.

Burns J. A., Lamy P. L., and Soter S. (1979) Radiation forces on small particles in the solar system. *Icarus* **40**, 1–48.

Caffee M. W. and Nishiizumi K. (1997) Exposure ages of carbonaceous chondrites: II. *Meteorit. Planet. Sci.* **32**, A26.

Caffee M. W. and Nishiizumi K. (2001) Exposure history of separated phases from the Kapoeta meteorite. *Meteorit. Planet. Sci.* **36**, 429–437.

Caffee M. W., Goswami J. N., Hohenberg C. M., Marti K., and Reedy R. C. (1988) Irradiation records in meteorites. In *Meteorites and the Early Solar System* (eds. J. F. Kerridge and M. S. Matthews). University of Arizona Press, Tucson, pp. 205–245.

Clayton R. N. and Mayeda T. K. (1996) Oxygen isotope studies of achondrites. *Geochim. Cosmochim. Acta* **60**, 1999–2017.

Eberhardt P., Geiss J., and Lutz H. (1963) Neutrons in meteorites. In *Earth Science and Meteoritics* (eds. J. Geiss and E. D. Goldberg). North Holland, Amsterdam, pp. 143–168.

Eberhardt P., Eugster O., and Geiss J. (1965) Radiation ages of aubrites. *J. Geophys. Res.* **70**, 4427–4434.

Eberhardt P., Eugster O., Geiss J., and Marti K. (1966) Rare gas measurements in 30 stone meteorites. *Naturforsch* **21A**, 414–426.

Eugster O. (1988) Cosmic-ray production rates for ^{3}He, ^{21}Ne, ^{38}Ar, ^{83}Kr, and ^{126}Xe in chondrites based on ^{81}Kr–Kr exposure ages. *Geochim. Cosmochim. Acta* **52**, 1649–1659.

Eugster O. (1989) History of meteorites from the Moon collected in Antarctica. *Science* **245**, 1197–1202.

Eugster O. (2003) Cosmic-ray exposure ages of meteorite and lunar rocks and their significance. *Geochemistry* (in press).

Eugster O. and Niedermann S. (1990) Solar noble gases in the unique chondritic breccia Allan Hills 85085. *Earth Planet. Sci. Lett.* **101**, 139–147.

Eugster O. and Michel T. (1995) Common asteroid break-up events of eucrites, diogenites and howardites, and cosmic-ray production rates for noble gases in achondrites. *Geochim. Cosmochim. Acta* **59**, 177–199.

Eugster O. and Lorenzetti S. (2001) Exposure history of some differentiated and lunar meteorites. *Meteorit. Planet. Sci.* **36**, A54.

Eugster O., Michel Th., and Niedermann S. (1991a) ^{244}Pu–Xe formation and gas retention age, exposure history and terrestrial age of angrites LEW 86010 and LEW 87051: comparison with Angra dos Reis. *Geochim. Cosmochim. Acta* **55**, 2957–2964.

Eugster O., Beer J., Burger M., Finkel R. C., Hofmann H. J., Krähenbühl U., Michel Th., Synal H. A., and Wölfli W. (1991b) History of paired lunar meteorites MAC 88104 and MAC 88105 derived from noble gas isotopes, radionuclides and some chemical abundances. *Geochim. Cosmochim. Acta* **55**, 3139–3148.

Eugster O., Michel Th., and Niedermann S. (1992) Solar wind and cosmic ray exposure history of lunar meteorite Yamato-793274. *Proc. NIPR Symp. Antarct. Meteorit.* **5**, 23–35.

Eugster O., Weigel A., and Polnau E. (1997) Ejection times of Martian meteorites. *Geochim. Cosmochim. Acta* **61**, 2749–2757.

Eugster O., Polnau E., Salerno E., and Terribilini D. (2000) Lunar surface exposure models for meteorites Elephant Moraine 96008 and Dar al Gani 262 from the Moon. *Meteorit. Planet. Sci.* **35**, 1177–1181.

Farinella P., Vokrouhlický D., and Hartmann W. (1998) Meteorite delivery via Yarkovsky orbital drift. *Icarus* **132**, 378–387.

Fink D., Klein J., Middleton R., Dezfouly-Arjomandy B., Herzog G. F., and Albrecht A. (1992) ^{41}Ca in the Norton County aubrite. *Lunar Planet. Sci.* **XXIII**. Lunar and Planetary Institute, Houston, pp. 355–356.

Fink D., Ma P., Herzog G. F., Albrecht A., Garrison D. H., Bogard D. D., Reedy R. C., and Masarik J. (2002) ^{10}Be, ^{26}Al, ^{36}Cl, and non-spallogenic ^{36}Ar in the Norton County aubrite. *Meteorit. Planet. Sci.* **37** (suppl. A46), pp.

Folco L., Franchi I. A., D'Orazio M., Rocchi S., and Schultz L. (2000) A new Martian meteorite from the Sahara: the shergottite Dar al Gani 489. *Meteorit. Planet. Sci.* **35**, 827–839.

Freundel M., Schultz L., and Reedy R. C. (1986) ^{81}Kr–Kr ages of Antarctic meteorites. *Geochim. Cosmochim. Acta* **50**, 2663–2673.

Gaffey M. J., Reed K. L., and Kelley M. S. (1992) Relationship of E-type Apollo asteroid 3103 (1982 BB) to the enstatite achondrite meteorites and the Hungaria asteroids. *Icarus* **100**, 95–109.

Garrison D. H. and Bogard D. D. (1998) Isotopic composition of trapped and cosmogenic noble gases in several Martian meteorites. *Meteorit. Planet. Sci.* **33**, 721–726.

Garrison D. H. and Bogard D. D. (2000) Cosmogenic and trapped noble gases in the Los Angeles Martian meteorite. *Meteorit. Planet. Sci.* **35**, A58.

Garrison D. H., Rao M. N., and Bogard D. D. (1995) Solar-proton-produced neon in shergottite meteorites and implications for their origin. *Meteoritics* **30**, 738–747.

Gersonde R., Kyte F. T., Bleil U., Diekmann B., Flores J. A., Gohl K., Grahl G., Hagen R., Kuhn G., Sierro F. J., Völker D., Abelmann A., and Bostwick J. A. (1997) Geological record and reconstruction of the late Pliocene impact of the Eltanin asteroid in the Southern Ocean. *Nature* **390**, 357–363.

Gilabert E., Lavielle B., Michel R., Leya I., Neumann S., and Herpers U. (2002) Production of krypton and xenon isotopes in thick stony and iron targets isotropically irradiated with 1600 MeV protons. *Meteorit. Planet. Sci.* **37**, 951–976.

Gladman B. J., Burns J. A., Duncan M. J., and Levinson H. F. (1995) The dynamical evolution of lunar impact ejecta. *Icarus* **118**, 302–321.

Gladman B. J., Migliorini F., Morbidelli A., Zappalà V., Michel P., Cellino A., Froeschlé C., Levison H. F., Bailey M., and Duncan M. (1997) Dynamical lifetimes of objects injected into asteroid belt resonances. *Science* **277**, 197–201.

Göbel R., Begemann F., and Ott U. (1982) On neutron-induced and other noble gases in Allende inclusions. *Geochim. Cosmochim. Acta* **46**, 1777–1792.

Goodrich C. A., Krot A. N., Scott E. R. D., Taylor G. J., Fioretti A. M., and Keil K. (2002) Formation and evolution of the ureilite parent body and its offspring. *Lunar Planet. Sci.* **XXXIII**, #1379. Lunar and Planetary Institute, Houston (CD-ROM).

Goswami J. N., Sinha N., Nishiizumi K., Caffee M. W., Komura K., and Nakamura K. (2001) Cosmogenic records in Kobe (CK4) meteorite: implications of transport of meteorites from the asteroid belt. *Meteorit. Planet. Sci.* **36**, A70–A71.

Graf Th. and Marti K. (1994) Collisional records in LL-chondrites. *Meteoritics* **29**, 643–648.

Graf Th. and Marti K. (1995) Collisional history of H chondrites. *J. Geophys. Res. (Planets)* **100**, 21247–21263.

Graf Th., Signer P., Wieler R., Herpers U., Sarafin R., Vogt S., Fieni Ch., Pellas P., Bonani G., Suter M., and Wölfli W. (1990a) Cosmogenic nuclides and nuclear tracks in the chondrite Knyahinya. *Geochim. Cosmochim. Acta* **54**, 2511–2520.

Graf Th., Baur H., and Signer P. (1990b) A model for the production of cosmogenic nuclides in chondrites. *Geochim. Cosmochim. Acta* **54**, 2521–2534.

Graf Th., Caffee M. W., Marti K., Nishiizumi K., and Ponganis K. V. (2001) Dating collisional events: $^{36}Cl-^{36}Ar$ exposure ages of chondritic metals. *Icarus* **150**, 181–188.

Greenberg R. and Nolan M. C. (1989) Delivery of asteroids and meteorites to the inner solar system. In *Asteroids II* (eds. R. P. Binzel, T. Gehrels, and M. S. Matthews). University of Arizona Press, Tucson, pp. 778–804.

Greshake A., Schmitt R. T., Stöffler D., Pätsch M., and Schultz L. (2001) Dhofar 081: a new lunar highland meteorite. *Meteorit. Planet. Sci.* **36**, 459–470.

Gustafson B. A. S. (1994) Physics of zodiacal dust. *Ann. Rev. Earth Planet. Sci.* **22**, 553–595.

Herzog G. F. and Anders E. (1971) Absolute scale for radiation ages of stony meteorites. *Geochim. Cosmochim. Acta* **35**, 605–611.

Herzog G. F., Vogt S., Albrecht A., Xue S., Fink D., Klein J., Middleton R., Weber H. W., and Schultz L. (1997) Complex exposure histories for meteorites with "short" exposure ages. *Meteorit. Planet. Sci.* **32**, 413–422.

Heymann D., Mazor E., and Anders E. (1968) Ages of calcium-rich achondrites: I. Eucrites. *Geochim. Cosmochim. Acta* **32**, 1241–1268.

Hidaka H., Ebihara M., and Yoneda S. (1999) High fluences of neutrons determined from Sm and Gd isotopic compositions in aubrites. *Earth Planet. Sci. Lett.* **173**, 41–51.

Hidaka H., Yoneda S., and Nishiizumi K. (2001) Neutron capture effects on Sm and Gd isotopes in Martian meteorites. *Meteorit. Planet. Sci.* **36**, A80–A81.

Hohenberg C. M., Nichols R. H., Jr., Olinger C. T., and Goswami J. N. (1990) Cosmogenic neon from individual grains of CM meteorites: extremely long pre-compaction exposure histories or an enhanced early particle flux. *Geochim. Cosmochim. Acta* **54**, 2133–2140.

Honda M. and Arnold J. R. (1967) Effects of cosmic rays on meteorites. *Handbook Phys.* **46**, 613–632.

Honda M., Caffee M. W., Miura Y. N., Nagai H., Nagao K., and Nishiizumi K. (2002) Cosmogenic products in the Brenham pallasite. *Meteorit. Planet. Sci.* **37**, 1711–1728.

Hudson B. G. (1981) Noble gas retention chronologies for the St. Severin meteorite. PhD Thesis, Washington University, Saint Louis, MO, 248p.

Imae N., Okazaki R., Kojima H., and Nagao K. (2002) The first nakhlite from Antarctica. *Lunar Planet. Sci.* **XXXIII**, #1483. Lunar and Planetary Institute, Houston (CD-ROM).

Jull A. J. T. and Donahue D. J. (1988) Terrestrial ^{14}C age of the Antarctic shergottite EETA 79001. *Geochim. Cosmochim. Acta* **52**, 1309–1311.

Jull A. J. T., Cielaszyk E., Brown S. T., and Donahue D. J. (1994) ^{14}C terrestrial ages of achondrites from Victoria Land, Antarctica. *Lunar Planet. Sci.* **XXV**. Lunar and Planetary Institute, Houston, pp. 647–648.

Jull A. J. T., Clandrud S. E., Schnabel C., Herzog G. F., Nishiizumi K., and Caffee M. W. (1999) Cosmogenic radio-nuclide studies of the Nakhlites. *Lunar Planet. Sci.* **XXX**, #1004. Lunar and Planetary Institute, Houston (CD-ROM).

Kitts K. and Lodders K. (1998) Survey and evaluation of eucrite bulk compositions. *Meteorit. Planet. Sci.* **33**, A197–A213.

Kong P., Mori T., and Ebihara M. (1997) Compositional continuity of enstatite chondrites and implications for heterogeneous accretion of the enstatite chondrite parent body. *Geochim. Cosmochim. Acta* **61**, 4895–4914.

Kortenkamp S. J. and Dermott S. F. (1998) Accretion of interplanetary dust particles by the Earth. *Icarus* **135**, 469–495.

Kurat G., Brandstätter F., Clayton R., Nazarov M. A., Palme H., Schultz L., Varela M. E., Wäsch E., Weber H. W., and Weckwerth G. (2001) D'Orbigny: a new and unusual angrite. *Lunar Planet. Sci.* **XXXII**, #1753. Lunar and Planetary Institute, Houston (CD-ROM).

Kyte F. T., Langenhorst F., and Tepley III F. J. (2000) The Eltanin meteorite: Large messenger from the HED or mesosiderite parent body? *Lunar Planet. Sci.* **XXXI** (#1811, Lunar and Planetary Institute, Houston, CD-ROM).

Lavielle B., Marti K., and Regnier S. (1985) Ages d'exposition des meteorites de fer: histoires multiples et variations d'intensité du rayonnement cosmique. In *Isotopic Ratios in the Solar System*. Cepadues-Éditions, Toulouse, France, pp. 15–20.

Lavielle B., Gilabert E., Soares M. R., Vasconcellos M. A. Z., Poupeau G., Canut de Bon C., Cisternas M. E., and Scorzelli R. B. (1998) Noble-gas and metal studies in the Vaca Muerta mesosiderite. *Meteorit. Planet. Sci.* **33**, A91–A92.

Lavielle B., Marti K., Jeannot J.-P., Nishiizumi K., and Caffee M. (1999) The $^{36}Cl-^{36}Ar-^{40}K-^{41}K$ records and cosmic ray production rates in iron meteorites. *Earth Planet. Sci. Lett.* **170**, 93–104.

Leya I., Lange H.-J., Neumann S., Wieler R., and Michel R. (2000) The production of cosmogenic nuclides in stony meteoroids by galactic cosmic ray particles. *Meteorit. Planet. Sci.* **35**, 259–286.

Leya I., Wieler R., Aggrey K., Herzog G. F., Schnabel C., Metzler K., Hildebrand A. R., Bouchard M., Jull A. J. T., Andrews H. R., Wang M.-S., Ferko T. E., Lipschutz M. E., Wacker J. F., Neumann S., and Michel R. (2001) Exposure history of the St-Robert (H5) fall. *Meteorit. Planet. Sci.* **36**, 1479–1494.

Lodders K. and Fegley B., Jr. (1998) *The Planetary Scientist's Companion*. Oxford University Press, New York, 371p.

Lorenzetti S., Eugster O., Busemann H., Marti K., Burbine T., and McCoy T. (2003) History and origin of enstatite achondrites. *Geochim. Cosmochim. Acta* **67**, 557–571.

Lugmair G. and Marti K. (1977) Sm–Nd–Pu time pieces in the Angra dos Reis meteorite. *Earth Planet. Sci. Lett.* **35**, 273–284.

Mahajan R. R., Murty S. V. S., and Ghosh S. (2000) Exposure ages of Lohawat (howardite) and Vissannapeta (eucrite), the recent falls in India. *Meteorit. Planet. Sci.* **35**, A101.

Marti K. (1967) Mass spectrometric detection of cosmic-ray produced ^{81}Kr in meteorites and the possibility of Kr–Kr dating. *Phys. Rev. Lett.* **18**, 264–266.

Marti K. and Graf T. (1992) Cosmic-ray exposure history of ordinary chondrites. *Ann. Rev. Earth Planet. Sci.* **20**, 221–243.

Marti K. and Lugmair G. (1971) Kr^{81}–Kr and K–Ar^{40} ages, cosmic-ray spallation products and neutron effects in lunar samples from Oceanus Procellarum. *Proc. 2nd Lunar Sci. Conf.*, 1591–1605.

Marti K. and Matthew K. J. (2002) Near-Earth asteroid origin for the Farmington meteorite. *Lunar Planet. Sci.* **XXXIII**, #1132. Lunar and Planetary Institute, Houston (CD-ROM).

Marty B., Marti K., Barrat J. A., Birck J. L., Blichert-Toft J., Chaussidon M., Deloule E., Gillet P., Göpel C., Jambon A., Manhès G., and Sautter V. (2001) Noble gases in new SNC meteorites NWA 817 and NWA 480. *Meteorit. Planet. Sci.* **36**, A122–A123.

Masarik J. and Reedy R. C. (1994) Effects of bulk composition on nuclide production processes in meteorites. *Geochim. Cosmochim. Acta* **58**, 5307–5317.

Masarik J. and Reedy R. C. (1995) Terrestrial cosmogenic-nuclide production systematics calculated from numerical simulations. *Earth Planet. Sci. Lett.* **136**, 381–395.

Masarik J., Nishiizumi K., and Reedy R. C. (2001) Production rates of ^{3}He, ^{21}Ne and ^{22}Ne in ordinary chondrites and the lunar surface. *Meteorit. Planet. Sci.* **36**, 643–650.

Mathew K. J. and Marti K. (2001) Early evolution of Martian volatiles: nitrogen and noble gas components in ALH84001 and Chassigny. *J. Geophys. Res.* **106**, 1401–1422.

Mazor E., Heymann D., and Anders E. (1970) Noble gases in carbonaceous chondrites. *Geochim. Cosmochim. Acta* **34**, 781–824.

McCoy T. J., Keil K., Clayton R. N., Mayeda T. K., Bogard D. D., Garrison D. H., Huss G. R., Hutcheon I. D., and Wieler R. (1996) A petrologic, chemical, and isotopic study of Monument Draw and comparison with other acapulcoites:

evidence for formation by incipient partial melting. *Geochim. Cosmochim. Acta* **60**, 2681–2708.

McCoy T. J., Keil K., Clayton R. N., Mayeda T. K., Bogard D. D., Garrison D. H., and Wieler R. (1997) A petrologic and isotopic study of lodranites: evidence for early formation as partial melt residues from heterogeneous precursors. *Geochim. Cosmochim. Acta* **61**, 621–637.

Megrue G. H. (1966) Rare-gas chronology of calcium-rich achondrites. *J. Geophys. Res.* **71**, 4021–4027.

Megrue G. H. (1968) Rare gas chronology of hypersthene achondrites and pallasites. *J. Geophys. Res.* **73**, 2027–2033.

Melosh H. J. (1989) *Impact Cratering,* Oxford University Press, NY.

Michel R., Leya I., and Borges L. (1996) Production of cosmogenic nuclides in meteoroids: accelerator experiments and model calculations to decipher the cosmic ray record in extraterrestrial matter. *Nucl. Instr. Meth. Phys. Res. B* **113**, 434–444.

Michlovich E. S., Vogt S., Masarik J., Reedy R. C., Elmore D., and Lipschutz M. E. (1994) ^{26}Al, ^{10}Be, and ^{36}Cl depth profiles in the Canyon Diablo iron meteorite. *J. Geophys. Res.: Planets* **99**, 23187–23194.

Mileikowsky C., Cucinotta F. A., Wilson J. W., Gladman B., Horneck G., Lindegren L., Melosh J., Rickman H., Valtonen M., and Zheng J. Q. (2000) Natural transfer of viable microbes in space. *Icarus* **145**, 391–427.

Mittlefehldt D. W. and Berkley J. L. (2002) Petrology and geochemistry of paired brachinites EET 99402 and EET 99407. *Lunar Planet. Sci.* **XXXIII**, #1008. Lunar and Planetary Institute, Houston (CD-ROM).

Miura Y. and Nagao K. (2000) Noble gases in Y-791192, Y 75032-type diogenites and A-881838. *Antarct. Meteorit.* **XXV**, 85–87.

Miura Y., Nagao K., and Fujitani T. (1993) ^{81}Kr terrestrial ages and grouping of Yamato eucrites based on noble gas and chemical compositions. *Geochim. Cosmochim. Acta* **57**, 1857–1866.

Miura Y. N. (1995) Studies on differentiated meteorites: evidence from ^{244}Pu-derived fission Xe, ^{81}Kr, other noble gases and nitrogen. PhD Dissertation, University of Tokyo, Tokyo, Japan, 209p.

Miura Y. N., Nagao K., Sugiura N., Fujitani T., and Warren P. H. (1998) Noble gases, ^{81}Kr–Kr exposure ages and ^{244}Pu–Xe ages of six eucrites, Béréba, Binda, Camel Donga, Juvinas, Millbillillie, and Stannern. *Geochim. Cosmochim. Acta* **62**, 2369–2387.

Nagai H., Honda M., Imamura M., and Kobayashi K. (1993) Cosmogenic ^{10}Be and ^{26}Al in metal, carbon and silicate of meteorites. *Geochim. Cosmochim. Acta* **57**, 3705–3723.

Nagao K., Takaoka N., and Saito K. (1983) Rare gas studies on the Antarctic Meteorites. *Abstr. 8th Symp. Ant. Meteorit.*, Tokyo, pp. 83–84.

Nagao K., Nakamura T., Miura Y. N., and Takaoka N. (1997) Noble gases and mineralogy of primary igneous materials of the Yamato-793605 shergottite. *Antarct. Meteorit. Res.* **10**, 125–142.

Nagao K., Nakamura T., Okazaki R., Miura Y. R., and Takaoka N. (1998) Two-stage irradiation of the Yamato 793605 Martian meteorite. *Meteorit. Planet. Sci.* **33**, A114.

Nier A. O. (1995) Helium and neon in interplanetary dust particles. In *Analysis of Interplanetary Dust*, AIP Conference Proceedings (eds. M. E. Zolensky, T. L. Wilson, F. J. M. Rietmeijer, and G. J. Flynn), Springer, New York, vol. 310, pp. 115–126.

Nishiizumi K. (1987) ^{53}Mn, ^{26}Al, ^{10}Be, and ^{36}Cl in meteorites: data compilation. *Nucl. Tracks Radiat. Meas.* **13**, 209–273.

Nishiizumi K. and Caffee M. (1996a) Exposure history of shergottite Queen Alexandra Range 94201. *Lunar Planet. Sci.* **XXVII**. Lunar and Planetary Institute, Houston, pp. 961–962.

Nishiizumi K. and Caffee M. (1996b) Exposure histories of lunar meteorites Queen Alexandra Range 94281 and 94269. *Lunar Planet. Sci.* **XXVII**. Lunar and Planetary Institute, Houston, pp. 959–960.

Nishiizumi K. and Caffee M. (1997) Exposure history of shergottite Yamato 793605. *Antarct. Meteorit.* **XXII**, 149–151.

Nishiizumi K. and Caffee M. W. (2001a) Exposure histories of lunar meteorites Northwest Africa 032 and Dhofar 081. *Lunar Planet. Sci.* **XXXII**, #2101. Lunar and Planetary Institute, Houston (CD-ROM).

Nishiizumi K. and Caffee M. (2001b) Exposure histories of lunar meteorites Dhofar 025, 026 and Northwest Africa 482. *Meteorit. Planet. Sci.* **36**, A148–A149.

Nishiizumi K., Klein J., Middleton R., Elmore D., Kubik P. W., and Arnold J. R. (1986) Exposure history of shergottites. *Geochim. Cosmochim. Acta* **50**, 1017–1021.

Nishiizumi K., Arnold J. R., Klein J., Fink D., Middleton R., Kubik P. W., Sharma P., Elmore D., and Reedy R. C. (1991a) Exposure histories of lunar meteorites: ALH81005, MAC88104, and Y791197. *Geochim. Cosmochim. Acta* **55**, 3149–3155.

Nishiizumi K., Arnold J. R., Klein J., Fink D., Middleton R., Sharma P., and Kubik P. W. (1991b) Cosmic ray exposure history of lunar meteorite Yamato 793274. In *16th Symp. Antarctic Meteorites*. Natl. Inst. Polar Res., Tokyo, Japan, pp. 188–191.

Nishiizumi K., Arnold J. R., Fink D., Klein J., Middleton R., Brownlee D. E., and Maurette M. (1991c) Exposure history of individual cosmic particles. *Earth Planet. Sci. Lett.* **104**, 315–324.

Nishiizumi K., Arnold J. R., Caffee M. W., Finkel R. C., Southon J., and Reedy R. C. (1992a) Cosmic ray exposure histories of lunar meteorites Asuka 881757, Yamato 793169 and Calcalong Creek. In *17th Symp. Antarctic Meteorites*. Natl. Inst. Polar Res., Tokyo, Japan, pp. 129–132.

Nishiizumi K., Arnold J. R., Caffee M. W., Finkel R. C., and Southon J. (1992b) Exposure histories of Calcalong Creek and LEW 88516 meteorites. *Meteoritics* **27**, 270.

Nishiizumi K., Caffee M. W., and Finkel R. C. (1994) Exposure histories of ALH 84001 and ALHA 77005. *Meteoritics* **29**, 511.

Nishiizumi K., Arnold J. R., Brownlee D. E., Caffee M. W., Finkel R. C., and Harvey R. P. (1995) Beryllium-10 and aluminum-26 in individual cosmic spherules from Antarctica. *Meteoritics* **30**, 728–732.

Nishiizumi K., Caffee M. W., Jull A. J. T., and Reedy R. C. (1996a) Exposure history of lunar meteorites Queen Alexandra Range 93069 and 94269. *Meteorit. Planet. Sci.* **31**, 893–896.

Nishiizumi K., Caffee M. W., Nagai H., and Imamura M. (1996b) Multiple breakup of ALH 85085-like and CR group chondrites. *Meteorit. Planet. Sci.* **31**, A99–A100.

Nishiizumi K., Caffee M. W., and Jull A. J. T. (1998) Exposure histories of Dar al Gani 262 lunar meteorites. *Lunar Planet. Sci.* **XXIX**, #1957. Lunar and Planetary Institute, Houston (CD-ROM).

Nishiizumi K., Masarik J., Caffee M. W., and Jull A. J. T. (1999) Exposure histories of pair lunar meteorites EET 96008 and EET 87521. *Lunar Planet. Sci.* **XXX**, #1980. Lunar and Planetary Institute, Houston (CD-ROM).

Nishiizumi K., Caffee M. W., and Masarik J. (2000a) Cosmogenic radionuclides in the Los Angeles Martian meteorite. *Meteorit. Planet. Sci.* **35**, A120.

Nishiizumi K., Caffee M. W., Bogard D. D., Garrison D. H., and Kyte F. T. (2000b) Noble gases and cosmogenic radionuclides in the Eltanin meteorite. *Lunar Planet. Sci.* **XXXI**, #2070. Lunar and Planetary Institute, Houston (CD-ROM).

Nishiizumi K., Caffee M. W., Jull A. J. T., and Klandrud S. E. (2001) Exposure history of shergottites Dar al Gani 476/489/670/735 and Sayh al Uhaymir 005. *Lunar Planet. Sci.* **XXXII**, #2117. Lunar and Planetary Institute, Houston (CD-ROM).

Nishiizumi K., Okazaki R., Park J., Nagao K., Masarik J., and Finkel R. C. (2002) Exposure and terrestrial histories of

Dhofar 019 Martian meteorite. *Lunar Planet. Sci.* **XXXIII**, #1366. Lunar and Planetary Institute, Houston (CD-ROM).

Nyquist L. E., Bogard D. D., Garrison D. H., and Reese Y. (1998) A single crater origin for Martian shergottites: resolution of the age paradox? *Lunar Planet. Sci.* **XXIX**, #1688. Lunar and Planetary Institute, Houston (CD-ROM).

Nyquist L. E., Bogard D. D., Shih C.-Y., Greshake A., Stöffler D., and Eugster O. (2001) Ages and geologic histories of Martian meteorites. *Space Sci. Rev.* **96**, 105–164.

Okazaki R., Takaoka N., Nakamura T., and Nagao K. (2000) Cosmic-ray exposure ages of enstatite chondrites. *Antarct. Meteorit. Res.* **13**, 153–169.

Olinger C. T., Maurette M., Walker R. M., and Hohenberg C. M. (1990) Neon measurements of individual Greenland sediment particles: proof of an extraterrestrial origin and comparison with EDX and morphological analyses. *Earth Planet. Sci. Lett.* **100**, 77–93.

Ostertag R., Stöffler D., Bischoff A., Palme H., Schultz L., Spettel B., Weber H., Weckwerth G., and Wänke H. (1986) Lunar Meteorite Yamato-791197: Petrography, shock history and chemical composition. *Mem. Natl. Inst. Polar. Res. (Tokyo), Spec. Issue* **41**, 17–44.

Ott U. and Begemann F. (2000) Spallation recoil and age of presolar grains in meteorites. *Meteorit. Planet. Sci.* **35**, 53–63.

Ott U., Löhr H.-P., and Begemann F. (1985) Noble gases and the classification of Brachina. *Meteoritics* **20**, 69–78.

Ott U., Löhr H.-P., and Begemann F. (1987) Noble gases in ALH 84025: like Brachina, unlike Chassigny. *Meteoritics* **22**, 476–477.

Paetsch M., Altmaier M., Herpers U., Kosuch H., Michel R., and Schultz L. (2000) Exposure age of the new SNC meteorite Sayh al Uhaymir 005. *Meteorit. Planet. Sci.* **35**, A124–A125.

Paetsch M., Weber H. W., and Schultz L. (2001) Noble gas investigations of new meteorites from Africa. *Lunar Planet. Sci.* **XXXII**, #1526. Lunar and Planetary Institute, Houston (CD-ROM).

Park J., Okazaki R., and Nagao K. (2001) Noble gases in the SNC meteorites: Dar al Gani 489, Sayh al Uhaymir 005 and Dhofar 019. *Meteorit. Planet. Sci.* **36**, A157.

Patzer A. and Schultz L. (2001) Noble gases in enstatite chondrites: I. Exposure ages, pairing and weathering effects. *Meteorit. Planet. Sci.* **36**, 947–961.

Patzer A., Franke L., and Schultz L. (2001) New noble gas data of four enstatite chondrites and Zaklodzie. *Meteorit. Planet. Sci.* **36**, A157–A158.

Pepin R. O., Palma R. L., and Schlutter D. J. (2001) Noble gases in interplanetary dust particles: II. Excess helium-3 in cluster particles and modeling constraints on interplanetary dust particle exposures to cosmic-ray irradiation. *Meteorit. Planet. Sci.* **36**, 1515–1534.

Polnau E. and Eugster O. (1998) Cosmic-ray produced, radiogenic and solar noble gases in lunar meteorites Queen Alexandra Range 94269 and 94281. *Meteorit. Planet. Sci.* **33**, 313–319.

Polnau E., Eugster O., Burger M., Krähenbühl U., and Marti K. (2001) Precompaction exposure of chondrules and implications. *Geochim. Cosmochim. Acta* **65**, 1849–1866.

Raisbeck G. M., Yiou F., Klein J., Middleton R., and Brownlee D. (1985a) ^{26}Al/^{10}Be in deep sea spherules as evidence of cometary origin. In *Properties and Interactions of Interplanetary Dust* (eds. R. H. Giese and P. Lamy). D. Reidel, Dordrecht, The Netherlands, pp. 169–174.

Raisbeck G. M., Yiou F., and Brownlee D. (1985b) Unusually high concentration of ^{10}Be in a cosmic spherule: possible evidence for irradiation outside the planetary solar system. *Meteoritics* **20**, 734–735.

Rao M. N., Garrison D. H., Palma R. L., and Bogard D. D. (1997) Energetic proton irradiation history of the howardite parent body regolith and implications for ancient solar activity. *Meteorit. Planet. Sci.* **32**, 531–543.

Reedy R. C. (1987a) Nuclide production by primary cosmic-ray protons. *Proc. 17th Lunar Planet. Sci. Conf. Part 2: J. Geophys Res.* **92**, E697–E702.

Reedy R. C. (1987b) Cosmogenic nuclide production in small metallic spherules. *Lunar Planet. Sci.* **XVIII**. Lunar and Planetary Institute, Houston, pp. 820–821.

Reedy R. C. (1990) Cosmogenic-radionuclide production rates in mini-spherules. *Lunar Planet. Sci.* **XXI**. Lunar and Planetary Institute, Houston, pp. 1001–1002.

Reedy R. C. and Masarik J. (1994) Cosmogenic-nuclide depth profiles in the lunar surface. *Lunar Planet. Sci.* **XXV**. Lunar and Planetary Institute, Houston, pp. 1119–1120.

Reedy R. C., Arnold J. R., and Lal D. (1983) Cosmic-ray record in solar system matter. *Ann. Rev. Nucl. Part. Sci.* **33**, 505–537.

Reedy R. C., Masarik J., Nishiizumi K., Arnold J. R., Finkel R. C., Caffee M. W., Southon J., Jull A. J. T., and Donahue D. J. (1993) Cosmogenic-radionuclide profiles in Knyahinya. *Lunar Planet. Sci.* **XXIV**. Lunar and Planetary Institute, Houston, pp. 1195–1196.

Rubin A. E., Kallemeyn G. W., Wasson J. T., Clayton R. N., Mayeda T. K., Grady M., Verchovsky A. B., Eugster O., and Lorenzetti S. (2003) Formation of metal and silicate globules in Gujba: a new Bencubbin-like meteorite fall from Nigeria. *Geochim. Cosmochim. Acta* (in press).

Scherer P. and Schultz L. (1999) Noble gases in the SNC meteorite Dar al Gani 476. *Lunar Planet. Sci.* **XXX**, #1144. Lunar and Planetary Institute, Houston (CD-ROM).

Scherer P. and Schultz L. (2000) Noble gas record, collisional history and pairing of CV, CO, CK and other carbonaceous chondrites. *Meteorit. Planet. Sci.* **35**, 145–153.

Scherer P., Paetsch A., and Schultz L. (1998) Noble gas study of the new highland meteorite Dar al Gani 400. *Meteorit. Planet. Sci.* **32**, A135.

Schnabel C., Ma P., Herzog G. F., Faestermann T., Knie K., and Korschinek K. (2001) ^{10}Be, ^{26}Al, and ^{53}Mn in Martian meteorites. *Lunar Planet. Sci.* **XXXII**, #1353. Lunar and Planetary Institute, Houston (CD-ROM).

Schultz L. and Hintenberger H. (1967) Edelgasmessungen an Eisenmeteoriten. *Z. Naturforsch.* **22a**, 773–779.

Schultz L. and Freundel M. (1984) Terrestrial ages of Antarctic meteorites. *Meteoritics* **19**, 310.

Schultz L. and Franke L. (2002) Helium, neon, and argon in meteorites: A data collection. Update 2002. CD-ROM Max-Planck-Institut für Chemie, Mainz.

Shu F. H., Shang H., and Lee T. (1996) Toward an astrophysical theory of chondrites. *Science* **271**, 1545–1552.

Shukolyukov A. and Begemann F. (1996) Cosmogenic and fissiogenic noble gases and Kr-81-Kr exposure age clusters of eucrites. *Meteorit. Planet. Sci.* **31**, 60–72.

Shukolyukov Yu. A. and Petaev M. I. (1992) Noble gases in the Omolon pallasite. *Lunar Planet. Sci.* **XXIII**. Lunar and Planetary Institute, Houston, pp. 1297–1298.

Shukolyukov Yu. A., Nazarov M. A., and Schultz L. (2000) Dhofar 019: a shergottite with an approximately 20-million-year exposure age. *Meteorit. Planet. Sci.* **35**, A147.

Shukolyukov Yu. A., Nazarov M. A., Pätsch M., and Schultz L. (2001) Noble gases in three meteorites from Oman. *Lunar Planet. Sci.* **XXXII**, #1502. Lunar and Planetary Institute, Houston (CD-ROM).

Sisodia M. S., Shukla A. D., Suthar K. M., Mahajan R. R., Murty S. V. S., Shukla P. N., Bhandari N., and Natarajan R. (2001) Lohawat howardite: mineralogy, chemistry and cosmogenic effects. *Meteorit. Planet. Sci.* **36**, 1457–1466.

Spergel M. S., Reedy R. C., Lazareth O. W., Levy P. W., and Slatest A. (1986) Cosmogenic neutron-capture-produced nuclides in stony meteorites. *Proc. 16th Lunar Planet. Sci. Conf.: J. Geophys. Res. Suppl.* **91**, D484–D494.

Spitale J. and Greenberg R. (2002) Numerical evaluation of the general Yarkovsky effect: effects on eccentricity and longitude of periapse. *Icarus* **156**, 211–222.

Stauffer H. (1962) On the production ratios of rare gas isotopes in stone meteorites. *J. Geophys. Res.* **67**, 2023–2028.

Swindle T. D., Burkland M. K., and Grier J. A. (1995) Noble gases in the lunar meteorites Calcalong Creek and Queen Alexandra Range 93069. *Meteoritics* **30**, 584–585.

Swindle T. D., Kring D. A., Burkland M. K., Hill D. H., and Boynton W. F. (1998) Noble gases, bulk chemistry, and petrography of olivine-rich achondrites Eagles Nest and Lewis Cliff 88763: comparison to brachinites. *Meteorit. Planet. Sci.* **33**, 31–48.

Takaoka N. and Yoshida Y. (1992) Noble gases in Yamato-793274 and 86032 lunar meteorites. *Proc. Natl. Inst. Polar Res. Symp. Antarct. Meteorit.* **5**, 36–48.

Terribilini D., Eugster O., Burger M., Jakob A., and Krähenbühl U. (1998) Noble gases and chemical composition of Shergotty mineral fractions, Chassigny and Yamato 793605: the trapped argon-40/argon-36 ratio and ejection times of Martian meteorites. *Meteorit. Planet. Sci.* **33**, 677–684.

Terribilini D., Eugster O., Herzog G. F., and Schnabel C. (2000a) Evidence for common break-up events of the acapulcoites/lodranites and chondrites. *Meteorit. Planet. Sci.* **35**, 1043–1050.

Terribilini D., Busemann H., and Eugster O. (2000b) Krypton-81-krypton cosmic-ray exposure ages of the Martian meteorites including the new shergottite Los Angeles. *Meteorit. Planet. Sci.* **35**, A155–A156.

Terribilini D., Eugster O., Mittlefehldt D. W., Diamond L. W., Vogt S., and Wang D. (2000c) Mineralogical and chemical composition and cosmic-ray exposure history of two mesosiderites and two iron meteorites. *Meteorit. Planet. Sci.* **35**, 617–628.

Thalmann C., Eugster O., Herzog G. F., Klein J., Krähenbühl U., Vogt S., and Xue S. (1996) History of lunar meteorites Queen Alexandra Range 93069, Asuka 881757 and Yamato 793169 based on noble gas isotopic abundances, radionuclide concentrations and chemical composition. *Meteorit. Planet. Sci.* **31**, 857–868.

Tuniz C., Bird J., Fink D., and Herzog G. F. (1998) *Accelerator Mass Spectrometry: Ultrasensitive Analysis for Global Science*. CRC Press, Boca Raton, chap. 9, pp. 155–176.

Vokrouhlický D., Milani A., and Chesley S. R. (2000) Yarkovsky effect on small near-Earth asteroids: Mathematical formulation and examples. *Icarus* **148**, 118–138.

Vogt S., Herzog G. F., and Reedy R. C. (1990) Cosmogenic nuclides in extraterrestrial materials. *Rev. Geophys.* **28**, 253–275.

Vogt S., Herzog G. F., Fink D., Klein J., Middleton R., Dockhorn B., Korschinek G., and Nolte E. (1991) Exposure histories of the lunar meteorites MacAlpine Hills 88104, MacAlpine Hills 88105, Yamato 791197 and Yamato 86032. *Geochim. Cosmochim. Acta* **55**, 3157–3165.

Vogt S., Herzog G. F., Eugster O., Michel Th., Niedermann S., Krähenbühl U., Middleton R., Dezfouly-Arjomandy B., Fink D., and Klein J. (1993) Exposure history of the lunar meteorite Elephant Moraine 87521. *Geochim. Cosmochim. Acta* **57**, 3793–3799.

Voshage H. (1967) Bestrahlungsalter und Herkunft der Eisenmeteorite. *Z. Naturforsch* **22a**, 477–506.

Voshage H. and Feldmann H. (1979) Investigations on cosmic-ray-produced nuclides in iron meteorites: 3. Exposure ages, meteoroid sizes and sample depths determined by mass spectrometric analyses of potassium and rare gases. *Earth Planet. Sci. Lett.* **45**, 293–308.

Voshage H. and Hintenberger H. (1963) The cosmic-ray exposure ages of iron meteorites as derived from the isotopic composition of potassium and the production rates of cosmogenic nuclides in the past. In *Radioactive Dating*, International Atomic Energy Agency, Vienna, pp. 367–379.

Voshage H., Feldmann H., and Braun O. (1983) Investigations of cosmic-ray-produced nuclides in iron meteorites: 5. More data on the nuclides of potassium and noble gases, on exposure ages and meteoroid sizes. *Z. Naturforsch.* **38a**, 273–280.

Warren P. (1994) Lunar and Martian meteorite delivery services. *Icarus* **111**, 338–353.

Warren P. H. (2001) Porosities of lunar meteorites: strength, porosity and petrologic screening during the meteorite delivery process. *J. Geophys. Res.* **106**, 10101–10111.

Wasson J. T. (1974) *Meteorites*. Springer, Berlin, 316p.

Wasson J. T. and Kallemeyn G. W. (2002) The IAB iron-meteorite complex: a group, five subgroups, numerous grouplets, closely related, mainly formed by crystal segregation in rapidly cooling melts. *Geochim. Cosmochim. Acta* **66**, 2445–2473.

Weber H. W. and Schultz L. (2001) Noble gases in five new Rumuruti chondrites. *Lunar Planet Sci.* **XXXII**, #1500. Lunar and Planetary Institute, Houston (CD-ROM).

Weber H. W., Franke L., and Schultz L. (2001) Subsolar noble gases in metal-rich carbonaceous (CH) chondrites. *Meteorit. Planet. Sci.* **36**, A220–A221.

Weigel A., Eugster O., Koeberl C., and Krähenbühl U. (1997) Differentiated achondrites Asuka 881371, an angrite and Divnoe: Noble gases, ages, chemical composition and relation to other meteorites. *Geochim. Cosmochim. Acta* **61**, 239–248.

Weigel A., Eugster O., Koeberl C., Michel R., Krähenbühl U., and Neumann S. (1999) Relationships among lodranites and acapulcoites: Noble gas isotopic abundances, chemical composition, cosmic-ray exposure ages and solar cosmic ray effects. *Geochim. Cosmochim. Acta* **63**, 175–192.

Welten K. C., Lindner L., van der Borg K., Loeken T., Scherer P., and Schultz L. (1997) Cosmic-ray exposure ages of diogenites and the recent collisional history of the howardite, eucrite, and diogenite parent body/bodies. *Meteorit. Planet. Sci.* **32**, 891–902.

Welten K. C., Nishiizumi K., Caffee M. W., and Schultz L. (2001a) Update on exposure ages of diogenites: the impact history of the HED parent body and evidence of space erosion and/or collisional disruption of stony meteoroids. *Meteorit. Planet. Sci.* **36**, A223.

Welten K. C., Bland P. A., Russell S. S., Grady M. M., Caffee M. W., Masarik J., Jull A. J. T., Weber H. W., and Schultz L. (2001b) Exposure age, terrestrial age and pre-atmospheric radius of the Chinguetti mesosiderite: not part of a much larger mass. *Meteorit. Planet. Sci.* **36**, 939–946.

Wieler R. (2002) Cosmic-ray-produced noble gases in meteorites. In *Noble Gases in Geochemistry and Cosmochemistry*, Reviews in Mineralogy and Geochemistry (eds. D. Porcelli, C. J. Ballertine, and R. Wieler). *Min. Rev. Geochem.* **47**, 125–170.

Wieler R. and Graf T. (2001) Cosmic ray exposure history of meteorites. In *Accretion of Extraterrestrial Material on Earth over Time* (eds. B. Peucker-Ehrenbrink and B. Schmitz). Kluwer Academic/plenum, New York, pp. 221–240.

Wieler R., Pedroni A., and Leya I. (2000) Cosmogenic neon in mineral separates from Kapoeta: no evidence for an irradiation of its parent body regolith by an early active Sun. *Meteorit. Planet. Sci.* **35**, 251–257.

Wright R. J., Simms L. A., Reynolds M. A., and Bogard D. D. (1973) Depth variation of cosmogenic noble gases in the 120-kg Keyes chondrite. *J. Geophys. Res.* **78**, 1308–1318.

Xue S., Herzog G. F., Klein J., and Middleton R. (1994) ^{26}Al And ^{10}Be activities and exposure ages of lodranites, acapulcoites, Kakangari, and Pontlyfni. *Lunar Planet. Sci.* **XXV**. Lunar and Planetary Institute, Houston, pp. 1523–1524.

Yanai K. (2000) Achondrite polymict breccia 1153: a new lunar meteorite classified to anorthositic regolith breccia. *Lunar Planet. Sci.* **XXXI**, #1101. Lunar and Planetary Institute, Houston (CD-ROM).

Zipfel J., Scherer P., Spettel B., Dreibus G., and Schultz L. (2000) Petrology and chemistry of the new shergottite Dar al Gani 476. *Meteorit. Planet. Sci.* **35**, 95–106.

1.14
Noble Gases

F. A. Podosek

Washington University, St. Louis, MO, USA

1.14.1 INTRODUCTION	381
1.14.2 *IN SITU* COMPONENTS AND NUCLEAR COMPONENTS	383
1.14.2.1 Radiogenic Noble Gases	384
1.14.2.2 Spallation Noble Gases	386
1.14.3 SOLAR NOBLE GASES	388
1.14.3.1 Deuterium Burning and the Solar $^3He/^4He$ Ratio	389
1.14.3.2 Solar Wind in Lunar Soils	390
1.14.3.3 Solar-wind Composition and Solar Composition	392
1.14.4 EXOTIC COMPONENTS	393
1.14.4.1 Presolar Carriers	394
1.14.4.2 Radically Anomalous Exotic Noble-gas Components	395
1.14.4.3 Moderately Anomalous Exotic Noble-gas Components	397
1.14.5 NOBLE-GAS CHEMISTRY AND TRAPPED COMPONENTS	398
1.14.5.1 Planetary Noble Gases	398
1.14.5.2 The Q-Component	400
1.14.5.3 Trapping Mechanisms	401
1.14.5.4 Loose Ends?	402
1.14.5.5 U–xenon?	402
1.14.5.6 Planetary Atmospheres	403
REFERENCES	403

1.14.1 INTRODUCTION

The noble gases are the group of elements—helium, neon, argon, krypton, xenon—in the rightmost column of the periodic table of the elements, those which have "filled" outermost shells of electrons (two for helium, eight for the others). This configuration of electrons results in a neutral atom that has relatively low electron affinity and relatively high ionization energy. In consequence, in most natural circumstances these elements do not form chemical compounds, whence they are called "noble." Similarly, much more so than other elements in most circumstances, they partition strongly into a gas phase (as monatomic gas), so that they are called the "noble gases" (also, "inert gases"). (It should be noted, of course, that there is a sixth noble gas, radon, but all isotopes of radon are radioactive, with maximum half-life a few days, so that radon occurs in nature only because of recent production in the U–Th decay chains. The factors that govern the distribution of radon isotopes are thus quite different from those for the five gases cited. There are interesting stories about radon, but they are very different from those about the first five noble gases, and are thus outside the scope of this chapter.)

In the nuclear fires in which the elements are forged, the creation and destruction of a given nuclear species depends on its nuclear properties, not on whether it will have a filled outermost shell when things cool off and nuclei begin to gather electrons. The numerology of nuclear physics is different from that of chemistry, so that in the cosmos at large there is nothing systematically special about the abundances of the noble gases as compared to other elements. We live in a very nonrepresentative part of the cosmos, however. As is discussed elsewhere in this volume, the outstanding generalization about the geo-/cosmo-chemistry of the terrestrial planets is that at some point thermodynamic conditions dictated phase separation of solids from gases, and that the Earth and the rest of the inner solar were made by collecting the solids, to the rather efficient exclusion of the gases. In this grand separation the noble gases, because they are noble, were partitioned strongly into the gas phase. The resultant generalization is that the noble gases are very scarce in the materials of the inner solar system, whence their common synonym "rare gases."

This scarcity is probably the most important single feature to remember about noble-gas cosmochemistry. As illustration of the absolute quantities, for example, a meteorite that contains xenon at a concentration of order 10^{-10} cm^{3}STP g^{-1} (4×10^{-15} mol g^{-1}) would be considered relatively rich in xenon. Yet this is only 0.6 ppt (part per trillion, fractional abundance 10^{-12}) by mass. In most circumstances, an element would be considered efficiently excluded from some sample if its abundance, relative to cosmic proportions to some convenient reference element, were depleted by "several" orders of magnitude. But a noble gas would be considered to be present in quite high concentration if it were depleted by only four or five orders of magnitude (in the example above, 10^{-10} cm^{3}STP g^{-1} of xenon corresponds to depletion by seven orders of magnitude), and one not uncommonly encounters noble-gas depletion of more than 10 orders of magnitude.

The second most important feature to note about noble-gas cosmochemistry is that while a good deal of the attention given to noble gases really is about chemistry, traditionally a good deal of attention is also devoted to nuclear phenomena, much more so than for most other elements. This feature is a corollary of the first feature noted above, namely scarcity. A variety of nuclear transmutation processes—decay of natural radionuclides and energetic particle reactions—lead to the production of new nuclei that are often new elements. Most commonly, the quantity of new nuclei originating in nuclear transmutation is very small compared to the quantity already present in the sample in question, metaphorically a drop in the bucket. Thus, they are very difficult or impossible to detect and, therefore, in practical terms, attracting little or no interest. When the bucket is empty, or nearly so, however, the "drop" contributed by nuclear transmutations may become observable or even dominant. Traditionally there are two types of (nearly) empty buckets that are most suitable for revealing the effects of nuclear transmutations: short-lived radionuclides (e.g., ^{10}Be and ^{26}Al) which would be entirely absent except for recent nuclear reactions, and the noble gases, renowned for their scarcity.

Emphasis on nuclear processes explains what sometimes seems to be an obsession with isotopes in noble-gas geo- and cosmochemistry. Different nuclear processes will produce different isotopes, singly or in suites with well-defined proportions (i.e., "components"), different from one process to another. Much of the traditional agenda of noble-gas geochemistry, and especially cosmochemistry, thus consists of isotopic analysis, and deconvolution of an observed isotopic spectrum into constituent components. (In most geochemical investigations, noble gases are detected by mass spectrometry, a technique that is inherently sensitive to specific isotopes, not just the chemical element. Isotopic data thus emerge naturally in most studies. Noble-gas mass spectrometry can be a much more sensitive technique than other traditional types of mass spectrometry because the gases are "noble," and therefore relatively easy to separate from other elements, and because they are scarce, so that they can be analyzed in "static"-mode (no pumping during analysis) gas-source spectrometers, permitting relatively high detection efficiency without overwhelming blanks.) In realistic terms, it is very difficult to appreciate noble-gas geo-/cosmochemistry without a basic familiarity with noble-gas isotopes: which isotopes occur in nature (i.e., which are stable), in what approximate abundance they are found, how they relate to non-noble neighbors, and, to some extent, how they are associated with specific nuclear processes. Figure 1 provides assistance in this regard.

When the goal is to identify and quantify different noble-gas components that may be present in a sample or group of samples, a common approach to this goal is to try to unmix the components, at least partially, to provide some leverage. One path to this end, of course, is analysis of different samples that may contain the components in different proportions, and thus have different isotopic compositions. Another path, available in addition to or instead of the first, is stepwise heating analysis, which has traditionally been very extensively used in noble-gas studies. Noble gases may be released from solid samples by volume diffusion, or by reaction, recrystallization, melting, or even evaporation of their host phases. If different noble-gas components reside in physically distinct locations within a complex sample, they may be liberated, and thus become available for analysis, at different steps in a time–temperature heating sequence. Differential release of isotopically distinct components will then result in variation of the isotopic composition of gas released in different steps (e.g., see Figures 2 and 4).

A common tool for visualization of isotopic variations is the so-called "three-isotope diagram," in which two isotope ratios, each with the same reference (denominator) isotope, are

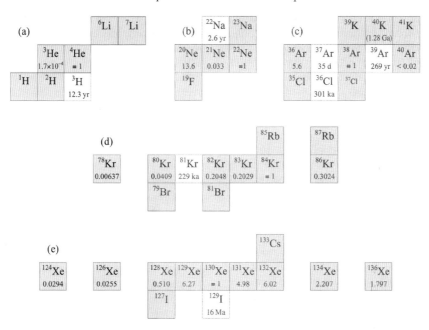

Figure 1 A display of the isotopes of the noble gases and neighboring isotopes in the familiar "chart of the nuclides" format. The abscissa is neutron number (N) and the ordinate is proton number (Z). The box corresponding to any pair (Z, N) represents an isotope; an element is represented by a horizontal row. Boxes for stable isotopes are shown with solid outline; for the noble gases, approximate solar (in the case of He, protosolar) isotope ratios are shown at the bottom of each box. Selected unstable isotopes are shown as boxes with broken line edges. The left-superscript isotope label is the atomic weight A ($= Z + N$). The five panels show regions around the five noble gases (excluding Rn).

displayed on abscissa and ordinate (e.g., Figure 2). Two isotopically distinct components will plot at distinct points on a three-isotope diagram, and an often-used feature is that mixtures of the two components will plot on the *straight line* joining those two points. A lever rule applies: the greater the proportion that one component contributes to a mixture, the closer the point representing the mixture will lie to the point representing that end-member component, and there is a *linear* relationship between fractional distance from one end-member to the other and the fraction that each component contributes to the mixture (specifically to the *reference isotope*). If observed isotopic data are variable but the variations in two ratios are correlated, so as to be consistent with a straight line on a three-isotope diagram, it can be inferred that at least two components are present and it will often be hypothesized that *only* two components are present, in which case their compositions can be constrained to lie on the line, one on either side of the data field. If three components are present, not coincidentally collinear on this diagram, mixtures will occupy the triangular field defined by the three compositions, and conversely if observed data are not consistent with linear correlation it can be inferred that at least three components are contributing to the mix. The concept of the three-isotope diagram is readily generalized. Four isotopes defining three ratios (all with the same reference isotope), for example, will define a three-dimensional space in which mixture of two components will produce compositions lying along a straight line, and mixture of three components will produce compositions lying in a plane, etc. Generalization to more dimensions is mathematically straightforward, even if difficult to envision.

1.14.2 IN SITU COMPONENTS AND NUCLEAR COMPONENTS

In noble-gas geo-/cosmochemistry the term "*in situ* component" is conventionally used to designate a component which is produced *in situ*, i.e., within any given (solid) sample, by nuclear transmutation, and whose constituent atoms are still identifiably in the places in which they were made. For the noble gases, there are many *in situ* components of interest in one context or another. As stressed above, this is because noble gases are very scarce in most solid materials. The small quantities of atoms produced in most nuclear transmutation are essentially lost and overwhelmed in the background sea of most elements, but they can be quite prominent when the product atoms are noble gases.

The term "nuclear component," meaning produced by nuclear transmutation, is sometimes used almost synonymously with "*in situ* component," but the two terms do not mean quite the same thing. (In this context it does not pay to be too fussy about definitions. Since all atomic nuclei are made in nuclear processes, by strict application of this definition everything is a nuclear component, and the term becomes useless. Instead, the term "nuclear component" is generally used when there is some specific, typically local, nuclear process in mind.)

An *in situ* component is necessarily a nuclear component, but the converse need not be true: an originally *in situ* component could be mobilized and transported, and in the process perhaps mixed with other components, in which case it will no longer be an *in situ* component, but it might still be a recognizable nuclear component. As one specific example, it is believed that radiogenic noble gases are exhaled from the interior of the Earth's Moon, and then partially implanted into lunar soils on the surface (see below), where they are readily distinguishable from other reservoirs of noble gases. Such a gas would still be called a nuclear component even though it is not an *in situ* component.

1.14.2.1 Radiogenic Noble Gases

One prominent and well-known kind of nuclear component is that which is produced by the decay of naturally occurring radionuclides (see Table 1). The best- and longest-known examples are ^{4}He, produced by alpha decay of the natural isotopes of uranium and Th, and ^{40}Ar, produced in one branch of the beta decay of ^{40}K. (There are several other natural radionuclides which produce ^{4}He by alpha decay, but whether because of low parent abundance and/or very slow decay, only in very unusual samples is the production of ^{4}He not strongly dominated by uranium and thorium.) Since radioactive decay laws are well known, the ratio of daughter to parent isotope(s) in a closed system is a simple function of time, whence this phenomenon has been long and extensively exploited as a geochronometer (e.g., see Chapter 1.16).

^{4}He and especially ^{40}Ar nuclei provide good specific examples of how scarcity affects the geo-/cosmochemistry of noble gases. It is very difficult to find or isolate samples in which the total inventory of either of these isotopes is not overwhelmingly dominated by the radiogenic component, and in the case of ^{40}Ar it is so despite the fact that it is produced in only a minor branch (~11%) of the decay of ^{40}K. The major branch leads to ^{40}Ca, which in most rocks is so abundant that radiogenic ^{40}Ca is difficult if not impossible to even detect.

Radionuclides can also decay by spontaneous fission. Fission of an actinide nucleus produces two main fragments, not always the same, so that the products of fission are not one or two specific isobars but rather a spectrum, whence a fission component typically consists of several isotopes in statistically well-defined proportions. The distribution of actinide fission fragments is characteristically asymmetric, one relatively heavy and one relatively light, so that the distribution of daughter masses is bimodal. By chance, xenon typically lies within the heavier mode, krypton within the lighter, so that xenon and krypton are prominent fission products (yields of a few to several percent). Actinide nuclei have higher neutron-to-proton ratios than lighter nuclei, so that their directly produced fission fragments characteristically lie far on the neutron-rich side of the valley of beta-stability at their respective isobars. Fission fragment nuclei thus typically beta-decay, more or less quickly, along an isobar until they reach a stable nuclide. This feature is often invoked to characterize a fission component, e.g., for xenon the heavier isotopes ^{136}Xe, ^{134}Xe, ^{132}Xe, ^{131}Xe, and ^{129}Xe can all appear in a fission component, whereas the lighter group ^{130}Xe, ^{128}Xe, ^{126}Xe, and ^{124}Xe cannot, because production of these lighter isobars yields stable isotopes of tellurium, not xenon. Sometimes the adjective "fissiogenic" is used to designate a component produced in fission, although this seems unnecessarily awkward when "radiogenic" will serve as well.

Table 1 Cosmochemically prominent radiogenic noble gases.

Isotope	Process	Parent	Remarks
^{4}He	α	^{232}Th, ^{238}U, ^{235}U, ...	Dominant in nearly all samples
^{40}Ar	β	^{40}K	Dominant in nearly all samples
^{129}Xe	β	^{129}I	Prominent in most meteorites
Xe, Kr	SF[a]	^{244}Pu	Observable in selected meteorites
Xe, Kr	SF[a]	^{238}U	Sometimes detectable in selected meteorites
^{36}Ar	β	^{36}Cl	Reported, but needs further study for verification
^{22}Ne	β	^{22}Na	Presolar component; see Table 2

[a] SF = spontaneous fission. Fission fragments are rich in neutrons and decay toward beta stability from the neutron-rich side. Fission produces all the heavier isotopes of Xe and Kr which are not "shielded" by the existence of more neutron-rich stable isobars, i.e., ^{136}Xe, ^{134}Xe, ^{132}Xe, ^{129}Xe, ^{86}Kr, ^{84}Kr, and ^{83}Kr. Each fissioning nuclide produces these isotopes in distinct and diagnostic proportions.

In practice, only one natural radionuclide, ^{238}U, produces observable *in situ* spontaneous-fission components in terrestrial (and most lunar) samples, and even then the fission components are observable only in samples unusually deficient in noble gases and/or rich in uranium. The spectra of ^{238}U fission components, and the yields at xenon and krypton, are well known (e.g., see Ozima and Podosek, 2002), so that radiogenic xenon and krypton from fission of ^{238}U are readily identifiable. In principle, a useful radiometric chronology scheme could be based on accumulation of radiogenic xenon (and krypton) from fission of ^{238}U, but this has not yet been much exploited.

Meteorites are characteristically old, many dating from the oldest times in the solar system, and at the time of their formation they incorporated several relatively short-lived natural radionuclides (e.g., see Podosek and Nichols, 1998), radionuclides which have not only decayed to effective extinction at present but which were also effectively extinct at the time of formation of the oldest terrestrial and most lunar samples. The abundances of these short-lived radionuclides were generally quite low, so that their presence may be inferred from excesses of their daughter isotopes primarily in cases where there are large fractionations between parent and daughter elements. Since noble gases are strongly depleted relative to nearly everything in solid materials, it is not surprising that the first two now-extinct radionuclides to be discovered, ^{129}I (half-life 16 Myr) and ^{244}Pu (half-life 78 Myr), have noble-gas daughters.

The presence of ^{129}I in the early solar system was inferred from excesses of its daughter ^{129}Xe (Reynolds, 1960), and it has subsequently been found to have been present in essentially all undifferentiated meteorites and even some differentiated meteorites. This discovery was important in establishing our present paradigm for solar-system chronology: it showed that undifferentiated meteorites were indeed the oldest solids in the solar system and also that they formed nearly simultaneously (differing in age by no more than 10–20 Ma). Moreover, on the premise that the ^{129}I must have been synthesized in some other star, it could be concluded that the solar system as a whole could not be much older (no more than 10^8 yr) than undifferentiated meteorites, else the abundance of ^{129}I would have decayed beyond observation. Tighter limits of both kinds are available at present, based on subsequently discovered shorter-lived radionuclides, but the general picture first based on ^{129}I persists. In addition, because radiogenic ^{129}Xe is so readily measurable as well as observable in most meteorites, I–Xe dating is arguably the most extensively applicable fine-scale chronometer for meteorite history in the early solar system (e.g., Swindle and Podosek, 1988).

The prior presence of ^{244}Pu, the only transuranic nuclide known to have been present in the early solar system, can be inferred from its spontaneous-fission decay branch, through production of fission tracks and, more diagnostically, by production of fission xenon and krypton. The identification of ^{244}Pu as the fissioning nuclide present in meteorites is unambiguous, since the meteoritic fission spectrum is distinct from that of ^{238}U but consistent with that of artificial ^{244}Pu (Alexander *et al.*, 1971). The demonstration of the existence of ^{244}Pu in the solar system reinforced the requirement (from the presence of ^{129}I) of a relatively short time between stellar nucleosynthesis and solar-system formation and made it incontrovertible, since while it might be possible to make ^{129}I in some models of early solar system development, the rapid capture of multiple neutrons (the r-process) needed to synthesize ^{244}Pu could not plausibly be supposed to have happened in the solar system.

Besides their presence due to *in situ* radioactive decay within a given solid sample, radiogenic ^{4}He, ^{40}Ar, ^{129}Xe, ^{244}Pu-fission xenon (and krypton), and likely also ^{238}U-fission xenon, are also prominent or observable constituents of planetary atmospheres, and their abundance is important in constraining models for planetary atmosphere evolution (see Chapter 4.12).

Besides these "known" cases of radiogenic noble gases, there are a few other cases of interest. One is ^{248}Cm (half-life 0.35 Myr), also a transuranic radionuclide that might plausibly be speculated to have been present in the early solar system. Evidence for fission xenon or krypton from ^{248}Cm has been long sought but never found. Another is ^{36}Cl (half-life 300 kyr), which decays to ^{36}Ar. There is indeed a positive report for the presence of ^{36}Cl (Murty *et al.*, 1997), but only in one meteorite, so until confirmed this must be considered to remain only a "hint." As discussed below, there is good evidence for generation of a pure ^{22}Ne component from decay of ^{22}Na (half-life 2.6 yr; see Figure 1(b)), but since the decay evidently occurred in the outflow of the star that made the ^{22}Na (see below), not within the solar system, this is not customarily listed among "radiogenic" noble gases. A particularly interesting case is that of CCF-Xe ("carbonaceous chondrite fission" xenon). Quite early in the study of noble-gas cosmochemistry, xenon isotopic variations revealed in stepwise heating of carbonaceous chondrites (Reynolds and Turner, 1964) suggested an interpretation in terms of fission, but the effect could not be successfully associated with any known fissioning radionuclide. A case was made (e.g., Anders *et al.*, 1975) that CCF-Xe was produced by fission of a superheavy element (in the vicinity of atomic number 115), but this was never substantiated.

The term CCF-Xe is no longer in use, and it is now understood that the effects that led to its use are attributable to Xe-HL (or Xe-H), an exotic nucleosynthetic component (see below), but not a fission product.

1.14.2.2 Spallation Noble Gases

Galactic cosmic rays (GCR) are very high energy particles, chiefly protons, which can induce a wide variety of nuclear reactions when they interact with some target nucleus. Often the products of a reaction induced by a primary GCR particle include secondary protons, neutrons, α-particles, etc., which are themselves sufficiently energetic to induce further reactions when they strike another target nucleus, and so on, so that one primary particle can generate a substantial cascade of secondaries. The characteristic attenuation depth for cosmic rays is some $100-150$ g cm^{-2}; the characteristic depth for secondaries, particularly neutrons, is greater, and the flux of thermalized neutrons may peak at depth $\sim 250-300$ g cm^{-2}.

In a strict sense, spallation is a nuclear fragmentation process in which the target nucleus loses several nucleons. As used in cosmochemistry, however, the term is used more broadly to designate the product of any nuclear transformation induced by cosmic rays, primary or secondary, whether produced by spallation in the strict sense or by more specific reaction channels involving fewer exiting particles (e.g., (p, pn) or (n, α) reactions).

In normal circumstances, the total quantity of nuclides transformed from one isotope to another through nuclear reactions ultimately induced by cosmic rays is small in absolute terms, and only in rare cases is depletion of target nuclides observable (the exceptions are a few cases of secondary neutron capture by isotopes with huge capture cross-sections). In most cases the products of cosmic-ray-induced spallation are also too sparse to be observed against the background of material already present in samples of interest. There are two prominent classes of exceptions to this latter generalization, however: one group is short-lived (tens of Ma and less) radioisotopes, which would not be present at all except for recent production by nuclear reaction; the second is noble-gas isotopes, whose background abundances are characteristically so low that the small quantities generated by spallation may be observable or even dominant. The study of spallation noble gases in meteorites has traditionally been an important part of the agenda of noble-gas cosmochemistry, having begun essentially as soon as the necessary analytical techniques were developed in the 1950s, and vigorous interest continues to the present. Since they became available through the Apollo program, lunar samples have also been studied extensively using similar techniques and goals.

In natural materials, the spallation production rate of a given isotope depends on the concentration of appropriate target nuclides, i.e., on the chemical composition of the sample. Typically, several target elements contribute significantly to the production of a given isotope, so the dependence on target chemistry can be complex. In addition, production rates depend on the fluxes (and their energy spectra) of both primary and secondary high-energy particles. Moreover, accounting for the dependence on flux/spectrum is not just a matter of once-and-for-all integrating over the GCR spectrum, because the energy spectra of various constituents of the cascade depend sensitively on position within the target sample. Position dependence is mostly a matter of distance from external surface, but not entirely, since the projectile fluxes/spectra at the center of a sphere (4π exposure geometry) of radius 50 cm, say, is different than at a depth of 50 cm below a plane surface (2π exposure geometry). All in all, the quantitative prediction of the production rate of a given isotope, for a given position within target material of arbitrary chemical composition, is a rather complicated affair. Nevertheless, quantitative understanding of spallation production rates has received a great deal of attention, involving theoretical calculations, measurements of natural samples, and empirical calibration experiments in particle accelerators, and in detail the prediction of production rates can be quite sophisticated (for a thorough and accessible review, see Wieler (2002a)).

Spallation is sometimes described as an egalitarian process, suggesting that all (neighboring) nuclides are produced (approximately) equally. This reflects the generalization that spallation involves mostly higher energies than stellar nucleosynthesis, so that the details of nuclear energy structure are less important. It is certainly not literally true that all nuclides are produced in equal abundance. Spallation products are characteristically in or on the proton-rich side of the valley of beta-stability, for example, and one of the staple assumptions invoked for heavy noble-gas isotopic deconvolution is that the production of neutron-rich isotopes (see Figure 1), particularly those such as ^{86}Kr and ^{136}Xe, which have more proton-rich stable isobars, is effectively nil. Also, production of ^{4}He, as another example, is generally several-fold greater than production of ^{3}He. Still, a ^{4}He/^{3}He spallation ratio of "several" is much nearer to unity than in other known sources of helium, in which this ratio is of order 10^{4} or greater. Perhaps the best example of spallation egalitarianism is in neon. There are significant variations in the composition of spallation neon according to target chemistry and especially shielding (indeed, variation in spallation ^{21}Ne/^{22}Ne is commonly used as a proxy for degree of shielding), but still, the diagnostic feature of spallation neon is that all three isotopes are produced in nearly equal proportions (cf. the GCR composition in Figure 2), which certainly distinguishes spallation neon from any other known neon component.

Because of this tendency toward egalitarian compositions, the accumulation of spallation products due to cosmic ray exposure of natural samples first becomes evident as an increased relative abundance of the scarcest isotopes. In practice, among the noble gases the most sensitive indicator for the presence of spallation products is usually ^{21}Ne, i.e., when spallation components are added to previously unexposed rock, the first observable shift in noble-gas composition is growth in the relative abundance of ^{21}Ne. The usual graphical representation is a three-isotope diagram such as Figure 2, in which the hallmark of cosmic ray exposure is deflection of plotted compositions to the right of reference compositions such as SW and in the direction of the GCR composition. Commonly, prominent enhancement of the ^{3}He/^{4}He ratio is not far behind. With further additions of spallation products the overall composition of neon can be dominated by spallation, and increases in the relative abundances of the scarcest isotopes of the heavier gases become observable or prominent, first ^{38}Ar, then ^{78}Kr, then ^{124}Xe and ^{126}Xe.

For near-surface exposure to GCR, spallation indicator isotopes such as ^{21}Ne and ^{38}Ar are produced at an approximate rate of order 10^{-8} cm^3 STP g^{-1} Ma^{-1} ($\sim 3 \times 10^{11}$ atom g^{-1} Ma^{-1}) from target elements a few amu heavier (magnesium for neon, calcium for argon). Production rates decrease with increasing mass difference between target and product, e.g., production of neon is down by a factor of a few for silicon targets, and by about two orders of magnitude for iron. ^{3}He and ^{4}He are among the small chips spalled off from a target nucleus, rather than the residue, so their production rate is higher and less sensitively dependent on target chemistry; in most meteorites the ^{4}He/^{21}Ne production ratio is between 10^2 and 10^3.

Besides GCR, which originate outside the solar system, spallation can also be induced by the so-called solar cosmic rays (SCR), i.e., an irregular flux of energetic particles (solar flares) from the Sun, sufficiently energetic to cause nuclear fragmentation. The energy spectrum of SCR is much steeper than that of GCR, i.e., the decline of flux with energy is greater, so that the effective penetration depth of SCR is much less than that of GCR, only a few g cm^{-2}, below which the production of spallation isotopes by SCR is small or negligible compared to that by GCR. Nevertheless, SCR flux is high, so that to depths of 1–2 cm in silicates the production of isotopes such as ^{21}Ne by SCR spallation is comparable to or several-fold greater than production by GCR (cf. Hohenberg et al., 1978). Integrated over depth, the SCR contribution to total spallation is thus small but not trivial compared to the GCR contribution. In most macroscopic meteorites

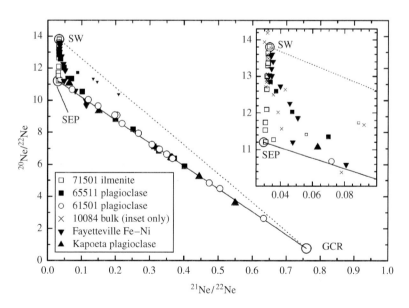

Figure 2 A three-isotope diagram illustrating compositional variations in lunar samples and meteorites, as observed in stepwise in vacuo etching and pyrolysis. Since the observed isotopic compositions do not lie on a single straight line, at least three isotopically distinct components must contribute in variable proportions. These data are interpreted as superposition of solar wind (SW), solar energetic particles (SEP), and galactic cosmic ray, i.e., spallation (GCR) Ne components (source Wieler, 1998).

SCR effects are small to negligible because the preatmospheric near-surface regions affected by SCR are ablated away during passage through the atmosphere, but SCR effects can be important near the surfaces of lunar samples. It is likely that the Sun was considerably more active early in its evolution than it is now, so the relative importance of SCR in inducing spallation may have been significantly greater early in solar-system history (e.g., Hohenberg et al., 1990).

Even in carbonaceous chondrites, which have higher noble-gas concentrations than most other meteorites, the concentration of trapped ^{21}Ne is generally less than 10^{-8} cm^3 STP g^{-1}. It follows that even in samples with comparatively high noble-gas concentrations, the effects of exposure to cosmic rays become observable, in the form of elevated relative abundance of ^{21}Ne, even for exposure times significantly less than a million years. In practice, essentially any extraterrestrial sample that falls to Earth as a meteorite (or which is brought from the Moon by spacecraft) has been exposed to cosmic rays long enough that its exposure age can be determined through noble gases, especially ^{21}Ne. The study of cosmic ray exposure ages, and their implications for sample history and provenance, provide a major impetus for the continuing great interest in spallation noble gases.

Cosmic ray exposure ages have been determined for many meteorites. The exposure ages of stony meteorites are found to be mostly a few Ma to a few tens of Ma, but there are a few lower and a few higher. The exposure ages of stony-iron and iron meteorites are characteristically higher than of stony meteorites, hundreds of Ma to Ga ages. No known meteorite (or lunar sample) has been exposed to cosmic rays for even approximately as long as the age of the solar system, whence it may be concluded that these extraterrestrial materials have spent most of the age of the solar system on larger bodies (parent bodies), shielded from cosmic rays by a few meters or more, so that the cumulative effects of cosmic ray exposure are very small or negligible. (This seems obvious today, but there was a time when it was not so.)

The simple one-stage model is that meteorites suddenly become unshielded in some event, generally an impact, beginning their exposure to cosmic rays as a relatively small body (with constant shielding) and presumably also launching them into orbits that will ultimately lead to capture by the Earth. This model seems adequate to account for much data. In some cases, more complicated and interesting histories are inferred, e.g., changes in shielding during exposure or differential exposure of some constituents, in parent-body regoliths or even in the solar nebula, prior to assembly of the present sample. A detailed description of cosmic ray exposure ages and their implications is given in Chapter 1.13 (also see review by Wieler (2002a) and references therein).

Because ^{21}Ne and ^{3}He are the most sensitive indicators of the presence of spallation products, it can be difficult to identify materials which can be confidently inferred to be *unexposed* to cosmic rays, i.e., in which we may confidently assign the abundance of these isotopes to a trapped component. Thus, the measured ^{21}Ne and/or ^{3}He abundances in some sample are upper limits to the trapped abundance, but it may be difficult to eliminate the possibility that some of the measured abundance was not produced in spallation. It is possible, for example, that some of the ^{21}Ne in prominent components such as air or planetary trapped gases (see below) is attributable to cosmic-ray-induced spallation.

1.14.3 SOLAR NOBLE GASES

Some of the noble gases found in extraterrestrial materials are from the Sun. This has been known for a long time (e.g., Signer and Suess, 1963), and there is no great mystery to it. Solar noble gases are samples of solar corpuscular emanation, mostly the "solar wind" (particles with velocities mostly in the range of several hundred km s^{-1}) but also including lesser amounts of more energetic particles. Solar gases are found in solid samples that have been exposed to the Sun on the surfaces of airless planetary bodies (or spacecraft), into which they are implanted by virtue of their kinetic energy, and they can be identified as "solar" by virtue of elemental abundances only modestly different from "cosmic" composition (see Figure 6) and of near-surface residence. Implantation depths are characteristically of order 0.1 μm or more, small in absolute terms but still hundreds of atom layers deep, in appropriate conditions deep enough to resist diffusive loss for geologically long times. Solar noble gases were first discovered in meteorites, specifically in types that are otherwise extremely gas poor. A few members of these classes contain abundant surface-correlated noble gases along with other manifestations of direct exposure to the Sun. It was and still is considered that they were exposed on the surfaces of their parent asteroids. In post-Apollo times the stereotypical solar-gas-bearing samples are lunar soils and breccias incorporating former soils, in which solar gases are commonly prominent because the Moon is apparently essentially devoid of intrinsic volatiles, including noble gases.

Solar noble gases have attracted much interest and analysis. In part, this is simply because the Sun has most of the mass of the solar system. A more focused reason, however, is the general belief that the noble gases in the Sun are the same as the noble gases that were present in the solar nebula, from which the terrestrial planets (including meteorites) were formed. For the noble gases this is a nontrivial proposition because there are so many different noble-gas compositions found in different kinds of planetary materials, which are often rather difficult to relate to each other. In particular, because noble gases are so scarce in planetary materials, and in the terrestrial planets more generally, there are apparently numerous specific cases of noble-gas isotopic compositions having been modified by the addition of nuclear components (see above). Because the current paradigm of solar-system formation implies that the *noble gases are not depleted in the Sun*, their compositions will not have been modified in this way, or at least modified much less. There is great attraction to the idea that solar noble gases are the same as nebular noble gases and, having preserved their compositions, can be taken to be the starting material for the gases now in planets and meteorites, anchoring the starting point for atmospheric and planetary volatile evolutionary models.

1.14.3.1 Deuterium Burning and the Solar ^{3}He/^{4}He Ratio

There is one prominent and well-known exception to the rule that the Sun preserves nebular noble-gas isotopic compositions better than planetary materials: helium. Before it began main-sequence hydrogen burning, the protosun experienced a nucleosynthetic phase known as deuterium burning, in which all the Sun's initial supply of deuterium (^{2}H, more commonly represented as D) was reacted ("burned") into ^{3}He. The result is that the Sun is essentially devoid of D, and its present-day ^{3}He/^{4}He ratio is higher than its primordial value; specifically, the post-deuterium-burning ^{3}He/^{4}He ratio is (D + ^{3}He)$_0$/^{4}He, where the "zero" subscript designates the primordial component, i.e., that of the interstellar medium from which the solar system formed. (The more direct quantity of interest is the D/H ratio, which can be translated from D/^{4}He by the standard atomic ratio ^{4}He/H = 0.10.)

The primordial D/H and ^{3}He/^{4}He ratios are both quite interesting quantities, not just in cosmochemistry and geochemistry, but also in astrophysics and cosmology. While the present-day ^{3}He/^{4}He ratio is readily enough measured in the solar wind, however, empirical determination of both D$_0$/^{4}He (i.e., D/H) and ^{3}He$_0$/^{4}He is rather challenging. Determination of D/H from terrestrial planetary material is complicated by the fact that D/H is evidently highly variable (several-fold) among different planetary materials, evidently because of very severe chemical mass-dependent fractionation at low temperatures, in the solar nebula or in its antecedent interstellar medium. The inference is that primordial D/H is much lower than D/H in any terrestrial planetary body or material, but it is hard to pin down an exact value. Analytical determination of primordial ^{3}He/^{4}He is complicated by the fact that in most natural planetary materials the ^{4}He is predominantly radiogenic and the ^{3}He is predominantly spallogenic. It is thus difficult to obtain a sample of suitable material in which helium is mostly primordial, enough so that plausibly small corrections for radiogenic and spallation components can be made. The coupling of these two important ratios into the present solar helium composition has often been approached in the hope that if an independent estimate of one of these ratios can be made, the other can be inferred through solar helium (cf. Geiss and Reeves, 1972). The game has been played in both directions.

Even before the advent of Apollo samples, for example, it was noted that ^{3}He/^{4}He in the solar wind, as observed in solar-gas-rich meteorites, was of order 4×10^{-4}. This is several-fold lower than it would be just from deuterium burning if primordial D/H were close to the terrestrial (seawater) value, 1.6×10^{-4}. This contrast provided the first solid argument that primordial D/H must have been several-fold lower than the seawater value (Geiss and Reeves, 1972).

Using the present best values (e.g., see review by Wieler, 2002b), the solar-wind ^{3}He/^{4}He ratio is $\sim 4.3 \times 10^{-4}$, as inferred from both lunar samples and spacecraft measurements. Because of isotopic fractionation in the solar wind (see below), the solar photospheric value is inferred to be somewhat lower, nominally 3.75×10^{-4}. Gloeckler and Geiss (2000) apply a further small correction for differential gravitational settling of the heavier isotope and conclude that the primordial ratio (D + ^{3}He)$_0$/^{4}He was $\sim 3.60 \times 10^{-4}$. Many investigators would consider that the present best estimate for the primordial ^{3}He/^{4}He is 1.66×10^{-4}, the spacecraft measurement for the atmosphere of Jupiter (Mahaffy *et al.*, 1998). By difference, the conclusion is that the primordial D/H ratio was 1.94×10^{-5}. An alternative interpretation based on meteorite samples, specifically the "Q" component (see below), is that primordial ^{3}He/^{4}He was 1.23×10^{-4} (Busemann *et al.*, 2001), whence primordial D/H was 2.47×10^{-5}. The difference between these two calculations is probably indicative of the uncertainty that should be assigned to this interpretation. Either way, however, it may be noted that this inferred D/H ratio is indeed nearly an order of magnitude below the terrestrial seawater value.

The primordial ^{3}He/^{4}He ratio has special relevance in models for planetary evolution, in that it constrains how much radiogenic ^{4}He must be added to produce a given mixture ratio, e.g., as in helium in the Earth's mantle. In addition, solar-wind ^{3}He/^{4}He should be a unique component within the solar system, and so might be decisive in distinguishing between hypotheses for component origins. As one example, there are reports (e.g., Ozima and Zashu, 1983) that some terrestrial diamonds contain trapped helium with ^{3}He/^{4}He ratio higher than the primordial ratio. This interpretation is disputed (the alternative interpretation is that the high ^{3}He/^{4}He indicates spallation in samples resident near the surface of the Earth) and cannot be considered well established, but if this or some similar future observation were more strongly established it would essentially demand that the source of the gas was the solar wind.

The prior discussion in this section has ignored the fact that for most of the age of the solar system the Sun has been conducting main-sequence nucleosynthesis of helium, not just the end product ^{4}He but also the p–p chain intermediary ^{3}He. This discussion is thus relevant only with a further assumption that the helium isotopes synthesized in the interior have not mixed into the outer convective zone, so that the photosphere can be taken to have preserved early (post-deuterium-burning) composition. Mixing from the inner nucleosynthetic zone to the outer convective zone would be manifested as a change in helium composition (specifically, an increase in ^{3}He/^{4}He) over the age of the Sun. Available data suggest that such mixing, i.e., evolution of solar-wind helium composition, has not occurred to a significant extent (see review by Wieler, 2002b).

1.14.3.2 Solar Wind in Lunar Soils

For several elements, including the noble gases, the best way to measure the composition of the solar wind is in lunar soils (and soil breccias). Lunar soils have been exposed to the Sun and have collected solar wind for geologically long times, and they provide greater quantities of solar-wind species than any other materials presently accessible. In addition, the Moon is very poor in volatile elements, so that for several elements—e.g., hydrogen, carbon, nitrogen, and noble gases—the solar-wind contribution is not overwhelmed by indigenous lunar materials. For the noble gases especially, the Moon is a good place to catch solar wind for study, since lunar primordial noble gases are so scarce that it is debatable whether any have ever been observed. Even so, however, analysis of solar-wind noble gases is not just a matter of analyzing the total noble-gas content of some lunar soil, and to some extent the accuracy to which we know solar-wind composition is limited by considerations of how it is measured in lunar samples.

One complicating factor is spallation: exposure to the Sun on the surface of the Moon also means exposure to cosmic rays, and for some isotopes production by spallation is not inconsequential compared to solar-wind accumulation (Figures 2 and 3). Spallation and solar-wind gas can largely be separated, however, because they are sited differently: spallation gas is volume correlated and solar-wind gas is surface correlated. (Spallation gases are not literally volume correlated. "Mass correlated" would be a better term, since their abundance will be proportional to their target elements, and they will be in the same places as the target elements, but "volume correlated" is the traditional term. Also, solar-wind gases are not literally surface correlated. Analyses of size-separated lunar soils indicate that solar-wind gas concentrations are proportional to a^{-n}, where a is grain radius, but the exponent n is less than the value unity which strict surface correlation demands (e.g., see Eberhardt et al., 1972). This can be understood in terms of solar-wind gases gradually changing from surface correlated to volume correlated through lunar regolith processes such as agglutination.)

Partial separation can thus be achieved by stepwise heating or etching (Figure 2), since they will be released at different temperatures and/or etch depths. Also, solar wind can be enhanced, relative to the spallation component, by selecting finer grain sizes. A particularly useful approach is an ordinate-intercept analysis, as illustrated in Figure 3. If two compositionally distinct reservoirs are mixed, and one of them (the spallation component, in the present instance) is volume correlated (actually, mass correlated), so that its concentration is constant, whereas the other (solar wind) is present in different concentrations (in different grain size fractions; note that strict surface correlation is not required), their mixing will yield a linear correlation between any given isotope ratio (plotted on the ordinate) and the inverse of the total concentration (plotted on the abscissa) of the reference isotope. The composition of the variable component can then be determined as the ordinate intercept of the correlation line, i.e., by extrapolating to infinite concentration of the reference isotope. Figure 3, which illustrates calculation of solar-wind xenon composition by Eberhardt et al. (1972), is actually a variation on this theme, in which the "constant" is not the concentration of the reference isotope (^{130}Xe) but its ratio to spallation target element barium.

Another complication is that some surface-correlated noble-gas isotopes in lunar soils are apparently indigenous to the Moon, not imported

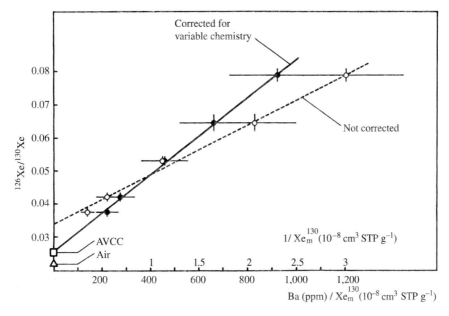

Figure 3 A modified ordinate-intercept diagram which illustrates mixing of surface-correlated (solar wind) and "volume"-correlated (spallation) Xe in lunar soil 12001. In such a diagram, if the abscissa were simply $1/^{130}$Xe, mixing of a (volume-, or mass-correlated) component present in *constant* concentration with another (surface-correlated) component present in *variable* concentration would produce a straight-line correlation, and extrapolation to infinite concentration (i.e., the ordinate intercept) would yield the composition of the variable component (i.e., the surface-correlated, or solar-wind component). This situation is illustrated by the open data symbols and broken-line correlation. A better assumption is that the more nearly constant "volume-correlated" component is the ratio of ^{130}Xe to Ba, the principal target element for xenon spallation. (The light rare-earth elements are also important targets, but these occur in nearly constant proportion to Ba.) This situation is illustrated by the solid data symbols and correlation line, whose ordinate intercept is a better determination of surface-correlated xenon composition. Note that this ^{126}Xe/^{130}Xe ratio is indistinguishable from the "average value carbonaceous chondrite" (AVCC) ratio. The term AVCC is no longer in common use; it is equivalent to "planetary" xenon (see Figure 6), which is a superposition of Q and other components (see Figure 7) (source Eberhardt et al., 1972).

in the solar wind. It is found that the concentration of surface-correlated ^{40}Ar in lunar soils can be up to several times the concentration of ^{36}Ar, and is in general much higher than could plausibly be attributed to solar-wind argon (Heymann and Yaniv, 1970). The total concentration of ^{40}Ar is in excess of what could be attributed to *in situ* decay, and so the excess came to be called orphan argon (i.e., parentless argon). It is widely believed to be ^{40}Ar, which is radiogenic but not *in situ*, formed by decay of ^{40}K in the lunar interior, degassed from the interior into the transient lunar atmosphere, and partially ionized and accelerated by solar-wind electromagnetic fields to impact on and become trapped in lunar soil grain surfaces (Heymann and Yaniv, 1970; Manka and Michel, 1971). More subtle, but apparently qualitatively similar, excesses of the heavy isotopes (and in some cases of ^{129}Xe and the heavy isotopes of krypton) have also been observed (Drozd et al., 1972, 1976), most prominently at the older landing sites (Apollo 14 and Apollo 16). The excess ^{129}Xe is plausibly ascribed to decay of ^{129}I, and the heavy isotopes to actinide fission, primarily from ^{244}Pu, as can be verified in a few cases by matching the spectrum of xenon isotopes to the composition of ^{244}Pu fission. It is commonly taken that these also are "orphan" components, similar in origin to the excess ^{40}Ar. From the early solar-system ratio of ^{244}Pu to ^{238}U it may be inferred that there is similarly excess fission xenon and krypton from ^{238}U present, likely even in soils at the younger landing sites, in quantities small enough that they are perhaps difficult to detect directly but still large enough to be significant. It may be noted that since orphan components are surface correlated, not volume correlated, they cannot be resolved from solar-wind characterization through the ordinate-intercept analysis (e.g., as in Figure 3), so there is some basis for concern that there may be non-negligible fission in surface-correlated xenon compositions derived from lunar soils.

Some elemental abundance patterns for solar gases in lunar soils (and one gas-rich meteorite, Pesyanoe, that was likely exposed to solar wind on the surface of an asteroid; see Marti, 1969) are illustrated in Figure 6 (left panel). Gases in true solar proportions would define a horizontal straight line in this diagram, and it is evident that the lunar (and meteoritic) regolith samples do not do this, but rather exhibit significant and variable fractionation patterns of progressive depletion of lighter gases (albeit still modest compared to the "planetary" pattern in the right panel of Figure 6). An easy inference is that this pattern reflects progressively greater (diffusive) loss of lighter gases from the lunar/meteoritic materials. For helium and neon this interpretation is supported by abundance ratios, relative to argon, which are not only less than those inferred for the Sun as a whole but also less than those directly measured by spacecraft and in foils exposed during Apollo missions (see reviews by Wieler, 1998, 2002b). In addition, although there are substantial uncertainties in estimating whole-regolith abundances, it can be estimated that the present regolith inventories of helium and neon are only 10^{-2}–10^{-1} of all the gas that has impinged on the lunar surface in the past ~4 Ga, i.e., there has been major loss of these light noble gases. In contrast, it can be argued that effectively the whole fluence of the heavier gases argon, krypton, and xenon that has ever hit the Moon (in the ~4 Ga age of surface formations) is still there. Primarily on these grounds it can be further argued that the relative proportions of the heavier gases trapped in lunar regolith materials represent actual solar-wind composition (cf. Wieler, 2002b), unmodified by fractionating loss, and thus that the solar wind itself is fractionated relative to solar composition.

If the heavy gases argon, krypton, and xenon in the solar wind are quantitatively retained in the lunar regolith, it can also be inferred that solar-wind isotopic compositions of these gases are preserved in the lunar regolith. For helium and neon, extensive loss leads to concern that the residual gases in the lunar soils may be isotopically fractionated with respect to true solar-wind composition. Wieler (1998, 2002b) argues that this is not the case, however, at least not for neon. The first gas released in closed-system etching, corresponding to the shallowest siting, is generally very close to a common composition, labeled SW in Figure 2, which is also the composition observed in spacecraft measurements and in the foils exposed to solar wind during the Apollo missions, and which is taken to be the true mean solar-wind neon composition.

When neon is released differentially, either by stepwise heating or by stepwise closed-system etching, the early release that indicates trapped neon of composition SW (Figure 2) is generally followed by trapped neon with lower ^{20}Ne/^{22}Ne, more nearly like the composition labeled solar energetic particles (SEP; solar corpuscular emanation at higher energies (100 keV amu^{-1} and higher)), and/or a correlation line between SEP and spallation (GCR in Figure 2). Comparable effects occur in the other gases. Although the term is entrenched in the noble-gas literature, it may be a misnomer, and the nature and origin of the SEP component is somewhat of a mystery (Wieler, 2002b). The principal problem is that there is too much of it: noble-gas SEP/SW in lunar soils is of order 10^{-1}, which is about three orders of magnitude too high compared to directly measured solar emanation. It has been suggested that SEP is not really a physical component but rather the result of differential diffusive migration of the noble-gas isotopes, or perhaps of a slightly greater implantation depth for heavier isotopes. It has also been suggested that it is not solar emanation at all, but interstellar gas entering the solar system at energies higher than the solar wind (Wimmer-Schweingruber and Bochsler, 2000). None of these possible interpretations is without problems, so the origin of SEP noble gases remains an open question.

1.14.3.3 Solar-wind Composition and Solar Composition

Even if lunar soils and regolithic meteorites preserve well the composition of the solar wind incident on them, the composition of "solar" noble gases cannot immediately be translated to the composition of the Sun (or the photosphere) because of the possibility of fractionation, both elemental and isotopic, arising in the mechanisms by which solar atmospheric atoms are accelerated into the solar wind. Indeed, spacecraft measurements indicate variability of both elemental and isotopic compositions in the solar wind, and in higher-energy emanation, at different times and at different energies, so fractionation mechanisms are clearly at work. In general, it may not be presumed that even time-averaged composition, even integrated over energy, will accurately yield the composition of the solar source.

Solar-wind ^{3}He/^{4}He is commonly some 10–20% higher than the underlying solar composition, varying slightly with speed (e.g., Gloeckler and Geiss, 2000; Wieler, 2002b). This ratio can be several-fold different and highly variable in higher-energy emanation, in the "solar energetic particles" component. Isotopic fractionation is expected to be smaller in heavier elements, it is of order a percent or two, varying with speed, for neon and neighboring elements (e.g., Kallenbach et al., 1998). Isotopic composition can be expected to be progressively still less for argon, krypton, and xenon, although isotopic data for real-time collection of solar-wind argon are insufficiently precise to rule on this issue. For krypton and xenon there are no data for direct collection, so only the time- and speed-integrated compositions recorded in lunar and meteoritic materials are available.

At the level of the best analytical errors it is difficult to determine whether averaged solar-wind isotopic compositions accurately represent true solar values because there is no other independent knowledge of solar compositions. Instead, solar-wind compositions can be compared with planetary compositions such as the Earth's atmosphere or the Q-component of "planetary" gases in meteorites (see below). This is a particularly problematic approach for noble gases, however, since in any given case there are generally ample grounds for suspecting that present compositions in planetary materials have been altered from true solar composition. In addition, at the finest levels of precision permitted by analytical errors it is not uncommon that different investigators disagree on what solar-wind composition really is (cf. Wieler, 2002b; Pepin and Porcelli, 2002).

There can also be substantial elemental fractionation, and fluctuation, in the solar wind (and still more in the solar energetic particles). The average value of solar-wind He/Ne observed directly (in spacecraft and Apollo foils) seems not quite a factor of 2 below the true solar value (inferred from spectroscopic data and stellar structure modeling), but Ne/Ar in the solar wind, within errors, may be consistent with the actual solar value. The lunar data evidently do not speak to this issue, because of apparent pervasive loss of the lighter gases (see above). Relative to their inferred solar abundances (thought reliably inferred from regularities in nucleosynthetic processes), krypton and xenon in the solar wind, as recorded in lunar and meteoritic materials, are significantly enhanced (see Figure 6, left panel), xenon by a factor of ~4. This is thought to be a manifestation of the well-known first ionization potential (FIP) effect, in which elements with ionization energies less than ~10 eV are present in the solar wind in greater abundance, relative to source solar abundances of other elements (von Steiger et al., 1997). Krypton and xenon actually have higher ionization energies than this threshold, but they are nevertheless enhanced in the solar wind in models that show that the governing parameter is really ionization time in the solar chromosphere (Geiss et al., 1994; Wieler, 1998).

1.14.4 EXOTIC COMPONENTS

Arguably the most significant development in cosmochemistry in the past generation is recognition and exploitation of the circumstance that some types of presolar solid materials have survived, more or less intact, throughout the formation and subsequent history of the solar system, that they are preserved in undifferentiated meteorites that have not been metamorphosed, and that they can be isolated in sufficient purity and quantity to support a wide variety of laboratory studies (e.g., see Bernatowicz and Zinner, 1997; Chapter 1.02). The isotopic compositions of these materials are radically different from solar-system normal, which is indeed the primary evidence that they are presolar. Since materials formed in the same interstellar medium from which the whole solar system formed would be expected to have isotopic compositions close to normal, it is further inferred that these materials are not just presolar but circumstellar, having formed in the atmospheres or ejecta of individual stars at specific stages in their evolution, and containing atoms whose nuclei were made in specific nucleosynthetic processes. Study of these materials has provided a wealth of empirical constraints on astrophysical theory. Among the elements identified and characterized in presolar materials are the noble gases, which are particularly prominent in the study of presolar materials for two reasons.

One reason is that, at least as far as we know at present, the abundance of identifiable preserved presolar materials in meteorites is low, so that known types of presolar materials do not supply a substantial fraction of the total meteoritic inventory of any elements *except for the noble gases*. This is another manifestation of the generalization that low noble-gas abundances render prominent some quantitatively small contributions that would be lost in the background sea of other elements.

The second reason is that noble gases have been intimately involved in the initial recognition and isolation of presolar materials. The first evidence for the presence of presolar materials in meteorites was isotopic variation among gases released in stepwise degassing, especially xenon and neon (e.g., Reynolds and Turner, 1964; Black and Pepin, 1969), reflecting differential thermal release of isotopically distinct components. At the time, however, these isotopic variations were not recognized as reflecting the presence of presolar materials. The main reason for this nonrecognition was likely that the host of isotopic variations in the noble gases that could be ascribed to nuclear components (radiogenic and spallation gases, in some cases including neutron-capture effects), plus the possibility of significant mass-dependent isotopic fractionation, which led to expectations that yet more isotopic effects would ultimately be explained in terms of yet more nuclear components or physical processes, effects operating within the solar system to modify the isotopic composition of once-uniform planetary materials. In addition, many cosmochemists were predisposed to the view that all previously existing solids in materials ancestral to the terrestrial planets were vaporized in the solar nebula, so that survival of presolar grains was unexpected. Even when larger isotopic effects (e.g., Figure 4), less interpretable in terms of solar-system processes, were explicitly suggested to represent presolar components (Black, 1972), this hypothesis did not gain much support or even attention. It was not until apparent isotopic anomalies, indicative of nonhomogenization of presolar components, were found in the major element oxygen (Clayton et al., 1973) that credence in this idea led to vigorous programs of study of isotopic anomalies.

1.14.4.1 Presolar Carriers

In hindsight, it is clear that stepwise heating of bulk meteorite samples, with the resultant partial separation of noble-gas components from host phases that release their gases at different temperatures (e.g., Figures 2 and 4), provided the essential clues pointing to the discovery of presolar materials preserved in meteorites. Still, stepwise heating of bulk meteorites provides only limited isotopic component separation, allows characterization of the noble-gas hosts only by temperature of gas release, and does not offer the opportunity to relate noble-gas components to most other elements which are not noble gases. Major advances followed the introduction of experimental efforts to physically separate the phases that carried the isotopically distinct noble gases, by properties such as grain size, density, and especially resistance to corrosive chemical attack. The effort to identify the carriers of anomalous noble gases provides a fascinating history, full of blind alleys and inspired detective work in which investigators were able to follow the isotopic anomalies in the noble gases to progressively more refined isolation of their hosts, eventually culminating in the discovery and identification of presolar phases that host exotic noble gases (see Figure 5). There is no need to recount this history here, however, and interested readers are referred to reviews such as given by Anders (1988) or Anders and Zinner (1993), various chapters in the book edited by Bernatowicz and Zinner (1997), or Chapter 1.02. Moreover, the discussion here focuses on aspects of presolar grains most relevant to noble gases; no attempt is made to describe the broader aspects of presolar grain studies, for which reference is made to Chapter 1.02.

There are three well-known types of presolar grains which were originally isolated and identified by following anomalous noble gases through complicated separation procedures: diamond, silicon carbide, and graphite. These phases account for only a very small fraction of the total mass of a meteorite, or of most elements within a meteorite, but they supply important parts of the noble-gas budget. To an extent, these materials can be separated by nondestructive techniques based on density, grain size, etc., but production of the cleanest and most abundant samples has generally involved the basic approach of dissolving away most of the other phases with which they were originally mixed in whole-rock meteorites.

There can be no doubt that at least the silicon carbide and graphite are presolar. Many elements have been analyzed in bulk collections of these grains, and in general all multi-isotope elements have isotopic compositions that are radically different from solar-system normal composition,

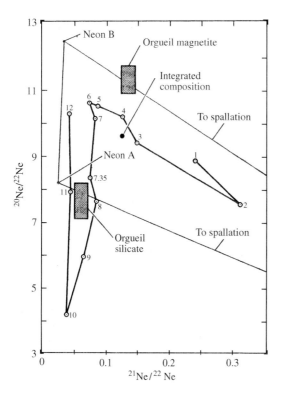

Figure 4 A three-isotope diagram for neon indicating compositional variations in neon in stepwise heating of the carbonaceous chondrite Orgueil. The numbers next to the data points are release temperatures in hundreds of degrees C. Also shown are the trapped component compositions for neon-A ("planetary") and neon-B ("solar"); the light lines connect neon-A, neon-B, and the composition of spallation Ne offscale to the right (cf. Figure 2). The shaded rectangles indicate the compositions of separated magnetite and silicate fractions found by earlier investigators. Much of the individual data points could be explained as mixtures of these three components, but the data excursions below this triangle demand another component, termed neon-E. It was later determined that neon-E is very nearly monisotopic ^{22}Ne, i.e. it would plot at the origin of this diagram (see Figure 5), and also that it is composite, made up of the G- and R-components (see Table 2 and Figure 7) (source Black, 1972).

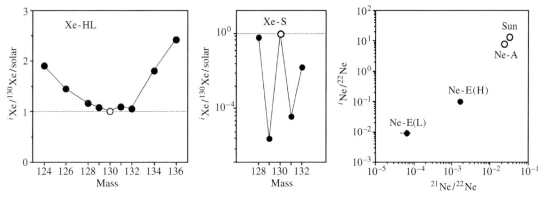

Figure 5 A display of prominent exotic (presolar) noble-gas compositions (from Anders and Zinner, 1993). In the left two panels, for each isotope on the abscissa the ordinate is the ratio (to ^{130}Xe) in the HL component (left panel) or the G (formerly termed Xe-S) component (center panel), divided by the equivalent ratio in solar xenon (i.e., solar xenon would plot with all isotopes at unity on the ordinate). The HL component shows the defining characteristics of enriched heavy and light isotopes. For the G-component, the pattern is that expected for s-process (slow neutron capture) nucleosynthesis. The right panel is a three-isotope diagram analogous to Figure 4, except that both scales are logarithmic. It shows experimental limits for the R-component (formerly Ne-E(L)) and the G-component (formerly Ne-E(H)).

compositions that could not plausibly have been produced within the solar system. Indeed, it is generally held that these compositions could not have been produced from materials in the interstellar medium, which mixes nucleosynthetic contributions from many stars; instead they are interpreted to be circumstellar grains, formed in the outflow of individual stars and incorporating nucleosynthetic products from those individual stars. There is also no doubt about carrier phase identification: these grains are mostly quite small (submicrometer), but their populations include representatives with sizes upwards of 1 μm, large enough for isotopic analysis of individual well-identified grains. Such analyses reveal radically anomalous isotopic compositions of the major elements as well as several minor and trace elements, including noble gases (e.g., Nichols et al., 2003).

Much the same can be said about the diamonds, in that most minor and trace elements are radically anomalous. But the major element, carbon, has an isotopic composition within the range of solar-system normal. Moreover, the individual diamond grains are extremely small, characteristically only a few nanometers, and too small to support analysis even of carbon in individual grains. It may be that diamond carbon appears isotopically normal only because any isotopic analysis is an average over many grains. But because of the normal carbon there is persistent suspicion that most the diamonds are not really circumstellar or even presolar after all, and that the real presolar grain carrier is a small subset of the diamonds or some other phase entirely, less abundant than the diamonds but which follows them in the separation procedures.

There are other forms of presolar/circumstellar grains in meteorites, such as silicon nitride, spinel, corundum, even a silicate (see Chapter 1.02), most of them isolated in the same or similar ways as the three phases noted above. Such phases have been discovered only more recently than diamond, silicon carbide, and graphite, and in smaller quantities, and so have not been studied as extensively. At present there is no evidence that any of these other phases bear any significant noble gases. Such evidence as exists is negative (cf. Lewis and Srinivasan, 1993), but there is precious little evidence bearing on this point. It is worth noting, however, that there appears to be no evidence in presently available meteorite data that requires yet another presolar noble-gas component (and another presolar grain carrier), at least not of the extremely anomalous variety.

1.14.4.2 Radically Anomalous Exotic Noble-gas Components

The known presolar/circumstellar phases diamond, silicon carbide, and graphite each contain a distinct noble-gas component which, like the major (except perhaps for carbon in diamonds), minor, and other trace elements in these phases, is radically anomalous compared to normal solar-system composition. These components are listed in Table 2 and illustrated in Figure 5. In the exploratory studies in which an understanding of these components was being developed, a variety of more-or-less complicated names, typically an acronym for some descriptive phrase or arbitrarily selected alphabetic characters (not all from the Latin alphabet), have been used. Some

Table 2 Exotic noble-gas components in presolar grains.

Component	Carrier	Remarks
Radically anomalous components		
HL	Diamond	r-Process (?) Xe and Kr; p-process (?) Xe; "planetary" Ne
G	Silicon carbide	From AGB stars; s-process Xe and Kr; Ne very rich in ^{22}Ne
R	Graphite	Monosiotopic ^{22}Ne from decay of ^{22}Na
Moderately anomalous components		
N	Silicon carbide	Probably initial composition of AGB stars
P3	Diamond	Very nearly normal, possibly not really exotic
P6	Diamond	Composition not well constrained; possibly quite anomalous

of the terms have been short lived, and there has been some confusion through the application of different names to the same thing. It appears that the highly anomalous components have now been substantially identified and characterized, and Ott (2002) has proposed adoption of a standard nomenclature, in some cases replacing older names in the interest of clarity. Table 2 and the discussion below use Ott's suggested nomenclature. In addition, Ott (2002) provides a detailed overview of the occurrence of these components and tabulation of the best present determinations of their compositions.

The radically anomalous component carried by silicon carbide is the G-component (Table 2). Isotopic analyses of the major elements in individual grains indicate that they come from a variety of different astrophysical sources, but most grains evidently formed in the outflow of AGB stars (on the asymptotic giant branch of the Hertzsprung–Russell diagram); the G in the name stands for "giant." The isotopic compositions characteristic of the G-component can be understood quite well in terms of nucleosynthesis in the helium-burning shells of carbon-rich AGB stars (e.g., Gallino *et al.*, 1990; Lewis *et al.*, 1994; Hoppe and Ott, 1997). For the heavy elements (like krypton and xenon), this means s-process nucleosynthesis (slow neutron capture, such that radioactive isotopes with lifetimes up to several years or decades are likely to decay before capturing another neutron), which produces only a characteristic set of isotopes in the central part of the valley of beta stability and an even–odd abundance variation pattern paralleling inverse neutron-capture cross-sections (Figure 5, center panel). End-member s-process isotopic compositions can often be determined relatively robustly, even when mixed with more nearly normal compositions, by applying the constraint that some of the lightest and heaviest isotopes (e.g., ^{78}Kr, ^{124}Xe, and ^{136}Xe; see Figure 1) are not made in the s-process. G-component krypton is of particular interest because it contains an s-process branch point, i.e., an isotope (^{85}Kr) where the s-process synthesis path divides into two branches in proportions sensitively dependent on astrophysical parameters during the synthesis.

G-component isotopic composition is not entirely uniform, but is variable in ways that can be understood to reflect minor variation in astrophysical conditions during nucleosynthesis. For the s-process elements, the compositional variation is readily attributable to differences in neutron exposure. For krypton and xenon the isotopic variations correlate with silicon carbide grain size (greater exposure for bigger grains), although the astrophysical reasons for such a correlation remain elusive. Not surprisingly, s-process compositions also vary in other G-component heavy elements (e.g., strontium and barium), also in ways attributable to neutron exposure. In addition, these variations also correlate with grain size, but in the opposite sense of the noble-gas correlation. The explanation for this is not known, but must involve different mechanisms for the incorporation of the noble gases and the incorporation of chemically reactive elements into the silicon carbide grains.

The G-component also includes neon that is strongly enriched in ^{22}Ne (Figure 5, right panel), the defining feature for a component previously termed neon-E (Figure 4). Further investigation showed that there were actually two kinds of neon-E (Jungck and Eberhardt, 1979). One was hosted by a relatively low-density phase and was released at relatively low pyrolysis temperatures, and was accordingly termed neon-E(L); the other, in a higher-density phase and released at higher temperatures, was termed neon-E(H). The G-component neon is neon-E(H). It is highly enriched in ^{22}Ne, but it is not monosiotopic: it contains low but measurable ^{20}Ne and is clearly associated with ^{4}He, and both features are consistent with helium burning in AGB stars.

As inferred from single-grain isotopic analyses, the origins of interstellar graphite are rather diverse, without evident preponderance of any one source. Among the sources of graphite are

AGB stars. The heavy noble gases in graphite (cf. Amari et al., 1995), like those in silicon carbide, are evidently of s-process type, but with a wider range of neutron exposure.

Graphite also contains a ^{22}Ne-rich component, not obviously associated with other gases and not the same as the ^{22}Ne-rich G-component in silicon carbide, but which can be identified as the complimentary form neon-E(L) (Amari et al., 1995), renamed the R-component (Table 2). It appears to contain no detectable amounts of the other neon isotopes, and is generally thought to result from in situ decay of ^{22}Na (half-life 2.6 yr).

Diamonds host the HL component, which is the most abundant of the three radically anomalous components in Table 2. The name reflects the great enrichment of both the heavy and the light isotopes of xenon (Figure 5, left panel; also see Huss and Lewis, 1994). The partial separation of HL-Xe from other components in stepwise heating accounts for the isotopic variations first observed by Reynolds and Turner (1964), and the heavy-isotope enrichments for a time supported the idea that it was a fission component, CCF-Xe (see above). Heavy-isotope enrichment is suggestive of the nucleosynthetic r-process, and light isotope enrichment is suggestive of the p-process, and the astrophysical provenance of the HL component is commonly thought to be supernovae. The compositions are not quite right for either of these "classical" nucleosynthetic processes, however. It may be that the heavy-isotope ratios in HL-Xe reflect r-process composition modified by chemical separations in the beta-unstable progenitors of the heavy xenon isotopes (Ott, 1996) or that the heavy isotopes were produced in a "neutron burst" (Howard et al., 1992) involving neutron exposure intermediate between the classical s-process and r-process. The heavy isotopes and the light isotopes must be synthesized in different nuclear processes, and on astrophysical grounds it is not evident why H- and L-components should be coupled into a single HL component, as they appear to be. This has led to suspicion that they are indeed not coupled, and that there are distinct H- and L-components hosted in distinct subpopulations of the diamond samples, and thus to analytical efforts to separate H from L. At present there is some evidence that H and L are separable, but it is not yet definitive (see Meshik et al., 2001).

The HL component includes all the noble gases. It is noteworthy that the krypton is enriched in the heavy isotopes but not the light (i.e., H but not L). The relative elemental abundance of the light gases neon and helium in HL is higher than in other prominent trapped components, so that while HL makes a small, albeit still significant contribution to ordinary "planetary" gases (see below), it provides the dominant contribution to neon and helium (Figure 7).

1.14.4.3 Moderately Anomalous Exotic Noble-gas Components

The isotopic compositions that are actually observed in presolar grain analyses, particularly in the many-grain ensembles needed for analysis of heavy trace elements, in general are not pure, radically anomalous nucleosynthetic end-member compositions, e.g., the components which are the focus in the prior section. More generally, for both noble gases and other elements alike, the nucleosynthetic components are diluted by compositions that are more nearly normal. In some cases it might be argued that these diluents really are normal, in the strict sense, because they are actually laboratory blank or bulk meteorite contamination. More generally, however, the evidence favors the conclusion that these components really are in the presolar grains, so that they really are exotic to the solar system, especially when they can be inferred to be isotopically anomalous. There are three such components that are reasonably well established (Table 2): the N-component in silicon carbide, and two components, P3 and P6, in the diamonds. These components would likely not have been noted except for their association with the radically anomalous components.

It is important to appreciate that even though these components may be anomalous, the anomalies in question are much more modest than those described in the preceding section, commonly expressed in percent deviations from normal rather than as multiples or even orders of magnitude. More fundamentally, these components do not appear to be the results of specific nucleosynthetic processes in specific stars, but rather a mixture of many different stellar nucleosynthetic contributions, as would be found in the interstellar medium, and so could be remnants of initial stellar compositions. In other words, these components are like our own normal composition, except that they are the "normal" of a different time, elsewhere in the galaxy.

It is difficult to doubt that the N-component is actually in the silicon carbide, and it is commonly held that the N-component represents the outer envelope of the same AGB stars in which helium-burning generates the G-component, i.e., the initial stellar composition. Since the diamonds are too small for single-grain analyses, however, it cannot be precluded that the samples prepared in the laboratory are not assemblages of different materials of fundamentally different origin. Thus, it cannot be concluded that the P3 and/or P6 components are in the same population of grains, or are closely related to the HL component or each other.

1.14.5 NOBLE-GAS CHEMISTRY AND TRAPPED COMPONENTS

Strictly speaking, the title "noble-gas chemistry" should be an oxymoron. But the noble gases are not literally and completely noble in the sense that they fail entirely to interact chemically with other forms of matter. Under appropriate conditions in the laboratory they can form real compounds with other elements, although there is no evidence that actual noble-gas compounds are relevant in cosmochemistry (possibly excepting ice clathrates in comets). Still, planetary materials do contain noble gases that were somehow incorporated into them, and at least some of these appear to have involved some form of chemical interaction. The issue of chemical interactions is a venerable topic in noble-gas cosmochemistry, but there are still questions that have been unanswered for a long time.

1.14.5.1 Planetary Noble Gases

As the noble gases in planetary materials were first being explored, it became evident that there were three broad categories of noble-gas occurrences: *in situ* gases and two kinds of trapped gases, solar and planetary. *In situ* gases are those produced by nuclear transformation—radioactive decay and spallation—in the samples in question, such that the gases are still in the same places where they were made. Solar gases are samples of solar corpuscular radiation, implanted in the near-surface layers of solids exposed directly to the Sun, as freely orbiting small particles or on the surfaces of larger but airless parent bodies such as the Moon or asteroids. Planetary gases are everything else, gases which are neither *in situ* nor solar.

Despite definition as a garbage-can category, planetary noble gases in most cases occur in a reasonably well-defined pattern. The elemental abundances in different meteorites occur in similar proportions, in a pattern defined by progressively greater depletion of the lighter gases, as illustrated in Figure 6 (right panel). This pattern is clearly distinct from the solar elemental pattern (Figure 6, left panel). The term "planetary" reflects the similarity of the nonsolar meteoritic abundance pattern to the relative abundances of the noble gases in the Earth's atmosphere (except that no comparison can be made for helium, which escapes from the atmosphere in geologically short times). The quantities are about right too, i.e., if the inventories of atmospheric gases divided by the mass of the Earth are taken as a reasonable measure of the initial concentrations in the solids that formed the Earth, these concentrations fall within the range seen in meteorites (mostly between Allende and Bruderheim in Figure 6, right). Accounting for the origin of planetary noble gases is another traditional problem that has long resisted solution, but there is a clear suggestion that whatever explanation there is for planetary gases in meteorites should also suffice for Earth (and perhaps Venus and Mars as well).

Besides the elemental abundance pattern by which they are defined, planetary gases also display characteristic isotopic features. The isotopic compositions of krypton and nonradiogenic argon are very similar in solar, planetary, and atmospheric gases, varying perhaps by modest isotopic fractionation. Planetary xenon is like solar xenon at the light isotopes, but has small (a few per mil to a few percent) excesses in the heavy isotopes (atmospheric xenon constitutes a special challenge, as described below). The largest distinctions occur in neon. Neon in carbonaceous chondrites characteristically defines a trapped composition called neon-A (Figure 4) with much lower $^{20}Ne/^{22}Ne$ than "solar" neon (the B composition in Figure 4, which actually has $^{20}Ne/^{22}Ne$ somewhat lower than the solar wind). (Atmospheric neon, with $^{20}Ne/^{22}Ne = 9.8$, is intermediate between the A and B compositions, closer to A.) Because of the prevalence of *in situ* components, planetary helium can be observed only in very few samples. Where observable, planetary $^{3}He/^{4}He$ is considerably lower than solar $^{3}He/^{4}He$, but this is a special case attributable to deuterium burning in the Sun (see above).

In the "classical" picture, i.e., before appreciation that presolar components were an important part of the total inventory of trapped gases in meteorites, neither the isotopic effects nor the generation of the elemental abundance pattern were ever explained satisfactorily in terms of quantitative models that gained consensus acceptance. Some aspects of this problem have become moot, however, since it is now recognized that "planetary" gas is composite: planetary gas includes the contributions of the exotic noble-gas components (Table 2) imported into the solar system by presolar grains. These contributions, especially of the HL component, can be substantial (Huss and Lewis, 1995; also see Figure 7).

For the heavy gases, the diamond-borne components commonly provide a few to several percent of observed gas (Figure 7, left), the other exotic components contribute only smaller amounts, and the major contribution is assigned to the component called "Q" (sometimes also called "P1," where the P stands for "planetary"). Some of the main characteristics of "planetary" gas are now attributed to the Q-component: it accounts for most of the heavy gases, it is what is left over after other

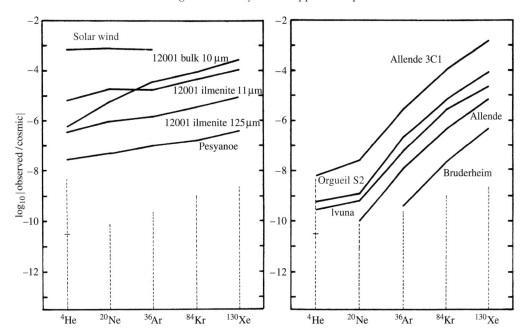

Figure 6 Elemental abundance patterns for trapped noble gases in various planetary materials. For each gas identified on the abscissa, the ordinate shows the depletion factor in a given sample, i.e., the gas concentration in the sample divided by what the concentration would be if the gas were present in undepleted cosmic proportion (normalized for a nominal rock with 17% Si). The relative elemental abundances in the left panel illustrate the "solar" pattern, those in the right panel the "planetary" pattern. The vertical broken lines for each gas illustrate typical *in situ* gas concentrations (the radiogenic component for ^{4}He, spallation for the others), below which it becomes progressively more difficult to characterize or even identify trapped components (source Ozima and Podosek, 2002).

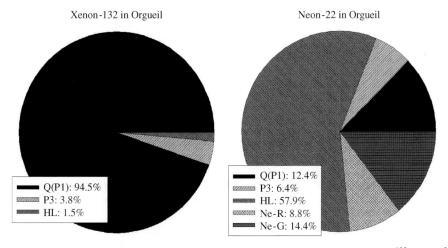

Figure 7 Pie diagrams illustrating the inferred superposition of components to total trapped ^{132}Xe and ^{22}Ne in the carbonaceous chondrite Orgueil. (Only components contributing at least 1% are shown.) The Q-component dominates ^{132}Xe but is only a minor contributor to ^{22}Ne; the main contribution to ^{22}Ne is the HL component (source Ott, 2002).

sources (*in situ*, solar and exotic components) have been accounted for, and it is often regarded as a (the?) local "background" component, derived from the ambient noble gases in the nebula from which planetary materials formed. An important difference, however, is that the ratio of light to heavy gas elemental abundances in Q is less than in "planetary" gas and less than in the exotic components. It is thus inferred that Q makes a progressively smaller contribution to the lighter gases. For ^{22}Ne in Orgueil (Figure 7, right), for example, Q makes only a minor contribution. Most of planetary neon (and helium), for which explanation was once sought in terms of how it could be derived from the solar nebula, is now inferred to be exotic, a superposition of components (chiefly HL) whose properties were established outside the solar system.

It is sometimes argued that the term "planetary" is obsolete and its use should be discontinued because it is now seen that it is not a monolithic component, and because its dissection into mostly extrasolar components weakens its association with the major terrestrial planets. This last viewpoint seems questionable, however, and no good alternative to the term "planetary" has emerged. The term seems still useful in denoting this ensemble of components, and also in suggesting some of its salient features. When it was once thought that planetary neon and xenon were parts of the same component extracted from the nebula, for example, the question was what process could establish such an Ne/Xe ratio and what process (or added component) could account for such a different isotopic composition for neon. In the light of present understanding, the isotopic question seems more relaxed, since it is quite reasonable that an extrasolar component could have quite different composition from solar gas, but there is now the question of how it is that independent components, hosted in different carriers, nevertheless occur in nearly the same proportions in many samples.

1.14.5.2 The Q-Component

As discussed above, there are a number of exotic noble-gas components that are important to reckon with in cosmochemistry, i.e., components whose identity was established outside the solar system, prior to its formation, which were imported into the solar system in presolar grains, and which have maintained their identities throughout the solar-system history. Some of these exotic components have radically anomalous isotopic compositions, and others, while not radically anomalous, are clearly distinct from normal composition. It is often held that the complement to these exotic components is the component most commonly designated "Q" (an alternative name is "P1"), with "normal" isotopic composition and thought to be of local rather than exotic origin.

It is frequently said that the Q-component is defined operationally. As initially reported by Lewis et al. (1975) as part of their program to identify the carrier phases of the exotic noble gases, suitable treatment of whole-rock carbonaceous chondrites with HF and HCl dissolves most of the mass, leaving a residue of order 1% or less of the starting mass, but nevertheless containing nearly all the initial trapped noble gases, i.e., the major phases removed by this treatment carried little to no trapped noble gases. When the residue is further treated with HNO_3 (or other oxidizing agent), there is only small additional mass loss, but most of the trapped noble gas is removed, and the remaining gases in the residue are markedly anomalous. The inference is that the normal noble gases are in a very minor phase, and the name Q was applied to both the (unknown) phase and the gases. This is not exactly the same definition as the one implied in the previous section, i.e., that Q is planetary noble gas, less the components listed in Table 2, but there seems to be no discrepancy arising in the difference. In both of these definitions, Q is defined by subtraction, the difference between two occurrences. In a third approach, more straightforward and much more amenable to studying its properties, Q gases are liberated by the closed-system stepwise etching technique, in which gases released in very superficial chemical attack can be directly collected for analysis (Wieler et al., 1991; Busemann et al., 2000). The Q-component thus obtained is consistent with the definitions by difference. There was initially some controversy about the nature of the carrier phase, but it seems now generally agreed that it is carbonaceous (Ott et al., 1981).

The Q-component contains all five noble gases, in an elemental abundance pattern defined by progressive depletion of lighter gases relative to heavier gases, like the nominal planetary pattern (Figure 6) but steeper, the depletion of helium and neon relative to xenon being some two orders of magnitude greater. As already noted in the prior section, this results in Q accounting for a smaller fraction of the total planetary abundance of the lighter gases than of the heavier gases (Figure 7). There is some variation in elemental abundance ratios in different instances of Q, amounting to a factor of several, significant but still small compared to more than six orders of magnitude depletion of helium and neon relative to xenon. Busemann et al. (2000) argue that Q is actually composite, consisting of at least two subcomponents that behave differently in planetary parent body processes. Q isotopic compositions are distinct from solar-wind compositions, but could be related to solar-wind compositions by isotopic fractionation. The degree of fractionation, which is progressively less in the heavier elements, is substantial, ~30% for $^{20}Ne/^{22}Ne$, ~1% amu^{-1} in xenon. The sense of the isotopic fractionation is depletion of light relative to heavy, the same as the elemental fractionation. Busemann et al. (2000) report distinct variations, of order 6% and bimodally distributed, in $^{20}Ne/^{22}Ne$ in Q-Ne. Assuming that Q-Ne really is derived from neon of solar-wind composition by fractionation, this isotopic difference corresponds to modest difference in degree of fractionation. The more profound generalization is that at least at present there is no evidence that Q itself is isotopically complex, i.e., its isotopic structures can be related to the composition of the Sun, and presumably of the solar nebula, by no more than fractionation, with no embedded isotopic anomalies.

Q gases are found in a variety of meteorite types, with just relatively modest compositional variation, from which it can be argued that Q was somehow established as a component, in some physical reservoir, prior to being distributed to and incorporated into diverse meteorite parents (cf. Ozima et al., 1998). In addition, the classical observation that planetary gases behave like a "component," e.g., in having a characteristic ratio of light to heavy gases, is now equivalent to stipulating that Q occurs in a characteristic ratio to exotic components of entirely separate origin. This feature is easier to understand if Q gases were incorporated into their solid carrier phase(s), and then the Q carrier phase mixed with the presolar grain carriers, such that meteorite precursors could sample them in the same proportions.

It is not clear how the defining characteristics of Q were generated. Ozima et al. (1998) advocate a model based on mass, suggesting Rayleigh distillation to generate both the elemental and isotopic fractionation, but it is difficult to reconcile both types of fractionation with the same process, and there is no known astrophysical scenario for the postulated Rayleigh distillation. Ott (2002) notes that Q gas abundances can also be correlated with ionization potential, but this would not provide a natural explanation for the isotopic fractionation, and again a complete model would require an appropriate astrophysical scenario. Besides the attempt to account for elemental and isotopic composition, these and any other models for generation of Q, especially those involving gas-phase separation, face the additional hurdle of accounting for incorporation of Q gases into the solid Q-phase carrier (see following section).

Huss and Alexander (1987) advocate a model in which Q is actually a presolar component, in which the Q-phase carrier is made, and acquires noble gases, not in the solar nebula but in the interstellar medium, in the molecular cloud from which the Sun later forms. This scenario is consistent with the observation that Q gas and solar gas have the same isotopic composition, except for fractionation, since Q gases would be drawn from the same reservoir that will later provide the Sun's and the nebula's noble gases. It would also account in a natural way for an essentially constant proportion of Q gases to other presolar components, since they would all enter the solar system in grains that would be mixed even before the solar system forms. In this scenario Q would be a presolar component, but whether it should also be termed "local" (made from the same material as the solar system) or "exotic" (made elsewhere, and imported into the solar system) seems more a matter of idiosyncratic taste than scientific substance. Postulating an interstellar medium origin for Q gas does not by itself solve the problem of how gases get incorporated into the solids, but at least it expands the parameter space in which a solution may be sought (see below).

1.14.5.3 Trapping Mechanisms

Elsewhere in this chapter, discussion of how a component is to be accounted for is generally a matter of how an isotopic ratio or an elemental abundance pattern is generated. But most noble-gas components of interest in cosmochemistry are in or once were contained in planetary solids, so that a model to account for a noble-gas component is incomplete without consideration of how the gases get into the solids. This is no great mystery if the nuclei of the gas atoms were made there by nuclear transformation, i.e., an *in situ* component, or if noble-gas atoms impinge on the surface of a solid with enough energy to penetrate below at least several atom layers, e.g., solar wind or more energetic particles. Otherwise, however, the means by which noble-gas atoms entered and became trapped in planetary material has been one of the enduring mysteries of noble-gas cosmochemistry. In particular, the traditional problem has been how to account for the origin of planetary noble gases (Figure 6). This traditional problem has been more or less inherited by the Q-component, and in fact amplified, since some accounting must also be made for the exotic components (Table 2).

The problem with planetary gases, or the Q-component, is not that the noble gases are so scarce, it is that in any simple quantitatively predictive model the content of noble gases, at least any gases extracted from the ambient solar nebula, should be much lower still. In thermodynamic equilibrium between solid and gas phases, the gases can be dissolved in the solids or adsorbed on their surfaces. For any materials with known or plausible thermodynamic parameters, solution is simply not a viable mechanism to account for trapped noble gases: it does not produce the right compositions, and quantities are orders of magnitude too low. Adsorption is not quantitatively adequate either, but at least it produces gases with the right qualitative elemental abundance features, i.e., progressive depletion of lighter gases and much higher effective concentrations than are possible in solution (e.g., Fanale and Cannon, 1972; Podosek et al., 1981). In practice, though, equilibrium adsorption does not really provide a feasible simulation of planetary or Q gases except at quite low temperatures, certainly much lower than the temperatures at which terrestrial–planetary materials seem to have last substantially interacted with nebular gas (the "accretion temperature") and plausible only in the outer reaches of the solar system (comets?) or in the interstellar medium. In addition, gases adsorbed in equilibrium will be desorbed as soon as the ambient gas phase is removed, so that models involving adsorption in the generation of planetary or Q gases are incomplete without also specifying some mechanism for fixing adsorbed gases, so that they really become "trapped."

There has been no shortage of ideas, or of laboratory simulations, for nonequilibrium conditions or processes that may have produced planetary or Q gases (e.g., Frick et al., 1979; Yang and Anders, 1982; Zadnik et al., 1985; Wacker, 1989; Nichols et al., 1992; Sandford et al., 1998; Hohenberg et al., 2002). Generally, such models/simulations involve the synthesis of some prospective host phase, invoke adsorption to concentrate gas in/on the relevant material, and continued growth to occlude or trap the gas so that it will not later desorb. Often some energetic phenomenon is involved, such as plasma discharge, ultraviolet irradiation, hypervelocity shock, etc. In the Huss and Alexander (1987) model for generation of the Q-component in the presolar interstellar medium, for example, adsorption at very low temperature is called upon to concentrate gas in icy grain mantles, where they become trapped as UV irradiation and incidence of energetic ions lead to polymerization of a carbonaceous phase around the gas atoms.

There is no doubt that laboratory simulations can effectively "trap" noble gases beyond equilibrium concentrations, sometimes in qualitatively appropriate elemental patterns, but evidently none have yet achieved the distribution coefficients implied by the gas contents of the Q-component. A common problem with such simulations is that they are difficult to control in detail, and they are generally rather complicated, so that it is unclear which parameters are really controlling gas trapping. It is certainly possible that one (or more) of the processes studied in laboratory simulations is responsible for generation of Q, but there is no consensus on which, or in what circumstances.

Most of the attention given to production of the exotic gas components (Table 2) has focused on their isotopic patterns, but except for the R-component, believed to be formed by *in situ* decay of ^{22}Na, they also must be trapped components in their host phases, so models for their origin are incomplete without a trapping mechanism. There is significant evidence, including absence of fractionation from neighboring elements and comparison with laboratory simulation, that at least some of the components carried in diamonds and silicon carbide were trapped by ion implantation (e.g., Koscheev et al., 2001;

Ott, 2002) in stellar winds, and by extension perhaps all of them were thus trapped (except for the R-component). The situation is complicated, though, and multiple types of winds may be necessary to get elemental ratios right. If not all the exotic components were implanted as ions, it is not clear whether any of the mechanisms advanced for Q will suffice.

1.14.5.4 Loose Ends?

Noble-gas geochemistry is a data-rich field, and the discussion above does not do justice to all the data. There are additional occurrences of elemental and/or isotopic patterns which may be indicative of still other significant noble-gas components. Except for one particularly significant case, U–Xe, which is discussed in the following section, the scope of this chapter does not permit adequate description (but see Ott (2002), for a more detailed review).

To only mention two of the more prominent cases, ureilites and enstatite chondrites exhibit noble gases in different elemental patterns and in different carriers than those in most meteorites. Ureilite elemental abundance patterns are variable but overall more strongly fractionated than Q, and the gases are in diamonds (produced by shock in the solar system, and not to be confused with presolar nanodiamonds). Isotopic patterns are similar to those in Q, however, so it is not clear whether ureilite noble gases are better viewed as a distinct component or as a variant of Q. Gases in enstatite chondrites have been termed a "subsolar" component, marked by less steeply fractionated elemental abundances, intermediate between the solar and planetary patterns, and are hosted in the main silicates, chiefly enstatite (Crabb and Anders, 1982). It is not clear whether subsolar gases reflect some variant of solar-wind implantation or a component fundamentally different from those known in other classes of meteorites.

There are also a handful of other occurrences (see Ott, 2002) of unusual elemental or isotopic patterns that are not very well characterized and/or occur only in very restricted samples. These may reflect unusual circumstances not often encountered or explored (e.g., recoil; see Marti et al., 1989), or possibly more anomalous components yet to be discovered.

1.14.5.5 U–xenon?

One of the major vexations of noble-gas geo-/cosmochemistry, and even of cosmochemistry more broadly, is the difficulty in accounting for the isotopic composition of terrestrial atmospheric xenon in terms of plausibly known starting materials and plausible planetary processes.

This problem was formulated in the earliest days of surveying the content of noble gases in planetary materials and has attracted the interest of multiple generations of cosmochemists, but a consensus resolution still remains elusive.

To first order, air xenon can be related to xenon known elsewhere in the solar system (solar wind, meteorites) by mass-dependent isotopic fractionation. The fractionation is severe, $\sim 4\%$ amu^{-1}, in the direction that lighter isotopes in air are depleted relative to, say, solar-wind xenon, but it is not impossible to construct a planetary history to account for this, for example, by stipulating that air xenon is the residue of extensive hydrodynamic escape of an early atmosphere (e.g., Pepin, 1991). In closer detail, the relative abundances of the lighter isotopes (^{124}Xe, ^{126}Xe, ^{128}Xe, and ^{130}Xe) can be quantitatively accounted for as fractionated solar-wind xenon. Relative to this pattern, air xenon has $\sim 7\%$ too much ^{129}Xe, which is quite plausibly attributed to a radiogenic component from decay of ^{129}I. However, there is no way to extend this fractionation pattern, assuming solar wind or any known bulk meteoritic component as the underlying composition, such as to give a quantitatively satisfactory accounting of the heavier isotopes (^{131}Xe, ^{132}Xe, ^{134}Xe, and ^{136}Xe), with or without addition of a plausible nuclear component (fission of ^{244}Pu or ^{238}U).

U–Xe is a mathematical construct designed to address this problem. It is a composition (not counting ^{129}Xe) constrained to lie on the multidimensional correlation surfaces (generalizations of three-isotope correlation diagrams) obtained in stepwise heating of carbonaceous chondrites and also to match fractionated air xenon less some fraction of ^{244}Pu-fission xenon (Pepin and Phinney, 1976; Pepin, 2000). A match can be made for $\sim 4\%$ fission contribution to ^{136}Xe. The resultant U–Xe composition is equivalent to solar-wind xenon at the light isotopes but has a few to several percent less of the heavy isotopes. This model is specific to xenon and does not involve other gases. U–Xe is hypothesized to be original terrestrial xenon, supplied by the solar nebula.

Factors that are favorable to this model are that the mathematics can be made to work at all, and that it has no quantitative competitor to explain terrestrial xenon. One unfavorable factor is that this model presupposes separation of H-Xe from L-Xe in carbonaceous chondrites. Another is that the existence of U–Xe as a physical entity remains speculative. It is supposed that U–Xe is the prevalent gas in the nebula, but evidence for its actual presence in meteorites is sparse at best and apparently not reproducible (Busemann and Eugster, 2000), and U–Xe is evidently not invoked as the underlying primitive composition for xenon on Mars (cf. Pepin and Porcelli, 2002). Perhaps the most compelling argument is that if U–Xe is primitive nebular xenon, then the Sun is enriched in the heavy isotopes of xenon, e.g., several percent H-Xe without L-Xe, and it is difficult to see how this can be accommodated without some significant change to the current picture of solar-system formation. All in all, the issue is central but seemingly remains unresolved.

1.14.5.6 Planetary Atmospheres

Traditionally, accounting for terrestrial planetary atmospheres has been part of the practice of noble-gas cosmochemistry. The overall program is to try to deduce the character of the initial gases in the planetary inventory—as they may have been acquired from gas in the nebula, through solar-wind irradiation, or by capturing components such as are observed to exist in planetary materials—and how they may have been modified through planetary processes. Pursuit of this program has been fruitful in establishing constraints on the origin and evolution of the planets, but this topic is beyond the scope of this chapter. The major issue in which consideration of planetary atmospheres has raised fundamental questions about the character and distribution of noble-gas components in the preplanetary solar system is that of U–Xe and the nature of the primordial xenon in the Earth's atmosphere, discussed above. For broader discussion of planetary atmospheres the reader is referred to Chapter 4.12.

REFERENCES

Alexander E. C., Jr., Lewis R. S., Reynolds J. H., and Michel M. (1971) Plutonium-244: confirmation as an extinct radioactivity. *Science* **172**, 837–840.

Amari S., Lewis R. S., and Anders E. (1995) Interstellar grains in meteorites: III. Graphite and its noble gases. *Geochim. Cosmochim. Acta* **59**, 1141–1426.

Anders E. (1988) Circumstellar material in meteorites: noble gases, carbon and nitrogen. In *Meteorites and the Early Solar System* (eds. J. F. Kerridge and M. S. Matthews). University of Arizona Press, Tucson, pp. 927–955.

Anders E. and Zinner E. K. (1993) Interstellar grains in meteorites: diamond, silicon carbide and graphite. *Meteoritics* **28**, 490–514.

Anders E., Higuchi H., Gros J., Takahashi H., and Morgan J. W. (1975) Extinct superheavy element in the Allende meteorite. *Science* **190**, 1261–1271.

Bernatowicz T. J. and Zinner E. K. (eds.) (1997) *Astrophysical Implications of the Laboratory Study of Presolar Materials* (AIP Conference Proceedings 402). AIP, New York, pp. 750.

Black D. C. (1972) On the origin of trapped helium, neon and argon isotopic variations in meteorites: II. Carbonaceous chondrites. *Geochim. Cosmochim. Acta* **36**, 377–394.

Black D. C. and Pepin R. O. (1969) Trapped neon in meteorites: II. *Earth Planet. Sci. Lett.* **6**, 395–405.

Busemann H. and Eugster O. (2000) Primordial noble gases in Lodran metal separates and the Tatahouine diogenite. *Lunar Planet. Sci.* **XXXI**, #1642. Lunar and Planetary Institute, Houston (CD-ROM).

Busemann H., Baur H., and Wieler R. (2000) Primordial noble gases in "phase Q" in carbonaceous and ordinary chondrites studied by closed-system stepped etching. *Meteorit. Planet. Sci.* **35**, 949–973.

Busemann H., Baur H., and Wieler R. (2001) Helium isotopic ratios in carbonaceous chondrites: significant for the early solar nebula and circumstellar diamonds? *Lunar Planet. Sci.* **XXXII**, #1598. Lunar and Planetary Institute, Houston (CD-ROM).

Clayton R. N., Grossman L., and Mayeda T. K. (1973) A component of primitive nuclear composition in carbonaceous meteorites. *Science* **182**, 495–488.

Crabb J. and Anders E. (1982) On the siting of noble gases in E-chondrites. *Geochim. Cosmochim. Acta* **46**, 2351–2361.

Drozd R. J., Hohenberg C. M., and Ragan D. (1972) Fission xenon from extinct ^{244}Pu in 14301. *Earth Planet. Sci. Lett.* **15**, 338–346.

Drozd R. J., Kennedy B. M., Morgan C. J., Podosek F. A., and Taylor G. J. (1976) The excess fission Xe problem in lunar samples. *Proc. 7th Lunar Sci. Conf.* 599–623.

Eberhardt P., Geiss J., Graf H., Grögler N., Mendia M. D., Mörgeli M., Schwaller H., and Stettler A. (1972) Trapped solar wind noble gases in Apollo 12 lunar fines 12001 and Apollo 11 breccia 10046. *Proc. 3rd Lunar Sci. Conf.* 1821–1856.

Fanale F. and Cannon W. A. (1972) Origin of planetary primordial rare gas: the possible role of adsorption. *Geochim Cosmochim. Acta* **36**, 319–328.

Frick U., Mack R., and Chang S. (1979) Solar gas trapping and fractionation during synthesis of carbonaceous matter. *Proc. 10th Lunar Planet. Sci. Conf.* 1961–1973.

Gallino R., Busso M., Picchio G., and Raiteri C. M. (1990) On the astrophysical interpretation of isotope anomalies in meteoritic SiC grains. *Nature* **348**, 298–302.

Geiss J. and Reeves H. (1972) Cosmic and solar system abundances of deuterium and helium-3. *Astron. Astrophys.* **18**, 126–132.

Geiss J., Gloeckler G., and von Steiger R. (1994) Solar and heliospheric processes from solar wind composition measurements. *Phil. Trans. Roy. Soc. London A* **349**, 213–226.

Gloeckler G. and Geiss J. (2000) Deuterium and helium-3 in the protosolar cloud. In *The Light Elements and their Evolution*, IUA Symposium, 198 (eds. L. da Silva, M. Spite, and J. R. Medeiros), pp. 224–233.

Heymann D. and Yaniv A. (1970) Ar^{40} anomaly in lunar samples from Apollo 11. *Proc. Apollo 11 Lunar Sci. Conf.* 1261–1267.

Hohenberg C. M., Marti K., Podosek F. A., Reedy R. C., and Shirck J. R. (1978) Comparisons between observed and predicted cosmogenic noble gases in lunar samples. *Proc. 9th Lunar Planet. Sci. Conf.* 2311–2344.

Hohenberg C. M., Nichols R. H., Jr., Olinger C. T., and Goswami J. N. (1990) Cosmogenic neon from individual grains of CM meteorites—extremely long pre-compaction exposure histories or an enhanced early particle flux. *Geochim. Cosmochim. Acta* **54**, 2133–2140.

Hohenberg C. M., Thonnard N., and Meshik A. (2002) Active capture and anomalous adsorption: new mechanisms for incorporation of heavy noble gases. *Meteorit. Planet. Sci.* **37**, 257–267.

Hoppe P. and Ott U. (1997) Mainstream silicon carbide grains from meteorites. In *Astrophysical Implications of the Laboratory Study of Presolar Materials*, AIP Proc. Conf. 402 (eds. T. J. Bernatowicz and E. K. Zinner), AIP, New York, pp. 27–58.

Howard W. M., Meyer B. S., and Clayton D. D. (1992) Heavy-element abundances from a neutron burst that produces Xe-X. *Meteoritics* **27**, 404–412.

Huss G. R. and Alexander E. C., Jr. (1987) On the pre-solar origin of the normal planetary noble gas component in meteorites. *Proc. 17th Lunar Planet. Sci. Conf.: J. Geophys. Res.* **92**, E710–E716.

Huss G. R. and Lewis R. S. (1994) Noble gases in presolar diamonds: I. Three distinct components and their implications for diamond origins. *Meteoritics* **29**, 791–810.

Huss G. R. and Lewis R. S. (1995) Presolar diamond, SiC, and graphite in primitive chondrites: abundances as a function of meteorite class and petrologic type. *Geochim. Cosmochim. Acta* **59**, 115–160.

Jungck M. H. A. and Eberhardt P. (1979) Neon-E in Orgueil density separates. *Meteoritics* **14**, 439–440.

Kallenbach R., Ipavich F. M., Kucharek H., Bochsler P., Galvin A. B., Geiss J., Gliem F., Glöeckler G., Grünwaldt H., Hefti H., Hilchenbach M., and Hovestadt D. (1998) Fractionation of Si, Ne, and Mg isotopes in the solar wind as measured by SOHO/CELIAS/MTOF. *Space Sci. Rev.* **85**, 357–370.

Koscheev A. P., Gromov M. D., Mohapatra M. K., and Ott U. (2001) History of trace gases in presolar diamonds as inferred from ion implantation experiments. *Nature* **412**, 615–617.

Lewis R. S. and Srinivasan B. (1993) A search for noble gas evidence for presolar oxide grains. *Lunar Planet. Sci.* **XXIV**. Lunar and Planetary Institute, Houston, pp. 873–874.

Lewis R. S., Srinivasan B., and Anders E. (1975) Host phase of a strange xenon component in Allende. *Science* **190**, 1251–1262.

Lewis R. S., Amari S., and Anders E. (1994) Interstellar grains in meteorites: II. SiC and its noble gases. *Geochim. Cosmochim. Acta* **58**, 471–494.

Mahaffy P. R., Donahue T. M., Atreya S. K., Owen T. C., and Niemann H. B. (1998) Galileo probe measurements of D/H and ^{3}He/^{4}He in Jupiter's atmosphere. *Space Sci. Rev.* **84**, 251–263.

Manka R. H. and Michel F. C. (1971) Lunar atmosphere as a source of lunar surface elements. *Proc. 2nd Lunar Sci. Conf.* 1717–1728.

Marti K. (1969) A new isotopic composition of xenon in the Pesyanoe meteorite. *Earth Planet. Sci. Lett.* **3**, 243–248.

Marti K., Kim J. S., Lavielle B., Pellas P., and Perron C. (1989) Xenon in chondritic metal. *Z. Naturforsch* **44a**, 963–967.

Meshik A. P., Pravdivtseva O. V., and Hohenberg C. M. (2001) Selective laser extraction of Xe-H from Xe-HL in meteoritic nanodiamonds: real effect or experimental artifact? *Lunar Planet. Sci.* **XXXII**, #2158. Lunar and Planetary Institute, Houston (CD-ROM).

Murty S. V. S., Goswami J. N., and Shukolyukov Y. A. (1997) Excess ^{36}Ar in the Efremovka meteorite: a strong hint for the presence of ^{36}Cl in the solar system. *Astrophys. J.* **475**, L65–L68.

Nichols R. H., Jr., Nuth J. A., III, Hohenberg C. M., Olinger C. T., and Moore M. H. (1992) Trapping of noble gases in proton-irradiated silicate smokes. *Meteoritics* **27**, 555–559.

Nichols R. H., Jr., Kehm K., Hohenberg C. M., Amari S., and Lewis R. S. (2003) Neon and helium in single interstellar SiC and graphite grains: asymptotic Giant Branch, Wolf-Rayet, supernova and nova sources. *Geochim. Cosmochim. Acta* (submitted).

Ott U. (1996) Interstellar diamond xenon and timescales of supernova ejecta. *Astrophys. J.* **463**, 344–348.

Ott U. (2002) Noble gases in meteorites—trapped components. *Rev. Mineral. Geochem.* **47**, 71–100.

Ott U., Mack R., and Chang S. (1981) Noble-gas-rich separates from the Allende meteorite. *Geochim. Cosmochim. Acta* **45**, 1751–1788.

Ozima M. and Podosek F. (2002) *Noble Gas Geochemistry*, 2nd edn. Cambridge University Press, Cambridge, 286p.

Ozima M. and Zashu S. (1983) Primitive helium in diamonds. *Science* **219**, 1067–1068.

Ozima M., Wieler R., Marty B., and Podosek F. A. (1998) Comparative studies of solar, Q-gases and terrestrial noble gases, and implications on the evolution of the solar nebula. *Geochim. Cosmochim. Acta* **62**, 301–314.

Pepin R. O. (1991) On the origin and early evolution of terrestrial planet atmospheres and meteoritic volatiles. *Icarus* **92**, 2–79.

Pepin R. O. (2000) On the isotopic composition of primordial xenon in terrestrial planet atmospheres. *Space Sci. Rev.* **92**, 371–395.

Pepin R. O. and Phinney D. (1976) The formation interval of the Earth. *Lunar Planet. Sci.* **VII**. Lunar and Planetary Institute, Houston, pp. 682–684.

Pepin R. O. and Porcelli D. (2002) Origin of noble gases in the terrestrial planets. *Rev. Mineral. Geochem.* **47**, 191–246.

Podosek F. A. and Nichols R. H., Jr. (1998) Short-lived radionuclides in the solar nebula. In *Astrophysical Implications of the Laboratory Study of Presolar Materials*, AIP Proc. Conf. 402 (eds. T. J. Bernatowicz and E. K. Zinner), pp. 617–647.

Podosek F. A., Bernatowicz T. J., and Kramer F. E. (1981) Adsorption of xenon and krypton on shales. *Geochim. Cosmochim. Acta* **45**, 2401–2415.

Reynolds J. H. (1960) Determination of the age of the elements. *Phys. Rev. Lett.* **4**, 8–10.

Reynolds J. H. and Turner G. (1964) Rare gases in the chondrite Renazzo. *J. Geophys. Res.* **69**, 3263–3281.

Sandford S. A., Bernstein M. P., and Swindle T. D. (1998) The trapping of noble gases by by the irradiation and warming of interstellar ice analogs. *Meteorit. Planet. Sci.* **33**, A135.

Signer P. and Suess H. E. (1963) Rare gases in the Sun, in the atmosphere, and in meteorites. In *Earth Science and Meteorites* (eds. J. Geiss and E. D. Goldberg). North Holland, Amsterdam, pp. 241–272.

Swindle T. D. and Podosek F. A. (1988) Iodine-xenon dating. In *Meteorites and the Early Solar System* (eds. J. F. Kerridge and M. S. Matthews). University of Arizona Press, Tucson, pp. 1127–1146.

von Steiger R., Geiss J., and Gloeckler G. (1997) Composition of the solar wind. In *Cosmic Winds and the Heliosphere* (eds. J. R. Jokipii, C. P. Sonnett, and M. S. Giampapa). University of Arizona Press, Tucson, pp. 591–616.

Wacker J. F. (1989) Laboratory simulation of meteoritic noble gases: III. Sorption of neon, argon, krypton, and xenon on carbon: elemental fractionation. *Geochim. Cosmochim. Acta* **53**, 1421–1433.

Wieler R. (1998) The solar noble gas record in lunar samples and meteorites. *Space Sci. Rev.* **85**, 303–314.

Wieler R. (2002a) Cosmic-ray-produced noble gases in meteorites. *Rev. Mineral. Geochem.* **47**, 125–170.

Wieler R. (2002b) Noble gases in the solar system. *Rev. Mineral. Geochem.* **47**, 21–70.

Wieler R., Anders E., Baur H., Lewis R. S., and Signer P. (1991) Noble gases in "phase-Q": closed-system etching of an Allende residue. *Geochim. Cosmochim. Acta* **55**, 1709–1722.

Wimmer-Schweingruber R. F. and Bochsler P. (2000) Is there a record of interstellar pick-up ions in lunar soils? In *Acceleration and Transport of Energetic Particles Observed in the Heliosphere,* ACE 2000 Symposium (eds. R. A. Mewaldt, J. R. Jokipii, M. A. Lee, E. Möbius, and T. H. Zurbuchen). Am. Inst. Phys. Conf. Proc. 528, pp. 270–273.

Yang J. and Anders E. (1982) Sorption of noble gases by solids, with reference to meteorites: III. Sulfides, spinels, and other substances: on the origin of planetary gases. *Geochim. Cosmochim. Acta* **46**, 877–892.

Zadnik G., Wacker J. F., and Lewis R. S. (1985) Laboratory simulation of meteoritic noble gases: II. Sorption of xenon on carbon: etching and heating experiments. *Geochim. Cosmochim. Acta* **49**, 1049–1059.

1.15
Condensation and Evaporation of Solar System Materials

A. M. Davis and F. M. Richter

University of Chicago, IL, USA

1.15.1 INTRODUCTION	407
1.15.2 THEORETICAL FRAMEWORK	409
1.15.2.1 Thermodynamic Equilibrium	409
1.15.2.1.1 Condensation of the major elements	409
1.15.2.1.2 Condensation of trace elements	411
1.15.2.1.3 High-temperature equilibrium isotopic fractionation	412
1.15.2.2 Kinetic Effects	412
1.15.3 LABORATORY EXPERIMENTS	415
1.15.3.1 Evaporation of Simple Oxides	416
1.15.3.2 Evaporation of Olivine	416
1.15.3.3 Evaporation of CMAS Melts	416
1.15.3.4 Evaporation of Chondritic Meteorites and Chondritic Melts	421
1.15.4 APPLICATIONS	421
1.15.4.1 Bulk Compositions of Planets and Meteorite Parent Bodies	421
1.15.4.2 Calcium-, Aluminum-rich Inclusions	422
1.15.4.2.1 Ultrarefractory inclusions	422
1.15.4.2.2 Textural evidence	422
1.15.4.2.3 Bulk elemental and isotopic compositions of type B CAIs	423
1.15.4.2.4 Refractory metal inclusions	424
1.15.4.2.5 FUN CAIs	425
1.15.4.2.6 Hibonites with cerium anomalies	425
1.15.4.3 Chondrites	425
1.15.4.3.1 Chondrules	425
1.15.4.3.2 Metal grains	426
1.15.4.4 Deep-sea Spherules	426
1.15.5 OUTLOOK	426
ACKNOWLEDGMENTS	427
REFERENCES	427

1.15.1 INTRODUCTION

It is widely believed that the materials making up the solar system were derived from a nebular gas and dust cloud that went through an early high-temperature stage during which virtually all of the material was in the gas phase. At one time, it was thought that the entire inner solar nebula was hot, but it is now believed that most material was processed through regions where high temperatures were achieved. Certainly some material, such as presolar grains (cf., Mendybaev et al., 2002a), has never been exposed to high temperatures. As the system cooled, solids and perhaps liquids began to condense, but at some point the partially condensed materials became isolated from the remaining gas. Various lines of evidence support this view. At the largest scale, there is the observation that the Earth, Moon, Mars, and all chondritic meteorites except for the CI chondrites are depleted to varying degrees in the abundances of moderately volatile elements relative to bulk solar system composition. The CI chondrites reflect the bulk composition of the solar system for all but hydrogen, carbon, nitrogen, oxygen, and the rare gases, the most volatile elements (see Chapter 1.03; Palme et al., 1988; McDonough and Sun, 1995; Humayun and Cassen, 2000). The depletions in moderately volatile elements are, to a significant degree, correlated with condensation temperature, suggesting progressive removal of gas as condensation proceeded (Cassen, 1996). Additional observations that can be explained by partial condensation are that various particularly primitive components of meteorites (e.g., calcium-, aluminum-rich refractory inclusions, and certain metal grains) have mineralogy and/or details of their chemical composition that are remarkably similar to what is calculated for equilibrium condensates from a solar composition gas. For example, the calcium-, aluminum-rich inclusions (CAIs) in chondritic meteorites have compositions very similar to that calculated for the first 5% of total condensable matter (see Chapter 1.08; Grossman, 1973; Wänke et al., 1974; Grossman and Ganapathy, 1976; Grossman et al., 1977), where CI chondrites are taken to represent total condensable matter.

Elemental abundance patterns ordered by volatility certainly could have been produced by partial condensation, but they could also have been caused by partial evaporation. The relative importance of these opposite processes is still subject to debate and uncertainty. It should be remembered that condensation calculations typically assume chemical equilibrium in a closed system, in which case the system has no memory of the path by which it arrived at a given state, and thus the chemical and isotopic composition of the condensed phase cannot be used to distinguish between partial condensation and partial evaporation. Humayun and Clayton (1995) have taken a somewhat different view by arguing that condensation and evaporation are distinguishable, in that evaporation, but not condensation, will produce isotopically fractionated residues. With this idea in mind, they carefully measured the potassium isotopic compositions of a broad range of solar system materials with different degrees of potassium depletion and found them to be indistinguishable. This they took as evidence that evaporation could not have been a significant process in determining the diverse elemental abundance patterns of the various solar system materials they measured, because had evaporation been important in fractionating potassium it would have also fractionated the potassium isotopes. We will qualify this line of reasoning by arguing that evaporation and condensation can under certain conditions produce isotopically fractionated condensed phases (i.e., that partial evaporation can produce isotopically heavy residues and that partial condensation can produce isotopically light condensates) but that under other conditions both can produce elemental fractionations without significant isotopic fractionation. The absence of isotopic fractionation in a volatile element-depleted condensed phase is more a measure of the degree to which the system maintained thermodynamic equilibrium than a diagnostic of whether the path involved condensation or evaporation.

The pervasive volatile element depletion of inner solar system planets and the asteroidal parent bodies of most meteorites is a major, but by no means the only reason to consider evaporation and condensation processes in the early history of the solar system. Chondrules appear to have been rapidly heated and then cooled over a period of minutes to hours (see Chapter 1.07). If this occurred in a gas of solar composition under nonequilibrium conditions, chondrules should have partially evaporated and an isotopic fractionation record should remain. The absence of such effects can be used to chonstrain the conditions of chondrule formation (e.g., Alexander et al., 2000; Alexander and Wang, 2001). There is good petrologic, chemical, and isotopic evidence suggesting that certain solar system materials such as the coarse-grained CAIs are likely evaporation residues. For example, the type B CAIs are often found to have correlated enrichments in the heavy isotopes of silicon and magnesium (Figure 1), and these isotopic fractionations are very much like those of evaporation residues produced in laboratory experiments. Condensation also appears to be a major control of elemental zoning patterns in metal grains in CH chondrites (Meibom et al., 1999, 2001; Campbell et al., 2001; Petaev et al., 2001; Campbell et al., 2002). A more contemporary example is the isotopic and chemical compositions of deep-sea spherules that have been significantly affected by evaporative loss during atmospheric entry (Davis et al., 1991a; Davis and Brownlee, 1993; Herzog et al., 1994, 1999; Xue et al., 1995; Alexander et al., 2002).

The volatile element depletion patterns of planetary size objects and the chemical and isotopic composition of numerous smaller objects such as chondrules and CAIs provide the motivation to consider evaporation and condensation process in the early solar system. The key point is that the processes that led to chondrules and planets appear to have occurred under conditions very close to equilibrium, whereas the processes that led to CAIs involved significant departures from equilibrium.

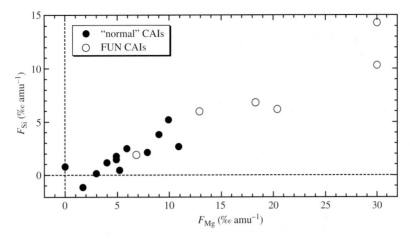

Figure 1 Isotopic mass fractionation effects in CAIs. Most coarse-grained CAIs have enrichments of a few ‰ amu^{-1} in magnesium and silicon, whereas "fractionation and unknown nuclear" (FUN) CAIs are isotopically heavier.

1.15.2 THEORETICAL FRAMEWORK

Volatility fractionation of elemental abundances and isotopic composition has traditionally been considered in terms of condensation and evaporation, under the assumption that condensation is a slow process controlled by thermodynamic equilibrium, and evaporation is a rapid process controlled by kinetic factors. We see no compelling reason to accept this distinction between evaporation and condensation. However, distinguishing between equilibrium and nonequilibrium conditions is crucial, and thus we have chosen to organize this section in terms of equilibrium and kinetically controlled processes.

1.15.2.1 Thermodynamic Equilibrium

The theory of equilibrium calculations is well established and the techniques employed can be found in the literature cited in the following subsections.

1.15.2.1.1 Condensation of the major elements

"Condensation calculations" in which the thermodynamic equilibrium condensate mineral assemblage is calculated as a function of temperature in a gas of fixed pressure and composition (usually solar) provide an extremely useful framework for interpreting bulk compositions of CAIs, chondrules, meteorites, and planets. Calculations of this sort date back to Urey (1952). Lord (1965) and Larimer (1967) calculated gas–solid equilibria for solar system composition, but did not correct the gas for removal of condensed material. The first detailed high-temperature condensation sequence was defined by Grossman (1972). The results of a calculation of Yoneda and Grossman (1995) are shown in Figure 2, for a total nebular pressure of 10^{-3} atm (100 Pa), and are not so different from the Grossman (1972) calculation. The pressure used in Figure 2 is the same as Grossman used and is within the range of 10^{-6}–10^{-2} atm that are thought to have prevailed in the solar nebula. This calculation assumed the solar system abundances of Anders and Grevesse (1989) and made use of the best currently available thermodynamic data for solid phases. It includes nonideal solid solution models for melilite [Ca$_2$(MgSi, Al$_2$)SiO$_7$], calcium-rich pyroxene [Ca(MgSi$_2$, Al$_2$Si, Ti^{+3}AlSi, Ti^{+4}Al$_2$)O$_6$], spinel [(Mg, Fe)(Al, Cr)$_2$O$_4$], and metal [Fe–Ni–Co–Cr]. The only significant change to solar system abundances of major elements since Anders and Grevesse (1989) is that the abundances of carbon and oxygen are nearly 50% lower (Allende Prieto et al., 2001, 2002; Lodders, 2003; see Chapter 1.03). Use of the new carbon and oxygen abundances has the effect of lowering condensation temperatures by ~30° relative to the temperatures shown in Figure 1, but does not change the sequence (L. Grossman, personal communication).

Other calculations have explored a wider range of parameters. Wood and Hashimoto (1993) did the first calculations exploring the effects of varying the proportions of four volatility components of solar material: refractory dust, carbonaceous matter, ice, and hydrogen gas. They found that dust enrichment increases the overall refractoriness of dust and the iron content of mafic minerals and permits melts to be in equilibrium with gas. Yoneda and Grossman (1995) explored condensation of melts in the CaO–MgO–Al$_2$O$_3$–SiO$_2$ (CMAS) system, using the more sophisticated Berman (1983) model for CMAS melts, and found that melts can coexist with gas at

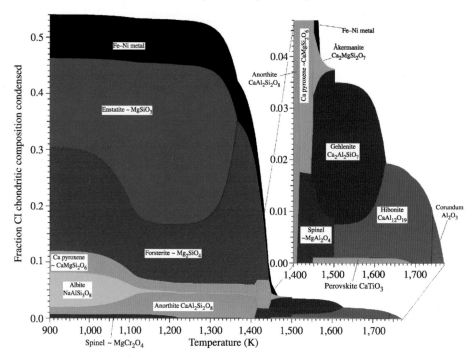

Figure 2 Condensation of major rock-forming phases from a gas of solar composition (Anders and Grevesse, 1989) at a total pressure of 10^{-3} atm. This calculation was done with the best currently available internally consistent thermodynamic data for solid and gaseous phases and includes nonideal solid solution models for melilite, Ca-rich pyroxene, feldspar, and metal. This calculation is the same as the one shown in Yoneda and Grossman (1995, table 1 and figure 1).

pressures as low as 10^{-2} atm. Ebel and Grossman (2000) extended such calculations to chondritic compositions using the well-known MELTS model of Ghiorso and co-workers. With nonideal melts and crystalline phases, sophisticated free-energy minimization techniques are needed (Ebel et al., 2000). Dust-enriched systems have two important differences from solar system composition: condensation temperatures are higher, so melts can be stabilized; and conditions can become oxidizing enough at high dust enrichments to allow condensation of a significant fraction of total iron into ferromagnesian minerals rather than metal (Wood and Hashimoto, 1993; Ebel and Grossman, 2000). The oxidation state of iron in chondritic meteorites and terrestrial planets is difficult to understand in the context of equilibrium condensation in a solar composition system, as iron does not become oxidized until below 800 K at a total pressure of 10^{-3} atm; thus, dust enrichment may play an important role in maintaining locally oxidizing conditions.

It must be recognized that the availability of thermodynamic data requires some compromises in condensation calculations. For condensation calculations at pressures where no melt condensation needs to be considered, the best thermodynamic data for mineral phases can be used, although there is still no model for solid solution of magnesium and titanium into hibonite. The plot shown in Figure 2 is an example of such a calculation. When the Berman (1983) model for CMAS liquids or MELTS model for more chondritic composition melts are used, sets of thermodynamic data for solids that are compatible with each melt model must be used. Furthermore, the MELTS model that is widely used in petrologic modeling and works well for chondritic melts (Ebel and Grossman, 2000; Alexander, 2002) fails for the CMAS system and calculations must make a transition to the Berman (1983) CMAS melt model when melts change from chondrule-like to CAI-like. For compatibility with melt models, solid data must also change at this transition. Thermodynamic data for solids compatible with MELTS were used in the calculations of Ebel (2000) and Grossman et al. (2000, 2002). When used at low pressures where melts are not stable, these data result in the condensation of several calcium–aluminum-oxide minerals that are seldom or never observed in CAIs (cf. figure 2 of Ebel, 2000). However, grossite ($CaAl_4O_7$) is common in CAIs in CH chondrites (Grossman et al., 1988; Weber and Bischoff, 1994) and $CaAl_2O_4$ has been observed (Ivanova et al., 2002), reflecting the fact that these phases are on the verge of stability and minor changes in conditions may affect whether or not they appear.

Further variations on equilibrium condensation calculations have been explored. Petaev and Wood (1998) investigated the effects of isolating early condensates from equilibrium with gas as temperature falls. They used the simple assumption that some fraction of condensate is isolated from the gas at each temperature step. Small degrees of isolation had little effect, but large degrees of isolation resulted in condensation of new phases such as silica, larnite, and merwinite. The latter two phases are not observed in CAIs or chondrules, but some aspects of high silica chondrules may be explained by constrained equilibrium calculations such as those of Petaev and Wood (1998).

Condensation calculations can be carried out to relatively low temperatures: Grossman and Larimer (1974) extended calculations to below 400 K, and Lodders (2003) lists 50% condensation temperatures as low as 9 K!, but the meaning of calculations below ~1,000 K is unclear as gas–solid equilibrium is unlikely to be reached.

1.15.2.1.2 Condensation of trace elements

The Grossman (1972) condensation calculations were restricted to the major elements, but such calculations were soon extended to refractory trace elements (Grossman, 1973; Grossman and Larimer, 1974; Wänke et al., 1974; Boynton, 1975; Grossman and Ganapathy, 1976). These calculations were motivated by trace-element abundances in CAIs, which were found by Grossman (1973) to be enriched in all refractory elements, regardless of lithophile or siderophile character. Condensation calculations of moderately and highly volatile trace elements were also done at that time to explain elemental abundances in chondrites (Larimer and Anders, 1967; Keays et al., 1971; Laul et al., 1973). The most commonly used table of trace-element condensation temperatures is that of Wasson (1984), although Lodders (2003) has recently provided an updated list of 50% condensation temperatures of all elements except hydrogen and helium. A common problem for trace-element condensation calculations is that the solid solution behavior of trace elements in most solids is not known. For this reason, ideal solid solution into mineral sites of appropriate ionic radius is often assumed. Since ideal solid solution is the exception rather than the rule, calculated condensation temperatures of trace elements are not as accurate as those for major elements.

Although the rare-earth elements (REEs) have similar geochemical behavior, since they are all large-ion lithophile elements and most of them partition among melts and mineral phases as a smooth function of ionic radius (with the exception of europium, which, commonly being divalent rather than trivalent, partitions differently from the other REEs), their volatilities are not smooth functions of ionic radius. The REEs exhibit a significant range in volatility (Boynton, 1975; Davis and Grossman, 1979). Rather than condensing into a separate phase, it is likely that the REEs condense in solid solution in refractory oxide phases, most likely initially into hibonite. Figure 3 shows REE patterns predicted for the equilibrium condensate assemblage at 10^{-3} atm total pressure (Figure 2) as a function of the temperature below the initial appearance of hibonite, the first likely host for REEs. The REEs can be divided into three groups: (i) most of the heavy REEs are highly refractory, with refractoriness increasing in the order gadolinium, terbium, dysprosium, holmium, erbium, and lutetium; (ii) the light REEs and thulium are less refractory and have similar volatilities; and (iii) europium and ytterbium are the most volatile of the REEs. The volatilities of a few of the REEs are also sensitive to oxidation state. Most of the REEs are trivalent under most natural conditions, but europium can be divalent under reducing conditions and cerium can be tetravalent under oxidizing conditions. Under oxidizing conditions, cerium becomes much more volatile than the other REEs and europium and ytterbium become more refractory (Davis et al., 1982).

The siderophile elements show a wide range of volatilities, ranging from rhenium, osmium, and

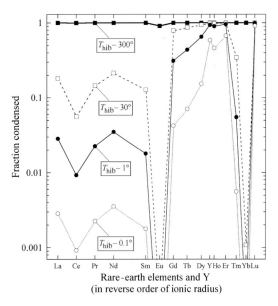

Figure 3 Condensation of REE into hibonite, perovskite, melilite, and fassaite. In a gas of solar composition at 10^{-3} atm total pressure, these phases condense at temperatures of 1,742 K, 1,688 K, 1,628 K, and 1,449 K, respectively. Nonideal solid solution of REE into hibonite makes thulium as volatile as the light REE, as seems to be required to explain REE patterns in group II CAIs.

tungsten, which can condense as metal grains 150° higher than the first major-element oxide, Al_2O_3, to cadmium, indium, thallium, and lead, which condense below 750 K at 10^{-4} bar total pressure. A calculation of siderophile elements in alloys (Campbell et al., 2001) compares well with earlier calculations of Grossman and Larimer (1974), Palme and Wlotzka (1976), Fegley and Palme (1985), and Sylvester et al. (1990). Fegley and Palme (1985) considered redox effects on condensation of the moderately siderophile elements molybdenum and tungsten, but further calculations on condensation of these and other moderately siderophile elements, such as rhenium and gallium into silicates and oxides, might be useful for explaining some elemental abundance features in chondrules and CAIs.

1.15.2.1.3 High-temperature equilibrium isotopic fractionation

Isotopic mass fractionation due to equilibrium partitioning of isotopes between gas and condensed phases at the temperatures of interest here is assumed to be negligible (less than 1–2‰ amu^{-1}), because the equilibrium isotope partition coefficients approach 1 at high temperature (Urey, 1947; Richter et al., 2002). Some kinetic fractionation processes, such as thermal evaporation, are independent of temperature and thus capable of significant isotopic fractionation even at high temperature, are discussed in the next section. However, not all kinetic isotopic fractionations are temperature independent. Any process with an activation energy, such as thermal decomposition, has a temperature dependence just like an equilibrium process, and isotopic fractionation becomes negligible at high temperature.

1.15.2.2 Kinetic Effects

Kinetic effects have been of great interest in cosmochemistry since the mid-1990s and we now review the generally accepted theory of evaporation and condensation in some detail. This discussion focuses on isotope fractionations, but is equally applicable to elemental fractionations.

The evaporative flux from a solid or liquid surface is given by the Hertz–Knudsen equation (see Hirth and Pound, 1963)

$$J_i^e = \frac{\sum_j \gamma_{ij}^e P_{ij,\text{sat}}}{\sqrt{2\pi m_{ij} RT}} \quad (1)$$

where J_i^e is the evaporation flux of element or isotope i per unit area, γ_{ij}^e is the evaporation coefficient of the jth gas species containing i, $P_{ij,\text{sat}}$ is the saturation vapor pressure of j, m_{ij} is the molecular weight of j, R is the gas constant, and T is absolute temperature. The evaporation coefficient is essentially a measure of kinetic hindrance of the evaporation reaction and has values ≤ 1. It should be noted that both γ_{ij}^e and $P_{ij,\text{sat}}$ are functions of temperature. Equation (1) can be simplified when the gas phase is dominated by a single species of an element of interest (e.g., Mg or SiO), in which case

$$J_i^e = \frac{\gamma_i^e P_{i,\text{sat}}}{\sqrt{2\pi m_i RT}} \quad (2)$$

Equation (1), or (2) if appropriate, is taken to represent the evaporation rate regardless of whether condensation is also taking place. The condensation flux, J_i^c, can in a similar way be represented by

$$J_i^c = \frac{\gamma_i^c P_i}{\sqrt{2\pi m_i RT}} \quad (3)$$

where γ_i^c is the condensation coefficient and P_i is the pressure of species i at the surface where condensation is taking place. The net evaporation or condensation rate, $J_{i,\text{net}}$, is the difference between the evaporation flux given by Equation (2) and the condensation flux given by Equation (3), thus

$$J_{i,\text{net}} = J_i^e \left[1 - \left(\frac{\gamma_i^c}{\gamma_i^e}\right)\left(\frac{P_i}{P_{i,\text{sat}}}\right)\right] \quad (4)$$

We will assume here for simplicity that $\gamma_i^c = \gamma_i^e$, so Equation (4) reduces to

$$J_{i,\text{net}} = J_i^e \left(1 - \frac{P_i}{P_{i,\text{sat}}}\right) \quad (5)$$

$J_{i,\text{net}}$ must approach 0 as $P_i/P_{i,\text{sat}}$ approaches 1 (i.e., no net mass loss once the pressure of the evaporating species at the surface equals the saturation vapor pressure). Thus, γ_i^c must equal γ_i^e as $P_i/P_{i,\text{sat}} \rightarrow 1$, because of the very definition of $P_{i,\text{sat}}$. Furthermore, experiments on evaporation and condensation of iron show no evidence that γ_i^c differs from γ_i^e away from equilibrium (Tachibana et al., 2001). Net evaporation will be the case whenever $P_i < P_{i,\text{sat}}$, whereas net condensation will take place when $P_i > P_{i,\text{sat}}$.

We can illustrate some of the effects associated with evaporation and condensation by considering a commonly invoked special case, which we will call free evaporation, and which arises when the pressure of the evaporating species in the gas is assumed to be negligible compared to their saturation vapor pressure. Equation (2) can then be used to represent the evaporative flux of elements or isotopes, and the flux will fractionate these in proportion to the ratio

$$\frac{J_2^e}{J_1^e} = \frac{P_{2,\text{sat}}}{P_{1,\text{sat}}} \frac{\gamma_2^e}{\gamma_1^e} \sqrt{\frac{m_1}{m_2}} \quad (6)$$

The effect of volatility in fractionating elements is due to the ratio of the saturation vapor pressures, but as shown by Equation (6), the relative masses of the gas species and possible differences in the evaporation coefficients also affect the degree of chemical fractionation produced by evaporation. When Equation (6) is used in connection with isotope fractionation, it is generally assumed that isotopes of the same element have the same evaporation coefficient and that the ratio of the saturation vapor pressures is equal to the isotopic ratio at the surface of the evaporating material (i.e., no equilibrium fractionation). This results in the following equation for the relative flux of the isotopes:

$$\frac{J_2^e}{J_1^e} = R_{2,1} \sqrt{\frac{m_1}{m_2}} \quad (7)$$

where $R_{2,1}$ is the isotopic ratio at the surface. The isotopic composition of the evaporative flux differs from that at the surface of the evaporating material by a fractionation factor $\alpha = \sqrt{m_1/m_2}$.

We can consider an equivalently idealized situation for condensation if we assume that the pressure of condensable species in the gas is large compared to their saturation vapor pressure relative to the substrate on which condensation takes place. It follows from Equation (3) that

$$\frac{J_2^c}{J_1^c} = \frac{P_{2,i}}{P_{1,i}} \frac{\gamma_2^c}{\gamma_1^c} \sqrt{\frac{m_1}{m_2}} \quad (8)$$

which, when applied to isotopes, gives

$$\frac{J_2^c}{J_1^c} = R_{2,1} \sqrt{\frac{m_1}{m_2}} \quad (9)$$

where $R_{2,1}$ now refers to the isotopic composition of the gas.

The cumulative effect of the instantaneous fractionations given by Equations (6)–(9) is easily calculated if it is further assumed that mass transport processes (e.g., chemical diffusion) are sufficiently fast to maintain chemical and isotopic homogeneity in both the gas and in the condensed phase. There are cases where diffusion in the residue or gas limits mass transport and these effects on isotopic and chemical fractionation have been explored by Richter et al. (2002). Let us consider first the isotopic fractionations associated with condensation in a supersaturated closed system. The change in the moles of isotope 1 of element k in the gas can be written as

$$dM_1 = -J_1 A = \frac{-\gamma_k^c X_1 P_k}{\sqrt{2\pi m_1 RT}} A \quad (10)$$

and for isotope 2

$$dM_2 = -J_2 A = \frac{-\gamma_k^c X_2 P_k}{\sqrt{2\pi m_2 RT}} A \quad (11)$$

where dM_i is the change in moles of i in the gas, A is the surface area onto which material is condensing, and X_i is the mole fraction of isotope i in the gas. Dividing Equation (11) by (10) gives

$$\frac{dM_2}{dM_1} = \frac{dX_2}{dX_1} = \frac{X_2}{X_1} \sqrt{\frac{m_1}{m_2}} \quad (12)$$

which we rewrite as

$$\frac{dX_2}{X_2} = \alpha \frac{dX_1}{X_1} \quad (13)$$

where $\alpha = \sqrt{m_1/m_2}$ is the kinetic isotopic fractionation factor. Equation (13) is then integrated from the starting isotopic fractions $X_{0,1}$ and $X_{0,2}$ to that at any later time, yielding

$$\ln\left(\frac{X_2}{X_{0,2}}\right) = \alpha \ln\left(\frac{X_1}{X_{0,1}}\right) \quad (14)$$

or equivalently

$$\frac{X_2}{X_{0,2}} = \left(\frac{X_1}{X_{0,1}}\right)^\alpha \quad (15)$$

Dividing Equation (15) by $X_1/X_{0,1}$ and writing the isotope ratios as $R_{2,1} = X_2/X_1$ and $R_0 = X_{0,2}/X_{0,1}$, we finally arrive at the classic relationship for Rayleigh fractionation

$$\frac{R_{2,1}}{R_0} = \left(\frac{X_1}{X_{0,1}}\right)^{\alpha-1} \quad (16)$$

$(X_1/X_{0,1})$ represents the fraction of isotope 1 remaining in the gas, which in most cases will be a very good approximation to the fraction of element k remaining in the gas. The isotopic composition of the condensate as a function of the fraction of gas remaining is given by

$$\frac{R_{2,1}}{R_0} = \frac{1 - F^\alpha}{1 - F} \quad (17)$$

Equations (16) and (17) were derived for the isotopic effects associated with condensation in the idealized limit of no re-evaporation of the condensed material. Such a situation can arise for highly supersaturated systems and then Equation (16) gives the isotopic fractionation of the gas and Equation (17) that of the condensate as a function of the amount condensed. The same result obtains if we consider evaporation when recondensation is negligible, and then Equation (16) applies to the evaporation residue and Equation (17) to the evaporated gas. The magnitude of fractionations given by Equations (16) and (17) are illustrated in Figure 4 in the specific case of magnesium isotopes. The key point is that for systems not in

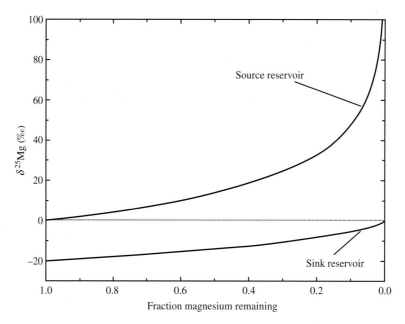

Figure 4 Magnesium isotopic fractionation of the source and sink reservoirs relative to the isotopic composition of the bulk system as a function of the magnesium fraction remaining in the source reservoir. In the case of condensation, the source reservoir is the gas and the sink the condensed phase. In the case of evaporation, the condensed phase is the source and the gas is the sink. $\delta^{25}\text{Mg} = [(^{25}\text{Mg}/^{24}\text{Mg})_{\text{reservoir}}/(^{25}\text{Mg}/^{24}\text{Mg})_{\text{bulk}} - 1] \times 1{,}000$.

equilibrium, both evaporation and condensation can produce large isotope fractionations.

The isotopic fractionations given by Equations (16) and (17) can be thought of as upper bounds given that they are derived assuming unidirectional fluxes (i.e., no recondensation or re-evaporation). Equation (5) shows that these unidirectional fluxes correspond to $P_i/P_{i,\text{sat}} \ll 1$ for free evaporation, while $P_i/P_{i,\text{sat}} \gg 1$ is the limit of condensation without re-evaporation. In a natural system it will often be the case that evaporation and condensation are taking place simultaneously, and we need to consider how this affects isotope fractionation. Richter et al. (2002) derive an equation for the effective kinetic isotope fractionation factor α' for use in place of $\alpha = \sqrt{m_1/m_2}$ in the Rayleigh fractionation equations (Equations (16) and (17)) for evaporation into a gas of finite total pressure. A slightly simplified (but entirely sufficient for our purposes) form of the relationship between α', α, and $P_i/P_{i,\text{sat}}$ is

$$\alpha' - 1 = (\alpha - 1)\left(1 - \frac{P_i}{P_{i,\text{sat}}}\right) \quad (18)$$

valid for $P_i/P_{i,\text{sat}} \leq 1$, which implies net evaporation (see Richter et al. (2002) for the exact result). The analogous result for net condensation (i.e., $P_{i,\text{sat}}/P_i \leq 1$) is

$$\alpha' - 1 = (\alpha - 1)\left(1 - \frac{P_{i,\text{sat}}}{P_i}\right) \quad (19)$$

Note that in the equilibrium limit $P_i/P_{i,\text{sat}} = 1$, isotopic fractionations vanish but the elemental abundances in the gas will still be fractionated in proportion to their relative saturation vapor pressures. The degree of equilibrium elemental fractionation of the condensed phase will depend on the volume of gas being sufficiently large that a substantial fraction of the elements of interest are in the gas phase.

There are various ways in which the different limits of $P_i/P_{i,\text{sat}}$ can be realized in an experiment or in a natural situation. If the evaporated gas is continuously removed from the system on a timescale fast compared to the rate of evaporation, then $P_i/P_{i,\text{sat}} \ll 1$ and the isotopic composition of the residue would be expected to evolve according to Equation (16). This is the case approached by vacuum evaporation experiments. At the other extreme is the case where the gas is removed slowly compared to the free evaporation rate (i.e., that measured in vacuum experiments) in which case $P_i/P_{i,\text{sat}} \to 1$ and elements will be fractionated by volatility but not their isotopes. In the following section, we discuss results from evaporations in a gas-mixing furnace that correspond to this limit. The ratio $P_i/P_{i,\text{sat}}$ will also reflect the rate at which environmental conditions (i.e., temperature and/or pressure) change compared to evaporation or condensation rates. If the environmental conditions change sufficiently fast, the system will not be able to maintain chemical and isotopic equilibrium and large fractionations can result even in a closed system that is only subsequently subject to a separation of the gas from partially condensed phases. It should also be kept in mind that the fractionations given by Equations (16) and (17) assume that both the gas and the condensed phase are well mixed and reasonably homogeneous. Especially in the case of the condensed phase, this will require that the timescale for diffusion to maintain a homogeneous liquid or solid be faster than the timescale of elemental and isotopic fractionation at the surface of the condensed phase. Liquids have sufficiently fast diffusion that homogeneity will often prevail; however, fractionation effects in solids will often be severely limited by the slowness of diffusion. Wang et al. (1999), in connection with their evaporation experiments of solid forsterite, give analytical solutions for the isotopic fractionation by diffusion-limited evaporation from a semi-infinite slab. Young et al. (1998) and Richter et al. (2002) used numerical methods to calculate the effect of finite diffusion in a molten silicate sphere on the isotopic fractionation associated with evaporation.

Before closing this section we should mention several relatively recent papers that provide additional details relevant to the overview given above. The effects of evaporation, back-reaction, diffusion, and dust enrichment on isotopic fractionation in forsterite have been discussed in great detail by Tsuchiyama et al. (1999) and Nagahara and Ozawa (2000) and extended to multicomponent systems in Ozawa and Nagahara (2001). Richter et al. (2002) combined theoretical and experimental approaches to study elemental and isotopic fractionation effects due to evaporation from CMAS liquids and included consideration of the effects of temperature, gas composition, and diffusion in both the residue and in the surrounding gas.

Elemental and isotopic fractionations by evaporation of silicate liquids, in particular limiting circumstances, can be simulated by equilibrium calculations, provided that an adequate thermodynamic model of the melt is available. In this approach, a particular starting temperature, pressure, and initial composition of condensed material are chosen and the gas in equilibrium with the melt is calculated from thermodynamic data. The gas is then removed from the system and equilibrium is recalculated. Repeated small steps of this sort can simulate the kinetic behavior during vacuum evaporation (i.e., the limit of fast removal of the gas relative to the rate it is generated by evaporation). This approach has been taken by Grossman et al. (2000, 2002) and Alexander (2001, 2002).

Even though much of the theoretical framework outlined above for describing kinetic evaporation and condensation has been around since the 1880s (Hertz, 1882; Knudsen, 1909), laboratory experiments are still very much needed. To begin with, the solution properties of molten silicates are sufficiently nonideal that one often needs to validate the thermodynamic model used to calculate the saturation vapor pressure in connection with Equations (1) and (2). One also needs experimental determinations of the evaporation coefficients, which, as we will show in the next section, are often very significantly less than one and not calculable from first principles. Finally, measurements of the isotopic composition of experimental evaporation residues are used to determine the fractionation factor α, and in this way confirm or modify the expectation given by Equation (16). In the next section, we discuss some of the relevant experimental data for the rates and consequences of evaporation from silicate systems.

1.15.3 LABORATORY EXPERIMENTS

Quantitative evaporation experiments using solar system and solar system-like materials were pioneered by Hashimoto (1983, 1990) in terms of chemical fractionations and by Davis et al. (1990) in terms of the isotopic fractionation of the evaporation residues. Quantitative condensation experiments are much more difficult, and there are no well-controlled condensation experiments available on chondritic or CAI compositions. The compositions that have been most extensively studied in evaporation experiments intended for application to the solar system are pure forsterite (Mg_2SiO_4), a type B CAI-like liquid consisting of CaO, MgO, Al_2O_3, and SiO_2, and a liquid in which major oxides including those of iron are initially in solar proportions.

We will describe several relatively recent experiments that used three different types of furnace, each designed to study evaporation in a particular pressure regime. What we will call vacuum experiments were carried out in a furnace especially designed for free evaporation, in which a sample is hung from a wire in the center of a resistance-heated chamber. The vacuum furnace is pumped continuously and the pressure at run temperatures (1,600–2,150 °C) is $P \leq 10^{-6}$ torr (for details, see Hashimoto, 1990; Wang et al., 1999). Evaporation experiments run in hydrogen gas at controlled pressures ($P = 6 \times 10^{-4}$ – 6×10^{-3} bar) were carried out in a highly modified vertical tube furnace (for details, see Kuroda and Hashimoto, 2002). An important consideration in the design of the latter furnace was to insure that the hydrogen gas, before reaching the sample, equilibrates both in temperature and speciation. Some of the quantitative discrepancies between the experimental results reported by different groups that measured evaporation rates in the presence of hydrogen could be a reflection of departures from equilibrium of the low-pressure hydrogen gas. In such a departure, reaction rates could be controlled by over- or underabundances of reactive monatomic hydrogen. Evaporation experiments have also been carried out in a 1 atm gas-mixing vertical tube furnace. The oxygen fugacity in the vicinity of the sample was controlled by flowing $H_2 + CO_2$ mixtures through the furnace tube and continuously monitored by a zirconia oxygen fugacity probe (Mendybaev et al., 1998).

1.15.3.1 Evaporation of Simple Oxides

Hashimoto (1988, 1989) and Hashimoto et al. (1989) used a vacuum furnace to perform evaporation experiments on a number of simple oxides and found that evaporation coefficients (Equation (3)) ranged from 1 to 0.03. Experiments on simple oxides have also been done by Wang et al. (1994), who showed that the evaporation coefficient for molten FeO is ~1, i.e., that there is no kinetic inhibition of evaporation, and that the iron isotopic fractionation factor, α, is that expected for evaporation of iron atoms. Similar experiments on other compositions have been done (e.g., Young et al., 1998), but it is difficult to derive evaporation coefficients from the limited data given in these reports.

1.15.3.2 Evaporation of Olivine

In addition to being the most abundant mineral in chondrites (thus being of great interest for studies of volatility effects on elemental and isotopic composition), olivine has the remarkable property that it evaporates congruently, both from the solid and liquid states. Davis et al. (1990) and Hashimoto (1990) carried out the first quantitative forsterite evaporation experiments, with Davis et al. (1990) emphasizing the isotopic fractionation and Hashimoto (1990) focusing on evaporation kinetics. Forsterite has a rather small evaporation coefficient, $\gamma = 0.03-0.06$ (Hashimoto, 1989; Davis et al., 1990; Wang et al., 1999); thus, its evaporation is significantly kinetically inhibited, both from the solid and liquid states. The isotopic compositions of magnesium, silicon, and oxygen in forsterite evaporation residues nicely follow Rayleigh fractionation curves, but the isotopic fractionation was significantly less than expected when the dominant evaporating species are Mg, SiO, O, and O_2 (Figure 5). The fractionation factor did not correspond to the inverse square root of mass of the isotopes of these gas species even though these species are those calculated from thermodynamics for forsterite in equilibrium with gas (Wang et al., 1999) and were also the dominant species measured by Nichols et al. (1998) in the gas above evaporating forsterite. Evaporation of solid forsterite led to negligible bulk isotopic fractionation, because diffusion in the residue is far too slow to allow internal equilibrium to be achieved and thus the isotopic fractionation was limited to a very small boundary at the surface (Wang et al., 1999). The evaporation rate of forsterite is greatly accelerated in the presence of hydrogen (Nagahara and Ozawa, 1996; Tsuchiyama, 1998; Kuroda and Hashimoto, 2002). We will return to the topic of evaporation in reducing gases later.

The next step in complexity involves evaporation studies of iron-bearing olivine, in which the residue becoming increasingly magnesium-rich as evaporation proceeds, but the residual phase remains olivine (i.e., the fayalite component of the olivine solid solution evaporates faster than the forsterite component). The evaporation behavior of iron-bearing olivine as reported by Ozawa and Nagahara (2000) is quite complicated in that the kinetic fractionation of the Fe/Mg ratio is found to be significantly less than the values expected based on calculated equilibrium Fe/Mg distribution between $Fo_{90}-Fo_{99}$ olivine and gas (i.e., the ratio of saturation vapor pressures of $Fe_{(g)}$ and $Mg_{(g)}$ over olivine). This implies that the evaporation coefficient for evaporation of iron from olivine must be significantly lower than that of magnesium. Recall that the evaporation coefficient for evaporation of magnesium from forsterite is of the order of 0.03–0.06 while that of iron from molten FeO is ~1. The clear implication is that the relative volatility of elements during kinetic evaporation depends on both saturation vapor pressures (calculable in most cases from thermodynamic data) and evaporation coefficients. The latter are neither calculable from first principles nor can they be reliably estimated from their values in end-member silicates or oxides.

1.15.3.3 Evaporation of CMAS Melts

The first set of experiments we discuss for this system involved partially evaporating small (3 mm typical dimension) CMAS liquid drops held at 1,800 °C in a vacuum furnace for times ranging from 15 min to 90 min (for details, see Richter et al., 2002). One purpose of these experiments was to validate a thermodynamic model for calculating saturation vapor pressures. Figure 6 shows how the SiO_2 and MgO contents of the liquid evolve as evaporation proceeds. The composition of the evaporation residues is compared to composition trajectories calculated using the thermodynamic model described by Grossman et al. (2000). The trajectories are

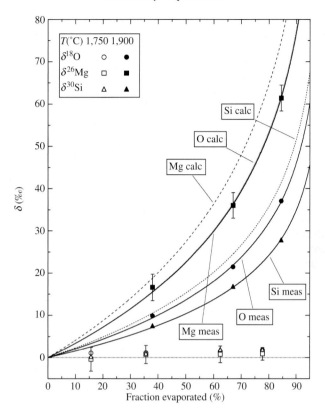

Figure 5 Isotopic mass fractionation of oxygen, magnesium, and silicon as a function of fraction evaporated (after Davis et al., 1990). Thermodynamic calculations predict that at 1,900 °C forsterite evaporates via the reaction $Mg_2SiO_{4(l)} \rightarrow 1.974Mg_{(g)} + 0.026MgO_{(g)} + 0.981SiO_{(g)} + 0.019SiO_{2(g)} + 0.659O_{(g)} + 1.148O_{2(g)}$. The figure shows two Rayleigh curves for each element: "meas," a best fit through the data; and "calc," an average weighted by the stoichiometry of the evaporation reaction of all gas phase species for each element, assuming that the gas/solid fractionation factor is the square root of the mass ratios of the isotopomers of the evaporating species. For evaporation of melts of forsterite composition, all three elements follow Rayleigh curves, but show less fractionation than expected from the thermodynamically calculated evaporating species. Note that solid forsterite shows no significant isotopic fractionation, despite significant mass loss.

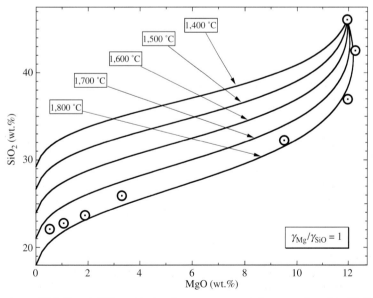

Figure 6 Comparison of MgO and SiO_2 contents of vacuum evaporation experiments with those calculated for an evaporating CMAS melt. Note how well the data match the curve for 1,800 °C, the temperature at which the experiments were performed (after Richter et al., 2002).

calculated using the ratio of the molar fluxes of silicon and magnesium as given by

$$\frac{J_{Si}}{J_{Mg}} = \frac{\gamma_{SiO}}{\gamma_{Mg}} \frac{P_{SiO}}{P_{Mg}} \sqrt{\frac{m_{Mg}}{m_{SiO}}} \quad (20)$$

Equation (20) follows from Equation (2) when the gas is dominated by Mg and SiO, and recondensation is negligible. The saturation vapor pressure of aluminum and calcium species in the gas is extremely low and thus the Al_2O_3 and CaO components of the liquid do not evaporate to any significant degree under the conditions of the experiments. The composition trajectory in $MgO-SiO_2$ space is surprisingly complicated, reflecting the thermodynamic nonideality of these liquids. The excellent agreement between the calculated trajectory and the experimental data shown in Figure 6 is an important indication that the activity model for CMAS liquids (Berman, 1983) used by Grossman et al. (2000) gives a very good representation of the relative activities of SiO_2 and MgO in the CMAS liquids of these experiments. Similarly, good results are obtained for residues evaporated in hydrogen at low pressures of the order of 10^{-4} bar. This is not really an independent test, since the chemical activities of the various components of the liquid are not affected by the surrounding hydrogen gas; what is affected by the surrounding gas is the saturation vapor pressures at the surface.

We now turn to the question of the evaporation kinetics of silicate systems. Figure 7 compares the evaporation rates in nominal vacuum $P \approx 10^{-6}$ torr) of forsterite (from Wang et al., 1999) and the CMAS liquid shown in Figure 6. Forsterite evaporates congruently, while in the case of CMAS liquids all the various oxide components evaporate at different rates. The silicon and magnesium evaporation rates plotted in Figure 7 are typical of the results for MgO contents from 1–10 wt.% (see figure 4(a) in Richter et al., 2002). The effect of a finite pressure of hydrogen is illustrated by the results shown in Figure 8, showing the evaporation rates of forsterite (Hashimoto, 1990) and a type B CAI-like liquid (Richter et al., 2002) at 1,500 °C and hydrogen pressures of $\sim 10^{-4}$ bar. Comparing Figures 7 and 8 shows that the evaporation rate of forsterite at $T = 1,500$ °C in 2×10^{-4} bar hydrogen is more than 300 times faster than in vacuum at the same temperature.

Given a thermodynamic model for calculating the saturation vapor pressures of the dominant gas species that would be in equilibrium with an evaporating surface and experimental data for the actual rate of evaporation at a given pressure, temperature, and composition, a determination of the evaporation coefficient can be made using Equation (2). The results for both forsterite and type B CAI-like liquids are shown in Figure 9. The evaporation coefficients for these two silicate systems have a common trend with respect to temperature, with the evaporation coefficient going from slightly larger than 0.1 at 1,800 °C to less than 0.01 at 1,200 °C. The differences in composition and state between solid forsterite and CMAS liquid do not seem to make much difference, nor does it seem to matter whether the evaporation takes place in vacuum or in a finite pressure of hydrogen.

A particularly important function of the laboratory studies of evaporation is to confirm the

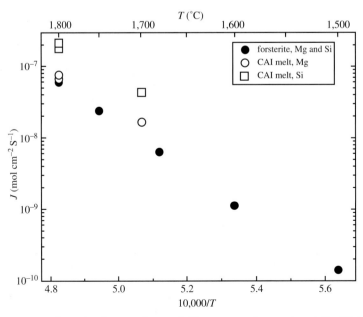

Figure 7 Vacuum evaporation rate of magnesium and silicon versus temperature (after Richter et al., 2002).

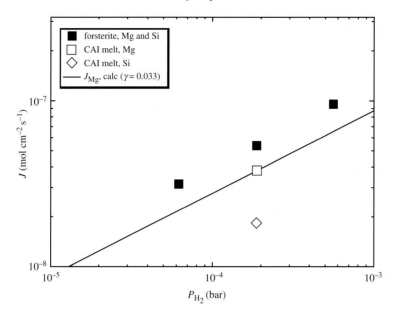

Figure 8 Evaporation rate versus hydrogen pressure (after Richter *et al.*, 2002).

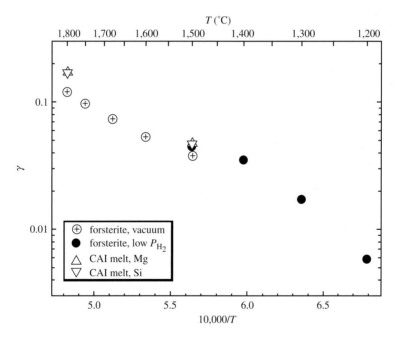

Figure 9 Evaporation coefficients for magnesium and silicon for forsterite and CAI-like liquids. The evaporation coefficients depend strongly on temperature, but are not strongly dependent on evaporating composition or ambient atmosphere (after Richter *et al.*, 2002).

isotopic fractionation predicted by Equation (15), which can be thought of as providing the "fingerprint" of evaporation having taken place. A special virtue of the Davis *et al.* (1990) paper is that the isotopic compositions of all the elements making up the melt of forsterite composition were measured, with the result that given sufficient evaporation, all of them were found to be significantly isotopically heavy. The isotopic fractionations follow a Rayleigh-like behavior, but the fractionation factors are in every case closer to 1 (i.e., less fractionating) than what one calculates from the inverse square root of the mass of rate-controlling gas species of different isotopic composition (Figure 5). This diminished isotopic fractionation could result from partial recondensation, from diffusion not being fast enough to maintain isotopic homogeneity in the molten forsterite and/or from the dominant gas species being heavier than those calculated from thermodynamics (Mg, O, O_2, and SiO).

Similar heavy-isotope enrichments were found in residues from evaporation experiments involving CAI-like liquids. Figure 10 shows the isotopic fractionation of magnesium from various sets of experiments involving the evaporation of magnesium-bearing silicate melts. The results for vacuum evaporation and for evaporation into $\sim 2 \times 10^{-4}$ bar hydrogen of CAI-like melts give very similar fractionations as a function of the amount of magnesium evaporated, but the apparent fractionation factor is somewhat closer to one than the expected value of $\alpha = \sqrt{24/25}$. There is, to our knowledge, no explanation that can account for all the experiments to date on the kinetic isotopic fractionation of liquid and solid forsterite (Davis *et al.*, 1990; Wang *et al.*, 1999), a solar composition liquid (Wang *et al.*, 2001) and type B CAI-like liquids in vacuum and in hydrogen (Richter *et al.*, 2002) find residues less fractionated than the generally accepted theory suggests. In discussing evaporation of solid SiO_2, Young *et al.* (1998) suggested that the kinetic fractionation factor might reflect bond breaking rather than the mean velocity of the volatilizing molecules. It seems unlikely that bond breaking can explain our results given that they involve liquids, solids, and different materials with vastly different evaporation rates depending on temperature and surrounding pressure. Furthermore, some materials such as FeO do have a kinetic isotope fractionation factor equal to the inverse square root of the evaporating species (Wang *et al.*, 1994), which would be a surprising coincidence if the fractionation were controlled by the sort of bond breaking suggested by Young *et al.* (1998). Some of the more obvious explanations for reduced isotopic fractionation, such as recondensation or lack of homogeneity of the evaporating material, have been tested and found not to be the cause (Mendybaev *et al.*, 2002b).

Figure 10 also includes data from evaporation experiments run in a 1 bar gas-mixing furnace through which a gas of solar f_{O_2} was flowing at a rate corresponding to a residence time in the furnace of ~ 1 min. In the 1 bar experiments no measurable isotopic fractionation was associated with the loss of magnesium from the CMAS liquid.

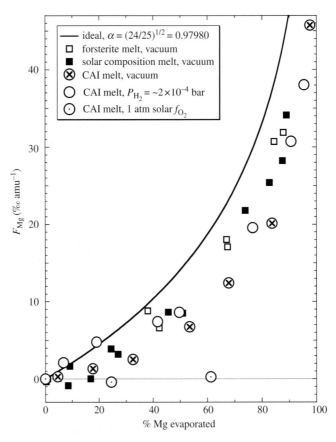

Figure 10 Magnesium isotopic composition versus % magnesium evaporated for a variety of initial compositions and evaporation conditions. All experiments conducted in low-pressure hydrogen or vacuum show similar behavior, with less isotopic fractionation than that expected for evaporation of Mg atoms (the major gas phase species of magnesium). Experiments run at 1 atm in a solar f_{O_2} gas show elemental without isotopic fractionation.

The lack of isotopic fractionation is the result of these experiments being done in an open system in which the gas was being removed at a slow rate compared to the time it takes evaporation to saturate the gas. Thus the experiments are in the limit $P_i/P_{i,\text{sat}} \to 1$ for which isotopic fractionation is not expected. In this limit, the gas that is continuously being removed from the system contains silicon and magnesium in proportions equal to the saturation vapor pressures of SiO and Mg, and negligible amounts of calcium and aluminum. Given enough time, the preferential loss of silicon and magnesium from the system will inevitably appear as an elemental fractionation of the condensed phase in the system. The vacuum experiments represent the opposite extreme of a open system in which the evaporated gas is removed much too quickly for the pressure of the various gas species to become a significant fraction of the saturation vapor pressures (i.e., $P_i/P_{i,\text{sat}} \ll 1$). As noted earlier, in this limit we expect, and indeed find, that the evaporation residue is fractionated in both its elemental and isotopic composition. The contrasting behavior of these two limiting cases is a good illustration of how the relative magnitude of the evaporation timescale and residence time of gas species in the system of interest can determine the isotopic consequences of evaporation.

We close this section by describing an experimentally calibrated parametrization for the evaporation kinetics of CMAS liquids. A similar parametrization for forsterite could be constructed from data and thermodynamic considerations found in the papers by Nagahara and Ozawa (1996), Tsuchiyama et al. (1998, 1999), Wang et al. (1999), and the more recent experimental data by Kuroda and Hashimoto (2002). Because of their use in constraining the thermal history of the type B CAIs, we will focus on the parametrizations for CMAS liquids. The following formula for the dependence of the evaporation rate of a type B CAI-like liquid is taken from Richter et al. (2002):

$$J_i(T, P) = \frac{J_i(T_0, P_0) e^{\frac{-E}{R}\left(\frac{1}{T} - \frac{1}{T_0}\right)} \sqrt{P/P_0}}{1 + (\gamma_a/D)\sqrt{RT/(2\pi m_i)}} \quad (21)$$

$J_i(T_0, P_0)$ is the experimentally determined evaporation rate at temperature T_0 and H_2 pressure P_0. The effect of hydrogen pressure is assumed proportional to $\sqrt{P/P_0}$, which is consistent with the forsterite evaporation rate data shown in Figure 8 as well as with the thermodynamically calculated dependence of the saturation vapor pressures of Mg and SiO over a CMAS liquid. A reasonable choice for the effective activation energy E for a type B CAI-like liquid is 300 kJ mol^{-1}, which is ~100 kJ mol^{-1} larger than the value given by a thermodynamic calculation for the saturation vapor pressures in order to reflect the dependence of the evaporation coefficients on temperature (see Figure 9). The evaporation coefficient does not appear explicitly in the numerator of Equation (21) because the parametrization is based on an experimentally determined evaporation rate. The term in the denominator of Equation (21) takes account of recondensation based on the pressure at the surface of a sphere of radius a due to the evaporating species having to diffuse away from the surface through a gas with a finite diffusivity D (for details, see Richter et al., 2002). The temperature and pressure dependence of the magnesium evaporation rates given by Equation (21) are illustrated in Figure 11.

1.15.3.4 Evaporation of Chondritic Meteorites and Chondritic Melts

A number of vacuum evaporation experiments have been done on chondritic meteorites (Hashimoto et al., 1979; Floss et al., 1996, 1998) and on synthetic melts of chondritic composition (Hashimoto, 1983; Wang et al., 2001). These experiments confirm that the major elements can be ordered by volatility, from volatile to refractory: iron, magnesium ≈ silicon, calcium, aluminum. The isotopic compositions of magnesium and silicon follow Rayleigh curves, but as for forsterite and CAI melts, the isotopic fractionation factor is closer to one (less fractionating) than expected from the square root of the ratio of the masses of the evaporating species.

Vacuum evaporation of REE-bearing melts leads to residues with large negative cerium anomalies (Wang et al., 2001). Evaporation of REE-bearing CAI melts in the presence of hydrogen does not produce cerium anomalies, because conditions remain reducing (Davis et al., 1999). The reason for this difference is as follows. For most REEs, such as lanthanum, the evaporation reaction is $H_2 + La_2O_3 \to 2LaO + H_2O$; for cerium, it is $H_2O + Ce_2O_3 \to 2CeO_2 + H_2$. Thus, as hydrogen pressure is increased, lanthanum and most other REEs become more volatile and cerium becomes more refractory; under very low pressures (as in vacuum experiments), cerium is significantly more volatile than the other REEs (Davis et al., 1982).

1.15.4 APPLICATIONS

1.15.4.1 Bulk Compositions of Planets and Meteorite Parent Bodies

The bulk compositions of the terrestrial planets and all meteorite parent bodies except that of the CI chondrites are depleted in volatile elements to various degrees. These depletions are reasonably smooth functions of 50% condensation

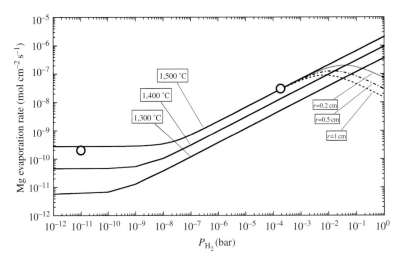

Figure 11 Magnesium evaporation rate versus hydrogen pressure. Heavy lines show the Mg evaporation rate from a type B CAI-like liquid as a function of hydrogen pressure and temperature calculated using Equation (2) with an evaporation coefficient $\gamma = 0.03$. The evaporation rates become independent of the hydrogen pressure at sufficiently low pressure in a regime corresponding to the vacuum limit. In the hydrogen-dominated regime where the evaporation rate increases with pressure, the evaporation rate is well characterized by the parametrized Equation (20). The effect of recondensation on the evaporation rate is illustrated by comparing the heavy curve for $T = 1,400\ °C$ (no recondensation assuming radius $r \rightarrow 0$) to the dashed curves which represent the evaporation rate from spheres of radius 0.1 cm and 0.5 cm. The experimentally determined evaporation rate in vacuum (plotted at $P = 10^{-11}$ bar) and for $P_{H_2} = 1.9 \times 10^{-4}$ bar are shown by the open circles. An evaporation rate of 3×10^{-8} mol cm^{-2} s^{-1} for $T = 1,500\ °C$ and $P_{H_2} = 1.9 \times 10^{-4}$ bar was used in connection with Equation (2) to determine the evaporation coefficient and as the reference evaporation rate for Equation (18).

temperature, but differ from body to body (e.g., Chapter 1.03, figure 2; Chapter 1.11, figure 2; Chapter 1.20, figure 1; Chapter 1.22, figure 8). The lack of any significant isotopic fractionation in volatile elements in these bodies (e.g., Humayun and Clayton, 1995) implies that the gas–solid fractionation occurred in near-equilibrium conditions. Cassen (1996) has constructed models of the solar nebula combining a slowly cooling solar nebula with loss of gas by stellar winds that provide a reasonable explanation of the observed patterns of increasing depletion with decreasing condensation temperature.

Most iron meteorites likely represent cores of asteroids (Chapter 1.12). The remarkably large depletions in gallium and germanium in some iron meteorite groups (groups IVA and IVB are depleted in Ga/Ni and Ge/Ni by a factor of ~1,000 compared to bulk solar system values) would provide a sensitive isotopic test for mechanisms of volatility fractionation.

1.15.4.2 Calcium-, Aluminum-rich Inclusions

1.15.4.2.1 Ultrarefractory inclusions

A very minor population of CAIs have ultra-refractory compositions, with enrichments in the most refractory trace elements of over a factor 1,000 times CI chondritic composition. The most spectacular of these is OSCAR, which consists of Sc-fassaite with up to 20 wt.% Sc_2O_3, and Y-rich perovskite. The REEs show clear evidence of volatility fractionation, with the heavy REEs enriched by a factor of 1,000 compared to the light REEs (Davis, 1984, 1991; Davis and Hinton, 1985). Several additional ultrarefractory inclusions have been described more recently (Davis, 1991; Simon et al., 1996). These inclusions appear to be extremely high-temperature condensates. No isotopic mass fractionations have been reported for these objects; thus, there is no positive evidence that they are evaporation residues. Furthermore, vacuum evaporation experiments done to high degrees of evaporation at temperatures as high as 2,150 °C have failed to reproduce the REE patterns seen in ultrarefractory inclusions (Davis and Hashimoto, 1995; Davis et al., 1995). Although the conditions necessary for partial evaporation or condensation to explain the REE patterns seen in ultrarefractory CAIs have not yet been reproduced in the laboratory, equilibrium thermodynamic calculations have successfully matched the observations (Simon et al., 1996).

1.15.4.2.2 Textural evidence

Clear textural evidence for evaporation is surprisingly rare. In the case of most CAIs, for which evaporation has been suggested to modify elemental and isotopic compositions (Grossman et al., 2000), the CAI is believed to have been molten during evaporation, erasing any textural evidence of the evaporation. One notable exception to this is an unusual forsterite-rich CAI from the Vigarano CV chondrite, 1623-5 (Davis et al., 1991b). In this CAI, there are clear isotopic and petrologic indications of subsolidus evaporation. In most of the unevaporated portion of this inclusion, forsterite occurs within a matrix of Åk$_{89}$ melilite. Where evaporation has occurred, forsterite has been pseudomorphically replaced by pure åkermanite melilite, just as predicted from analysis of the gehlenite–forsterite–anorthite (projected from spinel) ternary phase diagram widely used in CAI petrology (Stolper, 1982; Davis et al., 1991b). In places, fassaite is replaced by gehlenite and perovskite, again as expected from evaporative breakdown. The case for evaporation is made all the more compelling by the fact that the outer portions of the CAI, where there are petrologic indications of evaporation, are also found to be isotopically heavy relative to the interior of the inclusion.

1.15.4.2.3 Bulk elemental and isotopic compositions of type B CAIs

What are the conditions in the early solar system that could produce significant evaporation of the CAIs? The most studied CAIs are the type Bs. The characteristic coarse-grained igneous textures of these inclusions indicate a high degree of partial melting by a reheating event that raised their temperature to ~1,700 K, which is ~400 K higher than their condensation temperature. This peak temperature was short lived (i.e., a few hours) and must have been followed by cooling at rates of the order of $1-10\,°C\,h^{-1}$ in order to account for the large euhedral crystals of melilite in the type B CAIs (Stolper and Paque, 1986). There are various lines of evidence that taken together make a strong case that significant evaporation must have occurred during the one or more reheating events that partially melted the precursor materials of the type B CAIs. To begin with, as shown in Figure 1, the type B CAIs are often found to have correlated enrichment in the heavy isotopes of silicon and magnesium (Clayton et al., 1988) in much the same way as those found in laboratory produced evaporation residues by Davis et al. (1990). It was subsequently noted by Grossman et al. (2000) that the amount of silicon and magnesium that would have to evaporate from the type B CAIs to produce the range of observed isotopic fractionations of silicon and magnesium shown in Figure 1 corresponds to the range of their silicon and magnesium depletion relative to compositions along the condensation curve (Figure 12; see Grossman et al., 2000, figure 10). This suggests that the precursor materials of the type B CAIs were in equilibrium with a solar composition gas at temperatures of the order of $1,300-1,450$ K (for total pressures of $10^{-5}-10^{-3}$ atm), and then reheated to temperatures of the order of 1,700 K for a sufficient time for them to lose significant amounts of silicon and

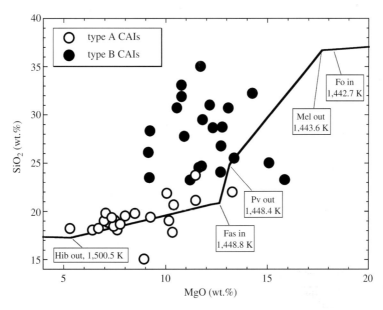

Figure 12 SiO$_2$ versus MgO in coarse-grained CAIs from CV chondrites, compared with the condensation trajectory for a gas of solar composition at 10^{-3} atm. Note that type A CAIs lie along the condensation trajectory, but type B CAIs lie to the left of the trajectory. For more details on type A and B CAIs, see Chapter 1.08, by MacPherson. The CAI data have been corrected for nonrepresentative sampling of different minerals by correcting to the solar system Ca/Al ratio (sources Grossman et al., 2000; Simon et al., 2002).

magnesium by evaporation. This interpretation is consistent with the experimentally determined evaporation rates of type B CAI-like liquids (Richter et al., 2002), which are such that significant evaporation would be expected during the reheating and subsequent cooling rates required by the textural arguments of Stolper and Paque (1986). Evaporation rates are significantly faster in reducing conditions (see Figure 11) and the Ti^{3+}/Ti^{4+} of pyroxenes in the type B CAIs indicate that very reducing conditions prevailed during their reheating and subsequent cooling (Beckett, 1986).

Figure 13 illustrates what we believe are the various stages in the formation and evolution of a type B CAI, in which a CAI condenses from a nebular gas under near equilibrium conditions along the condensation trajectory and is then melted and evaporated in a hydrogen-rich gas. The experimentally determined evaporation kinetics of type B CAI-like liquids as a function of temperature, hydrogen pressure, and size are summarized by the parametrization given in Equation (20). This parametrization can be used to calculate the amount of magnesium or silicon that would be lost from a molten droplet as a function of the duration of a high-temperature event such as we believe was responsible for the igneous texture of the type B CAIs. Given a relationship between elemental fractionation and the associated isotopic fractionation such as shown in Figure 10, one can also calculate the degree of heavy isotope enrichment of an evaporation residue subject to a particular thermal history. Figure 14 shows the results of such a calculation for particularly simple thermal histories involving a constant cooling rate starting from 1,700 K. This figure shows the cooling rate as a function of hydrogen pressure that would produce a 1‰ amu^{-1} or 5‰ amu^{-1} enrichment of the magnesium isotopic composition of a 1 cm diameter model CAI. We showed in Figure 1 that many CAIs are enriched in the heavy isotopes of magnesium in the range 1–5‰ amu^{-1}, which assuming a single reheating event would require cooling rates of the order of 1–100 °C h^{-1} for a reasonable range of hydrogen pressure (10^{-2}–10^{-6} bar). The fact that two independent estimators of the thermal history of the type B CAIs, one based on textures (Stolper and Paque, 1986), the other based on evaporation kinetics, arrive at the same order of magnitude for the cooling rates suggests that we are on the right track in terms of understanding the origin and evolution of these earliest materials to have formed in the solar system.

1.15.4.2.4 Refractory metal inclusions

CAIs often contain tiny metal nuggets consisting largely of highly refractory siderophile elements such as iridium, osmium, ruthenium, etc. The chemical compositions of these can be understood in terms of high-temperature condensation from a gas of solar composition (e.g., Sylvester et al., 1990). However, nuggets appear to be more complicated. It appears that the high-temperature condensation model sees an earlier stage prior to subsequent condensation of iron and nickel and then loss of the latter elements by oxidation and sulfidization (Blum et al., 1989; Campbell et al., 2003).

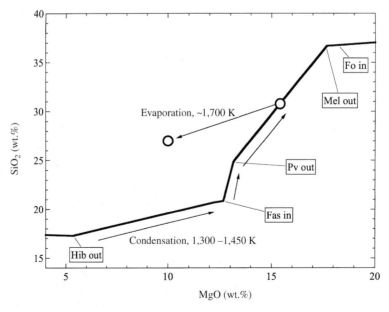

Figure 13 Cartoon for making a type B CAI. The temperature range for condensation is for total pressures of 10^{-5} atm (~1,300 K) to 10^{-3} atm (~1,450 K).

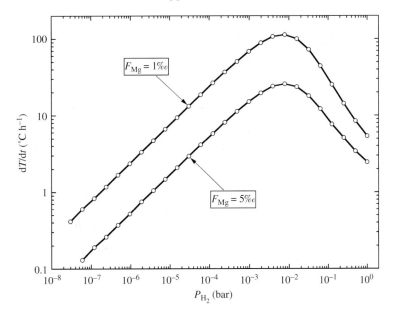

Figure 14 Cooling rate versus hydrogen pressure, showing the required cooling rates to produce residue compositions with 1‰ amu^{-1} and 5‰ amu^{-1} magnesium isotopic fractionation.

1.15.4.2.5 FUN CAIs

A few CAIs, the FUN (for fractionated and unknown nuclear CAIs), have large isotopic mass fractionation effects coupled with nuclear anomalies in many isotopes (Clayton and Mayeda, 1977; Wasserburg *et al.*, 1977; Davis *et al.*, 1991b; Loss *et al.*, 1994). These CAIs show much larger mass fractionation effects than normal CAIs. It remains unclear why the nuclear anomalies are associated with the large mass fractionations. The isotopic fractionation of silicon and magnesium in many FUN inclusions falls along the same trend as that of the normal CAIs, but with much larger fractionations (Figure 1). The trouble with any simple evaporation explanation for this much fractionation of the silicon and magnesium isotopes is that the presumed evaporation residue would contain only about half the original silicon and a few tens of percent of the original magnesium. The precursor would have had to be so rich in silicon and magnesium as to be very unlikely for any solar nebular condensate.

1.15.4.2.6 Hibonites with cerium anomalies

A few hibonite-rich CAIs have been found with significant isotopic mass fractionation effects and large negative cerium anomalies. Since both features result from evaporation in vacuum or under oxidizing conditions, it is likely that such CAIs are evaporation residues produced under these special conditions (Davis *et al.*, 1982; Hinton *et al.*, 1988; Ireland *et al.*, 1992). We should note that this is different from the very reducing conditions in which we believe the type B CAIs were evaporated.

1.15.4.3 Chondrites

1.15.4.3.1 Chondrules

Chondrules have long fascinated and frustrated meteoriticists, but it seems clear that they were heated to high temperatures and cooled fairly rapidly. Experiments and modeling indicate that if the heating event occurred in a gas of solar composition, significant evaporation of volatile elements would occur. Recent measurements show that many chondrules are depleted in volatile elements but that the volatile elements potassium and iron are not isotopically fractionated (Alexander *et al.*, 2000; Alexander and Wang, 2001; Yu *et al.*, 2003). These authors have suggested that evaporation of chondrules may have occurred in a region with an enhanced gas/dust ratio, such that there was significant back-reaction during evaporation (Alexander and Wang, 2001; Yu *et al.*, 2003). They argue that under such conditions, elemental fractionation is still possible, but isotopic fractionation will be controlled by equilibrium isotopic partitioning between gas and dust, which is small at high temperature. Galy *et al.* (2000) argued that chondrules formed at high nebular pressures, on the basis of the relatively small magnesium isotopic fractionations observed.

We can illustrate how evaporation kinetics provides constraint on the thermal history of chondrules by discussing a set of Chainpur chondrules recently studied by Alexander and Wang (2001).

These chondrules have textures indicating reheating to liquidus (barred) or near liquidus (porphyritic) temperatures of 1,800 K or more, and subsequent cooling at rates of 100–1,000 °C h^{-1} (Lofgren and Lanier, 1990; Radomsky and Hewins, 1990). Olivine from these chondrules was studied by Alexander and Wang (2001) and was found to have a large range of forsterite contents, from Fo$_{78}$ to Fo$_{99.9}$. Despite these large differences in iron content of the olivine, the iron isotopic ratios showed no evidence of fractionation greater than the analytical precision of the isotopic measurements of ±1–2‰. A natural question to ask is how much evaporation of iron would one expect given the peak temperatures and cooling rates inferred from the texture of the Chainpur chondrules? Alexander and Wang (2001) have addressed this and concluded that if the reheating and subsequent cooling took place in vacuum (i.e., using the evaporation kinetics measured by Wang et al., 2001), then a negligible amount of iron would have evaporated and no significant fractionation of the iron isotopes would have occurred. They also considered the case of enhanced evaporation due to hydrogen, and concluded that had hydrogen been present during the reheating of the chondrules, cooling rates several orders of magnitude faster than those suggested by the textures would be required to avoid detectable fractionation of the iron isotopes. Does this mean that the Chainpur chondrules must have been reheated in vacuum? Not necessarily. A plausible scenario for which there is some corroborating evidence is that chondrules formed and were reheated in regions of high dust to gas ratios. Ebel and Grossman (2000) have argued that condensing chondrules require a high dust to gas ratio for conditions to be sufficiently oxidizing to stabilize significant amounts of FeO in melts. Reheating chondrules in a region of high dust to gas ratio is very likely to significantly reduce evaporation rates. One reason is that the saturation vapor pressure of iron in the gas is expected to decrease as $\sqrt{f_{O_2}}$ (see discussion of this in Richter et al., 2002), and according to Equation (1) the evaporation rate of iron would be reduced by a similar amount. In addition, if there is a sufficiently large dust to gas ratio, an amount of evaporation too small to significantly fractionate the condensed material might be sufficient to saturate the gas. The effect would be to reduce the net evaporation to zero and eliminate isotopic fractionation. Additional reasons why the Chainpur chondrules show no significant isotopic fractionation of iron can be found in Alexander and Wang (2001). One should not be surprised that considerations of evaporation and condensation are less certain indicator of the conditions for forming and evolving chondrules compared to the more confident tone we adopted regarding the type B CAIs. This is because there are likely to be many more ways to suppress kinetic isotopic fractionations than there are of arriving at a finite amount of fractionation.

1.15.4.3.2 Metal grains

CH chondrites and some bencubbinites contain metal grains that are zoned in major and trace siderophile elements. It has been argued that the zoning patterns are consistent with condensation from a gas of solar composition (Meibom et al., 1999, 2000, 2001; Campbell et al., 2001; Petaev et al., 2001, 2003). However, this would require significant spreading of the siderophile element profiles by some process such as diffusion (Campbell et al., 2001). An alternative explanation for why the most refractory siderophile elements are not exclusively concentrated in the very first metal to condense (and thus found as a spike in the center of the grain) is that these metal grains are the result of nonequilibrium condensation from a supersaturated gas (Campbell et al., 2001). The nonequilibrium explanation should be testable in that materials condensing from a supersaturated gas (i.e. $P_i/P_{i,\text{sat}} > 1$) should be isotopically fractionated. Condensation of metal from a supersaturated gas would provide strong constraints on thermal history of the gas (Campbell et al., 2001; Meibom et al., 2001).

1.15.4.4 Deep-sea Spherules

Magnetite–wüstite spherules recovered from deep-sea sediments are of extraterrestrial origin and result from oxidation of metal particles during atmospheric entry. They have long been known to have isotopically heavy oxygen, but this was attributed to exchange with isotopically heavy oxygen in the stratosphere (Clayton et al., 1986). Subsequent measurement of the isotopic compositions of iron, nickel, and chromium in the spherules showed large isotopic mass fractionation effects leading one to conclude that the isotopic effects are most likely the result of evaporation during atmospheric heating (Davis et al., 1991a; Davis and Brownlee, 1993; Herzog et al., 1994, 1999; Xue et al., 1995).

1.15.5 OUTLOOK

The combination of volatility controlled fractionations found in virtually all solar system materials with experimental and theoretical characterizations of the conditions and consequences of evaporation and condensation allows powerful constraints to be placed on physicochemical processes and the timescales on which they operated in the early solar system. To the extent that isotopic fractionations can be used to infer volatilization processes and the timescales on which they operated, one measure of where the emphasis needs to be placed in future is the fact that more isotopic measurements have been made on synthetic evaporation residues than on the natural materials that motivated the evaporation experiments. Certainly, more high precision spatially resolved isotopic analyses of CAIs and chondrules would be very desirable. In the case of the type B CAIs, for example, if they crystallized as they evaporated, growing melilite crystals should preserve the isotopic record of evaporation, and the degree of magnesium isotopic fractionation should vary as a function of distance in the direction of growth. Were one to find that the melilite crystals are uniformly isotopically fractionated one would have to conclude that the fractionation, still most likely due to evaporation, occurred in some earlier high-temperature event prior to the one responsible for the present igneous texture of the type B CAIs. The advent of high-precision microbeam analyses of isotopic fractionation distributions will make for more compelling tests of the role of evaporation and condensation of early solar system materials.

There are experimental issues that also need to be addressed in order to use theoretical models more effectively. The thermodynamic data used for condensation, evaporation, and crystallization calculations still need to be improved. Coming to a better understanding of what determines the 100-fold range in the evaporation coefficients of elements evaporating from oxides or silicates would allow for more confident extrapolation of the experimental data to new systems. It remains unclear why magnesium, silicon, and oxygen do not fractionate in evaporation experiments by as much as expected from the kinetic theory of gases. Many tests have been made and it now seems unlikely that this discrepancy is an experimental artifact. Sophisticated surface analysis techniques that probe the first few atomic layers of the evaporating surface during high-temperature evaporation might help to solve this problem and also questions related to what determines the evaporation coefficients for a given system.

Perhaps the single most important issue that has not yet been sufficiently addressed, let alone resolved, involves how are partial condensates and/or partial evaporation residues are preserved. When we make a comparison between some component found in a meteorite (e.g., a type B CAI) and a condensate from a solar gas at some high temperature, there is the presumption that the condensate was for some (usually unstated) reason removed from any further interaction with the gas. This problem becomes compounded when one considers that CAIs are believed to be ~2 Myr older than chondrules, but they are often found in the same meteorite. Where were the CAIs for the 2 Myr they had to wait for the chondrules and then be incorporated together in a meteorite? How did the CAIs avoid chemical interaction with their environment over this time interval? Similar, and in many ways far more important, questions could be asked about the processes that gave rise to the volatility fractionated components that in various proportions gave planetary bodies their distinctive depletion patterns. Cassen's (1996) work is exemplary of the type of combined dynamical and chemical modeling that will be required in order to understand how each planetary body got its distinctive composition.

ACKNOWLEDGMENTS

This work was supported in part by the National Aeronautics and Space Administration, through grants NAG5-9378 and -9510. We thank S. B. Simon for unpublished data on bulk compositions of CAIs and A. J. Campbell, R. N. Clayton, N. Dauphas, and E. D. Young for helpful comments.

REFERENCES

Alexander C. M. O'D. (2001) Exploration of quantitative kinetic models for the evaporation of silicate melts in vacuum and in hydrogen. *Meteorit. Planet. Sci.* **36**, 255–283.

Alexander C. M. O'D. (2002) Application of MELTS to kinetic evaporation models of FeO-bearing silicate melts. *Meteorit. Planet. Sci.* **37**, 245–256.

Alexander C. M. O'D. and Wang J. (2001) Iron isotopes in chondrules: implications for the role of evaporation during chondrule formation. *Meteorit. Planet. Sci.* **36**, 419–428.

Alexander C. M. O'D., Grossman J. N., Wang J., Zanda B., Bourot-Denise M., and Hewins R. H. (2000) The lack of potassium-isotopic fractionation in Bishunpur chondrules. *Meteorit. Planet. Sci.* **35**, 859–868.

Alexander C. M. O'D., Taylor S., Delaney J. S., Ma P., and Herzog G. F. (2002) Mass-dependent fractionation of Mg, Si, and Fe isotopes in five stony cosmic spherules. *Geochim. Cosmochim. Acta* **66**, 173–183.

Allende Prieto C., Lambert D. L., and Asplund M. (2001) The forbidden abundance of oxygen in the sun. *Astrophys. J.* **556**, L63–L66.

Allende Prieto C., Lambert D. L., and Asplund M. (2002) A reappraisal of the solar photospheric C/O ratio. *Astrophys. J.* **573**, L137–L140.

Anders E. and Grevesse N. (1989) Abundances of the elements: meteoritic and solar. *Geochim. Cosmochim. Acta* **53**, 197–214.

Beckett J. R. (1986) The origin of calcium-, aluminum-rich inclusions from carbonaceous chondrites: an experimental study. PhD Thesis, University of Chicago.

Berman R. G. (1983) A thermodynamic model for multicomponent melts, with application to the system calcium oxide-magnesium oxide-alumina-silica. PhD Thesis, University of British Columbia.

Blum J. D., Wasserburg G. J., Hutcheon I. D., Beckett J. R., and Stolper E. M. (1989) Origin of opaque assemblages in C3V meteorites: implications for nebular and planetary processes. *Geochim. Cosmochim. Acta* **53**, 543–556.

Boynton W. V. (1975) Fractionation in the solar nebula: condensation of yttrium and the rare earth elements. *Geochim. Cosmochim. Acta* **39**, 569–584.

Campbell A. J., Humayun M., Meibom A., Krot A. N., and Keil K. (2001) Origin of zoned metal grains in the QUE94411 chondrite. *Geochim. Cosmochim. Acta* **65**, 163–180.

Campbell A. J., Humayun M., and Weisberg M. K. (2002) Siderophile element constraints on the formation of metal in the metal-rich chondrites Bencubbin, Weatherford, and Gujba. *Geochim. Cosmochim. Acta* **66**, 647–660.

Campbell A. J., Simon S. B., Humayun M., and Grossman L. (2003) Chemical evolution of metal in refractory inclusions in CV3 chondrites. *Geochim. Cosmochim. Acta* **67**, 3119–3134.

Cassen P. (1996) Models for the fractionation of moderately volatile elements in the solar nebula. *Meteorit. Planet. Sci.* **31**, 793–806.

Clayton R. N. and Mayeda T. K. (1977) Correlated oxygen and magnesium isotope anomalies in Allende inclusions: I. Oxygen. *Geophys. Res. Lett.* **4**, 295–298.

Clayton R. N., Mayeda T. K., and Brownlee D. E. (1986) Oxygen isotopes in deep-sea spherules. *Earth Planet. Sci. Lett.* **79**, 235–240.

Clayton R. N., Hinton R. W., and Davis A. M. (1988) Isotopic variations in the rock-forming elements in meteorites. *Phil. Trans. Roy. Soc. London A* **325**, 483–501.

Davis A. M. (1984) A scandalously refractory inclusion in Ornans. *Meteoritics* **19**, 214.

Davis A. M. (1991) Ultrarefractory inclusions and the nature of the group II REE fractionation. *Meteoritics* **26**, 330.

Davis A. M. and Brownlee D. E. (1993) Iron and nickel isotopic mass fractionation in deep-sea spherules. *Lunar Planet. Sci.* **XXIV**. Lunar and Planetary Institute, Houston, pp. 373–374.

Davis A. M. and Grossman L. (1979) Condensation and fractionation of rare earths in the solar nebula. *Geochim. Cosmochim. Acta* **43**, 1611–1632.

Davis A. M. and Hashimoto A. (1995) Volatility fractionation of REE and other trace elements during vacuum evaporation. *Meteoritics* **30**, 500–501.

Davis A. M. and Hinton R. W. (1985) Trace element abundances in OSCAR, a scandium-rich refractory inclusion from the Ornans meteorite. *Meteoritics* **20**, 633–634.

Davis A. M., Tanaka T., Grossman L., Lee T., and Wasserburg G. J. (1982) Chemical composition of HAL, an isotopically-unusual Allende inclusion. *Geochim. Cosmochim. Acta* **46**, 1627–1651.

Davis A. M., Hashimoto A., Clayton R. N., and Mayeda T. K. (1990) Isotope mass fractionation during evaporation of forsterite (Mg_2SiO_4). *Nature* **347**, 655–658.

Davis A. M., Clayton R. N., Mayeda T. K., and Brownlee D. E. (1991a) Large mass fractionation of iron isotopes in cosmic spherules collected from deep-sea sediments. *Lunar Planet. Sci.* **XXII**. Lunar and Planetary Institute, Houston, pp. 281–282.

Davis A. M., MacPherson G. J., Clayton R. N., Mayeda T. K., Sylvester P. J., Grossman L., Hinton R. W., and Laughlin J. R. (1991b) Melt solidification and late-stage evaporation in the evolution of a FUN inclusion from the Vigarano C3V chondrite. *Geochim. Cosmochim. Acta* **55**, 621–637.

Davis A. M., Hashimoto A., Clayton R. N., and Mayeda T. K. (1995) Isotopic and chemical fractionation during evaporation of $CaTiO_3$. *Lunar Planet. Sci.* **XXVI**. Lunar and Planetary Institute, Houston, pp. 317–318.

Davis A. M., Hashimoto A., and Parsad N. M. (1999) Trace element fractionation during evaporation in reducing atmospheres. *Lunar Planet. Sci.* **XXX**, #2023. Lunar and Planetary Institute, Houston (CD-ROM).

Ebel D. S. (2000) Variations on solar condensation: sources of interstellar dust nuclei. *J. Geophys. Res.* **105**, 10363–10370.

Ebel D. S. and Grossman L. (2000) Condensation in dust-enriched systems. *Geochim. Cosmochim. Acta* **64**, 339–366.

Fegley B., Jr. and Palme H. (1985) Evidence for oxidizing conditions in the solar nebula from molybdenum and tungsten depletions in refractory inclusions in carbonaceous chondrites. *Earth Planet. Sci. Lett.* **72**, 311–326.

Ebel D. S., Ghiorso M. S., Sack R. O., and Grossman L. (2000) Gibbs energy minimization in gas + liquid + solid systems. *J. Comput. Chem.* **21**, 247–256.

Floss C., El Goresy A., Zinner E., Kransel G., Rammensee W., and Palme H. (1996) Elemental and isotopic fractionations produced through evaporation of the Allende CV chondrite: implications for the origin of HAL-type hibonite inclusions. *Geochim. Cosmochim. Acta* **60**, 1975–1997.

Floss C., El Goresy A., Zinner E., Palme H., Weckwerth G., and Rammensee W. (1998) Corundum-bearing residues produced through the evaporation of natural and synthetic hibonite. *Meteorit. Planet. Sci.* **33**, 191–206.

Galy A., Young E. D., Ash R. D., and O'Nions R. K. (2000) The formation of chondrules at high gas pressures in the solar nebula. *Science* **290**, 1751–1753.

Grossman J. N., Rubin A. E., and MacPherson G. J. (1988) A unique volatile-poor carbonaceous chondrite with possible implications for nebular fractionation processes. *Earth Planet. Sci. Lett.* **91**, 33–54.

Grossman L. (1972) Condensation in the primitive solar nebula. *Geochim. Cosmochim. Acta* **36**, 597–619.

Grossman L. (1973) Refractory trace elements in calcium–aluminum-rich inclusions in the Allende meteorite. *Geochim. Cosmochim. Acta* **37**, 1119–1140.

Grossman L. and Ganapathy R. (1976) Trace elements in the Allende meteorite: I. Coarse-grained, calcium-rich inclusions. *Geochim. Cosmochim. Acta* **40**, 331–344.

Grossman L. and Larimer J. W. (1974) Early chemical history of the solar system. *Rev. Geophys. Space Phys.* **12**, 71–101.

Grossman L., Ganapathy R., and Davis A. M. (1977) Trace elements in the Allende meteorite: III. Coarse-grained inclusions revisited. *Geochim. Cosmochim. Acta* **41**, 1647–1664.

Grossman L., Ebel D. S., Simon S. B., Davis A. M., Richter F. M., and Parsad N. M. (2000) Major element chemical and isotopic compositions of refractory inclusions in C3 chondrites: the separate roles of condensation and evaporation. *Geochim. Cosmochim. Acta* **64**, 2879–2894.

Grossman L., Ebel D. S., and Simon S. B. (2002) Formation of refractory inclusions by evaporation of condensate precursors. *Geochim. Cosmochim. Acta* **66**, 145–161.

Hashimoto A. (1983) Evaporation metamorphism in the early solar nebula—evaporation experiments on the melt iron(II) oxide–magnesium oxide–silicon dioxide–calcium oxide–aluminum oxide and chemical fractionations of primitive materials. *Geochem. J.* **17**, 111–145.

Hashimoto A. (1988) Evaporation kinetics of REE oxides. *Lunar Planet. Sci.* **XIX**. Lunar and Planetary Institute, Houston, pp. 459–460.

Hashimoto A. (1989) Kinetics of evaporation of MgO, SiO_2, and Mg_2SiO_4, and their effect on isotope mass fractionations. *Lunar Planet. Sci.* **XX**. Lunar and Planetary Institute, Houston, pp. 385–386.

Hashimoto A. (1990) Evaporation kinetics of forsterite and implications for the early solar nebula. *Nature* **347**, 53–55.

Hashimoto A., Kumazawa M., and Onuma N. (1979) Evaporation metamorphism of primitive dust material in the early solar nebula. *Earth Planet. Sci. Lett.* **43**, 13–21.

Hashimoto A., Holmberg B. B., and Wood J. A. (1989) Effects of melting on evaporation kinetics. *Meteoritics* **24**, 276.

Hertz H. (1882) Ueber die Verdunstung der Flüssigkeiten, insbesondere des Quecksilbers, in luftleeren Raume. *Ann. d. Phys. u. Chem. Ser. 3* **17**, 177–193.

Herzog G. F., Hall G. S., and Brownlee D. E. (1994) Mass fractionation of nickel isotopes in metallic cosmic spheres. *Geochim. Cosmochim. Acta* **58**, 5319–5323.

Herzog G. F., Xue S., Hall G. S., Nyquist L. E., Shih C. Y., Wiesmann H., and Brownlee D. E. (1999) Isotopic and elemental composition of iron, nickel, and chromium in type I deep-sea spherules: implications for origin and composition of the parent micrometeoroids. *Geochim. Cosmochim. Acta* 63, 1443–1457.

Hinton R. W., Davis A. M., Scatena-Wachel D. E., Grossman L., and Draus R. J. (1988) A chemical and isotopic study of hibonite-rich refractory inclusions in primitive meteorites. *Geochim. Cosmochim. Acta* 52, 2573–2598.

Hirth J. P. and Pound G. M. (1963) *Condensation and Evaporation*. Pergamon, Oxford, 192pp.

Humayun M. and Cassen P. (2000) Processes determining the volatile abundances of the meteorites and terrestrial planets. In *Origin of the Earth and Moon* (eds. R. M. Canup and K. Righter). University of Arizona Press, Tucson, pp. 3–23.

Humayun M. and Clayton R. N. (1995) Potassium isotope cosmochemistry: genetic implications of volatile element depletion. *Geochim. Cosmochim. Acta* 59, 2131–2148.

Ireland T. R., Zinner E. K., Fahey A. J., and Esat T. M. (1992) Evidence for distillation in the formation of HAL and related hibonite inclusions. *Geochim. Cosmochim. Acta* 56, 2503–2520.

Ivanova M. A., Petaev M. I., MacPherson G. J., Nazarov M. A., Taylor L. A., and Wood J. A. (2002) The first known natural occurrence of calcium monoaluminate, in a calcium–aluminum-rich inclusion from the CH chondrite northwest Africa 470. *Meteorit. Planet. Sci.* 37, 1337–1344.

Keays R. R., Ganapathy R., and Anders E. (1971) Chemical fractionations in meteorites: IV. Abundances of fourteen trace elements in L-chondrites: implications for cosmothermometry. *Geochim. Cosmochim. Acta* 35, 337–363.

Knudsen M. (1909) Experimentelle Bestimmung des Druckes gesättigter Quecksilberdümpfe bei 0° und höheren Temperaturen. *Ann. d. Phys.* 29, 179–193.

Kuroda D. and Hashimoto A. (2002) The reaction of forsterite with hydrogen—its apparent and real temperature dependences. *Antarct. Meteorit. Res.* 15, 152–164.

Larimer (1967) Chemical fractionations in meteorites: I. Condensation of the elements. *Geochim. Cosmochim. Acta* 31, 1215–1238.

Larimer J. W. and Anders E. (1967) Chemical fractionations in meteorites: II. Abundance patterns and their interpretation. *Geochim. Cosmochim. Acta* 31, 1239–1270.

Laul J. C., Ganapathy R., Anders E., and Morgan J. W. (1973) Chemical fractionations in meteorites: VI. Accretion temperatures of H-, LL-, and E-chondrites, from abundance of volatile trace elements. *Geochim. Cosmochim. Acta* 37, 329–357.

Lodders K. (2003) Solar system abundances and condensation temperatures of the elements. *Astrophys. J.* 591, 1220–1247.

Lofgren G. E. and Lanier A. B. (1990) Dynamic crystallization study of barred olivine chondrules. *Geochim. Cosmochim. Acta* 54, 3537–3551.

Lord H. C., III (1965) Molecular equilibria and condensation in a solar nebula and cool stellar atmospheres. *Icarus* 4, 279–288.

Loss R. D., Lugmair G. W., Davis A. M., and MacPherson G. J. (1994) Isotopically distinct reservoirs in the solar nebula: isotope anomalies in Vigarano meteorite inclusions. *Astrophys. J.* 436, L193–L196.

McDonough W. F. and Sun S.-S. (1995) The composition of the Earth. *Chem. Geol.* 120, 223–253.

Meibom A., Petaev M. I., Krot A. N., Wood J. A., and Keil K. (1999) Primitive FeNi metal grains in CH carbonaceous chondrites formed by condensation from a gas of solar composition. *J. Geophys. Res.* 104, 22053–22059.

Meibom A., Desch S. J., Krot A. N., Cuzzi J. N., Petaev M. I., Wilson L., and Keil K. (2000) Large-scale thermal events in the solar nebula: evidence from Fe, Ni metal grains in primitive meteorites. *Science* 288, 839–841.

Meibom A., Petaev M. I., Krot A. N., Keil K., and Wood J. A. (2001) Growth mechanism and additional constraints on FeNi metal condensation in the solar nebula. *J. Geophys. Res.* 106, 32797–32801.

Mendybaev R. A., Beckett J. R., Stolper E., and Grossman L. (1998) Measurement of oxygen fugacities under reducing conditions: non-Nernstian behavior of Y_2O_3-doped zirconia oxygen sensors. *Geochim. Cosmochim. Acta* 62, 3131–3139.

Mendybaev R. A., Beckett J. R., Grossman L., Stolper E., Cooper R. F., and Bradley J. P. (2002a) Volatilization kinetics of silicon carbide in reducing gases: an experimental study with applications to the survival of presolar grains in the solar nebula. *Geochim. Cosmochim. Acta* 66, 661–682.

Mendybaev R. A., Davis A. M., and Richter F. M. (2002b) The effect of sample size on experimental evaporation of Type B CAIs. *Lunar Planet. Sci.* XXXIII, #2040. Lunar and Planetary Institute, Houston (CD-ROM).

Nagahara H. and Ozawa K. (1996) Evaporation of forsterite in H_2 gas. *Geochim. Cosmochim. Acta* 60, 1445–1459.

Nagahara H. and Ozawa K. (2000) Isotopic fractionation as a probe of heating processes in the solar nebula. *Chem. Geol.* 169, 45–68.

Nichols R. H., Jr., Grimley R. T., and Wasserburg G. J. (1998) Measurement of gas-phase species during Langmuir evaporation of forsterite. *Meteorit. Planet. Sci.* 33, A115–A116.

Ozawa K. and Nagahara H. (2000) Kinetics of diffusion-controlled evaporation of Fe–Mg olivine: experimental study and implication for stability of Fe-rich olivine in the solar nebula. *Geochim. Cosmochim. Acta* 64, 939–955.

Ozawa K. and Nagahara H. (2001) Chemical and isotopic fractionations by evaporation and their cosmochemical implications. *Geochim. Cosmochim. Acta* 65, 2171–2199.

Palme H. and Wlotzka F. (1976) A metal particle from a calcium, aluminum-rich inclusion from the meteorite Allende, and the condensation of refractory siderophile elements. *Earth Planet. Sci. Lett.* 33, 45–60.

Palme H., Larimer J. W., and Lipschutz M. E. (1988) Moderately volatile elements. In *Meteorites and the Early Solar System* (eds. J. F. Kerridge and M. S. Matthews). University of Arizona Press, Tucson, pp. 436–461.

Petaev M. I. and Wood J. A. (1998) The condensation with partial isolation (CWPI) model of condensation in the solar nebula. *Meteorit. Planet. Sci.* 33, 1123–1137.

Petaev M. I., Meibom A., Krot A. N., Wood J. A., and Keil K. (2001) The condensation origin of zoned metal grains in Queen Alexandra Range 94411: implications for the formation of the Bencubbin-like chondrites. *Meteorit. Planet. Sci.* 36, 93–106.

Petaev M. I., Wood J. A., Meibom A., Krot A. N., and Keil K. (2003) The ZONMET thermodynamic and kinetic model of metal condensation. *Geochim. Cosmochim. Acta* 67, 1737–1751.

Radomsky P. M. and Hewins R. H. (1990) Formation conditions of pyroxene-olivine and magnesian olivine chondrules. *Geochim. Cosmochim. Acta* 54, 3475–3490.

Richter F. M., Davis A. M., Ebel D. S., and Hashimoto A. (2002) Elemental and isotopic fractionation of Type B calcium-, aluminum-rich inclusions: experiments, theoretical considerations, and constraints on their thermal evolution. *Geochim. Cosmochim. Acta* 66, 521–540.

Simon S. B., Davis A. M., and Grossman L. (1996) A unique ultrarefractory inclusion from the Murchison meteorite. *Meteorit. Planet. Sci.* 31, 106–115.

Simon S. B., Grossman L., Krot A. N., and Ulyanov A. A. (2002) Bulk chemical compositions of Type B refractory inclusions. *Lunar Planet. Sci.* XXXIII, #1620. Lunar and Planetary Institute, Houston (CD-ROM).

Stolper E. (1982) Crystallization sequences of calcium–aluminum-rich inclusions from Allende: an experimental study. *Geochim. Cosmochim. Acta* 46, 2159–2180.

Stolper E. and Paque J. M. (1986) Crystallization sequences of calcium–aluminum-rich inclusions from Allende: the effects of cooling rate and maximum temperature. *Geochim. Cosmochim. Acta* **50**, 1785–1806.

Sylvester P. J., Ward B. J., Grossman L., and Hutcheon I. D. (1990) Chemical compositions of siderophile element-rich opaque assemblages in an Allende inclusion. *Geochim. Cosmochim. Acta* **54**, 3491–3508.

Tachibana S., Nagahara H., and Ozawa K. (2001) Condensation kinetics of metallic iron and its application to condensation in the solar nebula. *Lunar Planet. Sci.* **XXXII**, #1767. Lunar and Planetary Institute, Houston (CD-ROM).

Tsuchiyama A. (1998) Condensation experiments using a forsterite evaporation source in H_2 at low pressures. *Mineral. J.* **20**, 59–80.

Tsuchiyama A., Takahashi T., and Tachibana S. (1998) Evaporation rates of forsterite in the system Mg_2SiO_4–H_2. *Mineral. J.* **20**, 113–126.

Tsuchiyama A., Tachibana S., and Takahashi T. (1999) Evaporation of forsterite in the primordial solar nebula; rates and accompanied isotopic fractionation. *Geochim. Cosmochim. Acta* **63**, 2451–2466.

Urey H. C. (1947) The thermodynamic properties of isotopic substances. *J. Chem. Soc.* 562–581.

Urey H. C. (1952) *The Planets*. Yale University Press, New Haven.

Wang J., Davis A. M., Clayton R. N., and Mayeda T. K. (1994) Kinetic isotope fractionation during the evaporation of iron oxide from the liquid state. *Lunar Planet. Sci.* **XXV**. Lunar and Planetary Institute, Houston, pp. 1459–1460.

Wang J., Davis A. M., Clayton R. N., and Hashimoto A. (1999) Evaporation of single crystal forsterite: evaporation kinetics, magnesium isotope fractionation, and implications of mass-dependent isotopic fractionation of a diffusion-controlled reservoir. *Geochim. Cosmochim. Acta* **63**, 953–966.

Wang J., Davis A. M., Clayton R. N., Mayeda T. K., and Hashimoto A. (2001) Chemical and isotopic fractionation during the evaporation of the FeO–MgO–SiO_2–CaO–Al_2O_3–TiO_2 rare earth element melt system. *Geochim. Cosmochim. Acta* **65**, 479–494.

Wänke H., Baddenhausen H., Palme H., and Spettel B. (1974) Chemistry of the Allende inclusions and their origin as high temperature condensates. *Earth Planet. Sci. Lett.* **23**, 1–7.

Wasserburg G. J., Lee T., and Papanastassiou D. A. (1977) Correlated oxygen and magnesium isotopic anomalies in Allende inclusions: II. Magnesium. *Geophys. Res. Lett.* **4**, 299–302.

Wasson J. T. (1984) *Meteorites: Their Record of Early Solar-system History*. W. H. Freeman, New York, 267p.

Weber D. and Bischoff A. (1994) The occurrence of grossite ($CaAl_4O_7$) in chondrites. *Geochim. Cosmochim. Acta* **58**, 3855–3877.

Wood J. A. and Hashimoto A. (1993) Mineral equilibrium in fractionated nebular systems. *Geochim. Cosmochim. Acta* **57**, 2377–2388.

Xue S., Herzog G. F., Hall G. S., Bi D., and Brownlee D. E. (1995) Nickel isotope abundances of type I deep-sea spheres and of iron-nickel spherules from sediments in Alberta, Canada. *Geochim. Cosmochim. Acta* **59**, 4975–4981.

Yoneda S. and Grossman L. (1995) Condensation of CaO–MgO–Al_2O_3–SiO_2 liquids from cosmic gases. *Geochim. Cosmochim. Acta* **59**, 3413–3444.

Young E. D., Nagahara H., Mysen B. O., and Audet D. M. (1998) Non-Rayleigh oxygen isotope fractionation by mineral evaporation: theory and experiments in the system SiO_2. *Geochim. Cosmochim. Acta* **62**, 3109–3116.

Yu Y., Hewins R. H., Alexander C. M. O'D., and Wang J. (2003) Experimental study of evaporation and isotopic mass fractionation of potassium in silicate melts. *Geochim. Cosmochim. Acta* **67**, 773–786.

1.16
Early Solar System Chronology

K. D. McKeegan

University of California, Los Angeles, CA, USA

and

A. M. Davis

The University of Chicago, IL, USA

1.16.1 INTRODUCTION	431
1.16.1.1 Chondritic Meteorites as Probes of Early Solar System Evolution	431
1.16.1.2 Short-lived Radioactivity at the Origin of the Solar System	432
1.16.1.3 A Brief History and the Scope of the Present Review	433
1.16.2 DATING WITH ANCIENT RADIOACTIVITY	434
1.16.3 "ABSOLUTE" AND "RELATIVE" TIMESCALES	435
1.16.3.1 An Absolute Timescale for Solar System Formation	435
1.16.3.2 An Absolute Timescale for Chondrule Formation	437
1.16.3.3 An Absolute Timescale for Early Differentiation of Planetesimals	437
1.16.4 THE RECORD OF SHORT-LIVED RADIONUCLIDES IN EARLY SOLAR SYSTEM MATERIALS	438
1.16.4.1 Calcium-41	438
1.16.4.2 Aluminum-26	439
1.16.4.3 Beryllium-10	442
1.16.4.4 Manganese-53	444
1.16.4.5 Iron-60	446
1.16.4.6 Palladium-107	447
1.16.4.7 Hafnium-182	448
1.16.4.8 Iodine-129	448
1.16.4.9 Niobium-92	448
1.16.4.10 Plutonium-244 and Samarium-146	449
1.16.5 ORIGINS OF THE SHORT-LIVED NUCLIDES IN THE EARLY SOLAR SYSTEM	449
1.16.6 IMPLICATIONS FOR CHRONOLOGY	450
1.16.6.1 Formation Timescales of Nebular Materials	451
1.16.6.2 Timescales of Planetesimal Accretion and Early Chemical Differentiation	453
1.16.7 CONCLUSIONS	454
1.16.7.1 Implications for Solar Nebula Origin and Evolution	454
1.16.7.2 Future Directions	456
ACKNOWLEDGMENTS	456
REFERENCES	456

1.16.1 INTRODUCTION

1.16.1.1 Chondritic Meteorites as Probes of Early Solar System Evolution

The evolutionary sequence involved in the formation of relatively low-mass stars, such as the Sun, has been delineated in recent years through impressive advances in astronomical observations at a variety of wavelengths, combined with improved numerical and theoretical models of the physical processes thought to occur during each stage. From the models and the observational statistics, it is possible to infer in a general way how our solar system ought to have evolved through the various stages from gravitational collapse of a fragment of a molecular cloud to the accretion of planetary-sized bodies (e.g., Cameron, 1995; Alexander *et al.*, 2001; Shu *et al.*, 1987; André *et al.*, 2000; see Chapters 1.04, 1.17, and 1.20). However, the details of these processes remain obscured, literally from an astronomical perspective, and the dependence of such models on various parameters requires data to constrain the specific case of our solar system's origin.

Fortunately, the chondritic meteorites sample aspects of this evolution. The term "chondrite" (or chondritic) was originally applied to meteorites bearing chondrules, which are approximately millimeter-sized solidified melt droplets consisting largely of mafic silicate minerals and glass commonly with included metal or sulfide. However, the meaning of chondritic has been expanded to encompass all extraterrestrial materials that are "primitive," i.e., are undifferentiated samples having nearly solar elemental composition. Thus, the chondrites represent a type of cosmic sediment, and to a first approximation can be thought of as "hand samples" of the condensable portion of the solar nebula. The latter is a general term referring to the phase(s) of solar system evolution intermediate between molecular cloud collapse and planet formation. During the nebular phase, the still-forming Sun was an embedded young-stellar object (YSO) enshrouded by gas and dust, which was distributed first in an extended envelope which later evolved into an accretion disk that ultimately defined the ecliptic plane. The chondrites agglomerated within this accretion disk, most likely close to the position of the present asteroid belt from whence meteorites are currently derived. In addition to chondrules, an important component of some chondrites are inclusions containing refractory oxide and silicate minerals, so-called calcium- and aluminum-rich inclusions (CAIs) that also formed as free-floating objects within the solar nebula. These constituents are bound together by a "matrix" of chondrule fragments and fine-grained dust (which includes a tiny fraction of dust grains that predate the solar nebula; see Chapter 1.02). It is important to realize that, although these materials accreted together at a specific time in some planetesimal, the individual components of a given chondrite can, and probably do, sample different places and/or times during the nebular phase of solar system formation. Thus, each grain in one of these cosmic sedimentary rocks potentially has a story to tell regarding aspects of the early evolution of the solar system.

Time is a crucial parameter in constructing any story. Understanding of relative ages allows placing events in their proper sequence, and measures of the duration of events are critical to developing an understanding of process. If disparate observations can be related temporally, then structure (at any one time) and evolution of the solar system can be better modeled; or, if a rapid succession of events can be inferred, it can dictate a cause and effect relationship. This chapter is concerned with understanding the timing of different physical and chemical processes that occurred in the solar nebula and possibly on early accreted planetesimals that existed during the nebula stage. These events are "remembered" by the components of chondrites and recorded in the chemical, and especially, isotopic compositions of the host mineral assemblages; the goal is to decide which events were witnessed by these ancient messengers and to decipher those memories recorded long ago.

1.16.1.2 Short-lived Radioactivity at the Origin of the Solar System

The elements of the chondritic meteorites, and hence of the terrestrial planets, were formed in previous generations of stars. Their relative abundances represent the result of the general chemical evolution of the galaxy, possibly enhanced by recent local additions from one or more specific sources just prior to collapse of the solar nebula ~4.56 Gyr ago. A volumetrically minor, but nevertheless highly significant part of this chemical inventory, is comprised of radioactive elements, from which this age estimate is derived. The familiar long-lived radionuclides, such as ^{238}U, ^{235}U, ^{232}Th, ^{87}Rb, ^{40}K, and others, provide the basis for geochronology and the study of large-scale differentiation amongst geochemical reservoirs over time. They also provide a major heat source to drive chemical differentiation on a planetary scale (e.g., terrestrial plate tectonics).

A number of short-lived radionuclides also existed at the time that the Sun and the rocky bits of the solar system were forming (Table 1). These nuclides are sufficiently long-lived that they could exist in appreciable quantities in the earliest solar system rocks, but their mean lives are short enough that they are now completely decayed from their primordial abundances. In this sense they are referred to as extinct nuclides. Although less familiar than the still-extant radionuclides, these short-lived isotopes potentially play similar roles: their relative abundances can, in principle, form the basis of various chronometers that constrain the timing of early chemical fractionations, and the more abundant radio-isotopes can possibly provide sufficient heat to drive differentiation (i.e., melting) of early accreted planetesimals. The very rapid rate of decay of the short-lived isotopes, however, means that inferred isotopic differences translate

Table 1 Short-lived radioactive nuclides once existing in solar system objects.[a]

Fractionation[b]	Parent nuclide	Half-life (Myr)	Daughter nuclide	Estimated initial solar system abundance	Objects found in	References
Nebular	^{41}Ca	0.1	^{41}K	$10^{-8} \times ^{40}$Ca	CAIs	(1)
	^{26}Al	0.7	^{26}Mg	$(4.5 \times 10^{-5}) \times ^{27}$Al	CAIs, chondrules, achondrite	(2)
	^{10}Be	1.5	^{10}B	$(\sim 6 \times 10^{-4}) \times ^{9}$Be	CAIs	(3)
	^{53}Mn	3.7	^{53}Cr	$(\sim 2.4 \times 10^{-5}) \times ^{55}$Mn	CAIs, chondrules, carbonates, achondrites	(4)
	^{60}Fe	1.5	^{60}Ni	$(\sim 3 \times 10^{-7}) \times ^{56}$Fe	achondrites, chondrites	(5)
Planetary	^{107}Pd	6.5	^{107}Ag	$(\sim 5 \times 10^{-5}) \times ^{108}$Pd	iron meteorites, pallasites	(6)
	^{182}Hf	9	^{182}W	$10^{-4} \times ^{180}$Hf	planetary differentiates	(7)
	^{129}I	15.7	^{129}Xe	$10^{-4} \times ^{127}$I	chondrules, secondary minerals	(8)
	^{92}Nb	36	^{92}Zr	$10^{-4} \times ^{93}$Nb	chondrules, mesosiderites	(9)
	^{244}Pu	82	Fission products	$(7 \times 10^{-3}) \times ^{238}$U	CAIs, chondrites	(10)
	^{146}Sm	103	^{142}Nd	$(9 \times 10^{-4}) \times ^{147}$Sm	chondrites	(11)

References: (1) Srinivasan et al. (1994, 1996), (2) Lee et al. (1977), MacPherson et al. (1995); (3) McKeegan et al. (2000); (4) Birck and Allègre (1985), Lugmair and Shukolyukov (1998); (5) Shukolyukov and Lugmair (1993a), Tachibana and Huss (2003); (6) Chen and Wasserburg (1990); (7) Kleine et al. (2002a), Yin et al. (2002); (8) Jeffery and Reynolds (1961); (9) Schönbacher et al. (2002); (10) Hudson et al. (1988); and (11) Lugmair et al. (1983).
[a] Some experimental evidence exists suggesting the presence of the following additional isotopes, but confirming evidence is needed (half-lives are given after each isotope): ^{7}Be—53 d (Chaussidon et al., 2002); ^{99}Tc—0.2 Myr (Yin et al., 2000); ^{36}Cl—0.3 Myr (Murty et al., 1997); ^{205}Pb—15 Myr (Chen and Wasserburg, 1987). [b] Environment in which most significant parent–daughter fractionation processes occur.

into relatively short amounts of time, i.e., these potential chronometers have inherently high precision (temporal resolution). The realization of these possibilities is predicated upon understanding the origin(s) and distributions of the now-extinct radioactivity. While this is a comparatively easy task for the long-lived, still existing radionuclides, it poses a significant challenge for studies of the early solar system. However, this represents the best chance at developing a quantitative high-resolution chronology for events in the solar nebula and, moreover, the question of the origins of the short-lived radioactivity has profound implications for the mechanisms of formation of the solar system (as being, possibly, quite different from that for solar-mass stars in general).

1.16.1.3 A Brief History and the Scope of the Present Review

That short-lived radioactive isotopes existed in the early solar system has been known since the 1960s, since ^{129}Xe excesses were first shown to be correlated with the relative abundance of iodine, implicating the former presence of its parent nuclide, ^{129}I (Jeffery and Reynolds, 1961). Because the half-life of ^{129}I (~ 16 Myr) is not so short, its presence in the solar system can be understood as primarily a result of the ambient, quasi-steady-state abundance of this nuclide in the parental molecular cloud due to continuous r-process nucleosynthesis in the galaxy (Wasserburg, 1985). The situation changed dramatically in the mid-1970s when it was discovered that CAIs from the Allende meteorite exhibited apparent excesses of ^{26}Mg (Gray and Compston, 1974; Lee and Papanastassiou, 1974) and that the degree of excess ^{26}Mg correlated with Al/Mg in CAI mineral separates (Lee et al., 1976) in a manner indicative of the *in situ* decay of ^{26}Al ($t_{1/2} = 0.73$ Myr).

The high abundance inferred for this short-lived isotope ($\sim 5 \times 10^{-5} \times ^{27}$Al) demanded that it had been produced within a few million years of CAI formation, possibly in a single stellar source which "contaminated" the nascent solar system with freshly synthesized nuclides (Wasserburg and Papanastassiou, 1982). Because of the close time constraints an attractively parsimonious idea arose, whereby the very same dying star that threw out new radioactivity into the interstellar medium may also have served to initiate gravitational collapse of the molecular cloud fragment that would become the solar system, through the shock wave created by its expanding ejecta (Cameron and Truran, 1977). An alternative possibility that the new radioactive elements were produced "locally" through nuclear reactions between energetic solar particles and the surrounding nebular material was also quickly recognized (Heymann and Dziczkaniec, 1976; Clayton et al., 1977; Lee, 1978). However, many of the early models were unable to produce sufficient amounts of ^{26}Al by irradiation within the constraints of locally available energy sources and the lack of correlated isotopic effects in other elements (see discussion in Wadhwa and Russell (2000)). Almost by default, "external seeding" scenarios and the implied supernova trigger became the preferred class of models for explaining the presence of ^{26}Al and its distribution in chondritic materials.

In the intervening quarter century, as indicated in Table 1, many other short-lived isotopes have been found to have existed in early solar system materials. Several of these have been discovered in recent years, and the record of the distribution of ^{26}Al and other nuclides in a variety of primitive and evolved materials has been documented with much greater clarity. Nevertheless, at the time of writing of this review many of the fundamental issues remain unresolved. In part due to improvements in mass spectrometry, new data are being generated at an increasing pace, and in some cases, interpretations that seemed solid only a short time ago are now being revised. Some of the new evidence supports the notion of an external seeding or late injection of new material, while other evidence, both meteoritic and astronomical, points to nuclear irradiation as a source for radioactivity of early solar system matter. For further details the reader is directed to several excellent reviews (Wasserburg, 1985; Swindle et al., 1996; Podosek and Nichols, 1997; Gilmour, 2000; Wadhwa and Russell, 2000; Russell et al., 2001).

Development of a quantitative understanding of the source, or sources, of now-extinct radionuclides is important for constraining the distribution of these radioactive species throughout the early solar system and, thus, is critical for chronology. For the major part of this review, we will tacitly adopt the prevailing point of view, namely that external seeding for the most important short-lived isotopes dominates over possible local additions from nuclear reactions with energetic particles associated with the accreting Sun. This approach permits examination of timescales for self-consistency with respect to major chemical or physical "events" in the evolution of the solar system; the issues of the scale of possible isotopic heterogeneity within the nebula and assessment of local irradiation effects will be explicitly addressed following an examination of the preserved record.

1.16.2 DATING WITH ANCIENT RADIOACTIVITY

In "normal" radioactive dating, the chemical fractionation of a parent isotope from its radiogenic daughter results, after some decay of the parent, in a linear correlation of excesses of the daughter isotope with the relative abundance of the parent. For a cogenetic assemblage, such a correlation is an isochron and its slope permits the calculation of the time since the attainment of isotopic closure, i.e., since all relative transport of parent or daughter isotopes effectively ceased. If the fractionation event is magmatic, and the rock quickly cooled, then this time corresponds to an absolute crystallization age.

In a manner similar to dating by long-lived radioisotopes, the former presence of short-lived radioactivity in a sample is demonstrated by excesses of the radiogenic daughter isotope that correlate with the inferred concentration of the parent. However, because the parent isotope is extinct, a stable isotope of the respective parent element must serve as a surrogate with the same geochemical behavior (see Wasserburg, 1985, figure 2). The correlation line yields the initial concentration of radioactive parent relative to its stable counterpart and may represent an isochron; however, its interpretation in terms of "age" for one sample relative to another requires an additional assumption. The initial concentrations of a short-lived radionuclide among a suite of samples can correspond to relative ages only if the samples are all derived from a reservoir that at one time had a uniform concentration of the radionuclide. Under these conditions, differences in concentration correspond to differences in time only. As before, if the fractionation event corresponds to mineral formation and isotopic closure is rapidly achieved and maintained, then relative crystallization ages are obtained.

One further complication potentially arises that is unique to the now-extinct nuclides. In principle, excesses of a radiogenic daughter isotope could be "inherited" from an interstellar (grain) component, in a manner similar to what is known to have occurred for some stable isotope anomalies in CAIs and other refractory phases of chondrites (e.g., Begemann, 1980; Niederer et al., 1980; Niemeyer and Lugmair, 1981; Fahey et al., 1987). In such a case, the correlation of excess daughter isotope with radioactive parent would represent a mixing line rather than in situ decay from the time of last chemical fractionation. Such "fossil" anomalies (in magnesium) have, in fact, been documented in bona fide presolar grains (Zinner, 1998; see Chapter 1.02). These grains of SiC, graphite, and corundum crystallized in the outflows of evolved stars, incorporating very high abundances of newly synthesized radioactivity with ^{26}Al/^{27}Al close to unity. However, because these grains did not form in the solar nebula from a uniform isotopic reservoir, there is no chronological constraint that can be derived. Probably, the radioactivity in such grains decayed during interstellar transit, and hence arrived in the solar nebula as a "fossil."

Even before the discovery of presolar materials, Clayton championed a fossil origin for the magnesium isotope anomalies in CAIs in a series of papers (e.g., Clayton, 1982, 1986). A significant motivation for proposing a fossil origin was, in fact, to obviate chronological constraints derived from Al–Mg systematics in CAIs that apparently required a late injection and fast collapse timescales along with a long (several Myr) duration of small dust grains in the nebula. Although some level of inheritance may be present, and can possibly even be the dominant signal in a few rare samples or for specific isotopes (discussed below), for the vast majority of early solar system materials it appears that most of the inventory of short-lived isotopes did indeed decay following mineral formation in the solar nebula. MacPherson et al. (1995) summarized the arguments against a fossil origin for the ^{26}Mg excesses in their comprehensive review of the Al–Mg systematics in early solar system materials. In addition to the evidence regarding chemical partitioning during igneous processing of CAIs, must now be added the number of short-lived isotopes known (Table 1) and a general consistency of the isotopic records in a wide variety of samples. The new observations buttress the previous conclusions of MacPherson et al. (1995) such that the overwhelming consensus of current opinion is that correlation lines indicative of the former presence of now-extinct isotopes are truly isochrons representing *in situ* radioactive decay. This is a necessary, but not sufficient, condition for developing a chronology based on these systems.

1.16.3 "ABSOLUTE" AND "RELATIVE" TIMESCALES

In order to tie high-resolution relative ages to an "absolute" chronology, a correlation must be established between the short-lived and long-lived chronometers, i.e., the ratio of the extinct nuclide to its stable partner isotope must be established at some known time (while it was still alive). This time could correspond to the "origin of the solar system," which, more precisely defined, means the crystallization age of the first rocks to have formed in the solar system, or it could refer to some subsequent well-defined fractionation event, e.g., large-scale isotopic homogenization and fractionation occurring during planetary melting and differentiation. Both approaches for reconciling relative and absolute chronologies have been investigated in recent years, e.g., utilizing the ^{26}Al–^{26}Mg and Pb–Pb systems in CAIs and chondrules for constraining the timing and duration of events in the nebula, and the ^{53}Mn–^{53}Cr and Pb–Pb systems in differentiated meteorites to pin the timing of early planetary melting. The consistency of the deduced chronologies may be evaluated to give confidence (or not) that the assumptions necessary for a temporal interpretation of the record of short-lived radioactivity are, indeed, fulfilled.

1.16.3.1 An Absolute Timescale for Solar System Formation

The early evolution of the solar system is characterized by significant thermal processing of original presolar materials. This processing typically results in chemical fractionation that may potentially be dated by isotopic means in appropriate samples, e.g., nebular events such as condensation or distillation fractionate parent and daughter elements according to differing volatility. Likewise, chemical differentiation during melting and segregation leads to unequal rates of radiogenic ingrowth in different planetary reservoirs (e.g., crust, mantle, and core) that can constrain the nature and timing of early planetary differentiation. Several long-lived and now-extinct radioisotope systems have been utilized to delineate these various nebular and parent-body processes; however, it is only the U–Pb system that can record the absolute ages of the earliest volatility-controlled fractionation events, corresponding to the formation of the first refractory minerals, as well as the timing of melt generation on early planetesimals with sufficiently high precision as to provide a quantitative link to the short-lived isotope systems.

The U–Pb system represents the premier geochronometer because it inherently contains two long-lived isotopic clocks that run at different rates: ^{238}U decays to ^{206}Pb with a half-life of 4,468 Myr, and ^{235}U decays to ^{207}Pb with a much shorter half-life of 704 Myr. This unique circumstance provides a method for checking for isotopic disturbance (by either gain or loss of uranium or lead) that it is revealed by discordance in the ages derived from the two independent isotopic clocks with the same geochemical behavior (Wetherill, 1956; Tera and Wasserburg, 1972). Such an approach is commonly used in evaluating the ages of magmatic or metamorphic events in terrestrial samples. For obtaining the highest precision ages of volatility-controlled fractionation events in the solar nebula, the U–Pb concordance approach is of limited utility, however, and instead one utilizes ^{207}Pb/^{206}Pb and ^{204}Pb/^{206}Pb variations in a suite of cogenetic samples to evaluate crystallization ages. The method has a significant analytical advantage since only isotope ratios need to be determined in the mass spectrometer, but equally important is the high probability that the age obtained represents a true crystallization age, because the system is relatively insensitive to recent gain or loss of lead (or, more generally, recent fractionation of U/Pb). Moreover, this age is fundamentally based on the isotopic evolution of uranium, a refractory element whose isotopic composition is thought to be invariant throughout the solar system (Chen and Wasserburg, 1980, 1981), and the radiogenic $^{207}Pb/^{206}Pb$ evolves rapidly at 4.5 Ga because of the relatively short half-life of ^{235}U. In principle, ancient lead loss or redistribution (e.g., due to early metamorphic or aqueous activity on asteroids, the parent bodies of meteorites) can confound the interpretation of lead isotopic ages as magmatic ages, but such closure effects are usually considered to be insignificant for the most primitive meteorite samples. Whether or not this is a valid assumption is an issue that is open to experimental assessment and interpretation (see discussions in Tilton (1988) and Tera and Carlson (1999)).

Absolute crystallization ages have been calculated for refractory samples, CAIs that formed with very high depletions of volatile lead, by modeling the evolution of $^{207}Pb/^{206}Pb$ from primordial common (i.e., unradiogenic) lead found in early formed sulfides from iron meteorites. Such "model ages" can be determined with good precision (typically a few Ma), but accuracy depends on the correctness of the assumption of the isotopic composition of initial lead. Sensitivity to this correction is relatively small for fairly radiogenic samples such as CAIs where almost all the lead is due to *in situ* decay, nevertheless, depending on the details of data reduction and sample selection, even the best early estimates of Pb–Pb model ages for CAI formation ranged over ~15 Ma, from 4,553 Ma to 4,568 Ma, with typical uncertainties in the range of 4–5 Ma (see discussions in Tilton (1988) and Tera and Carlson (1999)). By progressively leaching samples to remove contaminating lead (probably introduced from the meteorite matrix), Allègre et al. (1995) were able to produce highly radiogenic ($^{206}Pb/^{204}Pb > 150$) fractions from four CAIs from the Allende CV3 chondrite, which yielded Pb–Pb model ages of $4,566 \pm 2$ Ma. Accuracy problems associated with initial lead corrections can also be addressed by an isochron approach where no particular composition of common lead need be assumed, only that a suite of samples are cogenetic and incorporated varying amounts of the same initial lead on crystallization (Tera and Carlson, 1999). Utilizing this approach, Tera and Carlson (1999) reinterpreted previous lead isotopic data obtained on nine Allende coarse-grained CAIs that had indicated a spread of ages (Chen and Wasserburg, 1981) to instead fit a single lead isochron of age = $4,566 \pm 8$ Ma which, however, is evolved from an initial lead isotopic composition that is unique to CAIs. More recently, Amelin et al. (2002) used the isochron method to determine absolute ages of formation for two CAIs from the Efremovka CV3 carbonaceous chondrite. Both samples are consistent with a mean age of $4,567.2 \pm 0.6$ Myr (Figure 1), which is the most precise absolute age obtained on CAIs. Because the previous best ages on Allende CAIs are consistent, within their relatively larger errors, with this new lead isochron age of Efremovka CAIs (Amelin et al., 2002), we adopt this value of $4,567.2 \pm 0.6$ Ma as the best estimate for the absolute formation age for coarse-grained (igneous) CAIs from CV chondrites.

To the extent that this high precision, high accuracy result represents the absolute age of crystallization of CAIs generally, it provides a measure of the age of formation of the solar system since several lines of evidence, in addition to the absolute Pb–Pb ages, indicate that CAIs are the first solid materials to have formed in the solar nebula (for a review, see Podosek and Swindle (1988)). In fact, it is the relative abundances of the short-lived radionuclides, especially ^{26}Al, which provides the primary indication that CAIs are indeed these first local materials. Other evidence is more circumstantial, e.g., the prevalence of large stable isotope anomalies in CAIs compared to other material of solar system origin (see Chapter 1.08). We will return to the issue of antiquity of CAIs when we examine the distribution of short-lived isotopes among different CAI types.

Other volatility-controlled long-lived parent/daughter isotope systems (e.g., Rb–Sr) yield absolute ages that are compatible with the coupled U–Pb systems, albeit with poorer precision. Because the chondrites are unequilibrated assemblages of components that may not share a common history, whole-rock or even mineral separate "ages" are not very meaningful for providing a very useful constraint on accretion timescales. High precision age determinations, approaching 1 Ma resolution, can in principle be obtained from initial $^{87}Sr/^{86}Sr$ in low Rb/Sr phases, such as CAIs (e.g., Podosek et al., 1991). However, such ages depend on deriving an accurate model of the strontium isotopic evolution of the reservoir from which these materials formed. The latter is a very difficult requirement, because it is not likely that a strictly chondritic Rb/Sr ratio was always maintained in the nebular regions from which precursor materials that ultimately formed CAIs, chondrules, and other meteoritic components condensed. Thus, initial strontium "ages," while highly precise, may be of little use in terms of quantitatively constraining absolute ages of formation of individual nebular objects and are best interpreted as only providing a

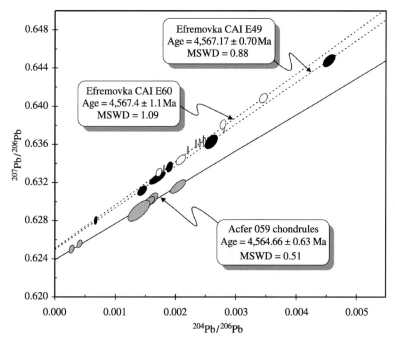

Figure 1 Pb–Pb isochrons for acid-washed fractions of two CAIs from CV3 Efremovka and for the six most radiogenic fractions of acid-washed chondrules from the CR chondrite Acfer 059. The $^{207}Pb/^{206}Pb$ data are not corrected for any assumed common lead composition; 2σ error ellipses are shown. Isochron ages for the two CAIs overlap with a weighted mean age of 4,567.2 ± 0.6 Ma, which is ~2.5 Myr older than the chondrules. Data and figure from Amelin *et al.* (2002) (reproduced with permission of the American Association for the Advancement of Science from *Science* **2002**, *297*, 1678–1683).

qualitative measure of antiquity (Podosek *et al.*, 1991). It is possible that initial $^{87}Sr/^{86}Sr$ ratios of similar nebular components, e.g., type B CAIs, could provide relative formation ages under the assumption that such objects share a common long-term Rb/Sr heritage; however, this has not yet been demonstrated.

1.16.3.2 An Absolute Timescale for Chondrule Formation

Although chondrule formation is thought to be one of the most significant thermal processes to have occurred in the solar nebula, in the sense of affecting the majority of planetary materials in the inner solar system (see Chapter 1.07), the mechanism(s) responsible remains hotly debated after many years of investigation. Similarly, it has long been recognized that obtaining good measurements of chondrule ages would be extremely useful for possibly constraining formation mechanisms and environments, as well as setting important limits on the duration of the solar nebula and, thus, on accretion timescales. However, determination of crystallization ages of chondrules is very difficult because their mineralogy is typically not amenable to large parent–daughter fractionation. Several short-lived isotope systems (discussed below) have been explored in recent years in order to try to delimit relative formation times for chondrules, e.g., compared to CAIs, but high precision absolute Pb–Pb ages have been measured for only a single meteorite. Amelin *et al.* (2002) used aggressive acid washing of a suite of chondrules from the unequilibrated CR chondrite Acfer 059 to remove unradiogenic lead (from both meteorite matrix and terrestrial contamination). Isochron ages ranged from 4,563 Ma to nearly 4,565 Ma, with a preferred value of 4,564.7 ± 0.6 Ma (Figure 1) for six of the most radiogenic samples ($^{206}Pb/^{204}Pb > 395$). It is argued that this result dates chondrule formation because lead closure effects are thought to be insignificant for these pristine samples. If these CR chondrules are representative of chondrules generally, then the data of Amelin *et al.* (2002) imply an interval of ~2.5 Ma between the formation of CV CAIs and chondrules in the nebula.

1.16.3.3 An Absolute Timescale for Early Differentiation of Planetesimals

Time-markers for tying short-lived chronometers to an absolute timescale can potentially be provided by early planetary differentiates. The basic requirements are that appropriately ancient samples would have to have evolved from a reservoir (magma) that had achieved isotopic equilibrium with respect to daughter elements of both long-lived and short-lived systems (i.e., lead, and chromium or magnesium, respectively), then cooled rapidly following crystallization and remained isotopically closed until analysis in the laboratory. In practice, the latter requirement means that samples should be undisturbed by shock and free of terrestrial contamination. No sample is perfect in all these respects, but the angrites are considered to be nearly ideal (the major problem being terrestrial lead contamination). By careful cleaning, Lugmair and Galer (1992) determined high precision Pb–Pb model ages for the angrites Lewis Cliff 86010 (LEW) and Angra dos Reis (ADOR). The results are concordant in U/Pb and with other isotopic systems as well as with each other, and provide an absolute crystallization age of $4,557.8 \pm 0.5$ Ma for the angrites (Lugmair and Galer, 1992). This is a significant time-marker ("event") because angrite mineralogy also provides large Mn/Cr fractionation that is useful for accurate $^{53}Mn/^{55}Mn$ determination.

The eucrites are highly differentiated (basaltic) achondrites that, along with the related howardites and diogenites, may have originated from the asteroid 4 Vesta (Binzel and Xu, 1993; see Chapter 1.11). Unfortunately, the U/Pb systematics of eucrites appear to be disturbed, yielding Pb–Pb ages up to ~220 Myr younger than angrites (Galer and Lugmair, 1996). This compromises the utility of the eucrites as providing independent tie points between long- and short-lived chronometers.

Evidence for an extended thermal history of equilibrated ordinary chondrites is provided by U–Pb analyses of phosphates (Göpel et al., 1994). The phosphates (merrillite and apatite) are metamorphic minerals produced by the oxidation of phosphorus originally present in metal grains. Phosphate mineral separates obtained from chondrites of metamorphic grade 4 and greater have Pb–Pb model ages (Göpel et al., 1994) from 4,563 Ma (for H4, Ste. Marguerite) to 4,502 Ma (for H6, Guareña). The oldest ages are nearly equivalent to Pb–Pb ages from CR chondrules (Amelin et al., 2002) and only a few million years younger than CAIs, indicating that accretion and thermal processing was rapid for the H4 chondrite parent body. The relatively long time interval of ~60 Myr has implications for the nature of the H chondrite parent body and the heat sources responsible for long-lived metamorphism (Göpel et al., 1994).

1.16.4 THE RECORD OF SHORT-LIVED RADIONUCLIDES IN EARLY SOLAR SYSTEM MATERIALS

Here, we discuss the evidence for the prior existence of now-extinct isotopes in meteoritic materials and, in the better-studied cases, what is known about the distribution of that isotope in the early solar system. Table 1 summarizes the basic facts regarding those short-lived radioisotopes that are unequivocally known to have existed as live radioactivity in rocks formed in the early solar system and provides an estimate of their initial abundances compared to a reference isotope. The table is organized in terms of increasing half-life and according to the main environment for parent–daughter chemical fractionation. The latter property indicates what types of events can potentially be dated and largely dictates what types of samples record evidence that a certain radioisotope once existed. Note that there is only a small degree of overlap demonstrated thus far for a few of the isotope systems. For example, it is well-documented that the Mn–Cr system is sensitive to fractionation in both nebular and parent-body environments, but other systems which might similarly provide linkages from the nebula through accretion to early differentiation have not been fully developed due to either analytical difficulties (e.g., Al–Mg, Fe–Ni) and/or difficulties in constraining mineral hosts and closure effects (e.g., I–Xe, ^{244}Pu). The initial abundances refer to the origin of the solar system, which, as discussed previously, means the time of CAI formation, and hence these can only be measured directly in nebular samples. The initial abundances of those isotopes that are found only in differentiated meteorites also refer back to the time of CAI formation, but such a calculation necessarily requires a chronological framework and is underpinned by the absolute time-markers provided by the Pb–Pb system.

1.16.4.1 Calcium-41

Calcium-41 decays by electron capture to ^{41}K with a half-life of only 103 kyr. It has the distinction of being the shortest-lived isotope for which firm evidence exists in early solar system materials, and this fact makes it key for constraining the timescale of last nucleosynthetic addition to solar system matter (in the external seeding scenario). It also makes ^{41}Ca exceedingly difficult to detect experimentally, because it can only be found to have existed in the oldest materials and then in only very small concentrations. Fortunately, its daughter potassium is rather volatile and calcium is concentrated in refractory minerals (the "C" in CAI) leading to large fractionations. Hutcheon et al. (1984) found hints for ^{41}Ca in Allende refractory inclusions, but could not clearly resolve ^{41}K excesses above measurement uncertainties.

The first unambiguous evidence of live ^{41}Ca came with the demonstration of correlated excesses of $^{41}K/^{39}K$ with Ca/K in Efremovka

CAIs by Srinivasan et al. (1994, 1996). Subsequent measurements by the PRL group have established that ^{41}Ca was also present in refractory oxide phases (hibonite) of CM and CV chondrites (Sahijpal et al., 1998, 2000). The CM hibonite grains are generally too small to permit enough multiple measurements to define an isochron on individual objects, even by ion probe; however, hibonite crystals from Allende CAIs show good correlation lines (Sahijpal et al., 2000) consistent with that found for Efremovka and indicating that ^{41}Ca decayed *in situ*. Most of the isolated CM hibonite grains also show ^{41}K/^{39}K excesses that are consistent with the isochrons obtained on silicate minerals of CAIs, except ~1/3 of the hibonite grains appear to have crystallized with "dead" calcium (i.e., they have normal ^{41}K/^{39}K compositions). The ensemble isochron (Figure 2) yields an initial value of ^{41}Ca/^{40}Ca = 1.4×10^{-8} with a formal error of ~10% relative and a statistical scatter that is commensurate with the measurement uncertainties. Such a small uncertainty would correspond to a very tight timescale (~15 kyr) for the duration of formation of these objects; however, possible systematic uncertainties in the mass spectrometry may increase this interval somewhat. The hibonite grains that contain no excess ^{41}K/^{39}K are unlikely to have lost that signal and, thus, must either have formed well after the other samples, or else they never incorporated live ^{41}Ca during their crystallization.

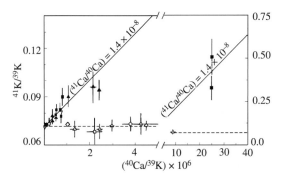

Figure 2 Potassium isotopic compositions measured in individual hibonite grains (Sahijpal et al., 1998) plotted as a function of Ca/K ratio. Hibonite grains from the carbonaceous chondrites Murchison, Allende, and Efremovka which formed with close to canonical levels of ^{26}Al are indicated as filled symbols, whereas hibonite grains that crystallized with no ^{26}Al are open circles and triangles. Terrestrial standards are plotted as open diamonds; error bars are 1σ. The isochron corresponding to live ^{41}Ca at the level ^{41}Ca/^{40}Ca = 1.4×10^{-8}, determined for Efremovka CAIs (Srinivasan et al., 1996), is also shown. Those hibonite grains that contained ^{26}Al are seen to plot on the same ^{41}Ca isochron as the CAIs, but grains lacking ^{26}Al are also lacking ^{41}Ca and plot on the horizontal dashed line corresponding to terrestrial ^{41}K/^{39}K. Data from Sahijpal et al. (1998); figure adapted from same.

An important clue is that these same grains also never contained ^{26}Al (Sahijpal and Goswami, 1998; Sahijpal et al., 1998, 2000); we will return to the significance of this correlation in discussing the scale of isotopic heterogeneity in the nebula and the source of ^{41}Ca and ^{26}Al.

1.16.4.2 Aluminum-26

Aluminum-26 decays by positron emission and electron capture to ^{26}Mg with a half-life of ~730 kyr. The discovery circumstances of ^{26}Al have already been discussed (Section 1.16.1.3) and since those early measurements a large body of data has grown to include analyses of CAIs from all major meteorite classes (carbonaceous, ordinary, enstatite) as well as important groups within these classes (e.g., CM, CV, CH, CR, CO, etc.); sparse data also exist for aluminum-rich phases from several differentiated meteorites and in chondrules. Data obtained prior to 1995 were the subject of a comprehensive review by MacPherson et al. (1995); for the most part, their analysis relied heavily on the extensive record in the large, abundant CAIs from CV chondrites, although significant numbers of refractory phases from other carbonaceous chondrite groups were also considered. Since that time, work has generally concentrated on extending the database to include smaller CAIs from underrepresented meteorite groups and, especially, chondrules (mostly from ordinary chondrites). Most measurements continue to be performed by ion microprobe because of the need to localize analysis on mineral phases with high Al/Mg ratios in order to resolve the addition of radiogenic ^{26}Mg*; this capability is particularly important for revealing internal Al–Mg isochrons in chondrules by examining small regions of trapped melt or glassy mesostasis in between the larger ferromagnesian minerals that dominate chondrules (Russell et al., 1996; Kita et al., 2000; McKeegan et al., 2000b; Mostefaoui et al., 2002). Inductively coupled mass spectrometry (ICPMS) analysis has produced the first high precision data that allow detection of very small levels of ^{26}Mg* in whole CAIs and chondrules (Galy et al., 2000); however, the technique has not been widely applied thus far.

To first order, the larger data set now available extends and confirms the general assessments of MacPherson et al. (1995), albeit with some modifications and enhancements. The distribution of inferred initial ^{26}Al/^{27}Al in CAIs is bimodal (Figure 3(a)), with the dominant peak at the so-called "canonical value" of 4.5×10^{-5}, and a second peak at "dead" aluminum (i.e., ^{26}Al/^{27}Al = 0). MacPherson et al. (1995) demonstrated that this pattern applied to all classes of carbonaceous chondrites, although the relative

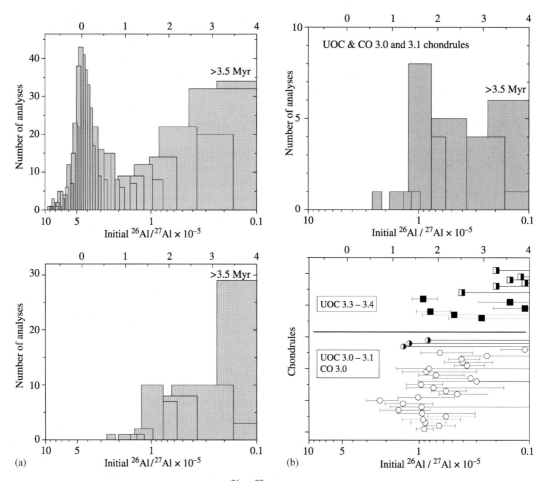

Figure 3 (a) Top panel: Histogram of initial $^{26}Al/^{27}Al$ inferred for CAIs; the number of analyses (taken to be representative of the number of samples) is plotted versus time after CAI formation (top axis), where time zero is taken as the "canonical" $^{26}Al/^{27}Al = 4.5 \times 10^{-5}$ peak of the distribution for CAIs. In addition to the canonical value, a significant number of CAIs do not preserve any evidence for having formed with live ^{26}Al; samples with only upper limits are summed in the last bin, indicating the achievement of isotopic closure at least 3.5 Myr after time zero, or alternatively, never having incorporated ^{26}Al at all (see text). Data sources are summarized by MacPherson et al. (1995). Bottom panel: Similar histogram summarizing data on plagioclase-olivine-inclusions (POIs) and chondrules (both aluminum-rich and ferromagnesian). In contrast to CAIs, there is no peak at $\sim 5 \times 10^{-5}$ and most chondrules show no evidence for having incorporated ^{26}Al. Some chondrules do show evidence for $^{26}Al/^{27}Al$ initial values at the level of $\sim 1 \times 10^{-5}$ or lower, indicating formation 1.5 to several million years after CAIs. Data sources are those summarized by MacPherson et al. (1995), supplemented by more recent data (Russell et al., 1996; Kita et al., 2000; McKeegan et al., 2000b; Huss et al., 2001; Mostefaoui et al., 2002; Hsu et al., 2003; Kunihiro et al., 2003). (b) Top panel: Histogram similar to 3(a)-bottom, except showing the inferred $^{26}Al/^{27}Al$ distribution for only those chondrules from the most unequilibrated meteorites, i.e., POIs and chondrules from metamorphic grades >3.1 have been removed from the plot. Also, this plot now shows the number of chondrules with that distribution, as opposed to the number of analyses considering each datum as a model isochron. Chondrules for which ^{26}Mg excesses are not well resolved (i.e., only upper limits are obtained or Al–Mg isochron slopes are within 2σ error of zero) are accumulated in the last histogram bin. A peak in the distribution may be discerned at $^{26}Al/^{27}Al \sim 1 \times 10^{-5}$, which corresponds to 1.5–2 Myr after time zero. Bottom panel: Inferred $^{26}Al/^{27}Al$ ratios for individual ferromagnesian and aluminum-rich chondrules with 2σ errors. Chondrules from the lowest metamorphic grades (3.0, 3.1) of unequilibrated ordinary (LL) and carbonaceous (CO) chondrites are shown in open circles, those from metamorphic grades 3.3 and above are shown in filled squares. Chondrules for which only upper limits are obtained are shown in half-open/half-filled symbols. It is apparent that chondrules from more intensely metamorphosed meteorites display apparently lower $^{26}Al/^{27}Al$ initial values. Among the most unequilibrated samples, an interval of >1 Myr is implied for the duration of chondrule formation. Data sources as in Figure 3(a).

heights of the two peaks varied among different meteorites (mostly reflecting a difference in CAI types; see Chapter 1.08). The dispersion of the canonical peak (amounting to $\sim 1 \times 10^{-5}$, FWHM) was considered to represent a convolution of measurement error and geologic noise, with no robust data indicating that any CAIs formed with $(^{26}Al/^{27}Al)_0$ significantly above the canonical ratio. The $\sim 5 \times 10^{-5}$ limit for $(^{26}\text{Al}/^{27}\text{Al})_0$ still holds, although Galy et al. (2000) compute a model isochron for one Allende CAI that yields $(^{26}\text{Al}/^{27}\text{Al})_0 = (6.24 \pm 0.23) \times 10^{-5}$, which is marginally higher than any previously determined value. It is noteworthy that all other measurements were by ion microprobe, where the slope of the Al–Mg correlation line is frequently set by analyses of anorthitic plagioclase which is known to be susceptible to mobilization of magnesium during metamorphism (LaTourrette and Wasserburg, 1998) or, possibly, during nebular events (Podosek et al., 1991). The high precision result of Galy et al. (2000) is based on a whole CAI, and thus is less sensitive to postcrystallization redistribution of radiogenic ^{26}Mg; however, the inferred $(^{26}\text{Al}/^{27}\text{Al})_0$ is not based on a measured isochron and may be susceptible to other systematic errors. Clearly, more high precision data are required before any modification of the canonical ratio would be warranted.

The existence of a canonical $(^{26}\text{Al}/^{27}\text{Al})_0$ value was previously based on analyses of CAIs only from carbonaceous chondrites; refractory inclusions from ordinary and enstatite chondrites are rare and often very small, and thus few had been discovered and none analyzed. There are now data for four CAIs from unequilibrated ordinary chondrites (Russell et al., 1996; Huss et al., 2001) and for 11 hibonite-bearing inclusions from enstatite chondrites (Guan et al., 2000); all are consistent with $(^{26}\text{Al}/^{27}\text{Al})_0$ in the range $\sim(3.5-5.5) \times 10^{-5}$, except for 4 of the (very small) hibonite grains for which $^{26}\text{Mg}^*$ could not be resolved. Thus, the same canonical value characterizes CAIs from all major meteorite classes. The possible meaning of this confirmation in terms of nebular chronology based on ^{26}Al is not completely straightforward, however.

The idea that many CAIs, whether they originally formed by melt crystallization or by condensation, have suffered some degree of disturbance to their Al–Mg isotopic system is well documented via correlated petrographic and isotopic evidence (MacPherson et al., 1995 and references therein). For example, in situ isotopic measurements have demonstrated that certain anorthite crystals within a CAI can record resetting events ~ 1 Myr or more following CAI formation (see figure 28 of Chapter 1.08). In general, it seems to be the large type B CAIs from CV chondrites that are the most prone to have suffered multiple thermal events capable of at least partially resetting the Al–Mg system (Podosek et al., 1991; Caillet et al., 1993; MacPherson and Davis, 1993; MacPherson et al., 1995); the protracted and complex thermal histories of type B CAIs are also evident in other chemical and isotopic systems, particularly the microdistribution of oxygen isotopes within individual inclusions (Clayton and Mayeda, 1984; Young and Russell, 1998; Yurimoto et al., 1998; McKeegan and Leshin, 2001). MacPherson et al. (1995) have argued that the trailing distribution of $^{26}\text{Al}/^{27}\text{Al}$ values downward from the canonical peak primarily represents a protracted period of thermal processing of CAIs, possibly accompanied by secondary mineral formation, over a few million years residence time in the solar nebula. Recently, Hsu et al. (2000) documented multiple isochrons within a single type B Allende CAI that they interpreted as signifying three discrete melting events separated in time by a few hundred thousand years. Such observations set lower bounds on the duration of the lifetime of the nebula and of significant heat sources, capable of producing CAIs, within regions of the nebula.

The duration of high-temperature processes in the solar nebula is closely related to the age difference between CAIs and chondrules, and it is in this area that some of the most significant new data have been developed since the review by MacPherson et al. (1995). The first evidence for radiogenic $^{26}\text{Mg}^*$ in non-CAI material was found in a plagioclase-bearing chondrule from the highly unequilibrated ordinary chondrite Semarkona (Hutcheon and Hutchison, 1989); the isochron implies an initial abundance of $(^{26}\text{Al}/^{27}\text{Al})_0 = (7.7 \pm 2.1) \times 10^{-6}$. In most cases, however, only upper limits on ^{26}Al abundances could be determined in a handful of plagioclase grains from chondrules in ordinary chondrites (Hutcheon et al., 1994; Hutcheon and Jones, 1995). Today, initial $^{26}\text{Al}/^{27}\text{Al}$ ratios have been determined in ~ 50 chondrules from several unequilibrated ordinary and carbonaceous chondrites. Chondrules with abundant aluminum-rich minerals (plagioclase-rich chondrules) and those with "normal" ferromagnesian mineralogy have been analyzed (Figure 3(a), bottom panel). Chondrules have distinctly lower $(^{26}\text{Al}/^{27}\text{Al})_0$ than CAIs, most by a factor of 5 or more. A significant number of chondrules show no resolvable $^{26}\text{Mg}^*$, implying that if they evolved from the same canonical $(^{26}\text{Al}/^{27}\text{Al})_0$ that characterized the nebular regions where many CAIs formed, then chondrules achieved isotopic closure of the Al–Mg system at least 3–4 Myr (and possibly significantly more) after CAI formation. A closer inspection of the record, however, indicates that those chondrules from meteorites that are more extensively metamorphosed tend to have lower $(^{26}\text{Al}/^{27}\text{Al})_0$ values (Figure 3(b)). This would indicate that metamorphic redistribution, on an asteroid, could be obscuring the nebular record of $^{26}\text{Mg}^*$ in these meteorites.

Chondrules that have been analyzed from the some of the most pristine meteorites (e.g., Semarkona, Bishunpur, Yamato 81020) tend to show detectable ^{26}Mg excesses that imply

$(^{26}Al/^{27}Al)_0$ values $\sim 1 \times 10^{-5}$, with some significant spread in this peak of the distribution (Russell *et al.*, 1996; Kita *et al.*, 2000; McKeegan *et al.*, 2000b; Huss *et al.*, 2001; Mostefaoui *et al.*, 2002; Hsu *et al.*, 2003; Kunihiro *et al.*, 2003; Hutcheon and Hutchison, 1989). A couple of chondrules have $(^{26}Al/^{27}Al)_0$ values that approach the range seen in some CAIs, and Galy *et al.* (2000) report one chondrule (not plotted on Figure 3(b)) with $(^{26}Al/^{27}Al)_0 = (3.7 \pm 1.2) \times 10^{-5}$, which overlaps the canonical CAI value within uncertainty. However, that datum is for ICPMS measurement of a whole chondrule, and there are currently no data showing internal Al–Mg isochrons for chondrules that fall within error of the CAI value. It is not possible to rule out mixing of CAI-like material as the cause of the ^{26}Mg excess in this one case, and, given that Galy *et al.* (2000) also measured a high $(^{26}Al/^{27}Al)_0$ for a CAI (see above), CAIs and chondrules measured by the same technique do not overlap in initial $^{26}Al/^{27}Al$. Thus, there are no data that unequivocally point toward coeval CAI and chondrule formation. Instead, if ^{26}Al chronology is valid for CAIs and chondrules, the overall data imply that chondrule formation began ~ 1 Myr after the formation of most CAIs and then continued for another ~ 2 Myr or more. Some chondrules may have formed later still, or more likely, only achieved closure temperatures for magnesium diffusion following parent body cooling at times exceeding ~ 4 Myr after CAIs. That mild metamorphism in chondrites could delay isotopic closure of the Al–Mg system is further evidenced by analyses of plagioclase grains from the H4 chondrites Ste. Marguerite and Forest Vale (Zinner and Göpel, 2002). The inferred $^{26}Al/^{27}Al$ ratios indicate retention of $^{26}Mg^*$ by $\sim 5-6$ Myr following CAIs, which is consistent with timescales of parent body metamorphism implied by absolute Pb–Pb ages of (secondary) phosphates in these meteorites.

A similar temporal interpretation is generally not invoked for those CAIs that exhibit an apparent lack of initial ^{26}Al (Figure 3(a)). As pointed out by MacPherson *et al.* (1995), many of the inclusions in the low $(^{26}Al/^{27}Al)_0$ peak are not mineralogically altered, which argues against late metamorphism. Moreover, these inclusions are typically hosts for very significant isotopic anomalies in a variety of elements, which argues strongly for their antiquity. Included in this group are the so-called FUN (fractionated and unknown nuclear isotopic effects) inclusions (e.g., Lee *et al.*, 1977; Lee *et al.*, 1980) and the platelet hibonite crystals, which are extremely refractory grains from CM chondrites that are characterized by huge isotopic anomalies in the sub-iron group elements like titanium and calcium (Fahey *et al.*, 1987; Ireland, 1988). Because of their preservation of extreme stable isotope anomalies, these refractory phases are best understood as having formed at an early time in the nebula, but from an isotopic reservoir (or precursor minerals) that was missing the ^{26}Al inventory sampled by other "normal" refractory materials. The scope of this heterogeneity, both spatially and temporally, is the focus of much conjecture and research, as this is a key issue for the utility of ^{26}Al as a high-resolution chronometer for nebular events (see discussion in Section 1.16.6).

Relatively few data exist for the former presence of ^{26}Al in differentiated (i.e., melted) meteorites, even though there is a widespread assumption that ^{26}Al provided a significant, if not the dominant, heat source for melting of early accreted planetesimals (e.g., Grimm and McSween, 1994; Schramm *et al.*, 1970). Plagioclase crystals in the eucrite Piplia Kalan have significant excess ^{26}Mg (Srinivasan *et al.*, 1999); however, the correlation of $^{26}Mg^*$ with Al/Mg in the plagioclase is poor, indicating that the system has suffered partial reequilibration of magnesium isotopes following crystallization. A best-fit correlation through plagioclase and pyroxene yields an apparent $(^{26}Al/^{27}Al)_0 = (7.5 \pm 0.9) \times 10^{-7}$, which would correspond to ~ 4 Myr after the CAI canonical value.

Recently, several abstracts have reported Al–Mg data for achondrites, which can potentially be tied to the $^{53}Mn-^{53}Cr$ system. The petrographically unique eucrite Asuka 881394 exhibits a good Al–Mg isochron with well-resolved $^{26}Mg^*$ in its anorthitic plagioclase that yields $^{26}Al/^{27}Al = (1.19 \pm 0.13) \times 10^{-6}$, corresponding to ~ 4 Myr after CAIs (Nyquist *et al.*, 2001b). In contrast, the eucrite Juvinas shows only an upper limit of $^{26}Al/^{27}Al \sim 10^{-7}$ (Wadhwa *et al.*, 2003). Basaltic clasts in the ultramafic ureilite DaG-319 all lie on a single Al–Mg isochron with slope $^{26}Al/^{27}Al = (3.95 \pm 0.59) \times 10^{-7}$ indicating that they achieved isotopic closure ~ 5 Myr after CAI formation (Kita *et al.*, 2003). The data for two angrites (Nyquist *et al.*, 2003) yield a two-point isochron with somewhat lower slope, corresponding to $^{26}Al/^{27}Al = (2.3 \pm 0.8) \times 10^{-7}$.

1.16.4.3 Beryllium-10

^{10}Be β-decays to ^{10}B with a half-life of 1.5 Myr. Evidence for its former existence in the solar system is provided by excesses of $^{10}B/^{11}B$ correlated with Be/B ratio (Figure 4), first found within coarse-grained (type B) CAIs from Allende (McKeegan *et al.*, 2000a). From the slope of the correlation line, McKeegan *et al.* calculated an initial $^{10}Be/^9Be = (9.5 \pm 1.9) \times 10^{-4}$ at the time corresponding to isotopic closure of the Be–B system. This discovery was rapidly confirmed and extended by analyses of a variety of CAIs of types A and B, and a FUN inclusion from various

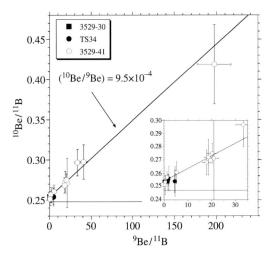

Figure 4 Boron isotopic composition of individual minerals from Allende CAIs as a function of Be/B ratio in the same material; error bars are 2σ. The $^{10}B/^{11}B$ values from various spots of CAI 3529-41 show ^{10}B excesses that are correlated with the Be/B ratio in a manner indicative of the *in situ* decay of ^{10}Be. The slope of the correlation line corresponds to an initial $^{10}Be/^{9}Be = (9.5 \pm 1.9) \times 10^{-4}$ at the time of crystallization. The intercept indicates $^{10}B/^{11}B = 0.254 \pm 0.002$, which is higher than $^{10}B/^{11}B$ for CI chondrites (shown by the horizontal line). Inset figure shows the same data at an expanded scale; data for CAIs 3529-30 and TS-34 are consistent with the Be–B isotope systematics of 3529-41. Data and figure from McKeegan et al. (2000a) (reproduced by permission of the American Association for the Advancement of Science from *Science* **2000**, 289, 1334–1337).

CV3 chondrites, including Allende, Efremovka, Vigarano, Leoville, and Axtell (MacPherson and Huss, 2001; McKeegan et al., 2001; Sugiura et al., 2001; MacPherson et al., 2003). Of the nearly two dozen CAIs that have been examined so far, in every case for which high Be/B ratios could be found in a sample (i.e., except where boron contamination is prevalent), excesses of $^{10}B/^{11}B$ are measured, implying that the existence of live ^{10}Be was rather widespread in the solar nebula, at least at the locale of CAI formation. Some spread in initial $^{10}Be/^{9}Be$ ratios is apparent, but overall it is remarkably uniform, especially considering the difficulties of the measurements and the susceptibility of samples to contamination by trace amounts of boron (cf. Chaussidon et al., 1997). Calculated initial $^{10}Be/^{9}Be$ ratios for "normal" CV CAIs range only over a factor of 2 from $(\sim 4.5-9.5) \times 10^{-4}$, with no difference seen between type B CAIs (mean of 12 samples: $^{10}Be/^{9}Be = (6.3 \pm 0.4) \times 10^{-4}$) and type A CAIs (mean of five samples, $^{10}Be/^{9}Be = (6.7 \pm 0.6) \times 10^{-4}$). The one FUN inclusion measured, a type A from Axtell (MacPherson et al., 2003), has the lowest initial $^{10}Be/^{9}Be = (3.6 \pm 0.9) \times 10^{-4}$, but even this value is within

error of the lower values measured on "normal" (i.e., non-FUN) CAIs. One CAI, Efremovka E44, has been measured independently in two laboratories with excellent agreement (McKeegan et al., 2001; Sugiura et al., 2001), indicating that potential systematic uncertainties are not significant compared to statistical errors. The initial boron isotopic composition (prior to any ^{10}Be decay) is the same among these various CAIs, with a small degree of relative scatter. However, the mean value, $^{10}B/^{11}B = 0.250 \pm 0.001$, is distinct from a chondritic value ($= 0.248$) measured for CI chondrites (Zhai et al., 1996).

The former presence of ^{10}Be was extended to another important class of refractory objects, hibonite from the CM2 Murchison meteorite (Marhas et al., 2002). Hibonite $[CaAl_{12-2x}(Mg_xTi_x)O_{19}]$ is one of the most refractory minerals calculated to condense from a gas of solar composition, and is known to host numerous isotopic anomalies, especially in the heavy isotopes of calcium and titanium (Ireland et al., 1985; Zinner et al., 1986; Fahey et al., 1987). Curiously, when these anomalies are of an exceptionally large magnitude (in the ~several to 10% range), the hibonite grains show a distinct lack of evidence for having formed with ^{26}Al (e.g., Ireland, 1988, 1990) or ^{41}Ca (Sahijpal et al., 1998, 2000). Marhas et al. (2002) found excesses of $^{10}B/^{11}B$ in three such hibonite grains that are each devoid of either $^{26}Mg^*$ or $^{41}K^*$ from the decay of ^{26}Al and ^{41}Ca, respectively. Collectively, the Be–B data imply $^{10}Be/^{9}Be = (5.2 \pm 2.8) \times 10^{-4}$ when these hibonites formed. This initial $^{10}Be/^{9}Be$ is in the same range as for other refractory inclusions and indicates that existence of ^{10}Be is decoupled from the other two short-lived nuclides that partition into refractory objects, namely ^{26}Al and ^{41}Ca. Even more striking evidence for decoupling of the $^{26}Al-^{26}Mg$ and $^{10}Be-^{10}B$ systems came with the report of Marhas and Goswami (2003) that hibonite in the well-known FUN CAI HAL had an initial $^{10}Be/^{9}Be$ ratio in the same range as other CAIs, yet had an initial $^{26}Al/^{27}Al$ ratio three orders of magnitude lower than the canonical early solar system ratio. The significance of this lack of correlation, for both chronology and source of radionuclides, is discussed further below.

Convincing evidence of live ^{10}Be has so far only been found in refractory inclusions because these samples exhibit large volatility controlled Be–B fractionation. A tantalizing hint for ^{10}Be was found in one anorthite-rich chondrule from a highly unequilibrated (CO3) chondrite: the Be–B correlation diagram displays a large amount of scatter, but an initial $^{10}Be/^{9}Be$ ratio of $7.2 \pm 2.9 \times 10^{-4}$ may be calculated (Sugiura, 2001). This value is similar to that seen in CAIs, but needs to be confirmed by further measurements. Finally, a possible hint for the existence of extremely short-lived ^{7}Be (half-life = 53 d), evidenced by ^{7}Li/^{6}Li anomalies in an Allende CAI (Chaussidon et al., 2002), also needs confirmation.

1.16.4.4 Manganese-53

^{53}Mn decays by electron capture to ^{53}Cr with a half-life of 3.7 Myr. This relatively long half-life, and the fact that manganese and chromium are reasonably abundant elements that undergo relative fractionation in evaporation/condensation processes as well as magmatic processes, make the ^{53}Mn–^{53}Cr system particularly interesting for bridging the time period from nebular events to accretion and differentiation of early-formed planetesimals. Accordingly, this system has been intensively investigated and evidence of live ^{53}Mn has now been found in nebular components such as (i) CAIs (Birck and Allègre, 1985; Birck and Allègre, 1988; Papanastassiou et al., 2002) and (ii) chondrules (Nyquist et al., 2001a), as well as (iii) bulk ordinary chondrites (Nyquist et al., 2001a; Lugmair and Shukolyukov, 1998), (iv) bulk carbonaceous chondrites (Birck et al., 1999), (v) CI carbonates (Endress et al., 1996; Hutcheon and Phinney, 1996; Hutcheon et al., 1999b), (vi) enstatite chondrite sulfides (Wadhwa et al., 1997), and (vii) various achondrites including angrites, eucrites, diogenites, pallasites, and SNC meteorites (Lugmair and Shukolyukov, 1998; Nyquist et al., 2001b, 2003). Due to the wealth of high-quality data, an impressively detailed high resolution relative chronometry can be developed (e.g., Lugmair and Shukolyukov, 2001), however interpretation of the ^{53}Mn–^{53}Cr system with respect to other chronometers is complex, particularly with respect to nebular events. The primary reasons for these complexities are difficulty in evaluating the initial ^{53}Mn/^{55}Mn of the solar system and in establishing its homogeneity in the nebula (see discussions in Birck et al., 1999; Lugmair and Shukolyukov, 2001; Nyquist et al., 2001a).

As with ^{26}Al, ^{41}Ca, and ^{10}Be, the obvious samples in which to try to establish the solar system initial value for ^{53}Mn/^{55}Mn are CAIs. However, in this case there are three factors which work against this goal: (i) volatility-controlled fractionation is not favorable when the parent (^{53}Mn) is more volatile than the daughter (^{53}Cr); (ii) both manganese and chromium are moderately volatile elements and significantly depleted in CAIs; and (iii) the daughter element is known to exhibit nucleogenetic anomalies in most CAIs (e.g., Papanastassiou, 1986). Together, these properties mean that there are no mineral phases with large Mn/Cr in CAIs, and it is not feasible to find large ^{53}Cr excesses that are uniquely and fully attributable to ^{53}Mn decay. Birck and Allègre (1988) first demonstrated the in situ decay of ^{53}Mn by correlating ^{53}Cr excesses with Mn/Cr in mineral separates of an Allende inclusion, deriving an initial ^{53}Mn/^{55}Mn = (3.7 ± 1.2) × 10^{-5}. Comparison to other Allende CAIs led these authors to estimate ∼4.4 × 10^{-5} as the best initial ^{53}Mn/^{55}Mn for CAIs; however, Nyquist et al. (2001a) prefer a somewhat lower value (2.8 ± 0.3) × 10^{-5} based on the same mineral separate analyses plus consideration of nonradiogenic chromium in a spinel separate from an Efremovka CAI. In recent work, Birck et al. (1999) have emphasized that refractory inclusions are inconsistent with solar system evolution of the ^{53}Mn–^{53}Cr system, noting that the inferred chronology is necessarily model dependent. Lugmair and Shukolyukov (1998) reach a similar assessment, describing the "chronological meaning of ^{53}Mn/^{55}Mn ratios in CAIs" as "tentative." Papanastassiou et al. (2002) also studied Mn–Cr systematics of CAIs and concluded that although spinel preserved the initial ^{53}Cr/^{52}Cr ratio, manganese with live ^{53}Mn was introduced during secondary alteration, so it was not clear what event was being dated in CAIs.

Whole chondrule Mn–Cr isochrons (Figure 5) have been reported for the ordinary chondrites Chainpur (LL3.4) and Bishunpur (LL3.1) by Nyquist et al. (2001a). The chondrules from both meteorites are consistent with a single isochron with $(^{53}$Mn/^{55}Mn$)_0$ = (8.8 ± 1.9) × 10^{-6} and an intercept $\varepsilon(^{53}$Cr) = −0.03 ± 0.06 (Figure 5). If the chondrule data are considered with Mn–Cr data for whole chondrites (Nyquist et al., 2001a), then the slope increases slightly to $(^{53}$Mn/^{55}Mn$)_0$ = (9.5 ± 1.7) × 10^{-6} which Nyquist and colleagues interpret as reflecting the time of Mn/Cr fractionation during the condensation of chondrule precursors. If this occurred in the same nebular environments as CAI mineral condensation characterized by the preferred ^{53}Mn/^{55}Mn initial = 2.8 × 10^{-5}, this implies a time difference of 5.8 ± 2.7 Myr. This is significantly longer than the CAI-chondrule timescale inferred from ^{26}Al/^{27}Al (also for Bishunpur chondrules); however it is not clear that the two chronometers are dating the same events (see discussion in Nyquist et al., 2001a).

A more straightforward interpretation of Mn–Cr ages can, in principle, be achieved for planetary differentiates since these certainly homogenized chromium isotopes during melting and also likely underwent Mn/Cr fractionation at a well-defined nebular locale (the asteroid belt). Although Lugmair and Shukolyukov (1998) have argued for heterogeneity of ^{53}Mn/^{55}Mn as a function of heliocentric distance, such effects would be negligible considered over the likely distances of

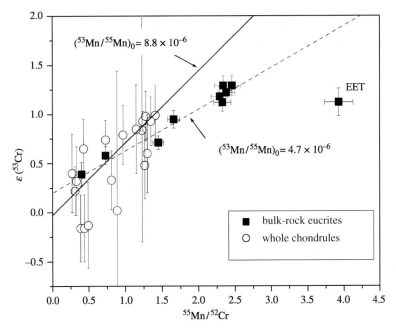

Figure 5 ^{53}Mn–^{53}Cr evolution diagram for nebular components (whole chondrules from ordinary chondrites Bishunpur and Chainpur; Nyquist et al., 2001) and for planetary differentiates (whole-rock eucrites; Lugmair and Shukolyukov, 1998). Plotted are measured values of $\varepsilon(^{53}\text{Cr})$, the deviation of $^{53}\text{Cr}/^{52}\text{Cr}$ in a sample from the terrestrial standard value in parts per 10^4, as a function of $^{55}\text{Mn}/^{52}\text{Cr}$. The correlation is interpreted as an isochron indicating the *in situ* decay of ^{53}Mn; the slope for the eucrites (dashed line) corresponds to an initial $^{53}\text{Mn}/^{55}\text{Mn} = (4.7 \pm 0.5) \times 10^{-6}$ and that for chondrules (solid line) indicates $(^{53}\text{Mn}/^{55}\text{Mn})_0 = (8.8 \pm 1.9) \times 10^{-6}$, implying that Mn/Cr fractionation in chondrule precursors preceded global fractionation of the eucrite parent body by approximately one half-life, or ~3.5 Myr. All data are replotted from Lugmair and Shukolyukov (1998) and Nyquist et al. (2001a); 2σ error bars are indicated and the datum for EET87520 is excluded from the fit for the eucrite whole-rock isochron.

formation for the asteroids (meteorite parent bodies). The rapidly cooled angrites provide the anchor point between ^{53}Mn–^{53}Cr and the absolute age determined by Pb–Pb since both isotopic systems should have closed contemporaneously (Lugmair and Shukolyukov, 1998). The olivine fraction of LEW has a high Mn/Cr and thus provides a good precision for the isochron, with $^{53}\text{Mn}/^{55}\text{Mn} = (1.25 \pm 0.07) \times 10^{-6}$ and $\varepsilon(^{53}\text{Cr}) = +0.40 \pm 0.16$ (Lugmair and Shukolyukov, 1998), which is tied to the Pb–Pb age of $4,557.8 \pm 0.5$ Ma (Lugmair and Galer, 1992).

As alluded to above in the discussion of absolute ages of differentiated objects, the eucrites have suffered a more prolonged and complex thermal and shock history, which is reflected in their internal ^{53}Mn–^{53}Cr systematics. Despite this, excesses of ^{53}Cr in *bulk* samples of eucrites are well correlated with Mn/Cr (Figure 5) indicating large-scale differentiation on the eucrite parent body prior to the decay of ^{53}Mn (Lugmair and Shukolyukov, 1998). The slope of the correlation line yields $^{53}\text{Mn}/^{55}\text{Mn} = (4.7 \pm 0.5) \times 10^{-6}$ which is nearly two half-lives of ^{53}Mn steeper (older) than the 1.25×10^{-6} value obtained for angrites. Thus, these data indicate that the parent asteroid of the eucrites (Vesta ?) was totally molten, probably during mantle-core differentiation, at 7.1 ± 0.8 Ma prior to the crystallization of angrite LEW. By calibration with the absolute Pb–Pb chronology of angrites, this indicates igneous differentiation of the eucrite parent body at $4,564.8 \pm 0.9$ Ma (Lugmair and Shukolyukov, 1998). It should be clear that this time does not necessarily represent the crystallization age of individual eucrite meteorites, but the last time of global chromium isotope equilibration and Mn/Cr fractionation. In fact, internal ^{53}Mn–^{53}Cr isochrons for individual cumulate and noncumulate eucrites show a range of apparent $^{53}\text{Mn}/^{55}\text{Mn}$ values, from close to the global fractionation event (e.g., 3.7×10^{-6} for Chervony Kut) to essentially "dead" ^{53}Mn (e.g., Caldera, Wadhwa and Lugmair, 1996). It is not certain whether these ages, especially the young ones, reflect prolonged igneous activity over a period of tens of millions of years, or cooling ages, or disturbance of the Mn–Cr system by impacts, or some combination of the above (Lugmair and Shukolyukov, 1998). The ^{53}Mn–^{53}Cr ages for individual eucrites do not correlate particularly well with Pb–Pb ages, for example Chervony Kut with a $^{53}\text{Mn}/^{55}\text{Mn}$ initial ratio indicating isotopic closure at ~4,564 Ma (almost contemporaneous with mantle differentiation) has a Pb–Pb age of

4,312.6 ± 1.6 Ma (Galer and Lugmair, 1996). This discrepancy can be attributed to the U–Pb system being more easily disturbed than Mn–Cr (Lugmair and Shukolyukov, 1998), however, as discussed in more detail by Tera and Carlson (1999), it also means that the eucrites cannot serve as an independent check on the validity of coupling ^{53}Mn–^{53}Cr model ages to an absolute timescale based on the Pb–Pb ages of angrites.

The ^{53}Mn–^{53}Cr system has also proved useful in constraining the timescales of earliest aqueous activity on the parent bodies of some carbonaceous chondrites by dating Mn/Cr fractionation associated with the formation of aqueously precipitated minerals. Carbonates from the CI chondrites Orgueil and Ivuna show very large ^{53}Cr excesses correlated with Mn/Cr; inferred initial ^{53}Mn/^{55}Mn ratios range from 1.42×10^{-6} to 1.99×10^{-6} (Endress et al., 1996). Carbonates from other carbonaceous chondrites show a wider range extending to significantly higher initial ^{53}Mn/^{55}Mn ratios: $(6.4 \pm 1.2) \times 10^{-6}$ in CM chondrites Nogoya and Y791198, and $(9.4 \pm 1.6) \times 10^{-6}$ in the unusual carbonaceous chondrite Kaidun (Hutcheon et al., 1999a; Hutcheon et al., 1999b). The latter values are similar to ^{53}Mn/^{55}Mn found in ordinary chondrite chondrules (Nyquist et al., 2001a). Fayalite (FeO-rich olivine) from the Mokoia oxidized and aqueously altered CV3 chondrite formed with very high ^{55}Mn/^{52}Cr ratios ($>10^4$) and exhibits (Hutcheon et al., 1998) ^{53}Mn/^{55}Mn = $(2.32 \pm 0.18) \times 10^{-6}$, similar to CI carbonates and eucrites. Mn–Cr data for fayalite from the Kaba chondrite yields the same ^{53}Mn/^{55}Mn within uncertainty (Hua et al., 2002).

1.16.4.5 Iron-60

^{60}Fe β-decays to ^{60}Ni with a half-life of 1.5 Myr. Unlike the other short-lived nuclides with half-lives of a few million years or less, and in particular contrast to ^{10}Be, ^{60}Fe is not produced by spallation because there are no suitable target elements, and therefore all of its solar system inventory must reflect recent stellar nucleosynthesis. The first plausible evidence for the existence of ^{60}Fe in the solar system was provided by ^{60}Ni excesses found in bulk samples of the eucrites Chervony Kut and Juvinas (Shukolyukov and Lugmair, 1993a,b). These are basaltic achondrites, the result of planetary-scale melting and differentiation (possibly on the asteroid Vesta; see Chapter 1.11) that fractionated nickel into the core. Thus, the excess ^{60}Ni cannot represent nucleogenetic isotope anomalies of the iron-group elements, as is seen in CAIs, and its presence in such a large volume material indicates wide scale occurrence of ^{60}Fe in the solar system (Shukolyukov and Lugmair, 1993a).

However, internal mineral isochrons could not be obtained on the eucrite samples because of element redistribution after the decay of ^{60}Fe (Shukolyukov and Lugmair, 1993b). Moreover, the inferred initial ^{60}Fe/^{56}Fe differs by an order a magnitude between these eucrites for which other isotopic systems (e.g., ^{53}Mn–^{53}Cr) indicate a similar formation age (Lugmair and Shukolyukov, 1998). These inconsistencies point out problems with interpreting eucrite ^{60}Fe/^{56}Fe abundances in chronologic terms and indicate that estimates of a solar system initial ^{60}Fe/^{56}Fe, based on an absolute age of eucrite formation, is likely subject to large systematic uncertainties.

Recent in situ measurements on high Fe/Ni phases in chondrites help to constrain this initial value. Tachibana and Huss (2003) found good correlations of excess ^{60}Ni with Fe/Ni ratios in sulfide minerals of the (LL3.1) unequilibrated ordinary chondrites Bishunpur and Krymka (Figure 6), which imply ^{60}Fe/^{56}Fe ratios of between 1.0×10^{-7} and 1.8×10^{-7}. Although it is somewhat ambiguous whether these phases achieved isotopic closure in the solar nebula or on an asteroidal parent body, it is likely that these sulfides have suffered significantly less disturbance of their Fe–Ni isotopic system than have eucrites, making an extrapolation back to the time of CAI formation more robust. With plausible assumptions, Tachibana and Huss (2003) estimate $(^{60}\text{Fe}/^{56}\text{Fe})_0$ for solar system formation of between 1×10^{-7} to 6×10^{-7} with a probable value (depending on the age of the sulfides relative to CAIs) of $(\sim 3-4) \times 10^{-7}$. This is consistent with an upper limit of $(^{60}\text{Fe}/^{56}\text{Fe})_0 \sim 3.5 \times 10^{-7}$ derived from analyses of nickel isotopes in FeO-rich olivine from a (LL3.0) Semarkona chondrule which exhibited $(^{26}\text{Al}/^{27}\text{Al})_0 = 0.9 \times 10^{-5}$ (Kita et al., 2000). These upper limits are significantly lower than a value of $(^{60}\text{Fe}/^{56}\text{Fe})_0 = (1.6 \pm 0.5) \times 10^{-6}$ inferred for an Allende CAI (Birck and Lugmair, 1988) indicating that the ^{60}Ni excesses in this sample are probably of a nucleosynthetic origin and are not due to in situ decay of ^{60}Fe. However, preliminary data on sulfides contained in Semarkona (Mostefaoui et al., 2003) indicate an Fe–Ni isochron with $(^{60}\text{Fe}/^{56}\text{Fe})_0 = (7.5 \pm 2.6) \times 10^{-7}$, which marginally exceeds the limit set by chondrule olivine (Kita et al., 2000) and may be compatible with the earlier CAI result. It would be desirable to have a direct measure of a ^{60}Fe/^{56}Fe isochron in a CAI; however, as a volatile element, iron is generally depleted in refractory inclusions and samples containing appropriate mineralogy for this determination may not be found. Clearly, over the next few years microanalytical techniques will be contributing more data to the question of the distribution of ^{60}Fe in early solar system objects.

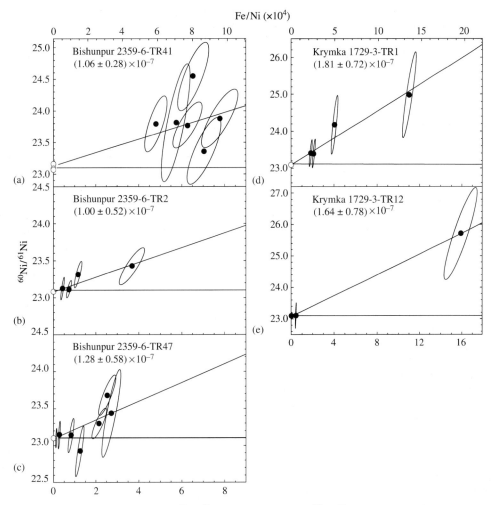

Figure 6 Ion microprobe analyses of ^{60}Ni/^{61}N as a function of ^{56}Fe/^{61}N in individual iron sulfide grains (filled circles) and metal grains (open circles) from the ordinary (LL3.1) chondrites Bishunpur and Krymka (with 2σ error ellipses). The correlation is interpreted as an isochron demonstrating the *in situ* decay of ^{60}Fe from initial ratios $(^{60}\text{Fe}/^{56}\text{Fe})_0 = (1.1 \pm 0.2) \times 10^{-7}$ and $(1.7 \pm 0.5) \times 10^{-7}$ for Bishunpur and Krymka, respectively. Data and figure from Tachibana and Huss (2003) (reproduced by permission of *The American Astronomical Society* from *Astrophys. J.* **2003**, *588*, L41–L44).

1.16.4.6 Palladium-107

^{107}Pd β-decays to ^{107}Ag with a half-life of 6.5 Myr. Evidence for this now-extinct nuclide is found in metallic phases of iron meteorites since large Pd/Ag fractionations occur during magmatic partitioning of metal (Kelly and Wasserburg, 1978; see also review by Wasserburg, 1985). Kaiser and Wasserburg (1983) demonstrated that a linear correlation exists between excess ^{107}Ag/^{109}Ag and Pd/Ag in different fractions of metal and sulfide from the group IIIB iron meteorite Grant and from the isochron inferred an initial ^{107}Pd/^{108}Pd = ~1.7×10^{-5} at the time of crystallization of this meteorite. Extrapolation back to the time of CAI formation would yield an initial ^{107}Pd/^{108}Pd of approximately twice this value for the solar system, though with considerable uncertainty. Further isochrons were determining in other many iron and stony-iron meteorites, showing that there is a wide range of initial ^{107}Pd/^{108}Pd ratios, but that many samples have ratios in the range $(1.5-2.5) \times 10^{-5}$ (Chen and Wasserburg, 1996; Chen *et al.*, 2002). Recently, Carlson and Hauri (2001) have developed ICPMS methods for determining silver isotope ratios with high precision, thus permitting the investigation of phases with more moderate Pd/Ag fractionation. They found good isochrons for the pallasite (stony-iron) Brenham and the IIIB iron Grant, both with inferred initial ^{107}Pd/^{108}Pd = 1.6×10^{-5}. A two-point correlation between metal and sulfide was also determined for Canyon Diablo (group IA iron), yielding an apparent initial ^{107}Pd/^{108}Pd essentially identical to that previously found for Gibeon (Chen and Wasserburg, 1990). Interpreted chronologically, the data imply that Brenham and Grant formed some 3.5 Myr following Canyon Diablo and Gibeon. Small (5ε) $^{107}Ag/^{109}Ag$ anomalies were also documented for the carbonaceous chondrite Allende (Carlson and Hauri, 2001), which, given its relatively low Pd/Ag content, would imply an enormous initial $^{107}Pd/^{108}Pd$ ($\sim 39 \times 10^{-5}$) if this anomaly had evolved from the most unradiogenic sample (Canyon Diablo sulfide) due to ^{107}Pd decay only. However, no internal isochron is obtained for Allende and considering its unequilibrated nature (i.e., it hosts many isotopic anomalies) there is no compelling reason to assume that this value represents a solar nebular abundance of live ^{107}Pd.

1.16.4.7 Hafnium-182

^{182}Hf β-decays to ^{182}W with a half-life of 9 Myr. This has been recognized as an extremely important isotopic system in recent years (e.g., Lee and Halliday, 1996; Halliday and Lee, 1999) because it is almost uniquely sensitive to metal–silicate fractionation and its rather long half-life makes it a useful probe for both nebular and planetary processes. Specifically, tungsten is highly siderophile, whereas hafnium is retained in silicates during melting and metal segregation. Thus, tungsten isotope compositions could be very different in silicates and metal from distinct planetary objects depending on whether or not metal/silicate fractionation in those objects predated significant decay of ^{182}Hf. Internal isochrons, demonstrating good correlations of $^{182}W/^{180}W$ with Hf/W, are found for several separates of ordinary chondrites (Kleine et al., 2002a,b; Yin et al., 2002); samples of whole-rock carbonaceous chondrites and a CAI from Allende also fall within error of these isochrons (Yin et al., 2002). The Pb–Pb ages of phosphates in the ordinary chondrites (Kleine et al., 2002a) and the coincidence of the CAI data (Yin et al., 2002) allow a robust estimate of the initial $^{182}Hf/^{180}Hf$ of the solar system of $1.0-1.1 \times 10^{-4}$ with an initial $^{182}W/^{180}W$ significantly ($\sim -3\varepsilon$) lower than terrestrial mantle samples. The meaning of these recent results with regard to timescales of accretion and core formation of the Earth and formation of the Moon is discussed in Chapter 1.20.

1.16.4.8 Iodine-129

^{129}I β-decays to ^{129}Xe with a half-life of 15.7 Myr. As mentioned in the historical introduction (Section 1.16.1.3), ^{129}I was the first extinct isotope whose presence in the early solar system was inferred from excesses of its daughter ^{129}Xe in meteorites (Jeffery and Reynolds, 1961). Both parent and daughter are mobile elements, and coupled with the relatively long half-life, this means that closure effects on the I–Xe system likely limit its utility to parent–body processes (e.g., Swindle et al., 1996), although arguments have been advanced that I–Xe can date nebular events in favorable circumstances (Whitby et al., 2001). New analytical techniques that enable the investigation of single mineral phases (Gilmour, 2000; Gilmour and Saxton, 2001) have helped in the understanding of apparent I–Xe isochrons (as differentiated from mixing lines of multiple phases) and enabled more confident chronological interpretations, particularly of secondary mineral phases formed on asteroidal parent bodies. Brazzle et al. (1999) demonstrated concordancy between I–Xe and Pb–Pb chronometers for chondrite phosphates over a timescale of tens of millions of years. At another extreme, Whitby et al. (2000) found an initial ratio of $^{129}I/^{127}I = (1.35 \pm 0.05) \times 10^{-4}$ in halite from a relatively unequilibrated ordinary chondrite. This result is close to the estimated initial value for the solar system ($\sim 10^{-4}$), implying that the aqueous activity responsible for precipitating the halite occurred immediately upon accretion, probably within a few million years of CAI formation (Whitby et al., 2000).

1.16.4.9 Niobium-92

^{92}Nb decays by electron capture to ^{92}Zr with a half-life of 36 Ma. ^{92}Nb is a p-process nuclide (see Chapter 1.01). The first hint that this isotope was present in the early solar system was based on an $8.8 \pm 1.7\varepsilon$ excess in ^{92}Zr in a niobium-rich rutile grain from the Toluca IAB iron meteorite (Harper, 1996). This corresponded to an initial $^{92}Nb/^{93}Nb$ ratio of $(1.6 \pm 0.3) \times 10^{-5}$, but the time of formation of Toluca rutile is not known. Three subsequent studies that used MC-ICPMS to measure zirconium isotopic composition reported that the initial solar system $^{92}Nb/^{93}Nb$ was $\sim 10^{-3}$, higher by two orders of magnitude (Yin et al., 2000; Münker et al., 2000; Sanloup et al., 2000). This initial $^{92}Nb/^{93}Nb$ was nearly one quarter of the p-process production ratio (Harper, 1996) and was difficult to understand, as most ^{93}Nb is made by the s-process. The situation was resolved with the work of Schönbachler et al. (2002), who reported internal Nb–Zr isochrons for the Estacado H6 chondrite and for a clast from the Vaca Muerta mesosiderite, both of which give an initial solar system $^{92}Nb/^{93}Nb$ of $\sim 10^{-5}$, a much more plausible value in terms of nucleosynthetic considerations. This lower initial ratio limits the utility of the $^{92}Nb-^{92}Zr$ for chronometry (see Chapter 1.20 for further discussion).

1.16.4.10 Plutonium-244 and Samarium-146

These relatively long-lived isotopes are mentioned here for completeness since both have been shown to have existed in the early solar system. However, neither ^{244}Pu nor ^{146}Sm have been developed for chronological applications, for very practical reasons. ^{244}Pu suffers from the fact that there are no long-lived isotopes of plutonium against which to normalize its abundance, and its primary application in meteorite studies is for obtaining cooling rates from the annealing of fission tracks in appropriate minerals. The half-life of ^{146}Sm (103 Myr) is too long and its abundance and relative fractionation from daughter ^{142}Nd are insufficient for it to constitute a useful chronometer for early solar system processes. Its primary interest is for nuclear astrophysics (e.g., Prinzhofer et al., 1989), because this isotope is on the neutron-deficient side of the valley of β-stability. Interested readers are referred to Stewart et al. (1994) and review by Podosek and Swindle (1988) and Wasserburg (1985) for more information.

1.16.5 ORIGINS OF THE SHORT-LIVED NUCLIDES IN THE EARLY SOLAR SYSTEM

The ability of short-lived radioisotopes to function as chronometers for the early solar system is critically dependent on there having been an initially uniform distribution of the radioactivity throughout the nebula, or at least in those regions from which meteoritic components are derived. Only in this circumstance can differences in initial abundances of a radionuclide compared to a stable counterpart, as inferred by the excesses of the respective daughter isotope, be interpreted as due to radioactive decay from the initial inventory. The homogeneity of the distribution of radionuclides in the solar nebula depends, in turn, on the processes that created those isotopes some time before the formation of early solar system materials. For the longer-lived isotopes listed in Table 1 (e.g., ^{182}Hf, ^{129}I, ^{92}Nb, ^{146}Sm, ^{244}Pu), continuous nucleosynthesis may have been sufficient to produce a quasi-equilibrium abundance of these species that was inherited by the solar nebula. However, the shorter half-life isotopes require a more immediate source (e.g., Meyer and Clayton, 2000; Wasserburg et al., 1996).

In principle, new (radioactive) isotopes could have been created by nuclear processes within the solar nebula itself, or they could have originated from sources external to the nebula. In the latter case, the most likely source is stellar nucleosynthesis in the interiors of nearby mass-losing stars (e.g., Cameron, 2001a,b; Cameron et al., 1995; Wasserburg et al., 1994, 1996, 1998), although spallation reactions in the molecular cloud parental to the solar nebula are also a possibility. If short-lived radioactivity is produced locally, for example by spallation reactions with nuclear particles (protons, alphas) accelerated by interaction with an active young Sun (e.g., Gounelle et al., 2001; Lee et al., 1998), then it is unlikely that the products of those reactions will be distributed uniformly throughout the accretion disk. Homogeneity over nebular scale-lengths is much more likely for an "external seeding" scenario, although even in this case strong isotopic heterogeneity is possible at the very early stages following injection, before local mixing can act to smooth out the memory of the particular mechanism for "contamination" of the nebula by the new isotopes. The injection of radioactive stellar debris in a "triggered" collapse scenario for solar system formation is reviewed by Boss and Vanhala (2001); later we consider the possible implications of this model for understanding isotopic heterogeneities in certain refractory inclusions.

The possible stellar sources of the short-lived isotopes, as well as constraints on nuclear spallation processes that could have produced them, are reviewed in detail by Goswami and Vanhala (2000). Since that work, two new developments have occurred: the discovery of evidence for live ^{10}Be in CAIs (McKeegan et al., 2000a) and the observation of in situ ^{60}Fe decay in chondrites (Tachibana and Huss, 2003) that leads to a factor of ~20 increase in the estimated (^{60}Fe/^{56}Fe)$_0$ for the solar system initial. These isotopes are particularly significant because their respective modes of origin are much more tightly constrained than those of the other extinct nuclides. ^{10}Be is not produced by stellar nucleosynthesis, thus its existence in the early solar system is strong evidence for a spallogenic source of some short-lived nuclides. However, ^{60}Fe is not produced by spallation reactions, but it is produced in core-collapse supernovae and in asymptotic giant branch (AGB) stars (Wasserburg et al., 1994). The existence of ^{60}Fe in the relatively high abundance of $\sim 3 \times 10^{-7}$ is therefore compelling evidence that stellar debris seeded the early solar system with new radioactivity. A recently proposed hypothesis considers that the source of spallogenic ^{10}Be is actually magnetically trapped cosmic rays in the interstellar medium prior to the collapse of a molecular cloud to form the solar system (Desch et al., 2003), but a detailed model has not yet been published. An alternative model considers ^{10}Be to be produced during supernova explosions (Cameron, 2001a,b), but there are problems in co-producing ^{10}Be with other short-lived isotopes (see below). The abundance of ^{10}Be in CAIs seems consistent with expectations based on observations of X-ray luminosity in young, solar-like stars (Feigelson et al., 2002a,b) and models of particle acceleration due to magnetic flare activity near the protosun (Lee et al., 1998). In summary, the most likely scenario implied by the new meteoritic data is that the overall inventory of extinct nuclides contained both a spallogenic component, probably produced locally, and a nucleogenetic component, probably produced in a supernova, although contributions from AGB and other rapidly evolving mass-losing stars are also possible.

Although ^{10}Be and ^{60}Fe are interesting isotopes for delimiting possible origins of short-lived radioactivity, it is ^{41}Ca, ^{26}Al, and ^{53}Mn that are potentially useful for chronology. Thus, a key task is to sort out, quantitatively, what sources are responsible for these isotopes in the early solar system. This can be addressed theoretically for both stellar and spallogenic sources; however a clear consensus is lacking (e.g., Goswami et al., 2001; Gounelle et al., 2001; Leya et al., 2002) since production models can be tweaked by adjustable parameters (e.g., energy spectrum and target compositions) that are poorly constrained by observation. Another approach is to examine the isotopic record in meteoritic components for correlations that may indicate common sources (and distributions) for these nuclides.

The refractory inclusions provide the best samples since they incorporated all three of these radioisotopes, as well as ^{10}Be. It has already been mentioned that ^{41}Ca and ^{26}Al are highly correlated in CAIs and hibonite grains (Figure 2). At face value, this would imply the same source for both these refractory elements. A problem with ^{41}Ca, however, is that its abundance is only marginally above detection limits and it decays very quickly, so that there is essentially no chance to test for concordant decay between the ^{41}Ca and ^{26}Al systems. This is not the case for ^{26}Al and ^{10}Be, which exist in much higher abundances and which have half-lives that differ by only a factor of two.

The initial ^{26}Al/^{27}Al and ^{10}Be/^{9}Be values have been measured in a variety of refractory phases from both CV and CM carbonaceous chondrites (Figure 7). "Normal" CAIs of both petrologic types A and B have inferred ^{26}Al/^{27}Al values that plot within error of the "canonical" solar system initial; even for cases where the Al–Mg system is disturbed in anorthite, other phases in the inclusion plot near the canonical value (e.g., Sugiura et al., 2001). As noted above, initial ^{10}Be/^{9}Be ratios for "normal" CV CAIs also show no discrimination based on petrology and the total range covered is approximately a factor of 2, which is only marginally outside of experimental uncertainty. Thus, for normal CAIs it is difficult to claim that the two isotopic systems are definitively discordant since the resolution of the data is not quite good enough.

However, the situation is different when one considers hibonites and FUN inclusions (Figure 7). For most of these objects only an upper limit on initial ^{26}Al/^{27}Al ($< \sim 10^{-5}$) is obtained, yet they have initial ^{10}Be/^{9}Be similar to most of the other refractory inclusions (MacPherson et al., 2003; Marhas et al., 2002; Marhas and Goswami, 2003). The data are still not completely convincing until one includes the famous FUN inclusion "HAL" (Lee et al., 1978, 1980). Recent analyses by Marhas and Goswami (2003) demonstrate that this hibonite-rich Allende CAI has ^{10}B/^{11}B excesses that imply ^{10}Be/^{9}Be = $\sim 4 \times 10^{-4}$, close to that of other CAIs, yet HAL has a well-resolved, but exceedingly low, initial ^{26}Al/^{27}Al = 5×10^{-8} (Fahey et al., 1987). These data clearly demonstrate that HAL formed from a reservoir with a characteristic ^{10}Be/^{9}Be similar to that of other refractory materials, but that it was almost completely lacking in ^{26}Al/^{27}Al. The low value of ^{26}Al/^{27}Al that it does have may, in fact, be commensurate with ambient background in the molecular cloud, i.e., independent of any specific additional source of ^{26}Al that spiked the CAI-forming regions of the solar nebula (Marhas and Goswami, 2003). Because the ^{10}Be is clearly spallogenic, this provides strong evidence that the vast majority of the ^{26}Al cannot have been produced that way and therefore that essentially all ^{26}Al is derived from external seeding of the nebula. The correlation of ^{26}Al with ^{41}Ca, even though it is not temporally quantitative, is then further evidence for the coproduction and injection of these nuclides into the solar nebula as freshly synthesized stellar debris.

Unfortunately, similar arguments cannot be advanced for ^{53}Mn, primarily because of the poor constraints on initial ^{53}Mn/^{55}Mn in CAIs. As discussed further below, the Mn–Cr systematics of nebular components are difficult to interpret in terms of a reasonable chronology, and one possible reason for this could be a not-insignificant contribution to the ^{53}Mn inventory by local production processes.

1.16.6 IMPLICATIONS FOR CHRONOLOGY

In principle, the record of each of the now-extinct isotopes can be interpreted to infer a chronology for various events that caused chemical fractionations in early solar system materials. Here we evaluate the consistency of these records, both internally and with each other, as well as with the Pb–Pb chronometer, to determine what quantitative constraints can be confidently

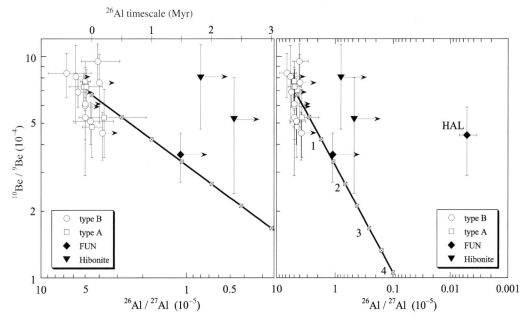

Figure 7 Inferred initial ^{26}Al/^{27}Al versus initial ^{10}Be/^{9}Be for refractory inclusions in CM and CV carbonaceous chondrites. All data are plotted with 2σ errors; upper limits are indicated by arrows. The locus of concordant ages by free decay from assumed solar system initial values of ^{26}Al/^{27}Al = 4.5×10^{-5} and ^{10}Be/^{9}Be = 6.7×10^{-4} is shown by the heavy line with 0.5 Myr tick marks. Left panel: the ^{26}Al/^{27}Al timescale is also shown on the top axis. It may be seen that "normal" CAIs of both petrologic types A and B have maximal ^{26}Al/^{27}Al values that plot within error of the "canonical" solar system initial, but the FUN inclusion (Axtell 2771; MacPherson et al., 2003) and the CM hibonite grains (Marhas et al., 2002) are depleted in ^{26}Al, with upper limits $< \sim 10^{-5}$. For nearly half of the type B CAIs, the Al–Mg system shows evidence of secondary disturbance; in these cases the maximum inferred ^{26}Al/^{27}Al is plotted as an upper limit (i.e., the inclusions are assumed to have formed with close to these values). With this approximation, the normal CAIs are relatively tightly clustered in ^{26}Al/^{27}Al, but show a range of approximately a factor 2 in ^{10}Be/^{9}Be, which is resolved at the 2σ level for several cases. Right panel: expanded scale showing new data from Marhas and Goswami (2003). In contrast to other FUN inclusions, HAL shows resolved ^{26}Mg excesses (Fahey et al., 1987) implying a very low initial ^{26}Al/^{27}Al = $(5.2 \pm 1.7) \times 10^{-8}$ but it also has ^{10}Be/^{9}Be similar to other refractory inclusions (Marhas and Goswami, 2003), demonstrating that ^{10}Be and ^{26}Al are decoupled. Data sources: Fahey et al. (1987), Podosek et al. (1991), McKeegan et al. (2000a); McKeegan et al. (2001), Srinivasan (2001), Sugiura et al. (2001), MacPherson et al. (2003), Marhas et al. (2002), Marhas and Goswami (2003).

inferred for the sequence and duration of processes in the solar nebula and on earliest planetesimals (planetary-scale differentiation, e.g., relative to the Earth, is considered in Chapter 1.20). To obtain reference points for cross-calibrating relative and absolute chronologies, we require samples which achieved rapid isotopic closure following a well-defined fractionation event and for which a robust and high-precision data set exists. By these criteria, only two anchor points are possible for the cross-calibration: (i) the Pb–Pb and Al–Mg records in CAIs and (ii) the Pb–Pb and Mn–Cr records in angrites. As demonstrated in Figure 8, the former provides a reasonably self-consistent, high-resolution record for nebular events, and the latter yields unique temporal information regarding early planetary differentiation processes, but that global consistency between the Al–Mg and Mn–Cr systems is problematic. The existing record for the other short-lived radionuclides is either not well preserved across different types of samples (e.g., ^{41}Ca, ^{10}Be, ^{182}Hf), or is insufficiently precise (e.g., ^{60}Fe), or uncertain as to the nature of isotopic closure (e.g., ^{129}I) so that cross-calibrations spanning the nebular and planetary accretion timescales are not yet possible.

1.16.6.1 Formation Timescales of Nebular Materials

A consistent timescale for fractionation events that occurred during high-temperature processing of nebular materials is obtained (Figure 8) by fixing the canonical ^{26}Al/^{27}Al value (4.5×10^{-5}) measured in CAIs to the absolute timescale provided by the recent high-precision Pb–Pb isochron age of $4,567.2 \pm 0.6$ Ma (Amelin et al., 2002). By this calibration, the initial ^{26}Al/^{27}Al values inferred for chondrules from the most unequilibrated chondrites ($\sim 1 \times 10^{-5}$; Figure 3) indicate that chondrule formation began by at least $\sim 4,565$ Ma and continued probably for

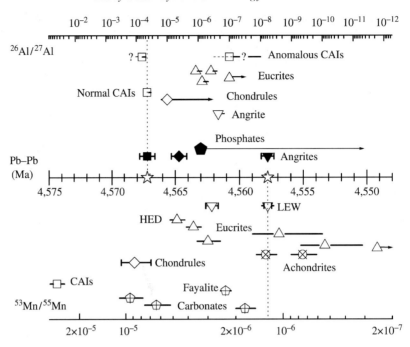

Figure 8 Timeline for early solar system events integrating the ^{26}Al–^{26}Mg and ^{53}Mn–^{53}Cr short-lived chronometers with the absolute timescale provided by the Pb–Pb chronometer. The anchor points (vertical dashed lines) are (1) the Pb–Pb age of CAIs (Amelin et al., 2002) with "canonical" ^{26}Al/^{27}Al and (ii) the Pb–Pb age of angrites (Lugmair and Galer, 1992) with the ^{53}Mn/^{55}Mn ratio in LEW (Lugmair and Shukolyukov, 1998). Pb–Pb ages are indicated for the filled symbols read against the absolute timescale (central axis); the top axis shows the initial ^{26}Al/^{27}Al values measured in various phases (open symbols) and the bottom axis refers to initial ^{53}Mn/^{55}Mn for the open symbols in the bottom panel. Squares represent CAIs, diamonds—chondrules, upright triangles—eucrites (basaltic achondrites), inverted triangles—angrites, crossed circles—pallasites and Acapulco, pentagons—secondary minerals in chondrites (phosphates, carbonates, fayalite). The datum labeled "HED" represents the Mn–Cr correlation line for bulk eucrites. "LEW" refers to the anchor point for ^{53}Mn/^{55}Mn and Pb–Pb; the remaining angrite datum represents Mn–Cr and Al–Mg analyses of D'Orbigny (Nyquist et al., 2003). "Anomalous CAIs" refers to those which apparently formed with no live ^{26}Al—see text for discussion.

another ~1–2 Myr. This time frame fits with the only high-precision Pb–Pb isochron for chondrules (from CR chondrites) which yields 4,564.7 ± 0.6 Ma (Figure 1). Chondrule ages which appear younger than ~4,563 by Al–Mg probably reflect metamorphic cooling rather than nebular formation.

The same is not true for the majority of "anomalous CAIs", those that apparently formed lacking any significant live ^{26}Al. These refractory inclusions, which are often hibonite-rich, typically exhibit very large anomalies in "stable" isotopes (e.g., calcium or titanium) that are most readily interpreted as indicating a lack of mixing with average solar nebula materials. Because isotopic homogenization is expected to be an ongoing process during nebular evolution, the preservation of these anomalies argues strongly for a very "primitive" nature of these materials, i.e., they probably formed early (not late) and also they escaped any significant isotopic reequilibration from later heating (MacPherson et al., 1995; Sahijpal and Goswami, 1998). Sahijpal and Goswami (1998) suggested that the highly anomalous CM hibonite grains might have formed in a triggered collapse scenario just prior to injection of the radionuclides (^{41}Ca and ^{26}Al), which could theoretically trail the shock front (Foster and Boss, 1997). It would be useful to demonstrate the plausibility of this scenario by measuring an absolute Pb–Pb age on a suite of these objects; even if such a measurement might lack the precision to resolve the prearrival interval, it could at least demonstrate that the samples were not anomalously young.

There are other refractory inclusions, e.g., grossite-bearing CAIs from CH chondrites (Weber et al., 1995), that do not fit this model since they lack calcium and titanium isotopic anomalies as well as ^{26}Al. One interpretation of such objects could be that CAI formation lasted several million years, but this is not supported by any independent evidence and there could well be other reasons for the lack of both short-lived radioactivity and large isotopic anomalies (aside from ^{16}O excesses; Sahijpal et al., 1999) in these inclusions. Circumstantial arguments against a long time period for CAI formation are that it leads to problems with understanding the distribution of the oxygen isotope anomalies in nebular components (see Chapters 1.06–1.08; also McKeegan and Leshin, 2001) and with calculations of dynamical lifetimes of CAIs as independent objects in the nebula (Weidenschilling, 1977). Alternative explanations must invoke spatial heterogeneity within the nebula, either with respect to radionuclide distribution or CAI distribution, or both. It is beyond the scope of this review to critically assess models of turbulence and mixing in the solar nebula or evidence regarding the provenance of various CAI types; see Shu et al. (2001), Cuzzi et al. (2003), McKeegan et al. (2000a), Krot et al. (2002) and Alexander et al. (2001) for discussions.

Difficulties in interpreting an absence of ^{26}Al in some samples notwithstanding, on the basis of the good concordance of the Al–Mg and Pb–Pb systems the first order conclusion is that ^{26}Al/^{27}Al records do have chronological significance for most CAIs and chondrules. Taking the conventional (and reasonable) point of view that chondrules are nebular products, their formation ages relative to normal CAIs imply a duration of at least ~2–3 Myr for the solar nebula. Such a duration is plausible from an astrophysical viewpoint (Podosek and Cassen, 1994; Cameron, 1995), and it has interesting implications for timescales of accretion and radioactive heating of early-formed planetary bodies.

The nebular chronology inferred from initial ^{53}Mn/^{55}Mn (Figure 8) is not consistent with the Al–Mg and Pb–Pb systems either in terms of intervals or absolute ages (when Mn–Cr is anchored by the absolute Pb–Pb age of angrites). Because (^{53}Mn/^{55}Mn)$_0$ is poorly defined for CAIs (see above and discussion in Nyquist et al., 2001a), the inferred interval between CAI and chondrule formation is rather uncertain, but is at least 4 Myr, with a more likely minimum value of ~6 Myr (Nyquist et al., 2001a). The angrite-calibrated ^{53}Mn/^{55}Mn age of CAIs is too old by a minimum of 7 Myr compared to the measured Pb–Pb age, and chondrules are calculated to be ~1–2 Myr older than their measured Pb–Pb absolute age.

The discrepancies due to aberrantly old Mn–Cr ages of CAIs and chondrules were recognized by Lugmair and Shukolyukov (2001), who argued that a 4,571 Ma absolute age of the solar system, with (^{53}Mn/^{55}Mn)$_0$ = 1.4 × 10^{-5}, would resolve the difficulties. In this case, ^{53}Mn/^{55}Mn could not be used to date CAI formation. More significantly, this would imply that Pb–Pb ages of CAIs could not be crystallization ages but must (based on the time interval) represent metamorphic cooling times. A problem with such an interpretation is the apparently unique composition of initial lead in CAIs (Tera and Carlson, 1999), which could not be maintained in a parent body setting above the closure temperature for lead diffusion. Additionally, this interpretation (based on a model of chromium isotopic evolution in the solar nebula) runs counter to the good concordance of the Al–Mg system with Pb–Pb. At this time, it seems more reasonable to conclude that Mn–Cr does not provide a consistent high-resolution chronology for nebular events because one or more of the assumptions (initial homogeneity, isotopic closure, etc.) regarding the behavior of this short-lived chronometer is not satisfied within nebular components of chondrites.

A relatively long interval (>4 Myr) between CAIs and chondrules can be inferred on the basis of I–Xe dating (see Swindle et al., 1996 for a review). At face value, this might be seen as support for a Mn–Cr age for chondrule formation; however, in detail it does not work. The siting of ^{129}I is uncertain in both CAIs and chondrules and isotopic closure effects are evidenced by I–Xe apparent ages of chondrules that span an interval of up to several tens of millions of years, implicating asteroidal rather than nebular processes (e.g., Swindle et al., 1991).

1.16.6.2 Timescales of Planetesimal Accretion and Early Chemical Differentiation

Although the interpretation of apparent initial ^{53}Mn/^{55}Mn values in terms of a chronology for nebular fractionation events is problematic, the Mn–Cr system seems amenable to timing chemical fractionations associated with "geologic" activity on early-formed planetary bodies. A timescale is presented in the bottom panel of Figure 8, following the suggestion of Lugmair and Shukolyukov (1998) to utilize the angrites as a reference point to cross-calibrate the Mn–Cr and Pb–Pb systems. Thus, ^{53}Mn/^{55}Mn = 1.25 × 10^{-6} is tied to an absolute Pb–Pb age of 4,557.8 Ma. By this reckoning, the "global" differentiation of the HED parent body is pinned by the ensemble eucrite Mn–Cr isochron to 4,565 Ma. As mentioned previously, individual eucrites show internal Mn–Cr isochrons that indicate attainment of isotopic closure from just slightly after this time to significantly later, implying an extended (>10^7 yr) history of thermal activity on the HED asteroid. This is qualitatively in agreement with the young U–Pb ages of eucrites; however, a quantitative correlation between Mn–Cr and U–Pb ages is lacking (Tera and Carlson, 1999).

The ^{53}Mn–^{53}Cr isochron for the HED parent body is generally consistent with the timing of other indicators of early planetary processes. The Pb–Pb age for the oldest phosphates, from the least metamorphosed (H4) chondrites studied, postdates HED differentiation by ~2 Myr.

This is approximately equivalent to the Mn–Cr closure age for Chervony Kut, the noncumulate eucrite with the highest individual ^{53}Mn/^{55}Mn initial ratio. Other achondrites, including a pallasite and the unusual basaltic achondrite Acapulco, have Mn–Cr ages ~8–10 Myr after the HED differentiation event. These timescales are consistent with the notion that a variety of differentiated meteorites sample various depths in asteroids of various sizes during this early epoch following accretion.

A problem arises with the apparent chronology of aqueous activity on carbonaceous chondrite parent bodies. The formation time of fayalite is reasonable from the Mn–Cr point of view; however, carbonates from CM chondrites and from the unique chondrite Kaidun have ^{53}Mn/^{55}Mn initial values commensurate with those of chondrules. Although we have argued above that there are problems in understanding the temporal meaning of Mn–Cr systematics in CAIs and chondrules, we note that if ^{53}Mn–^{53}Cr can serve as an accurate chronometer at least for chondrules, then it implies that aqueous activity on some chondrite parent bodies was contemporaneous with chondrule formation elsewhere. The data that exist thus far are for carbonates from carbonaceous chondrites and for chondrules from ordinary chondrites. There are no data to suggest that chondrules from carbonaceous chondrites might be older than those from ordinary chondrites; in fact limited Al–Mg data could be interpreted to suggest the opposite (Kunihiro et al., 2003). Clearly, carbonates formed very early, but whether chondrule formation was still ongoing during this aqueous activity will have be decided by further study, preferably of chondrules and secondary minerals from the same meteorites.

The experimental record documenting the prior existence of ^{26}Al in differentiated meteorites is mostly based on very recent data, the majority of which have only been published in abstract form. Certainly the finding of ^{26}Mg* and of good Al–Mg isochrons in eucrites is consistent with the early age of igneous activity inferred from Mn–Cr systematics of bulk eucrites. This is strong confirmation that planetary scale melting began not more than a few million years following CAI crystallization, and quite possibly, while chondrule formation was still ongoing.

One very important issue has to do with the uniqueness of the timescale for formation of angrites. Recently, Nyquist et al. (2003) reported Al–Mg and Mn–Cr ages for two new angrites, D'Orbigny and Sahara 99555, which could represent different parent magmas from LEW and ADOR. Analyses of D'Orbigny yielded a good ^{53}Mn–^{53}Cr isochron with a slope higher by a factor ~2 than that of LEW 86010, implying formation at 4,561.1 ± 0.5 Ma based on the Pb–Pb age of LEW and ADOR (Nyquist et al., 2003). The same authors found ^{26}Mg* in plagioclase and constructed a two-point Al–Mg isochron that implies ^{26}Al/^{27}Al = (2.3 ± 0.8) × 10^{-7} at crystallization. This low value is reasonably close to that expected based on ^{26}Al decay from the canonical ratio given an absolute age 4,561 Ma. If these preliminary results hold, they will represent a new point of concordance between the Al–Mg and Mn–Cr systems in igneously differentiated meteorites.

1.16.7 CONCLUSIONS

Both chondrites and differentiated meteorites preserve records of short-lived radionuclides which are now extinct, but which were present when the solar system formed (Table 1). These isotopic records yield information on the amount of radioactivity contained by ancient solar system minerals, from which the relative timing of chemical fractionations between parent and daughter elements can be inferred (assuming that the short-lived radionuclides were originally distributed homogenously). The fractionation events can often be related to thermal processes occurring in the solar nebula or on early-accreted planetesimals, thus allowing a high-resolution relative chronology to be delineated (Figure 8).

The existence of both ^{10}Be and ^{60}Fe in various early solar system materials provides strong evidence for a multiplicity of sources for short-lived isotopes. The former is most probably a result of local production by energetic particle irradiation, perhaps near the forming Sun, whereas the latter is evidence for seeding of the solar nebula by freshly synthesized stellar ejecta. In principle, the inventory of other radioisotopes may contain contributions from both these sources in addition to other nondiscrete ("background") sources such as galactic stellar nucleosynthesis or spallogenic nuclear reactions in the protosolar molecular cloud. However, correlations of radiogenic isotope signatures in CAIs and hibonite grains indicate that spallogenic contributions to the abundances of the shortest-lived isotopes, ^{41}Ca and ^{26}Al, are minor and that these refractory isotopes arrived together in the solar nebula.

1.16.7.1 Implications for Solar Nebula Origin and Evolution

The short lifetimes of ^{26}Al and, especially, ^{41}Ca, coupled with the evidence for an external origin of these nuclides, have important implications for the origin of the solar system. Based on estimated production rates and isotope mixing during interstellar transit and injection into the solar system, a duration of at most ~1 Myr can be accommodated for the total time between nucleosynthetic production and incorporation of these isotopes into crystalline solids in the early solar system. Such a rapid timescale implies a triggering mechanism for fragmentation and collapse of a portion of the presolar molecular cloud to the form the early Sun and its accretion disk. Although it is known that many AGB stars contributed dust to the early solar nebula (see Chapter 1.02) and that a wind from such a star could theoretically provide a sufficient shock to initiate collapse, astrophysical considerations of stellar lifetimes suggest a nearby type II supernova as a more likely trigger.

Supernovae can be the source of most of the short-lived radionuclides (except ^{10}Be); however there are difficulties in reconciling relative abundances of all species with a single event (see review by Goswami and Vanhala, 2000). While this may be aesthetically desirable, it is not required, especially for the longer-lived isotopes of Table 1. Other evidence indicates that it is probably not correct and that the truth is more complex than a single supernova triggering and injection. The "last" supernova is not the source of large stable isotope anomalies in oxygen, calcium, or titanium, demonstrating that isotopic memories of other presolar components survived to be incorporated into early solar system minerals. Additionally, the evidence for pervasive ^{10}Be signatures in CAIs and the abundant astronomical evidence for copious X-ray activity of YSO indicates that early-formed solar system materials were most likely strongly irradiated if they were not shielded. Further work is required to quantitatively assess the proportion of those radionuclides (besides ^{26}Al) that were produced locally by solar energetic particles.

Cross-calibration of the initial ^{26}Al/^{27}Al records inferred for nebular components of chondrites with the absolute Pb–Pb ages of CAIs results in a self-consistent high-resolution chronology for the high temperature phases of solar nebula evolution. A plausible scenario and timeline can be constructed:

(i) at nearly 4,568 Ma, a shock wave, probably initiated by a "nearby" supernova, triggers fragmentation and gravitational collapse of a portion of a molecular cloud;

(ii) near the central, hot regions of the nebula the first refractory minerals form by evaporation and/or recondensation and melting of mixtures of presolar dust grains from various interstellar heritages; these hibonite grains and FUN inclusions incorporate ^{10}Be produced by irradiation of the dust grains by solar energetic particles, but they do not sample the radioactivity accompanying the supernova shock wave;

(iii) shortly afterward, at ~4,567 Ma, the fresh radioactivity arrives in the inner nebula and most CAIs form over a short interval incorporating ^{26}Al and ^{41}Ca at the "canonical" levels;

(iv) high temperature processing of some CAIs continues for a few hundred thousand years, but most of those that do not accrete to the Sun are removed from high temperature regions of the nebula, perhaps by entrainment in bipolar outflows, and survive for a long period of time in undetermined nebular locations;

(v) at ~4,566 Ma, chondrule formation begins and continues for ~1–2 Myr; CAIs are largely absent from the nebular regions where chondrule melting occurs; and

(vi) at ~4,565–4,564 CAIs have joined chondrules and nebular dust in accreting to planetesimals in the asteroid belt. If the latter process is considered as the termination of the nebular phase of solar system evolution, then its lifetime is ~4 Ma as recorded by radionuclides in nebular materials.

The timescales for accretion and early evolution of these planetesimals are also constrained by short-lived radioactivity. This record is best elucidated with the ^{53}Mn–^{53}Cr isotopic system, even though as discussed previously the record of ^{53}Mn/^{55}Mn in solar nebula objects does not yield a consistently interpretable chronology. Accretion of some planetesimals started very early, perhaps even while chondrule formation was ongoing. By ~4,565–4,564, large scale melting and differentiation occurred on the HED parent body, most likely the asteroid 4 Vesta. Some eucrites crystallized soon after mantle differentiation, quickly cooling through isotopic closure for magnesium and chromium by ~4,564–4,563 Ma. Energy from ^{26}Al and ^{60}Fe decay probably contributed substantially to the heat required for melting, but the asteroid was large enough that igneous activity continued for several tens of million years. Some angrites appear to have erupted early, cooling by ~4,561 Ma, but ADOR and LEW did not crystallize until 4,558 Ma. Other asteroidal bodies, from which chondrites are derived, either accreted somewhat later than Vesta or remained as relatively small bodies for several million years. Absolute Pb–Pb ages of phosphates indicate that metamorphic temperatures were reached on some ordinary chondrite asteroids by ~4,563 Ma; this timescale is consistent with the ^{26}Al/^{27}Al records of chondrules. Metamorphism on chondrite parent bodies continued for up to tens of millions of years as indicated by Pb–Pb and I–Xe dating. Aqueous activity (formation of carbonate) happened very early, perhaps "too" early, on the parent asteroids of some carbonaceous chondrites. Calibration of the ^{53}Mn–^{53}Cr chronometer by the Pb–Pb age of angrites implies formation of the earliest of these carbonates by ~4,567 Ma, which is not compatible with the nebular chronology discussed above. Accretion and differentiation of planetary embryos continued from this early epoch for a period of several tens of millions of years (see Chapters 1.17 and 1.20).

1.16.7.2 Future Directions

The quantitative comparison of various short-lived radionuclide systems with each other and with Pb–Pb chronology has only been made possible by new data obtained during the last decade, or in many cases, the last few years. Over this same time period, evidence for the decay of several important new short-lived isotopes in the early solar system has been discovered. The record of now-extinct isotopes in early solar system materials is becoming sufficiently well defined to allow construction of a plausible timeline and scenario for solar system origin. However, even though broad areas of consistency have been revealed, there are significant problems that will require further investigation. One of the most important is trying to understand the role of energetic particle irradiation in the early solar system. Energetic processes associated with magnetic flare activity of the young Sun almost certainly occurred; the question is what effect these had on isotopic and mineralogical records of early-formed solar system rocks. Could solar system irradiation be responsible for some of the confusion of the nebular record of ^{53}Mn/^{55}Mn? There appears to be large-scale inhomogeneity in the ^{53}Mn–^{53}Cr systematics: could some of this be explicable in terms of solar system production and/or large-scale radial transport of nebular components?

It has been recently hypothesized (Desch et al., 2003) that ^{10}Be may result from magnetic trapping of cosmic radiation in molecular cloud material, such that all short-lived nuclides predate solar system formation. However, little attention has so far been paid to the role of magnetic fields in triggered collapse mechanisms. It is clear that magnetic pressure cannot substantially inhibit collapse, otherwise the delay would cause extinction of the signal of ^{41}Ca in CAIs. The correlation of ^{41}Ca with ^{26}Al needs to be better quantified, and even the canonical ^{26}Al record more closely examined to sort out the intrinsic dispersion in the distribution from the effects of secondary heating and alteration of CAI minerals. As it stands, the duration of CAI production seems implausibly short compared to CAI longevity in the nebula, but this is largely a model dependent result. A better understanding of the locales and formation mechanisms of CAIs and chondrules, and their relationships to each other, will help in constraining such models. Finally, it can be anticipated that in the near future much more data will be gathered by *in situ* methods and high-precision bulk methods that will greatly improve our knowledge of the distributions of ^{10}Be and ^{60}Fe in a wide of early materials. So far, these isotopes have been primarily exploited as semiquantitative indicators of process; perhaps with a more robust data set, it will be possible to employ them as further chronological tools for understanding solar nebula origin and evolution.

ACKNOWLEDGMENTS

We have benefited from intensely fun discussions about this subject with many colleagues, and although it is impossible to mention everyone we would especially like to thank M. Chaussidon, D. Clayton, J. Goswami, M. Gounelle, I. Hutcheon, T. Ireland, A. Krot, T. Lee, L. Leshin, G. Lugmair, G. MacPherson, S. Russell, S. Sahijpal, F. Shu, G. Srinivasan, M. Wadhwa, G. J. Wasserburg, E. Young, and E. Zinner. We acknowledge them and many other colleagues working in this area for their generosity and insight, and would especially like to thank Gary Huss, Ian Hutcheon, and Jerry Wasserburg for critically helpful reviews that significantly improved this manuscript. We also thank Y. Amelin, G. Huss, G. J. MacPherson, and S. Tachibana for providing figures and data used herein.

REFERENCES

Alexander C. M. O'D., Boss A. P., and Carlson R. W. (2001) The early evolution of the inner solar system: a meteoritic perspective. *Science* **293**, 64–68.

Allègre C. J., Manhès G., and Göpel C. (1995) The age of the Earth. *Geochim. Cosmochim. Acta* **59**, 1445–1456.

Amelin Y., Krot A. N., Hutcheon I. D., and Ulyanov A. A. (2002) Lead isotopic ages of chondrules and calcium–aluminum-rich inclusions. *Science* **297**, 1678–1683.

André P., Ward-Thompson W., and Barsony M. (2000) From prestellar cores to protostars: the initial conditions of star formation. In *Protostars and Planets IV* (eds. V. Mannings, A. P. Boss, and S. S. Russell). University of Arizona Press, Tucson, pp. 59–96.

Begemann F. (1980) Isotopic anomalies in meteorites. *Rep. Prog. Phys.* **43**, 1309–1356.

Binzel R. P. and Xu S. (1993) Chips off of asteroid 4 Vesta: evidence for the parent body of basaltic achondrite meteorites. *Science* **260**, 186–191.

Birck J. L. and Allègre C. J. (1985) Evidence for the presence of Mn-53 in the early solar-system. *Geophys. Res. Lett.* **12**, 745–748.

Birck J. L. and Allègre C. J. (1988) Manganese chromium isotope systematics and the development of the early solar-system. *Nature* **331**, 579–584.

Birck J. L. and Lugmair G. W. (1988) Nickel and chromium isotopes in Allende inclusions. *Earth Planet. Sci. Lett.* **90**, 131–143.

Birck J. L., Rotaru M., and Allègre C. J. (1999) ^{53}Mn–^{53}Cr evolution of the early solar system. *Geochim. Cosmochim. Acta* **63**, 4111–4117.

Boss A. P. and Vanhala H. A. T. (2001) Injection of newly synthesized elements into the protosolar cloud. *Phil. Trans. Roy. Soc. London* **A359**, 2005–2016.

Brazzle R. H., Pravdivtseva O. V., Meshik A. P., and Hohenberg C. M. (1999) Verification and interpretation of the I–Xe chronometer. *Geochim. Cosmochim. Acta* **63**, 739–760.

Caillet C., MacPherson G. J., and Zinner E. K. (1993) Petrologic and Al–Mg isotopic clues to the accretion of two refractory inclusions onto the Leoville parent body: one was hot, the other wasn't. *Geochim. Cosmochim. Acta* **57**, 4725–4743.

Cameron A. G. W. (1995) The first ten million years in the solar nebula. *Meteorit. Planet. Sci.* **30**, 133–161.

Cameron A. G. W. (2001a) Extinct radioactivities, core-collapse supernovas, jets, and the r-process. *Nucl. Phys. A* **688**, 289C–296C.

Cameron A. G. W. (2001b) Some properties of r-process accretion disks and jets. *Astrophys. J.* **562**, 456–469.

Cameron A. G. W. and Truran J. W. (1977) The supernova trigger for formation of the solar system. *Icarus* **30**, 447–461.

Cameron A. G. W., Höflich P., Myers P. C., and Clayton D. D. (1995) Massive supernovae, Orion gamma rays, and the formation of the solar system. *Astrophys. J.* **447**, L53–L57.

Carlson R. W. and Hauri E. (2001) Extending the ^{107}Pd–^{107}Ag chronometer to low Pd/Ag meteorites with MC–ICP–MS. *Geochim. Cosmochim. Acta* **65**, 1839–1848.

Chaussidon M., Robert F., Mangin D., Hanon P., and Rose E. F. (1997) Analytical procedures for the measurement of boron isotope compositions by ion microprobe in meteorites and mantle rocks. *Geostand. Newslett.* **21**, 7–17.

Chaussidon M., Robert F., and McKeegan K. D. (2002) Incorporation of short-lived ^{7}Be in one CAI from the Allende meteorite. In *Lunar Planet. Sci.* **XXXIII**, #1563. The Lunar and Planetary Institute, Houston (CD-ROM).

Chen J. H. and Wasserburg G. J. (1980) A search for isotopic anomalies in uranium. *Geophys. Res. Lett.* **7**, 275–278.

Chen J. H. and Wasserburg G. J. (1981) The isotopic composition of uranium and lead in Allende inclusions and meteorite phosphates. *Earth Planet. Sci. Lett.* **52**, 1–15.

Chen J. H. and Wasserburg G. J. (1987) A search for evidence of extinct lead 205 in iron meteorites. In *Lunar Planet. Sci.* **XVIII**, 165–166. The Lunar and Planetary Institute, Houston (CD-ROM).

Chen J. H. and Wasserburg G. J. (1996) Live ^{107}Pd in the early solar system and implications for planetary evolution. In *Earth Processes: Reading the Isotopic Code*, Geophysical Monograph 95 (eds. A. Basu and S. Hart). American Geophysical Union, Washington, pp. 1–20.

Chen J. H. and Wasserburg G. J. (1990) The isotopic composition of Ag in meteorites and the presence of ^{107}Pd in protoplanets. *Geochim. Cosmochim. Acta* **54**, 1729–1743.

Chen J. H., Papanastassiou D. A., and Wasserburg G. J. (2002) Re–Os and Pd–Ag systematics in group IIIAB irons and in pallasites. *Geochem. Cosmochim. Acta* **66**, 3793–3810.

Clayton D. D. (1982) Cosmic chemical memory—a new astronomy. *Quart. J. Roy. Astron. Soc.* **23**, 174–212.

Clayton D. D. (1986) Interstellar fossil ^{26}Mg and its possible relation to excess meteoritic ^{26}Mg. *Astrophys. J.* **310**, 490–498.

Clayton D. D., Dwek E., and Woosley S. E. (1977) Isotopic anomalies and proton irradiation in the early solar system. *Astrophys. J.* **214**, 300–315.

Clayton R. N. and Mayeda T. K. (1984) The oxygen isotope record in Murchison and other carbonaceous chondrites. *Earth Planet. Sci. Lett.* **67**, 151–161.

Cuzzi J. N., Davis S. S., and Dobrovolskis A. R. (2003) Creation and distribution of CAIs in the protoplanetary nebula. In *Lunar Planet. Sci.* **XXXIV**, #1749. The Lunar and Planetary Institute, Houston (CD-ROM).

Desch S. J., Srinivasan G., and Connolly H. C. (2003) An interstellar origin for the beryllium 10 in CAIs. In *Lunar Planet. Sci.* **XXXIV**, #1394. The Lunar and Planetary Institute, Houston (CD-ROM).

Endress M., Zinner E., and Bischoff A. (1996) Early aqueous activity on primitive meteorite parent bodies. *Nature* **379**, 701–703.

Fahey A. J., Goswami J. N., McKeegan K. D., and Zinner E. (1987) ^{26}Al, ^{244}Pu, ^{50}Ti, REE, and trace element abundances in hibonite grains from CM and CV meteorites. *Geochim. Cosmochim. Acta* **51**, 329–350.

Feigelson E. D., Broos P., Gaffney J. A., Garmire G., Hillenbrand L. A., Pravdo S. H., Townsley L., and Tsuboi Y. (2002a) X-ray emitting young stars in the Orion Nebula. *Astrophys. J.* **574**, 258–292.

Feigelson E. D., Garmire G. P., and Pravdo S. H. (2002b) Magnetic flaring in the pre main-sequence sun and implications for the early solar system. *Astrophys. J.* **572**, 335–349.

Foster P. N. and Boss A. P. (1997) Injection of radioactive nuclides from the stellar source that triggered the collapse of the presolar nebula. *Astrophys. J.* **489**, 346–357.

Galer S. J. G. and Lugmair G. W. (1996) Lead isotope systematics of non-cumulate eucrites. *Meteoritics* **31**, A47–A48.

Galy A., Young E. D., Ash R. D., and O'Nions R. K. (2000) The formation of chondrules at high gas pressures in the solar nebula. *Science* **290**, 1751–1753.

Gilmour J. D. (2000) The extinct radionuclide timescale of the solar system. *Space Sci. Rev.* **192**, 123–132.

Gilmour J. D. and Saxton J. M. (2001) A time-scale of formation of the first solids. *Phil. Trans. Roy. Soc. London* **A359**, 2037–2048.

Göpel C., Manhès G., and Allègre C. J. (1994) U–Pb systematics of phosphates from equilibrated ordinary chondrites. *Earth Planet. Sci. Lett.* **121**, 153–171.

Goswami J. N. and Vanhala H. A. T. (2000) Extinct radionuclides and the origin of the solar system. In *Protostars and Planets IV* (eds. V. Mannings, A. P. Boss, and S. S. Russell). University of Arizona Press, Tucson, pp. 963–994.

Goswami J. N., Marhas K. K., and Sahijpal S. (2001) Did solar energetic particles produce the short-lived nuclides present in the early solar system? *Astrophys. J.* **549**, 1151–1159.

Gounelle M., Shu F. H., Shang H., Glassgold A. E., Rehm K. E., and Lee T. (2001) Extinct radioactivities and protosolar cosmic rays: self-shielding and light elements. *Astrophys. J.* **548**, 1051–1070.

Gray C. M. and Compston W. (1974) Excess ^{26}Mg in the Allende meteorite. *Nature* **251**, 495–497.

Grimm R. E. and McSween H. Y. (1994) Heliocentric zoning of the asteroid belt by aluminum-26 heating. *Science* **259**, 653–655.

Guan Y., Huss G. R., MacPherson G. J., and Wasserburg G. J. (2000) Calcium–aluminum-rich inclusions from enstatite chondrites: indigenous or foreign? *Science* **289**, 1330–1333.

Halliday A. and Lee D. C. (1999) Tungsten isotopes and the early development of the Earth and Moon. *Geochim. Cosmochim. Acta* **63**, 4157–4179.

Harper C. L., Jr. (1996) Evidence for ^{92g}Nb in the early solar system and evaluation of a new p-process cosmochronometer from ^{92g}Nb/^{92}Mo. *Astrophys. J.* **466**, 437–456.

Heymann D. and Dziczkaniec M. (1976) Early irradiation of matter in the solar system: magnesium (proton, neutron) scheme. *Science* **191**, 79–81.

Hsu W., Huss G. R., and Wasserburg G. J. (2003) Al–Mg systematics of CAIs, POI, and ferromagnesian chondrules from Ningqiang. *Meteorit. Planet. Sci.* **38**, 35–48.

Hsu W. B., Wasserburg G. J., and Huss G. R. (2000) High time resolution by use of the Al-26 chronometer in the multistage formation of a CAI. *Earth Planet. Sci. Lett.* **182**, 15–29.

Hua J., Huss G. R., and Sharp T. G. (2002) ^{53}Mn–^{53}Cr dating of fayalite formation in the Kaba CV3 carbonaceous chondrite. In *Lunar Planet. Sci.* **XXXIII**, #1660. The Lunar and Planetary Institute, Houston (CD-ROM).

Hudson G. B., Kennedy B. M., Podosek F. A., and Hohenberg C. M. (1988) The early solar system abundance of ^{244}Pu as inferred from the St. Severin chondrite. *Proc. 19th Lunar Planet. Sci. Conf.* 547–557.

Huss G. R., MacPherson G. J., Wasserburg G. J., Russell S. S., and Srinivasan G. (2001) Aluminum-26 in calcium–aluminum-rich inclusions and chondrules from unequilibrated ordinary chondrites. *Meteorit. Planet. Sci.* **36**, 975–997.

Hutcheon I. D. and Hutchison R. (1989) Evidence from the Semarkona ordinary chondrite for ^{26}Al heating of small planets. *Nature* **337**, 238–241.

Hutcheon I. D. and Jones R. H. (1995) The ^{26}Al–^{26}Mg record of chondrules: clues to nebular chronology. In *Lunar Planet. Sci.* **XXVI**. The Lunar and Planetary Institute, Houston, pp. 647–648.

Hutcheon I. D. and Phinney D. L. (1996) Radiogenic ^{53}Cr* in Orgueil carbonates: chronology of aqueous activity on the CI parent body. In *Lunar Planet. Sci.* **XXVII**. The Lunar and Planetary Institute, Houston, pp. 577–578.

Hutcheon I. D., Armstrong J. T., and Wasserburg G. J. (1984) Excess ^{41}K in Allende CAI: confirmation of a hint. In *Lunar Planet. Sci.* **XV**. The Lunar and Planetary Institute, Houston, pp. 387–388.

Hutcheon I. D., Huss G. R., and Wasserburg G. J. (1994) A search for ^{26}Al in chondrites: chondrule formation times. In *Lunar Planet. Sci.* **XVII**. The Lunar and Planetary Institute, Houston, pp. 587–588.

Hutcheon I. D., Krot A. N., Keil K., Phinney D. L., and Scott E. R. D. (1998) ^{53}Mn–^{53}Cr dating of fayalite formation in the CV3 chondrite Mokoia: evidence for asteroidal alteration. *Science* **282**, 1865–1867.

Hutcheon I. D., Browning L., Keil K., Krot A. N., Phinney D. L., Prinz M., and Weisberg M. K. (1999a) Timescale of aqueous activity in the early solar system. In *Ninth Goldschmidt Conf. Abstr.*, #971, Lunar and Planetary Institute, Houston (CD-ROM).

Hutcheon I. D., Weisberg M. K., Phinney D. L., Zolensky M. E., Prinz M., and Ivanov A. V. (1999b) Radiogenic ^{53}Cr in Kaidun carbonates: evidence for very early aqueous activity. In *Lunar Planet. Sci.* **XXX**, #1722. The Lunar and Planetary Institute, Houston (CD-ROM).

Ireland T. R. (1988) Correlated morphological, chemical, and isotopic characteristics of hibonites from the Murchison carbonaceous chondrite. *Geochim. Cosmochim. Acta* **52**, 2827–2839.

Ireland T. R. (1990) Presolar isotopic and chemical signatures in hibonite-bearing refractory inclusions from the Murchison carbonaceous chondrite. *Geochim. Cosmochim. Acta* **54**, 3219–3237.

Ireland T. R., Compston W., and Heydegger H. R. (1985) Titanium isotopic anomalies in hibonites from the Murchison carbonaceous chondrite. *Geochim. Cosmochim. Acta* **49**, 1989–1993.

Jeffery P. M. and Reynolds J. H. (1961) Origin of excess Xe^{129} in stone meteorites. *J. Geophys. Res.* **66**, 3582–3583.

Kaiser T. and Wasserburg G. J. (1983) The isotopic composition and concentration of Ag in iron meteorites. *Geochim. Cosmochim. Acta* **47**, 43–58.

Kelly W. R. and Wasserburg G. J. (1978) Evidence for the existence of ^{107}Pd in the early solar system. *Geophys. Res. Lett.* **5**, 1079–1082.

Kita N. T., Nagahara H., Togashi S., and Morshita Y. (2000) A short duration of chondrule formation in the solar nebula: evidence from ^{26}Al in Semarkona ferromagnesian chondrules. *Geochim. Cosmochim. Acta* **64**, 3913–3922.

Kita N. T., Ikeda Y., Shimoda H., Morshita Y., and Togashi S. (2003) Timing of basaltic volcanism in ureilite parent body inferred from the ^{26}Al ages of plagioclase-bearing clasts in DAG-319 polymict ureilite. In *Lunar Planet. Sci.* **XXXIV**, #1557. The Lunar and Planetary Institute, Houston (CD-ROM).

Kleine T., Munker C., Mezger K., and Palme H. (2002a) Rapid accretion and early core formation on asteroids and the terrestrial planets from Hf–W chronometry. *Nature* **418**, 952–955.

Kleine T., Munker C., Mezger K., Palme H., and Bischoff A. (2002b) Revised Hf–W ages for core formation in planetary bodies. *Geochim. Cosmochim. Acta* **66**, A404–A404.

Krot A. N., McKeegan K. D., Leshin L. A., MacPherson G. J., and Scott E. R. D. (2002) Existence of an ^{16}O-rich gaseous reservoir in the solar nebula. *Science* **295**, 1051–1054.

Kunihiro T., Rubin A. E., McKeegan K. D., and Wasson J. T. (2003) Initial ^{26}Al/^{27}Al ratios in carbonaceous chondrite chondrules. In *Lunar Planet. Sci.* **XXXIV**, #2124. The Lunar and Planetary Institute, Houston (CD-ROM).

LaTourrette T. and Wasserburg G. J. (1998) Mg diffusion in anorthite: implications for the formation of early solar system planetesimals. *Earth Planet. Sci. Lett.* **158**, 91–108.

Lee D. C. and Halliday A. N. (1996) Hf–W isotopic evidence for rapid accretion and differentiation in the early solar system. *Science* **274**, 1876–1879.

Lee T. (1978) A local proton irradiation model for isotopic anomalies in the solar system. *Astrophys. J.* **224**, 217–226.

Lee T. and Papanastassiou D. A. (1974) Mg isotopic anomalies in the Allende meteorite and correlation with O and Sr effects. *Geophys. Res. Lett.* **1**, 225–228.

Lee T., Papanastassiou D. A., and Wasserburg G. J. (1976) Demonstration of ^{26}Mg excess in Allende and evidence for ^{26}Al. *Geophys. Res. Lett.* **3**, 109–112.

Lee T., Papanastassiou D. A., and Wasserburg G. J. (1977) Aluminum-26 in the early solar system: fossil or fuel? *Astrophys. J.* **211**, L107–L110.

Lee T., Russell W. A., and Wasserburg G. J. (1978) Calcium isotopic anomalies and the lack of aluminum-26 in an unusual Allende inclusion. *Astrophys. J.* **228**, L93–L98.

Lee T., Mayeda T. K., and Clayton R. N. (1980) Oxygen isotopic anomalies in Allende inclusion HAL. *Geophys. Res. Lett.* **7**, 493–496.

Lee T., Shu F. H., Shang H., Glassgold A. E., and Rehm K. E. (1998) Protostellar cosmic rays and extinct radioactivities in meteorites. *Astrophys. J.* **506**, 898–912.

Leya I., Wieler R., and Halliday A. N. (2002) Nucleosynthesis by spallation reactions in the early solar system. *Meteorit. Planet. Sci.* **37**, A86.

Lugmair G. W. and Galer S. J. G. (1992) Age and isotopic relationships among the angrites Lewis Cliff 86010 and Angra dos Reis. *Geochim. Cosmochim. Acta* **56**, 1673–1694.

Lugmair G. W. and Shukolyukov A. (1998) Early solar system timescales according to ^{53}Mn–^{53}Cr systematics. *Geochim. Cosmochim. Acta* **62**, 2863–2886.

Lugmair G. W. and Shukolyukov A. (2001) Early solar system events and timescales. *Meteorit. Planet. Sci.* **36**, 1017–1026.

Lugmair G. W., Shimamura T., Lewis R. S., and Anders E. (1983) Samarium-146 in the early solar system: evidence from neodymium in the Allende meteorite. *Science* **222**, 1015–1018.

MacPherson G. J. and Davis A. M. (1993) A petrologic and ion microprobe study of a Vigarano type B refractory inclusion: evolution by multiple stages of alteration and melting. *Geochim. Cosmochim. Acta* **57**, 231–243.

MacPherson G. J. and Huss G. R. (2001) Extinct ^{10}Be in CAIs from Vigarano, Leoville, and Axtell. In *Lunar Planet. Sci.* **XXXII**, #1882. The Lunar and Planetary Institute, Houston (CD-ROM).

MacPherson G. J., Davis A. M., and Zinner E. K. (1995) The distribution of aluminum-26 in the early Solar System—a reappraisal. *Meteoritics* **30**, 365–386.

MacPherson G. J., Huss G. R., and Davis A. M. (2003) Extinct ^{10}Be in type A calcium–aluminum-rich inclusions from CV chondrites. *Geochim. Cosmochim. Acta* **67**, 3165–3179.

Marhas K. K. and Goswami J. N. (2003) Be–B systematics in CM and CV hibonites: implications for solar energetic particle production of short-lived nuclides in the early solar system. In *Lunar Planet. Sci.* **XXXIV**, #1303. The Lunar and Planetary Institute, Houston (CD-ROM).

Marhas K. K., Goswami J. N., and Davis A. M. (2002) Short-lived nuclides in hibonite grains from Murchison: evidence for solar system evolution. *Science* **298**, 2182–2185.

McKeegan K. D. and Leshin L. A. (2001) Stable isotope variations in extraterrestrial materials. *Rev. Mineral. Geochem.* **43**, 279–318.

McKeegan K. D., Chaussidon M., and Robert F. (2000a) Incorporation of short-lived Be-10 in a calcium–aluminum-rich inclusion from the Allende meteorite. *Science* **289**, 1334–1337.

McKeegan K. D., Greenwood J. P., Leshin L. A., and Cosarinsky M. (2000b) Abundance of ^{26}Al in ferromagnesian chondrules of unequilibrated ordinary chondrites. In *Lunar Planet. Sci.* **XXXI**, #2009. The Lunar and Planetary Institute, Houston (CD-ROM).

McKeegan K. D., Chaussidon M., Krot A. N., Robert F., Goswami J. N., and Hutcheon I. D. (2001) Extinct radionuclide abundances in Ca, Al-rich inclusions from the CV chondrites Allende and Efremovka: a search for synchronicity. In *Lunar Planet. Sci.* **XXXII**, #2175. The Lunar and Planetary Institute, Houston (CD-ROM).

Meyer B. and Clayton D. D. (2000) Short-lived radioactivities and the birth of the Sun. *Space Sci. Rev.* **92**, 133–152.

Mostefaoui S., Kita N. T., Togashi S., Tachibana S., Nagahara H., and Morishita Y. (2002) The relative formation ages of ferromagnesian chondrules inferred from their initial aluminum-26/aluminum-27 ratios. *Meteorit. Planet. Sci.* **37**, 421–438.

Mostefaoui S., Lugmair G. W., Hoppe P., and El Goresy A. (2003) Evidence for live iron-60 in Semarkona and Chervony Kut: a NanoSIMS study. In *Lunar Planet. Sci.* **XXXIV**, #1585. The Lunar and Planetary Institute, Houston (CD-ROM).

Münker C., Weyer S., Mezger K., Rehkämper M., Wombacher F., and Bischoff A. (2000) ^{92}Nb–^{92}Zr and the early differentiation history of planetary bodies. *Science* **289**, 1538–1542.

Murty S. V. S., Goswami J. N., and Shukolyukov Y. A. (1997) Excess Ar-36 in the Efremovka meteorite: a strong hint for the presence of Cl-36 in the early solar system. *Astrophys. J.* **475**, L65–L68.

Niederer F. R., Papanastassiou D. A., and Wasserburg G. J. (1980) Endemic isotopic anomalies in titanium. *Astrophys. J.* **240**, L73–L77.

Niemeyer S. and Lugmair G. W. (1981) Ubiquitous isotopic anomalies in Ti from normal Allende inclusions. *Earth Planet. Sci. Lett.* **53**, 211–225.

Nyquist L., Lindstrom D., Mittlefehldt D., Shih C. Y., Wiesmann H., Wentworth S., and Martinez R. (2001a) Manganese–chromium formation intervals for chondrules from the Bishunpur and Chainpur meteorites. *Meteorit. Planet. Sci.* **36**, 911–938.

Nyquist L. E., Reese Y., Wiesmann H., Shih C.-Y., and Takeda H. (2001b) Live ^{53}Mn and ^{26}Al in an unique cumulate eucrite with very calcic feldspar (An-98). *Meteorit. Planet. Sci.* **36**, A151–A152.

Nyquist L. E., Shih C.-Y., Wiesmann H., and Mikouchi T. (2003) Fossil ^{26}Mg and ^{53}Mn in D'Orbigny and Sahara 99555 and the timescale for angrite magmatism. In *Lunar Planet. Sci.* **XXXIV**, #1338. The Lunar and Planetary Institute, Houston (CD-ROM).

Papanastassiou D. A. (1986) Chromium isotopic anomalies in the Allende meteorite. *Astrophys. J.* **308**, L27–L30.

Papanastassiou D. A., Bogdanovski O., and Wasserburg G. J. (2002) ^{53}Mn–^{53}Cr systematics in Allende refractory inclusions. *Meteorit. Planet. Sci.* **37**, A114.

Podosek and Cassen (1994) Theoretical, observational, and isotopic estimates of the lifetime of the solar nebula. *Meteoritics* **29**, 6–25.

Podosek F. A. and Nichols R. H., Jr. (1997) Short-lived radionuclides in the solar nebula. In *Astrophysical Implications of the Laboratory Study of Presolar Materials* (eds. T. J. Bernatowicz and E. Zinner). American Institute of Physics, Woodbury, pp. 617–647.

Podosek F. A. and Swindle T. D. (1988) Extinct radionuclides. In *Meteorites and the Early Solar System* (eds. J. F. Kerridge and M. S. Matthews). University of Arizona Press, Tucson, pp. 1093–1113.

Podosek F. A., Zinner E., MacPherson G. J., Lundberg L. L., Brannon J. C., and Fahey A. J. (1991) Correlated study of initial ^{87}Sr/^{86}Sr and Al–Mg isotopic systematics and petrologic properties in a suite of refractory inclusions from the Allende meteorite. *Geochim. Cosmochim. Acta* **55**, 1083–1110.

Prinzhofer A., Papanastassiou D. A., and Wasserburg G. J. (1989) The presence of ^{146}Sm in the early solar system and implications for its nucleosynthesis. *Astrophys. J.* **344**, 81–84.

Russell S. S., Srinivasan G., Huss G. R., Wasserburg G. J., and MacPherson G. J. (1996) Evidence for widespread ^{26}Al in the solar nebula and constraints for nebula timescales. *Science* **273**, 757–762.

Russell S. S., Gounelle M., and Hutchison R. (2001) Origin of short-lived radionuclides. *Phil. Trans. Roy. Soc. London* **359**, 1991–2004.

Sahijpal S. and Goswami J. N. (1998) Refractory phases in primitive meteorites devoid of ^{26}Al and ^{41}Ca: representative samples of first solar system solids? *Astrophys. J.* **509**, L137–L140.

Sahijpal S., Goswami J. N., Davis A. M., Grossman L., and Lewis R. S. (1998) A stellar origin for the short-lived nuclides in the early solar system. *Nature* **391**, 559.

Sahijpal S., McKeegan K. D., Krot A. N., Weber D., and Ulyanov A. A. (1999) Oxygen isotopic compositions of Ca–Al-rich inclusions from the CH chondrites, Acfer 182 and Pat91546. *Meteorit. Planet. Sci.* **34**, A101.

Sahijpal S., Goswami J. N., and Davis A. M. (2000) K, Mg, Ti, and Ca isotopic compositions and refractory trace element abundances in hibonites from CM and CV meteorites: implications for early solar system processes. *Geochim. Cosmochim. Acta* **64**, 1989–2005.

Sanloup C., Blicher-Toft J., Télouk P., Gillet P., and Albarède F. (2000) Zr isotope anomalies in chondrites and the presence of ^{92}Nb in the early solar system. *Earth Planet. Sci. Lett.* **184**, 75–81.

Schönbachler M., Rehkämper M., Halliday A. N., Lee D.-C., Bourot-Denise M., Zanda B., Hattendorf B., and Günther D. (2002) Niobium–zirconium chronometry and early solar system development. *Science* **295**, 1705–1708.

Shu F. H., Adams F. C., and Lizano S. (1987) Star formation in molecular clouds: observation and theory. *Ann. Rev. Astron. Astrophys.* **25**, 23–81.

Schramm D. N., Tera F., and Wasserburg G. J. (1970) The isotopic abundance of ^{26}Mg and limits on ^{26}Al in the early solar system. *Earth Planet. Sci. Lett.* **10**, 44–59.

Shu F. H., Shang H., Gounelle M., Glassgold A. E., and Lee T. (2001) The origin of chondrules and refractory inclusions in chondritic meteorites. *Astrophys. J.* **548**, 1029–1050.

Shukolyukov A. and Lugmair G. W. (1993a) Fe-60 in eucrites. *Earth Planet. Sci. Lett.* **119**, 159–166.

Shukolyukov A. and Lugmair G. W. (1993b) Live iron-60 in the early solar system. *Science* **259**, 1138–1142.

Srinivasan G. (2001) Be–B isotope systematics in CAI E65 from Efremovka CV3 chondrite. *Meteorit. Planet. Sci.* **36**, A195–A196.

Srinivasan G., Ulyanov A. A., and Goswami J. N. (1994) ^{41}Ca in the early solar system. *Astrophys. J.* **431**, L67–L70.

Srinivasan G., Sahijpal S., Ulyanov A. A., and Goswami J. N. (1996) Ion microprobe studies of Efremovka CAIs: potassium isotope composition and ^{41}Ca in the early solar system. *Geochim. Cosmochim. Acta* **60**, 1823–1835.

Srinivasan G., Goswami J. N., and Bhandari N. (1999) Al-26 in eucrite Piplia Kalan: plausible heat source and formation chronology. *Science* **284**, 1348–1350.

Stewart B. W., Papanastassiou D. A., and Wasserburg G. J. (1994) Sm–Nd chronology and petrogenesis of mesosiderites. *Geochim. Cosmochim. Acta* **58**, 3487–3509.

Sugiura N. (2001) Boron isotopic compositions in chondrules: anorthite-rich chondrules in the Yamato 82094 (CO3) chondrite. In *Lunar Planet. Sci.* **XXXII**, #1277. The Lunar and Planetary Institute, Houston (CD-ROM).

Sugiura N., Shuzou Y., and Ulyanov A. (2001) Beryllium–boron and aluminum–magnesium chronology of calcium–aluminum-rich inclusions in CV chondrites. *Meteorit. Planet. Sci.* **36**, 1397–1408.

Swindle T. D., Caffee M. W., Hohenberg C. M., Lindstrom M. M., and Taylor G. J. (1991) Iodine-xenon studies of petrographically and chemically characterized Chainpur chondrules. *Geochim. Cosmochim. Acta* **55**, 861–880.

Swindle T. D., Davis A. M., Hohenberg C. M., MacPherson G. J., and Nyquist L. E. (1996) Formation times of chondrules and Ca–Al-rich inclusions: constraints from short-lived radionuclides. In *Chondrules and the Protoplanetary Disk* (eds. R. H. Hewins, R. H. Jones, and E. R. D. Scott). Cambridge University Press, New York, pp. 77–86.

Tachibana S. and Huss G. R. (2003) The initial abundance of ^{60}Fe in the solar system. *Astrophys. J.* **588**, L41–L44.

Tera F. and Carlson R. W. (1999) Assessment of the Pb–Pb and U–Pb chronometry of the early solar system. *Geochim. Cosmochim. Acta* **63**, 1877–1889.

Tera F. and Wasserburg G. J. (1972) U–Th–Pb systematics in three Apollo 14 basalts and the problem of initial lead in lunar rocks. *Earth Planet. Sci. Lett.* **14**, 281–304.

Tilton G. R. (1988) Age of the solar system. In *Meteorites and the Early Solar System* (eds. J. F. Kerridge and M. S. Matthews). University of Arizona Press, Tucson, pp. 259–275.

Wadhwa M. and Lugmair G. W. (1996) Age of the eucrite "Caldera" from convergence of long-lived and short-lived chronometers. *Geochim. Cosmochim. Acta* **60**, 4889–4893.

Wadhwa M. and Russell S. S. (2000) Timescales of accretion and differentiation in the early solar system: the meteoritic evidence. In *Protostars and Planets IV* (eds. V. Mannings, A. P. Boss, and S. S. Russell). University of Arizona Press, Tucson, pp. 995–1018.

Wadhwa M., Zinner E. K., and Crozaz G. (1997) Manganese–chromium systematics in sulfides of unequilibrated enstatite chondrites. *Meteorit. Planet. Sci.* **32**, 281–292.

Wadhwa M., Foley C. N., Janney P., and Beecher N. A. (2003) Magnesium isotopic composition of the Juvinas eucrite: implications for concordance of the Al–Mg and Mn–Cr chronometers and timing of basaltic volcanism on asteroids. In *Lunar Planet. Sci.* **XXXIV**, #2055. The Lunar and Planetary Institute, Houston (CD-ROM).

Wasserburg G. J. (1985) Short-lived nuclei in the early solar system. In *Protostars and Planets II* (eds. D. C. Black and M. S. Matthews). University of Arizona Press, Tucson, pp. 703–737.

Wasserburg G. J. and Papanastassiou D. A. (1982) Some short-lived nuclides in the early solar system—a connection with the placental ISM. In *Essays in Nuclear Astrophysics* (ed. D. N. Schramm). Cambridge University Press, Cambridge, pp. 77–140.

Wasserburg G. J., Busso M., Gallino R., and Raiteri C. M. (1994) Asymptotic giant branch stars as a source of short-lived radioactive nuclei in the solar nebula. *Astrophys. J.* **424**, 412–428.

Wasserburg G. J., Busso M., and Gallino R. (1996) Abundances of actinides and short-lived nonactinides in the interstellar medium: diverse supernova sources for the r-processes. *Astrophys. J.* **466**, L109–L113.

Wasserburg G. J., Gallino R., and Busso M. (1998) A test of the supernova trigger hypothesis with Fe-60 and Al–26. *Astrophys. J.* **500**, L189–L193.

Weber D., Zinner E., and Bischoff A. (1995) Trace element abundances and magnesium, calcium, and titanium isotopic compositions of grossite-containing inclusions from the carbonaceous chondrite Acfer 182. *Geochim. Cosmochim. Acta* **59**, 803–823.

Weidenschilling S. J. (1977) Aerodynamics of solid bodies in the solar nebula. *Mon. Not. Roy. Astron. Soc.* **180**, 57–70.

Wetherill G. W. (1956) Discordant uranium-lead ages. *Trans. Am. Geophys. Union* **37**, 320–326.

Whitby J., Burgess R., Turner G., Gilmour J., and Bridges J. (2000) Extinct ^{129}I in halite from a primitive meteorite: evidence for evaporite formation in the early solar system. *Science* **288**, 1819–1821.

Whitby J., Gilmour J., Turner G., Prinz M., and Ash R. D. (2001) Iodine–xenon dating of chondrules from the Quinzhen and Kota Kota enstatite chondrites. *Geochim. Cosmochim. Acta* **66**, 347–359.

Yin Q. Z., Jacobsen S. B., McDonough W. F., and Horn I. (2000) Supernova sources and the ^{92}Nb–^{92}Zr p-process chronometer. *Astrophys. J.* **535**, L49–L53.

Yin Q. Z., Jacobsen S. B., Yamashita K., Blichert-Toft J., Telouk P., and Albarede F. (2002) A short timescale for terrestrial planet formation from Hf–W chronometry of meteorites. *Nature* **418**, 949–952.

Young E. D. and Russell S. S. (1998) Oxygen reservoirs in the early solar nebula inferred from an Allende CAI. *Science* **282**, 452–455.

Yurimoto H., Ito M., and Nagasawa H. (1998) Oxygen isotope exchange between refractory inclusion in Allende and solar nebula gas. *Science* **282**, 1874–1877.

Zhai M., Nakamura E., Shaw D. M., and Nakano T. (1996) Boron isotope ratios in meteorites and lunar rocks. *Geochim. Cosmochim. Acta* **60**, 4877–4881.

Zinner E. (1998) Stellar nucleosynthesis and the isotopic composition of presolar grains from primitive meteorites. *Ann. Rev. Earth Planet. Sci.* **26**, 147–188.

Zinner E. and Göpel C. (2002) Aluminum-26 in H4 chondrites: implications for its production and its usefulness as a fine-scale chronometer for early solar system events. *Meteorit. Planet. Sci.* **37**, 1001–1013.

Zinner E., Fahey A. J., Goswami J. N., Ireland T. R., and McKeegan K. D. (1986) Large ^{48}Ca anomalies are associated with ^{50}Ti anomalies in Murchison and Murray hibonites. *Astrophys. J.* **311**, L103–L107.

1.17
Planet Formation

J. E. Chambers

The SETI Institute, Mountain View, CA, USA

1.17.1	THE OBSERVATIONAL EVIDENCE	461
1.17.2	THE PROTOPLANETARY NEBULA AND THE FIRST SOLIDS	462
1.17.2.1	Circumstellar Disks	462
1.17.2.2	Viscous Accretion and Nebula Evolution	462
1.17.2.3	Disk Temperatures and Chemical Fractionation	463
1.17.2.4	Chondrules and Refractory Inclusions	463
1.17.2.5	Short-lived Isotopes	464
1.17.3	THE ORIGIN OF PLANETESIMALS	465
1.17.3.1	Dust Grain Compositions and the Ice Line	465
1.17.3.2	Gravitational Instability	465
1.17.3.3	Dust Grain Sticking	466
1.17.3.4	Larger Bodies and Turbulent Concentration	466
1.17.4	TERRESTRIAL PLANET FORMATION	466
1.17.4.1	Random Velocities and Gravitational Focusing	466
1.17.4.2	Runaway Growth	467
1.17.4.3	Oligarchic Growth	467
1.17.4.4	Late-stage Accretion and Giant Impacts	467
1.17.4.5	Core Formation and the Late Veneer	468
1.17.4.6	Formation of the Moon	468
1.17.4.7	Terrestrial-planet Volatiles	469
1.17.5	THE ASTEROID BELT	469
1.17.5.1	Formation and Mass Depletion	469
1.17.5.2	Thermal Processing and Composition of Asteroids	470
1.17.6	GIANT-PLANET FORMATION	470
1.17.6.1	Giant-planet Compositions	470
1.17.6.2	Core Accretion	471
1.17.6.3	Disk Instability	471
1.17.6.4	Planetary Migration and Disk Gap Formation	472
1.17.6.5	Formation of Uranus and Neptune	472
ACKNOWLEDGMENTS		473
REFERENCES		473

1.17.1 THE OBSERVATIONAL EVIDENCE

Modern theories for the origin of the planets are based on observations of the solar system and star-forming regions elsewhere in the galaxy, together with the results of numerical models. Some key observations are:

- The solar system contains eight large planets with roughly circular, coplanar orbits lying 0.4–30 AU from the Sun. There are few locations between the planets where additional large objects could exist on stable orbits.

- The major planets are grouped: small volatile-poor planets lie close to the Sun, with large volatile-rich planets further out. The main asteroid belt (2–4 AU from the Sun) is substantially depleted in mass with respect to other regions.

- The planets and asteroids are depleted in volatile elements compared to the Sun. The degree of fractionation decreases with distance: the terrestrial planets and inner-belt asteroids are highly depleted in volatiles, the outer-belt asteroids are less so, while many satellites in the outer solar

system are ice rich. Primitive CI meteorites (probably from the outer asteroid belt) have elemental abundances very similar to the Sun except for highly volatile elements.
- Ancient solid surfaces throughout the solar system are covered in impact craters (e.g., the Moon, Mercury, Mars, Callisto). Most of the planets have large axial tilts with respect to their orbits. Earth possesses a large companion with a mass ~1% that of the planet itself.
- The terrestrial planets and many asteroids have undergone differentiation. There is strong evidence that Saturn is highly centrally condensed, with a core of mass ~$10M_\oplus$, and weaker evidence that Jupiter has a core of similar mass. These cores have masses comparable to Uranus and Neptune.
- Meteorites from the main asteroid belt show evidence that they once contained short-lived radioactive isotopes with half-lives <10 Myr. The main components of primitive meteorites (chondrules and refractory inclusions) have sizes clustered around 1 mm. These components appear to have undergone rapid melting and cooling.
- Young stars generally exist in gas- and dust-rich environments. Many young stars possess massive, optically thick disks with diameters of 10–1,000 AU. These disks are inferred to have lifetimes of ~1–10 Myr.
- At least 4% of main sequence (ordinary) stars have planetary-mass companions. The companions have masses of 0.1–10 Jupiter masses (the lower limit is the current detection threshhold), and orbital distances from 0.05 AU to 5 AU (the upper limit is the current detection threshhold).

These observations have led to the development and refinement of a theory in which the planets formed from a disk-shaped protoplanetary nebula (Laplace) by pairwise accretion of small solid bodies (Safranov, 1969). A variant of the standard model invokes the gravitational collapse of portions of this disk to form gas giant planets directly. It should be pointed out that the standard model is designed to explain the planets observed in the solar system. Attempts to account for planetary systems recently discovered orbiting other stars suggest that planet formation is likely to differ in several respects from one system to another.

1.17.2 THE PROTOPLANETARY NEBULA AND THE FIRST SOLIDS

1.17.2.1 Circumstellar Disks

The solar system probably formed from the collapse of a fragment of a molecular cloud—a cold, dense portion of the interstellar medium containing gas and dust with a temperature of 10–20 K (Taylor, 2001). Collapse may have occurred spontaneously or been triggered externally, for example, by a supernova (e.g., Cameron, 1962). As the cloud fragment collapsed, the bulk of its mass fell to the center to form a protostar, while the remaining material formed a rotationally supported disk destined to become the protoplanetary nebula (see Chapter 1.04).

Roughly half of young "T Tauri" stars with ages <10 Myr are observed to have optically thick disks of gas and dust with masses of 0.001–1$M_\odot$ (Beckwith et al., 1990; Strom, 1994). These disks have spectra containing absorption features caused by the presence of water ice and silicates. Ultraviolet and visible emission lines indicate that the central stars are accreting mass from their disks at rates of 10^{-7}–$10^{-9} M_\odot$ yr^{-1} (Hartmann et al., 1998). Optically thick circumstellar disks are not observed around stars older than ~10 Myr (Strom, 1995), which provides an approximate upper limit for the lifetime of the Sun's protoplanetary nebula.

The protoplanetary nebula initially had a mass of at least $0.01M_\odot$. This "minimum mass" is obtained by estimating the total amount of rocky and icy material in all the planets, and adding hydrogen and helium to give a nebula of solar composition (Weidenschilling, 1977a). However, planet formation is probably an inefficient process, suggesting that the protoplanetary nebula was initially more massive than this.

Stars typically form in clusters. If the Sun formed in a large group, such as the Trapezium cluster in Orion, it is likely that one or more massive "OB stars" would have been present. These stars produce large amounts of ultraviolet radiation, which rapidly erodes the outer parts of nearby circumstellar disks by photoevaporation. Smaller clusters, such as the Taurus star-forming region, generally do not contain OB stars. Ultraviolet radiation from the Sun, together with any nearby external sources, would have slowly photoevaporated gas in the protoplanetary nebula beyond ~10 AU from the Sun, and this gas may have been removed entirely within ~10 Myr (Hollenbach et al., 2000).

1.17.2.2 Viscous Accretion and Nebula Evolution

The finite lifetime of circumstellar disks, and the fact that young stars are observed to be accreting material, has given rise to a model which views the protoplanetary nebula as a viscous accretion disk in which material was transported radially inwards, ultimately falling onto the Sun (Lynden-Bell and Pringle, 1974). As a consequence of this accretion, the mass of the nebula declined over time. However, the source of viscosity in the disk is unclear at present (see Chapter 1.04). If the disk was very massive ($\sim 0.1 M_\odot$), the cooler outer regions would have been susceptible to gravitational instabilities (e.g., Laughlin and Bodenheimer, 1994), causing mass to flow inwards and angular momentum to be transported outwards. The innermost part of the disk ($\ll 1$ AU) and the surface layers would have been partially ionized, and in these regions interaction with the Sun's rotating magnetic field generated magnetorotational instabilities, which transported material inwards (Balbus and Hawley, 1991). The absence of viable sources of viscosity elsewhere in the nebula suggests that viscous accretion may have been modest in these regions for most of the disk's lifetime.

Magnetic interactions between the Sun and the disk gave rise to powerful winds which ejected material in jets directed along open magnetic field lines out of the plane of the disk. The amount of material lost in this wind was perhaps 30–50% of the mass accreted by the Sun. Some solid particles entrained in the wind would have fallen back into the disk, possibly many AU from the Sun (Shu et al., 1988).

1.17.2.3 Disk Temperatures and Chemical Fractionation

Currently, there is considerable debate about the maximum temperatures reached in the Sun's protoplanetary disk (see Chapter 1.20). The approximate isotopic homogeneity of planetary material in the inner solar system argues that, at some point, most of the material within a few AU of the Sun was vaporized and mixed. In addition, many primitive meteorites are progressively depleted in volatile and moderately volatile elements with respect to the Sun (Palme et al., 1988), independent of cosmochemical behavior (lithophile, siderophile, chalcophile). This fractionation can be understood if elements more volatile than magnesium and iron were initially vaporized in the inner part of the disk. As the nebula cooled and refractory material began to condense, the gaseous phase continued to accrete onto the Sun. Hence, volatile elements were preferentially lost while refractory elements survived to coagulate into planetary bodies (e.g., Wasson and Chou, 1974; Cassen, 2001). The absence of *isotopic* fractionation of potassium in meteorites is consistent with the idea that chemical fractionation occurred by preferential condensation of refractory elements rather than gradual escape of volatile elements from solid bodies (Humayun and Clayton, 1995). However, the observed fractionation could also be the result of rapid escape of volatile elements due to collisions (Halliday and Porcelli, 2001), or be caused by the addition of a refractory substance to primitive material of solar composition (Hutchison, 2002).

The disks of mature T Tauri stars (ages >0.3 Myr) are observed to have surface temperatures of 50–300 K at 1 AU from the star (Beckwith et al., 1990). Temperatures in the midplanes of these disks are likely to be higher, in the range 200–750 K at 1 AU, and hot enough to vaporize silicates within 0.2–0.5 AU (Woolum and Cassen, 1999). Models of viscous accretion disks yield midplane temperatures that are substantially higher than this early in their history, due to the release of gravitational energy as material flows through the disk towards the central star. These models have disk mass accretion rates ~ 100 times greater than those observed in mature T Tauri stars (Boss, 1996; Bell et al., 1997), and similar mass accretion rates have been inferred for young stars that are still embedded in envelopes of gas from their molecular cloud (Kenyon et al., 1993). Mass accretion rates this high imply that silicates would have been vaporized out to several AU in the disk for a period of $\sim 10^5$ yr. However, the ubiquity of presolar grains in meteorites (identified by their unusual isotopic compositions) implies that at least some material in the inner solar system survived vaporization or that such material was added during the accretion of the planets. The variety of oxygen isotope abundances seen in solar system bodies suggests incomplete mixing of nebula material prior to the formation of the planets and meteorite parent bodies (Clayton, 1993), although iron isotopes *do* seem to have been homogenized (Zhu et al., 2001).

Disk temperatures would have decreased rapidly with distance from the Sun as accretional energy release, optical depth, and solar radiation all declined. For example, some meteorite samples from main-belt asteroids contain hydrated silicates, formed by reactions between anhydrous rock and water ice. This implies that temperatures at 2–3 AU became low enough for ice to condense while the asteroids were forming.

1.17.2.4 Chondrules and Refractory Inclusions

Chondrites—meteorites from parent bodies in the asteroid belt that never melted—represent the most primitive samples available of material that formed in the protoplanetary nebula. Chondrites are mainly composed of chondrules, with smaller amounts of refractory inclusions and a fine grained matrix of silicate, metal, and sulfide. Chondrules are roughly spherical objects, typically ~ 1 mm in size, and largely composed of olivine and pyroxene (Taylor, 2001). They appear to have formed from melt droplets that cooled on timescales of hours according to their texture (Jones et al., 2000). At least 25% of chondrules show signs that they were melted more than once (Rubin and Krot, 1996). The chondrules and matrix grains tend to have complementary chemical compositions (underabundant elements in one correspond to overabundant elements in the other), suggesting that they originated in the same part of the nebula.

As of early 2000s, it is not known how chondrules formed. The great abundance of chondrules in chondrites (up to 80% by mass; Jones et al., 2000) indicates either that chondrule formation was an efficient process, or that they were preferentially retained in objects that grew larger. The presence of moderately volatile elements (e.g., sulfur and sodium) and unmelted relict grains in chondrules implies that chondrule precursors existed in a cool environment with $T < 650$ K, and that chondrules remained molten for only a few minutes (Connolly et al., 1988). These characteristics suggest that chondrules formed in the protoplanetary nebula rather than within a parent body. Plausible formation mechanisms include melting due to lightning (Desch and Cuzzi, 2000) or shocks in the nebula. Models of shock melting have proved the most successful in terms of reproducing the observed properties of chondrules (Desch and Connolly, 2002). Such shocks may have been generated by large-scale gravitational instabilities in the disk if the nebula was still sufficiently massive when chondrules formed (Desch and Connolly, 2002).

Refractory inclusions (also called calcium–aluminum-rich inclusions or CAIs) are an order of magnitude less abundant than chondrules. These objects contain mostly calcium–aluminum silicates and oxides (Jones et al., 2000), and have undergone melting. However, they appear to have remained molten for longer than chondrules—hours rather than minutes—and cooled less rapidly (Jones et al., 2000). In addition, CAIs have uniform oxygen isotope ratios, suggesting that they all formed in the same region (McKeegan et al., 1998). It is possible that chondrules and CAIs both formed in situ, a few AU from the Sun. Models suggest that CAIs are likely to have formed under a narrower range of nebula conditions than chondrules, which would explain their lower abundance in chondrites (Alexander, 2003).

A second possibility is that CAIs formed elsewhere in the nebula and were subsequently transported to the asteroid belt. In the "X-wind" model, solids in the disk drifted inwards, emerged from a partially shielded environment and were melted by solar radiation (Shu et al., 1996). Some of these objects were entrained in the wind of material flowing away from the Sun, and millimeter-sized particles would have subsequently fallen back to the disk at distances of several AU. It seems unlikely that chondrules formed in an X-wind since their volatile components would have been lost due to prolonged heating, while the smallest observed chondrules should have been carried out of the solar system rather than falling back to the disk (Hutchison, 2002).

1.17.2.5 Short-lived Isotopes

Many CAIs, together with some chondrules and samples of differentiated asteroids, contained short-lived radioactive isotopes at the time they formed. This is deduced from the abundances of the daughter isotopes seen in modern meteorites. The short-lived isotopes include ^{41}Ca, ^{26}Al, ^{10}Be, ^{60}Fe, ^{53}Mn, and ^{107}Pd, with half-lives (in units of Myr) 0.13, 0.7, 1.5, 1.5, 3.7, and 6.5, respectively. Many of these isotopes could have been produced from stable ones by absorption of neutrons in a supernova or the outer layers of a giant star. In particular, ^{60}Fe can only be produced efficiently by stellar nucleosynthesis and so must have come from an external source (Shukolyukov and Lugmair, 1993). Conversely, some isotopes such as ^{10}Be almost certainly formed in the protoplanetary nebula when material was bombarded by solar cosmic rays (McKeegan et al., 2000). Multiple sources are possible for some short-lived isotopes. The abundances of the decay products of ^{26}Al and ^{41}Ca in CAIs from carbonaceous chondrites are correlated, suggesting that these isotopes come from a single stellar source (Sahijpal et al., 1995). Models indicate that no more than ~10% of the ^{26}Al was produced by cosmic ray bombardment, since otherwise ^{41}Ca and ^{53}Mn abundances would be higher than observed (Goswami and Vanhala, 2000). The short half-lives of these isotopes favor a scenario in which they were generated by the same supernova or stellar wind that triggered the collapse of the molecular-cloud fragment that went on to form the solar system (Vanhala and Boss, 2000).

The source of short-lived isotopes is important since if these isotopes were homogeneously mixed in the nebula they can be used as chronometers—the relative ages of materials can be obtained by measuring the abundance of the daughter products of the isotopes. CAIs appear to be the oldest surviving material in the solar system (with ages of 4.566 ± 0.002 Gyr, measured using the lead–lead chronometer; Allègre et al., 1995). Most CAIs, for which accurate measurements are available, formed with ^{26}Al/^{27}Al ratios of $(4-5) \times 10^{-5}$. The uniformity of these values suggests that ^{26}Al was thoroughly mixed at the time CAIs formed, and that they formed within a few hundreds of thousands of years of one another. A few refractory inclusions, referred to as FUN inclusions (fractionated and unidentified nuclear anomalies), had different initial amounts of ^{26}Al. These CAIs exhibit several isotopic anomalies, suggesting that they formed before the nebula was homogeneously mixed (Wadhwa and Russell, 2000).

Chondrules formed with a variety of ^{26}Al/^{27}Al ratios in the range $(0-2) \times 10^{-5}$ (Huss et al., 2001). These abundances are lower than those measured in most CAIs, suggesting that chondrules formed 2–5 Myr after CAIs. The initial abundances of other short-lived isotopes are also lower in chondrules than CAIs, supporting the interpretation that chondrules are the younger of the two (Alexander et al., 2001). This conclusion may be hard to reconcile with the short gas-drag lifetimes of millimeter-sized bodies ($\sim 10^4$ yr at 1 AU, see Section 1.17.3). It is possible that CAIs were incorporated into larger bodies, with longer gas-drag lifetimes, for a few million years, before being returned to the nebula in disruptive collisions. However, some CAIs show signs of thermal alteration and remelting over periods of up to 2 Myr, implying that these objects remained in the nebula until chondrules formed (Huss et al., 2001).

1.17.3 THE ORIGIN OF PLANETESIMALS

1.17.3.1 Dust Grain Compositions and the Ice Line

The earliest stage of planetary accretion is the most poorly understood at present. If planets did not form directly via instabilities in the disk (see Section 1.17.6), they would have formed by coagulation of solid material in the nebula. In a nebula with a similar elemental composition to the Sun, $\sim 0.5\%$ of the mass in the inner region would have formed solids following any initial hot phase of nebula evolution. These solids primarily consisted of silicates, metal, and sulfides. In the outer nebula, where temperatures were colder, icy materials also condensed, with water ice being the most abundant. In this region, up to 2% of material with a solar composition would have formed solids. The boundary between these two regions was marked by a discontinuity in the surface density of solid material called the "ice line" (or snow line). The magnitude of this discontinuity may have been enhanced by cold trapping of ice as gas was recycled across the ice line (Stevenson and Lunine, 1988). However, it is also likely that radial drifting of solid material smeared out the boundary to some degree. The ice line would have moved inwards over time as the nebula cooled.

Lacking gas pressure support, dust grains, and aggregates would have settled towards the midplane of the nebula, increasing the solid-to-gas ratio in this region. The rate of sedimentation depended on the particle size, with larger objects falling faster than small ones. As dust grains fell, they would have coagulated to form larger objects, increasing the rate of sedimentation. Calculations suggest that much of the solid material would have accumulated near the midplane in 10^3-10^4 yr (Weidenschilling, 1980).

1.17.3.2 Gravitational Instability

What happened next depends sensitively on the vertical thickness of this solid-rich layer and on the relative velocities of the solid particles. Pioneering studies of planetary accretion showed that if the solid-rich layer was very thin, portions of it would become gravitationally unstable, collapsing to form solid bodies ~ 1 km in diameter called "planetesimals" (Safranov, 1969; Goldreich and Ward, 1973). Note that this process is different from the large-scale disk instabilities that may have formed Jupiter-mass bodies.

Later studies suggested that things would not be so simple. Gravitational instability is most likely to occur when the volume density of solid particles is high and their relative velocities are low. However, a dense layer containing mostly solid material would revolve about the Sun with a velocity given by Kepler's third law. In the gas-rich layers above and below the midplane, the Sun's gravity was partially offset by gas pressure, so that the gas disk revolved at less than Keplerian velocity. This differential velocity (~ 100 m s^{-1} for objects at 1 AU; Adachi et al., 1976) generated turbulence which would have stirred up the solid-rich layer, rendering it less prone to gravitational instability (Weidenschilling, 1980). In addition, particles of differing sizes would have drifted through the nebula at different rates due to gas drag. This increased the relative velocities of the particles, also frustrating gravitational instability (Weidenschilling, 1988).

Ward (2001) has suggested that the onset of gravitational instability depends sensitively on disk parameters that are poorly constrained at present. There appears to be a critical solid-to-gas surface density ratio necessary for the onset of gravitational instability, which requires enhancement of solids by a factor of 2–10 times above that expected for material of solar composition (Youdin and Shu, 2002). In a nonturbulent disk, particles would have migrated at different rates in different regions, leading to a pileup of solid material at certain points (Youdin and Shu, 2002). In addition, particles are likely to migrate more slowly through regions where the volume density of particles is high, and this can also lead to a local increase in the solid-to-gas ratio (Goodman and Pindor, 1999). These factors may have favored the formation of planetesimals by gravitational instability.

1.17.3.3 Dust Grain Sticking

If planetesimals did not form via gravitational instability, such bodies must have formed as a result of the sticking together of dust grains and aggregates during collisions. Only when bodies reached ~1 km in size could gravity have played a significant role in further accretion. Some grain growth probably occurred in the Sun's molecular cloud fragment, forming objects up to ~0.1 mm in size (Weidenschilling and Ruzmaikina, 1994). However, many of these grains would have evaporated subsequently when they entered the protoplanetary disk, at least in the inner solar system.

Grain growth in the disk itself must have occurred quickly in order for the objects to survive against gas drag due to the differential velocity of solid objects relative to the nebula gas. Objects <1 m in size were somewhat coupled to the motion of the gas, reducing the effects of gas drag, but larger bodies had no such protection. These objects drifted towards the Sun as they lost angular momentum to the nebula gas, with drift timescales inversely proportional to their size. Meter-sized objects were especially vulnerable, with drift lifetimes ~100 yr at 1 AU (Weidenschilling, 1977b). Differential drift velocities for objects of different size may have aided the accretion of bodies smaller than 1 km, provided that impact speeds were not too high (Cuzzi et al., 1993).

Experiments suggest that small dust grains will stick together if they collide at velocities of less than a few meters per second (Poppe et al., 2000). The probability of sticking decreases with increasing collision speed. Spherical silica grains stick at collision speeds below ~1 m s^{-1}, while irregularly shaped particles can stick at higher velocities, up to 50 m s^{-1} (Poppe et al., 2000). Sticking is mainly due to van der Waals forces, which are weak. Coagulation of iron grains in the nebula would have been enhanced by the presence of the Sun's magnetic field (Beckwith et al., 2000), while in cool regions, frost coatings probably aided sticking of centimeter-to-meter size bodies at low collision speeds (Bridges et al., 1996). Collisions between small particles typically lead to charge exchange and formation of dipoles, and this could have enhanced sticking forces between chondrule-sized particles by a factor of ~1,000 (Marshall and Cuzzi, 2001).

1.17.3.4 Larger Bodies and Turbulent Concentration

It is unclear how aggregation continued for boulder-sized and larger objects. The presence of gas may have helped. This is because small fragments, formed by disruptive impacts onto large bodies, would have been blown back onto the body as the fragments became coupled to the gas, itself moving at less than Keplerian velocity (Wurm et al., 2001). Recent attention has also focused on processes that increased the concentration of solid bodies and may have aided further accretion. If the nebula was turbulent, solid objects would have been concentrated in the convergence zones of eddies. Numerical simulations show that particle concentration could be increased by a factor ~100 within 100 yr (Klahr and Henning, 1997), and much higher concentrations may have occurred on longer timescales (Cuzzi et al., 2001). Turbulent concentration is size dependent, working most efficiently for particles trapped in the smallest eddies. For plausible disk viscosities, the mean particle size and size distribution would have been similar to those observed for chondrules (Cuzzi et al., 2001), suggesting that turbulence played an important role in the early stages of accretion if gravitational instability was ineffective.

1.17.4 TERRESTRIAL PLANET FORMATION

1.17.4.1 Random Velocities and Gravitational Focusing

Once solid bodies reached sizes ~1 km, mutual gravitational interactions became significant. Objects of this size are traditionally referred to as planetesimals even if they did not form by gravitational instability. Close passages between planetesimals tended to increase their "random velocities" v (the radial and out-of-plane components of their motion) as a result of their mutual gravitational attraction. Conversely, drag caused by nebular gas and physical collisions between planetesimals damped their random velocities, making their orbits more circular and coplanar. Competition between these excitation and damping mechanisms established an equilibrium distribution of random velocities, which gradually changed over time as the planetesimals gained mass and their gravitational interactions grew stronger.

The frequency of collisions depended sensitively on the random velocities. When v was small, a pair of planetesimals undergoing a close encounter would have remained close to each other for some time, allowing their mutual gravity to "focus" their trajectories towards each other, increasing the chance of a collision. When v was large, close encounters were brief affairs rendering gravitational focusing ineffective, thus reducing the collision probability and the growth rate of planetesimals.

This intimate coupling between the planetesimals' masses and random velocities can give rise to three different growth modes, each with quite different characteristics and accretion timescales. It seems likely that each of these growth modes operated at different times in the solar nebula.

1.17.4.2 Runaway Growth

The early stages of accretion were marked by "runaway growth." Over the course of many close encounters, the largest planetesimals tended to acquire the smallest random velocities, a process that is often referred to as "dynamical friction." (A similar equipartition of energy is observed in gas molecules.) As a result, the largest bodies experienced the strongest gravitational focusing of their trajectories, and they grew most rapidly. Most of the solid material remained in smaller planetesimals that grew slowly if at all (Stewart and Kaula, 1980; Kokubo and Ida, 1996).

Initially at least, it seems likely that most collisions led to accretion rather than fragmentation (Leinhardt and Richardson, 2002). However, as the planetesimals grew larger, their random velocities increased, and collisions between small bodies were increasingly likely to result in disruption. Small collision fragments were strongly affected by gas drag, and these rapidly acquired circular, coplanar orbits. As a result, collision fragments were quickly accreted by the largest planetesimals due to their high mutual collision probability, and this allowed growth to proceed more rapidly (Wetherill and Stewart, 1993).

1.17.4.3 Oligarchic Growth

Once the largest objects, dubbed "planetary embryos," become more than ~100 times more massive than a typical planetesimal, the random velocities of the planetesimals became largely determined by gravitational perturbations from the embryos rather than by other planetesimals (Ida and Makino, 1993). As a result, gravitational focusing became less effective and runaway growth slowed down. The larger an embryo, the more it stirred up the velocities of nearby planetesimals, and the slower it grew, allowing neighboring embryos to catch up. This state of affairs is known as "oligarchic growth" (Kokubo and Ida, 1998). At the same time, gravitational interactions between embryos tended to keep them apart, so that each carved out its own niche, or "feeding zone" in the protoplanetary disk.

Numerical simulations of runaway growth suggest that bodies the size of the Moon or Mars could have formed in $\sim 10^5$ yr at 1 AU (Wetherill and Stewart, 1993). However, it is likely that oligarchic growth began at much lower masses than this, in the range 10^{-5}–$10^{-3} M_\oplus$ (Rafikov, 2003; Thommes et al., 2003). As a result, the formation of lunar-to-Mars size bodies took place in the oligarchic growth regime, requiring $\sim 10^6$ yr at 1 AU in a minimum mass nebula (Weidenschilling et al., 1997).

Planetary embryos accreted most of their mass from their feeding zones, which had widths of $\sim$10–12 Hill radii, where the Hill radius $R_H = r(m/3M_\odot)^{1/3}$, where r is the distance from the Sun (Kokubo and Ida, 1998). Hence, embryos probably had different compositions depending on where they formed in the protoplanetary disk, although radial drift of small bodies due to gas drag means that some radial mixing of material would have occurred.

1.17.4.4 Late-stage Accretion and Giant Impacts

The final stage of accretion began when the remaining planetesimals had too little mass to damp the random velocities of the embryos. As the random velocities of the embryos increased, growth slowed dramatically, and embryos' orbits began to cross those of their neighbors. This growth mode, known as "orderly growth," began when roughly half of the total solid mass was contained in embryos (Kokubo and Ida, 2000). Despite its name, the last stage of accretion was the most violent, marked by giant collisions between bodies the size of the Moon or Mars. It is unclear how efficient these collisions were, although numerical simulations suggest that growth continued even if there was significant fragmentation (Alexander and Agnor, 1998). The remaining planetesimals were swept up or lost, either by falling into the Sun or being ejected from the solar system. Numerical models suggest that the formation of a fully formed Earth required $\sim$100 Myr (Chambers and Wetherill, 1998).

The highly noncircular orbits of embryos and the long accretion timescales allowed considerable radial mixing of material over distances of 0.5–1.0 AU (Wetherill, 1994). It is likely that each of the inner planets accreted material from throughout the inner solar system, although the degree of radial mixing depends sensitively on the mass distribution of the embryos at this time (Chambers, 2001). The relative contributions from each part of the disk would have been different for each planet, however, producing somewhat different chemical compositions. Earth and Mars have similar but distinct oxygen isotopic compositions, which implies that embryos and planetesimals in the inner solar system were not thoroughly mixed before they accreted to form these planets (Drake and Righter, 2001). Alternatively, Mars may represent a surviving planetary embryo, with a unique chemical and isotopic signature, while the more massive Earth is a composite of many embryos.

Giant impacts were common during the final stages of accretion (Agnor et al., 1999). Several impacts would have been energetic enough to completely melt each of the inner planets, forming a magma ocean which homogenized existing material and erased chemical signatures of earlier stages of accretion. The presence of a massive atmosphere containing captured nebular gas would also have melted the surfaces of Earth and Venus (Sasaki, 1990). Impacts would have been particularly energetic in the innermost part of the disk where orbital, and hence collision, speeds were highest. The high density of Mercury compared to the other inner planets can best be explained as the result of a catastrophic collision occurring after the planet had differentiated (Wetherill, 1988). Numerical simulations suggest that a high-velocity impact onto proto-Mercury would have removed much of the planet's mantle, leaving an intact metal core encased in a thin layer of silicates (Benz et al., 1988).

1.17.4.5 Core Formation and the Late Veneer

Energy released from impacts, together with heat from the decay of radioactive isotopes, led to differentiation in planetary embryos, once these objects became partially molten (Tonks and Melosh, 1992). Iron and siderophile elements (e.g., platinum, palladium, and gold) preferentially sank to the center to form a core, while the lighter silicates and lithophile elements formed a mantle. Differentiation was probably a continuous process rather than a single event, so that large planets like Earth accreted from embryos that were already partially or wholly differentiated.

The time of core formation on the Earth and Mars can be constrained using chronometers based on short-lived isotopes. The W–Hf isotope system is particularly useful in this respect, consisting of a lithophile parent nucleus, ^{182}Hf, which decays to a siderophile daughter isotope with a half-life of 9 Myr. Recent measurements of an excess of the daughter isotope indicate that the cores of Mars and Earth formed in less than 13 Myr and 30 Myr, respectively (Kleine et al., 2002; Yin et al., 2002; see Chapter 1.20 by Halliday for a more detailed discussion).

This estimate for Mars is consistent with the 4.5 Gyr age for ALH 84001, the oldest known martian meteorite.

The affinity for iron of elements such as palladium and platinum, even at the high pressures present in planetary mantles, means that these elements should be essentially absent from the mantles of the inner planets (Holzheid et al., 2000). The fact that siderophiles *are* present in Earth's mantle and crust implies that some material was accreted as a "late veneer" when Earth's differentiation was largely complete. This material could have originated in the terrestrial-planet region or the asteroid belt. Earth's ^{187}Os/^{188}Os ratio appears to rule out a late veneer consisting primarily of enstatite or carbonaceous chondrites, leaving ordinary chondrites as the most likely source if the material was predominantly from the asteroid belt (Drake and Righter, 2001).

1.17.4.6 Formation of the Moon

The leading model for the origin of Earth's moon is an oblique impact between proto-Earth and a Mars-sized embryo (Cameron and Ward, 1976). Such an impact would have formed an accretion disk in orbit around Earth, consisting mostly of mantle material from the impactor, while the impactor's core coalesced with that of the Earth (Canup and Asphaug, 2001). The small size of the lunar core, and the Moon's extreme depletion in volatile and moderately volatile elements, arose as the result of its accretion from a hot circumplanetary disk containing mostly silicates. Today, the Earth and the Moon have essentially identical oxygen isotopic compositions, which suggests that the lunar impactor originated in the same region of the Sun's protoplanetary disk as proto-Earth (Wiechert et al., 2001). The fact that the Moon's core has remained small since its formation suggests that the moon-forming impact happened in the closing stages of planetary accretion, since subsequent collisions on the Moon would have increased its metal fraction (Canup and Asphaug, 2001).

The oldest known lunar rocks have ages of 4.4–4.5 Gyr (Carlson and Lugmair, 1988; see Chapter 1.21), and hence the Moon must have formed within the first 150 Myr of the solar system. This is consistent with the age of the oldest known terrestrial samples—zircon grains, which formed in crustal rocks ~4.4 Gyr ago (Wilde et al., 2001). However, the age of Earth's core, as determined using the W–Hf isotope system, suggests that the moon formed significantly earlier than this, since the moon-forming impact would probably have strongly affected the W–Hf chronometer.

1.17.4.7 Terrestrial-planet Volatiles

The presence of water and volatile elements (carbon, nitrogen, and the noble gases) on Earth, Mars, and Venus poses a problem for theories of planet formation. It is likely that the inner part of the protoplanetary nebula was too hot for these materials to condense at the time when planetesimals were forming (see Section 1.17.2). Enstatite chondrites, from undifferentiated bodies in the inner asteroid belt, are very dry, and ordinary chondrites also contain little water (Taylor, 2001). This suggests that planetesimals which formed <2.5 AU from the Sun were almost free of volatiles. If true, this implies that Earth acquired its volatiles by accreting material that originally formed beyond 2.5 AU, in regions of the nebula that were cold enough for ices to condense. In fact, Earth probably acquired more water than currently exists in the oceans and mantle, since some water would have been lost by reacting with iron (Righter and Drake, 1999).

Comets are rich in volatile elements, but they probably delivered no more than 10% of Earth's volatile inventory. There are several reasons for this. Comets have a very low impact probability with Earth over their dynamical lifetime ($\sim 10^{-6}$; Levison et al., 2000), limiting the amount of cometary material that Earth could have accreted. In addition, if most of Earth's water was acquired from comets, it seems likely that Earth's noble gas abundances would be higher than observed by several orders of magnitude (Zahnle, 1998). Finally, water measured spectroscopically in comets differs isotopically from that of seawater on Earth, with the cometary D/H ratio being greater by a factor of 2 (Lunine et al., 2000).

An asteroidal source of volatiles is more promising. Carbonaceous chondrites contain up to 10% water by mass, in the form of hydrated minerals (Taylor, 2001), while some ordinary chondrites also contain water. The hydrogen in many chondritic hydrated silicates has a similar D/H ratio to that of Earth's oceans. Numerical simulations show that Earth could have accreted several oceans worth of water from the asteroid belt, especially if lunar-to-Mars size planetary embryos formed in this region (Morbidelli et al., 2000). However, the amount of mass accreted from the asteroid belt depends sensitively on the early orbital evolutions of the giant planets, and these are poorly known at present (Chambers and Cassen, 2002). Oxygen isotope differences between Earth and carbonaceous chondrites imply that the latter contributed no more than a few percent of Earth's total mass (Drake and Righter, 2001), although this is enough to supply Earth's water. In addition, the relatively low abundance of siderophile elements in the terrestrial mantle argues that Earth acquired most of its asteriodal material before core formation was complete.

Mars is depleted in highly volatiles relative to Earth and contains water with a higher D/H ratio. Models suggest that it acquired a significant fraction of its volatiles from both comets and asteroids (Lunine et al., 2002). It is possible that Mars has lost much of its initial volatile inventory during large impacts. Giant impacts (including the moon-forming event) may have removed the early atmospheres of the terrestrial planets, leading to depletion of atmosphere-forming elements with respect to geochemical volatiles such as thalium (Zahnle, 1998). Hence, the modern atmospheres of the inner planets were produced by outgassing of material incorporated into the mantle during accretion (see Chapter 1.20 for a detailed discussion).

1.17.5 THE ASTEROID BELT

1.17.5.1 Formation and Mass Depletion

The primary feature of the main asteroid belt is its great depletion in mass relative to other regions of the planetary system. The present mass of the main belt is $\sim 5 \times 10^{-4} M_\oplus$, which represents 0.1–0.01% of the solid mass that existed at the time planetesimals were forming. There are several ways the main asteroid belt could have lost most of its primordial mass. Substantial loss by collisional erosion appears to be ruled out by the preservation of asteroid Vesta's basaltic crust, which formed in the first few million years of the solar system (Davis et al., 1994). More plausible models are based on the existence of orbital "resonances" associated with the giant planets.

Half a dozen strong resonances exist at particular heliocentric distances in the region 2–4 AU from the Sun. The orbit of an asteroid in one of these resonances becomes unstable on a timescale ~ 1 Myr, such that the asteroid ultimately falls into the Sun or is ejected from the solar system (Gladman et al., 1997). At present, the resonance locations are almost devoid of asteroids. However, resonances currently occupy only a small fraction of the orbital phase space in the main belt, so an additional mechanism must have operated to make them more effective at removing mass in the past. Some of the resonances existed in a different location when the protoplanetary nebula was present. As the nebula dispersed, these resonances swept across the asteroid belt and could have removed a substantial amount of mass in the process (Ward et al., 1976; Nagasawa et al., 2000). However, the clearing efficiency depends sensitively on the timescale for nebula removal, such that rapid nebula dispersal corresponds to less efficient clearing. Prior to the dispersal of the nebula, a combination of gas drag and resonances is likely to have caused many asteroids with diameters in the range 10–100 km to drift into the region now occupied by the terrestrial planets (Franklin and Lecar, 2000).

The asteroid belt would have also experienced rapid mass loss if planetary embryos accreted in this region via runaway and oligarchic growth. Close encounters between embryos would have caused frequent changes in their orbits until objects entered a resonance and were removed. Numerical simulations show that the most likely outcome is that *all* planetary embryos would have been lost from the asteroid region by this process (Chambers and Wetherill, 2001). The same mechanism would have removed ~99% of planetesimals and asteroid-sized bodies at the same time (Petit *et al.*, 2001). At present, it is unclear whether embryos did form in the main belt. Accretion simulations suggest that embryos ought to have formed within $\sim 10^6$ yr (Wetherill and Stewart, 1993), provided that Jupiter formed later than this. However, if Jupiter formed rapidly, its gravitational perturbations would have increased the random velocities of planetesimals in the asteroid belt and prevented runaway growth from taking place (Kortenkamp and Wetherill, 2000).

The disruptive effect of Jupiter's gravity suggests that asteroids accreted before the giant planets were fully formed. The fact that Jupiter and Saturn are mostly composed of nebular gas sets an upper limit of ~10 Myr for asteroid formation, based on estimates for the lifetimes of circumstellar disks (Strom, 1995). Rapid formation of some asteroids is confirmed by isotopic chronometers. The initial abundances of ^{53}Mn and ^{182}Hf in HED (howardite, eucrite, diogenite) meteorites, which probably originated on Vesta, suggest that this asteroid formed and differentiated within 4 Myr of the formation of CAIs (Lugmair and Shukolyukov, 1998; Kleine *et al.*, 2002; Yin *et al.*, 2002). This agrees with a formation time of ~5 Myr derived using the ^{26}Al chronometer (Srinivasan *et al.*, 1999). Similarly, the parent bodies of many iron and stony-iron meteorites and ordinary chondrites formed within ~10 Myr, according to the ^{107}Pd and Hf–W systems (Wadhwa and Russell, 2000; Lee and Halliday, 1996).

1.17.5.2 Thermal Processing and Composition of Asteroids

The existence of iron, stony-iron, and achondrite meteorites implies that many asteroids were heated sufficiently to cause melting and differentiation. The source of this heating is still a matter of debate. Impact melting is inefficient for asteroid-sized bodies, since collisions violent enough to cause melting tend to eject the melted material at greater than the asteroid's escape velocity (Tonks and Melosh, 1992). A more likely heating process was decay of short-lived isotopes, especially ^{26}Al. Calculations suggest that asteroids larger than ~10 km would have melted due to decay of ^{26}Al if they formed within 2 Myr (Woolum and Cassen, 1999). Asteroids that formed at later times would have escaped melting, since most ^{26}Al would have decayed, and this may explain why some primitive bodies exist in the asteroid belt. Some chondrites contain hydrated silicates, which suggests that their parent bodies were heated sufficiently for ice to melt in their interiors, and react with dry rock, but not enough to cause differentiation (Alexander *et al.*, 2001).

The short formation timescale of asteroids implies that they accreted from material in their immediate vicinity, unlike the terrestrial planets which probably accreted material from throughout the region <3 AU from the Sun. This conclusion is supported by differences in the magnesium, aluminum, and silicon abundances and oxidation states of different classes of chondrite (Drake and Righter, 2001). In addition, main-belt asteroids have spectral characteristics that vary with heliocentric distance (Gradie and Tedesco, 1982). Hence, solid material in the asteroid belt was not thoroughly mixed during accretion. The degree of thermal processing apparently decreases with distance from the Sun: M-type asteroids (probably associated with iron meteorites) are confined to the inner belt, C-types (many with spectral features associated with hydrated silicates) dominate in the middle belt, while P-type asteroids (primitive, unprocessed bodies?) exist in the outer belt. This presumably reflects the fact that accretion timescales would have increased with distance from the Sun, reducing the amount of heating caused by short-lived isotopes.

1.17.6 GIANT-PLANET FORMATION

1.17.6.1 Giant-planet Compositions

The dominant components of Jupiter and Saturn are hydrogen and helium (see Chapter 1.23). These elements would not have condensed or become trapped in solids at temperatures present in the protoplanetary nebula (Lunine *et al.*, 2000), so they must have been gravitationally captured as gases. Uranus and Neptune contain ~10% hydrogen and helium, and presumably these gases were captured from the nebula too. Protoplanetary disks orbiting young stars are observed to disperse in $\leq$10 Myr (Strom, 1995), which strongly suggests that the giant planets in the solar system took no more than 10 Myr to form. The gravitational fields of Jupiter and Saturn indicate that they possess dense cores, of unknown composition, with masses of $\sim 10 M_\oplus$ (Wuchterl et al., 2000). Jupiter's gaseous envelope is enriched in carbon and sulfur by a factor ~ 3 relative to the Sun, while nitrogen, argon, krypton, and xenon are also enriched (Owen et al., 1999). Elements heavier than helium are also enriched in Saturn relative to a solar composition.

The fact that Jupiter and Saturn have atmospheres enriched in elements heavier than helium suggests that they accreted a large amount of mass in the form of planetesimals. However, the similar enrichments of highly volatile argon and nitrogen compared to less volatile carbon, observed in Jupiter, imply that these planetesimals formed at temperatures ≤ 30 K (Owen et al., 1999). In contrast, temperatures at 5 AU are commonly thought to have been ~ 160 K at the time when Jupiter was forming. This suggests either that Jupiter formed much further from the Sun than its present location, which is hard to reconcile with the known models for planet formation, or that it was efficient at accreting planetesimals from the region now occupied by the Edgeworth–Kuiper belt. Alternatively, argon and nitrogen may have been trapped at higher temperatures in planetesimals composed of crystalline rather than amorphous ice (Gautier et al., 2001), in which case Jupiter could have formed and accreted planetesimals at its current location.

1.17.6.2 Core Accretion

As of early 2000s, two models for the origin of Jupiter and Saturn are being actively pursued. In the "core accretion" model, planetary accretion in the outer solar system initially proceeded through the same stages as in the inner solar system: the formation of planetesimals followed by runaway and oligarchic growth. The presence of additional solid material in the form of water ice, and the fact that Hill radii (and hence feeding zone widths) increase with distance from the Sun, means that, in principle, larger bodies would have formed here than in the inner solar system. As they grew, the largest planetary embryos would have acquired thick atmospheres of nebula gas. Numerical models indicate that once an object grew to $\sim 10 M_\oplus$, it could no longer support a static atmosphere, and it steadily accreted gas from the nebula (Pollack et al., 1996), eventually forming a gas-giant planet.

Analytic estimates suggest that $10 M_\oplus$ cores were unlikely to accrete in a minimum mass nebula on a timescale comparable to the lifetime of circumstellar disks. However, such cores could have formed if the surface density of solids at 5 AU was 5–10 times that of a minimum mass nebula (Lissauer, 1987). Numerical models of planetary accretion support this conclusion (Thommes et al., 2003), although it is possible that growth could have stalled when the largest bodies were a few Earth masses, due to loss of solid material via collisional fragmentation (Inaba and Wetherill, 2001). The surface density of solids at 5 AU may have been greater than that in a minimum mass nebula because the nebula itself was more massive than this, implying that planet formation was inefficient. In addition, solid material could have accumulated near 5 AU, either because of cold trapping of ice (Stevenson and Lunine, 1988) or drift of small solids to a local maximum in the gas surface density (Haghighipour and Boss, 2003).

The rate at which the planetary cores accreted gas increased slowly until their masses reached $\sim 30 M_\oplus$. After this, gas accretion was very rapid (Pollack et al., 1996). The growth timescale depends on the opacity of the planet's gas envelope, since this determines the rate at which the energy of accretion could be radiated away. For interstellar dust opacities, a $10 M_\oplus$ core would require ~ 10 Myr to grow to Jupiter's mass (Pollack et al., 1996). However, growth would have been quicker if the opacity was lower due to coagulation of grains in the envelope (Ikoma et al., 2000).

1.17.6.3 Disk Instability

A protoplanetary nebula containing enough mass to rapidly form a $10 M_\oplus$ giant-planet core would have been quite close to the limit of gravitational stability beyond ~ 5 AU (Boss, 2001). This suggests that the giant planets may have formed in a single step via "disk instability." This would occur on a timescale that is orders of magnitude shorter than the timescale for core accretion. Numerical simulations show that a marginally unstable disk gives rise to gravitationally bound clumps comparable to the mass of Jupiter on timescales of ~ 100 yr (Boss, 2001). High-resolution simulations show that these objects would remain bound and continue to collapse for at least 1,000 years (Mayer et al., 2002). The question of whether solids in these clumps would settle to the center to form a core similar to those thought to exist in Jupiter and Saturn has not been explored to date.

At present, it is unclear whether the protoplanetary nebula could have evolved to the point where it was marginally unstable, or whether disk instabilities would have redistributed mass in the disk prior to the formation of gravitationally bound clumps. If marginally unstable disks do develop, then giant-planet formation by disk instability seems unavoidable. It is worth noting that current simulations of disk instability tend to generate planets with masses greater than

Jupiter and Saturn, but comparable to some observed extrasolar planets.

1.17.6.4 Planetary Migration and Disk Gap Formation

The growth of the giant planets took place while the nebula gas was still present. This complicates the picture of planet formation because of gravitational interactions between the gas and the giant planets. In particular, a planet will strongly interact with the disk at "Lindblad resonances," i.e., at certain locations where the planet's gravitational perturbations build up constructively over time. A planet launches spiral density waves so that the distribution of gas in the disk becomes nonaxisymmetric. The gravitational attraction of overdense regions exerts a torque on the planet which changes its orbit. Torques due to gas orbiting inside and outside the planet have opposite signs but unequal magnitudes in general, such that a planet is likely to lose angular momentum and drift inwards (Goldreich and Tremaine, 1980). This is referred to as "type-I migration." The migration speed is proportional to the mass of the planet, and can be very rapid: a $10 M_\oplus$ core at 5 AU would have migrated into the Sun in $\sim 10^5$ yr (Tanaka et al., 2002), while the migration timescale for a Jupiter mass planet is 30 times shorter!

Clearly, type-I migration presents a problem for models of planet formation, both in terms of accreting fully formed planets before they migrate into the Sun and in terms of their survival once fully formed. However, it is likely that a sufficiently massive planet would have cleared a gap in the disk gas. Once this gap extended beyond the Lindblad resonances, type-I migration ceased. At present, there is considerable uncertainty about how massive a planet must be to clear a gap in the disk. This depends sensitively on the way in which waves damped in the nebula, and on the disk viscosity, both of which are poorly constrained. A recent estimate is that a body with a mass of $2-3 M_\oplus$ would have cleared a gap at 1 AU, while at 5 AU, a body $\sim 15 M_\oplus$ would do the job (Rafikov, 2002).

Once a planet clears a gap in the disk, its orbital evolution becomes tied to that of the nebula. In a viscous disk, material would have flowed towards the Sun and a giant planet would have migrated with this material, maintaining a gap in the disk as it moved. This is known as "type-II migration" (Lin and Papaloizou, 1986; Ward, 1997). Type-II migration rates depend on the nebula's viscosity, and so are poorly constrained at present. For plausible viscosities, the migration timescale would have been at least 1–2 orders of magnitude longer than for type-I migration (Ward and Hahn, 2000). Type-II migration provides a plausible explanation for the small sizes of the orbits of many observed extrasolar planets.

The existence of type-I migration implies that the cores of the giant planets must have accreted quickly if these planets formed by core accretion. It is likely that migration would have speeded up accretion as a giant-planet core drifted into regions containing additional planetesimals. However, simulations suggest that a migrating planet would accrete <10% of these planetesimals, so the need for a massive nebula remains (Tanaka and Ida, 1999).

The giant planets ceased growing when the flow of gas onto their envelopes was cut off. This may have been the result of gap formation or because the nebula dispersed. The latter seems unlikely, since the timescale for gas accretion onto a Jupiter-size planet is small compared to the lifetime of the nebula. However, hydrodynamical simulations suggest that gas would continue to flow onto Jupiter after it cleared a gap in the disk (Lubow et al., 1999), so this explanation is problematical too. In addition, it has been suggested that some gas would remain at the same orbital distance as the planet after it cleared a gap if the disk viscosity was low (Rafikov, 2002), and this would also be accreted by the planet eventually.

1.17.6.5 Formation of Uranus and Neptune

Attempts to model the accretion of Uranus and Neptune from planetesimals orbiting 20–30 AU from the Sun (the current locations of these planets) have met with severe difficulties. Long orbital periods in the outer solar system mean that accretion occurs very slowly. In addition, solar gravity is sufficiently weak here that gravitational interactions between planetary embryos would have ejected a substantial amount of mass from this region of the disk (Levison and Stewart, 2001). Numerical simulations show that it is unlikely that bodies larger than Earth could have accreted *in situ* at the locations of Uranus and Neptune, even if the nebula was substantially more massive than the minimum-mass nebula (Thommes et al., 2003).

A more likely scenario is that Uranus and Neptune, along with the cores of Jupiter and Saturn, formed in the region 5–10 AU from the Sun. Such a system would have remained dynamically stable until one object (Jupiter) accreted a large H/He-rich atmosphere. At this point at least two of the other bodies would have been perturbed into the region beyond 15 AU. Gravitational interactions with planetesimals in the outer solar system would then have circularized the orbits of Uranus and Neptune by dynamical friction, while at the same time scattering most of these planetesimals onto unstable orbits that crossed those of Jupiter and Saturn. Numerical simulations have shown that this model is robust provided that cores with masses $\sim 10 M_\oplus$ were able to survive type-I migration (Thommes et al., 1999). The failure of Uranus and Neptune to capture more than a small amount of nebula gas indicates that the nebula dispersed before they could accumulate massive atmospheres similar to Jupiter and Saturn. This may be a consequence of photoevaporation of nebula gas by ultraviolet radiation from the Sun, since this is most effective at distances beyond ~ 10 AU (Hollenbach et al., 2000).

ACKNOWLEDGMENTS

I am most grateful to Robbins Bell, Lindsey Bruesch, Jeff Cuzzi, Jack Lissauer, Ignacio Mosqueira, and Kevin Zahnle for helpful comments and discussion during the preparation of this chapter.

REFERENCES

Adachi I., Hayashi C., and Nakazawa K. (1976) The gas drag effect on the elliptical motion of a solid body in the primordial Solar Nebula. *Prog. Theor. Phys.* **56**, 1756–1771.

Agnor C. B., Canup R. M., and Levison H. F. (1999) On the character and consequences of large impacts in the late stage of terresrial planet formation. *Icarus* **142**, 219–237.

Alexander C. M. O. (2003) Making CAIs and chondrules from CI dust in a canonical nebula. *Lunar Planet. Sci.*, abstract 1391.

Alexander C. M. O., Boss A. P., and Carlson R. W. (2001) The early evolution of the inner solar system: a meteoritic perspective. *Science* **293**, 64–69.

Alexander S. G. and Agnor C. B. (1998) N-body simulations of late stage planetary formation with a simple fragmentation model. *Icarus* **132**, 113–124.

Allègre C. J., Manhès G., and Göpel C. (1995) The age of the Earth. *Geochim. Cosmochim. Acta* **59**, 1445–1456.

Balbus S. A. and Hawley J. F. (1991) A powerful local shear instability in weakly magnetized disks. *Astrophys. J.* **376**, 214–233.

Beckwith S. V. W., Sargent A. I., Chini R. S., and Guesten R. (1990) A survey for circumstellar disks around young stellar objects. *Astron. J.* **99**, 924–945.

Beckwith S. V. W., Henning T., and Nakagawa Y. (2000) Dust properties and assembly of large particles in protoplanetary disks. In *Protostars and Planets IV* (eds. V. Mannings, A. P. Boss, and S. S. Russell). University of Arizona Press, Tucson, pp. 533–558.

Bell K. R., Cassen P. M., Klahr H. H., and Henning Th. (1997) The structure and appearance of protostellar accretion disks: limits on disk flaring. *Astrophys. J.* **486**, 372–387.

Benz W., Slattery W. L., and Cameron A. G. W. (1988) Collisional stripping of Mercury's mantle. *Icarus* **74**, 516–528.

Boss A. P. (1996) Evolution of the solar nebula III: protoplanetary disks undergoing mass accretion. *Astrophys. J.* **469**, 906–920.

Boss A. P. (2001) Gas giant protoplanet formation: disk instability models with thermodynamics and radiative transfer. *Astrophys. J.* **563**, 367–373.

Bridges F. G., Supulver K. D., Lin D. N. C., Knight R., and Zafra M. (1996) Energy loss and sticking mechanisms in particle aggregation in planetesimal formation. *Icarus* **123**, 422–435.

Cameron A. G. W. (1962) The formation of the Sun and planets. *Icarus* **1**, 13–69.

Cameron A. G. W. and Ward W. R. (1976) The origin of the Moon. *Lunar Planet. Sci.* **7**, 120–122.

Canup R. M. and Asphaug E. (2001) Origin of the Moon in a giant impact near the end of the Earth's formation. *Nature* **412**, 708–712.

Carlson R. W. and Lugmair G. W. (1988) The age of Ferroan Anorthosite 60025: Oldest crust on a young Moon? *Earth Planet. Sci. Lett.* **90**, 119–130.

Cassen P. (2001) Nebular thermal evolution and properties of primitive planetary materials. *Meteoritics* **36**, 671–700.

Chambers J. E. (2001) Making more terrestrial planets. *Icarus* **152**, 205–224.

Chambers J. E. and Cassen P. (2002) Planetary accretion in the inner solar system: dependence on nebula surface density profile and giant planet eccentricities. *Lunar Planet. Sci.*, abstract 1049.

Chambers J. E. and Wetherill G. W. (1998) Making the terrestrial planets: N-body integrations of planetary embryos in three dimensions. *Icarus* **136**, 304–327.

Chambers J. E. and Wetherill G. W. (2001) Planets in the asteroid belt. *Meteoritics* **36**, 381–399.

Clayton R. N. (1993) Oxygen isotopes in meteorites. *Ann. Rev. Earth. Planet. Sci.* **21**, 115–149.

Connolly H. C., Jones B. D., and Hewins R. H. (1988) The flash melting of chondrules: an experimental investigation into the melting history and physical nature of chondrule precursors. *Geochim. Cosmochim. Acta* **62**, 2725–2735.

Cuzzi J. N., Dobrovolskis A. R., and Champney J. M. (1993) Particle-gas dynamics in the midplane of a protoplanetary nebula. *Icarus* **106**, 102–134.

Cuzzi J. N., Hogan R. C., Paque J. M., and Dobrovolskis A. R. (2001) Size-selective concentration of chondrules and other small particles in protoplanetary nebula turbulence. *Astrophys. J.* **546**, 496–508.

Davis D. R., Ryan E. V., and Farinella P. (1994) Asteroid collisional evolution: results from current scaling algorithms. *Planet. Space Sci.* **42**, 599–610.

Desch S. J. and Connolly H. C. (2002) A model of the thermal processing of particles in solar nebula shocks: application to the cooling rates of chondrules. *Meteoritics* **37**, 183–207.

Desch S. J. and Cuzzi J. N. (2000) The generation of lightning in the solar nebula. *Icarus* **143**, 87–105.

Drake M. J. and Righter K. (2001) Determining the composition of the Earth. *Nature* **416**, 39–44.

Franklin F. and Lecar M. (2000) On the transport of bodies within and from the asteroid belt. *Meteoritics* **35**, 331–340.

Gautier D., Hersant F., and Mousis O. (2001) Enrichments in volatiles in Jupiter: a new interpretation of the Galileo measurements. *Astrophys. J.* **550**, L227–L230.

Gladman B. J., Migliorini F., Morbidelli A., Zappala V., Michel P., Cellion A., Froeschle Ch., Levison H. F., Bailey M., and Duncan M. (1997) Dynamical lifetimes of objects injected into asteroid belt resonances. *Science* **277**, 197–201.

Goldreich P. and Tremaine S. (1980) Disk–satellite interactions. *Astrophys. J.* **241**, 425–441.

Goldreich P. and Ward W. R. (1973) The formation of planetesimals. *Astrophys. J.* **183**, 1051–1062.

Goodman J. and Pindor B. (1999) Secular instability and planetesimal formation in the dust layer. *Icarus* **148**, 537–549.

Goswami J. N. and Vanhala H. A. T. (2000) Extinct radionuclides and the origin of the solar system. In *Protostars and Planets IV* (eds. V. Mannings, A. P. Boss, and S. S. Russell). University of Arizona Press, Tucson, pp. 963–994.

Gradie J. and Tedesco E. (1982) Compositional structure of the asteroid belt. *Science* **216**, 1405–1407.

Haghighipour N. and Boss A. P. (2003) On pressure gradients and rapid migration of solids in a nonuniform solar nebula. *Astrophys. J.* **583**, 996–1003.

Halliday A. N. and Porcelli D. (2001) In search of lost planets—the paleocosmochemistry of the inner solar system. *Earth Planet. Sci. Lett.* **192**, 545–559.

Hartmann L., Calvet N., Gullbring E., and D'Alessio P. (1998) Accretion and the evolution of T Tauri disks. *Astrophys. J.* **495**, 385–400.

Hollenbach D. J., Yorke H. W., and Johnstone D. (2000) Disk dispersal around young stars. In *Protostars and Planets IV* (eds. V. Mannings, A. P. Boss, and S. S. Russell). University of Arizona Press, Tucson, pp. 401–428.

Holzheid A., Sylvester P., O'Neill H., Rubie D. C., and Palme H. (2000) Evidence for a late chondritic veneer in the Earth's mantle from high-pressure partitioning of palladium and platinum. *Nature* **406**, 396–399.

Humayun M. and Clayton R. N. (1995) Potassium isotope chemistry: genetic implications of volatile element depletion. *Geochim. Cosmochim. Acta.* **59**, 2131–2148.

Huss G. R., MacPherson G. J., Wasserburg G. J., Russell S. S., and Srinivasan G. (2001) Aluminum 26 in calcium–aluminum-rich inclusions and chondrules from unequilibrated ordinary chondrites. *Meteoritics* **36**, 975–997.

Hutchison R. (2002) Major element fractionation in chondrites by distillation in the accretion disk of a T Tauri Sun? *Meteoritics* **37**, 113–124.

Ida S. and Makino J. (1993) Scattering of planetesimals by a protoplanet: slowing down of runaway growth. *Icarus* **106**, 210–227.

Ikoma M., Nakazawa K., and Emori H. (2000) Formation of giant planets: dependences on core accretion rate and grain opacity. *Astrophys. J.* **537**, 1013–1025.

Inaba S. and Wetherill G. W. (2001) Formation of Jupiter: core accretion model with fragmentation. *Lunar Planet. Sci.*, abstract 1384.

Jones R. H., Lee T., Connolly H. C., Love S. G., and Shang H. (2000) Formation of chondrules and CAIs: theory vs. observation. In *Protostars and Planets IV* (eds. V. Mannings, A. P. Boss, and S. S. Russell). University of Arizona Press, Tucson, pp. 927–961.

Kenyon S. J., Calvet N., and Hartmann L. (1993) The embedded young stars in the Taurus Auriga molecular cloud: I. Models for the spectral energy distributions. *Astrophys. J.* **414**, 676–694.

Klahr H. H. and Henning T. (1997) Particle-trapping eddies in protoplanetary accretion disks. *Icarus* **128**, 213–229.

Kleine T., Münker C., Mezger K., and Palme H. (2002) Rapid accretion and early core formation on asteroids and the terrestrial planets from Hf–W chronometry. *Nature* **418**, 952–955.

Kokubo E. and Ida S. (1996) On runaway growth of planetesimals. *Icarus* **123**, 180–191.

Kokubo E. and Ida S. (1998) Oligarchic growth of protoplanets. *Icarus* **131**, 171–187.

Kokubo E. and Ida S. (2000) Formation of protoplanets from planetesimals in the solar nebula. *Icarus* **143**, 15–27.

Kortenkamp S. J. and Wetherill G. W. (2000) Terrestrial planet and asteroid formation in the presence of giant planets. *Icarus* **143**, 60–73.

Laughlin G. and Bodenheimer P. (1994) Nonaxisymmetric evolution in protostellar disks. *Astrophys. J.* **436**, 335–354.

Lee D. C. and Halliday A. N. (1996) Hf–W isotopic evidence for rapid accretion and differentiation in the early solar system. *Science* **274**, 1876–1879.

Leinhardt Z. M. and Richardson D. C. (2002) *N*-body simulations of planetesimal evolution: effect of varying impactor mass ratio. *Icarus* **159**, 306–313.

Levison H. F. and Stewart G. R. (2001) Remarks on modeling the formation of Uranus and Neptune. *Icarus* **153**, 224–228.

Levison H. F., Duncan M., Zahnle K., Holman M., and Dones L. (2000) Planetary impact rates from ecliptic comets. *Icarus* **143**, 415–420.

Lin D. N. C. and Papaloizou J. (1986) On the tidal interaction between protoplanets and the protoplanetary disk: III. Orbital migration of protoplanets. *Astrophys. J.* **309**, 846–857.

Lissauer J. J. (1987) Timescales for planetary accretion and the structure of the protoplanetary disk. *Icarus* **69**, 249–265.

Lubow S. H., Seibert M., and Artymowicz P. (1999) Disk accretion onto high-mass planets. *Astrophys. J.* **526**, 1001–1012.

Lugmair G. W. and Shukolyukov A. (1998) Early solar system timescales according to 53Mn–53Cr systematics. *Geochim. Cosmochim. Acta* **62**, 2863–2886.

Lunine J. I., Owen T. C., and Brown R. H. (2000) The outer solar system: chemical constraints at low temperatures on planet formation. In *Protostars and Planets IV* (eds. V. Mannings, A. P. Boss, and S. S. Russell). University of Arizona Press, Tucson, pp. 1055–1080.

Lunine J. L., Morbidelli A., and Chambers J. E. (2002) Origin of water on Mars. *Lunar Planet. Sci.*, abstract 1791.

Lynden-Bell D. and Pringle J. E. (1974) The evolution of viscous discs and the origin of the nebula variables. *Mon. Not. Roy. Astron. Soc.* **168**, 603–637.

Marshall J. and Cuzzi J. (2001) Electrostatic enhancement of coagulation in protoplanetary nebulae. *Lunar Planet Sci.*, abstract 1262.

Mayer L., Quinn T., Wadsley J., and Stadel J. (2002) Formation of giant planets by fragmentation of protoplanetary disks. *Science* **298**, 1756–1759.

McKeegan K. D., Leshin L. A., Russell S. S., and MacPherson G. J. (1998) Oxygen isotopic abundances in calcium–aluminum rich inclusions from ordinary chondrites: implications for nebula heterogeneity. *Science* **280**, 414–418.

McKeegan K. D., Chaussidon M., and Robert F. (2000) Evidence for the *in situ* decay of ^{10}Be in an Allende CAI and implications for short-lived radioactivity in the early Solar System. *Lunar Planet. Sci.*, abstract 1999.

Morbidelli A., Chambers J. E., Lunine J. I., Petti J.-M., Robert F., Valsecchi G. B., and Cyr K. E. (2000) Source regions and timescales for the delivery of water on Earth. *Meteoritics Planet. Sci.* **35**, 1309–1320.

Nagasawa M., Tanaka M., and Ida S. (2000) Orbital evolution of asteroids due to sweeping secular resonances. *Astron. J.* **119**, 1480–1497.

Owen T., Mahaffy P., Niemann H. B., Sushil A., Donahue T., Bar-Nun A., and de Pater I. (1999) A low-temperature origin for the planetesimals that formed Jupiter. *Nature* **402**, 269–270.

Palme H., Larimer J. W., and Lipschutz M. E. (1988) Moderately volatile elements. In *Meteorites and the Early Solar System* (eds. J. F. Kerridge and M. S. Matthews). University of Arizona Press, Tucson, pp. 436–461.

Petit J. M., Morbidelli A., and Chambers J. E. (2001) The primordial excitation and clearing of the asteroid belt. *Icarus* **153**, 338–347.

Pollack J. B., Hubickyj O., Bodenheimer P., Lissauer J. J., Podolak M., and Greenzweig Y. (1996) Formation of giant planets by concurrent accretion of solids and gases. *Icarus* **124**, 62–85.

Poppe T., Blum J., and Henning Th. (2000) Analogous experiments on the stickiness of micron-sized preplanetary dust. *Astrophys. J.* **533**, 454–471.

Rafikov R. R. (2002) Planet migration and gap formation by tidally induced shocks. *Astrophys. J.* **572**, 566–579.

Rafikov R. R. (2003) The growth of planetary embryos: orderly, runaway or oligarchic? *Astron. J.* **125**, 942–961.

Righter K. and Drake M. J. (1999) Effect of water on metal–silicate partitioning of siderophile elements: a high pressure and temperature magma ocean and core formation. *Earth Planet. Sci. Lett.* **171**, 383–399.

Rubin A. E. and Krot A. N. (1996) Multiple heating of chondrules. In *Chondrules and the Protoplanetary Disk* (eds. R. H. Hewins, R. H. Jones, and E. R. D. Scott). Cambridge University Press, Cambridge, pp. 173–180.

Safranov V. S. (1969) *Evolution of the Protoplanetary Cloud and Formation of the Earth and Planets* (English translation NASA TTF-677, 1972).

Sahijpal S., Srinivasan G., Wasserburg G. J., and Goswami J. N. (1995) Observation of correlated 41Ca and 26Al in CV3 hibonites. *Meteoritics* **30**, 570–571.

Sasaki S. (1990) Heating of an accreting protoplanet by blanketing effect of a primary solar-composition atmosphere. *Lunar Planet. Sci.*, abstract 1067.

Shu F. H., Shang H., and Lee T. (1996) Toward an astrophysical model of chondrites. *Science* **271**, 1545–1552.

Shu F. H., Lizano S., Ruden S. P., and Najita J. (1988) Mass loss from rapidly rotating magnetic protostars. *Astrophys. J.* **328**, L19–L23.

Shukolyukov A. and Lugmair G. W. (1993) Live iron-60 in the early solar system. *Science* **259**, 1138–1142.

Srinivasan G., Goswami J. N., and Bhandari N. (1999) ^{26}Al in eucrite Piplia Kalan: plausible heat source and formation chronology. *Science* **284**, 1348–1350.

Stevenson D. J. and Lunine J. I. (1988) Rapid formation of Jupiter by diffuse redistribution of water vapor in the Solar Nebula. *Icarus* **75**, 146–155.

Stewart G. R. and Kaula W. M. (1980) Gravitational kinetic theory for planetesimals. *Icarus* **44**, 154–171.

Strom S. E. (1994) The early evolution of stars. *Rev. Mex. Astron. Astrophys.* **29**, 23–29.

Strom S. E. (1995) Initial frequency, lifetime and evolution of YSO disks. *Rev. Mex. Astron. Astrophys.* **1**, 317–328.

Tanaka H. and Ida S. (1999) Growth of a migrating protoplanet. *Icarus* **139**, 350–366.

Tanaka H., Takeuchi T., and Ward W. R. (2002) Three-dimensional interaction between a planet and an isothermal gaseous disk: I. Corotation and Lindblad torques and planet migration. *Astrophys. J.* **565**, 1257–1274.

Taylor S. R. (2001) *Solar System Evolution, A New Perspective*, 2nd edn. Cambridge University Press, Cambridge.

Thommes E. W., Duncan M. J., and Levison H. F. (1999) The formation of Uranus and Neptune in the Jupiter–Saturn region of the solar system. *Nature* **402**, 635–638.

Thommes E. W., Duncan M. J., and Levison H. F. (2003) Oligarchic growth of giant planets. *Icarus* **161**, 431–455.

Tonks W. B. and Melosh H. J. (1992) Core formation by giant impacts. *Icarus* **100**, 326–346.

Vanhala H. A. T. and Boss A. P. (2000) Injection of radioactivities into the presolar cloud: convergence testing. *Astrophys. J.* **538**, 911–921.

Wadhwa M. and Russell S. S. (2000) Timescales of accretion and differentiation in the early solar system: the meteoritic evidence. In *Protostars and Planets IV* (eds. V. Mannings, A. P. Boss, and S. S. Russell). University of Arizona Press, Tucson, pp. 995–1018.

Ward W. R. (1997) Survival of planetary systems. *Astrophys. J.* **482**, L211–L214.

Ward W. R. (2001) On planetesimal formation: the role of collective particle behaviour. *Origin of the Earth and Moon* (eds. R. M. Canup and K. Righter). University of Arizona Press, Tucson, pp. 75–84.

Ward W. R., Colombo G., and Franklin F. A. (1976) Secular resonance, solar spin down and the orbit of Mercury. *Icarus* **28**, 441–452.

Ward W. R. and Hahn J. M. (2000) Disk–planet interactions and the formation of planetary systems. In *Protostars and Planets IV* (eds. V. Mannings, A. P. Boss, and S. S. Russell). University of Arizona Press, Tucson, pp. 1135–1155.

Wasson J. T. and Chou C. L. (1974) Fractionation of moderately volatile elements in ordinary chondrites. *Meteoritics* **9**, 69–84.

Weidenschilling S. J. (1977a) The distribution of mass in the planetary system and Solar Nebula. *Astrophys. Space Sci.* **51**, 153–158.

Weidenschilling S. J. (1977b) Aerodynamics of solid bodies in the solar nebula. *Mon. Not. Roy. Astron. Soc.* **180**, 57–70.

Weidenschilling S. J. (1980) Dust to planetesimals. *Icarus* **44**, 172–189.

Weidenschilling S. J. (1988) Formation processes and timescales for meteorite parent bodies. In *Meteorites and the Early Solar System* (eds. J. F. Kerridge and M. S. Matthews). University of Arizona Press, Tucson.

Weidenschilling S. J. and Ruzmaikina T. V. (1994) Coagulation of grains in static and collapsing protostellar clouds. *Astrophys. J.* **430**, 713–726.

Weidenschilling S. J., Spaute D., Davis D. R., Marzari F., and Ohtsuki K. (1997) Accretional evolution of a planetesimal swarm. *Icarus* **128**, 429–455.

Wetherill G. W. (1988) Accumulation of Mercury from planetesimals. The cratering record on Mercury and the origin of impacting objects. In *Mercury* (eds. F. Vilas, C. R. Chapman, and M. S. Matthews). University of Arizona Press, Tucson, pp. 670–691.

Wetherill G. W. (1994) Provenance of the terrestrial planets. *Geochim. Cosmochim. Acta* **58**, 4513–4520.

Wetherill G. W. and Stewart G. R. (1993) Formation of planetary embryos: effects of fragmentation, low relative velocity, and independent variation of eccentricity and inclination. *Icarus* **106**, 190–209.

Wiechert U., Halliday A. N., Lee D. C., Snyder G. A., Taylor L. A., and Rumble D. (2001) Oxygen isotopes and the Moon-forming giant impact. *Science* **294**, 345–348.

Wilde S. A., Valley J. W., Peck W. H., and Graham C. M. (2001) Evidence from detrital zircons for the existence of continental crust and oceans on the Earth 4.4 Gyr ago. *Nature* **409**, 175–178.

Woolum D. S. and Cassen P. (1999) Astronomical constraints on nebular temperatures: implications for planetesimal formation. *Meteoritics* **34**, 897–907.

WuchterI G., Guillot T., and Lissauer J. J. (2000) Giant planet formation. In *Protostars and Planets IV* (eds. V. Mannings, A. P. Boss, and S. S. Russell). University of Arizona Press, Tucson, pp. 1081–1109.

Wurm G., Blum J., and Colwell J. E. (2001) A new mechanism relevant to the formation of planetesimals in the Solar Nebula. *Icarus* **151**, 318–321.

Yin Q. Z., Jacobsen S. B., Yamashita K., Blicher-Toft J., Télouk P., and Albarède F. (2002) A short timescale for terrestrial planet formation from Hf–W chronometry of meteorites. *Nature* **418**, 949–952.

Youdin A. N. and Shu F. H. (2002) Planetesimal formation by gravitational instability. *Astrophys. J.* **580**, 494–505.

Zahnle K. (1998) Origins of atmospheres. *ASP Conf. Ser.* **148**, 364–391.

Zhu X. K., Guo Y., O'Nions R. K., Young E. D., and Ash R. D. (2001) Isotopic homogeneity of iron in the early solar nebula. *Nature* **412**, 311–312.

1.18
Mercury

G. J. Taylor and E. R. D. Scott

University of Hawaii, Honolulu, HI, USA

1.18.1 INTRODUCTION: THE IMPORTANCE OF MERCURY	477
1.18.2 WHAT WE KNOW ABOUT THE CHEMICAL COMPOSITION OF MERCURY	478
1.18.2.1 Density Implies High Metal/Silicate Ratio	478
1.18.2.2 Spectral Properties Imply Low FeO	478
1.18.2.3 Small Inventory of Volatile Elements	480
1.18.3 MODELS FOR THE COMPOSITION OF MERCURY	480
1.18.3.1 Metal Enrichment	480
1.18.3.2 Evaporation Models	481
1.18.3.3 Refractory Condensation Models	481
1.18.3.4 Refractory–Volatile Mixtures	482
1.18.3.5 Metal-rich Chondrite Model	482
1.18.3.6 Enstatite Meteorite Model	483
1.18.4 IMPLICATIONS FOR PLANETARY ACCRETION	483
ACKNOWLEDGMENTS	484
REFERENCES	484

1.18.1 INTRODUCTION: THE IMPORTANCE OF MERCURY

Mercury is an important part of the solar system puzzle, yet we know less about it than any other planet, except Pluto. Mercury is the smallest of the terrestrial planets (0.05 Earth masses) and the closest to the Sun. Its relatively high density (5.4 g cm^{-3}) indicates that it has a large metallic core (~3/4 of the planet's radius) compared to its silicate mantle and crust. The existence of a magnetic field implies that the metallic core is still partly molten. The surface is heavily cratered like the highlands of the Moon, but some areas are smooth and less cratered, possibly like the lunar maria (but not as dark). Its surface composition, as explained in the next section, appears to be low in FeO (only ~3 wt.%), which implies that either its crust is anorthositic (Jeanloz et al., 1995) or its mantle is similarly low in FeO (Robinson and Taylor, 2001).

The proximity of Mercury to the Sun is particularly important. In one somewhat outmoded view of how the solar system formed, Mercury was assembled in the hottest region close to the Sun so that virtually all of the iron was in the metallic state, rather than oxidized to FeO (e.g., Lewis, 1972, 1974). If correct, Mercury ought to have relatively a low content of FeO. This hypothesis also predicts that Mercury should have high concentrations of refractory elements, such as calcium, aluminum, and thorium, and low concentrations of volatile elements, such as sodium and potassium, compared to the other terrestrial planets.

Alternative hypotheses tell a much more nomadic and dramatic story of Mercury's birth. In one alternative view, wandering planetesimals that might have come from as far away as Mars or the inner asteroid belt accreted to form Mercury (Wetherill, 1994). This model predicts higher FeO and volatile elements than does the high-temperature model, and similar compositions among the terrestrial planets. The accretion process might have been accompanied by a monumental impact that stripped away much of the young planet's rocky mantle, accounting for the high density of the planet (Benz et al., 1988). Most planetary scientists consider such a giant impact as the most likely hypothesis for the origin of the Moon. A giant impact model could explain the high density of Mercury if much of the silicate material failed to reaccrete, but it would not explain the low FeO concentration of the planet. Thus, knowing the composition of Mercury is crucial to testing models of planetary accretion.

In this chapter we summarize what we know about the chemical composition of Mercury, with emphasis on assessing the amount of FeO in the bulk planet. FeO is a particularly useful quantity to evaluate the extent to which Mercury is enriched in refractory elements, because its concentration increases with decreasing temperature in a cooling gas of solar composition (e.g., Goettel, 1988). We then examine models for the composition of Mercury and outline tests that future orbital missions to Mercury will be able to make.

1.18.2 WHAT WE KNOW ABOUT THE CHEMICAL COMPOSITION OF MERCURY

Not much is known empirically about the bulk chemical composition of Mercury. We emphasize the three compositional features that are reasonably well determined: a large metallic core, a low FeO content, and the presence of at least a little sodium. This skimpy list gives us surprisingly useful information.

1.18.2.1 Density Implies High Metal/Silicate Ratio

The mean density of Mercury is 5.43 ± 0.01 g cm^{-3} (Anderson et al., 1987). This corresponds to a reduced (or uncompressed) density of 5.31 (Kaula, 1986). The reduced density is the density a planet would have if it was not compressed, but was at high enough pressure to squeeze out all the pore spaces, ~ 10 kbar (Kaula, 1986). Calculation of the reduced densities of the terrestrial planets assumes that the core is composed of iron and nickel, and that all the metal is in the core. The reduced density of Mercury is by far the largest of the terrestrial planets. Earth is the closest, with a reduced density of only 4.03 g cm^{-3}.

This high density has been interpreted to mean that Mercury has a large metallic FeNi core. Such a core would make up $\sim 70\%$ of the planet's mass and comprise $\sim 40\%$ of its volume, assuming a silicate density of ~ 3 g cm^{-3}. If this is correct, Mercury has a very high Fe/Si ratio. However, as Goettel (1988) points out, we have no direct evidence that the core is composed of iron and nickel only. If the core is made of pure FeS, it would make up an even larger percentage of the planet. Nevertheless, the most straightforward interpretation of the density value is that Mercury has a very large metallic core. Measurements of the moment of inertia, topography, and gravity field of Mercury by future missions will be able to place strong constraints on the size of the core and the thickness of the crust (e.g., Spohn et al., 2001).

1.18.2.2 Spectral Properties Imply Low FeO

A variety of spectral observations using Earth-based telescopes show that the surface of Mercury is low in FeO + TiO$_2$. Reflectance observations are summarized by Vilas (1988). Observations of Mercury in the mid-IR are reported by Sprague et al. (1994) and in the microwave and mid-IR by Jeanloz et al. (1995). Such observations are difficult to make because of the small angular separation between Mercury and the Sun. This results in a long path through the atmosphere and concomitantly large corrections. Nevertheless, the spectral observations provide clear evidence that the surface of Mercury is low in FeO.

The Moon has been used as a useful comparison to Mercury. Both bodies are airless, so space weathering effects ought to be similar, though probably more intense on Mercury: velocities of micrometeorites are higher and solar-wind implantation is more intense (Cintala, 1992), although Mercury's magnetic field may deflect some solar-wind gases. Blewett et al. (2002) and Warrell (2003) made detailed comparisons between spectra of Mercury and those of heavily space weathered ("mature") pure anorthosite regions on the Moon. Such regions (whether mature or not) have <3 wt.% FeO. Lunar anorthosite and mercurian spectra are quite similar. This implies that most of Mercury's surface has an FeO content like that of mature lunar anorthosites, <3 wt.%. This is consistent with an assessment of the FeO content of the mercurian surface (Blewett et al., 1997) by a different spectral technique developed for the Moon by Lucey et al. (1995). This approach allows both maturity and FeO to be deduced and indicates an FeO content of ~ 3 wt.%.

Burbine et al. (2002) tested an extreme case of a possible composition for the surface of Mercury. They made spectral observations of enstatite achondrites (igneous meteorites composed almost entirely of pure MgSiO$_3$, with some accessory minerals and essentially no FeO). The spectral features of enstatite achondrites (aubrites; see Chapter 1.05) are similar to those for Mercury, but lack the spectral reddening observed in spectra of Mercury and have an additional feature at 0.5 μm caused by troilite (FeS). This reddening (visible to UV ratio) is the result of space weathering, in which FeO is reduced to very small grains of metallic iron. Thus, the reddening indicates that some FeO must be present on Mercury to produce the nanophase iron. Alternatively,

Noble and Pieters (2001) suggest that metallic iron might have been added to a FeO-free surface by meteorite impact.

All these observations of Mercury indicate a small concentration of FeO on its surface, probably ~3 wt.%, though possibly even less. Is the interior similarly depleted in FeO? The compositions of volcanic deposits on planets can be used to probe the composition of the interior. There has always been some debate about whether smooth plains on Mercury are volcanic or impact products, but the balance of evidence from morphology and color measurements favors the presence of volcanic plains on Mercury (Murray, 1975; Murray et al., 1975; Strom, 1977; Dzurisin, 1978; Hapke et al., 1980; Rava and Hapke, 1987; Keiffer and Murray, 1987; Spudis and Guest, 1988). Robinson and Lucey (1997) recalibrated the Mariner 10 data for Mercury and concluded that there were compositional heterogeneities in the crust, though all of it is low in FeO. Using UV (375 nm) and orange (575 nm) mosaics of Mercury, they analyzed the data according to two trends established from lunar studies and theory. One shows decreasing albedo and UV/visible ratio with increasing FeO and increasing maturity. The second trend, which is due to addition of a spectrally neutral opaque mineral (e.g., ilmenite), causes the albedo to decrease as the UV/visible ratio increases. These trends can be separated and image maps of Mercury made of the resulting spectral parameters. The images show that the opaque parameter varies, while the FeO-maturity parameter remains relatively constant.

Many smooth plains, such as those associated with Rudaki crater and Tolstoj basin (Figure 1), have morphological features consistent with their origin as lava flows. The embayments indicated by the arrows in Figure 1 are similar to those of the lunar maria. More important, the smooth plains in Rudaki and Tolstoj have lower concentrations of opaque minerals (hence of TiO_2) than the highlands that surround them. This indicates that they are not simply impact deposits of average crustal material; they are partial melts of the interior. The plains do not differ in FeO and maturity from the global median on Mercury (Robinson and Lucey, 1997; Robinson and Taylor, 2001), indicating that these lava flows have the same FeO content as the global mean, ~3 wt.%.

Lava flows can be used as probes of a planet's interior, assuming that they have not fractionated dramatically as they migrated to the surface. It turns out that FeO is not greatly fractionated during partial melting and modest fractional crystallization, though the Fe/Mg ratio varies dramatically because of magnesium fractionation. Robinson and Taylor (2001) assembled experimental and observational data to show this quantitatively. For example, the average FeO concentration of terrestrial mid-ocean ridge basalts divided by the composition of the primitive terrestrial mantle is only 1.3. Thus, the FeO concentration of a lava is only ~30% higher than the concentration of its mantle source region. This led Robinson and Taylor (2001) to conclude that Mercury contained 2–3 wt.% FeO in its interior. Table 1 summarizes the inferred FeO contents of surface basalts and the interiors of the terrestrial planets and asteroid 4 Vesta. There appears to be a clear gradation in FeO with heliocentric distance, with Mercury lowest (2–3 wt.%), Earth and Venus intermediate (~8 wt.%), and Mars and Vesta highest (18–20 wt.%).

The morphological and spectral arguments for the existence of lava flows on Mercury are strong, but not unassailable. Jeanloz et al. (1995) argue on the basis of low FeO and mid-IR emission spectral data (Sprague et al., 1994) that the surface is dominated by plagioclase. The similarity of

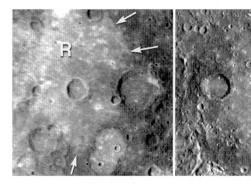

Figure 1 Mercurian smooth plains exhibit embaying boundaries similar to those found on the lunar maria, but because of their much lower FeO, they are much brighter than the lunar maria. The image on the left is of Rudaki plains (R), 3° S, 56° W; that on the right is of smooth plains in Tolstoj (T), 16° S, 164° W. The images were acquired with the Mariner 10 cameras. Scale of both images is 1 km per pixel at the center, orthographic projection, 500 × 500 pixels, north to the top.

Table 1 FeO (wt.%) in planetary basalts and interiors.

Planet	FeO in basalts (wt.%)	Refs.	Bulk planet FeO[a] (wt.%)	Refs.
Mercury	≤6 (conservative) 3 (preferred)	a	2–3	a
Venus	8.6	b, c	7–8	a
Earth	10.5 (MORB)	d	8	g, h, i
Mars	20	e	18	j
Vesta	18.5	f	20	k

After Robinson and Taylor (2001).
[a] These are the most reasonable estimates of the bulk compositions of the planets made on the basis of the compositions of rocks from their surfaces or, for Earth, surface and mantle.
References: (a) Robinson and Taylor (2001); (b) Surkov et al. (1984); (c) Surkov et al. (1986); (d) Melson et al. (1976); (e) Meyer (2003) and references therein; (f) Warren (1997); (g) Jagoutz et al. (1979); (h) Taylor and McLennan (1985); (i) McDonough and Sun (1995); (j) Longhi et al. (1992); and (k) Average of estimates in table 1 of Warren (1997).

the reflectance spectra of Mercury to the lunar highlands (Blewett et al., 2002) is consistent with this interpretation. This suggests that Mercury's crust might be much like the lunar highlands and might have formed in a magma ocean. Jeanloz et al. (1995) conclude that after the initial differentiation of Mercury, very little magma managed to reach the surface of Mercury. In this view, the smooth plains are impact products, not lava flows, and the smooth plains are compositionally similar to the rest of the highlands. If Jeanloz et al. are correct, then it is very difficult to deduce much about the interior composition of the planet. This highlights the basic problem that we do not know anything about the differentiation history of Mercury, which currently limits our understanding of its interior. Did Mercury have a magma ocean? Or did its ancient crust form by serial magmatism early in its history? We hope that data from NASA's MESSENGER and ESA's BepiColumbo will answer these questions.

1.18.2.3 Small Inventory of Volatile Elements

Mercury has a tenuous atmosphere that contains sodium and potassium (Potter and Morgan, 1985, 1986). These elements are liberated from the surface by a variety of loss mechanisms, apparently dominated by sputtering (e.g., Killen and Ip, 1999). Although it is not possible to determine the sodium and potassium abundance on the surface from their abundance in the atmosphere, their presence shows that some sodium and potassium must be present. This is very important in assessing the volatile inventory of the planet. Similarly, Sprague et al. (1995) suggested on the basis of mid-IR spectral observations that the surface of Mercury is relatively rich in iron sulfides. They also suggest that the radar anomalies in polar regions, which are generally attributed to cometary water ice (Slade et al., 1992; Harmon and Slade, 1992), might be caused by deposits of elemental sulfur in cold traps. So, although not as quantitative as the limit on the abundance of FeO, these observations suggest that Mercury is not totally devoid of volatile constituents.

1.18.3 MODELS FOR THE COMPOSITION OF MERCURY

There are several classes of models for the composition of Mercury: (i) either differential accumulation of metal and silicate or blasting away the silicate from a differentiated Mercury; (ii) high temperatures in the region where Mercury formed; (iii) mixing high-temperature ingredients with more volatile ones; and (iv) comparison of Mercury to appropriate chondritic meteorites. We discuss these here and point out what chemical and mineralogical tests the Messenger and BepiColumbo missions to orbit Mercury in 2009–2015 will be able to make in order to distinguish among the models. Both missions will carry sensors to measure the concentrations of several elements from orbit (γ-ray and X-ray spectroscopy) and will be able to assess the mineralogy of the surface through reflectance spectra at visible and near-IR wavelengths.

1.18.3.1 Metal Enrichment

Some authors have focused on physical mechanisms to account for the high density of Mercury, searching for ways to enrich metal compared to silicate. Weidenschilling (1976) appealed to aerodynamic fractionation to separate metal from silicate as Mercury was accreting. A more dramatic model involves a giant impact stripping the silicate mantle off Mercury after it has differentiated (Benz et al., 1988; Cameron et al., 1988; Wetherill, 1988). Both models account for the high metal/silicate ratio of Mercury, but neither explicitly predicts the FeO content of the planet. However, the impact model implies scrambling of planetesimals from a wide range of heliocentric distances (to account for the high relative velocities of impacting planetesimals). This implies that Mercury ought to have a relatively high FeO content like the other terrestrial planets (Table 1), but it does not. In addition, successful impact models required the following extreme conditions (Benz et al., 1988): (i) the bodies contain exceptionally large (32 wt.%) metal concentrations initially; (ii) the collisions were either head-on at 20 km s^{-1} or oblique at implausibly high collisions speeds of 35 km s^{-1}; and (iii) all the silicate ejecta is removed as sub-centimeter-sized particles by the Poynting–Robertson effect; otherwise silicates would reaccrete onto Mercury.

Tests. These models are almost completely unconstrained. The best bet is that concentrations of FeO and volatile elements will be larger than for other models. However, it is important to keep in mind that these metal-enhancing processes could have operated in consort with other processes that produced Mercury.

1.18.3.2 Evaporation Models

Cameron (1985) estimated that temperatures in the solar nebula could have reached 2,500–3,500 K at Mercury's distance from the Sun. Recent models of the solar nebula do not predict such high temperatures at 0.4 AU, but it is an interesting idea worth considering. If Mercury had formed very early, perhaps by gravitational instabilities in the gas phase, then its composition could be affected by these high temperatures. There would be two main effects. First, much of the silicate would be vaporized and lost, thereby increasing the ratio of metal to silicate. Cameron (1985) assumed that the original Mercury was ~2.25 times the mass of the present Mercury. Second, the composition of the silicate would become more refractory. Fegley and Cameron (1987) calculated the compositions of the silicate portion of the hypothetical Mercury, using both ideal and nonideal silicate magmas, assuming it began with oxides in solar proportions. They report four cases, and we reproduce in Table 2 (column 1) the one with the closest FeO concentration to our estimate of the FeO concentration in Mercury. This composition is very distinctive, characterized by very low SiO$_2$, high MgO, and no K. Thorium is greatly fractionated from uranium, due to the formation of volatile UO$_3$ gas.

Tests. Orbital measurements will see very low SiO$_2$ and high MgO, modest Al$_2$O$_3$, Al/Th roughly of chondritic proportions, and highly fractionated Th/U. The mineralogy is difficult to predict.

1.18.3.3 Refractory Condensation Models

Theoretical models for the solar nebula indicate a decreasing temperature with increasing distance from the Sun. If the thermal gradient were steep enough, the composition of condensates from the nebula, if at equilibrium, would vary as a function of distance and temperature (Lewis, 1972, 1974). Mercury might have accreted from

Table 2 Models for the bulk chemical composition (wt.%) of the silicate portion of Mercury (columns 1–6) and three possible surface magma compositions (columns 7–9).

	Suggested bulk compositions						Surface lava flow compositions		
	1	2	3	4	5	6	7	8	9
SiO$_2$	19.9	32.6	38–48	47.1	37.6	50.8	49.0	44.6	52.7
TiO$_2$	0.6	0.7	0.15–0.30	0.3	0.5	0.2	1.7	0.9	0.2
Al$_2$O$_3$	14.4	16.6	3.5–7	6.4	11.7	5.2	14.3	13.1	13.5
Cr$_2$O$_3$	0.0	0.0	0.0	3.3	0.2	0.7	1.6	0.3	
FeO	3.4	0.0	0.5–5	3.7	3.0	2.0	3.7	3.2	0.25
MnO	0.0	0.0	0.0	0.06	0.05	0.1	0.06	0.13	0.12
MgO	50.0	34.6	32–38	33.7	36.1	37.4	10.8	18.9	19.1
CaO	11.7	15.2	3.5–7	5.2	10.7	3.6	18.4	18.7	11.7
Na$_2$O	0.0	0.0	0.2–1	0.08	0.1	<0.04	0.3		
K$_2$O	0.0	0.0	a little	0.01	0.01	<0.04	0.08		
Th (ng g^{-1})	308	293[a]	60–120[a]	122	207[a]	93[a]	1,220[a]	930[a]	
U (ng g^{-1})	0.2	81[a]	17–33[a]	34	58[a]	26[a]	340[a]	258[a]	
Th/U	1.5 × 10^3	3.6[a]	3.6[a]	3.6	3.6[a]	3.6[a]	3.6[a]	3.6[a]	

[a] These estimates are from this work. They were made by assuming a chondritic Al/Th and Th/U ratios (columns 2–5). For column 6 the assumption is that U and Th have identical partition coefficients of 0.001, so the magma retains the same Th/U as the mantle source area. The estimate assumes 10% partial melting (hence Th is 10 times higher in the magma than in the original source, given in column 5).
Columns: (1) Vaporization model (model 4 of Fegley and Cameron, 1987). (2) Refractory end-member (Goettel, 1988). (3) Preferred model of Goettel (1988). (4) Model composition of Mercury (Morgan and Anders, 1980). (5) Mixture of Goettel's (1988) refractory end-member (column 2) and bulk silicate earth (Jagoutz et al., 1979; Taylor and McLennan, 1985; McDonough and Sun, 1995), to make FeO = 3 wt.% (61% refractory end-member and 39% Earth). (6) Average of skeletal olivine and cryptocrystalline chondrules in metal-rich chondrites (Krot et al., 2001). (7) Calculated 10% partial melt at 10 kbar of the bulk composition shown in column 4. (8) Calculated 10% partial melt at 10 kbar of bulk composition shown in column 6. (9) Experimental partial melt of the Indarch enstatite chondrite at 1,425 °C (29% partial melt).

high-temperature condensates. These would be composed of metallic iron and FeO-free silicates. This could explain why Mercury has a large metallic core: iron condenses before magnesium silicates, so could accrete first. However, the condensation temperatures of metallic iron and magnesium silicates (enstatite and forsterite) are really so close that it would require very rapid accretion of Mercury. Furthermore, the order of metal and magnesian silicate condensation changes as a function of pressure. Above 10^{-4} atm, metal condenses first; below 10^{-4} atm olivine condenses first, as first pointed out by Grossman (1972). This led to imaginative models for separating metal from silicate, such as Weidenschilling's (1976) sorting mechanism. The model also fell victim to the view that there was widespread mixing of planetesimals during planetary accretion, which would have erased much of the variation caused by condensation temperature (though Table 1 suggests that mixing was far from complete).

The refractory condensate model has fallen out of favor, including with Lewis (1988). Nevertheless, it is a useful end-member case. Goettel (1988) calculated the composition of the silicate portion of an ultrarefractory Mercury (Table 2, column 2). This model composition contains no FeO or volatiles, and has large concentrations of the refractory elements—aluminum, calcium, and magnesium. We calculated the thorium and uranium contents of such refractory condensates by assuming chondritic Al/Th and Al/U ratios. A surface of this composition will contain many of the phases in calcium–aluminum-rich inclusions (CAIs), such as forsterite, anorthite, spinel, perovskite, hibonite, and melilite.

Tests. If Mercury is this refractory, its surface will have high concentrations of aluminum, calcium, magnesium, no FeO, sodium, or potassium, and chondritic abundances of thorium and uranium. The surface mineral assemblage would be like that of CAIs. (Jeanloz et al. (1995) note that there is no spectral evidence for perovskite on Mercury. This observation, combined with the presence of at least some FeO, dampens enthusiasm for this type of extreme model.)

1.18.3.4 Refractory–Volatile Mixtures

The presence of FeO on Mercury and the notion that planetary accretion involved mixing throughout the inner solar system motivated models involving mixing of refractory and volatile materials. Goettel (1988) made broad estimates of Mercury's composition by combining his calculated refractory and volatile end-members (Table 2, column 3). Morgan and Anders (1980) used an elaborate seven-component model to calculate the composition of Mercury (Table 2, column 4). We have approached the problem in a slightly different way, devising a simple two-component model. We assume that Mercury contains a higher percentage of refractory materials than do the other planets because the planetesimals that accreted to form it probably formed, on average, closer to the Sun. Thus, we use Goettel's refractory end-member for one component (Table 2, column 2). We then assume that the Earth represents a good average of the bulk of the remaining planetesimals that formed Mercury. This leads us to use the Earth's primitive mantle as the second end-member. We mix the two end-members to give us 3 wt.% FeO in the model Mercury. The result is shown in Table 2, column 5.

Goettel's preferred composition and the composition calculated by Morgan and Anders are quite similar. The abundances of refractory elements are modest, but larger than those for the Earth (Jagoutz et al., 1979; Taylor and McLennan, 1985; McDonough and Sun, 1995). Both compositions have small amounts of sodium and potassium, as required by the presence of these elements in the mercurian atmosphere. The Morgan and Anders composition has surprisingly high Cr_2O_3. Our refractory + Earth mixture is much richer in refractory elements (aluminum, magnesium, calcium, and thorium). This causes a low amount of SiO_2 and perhaps too low a potassium content.

Orbiting spacecraft will be able to determine the compositions of smooth plains thought to be lava flows on Mercury. So, we calculated the compositions of a partial melt of the Morgan and Anders composition (Table 2, column 7). The calculation was done using MAGPOX (a phase-equilibria program written by John Longhi of Lamont–Doherty Earth Observatory) assuming 10% melting at a pressure of 10 kbar. We estimated the amount of thorium and uranium in the magma by assuming that their distribution coefficients were 0.001. The calculated magmas are still low in FeO, but have normal basaltic levels of aluminum, calcium, and magnesium. Sodium and potassium are enhanced compared to the assumed source rock.

Tests. These models have chondritic Th/U, subchondritic K/Th, and relatively low contents of refractory components compared to the more extreme cases described above. A typical magma produced from these compositions would have a low-FeO basaltic composition. The mineralogy of the basalt would be quartz normative for the Morgan and Anders bulk composition and consist of plagioclase and high-calcium pyroxene.

1.18.3.5 Metal-rich Chondrite Model

The notion that chondrites were the building blocks of the planets has a long history, but enthusiasm for the idea has waned, largely because there is no match between any chondrite group and the bulk Earth (Taylor, 1991; Drake and Righter, 2002). However, the discovery of three new chondrite groups (CR, CH, and Bencubbin-like) among Antarctic meteorites has greatly expanded the compositional range of chondrites (Scott, 1988; Grossman et al., 1988; Weisberg et al., 1993, 1995; Krot et al., 2001). The new chondrites have normal levels of refractory elements but are richer in metallic iron (some were classed as iron meteorites) and poorer in volatile elements like sodium, potassium, and sulfur than other chondrites. Bencubbin-like chondrites have more metal (~80 wt.%) than Mercury (~70 wt.%) and comparable FeO concentrations (Figure 2). We have suggested that metal-rich chondrites might be suitable Mercury building blocks (Scott and Taylor, 2000; Taylor and Scott, 2001; Scott, 2002).

We have modeled the silicate portion of Mercury by averaging the bulk compositions of primitive chondrules in two metal-rich chondrites (Krot et al., 2001); the result is shown in Table 2, column 6. This composition is actually quite similar to the one calculated by Morgan and Anders (1980), but has the virtue of being based on the compositions of real objects. It is somewhat lower in FeO, but still in the range of acceptable FeO for Mercury. Most importantly, it is lower in refractory elements (calcium and aluminum) and, ironically, lower in volatiles (sodium and potassium). A magma produced by 10% partial melting at 10 kbar (Table 2, column 8) is likewise similar to the partial melt of the Morgan and Anders composition (Table 2, column 7), except for higher MgO and lower sodium and potassium.

Tests. The compositional tests are similar to those for the refractory–volatile mixtures, though the metal-rich chondrite composition has lower concentrations of volatiles. However, the partial melt composition is olivine normative, so the surface mineralogy will consist of olivine (~20%), high-calcium pyroxene (~40%), and plagioclase (35%), plus minor phases (5%).

1.18.3.6 Enstatite Meteorite Model

Enstatite chondrites have also been considered as possible analogues for Mercury (Wasson, 1988). Like the hypotheses depicting Mercury as refractory rich, this idea has been advanced because of Mercury's location close to the Sun, where Wasson (1988) suggests enstatite chondrites formed. Burbine et al. (2002) studied the spectral properties of enstatite achondrites and concluded that they could fit observed spectra, though there are features due to sulfides at ~0.6 μm and less that have not yet been observed in spectra from Mercury. Features attributable to sulfides have been observed in the mid-IR (Sprague et al., 1995). We used partial melting experiments of an enstatite chondrite (McCoy et al., 1999) to estimate what the surface of Mercury might be like if its bulk composition were like a typical enstatite chondrite. This composition is shown in Table 2, column 9. It is lower in FeO and CaO compared to the calculated partial melts of other prospective mercurian bulk compositions. Its normative mineralogy is dominated by plagioclase (34%), enstatite (42%), high-calcium pyroxene (15%), and sulfides (6%).

Tests. If Mercury has a bulk composition as that of enstatite chondrites, the FeO content of the surface will be very low. Magmas produced by partial melting will be low in FeO and CaO, and the mineralogy will be dominated by low-calcium pyroxene, not diopsidic pyroxene.

Figure 2 Reflected-light photograph of the Weatherford meteorite, a metal-rich chondrite. Metal grains are bright, silicate is dark. Metal-rich chondrites contain up to 80% metal coexisting with silicates with low concentrations of FeO, suggesting that they might represent Mercury's bulk composition (photo courtesy of the Smithsonian Institution).

1.18.4 IMPLICATIONS FOR PLANETARY ACCRETION

Both observations and estimated compositions of Mercury indicate that the planet has a low concentration of FeO, and there appears to be a diversity of compositions among the terrestrial planets (Table 1). On the basis of chondrite compositions, the maximum range in FeO is from 0 wt.% to 30 wt.% (enstatite to carbonaceous chondrites). The range in planet compositions covers more than half this range, 3–18 wt.% (20 wt.% if we include Vesta). The feeding zone for Mercury might have included planetesimals from further out in the nebula, as argued by Wetherill (1988, 1994), but it seems to have been dominated by planetesimals formed near its current location. The FeO differences would not have survived if the Wetherill model is completely correct. In fact, Wetherill (1994) acknowledges that the concept of narrow feeding zones is probably satisfactory up to the stage of planetary accretion at which bodies between the sizes of the Moon and Mercury ($\sim 10^{26}$ g) formed rapidly by runaway growth. Safronov (1972) called such bodies planetary embryos. Thus, Mercury might well be a planetary embryo. Venus and Earth are also quite distinct from Mars, shown strikingly by their low FeO ($\sim$8 wt.%, versus 18 wt.% for Mars). This suggests that they might also have formed mostly from metal-rich chondrites (including enstatite chondrites), and possibly from planetesimals formed well within the orbit of Mars, again suggesting limited mixing (though perhaps much more than in the case of Mercury, if it is an embryo). Recent simulations of planetary accretion using new computer techniques also indicate more narrow feeding zones (Chambers, 2001). The amount of calculated mixing is low enough to preserve primary compositional differences among the terrestrial planets. Drake and Righter (2002) have reached the same conclusion on the basis of comparison of the Earth's bulk composition with chondrites.

Mercury is an important object. It is closest to the Sun, might be a planetary embryo, and may record the extent of mixing during accretion. Its differentiation, which we have not discussed in detail in this chapter, tells part of the story of planetary melting and crustal evolution. Ideas for its origin abound and lead to distinct chemical compositions and surface mineralogy. Data from the MESSENGER and BepiColumbo missions will allow us to distinguish among these models for the composition and formation of Mercury.

ACKNOWLEDGMENTS

The authors thank John Longhi for sharing his very useful programs with us, and Faith Vilas for a helpful review. This is SOEST publication number 6156 and HIGP publication number 1281.

REFERENCES

Anderson J. D., Columbo G., Esposito P. B., Lau P. B., and Trager G. B. (1987) The mass, gravity field and ephemeris of Mercury. *Icarus* **71**, 337–349.

Benz W., Slattery W. L., and Cameron A. G. W. (1988) Collisional stripping of Mercury's mantle. *Icarus* **74**, 516–528.

Blewett D. T., Lucey P. G., Hawke B. R., Ling G. G., and Robinson M. S. (1997) A comparison of Mercurian reflectance and spectral quantities with those of the Moon. *Icarus* **129**, 217–231.

Blewett D. T., Hawke B. R., and Lucey P. G. (2002) Lunar pure anorthosite as a spectral analog for Mercury. *Meteorit. Planet. Sci.* **37**, 1245–1254.

Burbine T. H., McCoy T. J., Nittler L. R., Benedix G. K., and Cloutis E. A. (2002) Spectra of extremely reduced assemblages: implications for Mercury. *Meteorit. Planet. Sci.* **37**, 1233–1244.

Cameron A. G. W. (1985) The partial volatilization of Mercury. *Icarus* **64**, 285–294.

Cameron A. G. W., Fegley B., Jr., Benz W., and Slattery W. L. (1988) The strange density of Mercury: theoretical considerations. In *Mercury* (eds. F. Vilas, C. R. Chapman, and M. S. Matthews). University of Arizona Press, Tucson, pp. 692–708.

Chambers J. E. (2001) Making more terrestrial planets. *Icarus* **152**, 205–224.

Cintala M. J. (1992) Impact-induced thermal effects in the lunar and mercurian regoliths. *J. Geophys. Res.* **97**, 947–973.

Drake M. J. and Righter K. (2002) Determining the composition of the Earth. *Nature* **416**, 39–44.

Dzurisin D. (1978) The tectonic and volcanic history of Mercury as inferred from studies of scarps, ridges, troughs, and other lineaments. *J. Geophys. Res.* **83**, 4883–4906.

Fegley B., Jr. and Cameron A. G. W. (1987) A vaporization model for iron/silicate fractionation in the Mercury protoplanet. *Earth Planet. Sci. Lett.* **82**, 207–222.

Goettel K. A. (1988) Present bounds on the bulk composition of Mercury: implications for planetary formation processes. In *Mercury* (eds. F. Vilas, C. R. Chapman, and M. S. Matthews). University of Arizona Press, Tucson, pp. 613–621.

Grossman L. (1972) Condensation in the primitive solar nebula. *Geochim. Cosmochim. Acta* **36**, 597–619.

Grossman J. N., Rubin A. E., and MacPherson G. J. (1988) ALH85085: a unique volatile-poor carbonaceous chondrite with possible implications for nebular fractionation processes. *Earth Planet. Sci. Lett.* **91**, 33–54.

Hapke B., Christman C., Rava B. B., and Mosher J. (1980) A color-ratio map of Mercury. *Proc. 11th Lunar Planet. Sci. Conf.* **1**, 817–821.

Harmon J. K. and Slade M. A. (1992) Radar mapping of Mercury: full-disk images and polar anomalies. *Science* **258**, 640–642.

Jagoutz E., Palme H., Baddenhausen H., Blum K., Cendales M., Dreibus G., Spettel B., Lorenz B., and Wänke H. (1979) The abundance of major, minor, and trace elements in the earth's mantle as derived from primitive ultramafic nodules. *Proc. 9th Lunar Planet. Sci. Conf.*, 2031–2050.

Jeanloz R., Mitchell D. L., Sprague A. L., and DePater I. (1995) Evidence for a basalt-free surface on Mercury and implications for internal heat. *Science* **268**, 1455–1457.

Kaula W. M. (1986) The interiors of the terrestrial planets: their structure and evolution. In *The Solar System* (ed. M. G. Kivelson). Prentice-Hall, Englewood Cliffs, NJ, pp. 78–93.

Keiffer W. S. and Murray B. C. (1987) The formation of Mercury's smooth plains. *Icarus* **72**, 477–491.

Killen R. M. and Ip W. H. (1999) The surface-bound atmospheres of Mercury and the Moon. *Rev. Geophys.* **37**, 361–406.

Krot A. N., Meibom A., Russell S. S., Alexander C. M. O'D, Jeffries T. E., and Keil K. (2001) A new astrophysical setting for chondrule formation. *Science* **291**, 1776–1779.

Lewis J. S. (1972) Metal/silicate fractionation in the solar system. *Earth Planet. Sci. Lett.* **15**, 286–290.

Lewis J. S. (1974) Chemistry of the planets. *Ann. Rev. Phys. Chem.* **24**, 339–351.

Lewis J. S. (1988) Origin and composition of Mercury. In *Mercury* (eds. F. Vilas, C. R. Chapman, and M. S. Matthews). University of Arizona Press, Tucson, pp. 651–666.

Longhi J., Knittle E., Holloway J. R., and Wänke H. (1992) The bulk composition, mineralogy and internal structure of Mars. In *Mars* (eds. H. H. Kieffer, B. M. Jakosky, C. W. Snyder, and M. S. Matthews). University of Arizona Press, Tucson, pp. 184–208.

Lucey P. G., Taylor G. J., and Malarete E. (1995) Abundance and distribution of iron on the Moon. *Science* **268**, 1150–1153.

McCoy T. J., Dickinson T. L., and Lofgren G. E. (1999) Partial melting of the Indarch (EH4) meteorite: a textural, chemical, and phase relations view of melting and melt migration. *Meteorit. Planet. Sci.* **34**, 735–746.

McDonough W. F. and Sun S.-S. (1995) The composition of the Earth. *Chem. Geol.* **120**, 223–253.

Melson W. G., Vallier T. L., Wright T. L., Byerly G., and Nelen J. (1976) Chemical diversity of abyssal volcanic glass erupted along Pacific, Atlantic, and Indian Ocean sea-floor spreading centers. In *The Geophysics of the Pacific Ocean Basin and its Margin* (eds. G. P. Woollard, G. H. Sutton, M. H. Manghnani, and R. Moberly). American Geophysics Union, Washington, DC, pp. 351–367.

Meyer C. (2003) *Mars Meteorite Compendium—2003*. http://www-curator.jsc.nasa.gov/curator/antmet/mmc/mmc.htm

Morgan J. W. and Anders E. (1980) Chemical composition of Earth, Venus, and Mercury. *Proc. Natl. Acad. Sci.* **77**, 6973–6977.

Murray B. C. (1975) The Mariner 10 pictures of Mercury. *J. Geophys. Res.* **80**, 2342–2344.

Murray B. C., Strom R. G., Trask N. J., and Gault D. E. (1975) Surface history of Mercury: implications for terrestrial planets. *J. Geophys. Res.* **80**, 2508–2514.

Noble S. K. and Pieters C. M. (2001) Space weathering in the mercurian environment. In *Workshop on Mercury: Space Environment, Surface, and Interior*, LPI Contrib. No. 1097 (eds. M. Robinson and G. J. Taylor). Lunar and Planetary Institute, Houston, pp. 68–69.

Potter A. and Morgan T. H. (1985) Discovery of sodium in the atmosphere of Mercury. *Science* **229**, 651–653.

Potter A. and Morgan T. H. (1986) Potassium in the atmosphere of Mercury. *Icarus* **67**, 336–340.

Rava B. and Hapke B. (1987) Analysis of the Mariner 10 color ratio map of Mercury. *Icarus* **71**, 397–429.

Robinson M. S. and Lucey P. G. (1997) Recalibrated Mariner 10 color mosaics: implications for mercurian volcanism. *Science* **275**, 197–200.

Robinson M. S. and Taylor G. J. (2001) Ferrous oxide in Mercury's crust and mantle. *Meteorit. Planet. Sci.* **36**, 841–847.

Safronov V. S. (1969) *Evolution of the Protoplanetary Cloud and Formation of the Earth and Planets*. (Transl. 1972 NASA TT F-677) Nauka, Moscow.

Scott E. R. D. (1988) A new kind of primitive chondrite, Allan Hills 85085. *Earth Planet. Sci. Lett.* **91**, 1–18.

Scott E. R. D. (2002) What chondrites can tell us about accretion in the solar nebula. In *Lunar Planet. Sci.* **XXXIII**, #1453. The Lunar and Planetary Institute, Houston (CD-ROM).

Scott E. R. D. and Taylor G. J. (2000) Composition and accretion of the terrestrial planets. The *Lunar Planet. Sci.* **XXXI**, #1546. The Lunar and Planetary Institute, Houston (CD-ROM).

Slade M., Butler B., and Muhleman D. (1992) Mercury radar imaging: evidence for polar ice. *Science* **258**, 635–640.

Spohn T., Sohl F., Wieczerkowski K., and Conzelmann V. (2001) The interior structure of Mercury: what we know, what we expect from BepiColombo. *Planet. Space Sci.* **49**, 1561–1570.

Sprague A. L., Kozlowski R. W. H., Witteborn F. C., Cruikshank D. P., and Wooden D. (1994) Mercury: evidence for anorthosite and basalt from mid-infrared (7.5–13.5 μm) spectroscopy. *Icarus* **109**, 156–167.

Sprague A. L., Hunten D. M., and Lodders K. (1995) Sulfur at Mercury, elemental at the poles and sulfides in the regolith. *Icarus* **118**, 211–215.

Spudis P. D. and Guest J. E. (1988) Stratigraphy and geologic history of Mercury. In *Mercury* (eds. F. Vilas, C. R. Chapman, and M. S. Matthews). University of Arizona Press, Tucson, pp. 118–164.

Strom R. G. (1977) Origin and relative age of lunar and mercurian intercrater plains. *Phys. Earth Planet. Sci.* **15**, 156–172.

Surkov Yu. A., Barsukov V. L., Moskalyeva L. P., Kharyukova V. P., and Kemurdzhian A. L. (1984) New data on the composition, structure, and properties of Venus rock obtained by Venera 13 and Venera 14. *Proc. 15th Lunar Planet. Sci. Conf.: J. Geophys. Res.* **89**, 393–402.

Surkov Yu. A., Moskalyova L. P., Kharyukova V. P., Dudin A. D., Smirnov G. G., and Zaitseva S. Ye. (1986) Venus rock composition at the Vega 2 landing site. *Proc. 17th Lunar Planet. Sci. Conf.: J. Geophys. Res.* **91**, 215–218.

Taylor G. J. and Scott E. R. D. (2001) Mercury: an end-member planet or a cosmic accident. In *Workshop on Mercury: Space Environment, Surface, and Interior*, LPI Contrib. No. 1097 (eds. M. Robinson and G. J. Taylor). Lunar and Planetary Institute, Houston, pp. 104–105.

Taylor S. R. (1991) Accretion in the inner nebula: the relationship between terrestrial planetary compositions and meteorites. *Meteoritics* **26**, 267–277.

Taylor S. R. and McLennan S. (1985) *The Continental Crust: Its Composition and Evolution*. Blackwell, Malden, MA.

Vilas F. (1988) Surface composition of Mercury from reflectance spectrophotometry. In *Mercury* (eds. F. Vilas, C. R. Chapman, and M. S. Matthews). University of Arizona Press, Tucson, pp. 59–76.

Warrell J. (2003) Properties of the Hermean regolith: III. Disk-resolved vis-NIR reflectance spectra and implications for the abundance of iron. *Icarus* **161**, 199–222.

Warren P. H. (1997) MgO–FeO mass balance constraints and a more detailed model for the relationship between eucrites and diogenites. *Meteorit. Planet. Sci.* **32**, 945–963.

Wasson J. T. (1988) The building stones of the planets. In *Mercury* (eds. F. Vilas, C. R. Chapman, and M. S. Matthews). University of Arizona Press, Tucson, pp. 622–650.

Weidenschilling S. J. (1976) Accretion of the terrestrial planets: II. *Icarus* **27**, 161–170.

Weisberg M. K., Prinz M., Clayton R. N., and Mayeda T. K. (1993) The CR (Renazzo-type) carbonaceous chondrite group and its implications. *Geochim. Cosmochim. Acta* **57**, 1567–1586.

Weisberg M. K., Prinz M., Clayton R. N., Mayeda T. K., Grady M. M., and Pillinger C. T. (1995) The CR chondrite clan. *Proc. NIPR Symp. Antarct. Meteorit.* **8**, 11–32.

Wetherill G. W. (1988) Accumulation of Mercury from planetesimals. In *Mercury* (eds. F. Vilas, C. R. Chapman, and M. S. Matthews). University of Arizona Press, Tucson, pp. 670–691.

Wetherill G. W. (1994) Provenance of the terrestrial planets. *Geochim. Cosmochim. Acta* **58**, 4513–4520.

1.19
Venus

B. Fegley, Jr.

Washington University, St. Louis, MO, USA

1.19.1	BRIEF HISTORY OF OBSERVATIONS	487
	1.19.1.1 Pre-twentieth Century	487
	1.19.1.2 The Twentieth Century to the Present Day	488
1.19.2	OVERVIEW OF IMPORTANT ORBITAL PROPERTIES	490
1.19.3	ATMOSPHERE	491
	1.19.3.1 Composition	491
	1.19.3.1.1 Basic definitions and general remarks	491
	1.19.3.1.2 Carbon, sulfur, and halogen gases	491
	1.19.3.1.3 Water vapor	493
	1.19.3.1.4 Nitrogen and noble gases	494
	1.19.3.1.5 Isotopic composition	494
	1.19.3.2 Thermal Structure and Greenhouse Effect	495
	1.19.3.3 Clouds and Photochemical Cycles	496
	1.19.3.4 Atmospheric Dynamics	497
	1.19.3.5 Upper Atmosphere and Solar-wind Interactions	497
1.19.4	SURFACE AND INTERIOR	497
	1.19.4.1 Geochemistry and Mineralogy	497
	1.19.4.2 Atmosphere–Surface Interactions	500
	1.19.4.2.1 Carbonate equilibria	500
	1.19.4.2.2 Equilibria involving HCl and HF	501
	1.19.4.2.3 Redox reactions involving iron-bearing minerals	501
	1.19.4.2.4 Minerals present in low radar emissivity regions	502
	1.19.4.3 The Venus Sulfur Cycle and Climate Change	502
	1.19.4.4 Topography and Geology	503
	1.19.4.5 Interior	504
1.19.5	SUMMARY OF KEY QUESTIONS	505
ACKNOWLEDGMENTS		506
REFERENCES		506

1.19.1 BRIEF HISTORY OF OBSERVATIONS

1.19.1.1 Pre-twentieth Century

Venus is Earth's nearest planetary neighbor, and has fascinated mankind since the dawn of history. Venus' clouds reflect most of the sunlight shining on the planet and make it the brightest object in the sky after the Sun and Moon. Venus is visible with the naked eye as an evening star until a few hours after sunset, or as a morning star shortly before sunrise. Many ancient civilizations observed and worshipped Venus, which had a different name in each society, e.g., Ishtar to the Babylonians, Aphrodite to the Greeks, Tai'pei to the Chinese, and Venus to the Romans (Hunt and Moore, 1982). Venus has continued to play an important role in myth, literature, and science throughout history.

In the early seventeenth century, Galileo's observations of the phases of Venus showed that the geocentric (Ptolemaic) model of the solar system was wrong and that the heliocentric (Copernican) model was correct. About a century later, Edmund Halley proposed that the distance from the Earth to the Sun (which was then unknown and is defined as one astronomical unit, AU) could be measured by observing transits of Venus across the Sun. These transits occur in pairs separated by eight years at intervals of 105.5 yr and 121.5 yr in an overall cycle of 243 yr, e.g., June 6, 1761, June 3, 1769; December 9, 1874, December 6, 1882, June 8, 2004, June 6, 2012, December 11, 2117, and December 8, 2125.

The first attempted measurements of the astronomical unit during the 1761 transit were unsuccessful. However, several observers reported a halo around Venus as it entered and exited the Sun's disk. Thomas Bergman in Uppsala and Mikhail Lomonosov in St. Petersburg, independently speculated that the halo was due to an atmosphere on Venus. Eight years later observations of the 1769 solar transit (including those made by Captain Cook's expedition to Tahiti) gave a value of 1 AU = 153 million kilometers, ~2.3% larger than the actual size (149.6 million kilometers) of the astronomical unit (Woolf, 1959; Maor, 2000).

1.19.1.2 The Twentieth Century to the Present Day

Modern observations of Venus date to the 1920s. During this decade Wright and Ross took the first UV photographs and discovered the UV dark markings in the clouds. Also in the 1920s, Pettit and Nicholson measured Venus' temperature as 240 K. Venus' atmospheric composition remained unknown until 1932, when Adams and Dunham serendipitously discovered CO_2 during an unsuccessful spectroscopic search for water vapor. Subsequent laboratory spectroscopy by Adel and Slipher in 1934 showed that Venus' atmosphere contains much more CO_2 than Earth's atmosphere. Shortly thereafter, in 1940, Wildt proposed that the large amount of CO_2 in Venus' atmosphere caused greenhouse heating and concluded that Venus' surface temperature "appears to be higher than the terrestrial boiling point of water." Wildt's idea, which was explored in more detail by Sagan in the early 1960s, was verified by microwave observations from Earth in the late 1950s and from Mariner 2 during its historic 1962 flyby (Barath et al., 1964).

From the late 1950s onward, Venus has been subjected to a variety of increasingly sophisticated Earth-based, Earth-orbital, and spacecraft observations. Spectroscopic observations of Venus were carried out using high-altitude telescopes carried on balloons or on airplanes. Fourier transform infrared (FTIR) spectrometers were applied to planetary spectroscopy and were used to observe Venus. As a result, the $^{13}C/^{12}C$ and $^{18}O/^{16}O$ isotopic ratios in CO_2 on Venus were measured and H_2O, CO, HCl, and HF were discovered in Venus' atmosphere (Connes et al., 1967, 1968; Bézard et al., 1987). During the 1960s, Earth-based radar observations measured Venus' radius, rotation rate, pole position, proved that Venus rotates in a retrograde (i.e., east to west) direction, and gave the first radar "images" of its surface, which cannot be seen through the thick atmosphere and clouds (Pettengill, 1968; Shapiro, 1968).

In the early 1970s, Earth-based measurements of the polarization and refractive index of the cloud particles led to their identification as droplets of concentrated (~75% by mass) sulfuric acid (Esposito et al., 1983). Several years later, Barker (1979) discovered SO_2 at Venus' cloud tops. Almost simultaneously, instruments on the Pioneer Venus and Venera 11–12 missions also observed SO_2.

The Pioneer Venus mission provided the first radar imaging and altimetry of Venus' surface from synthetic aperture radar on an orbiting spacecraft. Subsequently, the Venera 15 and 16 orbiters also carried out radar imaging and altimetry of part of Venus' northern hemisphere. Orbital spacecraft radar observations of Venus culminated with the very successful Magellan mission in the early 1990s.

In the 1980s, the discovery of spectral windows allowed Earth-based IR observations of the subcloud atmosphere on Venus' nightside (Allen and Crawford, 1984). High-resolution IR spectroscopy in these windows led to the discovery of carbonyl sulfide (OCS) and the first measurements of HCl and HF below the clouds (Bézard et al., 1990). The Galileo and CASSINI flybys of Venus (see Table 1) utilized these spectral windows to image the surface at near IR wavelengths.

Starting in the 1960s, Venus was the target of numerous spacecraft missions by the United States and the former Soviet Union. The major features of the successful missions are summarized in Table 1. Mission results, which have withstood the test of time and are relevant to Venusian geochemistry, are discussed in subsequent sections of this chapter.

One important point should be emphasized here. This is the paucity of spacecraft data on the chemical composition and thermal structure of Venus' lower atmosphere below ~22 km altitude (von Zahn et al., 1983). About 80% of Venus' atmospheric mass is below this altitude. Furthermore, altitudes of 0–12 km span the region where the atmosphere is interacting with the surface. However, with three exceptions we have no data on the chemical composition of Venus' near-surface atmosphere. First is the older measurements of CO_2 and N_2 from crude chemical experiments on the Venera 4–6 landers. Second, the water-vapor profile measured by the Pioneer Venus large probe neutral mass spectrometer. Third, the measurements of water-vapor and gaseous sulfur by spectrophotometer experiments on the Venera 11–14 landers. The gas chromatograph and mass spectrometer experiments on

Table 1 Spacecraft missions to Venus.

Launch date	Spacecraft[a]	Comments
Aug. 27, 1962	Mariner 2, flyby	Dec. 14, 1962 flyby (36,000 km), confirmed high surface temp., 1st USA success; see *Space Sci. Rev.* **2**(6), 750–777, Dec. 1963
June 12, 1967	Venera 4, atm. probe	First successful atmospheric probe on Oct. 18, 1967 showed CO_2 is major gas; see Jastrow and Rasool (1969)
June 14, 1967	Mariner 5, flyby	Oct. 19, 1967 flyby (3,900 km), atmospheric structure & composition expts.; see Jastrow and Rasool (1969)
Jan. 5, 1969	Venera 5, probe	Venera 5 entry on May 16, 1969, failed at 26 km altitude
Jan. 10, 1969	Venera 6, probe	Venera 6 entry on May 17, 1969, failed at 11 km altitude; see *J. Atm. Sci.* **27**(7), July 1970, and Marov (1972)
Aug. 17, 1970	Venera 7, lander	Dec. 15, 1970, first soft landing on Venus, measured atmospheric composition, pressure, and temperature, survived 23 min (Marov, 1972)
Mar. 27, 1972	Venera 8, lander	July 22, 1972 landing, first analysis of surface: K, U, Th measured by γ-ray analysis, survived for 50 min
Nov. 3, 1973	Mariner 10, flyby	4,200 km flyby en route to Mercury on Feb. 5, 1974, IR, UV spectra, imaging of clouds, see *J. Atm. Sci.* **32**(6), June 1975, *Science* **183**(4131) 29 March 1974)
June 8, 1975	Venera 9, orbiter & lander	Orbit insertion on Oct. 20, 1975, landed Oct. 22, 1975, survived 53 min, first TV images of surface, γ-ray analysis of K, U, Th; see Hunten *et al.* (1983)
June 14, 1975	Venera 10, orbiter & lander	Orbit insertion on Oct. 23, 1975, landed Oct. 25, 1975, survived 65 min; see references for Venera 9.
May 20, 1978	Pioneer Venus 1,[b] orbiter	Orbit insertion Dec. 4, 1978, first radar mapping of another planetary surface, Venus atm. entry Aug. 1992
Aug. 8, 1978	Pioneer Venus 2,[b] bus & probes	Atm. entry Dec. 9, 1978 (bus, large probe, 3 small probes), first successful gc and ms analyses of atm
Sep. 9, 1978	Venera 11	Venera 11 flyby & probe entry Dec. 25, 1978,
Sep. 14, 1978	Venera 12, flybys & probes	Venera 12 flyby & probe entry Dec. 21, 1978, atmospheric science from 2 probes, no TV or surface analyses (Hunten *et al.*, 1983; Krasnopolsky 1986, papers in *Cosmic Res.* **17**(5) Sept./Oct. 1979)
Oct. 30, 1981	Venera 13, flyby & probe	Landed March 1, 1982, first color TV images and X-ray fluorescence analyses of surface, survived 127 min
Nov. 4, 1981	Venera 14, flyby & probe	Landed March 5, 1982, Venera 13 twin, survived 57 min; see *Cosmic Res.* **21**(2), Mar./Apr. 1983, Hunten *et al.* (1983), Krasnopolsky (1986), and Barsukov *et al.* (1992)
June 2, 1983	Venera 15, orbiter	Oct. 10, 1983, orbit entry.
June 7, 1983	Venera 16, orbiter	Oct. 14, 1983, orbit entry. Radar imaging from N. pole to 30° N, radar altimetry, and atm. spectroscopy expts. (Barsukov *et al.*, 1992; Bougher *et al.*, 1997)
Dec. 15, 1984	Vega 1, lander & balloon	Landed June 11, 1985, balloon flew 11,500 km at ~54 km in 46 h, lander survived 20 min
Dec. 21, 1984	Vega 2, lander & balloon	Landed June 15, 1985, balloon flew 11,000 km at ~54 km in 46 h, lander survived 20 min. Atmospheric science, XRF & γ-ray analyses of the surface; see *Cosmic Res.* **25**(5), Sept./Oct. 1987, Bougher *et al.* (1997), balloon papers in Science **231**, 21 March 1986.
May 4, 1989	Magellan, orbiter	Orbit insertion Aug. 10, 1990, radar mapping, altimetry, emissivity data for surface; radio occultation expts. atm. science, end of mission and loss of orbiter Oct. 12/13, 1994; see *J. Geophys. Res.* **97**, Aug. 25 & Oct. 25, 1992.
Oct. 18, 1989	Galileo, flyby	Feb. 10, 1990 flyby. Imaging & spectroscopy of atmosphere (Bougher *et al.*, 1997, papers in *Planet. Space Sci.* **41**(7), July 1993).
Oct. 15, 1997	CASSINI, flyby	First flyby Apr. 26, 1998, second flyby June 24, 1999, imaging & atmospheric science (Baines *et al.*, 2000).

Sources: Lodders and Fegley (1998) and National Space Science Data Center, Greenbelt, MD.
[a] US spacecraft: Mariner, Pioneer Venus, Magellan, Galileo, CASSINI. USSR spacecraft: Venera, Vega. [b] Pioneer Venus papers in *J. Geophys. Res.* **85**(A13), Dec. 30, 1980, *Icarus* **51**(2), Aug. 1982, *Icarus* **52**(2), Nov. 1982, Hunten *et al.* (1983) and Bougher *et al.* (1997).

the *Venera 11–14* and *Vega 1–2* landers did not return data from altitudes below 12 km (apparently by design). The atmospheric structure and net flux radiometer instruments on the *Pioneer Venus* large probe mysteriously failed at 12.5 km altitude. Current knowledge of the thermal structure of Venus' near-surface atmosphere is based upon crude pressure and temperature measurements from several of the early *Venera* landers, extrapolation of the *Pioneer Venus* measurements below 12.5 km, and more accurate data from the *Vega* 2 lander (Seiff, 1983; Crisp and Titov, 1997).

Table 1 lists key references that summarize results of the space missions. Jastrow and Rasool (1969) is a collection of *Mariner 5* and *Venera 4* papers. *Venus*, edited by Hunten *et al.* (1983), gives a detailed picture of Venus after the *Pioneer Venus* and *Venera 11–12* missions. Barsukov *et al.* (1992) report results of Soviet missions including the *Venera 13–14* probes, the *Venera 15–16* orbiters, and the *Vega 1–2* probes. *Venus II*, edited by Bougher *et al.* (1997), focuses on *Magellan* results relevant to geology, geophysics, and geochemistry of Venus. *Venus II* also describes results from the more recent Soviet missions (*Venera 13–14* onward) for atmospheric and surface chemistry.

Other books of interest include Lewis and Prinn (1984), which emphasizes the use of observational data for understanding the origin, evolution, and present-day chemistry of planetary atmospheres. Krasnopolsky (1986) focuses on chemistry of the atmospheres of Mars and Venus. He also reviews the atmospheric composition, thermal structure, and cloud measurements by the Soviet *Venera* and *Vega* missions. Chamberlain and Hunten (1987) is the classic text about chemistry and physics of planetary atmospheres. It describes results of the US and Soviet missions to Venus that are relevant to the composition, dynamics, and structure of Venus' atmosphere and ionosphere.

1.19.2 OVERVIEW OF IMPORTANT ORBITAL PROPERTIES

Several of Venus' orbital properties (summarized in Table 2) are unique and were not fully appreciated until the space age. Radar observations revealed that Venus' orbital period is ~224.70 d, and that its rotation period is 243.02 d (retrograde). Thus, a Venusian "day" is 116.75 Earth days long (1/day = 1/243.02 + 1/224.70), with the Sun rising in the west and setting in the east after 58.375 Earth days of daylight and rising again after another 58.375 Earth days of night. As shown by Shapiro *et al.* (1979), Venus' rotation period is close to, but clearly different from 243.16 d, which would be in 3 : 2 resonance with the Earth's orbital period. The orientation of Venus' spin axis is almost normal to the ecliptic with a tilt of ~177°. The relationship between the orbital periods of Venus and Earth is such that inferior conjunction (i.e., closest approach of the two planets) occurs once every 583.92 Earth days ($1/t = 1/224.70 - 1/365.25$), at ~19 month intervals. As can be seen from Table 1, Soviet space missions to Venus during the 1960s were launched at similar intervals corresponding to successive inferior conjunctions. Finally, one should realize that Earth-based radar imaging and telescopic observations of Venus at inferior conjunction are of the same region of the planet because the 583.92 day synodic period is 5.001 Venusian days.

Table 2 Orbital and physical properties of Venus.

Property	Value	Property	Value
Semimajor axis (a, 10^6 km)	108.21	Mass (10^{24} kg)	4.8685
(AU)	0.7233	Modal radius (km)[d]	6,051.37
Perihelion (10^6 km)	107.48	Oblateness	0.0
Aphelion (10^6 km)	108.94	Volume (10^{10} km^3)	92.857
Orbital eccentricity (e)	0.0067	Mean density (kg m^{-3})	5,243
Inclination to the ecliptic (deg)	3.395	GM (10^{14} m^3 s^{-2})	3.2486
Sidereal orbital period (d)[a]	224.701	Mean gravity (m s^{-2})	8.870
Orbital obliquity (deg)	177.36	J_2 ($\times 10^6$)	4.4192
Tropical orbital period (d)[a]	224.695	Moment of inertia[e]	$0.331 \leq C/MR^2 \leq 0.341$
Mean orbital velocity (km s^{-1})	35.02	Solar constant (W m^{-2})	2,613.9
Synodic period (d)[b]	583.92	Bond albedo	0.750
Sidereal rotation period (d)[c]	−243.0185	Temperature at modal radius (K)[f]	740
Escape velocity (km s^{-1})	10.361	Pressure at modal radius (bar)[f]	95.6

After Lodders and Fegley (1998).
[a] The sidereal orbital period is the mean time for one revolution about the Sun relative to the fixed stars. The tropical orbital period is the mean time for one orbit from the same point to itself (such as equinox to equinox). [b] The time between Venus–Earth oppositions. [c] Retrograde rotation. [d] Median radius = 6,051.64 km, mean radius = 6,051.84 km. [e] Model-dependent constraint from Yoder (1997). [f] The temperature and pressure at the median and mean radii are 738.4 K, 94.5 bar and 736.8 K, 93.3 bar, respectively.

1.19.3 ATMOSPHERE

1.19.3.1 Composition

1.19.3.1.1 Basic definitions and general remarks

Throughout the rest of this chapter we discuss gas abundances using these three terms: volume-mixing ratio (henceforth mixing ratio), number density, and column density (or column abundance). The volume-mixing ratio is a dimensionless quantity also called the mole (or volume) fraction of a gas, and is the gas partial pressure (P_i) divided by the total pressure (P_T). Mixing ratios are given in percent for major gases and as parts per million, per billion, or per trillion by volume (ppmv, ppbv, or pptv) for trace gases.

The number density of a gas i is denoted by square brackets [i] and has the dimensions of particles (atoms plus molecules) per unit volume, e.g., particles cm^{-3}. It is equal to either $P_i N_A/RT$ or P_i/kT, where N_A is Avogadro's number, R is the ideal gas constant, k is Boltzmann's constant, and T is temperature in kelvin. For reference, the number density of Earth's atmosphere is 2.55×10^{19} particles cm^{-3} at sea level, where $T = 288.15$ K and $P = 1$ atm. The number density of Venus' atmosphere is 9.36×10^{20} particles cm^{-3} at 0 km, where $T = 740$ K and $P = 95.6$ bar. The 0 km level on Venus is the modal radius (see Table 2).

The column abundance of a gas is the number of gas particles throughout an atmospheric column and has the dimensions of particles per unit area, e.g., particles cm^{-2}. The column abundance is thus the integral of [i]dz from a specified altitude z_0, such as the planetary surface, to the top of the atmosphere ($z = $ infinity). The column density can also be calculated from $P_i N_A/gM$, where M is the formula weight of the gas and g is the gravitational acceleration as a function of altitude. Note that the total atmospheric mass per unit area is simply P_T/g. The mean molecular weight, column density, and column mass of the terrestrial atmosphere are 28.97 g mol^{-1}, 2.15×10^{25} particles cm^{-2}, and 1,034.2 g cm^{-2} at sea level. The corresponding values for Venus' atmosphere are 43.46 g mol^{-1}, 1.49×10^{27} particles cm^{-2}, and 107,531 g cm^{-2} at 0 km.

The chemical composition of Venus' atmosphere is described below. This discussion is based on sources listed in Table 3, Fegley and Treiman (1992), and Warneck (1988).

As shown in Table 3, Venus' atmosphere is dominantly CO_2 (96.5%) and N_2 (3.5%), with smaller amounts of SO_2, H_2O, CO, OCS, HCl, HF, the noble gases, and reactive species such as SO that are produced photochemically. The abundances of CO_2, N_2, the noble gases, HCl, and HF are apparently constant throughout most of Venus' atmosphere, but other gases such as SO_2, H_2O, CO, OCS, and SO have spatially and temporally variable abundances. Variations in the abundances of water vapor, CO, and sulfur gases are of particular interest because they result from the solar UV-driven photochemistry that maintains the global sulfuric acid cloud cover.

In addition to the gases listed in Table 3, Earth-based and spacecraft microwave spectroscopy indicates that H_2SO_4 vapor (with a mixing ratio of several tens of ppmv) is present below the clouds. Sulfur trioxide, as yet unobserved, is also expected to be present below the clouds in equilibrium with H_2SO_4 vapor. Spectrophotometers on *Venera 11–14* found absorption of blue sunlight in Venus' lower atmosphere. This is attributed to elemental sulfur vapor with a total mixing ratio (for all allotropes) of ~20 ppbv in Venus' lower atmosphere.

1.19.3.1.2 Carbon, sulfur, and halogen gases

The high atmospheric abundances of CO_2, SO_2, OCS, HCl, and HF on Venus are due to the high temperatures at Venus' surface (Fegley and Treiman, 1992). All these gases are present at much lower abundances in the Earth's atmosphere. For example, average mixing ratios in the terrestrial troposphere of CO_2, SO_2, OCS, HCl, and HF are 360 ppmv, 20–90 pptv, 500 pptv, ~1 ppbv, and ~25 pptv, respectively (cf. Table 3). Also the major sources and sinks for these gases on Earth are different from their probable sources and sinks on Venus.

Volcanic outgassing is probably the major source of CO_2 in Venus' atmosphere. Above the clouds CO_2 is converted to CO and O_2 by photolysis (see Section 1.19.3.3), while carbonate formation may be an important sink at the surface. In contrast, CO_2 in Earth's atmosphere has anthropogenic, biogenic, and geological sources with estimates indicating pre-industrial CO_2 levels of 290 ppmv or less. The major sinks are biogenic (i.e., consumption by photoautotrophs), dissolution in the oceans, and rock weathering (acting over longer timescales than the first two sinks). The column density of CO_2 in Earth's atmosphere is ~5.1×10^{21} molecules cm^{-2}, which is ~2.75×10^5 times lower than on Venus. However, the crustal carbon inventory on Earth is about the same as the CO_2 column density on Venus (~0.7×10^{27} C atoms cm^{-2} for crustal carbon on Earth versus 1.4×10^{27} CO_2 molecules cm^{-2} on Venus). Does this similarity mean that all CO_2 on Venus has been degassed and is now in the atmosphere? The short answer is that we do not know. We consider the related questions of whether or not carbonates exist on Venus' surface and whether or not the atmospheric abundances of CO_2, SO_2, OCS, HCl,

Table 3 Chemical composition of the atmosphere of Venus.

Gas	Abundance[a]	Source(s)	Sink(s)
CO_2	$96.5 \pm 0.8\%$	Outgassing	Carbonate formation
N_2	$3.5 \pm 0.8\%$	Outgassing	
SO_2[b]	150 ± 30 ppm (22–42 km)	Outgassing and	H_2SO_4 formation and
	25–150 ppm (12–22 km)	reduction of OCS, H_2S	$CaSO_4$ formation
H_2O[b]	30 ± 15 ppm (0–45 km)	Outgassing	H escape and
	30–70 ppm (0–5 km)		Fe^{2+} oxidation
^{40}Ar	31^{+20}_{-10} ppm	Outgassing (^{40}K)	
^{36}Ar	30^{+20}_{-10} ppm	Primordial	
CO[b]	45 ± 10 ppm (cloud top)	CO_2 photolysis	Photo-oxidation to CO_2
	30 ± 18 ppm (42 km)		
	28 ± 7 ppm (36–42 km)		
	20 ± 3 ppm (22 km)		
	17 ± 1 ppm (12 km)		
^4He[c]	0.6–12 ppm	Outgassing (U, Th)	Escape
Ne	7 ± 3 ppm	Outgassing, primordial	
^{38}Ar	5.5 ppm	Outgassing, primordial	
OCS[b]	4.4 ± 1 ppm (33 km)	Outgassing and sulfide weathering	Conversion to SO_2
H_2S[b]	3 ± 2 ppm (<20 km)	Outgassing and sulfide weathering	Conversion to SO_2
HDO[b]	1.3 ± 0.2 ppm (subcloud)	Outgassing	H escape
HCl	0.6 ± 0.12 ppm (cloud top)	Outgassing	Cl-mineral formation
	0.5 ppm (35–45 km)		
^{84}Kr	25^{+24}_{-18} ppb	Outgassing, primordial	
SO[b]	20 ± 10 ppb (cloud top)	Photochemistry	Photochemistry
S_{1-8}[b]	20 ppb (<50 km)	Sulfide weathering	Conversion to SO_2
HF	$5^{+5}_{-2.5}$ ppb (cloud top)	Outgassing	F-mineral formation
	4.5 ppb (35–45 km)		
^{132}Xe	<10 ppb	Outgassing, primordial	
^{129}Xe	<9.5 ppb	Outgassing (^{129}I)	

Sources: Lodders and Fegley (1998) and Wieler (2002).
[a] Abundance by volume, ppm = parts per million, ppb = parts per billion, e.g., 4.5 ppb is a volume mole fraction of 4.5×10^{-9}.
[b] Abundances of these species are altitude dependent (see text). [c] The He abundance in Venus' upper atmosphere where diffusive separation occurs is 12^{+24}_{-6} ppm by volume (von Zahn et al., 1983). The lower atmospheric value listed above is a model-dependent extrapolation.

and HF are regulated or buffered by mineral assemblages on Venus' surface in Section 1.19.4.2.

Carbon monoxide is the second most abundant carbon-bearing gas in Venus' atmosphere. The CO abundance in Venus' lower atmosphere is altitude dependent and decreases toward the surface as follows: 45 ± 10 ppmv (~64 km), 30 ± 18 ppmv (42 km), 20 ± 3 ppmv (22 km), and 17 ± 1 ppmv (12 km). This gradient is consistent with photochemical production of CO from CO_2 in Venus' upper atmosphere and CO consumption by thermochemical reactions with sulfur gases in Venus' lower atmosphere and with minerals at its surface. Carbon monoxide is also photo-oxidized back to CO_2 via catalytic cycles that are described in Section 1.19.3.3.

By comparison, the average CO mixing ratio in Earth's troposphere is ~0.12 ppmv and it is produced from a variety of anthropogenic and biogenic sources such as fossil fuel combustion, biomass burning, and oxidation of methane and other hydrocarbons. Most of the CO in Earth's troposphere is destroyed by reaction with OH radicals, which are also important for the catalytic reformation of CO_2 on Venus under some conditions.

Sulfur dioxide is the most abundant sulfur-bearing gas, and the third most abundant gas after CO_2 and N_2 in Venus' lower atmosphere. The SO_2 mixing ratio below the clouds is ~150 ppmv, as measured by gas chromatographs on the *Pioneer Venus* and *Venera 11–12* atmospheric entry probes. Its abundance varies with altitude and may also be temporally variable. Volcanic outgassing of SO_2 and outgassing and subsequent oxidation of reduced sulfur gases (OCS and H_2S) are probably the major sources of SO_2 on Venus. Photochemical oxidation to aqueous sulfuric acid droplets efficiently removes SO_2 from Venus' upper atmosphere (cf. ~150 ppmv below the clouds versus ~10 ppbv above the clouds). Reduction of SO_2 to OCS and reaction of SO_2 with calcium-bearing minerals to form anhydrite ($CaSO_4$) may be important sinks for SO_2 in the near-surface atmosphere.

In contrast to Venus, SO_2 is only a trace gas in Earth's atmosphere and it is generally less abundant than OCS and other reduced sulfur gases in Earth's troposphere. The average SO_2

column abundance in the terrestrial atmosphere is $\sim 4 \times 10^{15}$ molecules cm^{-2} versus a column abundance of $\sim 2.2 \times 10^{23}$ molecules cm^{-2} in Venus' atmosphere. The total amount of sulfur in Earth's oceans and crust is ~ 600 times larger than the SO$_2$ column abundance on Venus, which probably does not represent the planet's total sulfur inventory. Most of the SO$_2$ in the terrestrial troposphere is anthropogenic with volcanic emissions being one to two orders of magnitude less important. Sulfur dioxide is removed from Earth's troposphere by oxidation to sulfate, which occurs via photochemical and thermochemical processes in the gas phase, in cloud droplets, and on particulates.

Carbonyl sulfide is the most abundant reduced sulfur gas in Venus' subcloud atmosphere. The OCS mixing ratio is 4.4 ± 1.0 ppmv at 33 km altitude. Earth-based IR spectroscopy shows that the OCS mixing ratio increases with decreasing altitude in the 26–45 km range. Extrapolation of this gradient indicates that the OCS abundance may reach tens of ppmv at Venus' surface. The increase in OCS may be balanced by a decrease in SO$_2$ at low altitudes, such as that reported by UV spectrometers on the *Vega 1* and *2* probes. However, the 2.3 μm spectroscopic window used to observe OCS only probes altitudes in the 26–45 km range, and the *Vega* UV spectrometer data are difficult to understand. Thus, the gradients in the OCS and SO$_2$ abundances in the lowest 20 km of Venus' atmosphere require confirmation. Theoretical models and laboratory studies indicate that the major sources of OCS on Venus are probably volcanic outgassing and chemical weathering of iron sulfide minerals such as pyrrhotite (ranging in composition from FeS to Fe$_7$S$_8$). The major sink for OCS is photochemical oxidation to SO$_2$. Some OCS may also be lost by reaction with monatomic S to form CO and S$_2$ vapor.

Carbonyl sulfide is also the most abundant reduced sulfur gas in Earth's troposphere, but for completely different reasons. Volcanic sources of OCS are negligible by comparison with biogenic emissions, which are important sources of several reduced sulfur gases (e.g., OCS, H$_2$S, (CH$_3$)$_2$S, (CH$_3$)$_2$S$_2$, and CH$_3$SH) in the terrestrial troposphere. Many of these gases are ultimately converted into sulfate aerosols in the troposphere, but OCS is mainly lost by transport into the stratosphere, where it is photochemically oxidized to SO$_2$ and then to sulfuric acid aerosols, which form the Junge layer at ~ 20 km in Earth's stratosphere.

Volcanic outgassing is plausibly the major source of HCl and HF in Venus' atmosphere. Thermochemical equilibrium calculations suggest that formation of chlorine- and fluorine-bearing minerals are important sinks for these two gases. Observations by Connes *et al.* (1967) and Bézard *et al.* (1990) give the same HCl and HF mixing ratios, within error, of ~ 0.5 ppmv and ~ 5 ppbv above and below the clouds, respectively. The column densities of $\sim 7 \times 10^{20}$ HCl molecules cm^{-2} and $\sim 7 \times 10^{18}$ HF molecules cm^{-2} in Venus' atmosphere are many orders of magnitude greater than the column densities of $\sim 2 \times 10^{16}$ HCl molecules cm^{-2} and $\sim 5 \times 10^{14}$ HF molecules cm^{-2} in Earth's troposphere. Furthermore, most of the 25 pptv HF, and some of the 1 ppbv HCl in the terrestrial troposphere are from anthropogenic sources, and not from volcanic emissions. Tropospheric HF is mainly from industrial emissions and downward mixing from the stratosphere, where it forms as a result of photolytic destruction of synthetic chlorofluorocarbon (CFC) gases. The HCl originates from sea salt, volcanic gases, and from HCl mixed downward from the stratosphere, where it is produced from photolysis of CFC gases. However, most of the fluorine and chlorine in Earth's troposphere are in CFC gases with only minor amounts in HF and HCl.

1.19.3.1.3 Water vapor

Venus' atmosphere is so dry that Earth-based and spacecraft measurements of the water-vapor abundance are extremely difficult. Historically, many of the *in situ* water-vapor measurements gave values much higher than the actual water-vapor content. However, reliable values are now available from several sources including the *Pioneer Venus* mass spectrometer, spectrophotometer experiments on *Venera 11–14*, Earth-based FTIR spectroscopy of Venus' lower atmosphere on the nightside, and IR observations during the *Galileo* and *Cassini* flybys of Venus.

The average water-vapor mixing ratio below the clouds is ~ 30 ppmv. In comparison, the terrestrial troposphere has water-vapor mixing ratios of 1–4%. Although it is a trace gas, water vapor is the major reservoir of hydrogen in Venus' subcloud atmosphere, and is an important reactant in chemical reactions that are hypothesized to buffer or regulate the atmospheric abundances of HCl and HF. The loss of water from Venus, via oxidation of the surface and hydrogen escape to space, ultimately regulates the oxidation state of the atmosphere and surface. Furthermore, the high D/H ratio in atmospheric water vapor suggests that Venus was once wet, with the equivalent of a global ocean at least 4 m and possibly 530 m deep (Donahue *et al.*, 1997).

A long-standing question is whether or not the water-vapor abundance in Venus' lower atmosphere varies with altitude. Initial interpretation of spectrophotometer experiments on the *Venera 11–14* spacecraft suggested a monotonic decrease from ~ 200 ppmv at 50 km to ~ 20 ppmv at the surface. However, most of the measurements mentioned above now indicate a constant mixing ratio of 30 ± 15 ppmv throughout the lower atmosphere. Even though Venus is drier than Earth, volcanic emissions are the plausible major source of atmospheric water vapor. A major (reversible) sink is formation of the aqueous sulfuric acid clouds. Another sink, that is irreversible, and occurs on longer timescales, is reaction of water vapor with ferrous iron minerals on Venus' surface to form ferric iron minerals and hydrogen gas. The H_2 gas is then dissociated to hydrogen atoms in Venus' upper atmosphere and lost to space.

Venus' upper atmosphere is even drier than the lower atmosphere, and the average water-vapor mixing ratio above the clouds is only a few ppmv. The very low H_2O mixing ratios were hard to explain until it was realized that Venus' clouds are 75% sulfuric acid, which is a powerful drying agent. When dissolved in the acid, most of the water reacts with H_2SO_4 to form hydronium (H_3O^+) and bisulfate (HSO_4^-) ions. As a result, the concentrations of "free" H_2O in the acid solution and in the vapor over the acid are extremely low. The partial pressure of water at Venus' cloud tops is lower than that over water ice at the same temperature. Thus, the clouds are responsible for the extreme dryness of Venus' upper atmosphere, and play an important role in the photochemical stability of Venus' atmosphere (see Section 1.19.3.3).

1.19.3.1.4 Nitrogen and noble gases

Although N_2 is the second-most abundant gas in Venus' atmosphere, measurements of its abundance are difficult. The mass spectrometers on *Pioneer Venus* and *Venera 11–12* gave N_2 mixing ratios of 4% at ~22 km, but the gas chromatograph (GC) experiments on the same spacecraft gave different values. For example, the *Venera 11–12* GC reported 2.5 ± 0.3% in the 22–42 km region, while the *Pioneer Venus* GC reported an altitude-dependent N_2 mixing ratio of 3.4% at 22 km, 3.54% at 42 km, and 4.6% at 52 km. The GC data are difficult to explain. The recommended value of 3.5 ± 0.8% for the N_2 abundance reflects the disagreements between the mass spectrometer and GC results (von Zahn et al., 1983).

Volcanic outgassing is probably the major source of N_2, and formation of nitrogen oxides (NO_x) by lightning may be a sink for N_2 on Venus. The timescales for both of these processes are uncertain, but are plausibly very long.

Ultimately, Earth's N_2 is probably also due to volcanic emissions. At present, the major N_2 sources are denitrifying bacteria in soils and in the oceans. The major N_2 sinks on Earth are nitrogen-fixing bacteria in soils and the oceans.

These sources and sinks result in a lifetime of ~17 Myr for atmospheric N_2 on Earth. If the biological sources and sinks were removed, while lightning and forest fires continued at their present rates, the lifetime for atmospheric N_2 would increase to ~80 Myr. However, in the absence of biology, Earth's atmosphere would contain much less O_2, so combustion and lightning would be much less efficient sinks for nitrogen. In this case N_2 would have a lifetime of ~1 Gyr.

The observed noble-gas abundances and isotopic ratios on Venus are summarized in Tables 3 and 4. The helium mixing ratio is a model-dependent extrapolation of the value measured in Venus' upper atmosphere, where diffusive separation of gases occurs. The main differences between Venus and Earth are that Venus is apparently richer in ^4He, ^{36}Ar, and ^{84}Kr than the Earth, and the low $^{40}Ar/^{36}Ar$ ratio of ~1.1 on Venus, which is ~270 times smaller than on Earth. The low $^{40}Ar/^{36}Ar$ ratio may reflect more efficient solar-wind implantation of ^{36}Ar in solid grains accreted by Venus and/or efficient early outgassing that then stopped due to the lack of plate tectonics. Wieler (2002) discusses the noble-gas data. Volkov and Frenkel (1993) and Kaula (1999) describe implications of the $^{40}Ar/^{36}Ar$ ratio for outgassing of Venus.

1.19.3.1.5 Isotopic composition

Table 4 summarizes the data on the isotopic composition of Venus' atmosphere. Aside from the noble gases, the most important difference between Venus and Earth is the high D/H ratio, which is ~150 times greater than the D/H ratio of 1.558×10^{-4} in standard mean ocean water (SMOW). The high D/H ratio strongly suggests,

Table 4 Isotopic composition of Venus' atmosphere.

Isotopic ratio[a]	Observed value	Method
D/H	0.016 ± 0.002	Pioneer Venus (PV) MS[b]
	0.019 ± 0.006	IR spectroscopy
$^3He/^4He$	<3 × 10^{-4}	PV MS
$^{12}C/^{13}C$	86 ± 12	IR spectroscopy
	88.3 ± 1.6	Venera 11/12 MS
$^{14}N/^{15}N$	273 ± 56	PV MS
$^{16}O/^{18}O$	500 ± 25	PV MS
	500 ± 80	IR spectroscopy
$^{20}Ne/^{22}Ne$	11.8 ± 0.6	Venera 11/12 MS
$^{21}Ne/^{22}Ne$	<0.067	Venera 11/12 MS
$^{35}Cl/^{37}Cl$	2.9 ± 0.3	IR spectroscopy
$^{36}Ar/^{38}Ar$	5.45 ± 0.1	PV, Venera 11/12 MS
$^{40}Ar/^{36}Ar$	1.11 ± 0.02	PV, Venera 11/12 MS

Sources: Lodders and Fegley (1998) and Wieler (2002).
[a] No isotopic compositions are available for Kr and Xe on Venus.
[b] MS = mass spectrometer.

but does not prove, that Venus had more water and has lost most of it over time. This deduction is based on the assumption that Venus initially had the same D/H ratio as Earth. However, a wide range of D/H ratios are observed in other planets, meteorites, interplanetary dust particles (IDPs), and in comet P/Halley, so this assumption may be incorrect.

The carbon, nitrogen, oxygen, and chlorine isotopic ratios are the same as the terrestrial values, within rather large uncertainties. However, any difference in the oxygen isotopic composition of Venus and Earth is probably only a few parts per thousand (or less) and cannot be resolved from the present data. The isotopic ratios of the three stable oxygen isotopes (16, 17, 18) are of most interest because of the known variations in meteorites and IDPs and because oxygen is either the first or second (after iron) most abundant element in a rocky planet. However, the ^{17}O isotopic abundance has not been measured on Venus. This could be done using isotopic bands of CO_2 (e.g., see Bézard et al., 1987), although probably with uncertainties of ~10%. It is important to measure abundances of the three oxygen isotopes (with an uncertainty of <0.1‰) on Venus to see if it has the same oxygen isotopic composition as the Earth and Moon. Accurate measurements of oxygen, hydrogen, carbon, and sulfur isotopic ratios in gases in Venus' upper atmosphere are desirable, because photochemical reactions plausibly cause variations in these ratios, e.g., as seen with mass-independent variations in oxygen-bearing gases in Earth's atmosphere and in some nitrate and sulfate minerals on Earth.

1.19.3.2 Thermal Structure and Greenhouse Effect

Venus has the highest albedo of any planet (e.g., 0.75 versus 0.29 for Earth). Even though the solar constant at Venus (2613.9 W m^{-2}) is ~1.9 times larger (=1/0.723^2) than that at Earth, Venus absorbs only ~66% as much solar energy, i.e., ~160 W m^{-2} versus 243 W m^{-2}, as Earth. The energy deposition is dramatically different from that on Earth, where ~66% of the absorbed solar energy is deposited at the surface. In contrast, ~70% of the absorbed sunlight is deposited in Venus' upper atmosphere and clouds, another 19% is deposited in the lower atmosphere, and only ~11% reaches the surface. The "sunlight" at Venus' surface is ~5 times dimmer than that on Earth.

The high temperature and pressure at Venus' surface (740 K and 95.6 bar at the modal radius of 6,051.4 km) are due to a super-greenhouse effect maintained by the high IR opacity of CO_2, SO_2, and H_2O in its lower atmosphere. The origin, duration, and present stability of the Venusian super-greenhouse remain somewhat of a mystery. One problem is finding enough IR opacity to maintain the greenhouse. The earlier greenhouse models used the large water-vapor mixing ratios (1,000 ppmv or higher) indicated by the Venera 4–6 probes and Earth-based microwave sounding of the subcloud atmosphere. However, we now know that the water-vapor mixing ratio below Venus' clouds is ~30 ppmv. Another problem is an incomplete knowledge of IR opacities of CO_2, SO_2, H_2O, and other possible greenhouse gases at the high temperatures and high pressures prevailing in Venus' lower atmosphere. Finally, proper incorporation of the clouds into greenhouse modeling is yet another problem. Crisp and Titov (1997) review the state of the art in modeling the Venusian greenhouse and present current ideas for addressing these problems.

Figure 1 shows the temperature and pressure profiles from 0 km to 100 km altitude in Venus' atmosphere. The 0–60 km region is the troposphere, the 60–100 km region on the dayside or the 60–160 km region on the nightside is the mesosphere (sometimes also called the stratomesosphere), the 100–160 km region on the dayside is the thermosphere, and the exosphere is above 160 km. The latter regions are discussed in Section 1.19.3.5.

There are several key differences between the atmospheres of Venus and Earth. The terrestrial atmosphere has a pronounced temperature inversion at the tropopause, which is the boundary between the troposphere and the stratosphere. This inversion is due to absorption of UV sunlight by ozone, but is absent on Venus, which has too little O_2 (<0.3 ppmv) to form an ozone layer. The temperature gradient of ~8 K km^{-1} in Venus' troposphere is very close to the dry (i.e., condensation cloud free) adiabatic gradient. This is

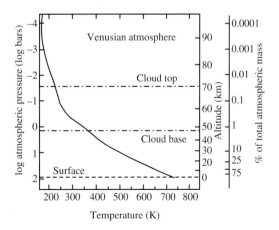

Figure 1 Temperature and pressure as a function of altitude in Venus' atmosphere. The position of the global clouds is also indicated.

given by $dT/dz = -gM/C_P = -g/c_p$, where C_P and c_p are the average molar heat capacity and specific heat, respectively, of Venus' atmosphere (96.5% CO_2 + 3.5% N_2). In contrast, the average temperature gradient in Earth's troposphere is ~ 6.5 K km^{-1}, significantly subadiabatic compared to the dry adiabatic gradient of ~ 10 K km^{-1} for air. The discrepancy is caused by water cloud condensation in Earth's troposphere, which releases the latent heat of vaporization (or sublimation) and thus warms the surrounding gas. The average surface temperature on Venus is ~ 450 K higher than the average surface temperature of 288 K on Earth. Analysis of the available temperature and wind-speed measurements indicates that meridional temperature gradients on Venus' surface are only a few degrees compared to ~ 50 K on Earth. However, *in situ* temperature measurements only extend to 60° N latitude on Venus. Temperatures on Venus' surface vary with altitude. The temperature is ~ 648 K and pressure is ~ 43 bar at the top of Maxwell Montes, which is ~ 12 km above the modal radius of 6,051.4 km and is the highest point on the planet.

1.19.3.3 Clouds and Photochemical Cycles

The most prominent feature of Venus' middle atmosphere is the global cloud layer that begins at ~ 45 km altitude and extends to ~ 70 km altitude, with thinner hazes ~ 20 km above and below these altitudes. Venus appears yellow-white in visible light, but the first UV images of Venus in the 1920s showed dark $\mathscr{Y}$- or $\mathscr{V}$-shaped cloud features. The features are formed by an unknown UV absorber that absorbs about half of all sunlight deposited in the clouds at wavelengths ≤ 500 nm. The UV absorber may be elemental sulfur, Cl_2, an S–Cl gas, chlorine compounds dissolved in cloud droplets, or another sulfur gas.

All of the clouds are low density, because the visibility inside the densest region of the clouds is a few kilometers. The average and maximum optical depths (τ) in visible light of all cloud layers are 29 and 40, respectively, versus average and maximum τ values of 6 and ~ 350 for terrestrial clouds. Average mass densities for Venus' clouds are 0.01–0.02 g m^{-3} versus an average mass density of 0.1–0.5 g m^{-3} for fog clouds on Earth. Venus' cloud layers are typically divided into the subcloud haze (32–48 km), the lower cloud (48–51 km), middle cloud (51–57 km), upper cloud (57–70 km), and upper haze (70–90 km).

Nephelometers, which use scattered light to measure particle size and number density, on *Venera 9–11* and *Pioneer Venus* showed that the cloud layers are composed of three different types of particles. The first type are aerosols of ~ 0.3 μm diameter (mode-1 particles), which occur in the upper and middle clouds. The second type are spherical droplets of ~ 2 μm diameter (mode-2 particles) composed of 75% sulfuric acid ($H_2SO_4 \cdot 2H_2O$), which occur throughout the clouds. The third type are the mode-3 particles with ~ 7 μm diameter, which have an unknown composition, and occur in the middle to lower clouds. By comparison, fog clouds on the Earth are composed of droplets with diameters of 0.5–30 μm. Aqueous sulfuric acid droplets comprise the visible clouds seen from Earth. The mode-1 and mode-3 particles may also be sulfuric acid particles. Some data suggest that the mode-3 particles may be crystalline and could be composed of iron or aluminum chlorides, solid perchloric acid hydrates, or phosphorus oxides.

The aqueous sulfuric acid droplets in the clouds result from UV sunlight photolysis of SO_2, which reduces the SO_2 abundance from ~ 150 ppmv below the clouds to ~ 10 ppbv at the cloud tops. The photochemistry of SO_2 and CO_2 is closely coupled. The O_2 produced from CO_2 photolysis is used to convert SO_2 to SO_3, which then forms sulfuric acid. Spectroscopic observations show temporal trends in the SO_2 abundance at the cloud tops. These observed variations are probably due to atmospheric dynamics.

The CO_2 in Venus' atmosphere (like CO_2 on Mars) is continually converted by UV sunlight to oxygen and CO:

$$CO_2 + h\nu \rightarrow CO + O(^3P) \quad (\lambda < 227.5 \text{ nm}) \quad (1)$$

$$\rightarrow CO + O(^1D) \quad (\lambda < 167.0 \text{ nm}) \quad (2)$$

The electronically excited (^1D) oxygen atoms formed in reaction (2) are rapidly converted to the ground state (^3P) by collisions with other molecules. The direct recombination of oxygen atoms and CO

$$CO + O(^3P) + M \rightarrow CO_2 \quad (3)$$

where M is any gas, is forbidden by quantum mechanical spin selection rules, and is much slower than oxygen atom recombination to form O_2:

$$O(^3P) + O(^3P) + M \rightarrow O_2 + M \quad (4)$$

Photolysis would completely destroy all CO_2 above the clouds in $\sim 1.4 \times 10^4$ yr and all CO_2 in Venus' atmosphere in ~ 5 Myr. In addition, CO_2 photolysis would produce observable amounts of O_2 in ~ 5 yr unless CO_2 is reformed by another route. However, O_2 is not seen in Venus' atmosphere and the spectroscopic upper limit is <0.3 ppmv. Gas-phase catalytic reformation of CO_2 by hydrogen, chlorine, or nitrogen gases has been proposed to solve this problem. The relative importance of the catalytic schemes depends on the H_2 abundance in Venus' strato-mesosphere, which is unknown, because H_2 is involved in chemistry forming OH radicals. For example, the reaction

$$CO + OH \rightarrow CO_2 + H \quad (5)$$

is important at H_2 levels of tens of ppmv. At very low H_2 levels of ~0.1 ppbv, reaction (5) is no longer important and catalytic cycles such as

$$CO + Cl + M \rightarrow COCl + M \quad (6)$$

$$COCl + O_2 + M \rightarrow ClCO_3 + M \quad (7)$$

$$ClCO_3 + O \rightarrow CO_2 + O_2 + Cl \quad (8)$$

$$\text{Net: } CO + O \rightarrow CO_2 \quad (9)$$

$$CO + Cl + M \rightarrow COCl + M \quad (6)$$

$$COCl + O_2 + M \rightarrow ClCO_3 + M \quad (7)$$

$$ClCO_3 + Cl \rightarrow CO_2 + ClO + O \quad (10)$$

$$ClCO_3 + O \rightarrow CO_2 + O_2 + Cl \quad (8)$$

$$\text{Net: } CO + O \rightarrow CO_2 \quad (9)$$

recycle CO to CO_2. At intermediate H_2 levels of ~0.1 ppmv, the reaction

$$NO + HO_2 \rightarrow NO_2 + OH \quad (11)$$

precedes reaction (5), which then recycles CO to CO_2. Yung and DeMore (1982) and Krasnopolsky (1986) give detailed discussions of Venusian atmospheric photochemistry. More recent work is described by Esposito et al. (1997) and by Mills (1998).

1.19.3.4 Atmospheric Dynamics

The major feature of atmospheric dynamics on Venus is that above ~16 km, i.e., above the first scale height ($H = RT/gM$), the atmosphere rotates much faster than the planet itself. The atmospheric zonal rotation is retrograde and reaches a maximum velocity of ~100 m s^{-1} at the cloud tops (70 km). Tracking of the UV-dark features in the clouds, first from Earth-based observations by Boyer and Carmichel in the 1960s and later from observations by *Pioneer Venus* and *Mariner 10*, showed the high-speed retrograde zonal winds known as the four-day super-rotation. *In situ* measurements by Doppler tracking of the *Pioneer Venus*, *Venera*, and *Vega* entry probes show that the zonal winds decrease with decreasing altitude and are ~1 m s^{-1} or less at the surface. The origin of the four-day super-rotation is still incompletely understood, and our present understanding is reviewed by Gierasch et al. (1997).

1.19.3.5 Upper Atmosphere and Solar-wind Interactions

Venus' ionosphere is the upper atmospheric region, where significant ion and electron densities exist. It coexists with the thermosphere and exosphere. The electrons and ions in Venus' dayside ionosphere are primarily formed by extreme solar UV photoionization of neutral gases in the thermosphere. However, CO_2^+, like N_2^+ on Earth, is only a minor ion in the ionosphere. The major species in Venus' lower ionosphere is O_2^+ with a maximum number density of ~10^6 cm^{-3}. The CO_2^+/O_2^+ ratio is less than 10%. Above 200 km O^+ is the major ion. The maximum electron density is ~3×10^5 cm^{-3} at ~140 km altitude. The solar wind directly interacts with the upper atmosphere, because Venus does not have a magnetic field. This interaction terminates the ionosphere at the ionopause, which is typically at 400–500 km. However, the ionopause level is variable. The nightside ionosphere is formed by flow of ionospheric plasma from the dayside of the planet. The maximum ion densities on the nightside are 10^4 cm^{-3}. Fox and Kliore (1997), Nagy and Cravens (1997), and Kasprzak et al. (1997) discuss Venus' ionosphere and upper atmospheric interactions with the solar wind in more detail.

1.19.4 SURFACE AND INTERIOR

1.19.4.1 Geochemistry and Mineralogy

Our knowledge of the geochemistry and mineralogy of Venus' surface primarily comes from six types of information: (i) elemental analyses of several major elements by X-ray fluorescence (XRF) spectroscopy; (ii) analyses of potassium, uranium, and thorium by γ-ray spectroscopy; (iii) TV imaging from several *Venera* landers; (iv) Earth-based and spacecraft radar observations of the dielectric constant and morphology of the surface (Pettengill et al., 1997); (v) models of thermochemical equilibria between atmospheric gases and presumed crustal minerals; and (vi) laboratory studies of the rates of reactions that are sinks or sources of sulfur-bearing gases. However, it is important to emphasize that we have no direct knowledge of the mineralogy of Venus' surface because there are no X-ray diffraction (XRD) measurements or other data that would unambiguously identify the minerals present.

Table 5 lists the seven *Venera* and *Vega* space probes that made elemental analyses of the surface of Venus. In addition to elemental analyses, several of the *Venera* and *Vega* landers measured density, bearing capacity, electrical resistivity, and atmospheric redox state (via the reaction of chemically impregnated asbestos with atmospheric CO). The *Venera 8–10*, and *Vega 1–2* probes analyzed

Table 5 Geochemical analyses and imaging on the surface of Venus.

Probe	°Lat[a]	°Long[a]	Alt. (km)[b]	Location and suggested rock types[c]	Experiment
Venera 8	−10.7	335.25	0.5 ± 0.2	Mottled volcanic plains E. of Navka Planitia; leucitite, lamprophyres, rhyolite, γ-ray density = 2.8 ± 0.1 g cm^{-3}	γ-ray
Venera 9	31.0	291.64	1.3 ± 0.5	N.E. slope of Beta Regio, lander site has 15° to 20° slope with decimeter-size rock fragments with soil between them, E-MORB-like basaltic tholeiite	γ-ray, TV image,[d] photometry[e]
Venera 10	15.42	291.51	0.8 ± 0.6	Lowlands near SE edge of Beta Regio, lander site has soil between 10–15 cm high outcrops of bedrock, N-MORB-like basaltic tholeiite	γ-ray, TV image,[d] photometry[e]
Venera 13	−7.55	303.69	0.8 ± 0.3	Navka Planitia at E. end of Phoebe Regio rise, landscape similar to Venera 10 site, mafic alkaline rocks such as weathered olivine leucitite, nephelinite, volume density = 1.4–1.5 g cm^{-3} from impact loading	XRF, redox expt., TV imaging,[d,e] impact load
Venera 14	−13.05	310.19	0.9 ± 0.3	S. Navka Planitia on flank of a volcano, landing site is a plain dominated by layered bedrock and minor amount of soil, weathered N-MORB-like basaltic tholeiite, volume density = 1.15–1.20 g cm^{-3} from impact loading	XRF, redox expt., TV imaging,[d,e] impact load
Vega 1	8.10	175.85	−0.1 ± 0.1	Rusalka Planitia, N. of Aphrodite Terra, No TV panoramas, MORB-like tholeiite	γ-ray, (XRF failed)
Vega 2	−7.14	177.67	1.2 ± 0.2	Transitional zone between Rusalka Planitia and E. edge of Aphrodite Terra rise, No TV panoramas, N-MORB-like basaltic tholeiite	γ-ray, XRF

After Fegley et al. (1997) (reproduced by permission of the University of Arizona Press from Venus II 1997, Table III).
[a] Typical uncertainties on latitude and longitude are ±1.5° (Basilevsky et al. 1992).   [b] Relative to modal radius of 6,051.4 km with one standard deviation uncertainties.   [c] Suggested rock types are taken from Kargel et al. (1993) and references therein.   [d] Black and white TV image for Venera 9 and 10; Red-green-blue TV imaging for Venera 13 and 14.   [e] See Florensky et al. (1983); Garvin et al. (1984), and Pieters et al. (1986) for interpretations of the CONTRAST redox experiment, photometric data, and imaging.

potassium, uranium, and thorium by γ-ray spectroscopy, and the *Venera 13*, *14*, and *Vega 2* probes analyzed silicon, titanium, aluminum, iron, manganese, magnesium, calcium, potassium, sulfur, and chlorine by XRF spectroscopy. The γ-ray and XRF analyses are summarized in Tables 6 and 7, and are taken from Surkov *et al.* (1984, 1986, 1987).

The samples analyzed were small drill cores ~1 cm³ from ~3 cm depth, and are probably mixtures of rock and soil, as indicated by the volume density measurements made by *Venera 13* and *14*. Elements lighter than magnesium could not be detected by XRF, and the sodium content was estimated using geochemical methods. At least some of the sulfur and chlorine in the XRF analyses could be due to atmospheric weathering instead of being primary. It is unlikely that the concentrations of any of the other elements were affected by weathering. The oxidation state of iron is not determined by XRF spectroscopy and the amounts of ferrous (Fe^{2+}) and ferric (Fe^{3+}) iron in the samples are unknown. The CONTRAST oxidation state experiments on the *Venera 13* and *14* landers gave lower limits of 0.6–7 ppmv for CO, which overlap the magnetite–hematite phase boundary at the two landing sites. Thus, both ferrous and ferric iron may be present on the surface. The imaging carried out by *Venera 13* and *14* supports this conclusion.

Several groups have performed comparisons with elemental analyses of terrestrial rocks, calculated normative mineralogical compositions, made geochemical correlations, and made geological interpretations of Magellan radar images of the *Venera* and *Vega* landing sites. Unique conclusions about the rocks present at the different landing sites are not possible, but some conclusions are broadly accepted. The *Venera 8* K/U data showed that Venus is differentiated. This ratio is higher than that for most terrestrial basalts and is closer to that for high-potassium alkaline basalts. Some authors think the *Venera 8* K/U data indicate rhyolites on Venus, but this is hard to understand on such a dry planet. The K/U ratios at the *Venera 9* and *10* landing sites are similar to those for terrestrial alkaline basalts and oceanic tholeiites. The *Venera 13* XRF analysis suggests high-potassium alkaline basalts. The rocks at the *Venera 8* and *Venera 13* sites are generally thought to be similar, but this is not certain. The *Venera 14* XRF analysis suggests rocks like normal mid-ocean ridge basalt (N-MORB) that have been

Table 6 XRF elemental analyses of Venus' surface.

Oxide	Mass percent[a]				
	Venera 13[b]	Leucitic basalt[c]	Venera 14[b]	Vega 2[d,e]	N-MORB[f]
SiO_2	45.1 ± 3.0	46.18	48.7 ± 3.6	45.6 ± 3.2	48.77
TiO_2	1.59 ± 0.45	2.13	1.25 ± 0.41	0.2 ± 0.1	1.15
Al_2O_3	15.8 ± 3.0	12.74	17.9 ± 2.6	16 ± 1.8	15.9
FeO[g]	9.3 ± 2.2	9.86	8.8 ± 1.8	7.7 ± 1.1	9.82
MnO	0.2 ± 0.1	0.19	0.16 ± 0.08	0.14 ± 0.12	0.17
MgO	11.4 ± 6.2	8.36	8.1 ± 3.3	11.5 ± 3.7	9.67
CaO	7.1 ± 0.96	8.16	10.3 ± 1.2	7.5 ± 0.7	11.16
Na_2O[h]	2 ± 0.5	2.36	2.4 ± 0.4	2.0	2.43
K_2O	4.0 ± 0.63	6.18	0.2 ± 0.07	0.1 ± 0.08	0.08
SO_3	1.62 ± 1.0	0.09	0.88 ± 0.77	4.7 ± 1.5	
Cl	<0.3		<0.4	<0.3	
Total	98.1	96.16	98.7	95.4	99.15

[a] Error values for *Venera 13, 14*, and *Vega 2* data are ±1σ. [b] Surkov *et al.* (1984). [c] Volkov *et al.* (1986). [d] Surkov *et al.* (1986). [e] In addition to Cl, Surkov *et al.* (1986) also report the following upper limits (in mass %): Cu, Pb <0.3; Zn < 0.2; Sr, Y, Zr, Nb, Mo < 0.1; As, Se, Br < 0.08. [f] Wilson (1989). [g] All Fe reported as FeO for all analyses. [h] Calculated by Surkov *et al.* (1984, 1986).

Table 7 Gamma ray analyses of Venus' surface.

Space probe	K (%)	U (ppm)	Th (ppm)	K/U ratio
Venera 8	4.0 ± 1.2	2.2 ± 0.7	6.5 ± 2.2	$18,200^{+16,500}_{-8,500}$
Venera 9	0.47 ± 0.08	0.60 ± 0.16	3.65 ± 0.42	$7,800^{+4,700}_{-2,700}$
Venera 10	0.30 ± 0.16	0.46 ± 0.26	0.70 ± 0.34	$6,500^{+16,500}_{-4,600}$
Vega 1	0.45 ± 0.22	0.64 ± 0.47	1.5 ± 1.2	$7,000^{+32,400}_{-4,900}$
Vega 2	0.40 ± 0.20	0.68 ± 0.38	2.0 ± 1.0	$5,900^{+14,100}_{-4,000}$

Source: Surkov *et al.* (1987).

weathered by atmospheric SO_2. The *Vega 1* XRF spectrometer failed and only γ-ray data are available. The K/U ratios are similar to those at the *Venera 9* and *10* sites. The *Vega 2* XRF and γ-ray data suggest rocks like those at the *Venera 14* site, namely, weathered N-MORB-like basalts. The difference between the potassium content determined by the XRF and γ-ray instruments may reflect a difference between two different analyzed samples. The high sulfur contents in the three XRF analyses may be due to chemical weathering by atmospheric SO_2 to form anhydrite ($CaSO_4$) or sulfate-bearing scapolite. Conversely, the sulfur could be primary anhydrite or iron sulfide as in some terrestrial lavas (Carroll and Rutherford, 1985).

Normative compositions for the *Venera 13*, *14*, and *Vega 2* XRF analyses are given in Table 8, along with comparisons to the norms for terrestrial rocks that are possible analogues to the samples analyzed on Venus. Sulfur and chlorine were excluded from the norms in Table 8. If all sulfur is assumed present as anhydrite, then anhydrite comprises 2.8 ± 1.7 mass%, 1.5 ± 1.3 mass%, and 8.2 ± 2.6 mass%, of the *Venera 13*, *14*, and *Vega 2* samples, respectively. Table 8 also shows that calculated liquidus temperatures for the Venusian samples are similar to the liquidus temperatures of the possible terrestrial analogues.

1.19.4.2 Atmosphere–Surface Interactions

The high temperatures and pressures drive chemical reactions of CO_2, SO_2, OCS, H_2S, HCl, and HF with rocks and minerals on Venus' surface. A possible exception is sulfur vapor chemistry initiated by absorption of blue sunlight. After the *Mariner 2* flyby Mueller (1963) wrote that Venus' surface temperature "corresponds with those [temperatures] attained during moderately high degrees of metamorphism on Earth. It is therefore possible that large parts of the atmosphere of Venus are partially equilibrated with the surface rocks. From this assumption, it follows that the composition of the atmosphere should reflect the mineralogical character of the rocks." However, as of early 2003, Mueller's assumption remains controversial and unproven.

During the 1960s and 1970s Mueller and J. S. Lewis in the US, and scientists with the Soviet Venus exploration program modeled thermochemical equilibria between gases and minerals expected on Venus' surface. The models predicted that the abundances of CO_2, SO_2, OCS, H_2O, HCl, and HF are controlled by reactions with reactive minerals on Venus' surface. From today's perspective it now seems that only the abundances of CO_2, HCl, and HF are regulated, or buffered by the surface mineralogy, while the abundances of water vapor, CO, and the sulfur gases are kinetically controlled by combinations of surface–atmosphere reactions and gas-phase chemistry.

1.19.4.2.1 Carbonate equilibria

The CO_2 pressure on Venus is plausibly regulated by the "Urey reaction,"

$$CaCO_3(\text{calcite}) + SiO_2(\text{silica})$$
$$= CaSiO_3(\text{wollastonite}) + CO_2 \quad (12)$$

$$\log_{10} P_{CO_2} = 7.97 - 4{,}456/T \quad (13)$$

because the CO_2 pressure of ~92 bar at 740 K is virtually identical to the equilibrium CO_2 pressure from reaction (12) at that temperature (Fegley and Treiman, 1992). But, are carbonates present on Venus?

Arguments in favor of carbonates include the following. Some of the flow features on Venus' surface look like they were made by magmas, such as carbonatites, with water-like rheologies. If correct, these geomorphological interpretations

Table 8 Normative compositions of Venusian samples and possible terrestrial analogues.

CIPW norms	*Venera 13*[a]	Leucitic basalt[a]	*Venera 14*[a]	Tholeiitic basalt[a]	*Vega 2*[b]
Hypersthene			18.2	14.2	25.4
Olivine	26.6	16.6	9.1	8.1	13.9
Diopside	10.2	29.4	9.9	21.2	2.5[d]
Anorthite	24.2	6.2	38.6	33.6	38.3
Albite	3.0		20.7	20.3	18.9
Orthoclase	25.0	11.8	1.2	0.3	0.5
Nepheline	8.0	11.2			
Leucite		20.6			
Ilmenite	3.0	4.2	2.3	2.3	0.5
Total	100.0	100.0	100.0	100.0	100.0
Liquidus T (°C)[c]	1,249	1,176	1,153	1,196	1,265

[a] On a volatile free basis, from Volkov *et al.* (1986). [b] Barsukov *et al.* (1986). [c] Liquidus temperatures calculated with the MAGPOX3 code of John Longhi. [d] Clinopyroxene.

mean that carbonates are present on Venus. Also, as done with the Viking XRF analyses, the mass deficits in the *Venera* and *Vega* XRF analyses can be attributed to carbonates. The calculated calcite abundances (by mass) are 4% at the *Venera 13* site, 3% at the *Venera 14* site, and 10% at the *Vega 2* site. Carbonate-bearing rocks would also provide chlorine- and fluorine-bearing minerals necessary for regulating the atmospheric abundances of the reactive hydrogen halides.

There are two major arguments against carbonates. Most terrestrial carbonates are sedimentary, while Venus is dry without liquid water. However, the high D/H ratio suggests that Venus was wet in the past and carbonates may have formed at that time. Second, calculations and experiments predict that carbonates on Venus will react with atmospheric SO_2 to form anhydrite ($CaSO_4$). However, the *Venera* XRF analyses show CaO/SO_3 ratios less than unity, so Venus' surface is not anhydrite saturated. Also, wind-driven erosion may abrade anhydrite layers and expose the underlying carbonate. But at present the questions of whether or not carbonates are present on Venus and are buffering atmospheric CO_2 remain unresolved. Sample return or *in situ* analyses that are sensitive to calcite and other carbonates are needed to answer these questions.

1.19.4.2.2 Equilibria involving HCl and HF

Hydrogen chloride and HF were discovered in Venus' atmosphere in the late 1960s. Shortly thereafter, it was proposed that the atmospheric abundances HCl and HF are regulated by equilibria involving chlorine- and fluorine-bearing minerals, such as sodalite or fluorphlogopite. For example, a reaction involving nepheline, albite, and sodalite

$$2HCl(g) + 9NaAlSiO_4(nepheline)$$
$$= Al_2O_3(corundum) + NaAlSi_3O_8(albite)$$
$$+ 2Na_4[AlSiO_4]_3Cl(sodalite)$$
$$+ H_2O(g) \quad (14)$$

may buffer HCl, while a reaction involving potassium feldspar and fluorphlogopite

$$2HF(g) + KAlSi_3O_8(K\text{-spar})$$
$$+ 3MgSiO_3(enstatite)$$
$$= KMg_3AlSi_3O_{10}F_2(fluorphlogopite)$$
$$+ 3SiO_2(silica) + H_2O(g) \quad (15)$$

may buffer HF (Fegley *et al.*, 1997). These reactions as well as other possible buffers for HCl and HF involve phases that are common in alkaline rocks on Earth and by analogy also on Venus. This is an interesting point because alkaline rocks, which are rare on Earth, seem to be present at several of the *Venera* and *Vega* landing sites. As argued by Kargel *et al.* (1993), Venus may have a more mafic alkaline crust than the Earth.

1.19.4.2.3 Redox reactions involving iron-bearing minerals

The oxidation of Fe^{2+}-bearing minerals in basalt (and other volcanic rocks on Venus' surface) is potentially very important for water loss via oxidation of the surface and hydrogen escape to space. The overall process is schematically represented by

$$H_2O(gas) + 2FeO(in\ rock)$$
$$= Fe_2O_3(hematite) + H_2(gas) \quad (16)$$

and

$$H_2O(gas) + 3FeO(in\ rock)$$
$$= Fe_3O_4(magnetite) + H_2(gas) \quad (17)$$

where FeO represents the Fe^{2+}-bearing pyroxenes, olivines, etc., in a rock. Several spacecraft observations indicate that oxidation of Fe^{2+}-bearing basalts occurs on Venus today. Wide-angle photometry at the *Venera 9* and *10* sites and color TV imaging at the *Venera 13* and *14* sites suggest the presence of hematite or another ferric iron mineral on Venus' surface (Pieters *et al.*, 1986). The electrical resistivity of the soil at the *Venera 13* and *14* landing sites was measured and found to be one to two orders of magnitude lower than for basalt at the same temperature. Electrically conductive minerals such as magnetite and/or hematite may be present. As noted above, the CONTRAST oxidation state experiments on these landers gave lower limits of 0.6–7 ppmv for CO, which overlap the magnetite–hematite phase boundary. The PV large probe neutral mass spectrometer reported 0–10 ppmv H_2 below 25 km. The PV Orbiter ion mass spectrometer detected a mass-2 peak that has been interpreted in terms of contributions from D^+ and H_2^+ (Donahue *et al.*, 1997). The derived H_2 mixing ratio is <0.1–10 ppmv below 140 km altitude. The *Venera 13* and *14* GC reported 25 ± 10 ppmv H_2 at 49–58 km (Krasnopolsky, 1986). The latter detection is controversial. In any case, the reported H_2 mixing ratio in Venus' lower atmosphere ranges from <0.1 ppmv to 10–25 ppmv. These H_2 mixing ratios are larger than those from the water–gas reaction

$$CO + H_2O = CO_2 + H_2 \quad (18)$$

at Venus surface temperatures and indicate that oxidation of Fe^{2+}-bearing minerals by H_2O may be involved in producing the observed hydrogen.

1.19.4.2.4 Minerals present in low radar emissivity regions

Radar observations from the *Pioneer Venus* and *Magellan* spacecraft show global variations in the radar emissivity of Venus' surface (Pettengill *et al.*, 1996, 1997). The variations are apparently mainly related to the types of rocks present. Typically, regions below ~2.4 km elevation (i.e., below 6,054 km radius) have radar properties characteristic of anhydrous rocks, such as dry basalt. However, higher elevation regions have lower radar emissivity, indicating the presence of semiconducting minerals with high dielectric constants. The critical level also shifts upward by ~1 km from the equator to high northern latitudes. The sharp boundary between normal rock at lower elevations and lower-emissivity rock at higher elevations indicates that temperature changes play a role in the unusual chemistry.

Several proposals have been made for the chemistry and minerals involved. The loaded dielectric model uses inclusions of iron sulfides, or other high dielectric minerals, in rocks. Higher amounts of the dielectric minerals produce lower radar emissivities and higher dielectric constants. The iron sulfides, or other dielectric minerals, are destroyed by chemical reactions with Venus' atmosphere. The reactions proceed faster at lower elevations, where the temperatures are higher, so the lowest radar emissivity regions are predicted at the highest elevations, or in areas with the youngest volcanic rocks.

The metallic frost model postulates volcanic outgassing and condensation of volatile elements and their compounds, such as elemental tellurium or galena (PbS), on mountain tops and other high-elevation areas. Many volatile elements (e.g., copper, zinc, tin, lead, arsenic, antimony, and bismuth) form halides or chalcogenides with high vapor pressures. Compounds of these metals are typically found around terrestrial volcanic vents and fumaroles, or are present in volcanic gases (Brackett *et al.*, 1995). Many of these elements and their compounds are semiconductors and have high dielectric constants. Volcanic eruptions release these volatile elements and their compounds, which then condense as metallic frosts on Venus' surface. The metallic frosts concentrate in the higher-elevation regions, which are cold traps for them (Brackett *et al.*, 1995), like snow on Mount Kilimanjaro. Thin layers of such metallic frosts can account for the observed radar properties, e.g., a 5 μm thick layer of elemental tellurium (Pettengill *et al.*, 1996). The metallic frost layers can be covered and/or vaporized by fresh volcanic flows, which should display normal radar emissivity. The radar properties do not uniquely constrain the semiconductor(s) present, which will remain unknown until elemental and mineralogical analyses are carried out *in situ* or on samples returned from low-emissivity highlands regions.

1.19.4.3 The Venus Sulfur Cycle and Climate Change

Current knowledge suggests that volcanic outgassing on Venus produces SO_2 and the reduced sulfur gases S_2, OCS, and H_2S. The relative proportions of these species in Venusian volcanic gases are unknown. Sulfur dioxide is generally the major sulfur gas in terrestrial basaltic volcanic gases, which have oxygen fugacities between those of the quartz–fayalite–magnetite (QFM) and nickel–nickel oxide (NNO) buffers (Symonds *et al.*, 1994). Hydrogen sulfide is often the next most abundant sulfur gas, and is sometimes as or more abundant than SO_2. Carbonyl sulfide and sulfur vapor are less abundant than SO_2 or H_2S. If Venusian basalts erupt at temperatures and oxygen fugacities similar to those for terrestrial basaltic volcanoes, then SO_2 should be more abundant than S_2, OCS, or H_2S. The very low H_2O abundance in Venus' atmosphere and considerations of chemical equilibria also imply that S_2 and OCS should be more abundant than H_2S in Venusian volcanic gases.

At present, the SO_2 abundance in Venus' atmosphere is 35–110 times larger than the equilibrium SO_2 pressure for the reaction

$$SO_2 + CaCO_3(\text{calcite}) = CaSO_4(\text{anhydrite}) + CO \quad (19)$$

which is thus a sink for SO_2. The *Vega* UV spectrometer data give the lower factor of 35, while *Pioneer Venus*, *Venera 11–12*, and Earth-based IR spectroscopic data give the higher factor of 110 (cf. Fegley *et al.*, 1997, figure 7). The rate of reaction (19) is known and it is sufficient to remove all SO_2 (and thus the sulfuric acid clouds) from Venus' atmosphere in ~1.9 Myr in the absence of a volcanic source (Fegley and Prinn, 1989). Similar, but slower, reactions of SO_2 occur with other calcium-bearing minerals such as anorthite, diopside, and wollastonite. The measured Ca/S ratios are greater than unity at the *Venera 13, 14*, and *Vega 2* sites. These ratios are larger than expected (i.e., unity) if all calcium were combined with sulfur in anhydrite. Thus, loss of atmospheric SO_2 via chemical weathering of calcium-bearing minerals on Venus' surface is probably an ongoing process.

Maintenance of atmospheric SO_2 at its current concentration requires eruption of ~1 km^3 yr of lava with the average composition of the *Venera 13, 14*, and *Vega 2* landing sites.

This volcanism rate is the same as the average rate of subaerial volcanism on Earth and is ~5% of the terrestrial plate creation rate of ~20 km^3 yr^{-1}. The required sulfur eruption rate to maintain SO$_2$ on Venus at steady state is ~28 Tg yr^{-1}. This is similar to estimates of 9 (subaerial), 19 (submarine), and 28 (total) Tg yr^{-1} for SO$_2$ emissions from terrestrial volcanism (Charlson et al., 1992).

Volcanism on Earth and on Io is episodic. By analogy, Venusian volcanism should be episodic, which may be one reason why active volcanism has not yet been seen on Venus. However, a volcanic source for SO$_2$ is required at present. What may happen if the volcanic source and anhydrite sink for SO$_2$ are not balanced? If less SO$_2$ is erupted than is lost by anhydrite formation, less SO$_2$ will be left in the atmosphere, less H$_2$SO$_4$ will be produced, and fewer clouds will form. Temperatures in Venus' atmosphere and at the surface may decrease, because SO$_2$ and volcanic volatiles such as CO$_2$ and H$_2$O are greenhouse gases. Magnesite (MgCO$_3$) and other carbonates unstable on Venus today may form and consume atmospheric CO$_2$ as temperatures drop. Conversely, if more SO$_2$ is erupted than is lost by anhydrite formation, more SO$_2$ will be added to the atmosphere, more H$_2$SO$_4$ will be produced, and more clouds will form. In this case atmospheric and surface temperatures may rise as more greenhouse gases enter the atmosphere. Minerals now stable at 740 K on Venus' surface may decompose as temperatures increase.

Some of these effects, which could operate in the future and may have done so in the past, have been studied in climate models that incorporate variations of SO$_2$ and H$_2$O abundances on the clouds and temperatures on Venus (e.g., Hashimoto and Abe, 2001; Bullock and Grinspoon, 2001). In particular, large temperature changes are predicted to result from the putative global resurfacing of Venus 500 ± 200 Ma. Solomon et al. (1999) modeled the atmospheric and geophysical consequences of this event. They propose that higher surface temperatures diffuse into Venus' interior and cause significant thermal stresses that influence tectonic deformation on a global scale. Their model may explain formation of some of the wrinkled ridge plains that cover 60–65% of Venus' surface today.

Finally, the role of OCS and H$_2$S in the Venus sulfur cycle needs to be considered. In addition to their possible volcanic sources, chemical weathering of iron sulfides may produce OCS and H$_2$S. These sources are exemplified by the following reactions:

$$4CO_2 + 3FeS(\text{pyrrhotite})$$
$$= 3OCS + CO + Fe_3O_4(\text{magnetite}) \quad (20)$$

$$4H_2O + 3FeS(\text{pyrrhotite})$$
$$= 3H_2S + H_2 + Fe_3O_4(\text{magnetite}) \quad (21)$$

The inferred gradient in OCS and observation by the *Pioneer Venus* mass spectrometer of 3 ± 2 ppmv H$_2$S below 22 km are consistent with these ideas. At higher altitudes, OCS and H$_2$S are converted to SO$_2$ via photochemical reactions that result in the net transformations:

$$2H_2S + 6CO_2 \rightarrow 2H_2O + 2SO_2 + 6CO \quad (22)$$

$$2OCS + 4CO_2 \rightarrow 2SO_2 + 6CO \quad (23)$$

The SO$_2$ is then photo-oxidized to aqueous sulfuric acid droplets as described earlier.

The sulfur cycle is closed as follows. Sulfuric acid cloud droplets vaporize to a gas mixture of H$_2$SO$_4$, SO$_3$, and H$_2$O at the cloud base. Reduction of SO$_3$ to SO$_2$ occurs via the reaction

$$CO + SO_3 \rightarrow SO_2 + CO_2 \quad (24)$$

The SO$_2$ formed via reaction (24) eventually reacts with calcium-bearing minerals on Venus's surface and the cycle repeats when sulfur gases are volcanically outgassed again.

1.19.4.4 Topography and Geology

Venus is often regarded as Earth's "twin planet" because of its similar size (~95%), mass (~82%), and gravity (~90%) compared to Earth. However, radar images from Earth-based observatories and from the *Pioneer Venus*, *Venera 15/16*, and *Magellan* spacecraft reveal both important similarities and differences to the Earth. Tanaka et al. (1997) distinguish three major terrain types on Venus: (i) lowlands, which comprise ~27% of Venus' surface that lies ~0–2 km below the modal radius (6,051.4 km); (ii) upland rolling plains, which comprise ~65% of the surface at an elevation of ~0–2 km; and (iii) highlands that are ~8% of the surface and are >2 km above the modal radius. The total range of elevations on Venus is ~14 km from Diana Chasma (−2 km) to the top of Maxwell Montes (12 km). However, ~80% of the surface is within ±1 km, and ~90% is between −1 km and +2.5 km of the modal radius. Venus' unimodal topographic distribution is in contrast to Earth, which has a bimodal hypsometric curve. Figure 2 shows the hypsometric curve for Venus.

The two major highland regions on Venus are Ishtar Terra at high northern latitudes and Aphrodite Terra in the equatorial regions. Both regions are continental in size. Ishtar is about the same size as Australia, while Aphrodite is roughly the size of South America. The western part of Ishtar is dominated by the Lakshmi Planum plateau that resembles, but is larger

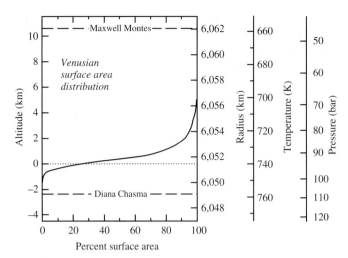

Figure 2 Distribution of surface area as a function of altitude for Venus, i.e., the hypsometric curve (after Fegley and Treiman, 1992) (reproduced by permission of American Geophysical Union from *Geophysical Monograph* **1992**, 66).

than, the Tibetan plateau on Earth. Maxwell Montes, which is higher than Mount Everest, is in eastern Ishtar. Aphrodite is rougher and more complex than Ishtar and is characterized by several deep narrow valleys, such as Diana Chasma, and by several distinct mountain ranges that reach up to 6 km high. Alpha, Beta, and Phoebe Regio are three smaller highland regions.

Venus' surface shows extensive evidence of widespread volcanism including large shield volcanoes such as Sif Mons, that are similar to the shield volcanoes of the Hawaiian islands, volcanic plains, volcanic calderas, smaller volcanic landforms such as cones and pancake domes, and long sinuous channels (such as Baltis Vallis) that can meander for several thousand kilometers across the surface. In some cases, the different landforms indicate different types of magmas, for example, the pancake domes were apparently formed by viscous SiO_2-rich magmas, whereas the long sinuous channels were apparently formed by fluid magmas, such as carbonatites or even liquid sulfur.

Tectonic features are also present on the surface. Tesserae, which are tectonically deformed regions formed by piling up blocks of crust, are common in the highlands. As noted earlier, ~65% of Venus' surface is covered by tectonically deformed lowlands and rolling plains containing wrinkle ridges formed by buckling of the crust. Other features formed by volcanism and tectonism are coronae, circular- or oval-shaped features a few hundred kilometers in diameter that may have raised outer rims and arachnoids, which are caldera-like collapse features surrounded by fractures. The topography of coronae is highly variable. Most of the coronae occur in chains associated with large scale rifting (chasmata), some occur in association with volcanic rises, and a few are isolated features in the plains.

Venus' surface has ~940 impact craters ranging in diameter from ~3 km to a few hundred kilometers. The small size cutoff is probably due to atmospheric disruption of small impactors. The smaller craters are more irregularly shaped, indicating the impact of several fragments instead of one object. The crater ejecta patterns are unlike those on other solar system bodies and probably have been affected by the dense atmosphere and prevailing winds. The number and distribution of craters was used to argue for global volcanism resurfacing ~500 ± 200 Ma (Schaber et al., 1992). However, this interpretation has been questioned by other workers, who argue that two large-scale plain units of distinct ages spread over 400 Ma exist instead (Hauck et al., 1998).

1.19.4.5 Interior

Venus' interior structure is unknown, but spacecraft data allow several inferences. No intrinsic magnetic field has been detected, and any dipole field is $<10^{-4}$ that of Earth. The ~243 d rotation rate may be too slow to generate a field by dynamo action in a core. Radar imaging does not show evidence of plate tectonics, which may be due to a lack of water and/or to difficulty in subducting the lithosphere, which is hotter and perhaps more buoyant than on Earth because of Venus' high surface temperature. The *Magellan* gravity data provide the best probe of Venus' interior. Unlike Earth, gravity is strongly correlated with topography on Venus, suggesting that higher regions are above regions of mantle upwelling. Figure 3 is a cartoon illustrating Venus' internal structure. Three models of Venus' (unknown) bulk composition are given in Table 9. The chondritic meteorite model assumes that the processes that affected chondritic meteorites in the solar nebula also affected the material accreted by terrestrial planets. Planetary bulk compositions are predicted based on abundances of key elements such as potassium and uranium that have been determined by sample return or *in situ* analyses. The equilibrium condensation model predicts planetary compositions based on temperature- and pressure-dependent thermodynamic equilibrium in a solar nebula model with T and P known as a function of radial distance. The pyrolite model is similar to the pyrolite model for the Earth and uses analyses of basalts and geochemical constraints to predict planetary compositions.

1.19.5 SUMMARY OF KEY QUESTIONS

Some of the key questions about Venus that are unresolved include the following. What is the oxygen isotopic composition (all three stable isotopes) of the silicate portion of Venus? Oxygen is either the first or second most abundant element in Venus, and its isotopic composition can help constrain accretion models for the terrestrial planets. The Earth and Moon have the same oxygen isotopic composition, but the oxygen isotopic composition of the SNC meteorite parent body, presumably Mars, of the eucrite parent body, presumably 4 Vesta, and of other meteorites, and IDPs are different. It is important to measure abundances of the three oxygen isotopes (with an uncertainty of <0.1‰) on Venus to see if it has the same or a different oxygen isotopic composition as the Earth and Moon. These measurements would have to be carried out on samples returned from Venus, which is technologically feasible at present.

Another important question is which, if any, gases in Venus' atmosphere are buffered by mineral assemblages on Venus' surface. Interpretations of data from *Pioneer Venus* and the *Venera* and *Vega* probes about the chemistry of Venus' lower atmosphere and surface are based upon this assumption, which has not been verified experimentally. Instead, research shows kinetic control of reactions of SO_2 and water vapor with minerals at Venus surface temperatures (Fegley and Prinn, 1989; Johnson and Fegley, 2000).

Finally, it is also important to understand the origin, duration, and stability of the Venusian super-greenhouse. The Venus sulfur cycle demonstrates the close connection between chemistry and climate on Venus, because SO_2 is one of the three key greenhouse gases sustaining the super-greenhouse, is involved in chemical reactions with abundant calcium-bearing minerals on the surface, is the feedstock for the global sulfuric acid clouds, and is probably also an important volcanic volatile. Current knowledge of the super-greenhouse and of climate change and stability on Venus requires new laboratory studies of IR absorption bands in hot, dense CO_2, SO_2, and H_2O and improved computer models of the influence of clouds on climate.

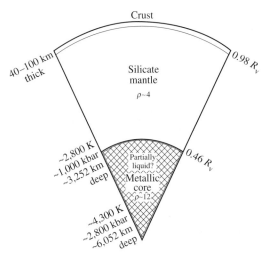

Figure 3 A cartoon illustrating Venus' interior structure based on current bulk composition models (see Table 9) (after Lodders and Fegley, 1998) (reproduced by permission of Oxford University Press from the *Planetary Scientist's Companion* **1998**, p. 118).

Table 9 Bulk composition models for Venus.

Component	Chondritic meteorite[a]	Equilibrium condensation[b]	Pyrolite[c]
Mantle/Crust			
SiO_2	49.8	52.9	40.4
TiO_2	0.21	0.20	0.24
Al_2O_3	4.1	3.8	3.4
FeO	5.4	0.24	18.7
Cr_2O_3	0.87		0.3
MnO	0.09		0.2
MgO	35.5	37.6	33.3
CaO	3.3	3.6	3.4
Na_2O	0.28	1.6	0.15
K_2O	0.027	0.174	0.018
Core			
Fe	88.6	94.4	78.7
Ni	5.5	5.6	6.6
Co	0.26		
S	5.1	0	4.9
O	0		9.8
Mass fractions			
Core	0.32	0.30	0.24
Mantle + Crust	0.68	0.70	0.76

After Lodders and Fegley (1998).
[a] Morgan and Anders (1980). [b] Basaltic Volcanism Study Project model Ve1. [c] BVSP model Ve4.

ACKNOWLEDGMENTS

I thank K. Baines, B. Bézard, D. Cahn, D. Crisp, A. Davis, K. Lodders, V. Moroz, R. Pepin, R. Wieler for discussions and R. Osborne for assistance with preparing the manuscript. This work was supported by Grant NAG5-11037 from the NASA Planetary Atmospheres Program.

REFERENCES

Allen D. A. and Crawford J. W. (1984) Cloud structure on the dark side of Venus. *Nature* **307**, 222–224.

Baines K. H. and 24 others (2000) Detection of sub-micron radiation from the surface of Venus by Cassini/VIMS. *Icarus* **148**, 307–311.

Barath F. T., Barrett A. H., Copeland J., Jones D. E., and Lilley A. E. (1964) Mariner 2 microwave radiometer experiment and results. *Astron. J.* **69**, 49–58.

Barker E. S. (1979) Detection of SO_2 in the UV spectrum of Venus. *Geophys. Res. Lett.* **6**, 117–120.

Barsukov V. L., Surkov Yu. A., Dmitriyev L. V., and Khodakovsky I. L. (1986) Geochemical studies on Venus with the landers from the Vega-1 and Vega-2 probes. *Geochem. Int.* **23**(7), 53–65.

Barsukov V. L., Basilevsky A. T., Volkov V. P., and Zharkov V. N. (eds.) (1992) *Venus Geology, Geochemistry, and Geophysics.* University of Arizona Press, Tucson.

Basilevsky A. T., Nikolaeva O. V., and Weitz C. M. (1992) Geology of the Venera 8 landing site region from Magellan data: morphological and geochemical considerations. *J. Geophys. Res.* **97**, 16315–16335.

Bézard B., Baluteau J. P., Marten A., and Coron N. (1987) The $^{12}C/^{13}C$ and $^{16}O/^{18}O$ ratios in the atmosphere on Venus from high-resolution 10-µm spectroscopy. *Icarus* **72**, 623–634.

Bézard B., DeBergh C., Crisp D., and Maillard J. P. (1990) The deep atmosphere of Venus revealed by high-resolution nightside spectra. *Nature* **345**, 508–511.

Bougher S. W., Hunten D. M., and Phillips R. J. (eds.) (1997) *Venus II.* University of Arizona Press, Tucson, 1362pp.

Brackett R. A., Fegley B., Jr., and Arvidson R. E. (1995) Volatile transport on Venus and implications for surface geochemistry and geology. *J. Geophys. Res.* **100**, 1553–1563.

Bullock M. A. and Grinspoon D. H. (2001) The recent evolution of climate on Venus. *Icarus* **150**, 19–37.

Carroll M. R. and Rutherford J. J. (1985) Sulfide and sulfate saturation in hydrous silicate melts. *Proc. 15th Lunar Planet Sci. Conf. J. Geophys. Res. Suppl.* **90**, C601–C612.

Chamberlain J. W. and Hunten D. M. (1987) *Theory of Planetary Atmospheres.* Academic Press, San Diego.

Charlson R. J., Anderson T. L., and McDuff R. E. (1992) The sulfur cycle. In *Global Biogeochemical Cycles* (eds. S. S. Butcher, R. J. Charlson, G. H. Orians, and G. V. Wolfe). Academic Press, London, pp. 285–300.

Connes P., Connes J., Benedict W. S., and Kaplan L. D. (1967) Traces of HCl and HF in the atmosphere of Venus. *Astrophys. J.* **147**, 1230–1237.

Connes P., Connes J., Kaplan L. D., and Benedict W. S. (1968) Carbon monoxide in the Venus atmosphere. *Astrophys. J.* **152**, 731–743.

Crisp D. and Titov D. (1997) The thermal balance of the Venus atmosphere. In *Venus II* (eds. S. W. Bougher, D. M. Hunten, and R. J. Phillips). University of Arizona Press, Tucson, pp. 353–384.

Donahue T. M., Grinspoon D. H., Hartle R. E., and Hodges R. R., Jr. (1997) Ion/neutral escape of hydrogen and deuterium: evolution of water. In *Venus II* (eds. S. W. Bougher, D. M. Hunten, and R. J. Phillips). University of Arizona Press, Tucson, pp. 385–414.

Esposito L. W., Knollenberg R. G., Marov M. Ya., Toon O. B., and Turco R. P. (1983) The clouds and hazes of Venus. In *Venus* (eds. D. M. Hunten, L. Colin, T. M. Donahue, and V. I. Moroz). University of Arizona Press, Tucson, pp. 484–564.

Esposito L. W., Burtaux J.-L., Krasnopolsky V., Moroz V. I., and Zasova L. V. (1997) Chemistry of the lower atmosphere and clouds. In *Venus II* (eds. S. W. Bougher, D. M. Hunter, and R. J. Phillips). University of Arizona Press, Tucson, pp. 415–458.

Fegley B., Jr. and Prinn R. G. (1989) Estimation of the rate of volcanism on Venus from reaction rate measurements. *Nature* **337**, 55–58.

Fegley B. and Treiman A. H. (1992) Chemistry of atmosphere-surface interactions on Venus and Mars. In *Venus and Mars: Atmospheres, Ionospheres, and Solar Wind Interactions*, Geophysical Monograph 66 (eds. J. G. Luhmann, M. Tatrallyay, and R. O. Pepin). American Geophysical Union, Washington, DC, pp. 7–71.

Fegley B., Jr., Klingelhöfer G., Lodders K., and Widemann T. (1997) Geochemistry of surface-atmosphere interactions on Venus. In *Venus II* (eds. S. W. Bougher, D. M. Hunten, and R. J. Phillips). University of Arizona Press, Tucson, pp. 591–636.

Florensky C. P., Nikolaeva O. V., Volkov V. P., Kudryashova A. F., Pronin A. A., Geektin Yu. M., Tchaikina E. A., and Bashkirova A. S. (1983) Redox indicator "CONTRAST" on the surface of Venus. *Lunar Planet. Sci.* **XIV**. (abstract). The Lunar and Planetary Institute, Houston, pp. 203–204.

Fox J. L. and Kliore A. J. (1997) Ionosphere: solar cycle variations. In *Venus II* (eds. S. W. Bougher, D. M. Hunten, and R. J. Phillips). University of Arizona Press, Tucson, pp. 161–188.

Garvin J. B., Head J. W., Zuber M. T., and Helfenstein P. (1984) Venus: the nature of the surface from Venus panoramas. *J. Geophys. Res.* **89**, 3381–3399.

Gierasch P. J., Goody R. M., Young R. E., Crisp D., Edwards C., Kahn R., Rider D., Del Genio A., Greeley R., Hou A., Leovy C. B., McCleese D., and Newman M. (1997) The general circulation of the Venus atmosphere: an assessment. In *Venus II* (eds. S. W. Bougher, D. M. Hunten, and R. J. Phillips). University of Arizona Press, pp. 459–500.

Hashimoto G. L. and Abe Y. (2001) Predictions of a simple cloud model for water vapor cloud albedo feedback on Venus. *J. Geophys. Res.* **106**, 14675–14690.

Hauck S. A., Phillips R. J., and Price M. H. (1998) Venus: crater distribution and plains resurfacing models. *J. Geophys. Res.* **103**, 13635–13642.

Hunt G. E. and Moore P. (1982) *The Planet Venus.* Faber and Faber, London.

Hunten D. M., Colin L., Donahue T. M., and Moroz V. I. (eds.) (1983) *Venus.* University of Arizona Press, Tucson.

Jastrow R. and Rasool S. I. (eds.) (1969) *The Venus Atmosphere.* Gordon and Breach, New York.

Johnson N. M. and Fegley B., Jr. (2000) Water on Venus: new insights from tremolite decomposition. *Icarus* **146**, 301–306.

Kargel J. S., Komatsu G., Baker V. R., and Strom R. G. (1993) The volcanology of Venera and VEGA landing sites and the geochemistry of Venus. *Icarus* **103**, 253–275.

Kasprzak W. T., Keating G. M., Hsu N. C., Stewart A. I. F., Colwell W. B., and Bougher S. W. (1997) Solar activity behavior of the thermosphere. In *Venus II* (eds. S. W. Bougher, D. M. Hunten, and R. J. Phillips). University of Arizona Press, Tucson, pp. 225–257.

Kaula W. M. (1999) Constraints on Venus evolution from radiogenic argon. *Icarus* **139**, 32–39.

Krasnopolsky V. A. (1986) *Photochemistry of the Atmospheres of Mars and Venus.* Springer, Berlin.

Lewis J. S. and Prinn R. G. (1984) *Planets and their Atmospheres.* Academic Press, Orlando.

Lodders K. and Fegley B., Jr. (1998) *The Planetary Scientist's Companion.* Oxford University Press, New York.

Maor E. (2000) *June 8, 2004 Venus in Transit.* Princeton University Press, Princeton.

Marov M. Ya. (1972) Venus: a perspective at the beginning of planetary exploration. *Icarus* **16**, 415–461.

Mills F. P. (1998) I. Observations and photochemical modeling of the Venus middle atmosphere. II. Thermal infrared spectroscopy of Europa and Callisto. Doctoral Thesis, California Institute of Technology, Pasadena, CA.

Morgan J. W. and Anders E. (1980) Chemical composition of the Earth, Venus, and Mercury. *Proc. Natl. Acad. Sci.* **77**, 6973–6977.

Mueller R. F. (1963) Chemistry and petrology of Venus: preliminary deductions. *Science* **141**, 1046–1047.

Nagy A. F. and Cravens T. E. (1997) Ionosphere: energetics. In *Venus II* (eds. S. W. Bougher, D. M. Hunten, and R. J. Phillips). University of Arizona Press, Tucson, pp. 189–223.

Pettengill G. H. (1968) Radar studies of the planets. In *Radar Astronomy* (ed. J. V. Evans). McGraw-Hill, NY, pp. 275–321.

Pettengill G. H., Ford P. G., and Simpson R. A. (1996) Electrical properties of the Venus surface from bistatic radar observations. *Science* **272**, 1628–1631.

Pettengill G. H., Campbell B. A., Campbell D. B., and Simpson R. A. (1997) Surface scattering and dielectrical properties. In *Venus II* (eds. S. W. Bougher, D. M. Hunten, and R. J. Phillips). University of Arizona Press, Tucson, pp. 527–546.

Pieters C. M., Head J. W., Patterson W., Pratt S., Garvin J., Barsukov V. L., Basilevsky A. T., Khodakovsky I. L., Selivanov A. S., Panfilov A. S., Gektin Yu. M., and Narayeva Y. M. (1986) The color of the surface of Venus. *Science* **234**, 1379–1383.

Schaber G. G., Strom R. G., Moore H. J., Soderblom L. A., Kirk R. L., Chadwick D. J., Davson D. D., Gaddis L. R., Boyce J. M., and Russell J. (1992) Geology and distribution of impact craters on Venus: what are they telling us? *J. Geophys. Res.* **97**, 13257–13301.

Seiff A. (1983) Thermal structure of the atmosphere of Venus. In *Venus* (eds. D. M. Hunten, L. Colin, T. M. Donahue, and V. I. Moroz). University of Arizona Press, Tucson, pp. 215–279.

Shapiro I. I. (1968) Spin and orbital motions of planets. In *Radar Astronomy* (ed. J. V. Evans). McGraw-Hill, NY, pp. 143–185.

Shapiro I. I., Campbell D. B., and DeCampli W. M. (1979) Nonresonance rotation of Venus? *Astrophys. J.* **230**, L123–L126.

Solomon S. C., Bullock M. A., and Grinspoon D. H. (1999) Climate change as a regulator of tectonics on Venus. *Science* **286**, 87–90.

Surkov Yu. A., Barsukov V. L., Moskalyeva L. P., Kharyukova V. P., and Kemurdzhian A. L. (1984) New data on the composition, structure, and properties of Venus rock obtained by Venera-13 and Venera-14. *J. Geophys. Res.: Proc. 15th LPSC* **89**, 393–402.

Surkov Yu. A., Moskalyova L. P., Kharyukova V. P., Dudin A. D., Smirnov G. G., and Zaitseva S. Ye. (1986) Venus rock composition at the Vega-2 landing site. *J. Geophys. Res. Proc. 17th LPSC* **91**, 215–218.

Surkov Yu. A., Kirnozov F. F., Glazov V. N., Dunchenko A. G., Tatsy L. P., and Sobornov O. P. (1987) Uranium, thorium, and potassium in the Venusian rocks at the landing sites of Vega-1 and Vega-2. *J. Geophys. Res.: Proc. 17th LPSC* **92**, 537–540.

Symonds R. B., Rose W. I., Bouth G. J. S., and Gerlach T. M. (1994) Volcanic-gas studies: methods, results, and applications. In *Volatiles in Magmas* (eds. M. R. Carroll and J. R. Holloway). Mineralogical Society of America, Washington, DC, pp. 1–66.

Tanaka K. L., Senske D. A., Price M., and Kirk R. L. (1997) Physiography, geomorphic/geologic mapping, and stratigraphy of Venus. In *Venus II* (eds. S. W. Bougher, D. M. Hunten, and R. J. Phillips). University of Arizona Press, Tucson, pp. 667–694.

Volkov V. P. and Frenkel M. Ya. (1993) The modeling of Venus' degassing in terms of potassium–argon system. *Earth Moon Planet.* **62**, 117–129.

Volkov V. P., Zolotov M. Yu., and Khodakovsky I. L. (1986) Lithospheric-atmospheric interaction on Venus. In *Chemistry and Physics of Terrestrial Planets* (ed. S. K. Saxena). Springer, New York, pp. 136–190.

Von Zahn U., Kumar S., Niemann H., and Prinn R. (1983) Composition of the Venus atmosphere. In *Venus* (eds. D. M. Hunten, L. Colin, T. M. Donahue, and V. I. Moroz). University of Arizona Press, Tucson, pp. 299–430.

Warneck P. (1988) *Chemistry of the Natural Atmosphere*. Academic Press, San Diego.

Wieler R. (2002) Noble gases in the solar system. In *Noble Gases in Geochemistry and Cosmochemistry* (eds. D. Porcelli, C. J. Ballentine, and R. Wieler). *Rev. Mineral. Geochem.* **47**, 21–70.

Wilson M. (1989) *Igneous Petrogenesis*. Unwin Hyman, London.

Woolf H. (1959) *The Transits of Venus a Study of Eighteenth-Century Science*. Princeton University Press, Princeton, 258pp.

Yung Y. L. and DeMore W. B. (1982) Photochemistry of the stratosphere of Venus: implications for atmospheric evolution. *Icarus* **51**, 199–247.

1.20
The Origin and Earliest History of the Earth

A. N. Halliday

Eidgenössische Technische Hochschule, Zürich, Switzerland

1.20.1	INTRODUCTION	510
1.20.2	OBSERVATIONAL EVIDENCE AND THEORETICAL CONSTRAINTS PERTAINING TO THE NEBULAR ENVIRONMENT FROM WHICH EARTH ORIGINATED	510
1.20.2.1	*Introduction*	510
1.20.2.2	*Nebular Gases and Earth-like versus Jupiter-like Planets*	511
1.20.2.3	*Depletion in Moderately Volatile Elements*	511
1.20.2.4	*Solar Mass Stars and Heating of the Inner Disk*	512
1.20.2.5	*The "Hot Nebula" Model*	514
1.20.2.6	*The "Hot Nebula" Model and Heterogeneous Accretion*	515
1.20.3	THE DYNAMICS OF ACCRETION OF THE EARTH	516
1.20.3.1	*Introduction*	516
1.20.3.2	*Starting Accretion: Settling and Sticking of Dust at 1 AU*	516
1.20.3.3	*Starting Accretion: Migration*	517
1.20.3.4	*Starting Accretion: Gravitational Instabilities*	517
1.20.3.5	*Runaway Growth*	518
1.20.3.6	*Larger Collisions*	518
1.20.4	CONSTRAINTS FROM LEAD AND TUNGSTEN ISOTOPES ON THE OVERALL TIMING, RATES, AND MECHANISMS OF TERRESTRIAL ACCRETION	519
1.20.4.1	*Introduction: Uses and Abuses of Isotopic Models*	519
1.20.4.2	*Lead Isotopes*	521
1.20.4.3	*Tungsten Isotopes*	522
1.20.5	CHEMICAL AND ISOTOPIC CONSTRAINTS ON THE NATURE OF THE COMPONENTS THAT ACCRETED TO FORM THE EARTH	527
1.20.5.1	*Chondrites and the Composition of the Disk from Which Earth Accreted*	527
1.20.5.2	*Chondritic Component Models*	528
1.20.5.3	*Simple Theoretical Components*	529
1.20.5.4	*The Nonchondritic Mg/Si of the Earth's Primitive Upper Mantle*	529
1.20.5.5	*Oxygen Isotopic Models and Volatile Losses*	530
1.20.6	EARTH'S EARLIEST ATMOSPHERES AND HYDROSPHERES	530
1.20.6.1	*Introduction*	530
1.20.6.2	*Did the Earth Have a Nebular Protoatmosphere?*	530
1.20.6.3	*Earth's Degassed Protoatmosphere*	532
1.20.6.4	*Loss of Earth's Earliest Atmosphere(s)*	533
1.20.7	MAGMA OCEANS AND CORE FORMATION	534
1.20.8	THE FORMATION OF THE MOON	535
1.20.9	MASS LOSS AND COMPOSITIONAL CHANGES DURING ACCRETION	541
1.20.10	EVIDENCE FOR LATE ACCRETION, CORE FORMATION, AND CHANGES IN VOLATILES AFTER THE GIANT IMPACT	542
1.20.11	THE HADEAN	543
1.20.11.1	*Early Mantle Depletion*	543
1.20.11.1.1	*Introduction*	543
1.20.11.1.2	$^{92}Nb-^{92}Zr$	543
1.20.11.1.3	$^{146}Sm-^{142}Nd$	543
1.20.11.1.4	$^{176}Lu-^{176}Hf$	544

1.20.11.2 Hadean Continents	544
1.20.11.3 The Hadean Atmosphere, Hydrosphere, and Biosphere	546
1.20.12 CONCLUDING REMARKS—THE PROGNOSIS	546
ACKNOWLEDGMENTS	547
REFERENCES	547

1.20.1 INTRODUCTION

The purpose of this chapter is to explain the various lines of geochemical evidence relating to the origin and earliest development of the Earth, while at the same time clarifying current limitations on these constraints. The Earth's origins are to some extent shrouded in greater uncertainty than those of Mars or the Moon because, while vastly more accessible and extensively studied, the geological record of the first 500 Myr is almost entirely missing. This means that we have to rely heavily on theoretical modeling and geochemistry to determine the mechanisms and timescales involved. Both of these approaches have yielded a series of, sometimes strikingly different, views about Earth's origin and early evolution that have seen significant change every few years. There has been a great deal of discussion and debate in the past few years in particular, fueled by new kinds of data and more powerful computational codes.

The major issues to address in discussing the origin and early development of the Earth are as follows:

(i) What is the theoretical basis for our understanding of the mechanisms by which the Earth accreted?

(ii) What do the isotopic and bulk chemical compositions of the Earth tell us about the Earth's accretion?

(iii) How are the chemical compositions of the early Earth and the Moon linked? Did the formation of the Moon affect the Earth's composition?

(iv) Did magma oceans exist on Earth and how can we constrain this from geochemistry?

(v) How did the Earth's core form?

(vi) How did the Earth acquire its atmosphere and hydrosphere and how have these changed?

(vii) What kind of crust might have formed in the earliest stages of the Earth's development?

(viii) How do we think life first developed and how might geochemical signatures be used in the future to identify early biological processes?

Although these issues could, in principle, all be covered in this chapter, some are dealt with in more detail in other chapters and, therefore, are given only cursory treatment here. Furthermore, there are major gaps in our knowledge that render a comprehensive overview unworkable. The nature of the early crust (item (vii)) is poorly constrained, although some lines of evidence will be mentioned. The nature of the earliest life forms (item (viii)) is so loaded with projections into underconstrained hypothetical environments that not a great deal can be described as providing a factual basis suitable for inclusion in a reference volume at this time. Even in those areas in which geochemical constraints are more plentiful, it is essential to integrate them with astronomical observations and dynamic (physical) models of planetary growth and primary differentiation. In some cases, the various theoretical dynamic models can be tested with isotopic and geochemical methods. In other cases, it is the Earth's composition itself that has been used to erect specific accretion paradigms. Therefore, much of this background is provided in this chapter.

All these models and interpretations of geochemical data involve some level of assumption in scaling the results to the big picture of the Earth. Without this, one cannot erect useful concepts that address the above issues. It is one of the main goals of this chapter to explain what these underlying assumptions are. As a consequence, this chapter focuses on the range of interpretations and uncertainties, leaving many issues "open." The chapter finishes by indicating where the main sources of uncertainty remain and what might be done about these in the future.

1.20.2 OBSERVATIONAL EVIDENCE AND THEORETICAL CONSTRAINTS PERTAINING TO THE NEBULAR ENVIRONMENT FROM WHICH EARTH ORIGINATED

1.20.2.1 Introduction

The starting place for all accretion modeling is the circumstellar disk of gas and dust that formed during the collapse of the solar nebula. It has been theorized for a long time that a disk of rotating circumstellar material will form as a normal consequence of transferring angular momentum during cloud collapse and star formation. Such disks now are plainly visible around young stars in the Orion nebula, thanks to the Hubble Space Telescope (McCaughrean and O'Dell, 1996). However, circumstellar disks became clearly detectable before this by using ground-based interferometry. If the light of the star is canceled out, excess infrared can be seen being emitted from the dust around the disk. This probably is caused by radiation from the star itself heating the disk.

Most astronomers consider nebular timescales to be of the order of a few million years (Podosek and Cassen, 1994). However, this is poorly constrained because unlike dust, gas is very difficult to detect around other stars. It may be acceptable to assume that gas and dust stay together for a portion of nebular history. However, the dust in some of these disks is assumed to be the secondary product of planetary accretion. Colliding planetesimals and planets are predicted to form at an early stage, embedded in the midplane of such optically thick disks (Wetherill and Stewart, 1993; Weidenschilling, 2000). The age of Beta Pictoris (Artymowicz, 1997; Vidal-Madjar et al., 1998) is rather unclear but it is probably more than ~20 Myr old (Hartmann, 2000) and the dust in this case probably is secondary, produced as a consequence of collisions. Some disks around younger (<10 Myr) stars like HR 4796A appear to show evidence of large inner regions entirely swept clear of dust. It has been proposed that in these regions the dust already may be incorporated into planetary objects (Schneider et al., 1999). The Earth probably formed by aggregating planetesimals and small planets that had formed in the midplane within such a dusty disk.

1.20.2.2 Nebular Gases and Earth-like versus Jupiter-like Planets

What features of Earth's composition provide information on this early circumstellar disk of dust that formed after the collapse of the solar nebula? The first and foremost feature of the Earth that relates to its composition and accretion is its size and density. Without any other information, this immediately raises questions about how Earth could have formed from the same disk as Jupiter and Saturn. The uncompressed density of the terrestrial planets is far higher than that of the outer gas and ice giant planets. The four most abundant elements making up ~90% of the Earth are oxygen, magnesium, silicon, and iron. Any model of the Earth's accretion has to account for this. The general explanation is that most of the growth of terrestrial planets postdated the loss of nebular gases from the disk. However, this is far from certain. Some solar-like noble gases were trapped in the Earth and although other explanations are considered (Trieloff et al., 2000; Podosek et al., 2003), the one that is most widely accepted is that the nebula was still present at the time of Earth's accretion (Harper and Jacobsen, 1996a). How much nebular gas was originally present is unclear. There is xenon-isotopic evidence that the vast majority (>99%) of Earth's noble gases were lost subsequently (Ozima and Podosekm, 1999; Porcelli and Pepin, 2000). A detailed discussion of this is provided in Chapter 4.11. The dynamics and timescales for accretion will be very different in the presence or absence of nebular gas. In fact, one needs to consider the possibility that even Jupiter-sized gas giant planets may have formed in the terrestrial planet-forming region and were subsequently lost by being ejected from the solar system or by migrating into the Sun (Lin et al., 1996). More than half the extrasolar planets detected are within the terrestrial planet-forming region of their stars, and these all are, broadly speaking, Jupiter-sized objects (Mayor and Queloz, 1995; Boss, 1998; Lissauer, 1999). There is, of course, a strong observational bias: we are unable to detect Earth-sized planets, which are not massive enough to induce a periodicity in the observed Doppler movement of the associated star or large enough to significantly occult the associated star (Boss, 1998; Seager, 2003).

For many years, it had been assumed that gas dissipation is a predictable response to radiative effects from an energetic young Sun. For example, it was theorized that the solar wind would have been ~100 times stronger than today and this, together with powerful ultraviolet radiation and magnetic fields, would have driven gases away from the disk (e.g., Hayashi et al., 1985). However, we now view disks more as dynamic "conveyor belts" that transport mass *into* the star. The radiative effects on the materials that form the terrestrial planets may in fact be smaller than previously considered. Far from being "blown off" or "dissipated," the gas may well have been lost largely by being swept into the Sun or incorporated into planetary objects—some of which were themselves consumed by the Sun or ejected (Murray et al., 1998; Murray and Chaboyer, 2001). Regardless of how the solar nebula was lost, its former presence, its mass, and the timing of Earth's accretion relative to that of gas loss from the disk will have a profound effect on the rate of accretion, as well as the composition and physical environment of the early Earth.

1.20.2.3 Depletion in Moderately Volatile Elements

Not only is there a shortage of nebular gas in the Earth and terrestrial planets today but the moderately volatile elements also are depleted (Figure 1) (Gast, 1960; Wasserburg et al., 1964; Cassen, 1996). As can be seen from Figure 2, the depletion in the moderately volatile alkali elements, potassium and rubidium in particular, is far greater than that found in any class of chondritic meteorites (Taylor and Norman, 1990; Humayun and Clayton, 1995; Halliday and Porcelli, 2001; Drake and Righter, 2002). The traditional explanation is that the inner "terrestrial" planets accreted where it was hotter,

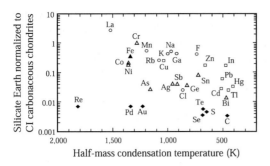

Figure 1 The estimated composition of the silicate portion of the Earth as a function of condensation temperature normalized to CI values in Anders and Grevesse (1989). Open circles: lithophile elements; shaded squares: chalcophile elements; shaded triangles: moderately siderophile elements; solid diamonds: highly siderophile elements. The spread in concentration for a given temperature is thought to be due to core formation. The highly siderophile element abundances may reflect a volatile depleted late veneer. Condensation temperatures are from Newsom (1995).

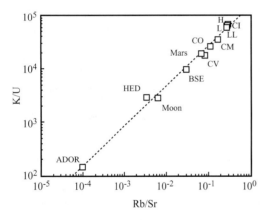

Figure 2 Comparison between the K/U and Rb/Sr ratios of the silicate Earth compared with other solar system objects. ADOR: Angra dos Reis; HED: howardite–eucrite–diogenite parent body; BSE: bulk silicate Earth; CI, CM, CV, CO, H, L, and LL are all classes of chondrites (source Halliday and Porcelli, 2001).

within the so-called "ice line" (Cassen, 1996; Humayun and Cassen, 2000). For several reasons, it has long been assumed that the solar nebula in the terrestrial planet-forming region started as a very hot, well-mixed gas from which all of the solid and liquid Earth materials condensed. The geochemistry literature contains many references to this hot nebula, as well as major *T-Tauri* heating events that may have further depleted the inner solar system in moderately volatile elements (e.g., Lugmair and Galer, 1992). Some nebula models predict early temperatures that were sufficiently high to prevent condensation of moderately volatile elements (Humayun and Cassen, 2000), which somehow were lost subsequently. To what extent these volatile elements condensed on grains that are now in the outer solar system and may be represented by interplanetary dust particles (Jessberger *et al.*, 1992) is unclear.

Nowadays, inner solar system models are undergoing major rethinking because of new observations of stars, theoretical models, and data from meteorites. It is important to keep track of the models and observational evidence on stars and disks as this continuously changes with greater resolution and detectability. The new data provide important insights into how our solar system may have developed. As of early 2000s, the linkage between temperature in the disk and accretion dynamics is anything but clear. There is no question that transient heating was important on some scale. But a large-scale hot nebula now is more difficult to accommodate. The depletion in volatile elements in the Earth is probably the result of several different processes and the latest astronomical evidence for these is summarized below. To understand these processes one has to have some idea of how solar mass stars and their disks are thought to "work."

1.20.2.4 Solar Mass Stars and Heating of the Inner Disk

Solar mass stars are thought to accrete rapidly. The pre-main-sequence solar mass *protostar* probably forms from collapse of a portion of a molecular cloud onto a "cloud core" in something like 10^5 yr (Hartmann, 2000). Strong outflows and jets are sometimes observable. Within a few hundred thousand years such protostars have developed into class I young stellar objects, as can be seen in the Orion nebula. These objects already have disks and are called *proplyds*. Remaining material from the cloud will accrete onto both the disk and onto the star itself. The disk also accretes onto the star and, as it does so, astronomers can track the accretion rate from the radiation produced at the innermost margin of the disk. In general terms, the accretion rate shows a very rough decrease with age of the star. From this, it can be shown that the mass of material being accreted from the disk onto the star is about the same as the minimum mass solar nebula estimated for our solar system (Hartmann, 2000).

This "minimum mass solar nebula" is defined to be the minimum amount of hydrogen–helium gas with dust, in bulk solar system proportions, that is needed in order to form our solar system's planets (Hoyle, 1960; Weidenschilling, 1977a). It is calculated by summing the assumed amount of metal (in the astronomical sense, i.e., elements heavier than hydrogen and helium) in all planets and adding enough hydrogen and helium to bring it up to solar composition. Usually, a value of 0.01 solar masses is taken to be the "minimum mass" (Boss, 1990). The strongest constraint on the value is the abundance of heavy elements in Jupiter and Saturn. This is at least partially independent of the uncertainty of whether these elements are hosted in planetary cores. Such estimates for the minimum mass solar nebula indicate that the disk was at least a factor of 10 more massive than the total mass of the current planets. However, the mass may have been much higher and because of this the loss of metals during the planet-forming process sometimes is factored in. There certainly is no doubt that some solids were consumed by the protosun or ejected into interstellar space. This may well have included entire planets. Therefore, a range of estimates for the minimum mass of 0.01–0.1 solar masses can be found. The range reflects uncertainties that can include the bulk compositions of the gas and ice giant planets (Boss, 2002), and the amount of mass loss from, for example, the asteroid belt (Chambers and Wetherill, 2001).

Some very young stars show enormous rapid changes in luminosity with time. These are called *FU Orionis* objects. They are young Sun-like stars that probably are temporarily accreting material at rapid rates from their surrounding disks of gas and dust; they might be consuming planets, for example (Murray and Chaboyer, 2001). Over a year to a decade, they brighten by a hundred times, then stay bright for a century or so before fading again (Hartmann and Kenyon, 1996). A protostar may go through this sequence many times before the accretion disk and surrounding cloud are dispersed. Radiation from the star on to the disk during this intense stage of activity could be partially responsible for volatile depletions in the inner solar system (Bell *et al.*, 2000), but the relative importance of this versus other heating processes has not been evaluated. Nor is it known if our Sun experienced such dramatic behavior.

T-Tauri stars also are pre-main-sequence stars. They are a few times 10^5 yr to a few million years in age and the *T-Tauri* effect appears to develop after the stages described above. They have many of the characteristics of our Sun but are much brighter. Some have outflows and produce strong stellar winds. Many have disks. The *T-Tauri* effect itself is poorly understood. It has long been argued that this is an early phase of heating of the inner portions of the disk. However, such disks are generally thought to have inclined surfaces that dip in toward the star (Chiang and Goldreich, 1997, 1999). It is these surfaces that receive direct radiation from the star and produce the infrared excess observed from the dust. The *T-Tauri* stage may last a few million years. Because it heats the disk surface it may not have any great effect on the composition of the gas and dust in the accretionary midplane of the disk, where planetesimal accretion is dominant.

Heating of inner solar system material in the midplane of the disk *will* be produced from compressional effects. The thermal effects can be calculated for material in the disk being swept into an increasingly dense region during migration toward the Sun during the early stages of disk development. Boss (1990) included compressional heating and grain opacity in his modeling and showed that temperatures in excess of 1,500 K could be expected in the terrestrial planet-forming region. The main heating takes place at the midplane, because that is where most of the mass is concentrated. The surface of the disk is much cooler. More recent modeling includes the detailed studies by Nelson *et al.* (1998, 2000), which provide a very similar overall picture. Of course, if the material is being swept into the Sun, one has to ask how much of the gas and dust would be retained from this portion of the disk. This process would certainly be very early. The timescales for subsequent cooling at 1 AU would have been very short (10^5 yr). Boss (1990), Cassen (2001), and Chiang *et al.* (2001) have independently modeled the thermal evolution of such a disk and conclude that in the midplane, where planetesimals are likely to accrete, temperatures will drop rapidly. Even at 1 AU, temperatures will be ~300 K after only 10^5 yr (Chiang *et al.*, 2001). Most of the dust settles to the midplane and accretes to form planetesimals over these same short timescales (Hayashi *et al.*, 1985; Lissauer, 1987; Weidenschilling, 2000); the major portion of the solid material may not be heated externally strongly after 10^5 yr.

Pre-main-sequence solar mass stars can be vastly (10^4 times) more energetic in terms of X-ray emissions from solar flare activity in their earliest stages compared with the most energetic flare activity of the present Sun (Feigelson *et al.*, 2002a). With careful sampling of large populations of young solar mass stars in the Orion nebula it appears that this is the normal behavior of stars like our Sun. This energetic solar flare activity is very important in the first million years or so, then decreases (Feigelson *et al.*, 2002a). From this it has been concluded that the early Sun had a 10^5-fold enhancement in energetic protons which may have contributed to short-lived nuclides (Lee *et al.*, 1998; McKeegan *et al.*, 2000; Gounelle *et al.*, 2001; Feigelson *et al.*, 2002b; Leya *et al.*, 2003).

Outflows, jets, and X-winds may produce a flux of material that is scattered across the disk from the star itself or the inner regions of the disk (Shu *et al.*, 1997). The region between the outflows and jets and the disk may be subject to strong magnetic fields that focus the flow of incoming material from the disk as it is being accreted onto the star and then project it back across the disk. These "X-winds" then produce a conveyor belt that cycle material through a zone where it is vaporized before being condensed and dispersed as grains of high-temperature condensates across the disk. If material from areas close to the Sun is scattered across the disk as proposed by Shu et al. (1997) it could provide a source for early heated and volatile depleted objects such as calcium-, aluminum-rich refractory inclusions (CAIs) and chondrules, as well as short-lived nuclides, regardless of any direct heating of the disk at 1 AU.

Therefore, from all of the recent examples of modeling and observations of circumstellar disks a number of mechanisms can be considered that might contribute to very early heating and depletion of moderately volatile elements at 1 AU. However, some of these are localized processes and the timescales for heating are expected to be short in the midplane.

It is unclear to what extent one can relate the geochemical evidence of extreme volatile depletion in the inner solar system (Figure 2) to these observations of processes active in other disks. It has been argued that the condensation of iron grains would act as a thermostat, controlling temperatures and evening out gradients within the inner regions of the solar nebula (Wood and Morfill, 1988; Boss, 1990; Wood, 2000). Yet the depletion in moderately volatile elements between different planetary objects is highly variable and does not even vary systematically with heliocentric distance (Palme, 2000). The most striking example of this is the Earth and Moon, which have very different budgets of moderately volatile elements. Yet they are at the same heliocentric distance and appear to have originated from an identical mix of solar system materials as judged from their oxygen isotopic composition (Figure 3) (Clayton and Mayeda, 1975; Wiechert et al., 2001). Oxygen isotopic compositions are highly heterogeneous among inner solar system objects (Clayton et al., 1973; Clayton, 1986, 1993; see Chapter 1.06). Therefore, the close agreement in oxygen isotopic composition between the Earth and Moon (Clayton and Mayeda, 1975), recently demonstrated to persist to extremely high precision (Figure 3), is a striking finding that provides good evidence that the Earth and Moon were formed from material of similar origin and presumably similar composition (Wiechert et al., 2001). The very fact that chondritic materials are not as heavily depleted in moderately volatile elements as the Earth and Moon provides evidence that other mechanisms of volatile loss must exist. Even the enstatite chondrites, with exactly the same oxygen isotopic composition as the Earth and Moon, are not as depleted in moderately volatile alkali elements (Newsom, 1995). The geochemical constraints on the origins of the components that formed the Earth are discussed below. But first it is necessary to review some of the history of the theories about how the Earth's chemical constituents were first incorporated into planetary building material.

1.20.2.5 The "Hot Nebula" Model

The current picture of the early solar system outlined above, with a dynamic dusty disk, enormous gradients in temperature, and a rapidly cooling midplane, is different from that prevalent in geochemistry literature 30 yr ago. The chemical condensation sequences modeled thermodynamically for a nebular gas cooling slowly and perhaps statically from 2,000 K were long considered a starting point for understanding the basic chemistry of the material accreting in the inner solar system (Grossman and Larimer, 1974). These traditional standard hot solar nebula models assumed that practically speaking *all* of the material in the terrestrial planet-forming region resulted from gradual condensation of such a nebular gas. Because so many of the concepts in the cosmochemistry literature relate to this hot nebula model, it is important to go through the implications of the newer ways of thinking about accretion of chemical components in order to better understand how the Earth was built.

Here are some of the lines of evidence previously used to support the theory of a large-scale hot nebula that now are being reconsidered.

(i) The isotopic compositions of a wide range of elements have long been known to be broadly similar in meteorites thought to come from Mars and the asteroid belt on the one hand and the Earth and Moon on the other. Given that stars produce huge degrees of isotopic heterogeneity it was assumed that the best way to achieve this homogenization was via a well-mixed gas from which all solids and liquids condensed

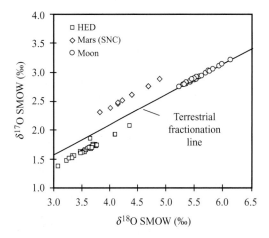

Figure 3 The oxygen isotopic compositions of the Earth and Moon are identical to extremely high precision and well resolved from the compositions of meteorites thought to come from Mars and Asteroid 4 Vesta (sources Wiechert et al., 2001, 2003).

(Suess, 1965; Reynolds, 1967). However, we know now that chondrites contain presolar grains that cannot have undergone the heating experienced by some of the other components in these meteorites, namely CAIs and chondrules. Presolar grains are unstable in a silicate matrix above a few hundred degrees Celsius (e.g., Mendybaev et al., 2002). The ubiquitous former presence of presolar grains (Huss, 1997; Huss and Lewis, 1995; Nittler, 2003; Nittler et al., 1994) provides unequivocal evidence of dust that has been physically admixed after the formation of the other components (CAIs and chondrules). It is this well-mixed cold dust that forms the starting point for the accretion of chondrite parent bodies, and probably the planets.

(ii) The models of Cameron (1978) using a 1 solar mass disk produced extremely high temperatures ($T > 2,000$ K) throughout most of the nebular disk. Such models fueled the hot nebula model but have been abandoned in favor of minimum mass nebula models. Some such viscous accretion disk models produced very low temperatures at 1 AU because these did not include compressional heating. However, Boss (1990) provided the first comprehensive thermal model including compressional heating and grain opacity, and this model does produce temperatures in excess of 1,500 K in the terrestrial planet region.

(iii) CAIs were found to have the composition of objects that condensed at high temperatures from a gas of solar composition (Grossman, 1972; Grossman and Larimer, 1974). Their old age confirmed that they were the earliest objects to form in the solar system (Göpel et al., 1991, 1994; Amelin et al., 2002). Although most CAIs have bulk compositions broadly consistent with high-temperature condensation (Wänke et al., 1974), nearly all of them have been melted and recrystallized, destroying any textural record of condensation. It now appears that they condensed and then were reheated and possibly partially evaporated, all within a short time. It is suspected by some that these objects condensed at very high temperatures close to the Sun and that they were scattered across the disk to be admixed with other components (Gounelle et al., 2001; Shu et al., 1997). This is far from certain and some "FUN" CAIs have isotopic compositions that cannot be easily reconciled with such a model (MacPherson et al., 1988; see Chapter 1.06). However, the important point is that their old age and refractory nature can be explained in ways other than just with a large-scale hot solar nebula.

(iv) The overall composition of the Earth is volatile-element depleted and this depletion is broadly consistent with that predicted from condensation theory (Cassen, 1996; Humayun and Cassen, 2000; Allègre et al., 2001). However, this agreement has rather little genetic significance. Why should chondrites be less depleted in volatiles like potassium and rubidium than the terrestrial planets and asteroids (Figure 2) if this is a nebular phenomenon? One explanation is that the chondrites accreted at 2–3 AU, where Boss (1990) shows that the nebula was cooler ($<1,000$ K). However, this provides no explanation for the extreme depletion in alkalis in eucrites and the Moon. The latter could be related to impact-induced losses (Halliday and Porcelli, 2001) but then the question arises as to whether the Earth's depletion in alkalis also relates to this in part. There is as yet no basis for distinguishing the volatile depletion that might be produced in planetary collisions (O'Neill, 1991a,b; Halliday and Porcelli, 2001) from that predicted to occur as a result of incomplete condensation of nebular gas.

(v) Strontium isotope differences between early very rubidium-depleted objects and planetesimals such as CAIs, eucrites and angrites have long been thought to provide evidence that they must have been created within a high-Rb/Sr environment such as the solar nebula but at a temperature above the condensation of rubidium (Gray et al., 1973; Wasserburg et al., 1977b; Lugmair and Galer, 1992; Podosek et al., 1991). The timescales over which the solar nebula has to be maintained above the condensation temperature of rubidium for this to work are a few million years. However, there is growing evidence that both cooling of the inner nebula and planetesimal growth may be very fast. Excluding the thermal effects from dense planetary atmospheres and the effects of planetary collisions, the timescale for major direct heating of the inner disk itself may be rather short (10^5 yr), but this view could change again with new observational data.

1.20.2.6 The "Hot Nebula" Model and Heterogeneous Accretion

It was at one time thought that even the terrestrial planets themselves formed directly by condensation from a hot solar nebula. This led to a class of models called heterogeneous accretion models, in which the composition of the material accreting to form the Earth changed with time as the nebula cooled. Eucken (1944) proposed such a heterogeneous accretion model in which early condensed metal formed a core to the Earth around which silicate accreted after condensation at lower temperatures. In this context the silicate-depleted, iron-enriched nature of Mercury makes sense as a body that accreted in an area of the solar nebula that was kept too hot to condense the same proportion of silicate as is found in the Earth (Lewis, 1972; Grossman and Larimer, 1974). Conversely, the lower density of Mars could partly reflect collection of an excess of silicate in cooler reaches of the inner solar nebula. So the concept of heliocentric "feeding zones" for accretion fitted this nicely. The discovery that iron metal condenses at a lower temperature than some refractory silicates made these models harder to sustain (Levin 1972). Nevertheless, a series of models involving progressive heterogeneous accretion at successively lower condensation temperatures were developed for the Earth (e.g., Turekian and Clark, 1969; Smith, 1977, 1980).

These models "produced" a zoned Earth with an early metallic core surrounded by silicate, without the need for a separate later stage of core formation. The application of condensation theory to the striking variations in the densities and compositions of the terrestrial planets, and how metal and silicate form in distinct reservoirs has been seen as problematic for some time. Heterogeneous accretion models require fast accretion and core formation if these processes reflect condensation in the nebula and such timescales can be tested with isotopic systems. The timescales for planetary accretion now are known to be far too long for an origin by partial condensation from a hot nebular gas. Nevertheless, heterogeneous accretion models have become embedded in the textbooks in Earth sciences (e.g., Brown and Mussett, 1981) and astronomy (e.g., Seeds, 1996).

An important development stemming from heterogeneous accretion models is that they introduced the concept that the Earth was built from more than one component and that these may have been accreted in separate stages. This provided an apparent answer to the problem of how to build a planet with a reduced metallic core and an oxidized silicate mantle. However, heterogeneous accretion is hard to reconcile with modern models for the protracted dynamics of terrestrial planet accretion compared with the shortness of nebular timescales. Therefore, they have been abandoned by most scientists and are barely mentioned in modern geochemistry literature any more.

1.20.3 THE DYNAMICS OF ACCRETION OF THE EARTH

1.20.3.1 Introduction

Qualitatively speaking, all accretion involves several stages, although the relative importance must differ between planets and some mechanisms are only likely to work under certain conditions that currently are underconstrained. Although the exact mechanisms of accretion of the gas and ice giant planets are poorly understood (Boss, 2002), all such objects need to accrete very rapidly in order to trap large volumes of gas before dissipation of the solar nebula.

Probably this requires timescales of $<10^7$ yr (Podosek and Cassen, 1994). In contrast, the most widely accepted dynamic models advocated for the formation of the terrestrial planets (Wetherill, 1986), involve protracted timescales $\sim 10^7 - 10^8$ yr. Application of these same models to the outer planets would mean even longer timescales. In fact, some of the outermost planets would not have yet formed. Therefore, the bimodal distribution of planetary density and its striking spatial distribution appear to require different accretion mechanisms in these two portions of the solar system. However, one simply cannot divide the accretion dynamics into two zones. A range of rate-limiting processes probably controlled accretion of both the terrestrial and Jovian planets and the debates about which of these processes may have been common to both is far from resolved. There almost certainly was some level of commonality.

1.20.3.2 Starting Accretion: Settling and Sticking of Dust at 1 AU

In most models of accretion at 1 AU, the primary process being studied is the advanced stage of gravitationally driven accretion. However, one first has to consider how accretion got started and in many respects this is far more problematic. Having established that the disk was originally dominated by gas and dust, it must be possible to get these materials to combine and form larger objects on a scale where gravity can play a major role. The starting point is gravitational settling toward the midplane. The dust and grains literally will "rain" into the midplane. The timescales proposed for achieving an elevated concentration of dust in the midplane of the disk are rapid, $\sim 10^3$ yr (Hayashi et al., 1985; Weidenschilling, 2000). Therefore, within a very short time the disk will form a concentrated midplane from which the growth of the planets ultimately must be fed.

Laboratory experiments on sticking of dust have been reviewed by Blum (2000), who concluded that sticking microscopic grains together with static and Van der Waals forces to build millimeter-sized compact objects was entirely feasible. However, building larger objects (fist- to football-pitch-sized) is vastly more problematic. Yet it is only when the objects are roughly kilometer-sized that gravity plays a major role. Benz (2000) has reviewed the dynamics of accretion of the larger of such intermediate-sized objects. The accretion of smaller objects is unresolved.

One possibility is that there was a "glue" that made objects stick together. Beyond the ice line, this may indeed have been relatively easy. But in the terrestrial planet-forming region in which early nebular temperatures were >1,000 K such a cement would have been lacking in the earliest stages. Of course, it already has been pointed out that cooling probably was fast at 1 AU. However, even this may not help. The baseline temperature in the solar system was then, and is now, above 160 K (the condensation temperature of water ice), so that no matter how rapid the cooling rate, the temperature would not have fallen sufficiently. The "stickiness" required rather may have been provided by carbonaceous coatings on silicate grains which might be stable at temperatures of >500 K (Weidenschilling, 2000). Waiting for the inner solar nebula to cool before accretion proceeds may not provide an explanation, anyway, because dynamic simulations provide evidence that these processes must be completed extremely quickly. The early Sun was fed with material from the disk and Weidenschilling (1977b, 2000) has argued that unless the dust and small debris are incorporated into much larger objects very quickly (in periods of less than $\sim 10^5$ yr), they will be swept into the Sun. Using a relatively large disk, Cuzzi et al. (2003) propose a mechanism for keeping a small fraction of smaller CAIs and fine debris in the terrestrial planet-forming region for a few million years. Most of the dust is lost. Another way of keeping the solids dust from migrating would have been the formation of gaps in the disk, preventing transfer to the Sun. The most obvious way of making gaps in the disk is by planet formation. So there is a "chicken and egg problem." Planets cannot form without gaps. Gaps cannot form without planets. This is a fundamental unsolved problem of terrestrial planetary accretion dynamics that probably deserves far more attention than has been given so far. Some, as yet uncertain, mechanism must exist for sticking small bodies together at 1 AU.

1.20.3.3 Starting Accretion: Migration

One mechanism to consider might be planetary migration (Lin et al., 1996; Murray et al., 1998). Observations of extrasolar planets provide strong evidence that planets migrate after their formation (Lin et al., 1996). Resonances are observed in extrasolar planetary systems possessing multiple Jupiter-like planets. These resonances can only be explained if the planets migrated after their formation (Murray et al., 1998). Two kinds of models can be considered.

(i) If accretion could not have started in the inner solar system, it might be that early icy and gas rich planets formed in the outer solar system and then migrated in toward the Sun where they opened up gaps in the disk prior to being lost into the Sun. They then left isolated zones of material that had time to accrete into planetesimals and planets.

(ii) Another model to consider is that the terrestrial planets themselves first started forming early in the icy outer solar system and migrated in toward the Sun, where gaps opened in the disk and prevented further migration. There certainly is evidence from noble gases that Earth acquired volatile components from the solar nebula and this might be a good way to accomplish this.

Both of these models have difficulties, because of the evidence against migration in the inner solar system. First, it is hard to see why the migrating planets in model (i) would not accrete most of the material in the terrestrial planet-forming region, leaving nothing for subsequent formation of the terrestrial planets themselves. Therefore, the very existence of the terrestrial planets would imply that such migration did not happen. Furthermore, there is evidence against migration in general in the inner solar system, as follows. We know that Jupiter had to form fast (<10 Myr) in order to accrete sufficient nebular gas (see Chapter 1.04). Formation of Jupiter is thought to have had a big effect (Wetherill, 1992) causing the loss of >99% of the material from the asteroid belt (Chambers and Wetherill, 2001). Therefore, there are good reasons for believing that the relative positions of Jupiter and the asteroid belt have been maintained in some approximate sense at least since the earliest history of the solar system. Strong supporting evidence against inner solar system migration comes from the fact that the asteroid belt is zoned today (Gaffey, 1990; Taylor, 1992). ^{26}Al heating is a likely cause of this (Grimm and McSween, 1993; Ghosh and McSween, 1999). However, whatever the reason it must be an early feature, which cannot have been preserved if migration were important.

Therefore, large-scale migration from the outer solar system is not a good mechanism for initiating accretion in the terrestrial planet-forming region unless it predates formation of asteroid belt objects or the entire solar system has migrated relative to the Sun. The outer solar system provides some evidence of ejection of material and migration but the inner solar system appears to retain much of its original "structure."

1.20.3.4 Starting Accretion: Gravitational Instabilities

Sticking together of dust and small grains might be aided by differences between gas and dust velocities in the circumstellar disk (Weidenschilling, 2000). However, the differential velocities of the grains are calculated to be huge and nobody has been able to simulate this adequately. An early solution that was proposed by Goldreich and Ward (1973) is that gravitational instabilities built up in the disk. This means that sections of the swirling disk built up sufficient mass to establish an overall gravitational field that prevented the dust and gas in that region from moving away. With less internal differential movement there would have been more chance for clumping together and sticking. A similar kind of model has been advocated on a much larger scale for the rapid growth of Jupiter (Boss, 1997). Perhaps these earlier models need to be looked at again because they might provide the most likely explanation for the onset of terrestrial planet accretion. This mechanism has recently been reviewed by Ward (2000).

1.20.3.5 Runaway Growth

Whichever way the first stage of planetary accretion is accomplished, it should have been followed by *runaway gravitational growth* of these kilometer-scale planetesimals, leading to the formation of numerous Mercury- to Mars-sized planetary embryos. The end of this stage also should be reached very quickly according to dynamic simulations. Several important papers study this phase of planetary growth in detail (e.g., Lin and Papaloizou, 1985; Lissauer, 1987; Wetherill and Stewart, 1993; Weidenschilling, 2000; Kortenkamp et al., 2000). With runaway growth, it is thought that Moon-sized "planetary-embryos" are built over timescales $\sim 10^5$ yr (Wetherill, 1986; Lissauer, 1993; Wetherill and Stewart, 1993). Exhausting the supply of material in the immediate vicinity prevents further runaway growth. However, there are trade-offs between the catastrophic and constructive effects of planetesimal collisions. Benz and Asphaug (1999) calculate a range of "weakness" of objects with the weakest in the solar system being ~ 300 km in size. Runaway growth predicts that accretion will be completed faster, closer to the Sun where the "feeding zone" of material will be more confined. On this basis material in the vicinity of the Earth would accrete into Moon-sized objects more quickly than material in the neighborhood of Mars, for example.

1.20.3.6 Larger Collisions

Additional growth to form Earth-sized planets is thought to require collisions between these "planetary embryos." This is a stochastic process such that one cannot predict in any exact way the detailed growth histories for the terrestrial planets. However, with Monte Carlo simulations and more powerful computational codes the models have become quite sophisticated and yield similar and apparently robust results in terms of the kinds of timescales that must be involved. The mechanisms and timescales are strongly dependent on the amount of nebular gas. The presence of nebular gas has two important effects on accretion mechanisms. First, it provides added friction and pressure that speeds up accretion dramatically. Second, it can have the effect of reducing eccentricities in the orbits of the planets. Therefore, to a first approximation one can divide the models for the overall process of accretion into three possible types that have been proposed, each with vastly differing amounts of nebular gas and therefore accretion rates:

(i) *Very rapid accretion in the presence of a huge nebula.* Cameron (1978) argued that the Earth formed with a solar mass of nebular gas in the disk. This results in very short timescales of $<10^6$ yr for Earth's accretion.

(ii) *Protracted accretion in the presence of a minimum mass solar nebula.* This is known as the Kyoto model and is summarized nicely in the paper by Hayashi et al. (1985). The timescales are $10^6 - 10^7$ yr for accretion of all the terrestrial planets. The timescales increase with heliocentric distance. The Earth was calculated to form in ~ 5 Myr.

(iii) *Protracted accretion in the absence of a gaseous disk.* This model simulates the effects of accretion via planetesimal collisions assuming that all of the nebular gas has been lost. Safronov (1954) first proposed this model. He argued that the timescales for accretion of all of the terrestrial planets then would be very long, in the range of $10^7 - 10^8$ yr.

Safronov's model was confirmed with the Monte Carlo simulations of Wetherill (1980), who showed that the provenance of material would be very broad and only slightly different for each of the terrestrial planets (Wetherill, 1994). The timescales for accretion of each planet also would be very similar. By focusing on the solutions that result in terrestrial planets with the correct (broadly speaking) size and distribution and tracking the growth of these objects, Wetherill (1986) noted that the terrestrial planets would accrete at something approaching exponentially decreasing rates. The half-mass accretion time (time for half of the present mass to accumulate) was comparable ($\sim 5-7$ Myr), and in reality indistinguishable, for Mercury, Venus, Earth, and Mars using such simulations. Of course these objects, being of different size, would have had different absolute growth rates.

These models did not consider the effects of the growth of gas giant planets on the terrestrial planet-forming region. However, the growth of Jupiter is unlikely to have slowed down accretion at 1 AU (Kortenkamp and Wetherill, 2000). Furthermore, if there were former gas giant planets in the terrestrial planet-forming region they probably would have caused the terrestrial planets to be ejected from their orbits and lost.

In order to distinguish between these models one has to know the amount of nebular gas that was present at the time of accretion. For the terrestrial planets this is relatively difficult to estimate. Although the Safronov–Wetherill model, which specifically assumes no nebular gas, has become the main textbook paradigm for Earth accretion, the discovery that gas giant planets are found in the terrestrial planet-forming regions of other stars (Mayor and Queloz, 1995; Boss, 1998; Lissauer, 1999; Seager, 2003) has fueled re-examination of this issue. Furthermore, recent attempts of accretion modeling have revealed that terrestrial planets can indeed be formed in the manner predicted by Wetherill but that they have high eccentricities (Canup and Agnor, 1998). Thus, they depart strongly from circular orbits. The presence of even a small amount of nebular gas during accretion has the effect of reducing this eccentricity (Agnor and Ward, 2002). This, in turn, would have sped up accretion. As explained below, geochemical data provide strong support for a component of nebular-like gases during earth accretion.

The above models differ with respect to timing and therefore can be tested with isotopic techniques. However, not only are the models very different in terms of timescales, they also differ with respect to the environment that would be created on Earth. In the first two cases the Earth would form with a hot dense atmosphere of nebular gas that would provide a ready source of solar noble gases in the Earth. This atmosphere would have blanketed the Earth and could have caused a dramatic buildup of heat leading to magma oceans (Sasaki, 1990). Therefore, the evidence from dynamic models can also be tested with compositional data for the Earth, which provide information on the nature of early atmospheres and melting.

1.20.4 CONSTRAINTS FROM LEAD AND TUNGSTEN ISOTOPES ON THE OVERALL TIMING, RATES, AND MECHANISMS OF TERRESTRIAL ACCRETION

1.20.4.1 Introduction: Uses and Abuses of Isotopic Models

Radiogenic isotope geochemistry can help with the evaluation of the above models for accretion by determining the rates of growth of the silicate reservoirs that are residual from core formation. By far the most useful systems in this regard have been the $^{235}U/^{238}U-^{207}Pb/^{206}Pb$ and $^{182}Hf-^{182}W$ systems. These are discussed in detail below.

Other long-lived systems, such as $^{87}Rb-^{87}Sr$, $^{147}Sm-^{143}Nd$, $^{176}Lu-^{176}Hf$, and $^{187}Re-^{187}Os$, have provided more limited constraints (Tilton, 1988; Carlson and Lugmair, 2000), although in a fascinating piece of work, McCulloch (1994) did attempt to place model age constraints on the age of the earth using strontium isotope data for Archean rocks (Jahn and Shih, 1974; McCulloch, 1994). The short-lived systems $^{129}I-^{129}Xe$ and $^{244}Pu-^{136}Xe$ have provided additional constraints (Wetherill, 1975a; Allègre et al., 1995a; Ozima and Podosek, 1999; Pepin and Porcelli, 2002). Other short-lived systems that have been used to address the timescales of terrestrial accretion and differentiation are $^{53}Mn-^{53}Cr$ (Birck et al., 1999), $^{92}Nb-^{92}Zr$ (Münker et al., 2000; Jacobsen and Yin, 2001), $^{97}Tc-^{97}Mo$ (Yin and Jacobsen, 1998), and $^{107}Pd-^{107}Ag$ (Carlson and Hauri, 2001). None of these now appear to provide useful constraints. Either the model deployed currently is underconstrained (as with Mn–Cr) or the isotopic effects subsequently have been shown to be incorrect or better explained in other ways.

Hf–W and U–Pb methods both work well because the mechanisms and rates of accretion are intimately associated with the timing of core formation and this fractionates the parent/daughter ratio strongly. For a long while, however, it was assumed that accretion and core formation were completely distinct events. It was thought that the Earth formed as a cold object in less than a million years (e.g., Hanks and Anderson, 1969) but that it then heated up as a result of radioactive decay and later energetic impacts. On this basis, it was calculated that the Earth's core formed rather gradually after tens or even hundreds of millions of years following this buildup of heat and the onset of melting (Hsui and Toksöz, 1977; Solomon, 1979).

In a similar manner isotope geochemists have at various times treated core formation as a process that was distinctly later than accretion and erected relatively simple lead, tungsten and, most recently, zirconium isotopic model ages that "date" this event (e.g., Oversby and Ringwood, 1971; Allègre et al., 1995a; Lee and Halliday, 1995; Galer and Goldstein, 1996; Harper and Jacobsen, 1996b; Jacobsen and Yin, 2001; Dauphas et al., 2002; Kleine et al., 2002; Schöenberg et al., 2002). A more complex model was presented by Kramers (1998). Detailed discussions of U–Pb, Hf–W, and Nb–Zr systems are presented later in this chapter. However, some generalities should be mentioned first.

In looking at these models the following "rules" apply:

(i) Both U–Pb and Hf–W chronometry are unable to distinguish between early accretion with late core formation, and late accretion with concurrent late core formation because it is dominantly core formation that fractionates the parent/daughter ratio.

(ii) If accretion or core formation or both are protracted, the isotopic model age does not define any particular event. In the case of U–Pb it could define a kind of weighted average. In the case of short-lived nuclides, such as the ^{182}Hf–^{182}W system with a half-life of ~9 Myr, it cannot even provide this. Clearly, if a portion of the core formation were delayed until after ^{182}Hf had become effectively extinct, the tungsten isotopic composition of the residual silicate Earth would not be changed. Even if >50% of the mass of the core formed yesterday it would not change the tungsten isotopic composition of the silicate portion of the Earth! Therefore, the issue of how long core formation persisted is completely underconstrained by Hf–W but is constrained by U–Pb data. It also is constrained by trace element data (Newsom et al., 1986).

(iii) Isotopic approaches such as those using Hf–W can only provide an indication of how quickly core formation may have started if accretion was early and very rapid. Clearly this is not a safe assumption for the Earth. If accretion were protracted, tungsten isotopes would provide only minimal constraints on when core formation started.

Tungsten and lead isotopic data can, however, be used to define the timescales for accretion, simply by assuming that core formation, the primary process that fractionates the parent/daughter ratio, started very early and that the core grew in constant proportion to the Earth (Halliday et al., 1996, 2000; Harper and Jacobsen, 1996b; Jacobsen and Harper, 1996; Halliday and Lee, 1999; Halliday, 2000). There is a sound basis for the validity of this assumption, as follows.

(i) The rapid conversion of kinetic energy to heat in a planet growing by accretion of planetesimals and other planets means that it is inescapable that silicate and metal melting temperatures are achieved (Sasaki and Nakazawa, 1986; Benz and Cameron, 1990; Melosh, 1990). This energy of accretion would be sufficient to melt the entire Earth such that in all likelihood one would have magma oceans permitting rapid core formation.

(ii) There is strong observational support for this view that core formation was quasicontinuous during accretion. Iron meteorites and basaltic achondrites represent samples of small planetesimals that underwent core formation early. A strong theoretical basis for this was recently provided by Yoshino et al. (2003). Similarly, Mars only reached one-eighth of the mass of the Earth but clearly its size did not limit the opportunity for core formation. Also, most of the Moon is thought to come from the silicate-rich portion of a Mars-sized impacting planet, known as "Theia" (Cameron and Benz, 1991; Canup and Asphaug, 2001; Halliday, 2000), which also was already differentiated into core and silicate. The amount of depletion in iron in eucrites, martian meteorites, and lunar samples provides support for the view that the cores of all the planetesimals and planets represented were broadly similar in their proportions to Earth's, regardless of absolute size. The slightly more extensive depletion of iron in the silicate Earth provides evidence that core formation was more efficient or protracted, but not that it was delayed.

There is no evidence that planetary objects have to achieve an Earth-sized mass or evolve to a particular state (other than melting), before core formation will commence. It is more reasonable to assume that the core grew with the accretion of the Earth in roughly the same proportion as today. If accretion were protracted, the rate-limiting parameter affecting the isotopic composition of lead and tungsten in the silicate Earth would be the timescale for accretion. As such, the "age of the core" is an average time of formation of the Earth itself. Therefore, simple tungsten and lead isotopic model ages do not define an event as such. The isotopic data instead need to be integrated with models for the growth of the planet itself to place modeled limits for the rate of growth.

The first papers exploring this approach were by Harper and Jacobsen (1996b) and Jacobsen and Harper (1996). They pointed out that the Monte Carlo simulations produced by Wetherill (1986) showed a trend of exponentially decreasing planetary growth with time. They emulated this with a simple expression for the accretionary mean life of the Earth, where the mean life is used in the same way as in nuclear literature as the inverse of a time constant. This model is an extension of the earlier model of Jacobsen and Wasserburg (1979) evaluating the mean age of the continents using Sm–Nd. Jacobsen and Harper applied the model to the determination of the age of the Earth based on (then very limited) tungsten isotope data. Subsequent studies (Halliday et al., 1996, 2000; Halliday and Lee, 1999; Halliday, 2000; Yin et al., 2002), including more exhaustive tungsten as well as lead isotope modeling, are all based on this same concept. However, the data and our understanding of the critical parameters have undergone major development.

Nearly all of these models assume that:

(i) accretion proceeded at an exponentially decreasing rate from the start of the solar system;

(ii) core formation and its associated fractionation of radioactive parent/radiogenic daughter ratios was coeval with accretion;

(iii) the core has always existed in its present proportion relative to the total Earth;

(iv) the composition of the accreting material did not change with time;

(v) the accreting material equilibrated fully with the silicate portion of the Earth just prior to fractionation during core formation; and

(vi) the partitioning of the parent and daughter elements between mantle and core remained constant.

The relative importance of these assumptions and the effects of introducing changes during accretion have been partially explored in several studies (Halliday et al., 1996, 2000; Halliday and Lee, 1999; Halliday, 2000). The issue of metal–silicate equilibration has been investigated recently by Yoshino et al. (2003). However, the data upon which many of the fundamental isotopic and chemical parameters are based are in a state of considerable uncertainty.

1.20.4.2 Lead Isotopes

Until recently, the most widely utilized approach for determining the rate of formation of the Earth was U–Pb geochronology. The beauty of using this system is that one can deploy the combined constraints from both ^{238}U–^{206}Pb ($T_{1/2} = 4,468$ Myr) and ^{235}U–^{207}Pb ($T_{1/2} = 704$ Myr) decay. Although the atomic abundance of both of the daughter isotopes is a function of the U/Pb ratio and age, combining the age equations allows one to cancel out the U/Pb ratio. The relative abundance of ^{207}Pb and ^{206}Pb indicates when the fractionation took place. Patterson (1956) adopted this approach in his classic experiment to determine the age of the Earth. Prior to his work, there were a number of estimates of the age of the Earth based on lead isotopic data for terrestrial galenas. However, Patterson was the first to obtain lead isotopic data for early low-U/Pb objects (iron meteorites) and this defined the initial lead isotopic composition of the solar system. From this, it was clear that the silicate Earth's lead isotopic composition required between 4.5 Gyr and 4.6 Gyr of evolution as a high-U/Pb reservoir. Measurements of the lead isotopic compositions of other high-U/Pb objects such as basaltic achondrites and lunar samples confirmed this age for the solar system.

In detail it is now clear that most U–Pb model ages of the Earth (Allègre et al., 1995a) are significantly younger than the age of early solar system materials such as chondrites (Göpel et al., 1991) and angrites (Wasserburg et al., 1977b; Lugmair and Galer, 1992). Such a conclusion has been reached repeatedly from consideration of the lead isotope compositions of early Archean rocks (Gancarz and Wasserburg, 1977; Vervoort et al., 1994), conformable ore deposits (Doe and Stacey, 1974; Manhès et al., 1979; Tera, 1980; Albarède and Juteau, 1984), average bulk silicate Earth (BSE) (Galer and Goldstein, 1996) and mid-ocean ridge basalts (MORBs) (Allègre et al., 1995a), all of which usually yield model ages of <4.5 Ga. Tera (1980) obtained an age of 4.53 Ga using Pb–Pb data for old rocks but even this postdates the canonical start of the solar system by over 30 Myr.

The reason why nearly all such approaches yield similar apparent ages that postdate the start of the solar system by a few tens of millions of years is that there was a very strong U/Pb fractionation that took place during the protracted history of accretion. The U–Pb model age of the Earth can only be young if U/Pb is fractionated at a late stage. This fractionation was of far greater magnitude than that associated with any later processes. Thus it has left a clear and irreversible imprint on the ^{207}Pb/^{206}Pb and ^{207}Pb/^{204}Pb isotope ratios of the silicate portion of the Earth. Uranium, being lithophile, is largely confined to the silicate portion of the Earth. Lead is partly siderophile and chalcophile such that >90% of it is thought to be in the core (Allègre et al., 1995a). Therefore, it was long considered that the lead isotopic "age of the Earth" dates core formation (Oversby and Ringwood, 1971; Allègre et al., 1982).

The exact value of this fractionation is poorly constrained, because lead also is moderately volatile, so that the U/Pb ratio of the total Earth (mantle, crust, and core combined) also is higher than chondritic. In fact, some authors even have argued that the dominant fractionation in U/Pb in the BSE was caused by volatile loss (Jacobsen and Harper, 1996b; Harper and Jacobsen, 1996b; Azbel et al., 1993). This is consistent with some compilations of data for the Earth, which show that lead is barely more depleted in the bulk silicate portion of the Earth than lithophile elements of similar volatility (McDonough and Sun, 1995). It is, therefore, important to know how much of the lead depletion is caused by accretion of material that was depleted in volatile elements at an early (nebular) stage. Galer and Goldstein (1996) and Allègre et al. (1995a) have compellingly argued that the major late-stage U/Pb fractionation was the result of core formation. However, the uncertainty over the U/Pb of the total Earth and whether it changed with accretion time remains a primary issue limiting precise application of lead isotopes.

Using exponentially decreasing growth rates and continuous core formation one can deduce an accretionary mean life assuming a ^{238}U/^{204}Pb for the total Earth of 0.7 (Halliday, 2000). This value is based on the degree of depletion of moderately volatile lithophile elements, as judged from the K/U ratio of the BSE (Allègre et al., 1995a). Application of this approach to the lead isotopic compositions of the Earth (Halliday, 2000) provides evidence that the Earth accreted with an accretionary mean life of between 15 Myr and

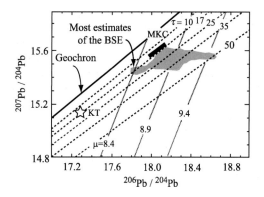

Figure 4 Lead isotopic modeling of the composition of the silicate Earth using continuous core formation. The principles behind the modeling are as in Halliday (2000). See text for explanation. The field for the BSE encompasses all of the estimates in Galer and Goldstein (1996). The values suggested by Kramers and Tolstikhin (1997) and Murphy et al. (2003) also are shown. The mean life (τ) is the time required to achieve 63% of the growth of the Earth with exponentially decreasing rates of accretion. The μ values are the ^{238}U/^{204}Pb of the BSE. It is assumed that the μ of the total Earth is 0.7 (Allègre et al., 1995a). It can be seen that the lead isotopic composition of the BSE is consistent with protracted accretion over periods of 10^7–10^8 yr.

50 Myr, depending on which composition of the BSE is deployed (Figure 4). A similar figure to this in Halliday (2000) is slightly different (i.e., incorrect) because of a scaling error. The shaded region covers the field defined by eight estimates for the composition of the BSE as summarized by Galer and Goldstein (1996). The star shows the estimate of Kramers and Tolstikhin (1997). The thick bar shows the recent estimate provided by Murphy et al. (2003). Regardless of which of these 10 estimates of the lead isotopic composition is used, the mean life for accretion is at least 15 Myr. Therefore, there is no question that the lead isotopic data for the Earth provide evidence of a protracted history of accretion and concomitant core formation as envisaged by Wetherill (1986).

1.20.4.3 Tungsten Isotopes

While lead isotopes have been useful, the ^{182}Hf–^{182}W chronometer ($T_{1/2} = 9$ Myr) has been at least as effective for defining rates of accretion (Halliday, 2000; Halliday and Lee, 1999; Harper and Jacobsen, 1996b; Jacobsen and Harper, 1996; Lee and Halliday, 1996, 1997; Yin et al., 2002). Like U–Pb, the Hf–W system has been used more for defining a model age of core formation (Kramers, 1998; Horan et al., 1998; Kleine et al., 2002; Lee and Halliday, 1995, 1996, 1997; Quitté et al., 2000; Dauphas et al., 2002; Schöenberg et al., 2002). As explained above this is not useful for an object like the Earth.

The half-life renders ^{182}Hf as ideal among the various short-lived chronometers for studying accretionary timescales. Moreover, there are two other major advantages of this method.

(i) Both parent and daughter elements (hafnium and tungsten) are refractory and, therefore, are in chondritic proportions in most accreting objects. Therefore, unlike U–Pb, we think we know the isotopic composition and parent/daughter ratio of the entire Earth relatively well.

(ii) Core formation, which fractionates hafnium from tungsten, is thought to be a very early process as discussed above. Therefore, the rate-limiting process is simply the accretion of the Earth.

There are several recent reviews of Hf–W (e.g., Halliday and Lee, 1999; Halliday et al., 2000), to which the reader can refer for a comprehensive overview of the data and systematics. However, since these were written it has been shown that chondrites, and by inference the average solar system, have tungsten isotopic compositions that are resolvable from that of the silicate Earth (Kleine et al., 2002; Lee and Halliday, 2000a; Schoenberg et al., 2002; Yin et al., 2002). Although the systematics, equations, and arguments have not changed greatly, this has led to considerable uncertainty over the exact initial ^{182}Hf abundance in the early solar system. Because this is of such central importance to our understanding of the timescales of accretion that follow from the data, it is discussed in detail below. Similarly, some of the tungsten isotopic effects that were once considered to reflect radioactive decay within the Moon (Lee et al., 1997; Halliday and Lee, 1999) are now thought to *partly* be caused by production of cosmogenic ^{182}Ta (Leya et al., 2000; Lee et al., 2002).

The differences in tungsten isotopic composition are most conveniently expressed as deviations in parts per 10,000, as follows:

$$\varepsilon_W = \left[\frac{(^{182}W/^{184}W)_{\text{sample}}}{(^{182}W/^{184}W)_{\text{BSE}}} - 1 \right] \times 10^4$$

where the BSE value $(^{182}W/^{184}W)_{\text{BSE}}$ is the measured value for an NIST tungsten standard. This should be representative of the BSE as found by comparison with the values for terrestrial standard rocks (Lee and Halliday, 1996; Kleine et al., 2002; Schoenberg et al., 2002). If ^{182}Hf was sufficiently abundant at the time of formation (i.e., at an early age), then minerals, rocks, and reservoirs with higher Hf/W ratios will produce tungsten that is significantly more radiogenic (higher ^{182}W/^{184}W or ε_W) compared with the initial tungsten isotopic composition of the solar system. Conversely, metals with low Hf/W that segregate at an early stage from bodies with chondritic Hf/W (as expected for most early planets and planetesimals) will sample unradiogenic tungsten.

Harper et al. (1991a) were the first to provide a hint of a tungsten isotopic difference between the iron meteorite Toluca and the silicate Earth. It is now clear that there exists a ubiquitous clearly resolvable deficit in ^{182}W in iron meteorites and the metals of ordinary chondrites, relative to the atomic abundance found in the silicate Earth (Lee and Halliday, 1995, 1996; Harper and Jacobsen, 1996b; Jacobsen and Harper, 1996; Horan et al., 1998). A summary of most of the published data for iron meteorites is given in Figure 5. It can be seen that most early segregated metals are deficient by $\sim(3-4)\varepsilon_W$ units (300–400 ppm) relative to the silicate Earth. Some appear to be even more negative, but the results are not well resolved. The simplest explanation for this difference is that the metals, or the silicate Earth, or both, sampled early solar system tungsten before live ^{182}Hf had decayed.

The tungsten isotopic difference between early metals and the silicate Earth reflects the time integrated Hf/W of the material that formed the Earth and its reservoirs, during the lifetime of ^{182}Hf. The Hf/W ratio of the silicate Earth is considered to be in the range of 10–40 as a result of an intensive study by Newsom et al. (1996). This is an order of magnitude higher than in carbonaceous and ordinary chondrites and a consequence of terrestrial core formation. More recent estimates are provided in Walter et al. (2000).

If accretion and core formation were early, an excess of ^{182}W would be found in the silicate Earth, relative to average solar system (chondrites). However, the tungsten isotopic difference between early metals and the silicate Earth on its own does not provide constraints on timing. One needs to know the atomic abundance of ^{182}Hf at the start of the solar system (or the (^{182}Hf/^{180}Hf)$_{BSSI}$, the "bulk solar system initial") and the composition of the chondritic reservoirs from which most metal and silicate reservoirs were segregated. In other words, it is essential to know to what extent the "extra" ^{182}W in the silicate Earth relative to iron meteorites accumulated in the accreted chondritic precursor materials or proto-Earth with an Hf/W ~ 1 prior to core formation, and to what extent it reflects an accelerated change in isotopic composition because of the high Hf/W (~ 15) in the silicate Earth.

For this reason some of the first attempts to use Hf–W (Harper and Jacobsen, 1996b; Jacobsen and Harper, 1996) gave interpretations that are now known to be incorrect because the (^{182}Hf/^{180}Hf)$_{BSSI}$ was underconstrained. This is a central concern in Hf–W chronometry that does not apply to U–Pb; for the latter system, parent abundances can still be measured today. In order to determine the (^{182}Hf/^{180}Hf)$_{BSSI}$ correctly one can use several approaches with varying degrees of reliability:

(i) The first approach is to model the expected (^{182}Hf/^{180}Hf)$_{BSSI}$ in terms of nucleosynthetic processes. Wasserburg et al. (1994) successfully predicted the initial abundances of many of the short-lived nuclides using a model of nucleosynthesis in AGB stars. Extrapolation of their model predicted a low (^{182}Hf/^{180}Hf)$_{BSSI}$ of $<10^{-5}$, assuming that ^{182}Hf was indeed produced in this manner. Subsequent to the discovery that the (^{182}Hf/^{180}Hf)$_{BSSI}$ was $>10^{-4}$ (Lee and Halliday, 1995, 1996), a number of new models were developed based on the assumption that ^{182}Hf is produced in the same kind of r-process site as the actinides (Wasserburg et al., 1996; Qian et al., 1998; Qian and Wasserburg, 2000).

(ii) The second approach is to measure the tungsten isotopic composition of an early high-Hf/W phase. Ireland (1991) attempted to measure the amount of ^{182}W in zircons (with very high hafnium content) from the mesosiderite Vaca Muerta, using an ion probe, and from this deduced

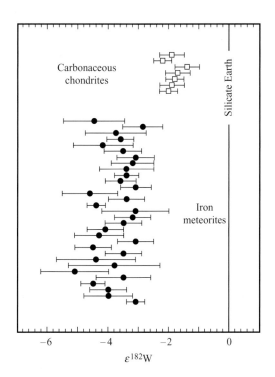

Figure 5 Well-defined deficiency in ^{182}W in early metals and carbonaceous chondrites relative to the silicate Earth (source Lee and Halliday, 1996; Horan et al., 1998; Kleine et al., 2002).

that the $(^{182}Hf/^{180}Hf)_{BSSI}$ was $<10^{-4}$. Unfortunately, these zircons are not dated with sufficient precision (Ireland and Wlotzka, 1992) to be very certain about the time extrapolation of the exact hafnium abundances. Nevertheless, on the basis of this work and the model of Wasserburg et al. (1994), Jacobsen and Harper (1996) assumed that the $(^{182}Hf/^{180}Hf)_{BSSI}$ was indeed low ($\sim 10^{-5}$). It was concluded that the difference in tungsten isotopic composition between the iron meteorite Toluca and the terrestrial value could only have been produced by radioactive decay within the silicate Earth with high Hf/W. Therefore, the fractionation of Hf/W produced by terrestrial core formation had to be early. They predicted that the Earth accreted very rapidly with a two-stage model age of core formation of <15 Myr after the start of the solar system.

(iii) The third approach is to simply assume that the consistent unradiogenic tungsten found in iron meteorites and metals from ordinary chondrites represents the initial tungsten isotopic composition of the solar system. This is analogous to the way in which the lead isotopic composition of iron meteorites has been used for decades. The difference between this and the present-day value of carbonaceous chondrites represents the effects of radiogenic ^{182}W growth *in situ* with chondritic Hf/W ratios. This in turn will indicate the $(^{182}Hf/^{180}Hf)_{BSSI}$. The difficulty with this approach has been to correctly determine the tungsten isotopic composition of chondrites. Using multiple collector ICPMS, Lee and Halliday (1995) were the first to publish such results and they reported that the tungsten isotopic compositions of the carbonaceous chondrites Allende and Murchison could not be resolved from that of the silicate Earth. The $(^{182}Hf/^{180}Hf)_{BSSI}$, far from being low, appeared to be surprisingly high at $\sim 2 \times 10^{-4}$ (Lee and Halliday, 1995, 1996). Subsequently, it has been shown that these data must be incorrect (Figure 5). Three groups have independently shown (Kleine et al., 2002; Schoenberg et al., 2002; Yin et al., 2002) that there is a small but clear deficiency in ^{182}W ($\varepsilon_W = -1.5$ to -2.0) in carbonaceous chondrites similar to that found in enstatite chondrites (Lee and Halliday, 2000a) relative to the BSE. Kleine et al. (2002), Schoenberg et al. (2002), and Yin et al. (2002) all proposed a somewhat lower $(^{182}Hf/^{180}Hf)_{BSSI}$ of $\sim 1.0 \times 10^{-4}$. However, the same authors based this result on a solar system initial tungsten isotopic composition of $\varepsilon_W = -3.5$, which was derived from their own, rather limited, measurements. Schoenberg et al. (2002) pointed out that if instead one uses the full range of tungsten isotopic composition previously reported for iron meteorites (Horan et al., 1998; Jacobsen and Harper, 1996; Lee and Halliday, 1995, 1996) one obtains a $(^{182}Hf/^{180}Hf)_{BSSI}$ of $>1.3 \times 10^{-4}$ (Table 1).

(iv) The fourth approach is to determine an internal isochron for an early solar system object with a well-defined absolute age (Swindle, 1993). The first such isochron was for the H4 ordinary chondrite Forest Vale (Lee and Halliday, 2000a). The best-fit line regressed through these data corresponds to a slope ($=^{182}Hf/^{180}Hf$) of $(1.87 \pm 0.16) \times 10^{-4}$. The absolute age of tungsten equilibration in Forest Vale is unknown but may be 5 Myr younger than the CAI inclusions of Allende (Göpel et al., 1994). Kleine et al. (2002) and Yin et al. (2002) both obtained lower initial $^{182}Hf/^{180}Hf$ values from internal isochrons, some of which are relatively precise. The two meteorites studied by Yin et al. (2002) are poorly characterized, thoroughly equilibrated meteorites of unknown equilibration age. The data for Ste. Marguerite obtained by Kleine et al. (2002) were obtained by separating a range of unknown phases with very high Hf/W. They obtained a value closer to 1.0×10^{-4}, but it is not clear whether the phases studied are the same as those analyzed from Forest Vale by Lee and Halliday (2000b) with lower Hf/W. Nevertheless, they estimated that their isochron value was closer to the true $(^{182}Hf/^{180}Hf)_{BSSI}$.

Both, the uncertainty over $(^{182}Hf/^{180}Hf)_{BSSI}$ and the fact that the tungsten isotopic composition of the silicate Earth is now unequivocally resolvable from a now well-defined chondritic composition (Kleine et al., 2002; Lee and Halliday, 2000a; Schoenberg et al., 2002; Yin et al., 2002), affect the calculated timescales for terrestrial accretion. It had been argued that accretion and core formation were fairly protracted and characterized by equilibration between accreting materials and the silicate Earth (Halliday, 2000; Halliday et al., 1996, 2000; Halliday and Lee, 1999). In other words, the tungsten isotope data provide very strong confirmation of the models of Safronov (1954) and Wetherill (1986). This general

Table 1 Selected W-isotope data for iron meteorites.

Iron meteorite	$\varepsilon^{182}W$	$^{182}Hf/^{180}Hf$
Bennett Co.	-4.6 ± 0.9	$(1.74 \pm 0.57) \times 10^{-4}$
Lombard	-4.3 ± 0.3	$(1.55 \pm 0.37) \times 10^{-4}$
Mt. Edith	-4.5 ± 0.6	$(1.68 \pm 0.46) \times 10^{-4}$
Duel Hill-1854	-5.1 ± 1.1	$(2.07 \pm 0.68) \times 10^{-4}$
Tlacotopec	-4.4 ± 0.4	$(1.68 \pm 0.41) \times 10^{-4}$

Source: Horan et al. (1998). The calculated difference between the initial and present-day W-isotopic composition of the solar system is equal to the $^{180}Hf/^{184}W$ of the solar system multiplied by the $(^{182}Hf/^{180}Hf)_{BSSI}$. The W-isotopic compositions of iron meteorites are maxima for the $(\varepsilon^{182}W)_{BSSI}$, and therefore provide a limit on the minimum $(^{182}Hf/^{180}Hf)_{BSSI}$. The $^{182}Hf/^{180}Hf$ at the time of formation of these early metals shown here is calculated from the W-isotopic composition of the metal, the W-isotopic composition of carbonaceous chondrites (Kleine et al., 2002), and the average $^{180}Hf/^{184}W$ for carbonaceous chondrites of 1.34 (Newsom et al., 1996).

scenario remains the same with the new data but in detail there are changes to the exact timescales.

Previously, Halliday (2000) estimated that the mean life, the time required to accumulate 63% of the Earth's mass with exponentially decreasing accretion rates, must lie in the range of 25–40 Myr based on the combined constraints imposed by the tungsten and lead isotope data for the Earth. Yin et al. (2002) have argued that the mean life for Earth accretion is more like ~11 Myr based on their new data for chondrites. The lead isotope data for the Earth are hard to reconcile with such rapid accretion rates as already discussed (Figure 4). Therefore, at present there is an unresolved apparent discrepancy between the models based on tungsten and those based on lead isotope data. Resolving this discrepancy highlights the limitations in both the tungsten and the lead isotope modeling. Here are some of the most important weaknesses to be aware of:

(i) The U/Pb ratio of the total Earth is poorly known.

(ii) In all of these models it is assumed that the Earth accretes at exponentially decreasing rates. Although the exponentially decreasing rate of growth of the Earth is based on Monte Carlo simulations and makes intuitive sense given the ever decreasing probability of collisions, the reality cannot be this simple. As planets get bigger, the average size of the objects with which they collide also must increase. As such, the later stages of planetary accretion are thought to involve major collisions. This is a stochastic process that is hard to predict and model. It means that the current modeling can only provide, at best, a rough description of the accretion history.

(iii) The Moon is thought to be the product of such a collision. The Earth's U/Pb ratio conceivably might have increased during accretion if a fraction of the moderately volatile elements were lost during very energetic events like the Moon-forming giant impact (Figure 6).

(iv) Similarly, as the objects get larger, the chances for equilibration of metal and silicate would seem to be less likely. This being the case, the tungsten and lead isotopic composition of the silicate Earth could reflect only partial equilibration with incoming material such that the tungsten and lead isotopic composition is partly inherited. This has been modeled in detail by Halliday (2000) in the context of the giant impact and more recently has been studied by Vityazev et al. (2003) and Yoshino et al. (2003) in the context of equilibration of asteroidal-sized objects. If correct, it would mean accretion was even slower than can be deduced from tungsten or lead isotopes. If lead equilibrated more readily than tungsten did, for whatever reason, it might help explain some of the discrepancy. One possible way to decouple lead from tungsten

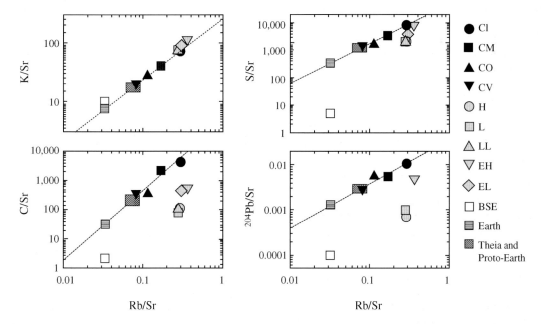

Figure 6 Volatile/refractory element ratio–ratio plots for chondrites and the silicate Earth. The correlations for carbonaceous chondrites can be used to define the composition of the Earth, the Rb/Sr ratio of which is well known, because the strontium isotopic composition of the BSE represents the time-integrated Rb/Sr. The BSE inventories of volatile siderophile elements carbon, sulfur, and lead are depleted by more than one order of magnitude because of core formation. The values for Theia are time-integrated compositions, assuming time-integrated Rb/Sr deduced from the strontium isotopic composition of the Moon (Figure 8) can be used to calculate other chemical compositions from the correlations in carbonaceous chondrites (Halliday and Porcelli, 2001). Other data are from Newsom (1995).

would be by their relative volatility. Lead could have been equilibrated by vapor-phase exchange, while tungsten would not have been able to do this and would require intimate physical mixing and reduction to achieve equilibrium.

(v) $(^{182}Hf/^{180}Hf)_{BSSI}$ is, at present, very poorly defined and could be significantly higher (Table 1). This would result in more protracted accretion timescales deduced from Hf–W (Table 2), which would be more consistent with the results obtained from U–Pb. An improved and more reliable Hf–W chronometry will depend on the degree to which the initial hafnium and tungsten isotopic compositions at the start of the solar system can be accurately defined (Tables 1 and 2). Techniques must be developed for studying very early objects like CAIs.

(vi) There is also a huge range of uncertainty in the Hf/W ratio of the silicate Earth (Newsom et al., 1996), with values ranging between 10 and 40. The tungsten isotope age calculations presented in the literature tend to assume a value at the lower end of this range. Adoption of higher values also would result in more protracted accretion timescales based on Hf–W. The most recent independent estimates (Walter et al., 2000) are significantly higher than those used by Yin et al. (2002) and Kleine et al. (2002) in support of their proposed timescales.

(vii) The lead isotopic composition of the BSE is not well defined (Figure 4). If the correct value lies closer to the Geochron than previously recognized, then the apparent accretion timescales for the Earth would be shortened.

(viii) Lastly, the decay constant for ^{182}Hf has a reported uncertainty of $\pm 22\%$ and really accurate determinations of accretion timescales require a significant reduction in this uncertainty.

Having made all of these cautionary statements, one still can state something useful about the overall accretion timescales. All recent combined accretion/continuous core formation models (Halliday, 2000; Halliday et al., 2000; Yin et al., 2002) are in agreement that the timescales are in the range $10^7 - 10^8$ yr, as predicted by Wetherill (1986). Therefore, we can specifically evaluate the models of planetary accretion proposed earlier as follows.

If the Earth accreted very fast, in $<10^6$ yr, as proposed by Cameron (1978), the silicate Earth would have a tungsten isotopic composition that is vastly more radiogenic than that observed today (Figure 5). Such objects would have $\varepsilon_W > +10$, rather than 0 (just two ε units above average solar system). Therefore, we can say with some confidence that this model does not describe the accretion of the Earth. Protracted accretion in the absence of nebular gas, as proposed by Safronov and Wetherill, is very consistent with the close agreement between chondrites and the silicate Earth (Figure 4). To what extent the Kyoto model, which involves a significant amount of nebular gas (Hayashi et al., 1985), can be confirmed or discounted is unclear at present. However, even the timescales presented by Yin et al. (2002) are long compared with the 5 Myr for accretion of the Earth predicted by the Kyoto model. This could change somewhat with further tungsten and lead isotope data and the definition of the critical parameters and modeling. However, the first-order conclusion is that nebular gas was at most somewhat limited during accretion and that there must have been much less gas than that implied by the minimum mass solar nebula scenario proposed in the Kyoto model.

Table 2 The apparent mean life of formation of the Earth, as well as the predicted W-isotopic compositions of the lunar mantle and silicate Earth.

$\varepsilon^{182}W_{BSSI}$	-5.0	-4.5	-4.0	-3.5
$(^{182}Hf/^{180}Hf)_{BSSI}$	2.0×10^{-4}	1.7×10^{-4}	1.4×10^{-4}	1.0×10^{-4}
Accretionary mean life of the earth (yr)	15×10^6	14×10^6	13×10^6	12×10^6
$\varepsilon^{182}W$ for: lunar initial/present-day lunar mantle/present-day silicate Earth				
Giant impact at 30×10^6 yr	$+0.3/+7.1/+0.4$	$-0.1/+5.6/0.0$	$-0.4/+4.2/-0.4$	$-0.8/+2.8/-0.7$
Giant impact at 40×10^6 yr	$-0.3/+2.9/+0.8$	$-0.6/+2.0/+0.3$	$-0.8/+1.3/-0.1$	$-1.1/+0.6/-0.5$
Giant impact at 45×10^6 yr	$-0.6/+1.6/0.0$	$-0.8/+1.0/-0.3$	$-1.0/+0.5/-0.6$	$-1.2/-0.1/-0.9$
Giant impact at 50×10^6 yr	$-0.8/+0.7/-0.6$	$-1.0/+2.0/-0.8$	$-1.2/-0.2/-1.0$	$-1.3/-0.6/-1.2$
Giant impact at 60×10^6 yr	$-1.2/-0.9/-1.2$	$-1.3/-1.0/-1.3$	$-1.4/-1.0/-1.4$	$-1.5/-1.2/-1.6$
Giant impact at 70×10^6 yr	$-1.3/-1.2/-1.5$	$-1.4/-1.3/-1.6$	$-1.5/-1.4/-1.7$	$-1.6/-1.5/-1.7$

All parameters are critically dependent on the initial W- and Hf- isotopic composition of the solar system, which at present are poorly known (see text). All calculated values assume that the W depletion in the silicate Earth and the lunar mantle are as given in Walter et al. (2000). The $(^{182}Hf/^{180}Hf)_{BSSI}$, model ages, and Earth and Moon compositions are all calculated using the W-isotopic composition of carbonaceous chondrites (Kleine et al., 2002) and the average $^{180}Hf/^{184}W$ for carbonaceous chondrites of 1.34 (Newsom, 1990). The principles behind the modeling are as in Halliday (2000). The giant impact is assumed to have occurred when the Earth had reached 90% of its current mass, the impactor adding a further 9%. Exponentially decreasing accretion rates are assumed before and after (Halliday, 2000). The solutions in bold type provide the best match with the current data for the W-isotopic compositions of the initial Moon (~ 0), and the present-day lunar mantle (<2) and silicate Earth (0).

1.20.5 CHEMICAL AND ISOTOPIC CONSTRAINTS ON THE NATURE OF THE COMPONENTS THAT ACCRETED TO FORM THE EARTH

1.20.5.1 Chondrites and the Composition of the Disk from Which Earth Accreted

A widespread current view is that chondrites represent primitive undifferentiated material from which the Earth accreted. In reality, chondrites are anything but simple and, although they contain early and presolar objects, how primitive and how representative of the range of very early planetesimals they are is completely unclear. In all probability the Earth was built largely from more extensively differentiated materials (Taylor and Norman, 1990). Nevertheless, chondrites do represent a useful set of reference reservoirs in chemical and isotopic terms from which one can draw some conclusions about the "average stuff" from which the Earth may have been built (e.g., Ganapathy and Anders, 1974; Anders, 1977; Wolf et al., 1980). Indeed, if geochemists did not have chondrite samples to provide a reference, many geochemical arguments about Earth's origin, not to say present-day interior structure, would be far weaker.

The degree to which different kinds of chondrites reflect bulk Earth composition has been extensively debated, since none of them provide a good match. CI carbonaceous chondrites define a reference reservoir for *undepleted* solar system compositions because they are strikingly similar in composition to the Sun when normalized to an element such as silicon (Grevesse and Sauval, 1998; see Chapter 1.03). However, they are not at all similar to the Earth's volatile-depleted composition. Nor are they much like the vast majority of other chondrites! Confirmation of the primitive nature of CIs is most readily demonstrated by comparing the relative concentrations of a well-determined moderately volatile major element to a well-determined refractory major element with the values in the Sun's photosphere. For example, the concentrations of sodium and calcium are both known to better than a percent in the photosphere (Grevesse and Sauval, 1998) and are identical to within a percent with the values found in CIs despite a big difference in volatility between the two elements (Anders and Grevesse, 1989; Newsom, 1995). It is tempting to ascribe CIs to a complete condensation sequence. However, they also contain dispersed isotopic anomalies in Cr, for example, that indicate a lack of homogenization on a fine scale (Podosek et al., 1997). This cannot be reconciled with nebular condensation from high temperatures.

Volatile-element depletion patterns in other (e.g., CV, CM, or CO) carbonaceous chondrites (Larimer and Anders, 1970; Palme et al., 1988; Takahashi et al., 1978; Wolf et al., 1980) and ordinary chondrites (Wasson and Chou, 1974) have long been considered to partly reflect incomplete condensation (Wasson, 1985), with the more volatile elements removed from the meteorite formation regions before cooling and the termination of condensation. However, this depletion must at least partially reflect the incorporation of volatile-depleted CAIs (Guan et al., 2000) and chondrules (Grossman, 1996; Meibom et al., 2000). The origins of CAIs have long been enigmatic, but recently it has been proposed that CAIs may be the product of rather localized heating close to the young Sun (Shu et al., 1997) following which they were scattered across the disk. Regardless of whether this model is exactly correct, the incorporation of volatile-depleted CAIs must result in some degree of volatile-depleted composition for chondritic meteorites. Chondrules are the product of rapid melting of chondritic materials (Connolly et al., 1998; Connolly and Love, 1998; Desch and Cuzzi, 2000; Jones et al., 2000) and such events also may have been responsible for some degree of volatile-element depletion (Vogel et al., 2002).

The discovery that pristine presolar grains are preserved in chondrites indicates that they were admixed at a (relatively) late stage. Many of the presolar grain types (Mendybaev et al., 2002; Nittler, 2003) could not have survived the high temperatures associated with CAI and chondrule formation. Therefore, chondrites must include a widely dispersed presolar component (Huss, 1988, 1997; Huss and Lewis, 1995) that either settled or was swept into the chondrule-forming and/or chondrule-accumulating region and would have brought with it nebular material that was not depleted in volatiles. Therefore, volatile depletion in chondrites must reflect, to some degree at least, the mixing of more refractory material (Guan et al., 2000; Kornacki and Fegley, 1986) with CI-like matrix of finer grained volatile-rich material (Grossman, 1996). CIs almost certainly represent accumulations of mixed dust inherited from the portion of the protostellar molecular cloud that collapsed to form the Sun within a region of the nebula that was never greatly heated and did not accumulate chondrules and CAIs (for whatever reason). Though undifferentiated, chondrites, with the possible exception of CIs, are not primitive and certainly do not represent the first stages of accretion of the Earth.

Perhaps the best current way to view the disk of debris from which the Earth accreted was as an environment of vigorous mixing. Volatile-depleted material that had witnessed very high temperatures at an early stage (CAIs) mixed with material that had been flash melted (chondrules) a few million years later. Then presolar grains that had escaped these processes rained into the midplane or more likely were swept in from the outer regions of the solar system. It is within these environments that dust, chondrules and CAIs mixed and accumulated into planetesimals. There also would have been earlier formed primary planetesimals (planetary embryos) that had accreted rapidly by runaway growth but were subsequently consumed. The earliest most primitive objects probably differentiated extremely quickly aided by the heat from runaway growth and live ^{26}Al (see Chapter 1.17).

It is not clear whether we have *any* meteorites that are samples of these very earliest planetesimals. Some iron meteorites with very unradiogenic tungsten (Horan et al., 1998) might be candidates (Table 1). Nearly all silicate-rich basaltic achondrites that are precisely dated appear to have formed later (more than a million years after the start of the solar system). Some of the disk debris will have been the secondary product of collisions between these planetesimals, rather like the dust in Beta Pictoris. Some have proposed that chondrules formed in this way (Sanders, 1996; Lugmair and Shukolyukov, 2001), but although the timescales appear right this is not generally accepted (Jones et al., 2000). A working model of the disk is of a conveyor belt of material rapidly accreting into early planetesimals, spiraling in toward the Sun and being fed by heated volatile-depleted material (Krot et al., 2001) scattered outwards and pristine volatile-rich material being dragged in from the far reaches of the disk (Shu et al., 1996). This mixture of materials provided the raw ingredients for early planetesimals and planets that have largely been destroyed or ejected. One of the paths of destruction was mutual collisions, and it was these events that ultimately led to the growth of terrestrial planets like the Earth.

1.20.5.2 Chondritic Component Models

Although it is now understood quite well that chondrites are complicated objects with a significant formation history, many have proposed that they should be used to define Earth's bulk composition. More detailed discussion of the Earth's composition is provided elsewhere in this treatise (e.g., see Chapter 2.01) and only the essentials for this chapter are covered here. A recent overview of some of the issues is provided by Drake and Righter (2002). The carbonaceous chondrites often are considered to provide the best estimates of the basic building blocks (Agee and Walker, 1988; Allègre et al., 1995a,b, 2001; Anders, 1977; Ganapathy and Anders, 1974; Herzberg, 1984; Herzberg et al., 1988; Jagoutz et al., 1979; Jones and Palme, 2000; Kato et al., 1988a,b; Newsom, 1990). Because the Earth is formed from collisions between differentiated material, it is clear that one is simply using these reference chondrites to provide little more than model estimates of the total Earth's composition. Allègre et al. (1995a,b, 2001), for example, have plotted the various compositions of chondrites using major and trace element ratios and have shown that the refractory lithophile elements are most closely approximated by certain specific kinds of chondrites. Relative abundances of the platinum group elements and osmium isotopic composition of the silicate Earth have been used extensively to evaluate which class of chondrites contributed the late veneer (Newsom, 1990; Rehkämper et al., 1997; Meisel et al., 2001; Drake and Righter, 2002), a late addition of volatile rich material. Javoy (1998, 1999) has developed another class of models altogether based on enstatite chondrites.

If it is assumed that one can use chondrites as a reference, one can calculate the composition of the total Earth and predict the concentrations of elements that are poorly known. For example, in theory one can predict the amount of silicon in the total Earth and determine from this how much must have gone into the core. The Mainz group produced many of the classic papers pursuing this approach (Jagoutz et al., 1979). However, it was quickly realized that the Earth does not fit any class of chondrite. In particular, Jagoutz presented the idea of using element ratios to show that the Earth's upper mantle had Mg/Si that was non-chondritic. Jagoutz et al. (1979) used CI, ordinary and enstatite chondrites to define the bulk Earth. Anders always emphasized that all chondrites were fractionated relative to CIs in his series of papers on "Chemical fractionations in meteorites" (e.g., Wolf et al., 1980). That is, CMs and CVs, etc., are also fractionated, as are planets. Jagoutz et al. (1979) recognized that the fractionations of magnesium, silicon, and aluminum were found in both the Earth and in CI, O, and E chondrites, but not in C2–C3 chondrites, which exhibit a fixed Mg/Si but variable Al/Si ratio, probably because of addition of CAIs.

One also can use chondrites to determine the abundances of moderately volatile elements such as potassium. Allègre et al. (2001) and Halliday and Porcelli (2001) pursued this approach to show that the Earth's K/U is $\sim 10^4$ (Figure 2). One can calculate how much of the volatile chalcophile and siderophile elements such as sulfur, cadmium, tellurium, and lead may have gone in to the Earth's core (Figure 6; Yi et al., 2000; Allègre et al., 1995b; Halliday and Porcelli, 2001). Allègre et al. (1995a) also have used this approach to estimate the total Earth's ^{238}U/^{204}Pb.

From the budgets of potassium, silicon, carbon, and sulfur extrapolated from carbonaceous chondrite compositions, one can evaluate the amounts of various light elements possibly incorporated in the core (Allègre et al., 1995b; Halliday and Porcelli, 2001) and see if this explains the deficiency in density relative to pure iron (Ahrens and Jeanloz, 1987). Such approaches complement the results of experimental solubility measurements (e.g., Wood, 1993; Gessmann et al., 2001) and more detailed comparisons with the geochemistry of meteorites (Dreibus and Palme, 1996).

It should be noted that the Jagoutz et al. (1979) approach and the later studies by Allègre and co-workers (Allègre et al., 1995a,b, 2001) are mutually contradictory, a fact that is sometimes not recognized. Whereas Jagoutz et al. (1979) used CI, ordinary, and enstatite chondrites, Allègre and co-workers used CI, CM, CV, and CO chondrites (the carbonaceous chondrite mixing line) to define the bulk earth, ignoring the others.

These extrapolations and predictions are based on two assumptions that some would regard as rendering the approach flawed:

(i) The carbonaceous chondrites define very nice trends for many elements that can be used to define the Earth's composition, but the ordinary and enstatite chondrites often lie off these trends (Figure 6). There is no obvious basis for simply ignoring the chemical compositions of the ordinary and enstatite chondrites unless they have undergone some parent body process of loss or redistribution in their compositions. This is indeed likely, particularly for volatile chalcophile elements but is not well understood at present.

(ii) The Earth as well as the Moon, Mars, and basaltic achondrite parent bodies are depleted in moderately volatile elements, in particular the alkali elements potassium and rubidium, relative to *most* classes of chondrite (Figure 2). This needs to be explained. Humayun and Clayton (1995) performed a similar exercise, showing that since chondrites are all more volatile rich than terrestrial planets, the only way to build planets from chondrite precursors was to volatilize alkalis and other volatile elements. This was not considered feasible in view of the identical potassium isotope compositions of chondrites, the Earth and the Moon. Taylor and Norman (1990) made a similar observation that planets formed from volatile-depleted and differentiated precursors and not from chondrites. Clearly, there must have been other loss mechanisms (Halliday and Porcelli, 2001) beyond those responsible for the volatile depletion in chondrites unless the Earth, Moon, angrite parent body and Vesta simply were accreted from parts of the disk that were especially enriched in some highly refractory component, like CAIs (Longhi, 1999).

This latter point and other similar concerns represent such a major problem that some scientists abandoned the idea of just using chondrites as a starting point and instead invoked the former presence of completely hypothetical components in the inner solar system. This hypothesis has been explored in detail as described below.

1.20.5.3 Simple Theoretical Components

In addition to making comparisons with chondrites, the bulk composition of the Earth also has been defined in terms of a "model" mixture of highly reduced, refractory material combined with a much smaller proportion of a more oxidized volatile-rich component (Wänke, 1981). These models follow on from the ideas behind earlier heterogeneous accretion models. According to these models, the Earth was formed from two components. Component A was highly reduced and free of all elements with equal or higher volatility than sodium. All other elements were in CI relative abundance. The iron and siderophile elements were in metallic form, as was part of the silicon. Component B was oxidized and contained all elements, including those more volatile than sodium in CI relative abundance. Iron and all siderophile and lithophile elements were mainly in the form of oxides.

Ringwood (1979) first proposed these models but the concept was more fully developed by Wänke (1981). In Wänke's model, the Earth accretes by heterogeneous accretion with a mixing ratio A:B ~ 85:15. Most of component B would be added after the Earth had reached about two thirds of its present mass. The oxidized volatile-rich component would be equivalent to CI carbonaceous chondrites. However, the reduced refractory rich component is hypothetical and never has been identified in terms of meteorite components.

Eventually, models that involved successive changes in accretion and core formation replaced these. How volatiles played into this was not explained except that changes in oxidation state were incorporated. An advanced example of such a model is that presented by Newsom (1990). He envisaged the history of accretion as involving stages that included concomitant core formation stages (discussed under core formation).

1.20.5.4 The Nonchondritic Mg/Si of the Earth's Primitive Upper Mantle

It has long been unclear why the primitive upper mantle of the Earth has a nonchondritic proportion of silicon to magnesium. Anderson (1979) proposed that the mantle was layered with the lower mantle having higher Si/Mg. Although a number of coeval papers presented a similar view (Herzberg, 1984; Jackson, 1983) it is thought by many geochemists these days that the major element composition of the mantle is, broadly speaking, well mixed and homogeneous as a result of 4.5 Gyr of mantle convection (e.g., Hofmann, 1988; Ringwood, 1990).

A notable exception is the enstatite chondrite model of Javoy (1999).

Wänke (1981) and Allègre et al. (1995b) have proposed that a significant fraction of the Earth's silicon is in the core. However, this is not well supported by experimental data. To explain the silicon deficiency this way conditions have to be so reducing that niobium would be siderophile and very little would be left in the Earth's mantle (Wade and Wood, 2001).

Ringwood (1989a) proposed that the nonchondritic Mg/Si reflected a radial zonation in the solar system caused by the *addition* of more volatile silicon to the outer portions of the solar system. That is, he viewed the Earth as possibly more representative of the solar system than chondrites.

Hewins and Herzberg (1996) proposed that the midplane of the disk from which the Earth accreted was dominated by chondrules. Chondrules have higher Mg/Si than chondrites and if the Earth accreted in a particular chondrule rich area (as it may well have done) the sorting effect could dominate planetary compositions.

1.20.5.5 Oxygen Isotopic Models and Volatile Losses

Some have proposed that one can use the oxygen isotopic composition of the Earth to identify the proportions of different kinds of chondritic components (e.g., Lodders, 2000). The isotopic composition of oxygen is variable, chiefly in a mass-dependent way, in terrestrial materials. However, meteorites show mass-independent variations as well (Clayton, 1986, 1993). Oxygen has three isotopes and a three-isotope system allows discrimination between mass-dependent planetary processes and mass-independent primordial nebular heterogeneities inherited during planet formation (Clayton and Mayeda, 1996; Franchi et al., 1999; Wiechert et al., 2001, 2003). Clayton pursued this approach and showed that the mix of meteorite types based on oxygen isotopes would not provide a chemical composition that was similar to that of the Earth. In particular, the alkalis would be too abundant. In fact, the implied relative proportions of volatile-rich and volatile-poor constituents are in the opposite sense of those derived in the Wänke-Ringwood mixing models mentioned above (Clayton and Mayeda, 1996). Even the differences in composition between the planets do not make sense. Earth would need to be less volatile-depleted than Mars, whereas the opposite is true (Clayton, 1993; Halliday et al., 2001). The oxygen isotope compositions of Mars (based on analyses of SNC meteorites) show a *smaller* proportion of carbonaceous-chondrite-like material in Mars than in Earth (Clayton, 1993; Halliday et al., 2001). It thus appears that, unless the Earth lost a significant proportion of its moderately volatile elements after it formed, the principal carrier of moderately volatile elements involved in the formation of the terrestrial planets was not of carbonaceous chondrite composition. This, of course, leads to the viewpoint that the Earth was built from volatile-depleted material, such as differentiated planetesimals (Taylor and Norman, 1990), but it still begs the question of how volatile depletion occurred (Halliday and Porcelli, 2001).

Whether it is plausible, given the dynamics of planetary accretion discussed above, that the Earth lost a major fraction of its moderately volatile elements during its accretion history is unclear. This has been advanced as an explanation by Lodders (2000) and is supported by strontium isotope data for early solar system objects discussed below (Halliday and Porcelli, 2001). The difficulty is to come up with a mechanism that does not fractionate K isotopes (Humayun and Clayton, 1995) and permits loss of heavy volatile elements. The degree to which lack of fractionation of potassium isotopes offers a real constraint depends on the mechanisms involved (Esat, 1996; Young, 2000). There is no question that as the Earth became larger, the accretion dynamics would have become more energetic and the temperatures associated with accretion would become greater (Melosh and Sonett, 1986; Melosh and Vickery, 1989; Ahrens, 1990; Benz and Cameron, 1990; Melosh, 1990; Melosh et al., 1993). However, the gravitational pull of the Earth would have become so large that it would be difficult for the Earth to lose these elements even if they were degassed into a hot protoatmosphere. This is discussed further below.

1.20.6 EARTH'S EARLIEST ATMOSPHERES AND HYDROSPHERES

1.20.6.1 Introduction

The range of possibilities to be considered for the nature of the earliest atmosphere provides such a broad spectrum of consequences for thermal and magmatic evolution that it is better to consider the atmospheres first before discussing other aspects of Earth's evolution. Therefore, in this section a brief explanation of the different kinds of early atmospheres and their likely effects on the Earth are given in cursory terms. More comprehensive information on atmospheric components is found elsewhere in Volume 4 of this treatise.

1.20.6.2 Did the Earth Have a Nebular Protoatmosphere?

Large nebular atmospheres have at various times been considered a fundamental feature of the early Earth by geochemists (e.g., Sasaki and Nakazawa, 1988; Pepin, 2000; Porcelli *et al.*, 1998). Large amounts of nebular gases readily explain why the Earth has primordial ^{3}He (Clarke *et al.*, 1969; Mamyrin *et al.*, 1969; Lupton and Craig, 1975; Craig and Lupton, 1976) and why a component of solar-like neon with a solar He/Ne ratio can be found in some plume basalts (Honda *et al.*, 1991; Dixon *et al.*, 2000; Moreira *et al.*, 2001). Evidence for a solar component among the heavier noble gases has been more scant (Moreira and Allègre, 1998; Moreira *et al.*, 1998). There is a hint of a component with different ^{38}Ar/^{36}Ar in some basalts (Niedermann *et al.*, 1997) and this may reflect a solar argon component (Pepin and Porcelli, 2002). Also, Caffee *et al.* (1988, 1999) have made the case that a solar component of xenon can be found in well gases.

The consequences for the early behavior of the Earth are anticipated to be considerable if there was a large nebular atmosphere. Huge reducing protoatmospheres, be they nebular or impact-induced can facilitate thermal blanketing, magma oceans and core formation. For example, based on the evidence from helium and neon, Harper and Jacobsen (1996b) suggested that iron was reduced to form the core during a stage with a massive early H_2–He atmosphere. A variety of authors had already proposed that the Earth accreted with a large solar nebular atmosphere. Harper and Jacobsen's model builds upon many earlier such ideas primarily put forward by the Kyoto school (Hayashi *et al.*, 1979, 1985; Mizuno *et al.*, 1980; Sasaki and Nakazawa, 1988; Sasaki, 1990). The thermal effects of a very large protoatmosphere have been modeled by Hayashi *et al.* (1979), who showed that surface temperatures might reach >4,000 K. Sasaki (1990) also showed that incredibly high temperatures might build up in the outer portions of the mantle, leading to widespread magma oceans. Dissolving volatiles like noble gases into early silicate and metal liquids may have been quite easy under these circumstances (Mizuno *et al.*, 1980; Porcelli *et al.*, 2001; Porcelli and Halliday, 2001).

Porcelli and Pepin (2000) and Porcelli *et al.* (2001) recently summarized the noble gas arguments pointing out that first-order calculations indicate that significant amounts of noble gases with a solar composition are left within the Earth's interior but orders of magnitude more have been lost, based on xenon isotopic evidence (see Chapter 4.11). Therefore, one requires a relatively large amount of nebular gas during Earth's accretion. To make the Earth this way, the timescales for accretion need to be characteristically short ($\sim 10^6$–10^7 yr) in order to trap such amounts of gas before the remains of the solar nebula are accreted into the Sun or other planets. Therefore, a problem may exist reconciling the apparent need to acquire a large nebular atmosphere with the longer timescales ($\sim 10^7$–10^8 yr) for accretion implied by Safronov–Wetherill models and by tungsten and lead isotopic data. It is hard to get around this problem because the nebular model is predicated on the assumption that the Earth grows extremely fast such that it can retain a large atmosphere (Pepin and Porcelli, 2002).

Therefore, other models for explaining the incorporation of solar-like noble gases should be considered. The most widely voiced alternative is that of accreting material that formed earlier elsewhere that already had acquired solar-like noble gases. For example, it has been argued that the neon is acquired as "Ne–B" from accretion of chondritic material (Trieloff *et al.*, 2000). With such a component in meteorites, one apparently could readily explain the Earth's noble gas composition, which may have a ^{20}Ne/^{22}Ne ratio that is lower than solar (Farley and Poreda, 1992). However, this component is no longer well defined in meteorites and the argument is less than certain because of this. In fact, Ne–B is probably just a fractionated version of solar (Ballentine *et al.*, 2001). This problem apart, the idea that the Earth acquired its solar-like noble gases from accreting earlier-formed objects is an alternative that is worth considering. Chondrites, however, mainly contain very different noble gases that are dominated by the so-called "Planetary" component (Ozima and Podosek, 2002). This component is nowadays more precisely identified as "Phase Q" (Wieler *et al.*, 1992; Busemann *et al.*, 2000). In detail, the components within chondrites show little sign of having incorporated large amounts of solar-like noble gases. Early melted objects like CAIs and chondrules are, generally speaking, strongly degassed (Vogel *et al.*, 2002, 2003).

Podosek *et al.* (2003) recently have proposed that the noble gases are incorporated into early formed planetesimals that are irradiated by intense solar-wind activity from the vigorous early Sun. With the new evidence from young solar mass stars of vastly greater flare activity (Feigelson *et al.*, 2002a,b), there is strong support for the notion that inner solar system objects would have incorporated a lot of solar-wind implanted noble gases. The beauty of this model is that small objects with larger surface-to-volume ratio trap more noble gases. Unlike the nebular model it works best if large objects take a long time to form. As such, the model is easier to reconcile with the kinds of long timescales for accretion implied by tungsten and lead isotopes. Podosek *et al.* (2003) present detailed calculations to illustrate the feasibility of such a scenario.

To summarize, although large nebular atmospheres have long been considered the most likely explanation for primordial solar-like noble gases in the Earth, the implications of such models appear hard to reconcile with the accretionary timescales determined from tungsten and lead isotopes. Irradiating planetesimals with solar wind currently appears to be the most promising alternative. If this is correct, the Earth may still at one time have had a relatively large atmosphere, but it would have formed by degassing of the Earth's interior.

1.20.6.3 Earth's Degassed Protoatmosphere

The discovery that primordial ^{3}He still is being released from the Earth's interior (Clarke et al., 1969; Mamyrin et al., 1969; Lupton and Craig, 1975; Craig and Lupton, 1976) is one of the greatest scientific contributions made by noble gas geochemistry. Far from being totally degassed, the Earth has deep reservoirs that must supply ^{3}He to the upper mantle and thence to the atmosphere. However, the idea that the majority of the components in the present-day atmosphere formed by degassing of the Earth's interior is much older than this. Brown (1949) and Rubey (1951) proposed this on the basis of the similarity in chemical composition between the atmosphere and hydrosphere on the one hand and the compositions of volcanic gases on the other. In more recent years, a variety of models for the history of degassing of the Earth have been developed based on the idea that a relatively undegassed lower mantle supplied the upper mantle with volatiles, which then supplied the atmosphere (Allègre et al., 1983, 1996; O'Nions and Tolstikhin, 1994; Porcelli and Wasserburg, 1995).

Some of the elements in the atmosphere provide specific "time information" on the degassing of the Earth. Of course, oxygen was added to the atmosphere gradually by photosynthesis with a major increase in the early Proterozoic (Kasting, 2001). However, the isotopic composition and concentration of argon provides powerful evidence that other gases have been added to the atmosphere from the mantle over geological time (O'Nions and Tolstikhin, 1994; Porcelli and Wasserburg, 1995). In the case of argon, a mass balance can be determined because it is dominantly (>99%) composed of ^{40}Ar, formed by radioactive decay of ^{40}K. From the Earth's potassium concentration and the atmosphere's argon concentration it can be shown that roughly half the ^{40}Ar is in the atmosphere, the remainder presumably being still stored in the Earth's mantle (Allègre et al., 1996). Some have argued that this cannot be correct on geophysical grounds, leading to the proposal that the Earth's K/U ratio has been overestimated (Davies, 1999). However, the relationship with Rb/Sr (Figure 2) provides strong evidence that the Earth's K/U ratio is ~10^4 (Allègre et al., 2001; Halliday and Porcelli, 2001). Furthermore, support for mantle reservoirs that are relatively undegassed can be found in both helium and neon isotopes (Allègre et al., 1983, 1987; Moreira et al., 1998, 2001; Moreira and Allègre, 1998; O'Nions and Tolstikhin, 1994; Niedermann et al., 1997; Porcelli and Wasserburg, 1995). The average timing of this loss is poorly constrained from argon data. In principle, this amount of argon could have been supplied catastrophically in the recent past. However, it is far more reasonable to assume that because the Earth's radioactive heat production is decaying exponentially, the amount of degassing has been decreasing with time. The argon in the atmosphere is the time-integrated effect of this degassing.

Xenon isotopes provide strong evidence that the Earth's interior may have undergone an early and catastrophic degassing (Allègre et al., 1983, 1987). Allègre et al. discovered that the MORB-source mantle had elevated ^{129}Xe relative to atmospheric xenon, indicating that the Earth and atmosphere separated from each other at an early stage. These models are hampered by a lack of constraints on the xenon budgets of the mantle and by atmospheric contamination that pervades many mantle-derived samples. Furthermore, the model assumes a closed system. If a portion of the xenon in the Earth's atmosphere was added after degassing (Javoy, 1998, 1999) by cometary or asteroidal impacts (Owen and Bar-Nun, 1995, 2000; Morbidelli et al., 2000), the model becomes underconstrained. The differences that have been reported between the elemental and atomic abundances of the noble gases in the mantle relative to the atmosphere may indeed be explained by heterogeneous accretion of the atmosphere (Marty, 1989; Caffee et al., 1988, 1999).

Isotopic evidence aside, many theoretical and experimental papers have focused on the production of a steam atmosphere by impact-induced degassing of the Earth's interior (Abe and Matsui, 1985, 1986, 1988; O'Keefe and Ahrens, 1977; Lange and Ahrens, 1982, 1984; Matsui and Abe, 1986a,b; Sasaki, 1990; Tyburczy et al., 1986; Zahnle et al., 1988). Water is highly soluble in silicate melts at high pressures (Righter and Drake, 1999; Abe et al., 2000). As such, a large amount could have been stored in the Earth's mantle and then released during volcanic degassing.

Ahrens (1990) has modeled the effects of impact-induced degassing on the Earth. He considers that the Earth probably alternated between two extreme states as accretion proceeded. When the Earth was degassed it would accumulate a large reducing atmosphere. This would provide a blanket that also allowed enormous surface temperatures to be reached:

Ahrens estimates ~1,500 K. However, when an impact occurred this atmosphere would be blown off. The surface of the Earth would become cool and oxidizing again (Ahrens, 1990). There is strong isotopic evidence for such early losses of the early atmospheres as explained in the next section.

1.20.6.4 Loss of Earth's Earliest Atmosphere(s)

The xenon isotope data provide evidence that much (>99%) of the Earth's early atmosphere was lost within the first 100 Myr. Several papers on this can be found in the literature and the most recent ones by Ozima and Podosek (1999) and Porcelli et al. (2001) are particularly useful. The basic argument for the loss is fairly simple and is not so different from the original idea of using xenon isotopes to date the Earth (Wetherill, 1975a). We have a rough idea of how much iodine exists in the Earth's mantle. The Earth's current inventory is not well constrained but we know enough (Déruelle et al., 1992) to estimate the approximate level of depletion of this volatile element. It is clear from the degassing of noble gases that the present ratio of I/Xe in the Earth is orders of magnitude higher than chondritic values. We know that at the start of the solar system ^{129}I was present with an atomic abundance of ~10^{-4} relative to stable ^{127}I (Swindle and Podosek, 1988). All of this ^{129}I formed ^{129}Xe, and should have produced xenon that was highly enriched in ^{129}Xe given the Earth's I/Xe ratio. Yet instead, the Earth has xenon that is only slightly more radiogenic than is found in meteorites rich in primordial noble gases; the ^{129}Xe excesses in the atmosphere and the mantle are both minute by comparison with that expected from the Earth's I/Xe. This provides evidence that the Earth had a low I/Xe ratio that kept its xenon isotopic compositions close to chondritic. At some point xenon was lost and by this time ^{129}I was close to being extinct such that the xenon did not become very radiogenic despite a very high I/Xe.

The xenon isotopic arguments can be extended to fissionogenic xenon. The use of the combined ^{129}I–^{129}Xe and ^{244}Pu–^{132,134,136}Xe (spontaneous fission half-life ~80 Myr) systems provides estimates of ~100 Myr for loss of xenon from the Earth (Ozima and Podosek, 1999). The fission-based models are hampered by the difficulties with resolving the heavy xenon that is formed from fission of ^{244}Pu as opposed to longer-lived uranium. The amount of ^{136}Xe that is expected to have formed from ^{244}Pu within the Earth should exceed that produced from uranium as found in well gases (Phinney et al., 1978) and this has been confirmed with measurements of MORBs (Kunz et al., 1998). However, the relative amount of plutonogenic Xe/uranogenic Xe will be a function of the history of degassing of the mantle. Estimating these amounts accurately is very hard.

This problem is exacerbated by the lack of constraint that exists on the initial xenon isotopic composition of the Earth. The xenon isotopic composition of the atmosphere is strikingly different from that found in meteorites (Wieler et al., 1992; Busemann et al., 2000) or the solar wind (Wieler et al., 1996). It is fractionated relative to solar, the light isotopes being strongly depleted (Figure 7). One can estimate the initial xenon isotopic composition of the Earth by assuming it was strongly fractionated from something more like the composition found in meteorites and the solar wind. By using meteorite data to determine a best fit to atmospheric Xe one obtains a composition called "U–Xe" (Pepin, 1997, 2000). However, this is based on finding a composition that is' consistent with the present-day atmosphere. Therefore caution is needed when using the fissionogenic xenon components to estimate an accretion age for the Earth because the arguments become circular (Zhang, 1998).

The strong mass-dependent fractionation of xenon has long been thought to be caused by hydrodynamic escape (Hunten et al., 1987; Walker, 1986) of the atmosphere. Xenon probably was entrained in a massive atmosphere of light gases presumably dominated by hydrogen and helium that was lost (Sasaki and Nakazawa, 1988). This is consistent with the view based on radiogenic and fissionogenic xenon that a large fraction of the Earth's atmosphere was lost during the lifetime of ^{129}I.

Support for loss of light gases from the atmosphere via hydrodynamic escape can be found in other "atmophile" isotopic systems. The ^{20}Ne/^{22}Ne ratio in the mantle is elevated

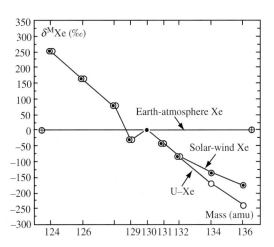

Figure 7 Mass fractionation of xenon in the atmosphere relative to the solar value (Pepin and Porcelli, 2002) (reproduced by permission of Mineralogical Society of America from *Rev. Mineral. Geochem.* **2002**, *47*, 191–246).

relative to the atmosphere (Trieloff et al., 2000). The mantle value is close to solar (Farley and Poreda, 1992), but the atmosphere plots almost exactly according to a heavy mass-dependent fractionated composition. Similarly, the hydrosphere has a D/H ratio that is heavy relative to the mantle. However, argon and krypton isotopes provide no support for the theory (Pepin and Porcelli, 2002). Therefore, this model cannot be applied in any simple way to all of the atmophile elements.

It is possible that the atmosphere was blown off by a major impact like the Moon-forming giant impact, but this is far from clear at this stage. Another mechanism that often is considered is the effect of strong ultraviolet wavelength radiation from the early Sun (Zahnle and Walker, 1982). This might affect Xe preferentially because of the lower ionization potential. It is of course possible that the Earth simply acquired an atmosphere, with xenon, like today's (Marty, 1989; Caffee et al., 1999). However, then it is not clear how to explain the strong isotopic fractionation relative to solar and meteorite compositions.

Taken together, the noble gas data provide evidence that Earth once had an atmosphere that was far more massive than today's. If true, this would have had two important geochemical consequences. First, there would have been a blanketing effect from such an atmosphere. This being the case, the temperature at the surface of the Earth would have been very high. There may well have been magma oceans, rock vapor in the atmosphere, and extreme degassing of moderately volatile as well as volatile elements. Blow-off of this atmosphere may have been related to the apparent loss of moderately volatile elements (Halliday and Porcelli, 2001). With magma oceans there would be, at best, a weak crust, mantle mixing would have been very efficient, and core formation would have proceeded quickly.

The second consequence is that it would have been relatively easy to dissolve a small amount of this "solar" gas at high pressure into the basaltic melts at the Earth's surface (Mizuno et al., 1980). This "ingassing" provides a mechanism for transporting nebular gases into the Earth's interior. The ultimate source of solar helium and neon in the mantle is unknown. At one time, it was thought to be the mantle (Allègre et al., 1983) and this still seems most likely (Porcelli and Ballentine, 2002) but the core also has been explored (Porcelli and Halliday, 2001) as a possible alternative.

1.20.7 MAGMA OCEANS AND CORE FORMATION

Core formation is the biggest differentiation process that has affected the planet, resulting in a large-scale change of the distribution of density and heat production. One would think that such a basic feature would be well understood. However, the very existence of large amounts of iron metal at the center of an Earth with an oxidized mantle is problematic (Ringwood, 1977). Large reducing atmospheres and magma oceans together provide a nice explanation. For example, Ringwood (1966) considered that the iron metal in the Earth's core formed by reduction of iron in silicates and oxides and thereby suggested a huge CO atmosphere. Clearly, if the Earth's core formed by reduction of iron in a large atmosphere, the process of core formation would occur early and easily.

However, there is very little independent evidence from mantle or crustal geochemistry to substantiate the former presence of magma oceans. There has been no shortage of proposals and arguments for and against on the basis of petrological data for the Earth's upper mantle. The key problem is the uncertainty that exists regarding the relationship between the present day upper mantle and that that may have existed in the early Earth. Some have argued that the present day lower mantle is compositionally distinct in terms of major elements (e.g., Herzberg, 1984; Jackson, 1983), whereas others (e.g., Hofmann, 1988; Hofmann et al., 1986; Davies, 1999) have presented strong evidence in favor of large-scale overall convective interchange. Although several papers have used the present-day major, trace, and isotopic compositions of the upper mantle to provide constraints on the earliest history of the Earth (Allègre et al., 1983; Agee and Walker, 1988; Kato et al., 1988a,b; McFarlane and Drake, 1990; Jones et al., 1992; Drake and McFarlane, 1993; Porcelli and Wasserburg, 1995; O'Nions and Tolstikhin, 1994; Righter and Drake, 1996), it is unclear whether or not this is valid, given 4.5 Gyr of mantle convection.

For example, the trace element and hafnium isotopic compositions of the upper mantle provide no sign of perovskite fractionation in a magma ocean (Ringwood, 1990; Halliday et al., 1995) and even the heavy REE pattern of the upper mantle appears to be essentially flat (e.g., Lee et al., 1996). This is not to be expected if majorite garnet was a liquidus phase (Herzberg et al., 1988). One explanation for the lack of such evidence is that the magma ocean was itself an efficiently mixed system with little if any crystal settling (Tonks and Melosh, 1990). Another possibility is that the entire mantle has been rehomogenized since this time. Ringwood (1990) suggested this possibility, which leaves some arguments regarding the relevance of the composition of the upper mantle in doubt. Of course, the subsequent introduction of heterogeneities by entrainment and radioactive decay and the development of an isotopically stratified mantle (Hofmann et al., 1986, Moreira and Allègre, 1998) are not inconsistent with this.

Whether or not magma oceans are a necessary prerequisite for core formation is unclear. It is necessary to understand how it is possible for metallic iron to migrate through the silicate mantle (Shaw, 1978). Many have assumed that a part of the mantle at least was solid during core formation. A variety of mechanisms have been studied including grain boundary percolation (Minarik et al., 1996; Rushmer et al., 2000; Yoshino et al., 2003) and the formation of large-scale metal structures in the upper mantle that sink like diapirs into the center of the Earth (Stevenson, 1981, 1990). Under some circumstances these may break up into small droplets of metal (Rubie et al., 2003). To evaluate these models, it is essential to have some idea of the physical state of the early Earth at the time of core formation. All of these issues are addressed by modern geochemistry but are not yet well constrained. Most effort has been focused on using the composition of the silicate Earth itself to provide constraints on models of core formation.

The major problem presented by the Earth's chemical composition and core formation models is providing mechanisms that predict correctly the siderophile element abundances in the Earth's upper mantle. It long has been recognized that siderophile elements are more abundant in the mantle than expected if the silicate Earth and the core were segregated under low-pressure and moderate-temperature equilibrium conditions (Chou, 1978; Jagoutz et al., 1979). Several explanations for this siderophile "excess" have been proposed, including:

(i) partitioning into liquid metal alloy at high pressure (Ringwood, 1979);

(ii) equilibrium partitioning between sulfur-rich liquid metal and silicate (Brett, 1984);

(iii) inefficient core formation (Arculus and Delano, 1981; Jones and Drake, 1986);

(iv) heterogeneous accretion and late veneers (Eucken, 1944; Turekian and Clark, 1969; Clark et al., 1972; Smith, 1977, 1980; Wänke et al., 1984; Newsom, 1990);

(v) addition of material to the silicate Earth from the core of a Moon-forming impactor (Newsom and Taylor, 1986);

(vi) very high temperature equilibration (Murthy, 1991); and

(vii) high-temperature equilibrium partitioning in a magma ocean at the upper/lower mantle boundary (Li and Agee, 1996; Righter et al., 1997).

Although the abundances and partition coefficients of some of the elements used to test these models are not well established, sufficient knowledge exists to render all of them problematic. Model (vii) appears to work well for some moderately siderophile elements. Righter and Drake (1999) make the case that the fit of the siderophile element metal/silicate partion coefficient data is best achieved with a high water content (per cent level) in the mantle. This would have assisted the formation of a magma ocean and provided a ready source of volatiles in the Earth. Walter et al. (2000) reviewed the state of the art in this area. However, the number of elements with well-established high-pressure partition coefficients for testing this model is still extremely small.

To complicate chemical models further, there is some osmium isotopic evidence that a small flux of highly siderophile elements from the core could be affecting the abundances in the mantle (Walker et al., 1995; Brandon et al., 1998). This model has been extended to the interpretation of PGE abundances in abyssal peridotites (Snow and Schmidt, 1998). The inventories of many of these highly siderophile elements are not that well established and may be extremely variable (Rehkämper et al., 1997, 1999b). In particular, the use of abyssal peridotites to assess siderophile element abundances in the upper mantle appears to be problematic (Rehkämper et al., 1999a). Puchtel and Humayun (2000) argued that if PGEs are being fluxed from the core to the mantle then this is not via the Walker et al. (1995) mechanism of physical admixture by entrainment of outer core, but must proceed via an osmium isotopic exchange, since the excess siderophiles were not found in komatiite source regions with radiogenic ^{187}Os. Taken together, the status of core to mantle fluxes is very vague at the present time.

1.20.8 THE FORMATION OF THE MOON

The origin of the Moon has been the subject of intense scientific interest for over a century but particularly since the Apollo missions provided samples to study. The most widely accepted current theory is the giant impact theory but this idea has evolved from others and alternative hypotheses have been variously considered. Wood (1984) provides a very useful review. The main theories that have been considered are as follows:

Co-accretion. This theory proposes that the Earth and Moon simply accreted side by side. The difficulty with this model is that it does not explain the angular momentum of the Earth–Moon system, nor the difference in density, nor the difference in volatile depletion (Taylor, 1992).

Capture. This theory (Urey, 1966) proposes that the Moon was a body captured into Earth's orbit. It is dynamically difficult to do this without the Moon spiraling into the Earth and colliding. Also the Earth and Moon have indistinguishable oxygen isotope compositions (Wiechert et al., 2001) in a solar system that appears to be highly heterogeneous in this respect (Clayton, 1986).

Fission. This theory proposes that the Moon split off as a blob during rapid rotation of a molten Earth. George Howard Darwin, the son of Charles Darwin originally championed this idea (Darwin, 1878, 1879). At one time (before the young age of the oceanfloor was known) it was thought by some that the Pacific Ocean might have been the residual space vacated by the loss of material. This theory is also dynamically difficult. Detailed discussions of the mechanisms can be found in Binder (1986). However, this model does have certain features that are attractive. It explains why Earth and Moon have identical oxygen isotope compositions. It explains why the Moon has a lower density because the outer part of the Earth would be deficient in iron due to core formation. It explains why so much of the angular momentum of the Earth–Moon system is in the Moon's motion. These are key features of any successful explanation for the origin of the Moon.

Impact models. Mainly because of the difficulties with the above models, alternatives were considered following the Apollo missions. Hartmann and Davis (1975) made the proposal that the Moon formed as a result of major impacts that propelled sufficient debris into orbit to produce the Moon. However, an important new facet that came from sample return was the discovery that the Moon had an anorthositic crust implying a very hot magma ocean. Also it was necessary to link the dynamics of the Moon with that of the Earth's spin. If an impact produced the Moon it would be easier to explain these features if it was highly energetic. This led to a series of single giant impact models in which the Moon was the product of a glancing blow collision with another differentiated planet (Cameron and Benz, 1991). A ring of debris would have been produced from the outer silicate portions of the Earth and the impactor planet (named "Theia," the mother of "Selene," the goddess of the Moon). Wetherill (1986) calculated that there was a realistic chance of such a collision. This model explains the angular momentum, the "fiery start," the isotopic similarities and the density difference.

The giant-impact theory has been confirmed by a number of important observations. Perhaps most importantly, we know now that the Moon must have formed tens of millions of years after the start of the solar system (Lee et al., 1997; Halliday, 2000). This is consistent with a collision between already formed planets. The masses of the Earth and the impactor at the time of the giant impact have been the subject of major uncertainty. Two main classes of models are usually considered. In the first, the Earth was largely (90%) formed at the time of the impact and the impacting planet Theia was roughly Mars-sized (Cameron and Benz, 1991). A recent class of models considers the Earth to be only half-formed at the time of the impact, and the mass ratio Theia/proto-Earth to be 3:7 (Cameron, 2000). The latter model is no longer considered likely; the most recent simulations have reverted to a Mars-sized impactor at the end of Earth accretion (Canup and Asphaug, 2001). The tungsten isotope data for the Earth and Moon do not provide a unique test (Halliday et al., 2000).

The giant-impact model, though widely accepted, has not been without its critics. Geochemical arguments have been particularly important in this regard. The biggest concern has been the similarities between chemical and isotopic features of the Earth and Moon. Most of the dynamic simulations (Cameron and Benz, 1991; Cameron, 2000; Canup and Asphaug, 2001) predict that the material that forms the Moon is derived from Theia, rather than the Earth. Yet it became very clear at an early stage of study that samples from the Moon and Earth shared many common features that would be most readily explained if the Moon was formed from material derived from the Earth (Wänke et al., 1983; Wänke and Dreibus, 1986; Ringwood, 1989b, 1992). These include the striking similarity in tungsten depletion despite a strong sensitivity to the oxidation state of the mantle (Rammensee and Wänke, 1977; Schmitt et al., 1989). Other basaltic objects such as eucrites and martian meteorites exhibit very different siderophile element depletion (e.g., Treiman et al., 1986, 1987; Wänke and Dreibus, 1988, 1994). Therefore, why should the Earth and Moon be identical if the Moon came from Theia (Ringwood, 1989b, 1992)? In a similar manner, the striking similarity in oxygen isotopic composition (Clayton and Mayeda, 1975), still unresolvable to extremely high precision (Wiechert et al., 2001), despite enormous heterogeneity in the solar system (Clayton et al., 1973; Clayton, 1986, 1993; Clayton and Mayeda, 1996), provides support for the view that the Moon was derived from the Earth (Figure 3).

One can turn these arguments around, however, and use the compositions of lunar samples to define the composition of Theia, assuming the impactor produced most of the material in the Moon (MacFarlane, 1989). Accordingly, the similarity in oxygen isotopes and trace siderophile abundances between the Earth and Moon provides evidence that Earth and Theia were neighboring planets made of an identical mix of materials with similar differentiation histories (Halliday and Porcelli, 2001). Their similarities could relate to proximity in the early solar system, increasing the probability of collision.

Certain features of the Moon may be a consequence of the giant impact itself. The volatile-depleted composition of the Moon, in particular, has been explained as a consequence of the giant impact (O'Neill, 1991a; Jones and Palme, 2000). It has been argued (Kreutzberger et al., 1986; Jones and Drake, 1993) that the Moon could not have formed as a volatile depleted residue of material from the Earth because it has Rb/Cs that is lower than that of the Earth and caesium supposedly is more volatile. However, the assumptions regarding the Earth's Rb/Cs upon which this is based are rather weak (McDonough et al., 1992). Furthermore, the exact relative volatilities of the alkalis are poorly known. Using the canonical numbers, the Earth, Moon, and Mars are all more depleted in less volatile rubidium than more volatile potassium (Figure 2). From the slope of the correlation, it can be seen that the terrestrial depletion in rubidium (50% condensation temperature ~1,080 K) is ~80% greater than that in potassium (50% condensation temperature ~1,000 K) (Wasson, 1985). Similar problems are found if one compares sodium depletion, or alkali concentrations more generally for other early objects, including chondrites.

Attempts to date the Moon were initially focused on determining the ages of the oldest rocks and therefore providing a lower limit. These studies emphasized precise strontium, neodymium, and lead isotopic constraints (Tera et al., 1973; Wasserburg et al., 1977a; Hanan and Tilton, 1987; Carlson and Lugmair, 1988; Shih et al., 1993; Alibert et al., 1994). At the end of the Apollo era, Wasserburg et al. (1977a) wrote "The actual time of aggregation of the Moon is not precisely known, but the Moon existed as a planetary body at 4.45 Ga, based on mutually consistent Rb–Sr and U–Pb data. This is remarkably close to the ^{207}Pb–^{206}Pb age of the Earth and suggests that the Moon and the Earth were formed or differentiated at the same time." Although these collective efforts made a monumental contribution, such constraints on the age of the Moon still leave considerable scope (>100 Myr) for an exact age.

Some of the most precise and reliable early ages for lunar rocks are given in Table 3. They provide considerable support for an age of >4.42 Ga. Probably the most compelling evidence comes from the early ferroan anorthosite 60025, which defines a relatively low first-stage μ (or ^{238}U/^{204}Pb) and an age of ~4.5 Ga. Of course, the ages of the oldest lunar rocks only date igneous events. Carlson and Lugmair (1988) reviewed all of the most precise and concordant data and concluded that the Moon had to have formed in the time interval 4.44–4.51 Ga. This is consistent with the estimate of 4.47 ± 0.02 Ga of Tera et al. (1973).

Model ages can provide upper and lower limits on the age of the Moon. Halliday and Porcelli (2001) reviewed the strontium isotope data for early solar system objects and showed that the initial strontium isotopic compositions of early lunar highlands samples (Papanastassiou and Wasserburg, 1976; Carlson and Lugmair, 1988) are all slightly higher than the best estimates of the solar system initial ratio (Figure 8). The conservative estimates of the strontium isotope data indicate that the difference between the ^{87}Sr/^{86}Sr of the bulk solar system initial at 4.566 Ga is 0.69891 ± 2 and the Moon at ~4.515 Ga is 0.69906 ± 2 is fully resolvable. An Rb–Sr model age for the Moon can be calculated by assuming that objects formed from material that separated from a solar nebula reservoir with the Moon's current Rb/Sr ratio. Because the Rb/Sr ratios of the lunar samples are extremely low, the uncertainty in formation age does not affect the calculated initial strontium isotopic composition, hence the model age, significantly. The CI chondritic Rb/Sr ratio (^{87}Rb/^{86}Sr = 0.92) is assumed to represent the solar nebula. This model provides an upper limit on the formation age of the object, because the solar nebula is thought to represent the most extreme Rb/Sr reservoir in which the increase in strontium isotopic composition could have been accomplished. In reality, the strontium isotopic composition probably evolved in a more complex manner over a longer time. The calculated time required to generate the difference in strontium isotopic composition in a primitive solar nebula environment is 11 ± 3 Myr. This is, therefore, the earliest point in time at which the Moon could have formed (Halliday and Porcelli, 2001).

A similar model-age approach can be used with the Hf–W system. In fact, Hf–W data provide the most powerful current constraints on the exact age of the Moon. The tungsten isotopic compositions of bulk rock lunar samples range from ε_W ~ 0 like the silicate Earth to ε_W > 10 (Lee et al., 1997, 2002). This was originally interpreted as the result of radioactive decay of formerly live ^{182}Hf within the Moon, which has a variable but generally high Hf/W ratio in its mantle (Lee et al., 1997). Now we know that a major portion of the ^{182}W excess in lunar samples is cosmogenic and the result of the reaction ^{181}Ta(n,γ)^{182}Ta(β$^-$)^{182}W while these rocks were exposed on the surface of the Moon (Leya et al., 2000; Lee et al., 2002). This can be corrected using (i) estimates of the cosmic ray flux from samarium and gadolinium compositions, (ii) the exposure age and Ta/W ratio, or (iii) internal isochrons of tungsten isotopic composition against Ta/W (Lee et al., 2002). The best current estimates for the corrected compositions are shown in Figure 9. The spread in the data is reduced and the stated uncertainties are greater relative to the raw tungsten isotopic compositions (Lee et al., 1997). Most data are within error of the Earth. A small excess ^{182}W is still resolvable for

Table 3 Recent estimates of the ages of early solar system objects and the age of the Moon.

Object	Sample(s)	Method	References	Age (Ga)
Earliest solar system	Allende CAIs	U–Pb	Göpel et al. (1991)	4.566 ± 0.002
Earliest solar system	Efremovka CAIs	U–Pb	Amelin et al. (2002)	4.5672 ± 0.0006
Chondrule formation	Acfer chondrules	U–Pb	Amelin et al. (2002)	4.5647 ± 0.0006
Angrites	Angra dos Reis and LEW 86010	U–Pb	Lugmair and Galer (1992)	4.5578 ± 0.0005
Early eucrites	Chervony Kut	Mn–Cr	Lugmair and Shukolyukov (1998)	4.563 ± 0.001
Earth accretion	Mean age	U–Pb	Halliday (2000)	≤4.55
Earth accretion	Mean age	U–Pb	Halliday (2000)	≥4.49
Earth accretion	Mean age	Hf–W	Yin et al. (2002)	≥4.55
Lunar highlands	Ferroan anorthosite 60025	U–Pb	Hanan and Tilton (1987)	4.50 ± 0.01
Lunar highlands	Ferroan anorthosite 60025	Sm–Nd	Carlson and Lugmair (1988)	4.44 ± 0.02
Lunar highlands	Norite from breccia 15445	Sm–Nd	Shih et al. (1993)	4.46 ± 0.07
Lunar highlands	Ferroan noritic anorthosite in breccia 67016	Sm–Nd	Alibert et al. (1994)	4.56 ± 0.07
Moon	Best estimate of age	U–Pb	Tera et al. (1973)	4.47 ± 0.02
Moon	Best estimate of age	U–Pb, Sm–Nd	Carlson and Lugmair (1988)	4.44–4.51
Moon	Best estimate of age	Hf–W	Halliday et al. (1996)	4.47 ± 0.04
Moon	Best estimate of age	Hf–W	Lee et al. (1997)	4.51 ± 0.01
Moon	Maximum age	Hf–W	Halliday (2000)	≤4.52
Moon	Maximum age	Rb–Sr	Halliday and Porcelli (2001)	≤4.55
Moon	Best estimate of age	Hf–W	Lee et al. (2002)	4.51 ± 0.01
Moon	Best estimate of age	Hf–W	Kleine et al. (2002)	4.54 ± 0.01

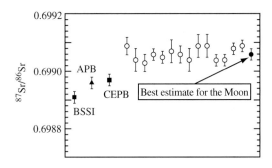

Figure 8 Initial strontium isotope composition of early lunar highland rocks relative to other early solar system objects. APB: Angrite Parent Body; CEPB: Cumulate Eucrite Parent Body; BSSI: Bulk Solar System Initial (source Halliday and Porcelli, 2001).

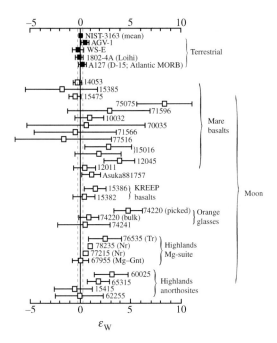

Figure 9 The tungsten isotopic compositions of lunar samples after calculated corrections for cosmogenic contributions (source Leya et al., 2000).

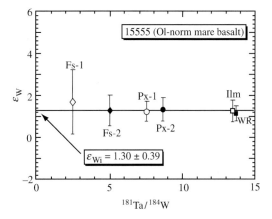

Figure 10 The tungsten isotopic composition of Apollo 15 basalt 15555 shows no internal variation as a function of Ta/W, consistent with its low exposure age (source Lee et al., 2002).

some samples, but these should be treated with caution.

The most obvious and clear implication of these data is that the Moon, a high-Hf/W object, must have formed late (Lee et al., 1997). Halliday (2000) argued that the tungsten isotopic composition was hard to explain if the Moon formed before ~50 Myr after the start of the solar system. The revised parameters for the average solar system (Kleine et al., 2002) mean that the Moon, like the Earth, has a well-defined excess of ^{182}W. This may have been inherited from the protolith silicate reservoirs from which the Moon formed. Alternatively, a portion might reflect ^{182}Hf decay within the Moon itself. Assuming that the Moon started as an isotopically homogeneous reservoir the most likely explanation for the small but well-defined excess ^{182}W found in Apollo basalts such as 15555 (Figure 10) relative to some of the other lunar rocks is that the Moon formed at a time when there was still a small amount of live ^{182}Hf in the lunar interior. This means that the Moon had to have formed within the first 60 Myr of the solar system (Halliday, 2000; Lee et al., 2002). The Moon must also have formed by the time defined by the earliest lunar rocks. The earliest most precisely determined crystallization age of a lunar rock is that of 60025 which has a Pb–Pb age of close to 4.50 Ga (Hanan and Tilton, 1987). Therefore, the Moon appears to have formed before ~4.50 Ga.

Defining the age more precisely is proving difficult at this stage. First, more precise estimates of the Hf–W systematics of lunar rocks are needed. The amount of data for which the cosmogenically produced ^{182}W effects are well resolved is very limited (Lee et al., 2002) and analysis is time consuming and difficult. Second, the (^{182}Hf/^{180}Hf)$_{BSSI}$ is poorly defined, as described above. If one uses a value of 1.0×10^{-4} (Kleine et al., 2002; Schöenberg et al., 2002; Yin et al., 2002), the small tungsten isotopic effects of the Moon probably were produced ≥30 Myr after the start of the solar system (Kleine et al., 2002). If, however, the (^{182}Hf/^{180}Hf)$_{BSSI}$ is slightly higher, as discussed above, the model age would be closer to 40–45 Myr (Table 3). The uncertainty in the ^{182}Hf decay constant (~±22%) also limits more precise constraints.

Either way, these Hf–W data provide very strong support for the giant-impact theory of lunar origin (Cameron and Benz, 1991). It is hard to explain how the Moon could have formed at such a late stage unless it was the result of a planetary collision. The giant-impact theory predicts that the age of the Moon should postdate the origin of the solar system by some considerable amount of time, probably tens of millions of years if Wetherill's predictions are correct. This is consistent with the evidence from tungsten isotopes.

The giant impact can also be integrated into modeling of the lead isotopic composition of the Earth (Halliday, 2000). Doing so, one can constrain the timing. Assuming a Mars-sized impactor was added to the Earth in the final stages of accretion (Canup and Asphaug, 2001) and that, prior to this, Earth's accretion could be approximated by exponentially decreasing rates (Halliday, 2000), one can calibrate the predicted lead isotopic composition of the Earth in terms of the time of the giant impact (Figure 11). It can be seen that all of the many estimates for the lead isotopic composition of the BSE compiled by Galer and Goldstein (1996) plus the more recent estimates of Kramers and Tolstikhin (1997) and Murphy et al. (2003) appear inconsistent with a giant impact that is earlier than ~45 Myr after the start of the solar system.

Xenon isotope data have also been used to argue specifically that the Earth lost its inventory of noble gases as a consequence of the giant impact (Pepin, 1997; Porcelli and Pepin, 2000). The timing of the "xenon loss event" looks more like 50–80 Myr on the basis of the most recent estimates of fissionogenic components (Porcelli and Pepin, 2000) (Figure 12). This agrees nicely with some estimates for the timing of the giant impact (Tables 2 and 3, and Figure 11) based on tungsten and lead isotopes. If the value of ~30 Myr is correct (Kleine et al., 2002), there appears to be a problem linking the xenon loss event with the Moon forming giant impact, yet it is hard to decouple these. If the tungsten chronometry recently proposed by Yin et al. (2002) and Kleine et al. (2002) is correct it would seem that the giant impact cannot have been the last big event that blew off a substantial fraction of the Earth's atmosphere. On the basis of dynamic simulations, however, it is thought that subsequent events cannot have been anything like as severe as the giant impact (Canup and Agnor, 2000; Canup and Asphaug, 2001). It is, of course, conceivable that the protoatmosphere was lost by a different mechanism such as strong UV radiation (Zahnle and Walker, 1982). However, this begs the question of how the (earlier) giant impact could have still resulted in retention of noble gases. It certainly is hard to imagine the giant impact without loss of the Earth's primordial atmosphere (Melosh and Vickery, 1989; Ahrens, 1990; Benz and Cameron, 1990; Zahnle, 1993).

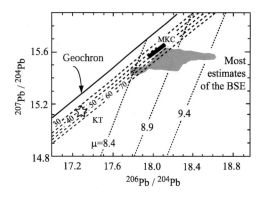

Figure 11 Lead isotopic modeling of the composition of the silicate Earth using continuous core formation and a sudden giant impact when the Earth is 90% formed. The impactor adds a further 9% to the mass of the Earth. The principles behind the modeling are as in Halliday (2000). See text for explanation. The field for the BSE encompasses all of the estimates in Galer and Goldstein (1996). The values suggested by Kramers and Tolstikhin (1997) and Murphy et al. (2002) are also shown. The figure is calibrated with the time of the giant impact (Myr). The μ values are the $^{238}U/^{204}Pb$ of the BSE. It is assumed that the μ of the total Earth is 0.7 (Allègre et al., 1995a). It can be seen that the lead isotopic composition of the BSE is hard to reconcile with formation of the Moon before ~45 Myr after the start of the solar system.

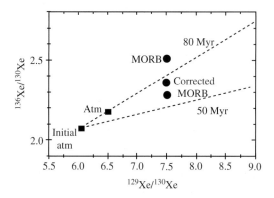

Figure 12 The relationship between the amounts of radiogenic ^{129}Xe inferred to come from ^{129}I decay within the Earth and fissionogenic ^{136}Xe thought to be dominated by decay of ^{244}Pu within the Earth. The differences in composition between the atmosphere and upper mantle relate to the timing of atmosphere formation. The compositions of both reservoirs are not very different from solar or initial U–Xe. This provides evidence that the strong depletion of xenon, leading to very high I/Xe, for example, was late. The data are shown modeled as a major loss event 50–80 Myr after the start of the solar system (Porcelli and Ballentine, 2002). The exact correction for uranium-derived fission ^{136}Xe in the MORB-mantle is unclear. Two values are shown from Phinney et al. (1978) and Kunz et al. (1998).

1.20.9 MASS LOSS AND COMPOSITIONAL CHANGES DURING ACCRETION

Collisions like the giant impact may well represent an important growth mechanism for terrestrial planets in general (Cameron and Benz, 1991; Canup and Agnor, 2000). These collisions are extraordinarily energetic and the question arises as to whether it is to be expected that accretion itself will lead to losses ("erosion") of material from the combined planetary masses. If this is the case it is only to be expected that the composition of the Earth does not add up to what one might expect from "chondrite building blocks" and, for example, the oxygen isotope composition (Clayton and Mayeda, 1996). There are several indications that the Earth may have lost a significant fraction of some elements during accretion and earliest development; all are circumstantial lines of evidence:

(i) As discussed above it has been argued that impact processes in particular are responsible for eroding protoatmospheres (Melosh and Vickery, 1989; Ahrens, 1990; Benz and Cameron, 1990; Zahnle, 1993). If these atmospheres were very dense and hot they may have contained a significant fraction of Earth's moderately volatile elements.

(ii) "Glancing blow" collisions between already differentiated planets, as during the giant impact, might be expected to preferentially remove major portions of the outer silicate portions of the planet as it grows. The Fe/Mg ratio of a planet is the simplest chemical parameter relating to planetary density and indicates the approximate size of the core relative to the silicate mantle. Mercury, with its high density, is a prime candidate for a body that lost a great deal of its outer silicate material by giant impacts (Benz et al., 1987). Therefore, by analogy the proportional size of the Earth's core may have increased as a consequence of such impact erosion. Conversely, Mars (Halliday et al., 2001) with a density lower than that of the Earth, may actually be a closer approximation of the material from which Earth accreted than Earth itself is (Halliday et al., 2001; Halliday and Porcelli, 2001).

(iii) Such late collisional loss of Earth's silicate is clearly evident from the low density of the Moon. The disk of material from which the Moon accreted during the giant impact was silicate-rich. Most simulations predict that Theia provided the major source of lunar material. The density difference between silicate and metal leads to a loss of silicate from the combined Theia–proto-Earth system, even when the loss primarily is from the impactor. Note, however, that the giant-impact simulations retain most of the mass overall (Benz and Cameron, 1990). Very little is lost to space from a body as large as the Earth.

(iv) If there was early basaltic crust on the Earth (Chase and Patchett, 1988; Galer and Goldstein, 1991) or Theia or other impacting planets, repeated impact erosion could have had an effect on the Si/Mg ratio of the primitive mantle. However, the maximum effect will be very small because silicon is a major element in the mantle. Earth's Si/Mg ratio is indeed low, as discussed above, but this may instead represent other loss processes. Nevertheless, the erosion effects could have been accentuated by the fact that silicon is relatively volatile and there probably was a magma ocean. With extremely high temperatures and a heavy protoatmosphere (Ahrens, 1990; Sasaki, 1990; Abe et al., 2000), it seems possible that one could form a "rock atmosphere" by boiling the surface of the magma ocean. This atmosphere would in turn be very vulnerable to impact-induced blow-off.

(v) The budgets of other elements that are more heavily concentrated in the outer portions of the Earth, such as the highly incompatible lithophile elements caesium, barium, rubidium, thorium, uranium, niobium, and potassium and the light rare earths, were possibly also depleted by impact erosion. Indeed some people have argued that the primitive mantle is slightly higher than chondritic in Sm/Nd (Nägler and Kramers, 1998). However, it is worth noting that there is no evidence for europium anomalies in the BSE as might be expected from impact loss of feldspathic flotation cumulates. These issues are further complicated by the fact that the early Earth (or impacting planets) may have had magma oceans with liquidus phases such as majorite or calcium- or magnesium-perovskite (Kato et al., 1988a,b). This in turn could have led to a very variable element distribution in a stratified magma ocean. There is no compelling evidence in hafnium isotopes, yet, for loss of a portion of the partially molten mantle fractionated in the presence of perovskite (Ringwood, 1990).

(vi) The oxygen isotopic compositions of lunar samples have been measured repeatedly to extremely high precision (Wiechert et al., 2001) using new laser fluorination techniques. The $\Delta^{17}O$ of the Moon relative to the terrestrial fractionation line can be shown to be zero (Figure 3). Based on a 99.7% confidence interval (triple standard error of the mean) the lunar fractionation line is, within ±0.005‰, identical to that of the Earth. There now is no doubt that the mix of material that accreted to form the Earth and the Moon was effectively identical in its provenance. Yet there is a big difference in the budgets of moderately volatile elements (e.g., K/U or Rb/Sr). One explanation is that there were major losses of moderately volatile elements during the giant impact (O'Neill, 1991a,b; Halliday and Porcelli, 2001).

(vii) The strontium isotopic compositions of early lunar highland rocks provide powerful support for late losses of alkalis (Figure 8). Theia had a time integrated Rb/Sr that was more than an order of magnitude higher than the actual Rb/Sr of the Moon, providing evidence that the processes of accretion resulted in substantial loss of alkalis (Halliday and Porcelli, 2001). Some of the calculated time-integrated compositions of the precursors to the present Earth and Moon are shown in Figure 6.

(viii) The abundances of volatile highly siderophile elements in the Earth are slightly depleted relative to refractory highly siderophile elements (Yi et al., 2000). This appears to reflect the composition of a late veneer. If this is representative of the composition of material that accreted to form the Earth as a whole it implies that there were substantial losses of volatile elements from the protoplanets that built the Earth (Yi et al., 2000).

Therefore, there exist several of lines of evidence to support the view that impact erosion may have had a significant effect on Earth's composition. However, in most cases the evidence is suggestive rather than strongly compelling. Furthermore, we have a very poor idea of how this is possible without fractionating potassium isotopes (Humayun and Clayton, 1995), unless the entire inventory of potassium is vaporized (O'Neill, 1991a,b; Halliday et al., 1996; Halliday and Porcelli, 2001). We also do not understand how to lose heavy elements except via hydrodynamic escape of a large protoatmosphere (Hunten et al., 1987; Walker, 1986). Some of the loss may have been from the proto-planets that built the Earth.

1.20.10 EVIDENCE FOR LATE ACCRETION, CORE FORMATION, AND CHANGES IN VOLATILES AFTER THE GIANT IMPACT

There are a number of lines of evidence that the Earth may have been affected by additions of further material subsequent to the giant impact. Similarly, there is limited evidence that there was additional core formation. Alternatively, there also are geochemical and dynamic constraints that strongly limit the amount of core formation and accretion since the giant impact. This is a very interesting area of research that is ripe for further development. Here are some of the key observations:

(i) It has long been recognized that there is an apparent excess of highly siderophile elements in the silicate Earth (Chou, 1978; Jagoutz et al., 1979). These excess siderophiles already have been discussed above in the context of core formation. However, until Murthy's (1991) paper, the most widely accepted explanation was that there was a "late veneer" of material accreted after core formation and corresponding to the final one percent or less of the Earth's mass. Nowadays the effects of high temperatures and pressures on partitioning can be investigated and it seems clear that some of the excess siderophile signature reflects silicate-metal equilibration at depth. The volatile highly siderophile elements carbon, sulfur, selenium, and tellurium are more depleted in the silicate Earth than the refractory siderophiles (Figure 1; Yi et al., 2000). Therefore, if there was a late veneer it probably was material that was on average depleted in volatiles, but not as depleted as would be deduced from the lithophile volatile elements (Yi et al., 2000).

(ii) The light xenon isotopic compositions of well gases may be slightly different from those of the atmosphere in a manner that cannot be easily related to mass-dependent fractionation (Caffee et al., 1988). These isotopes are not affected by radiogenic or fissionogenic additions. The effect is small and currently is one of the most important measurements that need to be made at higher precision. Any resolvable differences need to then be found in other reservoirs (e.g., MORBs) in order to establish that this is a fundamental difference indicating that a fraction of atmospheric xenon is not acquired by outgassing from the interior of the Earth. It could be that the Earth's atmospheric xenon was simply added later and that the isotopic compositions have no genetic link with those found in the Earth's interior.

(iii) Support for this possibility has come from the commonly held view that the Earth's water was added after the giant impact. Having lost so much of its volatiles by early degassing, hydrodynamic escape and impacts the Earth still has a substantial amount of water. Some have proposed that comets may have added a component of water to the Earth, but the D/H ratio would appear to be incorrect for this unless the component represented a minor fraction (Owen and Bar-Nun, 1995, 2000). An alternative set of proposals has been built around volatile-rich chondritic planetary embryos (Morbidelli et al., 2000).

(iv) Further support for this latter "asteroidal" solution comes from the conclusion that the asteroid belt was at one time relatively massive. More than 99% of the mass of the asteroid belt has been ejected or added to other objects (Chambers and Wetherill, 2001). The Earth, being en route as some of the material preferentially travels toward the Sun may well have picked up a fraction of its volatiles in this way.

(v) The Moon itself provides a useful monitor of the amount of late material that could have been added to the Earth (Ryder et al., 2000). The Moon provides an impact history (Hartmann et al., 2000) that can be scaled to the Earth (Ryder et al., 2000). In particular, there is evidence of widespread and intense bombardment of the Moon during the

Hadean and this can be scaled up, largely in terms of relative cross-sectional area, to yield an impact curve for the Earth (Sleep et al., 1989).

(vi) However, the Moon also is highly depleted in volatiles and its surface is very depleted in highly siderophile elements. Therefore, the Moon also provides a limit on how much can be added to the Earth. This is one reason why the more recent models of Cameron (2000) involving a giant impact that left an additional third are hard to accommodate. However, the database for this currently is poor (Righter et al., 2000).

(vii) It has been argued that the greater depletion in iron and in tellurium in the silicate Earth relative to the Moon reflects an additional small amount of terrestrial core formation following the giant impact (Halliday et al., 1996; Yi et al., 2000). It could also simply reflect differences between Theia and the Earth. If there was further post-giant-impact core formation on Earth, it must have occurred prior to the addition of the late veneer.

1.20.11 THE HADEAN

1.20.11.1 Early Mantle Depletion

1.20.11.1.1 Introduction

Just as the I–Pu–Xe system is useful for studying the rate of formation of the atmosphere and U–Pb and Hf–W are ideal for studying the rates of accretion and core formation, lithophile element isotopic systems are useful for studying the history of melting of the silicate Earth. Two in particular, ^{92}Nb ($T_{1/2} = 36$ Myr) and ^{146}Sm ($T_{1/2} = 106$ Myr), have sufficiently long half-lives to be viable but have been explored with only limited success. Another chronometer of use is the long-lived chronometer ^{176}Lu ($T_{1/2} = 34$–38 Gyr).

1.20.11.1.2 ^{92}Nb–^{92}Zr

^{92}Nb decays by electron capture to ^{92}Zr with a half-life of 36 ± 3 Myr. At one time it was thought to offer the potential to obtain an age for the Moon by dating early lunar ilmenites and the formation of ilmenite-rich layers in the lunar mantle. Others proposed that it provided constraints on the timescales for the earliest formation of continents on Earth (Münker et al., 2000). In addition, it was argued that it would date terrestrial core formation (Jacobsen and Yin, 2001). There have been many attempts to utilize this isotopic system over the past few years. To do so, it is necessary to first determine the initial ^{92}Nb abundance in early solar system objects accurately and various authors have made claims that differ by two orders of magnitude.

Harper et al. (1991b) analyzed a single niobium-rutile found in the Toluca iron meteorite and presented the first evidence for the former existence of ^{92}Nb from which an initial ^{92}Nb/^{93}Nb of $(1.6 \pm 0.3) \times 10^{-5}$ was inferred. However, the blank correction was very large. Subsequently, three studies using multiple collector inductively coupled plasma mass spectrometry proposed that the initial ^{92}Nb/^{93}Nb ratio of the solar system was more than two orders of magnitude higher (Münker et al., 2000; Sanloup et al., 2000; Yin et al., 2000). Early processes that should fractionate Nb/Zr include silicate partial melting because niobium is more incompatible than zirconium (Hofmann et al., 1986). Other processes relate to formation of titanium-rich (hence niobium-rich) and zirconium-rich minerals, the production of continental crust, terrestrial core formation (Wade and Wood, 2001) and the differentiation of the Moon. Therefore, on the basis of the very high ^{92}Nb abundance proposed, it was argued that, because there was no difference between the zirconium isotopic compositions of early terrestrial zircons and chondrites, the Earth's crust must have formed relatively late (Münker et al., 2000). Similarly, because it is likely that a considerable amount of the Earth's niobium went into the core it was argued that core formation must have been protracted or delayed (Jacobsen and Yin, 2001). We now know that these arguments are incorrect. Precise internal isochrons have provided evidence that the initial abundance of ^{92}Nb in the early solar system is indeed low and close to 10^{-5} (Schönbächler et al., 2002).

Therefore, rather than proving useful, the ^{92}Nb–^{92}Zr has no prospect of being able to provide constraints on these issues because the initial ^{92}Nb abundance is too low.

1.20.11.1.3 ^{146}Sm–^{142}Nd

High-quality terrestrial data now have been generated for the ^{146}Sm–^{142}Nd (half-life = 106 Myr) chronometer (Goldstein and Galer, 1992; Harper and Jacobsen, 1992; McCulloch and Bennett, 1993; Sharma et al., 1996). Differences in ^{142}Nd/^{144}Nd in early Archean rocks would indicate that the development of a crust on Earth was an early process and that subsequent recycling had failed to eradicate these effects. For many years, only one sample provided a hint of such an effect (Harper and Jacobsen, 1992) although these data have been questioned (Sharma et al., 1996). Recently very high precision measurements of Isua sediments have resolved a 15 ± 4 ppm effect (Caro et al., 2003).

Any such anomalies are clearly small and far less than might be expected from extensive, repeated depletion of the mantle by partial melting in the Hadean. It seems inescapable that there was melting on the early Earth. Therefore, the interesting and important result of these studies is that such isotopic effects must largely have been eliminated. The most likely mechanism is very efficient mantle convection. In the earliest Earth convection may have been much more vigorous (Chase and Patchett, 1988; Galer and Goldstein, 1991) because of the large amount of heat left from the secular cooling and the greater radioactive heat production.

1.20.11.1.4 $^{176}Lu-^{176}Hf$

A similar view is obtained from hafnium isotopic analyses of very early zircons. The ^{176}Lu–^{176}Hf isotopic system ($T_{1/2} = 34$–38 Gyr) is ideally suited for studying early crustal evolution, because hafnium behaves in an almost identical fashion to zirconium. As a result, the highly resistant and easily dated mineral zircon typically contains ~1 wt% Hf, sufficient to render hafnium isotopic analyses of single zircons feasible using modern methods (Amelin et al., 1999). The concentration of lutetium in zircon is almost negligible by comparison. As a result, the initial hafnium isotopic composition is relatively insensitive to the exact age of the grain and there is no error magnification involved in extrapolating back to the early Archean. Furthermore, one can determine the age of the single zircon grain very precisely using modern U–Pb methods. One can obtain an extremely precise initial hafnium isotopic composition for a particular point in time on a single grain, thereby avoiding the problems of mixed populations. The U–Pb age and hafnium isotopic compositions of zircons also are extremely resistant to resetting and define a reliable composition at a well-defined time in the early Earth. The hafnium isotopic composition that zircon had when it grew depends on whether the magma formed from a reservoir with a time-integrated history of melt depletion or enrichment. Therefore, one can use these early zircons to search for traces of early mantle depletion.

Note that this is similar to the approach adopted earlier with ^{147}Sm–^{143}Nd upon which many ideas of Hadean mantle depletion, melting processes and early crust were based (Chase and Patchett, 1988; Galer and Goldstein, 1991). However, the difficulty with insuring closed-system behavior with bulk rock Sm–Nd in metamorphic rocks and achieving a robust age correction of long-lived ^{147}Sm over four billion years has meant that this approach is now viewed as suspect (Nägler and Kramers, 1998). The ^{176}Lu–^{176}Hf isotopic system and use of low-Lu/Hf zircons is far more reliable in this respect (Amelin et al., 1999, 2000). In practice, however, the interpretation is not that simple, for two reasons:

(i) The hafnium isotopic composition and Lu/Hf ratio of the Earth's primitive mantle is poorly known. It is assumed that it is broadly chondritic (Blichert-Toft and Albarède, 1997), but which exact kind of chondrite class best defines the isotopic composition of the primitive mantle is unclear. Without this information, one cannot extrapolate back in time to the early Earth and state with certainty what the composition of the primitive mantle reservoir was. Therefore, one cannot be sure what a certain isotopic composition means in terms of the level of time-integrated depletion.

(ii) The half-life of ^{176}Lu is *not* well established and is the subject of current debate and research (Scherer et al., 2001). Although the determination of the initial hafnium isotopic composition of zircon is not greatly affected by this, because the Lu/Hf ratio is so low that the age correction is tiny, the correction to the value for the primitive mantle is very sensitive to this uncertainty.

With these caveats, one can deduce the following. Early single grains appear to have recorded hafnium isotopic compositions that provide evidence for chondritic or enriched reservoirs. There is no evidence of depleted reservoirs in the earliest (Hadean) zircons dated thus far (Amelin et al., 1999). Use of alternative values for the decay constants or values for the primitive mantle parameters increases the proportion of hafnium with an enriched signature (Amelin et al., 2000), but does not provide evidence for early mantle depletion events. Therefore, there is little doubt that the Hadean mantle was extremely well mixed. Why this should be is unclear, but it probably relates in some way to the lack of preserved continental material from prior to 4.0 Ga.

1.20.11.2 Hadean Continents

Except for the small amount of evidence for early mantle melting we are in the dark about how and when Earth's continents first formed (Figure 13). We already have pointed out that in its early stages Earth may have had a magma ocean, sustained by heat from accretion and the blanketing effects of a dense early atmosphere. With the loss of the early atmosphere during planetary collisions, the Earth would have cooled quickly, the outer portions would have solidified and it would thereby have developed its first primitive crust.

We have little evidence of what such a crust might have looked like. Unlike on the Moon and Mars, Earth appears to have no rock preserved that is more than 4.0 Gyr old. There was intense bombardment of the Moon until ~3.9 Gyr ago (Wetherill, 1975b; Hartmann et al., 2000; Ryder et al., 2000). Earth's earlier crust may therefore have been decimated by concomitant impacts.

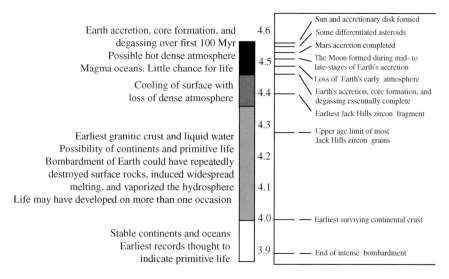

Figure 13 Schematic showing the timescales for various events through the "Dark Ages" of the Hadean.

It may also be that a hotter Earth had a surface that was inherently unstable. Some argued that the earliest crust was like the lunar highlands—made from a welded mush of crystals that had previously floated on the magma ocean. Others have suggested that it was made of denser rocks more like those of the Earth's present oceanfloor (Galer and Goldstein, 1991). But firm evidence has so far been sparse.

Froude et al. (1983) reported the exciting discovery of pre-4.0 Ga zircon grains formed on Earth. The host rock from which these grains were recovered is not so old. The pre-4.0 Gyr rocks were largely destroyed but the zircons survived by becoming incorporated in sands that formed a sedimentary rock that now is exposed in Australia as the Jack Hills Metaconglomerate. By measuring uranium-lead ages a sizeable population of grains between 4.1 Gyr and 4.2 Gyr was discovered (Froude et al., 1983). Subsequently, Wilde et al. (2001) and Mojzsis et al. (2001) reported uranium–lead ages and oxygen isotopic compositions of further old zircon grains. A portion of one grain appears to have formed 4.40 Gyr ago and this is the oldest terrestrial solid yet identified. More recent work has been published by Peck et al. (2001) and Valley et al. (2002).

These zircons provide powerful evidence for the former existence of some unknown amount of continental crust in the Hadean. Nearly all zircons grew from granite magmas, not similar at all to those forming the oceanfloor or the lunar highlands. Granite magmas usually form at >700 °C and >20 km depth, mainly by melting preexisting crust above subduction zones. Being buoyant they are typical of continental mountain regions such as the Andes—sites of very active erosion. The existence of such zircons would be consistent with continental crust as far back as 4.40 Gyr–~100 Myr after the formation of the Moon and the earliest atmosphere. It would be nice to have more data and one is extrapolating through many orders of magnitude in mass when inferring extensive continents from single zircons. Large ocean islands like Iceland have small volumes of granitic magma and so one can conceive of protocontinents that started from the accumulation of such basaltic nuclei, the overall mass of silicic crustal material increasing gradually.

Oxygen isotopic measurements can be used to infer the presence of liquid water on the Earth's early surface. The oxygen isotopic composition reflects that of the magma from which the zircon crystallized, which in turn reflects that of the rocks that were melted to form the magma. Heavy oxygen (with a high proportion of ^{18}O) is produced by low-temperature interactions between a rock and liquid water such as those that form clay by weathering. The somewhat heavy oxygen of these zircons provides evidence that the rocks that were melted to form the magma included components that had earlier been at the surface in the presence of liquid water. Early on, when Earth was hotter, they might also have formed by melting of wet oceanfloor basalt that was taken back into the mantle by a process potentially comparable to modern subduction. Either way, the data indicate that surface rocks affected by low-temperature fluids were probably being transported to significant depths and melted, as occurs today.

These grains represent a unique archive for information on the early Earth. The potential is considerable. For example, Wilde et al. (2001) and Peck et al. (2001) also used trace elements and tiny inclusions to reconstruct the composition of the parent magma. In all of this work there is a need to be very aware that the grains may have been disturbed after they formed. Wet diffusion of oxygen could lead to an ^{18}O-rich composition that was acquired during subsequent metamorphic or magmatic histories. Thus, ancillary information on diffusivities and the degree to which compositions might have been perturbed by later metamorphism must be acquired. Most importantly, however, there exists an urgent need of many more grains.

1.20.11.3 The Hadean Atmosphere, Hydrosphere, and Biosphere

The nature of the Earth's atmosphere and hydrosphere after the Earth had cooled, following the cessation of the main stages of accretion, has been the subject of a fair amount of research (Wiechert, 2002), particularly given the *lack* of data upon which to base firm conclusions. An interesting and important question is to what extent it may have been possible for life to develop during this period (Mojzsis et al., 1996; Sleep et al., 2001; Zahnle and Sleep, 1996). A great deal has been written on this and because it is covered elsewhere in this treatise only cursory background is provided here.

The first-order constraints that exist on the nature of Earth's early exosphere (Sleep et al., 1989; Sleep and Zahnle, 2001) are as follows:

(i) The Sun was fainter and cooler than today because of the natural start-up of fusion reactions that set it on the Main Sequence (Kasting and Grinspoon, 1991; Sagan and Chyba, 1997; Pavlov et al., 2000). Therefore, the level of radiation will have been less.

(ii) Earth's interior was a few 100 K hotter because of secular cooling from accretion and far greater radiogenic heat production. The Earth's heat flow was 2–3 times higher. Therefore, one can assume that more heat was escaping via mantle melting and production of oceanic crust.

(iii) This, in turn, means more mantle-derived volatiles such as CO_2 were being released.

(iv) It also means that there was more hydrothermal alteration of the oceanfloor. Therefore, CO_2 was converted to carbonate in altered basalt and returned to the mantle at subduction zones (if those really existed already).

(v) There may have been far less marine carbonate. We can infer this from the geological record for the Archean. Therefore, it appears that atmospheric CO_2 levels were low—most of the CO_2 was being recycled to the mantle.

(vi) Because atmospheric CO_2 exerts a profound effect on temperature as a greenhouse gas, low concentrations of CO_2 imply that atmospheric temperatures were cold, unless another greenhouse gas such as methane (CH_4) was very abundant (Pavlov et al., 2000). However, clear geochemical evidence for a strong role of methane in the Archean currently is lacking.

(vii) Impacts, depending on their number and magnitude (Hartmann et al., 2000; Ryder et al., 2000), may have had a devastating effect on the early biosphere. Impact ejecta will react with atmospheric and oceanic CO_2 and thereby lower atmospheric CO_2 levels reducing atmospheric temperatures still further.

The Hadean is fast becoming one of the most interesting areas of geochemical research. With so little hard evidence (Figure 13), much of the progress probably will come from theoretical modeling and comparative planetology.

1.20.12 CONCLUDING REMARKS—THE PROGNOSIS

Since the early 1990s, there has been great progress in understanding the origins and early development of the Earth. In some cases, this has been a function of improved modeling. This is particularly true for noble gases. However, in most cases it has been the acquisition of new kinds of data that has proven invaluable. The most obvious examples are in isotope geochemistry and cosmochemistry.

It is perhaps worth finishing by pointing out the kinds of developments that can be expected to have an impact on our understanding of the early Earth. One can ask a question like "what if we could measure...?" Here are some things that would be very interesting and useful to explore:

(i) If planets such as the Earth formed by very energetic collisions that were sufficient to cause vaporization of elements and compounds that normally are solid, it may be possible to find evidence for the kinetic effects of boiling in isotopic fractionations. Humayun and Clayton (1995) explored potassium at per mil levels of precision and found no significant evidence of fractionation. Poitrasson et al. (2003) have found evidence that the iron on the Moon may be very slightly heavy relative to other planetary objects. Note that this has nothing to do with the fractionations produced in the lunar surface during irradiation and implantation (Wiesli et al., 2003). Perhaps boiling during the giant impact caused this. This is just a preliminary inference at this stage. But if so, small fractionations also should be found in other elements of similar volatility such as lithium, magnesium, silicon, and nickel. There is much to be done to explore these effects at very high precision.

(ii) A vast amount of work still is needed on core formation, understanding the depletion of siderophile elements in the Earth's mantle (Walter et al., 2000), and determining the abundances of light elements in the core (Gessmann et al., 2001). Much of this depends on using proxy elements that are very sensitive to pressure, temperature, water content or f_{O_2} (e.g., Wade and Wood, 2001). A problem at present is that too many of the elements of interest are sensitive to more than one parameter and, therefore, the solutions are under constrained. Also, new experiments are needed at very high pressures—close to those of the core–mantle boundary.

(iii) The origins of Earth's water and volatiles more generally are the subject of considerable debate (e.g., Anders and Owen, 1977; Carr and Wänke, 1992; Owen and Bar-Nun, 1995, 2000; Javoy, 1998; Caffee et al., 1999; Righter and Drake, 1999; Abe et al., 2000; Morbidelli et al., 2000). A great deal probably will be learned from further modeling. For example, it is very important to understand what kinds of processes in the early Earth could have caused loss of early atmospheres. We do not understand how volatiles are retained during a Moon-forming giant impact (Melosh and Sonett, 1986; Melosh, 1990; Melosh et al., 1993) or what the early Sun might have done to the atmosphere. We need to acquire more reliable data on the isotopic compositions of volatiles in the deep Earth. We might be able to learn much more about Earth's volatile history from more precise measurements of the volatile components in other planets. In this respect more detailed studies of Mars and how closely it resembles the Earth (Carr and Wänke, 1992) could prove critical.

(iv) Mojzsis et al. (2001) and Wilde et al. (2001) have made major advances in studying the Hadean using single zircons. Apart from needing many more such samples, one has to ask what other kinds of information might be extractable from such zircons. The oxygen isotopic composition provides evidence for early low-temperature water. Exploring the melt inclusions and the trace-element concentrations also has been shown to have potential (Peck et al., 2001). Zircons are also iron-rich and conceivably could eventually be used to provide evidence for biological processes in the Hadean. However, the sensitivities of the techniques need to be improved vastly for this to be achieved with single grains. Furthermore, the current status of the rapidly expanding field of iron isotope geochemistry provides no clear basis for assuming a distinctive signal of biotic effects will be realizable. Also, the really interesting zircons are so precious that one should use minimally destructive techniques like SIMS on single grains. However, the required precision for measuring isotopic ratios in a useful manner is not available for trace elements in minerals using this method at present. Similarly, hafnium isotopes on single zircon grains provide the most reliable and powerful constraints on the extent of mantle depletion (Amelin et al., 1999) but require destruction of a part of the grain. Developing improved methods that achieve far higher overall sensitivity is critical.

(v) Determining the rates of accretion of the Earth and Moon more reliably will be critically dependent on the correct and precise determination of the initial tungsten and hafnium isotopic compositions of the solar system, and the ^{182}Hf decay constant. The initial isotopic compositions really require the more widespread application of negative ion thermal ionization mass spectrometry (N-TIMS) (Quitté et al., 2000). The decay constant work is going to require the acquisition of ^{182}Hf, probably from neutron-irradiated ^{180}Hf.

(vi) Similarly, progress in using hafnium isotopes to study the degree of early mantle depletion is being thwarted by the uncertainties associated with the ^{176}Lu decay constant (Scherer et al., 2001), and some new experimental work is needed in this area.

(vii) A major need is for a closer integration of the modeling of different isotopic systems. As of early 2000s, this has really only been attempted for tungsten and lead (Halliday, 2000). In the future it will be essential to integrate xenon isotopes in with these and other accretion models.

(viii) Finally, another major area of modeling has to occur in the area of "early Earth system science." There needs to be integrated modeling of the evolution of the atmosphere, oceans, surface temperature, mantle convection, and magma oceans. This is now being attempted. For example, the studies by Sleep and Zahnle (2001) and Sleep et al. (1989, 2001) are paving the way for more comprehensive models that might involve the fluid dynamics of mantle convection.

ACKNOWLEDGMENTS

This chapter benefited enormously from discussion with, and comments and criticism received from Tom Ahrens, Alan Boss, Pat Cassen, Andy Davis, Martin Frank, Tim Grove, Munir Humayun, Don Porcelli, Norm Sleep, Mike Walter, Uwe Wiechert, Rainer Wieler, Kevin Zahnle, and two anonymous reviewers.

REFERENCES

Abe Y. and Matsui T. (1985) The formation of an impact-generated H$_2$O atmosphere and its implications for the early thermal history of the Earth. *Proc. 15th Lunar Planet. Sci. Conf.: J. Geophys. Res.* **90**(suppl.), C545–C559.

Abe Y. and Matsui T. (1986) Early evolution of the earth: accretion, atmosphere formation, and thermal history. *Proc. 17th Lunar Planet. Sci. Conf.: J. Geophys. Res.* **91**(no. B13), E291–E302.

Abe Y. and Matsui T. (1988) Evolution of an impact-generated H$_2$O–CO$_2$ atmosphere and formation of a hot proto-ocean on Earth. *Am. Meteorol. Soc.* **45**, 3081–3101.

Abe Y., Ohtani E., and Okuchi T. (2000) Water in the early earth. In *Origin of the Earth and Moon* (eds. R. M. Canup and K. Righter). University of Arizona Press, Tucson, pp. 413–433.

Agee C. B. and Walker D. (1988) Mass balance and phase density constraints on early differentiation of chondritic mantle. *Earth Planet. Sci. Lett.* **90**, 144–156.

Agnor C. B. and Ward W. R. (2002) Damping of terrestrial-planet eccentricities by density-wave interactions with a remnant gas disk. *Astrophys. J.* **567**, 579–586.

Ahrens T. J. (1990) Earth accretion. In *Origin of the Earth* (eds. H. E. Newsom and J. H. Jones). Oxford University Press, Oxford, pp. 211–217.

Ahrens T. J. and Jeanloz R. (1987) Pyrite shock compression, isentropic release and composition of the Earth's core. *J. Geophys. Res.* **92**, 10363–10375.

Albarède F. and Juteau M. (1984) Unscrambling the lead model ages. *Geochim. Cosmochim. Acta* **48**, 207–212.

Alibert C., Norman M. D., and McCulloch M. T. (1994) An ancient Sm–Nd age for a ferroan noritic anorthosite clast from lunar breccia 67016. *Geochim. Cosmochim. Acta* **58**, 2921–2926.

Allègre C. J., Dupré B., and Brévart O. (1982) Chemical aspects of formation of the core. *Phil. Trans. Roy. Soc. London* **A306**, 49–59.

Allègre C. J., Staudacher T., Sarda P., and Kurz M. (1983) Constraints on evolution of Earth's mantle from rare gas systematics. *Nature* **303**, 762–766.

Allègre C. J., Staudacher T., and Sarda P. (1987) Rare gas systematics: formation of the atmosphere, evolution, and structure of the Earth's mantle. *Earth Planet. Sci. Lett.* **81**, 127–150.

Allègre C. J., Manhès G., and Göpel C. (1995a) The age of the Earth. *Geochim. Cosmochim. Acta* **59**, 1445–1456.

Allègre C. J., Poirier J.-P., Humler E., and Hofmann A. W. (1995b) The chemical composition of the Earth. *Earth Planet. Sci. Lett.* **134**, 515–526.

Allègre C. J., Hofmann A. W., and O'Nions R. K. (1996) The argon constraints on mantle structure. *Geophys. Res. Lett.* **23**, 3555–3557.

Allègre C. J., Manhès G., and Lewin E. (2001) Chemical composition of the Earth and the volatility control on planetary genetics. *Earth Planet. Sci. Lett.* **185**, 49–69.

Amelin Y., Lee D.-C., Halliday A. N., and Pidgeon R. T. (1999) Nature of the Earth's earliest crust from hafnium isotopes in single detrital zircons. *Nature* **399**, 252–255.

Amelin Y., Lee D.-C., and Halliday A. N. (2000) Early-middle Archean crustal evolution deduced from Lu–Hf and U–Pb isotopic studies of single zircon grains. *Geochim. Cosmochim. Acta* **64**, 4205–4225.

Amelin Y., Krot A. N., Hutcheon I. D., and Ulyanov A. A. (2002) Lead isotopic ages of chondrules and calcium-aluminum-rich inclusions. *Science* **297**, 1678–1683.

Anders E. (1977) Chemical compositions of the Moon, Earth, and eucrite parent body. *Phil. Trans. Roy. Soc. London* **A295**, 23–40.

Anders E. and Grevesse N. (1989) Abundances of the elements: meteoritic and solar. *Geochim. Cosmochim. Acta* **53**, 197–214.

Anders E. and Owen T. (1977) Mars and Earth: origin and abundance of volatiles. *Science* **198**, 453–465.

Anderson D. L. (1979) Chemical stratification of the mantle. *J. Geophys. Res.* **84**, 6297–6298.

Arculus R. J. and Delano J. W. (1981) Siderophile element abundances in the upper mantle: evidence for a sulfide signature and equilibrium with the core. *Geochim. Cosmochim. Acta* **45**, 1331–1344.

Artymowicz P. (1997) Beta Pictoris: an early solar system? *Ann. Rev. Earth Planet. Sci.* **25**, 175–219.

Azbel I. Y., Tolstikhin I. N., Kramers J. D., Pechernikova G. V., and Vityazev A. V. (1993) Core growth and siderophile element depletion of the mantle during homogeneous Earth accretion. *Geochim. Cosmochim. Acta* **57**, 2889–2898.

Ballentine C. J., Porcelli D., and Wieler R. (2001) A critical comment on Trieloff et al.. *Science* **291**, 2269 (online).

Bell K. R., Cassen P. M., Wasson J. T., and Woolum D. S. (2000) The FU Orionis phenomenon and solar nebular material. In *Protostars and Planets IV* (eds. V. Mannings, A. P. Boss, and S. S. Russell). University of Arizona Press, Tucson, pp. 897–926.

Benz W. (2000) Low velocity collisions and the growth of planetesimals. *Space Sci. Rev.* **92**, 279–294.

Benz W. and Asphaug E. (1999) Catastrophic disruptions revisited. *Icarus* **142**, 5–20.

Benz W. and Cameron A. G. W. (1990) Terrestrial effects of the giant impact. In *Origin of the Earth* (eds. H. E. Newsom and J. H. Jones). Oxford University Press, Oxford, pp. 61–67.

Benz W., Cameron A. G. W., and Slattery W. L. (1987) Collisional stripping of Mercury's mantle. *Icarus* **74**, 516–528.

Binder A. B. (1986) The binary fission origin of the Moon. In *Origin of the Moon* (eds. W. K. Hartmann, R. J. Phillips, and G. J. Taylor). Lunar and Planetary Institute, Houston, pp. 499–516.

Birck J.-L., Rotaru M., and Allègre C. J. (1999) ^{53}Mn–^{53}Cr evolution of the early solar system. *Geochim. Cosmochim. Acta* **63**, 4111–4117.

Blichert-Toft J. and Albarède F. (1997) The Lu–Hf isotope geochemistry of chondrites and the evolution of the mantle–crust system. *Earth Planet. Sci. Lett.* **148**, 243–258.

Blum J. (2000) Laboratory experiments on preplanetary dust aggregation. *Space Sci. Rev.* **92**, 265–278.

Boss A. P. (1990) 3D Solar nebula models: implications for Earth origin. In *Origin of the Earth* (eds. H. E. Newsom and J. H. Jones). Oxford University Press, Oxford, pp. 3–15.

Boss A. P. (1997) Giant planet formation by gravitational instability. *Science* **276**, 1836–1839.

Boss A. P. (1998) *Looking for Earths*. Wiley, New York, 240pp.

Boss A. P. (2002) The formation of gas and ice giant planets. *Earth Planet. Sci. Lett.* **202**, 513–523.

Brandon A. D., Walker R. J., Morgan J. W., Norman M. D., and Prichard H. M. (1998) Coupled ^{186}Os and ^{187}Os evidence for core–mantle interaction. *Science* **280**, 1570–1573.

Brett R. (1984) Chemical equilibrium between Earth's core and upper mantle. *Geochim. Cosmochim. Acta* **48**, 1183–1188.

Brown G. C. and Mussett A. E. (1981) *The Inaccessible Earth*. Allen and Unwin, London.

Brown H. (1949) Rare gases and the formation of the Earth's atmosphere. In *The Atmospheres of the Earth and Planets* (ed. G. P. Kuiper). University of Chicago Press, Chicago, pp. 258–266.

Busemann H., Baur H., and Wieler R. (2000) Primordial noble gases in "phase Q" in carbonaceous and ordinary chondrites studied by closed-system stepped etching. *Meteorit. Planet. Sci.* **35**, 949–973.

Caffee M. W., Hudson G. B., Velsko C., Alexander E. C., Jr., Huss G. R., and Chivas A. R. (1988) Non-atmospheric noble gases from CO_2 well gases. In *Lunar Planet. Sci.* **XIX**. The Lunar and Planetary Institute, Houston, pp. 154–155.

Caffee M. W., Hudson G. B., Velsko C., Huss G. R., Alexander E. C., Jr., and Chivas A. R. (1999) Primordial noble gases from Earth's mantle: identification of a primitive volatile component. *Science* **285**, 2115–2118.

Cameron A. G. W. (1978) Physics of the primitive solar accretion disk. *Moons and Planets* **18**, 5–40.

Cameron A. G. W. (2000) Higher-resolution simulations of the Giant Impact. In *Origin of the Earth and Moon* (eds. K. Righter and R. Canup). University of Arizona Press, Tucson, pp. 133–144.

Cameron A. G. W. and Benz W. (1991) Origin of the Moon and the single impact hypothesis: IV. *Icarus* **92**, 204–216.

Canup R. M. and Agnor C. (1998) Accretion of terrestrial planets and the earth–moon system. In *Origin of the Earth and Moon*, LPI Contribution No. **597** Lunar and Planetary Institute, Houston, pp. 4–7.

Canup R. M. and Asphaug E. (2001) Origin of the Moon in a giant impact near the end of the Earth's formation. *Nature* **412**, 708–712.

Carlson R. W. and Hauri E. H. (2001) Extending the ^{107}Pd–^{107}Ag chronometer to Low Pd/Ag meteorites with the MC-ICPMS. *Geochim. Cosmochim. Acta* **65**, 1839–1848.

Carlson R. W. and Lugmair G. W. (1988) The age of ferroan anorthosite 60025: oldest crust on a young Moon? *Earth Planet. Sci. Lett.* **90**, 119–130.

Carlson R. W. and Lugmair G. W. (2000) Timescales of planetesimal formation and differentiation based on extinct and extant radioisotopes. In *Origin of the Earth and Moon* (eds. K. Righter and R. Canup). University of Arizona Press, Tucson, pp. 25–44.

Caro G., Bourdon B., Birck J.-L., and Moorbath S. (2003) ^{146}Sm–^{142}Nd evidence from Isua metamorphosed sediments for early differentiation of the Earth's mantle. *Nature* **423**, 428–431.

Carr M. H. and Wänke H. (1992) Earth and Mars: water inventories as clues to accretional histories. *Icarus* **98**, 61–71.

Cassen P. (1996) Models for the fractionation of moderately volatile elements in the solar nebula. *Meteorit. Planet. Sci.* **31**, 793–806.

Cassen P. (2001) Nebular thermal evolution and the properties of primitive planetary materials. *Meteorit. Planet. Sci.* **36**, 671–700.

Chambers J. E. and Wetherill G. W. (2001) Planets in the asteroid belt. *Meteorit. Planet. Sci.* **36**, 381–399.

Chase C. G. and Patchett P. J. (1988) Stored mafic/ultramafic crust and Early Archean mantle depletion. *Earth Planet. Sci. Lett.* **91**, 66–72.

Chiang E. I. and Goldreich P. (1997) Spectral energy distributions of T Tauri stars with passive circumstellar disks. *Astrophys. J.* **490**, 368.

Chiang E. I. and Goldreich P. (1999) Spectral energy distributions of passive T Tauri disks: inclination. *Astrophys. J.* **519**, 279.

Chiang E. I., Joung M. K., Creech-Eakman M. J., Qi C., Kessler J. E., Blake G. A., and van Dishoeck E. F. (2001) Spectral energy distributions of passive T Tauri and Herbig Ae disks: grain mineralogy, parameter dependences, and comparison with ISO LWS observations. *Astrophys. J.* **547**, 1077.

Chou C. L. (1978) Fractionation of siderophile elements in the Earth's upper mantle. *Proc. 9th Lunar Sci. Conf.* 219–230.

Clark S. P., Jr., Turekian K. K., and Grossman L. (1972) Model for the early history of the Earth. In *The Nature of the Solid Earth* (ed. E. C. Robertson). McGraw-Hill, New York, pp. 3–18.

Clarke W. B., Beg M. A., and Craig H. (1969) Excess ^{3}He in the sea: evidence for terrestrial primordial helium. *Earth Planet. Sci. Lett.* **6**, 213–220.

Clayton R. N. (1986) High temperature isotope effects in the early solar system. In *Stable Isotopes in High Temperature Geological Processes* (eds. J. W. Valley, H. P. Taylor, and J. R. O'Neil). Mineralogical Society of America, Washington, DC, pp. 129–140.

Clayton R. N. (1993) Oxygen isotopes in meteorites. *Ann. Rev. Earth Planet. Sci.* **21**, 115–149.

Clayton R. N. and Mayeda T. K. (1975) Genetic relations between the moon and meteorites. *Proc. 11th Lunar Sci. Conf.* 1761–1769.

Clayton R. N. and Mayeda T. K. (1996) Oxygen isotope studies of achondrites. *Geochim. Cosmochim. Acta* **60**, 1999–2017.

Clayton R. N., Grossman L., and Mayeda T. K. (1973) A component of primitive nuclear composition in carbonaceous meteorites. *Science* **182**, 485–487.

Connolly H. C., Jr. and Love S. G. (1998) The formation of chondrules: petrologic tests of the shock wave model. *Science* **280**, 62–67.

Connolly H. C., Jr., Jones B. D., and Hewins R. G. (1998) The flash melting of chondrules: an experimental investigation into the melting history and physical nature of chondrules. *Geochim. Cosmochim. Acta* **62**, 2725–2735.

Craig H. and Lupton J. E. (1976) Primordial neon, helium, and hydrogen in oceanic basalts. *Earth Planet. Sci. Lett.* **31**, 369–385.

Cuzzi J. N., Davis S. S., and Dobrovolskis A. R. (2003) Creation and distribution of CAIs in the protoplanetry nebula. In *Lunar Planet. Sci.* **XXXIV**, #1749. The Lunar and Planetary Institute, Houston (CD-ROM).

Darwin G. H. (1878) On the precession of a viscous spheroid. *Nature* **18**, 580–582.

Darwin G. H. (1879) On the precession of a viscous spheroid and on the remote history of the Earth. *Phil Trans. Roy. Soc. London* **170**, 447–538.

Dauphas N., Marty B., and Reisberg L. (2002) Inference on terrestrial genesis from molybdenum isotope systematics. *Geophys. Res. Lett.* **29**, 1084 doi:10.1029/2001GL014237.

Davies G. F. (1999) Geophysically constrained mass flows and the ^{40}Ar budget: a degassed lower mantle? *Earth Planet. Sci. Lett.* **166**, 149–162.

Déruelle B., Dreibus G., and Jambon A. (1992) Iodine abundances in oceanic basalts: implications for Earth dynamics. *Earth Planet. Sci. Lett.* **108**, 217–227.

Desch S. J. and Cuzzi J. N. (2000) The generation of lightning in the solar nebula. *Icarus* **143**, 87–105.

Dixon E. T., Honda M., McDougall I., Campbell I. H., and Sigurdsson I. (2000) Preservation of near-solar neon isotopic ratios in Icelandic basalts. *Earth Planet. Sci. Lett.* **180**, 309–324.

Doe B. R. and Stacey J. S. (1974) The application of lead isotopes to the problems of ore genesis and ore prospect evaluation: a review. *Econ. Geol.* **69**, 755–776.

Drake M. J. and McFarlane E. A. (1993) Mg-perovskite/silicate melt and majorite garnet/silicate melt partition coefficients in the system CaO–MgO–SiO$_2$ at high temperatures and pressures. *J. Geophys. Res.* **98**, 5427–5431.

Drake M. J. and Righter K. (2002) Determining the composition of the Earth. *Nature* **416**, 39–44.

Dreibus G. and Palme H. (1996) Cosmochemical constraints on the sulfur content in the Earth's core. *Geochim. Cosmochim. Acta* **60**, 1125–1130.

Esat T. M. (1996) Comment on Humayun and Clayton (1995). *Geochim. Cosmochim. Acta* **60**, 2755–2758.

Eucken A. (1944) Physikalisch-Chemische Betrachtungen über die früheste Entwicklungsgeschichte der Erde. *Nachr. Akad. Wiss. Göttingen, Math-Phys. Kl.* **Heft 1**, 1–25.

Farley K. A. and Poreda R. (1992) Mantle neon and atmospheric contamination. *Earth Planet. Sci. Lett.* **114**, 325–339.

Feigelson E. D., Broos P., Gaffney J. A., III, Garmire G., Hillenbrand L. A., Pravdo S. H., Townsley L., and Tsuboi Y. (2002a) X-ray-emitting young stars in the Orion nebula. *Astrophys. J.* **574**, 258–292.

Feigelson E. D., Garmire G. P., and Pravdo S. H. (2002b) Magnetic flaring in the pre-main-sequence Sun and implications for the early solar system. *Astrophys. J.* **572**, 335–349.

Franchi I. A., Wright I. P., Sexton A. S., and Pillinger C. T. (1999) The oxygen isotopic composition of Earth and Mars. *Meteorit. Planet. Sci.* **34**, 657–661.

Froude D. O., Ireland T. R., Kinny P. D., Williams I. S., Compston W., Williams I. R., and Myers J. S. (1983) Ion microprobe identification of 4,100–4,200 Myr-old terrestrial zircons. *Nature* **304**, 616–618.

Gaffey M. J. (1990) Thermal history of the asteroid belt: implications for accretion of the terrestrial planets. In *Origin of the Earth* (eds. H. E. Newsom and J. H. Jones). Oxford University Press, Oxford, pp. 17–28.

Galer S. J. G. and Goldstein S. L. (1991) Early mantle differentiation and its thermal consequences. *Geochim. Cosmochim. Acta* **55**, 227–239.

Galer S. J. G. and Goldstein S. L. (1996) Influence of accretion on lead in the Earth. In *Isotopic Studies of Crust–Mantle*

Evolution (eds. A. R. Basu and S. R. Hart). American Geophysical Union, Washington, DC, pp. 75–98.

Ganapathy R. and Anders E. (1974) Bulk compositions of the Moon and Earth, estimated from meteorites. *Proc. 5th Lunar Conf.* 1181–1206.

Gancarz A. J. and Wasserburg G. J. (1977) Initial Pb of the Amîtsoq Gneiss, West Greenland, and implications for the age of the Earth. *Geochim. Cosmochim. Acta* **41**, 1283–1301.

Gast P. W. (1960) Limitations on the composition of the upper mantle. *J. Geophys. Res.* **65**, 1287–1297.

Gessmann C. K., Wood B. J., Rubie D. C., and Kilburn M. R. (2001) Solubility of silicon in liquid metal at high pressure: implications for the composition of the Earth's core. *Earth Planet. Sci. Lett.* **184**, 367–376.

Ghosh A. and McSween H. Y., Jr. (1999) Temperature dependence of specific heat capacity and its effect on asteroid thermal models. *Meteorit. Planet. Sci.* **34**, 121–127.

Goldreich P. and Ward W. R. (1973) The formation of planetesimals. *Astrophys. J.* **183**, 1051–1060.

Goldstein S. L. and Galer S. J. G. (1992) On the trail of early mantle differentiation: $^{142}Nd/^{144}Nd$ ratios of early Archean rocks. *Trans. Am. Geophys. Union*, 323.

Göpel C., Manhès G., and Allègre C. J. (1991) Constraints on the time of accretion and thermal evolution of chondrite parent bodies by precise U–Pb dating of phosphates. *Meteoritics* **26**, 73.

Göpel C., Manhès G., and Allègre C. J. (1994) U–Pb systematics of phosphates from equilibrated ordinary chondrites. *Earth Planet. Sci. Lett.* **121**, 153–171.

Gounelle M., Shu F. H., Shang H., Glassgold A. E., Rehm K. E., and Lee T. (2001) Extinct radioactivities and protosolar cosmic-rays: self-shielding and light elements. *Astrophys. J.* **548**, 1051–1070.

Gray C. M., Papanastassiou D. A., and Wasserburg G. J. (1973) The identification of early condensates from the solar nebula. *Icarus* **20**, 213–239.

Grevesse N. and Sauval A. J. (1998) Standard solar composition. *Space Sci. Rev.* **85**, 161–174.

Grimm R. E. and McSween H. Y., Jr. (1993) Heliocentric zoning of the asteroid belt by aluminum-26 heating. *Science* **259**, 653–655.

Grossman J. N. (1996) Chemical fractionations of chondrites: signatures of events before chondrule formation. In *Chondrules and the Protoplanetary Disk* (eds. R. H. Hewins, R. H. Jones, and E. R. D. Scott). Cambridge University of Press, Cambridge, pp. 243–253.

Grossman L. (1972) Condensation in the primitive solar nebula. *Geochim. Cosmochim. Acta* **36**, 597–619.

Grossman L. and Larimer J. W. (1974) Early chemical history of the solar system. *Rev. Geophys. Space Phys.* **12**, 71–101.

Guan Y., McKeegan K. D., and MacPherson G. J. (2000) Oxygen isotopes in calcium–aluminum-rich inclusions from enstatite chondrites: new evidence for a single CAI source in the solar nebula. *Earth Planet. Sci. Lett.* **183**, 557–558.

Halliday A. N. (2000) Terrestrial accretion rates and the origin of the Moon. *Earth Planet. Sci. Lett.* **176**, 17–30.

Halliday A. N. and Lee D.-C. (1999) Tungsten isotopes and the early development of the Earth and Moon. *Geochim. Cosmochim. Acta* **63**, 4157–4179.

Halliday A. N. and Porcelli D. (2001) In search of lost planets—the paleocosmochemistry of the inner solar system. *Earth Planet. Sci. Lett.* **192**, 545–559.

Halliday A. N., Lee D.-C., Tomassini S., Davies G. R., Paslick C. R., Fitton J. G., and James D. (1995) Incompatible trace elements in OIB and MORB and source enrichment in the sub-oceanic mantle. *Earth Planet. Sci. Lett.* **133**, 379–395.

Halliday A. N., Rehkämper M., Lee D.-C., and Yi W. (1996) Early evolution of the Earth and Moon: new constraints from Hf–W isotope geochemistry. *Earth Planet. Sci. Lett.* **142**, 75–89.

Halliday A. N., Lee D.-C., and Jacobsen S. B. (2000) Tungsten isotopes, the timing of metal-silicate fractionation and the origin of the Earth and Moon. In *Origin of the Earth and Moon* (eds. R. M. Canup and K. Righter). University of Arizona Press, Tucson, pp. 45–62.

Halliday A. N., Wänke H., Birck J.-L., and Clayton R. N. (2001) The accretion, bulk composition and early differentiation of Mars. *Space Sci. Rev.* **96**, 197–230.

Hanan B. B. and Tilton G. R. (1987) 60025: Relict of primitive lunar crust? *Earth Planet. Sci. Lett.* **84**, 15–21.

Hanks T. C. and Anderson D. L. (1969) The early thermal history of the earth. *Phys. Earth Planet. Int.* **2**, 19–29.

Harper C. L. and Jacobsen S. B. (1992) Evidence from coupled $^{147}Sm-^{143}Nd$ and $^{146}Sm-^{142}Nd$ systematics for very early (4.5-Gyr) differentiation of the Earth's mantle. *Nature* **360**, 728–732.

Harper C. L. and Jacobsen S. B. (1996a) Noble gases and Earth's accretion. *Science* **273**, 1814–1818.

Harper C. L. and Jacobsen S. B. (1996b) Evidence for ^{182}Hf in the early solar system and constraints on the timescale of terrestrial core formation. *Geochim. Cosmochim. Acta* **60**, 1131–1153.

Harper C. L., Völkening J., Heumann K. G., Shih C.-Y., and Wiesmann H. (1991a) $^{182}Hf-^{182}W$: new cosmochronometric constraints on terrestrial accretion, core formation, the astrophysical site of the r-process, and the origin of the solar system. In *Lunar Planet. Sci.* **XXII**. The Lunar and Planetary Science, Houston, pp. 515–516.

Harper C. L., Wiesmann H., Nyquist L. E., Howard W. M., Meyer B., Yokoyama Y., Rayet M., Arnould M., Palme H., Spettel B., and Jochum K. P. (1991b) $^{92}Nb/^{93}Nb$ and $^{92}Nb/^{146}Sm$ ratios of the early solar system: observations and comparison of p-process and spallation models. In *Lunar Planet. Sci.* **XXII**. The Lunar and Planetary Institute, Houston, pp. 519–520.

Hartmann L. (2000) Observational constraints on transport (and mixing) in pre-main sequence disks. *Space Sci. Rev.* **92**, 55–68.

Hartmann L. and Kenyon S. J. (1996) The FU Orionis phenomena. *Ann. Rev. Astron. Astrophys.* **34**, 207–240.

Hartmann W. K. and Davis D. R. (1975) Satellite-sized planetesimals and lunar origin. *Icarus* **24**, 505–515.

Hartmann W. K., Ryder G., Dones L., and Grinspoon D. (2000) The time-dependent intense bombardment of the primordial Earth/Moon system. In *Origin of the Earth and Moon* (eds. R. Canup and K. Righter). University of Arizona Press, Tucson, pp. 493–512.

Hayashi C., Nakazawa K., and Mizuno H. (1979) Earth's melting due to the blanketing effect of the primordial dense atmosphere. *Earth Planet. Sci. Lett.* **43**, 22–28.

Hayashi C., Nakazawa K., and Nakagawa Y. (1985) Formation of the solar system. In *Protostars and Planets II* (eds. D. C. Black and D. S. Matthews). University of Arizona Press, Tucson, pp. 1100–1153.

Herzberg C. (1984) Chemical stratification in the silicate Earth. *Earth Planet. Sci. Lett.* **67**, 249–260.

Herzberg C., Jeigenson M., Skuba C., and Ohtani E. (1988) Majorite fractionation recorded in the geochemistry of peridotites from South Africa. *Nature* **332**, 823–826.

Hewins R. H. and Herzberg C. (1996) Nebular turbulence, chondrule formation, and the composition of the Earth. *Earth Planet. Sci. Lett.* **144**, 1–7.

Hofmann A. W. (1988) Chemical differentiation of the Earth: the relationship between mantle, continental crust and oceanic crust. *Earth Planet. Sci. Lett.* **90**, 297–314.

Hofmann A. W., Jochum K. P., Seufert M., and White W. M. (1986) Nb and Pb in oceanic basalts: new constraints on mantle evolution. *Earth Planet. Sci. Lett.* **79**, 33–45.

Honda M., McDougall I., Patterson D. B., Doulgeris A., and Clague D. A. (1991) Possible solar noble-gas component in Hawaiian basalts. *Nature* **349**, 149–151.

Horan M. F., Smoliar M. I., and Walker R. J. (1998) ^{182}W and $^{187}Re-^{187}Os$ systematics of iron meteorites: chronology for melting, differentiation, and crystallization in asteroids. *Geochim. Cosmochim. Acta* **62**, 545–554.

Hoyle F. (1960) On the origin of the solar nebula. *Quart. J. Roy. Astron. Soc.* **1**, 28–55.

Hsui A. T. and Toksöz M. N. (1977) Thermal evolution of planetary size bodies. *Proc. 8th Lunar Sci. Conf.* 447–461.

Humayun M. and Cassen P. (2000) Processes determining the volatile abundances of the meteorites and terrestrial planets. In *Origin of the Earth and Moon* (eds. R. M. Canup and K. Righter). University of Arizona Press, Tucson, pp. 3–23.

Humayun M. and Clayton R. N. (1995) Potassium isotope cosmochemistry: genetic implications of volatile element depletion. *Geochim. Cosmochim. Acta* **59**, 2131–2151.

Hunten D. M., Pepin R. O., and Walker J. C. G. (1987) Mass fractionation in hydrodynamic escape. *Icarus* **69**, 532–549.

Huss G. R. (1988) The role of presolar dust in the formation of the solar system. *Earth Moon Planet.* **40**, 165–211.

Huss G. R. (1997) The survival of presolar grains during the formation of the solar system. In *Astrophysical Implications of the Laboratory Study of Presolar Materials*, American Institute of Physics Conf. Proc. 402, Woodbury, New York (eds. T. J. Bernatowicz and E. Zinner), pp. 721–748.

Huss G. R. and Lewis R. S. (1995) Presolar diamond, SiC, and graphite in primitive chondrites: abundances as a function of meteorite class and petrologic type. *Geochim. Cosmochim. Acta* **59**, 115–160.

Ireland T. R. (1991) The abundance of ^{182}Hf in the early solar system. In *Lunar Planet. Sci.* **XXII**. The Lunar and Planetary Institute, Houston, pp. 609–610.

Ireland T. R. and Wlotzka F. (1992) The oldest zircons in the solar system. *Earth Planet. Sci. Lett.* **109**, 1–10.

Jackson I. (1983) Some geophysical constraints on the chemical composition of the Earth's lower mantle. *Earth Planet. Sci. Lett.* **62**, 91–103.

Jacobsen S. B. and Harper C. L., Jr. (1996) Accretion and early differentiation history of the Earth based on extinct radionuclides. In *Earth Processes: Reading the Isotope Code* (eds. E. Basu and S. Hart). American Geophysical Union, Washington, DC, pp. 47–74.

Jacobsen S. B. and Wasserburg G. J. (1979) The mean age of mantle and crust reservoirs. *J. Geophys. Res.* **84**, 7411–7427.

Jacobsen S. B. and Yin Q. Z. (2001) Core formation models and extinct nuclides. In *Lunar Planet. Sci.* **XXXII**, #1961. The Lunar and Planetary Institute, Houston (CD-ROM).

Jagoutz E., Palme J., Baddenhausen H., Blum K., Cendales M., Drebus G., Spettel B., Lorenz V., and Wänke H. (1979) The abundances of major, minor, and trace elements in the Earth's mantle as derived from primitive ultramafic nodules. *Proc. 10th Lunar Sci. Conf.*, 2031–2050.

Jahn B.-M. and Shih C. (1974) On the age of the Onverwacht group, Swaziland sequence, South Africa. *Geochim. Cosmochim. Acta* **38**, 873–885.

Javoy M. (1998) The birth of the Earth's atmosphere: the behaviour and fate of its major elements. *Chem. Geol.* **147**, 11–25.

Javoy M. (1999) Chemical earth models. *C.R. Acad. Sci., Ed. Sci. Méd. Elseviers SAS* **329**, 537–555.

Jessberger E. K., Bohsung J., Chakaveh S., and Traxel K. (1992) The volatile enrichment of chondritic interplanetary dust particles. *Earth Planet. Sci. Lett.* **112**, 91–99.

Jones J. H. and Drake M. J. (1986) Geochemical constraints on core formation in the Earth. *Nature* **322**, 221–228.

Jones J. H. and Drake M. J. (1993) Rubidium and cesium in the Earth and Moon. *Geochim. Cosmochim. Acta* **57**, 3785–3792.

Jones J. H. and Palme H. (2000) Geochemical constraints on the origin of the Earth and Moon. In *Origin of the Earth and Moon* (eds. R. M. Canup and K. Righter). University of Arizona Press, Tucson, pp. 197–216.

Jones J. H., Capobianco C. J., and Drake M. J. (1992) Siderophile elements and the Earth's formation. *Science* **257**, 1281–1282.

Jones R. H., Lee T., Connolly H. C., Jr., Love S. G., and Shang H. (2000) Formation of chondrules and CAIs: theory vs. observation. In *Protostars and Planets IV* (eds. V. Mannings, A. P. Boss, and S. S. Russell). University of Arizona Press, Tucson, pp. 927–962.

Kasting J. F. (2001) The rise of atmospheric oxygen. *Science* **293**, 819–820.

Kasting J. F. and Grinspoon D. H. (1991) The faint young Sun problem. In *The Sun in Time* (eds. C. P. Sonett, M. S. Gimpapa, and M. S. Matthews). University of Arizona Press, Tucson, pp. 447–462.

Kato T., Ringwood A. E., and Irifune T. (1988a) Experimental determination of element partitioning between silicate perovskites, garnets and liquids: constraints on early differentiation of the mantle. *Earth Planet. Sci. Lett.* **89**, 123–145.

Kato T., Ringwood A. E., and Irifune T. (1988b) Constraints on element partition coefficients between MgSiO$_3$ perovskite and liquid determined by direct measurements. *Earth Planet. Sci. Lett.* **90**, 65–68.

Kleine T., Münker C., Mezger K., and Palme H. (2002) Rapid accretion and early core formation on asteroids and the terrestrial planets from Hf–W chronometry. *Nature* **418**, 952–955.

Kornacki A. S. and Fegley B., Jr. (1986) The abundance and relative volatility of refractory trace elements in Allende Ca, Al-rich inclusions: implications for chemical and physical processes in the solar nebula. *Earth Planet. Sci. Lett.* **79**, 217–234.

Kortenkamp S. J., Kokubo E., and Weidenschilling S. J. (2000) Formation of planetary embryos. In *Origin of the Earth and Moon* (eds. R. M. Canup and K. Righter). University of Arizona Press, Tucson, pp. 85–100.

Kortenkamp S. J. and Wetherill G. W. (2000) Terrestrial planet and asteroid formation in the presence of giant planets. *Icarus* **143**, 60–73.

Kramers J. D. (1998) Reconciling siderophile element data in the Earth and Moon, W isotopes and the upper lunar age limit in a simple model of homogeneous accretion. *Chem. Geol.* **145**, 461–478.

Kramers J. D. and Tolstikhin I. N. (1997) Two terrestrial lead isotope paradoxes, forward transport modeling, core formation and the history of the continental crust. *Chem. Geol.* **139**, 75–110.

Kreutzberger M. E., Drake M. J., and Jones J. H. (1986) Origin of the Earth's Moon: constraints from alkali volatile trace elements. *Geochim. Cosmochim. Acta* **50**, 91–98.

Krot A. N., Meibom A., Russell S. S., Alexander C. M. O'D., Jeffries T. E., and Keil K. (2001) A new astrophysical setting for chondrule formation. *Science* **291**, 1776–1779.

Kunz J., Staudacher T., and Allègre C. J. (1998) Plutonium-fission xenon found in Earth's mantle. *Science* **280**, 877–880.

Lange M. A. and Ahrens T. J. (1982) The evolution of an impact-generated atmosphere. *Icarus* **51**, 96–120.

Lange M. A. and Ahrens T. J. (1984) FeO and H$_2$O and the homogeneous accretion of the earth. *Earth Planet. Sci. Lett.* **71**, 111–119.

Larimer J. W. and Anders E. (1970) Chemical fractionations in meteorites: III. Major element fractionation in chondrites. *Geochim. Cosmochim. Acta* **34**, 367–387.

Lee D.-C. and Halliday A. N. (1995) Hafnium–tungsten chronometry and the timing of terrestrial core-formation. *Nature* **378**, 771–774.

Lee D.-C. and Halliday A. N. (1996) Hf–W isotopic evidence for rapid accretion and differentiation in the early solar system. *Science* **274**, 1876–1879.

Lee D.-C. and Halliday A. N. (1997) Core formation on Mars and differentiated asteroids. *Nature* **388**, 854–857.

Lee D-C. and Halliday A. N. (2000a) Accretion of primitive planetesimals: Hf–W isotopic evidence from enstatite chondrites. *Science* **288**, 1629–1631.

Lee D.-C. and Halliday A. N. (2000b) Hf–W isotopic systematics of ordinary chondrites and the initial ^{182}Hf/^{180}Hf of the solar system. *Chem. Geol.* **169**, 35–43.

Lee D.-C., Halliday A. N., Davies G. R., Essene E. J., Fitton J. G., and Temdjim R. (1996) Melt enrichment of shallow depleted mantle: a detailed petrological, trace element and isotopic study of mantle derived xenoliths and megacrysts from the Cameroon line. *J. Petrol.* **37**, 415–441.

Lee D.-C., Halliday A. N., Snyder G. A., and Taylor L. A. (1997) Age and origin of the Moon. *Science* **278**, 1098–1103.

Lee D.-C., Halliday A. N., Leya I., Wieler R., and Wiechert U. (2002) Cosmogenic tungsten and the origin and earliest differentiation of the Moon. *Earth Planet. Sci. Lett.* **198**, 267–274.

Lee T., Shu F. H., Shang H., Glassgold A. E., and Rehm K. E. (1998) Protostellar cosmic rays and extinct radioactivities in meteorites. *Astrophys. J.* **506**, 898–912.

Levin B. J. (1972) Origin of the earth. *Tectonophysics* **13**, 7–29.

Lewis J. S. (1972) Metal/silicate fractionation in the solar system. *Earth Planet. Sci. Lett.* **15**, 286–290.

Leya I., Wieler R., and Halliday A. N. (2000) Cosmic-ray production of tungsten isotopes in lunar samples and meteorites and its implications for Hf–W cosmochemistry. *Earth Planet. Sci. Lett.* **175**, 1–12.

Leya I., Wieler R., and Halliday A. N. (2003) The influence of cosmic-ray production on extinct nuclide systems. *Geochim. Cosmochim. Acta* **67**, 527–541.

Li J. and Agee C. B. (1996) Geochemistry of mantle–core differentiation at high pressure. *Nature* **381**, 686–689.

Lin D. N. C. and Papaloizou J. (1985) On the dynamical origin of the solar system. In *Protostars and Planets II* (eds. D. C. Black and M. S. Matthews). University of Arizona Press, Tucson, pp. 981–1072.

Lin D. N. C., Bodenheimer P., and Richardson D. C. (1996) Orbital migration of the planetary companion of 51 Pegasi to its present location. *Nature* **380**, 606–607.

Lissauer J. J. (1987) Time-scales for planetary accretion and the structure of the protoplanetry disk. *Icarus* **69**, 249–265.

Lissauer J. J. (1993) Planet formation. *Ann. Rev. Astron. Astrophys.* **31**, 129–174.

Lissauer J. J. (1999) How common are habitable planets? *Nature* **402**(suppl.2), C11–C14.

Lodders K. (2000) An oxygen isotope mixing model for the accretion and composition of rocky planets. *Space Sci. Rev.* **92**, 341–354.

Longhi J. (1999) Phase equilibrium constraints on angrite petrogenesis. *Geochim. Cosmochim. Acta* **63**, 573–585.

Lugmair G. W. and Galer S. J. G. (1992) Age and isotopic relationships between the angrites Lewis Cliff 86010 and Angra dos Reis. *Geochim. Cosmochim. Acta* **56**, 1673–1694.

Lugmair G. W. and Shukolyukov A. (1998) Early solar system timescales according to ^{53}Mn–^{53}Cr systematics. *Geochim. Cosmochim. Acta* **62**, 2863–2886.

Lugmair G. W. and Shukolyukov A. (2001) Early solar system events and timescales. *Meteorit. Planet. Sci.* **36**, C17–C26.

Lupton J. E. and Craig H. (1975) Excess ^{3}He in oceanic basalts: evidence for terrestrial primordial helium, *Earth Planet. Sci. Lett.* **26**, 133–139.

MacFarlane E. A. (1989) Formation of the Moon in a giant impact: composition of impactor. *Proc. 19th Lunar Planet. Sci. Conf.* 593–605.

MacPherson G. J., Wark D. A., and Armstrong J. T. (1988) Primitive material surviving in chondrites: refractory inclusions. In *Meteorites and the Early Solar System* (eds. J. F. Kerridge and M. S. Matthews). University of Arizona Press, Tucson, pp. 746–807.

Mamyrin B. A., Tolstikhin I. N., Anufriev G. S., and Kamensky I. L. (1969) Anomalous isotopic composition of helium in volcanic gases. *Dokl. Akad. Nauk. SSSR* **184**, 1197–1199 (in Russian).

Manhès G., Allègre C. J., Dupré B., and Hamelin B. (1979) Lead–lead systematics, the "age of the Earth" and the chemical evolution of our planet in a new representation space. *Earth Planet. Sci. Lett.* **44**, 91–104.

Marty B. (1989) Neon and xenon isotopes in MORB: implications for the earth-atmosphere evolution. *Earth Planet. Sci. Lett.* **94**, 45–56.

Matsui T. and Abe Y. (1986a) Evolution of an impact-induced atmosphere and magma ocean on the accreting Earth. *Nature* **319**, 303–305.

Matsui T. and Abe Y. (1986b) Impact-induced atmospheres and oceans on Earth and Venus. *Nature* **322**, 526–528.

Mayor M. and Queloz D. (1995) A Jupiter-mass companion to a solar-type star. *Nature* **378**, 355–359.

McCaughrean M. J. and O'Dell C. R. (1996) Direct imaging of circumstellar disks in the Orion nebula. *Astronom. J.* **111**, 1977–1986.

McCulloch M. T. (1994) Primitive ^{87}Sr/^{86}Sr from an Archean barite and conjecture on the Earth's age and origin, *Earth Planet. Sci. Lett.* **126**, 1–13.

McCulloch M. T. and Bennett V. C. (1993) Evolution of the early Earth: constraints from ^{143}Nd–^{142}Nd isotopic systematics. *Lithos* **30**, 237–255.

McDonough W. F. and Sun S.-S. (1995) The composition of the Earth. *Chem. Geol.* **120**, 223–253.

McDonough W. F., Sun S.-S., Ringwood A. E., Jagoutz E., and Hofmann A. W. (1992) Potassium, rubidium, and cesium in the Earth and Moon and the evolution of the mantle of the Earth, *Geochim. Cosmochim. Acta* **53**, 1001–1012.

McFarlane E. A. and Drake M. J. (1990) Element partitioning and the early thermal history of the Earth. In *Origin of the Earth* (eds. H. E. Newsom and J. H. Jones). Lunar and Planetary Institute, Houston, pp. 135–150.

McKeegan K. D., Chaussidon M., and Robert F. (2000) Incorporation of short-lived Be-10 in a calcium–aluminum-rich inclusion from the Allende meteorite. *Science* **289**, 1334–1337.

Meibom A., Desch S. J., Krot A. N., Cuzzi J. N., Petaev M. I., Wilson L., and Keil K. (2000) Large-scale thermal events in the solar nebula: evidence from Fe, Ni metal grains in primitive meteorites. *Science* **288**, 839–841.

Meisel T., Walker R. J., Irving A. J., and Lorand J.-P. (2001) Osmium isotopic compositions of mantle xenoliths: a global perspective. *Geochim. Cosmochim. Acta* **65**, 1311–1323.

Melosh H. J. (1990) Giant impacts and the thermal state of the early Earth. In *Origin of the Earth* (eds. H. E. Newsom and J. H. Jones). Oxford University Press, Oxford, pp. 69–83.

Melosh H. J. and Sonett C. P. (1986) When worlds collide: jetted vapor plumes and the Moon's origin. In *Origin of the Moon* (eds. W. K. Hartmann, R. J. Phillips, and G. J. Taylor). Lunar and Planetary Institute, Houston, pp. 621–642.

Melosh H. J. and Vickery A. M. (1989) Impact erosion of the primordial atmosphere of Mars. *Nature* **338**, 487–489.

Melosh H. J., Vickery A. M., and Tonks W. B. (1993) Impacts and the early environment and evolution of the terrestrial planets. In *Protostars and Planets III* (eds. E. H. Levy and J. I. Lunine). University of Arizona Press, Tucson, pp. 1339–1370.

Mendybaev R. A., Beckett J. R., Grossman L., Stolper E., Cooper R. F., and Bradley J. P. (2002) Volatilization kinetics of silicon carbide in reducing gases: an experimental study with applications to the survival of presolar grains in the solar nebula. *Geochim. Cosmochim. Acta* **66**, 661–682.

Minarik W. G., Ryerson F. J., and Watson E. B. (1996) Textural entrapment of core-forming melts. *Science* **272**, 530–533.

Mizuno H., Nakazawa K., and Hayashi C. (1980) Dissolution of the primordial rare gases into the molten Earth's material. *Earth Planet. Sci. Lett.* **50**, 202–210.

Mojzsis S. J., Arrhenius G., McKeegan K. D., Harrison T. M., Nutman A. P., and Friend C. R. L. (1996) Evidence for life on Earth before 3,800 million years ago. *Nature* **384**, 55–59.

Mojzsis S. J., Harrison T. M., and Pidgeon R. T. (2001) Oxygen isotope evidence from ancient zircons for liquid water at the Earth's surface 4300 Myr ago Jack Hills, evidence for more very old detrital zircons in Western Australia. *Nature* **409**, 178–181.

Morbidelli A., Chambers J., Lunine J. I., Petit J. M., Robert F., Valsecchi G. B., and Cyr K. E. (2000) Source regions and time-scales for the delivery of water to the Earth. *Meteorit. Planet. Sci.* **35**, 1309–1320.

Moreira M. and Allègre C. J. (1998) Helium–neon systematics and the structure of the mantle. *Chem. Geol* **147**, 53–59.

Moreira M., Kunz J., and Allègre C. J. (1998) Rare gas systematics in Popping Rock: isotopic and elemental compositions in the upper mantle. *Science* **279**, 1178–1181.

Moreira M., Breddam K., Curtice J., and Kurz M. D. (2001) Solar neon in the Icelandic mantle: new evidence for an undegassed lower mantle. *Earth Planet. Sci. Lett.* **185**, 15–23.

Münker C., Weyer S., Mezger K., Rehkämper M., Wombacher F., and Bischoff A. (2000) ^{92}Nb–^{92}Zr and the early differentiation history of planetary bodies. *Science* **289**, 1538–1542.

Murphy D. T., Kamber B. S., and Collerson K. D. (2003) A refined solution to the first terrestrial Pb-isotope paradox. *J. Petrol.* **44**, 39–53.

Murray N. and Chaboyer B. (2001) Are stars with planets polluted? *Ap. J.* **566**, 442–451.

Murray N., Hansen B., Holman M., and Tremaine S. (1998) Migrating planets. *Science* **279**, 69–72.

Murthy V. R. (1991) Early differentiation of the Earth and the problem of mantle siderophile elements: a new approach. *Science* **253**, 303–306.

Nägler Th. F. and Kramers J. D. (1998) Nd isotopic evolution of the upper mantle during the Precambrian: models, data and the uncertainty of both. *Precambrian Res.* **91**, 233–252.

Nelson A. F., Benz W., Adams F. C., and Arnett D. (1998) Dynamics of circumstellar disks. *Ap. J.* **502**, 342–371.

Nelson A. F., Benz W., and Ruzmaikina T. V. (2000) Dynamics of circumstellar disks: II. Heating and cooling. *Ap. J.* **529**, 357–390.

Newsom H. E. (1990) Accretion and core formation in the Earth: evidence from siderophile elements. In *Origin of the Earth* (eds. H. E. Newsom and J. H. Jones). Oxford University Press, Oxford, pp. 273–288.

Newsom H. E. (1995) Composition of the solar system, planets, meteorites, and major terrestrial reservoirs. In *Global Earth Physics: A Handbook of Physical Constants*, AGU Reference Shelf 1 (ed. T. J. Ahrens). American Geophysical Union, Washington, DC.

Newsom H. E. and Taylor S. R. (1986) The single impact origin of the Moon. *Nature* **338**, 29–34.

Newsom H. E., White W. M., Jochum K. P., and Hofmann A. W. (1986) Siderophile and chalcophile element abundances in oceanic basalts, Pb isotope evolution and growth of the Earth's core. *Earth Planet. Sci. Lett.* **80**, 299–313.

Newsom H. E., Sims K. W. W., Noll P. D., Jr., Jaeger W. L., Maehr S. A., and Bessera T. B. (1996) The depletion of W in the bulk silicate Earth. *Geochim. Cosmochim. Acta* **60**, 1155–1169.

Niedermann S., Bach W., and Erzinger J. (1997) Noble gas evidence for a lower mantle component in MORBs from the southern East Pacific Rise: decoupling of helium and neon isotope systematics. *Geochim. Cosmochim. Acta* **61**, 2697–2715.

Nittler L. R. (2003) Presolar stardust in meteorites: recent advances and scientific frontiers. *Earth Planet. Sci. Lett* **209**, 259–273.

Nittler L. R., Alexander C. M. O'D., Gao X., Walker R. M., and Zinner E. K. (1994) Interstellar oxide grains from the Tieschitz ordinary chondrite. *Nature* **370**, 443–446.

O'Keefe J. D. and Ahrens T. J. (1977) Impact-induced energy partitioning, melting, and vaporization on terrestrial planets. *Proc. 8th Lunar Sci. Conf.* 3357–3374.

O'Neill H. St. C. (1991a) The origin of the Moon and the early history of the Earth—a chemical model: Part I. The Moon. *Geochim. Cosmochim. Acta* **55**, 1135–1158.

O'Neill H. St. C. (1991b) The origin of the Moon and the early history of the Earth—a chemical model: Part II. The Earth. *Geochim. Cosmochim. Acta* **55**, 1159–1172.

O'Nions R. K. and Tolstikhin I. N. (1994) Behaviour and residence times of lithophile and rare gas tracers in the upper mantle. *Earth Planet. Sci. Lett.* **124**, 131–138.

Oversby V. M. and Ringwood A. E. (1971) Time of formation of the Earth's core. *Nature* **234**, 463–465.

Owen T. and Bar-Nun A. (1995) Comets, impacts and atmosphere. *Icarus* **116**, 215–226.

Owen T. and Bar-Nun A. (2000) Volatile contributions from icy planetesimals. In *Origin of the Earth and Moon* (eds. R. M. Canup and K. Righter). University of Arizona Press, Tucson, pp. 459–471.

Ozima M. and Podosek F. A. (1999) Formation age of Earth from ^{129}I/^{127}I and ^{244}Pu/^{238}U systematics and the missing Xe. *J. Geophys. Res.* **104**, 25493–25499.

Ozima M. and Podosek F. A. (2002) *Noble Gas Geochemistry*, 2nd edn. Cambridge University Press, Cambridge, 286p.

Palme H. (2000) Are there chemical gradients in the inner solar system? *Space Sci. Rev.* **92**, 237–262.

Palme H., Larimer J. W., and Lipschultz M. E. (1988) Moderately volatile elements. In *Meteorites and the Early Solar System* (eds. J. F. Kerridge and M. S. Matthews). University of Arizona Press, Tucson, pp. 436–461.

Papanastassiou D. A. and Wasserburg G. J. (1976) Early lunar differentiates and lunar initial ^{87}Sr/^{86}Sr. In *Lunar Sci.* **VII**. The Lunar Science Institute, Houston, pp. 665–667.

Patterson C. C. (1956) Age of meteorites and the Earth. *Geochim. Cosmochim. Acta* **10**, 230–237.

Pavlov A. A., Kasting J. F., Brown L. L., Rages K. A., and Freedman R. (2000) Greenhouse warming by CH_4 in the atmosphere of early Earth. *J. Geophys. Res.* **105**, 11981–11990.

Peck W. H., Valley J. W., Wilde S. A., and Graham C. M. (2001) Oxygen isotope ratios and rare earth elements in 3.3 to 4.4 Ga zircons: ion microprobe evidence for high δ^{18}O continental crust and oceans in the Early Archean. *Geochim Cosmochim. Acta* **65**, 4215–4229.

Pepin R. O. (1997) Evolution of Earth's noble gases: consequences of assuming hydrodynamic loss driven by giant impact. *Icarus* **126**, 148–156.

Pepin R. O. (2000) On the isotopic composition of primordial xenon in terrestrial planet atmospheres. *Space Sci. Rev.* **92**, 371–395.

Pepin R. O. and Porcelli D. (2002) Origin of noble gases in the terrestrial planets. In *Noble Gases in Geochemistry and Cosmochemistry*, Rev. Mineral. Geochem. 47 (eds. D. Porcelli, C. J. Ballentine, and R. Wieler). Mineralogical Society of America, Washington, DC, pp. 191–246.

Phinney D., Tennyson J., and Frick U. (1978) Xenon in CO_2 well gas revisited. *J. Geophys. Res.* **83**, 2313–2319.

Podosek F. A. and Cassen P. (1994) Theoretical, observational, and isotopic estimates of the lifetime of the solar nebula. *Meteoritics* **29**, 6–25.

Podosek F. A., Zinner E. K., MacPherson G. J., Lundberg L. L., Brannon J. C., and Fahey A. J. (1991) Correlated study of initial Sr-87/Sr-86 and Al–Mg isotopic systematics and petrologic properties in a suite of refractory inclusions from the Allende meteorite. *Geochim. Cosmochim. Acta* **55**, 1083–1110.

Podosek F. A., Ott U., Brannon J. C., Neal C. R., Bernatowicz T. J., Swan P., and Mahan S. E. (1997) Thoroughly anomalous chromium in Orgueil. *Meteorit. Planet. Sci.* **32**, 617–627.

Podosek F. A., Woolum D. S., Cassen P., Nicholls R. H., Jr., and Weidenschilling S. J. (2003) Solar wind as a source of terrestrial light noble gases. *Geochim. Cosmochim. Acta* (in press).

Poitrasson F., Halliday A. N., Lee D.-C., Levasseur S., and Teutsch N. (2003) Iron isotope evidence for formation of the Moon through partial vaporisation. In *Lunar Planet. Sci.* **XXXIV**, #1433. The Lunar and Planetary Institute, Houston (CD-ROM).

Porcelli D. and Ballentine C. J. (2002) Models for the distribution of terrestrial noble gases and evolution of the atmosphere. In *Noble Gases in Geochemistry and Cosmochemistry*, Rev. Mineral. Geochem. 47 (eds. D. Porcelli, C. J. Ballentine, and R. Wieler). Mineralogical Society of America, Washington, DC. pp. 411–480.

Porcelli D. and Halliday A. N. (2001) The possibility of the core as a source of mantle helium. *Earth Planet. Sci. Lett.* **192**, 45–56.

Porcelli D. and Pepin R. O. (2000) Rare gas constraints on early earth history. In *Origin of the Earth and Moon* (eds. R. M. Canup and K. Righter). University of Arizona Press, Tucson, pp. 435–458.

Porcelli D. and Wasserburg G. J. (1995) Mass transfer of helium, neon, argon and xenon through a steady-state upper mantle. *Geochim. Cosmochim. Acta* **59**, 4921–4937.

Porcelli D., Cassen P., Woolum D., and Wasserburg G. J. (1998) Acquisition and early losses of rare gases from the deep Earth. In *Origin of the Earth and Moon*, LPI Contribution No. 597. Lunar and Planetary Institute, Houston, pp. 35–36.

Porcelli D., Cassen P., and Woolum D. (2001) Deep Earth rare gases: initial inventories, capture from the solar nebula and losses during Moon formation. *Earth Planet. Sci. Lett* **193**, 237–251.

Puchtel I. and Humayun M. (2000) Platinum group elements in Kostomuksha komatiites and basalts: implications for oceanic crust recycling and core–mantle interaction. *Geochim. Cosmochim. Acta* **64**, 4227–4242.

Qian Y. Z. and Wasserburg G. J. (2000) Stellar abundances in the early galaxy and two r-process components. *Phys. Rep.* **333–334**, 77–108.

Qian Y. Z., Vogel P., and Wasserburg G. J. (1998) Diverse supernova sources for the r-process. *Astrophys. J.* **494**, 285–296.

Quitté G., Birck J.-L., and Allègre C. J. (2000) ^{182}Hf–^{182}W systematics in eucrites: the puzzle of iron segregation in the early solar system. *Earth Planet. Sci. Lett.* **184**, 83–94.

Rammensee W. and Wänke H. (1977) On the partition coefficient of tungsten between metal and silicate and its bearing on the origin of the Moon. *Proc. 8th Lunar Sci. Conf.* 399–409.

Rehkämper M., Halliday A. N., Barfod D., Fitton J. G., and Dawson J. B. (1997) Platinum group element abundance patterns in different mantle environments. *Science* **278**, 1595–1598.

Rehkämper M., Halliday A. N., Alt J., Fitton J. G., Zipfel J., and Takazawa E. (1999a) Non-chondritic platinum group element ratios in abyssal peridotites: petrogenetic signature of melt percolation? *Earth Planet. Sci. Lett.* **172**, 65–81.

Rehkämper M., Halliday A. N., Fitton J. G., Lee D.-C., and Wieneke M. (1999b) Ir, Ru, Pt, and Pd in basalts and komatiites: new constraints for the geochemical behavior of the platinum-group elements in the mantle. *Geochim. Cosmochim. Acta* **63**, 3915–3934.

Reynolds J. H. (1967) Isotopic abundance anomalies in the solar system. *Ann. Rev. Nuclear Sci.* **17**, 253–316.

Righter K. and Drake M. J. (1996) Core formation in Earth's Moon, Mars, and Vesta. *Icarus* **124**, 513–529.

Righter K. and Drake M. J. (1999) Effect of water on metal-silicate partitioning of siderophile elements: a high pressure and temperature terrestrial magma ocean and core formation. *Earth Planet. Sci. Lett.* **171**, 383–399.

Righter K., Drake M. J., and Yaxley G. (1997) Prediction of siderophile element metal/silicate partition coefficients to 20 GPa and 2800 °C: the effects of pressure, temperature, oxygen fugacity, and silicate and metallic melt compositions. *Phys. Earth Planet. Int.* **100**, 115–142.

Righter K., Walker R. J., and Warren P. W. (2000) The origin and significance of highly siderophile elements in the lunar and terrestrial mantles. In *Origin of the Earth and Moon* (eds. R. M. Canup and K. Righter). University of Arizona Press, Tucson, pp. 291–322.

Ringwood A. E. (1966) The chemical composition and origin of the Earth. In *Advances in Earth Sciences* (ed. P. M. Hurley). MIT Press, Cambridge, MA, pp. 287–356.

Ringwood A. E. (1977) Composition of the core and implications for origin of the Earth. *Geochem. J.* **11**, 111–135.

Ringwood A. E. (1979) *Origin of the Earth and Moon.* Springer, New York, 295p.

Ringwood A. E. (1989a) Significance of the terrestrial Mg/Si ratio. *Earth Planet. Sci. Lett.* **95**, 1–7.

Ringwood A. E. (1989b) Flaws in the giant impact hypothesis of lunar origin. *Earth Planet. Sci. Lett.* **95**, 208–214.

Ringwood A. E. (1990) Earliest history of the Earth–Moon system. In *Origin of the Earth* (eds. A. E. Newsom and J. H. Jones). Oxford University Press, Oxford, pp. 101–134.

Ringwood A. E. (1992) Volatile and siderophile element geochemistry of the Moon: a reappraisal. *Earth Planet. Sci. Lett.* **111**, 537–555.

Rubey W. W. (1951) Geological history of seawater. *Bull. Geol. Soc. Am.* **62**, 1111–1148.

Rubie D. C., Melosh H. J., Reid J. E., Liebske C., and Righter K. (2003) Mechanisms of metal-silicate equilibration in the terrestrial magma ocean. *Earth Planet. Sci. Lett.* **205**, 239–255.

Rushmer T., Minarik W. G., and Taylor G. J. (2000) Physical processes of core formation. In *Origin of the Earth and Moon* (eds. R. M. Canup and K. Righter). University of Arizona Press, Tucson, pp. 227–243.

Ryder G., Koeberl C., and Mojzsis S. J. (2000) Heavy bombardment of the Earth at ~3.85 Ga: the search for petrographic and geochemical evidence. In *Origin of the Earth and Moon* (eds. R. M. Canup and K. Righter). University of Arizona Press, Tucson, pp. 475–492.

Safronov V. S. (1954) On the growth of planets in the protoplanetary cloud. *Astron. Zh.* **31**, 499–510.

Sagan C. and Chyba C. (1997) The early faint Sun paradox: organic shielding of ultraviolet-labile greenhouse gases. *Science* **276**, 1217–1221.

Sanders I. S. (1996) A chondrule-forming scenario involving molten planetesimals. In *Chondrules and the Protoplanetary Disk* (eds. R. H. Hewins, R. H. Jones, and E. R. D. Scott). Cambridge University Press, Cambridge, UK, pp. 327–334.

Sanloup C., Blichert-Toft J., Télouk P., Gillet P., and Albarède F. (2000) Zr isotope anomalies in chondrites and the presence of live ^{92}Nb in the early solar system. *Earth Planet. Sci. Lett.* **184**, 75–81.

Sasaki S. (1990) The primary solar-type atmosphere surrounding the accreting Earth: H$_2$O-induced high surface temperature. In *Origin of the Earth* (eds. H. E. Newsom and J. H. Jones). Oxford University Press, Oxford, pp. 195–209.

Sasaki S. and Nakazawa K. (1986) Metal-silicate fractionation in the growing Earth: energy source for the terrestrial magma ocean. *J. Geophys. Res.* **91**, B9231–B9238.

Sasaki S. and Nakazawa K. (1988) Origin of isotopic fractionation of terrestrial Xe: hydrodynamic fractionation during escape of the primordial H$_2$–He atmosphere. *Earth Planet. Sci. Lett.* **89**, 323–334.

Scherer E. E., Münker C., and Mezger K. (2001) Calibration of the lutetium–hafnium clock. *Science* **293**, 683–687.

Schmitt W., Palme H., and Wänke H. (1989) Experimental determination of metal/silicate partition coefficients for P, Co, Ni, Cu, Ga, Ge, Mo, and W and some implications for the early evolution of the Earth. *Geochim. Cosmochim. Acta* **53**, 173–185.

Schneider G., Smith B. A., Becklin E. E., Koerner D. W., Meier R., Hines D. C., Lowrance P. J., Terrile R. I., and Rieke M.

(1999) Nicmos imaging of the HR 4796A circumstellar disk. *Astrophys. J.* **513**, L127–L130.

Schönberg R., Kamber B. S., Collerson K. D., and Eugster O. (2002) New W isotope evidence for rapid terrestrial accretion and very early core formation. *Geochim. Cosmochim. Acta* **66**, 3151–3160.

Schönbächler M., Rehkämper M., Halliday A. N., Lee D. C., Bourot-Denise M., Zanda B., Hattendorf B., and Günther D. (2002) Niobium–zirconium chronometry and early solar system development. *Science* **295**, 1705–1708.

Seager S. (2003) The search for Earth-like extrasolar planets. *Earth Planet. Sci. Lett.* **208**, 113–124.

Seeds M. A. (1996) *Foundations of Astronomy*. Wadsworth Publishing Company, Belmont, California, USA.

Sharma M., Papanastassiou D. A., and Wasserburg G. J. (1996) The issue of the terrestrial record of ^{146}Sm. *Geochim. Cosmochim. Acta* **60**, 2037–2047.

Shaw G. H. (1978) Effects of core formation. *Phys. Earth Planet. Int.* **16**, 361–369.

Shih C.-Y., Nyquist L. E., Dasch E. J., Bogard D. D., Bansal B. M., and Wiesmann H. (1993) Age of pristine noritic clasts from lunar breccias 15445 and 15455. *Geochim. Cosmochim. Acta* **57**, 915–931.

Shu F. H., Shang H., and Lee T. (1996) Toward an astrophysical theory of chondrites. *Science* **271**, 1545–1552.

Shu F. H., Shang H., Glassgold A. E., and Lee T. (1997) X-rays and fluctuating x-winds from protostars. *Science* **277**, 1475–1479.

Sleep N. H. and Zahnle K. (2001) Carbon dioxide cycling and implications for climate on ancient Earth. *J. Geophys. Res.* **106**, 1373–1399.

Sleep N. H., Zahnle K. J., Kasting J. F., and Morowitz H. J. (1989) Annihilation of ecosystems by large asteroid impacts on the early Earth. *Nature* **342**, 139–142.

Sleep N. H., Zahnle K., and Neuhoff P. S. (2001) Initiation of clement surface conditions on the earliest Earth. *Proc. Natl. Acad. Sci.* **98**, 3666–3672.

Smith J. V. (1977) Possible controls on the bulk composition of the Earth: implications for the origin of the earth and moon. *Proc. 8th Lunar Sci. Conf.* 333–369.

Smith J. V. (1980) The relation of mantle heterogeneity to the bulk composition and origin of the Earth. *Phil. Trans. Roy. Soc. London A* **297**, 139–146.

Snow J. E. and Schmidt G. (1998) Constraints on Earth accretion deduced from noble metals in the oceanic mantle. *Nature* **391**, 166–169.

Solomon S. C. (1979) Formation, history and energetics of cores in the terrestrial planets. *Earth Planet. Sci. Lett.* **19**, 168–182.

Stevenson D. J. (1981) Models of the Earth's core. *Science* **214**, 611–619.

Stevenson D. J. (1990) Fluid dynamics of core formation. In *Origin of the Earth* (eds. H. E. Newsom and J. H. Jones). Oxford University Press, Oxford, pp. 231–249.

Suess (1965) Chemical evidence bearing on the origin of the solar system. *Rev. Astron. Astrophys.* **3**, 217–234.

Swindle T. D. (1993) Extinct radionuclides and evolutionary timescales. In *Protostars and Planets III* (eds. E. H. Levy and J. I. Lunine). University of Arirona Press, Tucson, pp. 867–881.

Swindle T. D. and Podosek (1988) Iodine–xenon dating. In *Meteorites and the Early Solar System* (eds. J. F. Kerridge and M. S. Matthews). University of Arizona Press, Tucson, pp. 1127–1146.

Takahashi H., Janssens M.-J., Morgan J. W., and Anders E. (1978) Further studies of trace elements in C3 chondrites. *Geochim. Cosmochim. Acta* **42**, 97–106.

Taylor S. R. (1992) *Solar System Evolution: A New Perspective*. Cambridge University Press, New York.

Taylor S. R. and Norman M. D. (1990) Accretion of differentiated planetesimals to the Earth. In *Origin of the Earth* (eds. H. E. Newsom and J. H. Jones). Oxford University Press, Oxford, pp. 29–43.

Tera F. (1980) Reassessment of the "age of the Earth." *Carnegie Inst. Wash. Yearbook* **79**, 524–531.

Tera F., Papanastassiou D. A., and Wasserburg G. J. (1973) A lunar cataclysm at ~3.95 AE and the structure of the lunar crust. In *Lunar Sci.* **IV**. The Lunar Science Institute, Houston, pp. 723–725.

Tilton G. R. (1988) Age of the solar system. In *Meteorites and the Early Solar System* (eds. J. F. Kerridge and M. S. Matthews). University of Arizona Press, Tucson, pp. 259–275.

Tonks W. B. and Melosh H. J. (1990) The physics of crystal settling and suspension in a turbulent magma ocean. In *Origin of the Earth* (eds. H. E. Newsom and J. H. Jones). Oxford University Press, Oxford, pp. 17–174.

Treiman A. H., Drake M. J., Janssens M.-J., Wolf R., and Ebihara M. (1986) Core formation in the Earth and shergottite parent body (SPB): chemical evidence from basalts. *Geochim. Cosmochim. Acta* **50**, 1071–1091.

Treiman A. H., Jones J. H., and Drake M. J. (1987) Core formation in the shergottite parent body and comparison with the Earth. *J. Geophys. Res.* **92**, E627–E632.

Trieloff M., Kunz J., Clague D. A., Harrison D., and Allègre C. J. (2000) The nature of pristine noble gases in mantle plumes. *Science* **288**, 1036–1038.

Turekian K. K. and Clark S. P., Jr. (1969) Inhomogeneous accumulation of the earth from the primitive solar nebula. *Earth Planet. Sci. Lett.* **6**, 346–348.

Tyburczy J. A., Frisch B., and Ahrens T. J. (1986) Shock-induced volatile loss from a carbonaceous chondrite: implications for planetary accretion. *Earth Planet. Sci. Lett.* **80**, 201–207.

Urey H. C. (1966) The capture hypothesis of the origin of the Moon. In *The Earth–Moon System* (eds. B. G. Marsden and A. G. W. Cameron). Plenum, New York, pp. 210–212.

Valley J. W., Peck W. H., King E. M., and Wilde S. A. (2002) A cool early Earth. *Geology* **30**, 351–354.

Vervoort J. D., White W. M., and Thorpe R. I. (1994) Nd and Pb isotope ratios of the Abitibi greenstone belt: new evidence for very early differentiation of the Earth. *Earth Planet. Sci. Lett.* **128**, 215–229.

Vidal-Madjar A., Lecavelier des Etangs A., and Ferlet R. (1998) β Pictoris, a young planetary system? A review. *Planet. Space Sci.* **46**, 629–648.

Vityazev A. V., Pechernikova A. G., Bashkirov A. G. (2003) Accretion and differentiation of terrestrial protoplanetary bodies and Hf–W chronometry. In *Lunar Planet. Sci.* **XXXIV**, #1656. The Lunar and Planetary Institute, Houston (CD-ROM).

Vogel N., Baur H., Bischoff A., and Wieler R. (2002) Noble gases in chondrules and metal-sulfide rims of primitive chondrites—clues on chondrule formation. *Geochim. Cosmochim. Acta* **66**, A809.

Vogel N., Baur H., Leya I., and Wieler R. (2003) No evidence for primordial noble gases in CAIs. *Meteorit. Planet. Sci.* **38**(suppl.), A75.

Wade J. and Wood B. J. (2001) The Earth's "missing" niobium may be in the core. *Nature* **409**, 75–78.

Walker J. C. G. (1986) Impact erosion of planetary atmospheres. *Icarus* **68**, 87–98.

Walker R. J., Morgan J. W., and Horan M. F. (1995) Osmium-187 enrichment in some plumes: evidence for core–mantle interaction? *Science* **269**, 819–822.

Walter M. J., Newsom H. E., Ertel W., and Holzheid A. (2000) Siderophile elements in the Earth and Moon: Metal/silicate partitioning and implications for core formation. In *Origin of the Earth and Moon* (eds. R. M. Canup and K. Righter). University of Arizona Press, Tucson, pp. 265–289.

Wänke H. (1981) Constitution of terrestrial planets. *Phil. Trans. Roy. Soc. London* **A303**, 287–302.

Wänke H. and Dreibus G. (1986) Geochemical evidence for formation of the Moon by impact induced fission of the proto-Earth. In *Origin of the Moon* (eds. W. K. Hartmann,

R. J. Phillips, and G. J. Taylor). Lunar and Planetary Institute, Houston, pp. 649–672.

Wänke H. and Dreibus G. (1988) Chemical composition and accretion history of terrestrial planets. *Phil. Trans. Roy. Soc. London A* **325**, 545–557.

Wänke H. and Dreibus G. (1994) Chemistry and accretion of Mars. *Phil. Trans. Roy. Soc. London A* **349**, 285–293.

Wänke H., Baddenhausen H., Palme H., and Spettel B. (1974) On the chemistry of the Allende inclusions and their origin as high temperature condensates. *Earth Planet. Sci. Lett.* **23**, 1–7.

Wänke H., Dreibus G., Palme H., Rammensee W., and Weckwerth G. (1983) Geochemical evidence for the formation of the Moon from material of the Earth's mantle. In *Lunar Planet. Sci.* **XIV**. The Lunar and Planetary Institute, Houston, pp. 818–819.

Wänke H., Dreibus G., and Jagoutz E. (1984) Mantle chemistry and accretion history of the Earth. In *Archaean Geochemistry* (eds. A. Kroner, G. N. Hanson, and A. M. Goodwin). Springer, New York, pp. 1–24.

Ward W. R. (2000) On planetesimal formation: the role of collective particle behavior. In *Origin of the Earth and Moon* (eds. R. M. Canup and K. Righter). University of Arizona Press, Tucson, pp. 75–84.

Wasserburg G. J., MacDonald F., Hoyle F., and Fowler W. A. (1964) Relative contributions of uranium, thorium, and potassium to heat production in the Earth. *Science* **143**, 465–467.

Wasserburg G. J., Papanastassiou D. A., Tera F., and Huneke J. C. (1977a) Outline of a lunar chronology. *Phil. Trans. Roy. Soc. London A* **285**, 7–22.

Wasserburg G. J., Tera F., Papanastassiou D. A., and Huneke J. C. (1977b) Isotopic and chemical investigations on Angra dos Reis. *Earth Planet. Sci. Lett.* **35**, 294–316.

Wasserburg G. J., Busso M., Gallino R., and Raiteri C. M. (1994) Asymptotic giant branch stars as a source of short-lived radioactive nuclei in the solar nebula. *Astrophys. J.* **424**, 412–428.

Wasserburg G. J., Busso M., and Gallino R. (1996) Abundances of actinides and short-lived nonactinides in the interstellar medium: diverse supernova sources for the r-processes. *Astrophys. J.* **466**, L109–L113.

Wasson J. T. (1985) *Meteorites: Their Record of Early Solar-system History*. W. H. Freeman, New York, 251p.

Wasson J. T. and Chou C.-L. (1974) Fractionation of moderately volatile elements in ordinary chondrites. *Meteoritics* **9**, 69–84.

Weidenschilling S. J. (1977a) The distribution of mass in the planetary system and solar nebula. *Astrophys. Space Sci.* **51**, 153–158.

Weidenschilling S. J. (1977b) Aerodynamics of solid bodies in the solar nebula. *Mon. Not. Roy. Astron. Soc.* **180**, 57–70.

Weidenschilling S. J. (2000) Formation of planetesimals and accretion of the terrestrial planets. *Space Sci. Rev.* **92**, 295–310.

Wetherill G. W. (1975a) Radiometric chronology of the early solar system. *Ann. Rev. Nuclear Sci.* **25**, 283–328.

Wetherill G. W. (1975b) Late heavy bombardment of the moon and terrestrial planets. *Proc. 6th Lunar Sci. Conf.* 1539–1561.

Wetherill G. W. (1980) Formation of the terrestrial planets. *Ann. Rev. Astron. Astrophys.* **18**, 77–113.

Wetherill G. W. (1986) Accumulation of the terrestrial planets and implications concerning lunar origin. In *Origin of the Moon* (eds. W. K. Hartmann, R. J. Phillips, and G. J. Taylor). Lunar and Planetary Institute, Houston, pp. 519–551.

Wetherill G. W. (1992) An alternative model for the formation of the Asteroids. *Icarus* **100**, 307–325.

Wetherill G. W. (1994) Provenance of the terrestrial planets. *Geochim. Cosmochim. Acta* **58**, 4513–4520.

Wetherill G. W. and Stewart G. R. (1993) Formation of planetary embryos: effects of fragmentation, low relative velocity, and independent variation of eccentricity and inclination. *Icarus* **106**, 190–209.

Wiechert U. (2002) Earth's early atmosphere. *Science* **298**, 2341–2342.

Wiechert U., Halliday A. N., Lee D.-C., Snyder G. A., Taylor L. A., and Rumble D. A. (2001) Oxygen isotopes and the Moon-forming giant impact. *Science* **294**, 345–348.

Wiechert U., Halliday A. N., Palme H., and Rumble D. (2003) Oxygen isotopes in HED meteorites and evidence for rapid mixing in planetary embryos. *Earth Planet. Sci. Lett.* (in press).

Wieler R., Anders E., Baur H., Lewis R. S., and Signer P. (1992) Characterisation of Q-gases and other noble gas components in the Murchison meteorite. *Geochim. Cosmochim. Acta* **56**, 2907–2921.

Wieler R., Kehm K., Meshik A. P., and Hohenberg C. M. (1996) Secular changes in the xenon and krypton abundances in the solar wind recorded in single lunar grains. *Nature* **384**, 46–49.

Wiesli R. A., Beard B. L., Taylor L. A., Welch S. A., and Johnson C. M. (2003) Iron isotope composition of the lunar mare regolith: implications for isotopic fractionation during production of single domain iron metal. In *Lunar Planet Sci.* **XXXIV**, #1500. The Lunar and Planetary Institute, Houston (CD-ROM).

Wilde S. A., Valley J. W., Peck W. H., and Graham C. M. (2001) Evidence from detrital zircons for the existence of continental crust and oceans on the Earth 4.4 Gyr ago. *Nature* **409**, 175–178.

Wolf R., Richter G. R., Woodrow A. B., and Anders E. (1980) Chemical fractionations in meteorites: XI. C2 chondrites. *Geochim. Cosmochim. Acta* **44**, 711–717.

Wood B. J. (1993) Carbon in the core. *Earth Planet. Sci. Lett.* **117**, 593–607.

Wood J. A. (1984) Moon over Mauna Loa: a review of hypotheses of formation of Earth's moon. In *Origin of the Moon* (eds. W. K. Hartmann, R. J. Phillips, and G. J. Taylor). Lunar and Planetary Institute, Houston, pp. 17–55.

Wood J. A. (2000) Pressure ands temperature profiles in the solar nebula. *Space Sci. Rev.* **92**, 87–93.

Wood J. A. and Morfill G. E. (1988) A review of solar nebula models. In *Meteorites and the Early Solar System* (eds. J. F. Kerridge and M. S. Matthews). University of Arizona Press, Tucson, pp. 329–347.

Yi W., Halliday A. N., Alt J. C., Lee D.-C., Rehkämper M., Garcia M., Langmuir C., and Su Y. (2000) Cadmium, indium, tin, tellurium and sulfur in oceanic basalts: implications for chalcophile element fractionation in the earth. *J. Geophys. Res.* **105**, 18927–18948.

Yin Q. Z. and Jacobsen S. B. (1998) The ^{97}Tc–^{97}Mo chronometer and its implications for timing of terrestrial accretion and core formation. In *Lunar Planet Sci.* **XXIX**, #1802. The Lunar and Planetary Institute, Houston (CD-ROM).

Yin Q. Z., Jacobsen S. B., McDonough W. F., Horn I., Petaev M. I., and Zipfel J. (2000) Supernova sources and the ^{92}Nb–^{92}Zr p-process chronometer: *Astrophys. J.* **535**, L49–L53.

Yin Q. Z., Jacobsen S. B., Yamashita K., Blicher-Toft J., Télouk P., and Albarède F. (2002) A short timescale for terrestrial planet formation from Hf–W chronometry of meteorites. *Nature* **418**, 949–952.

Yoshino T., Walter M. J., and Katsura T. (2003) Core formation in planetesimals triggered by permeable flow. *Nature* **422**, 154–157.

Young E. D. (2000) Assessing the implications of K isotope cosmochemistry for evaporation in the preplanetary solar nebula. *Earth Planet. Sci. Lett.* **183**, 321–333.

Zahnle K. J. (1993) Xenological constraints on the impact erosion of the early Martian atmosphere. *J. Geophys. Res.* **98**, 10899–10913.

Zahnle K. J. and Sleep N. H. (1996) Impacts and the early evolution of life. In *Comets and the Origin and Evolution of Life* (eds. P. J. Thomas, C. F. Chyba, and C. P. McKay). Springer, Heidelberg, pp. 175–208.

Zahnle K. J. and Walker J. C. G. (1982) The evolution of solar ultraviolet luminosity. *Rev. Geophys.* **20**, 280.

Zahnle K. J., Kasting J. F., and Pollack J. B. (1988) Evolution of a steam atmosphere during Earth's accretion. *Icarus* **74**, 62–97.

Zhang Y. (1998) The young age of the Earth. *Geochim. Cosmochim. Acta* **62**, 3185–3189.

1.21
The Moon

P. H. Warren

University of California, Los Angeles, CA, USA

1.21.1 INTRODUCTION: THE LUNAR CONTEXT	559
1.21.2 THE LUNAR GEOCHEMICAL DATABASE	561
1.21.2.1 *Artificially Acquired Samples*	561
1.21.2.2 *Lunar Meteorites*	561
1.21.2.3 *Remote-sensing Data*	562
1.21.3 MARE VOLCANISM	564
1.21.3.1 *Classification of Mare Rocks*	564
1.21.3.2 *Chronology and Styles of Mare Volcanism*	569
1.21.3.3 *Mare Basalt Trace-element and Isotopic Trends*	573
1.21.4 THE HIGHLAND CRUST: IMPACT BOMBARDMENT AND EARLY DIFFERENTIATION	577
1.21.4.1 *Polymict Breccias and the KREEP Component*	577
1.21.4.2 *Bombardment History of the Moon*	578
1.21.4.3 *Impactite and Regolith Siderophile Signatures*	581
1.21.4.4 *Pristine Highland Rocks: Distinctiveness of the Ferroan Anorthositic Suite*	583
1.21.4.5 *The Magma Ocean Hypothesis*	587
1.21.4.6 *Alternative Models*	590
1.21.5 THE BULK COMPOSITION AND ORIGIN OF THE MOON	591
ACKNOWLEDGMENTS	592
REFERENCES	592

1.21.1 INTRODUCTION: THE LUNAR CONTEXT

Oxygen isotopic data suggest that there is a genetic relationship between the constituent matter of the Moon and Earth (Wiechert *et al.*, 2001). Yet lunar materials are obviously different from those of the Earth. The Moon has no hydrosphere, virtually no atmosphere, and compared to the Earth, lunar materials uniformly show strong depletions of even mildly volatile constituents such as potassium, in addition to N_2, O_2, and H_2O (e.g., Wolf and Anders, 1980). Oxygen fugacity is uniformly very low (BVSP, 1981) and even the earliest lunar magmas seem to have been virtually anhydrous. These features have direct and far-reaching implications for mineralogical and geochemical processes. Basically, they imply that mineralogical diversity and thus variety of geochemical processes are subdued; a factor that to some extent offsets the comparative dearth of available data for lunar geochemistry.

The Moon's gross physical characteristics play an important role in the more limited range of selenochemical compared to terrestrial geochemical processes. Although exceptionally large (radius = 1,738 km) in relation to its parent planet, the Moon is only 0.012 times as massive as Earth. By terrestrial standards, pressures inside the Moon are feeble: the upper mantle gradient is 0.005 GPa km^{-1} (versus 0.033 GPa km^{-1} in Earth) and the central pressure is slightly less than 5 GPa. However, lunar interior pressures are sufficient to profoundly influence igneous processes (e.g., Warren and Wasson, 1979b; Longhi, 1992, 2002), and in this sense the Moon more resembles a planet than an asteroid.

Another direct consequence of the Moon's comparatively small size was early, rapid decay of its internal heat engine. But the Moon's thermal disadvantage has resulted in one great advantage for planetology. Lunar surface terrains, and many of the rock samples acquired from them, retain for the most part characteristics acquired during the first few hundred million years of solar system existence. The Moon can thus provide crucial insight into the early development of the Earth, where the direct record of early evolution was effectively destroyed by billions of years of geological activity. Lunar samples show that the vast majority of the craters that pervade the Moon's surface are at least 3.9 Gyr old (Dalrymple and Ryder, 1996). Impact cratering has been a key influence on the geochemical evolution of the Moon, and especially the shallow Moon.

The uppermost few meters of the lunar crust, from which all lunar samples are derived, is a layer of loose, highly porous, fine impact-generated debris—regolith or lunar "soil." Processes peculiar to the surface of an atmosphereless body, i.e., effects of exposure to solar wind, cosmic rays, and micrometeorite bombardment, plus spheroidal glasses formed by in-flight quenching of pyroclastic or impact-generated melt splashes, all are evident in any reasonably large sample of lunar soil (Lindsay, 1992; Keller and McKay, 1997; Eugster et al., 2000). The lunar regolith is conventionally envisaged as having a well-defined lower boundary, typically 5–10 m below the surface (McKay et al., 1991); below the regolith is either (basically) intact rock, or else a somewhat vaguely defined "megaregolith" of loose but not so finely ground material. Ancient highland terrains tend to have a regolith roughly 2–3 times than that of the maria (Taylor, 1982). However, in much of the highlands the regolith/megaregolith "boundary" may be gradational. The growth of a regolith can approach a steady-state thickness by shielding its substrate against further impacts (Quaide and Oberbeck, 1975), but there is no reason to believe that the size–frequency spectrum of impactors bombarding the Moon (Melosh, 1989; Neukum et al., 2001) features a discontinuity at whatever size (of order 1–10 m) would be necessary to limit disintegration to ~10 m.

All lunar samples are from the regolith, so the detailed provenance of any individual lunar sample is rarely obvious; and for ancient highland samples, never obvious. The closest approach to *in situ* sampling of bedrock came on the Apollo 15 mission. The regolith is very thin near the edge of the Hadley Rille, and many samples of clearly comagmatic basalts were acquired within meters of their 3.3 Ga "young," nearly intact, lava flow, so that their collective provenance is certain (Ryder and Cox, 1996). Even the regional provenance of any individual lunar sample is potentially allocthonous. However, most lunar rocks, even ancient highland rocks, are found within a few hundred kilometers of their original locations. This conclusion stems from theoretical modeling of cratered landscapes (Shoemaker et al., 1970; Melosh, 1989), plus observational evidence such as the sharpness of geochemical boundaries between lava-flooded maria and adjacent highlands (e.g., Li and Mustard, 2000).

Besides breaking up rock into loose debris, impacts create melt. Traces of melt along grain boundaries may suffice to produce new rock out of formerly loose debris; the resultant rock would be classified as either regolith breccia or fragmental breccia, depending upon whether surface fines were important, or not, respectively, in the precursor matter (Stöffler et al., 1980). Features diagnostic of a surface component include the presence of glass spherules (typically a mix of endogenous mare-pyroclastic glasses and impact-splash glasses) or abundant solar-wind-implanted noble gases (e.g., Eugster et al., 2000).

Elsewhere, especially in the largest events in which a planet's gravitational strength limits displacement and the kinetic energy of impact is mainly partitioned into heat (Melosh, 1989), impact melt may constitute a major fraction of the volume of the material that becomes new rock. Rocks formed in this manner are classified as impact-melt breccias and subclassified based on whether they are clast-poor or clast-rich, and whether their matrix is crystalline or glassy (Stöffler et al., 1980). Obvious lithic and mineral clasts are very common in impact-melt breccias, although the full initial proportion of clasts may not be evident in the final breccia. Some of the clasts may be so pulverized, especially in large impact events (Schultz and Mendell, 1978), that they are "lost" by digestion into comingled superheated impact melt (Simonds et al., 1976). By some definitions, the term impact-melt breccia may be applied to products of melt plus clast mixtures with initial melt proportion as low as 10 wt.% (Simonds et al., 1976; Papike et al., 1998).

A few impactites feature a recrystallized texture, i.e., they consist dominantly of a mosaic of grains meeting at ~120° triple junctions. These metamorphic rocks, termed granulitic breccias, may form from various precursor igneous or impactite rocks, and the heat source may be regional (burial) or local, such as a nearby impact melt (Stöffler et al., 1980). But lunar granulitic breccias are almost invariably fine grained, and they tend to be "contaminated" with meteoritic siderophile elements (e.g., M. M. Lindstrom and D. J. Lindstrom, 1986; Warren et al., 1991; Cushing et al., 1999), implying that the precursor rocks were probably mostly shallow impact breccias (brecciation and siderophile-element contamination being concentrated near the surface), and the heat source was probably most often a proximal mass of impact melt.

Besides impactites, which are predominant near the bombarded surface, virtually all other lunar crustal rocks are igneous or annealed-igneous. The super-arid Moon has never produced (by any conventional definition) sedimentary rock, and most assuredly has never hosted life. Even metamorphism is of reduced scope, with scant potential for fluid-driven metasomatism. Evidence for metamorphism among returned lunar samples is mostly confined to impact shock and thermal effects. Although regional burial metamorphism may occur (Stewart, 1975), deeply buried materials seldom find their way into the surface regolith, whence all samples come. Annealing of lunar rocks is more likely a product of simple post-igneous slow cooling (at significant original depth), dry baking in proximity to an intrusion, or baking within a zone of impact heating.

The Moon's repertoire of geochemical processes may seem limited, but it represents a key link between the sampled asteroids (see Chapters 1.05 and 1.11) and the terrestrial planets. Four billion years ago, at a time when all but monocrystalline bits of Earth's dynamic crust were fated for destruction, most of the Moon's crust had already achieved its final configuration. The Moon thus represents a unique window into the early thermal and geochemical state of a moderately large object that underwent igneous differentiation in the inner solar system, and into the cratering history of near-Earth space.

1.21.2 THE LUNAR GEOCHEMICAL DATABASE

1.21.2.1 Artificially Acquired Samples

Six Apollo missions acquired a total of 382 kg of rocks and soil. Sampling was mostly by either simple scooping of bulk regolith, or by collection of large individual samples, mostly much bigger than 0.1 kg (four of the six missions collected individual rocks >8 kg). As a result, the particle size distribution of the overall sample is strongly bimodal. Of the total Apollo collection, rocks big enough to not pass a 10 mm sieve comprise 70 wt.%, yet the fraction between 1 nm and 10 mm adds only 2–3 wt.%, and the remaining 27–28 wt.% of the material is <1 mm fines (including core fines) (Vaniman et al., 1991). Three Russian unmanned Luna missions added a total of 0.20 kg of bulk regolith. All nine of the lunar sample-return sites are tightly clustered within the central-eastern region of the Moon's nearside hemisphere. These sites can be encompassed within a polygon covering just 4.4% of the Moon's surface; if limited to rock-sampling (Apollo) sites, the polygon's coverage is merely 2.7%.

1.21.2.2 Lunar Meteorites

In contrast with the Apollo and Luna sites, the distribution of source craters for lunar meteorites is virtually random. This conclusion is based on the randomness of overall cratering (Bandermann and Singer, 1973), plus celestial-mechanical constraints (Gault, 1983; Gladman et al., 1996) indicating that the vast majority of Moon–Earth journeys are not direct, but involve a phase of geocentric or even heliocentric orbit. All of the lunar meteorites are finds (in some cases significantly weathered), and their masses tend to be curiously low in comparison to other achondrites, including martian meteorites (Table 1). Lunar provenance is proven for these meteorites by a variety of observations, but the single most useful constraint comes from oxygen isotopes (Clayton and Mayeda, 1996).

In addition to the usual potential for pairing among meteorites, lunar meteorites have an important potential for launch (source-crater) pairing. To date, with only ~20 lunar meteorites to cross-compare, launch pairing can often be ruled out based on cosmic ray exposure (CRE) constraints (e.g., Benoit et al., 1996; Thalmann et al., 1996; Nishiizumi et al., 2002; Chapter 1.13). The launch age is the sum of the terrestrial age (usually brief, of order $10^3 - 10^4$ yr) plus the "4π" (all-directional exposure, Moon-to-Earth) CRE age. Unfortunately, the most unambiguous 4π CRE constraints, based on radioisotopic methods, become increasingly imprecise for CRE ages much lower than 10^6 yr, and hopelessly imprecise for CRE ages much less than 10^5 yr, whereas physical modeling shows that, over time, most lunar meteorites have Moon-to-Earth journeys that take less than 10^5 yr (Gault, 1983; Gladman et al., 1996). When CRE does not resolve the launch pairing question, it may be possible to virtually rule out launch pairing on the basis that materials from a single source crater will generally show a degree of overall geochemical similarity. This approach is most useful if one or both of the samples is a thoroughly polymict breccia, as exemplified by regolith breccia, i.e., a type of material that is usually representative of its region. In this connection, it is important to realize that the scale of the launch zone is far smaller than the full diameter of a crater, because ejection velocity is a strong function of proximity to the point of impact (Warren, 1993). Unfortunately, uncertainty about launch pairing among lunar meteorites is bound to increase as more and more samples are acquired.

In principle, a thorough sampling of the lunar surface might yield pieces of Earth impact-transported to the Moon (these might be dubbed "retrometeorites"), a large proportion of which would presumably date from the era of heavy bombardment, 3.9 Ga and before (see below). However, Earth's escape velocity is high even compared to the Moon and Mars, and rocks of terrestrial provenance have not been discovered among

Table 1 Known lunar meteorites, as of early 2003.

	Meteorite	Mass (g)	Pair total[a] (g)	Rock type[b]	Year(s)[c] discovered	wt.% Al_2O_3	mol.% mg
	Highland (and mostly highland) meteorites						
1	**ALH81005**[d]	31.4	31.4	Rego B, highland	1982	26	73
2	Calcalong Creek	18	18	Rego B, mainly highland	1991	21	?
3	Dar al Gani 262	513	513	Rego B, highland	1997	28	66
4	Dar al Gani 400	1,425	1,425	Polymict B, highland	1998	28	71
5	Dhofar 025	751	772	Rego B, highland	2000–2001	27	71
	paired samples		Dhofars 301 (2 g), 304 (10 g), 308 (9 g)				
6	Dhofar 026 (13 stones)	636	636	Polymict B, highland	2000	27	71
7	Dhofar 081	174	564	Polymict B, highland	1999–2001	31	61
	paired samples		Dhofars 280 (251 g), 302 (3.8 g), 303 (4.1 g), 305 (34), 306 (13), 307 (50), 490 (34)				
8	Dhofar 489	34.4	34.4	Polymict B, highland	2001	30 (est.)	?
9	**MAC88104**	61	723	Rego B, highland	1989	28	63
	paired sample		MAC88105 (662 g)				
10	NWA482	1,015	1,015	IMB, highland	2001	29	66
11	**QUE93069**	21.4	24.5	Rego B, highland	1994–1995	28	66
	paired sample		QUE94269 (3.1 g)				
12	**Yamato-791197**	52	341	Rego B, highland	1983–2002	26	63
	paired (???) sample		Yamato-983885 (289 g)				
13	**Yamato-82192**	37	712	Polymict B, highland	1984–1987	28	66
	paired samples		Yamato-82193 (27 g), Yamato-86302 (648 g)				
	Mare (and mostly mare) meteorites						
1	**Asuka-881757**	442	442	Gabbro, mare	1990	11	33
2	Dhofar 287	154	154	Basalt + Rego B, mare	2001	8	50
3	**EET87521**	31	84	Polymict B, mare	1989–1998	13	42
	paired sample		EET96008 (53 g)				
4	NWA032	300	456	Basalt, mare	2000	9	40
	paired sample		NWA479 (156 g)				
5	NWA773 (3 stones)	633	633	Gabbro + Rego B, mare	2001	6	70
6	**QUE94281**	23.4	23.4	Rego B, mainly mare	1995	16	52
7	**Yamato-793169**	6.1	6.1	Basalt, mare	1990	12	31
8	**Yamato-793274**	8.7	195	Rego B, mainly mare	1987–1999	15	53
	paired sample		Yamato-981031 (186 g)				

[a] Aside from the conventional pairings noted in the table, several launch pairs have been inferred: Yamato-793169 with Asuka-881757, and QUE94281 with Yamato-793274 (and possibly also with EET87521) (Arai and Warren, 1999; Korotev et al., 2003). There may be other launch pairs that are not manifested (i.e., the rocks are not sufficiently ideosyncratic). [b] Abbreviations: Rego, regolith; B, breccia; IMB, impact-melt breccia. The classification "polymict breccia" for six of the samples may seem vague, but most of these six samples are either difficult to classify or simply little-studied. EET87521 is clearly a fragmental breccia. [c] Listed "discovery" dates refer to year of discovery of lunar provenance, not year sample was first collected as a meteorite. [d] Samples shown in **bold** font are from Antarctica, and thus generally less weathered than the other (hot-desert) finds.

the meteorites hitting Earth in modern times; only glasses (tektites) are known to have re-entered from space. In any case, to date no "retrometeorite" has been found among the lunar samples.

1.21.2.3 Remote-sensing Data

The period since the early 1990s has been a golden age for lunar remote sensing, thanks to the 1994 Clementine mission and the 1997–1998 Lunar Prospector mission. Among the pre-1994 remote-sensing geochemical databases, the most notable (i.e., still not completely superceded) is probably the X-ray spectrometry data obtained for ~10% of the lunar surface on the Apollo 15 and 16 missions. These data were reported as Al/Si, Mg/Al, and Mg/Si ratios, at a spatial resolution of ~50 km (e.g., Adler et al., 1973; Bielefeld et al., 1977; Andre and El-Baz, 1981). The more precisely determined Al/Si ratio data show trends that could be gleaned almost as well from simple albedo variations. Among lunar rocks, aluminum is strongly linked with light-colored feldspar, while silicon is relatively constant (probably as a consequence of the lesser potential to form silicon-free oxides, except "late" ilmenite, in the reducing lunar environment). The Mg/Si and Mg/Al data suffer from poor precision. For example, the average 1σ error in Mg/Si reported for 22 large regions by Adler et al. (1973) is 26%.

Clementine used four different cameras to map the global surface reflectance of the Moon at eleven different wavelengths, from the near-ultraviolet (415 nm) to the near-infrared (2,800 nm), in roughly one million images.

The workhorse UV/visible camera had filter center wavelengths and bandpass widths (FWHM) of 415 (40) nm, 750 (10) nm, 900 (30) nm, 950 (30) nm, and 1,000 (30) nm; and a broadband filter covering 400–950 nm. Depending upon orbit parameters and camera, pixel resolution generally varied from 100 m to 300 m. Clementine's highest resolution camera took 600,000 images at typical pixel resolution of 7–20 m, for four different wavelengths (415–750 nm) but covering only selected areas of the Moon.

Exploitation of the vast treasure trove of Clementine data is still at a fairly basic stage. Clementine multispectral images have been used to map surface mineralogy and, for a few elements, chemical composition, and even regolith maturity (i.e., average extent of exposure to surface-regolithic processing) (e.g., Lucey et al., 2000a,b; Staid and Pieters, 2001; Shkuratov et al., 2003). Translation from spectral data into concentration is most straightforward for iron, using primarily data from 900 m to 1,000 nm, the region of a major Fe^{2+} absorption band for pyroxene (Lucey et al., 2000a). The technique for titanium is less direct, relying on the spectrum slope as determined by the ratio between the 415 nm and longer wavelength (especially 750 nm) reflectances. The technique for soil maturity depends on an even more subtle analysis of slopes in different portions of the spectrum (Pieters et al., 2002). Plagioclase (or aluminum) is gauged mainly from simple albedo (Shkuratov et al., 2003). There are a number of complications: e.g., olivine/pyroxene ratio, major opaque phases other than the presumed dominant opaque (ilmenite), the shock state of plagioclase, sunlight phase angle (mainly a function of latitude, but also affected by local slopes), and the maturity grain size and glass abundance (three related parameters) of the regolith (Lucey et al., 1998). The important database for TiO_2 has been somewhat controversial, as the early calibration showed poor agreement for a few of the sampled landing sites; Gillis et al. (2003) have proposed an *ad hoc* remedy. The high spatial resolution of the Clementine data can be exploited to address specific lunar geology issues, for example, the petrology of individual cryptomaria (Antonenko et al., 1995; Hawke et al., 2003), the petrology of uplifted central peaks of moderately large impact craters (Tompkins and Pieters, 1999), and compositional variations within the colossal and yet remote (in relation to all Apollo and Luna sampling sites) South Pole Aitken basin (Pieters et al., 2001).

Lunar Prospector's two most important geochemical mapping sensors were designed for gamma-ray spectrometry (GRS) and neutron spectrometry. The neutron spectrometer, a matched pair of ^{3}He gas proportional counters, was mainly designed to map the global distribution of regolith hydrogen (Feldman et al., 2001). One of the counters was covered by a 0.63 mm sheet of cadmium, making it responsive only to epithermal neutrons. The difference in counting rates between the cadmium-covered counter and the uncovered counter gave a measure of the thermal neutron flux. Locally very high hydrogen concentrations were found near the poles, and Feldman et al. (2001) argue that these must be more than mere accumulations of solar-wind-implanted hydrogen, i.e., that "a significant proportion of the enhanced hydrogen near both poles is most likely in the form of water [ice] molecules." Long ago, Arnold (1979) had conjectured that water liberated in impacts between comets and the Moon might have accumulated in cold traps within permanently shadowed regions near the lunar poles. However, the high hydrogen concentrations found by Feldman et al. (2001) extend over regions far larger than the minor, scattered areas of permanent shadow. As should surprise no one, the water–ice interpretation has been controversial (e.g., Crider and Vrondak, 2000; Starukhina and Shkuratov, 2000).

The GRS technique measures the upper few decimeters of the regolith, whereas reflectance spectrometry (e.g., Clementine) measures the very surface, and the Apollo orbital XRF data measured the uppermost ~10 μm. The difference is not very important, because the lunar regolith is so extensively turned over by repeated impacts, or in lunar terminology, gardened. Prospector's GRS detector was of bismuth germanate (BGO) type, with similar resolution to the NaI(Tl) detectors flown on the Apollo 15 and 16 missions. The Prospector GRS database is far superior, however, because of vastly longer detector acquisition times (i.e., better counting statistics), and because Prospector's coverage was global, whereas the Apollo data covered no more than 20% of the surface in two near-equatorial bands.

The two elements most amenable to orbital GRS are iron and thorium. Although initial thorium mapping was done at a spatial resolution of 150 km (Lawrence et al., 1998), iron and thorium have recently been mapped to a remarkable 15 km (Lawrence et al., 2002a,b). Spatial resolution of 60 km has been achieved for potassium and titanium (Prettyman et al., 2002), and also, based on a complex technique involving comparison with data from Prospector's neutron spectrometer, for the rare earth element (REE) samarium (Elphic et al., 2000). The GRS data have also yielded maps at spatial resolution of 150 km for oxygen, magnesium, aluminum, silicon, calcium, and uranium (Prettyman et al., 2002). Establishing the full scientific import of all these new data will take some years, but one of the most striking implications is a remarkable degree of global geochemical asymmetry. For example, average surface concentration of thorium (an exemplary incompatible trace element; on the Moon such elements are strongly concentrated into a potassium, REE- and phosphorus-rich component known as KREEP) is 3.5 times higher on the hemisphere centered over Oceanus Procellarum compared to the hemisphere antipodal to Procellarum, although it should be noted that this conclusion is based on a correction to the initial (e.g., Lawrence et al., 1998) calibration for the Prospector thorium data (Warren, 2003).

1.21.3 MARE VOLCANISM

1.21.3.1 Classification of Mare Rocks

The dark basalts that erupted to veneer the Moon's flat, low-lying "seas" (maria) during the waning of lunar magmatism are compositionally distinctive, even compared to other lunar samples. Mare basalts have high (mostly >16 wt.%) FeO, low to moderate mg (= MgO/[MgO + FeO]), low to high TiO_2, and low Al_2O_3 contents (Figure 1 and Table 2). Calling these rocks "basalt" is potentially misleading. Most mare basalts are far more melanocratic (mafic silicate-rich feldspar-poor) than terrestrial basalt, and arguably more analogous to a relatively leucocratic and low-mg komatiite. The contrast with the rest of the Moon's crust is stark. The typical composition of highland crust, as represented by lunar-meteoritic regolith breccias and regolith averages for the Apollo 14, Apollo 16, and Luna 20 sites (Table 3), features ~5 wt.% FeO, ~0.4 wt.% TiO_2, and ~27 wt.% Al_2O_3. The distinction between mare and highland materials is seldom controversial, but for quick-and-dirty mare versus highland classification based exclusively on bulk-analysis data, Wood (1975) proposed using a combination of wt.% TiO_2 and Ca/Al ratio (Figure 2).

The various subclassifications for mare basalt we describe below may seem arbitrary but, as will be discussed in the next section, mare basalt compositional diversity is in general not systematic, but haphazard. On Earth, volcanic diversity is largely a function of systematic global tectonics, such as upwelling at mid-ocean ridges to generate MORB. But plate tectonics probably never occurred on the Moon; and it surely never occurred during the comparatively late era of mare volcanism.

Mare basalts are classified primarily on the basis of their highly diverse bulk TiO_2 contents. Those with <1.5 wt.% TiO_2 are termed very-low-Ti (VLT), those with $1.5 < TiO_2 < 6$ wt.% as medium-Ti, and those with >6 wt.% TiO_2 as high-Ti. In pre-2001 literature (e.g., Neal and Taylor, 1992), these same classes were called "VLT," low-Ti, and high-Ti, respectively. The bizarre usage of "low-Ti" for basalts with up to 6 wt.% TiO_2 was a historical accident: in the first studies of lunar rocks from Apollo 11, the only large igneous rocks happened to all be "high-Ti." Le Bas (2001) has suggested calling these three groups titanium-poor, medium-Ti, and titanium-rich, but judging from more recent publications (e.g., Hiesinger et al., 2002; Gillis et al., 2003; Jolliff et al., 2003), this terminology has not been widely adopted. To minimize confusion, I choose to adopt the Le Bas "medium-Ti" replacement for "low-Ti," but continue using the old and logical terms VLT and high-Ti.

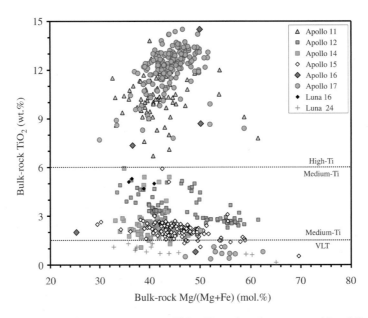

Figure 1 Mare basalt major elements: mg versus TiO_2. Note that the apparent bimodality in TiO_2 may be misleading (see text) (data are mainly from the compilation of Haskin and Warren, 1991).

Table 2 Summary of literature compositional data for lunar mare meteorites[a] and other major varieties of mare basalt (see text).

Sample		Na_2O (wt.%)	MgO (wt.%)	Al_2O_3 (wt.%)	SiO_2 (wt.%)	K_2O (wt.%)	CaO (wt.%)	TiO_2 (wt.%)	FeO (wt.%)	Sum (wt.%)	Sc (µg g⁻¹)	V (µg g⁻¹)	Cr (mg g⁻¹)	Co (µg g⁻¹)	Ni (µg g⁻¹)	Ga (µg g⁻¹)	Rb (µg g⁻¹)	Sr (µg g⁻¹)	Zr (µg g⁻¹)	Ba (µg g⁻¹)	La (µg g⁻¹)	Sm (µg g⁻¹)	Eu (µg g⁻¹)	Tb (µg g⁻¹)	Lu (µg g⁻¹)	Hf (µg g⁻¹)	Ir (ng g⁻¹)	Th (µg g⁻¹)
Dhofar 287	avg.	0.52	12.8	8.1	43.9	0.15	8.4	2.86	22.2	98.9	56		0.62	42	50			142	175		13		1.2		0.5	5.0		1.90
NWA 032	avg.	0.36	8.5	8.7	44.7	0.11	10.9	3.08	22.8	99.1			0.40	88	210						11.2	6.6	1.10	1.56	0.80			
NWA 773	avg.	0.03	27	4	43	0.02	5.9	0.3	19	99.3			2.7								10		0.4		0.4	2.6		
YA meteorite[b]	avg.	0.32	6.0	11.0	45.5	0.05	12.0	2.1	22.3	99.2	93	81	1.59	24	11	2.8	1.1	146	92	66	4.0	3.5	1.14	0.93	0.59		0.05	0.53
All Apollo 11 mare basalts	avg.	0.48	7.6	9.6	40.5	0.16	11.1	10.5	19.5	100.0	83	78	2.1	21	5.4	5.8	2.8	177	433	208	18	16	2.1	3.7	2.0	12.6	0.027	2.1
	SD	0.11	1.0	1.5	1.3	0.11	0.9	1.3	1.2		8	19	0.5	7		2.2	2.3	32	199	107	9	5	0.4	1.1	0.5	4.1		1.2
	N	97	89	89	29	100	89	90	89		74	62	86	78	6	18	22	17	35	74	89	85	85	82	80	73	7	65
All Apollo 12 mare basalts	avg.	0.27	10.8	9.3	45.1	0.06	10.0	3.4	20.4	100.0	50	158	3.6	46	47	3.6	1.0	120	133	70	6.8	5.0	1.13	1.35	0.65	4.0	0.036	0.78
	SD	0.10	3.5	1.6	1.6	0.01	1.6	0.8	1.2		7	30	1.0	13		1.2	0.3	30	62	23	2.3	1.2	0.38	0.398	0.18	1.0		0.23
	N	90	74	78	74	96	79	83	84		68	30	91	79	23	33	44	60	52	82	58	87	70	82	78	56	25	30
All Apollo 14 mare basalts	avg.	0.49	9.5	12.7	48.2	0.30	10.7	2.5	16.2	101.2	56	114	3.1	32	22	3.7	19	99	314	273	17	9	1.29	1.98	1.0	7.0	0.021	2.3
	SD	0.11	1.5	1.3	1.4	0.37	0.8	0.9	1.9		8	18	0.8	6	2	0.4	12	39	162	354	11	5	0.90	1.02	0.5	4.0		2.2
	N	94	91	90	24	87	94	91	93		93	59	93	93		2	24	39	46	81	95	93	94	93	93	93	3	90
All Apollo 15 mare basalts	avg.	0.27	10.4	9.1	46.3	0.05	9.8	2.1	21.2	100.1	40	202	3.9	51	58	3.8	1.0	116	90	72	5.7	3.4	0.85	0.75	0.34	2.34	0.043	0.57
	SD	0.05	2.6	1.3	1.7	0.01	1.1	0.5	1.5		5	29	1.2	12		1.0	0.9	21	34	25	2.6	0.8	0.18	0.19	0.08	0.84		0.18
	N	129	133	132	88	95	133	135	132		99	45	119	99	23	33	33	49	38	58	98	93	93	93	92	85	29	43
All Luna 16 mare basalts	avg.	0.52	6.4	13.4	43.8	0.18	11.7	5.0	18.7	100.2	64	75	1.56	17.5	79	3.7	1.9	404	300	335	18	13	3.27	2.53	1.13	0.60		1.9
	SD	0.02	0.5	0.2		0.04	0.3	0.2	0.4		6	10	0.12	0.9						80	2	2	0.32	0.16	0.10			0.3
	N	4	4	4	1	4	4	4	4		4	4	4	4	1	1	1	1	1	4	4	4	4	4	4	1		4
All Luna 24 mare basalts	avg.	0.27	9.4	11.2	45.7	0.03	11.6	0.8	20.1	99.7	46	163	2.1	40	43	1.8		103	50	43	2.0	1.6	0.66	0.36	0.23	1.4		0.19
	SD	0.07	4.3	1.9	1.2	0.01	2.1	0.3	1.6		8	13	0.8	7				8		5	1.1	0.7	0.08	0.16	0.11	1.4		0.04
	N	19	17	17	9	12	17	17	19		13	11	19	13	5	1		4	1	3	13	13	11	12	13	12		5
Large (high-Ti) Apollo 17 mare basalts	avg.	0.39	8.4	8.9	38.9	0.06	10.5	12.1	18.9	99.0	80	105	2.95	20.3	2.0	5.1	0.70	173	216	106	5.8	8.3	1.75	2.19	1.16	7.5	0.05	0.26
	SD	0.04	1.2	0.7	1.1	0.02	0.8	1.3	0.9		5	29	0.66	4.7		1.8	0.47	33	63	80	1.3	2.5	0.44	0.66	0.31	2.2		0.12
	N	170	169	169	50	170	169	170	169		165	115	170	165	6	18	45	72	50	79	159	158	158	158	154	143	5	38
Apollo 17 VLT mare basalts	avg.	0.15	12.5	10.4	48.2	0.02	9.4	0.7	18.3	100.7	47	221	5.3	46							1.7	8.8	1.81	2.24	1.23	8.0		0.40
	SD	0.04	3.0	1.2	0.7	0.01	1.1	0.2	1.7		11	30	1.0	15							1.3	2.0	0.42	0.56	0.28	1.9		0.40
	N	10	10	10	4	10	10	10	10		6	6	10	6							6	9	9	9	9	8		1
All Apollo/Luna mare basalts[c]	avg.	0.35	9.4	10.6	44.6	0.11	10.6	4.7	19.1	100.1	58	140	3.1	34	36	3.9	4	170	219	158	9.5	8.2	1.6	1.9	0.96	5.4	0.036	1.1
	SD	0.12	1.8	1.6	3.1	0.09	0.8		1.4		15	52	1.1	13		1.2												

Source: Data mainly from the compilation of Haskin and Warren (1991). A large set of additional trace element data have been reported by Neal (2001), but for a few elements (Mo, Sb) his data appear spuriously high. Abbreviations: avg. = average, SD = standard deviation, N = number of data averaged.
[a] Mass-weighted means of literature data; only pristine basalts are shown (mare-dominated polymict breccias are excluded). [b] Mass-weighted mean of literature data for Y-793169 and Asuka-881757 probably paired mare-basaltic meteorites. [c] Average of the eight major Apollo/Luna varieties shown above.

Table 3 The diversity of sampled lunar regolith compositions.

Sample	Sample type		Na_2O (wt.%)	MgO (wt.%)	Al_2O_3 (wt.%)	SiO_2 (wt.%)	CaO (wt.%)	TiO_2 (wt.%)	FeO (wt.%)	mg (mol.%)	K (mg g^{-1})	Sc (µg g^{-1})	V (µg g^{-1})
	Physical	*Geochemical*											
Ap 11 average	Soils (averaged)	Mare	0.47	7.9	12.6	42.0	11.7	7.9	16.4	46	1.37	62	70
Ap 12 average	Soils (averaged)	Mare	0.41	10.4	12.1	46.2	9.9	2.6	17.2	52	2.1	37	114
Ap 15, mare floor	Soils (averaged)	Mare	0.40	11.0	14.3	47.2	10.5	1.46	15.0	57	1.30	27	110
Ap 17, mare floor	Soils (averaged)	Mare	0.39	9.8	12.1	40.8	11.1	8.5	16.6	51	0.66	60	76
Luna 16	Core soil	Mare	0.38	8.8	15.6	44.4	11.7	3.30	16.4	49	0.95	52	80
Luna 24	Core soil	Mare	0.29	9.9	11.9	44.6	11.4	1.04	19.2	48	0.27	40	140
NWA773	Regolith breccia	Mare	0.23	13.2	10.6	46.2	10.8	0.78	17.3	58	0.83	ND	ND
QUE94281	Regolith breccia	Mainly mare	0.37	8.5	15.8	47.5	12.5	0.71	14.1	52	0.50	33	116
Yamato-793274	Regolith breccia	Mainly mare	0.38	9.1	15.3	47.9	12.2	0.61	14.2	53	0.67	33	99
Ap 14 average	Soils (averaged)	Mainly highland	0.69	9.2	17.7	48.2	11.0	1.72	10.4	61	4.3	23	51
Ap 15, Apennine Front	Core soil	Mainly highland	0.47	10.3	20.0	46.2	11.4	1.24	10.0	65	ND	18.5	65
Ap 17, South Massif	Soil 73141	Mainly highland	0.42	9.7	21.3	45.2	13.0	1.22	8.0	68	1.17	16.2	37
Luna 20	Core soil	Mainly highland	0.38	9.2	23.0	45.0	14.4	0.46	7.3	69	0.62	16.4	39
Calcalong Creek	Regolith breccia	Mainly highland	0.49	ND	20.9	ND	16.1	0.80	10.9	ND	2.0	23	57
ALH81005	Regolith breccia	Highland	0.30	8.2	25.6	45.7	15.0	0.25	5.4	73	0.19	9.1	25
Ap 14, 14076	Regolith breccia	Highland	0.44	3.3	30.5	44.1	16.8	0.33	3.8	61	0.73	7.8	17
Ap 14, 14315	Regolith breccia	Highland	0.57	7.9	22.1	47.1	13.0	0.85	7.4	65	2.6	15.6	ND
Ap 16 average	Soils (averaged)	Highland	0.48	5.8	27.1	44.7	15.5	0.58	5.2	67	1.07	10.4	20
Dar al Gani 262	Regolith breccia	Highland	0.36	4.7	27.7	44.6	16.1	0.20	4.3	66	0.42	7.6	26
Dhofar 025	Regolith breccia	Highland	0.31	6.5	26.8	44.5	15.8	0.29	4.8	71	0.48	10.0	<30
MAC88104/5	Regolith breccia	Highland	0.33	4.0	28.1	45.0	16.6	0.24	4.2	63	0.23	8.5	18
QUE93069	Regolith breccia	Highland	0.34	4.6	28.5	44.9	16.3	0.27	4.3	66	0.30	7.5	22
Yamato-791197	Regolith breccia	Highland	0.33	6.3	26.0	43.7	15.5	0.34	6.5	63	0.24	13.1	32
Average highland regolith		*Conceptual*	0.39	5.7	26.9	44.9	15.6	0.37	5.1	67	0.70	10.0	23

Sample	Cr (mg g^{-1})	Mn (mg g^{-1})	Co (µg g^{-1})	Ni (µg g^{-1})	Zn (µg g^{-1})	Ga (µg g^{-1})	Sr (µg g^{-1})	Zr (µg g^{-1})	Cs (µg g^{-1})	Ba (µg g^{-1})	La (µg g^{-1})	Ce (µg g^{-1})	Nd (µg g^{-1})
Ap 11 average	1.99	1.66	31	199	25	5.0	163	330	130	220	18.0	55	54
Ap 12 average	2.47	1.60	41	260	6	4.7	138	560	310	360	31	87	67
Ap 15, mare floor	2.53	1.45	45	216	14	4.0	138	320	170	200	21	50	30
Ap 17, mare floor	3.1	1.78	34	170	35	5.0	168	229	ND	98	8.0	24	21.8
Luna 16	2.20	1.64	32	167	26	5.1	275	253	63	185	11.2	34	24.6
Luna 24	2.91	1.97	47	129	13	1.2	91	78	52	41	3.2	8.3	6.4
NWA773	2.74	2.01	ND	ND	ND	ND	ND	ND	ND	ND	16	41	24
QUE94281	1.81	1.55	45	208	5	4.5	106	105	89	72	6.4	15.7	9.6
Yamato-793274	2.08	1.62	42	101	7	4.7	110	92	46	78	5.8	15.2	9.6
Ap 14 average	1.41	1.08	36	350	25	5.9	189	810	680	840	67	181	106
Ap 15, Apennine Front	2.01	1.11	30	162	31	5.1	145	363	260	256	99	59	34
Ap 17, South Massif	1.48	0.84	27	239	16	2.6	137	219	236	154	15.5	38	24.9
Luna 20	1.27	0.85	30	246	20	3.7	165	192	76	104	6.2	16.8	11.3
Calcalong Creek	1.30	1.12	24	360	ND	ND	ND	ND	ND	ND	20.1	52	28.8
ALH81005	0.89	0.58	21	202	9	2.7	135	27	24	28	1.98	5.2	3.2
Ap 14, 14076	0.50	0.45	15.8	231	7	10.8	175	108	146	90	8.0	18.2	11.9
Ap 14, 14315	1.24	0.78	33	430	ND	6.6	140	520	420	380	36	91	53
Ap 16 average	0.71	0.52	26	360	18	4.2	162	162	155	127	11.6	30	20.8
Dar al Gani 262	0.61	0.50	18	241	<40	4.0	210^a	33	90	120^a	2.2	6.1	3.4
Dhofar 025	0.77	0.58	15.7	141	<30	3.1	1,700^a	52	ND	200^a	3.3	8.0	5.0
MAC88104/5	0.63	0.48	14.6	148	7	3.5	151	34	38	32	2.52	6.3	4.0
QUE93069	0.57	0.47	22	295	15	3.2	149	46	41	42	3.4	8.5	5.0
Yamato-791197	0.90	0.67	19.4	178	21	5.5	135	32	67	32	2.13	5.5	3.5
Average highland regolith	0.76	0.56	20.6	247	13	4.8	149^a	113	123	100^a	7.87	19.9	12.2

Sample	Sm (µg g⁻¹)	Eu (µg g⁻¹)	Tb (µg g⁻¹)	Dy (µg g⁻¹)	Yb (µg g⁻¹)	Lu (µg g⁻¹)	Hf (µg g⁻¹)	Ta (µg g⁻¹)	Os (µg g⁻¹)	Ir (µg g⁻¹)	Au (µg g⁻¹)	Th (µg g⁻¹)	U (µg g⁻¹)
Ap 11 average	13.8	1.71	3.2	17	10.8	1.59	11.6	1.59	7.8	8.6	2.9	2.34	0.52
Ap 12 average	15.1	1.89	3.6	20	10.7	1.54	12.8	1.35	5.2	5.6	2.4	5.2	1.68
Ap 15, Mare	9	1.27	2.1	13	6.6	0.98	7.1	0.9	6.3	6.4	3.0	3.0	0.8
Ap 17, Mare	8.1	1.72	1.99	ND	7.2	0.99	6.8	1.18	ND	4.8	3.6	0.88	0.31
Luna 16	8.0	3.2	1.43	9.7	5.6	0.77	7.0	0.65	ND	10.3	2.4	1.12	0.30
Luna 24	2.02	0.66	0.49	3.0	1.74	0.27	1.69	0.25	7.7	5.7	6.9	0.45	0.10
NWA773	7	0.7	1.5	ND	5	0.7	ND	ND	ND	ND	ND	ND	ND
QUE94281	3.1	0.83	0.64	4.2	2.42	0.35	2.36	0.29	6.2	6.5	2.2	0.93	0.23
Yamato-793274	2.73	0.82	0.58	4.0	2.29	0.32	2.26	0.26	4.5	4.5	3.0	0.81	0.21
Ap 14 average	29.8	2.7	6.3	38	23	3.2	22	2.8	ND	17.5	5.8	13.8	3.3
Ap 15, Apennine Front	10.8	1.36	2.02	ND	7.7	1.04	8.7	1.00	ND	4.4	1.9	3.8	1.24
Ap 17, South Massif	7.0	1.2	1.50	9.3	5.4	0.80	5.2	0.75	ND	12	6	2.4	0.70
Luna 20	3.2	0.90	0.59	4.2	2.22	0.36	2.57	0.31	ND	9.5	3.6	1.06	0.37
Calcalong Creek	8.7	1.20	2.16	14.2	6.4	0.98	6.8	1.12	ND	ND	ND	4.2	1.13
ALH81005	0.95	0.69	0.21	1.33	0.84	0.12	0.72	0.09	8.4	6.8	2.3	0.29	0.10
Ap 14, 14076	3.2	1.09	0.67	4.2	2.5	0.34	2.2	0.29	ND	7.9	7.4	1.34	0.34
Ap 14, 14315	14.3	1.56	3.1	19.9	10.4	1.55	11.3	1.32	ND	18.8	12.7	5.6	1.38
Ap 16 average	5.5	1.22	1.12	7.2	3.8	0.58	4.1	0.52	ND	12.2	8.0	1.97	0.57
Dar al Gani 262	1.03	0.75	0.22	1.59	0.84	0.12	0.77	0.10	ND	10.1	3.9	0.37	0.16[a]
Dhofar 025	1.43	1.02	0.30	1.57	1.11	0.18	1.08	0.13	ND	5.4	4.3	0.57	0.22[a]
MAC88104/5	1.15	0.79	0.24	1.49	0.97	0.14	0.86	0.11	7.3	7.2	2.7	0.39	0.10
QUE93069	1.56	0.82	0.33	1.99	1.21	0.17	1.14	0.15	21	15.6	4.5	0.53	0.13
Yamato-791197	1.06	0.77	0.25	1.56	1.01	0.15	0.88	0.10	8.2	6.7	1.4	0.33	0.11
Average highland regolith	3.34	0.97	0.71	4.54	2.51	0.37	2.56	0.31	11.3	10.1	5.2	1.27	0.4[a]

[a] Hot-desert meteorites tend to be grossly contaminated with Ba, U, and especially Sr; these data are excluded from the "average highland regolith" calculation. ND: not determined. Data sources are too numerous to enumerate in full. Among the most noteworthy are previous compilations by Haskin and Warren (1991); Korotev (1987a), for Apollo 14 samples; Korotev (1987a), for Apennine Front core soil 15007; Warren and Kallemeyn (1986) for ALH81005, Luna 20 and South Massif soil (most Al-rich from Apollo 17) 73141; Korotev and Kremser (1992) for an average of Apollo 17 mare soils near the center of the Taurus-Littrow Valley; Jolliff et al. (2003) for NWA773; and Warren et al. (2003) for other recently discovered lunar meteorites.

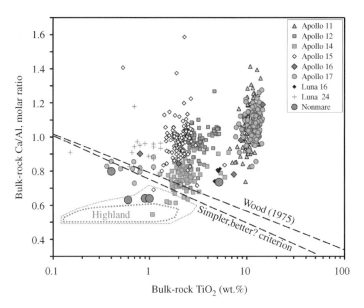

Figure 2 On a plot of TiO$_2$ versus Ca/Al, nearly all mare basalts plot above and to the right of a criterion proposed by Wood (1975) for distinguishing between mare and highland rocks, and well apart from the vast majority of highland rocks, which plot at Ca/Al = 0.5–0.6 and TiO$_2$ < 2 wt.% (0.5 is the stoichiometric plagioclase ratio; fields are based on Wood, 1975, figure 2). Five extraordinarily gabbronoritic pristine highland samples are plotted as individual filled circles that straddle Wood's line: sodic ferrogabbro clast 67915c (Marti et al., 1983), 14434 (Arai and Warren, 1997), 61224,11 (Marvin and Warren, 1980), 67667 (Warren and Wasson, 1978), and 73255c27 (James and McGee, 1979). Wood's (1975) criterion is Ca/Al = 0.786 − 0.229 log$_{10}$ TiO$_2$ (for Ca/Al as a molar ratio and TiO$_2$ in wt.%). Also plotted is a simpler variant proposed here, Ca/Al = 0.74 − 0.26 log$_{10}$ TiO$_2$, that seems better at including VLT mare basalts as mare samples, without significantly changing the situation for the nonmare samples.

Mare basalts are also classified on the basis of secondary compositional criteria, such as bulk-rock Al$_2$O$_3$ and potassium. In the most recent and systematic variant of this scheme (Neal and Taylor, 1992), samples with >11 wt.% Al$_2$O$_3$ are distinguished as aluminous (or high-Al), and samples with >2,000 µg/g K are distinguished as high-K. Thus, for example, the Apollo 14 samples include a few pieces of aluminous (or more specifically aluminous, high-K, medium-Ti) mare basalt (Dasch et al., 1987; Neal and Taylor, 1989). The Luna 24 VLT basalts as well as the two "YA" lunar meteorite basalts are borderline aluminous (Table 2), and the Luna 16 basalts are clearly aluminous. However, for the Luna 16 basalts, which were sampled strictly as tiny regolith particles, a caveat is in order. Most of the "bulk" analyses in the literature are uncorrected defocused-beam microprobe analyses, which tend to give spuriously high Al$_2$O$_3$ (e.g., 12 such analyses by Kurat et al. (1976) average 16.1 wt.% Al$_2$O$_3$, whereas analyses by true bulk methods such as wet chemistry, neutron activation analysis and X-ray fluorescence are consistently close to 13.4 wt.%). The vast majority of known mare basalt types are "low-Al" and (thus) undersaturated with respect to plagioclase (Longhi, 1992).

Aside from this overall classification, mare basalts are also classified based on more subtle distinctions among samples from a given landing site. For example, among the Apollo 12 medium-Ti mare basalts, three major groupings are distinguished, named for minerals that are especially abundant in each type: olivine, pigeonite, and ilmenite. Among Apollo 15 medium-Ti basalts, only two main types are recognized: olivine-normative and quartz-normative, which probably correspond to two distinct flows, with the olivine-normative atop the quartz-normative (Ryder and Cox, 1996). The most elaborate subclassification of a suite of basically similar mare basalts is for the high-Ti samples from Apollo 17 (Neal et al., 1990). These site-specific classifications are intended to link samples with specific magma types, or even specific lava flows. It is natural to ask whether differences between the Apollo 12 olivine and Apollo 12 pigeonite basalts indicate heterogeneity within a single lava (Rhodes et al., 1977) or require separate multiple lavas (Neal et al., 1994). However, considering the great diversity of mare materials as a whole, the big-picture observation is that Apollo 12 pigeonite and Apollo 12 olivine basalts are only subtly different from one another.

Besides basalts, mare volcanism also produced a variety of distinctive pyroclastic glasses. Sampled pyroclastic glasses are spherules with diameters generally between 0.03 mm and 0.3 mm. Distinguishing between these and impact-splash glasses is rarely difficult (Ryder et al., 1996). Remote observations indicate that scattered large regions are veneered with pyroclastic matter. Although the largest of these "dark mantle deposits" (~10 have area >2,500 km^2) occur near the rims of circular mare basins, the deposits are typically irregular-oval in shape, not arcuate. Clementine data have enabled identification of ~90 much smaller pyroclastic deposits that show similar spectral diversity but are much more widely distributed across the Moon (Gaddis et al., 2000). The mare pyroclastic glasses are classified by the same titanium-, aluminum-, and potassium-based criteria as the crystalline mare basalts, and also on the expedient basis of color of the glass. High-Ti glasses tend to be orange or red, VLT glasses green, and medium-Ti glasses tan or brown; devitrification of the glass can make rapidly accumulated spherules black (which may account for the "dark" mantle deposits).

1.21.3.2 Chronology and Styles of Mare Volcanism

The chronology of mare volcanism is constrained primarily by isotopic ages for samples (Table 4 and Figure 3), and secondarily, but with far broader application, by photogeologic (crater-counting) methods, calibrated by extrapolation from the few isotopically dated surfaces. The oldest isotopically dated mare-type samples are high-aluminum, medium-Ti cumulate clasts, found in Apollo 14 highland breccias, whose ages extend up to 4.23 Ga (Taylor et al., 1983; Shih et al., 1986; Dasch et al., 1987; Neal and Taylor, 1989). But in general, mare clasts within highland breccias are rare. The overall scarcity of mare basalt among the tens of thousands of diverse clasts studied from highland breccias, the scarcity of cryptomaria (regions where mare surfaces have been covered, but also sufficiently gardened to become detectible, by more recent impact cratering, Antonenko et al. (1995)), the sharpness of most mare–highland (and mare–mare) boundaries based on various remote-sensing data (e.g., Li and Mustard, 2000; Lucey and Steutel, 2003), and the low abundance of craters in the maria, all indicate that mare volcanism occurred mainly after the period of heavy bombardment that ended at 3.9 Ga (see below).

Most of the individual age data for mare samples are from Apollo 11, 12, 15, and 17, and these data gave rise to an early misconception (which still to some extent persists: e.g., Snyder et al., 2000b) that among mare basalts, titanium content correlates with antiquity. If anything, the overall database, including photogeologic inferences, suggests an anticorrelation between titanium and antiquity. The youngest isotopically dated mare rocks are a VLT cumulate, lunar meteorite NWA773, ~2.7 Ga, and a medium-titanium basalt, lunar meteorite NWA032, ~2.8 Ga (dated with ^{40}Ar–^{39}Ar measurements by Fernandes et al., 2002b and Fagan et al., 2002; there is a caveat: these could be shock-reset ages). But the very last gasp of mare volcanism, the least cratered mare lava, is believed to be a high-Ti flow in Oceanus Procellarum (Hiesinger et al., 2000). Assigning an absolute age to this surface requires a long extrapolation from rigorously dated surfaces, but Hiesinger et al. (2000) suggest that the age of the flow in question is ~1.3 Ga. In general, the youngest mare basalts in Procellarum tend to be high-Ti (Staid and Pieters, 2001). At the opposite extreme, the cryptomaria presumably tend to be uncommonly old, by mare standards, and remote-sensing data indicate that mare basalt in cryptomaria tends to be VLT, or at most medium-Ti (Hawke et al., 2003). In short, age seems a very poor predictor of mare lava composition.

The majority of mare rock samples are noncumulate basalts, formed by melts of generally very low viscosity, of order 1 poise (calculated by method of Persikov et al. (1987)) in flows whose thicknesses (assuming the Imbrium region is typical) were mostly less than 10 m (Gifford and El-Baz, 1981). By one recent estimate (Hiesinger et al. (2002), however, they averaged 30–60 m and ranged up to 220 m. The most widely cited method for estimating total accumulated thickness of mare lavas, based on crater submersion statistics (small, shallow craters are submerged sooner than big, deep craters), indicates that the maria are generally <0.5 km thick, and average ~1 km (DeHon, 1979). However, Hörz (1978) argued these estimates ignore prevolcanic erosion of crater depth/diameter ratio, and thus are too high by a factor of ~4. The abundance of highland component within regolith at mare-interior locales (where the highland component is probably mainly added by vertical mixing) also favors relatively thin maria (e.g., Ferrand, 1988). Even assuming DeHon's calibration as correct, the total volume of mare basalt is only ~7 × 10^6 km^3, or 0.03% of the lunar volume. However, it is conceivable that a larger volume of intrusive mare-like gabbros formed during the era of mare volcanism but remained unsampled, because the large post-3.9 Ga craters required to excavate them are rare.

Phenocrysts are common, especially in the Apollo 15 mare basalts, but they appear to have grown strictly during the main stage of cooling, i.e., during posteruptive flow across the surface, and not in a deep magma chamber (Lofgren et al., 1975; Walker et al., 1977). The phenocrysts are

Table 4 Summary of age constraints for mare basalts and pyroclastic glasses.

Sample	TiO$_2$ (wt.%)	Plotted age (Ga)	Ar age (Ga)	Sr age (Ga)	Nd age (Ga)	Pb age (Ga)	References
Lunar meteorites							
Dhofar 287	2.86	3.46			3.46 ± 0.03		Shih et al. (2002)
NWA 032	3.08	2.80	2.80 ± 0.02				Fagan et al. (2002)
NWA 773	0.30	2.7	2.7				Fernandes et al. (2002b)
YA meteorite	2.08	3.90	3.80 ± 0.01	3.89 ± 0.03	3.87 ± 0.06	3.94	Misawa et al. (1993)
Apollo 11							
10017	12.2	3.51		3.51 ± 0.05			BVSP (1981)
10024	12.6	3.53		3.53 ± 0.07			BVSP (1981)
10029	10.9	3.83	3.83 ± 0.03				BVSP (1981)
10062	10.5	3.86	3.79 ± 0.04	3.92 ± 0.11	3.88 ± 0.06		BVSP (1981)
Apollo 12							
12022	4.9	3.08	3.08 ± 0.06				BVSP (1981)
12020	2.83	3.09	3.09 ± 0.06				BVSP (1981)
12002	2.62	3.25	3.21 ± 0.05	3.29 ± 0.10			BVSP (1981)
12021	3.52	3.26		3.26 ± 0.03			BVSP (1981)
Apollo 14							
14053	2.72	3.93	3.93 ± 0.04				Ryder and Spudis (1980)
14072	2.57	4.04	4.04 ± 0.05	4.05 ± 0.08			Ryder and Spudis (1980)
14168	1.70	3.85	3.85 ± 0.02	3.82 ± 0.12	3.95 ± 0.17		BVSP (1981) and Shih et al. (1986)
14304, 113c	3[a]	3.95	3.85 ± 0.04	3.95 ± 0.04			Shih et al. (1987)
14304, 108c	3[a]	4.00		3.99 ± 0.02			Shih et al. (1987)
14305, 122–92c	3.60	4.23		4.23 ± 0.05	4.04 ± 0.11		Taylor et al. (1983)
14305, 304c	2.40	3.85	3.85 ± 0.05	3.83 ± 0.08			Shih et al. (1986)
14321, 184c	2.41	3.95	3.95 ± 0.05	3.95 ± 0.04	3.91 ± 0.16		Ryder and Spudis (1980)
14321, 223c	2.41	4.1		4.1			Dasch et al. (1987)
Apollo 15							
15668	2.49	3.10	3.10 ± 0.05				BVSP (1981)
15065	1.70	3.21		3.21 ± 0.04			BVSP (1981)
Ap15 green glass	0.36	3.34	3.3			3.38	Snyder et al. (2000) and Shih et al. (2001)
15385	2.19	3.35	3.35 ± 0.05				BVSP (1981)
15682	2.27	3.37		3.37 ± 0.07			BVSP (1981)
15388	3.10	3.38		3.36 ± 0.04	3.42 ± 0.07		Dasch et al. (1989)
15475	1.81	3.39		3.43 ± 0.15	3.37 ± 0.05		Snyder et al. (2000b)
Apollo 17 high-Ti							
71055	13.9	3.56		3.56 ± 0.09			BVSP (1981)
70017	13.4	3.59		3.59 ± 0.18			BVSP (1981)
Ap17 orange glass	8.8	3.56	3.6			3.48	Snyder et al. (2000a) and Shih et al. (2001)
70215	13.1	3.79	3.79 ± 0.04				BVSP (1981)
70255	11.4	3.84	3.84 ± 0.02				BVSP (1981)
79001, 2144	9.9	3.90	3.49 ± 0.04	3.89 ± 0.04	3.92 ± 0.04		Shearer et al. (2001)
Other samples							
Apollo 17 VLT	0.71	≥4.01	?–4.01				Taylor et al. (1991) (p. 209)
Luna 16	5.0	Range	3.5–3.04				Snyder et al. (2000a) and Fernandes et al. (2002a)
Luna 24	0.84	Range	3.6–3.0				Snyder et al. (2000a) and Fernandes et al. (2002a)

Sources: For the sake of brevity, most of the individual data from the extensive compilation of BVSP (1981) are not reproduced here. Instead, only the two oldest and two youngest members of each major sample type are shown. This listing is intended to be nearly comprehensive for Apollo 14 (for which most data appeared after BVSP, 1981), but the samples include many shock-altered clasts, and some of the more ambiguous Sr results of Dasch et al. (1987) have been omitted.

[a] TiO$_2$ for the Shih et al. (1987) 14304 clasts estimated based on sum "difference" and analogy to other Apollo 14 high-Al mare basalts. Except as noted, TiO$_2$ data are from compilation of Haskin and Warren (1991) or Table 2.

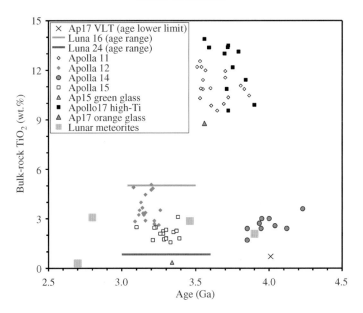

Figure 3 Ages of mare basalts and pyroclastic glasses show no correlation with TiO_2. Age data are from previous compilations by BVSP (1981), Ryder and Spudis (1980), Fernandes et al. (2002a), and (for pyroclastic glasses) Shih et al. (2001), plus a lower limit cited for Apollo 17 VLT basalts by Taylor et al. (1991). The TiO_2 data are averaged from the compilation of Haskin and Warren (1991). The five major Apollo basalt types are shown with small symbols because each point represents one of many available samples from the given locale, whereas each of the lunar meteorites represents (probably) our only sample from its locale.

invariably mafic, never (thus far) containing plagioclase. The absence of plagioclase in the phenocryst assemblage, and moreover its scarcity as a liquidus phase (Longhi, 1992), are powerful indications that mare magmas tended to transit directly from the mantle to the surface, and seldom paused in crustal magma chambers (cf. Wilson and Head, 2003). The aluminous mare basalts are plagioclase saturated; among those from Apollo 14, the best-documented suite of this type, many appear to have assimilated a crustal (or crust–mantle area) material, namely KREEP (Dickinson et al., 1985; Neal and Taylor, 1989). But, as noted above, aluminous mare basalts are rare. Mare cumulates managed to develop, as exemplified by rocks such as 12005 (Rhodes et al., 1977; Dungan and Brown, 1977), 15385 and 15387 (Ryder, 1985), 71597 (R. D. Warner et al., 1977), the Apollo 14 clast that Taylor et al. (1983) found to be exceptionally ancient, and the NWA773 meteorite (Jolliff et al., 2003). However, based on limited exsolution within their pigeonites (cf. Arai and Warren, 1999), these cumulates formed within rapidly cooling surface flows or lava ponds, and their trace-element compositions reflect relatively inefficient segregation of cumulus matter from "trapped" mare melt.

Many distinct mare melt types have been sampled via pyroclastic glasses. A compilation by Taylor et al. (1991) included 25 glass composition types. The process that produced the pyroclastic glasses (and dark mantle deposits) probably has a close analogue on present-day Io. The Moon and Io are nearly identical in size and density, and in their lack of atmosphere. On Io, ongoing volcanism produces steady, long-lived plumes that probably consist of tiny, diffuse droplets, akin to the lunar pyroclastic glass spherules; these plumes have been observed to rise as high as 400 km (briefly) and spread as far as 700 km from their volcanic vents (McEwen et al., 1998). The force that drives plumes like these is the explosive expansion of gas bubbles upon eruption into the near vacuum of the Ionian (or ancient lunar) atmosphere (Head and Wilson, 1992; McEwen et al., 1998). The explosive gas phase on Io is probably SO_2; the composition of the predominant gas phase during mare volcanism can only be reconstructed by inference, but it was probably dominated by CO and CO_2 (Fogel and Rutherford, 1995). The precursor of these gases would be graphite, which tends to undergo oxidation if entrained in a hot magma that depressurizes during the late stages of its ascent to the surface. Lunar pyroclastic spherules have volatile-rich coatings (Chou et al., 1975; Butler and Meyer, 1976; Fogel and Rutherford, 1995), and crystalline mare basalts can be highly vesicular (Ryder, 1985). Nonetheless, the smaller scale of the lunar dark mantle deposits compared to their Io analogues confirms a tendency toward relatively low proportions of volatile fuel in the lunar magmas.

Within each pyroclastic glass type, there tend to be systematic, relatively simple differentiation trends, encompassing in some cases wide ranges in chemical composition. The Apollo 14 orange glass array has MgO ranging from 14.5 wt.% down to 9.5 wt.% (Delano, 1986), and within an Apollo 14 green (VLT) glass suite, REE concentrations vary by a factor of 2 (Shearer et al., 1990). But the compositional vectors of these trends do not conform to expectations from any simple "batch" partial melting or fractional crystallization model. Instead, it appears that complex processes during the formation and migration of melt in the mantle, including assimilation and fractional crystallization (AFC) involving KREEP and ilmenite cumulates, are required (Shearer et al., 1996; Elkins et al., 2000). (KREEP is an important material mainly associated with highland rocks, discussed in a later section.)

Initial post-Apollo views of lunar evolution envisioned that the mantle is a "layer cake" stratigraphic sequence of compositionally distinct cumulate layers (deep, low-Ti, high-*mg* cumulates, grading upward to FeO-rich, high-titanium, low-*mg* cumulates), as a result of crystallization of a primordial magma ocean. The low-titanium mare basalts were assumed to come from the deep cumulates, and the high-titanium basalts from the shallow, late-stage cumulates (Taylor and Jakes, 1974). This view now appears oversimplified. There is little correlation between the titanium abundances of mare magmas and the experimentally estimated (from the pressure of multiple-saturation of phases) depths of their source regions (Taylor et al., 1991; Elkins et al., 2000). Ringwood and Kesson (1976) first articulated what has become a generally accepted feature of models for mare petrogenesis: ubiquitous modification of the initial deep mantle cumulates by varying degrees of hybridization and assimilation interactions with formerly shallow material, swept down by convective stirring of the mantle. Aside from regular convective motions, many authors (e.g., Hess and Parmentier, 1995; Zhong et al., 2000) have noted that the aftermath of magma ocean crystallization may have been a gravitationally unstable configuration (dense, FeO- and ilmenite-rich residual mush atop relatively FeO- and ilmenite-poor cumulates), which would imply an enhanced convective potential, and possibly a catastrophic overturn of the entire mantle. Moreover, modern theories of "polybaric" partial melting in a body as large as the Moon indicate that melting probably occurs over a range of depths, by a gradual increase in melt fraction within rising diapirs (Longhi, 1992, 2003; Shearer and Papike, 1999; Elkins et al., 2000). The depth of initial melting is difficult to constrain. Beard et al. (1998) have inferred from Lu–Hf isotopic systematics (cf. Neal, 2001) that at high pressure, garnet played a role in the residua of some mare basalts.

The picture that emerges from models like that of Ringwood and Kesson (1976) is an almost chaotic diversity of potential variations in the mixing between (already diverse) cumulates and "juicy" ingredients of the lunar mantle, such as downswept KREEP, during formation of precursors to mare volcanism. Nonetheless, it is often suggested that the distribution of TiO_2 among mare basalts is systematic, i.e., bimodal. Usually, the notion is that one mode is high-Ti, the other VLT plus medium-Ti (or low-Ti in the old nomenclature), with at gap at ~6–9 wt.%. Giguere et al. (2000) interpreted Clementine TiO_2 data as showing that the bimodality is not significant, although they acknowledged a low-Ti/high-Ti bimodality (interpreted as a fluke) among available samples. Gillis et al. (2003) have proposed an alternative calibration of the Clementine TiO_2 data, which restores the bimodality, with the gap shifted down to 4–7 wt.%, TiO_2 for the Apollo and Luna sampling locations, but not for the global lunar surface. Considering the tremendous range in the TiO_2 data (0.3–16 wt.%), it seems most appropriate, when weighing the bimodality issue, to plot the data on a log scale. When all known samples, including lunar meteoritic samples and ignoring subtle intrasite variations among basically similar basalts (e.g., averaging all Apollo 17 high-Ti basalts together), are plotted on a log scale (Figure 4), the only bimodality appears between the VLT and medium-Ti (plus high-Ti) types; i.e., the portion of the range lacking in samples is ~1–2 wt.%. (Unfortunately, TiO_2 concentrations <2 wt.% are not realistically resolvable by Clementine-style remote sensing.) Arai and Warren (1999) argue that a gap at 1–2 wt.% TiO_2 is more plausible, as a likely outcome of partial melting where absence versus presence of cumulus ilmenite plays a crucial role, than one at 6–9 wt.% TiO_2.

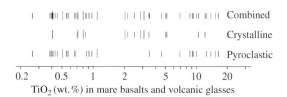

Figure 4 Plotted on a log scale, and with fundamentally related (artificially oversampled) suites averaged together (as in Table 2), mare volcanic samples exhibit a possible bimodality in TiO_2 content, but with the gap in the distribution at around 1–2 wt.%, not 6–9 wt.% as commonly claimed (see text). Lunar meteoritic data, including pyroclastic glass suites (Arai and Warren, 1999), are distinguished by longer symbols.

The pyroclastic glasses tend to be far more "primitive," in terms of MgO contents, than the crystalline basalts (Figure 5). The two types of material form a rough anticorrelation between MgO and Al_2O_3. But even the crystalline basalts are mostly picritic (far from plagioclase saturation), and detailed consideration of the major-element trends suggests that few, if any, of the sampled crystalline basalts are potentially related to spatially associated mare pyroclastic glasses (Longhi, 1987). Arai and Warren (1999) suggested that the glasses tend to be more MgO-rich because the fuel for explosive volcanism, graphite (density, $2.2\ g\ cm^{-3}$), must have come from a mantle component that remained (more or less) solid and intact through the extensive primordial (magmasphere) melting of the Moon, and thus remained MgO-rich.

1.21.3.3 Mare Basalt Trace-element and Isotopic Trends

Mare basalts, being young by lunar standards, are, as a rule, far less battered by impact processes than highland rocks; they are also generally closer in composition to their parent melts (i.e., the proportion of cumulates is far higher among highland rocks). Thus, mare basalts are our most suitable samples of lunar rocks that have an analogous terrestrial rock type (Figures 6 and 7). VLT and medium-Ti types are preferred for this purpose, because the high-Ti types are unlike any common variety of terrestrial basalt.

The most striking difference between mare basalt and terrestrial basalt is in their contents of volatile species. Besides H_2O (practically unmeasurable in most lunar rocks), this disparity is manifested by data for volatile alkalies (sodium, potassium, rubidium, and caesium) and by trace metals such as zinc, indium, bismuth, and cadmium (Figure 6). The same trend of enormous depletions in comparison to chondritic matter and even terrestrial basalt is also shown by highland samples, but the evidence is clearest and most definitive from the mare basalts. Gross overall volatile depletion is also demonstrated by the near absence of hydrated minerals among lunar rocks. Some Apollo rocks, most impressively Apollo 16 breccia 66095, do contain surface-associated "rust" patches (mainly FeOOH,Cl), but these are suspected to be products of terrestrial oxidation, not endogenous lunar volatile processing (Papike et al., 1991). Even extraordinarily "evolved" types of nonmare rock, such as a few tiny samples that are compositionally granite (a term not meant to imply the existence of lunar batholiths!) contain pyroxene, not amphibole or mica, as their mafic component (Warren et al., 1983).

Mare basalts indicate that the Moon's pattern of crust–mantle siderophile element depletions is roughly similar to that of the Earth (Figure 7). Unfortunately, for four siderophile elements, rubidium, rhodium, palladium, and platinum, no reliable data have been reported for lunar basalts. Neal et al. (1999, 2001) reported ICP-MS data, but their analyses show obvious artificial contamination, based on comparison with iridium data and palladium upper limits reported by Wolf et al. (1979). The most "noble" of the siderophile elements in Figure 7 are osmium and iridium. For noble siderophile elements, scatter among individual rocks is very great (e.g., Figure 8), but the factor-of-4 disparity between average mare and terrestrial basalt for osmium and iridium

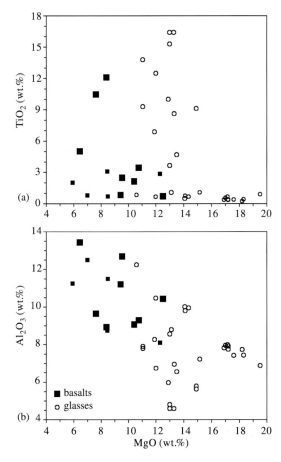

Figure 5 Mare pyroclastic glasses tend to have far higher MgO, yet a similar range in TiO_2, compared to crystalline mare basalts. For basalts, plotted data are averages of many literature analyses for basalt types (Table 2), except for lunar meteorites, which are individual rocks (shown with smaller filled squares). Other data sources are as cited by Arai and Warren (1999), i.e., primarily the compilation of Taylor et al. (1991); updated from Arai and Warren (1999) by adding lunar meteorites Dhofar 287 and NWA032 (but not the cumulate NWA773, 26 wt.% MgO, excluded because its composition is presumably unrepresentative of its magma type).

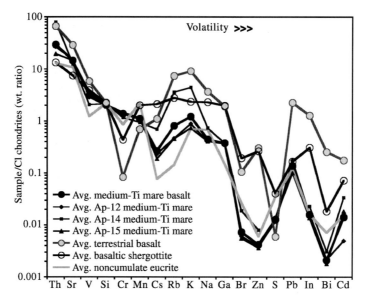

Figure 6 Volatile element concentrations in averaged medium-Ti mare basalts, normalized to CI chondrites (Wasson and Kallemeyn, 1988). Also shown for comparison are averages for terrestrial basalts, for basaltic shergottite (martian) meteorites, and for eucrite (asteroidal) meteorites. The elements are plotted in order of thermodynamically calculated solar nebula volatility (Wasson, 1985). The mare basalt data are mainly from the compilation of Haskin and Warren (1991); a noteworthy primary source is Wolf et al. (1979). The plotted overall average mare basalt composition is a 4:1:4 weighting of the Apollo-12, -14, and -15 types, respectively (the Apollo 14 type has been far less well studied, and moreover may be idiosyncratically enriched in volatile-incompatible elements: Dickinson et al., 1989). The average for terrestrial basalt is based on Govindaraju's (1994) compilation for USGS standards BCR-1, BHVO-1, BIR-1, and W-1, plus (given 1/5 weight) an average for MORB, based primarily on Hertogen et al. (1980). The average for basaltic shergottites is based on the up-to-date (internet) version of the Meyer (1998) compilation. The average for eucrites is based on a large number of references, most notably Chou et al. (1976), Morgan et al. (1978) and Paul and Lipschutz (1990).

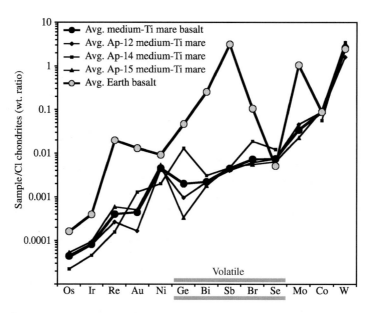

Figure 7 Siderophile element concentrations in averaged medium-Ti mare and terrestrial basalts, normalized to CI chondrites. The elements are plotted in order of CI-depletion factors in average medium-Ti mare basalt, but for some elements volatility may account in large part for the depletion. Data are from same sources as for Figure 6. To avoid an over-complex diagram, individual terrestrial compositions are not plotted, but all show similar patterns at the scale of this diagram; the most noteworthy exceptions being low iridium (9×10^{-6} times CI) and nickel (10^{-3} times CI) in BCR-1, and relatively low osmium and iridium (virtually identical to mare basalts) and antimony (0.11 times CI) in MORB.

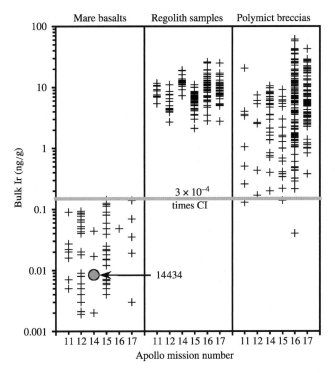

Figure 8 Despite great scatter in the data distribution, for a highly siderophile element such as iridium data reveal a strong contrast between compositionally "pristine" rocks, as exemplified by mare basalts, and lunar materials (regolith samples and polymict breccias) contaminated by a meteoritic component. This contrast can be exploited to help identify pristine nonmare rocks. A good example is Apollo 14 diabase 14434, which has an ambiguous texture and mineralogy (monomict-brecciated but not diagnostically coarse), yet is judged very probably pristine based on its depletion in iridium (Arai and Warren, 1997).

appears real. In lunar basalt, rhenium, gold, and germanium are more "nobly" depleted than in the terrestrial regime, where these elements, particularly rhenium, tend to exhibit incompatible-lithophile tendencies (probably linked to Earth's higher f_{O_2}). Also, some of the lunar depletion in elements such as germanium may reflect the Moon's general depletion of volatile elements. The same trend of strong siderophile depletions in comparison to chondritic matter, and even versus terrestrial igneous rocks, is also shown by highland samples. But for these, the evidence is far more complicated; a careful and, for siderophile elements, possibly biased selection of "pristine" rocks (see below) is required to obtain compositions representative of the endogenous igneous highland crust.

Among the highly siderophile elements, iridium and nickel contents in lunar rocks have been frequently and well determined, and both elements conform to a general pattern among planetary igneous rocks (Warren et al., 1999) by showing a correlation (albeit not linear) with MgO (Figure 9). These trends represent the strongest evidence that the lunar mantle was significantly siderophile-depleted in comparison to the terrestrial mantle. Figure 9 also includes martian meteorite data, and iridium and nickel are correlated with MgO in these samples as well.

The role of sulfide-driven (chalcophile) fractionations in lunar magmatism is difficult to constrain, but sulfides presumably play a lesser role on the Moon than on Earth, because the solubility of sulfide in mafic melt increases with decreasing f_{O_2} (Peach and Mathez, 1993). Compared to lithophile elements of similar volatility, sulfur is exceptionally depleted in terrestrial and martian basalts, but not so depleted in lunar basalts (Figure 6). Medium-Ti mare basalts are clearly unsaturated with sulfide, and although high-Ti mare basalts were originally believed to be sulfide-saturated (Gibson et al., 1977), Danckwerth et al. (1979) found them to be unsaturated, as well.

Chondrite-normalized REE patterns are shown in Figure 10. A comparable variety of patterns are found for mare pyroclastic glasses (Papike et al., 1998). Most mare basalts have significant negative europium anomalies, which increase in magnitude with increasing REE content and have important implications. In the reducing lunar environment, negative europium anomalies are only to be expected in basalts that are plagioclase-saturated, i.e., left mantle residua with plagioclase, with which europium is thoroughly compatible (distribution coefficient $D \sim 1.1-1.2$) when reduced to mostly Eu^{2+} (McKay and Weill, 1977). But most mare basalts (all those with Al_2O_3 less than ~ 12 wt.%) are not saturated with plagioclase at

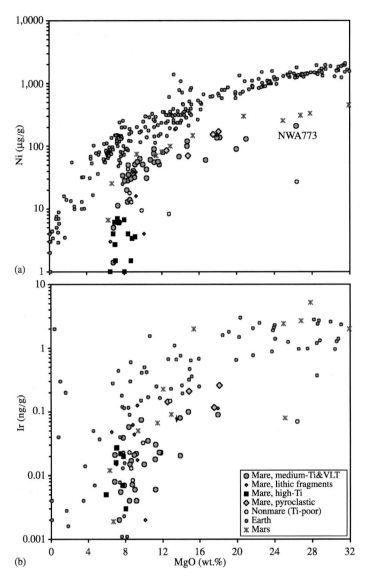

Figure 9 (a) Nickel and (b) iridium versus MgO, for igneous rocks from the Moon, Mars, and Earth. Lunar data are mainly from the compilation of Haskin and Warren (1991), the most noteworthy addition being lunar meteorite NWA773 (nickel only; Korotev et al., 2002a). Data for martian rocks are from the up-to-date (internet) version of the Meyer (1998) compilation. Data sources for terrestrial rocks are too numerous to list, but include Crocket and MacRae (1986), Brügmann et al. (1987), and the compilation of Govindaraju (1994).

any pressure (Taylor et al., 1991; Longhi, 1992; Papike et al., 1998). The inference, therefore, is that long before the mare magmas formed, their *source regions* (i.e., much of the lunar mantle) must have been *pre*-depleted in plagioclase, presumably by formation of the highland crust. This concept has long been viewed as a significant argument in favor of a primordial lunar magma ocean (Taylor and Jakes, 1974; Warren, 1985). Shearer and Papike (1989) suggested that the negative europium anomalies could conceivably have formed without major plagioclase fractionation, because pyroxene can impart a negative europium anomaly by excluding the relative large Eu^{2+} cation during crystallization. McKay et al. (1991) reported relevant D data for lunar pigeonite. Brophy and Basu (1990) applied these and earlier D results to show that accounting for the mare europium anomalies without appeal to prior plagioclase removal requires implausible assumptions about modal mineralogy and/or degree of melting in the source regions. Assimilation/mixing with KREEP swept down into the mantle may have affected many of the mare europium anomalies, but the mare basalts with the largest negative europium anomalies tend not to have KREEP-like enriched La/Sm ratios (Figure 10).

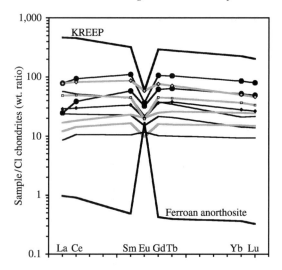

Figure 10 Chondrite-normalized REE concentrations in mare basalts, KREEP, and a representative ferroan anorthosite. Data for mare basalt types are from Table 2, with two deliberate omissions: the NWA773 cumulate, an individual rock presumably unrepresentative of its parent magma, and the Apollo 14 type, which is a suite with notoriously diverse REE abundances, although the patterns are generally parallel to KREEP (which is plotted). Individual mare basalt types are not labeled, but symbols on some of the patterns denote relatively titanium-rich varieties; the most titanium-rich basalts (largest symbols) tend to have the lowest (often subchondritic) La/Sm ratios. Data for high-K KREEP and ferroan anorthosite 15295c41 are from Table 5 and Warren et al. (1990), respectively.

Data for isotopic tracer ratios, I_{Sr}, ε_{Nd}, etc., in mare basalts have been reviewed by Papike et al. (1998) and Snyder et al. (2000b). Chapter 1.20 provides an extensive discussion of lunar ε_W data. Model ages based on strontium, neodymium, and lead for the mantle source regions typically cluster near 4.3–4.4 Ga, coinciding with a neodymium-based model age for urKREEP of 4.35 ± 0.03 Ga (Lugmair and Carlson, 1978). Most mare basalts have positive ε_{Nd}; the exceptions are some eastern Procellarum (Apollo 12 and Apollo 14) samples whose sources probably mixed to an unusual extent with KREEP, which is extremely abundant in the Procellarum region (Lawrence et al., 2002a). The Apollo 14 mare basalts also tend to have unusually high I_{Sr}, again consistent with KREEP assimilation. Low-Ti mare basalts show remarkably diverse ε_{Hf} (as high as 51), again with Apollo 14 and especially Apollo 12 samples accounting for much of the range (Unruh et al., 1984). Beard et al. (1998) inferred possible involvement of deep-mantle garnet, a suggestion also supported by Neal (2001) based on high Zr/Y and Zr/Yb ratios in some mare pyroclastic glasses. In any event, the highly diverse isotopic ratios of the mare basalts demonstrate that volcanism arose from a remarkably heterogeneous mantle, fundamentally different from the tectonically stirred, and thus relatively homogenized, terrestrial mantle.

1.21.4 THE HIGHLAND CRUST: IMPACT BOMBARDMENT AND EARLY DIFFERENTIATION

1.21.4.1 Polymict Breccias and the KREEP Component

The vast majority of rocks from the sampled near-surface portion of the ancient highland crust are impactites. The proportion of highland rocks that are completely unaltered by impact processing is so minor that in this context the term "pristine" is used to denote samples (in most cases monomict breccias) that preserve in effectively unaltered form the original chemical compositions of endogenous igneous lunar rocks (Warren and Wasson, 1977; Wolf et al., 1979; Ryder et al., 1980; Papike et al., 1998). The importance of pristine rocks in providing constraints on lunar crustal genesis will be discussed at length below, but the mixed-origin highland rocks (polymict breccias and pseudoigneous clast-poor impact-melt breccias) are important in their own way.

The polymict breccias, especially the regolith breccias, are useful as naturally produced composite samples (assembled from mostly local but otherwise random bits) of the lunar crust. The mixing involved in the genesis of these rocks diminishes their value as recorders of lunar igneous processes, but makes them ideal for constraining variations in the regional bulk composition of the crust (Table 3). The most dramatic compositional variations among these samples involve their contents of the unique lunar component, KREEP (Table 5). The chondrite-normalized REE pattern of KREEP is shown in Figure 10. Although named for its high contents of potassium, REE and potassium, KREEP is actually rich in all of the incompatible trace elements. Among the most notable are thorium and uranium, because these are the most extremely enriched (versus chondrites) in KREEP, and because (with potassium) these are the main sources of long-lived radiogenic heating within planets. Because the many hundreds of Apollo polymict impactites that are thorium- and REE-rich tend to contain all the incompatible elements in relatively constant (KREEP) ratios, Warren and Wasson (1979a) suggested that nearly all of the lunar crust's complement of incompatible elements was derived from a common reservoir, possibly a magma ocean residuum, which they dubbed urKREEP. Note, however, that urKREEP, like KREEP, is a strictly hypothetical material.

Table 5 Estimated composition of average "high-K" KREEP, as derived by Warren (1989).

		Warren and Wasson (1979b) KREEP	Warren (1989) KREEP (avg. high-K)	Strength of correlation with KR[a]	KREEP/CI chondrites wt. ratio	Uncertainty class[b] (see below)
Li	µg g^{-1}	56	40	Moderate +	25	III
Na	mg g^{-1}	6.4	7	Moderate +		II
Mg	mg g^{-1}	64	50[a]	Moderate −		II
Al	mg g^{-1}	88	80	Weak −	9.3	I
Si	mg g^{-1}	224	235	Weak + (?)		I
P	mg g^{-1}	3.4	3.5	Moderate +	3.4	III
K	mg g^{-1}	6.9	8	Weak +	14	III
Ca	mg g^{-1}	68	70	Weak −	7.6	II
Sc	µg g^{-1}	23	23	None	4.0	II
Ti	mg g^{-1}	10	12	Very weak +	29	III
V	µg g^{-1}	43	40	Weak −		II
Cr	mg g^{-1}	1.3	1.2	Weak −		II
Mn	mg g^{-1}	1.08	1.05	None		II
Fe	mg g^{-1}	82	80	Weak +		II
Co	µg g^{-1}	33	25	Weak −		IIII
Ga	µg g^{-1}	7.5	9	Weak +		(III)
Br	µg g^{-1}	Not estimated	120	Moderate +		IIII
Rb	µg g^{-1}	22	22	Weak +	9.9	III
Sr	µg g^{-1}	200	200	Very weak +	25	I
Y	µg g^{-1}	300	400	Strong +	278	(II)
Zr	µg g^{-1}	1700	1,400	Strong +	368	III
Nb	µg g^{-1}	80	100	Moderate +	370	(III)
Cs	ng g^{-1}	2000	1000	Weak +	5.46	III
Ba	µg g^{-1}	1200	1,300	Strong +	565	II
La	µg g^{-1}	110	110	Very strong +	466	
Ce	µg g^{-1}	270	280	Very strong +	455	I
Pr	µg g^{-1}	Not estimated	37	Very strong +	398	I
Nd	µg g^{-1}	180	178	Very strong +	389	
Sm	µg g^{-1}	49	48	Very strong +	322	
Eu	µg g^{-1}	3.0	3.3	Moderate +	59	I
Gd	µg g^{-1}	57	58	Very strong +	294	
Tb	µg g^{-1}	10	10.0	Very strong +	282	
Dy	µg g^{-1}	65	65	Very strong +	265	
Ho	µg g^{-1}	14	14	Strong +	256	I
Er	µg g^{-1}	39	40	Very strong +	250	
Tm	µg g^{-1}	Not estimated	5.7	Very strong +	231	I
Yb	µg g^{-1}	36	36	Very strong +	226	
Lu	µg g^{-1}	5.0	5.0	Very strong +	204	
Hf	µg g^{-1}	37	38	Strong +	317	I
Ta	µg g^{-1}	4.0	5.0	Very strong +	313	II
W	µg g^{-1}	2.0	3.0	Strong +	30	(III)
Th	µg g^{-1}	18	22	Very strong +	759	I
U	µg g^{-1}	5	6.1	Strong +	744	I
Molar Mg/(Mg + Fe)		0.64	0.59[c]	Moderate −		II

[a] Warren's (1989) parameter KR is the average of sample/KREEP ratios for a large set of incompatible trace elements. [b] Estimated uncertainties for the average high-K KREEP composition, expressed as maximum expected percentages of deviation between "true" average and estimates: blank = 5%, I = 10%, II = 20%, III = 30%, IIII = 40%. Parentheses denote elements for which extrapolation to high KR is required.
[c] Mg concentrations appear to be systematically higher in Apollo-14 KREEP versus KREEP from other locales.

It is expected/inferred that none of this material remains in its original form, since almost as soon as it formed it probably was massively involved in assimilative reactions with magnesium-rich magmas (Warren, 1988; Papike et al., 1996; Shervais and McGee, 1999). There may even have been a tendency for KREEP to differentiate (albeit temporarily and locally) by silicate liquid immiscibility (Neal and Taylor, 1991; Jolliff et al., 1999). The few KREEP-rich rocks that are pristine are mostly basalts, or in rare cases granitic (Papike et al., 1998). Despite the ephemeral nature of the hypothetical urKREEP, it is useful to compare all other lunar sample compositions to the average composition of the most concentrated "high-K" variety of polymict impactite KREEP, which occurs among Apollo 12 and especially Apollo 14 samples (Table 5).

Most of the lunar meteorite highland regolith breccias, which come from widely scattered random points, are remarkably KREEP-poor compared to the Apollo and Luna regolith samples (Table 3). The recent Lunar Prospector maps of the global distributions of thorium, uranium, potassium, and samarium (Lawrence et al., 2002a; Prettyman et al., 2002; Elphic et al., 2000) revealed that the Apollo/Luna sampling region happens to be atypically KREEP-rich.

1.21.4.2 Bombardment History of the Moon

The polymict impactites are extremely important as primary constraints on the impact bombardment of the Moon. This bombardment history can be directly extrapolated to Earth (in early times, the tidally recessing Moon was even closer to Earth), where pre-3.8 Ga crust has been virtually eliminated by vigorous geodynamism. The polymict impactite samples have yielded ages that are remarkably clustered near 3.9 Ga, especially for impact-melt breccias (Figure 11). This nearly unimodal age spectrum, which was first noted in the initial Apollo 16 sample investigations (e.g., Schaeffer and Husain, 1973; Kirsten et al., 1973), represents one of the most profound discoveries of planetary sample research. It revealed that the rate of cratering (i.e., collisions between the Moon and asteroids and comets) was vastly higher ~3.9 Gyr ago than it has been over the last 85% of solar system history. A few additional data for 3.1–3.7 Ga, derived from crater-abundance counts on mare lava terrains dated through Apollo and Luna mare basalt samples (BVSP, 1981), indicate that the cratering rate declined rapidly, by a factor of order 10, between 3.9 Ga and 3.1 Ga.

Based on an "Occam's Razor" preference for simplicity, but comparatively scant evidence (indirectly inferred ages of 1.0 Ga and 0.3 Ga for the Copernicus and Tycho craters, respectively, compared versus hugely uncertain crater count data for terrains resurfaced during the formation of these craters), for many years the consensus view was that the cratering rate had probably remained approximately constant since 3.1 Ga (Guinness and Arvidson, 1977; Young, 1977; BVSP, 1981; cf. Bogard et al., 1994). However, the $^{40}Ar-^{39}Ar$ ages obtained by Culler et al. (2000) for 155 Apollo 14 impact-glass spherules suggest that the cratering rate continued to decline by roughly a factor of two between 3.1 Ga and 2.0 Ga, and then remained roughly constant until ~0.4 Ga, when it abruptly increased by a factor of 3.7 ± 1.2. These results should be viewed with caution. The <0.4 Ga glass spherules conceivably are predominantly from one fresh crater (Cone?) near the Apollo 14 traverse area. Smaller samplings of 21 impact glasses from Apollo 16 and 17 (Zellner et al., 2003; Terada et al., 2002) show none younger than 0.6 Ga. The first reported ages for impact-melt clasts from lunar meteorites (Cohen et al., 2000) show a comparatively diffuse cluster, spread rather evenly from 2.6 Ga to 4.0 Ga (Figure 11). Most of the Cohen et al. (2000) lunar meteorite data are relatively imprecise (they were obtained from tiny clasts), but these data are important because they presumably represent several very widely separated regions of the Moon. The abundance of ages ≪3.9 Ga in the Cohen et al. (2000) data set supports the Culler et al. (2000) inference that the cratering rate was still declining between 3 Ga and 2 Ga.

The bombardment history before 3.9 Ga has been most controversial. The relative scarcity of breccia ages >3.9 Ga has led many (originally Tera et al., 1974; in recent years most notably Ryder, 1990) to infer a spike in the global lunar cratering rate at ~3.9 Ga; in other words, that the rate was considerably lower before 3.9 Ga. Following Tera (1974), this cratering spike concept is somewhat confusingly known as the lunar "cataclysm" hypothesis. A broader, generally accepted hypothesis known as "late heavy bombardment" simply postulates vastly higher, more destructive lunar cratering at ~3.9 Ga, without regard to the spike question. The controversy concerns the degree to which the clustered ~3.9 Ga ages reflect a large-factor and global spike, as opposed to a bump or inflection on a basically monotonic decline in the late-accretionary impact rate.

Skeptics of the cratering spike hypothesis have argued that the clustering of ages near 3.9 Ga may be a "stonewall" effect (Hartmann, 1975; Wetherill, 1981; Chapman et al., 2002). The stonewall model invokes a saturation of the

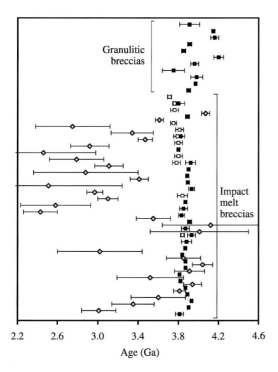

Figure 11 Ages of lunar polymict breccias, especially impact-melt breccias, show strong clustering near 3.9 Ga. Data from argon isotopes are shown by filled symbols; from strontium isotopes by open symbols. Squares: Apollo samples, data compiled by Papike et al. (1998), except excluding (in order to avoid complicating the plot) five much younger ages (0.4–2.3 Ga) for impact-melt glass samples. Diamonds: clasts in four different lunar meteorites (Cohen et al., 2000; Bogard et al., 2000). Error bars are generally 2σ (or unstated), but only 1σ for the Cohen et al. (2000) meteorite clast data.

pre-3.9 Ga crust with so many impacts that isotopic clocks were constantly being reset within a heated megaregolith, until the rocks that eventually became samples were excavated to the surface regolith (*sensu stricto*) by a few major impacts. This might seem a plausible explanation for the paucity of pre-3.9 Ga ages, if the 3.8–3.9 Ga clustering was limited to the easily reset argon age system. However, Papike et al. (1998) cite many strontium-based ages clustering near 3.8 Ga (Figure 11; the strontium-based cluster appears displaced by −0.1 Ga from the argon-based cluster, but this may be an analytical artifact). The relevance of the impact-melt age spectrum as portrayed by Ryder (1990) and Dalrymple and Ryder (1993, 1996) can still be questioned, however, on the more fundamental grounds that these authors always assumed that only samples with classic (near-subophitic) impact-melt breccia textures are appropriate for dating impact events. J. L. Warner et al. (1977) suggested that on the early (pre-3.9 Ga) Moon, where the ambient crustal temperature was still relatively close to the solidus, granulitic breccias tended to form in major impacts, in analogous proportion and position to where, on the post-3.9 Ga Moon, impact-melt breccias would form. This model, from a team of leading authorities on impactite genesis, has never been effectively refuted. As was already a point of emphasis to J. L. Warner et al. (1977), granulitic breccias are commonly older than 3.9 Ga (Figure 11).

Critics of the cratering spike hypothesis also note (e.g., Chapman et al., 2002) that the Apollo ∼3.9 Ga impact-melt samples come exclusively from sites clustered in the central nearside, and thus are likely from preponderantly just two or three impacts: Imbrium, Serenitatis, and with greater uncertainty, Nectaris. Cratering theory (Melosh, 1989; Grieve and Cintala, 1992) indicates that the proportional yield of impact melt, as opposed to solid ejecta, is vastly higher in large-basin-forming events than in smaller-scale impacts. Warren (1996) applied this principle to the particular case of the Moon.

On photogeologic–stratigraphic grounds, Nectaris is clearly older than Imbrium and Serenitatis, and two-thirds (30 out of 44) of the Moon's still-recognizable basins appear even older than Nectaris (Wilhelms, 1987). Impact-melt breccias of Nectaris origin are presumably present among the Apollo 16 samples, acquired ∼550 km from the center of Nectaris. ^{40}Ar–^{39}Ar ages for Apollo 16 impact-melt breccias mostly cluster in the range 3.87–3.92 Ga (Papike et al., 1998; Dalrymple et al., 2001), and 3.90–3.92 Ga has become almost canonical as the age of Nectaris (e.g., Wilhelms, 1987; Dalrymple et al., 2001). But, as Korotev et al. (2002b) have noted, the absolute age of Nectaris is unclear. A large group (at least 13) of Apollo 16 "light matrix breccias" (these rocks were *explicitly* classified as impact-melt breccias by the authoritative Stöffler et al. (1980) compilation) yielded argon ages in the range 4.12–4.26 Ga (Schaeffer and Husain, 1973; Mauer et al., 1978). Mauer et al. (1978) assumed these "group 1" samples cannot be basin-related, because they tend to be distinctly KREEP-poor compared to the other Apollo 16 impact melts; the assumption was that all impacts big enough to form basins would plumb into a KREEP layer in the lower crust. However, in light of what the Prospector data (Lawrence et al., 2002a) revealed about the extreme concentration of KREEP into the Procellarum (central-eastern nearside) region, a Nectaris provenance seems more plausible for the "group 1" samples than for the more typical, younger but unsuitably KREEP-rich, impact-melt breccias from Apollo 16. Also, recent basin ejecta thickness modeling (Haskin et al., 2002) suggests that ejecta from distant Imbrium and Serenitatis outweigh ejecta from the nearby but relatively small and older Nectaris, in the upper (sampled) Apollo 16 megaregolith. Considering all of these constraints, an age of ∼4.2 Ga for Nectaris seems entirely possible, if not probable.

The age of a fourth basin, Crisium, can, in principle, be constrained using Luna 20 samples. On photogeologic–stratigraphic grounds, Crisium appears similar in age to Serenitatis, i.e., older than Imbrium but younger than Nectaris (Wilhelms, 1987). Cohen et al. (2001) obtained argon ages for six Luna 20 rocklets and reviewed literature data (mainly from Swindle et al., 1991) for 12 others. Swindle et al. (1991) found a loose clustering of ages at 3.75–3.90 Ga and suggested that the oldest sample in this cluster, 3.895 ± 0.017 Ga, might date the Crisium impact. However, it is not even clear the sample in question is an impact-melt breccia (no thin section was made). The only two certain impact-melt breccias dated by Swindle et al. (1991) yielded ages of 0.52 ± 0.01 Ga and 4.09 ± 0.02 Ga. Among the samples dated by Cohen et al. (2001), most are, in this author's opinion, probably either impact-melt breccias or annealed impact-melt breccias (genuine, pristine "gabbros" seldom have grain sizes of <200 μm like rocklet 2004D; pristine troctolites seldom have grain sizes of <100 μm like rocklet 2004C). Considering all of the 13 or so Luna 20 rocklets that have yielded ^{40}Ar–^{39}Ar ages (Cohen et al., 2001) and are likely impact-melt breccias, the data (excluding the 0.52 Ga outlier) show an almost even distribution across the 3.75–4.19 Ga range. In other words, the age of the Crisium impact is not yet constrained beyond being probably within the range 3.8–4.2 Ga.

From a celestial-mechanical standpoint (e.g., Wetherill, 1981; Morbidelli et al., 2001), continued intense cratering as late as 3.9 Ga is plausible, but a major spike in the rate of inner solar system collisions 500 Ma after the origin of the solar system is hardly an obvious outcome from planetary accretion. Hartmann et al. (2000) discussed, in a conjectural way, several potential mechanisms to accomplish this.

By extrapolation to other solar system bodies, the Moon's bombardment history represents a crucial series of chronologic benchmarks for planetology. But even for application to today's mature solar system, this extrapolation is not straightforward (BVSP, 1981). Correcting for "gravitational cross-section" (i.e., the escape velocity) of the target body is straightforward, but also required is a far more complex correction for the flux and prevailing velocity of the asteroids and comets (i.e., the potential impactors) that are functions of, mainly, heliocentric distance (BVSP, 1981; Chyba, 1991). Going back in time, the extrapolation becomes even more uncertain. If the hypothesis of a 3.9 Ga cataclysm is correct, there is a slight possibility that the cataclysm was a local, ~1 AU phenomenon (Ryder, 1990). However, Bogard (e.g., 1995; cf. Kring and Cohen, 2002) has found that ages of numerous impactite meteorites from the HED asteroid (Vesta?), and many more from the (probably) separate mesosiderite asteroid, show a similar clustering near 3.9 Ga. The sole martian meteorite older than 1.3 Ga has also yielded an argon-based age of 3.9 Ga (Ash et al., 1996). Even if the cratering spike (cataclysm) hypothesis is incorrect, vastly higher inner solar system cratering rates at ~3.9 Ga, as demonstrated primarily and still most impressively from lunar samples, are a well-established fact. Implications for evolution of Earth, including the biosphere, are profound (e.g., Kring and Cohen, 2002).

1.21.4.3 Impactite and Regolith Siderophile Signatures

The siderophile elements in polymict lunar materials come almost entirely from meteoritic contamination added to the outer Moon in impacts. Siderophile elements thus may give clues to the nature of the materials that have bombarded the Moon. Unfortunately, however, the limited expanse of the Apollo/Luna sampling region, with samples dominated by just three or four major basins, again severely restricts the general applicability of this approach.

The largest basins clearly dominate the impact-melt inventory (Warren, 1996), but the degree to which they dominate the megaregolith siderophile element budget has been controversial. The Edward Anders (University of Chicago) group, in a series of papers reporting a wealth of excellent lunar siderophile analyses (e.g., Hertogen et al., 1977), interpreted some rather diffuse clustering in impactite siderophile element abundances as signatures of several different basins. For example, Wasson et al. (1975) and Korotev (1987a, 1994) argued that such an approach might be unreliable, because basin signatures tend to be obscured by significant siderophile contributions from smaller craters. Both views may be partly correct. The kinetic energy of impact is the main control over the basin (cavity, ejecta, and even melt) volume; and unfortunately for the aim of a simple, general answer to the basins-versus-craters controversy, the ratio of impact energy to projectile mass is sensitive to a wide-ranging and unrecoverable parameter: the collision velocity. The velocity of impact with the Moon is typically ~3 times faster for comets than for asteroids (e.g., Chyba, 1991).

Apart from impact velocity, the problem is partially constrained by the size–frequency spectrum of projectiles striking the Moon. This size distribution conforms to a power law with number of particles proportional to L^{-p}, where L is the projectile diameter and p is a constant >0. Numerous studies, reviewed by Melosh (1989), indicate that the slope p is 1.8 for $L > 10$ m, to 3.5 for $L < 10$ m. Thus (given that for large projectiles $p \ll 3$; and assuming that velocity, density, and siderophile concentrations all remain roughly constant across the size spectrum), most siderophile contamination, over time, should come from a comparatively few large collisions. Realistically, however, considering the stochastic potential for overlaps, and especially the great variability of collision velocity, it may be that in most regions no single basin dominates the megaregolith siderophile budget.

In any event, the steeper ($p > 3$) slope for impactor $L < 10$ m indicates that for the regolith (*sensu stricto*), a different regime from that of impact-melt rocks prevails, with micrometeorites important, if not dominant, in determining the siderophile budget. This is demonstrated by regolith sample data for iridium, the best-studied of the "noble" siderophile elements. Mare regoliths, which form by relatively small-scale impact gardening atop intact, siderophile-poor bedrock, show a strong correlation between iridium and regolith maturity (Figure 12) (cf. Wasson et al., 1975). Highland regoliths, however, show little correlation with maturity, probably because their siderophile enrichments are largely inherited from impact gardening at much larger scale, i.e., from megaregolith.

The Hertogen et al. (1977) interpretation of basin-dominated siderophile signatures has been refined in many subsequent works, usually emphasizing impact-melt breccias, most recently by Morgan et al. (2001), James (2002) and Norman et al. (2002a). The polymict highland

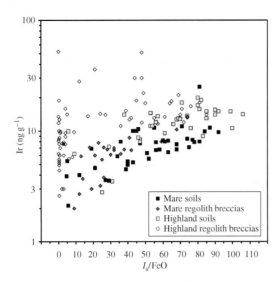

Figure 12 Variation of iridium concentration in lunar regolith samples as a function of the maturity index I_s/FeO, which is a magnetically determined measure of the abundance of ultra-fine-grained Fe^0 (which gradually accumulates through surface-regolithic processes) (Morris, 1978). The iridium data are from the compilation of Haskin and Warren (1991). A few soils of mixed mare–highland provenance (Apollo 17 stations 2 and 3) are excluded for the sake of clarity. Mare regolith samples show a strong positive correlation between iridium and I_s/FeO. Highland regolith samples feature generally higher iridium, and only a very weak correlation versus I_s/FeO.

impactites from the Apollo 14, 15, and 16 sites, where the siderophile components should be dominated by some combination of Imbrium, Serenitatis and (possibly, for Apollo 16 only) Nectaris ejecta, tend to feature remarkably nonchondritic siderophile patterns: Au/Ir is typically 3 times, and Ni/Ir 2 times, the CI-chondritic ratios. The polymict impactites from Apollo 17, presumably dominated by ejecta from adjacent Serenitatis, feature similar but less extremely high Au/Ir and Ni/Ir. Among chondrites (Wasson and Kallemeyn, 1988), only the EH enstatite type has Au/Ir approaching the typical Apollo 14, 15, and 16 ratio. Korotev (1987b) and Korotev et al. (1998) has conjectured that high-Au/Ir and high-Ni/Ir iron meteorites (akin to the IAB type) impacted at both Serenitatis and the Apollo 16 region. Morgan et al. (2001) argue that the high Au/Ir signature of late lunar accretion matches the composition they infer for the latest major accretion on Earth.

Ringwood et al. (1987) advocated a very different interpretation of the high Ni/Ir ratios in Apollo 16 impactites. These authors assumed that the meteoritic components have chondritic Ni/Ir, implying by mass balance (i.e., stripping away the meteoritic component based on iridium) a high nickel concentration for the indigenous highland crust. This model appears implausible in view of nickel and iridium systematics in lunar highland meteorites, which tend to feature chondritic Ni/Ir, implying that the Apollo 16 site happens to be unrepresentative (Warren et al., 1989, 2003).

Attempts to use siderophile element data to constrain the amount of material added to the Moon after origin of its crust (Ryder, 1999; Hartmann et al., 2000) are plagued with tremendous uncertainties. The most straightforward approach is to average breccia data in the hope that the samples are representative of some portion of the lunar crust. Unfortunately, siderophile elements are unevenly distributed, and available samples may be grossly unrepresentative. For example, a single Apollo 14 rock that consists of metal with noritic silicates (Albrecht et al., 1995) has 1,780 μg/g Ir, or 150 times the mean highland (Apollo 14 and 16) regolith concentration, which itself may be significantly enriched by surface-micrometeoritic processing (Figure 12). This metal-rich rock suggests that meteoritic impactor siderophile components have tended to accumulate in dense metal at the bottom of large impact melts. It seems unrealistic to expect subsequent gardening to stir the crust so well that the uppermost megaregolith is representative of the whole.

To utilize data for regolith samples, Ryder (1999) invoked a dubious "correction" of the observed iridium concentrations by assuming that, in each soil from a terrain older than the lavas atop which the Apollo 11 regolith formed (i.e., 3.6 Ga), 8.9 ng/g of iridium, the concentration found in the Apollo 11 regolith, was added by surface-micrometeoritic processing. This method ignored the high maturity ($I_s/\text{FeO} = 78$: Morris, 1978) of the sparsely sampled Apollo 11 regolith. A less-mature regolith atop the same Tranquillitatis lava would have a much lower iridium concentration. Ryder's method (cf. Hartmann et al., 2000) resulted in an estimate that an average highland regolith contains only 0.5 wt.% of CI-chondritic equivalent (i.e., 2.6 ng/g of iridium) as a pre-3.6 Ga meteoritic component. Considering that the mare trend on Figure 12 extends from ~3 to ~11 ng/g (the low end is not zero, because no mare soil is free of debris sprayed off the highlands), while the highland regoliths average ~15 ng/g (with no clear correlation with maturity), it seems a more realistic estimate for the contribution of pre-3.6 Ga Ir in average highland soil would be 8–12 ng/g, i.e., 2 wt.% of CI-chondritic equivalent. In any event, a long extrapolation would be needed to relate any surface regolith constraint to the content of meteoritic siderophiles in the crust as a whole.

1.21.4.4 Pristine Highland Rocks: Distinctiveness of the Ferroan Anorthositic Suite

Making the distinction between pristine rocks and polymict breccias is a crucial first step in any attempt to accurately gauge the original diversity, and thus the igneous processes involved in genesis, of the nonmare lunar crust (the terms nonmare and highland are essentially synonymous; arguably "highland" is an awkward term for virtually all of the Moon's crust, most of which is at levels lower than, including directly beneath, the veneers of "lowland" mare basalt). The vast majority of nonmare rocks are polymict breccias, and, unfortunately, identifying the pristine exceptions is seldom as easy as we would like it to be. In rare cases, vestiges of a coarse igneous cumulate texture, more consistent with an endogenous igneous origin than with an impact-melt origin, are preserved (e.g., Dymek et al., 1975; Warren and Wasson, 1980b). But the most straightforward and generally applicable criterion is based on siderophile elements. Nonpristine rocks, i.e., products of meteorite impact-induced mixing, generally contain enough meteoritic debris to impart high concentrations of siderophile elements such as iridium, in comparison to the levels characteristic of pristine rocks, as exemplified by mare basalts (Figure 8). Warren and Wasson (1977) suggested 3×10^{-4} times the CI chondritic concentration as the "cutoff" level.

It should be emphasized that this siderophile cutoff should never be construed as an upper limit, *sine qua non*, and nor should data below the cutoff be taken as complete proof of pristine composition. In evaluating suspected pristine samples all relevant traits, such as texture and mineralogy (silicates in gross disequilibrium, or more obviously FeNi metal with typical meteoritic kamacite composition, can be indicators of impact mixing), other aspects of bulk composition (absence of KREEP contamination can be a mildly favorable indicator of pristinity) and isotopic data (an extremely old age can be mildly favorable) should be assessed, if possible (cf. Ryder et al., 1980; Warren, 1993). Of course, a fine-grained texture, disequilibrium (zoned) silicates, and KREEP contamination, all are inevitable (but fortunately, high siderophile concentrations are not), if the pristine rock happens to be KREEP basalt.

The rate of discovery of pristine nonmare rocks has been limited by the scarcity of large lunar rock samples. Identification of a coarse cumulate or otherwise "plutonic" texture is difficult without a thin section at least several mm across, and determination of trace siderophile elements becomes increasingly difficult (and prone to the "nugget" effect) if available sample mass falls below ~ 0.1 g. Lunar geochemists must resist a temptation to overestimate the likelihood that new samples may be pristine. As a rather severe example, Snyder et al. (1995) classified 12 out of 18 small rock clasts from Apollo 14 as "probably pristine," even though "[n]one of the samples… are texturally pristine," and their siderophile data included no concentration (or limit) lower than 6×10^{-4} times CI, and even that level was approached for only one of the 12 samples in question. Rocklet 14286, which arguably might pass for an uncommonly fine-grained pristine norite, except that it contains a huge mass of meteoritic kamacite (Albrecht et al., 1995), indicates that potential pristine rocks should be evaluated with great caution.

When pristine rocks alone are considered, a remarkable geochemical bimodality is manifested within the nonmare crust. During the 1970s, petrologists gradually noticed that the most anorthositic pristine rocks tend to feature distinctively low mg in comparison to otherwise comparable nonmare rocks. Dowty et al. (1974) coined the term "ferroan anorthosite" and Warner et al. (1976) were the first to postulate a separate "magnesium-rich plutonic" suite to account for nearly all other pristine nonmare rocks. In more recent usage, the original terms may be replaced with "ferroan anorthositic suite" (or simply "ferroan suite") and "Mg-suite," but their basic meanings have not changed. The distinctiveness of the ferroan anorthositic (FA) suite was first noticed on the basis of mineral composition data, namely plots of plagioclase Ca/(Ca + Na) versus mafic silicate mg. Essentially the same pattern is found by simply plotting bulk-rock mg versus Na/(Na + Ca) (Figure 13(a)). The Mg-suite rocks distribute along a normal fractionation path, with Na/(Na + Ca) increasing as mg decreases, to form a diagonal trend on Figure 13(a). The FA suite forms a distinct cluster to the low Na/(Na + Ca), low-mg side of the Mg-suite trend; and there is a noticeable gap (or at least, a sparsely populated region) between the two groups. Warren and Kallemeyn (1984) found the same bimodality using other plagiophile element ratios, such as Ga/Al or Eu/Al, in place of Na/(Na + Ca). Most impressive is Eu/Al (Figure 13(b)), which results in a wide gap between the two pristine rock groups.

Besides having distinctive combinations of mg and plagiophile ratios, FA suite rocks tend to be more anorthositic than other types of pristine rocks (Figure 14). Not all of the ferroan pristine rocks are anorthosites, *sensu stricto* (>90 vol.% plagioclase, Stöffler et al., 1980), but prevalence of low mg among the most anorthositic components of the crust is borne out by major-element variations among regolith samples (Figure 15). The obvious yet far-reaching implication (Warren and Wasson, 1980a) is that the FA suite, and no other adequately

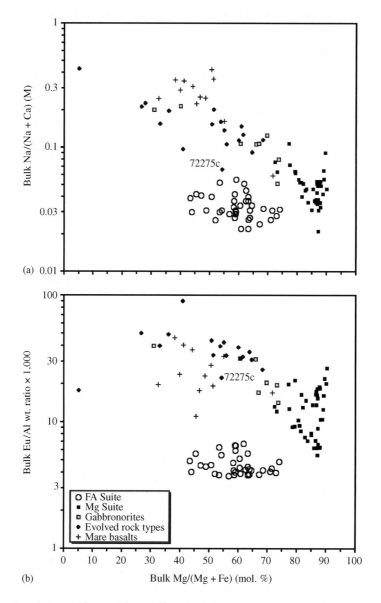

Figure 13 Examples of the geochemical bimodality of pristine nonmare rocks: (a) Na/(Na + Ca) and (b) Eu/Al versus bulk-rock *mg*. The category "evolved rocks" here includes such rock types as granite, felsite, quartz, monzodiorite, and KREEP basalt. The database used for these diagrams comprises all known nonmare rocks with "pristinity confidence index" ≥ 6 as compiled by Warren (1993); more recent data sources include Arai and Warren (1997), Jolliff *et al*. (1999), Zeigler *et al*. (2000) (gabbronorite 62283,7-15 only), Neal and Kramer (2003), plus a few unpublished analyses by the author. On the Eu/Al diagram (b), the single gray diamond represents quartz monzodiorite 14161; its actual Eu/Al ($\times 1,000$) is 126 (Jolliff *et al*., 1999), but it is plotted at a slightly lower value (90) to avoid compressing the *y*-axis of the diagram. For gabbronorite 62283,7-15, the *mg* ratio is inferred to be ~40 mol.% based on published mineralogy (source Zeigler *et al*., 2000).

sampled type of pristine lunar rock (note that Figure 14 includes no samples smaller than 1 g), would have been buoyant over its parental magma. If the magma ocean hypothesis (see below) has any validity, the FA suite represents the only rock type plausibly formed as magma ocean flotation crust; and the manifestly sunken cumulates of the Mg-suite must come from intrusions that were emplaced into an older, FA-dominated crust.

In the few FA suite rocks that (as sampled) contain more than 1–2% mafic silicates, the mafics are typically a mix of olivine and coarsely exsolved low-Ca pyroxene (igneous pigeonite) (Taylor *et al*., 1991; Papike *et al*., 1991, 1998). Another apparent geochemical discontinuity between the ferroan anorthositic and Mg-suites is manifested by Ni–Co systematics in olivine (Shearer *et al*., 2001b). Early models designed to

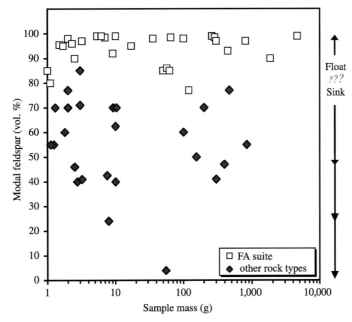

Figure 14 Ferroan suite rocks tend to be more anorthositic than any other type of pristine lunar rock. Most of the "other" large pristine rocks are troctolitic and noritic cumulates of the magnesium suite (source Warren, 1993).

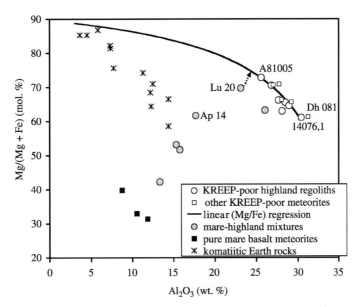

Figure 15 *mg* versus Al_2O_3 for lunar meteorites and averaged highland regolith samples from Apollo and Luna sites. The data set used for KREEP-poor highland regoliths comprises lunar meteorite breccias ALH81005, DaG 262, Dhofar 025, MAC88105, and QUE93069, Apollo 14 sample 14076,1 (Jerde et al., 1990), and average Apollo 16 station 11 regolith (Korotev, 1981). The other KREEP-poor lunar meteorites are polymict breccias DaG 400, Dhofar 026, Dhofar 081, NWA 482, and Y-82192. The mare–highland mixture lunar meteorites shown are EET87521, QUE94281, and Y-793274. The pure mare basalts are Asuka-881757, NWA 032, and Y-793169. Data for komatiitic Earth rocks are from BVSP (1981). Curve represents an extrapolated linear regression for all of the plotted KREEP-poor highland samples. Also shown is the position inferred for the pure highland component of the Luna 20 regolith sample.

account for all pristine nonmare rocks as magma ocean flotation cumulates (Wood, 1975; Longhi and Boudreau, 1979) assumed that the mafic silicates formed from locally high proportions of trapped melt. However, the more mafic members of the FA suite, such as 62236 (Borg et al., 1999) and 62237 (Warren and Wasson, 1978), have only slightly higher REE concentrations than the purest anorthosites, indicating that the mafics are almost purely cumulus (or geochemically equivalent "adcumulus") material; i.e., in FA rocks the trapped melt components are uniformly low.

FA suite rocks have distinctive but not entirely uniform compositions, and James et al. (1989) suggested a subclassification of the suite into four "subgroups" based on subtly different mg ratios of mafic silicates and alkali contents of plagioclase. Obviously, such terms can be useful for descriptive purposes, but our sampling of the FA suite (mainly from one site, Apollo 16) is probably inadequate for assessing whether the proposed subgroups reflect compositional discontinuities within the suite, or apparent divisions, due to nonrepresentative sampling, of a compositional continuum.

Most of the other pristine nonmare rocks are Mg-suite rocks: troctolitic, noritic, and gabbronoritic rocks that, based on evidence from a few texturally pristine samples (e.g., Dymek et al., 1975; Marvin and Warren, 1980), probably all formed as cumulates. The highest-mg cumulates include several with ultramafic modes, but only one of these, dunite 72415 (Dymek et al., 1975), has a mass >1 g (Figure 14), so many of them may actually be grossly unrepresentative samples of troctolites. High-Ca pyroxene apparently formed relatively late in the crystallization sequence of most Mg-suite magmas; gabbronorites are relatively rare, and they tend to have lower mg and higher Na/(Na + Ca) than norites (Figure 13(a)). They were first recognized as a distinctive subdivision of the Mg-suite (James and Flohr, 1983). As reviewed by Papike et al. (1998; cf. Shervais and McGee, 1999), the most evolved pristine rock types, alkali (high Na/Ca) suite rocks and rare granites (and similar quartz monzodiorites; also fine-grained felsites), may be extreme differentiates related to the Mg-suite and/or KREEP. Some of the best-sampled granitic rocks show clear millimeter-scale effects of silicate liquid immiscibility (Warren et al., 1987; Jolliff et al., 1999).

Isotopic studies have constrained the chronology of genesis of the various types of pristine rock. Results are summarized in Figure 16. Dating the REE- and rubidium-poor FA suite poses a severe analytical challenge, but argon isotopes seem far more prone than Sm–Nd and Rb–Sr to undergo resetting during the slow cooling and intense shock that ancient pristine rocks have typically endured. Considering the likelihood that the FA suite represents flotation crust from the

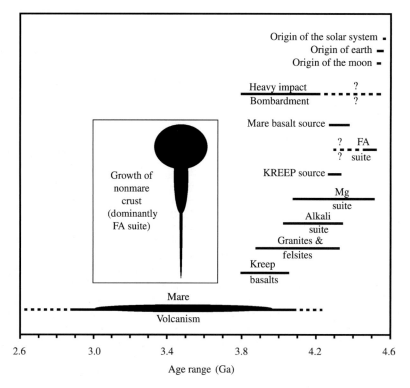

Figure 16 Summary of the chronology of lunar crustal genesis, based mainly on data previously reviewed by Papike et al. (1998) and in Chapter 1.20. Other noteworthy sources include Carlson and Lugmair (1979) and Shih et al. (1992) for KREEP and mare basalt ages, including source model ages; and Borg et al. (1999) for the lowest of arguably accurate/relevant ages for a ferroan anorthositic (FA) suite rock (~4.29 Ga, for 62236). The inset cartoon indicates (roughly) the relative volume of crust produced at the various stages; i.e., the main volume of the crust probably formed as FA suite flotation cumulate, whereas evolved rocks like granites and KREEP basalts are probably volumetrically minor.

putative magma ocean, Sm–Nd isochron ages obtained for FA samples have been surprisingly diverse, ranging from 4.56 ± 0.07 (Alibert et al., 1994) down to ~4.29 Ga (for 62236: Borg et al., 1999). However, Norman et al. (2002b) inferred that the plagioclase in these rocks might have been compositionally modified, most likely during a large-impact event many hundreds of millions of years after igneous crystallization. When Norman et al. (2002b) constructed an Sm–Nd isochron using mafic silicates only (from four different FA samples, including 62236), they found an age of 4.456 ± 0.040 Ga; they suggest that this may be closer to the true igneous crystallization age than any of the previous individual FA rock isochrons.

As reviewed by Papike et al. (1998) and Snyder et al. (2000b), Mg-suite rock ages distribute fairly evenly across a range from ~4.5 Ga (very close to the age of the Moon itself, but all of the oldest Mg-suite ages are strontium-based, and thus plagued by uncertainty associated with the decay constant) down to ~4.1 Ga. In a very general way, as illustrated in Figure 16, the sequence of typical or median age for the various major pristine rock types appears to be: FA suite, Mg-suite, gabbronorite (subset of the Mg-suite), alkali suite, granites and felsites, KREEP basalts, mare basalts.

Before leaving the topic of pristine rocks, it should be noted that not only might individual samples be misclassified as pristine, but conceivably, particularly in the light of 14286 (Albrecht et al., 1995), whole classes of putatively pristine rocks may actually be products of uncommonly large impact-melt ponds engendered by basin-forming impacts (Delano and Ringwood, 1978). The Mg-suite rocks are probably the most likely candidates for such an impact-melt genesis (A. E. Ringwood, 1993, personal communication; Hess, 1994). The diverse Mg-suite ages (Figure 16) imply that any impact-melt model is inconsistent with the bombardment cataclysm hypothesis, at least in its extreme form as advocated by Ryder (1990), because most ages should be the same in such a model. However, even if Ryder (1990) overstates the case for a cataclysm, the finite number of old basins is a major stumbling block for the Mg-suite-as-impact-melt-products hypothesis, because the hypothesis requires derivation of each Mg-suite sample through a combination of two major impacts. Impact 1 is required to form an exceptionally large pond of impact melt that yields the Mg-suite cumulates. Impact 2 is required to excavate the cumulates and add them to the surface regolith. Two overlapping basins are probably required for such a scenario. If Procellarum originally formed as an impact basin (a question that may never be resolved; see, e.g., Wilhelms, 1987), cumulates from its central "sheet" of impact melt, indistinguishable from truly pristine rocks, may well have been excavated by subsequent events like Imbrium. However, as noted by Warren et al. (1996), in an event as gigantic and early as the putative Procellarum impact, the distinction between pristine and nonpristine becomes blurred and potentially even misleading, because most of the energy in the impact melt is inherited from the very warm (and deep) target region, and the mantle beneath the point of impact probably does not respond in a strictly passive way to the aftermath of such a colossal near-surface heating event.

1.21.4.5 The Magma Ocean Hypothesis

The original inspiration for the lunar magma ocean hypothesis (Wood et al., 1970) was a discovery that among ~1,700 tiny (nearly all <2 mm) rocklets from the Apollo 11 Mare Tranquillitatis soil, a distinctive minority (5%) of exotic (i.e., highland) origin had anorthositic or "anorthositic gabbro" mineralogy or composition. Based on these few tiny samples, Wood et al. (1970) inferred that the Moon's main, highland crust formed as a globe-wide "surface layer of floating anorthite crystals" over a zone of "gabbroic liquid, of composition near the eutectic value of the Moon as a whole." Wood et al. (1970) admitted that their samples were "much too fine grained to have formed in such a grand event." They nonetheless claimed to discern consistent cumulate textures. From a modern perspective, one of their crystalline samples ("40–2") shows what is arguably a brecciated cumulate texture; most are fine-grained granulitic breccias. In any event, Wood et al. (1970) saw that "the cumulate interpretation must [also] be based on the specialized compositions of anorthositic rocks."

The most impressive argument for a lunar magma ocean is still that same straightforward observation: the bulk composition of the globe-wide highland crust is far too anorthositic to have formed by piecemeal aggregation of basaltic partial melts ascended from the deeper interior. The normative plagioclase contents of internally generated partial melts are constrained by phase equilibria (e.g., Longhi and Pan, 1988) to be less than ~55 wt.%, whereas the highly anorthositic (~75 wt.% normative plagioclase) composition of the global lunar crust has been amply confirmed in recent years by lunar meteorites (Table 3) and remote sensing (Lucey et al., 2000a; Prettyman et al., 2002).

Warren's (1985) review cited numerous other lines of evidence favoring the magma ocean hypothesis. Warren's (1985) second argument, the extreme antiquity of many pristine nonmare rocks, has been strengthened by recent work on

FA suite chronology (Norman et al., 2002b). But a more impressive argument, which I would now rank ahead of that one, is based on the geochemical bimodality of the pristine nonmare rocks (Figure 13). The singularly low-mg composition of the FA suite must reflect a special process of origin. Warren (1986) argued that a deep, originally ultramafic magma ocean is the only plausible scenario for engendering a global component of low-mg cumulate anorthosite while burying the comparable (or larger) volume of low-mg mafic (sunken) cumulates that must have formed simultaneously, but are so deep that they are absent among sampled lunar rocks. This scenario accounts for the peculiarly low mg of the flotation crust, *provided* that the initial magma ocean was ultramafic, so that onset of (copious) plagioclase crystallization was preceded by extensive fractional crystallization of mafic silicates, driving down the mg and driving up the density, yet not fractionating plagiophile ratios, such as Na/(Na + Ca), of the residual melt. A similarly low-mg, low-(Na/(Na + Ca)) composition was probably never reproduced by subsequent, post-magma-ocean magmatism, because these magmas originated by (more or less) conventional partial melting of the mantle, and thus were already at or near plagioclase saturation even as they ascended into the crust.

Even if the parent Mg-suite magmas of some of the highest-mg dunites and troctolites initially reached the crust undersaturated in plagioclase, assimilation of ferroan anorthositic country rock by superheated, adiabatically decompressed melt may have curtailed the extent of the fractional crystallization of mafic silicates prior to plagioclase saturation (Warren, 1996; Ryder, 1991). Warren (1986) constructed a rather detailed model of this proposed AFC process. Without explicitly finding fault with Warren's (1986) model, Hess (1994) argued that such assimilation could not be very extensive. The main difference between the two model approaches is probably that Warren (1986) assumed that real-world assimilation would seldom involve total melting. The assimilation could be mostly physical; i.e., plagioclase grains that became loosened by disaggregation of anorthositic wall rock (at, say, 30% melting), and after a while settled at the base of a Mg-suite intrusion, might be impossible to distinguish from plagioclase grown entirely from the parent melt. In any event, as Hess (1994) suggested, assimilation would not be needed, provided the initial melts were sufficiently magnesian; this might, however, require that the source regions included high-mg cumulates from the early stages of magma ocean crystallization.

Another key argument in support of the magma-ocean hypothesis, formulated by various authors in the mid-1970s (most notably Taylor and Jakes, 1974), holds that the prevalence of negative europium anomalies among plagioclase-poor mare basalts implies predepletion of mare basalt source regions in massive proportions of plagioclase with positive europium anomalies. As discussed above, the assumption that the mare basalt europium anomalies can only reflect plagioclase fractionation has been challenged, but it still appears sound (Brophy and Basu, 1990).

Warren (1985) also cited the uniformity of the trace-element pattern of KREEP, suggesting derivation from the residuum (urKREEP) of a single, global magma. This argument has been weakened slightly, because remote sensing (Lawrence et al., 2002a; cf. Warren, 2003) and lunar meteorites (Table 3) have revealed that KREEP is only abundant in one relatively small area, i.e., the Procellarum–Imbrium region. However, the same revelation about KREEP distribution reinforces another of Warren's (1985) arguments in favor of a magma ocean ("petrochemistry correlated with longitude"). The extremely heterogeneous distribution of KREEP stands in stark contrast to the almost uniform (anorthositic) major-element composition of the global crust. Both the Procellarum–Imbrium region and the giant, nearly antipodal South Pole Aitken basin have exceptionally mafic (noritic) compositions, yet strong KREEP enrichment is found only at Procellarum–Imbrium (Lawrence et al., 2002a; Pieters et al., 2001). Thus, the Procellarum–Imbrium KREEP anomaly is not simply a by-product of the large-scale cratering of that region. The uniqueness of the Procellarum–Imbrium region suggests that a globe-wide, extremely KREEP-concentrated magmatic plumbing system, such as a declining magma ocean, channeled and concentrated incompatible elements into this one region. The failure of KREEP to concentrate as well into the South Pole Aitken basin may be a consequence of origin of KREEP at a later stage, after the magma ocean had already nearly completely solidified.

Many authors have constructed detailed models for the crystallization of a lunar magma ocean (e.g., Longhi and Boudreau, 1979; Brophy and Basu, 1990; Snyder et al., 1992; cf. the review of Warren, 1985). One key constraint is that silicate adiabats have small dP/dT relative to melting curves, so crystallization occurs primarily along the base of the magma ocean (Thomson, 1864). This is important, because pressures of the order 0.1–4 GPa enhance stability of pyroxene at the expense of olivine. It is commonly assumed that the lunar magma ocean produced mainly olivine until late in its crystallization sequence. However, it is likely that even the early cumulates consisted largely of pyroxene (Warren and Wasson, 1979b), with important compositional implications for the initial magma ocean, for its early and deep cumulates, and for the bulk-Moon composition. All models indicate that the late magma ocean becomes (because of its low f_{O_2}) highly enriched in FeO, which gives the FA suite parent melt a high density, and, in turn, allowed a considerable proportion of mafic silicate to be rafted within the FA flotation crust, as modeled quantitatively by Warren (1990).

As reviewed by Shearer and Papike (1999), the origin of the other major class of ancient pristine rocks, the Mg-suite, is widely ascribed to more conventional layered intrusions, emplaced in the FA crust during and shortly after the magma ocean waned to the final dregs of urKREEP (e.g., Warren and Wasson, 1980a; Ryder, 1991; Papike et al., 1996). In a variant of this model, Hess (1994) suggested that the layered intrusions might have formed as impact melts in a few of the largest lunar accretionary events (cf. the previous section). The Mg-suite plutonism may have been triggered by the enhanced convective potential that many authors (e.g., Warren and Wasson, 1980a; Ryder, 1991; Hess, 1994; Hess and Parmentier, 1995; Zhong et al., 2000) have noted would likely ensue if the magma ocean crystallized into a gravitationally unstable configuration: dense, FeO- and ilmenite-rich residual mush atop relatively FeO- and ilmenite-poor early cumulates. The Mg-suite plutons would inevitably interact extensively with urKREEP stewing at the base of the FA crust; this seems a major advantage of the model, because pristine KREEP basalts have relatively high *mg* ratios (Warren, 1988), and Mg-suite cumulates appear to have formed from remarkably KREEP-rich parent melts (e.g., Papike et al., 1996; Shervais and McGee, 1999).

One obvious reason for skepticism regarding the magma ocean hypothesis is the colossal energy input that is required to engender a fully molten layer encompassing a large portion of the Moon (e.g., Wetherill, 1981). Another widely cited basis for skepticism stems from recent neodymium isotopic studies of ferroan anorthositic suite rocks. Until very recently, part of the problem lay in the young apparent ages (younger than Mg-suite rocks of manifest non-magma-ocean origin) for samples such as 62236, ~4.29 Ga (Borg et al., 1999). As discussed above, Norman et al. (2002b) seem to have largely eliminated this aspect of the neodymium-isotope based concern. These authors found that neodymium isotopic data from four FA suite rocks define a "robust" isochron, aged 4.456 ± 0.040 Ga. However, the neodymium isotopic data still pose a dilemma, because the same isochron implies a positive initial ε_{Nd} of $+0.8 \pm 1.4$. More significantly, Norman et al. (2002b) also report an initial ε_{Nd} of $+0.8 \pm 0.5$ for an individual FA suite sample 67215c, and endorse a value of $+0.9 \pm 0.5$ for sample 60025 (Carlson and Lugmair, 1988). These positive initial ε_{Nd} ratios imply derivation of the FA suite from material that had spent the prior few tens of Ma (between origin of the solar system and igneous crystallization) with a remarkably fractionated, light REE-depleted Sm/Nd ratio.

The potential seriousness of the problem is illustrated in Figure 17. The magnitude of the required light REE depletion is inversely related to the duration assumed for ^{147}Sm decay prior to isotopic closure. Assuming crystallization at 4.456 ± 0.040 Ga (Norman et al., 2002b), the maximum duration of the decay at fractionated Sm/Nd is 110 Ma, and ~80 Ma is more realistic, since the solar system formed at 4.566 Ga, and it probably took 20–30 Ma just to form the Moon (Chapter 1.20). A straightforward calculation (Dickin, 1995) yields the implied Sm/Nd ratio of the parent material as a function of assumed initial ε_{Nd} and duration of ^{147}Sm decay (Figure 17). The implied Sm/Nd for the FA parent material is ~1.3 times chondritic, and 1.14 times chondritic even at the extreme low end of the quoted 60025 ε_{Nd} uncertainty range. These may sound like minor fractionations, but samarium and neodymium are separated by just two atomic numbers. Extrapolation suggests that the ratio of the heaviest REE versus the lightest, Lu/La or even Yb/Ce, would be more strongly nonchondritic than Sm/Nd. Assuming log(REE concentration) scales linearly with atomic number, an Sm/Nd ratio of 1.3 times chondritic suggests Ce/Yb~0.21 times chondrites. The right side of Figure 17 indicates how the implied source Sm/Nd ratios translate into (approximate) source Ce/Yb based on more detailed modeling (see caption). Assuming 80 Ma of decay at fractionated Sm/Nd after the formation of the source, for the source to so quickly reach an ε_{Nd} of $+0.8$ implies the Sm/Nd of the source is 1.38 times chondritic, and its Ce/Yb is only ~0.14 times chondritic.

The severity of this ε_{Nd} conundrum appeared even worse before Norman et al. (2002b) adduced their revised interpretation of the neodymium isotopic data for the FA suite. Results for 62236 led Borg et al. (1999) to question the viability of the magma ocean hypothesis, where the positive initial ε_{Nd} of 3.1 ± 0.9 was seen as even more problematic than the young (~4.29 Ga) apparent age. While an age can be reset by shock and/or thermal processes, strongly positive ε_{Nd} ratio was claimed to be a firmly established trait of the FA suite. No conventional magmasphere model predicts light REE depletion at any point in the magma ocean's evolution, and least of all in its later stages. On the contrary, magma ocean crystallization models (e.g., Snyder et al., 1992)

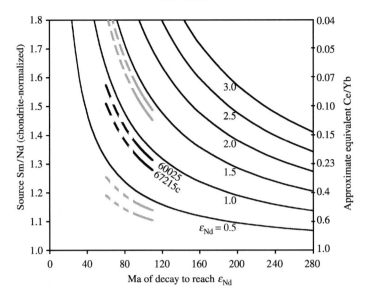

Figure 17 Source Sm/Nd required for development of a range of target ε_{Nd} versus time, where the former is calculated by rearrangement of equation 4.2 of Dickin (1995). As discussed in the text, these trends have important implications for origin of ferroan anorthosites with positive ε_{Nd}, such as 60025 and 67215c (Norman et al., 2002b). Black curves labeled 60025 and 67215c correspond to the nominal ε_{Nd} of these samples (Norman et al., 2002b) and the decay periods shown on the x-axis; unlabeled grey curves show limits implied by the quoted ε_{Nd} uncertainties. The curves for 60025 and 67915c end at 110 Ma, because if their age is 4.456 Ga (Norman et al., 2002b) they formed only 110 Ma after origin of the solar system. Shown at right are approximate Ce/Yb ratios, extrapolated from Sm/Nd based on modeling the FA source material as residual solids engendered by 0–6 wt.% "continuous" (1 wt.% porosity) partial melting (Albarède, 1995) of a plausible "primitive Moon mantle" with initially 60% olivine, 12% plagioclase, and 28% pigeonite, and applying distribution coefficients from Phinney and Morrison (1990), Kennedy et al. 1993, and McKay et al. 1991. Similar Ce/Yb would be implied by a variety of models; for example, under the simple assumption that log(REE concentration) scales linearly with atomic number, Sm/Nd ratios of 1.2, 1.4, and 1.6 would correspond to Ce/Yb = 0.33, 0.13, and 0.06, respectively.

indicate that crystallization of pyroxene tends to enrich light REE over heavy REE as the magma ocean evolves to approach plagioclase saturation (i.e., FA suite flotation). Based on the older age and less extreme ε_{Nd} results implied by their revised interpretation, Norman et al. (2002b) reverted toward acceptance of the hypothesis that the FA suite formed as a magma ocean flotation crust, and speculated that the positive initial ε_{Nd} ratio for the FA suite may have arisen because bulk Moon might have a subchondritic Nd/Sm ratio. However, the magnitude of the implied Ce/Yb fractionation (Figure 17) seems far in excess of what can be plausibly invoked without enormously depleting the Moon in light REE and other incompatible elements relative to cosmochemically similar refractory lithophile major elements (aluminum and calcium). Based on remote-sensing data (Lawrence et al., 2002a; Prettyman et al., 2002), incompatible elements are certainly not depleted by more than a factor of 2 relative to aluminum and calcium. As an alternative solution to the positive ε_{Nd} conundrum, Norman et al. (2002b) conceded that the canonical value for CHUR-Nd (i.e., the zero point on the ε_{Nd} scale) may not be "precisely representative" for lunar material.

1.21.4.6 Alternative Models

Models have been proposed that invoke no magma ocean, although not in any detail for many years. Walker (1983) reviewed the evidence for petrogenetic diversity among the various types of pristine rocks, and suggested that the lunar crust formed by "serial magmatism" as a series of flows and plutons. However, Walker offered little justification for his implicit assumption that the entire lunar magma ocean hypothesis should be rejected simply because important (yet subordinate to the FA suite) components of the crust formed by processes other than magma ocean cumulate flotation. Walker conceded that a magma ocean is implied "if the bulk crust... [is] significantly more feldspathic than internally generated partial melt." As discussed above, recent data amply confirm that the bulk composition of the crust (or at least its upper half) has ~75 wt.% plagioclase.

Wetherill (1981) argued that accretional heating could not melt a large fraction of the Moon; and yet, he suggested, it might have produced the Moon's anorthositic crust. In this model, the Moon is zone-refined by production of many regional magma chambers as the central impact melts of the biggest accretionary events.

As the Moon grows, these impact-melt plutons differentiate so that incompatible elements and low-density plagioclase concentrate upward. Melting is discontinuous in space and time, but the ultimate, cumulative effect is a global surface layer of anorthosite. Similar models were earlier (and briefly) mentioned by Wetherill himself, by Alfven and Arrhenius (1976), and by Hess et al. (1977).

The most detailed variant of this type of a model was proposed by Longhi and Ashwal (1985). In their model, the upward concentration of anorthosite is enhanced by diapiric detachment and ascent of feldspathic portions of the layered impact-melt plutons of the Wetherill (1981) model. Partial melting, partly due to pressure release, would lubricate the diapirs and enhance their potential for poking through the cool, mostly solid near-surface layer, and would be consistent with an argument (Haskin et al., 1981) that ferroan anorthosites may be residual solids from episodic partial melting. Through the years (e.g., Borg et al., 1999), this anorthosite diapirism model has been widely cited as the strongest competition for the magma ocean hypothesis. However, the diapirism model has some serious drawbacks. It implies that the anorthositic early crust should consist mainly of annealed (and possibly sheared and deformed) restite. But among the few pristine anorthositic (FA suite) samples that show little brecciation, only one, 15415, shows an annealed texture consistent with a restite origin (Taylor et al., 1991). The majority of the samples (e.g., 62237, Dymek et al., 1975, 66035c; Warren and Wasson, 1980b; 64435c, James et al., 1989) have igneous textures. Also, in common with all other published models for early lunar petrogenesis without a magma ocean, the diapirism model seems ill-suited to account for the geochemical bimodality of pristine rocks, i.e., the observation that nearly all of the pristine rocks with the distinctive "ferroan" geochemical signature (Figure 13) are anorthosites (Figure 14), and vice versa. Why should such a relationship have developed, and in such consistent way, if the crust formed by aggregation of diapirically mobilized components from many separate, isolated intrusions? Highly inconsistent geochemical data seem inevitable, considering that the original plutons are supposed to form and crystallize at variety of depths (i.e., pressures), and would be prone to diverse fates, due to regional remelting events during the prolonged but spotty accretional heating of the Moon. By the same token, the diapirism model also seems ill-suited to account for the characteristic uniformity of the KREEP component (Warren and Wasson, 1979a). Longhi (2002) recently advocated a deep, primordial magma ocean, citing evidence for major depletion of aluminum in the source region of VLT-mare pyroclastic glasses.

Shirley (1983) proposed a model that is, in some respects, intermediate between "serial" magmatic models and the magma ocean hypothesis. In this model, only a relatively thin zone near the top of the system is fully molten; the rest of the differentiating system, or at least a large portion immediately below the magma ocean, is a convective, mostly crystalline mush that Shirley (1983) called "magmifer." In a quasi-steady state, the magmifer continually bleeds partial melt into the magma ocean, which replenishes the mush layer by precipitating mafic crystals, while simultaneously floating buoyant anorthosite to form a FA suite crust. Warren (1985) suggested the term magmasphere for the combined (magmifer + thin magma ocean) partially molten system. This model has an obvious thermal advantage: most of the Moon can be differentiated by the magmasphere, without fully melting the differentiated volume. Realistically, any "magma ocean" must have some magmifer/magmasphere characteristics, but it is unclear whether a lunar magmifer could sustain itself against upward percolation of buoyant melt (McKenzie, 2000) over a thick enough depth range, with a large enough fraction of melt, to have any appreciable effect. A major role for equilibration between the magma ocean and a thick magmifer would have a major disadvantage: it would dampen, if not cancel, the potential for early fractional crystallization of ultramafic, plagioclase-undersaturated magma (see above) to engender the peculiarly low-mg ratio of the FA suite pristine rocks. It would also tend to reduce the tendency for heterogeneity to develop in the mantle (the magmifer is well-stirred, like a melt-lubricated variant of Earth's present-day asthenosphere), and thus come into conflict with evidence (e.g., Figure 1; hafnium isotopic data, Unruh et al., 1984; Beard et al., 1998) for extreme heterogeneity in the source mantle for mare basalts.

1.21.5 THE BULK COMPOSITION AND ORIGIN OF THE MOON

A vast literature exists concerning attempts to divine the bulk composition of the Moon, and to assess the implications for models of lunar origin of the Moon and Earth. The most obvious areas of compositional contrast versus Earth are the gross depletion in FeNi (i.e., core matter) implied by the Moon's low bulk density, and the lunar volatile depletions, which were discussed above (Figure 6). The FeNi depletion is a key advantage of the popular giant impact model, according to which the Moon originated as a form of giant impact spall after a collision between the protoearth and a doomed Mars-sized, or larger, body (e.g., Cameron and Ward, 1976; Canup and Asphaug, 2001).

More controversial (although sometimes cited as proven fact) have been claims (e.g., Taylor and Jakes, 1974; Taylor, 1982) that the bulk Moon is enriched roughly twofold in the cosmochemically refractory lithophile elements (a class that includes the REEs, the heat sources thorium and uranium, and the major elements aluminum, calcium, and titanium), and that compared to Earth's primitive mantle, the Moon's silicate mg ratio is much lower, i.e., its FeO concentration is much higher. Neither of these claims has been confirmed by recent lunar science developments, which include the advent of global thorium and samarium maps (Lawrence et al., 2002a; Prettyman et al., 2002), data from lunar meteorites, and some radically changed interpretations of the Apollo seismic database.

Starting with Cameron and Ward (1976), the putative refractory enrichments have been claimed as an advantage of the high temperatures implied by most variants of the giant impact model of Earth–Moon origin. It has often been claimed, or implied, that the well-documented lunar volatile depletions (Figure 6) are so great that they must represent a unique signature of origin by giant impact. However, an effectively identical overall volatile depletion pattern is found for the eucrite meteorites (Figure 6), which are products of an asteroid (or asteroids) that presumably was never involved in an Earth–Mars scale collision (Chapter 1.11). Another common misconception holds that the bulk-Moon volatile depletions imply a significant degree of refractory enrichment. In fact, total depletion of every constituent with cosmochemical volatility (solar nebular condensation temperature, Wasson, 1985) between those of SiO_2 and H_2O, from even the most volatile-rich (CI) type of chondrite, would increase the concentrations of all elements more refractory than silicon by a factor of only 1.22. For ten other types of chondrites (Jarosewich, 1990), the increase would be merely a factor of 1.02–1.09 (average 1.05).

Until recently, the canonical interpretation of the Apollo seismic data was that the thickness of the crust in the central nearside (Apollo 12 and Apollo 14) region is 60 km, and the global average thickness ~73 km (e.g., Taylor, 1982). Recent, more thorough processing of the Apollo lunar seismic data set suggests the central nearside crustal thickness is only ~30 km (Chenet et al., 2002) or perhaps ~38 km (Khan and Mosegaard, 2002). Wieczorek (2003) calculates from global mass distribution constraints that the thickness cannot be less than 33 km; but in any case, the required aluminum and calcium to account for the lunar crust now appears greatly reduced. Meanwhile, the global thorium map (Lawrence et al., 2002a; cf. Warren, 2003) and lunar meteorites (Table 3) have revealed that the Apollo (central nearside) sampling region is highly unrepresentative for KREEPy refractory lithophile elements, again relaxing the need to invoke refractory enrichments.

The anticorrelation between Al_2O_3 and mg shown by KREEP-poor highland regolith samples, including new lunar meteorites (Figure 15), extrapolates to a high, Earth-like mg for the mafic component of the highland crust. As noted by Warren (2003), this and other evidence (e.g., the extremely high mg of some Mg-suite rocks, Figure 13) is difficult to reconcile with a bulk Moon that is greatly enriched in FeO compared to Earth's primitive mantle. Warren (2003) argues that, apart from volatile and siderophile element constituents, plus possibly chromium and vanadium, available constraints imply a high degree of compositional similarity between the Moon and Earth's primitive mantle.

The precise oxygen isotopic match between Earth and the Moon (Wiechert et al., 2001) is a powerful constraint. Chromium isotopes also show a noteworthy match (Lugmair and Shukolyukov, 1998). It is not obvious, however, that such a precise oxygen isotopic match should be seen as a confirmation of the giant impact model. In many variants of the model, most of the material that ultimately forms the Moon is derived not from the protoearth, but from the Mars-sized intruder. In the estimation of Canup and Asphaug, 2001), the intruder's contribution was probably ~80 wt.%. Considering the easily measurable difference between the real Mars and Earth in terms of oxygen isotopes (Clayton and Mayeda, 1996), the giant impact model seems to require a remarkable degree of similarity between the putative giant impactor and the final Earth. Another cause for concern is the likelihood (Warren, 1992) that geochemical stratification in the impactor would have resulted in a Moon, i.e., the orbiting spall from the giant impact, with an idiosyncratic (e.g., thorium- and light REE-enriched) bulk composition.

ACKNOWLEDGMENTS

The author is indebted to Steve Simon and Andy Davis for helpful reviews. This research was supported by NASA grant NAG5-4215.

REFERENCES

Adler I., Trombka J. I., Schmadebeck R., Lowman P., Blodget H., Yin L., Eller E., Podwysocki M., Weidner J. R., Bickel A. L., Lum R. K. L., Gerard J., Gorenstein P., Bjorkholm P., and Harris B. (1973) Results of the Apollo 15 and 16 x-ray experiment. *Proc. 4th Lunar Sci. Conf.* **2**, 2783–2792.

Albarède F. (1995) *Introduction to Geochemical Modeling.* Cambridge University Press, Cambridge, 543pp.

Albrecht A., Herzog G. F., Klein J., Middleton R., Schultz L., Weber H. W., Kallemeyn G. W., and Warren P. H. (1995) Trace elements, ^{26}Al and ^{10}Be, and noble gases in lunar rock 14286. In *Lunar Planet. Sci.* **XXVI**. The Lunar and Planetary Institute, Houston, pp. 13–14.

Alfven H. and Arrhenius G. (1976) *Evolution of the Solar System.* NASA, 599p.

Alibert C., Norman M. D., and McCulloch M. T. (1994) An ancient age for a ferroan anorthosite clast from lunar breccia 67016. *Geochim. Cosmochim. Acta* **58**, 2921–2926.

Andre C. G. and El-Baz F. (1981) Regional chemical setting of the Apollo 16 landing site and the importance of the Kant Plateau. *Proc. 12th Lunar Planet. Sci. Conf.*, 767–779.

Antonenko I., Head J. W., Mustard J. F., and Hawke B. R. (1995) Criteria for the detection of lunar cryptomaria. *Earth Moon Planet.* **69**, 141–172.

Arai T. and Warren P. H. (1997) "Large" (1.7 g) compositionally pristine diabase 14434: a lithology transitional between Mg-gabbronorite and high-Al mare basalt. *Meteorit. Planet. Sci.* **32**, A7–A8.

Arai T. and Warren P. H. (1999) Lunar meteorite QUE94281: glass compositions and other evidence for launch pairing with Yamato-793274. *Meteorit. Planet. Sci.* **34**, 209–234.

Arnold (1979) Ice in the Lunar polar regions. *J. Geophys Res.* **84**, 5659–5668.

Ash R. D., Knott S. F., and Turner G. (1996) A 4-gyr shock age for a Martian meteorite and implications for the cratering history of Mars. *Nature* **380**, 57–59.

Bandermann L. W. and Singer S. F. (1973) Calculation of meteoroid impacts on Moon and Earth. *Icarus* **19**, 108–113.

Beard B. L., Taylor L. A., Scherer E. E., Johnson C. M., and Snyder G. A. (1998) The source region and melting mineralogy of high-titanium and low-titanium lunar basalts deduced from Lu–Hf isotope data. *Geochim. Cosmochim. Acta* **62**, 525–544.

Benoit P. H., Sears D. W. G., and Symes S. J. K. (1996) The thermal and radiation exposure history of lunar meteorites. *Meteorit. Planet. Sci.* **31**, 869–875.

Bielefeld M. J., Andre C. G., Eliason E. M., Clark P. E., Adler I., and Trombka J. I. (1977) Imaging of lunar surface chemistry from orbital X-ray data. *Proc. 8th Lunar Sci. Conf.*, 901–908.

Bogard D. D. (1995) Impact ages of meteorites: a synthesis. *Meteoritics* **30**, 244–268.

Bogard D. D., Garrison D. H., Shih C.-Y., and Nyquist L. E. (1994) ^{39}Ar–^{40}Ar dating of two lunar granites-the age of Copernicus. *Geochim. Cosmochim. Acta* **58**, 3093–3100.

Bogard D. D., Garrison D. H., and Nyquist L. E. (2000) Argon-39–Argon-40 ages of lunar highland rocks and meteorites. In *Lunar Planet. Sci.* **XXXI**, #1138. The Lunar and Planetary Institute, Houston, (CD-ROM).

Borg L., Norman M., Nyquist L., Bogard D., Snyder G., Taylor L., and Lindstrom M. (1999) Isotopic studies of ferroan anorthosite 62236: a young lunar crustal rock from a light rare-earth-element-depleted source. *Geochim. Cosmochim. Acta* **63**, 2679–2691.

Brophy J. G. and Basu A. (1990) Europium anomalies in mare basalts as a consequence of mafic cumulate fractionation from an initial lunar magma. *Proc. 20th Lunar Planet. Sci. Conf.*, 25–30.

Brügmann G. E., Arndt N. T., Hofmann A. W., and Tobschall H. J. (1987) Noble metal abundances in komatiite suites from Alexo, Ontario and Gorgona Island, Columbia. *Geochim. Cosmochim. Acta* **51**, 2159–2169.

Butler P., Jr. and Meyer C., Jr. (1976) Sulfur prevails in coatings on glass droplets: Apollo 15 green and brown glasses and Apollo 17 orange and black (devitrified) glasses. *Proc. 7th Lunar Sci. Conf.*, 1561–1582.

BVSP (1981) *Basaltic Volcanism on the Terrestrial Planets.* Pergamon, New York, 1286p.

Cameron A. G. W. and Ward W. R. (1976) The origin of the moon. In *Lunar Sci.* **VII**. The Lunar Science Institute, Houston, pp. 120–122.

Canup R. M. and Asphaug E. (2001) Origin of the Moon in a giant impact near the end of the Earth's formation. *Nature* **412**, 708–712.

Carlson R. W. and Lugmair G. W. (1979) Sm–Nd constraints on early lunar differentiation and the evolution of KREEP. *Earth Planet. Sci. Lett.* **45**, 123–132.

Carlson R. W. and Lugmair G. W. (1988) The age of ferroan anorthosite 60025: oldest crust on a young Moon? *Earth Planet. Sci. Lett.* **90**, 119–130.

Chapman C. R., Cohen B. A., and Grinspoon D. H. (2002) What are the real constraints on commencement of the late heavy bombardment? In *Lunar Planet. Sci.* **XXXIII**, #1627. The Lunar and Planetary Institute, Houston (CD-ROM).

Chenet H., Gagnepain-Beyneix J., and Lognonne P. (2002) A new geophysical view of the Moon. In *Lunar Planet. Sci.* **XXXIII**, #1684. The Lunar and Planetary Institute, Houston (CD-ROM).

Chou C.-L., Boynton W. V., Sundberg L. L., and Wasson J. T. (1975) Volatiles on the surface of Apollo 15 green glass and trace-element distributions among Apollo 15 soils. *Proc. 6th Lunar Sci. Conf.*, 1701–1727.

Chou C. L., Boynton W. V., Bild R. W., Kimberlin J., and Wasson J. T. (1976) Trace element evidence regarding a chondritic component in howardite meteorites. *Proc. 7th Lunar Sci. Conf.*, 3501–3518.

Chyba C. F. (1991) Terrestrial mantle siderophiles and the lunar impact record. *Icarus* **92**, 217–233.

Clayton R. N. and Mayeda T. K. (1996) Oxygen isotopic studies of achondrites. *Geochim. Cosmochim. Acta* **60**, 1999–2017.

Cohen B. A., Swindle T. D., and Kring D. A. (2000) Support for the lunar cataclysm hypothesis from lunar meteorite impact melt ages. *Science* **290**, 1754–1756.

Cohen B. A., Snyder G. A., Hall C. M., Taylor L. A., and Nazarov M. A. (2001) Argon-40–Argon-39 chronology and petrogenesis along the eastern limb of the Moon from Luna 16, 20, and 24 samples. *Meteorit. Planet. Sci.* **36**, 1345–1366.

Crider D. H. and Vondrak R. R. (2000) The solar wind as a possible source of lunar polar hydrogen deposits. *J. Geophys. Res.* **105**, 26773–26782.

Crocket J. H. and MacRae W. E. (1986) Platinum-group element distribution in komatiitic and tholeiitic volcanic rocks from Munro Township, Ontario. *Econ. Geol.* **81**, 1242–1251.

Culler T. S., Becker T. A., Muller R. A., and Renne P. R. (2000) Lunar impact history from ^{40}Ar/^{39}Ar dating of glass spherules. *Science* **287**, 1785–1788.

Cushing J. A., Taylor G. J., Norman M. D., and Keil K. (1999) The granulitic impactite suite-impact melts and metamorphic breccias of the earliest lunar crust. *Meteorit. Planet. Sci.* **34**, 185–195.

Dalrymple G. B. and Ryder G. (1993) ^{40}Ar/^{39}Ar age spectra of Apollo 15 impact melt rocks by laser step heating and their bearing on the history of lunar basin formation. *J. Geophys. Res.* **98**, 13085–13095.

Dalrymple G. B. and Ryder G. (1996) Argon-40/argon-39 age spectra of Apollo 17 highlands breccia samples by laser step heating and the age of the Serenitatis basin. *J. Geophys. Res.* **101**, 26069–26084.

Dalrymple G. B., Ryder G., Duncan R. A., and Huard J. J. (2001) ^{40}Ar–^{39}Ar ages of Apollo 16 impact melt rocks by laser step heating. In *Lunar Planet. Sci.* **XXXII**, #1225. The Lunar and Planetary Institute, Houston (CD-ROM).

Danckwerth P. A., Hess P. C., and Rutherford M. J. (1979) The solubility of sulfur in high-TiO_2 mare basalts. *Proc. 10th Lunar Planet. Sci. Conf.*, 517–530.

Dasch E. J., Shih C.-Y., Bansal B. M., Wiesmann H., and Nyquist L. E. (1987) Isotopic analysis of basaltic fragments from lunar breccia 14321: chronology and petrogenesis of pre-imbrium mare volcanism. *Geochim. Cosmochim. Acta* **51**, 3241–3254.

Dasch E. J., Ryder G., Shih C. Y., Wiesmann H., Bansal B. M., and Nyguist L. E. (1989) Time of crystallization of a unique A15 basalt. In *Lunar Planet. Sci.* **XX**. The Lunar and Planetary Institute, Houston, pp. 218–219.

DeHon R. A. (1979) Thickness of the western mare basalts. *Proc. 10th Lunar Planet. Sci. Conf.*, 2935–2955.

Delano J. W. (1986) Pristine lunar glasses: criteria, data, and implications. *Proc. 16th Lunar Planet. Sci. Conf.*, D201–D213.

Delano J. W. and Ringwood A. E. (1978) Siderophile elements in the lunar highlands: nature of the indigenous component and implications for the origin of the Moon. *Proc. 9th Lunar Planet. Sci. Conf.*, 111–159.

Dickin A. P. (1995) *Radiogenic Isotope Geology*. Cambridge University Press, Cambridge, p.78.

Dickinson T., Taylor G. J., Keil K., Schmitt R. A., Hughes S. S., and Smith M. R. (1985) Apollo 14 aluminous mare basalts and their possible relationship to KREEP. *Proc. 15th Lunar Planet. Sci. Conf.*, C365–C374.

Dickinson T., Taylor G. J., Keil K., and Bild R. W. (1989) Germanium abundances in lunar basalts: evidence of mantle metasomatism? *Proc. 19th Lunar Planet. Sci. Conf.*, 189–198.

Dowty E., Prinz M., and Keil K. (1974) Ferroan anorthosite: a widespread and distinctive lunar rock type. *Earth Planet. Sci. Lett.* **24**, 15–25.

Dungan M. A. and Brown R. W. (1977) The petrology of the Apollo 12 limonite basalt suite. *Proc. 8th Lunar Sci. Conf.*, 1339–1382.

Dymek R. F., Albee A. L., and Chodos A. A. (1975) Comparative petrology of lunar cumulate rocks of possible primary origin: dunite 72415, troctolite 76535, norite 78235, and anorthosite 62237. *Proc. 6th Lunar Sci. Conf.*, 301–341.

Elkins L. T., Fernandes V. A., Delano J. W., and Grove T. L. (2000) Origin of lunar ultramafic green glasses; constraints from phase equilibrium studies. *Geochim. Cosmochim. Acta* **64**, 2339–2350.

Elphic R. C., Lawrence D. J., Feldman W. C., Barraclough B. L., Maurice S., Binder A. B., and Lucey P. G. (2000) Lunar rare earth element distribution and ramifications for FeO and TiO_2: Lunar Prospector neutron spectrometer observations. *J. Geophys. Res.* **105**, 20333–20345.

Eugster O., Polnau E., Salerno E., and Terribilino D. (2000) Lunar surface exposure models for meteorites Elephant Moraine 96008 and Dar al Gani 262 from the Moon. *Meteorit. Planet. Sci.* **35**, 1177–1182.

Fagan T. J., Taylor G. J., Keil K., Bunch T. E., Wittke J. H., Korotev R. L., Jolliff B. L., Gillis J. J., Haskin L. A., Jarosewich E., Clayton R. N., Mayeda T. K., Fernandes V. A., Burgess R., Turner G., Eugster O., and Lorenzetti S. (2002) Northwest Africa 032: product of lunar volcanism. *Meteorit. Planet. Sci.* **37**, 371–394.

Feldman W. C., Maurice S., Lawrence D. J., Little R. C., Lawson S. L., Gasnault O., Wiens R. C., Barraclough B. L., Elphic R. C., Prettyman T. H., Steinberg J. T., and Binder A. B. (2001) Evidence for water ice near the lunar poles. *J. Geophys. Res.: Planets* **106**, 23231–23251.

Fernandes V. A., Burgess R., and Turner G. (2002a) Age determination of lunar regolith samples from the Luna 16 and 24 cores using IR step heating. In *Lunar Planet. Sci.* **XXXIII**, #1753. The Lunar and Planetary Institute, Houston (CD-ROM).

Fernandes V. A., Burgess R., and Turner G. (2002b) North West Africa 773 (NWA773): Ar–Ar studies of breccia and cumulate lithologies. In *The Moon Beyond 2002: Next Steps in Lunar Science and Exploration, #3033* (eds. D. J. Lawrence and M. B. Duke). Lunar and Planetary Institute, Houston.

Ferrand W. H. (1988) Highland contamination and minimum basalt thickness in northern Mare Fecunditatis. *Proc. 18th Lunar Planet. Sci. Conf.*, 319–329.

Fogel R. A. and Rutherford M. J. (1995) Magmatic volatiles in primitive lunar glasses: I. FTIR and EPMA analyses of Apollo 15 green and yellow glasses and revision of the volatile-assisted fire-fountain theory. *Geochim. Cosmochim. Acta* **59**, 201–215.

Gaddis L. R., Hawke B. R., Robinson M. S., and Coombs C. (2000) Compositional analyses of small lunar pyroclastic deposits using clementine multispectral data. *J. Geophys. Res.* **105**, 4245–4262.

Gault D. E. (1983) The terrestrial accretion of lunar material. In *Lunar Planet. Sci.* **XIV**. The Lunar and Planetary Institute, Houston, pp. 243–244.

Gibson E. K., Brett R., and Andrawes F. (1977) Sulfur in lunar mare basalts as a function of bulk composition. *Proc. 8th Lunar Sci. Conf.*, 1417–1428.

Gifford A. W. and El-Baz F. (1981) Thicknesses of mare flow fronts. *Moon Planets* **24**, 391–398.

Giguere T. A., Taylor G. J., Hawke B. R., and Lucey P. G. (2000) The titanium contents of lunar mare basalts. *Meteorit. Planet. Sci.* **35**, 193–200.

Gillis J. J., Jolliff B. L., and Elphic R. C. (2003) A revised algorithm for calculating TiO_2 from Clementine UVVIS data: a synthesis of rock, soil, and remotely sensed TiO_2 concentrations. *J. Geophys. Res.: Planets* **108** (E2), art. no. 5009.

Gladman B. J., Burns J. A., Duncan M., Lee P., and Levison H. F. (1996) The exchange of impact ejecta between terrestrial planets. *Science* **271**, 1387–1392.

Govindaraju K. (1994) 1994 compilation of working values and sample description for 383 geostandards. *Geostand. Newslett.* **18**, 1–158.

Grieve R. A. F. and Cintala M. J. (1992) An analysis of differential impact melt-crater scaling and implications for the terrestrial record. *Meteoritics* **27**, 526–538.

Guinness E. A. and Arvidson R. E. (1977) On the constancy of the lunar cratering flux over the past 3.3×10^9 yr. *Proc. 8th Lunar Sci. Conf.*, 3475–3494.

Hartmann W. K. (1975) "Lunar cataclysm": a misconception? *Icarus* **24**, 181–187.

Hartmann W. K., Ryder G., Dones L., and Grinspoon D. (2000) The time-dependent intense bombardment of the primordial Earth/Moon system. In *Origin of the Earth and Moon* (eds. R. M. Canup and K. Righter). University of Arizona Press, Tucson, pp. 493–512.

Haskin L. A. and Warren P. H. (1991) Chemistry. In *Lunar Sourcebook, A User's Guide to the Moon* (eds. G. Heiken, D. Vaniman, and B. M. French). Cambridge University Press, Cambridge, UK, pp. 357–474.

Haskin L. A., Lindstrom M. M., Salpas P., and Lindstrom D. J. (1981) On compositional variations among lunar anorthosites. *Proc. 12th Lunar Planet. Sci. Conf.*, 41–66.

Haskin L. A., Korotev R. L., Gillis J. J., and Jolliff B. L. (2002) Stratigraphies of Apollo and Luna highland landing sites and provenances of materials from the perspective of basin impact ejecta modelling. In *Lunar Planet. Sci.* **XXXIII**, #1364. The Lunar and Planetary Institute, Houston (CD-ROM).

Hawke B. R., Lawrence D. J., Blewett D. T., Lucey P. G., Smith G. J., Taylor G. J., and Spudis P. D. (2003) Remote sensing studies of geochemical and spectral anomalies on the nearside of the Moon. In *Lunar Planet. Sci.* **XXXIV**, #1598. The Lunar and Planetary Institute, Houston (CD-ROM).

Head J. W. and Wilson L. (1992) Lunar mare volcanism: stratigraphy, eruption conditions, and the evolution of secondary crusts. *Geochim. Cosmochim. Acta* **56**, 2155–2175.

Hertogen J., Janssens M. J., and Palme H. (1980) Trace elements in ocean ridge basalt glasses: implications for fractionations during mantle evolution and petrogenesis. *Geochim. Cosmochim. Acta* **44**, 2125–2143.

Hertogen J., Janssens M. J., Takahashi H., Palme H., and Anders E. (1977) Lunar basins and craters: evidence for compositional changes of bombarding population. *Proc. 8th Lunar Sci. Conf.*, 17–45.

Hess P. C. (1994) Petrogenesis of lunar troctolites. *J. Geophys. Res.* **99**, 19083–19093.

Hess P. C. and Parmentier E. M. (1995) A model for the thermal and chemical evolution of the Moon's interior: implications for the onset of mare volcanism. *Earth Planet. Sci. Lett.* **134**, 501–514.

Hess P. C., Rutherford M. J., and Campbell H. W. (1977) Origin and evolution of LKFM basalt. *Proc. 8th Lunar Sci. Conf.*, 2357–2374.

Hiesinger H., Head J. W., Wolf U., and Neukum G. (2000) Lunar mare basalts in oceanus procellarum: initial results on age and composition. In *Lunar Planet. Sci. XXXI*, #1278. The Lunar and Planetary Institute, Houston (CD-ROM).

Hiesinger H., Head J. W., III, Wolf U., Jaumann R., and Neukum G. (2002) Lunar mare basalt flow units: thicknesses determined from crater size-frequency distributions. *Geophys. Res. Lett.* **29**, 8891–8894.

Hörz F. (1978) How thick are lunar mare basalts? *Proc. 9th Lunar Planet. Sci. Conf.*, 3311–3331.

James O. B. (2002) Distinctive meteoritic components in lunar "cataclysm" impact-melt breccias. In *Lunar Planet. Sci. XXXIII*, #1210. The Lunar and Planetary Institute, Houston (CD-ROM).

James O. B. and Flohr M. K. (1983) Subdivision of the Mg-suite noritic rocks into Mg-gabbronorites and Mg-norites. *Proc. Lunar Planet. Sci. Conf.*, A603–A614.

James O. B. and McGee J. J. (1979) Consortium breccia 73255: genesis and history of two coarse-grained "norite" clasts. *Proc. 10th Lunar Sci. Conf.*, 713–743.

James O. B., Lindstrom M. M., and Flohr M. K. (1989) Ferroan anorthosite from lunar breccia 64435: implications for the origin and history of lunar ferroan anorthosites. *Proc. 19th Lunar Planet. Sci. Conf.*, 219–243.

Jarosewich E. (1990) Chemical analyses of meteorites: a compilation of stony and iron meteorite analyses. *Meteoritics* **25**, 323–337.

Jerde E. A., Morris R. V., and Warren P. H. (1990) In quest of lunar regolith breccias of exotic provenance—a uniquely anorthositic sample from the Fra Mauro (Apollo 14) highlands. *Earth Planet. Sci. Lett.* **98**, 90–108.

Jolliff B. L., Floss C., McCallum I. S., and Schwartz J. M. (1999) Geochemistry, petrology, and cooling history of 14161,7373: a plutonic lunar sample with textural evidence of granitic-fraction separation by silicate-liquid immiscibility. *Am. Mineral.* **84**, 821–837.

Jolliff B. L., Korotev R. L., Zeigler R. A., Floss C., and Haskin L. A. (2003) Northwest Africa 773: *lunar mare breccia with a shallow-formed olivine cumulate, very-low-Ti heritage, and a KREEP connection.* In *Lunar Planet. Sci. XXXIV*, #1935. The Lunar and Planetary Institute, Houston (CD-ROM).

Keller L. P. and McKay D. S. (1997) The nature and origin of rims on lunar grains. *Geochim. Cosmochim. Acta* **61**, 2331–2341.

Kennedy A. K., Lofgren G. E., and Wasserburg G. J. (1993) An experimental study of trace element partitioning between olivine, orthopyroxene, and melt in chondrules. *Earth Planet. Sci. Lett.* **115**, 177–195.

Khan A. and Mosegaard K. (2002) An inquiry into the lunar interior: a nonlinear inversion of the Apollo lunar seismic data. *J. Geophys. Res.* **107**, E63.1–E63.23.

Kirsten T., Horn P., and Kiko J. (1973) $^{39}Ar-^{40}Ar$ dating and rare gas analysis of Apollo 16 rocks and soils. *Proc. 4th Lunar Sci. Conf.*, 1757–1784. .

Korotev R. L. (1981) Compositional trends in Apollo 16 soils. *Proc. 12th Lunar Planet. Sci. Conf.*, 577–605.

Korotev R. L. (1987a) Mixing levels, the Apennine front soil component, and compositional trends in the Apollo 15 soils. *Proc. 17th Lunar Planet. Sci. Conf.*, E411–E431.

Korotev R. L. (1987b) The meteoritic component of Apollo 16 noritic impact melt breccias. *Proc. 17th Lunar Planet. Sci. Conf.*, E491–E512.

Korotev R. L. (1994) Compositional variation in Apollo 16 impact melt breccias and inferences for the geology and bombardment history of the central highlands of the Moon. *Geochim. Cosmochim. Acta* **58**, 3931–3969.

Korotev R. L. and Kremser D. T. (1992) Compositional variations in Apollo 17 soils and their relationship to the geology of the Taurus-Littrow site. *Proc. 22nd Lunar Planet. Sci. Conf.*, 275–301.

Korotev R. L., Haskin L. A., and Jolliff B. L. (1998) Ir/Au ratios and the origin of Apollo 17 and other Apollo impact-melt breccias. In *Lunar Planet. Sci. XXIX*, #1231. The Lunar and Planetary Institute, Houston (CD-ROM).

Korotev R. L., Gillis J. J., Haskin L. A., and Jolliff B. L. (2002a) On the age of the Nectaris basin. In *The Moon Beyond 2002: Next Steps in Lunar Science and Exploration*, #3029 (eds. D. J. Lawrence and M. B. Duke). The Lunar and Planetary Institute, Houston.

Korotev R. L., Zeigler R. A., Jolliff B. L., and Haskin L. A. (2002b) Northwest Africa 773—an unusual rock from the lunar maria. *Meteorit. Planet. Sci.* **37**, A81.

Korotev R. L., Jolliff B. L., Zeigler R. A., and Haskin L. A. (2003) Compositional evidence for launch pairing of the YQ and elephant moraine lunar meteorites. In *Lunar Planet. Sci. XXIV*, #1357. The Lunar and Planetary Institute, Houston.

Kring D. A. and Cohen B. A. (2002) Cataclysmic bombardment throughout the inner solar system. *J. Geophys. Res.* **107**, E2.4.1–E2.4.6.

Kurat G., Kracher A., Keil K., Warner R., and Prinz M. (1976) Composition and origin of Luna 16 aluminous mare basalts. *Proc. 7th Lunar Sci. Conf.* **2**, 1301–1322.

Lawrence D. J., Feldman W. C., Barraclough B. L., Binder A. B., Elphic R. C., Maurice S., and Thomsen D. R. (1998) Global elemental maps of the Moon: the Lunar Prospector gamma-ray spectrometer. *Science* **281**, 1484–1489.

Lawrence D. J., Elphic R. C., Feldman W. C., Gasnault O., Maurice S., and Prettyman T. H. (2002a) Small-area thorium enhancements on the lunar surface. In *Lunar Planet. Sci. XXXIII*, #1970. The Lunar and Planetary Institute, Houston (CD-ROM).

Lawrence D. J., Feldman W. C., Elphic R. C., Little R. C., Prettyman T. H., Maurice S., Lucey P. G., and Binder A. B. (2002b) Iron abundances on the Lunar surface as measured by the Lunar Prospector gamma-ray and neutron spectrometers. *J. Geophys. Res.* **107**, E12.13.1–E12.13.26.

Le Bas M. J. (2001) Report of the working party on the classification of the lunar igneous rocks. *Meteorit. Planet. Sci.* **36**, 1183–1188.

Li L. and Mustard J. F. (2000) Compositional gradients across mare–highland contacts: importance and geological implication of lateral transport. *J. Geophys. Res.* **105**, 20431–20450.

Lindsay J. F. (1992) Extraterrestrial soils: the lunar experience. *SDEDP* **2**, 41–70.

Lindstrom M. M. and Lindstrom D. J. (1986) Lunar granulites and their precursor anorthositic norites of the early lunar crust. *Proc. 16th Lunar Planet. Sci. Conf.*, D263–D276.

Lofgren G. E., Donaldson C. H., and Usselman T. M. (1975) Geology, petrology, and crystallization of Apollo 15 quartz-normative basalts. *Proc. 6th Lunar Sci. Conf.*, 79–100.

Longhi J. (1987) On the connection between mare basalts and picritic volcanic glasses. *Proc. Lunar 17th Planet. Sci. Conf.*, E349–E360.

Longhi J. (1992) Experimental petrology and petrogenesis of mare volcanics. *Geochim. Cosmochim. Acta* **56**, 2235–2251.

Longhi (2002) The extent of early lunar differentiation. In *Lunar Planet. Sci. XXXIII*, #2069. The Lunar and Planetary Institute, Houston

Longhi J. (2003) Green glasses: new pressure calibration, new ascent mechanism, new calculations, same story. In *Lunar Planet. Sci.* **XXXIV**, #1528. The Lunar and Planetary Institute, Houston (CD-ROM).

Longhi J. and Ashwal L. D. (1985) Two-stage models for lunar and terrestrial anorthosites: petrogenesis without a magma ocean. *Proc. 15th Lunar Planet. Sci. Conf.*, C571–C584.

Longhi J. and Boudreau A. E. (1979) Complex igneous processes and the formation of the primitive lunar crustal rocks. *Proc. 10th Lunar Planet. Sci. Conf.*, 2085–2105.

Longhi J. and Pan V. (1988) A reconnaissance study of phase boundaries in low-alkali basaltic liquids. *J. Petrol.* **29**, 115–147.

Lucey P. G. and Steutel D. (2003) Global mineral maps of the Moon. In *Lunar Planet. Sci.* **XXXIV**, #1051. The Lunar and Planetary Institute, Houston (CD-ROM).

Lucey P. G., Blewett D. T., and Hawke B. R. (1998) Mapping the FeO and TiO_2 content of the lunar surface with multispectral imagery. *J. Geophys. Res.* **103**, 3679–3699.

Lucey P. G., Blewett D. T., and Jolliff B. L. (2000a) Lunar iron and titanium abundance algorithms based on final processing of Clementine ultraviolet-visible images. *J. Geophys. Res.* **105**, 20297–20305.

Lucey P. G., Blewett D. T., Taylor G. J., and Hawke B. R. (2000b) Imaging of lunar surface maturity. *J. Geophys. Res.* **105**, 20377–20386.

Lugmair G. W. and Carlson R. W. (1978) The Sm–Nd history of KREEP. *Proc. 9th Lunar Planet. Sci. Conf.*, 689–704.

Lugmair G. W. and Shukolyukov A. (1998) Early solar system timescales according to ^{53}Mn–^{53}Cr systematics. *Geochim. Cosmochim. Acta* **62**, 2863–2886.

Marti K., Aeschlimann U., Eberhardt P., Geiss J., Grögler N., Jost D. T., Laul J. C., Ma M. S., Schmitt R. A., and Taylor G. J. (1983) Pieces of the ancient lunar crust: ages and composition of clasts in consortium breccia 67915. *Proc. 14th Lunar Planet. Sci. Conf.*, B165–B175.

Marvin U. B. and Warren P. H. (1980) A pristine eucrite-like gabbro from Descartes and its exotic kindred. *Proc. 11th Lunar Planet. Sci. Conf.*, 507–521.

Mauer P. P. E., Geiss J., Grögler N., Stettler A., Brown G. M., Peckett A., and Krähenbühl U. (1978) Pre-imbrian craters and basins: ages, compositions, and excavation depths of Apollo 16 breccias. *Geochim. Cosmochim. Acta* **42**, 1687–1720.

McEwen A. S., Keszthelyi L. P., Spencer J. R., Schubert G., Matson D. L., Lopes-Gautier R. M. C., Klaasen K. P., Johnson T. V., Head J. W., III, Geissler P. E., Fagents S., Davies A. G., Carr M. H., Breneman H. H., and Belton M. J. S. (1998) High-temperature silicate volcanism on Jupiter's moon Io. *Science* **281**, 87–90.

McKay G. A. and Weill D. F. (1977) KREEP petrogenesis revisited. *Proc. 8th Lunar Sci. Conf.*, 2339–2355.

McKay G., Le L., and Wagstaff J. (1991) Constraints on the origin of the rare basalt europium anomaly: REE partition coefficients for pigeonite. In *Lunar Planet. Sci.* **XXII**. The Lunar and Planetary Institute, Houston, (CD-ROM), pp. 883–884.

McKenzie D. (2000) Constraints on melt generation and transport from V-series activity ratios. *Chem. Geol.* **162**, 81–94.

Melosh H. J. (1989) *Impact Cratering: A Geologic Process*. Oxford University Press, New York, 245p.

Meyer C. (1998) *Mars Meteorite Compendium—1998*. NASA Johnson Space Center, 237p.

Misawa K., Tatsumoto M., Dalrymple G. B., and Yanai K. (1993) An extremely low U/Pb source in the Moon: U–Th–Pb, Sm–Nd, Rb–Sr, and ^{40}Ar–^{39}Ar systematics of lunar meteorite Asuka 881757. *Geochim. Cosmochim. Acta* **57**, 4687–4702.

Morbidelli A., Petit J. M., Gladman B., and Chambers J. E. (2001) A plausible cause of the late heavy bombardment. *Meteorit. Planet. Sci.* **36**, 371–380.

Morgan J. W., Higuchi H., Takahashi H., and Hertogen J. (1978) A "chondritic" eucrite parent body: inference from trace elements. *Geochim. Cosmochim. Acta* **42**, 27–38.

Morgan J. W., Walker R. J., Brandon A. D., and Horan M. F. (2001) Siderophile elements in Earth's upper mantle and lunar breccias: data synthesis suggests manifestations of the same late influx. *Meteorit. Planet. Sci.* **36**, 1257–1275.

Morris R. V. (1978) The surface exposure (maturity) of lunar soils: some concepts and I_s/FeO compilation. *Proc. Lunar 9th Planet. Sci. Conf.*, 2287–2297.

Neal C. R. (2001) Interior of the Moon: the presence of garnet in the primitive deep lunar mantle. *J. Geophys. Res.* **106**, 27865–27885.

Neal C. R. and Kramer G. (2003) The composition of KREEP: a detailed study of KREEP basalt 15386. In *Lunar Planet. Sci.* **XXXIV**, #2023. The Lunar and Planetary Institute, Houston (CD-ROM).

Neal C. R. and Taylor L. A. (1989) Metasomatic products of the lunar magma ocean: the role of KREEP dissemination. *Geochim. Cosmochim. Acta* **53**, 529–541.

Neal C. R. and Taylor L. A. (1991) Evidence for metasomatism of the lunar highlands and the origin of whitlockite. *Geochim. Cosmochim. Acta* **55**, 2965–2980.

Neal C. R. and Taylor L. A. (1992) Petrogenesis of mare basalts: a record of lunar volcanism. *Geochim. Cosmochim. Acta* **56**, 2177–2211.

Neal C. R., Taylor L. A., Hughes S. S., and Schmitt R. A. (1990) The significance of fractional crystallization in the petrogenesis of Apollo 17 type A and B high-Ti basalts. *Geochim. Cosmochim. Acta* **54**, 1817–1833.

Neal C. R., Hacker M. D., Snyder G. A., Taylor L. A., Liu Y.-G., and Schmitt R. A. (1994) Basalt generation at the Apollo 12 site: Part 1. New data, classification, and re-evaluation. *Meteoritics* **29**, 334–348.

Neal C. R., Jain J. C., Snyder G. A., and Taylor L. A. (1999) Platinum group elements from the ocean of storms: evidence of two cores forming? (abstract). In *Lunar Planet. Sci.* **XXX**, #1003. The Lunar and Planetary Institute, Houston (CD-ROM).

Neal C. R., Ely J. C., and Jain J. C. (2001) The siderophile element budget of the Moon: a revaluation, Part 1. In *Lunar Planet. Sci.* **XXXII**, #1658. The Lunar and Planetary Institute, Houston (CD-ROM).

Neukum G., Ivanov B. A., and Hartmann W. K. (2001) Cratering records in the inner solar system in relation to the lunar reference system. *Space Sci. Rev.* **96**, 55–86.

Nishiizumi K., Okazaki R., Park J., Nagao K., Masarik J., and Finkel R. C. (2002) Exposure and terrestrial histories of Dhofar 019 martian meteorite. In *Lunar Planet. Sci.* **XXXIII**, #1366. The Lunar and Planetary Institute, Houston (CD-ROM).

Norman M. D., Benne V. C., and Ryder G. (2002a) Incorporation of siderophile elements into impact melts from lunar basins: PGE@Serenitatis.nasa.org. In *Lunar Planet. Sci.* **XXXIII**, #1176. The Lunar and Planetary Institute, Houston (CD-ROM).

Norman M. D., Borg L. E., Nyquist L. E., and Bogard D. D. (2002b) Crystallization age and impact resetting of ancient lunar crust from the Descartes terrane. In *The Moon Beyond 2002: Next Steps in Lunar Science and Exploration*, #3028 (eds. D. J. Lawrence and M. B. Duke). The Lunar and Planetary Institute, Houston.

Papike J., Taylor L., and Simon S. (1991) Lunar minerals. In *Lunar Sourcebook: A User's Guide to the Moon* (eds. G. Heiken, D. Vaniman, and B. M. French). Cambridge University Press, Cambridge, pp. 121–181.

Papike J. J., Fowler G. W., Shearer C. K., and Layne G. D. (1996) Ion microprobe investigation of plagioclase and orthopyroxene from lunar Mg suite norites: implications for calculating parental melt REE concentrations and for assessing post-crystallization REE redistribution. *Geochim. Cosmochim. Acta* **60**, 3967–3978.

Papike J. J., Ryder G., and Shearer C. K. (1998) Lunar samples. *Rev. Mineral.* **36**, 5.1–5.234.

Paul R. L. and Lipschutz M. E. (1990) Chemical studies of differentiated meteorites: I. Labile trace elements in Antarctic and non-Antarctic eucrites. *Geochim. Cosmochim. Acta* **54**, 3185–3195.

Peach C. L. and Mathez E. A. (1993) Sulfide melt-silicate melt distribution coefficients for nickel and iron and implications for the distribution of other chalcophile elements. *Geochim. Cosmochim. Acta* **57**, 3013–3021.

Persikov E. S., Bukhtiyarov P. G., and Kalinicheva T. V. (1987) Effects of composition, temperature, and pressure on magma fluidity. *Geokhimiya* **3**(3), 483–498.

Phinney W. C. and Morrison D. A. (1990) Partition coefficients for calcic plagiolcase: implications for Archean anorthosites. *Geochim. Cosmochim. Acta* **54**, 2025–2043.

Pieters C. M., Head J. W., III, Gaddis L., Jolliff L., and Duke M. (2001) Rock types of South Pole-Aitken basin and extent of basaltic volcanism. *J. Geophys. Res.* **106**, 28001–28022.

Pieters C. M., Stankevich D. G., Shkuratov Y. G., and Taylor L. A. (2002) Statistical analysis of the links between lunar mare soil mineralogy, chemistry and reflectance. *Icarus* **155**, 285–298.

Prettyman T. H., Feldman W. C., McKinney G. W., Binder A. B., Elphic R. C., Gasnault O. M., Maurice S., and Moore K. R. (2002) Library least squares analysis of Lunar Prospector gamma-ray spectra. In *Lunar Planet. Sci.* **XXXIII**, #2012. The Lunar and Planetary Institute, Houston (CD-ROM).

Quaide W. L. and Oberbeck V. R. (1975) Development of the mare regolith: some model considerations. *Moon* **13**, 27–55.

Rhodes J. M., Blanchard D. P., Dungan M. A., Brannon J. C., and Rodgers K. V. (1977) Chemistry of Apollo 12 mare basalts: magma types and fractionation processes. *Proc. 8th Lunar Sci. Conf.*, 1305–1338.

Ringwood A. E. and Kesson S. E. (1976) A dynamic model for mare basalt petrogenesis. *Proc. 7th Lunar Sci. Conf.*, 1697–1722.

Ringwood A. E., Seifert S., and Wänke H. (1987) A komatiite component in Apollo 16 highland breccias: implications for the nickel–cobalt systematics and bulk composition of the Moon. *Earth Planet. Sci. Lett.* **81**, 105–117.

Ryder G. (1985) *Catalog of Apollo 15 Rocks* (Curatorial Facility Publication 20787). NASA Johnson Space Center, 1296p.

Ryder G. (1990) Lunar samples, lunar accretion and the early bombardment of the Moon. *EOS* **71**, 313–323.

Ryder G. (1991) Lunar ferroan anorthosites and mare basalt sources: the mixed connection. *Geophys. Res. Lett.* **18**, 2065–2068.

Ryder G. (1999) Meteoritic abundances in the ancient lunar crust. In *Lunar Planet. Sci.* **XXX**, #1848. The Lunar and Planetary Institute, Houston (CD-ROM).

Ryder G. and Cox B. T. (1996) An Apollo 15 mare basalt fragment and lunar mare provinces. *Meteorit. Planet. Sci.* **31**, 50–59.

Ryder G. and Spudis P. D. (1980) Volcanic rocks in the lunar highlands. In *Proceedings of the Conference on the Lunar Highlands Crust* (eds. J. J. Papike and R. B. Merrill). Pergamon, New York, pp. 353–375.

Ryder G., Norman M. D., and Score R. A. (1980) The distinction of pristine from meteorite-contaminated highlands rocks using metal compositions. *Proc. 11th Lunar Planet. Sci. Conf.*, 471–479.

Ryder G., Delano J. W., Warren P. H., Kallemeyn G. W., and Dalrymple G. B. (1996) A glass spherule of equivocal impact origin from the Apollo 15 landing site: unique target mare basalt. *Geochim. Cosmochim. Acta* **60**, 693–710.

Schaeffer O. A. and Husain L. (1973) Early lunar history: ages of 2 to 4 mm soil fragments from the lunar highlands. *Proc. 4th Lunar Sci. Conf.*, 1847–1864.

Schultz P. H. and Mendell W. (1978) Orbital infrared observations of lunar craters and possible implications for impact ejecta emplacement. *Proc. 9th Lunar Planet. Sci. Conf.*, 2857–2884.

Shearer C. K. and Papike J. J. (1989) Is plagioclase removal responsible for the negative Eu anomaly in the source regions of mare basalts. *Geochim. Cosmochim. Acta* **53**, 3331–3336.

Shearer C. K. and Papike J. J. (1999) Magmatic evolution of the Moon. *Am. Mineral.* **84**, 1469–1494.

Shearer C. K., Papike J. J., Simon S. B., Shimizu N., Yurimoto H., and Sueno S. (1990) Ion microprobe studies of trace elements in Apollo 14 volcanic glass beads: comparisons to Apollo 14 mare basalts and petrogenesis of picritic magmas. *Geochim. Cosmochim. Acta* **54**, 851–867.

Shearer C. K., Papike J. J., and Layne G. D. (1996) Deciphering basaltic magmatism on the Moon from the compositional variations in Apollo 15 very low-Ti picritic magmas. *Geochim. Cosmochim. Acta* **60**, 509–528.

Shearer C. K., Borg L., Ryder G., Papike J. J., and Nyquist L. (2001a) Deciphering ages of impacted basalts using a crystal chemical-ion microprobe approach. An example using the Apollo 17 Group D basalt. In *Lunar Planet. Sci.* **XXXII**, #1851. The Lunar and Planetary Institute, Houston (CD-ROM).

Shearer C. K., Papike J. J., and Hagert J. (2001b) Chemical dichotomy of the Mg-suite. Insights from a comparison of trace elements in silicates from a variety of lunar basalts. In *Lunar Planet. Sci.* **XXXII**, #1643. The Lunar and Planetary Institute, Houston (CD-ROM).

Shervais J. W. and McGee J. J. (1999) KREEP cumulates in the western lunar highlands: ion and electron microprobe study of alkali-suite anorthosites and norites from Apollo 12 and 14. *Am. Mineral.* **84**, 806–820.

Shih C.-Y., Nyquist L. E., Bogard D. D., Bansal B. M., Wiesmann H., Johnson P., Shervais J. W., and Taylor L. A. (1986) Geochronology and petrogenesis of Apollo 14 very high potassium basalts. *Proc. 16th Lunar Planet. Sci. Conf.*, D214–D228.

Shih C.-Y., Nyquist L. E., Bogard D. D., Dasch E. J., Bansal B. M., and Wiesmann H. (1987) Geochronology of high-K aluminous mare basalt clasts from Apollo 14 breccia 14304. *Geochim. Cosmochim. Acta* **51**, 3255–3271.

Shih C.-Y., Nyquist L. E., Bansal B. M., and Wiesmann H. (1992) Rb–Sr and Sm–Nd chronology of an Apollo 17 KREEP basalt. *Earth Planet. Sci. Lett.* **108**, 203–215.

Shih C.-Y., Nyquist L. E., Reese Y., Wiesmann H., and Schwandt C. (2001) Rb–Sr and Sm–Nd isotopic constraints on the genesis of lunar green and orange glasses. In *Lunar Planet. Sci.* **XXXII**, #1401. The Lunar and Planetary Institute, Houston (CD-ROM).

Shih C.-Y., Nyquist L. E., Reese Y., Wiesmann H., Nazarov M. A., and Taylor L. A. (2002) The chronology and petrogenesis of the mare basalt clast from lunar meteorite Dhofar 287: Rb–Sr and Sm–Nd isotopic studies. In *Lunar Planet. Sci.* **XXXIII**, #1344. The Lunar and Planetary Institute, Houston (CD-ROM).

Shirley D. N. (1983) A partially molten magma ocean model. *Proc. 13th Lunar Planet. Sci. Conf.* A519–A527.

Shkuratov Y., Pieters C., Omelchenko V., Stankevich D., Kaydash V., and Taylor L. (2003) Estimates of the lunar surface composition with clementine images and LSCC data. In *Lunar Planet. Sci.* **XXXIV**, #1258. The Lunar and Planetary Institute, Houston (CD-ROM).

Shoemaker E. M., Hait M. H., Swann G. A., Schleicher D. L., Schaber G. G., Sutton R. L., Dahlem D. H., Goddard E. N., and Waters A. C. (1970) Origin of the lunar regolith at Tranquillity base. *Proc. Apollo 11 Lunar Sci. Conf.*, 2399–2412.

Simonds C. H., Warner J. L., Phinney W. C., and McGee P. E. (1976) Thermal model for impact breccia lithification: Manicouagan and the moon. *Proc. 7th Lunar Sci. Conf.*, 2509–2528.

Snyder G. A., Taylor L. A., and Neal C. R. (1992) The sources of mare basalts: a model involving lunar magma ocean crystallization, plagioclase flotation and trapped instantaneous residual liquid. *Geochim. Cosmochim. Acta* **56**, 3809–3823.

Snyder G. A., Neal C. R., Taylor L. A., and Halliday A. N. (1995) Processes involved in the formation of magnesian-suite plutonic rocks from the highlands of Earth's Moon. *J. Geophys. Res.* **100**, 9365–9388.

Snyder G. A., Hall C. M., Taylor L. A., Nazarov M. A., and Semenova T. S. (2000a) ^{40}Ar–^{39}Ar geochronology of "new" basalts from Mare Fecunditatis and Mare Crisium. In *Lunar Planet. Sci.* **XXXI**, #1222. The Lunar and Planetary Institute, Houston (CD-ROM).

Snyder G. A., Borg L. E., Nyquist L. E., and Taylor L. A. (2000b) Chronology and isotopic constraints on lunar evolution. In *Origin of the Earth and Moon* (eds. R. Canup and K. Righter). University of Arizona Press, Tucson, pp. 361–395.

Staid M. I. and Pieters C. M. (2001) Mineralogy of the last lunar basalts: results from Clementine. *J. Geophys. Res.* **106**, 27887–27900.

Starukhina L. V. and Shkuratov Y. G. (2000) The lunar poles: water ice or chemically trapped hydrogen? *Icarus* **147**, 585–587.

Stewart D. B. (1975) Apollonian metamorphic rocks—the products of prolonged subsolidus equilibration. In *Lunar Sci.* **VI**. The Lunar Science Institute, Houston, pp. 774–776.

Stöffler D., Knöll H.-D., Marvin U. B., Simonds C. H., and Warren P. H. (1980) Recommended classification and nomenclature of lunar highland rocks—a committee report. In *Proceedings of the Conference on the Lunar Highlands Crust* (eds. R. B. Merrill and J. J. Papike). Pergamon, New York, pp. 51–70.

Swindle T. D., Spudis P. D., Taylor G. J., Korotev R. L., Nichols R. H., Jr., and Olinger C. T. (1991) Searching for Crisium basin ejecta: chemistry and ages of Luna 20 impact melts. *Proc. 21st Lunar Planet. Sci. Conf.*, 167–181.

Taylor G. J., Warren P., Ryder G., Delano J., Pieters C., and Lofgren G. (1991) Lunar rocks. In *Lunar Sourcebook: A User's Guide to the Moon* (eds. G. Heiken, D. Vaniman, and B. M. French). Cambridge University Press, Cambridge, pp. 183–284.

Taylor L. A., Shervais J. W., Hunter R. H., Shih C. Y., Nyquist L., Bansal B., Wooden J., and Laul J. C. (1983) Pre-4.2 AE mare basalt volcanism in the lunar highlands. *Earth Planet. Sci. Lett.* **66**, 33–47.

Taylor S. R. (1982) *Planetary Science: A Lunar Perspective*. Lunar and Planetary Institute, Houston, 512p.

Taylor S. R. and Jakes P. (1974) The geochemical evolution of the Moon. *Proc. 5th Lunar Planet. Sci. Conf.*, 1287–1305.

Tera F. (1974) Isotopic evidence for a terminal lunar cataclysm. *Earth Planet. Sci. Lett.* **22**, 1–21.

Tera F., Papanstassiou D. A., and Wasserburg G. J. (1974) Isotopic evidence for a terminal lunar cataclym. *Earth Planet. Sci. Lett.* **22**, 1–21.

Terada K., Saiki T., Hidaka H., Hashizume K., and Sano Y. (2002) *In-situ* ion microprobe U–Pb dating of glass spherules from Apollo 17 lunar soils. In *Lunar Planet. Sci.* **XXXIII**, #1481. The Lunar and Planetary Institute, Houston.

Thalmann C., Eugster O., Herzog G. F., Klein J., Krähenbühl U., Vogt S., and Xue S. (1996) History of lunar meteorites Queen Alexandra Range 93069, Asuka 881757, and Yamato 791639 based on noble gas abundances, radionuclide concentrations, and chemical composition. *Meteorit. Planet. Sci.* **31**, 857–868.

Thomson W. (1864) On the secular cooling of the Earth. *Trans. Roy. Soc. Edinburgh* **23**, 157–169.

Tompkins S. and Pieters C. M. (1999) Mineralogy of the lunar crust: results from clementine. *Meteorit. Planet. Sci.* **34**, 25–41.

Unruh D. M., Stille P., Patchett P. J., and Tatsumoto M. (1984) Lu–Hf and Sm–Nd evolution in lunar mare basalts. *Proc. 14th Lunar Planet. Sci. Conf.*, B459–B477.

Vaniman D., Dietrich J., Taylor G. J., and Heiken G. (1991) Exploration, samples, and recent concepts of the Moon. In *Lunar Sourcebook: A User's Guide to the Moon* (eds. G. Heiken, D. Vaniman, and B. M. French). Cambridge University Press, Cambridge, pp. 5–26.

Walker D. (1983) Lunar and terrestrial Crust Formation. *Proc. 14th Lunar Planet. Sci. Conf.*, B17–B25.

Walker D., Longhi J., Lasaga A. C., Stolper E. M., Grove T. L., and Hays J. F. (1977) Slowly cooled microgabbros 15555 and 15065. *Proc. 8th Lunar Sci. Conf.*, 1521–1548.

Warner J. L., Simonds C. H., and Phinney W. C. (1976) Genetic distinction between anorthosites and Mg-rich plutonic rocks: new data from 76255 (abstract). *Lunar Sci.* **VII**. The Lunar Science Institute, Houston, pp. 915–917.

Warner J. L., Phinney W. C., Bickel C. E., and Simonds C. H. (1977) Feldspathic granulitic impactites and pre-final bombardment lunar evolution. *Proc. 8th Lunar. Sci. Conf.* **2**, 2051–2066.

Warner R. D., Keli K., and Taylor G. J., (1977) Coarse-grained basalt 71597: a prodcut of partial olivine accumulation. *Proc. 8th Lunar Sci. Conf.*, 1429–1442.

Warren P. H. (1985) The magma ocean concept and lunar evolution. *Ann. Rev. Earth Planet. Sci.* **13**, 201–240.

Warren P. H. (1986) Anorthosite assimilation and the origin of the Mg/Fe-related bimodality of pristine Moon rocks: support for the magmasphere hypothesis. *Proc. 16th Lunar Planet. Sci. Conf.*, D331–D343.

Warren P. H. (1988) The origin of pristine KREEP: effects of mixing between urKREEP and the magmas parental to the Mg-rich cumulates. *Proc. 18th Lunar Planet. Sci. Conf.*, 233–241.

Warren P. H. (1989) KREEP: major-element diversity, trace-element uniformity (almost). In *Workshop on Moon in Transition: Apollo 14, KREEP, and Evolved Lunar Rocks* (Technical Report 89-03) (eds. G. J. Taylor and P. H. Warren). Lunar and Planetary Institute, Houston, pp. 149–153.

Warren P. H. (1990) Lunar anorthosites and the magma ocean hypothesis: importance of FeO enrichment in the parent magma. *Am. Mineral.* **75**, 46–58.

Warren P. H. (1992) Inheritance of silicate differentiation during lunar origin by giant impact. *Earth Planet. Sci. Lett.* **112**, 101–116.

Warren P. H. (1993) A concise compilation of key petrologic information on possibly pristine nonmare Moon rocks. *Am. Mineral.* **78**, 360–376.

Warren P. H. (1996) Global inventory of lunar impact melt as a function of parent crater size. In *Lunar Planet. Sci.* **XXVII**. The Lunar and Planetary Institute, Houston, pp. 1379–1380.

Warren P. H. (2003) "New" lunar meteorites: II. Implications for composition of the global lunar surface, of the lunar crust, and of the bulk Moon. *Meteorit. Planet. Sci.* **38** (submitted).

Warren P. H. and Kallemeyn G. W. (1984) Pristine rocks (8th foray): "Plagiophile" element ratios, crustal genesis, and the bulk composition of the Moon. *Proc. 15th Lunar Planet. Sci. Conf.* C16–C24.

Warren P. H. and Kallemeyn G. W. (1986) Geochemistry of lunar meteorite Yamato-791197: comparison with ALHA81005 and other lunar samples. *Proc. NIPR Symp. Antarct. Meteorit. (Tokyo)* **10**, 3–16.

Warren P. H. and Wasson J. T. (1977) Pristine nonmare rocks and the nature of the lunar crust. *Proc. 8th Lunar Sci. Conf.*, 2215–2235.

Warren P. H. and Wasson J. T. (1978) Compositional-petrographic investigation of pristine nonmare rocks. *Proc. 9th Lunar Planet. Sci. Conf.*, **1**, 185–217.

Warren P. H. and Wasson J. T. (1979a) Effects of pressure on the crystallization of a "chondritic" magma ocean and implications for the bulk composition of the Moon. *Proc. 10th Lunar Planet. Sci. Conf.*, **2**, 2051–2083.

Warren P. H. and Wasson J. T. (1979b) The origin of KREEP. *Rev. Geophys. Space Phys.* **17**, 73–88.

Warren P. H. and Wasson J. T. (1980a) Early lunar petrogenesis, oceanic and extraoceanic. In *Proceedings of the Conference on the Lunar Highlands Crust* (eds. R. B. Merrill and J. J. Papike). Pergamon, New York, pp. 81–99.

Warren P. H. and Wasson J. T. (1980b) Further foraging for pristine nonmare rocks: correlations between geochemistry and longitude. *Proc. 11th Lunar Planet. Sci. Conf.*, 431–470.

Warren P. H., Taylor G. J., Keil K., Shirley D. N., and Wasson J. T. (1983) Petrology and chemistry of two "large" granite clasts from the Moon. *Earth Planet. Sci. Lett.* **64**, 175–185.

Warren P. H., Jerde E. A., and Kallemeyn G. W. (1987) Pristine Moon rocks: a "large" felsite and a metal-rich ferroan anorthosite. *Proc. 17th Lunar Planet. Sci. Conf.*, E303–E313.

Warren P. H., Jerde E. A., and Kallemeyn G. W. (1989) Lunar meteorites: siderophile element contents, and implications for the composition and origin of the Moon. *Earth Planet. Sci. Lett.* **91**, 245–260.

Warren P. H., Jerde E. A., and Kallemeyn G. W. (1990) Pristine Moon rocks: an alkali anorthosite with coarse augite exsolution from plagioclase, a magnesian harzburgite, and other oddities. *Proc. 20th Lunar Planet. Sci. Conf.*, 31–59.

Warren P. H., Jerde E. A., and Kallemeyn G. W. (1991) Pristine Moon rocks: Apollo 17 anorthosites. *Proc. 21st Lunar Planet. Sci. Conf.*, 51–61.

Warren P. H., Claeys P., and Cedillo-Pardo E. (1996) Megaimpact melt petrology (Chicxulub, Sudbury, and the Moon): effects of scale and other factors on potential for fractional crystallization and development of cumulates. In *The Cretaceous-Tertiary Event and Other Catastrophes in Earth History*, Spec. Pap. 307 (eds. G. Ryder, D. Fastovsky, and S. Gartner). Geological Society of America, Boulder, CO, pp. 105–124.

Warren P. H., Kallemeyn G. W., and Kyte F. T. (1999) Origin of planetary cores: evidence from highly siderophile elements in martian meteorites. *Geochim. Cosmochim. Acta* **63**, 2105–2122.

Warren P. H., Ulff-Møller F., and Kallemeyn G. W. (2003) "New" lunar meteorites: I. Impact melt and regolith breccias and large-scale heterogeneities of the upper lunar crust. *Meteorit. Planet. Sci.* **38** (submitted).

Wasson J. T. (1985) *Meteorites: Their Record of Early Solar System History*. Freeman, New York, 267p.

Wasson J. T. and Kallemeyn G. W. (1988) Compositions of chondrites. *Phil. Trans. Roy. Soc. London* **A325**, 535–544.

Wasson J. T., Boynton W. V., Chou C.-L., and Baedecker P. A. (1975) Compositional evidence regarding the influx of interplanetary materials onto the lunar surface. *Moon* **13**, 121–141.

Wetherill G. W. (1981) Nature and origin of basin-forming projectiles. In *Multi-ring Basins* (eds. P. H. Schultz and R. B. Merrill). Pergamon, New York, pp. 1–18.

Wiechert U., Halliday A. N., Lee D.-C., Snyder G. A., Taylor L. A., and Rumble D. (2001) Oxygen isotopes and the Moon-forming giant impact. *Science* **294**, 345–348.

Wieczorek M. A. (2003) The thickness of the lunar crust: how low can you go? In *Lunar Planet. Sci.* **XXXIV**, #1330. The Lunar and Planetary Institute, Houston, CD-ROM.

Wilhelms D. E. (1987) *The Geologic History of the Moon (USGS Professional Paper 1348)*. US Geological Survey, 302p.

Wilson L. and Head J. W., III (2003) Depth generation of magmatic gas on the Moon and implications for pyroclastic eruptions. *Geophys. Res. Lett.* **30** (12), 1605.

Wolf R. and Anders E. (1980) Moon and Earth: compositional differences inferred from siderophiles, volatiles, and alkalis in basalts. *Geochim. Cosmochim. Acta* **44**, 2111–2124.

Wolf R., Woodrow A., and Anders E. (1979) Lunar basalts and pristine highland rocks: comparison of siderophile and volatile elements. *Proc. 10th Lunar Planet. Sci. Conf.*, 2107–2130.

Wood J. A. (1975) Lunar petrogenesis in a well-stirred magma ocean. *Proc. 6th Lunar Sci. Conf.*, 1087–1102.

Wood J. A., Dickey J. S., Marvin U. B., and Powell B. N. (1970) Lunar anorthosites and a geophysical model of the Moon. *Proc. Apollo 11 Lunar Sci. Conf.*, 965–988.

Young R. A. (1977) The lunar impact flux, radiometric age correlation, and dating of specific lunar features. *Proc. 8th Lunar Sci. Conf.*, 3457–3474.

Zeigler R. A., Jolliff B. L., Korotev R. L., and Haskin L. A. (2000) Petrology, geochemistry, and possible origin of monomict mafic lithologies of the Cayley plains. In *Lunar Planet. Sci.* **XXXI**, #1859. The Lunar and Planetary Institute, Houston (CD-ROM).

Zellner N. E. B., Spudis P. D., Delano J. W., Whittet D. C. B., and Swindle T. D. (2003) Geochemistry and impact history at the Apollo 16 landing site. In *Lunar Planet. Sci.* **XXXIV**, #1157. The Lunar and Planetary Institute, Houston (CD-ROM).

Zhong S., Parmentier E. M., and Zuber M. T. (2000) A dynamic origin for the global asymmetry of lunar mare basalts. *Earth Planet. Sci. Lett.* **177**, 131–140.

1.22
Mars

H. Y. McSween, Jr.

University of Tennessee, Knoxville, TN, USA

1.22.1	GEOCHEMICAL EXPLORATION OF MARS	601
1.22.2	SOURCES OF GEOCHEMICAL DATA	602
1.22.2.1	*Global Surface Chemistry from Orbiter Measurements*	602
1.22.2.2	*Geophysical Constraints on Geochemistry of the Bulk Planet*	603
1.22.2.3	In Situ *Analyses of Rocks, Soils, and Atmosphere by Landers*	603
1.22.2.4	*Martian Meteorites*	603
1.22.2.5	*Geologic Context and Relationships of Geochemical Data Sets*	605
1.22.3	GEOCHEMISTRY OF THE MANTLE AND CORE	607
1.22.3.1	*Geochemical Characteristics of Martian Meteorites*	607
1.22.3.2	*Geochemical Models for the Martian Mantle*	610
1.22.3.3	*Geochemical Models for the Martian Core*	610
1.22.3.4	*Consistency with Geophysical Constraints*	610
1.22.4	GEOCHEMISTRY OF THE CRUST	610
1.22.4.1	*Mafic versus Felsic Igneous Rocks*	610
1.22.4.2	*Geochemistry of Assimilated Crust in Martian Meteorites*	612
1.22.4.3	*Geochemistry of Crustal Sediments*	613
1.22.5	GEOCHEMISTRY OF VOLATILE RESERVOIRS	614
1.22.5.1	*Planet Volatile Inventory*	614
1.22.5.2	*Outgassing and Atmospheric Loss*	614
1.22.5.3	*Isotopic Evidence for Volatile Cycling*	615
1.22.5.4	*Organic Geochemistry or Biochemistry?*	615
1.22.6	GEOCHEMICAL COMPARISON OF MARS AND EARTH	616
1.22.6.1	*Differences in Bulk Planet Compositions and Major Reservoirs*	616
1.22.6.2	*Isotope Geochronology and Planetary Differentiation*	616
1.22.6.3	*Geochemical Cycles without Plate Tectonics*	617
1.22.6.4	*Major Unsolved Problems*	617
REFERENCES		618

1.22.1 GEOCHEMICAL EXPLORATION OF MARS

More than any other planet, Mars has captured our attention and fueled our speculations. Much of this interest relates to the possibility of martian life, as championed by Percival Lowell in the last century and subsequently in scientific papers and science fiction. Lowell's argument for life on Mars was based partly on geochemistry, in that his assessment of the planet's hospitable climate was dependent on the identification of H_2O ice rather than frozen CO_2 in the polar caps. Although this reasoning was refuted by Alfred Wallace in 1907, widespread belief in extant martian life persisted within the scientific community until the mid-twentieth century (Zahnle, 2001). In 1965 the Mariner 4 spacecraft flyby suddenly chilled this climate, by demonstrating that the martian atmosphere was thin and the surface was a cratered moonscape devoid of canals. This view of Mars was overturned again in 1971, when the Mariner 9 spacecraft discovered towering volcanoes and dry riverbeds, implying a complex geologic history. The first geochemical measurements on Mars, made by two Viking landers in 1976, revealed soils enriched in salts suggesting exposure to water, but lacking organic compounds which virtually ended discussion of martian life.

The suggestion that a small group of achondritic meteorites were martian samples (McSween and Stolper, 1979; Walker et al., 1979; Wasson and Wetherill, 1979) found widespread acceptance when trapped gases in them were demonstrated to be compositionally similar to the Mars atmosphere (Bogard and Johnson, 1983; Becker and Pepin, 1984). The ability to perform laboratory measurements of elements and isotopes present in trace quantities in meteorites has invigorated the subject of martian geochemistry. Indeed, because of these samples, we now know more about the geochemistry of Mars than of any other planet beyond the Earth–Moon system. Some studies of martian meteorites have prompted a renewed search for extraterrestrial life using chemical biomarkers.

Recent Mars spacecraft, including the Mars Pathfinder lander/rover in 1997 and Mars Global Surveyor and Mars Odyssey now orbiting the planet, have provided significant new geochemical findings. These missions have also generated geophysical data with which to constrain geochemical models of the martian interior.

1.22.2 SOURCES OF GEOCHEMICAL DATA

1.22.2.1 Global Surface Chemistry from Orbiter Measurements

Few geochemical measurements of Mars have been made from orbiting spacecraft. Gamma-ray spectroscopy on the Soviet Phobos-2 orbiter provided analyses of two broad areas (Surkov et al., 1989), and Mars Odyssey, now in orbit, carries a gamma-ray spectrometer (GRS) that may map the entire planet at somewhat higher spatial resolution. Trombka et al. (1992) compared element abundances for oxygen, silicon, iron, potassium, and thorium obtained by several independent analyses of data from the Phobos experiment (Table 1). Given the difficulties in analyzing these spectra and the huge footprints of the measurements (encompassing equatorial regions that span parts of the southern highlands, northern lowlands, and Tharsis volcanoes), the data are probably useful only as a comparison with globally homogenized soil compositions. Preliminary Mars Odyssey GRS measurements (Taylor et al., 2003) of potassium (0.3–0.5 wt.%) and thorium (0.5–2 ppm) of the same equatorial regions are consistent with these values. GRS has also provided a global map of hydrogen distribution (Boynton et al., 2002; Feldman et al., 2002).

Spectrometers on several spacecraft have provided information on the mineralogy of the martian surface, but only one set of spectral data is sufficiently quantitative to be recast into geochemistry. The thermal emission spectrometer (TES) on Mars Global Surveyor and the thermal emission spectrometer imaging system (THEMIS) on Mars Odyssey have provided global mineral mapping, and linear deconvolution allows the

Table 1 Mars surface chemistry from orbital spacecraft measurements.

	Phobos-2 gamma-ray spectra[a]	
Element	PC-3[b]	PC-4[b]
O	48 ± 5/40 ± 18%	46 ± 4/54 ± 27%
Si	19 ± 4/11 ± 6%	20 ± 3/15 ± 7%
Fe	9 ± 3/10 ± 4%	8 ± 4/4 ± 7%
K	0.3 ± 0.1/0.2 ± 0.1%	0.4 ± 0.1/0.3 ± 0.2%
Th	1.9 ± 0.6/3.1 ± 1.3 ppm	2.0 ± 0.4/2.2 ± 1.0 ppm

	MGS thermal emission spectra[c]	
Oxide	Surface type-1	Surface type-2
SiO_2	53.6 ± 1.4	58.4 ± 1.4
TiO_2	0.1 ± 0.9	0.1 ± 0.9
Al_2O_3	15.2 ± 1.5	15.0 ± 1.5
FeO(T)	6.9 ± 1.2	4.5 ± 1.2
MnO	0.1	0.1
MgO	8.7 ± 2.6	9.0 ± 2.6
CaO	10.3 ± 0.7	7.4 ± 0.7
Na_2O	2.8 ± 0.4	2.7 ± 0.4
K_2O	0.6 ± 0.4	1.2 ± 0.4
Cr_2O_3	0.1	0.1
Total	98.4	98.5

[a] Source: Trombka et al. (1992). [b] PC (for pericenter) -3 and -4 refer to the trajectories for two orbits; the two data sets for each orbit are USSR and USA science team analyses. [c] Source: Calculated from mineral abundances of Hamilton et al. (2001) using method of Wyatt et al. (2001).

identification of minerals and determination of their relative abundances. Using deconvolved TES modal mineralogy and the compositions of minerals in the spectral library, Wyatt *et al.* (2001) demonstrated that the concentrations of major element oxides can be estimated with an uncertainty of ±5 wt.% or less for most elements. Major and a few minor element abundances of the two major surface units on Mars, calculated from the deconvolved TES mineral modes of Hamilton *et al.* (2001), are given in Table 1.

1.22.2.2 Geophysical Constraints on Geochemistry of the Bulk Planet

The accepted mean density of Mars, based on its measured volume and determination of its mass from spacecraft orbits, is 3.9335 ± 0.0004 g cm^{-3} (Lodders and Fegley, 1998). The density of the elastic lithosphere (approximately equivalent to the crust), estimated from models of the relationship between gravity and topography from Mars Global Surveyor data, is 2.95–2.99 g cm^{-3} (McKenzie *et al.*, 2002), which is similar to the density of basalt. The planet's dimensionless moment of inertia (0.3662 ± 0.0017), calculated from Mars Pathfinder measurement of the rate at which its spin pole precesses (Folkner *et al.*, 1997), constrains the core radius to ~1,300–1,500 km, depending on core composition.

1.22.2.3 *In Situ* Analyses of Rocks, Soils, and Atmosphere by Landers

Viking landers carried out Rutherford backscattering and X-ray fluorescence (XRF) analyses of soils at two sites (Clark *et al.*, 1982). Rocks and soils were analyzed by an alpha-proton X-ray spectrometer (APXS) on the Mars Pathfinder rover (Rieder *et al.*, 1997) (Figure 1). Early APXS analyses (Rieder *et al.*, 1997; McSween *et al.*, 1999) have been superceded by new calibrations that take into account the effect of the CO_2 atmosphere and include alpha-proton data (Foley *et al.*, 2000; Wanke *et al.*, 2001). Viking analyses are absolute measurements because the landers carried a reference to calibrate X-ray interaction of the target. APXS analyses must be normalized to 100%, although they include more elements. Both kinds of measurements are surface analysis techniques (analyses represent the outer few tens or hundreds of micrometers, depending on the element), but some Viking soil samples were collected from several centimeters depth. McSween and Keil (2000) attempted to integrate the XRF and APXS data sets for martian soils. Elements in Mars Pathfinder rocks form nearly linear arrays when plotted versus sulfur, which has been interpreted to indicate variable coatings of sulfate-rich dust (Rieder *et al.*, 1997). Graphical extrapolation of these trends to zero sulfur (or low sulfur equal to that in martian meteorites) gives the dust-free rock composition (McSween *et al.*, 1999; Wanke *et al.*, 2001). Representative martian rock and soil compositions are given in Table 2.

The response of Viking surface fines to heating (Biemann *et al.*, 1977) is consistent with loss of <2 wt.% water of hydration, and spectral estimates of absorbed plus bound water in martian soils vary between <0.5% and 4% (Yen *et al.*, 1998). Mars Odyssey GRS measurements suggest that significant amounts of water ice are present at high latitudes, and mineralogically bound water is present in equatorial regions (Boynton *et al.*, 2002; Feldman *et al.*, 2002). Carbon in analyzed Mars Pathfinder soils is below the APXS detection limit of 0.8% (Wanke *et al.*, 2001), consistent with an apparent absence of organic compounds at Viking sites (Biemann *et al.*, 1977).

Martian atmospheric gases were analyzed by mass spectrometers on the Viking landers (Nier and McElroy, 1977; Owen *et al.*, 1977). Abundances of CO_2, H_2O (which is highly variable), N_2, O_2, and noble gases, as well as ratios of the isotopes of hydrogen, carbon, oxygen, nitrogen, argon, krypton, and xenon, have been determined (Table 3), although sometimes with large measurement uncertainties.

1.22.2.4 Martian Meteorites

The Shergotty, Nakhla, and Chassigny meteorites and their relatives, collectively called SNC meteorites (see Chapter 1.05), are thought to be martian igneous rocks (e.g., McSween, 1994; Treiman *et al.*, 2000). Shergottites (Figure 2) comprise the largest group of martian meteorites. Some shergottites are basalts (the basaltic shergottites), and others are ultramafic cumulate rocks

Figure 1 Mars Pathfinder rover performing an APXS analysis of a rock (Barnacle Bill).

Table 2 Mars rock and soil compositions from *in situ* measurements by landers.

Oxide (wt.%)	Mars Pathfinder rocks[a]						Unc[b] (%)
	A-3	A-7	A-16	A-17	A-18	Dust-free rock	
Na_2O	1.60	1.19	2.30	2.03	1.78	2.46	40
MgO	3.20	6.71	4.56	3.50	3.91	1.51	10
Al_2O_3	11.02	9.68	10.24	10.03	10.94	11.0	7
SiO_2	53.8	49.7	48.6	55.2	51.8	57.0	10
P_2O_5	1.42	0.99	1.00	0.98	0.97	0.95	20
SO_3	2.77	4.89	3.29	1.88	3.11	0.30	20
Cl	0.41	0.50	0.41	0.38	0.37	0.32	15
K_2O	1.29	0.87	0.96	1.14	1.10	1.36	10
CaO	6.03	7.35	8.14	8.80	6.62	8.09	10
TiO_2	0.92	0.91	0.95	0.65	0.82	0.69	20
Cr_2O_3	0.10	ND	ND	0.05	ND	ND	50
MnO	ND	0.47	0.65	0.49	0.52	0.55	25
Fe_2O_3	16.2	16.7	18.9	14.8	18.1	15.7	5

Oxide (wt.%)	Viking 1 soils[c]					Mars Pathfinder soils[a]			
	C-1	C-5	C-6	C-7	Unc[c]	A-4	A-5	A-10	A-15
Na_2O	ND	ND	ND	ND		1.00	1.05	1.32	0.97
MgO	6.4	7.3	6.4	5.3	$-3/+5$	9.95	9.20	8.16	7.46
Al_2O_3	8.4	7.2	7.8	7.8	± 4	8.22	8.71	7.41	7.59
SiO_2	46.0	44.1	47.1	46.6	± 6	42.5	41.6	41.8	44.0
P_2O_5	ND	ND	ND	ND		1.89	1.55	0.95	1.01
SO_3	7.5	10.0	7.2	7.2	$-2/+6$	7.58	6.38	7.09	6.09
Cl	0.8	0.9	0.9	0.6	$-0.5/+1.5$	0.57	0.55	0.53	0.54
K_2O	ND	ND	ND	ND		0.60	0.51	0.45	0.87
CaO	6.4	5.9	6.4	6.4	± 2	6.09	6.63	6.86	6.56
TiO_2	0.7	0.6	0.7	0.7	± 0.25	1.08	0.75	1.02	1.20
Cr_2O_3	ND	ND	ND	ND		0.2	0.4	0.3	0.3
MnO	ND	ND	ND	ND		0.76	0.34	0.51	0.46
Fe_2O_3	18.9	18.3	18.5	20.1	$-2/+5$	19.6	23.0	23.6	23.0

[a] Source: Wanke *et al.* (2001). [b] Average error, in relative %; applies to Mars Pathfinder rocks and soils. [c] Source: Clark *et al.* (1982).

(the lherzolitic shergottites). The nakhlites are clinopyroxenites, the Chassigny meteorite is a dunite, and ALH84001 is an orthopyroxenite (see McSween and Treiman, 1998, for mineralogic and petrographic descriptions). All SNCs except ALH84001 have crystallization ages that are too young (<1.3 billion years) for igneous activity on asteroids (Nyquist *et al.*, 2001), and the compositions of gases implanted during shock (probably as these rocks were ejected from their parent body) match those measured for the martian atmosphere (Bogard *et al.*, 2001), as shown in Figure 3.

Laboratory analyses of SNCs provide a wealth of trace element and isotopic data that is otherwise unobtainable from remote sensing measurements. Compilations of element abundances in these meteorites by Treiman *et al.* (1987), Warren and Kallemeyn (1997), and Lodders (1998) have been augmented by analyses of newly discovered meteorites (Folco *et al.*, 2000; Rubin *et al.*, 2000; Barrat *et al.*, 2001). Elemental abundances in a representative set of SNC meteorites are given in Table 4. Stable isotope data for martian meteorites are scattered throughout the literature, but the following references provide summaries for specific elements: hydrogen (Leshin *et al.*, 1996; Bridges *et al.*, 2001), carbon (Bridges *et al.*, 2001), oxygen (Karlsson *et al.*, 1992; Clayton and Mayeda, 1996; Franchi *et al.*, 1999; Bridges *et al.*, 2001), nitrogen (Mathew *et al.*, 1998), sulfur (Greenwood *et al.*, 1997, 2000; Farquhar *et al.*, 2000), and noble gases (Bogard *et al.*, 2001). Sources of radiogenic isotope measurements and their chronological interpretations have been summarized for the K–Ar, Rb–Sr, Sm–Nd, and U–Th–Pb systems by Nyquist *et al.* (2001) and for the Hf–W, Sm–Nd, Re–Os, Lu–Hf, and Mn–Cr systems by Halliday *et al.* (2001). The Hf–W chronology of Mars has been reinterpreted in light of a revision to the chondritic $\varepsilon^{182}W$ (Kleine *et al.*, 2002; Yin *et al.*, 2002). Sources for measured cosmogenic nuclide data and calculated cosmic-ray exposure ages for SNC meteorites are summarized by Eugster *et al.* (1997) and Nyquist *et al.* (2001).

Table 3 Composition of the Martian atmosphere.[a]

Component	Standard	Atmosphere-Viking	SNC impact glass
CO_2		~95%	
N_2		2.7%	
Ar		1.6%	
O_2		0.13%	
Ne		2.5 ppm	
Kr		0.3 ppm	
Xe		0.08 ppm	
$^{13}C/^{12}C$	%-terrestrial	0 ± 50	n.r.
$^{18}O/^{16}O$	%-terrestrial	0 ± 50	n.r.
$^{36}Ar/^{132}Xe$	ratio	350 ± 98	900 ± 100
$^{84}Kr/^{132}Xe$	ratio	11 ± 3	20.5 ± 1.5
$^{2}H/^{1}H$	%-terrestrial	~450[b]	~440[c]
$^{15}N/^{14}N$	%-terrestrial	63 ± 16	<50
$^{20}Ne/^{22}Ne$	ratio	n.r.	~10
$^{36}Ar/^{38}Ar$	ratio	5.5 ± 1.5	≤3.9
$^{40}Ar/^{36}Ar$	ratio	$3{,}000 \pm 500$	~1,800
$^{86}Kr/^{84}Kr$	%-solar	n.r.	~0
$^{129}Xe/^{132}Xe$	ratio	~2.5	2.4–2.6
$^{126}Xe/^{130}Xe$	%-solar	n.r.	~27

[a] Source: Bogard et al. (2001). [b] Measurement from ground-based spectra. [c] Measured in Zagami apatite; D/H ratios for impact glass have not been reported.

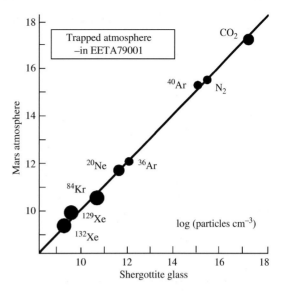

Figure 3 Comparison of molecular and isotopic abundances of the Mars atmosphere (Viking measurements) with trapped gases in impact-melted glass in the EETA79001 shergottite. This correspondence is evidence of the meteorite's martian origin (after Treiman et al., 2000).

Figure 2 An $Fe\,K_\alpha$ map of a thin section of the Shergotty meteorite. Gray phases are zoned pyroxenes, black phases are plagioclase (transformed to maskelynite by shock), and bright phases are Fe–Ti oxides.

The regularly updated Mars Meteorite Compendium (www-curator.jsc.nasa.gov/curator/antmet/antmet.htm) contains a wealth of geochemical information on SNCs.

1.22.2.5 Geologic Context and Relationships of Geochemical Data Sets

Martian stratigraphy is divided into the Noachian (>3.5–3.7 billion years old), Hesperian, and Amazonian (<2.9–3.3 billion years old) systems (Hartmann and Neukum, 2001). Distinct geologic terrains in the northern and southern hemispheres of Mars are separated by a marked topographic boundary (Figure 4). High densities of craters in the southern highlands of Mars, together with the existence of the 4.5-billion-year-old ALH84001 martian meteorite (Nyquist et al., 2001), indicate that some parts, perhaps most, of the crust are ancient. Noachian basement in the northern lowlands is covered by Hesperian sediments or volcanics. Mars Global Surveyor data indicate a lower limit of 40–50 km for the average crustal thickness, comprising more than 4% of the planetary volume (Zuber, 2001). Volcanic centers, resulting from plume magmatism, are represented by Tharsis and Elysium. The crust underlying Tharsis must be at least 100 km thick to provide isostatic support to its huge volcanoes. Although the Tharsis and Elysium bulges are the sites of fairly recent (Amazonian) volcanism, they have apparently existed for billions of years. Mars tectonics is dominated by Tharsis (Figure 4), which is surrounded by a flexural moat, radial rifts, and concentric compressional ridges. Mars lacks any observational evidence of plate tectonics or crustal recycling. However, magnetic lineations discovered by Mars Global Surveyor (Acuna et al., 1999) may imply ancient crustal spreading.

Mars is the only planet, besides Earth, that provides geomorphic evidence for the past operation of a hydrologic cycle. Today, the surface of Mars is cold and dry, but landforms generated by movement and ponding of water or ice indicate that brief but extensive episodes of aqueous activity punctuated the martian geologic record. These features include sinuous valleys,

Table 4 Representative elemental abundances in SNC meteorites.[a]

Element	Shergotty	QUE94201	ALH77005	ALH84001	Nakhla	Chassigny
Li (ppm)	4.5 ± 0.9		1.5 ± 0.2		3.9	1.4 ± 0.2
B (ppm)					4.6	63
C (ppm)	530 ± 130		140 ± 90	580	300 ± 100	847
F (ppm)	46 ± 6	40	22		57	15
Na (%)	1.03 ± 0.14	1.17 ± 0.19	0.35 ± 0.06	0.10 ± 0.02	0.34 ± 0.05	0.09 ± 0.01
Mg (%)	5.58 ± 0.11	3.77 ± 0.04	17.0 ± 0.8	15.1 ± 0.5	7.3 ± 0.2	19.2 ± 0.5
Al (%)	3.64 ± 0.27	5.81 ± 0.58	1.52 ± 0.15	0.68 ± 0.05	0.89 ± 0.11	0.42 ± 0.15
Si (%)	24.0	22.4	19.8 ± 0.4	24.7 ± 0.1	22.7 ± 0.8	17.5 ± 0.5
P (ppm)	3,230 ± 230		1,750 ± 150	61	500 ± 45	275 ± 35
S (ppm)	1,270 ± 760		510 ± 200	110	260 ± 80	260 ± 130
Cl (ppm)	108	91	14	8.0 ± 4.5	80	34
K (ppm)	1,440 ± 110	375 ± 80	250 ± 40	140 ± 50	1,070 ± 190	300 ± 110
Ca (%)	6.86 ± 0.39	8.14 ± 0.01	2.26 ± 0.23	1.30 ± 0.27	10.5 ± 0.5	0.47 ± 0.07
Sc (ppm)	52 ± 7	48 ± 2	21 ± 1	13 ± 1	51 ± 4	5.3 ± 0.4
Ti (ppm)	4,900 ± 430	1,1000 ± 900	2,340 ± 440	1,240 ± 70	2,020 ± 250	480 ± 90
V (ppm)	290 ± 40	113 ± 15	162 ± 6	201 ± 6	192	39 ± 9
Cr (ppm)	1,350 ± 100	950 ± 85	6,670 ± 520	7,760 ± 670	1,770 ± 280	5,240 ± 1,050
Mn (ppm)	4,110 ± 130	3,480 ± 150	3,470 ± 80	3,560 ± 100	3,820 ± 310	4,120 ± 840
Fe (%)	15.1 ± 0.5	14.4 ± 0.4	15.6 ± 0.3	13.6 ± 0.4	16.0 ± 1.2	21.2 ± 0.6
Co (ppm)	40 ± 7	24 ± 1	72 ± 4	47 ± 3	48 ± 5	123 ± 17
Ni (ppm)	79 ± 12	<20	290 ± 85	58	90	500 ± 70
Cu (ppm)	16 ± 9		5.1 ± 6		12 ± 6	2.6
Zn (ppm)	69 ± 9	110	60 ± 8	92 ± 9	54 ± 11	72 ± 4
Ga (ppm)	16.0 ± 1.3	27 ± 1	7.3 ± 1.2	2.9 ± 0.5	3 ± 0.5	0.7
Ge (ppm)	0.73 ± 0.06		0.58	1.8	3.0	0.011
As (ppb)	25	770	22	<30	15	8
Se (ppm)	0.38 ± 0.8		0.15	<0.16	0.08 ± 0.02	0.037
Br (ppm)	0.88 ± 0.2	0.35	0.077 ± 0.011		4.5 ± 0.2	0.088 ± 0.031
Rb (ppm)	6.4 ± 0.6		0.7 ± 0.08	0.83	3.8 ± 0.8	0.75 ± 0.5
Sr (ppm)	48 ± 8	70 ± 15	14 ± 3	4.5	59 ± 10	7.2
Y (ppm)	19	31	6.2	1.6	3.3 ± 1.1	0.64
Zr (ppm)	57 ± 14	100 ± 7	19.5	5.9	8.8 ± 1.0	2.1 ± 0.9
Nb (ppm)	4.6	0.68	0.65 ± 0.11	0.42	1.57	0.34 ± 0.03
Mo (ppm)	0.37		0.20		0.086	
Pd (ppb)	1.7				30 ± 12	0.15
Ag (ppb)	11.4 ± 5.1		4.4		40	2.6
Cd (ppb)	28 ± 16		2.1	77	93 ± 31	14
In (ppb)	26 ± 4		11		20 ± 7	3.9
Sn (ppm)	0.011		0.24		0.58	
Sb (ppb)	5.2 ± 3.6		69		40 ± 30	0.87
Te (ppb)	3.3 ± 0.9		0.5		<4.3	50
I (ppb)	43 ± 10	4,600	1,720		180	<10
Cs (ppb)	440 ± 50		53 ± 26	43 ± 4	390 ± 90	37
Ba (ppm)	34 ± 5	<15	4.2 ± 1.5	4.0	29 ± 6	7.6 ± 0.6
La (ppm)	2.16 ± 0.32	0.40 ± 0.06	0.34 ± 0.03	0.19 ± 0.06	2.06 ± 0.33	0.53 ± 0.12
Ce (ppm)	5.45 ± 0.86	1.47 ± 0.23	0.91 ± 0.15	0.59 ± 0.14	5.87 ± 0.37	1.12
Pr (ppm)	0.81 ± 0.10		0.13	0.06	0.67	0.13
Nd (ppm)	4.2 ± 0.5	2.2 ± 0.4	0.95 ± 0.17	0.265	3.23 ± 0.52	0.62 ± 0.11
Sm (ppm)	1.47 ± 0.14	2.3 ± 0.4	0.49 ± 0.07	0.12 ± 0.03	0.77 ± 0.08	0.14 ± 0.03
Eu (ppm)	0.60 ± 0.11	1.04 ± 0.07	0.22 ± 0.02	0.035 ± 0.008	0.235 ± 0.030	0.045 ± 0.007
Gd (ppm)	2.54 ± 0.26	4.3	0.92	0.14	0.86 ± 0.08	0.11
Tb (ppm)	0.48 ± 0.05	0.87 ± 0.09	0.17 ± 0.01	0.038 ± 0.005	0.12 ± 0.01	0.03 ± 0.02
Dy (ppm)	3.50 ± 0.64	5.8 ± 0.4	1.08 ± 0.08	0.28 ± 0.03	0.77 ± 0.06	0.20 ± 0.11
Ho (ppm)	0.71 ± 0.10	1.19	0.25 ± 0.03	0.076 ± 0.007	0.155 ± 0.020	0.044 ± 0.020
Er (ppm)	1.88 ± 0.02		0.66	0.21	0.37 ± 0.05	0.09
Tm (ppm)	0.30 ± 0.04		0.088 ± 0.010	0.036	0.047	
Yb (ppm)	1.52 ± 0.15	3.3 ± 0.3	0.59 ± 0.03	0.29 ± 0.03	0.39 ± 0.02	0.11 ± 0.01
Lu (ppm)	0.26 ± 0.03	0.50 ± 0.06	0.078 ± 0.004	0.049 ± 0.006	0.055 ± 0.007	0.015 ± 0.004
Hf (ppm)	2.3 ± 0.3	3.41 ± 0.02	0.62 ± 0.11	0.14 ± 0.03	0.27 ± 0.03	<0.1
Ta (ppb)	250 ± 40	23	33 ± 7	32	90	<20

Table 4 (continued).

Element	Shergotty	QUE94201	ALH77005	ALH84001	Nakhla	Chassigny
W (ppb)	460 ± 53		84	79	120 ± 80	46
Re (ppb)	0.44		0.102	0.002	0.036 ± 0.005	0.063 ± 0.012
Os (ppb)	0.4		4.4	0.010	0.007 ± 0.004	1.58 ± 0.31
Ir (ppb)	0.057 ± 0.025	<3	3.9 ± 0.3	0.08	0.22 ± 0.10	2.1 ± 0.4
Pt (ppb)	0.16				0.5	
Au (ppb)	0.92 ± 0.08	<0.5	0.21 ± 0.02	0.009	0.72 ± 0.20	0.73 ± 0.3
Hg (ppb)					0.7	
Tl (ppb)	12.8 ± 1.3		1.7		3.5 ± 0.5	3.7
Pb (ppm)				72		
Bi (ppb)	1.1 ± 0.6		<0.7		0.37	
Th (ppb)	380 ± 75	<50	57 ± 3	35	198 ± 13	57
U (ppb)	105 ± 20	<50	15 ± 3	11 ± 2	52 ± 9	18 ± 4
H_2O (ppm)	280 ± 120			550	570 ± 120	740 ± 70

[a] Source: Lodders (1998); Shergotty and QUE94201 are basaltic shergottites, ALH77005 is a lherzolitic shergottite, ALH84001 is an orthopyroxenite, Nakhla is a nakhlite, and Chassigny is a chassignite. Analytical uncertainties are 1σ deviations for multiple analyses.

outflow channels, and depositional basins (Baker, 2001; Masson et al., 2001).

Approximately half of the martian surface is covered by a veneer of fine-grained red dust, which precludes spectral measurements of underlying units. The Viking and Mars Pathfinder landing sites were within these dusty regions. In situ chemical analyses of soils from the Viking and Mars Pathfinder sites indicate similar compositions (Rieder et al., 1997; Bell et al., 2000), which suggests that surface sediments have been geochemically homogenized by wind. Thermal emission spectra of dark regions with high abundances of rock or sand indicate that two distinct units, with chemical compositions (calculated from deconvolved TES mineralogy) given in Table 1, dominate the planet's surface (Bandfield et al., 2000; Hamilton et al., 2001). The distribution of these units conforms approximately to the planetary dichotomy. The dust-free rock composition analyzed at the Mars Pathfinder site is andesitic (Rieder et al., 1997; McSween et al., 1999).

SNC meteorites, other than ALH84001, are generally thought to have come from the younger volcanic centers, which are spectrally obscured by dust. Cosmic-ray exposure ages for these meteorites indicate multiple ejection events (Eugster et al., 1997; see Chapter 1.13), implying that sampling of the martian crust is highly biased towards young igneous rocks. Older crust may be too pulverized by impacts to transmit the shock waves required to accelerate ejecta to martian escape velocity, or the older rocks may be too fragile to survive the ejection mechanism (McSween, 2002). Spacecraft imagery also shows that layered deposits, sometimes interpreted to be sediments, are widespread (Malin and Edgett, 2000). No sedimentary rocks, or igneous rocks having andesitic compositions, have been recognized among martian meteorites. However, salts and clays in martian meteorites demonstrate that some surface or near-surface rocks have been altered by interaction with fluids or brines (Bridges et al., 2001).

Several sources provide geologic context and overviews of geochemical data. Reviews by McSween (1985, 1994, 2002) and McSween and Treiman (1998) discuss what is known about the geochemistry and petrology of martian meteorites and how meteorite data constrain aspects of the geology and evolution of Mars. Two books provide especially useful geologic summaries of Mars: *Mars* (Kieffer et al., 1992) contains several geochemically relevant chapters, especially those by Banin et al. (1992), Longhi et al. (1992), and Owen (1992); and *Chronology and Evolution of Mars* (Kallenbach et al., 2001, reprinted from Space Science Reviews) contains excellent reviews of various aspects of martian geochemistry by Bogard et al., Bridges et al., Halliday et al., Nyquist et al., and Wanke et al. *Water on Mars* by Carr (1996) provides a comprehensive summary of martian hydrologic processes.

1.22.3 GEOCHEMISTRY OF THE MANTLE AND CORE

1.22.3.1 Geochemical Characteristics of Martian Meteorites

The geochemical characteristics of martian meteorites reflect those of their mantle source regions. However, their compositions have been modified by fractionation and, in some cases, assimilation, during ascent or emplacement.

SNC meteorites lack the depletion in iron and certain other siderophile elements commonly seen in terrestrial basalts. Manganese and phosphorus also have higher abundances in martian meteorites.

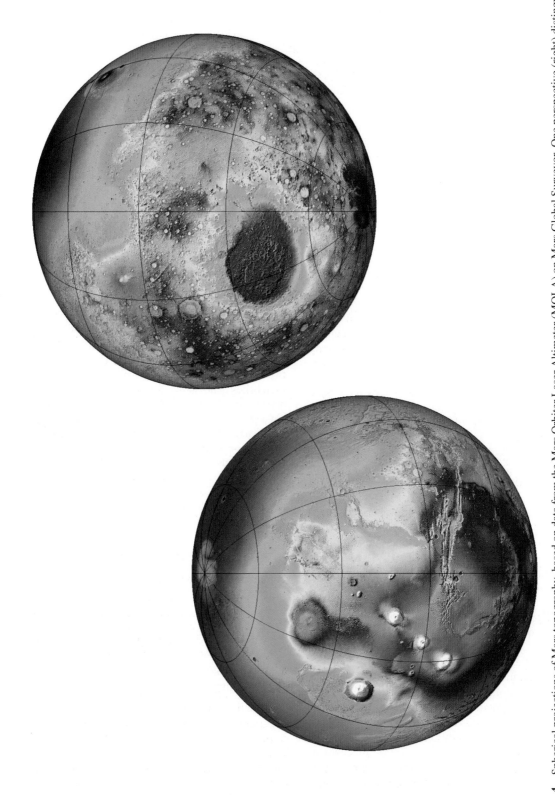

Figure 4 Spherical projections of Mars topography, based on data from the Mars Orbiter Laser Altimeter (MOLA) on Mars Global Surveyor. One perspective (right) distinguishes the heavily cratered southern highlands from the resurfaced northern lowlands. The Hellas impact basin is at the lower right edge. The other image (left) shows the Tharsis rise, containing several huge shield volcanoes.

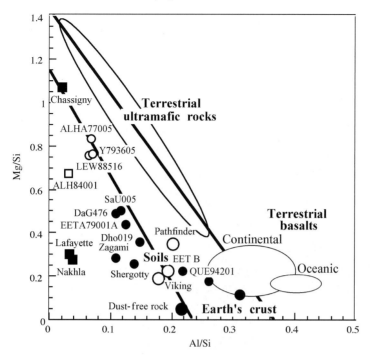

Figure 5 Mg/Si versus Al/Si weight ratios show that martian meteorites and Mars rocks and soils are depleted in Al, relative to terrestrial rocks. Filled circles are basaltic shergottites, open circles are Pherzolitic shergottites, filled squares are nakhlites and chassignites, open square is orthopyroxenite, and larger symbols are for Mars rocks and soils analyzed *in situ* (after Rieder et al., 1997; McSween, 2002).

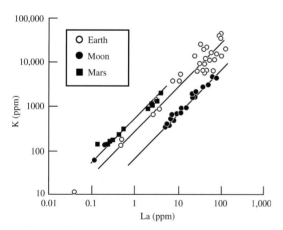

Figure 6 Volatile K versus refractory La in Mars (SNC meteorites), Earth, and Moon (after Wanke and Dreibus, 1988; Halliday et al., 2001).

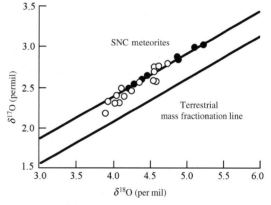

Figure 7 Oxygen isotopic compositions in martian meteorites define a distinct mass fractionation line from that of terrestrial geologic materials. Open circles are mean values of Clayton and Mayeda (1996), and filled circles are mean values of Franchi et al. (1999).

However, these meteorites are depleted in aluminum relative to terrestrial rocks, a characteristic shared by martian soils and the Mars Pathfinder dust-free rock (Figure 5). Ratios of volatile to refractory elements (e.g., K/La) are constant but higher than in terrestrial or lunar rocks (Figure 6).

Radiogenic isotopes in martian meteorites are remarkably heterogeneous. This apparently results from lack of mantle homogenization through plate tectonics and from assimilation of crustal materials (discussed below).

SNC meteorites define an oxygen isotope mass fractionation line displaced from that of terrestrial samples (Figure 7), indicating that oxygen in martian materials is isotopically distinct. In fact, oxygen isotopes are commonly used to classify a meteorite as martian. Some other stable isotopic systems in these meteorites suggest that Mars has interior reservoirs that are very different from the outgassed martian atmosphere.

1.22.3.2 Geochemical Models for the Martian Mantle

Geochemical models for the silicate portion of Mars (mantle plus crust, which is approximately equal to the mantle composition) are based on elemental and isotopic compositions of SNC meteorites. Wanke and Dreibus (1988) used SNC element correlations to estimate the composition of the martian mantle. Major and minor oxide abundances for their model are given in Table 5, and a more complete set of elemental abundances (relative to silicon and chondrite abundances) is shown in Figure 8. Most moderately volatile elements are higher in Mars than Earth. Some elements that are moderately siderophile in the Earth's mantle (manganese, chromium, tungsten, phosphorus) appear to have been more lithophile on Mars. This may relate to more volatile-rich, oxidizing and sulfur-rich conditions during martian core formation (Halliday et al., 2001). Wanke and Dreibus (1988) explained their mantle composition as having formed by accretional mixing of reduced and oxidized components, the latter containing volatiles in CI-chondrite proportions.

Two other compositional models for Mars are based on matching the oxygen isotopic composition of SNC meteorites by mixing various classes of chondrites. The models of Lodders and Fegley (1997) and Sanloup et al. (1999) combined ordinary chondrites with either carbonaceous or enstatite chondrites in the proportions required by oxygen isotopes, with the bulk elemental compositions following from mass balance (Table 5). Although all these models of mantle composition share important chemical characteristics, they differ in their high-pressure normative mineralogy (Table 5).

1.22.3.3 Geochemical Models for the Martian Core

The geochemical models of Wanke and Dreibus (1988), Lodders and Fegley (1997), and Sanloup et al. (1999) suggest that the core comprises 20.6–23.0% of the mass of Mars. All these model cores are sulfur rich, but differ significantly in core mass and sulfur abundance (Table 5). Measured siderophile element abundances in martian meteorites are consistent with equilibrium between sulfur-bearing metal and silicate at high temperature and pressure (Righter and Drake, 1996).

1.22.3.4 Consistency with Geophysical Constraints

The compressed mean densities and calculated moments of inertia for the three Mars models are compared in Table 5. The models are all in reasonable agreement with measured values, although the Sanloup et al. (1999) model has the most variance.

Bertka and Fei (1997) experimentally determined mantle mineral stabilities using the Wanke and Dreibus (1988) model composition. The mineral stability fields and resulting mantle density profile, as well as core densities and positions of the core–mantle boundary for a range of model core compositions, are illustrated in Figure 9. The moment of inertia calculated from these experimental data (0.354) is consistent with the Mars Pathfinder measurement (Bertka and Fei, 1998). However, this model requires an unrealistically thick crust.

Table 5 Composition of the Mars mantle + crust (wt.%), calculated densities, and moment of inertia.

Model[a]	WD88	LF97	S99
Mantle			
SiO_2	44.4	45.4	47.5
Al_2O_3	2.9	2.9	2.5
MgO	30.1	29.7	27.3
CaO	2.4	2.4	2.0
Na_2O	0.5	0.98	1.2
K_2O	0.04	0.11	ND
TiO_2	0.13	0.14	0.1
Cr_2O_3	0.8	0.68	0.7
MnO	0.5	0.37	0.4
FeO	17.9	17.2	17.7
P_2O_5	0.17	0.17	ND
High-pressure norm (wt.%)			
pyroxenes	37.8	42.6	63
olivine	51.9	50.9	26
garnet	8.6	4.8	11
other[b]	1.4	1.6	ND
High-pressure density (g cm^{-3})	3.52	3.50	3.46
Core			
Fe	53.1	61.5	48.4
Ni	8.0	7.7	7.2
FeS	38.9	29.0	44.4
High-pressure density (g cm^{-3})	7.04	7.27	7.02
Bulk planet			
Core (wt.%)	21.7	20.6	23.0
High-pressure density (g cm^{-3})	3.95	3.92	4.28
$C/(MR^2)$	0.367	0.367	0.361

[a] Models are WD88: Wanke and Dreibus (1988); LF97: Lodders and Fegley (1997); and S99: Sanloup et al. (1999). [b] Other normative minerals include ilmenite, chromite, and whitlockite.

1.22.4 GEOCHEMISTRY OF THE CRUST

1.22.4.1 Mafic versus Felsic Igneous Rocks

Orbital spectroscopy indicates that rocks having basaltic compositions are common

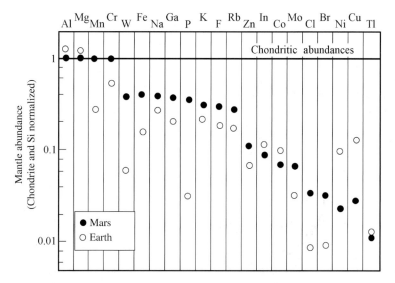

Figure 8 Comparison of elemental abundances in the mantles of Mars and Earth, based on the Mars geochemical model of Wanke and Dreibus (1988) (after Halliday et al., 2001).

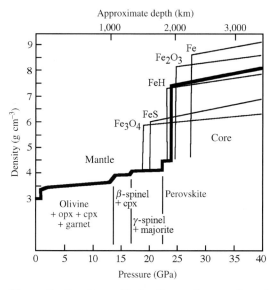

Figure 9 Density profile for the martian mantle and core, and stability fields for mantle minerals, based on experiments (Bertka and Fei, 1997) using the geochemical model of Wanke and Dreibus (1988). Positions of the core–mantle boundary for a range of model core compositions (Bertka and Fei, 1998) are also shown.

constituents of the martian surface (Mustard et al., 1997; Bandfield et al., 2000; Christensen et al., 2000a), especially in the southern highlands (Figure 10). The average Mars basalt composition (surface type-1) calculated from deconvolved TES data is given in Table 1. Thus, the discovery of rocks having andesitic composition at the Mars Pathfinder site (Rieder et al., 1997) came as a surprise. Subsequent analyses of TES spectra (Bandfield et al., 2000; Hamilton et al., 2001) suggest that the northern plains may be dominated by andesite (Figure 10). Silica and alkali contents calculated from deconvolved TES spectra for average andesite (surface type-2) are compared to the Pathfinder dust-free rock composition in Figure 11. The compositions of surface type-1 and several kinds of martian meteorites (basaltic shergottites and nakhlites) are also shown.

The formation of a hemisphere of andesite is not easily explained, especially in the absence of subduction. Fractionation of basaltic magma to produce andesitic residual melt is inefficient, although less fractionation is required if the magma is hydrous. Evidence has been presented that shergottite magmas contained water prior to eruption (McSween et al., 2001), although this hypothesis remains controversial.

Wyatt and McSween (2002) showed that TES spectra of Mars andesite and partially weathered basalt are virtually indistinguishable, and suggested that the andesite terrain might be equally well interpreted as a mixture of basaltic minerals and alteration phases (mostly clays). The region containing andesite has been interpreted as the site of an ancient ocean basin (Head et al., 1999), and the materials identified as weathered basalt are mostly bounded by topographic features previously mapped as shorelines. Therefore, the northern plains could be either andesite or weathered basalt. McSween et al. (1999) also discussed the possibility that the analyzed surfaces of Pathfinder andesitic rocks could be siliceous weathering rinds. APXS analyses of excess oxygen corresponding to ~2 wt.% H_2O in Pathfinder rocks (Foley et al., 2001) may support the idea that they have weathered rinds.

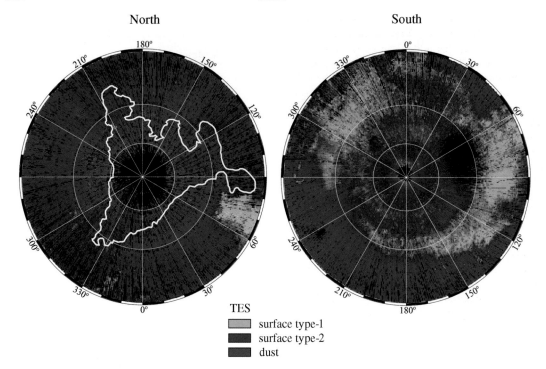

Figure 10 Polar projections showing the global distributions of materials having the composition of basalt (green), andesite (red), and dust (blue). Compositions are based on TES spectra from Mars Global Surveyor (after Wyatt and McSween, 2002).

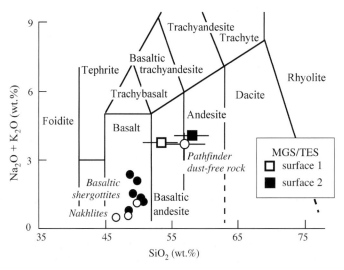

Figure 11 Chemical classification of martian volcanic rocks. Squares show basaltic materials in the southern highlands (surface 1) and andesitic materials in the northern lowlands (surface 2), derived from deconvolved TES spectra from Mars Global Surveyor (Hamilton et al., 2001). Analyzed compositions of the Mars Pathfinder dust-free rock (Wanke et al., 2001) and martian meteorites (basaltic shergottites are filled circles and nakhlites are open circles) are also shown (after McSween, 2002).

To the extent that SNC meteorites represent crustal lithologies, younger parts of the crust appear to be dominated by basalt and ultramafic cumulates from basaltic magmas. These meteorites probably are samples of volcanic centers like the Tharsis plume.

1.22.4.2 Geochemistry of Assimilated Crust in Martian Meteorites

The varying trace element and isotope geochemistry of basaltic shergottites is thought to result, in large part, from assimilation of an ancient crustal component (Jones, 1989; Longhi, 1991). Reconstructing the properties of this component provides additional geochemical data on the crust. The assimilant is characterized by enrichment of incompatible elements, reflected in high abundance of light versus heavy rare earth elements and high Rb/Sr and Nd/Sm. The latter are indicated by high initial $^{87}Sr/^{86}Sr$ and low initial $^{143}Nd/^{144}Nd$ ratios. The crust was also more oxidized than mantle-derived magmas, as revealed by Fe–Ti oxide compositions (Herd et al., 2001) and differences in the magnitude of europium anomalies in pyroxenes (Wadhwa, 2001) in meteorites that assimilated more crust. Variations in oxygen isotope data are minor (Clayton and Mayeda, 1996; Franchi et al., 1999) and uncorrelated with the degree of assimilation. Incorporation of this crustal component in varying amounts produced hybrid-shergottite magmas with varying trace element patterns, radiogenic isotopic compositions, and oxidation states. Correlations between some of these geochemical signatures are illustrated in Figure 12.

1.22.4.3 Geochemistry of Crustal Sediments

Compositions of the pervasive layered deposits seen in Mars Global Surveyor imagery (Malin and Edgett, 2000) have not been measured directly. However, TES spectra, which are dominated by sand-sized particles, indicate that igneous minerals (pyroxenes and plagioclase) are abundant on the martian surface. Quartz, which should be readily detectable in TES spectra, has not been observed. Deconvolutions of the spectra of soils and dust are dominated by framework silicates, either plagioclase or zeolites (Bandfield and Smith, 2003; McSween et al., 2003).

The geochemistry of soils analyzed at the Viking (Clark et al., 1982) and Mars Pathfinder (Wanke et al., 2001) sites is similar in many respects to basaltic shergottites. The compositional uniformity of soils at sites separated by thousands of kilometers has been explained by aeolian homogenization. Spectral identification of minerals in soils is hampered by the presence of fine-grained iron oxides, so the relative proportions of igneous versus alteration minerals remain controversial. However, the closest spectral match for Mars soils is palagonite (altered volcanic glass). Viking soils were originally interpreted as mixtures of clay minerals (Toulmin, 1977). More recent studies suggest that Mars soils are physical mixtures of basalt and andesite particles (Wanke et al., 2001; Larsen et al., 2000), or poorly crystalline materials derived from limited chemical weathering of basalt, perhaps also displaying some aeolian fractionation

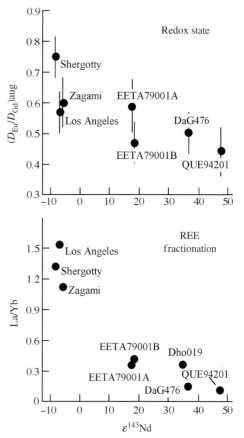

Figure 12 Chemical and isotopic correlations among shergottites, arising from assimilation of a crustal component. Decreasing values of $\delta^{143}Nd$ indicate increasing assimilation. This parameter correlates with magma redox state, indicated by size of the Eu anomaly in pyroxenes (Wadhwa, 2001), and ratio of light-to-heavy rare-earth elements (after McSween, 2002).

of heavy minerals (McLennan, 2000; McSween and Keil, 2000).

Although TES and THEMIS are sensitive to carbonates and sulfates, these minerals have not yet been detected unambiguously from orbit (Bandfield, 2002). The low carbon abundance in APXS-analyzed soils rules out much carbonate, although appreciable sulfur and chlorine are present in all soils. Thermodynamic stability considerations suggest that sulfates and iron carbonates should be present under martian conditions (Clark and Van Hart, 1981; Catling, 1999). It is unclear whether sulfate formed by reactions with acidic vapor from volcanic exhalations (Banin et al., 1997) or evaporation of surface brines (Warren, 1998; McSween and Harvey, 1998).

The existence of deposits of hematite in the Oxia Palus region, as revealed by TES (Christensen et al., 2000b), probably indicates processes requiring water. Magnetic measurements of dust by Mars landers (e.g., Madsen et al., 1999) indicate iron oxides, so the surface of Mars is highly oxidized.

Martian meteorites provide no direct information on the geochemistry of martian sediments. However, SNCs have higher Ca/Si ratios than martian soils. Rubin et al. (2000) suggested that the lower Ca/Si in soils might indicate removal of calcium into a global carbonate reservoir.

1.22.5 GEOCHEMISTRY OF VOLATILE RESERVOIRS

1.22.5.1 Planet Volatile Inventory

Compared to the Earth's mantle, the abundances of all moderately volatile elements (except manganese and chromium) are higher in Mars by a factor of 2. Wanke and Dreibus (1988) proposed that Mars is a volatile-rich planet, having inherited a high proportion of volatiles during accretion. According to this model, Mars also accreted with abundant water. However, the water reacted extensively with metallic iron to produce FeO, with concomitant loss of hydrogen, so the Mars mantle is now relatively dry despite having high abundances of other volatile elements. Based on the abundance of chlorine in martian meteorites and relative solubilities of chlorine and H_2O in basaltic magma, Wanke and Dreibus (1988) estimated that the mantle contains only 36 ppm water. Lodders and Fegley (1997) questioned this calculation, because chlorine has been sequestered in the martian crust. Their model suggests a bulk Mars chlorine abundance 8 times higher than previously modeled, with a correspondingly higher mantle water content.

The planetary CO_2 inventory has been estimated from C/N ratios of possible accreted volatile sources (chondrites or comets). Carbon dioxide estimates, expressed as equivalent global thicknesses of carbonate, vary from 3 m to 20 m (Bogard et al., 2001). Carbonates have not been detected spectroscopically on Mars as of early 2000s, although they are present in small quantities as martian weathering products in SNC meteorites. Carbon in soils at the Mars Pathfinder site is below detection limit, and the visible inventory of carbon on Mars is only 10^{-3} that of Earth or Venus (Bogard et al., 2001).

Noble gases and nitrogen in martian meteorites reveal several interior components having isotopic compositions different from those of the atmosphere. Xenon, krypton, and probably argon in the mantle components have solar isotopic compositions, rather than those measured in chondrites. However, ratios of these noble gas abundances are strongly fractionated relative to solar abundances. This decoupling of elemental and isotopic fractionation is not understood. The interior $^{15}N/^{14}N$ ratio in martian meteorites is similar to chondrites.

1.22.5.2 Outgassing and Atmospheric Loss

The isotopic compositions of atmospheric gases and trapped gases in martian meteorites provide information about planetary outgassing and atmospheric loss (Bogard et al., 2001). Isotopic fractionation in xenon has been modeled to reflect massive loss of the early atmosphere, where only xenon was retained (Pepin, 1991). Rejuvenation then occurred by outgassing or possibly by addition of a cometary veneer. Retention of a significant amount of primordial ^{36}Ar in the mantle component of martian meteorites (Bogard et al., 2001) suggests a relatively undegassed interior.

While there is compelling geomorphic evidence for water on Mars, uncertainties abound concerning the amount of outgassed water, when surface water existed, and where it went. Outgassing of water has been estimated by a variety of techniques. (This quantity is commonly expressed in terms of a globally distributed water layer of uniform depth; for comparison, the Earth has outgassed 2,700 m of water.) The present atmosphere contains only ~ 1 m of condensable water, and ice at the north pole, if melted, would amount to only ~ 10 m. Constraints from the isotopic compositions of atmospheric gases and extrapolated from the volume of eroded materials on Mars bracket the amount of outgassed water between a few tens to hundreds of meters (Jakosky and Phillips, 2001). Martian meteorites are very dry, leading to the idea that outgassing of water was limited. However, depletions of soluble light elements suggest that some shergottites contained water at depth and lost it on ascent (McSween et al., 2001), and models of the solidification of amphibole-bearing melt inclusions in martian meteorites also suggest that water has been outgassed (McSween and Harvey, 1993). Ground ice at high latitudes, as determined by Mars Odyssey GRS (Boynton et al., 2002; Feldman et al., 2002), may represent a significant reservoir, but its depth and hence abundance is unknown.

The mass ratio of atmosphere to planet is $\sim 10^{-4}$ for Earth and 5×10^{-8} for Mars. Even if degassing were less on Mars, this huge atmospheric depletion suggests that Mars has lost a substantial portion of its rejuvenated atmosphere. Enrichments in the heavy isotopes of hydrogen, nitrogen, argon, and xenon provide evidence for atmospheric loss over long timescales (Bogard et al., 2001). Escape of atmospheric gases from a planetary gravity field is often controlled by masses of the gaseous species. The very high atmospheric D/H and $^{15}N/^{14}N$ ratios on Mars are attributed to isotopic fractionation during escape, which requires that at least 90% of the original surface water and 99% of the original nitrogen have been lost. Loss of atmospheric gases is also favored by the high $^{129}Xe/^{132}Xe$ ratio of trapped gases in martian meteorites. Loss of argon has also occurred, as atmospheric ^{40}Ar represents less than 2% of that produced by radioactive decay of ^{40}K during the last four billion years.

Compared to nitrogen and noble gases, mass fractionation of carbon and oxygen isotopes during atmospheric loss has been mitigated by buffering and exchange with surface reservoirs of CO_2 and H_2O (e.g., the polar caps). Although precise atmospheric measurements of carbon and oxygen isotopes are not available, these isotopes have been measured in secondary alteration phases in martian meteorites. Carbonates are enriched in ^{13}C and ^{18}O (Figure 13), reflecting exchange with isotopically heavy atmospheric gases (Bridges et al., 2001). Based on such analyses, Jakosky (1991) estimated that 40–70% of carbon and 20–30% of oxygen has been lost to space.

1.22.5.3 Isotopic Evidence for Volatile Cycling

Significant differences exist in the isotopic compositions of hydrogen and oxygen between igneous silicate minerals and secondary alteration phases in the same martian meteorites (Karlsson et al., 1992; Leshin, 2000). This suggests that secondary phases in the lithosphere incorporated or exchanged with isotopically heavy water that was once in the atmosphere. Farquhar et al. (1998) also argued that ^{17}O enrichment of carbonates in ALH84001 resulted from mass-independent photochemical reactions of ozone in the atmosphere, which were then transferred to carbonate minerals. Another interpretation of these data is that repeated cycles of freezing and sublimation resulted in kinetic mass-dependent fractionation (Young et al., 2002).

Oxidation of sulfur-bearing gases in the atmosphere can lead to non-mass-dependent isotopic fractionation, manifested as excesses or deficiencies in ^{33}S. Farquhar et al. (2000) found small ^{33}S deficits in a nakhlite, and proposed that this was an isotopic signature of martian atmospheric chemistry. This subtle non-mass-dependent signal was not seen in other martian samples, but it bolsters the idea that outgassed volatiles are fractionated in the atmosphere and then returned to the lithosphere.

1.22.5.4 Organic Geochemistry or Biochemistry?

Viking landers failed to detect any organic compounds in martian soil (Biemann et al., 1977). Destruction of organic materials by superoxidants might make life on or near the surface improbable. Although organic compounds have been found in martian meteorites (Wright et al., 1989), they are generally terrestrial contaminants (McDonald and Bada, 1995; Jull et al., 1998).

Following the controversial hypothesis that ALH84001 provides evidence for extraterrestrial life (McKay et al., 1996), attempts have been made to discriminate between geochemical and biochemical signatures in martian meteorites. ALH84001 contains small amounts of polycyclic aromatic hydrocarbons (PAHs) that may be of martian origin, as judged from the carbon isotopic composition (Becker et al., 1999). However, PAHs commonly form abiotically, and there is nothing to link these specific PAHs to the decayed remains of organisms. Most of the PAHs in this meteorite (Becker et al., 1999), as well as other organic components such as amino acids (Bada et al., 1998), are demonstrably terrestrial.

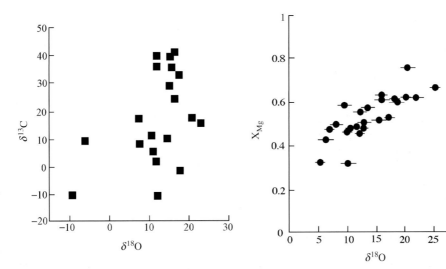

Figure 13 Isotopic composition of C (Jull et al., 1997) and O (Valley et al., 1997; Leshin et al., 1998) in carbonates of the ALH84001 martian meteorite. Heavy isotopes correlate with the Mg content of the carbonate.

Large sulfur isotope variations in terrestrial crustal rocks are commonly attributed to biological processes, because sulfate-respiring organisms preferentially utilize ^{32}S and produce isotopically light sulfides. However, similar characteristics can be caused by nonbiologic processes, e.g., fractionated sulfur isotopes in lunar soils due to micrometeorite impacts. Sulfides in ALH84001 show modest enrichment in ^{34}S (Greenwood et al., 1997), opposite to the expected trend for biogenic sulfides. Sulfides in other martian meteorites either show no fractionations or have small enrichments of ^{32}S that have been attributed to inorganic processes (Greenwood et al., 1997, 2000).

1.22.6 GEOCHEMICAL COMPARISON OF MARS AND EARTH

1.22.6.1 Differences in Bulk Planet Compositions and Major Reservoirs

Mars is more volatile rich than Earth, reflecting a higher component of accreted volatile-bearing planetesimals. It is also more highly oxidized, which produced a mantle with roughly twice the Fe^{2+} of the Earth's mantle. Oxidation reactions during accretion resulted in depletion of water but not other volatile components (Wanke and Dreibus, 1988). Core formation under more oxidized conditions affected the distribution of other elements. The martian mantle is not as depleted in most siderophile elements as the Earth's mantle, but it is substantially depleted in chalcophile elements. The mantle is also depleted in aluminum, possibly because of preferential melting of garnet or other aluminous phases during crust differentiation (Longhi et al., 1992).

It is likely that the ancient crust of Mars is more mafic than the Earth's continental crust. Pervasive andesite may signal crustal fractionation, but the identity and significance of andesitic rocks is disputed. The martian crust is relatively more voluminous than the Earth's crust, perhaps because it is not recycled. The crust is characterized by high concentrations of incompatible lithophile elements, but fractionations are not as extreme as in terrestrial continental crust, which has experienced repeated partial melting events over a protracted geologic history.

The martian hydrosphere, as represented by water of hydration in primary igneous minerals and secondary alteration minerals of SNC meteorites, has high D/H, as does the atmosphere. The compositions of atmospheric gases are significantly different from mantle gas components in martian meteorites. The martian atmosphere is enriched in ^{129}Xe relative to mantle xenon, and the atmosphere is enriched in ^{40}Ar relative to the mantle. These patterns are opposite to those observed on Earth, and the explanation is not clear. The isotopic compositions of most atmospheric species have been modified by loss of a significant portion of the atmosphere, although some species appear to have been buffered by exchange with surface reservoirs.

No discernable geochemical indication of an extant martian biosphere or unambiguous biomarkers indicating past martian life have been found as of the early 2000s.

1.22.6.2 Isotope Geochronology and Planetary Differentiation

Early fractionation of the Mars mantle at ~4.5 billion years is suggested by the whole-rock ^{87}Rb–^{87}Sr isochron (Borg et al., 1997) for martian meteorites and the close fit of ^{235}U–^{207}Pb and ^{238}U–^{206}Pb isotopic data to the geochron (Chen and Wasserburg, 1986). These data imply a remarkable absence of mantle remixing since its early differentiation.

Excess ^{142}Nd, formed by the rapid decay of now-extinct ^{146}Sm, has also been documented in SNC meteorites (Harper et al., 1995; Borg et al., 1997). This isotopic anomaly requires early differentiation of mantle and crust. Because hafnium and tungsten fractionated into silicate and metal, respectively, the short-lived ^{182}Hf–^{182}W system can be used to determine that martian core formation occurred within ~13 million years of the planet's accretion (Kleine et al., 2002; Yin et al., 2002). Correlation between ^{142}Nd and ^{182}W isotope anomalies, as well as the initial $^{187}Os/^{188}Os$ ratios for martian meteorites (Brandon et al., 2000), indicate synchronous differentiation of core, mantle, and crust (Figure 14). On Earth, core formation took substantially longer, convection has stirred the mantle sufficiently to erase any evidence of early isotopic heterogeneity, and crust formation continues throughout geologic history.

Various radiogenic isotope systems indicate that the nakhlites and Chassigny crystallized ~1.3 billion years ago (Nyquist et al., 2001). The radiometric ages for shergottites span an interval from 165 million years to 875 million years, with most samples clustering near the recent end of this range (Nyquist et al., 2001). The interpretation of the young ages remains controversial. Nyquist et al. (2001), following Jones (1986), argued that the ~175-million-year ages represent crystallization, whereas Blichert-Toft et al. (1999) favored the interpretation of Jagoutz and Wanke (1986) that the meteorites crystallized earlier and their ages were reset by fluid metasomatism.

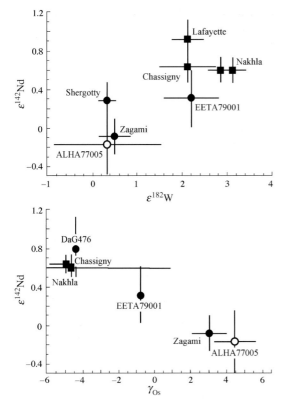

Figure 14 Correlation between Nd, W, and Os isotopic anomalies in martian meteorites. ^{142}Nd and ^{182}W are the decay products of short-lived radionuclides that partition into the crust and core, respectively. Initial ^{187}Os/^{188}Os ratios, expressed as γ_{Os} (percent deviation from chondrites at a specified time), increase by the slow decay of ^{187}Re, which should also reflect core differentiation. These correlations indicate early differentiation and chemical isolation of crust, mantle, and core, with minimal subsequent mixing of these reservoirs (after Lee and Halliday, 1997; Brandon et al., 2000).

Mittlefehldt et al. (1999) accepted the young ages as crystallization ages, but argued that a few of these meteorites solidified from impact melts. Young (<1.3 billion years) crystallization ages for most martian meteorites imply that continued mantle fractionation in their source regions (probably the Tharsis and Elysium plumes) is likely, but the geochemistry of these meteorites is not appreciably evolved.

1.22.6.3 Geochemical Cycles without Plate Tectonics

The geochemistry of martian meteorites confirms the conclusion from spacecraft imagery that crustal recycling has been minimal or absent. The mantle source regions for SNCs were differentiated early and then remained virtually unchanged until remelted to form the SNC parent magmas. Their heterogeneity in radiogenic isotopes and trace elements is much greater than in the Earth's mantle. Likewise, radiogenic isotopes in the ancient crust, as judged from assimilated crust in shergottites, suggest that it has been largely static since its formation. The pervasive tectonic recycling between crust and mantle that dominates terrestrial geochemistry is absent on Mars.

Atmospheric loss has depleted many volatile species in light isotopes, and atmospheric photochemical reactions have apparently produced non-mass-dependent fractionations in some species. These geochemical characteristics have been found in alteration minerals in martian meteorites. This implies that outgassed volatiles pass through the atmosphere and are then recycled as fluids into the lithosphere or otherwise exchange with groundwaters. It is not known how deeply volatile remixing might extend into the martian crust, but without subduction cycling into the mantle seems improbable. The details of critical geochemical cycles, such as the hydrologic (Baker, 2001) or carbon (Pollack et al., 1987; Catling, 1999) cycles, on Mars are speculative and relatively unconstrained.

1.22.6.4 Major Unsolved Problems

Among the significant but still unsolved issues related to Mars geochemistry are the following:

- Is the bulk composition of the crust basaltic, or does it contain significant quantities of silicic (andesitic) rocks?
- Have surface sediments experienced chemical weathering and, if so, what chemical changes resulted?
- What is the significance of differences in the compositions of relatively young martian meteorites and the ancient crust?
- How well do martian meteorites constrain the chemical and mineralogical composition and evolution of their mantle source regions?
- What is the global water inventory, how much was outgassed, and where did it go?
- How can the unique isotopic composition of the martian atmosphere be explained?
- Are there geochemical fingerprints of life on Mars?

These important questions will be addressed, at least in part, in future spacecraft missions. Two Mars Exploration Rovers, which are to arrive in early 2004, will carry a battery of instruments (Mössbauer, infrared, and visible spectrometers, APXS, microscopic imager, and a rock abrasion tool to remove dust and weathered coatings) that may analyze samples of the ancient crustal rocks and sediments that have been associated with water. A 2005 orbiter will carry instruments to analyze the atmosphere, and a visible near-infrared spectrometer that can map the surface at high spatial resolution. Other landers may analyze the mineralogy of surface dust and soil, and possibly make isotope measurements of the atmosphere or subsurface ices. Ultimately, these questions will be tackled by analyzing carefully selected martian samples that are returned to Earth. In the meantime, continued exploration in cold and hot deserts will likely yield new Mars meteorites. Given the apparent bias in sampling, it may be difficult to find additional pieces of the ancient crust, but ancient rocks and even sedimentary rocks should be high priorities. The arsenal of analytical instruments that can be applied in terrestrial laboratories makes any newly discovered martian meteorites of great geochemical value.

REFERENCES

Acuna M. H., Connerney J. E. P., Ness N. F., Lin R. P., Mitchell D., Carlson C. W., McFadden J., Anderson K. A., Reme H., Mazelle C., Vignes D., Wasilewski P., and Cloutier P. (1999) Global distribution of crustal magnetization discovered by the Mars Global Surveyor MAG/ER experiment. *Science* **284**, 790–793.

Bada J. L., Glavin D. P., McDonald G. D., and Becker L. (1998) A search for endogenous amino acids in martian meteorite ALH84001. *Science* **279**, 362–365.

Baker V. R. (2001) Water and the martian landscape. *Nature* **412**, 226–228.

Bandfield J. L. (2002) Global mineral distributions on Mars. *J. Geophys. Res.* **107**, 10.1029/2001JE001510.

Bandfield J. L. and Smith M. D. (2003) Multiple emission angle surface–atmosphere separations of Thermal Emission Spectrometer data. *Icarus* **161**, 47–65.

Bandfield J. L., Hamilton V. E., and Christensen P. R. (2000) A global view of martian surface compositions from MGS-TES. *Science* **287**, 1626–1630.

Banin A., Clark B. C., and Wanke H. (1992) Surface chemistry and mineralogy. In *Mars* (eds. H. H. Kieffer, B. M. Jakosky, C. W. Snyder, and M. S. Matthews). University of Arizona Press, Tucson, pp. 594–625.

Banin A., Han F. X., Kan I., and Cicelsky A. (1997) Acidic volatiles and the Mars soil. *J. Geophys. Res.* **102**, 13341–13356.

Barrat J. A., Blichert-Toft J., Nesbitt R. W., and Keller F. (2001) Bulk chemistry of Saharan shergottite Dar al Gani 476. *Meteorit. Planet. Sci.* **36**, 23–29.

Becker L., Popp B., Rust T., and Bada J. L. (1999) The origin of organic matter in the martian meteorite ALH84001. *Earth Planet. Sci. Lett.* **167**, 71–79.

Becker R. H. and Pepin R. O. (1984) The case for a martian origin of the shergottites: nitrogen and noble gases in EETA79001. *Earth Planet. Sci. Lett.* **69**, 225–242.

Bell J. F., McSween H. Y., Crisp J. A., Morris R. V., Murchie S. L., Bridges N. T., Johnson J. R., Britt D. T., Golombek M. P., Moore H. J., Ghosh A., Bishop J. L., Anderson R. C., Bruckner J., Economou T., Greenwood J. P., Gunnlaugsson H. P., Hargraves R. M., Hviid S., Knudsen J. M., Madsen M. B., Reid R., Rieder R., and Soderblom L. (2000) Mineralogic and compositional properties of martian soil and dust: results from Mars Pathfinder. *J. Geophys. Res.* **105**, 1721–1755.

Bertka C. M. and Fei Y. (1997) Mineralogy of the martian interior up to core–mantle pressures. *J. Geophys. Res.* **102**, 5251–5264.

Bertka C. M. and Fei Y. (1998) Implications of Mars Pathfinder data for the accretion history of the terrestrial planets. *Science* **281**, 1838–1840.

Biemann K., Oro J., Toulmin P., Orgel L. E., Nier A. O., Anderson D. M., Simmonds P. G., Flory D., Dinz A. V., Rushneck D. R., Biller J. E., and Lafleur A. L. (1977) The search for organic substances and inorganic volatile compounds on the surface of Mars. *J. Geophys. Res.* **82**, 4641–4658.

Blichert-Toft J., Gleason J. D., Telouk P., and Albarende F. (1999) The Lu–Hf isotope geochemistry of shergottites and the evolution of the martian mantle–crust system. *Earth Planet. Sci. Lett.* **173**, 25–39.

Bogard D. D. and Johnson P. (1983) Martian gases in an Antarctic meteorite? *Science* **221**, 651–654.

Bogard D. D., Clayton R. N., Marti K., Owen T., and Turner G. (2001) Martian volatiles: isotopic composition, origin, and evolution. In *Chronology and Evolution of Mars* (eds. R. Kallenbach, J. Geiss, and W. K. Hartmann). Kluwer Academic, Dordrecht, The Netherlands, pp. 425–458.

Borg L. E., Nyquist L. E., Taylor L. A., Wiesmann H., and Shih C.-Y. (1997) Constraints on martian differentiation processes from Rb–Sr and Sm–Nd isotopic analyses of the basaltic shergottite QUE94201. *Geochim. Cosmochim. Acta* **61**, 4915–4931.

Boynton W. V., Feldman W. C., Squyres S. W., Prettyman T. H., Bruckner J., Evans L. G., Reedy R. C., Starr R., Arnold J. R., Drake D. M., Englert P. A. J., Metzger A. E., Mitrofanov I., Trombka J. I., d'Uston C., Wanke H., Gasnault O., Hamara D. K., Janes D. M., Marcialis R. L., Maurice S., Mikheeva I., Taylor G. J., Tokar R., and Shinohara C. (2002) Distribution of hydrogen in the near surface of Mars: evidence for subsurface ice deposits. *Science* **297**, 81–85.

Brandon A. D., Walker R. J., Morgan J. W., and Goles G. G. (2000) Re–Os isotopic evidence for early differentiation of the martian mantle. *Geochim. Cosmochim. Acta* **64**, 4083–4095.

Bridges J. C., Catling D. C., Saxton J. M., Swindle T. D., Lyon I. C., and Grady M. M. (2001) Alteration assemblages in martian meteorites: implications for near-surface processes. *Space Sci. Rev.* **96**, 365–392.

Carr M. H. (1996) *Water on Mars*. Oxford University Press, New York, 229pp.

Catling D. C. (1999) A chemical model for evaporites on early Mars: possible sedimentary tracers of the early climate and implications for exploration. *J. Geophys. Res.* **104**, 16453–16469.

Chen J. H. and Wasserburg G. J. (1986) Formation ages and evolution of Shergotty and its parent planet from U–Th–Pb systematics. *Geochim. Cosmochim. Acta* **50**, 955–968.

Christensen P. R., Bandfield J. L., Smith M. D., Hamilton V. E., and Clark R. N. (2000a) Identification of a basaltic component on the martian surface from thermal emission spectrometer data. *J. Geophys. Res.* **105**, 9609–9622.

Christensen P. R., Bandfield J. L., Clark R. N., Edgett K. S., Hamilton V. E., Hoefen T., Kieffer H. H., Kuzmin R. O., Lane M. D., Malin M. C., Morris R. V., Pearl J. C., Pearson R., Roush T. L., Ruff S. W., and Smith M. D. (2000b) Detection of crystalline hematite mineralization on Mars by the thermal emission spectrometer: evidence for near-surface water. *J. Geophys. Res.* **105**, 9623–9642.

Clark B. C. and Van Hart D. C. (1981) The salts of Mars. *Icarus* **45**, 370–378.

Clark B. C., Baird A. K., Weldon R. J., Tsusaki D. M., Schabel L., and Candelaria M. P. (1982) Chemical composition of martian fines. *J. Geophys. Res.* **87**, 10059–10067.

Clayton R. N. and Mayeda T. K. (1996) Oxygen isotope studies of achondrites. *Geochim. Cosmochim. Acta* **60**, 1999–2017.

Eugster O., Weigel A., and Polnau E. (1997) Ejection times of martian meteorites. *Geochim. Cosmochim. Acta* **61**, 2749–2757.

Farquhar J., Thiemens M. H., and Jackson T. (1998) Atmosphere-surface interactions on Mars: ^{17}O measurements of carbonate from ALH84001. *Science* **280**, 1580–1582.

Farquhar J., Savarino J., Jackson T. L., and Thiemens M. H. (2000) Evidence of atmospheric sulphur in the martian regolith from sulphur isotopes in meteorites. *Nature* **404**, 50–52.

Feldman W. C., Boynton W. V., Rokar R. L., Prettyman T. H., Gasnault O., Squyres S. W., Elphic R. C., Lawrence D. J., Lawson S. L., Maurice S., McKinney G. W., Moore K. R., and Reedy R. C. (2002) Global distribution of neutrons from Mars: results from Mars Odyssey. *Science* **297**, 75–78.

Folco L., Franchi I. A., DíOrazio M., Rocchi S., and Schultz L. (2000) A new martian meteorite from the Sahara: the shergottite Dar al Gani 489. *Meteorit. Planet. Sci.* **35**, 827–839.

Foley C. N., Economou T., Dietrich W., and Clayton R. N. (2000) Chemical composition of martian soil and rocks: comparison of the results from the alpha, proton, and x-ray modes of the Mars Pathfinder Alpha-Proton-X-ray spectrometer (abstr.). *Meteorit. Planet. Sci.* **35** (suppl.), A55–A56.

Foley C. N., Economou T., and Clayton R. N. (2001) Chemistry of Mars Pathfinder samples determined by the APXS. *Lunar Planet. Sci.* **XXXII**, 1979. The Lunar and Planetary Institute, Houston (CD-ROM).

Folkner W. M., Yoder C. F., Yuan D. N., Standish E. M., and Preston R. A. (1997) Interior structure and seasonal mass redistribution of Mars from radio tracking of Mars Pathfinder. *Science* **278**, 1749–1752.

Franchi I. A., Wright I. P., Sexton A. S., and Pillinger C. T. (1999) The oxygen-isotopic composition of Earth and Mars. *Meteorit. Planet. Sci.* **34**, 657–661.

Greenwood J. P., Riciputi L. R., and McSween H. Y. (1997) Sulfide isotopic compositions in shergottites and ALH84001, and possible implications for life on Mars. *Geochim. Cosmochim. Acta* **61**, 4449–4453.

Greenwood J. P., Riciputi L. R., McSween H. Y., and Taylor L. A. (2000) Modified sulfur isotopic compositions of sulfides in the nakhlites and Chassigny. *Geochim. Cosmochim. Acta* **64**, 1121–1131.

Halliday A. N., Wanke H., Birck J.-L., and Clayton R. N. (2001) Accretion, composition and early differentiation of Mars. *Space Sci. Rev.* **96**, 197–230.

Hamilton V. E., Wyatt M. B., McSween H. Y., and Christensen P. R. (2001) Analysis of terrestrial and martian volcanic compositions using thermal emission spectroscopy: 2. Application to martian surface spectra from the Mars global surveyor thermal emission spectrometer. *J. Geophys. Res.* **106**, 14733–14746.

Harper C. L., Jr., Nyquist L. E., Bansal B. M., Wiesmann H., and Shih C.-Y. (1995) Rapid accretion and early differentiation of Mars indicated by $^{142}Nd/^{144}Nd$ in SNC meteorites. *Science* **267**, 213–216.

Hartmann W. K. and Neukum G. (2001) Cratering chronology and the evolution of Mars. *Space Sci. Rev.* **96**, 165–194.

Head J. W., Hiesinger H., Ivonov M. A., Kreslavsky M. A., Pratt S., and Thompson B. J. (1999) Possible ancient oceans on Mars: evidence from Mars orbiter laser altimeter data. *Science* **286**, 2134–2137.

Herd C. D. K., Papike J. J., and Brearley A. J. (2001) Oxygen fugacity of martian basalts from electron microprobe oxygen and TEM-EELS analyses of Fe–Ti oxides. *Am. Mineral.* **86**, 1015–1024.

Jagoutz E. and Wanke H. (1986) Sr and Nd isotopic systematics of Shergotty meteorite. *Geochim. Cosmochim. Acta* **50**, 939–953.

Jakosky B. M. (1991) Mars volatile evolution: evidence from stable isotopes. *Icarus* **94**, 14.

Jakosky B. M. and Phillips R. J. (2001) Mars' volatile and climate history. *Nature* **412**, 237–244.

Jones J. H. (1986) A discussion of isotopic systematics and mineral zoning in the shergottites: evidence for a 180 m.y. igneous crystallization age. *Geochim. Cosmochim. Acta* **50**, 969–977.

Jones J. H. (1989) Isotopic relationships among the Shergottites, the Nakhlites and Chassigny. *Proc. Lunar Planet. Sci. Conf.* **19**, 465–474.

Jull A. J. T., Eastoe C. J., and Cloudt S. (1997) Isotopic composition of carbonates in the SNC meteorites, Allan Hills 84001 and Zagami. *J. Geophys. Res.* **102**, 1663–1669.

Jull A. J. T., Coutney C., Jeffrey D. A., and Beck J. W. (1998) Isotopic evidence for a terrestrial source of organic compounds found in martian meteorites Allan Hills 84001 and Elephant Moraine 79001. *Science* **279**, 366–369.

Kallenbach R., Geiss J., and Hartmann W. K. (eds.) (2001) *Chronology and Evolution of Mars*. Kluwer Academic, Dordrecht, 498pp.

Karlsson H. R., Clayton R. N., Gibson E. K., and Mayeda T. K. (1992) Water in SNC meteorites: abundances and isotopic compositions in bulk samples. *Science* **255**, 1409–1411.

Kieffer H. H., Jakosky B. M., Snyder C. W., and Matthews M. S. (eds.) (1992) *Mars*. University of Arizona Press, Tucson, 1498pp.

Kleine T., Munker C., Mezger K., and Palmer H. (2002) Rapid accretion and early core formation on asteroids and the terrestrial planets from Hf–W chronometry. *Nature* **418**, 952–955.

Larsen K. W., Arvidson R. E., Jolliff B. L., and Clark B. C. (2000) Correspondence and least squares analyses of soil and rock compositions for the Viking Lander 1 and Pathfinder landing sites. *J. Geophys. Res.* **105**, 29207–29221.

Lee D.-C. and Halliday A. N. (1997) Core formation on Mars and differentiated asteroids. *Nature* **388**, 854–857.

Leshin L. A. (2000) Insights into martian water reservoirs from analysis of martian meteorite QUE94201. *Geophys. Res. Lett.* **27**, 2017–2020.

Leshin L. A., Epstein S., and Stolper E. M. (1996) Hydrogen isotope geochemistry of SNC meteorites. *Geochim. Cosmochim. Acta* **60**, 2635–2650.

Leshin L. A., McKeegan K. D., Carpenter P. K., and Harvey R. P. (1998) Oxygen isotopic constraints on the genesis of carbonates from martian meteorite ALH84001. *Geochim. Cosmochim. Acta* **62**, 3–13.

Lodders K. (1998) A survey of shergottite, nakhlite and Chassigny meteorites whole-rock compositions. *Meteorit. Planet. Sci.* **33** (suppl.), A183–A190.

Lodders K. and Fegley B., Jr. (1997) An oxygen isotope model for the composition of Mars. *Icarus* **126**, 373–394.

Lodders K. and Fegley B., Jr. (1998) *The Planetary Scientist's Companion*. Oxford University Press, New York.

Longhi J. (1991) Complex magmatic processes on Mars: inferences from the SNC meteorites. *Proc. Lunar Planet. Sci. Conf.* **21**, 695–709.

Longhi J., Knittle E., Holloway J. R., and Wanke H. (1992) The bulk composition, mineralogy and internal structure of Mars. In *Mars* (eds. H. H. Kieffer, B. M. Jakosky, C. W. Snyder, and M. S. Matthews). University of Arizona Press, Tucson, pp. 184–208.

Madsen M. B., Hviid S. R., Gunnlaugsson H. P., Goetz W., Pedersen C. T., Dinesen A. R., Morgensen C. T., Olsen M., Hargraves R. B., and Knudsen J. M. (1999) The magnetic properties experiments on Mars Pathfinder. *J. Geophys. Res.* **104**, 8761–8780.

Malin M. C. and Edgett K. S. (2000) Sedimentary rocks of early Mars. *Science* **290**, 927–937.

Masson P., Carr M. H., Costard F., Greeley R., Hauber E., and Jaumann R. (2001) Geomorphic evidence for liquid water. *Space Sci. Rev.* **96**, 333–364.

Mathew K. J., Kim J. S., and Marti K. (1998) Martian atmospheric and indigenous components of xenon and nitrogen in the Shergotty, Nakhla, and Chassigny group meteorites. *Meteorit. Planet. Sci.* **33**, 655–664.

McDonald G. D. and Bada J. L. (1995) A search for endogenous amino acids in the martian meteorite EETA79001. *Geochim. Cosmochim. Acta* **59**, 1179–1184.

McKay D. S., Gibson E. K., Thomas-Keprta K. L., Vali H., Romanek C. S., Clemett S. J., Chillier X. D. F., Maechling C. R., and Zare R. N. (1996) Search for past life on Mars: possible relic biogenic activity in martian meteorite ALH84001. *Science* **273**, 924–930.

McKenzie D., Barnett D. N., and Yuan D. N. (2002) The relationship between martian gravity and topography. *Earth Planet. Sci. Lett.* **195**, 1–16.

McLennan S. M. (2000) Chemical composition of martian soil and rocks: complex mixing and sedimentary transport. *Geophys. Res. Lett.* **27**, 1335–1338.

McSween H. Y. (1985) SNC meteorites: clues to martian petrologic evolution? *Rev. Geophys.* **23**, 391–416.

McSween H. Y. (1994) What we have learned about Mars from SNC meteorites. *Meteoritics* **29**, 757–779.

McSween H. Y. (2002) The rocks of Mars, from far and near. *Meteorit. Planet. Sci.* **37**, 7–25.

McSween H. Y. and Harvey R. P. (1993) Outgassed water on Mars: constraints from melt inclusions in SNC meteorites. *Science* **259**, 1890–1892.

McSween H. Y. and Harvey R. P. (1998) An evaporation model for formation of carbonates in the ALH84001 martian meteorite. *Int. Geol. Rev.* **9**, 840–853.

McSween H. Y. and Keil K. (2000) Mixing relationships in the martian regolith and the composition of globally homogeneous dust. *Geochim. Cosmochim. Acta* **64**, 2155–2166.

McSween H. Y. and Stolper E. M. (1979) Basaltic meteorites. *Sci. Am.* **242**, 54–63.

McSween H. Y. and Treiman A. H. (1998) Martian meteorites. In *Planetary Materials* (ed. J. J. Papike). Min. Soc. Am., Washington, DC, pp. 6–1 to 6–53.

McSween H. Y., Murchie S. L., Crisp J. A., Bridges N. T., Anderson T., Ghosh A., Golombek M. P., Greenwood J. P., Johnson J. R., Moore H. J., Morris R. V., Parker T. J., Rieder R., Singer R., and Wanke H. (1999) Chemical, multispectral, and textural constraints on the composition and origin of rocks at the Mars Pathfinder landing site. *J. Geophys. Res.* **104**, 8679–8715.

McSween H. Y., Grove T. L., Lentz R. C. F., Dann J. C., Holzheid A. H., Riciputi L. R., and Ryan J. G. (2001) Geochemical evidence for magmatic water within Mars from pyroxenes in the Shergotty meteorite. *Nature* **409**, 487–490.

McSween H. Y., Hamilton V. E., and Hapke B. W. (2003) Mineralogy of martian atmospheric dust inferred from spectral deconvolution of MGS TES and Mariner 9 IRIS data. *Lunar Planet Sci.* **XXXIV**, 1233. The Lunar and Planetary Institute, Houston (CD-ROM).

Mittlefehldt D. W., Lindstrom D. J., Lindstrom M. M., and Martinez R. R. (1999) An impact-melt origin for lithology A of martian meteorite Elephant Moraine A79001. *Meteorit. Planet. Sci.* **34**, 357–368.

Mustard J. F., Murchie S., Erard S., and Sunshine J. M. (1997) *In situ* compositions of martian volcanics: implications for the mantle. *J. Geophys. Res.* **102**, 25605–25615.

Nier A. and McElroy M. B. (1977) Composition and structure of Mars' upper atmosphere: results from the neutral mass spectrometers at Viking 1 and 2. *J. Geophys. Res.* **82**, 4341–4350.

Nyquist L. E., Bogard D. D., Shih C.-Y., Greshake A., Stöffler D., and Eugster E. (2001) Ages and geologic histories of martian meteorites. *Space Sci. Rev.* **96**, 105–164.

Owen T. (1992) The composition and early history of the atmosphere of Mars. In *Mars* (eds. H. H. Kieffer, B. M. Jakosky, C. W. Snyder, and M. S. Matthews). University of Arizona Press, Tucson, pp. 818–834.

Owen T., Biemann K., Rushneck D. R., Biller J. E., Howarth D. W., and Lafleur A. L. (1977) The composition of the atmosphere at the surface of Mars. *J. Geophys. Res.* **82**, 4635–4639.

Pepin R. O. (1991) On the origin and early evolution of terrestrial planet atmospheres and meteoritic volatiles. *Icarus* **92**, 2–79.

Pollack J. B., Kasting J. F., Richardson S. M., and Poliakoff K. (1987) The case for a wet, warm climate on early Mars. *Icarus* **71**, 203–224.

Rieder R., Economou T., Wanke H., Turkevich H., Crisp J., Bruckner J., Dreibus G., and McSween H. Y. (1997) The chemical composition of martian soil and rocks returned by the mobile Alpha Proton X-ray spectrometer: preliminary results from the x-ray mode. *Science* **278**, 1771–1774.

Righter K. and Drake M. J. (1996) Core formation in Earth's Moon, Mars, and Vesta. *Icarus* **124**, 513–529.

Rubin A. E., Warren P. H., Greenwood J. P., Verish R. S., Leshin L. A., Hervig R. L., Clayton R. N., and Mayeda T. K. (2000) Los Angeles: the most differentiated basaltic martian meteorite. *Geology* **28**, 1011–1014.

Sanloup C., Jambon A., and Gillet P. (1999) A simple chondritic model of Mars. *Earth Planet. Sci. Lett.* **112**, 43–54.

Surkov Y. A., Barsukov V. L., Moskaleva L. P., Kharyukova V. P., Zaitseva S. Y., Smirnov G. G., and Manvelyan O. S. (1989) Determination of the elemental composition of martian rocks from Phobos 2. *Nature* **341**, 595–598.

Taylor G. J., Boynton W., Hamara D., Kerry K., Janes D., Keller J., Feldman W., Prettyman T., Reedy R., Bruckner J., Wanke H., Evans L., Starr R., Squyres S., Karunatillake S., Gasnault O., and Odyssey GRS Team (2003) Evolution of the martian crust: evidence from preliminary potassium and thorium measurements by Mars Odyssey gamma-ray spectrometer. *Lunar Planet Sci.* **XXXIV**, 2004. The Lunar and Planetary Institute, Houston (CD-ROM).

Toulmin P., III, Baird A. K., Clark B. C., Keil K., Rose H., Christian R. P., Evans P. H., and Kelliher W. C. (1977) Geochemical and mineralogical interpretations of the Viking inorganic chemical results. *J. Geophys. Res.* **82**, 4625–4634.

Treiman A. H., Jones J. H., and Drake M. J. (1987) Core formation in the shergottite parent body and comparison with the Earth. *J. Geophys. Res.* **92**, E627–E632.

Treiman A. H., Gleason J. D., and Bogard D. D. (2000) The SNC meteorites are from Mars. *Planet. Space Sci.* **48**, 1213–1230.

Trombka J. I., Evans L. G., Starr R., Floyd S. R., Squyres S. W., Whelan J. T., Bamford G. J., Coldwell R. L., Rester A. C., Surkov Y. A., Moskaleva L. P., Kharyukova V. P., Manvelyan O. S., Zaitseva S. Y., and Smirnov G. G. (1992) Analysis of Phobos mission gamma ray spectra from Mars. *Proc. Lunar Planet. Sci. Conf.* **22**, 23–29. The Lunar and Planetary Institute, Houston.

Valley J. W., Eiler J. M., Graham C. M., Gibson E. K., Romanek C. S., and Stolper E. M. (1997) Low-temperature carbonate concretions in the martian meteorite ALH84001: evidence from stable isotopes and mineralogy. *Science* **275**, 1633–1638.

Wadhwa M. (2001) Redox state of Mars' upper mantle and crust from Eu anomalies in shergottite pyroxenes. *Science* **291**, 1527–1530.

Walker D., Stolper E. M., and Hays J. F. (1979) Basaltic volcanism: the importance of planet size. *Proc. Lunar Planet. Sci. Conf.* **10**, 1995–2015.

Wanke H. and Dreibus G. (1988) Chemical composition and accretion history of terrestrial planets. *Phil. Trans. Roy. Soc. London* **A325**, 545–557.

Wanke H., Bruckner J., Dreibus G., Rieder R., and Ryabchikov I. (2001) Chemical composition of rocks and soils at the Pathfinder site. *Space Sci. Rev.* **96**, 317–330.

Warren P. H. (1998) Petrologic evidence for low-temperature, possibly flood evaporitic origin of carbonates in the ALH84001 meteorite. *J. Geophys. Res.* **103**, 16759–16773.

Warren P. H. and Kallemeyn G. W. (1997) Yamato-793605, EET79001, and other presumed martian meteorites: compositional clues to their origins. *Proc. NIPR Symp. Antarctic Met.* **10**, 61–81.

Wasson J. T. and Wetherill G. W. (1979) Dynamical, chemical, and isotopic evidence regarding the formation location of asteroids and meteorites. In *Asteroids* (ed. T. Gehrels). University of Arizona Press, Tucson, pp. 926–974.

Wright I. P., Grady M. M., and Pillinger C. T. (1989) Organic materials in a martian meteorite. *Nature* **340**, 220–222.

Wyatt M. B. and McSween H. Y. (2002) Spectral evidence for weathered basalt as an alternative to andesite in the northern lowlands of Mars. *Nature* **417**, 263–266.

Wyatt M. B., Hamilton V. E., McSween H. Y., Christensen P. R., and Taylor L. R. (2001) Analysis of terrestrial and martian volcanic compositions using thermal emission spectroscopy: 1. Determination of mineralogy, chemistry, and classification strategies. *J. Geophys. Res.* **106**, 14711–14732.

Yen A. S., Murray B. C., and Rossman G. R. (1998) Water content of the martian soil: laboratory simulations of reflectance spectra. *J. Geophys. Res.* **103**, 11125–11133.

Yin Q., Jcobsen S. B., Yamashita K., Blichert-Toft J., Telouk P., and Albarede F. (2002) A short timescale for terrestrial planet formation from Hf–W chronometry of meteorites. *Nature* **418**, 949–952.

Young E. D., Galy A., and Nagahara H. (2002) Kinetic and equilibrium mass-dependent isotope fractionation laws in nature and their geochemical and cosmochemical consequences. *Geochim. Cosmochim. Acta* **66**, 1095–1104.

Zahnle K. (2001) Decline and fall of the martian empire. *Nature* **412**, 209–213.

Zuber M. T. (2001) The crust and mantle of Mars. *Nature* **412**, 220–227.

1.23
Giant Planets

J. I. Lunine

The University of Arizona, Tucson, AZ, USA

1.23.1	THE GIANT PLANETS IN RELATION TO THE SOLAR SYSTEM	623
1.23.1.1	Basic Physical and Orbital Parameters	623
1.23.1.2	Discovery and Historical Investigation of the Giant Planets	624
1.23.2	ESSENTIAL DETERMINANTS OF THE PHYSICAL PROPERTIES OF THE GIANT PLANETS	626
1.23.2.1	How We Know the Giant Planets Contain Hydrogen and Helium	626
1.23.2.2	The Equation of State of Hydrogen and Helium as a Determinant of the Structure	627
1.23.2.3	The Thermal Infrared Emission of the Giant Planets and Implications for Evolution	629
1.23.2.4	The Interior Structure of the Giant Planets	630
1.23.2.5	Elemental and Isotopic Abundances	630
1.23.2.6	Atmospheric Dynamics and Magnetic Fields	631
1.23.3	ORIGIN AND EVOLUTION OF THE GIANT PLANETS	633
1.23.3.1	Basic Model for the Formation of the Planets from a Disk of Gas and Dust	633
1.23.3.2	Constraints from the Composition of the Giant Planets	633
1.23.4	EXTRASOLAR GIANT PLANETS	634
1.23.5	MAJOR UNSOLVED PROBLEMS AND FUTURE PROGRESS	634
REFERENCES		635

1.23.1 THE GIANT PLANETS IN RELATION TO THE SOLAR SYSTEM

1.23.1.1 Basic Physical and Orbital Parameters

Beyond the inner solar system's terrestrial planets, with their compact orbits and rock–metal compositions, lies the realm of the outer solar system and the giant planets. Here the distance between planets jumps by an order of magnitude relative to the spacing of the terrestrial planets, and the masses of the giants are one to two orders of magnitude greater than Venus and Earth—the largest terrestrial bodies. Composition changes as well, since the giant planets are largely gaseous, with inferred admixtures of ice, rock, and metal, while the terrestrial planets are essentially pure rock and metal. The giant planets have many more moons than do the terrestrial planets, at last count 57 versus 3 for the terrestrial planets, and the range of magnetic field strengths is larger in the outer solar system as well. It is the giant planets that sport rings, ranging from the magnificent ones around Saturn to the variable ring arcs of Neptune. Were it not for the fact that only Earth supports abundant life (with life possibly existing, but not proved to exist, in the martian crust and liquid water regions underneath the ice of Jupiter's moon Europa), the terrestrial planets would pale in interest next to the giant planets for any extraterrestrial visitor.

Modern telescopic and spacecraft study of Jupiter, Saturn, Uranus, and Neptune, their properties, and their systems of rings, moons, and magnetospheres, has been the purview of the planetary scientist with little connection to the universe beyond until 1995, when the first extrasolar giant planet was discovered. Now the solar system's giants are the best-studied example of a class of some 100 objects which—while only one has been measured for size and hence density—may be present ~10% of Sun-like stars.

The basic physical and orbital parameters of the solar system's giant planets (Cox, 2000) are summarized in Table 1. Included as well in the table are data on the one giant planet beyond

Table 1 Basic physical parameters of measured giant planets.

Object	Orbital radius[a]	Period (yr)	Eccentricity of orbit	Inclination (deg)[b]	Planetary radius[c]	Planetary mass[d]	Magn. field[e]	Number of satellites
Jupiter	5.20	11.9	0.048	1.3	11.2	318	20,000	16
Saturn	9.54	29.4	0.054	2.5	9.45	95.2	590	18
Uranus	19.2	83.7	0.047	0.77	4.01	14.4	50	17
Neptune	30.1	164	0.009	1.77	3.88	17.1	28	8
HD209458b[f]	0.045	0.0096	0.0	?	16 ± 1	219 ± 6	?	?

Sources: Cox (2000) and Cody and Sasselov (2002).
[a] Semimajor axis, in AU, where 1 AU = Earth–Sun mean distance = 1.496×10^{13} cm. [b] Inclination of orbit relative to an invariable plane.
[c] In units of Earth radii, where 1 Earth radius = 6.378×10^8 cm at the equator. [d] In Earth masses, where 1 Earth mass = 5.974×10^{27} g.
[e] In units of the Earth's dipole field, where the centered magnetic dipole field of the Earth is 3×10^4 nT. [f] Orbital parameters refer to Charbonneau et al. (2000).

our solar system, the companion to the star HD209458, for which size and physical mass information exist in addition to orbital data (Cody and Sasselov, 2002). Notable about this object is its proximity to its parent star—which is much like the Sun—relative to the giant planets of our own solar system. This proximity is responsible for another interesting feature about the extrasolar giant planet, its low density compared to Jupiter and Saturn. Despite the low density, the basic properties of this object are fully consistent with its being primarily a hydrogen–helium planet, like Jupiter and Saturn (and, to a lesser extent, Uranus and Neptune). Hence, we must consider it very much of a class with our own system's giant planets—affording the first indication that such objects are potentially a common phenomenon in the cosmos.

In the context of Jovian-mass bodies being detected around other stars, the question often arises as to what might be considered the definition of a planets, and, in particular, where is the crossover between planets and "failed stars," or brown dwarfs, that do not undergo significant fusion reactions. There is a convenient, if approximate, hierarchy that can be assigned to self-gravitating objects from the most massive stars to Earth-sized planets, corresponding roughly to a progression of powers of 10 in mass. The mass of the Sun, 1.99×10^{33} g, and that of Jupiter, a bit more than three orders of magnitude smaller, are useful units of currency. Stars 10 times more massive than the Sun can be considered *high-mass* stars, with geologically and astronomically short lifetimes, $<10^7$ yr, defined by the "main-sequence" phase of stable hydrogen fusion (Kippenhahn and Weigert, 1991). Below 0.1 (strictly, 0.08) solar masses, the interiors of stars are not sufficiently hot to undergo self-sustained hydrogen fusion, and hence do not possess a stable main sequence phase of hydrogen burning. These so-called "brown dwarfs" evolve downward in luminosity and surface (photospheric) temperature over time.

Below ~0.01 (strictly 0.013) solar masses, interior temperatures are insufficient even for the fusion of deuterium to occur. This has been proposed as a convenient boundary between planets and brown dwarfs, with the virtue that it is unambiguously based on the cessation of a particular physical process (Hubbard, 1989). For those who prefer a distinction based on formation processes, motivated by the view that brown dwarfs form from direct gaseous collapse like stars, whereas planets form through accretion of solids and gas in a disk, the threshold may not be too different. There is some suggestion from theory that bodies larger than 10 Jupiter masses (0.01 solar masses) may form preferentially by direct collapse, though the threshold is likely to be uncertain by several factors, and significant overlap in masses generated by the two processes is likely (Bodenheimer et al., 2000). There may even be a weak preference for making bodies of the order of a Jupiter mass (0.001 solar masses) based on the current planet detection statistics (Mayor et al., 1998).

At 0.1 Jupiter masses, if Uranus and Neptune are any guide, giant planets are so noncosmic in composition (with a strong enrichment in elements heavier than hydrogen and helium) that a separate class of "ice giants" has been proposed. Whether 0.1 Jupiter mass objects that are more cosmic in composition exist around other stars is not known, but such objects could be difficult to generate without adding large amounts of rocky and icy elements. Finally, at 0.01 Jupiter masses (a few Earth masses), we enter the realm of the terrestrial planets—in essence, the rocky component of the giant planets bereft of icy and gaseous materials. Addition of volatiles to these bodies to form hydrospheres and organic carbon reservoirs is likely to vary significantly from one system to another, so that the possibility of an Earth-sized planet with the volatile complement of the Moon cannot be ruled out (Lunine, 2001).

1.23.1.2 Discovery and Historical Investigation of the Giant Planets

Jupiter and Saturn are naked-eye objects and were known to the ancients; Uranus is barely detectable by the unaided eye but has undoubtedly been seen by numerous keen-eyed individuals in the times before modern lighting systems dimmed the night sky over much of the settled Earth. The classical Greeks, one among many civilizations, that noticed the regular motions of a handful of the points of light in the sky, described these moving objects as planets, meaning wandering, and hence their modern name planet. The association of specific planets with particular deities was a practice established by non-Greeks and non-Romans, according to Plato. For example, it was in Babylon that one of the brightest planets was associated with Marduk, king of the gods, and by this was the planet eventually associated with Jupiter, the equivalent in the Roman pantheon (Krupp, 1983). The discovery of Uranus (so named to continue the classical tradition) is credited to W. Herschel, who first detected it with a telescope in 1781 and thought it a new comet. Further observations of its brightness and motion determined that it is a planet with an orbit almost twice the semimajor axis of that of Saturn. Tracking of Uranus over several decades suggested that the planet's orbital motion deviated from that prescribed by Kepler's laws, so that the position of a yet more distant planet was predicted independently by J. C. Adams in England and U. J. J. Le Verrier in France. G. Galle and H. L. D'Arrest found the planet, named Neptune, the evening they received the prediction, in 1846. (In fact, Galileo was the first to see Neptune, recording it as a star shifting in the field of view of his 1612 observations of Jupiter, but he did not recognize it as a previously unknown planet (Kowal and Drake, 1980). Apparent deviations from Keplerian motion in Neptune's orbit, now regarded largely as spurious, motivated the search for another giant, a search which came up empty handed but led to the twentieth-century discovery of Pluto and then the Kuiper Belt just beyond the realm of the giant planets. As described more thoroughly in Chapter 1.17, ~100 extrasolar giant planets have been indirectly detected between 1995 and 2002 via the gravitational pull on their parent stars. Only one of these, HD209458b, has been directly seen transiting in front of its parent star, an observation first made in 1999 (Charbonneau et al., 2000). Unlike Uranus, for which the evocative Olympian name soon replaced Herschel's initial proposal of "[King] George's star," no plans exist to do likewise for the only directly known extrasolar giant.

The discovery of the rings and moons of the giant planets constitutes a much more complex history beginning with Galileo's 1610 discovery of the four large moons of Jupiter and his glimpse of the rings of Saturn. Galileo himself did not understand the annular nature of the rings, thinking Jupiter might be oblate; C. Huygens first discerned this in 1659, and J. D. Cassini discovered the first major division, or gap, in the rings in 1676. As telescopes progressively increased in aperture and became optically more precise, more moons were progressively discovered, a process that continues today. The Pioneer and Voyager flyby spacecraft discovered a handful of the Jovian and Saturnian satellites, and the majority of the known satellites around Uranus and Neptune. The structure of Saturnian ring system was gradually revealed as well, while those of Uranus and Neptune discovered by stellar occultations observed from Earth-based aircraft and ground telescopes. The detailed structures of the three ring systems, and the discovery of the tenuous Jovian ring, were not discerned until the Voyager flybys of the four giant planets in the decade from 1979 to 1989, supplemented by studies in the 1990s using the Hubble Space Telescope (see Chapter 1.24 for a detailed treatment of the satellites of the outer planets).

The history of atmospheric and interior studies of the giant planets is likewise complex. The first atmospheric compounds to be detected in Jupiter were methane and ammonia, detected by telescopic spectroscopy (Wildt, 1932). Measurement of the overall density of Jupiter and Saturn essentially required that the bulk constituent of the interior be hydrogen, but this was not detected in Jupiter until 1957 (Kiess et al., 1960), and later in the other giant planets. The first reasonably accurate quantitative modeling of the composition of Jupiter based on the behavior of hydrogen and helium at high pressure was that of Demarcus (1958). This work established that Jupiter and Saturn are roughly of solar composition, an intriguing fact that was left unexplained by the star-formation models of the day (and, in fact, elicited little scientific interest). The discovery of an excess luminosity in the Jupiter and Saturn through pioneering thermal infrared work (Low, 1966) indicated that the two objects probably formed through some sort of collapse process. It would not be until the mid-1970s that serious work on the formation of these objects proceeded—along with ever more detailed compositional measurements and interior models. Much of what was learned about Jupiter and Saturn from ground-based studies required—for Uranus and Neptune—spacecraft observation by Voyager and then the European Space Agency Infrared Space Observatory (Atreya et al., 1999a). In 1995, Jupiter's atmosphere was studied *in situ* by the mass spectrometer aboard the Galileo probe, which transmitted down to a pressure level exceeding 20 bar (Niemann et al., 1996).

Information regarding the interiors of the giant planets from the presence of magnetic fields became available beginning in 1950 with the detection of radio-emissions from Jupiter at frequencies ranging from 10 MHz to 1,000 MHz. The highest-frequency radiation was interpreted, correctly, as synchrotron emission from electrons forced to gyrate along magnetic field lines generated by Jupiter. From the spatial pattern of the emission, observed from the Earth, the strength and dipolar nature of Jupiter's magnetic field were revealed (Dessler, 1983). Because of their greater distance from the Earth and smaller magnitudes, the magnetic fields of the three other giant planets awaited discovery by planetary spacecraft, specifically Pioneer 11 at Saturn in 1979, and Voyager 2 at Uranus and Neptune in 1986 and 1989, respectively. More directly tied to the modeling of the interior of the giant planets are data on the gravitational moments of the giant planets, derived from measuring the path of flyby spacecraft and optical tracking of the natural satellites. Also germane to interior models is the determination of the value of the internal heat emitted from the giant planets over and above that gained by the absorption of incident sunlight. While Earth-based studies were able to determine approximate values, it required the Voyager flybys to refine both the infrared emission and the angular dependence of the scattering of sunlight by the clouds to most accurately determine the energy balance on each of the giant planets (Pearl and Conrath, 1991).

1.23.2 ESSENTIAL DETERMINANTS OF THE PHYSICAL PROPERTIES OF THE GIANT PLANETS

1.23.2.1 How We Know the Giant Planets Contain Hydrogen and Helium

Hydrogen, the most abundant element in the cosmos, is difficult to measure spectroscopically in its molecular form, because it is a symmetric molecule with permitted features in the rotational infrared part of the spectrum only. Pressure-induced, also referred to as collision-induced, lines occur in the visible and near-infrared part of the spectrum for dense gases, in which frequent collisions among hydrogen molecules distort the symmetric electronic structures leading to transient electric moments. The fundamental rotation–vibration pressure-induced band for H_2 is at 2.4 μm, with overtones extending into the visible. Sharp quadrupole transitions of molecular hydrogen seen at 0.85 μm and 0.65 μm wavelengths, as well as the more diffuse dipolar features, establish that hydrogen is present and abundant in the atmospheres of all four of the giant planets of our solar system (Spinrad, 1963). While other molecules, such as methane, dominate the spectra of the giant planets, this is the result of the differing intrinsic strengths of the spectroscopic features; the contribution of hydrogen to their spectra is sufficient to suggest it as the dominant atmospheric constituent.

Helium, as the lightest noble gas, is so spectroscopically inactive that its presence can be detected only through the effect this atom has on the shape of the pressure-induced absorption features of hydrogen. The determination is difficult as the effect is subtle, and is most pronounced in the thermal infrared part of the spectrum (10–30 μm) where terrestrial atmospheric water vapor impedes the accuracy of ground-based studies. The Voyager infrared interferometric spectrometer (IRIS) made determinations for Jupiter and Saturn (Conrath and Gautier, 2000), but the atmospheres of Uranus and Neptune are so cold that insufficient sensitivity was available from IRIS. For all four planets, a second determination was obtained by deriving the temperature–pressure profile of the atmosphere through radio-occultation and infrared measurements. The refraction of a radio beacon sent through the atmosphere by Voyager, as it passed behind each of the giant planets, could be converted to a number density profile with altitude under the assumption of a hydrogen-rich atmosphere. This profile, to which is added the constraints of hydrostatic equilibrium and the equation of state of an ideal gas, yields the temperature divided by the atmospheric molecular weight. Infrared brightness temperatures at various altitudes obtained by IRIS then can be matched to the spectrum to extract the molecular weight of the atmosphere.

To the accuracy of the measurement only hydrogen and helium contribute to the molecular weight with any significance except perhaps in Neptune where N_2 may have a measurable effect (Gautier et al., 1995). Even for Jupiter and Saturn, the collision-induced determination is difficult, and subsequent in situ measurements of the helium abundance with the Galileo mass spectrometer and the helium abundance detector (Von Zahn et al., 1998) have led to reanalysis of the Voyager IRIS determination. The best-fit helium determination for each of the giant planets, expressed as a number (mole) fraction relative to molecular hydrogen, is 0.1359 ± 0.0027 for Jupiter, 0.135 ± 0.025 for Saturn, 0.152 ± 0.033 for Uranus, and 0.190 ± 0.032 for Neptune (Von Zahn et al., 1998; Conrath and Gautier, 2000; Fegley et al., 1991; Gautier et al., 1995). The solar abundance—that found in the atmosphere of the Sun—is virtually identical to the number for Jupiter, but the value obtained at the time the planets formed and derived by applying corrections associated with sedimentation of helium in the Sun is ~15% higher (Guillot et al., 2003). Therefore, helium in the upper layers of both Jupiter and Saturn is depleted relative to the solar primordial value, and hence perhaps relative to the total bulk abundance of this element in the interiors of the two planets.

1.23.2.2 The Equation of State of Hydrogen and Helium as a Determinant of the Structure

While both main sequence stars and giant planets are self-gravitating bodies composed mostly of hydrogen and helium, there is a fundamental difference between the two classes in the nature of the equation of state of these materials. Unlike stars, giant planets do not behave as ideal gases in their interiors. The difference is a result of the densities in the interiors of the two classes of bodies; stars are thermally expanded by virtue of hydrogen fusion, and hence thermal pressure is much larger than electronic or degeneracy pressure. The temperature–pressure profiles in giant planets pass through a complex region of the phase diagram of hydrogen and helium where the atoms and molecules are separated by distances comparable to the dimensions of the particles themselves. Hence, pressure dissociation of the molecules into atoms, and ionization compete with or are more important than ionization and dissociation due to thermal effects. The standard Saha relations for determining the fractional ionization and dissociation as a function of temperature, pressure, and composition do not work in this realm of giant planet interiors. For this reason, much experimental and theoretical effort has been expended to quantify the behavior of hydrogen–helium mixtures under conditions relevant to the giant planets.

This behavior of hydrogen and helium, coupled to its ideal gas behavior under much lower pressures, has a fundamental implication for the size of giant planets, as was shown in the classic paper by Zapolsky and Salpeter (1969). Imagine an object that is made of pure hydrogen, or an admixture by mass of 75% hydrogen, 25% helium, and is cold, so that thermal energy plays no role. At sufficiently low mass of such a self-gravitating object, the behavior of the material is that of an incompressible gas for which the addition of more mass leads primarily to an increase in the radius or size of the body. In the limit of high mass, the hydrogen–helium mixture behaves as a degenerate gas, in which electrons may occupy only the lowest energy states available and Fermi statistics apply. Addition of more mass, in this regime (in which relativistic effects that apply at irrelevantly high masses are neglected), leads to a reduction in the radius of the body. Extension of these two regimes toward each other leads to a maximum radius at a critical mass close to, but somewhat larger than, the mass of Jupiter (Zapolsky and Salpeter, 1969). The maximum radius is essentially that of Jupiter for a solar composition of hydrogen and helium; it is less for a superabundance of nonhydrogen elements, and is ~15% higher than that of Jupiter for a pure hydrogen body.

The observational implication of this analysis, which holds today despite the equation of state of hydrogen and helium having been refined many times since, is that giant planets are not much larger in surface area, hence in reflected brightness, than Jupiter. One cannot make a larger giant planet by simply adding mass; one must add thermal energy. Stars, defined as bodies 75–85 times the mass of Jupiter (a mass sufficient to ignite hydrogen fusion and hence sustain huge internal temperatures), are much larger because of the thermal pressure contribution to their structure. Young giant planets, or those so close to their parent stars that substantial energy is gained from stellar photons, may be larger, but except for the earliest million years or so of a giant planet's existence, radii more than a factor of 50%—surface areas more than a factor of 2—that of Jupiter are not possible. This limits the detectability of such objects in astronomical searches. Brown dwarfs are transitional, in that those close to the hydrogen-burning limit may remain hot and hence thermally expanded for astrophysically interesting timescales. However, others may not be, for example, the brown dwarf Gliese 229 B, while perhaps 40–50 times or so more massive than Jupiter, is in fact smaller in volume (Burrows et al., 2001). Conversely, Uranus and Neptune are substantially smaller than Jupiter both because of their lesser mass and their larger fractional complement of elements heavier than hydrogen and helium; the latter compositional effect yields a factor of 2 smaller radius over and above the mass effect (Zapolsky and Salpeter, 1969). The menagerie of giant planets and brown dwarfs can be shown conveniently on a single scale, at least as far as physical size is concerned (Figure 1).

Detailed modeling of the sizes of the giant planets, as indicators of their composition and thermal history, require equations of state that are more detailed and better tied to experimental measurements than that of Zapolsky and Salpeter (1969). Figure 2 illustrates a modern phase diagram for pure hydrogen along with some of the experimental constraints from which it is obtained, in the temperature–pressure regime appropriate for the interiors of Jupiter and Saturn. Most interesting about the diagram is the prediction in one heavily used equation-of-state model (Saumon et al., 1995) of a so-called "plasma phase transition" (PPT) by which the relative abundances of molecular and atomic hydrogen change in a discontinuous fashion.

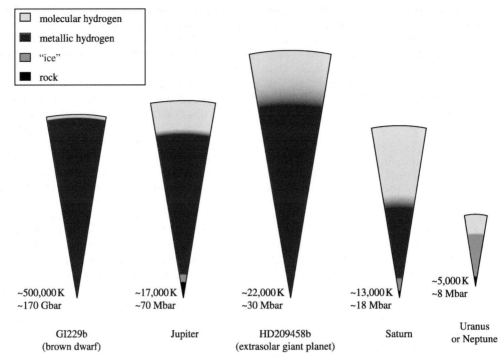

Figure 1 Some known giant planets and brown dwarfs, illustrated with a limited azimuthal slice (pie slice) to correct scale. The interiors are color coded according to the principal materials in each zone. Ice and rock refer to elements common in materials that are icy or rocky at normal pressures. Metallic hydrogen indicates ionization primarily through pressure effects. Modeled central temperature in K, and pressure in 10^9 bar, is shown. The radii of all but Gl229b are known directly; for Gl229b, modeling of the brightness versus wavelength must be used to derive the radius. From left to right, the masses (expressed relative to the mass of Jupiter) are 45, 1, 0.7, 0.3, and 0.05 (figure courtesy of W. B. Hubbard and T. Guillot).

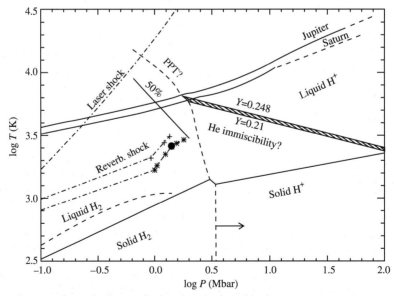

Figure 2 Phase diagram of pure hydrogen displayed as the logarithmic temperature versus pressure. Theoretical temperature–pressure profiles for Jupiter and Saturn are shown. Laser shocking of molecular hydrogen covers the temperature–pressure region indicated by the upper dashed-dot line, while the reverberation shock experiments, at cooler temperatures, measured the electrical conductivity of hydrogen at the points indicated. A possible discrete transition between molecular and atomic (metallic) hydrogen is marked as PPT, while the 50% line below it indicates where half the hydrogen is metallic in a competing model for which the transition is continuous (Ross, 1998). A possible boundary in T–P space below which helium is immiscible in hydrogen, for mass fractions of helium indicated by "Y," is shown with a hatched double line (source Hubbard et al., 2002).

This is based on a particular model of the interactions among protons and electrons in dense liquid hydrogen, wherein the free energy is minimized to identify the preferred phases. An alternative model that assumes that the PPT is replaced by a continuous transition from molecular to metallic hydrogen can better fit much of the shock data (Ross, 1998). However, this theory is *ad hoc* and some of the data it does fit have been called into question (Hubbard *et al.*, 1999).

The debate is not an academic one, because the different equations of state make different predictions for the solubility of helium in metallic hydrogen. The depleted mixing ratio of helium relative to hydrogen in the Jovian and Saturnian atmospheres suggests that helium is being extracted from the upper envelope of these planets and sequestered in the deep interior, perhaps because of immiscibility. This process should be revealed through the resulting conversion of gravitational potential energy into heat (Stevenson and Salpeter, 1977), and this appears to be the case at least for Saturn (Hubbard, 1989). Regardless of which equation-of-state model is correct, the existence of electrically conducting metallic hydrogen in the interiors of Jupiter and Saturn, combined with their rotation, provide a natural dynamo mechanism for their magnetic fields (Stevenson, 1998). These, in turn, have a pervasive impact on the nature and evolution of the surfaces and atmospheres of the natural satellites. The equation of state is also crucial in determining the radius of the extrasolar giant HD209458b under the extreme optical irradiation by its nearby parent star, as is discussed below. Finally, Uranus and Neptune represent another challenge in high-pressure equations of state, since their deep interiors are largely rocky and icy elements with no metallic hydrogen. The source of electrically conducting fluid in these bodies may instead be water ionized at high pressure (Podolak *et al.*, 1991).

1.23.2.3 The Thermal Infrared Emission of the Giant Planets and Implications for Evolution

Sources of energy leading to emission of heat, and hence thermal radiation, from planets include sunlight (or the equivalent from the parent star for extrasolar planets), virialized heat of collapse, gravitational differentiation, and radioisotope decay. The second of these sources refers to the conversion of gravitational potential energy into heat as distributed material accretes to form self-gravitating bodies. The third source, referred to as core formation in the evolution of the terrestrial planets, is due to the sinking of heavier material toward the center of the planet with release of gravitational energy. All four of these sources play important roles in the rocky planets during all or some part of their evolution. Except for the last source, the same holds for the giant planets.

Measurement of the net heat flow associated with formation of and internal processes within the giant planets requires measurement of both the thermal infrared emission and absorption of sunlight. The former is a difficult measurement to perform from Earth for Uranus and Neptune. For all four giant planets, measurement of the absorption of sunlight is also difficult, because it requires accounting for scattered sunlight off of clouds, at angles difficult or impossible to measure from the Earth. For these reasons, the energy balance within the four giant planets of our own solar system has required the Voyager spacecraft for definitive measurement. For the well-studied extrasolar giant planet HD209458b, no direct measurement is possible, but the inflation of its radius (see below) provides an indirect indicator of this process.

Define the energy balance of a planet as the ratio of the total thermal emission at present relative to the amount of sunlight re-emitted as thermal radiation, so that a value of unity obtains for an object whose sole source is sunlight. One finds a value of 1.67 ± 0.09 for Jupiter, 1.78 ± 0.09 for Saturn, 1.06 ± 0.08 for Uranus, and 2.61 ± 0.28 for Neptune (Hubbard *et al.*, 1995). That Uranus and Neptune are so different in regard to the energy balance can be seen also in the entirely coincidental correspondence in the effective radiating temperature of the two planets, which is 59 K despite Neptune being more than 50% further from the Sun. Interpretation of the lack of a signature of internal heat from Uranus is difficult, and some models have invoked different formation conditions while others the extreme tilt of Uranus on its axis, such that during the Voyager flyby one pole of Uranus was pointed toward the Sun (Lunine, 1993). The quiescence of the Uranian atmosphere relative to the Neptunian atmosphere was striking in Voyager images of the two bodies, and the subsequent apparent increase in convective activity on Uranus as imaged by Hubble Space Telescope suggests that indeed a seasonal effect plays a role in a possibly variable Uranian heat balance (Hammel *et al.*, 2001).

For Saturn versus Jupiter the story is a bit clearer. Models of the interior and evolution of Jupiter (described further below) produce the currently observed effective temperature with only sunlight and the original virialized energy of collapse; no additional differentiation is required at present. However, in the case of Saturn, additional energy is required to obtain a body of its effective temperature and mass at an age of 4.56 Gyr, implying that differentiation is taking place (Hubbard *et al.*, 2002). The main heavy constituent contributing to this additional source of energy as it sinks is most plausibly helium, on the basis of its observed abundance and the phase diagram discussed above, which implies the potential for immiscibility in the current epoch (Stevenson and Salpeter, 1977). The absence of evidence for the additional energy source on Jupiter may indicate that immiscibility began later on Jupiter, and has contributed proportionately less to the total energy balance than in the case of Saturn, or has not begun at all (Fortney and Hubbard, 2002).

1.23.2.4 The Interior Structure of the Giant Planets

Construction of a model of the interior structure of hydrogen–helium objects requires the imposition of an equation of state along with a thermal model, which together prescribes the temperature–pressure–density relationship at any given time in the planet's history. Integration of the equation of hydrostatic equilibrium then yields the distribution with radius of the mass (strictly mass density) in the interior of the planet. Since the composition and the internal distribution of the chemical elements are unknown, they must be constrained by an additional set of observations, namely, the gravitational field of the planet and its rotation rate (which, when coupled with the planetary oblateness, depends on the mass distribution). Ground-based studies can provide the required data, through the orbital precession of rings and satellites, measured shape and rotation rate, but with significant ambiguities. Satellite measurements, such as from Voyager, provide higher accuracy and—through observation of the magnetic field rotation rate—an indication of the spin rate of the bulk of the planet.

The temperature profiles within Jupiter and Saturn are thought to be essentially adiabatic, reflecting the high central temperatures and the dominant role of convection below the observable atmosphere where radiative processes become important. There may be deeper layers restricted in radial extent where the temperature profile becomes subadiabatic, due to a decrease in the total opacity, or by virtue of the behavior of the equation of state of hydrogen and helium. The same may hold for Uranus and Neptune, although with less certainty, because of the possibility that stable compositional gradients could exist and dominate the heat flow regime. In particular, Uranus' small heat flow, if primordial and not a function of seasonal insolation, could be the result of a stable compositional stratification and hence subadiabatic temperature profile in the interior (Podolak *et al.*, 1991).

The results of the modeling of the interior are summarized in rough pictorial fashion in Figure 1. It is clear that Uranus and Neptune possess multiple layers, and are similar despite the different heat flows. Neptune has a hydrogen-rich outer layer down to a radius 0.8 that of the planet and a mixed layer of ice, rock, and hydrogen and helium below that. Only two Earth masses, or less than 15%, of the planet is hydrogen and helium. Some of the rock might be separated out in the form of a core. Uranus may be more centrally condensed than Neptune, but the relative proportions of rocky, icy, and gaseous elements overall throughout the two bodies are the same. Jupiter and Saturn, alternatively, are dominated by hydrogen and helium, but the abundance of heavy elements—equivalent to some 20 Earth masses of material (Guillot *et al.*, 2003)—is enhanced over solar abundance. It is possible, but not required, that Jupiter and Saturn have small rocky cores. Most of the elements heavier than hydrogen and helium are distributed throughout the deep interior, rather than sequestered as a discrete layer. Unfortunately, the distribution of elements in the interiors of Jupiter and Saturn does not provide a tight constraint on the formation mechanisms described below, as such a distribution could be achieved either by initial seeding of the formation with heavy elements, or by later addition of this material, or both.

1.23.2.5 Elemental and Isotopic Abundances

Measurement of atmospheric abundances in the giant planets is another indicator of their bulk interior composition, although an indirect one, because the interiors are decidedly nonhomogeneous. However, the atmospheric abundances provide strong evidence for an overall enrichment of the elements heavier than hydrogen and helium relative to solar abundance, an enrichment that increases from Jupiter, to Saturn, and then to Uranus and Neptune. Because of the dominance of hydrogen, the elements are present in reduced molecular form—carbon as CH_4, nitrogen as NH_3, oxygen as H_2O, sulfur as H_2S, phosphorous as PH_3, germanium as GeH_4, etc. However, more oxidized molecular forms are not completely absent; for example, carbon monoxide has been detected on Jupiter, although whether its source is from deeper, hotter, and hence more oxidizing regions, or from photochemistry in the stratosphere, has remained an unsolved problem (Bezard *et al.*, 1999). The troposphere of Neptune appears, indirectly, to have nitrogen primarily in the form of N_2 rather than NH_3 (Gautier *et al.*, 1995). Photochemically produced species such as acetylene (C_2H_2) and ethane (C_2H_6) derived from methane are observed in the stratospheres of the giant planets.

Table 2 attempts to summarize the abundances of the major elements in the atmospheres of the giant planets, expressed relative to solar abundance. This is a difficult undertaking, and many abundances, particularly for Uranus and Neptune, are not well determined or are controversial. In addition, it is difficult to assign error bars that capture the diverse range of experimental sensitivities and observational wavelengths; hence, the number of significant figures or the range must serve to indicate the uncertainties. However, the general trend of increasing overabundances of the heavy elements relative to solar is well illustrated. For Jupiter, *in situ* measurement of noble gases and molecular species provides the most complete abundance determinations, but even here the condensation of ammonia and water as clouds leads to uncertainties in the abundances of these important species (Atreya *et al.*, 1999b). An outstanding uncertainty in the water abundance was measured by the Galileo probe to be below solar abundance down to 20 bar, but the atmospheric conditions indicated a zone that had been dried out by atmospheric circulation. Circumstantial evidence from observations by the Galileo orbiter of deep clouds that may be water (the most visible clouds are higher in the atmosphere and composed of ammonia) suggests abundance in excess of solar.

Isotopic abundance determinations are spotty for all but deuterium, which is seen in HD and CH_3D in the giant planets, as shown in Figure 3. The marginally consistent results from remote sensing and Galileo probe measurements for Jupiter illustrate the difficulty in making accurate chemical and isotopic measurements in giant planet atmospheres, but a trend of rising deuterium abundance seems to be present. Because comets, which are likely to be remnants of the planetesimals from which bodies in the outer solar system were built, contain elevated deuterium enrichments relative to protosolar, the trend in the giant planets may well reflect the increasing relative importance of the icy and rocky phases compared to the gas. The deuterium abundance in Jupiter, added to the abundance of ^{3}He, the light isotope of helium, measured by Galileo, is consistent with the same sum constructed for the local interstellar medium (ISM). However, relative to the local ISM, deuterium in Jupiter is elevated and ^{3}He is depleted. The ISM represents 4.5 Gyr of stellar hydrogen fusion since the birth of Jupiter—consuming deuterium, generating light helium, and expelling these into the interstellar medium. Hence, the trend of these isotopes is fully consistent with the gaseous (hydrogen–helium) component of Jupiter being a sample of protosolar gas that has been bottled up and hence isolated from hydrogen fusion for 4.5 Gyr (Lunine *et al.*, 2000).

1.23.2.6 Atmospheric Dynamics and Magnetic Fields

The wind systems on Jupiter and Saturn are strongly westward at the equator, with eastward jets at latitudes above and below the equator. Equatorial wind speeds on Jupiter approach 150 m s^{-1} in the direction of rotation, and are 3–4 times higher still on Saturn. Measurements by the Galileo probe indicate that the winds persist down to the deepest level, ~20 bar, measured by the probe. Uranus shows prograde (in the direction of rotation) winds of 150 m s^{-1} at midlatitudes, which decline in speed toward the equator. Because of the low contrast of the Uranian clouds at the time of the Voyager flyby, optical tracking

Table 2 Abundances of major element species in the atmospheres of the giant planets (expressed relative to solar abundance).

	Jupiter	Saturn	Uranus	Neptune
Helium	0.8	0.8	1	1.3
Methane, CH_4	2.5–3.5	6	30–60	30–60
Ammonia, NH_3	3–4 (>8 bar)	0.5–3	>1?	20–40 as N_2?
Water, H_2O	>1?	?	?	?
Phosphine, PH_3	>0.2	6	?	?
Hydr. sulf., H_2S	2.2–2.9	?	?	?
Arsine, AsH_3	0.6–3	2–8	?.	?
Germane, GeH_4	0.1	6	?	?
Neon	0.2	?	?	?
Argon	2.0–3.0	?	?	?
Krypton	2.2–3.2	?	?	?
Xenon	1.8–3.0	?	?	?

Notes: Units given in solar abundances as known at the time the cited works were published. Readers should consult those works for the abundance values in absolute units. Updated solar abundances are given in Chapter 1.03. Jupiter and Saturn data are from Atreya *et al.* (1999b), Gautier *et al.* (2001), Noll *et al.* (1989), and Fink *et al.* (1978); Uranus and Neptune from Gautier *et al.* (1995).

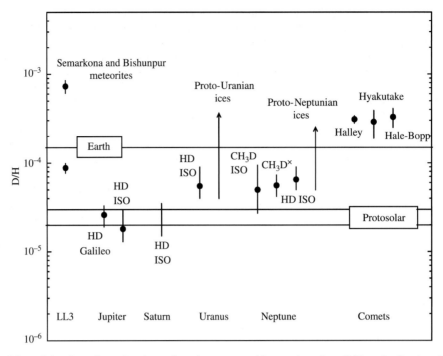

Figure 3 Map of the deuterium abundance in solar system objects, plotted as D/H mole fraction. The carrier molecules for which deuterium has been measured in each object are labeled. The protosolar value, derived from measurements of ^{3}He products of deuterium fusion in the Sun, and deuterium in the local region of the galaxy, is given, as are values for a carbonaceous chondrite meteorite, the Earth's oceans, and comets (source Hersant *et al.*, 2001).

of the winds was difficult. A radio-occultation experiment done as the Voyager spacecraft past behind Uranus as seen from the Earth indicated retrograde winds at the equator of 100 m s^{-1}. On Neptune, where cloud features at high contrast aided tracking by Voyager cameras, a narrow jet at 70° latitude was clocked at 300 m s^{-1}. Other than this feature, wind speeds seem comparable to that at Uranus. Cyclonic and anticyclonic systems are abundant on Jupiter, appear on Saturn, and at least one such feature (the Great Dark Spot) appears on Neptune (Lunine, 1993).

The winds on Jupiter are consistent with, but do not require, the presence of deep cylindrical flows, consistent with an efficiently convecting planet in which differential rotation is forced to be constant on cylinders (Hubbard *et al.*, 2002). However, other explanations, including shallow meteorological patterns, are possible. The rough increase in wind speed from Jupiter to Neptune may reflect the greater importance of convection impeding rotational momentum on the two more massive giant planets, the effects of declining insolation with distance, or more stochastic effects associated with formation.

The magnetic fields of the giant planets provide fiduciary spin rates of the interiors, create magnetospheres of charged and neutral particles that alter the appearance and composition of satellite surfaces and rings, and affect the energy and mass input into the ionospheres of the giant planet atmospheres. Direct measurement of the magnetic fields of the giant planets has revealed diverse geometries. The Jovian magnetic field has large quadrupole and octupole moments, while the Saturnian magnetic field is largely dipolar (Dessler, 1983). Both fields are opposite in polarity to that of the Earth, although this is probably not a significant difference since the Earth's field has flipped in polarity many times in its history. Uranus and Neptune both possess complex fields with surface intensities comparable to that of the Earth, but quite different geometries. The fields can be described by dipoles that are tilted with respect to the rotational axis and displaced relative to the center of the planet. Large quadrupolar and octupolar components are present as well (Ness *et al.*, 1995). The implication for the field generation within these planets is that, in contrast to Jupiter and Saturn, the dynamo generation is occurring at shallow levels in the interiors of Uranus and Neptune. This, in turn, suggests that an electrically conducting fluid, possibly high-pressure ionized water, must be invoked.

The chemical implications of the magnetic fields and associated magnetospheres for the giant planets are too numerous to even outline here (Dessler, 1983). One interesting example of the role of the magnetic field is the aurora, which is particularly observable on Jupiter with Hubble Space Telescope. The precipitation of charged particles into the Jovian atmosphere, guided along the field lines of the magnetic field, has a number of effects. They directly excite molecules, inducing photon emission at specific wavelengths, they heat the atmosphere and produce thermal emission, and they induce upper atmospheric chemistry. A high-altitude haze seen on Jupiter at latitudes exceeding 60° might be hydrocarbons generated at high altitudes by charged particle chemistry associated with the auroral precipitation.

1.23.3 ORIGIN AND EVOLUTION OF THE GIANT PLANETS

1.23.3.1 Basic Model for the Formation of the Planets from a Disk of Gas and Dust

Astronomical observations of disks around other stars, combined with theoretical modeling and meteorite studies, establish with little doubt that planetary systems are the result of the formation of a disk of gas and solids around a growing "proto-star"—the central mass that will become a star (Mannings et al., 2000). The disk is a natural consequence of the angular momentum possessed by molecular clouds, the conservation of which during collapse of a clump leads to the development of a disk. Dissipation in the disk caused by turbulent motions, gravitational, and magnetic torques leads to much of the disk mass being transported inward, to the star, while the angular momentum is moved outward (see Chapter 1.04 on the solar nebula). These processes, modeled in detail, lead to the observed fundamental properties of our own solar system: orbits of the major planets confined to a plane, most of the mass in the Sun, and most of the angular momentum in the orbits of the planets. The formation of the regular systems of moons around the giant planets may replicate some of these processes, but with different timescales and some important variations (Peale and Lee, 2002).

The decrease in accretion rate of material to the disk, in the terminal stages of star and planet formation, leads to a cooling of the disk, so that water ice becomes stable at distances of 5 AU or perhaps inward. The presence of large amounts of solids—silicates plus water ice—encourages the formation of large bodies by accretion (Lissauer, 1987; Stevenson and Lunine, 1988). Giant planets must form in no more than a several million-year timescale, which is the lifetime of the gaseous disk based on astronomical observations (Calvet et al., 2000), although some gas may remain for longer periods of time. Uranus and Neptune may be giant planets, whose accretion was truncated by the loss of gas, or interrupted by dynamical perturbations from Jupiter (Thommes et al., 2002). Terrestrial planets inward of the region of the giant planets will take longer to form, because of the smaller amount of solid material inwards of the ice condensation zone, and the presence of Jupiter plays a key role in stirring the orbits of the growing terrestrial planets and encouraging high velocity collisions (Morbidelli et al., 2000; see also Chapter 1.17 on planet formation). Delivery of water and organics to the Earth from the asteroid belt and from comets was thus strongly affected, if not determined, by the presence of Jupiter prior to the time of significant terrestrial planet growth (Lunine, 2001). Interaction of giant planets with the gaseous and even particulate disk may result in migration inward, disrupting terrestrial planet formation and leading to giant planets at small orbital distances (Lin et al., 1996; Ward and Hahn, 2000). Such migration did not occur in our own solar system, or occurred early, and the migrating giant planet was destroyed through falling into the Sun.

1.23.3.2 Constraints from the Composition of the Giant Planets

The specific mechanism for giant planet formation remains undetermined. Direct formation by collapse of gas within a relatively massive gas disk is one possibility (Boss, 2000); accretion of solids first to trigger the accelerated accretion or hydrodynamic collapse of gas is the other (Wuchterl et al., 2000). The substantial overabundance of heavy elements in Jupiter and Saturn, and the existence of Uranus and Neptune as heavy element "cores" surrounded by a much smaller amount of hydrogen–helium, would seem to support the latter model. One might conceive of direct collapse of gas in the disk followed by later accretion of solids, but this has not been examined quantitatively. It is conceivable that both processes work, leading to giant planets with differing abundances of heavy elements relative to the parent star, but in our own solar system accretion of solids first seems to best explain the compositions of the giant planets.

The noble gas, carbon, nitrogen, and sulfur abundances in Jupiter can be compared to the predicted compositions of icy planetesimals to provide details on when and how material was accreted during the formation of Jupiter. Unfortunately the oxygen abundance in Jupiter is unknown, and since water as the primary oxygen carrier was the dominant ice in planetesimals as well (based on observations of comets), one requires this abundance to decide among models. In its absence, the current heavy element inventory can be explained by a model in which Jupiter's ices were derived from very cold, possibly molecular cloud material (Owen et al., 1999), or by a model in which the ices were condensed directly from the nebula in the region of Jupiter formation (Gautier et al., 2001). Future microwave measurements from a flyby spacecraft, or a deep atmospheric entry probe capable of reaching 2–5 times deeper in pressure than Galileo, are required to determine the water and hence deep oxygen abundance.

1.23.4 EXTRASOLAR GIANT PLANETS

Of the 100 or so bodies in the Jupiter to 10 Jupiter mass range discovered indirectly by Doppler spectroscopy, only one can be directly studied, because it transits across the face of its parent star (Charbonneau et al., 2000). The fact that transits occur in the system HD209458 tightly constrains the orbital inclination of this planet relative to the line of sight to the Earth, removing the inclination ambiguity of the planet's mass that is inherent in the Doppler spectroscopic technique. The decrease in starlight as the planet passes across its parent star in its close orbit (less than a tenth the semimajor axis of Mercury's orbit around the Sun) provides a radius for the planet of 1.3–1.4 times the radius of Jupiter (Cody and Sasselov, 2002). With a mass of 0.69 times that of Jupiter, this giant planet is substantially less dense than Saturn.

The origin of this extended radius lies in the close proximity of the planet to its parent star. Although some controversy exists as to how stellar photons are transported to the interior of the planet (Showman and Guillot, 2002), the net effect is to slow the decline in the radius of the planet as it evolves over billions of years. Detailed modeling of this evolution coupled with determination of the age of the star leads to the observed radius, as long as the planet arrived in its present orbit within 10 Myr after formation (Burrows et al., 2000). Thus, the bulk properties of HD209458b are quite consistent with those of a primarily hydrogen–helium giant planet subjected to a degree of heating enormously larger than those experienced by Jupiter and Saturn.

More precise measurements of the transits of HD209458b across the face of its parent star with Hubble Space Telescope have permitted the change of radius with wavelength to be determined (Charbonneau et al., 2002). The radius variation is directly related to the opacity of the atmosphere as a function of wavelength, and hence the abundances of various absorbing species in the atmosphere (Hubbard et al., 2001). In particular, the signature of the D lines of sodium in the optical should be especially strong (Seager and Sasselov, 2000), and this is indeed what is seen. The strength of the sodium feature from the HST data is weaker than expected were the system to possess a heavy element abundance similar to the solar value, but the sodium signature may be reduced through the presence of clouds and charged-particle chemistry.

The transit studies of HD209458b have opened up the study of the atmospheric composition and dynamical processes in extrasolar giant planet atmospheres. However, only a small fraction, ~1%, of giant planets around other stars will exhibit transits. Future studies of giant planets around other stars will require direct detection and spectroscopy, and this in turn will demand huge (30 m mirror diameter) ground-based telescopes employing state-of-the-art adaptive optics, or space-borne telescopes with interferometers or coronagraphs to pull planets out from under the light of their parent stars. Such powerful systems will benefit study of our solar system's own giant planets as well, but additional visits by spacecraft (such as Cassini to Saturn beginning in 2004) will continue to be required for certain kinds of atmospheric, interior, and magnetospheric investigations.

1.23.5 MAJOR UNSOLVED PROBLEMS AND FUTURE PROGRESS

An exhaustive list of all unsolved problems regarding the giant planets would require much more space than this chapter itself occupies. Briefly, some of the key questions include the following.

- How do giant planets form? Two different models, disk instability versus core accretion followed by gas collapse, are viable. They require very different timescales, have very different implications for satellite formation and internal composition, and may have implications for the ubiquity of giant planets and terrestrial planets around other stars. The formation of Uranus and Neptune is even less well understood, and no agreement exists as to whether these are stillborn "Jupiters" or the product of a distinct kind of formation process.
- What are the detailed internal structures of the giant planets and how are the magnetic fields generated? While the separation of helium from hydrogen seems to be assured for Saturn, it is unclear to what extent this occurs in Jupiter. Further, the distribution of elements heavier than hydrogen and helium remains unclear in Jupiter and Saturn, in part because of equation-of-state uncertainties. The interiors of Uranus and Neptune are even less certain. For these reasons, and because of uncertainties in dynamo theory, the specific details of the magnetic field

generation in the giant planets remain uncertain.
- What is the relationship between the atmospheric circulation patterns and the deep circulation? Fluid planets are in many ways more complex than the solid planets, in that the atmospheric circulation patterns have some relationship—greater or lesser—to deep circulations. How the coupling occurs in the giant planets of our solar system, and the strength of the coupling, remain unresolved.
- What are the deep abundances of key elements in the giant planets? In spite of the Galileo mission, we do not know the deep oxygen abundance in Jupiter, which can help constrain formation models for the giant planets. Deep abundances in the other giant planets are even more poorly constrained.
- How do moons and rings form? The solid bodies around the giant planets formed as a consequence of the assembly of the giant planets, but stochastic events such as large collisions may have played crucial roles. For example, we do not know whether the massive Saturnian rings are as old as Saturn itself.

Future prospects for the study of the giant planets in our solar system from space probes center on the Cassini mission to Saturn, set to arrive in 2004, and possible future Jupiter orbiters and probes now in the planning stages. No missions to Uranus and Neptune are planned. The dramatic advances in large ground-based telescopes, detectors, adaptive optics systems, and the planned next generation of space-borne telescopes such as NASA's James Webb Space Telescope will benefit studies both of the giant planets of our own solar system and detection and characterization of giant planets around other stars.

REFERENCES

Atreya S. K., Edgington S. G., Encrenaz Th., and Feuchtgruber H. (1999a) ISO observations of C_2H_2 on Uranus and CH_3 on Saturn. In *The Universe as Seen by ISO* (eds. P. Cox and M. F. Kessler). ESA-SP 427, Paris, p. 149.

Atreya S. K., Wong M. H., Owen T. C., Mahaffy P. R., Niemann H. B., de Pater I., Drossart P., and Encrenaz Th. (1999b) A comparison of the atmospheres of Jupiter and Saturn: deep atmospheric composition, cloud structure, vertical mixing, and origin. *Planet. Space Sci.* **47**, 1243–1262.

Bezard B., Strobel D. F., Maillard J.-P., Drossart P., and Lellouch E. (1999) The origin of carbon monoxide on Jupiter. *Am. Astron. Soc. DPS Meet.* **31**, 69.02.

Bodenheimer P., Hubickyj O., and Lissauer J. J. (2000) Models of the *in situ* formation of detected extrasolar giant planets. *Icarus* **143**, 2–14.

Boss A. P. (2000) Possible rapid gas giant planet formation in the solar nebula and other protoplanetary disks. *Astrophys. J.* **536**, L101–L104.

Burrows A., Guillot T., Hubbard W. B., Marley M., Saumon D., Lunine J. I., and Sudarsky D. (2000) On the radii of close-in giant planets. *Astrophys. J.* **534**, L97–L100.

Burrows A., Hubbard W. B., Lunine J. I., and Liebert J. (2001) The theory of brown dwarfs and extrasolar planets. *Rev. Mod. Phys.* **73**, 719–765.

Calvet N., Hartmann L., and Strom S. E. (2000) Evolution of disk accretion. In *Protostars and Planets IV* (eds. V. Mannings, A. P. Boss, and S. S. Russell). University of Arizona Press, Tucson, pp. 377–399.

Charbonneau D., Brown T. M., Latham D. W., and Mayor M. (2000) Detection of planetary transits across a Sun-like star. *Astrophys. J.* **529**, L45–L48.

Charbonneau D., Brown T. M., Noyes R. W., and Gilliland R. L. (2002) Detection of an extrasolar planet atmosphere. *Astrophys. J.* **568**, 377–384.

Cody A. M. and Sasselov D. D. (2002) HD 209458: physical parameters of the parent star and the transiting planet. *Astrophys. J.* **569**, 451–458.

Conrath B. J. and Gautier D. (2000) Saturn helium abundance: a reanalysis of Voyager measurements. *Icarus* **144**, 124–134.

Cox A. N. (2000) *Allen's Astrophysical Quantities*. American Institute of Physics Press, New York.

Demarcus W. C. (1958) The constitution of Jupiter and Saturn. *Astronom. J.* **63**, 2–28.

Dessler A. J. (1983) *Physics of the Jovian Magnetosphere*. Cambridge University Press, Cambridge.

Fegley B., Jr., Gautier D., Owen T., and Prinn R. G. (1991) Spectroscopy and chemistry of the atmosphere of Uranus. In *Uranus* (eds. J. T. Bergstralh, E. D. Miner, and M. S. Matthews). The University of Arizona Press, Tucson, pp. 147–203.

Fink U., Larson H. P., and Treffers R. R. (1978) Germane in the atmosphere of Jupiter. *Icarus* **34**, 344–354.

Fortney J. J. and Hubbard W. B. (2002) Inhomogenous evolution of giant planets: Jupiter and Saturn. *Am. Astron. Soc. DPS Meet.* **34**, 10.03.

Gautier D., Conrath B., Owen T., de Pater I., and Atreya S. K. (1995) The troposphere of Neptune. In *Neptune and Triton* (ed. D. P. Cruikshank). University of Arizona Press, Tucson, pp. 547–611.

Gautier D., Hersant F., Mousis O., and Lunine J. I. (2001) Enrichments in volatiles in Jupiter: a new interpretation of the Galileo measurements. *Astrophys. J.* **550**, L227–L230 (erratum **559**, L183).

Guillot T., Stevenson D. J., and Hubbard W. B. (2003) The interior of Jupiter. In *Jupiter* (ed. F. Bagenal). Cambridge University Press, Cambridge (in press).

Hammel H. B., Rages K., Lockwood G. W., Karkoschka E., and de Pater I. (2001) New measurements of the winds of Uranus. *Icarus* **153**, 229–235.

Hersant F., Gautier D., and Huré J.-M. (2001) A two-dimensional model for the primordial nebula constrained by D/H in the solar system: implications for the formation of giant planets. *Astrophys. J.* **554**, 391–407.

Hubbard W. B. (1989) Structure and composition of giant planet interiors. In *Origin and Evolution of Planetary and Satellite Atmospheres* (eds. S. K. Atreya, J. B. Pollack, and M. S. Matthews). University of Arizona Press, Tucson, pp. 539–563.

Hubbard W. B., Podolak M., and Stevenson D. J. (1995) The interior of Neptune. In *Neptune* (ed. D. P. Cruikshank). University of Arizona Press, Tucson, pp. 109–138.

Hubbard W. B., Guillot T., Marley M. S., Burrows A. S., Lunine J. I., and Saumon D. (1999) Comparative evolution of Jupiter and Saturn. *Planet. Space Sci.* **47**, 1175–1182.

Hubbard W. B., Fortney J., Lunine J. I., Burrows A., Sudarsky D., and Pinto P. (2001) Theory of extrasolar giant planet transits. *Astrophys. J.* **560**, 413–419.

Hubbard W. B., Burrows A. S., and Lunine J. I. (2002) Theory of giant planets. *Ann. Rev. Astron. Astrophys.* **40**, 103–136.

Kiess C. C., Corliss C. H., and Kiess H. K. (1960) High-dispersion spectra of Jupiter. *Astrophys. J.* **132**, 221–231.

Kippenhahn R. and Weigert A. (1991) *Stellar Structure and Evolution*. Springer, Berlin.

Kowal C. T. and Drake S. (1980) Galileo's observations of Neptune. *Nature* **287**, 311–315.

Krupp E. C. (1983) *Echoes of the Ancient Skies: Astronomy of Lost Civilizations*. Harper and Row, New York.

Lin D. N. C., Bodenheimer P., and Richardson D. C. (1996) Orbital migration of the planetary companion of 51 Pegasi to its present location. *Nature* **380**, 606–607.

Lissauer J. J. (1987) Timescales for planetary accretion and the structure of the protoplanetary disk. *Icarus* **69**, 249–265.

Low F. J. (1966) Observations of Venus, Jupiter and Saturn at λ 20 μ. *Astron. J.* **71**, 391.

Lunine J. I. (1993) The atmospheres of Uranus and Neptune. *Ann. Rev. Astron. Astrophys.* **31**, 217–263.

Lunine J. I. (2001) The occurrence of Jovian planets and the habitability of planetary systems. *Proc. Natl. Acad. Sci. USA* **98**, 809–814.

Lunine J. I., Owen T. C., and Brown R. H. (2000) The outer solar system: chemical constraints at low temperatures on planet formation. In *Protostars and Planets IV* (eds. V. Mannings, A. P. Boss, and S. S. Russell). University of Arizona Press, Tucson, pp. 1055–1080.

Mannings V., Boss A. P., and Russell S. S. (2000) *Protostars and Planets IV*. University of Arizona Press, Tucson.

Mayor M., Queloz D., and Udry S. (1998) Mass function and orbital distributions of substellar companions. In *Brown Dwarfs and Extrasolar Planets* (eds. R. Rebolo, E. L. Martin, and M. R. Z. Osorio). ASP Conference Series, San Francisco, 140pp.

Morbidelli A., Chambers J., Lunine J. I., Petit J. M., Robert F., Valsecchi G. B., and Cyr K. E. (2000) Source regions and timescales for the delivery of water on Earth. *Meteoritics Planet. Sci.* **35**, 1309–1320.

Ness N. F., Acuña M. H., and Connerney J. E. P. (1995) Neptune's magnetic field and fluid-geometric properties. In *Neptune* (ed. D. P. Cruikshank). University of Arizona Press, Tucson, pp. 141–168.

Niemann H. B., Atreya S. K., Carignan G. R., Donahue T. M., Haberman J. A., Harpold D. N., Hartle R. E., Hunten D. M., Kasprzak W. T., Mahaffy P. R., Owen T. C., Spencer N. W., and Way S. H. (1996) The Galileo probe mass spectrometer: composition of Jupiter's atmosphere. *Science* **272**, 846–849.

Noll K. S., Geballe T. R., and Knacke R. F. (1989) Arsine in Saturn and Jupiter. *Astrophys. J.* **338**, L71–L74.

Owen T., Mahaffy P., Niemann H. B., Atreya S. K., Donahue T. M., Bar-Nun A., and de Pater I. (1999) A new constraint on the formation of giant planets. *Nature* **402**, 269–270.

Peale S. J. and Lee M. H. (2002) A primordial origin of the Laplace relation among the Galilean satellites. *Science* **298**, 593–597.

Pearl J. C. and Conrath B. J. (1991) The albedo, effective temperature, and energy balance of Neptune, as determined from Voyager data. *J. Geophys. Res.* **96**, 18921–18930.

Podolak M., Hubbard W. B., and Stevenson D. J. (1991) Model of Uranus' interior and magnetic field. In *Uranus* (eds. J. T. Bergstralh, E. D. Miner, and M. S. Matthews). University of Arizona Press, Tucson, pp. 29–61.

Ross M. (1998) Linear-mixing model for shock-compressed liquid deuterium. *Phys. Rev. B* **58**, 669–677.

Saumon D., Chabrier G., and Van Horn H. M. (1995) An equation of state for low-mass stars and giant planets. *Astrophys. J. Suppl.* **99**, 713–741.

Seager S. and Sasselov D. D. (2000) Theoretical transmission spectra during extrasolar planet transits. *Astrophys. J.* **537**, 916–921.

Showman A. P. and Guillot T. (2002) Atmospheric circulation and tides of "51 Pegasus b-like" planets. *Astron. Astrophys.* **385**, 166–180.

Spinrad H. (1963) Pressure-induced dipole lines of molecular hydrogen in the spectra of Uranus and Neptune. *Astrophys. J.* **138**, 1242–1245.

Stevenson D. J. (1998) States of basic matter in massive planets. *J. Phys.-Cond. Matt.* **10**, 11227–11234.

Stevenson D. J. and Lunine J. I. (1988) Rapid formation of Jupiter by diffusive redistribution of water vapor in the solar nebula. *Icarus* **75**, 146–155.

Stevenson D. J. and Salpeter E. E. (1977) The dynamics and helium distribution in hydrogen–helium planets. *Astrophys. J. Suppl.* **35**, 239–261.

Thommes E. W., Duncan M. J., and Levison H. F. (2002) The formation of Uranus and Neptune among Jupiter and Saturn. *Astron. J.* **123**, 2862–2883.

Von Zahn U., Hunten D. M., and Lehmacher G. (1998) Helium in Jupiter's atmosphere: results from the Galileo probe helium interferometer experiment. *J. Geophys. Res.* **103**, 22815–22830.

Ward W. R. and Hahn J. M. (2000) Disk–planet interactions and the formation of planetary systems. In *Protostars and Planets IV* (eds. V. Mannings, A. P. Boss, and S. S. Russell). University of Arizona Press, Tucson, pp. 1135–1155.

Wildt R. (1932) Absorptionsspektren und Atmosphären der grossen Planeten. *Veroeffentlichungen der Universitaets-Sternwarte zu Goettingen* **2**, 171–180.

Wuchterl G., Guillot T., and Lissauer J. J. (2000) Giant planet formation. In *Protostars and Planets IV* (eds. V. Mannings, A. P. Boss, and S. S. Russell). University of Arizona Press, Tucson, pp. 1081–1109.

Zapolsky H. S. and Salpeter E. E. (1969) The mass–radius relation for cold spheres of low mass. *Astrophys. J.* **158**, 809–813.

1.24
Major Satellites of the Giant Planets

T. V. Johnson

California Institute of Technology, Pasadena, CA, USA

1.24.1	INTRODUCTION	637
1.24.2	COSMOCHEMICAL CONTEXT	638
1.24.3	BULK COMPOSITION	639
1.24.3.1	*Bulk Density*	639
1.24.3.2	*Sources of Data*	639
1.24.4	SURFACE COMPOSITION	639
1.24.4.1	*Spectral Reflectance*	639
1.24.4.2	*Temperature and Atmospheres*	641
1.24.4.3	*Radiation Effects*	641
1.24.5	THE JUPITER SYSTEM	641
1.24.5.1	*General*	641
1.24.5.2	*Io*	641
1.24.5.2.1	*Surface composition and volcanism*	642
1.24.5.2.2	*Atmosphere and magnetospheric interactions*	643
1.24.5.3	*Ganymede and Callisto*	644
1.24.5.3.1	*Surface composition*	645
1.24.5.3.2	*Atmospheres and magnetospheric interactions*	647
1.24.5.4	*Europa*	647
1.24.5.4.1	*Surface composition*	647
1.24.5.4.2	*Atmosphere and magnetospheric interactions*	648
1.24.6	THE SATURN SYSTEM	649
1.24.6.1	*General*	649
1.24.6.2	*Iapetus*	649
1.24.6.3	*Titan*	649
1.24.6.4	*Magnetospheric Interactions*	650
1.24.7	THE URANUS SYSTEM	652
1.24.7.1	*General*	652
1.24.8	THE NEPTUNE SYSTEM—TRITON	653
1.24.8.1	*General*	653
1.24.8.2	*Composition*	654
1.24.8.3	*Atmosphere*	655
1.24.9	MAJOR ISSUES AND FUTURE DIRECTIONS	656
REFERENCES		656

1.24.1 INTRODUCTION

The geochemistry of the natural satellites of the outer solar system provides important clues and constraints on the formation of the planets about which they orbit, the formation and nature of the satellites themselves, and the various processes that have shaped their histories and evolution. Historically, little was known about the geochemical makeup of these worlds prior to the beginning of space exploration. Information about these objects ca. 1960 was limited to positional astronomical data, crude estimates of size and mass for only a few of the largest satellites, and in some cases measurements of color and albedo (reflectance). Informed speculation suggested the likely presence of rock and frozen volatiles (water, carbon dioxide, methane, and ammonia) in these cold, distant reaches of the solar system, but there were few hard data from which to deduce detailed composition. The only known chemical species related to satellite composition that had been positively confirmed was gaseous methane, identified from absorption features in the spectrum of Saturn's satellite, Titan (Kuiper, 1944).

Since the 1960s a combination of improved theoretical understanding of planetary formation, data from reconnaissance spacecraft and modern astronomical observations from earth and space-based platforms have revolutionized our knowledge of the geochemistry of these satellites. The most detailed information is available for the larger, relatively spherical satellites in equatorial orbits about their planets, which we will term the "major" satellites. Each system also contains numerous smaller bodies, associated with ring systems, or in swarms of distant, loosely bound orbits. For information on the characteristics of these smaller satellites, see Yoder (1995). As yet, we have little information on the geochemistry of these bodies, and this chapter will concentrate on the larger satellites.

1.24.2 COSMOCHEMICAL CONTEXT

One of the principal constraints on planetary chemistry, including the natural satellites, is the solar system abundance of elements. Lewis (1971, 1972, 1973) published the first detailed studies of the chemistry of the satellites based on solar system abundance and the likely conditions of temperature and pressure in the solar nebula. Two of his primary conclusions were: (i) beyond the asteroid belt, solid material available to form satellites should consist of approximately equal proportions (by mass) of "rock" (primarily silicates, iron, and other refractory components, including carbon compounds) and water ice, and (ii) more volatile ices, such as ammonia, methane, and clathrated/hydrated compounds of these material, might be present in smaller amounts and could play an important role by lowering the temperatures required for producing melting or partial melting in a satellite. He noted that at least some of the large satellites had densities consistent with his ice/rock models ($\sim 2,000$ kg m^{-3}) and suggested that the relatively low melting point of the icy portions of these bodies, combined with heat from natural radionuclides in the non-ice portions, might lead to more complex chemistry and geology for such satellites than one might otherwise expect on such relatively small, cold objects.

Subsequent developments of models for satellite chemistry have concentrated on modifications of Lewis' basic picture. The effect of the forming giant planets on the chemistry in their circumplanetary nebulae may be important in controlling differences in chemical makeup among the satellites of a given system. The most detailed calculations have been made for the Jupiter system, where several researchers have suggested that Io and Europa have basically rocky compositions due to proto-Jupiter's effects on the temperature and pressure in the inner portion of the circumplanetary nebula, preventing condensation of large amounts of water ice in their vicinity (Fanale *et al.*, 1977; Pollack and Fanale, 1982; Pollack and Reynolds, 1974). Canup and Ward (2002) have recently pointed out deficiencies in these and related "minimum mass" subnebula models, which have difficulty explaining the characteristics of the icy satellites without creating conditions in the modeled circumplanetary nebula that would have led to dynamical loss of the satellites. Their model involves slow accretion of the satellites from the inflow of gas and dust from the solar nebula during the late stages of Jupiter's accretion. In their models the satellites migrate inward from their original positions but survive. The inner satellites are still formed at temperatures above the water condensation conditions.

The other giant planet systems do not present such a clear case, either in variation of satellite densities within the system or in the theoretical treatment of the evolution of the circumplanetary nebula during formation. It may be that Jupiter is alone in exerting such obvious control over the local chemistry of its system. There is evidence that even the Jupiter system may be more complex than current models can explain. In 2002 Galileo made a close pass by Amalthea, the next satellite inward from Io. Analysis of radio tracking data resulted in a determination of mass yielding a density $<1,000$ kg m^{-3} (Anderson *et al.*, 2002). Even with allowance for significant bulk porosity within this relatively small satellite, a compositional mixture of rock and ice similar to the large icy satellites is suggested (Johnson and Anderson, 2003), which is difficult to reconcile with any of current formation models.

Another significant modification relates to carbon and nitrogen chemistry in the solar nebula, particularly in the outer solar system. Lewis' original models for equilibrium chemistry resulted in CH_4 being the primary carbon and NH_3 the main nitrogen species in the nebula. Lewis and Prinn (1980), using more refined nebula models and considering the kinetics of the relevant reactions in the outer nebula, suggested that CO and N_2 might dominate instead, leading to predictions of somewhat higher average densities for condensed rock and volatile mixtures in this region. Prinn and Fegley (1981) examined these arguments as they related to the conditions around the forming Jupiter and concluded that CH_4 and NH_3 should still dominate in the circumplanetary nebula. They later quantified in some detail the effects of these models on the rock/ice ratios expected for satellites (Prinn and Fegley, 1989). Whether these conclusions still obtain for the "gas-starved" accretion models of Canup and Ward remains to be seen. Further discussion of the conditions for planetary formation in the outer solar system and the chemistry of the giant planets themselves can be found in Chapter 1.23.

1.24.3 BULK COMPOSITION

1.24.3.1 Bulk Density

The most important observational constraint on the bulk chemistry of a satellite is its bulk density. For the purposes of modeling the interior structure and chemistry from mass, radius, and density, the outer planet satellites fall into two basic classes—those large enough for the effects of self-compression to be important and those small enough for bulk porosity to a major factor. Although there is no universally agreed upon definition of the size ranges for these categories, a useful working figure of merit can be obtained from considering shape or "degree of sphericity" of an object. Thomas et al. (1986) have considered the shapes of many small satellites and concluded that below a volume of $\sim 10^7$ km^3, objects have highly irregular shapes as defined by the ratio of their maximum to minimum axes, while above this threshold, the bodies are essentially spherical. This presumably results from a combination of their collisional histories and the ability of gravitational compression to eliminate major pore space below the outer layers of the larger bodies. Recent reports of extremely low densities found for irregularly shaped asteroids (Thomas et al., 1999; Veverka et al., 1999) and small icy Saturn satellites (Nicholson et al., 1992) below this volume threshold support this view. Bulk porosities in these cases, estimated from constraints on their composition, range from 10% to 40%.

For objects above the Thomas et al. volume threshold, bulk density should be closely related to the densities and therefore the compositions of the major constituents—"rock" and ice in most cases. The complications arising from the high-pressure phases of minerals under conditions found in the interiors of the larger terrestrial planets are much less severe for these objects. Interior pressure in even the largest outer planet satellites reaches only ~ 3.5 MPa, which will not affect the densities of minerals in the rock portion significantly (Schubert, 1986). The major pressure-related effect that must be taken into account is the phase diagram of the water–ice system, where temperatures and pressures in the larger icy satellites result in higher-density ice phases at depth. Ice phase behavior has been studied in detail for the Galilean satellites where interior pressures make it a relatively major effect, and most modeling of smaller icy satellites have included these effects as well where appropriate (Consolmagno and Lewis, 1977, 1978; Johnson et al., 1987; Lupo and Lewis, 1979).

1.24.3.2 Sources of Data

In the era before spacecraft, estimates of satellite densities were limited to the few cases where resonant orbital dynamics and mutual perturbations allowed a mass estimate and the object was large enough/close enough to measure its radius optically. The Galilean satellites of Jupiter were effectively the only outer planet satellites for which reasonably accurate estimates could be made, suggesting high, rock-like densities for Io and Europa and lower "icy" densities for Ganymede and Callisto. Accurate radii from stellar occultations improved these values somewhat for Io and Ganymede, but the major breakthrough came with the advent of spacecraft flybys of these satellite systems with the Pioneer and Voyager spacecrafts. These produced accurate data for satellite radii for essentially all the major satellites and direct measurements of many of their masses through analysis of their effects on spacecraft trajectories. Table 1 lists current values for radius, mass, and density as well as estimates for the approximate ratio of "rock" to "ice" implied by these bulk densities. The rock/ice estimates are based on a number of sources and models for satellite interior structures (Johnson et al., 1987; Schubert, 1986). In these models "rock" is a general term for the non-ice solid material in the object, including not only silicate material but also lighter carbonaceous components that are probably present in the non-ice fraction, especially at larger heliocentric distances. Detailed physical properties and satellite orbital data may be found in Yoder (1995).

1.24.4 SURFACE COMPOSITION

1.24.4.1 Spectral Reflectance

Most information about the surface geochemistry of outer planet satellites comes in one way or another from spectral analyses of reflected sunlight from their surfaces. The fraction of sunlight reflected compared with incident sunlight is generally referred to as the albedo (several somewhat different technical definitions exist in the astronomical literature depending on whether the value refers to a global planetary average or an

Table 1 Satellite properties.

Planet Satellites	Distance to planet (10^3 km)	Radius (km)	Mass (10^{19} kg)	Density (10^3 kg m^{-3})	Approx. silicate mass fraction
Jupiter					
Io	422	1,821	8,933	3.53	1.0
Europa	671	1,565	4,797	2.99	0.94
Ganymede	1,070	2,634	14,820	1.94	0.58
Callisto	1,880	2,403	10,760	1.85	0.52
Saturn					
Mimas	185	199	3.75	1.14	0.27
Enceladus	238	249	7.3	1.12	0.22
Tethys	295	530	62.2	1.00	
Dione	377	560	105.2	1.44	0.46
Rhea	527	764	231	1.24	0.40
Titan	1,222	2,575	13,455	1.88	0.55
Iapetus	3,561	718	159	1.02	
Uranus					
Miranda	130	236	6.59	1.20	0.30
Ariel	191	579	135	1.67	0.53
Umbriel	266	585	117	1.40	0.53
Titania	436	789	353	1.71	0.62
Oberon	583	761	301	1.63	0.60
Neptune					
Triton	355	1,353	2,147	2.05	0.66

element of the surface as well as over what spectral range the value is specified). For many objects, the albedo in the visible spectral range (~0.4–0.7 μm) is by itself a good first-order indicator of whether the surface composition is dominantly icy. This is so simply because most rocks and minerals, particularly the common iron- and magnesium-bearing materials or carbonaceous compounds, are relatively dark compared to the condensed ices of the common volatiles (e.g., water, carbon dioxide, ammonia, and methane).

More detailed information can be derived from higher-resolution spectra covering the extended spectral range dominated by reflected sunlight (~0.3–5.0 μm). Many common iron-bearing minerals have significant absorption signatures in the 1–2 μm region, while ices generally have relatively rich spectra with many features throughout the 1–5 μm spectral region (Clark *et al.*, 1986). Telescopic spectra from ground- and space-based observatories provide most of the existing information on satellite surface compositions. Spacecraft data from the Galileo mission have extended these global-scale studies to surface mapping for the Galilean satellites and the Cassini mission is expected to do the same for Saturn's satellites beginning in 2004.

The strong spectral absorptions due to water ice and frost dominate the spectra of most outer planet satellites. Water ice was first firmly identified in infrared spectra of Saturn's rings (Pilcher *et al.*, 1970) and then Europa and Ganymede in 1972 (Pilcher *et al.*, 1972). Since then water ice/frost has been reported as a major component of most outer planet satellite surfaces. Much more difficult has been the identification of the dark, non-ice materials present in varying quantities on these surface. Very low albedo materials appear to be ubiquitous on many objects in the outer solar system—from the asteroid belt to comets. Asteroids with much lower reflectance than common rocks are known to be common in the asteroid belt (Chapman *et al.*, 1975; Matson, 1971) as well as on cometary nuclei (Hanner *et al.*, 1985, 1987; Veeder *et al.*, 1987).

These low-albedo materials are also relatively spectrally featureless. Comparisons with the spectra of meteorites suggest that some of this material is similar to the opaque carbonaceous material found in carbonaceous chondrites (Gaffey *et al.*, 1993; Johnson and Fanale, 1973) and to material referred to as tholins (carbon-rich material formed in laboratory experiments containing complex solid hydrocarbons such as polycyclic aromatic hydrocarbons (Cruikshank *et al.*, 1991)). It is generally assumed that satellite surfaces contain a mixture of ices and silicate material, probably hydrated, mixed with similar dark carbon-rich material. However, detailed identification of these constituents is difficult due to the paucity of distinctive spectral features and to the fact that even relatively small amounts of these compounds mixed with brighter silicate materials and/or ices produce a dark mixture and reduce or eliminate the spectral contrast from absorptions in the higher-albedo materials.

1.24.4.2 Temperature and Atmospheres

Since condensed volatiles, particularly water, are major constituents of the outer planet satellites, temperature plays a key role in determining which species are stable as solids on their surface and in how the surface interacts with any atmosphere present. Water ice has extremely low vapor pressures at the surface temperatures of all of the satellites from Jupiter ($T \sim 130$ K) to Neptune ($T \sim 58$ K), so that in addition to its high abundance it is not surprising to find water ice a major surface and bulk constituent of these satellites. Other volatiles identified to date include sulfur dioxide on Io and methane and nitrogen on Triton. In both cases these more volatile species are also present in these satellites' tenuous atmospheres. In the case of Titan, the chemistry of the nitrogen–methane atmosphere and the surface temperature ($\sim$90 K) combine to strongly suggest the presence of condensed hydrocarbons (solid and/or liquid) on its surface even though we have yet no direct identification of these surface materials (Lorenz and Lunine, 1997; Lunine, 1993; Lunine et al., 1983). In the special case of volcanic Io, measurements of the temperature of erupting lava also place constraints on the composition of the lava (T. V. Johnson et al., 1988; Lopes et al., 2001; McEwen et al., 1998b).

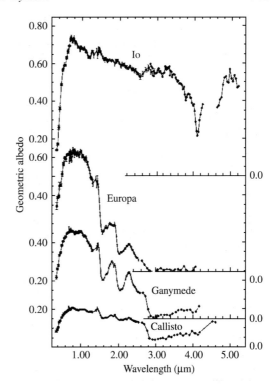

Figure 1 Spectral geometric albedo (1.0 = white disk of same radius) of the Galilean satellites. Hemispheric average values from telescopic observations (after Clark and McCord, 1980a).

1.24.4.3 Radiation Effects

Most outer planet satellites orbit within the magnetospheric environments of their primary planet. As a result their surfaces are constantly exposed to bombardment by particle radiation of varying types, ranging from low-energy (10–100 eV) plasma electrons and ions to very high-energy electrons and >10 MeV/nucleon heavy ions. Jupiter has the strongest magnetic field and the most severe environment, but all of the outer planets share these characteristics to some degree. The effects of this radiation exposure on surface chemistry include ion implantation, sputtering of surface material, and radiation mediated chemical reactions (referred to as "radiolysis" in the literature). See Cheng et al. (1986) for a review of magnetosphere-satellite interactions.

1.24.5 THE JUPITER SYSTEM

1.24.5.1 General

The Galilean satellites of Jupiter, discovered in 1610 by Galileo, are the most easily observed outer planet satellites because of their size and relative proximity to the Earth. Figure 1 shows the hemispheric-scale telescopic spectra of the satellites (Clark and McCord, 1980a). Immediately apparent are the high visible reflectance of all these objects. The darkest, Callisto, is still approximately twice as reflective as the Earth's moon. All also exhibit low ultraviolet reflectance, with Io having the most extreme spectral contrast. The other striking feature of these spectra is the strong contrast between Io's high infrared reflectance from 1 μm to 5 μm compared with the spectra of Europa, Ganymede, and Callisto, which all exhibit strong absorption features typical of water, ice, and frost in this spectral region. The following sections review the geochemical information for each of the major satellites.

1.24.5.2 Io

Io is unique among the outer planet satellites in being an essentially rocky, silicate world with no evidence for significant amounts of the water ice ubiquitously present in the other satellites. Gravitational data and models suggest a large iron and/or iron sulfide core with a silicate mantle and crust (Anderson et al., 2001a). Io's high albedo and striking color (due to very low ultraviolet and blue reflectance) seem at odds with this silicate-rich character, a puzzle only recently resolved with advent of detailed spectral and imaging data from the Galileo mission.

1.24.5.2.1 Surface composition and volcanism

Sulfur and sulfur compounds and their relationship to volcanic silicate rocks are at the heart of Io's surface chemistry. The satellite's low ultraviolet albedo combined with high visible and near-infrared reflectance suggested elemental sulfur to a number of researchers studying telescopic spectra (Wamsteker, 1973; Wamsteker et al., 1974), although laboratory measurements of pure sulfur differ somewhat from Io's average color. The presence of sulfur ions detected in Jupiter's magnetosphere near Io (Kupo et al., 1976) also pointed toward an Io source of sulfur.

A major advance in our understanding of Io came from the discovery of active volcanism by Voyager investigators (Morabito et al., 1979; Smith et al., 1979b), driven by tidal heating predicted theoretically by Peale et al. (1979) just prior to the Voyager encounters. At the same time, sulfur dioxide gas was identified in one of Io's geyser-like plumes (Pearl et al., 1979) and sulfur dioxide frost was identified as a major surface constituent (Fanale et al., 1979; Smythe et al., 1979). Two major types of volcanic features are recognized in the Voyager data set: volcanic landforms such as flows and volcanic calderas, frequently associated with elevated infrared signatures, also referred to as "hotspots" in the literature (Pearl and Sinton, 1982) and the striking eruption plumes of dust and gas typically rising 50–150 km above the surface.

Io's volcanic landscape combined with its striking coloration led to two opposed working hypotheses for the composition and origin of the volcanic flows and landforms: (i) sulfur volcanism—envisioned as occurring in a hundreds of meters to kilometers thick global sulfur-rich crust overlying a silicate base where molten silicate eruptions provided the heat input to mobilize sulfur eruptions on the surface (Sagan, 1979) and (ii) silicate volcanism—in this model the primary volcanic fluids and mechanisms are similar to terrestrial basaltic volcanism and the surface colors result from surficial deposits of sulfur compounds (Carr et al., 1979).

Voyager infrared data indicated maximum temperatures of 300–500 K (Pearl and Sinton, 1982), consistent with either liquid sulfur or cooling silicate lava. The absence of measurements suggesting the higher temperatures ($>\sim1,000$ K) that would be associated with active silicate eruptions tended to favor the sulfur hypothesis. The strongest argument in favor of sulfur volcanism came from differences in colors on flow features attributed to changes in sulfur optical properties and viscosity with temperature. Structures such as kilometers deep volcanic calderas, alternatively, suggested material strength greater than that expected for deep layers of sulfur (Clow and Carr, 1980), favoring molten silicates as the primary volcanic fluid (Carr, 1986). See Nash et al. (1986) for a review of both models.

Volcanic regions on Io can be detected from Earth since they produce intense infrared radiation in the 4–20 μm spectral region. This radiation can be measured directly by Earth and space-based telescopic instruments. The characteristics of Io's volcanically heated regions on a hemispheric scale can thus be studied remotely. Telescopic infrared observations have been used to measure Io's heat flow (Matson et al., 1981; Morrison and Telesco, 1980), study variations in volcanic emission with longitude (Johnson et al., 1984) and monitor temporal variations in activity (Sinton and Kaminski, 1988; Veeder et al., 1994). Telescopic imaging of Io in eclipse with increasing sophisticated instrumentation as well with the Hubble Space Telescope has allowed detailed study of Io's volcanism in space and time (McGrath et al., 2000; Spencer and Schneider, 1996; Spencer et al., 1997; Stansberry et al., 1997).

Temporal monitoring resulted in the first direct evidence for active silicate volcanism on Io, when a large increase in Io's infrared signal was detected one night in 1986 in multiple spectral bands. Analysis of this event suggested that material with a temperature of >900 K is necessary to explain the infrared spectral signature, ruling out molten sulfur (Johnson et al., 1988).

The Galileo mission provided over seven years of Io observations between 1995 and 2003, including several close flybys within a few hundred kilometers of the surface. Imaging and near-infrared observations under daytime, nighttime, and eclipsed conditions have yielded new insights and constraints to our understanding of this satellite's volcanism and surface chemistry. Sulfur dioxide absorption features dominate the near-infrared portion of the spectrum observed by the near-infrared mapping spectrometer (NIMS) (see Figure 2). Analyses of the relative strengths of these features allows mapping of SO_2 distribution and frost properties on Io's surface (Carlson et al., 1997; Doute et al., 2001).

Galileo observations of Io's volcanism have resoundingly confirmed high-temperature silicate driven volcanism and show that this is the dominant mode of volcanic activity on Io (Lopes et al., 2001; McEwen et al., 1998b). Imaging data sensitive to short-wavelength thermal radiation in the 1 μm region and NIMS data in the 1–5 μm range can both be used to constrain the eruption temperatures. Observations of numerous active eruptions indicate temperatures not only well above the range expected for sulfur volcanism but also significantly above the $\sim1,500$ K expected for lava of ordinary basaltic composition

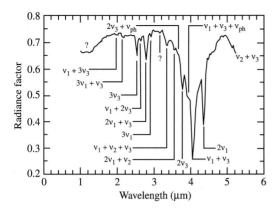

Figure 2 Reflectance spectrum of non-active regions on Io obtained by the Galileo NIMS instrument. Identified absorption features are all due to SO_2 frost (after Carlson et al., 1997).

(Davies et al., 2001; McEwen et al., 1998b). Since molten material will radiatively cool very quickly and form cooler crusts, such temperatures, even at active eruption sites, imply melt temperatures at least this high. If these temperatures are interpreted in terms of the composition of the lava, they must be magnesium-rich ultramafic melts. It has been suggested that volcanism on Io may resemble the conditions inferred for the emplacement of ultramafic, komatiitic lava on the early Earth (Keszthelyi et al., 2001; Williams et al., 2000, 2001a). There are no detailed reflectance spectra of unaltered silicate surfaces from NIMS, but multispectral images from the imaging system show evidence for absorption near 1 μm that is consistent with an ultramafic composition (Geissler et al., 1999b). Other mechanisms, such as superheating the melt under pressure through tidal forces might also produce high temperatures even for more normal composition, but these avenues have not yet been explored quantitatively.

Io's surface coloration is thus apparently controlled by mixtures of sulfur allotropes and sulfur dioxide frost deposited in relatively thin layers on the underlying silicate topography (Geissler et al., 1999b, 2001a; Moses et al., 2002). In addition, although sulfur volcanism is not the primary mode of volcanism there is evidence that localized sulfur flows may occur in some places (Williams et al., 2001b).

Another major Galileo result is the mapping of night-time surface temperatures by the photopolarimeter radiometer (PPR) experiment. These temperature distributions show large thermal anomalies, many of them associated with known volcanic features or high-temperature regions in NIMS maps. They also show lower (but still significantly elevated) temperatures in large areas surrounding and between the discrete anomalies that presumably result from previously emplaced, cooling lava. Significantly, there is little variation in these temperatures with latitude in the nighttime data (Spencer et al., 2000b). It has been suggested that these data imply an additional, mostly high-latitude component to Io's global heat flow (Matson et al., 2003).

The plume features were quickly recognized as being more similar to terrestrial geysers than to explosive volcanism (Smith et al., 1979a). Instead of water, as in the terrestrial case, the available working fluids for Io's geysers are sulfur and sulfur dioxide, both of which can exist as liquids under temperature and pressure conditions in the shallow crust. Heating from volcanic sources at depth or by newly emplaced lava flows provides the energy to drive phase-change volcanism, where hot gas flows isentropically up a vent, becoming mixed with the liquid and/or solid phases as it expands and cools. This process can result in a very rapidly moving mixture of gas and particulates reaching the surface with enough velocity to create an umbrella-shaped plume over a hundred kilometers high under the conditions of Io's low gravity (~1/6 the Earth's—similar to the Moon's) and negligible atmosphere.

Although first considered in the context of low-temperature sulfur volcanism, the geyser model for the plumes ultimately relied on heat from silicate lava, either to mobilize sulfur or as a direct intrusive heat source. A more extensive treatment of the thermodynamics of geyser plume systems on Io, decoupled from any particular assumption about the composition of the underlying volcanism, was given by Kieffer (1982). This work considered plume physics over a large range of potential reservoir conditions ranging from low energy ("sulfur-like") to high energy ("silicate-like"). The recognition that high-temperature silicate volcanism is the dominant mode on Io means that most plume activity will be of the high-energy variety, leading to higher flow velocities and the potential for nearly pure gas phase plumes without significant quantities of condensed particles (Johnson et al., 1995). A detailed high-temperature silicate model with a "rootless conduit" has been proposed to explain one long-lived plume, Prometheus (Kieffer et al., 2000).

1.24.5.2.2 Atmosphere and magnetospheric interactions

Io's atmosphere is essentially a transient, volcanically derived atmosphere, continually renewed by volcanic emissions while ~1 t s^{-1} of material (primarily sulfur and oxygen ions) is injected into the magnetosphere. Sulfur dioxide, existing in all three phases—as subsurface liquid, gas and condensed solid on the surface—is a key component of the atmosphere. Images of Io in eclipse taken by the Galileo imaging team vividly show the various elements of this complex system (Figure 3). In these images volcanic eruption centers are seen as intense spots due to blackbody radiation at 1 μm from hot lava. Diffuse glows, due to excitation by the impact of magnetospheric plasma electrons, show the distribution of the gaseous species. A thin global envelope is probably due to an extremely tenuous atomic oxygen atmosphere, derived from the dissociation of volcanic sulfur dioxide. Localized diffuse glows are seen in the vicinity of active volcanic centers and plume activity and are believed to result from sulfur dioxide emission (Geissler et al., 1999a; 2001b; McEwen et al., 1998a).

Io's interaction with Jupiter's magnetosphere, resulting in loss of material from Io and the formation of the "Io plasma torus," plays a key role in the energetics of the magnetosphere as a whole. In addition to sulfur and oxygen, Io-genic neutral and plasma species detected to date include sodium, potassium, and chlorine (Brown, 1974; Kuppers and Schneider, 2000; Trafton, 1975). Other species have been detected in plumes or atmosphere, including SO (de Pater et al., 2002; McGrath et al., 2000; Russell and Kivelson, 2001; Zolotov and Fegley, 1998) and S_2 (Spencer et al., 2000a). The physics controlling these processes is quite complex and a detailed discussion is beyond the scope of this chapter. Studies of the gases in volcanic plumes, Io's atmosphere and the plasma torus bear on the chemistry of Io, however. Sulfur species in the plumes have been used to estimate oxygen fugacity in the interior of Io (Zolotov and Fegley, 1999), and volcanically derived sodium chloride has been suggested as the source for sodium and chlorine in the system (Fegley and Zolotov, 2000; Lellouch et al., 2003).

1.24.5.3 Ganymede and Callisto

The "ice giants" Ganymede and Callisto are in many ways at the other end of the evolutionary spectrum from rocky, youthful, volcanic Io, having bulk compositions with approximately equal proportions of ice and rock, and ancient, cratered surfaces. Because of these similarities in some global properties, Ganymede and Callisto are usually considered together. The major

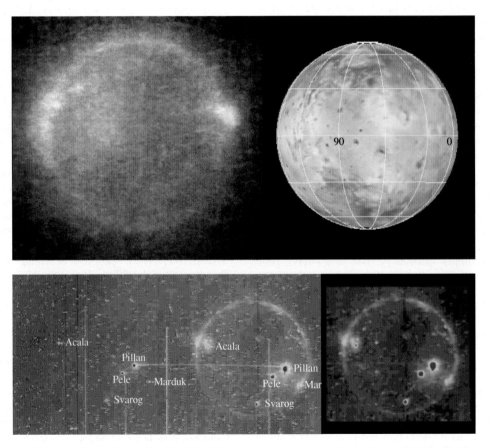

Figure 3 Images of Io in eclipse taken by the Galileo imaging team (photos courtesy of NASA/Jet Propulsion Laboratory).

differences between these satellites relate to Ganymede's generally more youthful or evolved character. Ganymede has apparently differentiated completely into a rock-iron rich mantle and/or core with a thick layer of ice overlying it. Callisto's moment of inertia suggests that it is not completely differentiated although it may have an outer layer that is depleted in heavier material (Anderson et al., 1996, 1997b, 1998a, 2001b). Ganymede also has an intrinsic magnetic field, presumably generated by a dynamo process in a molten iron–iron sulfide cone (Kivelson et al., 1996, 1997a, 1998). Callisto does not appear to have an intrinsic field, however (Khurana et al., 1997). In addition, Ganymede's surface exhibits a range of geologic provinces, including darker, older terrain similar in many ways to Callisto's battered surface as well as extensive regions where older terrains have been tectonically altered by fracturing and faulting (Collins et al., 1998; Pappalardo et al., 1998; Prockter et al., 2000).

Magnetic field observations from the Galileo mission indicate that both these satellites (as well as Europa) display a time-variable inductive response to Jupiter's imposed field, in addition to Ganymede's fixed internal field. The strongest inductive response is Callisto's (Khurana et al., 1998; Neubauer, 1998). Models of this interaction suggest that a global layer of moderate electrical conductivity is required to produce this response. The electrical properties of candidate crustal silicate and ice materials are insufficient to produce the observed signature. Liquid salty ocean layers lying perhaps 150 km below these icy surfaces have been proposed as the best explanation for the satellites' inductive response (Khurana et al., 1997, 1998; Kivelson et al., 2002; Zimmer et al., 2000).

1.24.5.3.1 Surface composition

The reflection spectra of Ganymede and Callisto (Figure 1) show many of the same features, dominated by absorptions due to water and hydrated silicates (Calvin and Clark, 1991; Clark et al., 1986; Clark and McCord, 1980a; Pilcher et al., 1972). Their relatively low ultraviolet reflectance has been attributed to implantation of sulfur ions from the magnetospheric plasma that interacts with their surfaces (Clark et al., 1986; Lane and Domingue, 1997; Nelson et al., 1987; Noll et al., 1997a).

In contrast with the relatively straightforward interpretation of the water ice/frost absorptions in Ganymede and Callisto's spectra, identification of the non-ice material on these satellites has proved more difficult. Although small amounts of dark, opaque material mixed with bright ices are sufficient to drastically lower the visible albedo of the mixture, strong water absorptions still dominate the spectral regions in the near infrared, where many silicates have diagnostic features (Clark, 1981a,b; Clark and Lucey, 1984). Also, dark material similar to the opaque phases in primitive meteorites is either spectrally neutral or lacks distinctive spectral features over most of the reflected spectral region. This type of material is believed to dominate the surfaces of low-albedo asteroids and comet nuclei and is a good candidate for the satellite non-ice component.

Galileo spectra from the NIMS experiment provided critical new information on the composition of this non-ice component on Ganymede and Callisto. In the 3–5 μm spectral region, water ice and frost is essentially black. Thus any non-ice material on the surface, even if it is low albedo, will dominate reflectance in this region. This critical spectral range had not been explored prior to Galileo due to strong Earth atmospheric absorptions limiting terrestrial observations and sensitivity limiting space-based observations from ISO (European Space Agency's (ESA's) Infrared Space Observatory).

Figure 4 shows details of both satellites' spectra in the 3–5 μm range from Galileo NIMS data (McCord et al., 1997, 1998a). The two are very similar with the exception of higher signal-to-noise ratios for the Callisto spectra. This results from the greater proportion of non-ice material included in the field of view for Callisto observations compared with Ganymede data. Several absorption features are evident; the strongest centered around 4.25 μm. This has been confidently identified as due to carbon dioxide (the opacity of the Earth's atmosphere in this whole spectral region results primarily from the absorption spectrum of gaseous carbon dioxide). The state of the carbon dioxide is less clear. The shape of the absorption and the location of the central frequency rule out a gaseous phase. The location of the absorption is also not compatible with pure solid carbon dioxide frost, being somewhere between the theoretical position for liquid and solid. McCord et al. (1997, 1998a) have analyzed these spectra in detail and suggest that the carbon dioxide exists as small deposits of molecules, possibly in radiation induced inclusions in the matrix water ice or silicate material.

The CO_2 feature at Callisto is strong enough in individual NIMS spectra that its distribution on the surface can be mapped. The observed distribution shows a strong asymmetry, suggesting higher concentrations on the trailing side, which is being impacted preferentially by magnetospheric plasma, lending general support to a role for radiation effects in creating the surface conditions for the condensed carbon dioxide inclusions. On a finer scale however, the CO_2 concentrations

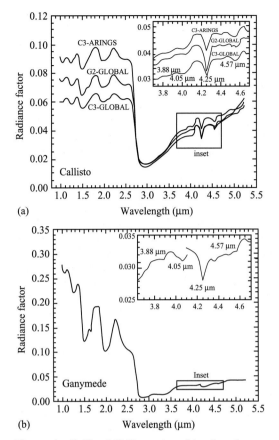

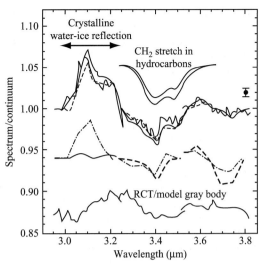

Figure 5 Callisto spectra from NIMS observations in the 3.0–3.8 μm spectral region (continuum removed). The complex absorption feature centered near 3.4 μm is attributed to CH$_2$. Note the similar features in the inset laboratory spectrum of hydrocarbons. The bottom curve shows a calibration target spectrum to assess systematic effects (after McCord et al., 1998a) (reproduced by permission of American Geophysical Union from *J. Geophys. Res.: Planets* **1988**, *103*, 8603–8626).

Figure 4 Galileo NIMS spectra of local regions on the surfaces of Ganymede and Callisto. The labels of individual curves for Callisto refer to individual spacecraft observation blocks during the second (G2) and third (C3) Galileo orbits. The insets show an expanded scale in the spectral region where features due to CO$_2$, S compounds and C≡N bearing materials have been identified (see text) (after McCord et al., 1998a) (reproduced by permission of American Geophysical Union from *J. Geophys. Res.: Planets* **1988**, *103*, 8603–8626).

are clearly influenced by the local geologic setting, particularly impact craters, indicating that the processes involved in producing this signature are more complex (Hibbitts et al., 2002, 2000).

In addition to the carbon dioxide feature, four other absorption features can be identified in these spectra. A weaker absorption identified around 3.4 μm (Figure 5) is attributed to a C–H stretch feature in CH$_2$, in a hydrocarbon of some variety. Three stronger features are centered at 3.88 μm, 4.05 μm, and 4.57 μm (Figure 4). Identification of these features is based primarily on their central frequencies (McCord et al., 1997, 1998a). The 3.88 μm feature is at the correct position for an S–H stretch and is consistent generally with the other evidence for S ion implantation from Io-genic plasma. The 4.05 μm feature is similar to Io's strong sulfur dioxide absorption and is attributed to sulfur dioxide also, although the position of the feature is slightly shifted from SO$_2$ frost, suggesting as in the case of CO$_2$ a more complex condensed state for this species. The feature at 4.57 μm has been identified as due to a C≡N bond based on its frequency. This feature is present in laboratory spectra of opaque hydrocarbon-rich "tholins" (McCord et al., 1997, 1998a) but the detailed nature of the carbon- and nitrogen-bearing molecules cannot be determined solely from these spectra.

These identifications are not as firm as for the stronger carbon dioxide feature. It is interesting to note in this context that there are similar features attributed to C–H, CO$_2$, and C≡N in ISO spectra of interstellar ice grains (Whittet et al., 1996). The interstellar ice spectra lack features due to sulfur, which may be specific to the satellites' magnetospheric environment and Io-genic sulfur. Carbon monoxide is also present in the interstellar grains but absent in the satellites' spectra, presumably because of its higher volatility at Jupiter system temperatures. This suggests that material similar to the interstellar grains may be incorporated in satellites during their formation and from subsequent delivery of primitive comet and asteroidal material.

Another potential non-ice component has been reported based on distortion of the usually symmetrical water absorptions in the 1–2.5 μm region in the spectra from some areas on Ganymede. Similar features on Europa have been attributed to heavily hydrated salt compounds or frozen sulfuric acid (see section on Europa) and their presence on Ganymede raises the possibility that similar processes involving briny fluids and radiolysis chemistry may be occurring here as well (McCord et al., 2001a).

1.24.5.3.2 Atmospheres and magnetospheric interactions

As for Io, the surfaces of Ganymede and Callisto are intimately connected to both radiation and atmospheric processes. Optical spectra have identified molecular oxygen in a "condensed state," trapped in the ices on both satellites, as well as Europa. The characteristics of this oxygen component have been attributed to interaction with the plasma environment (Calvin and Spencer, 1997; Spencer and Calvin, 2002; Spencer et al., 1995) and may be related to the processes controlling the formation and distribution of carbon dioxide discussed above. In addition, ozone has been detected on Ganymede and attributed to the presence of "micro-atmospheres" of O_2 and O_3 trapped in the ice (Noll et al., 1996).

In addition to trapped gases in their surfaces, Ganymede and Callisto both have tenuous extended atmospheres as well. Molecular oxygen has been identified as one constituent of Ganymede's atmosphere (Hall et al., 1998) with a column density of $\sim 5 \times 10^{14}$ cm^{-2}. Galileo observations also show a hydrogen exosphere (Barth et al., 1997) and evidence for proton outflow (Frank et al., 1997). Callisto's atmosphere is of similar density ($\sim 8 \times 10^{14}$ cm^{-2}) but the identified constituent is CO_2 (Carlson, 1999).

1.24.5.4 Europa

Europa is primarily a rocky satellite, but falls into the category of Jupiter's "icy" satellites by virtue of its icy outer layer. Gravity data suggests that Europa is differentiated with a core–mantle structure surmounted by 75–150 km of low-density water-rich crust (Anderson et al., 1997a, 1998b). Europa apparently does not possess an intrinsic magnetic field, but does exhibit the same inductive response to Jupiter's field that Ganymede and Callisto show (Khurana et al., 1998; Kivelson et al., 1997b, 1999). There are very few large impact craters evident on Europa's surface, indicating a geologically young surface, less than ~100–200 Myr. Combined with geological evidence for extensive resurfacing and fracturing of the crust, these data strongly suggest that a global liquid-water ocean lies only kilometers to tens of kilometers beneath the surface and that the ocean has communicated with the surface in the geologically recent past (Carr et al., 1998; Greeley et al., 1998, 2000).

1.24.5.4.1 Surface composition

Europa's surface reflection spectra are dominated almost exclusively by water absorptions. Figure 1 shows a typical telescopic spectrum of Europa. Analyses of ground-based telescopic spectra suggested that very pure water ice/frost of varying particle sizes was responsible for most if not all of the near-infrared spectral features (Calvin et al., 1995; Clark et al., 1986; Clark and McCord, 1980a; Pilcher et al., 1972). Low ultraviolet reflectance has been attributed to sulfur ion implantation as well as sulfur dioxide, creating a distinctive spectral signature below 0.3 μm (Lane et al., 1981; Nelson et al., 1987). The spatial distribution of the low ultraviolet reflectance signature in telescopic spectra and Voyager multispectral images strongly suggests a magnetospheric source, with the hemisphere exposed to overtaking magnetospheric plasma having a diffuse signature of low ultraviolet material (Johnson et al., 1983, 1988; McEwen, 1986; Nelson et al., 1986).

Hemispheric differences in the water absorption are seen in telescopic spectra, with distinctly different shapes to the absorption features in the 1–3 μm spectral region on the leading (centered on 90° W) and trailing hemispheres (270° W). Leading hemisphere spectra exhibit strong, symmetric absorptions characteristic of clean water ice/frost spectra, while the trailing hemisphere spectra have broader, asymmetric absorptions (Calvin et al., 1995; Clark et al., 1986). Galileo observations provided the first detailed mapping of these water absorption features on the surface with the NIMS experiment. The asymmetric spectral signature is closely associated with disrupted regions and the ridge systems that crisscross the satellite's surface. Figure 6 shows a comparison of ice-rich and ice-poor spectra for each of the icy satellites (McCord et al., 1999) where the asymmetric water features are evident in the ice-poor regions. These areas also typically have somewhat lower reflectance and redder color generally (Fanale et al., 1999; McCord et al., 1998b, 1999).

The cause of the Europa's asymmetrical water absorptions and the identity of any non-ice constituent on the surface have both been the subject of extensive analysis and debate. Large particle sizes and very clear ice with long optical path lengths were proposed to explain some but not all aspects of the anomalous telescopic spectra (Calvin et al., 1995; Clark et al., 1983). Analyses of Galileo NIMS data have produced two alternative explanations. Both propose that the asymmetric absorptions are due fundamentally

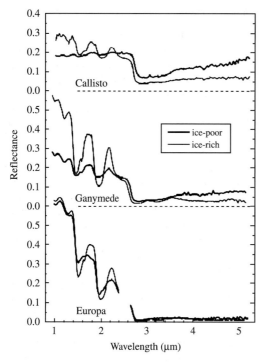

Figure 6 Examples of Galileo NIMS spectra from ice-rich (thin lines) and ice-poor (thick lines) regions from the three icy Galilean satellites. The ice-poor regions are interpreted as containing mixtures of water ice with darker hydrated minerals (Callisto and Ganymede) and heavily hydrated salts (Europa and Ganymede) (after McCord et al., 1999) (reproduced by permission of American Geophysical Union from *J. Geophys. Res.: Planets* **1999**, *104*, 11827–11851).

to water in very heavily hydrated compounds, but differ on the nature and origin of the hydrated species. McCord et al. (1998b, 1999) have suggested that hydrated sulfate salts, containing sodium and magnesium, are the best candidates. In their scenario these materials are brought to the surface in salty ice or by the extrusion of salty water from the subsurface ocean. Carlson et al. (2002) also invoke sulfate chemistry but as hydrogen sulfate (or sulfuric acid), produced by the interaction of magnetospheric radiation with sulfur compounds on the surface.

Spectrally there is little to distinguish between the two opposing hypotheses, since the distortion of the water absorptions in both cases is primarily due to the highly hydrated sulfate groups with little effect from the cation involved. Indeed, there are points of agreement between the two with respect to the likelihood of sulfuric acid production in the environment and the possibility that salts may be part of the proposed radiolytic cycle. Observations of neutral atoms in the vicinity of Europa tend to favor a sodium-bearing constituent. Europa has an extended atmosphere containing both sodium and potassium (Brown, 2001; Brown and Hill, 1996). Potassium-to-sodium ratios at Europa differ significantly from those at Io, further suggesting an indigenous source of sodium at Europa (Brown, 2001). Analyses of these observations and comparisons with the amount of Io-derived sodium delivered to the surface shows that Europa is a net source of sodium to the neutral atom clouds associated with the satellites (Leblanc et al., 2002).

Salts are not entirely unexpected from a geochemical point of view. Several studies of the interaction of liquid water with silicates in satellite interiors suggest that salty brines are a likely result (Fanale et al., 1977, 2001; Kargel, 1991). Further, these materials are relatively stable on Europa's surface even in the presence of radiation (McCord et al., 2001b; Zolotov and Shock, 2001).

Other evidence for non-ice materials comes from the 3–5 μm region where non-ice features show up as in Ganymede's and Callisto's spectra. However, even the higher spatial resolution Europa spectra from NIMS contain large amounts of ice surface in their fields of view and the signal-to-noise ratio in this spectral region is worse than for the larger icy satellites. Carbon dioxide absorptions, similar to those seen in the other icy satellite spectra, are detectable in some Europa spectra, suggesting that both carbon-bearing materials and surface radiation effects are present on this satellite also (McCord et al., 1999).

Other spectral features in the 3–5 μm region are more difficult to assess due to the low signal-to-noise mentioned above. A spectral feature near 3.4 μm appears different in detail from the feature attributed to C–H in the same spectral region on Ganymede and Callisto. Analyses of this feature by (Carlson et al., 1999) attribute it to hydrogen peroxide, H_2O_2. Radiolytic chemical processes resulting from charged particle irradiation of the surface water ice are proposed as the source of this constituent, once again highlighting the effects of the satellites' magnetospheric environment on surface chemistry.

Finally, ammonia and methane have both been proposed as minor constituents of the Galilean satellites in hydrated or clathrated form, possibly lowering the melting point in their interiors and promoting ice (or cryo-) volcanism (Hogenboom et al., 1997; Kargel, 1992; Lewis, 1973). However, as of the early 2000s, there has been no identification of these materials on the satellite surfaces.

1.24.5.4.2 Atmosphere and magnetospheric interactions

Observations have identified the same condensed oxygen signature as seen for the other icy satellites (Spencer and Calvin, 2002). In addition, a tenuous molecular oxygen atmosphere has been identified from Hubble Space Telescope spectra

(Hall et al., 1995). These data all suggest an array of atmospheric processes dominated by the evaporation and sputtering of water from Europa's surface (Shematovich and Johnson, 2001). The signature of these processes in the magnetosphere has been detected by measurements made by the neutral particle imaging experiment on Cassini during its flyby of Jupiter in 2000. These data show a gas torus of water products at Europa's orbit, similar to the sulfur and oxygen torus produced by Io's supply of material to the magnetosphere (Mauk et al., 2003).

1.24.6 THE SATURN SYSTEM

1.24.6.1 General

Saturn's retinue of satellites is qualitatively quite different from Jupiter's fellow travelers. The system contains just one large satellite, Titan, which is virtually identical in bulk properties to Ganymede and Callisto. Titan's density, determined from Voyager observations, suggests a ice/rock composition and a probable differentiated interior by analogy with the Jupiter satellites. Titan's atmosphere, the first discovered for a planetary satellite, was detected in 1944 through identification of methane gas absorptions in its spectrum (Kuiper, 1944).

The rest of the satellites are all significantly smaller than any of the Galilean satellites. Telescopic spectral evidence illustrated in Figure 7 shows that water ice/frost is the dominant, ubiquitous surface constituent in the system (Clark et al., 1986; Clark and Lucey, 1984; Fink et al., 1976; Johnson et al., 1975), including the ring (Clark and McCord, 1980b; Pilcher et al., 1970). Current knowledge of their densities comes from a combination of direct determinations from Voyager flybys and dynamical determinations from the numerous mutual perturbations and orbital resonant conditions in the system (see Table 1). These densities show that all the satellites are ice-rich, with no equivalents of rocky Io and Europa. Voyager images of the surfaces of the small icy satellites show mostly old cratered surfaces with some evidence for resurfacing and tectonic processes (Morrison et al., 1986; Smith, 1981; Smith et al., 1982). An outstanding exception is Enceladus, which has extensive crater-free areas and a nearly uniform high-albedo surface suggesting active resurfacing. Iapetus with its hemispheric dichotomy—a relatively bright icy surface on one hemisphere and very low albedo, red material on the other, and Titan with a thick chemically complex atmosphere are discussed below in more detail.

1.24.6.2 Iapetus

Although Iapetus' extreme hemispheric albedo contrast has been recognized essentially from the time of its discovery, the nature and origin of the dark material that dominates the leading hemisphere is still uncertain. Spectrally, the dark material resembles some low-albedo, red asteroids. This material is believed to be similar to hydrocarbon-rich materials produced in laboratory experiments, known as tholins, which are opaque, contain PAHs and are spectrally red (Cruikshank et al., 1991). Tholins are also candidates for at least some of the dark, non-ice constituents on Ganymede and Callisto. Unfortunately, spectra in the 3–5 μm region where these material have some spectral features are not yet available for Saturn satellites.

Voyager images are not of sufficiently high resolution to determine definitely the geologic relation of the dark material to the icy regions. Proposals for the origin of this material have ranged from exogenic (material from Phoebe swept up by Iapetus) to endogenic (dark hydrocarbon material extruded on the surface) to combined hypotheses (indigenous material modified by exogenic impact processes). More extensive discussions of possible composition of the dark material and its origins following the Voyager encounters can be found in the review by Morrison et al. (1986) and papers by Cruikshank et al. (1983) and Bell et al. (1985).

1.24.6.3 Titan

The complexity of Titan's atmospheric chemistry began to be apparent when telescopic infrared spectroscopy detected a number of hydrocarbons in addition to the methane discovered in 1944 (Danielson et al., 1973; Gillett, 1975; Kuiper, 1944). In 1981 Voyager observations determined that Titan's atmosphere is primarily nitrogen with methane as a minor constituent. The surface temperature is ~94 K and the surface pressure is high, ~1.5 times the Earth's (Hanel et al., 1981; Lindal et al., 1983; Tyler et al., 1981). Complex photochemistry in the atmosphere produces a rich array of hydrocarbons (Strobel, 1982; Yung et al., 1984), which have been identified in Voyager infrared spectra (Hanel et al., 1981; Kunde et al., 1981; Lutz et al., 1981, 1983a,b; Maguire et al., 1981; Samuelson et al., 1981). The currently identified atmospheric species are given in Table 2.

Optically the atmosphere is dominated by an opaque reddish aerosol haze produced by photochemical processes that masks the surface at visible wavelengths. As a result, little is known directly of the surface geology or composition.

Presumably, as with Callisto and Ganymede, the crust and mantle are primarily water ice. Models of the atmospheric chemistry however suggest that the surface should receive a continual "rain" of hydrocarbon aerosols, some of which may be liquid under Titan surface conditions (Lunine, 1993; Lunine *et al.*, 1983).

The haze layers are penetrable by radar (Muhleman *et al.*, 1990) and by infrared images made in "windows" between the strong methane absorptions features in the atmospheric spectra (Smith *et al.*, 1996). These data show that the surface is variegated in radar scattering properties and in near-infrared albedo (Griffith, 1993; Lorenz and Lunine, 1997; Smith *et al.*, 2002), mitigating against a uniform global layer of liquid hydrocarbons. Clouds in the atmosphere have also been detected in infrared images (Griffith *et al.*, 2000). Analyses of the relative albedoes in different spectral windows suggest water ice exposed in some regions (Coustenis *et al.*, 1995). Recent studies confirm the presence of water ice and suggest that some areas may resemble Ganymede's surface, with relatively high-albedo water ice exposed (Griffith *et al.*, 2003).

1.24.6.4 Magnetospheric Interactions

Although less intense than Jupiter's environment, the Saturn magnetosphere contains both low-energy plasma and high-energy radiation that interacts with the satellite surfaces. It is likely that oxygen and hydroxyl ions in the magnetosphere result from material sputtered from the

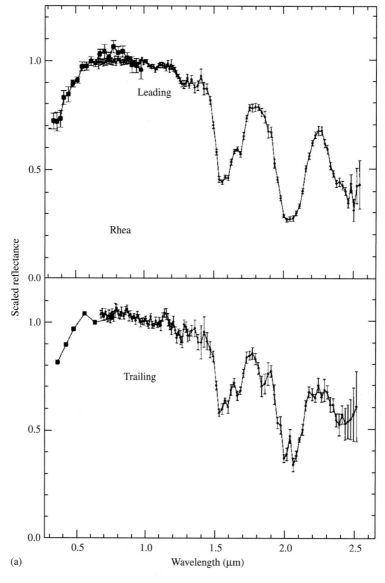

Figure 7 Spectra of the leading and trailing sides of (a) Rhea, (b) Iapetus and (c) one hemisphere of Hyperion (after Clark *et al.*, 1984) (reproduced by permission of Elsevier from *Icarus* **1984**, *58*, 265–281).

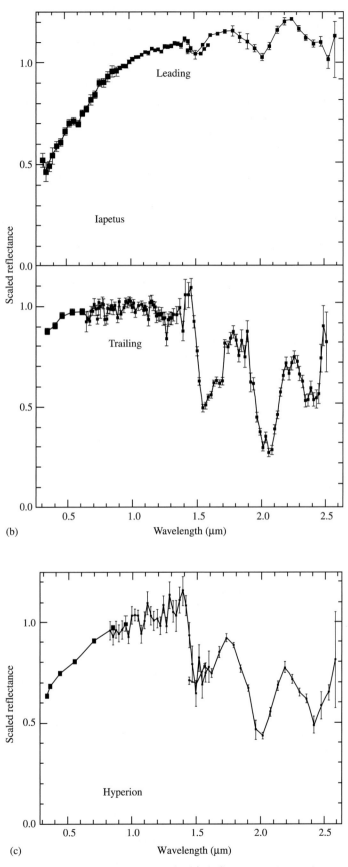

Figure 7 (continued).

Table 2 Satellite composition summary.

Planet Satellites	Surface composition (including "condensed" trapped species)		Atmospheric composition	
	Major	Minor	Major	Minor
Jupiter				
Io	Silicate, possibly ultramafic	SO_2, S_x, NaCl	SO_2	O, SO, S_2, Na, K, NaCl
Europa	H_2O	H_2O_2, $X_y SO_4 \cdot NH_2O$, SO_2, CO_2, O_2,	O_2	Na, K, H
Ganymede	H_2O	CO_2, CH_2, C≡N, H—S, $X_y SO_4 \cdot NH_2O$ SO_2, O_2, O_3	O_2	H
Callisto	H_2O, hydrated silicates	CO_2, CH_2, C≡N, H—S, $X_y SO_4 \cdot NH_2O$ SO_2, O_2	O_2?	CO_2
Saturn				
Mimas	H_2O			
Enceladus	H_2O			
Tethys	H_2O			
Dione	H_2O			
Rhea	H_2O			
Titan	H_2O, Hydrocarbons		N_2	Ar, CH_4, H_2, C_2H_6, C_2H_2, C_3H_8, C_2H_4, C_4H_2, HCN, CO, CO_2, H_2O
Iapetus	H_2O, dark material (?)			
Uranus				
Miranda	H_2O, dark material (?)			
Ariel	H_2O, dark material (?)			
Umbriel	H_2O, dark material (?)			
Titania	H_2O, dark material (?)			
Oberon	H_2O, dark material (?)			
Neptune				
Triton	H_2O, CH_4, N_2	CO_2, CO	N_2	CH_4, photochemical hydrocarbons

satellite surfaces. Ozone signatures similar to those seen on the icy satellites of Jupiter have also been reported for Rhea and Dione (Noll *et al.*, 1997b), and are presumably related to radiation mediated chemistry occurring in the surface ices.

1.24.7 THE URANUS SYSTEM

1.24.7.1 General

Due to the unique orientation of Uranus's spin axis, and therefore the satellites' orbits, only the southern hemispheres of these satellites have been sunlit and available for study from the Earth during the last few decades. Albedo and spectral information as well as visible imaging from Voyager are thus limited to only one half of the satellites' surfaces. Water-ice spectral features are seen in all the satellites' reflection spectra (Brown and Clark, 1984; Brown and Cruikshank, 1983, 1985; Cruikshank, 1980; Cruikshank and Brown, 1981, 1986). Figure 8 shows spectra of several of the Uranus satellites, along with Hyperion, as smaller Saturn satellite with similar optical properties. The moderately low albedo and the intensities of the absorptions indicate that the surface ice is

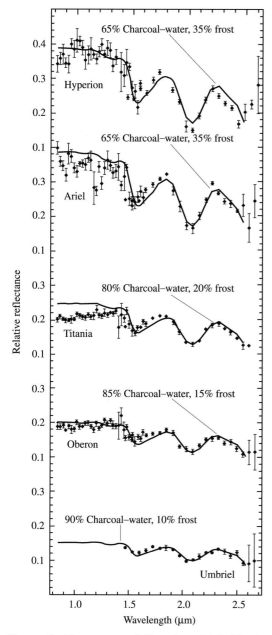

Figure 8 The spectra of Hyperion, Ariel, Titania, Oberon, and Umbriel (Brown and Cruikshank, 1983), compared with laboratory spectra of charcoal and ice mixtures (after Cruikshank and Brown, 1986) (reproduced by permission of the University of Arizona Press from *Satellites* **1986**).

mixed with some darker, spectrally neutral material as well, as shown by the comparison with charcoal–water-ice laboratory mixtures.

Voyager density determinations from optical navigation and radio tracking (Smith *et al.*, 1986; Tyler *et al.*, 1986) suggest bulk compositions with both rock and ice. The proportions of the rock (including carbon constituents) are relatively high compared with the Saturn small satellites, 45–65% (Brown *et al.*, 1991b; Johnson *et al.*, 1987).

The innermost of the larger satellites, Miranda is the only exception, having a low-silicate mass fraction more similar to the Saturn satellites Rhea and Mimas (Figure 9). As can be seen in Figure 9, the rock/ice ratios in this system do not agree with those expected from either a kinetically inhibited solar nebula model (CO-rich) (Lewis and Prinn, 1980), or a relatively massive circumplanetary sub-nebula (CH_4-rich) (Prinn and Fegley, 1989). Other, more complex, models seem to be required. Some of these were reviewed following the Voyager Uranus encounter in Pollack *et al.* (1991), with one possibility considered being formation from a disk blown out by a massive impact on Uranus.

1.24.8 THE NEPTUNE SYSTEM—TRITON

1.24.8.1 General

Neptune's largest moon, Triton, was discovered within weeks of the discovery of the planet itself. It is one of the most distant objects in the solar system. Even the "outermost" planet, Pluto, and its moon, Charon, spend considerable time on their eccentric orbits closer to the Sun than Triton. Its nature remained a mystery until the advent of new astronomical methods in the 1970s and 1980s and the flyby of the Voyager 2 spacecraft in 1989. In many ways, it is a planetary body "on the edge"—on the outer edge of the "main" part of the solar system, and the inner edge of the realm of comets and the recently discovered Kuiper belt objects. As such, it shares some of the characteristics of the icy satellites of the rest of the outer solar system with some of the nature of the colder, more distant, cometary bodies.

Triton's unusual orbit is a strong clue that its origin and history might be very complex. All of the other major moons of the outer solar system orbit their planets in the equatorial plane, with nearly circular orbits and revolving in the same direction as their planet's spin—these are known as "regular satellites." Triton's orbit, although circular, is inclined with respect to Neptune's equator, and the satellite goes around the planet opposite to its spin—known as a retrograde orbit. The other, smaller, "irregular satellites" of the solar system are generally believed to have been captured rather than having formed around their planet. Triton's unique characteristics–relatively large size, closeness to its primary (14.33 Neptune radii), inclination and very circular orbit—have led to a variety of ideas about its possible history. One of the first dynamical explanations for Triton's characteristics involved Pluto as well. The astronomer Raymond Lyttleton proposed in 1936 that both Pluto and Triton had started as satellites of Neptune and that Pluto's escape from the system had also resulted in perturbing Triton's

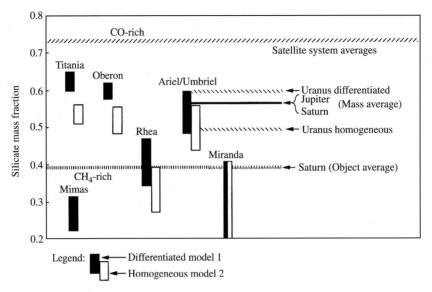

Figure 9 Comparison of silicate mass fractions. Two assumptions for interior structure are shown: (i) differentiated—rock core, ice mantle, and (ii) homogeneous—uniformly mixed ice and rock. Also shown are silicate mass fractions for the Jupiter and Saturn systems and expected values for two models of the early solar nebula carbon chemistry (see text) (after Johnson *et al.*, 1987) (reproduced by permission of American Geophysical Union from *J. Geophys. Res.: Space Phys.* **1987**, *92*, 14884–14894).

orbit into its present configuration. More recent research, however, has cast considerable doubt on this initially intriguing hypothesis and it is now deemed much more likely that Triton began as a body orbiting the Sun and was captured into orbit around Neptune.

It is not easy to devise a simple capture model that leads to all of Triton's characteristics however. There are two classes of capture models which have been extensively discussed: (i) capture by gas drag in a proto-Neptune gas/dust nebula and (ii) capture by collision with an existing regular satellite. Both types of capture can in principle explain Triton's current state, invoking considerable evolution of the initial capture orbit from gas drag and/or tidal dissipation. The two models each have their strengths and weaknesses and there is no conclusive evidence for one or the other, although a recent review of the problem concluded that the collisional capture model was "favored, but not overwhelmingly so" (McKinnon *et al.*, 1996).

1.24.8.2 Composition

Triton's bulk (or average) composition is constrained by its density. With a mean density of 2,000 kg m^{-3} it is one of the densest outer solar system satellites (exceeded only by volcanic Io and mostly rocky Europa). Triton's density suggests a rock-rich composition, ~65–70% silicate fraction. Interestingly, the only other denizen of these outer reaches of the solar system with a well-determined density is Pluto and it is very similar to Triton in this respect—perhaps even rockier, suggesting that the composition and formation conditions in the outer solar system differed somewhat from the circum-planetary nebulae from which the regular satellites formed and strengthening the argument that Triton formed in orbit about the Sun originally instead of around Neptune (see Figure 9 and discussion of Uranus satellites).

In contrast to its "rocky" nature as a whole, frozen ices of several types dominate Triton's surface. It is extremely bright, reflecting ~80% of the incident sunlight, and thus also very cold, even for an object so far from the Sun—only 58° above absolute zero (58 K). Telescopic spectra from Earth-based observatories gave the first hints as to how different Triton is from the satellites of the other outer planets. Instead of the typical absorption features due to water ice in the near-infrared part of the spectrum, observers recorded a very different spectrum, in which absorptions from methane were prominent (Cruikshank and Apt, 1984). Although ices of methane, ammonia, and carbon dioxide had been searched for on other outer satellites, Triton was the first satellite to show clear evidence for surface ices other than water. Even more intriguing, Dale P. Cruikshank and Robert H. Brown interpreted another absorption feature as due to nitrogen—either in liquid or solid form (Cruikshank *et al.*, 1984). Spectral studies of Triton have now detected, in order of volatility: N_2, CO, CH_4, and CO_2. Water ice, expected to be present, but mixed with and obscured by the more volatile frosts, has been detected as well (Cruikshank *et al.*, 2000). The presence of such volatile ices also implied at least a tenuous atmosphere resulting from sublimation of these materials. Figure 10 shows a recent, high-quality spectrum of Triton with the absorption features from the various ice constituents marked.

The most striking feature of Triton's surface is the impression of extreme youthfulness, geologically speaking (Smith et al., 1989). Impact craters, which dominate the landscape of most of the other satellites, are extremely rare on Triton, implying geological processes which have reshaped and/or covered the older surfaces in recent geological times. There is evidence for "cryovolcanic" activity creating flows of icy materials that have flooded and buried parts of the surface much like the mare lava flows on the Earth's moon. Tectonic faulting and fracturing of the surface has occurred in many places. The surface has also been modified by interactions with the atmosphere and Triton's extreme seasonal cycles, with a large southern polar "cap" reaching almost to the equator. Geyser-like eruptions were discovered in these regions by the Voyager cameras as well. The energy for all this activity must be a combination of the normal decay of radioactive materials in Triton's rocky portion, heating by the tides which have shaped its current orbit and solar energy absorbed in the surface ices (Brown et al., 1991a).

The geyser plumes discovered by Voyager make Triton the second outer planet satellite (after Io) with active volcanic eruptions (albeit extremely cold eruptions by terrestrial standards). They are relatively modest in size, rising ~8 km above the surface before winds blow the top of plume horizontal, leaving a long downwind trail of plume particles oddly resembling industrial smokestacks on Earth. Several models to explain the plumes have been suggested (Kirk et al., 1990), including the possibility that they are really dust devils and not geyser-like at all despite their appearance. The geyser models are similar to the sulfur dioxide models developed to explain Io's towering umbrella shaped plumes (Kirk et al., 1996). On Triton it is believed that the gas is either nitrogen or methane and the heat comes from deeply absorbed sunlight in the ice or from internal heat from the rocks and ice below the surface frosts (Duxbury and Brown, 1997).

1.24.8.3 Atmosphere

Voyager's radio occultations, the infrared spectrometer and the ultraviolet spectrometer experiments, all gave us information about the atmosphere. These data are all consistent with a nitrogen atmosphere in what is called "vapor pressure equilibrium." In vapor pressure equilibrium, the gas in the atmosphere comes from the sublimation of ice for the same material frozen on the surface. The amount of gas in the atmosphere is controlled by the temperature of the ice, and the atmosphere acts to keep the ice at a constant temperature by the transport and condensation of the gas from warm to colds areas. Mars' primarily carbon dioxide atmosphere is in a similar equilibrium with its polar carbon dioxide caps,

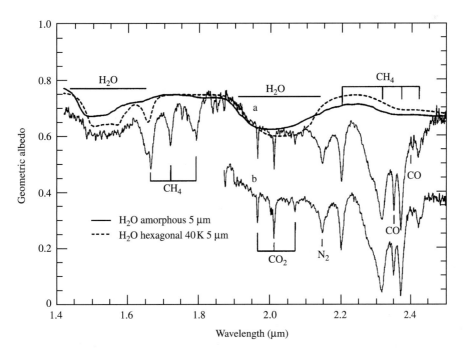

Figure 10 The reflectance spectrum of Triton in the 1.4–2.25 μm region. Locations of identified ice/frost species are marked and two laboratory spectra of water ice are shown (after Cruikshank et al., 2000) (reproduced by permission of Elsevier from *Icarus* **2000**, *147*, 309–316).

but Triton is an even more extreme case, with the frozen volatile being everywhere on the surface. The atmosphere that results from these processes is composed mostly of nitrogen and is in equilibrium with nitrogen ice on the surface at a temperature of 38 K with the atmospheric pressure being only 14 μbar. The atmosphere also contains thin clouds, probably nitrogen ice, and hazes, which are probably the result of the condensation of photochemically produced gases in the lower atmosphere (Strobel and Summers, 1996; Yelle et al., 1996).

The transport of volatile material from one region of Triton to another by sublimation and condensation makes Triton's seasons very important to what we see on the surface. Because of the nature of its orbit and Neptune's tilt from the plane of the ecliptic, Triton undergoes extreme and complicated seasonal changes, with the latitude of the Sun on the surface varying from $\sim 50°$ S to $50°$ N in a complex pattern over hundreds of years. Simple calculations based on the vapor pressure equilibrium described above suggest that the satellite's nitrogen should all be concentrated by now in small, thick polar caps at both poles. Instead, the southern hemisphere cap seen by Voyager—in high "summer" no less—extends almost to the equator. A number of theories have been developed to explain the distribution of frost on the surface with atmospheric models, including the somewhat counterintuitive idea that the freshest ices might be dark, not bright, but there is currently no completely satisfactory theory explaining all aspects of Triton's surface–atmosphere system (Brown and Kirk, 1994). One problem, of course, is that the only spacecraft to visit so far observed only about one-half of Triton. Future missions and advanced telescopic observations at other seasons will probably be necessary to fully understand this distant, cold and yet active and fascinating moon.

1.24.9 MAJOR ISSUES AND FUTURE DIRECTIONS

This survey of the current state of knowledge of satellite geochemistry also uncovers the many areas where there are unsolved problems and major issues still to be addressed by future observations and theoretical work. On the observational side, most of the chemical information we now have for objects beyond the Galilean satellites comes from bulk density and global telescopic spectral data. The Cassini/Huygens mission, which arrives at Saturn in July of 2004, will allow more detailed study of all the satellites of this major system. In particular, Titan is now known primarily through detailed study of its upper atmosphere from Voyager and Earth-based spectra, and we have tantalizing but incomplete information on its surface properties. The Huygens probe will address the chemistry and structure of the Titan atmosphere in unprecedented detail and hopefully reveal the nature of the surface at its landing site, while the Cassini orbiter will provide the first high-resolution information on its surface geology with a combination of radar, imaging, and infrared observations. New spacecraft data for the more distant reaches of the solar system await the New Horizons Pluto/Kuiper Belt mission and orbital missions to Uranus and Neptune still in the study phase.

Another major issue is the composition and nature of the dark non-ice material mixed in various quantities on the satellites. Galileo data has provided some clues for the Jupiter system and Cassini will do a similar job for Saturn. More powerful spectral techniques and in situ analyses (and possibly sample return) will be required for a full characterization and identification of what are believed to be mixtures of complex hydrocarbons and organic material. For Titan, missions that explore both the atmosphere and the surface are being studied as follow-on exploration after Cassini/Huygens.

In the Jovian system, the nature of the putative subsurface oceans of the icy satellites, particularly Europa, needs intensive future study. Issues include the thickness of liquid layers and ice crusts, the history of communication between liquid layers and the surface, and the brine composition of the liquid. The next mission planned to address these issues is the Jupiter Icy Moons Orbiter, which is designed to orbit each of the icy satellites and explore their surfaces and subsurfaces in detail.

Finally, on the theoretical front, the main issues are still those first raised in detail in the 1970s by Lewis: the relationship of the solar nebula, planet formation and satellite system formation to the composition of the satellites. The current information suggests that more complex models are required, including such factors as variations in the carbon chemistry in different systems and the effects of planetary and satellite migration during formation and accretion.

REFERENCES

Anderson J. D., Lau E. L., Sjogren W. L., Schubert G., and Moore W. B. (1996) Gravitational constraints on the internal structure of Ganymede. Nature **384**, 541–543.

Anderson J. D., Lau E. L., Sjogren W. L., Schubert G., and Moore W. B. (1997a) Europa's differentiated internal structure: inferences from two Galileo encounters. Science **276**, 1236–1239.

Anderson J. D., Lau E. L., Sjogren W. L., Schubert G., and Moore W. B. (1997b) Gravitational evidence for an undifferentiated Callisto. Nature **387**, 264–266.

Anderson J. D., Schubert G., Jacobson R. A., Lau E. L., Moore W. B., and Sjogren W. L. (1998a) Distribution of rock, metals, and ices in Callisto. Science **280**, 1573–1576.

Anderson J. D., Schubert G., Jacobson R. A., Lau E. L., Moore W. B., and Sjogren W. L. (1998b) Europa's differentiated internal structure: inferences from four Galileo encounters. *Science* **281**, 2019–2022.

Anderson J. D., Jacobson R. A., Lau E. L., Moore W. B., and Schubert G. (2001a) Io's gravity field and interior structure. *J. Geophys. Res.: Planets* **106**, 32963–32969.

Anderson J. D., Jacobson R. A., McElrath T. P., Moore W. B., Schubert G., and Thomas P. C. (2001b) Shape, mean radius, gravity field, and interior structure of Callisto. *Icarus* **153**, 157–161.

Anderson J. D., Anabtawi A., Asmar S., Jacobson R. A., Johnson T.V., Lau E. L., Lewis G. D., Moore W. B., Schubert G., Taylor A. H., Thomas P. C., and Weinwurm G. (2002) *Determination of the Mass and Density of Amalthea*. EOS, Trans., AGU, **83**, Fall Meeting Suppl., abstract P12C–13.

Barth C. A., Hord C. W., Stewart A. I. F., Pryor W. R., Simmons K. E., McClintock W. E., Aiello J. M., Naviaux K. L., and Aiello J. J. (1997) Galileo ultraviolet spectrometer observations of atomic hydrogen in the atmosphere of Ganymede. *Geophys. Res. Lett.* **24**, 2147–2150.

Bell J. F., Cruikshank D. P., and Gaffey M. J. (1985) The composition and origin of the Iapetus dark material. *Icarus* **61**, 192–207.

Brown M. E. (2001) Potassium in Europa's atmosphere. *Icarus* **151**, 190–195.

Brown M. E. and Hill R. E. (1996) Discovery of an extended sodium atmosphere around Europa. *Nature* **380**, 229–231.

Brown R. A. (1974) Optical line emission from Io. In *IAU Symposium No. 65 Exploration of the Planetary System* (eds. A. Woszczyk and C. Iwaniszewski). D. Reidel, Dordrecht, Holland, pp. 527–531.

Brown R. H. and Clark R. N. (1984) Surface of Miranda—identification of water ice. *Icarus* **58**, 288–292.

Brown R. H. and Cruikshank D. P. (1983) The Uranian satellites—surface compositions and opposition brightness surges. *Icarus* **55**, 83–92.

Brown R. H. and Cruikshank D. P. (1985) The moons of Uranus, Neptune, and Pluto. *Sci. Am.* **253**, 38–47.

Brown R. H. and Kirk R. L. (1994) Coupling of volatile transport and internal heat-flow on triton. *J. Geophys. Res.: Planets* **99**, 1965–1981.

Brown R. H., Johnson T. V., Goguen J. D., Schubert G., and Ross M. N. (1991a) Tritons global heat-budget. *Science* **251**, 1465–1467.

Brown R. H., Johnson T. V., Synnott S. P., Anderson J. D., Jacobson R. A., Dermott S. F., and Thomas P. C. (1991b) Physical properties of the Uranus satellites. In *Uranus* (eds. J. T. Bergstralh, E. D. Miner, and M. S. Matthews). University of Arizona Press, Tucson, pp. 513–527.

Calvin W. M. and Clark R. N. (1991) Modelling the reflectance spectrum of Callisto 0.25 to 4.1 μm. *Icarus* **89**, 305–317.

Calvin W. M. and Spencer J. R. (1997) Latitudinal distribution of O_2 on Ganymede: observations with the Hubble Space Telescope. *Icarus* **130**, 505–516.

Calvin W. M., Clark R. N., Brown R. H., and Spencer J. R. (1995) Spectra of the Icy Galilean satellites from 0.2 to 5 μm—a compilation, new observations, and a recent summary. *J. Geophys. Res.: Planets* **100**, 19041–19048.

Canup R. M. and Ward W. R. (2002) Formation of the Galilean satellites: conditions of accretion. *Astron. J.* **124**, 3404–3423.

Carlson R. W. (1999) A tenuous carbon dioxide atmosphere on Jupiter's moon Callisto. *Science* **283**, 820–821.

Carlson R. W., Smythe W. D., LopesGautier R. M. C., Davies A. G., Kamp L. W., Mosher J. A., Soderblom L. A., Leader F. E., Mehlman R., Clark R. N., and Fanale F. P. (1997) The distribution of sulfur dioxide and other infrared absorbers on the surface of Io. *Geophys. Res. Lett.* **24**, 2479–2482.

Carlson R. W., Anderson M. S., Johnson R. E., Smythe W. D., Hendrix A. R., Barth C. A., Soderblom L. A., Hansen G. B., McCord T. B., Dalton J. B., Clark R. N., Shirley J. H., Ocampo A. C., and Matson D. L. (1999) Hydrogen peroxide on the surface of Europa. *Science* **283**, 2062–2064.

Carlson R. W., Anderson M. S., Johnson R. E., Schulman M. B., and Yavrouian A. H. (2002) Sulfuric acid production on Europa: the radiolysis of sulfur in water ice. *Icarus* **157**, 456–463.

Carr M. H. (1986) Silicate volcanism on Io. *J. Geophys. Res.: Solid Earth Planets* **91**, 3521–3532.

Carr M. H., Masursky H., Strom R. G., and Terrile R. J. (1979) Volcanic features of Io. *Nature* **280**, 729–733.

Carr M. H., Belton M. J. S., Chapman C. R., Davies A. S., Geissler P., Greenberg R., McEwen A. S., Tufts B. R., Greeley R., Sullivan R., Head J. W., Pappalardo R. T., Klaasen K. P., Johnson T. V., Kaufman J., Senske D., Moore J., Neukum G., Schubert G., Burns J. A., Thomas P., and Veverka J. (1998) Evidence for a subsurface ocean on Europa. *Nature* **391**, 363–365.

Chapman C. R., Morrison D., and Zellner B. (1975) Surface properties of asteroids—synthesis of polarimetry, radiometry, and spectrophotometry. *Icarus* **25**(1), 104–130.

Cheng A. F., Haff P. K., Johnson R. E., and Lanzerotti L. J. (1986) Interactions of planetary magnetospheres with Icy satellite surfaces. In *Satellites* (eds. J. A. Burns and M. S. Matthews). University of Arizona Press, Tucson, pp. 342–402.

Clark R. N. (1981a) The spectral reflectance of water-mineral mixtures at low-temperatures. *J. Geophys. Res.* **86**, 3074–3086.

Clark R. N. (1981b) Water frost and ice—the near-infrared spectral reflectance 0.65–2.5 μm. *J. Geophys. Res.* **86**, 3087–3096.

Clark R. N. and Lucey P. G. (1984) Spectral properties of Ice-particulate mixtures and implications for remote-sensing: 1. Intimate mixtures. *J. Geophys. Res.* **89**, 6341–6348.

Clark R. N. and McCord T. B. (1980a) The Galilean satellites—new near-infrared spectral reflectance measurements (0.65–2.5 μm) and a 0.325–5 μm summary. *Icarus* **41**, 323–339.

Clark R. N. and McCord T. B. (1980b) The rings of Saturn—new near-infrared reflectance measurements and a 0.326–4.08 μm summary. *Icarus* **43**, 161–168.

Clark R. N., Fanale F. P., and Zent A. P. (1983) Frost grain-size metamorphism—implications for remote-sensing of planetary surfaces. *Icarus* **56**, 233–245.

Clark R. N., Brown R. H., Owensby P. D., and Steele A. (1984) Saturn's satellites—near-infrared spectrophotometry (0.65–2.5 μm) of the leading and trailing sides and compositional implications. *Icarus* **58**, 265–281.

Clark R. N., Fanale F. P., and Gaffey M. J. (1986) Surface composition of natural satellites. In *Satellites* (eds. J. A. Burns and M. S. Matthews). University of Arizona Press, Tucson, pp. 437–491.

Clow G. D. and Carr M. H. (1980) Stability of sulfur slopes on Io. *Icarus* **44**, 268–279.

Collins G. C., Head J. W., and Pappalardo R. T. (1998) Formation of Ganymede grooved terrain by sequential extensional episodes: implications of Galileo observations for regional stratigraphy. *Icarus* **135**, 345–359.

Consolmagno G. J. and Lewis J. S. (1977) Preliminary thermal history models of icy satellites. In *Planetary Satellites* (ed. J. A. Burns). University of Arizona Press, Tucson, pp. 492–500.

Consolmagno G. J. and Lewis J. S. (1978) Evolution of icy satellite interiors and surfaces. *Icarus* **34**, 280–293.

Coustenis A., Lellouch E., Maillard J. P., and McKay C. P. (1995) Titans surface-composition and variability from the near-infrared albedo. *Icarus* **118**, 87–104.

Cruikshank D. P. (1980) Near-infrared studies of the satellites of Saturn and Uranus. *Icarus* **41**, 246–258.

Cruikshank D. P. and Apt J. (1984) Methane on Triton—physical state and distribution. *Icarus* **58**, 306–311.

Cruikshank D. P. and Brown H. (1981) The Uranian satellites—water ice on Ariel and Umbriel. *Icarus* **45**, 607–611.

Cruikshank D. and Brown R. H. (1986) Satellites of Uranus and Neptune, and the Pluto–Charon System. In *Satellites* (eds. J. A. Burns and M. S. Matthews). University of Arizona Press, Tucson, pp. 836–873.

Cruikshank D. P., Bell J. F., Gaffey M. J., Brown R. H., Howell R., Beerman C., and Rognstad M. (1983) The dark side of Iapetus. *Icarus* **53**, 90–104.

Cruikshank D. P., Brown R. H., and Clark R. N. (1984) Nitrogen on Triton. *Icarus* **58**, 293–305.

Cruikshank D. P., Allamandola L. J., Hartmann W. K., Tholen D. J., Brown R. H., Matthews C. N., and Bell J. F. (1991) Solid CN bearing material on outer solar-system bodies. *Icarus* **94**, 345–353.

Cruikshank D. P., Schmitt B., Roush T. L., Owen T. C., Quirico E., Geballe T. R., de Bergh C., Bartholomew M. J., Ore C. M. D., Doute S., and Meier R. (2000) Water ice on Triton. *Icarus* **147**, 309–316.

Danielson R. E., Caldwell J. J., and Larach D. R. (1973) Inversion in atmosphere of Titan. *Icarus* **20**, 437–443.

Davies A. G., Keszthelyi L. P., Williams D. A., Phillips C. B., McEwen A. S., Lopes R. M. C., Smythe W. D., Kamp L. W., Soderblom L. A., and Carlson R. W. (2001) Thermal signature, eruption style, and eruption evolution at Pele and Pillan on Io. *J. Geophys. Res.: Planets* **106**, 33079–33103.

de Pater I., Roe H., Graham J. R., Strobel D. F., and Bernath P. (2002) Detection of the forbidden SO $a\,^1\Delta \rightarrow X\,^3\Sigma^{-1}$ rovibronic transition on Io at 1.7 μm. *Icarus* **156**, 296–301.

Doute S., Schmitt B., Lopes-Gautier R., Carlson R., Soderblom L., and Shirley J. (2001) Mapping SO_2 frost on Io by the modelling of NIMS hyperspectral images. *Icarus* **149**, 107–132.

Duxbury N. S. and Brown R. H. (1997) The role of an internal heat source for the eruptive plumes on Triton. *Icarus* **125**, 83–93.

Fanale F. P., Johnson T. V., and Matson D. L. (1977) Io's surface and the histories of the Galilean satellites. In *Planetary Satellites* (ed. J. A. Burns). University of Arizona Press, Tucson, pp. 379–405.

Fanale F. P., Brown R. H., Cruikshank D. P., and Clake R. N. (1979) Significance of absorption features in Io's IR reflectance spectrum. *Nature* **280**, 761–763.

Fanale F. P., Granahan J. C., McCord T. B., Hansen G., Hibbitts C. A., Carlson R., Matson D., Ocampo A., Kamp L., Smythe W., Leader F., Mehlman R., Greeley R., Sullivan R., Geissler P., Barth C., Hendrix A., Clark B., Helfenstein P., Veverka J., Belton M. J. S., Becker K., and Becker T. (1999) Galileo's multi-instrument spectral view of Europe's surface composition. *Icarus* **139**, 179–188.

Fanale F. P., Li Y. H., De Carlo E., Farley C., Sharma S. K., Horton K., and Granahan J. C. (2001) An experimental estimate of Europa's "ocean" composition independent of Galileo orbital remote sensing. *J. Geophys. Res.: Planets* **106**, 14595–14600.

Fegley B. and Zolotov M. Y. (2000) Chemistry of sodium, potassium, and chlorine in volcanic gases on Io. *Icarus* **148**, 193–210.

Fink U., Larson H. P., Gautier T. N., and Treffers R. R. (1976) Infrared-spectra of satellites of Saturn—identification of water ice on Iapetus, Rhea, Dione, and Tethys. *Astrophys. J.* **207**, L63–L67.

Frank L. A., Paterson W. R., Ackerson K. L., and Bolton S. J. (1997) Outflow of hydrogen ions from Ganymede. *Geophys. Res. Lett.* **24**, 2151–2154.

Gaffey M. J., Burbine T. H., and Binzel R. P. (1993) Asteroid spectroscopy—progress and perspectives. *Meteoritics* **28**, 161–187.

Geissler P. E., McEwen A. S., Ip W., Belton M. J. S., Johnson T. V., Smyth W. H., and Ingersoll A. P. (1999a) Galileo imaging of atmospheric emissions from Io. *Science* **285**, 870–874.

Geissler P. E., McEwen A. S., Keszthelyi L., Lopes-Gautier R., Granahan J., and Simonelli D. P. (1999b) Global color variations on Io. *Icarus* **140**, 265–282.

Geissler P., McEwen A., Phillips C., Simonelli D., Lopes R. M. C., and Doute S. (2001a) Galileo imaging of SO_2 frosts on Io. *J. Geophys. Res.: Planets* **106**, 33253–33266.

Geissler P. E., Smyth W. H., McEwen A. S., Ip W., Belton M. J. S., Johnson T. V., Ingersoll A. P., Rages K., Hubbard W., and Dessler A. J. (2001b) Morphology and time variability of Io's visible aurora. *J. Geophys. Res.: Space Phys.* **106**, 26137–26146.

Gillett F. C. (1975) Further observations of 8–13 micron spectrum of Titan. *Astrophys. J.* **201**, L41–L43.

Greeley R., Sullivan R., Klemaszewski J., Homan K., Head J. W., Pappalardo R. T., Veverka J., Clark B. E., Johnson T. V., Klaasen K. P., Belton M., Moore J., Asphaug E., Carr M. H., Neukum G., Denk T., Chapman C. R., Pilcher C. B., Geissler P. E., Greenberg R., and Tufts R. (1998) Europa: initial Galileo geological observations. *Icarus* **135**, 4–24.

Greeley R., Figueredo P. H., Williams D. A., Chuang F. C., Klemaszewski J. E., Kadel S. D., Prockter L. M., Pappalardo R. T., Head J. W., Collins G. C., Spaun N. A., Sullivan R. J., Moore J. M., Senske D. A., Tufts B. R., Johnson T. V., Belton M. J. S., and Tanaka K. L. (2000) Geologic mapping of Europa. *J. Geophys. Res.: Planets* **105**, 22559–22578.

Griffith C. A. (1993) Evidence for surface heterogeneity on Titan. *Nature* **364**, 511–514.

Griffith C. A., Hall J. L., and Geballe T. R. (2000) Detection of daily clouds on Titan. *Science* **290**, 509–513.

Griffith C. A., Owen T., Geballe T. R., Rayner J., and Rannou P. (2003) Evidence for the exposure of water ice on Titan's surface. *Science* **300**, 628–630.

Hall D. T., Strobel D. F., Feldman P. D., McGrath M. A., and Weaver H. A. (1995) Detection of an oxygen atmosphere on Jupiter's moon Europa. *Nature* **373**, 677–679.

Hall D. T., Feldman P. D., McGrath M. A., and Strobel D. F. (1998) The far-ultraviolet oxygen airglow of Europa and Ganymede. *Astrophys. J.* **499**, 475–481.

Hanel R., Conrath B., Flasar F. M., Kunde V., Maguire W., Pearl J., Pirraglia J., Samuelson R., Herath L., Allison M., Cruikshank D., Gautier D., Gierasch P., Horn L., and Koppany R. (1981) Infrared observations of the system from Voyager-1. *Science* **212**, 192–200.

Hanner M. S., Aitken D. K., Knacke R. F., McCorkle S., Roche P. F., and Tokunaga A. T. (1985) Infrared spectrophotometry of comet IRAS-Araki-Alcock (1983d): a bare nucleus revealed? *Icarus* **62**, 97–109.

Hanner M. S., Newburn R. L., Spinrad H., and Veeder G. J. (1987) Comet Sugano-Saigusa-Fujikawa (1983v) a small, puzzling comet. *Astron. J.* **94**, 1081–1087.

Hibbitts C. A., McCord T. B., and Hansen G. B. (2000) Distributions of CO_2 and SO_2 on the surface of Callisto. *J. Geophys. Res.: Planets* **105**, 22541–22557.

Hibbitts C. A., Klemaszewski J. E., McCord T. B., Hansen G. B., and Greeley R. (2002) CO_2-rich impact craters on Callisto. *J. Geophys. Res.: Planets* **107**, 5084.

Hogenboom D. L., Kargel J. S., Consolmagno G. J., Holden T. C., Lee L., and Buyyounouski M. (1997) The ammonia-water system and the chemical differentiation of icy satellites. *Icarus* **128**, 171–180.

Johnson R. E., Nelson M. L., McCord T. B., and Gradie J. C. (1988) Analysis of voyager images of Europa—plasma bombardment. *Icarus* **75**, 423–436.

Johnson T. V. and Anderson J. D. (2003) Galileo's encounter with Amalthea. *Geophys. Res. Abstr.* **5**, abstr. no. 07902.

Johnson T. V. and Fanale F. P. (1973) Optical-properties of carbonaceous chondrites and their relationship to Asteroids. *J. Geophys. Res.* **78**, 8507–8518.

Johnson T. V., Veeder G. J., and Matson D. L. (1975) Evidence for frost on Rhea's surface. *Icarus* **24**(4), 428–432.

Johnson T. V., Soderblom L. A., Mosher J. A., Danielson G. E., Cook A. F., and Kupferman P. (1983) Global multispectral mosaics of the icy Galilean satellites. *J. Geophys. Res.* **88**, 5789–5805.

Johnson T. V., Morrison D., Matson D. L., Veeder G. J., Brown R. H., and Nelson R. M. (1984) Volcanic hotspots on Io—stability and longitudinal distribution. *Science* **226**, 134–137.

Johnson T. V., Brown R. H., and Pollack J. B. (1987) Uranus satellites—densities and composition. *J. Geophys. Res.: Space Phys.* **92**, 14884–14894.

Johnson T. V., Veeder G. J., Matson D. L., Brown R. H., Nelson R. M., and Morrison D. (1988) Io—evidence for silicate volcanism in 1986. *Science* **242**, 1280–1283.

Johnson T. V., Matson D. L., Blaney D. L., Veeder G. J., and Davies A. (1995) Stealth plumes on Io. *Geophys. Res. Lett.* **22**, 3293–3296.

Kargel J. S. (1991) Brine volcanism and the interior structures of asteroids and Icy satellites. *Icarus* **94**, 368–390.

Kargel J. S. (1992) Ammonia-water volcanism on icy satellites: phase relationships at one atmosphere. *Icarus* **100**, 556–574.

Keszthelyi L., McEwen A. S., Phillips C. B., Milazzo M., Geissler P., Turtle E. P., Radebaugh J., Williams D. A., Simonelli D. P., Breneman H. H., Klaasen K. P., Levanas G., and Denk T. (2001) Imaging of volcanic activity on Jupiter's moon Io by Galileo during the Galileo Europa mission and the Galileo millennium mission. *J. Geophys. Res.: Planets* **106**, 33025–33052.

Khurana K. K., Kivelson M. G., Russell C. T., Walker R. J., and Southwood D. J. (1997) Absence of an internal magnetic field at Callisto. *Nature* **387**, 262–264.

Khurana K. K., Kivelson M. G., Stevenson D. J., Schubert G., Russell C. T., Walker R. J., and Polanskey C. (1998) Induced magnetic fields as evidence for subsurface oceans in Europa and Callisto. *Nature* **395**, 777–780.

Kieffer S. W. (1982) Dynamics and thermodynamics of volcanic eruptions: implications for the plumes on Io. In *Satellites of Jupiter* (ed. D. Morrison). University of Arizona Press, Tucson, pp. 647–723.

Kieffer S. W., Lopes-Gautier R., McEwen A., Smythe W., Keszthelyi L., and Carlson R. (2000) Prometheus: Io's wandering plume. *Science* **288**, 1204–1208.

Kirk R., Soderblom L., Brown R. H., Kieffer S. W., and Kargel J. (1996) Triton's plumes: discovery, characteristics and models. In *Neptune and Triton* (ed. D. Cruikshank). University of Arizona Press, Tucson, pp. 949–990.

Kirk R. L., Brown R. H., and Soderblom L. A. (1990) Subsurface energy-storage and transport for solar-powered geysers on Triton. *Science* **250**, 424–429.

Kivelson M. G., Khurana K. K., Russell C. T., Walker R. J., Warnecke J., Coroniti F. V., Polanskey C., Southwood D. J., and Schubert G. (1996) Discovery of Ganymede's magnetic field by the Galileo spacecraft. *Nature* **384**, 537–541.

Kivelson M. G., Khurana K. K., Coroniti F. V., Joy S., Russell C. T., Walker R. J., Warnecke J., Bennett L., and Polanskey C. (1997a) The magnetic field and magnetosphere of Ganymede. *Geophys. Res. Lett.* **24**, 2155–2158.

Kivelson M. G., Khurana K. K., Joy S., Russell C. T., Southwood D. J., Walker R. J., and Polanskey C. (1997b) Europa's magnetic signature: report from Galileo's pass on December 19, 1996. *Science* **276**, 1239–1241.

Kivelson M. G., Warnecke J., Bennett L., Joy S., Khurana K. K., Linker J. A., Russell C. T., Walker R. J., and Polanskey C. (1998) Ganymede's magnetosphere: magnetometer overview. *J. Geophys. Res.: Planets* **103**, 19963–19972.

Kivelson M. G., Khurana K. K., Stevenson D. J., Bennett L., Joy S., Russell C. T., Walker R. J., Zimmer C., and Polanskey C. (1999) Europa and Callisto: induced or intrinsic fields in a periodically varying plasma environment. *J. Geophys. Res.: Space Phys.* **104**, 4609–4625.

Kivelson M. G., Khurana K. K., and Volwerk M. (2002) The permanent and inductive magnetic moments of Ganymede. *Icarus* **157**, 507–522.

Kuiper G. P. (1944) Titan: a satellite with an atmosphere. *Astrophys. J.* **100**, 378–383.

Kunde V. G., Aikin A. C., Hanel R. A., Jennings D. E., Maguire W. C., and Samuelson R. E. (1981) C_4H_2, HC_3N, and C_2N_2 in Titans atmosphere. *Nature* **292**, 686–688.

Kupo I., Mekler Y., and Eviatar A. (1976) Detection of ionized sulfur in Jovian magnetosphere. *Astrophys. J.* **205**, L51–L53.

Kuppers M. and Schneider N. M. (2000) Discovery of chlorine in the Io torus. *Geophys. Res. Lett.* **27**, 513–516.

Lane A. L. and Domingue D. L. (1997) IUE's view of Callisto: detection of an SO_2 absorption correlated to possible torus neutral wind alterations. *Geophys. Res. Lett.* **24**, 1143–1146.

Lane A. L., Nelson R. M., and Matson D. L. (1981) Evidence for sulfur implantation in Europa's UV absorption-band. *Nature* **292**, 38–39.

Leblanc F., Johnson R. E., and Brown M. E. (2002) Europa's sodium atmosphere: an ocean source? *Icarus* **159**, 132–144.

Lellouch E., Paubert G., Moses J. I., Schneider N. M., and Strobel D. F. (2003) Volcanically emitted sodium chloride as a source for Io's neutral clouds and plasma torus. *Nature* **421**, 45–47.

Lewis J. S. (1971) Satellites of outer planets—their physical and chemical nature. *Icarus* **15**, 174–185.

Lewis J. S. (1972) Low-temperature condensation from solar nebula. *Icarus* **16**, 241–252.

Lewis J. S. (1973) Chemistry of outer solar system. *Space Sci. Rev.* **14**, 401–411.

Lewis J. S. and Prinn R. G. (1980) Kinetic inhibition of CO and N_2 reduction in the solar nebula. *Astrophys. J.* **238**, 357–364.

Lindal G. F., Wood G. E., Hotz H. B., Sweetnam D. N., Eshleman V. R., and Tyler G. L. (1983) The atmosphere of Titan—an analysis of the Voyager-1 radio occultation measurements. *Icarus* **53**, 348–363.

Lopes R. M. C., Kamp L. W., Doute S., Smythe W. D., Carlson R. W., McEwen A. S., Geissler P. E., Kieffer S. W., Leader F. E., Davies A. G., Barbinis E., Mehlman R., Segura M., Shirley J., and Soderblom L. A. (2001) Io in the near infrared: near-infrared mapping spectrometer (NIMS) results from the Galileo flybys in 1999 and 2000. *J. Geophys. Res.: Planets* **106**, 33053–33078.

Lorenz R. D. and Lunine J. I. (1997) Titan's surface reviewed: the nature of bright and dark terrain. *Planet. Space Sci.* **45**, 981–992.

Lunine J. I. (1993) Does Titan have an ocean—a review of current understanding of Titan's surface. *Rev. Geophys.* **31**, 133–149.

Lunine J. I., Stevenson D. J., and Yung Y. L. (1983) Ethane ocean on Titan. *Science* **222**, 1229–1230.

Lupo M. J. and Lewis J. S. (1979) Mass-radius relationships in icy satellites. *Icarus* **40**, 157–170.

Lutz B. L., Debergh C., Maillard J. P., Owen T., and Brault J. (1981) On the possible detection of CH_3D on Titan and Uranus. *Astrophys. J.* **248**, L141–L145.

Lutz B. L., Debergh C., and Owen T. (1983a) Carbon-monoxide in the atmosphere of Titan—search and discovery. *Publ. Astron. Soc. Pacific* **95**, 593–593.

Lutz B. L., Debergh C., and Owen T. (1983b) Titan—discovery of carbon-monoxide in its atmosphere. *Science* **220**, 1374–1375.

Maguire W. C., Hanel R. A., Jennings D. E., Kunde V. G., and Samuelson R. E. (1981) C_3H_8 and C_3H_4 in Titans atmosphere. *Nature* **292**, 683–686.

Matson D. (1971) Infrared observations of asteroids. In *Physical Studies of the Minor Planets*, NASA SP-267 (ed. T. Gehrels). US Government Printing Office, Washington, DC, pp. 41–44.

Matson D. L., Ransford G. A., and Johnson T. V. (1981) Heat-flow from Io (J1). *J. Geophys. Res.* **86**(NB3), 1664–1672.

Matson D. L., Johnson T. V., Davies A. G., Veeder G. J., and Blaney D. L. (2003) The conundrum posed by Io's minimum surface temperaures. *Geophys. Res. Abstr.* **5**, abstr. no. 07912.

Mauk B. H., Mitchell D. G., Krimigis S. M., Roelof E. C., and Paranicas C. P. (2003) Energetic neutral atoms from a trans-Europa gas torus at Jupiter. *Nature* **421**, 920–922.

McCord T. B., Carlson R. W., Smythe W. D., Hansen G. B., Clark R. N., Hibbitts C. A., Fanale F. P., Granahan J. C., Segura M., Matson D. L., Johnson T. V., and Martin P. D. (1997) Organics and other molecules in the surfaces of Callisto and Ganymede. *Science* **278**, 271–275.

McCord T. B., Hansen G. B., Clark R. N., Martin P. D., Hibbitts C. A., Fanale F. P., Granahan J. C., Segura M., Matson D. L., Johnson T. V., Carlson R. W., Smythe W. D., and Danielson G. E. (1998a) Non-water-ice constituents in the surface material of the icy Galilean satellites from the Galileo near-infrared mapping spectrometer investigation. *J. Geophys. Res.: Planets* **103**, 8603–8626.

McCord T. B., Hansen G. B., Fanale F. P., Carlson R. W., Matson D. L., Johnson T. V., Smythe W. D., Crowley J. K., Martin P. D., Ocampo A., Hibbitts C. A., and Granahan J. C. (1998b) Salts an Europa's surface detected by Galileo's near infrared mapping spectrometer. *Science* **280**, 1242–1245.

McCord T. B., Hansen G. B., Matson D. L., Johnson T. V., Crowley J. K., Fanale F. P., Carlson R. W., Smythe W. D., Martin P. D., Hibbitts C. A., Granahan J. C., and Ocampo A. (1999) Hydrated salt minerals on Europa's surface from the Galileo near-infrared mapping spectrometer (NIMS) investigation. *J. Geophys. Res.: Planets* **104**, 11827–11851.

McCord T. B., Hansen G. B., and Hibbitts C. A. (2001a) Hydrated salt minerals on Ganymede's surface: evidence of an ocean below. *Science* **292**, 1523–1525.

McCord T. B., Orlando T. M., Teeter G., Hansen G. B., Sieger M. T., Petrik N. G., and Van Keulen L. (2001b) Thermal and radiation stability of the hydrated salt minerals epsomite, mirabilite, and natron under Europa environmental conditions. *J. Geophys. Res.: Planets* **106**, 3311–3319.

McEwen A. S. (1986) Exogenic and endogenic Albedo and color patterns on Europa. *J. Geophys. Res.: Solid Earth Planets* **91**, 8077–8097.

McEwen A. S., Keszthelyi L., Geissler P., Simonelli D. P., Carr M. H., Johnson T. V., Klaasen K. P., Breneman H. H., Jones T. J., Kaufman J. M., Magee K. P., Senske D. A., Belton M. J. S., and Schubert G. (1998a) Active volcanism on Io as seen by Galileo SSI. *Icarus* **135**, 181–219.

McEwen A. S., Keszthelyi L., Spencer J. R., Schubert G., Matson D. L., Lopes-Gautier R., Klassen K. P., Johnson T. V., Head J. W., Geissler P., Fagents S., Davies A. G., Carr M. H., Breneman H. H., and Belton M. J. S. (1998b) High-temperature silicate volcanism on Jupiter's moon Io. *Science* **281**, 87–90.

McGrath M. A., Belton M. J. S., Spencer J. R., and Sartoretti P. (2000) Spatially resolved spectroscopy of Io's Pele plume and SO_2 atmosphere. *Icarus* **146**, 476–493.

McKinnon W. B., Lunine J. I., and Banfield D. (1996) Origin and evolution of Triton. In *Neptune and Triton* (ed. D. P. Cruikshank). University of Arizona Press, Tucson, pp. 807–878.

Morabito L. A., Synnott S. P., Kupferman P. N., and Collins S. A. (1979) Discovery of currently active extraterrestrial volcanism. *Science* **204**, 972.

Morrison D. and Telesco C. M. (1980) Io—observational constraints on internal energy and thermophysics of the surface. *Icarus* **44**, 226–233.

Morrison D., Owen T., and Soderblom L. A. (1986) The satellites of Saturn. In *Satellites* (eds. J. A. Burns and M. S. Matthews). University of Arizona Press, Tucson, pp. 764–801.

Moses J. I., Zolotov M. Y., and Fegley B. (2002) Photochemistry of a volcanically driven atmosphere on Io: sulfur and oxygen species from a Pele-type eruption. *Icarus* **156**, 76–106.

Muhleman D. O., Grossman A. W., Butler B. J., and Slade M. A. (1990) Radar reflectivity of Titan. *Science* **248**, 975–980.

Nash D. B., Carr M. H., Gradie J., Hunten D. M., and Yoder C. F. (1986) Io. In *Satellites* (eds. J. A. Burns and M. S. Matthews). University of Arizona Press, Tucson, pp. 629–688.

Nelson M. L., McCord T. B., Clark R. N., Johnson T. V., Matson D. L., Mosher J. A., and Soderblom L. A. (1986) Europa—characterization and interpretation of global spectral surface units. *Icarus* **65**, 129–151.

Nelson R. M., Lane A. L., Matson D. L., Veeder G. J., Buratti B. J., and Tedesco E. F. (1987) Spectral geometric Albedos of the Galilean satellites from 0.24 to 0.34 micrometers—observations with the international ultraviolet explorer. *Icarus* **72**, 358–380.

Neubauer F. M. (1998) The sub-Alfvenic interaction of the Galilean satellites with the Jovian magnetosphere. *J. Geophys. Res.: Planets* **103**, 19843–19866.

Nicholson P. D., Hamilton D. P., Matthews K., and Yoder C. F. (1992) New observations of Saturns coorbital satellites. *Icarus* **100**, 464–484.

Noll K. S., Johnson R. E., Lane A. L., Domingue D. L., and Weaver H. A. (1996) Detection of ozone on Ganymede. *Science* **273**, 341–343.

Noll K. S., Johnson R. E., McGrath M. A., and Caldwell J. J. (1997a) Detection of SO_2 on Callisto with the Hubble Space Telescope. *Geophys. Res. Lett.* **24**, 1139–1142.

Noll K. S., Roush T. L., Cruikshank D. P., Johnson R. E., and Pendleton Y. J. (1997b) Detection of ozone on Saturn's satellites Rhea and Dione. *Nature* **388**, 45–47.

Pappalardo R. T., Head J. W., Collins G. C., Kirk R. L., Neukum G., Oberst J., Giese B., Greeley R., Chapman C. R., Helfenstein P., Moore J. M., McEwen A., Tufts B. R., Senske D. A., Breneman H. H., and Klaasen K. (1998) Grooved terrain on Ganymede: first results from Galileo high—resolution imaging. *Icarus* **135**, 276–302.

Peale S. J., Cassen P., and Reynolds R. T. (1979) Melting of Io by tidal dissipation. *Science* **203**, 892–894.

Pearl J., Hanel R., Kunde V., Maguire W., Fox K., Gupta S., Ponnamperuma C., and Raulin F. (1979) Identification of gaseous SO_2 and new upper limits for other gases on Io. *Nature* **280**, 755–758.

Pearl J. C. and Sinton W. M. (1982) Hot spots of Io. In *Satellites of Jupiter* (ed. D. Morrison). University of Arizona Press, Tucson, pp. 724–755.

Pilcher C. B., Chapman C. R., Lebofsky L. A., and Kieffer H. H. (1970) Saturn's rings—identification of water frost. *Science* **167**, 1372–1373.

Pilcher C. B., Ridgway S. T., and McCord T. B. (1972) Galilean satellites—identification of water frost. *Science* **178**, 1087–1089.

Pollack J. B. and Fanale F. P. (1982) Origin and evolution of the Jupiter satellite system. In *Satellites of Jupiter* (ed. D. Morrison). University of Arizona Press, Tucson, pp. 872–910.

Pollack J. B. and Reynolds R. T. (1974) Implications of Jupiter's early contraction history for composition of Galilean satellites. *Icarus* **21**, 248–253.

Pollack J. B., Lunine J. I., and Tittemore W. C. (1991) Origin of the Uranus satellites. In *Uranus* (eds. J. T. Bergstralh, E. D. Miner, and M. S. Matthews). University of Arizona Press, Tucson, pp. 469–512.

Prinn R. G. and Fegley B. (1981) Kinetic inhibition of CO and N_2 reduction in circumplanetary nebulae—implications for satellite composition. *Astrophys. J.* **249**, 308–317.

Prinn R. G. and Fegley B. (1989) Solar nebula chemistry: origin of planetary, satellite, and commentary volatiles. In *Origin and Evolution of Planetary and Satellite Atmospheres* (ed. S. Atreya). University of Arizona Press, Tucson, pp. 78–136.

Prockter L. M., Figueredo P. H., Pappalardo R. T., Head J. W., and Collins G. C. (2000) Geology and mapping of dark terrain on Ganymede and implications for grooved terrain formation. *J. Geophys. Res., Planets* **105**, 22519–22540.

Russell C. T. and Kivelson M. G. (2001) Evidence for sulfur dioxide, sulfur monoxide, and hydrogen sulfide in the Io exosphere. *J. Geophys. Res., Planets* **106**, 33267–33272.

Sagan C. (1979) Sulfur flows on Io. *Nature* **280**, 750–753.

Samuelson R. E., Hanel R. A., Kunde V. G., and Maguire W. C. (1981) Mean molecular-weight and hydrogen abundance of Titan's atmosphere. *Nature* **292**, 688–693.

Schubert G. (1986) Thermal histories, compositions, and internal structures of the moons of the solar system. In *Satellites* (eds. J. A. Burns and M. S. Matthews). University of Arizona Press, Tucson, pp. 224–292.

Shematovich V. I. and Johnson R. E. (2001). Near-surface oxygen atmosphere at Europa. *Adv. Space Res., Planet. Ionosp.* **27**, 1881–1888.

Sinton W. M. and Kaminski C. (1988) Infrared observations of eclipses of Io, its thermo-physical parameters, and the thermal-radiation of the Loki Volcano and environs. *Icarus* **75**, 207–232.

Smith B. A. (1981) Voyager encounters with Saturn. *J. Opt. Soc. Am.* **71**, 1591–1591.

Smith B. A., Shoemaker E. M., Kieffer S. W., and Cook A. F. (1979a) Role of SO_2 in volcanism on Io. *Nature* **280**, 738–743.

Smith B. A., Soderblom L. A., Johnson T. V., Ingersoll A. P., Collins S. A., Shoemaker E. M., Hunt G. E., Masursky H., Carr M. H., Davies M. E., Cook A. F., Boyce J., Danielson G. E., Owen T., Sagan C., Beebe R. F., Veverka J., Strom R. G., McCauley J. F., Morrison D., Briggs G. A., and Suomi V. E. (1979b) Jupiter system through the eyes of Voyager-1. *Science* **204**, 951–972.

Smith B. A., Soderblom L., Batson R., Bridges P., Inge J., Masursky H., Shoemaker E., Beebe R., Boyce J., Briggs G., Bunker A., Collins S. A., Hansen C. J., Johnson T. V., Mitchell J. L., Terrile R. J., Cook A. F., Cuzzi J., Pollack J. B., Danielson G. E., Ingersoll A. P., Davies M. E., Hunt G. E., Morrison D., Owen T., Sagan C., Veverka J., Strom R., and Suomi V. E. (1982) A new look at the Saturn system—the Voyager-2 images. *Science* **215**, 504–537.

Smith B. A., Soderblom L. A., Beebe R., Bliss D., Boyce J. M., Brahic A., Briggs G. A., Brown R. H., Collins S. A., Cook A. F., Croft S. K., Cuzzi J. N., Danielson G. E., Davies M. E., Dowling T. E., Godfrey D., Hansen C. J., Harris C., Hunt G. E., Ingersoll A. P., Johnson T. V., Krauss R. J., Masursky H., Morrison D., Owen T., Plescia J. B., Pollack J. B., Porco C. C., Rages K., Sagan C., Shoemaker E. M., Sromovsky L. A., Stoker C., Strom R. G., Suomi V. E., Synnott S. P., Terrile R. J., Thomas P., Thompson W. R., and Veverka J. (1986) Voyager-2 in the Uranian system—imaging science results. *Science* **233**, 43–64.

Smith B. A., Soderblom L. A., Banfield D., Barnet C., Basilevksy A. T., Beebe R. F., Bollinger K., Boyce J. M., Brahic A., Briggs G. A., Brown R. H., Chyba C., Collins S. A., Colvin T., Cook A. F., Crisp D., Croft S. K., Cruikshank D., Cuzzi J. N., Danielson G. E., Davies M. E., Dejong E., Dones L., Godfrey D., Goguen J., Grenier I., Haemmerle V. R., Hammel H., Hansen C. J., Helfenstein C. P., Howell C., Hunt G. E., Ingersoll A. P., Johnson T. V., Kargel J., Kirk R., Kuehn D. I., Limaye S., Masursky H., McEwen A., Morrison D., Owen T., Owen W., Pollack J. B., Porco C. C., Rages K., Rogers P., Rudy D., Sagan C., Schwartz J., Shoemaker E. M., Showalter M., Sicardy B., Simonelli D., Spencer J., Sromovsky L. A., Stoker C., Strom R. G., Suomi V. E., Synott S. P., Terrile R. J., Thomas P., Thompson W. R., Verbiscer A., and Veverka J. (1989) Voyager-2 at Neptune—imaging science results. *Science* **246**, 1422–1449.

Smith P. H., Lemmon M., Tomasko M. (2002). The surface and lower atmosphere of Titan from HST observations. In *Highlights of Astronomy*. IAU, vol. 12, pp. 625–625.

Smith P. H., Lemmon M. T., Lorenz R. D., Sromovsky L. A., Caldwell J. J., and Allison M. D. (1996) Titan's surface, revealed by HST imaging. *Icarus* **119**(2), 336–349.

Smythe W. D., Nelson R. M., and Nash D. B. (1979) Spectral evidence for SO_2 frost or adsorbate on Io's surface. *Nature* **280**, 766–766.

Spencer J. R. and Calvin W. M. (2002) Condensed O_2 on Europa and Callisto. *Astron. J.* **124**, 3400–3403.

Spencer J. R. and Schneider N. M. (1996) Io on the eve of the Galileo mission. *Ann. Rev. Earth Planet. Sci.* **24**, 125–190.

Spencer J. R., Calvin W. M., and Person M. J. (1995) Charge-coupled-device spectra of the Galilean satellites—molecular-oxygen on Ganymede. *J. Geophys. Res.: Planets* **100**, 19049–19056.

Spencer J. R., Stansberry J. A., Dumas C., Vakil D., Pregler R., Hicks M., and Hege K. (1997) A history of high-temperature Io volcanism: February 1995 to May 1997. *Geophys. Res. Lett.* **24**, 2451–2454.

Spencer J. R., Jessup K. L., McGrath M. A., Ballester G. E., and Yelle R. (2000a) Discovery of gaseous S_2 in Io's Pele plume. *Science* **288**, 1208–1210.

Spencer J. R., Rathbun J. A., Travis L. D., Tamppari L. K., Barnard L., Martin T. Z., and McEwen A. S. (2000b) Io's thermal emission from the Galileo photopolarimeter-radiometer. *Science* **288**, 1198–1201.

Stansberry J. A., Spencer J. R., Howell R. R., Dumas C., and Vakil D. (1997) Violent silicate volcanism on Io in 1996. *Geophys. Res. Lett.* **24**, 2455–2458.

Strobel D. F. (1982) Chemistry and evolution of Titan's atmosphere. *Planet. Space Sci.* **30**, 839–848.

Strobel D. F. and Summers M. E. (1996) Triton's upper atmosphere and ionosphere. In *Neptune and Triton* (ed. D. Cruikshank). University of Arizona Press, Tucson, pp. 1107–1151.

Thomas P., Ververka J., and Dermott S. (1986) Small satellites. In *Satellites* (eds. J. A. Burns and M. S. Matthews). University of Arizona Press, Tucson, pp. 802–835.

Thomas P. C., Veverka J., Bell J. F., Clark B. E., Carcich B., Joseph J., Robinson M., McFadden L. A., Malin M. C., Chapman C. R., Merline W., and Murchie S. (1999) Mathilde: size, shape, and geology. *Icarus* **140**(1), 17–27.

Trafton L. (1975) Detection of a potassium cloud near Io. *Nature* **258**, 690–692.

Tyler G. L., Eshleman V. R., Anderson J. D., Levy G. S., Lindal G. F., Wood G. E., and Croft T. A. (1981) Radio science investigations of the Saturn system with Voyager-1—preliminary-results. *Science* **212**, 201–206.

Tyler G. L., Sweetnam D. N., Anderson J. D., Campbell J. K., Eshleman V. R., Hinson D. P., Levy G. S., Lindal G. F., Marouf E. A., and Simpson R. A. (1986) Voyager-2 radio science observations of the Uranian system—atmosphere, rings, and satellites. *Science* **233**, 79–84.

Veeder G. J., Hanner M. S., and Tholen D. J. (1987) The nucleus of comet P/Arend-Rigaux. *Astron. J.* **94**, 169–173.

Veeder G. J., Matson D. L., Johnson T. V., Blaney D. L., and Goguen J. D. (1994) Io's heat-flow from infrared radiometry—1983–1993. *J. Geophys. Res.: Planets* **99**, 17095–17162.

Veverka J., Thomas P., Harch A., Clark B., Bell J. F., Carcich B., Joseph J., Murchie S., Izenberg N., Chapman C., Merline W., Malin M., McFadden L., and Robinson M. (1999) NEAR encounter with asteroid 253 Mathilde: overview. *Icarus* **140**, 3–16.

Wamsteker W. (1973) Narrow-band photometry of the Galilean satellites. *Commun. Lunar Planet. Lab.* **9**, 171–177.

Wamsteker W., Kroes R. L., and Fountain J. A. (1974) Surface composition of Io. *Icarus* **23**, 417–424.

Whittet D. C. B., Schutte W. A., Tielens A., Boogert A. C. A., deGraauw T., Ehrenfreund P., Gerakines P. A., Helmich F. P., Prusti T., and vanDishoeck E. F. (1996) An ISO SWS view of interstellar ices: first results. *Astron. Astrophys.* **315**, L357–L360.

Williams D. A., Fagents S. A., and Greeley R. (2000) A reassessment of the emplacement and erosional potential of turbulent, low-viscosity lavas on the Moon. *J. Geophys. Res.: Planets* **105**, 20189–20205.

Williams D. A., Davies A. G., Keszthelyi L. P., and Greeley R. (2001a) The summer 1997 eruption at Pillan Patera on Io: implications for ultrabasic lava flow emplacement. *J. Geophys. Res.: Planets* **106**, 33105–33119.

Williams D. A., Greeley R., Lopes R. M. C., and Davies A. G. (2001b) Evaluation of sulfur flow emplacement on Io from Galileo data and numerical modelling. *J. Geophys. Res.: Planets* **106**, 33161–33174.

Yelle R. V., Lunine J. I., Pollack J. B., and Brown R. H. (1996) Lower atmospheric structure and surface-atmospheric interaction on Triton. In *Neptune and Triton* (ed. D. Cruikshank). University of Arizona Press, Tucson, pp. 1031–1106.

Yoder C. F. (1995) Astrometric and geodetic properties of Earth and the solar system. In *AGU Reference Shelf1: Global Earth Physics*, A Handbook of Physical Constants (ed. T. J. Ahrens). American Geophysical Union, Washington, DC, pp. 1–31.

Yung Y. L., Allen M., and Pinto J. P. (1984) Photochemistry of the atmosphere of Titan—comparison between model and observations. *Astrophys. J. Suppl. Ser.* **55**, 465–506.

Zimmer C., Khurana K. K., and Kivelson M. G. (2000) Subsurface oceans on Europa and Callisto: constraints from Galileo magnetometer observations. *Icarus* **147**, 329–347.

Zolotov M. Y. and Fegley B. (1998) Volcanic production of sulfur monoxide (SO) on Io. *Icarus* **132**, 431–434.

Zolotov M. Y. and Fegley B. (1999) Oxidation state of volcanic gases and the interior of Io. *Icarus* **141**, 40–52.

Zolotov M. Y. and Shock E. L. (2001) Composition and stability of salts on the surface of Europa and their oceanic origin. *J. Geophys. Res.: Planets* **106**, 32815–32827.

1.25
Comets

D. E. Brownlee

University of Washington, WA, USA

1.25.1 INTRODUCTION	663
1.25.2 COMET AND ASTEROID COMPARISONS	664
1.25.3 COMET ACTIVITY	665
1.25.4 COMET TYPES—ORBITAL DISTINCTION	666
1.25.4.1 Comet Source Regions	666
1.25.4.2 The Oort Cloud	666
1.25.4.3 The Kuiper Belt	667
1.25.4.3.1 KBO terminology	667
1.25.4.3.2 KBO orbital distribution	668
1.25.5 PHYSICAL EVOLUTION OF COMETS	669
1.25.5.1 Fragmentation	669
1.25.5.2 Crust/Mantle Formation	671
1.25.5.3 Strength and Structure of Cometary Materials	671
1.25.6 MAJOR COMPONENT COMPOSITION	672
1.25.6.1 Water Ice	672
1.25.6.2 CO and Very Volatile Compounds	674
1.25.6.2.1 Full volatile composition	674
1.25.6.3 Dust (and Rocks)	675
1.25.6.3.1 Information from astronomical observations	675
1.25.6.3.2 Information from collected samples	678
1.25.6.3.3 Information from in situ measurements	680
1.25.7 DIVERSITY AMONG COMETS	682
1.25.8 CONCLUSIONS	685
REFERENCES	685

1.25.1 INTRODUCTION

Comets are surviving members of a formerly vast distribution of solid bodies that formed in the cold regions of the solar nebula. Cometary bodies escaped incorporation into planets and ejection from the solar system and they have been stored in two distant reservoirs, the Oort cloud and the Kuiper Belt, for most of the age of the solar system. Observed comets appear to have formed between 5 AU and 55 AU. From a cosmochemical viewpoint, comets are particularly interesting bodies because they are preserved samples of the solar nebula's cold ice-bearing regions that occupied 99% of the areal extent of the solar nebula disk. All comets formed beyond the "snow line" of the nebula, where the conditions were cold enough for water ice to condense, but they formed from environments that significantly differed in temperature. Some formed in the comparatively "warm" regions near Jupiter where the nebular temperature may have been greater than 120 K and others clearly formed beyond Neptune where temperatures may have been less than 30 K (Bell *et al.*, 1997). Although comets are the best-preserved materials from the early solar system, they should be a mix of nebular and presolar materials that accreted over a vast range of distances from the Sun in environments that differed in temperature, pressure, and accretional conditions such as impact speed.

Comets, by conventional definition, are unstable near the Sun; they contain highly volatile ices that vigorously sublime within 2–3 AU of

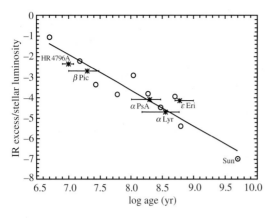

Figure 1 The ratio of infrared excess/stellar luminosity is a measure of the fraction of starlight absorbed by circumstellar dust and re-radiated in the infrared. The plot from Spangler et al. (2001) shows the temporal decline of dust around "Vega-like" stars (points) and stars in clusters with measured ages (circles). At least for the longer ages, the dust is most probably generated by comets.

the Sun. When heated, they release gas and solids due to "cometary activity," a series of processes usually detected from afar by the presence of a coma of gas and dust surrounding the cometary nucleus and or elongated tails composed of dust and gas. Active comets clearly have not been severely modified by the moderate to extreme heating that has affected all other solar system materials, including planets, moons, and even the asteroids that produced the most primitive meteorites. Comets have been widely described as the most primitive solar system materials, preserved at cryogenic temperature and low pressure since the formation of the Sun. This is likely to be true, in general, but there is a growing body of recent evidence suggesting that comets are both more physically complex and have had more complex histories than formerly believed. They formed over an order of magnitude range of distances from the Sun; some are fragments of relatively large bodies and collisional effects must have processed at least some comets, as they have processed asteroids (McSween and Weissman, 1989).

Comet-like materials are presumed to be the building blocks of Uranus and Neptune (the ice giants); they may have played a role in the formation of Jupiter and Saturn (the gas giants) and they also played some role in transporting outer solar system volatile materials to inner planets (Delsemme, 2000). The inner solar system flux of comets may have been much higher in the past and comets may have played a role in producing the late heavy bombardment on terrestrial planets (Levison et al., 2001). Comets also exist outside the solar system and there is good evidence that they orbit a major fraction of Sun-like stars. Circumstellar dust, which appears to have been generated by comets, is detected as thermal infrared emission and sometimes as scattered starlight (Backman et al., 1997; Weissman, 1984; Jewitt and Luu, 1995). It is particularly interesting that the amount of dust around stars declines with stellar age and is highest around stars younger than a few hundred million years. The common presence of what appears to be comet-generated dust around other stars suggests that comet formation is a normal and common consequence of star formation (Figure 1).

1.25.2 COMET AND ASTEROID COMPARISONS

Both comets and asteroids are surviving planetesimals from the solar nebula, preserved building blocks of planets. Comets are generally considered to be relic materials from the outer regions of the solar nebula, while asteroids are considered to be materials that formed or ended up near the asteroid Belt, mostly between 1.9 AU and 3.2 AU, or the Trojan asteroid region in synchronous orbit with Jupiter near 5.2 AU. The International Astronomical Union (IAU) defines the difference between comets and asteroids as due to the presence of cometary activity: the formation of a coma and/or tail. This is a practical working definition but the real distinction between comets and asteroids is sometimes arbitrary and uncertain. A limitation of the official terminology is that cometary activity, the official diagnostic property, only occurs relatively close to the Sun and not at the solar distances where the solar system's ice-bearing bodies normally reside. Another limitation is that an asteroid can become a comet but an observed comet can never become an asteroid. Comet Wilson–Harrington was discovered in 1949 with a short tail but has not shown cometary activity since. Although it is effectively now an asteroid, it must be considered a comet. Bodies are called comets if they have ever been observed to show cometary activity. Many asteroids are discovered and then later found to show comet activity and they are accordingly reclassified as comets. This has led to confusing terminology. Active comets can be perturbed to asteroid-like orbits and in time become inactive and therefore "asteroids," unless they were previously observed as comets. Bona-fide asteroids can occasionally be perturbed into the outer solar system with comet-like orbits. The regions of the solar system dominated by asteroids surely contain some "comets" and the regions dominated by "comets" must contain some asteroids. During the accretion disk phase of the solar nebula and continuing to the present day, there is mixing, to some degree, over the solar system and transfer of materials between regions.

The asteroids are generally considered to be relic planetesimals that formed too close to the center of the solar nebula to contain ices and comets to be ice-bearing planetesimals that formed beyond the nebular snow line. This is a convenient cosmochemical way to consider comets and asteroids but even it is misleading. Many asteroids would have been considered to be comets by the IAU had someone been around to observe them early in the history of the solar system. Meteorites are samples of asteroids, and the presence of hydrous minerals in some of them clearly indicates that a substantial fraction of the asteroids contained water, presumably due to the melting of ice. Hydrous alteration of CI and CM chondrites destroyed most of the primary phases of these meteorites, producing phyllosilicates such as smectite and serpentine as well as sulfates, magnetite and other phases (Bunch and Chang, 1980). The presence of an OH infrared feature and other features in the reflection spectra of asteroids reveals the presence of bound water in hydrous minerals (Rivkin et al., 2002). The more distant asteroids are dominated by anhydrous silicates (Jones et al., 1990), suggesting that ice that may have been in these bodies was not heated to melting and, if present, was lost by sublimation.

Decades ago, the presence of hydrated minerals in asteroids and meteorites was interpreted as the result of gas–grain reactions in the nebula, the result of low-temperature equilibrium between nebular water vapor and silicate grains that had previously condensed at higher temperature. A considerable amount of petrographic (Bunch and Chang, 1980) and isotopic (Clayton and Mayeda, 1983) data indicates that the hydrated minerals were formed by internal processes inside parent bodies and not preaccretionary nebula processes. Hydrated silicate formation was probably not prevalent by nebular gas–grain reactions, because the timescale of the reaction was longer than the time span of the gaseous nebula at the low temperatures where hydration reactions are thermodynamically possible (Fegley and Prinn, 1989). Alternatively, the work of Ganguly and Bose (1995) suggests that nebular hydration reactions could have occurred before the nebula dissipated. The preponderance of evidence from aqueous alteration products in primitive meteorites implies that many of the asteroids contained ice and they were heated to the melting point of ice, yielding internal liquid water, at least as surface films.

The heat source in the asteroids that produced the thermal metamorphism observed in meteorites may have come from the decay of short-lived ^{26}Al and ^{60}Fe (Tachibana and Huss, 2003). Whatever the heat source was, its effects were most prominent in the inner regions of the asteroid Belt (Bell et al., 1989) and are presumed to be the origin of the radial dependence of asteroid reflectance classes with increasing distance from the Sun. Once started, the internal heating process would have been enhanced by the exothermic heat from hydration processes. Some of the water must have escaped from the interiors of internally heated asteroids and perhaps a significant proportion of main Belt asteroids showed cometary activity before they turned into what are now considered to be asteroids. Had they been observed early in the history of the solar system all of the hydrated silicate-bearing asteroids would have been considered to be comets.

From a purely cosmochemical perspective, the present asteroid-comet terminology is a confusing and often misleading means of classifying primitive solar system bodies. An improvement would be to consider any primitive undifferentiated body that ever contained trace of higher amounts of ice to be an "ice-bearing planetesimal" or "true comet." By this scheme, most of the planetesimals in the solar nebula were ice-bearing particles or true comets.

While many asteroids may have previously been comet-like, even active comets are somewhat asteroid like. The traditional concept of comets as "dirty snowballs" has been modified by some authors to consider them to be more like "frosty rocks," because they contain more rock than ice. The dust gas production rate from the spectacularly active comet Hale-Bopp exceeded its gas production rate by a factor of 5 (Jewitt and Matthews, 1999). In considering the water contents of comets, it is interesting to consider that some comets may have ice abundances similar to the bound water content of hydrated silicate-rich asteroids. At least in some cases, comets and asteroids might have similar capacities for carrying water to other solar system bodies.

1.25.3 COMET ACTIVITY

As comets approach the Sun, the increasing intensity of sunlight drives cometary activity, the release of gas, dust, and rocks into space. Materials escaping from the comet nucleus form a coma of gas and dust that can reach 10^6 km in size. A roughly spheroidal coma of hydrogen, up to 10^7 km across, can be observed in the ultraviolet as Lyman-alpha radiation. Outflowing gas becomes ionized due to photoionization from solar ultraviolet radiation and charge exchange with the solar wind. The ions are then swept outward at the speed of the solar wind $(300-500$ km s$^{-1})$ to form the ion tail. Volatiles such as H_2O, CO, CH_4, C_3H_6, HCN, NH_3, CH_3OH, and CH_4CN sublime from the solar-heated nucleus as well as debris in the coma and the resulting gas expands into space at velocities of several kilometers per second, depending on the molecular weight of the gas and the temperature of the emitting surface. Volatiles are also produced by degradation of larger molecules. The dust particles, composed of less volatile material, are liberated from the nucleus and propelled into space by the expanding gas. Submicron particles (with high area/mass ratios) are accelerated nearly to the gas outflow speed, while larger and larger particles are proportionately less affected by the gas flow. Evidence from direct particle impacts detected at Halley and from meteor studies indicates that a major fraction of the particulate mass ejected from comets is in the millimeter and larger size range. Particles larger than a few centimeters should not be capable of being lifted off the nucleus by gas but bodies of meter and larger size are released and do exist in cometary meteor streams. Cometary meteor streams are thin Sun-encircling belts of solid comet debris that produce meteor showers when the Earth passes through them. Due to loss of volatiles, the timescale for comet activity by a comet orbiting inside Saturn's orbit is considerably shorter than its dynamical lifetime (Levison and Duncan, 1994).

1.25.4 COMET TYPES—ORBITAL DISTINCTION

Comets are traditionally classified into two groups on the basis of their orbital periods. Those with orbital periods longer than 200 yr are considered to be long-period (LP) comets and those with less than 200 yr to be short-period (SP) comets. SP comets with periods between 200 yr and 30 yr are considered to be Halley-like and those <30 yr to be Jupiter family comets because most have aphelia (greatest distance from the Sun) near Jupiter and were trapped into the short-period orbits as a result of close encounters with Jupiter. All comets close to the Sun and showing cometary activity are short-lived due to several effects. They are on unstable orbits that are quickly perturbed by planets, they rapidly lose volatiles on each perihelion passage (closest approach to the Sun), and they also fragment. The timescale for comet activity is short, and comets usually become either inactive or they totally disintegrate before being perturbed beyond the solar proximity needed to trigger cometary activity. Bodies on planet crossing orbits in the inner solar system have dynamical lifetimes $\sim 10^6 - 10^7$ yr before they have close encounters with planets and either collide with a planet and are ejected from the solar system or disintegrate near the Sun. The timescale of cometary activity is determined by the combined effects of loss of volatiles and buildup of inert lag deposits on the surface. Cometary activity may end as a comet becomes an inactive asteroid-like body or it disrupts leaving no detectable trace. Several comets have been observed to disintegrate at least into smaller fragments if not largely into dust. These include comets West, Ikea Seki, and C/1999 S4 (LINEAR).

1.25.4.1 Comet Source Regions

Comets are not stable near the Sun and they are short lived in regions of the solar system where they exhibit cometary activity. Active comets are derived from two major reservoirs where they can be stored in adequate long-term isolation from solar heating and planetary perturbations. These reservoirs are called the Oort cloud and the Kuiper Belt. It appears that virtually all comets with low-inclination orbits with orbital periods less than 30 yr are derived from the Kuiper Belt while others come from the Oort cloud.

1.25.4.2 The Oort Cloud

The Oort cloud is an extended distribution of comets orbiting the Sun in randomly inclined orbits. It extends out to roughly 5×10^4 AU (Figure 2) as estimated by the size of LP comet orbits (Weissman, 1996). The outer limits of the Oort cloud are believed to be determined by tidal stripping from encounters with other stars, giant molecular clouds, and gravitational effects when the Sun crosses the midplane of the galaxy. The encounters that limit the extent of the cloud also are the source of perturbations that alter Oort cloud comets into orbits that penetrate into the <5 AU region where cometary activity becomes important. Comets in the Oort cloud are too distant to be observed, and the presence of the cloud was proposed to explain the origin of LP comets by Oort (1950). The number of bodies needed in the cloud to produce the observed flux of LP comets is $>10^{12}$ and the total mass of the Oort cloud is on the order of an Earth mass (Stern and Weissman, 2001).

The size of Oort cloud is three orders of magnitude larger than the planetary region of the solar system and it does not seem plausible that Oort cloud comets assembled *in situ*. It is generally believed that the Oort cloud is populated with bodies that were scattered outwards from the formation of the giant planets. Numerical simulations indicate that roughly comparable amounts of planetesimals were scattered from the regions of all four of the giant planets (Weissman, 1999). It is also inevitable that some bodies from the terrestrial planet region were also ejected into the Oort cloud. The Oort cloud comets were exposed to severe impact regimes in the nebular disk

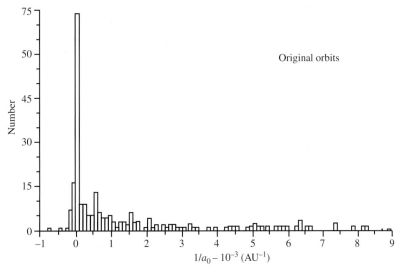

Figure 2 Distribution of $1/a$ in units of 10^{-3} AU^{-1}, the inverse of the semimajor axes (half length of the orbital ellipse) of 264 LP comets. The peak near zero contains dynamically "new" comets coming directly from the Oort cloud for the first time. Comets with larger values of $1/a$ have previously entered the planetary region of solar system and have been perturbed into smaller lower-energy orbits (source Marsden, 1989b).

before ejection into the protective isolation in the Oort cloud (Stern and Weissman, 2001). Oort cloud comets probably formed closer to the Sun than Kuiper Belt comets and many have been modified by substantial collisional processes. They also should have accreted in warmer and denser regions of the solar nebula disk relative to the formation region of the Kuiper Belt comets.

1.25.4.3 The Kuiper Belt

Originally it was thought that SP comets were Oort cloud comets that had been captured into short-period orbits following a close encounter with Jupiter. This process can occur and even explain why SP comets usually have prograde orbits with relatively low inclinations. However, Fernandez (1980) pointed out that this source could not produce the number of SP comets that are observed, and suggested that the SP comets were derived from a disk-like distribution of bodies beyond Neptune. It was shown by numerical simulations that this trans-Neptunian distribution of comets could quantitatively supply the SP flux and explain their inclinations (Duncan et al., 1988).

Like the Oort cloud, the Kuiper Belt was initially hypothetical but, due to its proximity, techniques were eventually developed so that the larger Kuiper Belt objects (KBOs) could be telescopically detected from Earth. In 1992 the first KBO was discovered by Jewitt and Luu (1993). It was a 23rd magnitude object with a diameter of ~320 km at an average solar distance of 44 AU. By the end of 2002, over 700 KBOs had been discovered, over 500 since the beginning of 1999. The dramatic rise in detection was due to heroic efforts of detecting faint, distant slow moving objects with rapidly improving imaging capabilities on large telescopes. Proposed telescopes have the potential of detecting several hundred thousand KBOs. The largest body found to date, since the discovery of Pluto and its moon Charon, is 2002 LM60 also named Quaoar. Quaoar is 1,300 km across and it is probably not the largest KBO. It is estimated that there are $\sim 1 \times 10^5$ KBOs larger than 50 km diameter. This is two orders of magnitude larger than number of similar size asteroids in the asteroid Belt.

1.25.4.3.1 KBO terminology

With the discovery of bodies near Neptune and beyond, there arose the difficulty of what to call them. They were initially called KBOs in honor of Gerald Kuiper, who in 1951 had discussed the possible presence of bodies just beyond Neptune, still in the planetary disk but not in Oort cloud. After Kuiper's name was attached to the Belt, it became generally recognized that Kenneth Edgeworth had previously discussed the bodies in 1949. The Belt is often called Edgeworth-KBO in honor of both scientists. Further complicating the issue is the fact the Belt was mentioned even earlier by F. C. Leonard. Fred Whipple and A. C. Cameron also discussed the possibility of these bodies in the early 1960s. Some have complained that naming the Belt to honor Kuiper is inappropriate because, although he mentioned bodies in this region, he specifically stated that they would not be small (comet-like bodies). To avoid misplaced honors some use the term trans-Neptune object (TNO). The first object discovered was

QB$_1$ and the objects have also been referred to as "cubewanos." All of these aliases are a source of considerable confusion and this review will simply refer to them as KBOs because this is the term that is most commonly used in 2002 literature.

1.25.4.3.2 KBO orbital distribution

There is a strong bias for discovery of KBOs that are close and large. Between their discovery in 1992 and 2003, over 700 KBOs were discovered beyond the inner edge of the Kuiper Belt at the 30 AU orbit of Neptune. With distances >30 AU the KBOs do not show comet activity and their brightness is due to sunlight reflected off bare black surface materials. All KBOs are very faint objects, and with their observed brightness declining as r^{-4} distance from the Sun there is a strong bias towards discovery of objects that are closer.

The observed KBOs orbits fit into three major groups: classical, resonant, and scattered (Jewitt and Luu, 2000; Luu and Jewitt, 2002). The classical KBOs have relatively circular orbits and low inclinations and they are the most common, accounting for two-thirds of the objects detected to date. They are characterized with orbital semi-major axes between 42 AU and 48 AU and they seem to be KBOs that suffered the least orbital perturbation since their formation. Orbital simulations have shown that this group is not strongly affected on Gyr timescales by gravitational perturbations from Neptune (Morbidelli et al., 1995). Although the classical KBOs seem to be the closest to their primordial distribution, their range in orbital inclination and eccentricity suggest that even the classical KBO orbits have been altered with time (Luu and Jewitt, 2002). Some classical orbits have inclinations as high as 30°. The distribution of observed classical KBOs ends at 50 AU, suggesting that the original KBOs were truncated at that distance. Dust apparently generated by Kuiper belts around other stars, seen as dust rings, has also been observed, in some cases, to also have outer truncations (Figure 3).

About 25% of the known KBOs have orbital resonances with Neptune, although considering selection effects, the actual abundance of this group may be ~10% (Figure 4). Resonant KBOs are closer and therefore easier to detect. Most of resonant KBOs orbit the Sun twice every time Neptune orbits three times—a 3:2 resonance. Some of the resonant KBOs have 2:1 resonant orbits. The resonant KBOs are believed to be have been trapped into orbital resonance by the outward migration of Neptune (Malhotra, 1995). As Neptune grew by accretion of solid bodies, it drifted outwards due to an imbalance in bodies it scattered inward into lower energy orbits and

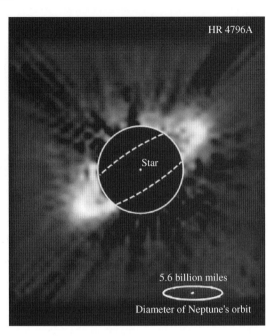

Figure 3 Light scattered off a ring of dust surrounding the star HR 4796A. Presumably the dust is generated by a Kuiper Belt distribution of comets with internal and external truncations. Image taken by NASA's Hubble Space Telescope. The figure is courtesy of B. Smith (University of Hawaii), G. Schneider (University of Arizona) and NASA.

outward into higher-energy orbits. The imbalance was caused by the presence of Jupiter, an effective sink of planetesimals scattered inwards into lower-energy orbits. Every time Neptune would scatter an object inwards, the planet itself would gain energy and drift into a slightly larger orbit. As Neptune migrated outward, its resonances moved with it and planetesimals in its path could be locked into 3:2 or 2:1 orbital resonance. The eccentricities of resonant KBOs imply that Neptune moved outwards by 8 AU, so the 3:2 KBOs represent bodies that may have originally been in a band from 32 AU to 40 AU.

The KBOs are clearly reservoirs of comets that have undergone past dynamical evolution. The scattered group have elliptical orbits that extend beyond the classical, some with aphelia beyond 400 AU. The scattered KBOs appear to be a consequence of an early history of stirring, perhaps related to a larger population of bodies that may have originally resided in the Kuiper Belt region. The orbital distribution of the KBOs is a rich source of information of the dynamics of accretion and scattering in the outer regions of the solar nebula.

A remarkable aspect of the Kuiper Belt is the number of large bodies that it contains. Pluto is in 3:2 resonance with Neptune and it is a member of the bodies swept up by Neptune. Due to orbital

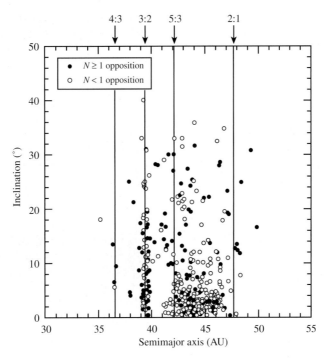

Figure 4 Orbital distribution of classical and resonant KBOs. The numbers at top are ratios of the orbital periods (time to complete an orbit) of the objects to the orbital period of Neptune. The objects clustered around the 3:2 resonance are Plutinos swept up by the outward migration of Neptune. The large range of orbital inclinations produces in relative collision speeds comparable to the orbital speed (~4 km s^{-1}) (source Luu and Jewitt, 2002).

similarities, the 3:2 KBOs are called Plutinos. Pluto is 2,300 km in diameter but there are other very large KBOs. Including Pluto and its moon there are six KBOs with diameters close to or larger than 1,000 km. The KBOs contain significantly more large bodies than are found in the asteroid Belt. The large size of the KBOs coupled with their dynamical evolution suggests that many comets from the Kuiper Belt may have been altered by internal parent-body process due either to radiogenic heat or from the kinetic energy of impacts. It is usually assumed that comets are pristine because of their volatile contents, but survival of volatiles does not mean that they were not heated far beyond nebular or black body temperatures at trans-Neptunian distances. Jewitt has suggested that the smaller bodies, such as usually seen in short-period orbits, are fragments of larger bodies. Davis and Farinella (1997) showed that the Kuiper Belt has undergone considerable collision evolution and that only the largest KBOs (>500 km) have survived the early solar system largely intact.

1.25.5 PHYSICAL EVOLUTION OF COMETS

Within a few AU of the Sun solar heating causes vigorous sublimation of volatiles yielding the formation of a ~10^5 km coma and gas and dust tails that can extend to more than 10^8 km. The active life of a comet can cease by either of three major fates (Keller and Jorda, 2002):

(i) it sublimes if it is dominated by ice, shrinks, and becomes invisible and ceases to exist;

(ii) it disintegrates by splitting and shedding of subnuclei and chunks finally ending up as dust within the planetary system; and

(iii) it becomes dormant when its surface becomes inert; it may remain as a body near its original size.

1.25.5.1 Fragmentation

One of the most remarkable aspects of active comets is that a major path to final destruction is fragmentation. Comets are observed to split into two or more major fragments or sometimes into dozens or more. At least 5% of active comets have been observed to split (Boehnhardt, 2002). The dynamic lifetime for Jupiter family comets is ~10^4–10^5 yr, while the typical chance of fragmentation of a comet in the inner solar system is ~1% yr^{-1} (Chen and Jewitt, 1994). Typical Jupiter family comets may fragment hundreds to thousands of times. Many comets have been observed to fragment. In some cases, fragmentation is due to the tidal forces of a close planetary flyby. This was the cause of fragmentation of Shoemaker-Levy 9 when it passed within 1.3 radii of Jupiter in 1992 and broke into 21 fragments. The weak comet body fragmented under tidal stress of only 10 Pa (10^{-4} atm). Numerous linear crater chains found on the surfaces of Ganymede and Callisto (Schenk et al., 1996) appear to have been created by impacts from tidally disrupted of comets similar to Shoemaker-Levy 9. The presence of crater chains suggests that tidal breakup of comets, into similar sized subfragments, is a relatively common process near Jupiter.

Tidal breakup is not, however, the major source of fragmentation of comets. Most fragmentation occurs somewhat mysteriously far from massive bodies in the absence of tidal stress. Many comets including S4 LINEAR, Hale-Bopp, Wilson, Kohoutek, West, and Ikeya-Seki have been seen to fragment far from the Sun and Jupiter. Remarkable insight into the disintegration of comets has come from the study of the Kreutz family of sungrazing comets, a family of comets that pass within a few solar radii of the Sun.

Sun grazing comets are remarkable in that they pass very close to the Sun but they are also remarkable because they are so numerous. Except for KBOs they are the most abundant observed comets. They pass the Sun at a rate of ~ 0.6 d^{-1} and a large number of observations have been made. Some are bright enough that they have been discovered by ground-based telescopic and even naked eye observations but most are faint and have been discovered by space-borne coronagraphic telescopes observing the solar corona. Most of these comets are considered to be members of the Kreutz group and are fragments of the breakup fragmentation of a single or just a few parents. They are associated with great comets observed as long ago as 372 BC (Marsden, 1967, 1989a,b). The odd circumstance of bodies having perihelia near the Sun is not unexpected. Comets with orbital inclinations near 90° can rapidly evolve to such orbits with perihelia near the Sun. Typically the Kreutz comets have orbital periods $\sim 1,000$ yr and their aphelia extend beyond Pluto. As of August 2002 the Solar Heliospheric observatory (SOHO) had discovered 500 such comets.

The pairing of Kreutz comets by Sekanina and Chodas (2002) has led to the implication that splitting of comets into major fragments occurs over the full orbit path, even beyond 5 AU. Sekanina (2003) has used this approach to explore the fragmentation processes that he suggests can be generalized to all comets. He provides evidence that comets fragment over all of their orbital path. Close to the Sun, the Kreutz comets brighten rapidly $B \sim r^{-5}$ but they then decline in seemingly chaotic fashion. Sekanina has modeled the rapid brightening, fading, and ultimate disappearance between ~ 40 and ~ 3 solar radii as due to processes of continuous, progressive bulk fragmentation, and sublimation of subfragments. Continuous fragmentation can produce the observed light curves of individual comets, and as a runaway process it can produce the entire stream. Kreutz comets smaller than 1–2 km are destroyed completely at perihelion and only larger ones survive solar passage. Fragmentation occurs close to the Sun but it also occurs over all of the orbital path. Sekanina suggests that most of the mass of the Kruetz system is still in large bodies and that the number of observed Sun grazers will increase in the future centuries and fragmentation continues. The origin of fragmentation and large-scale splitting can be caused by a number of factors including: tidal stress, sublimation, change of spin state, internal gas pressure, phase transition from amorphous to crystalline ice, thermal stress and impact with other bodies (Boehnhardt, 2002).

In addition to producing Kreutz family of comet fragments, fragmentation also plays a role in producing meteor streams. Cometary meteor streams consist of fragments of disintegrating comets. Particles larger than a millimeter stay with the stream. Particles smaller than 100 μm are influenced by light pressure drag, the Poynting Robertson effect, and are quickly perturbed from the stream. When Earth passes through the path of the larger particles, a meteor shower is seen. Small differences in orbital parameters, due to spread in ejection velocities from the nucleus and gravitational perturbations from the planets, cause the particles to eventually spread totally around the orbit. In special cases such as the Leonid stream, meteor storms can occur where the density of debris is high enough to produce visual meteor rates that can exceed 1×10^5 meteors per hour. For meteor storms, relatively dense substreams are preserved, each of which can be related to a particular perihelion passage (McNaught and Asher, 1999). In the case of the Leonids, storms often have occurred on a 33 yr timescale, the orbital period of Temple–Tuttle, the parent comet. The brightest meteors are called fireballs and are created by boulder-sized objects too large to be ejected from comets by the nominal flow of escaping gas. Typical surface conditions at comets cannot lift particles larger than centimeter size because of their low area to mass ratios. Large particles in comet debris streams are also indicated by infrared observations of emission from solid particles. The IRAS infrared telescope observed thin trails of emitting dust along the paths of several comets. Sykes et al. (1986) suggest that these particulate trails are made of rocks with typical sizes of 60 cm! It is clear that the much of the evolution of comets includes disintegration into very large chunks that subsequently disintegrate into smaller and smaller bodies, ultimately into dust.

1.25.5.2 Crust/Mantle Formation

All active comets are partly covered by less active or inactive crusts (Kuehrt and Keller, 1994) that commonly cover 90% or more of the surface. Volatile emission occurs from rare active regions when illuminated by sunlight (Keller and Thomas, 1997). The process of vacuum devolatilization of ices from particulate surface layers has been simulated by a number of extensive laboratory experiments (Gruen et al., 1993). Gas produced in confined active regions is believed to be responsible for the formation of jets observed on comets (Figure 5) and the production of concentric shells in the comas of comets produced as isolated "hot spots" revolve into sunlight and become active. The high-resolution images of comet Borelly show a highly complex surface with wide ranging surface topography and albedo variations from 1% to 4% (Figures 6–9). Jewitt (2002) has suggested that the crust is the result of the buildup of inactive lag deposits of fallback materials that were not ejected at speeds exceeding the escape velocity. He relates this process to the remarkable color differences between active comet nuclei, KBOs, Centaurs, and dead comets. All of these bodies are dynamically related and derived from KBOs and yet the KBOs systematically have spectral reflectance, with steeper slopes in the red. The KBOs are dark ultra-red objects and yet the populations of bodies that are derived from them, and travel closer to the Sun, are systematically less red (Jewitt, 2002; Jewitt and Luu, 2000) (Figure 8). Jewitt suggests that the active nuclei, Centaurs, and dead comets have either lost the original dark red KBO surface material or that this surface is obscured by events activated by cometary activity that lead to the buildup of less red surface rubble. The original ultra-red surface cover may be produced by irradiation of organic materials or some form of space weathering that occurs on cold organic-rich surfaces exposed to space for long periods of time.

Estimates of the percent of active areas on comets peak at very low values (Figure 9). A'Hearn et al. (1995) estimate that active area distribution histogram peaks at areas less than a square kilometer in size. Cometary surfaces are largely inert and it is clear that some comets can become totally inert dead comets (Figure 8).

1.25.5.3 Strength and Structure of Cometary Materials

There is abundant evidence from cometary meteors and tidal breakup, suggesting that comets are very weakly consolidated materials (Weissman, 1986). Sekanina (1982) found that

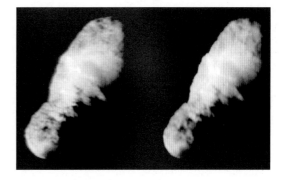

Figure 6 Stereo images of comet Borrelly taken by NASA's Deep Space 1 Mission.

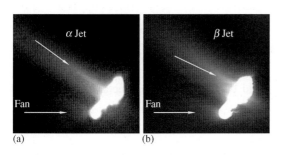

Figure 5 Active jets observed from comet Borrelly. This directed emission from sunlit areas is responsible for the "rocket effect" that measurably perturbs many comet from normal "Keplerian" orbits purely determined by gravity. Image taken by NASA's Deep Space 1 Mission.

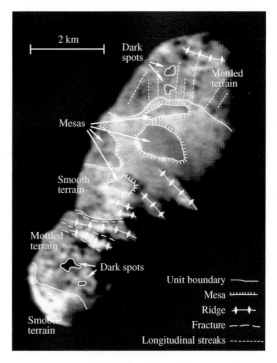

Figure 7 High-resolution image of comet Borrelly with topographic and albedo regions mapped (source Soderblom et al., 2002).

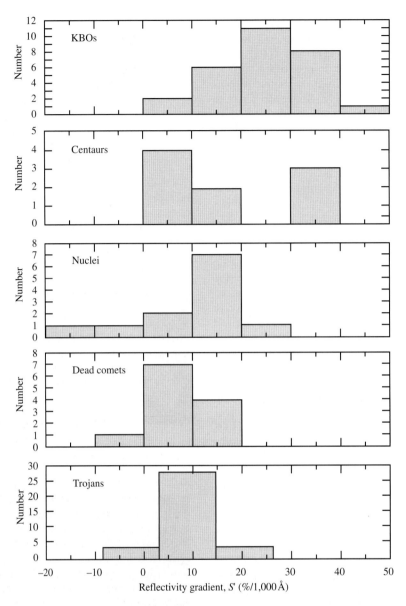

Figure 8 Histograms of the spectral reflectivity gradient of KBOs, Centaurs (bodies with perheilia between Jupiter and Neptune), active comet nuclei, dead comets, and Trojan asteroids. The similarity of dead comets and Trojan asteroids is remarkable (source Jewitt, 2002).

the sungrazing comets split with tidal stresses in the range of 10^2-10^4 Pa. Sirono and Greenberg (2000) deduce similar values for the strength of Shoemaker-Levy 9, tidally disrupted by a close Jovian flyby. The crushing strength of cometary meteors, deduced by the atmospheric ram pressure that causes fragmentation, is similarly low ~ 2 kPa (2×10^4 dyne cm^{-1}) (Verniani, 1969; Revelle 2001).

1.25.6 MAJOR COMPONENT COMPOSITION

Comets are composed of volatile and nonvolatile components that ultimately become dust and rocks or gas after ejection from the nucleus. A major fraction of the volatiles are ices. Volatileicy materials can exist as pure frozen materials, as materials trapped in amorphous ice or in clathrate hydrates. Table 1 lists the approximate temperature ranges where volatiles could be released from cometary ices (Prialnik, 2002).

1.25.6.1 Water Ice

Water ice is a major component of comets and one that dominates cometary activity and physical evolution. The phase of the ice is of considerable interest, because different phases have quite different properties. At temperatures

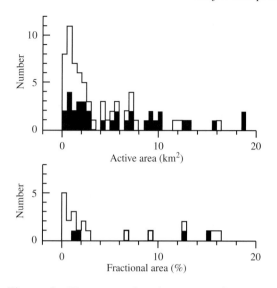

Figure 9 Histogram of active areas of comets showing peaking at low spatial and fractional areas. A few comets have larger active areas beyond the limits of the plot (source A'Hearn et al., 1995).

Table 1 Release of volatiles other than H_2O.

State	Released by	Temperature range
Frozen	Sublimation	~20–100 K
Trapped	Crystallization	~120–140 K
Clathrate–hydrate	Sublimation of H_2O ice	>180 K

Source: Prialnik (2002).

below 120 K ice can occur in a metastable amorphous state for timescales as long as the age of the solar system. If heated above 150 K, it should crystallize to cubic ice, which should transform to hexagonal ice at temperatures above 195–223 K. The presence of amorphous ice has important ramifications because of its remarkable properties including very low thermal conductivity, and ability to trap a wide range of volatile compounds including CO and Ar. It can accommodate ~5% of its mass as guest molecules. The transformation to crystalline ice releases heat—providing stimulus for limited thermal runaway effects and release of volatiles trapped at thermal conditions that exceed normal sublimation temperatures. Amorphous ice has often been considered in models of cometary activity.

The issue of amorphous ice is an important one for comet activity and evolution (Bar-Nun et al., 1987; Gronkowski, 2002). It is widely mentioned in the literature, but there is some debate on whether it exists in comets. Keller and Jorda (2002) outlines several arguments that it should not occur in comets. He uses the work of

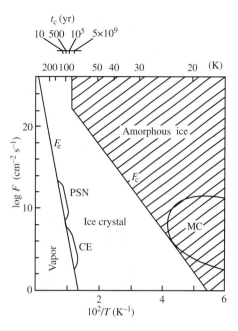

Figure 10 Phase diagram for ice as a function of temperature and F, the deposition rate of water vapor. PSN, Ce, and MC denote the primordial solar nebula, circumstellar envelope, and molecular cloud (source Kouchi et al., 1994).

Kouchi et al. (1994) to show that amorphous ice should not form in the solar nebula. As shown in Figure 10, the formation of amorphous ice in astrophysical conditions requires low temperatures and a deposition rate above a critical rate (F_c in Figure 10) to prevent crystallization. This work implies that these conditions could not be met in the solar nebula marked PSN in Figure 10. The work of Kouchi et al., suggests that while low enough temperatures may have existed in the trans-Neptunian region, the deposition rate was too low. Molecular clouds provide appropriate environments for amorphous ice formation and such presolar ice could have preserved in the trans-Neptunian region even if they could not have formed there. Keller and Jorda cautions that even short heating events such as by shocks during the collapse of the nebula might have initiated crystallization. Even if amorphous ice is incorporated into comets, there are heat sources that might initiate its crystallization. Thermal modeling of radiogenic heat from ^{26}Al, ^{40}K, ^{232}Th, ^{235}U, and ^{238}U has shown that interior of comets might be heated above the transition temperature of amorphous ice, even inside comets as small as 10 km radius. The role of ^{26}Al is an important question due to its short half-life of 0.74 Myr. Accretion time estimates in the Kuiper Belt range from <10^6 yr (Weidenschilling, 1997) to several million years (Farinella et al., 2000) or more. Even if the timescale of comet accretion is too slow for the survival of the short-lived isotopes

Table 2 Sublimation temperatures and ^{26}Al abundance needed for sublimation.

Molecule	T_{subl} (K)	$X_0(^{26}Al)$
H_2O	120	1×10^{-7}
CO	20	8.2×10^{-9}
CO_2	70	2.4×10^{-8}
NH_3	83	6.2×10^{-8}
CH_4	30	2.2×10^{-8}
H_2CO	73	3.1×10^{-8}
N_2	21	8.9×10^{-9}

Source: Choi et al. (2002).
Note: Initial $X_0 = 7 \times 10^{-7}$ (i.e., $^{26}Al/^{27}Al = 5 \times 10^{-5}$).

^{26}Al and ^{60}Fe, the models of Choi et al. (2002) indicate that internal heating by long-lived radiogenic isotopes may also heat interiors and cause loss of volatiles that sublime at temperatures of 40–50 K. Condensed CO, N_2, and CH_4 would be lost and even moderately volatile species such as CO_2 and H_2CO could be partially lost. The observation that CO and other volatiles are commonly seen flowing from comets is evidence that these species are trapped in less volatile materials like water ice. However, comet activity is sometimes seen far from the Sun and this must be driven by material much more volatile than water ice. Table 2 shows the approximate sublimation temperatures of selected volatiles and the abundance of ^{26}Al required to produce such internal heating.

1.25.6.2 CO and Very Volatile Compounds

Many comets show activity at great distance from the Sun. Halley had an outburst where its brightness increased by a factor of over 200 when it was 14 AU from the Sun. Hale-Bopp had a large coma and broad fan-shaped tail when it was nearly 13 AU from the Sun. Black body temperatures at this distance are <100 K and clearly activity is driven by materials much more volatile than H_2O ice. CO appears to be the major "super volatile" and its volatility far from the Sun can in principle explain the ejection of dust grains observed at great solar distance (Hanner, 1981). CO can be trapped in either amorphous or crystalline ice. The major idea for its release is either (i) by transformation of amorphous ice to crystalline ice or (ii) by sublimation of crystalline ice. Crystallization of amorphous ice would occur at low temperature deep in the interior and the resulting gas would have to diffuse through open spaces to reach the surface. Sublimation of crystalline ice would occur at the surface in warmer surface or near surface regions heated by sunlight.

High levels of CO emission from Hale-Bopp at 7 AU (Jewitt et al., 1996) as well as transient brightening of Schwassman-Wachmann 1 and Chiron at great solar distance from the Sun are anomalous emissions that would not be expected by a simple body outgassing highly volatile materials from its surface. These effects are often stated as evidence for internal instability such as can be caused by crystallization amorphous ice.

In addition to the nuclear source there is also clearly a "distributed" source above the nucleus, in the coma itself. Direct spacecraft measurements showed that two-thirds of the CO was released by Halley in the coma not directly from the nucleus (Eberhardt, 1999; Greenberg and Li, 1998). Detailed infrared observations of Hale-Bopp also showed that half of the CO was from a distributed source (Di Santi et al., 2001). These observations also showed that the distributed source turned on between 1.5 AU and 2 AU inbound and turned off at 2.2 AU outbound. The source turn-on was evidently triggered by a thermal threshold.

The UV discovery of argon in Hale-Bopp has been implicated as evidence that the interior of that comet had not been heated to ~35–40 K (Stern et al., 2000). Very low interior temperatures are also implied by the measured *ortho* to *para* ratios in NH_3 (Kawakita et al., 2001) and water (Mumma et al., 1987). If correctly interpreted, these data imply that the comets studied were never even moderately heated in their interiors.

1.25.6.2.1 Full volatile composition

A detailed summary of the present knowledge of the full volatile composition of comets from ground-based and space-based observations is beyond the scope of this geochemically inclined review. Detailed information can be found in Huebner (1990, 2002) and Huebner et al. (1987, 1989). A complicating factor in understanding the molecular composition of comets is compositional variation from comet to comet. It has been suggested that there are major groupings based on observed differences in the relative strengths of C_2, C_3, CN, and $NH_2:H_2O$ lines. For example, most comets have spectra similar to Halley with strong lines of all of these molecules while a lesser fraction are similar to comet Borrelly and are depleted in carbon chain molecules. In the extreme, the unique comet Yanaka shows no detected C_2 or CN emission. As will be discussed in Section 1.25.8, the most systematic study of comets concludes that the intrinsic compositional differences between comets are relatively modest.

One of the major goals of the study of cometary volatiles is to determine if the molecular composition of comets was determined primarily by nebular or prenebular processes. The most detailed information on this issue comes from

Table 3 Relative production rates of molecules in comet Hale-Bopp (C/195 O1).

Molecule	$[X]/[H_2O]$
H_2O	100
HDO	0.06
CO	23
CO_2	20
CH_4	0.6
C_2H_2	0.2
CH_3OH	2.4
H_2CO	1.1
HCOOH	0.08
NH_3	0.7
HCN	0.25
DCN	0.25
HNCO	0.10
HNC	0.25
CH_3CN	0.02
HC_3N	0.02
NH_2CHO	0.015
H_2S	1.5
OCS	0.4
SO	0.3
CS	0.2
SO_2	0.2
H_2CS	0.02
NS	0.02
H_2O_2	<0.03
CH_2CO	<0.032
C_2H_5OH	<0.05
HC_5N	<0.032
Glycine I	<0.5

Source: Bockelee-Morvan and Crovisier (2002).

the most comprehensive compositional study of any comet to date, the bright and massive Oort cloud comet Hale-Bopp. The data in Table 3, from Bockelee-Morvan and Crovisier (2002), list the measured production rates of a large number of molecules. The abundances are very similar to those inferred from interstellar ices, hot cores in molecular clouds, and in the bipolar outflows from protostars. This similarity suggests that the molecular composition of this comet was determined by presolar processes that occurred by gas grain reactions in interstellar environments rather than by processes in the solar nebula (Bockelee-Morran et al., 2000). Further disentangling of the effects of the solar nebula and presolar environments will be a major challenge of future work as both can involve a variety of gas grain reactions and radiation environments.

1.25.6.3 Dust (and Rocks)

1.25.6.3.1 Information from astronomical observations

Information on the properties of comet dust comes from astronomical observations in the infrared, *in situ* measurements on spacecraft, meteor observations, and analysis of collected interplanetary particles. Advances in astronomical observations of comets have been spectacular and have produced important information on mineralogical composition of cometary materials. The presence of silicates was first discovered by Maas et al. (1970), who detected the 10 μm "silicate" feature seen in the thermal infrared emission of outflowing coma dust. The feature is a small broad bump in the infrared emission by silicate grains. The emission feature is due to stretching vibrations in Si–O bonds and bending mode vibrations also produce a silicate feature near 20 μm but this is seldom observed. The silicate features are only seen in silicate grains less than 2 μm in size that are too small to be efficient black body emitters at 10 μm. The silicate feature cannot be seen in larger grains, because they radiate nearly as ideal blackbodies. The feature has been seen in many comets, but its strength and shape varies (Hanner, 1994, 1996; Hanner et al., 1994). Some of the variations are due to grain size and temperature variations. Strong 10 μm features have only been seen in long- or intermediate-period comets and although the silicate feature is observed in SP comets (Kuiper Belt bodies), it is weak. This difference could be due to intrinsic differences between grains in Oort cloud and Kuiper Belt comets or it could simply be that SP comets do not yield as many of the micron and smaller particles needed to provide strong emission at 10 μm.

Fine structure on the silicate feature was first seen in observations of comet Halley. It consisted of a small 11.2 μm bump on the silicate feature detected in high spectral resolution and good signal-to-noise ratio observations. This small bump has now been seen on the "silicate" feature of several comets and it is widely interpreted as evidence for the presence of olivine because IR studies of powdered olivine samples show fine structure at 11.2 μm. In the astronomical literature this feature is considered to be "crystalline olivine" as compared to amorphous silicates that cannot produce the pronounced 11.2 μm bump on the overall 10 μm silicate feature.

The 11.2 μm fine structure on the Si–O silicate feature has provided interesting insight into the relationship between cometary and interstellar materials, because IR observations of silicates in the diffuse interstellar medium and molecular clouds do not show the feature (Molster et al., 2002a,b). Searches for the 11.2 μm fine structure towards the Galactic center indicates that less than 0.5% of interstellar silicates are crystalline (Kemper and Tielens, 2003). The crystalline olivine feature is, however, seen in certain astronomical objects, stars surrounded with disks. It has been seen in Beta Pictoris (Knacke et al., 1993) and Herbig Ae/Be stars

(Waelkens *et al.*, 1996) massive pre-main sequence stars surrounded with disks of dust and gas. Herbig Ae/Be stars even show transient gas features in their spectra that have been interpreted as comets falling into the star (Beust *et al.*, 1994). The presence of the olivine feature in comets and circumstellar disk systems and the lack of it in interstellar and molecular clouds, the parental materials for star and planetary formation, is somewhat of a conundrum. A common astronomical interpretation is that interstellar grains are amorphous silicates and when warmed in a circumstellar disk environment, they anneal to produce crystalline materials. The other possibility is that olivine in comets and disks condenses from vapor produced by evaporation of original interstellar materials.

It is intriguing that the crystalline olivine fine structure at 11.2 μm has only been seen in LP comets. This is at least partially due to the facts that all recently observed bright comets have been LP comets and that only LP comets show strong 10 μm silicate features. As mentioned previously, the stronger silicate feature in LP comets may be due simply to the higher abundance of micron and smaller grains in these comets. If the lack of an olivine feature in SP comets is a true indicator of mineralogical difference, this would imply that Oort cloud and Kuiper Belt comets have different mineralogical compositions. In the extreme, it would imply that Kuiper Belt comets do not have any mineralogy that would suggest that they are composed of amorphous silicates. It would be quite remarkable if Kuiper Belt comets would be olivine free when Oort cloud comets and circumstellar disks are not. If true, it might result from olivine formation in warmer regions of nebular disks. In the case of the solar nebula disk, the most popular origin of Oort cloud comets is formation interior to the Kuiper Belt region and then ejection into the Oort cloud. The Oort cloud comets formed in warmer regions than Kuiper Belt comets and conceivably this could have initiated olivine formation. While the source region of the LP comets may have been warmer, it was still cold. The temperatures of these regions, at least at midplane, were certainly below 200 K (Bell *et al.*, 1997). Shocks in the solar nebula may have been a transient source of strong heating that could drive annealing processes that would have had a radial gradient in the solar system (Harker and Desch, 2002). Laboratory studies of condensation of amorphous silicates and moderately low-temperature annealing provide important insight into the annealing processes and they have demonstrated some complications, including the formation of tridymite associated with olivine formation (Rietmeijer *et al.*, 1986, 1999, 2002) and the difficulty of forming enstatite by annealing (Thompson *et al.*, 2002). As will be discussed shortly, some comets contain Fo (forsterite Mg_2SiO_4) and En (enstatite $MgSiO_3$). It is not at all clear how these magnesium-rich end-members could form from amorphous materials with near solar iron, magnesium, and silicon ratios. Another possible origin of particular phases in any region of the nebula is by transportation from other regions. If the Oort cloud and Kuiper Belt materials really are mineralogically different, it could be due to transport of materials from warmer and more active regions of the nebula. Oort cloud comets, having formed closer to the Sun, should incorporate more materials processed in the hottest regions of the solar nebula disk.

There is additional astronomical information on cometary silicates that provides far more information than simply the presence of olivine. High-resolution and good signal-to-noise ratio IR spectra show additional fine structure on the 10 μm silicate feature of bright LP comets. A small feature at 11.9 μm is also due to olivine and a slope change at 9.2 μm and 9.3 μm is attributed to pyroxene and amorphous silicate with pyroxene composition (Hanner and Bradley, 2003) (Figure 11).

Although olivine and probably pyroxene is present in LP comets, it is clear from the shape of the 10 μm silicate feature that the major fraction of the silicates are either poorly ordered or amorphous. The general shape of the silicate feature in LP comets can be reproduced by a mix of materials dominated by amorphous silicates with olivine as a minor (15–20%) component and some pyroxene (Hanner and Bradley, 2003; Wooden, 2002). A giant leap forward in the understanding of cometary silicates was obtained by the study of Comet Hale-Bopp, an LP comet that was the largest comet to enter the inner solar system in the last 500 yr. It provided a bonanza for comet science in many ways but it was also a breakthrough for understanding the mineralogy of comet dust. Spectral observations over the range from 16 μm to 45 μm (Figure 12) were made with the Infrared Space Observatory (ISO) (Crovisier *et al.*, 1997, 2000). This region is highly sensitive to mineralogical composition and provides much more diagnostic information than can be obtained by analysis of the 10 μm Si–O stretch feature. The spectra show sharp peaks that are beautifully matched with forsterite (Koike *et al.*, 1993). Five peaks match Fo 100 olivine and minor features have been attributed with enstatite (Wooden *et al.*, 1999; Crovisier *et al.*, 2000). The presence of magnesium-rich phases, the highest-temperature silicates that first condense from solar composition gas, is remarkable. Condensation of these reduced phases implies gas temperatures that exceeded 1,400 K. A complication for the proposed formation of crystalline phases by annealing is how these magnesium end-members could be produced by the annealing of amorphous

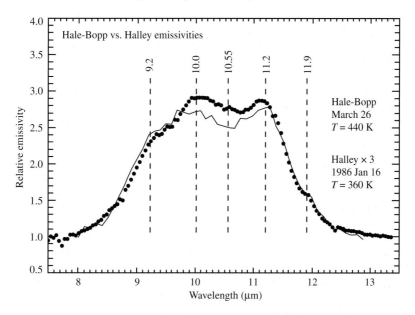

Figure 11 High-resolution, high signal-to-noise profile of the 10 μm silicate emission feature of Halley and Hale-Bopp. The peaks at 11.9 μm and 11.2 μm are due to olivine and the structure near 9.2 μm is attributed to pyroxene. This peak is broader than the diffuse interstellar 10 μm feature and also broader than and distinct from purely crystalline silicates (source Hanner and Bradley, 2003).

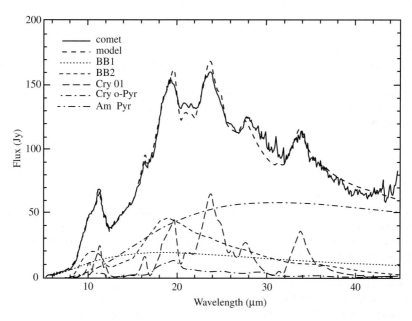

Figure 12 IR data from the ISO for comet Hale-Bopp compared with a five-component dust composition model. BB1 is a 280 K blackbody, BB2 is a 165 K black body, Cry Ol is 22% forsterite, Cry o-Pyr is 8% orthopyroxene, and Am Pyr is 70% amorphous silicate with pyroxene composition (source Crovisier et al., 2000).

silicates with magnesium, iron, and silicon composition that must have been close to solar where all three elements have similar abundances. It is difficult to see how the annealing hypothesis for formation of crystalline silicates could produce Fo and En composition from amorphous materials unless the pre-crystalline silicates also had Fo and En compositions. It is very difficult to produce olivine normative amorphous material in general and Fo in particular unless it is done by irradiation of precursors with olivine composition. Perhaps the only means of making amorphous silicate with olivine and particularly Fo elemental composition is by irradiation of olivine to make it metamict or by melting followed by extremely fast quenching. It is very difficult to make olivine composition glass by quenching unless it is quenched at a splat cooling timescale.

It is particularly intriguing that Fo and En are also seen in the IR spectra of young stars with disks and also in dust formed from gas outflows from evolved stars (Molster *et al.*, 2002a,b; Waters and Molster, 2002; Jaeger *et al.*, 1998). Forsterite and enstatite are common minerals around stars, in LP comets and in the most primitive meteorites even though they are not seen in the diffuse interstellar medium or in SP Kuiper Belt comets.

1.25.6.3.2 Information from collected samples

Laboratory analyses of cometary samples are the ultimate means of determining the nature, origin, and evolution of the nonvolatile components of comets. The fundamental nonvolatile building blocks of comets are micron size, and smaller materials and the most detailed information on them can be obtained with large complex instruments that are often not practical for use on spacecraft missions. The fine-grained nature of cometary solids is clearly evident from analysis of the structure of cometary dust tails (Sekanina, 1980) where particles are spread out in space, depending on their size, optical properties, and time of emission. The infrared emission of coma particles (Hanner, 1994; Maas *et al.*, 1970) and simply the presence of the 10 μm silicate feature requires micron and smaller grains. Direct detection of particle impacts at comet Halley also showed the presence of submicron particles (Maas *et al.*, 1989; Simpson *et al.*, 1987; Jessberger *et al.*, 1988). As will be discussed later, there is evidence that cometary solids are probably made of aggregate structures (Greenberg, 1982), where many of the larger components are aggregates of smaller grains. The fundamental building blocks of the nonvolatile fraction of comets are composed of submicron grains.

The obvious source of comet samples is by direct collection at a comet with Earth return. Stardust, the first comet sample mission (Brownlee *et al.*, 2000), will collect the positively identified particulate samples from a comet and return them to Earth. Stardust, a NASA Discovery mission, will collect thousands of particles from the coma of SP comet Wild 2 and return them in 2006. Hopefully future sample return missions will, in addition, recover subsurface samples of ice and dust and return them to Earth with cryogenic preservation.

In addition to sample return missions, meteoritic materials are available and provide important insight into comets even though it is not presently possible to positively identify any meteoritic sample as having a cometary origin. The meteoritic samples include conventional meteorites and samples of interplanetary dust. Conventional meteorites provide the largest specimens of meteoritic material, most are gram and larger bodies, but it is not evident that any single meteorite has a cometary origin. As relatively large bodies, all meteorites penetrate deeply into the atmosphere before losing their cosmic velocity. The atmospheric ram pressure acting on high-speed bodies is dependent on the ambient air density and the square of velocity. Meteorites retain high velocity deep into the atmosphere, where the density is high, and accordingly they are subjected to crushing ram pressures. All stony meteorites fragment during atmospheric entry, due to this effect, even though most meteorites are relatively strong rocks. Rocks that survive to become meteorites are much stronger than the rocks in cometary meteor showers that are seen to fragment at very high altitudes. Meteorite producing meteors are seen to penetrate deep into the atmosphere and decelerate, while most cometary meteors disintegrate at high altitude and low dynamic ram pressure without slowing down. Atmospheric entry is a barrier, a filter, that prevents typical cometary meteors from surviving atmospheric in rock-size objects, and large cometary meteors disintegrate into dust at high altitudes. Most light curves of bright cometary meteors do not show evidence of surviving bodies. Anders (1975) outlined a number of arguments why recovered meteorites could not be cometary. Among these, one is that all meteorite classes include gas-rich members that appear to have been exposed to solar-wind irradiation in the inner solar system.

It is clear that most primitive meteorites have asteroidal origins, but it remains an open question whether any recovered sample could be cometary. The atmospheric entry filter could only keep out all cometary materials only if comets never contain any strong rock-size components. Wetherill and Revelle (1982) pointed out, however, that while most cometary meteors are weak bodies, there are at least some that show atmospheric entry characteristics similar to strong meteorite producing events. The implication is that some comets contain a small fraction of rocks strong enough to survive atmospheric entry and should produce meteorites. Perhaps strong portions are formed by crust-formation processes on comets, internal formation of strong layers. Strong internal regions in comets could form by impact shock compression, localized heating, or even melting of ice to form water to drive silicate hydration reactions. All of the chemically primitive meteorites were modified and presumably compacted and strengthened by aqueous alteration processes that occurred inside asteroids early in solar system history. Perhaps analogous processes also occurred inside some comets and produced strong materials. This may be particularly true for comets that either are or contain fragments of larger bodies.

The Tagish Lake meteorite shower consisted of hundreds of fragile porous rocks resulting from the atmospheric breakup of a >100 t bolide. Only a minuscule fraction of the preatmospheric mass survived and was found. All of the meteorites fell on snow and many broke into fine powder upon impact, broken like highly friable fine-grained dirt clods. Tagish Lake is the most porous, lowest-density and most fragile meteorite ever found, although, if it had not been seen to fall, had not fallen on a snow-covered lake and had not had such a spectacular atmospheric entry, it surely would never have been found. Tagish Lake has D-type spectral reflectance (dark and reddish), a first for a meteorite providing a link with the most distant asteroid populations and with comets (Brown et al., 2001). Its orbit was consistent with an asteroidal origin although a cometary origin cannot be ruled out. Tagish Lake contains hydrated silicates, carbonates, and other signs of aqueous alteration. Tagish Lake experienced aqueous alteration inside a parent body, and if it were cometary it would be direct evidence for strong internal heating in at least one comet.

Meteoritic materials that enter the atmosphere as small particles do not have the atmospheric strength filter that hampers the survival of cometary rocks. Small cometary dust particles survive atmospheric entry, and studies of interplanetary dust collected in the stratosphere have provided significant insight into the nature of cometary materials. Particles of 10 μm size decelerate at altitudes near 100 km where the ram pressures are only ~0.02 kPa (Brownlee, 1985), many orders of magnitude lower than that experienced by conventional meteorites. Dust samples are also the largest source, by mass, of meteoritic materials. Over 3×10^4 t of interplanetary particles smaller than 300 μm diameter impact the Earth each year (Love and Brownlee, 1995).

The source of this material is the collisions of asteroids and asteroid debris as well as the disintegration of SP comets and the impact disintegration of cometary fragments. LP comets are not major sources of collected dust, because light pressure effects cause most dust particles to be ejected on solar escape orbits. LP comets are only weakly bound to the solar system, and slight effects of light pressure on dust give them positive total energy and puts them on hyperbolic orbits (Harwit, 1963). Some fraction of the dust reaching Earth has directly spiraled inwards from the Kuiper Belt due to the effects of Poynting–Robertson (light pressure) drag (Liou et al., 1996).

Particles less than a few hundred microns in diameter can enter the atmosphere without being heated to their melting points by atmospheric friction, depending on their speed and entry angle. Dust in the size range from ~2 μm to ~50 μm can be collected with high-altitude aircraft and balloons and particles in the micron to millimeter range have been collected from polar ice deposits (Maurette et al., 1994, 1995; Taylor et al., 1998). Particularly for particles smaller than 20 μm, there is a virtually certainty that the collected materials contain a cometary component. Particles in this size range are by convention called interplanetary dust particles (IDPs). There is a selection effect with size, because smaller particles are less heated and experience less ram pressure during atmospheric entry (Flynn, 1989, 1995; Greshake et al., 1998).

There are several classes of IDPs, but there are two distinctive groups that have quite different mineral contents although their elemental compositions are similar. The two groups both have elemental compositions close to primitive chondrites, but one class is composed entirely of anhydrous minerals and the other contains hydrated silicates (see Chapter 1.26). Some of the anhydrous particles are carbon-rich porous aggregates of largely submicron grains. Their porosity often exceeds 30%, and the preservation of such porous structure inside kilometer-sized parentbodies probably requires the presence of volatile phases to prevent compaction over the full lifetime of the solar system. Several groups of investigators have suggested that the porous anhydrous particles are the most likely to be cometary. Estimates of comparative entry speed based on IDP thermal indicators support but do not prove cometary origin for at least some of the particles (Brownlee et al., 1995; Joswiak et al., 2000). A major fraction of cometary particles intersect the Earth's orbit with comet-like orbits and they accordingly have higher entry velocities than asteroidal particles with more circular low-inclination orbits. Asteroidal dust reaches the Earth by a light pressure driven spiral inwards from the asteroid Belt. Many of the asteroid particles reaching Earth are expected to have fairly circular low-inclination orbits at 1 AU, while comet dust will often be on eccentric orbits similar to those of their parents and they have systematically higher entry speeds. Many asteroid particles can have entry speeds below 14 km s^{-1} while many comet particles can have entry speeds in the 16–72 km s^{-1} range. Larger meteorite-size asteroid fragments do not reach Earth at low speeds, because they are not driven to Earth crossing orbits by light pressure drag but by solely by gravitational perturbation.

Figure 13 shows an SEM picture of a typical IDP with entry speed consistent with cometary

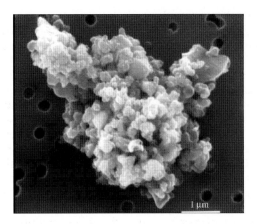

Figure 13 A 5 μm interplanetary dust particle. This is a carbon-rich particle with chondritic elemental composition. It is porous and entirely composed of anhydrous phases. This 10^{-10} g particle is an aggregate of $>10^4$ unrelated and unequilibrated grains. The smooth grains are typically single mineral grains such as Fo, En, or pyrrhotite or carbonaceous material, and the <0.5 μm lumpy grains are usually GEMSs. This is a relatively typical example of the particles that have entry speeds consistent with cometary origin.

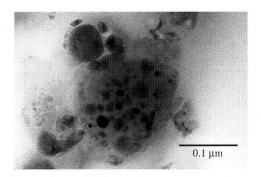

Figure 14 TEM image of a GEMS showing 10 nm and smaller dark beads of FeNi metal and sulfide imbedded in amorphous silicate. GEMS appear to be nonporous and usually range in size from 100 nm to 400 nm.

origin. Characteristic of this group has an aggregate structure, a loosely bound, uncompacted collection of submicron and micron grains. Few visible grains are less than 200 nm in diameter and the typical grain size is in the 200–400 nm range. There are smaller grains, down to nanometer size but they are subcomponents of solid grains usually larger than 100 nm in diameter. Larger components up to several microns in size are common but usually not major components. Normally, the larger grains are single minerals, commonly forsterite, enstatite, or pyrrhotite. Most of these particles are microporous with chondritic elemental composition particle with >10 wt.% carbon content and higher than CI abundance of trace volatile elements (Flynn et al., 1993, Kehm et al., 2002). Like many IDPs its major and minor element composition matches CI values within a factor of 2 or better (Schramm et al., 1989). It is largely composed of anhydrous silicate minerals, amorphous silicates, sulfides, and amorphous carbon-rich material.

An intriguing and unique component of the porous anhydrous IDPs are particles of glass with embedded metal and sulfides (GEMSs). These 200–500 nm diameter particles are composed of silicate glass and small embedded FeNi metal and FeNi sulfide beads (Figure 14). These odd reduced components are fundamental building blocks of many IDPs, but they have never been seen in meteorites. Their properties are consistent with intense radiation processing and they have most of the physical properties of interstellar grains, including the paradoxical ability of interstellar grains, composed largely of oxygen, carbon, magnesium, silicon, and iron, to have the superparamagnetic properties needed to allow alignment in weak interstellar magnetic fields (Bradley, 1994). Mixtures of GEMSs and minor amounts of crystalline silicates provide the best match for the shape of the 10 μm silicate feature observed in comets and circumstellar grains (Bradley et al., 1999).

One of the most important discoveries from the study of anhydrous "cometary" IDPs is the finding that some contain well-preserved presolar grains. Messenger (2000) describes the presence of subgrains that have enhanced D/H and $^{15}N/^{14}N$ ratios that are consistent with their being molecular cloud materials preserved intact within the aggregate particles. The abundance of these somewhat fragile materials is higher than found in the most primitive asteroidal meteorites. If the IDPs truly are cometary, this finding supports the concept that comets truly are storehouses of presolar materials. The effects are most common in "cluster particles" that are so fragile that they fragment into tens to thousands of micron and submicron fragments during the collection process. Messenger et al. (2003) also report the discovery that some IDPs contain silicate grains with highly anomalous oxygen-isotopic compositions. These are the first "stardust" silicates, formed around other stars, discovered inside meteoritic samples.

1.25.6.3.3 Information from in situ measurements

Time-of-flight dust impact mass spectrometers were flown to comet Halley on both of the Soviet Vega and the European Giotto flyby missions (Kissel et al., 1986a,b; Jessberger et al., 1988, 1989a,b, 1991; Schulze et al., 1997). Ions generated by the impact of micron-size dust particles were accelerated into a drift tube and masses determined by flight times in the tube.

These instruments provide the only direct elemental analyses of the composition of comet particles of proven origin. The particles were grouped into three groups: CHON (dominated by carbon, hydrogen, oxygen, and nitrogen), silicate (mainly magnesium, silicon, iron, and oxygen in roughly chondritic proportions), and mixed. There are a few particles near the end-point compositions, but nearly all of the silicates have a CHON component and all the CHON particles contain small peaks of magnesium, silicon, and iron. There is a continuum of compositions (Lawler and Brownlee, 1992). It is clear that the average Halley dust composition is close to chondritic elemental composition (except for enhanced carbon and nitrogen) and that carbonaceous and silicate matter is usually mixed at the submicron level. The high abundances of carbon and nitrogen showed that these elements are largely carried in dust at comet Halley. Combining dust and gas, the overall composition, is remarkably solar for condensable elements (Table 4).

The mineralogical composition of Halley dust is somewhat perplexing (Schulze and Kissel, 1992). Iron, magnesium, and silicon ratios from the highest-quality mass spectra show a wide dispersion of values that do not appear to be modulated by mineralogical compositions (Figure 15). No obvious correlation lines associated with major silicates are apparent. The Mg/Si ratios vary continuously over two orders of magnitude and Fe/Si ratios vary continuously over almost four orders of magnitude. By comparison (Lawler et al., 1989) analyses of similar microvolumes of the Orgueil CI chondrite show highly controlled Mg/Si ratios and constrained Fe/Si ratios. None of the hydrated silicate-bearing meteorites shows such compositional dispersion. Hydration effects tend to homogenize meteoritic materials at the micron spatial scale, although they can produce wide ranges of iron content with magnetite and chronstedite as high-iron end-members. The only meteoritic materials that are even similar to the Halley data are the anhydrous IDPs that also show large compositional variations at the submicron scale because of their GEMSs and glass content. But even IDPs do not show the vast scatter and seemingly impossible order-of-magnitude scatter in the Halley data.

Looking at individual Halley spectra, examples can be found that are consistent with minerals, but looking at the data as a whole there is no evidence of clustering at mineral compositions. This could be due to the lack of mineral grains at the micron size or instrumental effects that shift peak heights providing a level of "noise" that inhibits full quantitative use of the data. The dispersion of Fe/(Fe + Mg) in Figure 16 shows a peaking at magnesium-rich composition that would be evidence for Fo or En, if real, but these compositions do not appear in Figure 14. The presence of so many magnesium-rich silicates distinguishes Halley dust from hydrously altered meteorites. Fomenkova et al. (1992) examined a larger but less quality selected group of spectra than Lawler et al. (1989). In contrast with Lawler et al., they identified particles with compositions consistent with pyroxene, possibly with phyllosilicates, magnesium carbonate, FeNi metal, iron sulfide, iron oxide, and iron-rich silicates. They estimated the carbonate abundances ~10%. They also found a systematic effect that the Mg/Si ratio increases with decreasing particle size varying from 1 to >3. Mg/Si ratios greater than that for forsterite (i.e., 2) are exceedingly rare in meteoritic silicates and usually the presence of such a high ratio in a micron volume requires carbonates or other nonsilicate phases.

The Halley particle data provide remarkable insight into the composition of cometary solids, but unfortunately the precision and reliability of the data cannot be proven. The remarkable mass spectrometers determined compositions of micron and smaller particles impacting at 70 km s^{-1} but the full quantitative precision of the data must be considered to be uncertain. The fact that the mean of all spectra is generally consistent with chondritic values provides assurance that the results are moderately quantitative at the very least. Unfortunately, it remains controversial whether Halley has identifiable minerals or not. Infrared measurements of the 10 μm silicate feature imply the presence of an olivine content of at least several percent and yet olivine trend lines are not evident in Figure 15. The analysis technique used at Halley is not a standard method used for mineralogical analyses, a field where 1% accuracy is usually desired. The ion yields and even reducibility of ion yields from particles of different compositions, sizes, shapes, and porosity could not be tested or fully calibrated with standard minerals at the comet encounter impact speeds. Many spectra were ideal showing strong mass peaks of major and minor meteoritic elements,

Table 4 Relative atomic abundances in gas and dust at comet Halley.

	Geiss (1988)	Grün and Jessberger (1990)	Solar system
H/Mg	39	31	25,200
C/Mg	12	11.3	11.3
N/Mg	0.4–0.8	0.7	2.3
O/Mg	22.3	15	18.5
N/C	0.03–0.06	0.06	0.2
O/C	1.8	1.3	1.6

Source: Huebner (2002).

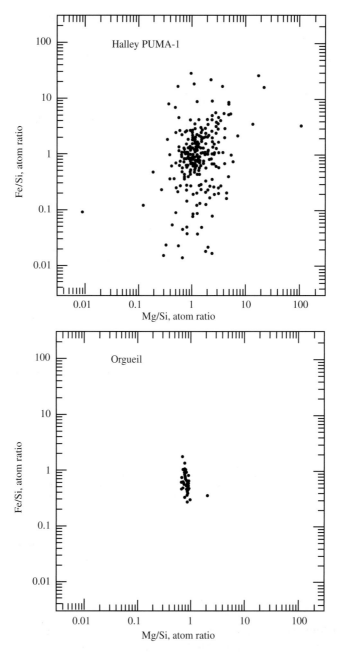

Figure 15 Fe/Si and Mg/Si ratios for micron-sized particles from comet Halley compared with analysis of similar sized volumes of the Orgueil CI chondrite. The vast scatter in the Halley data does not show evidence for mineralogical control of compositions. The tight highly constrained range of Mg/Si in Orgueil is common for phyllosilicate dominated chondrites. (source Lawler et al., 1989).

but there were also nonsensical spectra apparently affected by unknown instrumental effects. These include spectra with magnesium and silicon but no oxygen. Because there were such effects and the fact that good spectra had to be separated from bad, the Halley dust particle data must be used with some caution and not interpreted beyond the possibilities of the data. The chondritic nature, the fine-grained mix of organic and silicate components, and the high abundance of carbon and nitrogen are, however, quite secure.

1.25.7 DIVERSITY AMONG COMETS

Asteroids are a highly diverse group of bodies despite the fact that they formed in a very small region of the solar nebula. Spectral reflectance studies of asteroids show a variety of different

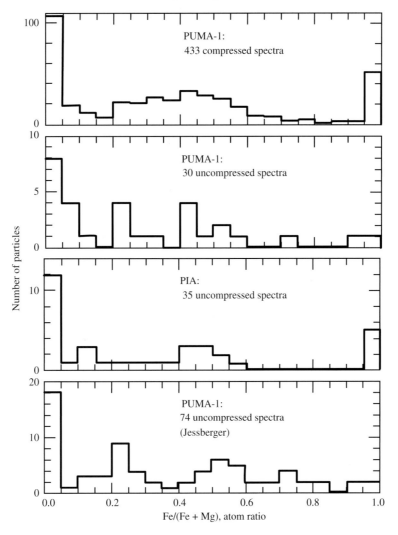

Figure 16 Histograms of Fe/(Fe + Mg) for micron-sized Halley particles. Data from the PUMA mass spectrometer flown to Halley on the Soviet VEGA missions (source Lawler et al., 1989).

types and asteroidal meteorites show remarkable differences ranging from highly oxidized and volatile-rich CI chondrites to the highly reduced E chondrites. Meteorites also show a vast range of metamorphic alteration due to internal parent-body heating. This ranges from the melting of ice, a temperature that required heating to twice the temperature typical asteroids are heated to by sunlight, to temperatures required to melt silicates. There are a variety of reasons to believe that comets might also show a diverse range of properties. Comets formed across a vast range of distances in the nebular accretion disk, from the snow line inside Jupiter's orbit at 5.2 AU to the edge of the Kuiper Belt just beyond 50 AU. They also must have formed over a range of time periods some of which included the lifetimes of internal heat generation from the short-lived extinct radioisotopes ^{26}Al and ^{60}Fe and some that extended beyond this time. Comets also have experienced storage in regions, ranging from the planetary portion of the solar system to the Oort cloud, and during the early history of the solar system they were exposed to wide ranging impact environments in the nebular accretion disk. An additional factor is that many comets may contain fragments of larger bodies ranging in size over 1,000 km in diameter. These large planetary-size bodies must have had some level of internal differentiation (Wallis, 1980). Comets also show some level of diversity simply due to loss of volatiles over time and the evolution of crustal materials.

In light of the above points it is remarkable how similar those comets seem to be. There are differences, but compared to the range of properties seen among asteroids the comets seem to be more similar to each other than they are different. The most complete study of compositional variation among comets was the study of

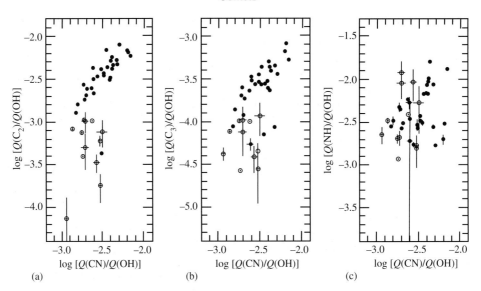

Figure 17 The two main comet compositional groups. Plots of C_2, C_3, and NH relative to CN emission from the A'Hearn *et al.* (1995) study. The filled points are Jupiter family comets, the group with depleted C_2 and C_3 and narrowed range of CN/OH emission.

A'Hearn *et al.* (1995), who did a narrow band photometry of 85 comets over the period from 1976 to 1992. They concluded that most comets have remarkably uniform compositions and that there was little evidence for variation with depth. Comets seen on various apparitions retained similar properties. Their study did detect two major compositional groups: one depleted in C_2 and C_3 and a normal undepleted group (Figure 17). The depleted groups are comets from the Jupiter family, derived from the Kuiper Belt but not all Kuiper Belt comets show this trend. Mumma *et al.* (2001, 2002) report studies of the remarkable LP comet C/LINEAR that appears to be on its first orbit in the inter solar system. This "new" comet is depleted in volatile compounds, suggesting that it formed in a different region than typical LP comets. Mumma *et al.* suggest that it may have formed in the comparatively warm Jupiter–Saturn region. They also suggest that comets of this type may have played a significant role delivering water to Earth. The enhanced D/H ratios observed in the few comets where D/H measurements have been possible yield values higher than seawater (Figure 18). Processes such as loss to space can make water isotopically lighter but not heavier. This implies that the measured comets cannot, in a straightforward way, be major contributors to Earth's oceans. The D/H ratio in ice condensates should be enhanced for ice that condensed in colder, more distant, regions of the solar nebula, and hence condensed outer solar ice should have higher D/H than inner solar system ice (Mousis *et al.*, 2000). D/H was not measured on C/LINEAR but the implication is that comets like it that accreted in the warm inner regions of

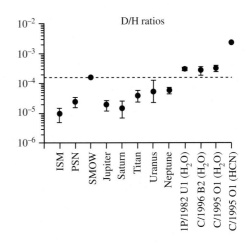

Figure 18 D/H ratios of several comets compared to the oceans (SMOW), planets, the solar nebula (PSN), and the interstellar medium. Low-temperature fractionation processes increase D/H. Jupiter and Saturn have compositions close to the original nebular composition, but low-temperature formation of ice caused the enhancements seen in Uranus and Neptune (the ice giants) and comets. The discrepancy between the plotted LP comets and SMOW argues against these comets providing Earth with a major fraction of its water. Other comets, formed in warmer environments, near Jupiter, could be more similar to SMOW (source Huebner, 2002).

the nebula probably near Jupiter, would not only have volatile depletion but also lower D/H, perhaps consistent with seawater. Other differences seen among comets also include the red spectral reflectivity discussed earlier and the differences in the olivine fine structure superimposed on the 10 μm silicate feature.

1.25.8 CONCLUSIONS

Comets are primitive but complex bodies that potentially contain solid materials from a wide sampling of the cold regions of the solar nebula disk. Comet formation appears to be a common consequence of star formation and studies of solar system comets provide important links between solar system studies and a broad range of astronomical investigations. Although the comets are undoubtedly the best preserved materials from the solar nebula, they have potentially been influenced and processed by a significant number of nebular and parent-body processes. The properties and mysteries of comets contain important clues to numerous materials, environments, and processes that occurred both in the cold regions of the solar nebula and in the environments that preceded it.

REFERENCES

A'Hearn M. F., Millis R. L., Schleicher D. G., Osip D. J., and Birch P. V. (1995) The ensemble properties of comets: results from narrowband photometry of 85 comets, 1976–1992. *Icarus* **118**, 223–270.

Anders E. (1975) Do stony meteorites come from comets. *Icarus* **24**, 363–371.

Backman D. E., Fajardo-Acosta S. B., Stencel R. E., and Stauffer J. R. (1997) Dust disks around main sequence stars. *Astrophys. Space Sci.* **255**, 91–101.

Bar-Nun A., Dror J., Kochavi E., and Laufer D. (1987) Amorphous water ice and its ability to trap gases. *Phys. Rev. B* **35**, 2427–2435.

Bell J. F., Davis D. R., Hartmann W. K., and Gaffey M. J. (1989) Asteroids—the big picture. Asteroids II (eds. R. P. Binzel, T. Gehrels, and M. S. Matthews). University of Arizona Press, Tucson, pp. 921–945.

Bell K. R., Cassen P. M., Klahr H. H., and Henning T. (1997) The structure and appearance of protostellar accretion disks: limits on disk flaring. *Astrophys. J.* **486**, 372.

Beust H., Vidal-Madjar A., Ferlet R., and Lagrange-Henri A. M. (1994) Cometary-like bodies in the protoplanetary disk around beta Pictoris. *Astrophys. Space Sci.* **212**, 147–157.

Bockelee-Morvan D. and Crovisier J. (2002) Lessons of comet Hale-Bopp for coma chemistry. *Earth, Moon and Planets* **89**, 53–71.

Bockelee-Morvan D., Lis D. C., Wink J. E., Despois D., Crovisier J., Bachiller R., Benford D. J., Biver N., Colom P., Davies J. K., Grard E., Germain B., Houde M., Mehringer D., Moreno R., Paubert G., Phillips T. G., and Rauer H. (2000) New molecules found in comet C/1995 O1 (Hale-Bopp): investigating the link between cometary and interstellar material. *Astron. Astrophys.* **353**, 1101–1114.

Boehnhardt H. (2002) Comet splitting—observations and model scenarios. *Earth Moon Planet.* **89**, 91–115.

Bradley J. P. (1994) Chemically anomalous, preaccretionally irradiated grains in interplanetary dust from comets. *Science* **265**, 925–929.

Bradley J. P., Keller L. P., Snow T. P., Hanner M. S., Flynn G. J., Gezo J. C., Clemett S. J., Brownlee D. E., and Bowey J. E. (1999) An infrared spectral match between GEMS and interstellar grains. *Science* **285**, 1716–1718.

Brown P. G., Hildebrand A. R., and Revelle D. O. (2001) Inferring the physical properties of a small D-class NEA: the Tagish Lake Meteoroid. AAS/Division for Planetary Sciences Meeting 33.

Brownlee D. E. (1985) Cosmic dust—collection and research. *Ann. Rev. Earth Planet. Sci.* **13**, 147–173.

Brownlee D. E., Joswiak D. J., Schlutter D. J., Pepin R. O., Bradley J. P., and Love S. G. (1995) Identification of individual cometary IDPs by thermally stepped He release. In *Lunar Planet. Sci.* **XXVI**. The Lunar and Planetary Institute, Houston, pp. 183–184.

Brownlee D. E., Tsou P., Clark B., Hanner M. S., Hörz F., Kissel J., McDonnell J. A. M., Newburn R. L., Sandford S., Sekanina Z., Tuzzolino A. J., and Zolensky M. (2000) Stardust: a comet sample return mission. *Meteorit. Planet. Sci.* **35**, A35.

Bunch T. E. and Chang S. (1980) Carbonaceous chondrites: II. Carbonaceous chondrite phyllosilicates and light element geochemistry as indicators of parent body processes and surface conditions. *Geochim. Cosmochim. Acta* **44**, 1543–1577.

Chen J. and Jewitt D. (1994) On the rate at which comets split. *Icarus* **108**, 265–271.

Choi Y., Cohen M., Merk R., and Prialnik D. (2002) Long-term evolution of objects in the Kuiper Belt zone-effects of insolation and radiogenic heating. *Icarus* **160**, 300–312.

Clayton R. N. and Mayeda T. K. (1983) The oxygen isotope record in Murchison and other carbonaceous chondrites. *Earth Planet. Sci. Lett.* **67**, 151–161.

Crovisier J., Leech K., Bockelee-Morvan D., Brooke T. Y., Hanner M. S., Altieri B., Keller H. U. and Lellouch (1997) The spectrum of Hale-Bopp observed with ISO 2.9 AU from the Sun. *Science* **275**, 1904–1907.

Crovisier J., Brooke T. Y., Leech K., Bockelee-Morvan D., Lellouch E., Hanner M. S., Altieri B., Keller H. U., Lim T., Encrenaz S., Griffin M., de Graauw T., van Dishoeck E., and Knacke R. F. (2000) The thermal infrared spectra of comets Hale-Bopp and 103 P/Hartley 2 observed with the Infrared Space Observatory. In *Thermal Emission Spectroscopy and Analysis of Dust, Disks, and Regoliths*. ASP Conf. Ser.196 (eds. M. L. Sitko, A. L. Sprague, and D. K. Lynch). Astronomical Society of the Pacific, San Francisco, pp. 109–117.

Davis D. R. and Farinella P. (1997) Collisional evolution of Edgeworth-Kuiper Belt objects. *Icarus* **125**, 50–60.

Delsemme A. H. (2000) 1999 Kuiper prize lecture: cometary origin of the biosphere. *Icarus* **146**, 313–325.

Di Santi M., Mumma M., Russo N., and Magee-Saver K. (2001) Carbon monoxide production and excitation in comet C/1995 O1 (Hale-Bopp): Isolation of nature and distributed CO sources. *Icarus* **153**, 361–390.

Duncan M., Quinn T., and Tremaine S. (1988) The origin of short-period comets. *Astrophys. J.* **328**, L69–L73.

Eberhardt P. (1999) Comet Halley's gas composition and extended sources: results from the neutral mass spectrometer on Giotto. *Space Sci. Rev.* **90**, 45–52.

Farinella P., Davis D. R., and Stern S. A. (2000) Formation and collisional evolution of the Edgeworth-Kuiper Belt. In *Protostars and Planet* **IV** (eds. V. Mannings, A. P. Boss, and S. S. Russell), University of Arizona Press, Tucson, pp. 1255–1282.

Fegley B., Jr. and Prinn R. (1989) Solar nebula chemistry: implications for volatiles in the solar system. In *The Formation and Evolution of Planetary Systems* (eds. H. Weaver and L. Danly). Cambridge University Press, Cambridge, pp. 171–211.

Fernandez J. A. (1980) On the existence of a comet Belt beyond Neptune. *Mon. Not. Roy. Astron. Soc.* **192**, 481–491.

Fomenkova M. N., Kerridge J. F., Marti K., and McFadden L.-A. (1992) Compositional trends in rock-forming elements of comet Halley dust. *Science* **258**, 266–269.

Flynn G. J. (1989) Atmospheric entry heating: a criterion to distinguish between asteroidal and cometary sources of interplanetary dust. *Icarus* **77**, 287–310.

Flynn G. J. (1995) Atmospheric entry heating of large interplanetary dust particles. *Meteoritics* **30**, 504–505.

Flynn G. J., Sutton S. R., Bajt S., Klöck W., Thomas K. L., and Keller L. P. (1993) The volatile content of anhydrous interplanetary dust. *Meteoritics* **28**, 349–350.

Ganguly J. and Bose K. (1995) Kinetics of formation of hydrous phyllosilicates in the solar nebula. In *Lunar Planet. Sci.* **XXVI**. The Lunar and Planetary Institute, Houston, pp. 441–442.

Geiss J. (1988) Composition in Halley's comet: clues to origin and history of commentary matter. *Rev. Mod. Astron.* **1**, 1–27.

Greenberg J. M. (1982) What are comets made of? A model based on interstellar grains. In *Comets* (ed. L. Wilkening). University of Arizona Press, Tucson, pp. 131–163.

Greenberg J. M. and Li A. (1998) From interstellar dust to comets: the extended CO source in comet Halley. *Astron. Astrophys.* **332**, 374–384.

Greshake A., Klöck W., Arndt P., Maetz M., Flynn G. J., Bajt S., and Bischoff A. (1998) Heating experiments simulating atmospheric entry heating of micrometeorites: clues to their parent body sources. *Meteorit. Planet. Sci.* **33**, 267–290.

Gronkowski P. (2002) Cometary outbursts—search of probable mechanisms—case of 29P/Schwassman-Wachmann 1. *Astronomische Nachrichten* **323**, 49–56.

Gruen E., Gebhard J., Bar-Nun A., Benkhoff J., Dueren H., Eich G., Hische R., Huebner W. F., Keller H. U., and Klees G. (1993) Development of a dust mantle on the surface of an insolated ice-dust mixture—results from the KOSI-9 experiment. *J. Geophys. Res.* **98**, 15091.

Grün E. and Jessberger E. (1990) *Physics and Chemistry of Comets* (ed. W. F. Huebner). Springer, Berlin, pp. 113–176.

Hanner M. S. (1981) On the detectability of icy grains in the comae of comets. *Icarus* **47**, 342–350.

Hanner M. S. (1994) Remote sensing of cometary dust and comparisons to IDPs. In *Analysis of Interplanetary Dust, Proceedings of the NASA/LPI Workshop held in Houston, TX, May 1993, New York*, AIP Conference Proceedings (eds. M. E. Zolensky, T. L. Wilson, F. J. M. Rietmeijer, and G. J. Flynn). American Institute of Physics Press, Woodbury, vol. 310, pp. 23–32.

Hanner M. S. (1996) Composition and optical properties of cometary dust. In *Physics, Chemistry, and Dynamics of Interplanetary Dust*, ASP Conf. Ser. 104, IAU Colloq. 150, (eds. B. Å. S. Gustatsun and M. S. Hanner). Astronomical Society of the Pacific, pp. 367–376.

Hanner M. S. and Bradley J. P. (2003) Composition and mineralogy of comet dust. In *Comets II* (eds. M. Festou, H. U. Keller, and H. A. Weaver). University of Arizona Press (in press).

Hanner M. S., Lynch D. K., and Russell R. W. (1994) The 8–13 micron spectra of comets and the composition of silicate grains. *Astrophys. J.* **425**, 274–285.

Harker D. E. and Desch S. J. (2002) Annealing of silicate dust by nebular shocks at 10 AU. *Astrophys. J.* **565**, L109–L112.

Harwit M. (1963) Origins of the zodiacal dust cloud. *J. Geophys. Res.* **68**, 2171–2180.

Huebner W. F. (1990) *Physics and Chemistry of Comets*. Springer, New York, 376p.

Huebner W. F. (2002) Composition of comets: observations and models. *Earth, Moon Planets.* **89**, 179–195.

Huebner W. F., Boice D. C., and Sharp C. M. (1987) Polyoxymethylene in comet Halley. *Astrophys. J.* **320**, L149–L152.

Huebner W. F., Boice D. C., and Korth A. (1989) Halley's polymeric organic molecules. *Adv. Space Res.* **9**, 29–34.

Jaeger C., Molster F. J., Dorschner J., Henning T., Mutschke H., and Waters L. B. F. M. (1998) Steps toward interstellar silicate mineralogy: IV. The crystalline revolution. *Astron. Astrophys.* **339**, 904–916.

Jessberger E. K., Christoforidis A., and Kissel J. (1988) Aspects of the major element composition of Halley's dust. *Nature* **332**, 691–695.

Jessberger E. K., Kissel J., Fechtig H., and Krueger F. R. (1989a) On the average chemical composition of cometary dust. In *Evolution of Interstellar Dust and Related Topics*, New Holland, New York, 455p.

Jessberger E. K., Kissel J., and Rahe J. (1989b) The composition of comets. In *Origin and Evolution of Planetary Satellite Atmospheres* (eds. J. B. Pollack, M. S. Matthews, and S. K. Atreya). University of Arizona Press, Tucson, AZ, pp. 167–191.

Jewitt D. C. (2002) From Kuiper Belt object to cometary nucleus: the missing ultrared matter. *Astron. J.* **123**, 1039–1049.

Jewitt D. and Luu J. (1993) Discovery of the candidate Kuiper Belt object 1992 QB1. *Nature* **362**, 730–732.

Jewitt D. and Luu J. (1995) Kuiper Belt: collisional production of dust around main-sequence stars. *Astrophys. Space Sci.* **223**, 164–165.

Jewitt D. C. and Luu J. X. (2000) Physical nature of the Kuiper Belt. *Protostar. Planet.* **IV**, 1201.

Jewitt D. and Matthews H. (1999) Particulate mass loss from comet Hale-Bopp. *Astron. J.* **117**, 1056–1062.

Jewitt D., Senay M., and Matthews H. (1996) Observations of carbon monoxide in comet Hale-Bopp. *Science* **271**, 1110–1113.

Jones T. D., Lebofsky L. A., Lewis J. S., and Marley M. S. (1990) The composition and origin of the C, P, and D asteroids—water as a tracer of thermal evolution in the outer Belt. *Icarus* **88**, 172–192.

Joswiak D. J., Brownlee D. E., Pepin R. O., and Schlutter D. J. (2000) Characteristics of asteroidal and cometary IDPs obtained from stratospheric collectors: summary of measured He release temperatures, velocities and descriptive mineralogy. In *Lunar Planet. Sci.* **XXXI**, 1500. The Lunar and Planetary Institute, Houston (CD-ROM).

Kawakita H., Watanabe J., Ando H., Aoki W., Fuse T., Honda S., Izumiura H., Kajino T., Kambe E., Kawanomoto S., Noguchi K., Okita K., Sadakane K., Sato B., Takada-Hidai M., Takeda Y., Usuda T., Watanabe E., and Yoshida M. (2001) The spin temperature of NH_3 in Comet C/1999S4 (LINEAR). *Science* **294**, 1089–1091.

Kehm K., Flynn G. J., Sutton S. R., and Hohenberg C. M. (2002) Combined noble gas and trace element measurements on individual stratospheric interplanetary dust particles. *Meteorit. Planet. Sci.* **37**, 1323–1335.

Keller H. U. L. and Jorda L. (2002) The morphology of cometary nuclei. In *The Century of Science.* (eds. J. A. M. Bleeker, J. Geiss, and M. C. E. Huber). Kluwer, New York, pp. 1323–1335.

Keller H. U. and Thomas N. (1997) Comet Halley's nucleus: a physical interpretation. *Adv. Space Res.* **19**, 187–194.

Kemper F. and Tielens A. G. (2003) The crystallinity of interstellar silicates. Astrophysics of Dust, Ester Park Co., May 26–30 (in press).

Kisseül J., Brownlee D. E., Buchler K., Clark B. C., Fechtig H., Grun E., Hornung K., Igenbergs E. B., Jessberger E. K., Krueger F. R., Kuczera H., McDonnell J. A. M., Morfill G. M., Rahe J., Schwehm G. H., Sekanina Z., Utterback N. G., Volk H. J., and Zook H. A. (1986a) Composition of comet Halley dust particles from Giotto observations. *Nature* **321**, 336.

Kissel J., Sagdeev R. Z., Bertaux J. L., Angarov V. N., Audouze J., Blamont J. E., Buchler K., Evlanov E. N., Fechtig H., Fomenkova M. N., von Hoerner H., Inogamov N. A., Khromov V. N., Knabe W., Krueger F. R., Langevin Y., Leonasv B., Levasseur-Regourd A. C., Managadze G. G., Podkolzin S. N., Shapiro V. D., Tabaldyev S. R., and Zubkov B. V. (1986b) Composition of comet Halley dust particles from VEGA observations. *Nature* **321**, 280–282.

Koike C., Shibai H., and Tuchiyama A. (1993) Extinction of olivine and pyroxene in the mid- and far-infrared. *Mon. Not. Roy. Astron. Soc.* **264**, 654–658.

Knacke R. F., Fajardo-Acosta S. B., Telesco C. M., Hackwell J. A., Lynch D. K., and Russell R. W. (1993) The silicates in the disk of beta Pictoris. *Astrophys. J.* **418**, 440.

Kouchi A., Yamamoto T., Kozasa T., Kuroda T., and Greenberg J. M. (1994) Conditions for condensation and preservation of amorphous ice and crystallinity of astrophysical ices. *Astron. Astrophys.* **290**, 1009–1018.

Kuehrt E. and Keller H. U. (1994) The formation of cometary surface crusts. *Icarus* **109**, 121–132.

Lawler M. E. and Brownlee D. E. (1992) CHON as a component of dust from comet Halley. *Nature* **359**, 810–812.

Lawler M. E., Brownlee D. E., Temple S., and Wheelock M. M. (1989) Iron, magnesium, and silicon in dust from comet Halley. *Icarus* **80**, 225–242.

Levison H. F. and Duncan M. J. (1994) The long-term dynamical behavior of short-period comets. *Icarus* **108**, 18–36.

Levison H. F., Dones L., Chapman C. R., Stern S. A., Duncan M. J., and Zahnle K. (2001) Could the lunar "late heavy bombardment" have been triggered by the formation of Uranus and Neptune? *Icarus* **151**, 286–306.

Liou J., Zook H. A., and Dermott S. F. (1996) Kuiper Belt dust grains as a source of interplanetary dust particles. *Icarus* **124**, 429–440.

Love S. G. and Brownlee D. E. (1995) A direct measurement of the terrestrial mass accretion rate of cosmic dust. *Science* **262**, 550–553.

Luu J. X. and Jewitt D. C. (2002) Kuiper Belt objects: relics from the accretion disk of the Sun. *Ann. Rev. Astron. Astrophys.* **40**, 63–101.

Maas R. W., Ney E. P., and Woolf N. J. (1970) The 10-micron emission peak of Comet Bennett 1969i. *Astrophys. J.* **160**, L101–L104.

Maas D., Göller J. R., Grün E., Lange G., McDonnell J. A. M., Nappo S., Perry C., and Zarnecki J. C. (1989) Cometary dust particles detected by the Didsy-IPM-P sensor on board Giotto. *Adv. Space Res.* **9**, 247–252.

Malhotra R. (1995) The origin of Pluto's orbit: implications for the solar system beyond Neptune. *Astron. J.* **110**, 420.

Marsden B. G. (1967) The sungrazing comet group. *Astron. J.* **72**, 1170.

Marsden B. G. (1989a) The sungrazing comet group: II. *Astron. J.* **98**, 2306–2321.

Marsden B. G. (1989b) *Catalogue of Cometary Orbits*, 6th edn., Smithsonian Astrophysical Observatory, Cambridge, 96p.

Maurette M., Immel G., Hammer R., Harvey R., Kurat G., and Taylor S. (1994) Collection and curation of IDPs from the Greenland and Antarctic Ice Sheets. In *Analysis of Interplanetary Dust: Proceedings of the NASA/LPI Workshop held in Houston, TX, May 1993, New York*, AIP Conf. Proc. (eds. E. Zolensky, T. L. Wilson, F. J. M. Rietmeijer, and G. J. Flynn). American Institute of Physics Press, Woodbury, vol. 310, pp. 277–290.

Maurette M., Engrand C., and Kurat G. (1995) "Chemical" search for cometary grains in Antarctic micrometeorites. In *Lunar Planet. Sci.* **XXVI**, The Lunar and Planetary Institute, Houston, pp. 915–916.

McNaught R. H. and Asher D. J. (1999) Leonid dust trails and meteor storms. *J. Int. Meteor Org.* **27**, 85–102.

McSween H. Y. and Weissman P. R. (1989) Cosmochemical implications of the physical processing of cometary nuclei. *Geochim. Cosmochim. Acta* **53**, 3263–3271.

Messenger S. (2000) Identification of molecular-cloud material in interplanetary dust particles. *Nature* **404**, 968–971.

Messenger S., Keller L. P., Stadermann F. J., Walker R. M., and Zinner E. (2003) Samples of stars beyond the solar system: silicate grains in interplanetary dust. *Science* **300**, 105–108.

Molster F. J., Waters L. B. F. M., and Tielens A. G. G. M. (2002a) Crystalline silicate dust around evolved stars: II. The crystalline silicate complexes. *Astron. Astrophys.* **382**, 222–240.

Molster F. J., Waters L. B. F. M., Tielens A. G. G. M., Koike C., and Chihara H. (2002b) Crystalline silicate dust around evolved stars: III. A correlations study of crystalline silicate features. *Astron. Astrophys.* **382**, 241–255.

Morbidelli A., Thomas F., and Moons M. (1995) The resonant structure of the Kuiper Belt and the dynamics of the first five trans-Neptunian objects. *Icarus* **118**, 322–340.

Mousis O., Gautier D., Bockelee-Morvan D., Robert F., Dubrulle B., and Drouart A. (2000) Constraints on the formation of comets from D/H Ratios measured in H_2O and HCN. *Icarus* **148**, 513–525.

Mumma M. J., Weaver H. A., and Larson H. P. (1987) The ortho-para ratio of water vapor in Comet p/Halley. *Astron. Astrophys.* **187**, 419–424.

Mumma M. J., Dello Russo N., DiSanti M. A., Magee-Sauer K., Novak R. E., Brittain S., Rettig T., McLean I. S., Reuter D. C., and Xu L. (2001) Organic composition of C/1999 S4 (LINEAR): a comet formed near Jupiter? *Science* **292**, 1334–1339.

Mumma, M. J., DiSanti, M. A., Dello Russo N., Magee-Sauer K., Gibb E., and Novak R. (2002) The organic volatile composition of Oort-cloud comets: evidence for chemical diversity in several Oort cloud comets. American Astronomical Society, DPS meeting #34.

Oort J. H. (1950) The structure of the cloud of comets surrounding the solar system and a hypothesis concerning its origin. *Bull. Astron. Inst. Neth.* **11**, 91–110.

Prialnik D. (2002) Modeling and comet nucleus interior. Application to comet C/1995 01 Hale-Bopp. *Earth, Moon Planets.* **89**, 27–52.

Revelle D. O. (2001) Bolide dynamics and luminosity modelling: comparisons between uniform bulk density and porous meteoroid models. In *Proceedings of the Meteoroids 2001 Conference, August 6–10, 2001, Kiruna, Sweden* (ed. B. Warmbein). ESA Publications Division, Noordwijk, ESA SP-495, ISBN 92-9092-805-0, pp. 513–517.

Rietmeijer F. J. M., Nuth J. A., and MacKinnon I. D. R. (1986) Analytical electron microscopy of Mg–SiO smokes—a comparison with infrared and XRD studies. *Icarus* **66**, 211–222.

Rietmeijer F. J. M., Nuth J. A., and Karner J. M. (1999) Metastable eutectic condensation in a $Mg-Fe-SiO-H_2O_2$ vapor: analogs to circumstellar dust. *Astrophys. J.* **527**, 395–404.

Rietmeijer F. J. M., Hallenbeck S. L., Nuth J. A., and Karner J. M. (2002) Amorphous magnesiosilicate smokes annealed in vacuum: the evolution of magnesium silicates in circumstellar and cometary dust. *Icarus* **156**, 269–286.

Rivkin A. S., Howell E. S., Vilas F., and Lebofsky L. A. (2002) Hydrated minerals on asteroids: the astronomical record. In *Asteroids III* (eds. W. F. Bottke, A. Cellino, P. Paolicehi, and R. P. Binzel). University of Arizona press, Tucson, pp. 235–253.

Schenk P. M., Asphaug E., McKinnon W. B., Melosh H. J., and Weissman P. R. (1996) Cometary nuclei and tidal disruption: the geologic record of crater chains on Callisto and Ganymede. *Icarus* **121**, 249–274.

Schramm L. S., Brownlee D. E., and Wheelock M. M. (1989) Major element composition of stratospheric micrometeorites. *Meteoritics* **24**, 99–112.

Schulze H. and Kissel J. (1992) Chemical heterogeneity and mineralogy of Halley's dust. *Meteoritics* **27**, 286.

Schulze H., Kissel J., and Jessberger E. K. (1997) Chemistry and mineralogy of Comet Halley's dust. In *From Stardust to Planetesimals*, ASP Conf. Ser. 122, 397p.

Sekanina Z. (1980) Physical characteristics of cometary dust from dynamical studies: a review. In *Solid Particles in the Solar System*, Proc. IAU Symp. 90 (eds. I. Halliday and B. A. Macintosh). Reidel, Dordrecht, pp. 237–250.

Sekanina Z. (1982) The problem of split comets in review. In *Comet* (ed. L. L. Wilkening). University of Arizona Press, Tucson, pp. 251–287.

Sekanina Z. (2002) Runaway fragmentation of sungrazing comets observed with the Solar and Heliospheric Observatory. *Astrophys. J.* **576**, 1085–1089.

Sekanina Z. (2003) Erosion model for the sungrazing comets observed with the Solar Heliospheric Observatory. *Astrophys. J.* (in press).

Sekanina Z. and Chodas P. W. (2002) Common origin of two major sungrazing comets. *Astrophys. J.* **581**, 760–769.

Soderblom L. A., Becker T. L., Bennett G., Boice D. C., Britt D. T., Brown R. H., Buratti B. J., Isbell C., Giese B., Hare T., Hicks M. D., Howington-Kraus E., Kirk R. L., Lee M., Nelson R. M., Oberst J., Owen T. C., Rayman M. D., Sandel B. R., Stern S. A., Thomas N., and Yelle R. V. (2002) Observations of Comet 19P/Borrelly by the miniature integrated camera and spectrometer aboard Deep Space 1. *Science* **296**, 1087–1091.

Simpson J. A., Rabinowitz D., Tuzzolino A. J., Ksanfomality L. V., and Sagdeev R. Z. (1987) The dust coma of Comet P/Halley: measurements on the Vega-1 and Vega-2 spacecraft. *Astron. Astrophys.* **187**, 742–752.

Sirono S. and Greenberg J. M. (2000) Do cometesimal collisions lead to bound rubble piles or to aggregates held together by gravity? *Icarus* **145**, 230–238.

Spangler C., Sargent A. I., Silverstone M. D., Becklin E. E., and Zuckerman B. (2001) Dusty debris around solar-type stars: temporal disk evolution. *Astrophys. J.* **555**, 932–944.

Stern S. A. and Weissman P. R. (2001) Rapid collisional evolution of comets during the formation of the Oort cloud. *Nature* **409**, 589–591.

Stern S. A., Slater D. C., Festou M. C., Parker J. W., Gladstone G. R., A'Hearn M. F., and Wilkinson E. (2000) The discovery of argon in Comet C/1995 O1 (Hale-Bopp). *Astrophys. J.* **544**, L169–L172.

Sykes M. V., Lebofsky L. A., Hunten D. M., and Low F. (1986) The discovery of dust trails in the orbits of periodic comets. *Science* **232**, 1115–1117.

Tachibana S. and Huss G. R. (2003) The initial abundance of ^{60}Fe in the solar nebula. *Astrophys. J.* **588**, L41–L44.

Taylor S., Lever J., and Harvey R. (1998) Accretion rate of cosmic spherules measured at the South Pole. *Nature* **392**, 899–903.

Thompson S. P., Fonti S., Verrienti C., Blanco A., Orofino V., and Tang C. C. (2002) Laboratory study of annealed amorphous $MgSiO_3$ silicate using IR spectroscopy and synchrotron X-ray diffraction. *Astron. Astrophys.* **395**, 705–717.

Verniani F. (1969) Structure and fragmentation of meteoroids. *Space Sci. Rev.* **10**, 230–261.

Waelkens C., Waters L. B. F. M., de Graauw M. S., Huygen E., Malfait K., Plets H., Vandenbussche B., Beintema D. A., Boxhoorn D. R., Habing H. J., Heras A. M., Kester D. J. M., Lahuis F., Morris P. W., Roelfsema P. R., Salama A., Siebenmorgen R., Trams N. R., van der Bliek N. R., Valentijn E. A., and Wesselius P. R. (1996) SWS observations of young main-sequence stars with dusty circumstellar disks. *Astron. Astrophys.* **315**, L245–L248.

Wallis M. K. (1980) Radiogenic melting of primordial comet interiors. *Nature* **284**, 431–433.

Waters L. B. F. M. and Molster F. J. (1980) Crystalline silicates in space. In *Highlights in Astronomy Volume 12* (ed. H. Rickman). Astronomical Society of the Pacific, San Francisco, pp. 48–51.

Weidenschilling S. J. (1997) The origin of comets in the solar nebula: a unified model. *Icarus* **127**, 290–306.

Weissman P. R. (1984) The VEGA particulate shell—comets or asteroids? *Science* **224**, 987–989.

Weissman P. R. (1986) Are cometary nuclei primordial rubble piles? *Nature* **320**, 242–244.

Weissman P. R. (1996) The Oort cloud. In *Completing the Inventory of the Solar System*, ASP Conf. Ser. 107, (eds. T. W. Rettig and J. M. Hahn). Astronomical Society of the Pacific, San Francisco, pp. 265–288.

Weissman P. R. (1999) Diversity of comets: formation zones and dynamical paths. *Space Sci. Rev.* **90**, 301–311.

Wetherill G. W. and Revelle D. O. (1982) Relationships between comets, large meteors, and meteorites. In *Comets* (ed. L. L. Wikening). University of Arizona Press, Tucson, pp. 297–319.

Wooden D. H. (2002) Comet grains: their IR emission and their relation to ISM grains. *Earth Moon Planets* **89**, 247–287.

Wooden D. H., Harker D. E., Woodward C. E., Butner H. M., Koike C., Wittebom F. C., and McMurtry C. W. (1999) Silicate mineralogy of the dust in the inner coma of comet C/1995 01, (Hale-Bopp) pre- and post-perihelion. *Astrophys J.* **517**, 1034–1058.

1.26
Interplanetary Dust Particles

J. P. Bradley

Institute of Geophysics and Planetary Physics, Lawrence Livermore National Laboratory, CA, USA

1.26.1	INTRODUCTION	689
1.26.2	PARTICLE SIZE, MORPHOLOGY, POROSITY, AND DENSITY	691
1.26.3	MINERALOGY	692
	1.26.3.1 CP IDPs	692
	1.26.3.2 Glass with Embedded Metal and Sulfides	697
	1.26.3.3 CS IDPs	699
1.26.4	OPTICAL PROPERTIES	701
1.26.5	COMPOSITIONS	702
	1.26.5.1 Major Elements	702
	1.26.5.2 Trace Elements	704
	1.26.5.3 Isotopes	704
	1.26.5.4 Noble Gases	706
1.26.6	CONCLUSIONS	707
ACKNOWLEDGMENTS		708
REFERENCES		708

How can you appreciate a castle if you don't cherish all the building blocks?

Stephen Jay Gould

1.26.1 INTRODUCTION

One of the fundamental goals of the study of meteorites is to understand how the solar system and planetary systems around other stars formed. It is known that the solar system formed from pre-existing (presolar) interstellar dust grains and gas. The grains originally formed in the circumstellar outflows of other stars. They were modified to various degrees, ranging from negligible modification to complete destruction and reformation during their $\sim 10^8$ yr lifetimes in the interstellar medium (ISM) (Seab, 1987; Mathis, 1993). Finally, they were incorporated into the solar system. Submicrometer-sized silicates and carbonaceous material are believed to be the most common grains in the ISM (Mathis, 1993; Sandford, 1996), but it is not known how much of this presolar particulate matter was incorporated into the solar system, to what extent it has survived, and how it might be distinguished from solar system grains. In order to better understand the process of solar system formation, it is important to identify and analyze these solid grains. Since all of the alteration processes that modified solids in the solar nebula presumably had strong radial gradients, the logical place to find presolar grains is in small primitive bodies like comets and asteroids that have undergone little, if any, parent-body alteration.

Trace quantities of refractory presolar grains (e.g., SiC and Al_2O_3) survive in the matrices of the most primitive carbon-rich chondritic meteorites (Anders and Zinner, 1993; Bernatowicz and Zinner, 1996; Bernatowicz and Walker, 1997; Hoppe and Zinner, 2000; see Chapter 1.02). Chondritic meteorites are believed to be from the asteroid belt, a narrow region between 2.5 and 3.5 astronomical units (AU) that marks the transition from the terrestrial planets to the giant gas-rich planets. The spectral properties of the asteroids suggest a gradation in properties with some inner and main belt C and S asteroids (the source region of most meteorites and polar micrometeorites) containing layer silicates indicative of parent-body aqueous alteration and the more distant anhydrous P and D asteroids exhibiting no evidence of (aqueous) alteration (Gradie and Tedesco, 1982). This gradation in spectral properties presumably extends several hundred AU out to the Kuiper belt, the source region of most short-period comets, where the distinction between comets and outer asteroids may simply be one of the orbital parameters (Luu, 1993; Brownlee, 1994; Jessberger et al., 2001). The mineralogy and petrography of meteorites provides direct confirmation of aqueous alteration, melting, fractionation, and thermal metamorphism among the inner asteroids (Zolensky and McSween, 1988; Farinella et al., 1993; Brearley and Jones, 1998). Because the most common grains in the ISM (silicates and carbonaceous matter) are not as refractory as those found in meteorites, it is unlikely that they have survived in significant quantities in meteorites. Despite a prolonged search, not a single presolar silicate grain has yet been identified in any meteorite.

Interplanetary dust particles (IDPs) are the smallest and most fine-grained meteoritic objects available for laboratory investigation (Figure 1). In contrast to meteorites, IDPs are derived from a broad range of dust-producing bodies extending from the inner main belt of the asteroids to the Kuiper belt (Flynn, 1996, 1990; Dermott et al., 1994; Liou et al., 1996). After release from their asteroidal or cometary parent bodies the orbits of IDPs evolve by Poynting–Robertson (PR) drag (the combined influence of light pressure and radiation drag) (Dermott et al., 2001). Irrespective of the location of their parent bodies nearly all IDPs under the influence of PR drag can eventually reach Earth-crossing orbits. IDPs are collected in the stratosphere at 20–25 km altitude using NASA ER2 aircraft (Sandford, 1987; Warren and Zolensky, 1994). Laboratory measurements of implanted rare gases, solar flare tracks (Figure 2), and isotope abundances have confirmed that the collected particles are indeed extraterrestrial and that, prior to atmospheric entry, they spent 10^4–10^5 yr as small particles orbiting the Sun (Rajan et al., 1977; Hudson et al., 1981; Bradley et al., 1984a; McKeegan et al., 1985; Messenger, 2000).

During atmospheric entry most IDPs are frictionally heated to within 100 °C of their peak heating temperature for ~1 s and, to a first-order approximation, the smallest particles are the least strongly heated. Although some IDPs may experience thermal pulses in excess of 1,000 °C for up to 10 s (depending on particle

Figure 1 (a)–(c) Secondary electron images. (a) Anhydrous CP IDP. (b) Hydrated CS IDP (RB12A44). (c) Single-mineral forsterite grain with adhering chondritic material. (d) Optical micrograph (transmitted light) of giant cluster IDP (U220GCA) in silicone oil on ER2 collection flag.

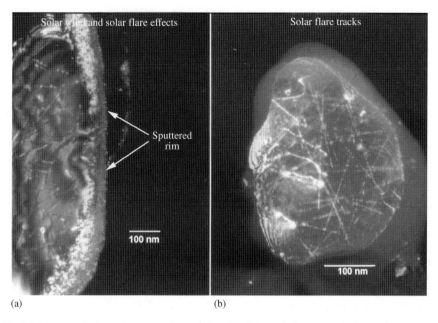

Figure 2 Darkfield transmission electron micrographs. (a) Solar-wind sputtered rim on exterior surface plus implanted solar flare tracks in chondritic IDP U220A19 (from Bradley and Brownlee, 1986). (b) Solar flare tracks in a forsterite crystal in chondritic IDP U220B11 (from Bradley et al., 1984a). The track densities in both IDPs are $\sim 10^{10}-10^{11}$ cm^2 corresponding to an orbital exposure age of $\sim 10^4$ yr.

size, mass, entry angle, and speed) (Love and Brownlee, 1991, 1996), the presence of solar flare tracks in an IDP establishes that it was not heated above ~ 650 °C (Bradley et al., 1984a). Since IDPs decelerate from cosmic velocities at altitudes >90 km, where the maximum aerodynamic ram pressure is a factor of $\sim 10^3$ less than that exerted on conventional meteorites, extremely fragile meteoritic materials that cannot survive as large objects can survive as small IDPs (Figure 1(a)) (Brownlee, 1994). Such fragile materials are suspected to be among the most primitive objects and potentially the most informative regarding early solar system and presolar processes. (Conventional meteorites penetrate deep into the atmosphere such that only relatively well-indurated rocks can survive.) Collected IDPs are briefly exposed to the terrestrial environment but since their residence time in the stratosphere is short (~ 2 weeks), they are not subjected to longer-term weathering that affects the surfaces of most meteorites (Flynn, 1994a).

This chapter examines the compositions, mineralogy, sources, and geochemical significance of IDPs. Additional reading can be found in reviews by Fraundorf (1981), Brownlee (1985), Sandford (1987), Bradley et al. (1988), Jessberger et al. (2001), Rietmeijer (1998), and the book edited by Zolensky et al. (1994). Despite their micrometer-scale dimensions and nanogram masses it is now possible, primarily as a result of advances in small particle handling techniques and analytical instrumentation, to examine IDPs at close to atomic-scale resolution. The most widely used instruments for IDP studies are presently the analytical electron microscope, synchrotron facilities, and the ion microprobe. These laboratory analytical techniques are providing fundamental insights about IDP origins, mechanisms of formation, and grain processing phenomena that were important in the early solar system and presolar environments. At the same time, laboratory data from IDPs are being compared with astronomical data from dust in comets, circumstellar disks, and the ISM. The direct comparison of grains in the laboratory with grains in astronomical environments defines the new discipline of "astromineralogy" (Jaeger et al., 1998; Bradley et al., 1999a,b; Molster et al., 2001; Keller et al., 2001; Flynn et al., 2002).

1.26.2 PARTICLE SIZE, MORPHOLOGY, POROSITY, AND DENSITY

Individual IDPs span the diameter range 1–50 μm, although most are between 5 μm and 15 μm (Figures 1(a)–(c)). Larger 50–500 μm diameter particles (10–20% of collected IDPs) that fragment into many pieces when they impact the flags are known as giant "cluster" particles (Figure 1(d)). Two principal morphological groups of IDPs are recognized, porous and smooth (Figures 1(a) and (b)). Since their bulk compositions are similar to chondritic meteorites (of types CI and CM), they are referred to as chondritic porous (CP) and chondritic smooth (CS)

IDPs. CP and CS IDPs are also mineralogically distinct classes of materials (see Section 1.26.3). The morphologies of CP particles resemble a bunch of grapes (Figure 1(a)). Porosities as high as 70% and densities ranging between 0.3 g cm^{-2} and 6.0 g cm^{-2} have been measured, although IDPs with densities above 3.5 g cm^{-2} typically contain a large FeNi sulfide grain (Fraundorf et al., 1982a; Love et al., 1994). Such low densities, high porosities, and fragile microstructures are consistent with the particulate matter in cometary meteors (Bradley and Brownlee, 1986). Most cluster particles belong to the CP class, presumably because their fragile microstructures predispose them to fragmentation during impact onto the collection substrates. Less common low-porosity CP IDPs are referred to as chondritic filled (CF) (Schramm et al., 1989). The CS IDPs are mostly solid objects with platy and/or fibrous surface textures (Figure 1(b)). It is important to note that although there are many particles that fall neatly into this category, some particles have characteristics of more than one particle type or they have unique characteristics. Some IDPs are composed mostly of refractory calcium-, aluminum-, silicon-rich minerals (Zolensky, 1987), and others are single mineral grains like olivines, pyroxenes, and iron-rich sulfides, some of which have adhering fine-grained chondritic material. Their morphologies are defined, to a large extent, by the shape of the mineral grain (Figure 1(c)).

1.26.3 MINERALOGY

It has been established using infrared (IR) and electron microscopic studies that there are three principal mineralogical classes of chondritic IDPs (Figure 3). They are referred to as "pyroxene," "olivine," and "layer silicate" after their most abundant silicate minerals. In a study of 26 IDPs, Sandford and Walker (1985) classified ~25% as "pyroxenes," ~25% as "olivines," and ~50% as "layer lattice silicate" IDPs. The pyroxene and olivine classes are usually porous CP IDPs and contain only anhydrous minerals, while the layer silicate classes are usually smooth CS IDPs and contain hydrous silicates (clays). Most IDPs fall into this mineralogical classification scheme, although IDPs with intermediate morphology and mineralogy are relatively common. Examples include anhydrous IDPs with similar amounts of pyroxene and olivine (Sandford and Walker, 1985; Bradley et al., 1989, 1992), porous IDPs containing minor amounts of hydrated layer silicates (Thomas et al., 1995), and smooth layer silicate particles containing large anhydrous silicate grains (Germani et al., 1990).

1.26.3.1 CP IDPs

This class of IDPs has been most intensively examined primarily because their high porosities and fluffy microstructures are unique among other known classes of extraterrestrial materials. IR spectra indicate that silicates are the most abundant minerals in CP IDPs and that the absence of hydrous silicates (clays) is a fundamental distinguishing property (Sandford and Walker, 1985; Bradley et al., 1992). Transmission electron microscopy studies of electron transparent thin sections confirm the absence of hydrous minerals and reveal that CP IDPs are heterogeneous aggregates of predominantly submicrometer-sized crystalline mineral grains (olivine, pyroxenes, iron-rich sulfides, FeNi metal), polycrystalline aggregates (e.g., glass with embedded metal and sulfides (GEMS), see Section 1.26.3.2), silicate glasses, and carbonaceous material (Bradley, 1994a) (Figures 4(a) and (b)). Enstatite and forsterite with <5 mol.% Fe are the most common crystalline silicates, although crystals with up to 30 mol.% Fe are also observed (Christoffersen and Buseck, 1986; Bradley et al., 1989; Bradley et al., 1999b). Enstatite-rich particles are typically more fine-grained than olivine-rich particles. Solar flare tracks have been observed in both enstatite-rich and forsterite-rich IDPs, confirming that they are indeed extraterrestrial and that they were not strongly heated during atmospheric entry. However, while tracks are observed in most enstatite-rich IDPs they are conspicuously absent in most but not all forsterite-rich IDPs. Some forsterite-rich IDPs exhibit melt textures, equilibrated silicate mineralogy, and surfaces that are decorated with magnetite that almost certainly formed from frictional heating during atmospheric entry (Germani et al., 1990). Therefore, it is likely that the olivine-rich subset of CP IDPs includes and is perhaps dominated by particles that were thermally modified during atmospheric entry.

The morphologies, crystal structures, and compositions of some enstatite and forsterite crystals in (track-rich) CP IDPs suggest that they formed by vapor phase growth. (Gas-to-solid condensation is the fundamental mechanism by which grains are formed from nebular gases throughout the galaxy.) Enstatite ($MgSiO_3$) crystals in IDPs have distinctive ultrathin platelet morphologies while others are whiskers with crystallographic screw dislocations characteristic of vapor phase growth (Figure 5) (Bradley et al., 1983). Forsterite and enstatite grains in some IDPs contain up to 5 wt.% MnO, in contrast to pyroxenes and olivines in meteorites that typically contain <0.5% MnO. Since iron and manganese are coupled during crystallization from a liquid melt and decoupled during vapor

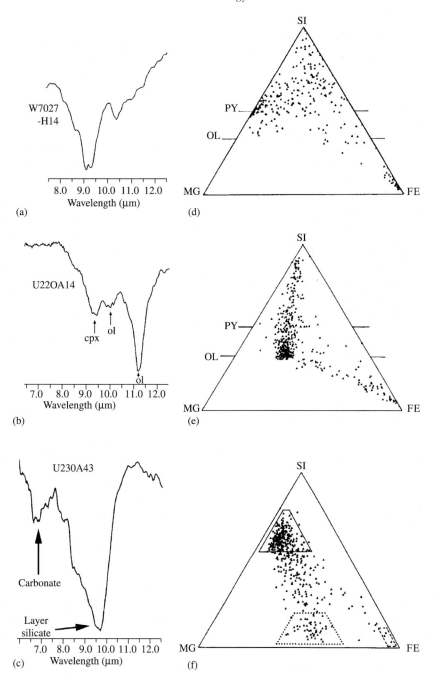

Figure 3 (a)–(c) IR spectra from thin sections of pyroxene-rich CP IDP W7027H14, olivine-rich CP IDP U220A14, and layer silicate-rich CS IDP U230A43. (d)–(f) Corresponding X-ray point count analyses obtained from the thin sections on a two-dimensional grid using a 200 keV electron probe with <50 nm spatial resolution at each point. Solid boxed area in (f) shows Mg–Fe–Si composition of the layer silicate and dotted boxed area shows carbonate Mg–Fe composition (source Bradley et al., 1992).

phase condensation, Klöck et al. (1989) propose that the (low-iron) manganese-enriched forsterite and enstatite grains in IDPs are vapor-phase condensates. CP IDPs contain more silicate glass than any other class of chondritic materials. Nonstoichiometric magnesium-rich silicate glass is most common although other glass compositions with highly variable amounts oxygen, magnesium, aluminum, silicon, calcium, titanium, and iron are observed. The glasses occur as discrete grains, rims on silicate crystals, and within GEMS (Section 1.26.3.2) (Bradley, 1994a, 1994b; Brownlee et al., 1999).

Sulfides are the second most abundant class of minerals in CP IDPs (Figure 6). The most common sulfide is pyrrhotite with up to ~20 at.%

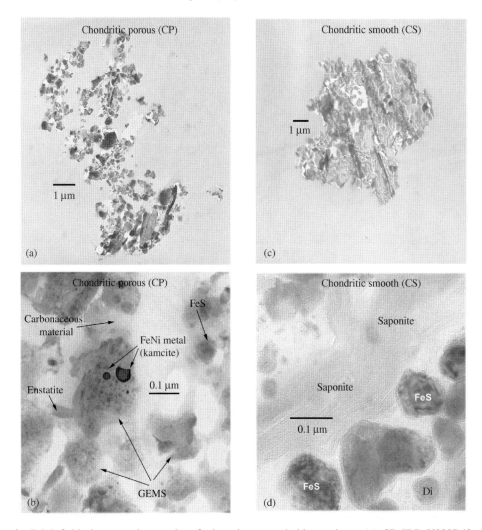

Figure 4 Brightfield electron micrographs of ultramictrotomed thin sections: (a) CP IDP U222B42, (b) CS IDP U230A43, (c) CP IDP U222B42 with carbonaceous material, GEMS, crystalline silicates (enstatite), and sulfides (FeS), (d) CS IDP U222C29 with fibrous layer lattice silicate (saponite), pyrrhotite (FeS), and the pyroxene diopside (Di).

Ni and a hexagonal unit cell (Fraundorf, 1981; Zolensky and Thomas, 1995; Dai and Bradley, 2001). Some crystals exhibit superlattice reflections. Grain sizes span a huge range from ~10 nm to ~5 μm. Rare troilite, pentlandite, sphalerite, and NiS crystals have also been observed or reported (Christoffersen and Buseck, 1986; Zolensky and Thomas, 1995). A cubic "spinel-like" sulfide with a composition similar to the hexagonal pyrrhotite has also been identified in CP IDPs (Dai and Bradley, 2001). Both polymorphs are sometimes coherently intergrown on a unit cell scale. The cubic polymorph is metastable and it transforms into hexagonal pyrrhotite when it is heated in the electron beam (Figures 6(c) and (d)). Since most IDPs are pulse heated above 200 °C during atmospheric entry (Sandford and Bradley, 1989; Love and Brownlee, 1991), it is possible that much of the hexagonal pyrrhotite in IDPs is a secondary thermal alteration product of the cubic sulfide. Although cubic spinel-like sulfides with pyrrhotite compositions have not been reported in nature, a crystallographically similar nickel-free pentlandite was synthesized in the laboratory by low-temperature (<200 °C), low-pressure vapor phase growth (Nakazawa et al., 1973). Therefore, it is possible that some or all of the sulfides in IDPs, like the pyroxene whiskers and platelets (Figure 5), formed by direct condensation from a nebular gas. Alternatively, the sulfides may have formed by gaseous sulfidization of pre-existing FeNi metal grains (Lauretta and Fegley, 1994, 1995; Lauretta et al., 1996; Zolensky and Thomas, 1995). Since the cubic sulfide has not been found in chondritic meteorites, it is likely that some sulfides in IDPs formed under conditions significantly different from those in

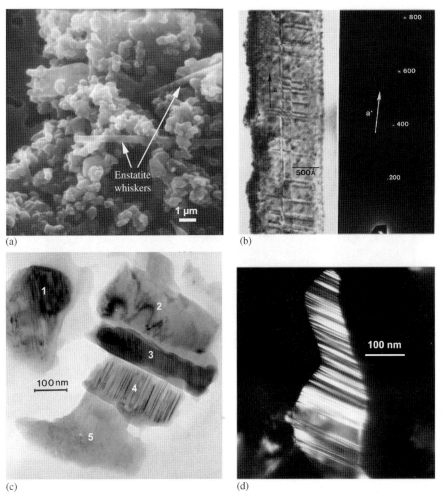

Figure 5 (a) Secondary electron image of a CP IDP with embedded enstatite whiskers (and a platelet, upper left). (b) (Left) Brightfield transmission electron micrograph of a segment of a clinoenstatite rod viewed down the *b* crystallographic axis. The two features parallel to the axis of the rod are screw dislocations and the features cutting across the rod are (100) stacking faults. (Right) Selected area electron diffraction pattern showing the (*h*00) reciprocal lattice row from the rod. The side bands (splitting) of the diffracted beams, especially visible for the 600 and 800, are related to the helical lattice distortions arising from the dislocations (see Bradley *et al.*, 1983). (c) Brightfield electron micrograph of five ultrathin enstatite ribbons and platelets. (d) Darkfield electron micrograph of an enstatite ribbon. Striations in the crystal result from extreme stacking disorder associated with unit-cell scale intergrowths of orthorhombic, orthoenstatite, and monoclinic clinoenstatite. All of enstatite crystals are likely condensates from a nebular gas.

conventional meteorites. Flynn (2000) observed selenium levels in IDPs 60% higher than those in meteoritic sulfides and concluded that IDP sulfides may have formed in a different environment than the sulfides in meteorites.

Carbonaceous material is widespread throughout CP IDPs both as discrete inclusions of noncrystalline material and as a semicontinuous matrix with embedded mineral grains. Often it has a vesiculated appearance consistent with an organic component (Figure 7(a)). The bulk abundance of carbon in CP IDPs varies from ~4% to ~45% with an average of ~13% (Keller *et al.*, 1994). In contrast to the fine-grained matrices of carbonaceous chondrites, ordered (graphitic) carbon exhibiting ~3.4 Å lattice fringes is rare in chondritic IDPs where it is rarely observed as rims on the surfaces of FeNi metal and FeNi carbide grains (e.g., Bradley *et al.*, 1984b). The carbonaceous material is present in clumps or as a semicontinuous matrix throughout which submicrometer-sized grains (e.g., GEMS and sulfides) are distributed. While elemental carbon is almost certainly a major component, IR spectra showing prominent C–H stretching resonances establish the presence of an aliphatic organic component in chondritic IDPs (Figure 7(b)) (Flynn *et al.*, 2000). Using two-step laser desorption mass spectroscopy ($\mu L^2 MS$), Clemett *et al.* (1993) have shown that both porous and

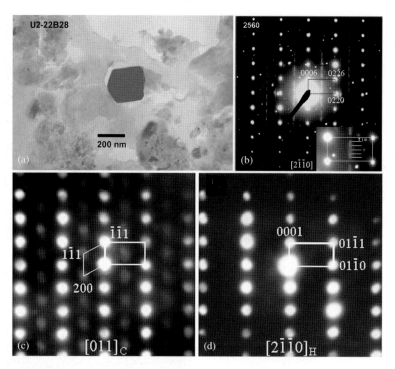

Figure 6 (a) Brightfield electron micrograph of a pyrrhotite crystal in a thin section of CP IDP U222B28. (b) Selected area electron diffraction pattern from the pyrrhotite crystal in (a) exhibiting prominent superlattice reflections consistent with the $a = 2A$, $c = 6C$ superstructure. Inset lower right is a magnified view of the central region of the pattern showing the 6C reciprocal lattice periodicity. (c)–(d) Electron micro-diffraction patterns from a pyrrhotite crystal in CP IDP W2070-8D before and after extended electron irradiation. The initial pattern (c) shows two reciprocal lattice nets, the stronger indexed as hexagonal pyrrhotite viewed along [010] (thick-line box) and the weaker as cubic spinel-like sulfide viewed along [011] (thin-line box). After the grain was illuminated in electron beam for several tens of seconds (c) only the strong reflections remain, indicating that the cubic sulfide has transformed into hexagonal pyrrhotite (source Dai and Bradley, 2001).

smooth IDPs contain polyaromatic hydrocarbons (PAHs). Nitrogen may be associated with the PAHs because the mass spectra of the PAHs are dominated by odd mass species in the intermediate molecular weight range from 200 amu to 300 amu. This is in contrast to the results previously obtained from PAHs in primitive meteorites using the same analytical technique. The odd mass peaks could be due to the substitution of functional groups containing odd numbers of nitrogen atoms such as cyano (–CN) and amino (–NH$_2$) groups. Spatially correlated ^{13}C depletion and ^{15}N enrichments were recently observed in an IDP (Floss and Stadermann, 2003).

Nanodiamonds have been identified in large cluster CP IDPs but not so far in smaller noncluster CP IDPs (Figures 7(c) and (d)) (Dai et al., 2002). The nanodiamonds are similar in size distribution and abundance to those found within carbonaceous chondrites although in one IDP (U220GCA, Figure 1(d)) their abundance appears to be ~10× higher. Defect structures in meteoritic nanodiamonds suggest that they formed by a vapor deposition process as opposed to shock metamorphism (Daulton et al., 1996).

The carrier of nanodiamonds in IDPs is a disordered (amorphous) carbonaceous material that may contain organic components, e.g., polyaromatic hydrocarbons (PAHs) (Dai et al., 2003). It is unclear why nanodiamonds are found in cluster IDPs and depleted or absent in smaller noncluster IDPs. Moreover, the relationship and distinction between large cluster and smaller noncluster IDPs is also unclear. It has been suggested that cluster IDPs are more isotopically primitive (Messenger, 2000), but their compositions, mineralogy, and petrography appear identical to those of smaller noncluster CP IDPs. Despite their large sizes, cluster IDPs appear less strongly heated than is predicted by atmospheric entry heating models, implying that they are captured from low-speed asteroidal or Kuiper belt orbits (Thomas et al., 1995; Liou et al., 1996; Flynn, 1996; Brownlee et al., 1995). Since (asteroidal) chondritic meteorites contain nanodiamonds, it is not unexpected that cluster IDPs also contain nanodiamonds, assuming they are from asteroids. Irrespective of the origins of cluster IDPs, the absence or depletion or nanodiamonds in smaller carbon-rich chondritic IDPs, some or all of which may be cometary, suggests

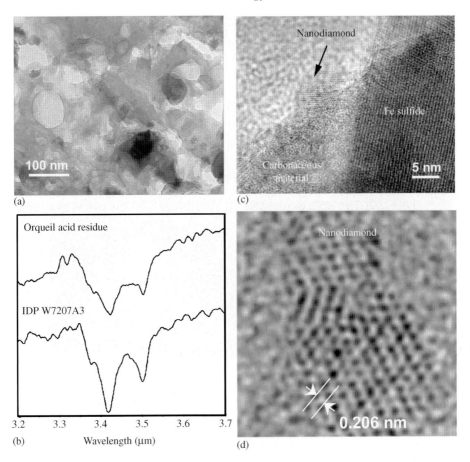

Figure 7 (a) Brightfield transmission electron micrograph of vesiculated amorphous carbonaceous material (in CP IDP W7027A1) that likely contains organic and inorganic carbon. (b) Comparative IR spectra from Orgueil (CI) meteorite acid residue and CP IDP W7207A3 showing prominent C–H stretch features at 3.4–3.5 μm indicative of aliphatic hydrocarbons (Brownlee et al., 2000). (c) Nanodiamond embedded in a carbonaceous mantle on the surface of a sulfide crystal at the edge of a GEMS (cluster IDP U220GCA). (d) A single nanodiamond within an acid-etched residue of cluster IDP W7110A-2E-D.

that nanodiamonds are heterogeneously distributed throughout the solar system and that they may actually be more abundant in asteroids than in comets. Therefore, it is possible that most meteoritic nanodiamonds formed within the inner solar system in the vicinity of asteroid accretion and not in a presolar environment as is widely believed, although xenon and tellurium isotopes indicate that at least a small fraction of them must be of presolar origin (Hoppe and Zinner, 2000).

A solar system origin could explain the anomalously high abundance of nanodiamonds relative to other types of presolar grains in meteorites (Hoppe and Zinner, 2000). The recent detection by the Infrared Space Observatory (ISO) of nanodiamonds formed *in situ* within the accretion disks of young stars confirms that nanodiamonds could indeed have formed in the inner solar system (Van Kerckhoven et al., 2002).

1.26.3.2 Glass with Embedded Metal and Sulfides

GEMS are perhaps the most exotic class of primitive meteoritic materials yet encountered (Figure 8). They are 0.1–0.5 μm spheroids that are ubiquitous throughout the matrices of CP IDPs. Because their bulk compositions are approximately chondritic they can be viewed as a discrete class of picogram-mass chondritic meteorites. Since they may also contain small amounts of carbon it is possible that they are carbonaceous chondrites (Bradley, 1994b; Brownlee et al., 2000). Despite their submicrometer dimensions, picogram masses, and unique nanometer-scale heterogeneity, their bulk compositions are similar to those of kilogram-mass chondritic meteorites and kiloton-mass asteroids. Terminology used to describe GEMS has evolved as increasingly sophisticated instruments have been used to analyze them, particularly those capable of highest spatial resolution and light

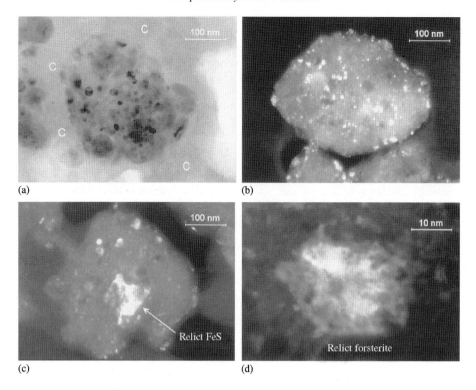

Figure 8 Transmission electron micrographs of GEMS within CP IDPs. (a) Brightfield image of GEMS embedded within amorphous carbonaceous material (labeled "C"). The inclusions are FeNi metal (kamacite) and Fe sulfides. (b) Darkfield image. Bright inclusions are metal and sulfides; matrix is magnesium silicate glass. (c, d) Darkfield images of GEMS with "relict" sulfide and forsterite (Mg_2SiO_4) inclusions (source Bradley et al., 1999a).

element (carbon and oxygen) analyses. GEMS have been described as "tar balls," "granular units," "microcrystalline aggregates," "unequilibrated aggregates," and "polyphase units" (Bradley, 1988; Rietmeijer, 1989, 1997; Klöck and Stadermann, 1994). They are mineralogically unequilibrated in that they contain nanometer-sized inclusions of FeNi alloy (kamacite) and iron-rich sulfide (pyrrhotite) embedded in oxygen-rich, low-iron magnesium silicate glass.

Figure 8 shows a brightfield and several darkfield images of GEMS. Typically, GEMS are found embedded within amorphous carbonaceous matrix (Figure 8(a)). In Figure 8(b) the bright inclusions are 2–100 nm diameter FeNi metal (kamacite) and pyrrhotite nanocrystals and the uniform gray matrix is magnesium-rich silicate glass. Some GEMS contain deeply eroded "relict" pyrrhotite, forsterite, or enstatite grains towards their cores (Figures 8(c) and (d)). Experiments with irradiated mineral standards and observations of the surfaces of lunar soils grains exposed to the solar wind indicate that the mineralogy and petrography of GEMS were shaped primarily by exposure to ionizing radiation and the exposure occurred *prior to* the accretion of the host CP IDPs (Bradley, 1994b; Bradley et al., 1996a).

The most intriguing property of GEMS is their similarity to "amorphous silicate" grains that are ubiquitous through interstellar space. The presence of silicate grains in the ISM is revealed by spectral features at $\sim 1,030$ cm^{-1} and ~ 525 cm^{-1} (9.7 μm and 19 μm) corresponding to the Si–O stretching and Si–O–Si bending modes of silicates (Millar and Duley, 1980, Aitken et al., 1989). The features are observed both in absorption and emission along various lines-of-sight and they generally lack fine structure, which suggests that the silicates are predominantly amorphous (i.e., glasses). The size range of the grains inferred from extinction is between 0.1 μm and 0.5 μm (Kim et al., 1994). These grains are believed to have originally formed in the atmospheres and outflows of oxygen-rich post-main-sequence AGB stars (Mathis, 1993; Henning, 1999). With the exception of carbon, most of the condensed rock-forming elements in ISM are associated with these silicate gains (Snow and Witt, 1996). Immediately prior to the collapse of the solar nebula, most of the condensed atoms in the solar system were carried within these interstellar amorphous silicate grains. The physical and chemical properties of interstellar silicates inferred from astronomical observations match the exotic properties of GEMS. For example, it has been proposed that polarization of starlight caused by alignment of IS amorphous silicates in the galactic magnetic can be explained by nanometer-sized inclusions of superparamagnetic FeNi metal (Jones and Spitzer, 1967). Other properties of GEMS including their bulk compositions and size distribution match those of IS amorphous silicates (Bradley et al., 1997).

The only way to prove that GEMS are indeed presolar interstellar amorphous silicates grains is to measure nonsolar isotope abundances. But if GEMS truly are interstellar amorphous silicates, it is not entirely clear that they should have nonsolar isotopic compositions, because grains undergo considerable processing during their 10^8-10^9 yr lifetimes in the ISM such that the chemical and isotopic compositions of most grains are likely homogenized (Seab, 1987; Mathis, 1993). It is significant that, without exception, all of the presolar grains so far identified in meteorites are highly refractory minerals capable of withstanding processing in the ISM and incorporation into asteroidal parent bodies. Although the glassy silicate structures, chondritic (solar) compositions, and irradiation effects in GEMS are consistent with considerable grain processing, relict grains indicate that some GEMS may retain a memory of their circumstellar origins (Figures 5(c) and (d)). The presolar interstellar origin of two GEMS was confirmed by measurements of their oxygen isotope compositions using the new-generation NanoSIMS ion microprobe (Messenger et al., 2003). Both exhibit ^{16}O abundances significantly different from those of solar system abundances, indicating that they contain preserved circumstellar silicate components, and confirming that they are interstellar "amorphous silicates." It is pointed out that the other 40 GEMS analyzed (even in the same IDP sections) have perfectly normal isotopic compositions and that it is therefore not possible to conclude whether they are solar system materials or isotopically homogenized interstellar grains. The observation that some GEMS are presolar indicates that at least some of them may indeed be the long-sought interstellar "amorphous silicates" (see Flynn, 1994b; Goodman and Whittet, 1995; Martin, 1995).

In addition to GEMS, there are other types of polycrystalline aggregates in CP IDPs. Rietmeijer (1997, 1998) described course-grained and ultra-fine-grained "polyphase units" composed of glass, Mg–Fe olivine and pyroxene, oxides, sulfides, and metal. Other submicrometer grains described as "equilibrated aggregates" also have bulk compositions that are approximately chondritic (Bradley, 1994a). But in contrast to GEMS that have highly unequilibrated compositions and mineralogy, "equilibrated aggregates" contain iron-bearing olivine and pyroxene grains with equilibrated Fe/Mg ratios and iron sulfides embedded in aluminosilicate glass. Their textures and petrography suggest that they have an igneous origin, e.g., formation by collisional shock melting. "Reduced aggregates" are yet another component of chondritic IDPs. These carbon-rich aggregates contain FeNi metal, FeNi carbide, and FeNi sulfide crystals embedded in carbonaceous matrix (Bradley, 1994a). Some of the metal grains are rimmed with a thin (<10 nm) rim of graphitic carbon (Bradley et al., 1984b). Since all of these aggregates have undergone an accretional event and processing *prior to* accretion of the IDPs in which they reside, they are older than the IDPs. Therefore, the study of aggregates in CP IDPs provides a window back to times predating the accretion of interplanetary dust. Whether the accretion occurred in the early solar system or presolar interstellar environments is unknown at this time.

1.26.3.3 CS IDPs

CS IDPs are low-porosity objects composed predominantly of hydrated layer lattice silicates (clays) (Figures 1(b) and 9). IR spectra indicate that silicates are the most abundant minerals in CS IDPs, some of them contain carbonates, and that the presence of hydrous silicates (clays) is a fundamental distinguishing property (Sandford and Walker, 1985; Germani et al., 1990; Bradley et al., 1992). Transmission electron microscopy studies of electron-transparent thin sections confirm the presence of hydrous layer lattice silicates and carbonates. Their mineralogical and petrographic similarity to the fine-grained matrices of type CI and CM carbonaceous chondrites is unmistakable but there are important differences. Whereas serpentine is the dominant layer lattice silicate in CI and CM chondrites, smectite is the dominant layer silicate in CS IDPs, suggesting that there were significant differences in the parent bodies of the IDPs and meteorites (Zolensky and McSween, 1988; Brearley and Jones, 1998; Germani et al., 1990). In contrast to the layer silicates in CI/CM chondrites, those in CS IDPs are poorly crystallized, the predominant basal spacing is ~10 Å, and they are compositionally more heterogeneous on a scale of ~50 nm than the fine-grained matrices of carbonaceous chondrites (Germani et al., 1990). Most of the layer silicates in CI and CM chondrites are serpentine with a ~7 Å basal spacing and they formed mostly from crystalline silicates like olivine and pyroxene. The layer silicates in CS IDPs formed *in situ* by aqueous alteration of silicate glasses and their compositions plus the ~10 Å spacing suggest that they are smectites.

A variety of other minerals have been reported in CS IDPs. They include: the anhydrous

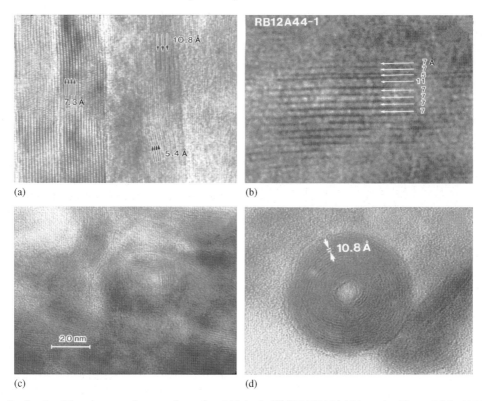

Figure 9 Lattice-fringe images of cronstedite and tochilinite in CS IDP RB12A44 (see also Figure 1(b)). (a) Mixture of iron-rich serpentine cronstedite (7.3 Å spacing) and tochilinite (5.4 Å and 10.8 Å spacings). (b) Unit cell intergrowth of cronstedite and tochilinite. (c) Pseudo-rectangular tochilinite "prototube" nucleated in a cronstedite plate. (d) Tochilinite tube (source Bradley and Brownlee, 1991).

crystalline silicates diopside (Figure 4(d)), enstatite, fassaite, and forsterite; amorphous silicates (glasses) with variable amounts of magnesium, calcium, aluminum, and iron; the sulfides pyrrhotite, troilite, and nickel sulfide; the oxides magnetite and chromite; and a phosphide, schreibersite. The sulfide mineralogy of CS IDPs differs from that of CP IDPs. Whereas low-nickel pyrrhotite ($[Fe, Ni]_{1-x}S$) is the dominant sulfide in CS IDPs (Dai and Bradley, 2001), nickel-rich sulfides with compositions ranging from low-nickel pyrrhotite compositions up to and including pentlandite ($[Fe,Ni]_9S_8$) are more abundant in CS IDPs (Zolensky and Thomas, 1995). A "low-nickel pentlandite" has been reported in a hydrated CS IDP (Tomeoka and Buseck, 1984). The composition and crystal structure of the low-nickel pentlandite are similar to those of the cubic spinel-like sulfide identified in CS IDPs (Dai and Bradley, 2001). To confirm whether they are one and the same mineral requires more study. CS IDPs are, on average, significantly enriched in carbon relative to CI chondrites and contain disordered carbonaceous material similar to that found in CP IDPs. Carbon abundances vary from 5% to >20% with an average of ~13% (Keller et al., 1994).

The hydrated silicate mineralogy of CS IDPs indicates that they are derived from parent bodies in which aqueous alteration has occurred. High-nickel sulfides (e.g., pentlandite) are also consistent with formation during asteroidal parent-body aqueous alteration because Ni-rich sulfides form at relatively high oxygen fugacities (Godlevskiy et al., 1971). Asteroids are the logical parent bodies, since it is well established that aqueous alteration is an important parent-body process within at least some regions of the asteroid belt (Brearley and Jones, 1998). The mineralogy of several CS IDPs provides a direct connection to the asteroids. Tochilinite, an ordered mixed-layer mineral containing magnesium, aluminum, iron, nickel, sulfur, and oxygen, identified in CS IDP RB12A44, has been found in only one other class of meteorites, the type CM carbonaceous chondrites (Bradley and Brownlee, 1991). Similarly, unit cell-scale intergrowths of serpentine and saponite observed in CS IDP W7013F5 are also found within the fine-grained matrices of type CI chondrites (Keller et al., 1992). The presence of these distinctive secondary mineral assemblages provides direct petrogenetic links between some IDPs and specific classes of meteorites and thus confirms that some IDPs collected in the stratosphere do indeed have an asteroidal origin (Rietmeijer, 1996). But the scarcity of these IDPs suggests that CS IDPs sample a broad range of hydrous parent bodies and that materials with CM and CI mineralogy are not abundant among the hydrous dust-producing asteroids.

1.26.4 OPTICAL PROPERTIES

The optical properties of chondritic IDPs have been measured in the IR and visible spectral regions (Figures 3, 10, and 11). Most IR measurements have been acquired in transmission over the 2–25 μm IR region using microscope spectrophotometers equipped with globar sources (Figure 3). More recently, high-brightness synchrotron light sources have proved to be ideal for spectral microanalysis of IDPs and even subcomponents of IDPs (e.g., GEMS and sulfides). Synchrotron sources can deliver an apertured beam as small as ~3 μm diameter spot >100× brighter than globar sources. The 2–25 μm region includes the important ~10 μm and ~20 μm silicate features. Almost all chondritic IDPs exhibit a dominant ~10 μm silicate feature, the position and shape of which have been used to classify particles as "pyroxene," "olivine," or "layer silicate" IDPs. The 10 μm feature of pyroxene-rich CP IDPs typically consists of principal bands at 9.1–9.4 μm (1,064–1,099 cm^{-1}) and 10.5–10.75 μm (930–953 cm^{-1}) that are consistent with monoclinic pyroxene. Olivine-rich CP IDPs exhibit an intense band at 11.2–11.3 μm (885–892 cm^{-1}) with less intense bands at 10.1 μm, 10.75 μm, and 11.9 μm (840 cm^{-1}, 930 cm^{-1}, 990 cm^{-1}). Hydrated CS IDPs usually produce a single featureless band at 9.7–9.8 μm (see Sandford and Walker, 1985; Bradley et al., 1992).

The 10 μm feature of chondritic IDPs has been compared with the 10 μm feature of astronomical silicates. No particular IDP IR class consistently matches the ~10 μm feature of solar system comets or silicate dust in the interstellar medium (Sandford and Walker, 1985). However, the ~10 μm features of CP IDPs composed mostly of GEMS and submicrometer enstatite and forsterite crystals generally resemble those of comets and late-stage Herbig Ae/Be stars in support of the hypothesis that some CP IDPs are of cometary origin (Figure 10).

The IR spectral features of individual subcomponents of chondritic IDPs have also been measured. A measurement of the ~10 μm silicate feature of GEMS produced a broad featureless band at ~9.7 μm that matches the spectra of interstellar molecular cloud dust,

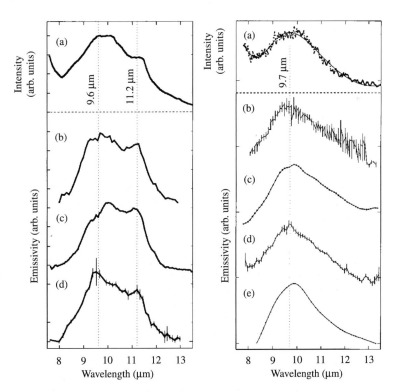

Figure 10 Comparison of the 10 μm Si–O stretch bands of a "GEMS-rich" IDP and astronomical silicates. (Left): (a) CP IDP L2008V42A, profile derived from transmittance spectrum; (b) Comet Halley (Campins and Ryan, 1989); (c) Comet Hale-Bopp (Hanner et al., 1997); (d) late stage Herbig Ae/Be star HD163296 (Sitko et al., 1999). (Right): (a) GEMS (in IDP L2011*B6); (b) Elias 16 molecular cloud (Bowey et al., 1998); (c) Trapezium molecular cloud (Hanner et al., 1995); (d) pre-main sequence T Tauri YSO DI Cephei (Hanner et al., 1998); (e) post-main sequence M-type supergiant μ-Cephei (Aitken et al. (1988) (source Bradley et al., 1999a).

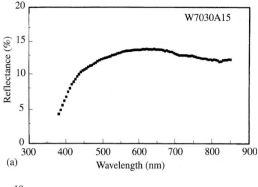

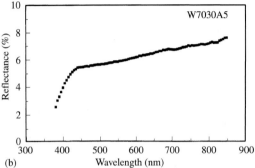

Figure 11 Reflectance spectra of (a) CS IDP W7030A15 and (b) CP IDP W7030A5 (source Bradley et al., 1996b).

providing further evidence in support of the controversial hypothesis that GEMS are interstellar amorphous silicates (Figure 10) (Bradley et al., 1999a). This hypothesis was recently confirmed by the discovery of GEMS with oxygen isotopic compositions indicating that they are indeed presolar silicates (Messenger et al., 2003). Some chondritic IDPs exhibit a broad feature at ~23.5 μm and a similar broad feature is seen in the IR spectra of young stellar objects. Detailed mineralogical analyses of the IDPs in conjunction with IR spectroscopy of mineral standards established that iron sulfides are responsible for the ~23.5 μm feature (Keller et al., 2001).

Reflectance spectra have been collected from chondritic IDPs over the visible 450–800 nm wavelength range (Figure 11). Interpretation of the indigenous reflectance characteristics of IDPs can be complicated because of spurious scattering effects from large mineral grains, secondary magnetite formed on the surfaces of some IDPs during atmospheric entry, and other small particle light scattering artifacts. In general, chondritic IDPs are spectrally dark objects with <15% reflectivity over the 400–800 nm range. Most anhydrous CP IDPs dominated by enstatite, forsterite, and GEMS exhibit spectral characteristics similar to those of smaller, more primitive solar system objects (e.g., P and D asteroids).

Carbon-rich CP IDPs are spectrally red with a redness comparable to the comet-like outer asteroid Pholus (Binzel, 1992). Hydrated CS IDPs that contain layer lattice silicates exhibit spectral characteristics similar to carbonaceous chondrites and main-belt C-type asteroids (Bradley et al., 1996b).

1.26.5 COMPOSITIONS

1.26.5.1 Major Elements

The bulk compositions of several hundred IDPs like those shown in Figure 1 have been measured using electron beam X-ray energy-dispersive spectroscopy (EDS), synchrotron X-ray fluorescence (SXRF), proton-induced X-ray emission (PIXE), and instrumental neutron activation analysis (INAA). The benchmark standard of comparison is with CI chondritic meteorites. The CI meteorites are considered to be the most chemically primitive class of meteoritic materials, because their bulk compositions, more than any other class of meteorites, closely correspond to the composition of the solar corona (see Chapter 1.03). Within a factor of 2–3, the element ratios for most chondritic CP and CS IDPs match those of the CI chondrites. (Carbon is an exception with abundances as high as 5 × higher than CI chondrites (Keller et al., 1994).) The compositions of CP IDPs are chondritic (solar) on a scale of less than 1 μm, indicating that they are mineralogically heterogeneous on a submicrometer scale (Bradley et al., 1989). CS IDPs are less heterogeneous, presumably as a result of in situ aqueous alteration (Germani et al., 1990). IDPs dominated by a single mineral grain (e.g., forsterite or pyrrhotite) typically have nonchondritic compositions reflecting the composition of the grain. Other stratospheric particles identified as IDPs include the so-called refractory IDPs rich in the elements calcium, aluminum, titanium (Zolensky, 1987).

The major-element compositions of 200 chondritic IDPs were measured by EDS (Table 1 and Figure 12). All of the particles were identified as extraterrestrial because they have approximately chondritic compositions or consist predominantly of a single mineral grain like forsterite or pyrrhotite (commonly found within chondritic IDPs): 37% of the particles are CS IDPs, 45% are CP IDPs, and 18% IDPs composed predominantly of a single mineral. Table 1 summarizes the compositions of the IDPs. Within a factor of 2 the abundances of oxygen, magnesium, aluminum, sulfur, calcium, chromium, manganese, iron, and nickel are approximately chondritic. CP IDPs are a closer match to CI carbonaceous chondrites than CS IDPs, and they are closer to CI bulk than to CI

Table 1 Mean atomic element/Si ratios for stratospheric micrometeorites versus those of various chondritic meteorite classes.

Type	C[a]	O	Na	Mg	Al	S	Ca	Cr	Fe	Ni
Chondritic IDPs[b]										
All	1.75	4.17	0.052	0.980	0.075	0.356	0.052	0.015	0.697	0.027
CS	1.32	4.49	0.051	0.824	0.082	0.341	0.021	0.014	0.742	0.032
CP	2.39	3.98	0.056	1.015	0.070	0.417	0.047	0.016	0.705	0.024
Coarse	1.31	3.81	0.043	1.203	0.075	0.231	0.125	0.013	0.585	0.019
Chondritic meteorites (bulk)										
CI[c]	0.70	7.64	0.057	1.040	0.083	0.444	0.061	0.013	0.868	0.048
CM[d]	0.35	4.38	0.029	1.023	0.088	0.201	0.070	0.012	0.804	0.048
L[d]	0.02	3.49	0.046	0.928	0.067	0.099	0.050	0.011	0.594	0.032
Chondritic meteorites (fine-grained matrices)										
CI[e]	NA	NA	0.016	0.920	0.094	0.129	0.011	0.012	0.539	0.047
CM[e]	NA	NA	0.038	0.957	0.121	0.194	0.029	0.010	0.935	0.057

[a] C and O analyses were done for only 30 IDPs. [b] IDP data from Schramm et al. (1989). [c] CI chondrite average; Palme and Jones (Chapter 1.03). [d] CM and L chondrite averages calculated from Jarosewich (1990). [e] CI and CM matrix compositions from McSween and Richardson (1977).

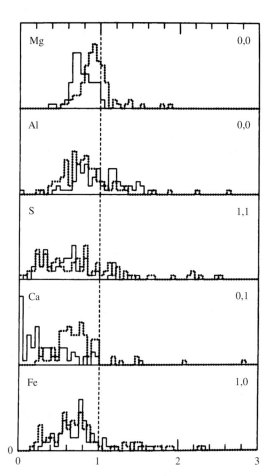

Figure 12 CI chondrite-normalized element to silicon ratios for CS and CP IDPs. The solid line represents frequency of CS IDPs and the dotted line frequency of CP IDPs. Numbers in upper right of each histogram are the number of CS and CP IDPs, respectively, with element to silicon ratios >3 CI. CS IDPs are systematically depleted in calcium and magnesium while CP IDPs are only slightly depleted in calcium, aluminum, sulfur, and iron relative to CI (vertical dotted line) (source Schramm et al., 1989).

or CM matrix. Anhydrous CP IDPs are the only known meteoritic materials that have a composition at the nanometer scale that is similar to CI bulk.

Despite the compositional similarities between CP and CS particles there are significant differences. While CP IDPs are a close match to CI abundances (they are a closer match to CI bulk than to CI or CM matrix), the CS group shows systematic magnesium and calcium depletions and a stoichiometric "excess" of oxygen. The mean Mg/Si ratio for CP IDPs is 6% below the CI mean but the Mg/Si ratio of CS IDPs is 25% below the CI mean. The Ca/Si ratio shows a large range with a mean for all IDPs that is depleted by 15% relative to CI. Like Mg/Si, there is a clear difference between the CP and CS particles, with the former containing normal calcium and the latter depleted in calcium (see also Fraundorf et al., 1982b). These element patterns are consistent with the presence of hydrous layer silicates in CS IDPs and the loss of magnesium and calcium by formation of secondary Mg–Ca carbonates on the parent bodies. Similar magnesium depletions in the fine-grained matrices of CI (and CM) meteorites also have been attributed to leaching during aqueous alteration. Thus, the magnesium compositions of the smooth group of chondritic IDPs suggest that they too have been processed by aqueous alteration, which is an important clue regarding their origin. Conversely CP IDPs are not significantly depleted in either magnesium or calcium suggesting that they have not been exposed to aqueous alteration. The lack of hydrous minerals in most CP IDPs supports this assertion.

The mean Al/Si ratio relative to CI varies among IDPs by an amount similar to that seen for Mg/Si. Again there is a systematic difference between CP and CS IDPs with the latter being enriched in aluminum. CS particles contain secondary layer lattice silicates (clays) with a high percentage of aluminum (Germani et al., 1990). The S/Si ratio shows a large range although there is no systematic difference between CS and CP particles. Although it is depleted from the CI ratio by 30%, it is still higher than any other chondrite group except CIs. Sulfur is the most volatile major element in IDPs and its measured abundance is complicated by the potential for sulfur loss by frictional heating during atmospheric entry and possible contamination of IDPs from stratospheric sulfate aerosols. Because pyrrhotite (FeS) is the major carrier of iron in chondritic IDPs, iron correlates strongly with sulfur (Figure 12). The correlated depletion of iron and sulfur is likely due to exclusion of single-mineral FeS-dominated grains from the data set. Fe/Si is depleted by 20% relative to CI chondrites and there is no significant difference in the Fe/Si ratio between CP and CS particles. Iron and aluminum are correlated in CS IDPs but not in CP IDPs. The average Fe/Al value for CS particles is 9.05, which is further from the solar system value of 10.50 (see Chapter 1.03) than the 10.13 Fe/Al mean value for CP IDPs. This same Fe–Al correlation was seen in point count areas analyses of a CS IDP that contains abundant layer lattice silicates. The correlation in CS IDPs almost certainly reflects the abundance of aluminum and iron-containing layer lattice silicates.

The C/Si ratio in chondritic IDPs is systematically higher than all classes of chondritic meteorites. The mean carbon abundance is ~10 wt.% versus 3.22 wt.% for CI (see Chapter 1.03). Nitrogen has been detected in chondritic IDPs but as yet not quantified, although Keller et al. (1995) report that the C/N ratio is approximately chondritic. Electron energy-loss spectra show that nitrogen is carried in amorphous carbonaceous material and that it is heterogeneously distributed as "hot spots." There is indirect evidence that the nitrogen is associated with polyaromatic hydrocarbons (Section 1.26.3.1).

1.26.5.2 Trace Elements

Most chondritic IDPs have "chondrite-like" trace-element compositions (Arndt et al., 1996). Abundances in individual chondritic IDPs generally scatter from ~0.3 × CI to ~3 × CI and that enrichments are more common than depletions (Flynn and Sutton, 1992a,b,c). Volatile elements tend to be enriched relative to CI meteorites (Ganapathy and Brownlee, 1979; Sutton, 1994). Enrichments of bromine measured in some IDPs probably reflect stratospheric contamination (Van der Stap et al., 1986; Flynn, 1994a; Flynn et al., 1996), and zinc depletions probably reflect loss of zinc by heating during atmospheric entry heating. Some low-zinc IDPs are also depleted in other volatile elements (copper, gallium, germanium, and selenium) (Flynn and Sutton, 1992a; Flynn et al., 1992). The most important trace-element trends in chondritic IDPs are illustrated in Figure 13. Element ratios for two different elements are plotted on the x- and y-axes and the reference lines are where CI-normalized element ratios are 1. Nickel and chromium do not show a trend and are scattered about the CI reference lines and average Cr/Fe and Ni/Fe are similar to CI (Figures 13(a) and (b)). Calcium is depleted in most of the IDPs in accordance with Schramm et al. (1989) (Figure 13(b)), and titanium appears to be enriched (Figure 2(c)). Volatile trace elements are plotted in Figures 13(d) and (e). There are both enrichments and depletions of zinc (Figure 13(d)). Figure 13(e) shows the relationship between selenium and zinc. Low-zinc IDPs tend to also have low selenium and selenium deficiencies, like zinc deficiencies, and are more common in chondritic IDPs than selenium enrichments.

1.26.5.3 Isotopes

Because a single 10 μm IDP can contain several tens of thousands of submicrometer grains, and the ion microprobe has traditionally measured isotopic composition on a scale of ~10 μm, large isotopic anomalies in *individual* grains within IDPs may not be recognized because they were averaged out on a scale of 10 μm. With this caveat in mind, the hydrogen, carbon, nitrogen, oxygen, magnesium, and silicon isotopic compositions of chondritic IDPs have been measured. Esat et al. (1979) first measured the magnesium isotopic compositions of four chondritic IDPs as well as calcium in one IDP. They found that the magnesium compositions were very close to normal isotopic composition but their normalized isotopic ratios appeared to show nonlinear effects of 3–4‰, which at that time was near the limit of detection. The isotopic composition of calcium was found to be normal (solar) within 2%. Esat et al. recommended that it might prove useful to measure individual ~1 μm components of IDPs. Significant enrichments and depletions in D/H have been found in IDPs (Zinner et al., 1983; McKeegan et al., 1985; McKeegan, 1987). The carrier of the D/H anomalies is believed to be the carbonaceous matrix within IDPs and, since the highest D/H enrichments (up to ~30,000‰) are found in large cluster IDPs that likely maintain a thermal gradient during atmospheric entry (i.e., their interiors remain cool), the carrier is presumably organic (Messenger, 2000). Enrichments of ^{15}N (up to δ^{15}N = 1,280‰) have also

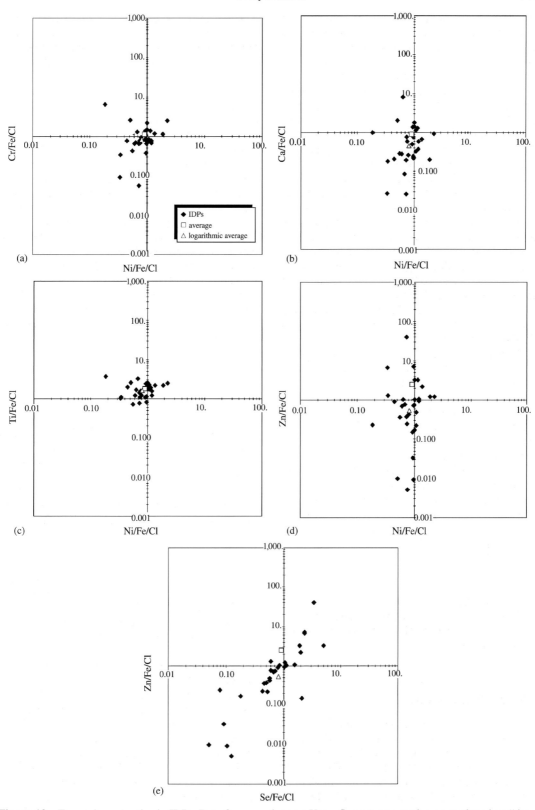

Figure 13 Trace-element ratios in IDPs. Data from synchrotron X-ray fluorescence analyses are plotted on "three element" diagrams. Element ratios are normalized to bulk CI abundances: (element/Fe)$_{sample}$/(element/Fe)$_{CI}$ also denoted "element/Fe/CI." CI composition lies at the point element/Fe/CI = 1 on each plot. Averages, assuming data are normally distributed (open squares) and assuming the data are log normally distributed (open diamonds), are also shown. Plots (a)–(c) exhibit the behavior of some more refractory elements chromium, calcium, and titanium with respect to nickel, while (d) and (e) show the behavior of zinc (relatively volatile) with respect to nickel (relatively refractory) and selenium (relatively volatile) (source Kehm *et al.*, 2002).

been found. Many but not all of the ^{15}N-enriched particles also show D/H enrichments, but the converse is not true (Stadermann et al., 1989). Both the D/H and ^{14}N/^{15}N often vary significantly within a given particle, in some cases displaying a pronounced "hot spot." The D and ^{15}N anomalies have been attributed to organic materials produced by ion–molecule reactions in cold interstellar molecular clouds (Messenger, 2000), although the same processes might equally work in the outer fringes of the solar nebula. PAHs, believed to be important constituents of interstellar molecular clouds, were found in only two (of seven measured) IDPs that had large D anomalies. D and N enrichments have been observed in both the CP and CS particles.

A new type of ion microprobe, the NanoSIMS, has made it possible to measure the isotopic compositions of chondritic IDPs on a scale of 0.5–1 μm (Floss and Stadermann, 2003; Messenger et al., 2003). Six silicate grains identified by Messenger et al. in nine chondritic IDPs have isotopic compositions confirming their presolar origins. Three of the grains exhibit elevated ^{17}O/^{16}O ratios and solar ^{18}O/^{16}O ratios consistent with origins in red giant and asymptotic giant branch stars, one is ^{16}O rich consistent with formation in a low-metallicity star, and two of uncertain stellar origin are ^{16}O depleted. One of the grains is a forsterite crystal and two others are GEMS. Floss and Stadermann (2003) measured carbon, nitrogen, and oxygen enrichments in two IDPs using the NanoSIMS (Figure 14). Two ^{17}O-enriched presolar grains (of unknown mineralogy) but with isotopic compositions similar to those of the presolar silicates were observed as well in a region with a modest but significant depletion in ^{13}C (δ^{13}C = −75‰) and spatially associated with a nitrogen "hot spot" with δ^{15}N = 1,280‰. Although hints of depletions of ^{13}C (with large errors) have previously been reported (McKeegan, 1987), the NanoSIMS measurements provide the first indication of correlated carbon and nitrogen isotope anomalies.

1.26.5.4 Noble Gases

Rajan et al. (1977) first measured the noble gases in chondritic IDPs and found solar-wind ^{4}He concentrations comparable to those observed in lunar soil grains. The measured concentrations were consistent with the ~10^{4} yr calculated exposure ages of small particles in solar orbit. Hudson et al. (1981) measured ^{20}Ne/^{22}Ne in 13 combined IDPs and observed a ratio of 13 ± 3, which is within the range of solar-wind neon. Nier and Schlutter (1990) measured ^{3}He/^{4}He and ^{20}Ne/^{22}Ne in 16 individual IDPs. The average helium content was 0.027 ± 0.01 cm^{3} STP g^{-1} (in the same range reported by Rajan et al., 1977), and the average ^{3}He/^{4}He ratio of 15 of the IDPs was (2.4 ± 0.3) × 10^{-4} (one IDP had a ^{3}He/^{4}He ratio of (1.45 ± 0.3) × 10^{-3}). But using stepwise heating, Pepin et al. (2000) measured ^{3}He/^{4}He ratios up to 40× the solar-wind ratio in several cluster particles, which they attribute to either cosmic-ray-induced spallogenic reactions during prolonged exposures of the IDPs in space or irradiation of the IDPs on their parent-body regoliths prior to their release into the interplanetary medium. The average ^{20}Ne/^{22}Ne value for 10 of the 16 IDPs measured by Nier and Schlutter (1990) was 12.0 ± 0.3 and the average

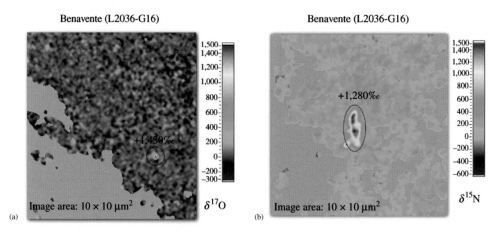

Figure 14 (a) δ^{17}O image of portion of Benavente (L2036-G16), showing a ^{17}O-rich subgrain within the IDP. The grain is ~300 nm^{2} in size. The extremely anomalous O isotopic composition indicates that this grain is of presolar origin. (b) δ^{15}N image of a portion of Benavente (L2036-G36) showing a strongly ^{15}N-enriched portion of the IDP. The "hotspot" is ~0.6 μm × 1.8 μm in size. The bulk IDP is also ^{15}N-enriched with an average δ^{15}N of about 230‰ (data courtesy of C. Floss and F. Stadermann, Washington University).

^{21}Ne/^{22}Ne value for three of the 16 IDPs was 0.035 ± 0.006.

Noble gas ratios (^{4}He/^{36}Ar versus ^{20}Ne/^{36}Ar) in 31 IDPs are plotted in Figure 15. Plotted uncertainties are 1σ. Also plotted are the noble gas elemental composition of the CI carbonaceous chondrite Orgueil ("planetary") and the solar wind ("SW"). The observed elemental ratios in Figure 15 indicate that the IDPs contain solar-wind noble gases diffusively fractionated either in solar orbit or by heating during atmospheric entry with helium and neon depleted with respect to argon. (An observed correlation between helium and zinc abundances in the IDPs suggests that it is more likely that helium is lost by frictional heating during atmospheric entry (Flynn and Sutton, 1992a; Kehm et al., 2002).)

Implanted helium is released from IDPs during pulsed stepwise heating in a furnace that mimics frictional heating during atmospheric entry. The helium release profile can be used to estimate the peak frictional heating temperatures and speeds experienced by individual IDPs during atmospheric entry (Nier, 1994). Stepwise heating has been used to distinguish high-speed cometary IDPs from low-speed asteroidal IDPs (Brownlee et al., 1995).

In summary, the abundances and isotopic compositions of noble gases in chondritic IDPs are consistent with a solar-wind origin, although fractionations due to cosmogenic spallation reactions and "secondary" diffusive processes are evident (e.g., heating during atmospheric entry). The solar-wind gases are implanted in IDPs during their lifetimes in solar orbit and they may also contain a primordial (pre-accretional) noble gas component.

1.26.6 CONCLUSIONS

Chondritic IDPs are an important resource of extraterrestrial materials because they sample a much broader range of primitive solar system bodies than do conventional meteorites and micrometeorites. Technical difficulties that limited interest in IDP research have been overcome as a result of rapid advances in microparticle handling and microanalytical instrumentation. The atmospheric entry speeds of IDPs suggest that some hydrated CS IDPs are from asteroids and some anhydrous CP IDPs are from comets (Brownlee et al., 1995). The mineralogy and petrography of CS IDPs clearly indicate that they were derived from hydrous objects where parent-body aqueous alteration occurred. Given their similarity to the fine-grained matrices of CI and CM meteorites, asteroids are the logical sources. In a few cases the mineralogy and petrography of chondritic CS IDPs link them directly to CI- or CM-like hydrous asteroids. The mineralogy and petrography of anhydrous CP IDPs suggest that they are from either anhydrous objects or very low-temperature hydrous objects where parent-body alteration was either minimal or nonexistent. Comets or "comet-like" outer asteroids are the likely sources of CP IDPs. But it is also likely that some CP IDPs are from asteroids and some CS IDPs are from comets. Since studies of IDPs are equivalent to a limited sample return,

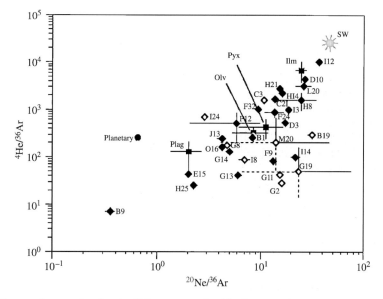

Figure 15 Noble-gas elemental ratios in IDPs compared with CI meteorites and solar-wind (star) noble gas compositions are also plotted. Closed and open diamonds represent "unheated" IDPs and Zn-depleted IDPs, respectively. Square and circle represent lunar mineral separates (Singer et al.1977) and planetary bulk CI chondrite (Jeffery and Anders, 1970), respectively (data courtesy of K. Kehm). See also Kehm et al. (2002).

clarification of the source(s) of the different classes of IDPs is a high priority of future research.

Anhydrous CP IDPs are unique among known natural geological materials in that they are mineralogically heterogeneous at the nanometer scale, and unique among known meteoritic materials in that they have *not* been subjected to significant post-accretional (parent-body) processing. They differ fundamentally from even the most primitive chondritic meteorites and micrometeorites. Some enstatite and forsterite crystals exhibit preserved evidence of condensation from nebular gases. Others have nonsolar isotopic compositions indicating that grain condensation occurred in the atmospheres of other stars. GEMS are perhaps the most enigmatic component of CP IDPs. Although they are cosmically primitive they have been extensively processed by ionizing radiation as free-floating objects. Material removed by sputtering has been thoroughly mixed and redeposited on grain surfaces producing GEMS with cosmic (chondritic) elemental compositions. The oxygen isotopic compositions of some GEMS establish that they are presolar grains. The ratio of presolar-to-solar system components in CP IDPs is unknown. It is possible that some GEMS-rich CP IDPs are relatively well-preserved aggregates of presolar circumstellar and interstellar materials, the "common stuff" of the ISM.

Emerging "nanobeam" analytical technologies will undoubtedly play a major role in future interplanetary dust research. Laboratory analytical data from IDPs will increasingly be compared with *in situ* spacecraft measurements as well as ground-based observational data from dust in space. Comet and asteroid sample return missions like STARDUST, MUSES-C, and Gulliver will undoubtedly provide new insight about stratospheric IDPs (Brownlee, 1996; Zolensky, 2000; Britt, 2003). The STARDUST mission is scheduled to return the first samples of contemporary interstellar dust and dust grains from comet Wild-2 to Earth in 2006. The samples will be a precious new resource of cosmic dust that will be examined using all of the analytical techniques that have been developed for IDPs. The STARDUST samples in particular will provide new perspectives about the scientific importance of IDPs collected in the stratosphere.

ACKNOWLEDGMENTS

This research is supported by NASA grants NAG5-10632 and NAG5-10696. I gratefully acknowledge discussions with and data from C. Floss, F. Stadermann, and K. Kehm and reviews by G. Flynn and A. Davis.

REFERENCES

Aitken D. K., Smith C. H., James S. D., Roche P. F., and Hough J. H. (1988) Infrared spectropolarimetry of AFGL 2591—evidence for an annealed grain component. *Mon. Not. Roy. Astron. Soc.* **230**, 629–638.

Aitken D. K., Smith C. H., and Roche P. F. (1989) 10 and 20 μm spectropolarimetry of the BN object. *Mon. Not. Roy. Astron. Soc.* **236**, 919–927.

Anders E. and Zinner E. (1993) Interstellar grains in primitive meteorites: diamond, silicon carbide, and graphite. *Meteoritics* **28**, 490–514.

Arndt P., Bohsung J., Maetz M., and Jessberger E. K. (1996) The elemental abundances in interplanetary dust particles. *Meteorit. Planet. Sci.* **31**, 817–834.

Bernatowicz T. J. and Walker R. M. (1997) Ancient stardust in the laboratory. *Phys. Today* **50**, 26–32.

Bernatowicz T. J. and Zinner E. (1996) *Astrophysical Implications of the Laboratory Study of Presolar Materials*, AIP Conf. Proc. 402. Am. Inst. Phys., New York, 750p.

Binzel R. P. (1992) The optical spectrum of 5145 Pholus. *Icarus* **99**, 238–240.

Bowey J. E., Adamson A. J., and Whittet D. C. B. (1998) The 10 μm profile of molecular-cloud and diffuse ISM silicate dust. *Mon. Not. Roy. Astron. Soc.* **298**, 131–138.

Bradley J. P. (1988) Analysis of chondritic interplanetary dust thin sections. *Geochim. Cosmochim. Acta* **52**, 889–900.

Bradley J. P. (1994a) Nanometer-scale mineralogy and petrography of fine-grained aggregates in anhydrous interplanetary dust particles. *Geochim. Cosmochim. Acta* **58**, 2123–2134.

Bradley J. P. (1994b) Chemically anomalous, preaccretionally irradiated grains in interplanetary dust from comets. *Science* **265**, 925–929.

Bradley J. P. and Brownlee D. E. (1986) Cometary particles: thin sectioning and electron beam analysis. *Science* **231**, 1542–1544.

Bradley J. P. and Brownlee D. E. (1991) An interplanetary dust particle linked directly to type CM meteorites and an asteroidal origin. *Science* **251**, 549–552.

Bradley J. P., Brownlee D. E., and Veblen D. R. (1983) Pyroxene whiskers and platelets in interplanetary dust particles: evidence of vapor phase growth. *Nature* **301**, 473–477.

Bradley J. P., Brownlee D. E., and Fraundorf P. (1984a) Discovery of nuclear tracks in interplanetary dust. *Science* **226**, 1432–1434.

Bradley J. P., Brownlee D. E., and Fraundorf P. (1984b) Carbon compounds in interplanetary dust particles: evidence for formation by heterogeneous catalysis. *Science* **223**, 56–58.

Bradley J. P., Sandford S. A., and Walker R. M. (1988) Interplanetary dust particles. In *Meteorites and the Early Solar System* (eds. J. F. Kerridge and M. S. Mathews). University of Arizona Press, Tucson, pp. 861–895.

Bradley J. P., Brownlee D. E., and Germani M. S. (1989) Automated thin-film analyses of anhydrous interplanetary dust particles in the analytical electron microscope. *Earth Planet. Sci. Lett.* **93**, 1–13.

Bradley J. P., Humecki H. J., and Germani M. S. (1992) Combined infrared and analytical electron microscope studies of interplanetary dust particles. *Astrophys. J.* **394**, 643–651.

Bradley J. P., Dukes C., Baragiola R., McFadden L., Johnson R. E., and Brownlee D. E. (1996a) Radiation processing and the origins of interplanetary dust. In *Lunar Planet. Sci.* XXVII. The Lunar and Planetary Institute, Houston, pp. 149–150.

Bradley J. P., Keller L. P., Brownlee D. E., and Thomas K. L. (1996b) Reflectance spectroscopy of interplanetary dust particles. *Meteorit. Planet. Sci.* **31**, 394–402.

Bradley J. P., Brownlee D. E., and Snow T. P. (1997) GEMS and other preaccretionally irradiated grains in interplanetary dust particles. In *From Stardust to Planetesimals* (eds. Y. J. Pendleton and A. G. G. M. Tielens). ASP Conf. Series vol. 122, pp. 217–225.

Bradley J. P., Keller L. P., Snow T. P., Hanner M. S., Flynn G. J., Gezo J. C., Clemett S. J., Brownlee D. E., and Bowey J. E. (1999a) An infrared spectral match between GEMS and interstellar grains. *Science* **285**, 1716–1718.

Bradley J. P., Snow T. P., Brownlee D. E., and Hanner M. S. (1999b) Mg-rich olivine and pyroxene grains in primitive meteoritic materials: comparison with crystalline silicate data from ISO. In *Solid Interstellar Matter: The ISO Revolution*, Les Houches No. 11 (eds. L. d'Hendecourt, C. Joblin, and A. Jones). EDP Sciences, Les Ullis, pp. 297–315.

Brearley A. J. and Jones R. H. (1998) Chondritic meteorites. In *Planetary Materials,* Rev. Mineral. **36**, (ed. J. J. Papike). Mineralogical Society of America, Washington, DC, 3-1–3-398.

Britt D. T. (2003) The Gulliver mission: sample return from the Martian moon Deimos. In *Lunar Planet. Sci.* XXXIV, #1841. The Lunar and Planetary Institute, Houston (CD-ROM).

Brownlee D. E. (1985) Cosmic dust—collection and research. *Ann. Rev. Earth Planet. Sci.* **13**, 147–173.

Brownlee D. E. (1994) The origin and role of dust in the early solar system. In *Analysis of Interplanetary Dust*, AIP Conf. Proc. 310 (eds. M. E. Zolensky, T. L. Wilson, F. J. M. Rietmeijer, and G. J. Flynn). Am. Inst. Phys., New York, pp. 5–8.

Brownlee D. E. (1996) STARDUST: comet and interstellar dust sample return mission. In *Physics, Chemistry, and Dynamics of Interplanetary Dust* (eds. B. Å. S. Gustafson and M. S. Hanner). Am. Inst. Phys., New York, pp. 223–226.

Brownlee D. E., Joswiak D. J., Schlutter D. J., Pepin R. O., Bradley J. P., and Love S. J. (1995) Identification of individual cometary IDPs by thermally stepped He release. In *Lunar Planet Sci.* XXVI. The Lunar and Planetary Institute, Houston, 183–184.

Brownlee D. E., Joswiak D. J., and Bradley J. P. (1999) High spatial resolution analyses of GEMS and other ultrafine grained IDP components. In *Lunar Planet Sci.* XXX, #2031. The Lunar and Planetary Institute, Houston (CD-ROM).

Brownlee D. E., Joswiak D. J., Bradley J. P., Gezo J. C., and Hill H. G. M. (2000) Spatially resolved acid dissolution of IDPs: the state of carbon and the abundance of diamonds in the dust. In *Lunar Planet. Sci.* XXXI, #1921. The Lunar and Planetary Institute, Houston (CD-ROM).

Campins H. and Ryan E. (1989) The identification of crystalline olivine in cometary silicates. *Astrophys. J.* **341**, 1059–1066.

Christoffersen R. and Buseck P. R. (1986) Mineralogy of interplanetary dust particles from the "olivine" infrared class. *Earth Planet. Sci. Lett.* **78**, 53–66.

Clemett S. J., Maechling C. R., Zare R. N., Swan P. D., and Walker R. M. (1993) Identification of complex aromatic molecules in individual: interplanetary dust particles. *Science* **262**, 721–725.

Dai Z. R. and Bradley J. P. (2001) Iron nickel sulfides in anhydrous interplanetary dust particles. *Geochim. Cosmochim. Acta* **65**, 3601–3612.

Dai Z. R., Bradley J. P., Joswiak D. J., Brownlee D. E., Hill H. G. M., and Genge M. J. (2002) Possible *in situ* formation of nanodiamonds in the early solar system. *Nature* **418**, 157–159.

Dai Z. R., Bradley J. P., Brownlee D. E., and Joswiak D. J. (2003) The petrography of meteoritic nano-diamonds. In *Lunar Planet. Sci.* XXXIV, #1121. The Lunar and Planetary Institute, Houston (CD-ROM).

Daulton T. L., Eisenhour D. D., Bernatowicz T. J., Lewis R. S., and Buseck P. R. (1996) Genesis of presolar diamonds: comparative high-resolution transmission electron microscopy study of meteoritic nano-diamonds. *Geochim. Cosmochim. Acta* **60**, 4853–4872.

Dermott S. F., Grogan K., Durda D. D., Jayaraman S., Kehoe T. J. J., Kortenkamp S. J., and Wyatt M. C. (2001) Orbital evolution of interplanetary dust. In *Interplanetary Dust* (eds. E. Grün, B. Å. S. Gustafson, S. F. Dermott, and H. Fechtig). Springer, Berlin, pp. 569–640.

Dermott S. F., Jayaraman S., Xu X.-L., Gustafson B.Å. S., and Liou J.-C. (1994) A circumsolar ring of asteroidal dust in resonant lock with the Earth. *Nature* **369**, 719–723.

Esat T. M., Brownlee D. E., Papanastassiou D. A., and Wasserburg G. J. (1979) Magnesium isotopic composition of interplanetary dust particles. *Science* **206**, 190–197.

Farinella P., Gonzi R., and Froeschl C. (1993) The injection of asteroid fragments into resonances. *Icarus* **101**, 174–187.

Floss C. and Stadermann F. J. (2003) Complimentary carbon, nitrogen, and oxygen isotopic imaging of interplanetary dust particles: presolar grains and an indication of a carbon isotope anomaly. In *Lunar Planet. Sci.* XXXIV, #1238. The Lunar and Planetary Institute, Houston (CD-ROM).

Flynn G. J. (1990) The near-Earth enhancement of asteroidal over cometary dust. *Proc. 20th Lunar Planet. Sci. Conf.* 363–371.

Flynn G. J. (1994a) Changes in the composition and mineralogy of interplanetary dust particles be terrestrial encounters. In *Analysis of Interplanetary Dust*, AIP Conf. Proc. 310 (eds. M. E. Zolensky, T. L. Wilson, F. J. M. Rietmeijer, and G. J. Flynn). Am. Inst. Phys., New York, pp. 127–143.

Flynn G. J. (1994b) Interplanetary dust—the common stuff of stardust. *Nature* **371**, 287–288.

Flynn G. J. (1996) Sources of 10 micron interplanetary dust: the contribution from the Kuiper belt. In *Physics, Chemistry, and Dynamics of Interplanetary Dust* (eds. B. Å. S. Gustavson and M. S. Hanner). ASP Conference Series, vol. 104, pp. 171–175.

Flynn G. J. (2000) A comparison of the selenium contents of sulfides from interplanetary dust particles and meteorites. *Meteorit. Planet. Sci.* **35**, A54.

Flynn G. J. and Sutton S. R. (1992a) Trace elements in chondritic stratospheric particles: zinc depletion as a possible as a possible indicator of atmospheric entry heating. *Proc. 22nd Lunar Planet. Sci. Conf.* 171–184.

Flynn G. J. and Sutton S. R. (1992b) Element abundances in stratospheric cosmic dust: indications for a new chemical type of chondritic material. In *Lunar Planet. Sci.* XXIII. The Lunar and Planetary Institute, Houston, 373–374.

Flynn G. J. and Sutton S. R. (1992c) Trace elements in chondritic cosmic dust: volatile correlation with Ca abundance. *Meteoritics* **27**, 220.

Flynn G. J., Sutton S. R., Thomas K. L., Keller L. P., and Klöck W. (1992) Zinc depletions and atmospheric entry heating in stratospheric cosmic dust particles. In *Lunar Planet. Sci.* XXIII. The Lunar and Planetary Institute, Houston, pp. 375–376.

Flynn G. J., Bajt S., and Sutton S. R. (1996) Evidence for weakly bound bromine in large interplanetary dust particles collected from the stratosphere. In *Lunar Planet. Sci.* XXVII. The Lunar and Planetary Institute, Houston, 367–368.

Flynn G. J., Keller L. P., Jacobsen C., Wirick S., and Miller M. A. (2000) Organic carbon in interplanetary dust particles. In *A New Era in Bioastronomy* (eds. G. Lemarchand and K. Meech). ASP Conf. Series, vol. 213, pp. 191–194.

Flynn G. J., Henning Th., Keller L. P., and Mutschke H. (2002) Infrared Spectroscopy of Cosmic Dust. In *NATO Science Series*. Optics of Cosmic Dust (eds. G. Videen and M. Kocifaj). Kluwer, Dordrecht, vol. 79, pp. 37–56.

Fraundorf P. (1981) Interplanetary dust in the transmission electron microscope: diverse materials from the early solar system. *Geochim. Cosmochim. Acta* **45**, 915–943.

Fraundorf P., Hintz C., Lowry O., McKeegan K. D., and Sandford S. A. (1982a) Determination of the mass, surface density, and volume density of individual interplanetary dust particles. In *Lunar Planet. Sci.* XIII. The Lunar and Planetary Institute, Houston, pp. 225–226.

Fraundorf P., Brownlee D. E., and Walker R. M. (1982b) Laboratory studies of interplanetary dust. In *Comets* (ed. L. L. Wilkening). University of Arizona Press, Tucson, Arizona, pp. 383–409.

Ganapathy R. and Brownlee D. E. (1979) Interplanetary dust: trace element analyses of individual particles by neutron activation. *Science* **206**, 1075–1077.

Germani M. S., Bradley J. P., and Brownlee D. E. (1990) Automated thin-film analyses of hydrated interplanetary dust particles in the analytical electron microscope. *Earth Planet. Sci. Lett.* **101**, 162–179.

Godlevskiy M. N., Likhachev A. P., Chuvikina N. G., and Andronov A. D. (1971) Hydrothermal synthesis of pentlandite. *Dokl. Akad. Nauk. SSSR* **196**, 146–149.

Goodman A. A. and Whittet D. C. B. (1995) A point in favor of the superparamagnetic grain hypothesis. *Astrophys. J.* **455**, L181–L184.

Gradie J. and Tedesco E. (1982) Compositional structure of the asteroid belt. *Science* **216**, 1405–1407.

Hanner M. S., Brooke T. Y., and Tokunaga A. T. (1995) 10 micron spectroscopy of younger stars in the rho Ophiuchi cloud. *Astrophys. J.* **502**, 250–258.

Hanner M. S., Brooke T. Y., and Tokunaga A. T. (1998) Micron spectroscopy of young stars. *Astrophys. J.* **502**, 871–882.

Hanner M. S., Gehrz R. D., Harker D. E., Hayward T. L., Lynch D. K., Mason C. C., Russell R. W., Williams D. M., Wooden D. H., and Woodward C. E. (1997) Thermal emission from the dust coma of Comet Hale-Bopp and the composition of the silicate grains. *Earth Moon Planets* **79**, 247–264.

Henning Th. (1999) Grain formation and evolution in the interstellar medium. In *Solid Interstellar Matter: The ISO Revolution*, Les Houches No 11 (eds. L. d'Hendecourt, C. Joblin, and A. Jones). EDP Sciences, Les Ullis, pp. 247–262.

Hoppe P. and Zinner E. (2000) Presolar dust grains from meteorites and their stellar sources. *J. Geophys. Res.* **105**, 10371–10398.

Hudson B., Flynn G. J., Thomas K. L., Keller L. P., Fraundorf C. M., and Shirck J. (1981) Noble gases in stratospheric dust particles: confirmation of extraterrestrial origin. *Science* **211**, 383–386.

Jaeger C., Molster F. J., Dorschner J., Henning Th., Mutschke H., and Waters L. B. F. M. (1998) Steps toward interstellar silicate mineralogy: IV. The crystalline revolution. *Astron. Astrophys.* **339**, 904–916.

Jarosewich E. (1990) Chemical analyses of meteorites: a compilation of stony and iron meteorite analyses. *Meteoritics* **25**, 323–337.

Jeffery P. M. and Anders E. (1970) Primordial noble gases in separated meteoritic minerals: I. *Geochim. Cosmochim. Acta* **34**, 1175–1198.

Jessberger E. K., Stephan T., Rost D., Arndt P., Maetz M., Stadermann F. J., Brownlee D. E., Bradley J. P., and Kurat G. (2001) Properties of interplanetary dust: information from collected samples. In *Interplanetary Dust* (eds. E. Grün, B. Å. S. Gustafson, S. F. Dermott, and H. Fechtig). Springer, Berlin, pp. 253–294.

Jones R. V. and Spitzer L., Jr. (1967) Magnetic alignment of interstellar grains. *Astrophys. J.* **147**, 943–964.

Kehm K., Flynn G. J., Sutton S. R., and Hohenberg C. M. (2002) Combined noble gas and trace element measurements on individual stratospheric interplanetary dust particles. *Meteorit. Planet. Sci.* **37**, 1323–1335.

Keller L. P., Hony S., Bradley J. P., Molster F. J., Waters L. B. F. M., Bouwman J., de Koter A., Brownlee D. E., Flynn G. J., Henning T., and Mutsche H. (2001) Sulfides in space: a possible match to the 23 μm feature detected by the Infrared Space Observatory. *Nature* **417**, 148–150.

Keller L. P., Thomas K. L., and McKay D. S. (1992) An interplanetary dust particle with links to CI chondrites. *Geochim. Cosmochim. Acta* **56**, 1409–1412.

Keller L. P., Thomas K. L., and McKay D. S. (1994) Carbon in primitive interplanetary dust particles. In *Analysis of Interplanetary Dust*, AIP Conf. Proc. 310 (eds. M. E. Zolensky, T. L. Wilson, F. J. M. Rietmeijer, and G. J. Flynn). Am. Inst. Phys., New York, pp. 51–87.

Keller L. P., Thomas K. L., Bradley J. P., and McKay D. S. (1995) Nitrogen in interplanetary dust particles. *Meteoritics* **30**, 526–527.

Kim S.-H., Martin P. G., and Hendry P. D. (1994) The size distribution of interstellar dust particles as determined from extinction. *Astrophys. J.* **422**, 164–175.

Klöck W. and Stadermann F. J. (1994) Mineralogical and chemical relationships of interplanetary dust particles, micrometeorites and meteorites. In *Analysis of Interplanetary Dust*, AIP Conf. Proc. 310 (eds. M. E. Zolensky, T. L. Wilson, and F. J. M. Rietmeijer). Am. Inst. Phys., New York, pp. 159–164.

Klöck W., Thomas K. L., McKay D. S., and Palme H. (1989) Unusual olivine and pyroxene composition in interplanetary dust and unequilibrated ordinary chondrites. *Nature* **339**, 126–128.

Lauretta D. S. and Fegley B., Jr. (1994) Troilite formation kinetics and growth mechanism in the solar nebula. *Meteoritics* **29**, 490.

Lauretta D. S. and Fegley B., Jr. (1995) Nickel fractionation during troilite formation in the solar nebula. In *Lunar Planet. Sci.* XXVI. The Lunar and Planetary Institute, Houston, pp. 831–832.

Lauretta D. S., Kremser D. T., and Fegley B., Jr. (1996) The rate of iron sulfide formation in the solar nebula. *Icarus* **122**, 288–315.

Liou J. C., Zook H. A., and Dermott S. F. (1996) Kuiper belt dust grains as a source of interplanetary dust. *Icarus* **124**, 429–440.

Love S. G. and Brownlee D. E. (1991) Heating and thermal transformation of micrometeoroids entering Earth's atmosphere. *Icarus* **89**, 26–43.

Love S. G. and Brownlee D. E. (1996) Peak atmospheric entry heating temperatures of micrometeorites. *Meteorit. Planet. Sci.* **31**, 394–402.

Love S. G., Joswiak D. J., and Brownlee D. E. (1994) Densities of stratospheric micrometeorites. *Icarus* **111**, 227–236.

Luu J. X. (1993) Spectral diversity among the nuclei of comets. *Icarus* **104**, 138–148.

Martin P. G. (1995) On the value of GEMS (glass with embedded metal and sulfides. *Astrophys. J.* **445**, L63–L66.

Mathis J. A. (1993) Observations and theories of interstellar dust. *Rep. Prog. Phys.* **56**, 605–652.

McKeegan K. D. (1987) Ion microprobe measurements of H, C, O, Mg, and Si isotopic abundances in individual interplanetary dust particles. PhD Thesis, Washington University.

McKeegan K. D., Zinner E., and Walker R. M. (1985) Ion microprobe isotopic measurements of individual interplanetary dust particles. *Geochim. Cosmochim. Acta* **49**, 1971–1987.

McSween H. Y., Jr. and Richardson S. M. (1977) The composition of carbonaceous chondrite matrix. *Geochim. Cosmochim. Acta* **41**, 1145–1161.

Messenger S. (2000) Identification of molecular-cloud material in interplanetary dust particles. *Nature* **404**, 968–971.

Messenger S., Keller L. P., Stadermann F. J., Walker R. M., and Zinner E. (2003) Samples of stars beyond the solar system: silicate grains in interplanetary dust. *Science* **300**, 105–108.

Millar T. J. and Duley W. W. (1980) Interstellar grains: constraints on composition from infrared observations. *Mon. Not. Roy. Astron. Soc.* **191**, 641–649.

Molster F. J., Bradley J. P., Sitko M. L., and Nuth J. A. (2001) Astromineralogy: the comparison of infrared spectra from astrophysical environments with those from interplanetary dust particles (IDPs). In *Lunar Planet Sci.* XXXII, #1391. The Lunar and Planetary Institute, Houston (CD-ROM).

Nakazawa H., Osaka T., and Sakaguchi K. (1973) A new cubic iron sulphide prepared by vacuum deposition. *Nature* **242**, 13–14.

Nier A. O. (1994) Helium and neon in interplanetary dust particles. In *Analysis of Interplanetary Dust*, AIP Conf. Proc. 310 (eds. M. E. Zolensky, T. L. Wilson, F. J. M. Rietmeijer, and G. J. Flynn). Am. Inst. Phys., New York, pp. 115–126.

Nier A. O. and Schlutter D. J. (1990) Helium and neon isotopes in stratospheric particles. *Meteoritics* **25**, 263–267.

Pepin R. O., Palma R. L., and Schlutter D. J. (2000) Noble gases in interplanetary dust particles: I. The excess helium-3 problem and estimates of the relative fluxes of solar wind and solar energetic particles in interplanetary space. *Meteoritics* **35**, 495–504.

Rajan R. S., Brownlee D. E., Tomandl D., Hodge P. W., Farrar H., and Britten R. A. (1977) Detection of ^{4}He in stratospheric particles gives evidence of extraterrestrial origin. *Nature* **267**, 133–134.

Rietmeijer F. J. M. (1989) Ultrafine-grained mineralogy and matrix chemistry of olivine-rich chondritic interplanetary dust particles. *Proc. 19th Lunar Planet. Sci. Conf.* 513–521.

Rietmeijer F. J. M. (1996) CM-like interplanetary dust particles in the lower stratosphere during 1989 October and 1991 June/July. *Meteorit. Planet. Sci.* **31**, 278–288.

Rietmeijer F. J. M. (1997) Interplanetary dust petrology, principal components analysis, chondrites, chemical composition, grain size, porosity, magnesium compounds, iron compounds, minerals. In *Lunar Planet Sci.* XXVIII, #1301. The Lunar and Planetary Institute, Houston (CD-ROM).

Rietmeijer F. J. M. (1998) Interplanetary dust particles. In *Planetary Materials*, Rev. Mineral. 36, (ed. J.J. Papike). Mineralogical Society of America, Washington, DC, 2-29–2-29 pp.

Sandford S. A. (1987) The collection and analysis of extraterrestrial dust particles. *Fund. Cosmic. Phys.* **12**, 1–73.

Sandford S. A. (1996) The inventory of interstellar materials available for the formation of the solar system. *Meteorit. Planet. Sci.* **31**, 449–476.

Sandford S. A. and Bradley J. P. (1989) Interplanetary dust particles collected in the stratosphere: observations of atmospheric heating and constraints on their inter-relationships and sources. *Icarus* **82**, 146–166.

Sandford S. A. and Walker R. M. (1985) Laboratory infrared transmission spectra of individual interplanetary dust particles from 2.5 to 25 microns. *Astrophys. J.* **291**, 838–851.

Schramm L. S., Brownlee D. E., and Wheelock M. M. (1989) Major element composition of stratospheric micrometeorites. *Meteoritics* **24**, 99–112.

Seab C. G. (1987) In *Interstellar Processes* (eds. D. J. Hollenbach and H. A. Thronson). Reidel, Dordrecht, 491 pp.

Signer P., Baur H., Derksen U., Etique P., Funk H., Horn P., and Wieler R. (1977) Helium, neon, and argon records of lunar soil evolution. *Proc. 8th Lunar Planet. Sci. Conf.* 3657–3683.

Sitko M. L., Grady C. A., Lynch D. K., Russell R. W., and Hanner M. S. (1999) Cometary dust in the debris disks of HD 31648 and HD 163296: two "baby" beta Pictoris stars. *Astrophys. J.* **510**, 408–412.

Snow T. P. and Witt A. N. (1996) Interstellar depletions updated: where all the atoms went. *Astrophys. J.* **468**, L65–L68.

Stadermann F. J., Walker R. M., and Zinner E. (1989) Ion microprobe measurements of nitrogen and carbon isotopic variations in individual IDPs. *Meteoritics* **24**, 327.

Sutton S. R. (1994) Chemical compositions of primitive solar system particles. In *Analysis of Interplanetary Dust*, AIP Conf. Proc. 310 (eds. M. E. Zolensky, T. L. Wilson, F. J. M. Rietmeijer, and G. J. Flynn). Am. Inst. Phys., New York, pp. 145–157.

Thomas K. L., Blanford G. E., Clemett S. J., Flynn G. J., Keller L. P., Klöck W., Maechling C. R., McKay D. S., Messenger S., Nier A. O., Schlutter D. J., Sutton S. R., Warren J. L., and Zare R. N. (1995) An asteroidal breccia: the anatomy of a cluster IDP. *Geochim. Cosmochim. Acta* **59**, 2797–2815.

Tomeoka K. and Buseck P. R. (1984) Transmission electron microscopy of the "LOW-CA" hydrated interplanetary dust particle. *Earth. Planet. Sci. Lett.* **69**, 243–254.

Van der Stap C. C. G. M., Vis R. D., and Verheul H. (1986) Interplanetary dust: arguments in favour of a late stage nebular origin of the chondritic aggregates. In *Lunar Planet. Sci.* XVII. The Lunar and Planetary Institute, Houston, 1013–1014.

Van Kerckhoven C., Tielens A. G. G. M., and Waelkens C. (2002) Nanodiamonds around HD 97048 and Elias: 1. *Astron. Astrophys.* **384**, 568–584.

Warren J. L. and Zolensky M. E. (1994) Collection and curation of interplanetary dust particles recovered from the stratosphere by NASA. In *Analysis of Interplanetary Dust*, AIP Conf. Proc. 310 (eds. M. E. Zolensky, T. L. Wilson, F. J. M. Rietmeijer, and G. J. Flynn). Am. Inst. Phys., New York, pp. 245–253.

Zinner E., McKeegan K. D., and Walker R. M. (1983) Laboratory measurements of D/H ratios in interplanetary dust. *Nature* **305**, 119–121.

Zolensky M. E. (1987) Refractory interplanetary dust particles. *Science* **237**, 1466–1468.

Zolensky M. E. (2000) The MUSES-C asteroid sample return mission. *Meteorit. Planet. Sci.* **35**, A178.

Zolensky M. E. and McSween H. Y., Jr. (1988) Aqueous alteration. In *Meteorites and the Early Solar System* (eds. J. F. Kerridge and M. S. Mathews). University of Arizona Press, pp. 114–143.

Zolensky M. E. and Thomas K. L. (1995) Iron and nickel sulfides in chondritic interplanetary dust. *Geochim. Cosmochim. Acta* **59**, 4707–4712.

Zolensky M. E., Wilson T. L., Rietmeijer F. J. M., and Flynn G. J. (eds.) (1994) *Analysis of Interplanetary Dust*. Am. Inst. Phys., New York.

Volume Subject Index

The index is in letter-by-letter order, whereby hyphens and spaces within index headings are ignored in the alphabetization (e.g. Arabian–Nubian Shield precedes Arabian Sea). Terms in parentheses are excluded from the initial alphabetization. In line with normal materials science practice, compound names are not inverted but are filed under substituent prefixes.

The index is arranged in set-out style, with a maximum of three levels of heading. Location references refer to the page number. Major discussion of a subject is indicated by bold page numbers. Page numbers suffixed by *f* or *t* refer to figures or tables.

acapulcoites
  age determination 301
  chemical composition *293t*
  cosmic-ray exposure ages *360f*, 369
  metal–sulfide system melts 298, *298f*
  mineralogy 104, *104f*, 294
  oxygen isotopic composition 140, *140f*
  petrologic characteristics *105t*, *292t*
  silicate-oxide system melts 298–300, *300f*
  trapped noble gases 300–301, *301f*
  veining 333, *333f*
accretion
  Mercury 477dash;478, 481–482, 483
  planetesimals
    giant impacts 467
    late-stage accretion 467
    Mercury 483
    oligarchic growth 467
    origins 465
    random velocities 466
    runaway growth 467
    short-lived radionuclides 453
    terrestrial planet formation 466
  planet formation **461–475**, 510, 516
  solar nebula 69, 72, 74, 79, 510
  star formation 69, 72, 74, 79
  viscous accretion disk models 74, 462
  volatile elements 469
acenaphthene *271f*
acetylene (C_2H_2) 630, *675t*
achondrites **291–324**
    see also chondrites; meteorites
  acapulcoites
    age determination 301
    chemical composition *293t*
    cosmic-ray exposure ages *360f*, 369
    elemental abundances *294f*
    metal–sulfide system melts 298, *298f*
    mineralogy 104, *104f*, 294
    oxygen isotopic composition 140, *140f*
    petrologic characteristics *105t*, *292t*
    silicate-oxide system melts 298–300, *300f*
    trapped noble gases 300–301, *301f*
    veining 333, *333f*
  angrites
    age determination 304–305, *538t*
    chemical composition *293t*
    manganese (Mn) isotopes 444, *445f*
    mineralogy 107, *107f*
    petrologic characteristics *105t*, *292t*
    solar system evolution *452f*
    strontium (Sr) isotopes 515, *539f*
    taxonomy *84f*, *85t*
  aubrites
    age determination 307
    chemical composition *293t*
    cosmic-ray exposure ages *360f*, *368f*, 369
    enstatite chondrites 305
    mineralogy 108, *108f*, 305
    oxygen isotopic composition 140, *140f*
    petrologic characteristics *105t*, *292t*
    siderophile elements 306–307
    taxonomy *84f*, *85t*
  brachinites
    age determination 308
    chemical composition *293t*
    cosmic-ray exposure ages 369
    lithophile elements 307–308
    mineralogy 108, *109f*, 307
    origins 308
    oxygen isotopic composition 140, *140f*
    petrologic characteristics *105t*, *292t*
    siderophile elements 308
    taxonomy *84f*, *85t*
    trapped noble gases 308
    volatile elements 307–308
  chemical composition *293t*
  differentiated achondrites
    age determination 304–305
    elemental abundances *304f*
    fractionation *305f*
    general discussion 291
    lithophile elements 304
    mineralogy 303
    oxygen isotopic composition 138
    petrologic characteristics *292t*
    volatile elements 304
  elemental abundances *294f*
  HED (howardite–eucrite–diogenite)
    age determination 310, *538t*
    chemical composition *293t*
    cosmic-ray exposure ages 369
    diogenites 111–112, *111f*, *365f*, 369
    eucrites 110, *111f*, *365f*, 369
    formation timescales 453
    fractionation 310–311
    howardites 110, *111f*, *366f*, 369
    iron (Fe) isotopes 446
    lithology 309
    lithophile elements 309–310
    manganese (Mn) isotopes 444, *445f*
    metallic melting process 333
    metamorphism 309
    mineralogy 109, *111f*, 309
    origins 309
    oxygen isotopic composition *118f*, 138, *139f*
    petrologic characteristics *105t*, *292t*
    solar system evolution *452f*
    strontium (Sr) isotopes 515, *539f*
    taxonomy *84f*, *85t*
    trace elements 310
    uranium/lead (U/Pb) isotopic ratios 438
    volatile elements 310, *574f*
Itqiy *292t*, 315
lodranites
  age determination 301
  chemical composition *293t*
  cosmic-ray exposure ages *360f*, 369
  metal–sulfide system melts 298, *298f*
  mineralogy 104, *104f*, 294
  oxygen isotopic composition 140, *140f*
  petrologic characteristics *105t*, *292t*
  silicate-oxide system melts 298–300, *300f*
  trapped noble gases 300–301, *301f*
  veining 333
iron silicates, IIE
  age determination 317
  bulk composition 316

achondrites (*continued*)
  chemical composition *293t*, 329
  formation conditions 317
  inclusion composition 315
  mineralogy 315
  petrologic characteristics *292t*
  volatile elements 316
    mesosiderites
  age determination 312
  chemical composition *293t*, 329
  cooling rates *337t*
  cosmic-ray exposure ages 369, *371f*
  formation conditions 312
  lithology 311
  matrix 311
  mineralogy 112, *112f*, *113f*, 311
  parent body size *337t*
  petrologic characteristics *105t*, *113t*, *292t*
  taxonomy *84f*, *85t*
    Northwest Africa 011
  bulk composition 315
  mineralogy 315
  oxygen (O) isotope composition 315
  petrologic characteristics *292t*
  primitive achondrites 291, 292, *292t*
  short-lived radionuclides *433t*
  sodium/aluminum (Na/Al) versus FeO/MnO plot *294f*
  uncategorized achondrites 292, *292t*, 315
  undifferentiated achondrites *84f*, *85t*, 140, *140f*
  ureilites
    age determination 314
    chemical composition *293t*
    core compositions 313
    diamonds 313
    graphite 313
    lithology 313
    lithophile elements 313
    mineralogy 108, *110f*, 312
    noble gas abundances 402
    origins 314
    oxygen isotopic composition 140, *140f*, 314
    petrologic characteristics *105t*, *292t*
    shock effects 313
    siderophile elements 313
    taxonomy *84f*, *85t*
    texture 313
    trapped noble gases 314
  winonaites
    age determination 302
    chemical composition *293t*
    formation conditions 302
    inclusion composition 301
    mineralogy 106, 301
    oxygen isotopic composition 139–140, 140, *140f*
    petrologic characteristics *292t*
  Zag (b) *292t*, 302
Adams, J. C. 624–625
adenine *271f*
AGB *see* asymptotic giant branch (AGB) evolutionary phases
åkermanite *410f*, 422
alabandite (MnS) 301, 305
alanine ($C_3H_7NO_2$) *273t*
albite *410f*, *500t*
alcohols *272t*, 276
aldehydes *272t*, 276
ALH84001
  cosmic-ray exposure ages 369, *370f*
  elemental abundances 603–604, *606t*
  isotopic composition 615, *615f*
  mineralogy 120
  oxygen isotopic composition 138, *139f*
  taxonomy *84f*, *85t*
ALH84025 307
ALHA77081 *293t*
alkanes *279f*
alpha-proton X-ray spectrometer (APXS) 603, *603f*, 611
aluminum (Al)
  achondrites *293t*, *294f*, 333
  aluminum oxide (Al_2O_3)
    chondrite matrix composition *182t*
    lunar crust *585f*
    lunar mare basalts 564, *565t*
    lunar regolith *566t*
    Mars *602t*, *604t*, *610t*
    Mercury *481t*
    Venus *499t*, *505t*
  calcium–aluminum-rich inclusions (CAIs) 237, *237f*, *238f*, 241, 439, *439f*, *440f*
  chondrites *49t*, *89f*, *181f*
  chondrules *239f*
  comets 673–674, *674t*
  cosmic-ray exposure ages *348t*, 369
  cosmochemical classification *46t*
  formation timescales 451
  interplanetary dust particles (IDPs) *703f*, *703t*
  KREEP composition *578t*
  Martian mantle composition *611f*
  Martian meteorites *606t*
  measurement units *348t*
  Moon 562, 563–564
  presolar grains *22f*, 23, 28, 30
  refractory inclusions 450, *451f*
  short-lived radionuclides 433, *433t*, 439
  solar nebula 78, 464
  solar photosphere *44t*
  solar system 439
Amazonian system 605
amides 275
amines *272t*, 275
amino acids *271f*, 272, *272t*, *273t*, *279f*
aminoisobutyric acid ($C_4H_9NO_2$) *273t*
amoeboid olivine aggregates (AOAs)
  chondrites
    abundances *147t*, 160, 219
    composition 144
    isotopic composition 162, *165f*
    mineralogy 161, *163f*, *164f*
    origins 162
    trace elements 162
  mineralogy *94f*, *99f*, *100f*, 219
Anders and Grevesse abundance table *49t*, 53
andesites 607, 610, *612f*
andradite 204–205, *208t*, 257
angrites
  age determination 304–305, *538t*
  chemical composition *293t*
  cosmic-ray exposure ages *367f*, 369
  elemental abundances *305f*
  fractionation *305f*
  lithophile elements 304
  manganese (Mn) isotopes 444, *445f*
  mineralogy 107, *107f*, 303
  petrologic characteristics *105t*, *292t*
  solar system evolution *452f*
  strontium (Sr) isotope composition *539f*
  taxonomy *84f*, *85t*
  volatile elements 304
angular momentum, solar nebula 65, 70, 71, 72
anhydrite 502
anorthite
  calcium–aluminum-rich inclusions (CAIs) 203, 206, *207t*, 211, *240f*
  CO chondrites 261–262
  condensation/fractionation processes *410f*
  mineral chemistry 211
  Venus *500t*
anorthosite
  ferroan anorthosites 537, *538t*, 583, *585f*, 588
  lunar mare basalts *577f*
antimony (Sb)
  chondrites *49t*, *89f*
  cosmochemical classification *46t*
  iron meteorites 330–331
  lunar mare basalts *574f*
  martian meteorites *606t*
  solar photosphere *44t*
  volatile element depletions *512f*
apatite 296, 301
Apollo missions **559–599**
APXS *see* alpha-proton X-ray spectrometer (APXS)
argon (Ar)
  acapulcoite–lodranite achondrites 300–301, *301f*
  brachinites 308
  chondrites *49t*
  comets 674
  cosmic-ray exposure ages *348t*, 369
  cosmochemical classification *46t*
  giant planets *631t*
  interstellar medium (ISM) *58t*, *59f*
  isotopic abundances 131, *383f*, 398
  lunar mare basalts *570t*
  lunar soils 390–391
  Mars *605t*, 614
  mesosiderites 312
  protoatmosphere 532
  radiogenic isotopes 384, *384t*
  solar photosphere *44t*
  solar wind concentrations 390, 392
  spallation 386
  terrestrial planets 532
  Venus *492t*, 494, *494t*

arsenic (As)
  achondrites *293t*, *294f*
  arsine (AsH$_3$) *631t*
  chondrites *49t*, *89f*
  interstellar medium (ISM) 57, *58t*, *59f*
  Martian meteorites *606t*
  solar photosphere *44t*
  volatile element depletions *512f*
aspartic acid (C$_4$H$_7$NO$_4$) *273t*
asteroid belt
  accretion effects 470
  composition 470
  formation conditions 469
  heating processes 470
  mass depletion 469
  orbital resonances 469
  volatile elements 469
asteroids
  chemical composition 339–340
  comets 664
  core formation 330, 332, 340
  differentiated achondrites 309
  evolution 325
  future research 340
  hydrated silicate mineral formation 665
  interplanetary dust particles (IDPs) 690–691, 700–701
  metallic melting process 332, *332f*, *333f*
  planetesimals 665
  spectral reflectance 339, *339f*,
asymptotic giant branch (AGB)
  evolutionary phases 1–2, 21–23, 22, 130
atmophile elements 533–534
atmospheric chemistry
  molecular nitrogen (N$_2$) 494
  noble gases 403
  protoatmosphere 530
  radiogenic isotopes 385
  thermal structure 495
  xenon fractionation 533, *533f*, *540f*, 542
aubrites
  age determination 307
  chemical composition *293t*
  cosmic-ray exposure ages *360f*, *368f*, 369
  enstatite chondrites 305
  mineralogy 108, *108f*, 305
  oxygen isotopic composition 140, *140f*,
  petrologic characteristics *105t*, *292t*
  siderophile elements 306–307
  taxonomy *84f*, *85t*
augite
  CO chondrites 261–262
  differentiated achondrites 315
  mesosiderites 311
  ureilites 313

baddeleyite 303, 309, 315
Balbus–Hawley instability 72
barium (Ba)
  achondrites *293t*
  chondrites *49t*
  cosmochemical classification *46t*
  KREEP composition *578t*
  lunar mare basalts *565t*
  lunar regolith *566t*
  Martian meteorites *606t*
  nucleosynthesis 13–14, *13f*
  presolar grains 24, *25f*, 28
  solar photosphere *44t*
baroclinic instability 71
basalts
  iron oxide (FeO) concentrations *480t*
  Mars 610–611, *612f*
  Mercury 479, *481t*
  mid-ocean ridge basalts (MORBs) 499, *500t*
  shergottites 117–118, *574f*, 603–604
  siderophile elements *574f*
  trace elements 573
  Venus *498t*, 499, *500t*
  volatile elements *574f*
benz(a)anthracene *278t*
benzene (C$_6$H$_6$) *278t*
benzo(ghi)fluoranthene *278t*
benzothiophenes *272t*
beryllium (Be)
  calcium–aluminum-rich inclusions (CAIs) 239, 442, *443f*
  chondrites *49t*
  cosmic-ray exposure ages *348t*
  cosmochemical classification *46t*
  measurement units *348t*
  refractory inclusions 450, *451f*
  short-lived radionuclides *433t*
  solar nebula 77, 464
  solar photosphere *44t*
  solar system evolution 239, 442, *443f*
biphenyl *278t*
bismuth (Bi)
  chondrites *49t*
  cosmochemical classification *46t*
  lunar mare basalts 573, *574f*
  Martian meteorites *606t*
  solar photosphere *44t*
  volatile element depletions *512f*
boron (B)
  boric acid (B(OH)$_3$) *59f*
  calcium–aluminum-rich inclusions (CAIs) 239, 442, *443f*
  chondrites *49t*
  cosmochemical classification *46t*
  elemental abundances *44t*
  interstellar medium (ISM) *58t*, *59f*
  Martian meteorites *606t*
  solar photosphere *44t*
  solar system evolution 239, 442, *443f*
brachinites
  age determination 308
  chemical composition *293t*
  cosmic-ray exposure ages 369
  lithophile elements 307–308
  mineralogy 108, *109f*, 307
  origins 308
  oxygen isotopic composition 140, *140f*,
  petrologic characteristics *105t*, *292t*
  siderophile elements 308
  taxonomy *84f*, *85t*
  trapped noble gases 308
  volatile elements 307–308

bromine (Br)
  achondrites *293t*
  chondrites *49t*
  cosmochemical classification *46t*
  interplanetary dust particles (IDPs) 704, *705f*
  isotopic abundances *383f*
  KREEP composition *578t*
  lunar mare basalts *574f*
  Martian mantle composition *611f*
  Martian meteorites *606t*
  solar photosphere *44t*
brown dwarfs 624, 627, *628f*
butane (C$_4$H$_{10}$) *278t*

cadmium (Cd)
  chondrites *49t*
  interstellar medium (ISM) *58t*, *59f*
  lunar mare basalts 573, *574f*
  Martian meteorites *606t*
  solar photosphere *44t*
  volatile element depletions *512f*
caesium (Cs)
  achondrites *293t*
  chondrites *49t*
  cosmochemical classification *46t*
  isotopic abundances *383f*
  KREEP composition *578t*
  lunar mare basalts 573, *574f*
  lunar regolith *566t*
  Martian meteorites *606t*
  solar photosphere *44t*
calcite *208t*, 502
calcium (Ca)
  see also calcium carbonate (CaCO$_3$)
  achondrites *293t*, *294f*
  calcium (^{41}Ca) isotopes 438
  calcium oxide (CaO)
    chondrite matrix composition *182t*
    lunar mare basalts *565t*
    lunar regolith *566t*
    Mars *602t*, *604t*
    Martian mantle composition *610t*
    Mercury *481t*
    Venus *499t*, *505t*
calcium–aluminum-rich inclusions (CAIs)
  age determination 229
  anorthite 211
  bulk element composition 211, 239, *240f*, 423
  calcium monoaluminate (CaAl$_2$O$_4$) 210, 226–227, 228
  CB chondrites 234–235, *235f*
  CH chondrites 224, *227f*, 234–235, *235f*
  chondrites
    abundances *147t*
    accretionary rims 160
    characteristics 132–133
    composition 144
    distribution 222
    origins 222
    refractory inclusions 156
    short-lived isotopes 464
  chondrules 239, *239f*, *240f*
  CM chondrites 223, *224f*, 233–234, *234f*, 252

calcium–aluminum-rich inclusions
(CAIs) (continued)
  CO chondrites 223, *224f*,
    233–234, 234, *234f*, 262
  cooling rates 423–424, *425f*
  CR chondrites 224, *225f*, 235,
    236f
  CV chondrites 158, 222–223, 232,
    233f, 235
  dating techniques 435
  enstatite chondrites 228, *229f*,
    234–235, *235f*
  evaporation processes 418, *418f*,
    419f, *420f*, 422
  formation conditions 239, 241
  formation timescales 451, *452f*,
    514
  fractionation and unidentified
      nuclear (FUN) inclusions
    aluminum (Al) isotopes *238f*
    beryllium/boron (Be/B)
      isotopic ratios 442, *443f*
    characteristics 219, *220f*
    isotopic fractionation 221, 425
    Murchison chondrite 220
    oxygen (O) isotopes 233
    rare earth elements (REEs) 205
  fractionation effects *409f*
  Fremdlinge 205
  geikelite 229
  grossite 210
  hibonite ($CaAl_{12}O_{19}$) 210, *210f*,
    221f, 425, 438–439, *439f*,
    450
  historical background 15
  isotopes
    aluminum (Al) isotopes 237,
      237f, *238f*, 241, *440f*
    beryllium/boron (Be/B)
      isotopic ratios 239, 442, *443f*
    bulk composition 423
    chromium (Cr) isotopes 444,
      445f
    hafnium (Hf) isotopes 448
    iron (Fe) isotopes 446
    magnesium (Mg) isotopes 237,
      237f, *238f*, 241
    manganese (Mn) isotopes 444,
      445f
    mass-dependent isotopic
      fractionation 239
    nickel (Ni) isotopes 446
    palladium (Pd) isotopes 447
    uranium/lead (U/Pb) isotopic
      ratios 436, *437f*
  melilite 202, 203, *203f*, 206, *207t*,
    208f, 239
  melt distillation 239
  metal grains 205
  meteorites 132–133, 464
  mineralogy *94f*, *96f*, *99f*, 203, 206,
    207t, *208t*,
  ordinary chondrites 228
  oxygen (O) isotopes
    chondrites *233f*, *234f*
    concentrations *134f*
    CV chondrites 133, *133f*
    data analysis 232, 235
    general discussion 232
  perovskite 210
  phase diagrams *213f*
  pyroxenes 209
  rare earth elements (REEs) 205,
    230, *230f*
  refractory inclusions 156, 219,
    231, 424
  rim sequences 204
  short-lived radionuclides 237,
    241, *433t*
  solar evolution 239
  spinel–pyroxene inclusions
    217–218, 218, *218f*, *219f*
  structural components 202,
    203f
  textures 422
  trace element abundances 230
  troilite 229
  type A 211, *213f*, *214f*, *224f*,
    225f, *228f*, 239, *240f*, *451f*
  type B 211, *213f*, *215f*, *237f*,
    239, *240f*, 416, 423, *423f*,
    424f, *451f*
  type C 217, 239, *240f*
  ultrarefractory inclusions 422
  unequilibrated ordinary
    chondrites 228, *228f*, 234,
    235f, 256, 263
  volatile element depletions 527
  Wark–Lovering rim 202, *203f*,
    204, *204f*, 223
chondrites *49t*, *89f*, *181f*
cosmic-ray exposure ages *348t*
cosmochemical classification *46t*
elemental abundances *89f*
interplanetary dust particles
  (IDPs) *703f*, *703t*, 704, *705f*
interstellar medium (ISM) *58t*,
  59f
KREEP composition *578t*
Martian meteorites *606t*
measurement units *348t*
Moon 563–564
presolar grains 30–31
short-lived radionuclides *433t*,
  438
solar nebula 78, 464
solar photosphere *44t*
solar system 438
calcium monoaluminate ($CaAl_2O_4$)
  207t, 210, 226–227, 228
Cameron, A. C. 667
carbon (C)
  carbon (^{12}C) isotope production 4
  carbon dioxide (CO_2)
    comets 673–674, *675t*
    Mars *605t*, 614
    Murchison chondrite *272t*,
      278t
    sublimation temperatures *674t*
    Venus 491, *492t*, 496
    volatile inventory 614
  carbon monosulfide (CS) *675t*
  carbon monoxide (CO)
    comets 673–674, 674, *675t*
    Murchison chondrite *272t*,
      278t
    sublimation temperatures *674t*
    Venus 492, *492t*
  carbonyl sulfide (COS)
    comets *675t*
    Venus *492t*, 493, 502
  chondrites *49t*, 88, *91f*
  comet Halley 680, *681t*
  comets *675t*
  cosmochemical classification *46t*
diamonds
  anomalous exotic components
    18–19, 394, 395, *396t*, 397
  presolar grains 18–19, *19t*
graphite
  anomalous exotic components
    18–19, 394, 395, *396t*
  presolar grains 18, *19t*
interplanetary dust particles
  (IDPs) 695–696, *697f*, *703t*
interstellar medium (ISM) *58t*, *59f*
Mars 615, *615f*
  atmospheric loss 615, *615f*
  volatile inventory 614
Martian meteorites *606t*
Murchison chondrite 277, *278t*,
  279f
nucleosynthesis 2, 8
polyaromatic hydrocarbons
  (PAHs) 695–696
presolar grains **17–39**, *21f*, 22,
  22f, *23f*, 394
silicon carbide (SiC)
  anomalous exotic components
    18–19, 394, 395, *396t*, 397
  presolar grains 18–19, *19t*
solar isotopic abundances 131
solar photosphere 43, *44t*
solar system *132t*
Venus *494t*, 495, 500
volatile element depletions *512f*
carboxylic acids 270, *271f*, *272t*,
  279f
Cassini, J. D. 625
CASSINI spacecraft 488, *489t*
caswellsilverite 305
celsian 303
cerium (Ce)
  achondrites *293t*
  calcium–aluminum-rich
    inclusions (CAIs) *230f*, 425
  chondrites *49t*, *230f*
  differentiated achondrites *304f*,
    306f, *308f*
  KREEP composition *578t*
  lunar mare basalts *577f*
  lunar regolith *566t*
  Martian meteorites *606t*
  presolar grains 24
  solar photosphere *44t*
  volatility 411
chalcophile elements *294f*, 298,
  298f, *512f*
Chassigny meteorite
  cosmic-ray exposure ages 369,
    369t, *370f*
  crystallization 616–617
  elemental abundances *606t*
  mineralogy *118f*
  origins 603–604
  taxonomy *84f*, *85t*
chlorine (Cl)
  chondrites *49t*
  cosmic-ray exposure ages *348t*,
    369
  cosmochemical classification *46t*
  hydrochloric acid (HCl) *492t*, 493
  interstellar medium (ISM) *58t*, *59f*
  isotopic abundances *383f*
  Mars
    mantle composition *611f*
    Martian meteorites *606t*
    soil analysis *604t*

Index

volatile inventory 614
measurement units *348t*
radiogenic isotopes *384t*, 385–386
solar photosphere *44t*
Venus *492t*, 493, *494t*, 495, *499t*, 501
volatile element depletions *512f*
chondrites **143–200**
  see also achondrites; meteorites
  accretionary rims *94f*, 160
  alterations 152
  aluminum (Al) isotopic composition *440f*, 442
  aluminum-rich chondrules
    composition 163, *166f*, *167f*
    isotopic composition 165
    origins 166
    oxygen isotopic composition *168f*
  amoeboid olivine aggregates (AOAs)
    abundances *147t*, 160, 219
    composition 144
    isotopic composition 162, *165f*
    mineralogy *94f*, *99f*, *100f*, 161, *163f*, *164f*
    origins 162
    trace elements 162
  asteroidal aqueous alteration
    alteration chronology 249, 251, 261
    background information 248
    CI chondrites 249
    CM chondrites 250
    CO chondrites 254
    CR chondrites 254
    CV chondrites 255
    hydrated silicate mineral formation 665
    olivine formation 261
    oxygen isotopic composition 261
    pre-accretionary alteration 251, 256
  breccia classification 90, *93t*
  bulk chemical compositions 87, *88f*, 149
  calcium–aluminum-rich inclusions (CAIs)
    abundances *147t*
    accretionary rims 160, *161f*, *162f*
    age determination 229
    beryllium/boron (Be/B) isotopic ratios 442
    characteristics 132–133
    chromium (Cr) isotopes 444, *445f*
    cooling rates 423–424, *425f*
    distribution 157, *157t*, 222
    enstatite chondrites 228
    evaporation processes 418, *418f*, *419f*, *420f*, 422
    formation timescales 451, *452f*
    fractionation and unidentified nuclear (FUN) inclusions 425
    hafnium (Hf) isotopes 448
    hibonite ($CaAl_{12}O_{19}$) 425
    iron (Fe) isotopes 446
    isotopic measurements 438–439, *439f*
    manganese (Mn) isotopes 444, *445f*
    melilite 202, 203, *203f*, 206, *207t*, *208f*, 239
    nickel (Ni) isotopes 446
    origins 222
    oxygen isotopic composition 158, *158f*, *159f*, 232, *233f*, *234f*, 235
    palladium (Pd) isotopes 447
    perovskite 210
    refractory inclusions 156, 424
    short-lived isotopes 464
    short-lived radionuclides 237, 241, *433t*
    solar system evolution *452f*
    textures 422
    ultrarefractory inclusions 422
  carbide–magnetite assemblages 263
  carbon abundances 88, *91f*
  carbonaceous chondrites
    cosmic-ray exposure ages 369
    hafnium/tungsten (Hf/W) isotopic ratios 523, *523f*
    mineralogy 91
    as models for bulk Earth composition 527, 528
    parent bodies 148
    petrologic classification *153t*
    taxonomy 86
  CB chondrites
    calcium–aluminum-rich inclusions (CAIs) 158, *159f*, 224
    composition 135, 146, *147t*
    compositional models 482–483
    elemental abundances *89f*
    iron meteorites 331–332
    metal grains 426
    metallic FeNi 176, *177f*
    mineralogy *96f*, 97–98, *97f*
    oxygen isotopic composition 234–235, *235f*
    petrographic properties *90t*
    petrologic classification *153t*
  CH chondrites
    aluminum-rich chondrules *167f*
    calcium–aluminum-rich inclusions (CAIs) 158, *159f*, 224, *227f*
    composition 135, 146, *147t*
    compositional models 482–483
    cosmic-ray exposure ages *357t*
    elemental abundances *89f*
    evaporation and condensation processes 408
    metal grains 426
    metallic FeNi 176
    mineralogy 95–96, *96f*, *97f*
    oxygen isotopic composition 234–235, *235f*
    petrographic properties *90t*
    petrologic classification *153t*
  chemical composition
    chondrule formation 463
    isotopic composition 151
    lithophile elements 150, *150f*, *176f*
    magnesium silicates 150
    metallic FeNi 151
    metallic grains 175
    refractory elements 150
    refractory inclusions 463
    siderophile elements *177f*
    troilite 177
    volatile elements 151
  chondritic asteroids 155
  CI chondrites
    alteration chronology 249
    amino acids 273
    asteroidal aqueous alteration 249
    background information 51
    bulk composition *148f*
    carbon compounds *278t*
    chemical variations 47, *48f*
    composition 146, *147t*, 248
    cosmic-ray exposure ages *356f*, 369
    Earth accretion dynamics 527
    elemental abundances 47, *49t*, 51, 53, *89f*
    manganese/chromium (Mn/Cr) ratios 249, *250f*
    matrix material *178t*, 179
    mineralogy 91
    oxygen isotopic composition *138f*
    petrographic properties *90t*
    petrologic classification *153t*
    rare earth elements (REEs) 230, *230f*,
    refractory elements 51, *52f*
    siderophile elements *574f*
    thermal metamorphism 138
    volatile element depletions 527
    volatile elements 46, 47, *47f*, 51, *52f*, *574f*
  CK chondrites
    composition 135, 146, *147t*
    cosmic-ray exposure ages *356f*, 369
    elemental abundances *89f*
    matrix material 182
    mineralogy 98, *100f*
    petrographic properties *90t*
    petrologic classification *153t*
  classification *84f*, *85t*, 86, 135, 146
  CM chondrites
    alteration chronology 251
    amino acids 273
    asteroidal aqueous alteration 250, *250f*
    brecciation 250
    bulk composition *148f*, 253, *253f*
    calcium–aluminum-rich inclusions (CAIs) 223, *224f*, 252
    carbon compounds *278t*
    chondrule alterations 253
    chondrule rim alteration 250, *250f*, 252
    composition 135, 146, *147t*, 251
    cosmic-ray exposure ages *356f*, 369
    elemental abundances *89f*
    formation timescales 453
    iodine/xenon (I/Xe) isotopic ratios 250
    iron-rich aureoles 252–253
    manganese/chromium (Mn/Cr) ratios 250

chondrites (*continued*)
   matrix material *178t*, 180
   mineralogy 91, *94f*
   oxygen isotopic composition *137f*, *138f*, 233, *234f*, 250, *251f*, 253
   petrographic properties *90t*
   petrologic classification *153t*
   phyllosilicates 136–137
   pre-accretionary alteration 251
   refractory inclusions *451f*
   rim sequences 204–205, *204f*
   thermal metamorphism 138
   veining 252, *252f*
 CO chondrites
   asteroidal aqueous alteration 254
   bulk composition *148f*
   calcium–aluminum-rich inclusions (CAIs) 158, *158f*, 223, *224f*, 262
   composition 135, 146, *147t*
   cosmic-ray exposure ages *356f*, 369
   elemental abundances *89f*
   matrix material *178t*, 180, *181f*, *182t*
   metasomatism 261, *261f*
   mineralogy 94–95, *94f*
   nebular alteration 262
   oxidation 261
   oxygen isotopic composition *137f*, 233–234, 234, *234f*
   parent-body alteration 262
   petrographic properties *90t*
   petrologic classification *153t*
   thermal metamorphism 138
 components 155, *157t*
 composition
   calcium–aluminum-rich inclusions (CAIs) 144
   major elements 702, *703f*, *703t*
   as models for bulk Earth composition 527, 528
 CR chondrites
   aluminum-rich chondrules *167f*
   asteroidal aqueous alteration 254
   calcium–aluminum-rich inclusions (CAIs) 158, *159f*, 224, *225f*
   chondrules *145f*
   composition 135, 146, *147t*
   compositional models 482
   cosmic-ray exposure ages 369
   elemental abundances *89f*
   isotopic composition 235
   matrix material *178t*, 180
   metallic FeNi 176
   mineralogy 95, *96f*, *97f*
   oxygen isotopic composition *168f*, *236f*
   petrographic properties *90t*
   petrologic classification *153t*
   uranium/lead (U/Pb) isotopic ratios 437, *437f*
 CV chondrites
   asteroidal aqueous alteration 255
   bulk composition *148f*, 256, *256f*

   calcium–aluminum-rich inclusions (CAIs) 133, *133f*, 158, 222–223
   composition 135, 146, *147t*
   cosmic-ray exposure ages *356f*, 369
   dark inclusions 258, *258f*
   elemental abundances *89f*
   matrix material *178t*, 182, *183f*
   metasomatism 257
   mineralogy 98, *99f*
   nebular alteration 259
   oxidation 257
   oxygen isotopic composition 138, 232, *233f*, 255, *255f*, 258
   parent-body alteration 256, 259
   petrographic properties *90t*
   petrologic classification *153t*
   pre-accretionary alteration 256
   refractory inclusions 450, *451f*
   rim sequences 204–205
 deep-sea spherules 426
 definition 144
 E chondrites *359f*, 369
 EH chondrites
   bulk composition *148f*
   composition 146, *147t*
   elemental abundances *89f*
   mineralogy 101, *102f*
   petrographic properties *90t*
   petrologic classification *153t*
 EL chondrites
   bulk composition *148f*
   composition 146, *147t*
   elemental abundances *89f*
   mineralogy 101
   petrographic properties *90t*
   petrologic classification *153t*
 elemental abundances *89f*, 149, *181f*
 enstatite chondrites
   aubrites 305
   calcium–aluminum-rich inclusions (CAIs) 228, *229f*
   composition 135, 146, *147t*
   manganese (Mn) isotopes 444, *445f*
   Mercury 483
   mineralogy 101, *102f*
   noble gas abundances 402
   oxygen isotopic composition *132f*, 234–235, *235f*
   parent bodies 149
   petrographic properties *90t*
   petrologic classification *153t*
   taxonomy *84f*, *85t*, 86
   volatile elements 469
 evaporation processes 421
 extractable organic matter
   abundances 270, *272t*
   alcohols *272t*, 276
   aldehydes *272t*, 276
   aliphatic hydrocarbons *272t*, 275
   alkanes *279f*
   amides 275
   amines *272t*, 275
   amino acids *271f*, 272, *272t*, *273t*, *279f*
   aromatic hydrocarbons *272t*, 274, *278t*

   carbon compounds 277, *278t*, *279f*
   carboxylic acids 270, *271f*, *278t*, *279f*
   contaminants 274
   heterocyclic compounds *271f*, *272t*, 274
   hydrogen (δ^2H) isotope concentrations *277t*, 280
   ketones *272t*, 276
   nitrogen (N) isotope concentrations 280
   organic fractions *276t*
   phosphonic acids 272, *272t*
   polyaromatic hydrocarbons (PAHs) 274, *279f*, 284, *285f*
   purines *272t*, 274
   pyrimidines *272t*, 274
   quinolines 274
   stable isotope compounds 276
   sugar-related compounds *272t*, 276
   sulfonic acids 272, *272t*, *277t*
   sulfur (S) isotope concentrations 280
 ferromagnesian chondrules 168, *169f*
 formation conditions 187, 285, *285t*, 451, *452f*
 future research 263
 glass with embedded metal and sulfides (GEMS) 680, *680f*, *694f*, 697, *698f*
 H chondrites
   bulk composition *148f*
   composition 146, *147t*
   cosmic-ray exposure ages 146–148, *358f*, 369
   elemental abundances *89f*
   matrix material *178t*, 184
   mineralogy 100
   petrographic properties *90t*
   petrologic classification *153t*
 historical background 145
 historical investigations 269
 impact processing 152, 155
 importance 144
 in situ studies 284
 interplanetary dust particles (IDPs) 679, 689, 695–696, *697f*
 isotopic fractionation 421, 426
 K chondrites
   composition 146, *147t*
   elemental abundances *89f*
   matrix material *178t*, 184
   mineralogy 101
   petrographic properties *90t*
   petrologic classification *153t*
 L chondrites
   bulk composition *148f*
   composition 146, *147t*
   cosmic-ray exposure ages *358f*, 369
   elemental abundances *89f*
   matrix material *178t*, 184
   mineralogy 100
   petrographic properties *90t*
   petrologic classification *153t*
 LL chondrites
   bulk composition *148f*
   composition 146, *147t*

cosmic-ray exposure ages 146, *359f*, 369
elemental abundances *89f*
matrix material *178t*, 184
mineralogy 100
petrographic properties *90t*
petrologic classification *153t*
macromolecular material
  aliphatic hydrocarbons 281
  aromatic hydrocarbons 280, *284t*, *285f*
  electromagnetic resonance studies 282
  heterocyclic compounds 281
  NMR investigations 281, *282f*, *282t*
  stable isotope studies 282, *283f*, *284t*
  thiophenes 281
magnesium (Mg) concentrations *145f*
Mars 610, *611f*
matrix
  abundances *147t*
  mineralogy *94f*, *96f*, *99f*, 177, *178t*
  origins 186
matrix-rich lithic clasts 185
metal-rich chondrites 482, *483f*
metamorphism 152, *153f*
Murchison chondrite 42–43, *42f*, 252, *252f*, **269–290**
nebular alteration
  alteration chronology 260
  background information 248
  CO chondrites 262
  CV chondrites 259
  pyroxene formation 260
  veining 260
neon (Ne) isotope concentrations 388, *394f*, 398, *399f*
nitrogen abundances 88, *91f*
ordinary chondrites
  calcium–aluminum-rich inclusions (CAIs) 228
  chondrules *145f*
  composition 135, 146, *147t*
  elemental abundances *89f*
  hafnium (Hf) isotopes 448
  iron (Fe) isotopes 446, *447f*
  manganese (Mn) isotopes 444, *445f*
  matrix material *178t*
  metamorphic ages 155
  mineralogy 100, *101f*
  oxygen isotopic composition *132f*, *136f*
  parent bodies 148
  petrographic properties *90t*
  petrologic classification *153t*
  sizes 155
  taxonomy *84f*, *85t*, 86
  thermal metamorphism 137
  uranium/lead (U/Pb) isotopic ratios 438
  volatile elements 469
organic matter 270, *271f*
origins 145
oxidation states 88, *91f*
oxygen isotopic composition
  anomalies 151
  bulk composition 88, *89f*
  chondrules 132, *132f*
  general discussion 135
  Rumuruti-type chondrites *132f*
  whole-rock data *135f*
parent bodies 256, 259, 262, 330
petrographic properties *90t*
petrologic classification 89, *92t* 152, *153t*,
R chondrites
  composition 146, *147t*
  cosmic-ray exposure ages *360f*, 369
  elemental abundances *89f*
  mineralogy 103
  petrographic properties *90t*
  petrologic classification *153t*
refractory elements
  composition 175
  inclusions 450, *451f*
  metallic grains 175
  volatile/refractory ratio 524, *525f*
relict calcium–aluminum-rich inclusions 168, *169f*
rim sequences 204
rubidium/strontium (Rb/Sr) isotopic ratios *523f*
Rumuruti-type chondrites *132f*
serpentine 699
shock metamorphism 90, *93t*
short-lived radionuclides *433t*
in situ studies 284
solar nebula 146
solar system evolution 431
taxonomy 86
terrestrial weathering 90
thermal history 154
type 1–2 chondrites 152
type 3 chondrites 153, *153f*
type 4–6 chondrites 154
unequilibrated ordinary chondrites 228, *228f*, 234, *235f*, 256, 446
volatile elements *512f*
chondrules **143–200**
  abundances 169
  age determination *538t*
  aluminum (Al) isotopes *440f*, 441
  aluminum (Al) isotopic composition *239f*, *240f*
  aluminum-rich chondrules
    composition 163, *166f*, *167f*
    isotopic composition 165
    origins 166
    oxygen isotopic composition *168f*
  asteroidal aqueous alteration 253, 256
  asteroid origins 175
  calcium–aluminum-rich inclusions (CAIs) 239
  classification 170, *170f*
  closed-system crystallization 172
  composition 171
  compound chondrules 172
  condensation processes 425
  CR chondrites *145f*
  definition 144, 168
  elemental abundances *171f*, *173f*, *174f*, *175f*, *176f*
  evaporation processes 425
  ferromagnesian chondrules 168, *169f*
  formation conditions
    absolute timescales 437
    heating processes 514
    jets 189
    models *188t*
    nebular shocks 189
    outflows 189
    planetesimal impacts 189
    shock fronts 76
    solar nebula *76f*, 78, 80, 463
    solar system 188
    volatile element depletions 527
  magnesium (Mg) isotopic composition *240f*
  mineralogy *94f*, *96f*, *99f*
  open-system crystallization 173
  ordinary chondrites *145f*
  oxygen (O) isotopes 132, *132f*, 135, *136f*, *172f*
  relict calcium–aluminum-rich inclusions 168, *169f*, 172
  rims 172, 250, *250f*, 252
  short-lived radionuclides *433t*
  textures 169
  thermal history 169–170
chromite
  differentiated achondrites 309, 315
  interplanetary dust particles (IDPs) 699–700
  iron silicates, IIE 316
  mesosiderites 311
  mineralogy 307
  pallasites 338
  primitive achondrites 296, 301, 302
chromium (Cr)
  achondrites *293t*, *294f*
  calcium–aluminum-rich inclusions (CAIs) 444, *445f*
  chondrites *49t*, *89f*, *181f*, 249, 250, *250f*
  chromium oxide (Cr_2O_3)
    chondrite matrix composition *182t*
    Mars *602t*, *604t*, *610t*
    Mercury *481t*
    Venus *505t*
  cosmochemical classification *46t*
  formation timescales 451
  interplanetary dust particles (IDPs) *703t*, 704, *705f*
  interstellar medium (ISM) *58t*, *59f*
  KREEP composition *578t*
  lunar mare basalts *565t*, *566t*, *574f*
  manganese/chromium (Mn/Cr) ratios 249, 250, *250f*, 444, *445f*, 453
  Martian mantle composition *611f*
  Martian meteorites *606t*
  solar photosphere *44t*
  solar system evolution 444, *445f*
  volatile element depletions *512f*
chrysene *278t*
circumstellar disks 462, 510
clinopyroxenes 301, 302, 316, 338
clinopyroxenites
  cosmic-ray exposure ages 369, *369t*
  mineralogy 118
  shergottites 603–604
  taxonomy *84f*, *85t*

CMAS melts 409–410, 416, *417f*, *418f*
cobalt (Co)
  achondrites *293t*, *294f*
  chondrites *49t*, *89f*
  cosmochemical classification *46t*
  interstellar medium (ISM) *58t*, *59f*
  iron meteorites *328t*
  KREEP composition *578t*
  lunar mare basalts *565t*, *574f*
  lunar regolith *566t*
  Martian mantle composition *611f*
  Martian meteorites *606t*
  solar photosphere *44t*
  Venus *505t*
  volatile element depletions *512f*
comets **663–688**
  activity 665
  asteroids 664
  background information 663
  Centaurs 671, *672f*
  classification 666
  comet Borrelly 671, *671f*
  comet Halley
    elemental abundances 55, *681t*
    grain structure 675, *677f*
    iron/magnesium (Fe/Mg) ratios 681, *683f*
    iron/silicon (Fe/Si) ratios 681, *682f*
    magnesium/silicon (Mg/Si) ratios 681, *682f*
    mineralogy 681, *682f*
    in situ measurements 680
  composition
    aluminum (Al) isotopes 673–674, *674t*
    amorphous ice 673–674, *673f*
    argon (Ar) 674
    carbon monoxide (CO) 673, 674
    hydrogen (D/hydrogen (D/H) ratio 683, *684f*
    diversity 682, *684f*
    dust 55, 664, *664f*, 675
    olivine 675, *677f*
    pyroxenes 676, *677f*
    sampling issues 678
    silicate minerals 676, *677f*
    in situ measurements 680
    sublimation temperatures *674t*
    volatile elements 672–673, *673t*, 674, *675t*
    water ice 672, *673f*
  dead comets 671, *672f*
  Hale Bopp 670, 674, *675t*, 676–678, *677f*
  interplanetary dust particles (IDPs) 679, *680f*
  Kuiper Belt 667, *668f*
  Kuiper Belt Objects (KBOs) 667, 668, *669f*, 671, *672f*
  Leonids 670
  meteorites 678
  meteor streams 670
  nuclei 671, *672f*
  Oort cloud 666, *667f*
  physical evolution
    active areas *673f*
    crust/mantle formation 671
    fragmentation 669
    material strength 671
  presolar grains 680
  solar nebula 663
  solar system 664
  source regions 666
  spectral reflectance *672f*
  Stardust 678
  sun grazing (Kreutz family) comets 670
  Tagish Lake meteorite shower 679
  Trojan comets *672f*
condensation processes
  chondrules 425
  kinetic processes 412
  laboratory experiments 415
  major elements 409
  Mercury 481
  meteorites 330, 421
  nickel (Ni) 330, *331f*
  solar nebula 330
  solar system **407–430**
  terrestrial planets 421
  trace elements 411
  volatile elements 407
copper (Cu)
  chondrites *49t*, *181f*
  cosmochemical classification *46t*
  interstellar medium (ISM) *58t*, *59f*
  iron meteorites *328t*, 330–331
  Martian mantle composition *611f*
  Martian meteorites *606t*
  solar photosphere *44t*
  volatile element depletions *512f*
corundum
  calcium–aluminum-rich inclusions (CAIs) 206, *207t*, *240f*
  condensation/fractionation processes *410f*
  presolar grains 32, *32f*
cosmic-ray exposure ages **347–380**
  acapulcoites *360f*, 369
  ALH84001 369, *370f*
  angrites *367f*, 369
  aubrites *360f*, *368f*, 369
  brachinites 369
  calculations
    aluminum/neon (^{26}Al/^{21}Ne) isotopic calibration 369
    argon/neon (^{38}Ar–^{22}Ne/^{21}Ne) isotopic calibrations 369
    basic calculations 369
    chlorine/argon (^{36}Cl/^{36}Ar) isotopic calibrations 369
    half-lives 369
    helium (^{3}He) ages 369
    krypton/krypton (^{81}Kr/Kr) isotopic calibrations 369
    krypton/krypton (^{83}Kr/^{81}Kr) isotopic calibration 369
    neon/neon (^{21}Ne–^{22}Ne/^{21}Ne) isotopic calibrations 369
    potassium/potassium (^{40}K/K) isotopic calibrations 369
    production rate calibration 369
  carbonaceous chondrites *356f*, 369
  Chassigny meteorite 369, *369t*
  CH chondrites *357t*
  CI chondrites *356f*, 369
  CK chondrites *356f*, 369
  clinopyroxenites 369, *369t*
  CM chondrites *356f*, 369
  CO chondrites *356f*, 369
  cosmogenic nuclides *348t*, 369
  CV chondrites *356f*, 369
  E chondrites *359f*, 369
  H chondrites 146–148, *358f*, 369
  HED meteorites
    diogenites *365f*, 369
    eucrites *365f*, 369
    howardites *366f*, 369
  interplanetary dust particles (IDPs) 369
  interstellar grains 369
  iron meteorites 369, *372f*
  Kapoeta 369
  L chondrites *358f*, 369
  lherzolites 369, *369t*
  LL chondrites 146–148, *359f*, 369
  lodranites *360f*, 369
  lunar meteorites
    composition 369
    cosmic-ray exposure histories *362t*, 369, 561
    parent bodies *374f*
    production rates 369
  Martian meteorites 369, *369t*, *370f*, *374f*
  mesosiderites 369, *371f*
  meteorites 388
  micrometeorites 369, *373f*
  nakhlites 369, *369t*
  noble gases 386
  orthopyroxenites 369, *369t*
  pallasites 369, *371f*
  production rates 369
  R chondrites *360f*, 369
  shergottites 369, *369t*
  ureilites *367f*, 369
  Vesta *374f*
cronstedite *208t*
crystalline rocks 603–604, 611
cubewanos (QB$_1$) 667
curium (Cm) 385–386

D'Arrest, H. L. 624–625
Darwin, G. H. 536
dating techniques
  argon/argon (Ar/Ar) isotopic ratios *570t*, 580
  calcium–aluminum-rich inclusions (CAIs) 435
  hafnium/tungsten (Hf/W) isotopic ratios 519, 522, *523f*, *526t*, *538t*, *539f*
  iodine/xenon (I/Xe) isotopic ratios 249, 250, 448
  lead (Pb/Pb) isotopic ratios 435
  lead (Pb) isotopes *570t*
  lutetium/hafnium (Lu/Hf) isotopic ratios 544
  manganese/chromium (Mn/Cr) ratios *538t*
  neodymium (Nd) isotopes *570t*
  niobium/zirconium (Nb/Zr) isotopic ratios 543
  presolar grains 434–435
  radiogenic isotopes 434, 519
  radiometric dating techniques 519, 522, *522f*, *523f*, *526t*
  rubidium/strontium (Rb/Sr) isotopic ratios
    chondrites *523f*
    lunar age determination *538t*

lunar highland crust *523f*
lunar mare basalts *570t*
samarium/neodymium (Sm/Nd)
    isotopic ratios
  lunar age determination *538t*
  lunar crust 589, *590f*
  mantle (Earth) 543
terrestrial planet formation 519,
    522f
uranium/lead (U/Pb) isotopic
    ratios
  achondrites *538t*
  calcium–aluminum-rich
    inclusions (CAIs) 436, *437f*
  chondrules *538t*
  CR chondrites 437, *437f*
  isotopes 435–436
  lunar age determination *538t*
  ordinary chondrites 438
  solar system evolution
    435–436, *538t*
  terrestrial planet formation
    519
daubreélite 301, 305
hydrogen (D ^{2}H)
  chondrites *277t*, 280
  comets *675t*, 683–684, *684f*
  solar noble gases 389
  solar system 631, *632f*
  Venus *494t*
diamonds
  anomalous exotic components
    18–19, 394, 395, *396t*, 397
  presolar grains 18–19, *18f*, *19t*,
    33
diatoms 313
dicarboximides *272t*, *278t*
diogenites *see* achondrites
diopside
  calcium–aluminum-rich
    inclusions (CAIs) *240f*
  CO chondrites 261–262
  CV chondrites 257
  differentiated achondrites 307,
    309
  interplanetary dust particles
    (IDPs) 699–700
  iron silicates, IIE 316
  primitive achondrites 296
  Venus *500t*
djerfisherite 305
dunites
  lunar crust *584f*, 586
  meteorites 339–340, 603–604
  mineralogy *118f*
  taxonomy *84f*, *85t*
dust
  comets 55, 664, *664f*, 675
  interstellar medium (ISM) 56,
    57
  Mars 607, *612f*, 613
  solar nebula 75
dysprosium (Dy)
  achondrites *293t*, *294f*
  calcium–aluminum-rich
    inclusions (CAIs) *230f*
  chondrites *49t*, *230f*
  differentiated achondrites *304f*,
    306f, *308f*
  KREEP composition *578t*
  lunar regolith *566t*
  Martian meteorites *606t*
  presolar grains 24, *25f*

solar photosphere *44t*
volatility 411

Earth
  see also mantle (Earth)
  accretion dynamics
    age determination *538t*
    chondrite models 527, 528,
      529
    compositional changes 541
    dust stickiness 516
    general discussion 516
    gravitational instability 517
    impact theories 536, 541, 546
    Kyoto model 526
    large collisions 518
    magnesium/silicon (Mg/Si)
      ratios 529, 541
    mass loss 541
    migration 517
    oxygen (O) isotopic
      composition 530
    post-impact evidence 542
    runaway growth 518
  bombardment history 536
  bulk composition *480t*, 527, 528
  core
    formation conditions 534
    heterogeneous accretion 515
    magma oceans 534
    siderophile elements 535
    uranium/lead (U/Pb) isotopic
      ratios 519
  crustal composition 616
  degassing models 532
  early history
    atmospheric composition 530
    atmospheric loss 533, *540f*
    degassed protoatmosphere 532
    lead (Pb) isotope composition
      540, *540f*
    nebular protoatmosphere 530
    xenon (Xe) isotope depletions
      533, *533f*, *540f*
  formation conditions
    accretion dynamics 516
    composition 515
    density 511
    geochemical constraints 519,
      527
    Hadean 544, *545f*
    hafnium/tungsten (Hf/W)
      isotopic ratios 522, *523f*,
      526t
    heterogeneous accretion 515
    hot nebula model 514, 515
    nebular gases 511
    radiogenic isotope models 519
    rubidium/strontium (Rb/Sr)
      isotopic ratios *523f*
    siderophile elements 542
    solar nebula 510
    uranium/lead (U/Pb) isotopic
      ratios 521, *522f*
    volatile element depletions
      511, *512f*, 527, 541
    volatile/refractory ratio 524,
      525f
  geochemical comparison with
    Mars 616
  Hadean
    atmospheric conditions 546
    continent formation 544, *545f*

    formation conditions 544,
      545f
    hydrosphere 544
    impact theories 546
    lutetium/hafnium (Lu/Hf)
      isotopic ratios 544
    mantle depletion 543
    niobium/zirconium (Nb/Zr)
      isotopic ratios 543
    samarium/neodymium
      (Sm/Nd) isotopic ratios 543
  heat flow 546
  oxygen (δ^{18}O) isotope
    concentrations 514, *514f*
Edgeworth, K. 667
elemental abundances
  Anders and Grevesse compilation
    49t, 53
  chondrites *181f*
  chondrules *171f*, *173f*, *174f*, *175f*,
    176f
  CI chondrites 47, *49t*, 51
  cosmochemical classification 45,
    46t, 149
  giant planets
    argon (Ar) *631t*
    germanium (Ge) 630, *631t*
    helium (He) 626, *631t*
    hydrogen (H) 626, 630
    krypton (Kr) *631t*
    methane (CH_4) 630, *631t*
    neon (Ne) *631t*
    nitrogen (N) 630, *631t*
    oxygen (O) 630
    sulfur (S) 630, *631t*
    water (H_2O) 630, *631t*
    xenon (Xe) *631t*
  interstellar medium (ISM) 55, *58t*,
    59f
  Jupiter 630, *631t*, 633
  mantle (Earth) 610, *611f*
  Mars 610, *611f*
  meteorites 45
    comparison with solar
      system abundances *44t*, 53,
      54f
    Martian 603, *606t*
  Neptune 630, *631t*, 633
  nucleosynthesis **1–15**
  Saturn 630, *631t*, 633
  shergottites *605f*, *606t*
  solar photosphere 43, *44t*
  solar system **41–61**
    comet Halley dust 55
    comparison with meteorite
      abundances *44t*, 53, *54f*
    condensation/fractionation
      processes 409, *410f*
    interplanetary dust particles
      (IDPs) 55
    isotopic abundances 54
    isotopic anomalies 42
    major elements *132t*
    mass number 54, *55f*, *56f*
    solar corona 55
    solar energetic particles (SEP)
      55
    solar nebula 42, 131
    solar photosphere 43
    solar wind 55
    Sun 131
    volatility fractionation 407
  Uranus 630, *631t*, 633

Elysium 605
enstatite
    see also chondrites
    calcium–aluminum-rich
        inclusions (CAIs) 228
    comets 676
    condensation/fractionation
        processes *410f*
    CP IDPs 692, *695f*
    differentiated achondrites 305
    interplanetary dust particles
        (IDPs) 699–700
erbium (Er)
    achondrites *293t*
    calcium–aluminum-rich
        inclusions (CAIs) *230f*
    chondrites *49t, 230f*
    differentiated achondrites *304f, 306f, 308f*
    KREEP composition *578t*
    Martian meteorites *606t*
    solar photosphere *44t*
    volatility 411
ethane (C_2H_6) *278t*, 630
ethanol (C_2H_6O) *675t*
ethene *278t*
eucrites *see* achondrites
europium (Eu)
    achondrites *293t, 294f*
    calcium–aluminum-rich
        inclusions (CAIs) *230f*
    chondrites *49t, 89f, 230f*
    differentiated achondrites *304f, 306f, 308f*
    KREEP composition *578t*
    lunar mare basalts *565t*, 575–576, *577f*
    lunar regolith *566t*
    Martian meteorites *606t*
    nucleosynthesis 13–14, *13f*
    solar photosphere *44t*
    ureilites 313
    volatility 411
evaporation processes
    calcium–aluminum-rich
        inclusions (CAIs) 418, *418f, 419f, 420f,* 422, *423f, 424f*
    chondrites 421
    chondrules 425
    CMAS melts 416, *417f, 418f*
    forsterite 415, 416, *417f, 418f, 419f, 420f*
    kinetic processes 412
    laboratory experiments 415
    Mercury 481
    olivine 416, *417f*
    oxide compounds 416
    rare earth elements (REEs) 421
    solar system **407–430**
extrasolar giant planets 634

fassaite
    calcium–aluminum-rich
        inclusions (CAIs) 209
    condensation/fractionation
        processes *411f*
    differentiated achondrites 303
    interplanetary dust particles
        (IDPs) 699–700
fayalite 258, *452f*
feldspar 301, 316
fluoranthene *271f, 278t*
fluorine (F)
    chondrites *49t*
    cosmochemical classification *46t*
    hydrofluoric acid (HF) *492t*, 493
    interstellar medium (ISM) *58t, 59f*
    isotopic abundances *383f*
    Martian mantle composition *611f*
    Martian meteorites *606t*
    solar photosphere *44t*
    Venus *492t*, 493, 501
    volatile element depletions *512f*
fluorphlogopite 501
formaldehyde (CH_2O) 673–674, *674f, 675t*
forsterite
    amoeboid olivine aggregates
        (AOAs) 219
    calcium–aluminum-rich
        inclusions (CAIs) 204–205, *207t, 240f*
    comets 676
    condensation/fractionation
        processes *410f*
    CP IDPs 692
    differentiated achondrites 305
    evaporation processes 415, 416, *417f, 418f, 419f, 420f*
    interplanetary dust particles
        (IDPs) 699–700
    isotopic fractionation 416, *417f, 420f*
Fremdlinge 205

gabbronorites *584f*, 586
gadolinium (Gd)
    achondrites *293t*
    calcium–aluminum-rich
        inclusions (CAIs) *230f*
    chondrites *49t, 230f*
    differentiated achondrites *304f, 306f, 308f*
    KREEP composition *578t*
    lunar mare basalts *577f*
    Martian meteorites *606t*
    solar photosphere *44t*
    volatility 411
galaxies
    carbon isotopic abundances 4, 131
    elemental abundance patterns 12, *12f, 13f*
    evolutionary timescale 12
    galactic chemical evolution
        carbon (^{12}C) isotope production 4
        carbon isotopic abundances 131
        elemental abundance patterns 12, *12f, 13f*
        nucleosynthesis 12
        timescale 12
    nucleosynthesis 12
galena (PbS) 502
Galileo 624–625
Galileo spacecraft 488, *489t*
Galle, G. 624–625
gallium (Ga)
    achondrites *293t, 294f*
    chondrites *49t, 89f, 181f*
    cosmochemical classification *46t*
    interstellar medium (ISM) *58t, 59f*
    iron meteorites **325–345**
    KREEP composition *578t*
    lunar mare basalts *565t, 574f*
    lunar regolith *566t*
    Martian mantle composition *611f*
    Martian meteorites *606t*
    solar photosphere *44t*
    volatile element depletions *512f*
gamma-ray spectrometer (GRS) 602
garnets
    calcium–aluminum-rich
        inclusions (CAIs) 203, *208t*
    CV chondrites 257
    Mars *610t, 611f*
gases
    interstellar medium (ISM) 56, 57
    Martian atmosphere 603, *605f, 605t*, 614, 616
gehlenite *240f, 410f*, 422
geikelite 229
geochemistry, historical background ix
geochronometry 519, *522f*
germanium (Ge)
    chondrites *49t, 181f*
    cosmochemical classification *46t*
    germane (GeH_4) 630, *631t*
    giant planets 630, *631t*
    interstellar medium (ISM) *58t, 59f*
    iron meteorites **325–345**
    lunar mare basalts 573–575, *574f*
    Martian meteorites *606t*
    solar photosphere *44t*
    volatile element depletions *512f*
GHRS *see* Goddard High Resolution Spectrometer (GHRS)
giant planets **623–636, 637–662**
Giotto spacecraft 55
glass with embedded metal and sulfides (GEMS) 680, *680f, 694f*, 697, *698f*
glutamic acid ($C_5H_9NO_4$) *273t*
glycine ($C_2H_5NO_2$) *273t, 675t*
Goddard High Resolution Spectrometer (GHRS) 57
gold (Au)
    achondrites *293t, 294f*
    chondrites *49t, 89f*
    cosmochemical classification *46t*
    iron meteorites *328t*, 330–331
    lunar mare basalts 573–575, *574f*
    lunar regolith *566t*
    Martian meteorites *606t*
    solar photosphere *44t*
    volatile element depletions *512f*
Goldschmidt, V. M. 9, 41–42
graphite
    anomalous exotic components 18–19, 394, 395, *396t*
    presolar grains
        anomalous exotic components 18–19, *18f*
        interplanetary dust particles (IDPs) *19t*
        isotopic composition 30, *30f*
        physical properties *21f*, 29, *31f*
        primitive achondrites 296, 301
        ureilites 313
greenhouse gases 546
grossite
    calcium–aluminum-rich
        inclusions (CAIs) 206, *207t*, 210, 219, *219f, 240f*
    CH chondrites 224–225
    mineral chemistry 210

GRS *see* gamma-ray spectrometer (GRS)
guanine *271f*

Hadean
  atmospheric conditions 546
  Earth formation 544, *545f*
  greenhouse gases 546
  hydrosphere 544
  impact theories 546
  volatile element depletions
    lutetium/hafnium (Lu/Hf) isotopic ratios 544
    niobium/zirconium (Nb/Zr) isotopic ratios 543
    samarium/neodymium (Sm/Nd) isotopic ratios 543
  zircon grains 544
hafnium (Hf)
  achondrites *293t*
  chondrites *49t*
  cosmochemical classification *46t*
  differentiated achondrites *304f*
  isotopes
    calcium–aluminum-rich inclusions (CAIs) 448
    dating techniques 519, 522, *523f, 526t*
    Earth formation 544
    short-lived radionuclides *433t*
    solar system evolution 448
    terrestrial planet formation 522, *523f, 526t*
  KREEP composition *578t*
  lunar mare basalts *565t*
  lunar regolith *566t*
  Martian meteorites *606t*
  primitive mantle composition 468
  solar photosphere *44t*
Hale Bopp 670, 674, *675t*, 676–678, *677f*
halogen chemistry 348–349
HD209458b
  equations of state 629
  historical investigations 624
  physical parameters *624t*
  thermal infrared emissions 629
  transit studies 634
heavy element production *see* nucleosynthesis
hedenbergite 204–205, *208t*, 209, 257
heideite 305
helium (He)
  chondrites *49t*
  equations of state 627
  giant planets 626, *631t*
  interstellar medium (ISM) 57
  isotopes
    cosmic-ray exposure ages *348t*, 369
    interplanetary dust particles (IDPs) 706
    isotopic abundances *383f*
    isotopic composition 398
    nucleosynthesis 4, 7
    radiogenic isotopes *384t*
    solar ^{3}He/^{4}He ratio 389
    solar isotopic abundances 131
    Venus *492t*, 494, *494t*
  solar photosphere 43, *44t*
  solar system *132t*

solar wind concentrations 390, 392
  spallation 386
hematite 613
Herschel, W. 624–625
Hesperian system 605
hibonite ($CaAl_{12}O_{19}$)
  see also chondrites
  beryllium (Be) isotopes 443
  calcium–aluminum-rich inclusions (CAIs)
    cerium anomalies 425
    characteristics 210, *210f*
    isotopic composition *221f*
    isotopic measurements 438–439, *439f*
    mineralogy 203, 206, *207t*
    refractory inclusions 450
    structure 202, *203f*
  CH chondrites *217f*, 224–225
  CM chondrites 216, *216f*, 223, *224f*
  condensation/fractionation processes *410f*, 411, *411f*
  CV chondrites 257
  mineral chemistry 210, *210f*
  presolar grains 32, *32f*
holmium (Ho)
  achondrites *293t*
  calcium–aluminum-rich inclusions (CAIs) *230f*
  chondrites *49t, 230f*
  differentiated achondrites *304f, 306f, 308f*
  KREEP composition *578t*
  Martian meteorites *606t*
  solar photosphere *44t*
  volatility 411
howardites *see* achondrites
Hubble Space Telescope (HST) 57
Huygens, C. 625
hydrocarbons
  acenaphthene *271f*
  aliphatic *272t*, 275, 281
  alkanes *279f*
  aromatic *271f, 272t*, 274, *278t*, 280, *284t, 285f*
  fluoranthene *271f*
  naphthalene *271f*
  phenanthrene *271f*
  polycyclic aromatic hydrocarbons (PAHs)
    interplanetary dust particles (IDPs) 695–696
    meteorites 615
    Murchison chondrite 274, *279f*, 284, *285f*
  pyrene *271f*
  thiophene *271f*
hydrogen (H)
  chondrites *49t*
  CI chondrites *49t*
  comet Halley 680, *681t*
  comets *675t*
  cosmochemical classification *46t*
  equations of state 627
  giant planets 626, 630
  hydrogen peroxide (H_2O_2) *675t*
  interstellar medium (ISM) 57
  isotopic abundances *383f*
  nucleosynthesis 7
  solar isotopic abundances 131
  solar photosphere *44t*

solar system *132t*
  temperature–pressure profiles 627–629, *628f*
hydrothermal vents 546
hypersthene *500t*

ice giants 624
ilmenite
  CO chondrites 262
  differentiated achondrites 309, 315
  mesosiderites 311
  Moon 563
  Venus *500t*
indium (In)
  chondrites *49t*
  cosmochemical classification *46t*
  lunar mare basalts 573, *574f*
  Martian mantle composition *611f*
  Martian meteorites *606t*
  solar photosphere *44t*
  volatile element depletions *512f*
infrared interferometric spectrometer (IRIS) 626
in situ measurements
  alpha-proton X-ray spectrometer (APXS) 603, *603f*
  Mars 603, 607
  X-ray fluorescence (XRF) 603
International Ultraviolet Explorer (IUE) 57
interplanetary dust particles (IDPs) **689–711**
  background information 84
  chondrites 679
  classification 103
  comets 679, *680f*
  composition
    isotopes 704, *706f*
    major elements 702, *703f, 703t*
    noble gases 706, *707f*
    solar ^{3}He/^{4}He ratio 706
    trace elements 704, *705f*
  cosmic-ray exposure ages 369
  CP IDPs
    carbon compounds *680f*, 695–696, *697f*
    enstatite crystals *695f*
    infrared spectra *693f*
    mineralogy 692, *694f*
    nanodiamonds 696–697, *697f*
    pyrrhotite crystals 693–695, *696f*
    solar flare tracks 692
    sulfide minerals 693
  CS IDPs
    infrared spectra *693f*
    latticed silicates 699, *700f*
    mineralogy *693f*, 699
    sulfide minerals 699–700
  definition 369
  density 691
  electron micrographs *690f, 691f*
  elemental abundances 55
  general discussion 689
  mineralogy
    CP IDPs 692
    CS IDPs 699
    equilibrated aggregates 699
    glass with embedded metal and sulfides (GEMS) *694f*, 697, *698f*
    olivine 692, *693f*

interplanetary dust particles (IDPs)
(*continued*)
  polyaromatic hydrocarbons
    (PAHs) 695–696
  pyroxenes 692, *693f*
  reduced aggregates 699
  morphology 691
  optical properties *693f*, 701, *701f*,
    702f
  particle size 691
  polyaromatic hydrocarbons
    (PAHs) 695–696
  porosity 691
  Poynting–Robertson (PR) drag
    690
  presolar grains *19t*, 680
  solar flare tracks 690, *691f*, 692
interstellar medium (ISM)
  chemical composition 57
  hydrogen (D/hydrogen (D/H)
    ratio *684f*
  dust composition 56, 57
  elemental abundances 17–18, 55
  elemental depletions 57, 59
  gas phase
    magnesium silicates *58t*
    metallic iron (Fe) *58t*
    refractory elements *58t*
    volatile elements 57, *58t*, 59,
      59f
  oxygen abundances 59
iodine (I)
  chondrites *49t*, 249, 250
  cosmic-ray exposure ages *348t*
  cosmochemical classification *46t*
  iodine/xenon (I/Xe) isotopic
    ratios 249, 250, 448
  isotopic abundances *383f*
  Martian meteorites *606t*
  measurement units *348t*
  radiogenic isotopes *384t*, 385
  short-lived radionuclides *433t*
  solar photosphere *44t*
  solar system evolution 448
iridium (Ir)
  achondrites *293t*, *294f*, 306–307,
    310
  chondrites *49t*, *89f*
  cosmochemical classification *46t*
  igneous rocks *576f*
  iron meteorites 328, *328t*,
    330–331, *331f*
  lunar mare basalts *565t*, 573–575,
    574f, *575f*, *576f*
  lunar regolith *566t*, *582f*
  Martian meteorites *606t*
  Martian rocks *576f*
  solar photosphere *44t*
iron (Fe)
  achondrites *293t*, *294f*
  calcium–aluminum-rich
    inclusions (CAIs) 446
  chondrites *49t*, *89f*, *181f*
  cosmochemical classification *46t*
  interplanetary dust particles
    (IDPs) *703f*, *703t*
  interstellar medium (ISM) *58t*, *59f*
  iron oxide (Fe$_2$O$_3$) *604t*
  iron oxide (FeO)
    chondrite matrix composition
      182t
    lunar mare basalts 564, *565t*
    lunar regolith *566t*

Mars *602t*
Martian mantle composition
  610t
Mercury 478, *481t*
Venus *499t*, 501, *505t*
iron sulfide (FeS) 46, *610t*
KREEP composition *578t*
magnetite (Fe$_3$O$_4$)
  chondrites 249
  CO chondrites 262
  CV chondrites 258
  interplanetary dust particles
    (IDPs) 699–700
  Mars *602t*, *604t*, *610t*, *611f*
Martian meteorites *606t*
meteorite formation 46
Moon 563–564
nucleosynthesis 7, 12–13, *12f*
pallasites 338
short-lived radionuclides *433t*
solar nebula 464
solar photosphere *44t*
solar system *132t*
solar system evolution 446, *447f*
troilite 333
Venus *505t*
volatile element depletions *512f*
isobutane *278t*
isoquinoline *271f*
isovaline (C$_5$H$_{11}$NO$_2$) *273t*
Itqiy achondrite *292t*, 315
IUE *see* International Ultraviolet
  Explorer (IUE)

Jupiter
  atmospheric chemistry 625
  Callisto
    atmospheric chemistry 647,
      652t
    carbon dioxide (CO$_2$)
      concentrations 645–646
    comet fragmentation 669–670
    physical properties *640t*
    spectral reflectance *641f*, *646f*
    surface composition 645, *652t*
    water/ice absorption spectra
      648f
  composition 470
  core accretion 471
  hydrogen (D/hydrogen (D/H)
    ratio *684f*
  elemental abundances 630, *631t*,
    633
  equations of state 629
  Europa
    atmospheric chemistry 648,
      652t
    magnetic field 647
    magnetospheric interactions
      648
    physical properties *640t*
    spectral reflectance *641f*
    surface composition 647, *652t*
    water/ice absorption spectra
      647, *648f*
  future research 634
  Ganymede
    atmospheric chemistry 647,
      652t
    comet fragmentation 669–670
    magnetic field 644–645
    physical properties *640t*
    spectral reflectance *641f*, *646f*

    surface composition 645, *652t*
    water/ice absorption spectra
      648f
  helium abundances 626
  historical investigations 624
  hydrogen (H) abundances 626
  interior models 625–626
  interior structure 630
  Io
    atmospheric chemistry 643,
      652t
    geyser plume systems 643
    magnetospheric interactions
      643
    physical properties *640t*
    spectral reflectance *641f*, *643f*
    sulfur dioxide (SO$_2$)
      concentrations 642, 643
    surface composition 642, *652t*
    temperature variations 643
    volatile elements 641
    volcanism 571, 642, 643, *644f*
  isotopic abundances 631
  magnetic field 632
  origins 633
  oxygen (O) abundances 633–634
  physical parameters *624t*
  planetary satellites
    atmospheric chemistry *652t*
    physical properties *640t*
    radiation effects 641
    silicate mass fractions *654f*
    spectral reflectance 639, 641,
      641f
    surface composition *652t*
    volatile elements 641
  size 627, *628f*
  temperature–pressure profiles
    628f
  thermal infrared emissions 629
  wind systems 631–632

kamacite
  CO chondrites 262
  CV chondrites 258
  differentiated achondrites 305,
    315
  iron meteorites 327, 336
Kapoeta 369
ketene (C$_2$H$_2$O) *675t*
ketones *272t*, 276
kirschsteinite 257, 303
krypton (Kr)
  anomalous exotic components
    396
  chondrites *49t*, 131, *383f*
  cosmic-ray exposure ages *348t*,
    369
  cosmochemical classification *46t*
  giant planets *631t*
  interstellar medium (ISM) *58t*, *59f*
  isotopic abundances 131, 398
  Mars *605t*, 614
  measurement units *348t*
  presolar grains 24, *25f*, 28
  radiogenic isotopes 384, *384t*
  solar photosphere *44t*, 45
  solar wind concentrations 390,
    392
  spallation 386
  Venus *492t*, 494
Kuiper Belt 624–625, 667, *668f*

Index

Kuiper Belt Objects (KBOs) 667, 668, *669f*, 671, *672f*
Kuiper, Gerald 667

lanthanum (La)
  achondrites *293t*, *294f*
  calcium–aluminum-rich inclusions (CAIs) *230f*
  chondrites *49t*, *89f*, *230f*
  differentiated achondrites *304f*, *306f*, *308f*
  KREEP composition *578t*
  lunar mare basalts *565t*, *577f*
  lunar regolith *566t*
  Martian meteorites *606t*
  solar photosphere *44t*
  ureilites 313
  volatile element depletions *512f*
lead (Pb)
  chondrites *49t*
  cosmochemical classification *46t*
  early Earth history 540, *540f*
  formation timescales 451
  interstellar medium (ISM) *58t*, *59f*
  lunar mare basalts *570t*, *574f*
  Martian meteorites *606t*
  solar photosphere *44t*
  uranium/lead (U/Pb) isotopic ratios 435–436
  volatile element depletions *512f*
Leonard, F. C. 667
Leonids 670
leucine ($C_6H_{13}NO_2$) *273t*
leucite *500t*
Le Verrier, U. J. J. 624–625
lherzolites 118, 369, *369t*, *370f*
lithium (Li)
  chondrites *49t*
  cosmochemical classification *46t*
  interstellar medium (ISM) *58t*, *59f*
  isotopic abundances *383f*
  KREEP composition *578t*
  Martian meteorites *606t*
  solar photosphere *44t*
lithophile elements
  brachinites 108
  chondrites *89f*, 150, *150f*, *176f*
  differentiated achondrites 304, 307–308, 309–310
  iron silicates, IIE 316
  primitive achondrites *294f*, 298–300, *300f*
  ureilites 313
  volatile elements *512f*
lodranites
  age determination 301
  chemical composition *293t*
  cosmic-ray exposure ages *360f*, 369
  metal–sulfide system melts 298, *298f*
  mineralogy 104, *104f*, 294
  oxygen isotopic composition 140, *140f*
  petrologic characteristics *105t*, *292t*
  silicate-oxide system melts 298–300, *300f*
  trapped noble gases 300–301, *301f*
  veining 333
Lowell, P. 601
Luna missions **559–599**

lutetium (Lu)
  achondrites *293t*, *294f*
  calcium–aluminum-rich inclusions (CAIs) *230f*
  chondrites *49t*, *89f*, *230f*
  differentiated achondrites *304f*, *306f*, *308f*
  Earth formation 544
  KREEP composition *578t*
  lunar mare basalts *565t*, *577f*
  lunar regolith *566t*
  Martian meteorites *606t*
  solar photosphere *44t*
  volatility 411
Lyttleton, R. 653–654

Magellan spacecraft 488, *489t*
magnesite 503
magnesium (Mg)
  achondrites *293t*, *294f*, 333
  calcium–aluminum-rich inclusions (CAIs) 237, *237f*, *238f*, 241
  chondrites *49t*, *89f*, *181f*
  chondrules *240f*
  comet Halley 680, *681t*
  cosmochemical classification *46t*
  formation timescales 451
  interplanetary dust particles (IDPs) *703f*, *703t*, 704
  interstellar medium (ISM) *58t*, *59f*
  isotopic fractionation *414f*
  KREEP composition *578t*
  magnesium oxide (MgO)
    chondrite matrix composition *182t*
    evaporation processes 416, *417f*, *418f*, *419f*, *420f*, *422f*, *423f*
    igneous rocks *576f*
    lunar mare basalts 564, *565t*, *576f*
    lunar mare pyroclastic glasses 573, *573f*
    lunar regolith *566t*
    Mars *576f*, *602t*, *604t*, *610t*
    Martian mantle composition *610t*
    Mercury *481t*
  Martian mantle composition *611f*
  Martian meteorites *606t*
  meteorite formation 46
  Moon 563–564
  nucleosynthesis 7
  refractory inclusions *451f*
  short-lived radionuclides 433
  solar photosphere *44t*
  solar system *132t*
  Venus *499t*, *505t*
magnetite (Fe_3O_4)
  chondrites 249
  CO chondrites 262
  CV chondrites 258
  interplanetary dust particles (IDPs) 699–700
  Venus 503
majorite *611f*
manganese (Mn)
  achondrites *293t*, *294f*
  calcium–aluminum-rich inclusions (CAIs) 444, *445f*
  chondrites *49t*, *89f*, *181f*, 249, 250, *250f*

cosmic-ray exposure ages *348t*
cosmochemical classification *46t*
formation timescales 451
interstellar medium (ISM) *58t*, *59f*
KREEP composition *578t*
lunar mare basalts *574f*
lunar regolith *566t*
manganese/chromium (Mn/Cr) ratios 249, 250, *250f*, 444, *445f*, 453
manganese oxide (MnO)
  chondrite matrix composition *182t*
  Mars *602t*, *604t*, *610t*
  Martian mantle composition *610t*
  Mercury *481t*
  Venus *499t*, *505t*
Martian mantle composition *611f*
Martian meteorites *606t*
measurement units *348t*
short-lived radionuclides *433t*
solar nebula 464
solar photosphere *44t*
solar system evolution 444, *445f*
volatile element depletions *512f*
mantle (Earth)
  elemental abundances 610, *611f*
  magma oceans 534
  magnesium/silicon (Mg/Si) ratios 529
  melt extraction 543
  planetary differentiation 534
  primitive mantle composition 529
  volatile elements 616
marine carbonate sediments 546
Mariner 10 spacecraft *489t*
Mariner 4 spacecraft 601
Mariner 5 spacecraft *489t*
Mariner 9 spacecraft 601
Mars
  Amazonian system 605
  atmospheric loss 614–615, *615f*, 617
  chondrites 610, *611f*
  core
    geochemistry 607, 610, *610t*, *611f*
    geochronology 616
    planetary differentiation 616, *617f*
  crust
    composition 616
    geochemistry 610, 613
    geochronology 616
    planetary differentiation 616, *617f*
    stratigraphy 605
  density 603, 610, *610t*, *611f*
  dust 607, *612f*, 613
  Elysium 605
  geochemistry
    atmospheric gases 603, *605t*, *605f*, 614, 616
    background information 601
    comparison with Earth 616
    core 607, 610, *610t*, *611f*
    crust 610, 613, 616
    data sources 602
    iron oxide (FeO) concentrations *480t*
    mantle 607, 610, *610t*, *611f*, 616

Mars (continued)
   meteorites 607
      soil analysis 603, *604t*, 607, 613
      unsolved problems 617
      volatile cycling 615
      volatile reservoirs 614
   hematite 613
   Hesperian system 605
   hydrologic cycle 605–607
   hydrosphere 616
   lander analyses 603, *604t*
   mantle
      composition *610t*, 616
      elemental abundances 610, *611f*
      geochemistry 607, 610
      geochronology 616
      planetary differentiation 616, *617f*
      volatile inventory 614
   Mars Pathfinder lander *603f*, *604t*
   Martian meteorites
      aluminum depletion 607–609, *609f*
      assimilated crust 612, *613f*
      background information 602
      cosmic-ray exposure ages 369, *369t*, *370f*
      crystallization 616–617
      elemental abundances 603, *606t*
      geochemistry 607
      historical background 16
      iron oxide/magnesium oxide (FeO/MgO) ratio *118f*
      manganese (Mn) isotopes 444, *445f*
      mineralogy 116, *119f*
      organic compounds 615
      outgassing 614
      oxygen isotopic composition *118f*, 138, *139f*, 609, *609f*
      parent bodies *374f*
      sources 607
      taxonomy *84f*, *85t*
      volatile cycling 615
      volatile elements *574f*
      volatile inventory 614
      volatile/refractory ratio 607–609, *609f*
      volcanic rock classification *612f*
   mineralogy 602–603
   moment of inertia 603, 610, *610t*
   Noachian system 605
   Oxia Palus region 613
   oxygen (δ^{18}O) isotope concentrations *514f*
   in situ measurements 603, *603f*
   spacecraft measurements 602, *602t*
   stratigraphy 605
   tectonics 605, 617
   Tharsis 605
   topography *608f*
   volatile elements 469
   volatile reservoirs 614
Mars Global Surveyor 602
Mars Odyssey spacecraft 602
Mars Pathfinder lander 602, 603, *603f*, *604t*, 610

melilite
   see also chondrites
   calcium–aluminum-rich inclusions (CAIs) 202, 203, *203f*, 206, *207t*, *208f*, 239
CH chondrites *227f*
CO chondrites 262
condensation/fractionation processes *410f*, *411f*
CV chondrites 257
evaporation processes 422
mineral chemistry 206, *208f*
Mercury **477–485**
   background information 477
   chemical composition
      general discussion 478
      iron oxide (FeO) concentrations 478, *480t*
      iron/silicon (Fe/Si) ratios 478
      models *481t*
      refractory elements 481, 482
      spectral analysis 478
      volatile elements 480, 482
   compositional models
      enstatite chondrites 483
      evaporation models 481
      metal enrichment 480
      metal-rich chondrites 482, *483f*
      refractory condensation models 481
   density 478
   formation proposals 480–481, 483
   lava flows 479, *481t*
   smooth plains 479, *479f*
mercury (Hg)
   CI chondrites *49t*
   cosmochemical classification *46t*
   Martian meteorites *606t*
   solar photosphere *44t*
   volatile element depletions *512f*
mesosiderites
   age determination 312
   chemical composition *293t*, 329
   cooling rates *337t*
   cosmic-ray exposure ages 369, *371f*
   formation conditions 312
   lithology 311
   matrix 311
   mineralogy 112, *112f*, *113f*, 311
   parent body size *337t*
   petrologic characteristics *105t*, *113t*, *292t*
   petrology *105t*, *113t*, *292t*
   taxonomy *84f*, *85t*
meteorites
   see also achondrites; chondrites; Mars
   ALH84001
      cosmic-ray exposure ages 369, *370f*
      elemental abundances 603–604, *606t*
      mineralogy 120
      oxygen isotopic composition 138, *139f*
      taxonomy *84f*, *85t*
   ALH84025 307
   ALHA77081 *293t*
   background information 84
   calcium–aluminum-rich inclusions (CAIs) 132–133, 464

Chassigny meteorite
   cosmic-ray exposure ages 369, *369t*, *370f*
   elemental abundances *606t*
   mineralogy *118f*
   origins 603
classification *84f*, *85t*, 86
comets 678
composition 462
cosmic-ray exposure ages **347–380**, 388
   basic calculations 369
   measurement units 369
   production rate calibration 369
differentiated achondrites
   elemental abundances 45
   mineralogy 107
   oxygen isotopic composition 138, *139f*
differentiated meteorites
   chemical composition 329
   oxygen isotopic composition *106f*
   taxonomy *84f*, *85t*
dunites *84f*, *85t*, 339, 603
elemental abundances 45
   comparison with solar system abundances *44t*, 53, *54f*
   historical background 41–42
   isotopic anomalies 42
   trapped noble gases 398, *399f*
formation conditions
   cosmochemical classification 45, *46t*
   magnesium silicates 46
   metallic iron (Fe) 46
   oxygen fugacity 47
   refractory elements 46, 47
   volatile elements 46, *47f*
historical background xv
interplanetary dust particles (IDPs) 689
iron meteorites **325–345**
   age determination 302
   CB chondrites 331–332
   chemical composition 327
   chemical groups 114
   classification 327
   cosmic-ray exposure ages 369, *372f*
   IVA group 328, *328t*, *331f*, *337t*
   IVB group 328, *328t*, *331f*, *337t*
   fractional crystallization 333, *334f*, 335, *335f*, *336f*
   future research 340
   hafnium/tungsten (Hf/W) isotopic ratios 524, *524t*
   IAB group 329, *337t*
   inclusion composition 301
   metallic melting process 332, *332f*, *333f*
   mineralogy *104f*, 106, 107, 114, *115f*, *117f*, 301
   IAB group *328t*
   oxygen isotopic composition 138, *139f*
   palladium (Pd) isotopes 447
   parent bodies 330, 332, 333, 339
   petrologic classification *105t*
   short-lived radionuclides *433t*
   silicate-bearing groups 116, 329

Index

spectral reflectance 339, *339f*
structural groups 115
sulfur-rich meteorites 336
taxonomy *84f, 85t*
 IIIAB group 327, *328t*, 333, *334f, 336f, 337f*
 IIICD group 329
 IIAB group 327, *328t*
 IIE group *292t*, 315, 329
 ungrouped iron meteorites 329, *331f*, 340
 volatile elements 330
 Widmanstätten pattern 336, *337f*
lunar meteorites
 chronology *570t*
 composition 369
 cosmic-ray exposure histories *362t*, 369, 561
 known meteorites *562t*
 mineralogy 120, *120f*
 oxygen isotopic composition *118f*
 parent bodies *374f*
 production rates 369
 taxonomy *84f, 85t*
Murchison chondrite **269–290**
nonchondritic 103, *116t*, 184
oxygen (O) isotopic abundances **129–142**
pallasites
 chemical composition 338, *338f*
 cooling rates *337t*
 cosmic-ray exposure ages 369, *371f*
 lithology 338
 metallic melting process 338–339
 mineralogy 113, *114f*
 origins 338–339
 oxygen isotopic composition 138, *139f*
 palladium (Pd) isotopes 447
 parent body size *337t*
 petrologic classification *105t*
 short-lived radionuclides *433t*
 taxonomy *84f, 85t*
parent bodies
 chemical composition 339–340
 condensation/fractionation processes 330
 cooling rates 336, *337t*
 core formation 332
 fractional crystallization 333
 iron meteorites 339
 size 336, *337t*
 spectral reflectance 339, *339f*
 stony iron meteorites 339
 volatile element depletions 421
polycyclic aromatic hydrocarbons (PAHs) 615
presolar grains **17–39**, 56, 394
primitive achondrites
 classification 104
 mineralogy *104f*
 oxygen isotopic composition *106f*
Q-component 400
radiogenic isotope concentrations 385
refractory elements
 formation conditions 46, 47

volatile/refractory ratio 524, *525f*, 607–609, *609f*
solar noble gases 388, 390
stony iron meteorites *84f, 85t*, **325–345**, 338, 339
Tagish Lake meteorite shower 679
IID meteorties *85t*
iron silicates, IIE
 age determination 317
 bulk composition 316
 chemical composition *293t*
 formation conditions 317
 inclusion composition 315
 mineralogy 315
 petrologic characteristics *292t*
 texture 316
 volatile elements 316
undifferentiated achondrites 45
 elemental abundances 45
 oxygen isotopic composition 140, *140f*
meteor streams 670
methane (CH_4)
 comets 673–674, *675t*
 giant planets 630, *631t*
 Murchison chondrite *272t, 278t*
 sublimation temperatures *674t*
methyl alcohol (CH_4O) *675t*
Michel, R. 350
micrometeorites 369, *373f*
mid-ocean ridge basalts (MORBs) 499, *500t*
models
 chondrite formation 463
 chondrule formation 188, *188t*
 lunar crust formation 587, 590
 Martian core composition 610, *610t, 611f*
 Martian mantle composition 610, *610t*
 Mercury compositional models 480
 terrestrial planet formation models 519
molecular clouds *see* precollapse clouds
molybdenum (Mo)
 calcium–aluminum-rich inclusions (CAIs) 231
 chondrites *49t*
 cosmochemical classification *46t*
 lunar mare basalts *574f*
 Martian mantle composition *611f*
 Martian meteorites *606t*
 meteorites 42–43, 130
 presolar grains 24, *28f*
 solar photosphere *44t*
monticellite *208t*, 262
montmorillonite 255–256
monzodiorites *584f*, 586
Moon
 age determination 537, *538t*
 Apollo missions **559–599**
 bombardment history 536, 578
 bulk composition 591
 Clementine mission 562
 Copernicus crater 579
 Crisium 580
 formation conditions 468, 535
 general discussion 559
 granulitic breccias 560, *579f*

hafnium/tungsten (Hf/W) isotopic ratios *526t*
highland crust
 age determination 537, *538t*, 578
 cratering rate 579
 dunites 586
 felsite *584f*
 ferroan anorthosites 583, *585f*, 588
 formation models 590
 gabbronorites *584f*, 586
 genesis chronology 586–587, *586f*
 hafnium/tungsten (Hf/W) isotopic ratios 537–539, *539f*
 KREEP 577, *578t*, 583, *584f*, *585f*, 588, 592
 magma ocean hypothesis 587
 monzodiorites *584f*, 586
 neodymium (Nd) isotopes 589, *590f*
 nickel/iridium (Ni/Ir) isotopic ratios 581–582
 norites *584f*, 586
 polymict breccias 577, *579f*, 581, 583
 pristine rocks 583, *584f*
 rubidium/strontium (Rb/Sr) isotopic ratios *523f*, 537, 542
 siderophile elements 581
 strontium (Sr) isotope composition *539f*
 thickness 592
 troctolites *584f*, 586
Imbrium 580
impact-melt breccias 560, 578, *579f*, 581
Luna missions **559–599**
lunar geochemical database
 artificially acquired samples 561
 lunar meteorites 561
 remote sensing data 562
 spectral analysis 562–563
lunar meteorites
 chemical composition *565t*
 chronology *570t*
 composition 369
 cosmic-ray exposure histories *362t*, 369, 561
 known meteorites *562t*
 mineralogy 120, *120f*
 oxygen isotopic composition *118f*
 parent bodies *374f*
 production rates 369
 taxonomy *84f, 85t*
mantle *526t*, 572
mare volcanism
 chemical composition *564f*, *565t, 566t, 568f, 572f*
 chronology 569, *570t, 571f*
 KREEP 563–564, 569–571, 575–576, *577f*
 lunar mare basalts *565t*
 mare basalts 564, *564f, 568f*, *570t, 571f*
 phenocrysts 569–571
 pyroclastic glasses 568–569, *570t*, 571, *571f, 573f*
 rare earth elements (REEs) 575–576, *577f*

728 Index

Moon (*continued*)
  rock types 564
    siderophile elements 573–575, *574f*, *575f*
    trace elements 573
    volatile elements *574f*
  Nectaris 580
  oxygen ($\delta^{18}O$) isotope concentrations 514, *514f*, 541
  Procellarum 577, 580
  regolith 560, *566t*, 581, *582f*, *585f*
  Serenitatis 580
  Tycho crater 579
  volatile element depletions 536
  volatile/refractory ratio 524, *525f*

nakhlites
  composition 603–604, *612f*
  cosmic-ray exposure ages 369, *369t*, *370f*
  crystallization 616–617
  elemental abundances *606t*
  mineralogy 118
  taxonomy *84f*, *85t*
nanodiamonds 33, 696–697, *697f*
NanoSIMS 19–20, 34, 706
naphthalene *271f*, *278t*
neodymium (Nd)
  achondrites *293t*, *294f*
  calcium–aluminum-rich inclusions (CAIs) *230f*
  chondrites *49t*, *230f*
  differentiated achondrites *304f*, *306f*, *308f*
  isotopes
    Earth formation 543
    lunar mare basalts *570t*
    presolar grains 24, *25f*
  KREEP composition *578t*
  lunar regolith *566t*
  Martian meteorites *606t*
  Murchison chondrite 42–43, *42f*
  solar photosphere *44t*
neon (Ne)
  anomalous exotic components 18–19, *18f*, 396
  chondrites *49t*, 388, *394f*, 398, *399f*
  cosmic-ray exposure ages *348t*, 350, *356f*, 369
  cosmochemical classification *46t*
  fractionation 533–534
  giant planets *631t*
  interplanetary dust particles (IDPs) 706
  isotopic abundances 131, *383f*, 398
  mantle (Earth) 533–534
  Mars *605t*
  nebular protoatmosphere 530
  nucleosynthesis 7, 8
  plagioclase *387f*
  presolar grains 23, 28
  radiogenic isotopes *384t*
  solar photosphere 43–45, *44t*
  solar wind concentrations *387f*, 390, 392
  spallation 386
  trapped components *399f*
  Venus *492t*, *494t*

nepheline
  calcium–aluminum-rich inclusions (CAIs) 203, 204–205, *208t*, 228
  CO chondrites 261–262
  CV chondrites 257
  differentiated achondrites 304
  Venus *500t*
Neptune
  atmospheric chemistry 625
  composition 470
  core accretion 472
  hydrogen (D/hydrogen (D/H) ratio *684t*
  elemental abundances 630, *631t*, 633
  equations of state 629
  future research 634
  helium abundances 626
  historical investigations 624
  hydrogen (H) abundances 626
  interior models 625–626
  interior structure 630
  isotopic abundances 631
  magnetic field 632
  origins 633
  physical parameters *624t*
  planetary satellites
    atmospheric chemistry *652t*
    physical properties *640t*
    radiation effects 641
    spectral reflectance 639
    surface composition *652t*
    volatile elements 641
  size 627, *628f*
  thermal infrared emissions 629
  Triton
    atmospheric chemistry *652t*, 655
    density 654
    geological processes 655
    geyser plume systems 655
    historical background 653
    orbit 653–654
    physical properties *640t*
    seasons 656
    surface composition *652t*, 654
    volatile elements 641
    water/ice absorption spectra *655f*
  wind systems 631–632
nickel (Ni)
  achondrites *293t*, *294f*
  calcium–aluminum-rich inclusions (CAIs) 446
  chondrites *49t*, *89f*, *181f*
  condensation/fractionation processes 330, *331f*
  cosmic-ray exposure ages *348t*
  cosmochemical classification *46t*
  fractional crystallization 333
  igneous rocks *576f*
  interplanetary dust particles (IDPs) *703t*, 704, *705f*
  interstellar medium (ISM) *58t*, *59f*
  iron meteorites *328t*, *331f*, 333, 336
  lunar mare basalts *565t*, *574f*, *576f*
  lunar regolith *566t*
  Martian core composition *610t*
  Martian mantle composition *611f*
  Martian meteorites *606t*
  Martian rocks *576f*

nickel oxide (NiO) *182t*
pallasites 338
solar photosphere *44t*
solar system evolution 446, *447f*
Venus *505t*
volatile element depletions *512f*
niningerite 305
niobium (Nb)
  chondrites *49t*
  cosmochemical classification *46t*
  Earth formation 543
  KREEP composition *578t*
  Martian meteorites *606t*
  short-lived radionuclides *433t*
  solar photosphere *44t*
  solar system evolution 448
nitrogen (N)
  acetonitrile (CH$_3$CN) *675t*
  ammonia (NH$_3$)
    comets *675t*
    giant planets 630, *631t*
    Murchison chondrite *272t*
    sublimation temperatures *674t*
  chondrites *49t*, 88, *91f*
  comet Halley 680, *681t*
  comets 673–674, *675t*
  cosmochemical classification *46t*
  cyanic acid (HNCO) *675t*
  hydrogen (D cyanide (DCN) *675t*
  formamide (CH$_3$NO) *675t*
  giant planets 630, *631t*
  heterocyclic compounds *271f*, *272t*, 274, 281
  hydrogen cyanide (HCN) *675t*
  interplanetary dust particles (IDPs) 704, *706f*
  interstellar medium (ISM) *58t*, *59f*
  Mars *605t*, 614
  molecular nitrogen (N$_2$)
    comets 673–674
    sublimation temperatures *674t*
    Venus *492t*
  mononitrogen monosulfide (NS) *675t*
  Murchison chondrite 280
  nitrogen oxides (NO$_x$) 494
  nucleosynthesis 3, 7
  presolar grains *21f*, 22, 25, 26, 28, 30
  propiolonitrile (C$_3$HN) *675t*
  silicon nitride (Si$_3$N$_4$) *19t*, 29
  solar isotopic abundances 131
  solar photosphere *44t*
  solar system *132t*
  Venus *492t*, 494, *494t*, 495
Noachian system 605
noble gases **381–405**
  abundances 381–382
  chondrites 531
  cosmic ray exposure 386
  enstatite chondrites 402
  exotic components
    anomalous components 18–19, *18f*, 395, *395f*, *396t*, 397
    isotopic composition 393
    presolar grains 394, *395f*
  interplanetary dust particles (IDPs) 706, *707f*
  isotopic abundances 382, *383f*, *387f*
  measurement techniques 382
  nuclear components
    radiogenic isotopes 383, *384t*

spallation 386, 390, *391f*
nuclear processes 382
partitioning 381
planetary atmospheres 403
planetary noble gases 398
in situ components 383
solar noble gases
  hydrogen (D (^{2}H)) burning 389
  general discussion 388
  helium/helium (^{3}He/^{4}He) ratio 389
  lunar soils 390, *391f*
  meteorites 388
  solar wind concentrations 390, 392
terrestrial planets
  abundances 494
  degassed protoatmosphere 532
  nebular protoatmosphere 530
  planetary degassing 532
  protoatmosphere 532
trapped noble gases *399f*
  acapulcoite–lodranite achondrites 300–301, *301f*
  brachinites 308
  planetary noble gases 398
  Q-component 398–400, *399f*, 400
  trapping mechanisms 401
  ureilites 314
ureilites 402
Venus 494
norites *584f*, 586
Northwest Africa 011
  bulk composition 315
  mineralogy 315
  oxygen (O) isotope composition 315
  petrologic characteristics *292t*
nucleosynthesis **1–15**
  asymptotic giant branch (AGB) evolutionary phases 1–2, 130
  backgrund information 1
  elemental abundance patterns 12, *12f*, *13f*
  explosive phase 5, *7t*, 8
  intermediate-mass stars
    carbon (^{12}C) isotope production 4
    evolution 3
    helium burning 4
  interstellar medium (ISM) 17–18
  isotopic abundances 2, *3f*
  massive stars
    carbon burning 8
    evolution 5, *5f*
    helium burning 7
    hydrogen burning 7
    interior structure *6f*
    neon burning 8
    nuclear burning stages 5, *7t*, 8
    oxygen abundances 7, 129–130
    oxygen burning 8
    p-process 9
    production factor *10f*
    r-process patterns 9
    silicon burning 8
    s-process patterns 7
  mechanisms 2–3
  nuclear burning stages
    massive stars 5, *6t*
    supernovae 5, *7t*, 8

red giants
  carbon (^{12}C) isotope production 4
  evolution 3
  helium burning 4
  r-process patterns 4
  s-process patterns 4, 130

oceans, hydrogen (D/hydrogen (D/H) ratio *684f*
Oddo–Harkins rule 41–42
oldhamite 305, 315
olivine
  amoeboid olivine aggregates (AOAs) 219
  calcium–aluminum-rich inclusions (CAIs) 206, *207t*
  comets 675, *677f*
  CV chondrites 257
  differentiated achondrites 303, 307, 309, 315
  evaporation processes 416, *417f*
  interplanetary dust particles (IDPs) 692, *693f*
  iron silicates, IIE 316
  isotopic fractionation 416, *417f*
  Martian core composition *611f*
  Martian mantle composition *610t*
  mesosiderites 311
  pallasites 338
  presolar grains *19t*
  primitive achondrites 296, 301, 302
  ureilites 313
  Venus *500t*
Oort cloud 666, *667f*
orthoclase *500t*
orthopyroxenes
  CO chondrites 261–262
  differentiated achondrites 309
  iron silicates, IIE 316
  mesosiderites 311
  mineralogy 307
  pallasites 338
  primitive achondrites 296, 301, 302
  ureilites 313
orthopyroxenites
  ALH84001 603–604
  cosmic-ray exposure ages 369, *369t*
  differentiated achondrites 305, 309
  mineralogy 120
  taxonomy *84f*, *85t*
osmium (Os)
  achondrites *293t*, *294f*
  chondrites *49t*, *89f*
  cosmochemical classification *46t*
  iron meteorites 328, 330–331
  lunar mare basalts 573–575, *574f*
  lunar regolith *566t*
  Martian meteorites *606t*
  primitive mantle composition 468
  solar photosphere *44t*
  volatility 411–412
Oxia Palus region 613
oxygen (O)
  calcium–aluminum-rich inclusions (CAIs) 133, *133f*, 232, *233f*, *234f*
  chondrites *49t*, 88, *89f*, 133, *236f*, 250–251, *251f*, 255, *255f*
  comet Halley 680, *681t*

cosmochemical classification *46t*
fractionation 129, *130f*, 135
giant planets 630
interplanetary dust particles (IDPs) *703t*, 704, *706f*
interstellar medium (ISM) *58t*, *59f*
Jupiter 633–634
Mars *602t*, *605t*, 615, *615f*
meteorite formation 47
meteorites **129–142**
meteoritic composition 609, *609f*
Moon 563–564
nucleosynthesis 2, 7, 8, 129–130
oxygen (δ^{18}O) isotopes 235, *236f*
oxygen fugacity 47
presolar grains 30, 32, 232–233
production factor *10f*
solar isotopic abundances 131
solar nebula 131, 134
solar photosphere 43, *44t*
solar system *132t*
Venus *494t*, 495, *505t*

palagonite 613
palladium (Pd)
  calcium–aluminum-rich inclusions (CAIs) 447
  chondrites *49t*
  cosmochemical classification *46t*
  lunar mare basalts 573–575
  Martian meteorites *606t*
  primitive mantle composition 468
  short-lived radionuclides *433t*
  solar nebula 464
  solar photosphere *44t*
  solar system evolution 447
  volatile element depletions *512f*
pallasites
  chemical composition 338, *338f*
  cooling rates *337t*
  cosmic-ray exposure ages 369, *371f*
  lithology 338
  metallic melting process 338–339
  mineralogy
    Eagle Station grouplet 113
    general discussion 113
    main-group pallasites 113, *114f*
    pyroxene-pallasite grouplet 114
  origins 338–339
  oxygen isotopic composition 138, *139f*
  palladium (Pd) isotopes 447
  parent body size *337t*
  petrologic classification *105t*
  short-lived radionuclides *433t*
  taxonomy *84f*, *85t*
partitioning 381, 535
pentlandite 258, 262, 699–700
peridotites 535
perovskite
  calcium–aluminum-rich inclusions (CAIs) 206, *207t*, 210
  condensation/fractionation processes *410f*, *411f*
  CV chondrites 257
  evaporation processes 422
  Martian core composition *611f*
  planetary differentiation 534
perylene *278t*
phenanthrene *271f*, *278t*
Phobos-2 spacecraft 602, *602t*

phosphorus (P)
  achondrites *293t, 294f*
  chondrites *49t, 182t*
  cosmochemical classification *46t*
  interstellar medium (ISM) 57, *58t, 59f*
  iron meteorites 330–331, 335, *336f*
  KREEP composition *578t*
  Mars *604t, 610t*
  Martian mantle composition *611f*
  Martian meteorites *606t*
  phosphates
    differentiated achondrites 309
    iron silicates, IIE 316
    primitive achondrites 302
  phosphine (PH_3) 630, *631t*
  phosphonic acids *271f, 272t*
  phosphorus oxide (P_2O_5) *182t, 604t, 610t*
  solar photosphere *44t*
phyllosilicates
  calcium–aluminum-rich inclusions (CAIs) *208t*
  CM chondrites 136–137
  CV chondrites 255
pigeonite
  differentiated achondrites 309, 315
  mesosiderites 311
  ureilites 313
Pioneer Venus spacecraft 488, *489t*
plagioclase
  CO chondrites 261–262
  differentiated achondrites 303, 305, 307, 309, 315
  iron silicates, IIE 316
  Martian crust 613
  mesosiderites 311
  Moon 563
  neon (Ne) isotope concentrations *387f*
  primitive achondrites 296, 301
planetary satellites
  bulk composition 639
  density 639, *640t*
  elemental abundances 638
  future research 656
  radiation effects 641
  spectral reflectance 639
  volatile elements 641
planetesimals
  absolute timescales 437
  accretion 483
  asteroids 665
  chondrule formation 189
  formation timescales 453
  origins
    dust grain composition 465
    dust grain sticking 466
    gravitational instability 465
    ice line 465, 511–512
    turbulent concentrations 466
  terrestrial planet formation 466
  volatile elements 469
planet formation **461–475**
  accretion dynamics
    compositional changes 541
    dust stickiness 516
    general discussion 516
    gravitational instability 517
    impact theories 536, 541, 546
    Kyoto model 526
    large collisions 518
    mass loss 541
    migration 517
    post-impact evidence 542
    runaway growth 518
  geochemical constraints
    chondrite models 527, 528, 529
    hafnium/tungsten (Hf/W) isotopic ratios 522, *523f, 526t*
    magnesium/silicon (Mg/Si) ratios 529, 541
    oxygen (O) isotopic composition 530
    radiogenic isotope models 519
    rubidium/strontium (Rb/Sr) isotopic ratios *523f*
    siderophile elements 542
    uranium/lead (U/Pb) isotopic ratios 521, *522f*
    volatile element depletions 527, 541
    volatile/refractory ratio 524, *525f*
    xenon (Xe) isotope depletions 533, *533f, 540f*
  giant planet formation
    composition 470
    core accretion 471
    disk gap formation 472
    disk instability 471
    gravitational torques 472
    migration 472
  protoplanetary nebula
    accretion 462, 510
    chemical fractionation 463
    circumstellar disks 462, 510
    temperatures 463
    volatile elements 463
  terrestrial planets
    atmospheric loss 533
    core formation 468
    degassed protoatmosphere 532
    density 511
    giant impacts 467
    heterogeneous accretion 515
    late-stage accretion 467
    Moon 468
    nebular gases 511
    nebular protoatmosphere 530
    oligarchic growth 467
    planetary embryos 467
    random velocities 466
    runaway growth 467
    volatile elements 469, 511, *512f*
planets, definition 624
platinum (Pt)
  chondrites *49t*
  cosmochemical classification *46t*
  lunar mare basalts 573–575
  Martian meteorites *606t*
  primitive mantle composition 468
  solar photosphere *44t*
Pluto 624–625, 668–669
plutonium (Pu)
  cosmochemical classification *46t*
  radiogenic isotopes *384t*, 385
  short-lived radionuclides *433t*
  solar system evolution 449
polycyclic aromatic hydrocarbons (PAHs)
  interplanetary dust particles (IDPs) 695–696
  meteorites 615
  Murchison chondrite 274, *279f*, 284, *285f*
potassium (K)
  achondrites *293t, 294f*
  chondrites *49t, 89f, 181t*
  cosmic-ray exposure ages 369
  cosmochemical classification *46t*
  interstellar medium (ISM) *58t, 59f*
  KREEP composition *578t*
  lunar mare basalts 573, *574f*
  lunar regolith *566t*
  Mars *602t, 611f*
  Martian meteorites *606t*
  Mercury 480
  potassium oxide (K_2O)
    chondrite matrix composition *182t*
    lunar mare basalts *565t*
    Mars *602t, 604t*
    Martian mantle composition *610t*
    Mercury *481t*
    Venus *499t, 505t*
  radiogenic isotopes 384, *384t*
  solar photosphere *44t*
  Venus *499t, 505t*
  volatile element depletions *512f*
Poynting–Robertson effect 55
praseodymium (Pr)
  achondrites *293t*
  calcium–aluminum-rich inclusions (CAIs) *230f*
  chondrites *49t, 230f*
  differentiated achondrites *304f, 306f, 308f*
  KREEP composition *578t*
  Martian meteorites *606t*
  solar photosphere *44t*
precollapse clouds
  ambipolar diffusion 65, 66
  angular momentum distributions 65
  astronomical observations 64
  collapse phase 65, 66
  composition 64, *64f*
  density profiles 64–65
  magnetic field support 65, 66
  oxygen abundances 130, 134
  shape 65
  shock-triggered collapse 65–66, 66
presolar grains **17–39**
  analysis techniques 19
  comets 680
  dating techniques 434–435
  diamond(s) 18–19, *19t*, 33
  grain types 19, *19t*, 394
  graphite
    anomalous exotic components 18–19
    interplanetary dust particles (IDPs) *19t*
    isotopic composition 30, *30f*
    physical properties *21f*, 29, *31f*
  historical background 18
  importance 20
  interplanetary dust particles (IDPs) 680, 689
  meteorites 394
  Murchison chondrite *25f*

Index

noble gases 394, *395f*
oxide grains 32, *32f*
oxygen (O) isotopes 30, 32, 232–233
silicon carbide (SiC)
  A + B grains *21f*, *23f*, 26
  aluminum (Al) isotope concentrations *22f*, 23
  barium (Ba) isotope concentrations 24, *25f*
  dysprosium (Dy) isotope concentrations 24, *25f*
  general discussion 20
  grain morphology 20–21, *21f*
  isotopic composition 21, *21f*, *22f*, *23f*, *31f*, *32f*
  krypton (Kr) isotope concentrations 24, *25f*
  mainstream grains 22, *22f*
  molybdenum (Mo) isotope concentrations 24, *28f*
  neodymium (Nd) isotope concentrations 24, *25f*
  neon (Ne) isotope concentrations 23
  nova grains *21f*, *23f*, 28
  origins 21–23, *22f*
  samarium (Sm) isotope concentrations 24, *25f*
  silicon (Si) isotope concentrations 23–24, *23f*
  s-process isotopic patterns 24, *25f*
  strontium (Sr) isotope concentrations 24, *25f*
  titanium (Ti) isotope concentrations 24, *27f*
  trace elements 21
  xenon (Xe) isotope concentrations 24, *25f*
  X grains *21f*, *22f*, *23f*, 26, *27f*, *28f*
  Y grains *21f*, *23f*, 25
  Z grains *21f*, *23f*, 25
silicon nitride (Si_3N_4) *19t*, 29
stellar evolution 21–23, 514, 527
supernovae 28, 30, 32
primitive mantle, magnesium/silicon (Mg/Si) ratios 529
proline ($C_5H_9NO_2$) *273t*
propane (C_3H_8) *278t*
protoplanetary nebula 462
protoplanets 66–67, 68–69
protostars 66
purines *272t*
pyrene *271f*, *278t*
pyridine *271f*
pyridinecarboxylic acids *272t*
pyrimidines *272t*
pyroxenes
  calcium–aluminum-rich inclusions (CAIs) 203, 206, *207t*, 209, *240t*
  CO chondrites 261–262
  comets 676, *677f*
  condensation/fractionation processes *410f*
  differentiated achondrites 303, 305, 307, 315
  europium anomalies 612, *613f*
  interplanetary dust particles (IDPs) 692, *693f*
  Martian crust 613
  Martian mantle composition *610t*
  mineral chemistry 209
  Moon 563
  pallasites 338
  ureilites 313
pyrrhotite
  CV chondrites 258
  interplanetary dust particles (IDPs) 693–695, *696f*, 699–700
  Venus 503

QB_1 (cubewanos) 667
Q-component 398–400, *399f*, 400
Quaoar 667–668
quinolines *271f*, 274

radioactive environmental contamination 449
radiocarbon (^{14}C) *348t*
radiogenic isotopes
  calcium–aluminum-rich inclusions (CAIs) 464
  chondrites 464
  cosmogenic nuclides **347–380**
  dating techniques 434
  formation timescales 451, *452f*
  Mars
    assimilated crust 612, *613f*
    atmospheric composition *605t*
    geochronology 616
    planetary differentiation 616, *617f*
    volatile reservoirs 614
  Martian meteorites 604–605, 609, *609f*,
  meteorites 609, *609f*
  radionuclides 449
  solar nebula
    aluminum (Al) isotopes 78
    beryllium (Be) isotopes 77
    calcium (Ca) isotopes 78
    isotopic anomalies 42
    short-lived isotopes 464
  solar system evolution 432, *433t*
  terrestrial planet formation models 519
radon (^{222}Rn) 381
rare earth elements (REEs)
  calcium–aluminum-rich inclusions (CAIs) 205, 230, *230f*
  condensation/fractionation processes 411, *411f*
  cosmochemical classification *46t*
  differentiated achondrites *304f*, 305, *306f*, *308f*
  evaporation processes 421
  iron silicates, IIE 316–317
  lunar mare basalts 575–576, *577f*
  primitive achondrites 302
  ureilites 313
red giants 3
redox processes 330, 501
refractory elements
  amoeboid olivine aggregates (AOAs) 219
  calcium–aluminum-rich inclusions (CAIs) 231
  chondrites
    abundances *147t*
    chemical composition 150, 463
    composition 175

CI chondrites 51, *52f*
interstellar medium (ISM) *58t*
lunar crust 592
Mercury 481, 482
meteorites
  formation conditions 46, 47
  volatile/refractory ratio 524, *525f*, 607–609, *609f*
  refractory lithophile elements (RLEs) 46, *46t*
  refractory siderophile elements (RSEs) 46, *46t*, 231
rhenium (Re)
  achondrites *293t*
  chondrites *49t*
  cosmochemical classification *46t*
  iron meteorites 328, *328t*, 330–331
  lunar mare basalts 573–575, *574f*
  Martian meteorites *606t*
  solar photosphere *44t*
  volatile element depletions *512f*
  volatility 411–412
rhodium (Rh) *44t*, *46t*, *49t*
rhönite *207t*
Rossby waves 71–72
rubidium (Rb)
  achondrites *293t*, *294f*
  chondrites *49t*
  cosmochemical classification *46t*
  isotopic abundances *383f*
  KREEP composition *578t*
  lunar mare basalts *565t*, *570t*, 573–575, *574f*
  Martian mantle composition *611f*
  Martian meteorites *606t*
  solar photosphere *44t*
  volatile element depletions *512f*
ruthenium (Ru)
  chondrites *49t*, *89f*
  cosmochemical classification *46t*
  solar photosphere *44t*

samarium (Sm)
  achondrites *293t*, *294f*
  calcium–aluminum-rich inclusions (CAIs) *230f*
  chondrites *49t*, *89f*, *230f*
  differentiated achondrites *304f*, *306f*, *308f*
  Earth formation 543
  KREEP composition *578t*
  lunar mare basalts *565t*, *577f*
  lunar regolith *566t*
  Martian meteorites *606t*
  presolar grains 24, *25f*
  short-lived radionuclides *433t*
  solar photosphere *44t*
  solar system evolution 449
saponite 255
sarcosine ($C_3H_7NO_2$) *273t*
Saturn
  atmospheric chemistry 625
  composition 470
  core accretion 471
  hydrogen (D)/hydrogen (D/H) ratio *684f*
  Dione
    atmospheric chemistry *652t*
    physical properties *640t*
    surface composition *652t*
  elemental abundances 630, *631t*, 633

Saturn (*continued*)
  Enceladus
    atmospheric chemistry *652t*
    physical properties *640t*
    surface composition *652t*
  equations of state 629
  future research 634
  helium abundances 626
  historical investigations 624
  hydrogen (H) abundances 626
  Hyperion *650f, 653f*
  Iapetus
    atmospheric chemistry *652t*
    physical properties *640t*
    spectral reflectance *650f*
    surface composition 649–650, *652t*
  interior models 625–626
  interior structure 630
  isotopic abundances 631
  magnetic field 632
  Mimas
    atmospheric chemistry *652t*
    physical properties *640t*
    silicate mass fractions *654f*
    surface composition *652t*
  origins 633
  physical parameters *624t*
  planetary satellites
    atmospheric chemistry *652t*
    magnetospheric interactions 650
    physical properties *640t*, 649
    radiation effects 641
    silicate mass fractions *654f*
    spectral reflectance 639
    surface composition *652t*
    volatile elements 641
  Rhea
    atmospheric chemistry *652t*
    physical properties *640t*
    silicate mass fractions *654f*
    spectral reflectance *650f*
    surface composition *652t*
  size 627, *628f*
  temperature–pressure profiles *628f*
  Tethys
    atmospheric chemistry *652t*
    physical properties *640t*
    surface composition *652t*
  thermal infrared emissions 629
  Titan
    atmospheric chemistry 649, *652t*
    hydrogen (D/hydrogen (D/H) ratio *684f*
    physical properties *640t*
    surface composition 649, *652t*
    volatile elements 641
  wind systems 631–632
scandium (Sc)
  achondrites *293t, 294f*
  chondrites *49t, 89f*
  cosmochemical classification *46t*
  KREEP composition *578t*
  lunar mare basalts *565t*
  lunar regolith *566t*
  Martian meteorites *606t*
  solar photosphere *44t*
scapolite 263
schreibersite
  differentiated achondrites 305
  interplanetary dust particles (IDPs) 699–700
  iron meteorites 337
  primitive achondrites 296, 301
seawater, hydrogen (D/hydrogen (D/H) ratio *684f*
selenium (Se)
  achondrites *293t, 294f*
  chondrites *49t, 89f, 181f*
  cosmochemical classification *46t*
  interplanetary dust particles (IDPs) 704, *705f*
  interstellar medium (ISM) *58t, 59f*
  lunar mare basalts *574f*
  Martian meteorites *606t*
  solar photosphere *44t*
  volatile element depletions *512f*
serpentine *208t*, 255, 699
shergottites
  see also Mars
  assimilated crust 612, *613f*
  basalts 117–118
  composition 603–604, *612f*
  cosmic-ray exposure ages 369, *369f, 370f*
  crystallization 616–617
  elemental abundances *605f, 606t*
  lherzolites 118
  mineralogy 116, 117–118
    taxonomy *84f, 85t*
    thin section image *605f*
  volatile elements *574f*
Shoemaker Levy 9 669–670
siderophile elements
  brachinites 108
  calcium–aluminum-rich inclusions (CAIs) 231
  chondrites *89f, 177f*
  condensation/fractionation processes 411–412
  differentiated achondrites 306–307, 308
  early Earth history 542
  Earth's core 535
  iron meteorites **325–345**
  iron silicates, IIE 316
  lunar breccias 560, 581
  lunar mare basalts 573–575, *574f, 575f*
  partitioning 542
  primitive achondrites *294f*, 298, *298f*
  ureilites 313
  volatile elements *512f*
silicon (Si)
  achondrites *293t, 294f*
  chondrites *49t, 181f*
  cosmochemical classification *46t*
  interstellar medium (ISM) *58t, 59f*
  KREEP composition *578t*
  lunar mare basalts *574f*
  Mars *602t*
  Martian meteorites *606t*
  Moon 563–564
  nucleosynthesis 8
  presolar grains *19t*, 23–24, *23f*, 30
  silicon carbide (SiC)
    anomalous exotic components 18–19, 395, *396t*, 397
    meteorites 42–43, *42f*, 130
    presolar grains 18–19, *19t*, 20, 394
  silicon dioxide (SiO_2)
    chondrite matrix composition *182t*
    evaporation processes 416, *417f, 418f, 419f, 420f, 423f*
    lunar mare basalts *565t*
    lunar regolith *566t*
    Mars *602t, 604t*
    Martian mantle composition *610t*
    Mercury *481t*
    Venus *499t, 505t*
  silicon nitride (Si_3N_4) *19t*, 29
  solar photosphere *44t*
  solar system *132t*
silver (Ag)
  chondrites *49t*
  cosmochemical classification *46t*
  Martian meteorites *606t*
  solar photosphere *44t*
  volatile element depletions *512f*
Sisterson, J. 350
smectite 699
sodalite
  calcium–aluminum-rich inclusions (CAIs) 203, *208t*, 228
  CO chondrites 262
  CV chondrites 257
  Venus 501
sodium (Na)
  achondrites *293t, 294f*
  chondrites *49t, 89f, 181f*
  cosmochemical classification *46t*
  interplanetary dust particles (IDPs) *703t*
  interstellar medium (ISM) *58t, 59f*
  isotopic abundances *383f*
  KREEP composition *578t*
  lunar mare basalts 573, *574f*
  Martian mantle composition *611f*
  Martian meteorites *606t*
  Mercury 480
  radiogenic isotopes *384t*, 385–386
  sodium oxide (Na_2O)
    chondrite matrix composition *182t*
    lunar mare basalts *565t*
    lunar regolith *566t*
    Mars *602t, 604t, 610t*
    Martian mantle composition *610t*
    Mercury *481t*
    Venus *499t, 505t*
  solar photosphere *44t*
  volatile element depletions *512f*
soils
  lunar soils 390, *391f*
  Mars 603, *604t*, 607, 613
solar corona 55
solar energetic particles (SEP) 55
solar nebula **63–82**
  see also calcium–aluminum-rich inclusions (CAIs)
  accretion 69, 72, 74, 79, 510
  aluminum (Al) isotopes 78
  angular momentum 65, 70, 71, 72
  astrophysical analogues 68
  Balbus–Hawley instability 72
  baroclinic instability 71
  beryllium (Be) isotopes 77
  beryllium/boron (Be/B) isotopic ratios 239, 442
  calcium (Ca) isotopes 78

chondrites 146, 248
chondrule formation 76, *76f*, 78, 80, 463
chronology 451, *452f*
clump formation 75
comets 663
condensation/fractionation processes 330
hydrogen (D/hydrogen (D/H) ratio *684f*
disk masses 70
dust 75
elemental abundances 42, 638–639
Eta Carina nebula *65f*
evolution 70, 73
formation 64, 451
gas drag transport 75, *76f*
gravitational torques 72, 76
hydrodynamic turbulence 71
inner disk heating 513
ionization structure 72
iron meteorites 326
isotopic composition 18–19
Keplerian rotation 70, 71
light elements 131
magnetorotational instability 72
manganese (Mn) isotopes 444, *445f*
minimum mass 512
mixing processes 76
nebular protoatmosphere 530
oxygen (O) isotopes 131, 134
precollapse clouds
  ambipolar diffusion 65, 66
  amorphous ice 673–674, *673f*
  angular momentum distributions 65, 70
  astronomical observations 64
  collapse phase 65, 66
  composition 64, *64f*
  density profiles 64–65
  magnetic field support 65, 66
  shape 65
  shock-triggered collapse 65–66
radionuclide origins 449
redox processes 330
removal
  accretion 79
  disk life times 78
  general discussion 78
  mechanisms 79
  stellar winds 79
  ultraviolet (UV) photoevaporation 79
Rossby waves 71–72
shock fronts 76, *76f*
short-lived radionuclides 454
star formation
  bipolar outflows 67, *68f*
  disk wind model 67
  infrared (IR) emissions 69
  luminosity 513
  mass accretion rates 69, 72, 463
  millimeter-wave emissions 70
  observations 67, 462
  outflows 513
  precollapse clouds 65, *65f*, 66
  protostellar phases 67, 512
  solar mass stars 512
  temperature profiles 69, *70f*, 79
  T Tauri 69, 462, 463, 511, 513
  ultraviolet (UV) emissions 69

X-wind model 67, 77, *77f*, 464, 513
transport mechanisms 70
viscosity 74
volatility patterns 74
solar photosphere 43, *44t*
solar system
  see also calcium–aluminum-rich inclusions (CAIs)
  accretion dynamics
    age determination *538t*
    dust stickiness 516
    general discussion 516
    gravitational instability 517
    large collisions 518
    planetary migration 517
    runaway growth 518
  chondrites 431
  chondrule formation 188, *188t*
  chronology **431–460**
  comets 664
  hydrogen (D (^{2}H) 389, 631, *632f*
  elemental abundances **41–61**
    comet Halley dust 55
    comparison with meteorite abundances *44t*, 53, *54f*
    giant planets 638
    interplanetary dust particles (IDPs) 55
    isotopic abundances 2, *3f*, *10f*, 54, 131
    isotopic anomalies 42
    major elements *132t*
    mass number 54, *55f*, *56f*
    solar corona 55
    solar energetic particles (SEP) 55
    solar nebula 42, 131
    solar photosphere 43
    solar wind 55
    Sun 131
    volatility fractionation 407
  evaporation and condensation processes **407–430**
  evolution
    absolute timescales 435
    aluminum (Al) isotopes 439
    beryllium/boron (Be/B) isotopic ratios 239, 442, *443f*
    calcium–aluminum-rich inclusions (CAIs) 239
    calcium (Ca) isotopes 438
    chondrites 431
    chromium (Cr) isotopes 444, *445f*
    chronology **431–460**
    hafnium (Hf) isotopes 448
    iodine (I) isotopes 448
    iron (Fe) isotopes 446, *447f*
    manganese (Mn) isotopes 444, *445f*
    nickel (Ni) isotopes 446, *447f*
    niobium (Nb) isotopes 448
    palladium (Pd) isotopes 447
    plutonium (Pu) isotopes 449
    radionuclide origins 449
    samarium (Sm) isotopes 449
    short-lived radionuclides 432, 433, *433t*
    uranium/lead (U/Pb) isotopic ratios 435
  extrasolar giant planets 634
  giant planets **623–636**

hot nebula model 514, 515
isotopic fractionation 426
neutron capture synthesis 2, *3f*
nucleosynthesis **1–15**
oxygen abundances 59
planetary differentiation 468
planet formation **461–475**
r-process patterns 2, 4, *10f*
short-lived radionuclides 432, 433, *433t*
solar ^{3}He/^{4}He ratio 389
s-process patterns 2, *3f*, 4
star formation 462
volatile elements
  condensation processes 511, *512f*
  depletions *512f*, 514
  fractionation 407
  isotopic fractionation 412, *414f*
  kinetic processes 412
  thermodynamic equilibrium 409
  volatility patterns 74
solar wind
  asteroidal core formation 333
  composition 390, 392
  elemental abundances 55
  interplanetary dust particles (IDPs) 706
  isotopic abundances 131
  lunar soils 390, *391f*
  nebula evolution 463
  oxygen isotopic abundances 135
  solar ^{3}He/^{4}He ratio 389
  solar noble gases 388
  Venus 497
spacecraft
  CASSINI 488, *489t*
  Galileo 488, *489t*
  geochemical measurements 602, *602t*
  Giotto 55
  Hubble Space Telescope (HST) 57
  International Ultraviolet Explorer (IUE) 57
  Magellan 488, *489t*
  Mariner 4 601
  Mariner 9 601
  Mars Global Surveyor 602
  Mars Odyssey 602
  Mars Pathfinder lander 602, 603, *604t*, 610
  Phobos-2 602, *602t*
  Pioneer Venus 488, *489t*
  unsolved geochemical problems 617
  Vega-I 55, *489t*, *498t*, *499t*
  Vega-II 55, *489t*, *498t*, *499t*
  Venera 488, *489t*, *498t*, *499t*
  Viking landers 601, *604t*, 613
Space Telescope Imaging Spectrograph (STIS) 57
spectrometers
  alpha-proton X-ray spectrometer (APXS) *603f*, 611
  gamma-ray spectrometer (GRS) 602
  Goddard High Resolution Spectrometer (GHRS) 57
  infrared interferometric spectrometer (IRIS) 626
  Lunar Prospector 563

spectrometers (*continued*)
  thermal emission spectrometer (TES) 602–603, 610–611, *612f*, 613
  thermal emission spectrometer imaging system (THEMIS) 602–603, 613
spinel
  calcium–aluminum-rich inclusions (CAIs) 202, *203f*, 204–205, 206, *207t*, *240f*
  condensation/fractionation processes *410f*
  CV chondrites 257
  differentiated achondrites 303
  Martian core composition *611f*
  presolar grains 32, *32f*
Stardust 678
star formation
  bipolar outflows 67, *68f*
  disk wind model 67–68
  elemental abundance patterns 12, *12f*, *13f*
  hot nebula model 514, 515
  infrared (IR) emissions 69
  intermediate-mass stars
    carbon (^{12}C) isotope production 4
    evolution 3
    helium burning 4
  luminosity 513
  mass accretion rates 69, 72, 463
  massive stars
    carbon burning 8
    evolution 5, *5f*
    helium burning 7
    hydrogen burning 7
    interior structure *6f*
    neon burning 8
    nuclear burning stages 5, *6t*, 8
    nucleosynthesis 7
    oxygen burning 8
    p-process 9
    production factor *10f*
    r-process patterns 9
    silicon burning 8
    s-process patterns 7
  millimeter-wave emissions 70
  neutron capture synthesis 2, *3f*
  nucleosynthesis **1–15**
  observations 67, 462
  outflows 513–514
  precollapse clouds 65–66, *65f*, 66
  protostellar phases 67, 512
  red giants
    carbon (^{12}C) isotope production 4
    evolution 3
    helium burning 4
    r-process patterns 4
    s-process patterns 4, 130
  r-process patterns 2, *3f*, 4
  solar mass stars 512
  s-process patterns 2, *3f*, 4
  stellar evolution
    asymptotic giant branch (AGB) phase 21–23
    presolar grains 21, 22, 514, 527
    supernovae 28
  temperature profiles 69, *70f*
  T Tauri 69, 462, 463, 511–512, 513
  ultraviolet (UV) emissions 69

X-wind model 67–68, 77, *77f*, 135, 464, 513–514
STIS *see* Space Telescope Imaging Spectrograph (STIS)
strontium (Sr)
  achondrites *293t*
  calcium–aluminum-rich inclusions (CAIs) age determination 229
  chondrites *49t*
  cosmochemical classification *46t*
  KREEP composition *578t*
  lunar age determination 537, *538t*, *539f*
  lunar mare basalts *565t*, *570t*, *574f*
  lunar regolith *566t*
  Martian meteorites *606t*
  presolar grains 24, *25f*, 28
  solar photosphere *44t*
sulfur (S)
  achondrites *293t*, *294f*
  carbon monosulfide (CS) *675t*
  carbonyl sulfide (COS) 502, *675t*
  chondrites *49t*, *181f*, *182t*
  comets *675t*
  cosmochemical classification *46t*
  differentiated achondrites 305, 307
  giant planets 630, *631t*
  hydrogen sulfide (H$_2$S)
    comets *675t*
    giant planets 630, *631t*
    Venus *492t*, 502
  interplanetary dust particles (IDPs) 693–695, 699–700, *703f*, *703t*
  interstellar medium (ISM) *58t*, *59f*
  iron meteorites *328t*, 330–331, 335, 336, *336f*
  iron sulfide (FeS) 46, *610t*
  lunar mare basalts *574f*
  Mars *604t*, 615
  Martian core composition *610t*
  Martian meteorites *606t*
  meteorite formation 46
  mononitrogen monosulfide (NS) *675t*
  Murchison chondrite *271f*, *272t*, *277t*, 280
  primitive achondrites 302
  solar photosphere *44t*
  sulfates *499t*
  sulfide minerals 302, 305, 307
  sulfonic acids *271f*, *272t*, *277t*
  sulfur dioxide (SO$_2$)
    comets *675t*
    Io 642, 643
    Venus 492, *492t*, 496, 502
  sulfur monoxide (SO) *492t*, *675t*
  sulfur trioxide (SO$_3$) *182t*, *604t*
  thioformaldehyde (H$_2$CS) *675t*
  Venus *492t*, *499t*, 502, *505t*
  volatile element depletions *512f*
Sun
  carbon abundances 2
  carbon (C) abundances 2
  isotopic abundances 131
  light elements 131
  oxygen abundances 2, 59
  solar ^{3}He/^{4}He ratio 389
  solar photosphere 43, *44t*
supernovae
  classification 9

elemental abundances 11, *11f*
explosive nucleosynthesis 5, *7t*, 8, 9
nuclear burning stages 5, *7t*, 8
presolar grains 28, 30, 32
short-lived radionuclides 454
structure *27f*

taenite 258, 326, 336
Tagish Lake meteorite shower 679
tantalum (Ta)
  achondrites *293t*
  chondrites *49t*
  cosmochemical classification *46t*
  differentiated achondrites *304f*
  KREEP composition *578t*
  lunar regolith *566t*
  Martian meteorites *606t*
  solar photosphere *44t*
technetium (Tc) 4
tellurium (Te)
  chondrites *49t*
  cosmochemical classification *46t*
  interstellar medium (ISM) *58t*, *59f*
  Martian meteorites *606t*
  solar photosphere *44t*
  Venus 502
  volatile element depletions *512f*
terbium (Tb)
  achondrites *293t*, *294f*
  calcium–aluminum-rich inclusions (CAIs) *230f*
  chondrites *49t*, *230f*
  differentiated achondrites *304f*, *306f*, *308f*
  KREEP composition *578t*
  lunar mare basalts *565t*, *577f*
  lunar regolith *566t*
  Martian meteorites *606t*
  solar photosphere *44t*
  volatility 411
terrestrial planet formation 466, 519, 522, *522f*, *523f*, *526t*
TES *see* thermal emission spectrometer (TES)
thallium (Tl)
  chondrites *49t*
  cosmochemical classification *46t*
  interstellar medium (ISM) *58t*, *59f*
  Martian mantle composition *611f*
  Martian meteorites *606t*
  solar photosphere *44t*
Tharsis 605
Theia 536
thermal emission spectrometer (TES) 602, 610, *612f*, 613
thermal emission spectrometer imaging system (THEMIS) 602–603, 613
thiophene *271f*
thorium (Th)
  achondrites *293t*
  chondrites *49t*
  cosmochemical classification *46t*
  KREEP composition *578t*
  lunar mare basalts *565t*, *574f*
  lunar regolith *566t*
  Mars *602t*
  Martian meteorites *606t*
  Mercury *481t*
  Moon 563–564
  radiogenic isotopes *384t*
  solar photosphere *44t*

Venus *499t*
thulium (Tm)
  calcium–aluminum-rich inclusions (CAIs) *230f*
  chondrites *49t, 230f*
  differentiated achondrites *306f*
  KREEP composition *578t*
  Martian meteorites *606t*
  solar photosphere *44t*
  volatility 411
tin (Sn)
  achondrites *293t, 294f*
  chondrites *49t*
  cosmochemical classification *46t*
  interstellar medium (ISM) *58t, 59f*
  Martian meteorites *606t*
  solar photosphere *44t*
  volatile element depletions *512f*
titanium (Ti)
  achondrites *293t, 294f*
  chondrites *49t, 181f*
  cosmochemical classification *46t*
  interplanetary dust particles (IDPs) 704, *705f*
  interstellar medium (ISM) *58t, 59f*
  KREEP composition *578t*
  Martian meteorites *606t*
  presolar grains 24, *27f*, 30–31
  solar photosphere *44t*
  titanium dioxide (TiO_2)
    chondrite matrix composition *182t*
    lunar mare basalts 564, *564f, 565t, 568f, 570t, 571f*
    lunar mare volcanism 569, *572f*
    lunar regolith *566t*
    Mars *602t, 604t*
    Martian mantle composition *610t*
    Mercury *481t*
    Venus *499t, 505t*
  volatile element depletions *512f*
tochilinite *208t*, 700–701
toluene (C_7H_8) *278t*
trace elements
  calcium–aluminum-rich inclusions (CAIs) 230
  condensation/fractionation processes 411
  differentiated achondrites 310
  interplanetary dust particles (IDPs) 704, *705f*
  lunar mare basalts 573
  Mars 612, *613f*
  Martian meteorites 604–605, *606t*
  presolar grain composition 21
  solar system **41–61**
Trans Neptune Objects (TNOs) 667
tridymite 311, 316, 328
troctolites *584f*, 586
troilite
  calcium–aluminum-rich inclusions (CAIs) 229
  CO chondrites 262
  CV chondrites 258
  differentiated achondrites 303, 305, 309, 315
  interplanetary dust particles (IDPs) 699–700
  iron (Fe) isotopes 333
  iron silicates, IIE 316
  mesosiderites 311
  primitive achondrites 296, 301

Trojan comets *672f*
troposphere 348–349
T Tauri 69, 333, 462, 463, 511–512, 513
tungsten (W)
  achondrites *293t*
  calcium–aluminum-rich inclusions (CAIs) 231
  chondrites *49t*
  cosmochemical classification *46t*
  isotopic dating techniques 519, 522, *523f, 526t*
  KREEP composition *578t*
  lunar age determination 537–539, *538t, 539f*
  lunar mare basalts *574f*
  Martian mantle composition *611f*
  Martian meteorites *606t*
  primitive mantle composition 468
  solar photosphere *44t*
  terrestrial planet formation 522, *523f, 526t*
  volatility 411–412
turbulence
  gas drag transport 75, *76f*
  hydrodynamic turbulence 71
  magnetorotational instability 72
  solar nebula 71, 72, 75
iron silicates, IIE
  age determination 317
  bulk composition 316
  chemical composition *293t*
  formation conditions 317
  inclusion composition 315
  mineralogy 315
  petrologic characteristics *292t*
  texture 316
  volatile elements 316

ultramafic rocks *see* crystalline rocks
ultraviolet (UV) wavelengths 69, 79
uracil *271f*
uranium (U)
  achondrites *293t*
  chondrites *49t*
  cosmochemical classification *46t*
  KREEP composition *578t*
  lunar regolith *566t*
  Martian meteorites *606t*
  Mercury *481t*
  Moon 563–564
  radiogenic isotopes *384t*, 385
  solar photosphere *44t*
  uranium/lead (U/Pb) isotopic ratios 435–436
  uranium/xenon (U/Xe) isotopic ratios 402
  Venus *499t*
Uranus
  Ariel
    atmospheric chemistry *652t*
    physical properties *640t*
    silicate mass fractions *654f*
    spectral reflectance *653f*
    surface composition *652t*
  atmospheric chemistry 625
  composition 470
  core accretion 472
  hydrogen (D/hydrogen (D/H) ratio *684f*
  elemental abundances 630, *631t*, 633

equations of state 629
future research 634
helium abundances 626
historical investigations 624
hydrogen (H) abundances 626
interior models 625–626
interior structure 630
isotopic abundances 631
magnetic field 632
Miranda
  atmospheric chemistry *652t*
  physical properties *640t*
  silicate mass fractions 643, *654f*
  surface composition *652t*
Oberon
  atmospheric chemistry *652t*
  physical properties *640t*
  silicate mass fractions *654f*
  spectral reflectance *653f*
  surface composition *652t*
origins 633
physical parameters *624t*
planetary satellites
  atmospheric chemistry *652t*
  physical properties *640t*
  radiation effects 641
  silicate mass fractions *654f*
  spectral reflectance 639, 652, *653f*
  surface composition 643, *652t*
  volatile elements 641
size 627, *628f*
thermal infrared emissions 629
Titania
  atmospheric chemistry *652t*
  physical properties *640t*
  silicate mass fractions *654f*
  spectral reflectance *653f*
  surface composition *652t*
Umbriel
  atmospheric chemistry *652t*
  physical properties *640t*
  silicate mass fractions *654f*
  surface composition *652t*
wind systems 631–632
ureas *272t*
ureilites
  age determination 314
  chemical composition *293t*
  core compositions 313
  cosmic-ray exposure ages *367f*, 369
  diamond(s) 313
  graphite 313
  lithology 313
  lithophile elements 313
  mineralogy 108, *110f*, 312
  noble gas abundances 402
  origins 314
  oxygen isotopic composition 140, *140f*, 314
  petrologic characteristics *105t, 292t*
  shock effects 313
  siderophile elements 313
  taxonomy *84f, 85t*
  texture 313
  trapped noble gases 314

vanadium (V)
  achondrites *293t*
  chondrites *49t, 89f*
  cosmochemical classification *46t*

vanadium (V) (*continued*)
  interstellar medium (ISM) *58t, 59f*
  KREEP composition *578t*
  lunar mare basalts *565t, 574f*
  lunar regolith *566t*
  Martian meteorites *606t*
  solar photosphere *44t*
Vega-I spacecraft 55, *489t, 498t, 499t*
Vega-II spacecraft 55, *489t, 498t, 499t*
Venera spacecraft 488, *489t, 498t, 499t*
Venus
  atmosphere–surface interactions
    carbonate equilibria 500
    chlorine (Cl) equilibria 501
    dielectric activity 502
    fluorine (F) equilibria 501
    general discussion 500
    radar emissivity 502
    redox processes 501
    volatile elements 502
  atmospheric chemistry
    atmospheric dynamics 497
    carbon (C) abundances 491, *492t*
    chemical composition 491, *492t*
    chlorine (Cl) abundances *492t*, 493
    clouds 496
    fluorine (F) abundances *492t*, 493
    greenhouse effects 495, 502
    ionosphere 497
    isotopic composition 494, *494t*
    nitrogen (N) abundances *492t*, 494
    noble gas abundances *492t*, 494
    photochemical cycles 496
    solar wind 497
    sulfur (S) abundances 491, *492t*
    thermal structure 495, *495f*
    volcanic emissions 491
    water (H_2O) *492t*, 493
  bulk composition 504, *505t*
  core composition *505t*
  galena (PbS) 502
  historical investigations 487
  interior structure 504, *505f*
  iron oxide (FeO) concentrations *480t*
  magnesite 503
  mantle/crust composition *505t*
  modern investigations 488
  orbital properties 490, *490t*
  spacecraft missions 488, *489t, 498t, 499t*
  sulfur cycle 502
  surface
    basalts *498t*, 499, *500t*
    climate effects 502
    gamma-ray analyses 497, *499t*
    geology 503
    mineralogy 497, *498t*
    surface area distribution *504f*
    tectonics 504
    topography 503
    volcanism 491, 502, 504
    X-ray fluorescence analysis 497, *499t*
Vernadsky, V. I. 9
Vesta
  cosmic-ray exposure ages *374f*
  differentiated achondrites 309
  formation conditions 470
  iron oxide (FeO) concentrations *480t*
  oxygen ($\delta^{18}O$) isotope concentrations *514f*
  oxygen isotopic composition 138, *139f*
Viking landers 601, 603, *604t*, 613
volatile elements
  accretion 469, *512f*
  achondrites *574f*
  chondrites 51, *52f*, 151
  comets 665, 672–673, *673t*, 674, *675t*
  depletions 511, 527, 541
  differentiated achondrites 304, 307–308, 310
  fractionation 407
  interplanetary dust particles (IDPs) 55
  interstellar medium (ISM) 57, *58t*, 59, *59f*
  iron meteorite concentrations 330
  iron silicates, IIE 316
  isotopic fractionation 412, *414f*
  kinetic processes 412
  lunar mare basalts *574f*
  Mars 469
  Mercury 480, 482
  meteorites 46, *46t, 47f, 574f*
  protoplanetary nebula 463
  shergottites *574f*
  solar nebula 74
  terrestrial planets 469, *512f*, 533, *533f, 540f*, 542
  thermodynamic equilibrium 409
  Venus 502
  xenon fractionation 533, *533f, 540f*, 542
volcanism
  Io 642, 643, *644f*
  Venus 491, 502, 504

Wallace, Alfred 601
Wark–Lovering rim 202, *203f*, 204, *204f*, 223
water (H_2O)
  comets 672, *675t*
  giant planets 630, *631t*
  Mars 614
  Martian meteorites *606t*
  solar nebula 75
  sublimation temperatures *674t*
  Venus *492t*, 493
weathering 90
wehrlites *84f, 85t*, 118
Whipple, F. 667
whitlockite 296, 303, 311
Widmanstätten pattern 336, *337f*
winonaites
  age determination 302
  chemical composition *293t*
  formation conditions 302
  inclusion composition 301
  mineralogy 106, 301
  oxygen isotopic composition 139–140, 140, *140f*
  petrologic characteristics *292t*
wollastonite *208t*, 257

xenon (Xe)
  anomalous exotic components 18–19, *18f*, 396
  brachinites 308
  chondrites *49t*, 249, 250
  cosmic-ray exposure ages *348t*
  cosmochemical classification *46t*
  degassing models 532, 533, *533f*
  fractionation 533, *533f, 540f*, 542
  giant planets *631t*
  iodine/xenon (I/Xe) isotopic ratios 249, 250, 448, 532, 533, *540f*
  isotopic abundances 131, *383f*, 398
  lunar soils 390–391, *391f*
  Mars *605t*, 614
  presolar grains 24, *25f*, 28
  radiogenic isotopes 384, *384t*
  solar photosphere *44t*, 45
  solar wind concentrations 390, 392
  spallation 386
  terrestrial atmospheric composition 402, 533, *533f, 540f*
  terrestrial planets 532
  trapped components *399f*
  uranium/xenon (U/Xe) isotopic ratios 402, 533, *533f, 540f*
  Venus *492t*
X-ray fluorescence (XRF) 603

Yarkovsky effect 374–375
ytterbium (Yb)
  achondrites *293t, 294f*
  calcium–aluminum-rich inclusions (CAIs) *230f*
  chondrites *49t, 89f, 230f*
  differentiated achondrites *304f, 306f, 308f*
  KREEP composition *578t*
  lunar mare basalts *577f*
  lunar regolith *566t*
  Martian meteorites *606t*
  solar photosphere *44t*
  volatility 411
yttrium (Y)
  chondrites *49t*
  cosmochemical classification *46t*
  KREEP composition *578t*
  Martian meteorites *606t*
  presolar grains 24
  solar photosphere *44t*

Zag (b) *292t*, 302
zeolites 613
zinc (Zn)
  achondrites *293t, 294f*
  chondrites *49t, 89f, 181f*
  cosmochemical classification *46t*
  interplanetary dust particles (IDPs) 704, *705f*
  interstellar medium (ISM) *58t, 59f*
  lunar mare basalts 573, *574f*
  lunar regolith *566t*
  Martian mantle composition *611f*
  Martian meteorites *606t*

solar photosphere *44t*
  volatile element depletions *512f*
zircon 544
zirconium (Zr)
  achondrites *293t*

chondrites *49t*
cosmochemical classification *46t*
Earth formation 543
KREEP composition *578t*
lunar mare basalts *565t*

lunar regolith *566t*
Martian meteorites *606t*
presolar grains 24, 30–31
solar photosphere *44t*
ζ Ophiucus 57, *58t*, *59f*